Separation of Molecules, Macromolecules and Particles

Providing chemical engineering undergraduate and graduate students with a basic understanding of how the separation of a mixture of molecules, macromolecules or particles is achieved, this textbook is a comprehensive introduction to the engineering science of separation.

- Students learn how to apply their knowledge to determine the separation achieved in a given device or process.
- Real-world examples are taken from biotechnology, chemical, food, petrochemical, pharmaceutical and pollution control industries.
- Worked examples, elementary separator designs and chapter-end problems give students a practical understanding of separation.

The textbook systematically develops different separation processes by considering the forces causing the separation, and how this separation is influenced by the patterns of bulk flow in the separation device. Readers will be able to take this knowledge and apply it to their own future studies and research in separation and purification.

Kamalesh K. Sirkar is a Distinguished Professor of Chemical Engineering and the Foundation Professor of Membrane Separations at New Jersey Institute of Technology (NJIT). His research areas are membranes and novel membrane based processes.

"The first comprehensive book that takes the fundamentals of separation on a molecular level as the starting point! The benefit of this approach is that it gives you a thorough insight in the mechanisms of separation, regardless of which separation is considered. This makes it remarkably easy to understand any separation process, and not only the classical ones. This textbook finally brings the walls down that divide separation processes in classical and non-classical."

Bart Van der Bruggen, University of Leuven, Belgium

"This strong text organizes separation processes as batch vs continuous and as staged vs differential. It sensibly includes coupled separation and chemical reaction. Supported by strong examples and problems, this non-conventional organization reinforces the more conventional picture of unit operations."

Ed Cussler, University of Minnesota

"This book fills the need by providing a very comprehensive approach to separation phenomena for both traditional and emerging fields. It is effectively organized and presents separations in a unique manner. This book presents the principles of a wide spectrum of separations from classical distillation to modern field-induced methods in a unifying way. This is an excellent book for academic use and as a professional resource."

C. Stewart Slater, Rowan University

"This book is an excellent resource for the topic of chemical separations. The text starts by using examples to clarify concepts. Then throughout the text, examples from many different technology areas and separation approaches are given. The book is framed around various fundamental approaches to chemical separations. This allows one to use this knowledge for both current and future needs."

Richard D. Noble, University of Colorado

"This book provides a unique and in depth coverage of separation processes. It is an essential reference for the practicing engineer. Unlike more conventional textbooks that focus on rate and equilibrium based separations, Prof Sirkar focuses on how a given separation takes place and how this is used in practical separation devices. Thus the book is not limited by application e.g. chemical or petrochemical separations.

"As chemical engineering becomes increasingly multidisciplinary, where the basic principles of separations are applied to new frontier areas, the book will become an essential guide for practitioners as well as students.

"The unique layout of the text book allows the instructor to tailor the content covered to a particular course. Undergraduate courses will benefit from the comprehensive and systematic coverage of the basics of separation processes. Whether the focus of a graduate course is traditional chemical separations, bioseparations, or separation processes for production of renewable resources the book is an essential text."

Ranil Wickramasinghe, University of Arkansas

"This advanced textbook provides students and professionals with a unique and thought-provoking approach to learning separation principles and processes. Prof. Sirkar has leveraged his years of experience as a separation scientist and membrane separation specialist, to provide the reader with a clearly written textbook full of multiple examples pulled from all applications of separations, including contemporary bioseparations. Compared to other separations textbooks, Prof. Sirkar's textbook is holistically different in its approach to teaching separations, yet provides the reader with a rich learning experience. Chemical engineering students and practicing professionals will find much to learn by reading this textbook."

Daniel Lepek, The Cooper Union

Separation of Molecules, Macromolecules and Particles

Principles, Phenomena and Processes

Kamalesh K. Sirkar
New Jersey Institute of Technology

CAMBRIDGE
UNIVERSITY PRESS

CAMBRIDGE
UNIVERSITY PRESS

Shaftesbury Road, Cambridge CB2 8EA, United Kingdom

One Liberty Plaza, 20th Floor, New York, NY 10006, USA

477 Williamstown Road, Port Melbourne, VIC 3207, Australia

314–321, 3rd Floor, Plot 3, Splendor Forum, Jasola District Centre, New Delhi – 110025, India

103 Penang Road, #05–06/07, Visioncrest Commercial, Singapore 238467

Cambridge University Press is part of Cambridge University Press & Assessment, a department of the University of Cambridge.

We share the University's mission to contribute to society through the pursuit of education, learning and research at the highest international levels of excellence.

www.cambridge.org
Information on this title: www.cambridge.org/9780521895736

First published 2014

A catalogue record for this publication is available from the British Library

Library of Congress Cataloging-in-Publication data
Sirkar, Kamalesh K., 1942- author.
Separation of molecules, macromolecules and particles : principles, phenomena and processes / Kamalesh Sirkar, New Jersey Institute of Technology.
pages cm. – (Cambridge series in chemical engineering)
ISBN 978-0-521-89573-6 (Hardback)
1. Separation (Technology)–Textbooks. 2. Molecules–Textbooks. I. Title.
TP156.S45S57 2013
541′.22–dc23 2012037018

ISBN 978-0-521-89573-6 Hardback

Additional resources for this publication at www.cambridge.org/sirkar

Contents

Preface

This is an introductory textbook for studying separation. Primarily, this book covers the separation of mixtures of molecules; in addition, it provides a significant treatment of particle separation methods. Separation of macromolecules has also received some attention. The treatment and coverage of topics are suitable for chemical engineering students at undergraduate and graduate levels. There is enough material here to cover a variety of introductory courses on separation processes at different levels.

This book is focused on developing a basic understanding of how separation takes place, and of how the resulting separation phenomenon is utilized in a separation device. The role of various forces driving molecules or particles from a feed mixture into separate phases/fractions/regions is basic to such an approach to studying separation. The separation achieved is then amplified in an open separator via different patterns of bulk-phase velocities vis-à-vis the direction(s) of the force(s). The forces are generated by chemical potential gradient, electrical field, rotational motion, gravity, magnetic field, etc. The resulting separation is studied under three broad categories of separation processes.

Separation processes driven by a negative chemical potential gradient are generally multiphase systems and are treated under the broad category of phase equilibrium driven processes. External force driven processes populate the second category, and include those operating under an electrical field, rotational motion, magnetic field or gravity; thermal diffusion processes are also briefly included here. The third category of membrane based processes studied is driven generally by a negative chemical potential gradient; however, electrical force is also relevant for some processes. The treatment of any external force driven processes will cover both separation of molecules and particle separations.

These physical separation methods are often reinforced by chemical reactions, which are usually reversible. An elementary treatment of the role of chemical reactions in enhancing separation across a broad spectrum of phase equilibrium driven processes and membrane based processes has been included. The level of treatment in this book assumes familiarity with elementary principles of chemical engineering thermodynamics and traditional undergraduate levels of knowledge of ordinary differential equations and elementary partial differential equations.

Specific aspects of a given separation process are studied in the chapter devoted to those aspects for all separation processes. To study a particular separation process in great detail, one therefore has to go to different chapters. The footprints of a given separation process are provided at the beginning of the book (Tables 1–7); there are quite a few tables to cover a variety of separation processes. The list of processes is large; however, it is far from being all inclusive. The introductory chapter, which provides additional details about various chapters, as well as about the book, is preceded by a notation section. All references appear at the end of the book.

The description of the extent of separation achieved in a closed vessel for a mixture of molecules is treated in Chapter 1. Chapter 2 illustrates how to describe the separation of molecules in open separators under steady and unsteady state operation; a description of separation for a size-distributed system of particles is also included. Chapter 3 introduces various forces developing species-specific velocities, fluxes and mass-transfer coefficients, and illustrates how the spatial variation of the potential of the force field can develop multicomponent separation ability. The criteria for chemical equilibrium are then specified for different types of multiphase separation systems, followed by an illustration of integrated flux expressions for two-phase and membrane based systems.

Chapter 4 develops the extent of separation achieved in a closed vessel to a variety of individual processes under each of the three broad categories of separation processes. Chapter 5 demonstrates how separation can be considerably enhanced by chemical reactions in phase equilibrium based and membrane based processes under both equilibrium- and rate-controlled conditions. For open separators having bulk flow in and out, including continuous stirred tank separators (CSTSs), Chapter 6 provides first the equations of change for molecular species concentration in single-phase and two-phase systems, the trajectory equation for a particle in a fluid and the general equation of change for a particle population. Chapter 6 then treats individual separation processes under each of the three broad categories of separation processes when the bulk flow is parallel to the direction of the force and in CSTS mode.

Chapter 7 follows this latter approach of treating individual separation processes under each of the three broad categories of separation processes when the bulk flow of feed-containing phase is perpendicular to the direction of the force. Chapter 8 follows the same approach when the bulk flows of two phases/regions in the separator are perpendicular to the direction(s) of the force(s). Chapter 9 briefly elaborates on cascades, which were already introduced in the countercurrent multistaged flow systems of Chapter 8. Chapter 10 introduces the energy required for a number of separation processes. Chapter 11 illustrates a few common separation sequences in a number of common industries involved in bioseparations, water treatment, chemical and petrochemical separations and hydrometallurgy. Conversion factors between various systems of units are provided in an Appendix.

Virtually all separation processes taught to chemical engineering students in a variety of courses have been covered via the approach illustrated in Chapters 3, 4, 6, 7 and 8; in addition, many particle separation methods have been treated. The structural similarity in the separation method between apparently unrelated separation processes becomes quite clear. A few basic principles equip the students with the capability to understand a wide variety of separation processes and techniques, including emerging ones. To aid the student, there are 118 worked examples, 300 problems, 340 figures, 100 tables and 1011 references. A website will provide guidance for computer simulations for a few selected problems.

The introductory chapter provides references to articles and books which influenced the development of various aspects of this book. I have benefitted considerably from the comments on selected chapters of the book by reviewers, anonymous or otherwise. Comments by Professors C. Stewart Slater, of Rowan University, Steven Cramer, of Rensselaer Polytechnic Institute, and Ranil Wickramasinghe, of Colorado State University (now at University of Arkansas), were particularly useful.

Many doctoral students and postdoctoral fellows were of invaluable help during the long gestation period of this book, either in formulating solutions of the problems or in developing illustrative drawings. I want to mention in particular Amit Sengupta, Theoharris Papadopoulos, Xiao-Ping Dai, Meredith Feins, Dimitrios Zarkadas, Quixi Fan, Praveen Kosaraju, Fei He, Atsawin Thongsukmak, Sagar Roy, Dhananjay Singh, John Tang and John Chau. The first two students helped me when we were at Stevens Institute of Technology. Sarah Matthews of Cambridge University Press patiently provided manuscript preparation guidelines and encouraging comments during an ever-shifting timetable. Irene Pizzie did an extraordinary job as the copy editor. Brenda Arthur of New Jersey Institute of Technology tirelessly typed the draft of the whole manuscript over a considerable length of time, while carrying out many other duties.

I must also mention at the end my wife, Keka, without whose patience, help and understanding this book would never have been finished.

Notation

Equation numbers identify where the symbols have been introduced or defined.

The following styles have been adopted.

Bold vector quantity

Overlines

—	quantity averaged over time or a specific coordinate direction, multicomponent system, Laplace transformed dependent variable
— =	averaged quantity
^	quantity in a mixture, per unit mass of bulk phase

Underlines

_	hypothetical binary system quantity (2.4.23), (2.4.24)
~, =	vector quantity, tensor quantity

Brackets

$\langle v_{tj} \rangle$ average value of v_{tj} over surface area $\mathbf{S}_j$

a	ellipsoid semiaxis dimension; see also (3.1.10a); constant in relations (3.1.49), (4.3.7) and (4.3.43a); interfacial area per unit volume, defined by (7.2.191), (7.3.25)
a_1; a_2, a_3,...	constant in (3.3.105), (4.1.42b), (4.3.29) and (5.2.147); constants in (7.2.73), (7.3.50) and (7.3.139)
a_1	constant in (7.2.198a)
a_{A}	stoichiometric coefficient for species A
a_H	Hamaker constant (3.1.16)
a_i; $\mathrm{a}_i(T)$; a_{ij}, a_{il}; $\hat{\mathrm{a}}_i$; a'_{m}; $a_{\mathrm{ms}\ell}$	activity of species i; equilibrium constant (7.1.63); atom fraction of ith isotope of an element in region j (1.3.6), value of a_i in region j and liquid phase, respectively; amplitude (7.1.72b); constant in (4.3.43c); membrane surface per unit channel length (7.2.70)
a_p	surface area of a particle
a_{s1}, a_{s2}; a_{sw}	activity of solvent s in regions 1 and 2, respectively; activity of salt in water
a_{sp}	pore surface area per unit volume of the porous medium of porosity ε
a_v, a_{vc}	surface area of a particle per unit particle volume, value of a_v in a cake (6.3.135j)
$a_\pm$	mean electrolyte activity (3.3.119d)
A; A_1, A_2, A_3	amplitude (7.3.18), pure-water permeability constant in reverse osmosis and diffusive ultrafiltration; constants in (7.1.90b), three surface areas in control volume of Figure 7.2.6(b)
$A(\mathrm{r})$	cross-sectional area of a cone in centrifugal elutriation
A_c	cross-sectional area of duct
A_{hex}; A_{ij}; A_m; A^t_m; $A_i(T)$; A^o_i	heat exchanger surface area; constant in equations for activity coefficients (4.1.34d); surface area of membrane; total membrane surface area; modified equilibrium constant (7.1.66); constant (7.1.72b)
$A_{1\rho}$	constant in crystal growth rate equation (6.4.27)
A_p	projected area of a particle (3.1.64); transport coefficient in solution-diffusion-imperfection model (3.4.60a)
$A_p^+(\mathrm{x})$	cumulative crystal surface area distribution fraction (6.4.17)
A_T	total particle surface area per unit volume of total mixture (Example 2.4.2); total crystal surface area per unit liquid volume (6.4.16)
Am^+, Am^-, $\mathrm{Am}^\pm$	three forms of amino acid
b	ellipsoid semiaxis dimension; proportionality constant in osmotic pressure relation (3.4.61b); half of channel gap; width of region of gas flow completely cleaned up by a fiber (6.3.42a); constant in crystal growth rate expression (6.4.35); parameter (7.2.18); liquid envelope radius (7.2.208)

b_1, b_2, b_3	constants in relations (3.1.143f)
b_1	constant in (3.3.105), (5.2.147)
b_f; b_i; b_{if}; $b_{i\ell}$; b_1^i	membrane feed channel height (Figure 7.2.3(a)); constant in equilibrium relation (3.3.81); constant in (7.1.73), constant in Freundlich isotherm (3.3.112c); constant in Langmuir isotherm (3.3.112a); constant in equation (4.1.42a)
b'_{i-k}, b'_k	proportionality constant in (5.2.154), equilibrium constant in (5.2.155)
b_m, b'_m	constants in (4.3.43a,c)
$\boldsymbol{B}$	magnetic induction vector (3.1.19)
B; B^o; B_i^o; B_p	constant, density function of the birth rate of new particles (6.2.50g), value of B as $r_p \to 0$; value of B^o for crystal growth rate gr_i; duct perimeter (7.3.41)
Bi	Biot number (3.4.35)
B_{1p}, B_{2p}	second virial coefficients for interaction between polymers 1 and 2 (4.1.34p)
c; c_C	velocity of light; gap between plates at entrance (Figure 7.3.10), stoichiometric coefficient for species C
C	clearance of a solute (8.1.390)
C_1, C_2; $C_2(x)$	integration constants (3.3.10b); molar concentration of species 2 at location x
C_{2f}	molar concentration of species 2 in initial mixture, mol/liter
$C^{(2)}$	defined in (6.3.75)
C_A^0, C_A^δ	molar concentration of A at $z = 0$, δ (3.1.124)
$C_{A\ell b}$, $C_{A\ell i}$	molar concentration of A in phase bulk and phase interface and in liquid at gas–liquid interface (3.4.1b)
C_{Awb}, C_{Awi}; C'_{Awb}	molar concentration of A in bulk water and in water at phase interface (3.4.45e); critical value of C_{Awb} for maximum enhancement (5.3.53)
C_{Bob}, C_{Boi}	molar concentration of species B in organic-phase bulk and interface, respectively
C_c; C'_{Cb}; C_D	slip (Cunningham) correction factor (3.1.215); critical value of bulk concentration of C for maximum enhancement (5.3.29b); particle drag coefficient (3.1.64), (6.3.4)
C_{FC}	molar concentration of fixed charges in ion exchange resin
C'_{H1}, C'_{H2}	dual mode sorption constants for species 1 and 2 (3.3.82a,b)
C_i; $\overline{C}_i$; C_i^t; C_i^*; C_{ig}^*; C_{i1}^i, C_{i2}^i	molar concentration of species i; an average of the molar concentration of species i in the feed and the permeate (5.4.74), (6.3.158b); total molar concentration of species i in the porous medium per unit volume (3.1.118b); nondimensional species i concentration (5.3.35j), hypothetical gas-phase species i concentration (8.1.47); initial bed concentration of solute i in phases 1 and 2
C_{i10}, C_{i20}	values of C_{i1}, C_{i2} at $z = 0$
$C_i^o(\pi)$	liquid-phase concentration of pure solute i at spreading pressure π and temperature T providing the same surface phase concentration of i as the mixture
C_{igel}	molar species i concentration in gel
C_{ij}	molar concentration of species i in region j or location j or stream j; $j = b$, bulk; $j = E$, extract; $j = f$, feed region; $j = g$, gas phase; $j = k$, kth phase; $j = \ell$, liquid; $j = \ell'$, adhering liquid phase on crystal; $j = m$, membrane; $j = o$, organic; $j = p$, permeate, product; $j = r$, raffinate; $j = R$, ion exchange resin phase, raffinate; $j = s$, solution, solid phase or pore surface; $j = w$, water, mol/liter
C_{ij}^t; C_{ijn}	total molar concentration of species i in region j including complexed or dissociated forms; value of C_{ij} on nth plate/stage
$C_{ij}^0, C_{ij}^\ell, C_{ij}^\delta$	species i concentrations in phase j at locations 0 (or initial concentration), ℓ and δ
$C_{i\ell b}, C_{i\ell e}, C_{i\ell i}$	molar concentrations of species i in bulk solution, at the end of concentration process and at the beginning of concentration process (6.3.173)
C'_{ig}, C'_{igf}	molar species i concentration per unit gas phase volume in a pore
$\overline{C}_{ik}, \overline{\overline{C}}_{ik}, \hat{C}_{ik}$; $\overline{C}_{it}$	intrinsic phase average, phase average and deviation in C_{ik} for species i concentration in phase k (6.2.24a,b), (6.2.28); defined by (7.1.94)
C_{im}^d, C_{im}^H; C_{im}^p	species i concentration in membrane: for Henry's law and Langmuir species, respectively, in dual sorption model (3.3.81); membrane pore liquid
$C_{imi}, C_{imo}, C_{iob}, C_{iwi}, C_{iwb}$	molar concentrations of species i at various locations in Figure 3.4.11
C_{pR}, C_{pw}	molar protein concentrations in resin phase and aqueous solution

C_{sf}, C_{sm}, C_{sp}	molar solvent concentration in feed, membrane and permeate, respectively
$C_t; C_{tj}; C_{vp}$	total molar concentration; C_t in a mixture in region j; volume of particles per unit fluid volume (7.2.176)
$d; d_h; d_i; d_{\ell m}$	diameter of tube/pipe/vessel; hydraulic diameter (Table 3.1.8); effective diameter of a molecule of gas species i (3.3.90a); logarithmic mean diameter (8.1.417)
$\boldsymbol{d}_i; d_{gr}; d_{\mathrm{imp}}$	force-type term (3.1.178), (3.1.181); grain diameter; diameter of an impeller
$d_{\mathrm{ion}}; d_p; d_w; d_{p1}, d_{p2}$	mean diameter of a molecular ion; mean diameter of a particle (6.1.4b); wire diameter (3.1.23); diameters of particles 1 and 2
d_{32}	Sauter mean diameter of a drop or particle (6.4.88), (6.4.89)
D	diffusion coefficient of species in countertransport through liquid membrane
$D_{\mathrm{A}}, D_{\mathrm{C}}; D_{\mathrm{B}}$	diffusion coefficient of species A and C, respectively; dialysance in hemodialysis (8.1.389)
D_{eff}	effective diffusion coefficient (5.4.64a)
D_{gr}	crystal growth diffusivity (6.4.45)
$D_p; D_p(\phi)$	diffusion coefficient for particle (3.1.68), (6.2.52); shear-induced particle diffusivity (3.1.74), (7.2.126), (7.2.131a)
D_r	desalination ratio (1.4.25), (2.2.1a)
$D_{i,\mathrm{eff}}; D_{i,\mathrm{eff},r}, D_{i,\mathrm{eff},z}; D_{i,\mathrm{eff},1}$	effective diffusion coefficient of i in liquid (6.2.18), (6.3.16b); value of $D_{i,\mathrm{eff}}$ in r-and z-directions; value of $D_{i,\mathrm{eff}}$ in phase/region 1
$D_{i,\mathrm{eff},k}; \underline{\underline{D}}_{i,\mathrm{eff},k}$	effective diffusion coefficient of i in phase k (6.2.33); dispersion tensor (6.2.31)
$D_{ij}; D_{il}; D_{is}$	diffusion coefficient of species i in region j; j = l, liquid; $j = s$, solvent
$D_{\mathrm{A}}^{\mathrm{T}}, D_{\mathrm{B}}^{\mathrm{T}}; D_{\mathrm{is}}^{T}$	thermal diffusion coefficient for species A and B (3.1.44); for species i in solution
$D_{\mathrm{AB}}; D_{\mathrm{BR}}$	binary diffusion coefficient for mixture of gases A and B; diffusion coefficient of particles due to Brownian motion (7.2.216)
D_{iD}, D_{iH}	diffusion coefficients in dual sorption-dual transport model (3.4.78)
D_{ie}	effective diffusion coefficient of i in a porous medium (3.1.112d)
$D_{im}; D_{imo}; D_{ip}$	diffusion coefficient of i in membrane (6.3.149); value of D_{im} for $C_{im} = 0$ (3.4.67b), D_i in a pore (3.4.89c), (6.3.145a); effective D_i in the pores of a particle
$\overline{D}_{ik}, \overline{D}_{im}$	multicomponent diffusion coefficient of species pair (i, k) and (i, m) in Maxwell–Stefan approach
D_{iK}	Knudsen diffusion coefficient for species i (3.1.115c)
D_{iM}	effective binary diffusivity of species i in a mixture (3.1.184), (3.1.185)
$D_{is}; D_{is}^{0}; D_{is}^{N}$	binary diffusion coefficient for solute i/solvent s; value of D_{is} at infinite dilution; Nernst-Planck binary diffusion coefficient for species i/solvent s
D_{12}	binary diffusion coefficient for species 1 and 2
$De; (De)_{mv}$	density function of particles which disappear (die) (6.2.50h); see (7.2.170a) and (7.2.172)
$Df; DF$	decontamination factor (2.2.1c); dilution factor (6.4.106), (7.2.91c)
e	charge of an electron, 1.60210×10^{-19} coulomb
e_i	constant in adsorption isotherm for solute i (3.3.113d)
$en_b, en_M, en_{MN}, en_{MP}, en_p$	molecular energy in the bulk, due to intramolecular interactions, due to intermolecular interactions, due to interaction between molecules and pores and total energy for molecules in the pore (3.3.89d)
$\boldsymbol{E}; E; E_c; E_{D_i}; E_e$	electrical force field; its magnitude (3.1.8), (6.1.22), (6.3.8f), extraction factor (8.1.281), stage efficiency (6.4.72); electrical field strength E_c (7.3.32a); activation energy for diffusion of species i in polymer (4.3.46b); extraction factor for extraction section (8.1.303)
$E_i; E_o; E_{oG}; E_s; E_y; E_{\mathrm{BRS}}, E_{\mathrm{GrS}}$	enrichment of species i by pervaporation (6.3.193b), particle collection efficiency (7.2.200b); overall column efficiency (8.1.195); point efficiency (8.3.13); extraction factor for the scrubbing section; electrical field strength in y-direction (7.3.48); particle collection efficiency (7.2.219), (7.2.214)
$E_{IS}; E_{\mathrm{IS}}$	inertial impaction based single fiber capture efficiency (6.3.42a); particle capture efficiency by interception (7.2.224)

$E_{ME}, E_{MR}; E_{MV}; E_N$	Murphree extract stage efficiency, Murphree raffinate stage efficiency (6.4.70), (6.4.71); Murphree vapor efficiency (8.1.198); Newton particle separation efficiency (2.4.14a)
$E_T; E_T^1; E_{T_i}$	total efficiency in solid–fluid separation (2.4.4a), overall filter efficiency (6.3.45), (7.2.201); reduced efficiency of Kelsall (2.4.16a); E_T for *i*th solid–fluid separator (2.4.17c,d)
f	friction factor (6.1.3a), fractional consumption of chemical adsorbent (5.2.19d)
f_2	fraction of the solute in ionized form ($i = 2$) in RO (5.4.4)
$f(r); f(r_p); f(\varepsilon)$	molar density function in a continuous/semi-continuous mixture with characteristic property r; particle size probability density function (2.4.1a), pore size distribution function in a membrane; defined by (7.2.222a,c)
$f_f(r_p), f_1(r_p), f_2(r_p)$	value of $f(r_p)$ for feed stream, overflow and underflow based on particle weight fraction in a given size range (2.4.1b)
$f_A, f_i; f_i^0$	fugacity of species A, species i; standard state fugacity of species i
$f_g(\text{gr})$	probability density function of crystal growth rate (6.4.41a)
f_{il}	fugacity of pure species i in liquid phase
f_m, f_m	quantities characteristic of a membrane polymer (4.3.46a,d)
$f_{ij}^0, f_{ig}^0, f_{il}^0$	standard state fugacity of species i in region j; $j = g$, gas phase; $j = \ell$, liquid phase
f_i^d, f_p^d	frictional coefficient for species i and spherical particle
$\hat{f}_{ig}, \hat{f}_{ij}, \hat{f}_{il}; \hat{f}_{ijpl}$	value of fugacity of i in a mixture in gas phase, phase j and liquid phase, respectively; value of $\hat{f}_{ij}$ for a planar surface
f_{io}^d	value of f_i^d for a sphere of equivalent volume (3.1.91e)
f_{im}^d, f_{sm}^d	frictional coefficient for solute i and solvent s in a membrane
$f_f(M), f_l(M), f_v(M); f_\infty, f_\lambda$	value of $f(r)$, where $r = M$, molecular weight, for the feed mixture, liquid fraction and the vapor fraction, respectively; defined by (7.1.59a), (7.2.187), respectively
$f_M; f_{Qm}, f_{yo}$	fraction of the total metal ion concentration in the aqueous phase present as M^{n+} (5.2.97); probability density functions (7.3.79), (7.3.80)
$\boldsymbol{F}; \hat{\boldsymbol{F}}, \hat{\boldsymbol{F}}_p; F$	force on a particle; value of $\boldsymbol{F}$ per unit particle mass; degrees of freedom (4.1.22)
$F(r_p)$	probability distribution function corresponding to $f(r_p)$ (2.4.1c), crystal size distribution function (6.4.11)
$\boldsymbol{F}_i; F_i^{\text{ext}}$	electrostatic force on 1 gmol of a charged species in solution (6.3.8a); magnitude of external force on 1 gmol of species i
F_{acrx}	acoustic radiation force in x-direction (3.1.48)
$\boldsymbol{F}_{\text{rad}}$	radiation pressure force (3.1.47), (7.3.267)
$\boldsymbol{F}^{\text{BR}}$	force on very small particle due to random Brownian motion (3.1.43)
$\boldsymbol{F}_i^{\text{ELK}}$	electrokinetic force on particle in double layer (3.1.17)
$\boldsymbol{F}_i^{\text{ELS}}$	electrostatic force on particle i, Coulomb's law (3.1.15)
$\boldsymbol{F}_i^{\text{Lret}}$	London attraction force (3.1.16)
$\boldsymbol{F}_k^{\text{m}}$	force on species k in mass flux $\boldsymbol{j}_i$, force relation (3.1.202)
$\boldsymbol{F}_{\text{net}}^{\text{ext}}$	net external force; for gravity see (3.1.5)
$\boldsymbol{F}_p^{\text{drag}}; \boldsymbol{F}_p^{\text{iner}}; F_{pz}^{\text{ext}}$	frictional force on a particle; inertial force on a particle (6.2.45); external force on a particle in z-direction
$\boldsymbol{F}_{ti}, \boldsymbol{F}_{ti}^{\text{ext}}$	total force and total external force on 1 gmol of species i (3.1.50)
$\boldsymbol{F}_{tp}^{\text{ext}}; F_{tpx}^{\text{ext}}, \boldsymbol{F}_{tpy}^{\text{ext}}; F_{tpz}^{ext}$	total external force on a particle (3.1.59), (6.2.45); components of F_{tp}^{ext} in x-, y-and z-directions
$\boldsymbol{F}_{\text{TA}}$	force on 1 gmol of species A due to a temperature gradient (3.1.44)
$\mathcal{F}$	Faraday's constant, 96 485 coulomb/gm-equivalent
$g; g_c; g_m$	acceleration due to gravity; conversion factor; a quantity characteristic of a membrane polymer (4.3.46a)
$\boldsymbol{g}^{\text{ext}}; g_x^{\text{ext}}, g_y^{\text{ext}}, g_z^{\text{ext}}$	external body force per unit mass; its components in x-, y-and z-directions
gr_i	intrinsic growth rate of *i*th crystal (6.4.41a)
$G; G_g$	superficial mass average velocity based on empty flow cross section, G for gas phase
$G; G_a, G_b; G_i; G_o$	growth rate of crystal (6.4.25), (6.4.3b); value of G under condition a, condition b; factor representing contribution of species i properties to Q_{im} (4.3.56a); constant

$\overline{G_c}$; $\overline{G_{Dr}}$; $\overline{G}_{D\mu}$	convective hindrance factor (3.1.113); drag factor reducing solute diffusion by hindrance (3.1.112e); function of particle volume fraction in hindered settling (4.2.61)
$\overline{G_i}$;$\overline{G_{ij}}$	partial molar Gibbs free energy of species *i*, ratio of solute *i* velocity to the averaged pore solvent velocity, convective hindrance factor (=$\overline{G_c}$) (3.1.113), (3.4.89b); value of $\overline{G_i}$ in region *j*
$G_r(r_p)$	grade efficiency function (2.4.4b)
G_{tj}; G_{crit}	total Gibbs free energy of all molecules in region *j* (3.3.1), (4.2.23); defined by (7.1.58e)
Gr; $Gr(\beta_r, \sigma_v)$; Gz	Grashof number (3.1.143e); function defined by (7.2.174); Graetz number (8.1.276)
h; h_o	membrane flow channel height, distance between particle and collector (3.1.17), constant in (4.1.9a) for Henry's law constant, $(1/h)$ is a characteristic thickness of double layer (6.3.31a), height of liquid in a capillary at any time *t*; value of *h* as $t \rightarrow \infty$ (6.1.11)
h_+, h_-, h_G	contributions of different species to *h* (4.1.9c)
h_{min}	minimum value of *h*
H; H_f; H_i; H_i^C, $[H_i^C]'$; H_i^P; H_{lf}, H_{vf}	plate height, stack height; molar enthalpy of feed; value of *H* for species *i* (6.3.22), Henry's law constants for species *i* in gas–liquid equilibrium; (3.3.59) (4.1.7); (3.4.1b), (5.2.6); (5.2.7); (8.1.49), (3.4.1a); molar enthalpy of liquid fraction and vapor fraction, respectively, of the feed
$H_{o\ell}$, $H_{o\ell P}$	height of a transfer unit defined by (8.1.96)
$\mathbf{H}^m$	magnetic field strength vector
H_A, H_A^o, H_A^c; H_1, H_2	Henry's law constants for species A (3.4.1a,b), (3.4.8); defined by (7.1.20b)
H_D, H_M, H_S, H_{SM}	components of plate height (7.1.107e–i)
$\overline{H}_i$	partial molar enthalpy of species *i*
HTU	height of a transfer unit (6.4.85), (8.1.54b), (8.1.57b), (8.1.65e), (8.1.245b), (8.1.247a), (8.1.357b)
i	unit vector in positive *x*-direction, current density (3.1.108c)
I; $\mathbf{I}$; I_j; $I(C^+_{isbL})$	ionic strength of the solution (3.1.10c), (4.1.9b); purity index (1.4.3b), current; value of purity index for region *j* (1.4.3b); integral (7.2.86)
j	unit vector in positive *y*-direction
$\boldsymbol{j}_i$; j_{ix}, j_{iy}, j_{iz}	mass flux vector of species *i*, $M_i\boldsymbol{J}_i$ (3.1.98), Tables 3.1.3A, 3.1.3B, (6.2.5n); components of $\boldsymbol{j}_i$ in *x*-, *y*- and *z*-directions
J_D	factor defined by (3.1.143g)
$\boldsymbol{J}_i$, J_i^*; $(J_i)_k$; J_1, J_1^T	molar flux vector of species *i* (3.1.98), (3.1.99), Tables 3.1.3A, 3.1.3B; value of $\boldsymbol{J}_i$ in region *k*; diffusive molar flux vector of species 1 (4.2.63); temperature gradient driven molar flux vector of species 1 (4.2.62)
J_{iz}, J_{sz}; J^*_{iy}	*z*-components of flux vectors $\boldsymbol{J}_i$ and $\boldsymbol{J}_s$; *y*-component of flux vector J_i^*
J^*_{iz}, J^*_{jz}, J^*_{sz}, J^*_{Az}	*z*-components of flux vectors J_i^*, J_j^*, J_s^* and J_A^*
J^{*t}_{Ay}	total molar flux of species A in *y*-direction (5.4.51)
J_{vz}	volume flux through membrane in *z*-direction (3.4.60c), (6.3.155a)
k; k	unit vector in positive *z*-direction; region or phase, constant in (2.2.8a–c), $(2\pi/\lambda)$ (3.1.48)
k^B	Boltzmann's constant (3.1.72), (3.3.90c)
k_a, k_d	rate constants for adsorption and dissociation, respectively (4.1.77a)
k_b, k_f	backward and forward reaction rate constants (5.4.42)
k_{Ao},k_{Aw}	mass-transfer coefficient of species A in organic or water phase
k_c, k_g, k_{xj}, k_y	mass-transfer coefficients for species *i* (3.1.139), (3.4.3)
k'_c, k'_g, k'_x, k'_y	values of k_c, k_g, k_x and k_y for equimolar counterdiffusion (3.1.124)
k_d, k_s; k_ℓ,	mass-transfer coefficients in crystallization (3.4.23a,b); liquid film mass transfer coefficient (5.3.3)
k_{igc}, k_{igx}, k_{igy}	mass-transfer coefficients for species *i* in gas phase when the concentration gradient is expressed in terms of *C*, molar concentration of species *i* in gas phase, *x*, mole fraction of species *i* in gas phase and similarly *y*, mole fraction in gas phase, respectively
k_T; k'_T	thermal diffusion ratio (3.1.45); thermal diffusion constant (4.2.64)
k_{gf}, k_{gs}	gas film mass-transfer coefficient on feed side and strip side of a liquid membrane (5.4.97a), (5.4.99a)
$k_{\ell f}$, $k_{\ell s}$; k_{pp}	liquid film mass-transfer coefficient on feed side and strip side of a liquid membrane (5.4.97b), (5.4.99b); particle mass-transfer coefficient (7.2.217b)

k_{s^+}, k_{s^-}	rate constants for forward and backward interfacial reactions (5.3.40)
$k'_{il}; k'_{i1}$	distribution ratio of species i between regions 1 and 2, also called capacity factor (1.4.1); distribution ratio defined by (2.2.19) for species i between streams 1 and 2
k_{cR}, k_{cE}	mass-transfer coefficient in the continuous phase, raffinate based, extract based (6.4.97a,b)
k_{gr}, k_{nu}	rate constant for crystal growth and nucleation, respectively (6.4.51)
$k_{il}; k_{il3}, k_{il4}, k_{ilo}; k_{imo}$	mass-transfer coefficients for species i in liquid phase; value of k_{il} for condition 3, condition 4, channel inlet; species i mass-transfer coefficient through organic filled membrane pore
k_1	first-order reaction rate constant (5.3.7)
k_{1m}	membrane mass-transfer coefficient for species 1 (4.3.1)
K	equilibrium constant for a chemical reaction (3.3.68), or an ion exchange process (3.3.121i), a constant (6.3.49)
$K^x; K^*$	mole fraction based K for a chemical reaction (5.2.35); defined by (5.4.100)
K_1, K_2, K_3	constants in membrane transport (6.3.155a,b)
$K_{Ao}, K_{Aw}; K_{AB}$	overall mass-transfer coefficient of species A based on organic or water phase; equilibrium constant (7.1.42c)
K_c, K_g, K_x, K_y	overall mass-transfer coefficients (3.4.5), (3.4.6)
K_{cE}	K_c based on extract phase (6.4.80)
K_C^A	equilibrium constant for ion exchange reaction (5.2.122)
K_d	equilibrium constant for protein–ion exchange resin binding (4.1.77c), ionization equilibrium constant (5.2.4)
K_{d1}, K_{d2}	dissociation constant for solutes 1 and 2, respectively (5.2.61a), (6.3.29)
$K_i; K_1, K_2, K_3; K_i^a; K_i^\infty$	equilibrium ratio of species i between regions 1 and 2 (1.4.1) or (3.3.61); value of K_i for species 1, 2 and 3; values of K_i in terms of activities (4.1.3); $f_{il}^o \gamma_{il}^\infty / P$ in dilute solution stripping (4.1.19b)
$K_{i\ell}; K_\ell$	overall liquid-phase mass-transfer coefficient for species i (7.1.5a); reaction equilibrium constant in the liquid phase based on molar concentrations (5.2.52a)

$K_{ijc}; k_{ijx}; k'_{ijx}$	molar concentration based overall mass-transfer coefficient for phase j (8.1.1c); j phase mass-transfer coefficient (8.1.60); value of k_{ijx} for equimolar counterdiffusion (8.1.62a)
K_o, K_w	overall mass-transfer coefficient based on organic or aqueous phase, ionization product for water (5.4.41c)
K_{is}, K_{ps}	values of K for ion i/protein (p)–salt (s) exchange on an ion exchange resin (7.1.109d), (3.3.122b)
K_{xE}, K_{xR}	overall mass-transfer coefficient K_x (3.4.5) based on extract phase and raffinate phase, respectively (6.4.77), (6.4.81)
$l; \ell$	length of a device, length of molten zone in zone melting (6.3.109b), characteristic dimension of the separator; constant in (2.2.8a–c), length
ℓ_{ik}, ℓ_{ki}	phenomenological coefficients (3.1.203)
ℓ_{loc}	characteristic length of a local volume corresponding to a point in volume averaging Section 6.2.1.1
ℓ_x, ℓ_y, ℓ_z	dimensions of a rectangular separator, Figure 3.2.1
$L; L_f; L^+$	length of a separator, dimension of length, characteristic crystal size; molar feed flow rate; nondimensional L (7.2.38)
L_{ii}, L_{is}, L_{ss}	phenomenological coefficients for binary system (i, s) (3.1.208), (3.1.209)
L_{ik}, L_{ki}, L_{iT}	phenomenological coefficients (3.1.205)
$L_p; L_p^a; L_p^b$	hydraulic transport parameter in Kedem–Katchalsky model (6.3.158a); value of L_p in perfect region; value of L_p in leaky region
$L_T; L_{min}; L_{MTZ}; LUB$	separator length (= L); (7.1.60); Figure 7.1.5(b); (7.1.21g)
$m; m_B; m_i; m_1; m_i^0$	velocity profile constant (7.3.134); moles of B; moles of species i in separator; moles of species 1 in separator; total number of moles of species i in separator
$m_{ij}; m_{ij}^o; m_{ij}(n); m_{ij}^a$	moles of species i in region j, total number of moles of i in region j at $t = 0$; moles of i in region j after nth contact; number of atoms of ith isotope of the element in region j
$m_{i\sigma}$	moles of species i in interfacial region σ

$m_{i,j}$	molality, moles of i per kilogram of solvent in region j: $j = R$, resin; $j = w$, aqueous phase
$m_{F,R}$	molality of fixed charges in the resin phase
m_p; m_{sl}	mass of particle; solvent moles in stationary liquid phase (7.1.104b)
m_{tj}	total moles of all species in region j ($j = f$, feed; j = 1, vapor phase; j = 2, ℓ, liquid phase)
$m_{11}(\mathrm{t})$, $m_{21}(\mathrm{t})$	moles of species 1 and 2 in region 1 at time t
$m^r_{11}(t), m^r_{21}(t)$	values of $m_{11}(t)$ and $m_{21}(t)$ in the case of a chemical reaction
$mo_{\mathrm{p}}^{\mathrm{mag}}$	magnetophoretic mobility (7.3.251)
M; M_i, M_s; M_{sl}	molecular weight, a metal species, number of stages in stripping section of a cascade; value of M for species i, for solvent s; M for coating liquid in stationary phase
M_t	average molecular weight of solution (3.1.56)
M_w	magnetization of wire (3.1.23)
M_{se}	seed mass density per unit liquid volume (6.4.40a)
M_T; M_{Ta}, M_{Tb}	suspension density of a crystal-containing solution (2.4.2f), (6.4.18); value of M_T for cases a and b
$Mo^{(n)}$; $Mo_f^{(i)}$, $Mo_\ell^{(i)}, Mo_v^{(i)}$	nth moment of the density function (2.4.1g); ith moment of molecular weight density function of feed, liquid and vapor, respectively (6.3.70)
Mo^j_{fr}, Mo^j_{gr}	jth moment of crystal size density functions $f(r_p)$ and $f_g(gr)$
MWCO	molecular weight cut off of a membrane
n; n_c	number of species/components in a system, number of contacts, stage/plate number, number of positive charges in a metal ion, number of unit bed elements; number of collectors, number of channels (7.3.109)
$n(r_p); n^o; \tilde{n}(r_p)$	population density function, particle number density function (2.4.2a); nucleation population density parameter (6.4.7); defined by (6.4.46b)
$\boldsymbol{n}_i$; n_{ix}, n_{iy}, n_{iz}; $\boldsymbol{n}_{ij}$	mass flux vector of species i, $M_i\boldsymbol{N}_i$; its components in x-, y- and z-directions; $\boldsymbol{n}_i$ through surface area $\mathbf{S}_j$
$\mathbf{n}_k$	outwardly directed unit normal to the k-phase surface (6.2.26b–d)
$n_{\max}$; n_{med}, n_{par}	peak capacity (3.2.32) (6.3.26a); refractive index of medium and particle
$\boldsymbol{n}_p$, n_p; n_{py}; n_t	particle number flux, (3.1.65), (3.1.66), (3.1.68); particle flux in y-direction; number of turns by gas in a cyclone (7.3.146b)
N; $N(r_p)$; $N(r_{\min}, r_p)$, $\mathrm{N}(r_{p_{\max}})$	total number of stages in a multistage device or in the enriching section of a cascade, anionic species in Donnan dialysis, a metal species, number concentration of molecules; numbers/cm[3], number of particles per unit fluid volume in the size range of $r_{\min}$ to r_p; value of $N(r_{\min}, r_p)$ for $r_{p_{\max}}$
$\tilde{\mathrm{N}}; N_i$	Avogadro's number (6.02×10^{23} molecules/gmol); plate number for i (6.3.27a)
$N_{\mathrm{A}}; N^r_{\mathrm{A}}, N^{\mathrm{t}}_{\mathrm{A}y}$; N_{dil}	molar flux of species A in a fixed reference frame without and with reaction; total molar flux of species A in facilitated transport or counter-transport or co-transport in y-direction; normality of diluate solution (8.1.404)
$\boldsymbol{N}_i$; $\boldsymbol{N}_{im}$, $\boldsymbol{N}_{jm}$; N_{ix}, N_{iy}, N_{iz}; N_{ir}; $\boldsymbol{N}_{ij}$; $\lvert N_{iy}\rvert$	species i flux; $\boldsymbol{N}_i$ through membrane, $\boldsymbol{N}_j$ through membrane; components of $\boldsymbol{N}_i$ in x-, y- and z-directions; radial component of $\boldsymbol{N}_i$; $\boldsymbol{N}_i$ through surface area $\boldsymbol{S}_j$; magnitude of N_{iy}
N^p_{iz}	z-component of species i flux, N_{iz}, based on unit pore cross-sectional area (3.1.112a)
N_{oj}; $N_{o\ell P}$; N_p	number of transfer units (8.1.92), (8.1.96); defined in (8.1.96); number of pores per unit area size r_p (6.3.135d)
N_R; N_{S}	flux ratio (3.1.129a); solvent flux
$N_i(r_p)$; N_{it}	number density of crystals having a size less than r_p and growth rate gr_i; total number of crystals per unit volume having a growth rate of gr_i (6.4.41c)
N_t	total number of particles per unit volume (6.4.10)
N_{toE}, N_{toR}	number of transfer units based on extract and raffinate phases, respectively (6.4.86a), (6.4.83)
No_i	number of particles of size r_{pi} (2.4.2k)
NTU	number of transfer units (8.1.54c), (8.1.57c), (8.1.66b), (8.1.67d), (8.1.338)
$\boldsymbol{p}$	dipole moment of a dielectric particle
p	stoichiometric coefficient for product P, kinetic order in the dependence of nucleation rate (6.4.30a)

$p_A; p_B; p_i, p_j$	partial pressures of species A, species B; species i and species j
$p_{Ab}; p_{Ai}; p_{B,\ell m}$	value of p_A in the gas bulk; value of p_A at gas–liquid interface; logarithmic mean of p_B (3.1.131b)
$p_{if}, p_{ij}, p_{ip}, p_{iv}$	value of p_i in feed gas, region/stream j, permeate gas, vapor phase
$\overline{p}_{ib}, \overline{p}_{ii}$	species i partial pressure in bulk and particle interface, averaged over bed cross section, Figures 3.4.4(a), (b)
$pH; pI$	indicator of hydrogen ion concentration (5.2.65a); isoelectric point for a protein/amino acid; at $pI = pH$, net charge is zero
pK_i	$-\log_{10} K_{di}$, (5.2.65b) for $i = 1$, (5.2.74b) for $i = 2$, (6.3.29)
$P; P^o; P_c$	total pressure, system pressure; standard state pressure; critical pressure
$\overline{P}; P_f; P_j, P_p; P_p'; P_\ell$	local solute permeability coefficient (6.3.157b); feed pressure; total pressure of jth region and permeate, respectively; gas pressure (Figure 7.2.1(b)); gas pressure at the end of a capillary of length l (6.1.5d)
$P_i^0(\pi); P_i^*$	equilibrium gas-phase pressure for pure i adsorption at spreading pressure π, which is the same for a mixture (3.3.111a); pressure at crossover point for solute i in supercritical extraction
$P_{atm}; P_{liq}; P_1, P_2$	atmospheric pressure; pressure in the liquid (6.1.12); purification factors (7.2.97)
$Pe; Pe_i; Pe_i^m; Pe_{z,\text{eff}}; Pe_{zj}$	Péclet number (3.1.143g), (7.3.34d); Pe number for dispersion of solute i (6.3.23a); pore Péclet number (6.3.145a); $(z\, v_z/D_{i,\text{eff},z})$ (7.1.18h); j phase Pe_z (8.1.92)
$P_i^{sat}, P_j^{sat}; P_{iP\ell}^{sat}; P_{i,\text{curved}}^{sat}$	vapor pressure of pure i and pure j, respectively, at system temperature; value of P_i^{sat} on a plane surface; value of P_i^{sat} on a curved surface
P_M^{sat}	value of P_i^{sat} for pure species i of molecular weight M
P_0	amplitude of pressure wave (3.1.48)
P_R, P_w	pressure of resin phase and external aqueous solution, respectively
Po, Pr	power number (6.4.976), power (3.1.47)
q	number of variables in a problem, stoichiometric coefficient for species Q, heat flux in a heat exchanger attached to a cooling crystallizer (6.4.47a), the power of M_T in expression (6.4.39a) for B^0, factor (8.1.150)
q_{fr}	fraction of light reflected (3.1.47)
$q_i(C_{i2}), q_{i1}(C_{i2}), \overline{q}_{i1}; q_{i1}^0, q_{i1}^s$	moles of species i in solid phase 1 per unit mass of solid phase, cross-sectional average of q_{i1}, initial value of q_{i1}; saturation value of q_{i1}
q_{iR}, q_{is}	moles of species i per unit mass of ion exchange resin (R) or solid adsorbent (s)
$q_{\max R}$	maximum molar fixed charge density per unit resin mass
$q(\boldsymbol{r}, \psi, \lambda)$	probability that a molecule having configuration $(\boldsymbol{r}, \psi, \lambda)$ does not intersect pore wall (3.3.89f)
Q	volumetric fluid flow rate, product species in reaction (5.3.5), hydraulic permeability in Darcy's law (6.1.4g,h), heat transfer rate (6.4.47a,d)
$Q_c; Q_d$	electrical charge of a collector; volume flow rate of dialysate solution
$Q_f; Q_g; Q_h; Q_i; Q_\ell, Q_o, Q_p, Q_R, Q_1, Q_2$	volumetric feed flow rate to separator; Darcy permeability for gas through packed bed; amount of heat supplied at a high temperature; electrical charge on 1 gmol of charged species i; amount of heat rejected at a low temperature; volumetric flow rate at membrane channel inlet; electrical charge of a particle; excess particle flux (7.2.123), volumetric flow rate of product stream from separator; heat supplied at the reboiler per mole of feed; volume flow rate of overflow; volumetric rate of underflow/concentrate
$Q_{Am}, Q_{Bm}, Q_{im}, Q_{jm}$	permeabilities of species A, B, i and j, respectively, through membrane in gas permeation and pervaporation, respectively
Q_{ij}	permeability of species i through region j (= A, B, C, D, 1, 2) of the membrane
Q_{im}^{ov}	overall permeability of species in membrane pervaporation for a membrane of thickness δ_m
Q_{crys}, Q_{sub}	heat transfer rate of a solution during crystallization and subcooling, respectively
Q_{im0}	value of Q_{im} for $C_{im} = 0$
Q_{sm}, Q_{sc}	solvent permeability through membrane and cake in cake filtration
$\mathbf{r}$	vector of molecular position, radius vector, unit vector in radially outward direction

r	radial coordinate, any characterizing property of a continuous mixture
$r_1, r_2; r_{1i}, r_{2o}; r_c; r_f$	radii of curvature of interface (3.3.47); liquid outlet radii in tubular centrifuge; critical size of a nucleus (3.3.100b), cyclone radius; free surface radius in tubular centrifuge
$r_g; r_h; r_i; r_{in}$	radius of gyration of a macromolecule; hydrodynamic viscosity based radius (3.3.90f); radius of spherical solute molecule (3.3.90a); radius of liquid–liquid interface in tubular centrifuge
r_o	radius of a sphere whose volume is equal to that of an ellipsoid (3.1.91g), radial location of the center of solute peak profile, radius of a cylindrical centrifuge
$r_p; r_{p,a}; r_{p1}, r_{p2}; r_{p_i}; r_s; r_t; r_w$	radius of particle, pore radius; analytical cut size (2.4.18); particles of two different sizes; dimension of particle of a certain size; particle size for classification (2.4.8); cyclone exit pipe radius; radius of wire
$r_{max}, r_{min}; r_{p,50}$	maximum and minimum radius of membrane pores or particle sizes; equiprobable size
$\sqrt{\overline{r_p^2}}$	hydraulic mean pore radius (3.4.87)
$\bar{r}_p, \bar{r}_{p_{1,0}}, \bar{r}_{p_{i+1,i}}$	mean of particle size distribution based on $f(r_p)$ and $n(r_p)$, respectively (2.4.1e), (2.4.2g), (2.4.2h)
$\bar{r}_{p_{3,2}}$	Sauter mean radius of a drop, bubble or particle (6.4.89)
$ra; re; re\|_{flow}$	rate of arrival of cells; fractional water recovery; value of re in a cell (6.3.172b)
R; R	universal gas constant; radius of a tube or capillary, reflux ratio (2.3.5), (2.3.7), (8.1.137), solute rejection in reverse osmosis, solute retention in ultrafiltration
$R_c; \hat{R}_{c\delta}, \hat{R}_{cw}$	cake resistance; specific cake resistance per unit cake thickness and unit cake mass, respectively (6.3.135l)
$R_h; R_i; R'_i; R_L$	hydraulic radius (6.1.4c); solute rejection/retention by membrane for species i, retention ratio for species i (7.3.211); fraction of solute i in mobile phase (7.1.16c); largest radius of a conical tube
$R_m; R_s; R_\infty$	membrane resistance; resolution between neighboring peaks (2.5.7) in chromatography; value of R_i at large Pe_i^m
R_A, R_i	molar rate of production of A per unit volume (5.3.7), (6.2.2d), for species i
R_1, R_2	intrinsic RO rejection of the unionized species 1 and ionized species 2
R^+ (R^-)	ion exchange resin with fixed positive (negative) charge; cation (anion) of a surface active solute
$R_{ij}; R_{im}; R_{i,reqd}$	permeation resistance of region j (j = A, B, C, D) in the membrane to species i; membrane permeation resistance of species i; solute i rejection required in a RO membrane
$R_{ik}, R_{ki}; R_{min}$	phenomenological coefficients (3.1.207); minimum reflux ratio (8.1.170)
$R_{obs}, R_{true}; Re; Re_L; Re_{imp}$	observed and true solute rejection in ultrafiltration and RO; Reynolds number; Re for a plate of length L (3.1.143a), Table 3.1.5; Re for an impeller (Table 3.1.7), (6.4.96)
s	fractional supersaturation (3.3.98b), power of the ΔP dependence of $\hat{R}_{c\delta}$ (6.3.138j), Laplace transform complex variable (5.4.35), solution volume fed to bed per unit empty bed cross section (7.1.17a), eluent/salt (7.1.109b), salinity (10.1.13)
$s_m; s_{mp}, s_{ms}; s_p$	pore surface area/membrane volume (6.3.135e); value of s_m for membrane pore volume or membrane solids volume (6.3.135f,g); sedimentation coefficient (4.2.16b)
s_1, s_2	two different solvents
$S; S_c$	solute transmission/sieving coefficient (6.3.141e), supersaturation ratio (3.3.98c), stripping factor (8.1.135), (8.1.189b); bed/column cross-sectional area
$\mathbf{S}_f, \mathbf{S}_k, \mathbf{S}_M; S_j; S_M, S_{d_p}$	cross-sectional area vector of feed entrance, kth feed exit and membrane surface area vector in a separator; cross-sectional area of flow for jth stream; selectivities (7.3.219a,b)
$S_{ij}; \overline{S_{ij}}$	molar entropy of species i; partial molar entropy of species i (3.3.17b)
$S_{im}; S_{im0}$	solubility coefficient of species i in membrane; value of S_{im} for $C_{im} = 0$
S_∞	value of S when $Pe_i^m >> 1$ (6.3.145b)
$S_\sigma; \overline{S_\sigma}; S_1, S_2$	surface area of interfacial region (Figure 3.3.2A); molar surface area in gas–solid adsorption (3.3.107); sieving coefficients for species 1 and 2
$S_{tj}; S_{mag}$	total entropy for region j (3.3.3); magnetic field force strength (7.3.251)

S_{obs}; S_{true}	observed solute transmission/sieving coefficient; true value of S
Sc; Sc_c	Schmidt number (3.1.143a); Sc for continuous phase
Sh; Sh_c, Sh_D; Sh_p; Sh_z	Sherwood number (3.1.143a); Sh for continuous phase and dispersed phase, respectively; Sh in a packed bed (7.2.218a); Sh at location z (7.2.64)
St	Stanton number (3.1.143g), Stokes' number (6.3.41)
t; t_{br}; t_c; t_{res}; t_σ	time; breakthrough time; time for cut point in the chromatographic separator output (2.5.1); residence time; thickness of interfacial region (Figure 3.3.2A)
t_i^+; t_1^+, t_2^+; $\bar{t}$; $\bar{t}_1$, $\bar{t}_2$, $\bar{t}_3$	nondimensional time variable for species i (6.3.12); value for t_i^+ for species 1, 2 (3.2.9), (3.2.20); breakthrough time (7.1.15c); value of $\bar{t}$ for species 1, 2 and 3
t_i^{in}, t_i^0	times when species i appears and disappears, respectively, from a chromatographic separator output
t_{im}, t_{is}	transport number of i in membrane or solution (3.1.108d)
t_{s_1}, t_{s_2}	time required for solvents, s_1, s_2 (3.2.24)
t_{R_i}; t_{R_M}; t_{R_o}	retention time for species i in capillary electrophoresis (6.3.18a), (7.1.99d); retention time for the mobile phase; retention time based on $v_{z,avg}$ (7.3.207)
T; T_1; T_2; T_c	absolute temperature; temperature of cooled plate; temperature of the heated plate; critical temperature
T_f; T_g; T_i; T_p	feed temperature; glass transition temperature of a glassy polymer; value of absolute temperature T of region i; product temperature
T_C; T_H, T_L; T_R	temperature of condenser; two temperatures in supercritical extraction; reboiler temperature
T_{cf}, T_{ci}; T_{mi}	temperature of cooling fluid, critical temperature of species i; melting temperature for species i
T_{sat}	temperature at which the solution is saturated
T_{sol}	temperature of the solution due to undergo crystallization
Th	dimensionless group (3.1.46b)
u	number of fundamental dimensions (Section 3.1.4.1)
u_c	mass fraction of solute in the crystallized solution on a solid-free basis
u_i; u_k; u_{ij}; u_{il}; u_{io}; u_{is}; u_{ilb}	mass fraction of impurity species i; mass fraction of species k; mass fraction of species i in region j (1.3.5) or jth stream (2.1.20); value of u_i in melt; value of u_i initially in the solid; value of u_i in the recrystallized solid; value of u_{il} in the bulk melt
u_o; u_{rc}; u_{ro}	mass fraction of solute in the solution charged for crystallization; $u_c/(1-u_c)$; $u_o/(1-u_o)$ (6.4.47b)
u_i^m	ionic mobility (3.1.108j)
u_{iEn}, u_{iRn}	mass fraction of species i in extract and raffinate streams, respectively, from stage n
ur_{ig}, ur_{is}	defined by (8.1.349) and (8.1.350)
ur_{ij}	weight of solute i per unit weight of phase j (9.1.32)
$\boldsymbol{U}_i$; $\boldsymbol{U}_p$	migration velocity vector for species i (3.1.84b); particle velocity vector
$\overline{U}_i$	averaged velocity vector of ith molecules due to all forces
$\overline{U}_i^{new}$	defined by (3.1.103)
U_{ix}, U_{iy}, U_{iz}	components of migration velocity of species i in x-, y- and z-directions, (3.1.82)
U_{ik}; $\langle U_{ik}\rangle^k$; $\hat{U}_{ik}$	value of $\boldsymbol{U}_i$ in the kth phase/region; average of $\boldsymbol{U}_{ik}$ in the kth phase/region; defined as a fluctuation by (6.2.28)
U_p^{int}; $U_{px^i}^{int}$; $U_{pr_p}^{int}$	internal particle velocity vector (6.2.50d); its component in the direction of the internal coordinate x^i; internal particle velocity vector for all particles of size r_p
$\boldsymbol{U}_{pt}$; U_{pr}; U_{prt}; U_{pzt}, U_{pyt}	terminal velocity vector of particle (3.1.62); radial particle velocity; terminal value of U_{pr}; value of $\boldsymbol{U}_{pt}$ in z-direction (6.3.1), (7.2.211), y-direction (7.3.154)
$\boldsymbol{v}$, $\boldsymbol{v}_t$; $\boldsymbol{v}^*$, $\boldsymbol{v}_t^*$; $\boldsymbol{v}^+$, $\boldsymbol{v}_t^{ref}$; $\boldsymbol{v}_{tj}$, $\boldsymbol{v}_{tj}^*$	mass averaged velocity vector of a fluid (also $\boldsymbol{v}_t$); molar average velocity vector of a fluid (also $\boldsymbol{v}_t^*$); nondimensional $\boldsymbol{v}$ (6.3.39); reference $\boldsymbol{v}_t$; values of $\boldsymbol{v}_t$ and $\boldsymbol{v}_t^*$ on surface area $\boldsymbol{S}_j$ (2.1.1), (2.1.2)
$\boldsymbol{v}_i$; $\boldsymbol{v}_{ij}$	averaged velocity vector of ith species; value of $\boldsymbol{v}_i$ on surface area $\boldsymbol{S}_j$ (2.1.1)
v_k; v_{kz}; v_k^{int}	velocity of region k; z-component of velocity $\boldsymbol{v}_k$ of region k; mass average velocity of the interface of two phases
$\boldsymbol{v}_p^r$	particle diffusion velocity relative to that of the fluid phase (3.1.43)

v_{ci}^*; v_{Ci}^*; v_o; v^o; v_{Ci}^{*+}; v_{og}	average velocity of liquid zone carrying species i (7.3.208); concentration wave velocity of species i (7.1.12a); velocity of micelles, superficial velocity (3.1.176), in a packed bed (6.1.4a); amplitude of square-wave velocity (7.1.72a); nondimensional concentration wave velocity (7.1.105o); superficial velocity for gas phase
v_r; v_s; v_{so}; v_{tw}; v_{tz}; v_{vz}	radial velocity; shell-side velocity (3.1.175), volume flux of solvent/permeate (7.2.71); value of v_s at membrane channel inlet; tangential fluid velocity in wall region (7.3.134); same as v_z; vapor velocity in z-direction (6.3.47a)
v_x, v_y; v_{yw}; v_{ywf}	local convective velocity in x-direction, y-direction; value of v_y at wall; value of v_{yw} at feed entrance (7.2.39)
v_z; $v_{z\max}$; $v_{z,\mathrm{avg}}$; $v_{z,\mathrm{avg},f}$	velocity of fluid in z-direction; maximum value of v_z, averaged value of v_z over flow cross section; value of $v_{z,\mathrm{avg}}$ at feed location (6.1.5f)
v_∞	uniform gas velocity far away from object
v_{fr}; $v_{i,\mathrm{eff}}^+$	velocity of freezing interface in zone melting (6.3.110c); nondimensional velocity of species i in capillary electrophoresis (6.3.12)
v_z^{AA}; v_{zH}, v_{zL}	velocity of interface AA between suspension and clarified liquid in z-direction (4.2.52); interstitial gas velocity at high-pressure feed step and low-pressure purge step, respectively (7.1.53a,b)
v_{EOF}; $v_{\mathrm{EOF},z}$	electroosmotic velocity (6.1.22); value in z-direction
vr	defined by (7.2.167)
V; V_b; V_d; V_f; V_h; $\hat{V}_a$; $\dot{V}$; $\overline{V}$	volume of a region, volume of separator, volume of feed solution/eluent passed, volume of a sphere (3.3.52a), voltage between electrodes; volume of buffer; volume of the dialysate solution; volume of feed solution; hydrodynamic volume of a macromolecule/protein (3.3.90f); adsorbed monolayer phase volume/weight of adsorbent (3.3.114b); radial volume flow rate of solution (Figure 7.1.6); defined by (7.1.18i)
$\tilde{V}_c$; V_i; $\overline{V_i}$; V_j	critical molar volume of a species; molar volume of pure species i; partial molar volume of species i; volume of region j: j = 1, stationary phase; j = 2, mobile phase
V_k; V_l; V_ℓ^p; V_o; V_o^t; V^o	volume of region k; liquid volume; specific pore volume of a microporous adsorbent (4.1.64a); volume of organic solvent; total amount of V_o (6.3.99); sample volume (7.1.101b)
V_M; V_p; V_s; $V_{S\ell}$; V_{sp}	mobile-phase volume in a column (7.1.99g); volume of a particle, cumulative permeate volume (7.2.87); stationary adsorbent-phase volume in a column; suspension volume; stationary liquid-phase volume; mobile-phase volume present inside the pores of a particle (7.1.110c)
$\overline{V}_s$; $\overline{V}_t$; V_w; V_σ	partial molar volume of solvent; averaged partial molar volume (3.1.56); volume of water; volume of interfacial region, Figure 3.3.2A
V_{fi}; V_{fe}	initial volume of feed solution; final volume of solution
V_{f0}; V_{fR}; V_{N_i}; V_{R_i}	volume of feed solution at time t = 0; volume of retentate; net retention volume for species i (7.1.99j), retention volume for species i (7.1.99e,f), (7.1.99h)
V_{cap}^{iex}; V_{eff}; V_{channel}	volumetric ion exchange capacity of packed bed; liquid filled centrifuge volume; volume of channel
$\overline{V}_{\mathrm{AY}}$	partial molar volume of electrolyte AY (3.3.119b)
$\overline{V}_{ic}$	value of $\overline{V}_i$ for a crystal of i
V_{im}	molar volume of species i at its normal boiling point
$\overline{V}_{ij}$	value of $\overline{V}_i$ in region j
VCR	volume concentration ratio (V_{f0}/V_{fR}) (6.4.98)
w; w_{crys}; w_i; w_{so}	mass of adsorbent, quantity defined by (5.2.54); mass of crystals formed at any time, mass of solute i charged; mass of solution charged to the crystallizer
w_{ij}; w_{tj}; w_{tz}; w_{tj}^s; w_{tE}, w_{ER}	mass flow rate of species i through $\boldsymbol{S}_j$ (2.1.4); total mass flow rate through $\boldsymbol{S}_j$ (2.1.5); total mass flow rate in z-direction; total solids flow rate in jth stream (2.4.3a); w_{tj} for j = extract, E, w_{tj} for j = raffinate, R
w_E, w_M, w_R	weight of mixture at E, M and R, respectively
w^l; w^s; $w_p^s(r_p)$; wa	mass of stationary liquid phase used as coating; mass of particles per unit of fluid volume (2.4.2e); mass of a solid particle of size r_p (2.4.2e); waist size of a light beam

W, W_A	width of rectangular channel walls, molar transfer rate of species A
$\underline{W}$	total reversible work done (3.1.25a,b)
W_{bi}, W_{ij}, W_{tj}	band width of the chromatographic output of species i (Figure 2.5.2), molar flow rate of species i through $\boldsymbol{S}_j$ (2.1.7), total molar flow rate through $\boldsymbol{S}_j$ (2.1.8)
W_{tAm}, W_{tBm}	molar rate of permeation of species A and B through membrane
W_{tf}, W_{tl}, W_{tv}	molar flow rate of feed, liquid fraction and vapor fraction, respectively
W_{tlb}, W_{tld}	total molar flow rate of bottoms product and distillate product from a distillation column
W_{t1}, W_{t2}, W_{tp}	total molar flow rate of streams $j = 1$, $j = 2$ and j = permeate
W^t_{tjn}	total molar flow rate of stream j for stage n in enriching section of cascade
We	Weber number, $(\rho_c v_c^2 d_p/\gamma)$ (6.4.91)
x, x_A	coordinate direction, mole fraction of species A in liquid phase
x^*_A	hypothetical liquid phase mole fraction of species A in equilibrium with y_{Ab}
x_{Ab}, x_{Ai}	bulk liquid mole fraction of species A, value of x_A at a two-phase interface
x_i, x_{ij}, $x_{i\sigma}$, $x_{i,j}$	mole fraction of species i, value of x_i (i = A,B, 1, 2, s(solvent), etc.) in region j or stream j (j =1, 2, f(feed), g = gas, l = liquid, p = permeate, R = resin, s = solution, v = vapor, w = water), x_{ij} where $j=\sigma$, the surface adsorbed phase, mole fraction of species i in region j when molecular formula is substituted for species i
x_{iE}, x^*_{iE}, x^e_{iE}	mole fraction of species i in extract phase, value of x_{iE} in equilibrium with x_{iR}, value of x_{iE} if both streams leave stage in equilibrium
x_{ilb}, x_{ild}	bulk value of x_{il}, mole fraction of i in liquid product from reboiler in a distillation column, x_{il} in liquid product from condenser at the top of a distillation column
x_{iR}, x^*_{iR}, x^e_{iR}	mole fraction of species i in raffinate phase, value of x_{iR} in equilibrium with x_{iE}, value of x_{iR} if both streams leave stage in equilibrium
x_{jf}, x_{jp}, x'_{ip}	mole fraction of species j in feed and permeate, respectively, mole fraction (Figure 7.2.1(b))
$x_{s,lm}$	defined by (3.1.137)
X	molar density of fixed charges in electrically charged system
X_{ij}, X_{ijn}	mole ratio of species i in region j (1.3.2) or stream j (2.2.2d), X_{ij} for state n
X^a_{ij}	abundance ratio of isotope i in region j (1.3.7)
y, y_f	coordinate direction, normal to gas-liquid, liquid–liquid or membrane-fluid interface, feed gas mole fraction (Figure 7.1.14)
y_i, y_A, y_H, y_L	mole fraction of species i and A in vapor/gas phase, respectively, fluid mole fraction (7.1.62b), gas-phase mole fractions defined by (7.1.54a)
y^*_A, $y^{*'}$, $y_{\ell im}$	hypothetical gas-phase mole fraction of species A in equilibrium with x_{Ab}, nondimensional y-coordinate (5.3.5j), y-coordinate of limiting trajectory (7.3.263a)
y_{Ab}, y_{Ai}	bulk gas mole fraction of species A, value of y_A at a two-phase interface
Y_{ij}, Y_{iR}, Y_2, $Y_{2,f}$	segregation fraction of species i in region j (1.3.8a), segregation fraction of solute i in retentate in a membrane process (6.4.107), defined by (7.2.19) and (7.2.20)
$Y_{11}(t)$, $Y_{21}(t)$	segregation fraction of species 1 and 2 in region 1 at time t
Y^a_{ij}, $\dot{Y}_{ij}$, $\dot{Y}_{rj}$	segregation fraction for an isotopic mixture (1.3.9a), of species i stream j (2.2.12), of particles of size r in stream j (2.4.5)
z, z_{cr} (or z_{crit}), z_H, z_L, z_{AA}, z_{BB}	coordinate direction, critical distance (7.2.133), characteristic locations at pressures P_H and P_L (7.1.55e), vertical coordinates of interfaces AA and BB, respectively
z_{HL}, z_{HS}	height of the liquid level of a dilute suspension, height of sludge layer
z^+, z^+_i	nondimensional z-coordinate (Figure 3.2.2), (6.3.12), value of z^+ for center of mass of concentration profile of species i (3.2.22a,b)
z_α, z_β, $z_o(\overline{C}_{i2})$	z-coordinate locations of regions α and β, respectively, defined by (7.1.17e,f)
Z_i, $Z_{i,\text{eff}}$, Z_p	electrochemical valence of species i, effective charge on ion i due to the diffuse double layer (< Z_i) (6.3.31a), value of Z_i for a protein

Greek letters

α	constant in (3.1.215), separation factor between two species 1 and 2 in capillary electrophoresis (6.3.27e)
$\alpha_f, \alpha_l, \alpha_v$	parameters for feed, liquid fraction, vapor fraction in a continuous chemical mixture described by a Γ distribution function (4.1.33f), (6.3.71)
$\alpha_{ij}; \alpha_{ij}^{hf}, \alpha_{ij}^{ft}, \alpha_{ij}^{ht}; \alpha_{ijn}$	separation factor between two species i (i = 1, A, s (solvent)) and regions j (j = 2, B (species), i), value of α_{ij} between two streams (2.2.3), (2.2.4), (2.2.2a); value of α_{ij} for nth plate
α_{in}	multicomponent separation factor between species i and n (1.6.6)
α'	separation factor of Sandell (1.4.15)
$\alpha_{AB}^*; \alpha_{ij}^{evap}; \alpha_{ij}^{perm*}$	value of α_{AB} under ideal condition of zero pressure ratio (6.3.198), (8.1.426); value of α_{ij} for evaporation only (6.3.180); ideal separation factor in vapor permeation (6.3.180)
β	defined by (2.2.8a,b), exponent in (3.3.89a), indicator of macromolecular shape for a given r_g (3.3.90e), constant in (3.1.215), defined by (7.1.52b), (7.2.58), (7.2.138)
$\beta_f, \beta_\ell, \beta_v$	compressibility of fluid (3.1.48), parameters for feed, liquid fraction and vapor fraction in a continuous chemical mixture (4.1.33f), (6.3.71)
$\beta_i; \beta_p; \beta_A, \beta_B; \bar{\beta}; \bar{\beta}_c$	exponent in (3.3.112c), parameter (7.2.18); compressibility of particle (3.1.48); defined by (7.1.52a); coefficient of volume expansion (6.1.9); a parameter (6.4.129); coefficient of thermal expansion of solution density (7.3.233)
$\gamma; \gamma_{cr}; \gamma_A; \gamma_R; \gamma_{R_2}$	surface tension, interfacial tension, pressure ratio (6.3.197), (6.4.123), (8.1.426), parameter (7.1.18f); critical surface tension for a polymer; Damkohler number $(k_b\delta_m^2/D_A)$ (5.4.100); $(D_A k_1/k_\ell^2)$ (5.3.11); defined by (5.3.35j)
$\gamma_f, \gamma_\ell, \gamma_v$	parameters for feed, liquid and vapor fraction (4.1.33f), (6.3.69)
$\gamma_i; \gamma_{if}; \gamma_{ij}; \gamma_{jf}$	activity coefficient for species i, dimensionless steric correction factor; value of γ_i in feed stream, in phase j; value of γ_j in feed stream (species j)
$\gamma_1, \gamma_2, \gamma_3, \gamma_4, \gamma^{12}; \gamma^{SL}, \gamma^{LG}, \gamma^{SG}$	factors/parameters in plate height (7.1.107f–i); interfacial tension between two bulk phases 1 and 2; interfacial tension between phases S and L, L and G, S and G, respectively (S = solid, L = liquid, G = gaseous)
$\dot{\gamma}; \dot{\gamma}_w$	shear rate (3.1.74), (6.1.31); wall shear rate
$\gamma_{il}^\infty; \gamma_{il,u}^\infty$	infinite dilution activity coefficient for species i in the liquid; activity coefficient of species i in liquid phase on a mass fraction basis at infinite dilution (6.3.83)
Γ_{is}^E	molar excess surface concentration of species i/cm^2 of pore surface area (3.1.117a)
$\Gamma_{i\sigma}; \Gamma_{i\sigma}^E$	surface concentration of species i (3.3.34); algebraic surface excess of species i (3.3.40b), (5.2.145a)
$\delta; \delta(x), \delta(z), \delta(z^+); \delta_c$	falling film thickness; delta function in x-, z- and z^+-directions (3.2.14b); cake layer thickness
$\delta_g; \delta_i; \delta_\ell; \delta_m; \delta_s; \delta'_\ell$	thickness of gas film; solubility parameter of species i, characteristic thickness of concentration profile (7.3.202); thickness of liquid film; membrane thickness, pore length; sorbed surface phase thickness (3.1.118b); thickness of liquid film from phase interface to reaction interface (5.3.22)
$\delta_A, \delta_B, \delta_C$	membrane thickness in regions A, B, C of a membrane (Figure 6.3.35(b))
$\delta_1, \delta_2, \delta_n$	thickness of layer 1, layer 2, layer n in a composite membrane
$\Delta C; \Delta E_s; \Delta G; \Delta G_t; \Delta H_s; \Delta P; \Delta S; \Delta T$	molar supersaturation (3.3.98a); activation energy for crystal growth rate (3.4.29); Gibbs free energy change, e.g. for forming a crystal (3.3.100a); change in Gibbs free energy for all molecules; enthalpy of solution of a gas species in a polymer; pressure difference $(P_f - P_p)$, etc.; selectivity index (7.2.93); extent of supercooling (3.3.98d), temperature difference $(T_2 - T_1)$
ΔH_{crys}	heat of crystallization
$\Delta G^0, \Delta H^0, \Delta S^0$	standard Gibbs free energy change, enthalpy change and entropy change for a reaction
$\Delta d_p; \Delta r_p; \Delta w^s$	particle diameter difference between two consecutive sieves; $(\Delta d_p/2)$ or the net growth in size of all seed crystals (6.4.40i); mass of crystals retained on a sieve of given size

$\Delta x, \Delta y, \Delta z, \Delta z_L, \Delta z_H$	lengths of a small rectangular volume element in three coordinate directions, defined by (7.1.53a,b)
$\Delta\mu_i^0; \Delta\nu_1, \Delta\nu_+, \Delta\nu_-, \Delta\pi$	$\mu_{i1}^0 - \mu_{i2}^0$, number of water molecules, cations and anions, respectively released during binding; osmotic pressure difference $(\pi_f - \pi_p)$
$\Delta\rho_i; \Delta\rho_{i,\text{sat}}; \Delta\phi$	mass concentration change of species i in crystallizer (6.4.24); change in saturation concentration due to ΔT (6.4.51); potential difference between two phases
Δ_R	defined by (8.1.328)
∇^i	del operator in internal coordinates x^i, y^i and z^i
ε	void volume fraction in a packed bed, electrical permittivity of the fluid ($= \varepsilon_0\varepsilon_d$)
$\varepsilon_a; \varepsilon_b$	phase angle; fractional cross-sectional area of membrane which is defective
$\varepsilon_d; \varepsilon_i; \varepsilon_k$	dielectric constant of a fluid, porosity of particle deposit on filter media; Lennard-Jones force constant for species i; volume fraction of phase k in a multiphase system
$\varepsilon_m; \varepsilon_p; \varepsilon_0; \varepsilon_0$	porosity of membrane, pellet, bead or particle; dielectrical constant of a particle; electrical permittivity of vacuum; porosity of spacer in spiral-wound module (3.1.170), porosity of clean filter
$\varepsilon_B; \varepsilon_C$	fractional area of defects in glassy skin region B (Figure 6.3.35(b)); value of fractional membrane cross section in region C
ε_{12}	enrichment factor for species 1 and 2 (1.4.10), (2.2.2e)
ζ	zeta (electrokinetic) potential (3.1.11a)
η	recycle ratio (2.2.22), nondimensional variable (3.2.11), (6.3.14b), (6.3.114), fraction of particles collected (7.3.37), (7.3.38), (7.3.46), intrinsic viscosity
η^+	variable defined by (3.2.17)
$\eta_i; \eta_j; \eta_{ij}; \eta_j'$, $[\eta]; [\eta]_i; \eta_{iF}$	variable (2.5.5), η for species i (3.2.25); impurity ratio for region j (1.4.3a); impurity ratio for ith species in region j (1.6.4); defined by (1.4.4); intrinsic viscosity (3.3.90f); its value for polymer i (7.1.110f); current utilization factor (8.1.406)
$\langle\eta_i\rangle; \langle\eta_i^2\rangle$	first moment of $C_i(\eta_i, t_i^+)$ (3.2.26); defined by (3.2.29)
θ	cut (2.2.10a), stage cut, fraction of adsorbent sites occupied (3.3.112a), contact angle, gas-liquid-solid system (Figure 3.3.16), contact angle between membrane surface and liquid (6.3.140)
θ_i	component cut for species i (2.2.10b), Langmuir isotherm (7.1.36a), nondimensional mobile phase concentration (7.1.18b)
θ_{id}, θ_{kd}	component cut for species i and k, respectively, in the distillation column top product (8.1.224)
$\kappa_{Ao}; \kappa_d; \kappa_i$	distribution coefficient of species A between aqueous and organic phases (3.4.16); reciprocal Debye length (3.1.17); distribution coefficient of species i
$\kappa_{i1}, \kappa_{i1}^N$	distribution coefficient of species i between regions 1 and 2 (1.4.1), its Nernst limit (3.3.78), (3.3.79)
$\kappa_{if}; \kappa_{ij}; \kappa_{im}$	partition coefficient of solute i between feed solution and feed side of the membrane; partition/distribution coefficient of solute i between two immiscible phases; κ_{ij} for feed solution and the membrane (3.3.87), (3.3.89a)
$\kappa_{ij}', \kappa_{ij}''$	effective value of κ_{ij} when there is no reaction or there is reaction
κ_{io}	partition coefficient of i between an organic and an aqueous phase (6.3.101a)
κ_{ip}	κ_{im} at permeate-membrane interface (3.4.58)
$\kappa_{i,\text{loc}}(\boldsymbol{r})$	local equilibrium partition constant (3.3.89g)
$\kappa_{is}, \kappa_{is}'$	impurity distribution coefficient (6.3.109a)
$\kappa_{is,\text{eff}}'$	effective partition coefficient in zone melting for species i (6.3.119a)
κ_{iE}	distribution (partition) coefficient for species i in solvent extraction
κ_{iR}	distribution coefficient of i in ion exchange system (3.3.115)
$\kappa_{p_1}; \kappa_{p1}^0$	protein partition coefficient in aqueous two-phase extraction; protein distribution coefficient in the absence of a charge gradient at interface
κ_{salt}^m	distribution coefficient of salt, molal basis (4.1.34o)

λ	Debye length (3.1.10b), mean free path of a gas molecule (3.1.114), filter coefficient (7.2.187), parameter for a dialyzer (8.1.399), parameter for a distillation plate/stage (8.3.38), latent heat of vaporization/condensation
λ; λ_x, λ_z; λ_i; λ_+; λ_-; λ_i^o	molecular conformation coordinate (3.3.89c); electrode spacings (7.3.18); retention parameter for species i (7.3.213), ionic equivalent conductance of ion i (3.1.108r); value of λ_i for a cation; value of λ_i for an anion; value of λ_i at infinite dilution (Table 3.A.8)
λ_1, λ_2	defined by (5.4.100)
Λ	equivalent conductance of a salt (an electrolyte) (3.1.108s)
μ; μ_0; μ_{dr}	dynamic fluid viscosity; value of μ at inlet; viscosity of drop fluid
μ_C, μ_D	fluid viscosity of continuous and dispersed phases, respectively
μ_{EOF}^m	electroosmotic mobility (6.3.10c)
μ_i, μ_i^0, μ_i^m; $\mu_{i,\text{eff}}^m$; $\mu_{\text{ion},g}^m$	species i chemical potential and its standard state value; ionic mobility of ion i (3.1.108m); effective value of μ_i^m (6.3.8d,g); ionic mobility of ion in gas phase
μ_{ij}; μ_{ijn}; $\mu_{ij}^{e\ell}$	value of μ_i in region j; value of μ_{ij} for plate/stage n; value of μ_{ij} in a system with electrical charge (3.3.27)
μ_{ijPl}	value of μ_{ij} for a planar interface (3.3.50)
$\mu_\ell(M)$, $\mu_v(M)$	chemical potential of species of molecular weight M in liquid and vapor phases, respectively, in continuous chemical mixtures
μ_0^m, μ_p^m, μ_s^m; μ_∞	magnetic permeabilities of vacuum, particle and solution, respectively; chemical potential of a crystal
υ; $\upsilon(z)$; υ_0	kinematic viscosity; υ at axial location z; υ at inlet
υ_i	stoichiometric coefficient for species i in any chemical reaction
υ_A, υ_Y	moles of ions A and Y produced by dissociation of 1 mole of electrolyte AY
ξ	extent of separation for a binary system (1.4.16), (2.2.11), (6.3.105)
ξ_{12}, $\xi_{r_{p_1}, r_{p_2}}$	extent of separation for components 1 and 2 in particle classification (2.4.9), for particles of size r_{p_1} and r_{p_2} (2.4.6)
π; π	constant (3.1416...); osmotic pressure, spreading pressure (3.3.41a)
π_f; π_i; π_p, π_w	osmotic pressure of feed solution; spreading pressure in adsorption of solute i from a solution (4.1.63a); osmotic pressures of permeate solution and the solution at the membrane wall, respectively
π_1, π_2	osmotic pressure of solution in regions 1 and 2
ρ; ρ_b; ρ_c; ρ_e; ρ_f; ρ_{fm}; ρ_i; ρ_{ij}	fluid density; bulk density of the packed bed; density of continuous phase (also fluid density at critical point); electric charge density per unit volume; density of feed fluid; moles/fluid volume; mass concentration of species i (also density of solute i (3.3.90b)); value of ρ_i in region j (1.3.4)
ρ_ℓ; ρ_p; ρ_{pg}; ρ_s; ρ_t; ρ_{tj}	density of liquid (melt); mass density of particle material; particle mass concentration in gas; density of solid material; total density of fluid; total mass density of mixture in region j (1.3.5)
ρ_v; ρ_{i1}, ρ_{if}	density of vapor phase; mass concentration of species i in crystallizer outlet solution and feed solution, respectively
$\rho_{i\ell}$; ρ_{ip}; ρ_{p,r_p}	mass density of species i in liquid (melt); mass density of crystals of species i; mass density contribution of particles of size r_p to $r_p + dr_p$ (6.2.55)
ρ_R; ρ_1, ρ_2	density of resin particle, reduced density of fluid ($= \rho/\rho_c$); density of liquids 1 and 2
ρ_{dR}; ρ_{pR}	fluid resistivity; particle resistivity
ρ_{avg}; ρ_{isat}	average gas density (6.1.5e); mass density of solute i in solution at saturation
$\overline{\rho}_{ik}$, $\overline{\overline{\rho}}_{ik}$	intrinsic phase average and phase average, respectively, of ρ_{ik} (6.2.24a,b)
$\rho_{ik,\text{avg}i}$	$\overline{\rho}_{ik}$ (6.2.24a)
$\langle\rho_i\rangle$	mass density of solute i in solution averaged over flow cross section (6.2.16b)
σ	electrical conductivity of the solution (3.1.108q), particle sticking probability
σ_i	standard deviation of any profile (3.2.21a,b), Lennard-Jones parameter, ionic equivalent conductance (3.1.108r), solute i reflection coefficient through membrane (6.3.157a,b)

$\sigma_p; \sigma_v, \sigma_v^i; \sigma_x, \sigma_y$	steric factor for protein SMA model (3.3.122d); specific volume based deposit in filter (7.2.172), (7.2.189); standard deviations (7.3.173)
$\sigma_{AB}; \sigma_{ix}$	average of σ_A and σ_B (3.1.91b); standard deviations (7.3.12a)
$\sigma_i^{\mid}; \overline{\sigma}_m$	nondimensional standard deviation of a profile (3.2.21a,b); average electrical conductivity of the ionic and electronic species in a mixed conducting membrane
$\sigma_{ti}, \sigma_{zi}; \sigma_{zi}^+; \sigma_{Vi}$	standard deviation in the output profile of species i in t-coordinate and z-coordinate, respectively (2.5.3), (6.3.18c,b); nondimensionalized σ_{zi} (6.3.16b); standard deviation in volume flow units (7.1.102b)
$\tau; \tau_m; \tau_w; \tau_{wc}$	tortuosity factor; value of τ for porous membrane; wall shear stress (7.2.118); τ_w in cake region (7.2.136a)
τ_{yx}, τ_{yz}	components of tangential stress τ_y in x- and z-directions (6.1.24)
$\phi; \phi_A, \phi_B$	potential of any external force field (also ϕ^{ext}), angle in spherical polar coordinate system, enhancement factor due to reaction, solids volume fraction profile in suspension boundary layer (7.2.109); enhancement factors for species A and B
$\phi_c; \phi_D; \phi_s$	volume fraction of solids in cake; volume fraction of continuous, dispersed phase (Table 3.1.7), (6.4.88); volume fraction of particles and solids in a suspension
$\phi_j; \phi_m$	electrical potential of phase j; defined in (7.1.107h)
$\phi_{i\ell}; \phi_{im}; \phi_m; \phi_p; \phi_{p\ell}$	volume fraction of species i in liquid phase; volume fraction of species i in membrane; polymer volume fraction in membrane; voltage drop over a cell pair in electrodialysis; volume fraction of polymer in liquid phase
ϕ_N	correction factor for nonequimolar counterdiffusion (3.1.136b)
ϕ_0	arbitrary value of centrifugal potential at $r = 0$ (3.1.6d)
ϕ_w	volume fraction of water in resin, particle volume fraction at wall
ϕ_{max}	maximum enhancement due to an instantaneous reaction, maximum particle volume fraction
$\phi_i^{ext}; \phi_{i\alpha}^{ext}, \phi_{i\beta}^{ext}$	value of ϕ or ϕ^{ext} for any species i; values of ϕ_i^{ext} in regions α and β
$\Phi; \hat{\Phi}_{ig}; \Phi_i^{sat}$	fugacity coefficient (3.3.56); fugacity coefficient of species i in a mixture in gas phase; value of $\hat{\Phi}_{ig}$ for pure i at P_i^{sat} at system temperature
$\phi_i^{tot}; \phi_i^*; (\phi_i^*)_{max}; \phi_i^+; (\phi_i^+)$	defined by (3.2.4); defined by (3.2.5b), defined by (3.2.6); defined by (3.2.9) and (6.3.14a)
ϕ_{ti}^{ext}	$\sum_j \phi_i^{ext}$, summation over different external forces (3.2.4)
χ, χ_p, χ_s	volume suspectibility of fluid, particle and the solution
χ_{ij}, χ_{ip}	Flory interaction parameter between species i and j and species i and polymer
$\psi; \psi$	molecular orientation coordinate vector (3.3.89c); extent of facilitation of flux (5.4.58) (5.4.59a), selectivity (7.2.92)
ψ_k	any property or characteristic of the kth phase
ψ_s, ψ_v	shape factor for particle surface and volume (Example 2.4.1), (2.4.2e) (3.4.26), (6.4.15), (6.4.19)
ψ_{As}	association factor (3.1.91a)
$\psi_{e\ell}$	electrical potential in the double layer
$\langle \psi_k \rangle$	phase average of ψ in the kth phase (6.2.25a) = $\psi_{k,avg}$
$\langle \psi_k \rangle^k$	intrinsic phase average of ψ in the kth phase (6.2.25b) = $\psi_{k,avgi}$
ω	angular velocity, solute permeation parameter (6.3.158b), sign of fixed charge in (3.3.30b)
$\Omega_{D,AB}$	quantity in diffusion coefficient expression (3.1.91b)

Superscripts

a; b; Br	activity based; bottom/stripping section of a column; Brownian motion
d, drag	drag
eff; eph; ext	effective; electrophoretic; external
ft; G	between the feed stream and the tails stream; gas phase
hf	between the heads stream and the feed stream
ht	between the heads stream and the tails stream
i; iner; int	species i; inertial; internal

L	liquid phase
m	magnetic, mobility, based on molal quantities, maximum number
N	related to Nernst–Plank equation (3.1.106)
0, o	location $z = 0$, standard state, infinite dilution, original quantity
oo, ov	(7.1.101c), overall
p	pore, pure species
P	mole fraction based in Henry's law
r	quantity in a system with recycle or reflux, for the case of a chemical reaction
s	related to solids only in solid–fluid separation, at saturation, scrubbing cascade
S	solid phase
t, T	total value, top/enriching section of a column, thermal diffusion
$v\ell$	vapor liquid
x, δ	mole fraction based, location $z = \delta$ or δ_m
∞	infinite dilution condition
$'$	equimolar counterdiffusion case, first derivative, feed side of membrane
$''$	permeate side of membrane
$+$	nondimensional quantity
*	hypothetical quantity

Subscripts

atm; b	atmospheric; bulk phase value, bottom product stream from a column, backward reaction
A,B,C	species A, B and C
c	based on molar concentration, continuous phase, cake, critical quantity
d	diffusive, dialysate phase, dissociation, distillate
D	drag related, Henry's law species related, dispersed phase
e	enriching section
eff; ex; ext	effective value; exit location; external
f; fr	feed side, feed based; during formation of drop, size based
g; gr	gas phase; growth based
H; hex	high; heat exchanger
i	species i, where i = 1, 2, 3, A, B, M, s, phase interface, ith module/tube/stage
ij	ith species in jth stream, $j = f$ for a single feed stream, $j = f_1, f_2$ for two feed streams entering separator, j = 1,2 for product streams rich in species 1 and species 2, respectively
imp; iner	impeller; inertial
in; ion	at inlet; ionic species
j	region/phase j, where j = 1, 2, E, f, g, l, ℓ' m, o, p, R, s, t, v, w, T, σ
k	species k
l, liq; loc; ℓ'; ℓ; L	liquid phase; local; adhering liquid phase on crystal; low; at end of separator of length L
m; mc; max; min	membrane phase; micellar; maximum; minimum
M	mixture, molecular weight based, metallic species
ME, MR	Murphree based on extract phase, raffinate phase
n, $(n+1)$, $(n-1)$	stage number n, $(n+1)$ and $(n-1)$, respectively
N; nu	metallic species; nucleation
o; obs; og; ol; out	organic phase; observed; overall gas phase based; overall liquid phase based; at outlet
p; pd; P	permeate side, product side, particle; dominant particle; planar interface
r; r_{p_1}, r_{p_2}; R	radial direction; related to particles of size r_1 or r_2; ion exchange resin phase, resistive, raffinate phase
s	solvent, surface integration step in crystallization, surface adsorption site, salt/eluent counterion, stripping section, stage number
s^+, s^-; se	forward and reverse surface reaction; seed crystal
S	location S, stationary phase
t; tOE; tOR	total, top product stream in a column; transfer units (overall) based on extract phase; transfer units (overall) based on raffinate phase
tj	total quantity in jth stream, $j = f$ for a single feed stream, $j = f_1, f_2$ for two feed streams entering separator, j = 1,2 for product streams rich in species 1 and 2, respectively
true; T	true value; thermal diffusive, total
v;w	vapor phase, volumetric; aqueous phase, water
x	liquid mole fraction based, coordinate direction
y	gas phase mole fraction based, coordinate direction
Y	as in species AY
z	coordinate direction
1, 2	species 1 and species 2, phase 1 and phase 2
3, 4	species 3 and species 4, phases 3 and 4
α, β	phase α and phase β
σ, s	surface phase

Abbreviations and acronyms

Å	angstrom
AEM	anion exchange membrane
atm	atmosphere
avg	average
AVLIS	atomic vapor laser isotope separation
BET	Brunauer–Emmet–Teller
bar	10^5 pascal
barrer	unit for permeability coefficient of gases through membrane, $1\,\text{barrer} = 10^{-10}\dfrac{\text{cm}^3(\text{STP})\text{-cm}}{\text{cm}^2\text{-s-cmHg}}$
Btu	British thermal unit
CAC	continuous annular chromatograph
CACE	counteracting chromatographic electrophoresis
CCEP	countercurrent electrophoresis
CD(s)	cyclodextrin(s)
CDI	capacitive deionization
CE	capillary electrophoresis
CEC	capillary electrochromatography
CEDI	continuous electrodeionization
CEM	cation exchange membrane
CFD	computational fluid dynamics
CFE	continuous free-flow electrophoresis
CGC	countercurrent gas centrifuge
CHO	Chinese hamster ovary
CHOPs	Chinese hamster ovary cell proteins
cm Hg	pressure indicated in the height of a column of mercury
CMC	critical micelle concentration
CMS	carbon molecular sieve
CSC	continuous-surface chromatography
CSTR	continuous stirred tank reactor
CSTS	continuous stirred tank separator
CV	control volume
CZE	capillary zone electrophoresis
ED	electrodialysis
ELM	emulsion liquid membrane
EOF	electroosmotic flow
ESA	energy-separating agent
FACS	fluorescence-activated cell sorting
FFE	free-flow electrophoresis
FFF	field-flow fractionation
FFM	free-flow magnetophoresis
GAC	granular activated carbon
GLC	gas–liquid chromatography
GPC	gel permeation chromatography
HDPs	high-density particles
HETP	height of an equivalent theoretical plate
HFCLM	hollow fiber contained liquid membrane
HGH	human growth hormone
HGMS	high-gradient magnetic separation
HK	heavy key
HPCE	high-performance capillary electrophoresis
HPLC	high-performance liquid chromatography
HPTFF	high-performance tangential-flow filtration
IEF	isoelectric focusing
ILM	immobilized liquid membrane
IMAC	immobilized metal affinity chromatography
ITM	ion transport membrane
ITP	isotachophoresis
LDF	linear driving force approximation
LK	light key
LLC	liquid–liquid chromatography
LM	logarithmic mean
LRV	log reduction value
LSC	liquid–solid adsorption chromatography
LTU	length of transfer unit
mAB	monoclonal antibody
MBE	moving boundary electrophoresis
ME	multiple effect
MEKC	micellar electrokinetic chromatography
MEUF	micellar enhanced ultrafiltration
MSA	mass-separating agent
MSC	molecular sieve carbon
MSF	multistage flash
MSMPR	mixed suspension, mixed product removal
MTZ	mass-transfer zone
NEA	nitrogen-enriched air
OEA	oxygen-enriched air
PAC	powdered activated carbon
PBE	population balance equation
PSA	pressure-swing adsorption
psia	pound force per square inch absolute
psig	pound force per square inch gauge
RO	reverse osmosis
SCF	supercritical fluid
SEC	size exclusion chromatography
SLM	supported liquid membrane
SMA	steric mass action
SMB	simulated moving bed
SPs	structured packings
TDS	total dissolved solids
TFF	tangential-flow filtration
TOC	total organic carbon

TSA	thermal-swing adsorption
UF	ultrafiltration
UPW	ultrapure water
VCR	volume concentration ratio
VLE	vapor–liquid equilibrium
VOCs	volatile organic compounds
WFI	water for injection
WPU	purified water
ZE	zone electrophoresis

Introduction to the book

The basic objective of this chapter is to describe the organization of this book vis-à-vis separations from a chemical engineering perspective. Separation, sometimes identified as concentration, enrichment or purification, is employed widely in large industrial-scale as well as small laboratory-scale processes. Here we refer primarily to physical separation methods. However, chemical reactions, especially reversible ones, can enhance separation and have therefore received significant attention in this book. Further, we have considered not only separation of mixtures of molecules, but also mixtures of particles and macromolecules.

The number of different separation processes, methods and techniques is very large. Further new techniques or variations of older techniques keep appearing in industries, old and new. The potential for the emergence of new techniques is very high. Therefore, the approach taken in this book is focused on understanding the basic concepts of separation. Such an approach is expected not only to help develop a better understanding of common separation processes, but also to lay the foundation for deciphering emerging separation processes/techniques. The level of treatment of an individual separation process is generally elementary. Traditional equilibrium based separation processes have received considerable but not overwhelming attention. Many other emerging processes, as well as established processes dealing with particles and external forces, are not usually taught to chemical engineering students; these are integral parts of this book. To facilitate the analysis of processes over such a broad canvas, a somewhat generalized structure has been provided. This includes a core set of equations of change for species concentration, particle population and particle trajectory. These equations are expected to be quite useful in general; however, separation systems are quite often very complicated, thereby limiting the direct utilization of such equations.

Separation and purification are two core activities of chemical engineers. The first wave of textbooks on separation/mass transfer (until the early 1980s) concentrated on distillation, absorption and extraction, with some attention to adsorption/ion exchange/chromatography. The second wave expanded the treatment of adsorption/ion exchange/chromatography and incorporated an introduction to membrane processes. The overwhelming emphasis in these books was on chemical separations. Simultaneously, a series of textbooks emerged focusing on bioseparations. In these textbooks, the treatment of particle based separations appears briefly under mechanical separations, or under the equilibrium based process of crystallization, or as special operations under bioseparations.

Fundamental principles that facilitate understanding of a variety of different separations have, however, been emerging in the literature for quite some time. It is useful to structure the learning of separation around these basic principles. Forces present in the separation system act on molecules, macromolecules or particles and make them migrate at different velocities and sometimes in different directions. When such velocities/forces interact with the bulk velocity of the individual phase(s)/region(s) present in the separation system, molecular species or particles follow different trajectories or concentrate in different phase(s)/region(s), leading to separation. This book will systematically develop this overall framework of a few important configurations of bulk flow direction vis-à-vis the direction of the force(s) for open separators. Chemical thermodynamics provides the local boundaries/limits in such configurations for chemical separations. The individual separation processes/techniques will then be illustrated in each such configuration of bulk flow direction vs. force direction in three categories of processes: phase equilibrium driven; external force driven; membrane processes.

These three basic categories of separation processes rely on three different types of separation phenomena. These different types of separation phenomena achieve different extents of separation when coupled with particular configurations of bulk flow vs. force pattern. The description of each process/technique generally includes its conventional treatment and often elementary process/equipment design considerations.

This illustrative framework is preceded by a few chapters that provide the basic tools for achieving this goal. In earlier literature, a broad category of separation techniques is identified as a mechanical separation process. These techniques are invariably restricted to the separation of particles in a fluid or drops in another fluid subjected generally to an external force. In this book, particle separations have been studied along with chemical separations when a particular external force is considered. Therefore the structure of this book is somewhat different. The following section provides a brief introduction to each chapter in the book.

Introduction to chapters

What happens to a perfectly mixed binary mixture of two species in a closed vessel as separation takes place is introduced in Chapter 1. How one describes the extent of separation achieved in the closed vessel is illustrated via a few common separation indices. The separation indices are based on the notion of different species-specific regions in the separation system, and their differing compositions and capacities. Double subscript based notation, with the first subscript i referring to a component and the second subscript j referring to a region/phase/fraction, is introduced here. This notation has been used throughout the book as often as possible. Use of these separation indices is illustrated for three basic classes of separation systems without any particles: immiscible phases; membrane-containing systems; and systems having the same phase throughout the separation system. A description of separation in multicomponent systems has been included along with the notion of a separating agent required for separation.

Chapter 2 presents the description of quantities needed to quantify separation in open systems with flow(s) in and out of single-entry and double-entry separators for binary, multicomponent and continuous chemical mixtures, as well as a size-distributed particle population. Separation indices useful for describing separation in open systems with or without recycle or reflux are illustrated for steady state operation (Sections 2.2 and 2.3); those for a particle population are provided in Section 2.4. At the end (Section 2.5), indices for description of separation in time-dependent systems, e.g. chromatography, have been introduced.

The physicochemical basis for separation is the primary focus of Chapter 3. Separation happens via species-specific force driven relative displacement of molecules of one species in relation to other species into species-specific region in the separation system. Particles of different sizes/properties similarly undergo relative displacements. To develop this perspective, Chapter 3 (Section 3.1) identifies various external forces and chemical potential gradient based force generating different migration/terminal velocities and fluxes for chemical species and particles. Integrated flux expressions for molecular diffusion and convection for single-phase systems, mass-transfer coefficients and empirical correlations for mass-transfer coefficients are introduced. Chapter 3 (Section 3.2) further points out the role of the spatial profile of the potential attributable to the force in developing a multicomponent separation capability. The criteria for achieving equilibrium between different phases and regions in the separation system with or without an external force and various types of phase equilibria are discussed in Section 3.3. The presentation of the partitioning of a species between two phases is at a phenomenological level. The molecular basis of this partitioning via intermolecular interactions has not been considered. This is followed by species flux expressions in interphase transport, including membrane transport (Section 3.4). The notion of an overall mass-transfer coefficient and its relation to single-phase mass-transfer coefficients are introduced here.

Chapter 4 provides a quantitative exposition of how much separation is achieved at equilibrium in a closed vessel for three broad classes of separation systems: phase equilibrium between two phases (Section 4.1); single phase or a particle suspension in an external force field (Section 4.2); two regions separated by a membrane (Section 4.3). The phase equilibrium systems considered are: gas–liquid, vapor–liquid, liquid–liquid, liquid–solid, interfacial adsorption systems, liquid–ion exchanger and the supercritical fluid–solid/liquid phase. The external force fields and configurations studied are: centrifuges, isopycnic sedimentation, isoelectric focusing, gravity (sedimentation, inclined settlers), acoustic forces and thermal diffusion. In the case of a membrane based system of dialysis and gas permeation if separation is to be achieved, we come across the need for an open system. Chapter 5 focuses on the beneficial effects of chemical reactions in phase equilibrium and membrane based separation systems. A few common types of reactions, such as ionizations, acid–base reactions and different types of complexation equilibria, are found to influence strongly the separation achieved across the whole spectrum of separations involving molecules and macromolecules. The phase equilibrium systems studied are: gas–liquid, vapor–liquid, liquid–liquid, liquid–solid, surface adsorption and Donnan equilibrium. Reaction based enhancement in the rates of interphase transport as well as membrane transport has been illustrated for a variety of systems.

Separation is most often implemented in open systems/devices with bulk flow(s) in and out. The treatment of separation achieved in such separators is carried out in Chapters 6, 7 and 8. Chapter 6 begins with the sources and the nature of bulk flow in separation systems in a multiscale context as well as the feed introduction mode vis-à-vis time (Section 6.1). The various equations of change for species concentration in a mixture, the equation of motion of a particle in a fluid and the general equation of change for a particle population, including that in a continuous stirred tank separator, are provided in Section 6.2. Section 6.3 covers the separation processes/techniques in which the direction of the bulk flow is parallel to the direction of the force(s). Figure 6.3.1 illustrates the widespread use of this flow vs. force configuration for three basic classes of separation systems. External force based processes of elutriation, capillary electrophoresis, centrifugal elutriation, inertial impaction and electrostatic separation of fine particles are introduced first. Chemical potential gradient driven processes of flash/vaporization/devolatilization, batch distillation, liquid–liquid extraction, zone melting, normal freezing and drying are studied next. The membrane based processes covered are: cake filtration/microfiltration, ultrafiltration, reverse osmosis, pervaporation and gas permeation. In the final section (Section 6.4), Chapter 6 considers the continuous stirred tank separator (CSTS) as a special category of bulk flow vs. force configurations; the separation processes studied are: crystallization (precipitation), solvent extraction, ultrafiltration and gas permeation.

The nature and extent of separation achieved when the direction of flow of the feed-containing fluid phase is perpendicular to the direction of the force(s) are studied in Chapter 7. This treatment illustrates the basic separation mechanism clearly, even though the particulars vary widely, as in, for example, free-flow electrophoresis, electrostatic precipitators, electrostatic separation of plastic mixtures, laser excitation of isotopes and flow cytometry (all of them driven by an electrical force field perpendicular to the bulk flow). Figure 7.0.1 provides this broad perspective across all three classes of separation processes: phase equilibrium driven, membrane based, external force driven. This chapter begins (Section 7.1) with the treatment of fixed-bed adsorption processes, pressure-swing adsorption, parametric pumping and chromatography. Crossflow membrane processes considered next (Section 7.2) are: gas permeation, reverse osmosis, ultrafiltration, microfiltration; this has been followed by granular filtration. The external force field based processes studied in Section 7.3 involve electrical force (mentioned earlier), centrifugal force (centrifuges, cyclones), gravity (gravity based settlers), magnetic force field (high-gradient magnetic separation) and optical force. Field-flow fractionation as a special case of a force perpendicular to bulk flow interacting with the velocity profile in a novel way has also been treated; a variety of forces may be used.

Chapter 8 deals with the configuration of bulk flows of two phases/regions (one of which may be solid) perpendicular to the direction of force(s). The directions of motion of the two phases may be parallel to each other in either countercurrent or cocurrent fashion, or they may be in crossflow. Figures 8.1.1–8.1.4 illustrate the countercurrent flow vs. force configuration for all three classes of separation systems. Conventional countercurrent devices/processes of gas absorption/stripping, column distillation with a condenser and reboiler, solvent extraction in columns, melt crystallization, adsorption and simulated moving beds, dialysis and electrodialysis, liquid membrane separation, gas permeation, gas centrifuge, thermal diffusion and mass (sweep) diffusion are studied in Section 8.1. How cocurrent flow of the two phases/flows changes the separation achieved is considered vis-à-vis a few systems in Section 8.2. Local multicomponent feed injection in a crossflow format in fluid–solid systems leads to the achievement of continuous chromatography. Overall crossflow of two phases is exemplified by a crossflow distillation plate (Section 8.3).

Although countercurrent multistaged processes of distillation, gas absorption, solvent extraction, etc. have been studied in some detail in Chapter 8, the subject of multistaging/cascades is considered briefly in Chapter 9. Ideal cascades and constant or variable cross-sectional area are introduced, as are cascades of multistage columns for non-binary systems. Chapter 10 describes at an elementary level the minimum energy required for separation by different separation processes. Additional topics discussed in Chapter 10 include the consideration of various concepts that reduce the energy required for separation, recovering the free energy of mixing via a dialytic battery and additional deliberations for treating dilute solutions in the context of bioseparations. In many real-life applications, sequences of different separation processes are employed with and without reaction processes. Chapter 11 illustrates such sequences of separation processes for bioseparations, water treatment, chemical and petrochemical industries and hydrometallurgical separations.

Each of these chapters provides a particular aspect/perspective of the broad subject of separations. One is often interested, however, in a particular separation process/technique in all of its aspects, beginning with the basic concept and ending with devices designed to implement the separation. Tables are therefore provided at the end of this chapter that identify the essential and important components located in different chapters for a given separation process. Obviously it is not possible to provide a comprehensive treatment of every process, and a few commonly used separation processes have received much more attention than others. However, the treatment of each such commonly used separation process is at a level illustrative of the basic principles relevant to the particular chapter. Furthermore, the treatments are not exhaustive. Readers interested in greater detail are encouraged to go to major texts on such separation processes identified along with their treatments.

Linked footprints of a separation process/technique

We provide in this section seven tables; they appear at the end of this chapter Each table has nine columns. The first column identifies the name of a particular separation process in a particular row (e.g. 'Absorption' in row 1 of Table 1). The second column focuses on Chapters 1 and 2. Six more columns are identified progressively with each of the Chapters 3–8. The final ninth column covers the much smaller Chapters 9–11. Each row in the tables is dedicated to a particular separation process. The entry in a box for a given row and a given column identifies sections in the chapter where that particular separation process or fundamental material needed to understand the transport and thermodynamics for that process has been presented. In Chapters 1 and 2 and in Sections 3.1, 3.2, 3.3.1–3.3.6, 6.1 and 6.2, general features or fundamental relations valid for a variety of separation processes are presented. Therefore, entries for a given separation process under columns 2, 3 and 6, specifically Sections 3.1, 3.2, 3.3.1–3.3.6, 6.1 and 6.2, providing fundamental information on the description of separation species/particle transport, thermodynamics relations, balance/conservation equations and equations of change, respectively, for species/particles are not tied in general specifically to that separation process; however, any entry will be useful for understanding that separation process.

Table 1 covers many of the common phase equilibrium based separation processes. The entries contain a few separation techniques/processes which are not employed on a large scale or illustrate important conceptual developments, e.g. cycling zone adsorption, foam fractionation, parametric pumping. Table 2 includes membrane separation processes, where different membrane transport rates of different species provide the selectivity in open systems. This table also includes membrane contactor based separation processes, where the basis for separation is the partitioning equilibrium between two fluid phases contacting each other at membrane pore mouths. Tables 3 and 4 identify separation processes driven by centrifugal force and electrical force, respectively. Table 5 is devoted to a few processes driven by magnetic force or gravity. A few separation processes/techniques driven by other forces, such as acoustic force, radiation pressure, inertial force and thermal gradient driven force, are listed in Table 6. Table 7 is devoted to additional separation processes such as field-flow fractionation and mass (sweep) diffusion.

It is useful now to illustrate how the descriptive treatment of a particular separation process, e.g. distillation, has been implemented in an evolutionary fashion via the different chapters as identified in row 7 of Table 1. In Section 1.1, Example I of Figure 1.1.2 illustrates the result of heat addition to an equimolar liquid mixture of benzene–toulene: a benzene-rich vapor phase and a toluene-rich liquid phase. Using definitions of compositions etc. introduced in Section 1.3, separation indices such as the separation factor α_{ij} (also the equilibrium ratio K_i) describe the separation achieved in a closed vessel for the benzene–toluene system and a methanol–water system for various liquid-phase compositions. Section 1.5 illustrates via Example 1.5.1 and the values of various separation indices, α_{12} and ξ, the separation achieved in the benzene–toluene system in a closed vessel. Section 1.6 describes multicomponent mixtures and develops the relations between the compositions of two phases in equilibrium, a result useful for distillation in later chapters.

Section 2.1 introduces various quantities describing flow rates and compositions in an open system; a sieve plate in a distillation column is used as one example, among others, of a double-entry separator. A flash distillation stage with liquid fraction recycle illustrates recycle in a single-entry separator (Section 2.2). Section 2.3 for double-entry separators provides a numerical example of benzene–toluene distillation in a countercurrent column without a condenser or reboiler. This and other examples provide a quantitative background on the separation achieved in a given device without discussing the separation mechanism. The same strategy of description of separation achieved via reflux to a column is pursued in this section to demonstrate that a higher reflux ratio leads to higher separation. Sections 2.4.2 and 2.4.3 introduce indices to describe continuous chemical mixtures and multicomponent mixtures vis-à-vis flash vaporization.

The introductory Section 3.1.2.5 in Chapter 3 identifies the negative chemical potential gradient as the driver of targeted separation, and the relevant species flux expression is developed in Section 3.1.3.2 (see Example 3.1.9 also). Section 3.1.4 introduces molecular diffusion and convection and basic mass-transfer coefficient based flux expressions essential to studies of distillation and other phase equilibrium based separation processes. Section 3.1-5.1 introduces the Maxwell–Stefan equations forming the basis of the rate based approach of analyzing distillation column operation. After these fundamental transport considerations (which are also valid for other phase equilibrium based separation processes), we encounter Section 3.3.1, where the equality of chemical potential of a species in all phases at equilibrium is illustrated as the thermodynamic basis for phase equilibrium (i.e. $\mu_{iv} = \mu_{il}$). Direct treatment of distillation then begins in Section 3.3.7.1, where Raoult's law is introduced. It is followed by Section 3.4.1.1, where individual phase based mass-transfer coefficients are related to an overall mass-transfer coefficient based on either the vapor or liquid phase.

Section 4.1 via Section 4.1.2 formally illustrates vapor–liquid equilibria vis-à-vis distillation in a closed vessel along with bubble-point and dew-point calculations for multicomponent systems. How vapor–liquid equilibrium is influenced by chemical reactions in the liquid phase is treated in Section 5.2.1.2, where two subsections, 5.2.1.2.1 and 5.2.1.2.2, deal with reactions influencing vapor–liquid equilibria in isotopic systems. We next encounter open systems in Chapter 6. The equations of change for any two-phase system (e.g. a vapor–liquid system) are provided in Section 6.2.1.1 based on the pseudo-continuum approach for the dependences of species concentrations on time and the main axial coordinate (i.e. z) direction. Section 6.3.2.1 starts with the simplest of open systems, a flash vaporizer, and illustrates isothermal flash calculations under the constraint of phase equilibrium and bulk flow parallel to (||) the force direction for multicomponent systems and continuous chemical mixtures. Batch distillation without any reflux is then studied as a particular illustration of this flow vs. force configuration for a fixed amount of feed liquid as well as for constant-level batch distillation employed for solvent exchange. Residue curves are introduced here.

Column distillation is the most common form of an open separation system in distillation. Here the two phases have, on an overall basis, bulk motions in parallel flow in the countercurrent direction with the forces causing separation being perpendicular ($\perp$) to the directions of bulk flows. The general characteristics of such a separation system are briefly identified in Section 8.1.1, specifically 8.1.1.1–8.1.1.3. We learn the structural consequences of this flow vs. force configuration, namely a distillation column cannot at steady state separate a ternary mixture; we need two columns for a ternary mixture. Further, the particular forms of equations of change for the two phases are obtained from the more general equations in Section 6.2.1.1 (as well as by a control volume analysis). Distillation columns with reflux and recycle are studied in detail in various parts of Section 8.1.3. The conventional approach of assuming ideal equilibrium stages (the stage may have crossflow in an overall countercurrent flow configuration) is adopted in the McCabe–Thiele graphical framework to study the following: operating lines in both sections of a column, q-line, total reflux, minimum reflux, partial/total reboiler, partial condenser, open steam introduction, Kremser equation and side stream.

The deviation from ideal equilibrium stages is studied next via stage efficiency in Section 8.1.3.4. Vapor–liquid contacting on a plate/tray in a column is considered in Section 8.1.3.5 vis-à-vis estimation of column diameter (with reference to Section 6.3.2.1). Topics such as the rate based approach for modeling distillation and separation of a multicomponent mixture in a column are briefly introduced, the latter via the Fenske equation, the Underwood equation and the Gilliland correlation. Distillation in a packed tower and in a batch vessel with reflux are studied next. Section 8.2.1 briefly touches on distillation in a cocurrent two-phase flow device. Section 8.3.2 studies separation in a crossflow distillation plate employing general equations from Section 6.2.1.1, ultimately yielding the American Institute of Chemical Engineers (AIChE) tray efficiency expression. The total number of worked examples involving distillation in one form or another in Chapters 1–8 is 19. Various other aspects of distillation are considered further in Chapters 9–11. Chapter 9 (Section 9.2) introduces briefly the methodology for multicolumn distillation for separating a mixture containing more than two species. Chapter 10 covers the minimum energy required for distillation, and the concepts of net work consumption, multieffect distillation and heat pump vis-à-vis distillation. Chapter 11.3 intro duces very briefly the important role of distillation in the chemical and petrochemical industries.

If the treatment of distillation in a given section of the book needs certain building blocks, it is most likely that those concepts/methods/building blocks have been introduced in an earlier chapter or section of the book. Furthermore, in whichever section distillation appears, it is studied as part of a specific pattern followed by many other separation processes based on phase equilibrium. Such patterns have been emphasized often throughout particular chapters.

A few pointers on phase equilibrium based separation processes are useful. Table 3.3.1 lists possible useful combinations of two bulk immiscible phases for separation such as gas–liquid (vapor–liquid included), gas–solid, liquid–liquid, etc. Quite a few of these combinations form the basis of existing separation processes. In this book, therefore, each chapter, from Section 3.3 onwards, focusing on a particular aspect of the subject of separation, has the subject of phase equilibrium driven separation processes organized along such two immiscible phase combinations. However, all such combinations in practical use do not appear in each chapter.

The treatment of membrane separation processes in this book merits some deliberation. The most commonly used driving force in membrane separation processes is negative chemical potential gradient; a few processes also employ electrical force. Figure 3.4.5 identifies the variety of feed phase–membrane type combinations with variations due to the nature of the permeate phase when negative chemical potential gradient is imposed across the membrane. Section 3.4.2 illustrates the interphase membrane transport aspects of many such configurations. The developments in later chapters follow these feed phase–membrane type permeate phase combinations as often as possible, subject to space limitations. Electrodialysis as an example of an application of electrical force appears in Sections 3.4.2 and 8.1.7. Membrane contactors appear with their phase equilibrium process counterparts in Sections 8.1.2 and 8.1.4, whereas the basic transport considerations in such membrane devices appear much earlier in Sections 3.4.3.1 and 3.4.3.2. A most important item in membrane separation processes is that such devices in the absence of external forces achieve separation when operated as an open system - Sections 4.3.1 and 4.3.3 demonstrate this feature via the processes of dialysis and gas permeation.

The descriptive treatment of the membrane process of reverse osmosis (RO) in the book as identified in Table 2 will be briefly illustrated here. Section 1.1 identifies the basic configuration of RO in Figure 1.1.3. Example 1.5.4 illustrates calculations of separation indices describing separation in RO shown in Figure 1.5.1. Sections 2.1 and 2.2 describe various quantities, as well as the separation indices relevant for RO; Example 2.2.1(c) is directly applicable to RO. Sections 3.1.2.5, 3.1.3.2 and 3.1.5.1 provide a general transport background. Section 3.1.5.2 is directly relevant to an irreversible thermodynamics based solute and solvent transport through RO membranes. Section 3.3.7.4 provides a membrane–liquid equilibrium relation from an osmotic equilibrium point of view. Section 3.4.2.1 formally introduces transport rates in RO membranes and flux expressions, along with issues of concentration polarization in a feed solution. A closed vessel of Chapter 4 has very limited relevance for RO (Section 4.3.4). Sections 5.4.1 and 5.4.1.1 describe how chemical reactions, such as ionization, in the solutions influence separation in RO processes.

Section 6.3.3.3 studies RO in bulk flow parallel to the force configuration and describes various membrane transport considerations and flux expressions. Practical RO membranes are employed in devices with bulk feed flow perpendicular to the force configuration, as illustrated in Section 7.2.1.2. A simplified solution for a spiral-wound RO membrane is developed: analytical expressions for the water flux as well as for salt rejection are obtained and illustrated through example problem solving. A total of six worked example problems have been provided up to Chapter 7. Chapter 9 (Figure 9.1.5) shows a RO cascade in a tapered configuration. Section 10.1.2 calculates the minimum energy required in reverse osmosis based desalination and compares it with that in evaporation. Section 11.2 covers the sequence of separation steps in a water treatment process for both desalination and ultrapure water production. The very important role played by RO in such plants is clearly illustrated.

The evolution of separation through different chapters due to an external force needs some discussion as well. Whereas negative chemical potential gradient driven distillation is utilized to separate low molecular weight liquids having different volatilities, an external force, such as electrical force arising from a negative gradient of electrical potential, can be used to separate small charged molecules, charged macromolecules, charged cells, charged particles, etc.; the medium may be liquid or gaseous. The canvas is large, and the variety of separation processes/techniques driven by electrical force is significant. Although there is considerable variety also in phase equilibrium processes resulting from a variety of two-phase systems, the separation systems are more often limited to smaller molecules. Separation of proteins/macromolecules via chromatography (Section 7.1.5.1) and biphasic/reverse micellar extraction (Sections 4.1.4 and 4.1.8) provide exceptions; flotation (Section 3.3.8) separates particles with the helping hand of an external force, gravity, as does a Venturi scrubber (Section 8.2.3) via inertial impaction.

Consider the electrophoretic motion of charged molecules/macromolecules/proteins in an aqueous solution/buffer subjected to an electrical force. Three separation techniques, isoelectric focusing, capillary electrophoresis and continuous free-flow electrophoresis, exploit, among others, electrophoretic transport under the constraints of a closed vessel, bulk flow parallel to force and bulk flow perpendicular to force, respectively. Correspondingly, in Table 4, isoelectric focusing does not appear in Chapters 6–8; capillary electrophoresis is absent from Chapter 8. However, each such technique benefits from relevant discussions in earlier chapters, even though the technique itself is treated in detail in a later chapter; therefore materials in Chapters 2 and (especially) Chapter 3 are identified for each of the three techniques. Capillary electrophoresis appears also in Section 7.1.7.1, where it has been coupled with chromatography where the bulk flow is perpendicular to the force.

Classification of separation processes

This book has not adopted a comprehensive classification scheme for all separation processes. Readers should go to the references, especially Figure 30 and Table 7 of Lee *et al.* (1977a) and Table 1-1 of King (1980), to that end. What has been adopted here is apparent from the titles of Tables 1–7. Separation processes are classified into three categories based on the three basic types of physiocochemical phenomena: (1) phase equilibrium based separation processes; (2) membrane separation processes; (3) external force based separation processes. There are a few processes where there is an overlap. For example, electrodialysis is a membrane-separation process driven primarily by an external force, the electrical potential gradient; most membrane-separation processes are driven by negative chemical potential gradient. There are a few others, e.g. mass diffusion/sweep diffusion, which cannot be neatly put into these three categories; they possess characteristics of different categories.

In this framework of three broad categories of separation processes, further separation development/classification comes about due to the nature of the interaction between the basic separation phenomena in each category and the directions of bulk flow vis-à-vis the direction of force(s) responsible for the basic separation mechanism. Considerable additional separation development is achieved by reflux, recycle, creation of an additional property gradient in an external force field, mode of feed introduction, etc. These aspects have been addressed in the following sections: reflux (Sections 2.3.2, 8.1.1, 8.1.4, 10.1.4.2, 10.2.2.1); recycle (Sections 2.2.2, 2.4.1, 7.2.1.1, 7.2.4, 8.1.1); development of an additional property gradient in an external force field (Sections 4.2.1.3, 4.2.2.1, 4.2.3.3, 7.1.7); mode of feed introduction (Sections 6.1.9, 7.1.5, 7.1.6, 8.1.1, 8.2.2.1, 8.2.2.2, 8.3.1).

An additional classification approach considers the nature of the mixture to be separated: mixtures of small molecules and/or ions in solution or gas phase; mixtures of macromolecules in solution; mixtures of particles, where particles in this book include biological cells (Tables 4.2.1, 7.3.1), cell debris, colloidal material and inorganic and organic particles of varying dimensions (submicron to visible particles, Figure 2.4.1(b)). Of the numerous separation techniques involving different types of macromolecules, the following have received some attention here: separation of proteins from each other/ one another or solvent via isoelectric focusing, etc. (Sections 4.2.2.1, 4.2.2.2), ultrafiltration (Sections 6.3.3.2, 6.4.2.1, 7.2.1.3), chromatography (Sections 4.1.6, 4.1.9.4, 7.1.5.1.6, 7.1.5.1.7, 7.1.5.1.8, 7.1.6, 7.1.7), electrophoresis (Section 7.3.1.1), field-flow fractionation (Section 7.3.4), aqueous biphasic extraction (Section 4.1.3) and reverse micelles (Section 4.1.9); separation of nonbiological macromolecules via size exclusion chromatography (Section 7.1.5.1.7), flash devolatilization (Section 6.3.2.1), sol-gel separation (Section 2.4.2); DNA separation via isopycnic sedimentation (Section 4.2.1.3).

It is useful to provide a list of the basic physical or physiochemical properties, each of which could be a basis for separation; it is also useful to list simultaneously the core phenomenon exploiting such a physical or physicochemical property for separation. It is to be noted that this list is not exhaustive; rather, it contains the more familiar properties. Table 8 identifies a variety of these basic properties and lists phenomena employing a particular basic property leading to separation. For each basic property and phenomenon in this table, there are three columns corresponding to three different types of basic separation processes: phase-equilibrium-based separation processes; membrane-separation processes; and external force based separation processes. An entry into these three columns identifies a separation process or processes where the particular basic property is key to separation. References to Tables 1–7, a section in the book or a separate reference have been provided to each entry in these three columns.

There are some items of interest here. A few basic properties are the basis for separation in two different types of basic separation processes. For example, condensability of a vapor/gas species is useful for vapor absorption as well as for membrane gas separation; geometrical partitioning (or partitioning by other means between a pore and an external solution) is useful both in adsorption/chromatography as well as in the membrane processes of dialysis and ultrafiltration, etc. Further, there are many cases where chemical reactions are extraordinarily useful for separation; these are not identified here since chemical reactions can enhance separation only if the basic mechanism for separation exists, especially in phase equilibrium based separations. However, there are a few cases where chemical reactions, especially complexations, provide the fundamental basis for separation, as in affinity chromatography, metal extractions and isotope exchange reactions.

Additional comments on using the book

This book has 118 separate numerical examples spread over Chapters 1–4 and 6–9. The numerical examples are not in finer print. Chapter 5 has sometimes employed numerical calculations to illustrate the effect of chemical reactions on separations without formal numerical examples. Chapter 10 follows this strategy as well to illustrate the amount of energy required for a particular separation. The total number of problems provided at the ends of all the chapters is 299. The specific separation process relevant for the problem is generally obvious from the introductory sentence in the problem. Further, the sequence of appearance of a problem on a given separation process reflects/follows the sequences of appearance of that separation process in the text.

Footnotes have been employed occasionally. All references used appear at one location in alphabetical order at the end of the book. The symbols and notation employed throughout the book are consistent; any local deviation has been identified. In a few locations, advanced material or additional information has been provided.

Textbooks, handbooks and major references on separation processes

There is an extraordinarily rich literature on separations. This book has freely drawn material from this literature consisting of textbooks, monographs or extended chapters in multiauthor edited volumes apart from numerous journal articles. Here we list these books and chapters (but no journal articles) under the following categories: separations; chemical separations; bioseparations; membrane separations; particle separations; other books. Such books and relevant journal articles have been cited through each section in each chapter. Occasionally some comments have been attached here to a given reference. Books devoted solely to a given separation process/ technique are not, in general, mentioned below. The following list is given in *chronological* order. At the end of each reference, its formal reference has been identified.

Separations

(1) Karger, B.L., L.R. Snyder and C. Horvath, *An Introduction to Separation Science*, Wiley, New York (1973). Chapter 18 devotes ~19 pages to particle separation; otherwise it covers primarily separations of chemicals and macromolecules. (Karger *et al.*, 1973.)

(2) Lee, H.L., E.N. Lightfoot, J.F.G. Reis and M.D. Waissbluth, "The systematic description and development of separation processes," in *Recent Developments in Separation Science*, Vol. III, N.N. Li (ed.), Part A, CRC

Press, Cleveland, OH (1977), pp.1–70. An important contribution to structuring separations from a morphological perspective with a distinct transport-based input. (Lee *et al.*, 1977a.)

(3) Giddings, J.C., "Principles of chemical separations," in *Treatise on Analytical Chemistry, Part I. Theory and Practice*, Vol. 5, P.J. Elving, E. Grushka and I.M. Kolthoff (eds.), Wiley-Interscience, New York (1982), chap.3. An early and useful contribution toward transport-based understanding of analytical separations. (Giddings, 1982.)

(4) Giddings, J.C., *Unified Separation Science*, John Wiley, New York (1991). An important contribution to separation science with an emphasis on methods used in analytical chemistry, especially chromatography. (Giddings, 1991.)

(5) Schweitzer, P.A., *Handbook of Separation Techniques for Chemical Engineers*, 3rd edn., McGraw-Hill, New York. (Schweitzer, 1997.)

Chemical separations

(1) Benedict, M. and T.H. Pigford, *Nuclear Chemical Engineering*, McGraw-Hill, New York (1957). Introduces isotope separations and cascades in a chemical engineering context for the nuclear industry. The second edition (1981), with added author H.W. Levi, substantially expands the treatment and coverage. (Benedict *et al.*, 1981.)

(2) Pratt, H.R.C., *Countercurrent Separation Processes*, Elsevier, Amsterdam (1967). Contains, among others, an introduction to cascade analysis for chemical separations and isotope separations. (Pratt, 1967.)

(3) Sherwood, T.K., R.L. Pigford and C.R. Wilke, *Mass Transfer*, McGraw-Hill, New York (1975). (Sherwood *et al.*, 1975.)

(4) King, C.J., *Separation Processes*, 2nd edn., McGraw-Hill, New York, (1980). An important textbook which analyzes chemical separations in a generalized framework with considerable emphasis on multistage separation processes. The first edition appeared in 1970. (King, 1980.)

(5) Treybal, R.E., *Mass-transfer Operations*, 3rd edn., McGraw-Hill, New York (1980). A textbook focusing primarily on conventional mass transfer operations. (Treybal, 1980.)

(6) Hines, A.L. and R.M. Maddox, *Mass Transfer: Fundamentals and Applications*, Prentice-Hall PTR, Upper Saddle River, NJ (1985). (Hines and Maddox, 1985.)

(7) Wankat, P.C., *Rate-Controlled Separations*, Elsevier Applied Science, New York (1990). Textbook providing an extensive treatment of adsorption, chromatography, crystallization, ion exchange and membrane separations. (Wankat, 1990.)

(8) Humphrey, J.L. and G.E. Keller II, *Separation Process Technology*, McGraw-Hill, New York (1997). Book oriented towards separation technology, useful for process design. (Humphrey and Keller, 1997.)

(9) Seader, J.D. and E.J. Henley, *Separation Process Principles*, John Wiley, New York (1998). Textbook with broad coverage of chemical separation processes including adsorption, crystallization and membrane separations. A second edition was published in 2006. (Seader and Henley, 1998.)

(10) Noble, R.D. and P.A. Terry, *Principles of Chemical Separations with Environmental Applications*, Cambridge University Press, Cambridge, UK (2004). Treats chemical separations in an environmental context. (Noble and Terry, 2004.)

(11) Wankat, P.C., *Separation Process Engineering*, 2nd edn., Prentice Hall, Upper Saddle River, NJ (2007) (formerly published as *Equilibrium Staged Separations*, Elsevier, New York, 1987). (Wankat, 2007.)

(12) Benitez, J., *Principles and Modern Applications of Mass Transfer Operations*, 2nd edn., John Wiley, Hoboken, NJ (2009). (Benitez, 2009.)

Bioseparations

(1) Belter, P.A., E.L. Cussler and W.-S. Hu, *Bioseparations: Downstream Processing in Biotechnology*, Wiley-Interscience, John Wiley, New York (1988). An introduction to bioseparations. (Belter *et al.*, 1988.)

(2) Garcia, A.A., M.R. Bonen, J. Ramirez-Vick, M. Sadaka and A. Vuppu, *Bioseparation Process Science*, Blackwell Science, Malden, MA (1999). (Garcia *et al.*, 1999.)

(3) Ladisch, M.R., *Bioseparations Engineering: Principles, Practice and Economics*, John Wiley, New York (2001). (Ladisch, 2001.)

(4) Harrison, R.G., P. Todd, S.R. Rudge and D.P. Petrides, *Bioseparations Science and Engineering*, Oxford University Press, New York (2003). (Harrison *et al.*, 2003.)

Membrane separations

(1) Hwang, S.T. and K. Kammermeyer, "Membranes in separations," Vol. VII in *Techniques of Chemistry*, A. Weissberger (ed.), Wiley-Interscience, New York (1975). Reprinted, Kreiger Publishing, Malabar, FL (1984). (Hwang and Kammermeyer, 1984.)

(2) Meares, P. (ed.), *Membrane Separation Processes*, Elsevier Scientific Publishing Co., Amsterdam (1976). (Meares, 1976.)

(3) Belfort, G., *Synthetic Membrane Processes*, Academic Press, New York (1984). (Belfort, 1984.)

(4) Ho, W.S.W. and N.N. Li, "Membrane processes," in *Perry's Chemical Engineers' Handbook*, R.H. Perry

and D.W. Green (eds.), 6th edn., McGraw-Hill, New York, (1984), pp. 17.14–17.35. (Ho and Li, 1984a.)

(5) Rautenbach, R. and R. Albrecht, *Membrane Processes*, John Wiley, New York (1989). (Rautenbach and Albrecht, 1989.)

(6) Mulder, M., *Basic Principles of Membrane Technology*, 2nd edn., Kluwer Academic Publishers, Dordrecht (1991); a second edition followed in 1996. (Mulder, 1991.)

(7) Ho, W.S.W. and K.K. Sirkar (eds.), *Membrane Handbook*, Van Nostrand Reinhold (1992). Reprinted, Kluwer Academic Publishers, Boston (2001). (Ho and Sirkar, 2001.)

(8) Noble, R.D. and S.A. Stern (eds.), *Membrane Separations Technology: Principles and Applications*, Elsevier, Amsterdam (1995). (Noble and Stern, 1995.)

(9) Baker, R.W., *Membrane Technology and Applications*, 2nd edn., John Wiley, Hoboken, NJ (2004). (Baker, 2004.)

Particle separations

(1) Wark, K. and D.F. Warner, *Air Pollution, Its Origin and Control*, IEP - Dun-Donnelley, Harper & Row, New York (1976). (Wark and Warner, 1976.)

(2) Friedlander, S.K., *Smoke, Dust and Haze: Fundamentals of Aerosol Behavior*, John Wiley, New York (1977).

(3) Svarovsky, L. (ed.), *Solid-Liquid Separation*, Butterworths, London (1977). (Svarovsky, 1977.)

(4) Svarovsky, L., *Solid-Gas Separation*, Elsevier Scientific Publishing, Amsterdam (1981). (Svarovsky, 1981.)

(5) Flagan, R.C. and J.H Seinfeld, *Fundamentals of Air Pollution Engineering*, Prentice Hall, Englewood Cliffs, NJ (1988). A useful book written from a fundamental perspective. (Flagan and Seinfeld, 1988.)

(6) Randolph, A.D. and M.A. Larson, *Theory of Particulate Processes: Analysis and Techniques of Continuous Crystallization*, 2nd edn., Academic Press, New York (1988). (Randolph and Larson, 1988.)

(7) Soo, S.L., *Particulates and Continuum: Multiphase Fluid Dynamics*, Hemisphere Publishing, New York (1989). (Soo, 1989.)

(8) Tien, C., *Granular Filtration of Aerosols and Hydrosols*, Butterworths, Boston, MA (1989). (Tien, 1989.)

Other books

(1) Bird, R.B., W.E. Stewart and E.N. Lightfoot, *Transport Phenomena*, John Wiley, New York (1960); 2nd edn., John Wiley, New York (2002). A key book for transport phenomena. Bird *et al.* (2002.)

(2) Foust, A.S., L.A. Wenzel, C.W. Clump, L. Maus and L. B. Anderson, *Principles of Unit Operations*, John Wiley, New York (1960). (Foust *et al.*, 1960.)

(3) McCabe, W.L. and J.C. Smith, *Unit Operations of Chemical Engineering*, 3rd edn., McGraw-Hill, New York (1976); 5th edn., (1993), with additional author P. Harriott. (McCabe *et al.*, 1993.)

(4) Perry, R.H. and D.W. Green (eds.), *Perry's Chemical Engineers' Handbook*, 6th edn., McGraw-Hill, New York (1984). (Perry and Green, 1984.)

(5) Cussler, E.L., *Diffusion: Mass Transfer in Fluid Systems*, 2nd edn., Cambridge University Press, Cambridge, UK (1997); 3rd edn. (2009).

(6) Geankoplis, C.J., *Transport Processes and Separation Process Principles*, 4th edn., Prentice-Hall PTR, Upper Saddle River, NJ (2003). (Geankoplis, 2003.)

Table 1. Relevant sections for each phase equilibrium based separation process

	Chapters 1 & 2 Describe separation	Chapter 3 Basis for separation	Chapter 4 Separation in a closed vessel	Chapter 5 Effect of chemical reaction	Chapter 6 Bulk flow ∥ to force(s) and CSTS	Chapter 7 Bulk flow ⊥ to force(s)	Chapter 8 Bulk flow of two phases ⊥ to force(s)	Chapters 9, 10 & 11 Other aspects
Absorption (and stripping)	1.1–1.4, 2.1, 2.3	3.3.7.1,[a] 3.4.1.1, 3.4.3.1	4.1, 4.1.1.1	5.1, 5.2.1.1, 5.3.1	6.2.1.1	7.1.2.1	8.1.1, 8.1.1.1–8.1.1.3, 8.1.2, 8.2.1.1, 8.2.2.2	
Adsorption (& simulated moving beds)	1.1–1.4, 2.1, 2.5	3.1.3.2.3, 3.1.3.2.4,[b] 3.3.5, 3.3.7.4, 3.3.7.6, 3.4.1.4	4.1, 4.1.5	5.2.3.2	6.2.1.1	7.1.1–7.1.7.1	8.1.1, 8.1.1.1–8.1.1.3, 8.1.6	10.1.4.4
Chromatography[c]	1.1–1.4, 2.1, 2.5	3.1.3.2.3, 3.1.3.2.4, 3.2.1, 3.2.2, 3.3.7.4, 3.3.7.6, 3.4.1.4, 3.4.1.5	4.1.3, 4.1.5–4.1.8, 4.1.9.1	5.2.3.2	6.2.1.1	7.1.1.1, 7.1.5–7.1.7.1	8.2.2, 8.3.1	11.1
Crystallization[d]	1.1–1.4, 2.1, 2.4.1	3.3.1, 3.3.7.5, 3.4.1.3	4.1.4, 4.1.9.1	5.2.4	6.2.3, 6.4.1.1		8.1.1, 8.1.1.1–8.1.1.3, 8.1.5	9.1.2.2, 9.1.2.3
Cycling zone[e] adsorption						7.1.4.4		
Devolatilization			4.1, 4.1.2		6.3.2.1			
Distillation	1.1–1.7, 2.1–2.3, 2.4.2, 2.4.3	3.1.5.1,[a] 3.3.7.1, 3.4.1.1, 3.4.3.1	4.1, 4.1.2	5.2.1.2	6.2.1.1, 6.3.2.1		8.1.1, 8.1.1.1–8.1.1.3, 8.1.3, 8.2.1, 8.3.2	10.1.1, 10.1.3,[f] 10.1.4.2, 10.2.1.1, 10.2.1.2, 10.2.1.3, 10.2.2
Drying (and freeze-drying)	1.1–1.4	3.1.4, 3.3.7.5			6.1.4, 6.3.2.4			11.1
Evaporation	1.1–1.4, 2.2, Ex. 2.2.2							10.1.1, 10.2.1
Extraction	1.1–1.6, 2.1, 2.3	3.2.2, 3.3.7.2, 3.3.7.9, 3.4.1.2, 3.4.3.2	4.1.3, 4.1.7, 4.1.8	5.2.2, 5.3.2	6.2.1.1, 6.3.2.2, 6.4.1.2		8.1.1, 8.1.1.1–8.1.1.3, 8.1.4	10.1.4.3, 11.4
Flotation		3.3.8						
Foam fractionation	Prob. 2.2.4	3.3.5, 3.3.7.6	4.1.5	5.2.5				
Ion exchange	1.1–1.4, 2.1, 2.5	3.1.3.2, 3.3.7.7, 3.4.1.5, 3.4.2.5	4.1.6	5.2.3.2		7.1.1.4, 7.1.5.1.6	8.1.1, 8.1.6	11.2
Leaching		3.3-7.4	4.1.4					
Melt crystallization	1.1–1.4, 2.1, 2.4.1	3.3.7.5	4.1.4		6.2.1, 6.3.2.3		8.1.1, 8.1.5	
Normal freezing	1.1–1.4, 2.1	3.3.7.5	4.1.4		6.3.2.3			
Parametric pumping (see adsorption)	1.1–1.4, 2.1, 2.5	3.3.7.4, 3.3.7.6, 3.4.1.4	4.1.5		6.2.1.1	7.1.4, 7.1.4.1, 7.1.4.2, 7.1.4.3		

Precipitation (see crystallization)		3.3.7.5			6.2.3, 6.4.1.1			
Pressure-swing adsorption (see adsorption)		3.1.3.2.3, 3.1.3.2.4, 3.3.5, 3.3.7.4, 3.3.7.6, 3.4.1.4	4.1, 4.1.5		6.2.1.1	7.1.1, 7.1.2		
Solvent extraction (see extraction)	1.1–1.6, 2.1, 2.3	3.3.7.2, 3.3.7.9, 3.4.1.2, 3.4.3.2	4.1.3, 4.1.7, 4.1.8	5.2.2, 5.3.2	6.2.1.1, 6.3.2.2, 6.4.1.2		8.1.1, 8.1.1.1–8.1.1.3, 8.1.4	10.1.4.3, 11.3, 11.4
Supercritical fluid extraction	1.1–1.6, 2.1, 2.3	3.3.7.2, 3.3.7.9, 3.4.1.2, 3.4.1.6, 3.4.3.2	4.1.3, 4.1.7	5.2.2	6.2.1.1, 6.3.2.2, 6.4.1.2		8.1.1, 8.1.4	
Zone melting (see crystallization)	1.1–1.4, 2.1	3.3.7.5	4.1.4, 4.1.9.1		6.2.1, 6.3.2.3		8.1.5	

[a] See also Sections 3.1.2.5, 3.1.3.2, 3.1.4, 3.3.1 and Example 3.1.9.
[b] See also Sections 3.1.2.5, 3.1.4, 3.2.1, 3.2.2, 3.3.1.
[c] See also those listed under "Adsorption".
[d] See also adductive crystallization, clathration, Sections 4.1.9.1.1, 4.1.9.1.2.
[e] See also those listed under "Adsorption" and "Parametric pumping".
[f] See also Chapter 9 and Sections 10.2.4, 11.3.

Table 2. Relevant sections for each membrane based separation process

	Chapters 1 & 2 Describe separation	Chapter 3 Basis for separation	Chapter 4 Separation in a closed vessel	Chapter 5 Effect of chemical reaction	Chapter 6 Bulk flow ‖ to force(s) and CSTS	Chapter 7 Bulk flow ⊥ to force(s)	Chapter 8 Bulk flow of two phases ⊥ to force(s)	Chapters 9, 10 & 11 Other aspects
Cake filtration	2.1, 2.2, 2.4	3.4.2.3	4.2.1.2		6.1.6, 6.3.1.4, 6.3.3.1	7.2.1.4, 7.2.1.5		11.1, 11.4
Dialysis	1.1–1.4, 2.1, 2.3	3.1.3.2.3, 3.1.4, 3.3.7.4, 3.4.2.3.1	4.3.1	5.4.3			8.1.1, 8.1.7, 8.2.4.1	11.1
Donnan dialysis	2.1, 2.3	3.1.3, 3.3.7.7, 3.4.2.5	4.3.1, 4.3.2	5.2.6				11.4
Electrodialysis	1.1–1.4, 2.1, 2.3	3.1.3.2, 3.3.7.7, 3.4.2.5	4.3.2				8.1.1, 8.1.7	10.2.3, 11.2
Emulsion liquid membrane	1.1–1.4, 2.1, 2.3	3.1.4, 3.3.7.2, 3.4.1.2	4.1.3, 4.1.8	5.2.2, 5.3.2, 5.4.4, 5.4.4.1–5.4.4.3, 5.4.4.5	6.2.1.1, 6.3.2.2, 6.4.2.2		8.1.1, 8.1.8	
Filtration (see cake filtration)	2.1, 2.2, 2.4	3.4.2.3			6.1.6, 6.3.1.4, 6.3.3.1	7.2.1.4, 7.2.1.5 7.2.2		11.2, 11.4
Gas permeation (membrane gas separation)	1.1–1.4, 2.1, 2.2	3.1.3.2, 3.3.1, 3.3.7.3, 3.4.2.2	4.3.3	5.4.4.1, 5.4.4.5, 5.4.4.6	6.3.3.5, 6.4.2.2	7.2.1.1	8.1.1, 8.1.9, 8.2.4.2	10.1.4.1
Gaseous diffusion	2.1, 2.2	3.1.3.2.4, 3.4.2.4	Prob. 4.3.9					9.1.1
Hollow fiber contained liquid membrane	1.1–1.4, 2.1–2.3	3.1.4, 3.3.7.2, 3.3.7.4, 3.4.3	4.1.3, 4.1.8	5.2.2, 5.3.2, 5.4.4, 5.4.4.1–5.4.4.4	6.2.1.1, 6.3.3.5, 6.4.2.2		8.1.8, 8.1.9	
Immobilized liquid membrane	1.1–1.4, 2.1–2.3	3.1.4, 3.3.7.2, 3.3.7.4, 3.4.3	4.1.3, 4.1.8	5.2.2, 5.3.2, 5.4.4, 5.4.4.1–5.4.4.3, 5.4.4.5	6.2.1.1, 6.3.3.5, 6.4.2.2	7.2.1.1	8.1.8, 8.1.9	
Ion transport membrane				5.4-5.2				
Liquid membrane processes	1.1–1.4, 2.1–2.3	3.1.4, 3.3.7.2, 3.3.7.4	4.1.3, 4.1.8	5.2.2, 5.3.2, 5.4.4 5.4.4.1–5.4.4.3, 5.4.4.5	6.4.1.2		8.1.8, 8.1.9	
Membrane contactor gas absorption/ stripping See reverse	1.1–1.4, 2.1–2.3	3.1.2.5, 3.1.3.2.3, 3.1.3.2.4, 3.1.4, 3.3.7.1, 3.4.1.1, 3.4.3.1	4.1.1.2, 4.1.7	5.2.1, 5.3.1	6.2.1.1		8.1.1.1, 8.1.1.2, 8.1.2.1, 8.1.2.2, 8.1.2.2.1	11.2
Microfiltration	2.1, 2.4	3.1.2, 3.1.3.1, 3.3.6, 3.4.2.3	4.2.3.1–4.2.3.3		6.2.2, 6.2.3, 6.3.1.4, 6.3.3.1	7.2.1.4, 7.2.1.5		11.1, 11.2
Pervaporation	1.1–1.4, 2.1, 2.2	3.1.3.2, 3.3.7.3, 3.3.7.4, 3.4.1.1, 3.4.2.1.1	4.1.2, 4.3.3		6.3.3.4			
Reverse osmosis[a] (Hyperfiltration)	1.1–1.4, 1.5, 1.7, 2.1, 2.2	3.1.2.5, 3.1.3.2, 3.1.5.1, 3.1.5.2, 3.3.7.4, 3.4.2.1	4.3.4	5.4.1, 5.4.1.1	6.3.3.3	7.2.1.2		9.1.2.1, 10.1.2, 11.2
Ultrafiltration	1.1–1.4, 1.5, 2.1, 2.2, 2.4.2, 2.5	3.1.3.2, 3.1.3.2.3, 3.1.5, 3.3.7.4, 3.4.2.3	4.3.4	5.4.2, 5.4.2.1–5.4.2.3	6.3.3.2	7.2.1.3		9.1.1, 9.1.2.1, 11.1, 11.2

[a] Forward osmosis, Prob. 4.3.12.

Table 3. Relevant sections for each centrifugal force driven separation process

	Chapters 1 & 2 Describe separation	Chapter 3 Basis for separation	Chapter 4 Separation in a closed vessel	Chapter 5 Effect of chemical reaction	Chapter 6 Bulk flow $\parallel$ to force(s) and CSTS	Chapter 7 Bulk flow $\perp$ to force(s)	Chapter 8 Bulk flow of two phases $\perp$ to force(s)	Chapters 9, 10 & 11 Other aspects
Centrifugal elutriation	2.1, 2.4, 2.5	3.1.2.2, 3.1.2.7, 3.1.3.1, 3.1.3.2, 3.2.1			6.2.2, 6.3.1.3			11.1
Centrifugal separations (centrifugal filtration)	2.1, 2.4, 2.5	3.1.2.2, 3.1.2.7, 3.1.3.1, 3.1.3.2, 3.2.1	4.2.1.1, 4.2.1.2, 4.2.1.3		6.1.3, 6.1.6, 6.2.2, 6.3.1.3	7.3.2.1–7.3.2.4	8.2.3	11.1
Cyclone dust separator	2.1, 2.4, 2.5	3.1.2.2, 3.1.2.7, 3.1.3.1, 3.1.3.2, 3.2.1	4.2.1.1, 4.2.1.3		6.2.2	7.3.2.3	8.2.3	
Gas centrifuge	2.1, 2.2, 2.3	3.1.2.2, 3.1.2.7, 3.1.3.1, 3.1.3.2, 3.2.1, 3.3.3	4.2.1.1		6.1.3, 6.2.1	7.3.2.4	8.1.1, 8.1.10	9.1.1
Hydroclone	2.1, 2.2, 2.4	3.1.2.2, 3.1.2.7, 3.1.3.1, 3.1.3.2	4.2.1.2		6.2.1, 6.2.2	7.3.2.3.1		
Isopycnic sedimentation	1.1–1.4, 2.1, 2.2, 2.4	3.1.2.2, 3.1.2.7, 3.1.3.1, 3.1.3.2, 3.2.1, 3.3.3	4.2.1.3					
Separation nozzle	1.1–1.4, 2.1 2.2, 2.4	3.1.2.2, 3.1.2.7, 3.1.3.1, 3.1.3.2, 3.2.1	4.2.1.1			7.3.2.4		

Table 4. Relevant sections for each electrical force driven separation process

	Chapters 1 & 2 Describe separation	Chapter 3 Basis for separation	Chapter 4 Separation in a closed vessel	Chapter 5 Effect of chemical reaction	Chapter 6 Bulk flow ∥ to force(s) and CSTS	Chapter 7 Bulk flow ⊥ to force(s)	Chapter 8 Bulk flow of two phases ⊥ to force(s)	Chapters 9, 10 & 11 Other aspects
Capillary electrophoresis	1.1–1.4, 2.1, 2.4.2, 2.4.3, 2.5	3.1.2.2, 3.1.2.7, 3.1.3.2, 3.2.1	4.2.2.1, 4.2.2.2	5.2.2.1, 5.2.2.2	6.1.5, 6.2.1 6.3.1.2	7.1.7, 7.1.7.1	8.2.2.1, 8.2.2.3	11.1
Continuous free-flow electrophoresis[a]	1.1–1.4, 2.1, 2.2, 2.4, 2.5	3.1.2.2, 3.1.2.7, 3.1.3.2, 3.2.1	4.2.2.1, 4.2.2.2	5.2.2.1, 5.2.2.2	6.1.1, 6.1.5, 6.2.1, 6.3.1.2	7.3.1.1		
Corona-discharge reactor		3.1.6.1, 3.1.6.2				7.3.1.2		
Dielectrophoresis		3.1.2.2, 3.1.2.7				7.3.1.1.1		11.1
Electrochemical membrane gas separation				5.4.4.6			8.1.8	
Electrochromatography		3.1.2.2, 3.1.2.7, 3.1.3.2, 3.2.1	4.2.2.1			7.1.7, 7.1.7.1	8.1.1, 8.1.6	11.2
Electrodialysis		see Table 2						
Electrosettler		3.1.2.2, 3.1.2.7	4.2.2, 4.2.3.4.1					
Electrostatic particle separator	2.1, 2.2, 2.4	3.1.2.2, 3.1.2.7, 3.1.3.1			6.2.2, 6.3.1.5	7.2.2, 7.3.1.4		
Electrostatic precipitator	2.1, 2.2, 2.4	3.1.2.2, 3.1.2.7, 3.1.3.1			6.2.2, 6.3.1.5	7.2.2, 7.3.1.3		
Flow cytometry	2.1, 2.2, 2.4	3.1.2.2, 3.1.2.7, 3.1.3.1			6.2.2	7.3.1.5		
Ion mobility spectrometry	2.1, 2.2, 2.4	3.1.2.2, 3.1.2.7, 3.1.6.1, 3.1.6.2				7.3.1.2		
Isoelectric focusing	1.1–1.4, 2.1, 2.5	3.1.2.2, 3.1.2.7, 3.1.3.2, 3.2.1	4.2.2.1, 4.2.2.2					
Laser isotope separation	2.1, 2.2, 2.4	3.1.2.2 3.1.2.7, 3.1.6.1/2				7.3.1.2		
Micellar electro-kinetic chromatography	1.1–1.4, 2.1, 2.4.2, 2.4.3, 2.5	3.1.2.2, 3.1.2.7, 3.1.3.2, 3.2.1	4.2.2.1, 4.2.2.2	5.2.3.2	6.1.5, 6.2.1, 6.3.1.2		8.2.2.1	
Potential swing adsorption (Electrosorption, capacitive deionization)						7.1.3		

[a] Moving boundary electrophoresis, zone electrophoresis: Section 4.2.2.2.

Table 5. Relevant sections for gravitational or magnetic force driven separation processes

	Chapters 1 & 2 Describe separation	Chapter 3 Basis for separation	Chapter 4 Separation in a closed vessel	Chapter 5 Effect of chemical reaction	Chapter 6 Bulk flow ‖ to force(s) and CSTS	Chapter 7 Bulk flow ⊥ to force(s)	Chapter 8 Bulk flow of two phases ⊥ to force(s)	Chapters 9, 10 & 11 Other aspects
Gravity driven separation (elutriation, jigging)		3.1.2.1, 3.1.2.7, 3.1.3.1, 3.1.3.2, 3.2.3	4.1.10, 4.2.3, 4.2.3.1, 4.2.3.2–4.2.3.4		6.3.1.1, 6.3.1.5	7.2.2, 7.3.3, 7.3.3.1, 7.3.4		
Magnetic force driven separation		3.1.2.4, 3.1.2.7, 3.1.3.1, 3.1.3.2				7.3.4, 7.3.5, 7.3.5.1, 7.3.5.2, 7.3.5.2.1		
Magnetophoresis		3.1.2.4, 3.1.2.7, 3.1.3.1, 3.1.3.2				7.3.5, 7.3.5.1		
High-gradient magnetic separation		3.1.2.4, 3.1.2.7, 3.1.3.1, 3.1.3.2				7.3.5, 7.3.5.1, 7.3.5.2, 7.3.5.2.1		

Table 6. Relevant sections for each separation process driven by other forces

	Chapters 1 & 2 Describe separation	Chapter 3 Basis for separation	Chapter 4 Separation in a closed vessel	Chapter 5 Effect of chemical reaction	Chapter 6 Bulk flow ‖ to force(s) and CSTS	Chapter 7 Bulk flow ⊥ to force(s)	Chapter 8 Bulk flow of two phases ⊥ to force(s)	Chapters 9, 10 & 11 Other aspects
Acoustic forces/waves		3.1.2.6	4.2.4, 4.2.4.1					
Inertial impaction–depth filtration		3.1.2.6.1			6.2.2, 6.3.1.4	7.2.2		11.2
Radiation pressure–photophoretic separation		3.1.2.6, Ex. 3.1.4			6.2.2	7.3.6		
Thermal diffusion	1.1–1.5	3.1.2.6	4.2.5		6.2.1.1		8.1.1, 8.1.11	
Thermophoresis		3.1.2.6						

Table 7. Relevant sections for additional separation processes

	Chapters 1 & 2 Describe separation	Chapter 3 Basis for separation	Chapter 4 Separation in a closed vessel	Chapter 5 Effect of chemical reaction	Chapter 6 Bulk flow ‖ to force(s) and CSTS	Chapter 7 Bulk flow ⊥ to force(s)	Chapter 8 Bulk flow of two phases ⊥ to force(s)	Chapters 9, 10 & 11 Other aspects
Field-flow fractionation	1.1–1.4, 2.1, 2.5	3.1.2.1–3.1.2.7, 3.1.3.1, 3.1.3.2			6.1.1	7.3.4		
Mass diffusion (sweep diffusion)	1.1–1.6, 2.1, 2.3, 2.4.3	3.1.2.5, 3.1.3.1, 3.1.3.2, 3.1.4, 3.1.5.1					8.1.11	

Table 8. Basic properties and phenomena underlying separation processes for molecules, macromolecules and particles

Basic property	Phenomenon causing separation	Phase equilibrium based separation processes	Membrane separation processes	External force based and other separation processes
Charge-to-mass ratio	different charge-to-mass ratios of different particles/molecules			electrostatic separation of particles (Section 7.3.1.4), mass spectrometry (Karger *et al.*, 1973)
Chelation of metal ions	selective chelation	solvent extraction of metals (Section 5.2.2.4, Table 1)	liquid membrane processes (Table 2, Section 8.1.8)	
Condensability of a vapor/gas species	different condensabilities of vapors/gases	vapor absorption, vapor–liquid chromatography (Table 1, Section 4.1.1.1)	membrane gas permeation (Section 4.3.3, Figure 4.3.4, Table 2)	
Density of liquid	heavier liquid is located further from the center for an immiscible liquid mixture in a centrifugal force field			centrifugal separation (Section 7.3.2.1.1)
Dielectric constant	particles/cells develop a dipole moment in a nonuniform electrical field			dieelectrophoresis (Table 4, Section 7.3.1.1.1)
Diffusivity	different diffusivities for different species		ultrafiltration, nanofiltration, dialysis, gas permeation, gaseous diffusion (Table 2)	thermal diffusion (Table 6), mass/sweep diffusion (Table 7, Section 7.3.4)
Electrical charge on an element, molecule, macromolecule, particle	attraction toward oppositely charged electrode		electrodialysis (Table 2)	element, molecule, particle in gas phase: laser isotope separation, corona-discharge reactor, ion mobility spectrometry, electrostatic precipitator, flow cytometry (Table 4)
	different ionic/electrophoretic mobilities and attraction toward oppositely charged electrode			capillary electrophoresis, continuous free-flow electrophoresis
	membrane surface has opposite charge		nanofiltration separation of divalent anions, ultrafiltration of charged proteins (Table 2)	
Geometrical partitioning of a solute between pore and a solution	size exclusion for different size solutes	gel permeation chromatography, size exclusion chromatography (Table 1, Sections 3.3.7.4, 3.4.2.3.1, 7.1.5.1.7)	dialysis, ultrafiltration (Table 2, Sections 3.3.7.4, 3.4.2.3, 4.3.1, 6.3.3.2)	

Magnetic permeability or susceptibility	different magnetic permeability/susceptibility between particles and the solution			magnetophoresis, high-gradient magnetic separation (Table 5)
Mass, density of particles	different terminal/settling velocities in gravitational/centrifugal force field			gravitational settlers, jigging, elutriational devices, centrifuges, cyclone separators (Tables 3, 5)
Molecular mass of species	heavier gas/isotope concentrates further from the center in a centrifugal force field			gas centrifuge, separation nozzle (Table 3)
Particle compressibility	positive and negative values of ϕ-factor in an acoustic force field			acoustic force based separation (Table 6)
pI, isoelectric point of a protein	proteins having different pI values equilibrate at different locations in a pH gradient			isoelectric focusing (Table 4)
Reversible complexation of solute/macrosolute with a ligand	selective complexation capability	affinity chromatography (Section 7.1.5.1.8)	membrane chromatography (Section 7.1.5.1.8), ultrafiltration (Section 5.4.2.1)	
Reversible electrostatic interaction between counterions in solution and porous charged solids	preferential interaction of a counterion in solution with porous charged solid	ion exchange, ion exchange chromatography (Table 1, Sections 3.3.7.7, 4.1.6)	electrodialysis, Donnan dialysis (Table 2, Sections 3.4.2.5, 4.3.2)	
Solubility in a solvent of a solute present in a solution or a solid	different solubilities for different solutes	solvent extraction, liquid–liquid chromatography, leaching, supercritical extraction (Table 1, Sections 4.1.3, 4.1.7, 3.3.7.5)	liquid membrane processes (Table 2, Section 8.1.8)	
Solubility in micellar core or selective complexation with micellar headgroups	selective partitioning into micellar core or complexation with charged headgroups	micellar extraction, reverse micellar extraction (Section 4.1.8)	micellar-enhanced ultrafiltration (Section 5.4.2.2)	micellar electrokinetic chromatography (Section 8.2.2.1)
Solubility of a gas/vapor in a liquid	different solubilities of gases/vapors in a liquid	absorption, gas–liquid chromatography (Table 1, Figure 3.3.3, Section 4.1.1.1)	membrane contactor (Table 2)	
Solubility of a gas/vapor in a polymer	different solubilities of gases/vapors		membrane gas permeation, pervaporation (Table 2)	
Species come out of solution or melt as a crystal	phase change leading to crystal formation	crystallization, zone melting (Table 1, Section 3.3.7.5)		
Surface active nature of solute	excess accumulation of a species at the interface of phase 1 and phase 2: gas–liquid, liquid–liquid	foam fractionation (Table 1, Sections 3.3.7.6, 4.1.5, 5.2.5)		

Table 8. (*cont.*)

Basic property	Phenomenon causing separation	Phase equilibrium based separation processes	Membrane separation processes	External force based and other separation processes
Surface adsorption potential	excess accumulation of a species at the interface of phase 1 and phase 2: gas–solid, liquid–solid	adsorption, chromatography (gas–solid, liquid–solid) (Table 1, Sections 3.3.7.6, 4.1.5)	gas separation by surface diffusion (Section 3.4.2.4), preferential sorption and capillary transport in reverse osmosis (Table 2)	
Vapor pressure of a liquid, volatility of a solute	different volatilities of bulk liquids and solutes	distillation, stripping (Table 1, Sections 4.1.1, 4.1.2)	pervaporation (Table 2, Section 6.3.3.4) (plus membrane permeability)	
Water removed by vaporization from a solution or moist solid	evaporation of water	evaporation, drying (Table 1)	membrane distillation (Song *et al.*, 2008)	
Water removed from a solid by sublimation	sublimation of water	freeze-drying (Table 1)		

1

Description of separation in a closed system

Separation is a major activity of chemical engineers and chemists. To separate a mixture of two or more substances, various operations called *separation processes* are utilized. Before we understand how a mixture can be separated using a given separation process, we should be able to describe the amount of separation obtained in any given operation. This chapter and Chapter 2 therefore deal with qualitative and quantitative descriptions of separation. Chapter 2 covers open systems; this chapter describes separations in a closed system.

In Section 1.1, we briefly illustrate the meaning of separation between two regions for a system of two components in a closed vessel. Section 1.2 extends this to a multicomponent system. In Section 1.3, various definitions of compositions and concentrations are given for a two-component system. In Section 1.4, we are concerned with describing the various indices of separation and their interrelationships for a two-region, two-component separation system. A number of such indices are compared with regard to their capacity to describe separation in Section 1.5 for a binary system. Next, Section 1.6 briefly considers the definitions of compositions and indices of separation for the description of separation in a multicomponent system between two regions in a closed vessel. Finally, Section 1.7 briefly describes some terms that are frequently encountered.

The separation of a mixture of chemical species 1 and 2 may involve separating either a mechanical mixture or a true solution of the two species. Examples of mechanical mixtures include water containing fine particles forming a slurry, air containing fine dust particles, etc. Here water or air comprise species 1, whereas the chemical substance present in the form of solid particles comprises species 2. (Although air is a mixture of various species, we consider it here as species 1.) Examples of true solutions of species 1 and 2 are: a solution of sugar in water, a liquid solution of benzene and toluene, a gas mixture of nitrogen and carbon dioxide, a solid solution of silver in gold, etc. In such true solutions, the two different chemical species are intimately mixed at the molecular level. In a mechanical mixture, solid particles, comprise aggregates of molecules of species 2, so the two species are not intimately mixed at the molecular level. Colloidal solutions and macromolecular solutions provide a spectrum of behavior in between these two limits. In this book we will primarily treat separation of true solutions, the so-called chemical separations. Separation of mechanical mixtures, as well as colloidal and macromolecular solutions, will also receive significant coverage.

1.1 Binary separation between two regions in a closed vessel

Consider a closed vessel A containing a true solution of species 1 and 2, as shown in Figure 1.1.1. Let the solution composition be uniform throughout the vessel. We will refer to this condition as the *perfectly mixed state*. The process of separation is now initiated somehow, either by the addition or extraction of some amount of energy without allowing any mass to enter or leave the system. The end result of any such process is shown by the condition existing in another vessel, B. Depending on the solution and the process employed, vessel B can have any number of conditions. If the condition is represented by I, we note that vessel B has two regions, each region being occupied by molecules of only one kind. Thus species 1 and 2 have been completely separated from each other. If the material in any one of the two regions is now removed from vessel B, we will have obtained the corresponding pure species. Such a condition is called *perfect separation*. It is the exact opposite of the perfectly mixed state present in vessel A. Further, the objective of all separation would be to achieve condition I in vessel B whatever the initial conditions. **Note that, henceforth, we would like to identify region 1 with species 1 and region 2 with species 2. This**

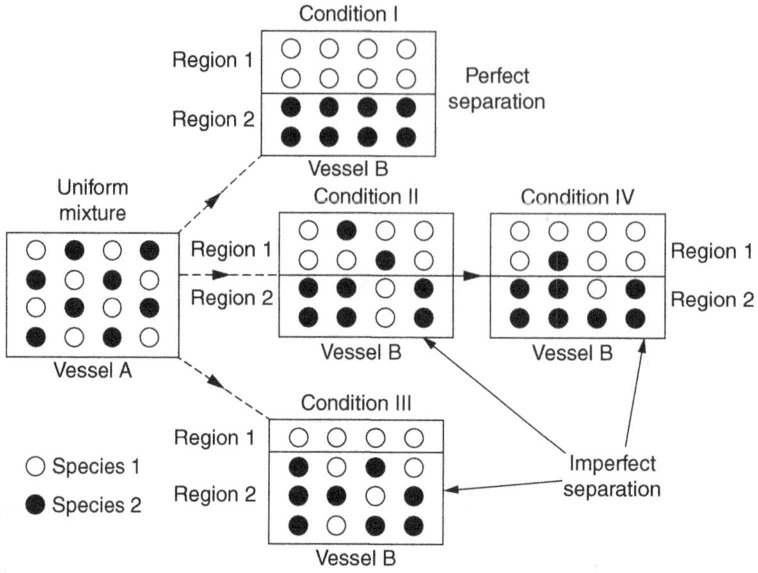

Figure 1.1.1. Binary separation between two regions in a closed vessel.

means that, for perfect separation, region 1 will have only molecules of species 1, and region 2 will have, only molecules of species 2.

The condition represented by II in Figure 1.1.1 is, however, quite different. Vessel B still has two regions. But region 1, instead of having only species 1, has some molecules of species 2, while region 2, instead of having only species 2, has some molecules of species 1. If the material in any one of the two regions is removed from vessel B, we will obtain a solution of two species instead of an amount of pure species. This condition illustrates *imperfect separation*. All real separation processes almost always result in an imperfect separation. Although perfect separation is therefore never attained, the condition of perfect separation is a valuable yardstick for measuring the amount of separation achieved by an actual separation process. If we refer to the separated material in any one of the two regions in condition II as a *fraction*, then the *purity level* or the *impurity level* of the fraction would provide one such yardstick. **By the convention adopted earlier, region 1 will have now species 2 as an impurity, whereas region 2 will have species 1 as an impurity.**

Another estimate of the separative efficiency of the process is provided by how far removed the compositions of the separated regions are from the original mixture in vessel A. Before we develop and use an *index of separation*, which serves as a criterion of merit to indicate the separative efficiency, we need to examine the quantities that describe the *composition* of a binary mixture. However, knowing the composition only can sometimes be misleading. For example, see the condition represented by III in Figure 1.1.1. We have, as before, two regions, region 1 and region 2. Region 1 has only pure species 1, whereas region 2 has both species 1 and 2. If we consider region 1 in III and region 1 in I and consider only the compositions of this region, we may mistakenly conclude that perfect separation has been achieved in both cases, since region 1 has only pure component 1 under both conditions. What we should also consider is how much of species 1 present in the original mixture in vessel A has been *recovered* in region 1 of vessel B in condition III. Since not all of the species 1 has been recovered and put in region 1, condition III represents another case of imperfect separation. Similarly, if region 2 had only species 2 and region 1 had a mixture of species 1 and 2, we again have imperfect separation.

If the previous paragraphs have led to the belief that a separation process always starts with a uniform mixture, as in vessel A of Figure 1.1.1, and ends up with any of the three types of conditions I, II or III in vessel B, this is not the case. One can start with a condition shown in II and carry out a separation process, perhaps by altering the physical state (e.g. temperature, pressure, etc.), and achieve greater separation. For example, consider condition IV in vessel B achieved by changing condition II. Obviously, region 1 in condition IV is purer in species 1, and similarly for species 2 in region 2. **Thus a separation operation can be carried out not only on a uniform mixture, but also on two mixtures of different compositions existing in two contiguous regions of a vessel or a container.**

We have used the notion of a *region* in a separation system without much explanation; it is worthwhile to

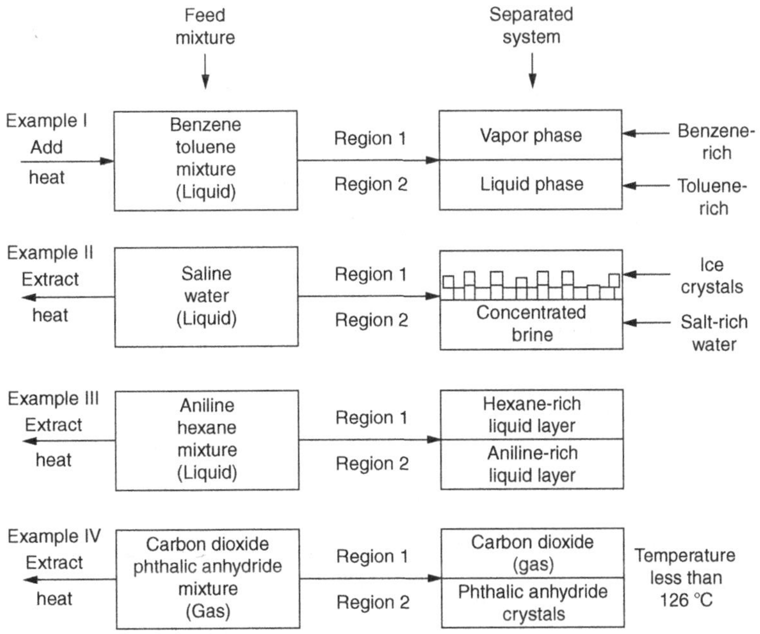

Figure 1.1.2. Binary separation systems with immiscible phases or regions.

explain its physical meaning. A region is a certain volume in space enclosed within given boundaries such that its contents have a composition different from that of an adjoining region (or those of adjoining regions if the separated system has more than two regions). The words *fraction* or *phase* may also be used to denote a region in a separation system. When the separation system contains two *immiscible phases* in a vessel, e.g. vapor (or gas) and liquid, solid and liquid, solid and vapor (or gas) or liquid and liquid, the use of "phase" or "fraction" is common.

Consider Figure 1.1.2, where we show examples of several practical separation processes. In example I, a liquid mixture containing the same number of moles of benzene and toluene is separated by the addition of heat into two immiscible phases - a vapor phase richer in benzene and a liquid phase richer in toluene. The space occupied by vapor is region 1, and the liquid phase is region 2.

In example II of Figure 1.1.2, a solution of salt in water can be separated by cooling the solution (extraction of heat) to form two distinct phases: ice crystals floating at the top (since they are significantly lighter than water or brine) and concentrated brine at the bottom. If the ice crystals are collected together without any brine sticking to the ice crystals (something which is almost impossible to carry out), then the concentrated brine is phase 2 and the ice crystals make up phase 1. If ice crystals cannot be collected together, then each ice crystal is considered to be part of region 1. Similarly, examples III and IV have two regions in a separated system; example III has a liquid–liquid system and example IV has a vapor–solid system[1] (Lowenheim and Moran, 1975).

In each of these four examples, the top phase, which is the lighter phase, is usually referred to as the *light fraction*, while the bottom phase is called the *heavy fraction*. Remember that for separation systems containing immiscible phases, *phase* or *fraction* is more commonly used than *region*.

On the other hand, consider the separation of saline water by a semipermeable membrane, as shown in example I of Figure 1.1.3. The uniform saline solution can be separated by the application of pressure energy on the feed saline to yield almost pure water on the low-pressure side of the membrane and concentrated brine on the high-pressure side. The membrane (or the barrier or partition) separates the two fractions in the two regions. However, both of these fractions are miscible with each other, unlike the immiscible phases of Figure 1.1.2. The words *fraction* or *region* are more appropriate here. Whereas in Figure 1.1.2, the two regions were separated only by the individual phase boundaries, in this case the membrane actually defines the boundaries of the two regions.

Another example of a membrane providing the boundaries between two regions is obtained in the separation of compressed air by a silicone membrane (example II of

[1] In the manufacture of phthalic anhydride by naphthalene oxidation, the gaseous products are CO_2, water vapor and phthalic anhydride. For illustration here we use a CO_2 and phthalic anhydride system, where phthalic anhydride crystals are formed from the vapor below 126 °C.

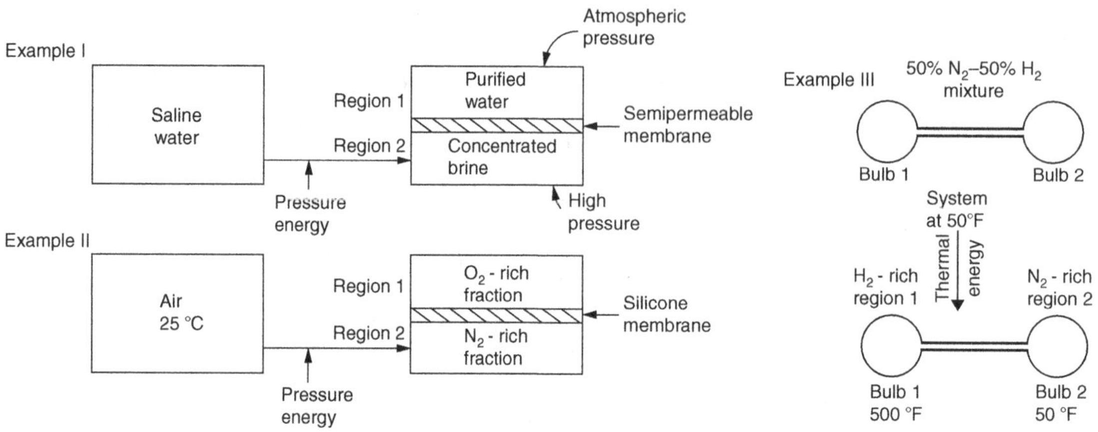

Figure 1.1.3. Binary separation systems with both regions miscible with each other.

Figure 1.1.3). Although separation in a closed vessel with membranes is often time-dependent, the conditions at any instant of time are sufficient for describing separation.

There can also be a separated system in which the two regions are not separated from each other by a membrane nor are they immiscible with each other. Such a separation system is shown in the form of a two-bulb cell in example III of Figure 1.1.3. Originally, the two-bulb cell contained a uniform gas mixture of nitrogen and hydrogen (say, 50% N_2–50% H_2, mole percent) at, say, 50 °F. If now the bulb 1 (region 1) has its temperature raised to and held at, say, 500 °F, while the temperature of bulb 2 (region 2) is maintained at 50 °F, we will observe that region 1 has become richer in the light component, hydrogen, whereas region 2 has become richer in the heavy component, nitrogen, due to the phenomenon of *thermal diffusion*.

What is important here is to recognize that the two regions having different gas compositions at steady state are not separated by either a membrane or a phase boundary.[2] In this case, the composition changes continuously through the capillary from one bulb to the other. However, since the capillary volume is very small compared to the volume of either of the bulbs, we may neglect it. So we have a separated system with two regions of different compositions such that not only are the materials in both regions completely miscible, but also there is no barrier separating the two regions. Although "light fraction" ("hydrogen-rich fraction") or "heavy fraction" ("nitrogen-rich fraction") are used to describe such a separation, the word *region* is more descriptive.

A few separated systems with continuously varying compositions are described in Problems 1.4.3 and 1.4.4. These involve systems without any barriers or immiscible phases.

[2]In any separated system without two immiscible phases or a barrier separating the two regions, the composition profile will be continuously varying.

1.2 Multicomponent separation between two regions in a closed vessel

If the separation system contains more than two species, strictly we are dealing with multicomponent separation. Thus, if we have species 1, 2 and 3 or 1, 2, 3 and 4 or 1, 2,. . ., n, etc., the separated system will have, in general, different compositions with respect to each species in the two regions. All situations described in Figures 1.1.1–1.1.3 are also valid here. The only difference is that the uniform solution we start with in Figure 1.1.1 is a multicomponent mixture. Further, whereas in Figure 1.1.1 one region has only pure component 1 present and the other region has a binary mixture of species 1 and 2, for multicomponent separation, if region 1 has only pure species 1, region 2 will have a mixture of all species (with or without species 1).

There is an additional point worth emphasizing. If there are only two regions in the closed vessel, it is not possible to have perfect separation since one requires as many regions as there are numbers of species present in the original mixture. Therefore, perfect separation for a three-component mixture requires three regions (or four regions for a four-component mixture). These requirements are illustrated by various alternative separation conditions in Figure 1.2.1 for a four-component mixture. For example, starting with a uniform mixture in vessel A, one can conceive of perfect separation in vessel B with four regions (conditions I or II). On the other hand, if two regions are available, only an imperfect separation is possible (conditions III or IV).

The case of multicomponent separation between two immiscible phases requires further consideration if each phase is primarily made up of one species and the remaining components are present in these phases in small amounts. Consider a three-component system consisting of, say, water, benzene and picric acid, the latter being present in very small amounts. Water and benzene

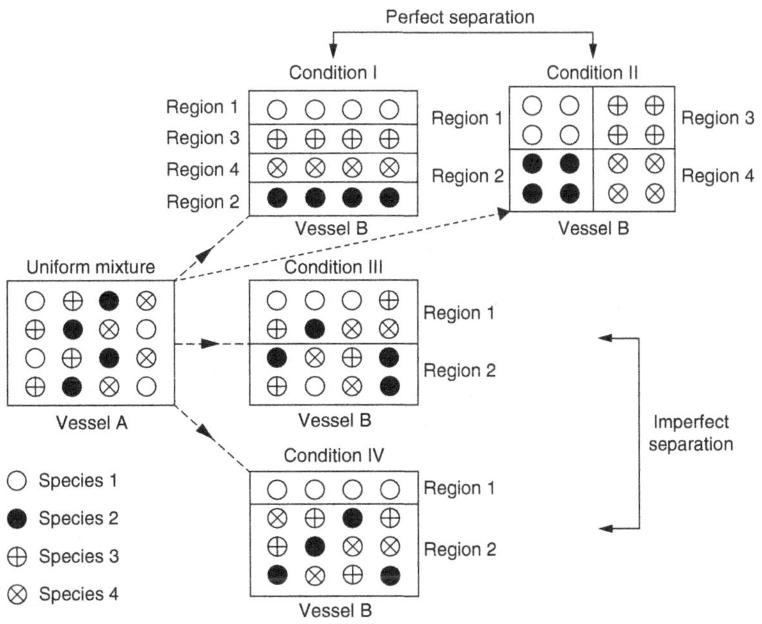

Figure 1.2.1. Separation of a four-component mixture in a closed vessel.

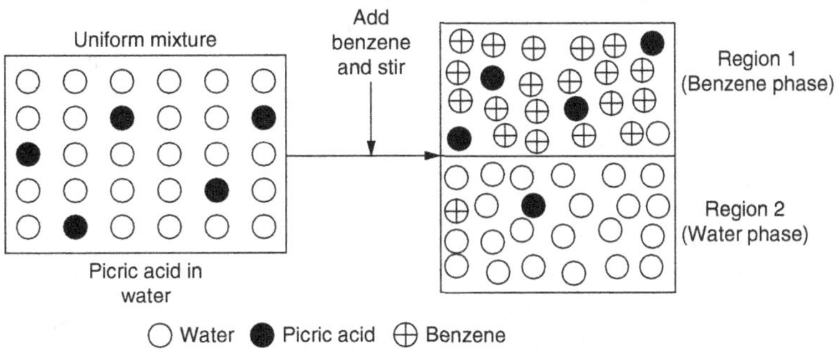

Figure 1.2.2. Multicomponent separation between two immiscible liquid phases.

form two immiscible liquid phases, with the lighter benzene layer above the heavier water layer. It is known that picric acid at low concentrations will be distributed (Hougen *et al.*, 1954) between the two phases such that its concentration in the benzene phase is higher than that in the water phase. Let the original mixture to be separated be picric acid in water. Benzene is added to this mixture, shown in Figure 1.2.2, since it is easier to recover picric acid from benzene than from its original mixture in water. Two fractions have been formed: a benzene based fraction, the *extract*, and the water based residue, the *raffinate*. Although this is a three-component system, the description of each phase can be given as if it were a binary system, i.e. a solution of picric acid in water or a solution of picric acid in benzene. Similarly, suppose two solutes, picric acid and benzoic acid, are distributed between two phases, one of which is a benzene phase and the other one is a water phase. Each phase can be described as if it were a ternary system since benzene and water are immiscible (in reality, very small amounts of water are soluble in benzene and vice versa).

In fact, one can go a step further by identifying the benzene extract phase as region 1 and the water raffinate phase as region 2 and working only with the concentrations of the two acids in each region. We then, in effect, have a binary system of benzoic acid and picric acid between two regions, 1 (benzene layer), and 2 (water layer), and one could proceed with the description in the manner of Section 1.1. Thus the description of a four-component separation system may be reduced to that of a two-component separation system provided each of the immiscible phases is made up of essentially one species only and the two components, in whose separation we are interested, are present in small amounts.

A variety of conditions are possible with multicomponent systems in general. **When one needs to separate particles of one size from particles of all other sizes present in a suspension, or when particles of different sizes are to be separated from one another as well as from the medium of suspension, a multicomponent separation problem exists. If a macromolecular substance or solution is to be fractionated, the nature of the description of the separation problem is similar**. Such problems are handled most conveniently by dealing with pseudo binary systems. We will come across such cases as we proceed further in this book, although binary systems will be encountered much more frequently.

1.3 Definitions of composition for a binary system in a closed vessel

So far we have described separation qualitatively. In order to describe separation quantitatively, first we have to define compositions and then use such compositions to define indices of separation. Only with the help of indices of separation can one describe separation quantitatively.

Refer to Figure 1.1.1 for a physical background behind the definitions of compositions. With both vessels A and B closed, we assume that the separation process which resulted in the condition of vessel B from that of vessel A conserved the total number of moles of each of the species 1 and 2. Assume further that within any region the composition is uniform everywhere (although this is not true for some of the representations actually used in Figure 1.1.1). Let m_{ij} denote the number of moles of the ith species in the jth region in vessel B under any particular condition. For a binary system, $i = 1, 2$, since there are two regions only, j has values 1 and 2. The *mole fraction* x_{ij} of species i in region j is defined by

$$x_{ij} = \frac{m_{ij}}{\sum_{i=1}^{2} m_{ij}}; \qquad i = 1,2; \qquad j = 1,2. \tag{1.3.1}$$

The *mole ratio* X_{ij} of species i in region j is given by

$$X_{ij} = \frac{m_{ij}}{m_{kj}}; \qquad i = 1,2; \qquad k = 1 \text{ or } 2. \tag{1.3.2}$$

Note that $(x_{ij} + x_{2j}) = 1$ but $(X_{ij} + X_{2j})$ need not equal 1. If the volume of the jth region is V_j, then the *molar concentration* C_{ij} of species i in region j of vessel B is given by

$$C_{ij} = \frac{m_{ij}}{V_j}. \tag{1.3.3}$$

It should be noted that $C_{1j} + C_{2j} = C_{tj}$, the total molar density of the mixture in region j. The *mass concentration* ρ_{ij} and the *mass fraction* u_{ij} of species i in region j are defined by

$$\rho_{ij} = \frac{m_{ij} M_i}{V_j} = C_{ij} M_i, \tag{1.3.4}$$

$$u_{ij} = \frac{\rho_{ij}}{\sum_{i=1}^{2} \rho_{ij}} = \frac{\rho_{ij}}{\rho_{tj}}, \tag{1.3.5}$$

where M_i is the molecular weight of the ith species and ρ_{tj} is the total mass density of the mixture in region j.

In dealing with mixtures of two isotopes (e.g. H_2 and D_2), one comes across instead of mole fraction the atom fraction a_{ij}. The *atom fraction* a_{ij} of the ith isotope of an element in region j is obtained as

$$a_{ij} = \frac{m^a_{ij}}{\sum_{i=1}^{2} m^a_{ij}}, \tag{1.3.6}$$

where m^a_{ij} is the number of atoms of the ith isotope of the element in the jth region. The sum $\left(m^a_{1j} + m^a_{2j}\right)$ represents the total number of atoms of the element in the jth region, with $i = 1, 2$ being the only two isotopic forms under consideration. For example, in a gas mixture of H_2 and D_2, if there are 0.186 moles of D_2 for 0.189 moles of H_2 and D_2, the atom fraction of deuterium in that gas mixture is $(0.186 \times 2/0.189 \times 2) = 0.984$.

For binary isotopic mixtures, the quantity similar to mole ratio X_{ij} is the *abundance ratio* X^a_{ij}, which is given by

$$X^a_{ij} = \frac{a_{ij}}{a_{kj}} = \frac{a_{ij}}{1 - a_{ij}}; \qquad i = 1,2; \quad k = 1 \text{ or } 2. \tag{1.3.7}$$

Thus the abundance ratio represents the number of atoms of the desired ith isotope per atom of the other isotope, where $a_{1j} + a_{2j} = 1$.

The quantities defined so far describe the composition of a given region, phase or fraction with respect to how much of one species is present amongst all the species in the given region, phase or fraction. It is also useful to know what fraction of the total amount of a species present in the total separation system is present in a given region. Such a quantity is the *segregation fraction* Y_{ij}:

$$Y_{ij} = \frac{m_{ij}}{\sum_{j=1}^{2} m_{ij}} = \frac{m_{ij}}{m^0_i}. \tag{1.3.8a}$$

Here m^0_i moles of the ith species are present in separation vessel B having two regions $j = 1$ and 2. Further, $Y_{i1} + Y_{i2} = 1$. For a two-component system with $j = 1, 2$, one may define a segregation matrix $[Y_{ij}]$ as well as the corresponding matrix for mole numbers $[m_{ij}]$:

$$[Y_{ij}] = \begin{vmatrix} Y_{11} & Y_{12} \\ Y_{21} & Y_{22} \end{vmatrix}; \; [m_{ij}] = \begin{vmatrix} m_{11} & m_{12} \\ m_{21} & m_{22} \end{vmatrix}. \qquad (1.3.8b)$$

See Problem 1.3.1 for related representations.

Following the segregation fraction Y_{ij} for nonisotopic systems, one can define the segregation fraction Y_{ij}^a of an isotopic mixture as

$$Y_{ij}^a = \frac{m_{ij}^a}{\sum_{j=1}^{2} m_{ij}^a} = \frac{m_{ij}^a}{m_i^{a0}}. \qquad (1.3.9a)$$

All three quantities, the mole fraction x_{ij}, the mass fraction u_{ij} and the atom fraction a_{ij}, vary between the limits of 0 and 1. When their value is 1, we have pure species i in region j. The closer their value is to 1, the greater is the purity of the region (fraction or phase) in the ith species.

On the other hand, the mole ratio X_{ij} and the abundance ratio X_{ij}^a vary between 0 and ∞, the latter value indicating only pure species i in region j. Thus, the upper limits of the two sets of quantities (x_{ij}, u_{ij}, a_{ij} and X_{ij}, X_{ij}^a) have radically different values, although all of these quantities indicate the level of purity of the ith species in the given jth region in their own ways.

The remaining quantities Y_{ij} and Y_{ij}^a have limits similar to those of x_{ij}, u_{ij} and a_{ij}, namely 0 and 1. But the upper limit has a different meaning. For example, $Y_{22} = 1$ signifies that region 2 has all of species 2 present in the separation system. But this does not mean that region 2 has pure species 2, since some species 1 may also be present in region 2. Thus if $Y_{22} = 1$, Y_{12} need not be zero (compare: if $x_{22} = 1$, $x_{12} = 0$). Further, Y_{11} need not be equal to 1. Condition III of Figure 1.1.1 illustrates such a separation. So **Y_{ij} provides an estimate of the extent of recovery of species i in region j with respect to the total amount of species i present in the whole separation system.** The goal of perfect separation is to segregate all of each species in its designated region in a pure form. For a separation system with $i = 1, 2$ and $j = 1, 2$, perfect separation may be represented as follows:

$$[Y_{ij}] = \begin{vmatrix} (Y_{11} = 1) & 0 \\ 0 & (Y_{22} = 1) \end{vmatrix};$$

$$[Y_{ij}^a] = \begin{vmatrix} (Y_{11}^a = 1) & 0 \\ 0 & (Y_{22}^a = 1) \end{vmatrix}. \qquad (1.3.9b)$$

It should be noted at this stage that some mass balance relations are necessary for relating conditions in vessel A to those in vessel B due to the law of conservation of mass. If the ith species mole fraction in the initial uniform mixture in vessel A is x_{if}, then, in the absence of chemical reactions, a balance of the total number of moles of both species leads to

$$\sum_{i=1}^{2} \sum_{j=1}^{2} m_{ij} = m_1^0 + m_2^0. \qquad (1.3.10)$$

A balance of the ith species only yields

$$(m_1^0 + m_2^0)\, x_{if} = \sum_{j=1}^{2} m_{ij}. \qquad (1.3.11)$$

Sometimes the initial binary mixture to be separated is not a uniform mixture as in vessel A, but instead has two regions whose compositions are characterized as x_{if_1} and x_{if_2} corresponding to $j = f_1, f_2$. Note that these regions are such that region f_1 usually has more of species 1, whereas region f_2 has more of species 2. The total and ith component mass balances in such a case lead to

$$m_1^0 + m_2^0 = \sum_{i=1}^{2} m_{if_1} + \sum_{i=1}^{2} m_{if_2};$$

$$\left(\sum_{i=1}^{2} m_{if_1}\right) x_{if_1} + \left(\sum_{i=1}^{2} m_{if_2}\right) x_{if_2} = m_i^0. \qquad (1.3.12)$$

1.4 Indices of separation for binary systems

A number of indices of separation may now be developed using the quantities defined in Section 1.3. A *separation index* is needed to indicate how much separation is obtained in a given separation process. Such a quantity enables one to determine quickly the separation capabilities of any particular process with respect to a given separation task. Since none of the quantities defined in Section 1.3 are restricted to any particular material or separation process (except a_{ij}, X_{ij}^a and Y_{ij}^a, which are restricted to isotopic mixtures), it is expected that the indices of separation given below for a two-component system will likewise be sufficiently general in their application. We restrict ourselves here to describing only the separation that has been achieved in vessel B (Figure 1.1.1) between regions 1 and 2. Further, the list of indices given below is not exhaustive.

The simplest separation indices are the *distribution ratio* (*or capacity factor*) k'_{i1}, the *distribution coefficient* κ_{i1} and the *equilibrium ratio* K_i:[3]

$$k'_{i1} = \frac{m_{i1}}{m_{i2}}; \qquad \kappa_{i1} = \frac{C_{i1}}{C_{i2}}; \qquad K_i = \frac{x_{i1}}{x_{i2}}. \qquad (1.4.1)$$

In general,

$$k'_{i1} = \frac{m_{i1}}{m_{i2}} = \frac{C_{i1}}{C_{i2}} \frac{V_1}{V_2} = \kappa_{i1} \frac{V_1}{V_2}, \qquad (1.4.2)$$

so that if $V_2 = V_1$, $k'_{i1} = \kappa_{i1}$. Note that the three indices in equation (1.4.1) are all defined with the light fraction, i.e. fraction 1, on the top. In this convention, then, for species 1 to concentrate more in region 1 implies that κ_{i1} is greater than 1, since $C_{i1} > C_{i2}$. This does not mean that $k'_{il} > 1$

[3] In chemical engineering literature k'_{il} is rarely used. But κ_{i1} is often used, especially in liquid extraction where the system, in general, is a multicomponent system (see Section 1.2). In distillation and flash separation processes, K_i is used frequently.

since for that to be true $V_1 \geq V_2$. However, the upper and lower limits for all three indices are, respectively, ∞ and 0. Further, an upper limit of ∞ for k'_{il} and κ_{il} as well as for K_i does not necessarily imply perfect separation. Similarly, the lower limit of zero does not necessarily imply either zero or perfect separation.

Before we present some complex indices of separation, some more indices apparently similar to those of equation (1.4.1) require further consideration. These are as follows.

Impurity ratios (Gleuckauf, 1955a):

$$\eta_1 = \frac{m_{21}}{m_{11}}; \qquad \eta_2 = \frac{m_{12}}{m_{22}}. \tag{1.4.3a}$$

Purity indices (de Clerk and Cloete, 1971):

$$I_j = -\log_{10}\eta_j; \qquad j = 1,2; \qquad I = \sum_{j=1}^{2} I_j. \tag{1.4.3b}$$

The subscript on the impurity ratio η_j refers to the phase (fraction or region) under consideration. Since phase 1 is supposed to contain primarily species 1 as a valid separation goal, species 2 in phase 1 is an impurity. Perfect separation therefore requires $\eta_1 = 0$ and $\eta_2 = 0$. For imperfect separation $\eta_j > 0$. Note, however, that $\eta_1 = 0$ does not imply $\eta_2 = 0$ or vice versa.

Since perfect separation requires $\eta_j = 0$ for both $j = 1, 2$ simultaneously, improved separation in a given problem implies decreasing values of η_j. If it is desired that the separation index value should increase as separation improves, one can define an *enrichment ratio* η'_j by utilizing the definition of η_j (Boyde, 1971):

$$\eta'_j = 1 - \eta_j. \tag{1.4.4}$$

For such a definition, the maximum value of $\eta'_j = 1$ implies no impurity in region j. One could add subscripts and superscripts to η'_j to indicate whether the separation of the desirable species 1 is being sought from the undesirable species 2 or vice versa (Boyde, 1971).

The most commonly used separation index in chemical engineering is the *separation factor* α_{12}, where the subscripts refer to the two species 1 and 2. It is defined by

$$\alpha_{12} = \frac{x_{11}\, x_{22}}{x_{21}\, x_{12}}. \tag{1.4.5}$$

For a binary system of species 1 and 2, since $(x_{1j} + x_{2j}) = 1$, we can express it also as

$$\alpha_{12} = \frac{x_{11}}{(1 - x_{11})}\,\frac{(1 - x_{12})}{x_{12}} = \frac{X_{11}}{X_{12}}, \tag{1.4.6}$$

where $k = 2$ in (1.3.2) so that

$$X_{11} = \alpha_{12} X_{12}. \tag{1.4.7a}$$

The mole ratios of any species between two regions are therefore related linearly through the separation factor α_{12}. Similarly, using relations (1.4.6) one can easily obtain the following two relations for the mole fractions of any species between the two regions:

$$x_{11} = \frac{\alpha_{12}\, x_{12}}{1 + x_{12}\,(\alpha_{12} - 1)}, \tag{1.4.7b}$$

$$x_{12} = \frac{x_{11}}{\alpha_{12} - x_{11}\,(\alpha_{12} - 1)}. \tag{1.4.7c}$$

The relationship between x_{11} and x_{12} is often presented graphically. For example, in systems having a vapor phase ($j = 1$) and a liquid phase (example I, Figure 1.1.2), x_{11} and x_{12} are plotted as the ordinate and the abscissa, respectively. Figure 1.4.1 displays this for the systems benzene–toluene and methanol–water. The straight line in each figure represents $x_{11} = x_{12}$, a condition where no separation is possible, since $x_{11} = x_{12}$, $x_{21} = x_{22}$ and $\alpha_{12} = 1$. The figure for each system has a dashed line representing the value of α_{12} as a function of x_{12}. In the benzene–toluene system, α_{12} is a constant; in the other system, α_{12} varies with composition. The nature of this variation can sometimes be very complex.

By using equation (1.4.1) in definition (1.4.5), we get a few relations between α_{12} and some other indices:[4]

$$\alpha_{12} = (K_1/K_2) = \left(k'_{11}/k'_{21}\right). \tag{1.4.8}$$

Similarly, definitions (1.3.1) and (1.4.3a) substituted into the definition of α_{12} yield

$$\alpha_{12} = (1/\eta_1\eta_2). \tag{1.4.9}$$

Note that if region 1 or region 2 has pure component 1 or 2, respectively, $\alpha_{12} = \infty$, a case of *infinite separation factor*. However, as indicated in Figure 1.1.1 (conditions I and III), unless both η_1 and η_2 are zero simultaneously, perfect separation is not achieved. Thus $\alpha_{12} = \infty$ need not imply perfect separation, although at least one region has a pure component only. An example of such a separation is the evaporative desalination of brine, where the vapor generated is pure water. Thus, with water as component 1 in the vapor region designated as 1, salt is absent in the vapor so that $\eta_1 = 0$ or $x_{21} = 0$ and $\alpha_{12} = \infty$. However, if during vapor generation some brine droplets are entrained by the rising vapor, $x_{21} \neq 0$ and $\alpha_{12} \neq \infty$. (See Example 1.4.2 for another problem of this type.)

Often instead of the separation factor α_{12}, one encounters the *enrichment factor*[5] ε_{12} defined by

$$\varepsilon_{12} = (\alpha_{12} - 1) = \frac{(x_{11} - x_{12})}{x_{12}(1 - x_{11})}. \tag{1.4.10}$$

[4] The ratio $\left(k'_{11}/k'_{21}\right)$ has been called the separation quotient (Rony, 1968a).

[5] In separation literature, subscripts 1 and 2 are usually dropped and only ε is used.

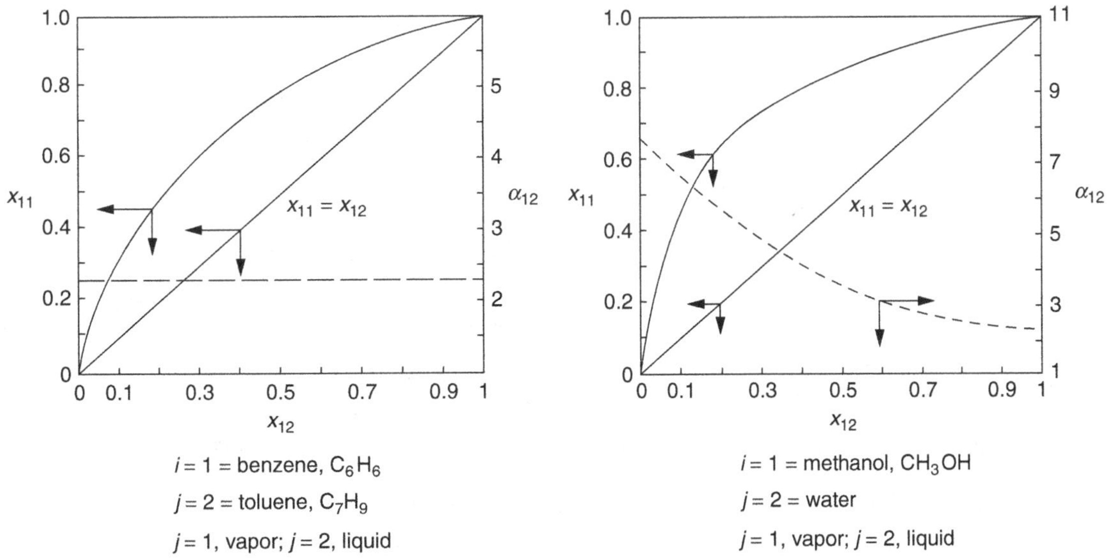

Figure 1.4.1. Relationships between x_{11} and x_{12} and α_{12} and x_{12} for benzene-toluene and methanol-water mixtures in a vapor-liquid system.

If ε_{12} is nonzero, α_{12} is greater than 1, indicating that separation is possible. If $\varepsilon_{12} \ll 1$, the situation corresponds to *close separation*. In such a case, the following approximation is valid (Pratt, 1967a, p. 10):

$$\ell n(\alpha_{12}) = (\alpha_{12} - 1) - \frac{(\alpha_{12} - 1)^2}{2} + \cdots \cong (\alpha_{12} - 1) = \varepsilon_{12}. \tag{1.4.11}$$

Therefore

$$\alpha_{12} = e^{\varepsilon_{12}} = \exp(\varepsilon_{12}). \tag{1.4.12}$$

Further, for close separation, $(x_{11} - x_{12}) = \varepsilon_{12}x_{12}(1-x_{11})$, a small quantity so that x_{11} is quite close to x_{12}. Therefore, often x_{12} can be interchanged with x_{11} in the term $\varepsilon_{12}x_{12}(1-x_{11})$ and vice versa, leading to the following two relations:

$$x_{11} = x_{12} + \varepsilon_{12}x_{12}(1 - x_{12}); \tag{1.4.13}$$

$$x_{12} = x_{11} - \varepsilon_{12}x_{11}(1 - x_{11}). \tag{1.4.14}$$

Some of the more recent indices of separation between two regions for a two-component system in a closed vessel are as follows.

Separation factor α' (Sandell, 1968):

$$\alpha' = \frac{Y_{11}}{Y_{22}}. \tag{1.4.15}$$

Extent of separation ξ (Rony, 1968a):

$$\xi = \text{abs}|Y_{11} - Y_{21}|. \tag{1.4.16}$$

One should consider the index ξ, the extent of separation, in some detail due to its versatility (Rony, 1972). Substituting definition (1.3.8a) of Y_{ij}, we can easily show that the following relations are valid:

$$\begin{aligned}\xi &= \text{abs}|Y_{11} - Y_{21}| = \text{abs}|1 - Y_{12} - Y_{21}| \\ &= \text{abs}|Y_{11} + Y_{22} - 1|; \quad \xi = \text{abs}|Y_{22} - Y_{12}| \\ &= \text{abs}|Y_{11}Y_{22} - Y_{12}Y_{21}| = |\det[Y_{ij}]|\end{aligned} \tag{1.4.17}$$

Therefore one may express ξ in general by

$$\xi = \text{abs}|Y_{ij} - Y_{kj}|, \quad i, k = 1, 2; \; j = 1, 2; \; i \neq k. \tag{1.4.18}$$

Since the largest value of Y_{ij} is 1 and the smallest value is 0, ξ varies between 1 and 0, with the absolute sign taking care of any negative sign if it arises. In terms of perfect and imperfect separation, since the maximum value of Y_{ij} is 1, $\xi = 1$ implies perfect separation (if Y_{ij} is 1 and Y_{kj} is zero; if Y_{kj} is 1 and Y_{ij} is zero). A uniform composition between the two regions is indicated by $\xi = 0$, and there is no separation (see equation (1.4.22)).

The relations between ξ and some of the other indices are going to have some use later, so we will obtain them now. By definition (1.4.16),

$$\xi = \text{abs}|Y_{11} - Y_{21}| = \text{abs}\left|\frac{m_{11}}{m_1^0} - \frac{m_{21}}{m_2^0}\right|. \tag{1.4.19}$$

Further, from equations (1.4.1)

$$1 + k'_{il} = \frac{m_i^0}{m_{i2}}; \qquad \frac{k'_{i1} + 1}{k'_{i1}} = \frac{m_i^0}{m_{i1}}, \tag{1.4.20}$$

yielding

$$\xi = \text{abs}\left|\frac{k'_{11}}{1 + k'_{11}} - \frac{k'_{21}}{1 + k'_{21}}\right| = \text{abs}\left|\frac{k'_{11}}{1 + k'_{11}} - \frac{k'_{11}}{(\alpha_{12} + k'_{11})}\right|, \tag{1.4.21}$$

where for the final relation we have utilized equation (1.4.8). An alternative relation between ξ and α_{12} can be obtained by expressing the last of the equalities (1.4.17) as

$$\xi = Y_{12}\, Y_{21} \text{ abs}\left|\frac{Y_{11} Y_{22}}{Y_{12} Y_{21}} - 1\right| = Y_{12} Y_{21} \text{ abs}\,|\alpha_{12} - 1|. \tag{1.4.22}$$

Similarly, the other separation factor α' is related to k'_{i1} by

$$\alpha' = \frac{Y_{11}}{Y_{22}} = \frac{m_{11}/m_1^0}{m_{22}/m_2^0} = \frac{k'_{11}(1 + k'_{21})}{(1 + k'_{11})}. \tag{1.4.23}$$

There is no direct relation between ξ and I or I_j, the purity indices. But both are different functions of two quantities, as can be observed from Problem 1.4.1.

Although indices of separation have been devised to indicate the nature of separation between two contiguous regions in a separated system, sometimes the nature of the composition difference between one of the separated regions and the initial mixture is of interest. The following indices, with their names originating from specialized separation processes, are of this type:

decontamination factor,

$$Df|_j = (C_{if}/C_{ij}); \tag{1.4.24}$$

desalination ratio,

$$D_r = (C_{if}/C_{i1}), \qquad i = 2. \tag{1.4.25}$$

In the above definitions, C_{if} is the molar concentration of i in the initial uniform mixture. For the desalination ratio, i is usually 2, corresponding to salt, which concentrates in region 2, whereas water concentrates in region 1 (ice in example II, Figure 1.1.2; purified water in example I, Figure 1.1.3). For the decontamination factor used with radioactive or surfactant impurities in solvents (or particles in air filtration), i is the impurity and j is the region for the purified solvent. These definitions need not be restricted to salt or a radioactive or surfactant impurity. They are applicable to any impurity to be removed from a solvent, especially in the case of a dilute solution. Note that, in a similar manner, one can define α_{12} and ε_{12} between the initial mixture and the separated region of interest.

Example 1.4.1 Close separation in thermal diffusion of an isotopic mixture Isotopic mixtures are difficult to separate since isotopes are very similar to one another. One method sometimes adopted is thermal diffusion. Consider a two-bulb cell as shown in example III of Figure 1.1.3, with one bulb at 300 °C and the other at 23 °C. Initially, both bulbs at the same temperature, 23 °C, contained an equimolar mixture of $C^{12}H_4$ and $C^{13}H_4$. After the temperature of bulb 1 was raised to 300 °C, the mole fractions of $C^{12}H_4$ in bulbs 1 and 2 were found to be 0.5006 and 0.4994, respectively. The separation factor α_{12} for $C^{12}H_4$ as species 1, with the hot bulb as region 1, is given by

$$\alpha_{12} = \frac{x_{11}\, x_{22}}{x_{21}\, x_{12}} = \frac{0.5006 \times 0.5006}{0.4994 \times 0.4994} = 1.00482;$$
$$\varepsilon_{12} = \alpha_{12} - 1 = 0.00482.$$

The expressions for some other indices of separation in two-bulb thermal diffusion are given in Example 1.5.3. Such a small separation would render the process of thermal diffusion useless unless a way is found to increase α_{12}.

Example 1.4.2 Description of freeze-concentration of fruit juices Large quantities of water are removed from freshly obtained fruit juices to produce fruit juice concentrates. This reduces, among others, the problem of shipping large amounts of liquid material from production centers to distribution centers. This concentration may be achieved by cooling the fruit juice to between −3 °C and −15 °C when highly pure ice crystals are formed as a suspension in the concentrated fruit juice (Figure 1.4.2). Next, the slurry can be separated in a filtering centrifuge or a filter press or wash column (Thijssen, 1979) to yield two products: almost pure ice and the fruit juice concentrate. Consider water as component 1 and the active constituents of fruit juice as component 2. Let the ice phase be region 1 and the concentrate be region 2. If the ice crystals formed were absolutely devoid of fruit juice active constituents, the separation factor α_{12} would be infinity since x_{21} would be zero. Actually when ice

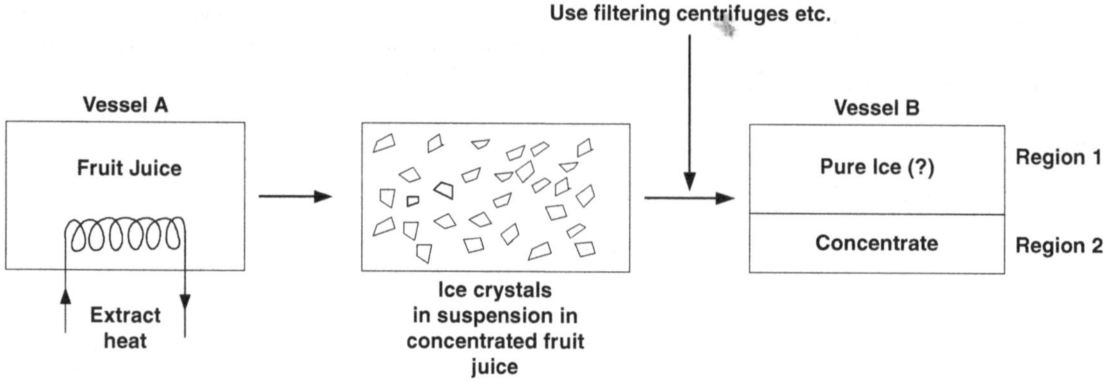

Figure 1.4.2. Freeze-concentration process for fruit juices.

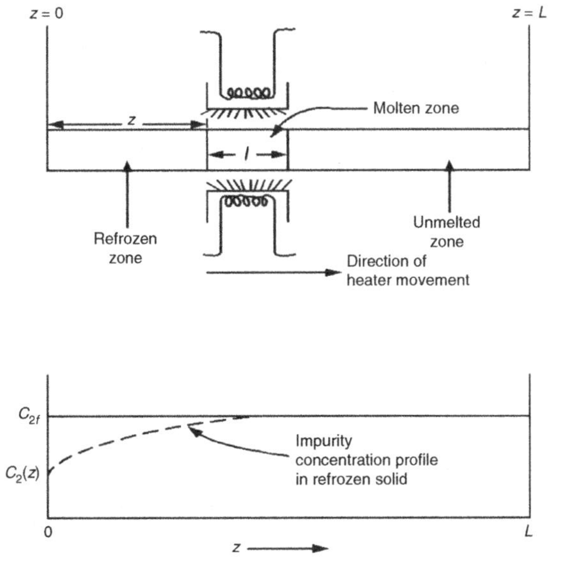

Figure 1.4.3. Zone refining of a rod.

crystals are formed, some fruit juice solids are frozen inside the crystals as impurities. In addition, some fruit juice concentrate will stick to the surfaces of ice crystals in the separator vessel B, which may be any one of the three types of equipment mentioned earlier. Therefore $x_{21} \neq 0$ and α_{12} has a high value, which is, however, less than infinity.

Example 1.4.3 Zone refining of solid materials Consider a solid rod of silicon of length L cm containing some impuity (species 2) at a low concentration level of C_{2f} gmol/cm^3. By a process called zone refining (Pfann, 1966) (see Section 6.3.2.3), a portion of the solid rod can be made substantially more pure by slowly moving a heater of length ℓ ($\ll L$) along the rod from one end to the other (Figure 1.4.3). The portion of the rod directly concentric with the heater remains molten, while that immediately to the left starts solidifying. This solidifying section rejects part of its impurity content into the molten zone. As the heater moves to the right, the molten zone also moves to the right. The impurity follows this movement so that the left end of the refrozen rod is substantially purer than the right end. Consider the whole rod as the separated system. The impurity concentration in the solid after the process is over is given as a function of distance z from the left end by (Pfann, 1966)

$$\left(C_2(z)/C_{2f}\right) = 1 - (1 - \kappa_{21})\exp\left[-\kappa_{21}\, z/\ell\right].$$

During the melting and refreezing process, $\kappa_{21} = C_2(z)/C_2^1(z)$, $C_2^1(z)$ being the impurity concentration in the molten rod at a distance z from the left end, which is in contact with the refrozen rod at z, having a concentration of $C_2(z)$. Determine the separation factor for this process for $\kappa_{21} = 0.5$, $L = 10\ell$ if region 1 is $z = 0$ and region 2 is $z = L$. If $\kappa_{21} = 0.1$, what happens to α_{12}?

Solution For $\kappa_{21} = 0.5$, $L = 10\ell$:

$$\alpha_{12} = \frac{x_{11}\, x_{22}}{x_{21}\, x_{12}} = \frac{m_{11}\, m_{22}}{m_{12}\, m_{21}}.$$

Since the impurity concentration is low, the density of the rod material may be assumed equal everywhere. Therefore taking two small and equal volumes at the two ends of the rod (effectively the same amount of material), we have

$$\alpha_{12} = \frac{m_{11}\, m_{22}}{m_{12}\, m_{21}} = \frac{C_{11}\, C_{22}}{C_{12}\, C_{21}}.$$

But the impurity concentration being very low, $C_{11} \cong C_{12}$. Therefore

$$\alpha_{12} \cong \frac{C_{22}}{C_{21}} = \frac{C_{2f}\left[1 - (1 - \kappa_{21})\exp(-10\,\kappa_{21})\right]}{C_{2f}\left[1 - (1 - \kappa_{21})\right]}$$

$$= \frac{1 - 0.5\exp(-5)}{0.5} = 1.9932.$$

For $\kappa_{21} = 0.1$,

$$\alpha_{12} = \frac{1 - 0.9\exp(-1)}{0.1} = 6.7.$$

Note here that κ_{21} is a distribution coefficient for species 2 during the melting and refreezing processes. Thus, the lower the distribution coefficient of the impurity between the solid and the melt, the greater is the separation factor. Further, such a separated system exists without a membrane (Example 1.5.4) or two immiscible phases.

1.5 Comparison of indices of separation for a closed system

There are many desirable properties of a versatile index of separation. Some of these are listed below (Rony, 1972):

(1) It should be dimensionless.
(2) It should be usable at all levels of concentrations of any species in the mixture as well as with any separation process.
(3) It should be easily calculable.
(4) It should preferably be normalized, varying between one (indicating perfect separation) and zero (implying no separation at all), with increasing number indicating improved separation.
(5) It should be unaffected if the component subscript (i = 1, 2) and the region subscript (j = 1, 2) are interchanged for a binary system.
(6) It should be sensitive.

One can, no doubt, list additional desirable features. In Table 1.5.1, we have indicated some of these features for all indices introduced in Section 1.4. Note that indices k'_{i1}, κ_{i1}, K_i, η_1, η_2, I_1 and I_2 are particularly inadequate when it comes to describing separation which involves both species and both regions, since each index either incorporates both regions or both components. It is also apparent that, of the remaining indices, e.g. I, α_{12}, ε_{12}, α' and ξ, only the extent of separation, ξ, is normalized. Further, except for the purity index I and ξ, the maximum values of the others, namely α_{12}, ε_{12} and α' do not represent perfect separation.

Table 1.5.1. Properties of various indices of separation[a]

Index of separation	Dimensionless	Applicable to both components	Applicable to both regions	Minimum value (= (?) zero separation)	Maximum value (= (?) perfect separation)	What value implies perfect separation?
k'_{i1}	yes	no	yes	0 (?)	∞ (?)	[c]
κ_{i1}	yes	no	yes	0 (?)	∞ (?)	[c]
K_{i1}	yes	no	yes	0 (?)	∞ (?)	[c]
η_1	yes	yes	no	0 (?)	[b]	[c]
η_2	yes	yes	no	0 (?)	[b]	[c]
I_1	yes	yes	no	[b]	[b]	[c]
I_2	yes	yes	no	[b]	[b]	[c]
I_1	yes	yes	yes	[b]	∞ (yes)	∞
α_{12}	yes	yes	yes	1 (yes)	∞ (?)	[c]
ε_{12}	yes	yes	yes	0 (yes)	∞ (?)	[c]
α'	yes	yes	yes	0 (?)	∞ (no)	[c]
ξ	yes	yes	yes	0 (yes)	1 (yes)	1

[a] For a binary system in a closed vessel with two regions only.
[b] Depends on further conventions about the limits of contents of each region. See Problem 1.4.1.
[c] No specific value is capable of describing perfect separation.

Next we give some examples and calculate the values of a selected few indices to obtain an estimate of their capabilities to describe the quality and amount of separation. Consider first the hypothetical cases of a binary separation given by the following four conditions.[6]

(a) Condition 1: $k'_{11} = 10^6$, $k'_{21} = 10^4$;
(b) condition 2: $k'_{11} = 10^{-4}$, $k'_{21} = 10^{-6}$;
(c) Condition 3: $k'_{11} = 10$, $k'_{21} = 0.1$;
(d) Condition 4: $k'_{11} = 20$, $k'_{21} = 0.05$.

In Table 1.5.2, we show the calculated values of Y_{ij} for all four conditions of separation. In condition 1, most of species 1 as well as species 2 are located in region 1, whereas in condition 2 they are almost totally located in region 2. Therefore, these two conditions represent poor separation. On the other hand, in condition 3, species 1 is located mostly in region 1, whereas species 2 is located mostly in region 2. Thus condition 3 represents much better separation than conditions 1 and 2. Condition 4 represents even better separation than condition 3 since the amounts of impurities in both regions are reduced considerably compared to those in condition 3.

We can now judge which one of the four indices of separation α_{12}, α', ξ or I accurately reflects the conditions of separation described above. Note that in Table 1.5.2 the two indices α_{12} and I do not discriminate between the three conditions of separation 1, 2 and 3. The index α' incorrectly indicates condition 1 as a much better separation than either of conditions 2 or 3. The extent of separation, ξ, however, correctly demonstrates that condition 3 represents far better separation than conditions 1 or 2. Thus ξ is a much better index of separation. However, with regard to indicating how much more improved the separation in condition 4 is with respect to that in condition 3, note that the relative changes in both α_{12} and I are much more than that of ξ. Therefore, as perfect separation is approached, the sensitivity of ξ becomes limited compared to either I or α_{12}.

[6] The first three conditions are based on an example provided by Rony (1972) in his discussion on the desirable properties of indices of separation. Note, however, the differences in our definition, k'_{i1}, and Rony's k_{12}.

Example 1.5.1 Consider the binary system of benzene (1) and toluene (2) distributed between a liquid and a vapor region within a closed vessel. Given the benzene mole fraction in the liquid and the vapor phase to be 0.780 and 0.90, respectively, determine the values of α_{12} and ξ for the following two cases: (a) 1 gmol of liquid, vapor volume 0.293 liters; (b) 1 gmol of liquid, vapor volume 2.93 liters. The total pressure is 1 atmosphere and the temperature is 85 °C. The vapor mixture may be assumed to be ideal.

Solution Since benzene (species 1) is present more in the vapor phase, region 1 is vapor phase. Therefore

$$x_{12} = 0.78, \quad x_{22} = 0.22, \quad x_{11} = 0.90, \quad x_{21} = 0.10.$$

So

$$\alpha_{12} = \frac{x_{11}\, x_{22}}{x_{12}\, x_{21}} = \frac{0.90 \times 0.22}{0.78 \times 0.1} = 2.54.$$

This is valid for both cases (a) and (b). For the calculation of ξ, the values of four quantities, m_{11}, m_{12}, m_{22} and m_{21}, have to be determined.

Case (a): 0.293 liters vapor volume. Assuming that the ideal gas law holds, the total number of gram moles present in the vapor is given by

$$(m_{11} + m_{21}) = \frac{0.293}{22.4} \times \frac{273}{358} = 0.01 \text{ gmol}.$$

Table 1.5.2. Comparative descriptions of separation by indices α_{12}, α', ξ and I

Description of separation		Condition 1[a]	Condition 2[b]	Condition 3[c]	Condition 4[d]
Y_{ij}	Y_{11}	0.999999	0.0001	0.909	0.9523
	Y_{21}	0.9999	0.000001	0.091	0.0477
	Y_{12}	0.000001	0.9999	0.091	0.0477
	Y_{22}	0.0001	0.999999	0.909	0.9523
α_{12}		100	100	100	400
α'		10^4	10^{-4}	1	1
ξ		0.000099	0.000099	0.818	0.9046
I		2	2	1.9991	2.601

[a] $k'_{11} = 10^6, k'_{21} = 10^4$.
[b] $k'_{11} = 10^{-4},\ k'_{21} = 10^{-6}$.
[c] $k'_{11} = 10,\ k'_{21} = 0.1$.
[d] $k'_{11} = 20, k'_{21} = 0.05$.

Therefore $m_{11} = 0.01 \times 0.9 = 0.009$ gmol; $m_{21} = 0.001$ gmol; $m_{12} = 0.78 \times 1 = 0.78$ gmol; $m_{22} = 0.22$ gmol, so

$$\xi = \left|\frac{m_{11}}{m_{11}+m_{12}} - \frac{m_{21}}{m_{21}+m_{22}}\right| = \left|\frac{0.009}{0.789} - \frac{0.001}{0.221}\right| = 0.00688,$$

a case of very poor separation.

Case (b): 2.93 liters vapor volume. Therefore $m_{11} + m_{21} = 0.10$ gmol $\Rightarrow m_{11} = 0.1 \times 0.9 = 0.09$ gmol; $m_{21} = 0.01$ gmol; $m_{12} = 0.78$ gmol; $m_{22} = 0.22$ gmol, and

$$\xi = \left|\frac{0.09}{0.09+0.78} - \frac{0.01}{0.01+0.22}\right| = 0.0597.$$

In both cases, ξ indicates poor separation, although case (b) represents somewhat better separation than case (a). However, α_{12}, based on mole fraction only, is insensitive to such changes. Further, $\alpha_{12} = 2.54$ mistakenly indicates reasonable separation since it is far away from $\alpha_{12} = 1.0$, corresponding to zero separation.

Example 1.5.2 Aniline and hexane are completely miscible with each other at all temperatures higher than 59.6 °C, below which they separate into two immiscible phases. A uniform mixture of aniline and hexane containing 52 mole percent aniline is cooled in a closed vessel from 60 °C to 42 °C. At 42 °C, the hexane-rich layer has 83 mole percent hexane while the heavier and immiscible aniline-rich layer has 88 mole percent aniline. Determine the values of α_{12} and ξ to indicate the separation that has been achieved compared to the original feed mixture.

Solution Basis: 100 gmol of feed mixture. Species 1 - hexane; species 2 - aniline; $j = 1$, hexane-rich top layer; $j = 2$, aniline-rich bottom layer. (Note: aniline is heavier than hexane.)

Given that

$$x_{11} = 0.83 = \frac{m_{11}}{m_{11}+m_{21}}, \tag{1.5.1}$$

$$x_{22} = 0.88 = \frac{m_{22}}{m_{22}+m_{12}} \tag{1.5.2}$$

From equation (1.3.11),

$$100\,x_{1f} = 100 \times 0.48 = m_{11} + m_{12} = 48; \tag{1.5.3}$$

$$100\,x_{2f} = 52 = m_{21} + m_{22}. \tag{1.5.4}$$

Solving the four equations (1.5.1)-(1.5.4) for m_{11}, m_{12}, m_{21} and m_{22}, we get $m_{11} = 42.09$, $m_{12} = 5.91$, $m_{21} = 8.62$ and $m_{22} = 43.38$ gmol. The extent of separation is given by

$$\xi = \left|\frac{m_{11}}{m_{11}+m_{12}} - \frac{m_{21}}{m_{21}+m_{22}}\right| = \left|\frac{42.09}{48} - \frac{8.62}{52}\right| = 0.7122.$$

The separation factor is given by

$$\alpha_{12} = \frac{x_{11}\,x_{22}}{x_{12}\,x_{21}} = \frac{0.83 \times 0.88}{0.12 \times 0.17} = 35.8.$$

Although the value of α_{12} indicates an extremely effective separation, the value of ξ indicates that, since the extent of recoveries is not sufficient, the separation is far from perfect.

Example 1.5.3 Consider the thermal diffusion separation of a gas mixture of H_2 and N_2 initially present with a hydrogen mole fraction x_{1f} in a two-bulb cell. The volumes of the two bulbs are V_1 and V_2. Let the uniform temperature of the two-bulb cell be changed so that the bulb of volume V_1 now has a temperature T_1 while the other one is at T_2 ($T_1 > T_2$) (Figure 1.1.3, example III). Assuming that this is a case of close separation such that $x_{11} \cong x_{1f}$, we obtain

$$\alpha_{12} - 1 = \varepsilon_{12} \cong \frac{x_{11} - x_{22}}{x_{1f}\,(1 - x_{1f})}. \tag{1.5.5}$$

Given that (Pratt, 1967b, p. 404)

$$\alpha_{12} - 1 = \varepsilon_{12} \cong \gamma_{12} \ln\left(\frac{T_1}{T_2}\right), \tag{1.5.6}$$

where γ_{12} is a property of the H_2-N_2 system, obtain a relation between ξ and ε_{12}. How can you increase separation in such a system for given T_1 and T_2? Assume the ideal gas law to be valid. Neglect the volume of connecting capillary.

Solution Since the ideal gas law is valid, with the light component hydrogen (species 1) concentrating more in region 1 (higher temperature)

$$m_{11} = \frac{PV_1}{RT_1}\,x_{11}; \qquad m_{12} = \frac{PV_2}{RT_2}\,x_{12}; \tag{1.5.7}$$

$$m_{22} = \frac{PV_2}{RT_2} x_{22}; \qquad m_{21} = \frac{PV_1}{RT_1} x_{21}, \tag{1.5.8}$$

where P is the total pressure of the gas mixture everywhere in the cell. Therefore

$$\xi = \text{abs}\left|\frac{m_{11}}{m_{11}+m_{12}} - \frac{m_{21}}{m_{21}+m_{22}}\right|$$
$$= \text{abs}\left|\frac{(x_{11}V_1/T_1)}{(x_{11}V_1/T_1) + (x_{12}V_2/T_2)} - \frac{(x_{21}V_1/T_1)}{(x_{21}V_1/T_1) + (x_{22}V_2/T_2)}\right|. \tag{1.5.9}$$

But

$$x_{1f} = \frac{m_{11}+m_{12}}{(m_{11}+m_{21}) + (m_{12}+m_{22})}$$
$$= \frac{(V_1 x_{11}/T_1) + (V_2 x_{12}/T_2)}{(V_1/T_1) + (V_2/T_2)}. \tag{1.5.10}$$

Similarly,

$$(1 - x_{1f}) = \frac{(V_1 x_{21}/T_1) + (V_2 x_{22}/T_2)}{(V_1/T_1) + (V_2/T_2)}. \tag{1.5.11}$$

Therefore

$$\xi = \frac{V_1 V_2}{\left(\frac{V_1}{T_1} + \frac{V_2}{T_2}\right)^2 T_1 T_2} \times \text{abs}\left|\frac{x_{11} - x_{12}}{x_{1f}(1 - x_{1f})}\right| \tag{1.5.12}$$

on introducing (1.5.10) and (1.5.11) in the expression for ξ obtained by simplifying expression (1.5.9). On rearrangement, the relation between ξ and ε_{12} is obtained as (Sirkar, 1977)

$$\xi = \frac{1}{\left(\frac{V_1}{V_2}\frac{T_2}{T_1} + \frac{V_2}{V_1}\frac{T_1}{T_2} + 2\right)} \times \varepsilon_{12}. \tag{1.5.13}$$

Suppose $T_1/T_2 = 10$. Let us take two cases: (a) $(V_1/V_2) = 1$; (b) $(V_1/V_2) = 10$.

Case (a) $\quad \xi = \dfrac{\varepsilon_{12}}{(0.1 + 10 + 2)} = \dfrac{\varepsilon_{12}}{12.1};$

Case (b) $\quad \xi = \dfrac{\varepsilon_{12}}{\left(\frac{10}{10} + 0.1 \times 10 + 2\right)} = \dfrac{\varepsilon_{12}}{4}.$

Obviously case (b) has much better separation since by providing the hotter bulb with a much larger volume, much more hydrogen can now be segregated in the hotter bulb which provides the light fraction. This is not evident from α_{12} or ε_{12}; both are independent of V_1 or V_2.

Example 1.5.4 In the vessel shown in Figure 1.5.1, a membrane which allows water to pass through easily and prevents salt from going through has on one side 1000 cm^3 of aqueous solution containing 0.1 gmol of salt per cm^3 and atmospheric air on the other side. If the saline water is pressurized by means of a piston to a constant but high enough pressure, water permeates through the membrane but salt does not.

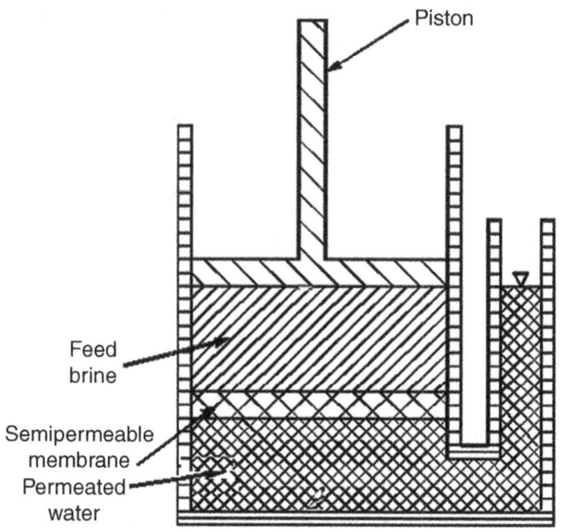

Figure 1.5.1. Separation of water from a salt solution under pressure through a membrane.

(a) After operation for some time, 500 cm^3 of pure water has come to the atmospheric air side through the membrane. Determine the separation factor α_{12} and the extent of separation ξ if the density of brine in the concentration ranges encountered is 1 g/cm^3.
(b) Suppose some salt also leaks through the membrane to the permeate side so that the salt concentration in the solution on the atmospheric pressure side is 0.026 gmol/cm^3. If 500 cm^3 of water has permeated through the membrane as before, what are the values of α_{12} and ξ?

Solution Region 1, permeated solution or water; region 2, high-pressure brine; component 1, water; component 2, salt.

Part (a): $\alpha_{12} = x_{11}x_{22}/x_{12}x_{21}$. But $x_{11} = 1$ (pure water), $x_{21} = 0$ (no salt in permeate) and $x_{12}, x_{22} \neq 0$. Therefore $\alpha_{12} = \infty$, a case of infinite separation factor.
By definition,

$$\xi = \left|\frac{m_{11}}{m_{11}+m_{12}} - \frac{m_{21}}{m_{21}+m_{22}}\right|.$$

Now $m_{21} = 0$, $m_{11} = (500/\overline{V}_1)$ and $m_{12} = (500/\overline{V}_1)$, where $\overline{V}_1$ is the partial molar volume of water in the brine solution, assumed to be constant in the range of pressures and concentrations (unit, cm^3/gmol). Therefore

$$\xi = \left|\frac{500}{500+500} - 0\right| = \frac{1}{2},$$

definitely not a case of perfect separation, inspite of an infinite separation factor.

Part (b): $m_{21} = 0.026 \times 500 = 13$ gmol; $m_{22} = (1000 \times 0.1) - 13 = 87$ gmol;

$$\xi = \left|\frac{500}{500+500} - \frac{13}{100}\right| = 0.5 - 0.13 = 0.37,$$

so separation is reduced from case (a).

$$\alpha_{12} = \frac{(m_{11})\,(m_{22})}{(m_{12})\,(m_{21})} = \frac{(500/\overline{V}_1)\,(m_{22})}{(500/\overline{V}_1)\,(m_{21})} = \frac{500 \times 87}{500 \times 13} = 6.7$$

Thus the change in α_{12} between cases (a) and (b) is much more dramatic than that in ξ.

From the above examples, it is clear that ξ, the extent of separation, often describes separation much better than α_{12}, the separation factor. This is due to the ability of ξ to take into account the extent of recovery of a given species in a given region. On the other hand, when one deals with two regions with small amounts of impurity, any change in the impurity levels are better recognized through the indices indicating the composition of any region, e.g. α_{12}, η_j, I, I_j, etc. Therefore, no single existing index describes all aspects of separation efficiently (see Problem 1.5.3).

1.6 Indices for separation of multicomponent systems between two regions

In Section 1.2, we introduced a brief description of separation for multicomponent systems. Although we learnt there that perfect separation in such a system requires as many regions as there are components, we will restrict ourselves here to separation systems with only two regions in a closed vessel. Thus perfect separation is, in general, ruled out from our considerations.

The familiar quantities used to define the composition of a mixture in a region have to be redefined since $i = 1, 2, 3, \ldots, n$ for the n-component system:

$$x_{ij} = \frac{m_{ij}}{\sum_{i=1}^{n} m_{ij}} \qquad i = 1, 2, \ldots, n; \qquad j = 1, 2; \tag{1.6.1a}$$

$$X_{ij} = \frac{m_{ij}}{m_{kj}} \qquad k \neq i; \qquad k, i = 1, 2, \ldots, n; \qquad j = 1, 2. \tag{1.6.1b}$$

Note

$$\sum_{i=1}^{n} x_{ij} = 1 \quad \text{for} \quad j = 1, 2; \tag{1.6.1c}$$

$$\sum_{i=1}^{n} C_{ij} = C_{tj}; \qquad a_{ij} = \frac{m_{ij}^a}{\sum_{i=1}^{n} m_{ij}^a}; \qquad u_{ij} = \frac{\rho_{ij}}{\sum_{i=1}^{n} \rho_{ij}}. \tag{1.6.1d}$$

All other basic definitions of concentrations remain the same, except for $i = 1, \ldots, n$. For multicomponent systems, it is worthwhile illustrating the value of a_{ij} by way of an example. Consider an isotopic mixture of water having H_2O, D_2O and HDO present in the amounts 0.6, 0.1, 0.3, respectively (in mole fractions, i.e. if H_2O is component 1, then $x_{1f} = 0.6$). The atom fraction of hydrogen is then $[(0.6 \times 2 + 0.3 \times 1)/2] = 0.75$.

A *graphical method* of describing compositions is often adopted for three-component systems. Consider an equilateral triangle with apexes identified as 1, 2 and 3

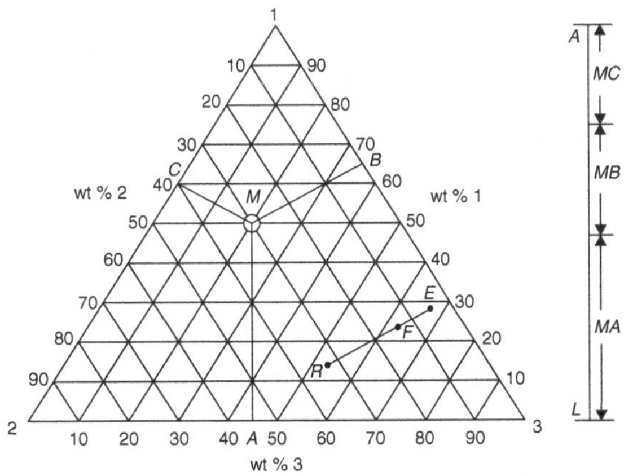

Figure 1.6.1. Compositions of a ternary system in equilateral triangular diagram.

(Figure 1.6.1) representing pure components 1, 2 and 3, respectively. From any point M inside the triangle representing a ternary mixture, draw perpendiculars MA, MB and MC to the sides 23, 31 and 12.

The distances MA, MB and MC represent weight percentages of species 1, 2 and 3, respectively, in the mixture. The sum of the distances MA, MB and MC is equal to the altitude of the triangle. Since the perpendicular distance from any one of the apexes to the opposing base is the same for an equilateral triangle, this distance represents 100 percent of the mixture. Therefore the weight fractions of species 1, 2 and 3 in the mixture are given by

$$u_{1j} = \frac{MA}{AL}, \; u_{2j} = \frac{MB}{AL}, \; u_{3j} = \frac{MC}{AL}, \tag{1.6.1e}$$

where AL is the altitude of the triangle. For easy determination of these fractions, triangle 123 has percentages marked on each side. Thus point M in Figure 1.6.1 has weight fractions of 0.50 of 1, 0.30 of 2 and 0.20 of 3.

Two different points inside the triangle represent two different mixtures. These two mixtures may or may not be immiscible. If these two mixtures are brought together, the composition of the overall mixture (subscript $j = M$) will be represented by the point F on the straight line joining the two points E (extract) and R (raffinate). The location of point F is determined by the *lever rule*:

$$\frac{\text{line } EF}{\text{line } RF} = \frac{u_{iE} - u_{iM}}{u_{iM} - u_{iR}} = \frac{w_R}{w_E}, \tag{1.6.2}$$

where w_R and w_E are the total weights of mixtures at R and E, respectively. This relation follows from a total mass balance,

$$w_R + w_E = w_M, \tag{1.6.3a}$$

and an ith-component balance,

$$w_R\, u_{iR} + w_E\, u_{iE} = w_M\, u_{im}. \tag{1.6.3b}$$

With regard to the various indices of separation, the definitions of k'_{i1}, κ_{i1} and K_i remain unaffected, regardless of whether $i = 1, 2$ or $i = 1, 2, \ldots, n$. The *impurity ratio* η_j defined earlier according to de Clerk and Cloete (1971) has to be modified for a multicomponent system to the impurity ratio for the ith species in the jth region:

$$\eta_{ij} = \frac{m_{ij}}{m_{jj}} \qquad j = 1, 2; \quad i = 1, 2, \ldots, n, \quad i \neq j. \tag{1.6.4}$$

One could define a *purity index* corresponding to such a definition of the impurity ratio as follows:

$$I = -\sum_{j=1}^{2} \sum_{i=1}^{n} \log_{10} \eta_{ij}, \tag{1.6.5}$$

following the suggestions of de Clerk and Cloete (1971). The *separation factor* α_{12} for two species 1 and 2, with species 1 being lighter than species 2, is usually changed for multicomponent separations to α_{in} (Pratt, 1967b, p. 451),

$$\alpha_{in} = \frac{x_{i1}\, x_{n2}}{x_{i2}\, x_{n1}}, \tag{1.6.6}$$

such that species n is the heaviest (with the highest molecular weight in general) and species 1 is the lightest with $i = 1, 2, \ldots, n$. Such a definition ensures that α_{in} is greater than 1 if there is any separation. Species n is often referred to as the *heavy key component*. The corresponding index of *enrichment factor* is defined as

$$\varepsilon_{in} = (\alpha_{in} - 1). \tag{1.6.7}$$

It is not necessary, however, to have the separation factor of species i always defined with respect to the heaviest species. One could define the separation factor α_{ij} as

$$\alpha_{ij} = \frac{x_{i1}\, x_{j2}}{x_{i2}\, x_{j1}}, \tag{1.6.8}$$

where the jth species is heavier than the ith species. This ensures that $\alpha_{ij} \geq 1$ since, by convention, the ith species concentrates more in region 1 and a species heavier than i will concentrate in region 2. The corresponding *enrichment factor* is defined as

$$\varepsilon_{ij} = (\alpha_{ij} - 1). \tag{1.6.9}$$

Note that if the separation factor is defined as

$$\alpha_{ji} = \frac{x_{j1}\, x_{i2}}{x_{j2}\, x_{i1}}, \tag{1.6.10}$$

where the jth component is heavier than the ith component, α_{ji} will have a value less than 1.

As in equation (1.4.7b), the relation between the mole fractions x_{i1} and x_{i2} is of interest for the n-component system. Now

$$K_i = \frac{x_{i1}}{x_{i2}}, \qquad i = 1, 2, \ldots, n. \tag{1.6.11a}$$

But

$$\sum_{i=1}^{n} x_{i1} = 1.0,$$

and therefore

$$\sum_{i=1}^{n} K_i\, x_{i2} = 1. \tag{1.6.11b}$$

If the separation factor α_{in} is defined by (1.6.6) with respect to the heavy species n, then

$$\sum_{i=1}^{n} \frac{K_i}{K_n} x_{i2} = \frac{1}{K_n} = \sum_{i=1}^{n} \alpha_{in}\, x_{i2}. \tag{1.6.11c}$$

But

$$x_{i1} = K_i\, x_{i2} = \left(\frac{K_i}{K_n}\right) K_n (x_{i2}),$$

and therefore

$$x_{i1} = \frac{\alpha_{in} x_{i2}}{\sum_{i=1}^{n} (\alpha_{in} x_{i2})}. \tag{1.6.12}$$

The inverse relation is

$$x_{i2} = \frac{x_{i1}/\alpha_{in}}{\sum_{i=1}^{n} (x_{i1}/\alpha_{in})}. \tag{1.6.13}$$

None of the definitions (1.6.4) and (1.6.6)–(1.6.9) provide a single number to indicate the quality of separation, as was the case for a binary system. If a single number is required to describe the separation of a particular ith component from the rest, one can lump all the other species together and treat them as the other component of a binary system. With this rearrangement, all the indices of separation defined earlier for binary systems become useful. This is especially true for ξ, the extent of separation, since Rony (1972) has shown that the automatic extension of $\xi = |\det[Y_{ij}]|$ from a binary system to an n-component system distributed between n regions is not of much use:

$$\xi = \left| \det \begin{vmatrix} Y_{11} Y_{12} Y_{13} \ldots Y_{1n} \\ Y_{21} Y_{22} Y_{23} \ldots Y_{2n} \\ Y_{31} Y_{32} Y_{33} \ldots Y_{3n} \\ Y_{n1} Y_{n2} Y_{n3} \ldots Y_{nn} \end{vmatrix} \right|. \tag{1.6.14}$$

Since a determinant is zero if two rows or columns of the determinant are identical element by element, ξ would be zero, even though $(n - 2)$ components (say) are completely separated and segregated in their respective regions. (See Problem 1.6.1 for other descriptions of multicomponent systems.)

As pointed out in Section 1.2, for a three- or four-component system in which two of the components form two immiscible phases while the other components are distributed between these phases in small quantities, indices of separation defined for a two-component system are quite useful. If we have a three-component system with species 1 distributing itself between phase 1 (essentially species 2, say) and phase 2 (essentially species 3, say) then the indices k'_{i1}, κ_i and K_i are particularly useful. In fact, other indices, e.g. η_j, I_j, I, α_{12}, ε, α' and ξ are not useful at all in such a case since $i = 1$ only. On the other hand, with a four-component system with species 1 and 2 distributing themselves between immiscible phases 1 and 2 (phase 1 is, say, essentially species 3 and phase 2 is essentially species 4), the indices η_j, I_j, I, α_{12}, ε, α' and ξ regain their usefulness. Here, the system is to be treated as if two components 1 and 2 are being separated between two regions. Therefore perfect separation is possible, with species 1 completely segregated in region 1 and species 2 in region 2.

In a five-component system with two immiscible phases, the above argument would suggest that we are effectively dealing with a three-component system. The indices of separation (1.6.4)–(1.6.7) valid for multicomponent systems are to be used in such a case.

There are systems where there may be thousands of species. Quantitative indices have been developed to describe composition and separation in such systems. Chapter 2 provides an introduction to this topic.

1.7 Some specialized nomenclature

Terms such as concentration, enrichment, purification and separation are commonly used in describing chemical separation processes. Based on the usage pattern for these terms, Rony (1972) has suggested that a specified range of composition should be associated with each term. For example, *purification* processes will increase the mole fraction of a species being purified, with the mole fraction of the species remaining above 0.90 always. In *concentration* processes, the species mole fraction being increased remains below 0.90. *Enrichment* processes, traditionally used for isotopic separations, rarely get the mole fraction of the species above 0.10. All processes achieving purification, concentration or enrichment are, however, separation processes.

When salt is removed from sea water by various desalination processes, we have purification of water. Germanium is purified to contain only 1 in 10^{10} electronically active impurities by the zone-refining process (Example 1.4.3). Cane juice containing 10 to 12 wt% sugar is concentrated to about 65 wt% sugar solution by multiple-effect evaporation (King, 1980). Orange juice solids are concentrated sometimes by the freeze-concentration process (Example 1.4.2). The amount of U^{235} isotope in natural uranium compounds (usually around 0.0075 atom fraction) are increased (with respect to the U^{238} isotope) by gaseous diffusion (Pratt, 1967b, p. 349), which becomes an enrichment process.

Chemical separation processes are often described broadly by means of the external agents introduced into the feed mixture to effect separation. In Figures 1.1.2 and 1.1.3, the external agent is energy addition or energy extraction. The form of energy is either heat or pressure energy. Many other forms of energy addition are also practiced. Such separation processes are called *energy-separating-agent* (ESA) processes. On the other hand, consider Figure 1.2.2 where benzene is added from outside to the feed mixture to create a separated system. Such a process is characterized as a *mass-separating-agent* (MSA) process; benzene is the mass-separating agent. Different materials and different phases have been used in mass-separating-agent processes. A comprehensive characterization of different separation processes in terms of the ESA process or the MSA process is available in King (1980).

It is wrong to assume that no energy is required in separation processes using a mass-separating agent. In the example of Figure 1.2.2, the extract phase containing the solute picric acid in benzene has to be subjected to an ESA process to separate benzene from picric acid. Only then is picric acid recovered in purer form, and the two initial species, water and picric acid, are separated.

Problems

1.3.1 One can characterize a binary separation system with two regions in terms of a 2×2 matrix of elements indicating the number of moles of species i in region j (Rony, 1972). Write the matrix S_1 for ideal separation.

(a) Obtain the matrix for the actual separation achieved, S.

(b) Determine the matrix S_f for the initial condition before the separation was started in terms of m_i^0 and Y_j^0. Here Y_j^0 is that fraction of the total number of moles present in the system which is present in region j before the separation was initiated at time $t = 0$.

1.4.1 (a) If the purity index $I = \sum_{j=1}^{2} I_j$ may be expressed as $I = \log_{10} A - \log_{10} B$, show that the extent of separation, ξ, is given by $\xi = \text{abs}\,[A - B]$ for a binary system distributed between two regions in a closed vessel. Indicate the expressions for each of A and B.

(b) Determine the minimum value of I corresponding to maximum impurity in both regions. For this purpose, recognize that since each region is identified with a species, the composition of any region may not be such that the region-specific component i has less than half of m_i^0 in its designated region. While determining the condition for minimum I with respect to a species, note that you have to assume that the distribution of the other species is unaffected. Avoid differentiation. (Ans. 0.)

1.4.2 Two enantiomers, species 1 and species 2, are present in a solution. By a process called resolution, two fractions are obtained. The optical purity of one of the fractions ($j = 1$) highly purified in species 1 is generally expressed in terms of an enantiomeric excess

$$ee(1) = \frac{C_{11} - C_{21}}{C_{11} + C_{21}}.$$

Relate this quantity to an appropriate index identifying the purity of the fraction in terms of species 1.

1.4.3 If a gas or gas mixture is enclosed within a cylindrical vessel of radius r_2 and height h, and the closed vessel is rotated about its axis at an angular velocity of ω radian/s, then, due to centrifugal forces, the partial pressure $p_i(r)$ of species i at any radial location at a distance r from the center is given by

$$p_i(r) = p_i(0) \exp\left[\frac{M_i \omega^2 r^2}{2RT}\right],$$

where M_i is the molecular weight of species i.

(a) Assuming the ideal gas law to be valid, obtain the following expression for the separation factor α_{12} between gas species 1 and 2 at radial locations r_1 and r_2, where $r_1 < r_2$ and $M_1 < M_2$:

$$\alpha_{12} = \frac{x_{11}\, x_{12}}{x_{12}\, x_{21}} = \exp\left[\frac{(M_2 - M_1)\,\omega^2\,(r_2^2 - r_1^2)}{2RT}\right],$$

where region 1 is located at radius r_1, etc. Remember that the total pressure varies with r in such a gas centrifuge.

(b) Consider now the regions 1 and 2 to be thin cylindrical shells of height h and thickness dr at radial locations r_1 and r_2, respectively. Obtain the following relation:

$$\xi = \left[\frac{1}{1 + \frac{r_1}{r_2}\exp\left[\frac{M_1\omega^2(r_1^2 - r_2^2)}{2RT}\right]}\right]\left[\frac{1}{1 + \frac{r_2}{r_1}\exp\left[\frac{M_2\omega^2(r_2^2 - r_1^2)}{2RT}\right]}\right]|\alpha_{12} - 1|.$$

1.4.4 For the thermal diffusion separation of solute 2 present in solvent 1, located between two flat plates as shown in Figure 1.P.1, it is known that the solute will concentrate near the cold plate (region 2) while the solution near the hot plate at the top gets depleted in solute. With y coordinate values of 0 at the top plate and ℓ at the bottom plate, it is known that the mole fraction distribution of the solute $x_2(y)$ is given by the equation

$$\frac{dx_2}{dy} = \frac{\alpha}{\overline{T}}\left[\frac{T_1 - T_2}{\ell}\right] x_2(1 - x_2).$$

Initially, the uniform solute mole fraction in the solution was x_{2f}, where the relation

$$x_{2f} = (1/\ell)\int_0^{\ell} x_2\, dy$$

is valid due to solute conservation.

(a) Show that the solute mole fraction profile is given by (Powers, 1962, p. 29)

$$x_2(y) = [1 + \psi_0 \exp(-2A\phi)]^{-1}, \text{ where}$$

$$A = [\alpha(T_1 - T_2)/2\overline{T}],\ \phi = (y/\ell) \text{ and}$$

$$\psi_0 = \frac{[\exp\{2A(1 - x_{2f})\} - 1]}{[1 - \exp\{-2A\, x_{2f}\}]}.$$

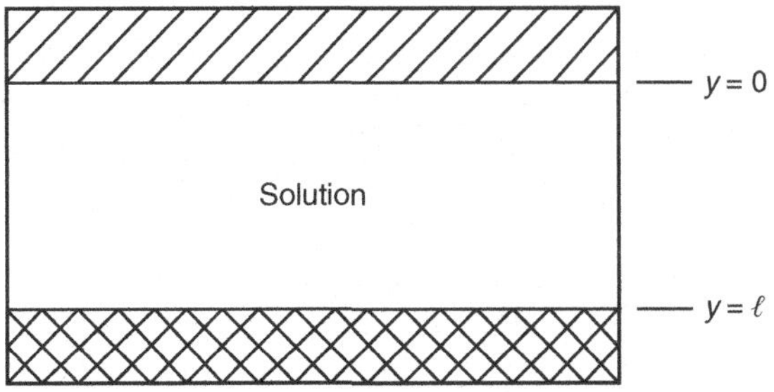

Figure 1.P.1. Static thermal diffusion cell with two plates for a solution.

(b) Obtain an expression for the separation factor α_{12}, with $y = 0$ being region 1 and $y = \ell$ being region 2. (Ans. $\alpha_{12} = \exp(2A)$.)
(c) Obtain the expression for ε_{12} if this is a case of close separation. (Ans. $\varepsilon_{12} = 2A$.)

1.5.1 (a) In Example 1.5.3, the index ξ was found to depend on V_2 and V_1, both of which can be varied to achieve an increased value of ξ. Thus, if (V_2/V_1) is considered to be a variable, ξ can be maximized with respect to (V_2/V_1). In general, the maximum value of ξ, ξ_{max}, with respect to a variable γ can be obtained for the separation of a two-component system between two regions from

$$\frac{\partial \xi}{\partial \gamma} = \frac{\partial}{\partial \gamma}\,[\text{abs}\,(Y_{11} - Y_{21})] = 0,$$

whose solution will yield a value of γ, which, when substituted in ξ, will give ξ_{max}. Consider a two-component system distributed between two immiscible phases such that α_{12} is constant. Show that for a value of the variable $k'_{11} = (\alpha_{12})^{1/2}$,

$$\xi_{max} = \text{abs}\left|\frac{\alpha_{12}^{1/2} - 1}{\alpha_{12}^{1/2} + 1}\right|.$$

(b) If the above system is such that $\varepsilon_{12} \to 0$ (i.e. close separation), show that ξ_{max} is given by $\xi_{max} = \varepsilon_{12}/4$.
(c) Consider relation (1.5.13) between ξ and ε_{12} for binary thermal diffusion separation. Show that, for constant values of T_1 and T_2 for a given two-component system (equation (1.5.6)), $\xi_{max} = (\varepsilon_{12}/4)$ and that the corresponding value of $(V_2/V_1) = (T_2/T_1)$.

1.5.2 (a) Consider the separation of a hexane–aniline liquid mixture as described in Example 1.5.2. Obtain the following relation between ξ, α_{12} and κ_{11}:

$$\xi = V_2 V_1 \left[\frac{C_{11} C_{21}}{m_1^0\, m_2^0}\right] \frac{1}{\kappa_{11}}\ \text{abs}\left|\alpha_{12} - 1\right|.$$

(b) For the separation of picric and benzoic acids between two immiscible layers of benzene and water, show that $\alpha_{12} = \kappa_{11}/\kappa_{21}$ if both liquid phases may be considered as dilute solutions in benzene and water.
(c) For the problem posed in part (b), it is known that both κ_{11} and κ_{21} are substantially independent of concentrations of either solutes in either phases. Utilizing the result in (a), suggest ways to increase the separation at constant temperature (this means κ_{11} and κ_{21} are constants).

1.5.3 Consider two different separators indicated by supercripts A and B. Each has a uniform binary mixture to start with. After separation of each mixture into two different regions, both separators have $k'_{11} = 10^6$ and $k'_{21} = 10^{-6}$. In separator A, $m_1^o = m_2^o = 1$ gmol. But in separator B, $m_1^o = 10^{-6}$ gmol and $m_2^o = 10^6$ gmol.
(a) Obtain the value of the extent of separation for each vessel and show that they are equal. (Ans. $\xi^A = \xi^B = 0.9999981$.)

(b) Define an extent of purification ξ^p by the absolute value of the determinant of the matrix whose elements are x_{ij}.

(c) Show that the extent of purification for separator A is much greater than that of separator B.

This problem illustrates the role of initial mole ratio (m_1^0/m_2^0) in developing a comparison between two different separation systems behaving identically with respect to distribution ratios (Rony, 1968b). Conversely, it illustrates the relative insensitivity of the extent of separation ξ to considerable degrees of purification (as pointed out in Section 1.5).

1.6.1 For a qualitative description of separation of a single multicomponent feed mixture of n species into k product regions or fractions, Lee *et al.* (1977a) have defined a relative molar fraction of species i in product region j by $\pi_{ij} = x_{ij}/x_{if}$, where each of the mole fractions refer to its averaged value in a given region. Further, r_j is defined as

$$r_j = \frac{m_j}{\sum_{j=1}^{k} m_j},$$

where m_j is the total number of moles in the jth product or region.

(a) Develop a relation between Y_{ij} and π_{ij}. $\left(\text{Ans.}\, Y_{ij} = \pi_{ij}\, r_j / \sum_{j=1}^{k} (\pi_{ij}\, r_j).\right)$

(b) Obtain a matrix of product compositions in terms of relative molar fractions.

(c) What criterion must be satisfied to indicate that there is some separation between two species in one of the product regions?

2

Description of separation in open separators

Separations for preparative or analytical purposes are often carried out batchwise in a closed vessel. On the other hand, industrial-scale separations are commonly achieved in a continuous manner with open vessels into which feed streams enter and from which product streams leave. In this chapter, we consider the available methods of describing separation in such open separators. The quantities, fluxes and mass balances necessary for such descriptions are presented first in Section 2.1. Section 2.2 describes the available indices of separation and their interrelationships for binary separation with a single feed stream entering the separator. In Section 2.3, we briefly introduce indices for binary separation with two feed streams entering a separator. The complications encountered in describing multicomponent separations with a single-entry or double-entry separator are presented in Section 2.4. This section provides also an introduction to the description of systems of continuous chemical mixtures and size-distributed population of particles. Separation by any of the separators considered in these sections presupposes that the output streams have different compositions. There are separation processes, e.g. chromatography, in which the separator has only one output stream, but that has a time-varying composition. The description of separation in such a separator with the help of various indices has been considered in Section 2.5. Triple-entry separators etc. have not been dealt with here. Further, except for Section 2.5, steady state operation is assumed throughout.

2.1 Preliminary quantitative considerations

Open separators can be broadly classified into single-entry and double-entry separators. A single feed stream enters the *single-entry separator*. Two feed streams are necessary for a *double-entry separator* (Figure 2.1.1). A feed stream enters a separator through a defined fraction of its surface area, just as the product streams (at least two in number in general) leave through some other defined surface areas of the separator.

The two product streams may be in contact with each other or they may be separated from each other by a barrier. Next we provide some examples of the nature of contact between various feed and product streams in open separators. Figure 2.1.2(a) shows a nitrogen-rich gas fraction leaving the single-entry separator on one side of the silicone membrane, whereas the oxygen-rich gas fraction is withdrawn from the low-pressure side of the silicone membrane permeator used for separating air. Figure 2.1.2(b) shows a single plate in a distillation column in which bubbles of vapor leave the plate (i.e. a double-entry separator) after contacting a liquid stream which flows along the plate. Only a small part of the vapor product stream is in contact with part of the liquid product stream leaving the plate separator. A product stream may also be in contact with a feed stream as it leaves the separator. Figure 2.1.2(c) shows how a gas stream leaves the top of a spray scrubber where a feed liquid stream is introduced to absorb an undesirable species from a gas mixture. Only part of the cross-sectional area of the tower is utilized by the falling drops of liquid feed, with the rest being utilized by the product gas. While the specification of the flow cross-sectional area of the separator for a given stream is quite difficult in this case, it is straightforward in Figure 2.1.2(a).

We now consider these separators to be fixed in space and focus attention on a particular surface area $\boldsymbol{S}_j$ of any separator shown in Figure 2.1.1. Let $\boldsymbol{v}_{ij}$ be the local average velocity vector of species i with respect to coordinate axes fixed on the surface area $\boldsymbol{S}_j$. If the mass concentration of species i at this location on such a surface area is indicated by ρ_{ij}, with ρ_{tj} being the local value of the total mass concentration, we define the local mass average velocity vector[1] $\boldsymbol{v}_{tj}$ by

[1] See Bird *et al.* (2002), pp. 533–535 for more details.

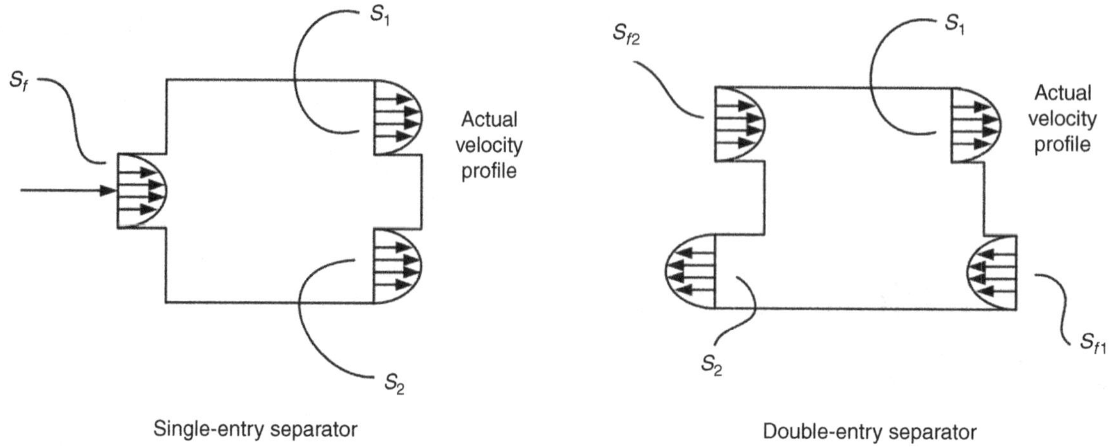

Figure 2.1.1. Single-entry and double-entry separator. The velocity profiles shown at the inlet(s) and exits of the separators need not be parabolic.

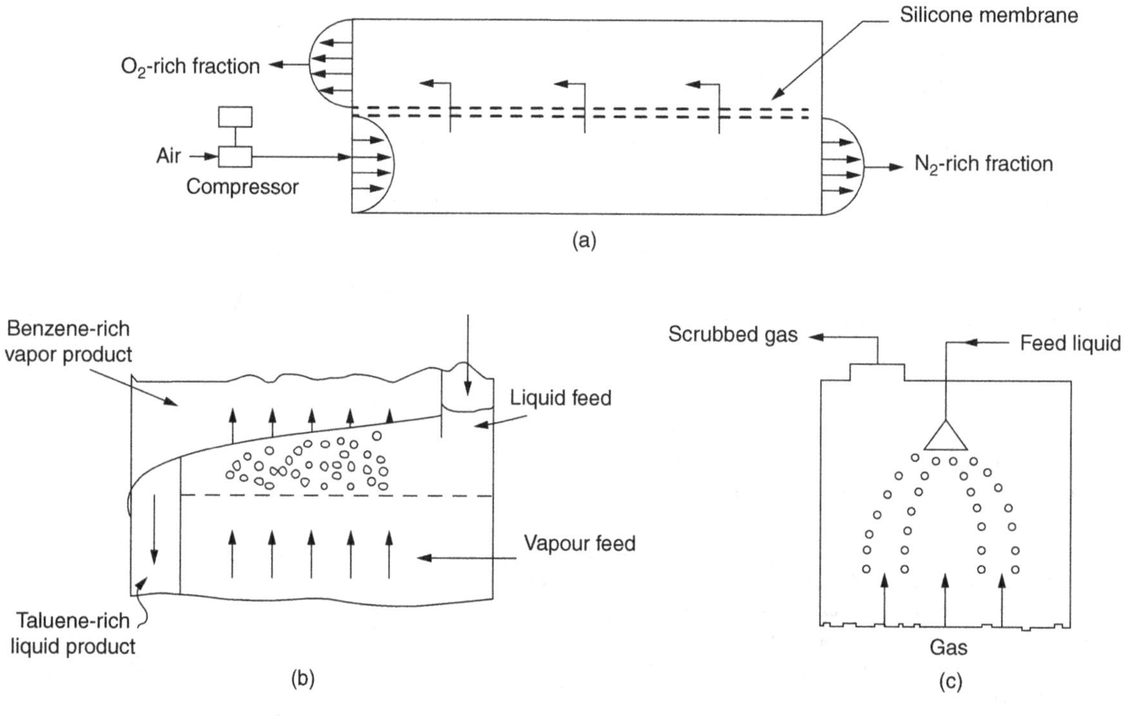

Figure 2.1.2. Some examples of the nature of contact between various feed and product streams. (a) Membrane separates air into two product streams in a single-entry separator, which is a silicone membrane permeator for air separation. (b) Limited contact between vapor and liquid products in a sieve plate (the separator) in a distillation column. (c) Complete contact between a feed scrubbing liquid stream introduced as drops in a spray and the product gas stream after scrubbing.

$$\boldsymbol{v}_{tj} = \frac{\sum_{i=1}^{n} \rho_{ij}\boldsymbol{v}_{ij}}{\sum_{i=1}^{n} \rho_{ij}} = \frac{\sum_{i=1}^{n} \rho_{ij}\boldsymbol{v}_{ij}}{\rho_{tj}} \tag{2.1.1}$$

for a system of n components. The magnitude of this velocity, v_{tj}, is, for example, measurable for liquids by use of a suitable pitot tube. The local molar average velocity vector $\boldsymbol{v}_{tj}^{*}$ is defined by

$$\boldsymbol{v}_{tj}^{*} = \frac{\sum_{i=1}^{n} C_{ij}\boldsymbol{v}_{ij}}{\sum_{i=1}^{n} C_{ij}} = \frac{\sum_{i=1}^{n} C_{ij}\boldsymbol{v}_{ij}}{C_{tj}}, \tag{2.1.2}$$

where C_{ij} is the local molar concentration of species i. In general, the quantities ρ_{ij}, ρ_{tj}, C_{ij}, C_{tj}, $\boldsymbol{v}_{ij}$, $\boldsymbol{v}_{tj}$, $\boldsymbol{v}_{tj}^*$, will vary with location on the surface area $\boldsymbol{S}_j$. The local mass flux vector $\boldsymbol{n}_{ij}$ of species i relative to the stationary surface area $\boldsymbol{S}_j$ is defined as

$$\boldsymbol{n}_{ij} = \rho_{ij}\boldsymbol{v}_{ij}. \tag{2.1.3}$$

The mass rate of inflow or outflow of species i through surface area $\boldsymbol{S}_j$ is given by

$$w_{ij} = \int_{S_j} \boldsymbol{n}_{ij} \cdot \mathrm{d}\boldsymbol{S}_j. \tag{2.1.4}$$

The rate at which mass is entering or leaving the separator through $\boldsymbol{S}_j$ is then given by

$$w_{tj} = \sum_{i=1}^{n} \left[\int_{S_j} \boldsymbol{n}_{ij} \cdot \mathrm{d}\boldsymbol{S}_j \right] = \sum_{i=1}^{n} w_{ij}. \tag{2.1.5}$$

We adopt the convention that when material is entering a separator, it is to be considered positive for mass balance purposes, whereas when material is leaving the separator, we consider it to be negative. Further, diffusive contributions to the mass rate of inflow or outflow of species i through $\boldsymbol{S}_j$ are not considered in this chapter.

Quite often the molar rate of inflow or outflow of mass with respect to the surface area $\boldsymbol{S}_j$ of a separator is of greater interest. In such a case, the molar concentration C_{ij} is used. The quantities corresponding to $\boldsymbol{n}_{ij}$, w_{ij} and w_{tj} are then $\boldsymbol{N}_{ij}$, W_{ij} and W_{tj}, the local molar flux vector of species i, the molar rate of inflow or outflow of species i and the total molar rate of inflow or outflow of all species, respectively:

$$\boldsymbol{N}_{ij} = C_{ij}\boldsymbol{v}_{ij}; \tag{2.1.6}$$

$$W_{ij} = \int_{S_j} \boldsymbol{N}_{ij} \cdot \mathrm{d}\boldsymbol{S}_j; \tag{2.1.7}$$

$$W_{tj} = \sum_{i=1}^{n} W_{ij} = \sum_{i=1}^{n} \left[\int_{S_j} \boldsymbol{N}_{ij} \cdot \mathrm{d}\boldsymbol{S}_j \right]. \tag{2.1.8}$$

If the quantities ρ_{ij}, C_{ij}, $\boldsymbol{v}_{ij}$, $\boldsymbol{v}_{tj}$ and $\boldsymbol{v}_{tj}^*$ do not vary across the surface area $\boldsymbol{S}_j$, we obtain the following relations from (2.1.4), (2.1.5), (2.1.7) and (2.1.8), respectively:

$$w_{ij} = \rho_{ij} v_{ij} S_j; \tag{2.1.9}$$

$$w_{tj} = \left(\sum_{i=1}^{n} \rho_{ij} v_{ij} \right) S_j = \rho_{tj} v_{tj} S_j; \tag{2.1.10}$$

$$W_{ij} = C_{ij} v_{ij} S_j; \tag{2.1.11}$$

$$W_{tj} = \left(\sum_{i=1}^{n} C_{ij} v_{ij} \right) S_j = C_{tj} v_{ij}^* S_j. \tag{2.1.12}$$

For these special cases, we have assumed that $\boldsymbol{v}_{ij}$ is always locally perpendicular to $\boldsymbol{S}_j$. Further, the quantities S_j, v_{ij}, v_{tj} and v_{tj}^* are the magnitudes of the respective vector quantities. If now $\boldsymbol{v}_{ij}$ and therefore $\boldsymbol{v}_{tj}$ and $\boldsymbol{v}_{tj}^*$ vary in magnitude only along $\boldsymbol{S}_j$, and $\boldsymbol{v}_{ij}$ is locally perpendicular to $\boldsymbol{S}_j$, the following average values $\langle v_{ij} \rangle$, $\langle v_{tj} \rangle$ and $\langle v_{tj}^* \rangle$ may be defined such that

$$w_{ij} = \rho_{ij} \langle v_{ij} \rangle S_j; \tag{2.1.13}$$

$$w_{tj} = \rho_{tj} \langle v_{tj} \rangle S_j; \tag{2.1.14}$$

$$W_{ij} = C_{ij} \langle v_{ij} \rangle S_j; \tag{2.1.15}$$

$$W_{tj} = C_{tj} \langle v_{tj}^* \rangle S_j. \tag{2.1.16}$$

Note that

$$\rho_{ij}/M_i = C_{ij}. \tag{2.1.17}$$

If the concentrations also vary across the surface area $\boldsymbol{S}_j$, we can define the various mass and molar flow rates with respect to averaged concentrations and averaged velocities.

The mole fraction x_{ij} of species i in the stream entering or leaving the separator through $\boldsymbol{S}_j$ is defined by

$$x_{ij} = (W_{ij}/W_{tj}). \tag{2.1.18}$$

If $v_{ij} = v_{tj}^*$ for the uniform velocity cases, (2.1.11) and (2.1.12), then only

$$x_{ij} = C_{ij}/C_{tj}. \tag{2.1.19}$$

The mass fraction u_{ij} of the ith species in the stream passing through area $\boldsymbol{S}_j$ is defined for an open separator by

$$u_{ij} = (w_{ij}/w_{tj}). \tag{2.1.20}$$

When uniform velocity profiles exist and $v_{ij} = v_{tj}$, this relation reduces to

$$u_{ij} = (\rho_{ij}/\rho_{tj}). \tag{2.1.21}$$

An open separator will have at least three streams coming in and out through three different surface areas $\boldsymbol{S}_j$ (see Section 2.5 for a different case). We adopt the convention that for a single-entry separator the subscript j will have values $1, 2, \ldots, k$ corresponding to k product streams, but $j = f$ for the single feed stream. For a double-entry separator, the two feed streams are to be denoted by $j = f_1$ and f_2, respectively, while the k product streams will continue to have $j = 1, 2, \ldots, k$.

Consider the steady state operation of the single-entry separator shown in Figure 2.1.1. Since there is no accumulation of mass inside the separator,

$$\text{rate of mass input} = \text{rate of mass outflow}$$

$$\Rightarrow w_{tf} = \sum_{j=1}^{k} w_{tj}. \tag{2.1.22}$$

Assuming further no accumulation of species i in the separator and no chemical reaction, we have the following mass balance for species i:

$$u_{if} w_{tf} = \sum_{j=1}^{k} u_{ij} w_{tj}, \quad (2.1.23a)$$

where we know that

$$\sum_{i=1}^{n} u_{ij} = 1, \qquad \sum_{i=1}^{n} u_{if} = 1. \quad (2.1.23b)$$

For a single-entry separator with two output streams, (2.1.22) and (2.1.23a) reduce to

$$w_{tf} = w_{t1} + w_{t2} \quad (2.1.24)$$

and

$$u_{if} w_{tf} = u_{i1} w_{t1} + u_{i2} w_{t2}. \quad (2.1.25)$$

The corresponding molar balances in the absence of a chemical reaction are:

$$W_{tf} = \sum_{j=1}^{k} W_{tj}; \quad (2.1.26)$$

$$x_{if} W_{tf} = \sum_{j=1}^{k} x_{ij} W_{tj}; \quad (2.1.27)$$

$$W_{tf} = W_{t1} + W_{t2}; \quad (2.1.28)$$

$$x_{if} W_{tf} = x_{i1} W_{t1} + x_{i2} W_{t2}. \quad (2.1.29a)$$

Further,

$$\sum_{i=1}^{n} x_{if} = 1, \qquad \sum_{i=1}^{n} x_{ij} = 1. \quad (2.1.29b)$$

With identical assumptions, the total mass balance and the ith species balance equations for a double-entry separator (see Figure 2.1.1) are:

$$w_{tf_1} + w_{tf_2} = \sum_{j=1}^{k} w_{tj}; \quad (2.1.30)$$

$$u_{if_1} w_{tf_1} + u_{if_2} w_{tf_2} = \sum_{j=1}^{k} u_{ij} w_{tj}. \quad (2.1.31a)$$

Note that in this case

$$\sum_{i=1}^{n} u_{ij} = 1, \quad \sum_{i=1}^{n} u_{if_1} = 1, \quad \sum_{i=1}^{n} u_{if_2} = 1. \quad (2.1.31b)$$

Like (2.1.23b) and (2.1.29b), these equations (2.1.31b) are needed as such to solve for the variables, the mass fractions and the mass flow rates. However, only $(n - 1)$ of equations (2.1.31a) are to be used.

The corresponding steady state molar balances,

$$W_{tf_1} + W_{tf_2} = \sum_{j=1}^{k} W_{tj} \quad (2.1.32)$$

and

$$x_{if_1} W_{tf_1} + x_{if_2} W_{tf_2} = \sum_{j=1}^{k} x_{ij} W_{tj}, \quad (2.1.33a)$$

are valid in the absence of any chemical reaction in the separator. Further,

$$\sum_{i=1}^{n} x_{if_1} = 1; \qquad \sum_{i=1}^{n} x_{if_2} = 1; \qquad \sum_{i=1}^{n} x_{ij} = 1. \quad (2.1.33b)$$

2.2 Binary separation in a single-entry separator with or without recycle

We consider now the description of binary separation in a single-entry separator with only two product streams. Assume steady state operation without any chemical reaction. As in Chapter 1 with region 1 and species 1, we assume here arbitrarily that species 1 is lighter than species 2 and that $j = 1$ refers to that product stream which is richer in the lighter species 1. If we had perfect separation, product stream $j = 1$ will have only species 1 and product stream $j = 2$ will have only species 2. To start with, consider only a nonrecycle separator. Such a separator is sometimes called a splitter (Figure 2.2.1a).

If $x_{i1} \neq x_{i2}$ and both are different from x_{if}, the single-entry separator has achieved some separation. How much separation has been attained can be estimated by suitable descriptors, indices of separation. The simplest of these indices bear the name of specific separation processes where they are extensively used: the desalination ratio D_r, the solute rejection R and the decontamination factor Df.

Desalination ratio:

$$D_r = \frac{C_{2f}}{C_{21}}. \quad (2.2.1a)$$

Solute rejection:

$$R = \left(1 - \frac{C_{21}}{C_{2f}}\right). \quad (2.2.1b)$$

Decontamination factor:

$$Df = \frac{C_{2f}}{C_{21}}. \quad (2.2.1c)$$

Note that all three definitions involve elimination of solute from a solvent (species 1). Further, the solvent is supposed to concentrate in exiting stream 1 and solute (species 2) in exiting stream 2. If there is separation, both D_r and Df will have values greater than 1, whereas $R \leq 1$. It is also clear that

$$R = \left(1 - \frac{1}{D_r}\right) = \left(1 - \frac{1}{Df}\right). \quad (2.2.1d)$$

For dilute solutions, the following simplifications are valid:

$$D_r \cong \frac{x_{2f}}{x_{21}}; \qquad R \cong \left(1 - \frac{x_{21}}{x_{2f}}\right); \qquad Df = \frac{x_{2f}}{x_{21}}. \quad (2.2.1e)$$

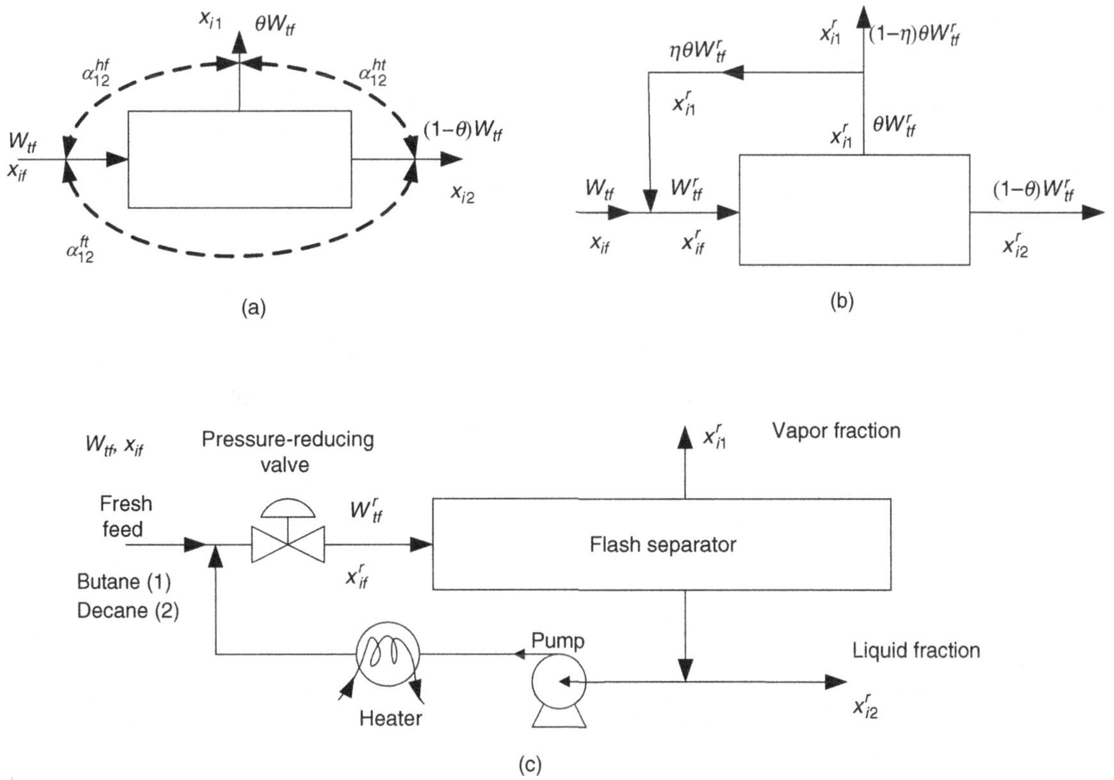

Figure 2.2.1. (a) Nonrecycle single-entry separator. (b) Single-entry separator with part of light fraction recycled to feed stream. (c) Single-entry separator with part of heavy fraction recycled to the feed.

All such indices therefore describe separation by indicating how different the exit stream 1 composition is from the feed stream with respect to the solute species. A somewhat similar index is the equilibrium ratio of species i (obtained under phase-equilibrium conditions),

$$K_i = (x_{i1}/x_{i2}), \tag{2.2.1f}$$

which relates the mole fraction of species 1 in product stream 1 to that in product stream 2. The most common index of separation for the single-entry separator with two product streams is the *separation factor* α_{12}^{ht} for species 1 and 2 between the product streams $j = 1$ (the heads or light fraction, indicated by the superscript h) and $j = 2$ (the tails or heavy fraction, indicated by the superscript t):

$$\alpha_{12}^{ht} = \frac{x_{11}x_{22}}{x_{12}x_{21}}. \tag{2.2.2a}$$

It is sometimes called the *stage separation factor*.

If $\alpha_{12}^{ht} > 1$, species 1 is preferentially present in the heads stream $j = 1$ and species 2 concentrates in the tails stream $j = 2$. Commonly conditions are chosen such that $\alpha_{12}^{ht} > 1$. A few common and useful relations involving the separation factor and the compositions of the exiting streams are given below:

$$x_{11} = \left[\frac{\alpha_{12}^{ht}x_{12}}{1 + x_{12}(\alpha_{12}^{ht} - 1)}\right]; \tag{2.2.2b}$$

$$x_{12} = \left[\frac{x_{11}}{\alpha_{12}^{ht} - x_{11}(\alpha_{12}^{ht} - 1)}\right]. \tag{2.2.2c}$$

Further, if one defines the *mole ratio* $X_{ij} = (x_{ij}/(1 - x_{ij}))$ for a binary system, then

$$X_{11} = \alpha_{12}^{ht} X_{12}. \tag{2.2.2d}$$

The *enrichment factor* ε_{12}^{ht}, defined by

$$\varepsilon_{12}^{ht} = (\alpha_{12}^{ht} - 1) = \frac{(x_{11} - x_{12})}{x_{12}(1 - x_{11})}, \tag{2.2.2e}$$

takes on very small values for close separations ($\varepsilon_{12}^{ht} <<< 1$); the following approximations are then valid:

$$\alpha_{12}^{ht} \cong \exp(\varepsilon_{12}^{ht}); \tag{2.2.2f}$$

$$x_{11} \cong x_{12} + \varepsilon_{12}^{ht}x_{12}(1 - x_{12}); \tag{2.2.2g}$$

$$x_{12} \cong x_{11} - \varepsilon_{12}^{ht}x_{11}(1 - x_{11}). \tag{2.2.2h}$$

For processes where K_i is useful,

$$\alpha_{12}^{ht} = (K_1/K_2). \quad (2.2.2i)$$

How far apart the compositions of the two product streams are is indicated by α_{12}^{ht}. But this index does not indicate how much richer the exiting stream 1 is in species 1 compared to the feed stream. That is expressed by the *heads separation factor* α_{12}^{hf} (Benedict *et al.*, 1981, chap. 12):

$$\alpha_{12}^{hf} = \frac{x_{11}x_{2f}}{x_{1f}x_{21}}, \quad (2.2.3)$$

where the superscript f refers to the feed stream. Similarly, the composition difference between the feed stream and the tails stream is characterized by the *tails separation factor* α_{12}^{ft} (Benedict *et al.*, 1981, chap. 12):

$$\alpha_{12}^{ft} = \frac{x_{1f}x_{22}}{x_{2f}x_{12}}. \quad (2.2.4)$$

Note that

$$\alpha_{12}^{ht} = \alpha_{12}^{hf}\alpha_{12}^{ft}. \quad (2.2.5)$$

Thus if $\alpha_{12}^{hf} > 1$ and $\alpha_{12}^{ft} > 1$, the stage separation factor α_{12}^{ht} is greater than both α_{12}^{hf} and α_{12}^{ft}. This need not be true if either α_{12}^{hf} or α_{12}^{ft} is less than one.

The single-entry separator is said to be *symmetric* if

$$\alpha_{12}^{hf} = \alpha_{12}^{ft}. \quad (2.2.6)$$

In such a case

$$\alpha_{12}^{ht} = \left(\alpha_{12}^{hf}\right)^2 = \left(\alpha_{12}^{ft}\right)^2. \quad (2.2.7)$$

When equation (2.2.6) is not valid, we have an *asymmetric separator*. If β is the highest common root of α_{12}^{hf} and α_{12}^{ft} such that (Wolf *et al.*, 1976)

$$\alpha_{12}^{hf} = \beta^k \quad \text{and} \quad \alpha_{12}^{ft} = \beta^\ell, \quad (2.2.8a)$$

we get

$$\alpha_{12}^{ht} = \beta^{k+\ell}, \quad (2.2.8b)$$

$$\alpha_{12}^{hf} = (\alpha_{12}^{ht})^{k/k+\ell}; \qquad \alpha_{12}^{ft} = (\alpha_{12}^{ht})^{\ell/k+\ell}. \quad (2.2.8c)$$

Although α_{12}^{ht} has been defined in terms of x_{ij} ($j = 1, 2$) only, it may be defined in terms of W_{ij} also. Use (2.1.18) in (2.2.2a) to obtain

$$\alpha_{12}^{ht} = \frac{W_{11}W_{22}}{W_{12}W_{21}}. \quad (2.2.9a)$$

Similarly,

$$\alpha_{12}^{hf} = \frac{W_{11}W_{2f}}{W_{1f}W_{21}}. \quad (2.2.9b)$$

and

$$\alpha_{12}^{ft} = \frac{W_{1f}W_{22}}{W_{2f}W_{12}}. \quad (2.2.9c)$$

In the foregoing treatment, separation was synonymous with the development of a composition difference between the two exiting streams. Since perfect separation requires all of species 1 in exiting stream 1 and all of species 2 in exiting stream 2, we need also some quantities to indicate how much of species 1 in the feed stream is present in the product stream $j = 1$. The earliest index of this kind is the *cut*, θ (sometimes called the *stage cut*):

$$\theta = \frac{W_{t1}}{W_{tf}} = \frac{W_{11} + W_{21}}{W_{1f} + W_{2f}} = \frac{x_{1f} - x_{12}}{x_{11} - x_{12}}, \quad (2.2.10a)$$

with $0 \leq \theta \leq 1$. For example, suppose $x_{11} >> x_{21}$ but $\theta << 1$. Let $x_{1f} = x_{2f}$, then $\alpha_{12}^{hf} >> 1$. But the separation is poor, since very little of species 1 present in the feed has been recovered in the heads fraction $j = 1$. Most of species 1 is present along with most of species 2 in the stream $j = 2$. Note, however, that θ does not directly indicate the extent of recovery in the light fraction of species 1 present in the feed stream. That is indicated by the *component cut*, θ_i, for the ith species, defined by

$$\theta_i = \frac{W_{t1}x_{i1}}{W_{tf}x_{if}}, \quad (2.2.10b)$$

the value of which ranges between 0 and 1.

The index ξ, the extent of separation, was shown in Chapter 1 to reflect both the quality of the separated regions in terms of composition and the amount of recovery of a species in its designated region. For a single-entry separator, an appropriate definition of ξ is as follows (see Rony (1970) and the efficiency formula 9 in Rietema (1957)):

$$\xi = |\dot{Y}_{11} - \dot{Y}_{21}| = |\theta_1 - \theta_2|, \quad (2.2.11)$$

where the quantities $\dot{Y}_{11}$ and $\dot{Y}_{21}$ are obtained from the following general definition of the segregation fraction $\dot{Y}_{ij}$ of the ith species in the jth stream:

$$\dot{Y}_{ij} = \frac{W_{ij}}{\sum_{j=1}^{k} W_{ij}} = \frac{x_{ij}W_{tj}}{W_{if}} = \frac{x_{ij}W_{tj}}{x_{if}W_{tf}}. \quad (2.2.12)$$

Note that $\dot{Y}_{11} = \theta_1$ and $\dot{Y}_{21} = \theta_2$, where θ_i is the ith component cut. For perfect separation in a single-entry separator and two product streams,

$$\dot{Y}_{11} = \frac{x_{11}W_{t1}}{x_{1f}W_{tf}} = 1 \quad \text{and} \quad \dot{Y}_{22} = \frac{x_{22}W_{t2}}{x_{2f}W_{tf}} = 1. \quad (2.2.13a)$$

If there is no separation at all (i.e. $x_{1f} = x_{11} = x_{12}$), then

$$\dot{Y}_{11} = \theta \quad \text{and} \quad \dot{Y}_{22} = 1 - \theta, \quad (2.2.13b)$$

since, for a binary system, $x_{22} = x_{2f} = x_{21}$ for no separation.

We will now obtain a more useful form of ξ for a single-entry separator. Substitution of definition (2.2.12) in (2.2.11) leads, after simplification, to (Sirkar, 1977)

$$\xi = \left(\frac{W_{t1}}{W_{tf}}\right)\left(\frac{x_{21}}{x_{2f}}\right)\left|\frac{x_{11}x_{2f}}{x_{21}x_{1f}} - 1\right|;$$

$$\xi = \left(\frac{W_{t1}}{W_{tf}}\right)\left(\frac{x_{11}}{x_{1f}}\right)\left|1 - \frac{x_{1f}x_{21}}{x_{11}x_{2f}}\right|; \tag{2.2.14}$$

$$\xi = \theta\left(\frac{x_{21}}{x_{2f}}\right)\left|\alpha_{12}^{hf} - 1\right| = \theta_2\left|\alpha_{12}^{hf} - 1\right| = \theta_2\varepsilon_{12}^{hf}; \qquad \xi = \theta_1\left[1 - \frac{1}{\alpha_{12}^{hf}}\right]. \tag{2.2.15}$$

The separation index ξ thus combines the descriptive features of the heads separation factor α_{12}^{hf} and the component cut θ_1 or θ_2. Note that α_{12}^{hf} indicates the degree of enrichment of the light fraction in species 1 with respect to the feed stream, with θ_1 providing an estimate of the extent of recovery of species i if there is separation. Some special cases of relation (2.2.15) are given in the following.

Example 2.2.1 Obtain simplified expressions of ξ for (a) close separation, (b) a close-separation symmetrical separator and (c) a dilute solution of species 2 in solvent 1.

Solution (a) **Close separation** In such a case, $x_{11} > x_{1f}$ and $x_{21} < x_{2f}$, but $x_{11} - x_{1f} \cong 0(\epsilon)$, where ϵ is a very small quantity $<<1$. Similarly, $x_{2f} - x_{21} \cong 0(\epsilon)$. Therefore

$$\xi = \theta\left|\alpha_{12}^{hf} - 1\right| = \theta\varepsilon_{12}^{hf} \tag{2.2.16}$$

since $(x_{21}/x_{2f}) \cong 1$.

(b) **Close-separation symmetrical separator** Due to symmetry, $\alpha_{12}^{hf} = (\alpha_{12}^{ht})^{1/2}$; further, with close separation

$$\ln\left(\alpha_{12}^{hf}\right) = \ln[1 + (\alpha_{12}^{hf} - 1)] \cong \left(\alpha_{12}^{hf} - 1\right);$$

but

$$\ln \alpha_{12}^{hf} = \frac{1}{2}\ln \alpha_{12}^{ht} = \frac{1}{2}\ln[1 + (\alpha_{12}^{ht} - 1)] \cong \frac{1}{2}(\alpha_{12}^{ht} - 1)$$

since $(\alpha_{12}^{ht} - 1)^2$ and higher-order terms (similarly for $(\alpha_{12}^{hf} - 1)^2$ terms, etc.) are negligibly small due to close separation; in addition, $(x_{21}/x_{2f}) \cong 1$. Therefore (Sirkar, 1977)

$$\xi = \frac{\theta}{2}\left|\alpha_{12}^{ht} - 1\right| = \frac{\theta}{2}\varepsilon_{12}^{ht}. \tag{2.2.17}$$

(c) **Dilute solution of species 2 in solvent 1:** $x_{1f} \cong x_{11}$, so

$$\xi = \theta\left|1 - \frac{x_{21}}{x_{2f}}\right| = \theta\left|1 - \frac{1}{D_r}\right| = \theta R = \theta\left|1 - \frac{1}{D_f}\right|. \tag{2.2.18}$$

The index ξ, as per expression (2.2.14) is such that it can have a maximum value of 1 and a minimum value of 0. A variety of conditions can lead to the zero value; the reader can easily verify this. An aspect of ξ that should be noted is that it is invariant to a permutation in the subscripts i or j in $\dot{Y}_{ij}$.

Since ξ combines the separation factor as well as the component cut, it is a composite index. Another such composite index would be the *distribution ratio* $\dot{k}_{i1}'$ defined as

$$\dot{k}_{i1}' = \frac{W_{i1}}{W_{i2}}. \tag{2.2.19}$$

Assuming properties as well as velocities to be invariant along the separator surfaces $\boldsymbol{S}_1$ and $\boldsymbol{S}_2$ as well as $v_{ij} = v_{tj}^*$, we can write (Rony, 1970)

$$\dot{k}_{i1}' = \frac{S_1 C_{i1} v_{i1}}{S_2 C_{i2} v_{i2}} = \frac{S_1 x_{i1} C_{t1}\langle v_{t1}^*\rangle}{S_2 x_{i2} C_{t2}\langle v_{t2}^*\rangle}. \tag{2.2.20}$$

By (1.4.1), $(C_{i1}/C_{i2}) = \kappa_{i1}$, the distribution coefficient. Thus

$$\dot{k}_{i1}' = \kappa_{i1} \times \text{area ratio} \times (\text{velocity ratio})_{i\text{th species}}. \tag{2.2.21}$$

If one further assumes that $v_{i1} = v_{t1}$, $v_{i2} = v_{t2}$ and $S_1 = S_2$, then the distribution ratio $\dot{k}_{i1}'$ is directly governed by κ_{i1} and the bulk velocity ratio of the two streams leaving the separator. Obviously $\dot{k}_{i1}'$ can have any value between 0 and ∞ due to the possible values of κ_{i1}, even if both the area ratio and the velocity ratio are each equal to 1. For the special case of $C_{t1} = C_{t2}$, $\langle v_{t1}^*\rangle = \langle v_{t2}^*\rangle$, and $S_1 = S_2$, $\dot{k}_{i1}' = K_i$.

2.2.1 Examples of separation in single-entry separators

Example 2.2.2 In the *flash expansion desalination* process (also called the *flash vaporization process*), cold sea water, heated to a temperature of 121 °C at the corresponding saturation pressure in a preheater, is allowed to enter a flash chamber where the temperature is 38 °C and the pressure is much lower at the corresponding saturation pressure. This produces some pure water vapor and a concentrated brine. In practical processes, the vapor mass flow rate produced can be at most 20% (Silver, 1966) of the feed brine mass flow rate. If the brine feed has 3.5 wt% salt, describe the separation achieved with the separation factors, extent of separation, distribution ratio and equilibrium ratio for two vapor flow rates expressed as a percentage of the feed brine mass flow rate: (a) 20 wt%, (b) 10 wt%.

Solution Basis: 100 g/s of feed brine flow rate $= w_{tf}$; species 1 is water, species 2 is NaCl, so

$$x_{1f} = \frac{(96.5/18)}{(96.5/18) + (3.5/58.5)} = 0.988, \qquad x_{2f} = 0.012.$$

(a) $w_{t1} = 20$ g/s, $w_{t2} = 80$ g/s, $u_{1f} = 0.965$, $u_{2f} = 0.035$, $u_{21} = 0$, $u_{11} = 1$, $x_{21} = 0$, $x_{11} = 1$ (pure H_2O). We have

$$w_{tf}u_{2f} = w_{t2}u_{22} + w_{t1}u_{21} = 80u_{22} + 0 = 100 \times 0.035;$$

$$u_{22} = \frac{3.5}{80} = 0.04375.$$

To determine x_{22} from u_{22}:

$$x_{22} = \frac{4.375/58.5}{\frac{4.375}{58.5} + \frac{95.625}{18}} = \frac{0.0747}{0.0747 + 5.3125} = \frac{0.0747}{5.3872}.$$

So $x_{22} = 0.01386$; $x_{12} = 0.9861$. The value of the indices may now be calculated as follows:

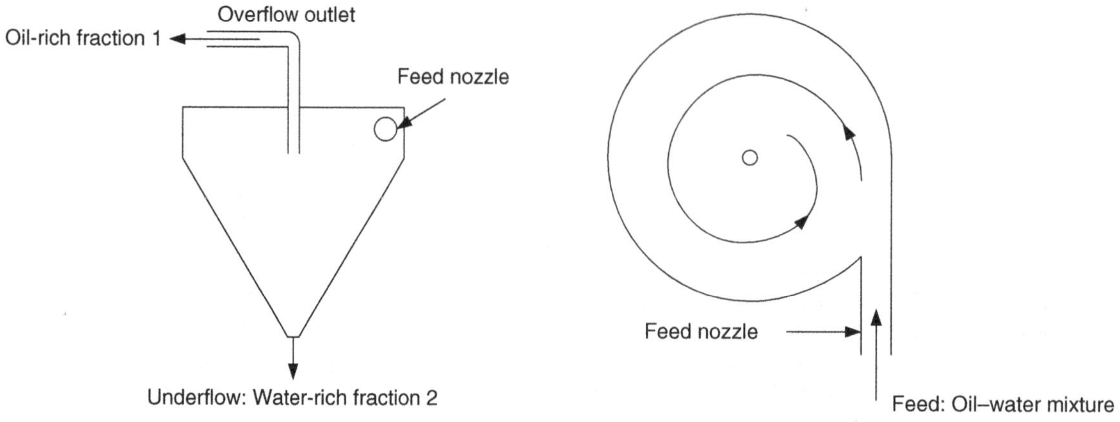

Figure 2.2.2. Hydrocyclone separating an oil–water mixture.

$$\alpha_{12}^{ht} = \frac{x_{11}x_{22}}{x_{21}x_{12}} = \frac{(1)(0.01386)}{(0)(0.9861)} = \infty;$$
$$\alpha_{12}^{hf} = \frac{x_{11}x_{2f}}{x_{21}x_{1f}} = \frac{(1)(0.012)}{(0)(0.988)} = \infty;$$
$$\alpha_{12}^{ft} = \frac{x_{1f}x_{22}}{x_{2f}x_{12}} = \frac{(0.988)(0.01386)}{(0.012)(0.9861)} = 1.1523.$$

Obviously, a flash desalinator is an asymmetric single-entry separator. We have

$$\xi = \theta\left|\frac{x_{11}}{x_{1f}} - \frac{x_{21}}{x_{2f}}\right| = \frac{(20/18)}{\left(\frac{96.5}{18}\right) + \left(\frac{3.5}{58.8}\right)}\left|\frac{1}{0.988} - \frac{0}{0.012}\right|$$
$$= 0.205\left|1.01 - 0\right| = 0.2075;$$

$$k'_{11} = \frac{W_{11}}{W_{12}} = \frac{(20/18)}{(76.5/18)} = 0.262; \qquad k'_{21} = \frac{W_{21}}{W_{22}} = 0;$$
$$K_1 = \frac{x_{11}}{x_{12}} = \frac{1}{0.9861} = 1.0141; K_2 = \frac{x_{21}}{x_{22}} = 0.$$

(b) $w_{t1} = 10$ g/s, $w_{t2} = 90$ g/s, $u_{11} = 1$, $u_{21} = 0$, $u_{1f} = 0.965$,

$$u_{2f} = 0.035, \qquad u_{22} = \frac{3.5}{90} = 0.0388;$$
$$x_{22} = 0.01226; \qquad x_{12} = 0.9877;$$

$$\alpha_{12}^{ht} = \text{infinity} = \alpha_{12}^{hf}; \qquad \alpha_{12}^{ft} = \frac{(0.988)(0.0122)}{(0.012)(0.9877)} = 1.0219;$$

$$\xi = \frac{(10/18)}{5.41}\left|\frac{x_{11}}{x_{1f}} - \frac{x_{21}}{x_{2f}}\right| = 0.1025\left|\frac{1}{0.988} - \frac{0}{0.012}\right|;$$

$$\xi = 0.1037; \qquad k'_{11} = \frac{(10/18)}{(86.5/18)} = 0.1155; \qquad k'_{21} = 0;$$

$$K_1 = \frac{x_{11}}{x_{12}} = \frac{1}{0.987} = 1.011; \qquad K_2 = 0.$$

The two components in binary mixtures dealt with so far are intimately mixed at the molecular level. Suppose, instead, we have to separate a mixture of water and oil such that the oil is made up of only one chemical species. Water and oil molecules are not intimately mixed at the molecular level; rather, there are drops of oil dispersed in water. However, we may treat the problem as if it were a binary separation problem.

Example 2.2.3 An oil and water mixture may be separated by a *hydrocyclone* (Figure 2.2.2), which is essentially a truncated hollow cone with a cylindrical section at the top. The feed mixture enters tangentially in the cylindrical section through the feed nozzle. A water-rich heavy fraction leaves the unit as an underflow at the bottom of the device. The oil-rich light fraction leaves the unit at the top center opening.

(a) Show that the conventional efficiency definition for a hydrocyclone (Van Ebbenhorst Tengbergen and Rietema, 1961),

$$\left|\frac{Q_1 x_1}{Q_f x_f} - \frac{Q_1 y_1}{Q_f y_f}\right| = \left|\frac{Q_2 x_2}{Q_f x_f} - \frac{Q_2 y_2}{Q_f y_f}\right|,$$

is exactly equal to definition (2.2.11) of ξ. Here Q_f, Q_1 and Q_2 are the volumetric flow rates of the feed, overflow and underflow, respectively; x_1 and y_1 are the volume fractions of oil and water in the overflow; x_f and y_f refer to similar quantities in the feed; and x_2 and y_2 refer to those in the underflow.

(b) Calculate the value of the extent of separation in a 35 mm hydrocyclone fed with a 70 cm^3/s of 50% oil–50% water mixture such that $x_1 = 0.85$ and $(Q_1/Q_2) = 0.815$.

Solution (a) $Q_1 x_1$ is the volumetric flow rate of oil only in the overflow, whereas $Q_f x_f$ refers to that in the feed. Therefore,

$$\frac{Q_1 x_1}{Q_f x_f} = \frac{(Q_1 x_1 \rho_{\text{oil}}/M_{\text{oil}})}{\left(Q_f x_f \rho_{\text{oil}}/M_{\text{oil}}\right)} = \frac{x_{11} W_{t1}}{x_{1f} W_{tf}},$$

where $i = 1$ refers to oil, $j = 1$ refers to the overflow in x_{ij}, W_{tj}; ρ_{oil} and M_{oil} are, respectively, the density and molecular weight of the oil. Similarly,

$$\frac{Q_1 y_1}{Q_f y_f} = \frac{(Q_1 y_1 \rho_{\text{water}}/M_{\text{water}})}{\left(Q_f y_f \rho_{\text{water}}/M_{\text{water}}\right)} = \frac{x_{21} W_{t1}}{x_{2f} W_{tf}};$$

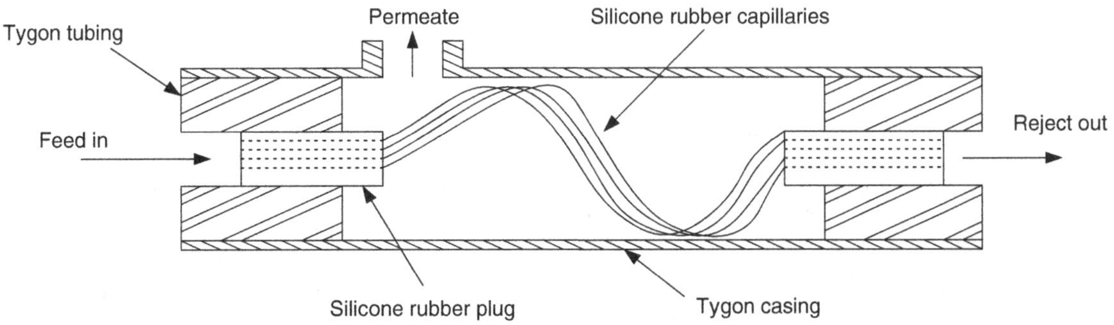

Figure 2.2.3. Gas separation by permeation through silicone rubber capillaries. Reprinted from Chem. Eng. Sci., **30***(7), 751 (1975), J.M. Thorman, S. Rhim and S.T. Hwang, "Gas separation by diffusion through silicone rubber capillaries." Copyright (1975), with permission from Elsevier.*

thus

$$\xi = \left|\dot{Y}_{11} - \dot{Y}_{21}\right| = \left|\frac{Q_1 x_1}{Q_f x_f} - \frac{Q_1 y_1}{Q_f y_f}\right|.$$

Note: Since component 1 is oil, perfect separation corresponds to only oil in the overflow (product stream $j = 1$) and only water in the underflow (product stream $j = 2$). The other equality follows using the following two mass balances:

$$Q_f x_f \rho_{oil} = Q_1 x_1 \rho_{oil} + Q_2 x_2 \rho_{oil};$$
$$Q_f y_f \rho_{water} = Q_1 y_1 \rho_{water} + Q_2 y_2 \rho_{water}.$$

For part (b),

$$Q_f = Q_1 + Q_2 = 70$$
$$= 1.815\, Q_2 \Rightarrow Q_2 = \frac{70}{1.815} = 38.5 \text{cm}^3/\text{s};$$

$$Q_1 = 31.5 \text{cm}^3/\text{s}.$$

We have

$$x_f = 0.5, \quad y_f = 0.5, \quad y_1 = 0.15;$$

$$\xi = \left|\frac{Q_1 x_1}{Q_f x_f} - \frac{Q_1 y_1}{Q_f y_f}\right| = \left|\frac{31.5 \times 0.85}{70 \times 0.5} - \frac{31.5 \times 0.15}{70 \times 0.5}\right| = 0.63.$$

Note: In Figure 2.2.2 we do not describe particular flow patterns or forces causing separation and yielding $\xi = 0.63$. It is sufficient here to recognize that two predominantly immiscible phase product streams are obtained from a dispersed phase feed through nothing but a suitable flow arrangement with tangential entry.

Example 2.2.3 is modeled after a practical example treated by Sheng (1977) on separation of liquids by a conventional hydrocyclone, and the value of $\xi = 0.63$ corresponds to the actual efficiency E used by Sheng. The above approach is also adaptable to cases where the oil itself is a multicomponent mixture of various hydrocarbons.

Example 2.2.4 (a) Thorman *et al.* (1975) investigated the separation of oxygen from air by preferential permeation of oxygen through silicone rubber capillaries kept in a permeator. Air containing 20.9 mole percent of oxygen was introduced at a pressure of 176 cm Hg into the inside of silicone rubber capillary tubings (0.245 mm inner diameter (i.d.) × 0.655 mm outer diameter (o.d)), a bundle of which was kept in a shell-and-tube arrangement inside a 12.5 mm i. d. tygon tubing shown in Figure 2.2.3. The gas that permeated through the walls of the silicone rubber capillary was withdrawn at atmospheric pressure through a port in the shell. The feed air flow rate is 1.80 std. cm^3/s and the permeated gas flow rate is 0.243 std. cm^3/s. The permeated gas contains 0.274 mole fraction oxygen. Determine the separation factor α_{12}^{ht}, the extent of separation and the cut.

(b) If instead of air, a mixture of O_2 and CO_2 containing 50.2 mole percent oxygen is introduced at 176 cm Hg as feed into the same capillary tubings in the separator at a flow rate of 1.91 std. cm^3/s, the permeated gas containing 0.315 mole fraction oxygen comes out at a rate of 0.97 std. m^3/s. Determine the separation factor α_{12}^{ht}, the extent of separation and the cut.

Solution (a) The composition and the flow rate of the reject stream are determined first. Here, $i = 1$ (O_2), $i = 2$ (N_2), $j = 1$ permeated stream, $j = 2$ reject stream, since the permeated stream is becoming enriched in oxygen. With the molar flow rate of a gas stream in gmol/s being directly proportional to its volumetric flow rate in std. cm^3/s, we get, using (2.1.29a),

$$x_{12} = \frac{x_{1f} W_{tf} - x_{11} W_{t1}}{W_{t2}};$$
$$x_{12} = \frac{0.209 \times 1.80 - 0.274 \times 0.243}{1.557} = 0.199,$$

since $x_{1f} = 0.209$, $x_{11} = 0.274$ and

$$W_{t2} = W_{tf} - W_{t1} = \left[(\text{conv. factor}) \text{ for } \frac{\text{gmol}}{\text{std. cm}^3}\right] \times (1.80 - 0.243)\frac{\text{std. cm}^3}{\text{s}}.$$

Therefore,

$$\alpha_{12}^{ht} = \frac{x_{11} x_{22}}{x_{21} x_{12}} = \frac{0.274 \times 0.801}{0.726 \times 0.199} = 1.516.$$

Further,

$$\xi = \theta_2 \left|\alpha_{12}^{hf} - 1\right| = \frac{0.243 \times 0.726}{1.80 \times 0.791}\left|\frac{0.274 \times 0.791}{0.726 \times 0.209} - 1\right| = 0.0545,$$

and the cut

$$\theta = \frac{W_{t1}}{W_{tf}} = \frac{0.243}{1.80} = 0.135.$$

(b) Here $i = 1$ (CO_2), $i = 2$ (O_2), $j = 1$ (permeated stream) and $j = 2$ (reject stream); the permeated stream is enriched in CO_2 compared to the feed stream. Given $x_{2f} = 0.502$, $x_{21} = 0.315$, and taking the volumetric flow rate of the reject stream to be 0.94 cm^3/s (= 1.91 − 0.97), we get

$$x_{22} = \frac{x_{2f} W_{tf} - x_{21} W_{t1}}{W_{t2}} = \frac{0.502 \times 1.91 - 0.315 \times 0.97}{0.94} = 0.694.$$

Therefore,

$$\alpha_{12}^{ht} = \frac{0.685 \times 0.694}{0.315 \times 0.306} = 4.83.$$

Further

$$\xi = \frac{W_{t1} x_{21}}{W_{tf} x_{2f}} \left| \alpha_{12}^{hf} - 1 \right| = \frac{0.97 \times 0.315}{1.91 \times 0.502} \left| \frac{0.685 \times 0.502}{0.315 \times 0.498} - 1 \right|$$
$$= 0.318 |2.19 - 1| = 0.378.$$

The cut $\theta = W_{t1}/W_{tf} = 0.506$. Note that $\alpha_{12}^{ft} = (4.83/2.19) = 2.20$; it is behaving almost as a symmetric separator.

It would appear from the preceding treatments that for a single-entry separator, the cut θ is independent of α_{12}^{ht}, α_{12}^{hf} or α_{12}^{ft}. For any given separation process, the relation between a separation factor and the cut, θ, can be derived only by detailed considerations of the separation mechanism operative in the separator. In general, the separation factor is likely to depend on θ. These dependencies will be considered when individual separation processes are discussed in later chapters.

2.2.2 Single-entry separator with a product recycle

Sometimes, to achieve a better separation, a fraction of one of the product streams is recycled to the feed end of the single entry separator. Consider Figure 2.2.1 where we show a nonrecycle traditional separator with an actual cut θ and actual heads and tails separation factors α_{12}^{hf} and α_{12}^{ft}. This figure also shows the schematic of a separator, where a fraction η of the actual light fraction molar output from the separator is recycled to the feed stream (Figure 2.2.1 (b)). Thus the actual molar feed rate to the *recycle separator* is higher than that to the nonrecycle separator. Let us now consider a special case of a binary mixture in which $x_{1f} << x_{2f}$ and x_{2f} is close to 1. Suppose further that the actual cut and the actual heads and tails separation factors of the recycle separator have the same values θ, α_{12}^{hf} and α_{12}^{ft}. What are the values of the apparent cut and the apparent heads and tails separation factors for this recycle separator? Can we compare the separating ability of these two arrangements with suitable descriptors?

The molar fresh feed supply rate to both separators is W_{tf}. While this is the actual feed flow rate into the nonrecycle separator, the actual feed flow rate into the recycle separator is $W_{tf}^r (> W_{tf})$. Similarly, the fresh feed composition is given by x_{1f}, whereas the actual feed composition to the recycle separator is x_{1f}^r ($> x_{1f}$ since $x_{11} > x_{1f}$ for $\alpha_{12}^{hf} > 1$). Since we have assumed that the actual values of θ, α_{12}^{hf} and α_{12}^{ft} for the recycle separator are the same as those of the simple separator, we obtain, by a total molar balance at steady state for a recycle ratio of η,

$$W_{tf}^r = \frac{W_{tf}}{(1 - \eta\theta)}. \tag{2.2.22}$$

Due to the binary mixture being dilute in species 1, the following relations are valid:

$$\alpha_{12}^{hf} = \frac{x_{11}^r x_{2f}^r}{x_{21}^r x_{1f}^r} \cong \frac{x_{11}^r}{x_{1f}^r}; \qquad \alpha_{12}^{hf} = \frac{x_{11} x_{2f}}{x_{1f} x_{21}} \cong \frac{x_{11}}{x_{1f}}; \tag{2.2.23}$$

$$\alpha_{12}^{ft} = \frac{x_{1f}^r x_{22}^r}{x_{2f}^r x_{12}^r} \cong \frac{x_{1f}^r}{x_{12}^r}; \qquad \alpha_{12}^{ft} = \frac{x_{1f} x_{22}}{x_{2f} x_{12}} \cong \frac{x_{1f}}{x_{12}}. \tag{2.2.24}$$

A light component mass balance for the actual feed stream in the recycle separator leads to

$$W_{tf} x_{1f} + \eta\theta W_{tf}^r x_{11}^r = x_{1f}^r W_{tf}^r, \tag{2.2.25}$$

which on rearrangement using (2.2.22) and (2.2.23) yields

$$x_{1f}^r = x_{1f} \frac{(1 - \eta\theta)}{[1 - \eta\theta\alpha_{12}^{hf}]}. \tag{2.2.26}$$

Since $\alpha_{12}^{hf} > 1$, obviously $x_{1f}^r > x_{1f}$, and therefore $x_{11}^r (= \alpha_{12}^{hf} x_{1f}^r)$ is greater than x_{11}. The *apparent cut* and the *apparent heads separation factor* for the recycle separator are given by

$$\theta_{app} = \frac{\theta W_{tf}^r (1 - \eta)}{W_{tf}} = \frac{\theta(1 - \eta)}{(1 - \eta\theta)} < \theta; \tag{2.2.27}$$

$$\left(\alpha_{12}^{hf}\right)_{app} \cong \frac{x_{11}^r}{x_{1f}} = \left(\alpha_{12}^{hf}\right)\left(\frac{x_{1f}^r}{x_{1f}}\right) > \alpha_{12}^{hf}. \tag{2.2.28}$$

Thus, the apparent values indicate a better quality of separation accompanied by a lesser amount of product stream 1. For a better picture, let us compute the extent of separation with recycle and without recycle:

$$\xi^r = \left| \dot{Y}_{11}^r - \dot{Y}_{21}^r \right| = \left| \frac{x_{11}^r (1 - \eta)\theta W_{tf}^r}{W_{tf} x_{1f}} - \frac{x_{21}^r (1 - \eta)\theta W_{tf}^r}{W_{tf}(1 - x_{1f})} \right|;$$

$$\xi^r = \frac{(1 - \eta)\theta x_{11}^r}{(1 - \eta\theta) x_{1f}} \left| 1 - \frac{x_{21}^r x_{1f}}{x_{2f} x_{11}^r} \right|; \tag{2.2.29}$$

$$\xi = \left| \dot{Y}_{11} - \dot{Y}_{21} \right| = \frac{\theta x_{11}}{x_{1f}} \left| 1 - \frac{x_{21} x_{1f}}{x_{2f} x_{11}} \right|. \tag{2.2.30}$$

Therefore,

$$\left[\frac{\xi^r}{\xi}\right] = \left[\frac{(1 - \eta)}{(1 - \eta\theta)} \frac{x_{11}^r}{x_{11}}\right] \left| \frac{1 - \dfrac{x_{21}^r x_{1f}}{x_{2f} x_{11}^r}}{1 - \dfrac{x_{21} x_{1f}}{x_{2f} x_{11}}} \right|. \tag{2.2.31a}$$

Since all the mixtures are dilute in species 1, we may assume that

$$\frac{x_{21}^r x_{1f}}{x_{2f} x_{11}^r} \simeq \frac{x_{1f}}{x_{11}^r}; \qquad \frac{x_{21} x_{1f}}{x_{2f} x_{11}} \simeq \frac{x_{1f}}{x_{11}}. \qquad (2.2.31b)$$

Utilizing relations (2.2.23) and (2.2.26) in (2.2.31a), we get, after various manipulations,

$$\xi^r/\xi = (1-\eta)/[(1-\eta\theta\alpha_{12}^{hf})(1-\eta\theta)], \qquad (2.2.31c)$$

from which we can conclude that if $\xi^r > \xi$, then the following inequality must hold:[2]

$$\alpha_{12}^{hf} > \frac{1-\theta}{\theta(1-\eta\theta)}. \qquad (2.2.31d)$$

For a binary mixture dilute in one species and being separated in a single-entry separator, there will therefore be a better separation with recycle of light fraction for constant θ, α_{12}^{hf} and α_{12}^{ft} if condition (2.2.31d) is satisfied. Obviously close-separation cases will not be helped by recycle unless θ is unrealistically close to 1. The description of separation in a recycle single-entry separator is now less ambiguous compared to that based on earlier results (2.2.27) and (2.2.28).[3] Note that ξ^r has been calculated by using the fresh feed and the net separated output streams to estimate the actual separation achieved.

A recycle single-entry separator may have part of its tails stream recycled to the feed (flash distillation, Figure 2.2.1(c)). Examples of radioactive rare gas separation with various recycle arrangements are given in Ohno *et al.* (1977, 1978). See Problem 2.2.7. One aspect to be noted while comparing separators with or without recycle is that since θ is constant, but the actual feed flow rate to the separator changes with recycle from that without recycle, the dimensions of the two separators will be different. See, however, Problem 2.2.8.

2.2.3 Separative power and value function

An additional index, often used for *isotope separation plants* and originally introduced by P.A.M Dirac (Cohen, 1951), is the *separative power* δU of a single-entry separator. Assuming that the value of a particular process stream increases with the increase in mole fraction of a given species, the separative power of a given separating unit could be determined from the net increase in the value of the streams coming into and out of the separator. Such an index may be defined, following Dirac, as

$$\delta U = W_{t1} Va(x_{11}) + W_{t2} Va(x_{12}) - W_{tf} Va(x_{1f}), \qquad (2.2.32a)$$

where $Va(x_{ij})$, the *value function*, is the value of one mole of the jth stream of composition x_{ij}. For the separator of cut θ, the following alternative expression is more useful:

$$\frac{\delta U}{W_{tf}} = \theta Va(x_{11}) + (1-\theta) Va(x_{12}) - Va(x_{1f}). \qquad (2.2.32b)$$

One demands $Va(x_{ij})$, the value function, to be such that δU is independent of feed and product compositions. The quantity δU then indicates the net increase in the value of products over that of the feed stream, and is therefore a measure of the separator capability or separation achieved.

There exists no general solution for $Va(x_{ij})$ to make δU independent of feed and product compositions. For a few special cases, solutions for $Va(x_{ij})$ satisfying such a requirement are available. One such case is close separation, for which both x_{i1} and x_{i2} are sufficiently close to x_{if} to permit a Taylor series expansion of $Va(x_{ij})$ around $Va(x_{if})$:

$$Va(x_{i1}) = Va(x_{if}) + \left.\frac{dVa(x_{i1})}{dx_{i1}}\right|_{x_{i1}=x_{if}} (x_{i1}-x_{if}) + \left.\frac{d^2 Va(x_{i1})}{dx^2_{i1}}\right|_{x_{i1}=x_{if}} \frac{(x_{i1}-x_{if})^2}{2} + \ldots; \qquad (2.2.33)$$

$$Va(x_{i2}) = Va(x_{if}) + \left.\frac{dVa(x_{i2})}{dx_{i2}}\right|_{x_{i2}=x_{if}} (x_{i2}-x_{if}) + \left.\frac{d^2 Va(x_{i2})}{dx^2_{i2}}\right|_{x_{i2}=x_{if}} \frac{(x_{i2}-x_{if})^2}{2} + \ldots. \qquad (2.2.34)$$

For a continuous and well-behaved function $Va(x_{ij})$,

$$\left.\frac{dVa(x_{i1})}{dx_{i1}}\right|_{x_{i1}=x_{if}} = \left.\frac{dVa(x_{i2})}{dx_{i2}}\right|_{x_{i2}=x_{if}} \quad \text{and} \quad \left.\frac{d^2 Va(x_{i1})}{dx^2_{i1}}\right|_{x_{i1}=x_{if}} = \left.\frac{d^2 Va(x_{i2})}{dx^2_{i2}}\right|_{x_{i2}=x_{if}}. \qquad (2.2.35)$$

Recognizing from equations (2.1.29a) and (2.2.10a) that

$$x_{if} = \theta x_{i1} + (1-\theta) x_{i2}, \qquad (2.2.36)$$

we obtain from definition (2.2.32b)

$$\frac{\delta U}{W_{tf}} = \frac{1}{2}\left[\theta(x_{i1}-x_{if})^2 + (1-\theta)(x_{i2}-x_{if})^2\right] \left.\frac{d^2 Va(x_{i1})}{dx^2_{i1}}\right|_{x_{i1}=x_{if}} \qquad (2.2.37a)$$

when relations (2.2.33), (2.2.34) and (2.2.35) are used. To simplify the above result further, rewrite the ith species balance equation (2.2.36) in the following two forms:

$$\theta(x_{i1}-x_{if}) = (1-\theta)(x_{if}-x_{i2}) \qquad (2.2.37b)$$

and

$$\theta(x_{i1}-x_{i2}) = (x_{if}-x_{i2}). \qquad (2.2.37c)$$

With these two relations, expression (2.2.37a) for $(\delta U/W_{tf})$ is simplified:

[2] This analysis is valid only if $\left(1-\eta\theta\alpha_{12}^{hf}\right)$ is nonzero and positive.

[3] K.K. Sirkar and S. Teslik, "Description of separation in separators with or without recycle," unpublished (1982).

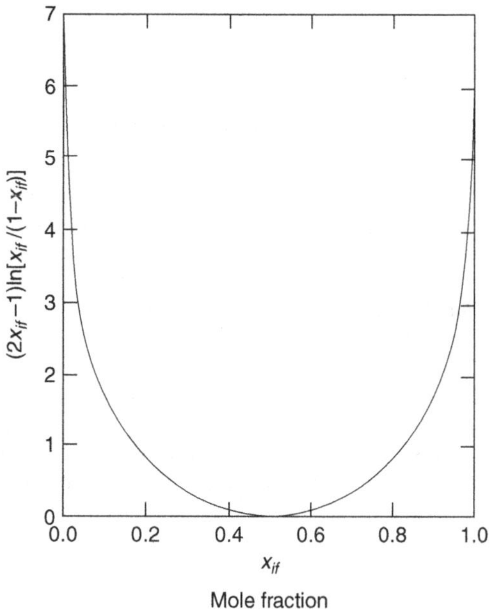

Figure 2.2.4. Value function of equation (2.2.42).

$$\delta U = \frac{W_{tf}}{2}\theta(1-\theta)(x_{i1}-x_{i2})^2 \left.\frac{d^2 Va(x_{i1})}{dx_{i1}^2}\right|_{x_{i1}=x_{if}}. \quad (2.2.37d)$$

For *close separations*, use relation (2.2.2h) with $x_{i1} \cong x_{i2} \cong x_{if}$ on the right-hand side and substitute in (2.2.37d):

$$\delta U = \frac{W_{tf}}{2}\theta(1-\theta)\left(\alpha_{12}^{ht}-1\right)^2\left[x_{if}\left(1-x_{if}\right)\right]^2 \frac{d^2 Va(x_{if})}{dx_{if}^2}. \quad (2.2.38)$$

To make δU independent of compositions of various streams, assume

$$\frac{d^2 Va(x_{if})}{dx_{if}^2} = \frac{1}{\left[x_{if}\left(1-x_{if}\right)\right]^2}. \quad (2.2.39)$$

Therefore,

$$\delta U = \frac{W_{tf}}{2}\theta(1-\theta)\left(\alpha_{12}^{ht}-1\right)^2 \quad (2.2.40)$$

for close separations. The corresponding value function can be obtained by specifying the following conditions:

$$Va(x_{if}=0.5)=0 \quad (2.2.41a)$$

and

$$\left.\frac{dVa(x_{if})}{dx_{if}}\right|_{x_{if}=0.5} = 0 \quad (2.2.41b)$$

for $Va(x_{if})$ satisfying (2.2.39). The solution is simply

$$Va(x_{if}) = \left(2x_{if}-1\right)\ln\left[x_{if}/\left(1-x_{if}\right)\right]. \quad (2.2.42)$$

The value function is plotted in Figure 2.2.4 against x_{if}. As has been pointed out (Pratt, 1967), a zero value for $Va(x_{if}=0.5)$ is reasonable, especially for nonisotopic mixtures. This is because the mixture is in the lowest energy state and will require the largest energy for separation (see Chapter 10). See Cohen (1951), Benedict *et al.* (1981, chap. 12) and Pratt (1967) for detailed treatments on separative power, value functions, etc. for isotope separation plants.

2.3 Binary separation in a double-entry separator

Unlike for a single-entry separator, *two* feed streams enter a double-entry separator. The number of product streams leaving such a separator in general is two. As shown in Figure 2.3.1, the relative orientations of the two feed streams with respect to each other as well as the separator may vary, resulting in crosscurrent, cocurrent or countercurrent configurations. Separation with such an arrangement is achieved when $x_{i1} \neq x_{i2}$, and either $x_{if1} \neq x_{i1}$ or $x_{if2} \neq x_{i2}$ or both. The most frequent situation encountered is $x_{if1} \neq x_{i1}$, $x_{if2} \neq x_{i2}$, $x_{i1} \neq x_{i2}$, with all three inequalities being valid simultaneously. Further, product stream 1, usually termed the *heads fraction*, the *light fraction* or the *vapor fraction*, will have only species 1 when pure, whereas product stream 2, usually identified as the *tails fraction*, the *heavy fraction* or the *liquid fraction*, will have only species 2 when pure.

The double-entry separator performance is often characterized for a two-component system by the equilibrium ratio K_i (definition (2.2.1f)) as well as by the separation factor α_{12}^{ht} (definition (2.2.2a)). Both definitions relate the two product stream compositions and are related to each other by (2.2.2i). The compositional differences between the light fraction and feed stream 1 may be indicated by a heads separation factor

$$\alpha_{12}^{hf} = \frac{x_{1l}\,x_{2f_1}}{x_{1f_1}\,x_{21}}. \quad (2.3.1a)$$

Similarly, the tails separation factor between feed stream 2 and the heavy fraction is

$$\alpha_{12}^{ft} = \frac{x_{1f_2}\,x_{22}}{x_{2f_2}\,x_{12}}. \quad (2.3.1b)$$

Unlike that for a single-entry separator, where $\alpha_{12}^{ht} = \alpha_{12}^{hf} \times \alpha_{12}^{ft}$ by definition, here,

$$\alpha_{12}^{ht} \neq \alpha_{12}^{hf} \times \alpha_{12}^{ft} \quad (2.3.2)$$

unless the two feed streams have the same composition, i.e. $x_{if_1} = x_{if_2}$. However, relations (2.2.2b)–(2.2.2d) *are* valid for a double-entry separator. One may also define an enrichment factor ε_{12}^{ht} in exactly the same way as in (2.2.2e). In addition, for close separations ($\varepsilon_{12}^{ht} << 1$), relations (2.2.2f)–(2.2.2h) also hold for a double-entry separator.

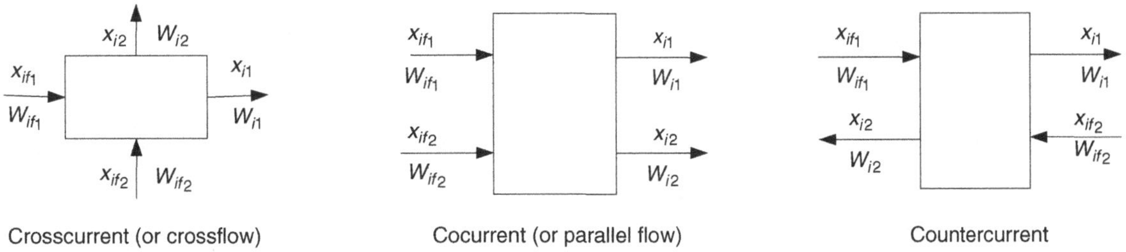

Figure 2.3.1. Double-entry separator: various orientations of feed streams.

The distribution ratio $\dot{k}'_{il}$ of species i between the two product streams 1 and 2 can be defined here in the same manner as in (2.2.19):

$$\dot{k}'_{i1} = \frac{W_{i1}}{W_{i2}} = \frac{x_{i1}\,W_{t1}}{x_{i2}\,W_{t2}} = K_i\frac{W_{t1}}{W_{t2}}. \qquad (2.3.3a)$$

For binary vapor–liquid systems used in distillation (see Example 2.3.1) if vapor phase $j = 1$ and liquid phase $j = 2$, then, for species i, $\dot{k}'_{i1}$ is equal to λ, a parameter frequently used in distillation analysis (see Problem 2.2.1). We see further from (2.2.2a) that

$$\alpha_{12}^{ht} = \left(\dot{k}'_{11}/\dot{k}'_{21}\right). \qquad (2.3.3b)$$

In terms of average velocities of species i across the surface areas S_1 and S_2 and constant properties across S_1 and S_2 (see relation (2.2.21))

$$\alpha_{12}^{ht} = \frac{\kappa_{1l}\,v_{1l}\,v_{22}}{\kappa_{21}v_{12}\,v_{21}}. \qquad (2.3.3c)$$

If $v_{i1} = v_{t1}$ and $v_{i2} = v_{t2}$, then only

$$\alpha_{12}^{ht} = (\kappa_{11}/\kappa_{21}), \qquad (2.3.3d)$$

which is also valid for closed separators when $V_1 = V_2$.

The index extent of separation for a double-entry separator may be expressed as (Sirkar, 1977)

$$\xi = \text{abs}\,|\dot{Y}_{11} - \dot{Y}_{21}| = \text{abs}\,|\dot{Y}_{11}\dot{Y}_{22} - \dot{Y}_{12}\dot{Y}_{21}|; \qquad (2.3.4a)$$

$$\xi = \dot{Y}_{12}\dot{Y}_{21}\text{abs}\left|\frac{\dot{Y}_{11}\dot{Y}_{22}}{\dot{Y}_{12}\dot{Y}_{21}} - 1\right|; \;\; \xi = \dot{Y}_{12}\dot{Y}_{21}\text{abs}\left|\frac{x_{11}x_{22}}{x_{12}x_{21}} - 1\right|;$$

$$\xi = \dot{Y}_{12}\dot{Y}_{21}\text{abs}\left|\alpha_{12}^{\text{ht}} - 1\right|. \qquad (2.3.4b)$$

The quantity $\dot{Y}_{21}$ may be interpreted as the component cut θ_2, i.e. that fraction of species 2 entering the separator which appears in the light fraction $j = 1$. Similarly, $\dot{Y}_{12}$ may be interpreted as $(1 - \theta_1)$, where θ_1 is the component cut for species 1. Note that both feed streams have to be lumped together here for determining θ_1 and θ_2. Thus

$$\xi = (1 - \theta_1)\theta_2\,\text{abs}\left|\alpha_{12}^{ht} - 1\right|. \qquad (2.3.4c)$$

2.3.1 Examples of separation in double-entry separators

Example 2.3.1 In a tall vertical column packed with Raschig rings, the separation of benzene from toluene is being carried out by distillation (Figure 2.3.2(a)). A vapor mixture of benzene and toluene containing 0.51 mole fraction benzene is introduced at the column bottom at a rate of 20 gmol/s. At the column top, a benzene-rich stream having the same molar flow rate (mole fraction of benzene $\cong$ 0.95) is introduced as a liquid through a distributor. As the vapor goes up the column with the liquid coming down, toluene from the vapor preferentially goes to the liquid stream, and the benzene from the liquid stream preferentially comes to the vapor stream. The vapor product stream flow rate at the top is 20 gmol/s and the benzene mole fraction in it is 0.98. Obtain the values of α_{12}^{ht} and ξ for this double-entry countercurrent separator. Determine the value of ξ for the two feed streams only, and compare it with that determined from the product streams. Comment thereafter on the utility of this separation device.

Solution $W_{tf_1} = 20$ gmol/s, $W_{tf_2} = 20$ gmol/s, $W_{t1} = 20$ gmol/s ($j = 1$, vapor stream). Since $W_{tf_1} + W_{tf_2} = W_{t1} + W_{t2}$, $W_{t2} = 20$ gmol/s ($j = 2$ liquid product stream). Making a component balance on benzene ($i = 1$), $W_{tf_1}\,x_{1f_1} + W_{tf_2}\,x_{1f_2} = W_{t1}\,x_{11} + W_{t2}\,x_{12}$. But $x_{1f_1} = 0.51$, $x_{1f_2} = 0.95$, $x_{11} = 0.98$ and $x_{12} = ?$ Therefore $x_{12} = [0.51 + 0.95 - 0.98] = 0.48$;

$$\alpha_{12}^{ht} = \frac{0.98 \times 0.52}{0.02 \times 0.48} = 53; \qquad \xi = \dot{Y}_{12}\dot{Y}_{21}\left|\alpha_{12}^{ht} - 1\right|;$$

$$\xi = \left\{\frac{W_{t2}x_{12}}{W_{tf_1}x_{1f_1} + W_{tf_2}x_{1f_2}}\right\}\left\{\frac{W_{t1}x_{21}}{W_{tf_1}x_{2f_1} + W_{tf_2}x_{2f_2}}\right\}|53 - 1|$$

$$= \left(\frac{0.48}{1.46}\right) \times \left(\frac{0.02}{0.54}\right) \times 52 = 0.632.$$

Concentrate now only on the two feed streams and assume them to be product streams whose extent of separation is to be determined. Assuming feed stream f_2 to be stream $j = 1$ and feed stream f_1 to be stream $j = 2$, we get (with $i = 1$ for benzene)

$$\xi = \left\{\frac{0.51}{0.51 + 0.95}\right\}\left\{\frac{0.05}{0.05 + 0.49}\right\}\left|\frac{0.95 \times 0.49}{0.05 \times 0.51} - 1\right|$$

$$= 0.35 \times 0.0927 \times 17.2 = 0.558.$$

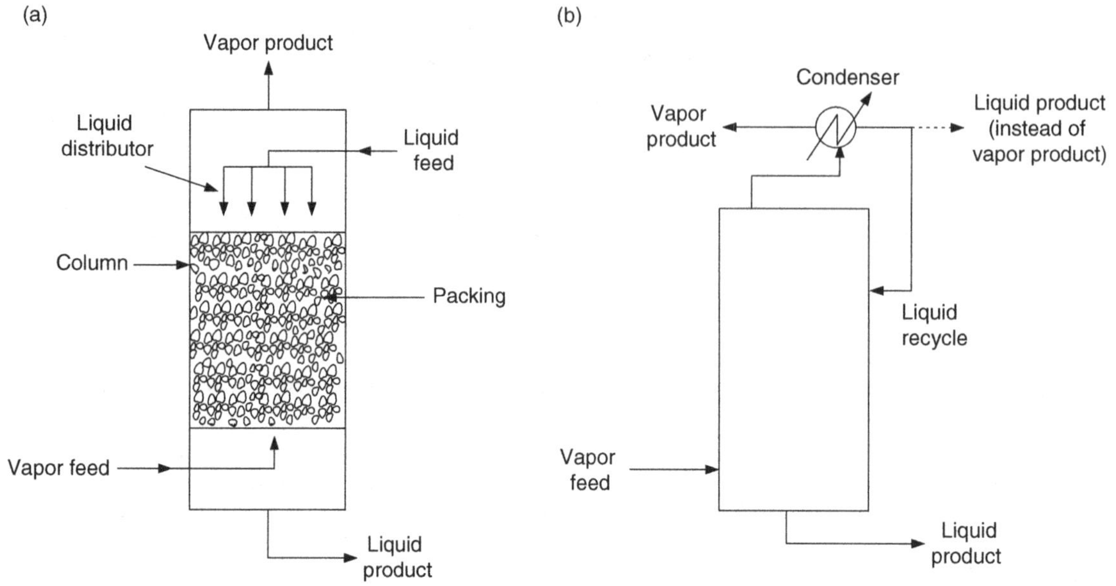

Figure 2.3.2. Packed column distillation of benzene–toluene mixture: (a) double-entry separator; (b) overhead condenser providing recycle liquid feed.

Obviously, little separation has been accomplished since this value of ξ is only marginally smaller than the actual ξ (= 0.632) for this separator. In fact, in exchange for a benzene-rich liquid stream (benzene mole fraction = 0.95) introduced as feed at the top, a benzene-rich vapor stream (benzene mole fraction = 0.98) is being obtained, although the objective is the separation of benzene and toluene introduced as vapor feed at the column bottom. In practice, the liquid feed at the top is obtained by condensing the vapor product stream in a partial or total condenser. Thus, the net separator setup appears as in Figure 2.3.2(b), which, is in effect, a single-entry separator. Note that the flow rates and compositions of the various product streams will differ from the values given in the example if this is the case.

Example 2.3.1 described separation in a double-entry separator having two distinct phases - a vapor feed and a liquid feed, with both phases existing throughout the column. Double-entry separators are used even when two product streams and the two feed streams are of the same phase.

Example 2.3.2 The permeator described in Example 2.2.4 has been used by Thorman *et al.* (1980) in a different way to obtain a much higher composition change in the separated streams. Consider Figure 2.3.3, where feed air is introduced at the bottom of the separator on the shell side and a compressed and O_2-rich stream is introduced as feed inside the tubes at the top. Part of the O_2-rich stream at the top is withdrawn as product, while the rest is sent back to the permeator to go through the silicone capillaries and finally emerge as another product stream. Thus this separator has two feed streams and two product streams.

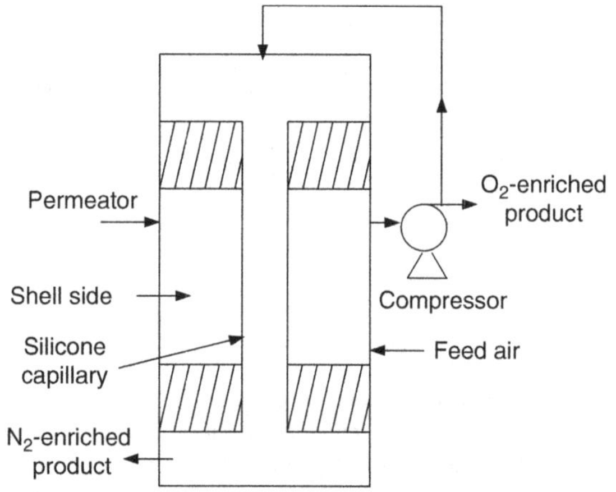

Figure 2.3.3. Double-entry membrane separator: continuous membrane column for O_2 separation from air.

However, the overall separation system incorporating the compressor has only one feed stream and two product streams. This arrangement is quite similar to the distillation column, and is called the *continuous membrane column*.

The wall of the silicone capillaries separates the two gas streams in the permeator described above. It is not necessary to have a barrier between two streams within a separator. Later in the book we will come across double-entry separators in which two gas streams are in contact with each other and yet separation is taking place.

2.3.2 Separation in a double-entry separator with recycle/reflux

We now briefly carry out a quantitative description of separation in a double-entry separator with recycle. A perfectly general double-entry separator with recycle would have two independent feed streams, and a part of one (or both) of the product streams is recycled to the feed stream closest to it (Figure 2.3.4). A more commonly encountered scheme would have only one independent feed stream entering the separator ($W_{tf2} = 0$). For such a case, the total and the ith component molar balances are given by

$$W_{tf_1} + W^r_{tf_2} = W^r_{t1} + W^r_{t2}, \tag{2.3.5}$$

$$W_{tf_1}x_{if_1} + W^r_{tf_2}x^r_{if_2} = W^r_{tl}x^r_{i1} + W^r_{t2}x^r_{i2}. \tag{2.3.6}$$

In general, $x^r_{i1} \neq x^r_{if_2}$ (for example, in Figure 2.3.2(b), where the vapor from the top of the column is only partially condensed, the composition of the product vapor and the product liquid will be different; the partial condenser is also a single-entry separator. However, with a total condenser, all vapor from the column top is condensed to liquid: $x^r_{if_2} = x^r_{il}$). But with a total condenser on the column, $W_{tf2}=0$, $x_{if_2} = 0, x^r_{if_2} = x^r_{i1}$ and R is the *reflux* (instead of the recycle) ratio (so that $(R/R+1)$ fraction of W^r_{t1} is recycled to the top of the separator). Equations (2.3.5) and (2.3.6) may now be simplified to

$$W_{tf_1} = \frac{W^r_{t1}}{R+1} + W^r_{t2}, \tag{2.3.7}$$

$$W_{tf_1}x_{if_1} = \frac{W^r_{t1}}{R+1}x^r_{i1} + W^r_{t2}x^r_{i2}. \tag{2.3.8}$$

To simplify the problem further, let us assume that $W_{tf_1} = W^r_{t1}$ and $W^r_{tf_2} = W^r_{t2}$ (the so-called *constant total molar overflow assumption* to be encountered in Sections 8.1-3.1/8.1-1.1). Further, let the following relation be valid regardless of the value of R:

$$x_{if_1} = K_i x^r_{i2}. \tag{2.3.9}$$

For given values of R, K_i and W_{tf1}, the following solutions are obtained:

$$W^r_{t2} = W^r_{tf_2} = \left(\frac{R}{R+1}\right)W_{tf_1}, \tag{2.3.10}$$

$$x^r_{i1} = \left[(R+1)x_{if_1} - \frac{R}{K_i}x_{if_1}\right]. \tag{2.3.11}$$

Could we now describe the relative performance of this separation system for two nonzero values of reflux ratio R_1 and R_2 ($R_1 > R_2$)? We opt to determine the relative values of the extents of separation for the two cases. But we recognize that, in effect, the separator of Figure 2.3.4 has one feed stream and two product streams. Therefore we use ξ defined by (2.2.14) to obtain the following:

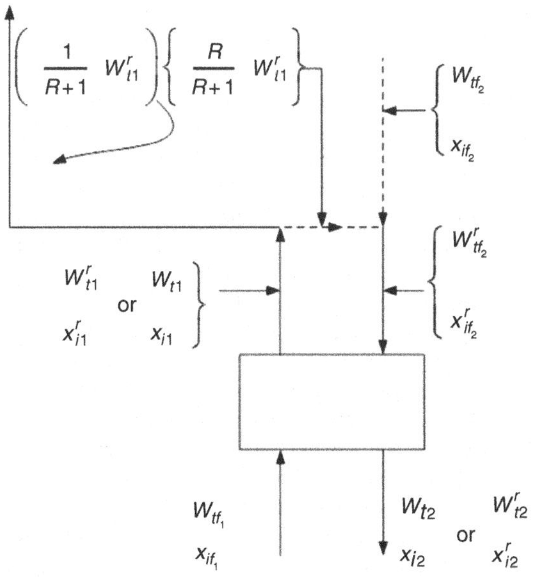

Figure 2.3.4. Double-entry separator with or without reflux.

$$\frac{\xi^r|_{R=R_1}}{\xi^r|_{R=R_2}} = \frac{\left|\dfrac{\left(\frac{W^r_{t1}}{R_1+1}\right)x^r_{11}\big|_{R_1}}{W_{tf_1}x_{1f_1}} - \dfrac{\left(\frac{W^r_{t1}}{R_1+1}\right)x^r_{21}\big|_{R_1}}{W_{tf_1}x_{2f_1}}\right|}{\left|\dfrac{\left(\frac{W^r_{t1}}{R_2+1}\right)x^r_{11}\big|_{R_2}}{W_{tf_1}x_{1f_1}} - \dfrac{\left(\frac{W^r_{t1}}{R_2+1}\right)x^r_{21}\big|_{R_2}}{W_{tf_1}x_{2f_1}}\right|}. \tag{2.3.12}$$

On simplification, it yields (Sirkar and Teslik, 1982; see footnote 3, p. 49) for R_1, $R_2 \neq 0$

$$\frac{\xi^r|_{R=R_1}}{\xi^r|_{R=R_2}} = \frac{(1+1/R_2)}{(1+1/R_1)}. \tag{2.3.13}$$

Thus $\xi^r|_{R=R_1} > \xi^r|_{R=R_2}$ if $R_1 > R_2$. That a higher reflux ratio achieves a better separation (which is a well-known fact, as we shall see in Section 8.1-3.2) can therefore be described with the help of the separation index ξ.

2.4 Multicomponent systems

Up to this point, we have considered only the description of separation of a binary mixture in a single- or double-entry separator. We now briefly turn to *multicomponent* mixtures.

In general, a multicomponent mixture will have, say, n different chemical species so that $i = 1, 2,\ldots,n$. An analogous problem is encountered when solid particles of the same material having a wide distribution of sizes are to be separated from one another or from a fluid in which they are suspended. If solid particles of a certain size are to be considered as one species or one component, then such a mixture, in general, will have an infinite number of species or components since the particle sizes are, in general, continuously distributed over a range. Similarly, a macromolecular mixture which is not monodisperse will have a molecular weight distribution and therefore an infinite number of species in general.

We should recognize now that whether the system is of n components or an infinite number of components, no special indices are available to accommodate this complexity. The indices developed for describing binary separation have to be adopted for this task. This is done by lumping a large number of components or a range of particle sizes (or molecular weights) into one component of a hypothetical binary system whose other component embraces the remaining components or the remaining range of particle sizes (or molecular weights). An alternative approach is to work with the heavy key component and the light key component, as pointed out in Section 1.6.

We will study first the description of separation of a *particle population* and then a *chemical solution* containing an infinite number of species. At the end of the section, we describe n-component chemical solutions using the approach of a *heavy key component* and a *light key component.*

2.4.1 Size-distributed particle population

Consider a single-entry separator used either for separating solid particles from a fluid or separating solid particles having sizes above a particular value from those having sizes below the particular value. Let w_{tf}^s be the total mass flow rate of solids in the feed fluid whose total volumetric flow rate is Q_f. In such a separator, there are only two product streams, the overflow ($j = 1$) and the underflow ($j = 2$). The total mass flow rate of solids in the overflow and the underflow are, respectively, w_{t1}^s and w_{t2}^s. The overflow is identified with essentially the carrier fluid and the finer particles, whereas the underflow is assumed to have most of the coarser particles and small amounts of carrier fluid. Figure 2.4.1 illustrates this for a hydrocyclone (Talbot, 1980). Sizes of various natural and industrial particles are shown in Figure 2.4.2.

If the objective is to obtain a particle-free fluid, then perfect separation means no solid particles in the overflow (equivalent to the heads stream of Section 2.2) and no carrier fluid in the underflow (the tails stream of Section 2.2). If particle classification is the goal, then perfect separation requires all particles above a given size to be in the underflow and all particles smaller than the given size in the overflow. In an imperfect separation, some particles are always present in the overflow (when the goal is to have a particle-free carrier fluid). Similarly, due to imperfections in the separator, some particles coarser than the given size are in the overflow, just as some finer particles are in the underflow from the separator functioning as a classifier.

The composition of the particle population is usually indicated by the *particle size density function* $f(r_p)$, where r_p is the characteristic particle dimension of importance. We denote the values of this function for the feed stream and the product streams by $f_f(r_p)$, $f_1(r_p)$ and $f_2(r_p)$, respectively. Since the fraction of particles in the size range r_p to $r_p + \mathrm{d}r_p$ is given by[4] $f(r_p)\,\mathrm{d}r_p$ when the particle size density function is $f(r_p)$, such that

$$\int_{r_{\min}}^{r_{\max}} f(r_p)\mathrm{d}r_p = 1, \tag{2.4.1a}$$

the following relations are also valid:

$$\int_{r_{\min}}^{r_{\max}} f_f(r_p)\mathrm{d}r_p = \int_{r_{\min}}^{r_{\max}} f_1(r_p)\mathrm{d}r_p = \int_{r_{\min}}^{r_{\max}} f_2(r_p)\mathrm{d}r_p = 1. \tag{2.4.1b}$$

Here, $r_{\max}$ and $r_{\min}$ refer to the maximum and minimum particle sizes in the feed stream. The nature of such a density function is illustrated in Figure 2.4.3(a). *The particle size distribution function* $F(r_p)$ is defined by

$$\mathrm{d}F = f(r_p)\mathrm{d}r_p \Rightarrow \int_{r_{\min}}^{r_p} f(r_p)\mathrm{d}r_p = \int_{r_{\min}}^{r_p} \mathrm{d}F(r_p) = F(r_p). \tag{2.4.1c}$$

The maximum value of $F(r_p)$, namely 1, is achieved when the upper limit of integration is $r_{\max}$.

A variety of particle size distributions are used in practice. The most widely used example is the Normal distribution (Gaussian distribution),

$$f(r_p) = \frac{1}{\sqrt{2\pi}\sigma_r}\exp\left[-\frac{(r_p - \bar{r}_p)^2}{2\sigma_r^2}\right], \tag{2.4.1d}$$

where $\bar{r}_p$ is the mean of the distribution defined by

$$\bar{r}_p = \int_{r_{\min}}^{r_{\max}} r_p f(r_p)\mathrm{d}r_p \tag{2.4.1e}$$

and σ_r^2, the variance, is defined by

$$\sigma_r^2 = \int_{r_{\min}}^{r_{\max}} (r_p - \bar{r}_p)^2 f(r_p)\mathrm{d}r_p. \tag{2.4.1f}$$

Other distributions employed are the log-normal distribution, the gamma distribution, the Rosin–Rammler distribution, etc. We will encounter some of them later in the book.

[4] One can also think of $f(r_p)\,\mathrm{d}r_p$ as the probability of finding a particle in the size range r_p to $r_p + \mathrm{d}r_p$.

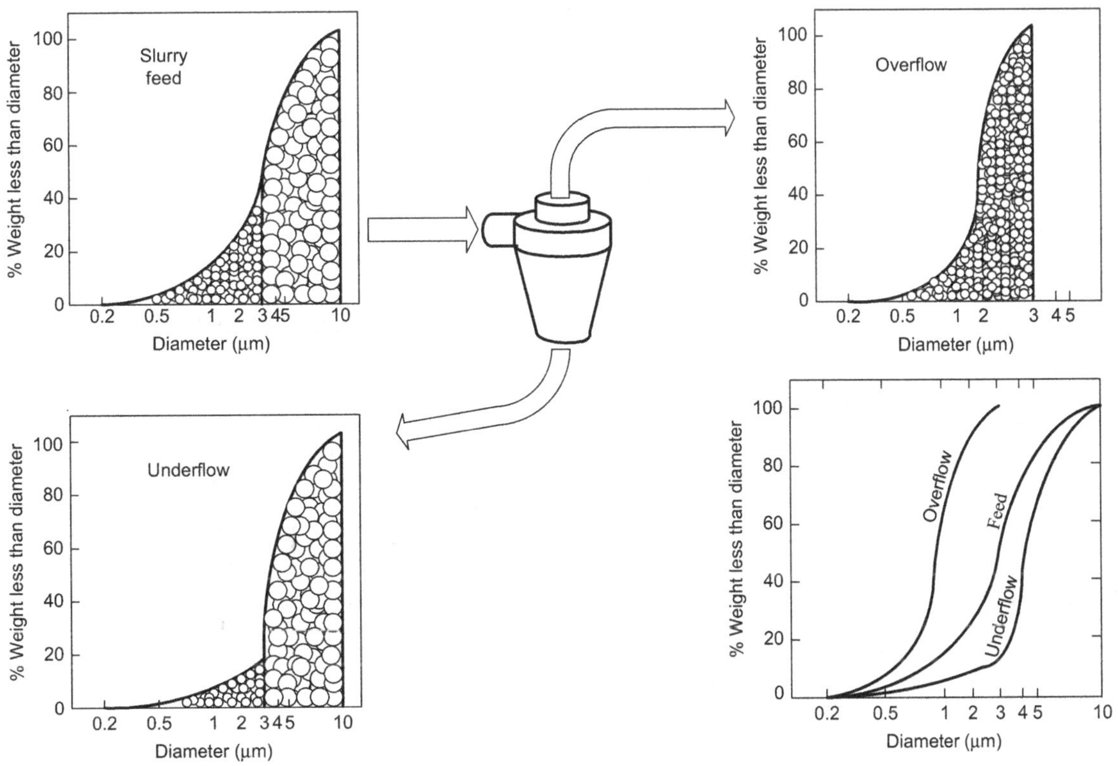

Figure 2.4.1 Typical particle size distribution curves for removal and separation of solids from an aqueous kaolin solution by a hydrocyclone. (After Talbot (1980).)

Often, *moments of distribution/density functions* are used in particulate systems. The nth moment of the particle size density function is defined by

$$Mo^{(n)} = \int_{r_{min}}^{r_{max}} r_p^n f(r_p) \mathrm{d}r_p \qquad (2.4.1g)$$

From definition (2.4.1e), we see that the mean of the particle size distribution, $\bar{r}_p$, is the first moment of $f(r_p)$, $Mo^{(1)}$. The zeroth moment of $f(r_p)$, $Mo^{(0)}$, is $F(r_{max}) = 1$. The variance σ_r^2 of the Gaussian distribution from (2.4.1f) may be related to

$$\sigma_r^2 = \int_{r_{min}}^{r_{max}} r_p^2 f(r_p) \mathrm{d}r_p - 2\bar{r}_p \int_{r_{min}}^{r_{max}} r_p f(r_p) \mathrm{d}r_p + (\bar{r}_p)^2 \int_{r_{min}}^{r_{max}} f(r_p) \mathrm{d}r_p$$
$$= Mo^{(2)} - 2\bar{r}_p Mo^{(1)} + (\bar{r}_p)^2 = Mo^{(2)} - 2\bar{r}_p^2 + (\bar{r}_p)^2,$$
$$Mo^{(2)} = \bar{r}_p^2 + \sigma_r^2. \qquad (2.4.1h)$$

Thus the variance, σ_r^2, is easily related to the second moment $Mo^{(2)}$ and the first moment $Mo^{(1)}$.

An item of significant interest is the actual number of particles N (r_{min}, r_p) in the size range r_{min} to r_p per unit fluid volume; the corresponding number is $\mathrm{d}N(r_p)$ in the size range r_p to $r_p + \mathrm{d}r_p$. This quantity is related to the *particle number density function, $n(r_p)$,* (also called the *population density function*) by

$$\mathrm{d}N(r_p) = n(r_p)\mathrm{d}r_p, \qquad (2.4.2a)$$

$$\int_{r_{min}}^{r_p} n(r_p)\mathrm{d}r_p = N(r_{min}, r_p). \qquad (2.4.2b)$$

The dimensions of $N(r_{min}, r_p)$ are number/volume, i.e. number/$[L]^3$; the dimensions of $n(r_p)$ are number /$[L]^4$. The total number of particles per unit fluid volume, N_t, is obtained from

$$N_t = \int_0^\infty \mathrm{d}N(\mathrm{r}_{min}, r_p) = \int_0^\infty \mathrm{d}N(r_p) = \int_0^\infty n(r_p)\mathrm{d}r_p, \qquad (2.4.2c)$$

by integrating over all possible dimensions of the particles, where we have assumed $r_{min} = 0$. For such a case, we write $N(r_p)$ instead of $N(0, r_p)$. Figure 2.4.3(b) illustrates the population density function $n(r_p)$ as well as the cumulative number of particles $N(r_{min}, r_p)$ per unit volume against the particle dimension, r_p. See Randolph and Larson (1988) for more details. Problem 2.4.1 is a useful exercise for determining $n(r_p)$. One can now relate the particle size distribution function $F(r_p)$ to the number density function $n(r_p)$ as follows:

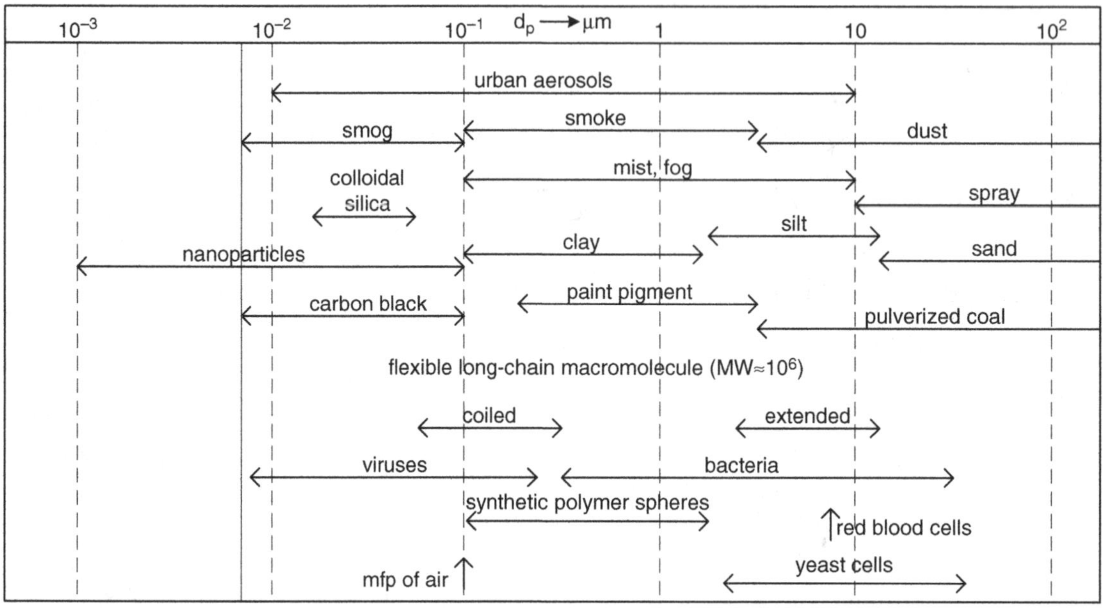

Figure 2.4.2. Sizes of various natural and industrial particles: d_p = particle diameter or length; 1 $\mu m = 10^{-4}$ cm; 1 nm = 10 Å = 10^{-3} μm; mfp = mean free path; MW = molecular weight. (After Davis (2001).)

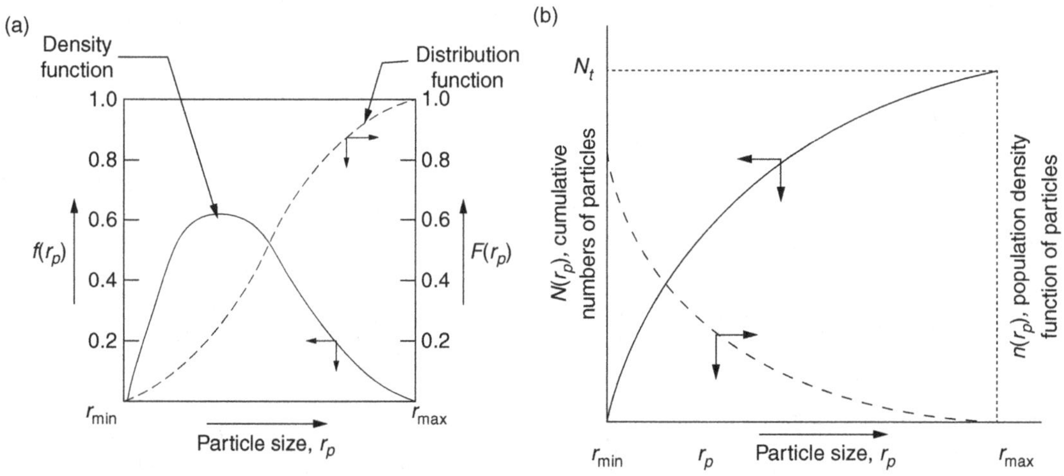

Figure 2.4.3. (a) Particle size density function and distribution function. (b) Typical plots of population density function and cumulative particle numbers against the particle size.

$$F(r_p) = \int_{r_{min}}^{r_p} n(r_p)\mathrm{d}r_p/N_t. \tag{2.4.2d}$$

If, in lieu of $f(r_p)$ based on mass fraction, we use $n(r_p)$, where the number of particles present per unit volume between the limits r_p and $r_p + \mathrm{d}r_p$ is $n(r_p)\mathrm{d}r_p$, then the differential change in mass of the particle population w^s per unit fluid volume for a differential change in r_p is given by

$$\mathrm{d}w^s = f(r_p)\mathrm{d}r_p w^s = [n(r_p)\mathrm{d}r_p]w_p^s(r_p) = [n(r_p)\mathrm{d}r_p]\rho_s \psi_v r_p^3, \tag{2.4.2e}$$

where $w_p^s(r_p)$ is the mass of a particle of size r_p, ρ_s is the density of the particle material and ψ_v is a shape factor (Foust *et al.*, 1960; Svarovsky, 1977) for particle volume assumed independent of particle size and having the value of $(4\pi/3)$ for a sphere (for example). The *total particle mass per unit volume*, w^s (also referred to as the *suspension density*, M_T) is

$$\int_{r_{min}}^{r_{max}} dw^s = w^s = M_T = \rho_s \psi_v \int_{r_{min}}^{r_{max}} r_p^3 n(r_p) dr_p = \rho_s \psi_v Mo^{(3)}. \tag{2.4.2f}$$

The number-based mean radius $\overline{r}_{p_{1,0}}$ of the particle population is obtained by using the zeroth and the first moment, $Mo^{(0)}$ and $Mo^{(1)}$, of $n(r_p)$:

$$\overline{r}_{p_{1,0}} = \left[\int_{r_{min}}^{r_{max}} r_p n(r_p) dr_p \Big/ \int_{r_{min}}^{r_{max}} n(r_p) dr_p \right]$$

$$= \frac{Mo^{(1)}}{Mo^{(0)}} = \int_0^\infty r_p n(r_p) dr_p / N_t, \tag{2.4.2g}$$

where we have replaced the limits r_{min} and r_{max} by 0 and ∞, respectively.

In general,

$$\overline{r}_{p_{i+1,i}} = \int_{r_{min}}^{r_{max}} r_p^{i+1} n(r_p) dr_p \Big/ \int_{r_{min}}^{r_{max}} r_p^i n(r_p) dr_p. \tag{2.4.2h}$$

Another commonly used radius is based on the mass of the particle (which varies with r_p^3; see (2.4.2f)), $\overline{r}_{p_{4,3}}$:

$$\overline{r}_{p_{4,3}} = \int_0^\infty r_p^4 n(r_p) dr_p \Big/ \int_0^\infty r_p^3 n(r_p) dr_p. \tag{2.4.2i}$$

Example 2.4.1 (a) We are going to consider the issue of volume shape factor ψ_v for different types of particles.

(i) What is the value of ψ_v for spherical particles of radius r_p?
(ii) Consider crystals shaped as rectangular parallelepipeds with dimensions of L, W, W, where $(L/W) = 3$. What is the value of ψ_v if ψ_v is defined with respect to L^3?

(b) When a mixture of different sized particles was fractionated by a sieve having openings of size 0.09 cm, the particles which passed through were fractionated next by a sieve of size 0.07 cm. Those particles which were retained by this sieve may be considered to have a characteristic dimension ($2r_p = d_p$) of 0.08 cm. The number of particles retained on the sieve is 50; their weight is 4×10^{-2} g. The particle density is 2 g/cm^3. What is the value of ψ_v? If ψ_v were defined with respect to d_p^3 what would be its value?

(c) A surface shape factor ψ_s of particles may be defined as follows:

$$\psi_s = \frac{\text{surface area}}{(\text{characteristic dimension})^2}.$$

Show that the sphericity ψ, defined by

$$\psi = \frac{(6\psi_v/\pi)^{2/3}}{\psi_s/\pi}$$

has the value of 1 for spherical particles when the characteristic dimension for ψ_v and ψ_s is the diameter.

Solution (a)(i) From equation (2.4.2e), for one spherical particle (i.e. $dN(r_p) = n(r_p)dr_p = 1$), the mass of one spherical particle is $dw^s = \rho_s \frac{4}{3}\pi r_p^3$; by definition, $dw^s = \rho_s \psi_v r_p^3$ which implies that $\psi_v = \frac{4}{3}\pi$. If ψ_v is defined as $dw^s = \rho_s \psi_v d_p^3$ this implies that $\psi_v = \pi/6$.

(ii) The volume of a rectangular parallelepiped of dimensions $L \times W \times W$ is LW^2. By definition, $dw^s = \rho_s \psi_v L^3 = \rho_s LW^2$. Since $W = L/3$, we get $\psi_v = W^2/L^2 = 1/9$.

(b) From equation (2.4.2e),

$$dw^s = dN(r_p)\rho_s \psi_v r_p^3 \Rightarrow 4 \times 10^{-2}\,\text{g}$$
$$= 50 \times 2\,(\text{g/cm}^3) \times \psi_v \times (0.04)^3 \text{cm}^3,$$

so

$$\psi_v = \frac{4 \times 10^{-2} \times 10^6}{100 \times 64} = \frac{4 \times 10^2}{64} = 6.25.$$

If ψ_v were defined via $dw^s = dN(r_p)\rho_s \psi_v d_p^3$, then

$$\psi_v = \frac{4 \times 10^{-2} \times 10^6}{100 \times 64 \times 8} = \frac{4 \times 10^2}{64 \times 8} = 0.78.$$

(c) For a spherical particle,

$$\psi_s = \frac{\text{surface area}}{(\text{characteristic dimension})^2} = \frac{4\pi r_p^2}{d_p^2} \Rightarrow \psi_s = \pi.$$

For a spherical particle,

$$\psi_v = \frac{\text{volume}}{(\text{characteristic dimension})^3} = \frac{(\pi/6)d_p^3}{d_p^3} = \pi/6.$$

Therefore

$$\psi = \frac{\left(\frac{6\psi_v}{\pi}\right)^{2/3}}{\psi_s/\pi} = \frac{\left(\frac{6}{\pi}\frac{\pi}{6}\right)^{2/3}}{\pi/\pi} = 1.$$

Generally, for other particle shapes, $0 \leq \psi < 1$.

Example 2.4.2 Consider a particle number density function $n(r_p)$ having the form $n(r_p) = n^0 \exp(-ar_p)$.

Determine expressions for N_t, $F(r_p)$, $f(r_p)$, $\overline{r}_{p_{1,0}}$, $\overline{d}_{32}$, M_T and A_T. Here $\overline{d}_{32}$ is the Sauter mean diameter equal to $2\overline{r}_{p_{3,2}}$; A_T is the total particle surface area per unit volume of the total mixture. Assume that r_p varies between 0 and ∞.

Solution From definition (2.4.2c),

$$N_t = \int_0^\infty n(r_p) dr_p = n^0 \int_0^\infty \exp(-ar_p) dr_p = (n^0/a).$$

From definition (2.4.2d),

$$F(r_p) = \int_{r_{min}}^{r_p} n(r_p) dr_p / N_t.$$

Assume now that $r_{min} = 0$. Then

$$F(r_p) = \frac{a}{n^0} \int_0^{r_p} n^0 \exp(-ar_p) dr_p = a\left[\frac{\exp(-ar_p)}{-a} - \frac{\exp(-0)}{-a}\right]$$
$$= -1\left[\exp(-ar_p) - 1\right] = \left[1 - \exp(-ar_p)\right].$$

By definition (2.4.1c),

$$\mathrm{d}F(r_p) = f(r_p)\mathrm{d}r_p \Rightarrow f(r_p) = \frac{\mathrm{d}F(r_p)}{\mathrm{d}r_p} = n(r_p)/N_t$$

(from definition (2.4.2d)), so

$$f(r_p) = \frac{n^0 \exp(-ar_p)}{n^0/a} = a\exp(-ar_p).$$

By definition (2.4.2g),

$$\bar{r}_{p_{1,0}} = \left(\int_0^\infty r_p n(r_p)\mathrm{d}r_p / N_t \right)$$
$$= \frac{n^0 a}{n^0} \int_0^\infty r_p \exp(-ar_p)\mathrm{d}r_p \Rightarrow \bar{r}_{p_{1,0}} = \frac{a \times 1}{a^2} = \frac{1}{a}.$$

By definition (2.4.2h), $\overline{d_{32}}$ = Sauter mean diameter $= 2\bar{r}_{p_{3,2}}$:

$$\overline{d_{32}} = \frac{2\int_0^\infty r_p^3 n(r_p)\mathrm{d}r_p}{\int_0^\infty r_p^2 n(r_p)\mathrm{d}r_p} = 2\frac{6/a^4}{2/a^3} = \frac{6}{a}.$$

By definition (2.4.2f),

$$M_T = \rho_s \psi_v \int_0^\infty r_p^3 n(r_p)\mathrm{d}r_p = \rho_s \psi_v n^0 \int_0^\infty r_p^3 \exp(-ar_p)\mathrm{d}r_p$$
$$= \rho_s \psi_v n^0 \frac{\Gamma(4)}{a^4} = \rho_s \psi_v n^0 \frac{3!}{a^4}; \quad M_T = 6\rho_s \psi_v n^0/a^4.$$

From the given definition of A_T, if a_p is the surface area of a particle of size r_p, then, from the previous example, $a_p = \psi_s r_p^2$. Therefore

$$A_T = \int_0^\infty a_p n(r_p)\mathrm{d}r_p = \psi_s n^0 \int_0^\infty r_p^2 \exp(-ar_p)\mathrm{d}r_p$$
$$= \frac{\psi_s n^0 2!}{a^3} = \frac{2\psi_s n^0}{a^3}.$$

The treatment of size-distributed particle populations has so far assumed that the number of particles in the population is quite large and there is almost a continuous distribution in particle sizes in the particle population under consideration. If the number of particles in the population is not large, then we have essentially a discrete distribution in particle sizes. Suppose the number of particles in the size range r_{p_i} to $r_{p_i} + \Delta r_{p_i}$ is ΔN_i. Then the population density function n_i of particles in this size range is

$$n_i = \Delta N_i / \Delta r_{p_i} \qquad (2.4.2j)$$

(*Note*: we are dealing with a discrete function not a continuous function, hence the subscript *i*.) This approach is needed to treat experimental information on different particle size fractions collected over different size sieves: that is, particles collected on a particular size sieve have sizes larger than this sieve opening but smaller than the opening size of the sieve immediately above it through which the particles fell. If we can characterize the number of particles of a certain size r_{p_i} as No_i, then the mean particle size $\bar{r}_{p_{1,0}}$ is

$$\bar{r}_{p_{1,0}} = \frac{\sum_{i=1} No_i r_{p_i}}{\sum_{i=1} No_i}, \qquad (2.4.2k)$$

where the denominator represents the total number of particles in the sample.

In many particle based separation processes, particles break, coalesce or grow (e.g. crystal growth). Common expectations for the mean particle size in small-sized populations can often be misleading. Consider the example provided by Neumann *et al.* (2003). Let there be ten particles of volume equivalent size of 1 unit and one large particle of size 100 units. The number mean size $\bar{r}_{p_{1,0}}$ from definition (2.4.2k)

$$\bar{r}_{p_{1,0}} = \frac{10 \times 1 + 1 \times 100}{11} = 10. \qquad (2.4.2l)$$

If the large particle breaks into two small particles of similar shape, then since the particle volume/mass, $\rho_s \psi_v r_p^3$, is conserved, we get two new particles of size r_{pn}:

$$\rho_s \psi_v (100)^3 = 2\rho_s \psi_v (r_{pn})^3; \qquad r_{pn} = \frac{100}{2^{1/3}} = 79.37. \qquad (2.4.2m)$$

The new value of $\bar{r}_{p_{1,0}}$ is

$$\bar{r}_{p_{1,0}} = \frac{10 \times 1 + 2 \times 79.37}{12} = 14.06,$$

which is larger than before. Since the breakage led to two smaller particles, one is tempted to think that the mean particle size will decrease. It did not because the sample size was small. Neumann *et al.* (2003) have suggested that the number mean size will increase due to halving breakage of the larger particles (each particle breaking into two equal fragments) provided the particles broken are more than 70% larger than the mean particle size of the initial distribution. We will now focus on describing separation in devices used for separation of particles.

The fraction of particles in the range r_p to $r_p + \mathrm{d}r_p$ could be based on the weight fraction of particles, or the number fraction of particles, etc. Suppose we assume that $f(r_p)\,\mathrm{d}r_p$ gives the weight fraction (and therefore the mass fraction) of particles in the size range r_p to $r_p + \mathrm{d}r_p$. Then a mass balance of all particles, as well as of particles in this size range, yields at steady state for the single-entry separator:

$$w_{tf}^s = w_{t1}^s + w_{t2}^s; \qquad (2.4.3a)$$

$$w_{tf}^s f_f(r_p)\,\mathrm{d}r_p = w_{t1}^s f_1(r_p)\,\mathrm{d}r_p + w_{t2}^s f_2(r_p)\,\mathrm{d}r_p. \qquad (2.4.3b)$$

Current practice utilizes two descriptors for such a separation problem, the *total efficiency* E_T and the grade efficiency G_r for particles of size r_p (Svarovsky, 1977, chap. 3):

$$E_T = \frac{w_{t2}^s}{w_{tf}^s}; \qquad (2.4.4a)$$

$$G_r = \frac{w^s_{t2} f_2(r_p)}{w^s_{tf} f_f(r_p)} = E_T \frac{f_2(r_p)}{f_f(r_p)}. \quad (2.4.4b)$$

The *total efficiency* E_T is the fraction of particles of the feed which are exiting through the underflow or the tails stream. When E_T equals 1, the separation between the solid and the overflow fluid is complete. If the carrier fluid stream is the preferred one containing finer particles, then obviously $E_T = (1-\theta^s)$, where θ^s is the cut defined by (2.2.10a) in terms of only the solid particle flow rates. The *grade efficiency* G_r of particles of size r_p is the ratio of the mass of particles of size r_p in the underflow to that in the feed. It depends on r_p and the separator characteristics. If the particle classifier or the solid–liquid separator is of any use, then at least all particles of size r_{max} should be in the underflow. To generalize, $G_r|_{r_p=\infty} = 1$. As $r_p \to 0$, G_r tends to a limiting value defined by (Svarovsky, 1977, chap. 3)

$$G_r|_{r_p=0} = \frac{Q_2}{Q_f}, \quad (2.4.4c)$$

where Q_2 is the total volumetric flow rate of the underflow. In other words, even if there is no separation per se of the finest solid by the separation mechanism from the fluid, the finest solids of the feed are split between exit streams 1 and 2 since some fluid inevitably leaves with the underflow. The behavior of such types of grade efficiency functions, along with those for which $G_r|_{r_p=0} = 0$, are shown in Figure 2.4.4. Such functions are also known as Tromp curves.

We now turn our attention to the description of separation of particles of only two sizes, r_{p1} and r_{p2}. Assume particles of size r_{p1} to be component 1 and those of size r_{p2} to be component 2. We can define the segregation fractions $\dot{Y}_{11}(=\dot{Y}_{r_{p_1}1})$ and $\dot{Y}_{21}(=\dot{Y}_{r_{p_2}1})$ as follows:

$$\dot{Y}_{r_{p_1}1} = \frac{w^s_{t1} f_1(r_{p_1})}{w^s_{tf} f_f(r_{p_1})}, \qquad \dot{Y}_{r_{p_2}1} = \frac{w^s_{t1} f_1(r_{p_2})}{w^s_{tf} f_f(r_{p_2})}, \quad (2.4.5)$$

where we have assumed that the mass fraction of the particles is equal to the mole fraction of the particles. Borrowing the definition of the extent of separation ξ from (2.2.11), we have for the separation of particles of two sizes r_{p1} and r_{p2}

$$\xi_{r_{p_1} r_{p_2}} = \left|\dot{Y}_{r_{p_1}1} - \dot{Y}_{r_{p_2}1}\right| = \left|\frac{w^s_{t1} f_1(r_{p_1})}{w^s_{tf} f_f(r_{p_1})} - \frac{w^s_{t1} f_1(r_{p_2})}{w^s_{tf} f_f(r_{p_2})}\right|. \quad (2.4.6)$$

Utilizing the relations (2.4.3b), (2.4.4a) and (2.4.4b) for two particle sizes r_{p_1} and r_{p_2}, we can easily show that

$$\xi_{r_{p_1} r_{p_2}} = \left|G_r(r_{p_1}) - G_r(r_{p_2})\right| = E_T \left|\frac{f_2(r_{p_1})}{f_f(r_{p_1})} - \frac{f_2(r_{p_2})}{f_f(r_{p_2})}\right|. \quad (2.4.7)$$

If $r_{p_1} = r_{max}$ and $r_{p_2} = r_{min}$, then $\xi_{r_{p_1} r_{p_2}} = 1 - (Q_2/Q_f)$, the maximum possible value. Since $Q_2 \neq 0$, $\xi_{r_{p_1} r_{p_2}}|_{max}$ is never equal to 1. On the other hand, as $r_{p_1} \to r_{p_2}$, $\xi_{r_{p_1} r_{p_2}} \to 0$. The familiar quantity grade efficiency may thus be easily related to the index, ξ, the extent of separation.

If we are interested in describing the separation of all particles above and below a particular size r_s in the overflow and the underflow, we may define component 1 to be all particles with $r_p < r_s$ and component 2 to be all particles with $r_p > r_s$. Continuing with the assumption of the equivalence of the mass fractions of particles with mole fractions of particles, we have in such a case

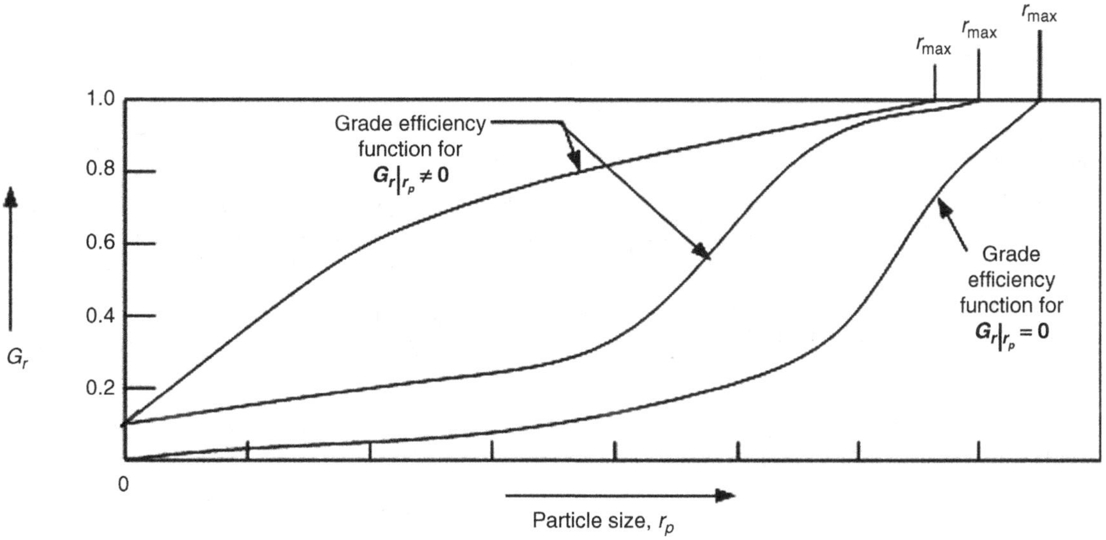

Figure 2.4.4. Nature of various grade efficiency functions.

$$\dot{Y}_{11} = \frac{w_{t1}^s \int_{r_{min}}^{r_s} f_1(r_p)\mathrm{d}r_p}{w_{tf}^s \int_{r_{min}}^{r_s} f_f(r_p)\mathrm{d}r_p}; \qquad \dot{Y}_{21} = \frac{w_{t1}^s \int_{r_s}^{r_{max}} f_1(r_p)\mathrm{d}r_p}{w_{tf}^s \int_{r_s}^{r_{max}} f_f(r_p)\mathrm{d}r_p}. \tag{2.4.8}$$

The extent of separation for components 1 and 2 defined in this manner is then

$$\xi_{12} = \frac{w_{t1}^s}{w_{tf}^s}\left| \frac{\int_{r_{min}}^{r_s} f_1(r_p)\mathrm{d}r_p}{\int_{r_{min}}^{r_s} f_f(r_p)\mathrm{d}r_p} - \frac{\int_{r_s}^{r_{max}} f_1(r_p)\mathrm{d}r_p}{\int_{r_s}^{r_{max}} f_f(r_p)\mathrm{d}r_p} \right|. \tag{2.4.9}$$

If we define the particle size distribution function $F(r_p)$ (see Figure 2.4.3(a) for illustration) by

$$F(r_p) = \int_{r_{min}}^{r_p} f(r_p)\mathrm{d}r_p, \tag{2.4.10}$$

we may rewrite (2.4.9) as

$$\xi_{12} = [1 - E_T]\left| \frac{F_1(r_s)}{F_f(r_s)} - \frac{1 - F_1(r_s)}{1 - F_f(r_s)} \right| \tag{2.4.11}$$

since

$$(w_{t1}^s / w_{tf}^s) = (1 - E_T) \tag{2.4.12a}$$

and

$$\int_{r_{min}}^{r_{max}} f(r_p)\mathrm{d}r_p = F(r_{max}) = 1. \tag{2.4.12b}$$

Note further that E_T in such a case is given by

$$E_T = \frac{F_f(r_s) - F_1(r_s)}{F_2(r_s) - F_1(r_s)} \tag{2.4.12c}$$

if one uses equations (2.4.3a) and (2.4.3b) and integrates over the size range.[5]

Instead of considering only the separation of particles from one another, consider now the separation of all particles (species 2) from the fluid (species 1) feed or the elimination of liquid from the solid-rich underflow (region 2). Instead of the segregation fractions defined by (2.4.8), we will have

$$\dot{Y}_{21} = \frac{w_{t1}^s \int_{r_{min}}^{r_{max}} f_1(r_p)\mathrm{d}r_p}{w_{tf}^s \int_{r_{min}}^{r_{max}} f_f(r_p)\mathrm{d}r_p} = \frac{w_{t1}^s}{w_{tf}^s} = \frac{w_{t1}u_{21}}{w_{tf}u_{2f}}. \tag{2.4.13a}$$

Similarly,

$$\dot{Y}_{11} = \frac{w_{t1}u_{11}}{w_{tf}u_{1f}}, \tag{2.4.13b}$$

leading to

$$\xi_{12} = \left| \frac{w_{t1}u_{11}}{w_{tf}u_{1f}} - \frac{w_{t1}u_{21}}{w_{tf}u_{2f}} \right|, \tag{2.4.13c}$$

where we recall that w_{tj} is the total mass flow rate of both species 1 and 2 in the jth stream. In solid–liquid separation practice, one comes across the *Newton efficiency* and the *reduced efficiency*. The *Newton efficiency* has been defined as follows (Van Ebbenhorst Tengbergen and Rietema, 1961; Svarovsky, 1979):

$$E_N = \frac{w_{t2}u_{22}}{w_{tf}u_{2f}} - \frac{w_{t2}u_{12}}{w_{tf}u_{1f}}. \tag{2.4.14a}$$

Using mass balance relations (2.1.24) and (2.1.25), one can easily show that ξ_{12} of (2.4.13c) is the same as E_N. Note that if the mass fraction of solids recovered in any stream (e.g. $w_{t2}\, u_{22}/w_{tf}\, u_{2f}$) can be replaced by the corresponding mole fraction ($W_{t2}\, x_{22}/W_{tf}\, x_{2f}$) and similarly for the fluid, we have

$$E_N \cong \frac{W_{t2}x_{22}}{W_{tf}x_{2f}} - \frac{W_{t2}x_{12}}{W_{tf}x_{1f}}, \tag{2.4.14b}$$

which is the extent of separation ξ for components 1 and 2 being separated between product streams 1 (overflow) and 2 (underflow). If we rewrite (2.4.14a) as

$$E_N^1 = \frac{w_{t2}u_{22}}{w_{tf}u_{2f}}\left(1 - \frac{u_{12}u_{2f}}{u_{22}u_{1f}}\right) = \frac{w_{t2}^s}{w_{tf}^s}\left(1 - \left(\frac{u_{22}u_{1f}}{u_{12}u_{2f}}\right)^{-1}\right) \tag{2.4.15}$$

to illustrate the composite nature of E_N, we see that (w_{t2}^s/w_{tf}^s) indicates how much of the solid in the feed is recovered in the underflow (and therefore is essentially separated from the feed liquid), whereas ($u_{22}\, u_{1f}/u_{12}\, u_{2f}$) indicates how free from liquid this solid in the underflow is.

The reduced efficiency of Kelsall (1966) (Svarovsky, 1977, chap. 3) is given by

$$E_T^1 = \frac{E_T - (Q_2/Q_f)}{1 - (Q_2/Q_f)} = \frac{Q_f\left(\dfrac{w_{t2}^s}{w_{tf}^s}\right) - Q_2}{Q_f - Q_2}. \tag{2.4.16a}$$

When $w_{t2}^s = w_{tf}^s$, $E_T^1 = 1$, the maximum value, since $w_{t2}^s \leq w_{tf}^s$. Further, note that, for dilute slurries or suspensions,

---

[5]From (2.4.4b), the grade efficiency may be related to E_T in general by (see Prob. 2.4.2b)

$$E_T = \int_0^{\infty} G_r f_f(r_p)\mathrm{d}r_p = \int_0^1 G_r \mathrm{d}F_f(r_p). \tag{2.4.12d}$$

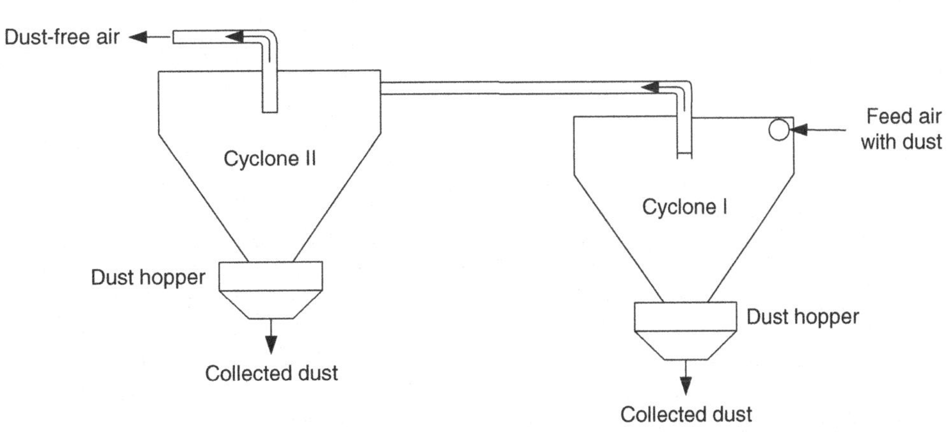

Figure 2.4.5. Two dust-collecting cyclones in series.

$$E_N \cong E_T - \left(Q_2/Q_f\right) \tag{2.4.16b}$$

since the solid volume is small compared to the liquid volume so that

$$E_T^1 \cong \frac{E_N}{1-(Q_2/Q_f)}. \tag{2.4.16c}$$

In solid–liquid separation practice, sometimes three or more phases are encountered. Each phase may be made up of only one chemical species or a number of chemical species. But the phases are immiscible. Such a system, if separated into two product streams, may be described by indices similar to E_N; one such is available in Sheng (1977).

Example 2.4.3. Consider the separation of dust particles from air by means of two cyclones connected in series (shown schematically in Figure 2.4.5). The particle size distribution function of the feed to cyclone 1 is given by (Van der Kolk, 1961)

$$F_{f_1}(r_p) = 1 - e^{-ar_p}. \tag{2.4.17a}$$

The types of cyclones being used are such that the grade efficiency function is given for the *i*th cyclone by

$$G_{r_i} = 1 - e^{-b_1 r_p}. \tag{2.4.17b}$$

(a) Show that the total efficiency E_{T1} for the first cyclone is given by

$$E_{T_1} = \frac{b_1}{b_1+a}. \tag{2.4.17c}$$

(b) Similarly, show that the total efficiency E_{T2} for the second cyclone, based on the quantity of dust entering cyclone 1, is

$$E_{T_2} = \left[\frac{a}{a+b_1} - \frac{a}{a+b_1+b_2}\right]. \tag{2.4.17d}$$

(c) What is the total efficiency of the two-cyclone system based on the material entering the first cyclone?

Solution (a) Use the results of Problem 2.4.2 (b), and obtain, in general,

$$\int_0^\infty G_r f_f(r_p)\,dr_p = E_T = \int_0^1 G_r\,dF_f(r_p).$$

For the first cyclone,

$$E_{T_1} = \int_0^\infty \left(1-e^{-b_1 r_p}\right)(-)(-a)e^{-ar_p}\,dr_p = \frac{b_1}{a+b_1}.$$

(b) The air that comes out through the top of cyclone 1 enters cyclone 2 as feed. This air has some dust in it. The amount of dust entering cyclone 2 is the difference between the total quantity of dust entering cyclone 1 and the amount of dust collected in cyclone 1. If we adopt the basis that a total of 1 kg of dust is entering cyclone 1, then

$$1 = \int_0^1 dF_{f_1}(r_p) = \int_0^\infty ae^{-ar_p}\,dr_p.$$

Further, the quantity of dust collected by cyclone 1 and appearing in the underflow stream 2 is then equal to

$$\int_0^1 G_{r_1}\,dF_{f_1}(r_p) = \int_0^\infty G_{r_1} f_{f1}(r_p)\,dr_p = E_{T_1} = \frac{b_1}{a+b_1}$$

$$= \int_0^\infty ae^{-ar_p}\left(1-e^{-b_1 r_p}\right)dr_p.$$

For the second cyclone, the total quantity of dust entering between size r_p and $r_p + dr_p$ is (based upon 1 kg of dust entering cyclone 1)

$$1 \times \left(ae^{-ar_p} - ae^{-ar_p}\left[1-e^{-b_1 r_p}\right]\right)dr_p.$$

If the grade efficiency function for cyclone 2 is $(1 - e^{-b_2 r_p})$, the value of total efficiency E_{T_2} in this case is

$$E_{T_2}\int_0^\infty 1\cdot\left[ae^{-ar_p} - ae^{-ar_p}(1-e^{-b_1 r_p})\right](1-e^{-b_2 r_p})\mathrm{d}r_p$$
$$= \left(\frac{a}{a+b_1} - \frac{a}{a+b_1+b_2}\right).$$

(c) The total efficiency of the series connected system is $E_{T1,2} = E_{T1} + E_{T2}$, where E_{T2} is based on the amount of dust entering cyclone 1. Therefore

$$E_{T_{1,2}} = \frac{b_1}{b_1+a} + \frac{a}{b_1+a} - \frac{a}{b_1+b_2+a} = 1 - \frac{a}{b_1+b_2+a}.$$

Obviously $E_{T_{1,2}} > E_{T_1}$. Other types of connections between two cyclones are also possible. See Problems 2.4.4 and 2.4.7 for two such arrangements.

Description of separation in solid particle–fluid separation devices by means of the grade efficiency G_r or related functions is often avoided in practice. Instead, a single number is used to describe the separation characteristics. Such a single number is provided by the "cut size," which is the size of the opening of a hypothetical and ideal screen achieving the same separation as the device under consideration. An ideal screen will generate a step function for G_r: all particles of size smaller than the screen opening will appear in the overflow, whereas all particles of size larger than the screen opening will appear in the underflow. Actual grade efficiency functions for separators in practice are hardly of the step function type; therefore a "cut size" approach cannot be a true substitute for G_r. However, since the "cut-size" concept is used frequently, we will indicate very briefly a few definitions of cut size.

One of the most commonly used cut sizes is the *equiprobable size* $r_{p,50}$, for which the value of the grade efficiency G_r is 0.50 (Figure 2.4.4). A particle having this size has an equal probability of appearing in both the overflow, as well as the underflow from the separator. A smaller particle will most likely be carried away by the fluid in the overflow whereas a larger particle will most likely be separated from the fluid and appear in the underflow (Svarovsky, 1979). Note that if G_r is plotted against a normalized particle radius (r_p/r_{max}), the equiprobable size will be independent of the actual values of the particle sizes for a given problem.

Since the grade efficiency function G_r is needed to know the equiprobable size $r_{p,50}$, and substantial information is necessary before G_r is known (e.g. E_T, $f_2(r_p)$, and $f_f(r_p)$ should be available according to equation (2.4.4b)), two other cut-size definitions are frequently used in industry: the analytical cut size $r_{p,a}$ and the cut size by curve intersection (Svarovsky, 1979). We will only touch upon the analytical cut size here.

The analytical cut size $r_{p,a}$ is defined such that a hypothetical and ideal screen having this size opening will give from the feed solid mixture the same value of total efficiency E_T as the actual separator. In terms of the feed particle size distribution function, $F_f(r_p)$,

$$1 - F_f(r_{p,a}) = E_T. \qquad (2.4.18)$$

The analytical cut size $r_{p,a}$ is not equal to the equiprobable cut size $r_{p,50}$. The relation between these two can be derived, for example, for a log-normal particle size distribution.

2.4.2 Continuous chemical mixtures

Coal-derived liquids, heavy petroleum fractions, vegetable oils and polymers are mixtures that have very large numbers of components. It is practically impossible to identify each component by ordinary chemical analysis. One can no longer use mole fractions of individual components. Traditionally, the pseudo-component approach, or the key component approach, has been used to handle such complex mixtures.

In the *pseudo-component approach*, the complex mixture is represented by a discontinuous distribution of pseudo-components, where the mole fraction x_i of pseudo-component i is represented by a bar in Figure 2.4.6(a) (which shows ten components). In the approach of *continuous chemical mixtures*, discrete components are not identified. Instead, the mixture is described by a property density function $f(r)$ of a single distribution variable r (Figure 2.4.6(b)). The variable r can be the normal boiling point, the molecular weight, the number of carbon atoms, the degree of polymerization, etc. (Cotterman *et al.*, 1985; Kehlen *et al.*, 1985). The fraction of molecules in the mixture characterized by the range r to $r + \mathrm{d}r$ is given by $f(r)\mathrm{d}r$ such that

$$\int_{r_{min}}^{r_{max}} f(r)\mathrm{d}r = 1. \qquad (2.4.19)$$

Figure 2.4.6(c) shows the cumulative distribution of components up to any component i. For continuous chemical mixtures, we have instead a property distribution function. The property distribution function $F(r)$ is defined by (Figure 2.4.6(d))

$$\int_{r_{min}}^{r} f(r)\mathrm{d}r = F(r). \qquad (2.4.20)$$

These quantities are thus analogous to those we have already defined for a size distributed particle population. Instead of particle size, we have a distribution variable r, which is intrinsic to a given chemical species. For example, for a flash vaporizer single-entry separator (Figure 2.4.7), the material balance for 1 mole of feed having a molecular weight density function of $f_f(M)$ is

$$f_f(M)\mathrm{d}M = \theta f_v(M)\mathrm{d}M + (1-\theta)f_l(M)\mathrm{d}M, \qquad (2.4.21a)$$

where θ is the fraction of feed vaporized, $f_v(M)$ is the molecular weight density function of the vapor and $f_l(M)$ is the molecular weight density function of the liquid product (Cotterman and Prausnitz, 1985). Again,

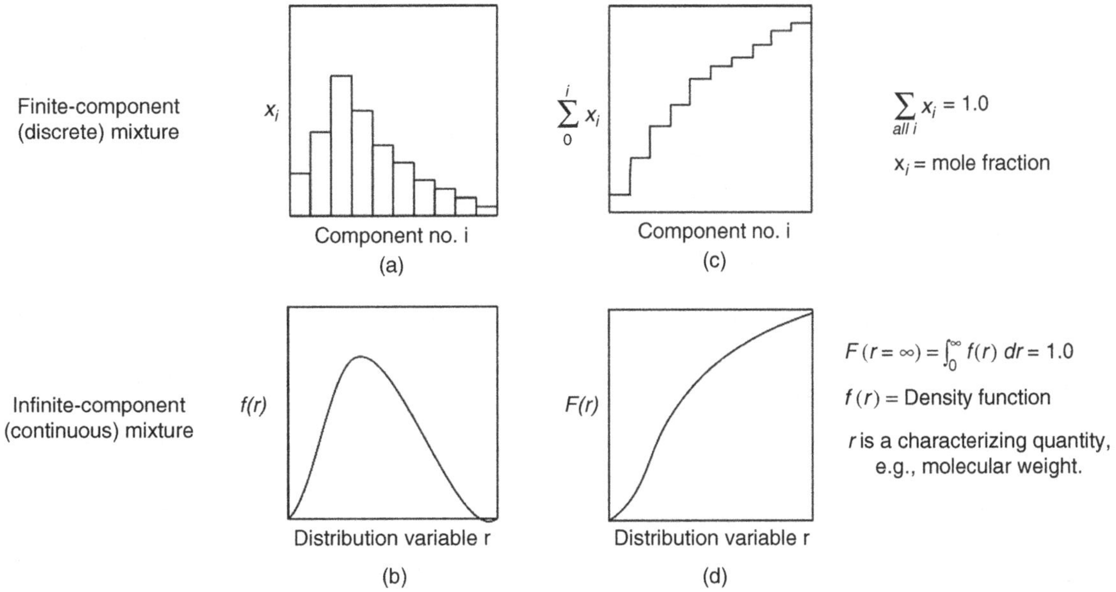

Figure 2.4.6. Discrete and continuous composition for a multicomponent mixture. Reprinted, with permission, from Ind. Eng. Chem. Proc. Des. Dev., **24***(1) (1985), 194. Copyright 1985 American Chemical Society.*

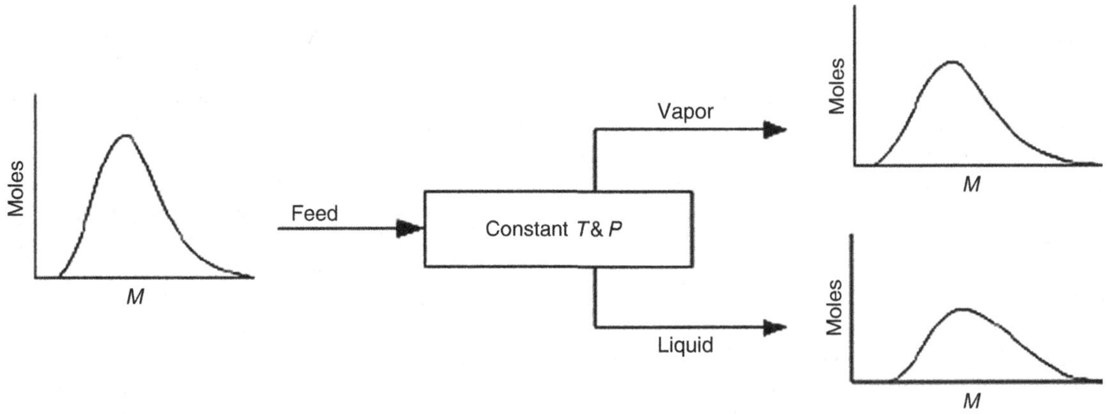

Figure 2.4.7. Flash vaporizer for a continuous chemical mixture. (After Cotterman and Prausnitz, 1985.)

$$\int_{M_{min}}^{M_{max}} f_v(M)\mathrm{d}M = 1; \qquad \int_{M_{min}}^{M_{max}} f_l(M)\mathrm{d}M = 1. \qquad (2.4.21b)$$

When a particular quantity, such as the pressure or volume, of a vapor–liquid mixture is to be determined for a continuous chemical mixture, the method of moments is to be adopted. Suppose the molar volume V of a pure chemical liquid species of molecular weight M can be expressed as a function of temperature T, pressure P and M, i.e. $V(M,T,P)$. If this mixture is an ideal mixture and we are dealing with a simple n-component system, then the molar volume V of the solution is given by

$$V = \sum_{i=1}^{n} x_i V_i(T, P, M_i). \qquad (2.4.22a)$$

In a continuous chemical mixture, we have instead

$$V = \int_{M_{min}}^{M_{max}} V(T, P, M) f_1(M)\mathrm{d}M. \qquad (2.4.22b)$$

Such a description of continuous chemical mixtures can also be adopted for polymer solutions containing a molecular weight distribution. Consider a solution of polydisperse polystyrene in cyclohexane. The mass fraction

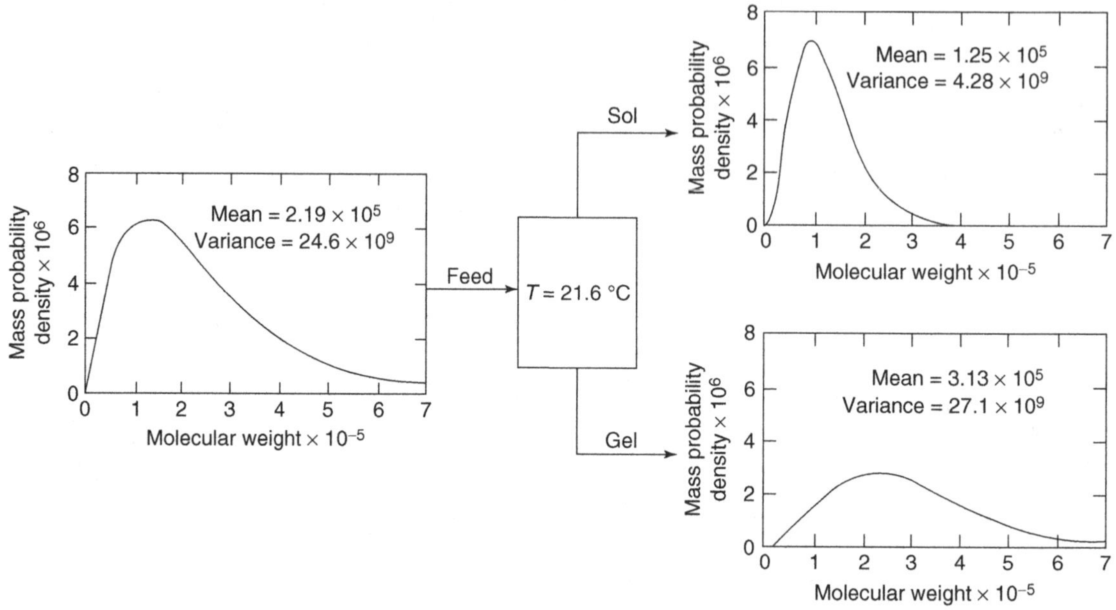

Figure 2.4.8. Phase equilibria in the system (polydisperse) polystyrene/cyclohexane. Flory parameter $\chi \approx$ 0.50–0.60. Reprinted, with permission, from Ind. Eng. Chem. Proc. Des. Dev., **24***(1) (1985), 194. Copyright 1985 American Chemical Society.*

of polymers having a molecular weight in the range M to $M + \mathrm{d}M$ in the solution is shown in Figure 2.4.8. The fraction of polymers in the solution is 0.67. When this solution is chilled to 21.6 °C, it forms two phases, a polymer-rich phase called the *gel* and a solvent-rich phase called the *sol*. Obviously, the density function of polymer molecular weight in the gel is different from that in the sol; the gel has much more of the higher molecular weight species than the sol as shown (Cotterman *et al.*, 1985).

Semi-continuous mixtures form a category in between the continuous chemical mixtures and the ordinary multicomponent mixtures. For example, solvents in a polymer solution or light hydrocarbons in a gas-condensate system can be described by discrete concentrations or mole fractions, whereas the continuous components are described by a density or distribution function approach as just outlined. For an introduction to such systems and the basics of calculation procedures needed to describe separation, consult Cotterman *et al.* (1985) and Cotterman and Prausnitz (1985).

2.4.3 Multicomponent chemical mixtures

For a *multicomponent chemical mixture* being fed into a single-entry separator with only two product streams, existing methods of description involve amongst others separation factors for selected components. Some of these separation factors, namely α_{in} (definition (1.6.6)), α_{ij} (definition (1.6.8)), etc., were introduced earlier with closed separators, but are equally applicable here if regions 1 and 2 are substituted by product streams 1 and 2.

Current practice regarding the definition of heavy key (Section 1.6) and light key is as follows. The component with the highest molecular weight appearing in the light fraction in significant amounts is termed the *heavy key*. The component with the lowest molecular weight appearing in the heavy fraction in significant amounts is the *light key*. Instead of the molecular weight, other criteria regarding the heaviness or lightness of a component may also be used.

Example 2.4.4 illustrates the selection of heavy key and light key components for the distillation process. Note also that one can analyze the separator as though there are only two components, the heavy key ($i = 2$) and the light key ($i = 1$). Of course, in such a case the total molar feed flow rate is to be reduced to $W_{tf}\,[x_{1f} + x_{2f}]$ for a single-entry separator, where x_{1f} and x_{2f} refer, respectively, to the actual mole fractions of species 1 and species 2 in the actual feed. Thus $x_{1f} + x_{2f} < 1$. But in the hypothetical binary system, define $\underline{x}_{1f}$ and $\underline{x}_{2f}$ such that (Hengstebeck, 1961)

$$\underline{x}_{1f} = \frac{x_{1f}}{x_{1f} + x_{2f}} \quad \text{and} \quad \underline{x}_{2f} = \frac{x_{2f}}{x_{1f} + x_{2f}}. \tag{2.4.23}$$

Furthermore,

$$\underline{W}_{tf} = W_{tf}[x_{1f} + x_{2f}] \quad \text{and} \quad \underline{W}_{tf}\underline{x}_{if} = W_{tf}x_{if}; \quad i = 1, 2, \tag{2.4.24}$$

where $\underline{W}_{tf}$ is the total molar feed flow rate in this hypothetical binary system.

If a multicomponent feed stream is separated by a single-entry separator into more than two product streams, the description acquires much greater complexity. Very little is available in the literature for such cases. The best approach appears to be to concentrate on two specific species and the respective species-specific product streams. Perfect separation will then consist of each product stream containing only a specified species. The various indices defined for binary systems in this chapter may be used with due provisions for the distribution of each species in each product stream.

Example 2.4.4 Consider a paraffinic hydrocarbon liquid feed having a composition in mole fractions as given in Table 2.4.1.

This liquid feed at a pressure of 15 atm and 90 °C is introduced into a tall distillation column, which produces a light and a heavy fraction having the compositions given in Table 2.4.2.

The light key component in this case is n-C_3H_8 because C_2H_6 is absent in the heavy fraction. Since n-hexane does not appear in the overhead vapor, n-pentane is the heavy key. The component n in α_{in} defined by (1.6.6) is then n-C_5H_{12}. In Table 2.4.3 we calculate α_{in} values for all the six components; here $i = 1, 2, 3$ and 4 correspond, respectively, to CH_4, C_2H_6, n-C_3H_8 and n-C_4H_{10}, whereas $i = 6$ corresponds to n-C_6H_{14}. Naturally $\alpha_{in} > 1$ for $i = 1, 2, 3$ and 4, but $\alpha_{in} < 1$ for $i = 6$.

Another type of multicomponent separation involves one or more solutes distributed between two solvents that are immiscible and which form two different phases. A third type of multicomponent separation involves a gas phase and a liquid phase such that the solvent, which is the dominant constituent of the liquid phase, does not appear to a significant extent in the gas phase. Although there are many other types of multicomponent, multiphase systems, we will describe one of the preceding types of separation now since the pseudo-binary approach is valid for both cases.

Table 2.4.1. Composition in mole fractions for liquid feed in Example 2.4.4

n-C_6H_{14}	n-C_5H_{12}	n-C_4H_{10}	n-C_3H_8	C_2H_6	CH_4
0.20	0.25	0.31	0.11	0.09	0.04

Table 2.4.2. Compositions of light and heavy fractions in Example 2.4.4

	n-C_6H_{14}	n-C_5H_{12}	n-C_4H_{10}	n-C_3H_8	C_2H_6	CH_4
Overhead vapor		0.01	0.32	0.33	0.22	0.12
Bottoms liquid	0.35	0.362	0.281	0.007	-	-

Table 2.4.3. Value of α_{in} for the six components in Example 2.4.4

	n-C_6H_{14}	n-C_5H_{12}	n-C_4H_{10}	n-C_3H_8	C_2H_6	CH_4
α_{in}	0	1	41.2	1710	∞	∞

Example 2.4.5 Benzene is to be absorbed in a heavy wash oil from a nitrogen stream which enters a countercurrent absorber at a total molar flow rate of 10 gmol/s. The mole fraction of benzene in this stream is 0.020. The heavy wash oil entering the absorber at 1.0 gmol/s contains no benzene or nitrogen. It is known that benzene is distributed between the gas and the wash oil according to the relation $(x_{\text{benzene}})\text{gas} = 0.125\ (x_{\text{benzene}})\text{oil}$ for the higher concentration levels near the gas inlet, but that nitrogen is not absorbed at all. Obtain values of the extent of separation and the separation factor. Consult the countercurrent schematic in Figure 2.3.1.

Solution Product stream $j = 1$ is the nitrogen stream stripped of benzene (almost). Product stream $j = 2$ is the wash oil stream leaving the absorber after absorbing benzene. Component $i = 1$ is nitrogen; component $i = 2$ is benzene. In a pseudo-binary calculation scheme, we neglect the wash oil content in $j = 2$. Similarly, feed stream f_1 is the nitrogen stream, whereas feed stream f_2 is the wash oil stream on a wash oil-free basis. The extent of separation ξ is

$$\xi = \left|\dot{Y}_{11} - \dot{Y}_{21}\right|$$
$$= \left|\frac{W_{t1}x_{11}}{W_{tf_1}x_{1f_1} + W_{tf_2}x_{1f_2}} - \frac{W_{t1}x_{21}}{W_{tf_1}x_{2f_1} + W_{tf_2}x_{2f_2}}\right|;$$

but $x_{1}f_2 = 0$, $x_{2}f_2 = 0$, and also $x_{12} = 0$. Furthermore, $W_{tf1}\ x_{1}f_1 = W_{t1}\ x_{11} + W_{t2}\ x_{12} = W_{t1}\ x_{11}$. Therefore

$$\xi = \left|1 - \frac{W_{t1}x_{21}}{W_{tf_1}x_{2f_1}}\right|.$$

So, the species 2 balance is given by

$$W_{tf_1}x_{2f_1} = W_{t1}x_{21} + W_{t2}x_{22} \Rightarrow \xi = \left|\frac{W_{t2}x_{22}}{W_{tf_1}x_{2f_1}}\right|.$$

Since some benzene always escapes with the nitrogen in product stream $j = 1$, $W_{t2}\ x_{22} < W_{tf1}\ x_{2f_1}$, so $\xi < 1$ unless we have complete absorption of benzene. Now,

$$W_{tf_1}x_{2f_1} = \left(10\frac{\text{gmol}}{s} \times 0.020\right) = 0.20\frac{\text{gmol benzene}}{\text{s}}.$$

Using the distribution relation at the absorber gas inlet, we have

$$(x_{\text{benzene}})_{\text{oil}} = \frac{(x_{\text{benzene}})_{\text{gas}}}{0.125} = \frac{0.020}{0.125} = 0.16.$$

Therefore

$$W_{t2}x_{22} = (\text{wash oil flow rate})\left(\frac{\text{moles benzene}}{\text{moles wash oil}}\right)$$
$$= 1.0\left\{\frac{(x_{\text{benzene}})_{\text{oil}}}{1 - (x_{\text{benzene}})_{\text{oil}}}\right\} = 1.0 \times \frac{0.16}{0.84} \cong 0.1905.$$

So,

$$\xi = \left| \frac{0.1905}{0.20} \right| = 0.954,$$

a reasonably good separation. But $\alpha_{12} = x_{11}x_{22}/x_{21}x_{12} = x_{11}x_{22}/x_{21}\, 0$ since $x_{12} \cong 0$ (no nitrogen in wash oil product stream (on a wash oil-free basis or otherwise)). Therefore $\alpha_{12} = \infty$, which is not helpful in determining the changes in the performance of the absorber (if any) since nitrogen is always absent in the $j = 2$ stream.

2.5 Separation in an output stream with time-varying concentration

In **some separation processes, only one product stream comes out of a single-entry separator**. Further, **the molar flow rates of various species in this product stream as measured by a detector** (Figure 2.5.1) **vary with time in a manner useful for separation.**

Consider the molar rates at which two chemical species 1 and 2 present in small amounts in a carrier liquid or gas stream emerge from the separator. Figure 2.5.1 shows two general types of output profile. In Figure 2.5.1 (a) between t_1^{in} and t_1^0 seconds, the exit gas or liquid stream contains only species 1, whereas between t_2^{in} and t_2^0 seconds, the exit stream has only species 2 with $t_2^0 > t_2^{in} > t_1^0 > t_1^{in}$. (Here the superscripts in and 0 refer, respectively, to the times when a particular species appears in the output for the first time and when it disappears completely.) Thus, if we assume the volumetric output of the separator between t_1^{in} and t_1^0 seconds to constitute region 1 and the volumetric output between t_2^{in} and t_2^0 to constitute region 2, we have a case of perfect separation of species 1 and 2 between the two regions. On the other hand, consider Figure 2.5.1(b), where the output is such that $t_2^0 > t_2^{in}$, $t_1^0 > t_1^{in}$ but $t_2^{in} < t_1^0$. In this case, the output stream between t_2^{in} and t_1^0 will have both species 1 and 2. If we now consider the purity of two regions obtained by collecting the outputs between, say, t_1^{in} and t_c and t_c and t_2^0, it is obvious that we are dealing with an imperfect separation. Both the regions now have some impurity.

It is now apparent that the description of separation between species 1 and 2 achieved with such an output stream can be carried out with the indices developed for a closed separator with two regions. There are also other indices which grew out of the practice of processes using such separators (e.g. in chromatographic separations). Before we introduce these indices, consider the variation in concentration $C_i(t)$ of the ith solute species in the output stream as a function of time. This is necessary to estimate the total number of moles of the given species in a given region obtained by collecting the output between specified times.

Let m_i^0 be the total number of moles of species i introduced into the separator by the feed stream. Assume further that all of m_i^0 has come out of the separator in the output stream between times t_1^{in} and t_i^0 seconds. Let the output volumetric flow rate be constant at Q_f cm^3/s. It is also assumed to be equal to the input volumetric flow rate. The number of moles of species i in the output between t_1^{in} and t_c (for $t_c > t_1^{in}$) is given by

$$m_{ii} = \int_{t_i^{in}}^{t_c} Q_f C_i(t)\mathrm{d}t \tag{2.5.1a}$$

if region i for $t_c > t_1^{in}$ is $t_1^{in} \le t \le t_c$. On the other hand, if region i for $t_c > t_1^{in}$ is $t_i^0 \ge t \ge t_c$, then

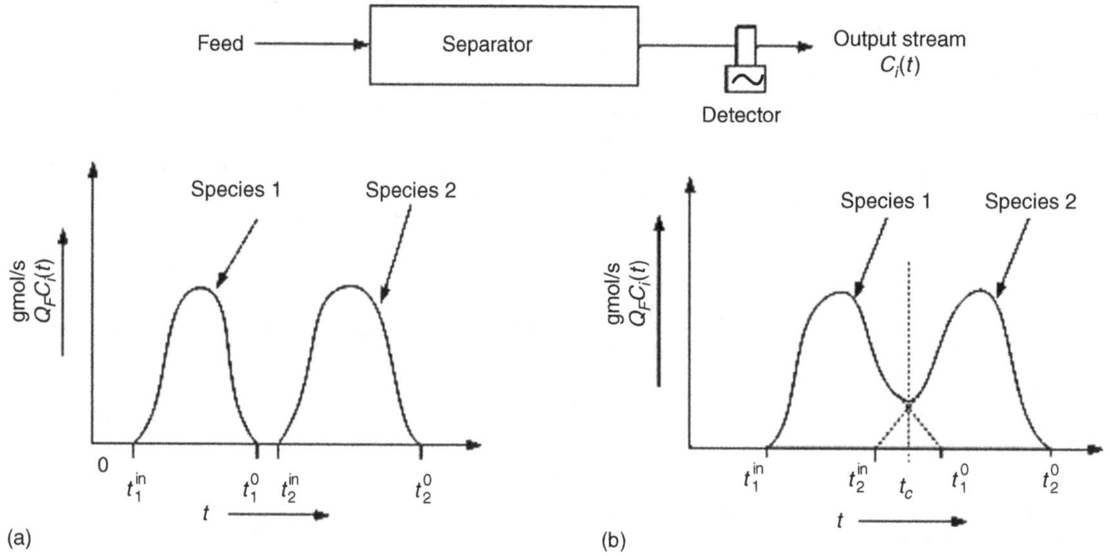

Figure 2.5.1. Time-varying molar output rate of solute species in the single product stream from a single-entry separator.

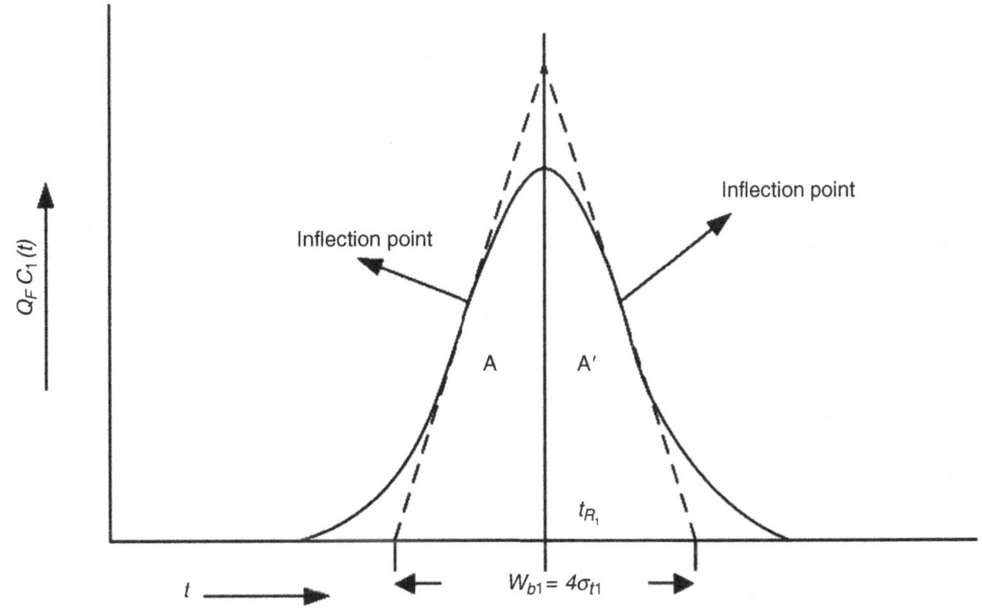

Figure 2.5.2. Molar output rate of species 1 for a Gaussian profile.

$$m_{ii} = \int_{t_c}^{t_i^0} Q_f C_i(t) \mathrm{d}t. \qquad (2.5.1b)$$

Note that here $C_i(t)$ is the number of moles of species i per unit volume of solution collected at any time t. Along with definition (2.5.1a), we have

$$m_i^0 = \int_{t_i^{\text{in}}}^{t_i^0} Q_f C_i(t) \mathrm{d}t, \qquad (2.5.2)$$

which is also valid for definition (2.5.1b).

Sometimes it is difficult to specify a value of t_i^0 where $C_i(t) = 0$; one then uses $t_i^0 = \infty$. Similarly, one uses $t_1^{\text{in}} = -\infty$. At this time, we require a functional description of $C_i(t)$ vs. t to make estimates of m_{ii} to be used in various indices. A **Gaussian profile is often close to what is observed in many separator outputs** (e.g. **the so-called chromatograms**, Figure 2.5.2, where t_{R1} is **called the retention time** for species 1):

$$\frac{Q_f C_i(t)}{m_i^0} = \frac{1}{\sigma_{ti}\sqrt{2\pi}} \exp\left[-\frac{1}{2}\left\{\frac{t_{R_i}-t}{\sigma_{ti}}\right\}^2\right]. \qquad (2.5.3)$$

Note that with $t_i^0 = \infty$ and $t_1^{\text{in}} = -\infty$, such a profile yields

$$\int_{-\infty}^{\infty} Q_F C_i(t) \mathrm{d}t = m_i^0 \int_{-\infty}^{\infty} \frac{1}{\sigma_{ti}\sqrt{2\pi}} \exp\left[-\frac{1}{2}\left\{\frac{t_{R_i}-t}{\sigma_{ti}}\right\}^2\right] \mathrm{d}t. \qquad (2.5.4)$$

On defining

$$\left[(t_{R_i}-t)/\sqrt{2}\sigma_{ti}\right] = -\eta_i, \qquad (2.5.5)$$

relation (2.5.4) becomes

$$\int_{-\infty}^{\infty} Q_F C_i(\text{t}) \mathrm{d}t = m_i^0 \left[\frac{1}{\sqrt{\pi}} \int_{-\infty}^{\infty} \mathrm{e}^{-\eta_i^2} \mathrm{d}\eta_i\right] = \frac{2m_i^0}{\sqrt{\pi}} \int_0^{\infty} \mathrm{e}^{-\eta_i^2} \mathrm{d}\eta_i = m_i^0,$$

as required by relation (2.5.2).

A brief look at the profile of the molar solute output rate as a function of time will help in specifying the role of t_{Ri} and σ_{ti} (appearing in the Gaussian profile (2.5.3)) in the separation system. In Figure 2.5.2, the Gaussian output flow rate of species 1 has been shown as gram moles of species 1 per unit time ($= Q_f \times C_1(t)$) against time. With Q_f constant, the concentration $C_1(t)$ quickly rises to a peak value at $t = t_{R1}$ and then decreases. If two tangents are drawn at the two inflection points A and A′ and extended in both directions, they intersect above the peak at $t = t_{R1}$. Furthermore, the distance between the intersections of these two lines with $Q_f C_1(t) = 0$ (the base line) is given as W_{b1}; for a Gaussian profile, $W_{b1} = 4\sigma_{t1}$, where σ_{t1} is the standard deviation of the species 1 profile. When $t = t_{R1}$, the height of the peak is obtained as follows:

$$Q_f C_1(t_{R_1}) = \frac{m_1^0}{\sigma_{t1}\sqrt{2\pi}}. \qquad (2.5.6a)$$

One can therefore rewrite the Gaussian profile (2.5.3) as

$$\frac{Q_f C_i(t)}{Q_f C_i(t_{R_i})} = \frac{Q_f C_i(t)}{Q_f C_i(t)|_{\max}} = \frac{C_i(t)}{C_i(t)|_{\max}} = \exp\left[-\frac{1}{2}\left\{\frac{t_{R_i}-t}{\sigma_{ti}}\right\}^2\right]. \qquad (2.5.6b)$$

An approximate measure of the time when species 1 appears in the output stream is therefore t_{R_1}. Similarly, for species 2, one obtains t_{R_2}. These time estimates would be exact only if $\sigma_{t1} = 0$ or $\sigma_{t2} = 0$. It is also evident from Figures 2.5.1 and 2.5.2 that if t_{R_1} and t_{R_2} are wide apart, then, unless σ_{t1} and σ_{t2} are very large, the two concentration profiles will not overlap.

A simple index commonly used for describing **separation in such an output stream** is the **resolution** R_s of **two adjacent profiles or bands or chromatograms**:

$$R_s = \frac{2(t_{R_2} - t_{R_1})}{W_{b1} + W_{b2}}. \tag{2.5.7}$$

It is merely the distance between the two concentration peaks divided by the arithmetic average value of W_{b1} and W_{b2}, *each referred to as a band width*. The larger the value of R_s, the greater the separation. Separation is perfect when $R_s \to \infty$. Further, the smaller the values of σ_{t1} and σ_{t2} for given t_{R1} and t_{R2}, the better the separation, since the overlap region becomes smaller. Usually, when $R_s = 1.5$, the separation between two chromatographic profiles is quite good, with only a small section of each region (less than 1%) contaminated by the other species.

When, $\sigma_{t1} = \sigma_{t2} = \sigma_t$ and both the outputs are Gaussian, the definition of resolution is simplified:

$$R_s = \frac{(t_{R_2} - t_{R_1})}{4\sigma_t}. \tag{2.5.8}$$

In general, however, the standard deviations for the two solutes are not equal. If we can express σ_{t1} and σ_{t2} thus

$$\sigma_{t1} = \sigma_t \qquad \text{and} \qquad \sigma_{t2} = \sigma_t + \Delta\sigma_t, \tag{2.5.9}$$

then

$$R_s = \frac{(t_{R_2} - t_{R_1})}{4\sigma_t\left[1 + \frac{\Delta\sigma_t}{2\sigma_t}\right]}. \tag{2.5.10}$$

Note that if $\Delta\sigma_t << 2\sigma_t$, expression (2.5.10) for R_s is reduced to (2.5.8). For additional considerations, see Vink (1972). Other measures describing the degree of overlap between non-Gaussian outputs have been considered by Dose and Guiochon (1990).

We now propose to calculate the extent of separation, ξ, for a typical chromatographic output. We need to calculate Y_{11} and Y_{21}. Substitution of profile (2.5.3) in definition (2.5.la) leads to

$$m_{ii} = \int_{-\infty}^{t_c} \frac{m_i^0}{\sigma_{ti}\sqrt{2\pi}} \exp\left[-\frac{1}{2}\left\{\frac{t_{R_i} - t}{\sigma_{ti}}\right\}^2\right] dt. \tag{2.5.11}$$

On using transformation (2.5.5), we obtain

$$\frac{m_{ii}}{m_i^0} = \int_{-\infty}^{\frac{t_c - t_{R_i}}{\sqrt{2}\sigma_{ti}}} \frac{\exp[-\eta_i^2]}{\sqrt{\pi}} d\eta_i, \tag{2.5.12}$$

which on rearrangement yields

$$\frac{m_{ii}}{m_i^0} = \frac{1}{\sqrt{\pi}} \int_0^{\frac{t_c - t_{R_i}}{\sqrt{2}\sigma_{ti}}} \exp[-\eta_i^2] d\eta_i + \frac{1}{\sqrt{\pi}} \int_0^{\infty} \exp[-\eta_i'^2] d\eta_i'$$
$$= \frac{1}{2} \text{erf}\left[\frac{t_c - t_{R_i}}{\sqrt{2}\sigma_{ti}}\right] + \frac{1}{2}, \tag{2.5.13}$$

where the value of the second integral with variable $\eta'_i = -\eta_i$ is $\sqrt{\pi}/2$. Note that (m_{ii}/m_i^0) is simply Y_{ii}, the extent of segregation of species i in region i. However, this derivation is valid for $t_c > t_i^{\text{in}}$ and $t_c \geq t \geq t_i^{\text{in}}$.

For the second type of region i with $t_c > t_i^{\text{in}}$ and $t_i^0 \geq t \geq t_c$, we use definition (2.5.1b) to obtain

$$m_{ii} = \int_{t_c}^{\infty} \frac{m_i^0}{\sigma_{ti}\sqrt{2\pi}} \exp\left[-\frac{1}{2}\left\{\frac{t_{R_i} - t}{\sigma_{ti}}\right\}^2\right] dt. \tag{2.5.14}$$

On using transformation (2.5.5) and splitting the integral, we get

$$\frac{m_{ii}}{m_i^0} = \frac{1}{\sqrt{\pi}} \int_0^{\infty} \exp[-\eta_i^2] d\eta + \int_{\frac{t_c - t_{R_i}}{\sqrt{2}\sigma_{ti}}}^{0} \frac{1}{\sqrt{\pi}} \exp[-\eta_i^2] d\eta_i \tag{2.5.15}$$

since $(t_c - t_{R_i}/\sqrt{2}\sigma_{ti})$ is negative here. Changing the variables in the second integral to $\eta_i'' = -\eta_i$, we get

$$\frac{m_{ii}}{m_i^0} = \frac{1}{2} + \frac{1}{\sqrt{\pi}} \int_0^{\frac{t_{R_i} - t_c}{\sqrt{2}\sigma_{ti}}} \exp[-\eta''^2_i] d\eta_i'' = \frac{1}{2} + \frac{1}{2} \text{erf}\left[\frac{t_{R_i} - t_c}{\sqrt{2}\sigma_{ti}}\right]. \tag{2.5.16}$$

With region 1 as $t_c > t_1^{\text{in}}$ and $t_c \geq t \geq t_1^{\text{in}}$, we can use relation (2.5.13) to determine Y_{11} and Y_{21}. The index ξ, the extent of separation for such a system of two solute species 1 and 2 distributed between regions 1 and 2, is therefore given by (Rony, 1968b)

$$\xi = \text{abs}|Y_{11} - Y_{21}| = \frac{1}{2} \text{abs}\left[\text{erf}\left\{\frac{t_c - t_{R_1}}{\sqrt{2}\sigma_{t1}}\right\} - \text{erf}\left\{\frac{t_c - t_{R_2}}{\sqrt{2}\sigma_{t2}}\right\}\right]. \tag{2.5.17}$$

Here t_{R_i} and σ_{ti} are characteristic parameters for the solute species i and the separation system. If $t_{R_2} - t_{R_1} = \infty$ and t_c halfway between t_{R_1} and t_{R_2}, both error functions will have an upper limit whose magnitude is infinitely large. Therefore, from (2.5.17) we have

$$\xi = \frac{1}{2} \text{abs}\left|\left[\frac{2}{\sqrt{\pi}} \int_0^{\frac{t_c - t_{R_1}}{\sqrt{2}\sigma_{t1}}} e^{-\eta^2} d\eta - \frac{2}{\sqrt{\pi}} \int_0^{\frac{t_c - t_{R_2}}{\sqrt{2}\sigma_{t2}}} e^{-\eta^2} d\eta\right]\right|$$
$$\cong \frac{1}{2} \text{abs}\left|\left[\frac{2}{\sqrt{\pi}} \int_0^{\infty} e^{-\eta^2} d\eta - \frac{2}{\sqrt{\pi}} \int_0^{-\infty} e^{-\eta^2} d\eta\right]\right|$$
$$\cong \frac{1}{2} \text{abs}\left|\left[1 + \frac{2}{\sqrt{\pi}} \int_0^{\infty} e^{-\eta'^2} d\eta'\right]\right| \cong \frac{1}{2} \text{abs}[1 + 1] = 1. \tag{2.5.18}$$

The maximum extent of separation of $\xi = 1$ is achieved even if t_c is not halfway between the two t_R values as long as each $|t_c - t_{R_i}|$ is large. Obviously, if $t_{R_1} = t_{R_2}$ and $\sigma_{t1} = \sigma_{t2}$, $\xi = 0$. Thus, there is no separation when the two concentration profiles have identical shape and come out at the same time. Separation under any other conditions then depends on the nature of the two profiles, their locations and the location of t_c, the cut point.

It is also possible to calculate the value of the separation factor α_{12} and some of the other types of indices defined in Chapter 1 for such an open system. The interrelationships been various separation indices (ξ and R_s, for example) may also be developed under certain conditions. See Example 2.5.1 and Problems 2.5.1 and 2.5.2.

For a description of multicomponent separation in chromatographic outputs, the semi-quantitative approach by Stewart (1978) involving the concept of a resolution matrix is likely to be useful to interested readers.

Example 2.5.1 Consider the time-varying molar outputs of two species 1 and 2 from a chromatographic system (Figure 2.5.1(b)). Assume both outputs, which are overlapping each other, to be Gaussian and $\sigma_{t1} = \sigma_{t2}$. If the cut point is located such that the impurity ratio in each region is the same, and if it is known that $m_1^0 = m_2^0$, develop a relation between this impurity ratio and the resolution. Calculate the values of this impurity ratio for $R_s = 0.2$, 0.6, 1.0 and 1.4.

Solution The impurity ratios for regions 1 and 2 are defined by relation (1.4.3), where region 1 is from, say, t_1^{in} to t_c and region 2 is from t_c to t_2^0. We have $\eta_1 = m_{21}/m_{11}$ and $\eta_2 = m_{12}/m_{22}$,

$$\eta_1 = \frac{m_{21}}{m_{11}} = \frac{m_{21}/m_2^0\, m_2^0}{m_{11}/m_1^0\, m_1^0} = \frac{m_{21}/m_2^0}{m_{11}/m_1^0} = \frac{Y_{21}}{Y_{11}}$$

since $m_2^0 = m_1^0$. Similarly for η_2:

$$\eta_2 = \frac{m_{12}}{m_{22}} = \frac{m_{12}/m_1^0\, m_1^0}{m_{22}/m_2^0\, m_2^0} = \frac{m_{12}/m_1^0}{m_{22}/m_2^0} = \frac{Y_{12}}{Y_{22}}.$$

From (2.5.13),

$$Y_{11} = \frac{1}{2}\left[\mathrm{erf}\left(\frac{t_c - t_{R_1}}{\sqrt{2}\sigma_{t1}}\right) + 1\right].$$

But

$$Y_{12} = 1 - Y_{11} = \frac{1}{2}\left[1 - \mathrm{erf}\left(\frac{t_c - t_{R_1}}{\sqrt{2}\sigma_{t1}}\right)\right].$$

This last result could also have been obtained from relation (2.5.16). Similarly,

$$Y_{22} = \frac{1}{2}\left[1 + \mathrm{erf}\left(\frac{t_{R_2} - t_c}{\sqrt{2}\sigma_{t2}}\right)\right] \quad \text{and}$$

$$Y_{21} = 1 - Y_{22} = \frac{1}{2}\left[1 - \mathrm{erf}\left(\frac{t_{R_2} - t_c}{\sqrt{2}\sigma_{t2}}\right)\right].$$

Therefore, as required in Example 2.5.1,

$$\eta_1 = \left[\frac{1 - \mathrm{erf}\left(\frac{t_{R_2} - t_c}{\sqrt{2}\sigma_{t2}}\right)}{1 + \mathrm{erf}\left(\frac{t_c - t_{R_1}}{\sqrt{2}\sigma_{t1}}\right)}\right] = \eta_2 = \left[\frac{1 - \mathrm{erf}\left(\frac{t_c - t_{R_1}}{\sqrt{2}\sigma_{t1}}\right)}{1 + \mathrm{erf}\left(\frac{t_{R_2} - t_c}{\sqrt{2}\sigma_{t2}}\right)}\right].$$

This leads to

$$\left[\mathrm{erf}\left(\frac{t_{R_2} - t_c}{\sqrt{2}\sigma_{t2}}\right)\right]^2 = \left[\mathrm{erf}\left(\frac{t_c - t_{R_1}}{\sqrt{2}\sigma_{t1}}\right)\right]^2.$$

Since $\sigma_{t1} = \sigma_{t2} = \sigma_t$ (say), this means *either*

$$\mathrm{erf}\left(\frac{t_{R_2} - t_c}{\sqrt{2}\sigma_t}\right) = \mathrm{erf}\left(\frac{t_c - t_{R_1}}{\sqrt{2}\sigma_t}\right),$$

which means $t_{R_2} - t_c = t_c - t_{R_1}$, i.e. $t_c = (t_{R_1} + t_{R_2})/2$, or

$$-\mathrm{erf}\left(\frac{t_{R_2} - t_c}{\sqrt{2}\sigma_t}\right) = +\,\mathrm{erf}\left(\frac{t_c - t_{R_1}}{\sqrt{2}\sigma_t}\right).$$

Since erf $(-x)$ = $-$erf (x), this means $t_c - t_{R_2} = t_c - t_{R_1}$ i.e. $t_{R_1} = t_{R_2}$, which is a trivial result since we know $t_{R_2} \neq t_{R_1}$. Therefore $t_c = ((t_{R_1} + t_{R_2})/2)$ and

$$\eta_1 = \eta_2 = \left.\left[\frac{1 - \mathrm{erf}\left(\frac{t_{R_2} - t_c}{\sqrt{2}\sigma_t}\right)}{1 + \mathrm{erf}\left(\frac{t_c - t_{R_1}}{\sqrt{2}\sigma_t}\right)}\right]\right|_{t_c = \frac{t_{R_1} + t_{R_2}}{2}} = \frac{1 - \mathrm{erf}\left(\frac{t_{R_2} - t_{R_1}}{2\sqrt{2}\sigma_t}\right)}{1 + \mathrm{erf}\left(\frac{t_{R_2} - t_{R_1}}{2\sqrt{2}\sigma_t}\right)}.$$

But the resolution in this case is $R_s = (t_{R_2} - t_{R_1})/4\sigma_t$. Thus

$$\eta_1 = \eta_2 = \frac{1 - \mathrm{erf}(\sqrt{2}R_s)}{1 + \mathrm{erf}(\sqrt{2}R_s)}$$

is the required relation. This η has been indicated by Glueckauf (1955a) as $\eta_{1:1}$.

The values of erf $(\sqrt{2}R_s)$ for $R_s = 0.2$, 0.6, 1.0 and 1.4 are, respectively, approximately 0.310, 0.7658, 0.9545 and 0.9949. The corresponding values of $\eta_{1:1}$ are 0.526, 0.133, 0.0233 and 0.00256. Thus a resolution value of 1.4 implies excellent separation with very low impurities in each band. For a graphical relation between $\eta_{1:1}$ and R_s over a wider range of R_s, see Said (1978).

The preceding treatment considered separation in a time-varying output from a separator of the chromatographic type. Similar methods of describing separation, namely R_s, ξ, η_1, η_2, etc., may also be adopted for describing the separation between two concentration profiles in Figure 2.5.1 if the abscissa is a distance coordinate z instead of a time coordinate t. One then replaces t_{R_i} by z_i, σ_{ti} by σ_{zi}, $C_i(t)$ by $C_i(z)$, t_c by z_c, etc. The mathematical formulae will be identical with the proper substitutions made, and we may conclude that no special treatment is needed to describe the separation between two concentration profiles spatially displaced but having some overlap.

Problems

2.2.1 The *absorption factor*, A, in the absorption of species i in a flowing gas stream by a flowing solvent stream is defined by

$$A = \left[\frac{\text{molar liquid flow rate}}{\text{molar gas flow rate}}\right] \times \left[\frac{\text{liquid mole fraction}}{\text{gas mole fraction}}\right]_{i,\text{equilibrium}}.$$

The stripping factor, S, which describes the stripping of a volatile species i into a flowing gas stream from a flowing liquid stream is the inverse of A. The extraction factor for any species is defined in solvent extraction by

$$E = \left[\frac{\text{mole fraction in extract}}{\text{mole fraction in raffinate}}\right]_{i,\text{equilibrium}} \times \left[\frac{\text{extract molar flow rate}}{\text{raffinate molar flow rate}}\right].$$

Relate A, S and E to $\dot{k}'_{i1}$.(Ans. $S = (1/\dot{k}'_{i1}); E = \dot{k}'_{i1}$.)

2.2.2 Obtain the following relation between the separation factor α_{12}^{ht}, the heads separation factor α_{12}^{hf}, the cut θ and x_{11} for a single-entry separator and a binary feed:

$$(\alpha_{12}^{hf} - 1) = \frac{(\alpha_{12}^{ht} - 1)(1 - \theta)}{1 + \theta(\alpha_{12}^{ht} - 1)(1 - x_{11})}.$$

2.2.3 Consider the separation of a binary mixture in two single-entry separators connected together as shown in Figure 2.P.1. The tails stream from separator 2 is recycled back to the feed stream to separator 1. The mole fraction of the ith species in the jth stream of separator n is indicated by $x_{ij}(n)$ (for n = 1,2). If these two separators are operated such that[6]

$$x_{i2}(2) = x_{if}(1)$$

and if the values of both α_{12}^{hf} and α_{12}^{ft} are the same for both separators, show that

$$\alpha_{12}^{hf} = \alpha_{12}^{ft} = (\alpha_{12}^{ht})^{1/2}.$$

(Hint: Use mole ratios to convert separation factor definitions to $X_{11}(n) = \alpha_{12}^{hf} X_{1f}(n)$.)

2.2.4 Consider *the elimination of a surface active impurity* from water by the foam fractionation process as shown in Figure 2.P.2. An inert gas is bubbled through the impure water generating a foam rich in the surface

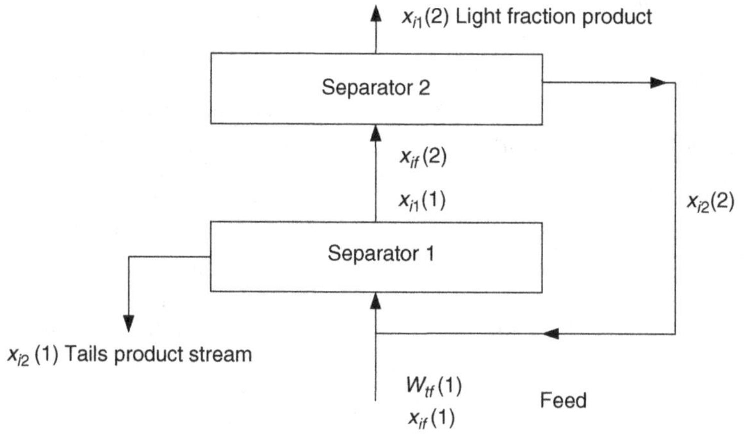

Figure 2.P.1. Two single-entry connected separators with tails recycle.

[6]When two streams to be mixed together are such that they have the same composition, we encounter the so-called "no-mixing" condition.

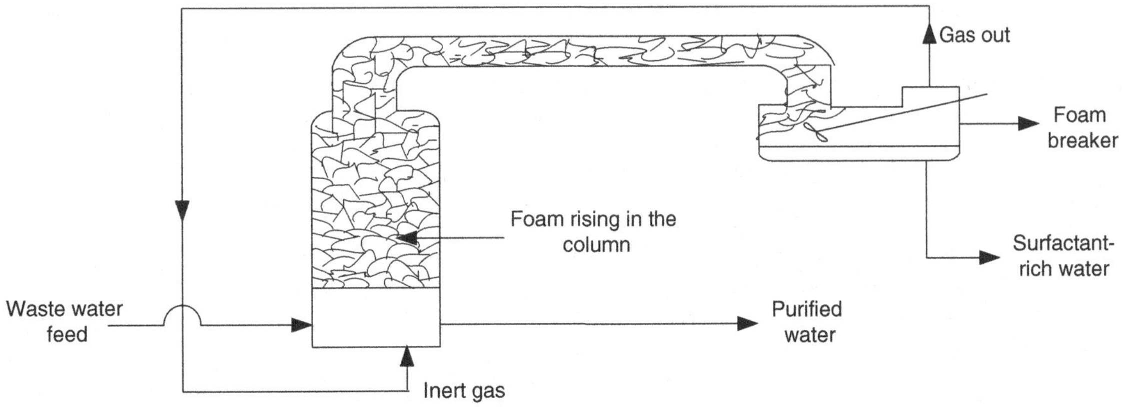

Figure 2.P.2. Water purification by foam fractionation removal of detergent impurities in a column.

active impurity. The foam rises up the column and is collapsed to obtain an impurity-rich water phase and the inert gas, which is recirculated to the column bottom for renewed foam making. The surfactant concentration in the feed is in the range 10^{-6} gmol/cm^3. The decontaminated water has a surfactant concentration 0.01 times that of the feed. The molar water flow rate in the impurity-rich collapsed foam is 3% of the feed molar flow rate, W_{tf}.

(a) Analyze this double-entry separator as a single-entry separator by drawing a suitable envelope in the figure and showing any other necessary arrangements.
(b) What is the value of the extent of separation?
(c) Show that the value obtained in (b) is very close to that obtained if the decontamination factor is 1000.
(d) Although the extent of purification of water in (c) is much larger than that in (b), ξ does not reflect it. Define impurity ratios for the flow system as $\dot{\eta}_1 = (W_{21}/W_{11})$ and $\dot{\eta}_2 = (W_{12}/W_{22})$. Is the new impurity ratio $\dot{\eta}_j$ or $I_j = -\log_{10}\dot{\eta}_j$ or $I = \sum_j I_j$ more sensitive for such a case? (*Note*: I_j or I is the purity index for the flow system.)

2.2.5 Thorman *et al.* (1975) have used two permeators of the type described in Example 2.2.4 in series to obtain O_2-enriched air containing 0.310 mole fraction O_2 from feed air (0.209 mole fraction O_2). The feed air flow rate to the first permeator is 0.665 std. cm^3/s. The O_2-enriched permeated stream containing 0.250 mole fraction O_2 and having a flow rate of 0.365 std. cm^3/s is introduced as feed after compression into the second permeator, which has a permeate composition of 0.310 mole fraction O_2. The permeate flow rate from the second permeator is 0.151 std. cm^3/s. The second permeator is somewhat smaller (silicone capillary length = 217 cm) than the first permeator (silicone capillary length = 691 cm). Obtain an estimate of the overall separation using α_{12}^{ht}, θ and ξ.

2.2.6 To recover as pure a C_8 hydrocarbon liquid as possible from a feed containing small amounts of a C_5 hydrocarbon, it is proposed to use a *flash separator* with a fraction of the bottoms liquids recycled to the feed. (See Figure 2.2.1(c) for a schematic; replace decane by C_8 hydrocarbon and butane by C_5 hydrocarbon.) A fraction η' of the bottoms liquid fraction is compressed and heated before it is mixed with fresh feed. It is proposed to compare the purification ability of the recycle system with the no-recycle system. Assume that the equilibrium ratio K_{i1} for the flash separator for the ith species is independent of recycle and depends only on the pressure and temperature of the flash, which do not change much with recycle. Assume further that the cut θ for the flash separator is independent of recycle (not strictly correct). Do not introduce any simplification based on the C_5 concentration in the feed.

(a) Develop a relation between x_{22}^r and x_{2f} containing only η', θ and K_{21}. Ans.

$$x_{22}^r = \frac{x_{2f}[1-\eta'(1-\theta)]}{[\theta K_{21} + (1-\eta')(1-\theta)]}.$$

(b) Show that x_{22}^r is always greater than x_{22} (corresponding to no recycle).

(c) Show further that the fraction of the fresh feed recovered as the liquid fraction is lower due to the recycle. *Note*: C_5 hydrocarbon goes much more into the vapor phase than C_8 hydrocarbon, which has a very low volatility.

2.2.7 Consider a single-entry separator with a dilute feed solution of heavy species 2. If a fraction η' of the tails stream is recycled to the feed stream and the separator operates with the same values of θ, α_{12}^{hf} and α_{12}^{ft} as does the non-recycle separator, show that the following are true.

(a) The apparent cut for the recycle separator based on the light fraction product flow rate from the system and the fresh feed flow rate is greater than θ, based on the light fraction actually leaving the separator and the feed rate actually entering the separator.

(b) The apparent value of $\alpha_{12}^{ft}(\cong x_{22}^{r}/x_{2f})$ is greater than the actual value of α_{12}^{ft} for the separator feed and tails stream.

(c) The condition under which the extent of separation based on the net product flow rates from the recycle system and the fresh feed is greater than that for a no-recycle system is (Sirkar and Teslik, 1982; see footnote 3, p. 49)

$$\eta' < \left[\frac{\alpha_{12}^{ft}+1}{(1-\theta)\alpha_{12}^{ft}} - \frac{1}{(1-\theta)^2\alpha_{12}^{ft}}\right].$$

Comment in general on the utility of such systems.

2.2.8 Consider a single-entry separator for the separation of a binary mixture of species 1 and 2 (Figure 2.2.1(a)). We wish to compare its separation performance with that of a single-entry separator having a fraction η of its light fraction recycled to the feed stream. The operational conditions are as follows:

(a) the heads and tails separation factors α_{12}^{hf} and α_{12}^{ft} have the same value for both the recycle and the non-recycle separator;

(b) the recycle separator operates such that its apparent cut (definition (2.2.27)) is equal to the cut of the non-recycle separator.

Show that under such conditions for any feed mixture $\xi^r > \xi$ holds and that the recycle separator is therefore a better separation device (Sirkar and Teslik, 1982; see footnote 3, p. 49).

2.2.9 The *separative power* δU of a single-entry separator for a binary feed stream is defined by Dirac in terms of a value function $Va(x_{ij})$ of composition x_{ij} as

$$\delta U = W_{t1}Va(x_{i1}) + W_{t2}Va(x_{i2}) - W_{tf}Va(x_{if}),$$

where the value function $Va(x_{ij})$ indicates the value of one mole of a given fluid stream of composition x_{ij}, and it changes as the composition of the fluid stream changes. Show that the above relation may also be expressed as follows (Lee *et al.*, 1977a):

$$\frac{\delta U}{W_{tf}} = \left(\frac{\alpha_{12}^{ft}-1}{\alpha_{12}^{ht}-1}\right)\left(\frac{1+\alpha_{12}^{hf}X_{1f}}{1+X_{1f}}\right)Va\left(\frac{\alpha_{12}^{hf}X_{1f}}{1+\alpha_{12}^{hf}X_{1f}}\right) + \left(\frac{\alpha_{12}^{ft}-\alpha_{12}^{ht}}{1-\alpha_{12}^{ht}}\right)\left(\frac{\alpha_{12}^{ft}+X_{1f}}{\alpha_{12}^{ft}+X_{1f}\alpha_{12}^{ft}}\right)Va\left(\frac{X_{1f}}{\alpha_{12}^{ft}+X_{1f}}\right) - Va(x_{1f}),$$

where the feed light component mole ratio X_{1f} is given by $X_{1f} = (x_{1f}/(1-x_{1f}))$, and α_{12}^{hf}, α_{12}^{ft} and α_{12}^{ht} are defined by (2.2.3)-(2.2.5).

2.2.10 Consider a single-entry separator separating two species g and h in the feed stream. The molar flow rates of species g and h in the feed stream are, respectively, G_0 and H_0. The corresponding molar flow rates of species g and h in stream 1 and stream 2 are, respectively, G_1, H_1 and G_2, H_2. Stream 1 is the light fraction. The following quantities are defined as follows:

a = mole percent of g in the feed;
b = mole percent of g in stream 1;
j = amount of stream 1 as a percentage of the feed.

(1) Define the extent of separation ξ in terms of the quantities defined here (G_0, H_0, G_1, H_1, G_2, H_2, a, b, j).

(2) A new separation efficiency η has been defined as

$$\eta = 100j\frac{b-a}{a(100-a)}.$$

Show that $\eta = 100\xi$.

2.4.1 Particle size distribution may be determined in practice for liquid suspensions by sieving, i.e. using screens/sieves having different openings and determining the weight of the particles retained by the sieve of a particular size. Part of the following information is adopted from Randolph and Larson (1988):
$w^s = M_T = 0.15$ g/cm^3 of particle free liquid volume; $\rho_s = 2$ g/cm^3;
particle volume shape factor = 0.6.
The sieving process yielded the following information for two sizes of particles: the fraction of particles $(\Delta w^s/w^s)$ around the two sizes, 100–120 μm (av. 100 μm), 200–220 μm (av. 210 μm), are 0.15 and 0.04, respectively. Calculate the values of the population density $n(r_p)$ for these two sizes in units of numbers/cm$^3\cdot$μm. (Hint: Employ (2.4.2e) over the size range.) (Ans. 704 numbers /cm$^3\cdot$μm; 27 numbers/cm$^3\cdot$μm.)

2.4.2 (a) Particle removal effectiveness of dead-end filters (Chapter 6.3-3.1) is determined by *fractional penetration*

$$P = \frac{\phi_{out}}{\phi_{in}},$$

where ϕ_{out} and ϕ_{in} are the particle volume fractions at the outlet and the inlet of the filter, respectively. An additional index is the log reduction value (LRV) of the filter:

$$\text{LRV} = \log_{10}(1/P).$$

Identify the indices in Chapters 1 and 2 which are closest to these two indices.
(b) Obtain the following relation for the total efficiency E_T in a solid–liquid separator operating as a classifier as a function of the grade efficiency G_r of particles of size r_p:

$$E_T = \int_0^1 G_r(r_p)\mathrm{d}F_f(r_p),$$

where $F_f(r_p)$ is the particle size distribution function of the feed ($0 \leq F_f(r_p) \leq 1$). Develop a relation to predict $f_1(r_p)$ from a knowledge of G_r, E_T and $f_f(r_p)$. (Ans. $f_1(r_p) = f_f(r_p) \times ((1-G_r)/(1-E_T)$.)

2.4.3 For classification of particles above size r_s and below size r_s, respectively, into the underflow and the overflow of a solid–liquid classifier, define component 1 to be the undersize particles and component 2 to be the oversize particles. Show that the extent of separation based on mass fractions in such a case is given by

$$\xi = \frac{w_{t1}^s}{w_{tf}^s}\frac{w_{t2}^s}{w_{tf}^s}\left|\frac{u_{22}-u_{21}}{u_{1f}u_{2f}}\right|.$$

Compare its numerical value with that of another efficiency E, defined in the literature (McCabe and Smith, 1976, p. 920) as

$$E = \frac{w_{t1}^s}{w_{tf}^s}\frac{w_{t2}^s}{w_{tf}^s}\left[\frac{u_{21}u_{12}}{u_{1f}u_{2f}}\right]$$

for the case where $u_{2f} = 0.540$, $u_{21} = 0.895$ and $u_{22} = 0.275$, obtained for a r_s value equal to 0.75 mm.
Assume that mass fractions of particles are equivalent to mole fractions. (Ans. $\xi_{12} = 0.61$; $E = 0.63$.)

2.4.4 Consider two cyclones in a series for dust cleaning of air, such that the dust collected in the underflow of cyclone 1 is withdrawn, along with a small amount of air, and introduced as feed to the smaller cyclone 2, whose overflow gas stream containing some dust is recycled to the fresh feed air stream entering cyclone 1. Assume that the grade efficiency functions for cyclones 1 and 2 are given, respectively, by (Van der Kolk, 1961)

$$G_{r1} = 1-\mathrm{e}^{-b_1 r} \quad \text{and} \quad G_{r2} = 1-\mathrm{e}^{-b_2 r}.$$

(a) Show that if the cleaned air from cyclone 1 and the dust from cyclone 2 are the net products, and if η_r kg of dust of certain size r is recycled from cyclone 2 for 1 kg of dust of the same size in the fresh air feed to cyclone 1, then

$$\eta_r = \frac{G_{r_1} - G_{r_1} G_{r_2}}{1 - G_{r_1} + G_{r_1} G_{r_2}}.$$

(b) Show that the total efficiency $E_{T_{1,2}}$. of this arrangement of two cyclones is given for a feed size distribution function $F_{f_1}(r) = 1 - e^{-ar}$ by

$$E_{T_{1,2}} = \int_0^\infty \frac{a e^{-ar}\{1 - e^{-b_1 r} - e^{-b_2 r} + e^{-(b_1+b_2)r}\}}{\{1 - e^{-b_2 r} + e^{-(b_1+b_2)r}\}} dr.$$

2.4.5 A disk centrifuge has the following grade efficiency function (Svarovsky, 1977, chap. 7)

$$G_r = \frac{r_p^2}{r_{max}^2} \quad \text{for } r_p \leq r_{max}; \qquad G_r = 1 \text{ for } r_p > r_{max.}$$

It is known that the feed particle size density function is Gaussian with an average value $\bar{r}_p$ and a standard deviation of σ_r. (For the limits of integration, a value of infinity may be substituted for r_{max}. However, in the integrand, retain r_{max}.) Obtain an expression for a total efficiency E_T as a function of $\bar{r}_p$, r_{max} and σ_r.

2.4.6 The feed solids in a slurry to a hydrocyclone obey the following log-normal law:

$$\frac{dF(r_p)}{d(\ell n r_p)} = \frac{1}{\ell n(\sigma_g)\sqrt{2\pi}} \exp\left[-\frac{(\ell n r_p - \ell n r_g)^2}{2(\ell n \sigma_g)^2}\right],$$

where $F(r_p)$ is the particle size distribution function, with σ_g being the standard deviation and r_g being the mean particle size. The feed volumetric flow rate is 6.35×10^{-4} m^3/s. The feed solids concentration is 30 kg/m^3. The underflow volumetric flow rate is 3.5×10^{-5} m^3/s and the underflow solids concentration is 314.2 kg/m^3. Determine

(1) the total efficiency E_T;
(2) the reduced efficiency of Kelsall;
(3) the particle size distribution function $F(r_p)$.

2.4.7 Consider the separation of dust particles from air by means of two cyclones connected in the following fashion. Feed air containing dust enters cyclone 1; the feed particle size distribution is $F_{f_1}(r_p) = 1 - e^{-ar_p}$. The underflow from cyclone 1 is introduced with a small amount of air as feed to a small cyclone 2. The underflow from cyclone 2 is collected as dust from the system. The overflow air from cyclone 2 is mixed with the overflow air from cyclone 1 to obtain the cleaned air. The grade efficiency of the ith cyclone is given by

$$G_{r_i}(r_p) = 1 - e^{-b_i r_p}.$$

(1) Show by actual calculation the total efficiency of the first cyclone, E_{T_1}.
(2) Based on 1 kg of dust entering cyclone 1, determine step-by-step the total efficiency $E_{T_{1,2}}$ of the two-cyclone system.

2.4.8 Consider the separation of a ternary gas mixture of species 1, 2 and 3 through a membrane separator with two different types of membranes M′ and M″ such that species 1 appears preferentially in product stream $j = 1$ from membrane M′, product stream $j = 2$ from membrane M″ is enriched in species 2, while the tails stream $j = 3$ is enriched in species 3. If the gas composition everywhere inside the separator is indicated by x_{i3}, the mole fraction of species i in tails stream 3, define the following separation factors (Sirkar, 1980):

(a) α'_{1n} and α''_{1n} for species 1 in the product streams 1 and 2, respectively, with respect to all other species $n = 2, 3$ and the reject stream $j = 3$;
(b) α_{1n} for species 1 and all other species ($n = 2, 3$) with respect to product streams $j = 1$ and $j = 2$. Show that $\alpha_{1n} = (\alpha'_{1n}/\alpha''_{1n})$.

2.5.1 It has been pointed out in Section 2.5 that separation in the output stream from a chromatographic separator depends on the location of t_c, the cut point. Assume $\sigma_{t1} = \sigma_{t2} = \sigma_t$ and Gaussian output streams for two species 1 and 2. Use the extent of separation to obtain the value of the optimum location of the cut point and the corresponding optimum value of the extent of separation. (See Problem 1.5.1 prior to solving this problem.) With this optimum value of ξ, show that (Rony, 1968b)

$$\xi_{\text{opt}} = \text{abs}\left|\text{erf}\left[\sqrt{2R_s}\right]\right|,$$

where R_s is the resolution for the given problem.

2.5.2 Obtain an expression for the separation factor α_{12} between two solute species coming out of a chromatographic separator.

3

Physicochemical basis for separation

The preceding chapters introduced first the notion of separation and then a variety of indices to describe separation. These indices were used to characterize quantitatively the amount of separation achieved in a closed or an open separation vessel. The quantitative description included systems at steady or unsteady state involving chemical or particulate systems. Systems studied were either binary or multicomponent or a continuous mixture. Not considered in these two chapters was the fundamental physicochemical basis for these separations; appropriately, this is the focus of our attention in this chapter.

In Section 3.1, we distinguish between bulk and relative displacements and describe the external and internal forces that cause separation-inducing displacements. This section then identifies species migration velocities and the resulting fluxes as a function of various potential gradients. Section 3.2 is devoted to a quantitative analysis of separation phenomena and multicomponent separation ability in a closed vessel as influenced by two basic types of forces. The criteria for equilibrium separation in a closed separator vessel and individual species equilibrium between immiscible phases are covered in Section 3.3. Section 3.4 treats flux expressions containing mass-transfer coefficients in multiphase systems. Flux expressions for transport through membranes are also introduced here.

3.1 Displacements, driving forces, velocities and fluxes

When a particular component in a mixture is displaced in a given direction, it moves with a certain velocity. This velocity leads to a flux of the species, which is the molar rate of species movement per unit area in any given frame of reference. The nature of the displacements and the forces that cause the displacements leading to species velocity and flux are considered first in this section. Expressions for species velocities and fluxes are then studied to provide the foundations for a quantitative analysis of separation later.

3.1.1 Nature of displacements

The separation of a mixture involves the setting apart of the mixture components present in a given region. When perfect separation is attained, each component occupies a separate region where no other component is present. The total volume of all such regions may or may not be equal to the volume of the region originally occupied by the mixture. To achieve this separation, each component must move selectively toward its own designated region. Therefore, molecules of each component undergo *displacement* toward their own region during a separation process.

The initial mixture, as well as the final separated state, may consist of either a single phase or a collection of immiscible phases. If separation is desired in a feed consisting of two immiscible phases, then each component has to be selectively displaced toward its designated phase. Such component-specific displacement and the separation achieved thereby may or may not lead to pure phases. It is, however, a prerequisite to any separation.

The direction and the rate of displacements of molecules of a component (or particles of a certain type) will in general depend on the nature of the chemical species (or particle), the type, magnitude and direction of forces acting on the chemical species (or particle) and the surrounding medium. Such species movements take place *spontaneously* at the microscopic level in response to conditions imposed on the separation system (Sweed, 1971, p. 175).

In most separation processes, *bulk displacements* also take place simultaneously. Generally, bulk displacements move molecules of all components at the same speed and in the same direction. On the other hand, the spontaneous microscopic species-specific movement in response to forces acting on a particular species (mentioned in the preceding paragraph) is identified as *relative displacement* (Giddings, 1978). These two types of displacements may occur simultaneously or consecutively (Sweed, 1971).

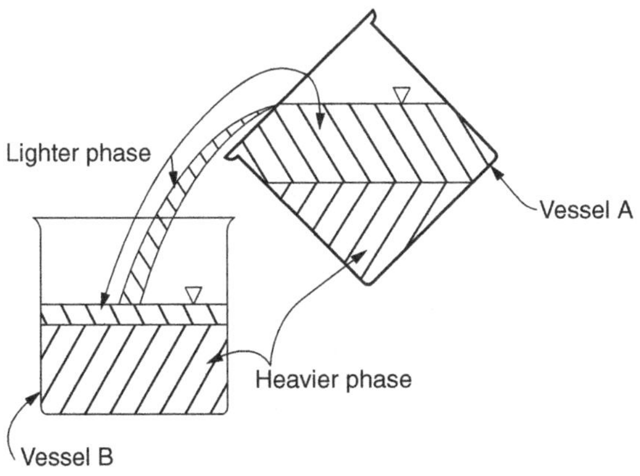

Figure 3.1.1. Decantation of upper phase from vessel A to vessel B.

Bulk displacement may be caused by *fluid flow* or *direct mechanical conveying* of a complete phase or fraction from one location or vessel to another. Consider, for example, the mechanical decantation of the upper phase from vessel A to vessel B. In vessel B, additional separation takes place because of the redistribution of components of the upper phase in vessel A between itself and the lower phase in vessel B (Figure 3.1.1). But all the components in the upper phase of vessel A are displaced to the new vessel B at the same rate and in a nonselective fashion. Such a procedure is followed in laboratory decantations as well as in industrial decanters.

Bulk displacement by fluid flow is much more common. When a multicomponent mixture moves in a vessel or in a region of a vessel at a certain bulk velocity, the fluid motion carries all the components, in general, at the same velocity. In the case of a flat velocity profile with convective transport being dominant over diffusive transport, the fluid flows like a "plug" (Froment and Bischoff, 1979), and every species has the same displacement and velocity. If no forces are present to impart different displacement rates to different components, all components will be non-selectively carried by bulk fluid motion and there would not be any separation.[1]

Relative displacement takes place when different forces acting on molecules of different species (or on different particles) cause molecules of one species to move relative to those of the other species (similarly for particles). Such motions lead, in general, to separation, whether the mixture as a whole has a bulk velocity or not. Examples of such species-specific forces are: chemical potential gradient, electrostatic potential gradient, centrifugal force field, gravitational force field, magnetic force field, thermal gradient, etc. Of these, only the chemical potential gradient is referred to as an *internal force*; the others are caused by devices or phenomena external to the separation system.

We have described above the notions of bulk displacement and relative displacement primarily for molecules. They are equally applicable to the separation of macromolecules, colloids or macroscopic particles from the continuous phase (or to fractionation of macromolecules, particles, etc.); however, the detailed mechanisms and methods of description may be different. For example, as we will see later in deep bed filtration and aerosol filtration, particles of different sizes and therefore different masses may have different inertial forces in an accelerating or decelerating flow field. With larger particles, certain types of bulk motion may, therefore, lead to a relative displacement. Even for gas molecules, the *persistence of velocity* phenomenon creates different displacements for molecules of differing masses moving at a given bulk velocity when an impingement on a target gas occurs (Anderson, 1980).

3.1.2 Forces on particles and molecules

A variety of external forces are of use in separation. Of course, the internal force of chemical potential gradient has great importance in chemical separations. We will begin, however, by treating external forces. Of these, the gravitational force due to the earth is perhaps the most familiar one and is therefore appropriate for consideration now.

3.1.2.1 *Gravitational force*

In the external *gravitational force field*, the work required to raise a particle of mass m_p from a location of height z_1 (where the gravitational potential is $\Phi_1 = gz_1$; Φ is a scalar quantity) to one of height z_2 (Figure 3.1.2A) in free space is $m_p(gz_2 - gz_1)$ for $z_2 > z_1$.[2] Here the positive z-coordinate is vertically upward. For a differential displacement dz, the vertically downward gravitational force $\boldsymbol{F}$ is related to $\mathrm{d}\underline{W}$, the differential amount of work done on the particle and $\mathrm{d}\Phi$, the change in gravitational potential, by a force $-\boldsymbol{F}$ in the positive z-direction acting over the distance dz:

$$\mathrm{d}\underline{W} = -\boldsymbol{F} \cdot \mathrm{d}\boldsymbol{z} = m_p\,\mathrm{d}\Phi = F\,\mathrm{d}z = m_p g\,\mathrm{d}z, \qquad (3.1.1)$$

where the magnitude of the force per unit mass of the particle, $\hat{F}$, is

[1] There can be separation processes in which bulk displacement could vary from location to location due to gradients in bulk velocity. Additional complexities and sometimes separation can be achieved by a time dependence in the bulk displacements. The effects of spatial and temporal dependence of bulk displacement will be treated in later chapters.

[2] g is acceleration due to gravity, 980 cm s^{-2}. The unit for force is the newton or kg m/s^2. The unit for work is the joule or newton-meter (N-m).

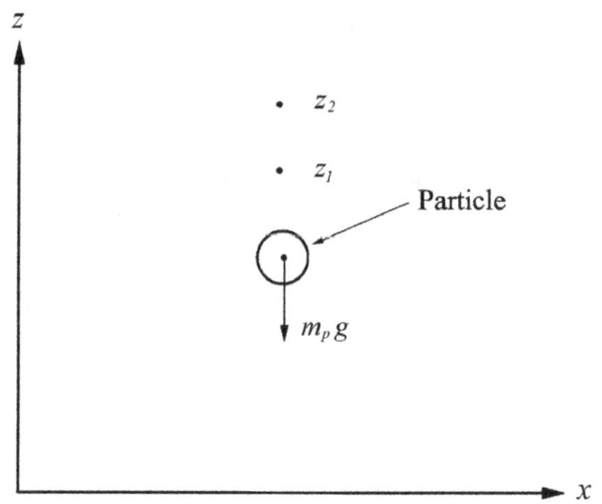

Figure 3.1.2A. Gravitational force of magnitude m_pg *acting on a particle of mass* m_p *vertically downward.*

$$\hat{F} = \frac{F}{m_p} = \frac{d\Phi}{dz} = g. \tag{3.1.2}$$

In vector notation, the vertically downward gravitational force per unit mass may be described as

$$\hat{\boldsymbol{F}} = \frac{\boldsymbol{F}}{m_p} = -\nabla\Phi. \tag{3.1.3a}$$

The relevant z-component in vector form is

$$\hat{F}_z \mathbf{k} = -(d\Phi/dz)\mathbf{k}. \tag{3.1.3b}$$

The vector gravitational driving force per unit mass of the particle due to the external force field is simply the negative of the gradient of the scalar potential of that force field. Here Φ increases with positive z; nature, i.e. gravity, spontaneously drives the particle to a lower Φ.

The movement of the particle of mass m_p under the action of the gravitational force considered above, however, assumed free space. If this particle were instead immersed in a fluid of density ρ, a buoyancy force would act on the particle in the vertically upward direction. The net external driving force on the particle acting vertically downward would be

$$F_{\text{net}}^{\text{ext}} = m_p g - \left(\frac{m_p}{\rho_p}\right)\rho g = m_p g\left(1 - \frac{\rho}{\rho_p}\right), \tag{3.1.4}$$

where ρ_p is the mass density of the particle. This force acts whether the particle moves or not. If, instead of a single particle, we have one mole of molecules of the ith species, the *total gravitational force on one mole* is

$$F_{\text{net}}^{\text{ext}} = M_i g\left(1 - \frac{\rho}{\rho_i}\right); \qquad \boldsymbol{F}_{\text{net}}^{\text{ext}} = -M_i g\left(1 - \frac{\rho}{\rho_i}\right)\mathbf{k}, \tag{3.1.5}$$

where M_i is the molecular weight and ρ_i is the density of the ith species. (To develop this expression, follow the procedure illustrated later for a centrifuge in (3.1.51).)

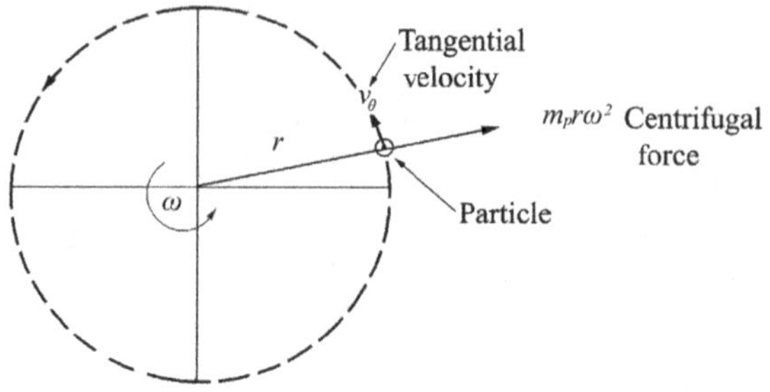

Figure 3.1.2B. Centrifugal force of magnitude $m_p r\omega^2$ $(= m_p v_\theta^2/r)$ *acting radially outward on a particle of mass* m_p *rotating in a circle of radius* r *with a tangential velocity* $v_\theta = r\omega$.

For a particle in the gravitational field, the net external force, $F_{\text{net}}^{\text{ext}}$, will remain constant if m_p, ρ_p and ρ are constant in the z-direction. In a separation system, one can create a condition such that the surrounding fluid composition, and therefore the fluid density ρ, change with the height z. The net external force on the particle will now depend on the vertical location of the particle. A vertical gradient in fluid density may then be considered as an additional source of an external driving force on a particle along with the gradient of gravitational potential. Clearly, such an additional external force requires the existence of the gravity force. Note that $F_{\text{net}}^{\text{ext}}$ will be zero if $\rho = \rho_p$ for a particle or $\rho = \rho_i$ for molecules of species i. Further, the net external force vector per unit particle mass is $-g(1 - \rho/\rho_p)\mathbf{k}$.

3.1.2.2 *Centrifugal force*

If a particle rotates at an angular velocity of ω radian/second (rad/s) in a circle of radius r (Figure 3.1.2B), the *centrifugal force*[3] on the particle per unit mass in the outward radial direction is

$$\hat{\boldsymbol{F}} = (\boldsymbol{F}/m_p) = r\omega^2\mathbf{r} = -(d\Phi/dr)\mathbf{r}. \tag{3.1.6a}$$

Here Φ is the scalar centrifugal force field potential and $\mathbf{r}$ is the unit vector along the outward radial direction. If a fluid particle of mass density ρ is rotated as a rigid body in a vessel such that ρ is constant, the centrifugal driving force on a unit mass of the fluid particle is also given by the above relation. The vector centrifugal force acting on one gram mole of molecules of species i in the radially outward direction is

$$\boldsymbol{F} = M_i r\omega^2\mathbf{r}. \tag{3.1.6b}$$

From relation (3.1.6a), we obtain

---

[3]The tangential velocity v_θ of the particle is $r\omega$. The magnitude of the centrifugal force is $m_p v_\theta^2/r = m_p r\omega^2$. The Coriolis force is also present under these conditions. However, it may be neglected in comparison to the $r\omega^2$ term.

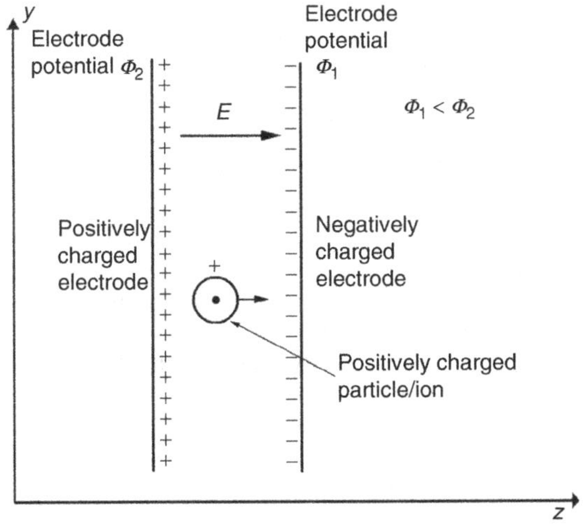

Figure 3.1.2C. Ion/charged particle placed in an electric force field of constant strength $E = -(d\phi/dz)$.

$$r\omega^2 = -(d\Phi/dr). \tag{3.1.6c}$$

Integrating, we can write

$$\Phi = \Phi_0 - (r^2\omega^2/2), \tag{3.1.6d}$$

where Φ_0 is an arbitrary value of the centrifugal potential at $r = 0$. As r increases, Φ decreases.

3.1.2.3 *Electrical force*

Consider next an *electrical force field* of constant strength E (volt/m) whose electrostatic scalar potential Φ (volt) is such that the potential difference $d\Phi$ over a distance dz in the direction of the field is $-E$ dz (Figure 3.1.2C). If molecular ions of the *i*th species, each having a valence of Z_i (the magnitude varies from 1 to 10), are exposed to this field in an aqueous solution in between the electrodes, the force exerted on 1 gmol of such charged molecules in the z-direction is

$$F_{iz} = Z_i\mathcal{F}E = -Z_i\mathcal{F}\frac{d\Phi}{dz} \tag{3.1.7}$$

or

$$\boldsymbol{F}_i = -Z_i\mathcal{F}\nabla\Phi \quad \text{and} \quad \boldsymbol{E} = -\nabla\Phi. \tag{3.1.8}$$

Here $\mathcal{F}$ is Faraday's constant having the value of 96 485 coulomb per g-equivalent (C/g-equiv.); the electrical field has the unit of newton/coulomb (N/C). The above expression for force assumes that the electrical field has not been disturbed by the ions and vice versa. The work done to move these charged molecular ions having a total charge of $Z_i\mathcal{F}$ coulomb along the z-coordinate from z_1 to z_2 is given by

$$\underline{W} = Z_i\,\mathcal{F}E(z_2 - z_1) \tag{3.1.9}$$

per gram mole or gram ion obtained from 1 gmol of the compound (the unit of work is the joule (=volt-coulomb)) with Φ decreasing in the positive z-direction. The electrical charge on a single ion is $(Z_i\mathcal{F}/\tilde{N})$, where $\tilde{N}$ is Avogadro's number. (See Section 3.1.6.2 for ions in a gaseous mixture.)

A binary electrolyte $A^{Z_+}_{\nu_+}Y^{Z_-}_{\nu_-}$ when dissociated fully in a solvent will produce ν_+ ions of cation A^{Z_+} (whose electrochemical valence is Z_+) and ν_- ions of anion Y^{Z_-} (whose electrochemical valence is Z_-). For example, for Na_2SO_4 in water, $\nu_+ = 2, Z_+ = 1, \nu_- = 1, Z_- = -2$ since the ions are Na^+ and SO_4^{--}. Thus $Z_i - s$ are positive for cations and negative for anions. If there is an electrical field in the solution, each species, positive and negative, will experience a force due to it according to (3.1.7).

In any electrolytic solution, any molecular ion of interest having a mean diameter d_{ion} and an algebraic valence Z_{ion} will attract ions of opposite charge, namely counterions. However, the centers of such counterions can approach the center of the ion of interest to a distance of "a" only due to short-range repulsive forces. For radial distances r (> a), the distribution of the net potential Φ_{net} due to the ion of interest and the counterions is given as (see Newman, 1973)

$$\Phi_{net}(r) = \frac{Z_{ion}e}{4\pi\varepsilon r}\,\frac{\exp\left(-\frac{(r-a)}{\lambda}\right)}{1+\frac{a}{\lambda}}, \tag{3.1.10a}$$

where the potential due to the ion of interest only at location r is $(Z_{ion}e/4\pi\varepsilon r)$ and e is the electronic charge, 1.6021 $\times$ 10^{-19} coulomb. Thus the net potential $\Phi_{net}(r)$ decays very rapidly with r. The extent of rapidity in decay is determined by the parameter λ called the *Debye length*

$$\lambda = \left(\frac{\varepsilon RT}{\mathcal{F}^2 I}\right)^{1/2}, \tag{3.1.10b}$$

which is influenced by the ionic strength I defined by

$$I = \frac{1}{2}\sum_{i=1}^{n} Z_i^2\,C_{il}. \tag{3.1.10c}$$

The higher the value of I, the smaller the value of λ. Here $\mathcal{F}$ is Faraday's constant, R is the universal gas constant, T is the absolute temperature and ε is the electrical permittivity of the fluid. This quantity ε is the product of the relative dielectric constant of the medium ε_d (for water $\varepsilon_d = 78.54$ at 25 °C) and the electrical permittivity ε_0 of vacuum (= 8.8542 $\times$ 10^{-14} farad/cm or coulomb/volt-cm = 8.854 $\times$ 10^{-12} coulomb^2/newton-m^2, where recall that 1 newton-m = 1 volt-coulomb = 1 joule). In a uni-univalent electrolyte solution of 0.1 M strength (of, say, NaCl) the value of λ at 25 °C is 9.6 $\times$ 10^{-8} cm, i.e. 0.96 nm (Newman, 1973). The ions of opposite charge shield the charge of the ion of interest, and the effect of the ion of interest decays very rapidly with distance. So the description of the electrical force on an ion in an applied field $\boldsymbol{E}$ by definition (3.1.8) is generally satisfactory.

The ion of interest, however, has in an aqueous solution a solvation shell of water molecules. Similarly, the counterions have water molecules around them. When the ion of interest moves in an applied electrical field $\boldsymbol{E}$ (as in electrophoresis), the water molecules solvating the counterions move in an opposite direction with the counterions whose charge is equal and opposite to the charge of interest. This motion of the solvent molecules located in a shell at a distance λ exerts a retardation force, $\boldsymbol{F}_i^{\text{ret}}$, on the main force on the ion of interest, $-Z_i\mathcal{F}\nabla\Phi$. The net driving force, $\boldsymbol{F}_i^{\text{net}}$, on the ion of interest becomes (Wieme, 1975)

$$\boldsymbol{F}_i + \boldsymbol{F}_i^{\text{ret}} = -Z_i\mathcal{F}\nabla\Phi - (Z_i\mathcal{F}\nabla\Phi)\left\{-\frac{d_{\text{ion}}}{2\lambda}\right\};$$

$$\boldsymbol{F}_i^{\text{net}} = -Z_i\mathcal{F}\nabla\Phi\left(1 - \frac{d_{\text{ion}}}{2\lambda}\right). \qquad (3.1.10\text{d})$$

Often, this correction is ignored for small ionic species where the description of the species-specific force (3.1.8) is reasonably accurate.

If the charged species are larger, for example proteins, or if we are dealing with colloidal particles in a solution, the retardation forces are regularly taken into account. Generally, small ions of charge opposite to that of the charged protein or particle will collect in a diffuse layer next to the protein or particle. An *electrical double layer* is created by the fixed charges of the protein or particle and the counterions collected from the solution (Figure 3.1.2D). The total charge in this double layer is zero. However, there is an electrical potential Ψ_{el} in this layer which decreases to zero with distance from the particle or protein surface. Note that the counterions are mobile and can be influenced by the external electrical field $\boldsymbol{E}$.

If the macroion, protein or particle moves in a given direction due to the external field $\boldsymbol{E}$, resulting in *electrophoretic* motion, the large number of counterions move in an opposite direction. Since these ions carry some solvent

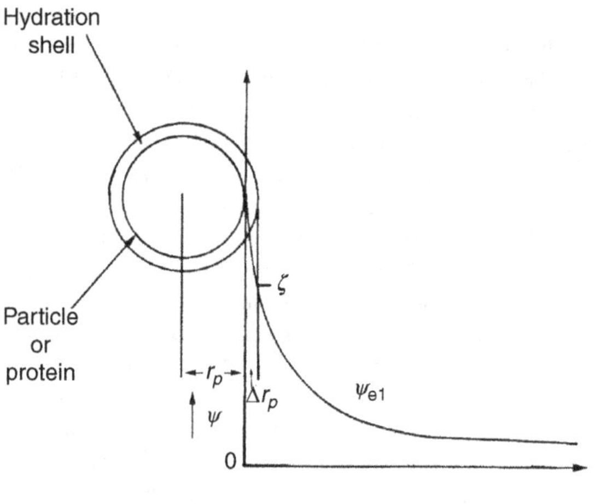

Figure 3.1.2D. Charged particle with its hydration shell and double-layer potential profile.

with them by the phenomenon of electro-osmosis, the macroion motion is retarded by this solvent motion. The electrophoretic retarding force is given by

$$\boldsymbol{F}_p^{\text{ret}} = (\varepsilon_d\zeta r_p - Q_p)\boldsymbol{E}. \qquad (3.1.11\text{a})$$

Here ε_d is the dielectric constant of the medium, r_p is the effective particle/macroion radius, Q_p is its charge and ζ is the *zeta* (or *electrokinetic*) *potential.* (This is the potential at the surface of shear around the particle; there are solvent molecules tightly bound to the particle of radius r_p up to the radius $r_p+\Delta r_p$ defining the hydration shell, and they move with the particle defining the shear surface.) Thus the net force on the particle is

$$\begin{aligned}\boldsymbol{F}_p + \boldsymbol{F}_p^{\text{ret}} &= Q_p\boldsymbol{E} + (\varepsilon_d\zeta r_p - Q_p)\boldsymbol{E}\\ &= \varepsilon_d\,\zeta\, r_p\boldsymbol{E}.\end{aligned} \qquad (3.1.11\text{b})$$

A more general analysis yields the following results (Wieme, 1975):

$$\boldsymbol{F}_p + \boldsymbol{F}_p^{\text{ret}} = \left[\varepsilon_d\zeta r_p\boldsymbol{E}\right]\frac{\text{f}(r_p/\lambda)}{(1+r_p/\lambda)}, \qquad (3.1.11\text{c})$$

where the function $\text{f}(r_p/\lambda)$ tends to 1 when $(r_p/\lambda) >> 1$ $(r_p/\lambda \to 1000)$. For proteins the range is $1 < r_p/\lambda < 300$. When $(r_p/\lambda) \to 0.1$, $\text{f}(r_p/\lambda)$ tends to 2/3.

We assume the electric field strength E to be constant here, independent of the z-coordinate. Therefore, the electrical force on 1 gmol of the charged ith species is also independent of the z-coordinate location (we can generalize this to all three coordinate directions). However, if the charge on the ith molecular ion species, Z_i, changes from one location to another, the electrical force will become location dependent. For example, it is known that the net charge on protein molecules in a solution depends on the solution *pH*. At the isoelectric *pH* (identified as *pI*), $Z_i = 0$, but $Z_i \neq 0$ for all other *pH* values (see Figure 4.2.5(c)). The net charge is positive at sufficiently low *pH* and negative at sufficiently high *pH*. Thus, the electrical force on protein molecules can vary even if the field strength E is constant.

It is possible to create a *pH* gradient in the solution of concern in the separation system by external means. At the location where *pH* has the isoelectric value for the ith protein species, Z_i will be zero, leading to a zero external electrical force on the ith species. At other locations, the force will be nonzero. The forces on molecules or ions in solution due to an externally imposed primary field (i.e. electrical field) can then be suitably altered by the imposition of additional property gradients in the solution by external means. If we have macroscopic particles instead of molecules or macromolecules, the driving force per particle may be obtained from definition (3.1.8) simply by replacing $Z_i\mathcal{F}$ by Q_p, the net particle charge in coulomb.

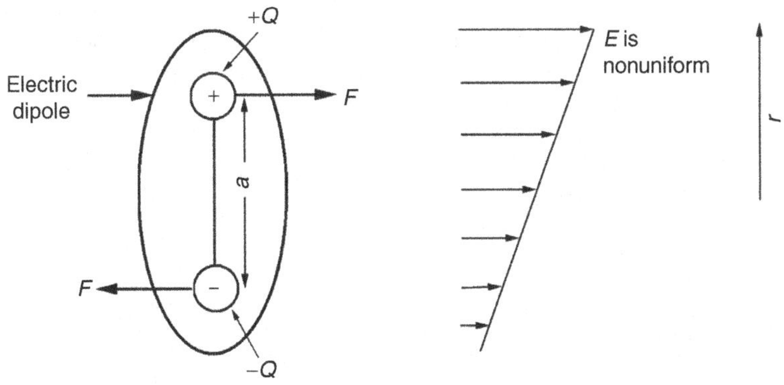

Figure 3.1.2E A dielectric uncharged particle placed in a nonuniform electric field develops an induced dipole of moment **p** = Q**a**, *where* Q *is the magnitude of the charge of each pole separted by a distance* **a** *where the vector coordinate of negative charge* -Q *is* **r**.

If the electrical force field **E** is *nonuniform* and a dielectric particle[4] is placed in such a field and develops a dipole moment $\boldsymbol{p}$, then the net electrical force on the particle with the induced dipole is (Figure 3.1.2E)

$$\boldsymbol{F} = (\boldsymbol{p} \cdot \nabla)\,\boldsymbol{E} = p\nabla E. \tag{3.1.12}$$

The motion of particles caused by polarization effects in the nonuniform electrical field is identified as *dielectrophoresis* (Pohl, 1978), whereas, as we have seen already, the motion of charged particles in a uniform electric field is termed *electrophoresis*. In the force expression (3.1.12), ∇E is the gradient of the nonuniform electric field at the particle center and p is the magnitude of the dipole moment given by the product of the charge and the distance between the charges (in coulomb-meter). The gradient of the electrical field is to be determined at the center of the particle (Halliday and Resnick, 1962).

If an uncharged particle is placed in a dielectric fluid having a dielectric constant ε_d and an electrical resistivity ρ_{dR}, the particle experiences a force only if the electrical field $\boldsymbol{E}$ is nonuniform in space. Assume the particle to be spherical with radius r_p. If the particle is not metallic and the fluid resistivity ρ_{dR} is very high, the force is given by (Jones, 1995)

$$\boldsymbol{F} = 2\pi r_p^3\, \varepsilon_d \left(\frac{\varepsilon_p - \varepsilon_d}{\varepsilon_p + 2\,\varepsilon_d}\right) \nabla \boldsymbol{E}^2, \tag{3.1.13}$$

where ε_p is the dielectric constant of the particle having a resistivity ρ_{pR}. The quantity within the brackets is called the Clausius-Mossoti function, and it indicates the strength of the effective polarization of the spherical particle (Jones, 1995). When the particle is metallic ($\rho_{pR} = 0$), with $\varepsilon_p \to \infty$,

$$\boldsymbol{F} = 2\pi r_p^3\, \varepsilon_d\, \nabla \boldsymbol{E}^2. \tag{3.1.14}$$

Note: $\nabla\, \boldsymbol{E}^2$ is $\nabla\, (\boldsymbol{E} \cdot \boldsymbol{E})$ so that $(\boldsymbol{E} \cdot \boldsymbol{E})$ is the local electrical field intensity.

Both of these forces are proportional to the particle volume as well as gradient of the square of the electrical field (Von Hippel, 1954; Lin and Benguigui, 1977). The above results can be obtained from (3.1.12) by introducing an expression for $\boldsymbol{p}$, the dipole moment, resulting from the polarization of the particle in the nonuniform electric field ($\boldsymbol{p}$ = (polarizability) × (particle volume) × ($\boldsymbol{E}$)).

Note that the dielectric force $\boldsymbol{F}$ is directed along the gradient of the electric field intensity $\nabla\, \boldsymbol{E}^2$. For the metallic particle, the force direction is always toward the direction of the largest field. On the other hand, the force on a nonmetallic particle will be toward the direction of the largest field only if $\varepsilon_p > \varepsilon_d$ (positive dielectrophoresis); it will be toward the lowest field if $\varepsilon_d > \varepsilon_p$ (negative dielectrophoresis). If $\varepsilon_p >> \varepsilon_d$ or $\varepsilon_d >> \varepsilon_p$, the magnitude of the force is not influenced, but the direction is. Obviously if there are two particles with $\varepsilon_{p_1} < \varepsilon_d < \varepsilon_{p_2}$, the forces experienced by the two particles will be in opposite directions (Lin and Benguigui, 1977).

The dielectrophoretic force expressions given above are proportional to the square of the field strength and are independent of the direction of the field. Therefore an alternating current (AC) field can be used here, unlike electrophoretic motions induced by a DC field (Pohl and Kaler, 1979). Such a time-varying nonuniform electrical field has been used to separate mixtures of whole cells.

In the cases considered above, there was an applied electrostatic potential gradient $\boldsymbol{E}$, uniform or nonuniform, acting on the particle with or without charge. If the particle had a charge, it was assumed (although we did not indicate it) that the field generated by this charge did not influence the external electrical force field generated by $\boldsymbol{E}$. On the

[4]The small dipole is characterized by two equal and opposite charges, +Q and −Q, located at a vector distance $\boldsymbol{a}$ apart (Jones, 1995). Since the electric field is nonuniform, the two charges are subjected to different values of the electric field, $\boldsymbol{E}(\boldsymbol{r} + \boldsymbol{a})$ and $\boldsymbol{E}(\boldsymbol{r})$; here $\boldsymbol{r}$ is the location of -Q. The force $\boldsymbol{F} = Q\boldsymbol{E}(\boldsymbol{r} + \boldsymbol{a}) - Q\boldsymbol{E}(\boldsymbol{r})$. However, since $|\boldsymbol{a}| << |\boldsymbol{r}|$, $\boldsymbol{E}(\boldsymbol{r} + \boldsymbol{a}) = \boldsymbol{E}(\boldsymbol{r}) + \boldsymbol{a} \cdot \nabla\boldsymbol{E}(\boldsymbol{r})$ by the Taylor series expansion. Therefore $\boldsymbol{F} = Q\boldsymbol{a} \cdot \nabla\boldsymbol{E}(\boldsymbol{r})$, where $Q\boldsymbol{a}$ is the dipole moment $\boldsymbol{p}$ of the particle.

other hand, if there is *no* externally applied field $\boldsymbol{E}$, and we have charged particles or molecules, there will be *coulombic attraction* between oppositely charged particles or *coulombic repulsion* between particles with similar charges. In aerosol or deep bed filtration, stationary collectors having charges attract particles with opposite charges. If both the charged stationary collector and the particle (with charge Q_p) in the fluid may be considered as point charges, the magnitude of the electrostatic force exerted on the particle by *Coulomb's law* is

$$F^{\text{ELS}} = \frac{Q_p Q_c}{4\pi\varepsilon r^2}. \tag{3.1.15}$$

Here Q_c is the charge on the collector whose center is located at a distance r from the particle center and ε is the electrical permittivity of the fluid. The electrical permittivity ε of the fluid is defined as $\varepsilon_d\,\varepsilon_0$ (where ε_0 is the electrical permittivity of vacuum and ε_d is the dielectric constant of the medium). The value of $1/4\,\pi\,\varepsilon_0$ commonly used is 9×10^9 newton-m^2/coulomb^2. The force is directed along the minimum distance r connecting the two charges. The magnitude of the force, F^{ELS}, on the particle and the collector is the same, but the forces point in opposite directions. The expression above is based on the point charge model, which is valid only at large values of r. For smaller distances, complicated procedures are followed using image methods (Kurrelmeyer and Mais, 1967).

In aqueous-suspended particles, or colloidal systems, there are two other forces between the particle and the collector. The unretarded London attraction force between a particle of radius r_p and a collector is given by (Spielman and Fitzpatrick, 1972)

$$\boldsymbol{F}^{\text{Lret}} = -\left(\frac{2\text{a}_H}{3r_p}\right)\left\{\frac{1}{\left[\left(\frac{h_{\min}}{r_p}\right)+2\right]^2 (h_{\min}/r_p)^2}\right\}\mathbf{i}_h, \tag{3.1.16}$$

where

$\text{a}_H =$ Hamaker constant;
$h_{\min} =$ the minimum separation between particle and collector;
$\mathbf{i}_h =$ unit vector in the direction $h_{\min}$ outward from the collector.

The electrokinetic force in the double layer (Spielman and Cukor, 1973) is given by

$$\boldsymbol{F}^{\text{ELK}} = \frac{\varepsilon_d r_p \kappa_d}{4}\left[(\zeta_1+\zeta_2)^2\frac{\exp(-\kappa_d h)}{1 \mp \exp(-\kappa_d h)} - (\zeta_1-\zeta_2)^2\frac{\exp(-\kappa_d h)}{1 \pm \exp(-\kappa_d h)}\right]\boldsymbol{i}_h, \tag{3.1.17}$$

where

$h =$ distance between particle and collector;
$\varepsilon_d =$ dielectric constant for the fluid;
$\kappa_d =$ reciprocal Debye length $(= 1/\lambda)$;
$\zeta_1, \zeta_2 =$ zeta potential of collector and particle, respectively.

The upper sign in this equation corresponds to approach at constant charge, while the lower sign corresponds to approach at constant potential.

3.1.2.4 *Magnetic field*

An ionic species i with a charge Z_i moving with a velocity $\boldsymbol{v}_i$ can essentially be considered to be a current flow with a current of magnitude $Z_i\mathcal{F}\,|\boldsymbol{v}_i|$ for 1 gmol of ionic species. A conductor carrying current in a *magnetic field of constant strength* $\boldsymbol{B}$ experiences a force. Since the motion of ionic species constitutes a current, the force on 1 gmol of the ith ionic species in the magnetic field is given by

$$\boldsymbol{F}_i^{\text{mag}} = Z_i\mathcal{F}[\boldsymbol{v}_i \times \boldsymbol{B}]. \tag{3.1.18}$$

The vector magnetic potential $\boldsymbol{\Phi}$ for the magnetic field $\boldsymbol{B}$ is given by

$$\boldsymbol{B} = \nabla \times \boldsymbol{\Phi}. \tag{3.1.19}$$

Vector $\boldsymbol{B}$ is also called the magnetic induction. The unit of $\boldsymbol{B}$ is (newton/coulomb)/(m/s) and is identified usually as weber/m^2, or tesla.

Consider now particles without charges in the magnetic field. Particles can be classified into three general classes depending on how much magnetization is induced in them in a magnetic field. *Ferromagnetic* materials, such as iron, cobalt and nickel, are strongly magnetized. In general, the induced magnetism in these materials become relatively independent of the applied magnetic field. The extent of magnetization induced in the second class of materials, *paramagnetics*, is far weaker than in ferromagnetics. However, 55 elements in the periodic table are paramagnetic, and a magnetic field is used to separate paramagnetic particles. The third class of materials, *diamagnetics*, have even weaker magnetization.

The *magnetic force* exerted on a small magnetizable paramagnetic particle of volume (m_p/ρ_p) in a magnetic field of strength $\boldsymbol{H}^m$ (unit, amp/m) and volume susceptibility χ_p in vacuo is

$$\boldsymbol{F} = \frac{1}{2}\mu_0^m\left(\frac{m_p}{\rho_p}\right)\chi_p\nabla(\boldsymbol{H}^m \cdot \boldsymbol{H}^m), \tag{3.1.20}$$

where μ_0^m is the magnetic permeability of vacuum (Watson, 1973; Birss and Parker, 1981). Note that the force is proportional to the volume of the particle (i.e. proportional to d_p^3). For paramagnetic and diamagnetic particles, H^m is linearly proportional to the magnetic induction vector $\boldsymbol{B}$. The magnitude of this magnetic force in the z-direction (say) is given by

$$F_z = \mu_0^m\left(\frac{m_p}{\rho_p}\right)\chi_p H^m\frac{\text{d}H^m}{\text{d}z}, \tag{3.1.21}$$

so that increasing H^m or $(\text{d}H^m/\text{d}z)$ or both will increase F_z. If the particle is in a fluid of volume susceptibility χ, then the force on the particle is modified to (Birss and Parker, 1981)

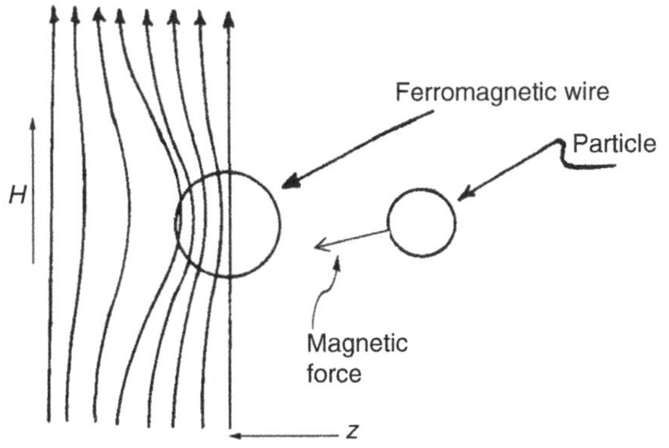

Figure 3.1.2F. Simplified cross-sectional view of the magnetic forces acting on a paramagnetic particle flowing past a magnetized ferromagnetic wire in a background field H. *(After Dobby and Finch (1977).)*

$$F = \frac{1}{2}\mu_0^m v_p(\chi_p - \chi)\nabla(\boldsymbol{H}^m \cdot \boldsymbol{H}^m). \tag{3.1.22}$$

The variation of H^m in the z-direction could be brought about in the following way even if a uniform H^m is created by a solenoid as a magnet. If the region where the magnetic field is applied has fine ferritic steel wires, the uniform field is grossly distorted near the wire (Figure 3.1.2F) and the particles are therefore forced toward the wire (Dobby and Finch, 1977) due to the positive gradient of the magnetic field. The background field H^m magnetizes the ferromagnetic wire of magnetic permeability μ_w. If the magnetization of the wire is M_w, the local field gradient becomes, approximately,

$$\frac{dH^m}{dz} = \frac{(H^m + M_w) - H^m}{d_w} = \frac{M_w}{d_w}\frac{\text{amp}}{\text{m}^2}, \tag{3.1.23}$$

where d_w is the wire diameter.

The magnetic force on a spherical paramagnetic (linearly polarizable) particle of radius r_p in a paramagnetic solution exposed to a nonuniform magnetic field having a local field intensity vector of $\boldsymbol{H}_0^m$ in vacuo is given by (Jones, 1995)

$$\boldsymbol{F} = 2\pi\mu_s^m r_p^3\left(\frac{\mu_p^m - \mu_s^m}{\mu_p^m + 2\mu_s^m}\right)\nabla H_0^{m^2}. \tag{3.1.24}$$

Here μ_s^m and μ_p^m are the magnetic permeabilities of the solution and the particle, respectively. The permeabilities may be related to the susceptibilities via $\mu^m = \mu_0^m(\chi + 1)$, where μ_0^m corresponds to that in vacuum. Further $\nabla H_0^{m^2}$ is related to the magnetic flux density $\boldsymbol{B}_0$ in the absence of matter via $\nabla H_0^{m^2} = \nabla(B_0/\mu_0^m)^2$. This phenomenon of particle motion is sometimes described as *magnetophoresis*. Note that the force is proportional to the particle volume through r_p^3, the magnetic permeability μ_s^m of the suspension medium and is directed along the gradient of the magnetic field intensity $\nabla H_0^{m^2}$. The quantity in brackets on the right-hand side of (3.1.24) is a Clausius–Mossotti function K of sorts, where ε has been replaced by μ^m–s. In positive magnetophoresis, $K > 0$, and particles are attracted to magnetic field intensity maxima and are repelled from the minima. In negative magnetophoresis, $K < 0$, and the phenomenon is reversed.

3.1.2.5 *Chemical potential gradient*

A number of external force fields have been described in the preceding subsections. Of these, the gravitational force field exists in nature regardless of our desire to require it or not. Other force fields, like centrifugal, electrical and magnetic, are created by engineers and scientists to achieve separation. Most of these external force fields are often described by the negative of the gradient of their respective scalar potentials, i.e. $-\nabla\Phi$ (exceptions are the magnetic and nonuniform electrical fields). But all such force fields originate *outside* the separation system.

The force of *chemical potential gradient*, on the other hand, originates *inside* the separation system due to the escaping tendency of molecules. For example, suppose we introduce two species into a closed vessel at constant temperature (T) and pressure (P) as a vapor-liquid system which is not in chemical equilibrium. As time progresses, we will observe that the species with the lower boiling point will preferentially escape the liquid phase and accumulate in the vapor phase; simultaneously, the higher boiling species will preferentially escape into the liquid phase. This directionally oriented escaping tendency disappears when the chemical potential of each species becomes uniform throughout the two phases and chemical equilibrium is attained (see Section 3.3). No external agency is involved in this separation phenomenon. We shall soon see that a large number of separation processes are governed by the gradient of this chemical potential, $\nabla\mu_i$, where μ_i is the chemical potential of the ith species per gmol of the ith species. For convenience, this chemical potential is sometimes referred to as μ_i^{int}, the *internal chemical potential*, the superscript indicating its origin inside the separation system (Giddings, 1982).

It is known from chemical thermodynamics (Guggenheim, 1967) that, at constant T and P, the reversible work $\underline{W}$ done to transfer m_i moles of species i from a state of partial molar Gibbs free energy $\overline{G}_i|_1$ to $\overline{G}_i|_2(> \overline{G}_i|_1)$ is given by

$$\underline{W} = m_i(\overline{G}_i|_2 - \overline{G}_i|_1). \tag{3.1.25a}$$

However, μ_i, the chemical potential, is simply $\overline{G}_i$. If states 1 and 2 refer to spatial locations z_1 and z_2, where $z_2 > z_1$, then we immediately conclude that

$$\underline{W} = m_i(\mu_i|_2 - \mu_i|_1). \tag{3.1.25b}$$

This work is needed to overcome a natural force which resists taking m_i moles from $\mu_i|_1$ to a higher value $\mu_i|_2$. This natural and spontaneous force on species i must then act in the opposite direction:

$$\underline{W} = m_i(\mu_i|_2 - \mu_i|_1) = -F_{iz}^{\text{total}}(z_2 - z_1). \tag{3.1.25c}$$

Per gram mole of species i, the force F_{iz} is

$$F_{iz} = \frac{F_{iz}^{\text{total}}}{m_i} = -\frac{\mu_i|_2 - \mu_i|_1}{z_2 - z_1}. \tag{3.1.25d}$$

Vectorially, we can now describe this force due to this chemical potential per gmol of ith species as

$$\boldsymbol{F}_i = \boldsymbol{F}|_{\text{gmol of }i\text{th species}} = -\frac{\mathrm{d}\overline{G}_i}{\mathrm{d}z}\mathbf{k} = -\frac{\mathrm{d}\mu_i}{\mathrm{d}z}\mathbf{k}. \tag{3.1.26}$$

In vector notation, the force on 1 gmol of the ith species due to the chemical potential gradient is $-\nabla\mu_i$. By analogy, it has often been indicated that all external driving forces per mole of the ith species may be represented as $-\nabla\mu_i^{\text{ext}}$. The shortcomings of such an approach (from the point of view of the fundamental property relation in chemical thermodynamics) for external forces have been demonstrated by Martin (1972, 1983). We therefore express the total driving force on 1 gmol of the ith species at constant temperature as

$$\boldsymbol{F}_{ti} = \boldsymbol{F}_t|_{\text{mole of }i\text{th species}} = -\nabla\mu_i + \boldsymbol{F}_{ti}^{\text{ext}}. \tag{3.1.27}$$

Of course, when possible, $\boldsymbol{F}_{ti}^{\text{ext}}$ may be represented by $-\Sigma\nabla\Phi_i^{\text{ext}}$ so that

$$\boldsymbol{F}_{ti} = \boldsymbol{F}_t|_{\text{mole of }i} = -\nabla(\mu_i + \Sigma\Phi_i^{\text{ext}}) = -\nabla(\mu_i + \Phi_{ti}^{\text{ext}}), \tag{3.1.28}$$

where the summation for the external force potentials indicates the sum of the available external forces representable as $-\nabla\Phi$.

The above representation of external forces combined with the chemical potential force for molecules or ions is quite useful for those external force fields representable by the negative of the gradient of their scalar potentials. We indicate in Table 3.1.1 the value of Φ_i^{ext} for 1 gmol of the ith species for a few cases. For magnetic and nonuniform electrical force fields, such representations are not possible for a system of molecules or ions. For macroscopic particles, where no chemical potential exists, only external force fields are important.

Table 3.1.1. The value of external force field Φ_i^{ext} for 1 gmol of species i

(1) Uniform electrical field of electrostatic potential Φ and constant Z_i	$Z_i\mathcal{F}\Phi$
(2) Gravity; species i in a solution of density ρ with z-axis vertically upwards	$M_i g\left(1 - \frac{\rho}{\rho_i}\right)z$
(3) Centrifugal force field	$\Phi_0 - \frac{1}{2}M_i\omega^2 r^2$ (3.1.29)

Just as we have an idea of how to calculate Φ_i^{ext} and therefore $-\nabla\Phi_i^{\text{ext}}$, we need to know more details about $-\nabla\mu_i$. Consider any region in the separation system where we assume equilibrium to exist (see Section 3.3). *For a binary system of i = 1, 2 at constant temperature,* if the pressure P and the mole fraction x_i in the region are changed by differential amounts, the corresponding change in μ_i is given by

$$\mathrm{d}\mu_i\Big|_T = \left(\frac{\partial\mu_i}{\partial P}\right)_{T,x_i}\mathrm{d}P + \left(\frac{\partial\mu_i}{\partial x_i}\right)_{T,P}\mathrm{d}x_i. \tag{3.1.30}$$

The partial molar volume $\overline{V}_i$ of the ith species in this region is defined as

$$\overline{V}_i = \left(\frac{\partial\mu_i}{\partial P}\right)_{T,x_i}. \tag{3.1.31}$$

Furthermore,

$$\left(\frac{\partial\mu_i}{\partial x_i}\right)_{T,P} = \mathrm{R}T\frac{\mathrm{d}\,\ell\mathrm{n}\,a_i}{\mathrm{d}x_i}, \tag{3.1.32}$$

where a_i is the activity of species i in this region. One can therefore write

$$\mathrm{d}\mu_i|_T = \overline{V}_i\,\mathrm{d}P + \mathrm{R}T\,\mathrm{d}\,\ell\mathrm{n}\,a_i. \tag{3.1.33}$$

Correspondingly, the gradient of μ_i is given by

$$\nabla\mu_i|_T = \overline{V}_i\,\nabla P + \mathrm{R}T\,\nabla\,\ell\mathrm{n}\,a_i \tag{3.1.34}$$

for a binary system at constant temperature.

The force due to a chemical potential gradient on species i at constant temperature, $-\nabla\mu_i|_T$, then arises due to the existence of a pressure gradient, an activity gradient, or both, in the separation system in a given region.

For liquid phases at not too high a pressure, the activity a_i may be related through an activity coefficient γ_i to the mole fraction x_i as follows:

$$a_i = \gamma_i x_i. \tag{3.1.35}$$

Thus

$$-\nabla\mu_i|_T = -\overline{V}_i\,\nabla P - \mathrm{R}T\,\nabla\,\ell\mathrm{n}\,(\gamma_i x_i),$$

which can be expressed as

$$-\nabla\mu_i|_T = -\overline{V}_i\,\nabla P - \mathrm{R}T\left\{\frac{\mathrm{d}\ell\mathrm{n}\,\gamma_i x_i}{\mathrm{d}\ell\mathrm{n}\,x_i}\right\}_T\nabla\,\ell\mathrm{n}\,x_i. \tag{3.1.36}$$

This expression shows that, ultimately, the mole fraction gradient can be used instead of the activity gradient. For an ideal solution, we have a simpler expression (since $\gamma_i \to 1$):

$$-\nabla\mu_i|_T = -\overline{V}_i\,\nabla P - \mathrm{R}T\,\nabla\,\ell\mathrm{n}\,x_i. \tag{3.1.37}$$

For gaseous mixtures, we can use

$$a_i = \frac{\hat{f}_{ig}}{f^0_{ig}}, \tag{3.1.38}$$

where f^0_{ig} is the standard state fugacity of the gas i at system temperature and specified pressure of 1 atmosphere and $\hat{f}_{ig}$ is the fugacity of species i at system temperature and pressure. For an ideal gas mixture, $\hat{f}_{ig} = p_{ig}$, the partial pressure of species i in the mixture; for such a case,

$$-\nabla\mu_i|_T = -\overline{V}_i\,\nabla P - RT\,\nabla\,\ell n\,p_{ig}. \tag{3.1.39}$$

For *isobaric* systems, we find, for ideal solutions,

$$\begin{aligned} &\textit{liquid solutions}: -\nabla\mu_i|_{T,p} = -RT\,\nabla\,\ell n\,x_i; \\ &\textit{gaseous mixtures}: -\nabla\mu_i|_T = -RT\,\nabla\,\ell n\,p_{ig}. \end{aligned} \tag{3.1.40}$$

Thus, mole fraction gradient, or partial pressure gradient, is the basis of the chemical potential gradient in a binary system at constant T and P. Recognize, however, that there is no specific force leading to a mole fraction or partial pressure gradient (see Section 3.1.3 after equation (3.1.76)).

It is important to qualify these characterizations of $\nabla\mu_i$ by indicating that they are valid within a phase, liquid, gaseous or solid. When the binary separation system has two phases, the variation of standard state chemical potential between the two phases also needs to be considered. Suppose ∇P=0 in the separation system. Integrate relation (3.1.33) at constant P and obtain, for any phase j,

$$\mu_{ij} = \mu^0_{ij} + RT_j\,\ell n\,a_{ij}. \tag{3.1.41}$$

Thus

$$-\nabla\mu_{ij}|_{T,P} = -\nabla\mu^0_{ij} - RT_j\,\nabla\,\ell n\,a_{ij}, \tag{3.1.42}$$

where μ^0_{ij} is the standard state chemical potential of the ith species in phase j at the standard state conditions used to calculate of a_{ij}. (In Section 3.3, Table 3.3.2 identifies standard states for various conditions.) In a two-phase system, μ^0_{ij} for species i is usually different for the two phases; thus $-\nabla\mu^0_{ij}$ is also a driving force for species movement and separation. The value of μ^0_{ij} in any phase may be calculated from standard thermodynamic relations (Denbigh, 1971).

Very small particles in liquid or gas streams have a *random Brownian motion* due to the thermal energy of the continuous phase molecules. If there is a concentration gradient of particles due to a particle sink, then there is a *Brownian motion force* on these particles:

$$\boldsymbol{F}^{BR} = 6\pi\mu\,\boldsymbol{v}^r_p\,r_p, \tag{3.1.43}$$

where $\boldsymbol{v}^r_p$ is the particle diffusion velocity relative to that of the liquid or gas phase, r_p is the particle radius and μ is the fluid viscosity.

3.1.2.6 *Other forces: thermal gradient, radiation pressure, acoustic force*

Until now, we have considered the temperature to be uniform in our separation system. The existence of a *temperature gradient* exerts unequal forces on molecules of different kinds. In a gas mixture subjected to a temperature gradient, it has been observed that the lighter gas species concentrates in the hot region and the heavier species concentrates in the cold region.[5] This phenomenon, known as *thermal diffusion*, is also observed in liquid mixtures. The force exerted on gas species A in a binary mixture of A and B subject to a temperature gradient is given by[6]

$$\boldsymbol{F}|_{\text{gmol of A}} = -\frac{RT\,D^T_A}{D_{AB}C_A M_A}\nabla\,\ell n\,T = \boldsymbol{F}_{TA}, \tag{3.1.44}$$

where D^T_A is the thermal diffusion coefficient for species A in a binary mixture of species A and B and D_{AB} is the ordinary diffusion coefficient. A thermal diffusion ratio k_T has been defined (Bird *et al.*, 1960) as

$$k_T = \frac{\rho_t D^T_A}{C^2_t M_A M_B D_{AB}}. \tag{3.1.45}$$

Furthermore $D^T_A = -D^T_B$. Thus if species A goes toward the hotter region, species B moves to the colder region.

No phenomenon analogous to thermal diffusion in chemical mixtures is encountered with particles. However, when placed in a temperature gradient, small particles in a stagnant liquid have been found to move in the direction of lower temperature. This phenomenon has been referred to as *thermophoresis* (Fuchs, 1964). It is said to arise in a gas because gas molecules originating in the hot regions impinge on the particles with greater momenta than molecules coming from the colder regions. The magnitude of the *force due to the thermophoresis*, $\boldsymbol{F}_{TP}$, for a spherical particle obeying Stokes' law in the z-direction is

$$F_{TPz} = \frac{6\pi\mu^2 r_p}{C_c\,\rho_t\,T}\left(-Th\frac{dT}{dz}\right), \tag{3.1.46a}$$

where Th is a dimensionless group defined for a gas phase of viscosity μ by

$$Th = -\frac{U_{pzt}\,\rho_t\,T}{\mu\,(dT/dz)}. \tag{3.1.46b}$$

The value of Th is said to range between 0.42 and 1.5 from theoretical predictions. Here U_{pzt} is the steady particle velocity due to thermophoresis (see Section 3.1.3.2 for its definition), r_p is the particle radius and C_c is a correction factor (see Section 3.1.6.1). For additional details, see Talbot *et al.* (1980). Explanations for the same phenomenon in a liquid are somewhat uncertain (McNab and Meisen, 1973). An alternative situation, in which particle motion is generated in a solute gradient and termed *diffusiophoresis*, is treated in Anderson *et al.* (1982).

[5] This is true for gases observing the inverse power law of repulsion: (force) = (constant) × (intermolecular distance)$^{\upsilon}$ and $\upsilon > 5$. See Present (1958).

[6] In a frame of reference where $\boldsymbol{v}_t = \boldsymbol{v}^*_t$ (see Section 3.1.3).

Radiation pressure from continuous wavelength (cw) visible laser light is known to accelerate freely suspended particles in the direction of the light. The magnitude of the *radiation pressure force*, F_{rad}, has been given as (Ashkin, 1970)

$$F_{rad} = \frac{2\,q_{fr}\,Pr}{c}. \tag{3.1.47}$$

Here q_{fr} is the fraction of light effectively reflected back (generally assumed to be ~0.1), Pr is the power of the laser light and c is the velocity of light. The medium of interest is a liquid.

The *acoustic radiation (acr) force* F_{acrx} in the x-direction due to an ultrasound of wavelength λ induced on a particle of volume V_p, density ρ_p and compressibility β_p suspended in a fluid medium of density ρ_f and compressibility β_f, is given by (Petersson *et al.*, 2005)

$$F_{acrx} = -\left(\frac{\pi P_0^2\,V_p\,\beta_f}{2\lambda}\right)\left\{\frac{5\rho_p - 2\rho_f}{2\rho_p + \rho_f} - \frac{\beta_p}{\beta_f}\right\}\sin(2kx), \tag{3.1.48}$$

where k is defined as $(2\pi/\lambda)$, x is the distance from the pressure node and P_0 is the amplitude of the pressure wave. The direction of the force F_{acrx} is dependent on the sign of the quantity within the brackets { }, which is sometimes called the ϕ-factor. Those particles which have a positive value for the ϕ-factor will move toward the pressure node; those with the reverse sign will move toward the pressure antinode. For an introduction, see Nyborg (1978), Ter Haar and Wyard (1978) and Weiser *et al.* (1984).

3.1.2.6.1 Inertial force So far, we have presented a variety of forces that can come into play and act on molecules of a given species and/or particles. When such forces act, the molecules or particles move in a given direction. At the beginning of this motion, the molecules/particles undergo acceleration. From Newton's second law, we know that the molecules/particles are subjected to an *inertial force*, $\boldsymbol{F}^{iner}$, in the same direction during this period of acceleration. The magnitude of this inertial force is simply the product of the mass of the molecules/species and the magnitude of the acceleration (see the beginning materials in Sections 3.1.3.1 and 3.1.3.2). In most circumstances encountered in separations, one can assume that the acceleration ceases to exist after some time and a steady velocity comes about; therefore the inertial force ceases to exist. However, if the fluid flow field is such that the direction and magnitude of its velocity continues to change, inertial forces will always be present. This is particularly true of particles flowing in a medium with many obstacles, or if there is a change in flow direction and flow cross-sectional area.

3.1.2.6.2 Lift force on a particle in shear flow When a particle is flowing in a shear flow field, it experiences a *lift force* normal to the fluid flow direction. The magnitude of this lift force normal to (i.e. in the y-direction) the flow direction (say, in the z-direction) is given by

$$F_{lift} = a\,\mu\,v_z\,(dv_z/dy)^{1/2}\,r_p^2/\upsilon^{1/2} \tag{3.1.49}$$

for a spherical particle of radius r_p in an axial fluid flow field (i.e. velocity v_z) having a simple shear ($(dv_z/dy) \neq 0$) (Saffman, 1965); the constant "a" has the value 6.46 and $\upsilon = (\mu/\rho)$ for the fluid in which the particle is suspended.

3.1.2.7 *A generalized expression for all forces*

In the preceding discussions, a large number of forces, both external and internal to the separation system, have been identified and described briefly. Note that any force so identified was, for example, specific to molecules of the *i*th species.[7] However, it is known that forces specific to the *j*th species can also affect the motion of molecules of the *i*th species. For the immediate objectives in the paragraphs that follow, these effects are ignored by assuming *uncoupled conditions*:[8] molecules of species i in a stagnant fluid move only due to forces specific to the *i*th species; similarly for the *jth* species. It is further assumed that the conditions are not too far removed from equilibrium (see the introduction to Section 3.3 and Sections 3.3.1–3.3.6 for descriptions of equilibrium conditions); therefore thermodynamic quantities (defined only for equilibrium conditions) can be used to described nonequilibrium conditions where a net transport of molecules of species i exist due to external and internal forces. For illustrative purposes, an expression for the *total driving force* $\boldsymbol{F}_{ti}$ on 1 gmol of species i or 1 gion of ion i in a solution or mixture due to a variety of forces identified earlier is given below (force is positive in the direction of the positive axis). Obviously, only one or a few of these forces exist at any time in a given separation system:

$$\boldsymbol{F}_{ti} = -\nabla\mu_i|_T + \boldsymbol{F}_{ti}^{ext} + \boldsymbol{F}_{Ti};$$
$$\boldsymbol{F}_{ti} = -\nabla\mu_i|_T + -\Sigma\nabla\Phi_i^{ext} + \boldsymbol{F}_i^{mag} + \boldsymbol{F}_{Ti};$$
$$\boldsymbol{F}_{ti} = -RT\nabla\ln a_i - \overline{V}_i\nabla P - \nabla\mu_i^0 + M_i\omega^2 r\mathbf{r} - M_i g\mathbf{k} - Z_i\mathcal{F}\nabla\Phi + Z_i\mathcal{F}(v_i \times \boldsymbol{B}) - \frac{RTD_i^T}{D_{ij}C_iM_i}\nabla\ln T. \tag{3.1.50}$$

This expression[9] does not include the effect of a nonuniform electrical field (see (3.1.12)) and the electrophoretic retardation. Note that buoyancy forces do not appear in $\boldsymbol{F}_{ti}$ as such. Lee *et al.* (1977a) have tabulated the magnitudes of each of these forces, i.e. gradients, which can create a value of $\nabla\ln a_i|_{T,P}$ equal to 1 cm^{-1}.

A clarification on the forces acting on solute molecules in a solvent undergoing rigid body rotation in a centrifuge

[7] Particles will be considered next.
[8] For an illustration of coupling, see Section 3.1.5.2.
[9] For alternate representations, see Lee *et al.* (1977a).

is useful here. The (non-Brownian) force exerted on the solute molecules is due not only to the $\omega^2 r\mathbf{r}$ term, but also to the pressure gradient in the solvent developed by rotation. To determine this pressure gradient, consider only the rotation of solvent species $i = s$ in the centrifuge (no solute present). At equilibrium, there is no net force acting on solvent molecules in the centrifuge:

$$\boldsymbol{F}_{ts} = 0 = -\overline{V}_s \frac{\mathrm{d}P}{\mathrm{d}r}\mathbf{r} + M_s \omega^2 r\,\mathbf{r}. \tag{3.1.51}$$

This provides an expression for $(\mathrm{d}P/\mathrm{d}r)$ due to rotation. Assume now that this pressure gradient generated by solvent rotation is unaffected by the presence of solute species i in the rotating centrifuge. The net force[10] on solute species i in the radial direction is then obtained as (*excluding the concentration gradient contribution*)

$$\begin{aligned}\boldsymbol{F}_{ti} &= -\overline{V}_i \frac{\mathrm{d}P}{\mathrm{d}r}\mathbf{r} + M_i\omega^2 r\,\mathbf{r}\\ &= M_i\omega^2 r\left[1 - \left(\frac{\overline{V}_i}{M_i}\right)\left(\frac{M_s}{\overline{V}_s}\right)\right]\mathbf{r},\end{aligned} \tag{3.1.52}$$

where $(\overline{V}_i/M_i)$ is the partial specific volume of solute i and $(M_s/\overline{V}_s)$ is essentially the solvent density in any region.

The assumption that the solute species will not affect the pressure gradient in a centrifuge is not generally valid. Consider the total force on any ith species in such a case. At equilibrium, $\boldsymbol{F}_{ti}$ is zero, leading to

$$\begin{aligned}\boldsymbol{F}_{ti} = 0 = -\left.\frac{\mathrm{d}\mu_i}{\mathrm{d}r}\right|_T \mathbf{r} + M_i\,\omega^2 r\,\mathbf{r};\\ -\overline{V}_i\frac{\mathrm{d}P}{\mathrm{d}r} - \left.\frac{\mathrm{d}\mu_i}{\mathrm{d}r}\right|_{T,P} + M_i\omega^2 r = 0.\end{aligned} \tag{3.1.53}$$

From the principles of chemical thermodynamics (Guggenheim, 1967), at constant T and P the Gibbs–Duhem equation is

$$\sum_{i=1}^{n} x_i \left.\mathrm{d}\mu_i\right|_{T,P} = 0. \tag{3.1.54}$$

Multiply equation (3.1.53) by x_i, sum over n species and use the Gibbs–Duhem equation (3.1.54) to obtain

$$-\sum_{i=1}^{n} x_i\overline{V}_i\,\mathrm{d}P + \sum_{i=1}^{n} x_i M_i\,\omega^2 r\,\mathrm{d}r = 0. \tag{3.1.55}$$

Define an average partial molar volume $\overline{V}_t$ and an average molecular weight of the solution M_t as follows:

$$\sum_{i=1}^{n} x_i\overline{V}_i = \overline{V}_t \quad\text{and}\quad \sum_{i=1}^{n} x_i M_i = M_t. \tag{3.1.56}$$

Rewrite equation (3.1.55) as

$$-\overline{V}_t\,\mathrm{d}P + (M_t\,\omega^2 r)\mathrm{d}r = 0. \tag{3.1.57}$$

With solution density ρ_t defined as $(M_t/\overline{V}_t)$, we can now simplify equation (3.1.53):

$$-\left.\mathrm{d}\mu_i\right|_{T,P} + (M_i - \rho_t\overline{V}_i)\,\omega^2 r\,\mathrm{d}r = 0. \tag{3.1.58}$$

Note that ρ_t and $\overline{V}_t$ will, in general, depend on radial location r since the pressure and composition vary with r.

We now focus on *macroscopic particles* and provide an expression for the *total external force* acting on a particle of mass m_p, density ρ_p, radius r_p, charge Q_p, velocity $\boldsymbol{v}_p$ and volume (m_p/ρ_p). We have not included here the Brownian motion force $\boldsymbol{F}^{\mathrm{Br}}$, nor any force due to thermophoresis, radiation pressure, acoustic force and the electrical force in a nonuniform electrical field given by (3.1.13). Although not generated by an external force field, coulombic types of interactions, London dispersion and electrokinetic forces in the double layer are included in the expression given below:

$$\begin{aligned}\boldsymbol{F}_{tp}^{\mathrm{ext}} &= -m_p g\left(1 - \frac{\rho_t}{\rho_p}\right)\mathbf{k} + m_p r\omega^2\left(1 - \frac{\rho_t}{\rho_p}\right)\boldsymbol{r} - Q_p\nabla\Phi\\ &\quad + (\varepsilon_d\zeta r_p - Q_p)\boldsymbol{E}\\ &\quad + Q_p[\boldsymbol{v}_p \times \boldsymbol{B}] + \frac{1}{2}\mu_0^m\left(\frac{m_p}{\rho_p}\right)\chi_p\nabla(\boldsymbol{H}^m\cdot\boldsymbol{H}^m)\\ &\quad + \boldsymbol{F}^{\mathrm{ELS}} + \boldsymbol{F}^{\mathrm{Lret}} + \boldsymbol{F}^{\mathrm{ELK}}.\end{aligned} \tag{3.1.59}$$

The expressions for $\boldsymbol{F}^{\mathrm{ELS}}$, $\boldsymbol{F}^{\mathrm{Lret}}$ and $\boldsymbol{F}^{\mathrm{ELK}}$ may be obtained, respectively, from (3.1.15), (3.1.16) and (3.1.17). Note that the pressure gradient generated in a centrifugal field has been replaced by means of equation (3.1.58) as a centrifugal buoyancy term in which $\overline{V}_i$ has been replaced by particle volume (m_p/ρ_p) and M_i by m_p to provide $(1 - [\rho_t/\rho_p])$.

Example 3.1.1 Calculate the gravitational force exerted on a particle for the following two cases: (1) particle diameter 10 μm, particle density 2 g/cm^3, fluid density 1.3 g/cm^3; (2) particle diameter 2 cm, particle density 2 g/cm^3, fluid density 1.3 g/cm^3.

Solution (1) From the force expression (3.1.59), the z-direction gravitational force on the particle is given by

$$\begin{aligned}F_{tz} &= -\frac{4}{3}\pi r_p^3\rho_p g(1 - (\rho_t/\rho_p))\\ &= -\frac{4\pi(5\times10^{-4})^3 \times 2 \times 980(1 - (1.3/2))}{3}\frac{\mathrm{g\,cm}}{\mathrm{s}^2}\\ &= -\frac{4\times\pi\times125\times10^{-12}\times2\times980\times0.35}{3}\\ &= -\frac{4\times\pi\times1.25\times2\times0.98\times0.35\times10^{-7}}{3}\frac{\mathrm{g\,cm}}{\mathrm{s}^2}\\ &= -3.59\times10^{-7}\frac{\mathrm{g\,cm}}{\mathrm{s}^2} = -3.59\times10^{-12}\ \text{newton}.\end{aligned}$$

[10] One can similarly develop (3.1.5) for species i in a solvent under gravity.

(2)
$$F_{tz} = -\frac{4}{3}\pi r_p^3 \rho_p g\,(1-(\rho_t/\rho_p))$$
$$= -\frac{4\pi\,(1)^3 \times 2 \times 980(1-(1.3/2))}{3}\frac{\text{g cm}}{\text{s}^2}$$
$$= \frac{-8\pi \times 980 \times 0.35}{3} = -2873.5\frac{\text{g cm}}{\text{s}^2}$$
$$= -2.873 \times 10^{-2}\ \text{newton}.$$

Example 3.1.2 Calculate the centrifugal force exerted for the following two cases.

Case (1) 1 gmol of ovalbumin molecules (molecular weight = 45 000) in an aqueous solution of density 1 g/cm^3 rotating in a centrifuge at 7000 radians/s at a radial coordinate of 1 cm, given ovalbumin density in solution = 1.34 g/cm^3.

Case (2) a particle of diameter 10 μm, density 2 g/cm^3 in a solution of density 1.3 g/cm^3 at a radial distance of 1 cm.

Solution *Case (1)* From the force expression (3.1.52), the magnitude of the radial force experienced by 1 gmol of ovalbumin molecules (species *i*) is given by

$$F_{tir}\Big|_{\text{mole}} = M_i\left[1-\left(\frac{\overline{V}_i}{M_i}\right)\left(\frac{M_s}{\overline{\overline{V}}_s}\right)\right]\omega^2 r$$
$$= \frac{45\,000\,\text{g}}{\text{gmol}}\left[1-\left(\frac{1}{1.34}\right)(1)\right]\frac{(7000)^2}{\text{s}^2}1\,\text{cm}$$
$$= 45000[1-0.746]49\times 10^6\frac{\text{g cm}}{\text{s}^2\,\text{gmol}}$$
$$= 560070 \times 10^6\frac{\text{g cm}}{\text{s}^2\,\text{gmol}}$$
$$= 5.60\times 10^6\,\text{newton/gmol}.$$

Case (2) From the generalized force expression (3.1.59), we get $F_{tr}|_{\text{particle}}$ experienced by the particle as follows:

$$F_{tr}\Big|_{\text{particle}} = m_p r\omega^2\left(1-\frac{\rho_t}{\rho_p}\right) = \frac{4}{3}\pi r_p^3\rho_p r\omega^2\left(1-\frac{\rho_t}{\rho_p}\right)$$
$$= \frac{4\pi}{3}\times(5\times 10^{-4})^3\times 2\times 1\times(7000)^2\times\left(1-\frac{1.3}{2}\right)$$
$$= \frac{4\pi\times 125\times 2\times 49\times 0.35\times 10^{-6}}{3}\ \text{g cm/s}^2$$
$$= 0.179\times 10^{-1}\ \text{g cm/s}^2.$$

Example 3.1.3 Calculate the force exerted on ovalbumin molecules exposed to an electrical field of constant strength 30 volts/cm, given $Z_i = 4.5$ for ovalbumin at the solution *pH*. Neglect the electrical double-layer effect.

Solution The electrical force per mole of ovalbumin is obtained from (3.1.8) in the absence of other information (e.g. zeta potential, ionic strength, etc.) It is given by

$$Z_i\mathcal{F}E = 4.5\times 96\,500\times 30\ \text{coulomb-volt/cm}$$
$$= 4.5\times 96\,500\times 3000\ \text{coulomb-volt/m}$$
$$= 1.3\times 10^9\ \text{coulomb-volt/m}.$$

But 1 coulomb-volt = 1 joule = 1 newton-m. Therefore, $Z_i\mathcal{F}E = 1.3\times 10^9$ newton, a force much stronger than the centrifugal force considered in Example 3.1.2, case (1).

Example 3.1.4 Calculate the magnitude of the force due to radiation pressure on a lossless dielectric spherical particle of radius 0.5145 μm and density $\rho_p = 1\text{g/cm}^3$ subjected to a cw argon laser light of power 1 watt at a wavelength $\lambda = 0.5145$ μm. Calculate also the instantaneous acceleration experienced by the particle.

Solution From equation (3.1.47),

$$F_{\text{rad}} = \frac{2\,q_{\text{fr}}\,Pr}{c}.$$

Assume that $q_{\text{fr}} = 0.1$; velocity of light, $c = 2.799\times 10^{10}$ cm/s; $Pr = 1$ watt. Then

$$F_{\text{rad}} = \frac{2\times 0.1\times 1\,\text{watt}}{2.799\times 10^{10}\text{cm/s}} = \frac{0.2\times 10^{-10}\,\text{watt-s}}{2.799\,\text{cm}}.$$
$$= \frac{0.2\times 10^{-10}\times 10^7\,\text{erg}}{2.799\,\text{cm}} = \frac{0.2\times 10^{-3}}{2.799\,\text{cm}}\frac{\text{g-cm}^2}{\text{s}^2}$$
$$= 7.145\times 10^{-5}\,\frac{\text{g-cm}}{\text{s}^2} = 7.145\times 10^{-5}\,\text{dyne}.$$

If the instantaneous acceleration is d^2z/dt^2, then $F_{\text{rad}} = m_p\, d^2z/dt^2$. Now, $m_p = (4/3)\pi r_p^3\rho_p$, so

$$m_p = \frac{4}{3}\pi(0.5145\times 10^{-4})^3\times 1\ \text{g} = \frac{4\pi}{3}\times 0.1362\times 10^{-12}$$
$$= 0.57\times 10^{-12}\text{g};$$
$$\frac{d^2z}{dt^2} = \text{acceleration} = \frac{7.145\times 10^{-5}\,\text{dyne}}{0.57\times 10^{-12}\,\text{g}} = 1.25\times 10^8\,\text{cm/s}^2.$$

The magnitude of this acceleration is quite high - much larger than that due to gravity (Ashkin, 1970).

Note that the following forces acting on the particle are proportional to the particle volume (and therefore r_p^3): gravity force; centrifugal force; dielectrophoretic force; magnetic force on a paramagnetic particle. The electrophoretic retardation force is proportional to the particle radius.

3.1.3 Particle velocity, molecular migration velocity and chemical species flux

It is now appropriate to calculate the velocities of particles or molecules resulting from the forces acting on them. For clarity, we assume first that velocities created by bulk flow are zero. Let the particle velocity vector due to external forces be represented by $\boldsymbol{U}_p$. Similarly, let $\overline{\boldsymbol{U}}_i$ be the average velocity vector of the ith chemical species in any region due to external as well as internal forces.

3.1.3.1 *Particle velocity and particle flux*

We focus on particle motion first. If a macroscopic particle (mass m_p) moves in the z-direction (positive z, vertically upward) under the action of forces acting in the z-direction, then, from the principles of mechanics (Newton's second law),

$$\left(\begin{array}{l}\text{magnitude of force on the}\\ \text{particle in the } z\text{-direction}\end{array}\right) = m_p\,\frac{\mathrm{d}^2 z}{\mathrm{d}t^2}. \tag{3.1.60}$$

If the particle was moving in free space, this force would equal the external force or forces we have identified earlier in magnitude and direction. If, however, the particle moves in a gaseous, liquid or solid (rarely) medium, the force on the particle consists of the external force, (or forces, including the buoyancy force) and a frictional force, $\boldsymbol{F}_p^{\text{drag}}$, which opposes the particle motion (and thus has a negative sign). This frictional force comes into play as soon as the particle moves.

It is known that, at small values of particle velocity U_{pz} (in the direction of the positive z-coordinate), the frictional resistive force is linearly proportional to the magnitude of the particle velocity. For example, according to Stokes' law, the resistive or drag force vector on a spherical particle of radius r_p moving at a velocity $U_{pz}\mathbf{k}$ through a fluid of viscosity μ is $-6\pi\mu\, r_p U_{pz}\mathbf{k}$. If this particle is falling under gravity in a fluid of density ρ_t and viscosity μ, then

$$m_p\,\frac{\mathrm{d}^2 z}{\mathrm{d}t^2} = F_{pz}^{\text{ext}} - F_{pz}^{\text{drag}} = -m_p g\left(1-\frac{\rho_t}{\rho_p}\right) - 6\,\pi\mu r_p\,\frac{\mathrm{d}z}{\mathrm{d}t}, \tag{3.1.61}$$

where $U_{pz} = \mathrm{d}z/\mathrm{d}t$.

Due to a nonzero acceleration ($\mathrm{d}^2z/\mathrm{d}t^2 \neq 0$), the velocity of the particle increases; the frictional resistive force (the drag force) also increases. In many systems, after some time the particle acceleration in the stagnant fluid becomes zero and the particle velocity becomes constant. This velocity is called the *terminal velocity*. We express the resistive force vector on the particle as $-f_p^d\,\boldsymbol{U}_p$; when $\boldsymbol{U}_p$ is the terminal velocity, for the z-component,

$$F_{pz}^{\text{ext}} = f_p^d\,U_{pz} \Rightarrow U_{pz} = \frac{F_{pz}^{\text{ext}}}{f_p^d}.$$

Thus

$$\boldsymbol{U}_p = \frac{\boldsymbol{F}_p^{\text{ext}}}{f_p^d}, \tag{3.1.62}$$

where f_p^d is the frictional coefficient of the spherical particle moving slowly in a stagnant medium of viscosity μ. For a particle resistive force described by Stokes' law, f_p^d is equal to $6\pi\mu r_p$. Often, this terminal velocity will be identified as $\boldsymbol{U}_{pt}$, for example in the z-direction as U_{pzt}. (See Section 3.1.6.2 for corresponding quantities for ions in a gas phase.)

Suppose the fluid medium is not stagnant but has a mass average velocity $\boldsymbol{v}_t$. If Stokes' law still determines the resistive force, then the resistive or frictional (or drag) force on the spherical particle is to be determined using the velocity of the particle relative to the fluid:

$$\boldsymbol{F}_p^{\text{drag}} = -6\pi\mu r_p\,(\boldsymbol{U}_p - \boldsymbol{v}_t). \tag{3.1.63}$$

These relations based on Stokes' law are valid only for small particles whose *particle Reynolds number*, defined by $(2r_p U_p \rho/\mu)$, is very small (<0.1). For larger particles and/or higher velocities, the magnitude of the resistive (drag) force is given by

$$F_{pz}^{\text{drag}} = C_D\,\frac{\rho\,U_{pz}^2}{2}\,A_p, \tag{3.1.64}$$

where C_D is a drag coefficient and A_p is the projected area of the particle measured in a plane perpendicular to the direction of particle motion (Bird *et al.*, 1960, 2002). The drag force varies with particle radius for fine particles in Stokes' law range, whereas it varies with a larger power of particle radius up to ${r_p}^2$ at higher particle Reynolds numbers.

The external force based particle velocity in a stagnant fluid medium leads to a number flux, $\boldsymbol{n}_p$, of the particles across any given cross-sectional area. If the total number of particles per unit volume of the fluid is N_t (see equation (2.4.2c)) and the terminal velocity of the particles is $\boldsymbol{U}_{pt}$, then the particle flux $\boldsymbol{n}_p$ in terms of numbers of particles crossing a surface area perpendicular to $\boldsymbol{U}_{pt}$ is given by (see Figure 3.1.3A)

$$\boldsymbol{n}_p = N_t\,\boldsymbol{U}_{pt}. \tag{3.1.65}$$

For any other surface area, indicated by $\boldsymbol{A}$, the particle flux across the surface area is obtained as

$$n_p = N_t\,\boldsymbol{U}_{pt}\cdot\boldsymbol{A}/|\boldsymbol{A}|. \tag{3.1.66}$$

If there is a particle size distribution indicated by $n(r_p)$, the particle number density function (equation (2.4.2a)), such that $\boldsymbol{U}_{pt}(r_p)$ is the terminal velocity of particles in the size range of r_p to $r_p + \mathrm{d}r_p$, the particle flux n_p across a surface area perpendicular to $\boldsymbol{U}_{pt}(r_p)$ is

$$\boldsymbol{n}_p = \int_{r_{\min}}^{r_{\max}} \boldsymbol{U}_{pt}(r_p)\,n(r_p)\mathrm{d}r_p. \tag{3.1.67}$$

When the size of the particles becomes quite small, Brownian motion becomes important. The flux of particles due to Brownian motion is generally indicated by (Flagan and Seinfeld, 1988)

$$\boldsymbol{n}_p = -D_p\,\nabla N_t, \tag{3.1.68}$$

where we have assumed that the particle diffusivity D_p has the same value regardless of the direction of diffusion. In a given direction (say z), this flux is obtained as

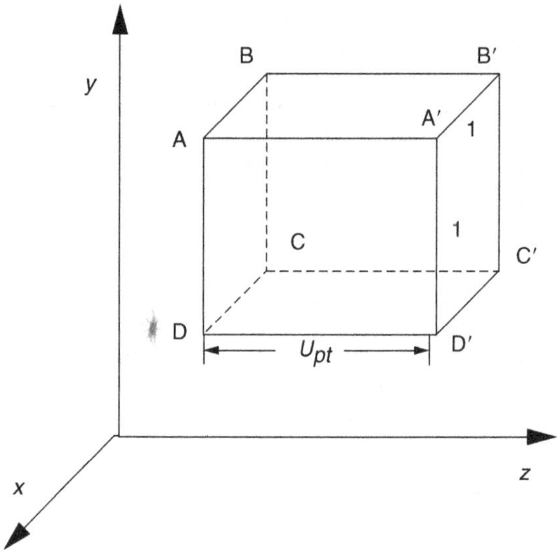

Figure 3.1.3A. Explanation for particle flux expression (3.1.65). Consider the rectangular parallelepiped shown. Let the length of side DD′ in the positive z-direction be U_{pt} cm, which is the distance crossed in one second by a particle having the terminal velocity U_{pt} of all particles in the positive z-direction. Let the lengths of A′B′ (or AB) and C′D′ (or CD) be 1 cm. The cross-sectional area of ABCD or A′B′C′D′ in the xy-plane is then 1 cm^2. Any particle in the volume ABCDA′B′C′D′ (= U_{pt} cm^3) will cross area A′B′C′D′ in one second since the particle furthest from A′B′C′D′ will be at a distance U_{pt} cm; others closer to A′B′C′D′ will surely cross this area. Now the number of particles in this volume ABCDA′B′C′D′ is particle number density × volume, i.e. N_t (particles/cm^3) × U_{pt} cm^3, N_tU_{pt}. All of these particles will cross the unit area of A′B′C′D′ in one second. Therefore the particle flux across A′B′C′D′ is $n_p = N_tU_{pt}$ particles/area-time.

$$n_{pz} = -D_p \frac{\partial N_t}{\partial z}. \tag{3.1.69}$$

When both Brownian motion as well as external forces act on the particles, the overall flux expression in, say, the z-direction is obtained as follows:

$$n_{pz} = -D_p \frac{\partial N_t}{\partial z} + U_{pzt} N_t. \tag{3.1.70}$$

If there is a particle size distribution given by $n(r_p)$, we obtain the particle flux as

$$n_{pz} = -D_p \int_{r_{min}}^{r_{max}} \frac{\partial n(r_p)}{\partial z} \mathrm{d}r_p + \int_{r_{min}}^{r_{max}} U_{pzt}(r_p)\, n(r_p) \mathrm{d}r_p. \tag{3.1.71}$$

The diffusion coefficient of small particles undergoing Brownian motion is of interest. For the diffusion coefficient of droplets/particles of very small dimensions in air, the Stokes–Einstein relation has been found to be valid:

$$D_p = \frac{k^B T}{f_p^d} \tag{3.1.72}$$

where k^B is Boltzmann's constant, 1.38066×10^{-23} J/K (1.38060×10^{-16} g-cm^2/s^2-K) and f_p^d is the frictional coefficient for the particle (see relation (3.1.86) for its molecular equivalent using molar concentrations). For such aerosols (suspensions of small particles in a gas/air), a more detailed result based on Stokes' law and the slip correction factor C_c (see expression (3.1.215)) for a particle of radius r_p is

$$D_p = \frac{k^B T C_c}{6\pi \mu r_p}, \tag{3.1.73}$$

where μ is the gas phase viscosity and T is the absolute temperature in kelvin. One can use the same equation for the liquid phase without the correction factor C_c.

If the particle is flowing in a shear flow of a liquid, the shear-induced diffusivity of a particle in a particle suspension, where $\dot{\gamma}$ is the local shear rate (e.g. (dv_z/dy) in relation (3.1.49)) has been experimentally measured by Eckstein *et al.* (1977) to be

$$\begin{aligned} D_p &\cong 0.1\phi_p r_p^2 \dot{\gamma} \quad \text{for } 0 < \phi_p < 0.2 \\ &\cong 0.25 r_p^2 \dot{\gamma} \quad \text{for } 0.2 < \phi_p < 0.5, \end{aligned} \tag{3.1.74}$$

where ϕ_p is the volume fraction of the particles.

3.1.3.2 *Molecular migration velocity and species flux*

Instead of macroscopic particles, we consider now the motion of molecules of the ith species in any region *having no bulk velocity*. The governing equations are similar in principle to equation (3.1.61). From a simplistic point of view, the major difference is that we will deal with a mole of ith species instead of one molecule of ith species. Therefore we will consider the averaged z-directional motion of a large number of molecules (since individual molecular motion is chaotic). For one mole, this motion may be simply modeled in the absence of a temperature gradient as (Giddings, 1982)

$$M_i \frac{\overline{d^2 z}}{dt^2} = +F_{tiz} - f_i^d \frac{\overline{dz}}{dt}, \tag{3.1.75}$$

where F_{tiz} is to be determined from equation (3.1.50) without F_{ti}. Further, $(\overline{dz/dt})$ is the averaged species i velocity in the z-direction and f_i^d is the frictional coefficient for one mole of the ith species in the region[11] under consideration, the overbar indicating an averaged quantity. When the force F_{tiz} is applied at time $t = 0$, the molecules experience an acceleration, i.e. d^2z/dt^2 is nonzero. However, unless we have vacuum in the region, the resistance encountered by the

[11] Sometimes the medium or the region in which species i moves will be explicitly identified via an additional subscript; the frictional coefficient of species i through a membrane, for example, will be described by f_{im}^d.

molecules of the ith species is so large that the period during which acceleration exists is very small (Tanford, 1961; Giddings, 1982). Thus, for all practical time scales of interest in separation, the equation given above is simplified to[12]

$$F_{tiz} = f_i^d \frac{\overline{dz}}{dt} = f_i^d \overline{U}_{iz}. \qquad (3.1.76)$$

Here, $\overline{U}_{iz}$ is the averaged velocity of the ith molecules in the z-direction due to all forces acting on the ith species. We have assumed that f_i^d has the same value in all directions (isotropic) and there is no bulk velocity.

At the beginning of this chapter, we identified the relative displacement of a species as the key to separation. What is the velocity of species i that would allow calculation of such a displacement? At first glance, $\overline{U}_{iz}$ may appear to be such a quantity. Note that F_{tiz} includes ordinary diffusion (molecular diffusion due to $-\nabla \ell n\, a_i$), which, arising from a random Brownian motion of molecules, is often detrimental to separation (see Section 3.2.1, in particular the text following equation (3.2.22)). We will separate its contribution to determine the most useful expression for displacement or separation velocity due to forces acting on individual molecules.

The key factor is consideration of the *forces acting on individual molecules*. In simple molecular diffusion, resulting from Brownian motion, an individual molecule is not subjected to any specific force, although we often tend to consider the concentration gradient as a force that causes the molecules to go from a region of higher concentration to a region of lower concentration. (Similarly, for gaseous phases, we have transport from a higher partial pressure region to a region with lower partial pressure.) That there is a flux of a species from a higher concentration (higher partial pressure) to a lower concentration (lower partial pressure) is simply a result of a larger number of molecules making random jumps across any surface area in the direction of decreasing concentration (partial pressure).

Calculate now the molar flux N_{iz} of the ith species (in a binary system) in the positive z-direction across a plane fixed with respect to stationary axes (Bird *et al.*, 1960, 2002). Assume *temporarily* that there can be bulk flow; the molar average and mass average velocities in the z-direction are v_{tz}^* and v_{tz}, respectively. If the averaged (over one mole, say) z-directional velocity of molecules of the ith species is v_{iz}, then the flux N_{iz} is defined by (see Figure 3.1.3B)

$$N_{iz} = C_i\, v_{iz}. \qquad (3.1.77)$$

Obviously[13]

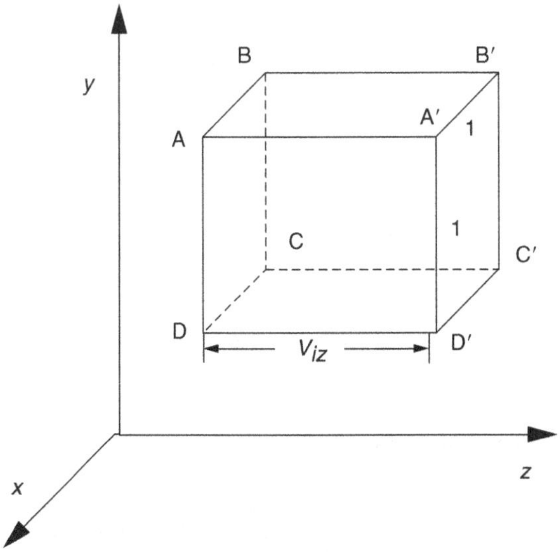

Figure 3.1.3B. Explanation for molar flux expression (3.1.77) of species i. Replace U_{pt} in Figure 3.1.3A by v_{iz}, particles by molecules of species i and particle number density N_t by C_i, the molar density of i in the volume element ABCDA′B′C′D′. We obtain $N_{iz} = C_i v_{iz}$, if we follow the explanation provided in the caption to Figure 3.1.3A.

$$v_{iz} = v_{tz}^* + \overline{U}_{iz}, \qquad (3.1.78)$$

so that

$$N_{iz} = C_i\, v_{tz}^* + C_i\, \overline{U}_{iz}. \qquad (3.1.79)$$

Using (3.1.76), we obtain

$$N_{iz} = C_i \left[v_{tz}^* + \frac{F_{tiz}}{f_i^d} \right]. \qquad (3.1.80)$$

From relations (3.1.27) and (3.1.42), we obtain *at constant T and P*

$$N_{iz} = C_i \left[v_{tz}^* + \left\{ -\frac{1}{f_i^d} \left(\frac{\partial \mu_i^o}{\partial z} - F_{tiz}^{\text{ext}} \right) \right\} \right] - \frac{RTC_i}{f_i^d} \frac{\partial \ell n\, a_i}{\partial z}. \qquad (3.1.81)$$

Now define the *displacement* or *migration velocity* of the ith species as U_{iz} such that, at *constant T and P*,

$$U_{iz} = -\frac{1}{f_i^d} \left(\frac{\partial \mu_i^o}{\partial z} - F_{tiz}^{\text{ext}} \right). \qquad (3.1.82)$$

The expression for N_{iz}, (3.1.81), may now be rewritten as follows:

$$N_{iz} = C_i\, (v_{tz}^* + U_{iz}) - C_i \left(\frac{RT}{f_i^d} \right) \frac{\partial\, \ell n\, a_i}{\partial z}. \qquad (3.1.83)$$

In vector notation,

$$\boldsymbol{N}_i = C_i\, (\boldsymbol{v}_t^* + \boldsymbol{U}_i) - C_i \left(\frac{RT}{f_i^d} \right) \nabla \ell n\, a_i \qquad (3.1.84a)$$

and

---

[12] See Lightfoot (1974, p. 54) for calculations of time scales for the transient behavior of red blood cells (about 10^{-4} seconds) and comments on protein molecules.

[13] The implication of $v_{iz} = v_{tz} + \overline{U}_{iz}$ (or $\boldsymbol{v}_i = \boldsymbol{v}_t + \overline{\boldsymbol{U}}_i$) is explored later in this section (see equation (3.1.102)). For an observed v_{iz}, this will lead to a different $\overline{U}_{iz}$.

$$\boldsymbol{U}_i = -\frac{1}{f_i^d}(\nabla\mu_i^0 - \boldsymbol{F}_{ti}^{\text{ext}}). \qquad (3.1.84\text{b})$$

The quantity $C_i \nabla \ell\text{n}\, a_i$ may be written as $\left(\frac{\text{d}\,\ell\text{n}\, a_i}{\text{d}\,\ell\text{n}\, C_i}\right)\nabla C_i$ in general. When $a_i = \gamma_i\, x_i$, then

$$C_i \nabla \ell\text{n}\, a_i = C_t\left(1 + \frac{\text{d}\,\ell\text{n}\,\gamma_i}{\text{d}\,\ell\text{n}\, x_i}\right)_{T,P} \nabla x_i. \qquad (3.1.84\text{c})$$

For *nonconstant P*,

$$\boldsymbol{U}_i = -\frac{1}{f_i^d}(\nabla\mu_i^0 - \boldsymbol{F}_{ti}^{\text{ext}} + \overline{V}_i \nabla P). \qquad (3.1.84\text{d})$$

We will now place the flux expression (3.1.84a) for $\boldsymbol{N}_i$ against our basic knowledge of mass flux expression in *molecular diffusion*. We have two objectives: first, to develop a simple way of estimating f_i^d, and second to show that we are familiar with the flux expression (3.1.84a) under a somewhat simpler situation. Consider an *ideal liquid solution* at constant T and P; therefore, $a_i = x_i$. Assume $C_i \cong C_t x_i$, with C_t being the constant total molar concentration. Further, let there be no external force in this single-phase liquid solution (i.e. $\boldsymbol{F}_{ti}^{\text{ext}} = 0; \nabla\mu_i^0 = 0$). Then there is only molecular diffusion, given by

$$\boldsymbol{N}_i = C_i \boldsymbol{v}_t^* - \frac{\text{R}T}{f_i^d}\nabla C_i. \qquad (3.1.85)$$

This equation is simply *Fick's first law of diffusion*. Note that $\boldsymbol{U}_i = 0$, but $\overline{\boldsymbol{U}}_i$ is nonzero. Furthermore, for this binary system of solute i in a liquid s, the ideal solution binary diffusion coefficient D_{is}^0 (valid at infinite dilution) is

$$D_{is}^0 = (\text{R}T/f_i^d). \qquad (3.1.86)$$

Using this identification for D_{is}^0, the general flux expression (3.1.83) may be rewritten as

$$N_{iz} = C_i(v_{tz}^* + U_{iz}) - C_t D_{is}^0\left(1 + \frac{\text{d}\,\ell\text{n}\,\gamma_i}{\text{d}\,\ell\text{n}\, x_i}\right)_{T,P}\frac{\partial x_i}{\partial z}, \qquad (3.1.87)$$

or

$$N_{iz} = C_i(v_{tz}^* + U_{iz}) - C_t D_{is}\frac{\partial x_i}{\partial z}, \qquad (3.1.88)$$

where the diffusion coefficient for a nonideal solution of solute i in solvent s, D_{is}, is defined by

$$D_{is} = D_{is}^0\left(1 + \frac{\text{d}\,\ell\text{n}\,\gamma_i}{\text{d}\,\ell\text{n}\, x_i}\right)_{T,P}. \qquad (3.1.89)$$

The flux expression (3.1.88) clearly shows how N_{iz} is increased by the presence of U_{iz}, *the displacement or migration velocity of species i in the z-direction due to forces that lead to separation.* The corresponding expression for U_{iz} in an isothermal isobaric system is given by

$$U_{iz} = \frac{D_{is}^0}{\text{R}T}\left[+F_{tiz}^{\text{ext}} - \frac{\partial\mu_i^0}{\partial z}\right], \qquad (3.1.90\text{a})$$

or

$$\boldsymbol{U}_i = \frac{D_{is}^0}{\text{R}T}[+\boldsymbol{F}_{ti}^{\text{ext}} - \nabla\mu_i^0], \qquad (3.1.90\text{b})$$

where we have assumed that D_{is}^0 is independent of the direction of diffusion.

Since U_{iz} is directly proportional to D_{is}^0, the higher the diffusion coefficient, the greater is the displacement or migration velocity. In the appendix to this chapter (see Tables 3.A.1–3.A.8), we list, the binary diffusion coefficients for a few gaseous, liquid and solid systems. The diffusion coefficients in solid systems are extremely low; those in liquid systems (D_{is}^0) are a couple of orders of magnitude larger, whereas the diffusivities D_{AB} for a gaseous system of species A and B are almost five orders of magnitude larger than those in liquid systems. For an introduction to diffusion in fluids, see Cussler (1997).

Values of diffusion coefficients are frequently needed in many separation calculations. Readers should refer to Reid *et al.* (1987) and Cussler (1997) for diffusion coefficients in liquids and gases. For immediate use, the following correlations may be used.

Dilute solution of i *in solvents* The Wilke-Chang (Wilke and Chang, 1955) correlation is given by

$$D_{is}^0 = 7.4 \times 10^{-8} \times (\psi_{\text{As}} M_S)^{1/2}\frac{T}{\mu_S V_{im}^{0.6}}\frac{\text{cm}^2}{\text{s}}, \qquad (3.1.91\text{a})$$

where

T is the absolute temperature in kelvin;
M_S is the molecular weight of the solvent;
μ_S is the solvent viscosity in centipoise;
V_{im} is the solute molar volume at boiling point, in cm^3/gmol (estimate from Reid *et al.* (1987));
ψ_{As} is the association factor for the solvent: 2.6 for water, 1.9 for methanol, 1.5 for ethanol and 1 for benzene, ether, heptane and other unassociated solvents.

Binary gaseous mixture of species A and B at low density The Chapman-Enskog formula (Bird *et al.*, 2002) is given by

$$D_{\text{AB}} = 0.0018583\,\frac{\sqrt{T^3\left(\frac{1}{M_{\text{A}}} + \frac{1}{M_{\text{B}}}\right)}}{P\sigma_{\text{AB}}^2\,\Omega_{D,\text{AB}}}\,\frac{\text{cm}^2}{\text{s}}, \qquad (3.1.91\text{b})$$

Here

T is the absolute temperature in kelvin;
P is the pressure in atmospheres;
$\sigma_{\text{AB}} = \frac{1}{2}(\sigma_{\text{A}} + \sigma_{\text{B}})$ Å, σ_{A} and σ_{B} are Lennard-Jones parameters for gases A and B (see Bird *et al.* (1960, table B-1, p. 744); When values of σ are not known, the following empirical relations (Bird *et al.*, 2002, p. 26) may be employed using critical point information:

$$\sigma(\text{Å}) = 0.841\,\tilde{V}_c^{1/3}; \qquad \sigma(\text{Å}) = 2.44\,(T_c/P_c)^{1/3}. \qquad (3.1.91\text{c})$$

Here, the critical molar volume $\tilde{V}_c$ has the units of cm^3/gmol, T_c is in kelvin and P_c is in atmospheres.

$\Omega_{D,AB}$ is a dimensionless function of temperature and intermolecular potential field (Bird *et al.*, 2002, table E.2).

The magnitudes of D^0_{is}, as shown in Table 3.A.1–3.A.8 (see the appendix to this chapter), vary widely. Such a variation has clear implications for the suitability of different phases in a separation system. For example, if the time required to achieve separation is to be determined from that required to traverse a given distance by a given species, then, other quantities remaining constant, this time for species i will be directly proportional to, say, U_{iz}. Solid systems will therefore require an extraordinarily large time compared to a liquid or gaseous system (see Table 3.3.1 for some practical consequences; namely, for a practical two-phase based separation system, one phase should be a fluid).

Of further interest here is the variation of f_i^d with the size of solute molecules i in a given system, liquid, gaseous or solid. Analogous to the value of $f_p^d = 6\pi\mu r_p$ (obtained from Stokes' law) for a single spherical particle of radius r_p falling in a medium of viscosity μ, f_i^d from equation (3.1.75) may be represented by $6\pi\mu r_i \tilde{N}$, where $\tilde{N}$ is Avogadro's number (6.02×10^{23}/gmol) and r_i is the hydrodynamic radius of an ith solute molecule. We can therefore write in general (for spherical molecules of hydrodynamic radius r_i) from equation (3.1.84d) that

$$\begin{aligned} \boldsymbol{U}_i &= -\frac{1}{f_i^d}(\nabla\mu_i^0 - \boldsymbol{F}_{ti}^{\text{ext}} + \overline{V}_i \nabla P) \\ &= -\frac{\nabla\mu_i^0 - \boldsymbol{F}_{ti}^{\text{ext}} + \overline{V}_i \nabla P}{6\pi\mu r_i \tilde{N}}. \end{aligned} \quad (3.1.91\text{d})$$

Estimates of r_i, the hydrodynamic radius, may be obtained from expressions (3.3.90a–c). (The basic assumption here is that Stokes' law holds reasonably well for the motion of solute molecules also. In fact, it is routinely used for proteins and macromolecules (Tanford, 1961).) Therefore for a given magnitude of the driving force, as solute size increases, the displacement or migration velocity U_{iz} decreases.

If the solute shape is nonspherical, a relation other than Stokes' law will apply. For the determination of the resistance of nonspherical macromolecules, the reader may consult pp. 356–364 of Tanford (1961). We will provide a very brief perspective on this effect here. Some cells, and especially many proteins, are ellipsoids of revolution. The drag force encountered by such an ellipsoid of revolution of species i is described in terms of the drag encountered by a sphere of equal volume whose radius is r_0 via an appropriate correction factor (f_i^d/f_{i0}^d) which is always greater than 1. This factor is called the Perrin factor. The magnitude of F_i^{drag} is enhanced by this factor, i.e.

$$F_i^{\text{drag}} = 6\pi\mu r_0 U_i (f_i^d/f_{i0}^d). \quad (3.1.91\text{e})$$

For prolate ellipsoids (semiaxes $a > b$),

Table 3.1.2. Estimation of Perrin factor for various proteins

Protein	Molecular weight	pI	r_0, radius, (nm)	f_i^d/f_{i0}^d
Bovine serum albumin	66 000	5.74	3.5	1.29
Ovalbumin	45 000	5.08	2.78	1.16
α-Lactalbumin	14 200	4.57	2.3	1.18
Myoglobin	16 900	7.1	2.4	1.18

$$\frac{f_i^d}{f_{i0}^d} = \frac{\sqrt{1-(b^2/a^2)}}{\left(\frac{b}{a}\right)^{2/3} \ln\left(\frac{1+\sqrt{1-(b^2/a^2)}}{b/a}\right)}, \quad (3.1.91\text{f})$$

where the volume of the prolate spheroid is equal to the volume of a sphere of radius r_0,

$$\frac{4}{3}\pi a b^2 = \frac{4}{3}\pi r_0^3. \quad (3.1.91\text{g})$$

For oblate ellipsoids (semiaxes $a < b$), correspondingly $r_0^3 = a^2 b$ and

$$\frac{f_i^d}{f_{i0}^d} = \frac{\sqrt{1-(a^2/b^2)}}{\left(\frac{a}{b}\right)^{1/3} \tan^{-1}\left[\left(1-\frac{a^2}{b^2}\right)^{1/2} \frac{b}{a}\right]}. \quad (3.1.91\text{h})$$

For a brief introduction, see Probstein (1989) for references to the work by Perrin; for greater details, see Happel and Brenner (1983). In Table 3.1.2, we list estimates of the Perrin factor for several proteins (information from Basak and Ladisch (1995)).

The increase in the frictional coefficient is natural since the surface area for the spheroid is larger than that of a sphere of equal volume. In any given medium, when the solute size increases, the frictional coefficient will increase. For a given solute, as the medium viscosity increases, f_i^d will increase, leading to a decrease in diffusion coefficients and therefore in the displacement velocity. For linear flexible macromolecules, especially charged ones such as polyelectrolytes, estimation of the drag force is complicated; see Viovy (2000) for an introduction. Note that $\boldsymbol{U}_i$ for a polyelectrolyte that is uniformly charged is likely to be independent of its length in free solution; the situation is different in gels.

Example 3.1.5 *Case (1)* Calculate the displacement or migration velocity of ovalbumin molecules subjected to the conditions in a centrifuge described in Example 3.1.2, case (1), where the solution viscosity is 1 cp.

Case (2) Calculate the terminal settling velocity of the particles in Example 3.1.2, case (2), where the solution viscosity in 1.5 cp. You are given that for ovalbumin $r_i = 2.78$ nm and $(f_i^d/f_{i0}^d) = 1.16$.

Solution *Case (1)* If we assume Stokes' law to be valid, then the radial displacement/migration velocity of ovalbumin molecules is given by

$$U_{ir} = (F_{tir/\text{gmol}})/f^d_{i|\text{gmol}} = \frac{5.60 \times 10^{11}\ \text{g-cm/s}^2}{6\pi\mu r_p \tilde{N}\,(f_i^d/f_{i0}^d)}$$

$$= \frac{5.60 \times 10^{11}\ \dfrac{\text{g-cm}}{\text{s}^2}}{6\pi \times 1 \times 10^{-2}\dfrac{\text{g}}{\text{cm-s}} \times 27.8 \times 10^{-8}\text{cm}}$$

$$\times 1.16 \times 6.02 \times 10^{23}$$

$$= \frac{5.60 \times 10^{11} \times 10^{-13}}{36.12 \times \pi \times 32.48}\ \frac{\text{cm}}{\text{s}} = 0.151 \times 10^{-4}\ \text{cm/s}.$$

Check now the validity of Stokes' law:

$$\text{Reynolds no.} = \frac{U_{ir} r_i \rho}{\mu}$$

$$= \frac{0.151 \times 10^{-4}\ \frac{\text{cm}}{\text{s}} \times 27.8 \times 10^{-8}\ \text{cm} \times 1\ \frac{\text{g}}{\text{cm}^3}}{1 \times 10^{-2} (\text{g/cm s})}$$

$$= 4.19 \times 10^{-10} <<< 0.1;$$

assumption okay.

Case (2) U_{prt}, the terminal settling velocity of the particle, is given by

$$U_{prt} = \frac{F_{tr}|_{\text{particle}}}{6\pi\mu r_p} = \frac{0.179 \times 10^{-3}\ \text{g-cm/s}^2}{6\pi \times 1.5 \times 10^{-2}\ \dfrac{\text{g}}{\text{cm-s}} \times 5 \times 10^{-4}\text{cm}}$$

$$= \frac{0.179 \times 10^{3}}{45\pi}\ \frac{\text{cm}}{\text{s}} = 1.26\ \text{cm/s}.$$

Check for validity of Stokes' law:

$$\text{Reynolds no.} = \frac{U_{prt} r_p \rho}{\mu} = \frac{1.26 \times 5 \times 10^{-4} \times 1}{1.5 \times 10^{-2}} = 0.042 < 0.1;$$

Stokes' law is valid.

Example 3.1.6 Compute the migration velocity of ovalbumin molecules subjected to an electrical field of uniform strength described in Example 3.1.3. The solution viscosity is 1 cp. Other required information is provided in Example 3.1.5, case (1).

Solution The migration velocity of ovalbumin molecules is

$$U_i = \frac{F_{ti}|_{\text{gmol}}}{6\pi\mu r_p \tilde{N}\,(f_i^d/f_{i0}^d)}$$

$$= \frac{1.3 \times 10^9\ \text{newton}}{6\pi \times 1 \times 10^{-2}\ \dfrac{\text{g}}{\text{cm-s}} \times 28 \times 10^{-8}\ \text{cm} \times 6.02 \times 10^{23} \times 1.16}$$

$$= \frac{1.3 \times 10^9 \times 10^5\ \dfrac{\text{g-cm}}{\text{s}^2}}{6\pi \times 28 \times 6.02 \times 1.16 \times 10^{13}\ \dfrac{\text{g}}{\text{s}}}$$

$$= \frac{1.3 \times 10^{14} \times 10^{-13}}{3685}\ \frac{\text{cm}}{\text{s}} = 0.35 \times 10^{-2}\ \text{cm/s}.$$

This migration velocity appears to be much larger than that generated in the centrifugal field of Example 3.1.5, case (1).

Example 3.1.7 Determine the gravitational terminal settling velocity of particles in the following two systems.

(1) Particle of diameter 10 μm, density 2 g/cm³ in a solution of density 1.3 g/cm³ and viscosity 1.5 cp.
(2) *E. coli* cells of diameter 1 μm, particle density 1.09 g/cm³, solution density 1 g/cm³ and viscosity 1.5 cp.

Solution *Case (1)*

$$U_{pzt} = -\frac{|F_{tz}|}{6\pi\mu r_p} = -\frac{\frac{4}{3}\pi r_p^3 \rho_p g\,(1 - (\rho_t/\rho_p))}{6\pi \times 1.5 \times 10^{-2}\ \dfrac{\text{g}}{\text{cm-s}} \times r_p\ \text{cm}}$$

$$= -\frac{2 r_p^2 g \rho_p (1 - (1.3/2))}{9 \times 1.5 \times 10^{-2}}\ \frac{\text{cm}}{\text{s}}$$

$$= -\frac{2 \times (5 \times 10^{-4})^2 \times 980 \times 2 \times 0.35}{13.5 \times 10^{-2}}$$

$$= -\frac{100 \times 980 \times 0.35 \times 10^{-8} \times 10^2}{13.5}\ \frac{\text{cm}}{\text{s}}$$

$$= -\frac{0.35 \times 0.98 \times 10^{-1}}{13.5} = -2.54 \times 10^{-3}\ \frac{\text{cm}}{\text{s}}.$$

The terminal velocity in the centrifugal field of Example 3.1.5, case (2), is much larger than that due to the gravitational field.

Case (2)

$$U_{pzt} = -\frac{2 r_p^2 g \rho_p (1 - (1/1.09))}{9 \times 1.5 \times 10^{-2}}\ \frac{\text{cm}}{\text{s}}$$

$$= -\frac{2 \times (0.5 \times 10^{-4})^2 \times 980 \times 1.09 \times (1 - 0.917)}{13.5 \times 10^{-2}}\ \frac{\text{cm}}{\text{s}}$$

$$= -\frac{2 \times 0.25 \times 10^{-8} \times 980 \times 1.09 \times 0.083}{13.5 \times 10^{-2}}$$

$$= -\frac{0.5 \times 0.98 \times 1.09 \times 0.83 \times 10^{-4}}{13.5}\ \frac{\text{cm}}{\text{s}}$$

$$= -0.32 \times 10^{-5}\text{cm/s}.$$

If we had used the centrifugal field of Example 3.1.2, case (2), we would have obtained a settling velocity of

$$U_{prt} = \frac{4\pi r_p^3 \rho_p r\omega^2 (1 - (\rho_t/\rho_p))}{3 \times 6\pi\mu r_p} = \frac{2}{9}\ \frac{r_p^2 r\omega^2 \rho_p (1 - (\rho_t/\rho_p))}{\mu}$$

$$= \frac{2 \times (0.5 \times 10^{-4})^2 \times 1 \times (7000)^2 \times 1.09 \times (1 - 0.917)}{9 \times 1.5 \times 10^{-2}}$$

$$= \frac{0.5 \times 10^{-8} \times 49 \times 10^6 \times 1.09 \times 0.083}{13.5 \times 10^{-2}}$$

$$= \frac{0.5 \times 49 \times 1.09 \times 0.083}{13.5}\ \frac{\text{cm}}{\text{s}} = 0.164\ \text{cm/s},$$

an orders of magnitude higher velocity than that due to gravity. Gravity is therefore not likely to be used for the separation of *E. coli* cells from water.

Example 3.1.8 Suppose you have a solution of a solute of molecular weight 100 such that the concentration of the solute on two sides of the solution are 10 gmol/cm^3 and 1 gmol/cm^3. Calculate the value of U_i from the high-concentration region to the low-concentration region.

Solution This is a case of random Brownian diffusion; U_i cannot be defined for this case.

Example 3.1.9 In Example 3.1.8, the phenomenon under consideration was ordinary molecular diffusion. Ordinary molecular diffusion due to a concentration gradient or partial pressure gradient (see relations (3.1.40) for example) does result in a molecular velocity v_i (we assume no bulk motion of the gas or liquid phase). Here we would like to calculate this velocity for arbitrary values of concentration difference or partial pressure difference over a diffusional path length.

(1) Ovalbumin molecules in an aqeuous solution (stagnant) diffuse due to a concentration ratio of 10 over a distance of 10 cm.
(2) Same as case (1) except the magnitude of $\nabla \ell n\, a_i$ is 1 cm^{-1}.

Solution

$$\mathbf{F}_{ti} = -RT\,\nabla \ell n\, a_i;$$

$$\mathbf{v}_i = \frac{\mathbf{F}_{ti}}{f_i^d} = -\frac{RT}{f_i^d}\nabla \ell n\, a_i = -D_{is}\nabla \ell n\, a_i.$$

Case (1)

$$v_{iz} = +D_{is}\frac{\ell n\,(a_{i1}/a_{i2})}{10} = +7\times 10^{-7}\,\frac{\text{cm}^2}{\text{s}}\,\frac{2.303\times \log 10}{10}$$
$$= 7\times 2.303\times 10^{-8}\text{cm/s} = 1.61\times 10^{-7}\text{ cm/s}.$$

Case (2)

$$v_{iz} = 7\times 10^{-7}\times 1 = 7\times 10^{-7}\text{cm/s}.$$

Note: These velocities are orders of magnitude smaller than the values of $\boldsymbol{U}_{iz}$ for the ovalbumin molecules in Examples 3.1.5 (centrifugal acceleration) and 3.1.6 (uniform electrical field). Therefore diffusion/Brownian motion is a small nuisance in the separation of proteins in these external force fields having appropriate values.

Example 3.1.10 If, however, we had considered the diffusion of a gas molecule in a gaseous medium, the diffusion coefficient would be orders of magnitude higher (e.g. $D_{\text{He-}N_2} = 0.687$ cm^2/s at 25 °C); therefore the values of v_{iz} will also be so much larger (for, say, $\nabla\, \ell n\, a_i$ of 1 cm^{-1}).

Example 3.1.11 Determine the terminal velocity achieved by a spherical particle of radius 1.34 μm in water when subjected to a cw argon laser of power Pr = 19 mW. The laser beam radius r_l is 6.2 μm; its wavelength $\lambda = 0.5145$ μm. The viscosity of the liquid medium (water) is 1 cp. Assume $q_{fr} = 0.1$.

Solution Since the laser beam cross section corresponds to a diameter (r_l) of 6.2 μm, only a small fraction of its total power of 10 mW is falling on the particle of radius 1.34 μm; therefore the value of power Pr in equation (3.1.47) is

$$Pr = \frac{\text{laser power}\times\text{particle cross-sectional area}}{\text{laser beam cross-sectional area}}$$
$$= \frac{19\text{ mW}\times \pi r_p^2}{\pi r_l^2} = 19\text{ mW}\times (r_p/r_l)^2.$$

Introduce it into equation (3.1.47) to obtain

$$F_{\text{rad}} = \frac{2q_{fr}\,Pr}{\text{c}} = \frac{2\times 0.1\times 19\text{mW}\times r_p^2}{\text{c}\ r_l^2} = 6\pi\mu r_p U_{pzt}$$

$\Rightarrow U_{pzt}$ = particle terminal velocity

$$= \frac{2\times 0.1\times 19\times 10^{-3}\text{watt}\times r_p^2}{\text{c}\times 6\pi\times(1\text{ cp})\times r_p\times r_l^2}$$

$$= \frac{2\times 0.1\times 1.34\times 10^{-4}\text{ cm}\times 19\times 10^{-3}\text{watt}\times 10^7\dfrac{\text{g-cm}^2}{\text{s}^2\text{-s}}\dfrac{1}{\text{watt}}}{6\times 2.799\times 10^{10}\dfrac{\text{cm}}{\text{s}}\times\pi\times 10^{-2}\dfrac{\text{g}}{\text{cm-s}}\times 6.2\times 6.2\times 10^{-8}\text{cm}^2}$$

$$= \frac{2\times 0.1\times 1.34\times 19}{6\times 2.799\times\pi\times 6.2\times 6.2}\text{cm/s} \cong 25\times 10^{-4}\text{ cm/s}$$

(Ashkin, 1970). *Note*: This radiation pressure driven motion has other complex dimensions. If the laser beam hits the spherical particle off-center, then the particle is drawn also to the laser beam axis.

We need to have additional considerations on the *i*th species flux expression (3.1.87). The vectorial form of this expression for a liquid solution is good for a binary system of species *i* and species *s*

$$\boldsymbol{N}_i = C_i\,(\boldsymbol{v}_t^* + \boldsymbol{U}_i) - C_t D_{is}^0\left(1 + \frac{\text{d}\,\ell n\,\gamma_i}{\text{d}\,\ell n\,x_i}\right)_{T,P}\nabla x_i \quad (3.1.92)$$

Here species *i* moves only (1) due to concentration gradient of species *i* (since $C_t\nabla x_i$ may be written as ∇C_i), (2) the external species *i*-specific forces and (3) the bulk motion (if any are considered via $\boldsymbol{v}_t^*$). If there is a concentration gradient of any other species or some external forces specific to that other species, they are not supposed to influence $\boldsymbol{N}_i$ as written above except through their contribution to the bulk velocity $\boldsymbol{v}_t^*$.

Equation (3.1.92) is most useful for liquid-phase systems. For a *binary gaseous phase mixture of species* A *and* B, use equations (3.1.81), (3.1.82), (3.1.86) and (3.1.88) to obtain the flux of species A as

$$N_{\text{A}z} = C_{\text{A}}\left[v_{tz}^* + U_{\text{A}z}\right] - D_{\text{AB}}\,C_{\text{A}}\frac{\partial\,\ell n\hat{f}_{\text{A}}}{\partial z}, \quad (3.1.93)$$

where we have assumed that $(\partial\,\ell n f_{\text{A}}^0/\partial z) = 0$.

For an *ideal gas mixture*, $\hat{f}_{\text{A}} = p_{\text{A}}$ and $p_{\text{A}} = C_{\text{A}}RT$; the above equation is then reduced to

$$N_{\text{A}z} = C_{\text{A}}\left[v_{tz}^* + U_{\text{A}z}\right] - \frac{D_{\text{AB}}}{RT}\frac{\text{d}p_{\text{A}}}{\text{d}z}. \quad (3.1.94)$$

Vectorially, this equation is written as

$$\boldsymbol{N}_{\text{A}} = C_{\text{A}}\left[\boldsymbol{v}_t^* + \boldsymbol{U}_{\text{A}}\right] - \frac{D_{\text{AB}}}{RT}\nabla p_{\text{A}}. \quad (3.1.95)$$

Since $D_{\text{AB}} = D_{\text{BA}}$, the flux expression for species B is, correspondingly,

Table 3.1.3A. Different mass and molar fluxes of a species and their relationships to various velocities[a]

Frame of reference	Mass flux of species i	Sum of mass fluxes	Molar flux of species i	Sum of molar fluxes
Flux in a fixed frame of reference (fixed coordinate axes)	$\boldsymbol{n}_i = \rho_i \boldsymbol{v}_i$ (a)	$\sum_{i=1}^{n} \boldsymbol{n}_i = \rho \boldsymbol{v}_t$ (b)	$\boldsymbol{N}_i = C_i \boldsymbol{v}_i$ (c)	$\sum_{i=1}^{n} \boldsymbol{N}_i = C_t \boldsymbol{v}_t^*$ (d)
Flux in a frame of reference moving with mass average velocity, $\boldsymbol{v}_t$	$\boldsymbol{j}_i = \rho_i(\boldsymbol{v}_i - \boldsymbol{v}_t)$ (e)	$\sum_{i=1}^{n} \boldsymbol{j}_i = 0$ (f)	$\boldsymbol{J}_i = C_i(\boldsymbol{v}_i - \boldsymbol{v}_t)$ (g)	$\sum_{i=1}^{n} \boldsymbol{J}_i - C_t(\boldsymbol{v}_t^* - \boldsymbol{v}_t)$ (h)
Flux in a frame of reference moving with molar average velocity, $\boldsymbol{v}_t^*$	$\boldsymbol{j}_i^* = \rho_i(\boldsymbol{v}_i - \boldsymbol{v}_t^*)$ (i)	$\sum_{i=1}^{n} \boldsymbol{j}_i^* = \rho(\boldsymbol{v} - \boldsymbol{v}_t^*)$ (j)	$\boldsymbol{J}_i^* = C_i(\boldsymbol{v}_i - \boldsymbol{v}_t^*)$ (k)	$\sum_{i=1}^{n} \boldsymbol{J}_i^* = 0$ (l)

[a] $\boldsymbol{v}_i$, averaged velocity vector of species i; $\boldsymbol{v}_t$, mass averaged velocity (equation (2.1.1), $\boldsymbol{v}_t^*$, molar averaged velocity (equation (2.1.2)).

Table 3.1.3B. Important relations between different fluxes

Frame of reference	Mass flux of species i	Molar flux of species i
Flux in a fixed frame of reference (fixed coordinate axes)	$\boldsymbol{n}_i$	$\boldsymbol{N}_i = (\boldsymbol{n}_i/M_i)$
Flux in a frame of reference moving with $\boldsymbol{v}_t$	$\boldsymbol{j}_i$	$\boldsymbol{J}_i = (\boldsymbol{j}_i/M_i)$
Flux in a frame of reference moving with $\boldsymbol{v}_t^*$	$\boldsymbol{j}_i^*$	$\boldsymbol{J}_i^* = (\boldsymbol{j}_i^*/M_i)$
Relation between $\boldsymbol{N}_i$ and $\boldsymbol{J}_i^*$	$\boldsymbol{N}_i = C_i \boldsymbol{v}_t^* + \boldsymbol{J}_i^*$ (*a*)	$\boldsymbol{N}_i = x_i \sum_{i=1}^{n} \boldsymbol{N}_i + \boldsymbol{J}_i^*$ (*b*)
Relation between $\boldsymbol{j}_i$ and $\boldsymbol{n}_i$	$\boldsymbol{n}_i = \rho_i \boldsymbol{v}_t + \boldsymbol{j}_i$ (c);	$\boldsymbol{n}_i = \omega_i \sum_{i=1}^{n} \boldsymbol{n}_i + \boldsymbol{j}_i$ (*d*)

Table 3.1.3C. Relations between important species fluxes, bulk velocities, migration velocities and components of chemical potential gradients

$$\boldsymbol{N}_i = C_i(\boldsymbol{v}_t^* + \boldsymbol{U}_i) - C_i\left(\tfrac{\mathrm{RT}}{f_i^d}\right)\nabla \ell\mathrm{n}\, a_i = C_i\left(\boldsymbol{v}_t^* - \tfrac{1}{f_i^d}\left(\nabla\mu_i^0 - \boldsymbol{F}_{ti}^{\mathrm{ext}} + \overline{V}_i \nabla P\right)\right) - C_i(\mathrm{R}T/f_i^d)\nabla \ell\mathrm{n}\, a_i \text{ (a)}$$

$$\boldsymbol{J}_i^* = C_i \boldsymbol{U}_i - C_i\left(\tfrac{\mathrm{R}T}{f_i^d}\right)\nabla \ell\mathrm{n}\, a_i = C_i\left(-\tfrac{1}{f_i^d}\left(\nabla\mu_i^0 - \boldsymbol{F}_{ti}^{\mathrm{ext}} + \overline{V}_i \nabla P\right)\right) - C_i(\mathrm{R}T/f_i^d)\nabla \ell\mathrm{n}\, a_i \text{ (b)}$$

$$\boldsymbol{J}_i = C_i\left((\boldsymbol{v}_t^* - \boldsymbol{v}_t) + \boldsymbol{U}_i\right) - C_i\left(\tfrac{\mathrm{R}T}{f_i^d}\right)\nabla \ell\mathrm{n}\, a_i \text{ (c)}$$

$$\boldsymbol{N}_\mathrm{B} = C_\mathrm{B}\left[\boldsymbol{v}_t^* + \boldsymbol{U}_\mathrm{B}\right] - \frac{D_\mathrm{AB}}{\mathrm{R}T}\nabla p_\mathrm{B}. \qquad (3.1.96)$$

In simple molecular diffusion in a binary system under isobaric condition,

$$\boldsymbol{U}_\mathrm{A} = 0 = \boldsymbol{U}_\mathrm{B}; \qquad \nabla(p_\mathrm{A} + p_\mathrm{B}) = 0.$$

Thus

$$\boldsymbol{N}_\mathrm{A} + \boldsymbol{N}_\mathrm{B} = \boldsymbol{v}_t^*(C_\mathrm{A} + C_\mathrm{B}) = C_t \boldsymbol{v}_t^*. \qquad (3.1.97)$$

Expression (3.1.92) describes the species i flux in a binary system across a cross section stationary with respect to the fixed coordinate axes. There can be a number of other flux expressions depending on the chosen nature of the motion of this cross section (across which the species flux takes place) with respect to the fixed coordinate axes. If the cross section moves with the mass average velocity of the mixture $\boldsymbol{v}_t$, the molar flux identified as $\boldsymbol{J}_i$ is given by

$$\boldsymbol{J}_i = C_i\{(\boldsymbol{v}_t^* - \boldsymbol{v}_t) + \boldsymbol{U}_i\} - C_t D_{is}^0\left(1 + \frac{d\,\ell\mathrm{n}\,\gamma_i}{d\,\ell\mathrm{n}\,x_i}\right)_{T,P}\nabla x_i. \qquad (3.1.98)$$

If the cross section moves with the molar average velocity of the mixture $\boldsymbol{v}_t^*$, the molar flux identified as $\boldsymbol{J}_i^*$ is expressed as

$$\boldsymbol{J}_i^* = C_i \boldsymbol{U}_i - C_t D_{is}^0\left(1 + \frac{d\,\ell\mathrm{n}\,\gamma_i}{d\,\ell\mathrm{n}\,x_i}\right)_{T,P}\nabla x_i. \qquad (3.1.99)$$

Various other mass and molar fluxes and some relations between the different flux expressions are explored in Section 3.1.3.2.1. The expressions for $\boldsymbol{J}_i$ and $\boldsymbol{J}_i^*$ for simple molecular diffusion when $\boldsymbol{U}_i$ is zero are given below:

$$\boldsymbol{J}_i = C_i(\boldsymbol{v}_t^* - \boldsymbol{v}_i) - C_t D_{is}^0\left(1 + \frac{d\,\ell\mathrm{n}\,\gamma_i}{d\,\ell\mathrm{n}\,x_i}\right)_{T,P}\nabla x_i; \quad (3.1.100)$$

$$\boldsymbol{J}_i^* = -C_t D_{is}^0\left(1 + \frac{d\,\ell\mathrm{n}\,\gamma_i}{d\,\ell\mathrm{n}\,x_i}\right)_{T,P}\nabla x_i. \qquad (3.1.101)$$

Tables 3.1.3A–C summarize three items regarding the various fluxes, species velocities and forces: Table 3.1.3A identifies the definitions of species fluxes with respect to different frames of references with respect to the individual species velocities; Table 3.1.3B provides the interrelationships between the different flux expressions in Table 3.1.3A; Table 3.1.3C illustrates the relation between the important fluxes, bulk velocities, migration velocities and components of chemical potential gradients.

The definition in (3.1.90b) of the migration velocity $\boldsymbol{U}_i$ is independent of any of the frames of reference used above to define the fluxes $\boldsymbol{J}_i$ and $\boldsymbol{J}_i^*$, $\boldsymbol{N}_i$, etc. However, if we define $\overline{\boldsymbol{U}}_i$ with respect to the observed species velocity $\boldsymbol{v}_i$ by

$$\boldsymbol{v}_i = \boldsymbol{v}_t + \overline{\boldsymbol{U}}_i^{\text{new}}, \qquad (3.1.102)$$

instead of the definition (3.1.78), we will get

$$\boldsymbol{N}_i = C_i\,\boldsymbol{v}_t \; + \; C_i\,\overline{\boldsymbol{U}}_i^{\text{new}} \qquad (3.1.103)$$

$$= C_i\,\boldsymbol{v}_t \; + \; C_i\,\boldsymbol{U}_i^{\text{new}} - \frac{RT\,C_i}{f_i^d}\,\nabla\,\ell n\,a_i. \qquad (3.1.104)$$

This suggests that the alternately defined migration velocity $\boldsymbol{U}_i^{\text{new}}$ is related to $\boldsymbol{U}_i$ of definition (3.1.90b) by

$$\boldsymbol{U}_i = \boldsymbol{U}_i^{\text{new}} \; + \; (\boldsymbol{v}_t - \boldsymbol{v}_t^*). \qquad (3.1.105)$$

These two definitions become identical only when $\boldsymbol{v}_t = \boldsymbol{v}_t^*$ (which is true, for example, for a dilute solution of species i in solvents s). The flux expressions described earlier for different frames of reference will also change if $\boldsymbol{U}_i^{\text{new}}$ is used.

It is appropriate to indicate now an approximate form of flux-force relation in a binary isothermal liquid mixture in the presence of an *electrical potential gradient*. It is called the *Nernst-Planck relation* and is used for systems containing ions:

$$\boldsymbol{J}_i^* = -C_t\,D_{is}^N\left(\nabla x_i + \frac{x_i\,Z_i\,\mathcal{F}}{RT}\,\nabla\boldsymbol{\Phi}\right). \qquad (3.1.106)$$

It has often been used for a dilute binary solution of species i and solvent s under the following conditions: in the presence of (i) a concentration gradient and (ii) an electrical potential gradient; there is no convection; $\nabla P = 0$. Here D_{is}^N is the Nernst-Planck binary diffusion coefficient for species i and solvent s. In the absence of any electrical field $\nabla\boldsymbol{\Phi}$ in the system, it reduces to the ordinary binary diffusion coefficient D_{is}^0 if the solution behaves ideally. Note that it is not necessary for $\nabla\boldsymbol{\Phi}$ to be generated by an external source; it can be generated within the system by diffusion of ionic species. An individual ion has no way of knowing the source of the electrical field (Helfferich, 1962).

In the solution of an electrolyte, the intrinsic migration velocities of ions (see equations (6.3.8c, e) and Table 6.3.1) are, in general, different. Thus, the velocity of the cation is different from that of the anion. In the absence of any external electric field, these differential migration velocities create a potential gradient, the *diffusion potential*, which reduces the speed of the faster ion and increases the speed of the slower ion. Ultimately, the rate of diffusion of each ionic species is equal to that of the binary electrolyte. The diffusion coefficient D of a binary electrolyte, e.g. $A_{\nu+}^{Z+}Y_{\nu-}^{Z-}$ in a dilute solution, given by (Newman, 1973, 1991)

$$D = \frac{D_+^0\,D_-^0\,(Z_+ - Z_-)}{(Z_+\,D_+^0 - Z_-\,D_-^0)}, \qquad (3.1.107)$$

reflects this phenomenon. Here D_+^0 and D_-^0 are the intrinsic diffusion coefficients of the cation and the anion, respectively (see illustrative values in Table 3.A.4). Section 3.1.3.2.2 is concerned with developing this result for dilute solutions with concentration-independent diffusion coefficients in the absence of any current and externally applied electrical field. The two basic relations needed for this development are given below.

For calculations in electrolytic solutions, which may or may not be subjected to an electrical field, the *principle of electroneutrality* holds (see Chapman (1969) or Newman (1973, 1991) for an introduction to ionic transport and electrochemical systems):

$$\sum_{i=1} Z_i\,C_i = 0. \qquad (3.1.108a)$$

This condition implies the absence of any net charge anywhere (except in the double layer (Figure 3.1.2D) and electrode surfaces). In the absence of any electrical current in the elelctrolytic solution,

$$\sum_{i=1} Z_i\,\boldsymbol{N}_i = 0, \qquad (3.1.108b)$$

since the current density $\boldsymbol{i}$ (amp/m^2) is defined by

$$\boldsymbol{i} = \mathcal{F}\sum_{i=1} Z_i\,\boldsymbol{N}_i. \qquad (3.1.108c)$$

The current in a solution is carried by all ionic species, $i = 1,\ldots,n$. The fraction of the total current carried by a particular ionic species i in a solution is defined as its *transport number*,

$$t_{is} = \frac{Z_i\mathcal{F}\,\boldsymbol{N}_i}{\sum_{k=1}^{n} Z_k\mathcal{F}\,\boldsymbol{N}_k}. \qquad (3.1.108d)$$

The transport number of any ion is always positive. Obviously,

$$\sum_{i=1}^{n} t_{is} = 1. \qquad (3.1.108e)$$

When electric current flows in an aqueous electrolytic solution, some amount of water is also transferred along with the ions in the current direction. The *transference number* of water, $\bar{t}_w$, is defined as the number of water moles transported with 1 faraday of electricity in the current direction. Substitute w by i, and $\bar{t}_{is}$ is the transference number of any species i in the solution. It is related to the transport number t_{is} by

$$t_{is} = \bar{t}_{is}\,Z_i. \qquad (3.1.108f)$$

Note that, unlike t_{is}, $\bar{t}_{is}$ may be positive or negative. But t_{ws} is zero for water always; the transport number for uncharged water molecules is always zero.

The Nernst-Planck equation (3.1.106) is valid in the absence of any convection. In the *presence of convection*, one can use the *extended Nernst-Planck equation*:

$$\boldsymbol{N}_i = -D_{is}^N\left(\nabla C_i + \frac{C_i\,Z_i\,\mathcal{F}}{RT}\,\nabla\boldsymbol{\Phi} \; + \; C_i\,\text{grad}\,\ell n\,\gamma_i\right) + C_i\,\boldsymbol{v}_t, \qquad (3.1.108g)$$

where γ_i is the activity coefficient of the ith ionic species. The activity coefficient term is often neglected, and the mass velocity $\boldsymbol{v}_t$ is used instead of molar average velocity.

At this time, one should ask what is the *migration velocity* $\boldsymbol{U}_i$ of the ith ionic species exposed to an electrical field $-\nabla\Phi$ and what is its contribution to the total ionic flux $\boldsymbol{N}_i$? From definitions (3.1.82) and (3.1.84b),

$$\boldsymbol{U}_i = -\frac{1}{f_i^d}\,(\nabla\mu_i^0 - \boldsymbol{F}_{ti}^{\text{ext}}).$$

In a solution, $\nabla\mu_i^0$ is zero since the standard state μ_i^0 exists everywhere. Per gmol of ith ionic species, the value of $\boldsymbol{F}_{ti}^{\text{ext}}$ is $-Z_i\,\mathcal{F}\,\nabla\,\Phi$. Therefore for an ionic species i in solution,

$$\boldsymbol{U}_i = -\frac{Z_i\mathcal{F}\nabla\Phi}{f_i^d} \tag{3.1.108h}$$

The flux due to this ionic velocity when the molar ion concentration is C_i is

$$C_i\,\boldsymbol{U}_i = -\frac{C_i\,Z_i\,\mathcal{F}\,\nabla\Phi}{f_i^d} = -Z_i\,u_i^m\,\mathcal{F}\,C_i\,\nabla\Phi. \tag{3.1.108i}$$

Here u_i^m is identified as the ionic mobility of the ith ion and is the average velocity of the ionic species in a solution when acted on by a force of 1 newton/mol. Therefore

$$\boldsymbol{U}_i = u_i^m(-Z_i\mathcal{F}\nabla\,\Phi). \tag{3.1.108j}$$

The units of the ionic mobility u_i^m are m/(s-newton/gmol).

From the extended Nernst–Planck equation (3.1.108g), we see that

$$C_i\,\boldsymbol{U}_i \;=\; -Z_i\,u_i^m\,\mathcal{F}\,C_i\,\nabla\Phi = \frac{-D_{is}^N\,Z_i\mathcal{F}}{RT}\,C_i\nabla\Phi$$

$$\Rightarrow\quad u_i^m = \frac{D_{is}^N}{RT} \Rightarrow D_{is}^N = \text{R}T\,u_i^m. \tag{3.1.108k}$$

This result is called the *Nernst-Einstein relation*; it relates the diffusivity of the ion to the ionic mobility and is valid only at infinite dilution. It is clear from relations (3.1.108h) and (3.1.108i) and the definition (3.1.86) of diffusion coefficient D_{is}^0 at infinite dilution that

$$u_i^m = \frac{1}{f_i^d} = \frac{D_{is}^0}{RT}, \tag{3.1.108l}$$

leading to relation (3.1.108k). However, in the conventional literature on capillary electrophoresis etc., ionic mobility, μ_i^m, is defined as

$$\boldsymbol{U}_i = \mu_i^m(-\nabla\,\Phi) = \mu_i^m\,\boldsymbol{E}. \tag{3.1.108m}$$

The units of this ionic mobility μ_i^m are either

$$\frac{\text{m}}{\text{s-volt/m}} \quad\text{or}\quad \frac{\text{cm}}{\text{s-volt/cm}}.$$

Comparing this definition of μ_i^m in the context of relation (3.1.108h), we see that

$$\mu_i^m = \frac{Z_i\,\mathcal{F}}{f_i^d} = Z_i\,\mathcal{F}\,u_i^m. \tag{3.1.108n}$$

Recall that f_i^d here is the frictional coefficient for 1 gmol of ions whose total charge is $Z_i\mathcal{F}$ coulombs; there are 6.022×10^{23} ionized molecules in 1 gmol, each having a charge of Z_i e, where e is the charge of an electron, namely 1.602×10^{-19} coulomb, leading to a total charge of

$$Z_i\text{e}\tilde{\text{N}} = Z_i\times1.602\times10^{-19}\times6.022\times10^{23}\,\frac{\text{coulomb}}{\text{gmol}}$$
$$= Z_i\times96500\,\text{coulomb/gmol}.$$

Equation (3.1.108c) has already defined the current density i (amp/m^2) in the presence of the electrical field. If there is no bulk fluid motion and no concentration gradients, the species in solution move due to the electrical field only; therefore

$$\boldsymbol{i} = \mathcal{F}\sum_{i=1} Z_i\,\boldsymbol{N}_i = \mathcal{F}^2\sum_{i=1} Z_i^2\,u_i^m\,C_i(-\nabla\Phi), \tag{3.1.108o}$$

which may be written in the manner of Ohm's law as

$$\boldsymbol{i} = \sigma(-\nabla\Phi) = \sigma\,\boldsymbol{E}, \tag{3.1.108p}$$

where σ, defined as

$$\sigma = \mathcal{F}^2\sum_{i=1} Z_i^2\,u_i^m\,C_i, \tag{3.1.108q}$$

is the *electrical conductivity* of the solution and has units of siemens/meter, S/m, where the unit of siemens is amp/volt. The quantity most often reported is the *ionic equivalent conductance* due to a specific ion i, λ_i; this is obtained by dividing σ_i, the contribution to σ by ionic species i, $(= \mathcal{F}^2\,Z_i^2\,u_i^m\,C_i)$, by C_i, the molar concentration of ions i and Z_i:

$$\lambda_i = \frac{\sigma_i}{Z_i\,C_i} = \left|Z_i\right|\,\mathcal{F}^2\,u_i^m. \tag{3.1.108r}$$

Here, $|Z_i|$ is used to indicate that conductance is a positive quantity. The units of σ_i/C_i are S-m^2/gmol, and those of λ_i are S-m^2/g-equiv.

(*Note*: The ohm is the unit for volt/amp; mho is the unit for amp/volt, therefore mho = siemens. Correspondingly, the units of λ_i are mho-m^2/g equiv.)

Sometimes the equivalent conductance Λ of a salt producing a cation and an anion is used:

$$\Lambda = \lambda_+ + \lambda. \tag{3.1.108s}$$

Table 3.A.8 lists the values of λ_i^0 for a few common ions corresponding to infinite dilution. Their diffusion coefficients (based on the Nernst-Einstein relation, (3.1.108k), namely $D_{is} = \text{R}T\,u_i^m = (\text{R}T\lambda_i/|Z_i|\,\mathcal{F}^2))$, are listed in Table 3.A.4.

One can now use the definition of ionic mobility u_i^m to redefine the transport number t_{is} in cases where there is no concentration gradient in the solution. Use equation (3.1.108g) to get

$$t_{is} = \frac{Z_i^2\,\mathcal{F}^2\,C_i\,u_i^m\,(-\nabla\Phi)}{\mathcal{F}^2\sum_{i=1} Z_i^2\,u_i^m\,C_i\,(-\nabla\Phi)} = \frac{Z_i^2\,C_i\,u_i^m}{\sum_{i=1} Z_i^2\,C_i\,u_i^m} \tag{3.1.108t}$$

since $v_t\sum_{i=1} Z_i\mathcal{F}C_i = v_t\,\mathcal{F}\sum_{i=1} Z_i\,C_i = 0$ due to the electroneutrality condition (3.1.108a).

So far, the medium in which the species were diffusing was either gaseous or liquid. Diffusion of a species i through a *solid* with no holes, defects or pores is also encountered in separation processes. In general, Fick's first law is assumed to be valid without any convective component:

$$\boldsymbol{N}_i = -D_i\,\nabla C_i. \tag{3.1.109a}$$

For one-dimensional transport in a planar geometry (say the z-direction),

$$N_{iz} = -D_i\,\frac{\text{d}C_i}{\text{d}z}. \tag{3.1.109b}$$

For radial transport in a cylindrical or spherical geometry,

$$N_{ir} = -D_i \frac{\mathrm{d}C_i}{\mathrm{d}r}. \tag{3.1.109c}$$

The diffusion coefficient D_i may or may not vary in the direction of diffusion. More on transport of a species through a solid material is provided in Section 3.4.2. The solid material considered there is in the form of a thin membrane with or without holes, pores or defects. Diffusion in liquid and gaseous phases through a porous solid material/membrane is considered in Sections 3.1.3.2.3 and 3.1.3.2.4. Evaporative flux of molecules from a free liquid surface under high vacuum is described in Section 3.1.3.2.5.

3.1.3.2.1 Some relations for different flux expressions

Two aspects are illustrated here. We consider a binary system of solute i and solvent s. First, we write down expressions for mass fluxes $\boldsymbol{n}_i$ and $\boldsymbol{j}_i$, since other relevant ones have been illustrated already. Second, and more important, the values of quantities like $\boldsymbol{J}_i^* + \boldsymbol{J}_s^*$ and $\boldsymbol{N}_i + \boldsymbol{N}_s$ are determined for two cases: (1) migration velocities $\boldsymbol{U}_i$ and $\boldsymbol{U}_s$ are nonzero; (2) $\boldsymbol{U}_i$ and $\boldsymbol{U}_s$ are both zero.

By definition, $\boldsymbol{n}_i$, the mass flux of species i across a cross section stationary with respect to the fixed coordinate axes, is related to $\boldsymbol{N}_i$ (from equation (3.1.92) and Table 3.1.3B) by

$$\boldsymbol{n}_i = \boldsymbol{N}_i M_i = M_i C_i(\boldsymbol{v}_t^* + \boldsymbol{U}_i) - M_i C_t D_{is}^0 \left[1 + \frac{\mathrm{d}\ln\gamma_i}{\mathrm{d}\ln x_i}\right]_{T,P} \nabla x_i. \tag{3.1.110a}$$

Similarly,

$$\boldsymbol{j}_i = M_i \boldsymbol{J}_i = M_i C_i\{(\boldsymbol{v}_t^* - \boldsymbol{v}_t) + \boldsymbol{U}_i\} - M_i C_t D_{is}^0 \left[1 + \frac{\mathrm{d}\ln\gamma_i}{\mathrm{d}\ln x_i}\right]_{T,P} \nabla x_i, \tag{3.1.110b}$$

where we have used equation (3.1.98) for $\boldsymbol{J}_i$.

When the migration velocities are nonzero, we obtain, from the flux expression (3.1.99) for $\boldsymbol{J}_i^*$, the following:

$$\boldsymbol{J}_i^* + \boldsymbol{J}_s^* = \boldsymbol{N}_i + \boldsymbol{N}_s - \boldsymbol{v}_t^*(C_i + C_s). \tag{3.1.110c}$$

By definition (3.1.79), $\boldsymbol{N}_i = C_i \boldsymbol{v}_t^* + C_i \overline{\boldsymbol{U}}_i$ and $\boldsymbol{N}_s = C_s \boldsymbol{v}_t^* + C_s \overline{\boldsymbol{U}}_s$. Therefore

$$\boldsymbol{N}_i + \boldsymbol{N}_s = \boldsymbol{v}_t^*(C_i + C_s) + C_i \overline{\boldsymbol{U}}_i + C_s \overline{\boldsymbol{U}}_s = C_t \boldsymbol{v}_t^* + C_i \overline{\boldsymbol{U}}_i + C_s \overline{\boldsymbol{U}}_s.$$

But by definition also $\boldsymbol{N}_i + \boldsymbol{N}_s = C_i \boldsymbol{v}_i + C_s \boldsymbol{v}_s = C_t \boldsymbol{v}_t^*$. Therefore $\boldsymbol{J}_i^* + \boldsymbol{J}_s^* = 0$. Also

$$C_i \overline{\boldsymbol{U}}_i + C_s \overline{\boldsymbol{U}}_s = 0. \tag{3.1.110d}$$

These relations are also valid when migration velocities are zero.

A similar result can be obtained for $\boldsymbol{j}_i + \boldsymbol{j}_s$. By definition (see Tables 3.1.3A and B),

$$\begin{aligned} \boldsymbol{j}_i + \boldsymbol{j}_s &= M_i \boldsymbol{J}_i + M_s \boldsymbol{J}_s = M_i(\boldsymbol{N}_i - C_i \boldsymbol{v}_t) + M_s(\boldsymbol{N}_s - C_s \boldsymbol{v}_t) \\ &= \rho_i \boldsymbol{v}_i - \rho_i \boldsymbol{v}_t + \rho_s \boldsymbol{v}_s \\ -\rho_s \boldsymbol{v}_t &= \{\rho_i \boldsymbol{v}_i + \rho_s \boldsymbol{v}_s\} - \boldsymbol{v}_t(\rho_i + \rho_s) = 0. \end{aligned} \tag{3.1.110e}$$

3.1.3.2.2 Diffusion of a binary electrolyte in a solution

Consider the diffusion of an electrolyte $A_{\nu+}^{Z+} Y_{\nu-}^{Z-}$ in a dilute solution without any applied electrical field. In order to show that the diffusion coefficient expression (3.1.107) is to be used for the electrolyte as a whole or for any of the ions, we start with the flux expressions for the two charged species in a stagnant medium. Here the *unknown diffusion potential* is to be eliminated. Define the molar concentrations of positively charged and negatively charged ions in solution as C_+ and C_- (gmol/cm^3), respectively.

From the electroneutrality relation, (3.1.108a),

$$Z_+ C_+ + Z_- C_- = 0 = Z_+ \nabla C_+ + Z_- \nabla C_-. \tag{3.1.111a}$$

From (3.1.108b,g), for $\boldsymbol{v}_t^* = 0$, we get

$$Z_+ \boldsymbol{N}_+ + Z_- \boldsymbol{N}_- = 0 = Z_+ \left[C_+ \frac{D_+^0}{RT} Z_+ \mathcal{F}(-\nabla\Phi) - D_+^0 \nabla C_+\right] + Z_- \left[C_- \frac{D_-^0}{RT} Z_- \mathcal{F}(-\nabla\Phi) - D_-^0 \nabla C_-\right] = 0,$$

where we have assumed ideal solution behavior. This last relation may be rearranged to yield

$$\frac{\mathcal{F}}{RT}(-\nabla\Phi) = \frac{(D_+^0 - D_-^0)\nabla C_+}{C_+(Z_+ D_+^0 - Z_- D_-^0)}, \tag{3.1.111b}$$

using result (3.1.111a). Now substitute this expression for diffusion potential $(-\nabla\Phi)$ into the Nernst-Planck expression (3.1.106) for

$$\begin{aligned} \boldsymbol{J}_+^*(=\boldsymbol{N}_+) &= -D_+^0 \nabla C_+ + \frac{C_+ D_+^0 Z_+}{RT} \mathcal{F}(-\nabla\Phi) \\ &= -D_+^0 \nabla C_+ + \frac{D_+^0 Z_+ (D_+^0 - D_-^0)}{(Z_+ D_+^0 - Z_- D_-^0)} \nabla C_+ \\ &= -\frac{D_+^0 D_-^0 (Z_+ - Z_-)}{(Z_+ D_+^0 - Z_- D_-^0)} \nabla C_+. \end{aligned} \tag{3.1.111c}$$

One can obtain the same diffusion coefficient in J_-^* and $\boldsymbol{J}_{AY}^*$.

3.1.3.2.3 Solute flux through porous liquid-filled material

We consider here the expression for solute i flux through a porous liquid-filled pellet/bead/membrane. Under conditions of *no convection* and solute radius, r_i, being at least two orders of magnitude smaller than the pore radius, r_p, $(r_i <<< r_p)$, the governing equation for solute flux N_{iz}^p along the pore length (z-coordinate) in the pore fluid is simply Fick's first law (3.1.85) with $\boldsymbol{v}_t^* = 0$:

$$N_{iz}^p = -D_{il} \frac{\mathrm{d}C_{im}^p}{\mathrm{d}z}. \tag{3.1.112a}$$

This expression for N_{iz}^p provides the moles of i being transported per unit pore cross-sectional area per unit time. This flux expression also assumes that there are no electrostatic interactions between the pore material and the solute. Here subscript m refers to the porous pellet/bead/membrane. The solute concentration in the liquid within the pore, C_{im}^p, is related to the overall solute concentration C_{im} in the pellet/bead/membrane having a porosity/void volume fraction ε_m by

$$C_{im} = \varepsilon_m C_{im}^p. \tag{3.1.112b}$$

The pore concentration of solute i, C_{im}^p, is related to the external solution concentration C_{il} via a partition coefficient κ_{im} (see Section 3.3.7.4). When there is no partitioning effect, $\kappa_{im} = 1$. In reality, the pores are tortuous so that solute molecules traverse a distance longer than the pellet/bead/membrane thickness δ_m; this is commonly taken into account by a correction factor, the "tortuosity factor," τ_m. Correspondingly, if the pellet/bead/membrane pore length is to be considered as δ_m, then the solute flux in moles per unit pellet/bead/membrane cross-sectional area per unit time N_{iz} is expressed as

$$N_{iz} = \varepsilon_m N_{iz}^p = -\frac{D_{il}\,\varepsilon_m}{\tau_m}\frac{\mathrm{d}\,C_{im}^p}{\mathrm{d}z} = -D_{im}\frac{\mathrm{d}\,C_{im}^p}{\mathrm{d}z}, \tag{3.1.112c}$$

where the effective diffusion coefficient of solute i for the porous pellet/bead/membrane D_{ie} (or D_{im}) is defined as

$$D_{ie} = D_{im} = \frac{D_{il}\,\varepsilon_m}{\tau_m}. \tag{3.1.112d}$$

Here D_{il} is the conventional diffusion coefficient of species i in the liquid (Tables 3.A.2 and 3.A.3). Such an expression is frequently used for pellets/beads/particles used in various separation techniques as well as for membranes. (Empirical estimates of τ_m made in a variety of catalyst particles, adsorbents and membranes are available: silica-alumina ~2.1–2.3 (Satterfield *et al.*, 1973); hollow fiber membranes ~2–14 (Prasad and Sirkar, 1988)).

When the solute dimension is no longer orders of magnitude smaller than the pore dimensions, the solute molecules experience an additional transport resistance due to the proximity of the pore wall. The effective diffusion coefficient is reduced further by a hindrance factor/drag factor $\overline{G}_{\mathrm{Dr}}(r_i,r_p)$. If the solute molecules are assumed spherical of radius r_i and diffusing through the centerline of the pore of radius, r_p, Faxen's expression may be used to estimate $\overline{G}_{\mathrm{Dr}}$ for $(r_i,r_p) \leq 0.5$ (Lane, 1950; Renkin, 1954):

$$\overline{G}_{\mathrm{Dr}}(r_i,r_p) = 1 - 2.1044\left(\frac{r_i}{r_p}\right) + 2.089\left(\frac{r_i}{r_p}\right)^3 - 0.948\left(\frac{r_i}{r_p}\right)^5 + \ldots \tag{3.1.112e}$$

The solute flux expression (3.1.112c) is now changed to

$$N_{iz} = -\frac{D_{il}\,\varepsilon_m\,\overline{G}_{\mathrm{Dr}}(r_i,r_p)}{\tau_m}\frac{\mathrm{d}\,C_{im}^p}{\mathrm{d}z}. \tag{3.1.112f}$$

For solute molecules having $r_i << r_p$, $\overline{G}_{\mathrm{Dr}}(r_i,r_p) = 1$. The flux of solute will vary with the radial location in the pore as well (Anderson and Quinn, 1974). If the local value of $\overline{G}_{\mathrm{Dr}}$ is available as a function of the radial location, r, then the solute flux averaged over the pore cross section, $\overline{N}_{iz}$, is

$$\overline{N}_{iz} = -\left(\frac{D_{il}\,\varepsilon_m}{\tau_m}\right)\left[\int_0^{1-r_i/r_p} 2\left(\frac{r}{r_p}\right)\overline{G}_{\mathrm{Dr}}(r_i,r_p,r)\,\mathrm{d}\left(\frac{r}{r_p}\right)\right]\frac{\mathrm{d}\,C_{im}^p}{\mathrm{d}z}, \tag{3.1.112g}$$

where the effective diffusion coefficient under no-convection conditions, D_{ie} (or D_{im}) may be expressed as

$$\overline{N}_{iz} = -D_{ie}\frac{\mathrm{d}\,C_{im}^p}{\mathrm{d}z}, \tag{3.1.112h}$$

where

$$D_{ie} = \left(\frac{D_{il}\,\varepsilon_m}{\tau_m}\right)\left[\int_0^{1-r_i/r_p} 2\left(\frac{r}{r_p}\right)\overline{G}_{\mathrm{Dr}}(r_i,r_p,r)\,\mathrm{d}\left(\frac{r}{r_p}\right)\right]. \tag{3.1.112i}$$

When there is *convection* through the pores along with diffusion, the solute flux N_{iz} has a *diffusive* as well as a *convective* component:

$$N_{iz} = \underbrace{-\frac{D_{il}\,\varepsilon_m\,\overline{G}_{\mathrm{Dr}}(r_i,r_p)}{\tau_m}\frac{\mathrm{d}\,C_{im}^p}{\mathrm{d}z}}_{\text{diffusive}} + \underbrace{\overline{G}_c\,C_{im}^p\,v_z\,\varepsilon_m}_{\text{convective}}. \tag{3.1.113}$$

Here, v_z is the local fluid velocity in the pore and $\overline{G}_c$ is a factor by which the solute velocity is reduced from the fluid velocity (Anderson and Quinn, 1974). One can also average this flux expression over the whole pore cross section in the manner of expression (3.1.112g). The quantity $\overline{G}_c$ is called the convective hindrance factor; a pore-average value of this may be developed as in expression (3.1.112i). The treatment provided assumes that there are no solute-pore wall interactions via adsorption or electrical potential effects.

3.1.3.2.4 Gas transport through porous beads/pellets/membranes In gas transport through porous materials/membranes, if the matrix openings are large enough to prevent molecular diffusion encountered in nonporous materials, gas convection occurs. Three types of gas transport in the gas phase through the porous/mesoporous/microporous material/membrane openings (pore or holes) have been postulated. The type of gas convection mechanism is determined by the relative values of the mean free path λ of the gas molecules (average distance traversed by a molecule before it suffers collision with another gas molecule) at the prevailing pressure and the matrix opening/pore diameter $2r_p$. When $(r_p/\lambda) < 0.05$, *Knudsen flow* or *gaseous diffusion* takes place (Liepmann, 1961). When $(r_p/\lambda) > 50$, *conventional Poiseuille* or *viscous flow* occurs. In the intermediate range, slip flow is observed. barrer (1963) has provided a review of the different types of flow in a microporous medium or membrane.

According to the rigid sphere model in the kinetic theory of gases, the mean free path λ is given by (Kennard, 1938)

$$\lambda = \frac{1.256\mu}{P}\sqrt{\frac{RT}{M_i}}. \tag{3.1.114}$$

As the total pressure decreases, λ increases. At very low pressures, then, Knudsen flow is likely, whereas at high pressures, viscous or slip flow will dominate.

The steady state *Knudsen diffusion flux* of a gas species i of molecular weight M_i through straight circular pores of radius r_p and length δ_m under the condition of total pressure being the same at both ends of the pore is given by

$$N_{iz}^p = -\frac{4r_p}{3}\left[\frac{2RT}{\pi M_i}\right]^{1/2}\frac{dC_i}{dz} = -\frac{4r_p}{3}\left[\frac{2RT}{\pi M_i}\right]^{1/2}\frac{1}{RT}\frac{dp_{ig}}{dz}, \tag{3.1.115a}$$

This flux expression is based on the pore cross section only; for the second expression on the right-hand side, we have assumed ideal gas behavior. For a pellet/bead/membrane of porosity/void volume fraction ε_m and pore tortuosity factor of τ_m, the gas species flux based on the total cross section of the pellet/bead/membrane is

$$N_{iz} = -\frac{4r_p\varepsilon_m}{3\tau_m}\left(\frac{2RT}{\pi M_i}\right)^{1/2}\frac{1}{RT}\frac{dp_{ig}}{dz}. \tag{3.1.115b}$$

An appropriate expression for the Knudsen diffusivity D_{ik} of species i through a straight circular pore is

$$D_{iK} = \frac{4r_p}{3}\left(\frac{2RT}{\pi M_i}\right)^{1/2} = 9.7\times 10^3\, r_p\left(\frac{T}{M_i}\right)^{1/2} \text{cm}^2/\text{s}, \tag{3.1.115c}$$

where r_p is the pore radius in centimeter. Since $r_p <<< \lambda$ in Knudsen flow, gas molecules of species i collide with the pore wall rather than with other gas molecules of species i or species j. Therefore Knudsen diffusion based flux of any species i is merely proportional to $(M_i)^{-1/2}$. When the pore dimensions are larger and/or the gas pressure is high enough so that $(r_p/\lambda) > 50$, the equations of molecular diffusion (3.1.94) should be used for a gas mixture when the total pressure is the same at both ends of the pore.

For conditions intermediate between Knudsen flow and Poiseuille flow, *slip flow* or *transitional flow* takes place. The molar flux of species A through the circular pores under such conditions has been shown for a binary gas mixture of species A and B to be (Scott and Dullien, 1962; Rothfeld, 1963)

$$N_{Az} = -\left[\frac{1}{\dfrac{1-\dfrac{x_{Ag}}{N_R}}{D_{AB}}+\dfrac{1}{D_{AK}}}\right]\frac{P}{RT}\frac{dx_{Ag}}{dz}, \tag{3.1.115d}$$

where $N_R = N_{Az}/[N_{Az}+N_{Bz}]$) is the flux ratio (see equation (3.1.129a)), D_{AB} is the ordinary diffusion coefficient of the binary gas mixture and D_{AK} is the Knudsen diffusivity of species A. An alternative way of writing this expression, namely

$$N_{Az} = 1 - \frac{1}{\left(1-\frac{x_{Ag}}{N_R}\right)\frac{D_{AK}}{D_{AB}}+1}D_{AK}\frac{P}{RT}\frac{dx_{Ag}}{dz}, \tag{3.1.115e}$$

shows that, at low pressures where $D_{AK}/D_{AB} <<1$, this expression reduces to the Knudsen flux expression (3.1.115a); at high pressures, $D_{AK}/D_{AB} >> 1$ (since (D_{AK}/D_{AB}) is proportional to P (from definition (3.1.115c), D_{iK} is independent of P and from definition (3.1.91b), D_{AB} is proportional to $(P)^{-1}$)) and the expression reduces to that due to molecular diffusion.

When the total pressures at the two ends of the pore are different, there will be bulk flow. If conditions for Knudsen diffusion exist, then the molar average bulk flow velocity (v_z^*) may be estimated from

$$C_{tg}\,v_z^* = N_{Az}+N_{Bz} = -\frac{4r_p\varepsilon_m}{3\tau_m}\left(\frac{2RT}{\pi}\right)^{1/2}\frac{1}{RT}\left\{\frac{1}{\sqrt{M_A}}\frac{dp_{Ag}}{dz}+\frac{1}{\sqrt{M_B}}\frac{dp_{Bg}}{dz}\right\}, \tag{3.1.116a}$$

where $C_{tg} = C_{Ag}+C_{Bg}$ for a binary system.

When Poiseuille flow conditions exist, we use the expression

$$-\frac{dP}{dz} = \frac{8\mu_i}{r_p^2}v_z = -\frac{dp_{Ag}}{dz}-\frac{dp_{Bg}}{dz} \tag{3.1.116b}$$

to estimate the mass average velocity v_z for the binary gas mixture flow. The molar flux of a gas species i under such conditions can be estimated from v_z or (v_z^*) and a mean density $(C_{\text{mean}})_i$ of species in the gas from one end of the pore to the other. Approximate values of the tortuosity factor τ_m for a variety of beads/pellets in gas transport/separation processes have been summarized by Yang (1987): alumina, 2–6; silica gel, 2–6; activated carbon, 5–65; micropores in zeolites, 1.7–4.5.

In gas transport through the pores of a pellet/bead/particle/membrane, the gas molecules may also be adsorbed on the pore wall or the solid surface. This leads to an additional type of gas transport, namely *surface diffusion*. Normally, the surface adsorbed species are assumed to be in equilibrium with the bulk gas phase being convected through the pores. The pore gas pressure (or partial pressure) usually decreases in the direction of transport. Therefore, the concentration of adsorbed gas species in the pore walls also decreases in the direction of transport, leading to a concentration gradient of surface-adsorbed molecules which diffuse downhill, generating what is often called the *surface flow*. The additional flux due to surface flow is expressed via a Fickian diffusion-type expression:

$$N_{iz} = -D_{is}\frac{d\Gamma_{is}^E}{dz}a_{sp}, \tag{3.1.117a}$$

where Γ_{is}^E is the molar excess surface concentration of species i in moles/cm^2 of the pore surface area, D_{is} is the surface diffusion coefficient of species i and a_{sp} is the pore surface area in cm^2/unit volume (cm^3) of the porous/microporous medium of porosity ε (pore volume/per unit volume of the porous medium) (Ash *et al.*, 1973a, b). Alternative approaches employ the molar surface concentration of i per unit weight of the adsorbent multiplied by the density of the adsorbent instead of the product $\Gamma_{is} a_{sp}$. For dilute adsorbed films, each species diffuses independently of the other (Ash *et al.*, 1973b).

Surface flow exists along with flow along the pore by Knudsen diffusion, molecular diffusion and/or convection. If Knudsen diffusion exists, then the fluxes due to the two different mechanisms are

$$N_{iz}\Big|_{\text{Knudsen}} = -D_{iK}\frac{\mathrm{d}C_{ig}}{\mathrm{d}z}; \quad N_{iz}\Big|_{\text{surface flow}} = -D_{is}\frac{\mathrm{d}(a_{sp}\Gamma_{is}^E)}{\mathrm{d}z} = -D_{is}\frac{\mathrm{d}C_{is}}{\mathrm{d}z}, \tag{3.1.117b}$$

where C_{is} is the molar concentration of the surface adsorbed species per unit volume, the porous medium.

The ratio of the two fluxes is

$$\frac{N_{iz}/_{\text{surface}}}{N_{iz}/_{\text{Knudsen}}} = \frac{D_{is}}{D_{iK}}\,a_{sp}\,\frac{\mathrm{d}\Gamma_{is}^E}{\mathrm{d}\,C_{ig}}. \tag{3.1.117c}$$

The total flux due to the two mechansims is given by

$$N_{iz} = -\left[\frac{D_{iK}}{RT}\frac{\mathrm{d}p_{ig}}{\mathrm{d}z} + D_{is}\,a_{sp}\frac{\mathrm{d}\Gamma_{is}^E}{\mathrm{d}z}\right]. \tag{3.1.118a}$$

However, since there is likely to be an equilibrium between p_i in the gas phase and Γ_{is}^E in the surface phase, one can reduce this expression to a relation between N_{iz} and $(\mathrm{d}p_i/\mathrm{d}z)$.

A more detailed description requires identification of the thickness of the sorbed surface phase δ_s and its volume $a_{sp}\delta_s$ so that the volume of the gas phase in the porous medium per unit volume of the porous medium is $(\varepsilon - a_{sp}\delta_s)$. Consider a pure gas i in the porous medium. If the total concentration of species i in the porous medium per unit volume, C_i^t, is considered as a sum of that in the gas phase C_{ig} and the adsorbed phase C_{is}, then

$$C_i^t = C_{ig} + C_{is} = C'_{ig}(\varepsilon - a_{sp}\delta_s) + \Gamma_{is}^E a_{sp}, \tag{3.1.118b}$$

where C'_{ig} is the number of molecules in the gas phase per unit gas phase volume in equilibrium with the surface-adsorbed phase (Ash *et al.*, 1967, 1973a). Rewriting (3.1.117c) using these expressions, one obtains

$$\frac{N_{iz}/_{\text{surface}}}{N_{iz}/_{\text{Knusden}}} = \frac{D_{is}}{D_{iK}}\,\frac{a_{sp}}{(\varepsilon - a_{sp}\delta_s)}\,\frac{\mathrm{d}\Gamma_{is}^E}{\mathrm{d}\,C'_{ig}}. \tag{3.1.118c}$$

Furthermore

$$\frac{\mathrm{d}\Gamma_{is}^E}{\mathrm{d}\,C'_{ig}} = \frac{\mathrm{d}\Gamma_{is}^E}{\mathrm{d}p_{ig}}RT, \tag{3.1.118d}$$

and this last ratio is obtainable from adsorption isotherm measurements. Hwang and Kammermeyer (1975) have provided an overview of surface diffusion in microporous membranes. Yang (1987) has summarized surface diffusion in pellets and beads used for adsorption processes.

3.1.3.2.5 Evaporative flux of molecules from a free liquid surface under high vacuum When molecules evaporate from a liquid surface, molecules also come back from the vapor phase into the liquid. Therefore, the net rate of evaporation from a liquid is determined by the net rate of heat transfer to the liquid. However, if conditions of high vacuum exist on the evaporating liquid surface and there is a cold condensing surface in the immediate vicinity, the molecules escaping the evaporating surface are condensed on the condensing surface; this process, called *molecular distillation*, is characterized by the following flux expression for a pure liquid (Langmuir, 1916):

$$N_{iz} = \frac{p_i^{\text{sat}}}{\sqrt{2\pi M_i RT}}. \tag{3.1.119a}$$

For a liquid mixture having a mole fraction x_{il} of species i,

$$N_{iz} = \frac{p_i^{\text{sat}}\,x_{il}}{\sqrt{2\pi M_i RT}}. \tag{3.1.119b}$$

Here, the vapor pressure units are dyne/cm^2 and R is in erg/gmol-K.

3.1.4 Integrated flux expressions for molecular diffusion and convection: single-phase systems

The flux expressions we have considered so far for a mixture of chemical species often relate the flux to a number of gradients of potentials. For the case of a gradient in chemical potential in an isobaric single-phase system, we find from expressions (3.1.84a), (3.1.84c), (3.1.94) and (3.1.95) that the species i flux is controlled by $C_t\,\nabla x_i$ (or ∇C_i) or ∇p_i. Gradients of quantities such as C_{ij} and p_{ij} in a given phase j are not directly available in practical separation systems. The value of C_{ij} or p_{ij} at the two ends of the diffusion path are much more likely to be available. We therefore integrate here the flux expressions along the diffusion/transport path, if possible, and express the flux in terms of differences in concentrations (or mole fractions) or partial pressures existing at the ends of the transport path. In this process, proportionality constants called *mass-transfer coefficients* arise: these are very important for a quantitative analysis of separation achieved in practical separation problems. We first consider a mixture of gases and then deal with diffusion in liquids (Cussler, 1997; Bird *et al.*, 2002).

Consider **simple molecular diffusion in a binary gaseous mixture of species A and B** under **isobaric condition**; so $\nabla P = 0$ and $\boldsymbol{U}_A = 0 = \boldsymbol{U}_B$. If the diffusion takes place in only the z-direction along the capillary length in Figure 3.1.4, then from flux expression (3.1.94)

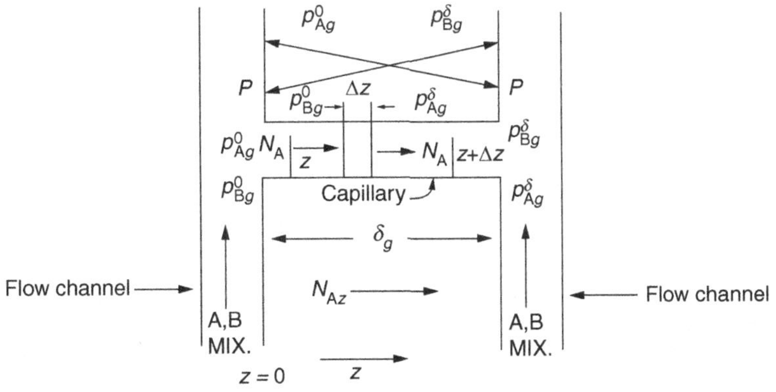

Figure 3.1.4. Two flow channels connected by a capillary having molecular diffusion; the total pressure in both channels at the capillary location is P. *Diffusion takes place in the capillary.*

$$N_{Az} = C_{Ag}\, v^*_{tz} - \frac{D_{AB}}{RT}\frac{dp_{Ag}}{dz}. \tag{3.1.120}$$

If the conditions are such that $\boldsymbol{v}^*_t = 0$, i.e. $N_{Az} + N_{Bz} = 0$ (by relation (3.1.97)), we have simply

$$N_{Az} = -\frac{D_{AB}}{RT}\frac{dp_{Ag}}{dz}. \tag{3.1.121}$$

This condition is identified as the *equimolar counterdiffusion* ($N_{Az} = -N_{Bz}$). For each mole of A transferred in the positive z-direction per unit area per unit time, one mole of B is transferred in the opposite direction. Assume steady state. Since the cross-sectional area through which diffusion occurs is constant, by making a balance on species A over a differential element of length Δz, we get

$$\frac{dN_{Az}}{dz} = 0. \tag{3.1.122a}$$

Integrating along the z-direction, we find that

$$N_{Az} = -\frac{D_{AB}}{RT}\frac{dp_{Ag}}{dz} = \text{constant}. \tag{3.1.122b}$$

Integrating again from 0 to δ_g we get

$$N_{Az} = \frac{D_{AB}}{RT\,\delta_g}\left(p^0_{Ag} - p^\delta_{Ag}\right) = k'_g\left(p^0_{Ag} - p^\delta_{Ag}\right), \tag{3.1.123}$$

where $k'_g = (D_{AB}/RT\delta_g)$ is the *mass-transfer coefficient* for equimolar counterdiffusion in the gas phase. Alternatively, $(1/k'_g)$ is the *mass-transfer resistance* since N_{Az} may be considered the *current* and the partial pressure difference, $(p^0_{Ag} - p^\delta_{Ag})$, the *potential difference* (from Ohm's law).

It is possible to express the preceding result in several other forms using the difference in concentrations or mole fractions at the ends of the diffusion path instead of a difference in partial pressures:

$$N_{Az} = k'_g\left(p^0_{Ag} - p^\delta_{Ag}\right) = k'_c\left(C^0_{Ag} - C^\delta_{Ag}\right) = k'_{xg}\left(x^0_{Ag} - x^\delta_{Ag}\right),$$

where

$$k'_g = \frac{D_{AB}}{RT\,\delta_g} = \frac{k'_c}{RT} = \frac{k'_{xg}}{P}; \qquad k'_{xg} = k'_c\,\frac{P}{RT} = k'_c\,C_{tg}. \tag{3.1.124}$$

Note that the prime indicates equimolar counterdiffusion with no bulk velocity in terms of $\boldsymbol{v}^*_t$.

Example 3.1.12 In Figure 3.1.4, on one side of the capillary, a gas containing 90% N_2–10% He is flowing. On the other side of the capillary, a gas containing 90% He–10% N_2 is flowing. The capillary length is 30 cm and its diameter is 0.1 mm. The pressure is atmospheric on both sides; the system temperature is 25 °C. Given $D_{AB} = 0.687\ \text{cm}^2/\text{s}$, determine (1) the rate at which N_2 gas is introduced into the helium-rich gas stream and the value of its k'_g; (2) the rate at which He gas is introduced into the N_2-rich gas stream.

Solution This is a case of equimolar counterdiffusion since the partial pressure difference of both species over the diffusion path is the same. This case of equimolar counterdiffusion can be treated via the flux expression (3.1.123); let species A = N_2, species B = He. Then we have

$$N_{Az} = \frac{D_{AB}}{RT\,\delta_g}\left(p^0_{Ag} - p^\delta_{Ag}\right) = \frac{D_{AB}P}{RT\,\delta_g}\left(x^0_{Ag} - x^\delta_{Ag}\right)$$

$$= -N_{Bz} = -N_{Hez} = k'_g\left(p^0_{Ag} - p^\delta_{Ag}\right)$$

$$= \frac{0.687(\text{cm}^2/\text{s})\ 1(\text{atm})}{82\ \dfrac{\text{cm}^3\text{-atm}}{\text{gmol-K}} \times 298\,\text{K} \times 30\,\text{cm}}(0.9 - 0.1)$$

$$= N_{N_2z}\,\frac{\text{gmol}}{\text{cm}^2\text{-s}}; \quad k'_g = \frac{0.687}{82 \times 298 \times 30}$$

$$= 0.937 \times 10^6\ \frac{\text{gmol}}{\text{cm}^2\text{-s-atm}}.$$

(In chemical engineering process equipments, δ_g is far smaller, ~ <0.01 cm, so k'_g is much larger.) We also have

$$W_{Az} = N_{Az} \times (\text{capillary cross-sectional area})$$

$$= \left(\frac{0.687 \times 1 \times 0.8}{82 \times 298 \times 30}\frac{\text{gmol}}{\text{cm}^2\text{-s}}\right) \times \left(\frac{\pi}{4}(10^{-2})^2\right)\text{cm}^2$$

$$= \frac{0.687 \times 0.8 \times \pi \times 10^{-4}}{82 \times 298 \times 30 \times 4}\frac{\text{gmol}}{\text{s}}$$

$$= \frac{0.687 \times 0.8 \times \pi \times 10^{-11}}{0.82 \times 0.298 \times 1.20}\frac{\text{gmol}}{\text{s}} = 5.88 \times 10^{-11}\frac{\text{gmol}}{\text{s}}$$

into the helium stream.

Since $|N_{Az}| = |N_{Bz}| = |N_{He}|$, $W_{He} = 5.88 \times 10^{-11}$gmol/s into the nitrogen stream in the opposite direction.

Note: $N_{Bz} = -N_{Az} = \dfrac{D_{AB}}{RT\delta_g}(p^{\delta}_{Bg} - p^0_{Bg})$.

In equimolar counterdiffusion, $N_{Az} + N_{Bz} = 0$ in a binary gas mixture; so is v^*_{tz}. Under other conditions, the bulk velocity $v^*_{tz} \neq 0$; the corresponding expressions of N_{Az} and the various mass-transfer coefficients will be developed now. The only restriction is that the magnitude of the transfer rate of species A is low. Consider again isobaric conditions and no external forces. From flux expression (3.1.94) and the relation (3.1.97),

$$N_{Az} = C_{Ag}\, v^*_{tz} - \frac{D_{AB}}{RT}\frac{dp_{Ag}}{dz} = \frac{C_{Ag}}{C_{tg}}(N_{Az} + N_{Bz}) - \frac{D_{AB}}{RT}\frac{dp_{Ag}}{dz}, \tag{3.1.125a}$$

where $(C_{Ag}/C_{tg}) = x_{Ag}$. Assume now a constant D_{AB} and a value of $(N_{Az} + N_{Bz})$ independent of the z-coordinate at steady state. Rearrange this relation (3.1.125a) as

$$-\int_{p^0_{Ag}}^{p^{\delta}_{Ag}} \frac{dp_{Ag}}{RT\, C_{tg}\frac{N_{Az}}{N_{Az}+N_{Bz}} - C_{Ag}RT} = \frac{(N_{Az} + N_{Bz})}{C_{tg}\, D_{AB}}\int_0^{\delta_g} dz, \tag{3.1.125b}$$

where $C_{Ag}RT = p_{Ag}$, $C_{tg}RT = P$ and the partial pressure of species A, p_{Ag}, has values of p^0_{Ag} and p^{δ}_{Ag} at $z = 0$ and δ_g, respectively. Simplifying, we find that

$$\int_{p^{\delta}_{Ag}}^{p^0_{Ag}} \frac{dp_{Ag}}{P\left(\frac{N_{Az}}{N_{Az} + N_{Bz}}\right) - p_{Ag}} = (N_{Az} + N_{Bz})\frac{\delta_g}{C_{tg}\, D_{AB}} \tag{3.1.125c}$$

On integration,

$$\frac{1}{(N_{Az} + N_{Bz})}\,\ell n\left[\frac{\left(\frac{N_{Az}}{N_{Az} + N_{Bz}}\right) - \frac{p^{\delta}_{Ag}}{P}}{\left(\frac{N_{Az}}{N_{Az} + N_{Bz}}\right) - \frac{p^0_{Ag}}{P}}\right] = \frac{\delta_g}{C_{tg}\, D_{AB}}, \tag{3.1.125d}$$

where we have assumed that $N_{Az}/(N_{Az} + N_{Bz})$ does not vary with z. Therefore

$$N_{Az} = \left(\frac{N_{Az}}{N_{Az} + N_{Bz}}\right)\frac{C_{tg}\, D_{AB}}{\delta_g}\,\ell n\left[\frac{\left(\frac{N_{Az}}{N_{Az} + N_{Bz}}\right) - x^{\delta}_{Ag}}{\left(\frac{N_{Az}}{N_{Az} + N_{Bz}}\right) - x^0_{Ag}}\right]. \tag{3.1.126}$$

From the definitions of k'_{xg}, k'_c and k'_g in relations (3.1.124), we obtain

$$N_{Az} = \frac{k'_{xg}}{\Phi_N}(x^0_{Ag} - x^{\delta}_{Ag}) = \frac{k'_c}{\Phi_N}\left(C^0_{Ag} - C^{\delta}_{Ag}\right) = \frac{k'_g}{\Phi_N}\left(p^0_{Ag} - p^{\delta}_{Ag}\right), \tag{3.1.127}$$

where the bulk flow correction factor Φ_N is given by

$$\Phi_N = \frac{(x^0_{Ag} - x^{\delta}_{Ag})}{\left(\frac{N_{Az}}{N_{Az} + N_{Bz}}\right)\ell n\left[\frac{\left(\frac{N_{Az}}{N_{Az} + N_{Bz}}\right) - x^{\delta}_{Ag}}{\left(\frac{N_{Az}}{N_{Az} + N_{Bz}}\right) - x^0_{Ag}}\right]}. \tag{3.1.128}$$

Defining the flux ratio, $(N_{Az}/\{N_{Az} + N_{Bz}\})$, as N_R (Geankoplis, 1972), Φ_N may be given a simpler representation as follows:

$$\Phi_N = \frac{(N_R - x_{Ag})_{\text{log mean}}}{N_R} = \frac{(N_R - x_{Ag})_{lm}}{N_R} \tag{3.1.129a}$$

since

$$\Phi_N = \frac{(N_R - x^{\delta}_{Ag}) - (N_R - x^0_{Ag})}{N_R\,\ell n\left(\frac{N_R - x^{\delta}_{Ag}}{N_R - x^0_{Ag}}\right)}. \tag{3.1.129b}$$

In equimolar counterdiffusion, knowledge of the mass-transfer coefficient k'_g (or k'_c or k'_{xg}) and the values of partial pressures (or concentrations or mole fractions) at the two ends of the diffusion path ($z = 0$ and $z = \delta_g$) were sufficient to determine N_{Az} (or N_{Bz}). In all other cases of molecular diffusion, the flux ratio N_R needs to be known; Φ_N can then be calculated to provide N_{Az} or N_{Bz} from definition (3.1.127).

A particular case of considerable use arises when species A *diffuses through nondiffusing or stagnant* B. This can happen, for example, when SO_2 (A) from air (B) is absorbed by an aqueous solution (Figure 3.1.5(a)) or the vapor of a volatile liquid, e.g. acetone (A), diffuses through stagnant air (B) (Figure 3.1.5(b)). Since species B is stagnant, N_{Bz}=0. Therefore,

$$N_R = \frac{N_{Az}}{N_{Az} + N_{Bz}} = 1. \tag{3.1.130}$$

From definitions (3.1.127) and (3.1.129),

$$N_{Az} = \frac{k'_g}{\Phi_N}(p^0_{Ag} - p^{\delta}_{Ag}) \tag{3.1.131a}$$

and

$$\Phi_N = \frac{(1 - x^{\delta}_{Ag}) - (1 - x^0_{Ag})}{\ell n\left(\frac{1 - x^{\delta}_{Ag}}{1 - x^0_{Ag}}\right)} = \frac{p^{\delta}_{Bg} - p^0_{Bg}}{P\,\ell n\left(\frac{p^{\delta}_{Bg}}{p^0_{Bg}}\right)} = \frac{p_{B,lm}}{P}, \tag{3.1.131b}$$

where $p_{B,\,lm}$ represents a logarithmic average of the partial pressure P_{Bg} over the gas film thickness δ_g, so that

$$N_{Az} = \frac{k'_g P\,(p^0_{Ag} - p^{\delta}_{Ag})}{p_{B,lm}} = \frac{k'_{xg}(x^0_{Ag} - x^{\delta}_{Ag})}{x_{B,lm}} = \frac{k'_c\, C_t(C^0_A - C^{\delta}_A)}{C_{B,lm}},$$

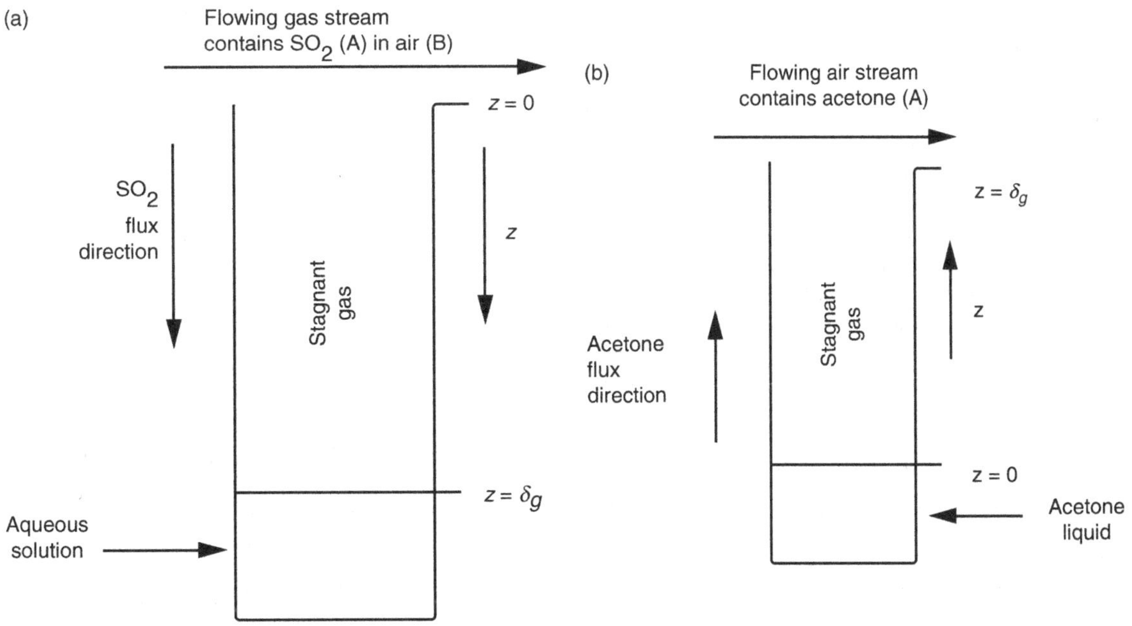

Figure 3.1.5. Diffusion of (a) SO_2 (A) through stagnant air (B) prior to absorption in an aqueous solution ($z = \delta_g$); (b) acetone (A) through stagnant air (B) after evaporation from liquid into flowing air stream at the top.

where $k'_g = (D_{AB}/RT\delta_g)$, $k'_{xg} = (D_{AB}\,C_t/\delta_g)$ and

$$x_{B,lm} = \frac{x^{\delta}_{Bg} - x^{0}_{Bg}}{\ell n\left(\dfrac{x^{\delta}_{Bg}}{x^{0}_{Bg}}\right)} = (1 - x_A)_{lm}, \tag{3.1.132}$$

where $x_{B,lm}$ represents a logarithmic average of the mole fraction of species B, x_{Bg}, over the gas film thickness δ_g. When neither equimolar counterdiffusion nor diffusion of A through a stagnant B describes the situation, other criteria have to be used to relate N_{Az} to N_{Bz}. This may involve reaction conditions, enthalpy balances, etc. Such cases will be treated as and when they arise.

Example 3.1.13 In Figure 3.1.5(b), a layer of ethanol is maintained at the bottom of a glass column 30 cm long and 1.5 cm in diameter at 25 °C. Air is blowing over the top of the open glass column; it sweeps away any ethanol evaporating from the bottom of the column. Air pressure, as well as the pressure in the glass column, may be assumed to be 1 atm. Determine (k'_g/Φ_N) for ethanol and its rate of loss into the air stream, given diffusion coefficient in ethanol–air system at 25 °C = 0.135 cm²/s; vapor pressure of ethanol at 25 °C = 0.075 atm.

Solution This is a case of ethanol (A) diffusing through a stagnant column of air (B). From equations (3.1.131) and (3.1.132), the steady state flux of ethanol is

$$N_{\text{ethanol}} = N_{Az} = \frac{D_{AB}\,P}{RT\,\delta_g\,p_{B,lm}}\left(p^0_{Ag} - p^{\delta}_{Ag}\right)$$

The partial pressure of ethanol at $z = 0$ corresponds to its vapor pressure, $p^0_{Ag} = 0.075$ atm. At the top of the column ($z = \delta_g$), $p^{\delta}_{Ag} \cong 0$ since the steady air flow rate is considerable. We determine $p_{B,lm}$, where $P = 1$ atm, as follows:

$$p_{B,lm} = \frac{p^{\delta}_{Bg} - p^{0}_{Bg}}{2.303\,\log\left(p^{\delta}_{Bg}/p^{0}_{Bg}\right)} = \frac{P - (1 - 0.075)}{2.303\log\left(\dfrac{P}{1 - 0.075}\right)}$$

$$= \frac{0.075}{2.303\log\left(\dfrac{1}{0.925}\right)} = \frac{0.075}{2.303 \times 0.0334} = 0.975\text{atm.};$$

$$\frac{k'_g}{\Phi_N} = \frac{(D_{AB})}{RT\,\delta_g}\,\frac{P}{p_{B,lm}} = \frac{0.135\,\dfrac{\text{cm}^2}{\text{s}}\,\dfrac{1\text{ atm}}{0.975\text{ atm}}}{82\,\dfrac{\text{cm}^3\text{-atm}}{\text{gmol-K}} \times 298\text{ K} \times 30\text{ cm}}$$

$$= 0.188\,\frac{\text{gmol}}{\text{cm}^2\text{-s-atm}};$$

$$N_{\text{ethanol}} = \frac{0.135\,\dfrac{\text{cm}^2}{\text{s}}\,1\text{ atm}\,(0.075 - 0)\text{ atm}}{82\,\dfrac{\text{cm}^3\text{-atm}}{\text{gmol-K}} \times 298\text{ K} \times 30 \times 0.975\text{ atm}}$$

$$= \frac{0.135 \times 0.075}{82 \times 298 \times 30 \times 0.975}\,\frac{\text{gmol}}{\text{cm}^2\text{-s}}$$

$$N_{\text{ethanol}} = \frac{0.135 \times 0.75 \times 10^{-7}}{0.82 \times 0.298 \times 3.0 \times 0.975} = 0.14 \times 10^{-7}\,\frac{\text{gmol}}{\text{cm}^2\text{-s}}.$$

The rate of loss of ethanol into the air stream

$$W_A = N_{\text{ethanol}} \times \pi r^2 = 0.14 \times 10^{-7} \times \pi \times (0.75)^2$$
$$= 0.14 \times \pi \times 0.562 \times 10^{-7} = 0.247 \times 10^{-7} \text{gmol/s}.$$

So far, the exposition has focused on molecular diffusion in a gaseous mixture of species A and B. Almost identical results may also be used for *molecular diffusion in liquids*. In gaseous systems, $\Delta P = 0$ implied that C_{tg} was constant along thc diffusion path. In a liquid mixture of solute i in solvent s, ΔP may be zero, but the molar solution density C_{tl} generally varies substantially along the diffusion path. Further, the diffusion coefficient D_{is} also varies with the solute concentration. The governing diffusion equation for species i in an i–s system is obtained from equations (3.1.87) and (3.1.89) for diffusion in the z-direction and $U_{iz} = 0$ as

$$N_{iz} = C_{il}(v^*_{tz}) - C_{tl} D_{is} \frac{dx_{il}}{dz}. \tag{3.1.133a}$$

Now,

$$N_{iz} + N_{sz} = v^*_{tz}\,(C_{il} + C_{sl}) - C_{tl} D_{is}\left(\frac{dx_{il}}{dz} + \frac{dx_{sl}}{dz}\right),$$

where we have used D_{is} for the diffusion of i in solvent s. By definition, locally, $x_{il} + x_{sl} = 1$; therefore, with $C_{il} + C_{sl} = C_{tl}$,

$$N_{iz} = \frac{C_{il}}{C_{tl}}\,(N_{iz} + N_{sz}) - C_{tl}\, D_{is} \frac{dx_{il}}{dz}. \tag{3.1.133b}$$

Assume that an averaged value may be used for C_{tl} and D_{is}. Then the result (3.1.126) or (3.1.127) with Φ_N defined by (3.1.128) may be used. Replacing, however, D_{AB} by D_{is} and δ_g by δ_l, the diffusion path length in the liquid is given by

$$N_{iz} = \left(\frac{N_{iz}}{N_{iz} + N_{sz}}\right) \frac{D_{is}\, C_{tl}}{\delta_l} \,\ell\text{n}\left[\frac{x^{\delta}_{il} - \left(\dfrac{N_{iz}}{N_{iz} + N_{sz}}\right)}{x^0_{il} - \left(\dfrac{N_{iz}}{N_{iz} + N_{sz}}\right)}\right]. \tag{3.1.134}$$

Define

$$k'_{cl} = \frac{D_{is}}{\delta_l} \quad \text{and} \quad k'_{xl} = k'_c\, C_{tl}. \tag{3.1.135}$$

Then

$$N_{iz} = \frac{k'_{xl}}{\Phi_N}\,(x^0_{il} - x^{\delta}_{il}) = \frac{k'_c}{\Phi_N}\,(C^0_{il} - C^{\delta}_{il}), \tag{3.1.136a}$$

where

$$\Phi_N = \frac{(N_R - x^{\delta}_{il}) - (N_R - x^0_{il})}{N_R\,\ell\text{n}\left(\dfrac{N_R - x^{\delta}_{il}}{N_R - x^0_{il}}\right)} \tag{3.1.136b}$$

and

$$N_R = \frac{N_{iz}}{N_{iz} + N_{sz}}. \tag{3.1.136c}$$

A special case of this arises as before: the diffusion of solute i through stagnant solvent s. Therefore put $N_{sz} = 0,\ N_R = 1$ and we get

$$N_{iz} = \frac{D_{is}}{\delta_l}\, C_{tl}\,\ell\text{n}\left(\frac{1 - x^{\delta}_{il}}{1 - x^0_{il}}\right) = \frac{D_{is}\, C_{tl}\,(x^0_{il} - x^{\delta}_{il})}{\delta_l\, x_{s,lm}},$$

where

$$x_{s,lm} = \frac{x^{\delta}_{sl} - x^0_{sl}}{\ell\text{n}\,(x^{\delta}_{sl}/x^0_{ls})}. \tag{3.1.137}$$

The case of equimolar counterdiffusion can also be obtained when $N_{iz} = -N_{sz}$, yielding

$$N_{iz} = -C_{tl}\, D_{is} \frac{dx_{il}}{dz}, \tag{3.1.138a}$$

where both C_{tl} and D_{iS} may vary with composition and therefore z-coordinate. Assume an averaged value of C_{tl} and D_{is}, integrate the above equation at steady state from $z = 0$ to $z = \delta_l$, the diffusion path length, to obtain

$$N_{iz} = \frac{C_{tl}\, D_{is}}{\delta_l}\,(x^0_{il} - x^{\delta}_{il}) = k'_{xl}(x^0_{il} - x^{\delta}_{il}) = k'_{cl}(C^0_{il} - C^{\delta}_{il}), \tag{3.1.138b}$$

where k'_{cl} and k'_{xl} have been defined by (3.1.135). Note that $\Phi_N = 1$ in this case.

The preceding treatments were limited to simple molecular diffusion. If the value of $\boldsymbol{v}_t$ or $\boldsymbol{v}^*_t$ was nonzero, it was directly a result of individual species velocity $\boldsymbol{v}_i$ arising from a concentration gradient or a partial pressure gradient. In practical separation processes, there is *fluid convection*; the flow may be *laminar* or *turbulent*. The rate of transfer of a species is enhanced considerably by such fluid motion. The fluid motion may or may not be in the direction of the intended species transport. More often than not, the desired direction of species transport is perpendicular to the direction of bulk motion of the fluid.

It is possible to predict theoretically the mass transfer rate (or flux $\boldsymbol{N}_i$) across any surface located in a fluid having laminar flow in many situations by solving the differential equation (or equations) for mass balance (Bird *et al.*, 1960, 2002; Skelland, 1974; Sherwood *et al.*, 1975). Our capacity to predict the mass transfer rates a priori in turbulent flow from first principles is, however, virtually nil. In practice, we follow the form of the integrated flux expressions in molecular diffusion. Thus, the flux of species i is expressed as the product of a mass-transfer coefficient in phase j and a concentration difference in the forms shown below:

$$N_{iz} = k_{cj}(C^0_{ij} - C^{\delta}_{ij}) = k_{xj}(x^0_{ij} - x^{\delta}_{ij}) = k_g(p^0_{ig} - p^{\delta}_{ig}). \tag{3.1.139}$$

However, the values of the mass-transfer coefficients are obtained from empirical correlations instead of the simple molecular diffusion based expressions (3.1.124) or

Table 3.1.4A. Relations between species flux and mass-transfer coefficients for two special cases

Flux expressions in gaseous systems - z-direction

Equimolar counterdiffusion	Diffusion of A through stagnant B	Units of transfer coefficient	Relation between the two transfer coefficients
$N_{Az} = k'_{xg}(x^0_{Ag} - x^\delta_{Ag})$	$N_{Az} = k_{xg}(x^0_{Ag} - x^\delta_{Ag})$	$\dfrac{\text{mol}}{\text{cm}^2\text{-s-mole fraction}}$	$k_{xg} = \dfrac{k'_{xg}}{x_{B,lm}} = \dfrac{k'_{xg}P}{p_{B,lm}}$
$N_{Az} = k'_{yg}(y^0_{Ag} - y^\delta_{Ag})$	$N_{Az} = k_{yg}(y^0_{Ag} - y^\delta_{Ag})$	$\dfrac{\text{mol}}{\text{cm}^2\text{-s-mole fraction}}$	$k_{yg} = \dfrac{k'_{yg}}{y_{B,lm}} = \dfrac{k'_{yg}P}{p_{B,lm}}$
$N_{Az} = k'_g(p^0_{Ag} - p^\delta_{Ag})$	$N_{Az} = k_g(p^0_{Ag} - p^\delta_{Ag})$	$\dfrac{\text{mol}}{\text{cm}^2\text{-s-atm}}$	$k_g = \dfrac{k'_g P}{p_{B,lm}}$
$N_{Az} = k'_{cg}(C^0_{Ag} - C^\delta_{Ag})$	$N_{Az} = k_{cg}(C^0_{Ag} - C^\delta_{Ag})$	$\dfrac{\text{mol}}{\text{cm}^2\text{-s-(mol/cm}^3)}$	$k_{cg} = \dfrac{k'_{cg}C_t}{C_{B,lm}} = \dfrac{k'_{cg}P}{p_{B,lm}}$

Flux expressions in liquid systems - z-direction

Equimolar counterdiffusion	Diffusion of i through stagnant s	Units of transfer coefficient	Relation between the two transfer coefficients
$N_{iz} = k'_{xl}(x^0_{il} - x^\delta_{il})$	$N_{iz} = \dfrac{k'_{xl}}{x_{s,lm}}(x^0_{il} - x^\delta_{il})$	$\dfrac{\text{mol}}{\text{cm}^2\text{-s-mole fraction}}$	$k_{xl} = \dfrac{k'_{xl}}{x_{s,lm}}$
$N_{iz} = k'_{cl}(C^0_{il} - C^\delta_{il})$	$N_{iz} = k_{cl}(C^0_{il} - C^\delta_{il})$	$\dfrac{\text{mol}}{\text{cm}^2\text{-s-(mol/cm}^3)}$	$k_{cl} = \dfrac{k'_{cl}}{x_{s,lm}}$

Relations for a gaseous system between k_g, k_c, k_x and k_y: $k_g = (k_y/P) = (k_c/RT) = (k_x/P)$.
Relation for a liquid system between k_c and k_x: $k_x = k_c C_t$.

(3.1.135). The mass-transfer coefficients are strong functions of the extent of convection. A few such correlations used frequently are given in Section 3.1.4.1. The quantities $C^0_i, C^\delta_i, x^0_{ij}, x^\delta_{ij}, p^0_{ig}$ and p^δ_{ig} are, as before, values at the two ends of the mass-transfer path.

There is a need to provide additional clarification about the above-mentioned quantities used to define flux $\boldsymbol{N}_{iz}$. Consider transfer of a species from the wall of a duct to the fluid flowing in the duct. The duct wall is obviously one end of the diffusion path. It is not clear where the other end is since material from the duct wall is coming to (and being convected away from) at all distances normal to the duct wall. Common practice, therefore, is to define a *bulk concentration* $\boldsymbol{C}_{ijb}$ (the *cup-mixing concentration*, (Bird *et al.*, 1960, 2002)) by

$$C_{ijb} = \frac{\int_{A_c} v_{xj} C_{ij}\, dA_c}{\int_{A_c} v_{xj}\, dA_c}, \tag{3.1.140}$$

where A_c is the cross-sectional area of the duct, v_{xj} is the local convective velocity in the mean flow direction (say the x-coordinate) along the duct length in phase j and C_{ij} is the value of the species i concentration at the location of velocity v_x. This C_{ijb} is then used as the concentration at the other end of the diffusion path in the flux expressions (3.1.139).

Some clarifications on the mass-transfer coefficients for use in the flux expressions (3.1.139) are also in order. If the mass-transfer coefficient under convection was measured or correlated under particular conditions of N_R $(= \{N_{iz}/(N_{iz} + N_{sz})\}$ or $\{N_{Az}/(N_{Az} + N_{Bz})\})$, then either of the basic relations (3.1.126) or (3.1.134) should be used to determine the value of $(D_{AB}C_t/\delta_g)$ or $(D_{is}C_t/\delta_l)$ as the case may be. These values are simply k'_{xj}, the mass-transfer coefficient for equimolar counterdiffusion in the presence of convection. For any other value of N_R, the value of k_{xj}, k_{cj} or k_g in equation (3.1.139) is obtained by using the bulk flow correction factor Φ_N defined by (3.1.129b) or (3.1.136b). Table 3.1.4A summarizes the flux vs. concentration difference relations in convective mass transfer for Δp_{ij}, ΔC_{ij}, Δx_{ij} under two conditions encountered frequently in separation processes using gases or liquids. Since the mole fraction of i in the literature is often expressed as x_i for a liquid phase and y_i for a gas phase, we have also incorporated Δy_i in this table. Further, k_c and k_l are interchangeable for the liquid phase. Sometimes subscript l may be replaced by w or o to indicate an aqueous or organic phase. For gaseous mixtures, species A and B have been used instead of i and s. Table 3.1.4B provides corresponding expressions where species i has been identified specifically in the mass-transfer coefficient terms as, for example, k_{ijc}, for species i in phase j and concentration units C.

In separation process devices/equipment having convection, the flows are often complex. Yet a simplified

Table 3.1.4B. Relations between species flux and mass-transfer coefficients for two special cases with species i *identified in mass-transfer coefficients*

Flux expressions in gaseous systems - *z*-direction			
Equimolar counterdiffusion	Diffusion of A through stagnant B, i = A	Units of transfer coefficient	Relation between the two transfer coefficients
$N_{iz} = k'_{igx}(x_i^0 - x_i^\delta)$	$N_{Az} = k_{Agx}(x_A^0 - x_A^\delta)$	$\dfrac{\text{mol}}{\text{cm}^2\text{-s-mole fraction}}$	$k_{ix} = \dfrac{k'_{ix}P}{(P-p_i)_{lm}}$
$N_{iz} = k'_{igy}(y_i^0 - y_i^\delta)$	$N_{Az} = k_{Agy}(y_A^0 - y_A^\delta)$	$\dfrac{\text{mol}}{\text{cm}^2\text{-s-mole fraction}}$	$k_{iy} = \dfrac{k'_{iy}P}{(P-p_i)_{lm}}$
$N_{iz} = k'_{ig}(p_i^0 - p_i^\delta)$	$N_{Az} = k_{Ag}(p_A^0 - p_A^\delta)$	$\dfrac{\text{mol}}{\text{cm}^2\text{-s-pressure}}$	$k_{ig} = \dfrac{k'_{ig}P}{(P-p_i)_{lm}}$
$N_{iz} = k'_{igc}(C_{ig}^0 - C_{ig}^\delta)$	$N_{Az} = k_{Agc}(C_{Ag}^0 - C_{Ag}^\delta)$	$\dfrac{\text{mol}}{\text{cm}^2\text{-s-(mol/cm}^3)}$	$k_{igc} = \dfrac{k'_{igc}C_t}{(C_t - C_i)_{lm}} = \dfrac{k'_{igc}P}{(P-p_i)_{lm}}$
Flux expressions in liquid systems - *z*-direction			
Equimolar counterdiffusion	Diffusion of i through stagnant s	Units of transfer coefficient	Relation between the two transfer coefficients
$N_{iz} = k'_{ilx}(x_{il}^o - x_{il}^\delta)$	$N_{iz} = \dfrac{k'_{ilx}}{x_{s,lm}}(x_{il}^o - x_{il}^\delta)$	$\dfrac{\text{mol}}{\text{cm}^2\text{-s-mole fraction}}$	$k_{ilc} = \dfrac{k'_{ilc}}{x_{s,lm}}$
$N_{iz} = k'_{ilc}(C_{il}^0 - C_{il}^\delta)$	$N_{iz} = k_{ilc}(C_{il}^0 - C_{il}^\delta)$	$\dfrac{\text{mol}}{\text{cm}^2\text{-s-(mol/cm}^3)}$	$k_{ilc} = \dfrac{k'_{ilc}}{x_{s,lm}}$

Relations for a gaseous system between k_{ig}, k_{ic}, k_{ix} and k_{iy}: $k_{ig} = (k_{iy}/P) = (k_{ic}/RT) = (k_{ix}/P)$.
Relation for a liquid system between k_{ilc} and k_{ilx} : $k_{ilx} = k_{ilc}C_{tl}$.

description of the mass-transfer coefficient (instead of a detailed correlation) is desirable in many cases, including those having a chemical reaction. A number of theories of increasing complexity, e.g. film theory, penetration theory, surface renewal theory, boundary layer theory (Bird *et al.*, 1960, 2002; Danckwerts, 1970; Sherwood *et al.*, 1975), have been developed to this end to describe the mass-transfer coefficient and the mass-transfer rate. Of these, film theory is the simplest and will be used in this book often to describe the mass-transfer coefficient and the mass-transport process at the interface of a gas or liquid phase for the following systems: gas-liquid; gas-solid; liquid-solid; liquid-liquid; liquid-membrane; liquid-ion exchange resin.

In the *film theory* description of the mass-transfer process occurring between two fluid phases or between a solid and a fluid phase, the complex mass-transfer phenomenon is substituted by the notion of simple molecular diffusion of the species through a stagnant fluid film of thickness δ. The actual concentration profiles of species A being transferred from phase 2 to phase 1 are shown in Figures 3.1.6 (a) and (b) in one phase only for a solid-liquid and a gas-liquid system, respectively. The concentration of A in the liquid phase at the solid-liquid or the gas-liquid interface is C_A^0. Far away from the interface it is reduced to a low value in the liquid phase. In turbulent flow, the curved profile of species A shown would correspond to the time-averaged value (Bird *et al.*, 1960, 2002). According to the film theory, the complex mass-transfer rate can be determined by postulating that simple molecular diffusion through a stagnant film of thickness δ_l next to the phase interface can describe the mass-transfer coefficient. Thus

$$N_{Az} = k_{cl}(C_{Al}^0 - C_{Alb}) = \frac{D_{As}}{\delta_l}(C_{Al}^0 - C_{Alb}), \tag{3.1.141}$$

so that

$$k_{cl} = (D_{As}/\delta_l), \tag{3.1.142}$$

where δ_l is the unknown effective film thickness. If k_{cl} is known, δ_l can be determined. As a predictive theory, film theory is rather poor, for example, in predicting the dependence of k_{cl} on D_{As} in most situations encountered. However, its use is frequent in complex geometries and flow conditions in the presence of chemical reactions. Additionally, the notion of a stagnant film has a vestigial effect; the **mass-transfer coefficient of the gas phase or the liquid phase is routinely identified as the gas film coefficient or the liquid film coefficient.**

3.1.4.1 *Empirical correlations of single-phase mass-transfer coefficients*

The analysis and estimation of separation in laboratory-scale techniques as well as large-scale processes generally require an estimate of the mass-transfer coefficient in a single-phase system which may be part of a multiphase

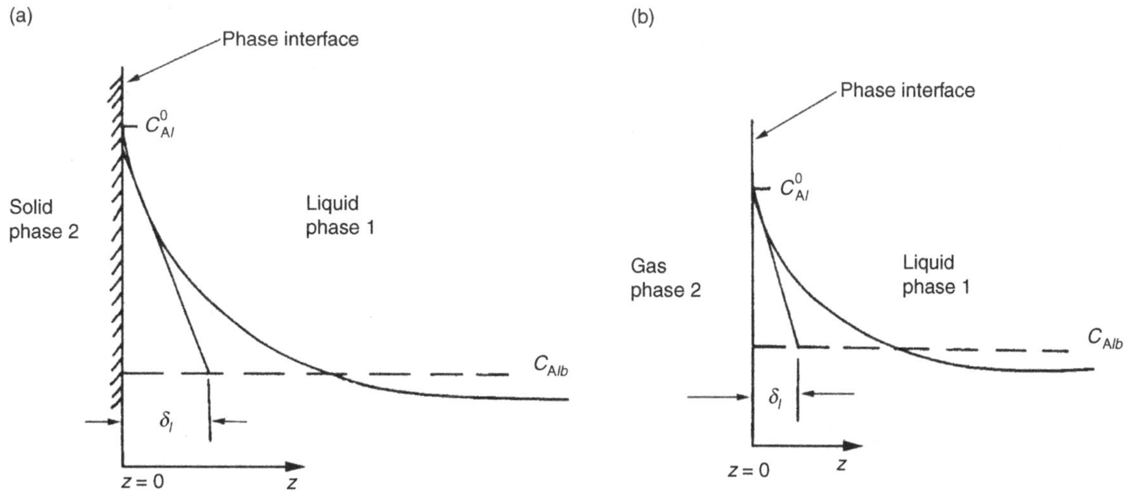

Figure 3.1.6. Stagnant film model for mass transfer near a phase interface. (a) Solid–liquid system. (b) Gas–liquid system.

separation system. Most practical separation systems have complex geometries and convective flow patterns[14] for which an exact mass-transfer analysis may not be available. Over the years, numerous experimental investigations of many such systems have been carried out, and the resulting mass-transfer coefficients have been correlated with a number of dimensionless numbers/groups formed out of the variables that affect the mass transfer. This correlation activity is guided by *Buckingham's pi theorem*: the functional relationship between q variables whose units may be described in terms of u fundamental units or dimensions may be written as a function of $q - u$ independent dimensionless groups (Langhaar, 1951).

There are three fundamental units or dimensions, mass M, length L and time t in these problems without any heat transfer (which will require T, the temperature, as the fourth one). The following is a list of the variables of importance: a characteristic physical dimension of the device, say, d, the diameter;[15] density, ρ; velocity, v; viscosity, μ; diffusion coefficient, for example, D_{AB} (or D_{is}): a total of six variables (including k). The number of independent dimensionless groups are 6−3, or 3. The three basic dimensionless groups in mass transfer are:

$$\text{Sherwood number} = Sh = \frac{k\,d}{D_{AB}};$$

$$\text{Reynolds number} = Re = \frac{dv\rho}{\mu};$$

$$\text{Schmidt number} = Sc = \frac{\mu}{\rho\, D_{AB}}$$

$$= \frac{\nu}{D_{AB}} (= \nu/D_{is} \text{ for solute } i \text{ in solvent } s). \quad (3.1.143a)$$

These three basic dimensionless groups are related functionally by

$$Sh = f(Re, Sc). \quad (3.1.143b)$$

Occasionally, an additional geometrical dimension of length, l, beside a diameter, d, is simultaneously involved. Then an additional basic dimensionless number is the (l/d) ratio:

$$Sh = f\left(Re, Sc, \frac{l}{d}\right). \quad (3.1.143c)$$

The functional relation between the Sherwood number and the other numbers has a general form of the following type:

$$Sh = \mathrm{a} + \mathrm{b}\, Re^{\mathrm{b_1}}\, Sc^{\mathrm{b_2}} \left(\frac{l}{d}\right)^{\mathrm{b_3}}, \quad (3.1.143d)$$

where the constants a, b, b_1, b_2 and b_3 reflect particular flow regimes, flow patterns, mass-transfer regimes, concentration profile development, etc.

In separation systems where natural convection is involved (see Section 6.1.3), T, the temperature, becomes the fourth fundamental unit ($u = 4$), and the density difference, $\Delta\rho$, due to a temperature difference and the acceleration due to gravity, g, become additional variables of

[14] See Section 6.1 for the sources and the nature of bulk flow commonly employed in separations.

[15] For systems containing particles or drops, it is d_p, the particle/drop diameter.

Table 3.1.5. Limiting analytical solutions and experimental correlations for convective mass transfer between a fluid and a solid wall

Configuration	Analytical solution/experimental correlation	Variable range	Reference	Equation number(s)
Laminar flow in a tube[a,b]	$Sh_z = \dfrac{k_z d}{D_{AB}} = 1.077\left(\dfrac{d}{z}\right)^{1/3}(Re\,Sc)^{1/3}$	$\dfrac{\pi}{4}\left(\dfrac{d}{z}\right) Re\,Sc = \dfrac{W}{\rho D_{AB} z} > 400$	LéVêque (1928); Skelland (1974)	(3.1.144), (3.1.145)
(1) Entrance region	$Sh_{avg} = \dfrac{k_{avg} d}{D_{AB}} = 1.615\left(\dfrac{d}{L}\right)^{1/3}(Re\,Sc)^{1/3}$			
(2) Fully developed concentration and velocity profiles	$Sh = (kd/D_{AB}) = 3.656$	uniform wall concentration	Skelland (1974)	(3.1.146)
	$Sh = (kd/D_{AB}) = 4.36$	uniform mass flux at wall	Skelland (1974)	(3.1.147)
Laminar flow between flat parallel plates: fully developed velocity and concentration profiles[c]	$Sh = (k2b/D_{AB}) = 7.6$	uniform wall concentration	Skelland (1974)	(3.1.148)
	$Sh = (k2b/D_{AB}) = 8.23$	uniform mass flux at wall	Skelland (1974)	(3.1.149)
Turbulent flow in a tube	$Sh = 0.023\,(Re)^{0.83}\,(Sc)^{0.33}$	$Re > 2100$; $0.6 < Sc < 3000$	Linton and Sherwood (1950)	(3.1.150)
Laminar boundary layer flow over a flat plate: integral analysis[d]	$J_D = 0.664\,Re_L^{-1/2}$	$Re_L < 15\,000$	Sherwood *et al.* (1975)	(3.1.151)
	$Sh_{avg} = 0.646\,Re_L^{1/2}\,Sc^{1/3}$	$Re_L < 3 \times 10^5$	Skelland (1974)	(3.1.152)

[a] z is the distance along the tube from the entrance.
[b] L is the length of the pipe; W is the total mass flow rate in the tube; see Skelland (1974) for solutions of the mass-transfer rate in developing concentration profiles for fully developed velocity profiles (plug or parabolic) in a tube.
[c] $2b$ is the gap between the parallel plates.
[d] $Re_L = (Lv_\infty/v)$, where L is the plate length and v_∞ is the free stream velocity outside the boundary layer.

importance: the larger the $\Delta\rho$, the higher the velocity due to the natural convective driving force. Correspondingly, the number of variables is now eight so the number of independent dimensionless groups is 8 - 4, or 4. The additional dimensionless group is the Grashof number,

$$Gr = \frac{d^3\rho^2 g}{\mu^2}\frac{\Delta\rho}{\rho}, \qquad (3.1.143e)$$

where $\Delta\rho$ is a density difference $(\rho_1 - \rho_2)$ in a given phase between two locations and ρ is an averaged density at the mean temperature. (Note that $\Delta\rho$ may be created by a concentration difference in a single phase as well without any temperature difference.) Correspondingly, the functional relation between the Sherwood number and the other dimensionless numbers may be presented as

$$Sh = a + b\,Re^{b_1}\,Sc^{b_2}\,Gr^{b_3} \qquad (3.1.143f)$$

in cases where only one length variable, d, is important.

Often special combinations of these dimensionless groups are employed to illustrate the correlations of mass-transfer coefficients. A few of them are as follows:

$$\text{Stanton number, } St = \frac{Sh}{Re\,Sc} = \frac{k}{v};$$

$$\text{Péclet number, } Pe = Re\,Sc = \frac{dv}{D_{AB}} = \frac{dv}{D_{is}}; \qquad (3.1.143g)$$

$$J_D \text{ factor, } J_D = \frac{Sh}{Re\,Sc^{1/3}}.$$

The correlations that have been developed for mass transfer are most often for system configurations of the following type.

Table 3.1.6. Experimental correlations and limiting analytical solutions for convective mass transfer between a fluid and a solid sphere

Configuration	Experimental correlation/analytical solution	Variable range	Reference	Equation number
One sphere in a stagnant fluid	$Sh = \frac{k_g d_p\, p_{B,lm}}{D_{AB}\, C_t} = 2$ (analytical solution)	stagnant gas around sphere	Cornish (1965)	(3.1.153)
Gas flowing over a single sphere	$Sh = 2 + 0.552\, Re^{0.53}\, Sc^{1/3}$	$1 < Re < 48\,000$ $0.6 < Sc < 2.7$	Geankoplis (1972)	(3.1.154)
Liquid flowing over a single sphere	$Sh = 2 + 0.95\, Re^{0.50}\, Sc^{1/3}$	$2 < Re < 2000$ $788 < Sc < 1680$	Geankoplis (1972)	(3.1.155)
	$Sh = \frac{k_c d_p}{D_{AB}} = 0.347\, Re^{0.62}\, Sc^{1/3}$	$2000 < Re < 16\,900$	Geankoplis (1972)	(3.1.156)
Gas flowing over a packed bed[a]	$J_D = (0.4548/\varepsilon)\, Re^{-0.4069}$	$10 < Re < 10\,000$	Dwivedi and Upadhyay (1977)	(3.1.157)
	$Sh = 2 + 1.1\, Sc^{1/3}\, Re^{0.6}$	$3 < Re < 10^4$	Wakao and Funazkri (1978)	(3.1.158)
Liquid flowing over a packed bed	$J_D = (1.09/\varepsilon)\, Re^{-(2/3)}$	$0.0016 < Re < 55$ $165 < Sc < 70\,600$	Wilson and Geankoplis (1966)	(3.1.159)
	$J_D = (0.25/\varepsilon)\, Re^{-0.31}$	$55 < Re < 1\,500$ $165 < Sc < 10\,690$		(3.1.160)
Fluidized bed of spheres (gases and liquids)	$\varepsilon J_D = 0.01 + [0.863/(Re^{0.58} - 0.483)]$	$1 \le Re \le 2140$	Gupta and Thodos (1962)	(3.1.161)

ε is the void volume fraction (porosity) of the packed bed; Re = $(d_p\, G/\nu)$, where G is the superficial mass average velocity in the empty flow section (before the single sphere or the empty bed without packing); d_p is the sphere diameter; for liquids, $Sh = (k_c\, d_p/D_{AB})$.

[a] The equation for *Sh* was developed by using an effective axial diffusion coefficient $D_{i,\text{eff},z}$, (equation (7.1.4)) called the axial dispersion coefficient in fixed-bed adsorbers via the following expression:

$$(\varepsilon D_{i,\text{eff},z}/D_{AB}) = (\varepsilon D_{i,\text{eff},z}/D_{is}) = 20 + 0.5\, Re\, Sc.$$

(1) Mass transfer between a fluid and a solid wall under the following conditions:
 (a) fluid is flowing inside a channel/pipe/tube;
 (b) fluid is flowing outside a channel/pipe/plate/tube.
(2) Mass transfer between a fluid and a single solid particle or a collection of particles, as in a packed or fluidized bed.
(3) Mass transfer between a fluid and a single immiscible drop or a collection of drops as, for example, in a liquid–liquid dispersion.
(4) Mass transfer between a liquid and a gas bubble or a collection of gas bubbles or particles in suspension.
(5) Mass transfer between a flowing thin liquid layer and another immiscible fluid layer flowing in parallel.

Many of these areas have a vast amount of literature devoted to them. What will be included here is as follows: a few important correlations will be provided, if possible, in each case, along with a few references. Often the geometries and the flow configurations are more complex than the five categories identified above. Further, there will be additional limitations depending on the flow regime, laminar or turbulent, as characterized by the Reynolds number, the extent of development of the concentration boundary layer (developing or fully developed), and any other range specified by the developer of the correlations. Sometimes, analytical solutions are available, and a few will be provided. For a detailed introduction to mass transfer and mass-transfer coefficients in such systems, consult Geankoplis (1972, 2003), Skelland (1974), Sherwood *et al.* (1975), Cussler (1997), Middleman (1998) and Bird *et al.* (2002). Table 3.1.5 summarizes limiting analytical solutions and experimental correlations for convective mass transfer between a fluid and a solid wall. Table 3.1.6 lists experimental correlations and some limiting analytical solutions for convective mass transfer between a fluid and a solid sphere, as in packed or fluidized beds. Table 3.1.7 provides the same information between a fluid and small particles, bubbles or drops. Table 3.1.8 illustrates a few experimental correlations (sometimes with their analytical basis) for convective mass transfer between a liquid and a permeable membrane (porous or nonporous) having a particular geometry, e.g. spiral-wound reverse osmosis unit, hollow fiber membrane contactor.

Table 3.1.7. Experimental correlations and limiting analytical solutions for convective mass transfer between a fluid and small particles, bubbles or drops

Configuration	Experimental correlation/Analytical solution	Variable range	Reference	Equation number
Small particles, gas bubbles or liquid drops in suspension	$Sh_c = \frac{k_c d_p}{D_{AB}} = 2.0 + 0.31\, Sc^{1/3}\left(\frac{d_p^3 \rho_c \Delta\rho g}{\mu_c^2}\right)$	$d_p < 0.6$ mm	Geankoplis (2003)	(3.1.162)
Particles in a turbulent stirred tank	$Sh_c = \frac{k_c d_T}{D_{AB}} = 0.267\,(Sc_c)^{1/4}\, Re_{imp}{}^{3/4} N_p{}^{1/4}\left(\frac{d_T}{V d_{imp}}\right)^{1/4}$	particle size independent	Kulov (1983)	(3.1.163)
	$Sh_c = \frac{k_c d_p}{D_{AB}} = 0.13\,(Sc_c)^{1/3}\left(\frac{(Pr/V)\, d_p^4 \rho_c^2}{\mu_c^3}\right)^{1/4}$		Calderbank and Moo-Young (1961); Blakebrough (1967)	(3.1.164)
Large gas bubbles	$k_c = 0.42\, Sc^{-0.5}\left(\frac{\Delta\rho\, \mu_c g}{\rho_c^2}\right)^{1/3}$	$d_p > 2.5$ mm	Geankoplis (2003)	(3.1.165)
Continuous phase in liquid-liquid dispersion	$Sh_c = \frac{k_c d_p}{D_{AB} C} = 0.725\, Pe^{0.57}\, Sc_c{}^{-0.15}(1 - \phi_D)$		Treybal (1963, p. 480)	(3.1.166)
Dispersed phase in liquid-liquid dispersion	$Sh_D = \frac{k_D d_p}{D_{ABD}} = \frac{0.00375\,(d_p v_t / D_{ABD})}{(1 + (\mu_D/\mu_c))}$	circulating liquid sphere	Treybal (1963, p. 187)	(3.1.167)
	$Sh_D = (2\pi^2/3)$	rigid sphere	Treybal (1963, p. 186)	(3.1.168)
Droplet formation in liquid-liquid dispersion where t_f is the time of formation of a single drop	$k_{Df} = 2.957(\rho/M)_{avg}(D_{ABD}/\pi t_f)^{0.5}$ $k_{Cf} = 4.6(\rho/M)_{avg}(D_{ABc}/\pi t_f)^{0.5}$	Based on total drop surface area at detachment	Skelland (1974)	(3.1.169)

Sh_c, Sc_c, k_c, ρ_c, μ_c refer to liquid phase for particles, bubbles, etc.
$\Delta\rho = \rho_p - \rho_c$, k_c, k_D, Sh_c, Sh_D, D_{ABc}, D_{ABD}, μ_c, μ_D refer to continuous phase (c) and dispersed phase (D) in liquid-liquid dispersion: v_t = drop terminal velocity; ϕ_D = dispersed phase volume fraction; d_T = vessel diameter; d_{imp} = impeller diameter; V = liquid volume; N_p = power number (= Pr (mixing power, watts)/$\rho_c\, d_{imp}^5\, N^3$), where N is the rotational speed in revolutions per minute, $Re_{imp} = N\, d_{imp}^2/\upsilon_c$, υ_c being the kinematic viscosity of the liquid.

3.1.5 Flux expressions in multicomponent systems

Chemical separation problems can involve a binary or a multicomponent mixture. The flux expressions commonly used for multicomponent mixtures are considered here.

3.1.5.1 *Maxwell-Stefan equations*

The Maxwell-Stefan equations for describing the diffusion of gases in a multicomponent gas mixture have been developed from the kinetic theory of gases. A highly simplified illustration may be pursued as follows. Consider a system of a gas mixture of n species at constant T and P. Focus first on molecules of species i. The net force exerted on species i in the absence of any external forces is $-\nabla\mu_{ig}$ / gmol of i. The net force exerted on species i per unit volume of the mixture is $-(\nabla\mu_{ig})\, x_{ig}\, C_{tg}$, where C_{tg} is the total molar density of the gas mixture; from equation (3.1.40) and an ideal gas mixture, the net force on species i per unit mixture volume is $-x_{ig}\, C_{tg}\, RT\, \nabla\, \ell n\, (Px_{ig})$, which is equivalent to

$$-x_{ig}\, C_{tg}\, RT\, \nabla\, (Px_{ig})/Px_{ig} = -\nabla(Px_{ig})$$

at constant P and T.

Now, these molecules of species i collide with molecules of species k, and momentum is transferred/lost by molecules of species i, resulting essentially in a frictional force on molecules of species i; this loss of momentum is proportional to the difference in velocities ($\boldsymbol{v}_{ig} - \boldsymbol{v}_{kg}$) between the two types of molecules. Further, the number of such collisions per unit time per unit volume of the gas mixture is proportional to $x_{ig}\, x_{kg}$. At steady state, the force on molecules of species i per unit volume of the mixture, $-C_{tg}\, RT\, \nabla x_{ig}$, should be equal to the frictional force on molecules of species i created by the loss of momentum of species i via collisions with species k molecules, namely $f_{ikm}\, x_{ig}\, x_{kg}(\boldsymbol{v}_{ig} - \boldsymbol{v}_{kg})$, where f_{ikm} is a frictional coefficient of sorts of species i due to species k in the unit volume:

$$-\nabla(Px_{ig}) = -P\, \nabla\, x_{ig} = f_{ikm}\, x_{ig}\, x_{kg}(\boldsymbol{v}_{ig} - \boldsymbol{v}_{kg}). \quad (3.1.177a)$$

This relation is written generally as

Table 3.1.8. Experimental correlations and limiting analytical solutions for convective mass transfer between a liquid and a membrane surface

Configuration	Experimental correlation/analytical solution	Variable range	Reference	Equation number(s)
Spiral-wound membrane channel in reverse osmosis[a]	$Sh_c = \frac{kd_h}{D_{il}} = 0.065\ Re^{0.875}\ Sc^{0.25}$ $Re = (vd_h/v)$		Schock and Miquel (1987)	(3.1.170)
Hollow fiber membrane module: Laminar tube-side flow				
(1) Gas absorption into liquid in hollow fiber lumen	$Sh_{avg} = \frac{k_{avg} d}{D_{AB}} = 1.615(d/L)^{1/3}(Re\,Sc)^{1/3}$ $Sh_{avg} = 1.64(d/L)^{1/3}(Re\,Sc)^{1/3}$	$(W/\rho D_{AB} z) > 400$ $Sc = 464$; $3 < (vd/D_{AB}L) < 500$	LéVêque (1928); Skelland (1974); Yang and Cussler (1986)	(3.1.145), (3.1.171)
(2) Membrane solvent extraction in hollow fiber lumen[b]	$Sh_{avg} = 0.5(d/L)\ Re\ Sc\ \frac{1-\varsigma}{1+\varsigma}$	$300 < Sc < 1000$ $0 < Re < 600$	Prasad and Sirkar (1988)	(3.1.172)
Hollow fiber membrane module: shell-side flow				
(1) Liquid in parallel flow[a]				
(i) in solvent extraction	$Sh_{avg} = \frac{k_{avg} d_h}{D_{il}} = 5.85\,(1-\phi)\left(\frac{d_h^*}{L}\right) Re^{0.66}\ Sc^{0.33}$	$0.08 < \phi \leq 0.4$	Prasad and Sirkar (1988)	(3.1.173)
(ii) in gas absorption	$Sh_{avg} = 1.25\,(d_h\,Re/L)^{0.93}\ Sc^{0.33}$	$\phi \leq 0.26$, $Gr < 60$	Yang and Cussler (1986)	(3.1.174)
(2) Liquid in cross flow in gas absorption[c,d]	$Sh = kd/D_{il} = 0.57\,Re^{0.31}\ Sc^{0.33}$	$0.01 < Re < 1$; no. of fiber layers 2 to 7	Bhaumik *et al.* (1998)	(3.1.175)
	$Sh = kd/D_{il} = 0.80\,Re^{0.47}\ Sc^{0.33}$	$0.1 < Re < 10$; $Sc = 480$	Wickramasinghe *et al.* (1993)	(3.1.176)

[a] d_h = hydraulic diameter = 4 R_h (hydraulic radius) = $[2\varepsilon_0/\{(2/b) + (1-\varepsilon_0)\, s_{v,sp}\}]$: b = feed channel height; ε_0 = porosity of spacer; $s_{v,sp}$ = specific surface of the spacer; $d_h = 4 \times$ (shell-side available flow cross-sectional area/wetted perimeter).
[b] For details on ς, see Skelland (1974, eq. (5.145)) and Prasad and Sirkar (1988, 2001, eqs. (41-30a) and (41-49)).
[c] $Re = d\, v_s/v$; d = outside diameter of fiber; v_s is based on the shell-side cross-sectional area available for flow.
[d] $Re = d\, v_o/v$; d = outside diameter of fiber; v_0 is based on the superficial velocity across the fibers.

$$\nabla x_{ig} = \frac{f_{ikm}}{P}\, x_{ig}\, x_{kg}(\boldsymbol{v}_{kg} - \boldsymbol{v}_{ig}) = \frac{C_{ig} C_{kg}}{C_{tg}^2\, \overline{D}_{ik}}\,(\boldsymbol{v}_{kg} - \boldsymbol{v}_{ig}), \tag{3.1.177b}$$

where the gas pair (i, k) based diffusion coefficient

$$\overline{D}_{ik} = (P/f_{ikm}) = \left(\frac{C_{tg}\,RT}{f_{ikm}}\right).$$

When we add up the frictional contributions of n species to the drag on species i in an n-component mixture, we get

$$\nabla x_{ig} = \boldsymbol{d}_{ig} = \sum_{k=1}^{n} \frac{C_{ig}\, C_{kg}}{C_{tg}^2\, \overline{D}_{ik}}\,(\boldsymbol{v}_{kg} - \boldsymbol{v}_{ig}). \tag{3.1.177c}$$

For multicomponent gas mixtures containing n species and *ordinary diffusion*, rigorous results from the kinetic theory of gases provide a set of $(n - 1)$ Maxwell–Stefan (MS) equations (Curtiss and Hirschfelder, 1949; Bird *et al.*, 1960, 2002; Taylor and Krishna, 1993):

$$\begin{aligned} \boldsymbol{d}_{ig} = \nabla x_{ig} &= \sum_{k=1}^{n} \frac{x_{ig}\,\boldsymbol{N}_{kg} - x_{kg}\,\boldsymbol{N}_{ig}}{C_{tg}\,\overline{D}_{ik}} \\ &= \sum_{k=1}^{n} \frac{C_{ig}\, C_{kg}}{C_{tg}^2\, \overline{D}_{ik}}\,(\boldsymbol{v}_{kg} - \boldsymbol{v}_{ig}) \\ &= \sum_{k=1}^{n} \frac{x_{ig}\, x_{kg}}{\overline{D}_{ik}}\,(\boldsymbol{v}_{kg} - \boldsymbol{v}_{ig}). \end{aligned} \tag{3.1.178}$$

Here $\overline{D}_{ik}$ is the multicomponent diffusion coefficient of the gas pair (i, k) in any region, and $\boldsymbol{d}_{ig}$ is a force-type term for the ith species. Note that

$$\overline{D}_{ik} \geqq 0; \qquad \overline{D}_{ik} = \overline{D}_{ki}; \qquad \sum_{i=1}^{n} \frac{x_{ig}}{\overline{D}_{ik}} = 0; \qquad \sum_{i=1}^{n} \boldsymbol{d}_{ig} = 0. \tag{3.1.179}$$

Equation (3.1.178) indicates that the force-type term $\boldsymbol{d}_{ig}$ is linearly proportional to the relative velocity of the ith species with respect to every other species in the system. Since $\overline{D}_{ik}$ is a diffusion coefficient, $(P/\overline{D}_{ik})$ is a frictional coefficient (see (3.1.86)) of sorts; therefore each term on the right-hand side of equation (3.1.178) is a force-type term (see (3.1.76)). Each term may be thought of as a frictional drag force experienced by the ith molecules as they move past the molecules of the kth type; further, all types of molecules contribute to the net force on the ith species, resulting in a concentration gradient, or mole fraction gradient, or partial pressure gradient, of the ith species. This equation also indicates that $\boldsymbol{N}_{ig}$ need not be zero when ∇x_{ig} is zero. Further, the flux of i is influenced by the fluxes of all other species, $\boldsymbol{N}_{kg}$, in the system.

For ordinary diffusion in liquids and dense gases, the same set of equations is often used, although there is no rigorous theoretical basis. The multicomponent diffusion coefficient $\overline{D}_{ik}$ for the species pair (i, k) has similar properties, namely it is the inverse of the frictional coefficient f^{d}_{ik} in the dense gas or liquid phase. In fact, such formulations are also used in denser phases such as membranes, etc.; Curtiss and Bird (1999), however, have shown that the Maxwell-Stefan formalism is applicable to dense gases, liquids and polymers provided the strongly concentration-dependent diffusivities $\overline{D}_{ik}$ are not the binary diffusivities. For solutions, an interpretation of $\boldsymbol{d}_{ij}$ by Lightfoot (1974) is useful: $(\boldsymbol{d}_{ij})RTC_t$ is the force moving species i relative to the solution per unit volume of solution in the region under consideration. If we estimate it per gmol of species i, i.e. $\{(\boldsymbol{d}_{ij})RTC_t/C_i\}$, we find it to be equal to $\{RT\,\nabla x_i/x_i\}$, i.e. $RT\,\nabla \ell n\, x_i$ (consider expression (3.1.50) for ordinary diffusion only).

For cases where external forces are present along with the chemical potential, the generalized Maxwell-Stefan type of equations in any region j for any i are as follows (Hirschfelder *et al.*, 1954; Lightfoot, 1974; Lee *et al.*, 1977a; Bird *et al.*, 2002):

$$\boldsymbol{d}_{ij} = \sum_{k=1}^{n} \frac{x_{ij}\boldsymbol{N}_{kj} - x_{kj}\boldsymbol{N}_{ij}}{C_{tj}\overline{D}_{ik}} = \sum_{k=1}^{n} \frac{x_{ij}x_{kj}}{\overline{D}_{ik}}(\boldsymbol{v}_{kj} - \boldsymbol{v}_{ij}) \tag{3.1.180}$$

and

$$\boldsymbol{d}_{ij} = x_{ij}\nabla \ell n\, a_i + \frac{\left(\overline{V}_i C_{ij} - u_{ij}\right)\nabla P}{C_{tj}RT} - \frac{\rho_{ij}\left(\boldsymbol{g}_{ij} - \sum_{k=1}^{n} u_{kj}\boldsymbol{g}_{kj}\right)}{C_{tj}RT} + \frac{Z_i \mathcal{F} x_{ij}}{RT_j}\nabla \Phi, \tag{3.1.181}$$

where u_{ij} is the mass fraction of species i and $\boldsymbol{g}_{kj}$ is the body force on species k per unit mass of species k. For example, for gravitational forces $\boldsymbol{g}_{kj} = \mathbf{g} = -g\mathbf{k}$. Centrifugal force is another body force, where $\boldsymbol{g}_{kj} = \omega^2 r\,\mathbf{r}$ at any radial location. We have not included forces due to thermal diffusion, magnetic forces, nonuniform electrical field and electrophoretic retardation in the expression (3.1.181) for $\boldsymbol{d}_{ij}$.

It is obvious that the relations (3.1.178), (3.1.180) are quite complicated. To provide a simpler picture, it is customary to reduce these complicated relations to the form of Fick's first law of diffusion (3.1.87) or (3.1.99) for a binary system. Thus species i is assumed to diffuse through a second species, which is, in effect, the multicomponent mixture. The manner in which this rearrangement is achieved is as follows. Rearrange (3.1.180) to obtain

$$\left[\sum_{\substack{k=1\\k\neq i}}^{n} \frac{x_{kj}}{\overline{D}_{ik}}\right] x_{ij}\boldsymbol{v}_{ij} = -\boldsymbol{d}_{ij} + x_{ij}\left[\sum_{k=1}^{n} \frac{x_{kj}}{\overline{D}_{ik}}\boldsymbol{v}_{kj}\right]. \tag{3.1.182}$$

Using any arbitrary constant reference velocity $\boldsymbol{v}_t^{\text{ref}}$, the above equation can be arranged as follows:

$$\left[\sum_{\substack{k=1\\k\neq i}}^{n} \frac{x_{kj}}{\overline{D}_{ik}}\right] x_{ij}(\boldsymbol{v}_{ij} - \boldsymbol{v}_t^{\text{ref}}) = -\boldsymbol{d}_{ij} + x_{ij}\left[\sum_{\substack{k=1\\k\neq i}}^{n} \frac{x_{kj}}{\overline{D}_{ik}}(\boldsymbol{v}_{kj} - \boldsymbol{v}_t^{\text{ref}})\right]. \tag{3.1.183}$$

It may be shown, after some manipulation, that this relation is equivalent to

$$x_{ij}(\boldsymbol{v}_{ij} - \boldsymbol{v}_{tj}^{\text{ref}}) = -D_{iM}\,\boldsymbol{d}_{ij} + x_{ij}\left[\sum_{k=1}^{n} \frac{x_{kj}D_{iM}}{\overline{D}_{ik}}(\boldsymbol{v}_{kj} - \boldsymbol{v}_t^{\text{ref}})\right], \tag{3.1.184}$$

where D_{iM}, an effective binary diffusivity of i through the mixture, is defined by

$$D_{iM} = \frac{1 - x_{ij}}{\left[\sum_{\substack{k=1\\k\neq i}}^{n} \left(\frac{x_{kj}}{\overline{D}_{ik}}\right)\right]}. \tag{3.1.185}$$

Multiplying (3.1.184) by C_{tj} will lead to familiar flux expressions, if we consider the three reference velocities of greatest interest, namely $\boldsymbol{v}_t^{\text{ref}} = 0$, $\boldsymbol{v}_{tj}^{*}$ and $\boldsymbol{v}_{tj}$. Following Lee *et al.* (1977a), consider a special case of all $\overline{D}_{ik}$ being equal. Then $D_{iM} = \overline{D}_{ik}$. Suppose $\boldsymbol{v}_t^{\text{ref}} = 0$, which implies that

$$\boldsymbol{N}_{ij} = C_{ij}\boldsymbol{v}_{ij} = -D_{iM}C_{tj}\boldsymbol{d}_{ij} + x_{ij}\left[\sum_{k=1}^{n} C_{kj}\boldsymbol{v}_{kj}\right]. \tag{3.1.186}$$

Since $C_{kj}\boldsymbol{v}_{kj} = \boldsymbol{N}_{kj}$, we get

$$\boldsymbol{N}_{ij} = x_{ij}\left[\sum_{k=1}^{n} \boldsymbol{N}_{kj}\right] - D_{iM}C_{tj}\boldsymbol{d}_{ij}. \tag{3.1.187}$$

When $\boldsymbol{v}_t^{\text{ref}} = \boldsymbol{v}_{tj}^{*}$,

$$\boldsymbol{J}_{ij}^{*} = C_{ij}(\boldsymbol{v}_{ij} - \boldsymbol{v}_{tj}^{*}) = -D_{iM}C_{tj}\boldsymbol{d}_{ij}. \tag{3.1.188a}$$

When $\boldsymbol{v}_t^{\text{ref}} = \boldsymbol{v}_{tj}$,

$$\boldsymbol{J}_{ij} = C_{ij}(\boldsymbol{v}_{ij} - \boldsymbol{v}_{tj}) = -D_{iM}C_{tj}\boldsymbol{d}_{ij} + C_{ij}[\boldsymbol{v}_t^{*} - \boldsymbol{v}_{tj}]. \tag{3.1.188b}$$

We should not fail to notice that, for a binary system, the Fick's law flux expressions for $\boldsymbol{J}_{ij}^{*}$ and $\boldsymbol{J}_{ij}$, namely (3.1.101) and (3.1.100), are identical to the above two expressions for ordinary diffusion. The same statement is also valid for the flux expression N_{ij}. We now consider the application of the MS equations for three different systems.

3.1.5.1.1 Diffusion of species *i* in a multicomponent stagnant gas mixture Consider flux expression (3.1.186) obtained under the condition of $\boldsymbol{v}_t^{\text{ref}} = 0$ (subscript g not used here):

$$C_i \boldsymbol{v}_i = \boldsymbol{N}_i = -D_{iM} C_t \, \boldsymbol{d}_i + x_i \left[\sum_{k=1}^{n} C_k \, \boldsymbol{v}_k \right].$$

For species i diffusing in a stagnant multicomponent gas mixture, however, $\boldsymbol{v}_k = 0$ for all $k \neq i$. Further, at $z = 0$, $p_i = p_i^0$ and at $z = \delta_g$, $p_i = p_i^\delta$. Therefore

$$\boldsymbol{N}_i = C_i \boldsymbol{v}_i = -D_{iM} C_t \, \boldsymbol{d}_i + x_i \, \boldsymbol{N}_i.$$

For ordinary diffusion, $\boldsymbol{d}_i = \nabla x_i$. Further, $C_t = (P/RT)$:

$$\boldsymbol{N}_i = \frac{D_{iM} \nabla p_i}{RT(1 - x_i)} \qquad (3.1.189)$$

if total pressure P is independent of the coordinate directions. For one-dimensional transport in the z-direction, use definition (3.1.185) to obtain

$$N_{iz} = -\frac{(1 - x_i)/(1 - x_i)}{RT\left[\sum\limits_{\substack{k=1 \\ k\neq i}}^{n} \frac{x_k}{\overline{D}_{ik}}\right]} \frac{dp_i}{dz}$$

$$= -\frac{\frac{dp_i}{dz}}{(1 - x_i)\left[\sum\limits_{\substack{k=1 \\ k\neq i}}^{n} \frac{x_k/(1 - x_i)}{\overline{D}_{ik}}\right] RT}. \qquad (3.1.190)$$

Now note that the $[x_k/(1 - x_i)]$ term for each $k \neq i$ is (moles of k/total moles of all species except i). Since all species except i are stagnant, then $\{x_k/(1 - x_i)\}$ is constant along the z-coordinate. Since $\overline{D}_{ik}$ is constant for each k, it is obvious that D_{iM} is constant along z. Rewrite (3.1.190) as

$$N_{iz} = -\frac{D_{iM}}{RT} \frac{1}{(1 - x_i)} \frac{dp_i}{dz} = -\frac{D_{iM}}{RT} \frac{P}{(P - p_i)} \frac{dp_i}{dz}.$$

Integrate from $z = 0$, $p_i = p_i^0$ to $z = \delta_g, p_i = p_i^\delta$ to obtain

$$N_{iz} = \frac{D_{iM}}{RT} \frac{(p_i^0 - p_i^\delta)}{\delta_g (P - p_i)_{LM}}, \qquad (3.1.191a)$$

where

$$(P - p_i)_{LM} = \frac{(P - p_i^\delta) - (P - p_i^0)}{\ln\left(\frac{P - p_i^\delta}{P - p_i^0}\right)}. \qquad (3.1.191b)$$

Alternatively,

$$N_{iz} = k_g (p_i^0 - p_i^\delta), \qquad (3.1.192a)$$

where

$$k_g = \frac{D_{iM}}{RT\,\delta_g} \frac{1}{(P - p_i)_{LM}}. \qquad (3.1.192b)$$

3.1.5.1.2 Mass diffusion in a ternary system In the mass diffusion or sweep diffusion separation process for isotopes 1 and 2 (Benedict *et al.*, 1981, chap. 14), an inert vapor, e.g. steam, diffuses radially outward while the mixture of two isotopes diffuses against the vapor. The two isotopes should have different rates of diffusion due to different diffusion coefficients (as well as different concentration differences). We use Maxwell–Stefan formalism to obtain the governing equations (subscript g not used here).

Identify two isotopes by subscripts $i = 1$ and 2 and steam by $i = s$. Use flux expression (3.1.178) for $i = 1$ and 2 and the radial coordinate r only:

$$\frac{\partial x_1}{\partial r} = \frac{x_1 N_{2r} - x_2 N_{1r}}{C_t \overline{D}_{12}} + \frac{x_1 N_{sr} - x_s N_{1r}}{C_t \overline{D}_{1s}}; \qquad (3.1.193)$$

$$\frac{\partial x_2}{\partial r} = \frac{x_2 N_{1r} - x_1 N_{2r}}{C_t \overline{D}_{21}} + \frac{x_2 N_{sr} - x_s N_{2r}}{C_t \overline{D}_{2s}}. \qquad (3.1.194)$$

Determine now $(\partial x_s / \partial r)$ by using one of the relations (3.1.179), namely, $\sum_{i=1}^{n} \boldsymbol{d}_i = 0$:

$$\frac{\partial x_s}{\partial r} = -\frac{\partial x_1}{\partial r} - \frac{\partial x_2}{\partial r}. \qquad (3.1.195)$$

Recall that $\overline{D}_{12} = \overline{D}_{21}$ but $\overline{D}_{1s} \neq \overline{D}_{2s}$; the latter is providing the primary basis for separation, resulting in different diffusional speeds for species 1 and 2 through steam. Therefore

$$\frac{\partial x_s}{\partial r} = -\frac{N_{sr}}{C_t}\left[\frac{x_1}{\overline{D}_{1s}} + \frac{x_2}{\overline{D}_{2s}}\right] + \frac{N_{1r} x_s}{C_t \overline{D}_{1s}} + \frac{N_{2r} x_s}{C_t \overline{D}_{2s}}. \qquad (3.1.196)$$

3.1.5.1.3 Diffusion of two species through a membrane When two species $i = 1,2$ diffuse through a membrane, we have a ternary system of $i = 1,2,m$, where m represents the membrane material. We will now obtain a Fick's first law type of expression for $\boldsymbol{N}_i$ to represent the diffusion of 1 and 2 through a membrane mechanically restrained and therefore having a zero velocity (Lightfoot, 1974) using Maxwell–Stefan formulation. Assume no external forces, no pressure gradients, no temperature gradients and ideal solution behavior. The two phases on two sides are liquids. The subscript m for the membrane phase has not been used with species subscripts.

Here $\boldsymbol{d}_i = x_i \nabla \ln a_i = x_i \nabla \ln x_i$ for ideal solutions. From equation (3.1.186),

$$\boldsymbol{N}_i = C_i \boldsymbol{v}_i = -D_{iM} \, C_t \, \boldsymbol{d}_i + x_i \left[C_1 \boldsymbol{v}_1 + C_2 \boldsymbol{v}_2 + C_m \boldsymbol{v}_m\right]. \qquad (3.1.197)$$

But $\boldsymbol{v}_m = 0$. Therefore, with $i = 1$,

$$\boldsymbol{N}_1 = -D_{1M} \, C_t x_1 \nabla \ln x_1 + x_1 [\boldsymbol{N}_1 + \boldsymbol{N}_2] = x_1 [\boldsymbol{N}_1 + \boldsymbol{N}_2] - D_{1M} \, C_t \nabla x_1. \qquad (3.1.198)$$

Similarly,

$$\boldsymbol{N}_2 = x_2 [\boldsymbol{N}_1 + \boldsymbol{N}_2] - D_{2M} \, C_t \nabla x_2. \qquad (3.1.199)$$

Here

$$D_{iM} = \left[\sum_{\substack{k=1 \\ k \neq i}}^{n} \left(\frac{x_k}{D_{ik}} \right) \frac{1}{(1 - x_i)} \right]^{-1}.$$

It follows that

$$D_{1M} = \frac{1 - x_1}{\frac{x_2}{D_{12}} + \frac{x_m}{D_{1m}}} \tag{3.1.200a}$$

and

$$D_{2M} = \frac{1 - x_2}{\frac{x_1}{D_{21}} + \frac{x_m}{D_{2m}}}. \tag{3.1.200b}$$

Thus D_{1M}, the effective binary diffusivity of species 1 through the mixture of 1, 2 and m is different from D_{2M}, the effective binary diffusivity of species 2 through the mixture of 1, 2 and m.

A ternary system of this type may be reduced to two pseudo-binary systems: solute 1 and, say, a solvent $2m$ of species 2 and membrane m; an alternative pseudo-binary system would be solute 2 and, say, a solvent $1m$ of species 1 and membrane m. The flux expression for species 1 in the first pseudo-binary system is

$$\boldsymbol{N}_1 = x_1 [\boldsymbol{N}_1 + \boldsymbol{N}_{2m}] - C_t D_{1-2m} \nabla x_1. \tag{3.1.201a}$$

The flux expression for species 2 in the second pseudo-binary system is

$$\boldsymbol{N}_2 = x_2 [\boldsymbol{N}_2 + \boldsymbol{N}_{1m}] - C_t D_{2-1m} \nabla x_2. \tag{3.1.201b}$$

Note that D_{1-2m} need not equal D_{2-1m} (Spriggs and Gainer, 1973).

3.1.5.2 *Irreversible thermodynamics approach*

The principles of irreversible thermodynamics (de Groot and Mazur, 1962) provide an alternative way of relating the flux of species i to all the forces that are present in a multicomponent multiforce system. The basic assumptions behind this approach are as follows.

(a) The nonequilibrium conditions leading to species transport depart from the equilibrium conditions by sufficiently small amounts. Thermodynamic equilibrium relations, valid locally under "thermostatic"[16] conditions, may be used for nonequilibrium conditions.
(b) The flux of any species is a function of all forces present in the system. The flux-driving force relations are linear as long as condition (a) holds.
(c) The phenomenological coefficients in (b) are symmetric.

[16] In conventional thermodynamics studied for developing equilibrium relations, the conditions are really static, with no change anywhere, except when going from one state to another; however, the process of change, or the dynamics of change, is not studied. Therefore it should be considered as "*thermostatics.*"

The mathematical representations of the flux-force relations in any given phase/region are:

$$\boldsymbol{j}_i = \sum_{k=1}^{n} l_{ik} \boldsymbol{F}_k^m; \tag{3.1.202}$$

$$l_{ik} = l_{ki}, \tag{3.1.203}$$

where the $l_{ik} - s$ are phenomenological coefficients. Here $\boldsymbol{j}_i$ is the mass flux of species i in a frame of reference moving with the mass average velocity $\boldsymbol{v}_t$ and $\boldsymbol{F}_k^m$ is the force on species k in an isothermal system of n components such that the fluxes and forces are proper and form the right pair. For a noniosthermal system, one of the forces would be due to a thermal gradient in addition to the n species-specific forces:

$$\boldsymbol{j}_i = \sum_{k=1}^{n} l_{ik} \boldsymbol{F}_k^m + l_{iT} \nabla T. \tag{3.1.204}$$

Under systems practically in mechanical equilibrium, it has been shown (de Groot and Mazur, 1962) that molar fluxes in frames of reference other than barycentric (i.e. mass-averaged velocity) can also be used. For example,

$$\boldsymbol{J}_i^* = \sum_{k=1}^{n} L_{ik} \boldsymbol{F}_k + L_{iT} \nabla T. \tag{3.1.205}$$

Here $\boldsymbol{F}_k$ is the force specific for species k. Such a force may be obtained from (3.1.50). The phenomenological coefficients are symmetrical; L_{ii} coefficients are always positive, whereas L_{ik} or L_{iT} may have any sign:

$$L_{ik} = L_{ki}; \quad L_{iT} = L_{Ti}; \quad L_{ii} \geq 0; \quad L_{ii} L_{kk} - L_{ik}^2 \geq 0. \tag{3.1.206}$$

For isothermal conditions, an inverted representation of flux-force relations would be

$$\boldsymbol{F}_i = \sum_{k=1}^{n} R_{ik} \boldsymbol{J}_k^*; \quad R_{ik} = R_{ki}. \tag{3.1.207}$$

But the phenomenological coefficients $L_{ik} - s$ are not necessarily reciprocals of the phenomenological coefficients $R_{ik} - s$. Conceptually, $R_{ik} - s$ are similar to resistances and $L_{ik} - s$ are similar to conductances in electrical circuits.

For a *binary system* of solute i and solvent s in an isothermal system, equation (3.1.205) is reduced to

$$\boldsymbol{J}_i^* = L_{ii} \boldsymbol{F}_i + L_{is} \boldsymbol{F}_s; \tag{3.1.208}$$

$$\boldsymbol{J}_s^* = L_{si} \boldsymbol{F}_i + L_{ss} \boldsymbol{F}_s. \tag{3.1.209}$$

Here L_{si} $(= L_{is})$ is the cross coefficient coupling the two fluxes $\boldsymbol{J}_i^*$ and $\boldsymbol{J}_s^*$; L_{ii} and L_{ss} are the straight coefficients. In most separation problems of interest, the coupling is essentially negligible, i.e. $L_{si} \cong 0$. For example, in ordinary diffusion with $\boldsymbol{U}_i = 0$, we observe from (3.1.50), (3.1.81), (3.1.89) and (3.1.101) that $L_{is} = 0$ and

$$L_{ii} = \frac{D_{is}\, C_i}{\mathrm{R}T}. \tag{3.1.210}$$

This gives us an idea about the nature of the L_{ii} coefficient in terms of other known transport coefficients. (Note that irreversible thermodynamics does not provide any such relation for determining L_{ii} - they have to be determined by experiment.) One can similarly show in this case that

$$R_{ii} = \frac{\mathrm{R}T}{D_{is}\, C_i}. \tag{3.1.211}$$

For a two-component system, a very brief excursion into the role of the coupling of fluxes is illustrative. For the system of solute i and solvent s and equations (3.1.208) and (3.1.209), eliminate the force $\boldsymbol{F}_s$ and rearrange the equations to get (Johnson *et al.*, 1966, 1980; Meares, 1976)

$$\boldsymbol{J}_i^* = L_{ii}\left[\left(1 - \left\{\frac{L_{is}^2}{L_{ii}\, L_{is}}\right\}\right)\boldsymbol{F}_i + \boldsymbol{J}_s^*\left\{\frac{L_{is}}{L_{ii}\, L_{ss}}\right\}\right]. \tag{3.1.212}$$

By (3.1.206)

$$L_{ii}L_{is} - L_{is}^2 \geq 0 \Rightarrow \frac{L_{is}^2}{L_{ii}L_{is}} \leq 1 \Rightarrow \left(1 - \frac{L_{is}^2}{L_{ii}\, L_{is}}\right) \tag{3.1.213}$$

is positive. From the first term within brackets in (3.1.212), we see that only a part of the force $\boldsymbol{F}_i$ is used to moves species i. The second term indicates the contribution of the movement of species s to the flux of species i (due to coupling). The higher the magnitude of L_{is} and $\boldsymbol{J}_s^*$, the larger the contribution of coupling to the observed flux $\boldsymbol{J}_i^*$ of species i.

3.1.6 Additional topics

A few special topics are covered here. First, slip correction factor for Stokes' law in a rarefied gaseous medium for smaller particles is considered. Second, ion transport due to a uniform electrical field is treated in the atmospheric/rarefied gas phase.

3.1.6.1 *Slip correction factor for Stokes' law in the gas phase*

We have seen in Section 3.1.3.2.4 that, depending on the value of $(r_p\ /\ \lambda)$, where r_p is the pore radius and λ is the value of the mean free path for the gas molecules, the gas transport regime in the porous medium changed. When λ is large (for example, at low gas pressures) and r_p is small, it is much more likely for the gas molecules to hit the pore wall rather than another gas molecule (Knudsen diffusion). Similarly, in the case of transport of small particles of radius r_p in a gas where λ is the mean free path for the gas molecules, if λ is large and r_p is small ($r_p/\lambda << 1$ or $\lambda/r_p >> 1$, (λ/r_p) is the **Knudsen number**) the particle appears to other gas molecules (which are far apart) as if it were a gas molecule. On the other hand, when $(\lambda/r_p) << 1$, the particle encounters a denser gas where gas molecules collide with one another much more often; correspondingly, the particle encounters much more resistance to motion due to a high density of gas molecules impinging on it.

The drag force $6\pi\mu r_p U_{pz}$ experienced by a particle moving with velocity U_{pz} in a fluid according to Stokes' law (3.1.61) and (3.1.63) is valid for a gaseous medium when $\lambda/r_p << 1$. When we have very small particles and a rarefied gas condition ($(\lambda/r_p) >> 1$), the drag force encountered by a small particle is lower than that given by Stokes' law since there are fewer gas molecules collisions that create the resistive force encountered by the particle in a gas moving with a velocity v_{tz}:

$$F_{pz}^{\mathrm{drag}} = \frac{6\pi\mu\, r_p\, U_{pz}}{C_c} \qquad (\text{for } v_{tz} = 0); \tag{3.1.214a}$$

$$F_{pz}^{\mathrm{drag}} = \frac{6\pi\mu\, r_p\, (U_{pz} - v_{tz})}{C_c}. \tag{3.1.214b}$$

Here the quantity C_c, which is greater than 1, is called the *slip correction factor*. It has the general form (Flagan and Seinfeld, 1988)

$$C_c = 1 + \frac{\lambda}{r_p}\left[\alpha + \beta \exp\left(-\frac{\gamma}{(\lambda/r_p)}\right)\right], \tag{3.1.215}$$

where $\alpha = 1.257$, $\beta = 0.4$, $\gamma = 1.10$. For small particles, the deviation from Stokes' law is substantial. For example, the values of C_c for atmospheric air at 25 °C for a few different particle sizes are (Flagan and Seinfeld, 1988) in the form of (r_p, C_c): (0.005 μm, 22.7); (0.025 μm, 5.06); (0.05 μm, 2.91); (0.5 μm, 1.168); (5 μm, 1.017).

3.1.6.2 *Motion of slow ions in a gas under an electrical field*

Chemicals (represented by M here) are sometimes separated in the gas phase by separating their ionized forms. The ionized product may have positive charge due to proton transfer (MH^+), dimerization ($\mathrm{MH}^+ + \mathrm{M} \rightarrow \mathrm{M_2H}^+$) etc.; or negative product ions are generated by, for example, electron attachment ($M + e^- \rightarrow M^-$) or dissociative electron attachment ($\mathrm{MX} + e^- \rightarrow \mathrm{M} + \mathrm{X}^-$), etc. Separation of the different ions can also allow their characterization (Eiceman and Karpas, 1994). Surrounded by neutral gas molecules (sometimes called buffer-gas molecules), an ion is acted on by a uniform electric field $\boldsymbol{E}$ (if it exists), which generates motion of the ion. The force on a single ion is e$\boldsymbol{E}$, where e is the electronic charge. This ion motion is resisted by the buffer-gas molecules in the surrounding gaseous medium. As we have considered earlier (see equation (3.1.62)), the initial acceleration encountered by the ion having a mass of m_{ion} due to the electrical field, namely $(\mathrm{e}\boldsymbol{E}/m_{\mathrm{ion}})$, will soon disappear, and a steady velocity, $\boldsymbol{U}_{\mathrm{ion}}$, called the *drift velocity* (similar to terminal velocity) is achieved on average by each ion:

$$\mathrm{e}\boldsymbol{E} = \frac{f^d_{\mathrm{ion}}}{\tilde{\mathrm{N}}}\boldsymbol{U}_{\mathrm{ion}}, \tag{3.1.216}$$

where the frictional coefficient for 1 gmol of ions is f^d_{ion} and $\tilde{\mathrm{N}}$ is Avogadro's number. Further, by Einstein's relation,

$$f^d_{\mathrm{ion}} = \mathrm{R}T/D_{\mathrm{ion}}, \tag{3.1.217}$$

where D_{ion} is the diffusion coefficient of the ion. Therefore the drift velocity of the ionized species is

$$\boldsymbol{U}_{\mathrm{ion}} = \frac{\mathrm{e}\,D_{\mathrm{ion}}\,\tilde{\mathrm{N}}}{\mathrm{R}T}\boldsymbol{E} = \frac{\mathrm{e}\,D_{\mathrm{ion}}}{\mathrm{k}^{\mathrm{B}}\,T}\boldsymbol{E}, \tag{3.1.218}$$

where k^{B} is Boltzmann's constant (1.380×10^{-23} joules/K). Since e is equal to 1.60210×10^{-19} coulomb, we obtain

$$\boldsymbol{U}_{\mathrm{ion}} = \frac{1.6021 \times 10^{-19}\,\mathrm{coulomb}}{1.380 \times 10^{-3}\,\frac{\mathrm{joule}}{\mathrm{K}}}\,\frac{D_{\mathrm{ion}}}{T}\boldsymbol{E} \Rightarrow$$

$$\boldsymbol{U}_{\mathrm{ion}} = \left(1.16 \times 10^{-4}\,\frac{D_{\mathrm{ion}}}{T}\right)\boldsymbol{E}; \; = \mu^m_{\mathrm{ion},g}\,\boldsymbol{E}, \tag{3.1.219}$$

where $\mu^m_{\mathrm{ion},g}$ is the ionic mobility in the gas phase having the units of cm^2/s-volt since 1 joule = 1 volt-coulomb. (The electrical field strength used in the separation of ions in ion mobility spectrometry is usually around $E = 250$ volt/cm (Eiceman *et al.*, 2004). See Section 7.3.1.2 for ion mobility spectrometry.)

This analysis is useful when the applied uniform electrical field (of strength $\boldsymbol{E}$) is weak. Further, the concentration of the ions in the gas is low enough so that the forces of repulsion from similar ions due to Coulomb's law (3.1.15) may be neglected. Also, the ionic drift velocity is not in too much excess over their thermal velocities. If the electric field strength is increased substantially, the ionic mobility $\mu^m_{\mathrm{ion},g}$ becomes dependent on the ratio E/C_g, where C_g is the gas concentration. Instead of C_g, the number concentration of molecules, N, is used:

$$\frac{E}{N}\,\frac{\mathrm{volt/cm}}{\mathrm{numbers/cm^3}} = \frac{E}{N}\,\mathrm{volt\text{-}cm^2}.$$

However the actual unit used is Td for "townsend," where $1\,\mathrm{Td} = 10^{-17}$ volt-cm^2 and the units of (E/N) are 10^{-17} volts-cm^2. Often instead of (E/N), (E/P) is used, where the pressure P is expressed in torr (1 mm Hg). The corresponding conversion is obtained as follows. For ideal gas behavior valid at low pressures,

$$P = C_t\,\mathrm{R}T = \frac{N}{\tilde{\mathrm{N}}}\mathrm{R}T = N\frac{\mathrm{R}}{\tilde{\mathrm{N}}}T = N\mathrm{k}^{\mathrm{B}}\,T. \tag{3.1.220}$$

Therefore

$$\frac{E}{N} = \frac{E}{P}\mathrm{k}^{\mathrm{B}}T = \frac{E}{P}\frac{\mathrm{R}}{\tilde{\mathrm{N}}}T; \quad \frac{E}{N} = \frac{E}{P}\left(\frac{\mathrm{volt/cm}}{\mathrm{torr}}\right)T(\mathrm{K})$$

$$\frac{82.05 \times 760}{6.022 \times 10^{23}}\,\frac{(\mathrm{cm^3\text{-}atm/gmol\text{-}K})}{\mathrm{no.of\ molecules/gmol}}\cdot\frac{\mathrm{torr}}{\mathrm{atm}}$$

$$= \left(\frac{E}{P}T\right)1.0354 \times 10^{-19}\,\mathrm{volt\text{-}cm^2}.$$

If the unit of 10^{-17} volts-cm^2 is employed for (E/N), then

$$\frac{E}{N} = \frac{E}{P}T \times 1.0354 \times 10^{-2}, \tag{3.1.221}$$

where P is in torr (Mason and McDaniel, 1988). When the electrical field is such that (E/N) is less than 10 Td, the ionic mobility $\mu^m_{\mathrm{ion},g}$ is almost independent of the field.

Consider a situation where ions present in a gas are not subjected to any external force field, temperature gradient or coulombic repulsion. Ions will be dispersed by the random motion of the ions, leading to a diffusion process. Fick's first law is used to describe the corresponding diffusive flux expression:

$$\boldsymbol{J}_{\mathrm{ion}} = -D_{\mathrm{ion}}\nabla C_{\mathrm{ion}}. \tag{3.1.222}$$

The total flux of the ionic species in an electrical field of uniform strength $\boldsymbol{E}$ is a sum of the two fluxes:

$$\boldsymbol{N}_{\mathrm{ion}} = -C_{\mathrm{ion}}\,\boldsymbol{U}_{\mathrm{ion}} - D_{\mathrm{ion}}\,\nabla C_{\mathrm{ion}}. \tag{3.1.223}$$

For an introduction to, and a detailed analysis of, motion of slow ions in gases, Mason and McDaniel (1988) should be consulted.

In a highly ionized gas, the number density (number per unit volume) of electrons is approximately equal to that of the positively charged gas ions, except very close to a boundary. Like the principle of electroneutrality in a solution (see equation (3.1.108a)), there is charge equality in an ionized gas. If there is a charge imbalance, it will produce forces to restore the balance: what it achieves is to slow down the diffusional rate of electrons whose diffusion coefficient (D_-) is much higher than that of the positively charged ions (D_+) whose diffusional rate is enhanced. The effective diffusional velocity of both species are the same. This diffusion is called *ambipolar* (Schottky, 1924).

3.2 Separation development and multicomponent separation capability

We have now developed a quantitative understanding of the movement of specific molecular species (or particles) in a given direction at a particular speed under the action of a negative chemical potential gradient and/or external forces. Suppose a mixture to be separated is located in a closed container identified as the separation system and subjected to a particular force (or forces). It should now be possible to follow the displacement characteristic of each species in the mixture. For *chemical solutions*, such displacements are conveniently studied by following the spatial and temporal variations of concentration of any species; we seek to find out how C_i varies with time t and the spatial coordinates x, y, z. Development of species-specific regions necessary for the separation of a mixture can then be observed first by considering the relative spatial locations of concentration maxima of each individual species.

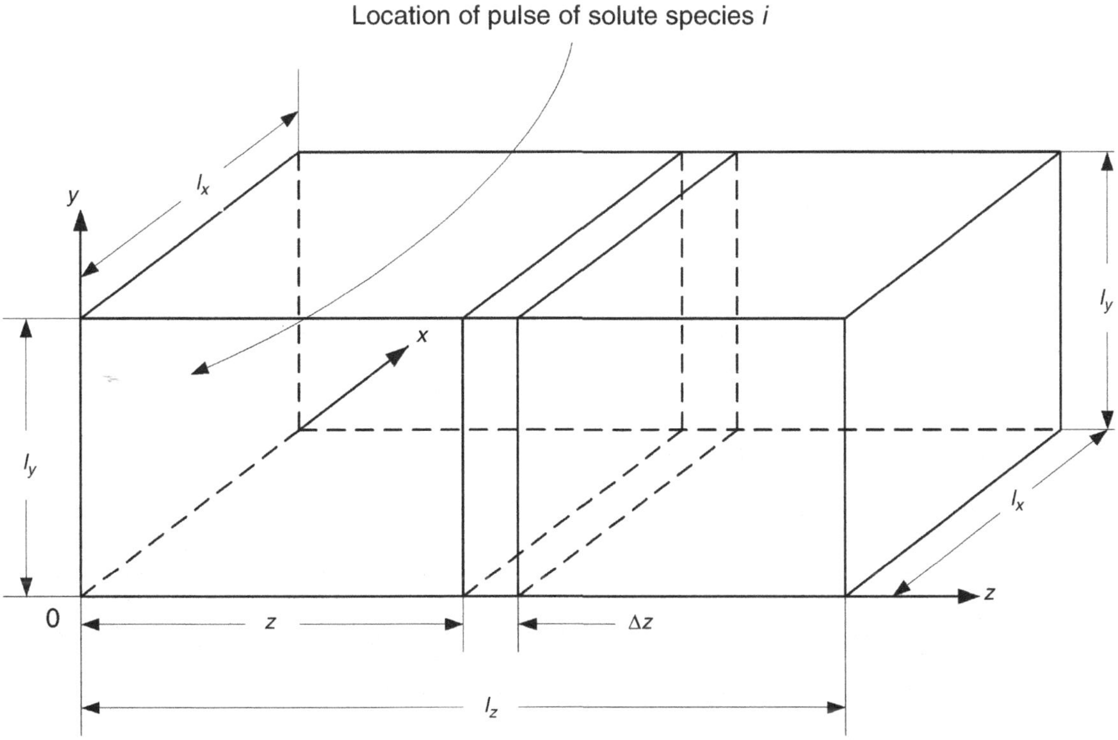

Figure 3.2.1. Rectangular separator vessel.

An associated question of fundamental importance (studied next) is how many different chemical species can be separated in a given separation system under ideal conditions. This multicomponent separation capability of a given separation technique could then be used as a criterion, amongst others, for developing a hierarchy of separation techniques. Concurrently, the crucial role of the nature of the force potential profile in developing the multicomponent separation capability will be illustrated. The corresponding treatments given below closely follow those by Giddings (1978).

3.2.1 Separation development in a closed system

Consider a rectangular separator vessel of dimensions (l_x, l_y, l_z) in the (x,y,z)-coordinate directions, respectively (Figure 3.2.1). Let the separator contain a single-phase system, say a solvent. At time $t = 0$, we subject the system to a pulse of solute species 1 and solute species 2, the total number of moles of each species being m_1 and m_2, respectively. The pulse is located at the vessel side characterized by $x = 0, l_x$; $y = 0, l_y$; $z = 0$. Assume the following.

(1) There is no bulk velocity of the solvent in the closed container; if there is any solvent motion at all, it is rigid body rotation, as in a centrifuge.
(2) The concentration of each species in the pulse at $t = 0$ is uniform in the xy-plane.
(3) The forces acting on the species make them move only in the positive z-direction for time $t > 0$; as molecules of the solute species move in the positive z-direction, the concentration of the solute remains uniform in the xy-plane (at all z).

How the concentration pulse of a given solute species moves as a function of z and t is best studied using the solution of the species conservation equation to be developed now. The species conservation equation can be obtained by a simple mass balance of solute i in a small volume element of cross-sectional area $l_x l_y$ and thickness Δz in Figure 3.2.1 at any z for $0 \le z \le l_z$. For species $i = 1$,

$$\begin{pmatrix}\text{rate of Species 1}\\ \text{accumulation}\end{pmatrix} = \begin{pmatrix}\text{rate of Species 1}\\ \text{coming in}\end{pmatrix} - \begin{pmatrix}\text{rate of Species 1}\\ \text{going out}\end{pmatrix}$$
$$\frac{\partial C_1}{\partial t}(l_x l_y \Delta z) \quad = N_{1z}\big|_z l_x l_y \quad - N_{1z}\big|_{z+\Delta z} l_x l_y \qquad (3.2.1)$$

Dividing both sides by $l_x l_y$ Δz and taking the limit $\Delta z \to 0$, we get

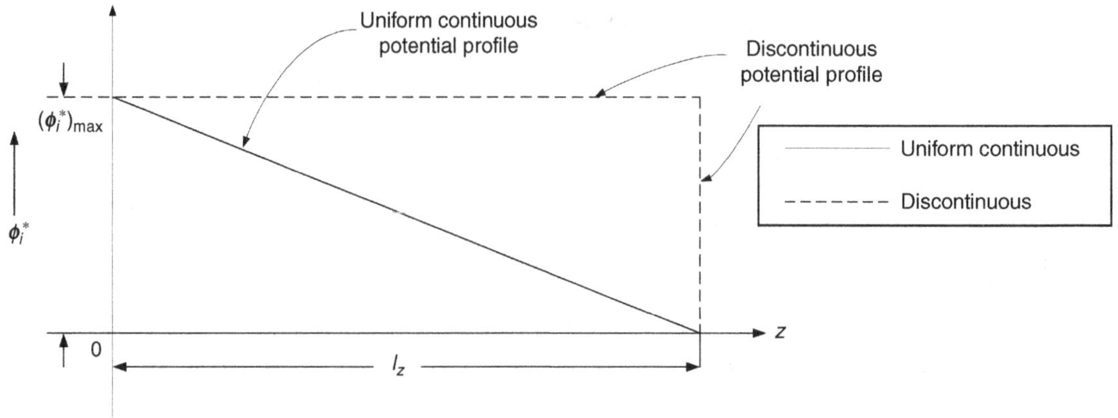

Figure 3.2.2. Profiles of ϕ_i^* analyzed. (After Giddings (1978).)

$$\frac{\partial C_1}{\partial t} = -\frac{\partial N_{1z}}{\partial z}. \tag{3.2.2}$$

Assume now that the movement of solute species 1 in the solvent does not affect the movement of solute species 2 and vice versa. Following the derivation of the species balance equation (3.2.2) for $i = 1$, we may obtain a similar species conservation equation for $i = 2$:

$$\frac{\partial C_2}{\partial t} = -\frac{\partial N_{2z}}{\partial z}. \tag{3.2.3}$$

The solution of these two partial differential equations, subject to appropriate boundary and initial conditions, will yield the concentration profiles for solutes i =1,2 as a function of z and t.

How each of these two concentration profiles develops along the z-coordinate with time is intimately connected with the nature of the species-specific forces, since the latter determine N_{iz}. Recall from expressions (3.1.27) and (3.1.42), at constant temperature,

$$\begin{aligned} \boldsymbol{F}_{ti} &= \boldsymbol{F}_{ti}^{\text{ext}} - \nabla\mu_i = -\nabla\left(\sum \phi_i^{\text{ext}} + \mu_i\right) \\ &= -\nabla(\phi_{ti}^{\text{ext}} + \mu_i^0) - RT\nabla\, \ell n\, a_i = -\nabla\phi_i^{\text{tot}}, \end{aligned}$$

where

$$\phi_i^{\text{tot}} = \phi_{ti}^{\text{ext}} + \mu_i^0 + RT\, \ell n\, a_i \tag{3.2.4}$$

is the total potential acting on species i (note: μ_i^0 includes here any pressure gradient effects also; otherwise employ the expression from (3.1.84d). Further, the $-\nabla\phi_i^{\text{ext}}$ representation for any external force is insufficient for magnetic and some other forces). Rewrite ϕ_i^{tot} as

$$\phi_i^{\text{tot}} = \phi_i^* + RT\ell n\, a_i, \tag{3.2.5a}$$

where

$$\phi_i^* = \phi_{ti}^{\text{ext}} + \mu_i^0; \qquad \phi_{ti}^{\text{ext}} = \sum \phi_i^{\text{ext}}. \tag{3.2.5b}$$

We will find out later that the nature of the profile ϕ_i^* in the z-direction along the separator is crucial. It can be either continuous or discontinuous, with all sorts of variation with z in either category.

We consider now a continuous profile of ϕ_i^*, specifically a *uniformly continuous profile* of ϕ_i^* (the discontinuous profile is studied in Section 3.2.2). In Figure 3.2.2, the uniform continuous profile of ϕ_i^* is represented as a straight line varying from $(\phi_i^*)_{max}$ at $z = 0$ to 0 at $z = l_z$. Two simple examples of such a profile are: a uniform electrical field of electrostatic potential; a gravitational potential along the vertical axis. A nondimensional representation

$$\phi_i^+ = \phi_i^*/(\phi_i^*)_{max} \tag{3.2.6}$$

will have a maximum value of 1 at $z = 0$ and a minimum value of 0 at $z = l_z$. Since both ϕ_i^+ and ϕ_i^* are functions of the z-coordinate only, we may write them as $\phi_i^+(z^+)$, where

$$z^+ = z/l_z. \tag{3.2.7}$$

Substitute now the expression (3.1.81) for N_{1z} in the differential equation (3.2.2), under the conditions of $v_{tz}^* = 0$ and an ideal solution, and obtain for species 1 in solvent s the following:

$$\frac{\partial C_1}{\partial t} = \frac{1}{f_1^d}\frac{\partial}{\partial z}\left[C_1\frac{\partial \mu_1^0}{\partial z} + C_1\frac{\partial \phi_{t1}^{\text{ext}}}{\partial z}\right] + \frac{RT}{f_1^d}\frac{\partial^2 C_1}{\partial z^2}.$$

Therefore

$$\frac{\partial C_1}{\partial t} = \frac{1}{f_1^d}\frac{\partial}{\partial z}\left[C_1\frac{\partial \phi_1^*}{\partial z}\right] + \frac{RT}{f_1^d}\frac{\partial^2 C_1}{\partial z^2}, \tag{3.2.8}$$

where $(RT/f_1^d) = D_{1s}^0$.

Nondimensionalize $(\partial\phi_1^*/\partial z)$ and the time t by

$$\frac{\partial\phi_1^*}{\partial z}=\frac{(\phi_1^*)_{\max}}{l_z}\frac{\partial\phi_1^+}{\partial z^+}=\frac{(\phi_1^*)_{\max}}{l_z}(\phi_1^+)';\quad t_1^+=\frac{t(\phi_1^*)_{\max}}{l_z^2 f_1^d},\tag{3.2.9}$$

where we note that $(\phi_1^+)'$ is a constant for the uniform continuous ϕ_1^* profile (Figure 3.2.2). Equation (3.2.8) may now be rewritten as

$$\frac{\partial C_1}{\partial t_1^+}-(\phi_1^+)'\frac{\partial C_1}{\partial z^+}=\frac{RT}{(\phi_1^*)_{\max}}\frac{\partial^2 C_1}{\partial z^{+2}},\tag{3.2.10}$$

where C_1 depends on the independent variables z^+ and t_1^+. To simplify equation (3.2.10) further, define a new set of independent variables η and t_1^+, where

$$\eta=z^+ +(\phi_1^+)' t_1^+.\tag{3.2.11}$$

In these two new independent variables, equation (3.2.10) is reduced to[17]

$$\frac{\partial C_1}{\partial t_1^+}=\frac{RT}{(\phi_1^*)_{\max}}\frac{\partial^2 C_1}{\partial \eta^2}.\tag{3.2.12}$$

This equation is far simpler than the earlier ones; in fact, it is essentially the *diffusion equation*.

We now need the solution of this equation for the given initial condition and suitable boundary conditions. Although the z-dimension of our separator is finite ($= l_z$), we may assume that l_z is sufficiently large for the values of t in our range of interest. Then the solution for $z\rightarrow\infty$, i.e. $\eta\rightarrow\infty$ will be usable for our purpose. Obviously, at all t_1^+,

$$\text{for } z^+=\infty \text{ or } \eta=\infty,\ C_1=0 \text{ and } \frac{\partial C_1}{\partial z^+}=0=\frac{\partial C_1}{\partial\eta};\tag{3.2.14a}$$

i.e. the value of C_1 as well as its gradient in z will be zero far away from the source of C_1 at $z=0$.

The initial condition is:

$$\text{at } t^+=0,\ C_1=(m_1/l_x l_y)\delta(z^+);\tag{3.2.14b}$$

i.e. at time $t=0$, all of the solute species 1 (m_1 moles) is contained in a very thin slab of cross-sectional area $l_x l_y$ at $z^+=0$. An alternative way of looking at this is as follows:

$$\int_0^\infty C_1 l_x l_y\, dz^+ = m_1\int_0^\infty \delta(z^+)dz^+ = m_1.\tag{3.2.14c}$$

A well-known solution of the diffusion equation (Carslaw and Jaeger, 1959),[18]

$$C_1(z,t)=C_1(z^+,t_1^+)=C_1(\eta,t_1^+)=\frac{A}{\sqrt{t_1^+}}\exp\left[-\frac{\eta^2}{\frac{4RT}{(\phi_1^*)_{\max}}t_1^+}\right],\tag{3.2.15}$$

is applicable here provided the z^+-coordinate varies between $+\infty$ and $-\infty$. To determine the constant A, note that solute species 1 is conserved. Therefore, the total number of moles of species 1 in a separator of dimensions $x=0,\ l_x;\ y=0,\ l_y;\ z=-\infty$ to $+\infty$ must be the sum of the number of moles in two separators, one from $z=0$ to $+\infty$ and the other from $z=-\infty$ to 0. However, the latter reservoir does not have any solute in this problem. Therefore

$$\begin{aligned}m_1+0=m_1 &= \int_{-\infty}^{\infty}\int_0^{l_y}\int_0^{l_x}\frac{A}{\sqrt{t_1^+}}\exp\left[-\frac{\eta^2}{\frac{4RT}{(\phi_1^*)_{\max}}t_1^+}\right]dx\,dy\,dz\\ &= A l_x l_y l_z\int_{-\infty}^{\infty}\frac{1}{\sqrt{t_1^+}}\exp\left[-\frac{\eta^2}{\frac{4RTt_1^+}{(\phi_1^*)_{\max}}}\right]dz^+.\end{aligned}\tag{3.2.16}$$

Define

$$\eta^+=\left(\eta\Big/\sqrt{\frac{4RTt_1^+}{(\phi_1^*)_{\max}}}\right).\tag{3.2.17}$$

Since $dz^+=d\eta$, we get

$$\begin{aligned}m_1 &= A l_x l_y l_z\sqrt{\frac{4RT}{(\phi_1^*)_{\max}}}\int_{-\infty}^{\infty}\exp\left[-(\eta^+)^2\right]d\eta^+\\ &= A l_x l_y l_z 4\sqrt{\frac{RT}{(\phi_1^*)_{\max}}}\int_0^{\infty}\exp\left[-(\eta^+)^2\right]d\eta^+.\end{aligned}$$

Therefore

---

[17] For $C_1=C_1(t_1^+,z^+)=C_1(t_1^+,\eta)$,

$$dC_1=\left(\frac{\partial C_1}{\partial t_1^+}\right)_\eta dt_1^+ + \left(\frac{\partial C_1}{\partial\eta}\right)_{t_1^+}d\eta.\tag{3.1.13a}$$

Therefore

$$\left(\frac{\partial C_1}{\partial t_1^+}\right)_{z^+}=\left(\frac{\partial C_1}{\partial t_1^+}\right)_\eta+\left(\frac{\partial C_1}{\partial\eta}\right)_{t_1^+}\frac{\partial\eta}{\partial t_1^+}=\left(\frac{\partial C_1}{\partial t_1^+}\right)_\eta+(\phi_1^+)'\left(\frac{\partial C_1}{\partial\eta}\right)_{t_1^+}.\tag{3.1.13b}$$

Further,

$$\left(\frac{\partial C_1}{\partial z_1^+}\right)_{t_1^+}=\left(\frac{\partial C_1}{\partial\eta}\right)_{t_1^+}\Rightarrow\left(\frac{\partial^2 C_1}{\partial z^{+2}}\right)_{t_1^+}=\left(\frac{\partial^2 C_1}{\partial\eta^2}\right)_{t_1^+}.\tag{3.1.13c}$$

[18] The solution of the heat conduction equation given in sect. 10.3.II of this reference for an instantaneous plane source of strength Q parallel to the plane $z=0$ can be used to solve the particular diffusion problem. The reader can verify it by substituting solution (3.2.15) into equation (3.2.12) and seeing that it is satisfied.

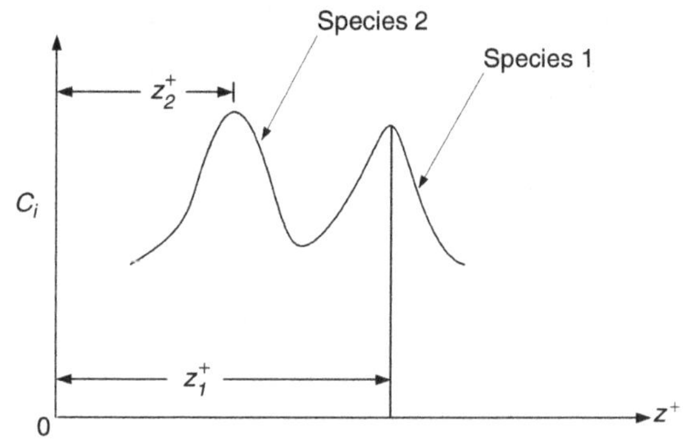

Figure 3.2.3. Concentration profiles of species 1 and 2 at any time t.

$$A = \frac{m_1}{2 l_x l_y l_z \sqrt{\frac{RT\pi}{(\phi_1^*)_{\max}}}} \tag{3.2.18}$$

since the value of the integral is $\sqrt{\pi}/2$.

The concentration profile for species 1 is therefore given by

$$C_1(z^+, t_1^+) = \left(\frac{m_1}{l_x l_y l_z}\right) \frac{1}{\sqrt{\frac{4\pi RT t_1^+}{(\phi_1^*)_{\max}}}} \exp\left[-\frac{\left\{z^+ + (\phi_1^+)' t_1^+\right\}^2}{4\left\{\frac{RT}{(\phi_1^*)_{\max}}\right\} t_1^+}\right]. \tag{3.2.19}$$

The concentration profile for species 2 may be similarly obtained as follows:

$$C_2(z^+, t_2^+) = \left(\frac{m_2}{l_x l_y l_z}\right) \frac{1}{\sqrt{\frac{4\pi RT t_2^+}{(\phi_2^*)_{\max}}}} \exp\left[-\frac{\left\{z^+ + (\phi_2^+)' t_2^+\right\}^2}{4\left\{\frac{RT}{(\phi_2^*)_{\max}}\right\} t_2^+}\right], \tag{3.2.20}$$

where

$$t_2^+ = \frac{t(\phi_2^*)_{\max}}{l_z^2 f_2^d}.$$

Here

$$C_1^{\max}(z^+, t_1^+) = \left(\frac{m_1}{l_x l_y l_z}\right) \sqrt{\frac{(\phi_1^+)_{\max}}{4\pi RT t_1^+}};$$

similarly for $C_2^{\max}(z^+, t_2^+)$.

Figure 3.2.3 shows the instantaneous concentration profiles for solutes 1 and 2 at any time t (only the z^+-coordinate is shown for simplicity). The *concentration profile* of species i $(i = 1,2)$ is located around a $z_i^+ = -(\phi_i^+)' t_i^+$ and *has its maxima there*. The profiles are *Gaussian* with a standard deviation σ_i^+ (nondimensional σ_i), where

$$\sigma_i^+ = \left\{2RT t_i^+ / (\phi_i^*)_{\max}\right\}^{1/2} = \sigma_i / l_z \tag{3.2.21a}$$

or

$$(\sigma_i / l_z) = \sigma_i^+ = \left\{2 D_{is}^0 t / l_z^2\right\}^{1/2}, \tag{3.2.21b}$$

since $(RT/f_i^d) = D_{is}^0$. Therefore, the larger the value of D_{is}^0 or the longer the time allowed for separation, the higher will be the value of σ_i as the species i concentration profile moves along the z-axis.

These results have several important implications. Consider first the instantaneous location of the center point of each of the two concentration profiles where $C_i^{\max}(z^+, t_i^+)$ occurs.

Species 1:

$$z_1^+ = -t_1^+ (\phi_1^+)' = -\frac{t(\phi_1^*)_{\max} l_z}{l_z^2 f_1^d (\phi_1^*)_{\max}} \left(\frac{\partial \phi_1^*}{\partial z}\right).$$

But

$$\frac{\partial \phi_1^*}{\partial z} = \frac{\partial \phi_{t1}^{\text{ext}}}{\partial z} + \frac{\partial \mu_1^0}{\partial z} = -f_1^d U_{1z},$$

so that

$$z_1^+ = \frac{t U_{1z}}{l_z}. \tag{3.2.22a}$$

Species 2: Similarly,

$$z_2^+ = \frac{t U_{2z}}{l_z}. \tag{3.2.22b}$$

Note that U_{1z} and U_{2z} are the z-components of the displacement or migration velocities of species 1 and 2, respectively. Thus, as long as $U_{1z} \neq U_{2z}$, the two concentration profiles (of 1 and 2) will have different locations of their centers. Although at $t = 0$, species 1 and 2 were present in a uniform mixture at $z = 0$ (and $x = 0$, l_x; $y = 0$, l_y), **different values of U_{1z} and U_{2z} have created a condition whereby species 1 and 2 are now spatially separated in the solvent present in the separator.**

From (3.2.21a,b) for σ_i, it is clear that, as the diffusion coefficient D_{is}^0 of solute species i increases, the concentration profile becomes broader at any time t. Therefore, the larger the value of D_{is}^0, the greater the extent of overlap between two contiguous solute profiles. **Thus, an increased value of D_{is}^0 leads to reduced separation. Although molecular diffusion due to a concentration gradient could be interpreted in terms of a concentration driving force that is part of the overall driving force F_{ti}, here it acts to reduce separation in the presence of external forces. Therefore it is useful to subtract it from F_{ti} in developing an expression for displacement or migration velocity U_i of species i (as was done in definition (3.1.82)).**

The effect of time on the separation achieved is an additional feature that can be investigated. First, the

z^+-coordinate of the center of mass of the concentration profile of the ith species, z_i^+, increases linearly with time (see the developments after equation (3.2.25)). The distance between the centers of two contiguous profiles, i.e. $z_1^+ - z_2^+$, for species 1 and 2 therefore increases as $[t(U_{1z} - U_{2z})/l_z]$. Second, the standard deviation σ_i of each profile increases as $t^{1/2}$. Thus, as time progresses, any given solute species disperses over a wider region of space around the center of the profile. This feature is often called *band broadening* or *dispersion*. The net effect of time on the separation between species 1 and 2 can be determined by using an index, e.g. the *resolution* R_s, where

$$R_s(t) = \frac{2(z_1^+ - z_2^+)}{4(\sigma_1^+ + \sigma_2^+)} = \frac{2(z_1 - z_2)}{4(\sigma_1 + \sigma_2)} \propto \frac{t}{t^{1/2}} = t^{1/2} \qquad (3.2.23)$$

Thus **separation between species 1 and 2 improves with increasing time, although there is increased dispersion in each profile. With time, then, we have better separation** but **increased dilution in each product region.** Obviously, the finite boundaries of the closed separator impose a limit on the time during which such a behavior is feasible with a uniform continuous ϕ_i^* potential profile. In practice, open separators are used and additional factors come into play.

We have studied here the one-dimensional migration of solutes introduced as a δ-function at one end of the vessel. Lee *et al.* (1977a) have illustrated the three-dimensional migration of a compact solute pulse.

If a certain state of separation of two solutes or the development of an individual solute profile is under consideration, the time lapsed, as we have seen above, is quite important. This time will depend amongst other things on the solvent used. For example, to arrive at a given nondimensional position in the separator with two different solvents s_1 and s_2, the times t_{s_1} and t_{s_2} required for a given $(\phi_i^+)'$ value may be obtained from

$$t_{s_1}^+ = t_{s_2}^+ \Rightarrow \frac{t_{s_1}}{t_{s_2}} = \frac{(l_z^2)_{s_1}(f_i^d)_{s_1}[(\phi_i^*)_{\max}]_{s_2}}{(l_z^2)_{s_2}(f_i^d)_{s_2}[(\phi_i^*)_{\max}]_{s_1}} \qquad (3.2.24)$$

for any solute i. Obviously, the frictional coefficient of solute i in the solvent and therefore solvent viscosity will be important for estimating the time required for achieving a specified state of solute profile development.

The previous conclusions on separation development were based on two primary characteristics of any solute profile: (1) the center point of each concentration profile is given by $z_i^+ = -(\phi_i^+)' t_i^+$; (2) the nondimensional standard deviation σ_i^+ of the Gaussian profile is equal to $\sqrt{(2D_{is}^0 t/l_z^2)}$. These characteristics of the concentration profiles (3.2.19) and (3.2.20) need to be justified.

To justify our locating of the center point of each concentration profile at $z_i^+ = (-\phi_i^+)' t_i^+$, take the first moment of the concentration profile (see the definition in (2.4.1g)) with respect to the variable η_i defined from (3.2.11) as

$$\eta_i = z_i^+ + (\phi_i^+)' t_i^+. \qquad (3.2.25)$$

Then

$$\langle \eta_i \rangle = \frac{\int_0^\infty \eta C_i(\eta_i, t_i^+) d\eta_i}{\int_0^\infty C_i(\eta_i, t_i^+) d\eta_i}; \qquad (3.2.26)$$

$$\langle \eta_i \rangle = \frac{\int_0^\infty \frac{\eta_i (m_i/l_x l_y l_z)}{\sqrt{4\pi RT t_i^+/(\phi_i^*)_{\max}}} \exp\left(- \frac{\eta_i^2}{4\left(\frac{RT}{(\phi_i^*)_{\max}}\right) t_i^+} \right) d\eta_i}{\int_0^\infty \frac{(m_i/l_x l_y l_z)}{\sqrt{4\pi RT t_i^+/(\phi_i^*)_{\max}}} \exp\left(- \frac{\eta_i^2}{4\left(\frac{RT}{(\phi_i^*)_{\max}}\right) t_i^+} \right) d\eta_i}. \qquad (3.2.27)$$

At any given t_i^+, the numerator has a zero value. Since $\langle \eta_i \rangle$ should provide the center of mass of the species i concentration profile,

$$\langle \eta_i \rangle = 0 = \langle z_i^+ + (\phi_i^+)' t_i^+ \rangle = 0 \qquad (3.2.28a)$$

$$\Rightarrow z_i^+ = -(\phi_i^+)' t_i^+. \qquad (3.2.28b)$$

To determine the standard deviation of the concentration profile, we take the second moment of the concentration profile with respect to the variable η_i, since

$$(\sigma_i^+)^2 = \langle (\eta_i - \langle \eta_i \rangle)^2 \rangle = \langle \eta_i^2 \rangle \text{ with } \langle \eta_i \rangle = 0;$$

$$\langle \eta_i^2 \rangle = \frac{\int_0^\infty \eta_i^2 C_i(\eta_i, t_i^+) d\eta_i}{\int_0^\infty C_i(\eta_i, t_i^+) d\eta_i}. \qquad (3.2.29)$$

Now, the integral in the numerator,

$$(m_i/l_x l_y l_z) \int_0^\infty \frac{\eta_i^2}{\sqrt{4\pi RT t_i^+/(\phi_i^*)_{\max}}} \exp\left(- \frac{\eta_i^2}{4\left\{\frac{RT}{(\phi_i^*)_{\max}}\right\} t_i^+} \right) d\eta_i,$$

can be changed to the integral below if we define

$$\eta_i^+ = \frac{\eta_i}{2\left\{\frac{RT t_i^+}{(\phi_i^*)_{\max}}\right\}^{1/2}}:$$

$$\frac{(m_i/l_x l_y l_z)}{\sqrt{\pi}} \frac{4RT t_i^+}{(\phi_i^*)_{\max}} \int_0^\infty (\eta_i^+)^2 \exp\left((\eta_i^+)^2\right) d\eta_i^+$$

$$= \left(\frac{m_i}{l_x l_y l_z} \right) \frac{4RT t_i^+}{\sqrt{\pi}(\phi_i^*)_{\max}} \frac{\sqrt{\pi}}{4}$$

at any t_i^+. The value of the integral in the denominator of (3.2.29) is

$$\left(\frac{m_i}{l_x l_y l_z}\right)\int_0^\infty \frac{1}{\sqrt{4\pi RTt_i^+/(\phi_i^*)_{\max}}}\exp\left(-\frac{\eta_i^2}{4\frac{RT}{(\phi_i^*)_{\max}}t_i^+}\right)d\eta_i.$$

This is considered the zeroth moment of the concentration profile C_i. Using η_i^+, it can be simplified to

$$\left(\frac{m_i}{l_x l_y l_z}\right)\frac{1}{\sqrt{\pi}}\int_0^\infty \exp(-\eta_i^{+2})d\eta_i^+ = \left(\frac{m_i}{l_x l_y l_z}\right)\frac{\sqrt{\pi}}{2\sqrt{\pi}}.$$

Therefore

$$(\sigma_i^+)^2 = \langle \eta_i^2 \rangle = \frac{(m_i/l_x l_y l_z)\{RTt_i^+/(\phi_i^*)_{\max}\}}{(m_i/l_x l_y l_z)\frac{1}{2}} = \frac{2RTt_i^+}{(\phi_i^*)_{\max}}$$
$$= \frac{2RTt_i^+}{(\phi_i^*)_{\max}} = \frac{2RTt(\phi_i^*)_{\max}}{l_z^2 f_i^d (\phi_i^*)_{\max}} = \frac{2D_{is}^0 t}{l_z^2} \tag{3.2.30}$$

at any time t.

Example 3.2.1 A pulse of ovalbumin molecules is introduced at the xy-plane at $z = 0$ at $t = 0$ in a uniform electrical field (in the z-direction) of strength 30 volt/cm (see Figure 3.2.1). The length of the vessel $l_z = 10$ cm; the solution viscosity is 1 cp; the molecular weight of ovalbumin is 45 000; $r_i = 2.78$ nm; $(f_i/f_{iw}) = 1.16$ (see Examples 3.1.2, 3.1.5 and 3.1.6). Calculate

(1) the time-dependent location of the center point of the ovalbumin concentration profile as it moves along the z-coordinate;
(2) the standard deviation of this profile ($D_{iw}^0 = 7.76 \times 10^{-7}$ cm^2/s; see Table 3.A.5), given $T = 20$ °C.
(3) How long would it take for the profile center point to reach the end of the vessel?

Solution (1) From the solution of Example 3.1.6, the migration velocity of ovalbumin molecules under the above-mentioned conditions is 0.35×10^{-2} cm/s. From (3.2.22a),

$$z_i^+ = \frac{U_{iz}t}{l_z} \Rightarrow z_i = U_{iz}t = 0.35 \times 10^{-2}t \text{ cm}.$$

(2) The dimensional standard deviation of the profile around the center point z_i^+ is equal to (from (3.2.21b))

$$\sigma_i^+ \times l_z = [(2D_{is}^0 t)^{1/2}/l_z] \times l_z = (2D_{is}^0 t)^{1/2}$$
$$= \sigma_i = (2 \times 7.76 \times 10^{-7})^{1/2} \times t^{1/2}$$
$$= 1.245 \times 10^{-3} t^{1/2} \text{ cm}.$$

(3)

$$z_i = l_z = 10 = 0.35 \times 10^{-2} t \Rightarrow t = \frac{10^3}{0.35} \text{ s};$$

t = 2857 s = 47.6 minutes. At this time, the standard deviation will be

$$1.245 \times 10^{-3} \times (2857)^{1/2}\text{cm} = 1.245 \times 53.45$$
$$\times 10^{-3} \text{ cm} = 0.066 \text{ cm}.$$

3.2.2 Multicomponent separation capability

What kind of separation processes or forces causing separation has the capacity to achieve multicomponent separation in a single separation vessel?

To answer this question, consider the solute concentration profiles (3.2.19) and (3.2.20) obtained earlier. If we assume that the solute mixture pulse at $z = 0$ and $t = 0$ contained more than two solute species, and if the movement of each species does not influence those of any other species, then

$$C_i(z^+, t_i^+) = \frac{(m_i/l_x l_y l_z)}{\sqrt{\frac{4\pi RTt_i^+}{(\phi_i^*)_{\max}}}}\exp\left[-\frac{\{z^+ + (\phi_i^+)' t_i^+\}^2}{4\left\{\frac{RT}{(\phi_i^*)_{\max}}\right\}t_i^+}\right] \tag{3.2.31}$$

for $i = 1, 2, \ldots, n$. Obviously if each solute species i has a unique value of U_{iz}, the center point of the instantaneous concentration profile for each solute will be at different locations in the separator. Further, if the standard deviations of the profiles are small compared to the distance between the center points of two neighboring concentration profiles, i.e. the z_i^+, we have succeeded in separating a multicomponent mixture at any given instant of time. For practical separations, one has to isolate physically the solvent in each region characteristic of a given species at any instant of time. Regardless of the practical difficulties in achieving this, it can be concluded that *uniform continuous profiles of* ϕ_i^*, where $i = 1, \ldots, n$, are inherently capable of separating an n-component solute mixture in a closed system with no convection since each i has a separate migration velocity $\boldsymbol{U}_i$.

The above analysis was carried out for a constant value of $(\phi_i^+)'$ found in systems with a uniform continuous ϕ_i^* profile. As long as $(\phi_i^+)'$ varies continuously along the separator and has a nonzero value, the multicomponent separation ability is retained. We will learn soon that multicomponent separation ability is absent in separation systems where $(\phi_i^+)' = 0$ and there is no bulk velocity.

What is the maximum number of components that can be separated in a closed separator of length l_z having uniform continuous profiles of ϕ_i^* ? This number, called the *peak capacity*, is defined by

$$n_{\max} = \frac{l_z}{4\sigma}, \tag{3.2.32}$$

where we have assumed an averaged value of standard deviations for all profiles and indicated it by σ. The quantity $n_{\max}$ is thus the maximum number of Gaussian solute concentration profiles that can exist in the separator of length l_z with contiguous profiles being separated, as shown in Figure 3.2.4, by a distance of 4σ (thus $R_s = 1$ between each neighboring peak). To calculate σ, choose a

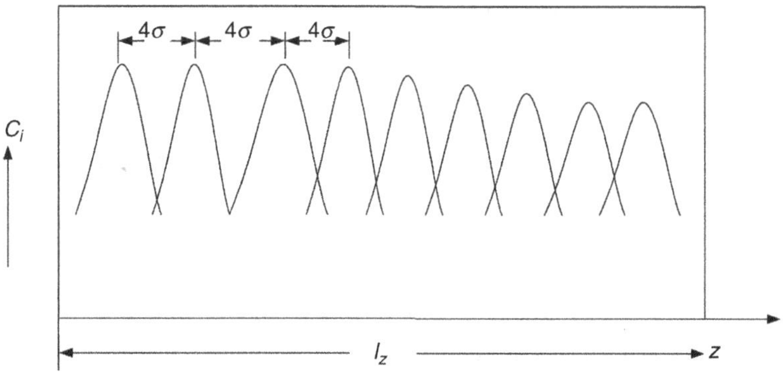

Figure 3.2.4. Maximum number of species separable in a separator of length l_z.

particular *i*th species and use[19] its $(\phi_i^*)_{max}$ and f_i^d. Remember, however, that the expression for σ_i indicated earlier by (3.2.21a/b) is nondimensionalized by l_z since η or z^+ is nondimensionalized with respect to l_z. Therefore

$$n_{max} = \frac{l_z}{4\sigma_i} = \frac{l_z}{4\left\{\frac{2RT}{(\phi_i^*)_{max}}\frac{t(\phi_i^*)_{max}}{l_z^2 f_i^d}\right\}^{1/2} l_z}.$$

For an estimate of the characteristic time t, use $t = (l_z/U_{iz})$,where U_{iz} is the migration velocity of species i and is given by $-(\partial\phi_i^*/\partial z)/f_i^d$. With these, and $-(\partial\phi_i^*/\partial z) = (\phi_i^*)_{max}/l_z$, we obtain

$$n_{max} = \frac{1}{4\left\{\frac{2RT}{l_z^2 f_i^d}\frac{l_z f_i^d}{\partial\phi_i^*/\partial z}(-)\right\}^{1/2}} = \sqrt{\frac{(\phi_i^*)_{max}}{32RT}} \tag{3.2.33a}$$

for uniform continuous ϕ_i^* profiles.

This result based on definition (3.2.32) assumes that an average value of 4σ may be defined. However, in practical situations, when long separation length or times are involved, 4σ will not be a constant. Suppose one is interested in finding out the maxmum number of components that can be separated in a given time window Δt at any given location in a column/vessel. Since the basic migration of peaks remains unchanged, n_{max} may be defined as

$$n_{max} = \frac{\Delta t}{4\sigma} \tag{3.2.33b}$$

for $R_s = 1$. Now 4σ is not a constant for different species/peaks. Expressing 4σ as the base width of the peak, W_b, we can rewrite the expression for n_{max}:

$$n_{max} = \frac{\Delta t}{W_b}. \tag{3.2.33c}$$

In a differential form, this expression may be written as (Giddings, 1969)

$$dn = \frac{dt}{W_b}. \tag{3.2.33d}$$

Integrating this over the time difference $\Delta t = t_2 - t_1$, we get

$$n_{max} = \int_0^{n_{max}} dn = \int_{t_1}^{t_2} (dt/W_b). \tag{3.2.33e}$$

For a homologous series of compounds, Medina *et al.* (2001) have proposed that W_b may be related to the time t when the peak appears by

$$W_b = a_1(a_2 t + a_3), \tag{3.2.33f}$$

where a_1, a_2, a_3 are empirical constants. Substituting into (3.2.33e), we get

$$n_{max} = \int_{t_1}^{t_2} \frac{dt}{a_1(a_2 t + a_3)} = \frac{1}{a_1 a_2}\ell n \frac{a_2 t_2 + a_3}{a_2 t_1 + a_3}. \tag{3.2.33g}$$

Example 3.2.2 A pulse of ovalbumin molecules (molecular weight, 45 000) in a mixture of a variety of proteins is exposed to an aqueous solution (density, 1 g/cm^3) subjected to a steady electrical field of strength 30 volt/cm. Given $Z_i = 4.5$, develop an estimate of the peak capacity when the electrodes are separated by a distance of 1 cm and the potential difference is 30 volt; ovalbumin may be used as the representative species.

Solution From equation (3.2.33a), the expression for peak capacity n_{max} for $R_s = 1.0$ is

$$n_{max} = \left((\phi_i^*)_{max}/32RT\right)^{1/2}.$$

Here,

$$(\phi_i^*)_{max} = (\phi_i^{ext})_{max} = Z_i \phi_{max}\mathcal{F},$$

where ϕ_{max} is the electrical potential, 30 volt. Therefore

$$Z_i\mathcal{F}\,\phi_{max} = 4.5 \times 96500 \times 30 \text{ volt-coulomb/gmol};$$

[19] As a representive value for obtaining σ.

R = 8.317 joule/gmol-K. Assume the temperature to be 25 °C, $T = 298$ K. We have

$$n_{max} = \left(4.5 \times 96500 \times 30\frac{\text{joule}}{\text{gmol}}/32 \times 8.317 \times 298\frac{\text{joule}}{\text{gmol}}\right)^{1/2}$$
$$= (164)^{1/2} = 12.8.$$

About 12 peaks may be separated. Compare this with the number of peaks in the centrifugal separation of Problem 3.2.1(c).

Example 3.2.3 The definition of peak capacity (3.2.32) was developed for the case of $R_s = 1$ where the two neighboring peaks were separated by the distance 4σ (where σ is an assumed average value of the standard deviations of all profiles). If poorer separation is acceptable, i.e. $R_s < 1$, what is an estimate of the peak capacity in a given system?

Solution From the definition of resolution,

$$R_s = \frac{z_1^+ - z_2^+}{2(\sigma_1^+ + \sigma_2^+)}.$$

Assuming $(\sigma_1^+ + \sigma_2^+) = 2\sigma^+$, where σ^+ is an averaged standard deviation, $R_s = (z_1^+ - z_2^+)/4\sigma^+$. Therefore for any value of R_s less than but close to 1, $(z_1^+ - z_2^+) \cong R_s 4\sigma^+$. We can then write

$$n_{max} = \frac{l_z}{(z_1^+ - z_2^+)l_z}$$
$$\Rightarrow \quad n_{max} = \frac{1}{R_s 4\sigma^+},$$

where σ^+ is nondimensional.

For dimensional standard deviation σ,

$$n_{max} = \frac{l_z}{R_s 4\sigma}.$$

We will now study the multicomponent separation capability of a *discontinuous* ϕ_i^* *profile* (shown in Figure 3.2.2 by the dashed line). The particular discontinuous profile we have chosen has a maximum value of $(\phi_i^*)_{max}$ at all values of $0 \le z \le l_z$, but at $z = l_z$, ϕ_i^* abruptly drops to the value of zero and stays at that level beyond l_z. It is thus a *step function*. Such a profile can be obtained in practice by having $\phi_{ti}^{ext} = 0$ or $\boldsymbol{F}_{ti}^{ext} = 0$ and maintaining two different phases or solvents in the two regions: $0 \le z \le l_z$ and $l_z \le z \le 2l_z$ (say). Recognize that, for all $z < l_z$, the value of $(\phi_i^+)'$ (equation (3.2.9)) is zero. The governing equation for the concentration distribution of species 1 is then obtained from equation (3.2.10) as the simpler

$$\frac{\partial C_1}{\partial t_1^+} = \frac{RT}{(\phi_i^*)_{max}}\frac{\partial^2 C_1}{\partial z^{+2}}. \tag{3.2.34}$$

Note that this is merely a special case of equation (3.2.12) obtained when $(\phi_i^+)'$ is zero so that $\eta = z^+$ from definition (3.2.11). The initial and boundary conditions of (3.2.14b) and (3.2.14a) for solute 1 may still be used here; the initial condition merely states that all solute is contained in a very

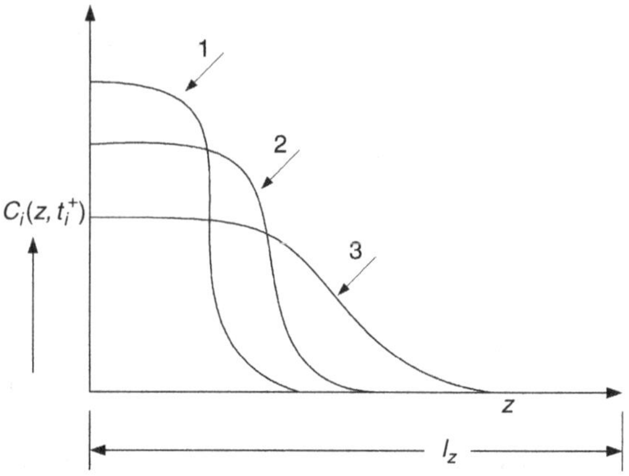

Figure 3.2.5A. Concentration profiles of species 1, 2 and 3 at any time t for a discontinuous ϕ_i^* profile in the vessel $0 \le z \le l_z$.

thin slab of cross-sectional area $l_x l_y$ at $z = 0$ for $t = 0$. The solution is obtained as a special case of solution (3.2.19) as

$$C_1(z^+, t^+) = \frac{(m_1/l_x l_y l_z)}{\sqrt{\frac{4\pi RT t_1^+}{(\phi_1^*)_{max}}}} \exp\left(-\frac{z^{+2}}{4\frac{RT}{(\phi_1^*)_{max}} t_1^+}\right). \tag{3.2.35}$$

The maximum value, as well as the center of mass of this profile, is located at $z^+ = 0$ (try calculating $\langle z_1^+ \rangle$ using the approach of (3.2.26) with z_1 instead of η_1). The nondimensional standard deviation of this profile will be given, as before, by $\sigma_1^+ = (2tD_{1s}^0/l_z^2)^{1/2}$. In Figure 3.2.5A, the profile of C_1 is shown at any time t.

Example 3.2.4 In the configuration shown in Figure 3.2.5A, a pulse of species i is injected in fluid j in the thin slab at $x = 0, l_x$; $y = 0, l_y$; $z = 0$. Calculate the standard deviation of the profile as a function of time for the following systems: system 1, n-propanol (species i)-water (fluid s); system 2, CO_2 (species i)-N_2 (fluid s). The system temperature is 25 °C. Obtain numerical values for $t = 100$ s.

Solution

System 1: Propanol-water, $\Rightarrow D_{is}^0 = 1.1 \times 10^{-5}\text{cm}^2/\text{s}$ (Table 3.A.3). From the formula for standard deviation (dimensional), $\sigma_i = (2 \times 1.1 \times 10^{-5} t)^{1/2} = 4.69 \times 10^{-3} t^{1/2}$cm.
System 2: CO_2-$N_2 \Rightarrow D_{is}^0 = 0.165\,\text{cm}^2/\text{s} = D_{CO_2-N_2}$; $\sigma_i = (2 \times 0.165 \times t)^{1/2} = 0.574 \times t^{1/2}$cm.

For $t = 100$ s we have (1) propanol-water, $\sigma_i = 0.0469$ cm; (2) CO_2-N_2, $\sigma_i = 5.74$ cm.

Assuming that the diffusion of each of the solute species $i = 1,2,3,\ldots,n$ contained in the solute mixture pulse at $z = 0$ takes place independently of one another, we have plotted the profiles of two more solutes $i = 2$ and 3 in Figure 3.2.5A. Obviously, at $z^+ = 0$, the solute whose diffusion coefficient is smallest will have the lowest σ_i and the highest peak

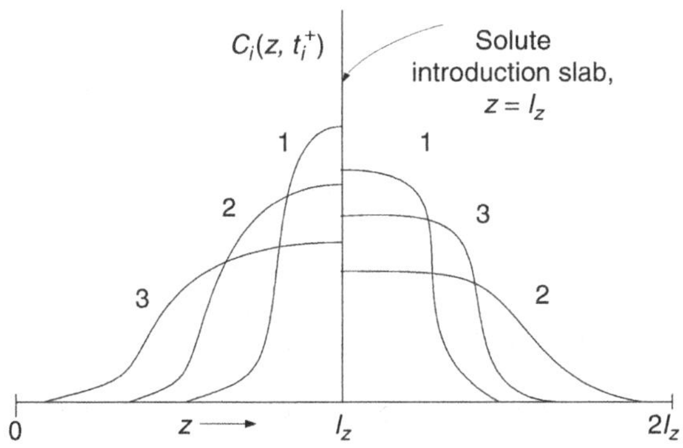

Figure 3.2.5B. Concentration profiles of species 1, 2 and 3 at any time t *for a discontinuous* ϕ_i^* *in the two adjoining vessels,* $0 \leq z \leq l_z$ *and* $l_z \leq z \leq 2l_z$, *with solute slab introduced at* $z = l_z$ *at time* $t = 0$.

for the same m_i; suppose $D_{1s}^0 < D_{2s}^0 < D_{3s}^0$ and $m_i = m_1$ (a constant) for all i, then

$$C_1(0,t_1^+) = \frac{(m_1/l_x l_y l_z)}{\sqrt{2\pi}\sqrt{2D_{1s}^0 t/l_z^2}} > \frac{(m_1/l_x l_y l_z)}{\sqrt{2\pi}\sqrt{2D_{2s}^0 t/l_z^2}} = C_2(0,t_2^+) > C_3(0,t_3^+). \qquad (3.2.36)$$

Therefore, solute 1 may be recovered at a high concentration around $z = 0$ if D_{2s}^0 and D_{3s}^0 are much larger than D_{1s}^0. However, this high concentration of species 1 will have impurities of species 2 and 3. On the other hand, solute 3 will have spread furthest; one can collect the solvent from this furthest region and get pure species 3 in the solvent. Obviously it will be a very dilute solution of pure species 3.

We observe then that no more than two species may be imperfectly separated when $(\phi_i^+)'$ is zero, and that only one pure species may be obtained with a poor recovery. To determine what will happen in the other solvent in the region $z > l_z$, it is better to start imagining what will happen if the initial solute mixture pulse were introduced at $z = l_z$ (Figure 3.2.5B). If we can assume that a different sequence[20] of profiles of species will be observed in each solvent vessel, then one can get at the most two pure species, one in each vessel from the furthest region from the solute introduction point. It is clear that this type of discontinuous potential profile can at the most separate two pure species and thus inherently lacks multicomponent separation capability. **When the diffusion process ceases in such a system after an adequate time has elapsed, the species are going to distribute themselves between the two solvents in a manner governed by the criterion for chemical equilibrium. This aspect is treated in Section 3.3.**

It is important to understand why a continuous ϕ_i^* profile can separate a multicomponent mixture when a discontinuous profile cannot. In the case of a continuous profile, the major feature is that the concentration profile of any species i appears to be translated bodily in the z-direction. In reality, the process may be understood by considering the C_i profile at any instant of time. There are two regions in this profile: the *trailing edge*, whose z^+-values are less than z_i^+, the center-of-mass location, and where the $\partial C_i/\partial z^+$ term is positive, and the *leading edge*, whose z^+ values are more than z_i^+ and where the $\partial C_i/\partial z^+$ term is negative. Consider now equation (3.2.10) without the diffusion term $\{RT/(\phi_1^*)_{max}\}(\partial^2 C_1/\partial z^2)$:

$$\frac{\partial C_1}{\partial t_1^+} = (\phi_1^+)' \frac{\partial C_1}{\partial z^+}. \qquad (3.2.37)$$

The value of $(\phi_1^+)'$ is negative for the continuous ϕ_i^* profile chosen. Therefore, in the *trailing edge* of the profile, C_1 is decreasing with time: the solute species 1 is being picked up and taken ahead (to larger z-values) to make room for another species coming from behind. In the *leading edge* of the profile, C_1 is increasing with time since solute is being transferred from the trailing edge into it (both $\partial C_1/\partial z^+$ and $(\phi_1^+)'$ negative). Thus, the existence of a nonzero $(\phi_1^+)'$ directly leads to the capacity of continuously evacuating solute from the trailing edge and bringing it to the leading edge to make room for another species profile to move in. Meanwhile, all of the species are moving forward in the positive z-direction. A discontinuous ϕ_i^* profile just does not have this capability.

We have already observed that there are two components in a ϕ_i^* profile: ϕ_{ti}^{ext} and μ_i^0. In developing the result (3.2.35), we had $\phi_{ti}^{ext} = 0$ and nonzero μ_i^0; but, as is true in two-phase systems (see Section 3.3), μ_i^0 does not vary within a phase at equilibrium. Therefore $(\partial\mu_i^0/\partial z) = 0$ within a phase. Thus, it appears that, unless we have external forces, $(\phi_i^+)'$ will be zero and there will be no multicomponent separation capability. We will learn in Section 7.1.5 (equation (7.1.100)) that this is not correct; the multicomponent separation capability can exist even though external forces are absent, e.g. in chromatographic processes in the presence of a *nonzero bulk velocity*.

3.2.3 Particulate systems

So far, we have studied the separation of chemical solutions including solutions of macromolecules, e.g. proteins. In macroscopic particulate systems without any Brownian forces of significance, for particles of size r_p, mass m_p, etc.,

$$\boldsymbol{F}_{tp} = \boldsymbol{F}_{tp}^{ext} \qquad (3.2.38)$$

since there is no chemical potential. We know that, in many cases, these external forces may be represented by

[20] Different solutes will have different solubilities and partitioning properties in the two solvents; thus the concentration difference driving the diffusion will be different in the two solvents for any species.

their scalar potentials. Such scalar potentials are usually continuous functions of the spatial coordinate, quite often uniformly continuous analogous to the cases studied in Section 3.2.1. Thus, it would appear that the external forces acting on particles have a multicomponent separation ability. Just as different chemical species had different migration velocities $\boldsymbol{U}_i$, similarly particles with different sizes or different densities or different charges will have different terminal velocities $\boldsymbol{U}_{pt}$. For example, in the gravitational force field, the terminal particle velocity vector is given by

$$\boldsymbol{U}_{pt} = \frac{\boldsymbol{F}_{tp}^{\text{ext}}}{f_p^d} = \frac{-m_p g\left(1 - \frac{\rho_t}{\rho_p}\right)\mathbf{k}}{6\pi\mu r_p} = -\frac{2}{9}\frac{r_p^2 g}{\mu}(\rho_p - \rho_t)\mathbf{k} \tag{3.2.39}$$

for spherical particles obeying Stokes' law. Thus particles of different radii or different densities (or both) will have different terminal velocities, providing the possibility of separating different particles. However, different particle speeds, by themselves, are insufficient for separation since all of them are moving in the same direction. Some other technique or condition has to be created to collect these particles at different locations of the separator (as we shall see later in Chapters 6, 7, etc.). However, different values of $\boldsymbol{U}_{pt}$ provide necessary conditions for separation. Note that identical comments can also be made about different molecular species i in a chemical mixture subjected to a uniform continuous potential profile.

3.3 Criteria for equilibrium separation in a closed separator

If two or more immiscible phases are kept in a closed container for a sufficient length of time, isolated from their surroundings, the phases come to equilibrium with one another. The amount of separation achieved at equilibrium is of considerable interest. We need to know the thermodynamic criteria for equilibrium to determine this separation. In this section, such criteria are specified for a variety of equilibrium conditions encountered in separation processes, including those where a single phase is exposed to an external force field in a closed vessel. Chapter 4 covers the extent of separation achieved under equilibrium conditions in a closed container.

Thermodynamic equilibrium between **two or more phases** or **two or more regions requires the existence** of **thermal**, **mechanical** and **chemical equilibrium**. The closed separator has *thermal equilibrium* if all phases and regions are at the same temperature T, which is constant. *Mechanical equilibrium* requires equality of pressure P in all phases or regions of the separator with plane phase interfaces. If the phase interfaces are curved, the pressures in the two phases under mechanical equilibrium will be different. If the two regions are separated by a semipermeable membrane, or if there is a swelling pressure in one phase (e.g. ion exchange resin), the constant pressure in each phase or region will also be different under mechanical equilibrium.

We are more interested in *chemical equilibrium*, achieved after transfer of species between two or more phases or regions. The criteria for equilibrium here will directly allow the calculation of different concentrations of a given species in different phases. This calculation presumes the existence of thermal and mechanical equilibrium. If the region is subjected to an *external force field*, the criterion for equilibrium separation is affected by the *external potential field*. This and other related criteria will be indicated in Section 3.3.1 without extensive and formal derivations (for which the reader should refer to different thermodynamics texts and references). The development of such criteria will be preceded by a brief illustration of the variety of two-phase systems encountered in separation processes. Our emphasis will be on two immiscible phase systems.

Mixtures in a separation system can exist in a number of different bulk phases. The most common adjectives used to characterize different phases are: gaseous, liquid, solid, supercritical fluid, membrane and ion exchange material.[21] The gaseous phase includes both gas and vapor, just as solid includes crystalline as well as amorphous materials. Some combinations, e.g. gas-gas, gas-supercritical fluid, are to be eliminated since they do not behave as two immiscible phases. From diffusional rate considerations, two-phase combinations of solid, membrane and ion exchange materials (a total of six, i.e. solid-membrane, solid-ion exchange material, solid-solid, membrane-membrane, membrane-ion exchange material, ion exchange material-ion exchange material) are not useful because separation would take forever![22] Thus one of the phases in any useful two-phase combination should be a fluid; i.e. gaseous, liquid or supercritical fluid. This eliminates a total of eight combinations. (Giddings (1982) has suggested that, in theory, there can be $[k(k+1)/2]$ combinations of two phases for k available phases, i.e. there are 21 possible combinations of six phases.) The remaining possible combinations are identified in Table 3.3.1.

Of the 13 two-phase combinations identified, two combinations, gas-ion exchanger and supercritical fluid-ion

[21] We ignore liquid crystals or mesophases, vesicles, caged molecules, etc. for the time being. They are, however, studied in Section 4.1.8. The membrane phase includes gels. Interfacial phases are ignored for now.

[22] Echoes of this may have mistakenly led to the formulation "Corpora non agunt nisi fluida sive soluta" - substances do not react unless in a liquid or a dissolved state (Helfferich, 1995).

Table 3.3.1 Possible useful combinations of two bulk immiscible phases[a]

Combinations having a particular phase	Possible useful combinations
Combinations having a gaseous phase	gas-liquid; gas-solid; gas-membrane; gas-ion exchanger
Combinations having a liquid phase	liquid-liquid; liquid-solid; liquid-membrane; liquid-ion exchanger; liquid-supercritical fluid
Combinations having a supercritical fluid phase	supercritical fluid-solid; supercritical fluid-membrane; supercritical fluid-supercritical fluid; supercritical fluid-ion exchanger

[a] Table 3.V in Giddings (1982) is based on seven phases, unlike the six phases used here.

exchanger, may not be fruitful since ions are not normally transferred between these phases. The rest are the basis of existing equilibrium based separation processes or have a significant potential for becoming practical techniques. These useful two immiscible phase combinations are to be kept in mind as we consider now the criteria for chemical equilibrium in such systems, as well as other related material.

3.3.1 Phase equilibrium with equal pressure in all phases

We choose for this purpose first a *two-region* (or *two-phase*) separator at constant and uniform T and P without any chemical reaction. If we focus on a particular region, it is obvious that it is open since molecules can be transferred from this to a contiguous region or vice versa. Impose now the restriction that a region can only have a single phase. For this particular open single-phase region chosen as a thermodynamic system, the total Gibbs free energy of all molecules in region j, G_{tj}, is a function of temperature T, pressure P and the mole numbers, m_{ij}, of all chemical species i in region j:

$$G_{tj} = G_{tj}\,(T,P, m_{1j}, m_{2j}, m_{3j},\ldots), \tag{3.3.1}$$

for the n-component system $i = 1, 2\ldots, n$. From standard thermodynamics texts (Van Ness and Abbott, 1982), we know that the total differential of G_{tj} following a small change in T,P and $m_{ij}-s$ is

$$\mathrm{d}G_{tj} = -S_{tj}\,\mathrm{d}T + V_j\,\mathrm{d}P + \sum_{i=1}^{n} \mu_{ij}\,\mathrm{d}m_{ij}. \tag{3.3.2}$$

Here S_{tj} is the total entropy of all molecules in region j, V_j is the volume of region j,

$$\begin{aligned}(\partial G_{tj}/\partial T)_{p,m_{ij},all\,i} &= -S_{tj};\\ (\partial G_{tj}/\partial P)_{T,m_{ij},all\,i} &= V_j;\\ (\partial G_{tj}/\partial m_{ij})_{T,P,m_{kj},k\neq i} &= \mu_{ij}.\end{aligned} \tag{3.3.3}$$

With the closed separator considered as a thermodynamic system, the species mole numbers, $m_i^0 = \sum_{j=1} m_{ij}$, are constant in the absence of any chemical reaction. The total Gibbs free energy of the whole separator as a system (obtained by summing G_{tj} over all j) is then only a function of T and P. At constant T and P, the total differential of this total Gibbs free energy is zero:

$$\sum_{j=1}^{2} \mathrm{d}G_{tj}\Big|_{T,P} = 0 \Rightarrow \sum_{j=1}^{2}\sum_{i=1}^{n} \mu_{ij}\,\mathrm{d}m_{ij} = 0, \tag{3.3.4}$$

where we have used (3.3.2) to obtain the second result. For a two-phase (or two-region) system, $\mathrm{d}m_{i1}=-\mathrm{d}m_{i2}$ due to conservation of species i. This simplifies (3.3.4) to

$$\sum_{i=1}^{n}(\mu_{i1} - \mu_{i2})\,\mathrm{d}m_{il} = 0, \tag{3.3.5}$$

which is valid for any arbitrary change $\mathrm{d}m_{i1}$. Therefore

$$\mu_{i1} = \mu_{i2}. \tag{3.3.6}$$

This criterion for chemical equilibrium in a two-phase system then requires that the chemical potential of any species i should be uniform and constant through the separator.

If the closed separator has more than two phases or two regions, a criterion for chemical equilibrium can be derived by considering any two of the phases or regions, say j_1 and j_2, at a time. The relation

$$\mu_{ij_1} = \mu_{ij_2} \tag{3.3.7}$$

can be easily shown to be valid for the total system of regions j_1 and j_2 by following the earlier procedure. Since this result is valid for any (j_1, j_2) pair of phases or regions, the general criterion for chemical equilibrium in a closed separator containing a total of k phases or regions is

$$\mu_{i1} = \mu_{i2} = \cdots = \mu_{ij} = \cdots = \mu_{ik} \qquad j = 1,2,\ldots,k. \tag{3.3.8}$$

A different form of the criterion for chemical equilibrium (3.3.6) or (3.3.8) is often quite useful since it uses the fugacities of a component i in solution instead of the chemical potential of species i. Let $\hat{f}_{ij}$ be the fugacity of species i in a homogeneous mixture in region or phase j. At a constant temperature, we know from the thermodynamic properties of homogeneous mixtures (Van Ness and Abbott, 1982) that any small change in the chemical potential of species i in region or phase j is related to the change in $\hat{f}_{ij}$ by

$$\mathrm{d}\mu_{ij}|_T = \mathrm{R}T\,\mathrm{d}\,\ell\mathrm{n}\,\hat{f}_{ij}|_T. \tag{3.3.9}$$

Integration yields

$$\mu_{ij} - \mu_{ij}(\text{initial}) = RT\,\ell n\,\hat{f}_{ij} - RT\,\ell n\Big[\hat{f}_{ij}(\text{initial})\Big]. \quad (3.3.10a)$$

We assume that

$$\mu_{ij}(\text{initial}) = C_1(\text{a constant}), \quad \hat{f}_{ij}(\text{initial}) = C_2(\text{a constant}), \quad (3.3.10b)$$

where the constants are dependent on the constant T and P of the separator and are assumed independent of j, the region or phase. Then, using the general criterion (3.3.8), we find

$$\ell n(\hat{f}_{i1}/C_2) = \ell n(\hat{f}_{i2}/C_2) = \cdots = \ell n(\hat{f}_{ij}/C_2) = \cdots = \ell n(\hat{f}_{ik}/C_2), \quad (3.3.10c)$$

which yields

$$\hat{f}_{i1} = \hat{f}_{i2} = \cdots = \hat{f}_{ij} = \cdots = \hat{f}_{ik} \qquad \text{for } j = 1,2\ldots,k, \quad (3.3.11)$$

as another form of the criterion for chemical equilibrium in a chemically nonreactive separator at constant T and P. (*Note*: The ^ over a quantity indicates, among other things, "valid for a mixture"; here, the fugacity of species i in a mixture.)

Equality of the chemical potential of a component/species in two contiguous immiscible phases constituting two regions rarely implies equality of concentration. In fact, the relation between the different concentrations of different species in two phases at equilibrium in a separator will be developed for various types of separation phenomena in Section 4.1 using these relations.

The relations given above are valid for any n-component system. As pointed out in Section 2.4, the mixture may be a *continuous chemical mixture*, where the composition is described by a molar density function $f(r)$ whose independent variable r is some characterizing property, e.g. molecular weight, carbon number, etc. The criterion for chemical equilibrium in a two-phase (or two-region) system of $j = 1, 2$ is, for all values of r, (Cotterman *et al.*, 1985)

$$\mu_1(r) = \mu_2(r), \quad (3.3.12)$$

where $\mu_1(r)$ and $\mu_2(r)$ refer, respectively, to the chemical potentials of species of property r in phases 1 and 2, respectively.

If the mixture is *semicontinuous*, i.e. the concentrations of some components ($i = 1,\ldots, n$) have discrete values in terms of mole fractions, whereas the concentrations of all others are described by a molar density function $f(r)$, there are two criteria for chemical equilibrium (Cotterman and Prausnitz, 1985),

$$\mu_{i1} = \mu_{i2}, \quad (3.3.13a)$$

for each $i = 1,\ldots, n$; for the continuous mixture part,

$$\mu_1(r) = \mu_2(r) \quad (3.3.13b)$$

for all values of r.

3.3.2 Phase equilibrium where different phases have different pressures

We consider next the requirements for chemical equilibrium between two regions or phases when the temperature is uniform and constant throughout the separator, but the pressure has different (but uniform) values in different phases (or regions). Assume the phase interface to be planar and assume no chemical reactions. Equation (3.3.2) is changed now to

$$dG_{tj} = -S_{tj}\,dT + V_j\,dP_j + \sum_{i=1}^{n} \mu_{ij}\,dm_{ij}, \quad (3.3.14)$$

where

$$(\partial G_{tj}/\partial T)_{P_j, m_{ij}, \text{all } i} = -S_{tj}; \quad (\partial G_{tj}/\partial P_j)_{T, m_{ij}, \text{all } i} = V_j; \quad (\partial G_{tj}/\partial m_{ij})_{T, P_j, m_{kj}, k\neq i} = \mu_{ij}. \quad (3.3.15)$$

The total Gibbs free energy of the system consisting of the two phases or regions of the separator as a whole depends only on T, P_1 and P_2 for $j = 1,2$ and not on the conserved mole number m_i^0 of any species i. At constant, T, P_1 and P_2, the total differential of the Gibbs total free energy of the two phases should be zero:

$$\sum_{j=1}^{2} dG_{tj}\Big|_{T,P_1,P_2} = 0 \Rightarrow \sum_{j=1}^{2}\sum_{i=1}^{n} \mu_{ij}\,dm_{ij} = 0.$$

Using the familiar argument of $dm_{i1} = -dm_{i2}$ and the fact that the magnitude of dm_{i1} or dm_{i2} is arbitrary, we obtain

$$\mu_{i1} = \mu_{i2}. \quad (3.3.16)$$

The chemical potential of a species i should then be uniform and constant throughout the separator under conditions of equilibrium that include different pressures in different phases.

In the *osmotic equilibrium* of two liquid solutions on either side of a *semipermeable membrane* (see Example 1.5.4), the semipermeable membrane, if perfect, prevents the exchange of the solute across the region boundary while the solvent passes through the membrane, the region boundary. In the development of the criteria for chemical equilibrium, it was always presumed that any species i could be exchanged between adjoining phases or regions. Therefore, the criteria for equality of chemical potential apply only to those species which are permeable through the membrane and not to impermeable species.

We will digress now and describe the dependence of $\mu_{ij}(P_j, T, x_{ij})$ on P_j, T and x_{ij}. A total differential of μ_{ij} for any phase j can be written as follows (Denbigh, 1971):

$$d\mu_{ij} = -\overline{S}_{ij}\,dT + \overline{V}_{ij}\,dP_j + \sum_{k=1}^{n-1}\left(\frac{\partial \mu_{ij}}{\partial x_{kj}}\right)_{T,P_j,x_{l,l\neq k}} dx_{kj}, \quad (3.3.17a)$$

where

$$\overline{S}_{ij} = -(\partial \mu_{ij}/\partial T)_{P_j,x_{nj},\text{all } n}; \qquad (\partial \mu_{ij}/\partial P_j)_{T,x_{nj},\text{all } n} = -\overline{V}_{ij} \tag{3.3.17b}$$

and $i = 1,2,\ldots, n$. Here, $\overline{S}_{ij}$ is the partial molar entropy of species i in region j.

Note that only $(n-1)$ x_{kj}s are independent. At any fixed composition $x_{ij,\text{all } i}$, we get

$$d\mu_{ij} = -\overline{S}_i\, dT + \overline{V}_{ij} dP_j. \tag{3.3.18a}$$

Integrating from standard state pressure P^0 to the pressure P_j of region j, we get, at constant T,

$$\mu_{ij}(P_j,T)_{x_{ij}} = \mu_{ij}(P^0,T)_{x_{ij}} + \int_{P^0}^{P_j} \overline{V}_{ij} dP_j. \tag{3.3.18b}$$

For liquid solutions, $\overline{V}_{ij}$ is generally a weak function of pressure. Therefore

$$\mu_{ij}(P_j,T)_{x_{ij}} = \mu_{ij}(P^0,T)_{x_{ij}} + \overline{V}_{ij}(P_j - P^0). \tag{3.3.19}$$

At any fixed pressure P_j, we can also integrate (3.3.17a) between the pure ith species and the composition x_{ij} to get, at constant T,

$$\mu_{ij}(P_j,T,x_{ij}) - \mu_{ij}(P_j,T)\Big|_{x_{ij}=1} = \int_{x_{ij}=1}^{x_{ij}} \sum_{k=1}^{n-1} \left(\frac{\partial \mu_{ij}}{\partial x_{kj}}\right)_{T,P_j,x_l l\neq k,} dx_{kj}. \tag{3.3.20a}$$

But we know from relation (3.3.10a) that, for the initial state being pure i,

$$\mu_{ij}(P_j,T,x_{ij}) - \mu_{ij}(P_j,T)|_{x_{ij}=1} = RT\ \ell n \left(\frac{\hat{f}_{ij}}{f^0_{ij}}\right) = RT\ \ell n\, a_{ij}, \tag{3.3.20b}$$

where a_{ij} is the activity of species i in region j. Use relation (3.3.19) for $x_{ij} = 1$ to obtain

$$\mu_{ij}(P_j,T,x_{ij}) = \mu_{ij}(P^0,T)|_{x_{ij}=1} + \overline{V}_{ij}(P_j - P^0) + RT\ \ell n\, a_{ij}. \tag{3.3.20c}$$

Usually, $\mu_{ij}(P^0,T)$ at standard state pressure P^0 and system temperature T for a pure substance i is identified as the standard state chemical potential $\mu^0_i(P^0,T) = \mu^0_i$. Then

$$\mu_{ij}(P_j,T,x_{ij}) = \mu^0_i + \overline{V}_{ij}(P_j - P^0) + RT\ \ell n\, a_{ij} \tag{3.3.21}$$

is the general relation linking μ_{ij}, P_j, T and x_{ij} (through a_{ij}). A special result for a pure substance is

$$\mu_{ij}(P_j,T) = \mu^0_i(P^0,T) + \overline{V}_{ij}(P_j - P^0). \tag{3.3.22}$$

One further obtains, at standard state conditions,

$$d\mu^0_{ij} = \overline{V}_{ij}\ dP^0 \tag{3.3.23}$$

for a differential change in the standard state pressure P^0. Since the standard state is important for many calculations, Table 3.3.2 summarizes the commonly employed standard states for the more frequently encountered conditions.

Table 3.3.2. Standard states

State or phase of system and/or species under consideration	Standard state defined by pressure, temperature or composition
(1) Gas or vapor	pure component gas or vapor behaving ideally at 1 atm and system temperature
(2) Liquid acting as a solvent	pure liquid component at the same temperature and pressure as the solution
(3) Solid acting as a solvent	pure solid component at the same temperature and pressure as the solid solution
(4) Solute in a liquid or solid solution such that the pure component solute is liquid or solid at the temperature and pressure of the solution	pure liquid or solid component at the same temperature and pressure as the solution
(5) Pure component solute does not exist in the same phase as the solution at the temperature and pressure of the solution	Infinite dilution standard state whereby the standard state value of fugacity or activity of a component at the temperature and pressure of the solution is given by the ratio of the fugacity or activity to the mole fraction[a] under conditions of infinite dilution

[a] Sometimes instead of the mole fraction x_{ij}, the molality, ($m_{i,j}$, moles of i/kg of solvent in region j) is used. Whether molality or mole fraction is used, this standard state is intimately connected with Henry's law:

$$\lim_{x_{il}\to 0} \hat{f}_{il} = x_{il} f^0_{il} = x_{il} H_i,$$

where H_i is Henry's law constant for species i in the solution and is the hypothetical fugacity f^0_{il} of the species at the infinite dilution standard state (see relation (3.3.59)).

3.3.3 Single-phase equilibrium in an external force field

We observed in Chapter 1 that separation is possible in a closed separator even if it contains only one phase (see Figure 1.1.3, example III; Examples 1.4.1 and 1.5.3; Problem 1.4.3; Section 3.2.1 analyses such a system in general). In such cases, there exists usually one (or more) external force field (or temperature gradient) which creates a difference in composition at different locations within the separator. These different locations then become different regions. A phase at equilibrium in our earlier results for criteria for equilibrium has the same properties everywhere. The presence of an external force field, however, imparts different values of the external potential ϕ_i^{ext} to different locations in a single-phase closed separator. Consequently, no two locations in such a separator are identical; one could presume the system to be composed of an infinite number of phases of differential thickness in the direction of the external force field.

Assume now two locations α and β a differential distance apart. At equilibrium, the net total force $\boldsymbol{F}_{ti}$ on species i should be zero. For the case of a single external force field $\boldsymbol{F}_i^{\text{ext}}$, we get from the general expression (3.1.50)

$$\nabla\mu_i\Big|_T = \boldsymbol{F}_i^{\text{ext}}. \tag{3.3.24}$$

For an external force field describable by $-\nabla\phi_i^{\text{ext}}$, the one-dimensional representation of the above relation along the coordinate direction z is

$$\frac{\mathrm{d}\mu_i}{\mathrm{d}z}\Big|_T \boldsymbol{k} = -\frac{\mathrm{d}\phi_i^{\text{ext}}}{\mathrm{d}z}\boldsymbol{k} \Rightarrow \mathrm{d}\mu_i\Big|_T = -\,\mathrm{d}\phi_i^{\text{ext}}, \tag{3.3.25}$$

relating the change in chemical potential of the ith species for a differential distance $\mathrm{d}z$ to the corresponding external-force-created potential difference. For a finite distance $z_\alpha - z_\beta$ between two regions α and β, we obtain, on integrating (3.3.25) (Denbigh, 1971),

$$\mu_i\Big|_{z_\alpha} - \mu_i\Big|_{z_\beta} = \phi_i^{\text{ext}}\Big|_{z_\beta} - \phi_i^{\text{ext}}\Big|_{z_\alpha} \Rightarrow \mu_{i\alpha} - \mu_{i\beta} = \phi_{i\beta}^{\text{ext}} - \phi_{i\alpha}^{\text{ext}}. \tag{3.3.26}$$

3.3.4 Equilibrium between phases with electrical charges

In systems using, say, ion exchange resins (see Section 3.3.7.7), the resin particles have fixed electrical charges. Similarly, ion exchange membranes (Section 3.4.2.5) have fixed electrical charges. If there is no externally applied electrical field, it is useful to enquire what criteria govern chemical equilibrium in two-phase systems containing such a phase.

It is common practice to define an *electrochemical potential* μ_{ij}^{el} in the jth region by

$$\mu_{ij}^{\text{el}} = \mu_{ij} + Z_i\mathcal{F}\phi_j, \tag{3.3.27}$$

where ϕ_j is the electrical potential of the jth phase and Z_i is the electrochemical valence of species i. It is defined as the total work done in bringing the species i from a vacuum into phase j (Adamson, 1967). The criterion for equilibrium for all ith species ($i = 1,\ldots, n$) is that the electrochemical potential of a species must be the same in all phases:

$$\mu_{i1}^{\text{el}} = \mu_{i2}^{\text{el}}; \qquad j = 1,2. \tag{3.3.28}$$

Using definition (3.3.27) in this relation leads to

$$\mu_{i1} - \mu_{i2} = Z_i\mathcal{F}(\phi_2 - \phi_1), \tag{3.3.29}$$

where the phase potentials ϕ_2 and ϕ_1 are independent of the ionic species under consideration. This potential difference ($\phi_2-\phi_1$) has been identified as the *Donnan potential*.

An additional relation has to be used in each phase containing charged species/ions. This is the *electroneutrality condition*, according to which there should be no net electrical charge at any bulk location (there may be deviations from it at phase boundaries; see Helfferich (1962, 1995)):

$$\sum Z_i\,C_{i1} = -\sum Z_k C_{k1}, \tag{3.3.30a}$$

where i = positive ions, k = negative ions, for phase $j = 1$ without any fixed charges;

$$\sum Z_i\,C_{i2} = -\sum Z_k C_{k2} - \omega X, \tag{3.3.30b}$$

for phase $j = 2$ with fixed charges, where X is the molar density of fixed charges and ω is the sign of fixed charges (+ for positive charges, - for negative charges).

It is useful to consider an explanation regarding electroneutrality and the potential difference $\phi_2-\phi_1$ in ion exchange systems provided by Helfferich (1962, 1995). *Ion exchange resin* particles have *fixed* positive or negative charges.[23] When they are placed in a solution of electrolytes, *counterions* (i.e. ions with charges opposite to the fixed charges) diffuse into the porous ion exchange resin particles and maintain electroneutrality. However, there then develop large concentration differences between the two phases. For a cation exchange resin system, say (with fixed negative charges), the cations are counterions; their concentrations are larger in the ion exchange resin particle, whereas the concentrations of anions (*coions*) are larger in the solution. Therefore anions tend to migrate to the resin particle and cations from the resin tend to migrate to the solution. The migration of a few ions in both directions immediately builds up an electrical potential difference between the two phases. This is the Donnan potential. The positive charge in the solution prevents cations from leaving the resin, whereas the negative charge on the resin prevents the migration of anions from the solution to the resin. An equilibrium is established between the repulsive

[23] See Figures 3.3.9, 3.3.10 and 3.3.12 for schematics of such particles and their charges.

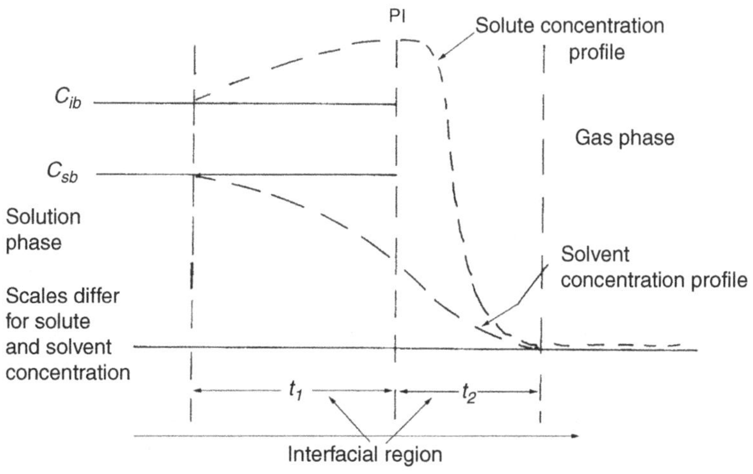

Figure 3.3.1. Concentration profiles across a gas–liquid interface: PI = phase interface.

forces of similar charges and the so-called driving force of a concentration difference. An identical condition is obtained with an anion exchange resin, except the Donnan potential has an opposite sign.

Regardless of the Donnan potential, however, electroneutrality is maintained in the ion exchange resin or the solution. Helfferich (1962) states that

> Migration of just a few ions is sufficient to build up so strong an electric field counteracting any further migration that deviations from electroneutrality remain far from the limit of accuracy of any method except for the measurement of the electric field itself.

3.3.5 Equilibrium between bulk and interfacial phases

In the absence of external force fields, we have so far assumed that, in a two-phase or multiphase system, the intensive properties of any phase such as pressure, temperature, composition, etc., were constant and uniform throughout the phase, including and up to the interface between the phase under consideration and the adjoining phase or phases. However, in many systems, the interfacial region demonstrates properties different from those of either of the bulk phases. For example, in a dilute aqueous solution of surfactants exposed to air, the surfactant concentration at the air–solution interface on the solution side is greater than that in the solution bulk. Skimming off the surface layer provides a way of obtaining a more concentrated surfactant solution, and is therefore a reasonable basis for a separation process or processes. A two-phase system has thus become a three-phase system: two bulk phases and one interfacial or surface phase. Figure 3.3.1 illustrates how the composition of a surface active solute i and the solvent s varies in the interfacial region of the surfactant solution ($j = 1$) - air ($j = 2$) system.

Since the interfacial phase is in addition to the two bulk phases, the criteria (3.3.8) for phase equilibrium should be applicable if there are no external force fields. Therefore

$$\mu_{i1} = \mu_{i2} = \mu_{i\sigma}, \tag{3.3.31}$$

where we have identified the surface phase by the subscript σ. In conventional phase equilibria, such a relation is generally sufficient to determine the equilibrium concentrations of any species in the two phases at equilibrium. But estimation of the surface phase concentration to determine equilibrium separation between a bulk phase and the surface phase requires consideration of surface forces.

The molecules of any species i present in the bulk of a liquid phase experience a net force if they are at the surface of this phase (i.e. at the interface of two contiguous phases, liquid–liquid or liquid–air). This force creates a tendency for the bulk phase surface to contract and is termed the *interfacial tension* γ^{12} (unit, dyne/cm), where the superscripts refer to the two bulk phases 1 and 2.

Consider a system with a planar interface between the phases. Let the volume, surface area and thickness of such a surface region shown in Figure 3.3.2A between two bulk phases 1 and 2 with planar interfaces 11′ and 22′ be, respectively, V_σ, S_σ and t_σ (Guggenheim, 1967).

Denote the uniform pressure in both bulk phases by P. The force in the direction parallel to planes 11′ and 22′ inside the surface phase is, however, given by $Pt_\sigma - \gamma^{12}$. This force acts on an area of height t_σ and unit width perpendicular to the plane of the paper in Figure 3.3.2A. Following Guggenheim (1967), it may be shown that, if the surface layer values V_σ, S_σ and t_σ are changed by a differential amount to $V_\sigma + dV_\sigma$, $S_\sigma + dS_\sigma$ and $t_\sigma + dt_\sigma$ through a

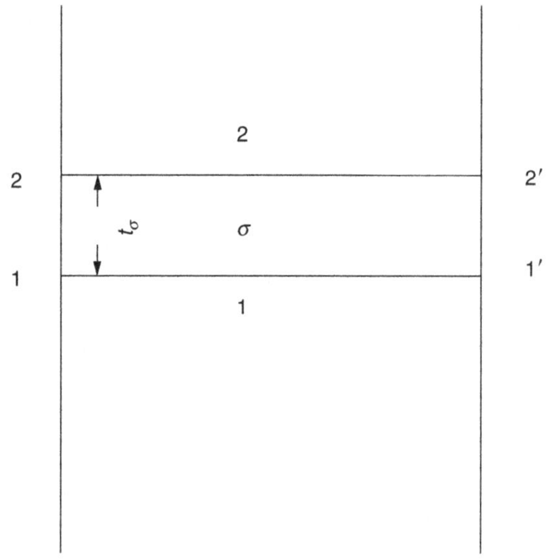

Figure 3.3.2A. Schematic of the interfacial region according to Guggenheim (1967).

reversible process, the reversible work done by the interfacial phase as a system is $P\,dV_\sigma - \gamma^{12}\,dS_\sigma$. This is in contrast to the reversible work $P\,dV$ done by a bulk phase when there is a differential change dV in volume V.

Incorporating such a departure in a total differential of the internal energy of the surface phase, and using Euler's theorem on homogeneous functions, the following relation can be shown to be valid for the surface phase (Guggenheim, 1967):

$$S_{t\sigma}\,dT - V_\sigma\,dP + S_\sigma\,d\gamma^{12} + \sum_{i=1}^{n} m_{i\sigma}\,d\mu_{i\sigma} = 0. \quad (3.3.32)$$

At constant T and P,

$$S_\sigma\,d\gamma^{12} = -\sum_{i=1}^{n} m_{i\sigma}\,d\mu_{i\sigma}. \quad (3.3.33)$$

Define the surface concentration $\Gamma_{i\sigma}$ of species i (in gmol/cm^2) by

$$\Gamma_{i\sigma} = \frac{m_{i\sigma}}{S_\sigma}. \quad (3.3.34)$$

The isotherm (3.3.33) may be rewritten as

$$-d\gamma^{12} = \sum_{i=1}^{n} \Gamma_{i\sigma}\,d\mu_{i\sigma} = \sum_{i=1}^{n} \Gamma_{i\sigma}\,d\mu_{i1} = \sum_{i=1}^{n} = \Gamma_{i\sigma}\,d\mu_{i2} \quad (3.3.35)$$

using relation (3.3.31). This result is, however, not used in this form.

Consider two species $i = 1$, the solute and $i = s$, the solvent. Equation (3.3.35) can be written for $i = 1, s$ as

$$-d\gamma^{12} = \Gamma_{1\sigma}\,d\mu_{1\sigma} + \Gamma_{s\sigma}\,d\mu_{s\sigma} = \Gamma_{1\sigma}\,d\mu_{11} + \Gamma_{s\sigma}\,d\mu_{s1} \quad (3.3.36)$$

where one of the bulk phases is $j = 1$. The Gibbs–Duhem equation for bulk phase 1 at constant T and P is

$$x_{11}\,d\mu_{11} + x_{s1}\,d\mu_{s1} = 0. \quad (3.3.37)$$

Substitution of this relation into (3.3.36) results in

$$\left(\Gamma_{1\sigma} - \frac{x_{11}}{x_{s1}}\Gamma_{s\sigma}\right) d\mu_{11} = -d\gamma^{12}. \quad (3.3.38)$$

The coefficient in brackets on the left-hand side may be interpreted as follows. Since species 1 is the solute and s is the solvent, $\Gamma_{s\sigma}(x_{11}/x_{s1})$ is the number of moles of solute 1 per unit area of the interface if the solvent and solute are present in the σ-phase in the same ratio as in bulk phase 1. The coefficient in brackets in (3.3.38), if positive, indicates the excess number of moles of solute 1 per unit interfacial area at the interface due to a deviation in behavior from the bulk phase. Denote this explicitly by rewriting (3.3.38) as

$$\Gamma^E_{1\sigma}\,d\mu_{11} = -d\gamma^{12} \quad (3.3.39a)$$

or

$$\Gamma^E_{1\sigma} = -\frac{d\gamma^{12}}{d\mu_{11}} \quad (3.3.39b)$$

for the two-component (solute $i = 1$) two-bulk-phase system with an interfacial region of distinct properties. (The superscript E denotes a surface excess quantity.) Generalizing such a procedure for an n-component solution with a solvent $i = s$, we get, corresponding to (3.3.39a),

$$\sum_{\substack{i=1\\ i\neq s}}^{n} \Gamma^E_{i\sigma}\,d\mu_{i1} = \sum_{\substack{i=1\\ i\neq s}}^{n} \Gamma^E_{i\sigma}\,d\mu_{i\sigma} = -d\gamma^{12}, \quad (3.3.40a)$$

where

$$\Gamma^E_{i\sigma} = \left(\Gamma_{i\sigma} - \frac{x_{i1}}{x_{s1}}\Gamma_{s\sigma}\right), \qquad i = 1, \underset{i\neq s}{2}, \ldots, n \quad (3.3.40b)$$

Here $\Gamma^E_{i\sigma}$ is called the *surface excess* of species i. Note that when $i = s$, $\Gamma^E_{s\sigma} = 0$, i.e. the surface excess of the solvent is zero. For surface active solutes, $(d\gamma^{12}/d\mu^{\text{int}}_{i1}) = (a_{i1}/RT)\,(d\gamma^{12}/da_{i1})$ is negative since γ^{12} decreases with increasing solute concentration: thus $\Gamma^E_{i\sigma}$ is positive. For electrolytic solutes (e.g. salts) in an aqueous solution at an air–water interface, $\Gamma^E_{i\sigma}$ could be negative since the electrolytic solute prefers the environment of the water molecules in the bulk compared to that of the air–water interface, raising the value of γ^{12} over that of pure water.

Consider now equation (3.3.39b). It relates the interfacial concentration of solute 1 to the bulk concentration of solute 1 through the dependence of interfacial tension on the bulk chemical potential of solute 1. Remember, we need the equilibrium criterion so that we can relate the solute concentrations in the two phases. Both results (3.3.39a) and (3.3.40a) are known as the *Gibbs equation* or *Gibbs adsorption isotherm*. The procedure normally used to derive the Gibbs equation, however, is different and is based on the assumptions that $V_\sigma = 0$ and the solvent surface excess is also zero (Davies and Rideal, 1963; Adamson, 1967). There is a sound reason for adopting the Gibbsian strategy of $V_\sigma = 0$. The quantities $\Gamma_{1\sigma}^E$ and $\Gamma_{s\sigma}^E$ in equation (3.3.38) are arbitrary since their values are dependent on the locations of the boundaries 11′ and 22′. However, the quantity $(\Gamma_{1\sigma} - \{x_{11}/x_{s1}\}\Gamma_{s\sigma})$ is independent of the volume V_σ and the locations of the boundaries. Gibbs therefore replaced the real situation with an imaginary one where bulk properties continued all the way to a hypothetical phase interface, a two-dimensional region which had the surface excess or surface deficiency of the solute only.

For surface-active solutes lowering the value of γ^{12}, a *film pressure* or *spreading pressure* π has been defined, for example for an air–water system, as

$$\pi = \gamma^{12}_{\text{solvent}} - \gamma^{12}_{\text{solution}}. \qquad (3.3.41a)$$

This is a sort of **two-dimensional pressure** for the two-dimensional **interfacial** region with **no volume** (**Gibbsian formulation** $V_\sigma = 0$) but a **surface area** S_σ. Thus instead of pressure, volume and temperature (P, V, T) of a conventional three-dimensional phase, the hypothetical interfacial phase has π, area and temperature (π, S_σ, T). Correspondingly, the ideal gas equation of state for a pure gas, $PV = RT$ for 1 mole of gas, is replaced by

$$\pi \overline{S}_\sigma = RT, \qquad (3.3.41b)$$

where $\overline{S}_\sigma$ is the surface area per mole of the gas. For $q_{i\sigma}$ moles of a gas i in the surface phase per unit adsorbent mass (see equation (3.3.43a) for a changed definition of S_σ)

$$\pi S_\sigma = q_{i\sigma} RT. \qquad (3.3.41c)$$

In the application of equation (3.3.40a), all independent ionic species are to be included along with nonionic solutes. Further, the electroneutrality condition requiring the absence of net charge anywhere suggests that the following should be satisfied:

$$\sum_{i=\text{positive ions}} Z_i C_{i1} = -\sum_{k=\text{negative ions}} Z_k C_{k1}; \qquad (3.3.42a)$$

$$\sum_{i=\text{positive ions}} Z_i \Gamma_{i\sigma}^E = -\sum_{k=\text{negative ions}} Z_k \Gamma_{k\sigma}^E. \qquad (3.3.42b)$$

If an applied electrical field is present, the surface adsorption equilibrium relation (3.3.40a) is modified by the inclusion of an electrical potential term (Davis and Rideal, 1963).

The criteria for equilibrium in *gas adsorption* on the surface of a *solid adsorbent* may also be obtained using the same thermodynamic equations which led to the Gibbs equation. The Gibbs adsorption isotherm in this case has been shown to be (Hill, 1949)

$$S_\sigma \,\mathrm{d}\pi = \sum_{i=1}^{n} q_{i\sigma}\,\mathrm{d}\mu_{i\sigma} = \sum_{i=1}^{n} q_{i\sigma}\,\mathrm{d}\mu_{ig}, \qquad (3.3.43a)$$

where $q_{i\sigma}$ is the number of moles of gas i adsorbed per unit mass of adsorbent, S_σ is the surface area per unit mass of the adsorbent, and π, the spreading pressure, is defined to be the lowering of surface tension at the gas–solid interface due to adsorption. Unlike gas–liquid interfaces where the interfacial tension γ^{12} is measurable, the spreading pressure π of a gas–solid surface is not measurable. The interfacial region is often identified as the *adsorbate*. Expressing (3.3.43a) for a binary gas mixture of species 1 and 2 as

$$S_\sigma \,\mathrm{d}\pi = \sum_{i=1}^{2} q_{i\sigma}\,\mathrm{d}\mu_{ig}, \qquad (3.3.43b)$$

$$\mathrm{d}\pi = \sum_{i=1}^{2} \frac{q_{i\sigma}}{S_\sigma} RT\,\mathrm{d}\ln\hat{f}_{ig} = \Gamma_{1\sigma}RT\,\mathrm{d}\ln\hat{f}_{1g} + \Gamma_{2\sigma}RT\,\mathrm{d}\ln\hat{f}_{2g},$$

$$\frac{\mathrm{d}\pi}{RT} = \Gamma_{1\sigma}\,\mathrm{d}\ln\hat{f}_{1g} + \Gamma_{2\sigma}\mathrm{d}\ln\hat{f}_{2g}. \qquad (3.3.44)$$

For a pure gas species i

$$\frac{\mathrm{d}\pi}{RT} = \Gamma_{i\sigma}\,\mathrm{d}\ln f_{ig} \qquad (3.3.45)$$

or

$$\frac{\mathrm{d}\pi}{\mathrm{d}\Gamma_{i\sigma}} = RT\,\Gamma_{i\sigma}\frac{\mathrm{d}\ln f_{ig}}{\mathrm{d}\Gamma_{i\sigma}}. \qquad (3.3.46a)$$

Since $(\mathrm{d}\ln f_{ig}/\mathrm{d}\Gamma_{i\sigma})$ is obtainable from an isotherm (experimental or otherwise), $\mathrm{d}\pi/\mathrm{d}\Gamma_{i\sigma}$ can be determined (Lewis and Randall, 1961). Note that for an ideal gas behavior with the pure species at pressure P,

$$\frac{\mathrm{d}\pi}{\mathrm{d}\Gamma_{i\sigma}} = RT\,\Gamma_{i\sigma}\frac{\mathrm{d}\ln P}{\mathrm{d}\Gamma_{i\sigma}}. \qquad (3.3.46b)$$

Expressing $\Gamma_{i\sigma}$ as $q_{i\sigma}$ / S_σ, we obtain, by integrating from 0 to P as π increases from 0 to π,

$$\frac{1}{RT}\int_0^{\pi} S_\sigma\,\mathrm{d}\pi = \frac{\pi S_\sigma}{RT} = \int_0^P \frac{q_{i\sigma}\,\mathrm{d}P}{P}, \qquad (3.3.46c)$$

an alternative form of (3.3.46a) for a pure ideal gas being adsorbed.

So far, the surface excess of a solute species was considered on the surface of a bulk phase, e.g. water, air, solid adsorbent, etc. The surfaces of macromolecules, especially proteins, have an *interfacial region*, sometimes called the *local domain*, which can have compositions different from the *bulk domain*, namely the bulk of the solution. For example, the bulk of an aqueous protein solution may have cosolvents (or cosolutes) such as urea, guanidine

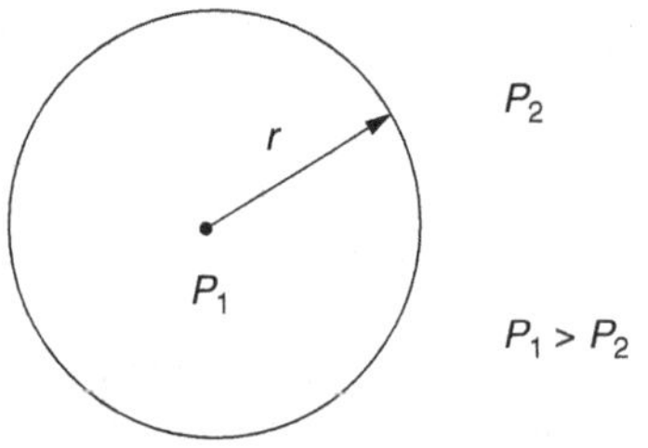

Figure 3.3.2B. Pressure in two regions (inside and outside) of a spherical bubble.

hydrochloride, sucrose, etc., added in large amounts (40–50% by weight). As Tang and Bloomfield (2002) state:

> These cosolute molecules bathe and solvate the macromolecular solute as water does.

The composition of the local domain is important from the point of view of how the macromolecular solute, for example protein, will behave. If the local domain is depleted in the cosolutes and enriched in water relative to that in the bulk solution, the macromolecule (protein) is said to undergo *preferential hydration*. If, however, the local domain, is depleted in water relative to the bulk, then either of two things can happen. The "cosolutes" may be preferentially accumulated in the local domain, or the local domains of neighboring macromolecules/proteins may interact with each other (for example via hydrophobic interaction; see Section 4.1.9.4) leading to, for example, precipitation (see Section 3.3.7.5).

3.3.6 Curved interfaces

The interfaces between two bulk phases considered so far have been planar, and the pressures in the two bulk phases under equilibrium have been equal for gas–liquid and liquid–liquid systems. When the interface has a curvature, mechanical equilibrium requires different values of the pressures in the two phases. The general relation (the *Young–Laplace equation*) governing the pressure difference between bulk phases 1 and 2 is as follows (Adamson, 1967; Guggenheim, 1967):

$$(P_1 - P_2) = \gamma^{12}\left(\frac{1}{r_1} + \frac{1}{r_2}\right), \tag{3.3.47}$$

where γ^{12} is the interfacial tension and r_1 and r_2 are the principal radii of curvature of the interfaces. By convention, r_1 and r_2 are positive if they are in phase 1 and ΔP is positive, with the interface convex going from phase 1 to phase 2. For a true spherical surface, as in a spherical bubble of radius r (Figure 3.3.2B),

$$(P_1 - P_2) = 2\gamma^{12}/r \tag{3.3.48}$$

since $r_1 = r_2 = r$. For both results, it has been assumed that γ^{12} is not affected by r_1, r_2 or r. For a planar interface, $r_1 = r_2 = \infty$ and $P_1 = P_2$ in relation (3.3.47).

A question of some importance is: **Does the curved surface of a drop or other geometries affect the chemical potential (or fugacity) of a species compared with that on a planar surface?** For gas–liquid, vapor–liquid, solid–liquid or liquid–liquid systems, this subject has been considered by Lewis and Randall (1961), starting from the general equation of a differential change in total Gibbs free energy of a phasc j (compare with relation (3.3.2)):

$$dG_{tj} = -S_{tj}\,dT + V_j\,dP_j + \gamma^{12}\,dS_\sigma + \sum \mu_{ij\text{Pl}}\,dm_{ij}, \tag{3.3.49}$$

where $\gamma^{12}\,dS_\sigma$ appears due to interfacial tension between phases $j = 1$ and $j = 2$. Note that the quantity μ_{ijPl} in (3.3.49) refers to any process of transfer of m_{ij} without any change in S_σ (planar interface, subscript Pl); to indicate this explicitly,

$$\mu_{ij\text{Pl}} = \left(\frac{\partial G_{tj}}{\partial m_{ij}}\right)_{T,\text{Pl},S_\sigma,m_{kj},k\neq i}, \tag{3.3.50}$$

where we have included a subscript Pl to refer to a planar interface. From Lewis and Randall (1961), for any curved interface system, we can express relation (3.3.49) in general as

$$dG_{tj} = -S_{tj}\,dT + V_j\,dP_j + \sum \mu_{ij}\,dm_{ij}. \tag{3.3.51}$$

For a sphere, any small change in the surface area of the sphere dS_σ due to a differential change dr in the radius will lead to the following change in volume dV of the sphere:

$$dV = 4\pi r^2\,dr; \quad dS_\sigma = 4\pi(r + dr)^2 - 4\pi r^2 \cong 8\pi r\,dr.$$

Correspondingly,

$$dS_\sigma = 2\frac{dV}{r}. \tag{3.3.52a}$$

If the volume change is due to the addition of dm_i moles of species i for all i, then

$$dV = \sum \overline{V}_{ij}\,dm_i, \tag{3.3.52b}$$

which leads to

$$\gamma^{12}\,dS_\sigma = 2\gamma^{12}\frac{dV}{r} = \frac{2\gamma^{12}}{r}\sum_{i=1} \overline{V}_{ij}\,dm_{ij} \tag{3.3.52c}$$

$$\Rightarrow \mu_{ij} - \mu_{ij\text{Pl}} = \frac{2\overline{V}_{ij}\,\gamma^{12}}{r} \tag{3.3.52d}$$

is a specific example for a spherical droplet of radius r (Figure 3.3.2B). Using the relation (3.3.9) between μ_{ij} and $\hat{f}_{ij}$, we get[24]

[24] Other considerations may come into play to prevent $\hat{f}_{ij}$ from increasing continually as r is reduced.

$$\ell n \frac{\hat{f}_{ij}}{\hat{f}_{ij\text{Pl}}} = \frac{2\overline{V}_{ij}\gamma^{12}}{RTr}. \qquad (3.3.53)$$

Here, $\hat{f}_{ij}$ is the fugacity in a system with a spherical surface, with the interface convex going from the droplet phase to the surroundings, and $\hat{f}_{ij\text{Pl}}$ corresponds to that with a planar surface. For ideal gas behavior, replace $\hat{f}_{ij}$ by p_{ij}.

This result has implications for droplets, which have their surface convex toward the vapor. Consider a pure liquid drop of radius r. Then $\hat{f}_{ij\text{Pl}}$ is equal to the vapor pressure of the pure liquid (on a flat surface), $P_{i\text{Pl}}^{\text{sat}}$, if the fugacity coefficient ϕ_i^{sat} under the conditions of P and T is assumed to be unity. However, equation (3.3.53) implies that the vapor pressure of the pure liquid in the space above the convex curved surface of the drop, P_i^{sat}, is higher than $P_{i\text{Pl}}^{\text{sat}}$ at equilibrium (Figure 3.3.2B):

$$P_i^{\text{sat}} = P_{i\text{Pl}}^{\text{sat}} \exp\left(\frac{2\, V_{ij}\,\gamma^{12}}{RTr}\right). \qquad (3.3.54)$$

Further, the smaller the drop diameter, the higher the observed vapor pressure of the liquid in the vapor space, leading to a higher rate of evaporation from the drop. Equation (3.3.54) is identified as the *Kelvin equation* and the phenomenon is called the *Kelvin effect*. This increased evaporation tendency appears because a molecule near the droplet surface is attracted to a lesser extent toward the interior by the surrounding molecules (since there are fewer of these surrounding molecules). However, this analysis breaks down as the drop size becomes too small and the number of molecules become much smaller. If the drop consists of a solution of a nonvolatile solute, the Kelvin equation applies to the solvent. Note that if the liquid surface is concave toward the vapor, the fugacity $\hat{f}_{ij}$ at the surface is less than that on a planar surface. See Section 3.3.7.5 as well as Problem 3.3.2 for the application to a solid-liquid system.

3.3.7 Solute distribution between phases at equilibrium: some examples

When two or more immiscible phases are at equilibrium, species *generally* distribute themselves between the phases such that the uniform and constant concentration of a species in one phase is different from those in other phases. Since there can be a large number of combinations of two-phase systems in equilibrium (see Table 3.3.1), the variety in distribution coefficient (κ_{il}) relations is enormous. Our objective here is somewhat limited. We pick a few common two-phase systems in equilibrium and illustrate the relations between the concentrations of any solute between the phases at equilibrium. This will not only allow us to develop integrated flux expressions in multiphase systems in Section 3.4, but will also facilitate calculating the separation achieved in a closed vessel in many systems considered in Chapter 4. We consider separately equilibrium in gas-liquid, liquid-liquid, gas-membrane, liquid-membrane/porous sorbent, liquid-solid, interfacial adsorption systems, ion exchange systems, as well as in supercritical fluid-solid/liquid systems. Only elementary and essential aspects of each equilibrium solute distribution will be our concern. Although we describe the equilibrium distribution of a solute[25] between two phases in general, there are cases here, e.g. vapor-liquid equilibrium, where the solute is a major constituent of the phases under consideration. The notion of a solute in a solvent is inappropriate in such cases.

3.3.7.1 *Gas-liquid equilibrium*

We discuss first the distribution of a solute i between a gas and a liquid at pressure P. At equilibrium, we obtain from the criterion (3.3.11)

$$\hat{f}_{ig} = \hat{f}_{il}. \qquad (3.3.55)$$

From standard thermodynamics texts, the fugacity of species i in each phase may be expressed as

$$\hat{f}_{ig} = x_{ig}\hat{\Phi}_{ig}P; \qquad (3.3.56)$$

$$\hat{f}_{il} = x_{il}\gamma_{il}f_{il}^0, \qquad (3.3.57)$$

where $\hat{\Phi}_{ig}$ is the fugacity coefficient of species i in the gas phase at P and system temperature T, and f_{il}^0 is the standard state fugacity of species i in the liquid phase at a standard temperature and pressure. Therefore, the mole fraction ratio of i between the two phases in equilibrium is given by

$$\frac{x_{il}}{x_{ig}} = \frac{\hat{\Phi}_{ig}P}{\gamma_{il}f_{il}^0}. \qquad (3.3.58)$$

If the gas phase behaves as an ideal gas and the liquid phase behaves as an ideal solution, we know that $\hat{\Phi}_{ig} \cong 1$, $\gamma_{il} \cong 1$ and $p_{ig} = x_{ig}P$. If the gas phase is distinctly in the nonvapor region (see item (5), Table 3.3.2), it is known from thermodynamics that

$$\lim_{x_{il}\to 0} \hat{f}_{il} = x_{il}f_{il}^0 = x_{il}H_i, \qquad (3.3.59)$$

where H_i is Henry's law constant. The relation (3.3.58) is then reduced to either

$$x_{ig} = (H_i/P)x_{il} \qquad (3.3.60a)$$

or

[25] Particle distribution between two immiscible phases has been considered separately at the end of this section.

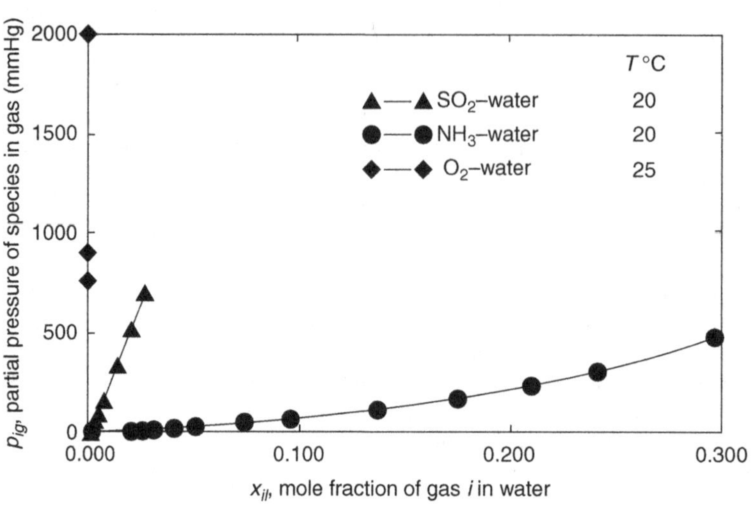

Figure 3.3.3. Solubility behavior of three different gas species in water. The value of the Henry's law constant may be calculated from (p_{ig}/x_{il}) = H_i. For example, for the SO_2-water system, at a p_{ig} = 450 mm Hg, $x_{il} \approx 0.017$. Therefore H_i = (450/0.017) mm Hg/mole fraction = 2.65×10^4 mm Hg/mole fraction = 35 atm/mole fraction.

$$p_{ig} = H_i x_{il}, \tag{3.3.60b}$$

which is *Henry's law* of solubility of gas species i in the liquid under consideration. Figure 3.3.3 illustrates the Henry's law behavior of a number of gases in water according to relation (3.3.60b). These are based on values of H_i available in the literature for three gases, NH_3, SO_2 and O_2 (Geankoplis, 1972; Perry *et al.*, 1984). Note that the solubility of O_2 is so low that its mole fraction in the liquid phase does not show up at the scale used. Further, the lower the solubility of a gas species, the higher the value of H_i. The liquid phase is often called *absorbent*.

If the gaseous phase can be considered as a vapor, then the distribution of a species between the vapor and liquid phase is of interest. Obviously, the species can be a major constituent of each phase here. The equilibrium ratio K_i of a component i is defined as (equation (1.4.1))

$$K_i = \frac{x_{i1}}{x_{i2}}.$$

Generally, phase 1 is the vapor phase ($j = v$) and phase 2 is the liquid phase ($j = l$). At equilibrium, use relations (3.3.56) and (3.3.57); then,

$$K_i = \frac{x_{iv}}{x_{il}} = \frac{\gamma_{il} f_{il}^0}{\hat{\Phi}_{iv} P}, \tag{3.3.61}$$

since $\hat{f}_{il} = \hat{f}_{iv}$. For low to moderate pressures, the standard state fugacity in the liquid phase is the fugacity of pure liquid at the temperature and pressure of the system (Smith and Van Ness, 1975):

$$f_{il}^0 = f_{il} = P_i^{\text{sat}} \Phi_i^{\text{sat}}, \tag{3.3.62}$$

where P_i^{sat} is the vapor pressure of pure component i at the system temperature and Φ_i^{sat} is the fugacity coefficient of pure i at P_i^{sat} at the system temperature. For pressures when the vapor phase is an ideal gas, $\Phi_i^{\text{sat}} \cong 1$. Further, $\hat{\Phi}_{iv} \cong 1$. If the liquid phase is an ideal solution, then $\gamma_{il} \cong 1$. Thus

$$\frac{x_{iv}}{x_{il}} = K_i = \frac{P_i^{\text{sat}}}{P}, \tag{3.3.63}$$

where the system pressure is P, and both the vapor and the liquid phase behave ideally. When written as

$$x_{iv} = (P_i^{\text{sat}} x_{il}/P), \tag{3.3.64}$$

we obtain the well-known *Raoult's law* of vapor-liquid equilibrium.

An alternative representation of Raoult's law is useful for the case of *continuous chemical mixtures*. Rewrite relation (3.3.63) as

$$P x_{iv} = p_{iv} = P_i^{\text{sat}} x_{il}, \tag{3.3.65}$$

where p_{iv} is the partial pressure of species i in the vapor phase behaving ideally. For a continuous chemical mixture of vapor and liquid phases at equilibrium, let $f_l(M)$ and $f_v(M)$ be the molecular weight density functions of the liquid and vapor phases, respectively. If the vapor pressure of a species of molecular weight M and T is indicated by P^{sat} $(T;M)$, then Raoult's law for species in the molecular weight range M to $M + \text{d}M$ is

$$P f_v(M)\,\text{d}M = P^{\text{sat}}(T;M) f_l(M)\,\text{d}M,$$

i.e.

$$Pf_v(M) = P^{sat}(T;M)f_l(M). \tag{3.3.66}$$

One can use density functions f_v and f_l as functions of quantities other than molecular weight, e.g. boiling point, carbon number.

In the vapor-liquid or gas-liquid equilibrium studied above, the vapor or gas species existed as molecules of that species in the liquid. With *diatomic gases*, e.g. H_2, O_2, N_2, etc., and *liquid metal*, it is, however, found that the gas is dissolved atomically in the molten metal. For example, in molten iron, the following equilibrium has been suggested for nitrogen gas (Darken and Gurry, 1953):

$$N_2(g) = 2N(l). \tag{3.3.67}$$

The equilibrium constant for this chemical reaction may be written in terms of activities of each species as

$$K = \frac{(a_{Nl})^2}{a_{N_2g}}. \tag{3.3.68}$$

For an ideal gas,

$$a_{N_2g} = x_{N_2g} = (p_{N_2g}/P). \tag{3.3.69a}$$

For the liquid phase, the activity of the atomic nitrogen species is related to its mole fraction:

$$a_{Nl} = x_{Nl}\,\gamma_{Nl}. \tag{3.3.69b}$$

Substituting these into (3.3.68), we get

$$x_{Nl} = \left(\frac{K}{P}\right)^{1/2} \frac{(p_{N_2g})^{1/2}}{\gamma_{Nl}}. \tag{3.3.70}$$

Thus, the mole fraction of atomic nitrogen in the liquid metal phase is proportional to the square root of the partial pressure of molecular nitrogen in the gas phase. This square-root dependency is known as *Sievert's law* and is found to be valid for diatomic gases like O_2, H_2, etc. The effect of a chemical reaction on solute distribution between two phases in equilibrium has been considered in great detail in Chapter 5.

3.3.7.2 *Liquid-liquid equilibrium*

Next, we consider the distribution of a solute in liquid-liquid equilibrium. Liquid-liquid systems can be of three main types: *aqueous-organic, nonpolar organic-polar organic; aqueous-aqueous.* Both phases in each case must be immiscible with each other. (See Section 4.1.3 for an illustration of each type of system.) In the general case, we use the equilibrium criterion (3.3.6) for solute i between two immiscible liquid phases $j = 1,2$, namely $\mu_{i1} = \mu_{i2}$. But at any P, T,

$$\mu_{il}(P) = \mu_{il}^0(P^0) + RT\ell n\, a_{il} + \overline{V}_{ij}(P - P^0) \tag{3.3.71}$$

from (3.3.20c) and (3.3.21) for μ_{ij}. Therefore

$$\overline{V}_{i1}(P - P^0) + RT\ell n\, a_{i1} + \mu_{i1}^0(P^0)$$
$$= RT\ell n\, a_{i2} + \mu_{i2}^0(P^0) + \overline{V}_{i2}(P - P^0). \tag{3.3.72}$$

Two cases arise depending on the solute, P and T if we assume $\overline{V}_{i1} = \overline{V}_{i2}$.

Case 1 If pure solute i exists as a liquid at P and T, then the standard state is independent of phase 1 or phase 2 (see item (4), Table 3.3.2):

$$\mu_{i1}^0(P^0) = \mu_{i2}^0(P^0). \tag{3.3.73}$$

This implies

$$(a_{i1}/a_{i2}) = 1, \tag{3.3.74}$$

i.e.

$$\frac{x_{i1}}{x_{i2}} = \frac{\gamma_{i2}}{\gamma_{i1}}. \tag{3.3.75}$$

In such a case, the mole fraction of solute i in the two liquid phases 1 and 2 will be different only if the solute behaves nonideally in both phases and $\gamma_{i2} \neq \gamma_{i1}$. Partitioning of, say, acetic acid at 25 °C between water and isopropyl ether will fall into this category since pure acetic acid is a liquid.

Case 2 Pure solute i does not exist as a liquid at P and T; an infinite dilution standard state is necessary (item (5), Table 3.3.2). In general, such standard state values $\mu_{ij}^0\,(P^0)$ are dependent on phase j. Therefore, from relation (3.3.72),

$$\frac{a_{i1}}{a_{i2}} = \exp\left[\frac{\mu_{i2}^0(P^0) - \mu_{i1}^0(P^0)}{RT}\right] = \exp\left[\frac{\Delta\mu_i^0}{RT}\right],$$

where

$$\Delta\mu_i^0 = \mu_{i2}^0(P^0) - \mu_{i1}^0(P^0). \tag{3.3.76}$$

Correspondingly,

$$\frac{x_{i1}}{x_{i2}} = \left(\frac{\gamma_{i2}}{\gamma_{i1}}\right) \exp\left[-\frac{\Delta\mu_i^0}{RT}\right]. \tag{3.3.77}$$

If the solutions in each phase are dilute enough for $\gamma_{i2} \cong 1$ and $\gamma_{i1} \cong 1$, then (x_{i1}/x_{i2}) is a constant:

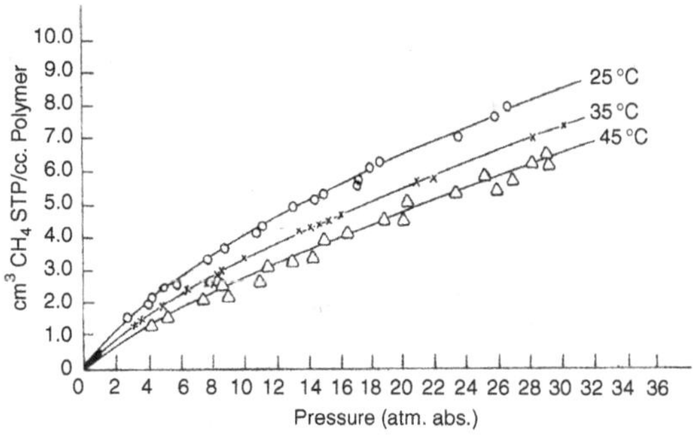

Figure 3.3.4. Solubility of methane in oriented polystyrene. Reprinted from Journal of Membrane Science, *Vol. 1, W. R. Vieth, J. M. Howell and J. H. Hsieh, "Dual sorption theory," pp. 177–220. Copyright (1976), with permission from Elsevier.*

$$(x_{i1}/x_{i2}) = \kappa_{i1}^N, \tag{3.3.78}$$

where κ_{i1}^N can be determined from $\mu_{ij}^0 - s$. This behavior is identified as the *Nernst distribution law*. Even if the amount of solute *i*, and therefore its mole fraction, is varied in one phase, the ratio of the solute mole fractions in the two phases remains constant. An alternative form is used for dilute solutions:

$$(C_{i1}/C_{i2}) = \kappa_{i1}, \tag{3.3.79}$$

where k_{il} is the constant distribution coefficient.

3.3.7.3 *Gas-membrane equilibrium*

The dissolution of a gas in a nonporous organic polymeric membrane is an example of *gas-membrane equilibrium*. Organic polymeric membranes can be amorphous, semicrystalline or crystalline. Crystalline regions are virtually impenetrable. Therefore practical polymer membranes are either amorphous or semicrystalline. Semicrystalline polymers are a combination of amorphous and crystalline regions. The amorphous region/membrane may be rubbery or glassy. The gas molecules dissolve in the amorphous region of the membrane as if it were a liquid. For permanent gases at temperatures $T > T_{ci}$ (= the critical temperature of gas species *i*) and nonporous rubbery polymeric membranes, this behavior has been observed as long as there is no interaction with the membrane material. Further, under these conditions, Henry's law is obeyed for gases such as H_2, He, Ne, Ar, O_2, N_2, CH_4 (Stern and Frisch, 1981). The form of Henry's law used in gas-membrane equilibrium with the gas phase behaving as an ideal gas is

$$C_{im} = S_{im}\, p_{ig}, \tag{3.3.80}$$

where S_{im} is the solubility coefficient of gas species *i*, C_{im} is the concentration of species *i* in the membrane exposed to a gas, with the species *i* partial pressure being p_{ig}.

The commonly used units of the solubility coefficient S_{im} are $(cm^3(STP))/(cm^3 \cdot cm\ Hg)$, with C_{im} and p_{ig} expressed, respectively, in $cm^3(STP)/cm^3$ and cm Hg. The gas species is often identified as the penetrant. Generally, the more condensible the gas, the higher the solubility coefficient. For a useful correlation of solubility coefficient with the normal boiling point, see Stannett *et al.* (1979). Sometimes, the solubility coefficient is also reported as the sorption coefficient. Figure 4.3.4 illustrates the values of the solubility coefficient for a number of gases and vapors in natural rubber membranes. Section 4.3.3, specifically, equations (4.3.43a–c) and (4.3.44a), illustrate how the solubility coefficient S_{im} depends on the polymer, the temperature as well as the critical temperature T_{ci} of the gas/vapor species for amorphous polymers.

When the organic polymer is glassy (i.e. the system temperature is lower than T_g, the polymer glass transition temperature), the solubility behavior of a pure gas species in the membrane is different from Henry's law (Figure 3.3.4):

$$\begin{aligned} C_{im} &= S_{im}\, P + \frac{C'_{Hi}\, b_i\, P}{[1 + b_i P]} \\ &= C_{im}^d + C_{im}^H. \end{aligned} \tag{3.3.81}$$

At very low pressures, where $b_i P << 1$, $C_{im} = (S_{im} + C'_{Hi} b_i)P$, a linear behavior as in Henry's law. At very high pressures, where $b_i P >> 1$, a linear behavior ($C_{im} = S_{im}P + C'_{Hi}$) is again observed. Figure 3.3.4 displays how a low-pressure linear region is connected by a nonlinear curved behavior to the high-pressure linear region for dissolution of CH_4 in a polystyrene film (Vieth *et al.*, 1976).

The first term in the above isotherm corresponds to species dissolving according to Henry's law. The second term is due to species dissolving according to the Langmuir isotherm (see Section 3.3.7.6). Specifically, the Langmuir species are assumed to be sorbed in microvoids in the membrane (regions having 0.5–0.6 nm diameter). The overall behavior is described as the *dual sorption*[26] *mode* for two different dissolution modes of gas molecules in the glassy membrane.

When a binary gas mixture ($i = 1, 2$) is in equilibrium with a glassy polymeric membrane, the following sorption behavior has been suggested (Koros, 1980):

---

[26] It has been suggested that there is no sharp boundary between absorption and adsorption. The term "sorption" was introduced by J. W. McBain in 1909 to describe situations where both are important (Bikerman, 1970).

$$C_{1m} = S_{1m} p_{1g} + \frac{C'_{H1}\, b_1\, p_{1g}}{\left[1 + b_1\, p_{1g} + b_2\, p_{2g}\right]}; \qquad (3.3.82a)$$

$$C_{2m} = S_{2m} p_{2g} + \frac{C'_{H2}\, b_2\, p_{2g}}{\left[1 + b_2\, p_{2g} + b_1\, p_{1g}\right]}. \qquad (3.3.82b)$$

Here, S_{1m} and S_{2m} are the solubility coefficients of those molecules of species 1 and 2 which dissolve according to Henry's law. The utility of such a mixture sorption isotherm model has been verified (see, for example, Sanders *et al.* (1984) for a mixture of CO_2 and C_2H_4 in poly(methyl methacrylate)).

The gas-membrane equilibrium is basically somewhat different if the membrane is made of a metal or an alloy and the gas is diatomic. We have already noted that a diatomic gas such as N_2, H_2 or O_2 dissolves in a molten metal according to Sievert's law, i.e. the mole fraction of the gas (in the atomic state) in the molten metal is proportional to the square root of the gas partial pressure. The same equilibrium behavior, relation (3.3.70), is also observed between a gas and a solid metallic membrane. It is therefore necessary to assume that the gas is present in the atomic state while developing any relation for membrane transport of diatomic gases.

3.3.7.4 *Liquid-membrane equilibrium and liquid-porous sorbent equilibrium*

There are *two* aspects to liquid-membrane equilibrium. The first one is concerned with the *osmotic equilibrium* between two solutions on two sides of a semipermeable membrane permeable to the solvent and impermeable to the solute; the second one covers partitioning of the solute between the solution and the membrane. Both porous and nonporous membranes are of interest. The second aspect is also useful for porous sorbent/gel particles.

In osmotic equilibrium between two regions 1 and 2 separated by a semipermeable membrane, the pressures P_1 and P_2 of the two regions are usually different (see Figure 1.5.1). For the solvent species *i* transferable between the two regions, we have, at equilibrium,

$$\mu_{i1}(P_1, T, x_{i1}) = \mu_{i2}(P_2, T, x_{i2}) \qquad (3.3.83)$$

from relation (3.3.16). Using expression (3.3.21) for $\mu_{ij}(P_j, T, x_{ij})$, we get

$$\mu^0_{i1}(P^0, T) + \overline{V}_{i1}(P_1 - P^0) + RT\ell n\, a_{i1} = \mu^0_{i2}(P^0, T) + \overline{V}_{i2}(P_2 - P^0) + RT\ell n\, a_{i2}. \qquad (3.3.84)$$

If the pure *i*th species exists in the same physical state in both phases at P^0 and T, then

$$\mu^0_{i1}(P^0, T) = \mu^0_{i2}(P^0, T) \Rightarrow \overline{V}_{i1}(P_1 - P_2) = RT\ell n\, \frac{a_{i2}}{a_{i1}}, \qquad (3.3.85)$$

where we have assumed $\overline{V}_{i1} = \overline{V}_{i2}$. This relation illustrates how the pressure difference between two phases under osmotic equilibrium can be tied to the composition difference between the two solutions. For a one solute-one solvent system with a semipermeable membrane impervious to solute *i* but permeable to solvent *s*, the above results can be reexpressed as follows:

$$\overline{V}_s(P_1 - P_2) = RT\ell n\, \frac{a_{s2}}{a_{s1}} = \overline{V}_s(\pi_1 - \pi_2), \qquad (3.3.86a)$$

where π_1 and π_2 are the osmotic pressures of the solutions in region 1 and region 2, respectively. If $\pi_1 > \pi_2$, $a_{S2} > a_{S1}$; further, $P_1 > P_2$ for osmotic equilibrium. The osmotic pressure of a dilute solution of small molecules of species *i* is often calculated from the *van 't Hoff equation*

$$\pi_i = C_i RT. \qquad (3.3.86b)$$

Consider now the second type of equilibrium when a solute *partitions* between the solution and the membrane, which is no longer semipermeable to the solute: the solute distribution is somewhat similar to that in liquid-liquid equilibrium of the type (3.3.78) or (3.3.79). Solute partitions into the membrane, which is assumed to be *nonporous*, and is dissolved in it as if the membrane were a liquid. For low solute concentrations in the membrane and no swelling, the following distribution equilibria generally hold:

$$(C_{im}/C_{il}) = \kappa_{im}, \qquad i = 1, \ldots, n. \qquad (3.3.87)$$

If the membrane happens to be *porous* (or microporous) and *uncharged*, the nature of the liquid-membrane equilibrium will be determined by the relative size of the solute molecules with respect to the pore dimensions in the absence of any specific solute-pore wall interaction. Similar considerations are also valid for liquid-porous sorbent equilibria. If the solute dimensions are at least two orders of magnitude smaller, then the solute concentration in the solution in the pore should be essentially equal to that in the external solution. However, the solute concentration in the porous membrane/porous sorbent/gel will be less than that in the external solution due to the porosity effect. Assuming that the solute exists only in the pores of the membrane/porous sorbent with a porosity ε_m, the value of k_{im} should be equal to the membrane or sorbent porosity ε_m if the solute characteristic dimensions are at least two orders of magnitude smaller than the radius of the pore.

When the solute dimensions are larger and there are no specific solute-pore wall interactions, the partitioning effect is indicated by a *geometrical partitioning factor* for a cylindrical pore as

$$\frac{C^p_{im}}{C_{il}} = \left(1 - \frac{r_i}{r_p}\right)^2, \qquad (3.3.88a)$$

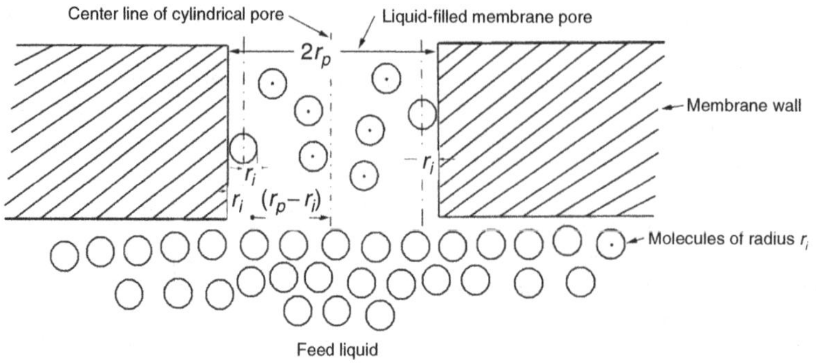

Figure 3.3.5A. Geometrical partitioning effect: only a distance of $r_p - r_i$ from the pore center is available for locating the center of the molecules of the solute.

where C^p_{im} is the concentration of solute i in the membrane/sorbent pore liquid, r_p is the pore radius and r_i is the radius of the solute molecule (assumed spherical). The center of the solute molecule cannot approach the wall beyond a radius of $r_p - r_i$ (Figure 3.3.5A). Thus only the volume fraction $\pi(r_p - r_i)^2/\pi r_p^2$ of the pore can have solute molecules at the same concentration as the external solution, C_{il}; yet the solvent molecules exist effectively throughout the pore, making the pore liquid (total volume $\pi r_p^2 \times 1$ for unit length) leaner in solute molecules compared to the liquid phase concentration C_{il} outside the pores:

$$C^p_{im}(\pi r_p^2) \times 1 = C_{il}\,\pi\,(r_p - r_i)^2 \times 1. \tag{3.3.88b}$$

This was originally suggested by Ferry (1936) as a reduction in cross-sectional area for diffusion. Giddings *et al.* (1968) have developed the same result theoretically by considering the limitations in orientations and structural configurations of macromolecules to avoid sterical overlap with the pore wall. If the pore has the shape of a slit ($\beta = 1$), cylinder ($\beta = 2$) or a cone ($\beta = 3$), the geometrical partitioning factor is given by

$$\kappa_{im} = (C^p_{im}/C_{il}) = \left(1 - \frac{r_i}{r_p}\right)^\beta, \tag{3.3.89a}$$

where r_p is the characteristic pore dimension. See Colton *et al.* (1975) for a brief review of the relevant literature. In most porous media being modeled as a collection of cylindrical capillaries, most pores are unlikely to be truly cylindrical. Therefore, r_p may be replaced by $(2/s)$, where s is the surface area of the wall of the capillary per unit pore volume:

$$\kappa_{im} = \left(1 - \frac{r_i\,s}{2}\right)^2. \tag{3.3.89b}$$

We will illustrate very briefly now a fundamental approach to calculating the partitioning factor of (3.3.88a) as developed by Giddings *et al.* (1968). In the formalism of statistical mechanics, the equilibrium partition constant k_{im} is the ratio of partition functions for molecules within the pores and within the bulk liquid:

$$\kappa_{im} = \frac{\iiint \exp\left[-en_p(\boldsymbol{r},\boldsymbol{\Psi},\lambda)/\mathrm{k}^B T\right]\ (\mathrm{d}\boldsymbol{r}\,\mathrm{d}\boldsymbol{\Psi}\,\mathrm{d}\lambda)}{\iiint \exp\left[-en_b(\lambda)/\mathrm{k}^B T\right]\ (\mathrm{d}\boldsymbol{r}\,\mathrm{d}\boldsymbol{\Psi}\,\mathrm{d}\lambda)}. \tag{3.3.89c}$$

Here the coordinates $\boldsymbol{r}$, $\boldsymbol{\Psi}$ and λ describe the molecular position, orientation and conformation, respectively. The energy en_p in the porous configuration consists of the en_M due to intramolecular interactions, an energy en_{MN} due to intermolecular interactions and an energy en_{Mp} due to interactions between the macromolecule and the pore; therefore

$$en_p = en_M + en_{MN} + en_{MP}. \tag{3.3.89d}$$

Correspondingly, for the bulk liquid,

$$en_b = en_M + en_{MN}. \tag{3.3.89e}$$

If we choose the condition of infinite dilution, en_{MN} will be zero since the macromolecules are far apart. We further assume that essentially there is only one conformation of the molecule/macromolecule (rigid) (alternatively all conformations have the same energy); therefore, $en_M = 0$. Further, we assume that there are no adsorptive forces between the macromolecule and the pore wall; in addition, the pore wall and macromolecule are distinct and discontinuous: consequently, $\exp(-en_{MP}/\mathrm{k}^B T)$ has the value of 1 for molecular configurations free from overlap with the wall and the value of 0 for configurations of overlap with the wall. In random-pore networks, an ensemble average of $\exp(-en_{MP}/\mathrm{k}^B T)$, $q(\boldsymbol{r}, \boldsymbol{\Psi}, \lambda)$, can replace it:

$$\kappa_{im} = \frac{\iiint q(\boldsymbol{r},\boldsymbol{\Psi},\lambda)\,\mathrm{d}\boldsymbol{r}\,\mathrm{d}\boldsymbol{\Psi}\,\mathrm{d}\lambda}{\iiint \mathrm{d}\boldsymbol{r}\,\mathrm{d}\boldsymbol{\Psi}\,\mathrm{d}\lambda}. \tag{3.3.89f}$$

This quantity $q(\boldsymbol{r}, \boldsymbol{\Psi}, \lambda)$ may be defined in general as the probability that a molecule having a given configuration does not intersect a pore wall. If we define a local equilibrium partition constant $k_{i,\mathrm{loc}}(\boldsymbol{r})$ such that

$$\kappa_{i,\mathrm{loc}}(\boldsymbol{r}) = \frac{\iiint q(\boldsymbol{r},\boldsymbol{\Psi},\lambda)\;\mathrm{d}\boldsymbol{\Psi}\,\mathrm{d}\lambda}{\iint \mathrm{d}\boldsymbol{\Psi}\,\mathrm{d}\lambda}, \tag{3.3.89g}$$

where $\kappa_{i,\mathrm{loc}}(\boldsymbol{r})$ depends only on the position coordinate $\boldsymbol{r}$ in the pore, then

$$\kappa_{im} = \frac{\int \kappa_{i,\mathrm{loc}}(\boldsymbol{r})\,\mathrm{d}\boldsymbol{r}}{\int \mathrm{d}\boldsymbol{r}}. \tag{3.3.89h}$$

For rigid macromolecules/molecules, dependence on λ disappears, and we get

$$\kappa_{im} = \frac{\iint q(\boldsymbol{r},\boldsymbol{\Psi})\,\mathrm{d}\boldsymbol{r}\,\mathrm{d}\boldsymbol{\Psi}}{\iint \mathrm{d}\boldsymbol{r}\,\mathrm{d}\boldsymbol{\Psi}} = \frac{\int \kappa_{i,\mathrm{loc}}(\boldsymbol{r})\,\mathrm{d}\boldsymbol{r}}{\int \mathrm{d}\boldsymbol{r}}. \tag{3.3.89i}$$

To illustrate, consider spherical molecules of radius r_i and pores of radius r_p in a rigid medium. For pores of infinite length, it is obvious that, for radial locations r of the center for the molecule,

$$0 < r \leq (r_p - r_i), \kappa_{im} = 1; \tag{3.3.89j}$$

$$(r_p - r_i) \leq r \leq r_p, \kappa_{im} = 0. \tag{3.3.89k}$$

Use now equation (3.3.89h); it is now the ratio of two circular areas, one with the diameter $(2(r_p - r_i))$ corresponding to (3.3.89j) in the numerator and the denominator corresponding to the pore area of diameter $2r_p$:

$$\kappa_{im} = \frac{\pi 4 (r_p - r_i)^2/4}{\pi 4 r_p^2/4} = \left(1 - \frac{r_i}{r_p}\right)^2, \tag{3.3.89l}$$

which is the relation (3.3.88a) described earlier.

The radius, r_i cm, of the solute molecule (which is assumed spherical) needs to be known in liquid-porous membrane/porous sorbent equilibrium. For relatively small molecular species, the specific molar volume of the solute will lead to

$$r_i = \left(\frac{3M_i}{4\pi \tilde{\mathrm{N}} \rho_i}\right)^{1/3}, \tag{3.3.90a}$$

where M_i is the molecular weight, ρ_i is the solute density and $\tilde{\mathrm{N}}$ is Avogadro's number (6.022×10^{23}). Spriggs and Li (1976) have therefore suggested that, for solids,

$$2r_i = 1.465 \times 10^{-8} (M_i/\rho_i)^{1/3}, \tag{3.3.90b}$$

where ρ_i is in g/cm^3 and r_i is in cm, whereas, for liquids,

$$2r_i = 10^{-8}\, V_{im}{}^{1/3},$$

where V_{im} is the molecular volume at the normal boiling point (MVNBP) in cm^3 (Reid *et al.*, 1977). Expression (3.3.90a) is also used for determining r_i values for larger molecules, like globular proteins, where one usually assumes that $(1/\rho_i) \cong 0.75\,\mathrm{cm^3/g}$ is the partial specific volume.

Since the solute molecule is in a solution, an alternative estimate of r_i is provided by the Stokes-Einstein equation:

$$r_i = \mathrm{k^B} T / 6\pi D_{is}^0 \mu, \tag{3.3.90c}$$

where $\mathrm{k^B}$ is Boltzmann's constant ($= \mathrm{R}/\tilde{\mathrm{N}}$), D_{is}^0 is the diffusivity of solute i in the solvent and μ is the solvent viscosity. (This r_i is the same hydrodynamic radius discussed in Section (3.1.3.2).) This solute radius based on the Stokes-Einstein equation and identified therefore as the *hydrodynamic radius* has been correlated with the molecular weight of globular proteins by (Tanford *et al.*, 1974)

$$\log(r_i \rho_i) = \log M_i - (0.147 \pm 0.041), \tag{3.3.90d}$$

where the recommended value of $(1/\rho_i)$ is 0.74 cm^3/g. For a macromolecule in general, whether it can enter a given pore of certain size or not depends on its radius of gyration r_g, which depends on the molecular weight M_i and the macromolecular shape β_1 via

$$r_g \propto M_i^{\beta_1}. \tag{3.3.90e}$$

For rod-shaped molecules, $\beta_1 = 1$, for spheres, $\beta_1 = (1/3)$, whereas for flexible chains, $\beta_1 = (1/2)$. The radius of gyration is proportional to the *hydrodynamic viscosity based radius* r_h determined from the intrinsic viscosity of the macrosolute via the hydrodynamic volume V_h (Ladisch, 2001, pp. 557-561):

$$r_h = \left(\frac{3\,V_h}{4\pi}\right)^{1/3}; \qquad V_h = \frac{[\eta]\,M_i}{\tilde{\mathrm{N}}\,\Psi_{\mathrm{sh}}}. \tag{3.3.90f}$$

Here $[\eta]$ is the intrinsic viscosity and Ψ_{sh} is a shape factor having the value of 2.5 for spheres; Ψ_{sh} is greater than 2.5 for ellipsoids. For compact globular proteins,

$$r_h = 0.718 M_i{}^{1/3}, \tag{3.3.90g}$$

which is very close to the estimate of r_i by equation (3.3.90b) for $(1/\rho_i) \cong 0.75\,\mathrm{cm^3/g}$. Hagel (1989) has indicated that the estimates of macromolecule/protein dimensions via the hydrodynamic radius, r_i, and the hydrodynamic viscosity based radius, r_h, follow the relation

$$r_h > r_i. \tag{3.3.90h}$$

The porous medium (sorbent, membrane, gel particle) can be made of soft spherical particles that swell quite a bit when immersed in a solvent; they are called *gels*. If they swell in water, they are called *hydrogels*; however, the polymers are crosslinked so that they retain their overall structure. Of course, the polymers are soft and compress under pressure; a typical example would be gels (which are crosslinked) based on agarose (Figure 3.3.5B). Such gel particles are widely used in a variety of separation techniques for separating macromolecules/proteins, etc.

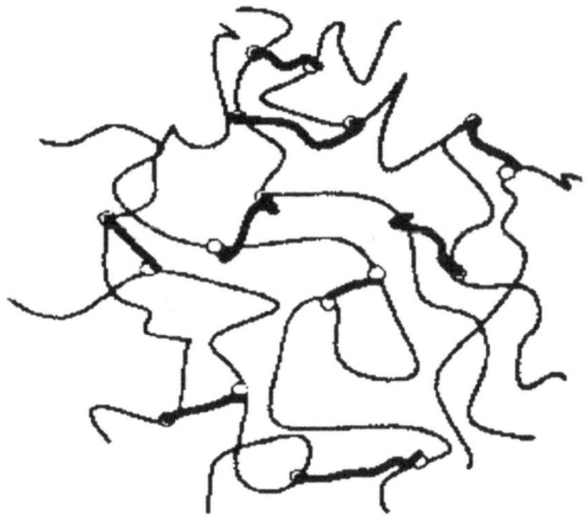

Figure 3.3.5B. Microporous polymer in the form of a gel particle formed by crosslinking linear chains of monomer; crosslinks are shown by heavy lines.

Various empirical expressions have been developed that describe the partition coefficient k_{im} of solute i as a function of the molecular weight M_i or hydrodynamic radius r_i. A few are illustrated in the following:

$$(\kappa_{im})^{1/3} = \mathrm{a} - \mathrm{b}\,(M_i^{1/2}); \qquad \kappa_{im} = \mathrm{a} + \mathrm{b} \ \log M_i;$$
$$\kappa_{im} = \mathrm{a} + \mathrm{b} r_i. \tag{3.3.90i}$$

(For more details, see Hagel (1989); for an introduction in a gel permeation chromatography context, see Ladisch (2001).) If the molecule cannot enter the pore, $k_{im} = 0$. From the relation (3.3.90e), for a given molecular weight M_i, the radius of gyration r_g is the highest for a rod-shaped molecule and lowest for a sphere. Therefore, for a given molecular weight, a rod-shaped molecule has the lowest k_{im} and a spherical molecule the highest k_{im}.

3.3.7.5 *Liquid–solid equilibria: leaching, crystallization, precipitation*

Liquid–solid equilibria in which a liquid and a solid phase (or solid phases) are coexisting can be of the following types:

(1) a solid solution in equilibrium with a molten mixture;
(2) a solute being leached from a solid mixture by a solvent;
(3) crystallization equilibria between a crystal and the mother liquor.

When components 1 and 2 are miscible in the solid phase over the whole composition range between pure 1 and pure 2, we have a *solid solution*. For example, Figure 3.3.6A shows the temperature vs. composition diagram for a naphthalene/β-naphthol system. The lower curve is

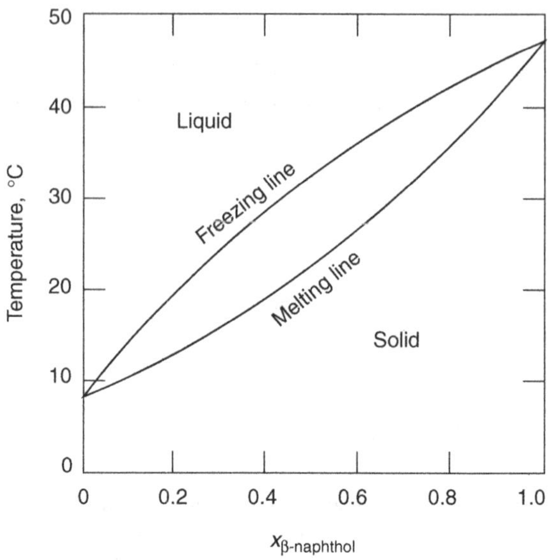

Figure 3.3.6A. Phase diagram for the system naphthalene/β-naphthol. (J. M. Prausnitz, Molecular Thermodynamics of Fluid-Phase Equilibria, *1st edn. © 1969, pp. 404. Reprinted by permission of Pearson Education, Inc., Upper Saddle River, NJ.)*

known as the *solidus* and the upper curve is known as the *liquidus*. The system is completely liquid above the liquidus and is completely solid below the solidus. In between, both liquid and solid phases exist. At any temperature, the solid and liquid phases at equilibrium have different compositions. Not all solid solutions span the whole composition. For example, in iron–copper alloys around 1000 °C, a homogeneous solid solution exists only below a composition of 10 wt% copper (Darken and Gurry, 1953).

To determine the equilibrium ratio of species i between the melt ($j = 1 = l$) and the solid solution ($j = 2 = s$), note that, at equilibrium,

$$\hat{f}_{i1} = \hat{f}_{i2} \Rightarrow \hat{f}_{il} = \hat{f}_{is} \tag{3.3.91}$$

But

$$\hat{f}_{il} = \gamma_{il}\, x_{il} f_{il}^0 \quad \text{and} \quad \hat{f}_{is} = \gamma_{is}\, x_{is} f_{is}^0, \tag{3.3.92}$$

leading to

$$\frac{x_{il}}{x_{is}} = \frac{x_{i1}}{x_{i2}} = K_i = \frac{\gamma_{is} f_{is}^0}{\gamma_{il} f_{il}^0} \cong |_{\text{ideal}} \frac{f_{is}^0}{f_{il}^0}. \tag{3.3.93}$$

The final simplification is valid only for ideal behavior. If experimental data on K_i are not available, the ratio f_{is}^0/f_{il}^0 has to be calculated by procedures that are elaborate and complicated (Prausnitz, 1969). If the behavior is nonideal, the liquidus and solidus behavior shown in Figure 3.3.6A is changed to more complicated ones similar to maximum or minimum boiling azeotropes (see Section 4.1.2).

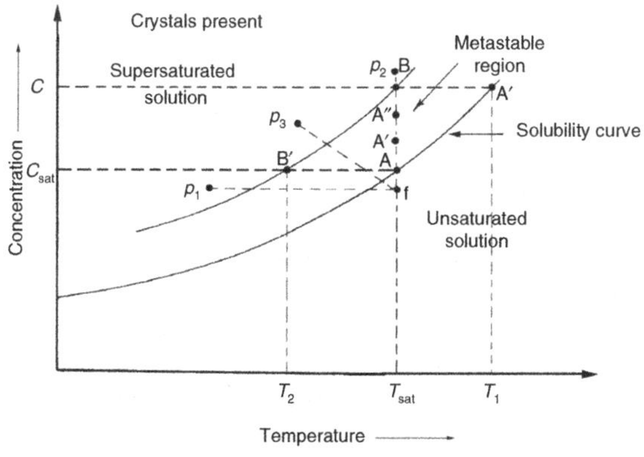

Figure 3.3.6B. Concentration vs. temperature behavior for saturation, supersaturation and crystallization.

Consider now the *leaching* of a solid mixture (which can be treated as a collection of aggregates of pure components) by a solvent. In the solid phase, each component fugacity is then equal to its pure component solid fugacity f_{is}^0. Let the leaching solvent be insoluble in the solid mixture. At equilibrium between the solid mixture and the solvent, for species i,

$$f_{is} = f_{is}^0 = \hat{f}_{il} = \gamma_{il}\, x_{il}\, f_{il}^0. \tag{3.3.94}$$

So

$$x_{il} = f_{is}^0/\gamma_{il}\, f_{il}^0. \tag{3.3.95}$$

If the solution temperature is lower than the melting point of solute i (which is usually the case), and if vapor pressures could be substituted for f_{is}^0 and f_{il}^0, we find

$$x_{il} = \frac{P_{is}^{\text{vap}}}{\gamma_{il}\, P_{il}^{\text{vap}}}. \tag{3.3.96}$$

If the system temperature is the normal melting temperature of the solid, then $f_{is}^0 = f_{il}^0$ and

$$x_{il} = (1/\gamma_{il}). \tag{3.3.97}$$

Refer to Prausnitz *et al.* (1999) for a more comprehensive treatment on standard state fugacity calculations in solid-liquid systems.

At any given temperature and pressure, a solvent has a certain capacity for a solute. When the limit is reached, the solution is said to be *saturated*. This limit generally increases with temperature and provides the solubility curve (Figure 3.3.6B). Conversely, if a solution is cooled to a temperature below the saturation temperature, there would be a tendency for the solute to come out of the solution. The solute may crystallize or precipitate. In reality, *crystallization* does not occur till the solution gets supersaturated to some extent, as shown in Figure 3.3.6B.

So there are three regions of importance in *crystallization* (Miers, 1927): an *unsaturated* region (with no crystals), a *supersaturated* region (with crystals) and an *intermediate* region, where crystal growth will occur if there are crystals, but there will be no nucleation (i.e. crystal formation without any crystals). This intermediate region is also called the *metastable* region. A feed solution represented by f in the unsaturated region can produce crystals if (1) it is cooled to the supersaturated region p_1, (2) the solvent is evaporated to reach the supersaturated region p_2 at constant T, or (3) a combination of cooling and evaporation is implemented to reach p_3. (An additional method employed to attain supersaturation involves the addition of an *antisolvent*). It is now believed that there are a number of other factors such as solution cooling rate, mechanical shock or friction which influence the locations of the curves. True equilibrium is rarely attained in crystallization processes.

Consider point A on the *solubility curve* (Figure 3.3.6B). The saturated solution concentration of solute at temperature T_{sat} is C_{sat}. When this solution loses some solvent at the same temperature T and reaches the curve for the supersaturated solution (point B), the solution concentration becomes $C > C_{sat}$. If the solution were saturated at this higher concentration C, its temperature would have to be higher, namely T_1, corresponding to point A′ on the solubility curve. At temperature T_{sat}, the molar supersaturation ΔC is defined as

$$\Delta C = C - C_{sat}. \tag{3.3.98a}$$

The fractional supersaturation, or relative supersaturation, s, is defined as

$$s = \frac{\Delta C}{C_{sat}} = \frac{C - C_{sat}}{C_{sat}} = \frac{C}{C_{sat}} - 1, \tag{3.3.98b}$$

whereas the supersaturation ratio S is defined as

$$S = \frac{C}{C_{sat}} = 1 + s. \tag{3.3.98c}$$

If instead of removing the solvent at T_{sat}, we lower the solution temperature and reach point B′ at T_2 ($< T_{sat}$), the saturated solution at A will become supercooled; the solution will also become supersaturated for the temperature T_2. The solution concentration remains the same as that of point A, but the supercooling undergone (leading to supersaturation) may be identified by

$$\Delta T = T_{sat} - T_2. \tag{3.3.98d}$$

The fractional supercooling may be defined by

$$\Delta T/T_{sat} = 1 - (T_2/T_{sat}). \tag{3.3.98e}$$

The distance AB′ (or AB) provides an estimate of the metastable zone width. Table 3.3.3 illustrates the metastable zone width for a number of salts in terms of ΔT values for three different cooling rates (Nývlt *et al.*, 1985).

Table 3.3.3. Metastable zone width in crystallization of various salts[a]

Salt	Equilibrium temperature, T_{sat} (°C)	Maximum supercooling prior to nucleation		
		2 °C/hr	5 °C/hr	20 °C/hr
$Ba(NO_3)_2$	30.8	1.65	2.17	3.27
$CuSO_4 \cdot 5H_2O$	33.6	5.37	6.82	9.77
$FeSO_4 \cdot 7H_2O$	30.0	0.89	1.21	1.90
KBr	61.0	1.69	2.41	4.11
KCl	29.8	1.62	1.86	2.30
$NaBr \cdot 2H_2O$	30.6	4.6	6.97	13.08

[a] Data from Nyvlt *et al.* (1985).

Focus now on two points A′ and A″ on the vertical line AB (see Figure 3.36B) at temperature T_{sat}. The solution is obviously less supersaturated at A′ than at A″. If, in fact, we take the solutions at A′ and A″ and allow them to sit without doing anything, namely stirring, etc., at temperature T_{sat}, we will observe that crystals may appear much sooner in the solution A″ having a higher fractional supersaturation. Further, if a crystal is introduced into any of these solutions (e.g. at A′, A″ or B), we will observe that the crystal will grow. To understand these phenomena better, consider the following brief introduction to the thermodynamics of nucleation and growth processes in crystallization. However, first, a brief discussion on different mechanisms of nucleation is useful.

When nucleation takes place without any crystal surfaces, we have *primary* nucleation. Primary nucleation is said to occur by *homogeneous* nucleation when no dissolved impurities are present. When primary nucleation occurs due to the presence of dissolved impurities, we encounter *heterogeneous* nucleation. When nuclei are formed due to the presence of existing macroscopic crystals, interaction with the crystallizer wall, rotary impellers, fluid shear, etc., we have *secondary* nucleation. Mechanisms of secondary nucleation are not sufficiently clear. An introduction to theories on secondary nucleation is provided by Myerson (1993). Here we will focus on homogeneous nucleation. Note that homogeneous nucleation is rarely achieved or desired in practical crystallization (McCabe and Smith, 1976; Myerson, 1993).

In the process of homogeneous nucleation taking place without any mechanical shock, friction, rapid cooling, etc., molecules, atoms or ions join together to form a crystal *nuclei* called a *kinetic unit* (McCabe and Smith, 1976). Individual kinetic units whose volumes are of the order of 10 nm^3 can come together or collide; combinations of a few such kinetic units lead to a *cluster* formation in the manner of a chain reaction (an addition mechanism):

$$\begin{array}{cccccc} A_1 & + & A_1 & \Leftrightarrow & A_2 \\ A_2 & + & A_1 & \Leftrightarrow & A_3 \\ \vdots & & \vdots & & \vdots \\ A_{m-1} & + & A_1 & \Leftrightarrow & A_m \end{array} \quad (3.3.99)$$

The subscript indicates the number of kinetic units in the cluster. With an increase of m, recognizable clusters become *embryos* and a few of them grow enough to be a *nucleus* (for liquid water, the number m is about 76 for a nucleus). Although such nuclei are not in a stable equilibrium (some nuclei become embryos via loss of kinetic units), some are able to gather kinetic units, attain a critical size r_c and become a crystal which will be in stable equilibrium with a saturated solution.

To determine this critical size, r_c, we have to have a situation where the Gibbs free energy change, ΔG, as a function of increase of nucleus radius r is negative. The Gibbs free energy change, ΔG, is needed to bring kinetic units together and expel the solvent; the ΔG change needed to form a nucleus of radius r has two contributions: (1) the energy needed to form the surface of the nucleus, $4\pi r^2 \gamma^{12}$, where γ^{12} is an interfacial tension (dyne/cm) or interfacial energy/area (ergs/cm^2); (2) the energy needed to build the crystal of volume $(4/3)\pi r^3$. This second energy term consists of the product of the molar chemical potential difference $\mu_\infty - \mu_{ij}$ (between the chemical potential of the crystal μ_∞ in equilibrium with a saturated solution and the chemical potential of the solute i in a supersaturated solution) and the number of gmols in the crystal mass of volume $(4/3)\pi r^3$, namely $(4/3)\,\pi r^3/\overline{V}_{ic}$:

$$\Delta G = 4\pi r^2 \gamma^{12} + \left(\mu_\infty - \mu_{ij}\right) \times (4/3)\,\pi\, r^3/\overline{V}_{ic}. \quad (3.3.100a)$$

When thermodynamic equilibrium is reached and the nucleus becomes stable, the nucleus dimension is called the *critical size*, r_c.

To obtain this value r_c achieved when $(d(\Delta G)/dr) = 0$, we get, after differentiation,

$$r_c = (2\gamma^{12}\,\overline{V}_{ic}/(\mu_{ij} - \mu_\infty)). \quad (3.3.100b)$$

Expressing $\mu_{ij} - \mu_\infty$ in terms of fugacities, we obtain

$$\ln\left(\frac{f_{ij}}{f_{ij|\text{sat}}}\right) = \frac{2\gamma^{12}\,\overline{V}_{ic}}{RT\,r_c}. \quad (3.3.100c)$$

Assuming that the activity coefficients of a nonelectrolytic solute i in the two solutions (f_{ij}, supersaturated; $f_{ij|\text{sat}}$, saturated) are nearly equal, we may replace the fugacity ratio by a concentration ratio, i.e. $(f_{ij}/f_{ij}|_{\text{sat}}) = (C_{ij}/C_{\text{sat}})$,

$$\ln\,(f_{ij}/f_{ij}|_{\text{sat}}) = \ln\,(C_{ij}/C_{\text{sat}})$$
$$= \ln\,(S) = \ln\,(1+s) = \frac{2\gamma^{12}\overline{V}_{ic}}{RT\,r_c}. \quad (3.3.101)$$

Beyond this critical size r_c, where ΔG achieves a critical value,

$$\Delta G_{\text{crit}} = 4\pi r_c^2 \gamma^{12} - (8/3)\gamma^{12} r_c^2 = (4/3)\pi r_c^2 \gamma^{12}, \quad (3.3.102)$$

the value of ΔG starts decreasing as r increases beyond r_c. Thus the crystal nucleus has become stable and the addition of kinetic units decreases the free energy, making it a spontaneous process. These results are useful for understanding the nucleus formation process. When the supersaturation ratio S increases, the critical size r_c decreases, and the solution becomes less stable since it can sustain smaller and smaller nuclei. The result (3.3.101) has another dimension. Since f_{ij} or C_{ij} refers to the supersaturated solution, one can argue also that the smaller the crystal size, the higher its solubility compared to the value C_{sat} for a large crystal or a flat surface, an example of the Kelvin equation (3.3.53) already studied. This leads to dissolution of smaller particles and growth of larger particles, a phenomenon called *Ostwald ripening*. For an introduction to these subjects, see Mullin (1972), McCabe and Smith (1976), Randolph and Larson (1988) and Myerson (1993).

The phenomenon of supersaturation illustrated in Figure 3.3.6B was created either by solvent evaporation or decreasing the temperature. In general, supersaturation may be developed via the following routes: (1) solvent evaporation; (2) temperature decrease (or increase in some cases); (3) addition of antisolvent; (4) precipitation via chemical reaction. We have already considered the results of solvent evaporation and temperature decrease in Figure 3.3.6B. For systems where solubility decreases with decreasing temperature (as in Figure 3.3.6B), evaporation-led cooling by applying vacuum (pathway fp_3) is also practiced with or without some heat addition. In case such steps are not practical, an alternative strategy involves the addition of another solvent into the solution, which drastically decreases the solute solubility in the mixed solvent environment. The solvent added is usually miscible, and this process of miscible nonsolvent addition is practiced quite often both for organic/inorganic solutes as well as amino acids.

There are many systems where the solubility-temperature curve is essentially flat. If one can carry out a chemical reaction between the solute and another added agent, one of the products may precipitate. For example, Na_2SO_4 in solution reacts with $CaCl_2$ in solution (both having essentially flat concentration vs. temperature behavior) to produce insoluble $CaSO_4$ (which has a low solubility and which precipitates) and NaCl (which remains in solution and which has a flat concentration vs. temperature profile):

$$\begin{aligned} &Na_2SO_4(\text{solution}) + CaCl_2(\text{solution}) \\ &\Leftrightarrow 2NaCl(\text{solution}) + CaSO_4(\text{precipitate}). \quad (3.3.103) \end{aligned}$$

For many inorganic compounds ($BaSO_4 \cdot H_2O$; silver halides; calcium phosphates; alumina trihydrate; titanium dioxide, etc.) considered for and produced by precipitation processes, excellent introductions are available in Nyvlt *et al.* (1985) as well as in Estrin (1993).

A major aspect of precipitation, however, does not involve chemical reactions as such. Instead, they are akin to the addition of an antisolvent in that somehow the solubility of the solute is reduced. Specifically, we focus on *proteins* and *macrosolutes*, where addition of neutral salts or an acid or base to an aqueous solution causes precipitation of the proteins or other macrosolutes. The reduction in solubility of the protein by the addition of neutral salts is generally due to a *hydrophobic effect* (see Section 4.1.9.4). A protein molecule in an aqueous solution has a *hydration layer* around it in the local domain (see Section 3.3.5). Such a layer prevents the hydrophobic patches on a protein molecule's surface (which consists of hydrophilic as well as hydrophobic patches; for example, nonpolar amino acids such as alanine, methionine, tryptophan and phenylalanine, if present in a protein, will provide hydrophobic patches with polar amino acids creating hydrophilic regions) from interacting with each other and forming aggregates leading to precipitation. Addition of neutral salt molecules into water (for example, $(NH_4)_2SO_4$) leads to the spontaneous withdrawal of water molecules from the protein's local domain to hydration of individual salt molecules. This leads to aggregation of protein molecules (via hydrophobic interaction) and their precipitation (salting-out effect).

The protein solubility C_2 (in g/liter) decreases as a logarithmic function of the solution's ionic strength (Cohn and Ferry, 1943):

$$\log C_2 = \log C_2|_0 - b\,C_3. \quad (3.3.104)$$

Here $C_2|_0$ is the hypothetical/extrapolated solubility at zero ionic strength (corresponding to $C_3 = 0$) and C_3 is the ionic strength of salt in gmol/liter ($= 1/2\sum Z_i^2 C_i^2 = I$, where C_i includes cationic and anionic species). There are some salts that may somewhat increase the solubility (salting-in effect). Both of these effects are schematically illustrated in Figure 3.3.6C. The logarithmic decrease in the protein solubility as a function of the ionic strength is an important basis for protein separation via precipitation (for an introduction, see Ladisch (2001)).

3.3.7.6 *Interfacial adsorption systems*

The previous three types of liquid-solid equilibria considered equilibrium between two bulk phases; we will now briefly look at two types of *interfacial adsorption* systems where the two bulk phases are either fluid-fluid or fluid-solid. Consider first the interfacial equilibrium relation for a *nonelectrolytic surface active solute* in an air-water system (a fluid-fluid system). If the surface active solute i is such that the interfacial tension γ^{12} decreases linearly with the surfactant concentration C_{i1} as

$$\gamma^{12} = a_1 - b_1 C_{i1} \qquad (3.3.105)$$

(a_1 and b_1 are constants in this linear relation) below the critical micelle concentration (CMC), then it is possible to get an expression for $(\mathrm{d}\gamma^{12}/\mathrm{d}\mu_{11})$ in relation (3.3.39b) and relate it to $\Gamma^{\mathrm{E}}_{1\sigma}$. In the present case, $i = 1$ is the solute and $\mathrm{d}\mu_{11} = RT\,\mathrm{d}\,\ell\mathrm{n}\,a_{11} \cong RT\,\mathrm{d}\,\ell\mathrm{n}\,C_{11}$. Further, $\mathrm{d}\gamma^{12} = -b_1 \mathrm{d}C_{11}$. Therefore

$$\Gamma^{E}_{1\sigma} = \frac{b_1}{RT}\frac{\mathrm{d}C_{11}}{\mathrm{d}\,\ell\mathrm{n}\,C_{11}} = \frac{b_1 C_{11}}{RT}, \qquad (3.3.106)$$

a linear equilibrium relation between the surface excess concentration of solute 1 and its bulk concentration C_{11} in water. The proportionality constant (b_1/RT) can be determined from the dependence of γ^{12} on the bulk solute concentration C_{11}.

Figure 3.3.7(a) shows the general behavior of γ^{12} with the surfactant solute concentration C_{11} in the bulk of water ($j = 1$). Beyond the CMC, γ^{12} is almost constant with respect to C_{11}; thus $\mathrm{d}\gamma^{12}=0$. Therefore a change in bulk concentration of species 1 no longer changes the excess concentration of solute 1 in the interface, which is saturated with a monolayer of surfactant molecules. The relation between $\Gamma^{\mathrm{E}}_{1\sigma}$ and C_{11} over the whole range of C_{11} is then of the type shown in Figure 3.3.7(b). It is of the *Langmuir adsorption* isotherm type (see equations (3.3.112a,b)).

We focus now on fluid–solid systems. This type of interfacial adsorption system is encountered, for example, in *gas–solid adsorption*. The solids on which the gas is adsorbed is called an *adsorbent* and the gas species adsorbed is called the *adsorbate*. More often than not, the adsorbents used in practice are porous solid particles; these particle types and their uses are identified in Table 7.1.1. The porosity values of such particles are high, in the range 0.4–0.85. The particle sizes can vary from quite small (e.g. 0.1 mm) to large (e.g. 2.54 cm). The surface area of these pores is very high and can range easily from 100–1000 m^2/g of adsorbent. The pore sizes can vary from

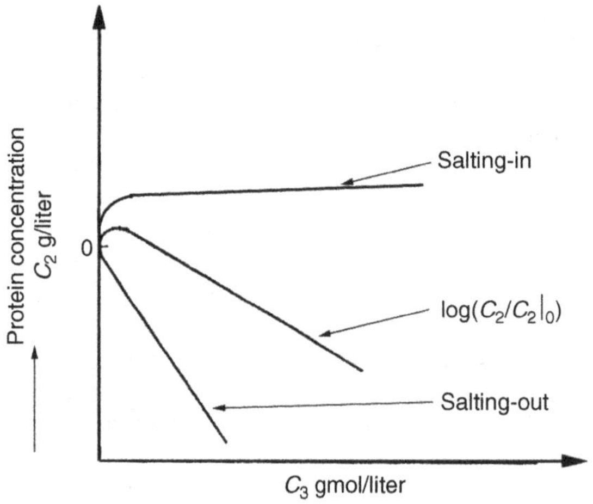

Figure 3.3.6C. Solubility of a protein as a function of salt concentration due to the salting-out effect. (After Ladisch (2001).)

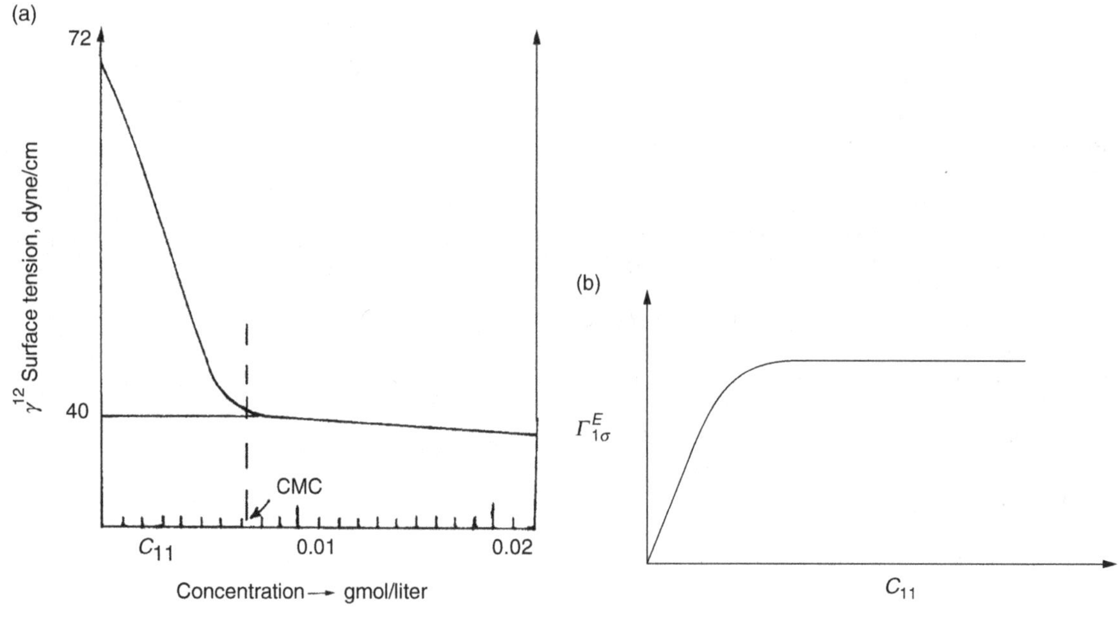

Figure 3.3.7. (a) Surface tension vs. concentration of a surfactant, sodium lauryl sulfate, in water. (b) Surface adsorption isotherm in a gas–liquid system.

large macropores of diameter > 50 nm to micropores < 2 nm, to pores of molecular dimensions (0.3–1 nm and above) in zeolites, which are crystalline aluminosilicates. The pores are invariably tortuous. Gas/vapor molecules are adsorbed on the pore surfaces generally by *physical adsorption*, involving relatively weak intermolecular forces, and rarely via *chemisorption*, involving bond formation. The physical adsorption due to these weak forces is assumed to take place on "sites" strewn throughout the pore surface area.

To understand the functional nature of gas–solid equilibria for physical adsorption processes, one may develop a relation between $\Gamma_{i\sigma}$ and the gas phase composition using equation (3.3.43b) and the illustrations thereafter. Two alternate approaches will be illustrated here using the same basic equation (3.3.43b) and a binary system of species $i = 1,2$. Recognize that S_σ is the total interfacial area per unit mass of adsorbent and therefore may be expressed as $\left(\sum_{i=1}^{2} q_{i\sigma}\right)\overline{S}_\sigma$, where $\overline{S}_\sigma$ is the molar surface area. Therefore (Van Ness, 1969)

$$\overline{S}_\sigma \,\mathrm{d}\pi = \sum_{i=1}^{2} \frac{q_{i\sigma}}{\sum q_{i\sigma}} \mathrm{d}\mu_{ig} = \sum_{i=1}^{2} x_{i\sigma}\,\mathrm{d}\mu_{ig}. \tag{3.3.107}$$

For ideal gas behavior in the gas phase,

$$\overline{S}_\sigma \,\mathrm{d}\pi = \sum_{i=1}^{2} x_{i\sigma} RT\,\mathrm{d}\ln x_{ig}P;$$

$$\frac{\overline{S}_\sigma}{RT}\mathrm{d}\pi = \frac{S_\sigma\,\mathrm{d}\pi}{\sum_{i=1}^{2} q_{i\sigma} RT} = \mathrm{d}\ln P + \frac{x_{1\sigma}\,\mathrm{d}x_{1g}}{x_{1g}} + \frac{x_{2\sigma}\,\mathrm{d}x_{2g}}{x_{2g}}, \tag{3.3.108}$$

since $x_{1\sigma} + x_{2\sigma} = 1$. Further, $x_{1g} + x_{2g} = 1$. It follows, therefore,

$$\frac{S_\sigma\,\mathrm{d}\pi}{RT\sum_{i=1}^{2} q_{i\sigma}} = \mathrm{d}\ln P + \frac{x_{1\sigma} - x_{1g}}{x_{1g}(1 - x_{1g})}\mathrm{d}x_{1g}. \tag{3.3.109}$$

For an ideal gas, this is a form of Gibbs adsorption isotherm for a binary mixture: it relates the species 1 mole fraction in gas phase, x_{1g}, to species 1 mole fraction, $x_{1\sigma}$, in the adsorbed phase, the adsorbate (Van Ness, 1969). At constant total gas pressure, we obtain

$$\left[\frac{\partial(\pi S_\sigma/RT)}{\partial x_{1g}}\right]_P = \left(\sum_{i=1}^{2} q_{i\sigma}\right) \frac{x_{1\sigma} - x_{1g}}{x_{1g}(1 - x_{1g})}. \tag{3.3.110a}$$

At constant gas phase composition, x_{1g}, we get

$$\left[\frac{\partial(\pi S_\sigma/RT)}{\partial P}\right]_{x_{1g}} = \left(\sum_{i=1}^{2} q_{i\sigma}\right)\Big/P. \tag{3.3.110b}$$

These two first-order partial differential equations can be solved numerically to express $x_{1\sigma}$ and $(\pi S_\sigma/RT)$ as a function of total pressure P and gas phase composition x_{1g} for observed data of $\sum_{i=1}^{2} q_{i\sigma}$. The experimental and numerical procedures are described in Myers *et al.* (1982).

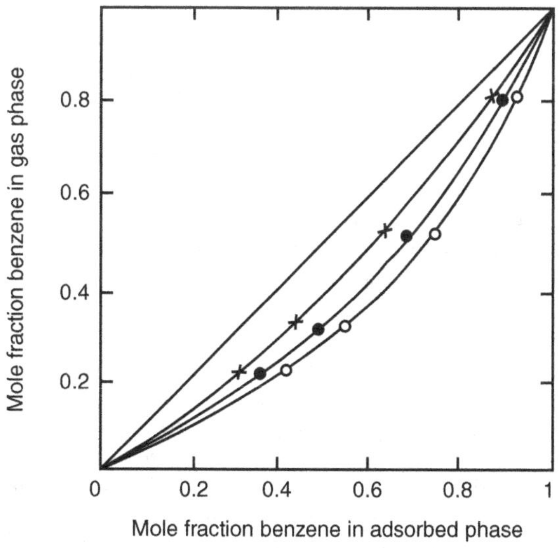

Figure 3.3.8A. Vapor-adsorbed phase equilibria of benzene and cyclohexane on activated charcoal at 30 °C. ×: P = 0.27 kPa; •: P = 1.33 kPa; o: P = 13.3 kPa. Reprinted with permission from A. L. Myers et al., *AIChE J., 28(1), 97 (1982). Copyright © [1982] American Institute of Chemical Engineers (AIChE).*

The adsorption behavior of benzene in a benzene-cyclohexane mixture for the adsorbent, activated charcoal, is shown for a particular temperature in Figure 3.3.8A. Note that $\sum_{i=1}^{2} q_{i\sigma}$, i.e. the total number of moles adsorbed on the solid adsorbent surface per unit adsorbent mass, is obtained from the experimental data of mass of gas adsorbed per mass of adsorbent using

$$\sum_{i=1}^{2} q_{i\sigma} = \frac{\text{total mass of gas adsorbed/mass of adsorbent}}{M_1 x_{1\sigma} + M_2 x_{2\sigma}}. \tag{3.3.110c}$$

For real gas mixtures, Ritter and Yang (1989) have developed simple numerical and graphical procedures to determine adsorbed phase compositions at elevated pressures.

An alternative procedure suggested by Myers and Prausnitz (1965) based on the molar Gibbs free energy of mixing provides

$$Px_{ig} = P_i^0(\pi)\,\gamma_{i\sigma}\,x_{i\sigma} \tag{3.3.111a}$$

as the relation between the gas phase mole fraction, x_{ig}, and the adsorbate phase mole fraction, $x_{i\sigma}$, for a total gas pressure P. Here $P_i^0(\pi)$ is the equilibrium gas phase pressure for pure i adsorption at spreading pressure π (which should be the same for the mixture). Further, the gas phase was assumed to behave ideally. If the adsorbate phase also behaves ideally, $\gamma_{i\sigma} = 1$. Under this condition,

$$x_{ig} = \frac{P_i^0(\pi)\, x_{i\sigma}}{P}; \qquad (3.3.111b)$$

i.e., the equilibrium behavior is similar to that of Raoult's law in vapor-liquid equilibrium (equation (3.3.64)) if $P_i^0(\pi)$ can be thought of as the pure adsorbate vapor pressure for component i at temperature T and spreading pressure π of the mixture.

The procedures for obtaining the *gas-solid adsorption isotherms* briefly outlined above are based on a solution thermodynamic Gibbs approach. There are two other approaches: the potential theory and the Langmuir approach. The latter approach is based on a dynamic equilibrium between the rates of adsorption and desorption of any species from adsorption sites on the solid surface. Since most data correlation in separation processes employs the Langmuir approach (Yang, 1987), with the adsorbate amount expressed as a function of species partial pressure, such isotherm types will be briefly identified here. For a comprehensive introduction to adsorption isotherms and phenomena, consult Ruthven (1984) and Yang (1987).

Consider a few types of *isotherms* observed in practice (Figure 3.3.8B); the first one is the common *Langmuir type*. For such an isotherm, initially the amount adsorbed is linearly proportional to the gas pressure P. After further increases in gas pressure, the amount adsorbed approaches a limit, which does not change with further increases in gas pressure. Under these conditions, all adsorption sites become covered with gas molecules, the adsorbate, in a monolayer fashion. The fraction of sites occupied, or the fraction of the maximum gas that can be adsorbed under other conditions, is given, for a *pure* gas i, by

$$\theta = \frac{q_{i\sigma}}{(q_{i\sigma})_{\max}} = \frac{b_{il} P}{1 + b_{il} P}, \qquad (3.3.112a)$$

where $(1/b_{il})$ is the Langmuir constant for species i. For a gas mixture of n species, θ_t is the total fraction of sites covered by all species and is equal to $\sum_{i=1}^{n} \theta_i$, where θ_i is the fraction of sites covered by species i:

$$\theta_i = \frac{b_{il}\, p_{ig}}{1 + \sum_{i=1}^{n} b_{il}\, p_{ig}}. \qquad (3.3.112b)$$

An additional isotherm type is the *Freundlich isotherm*. For a pure gas (species i), the Freundlich isotherm has a power law relation

$$\theta = b_{if}\, P^{1/\beta_i}, \qquad (3.3.112c)$$

where $\beta_i < 1$, $= 1$ or > 1. Figure 3.3.8B illustrates the case corresponding to $\beta_i < 1$. A log-log plot of the data satisfying this type of isotherm will yield from the slope $(1/\beta_i)$ and b_{if} from the intercept.

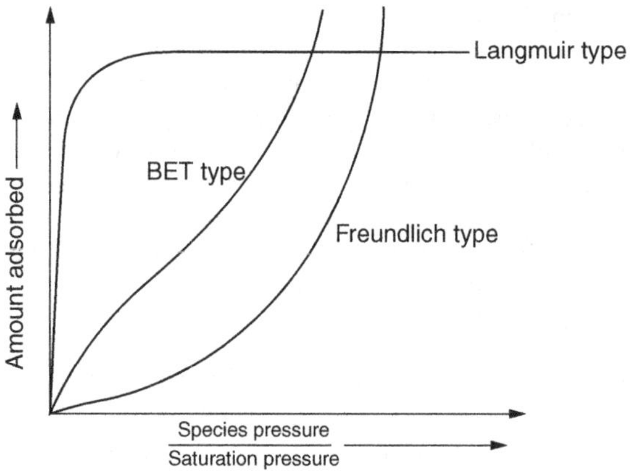

Figure 3.3.8B. Typical gas-solid adsorption isotherms.

For a mixture of n species,

$$\theta_i = \frac{b_{if}\, {p_{ig}}^{1/\beta_i}}{1 + \sum_{i=1}^{n} b_{if}\, {p_{ig}}^{1/\beta_i}}. \qquad (3.3.112d)$$

The similarities between the Langmuir gas-solid isotherm for a pure gas or a mixture and the Langmuir component of gas-membrane equilibrium in equation (3.3.81) or (3.3.82a) should be obvious.

The BET (Brunauer-Emmett-Teller)-type isotherm in Figure 3.3.8B reflects multilayer adsorption of the adsorbate. After a monolayer adsorbate coverage is achieved in the adsorbent pores, additional molecular layers are formed on top of the adsorbed monolayer by condensation of vapors. In adsorbents having small-diameter pores, multilayer condensation of the adsorbate vapor can fill the pore completely with the liquid adsorbate. This phenomenon is called *capillary condensation* (Figure 3.3.8C). Consider the curved interface between the vapor phase and the condensed phase of species i in the micropore in this figure. The vapor pressure of the condensed liquid above the concave curved liquid surface in the capillary $P^{\text{sat}}_{i,\text{curved}}$ is less than that over a plane condensed liquid surface $P^{\text{sat}}_{i,\text{Pl}}$:

$$(P^{\text{sat}}_{i,\text{curved}}/P^{\text{sat}}_{i,\text{Pl}}) = \exp\left(-\frac{2\gamma^{12}\, \overline{V}_{ij} \cos\theta}{R\,Tr}\right) \qquad (3.3.112e)$$

(see also relations (3.3.50)–(3.3.53)). This equation is also identified as the *Kelvin equation* (compare (3.3.54)). As a result, *capillary condensation* or *pore condensation* in a fine capillary or pore can take place at a lower value of equilibrium vapor pressure than the saturation vapor pressure $P^{\text{sat}}_{i,\text{Pl}}$ at that temperature (so far we have used this quantity as P^{sat}_i everywhere). The magnitude of this effect may be illustrated for pore condensation of benzene (Ruthven, 1984) (where γ^{12} for benzene $= 29$ dyne/cm; $\overline{V}_{ij} = \overline{V}_{\text{benzene}} = 89\ \text{cm}^3/\text{gmol}$; $T = 293$ K; $\theta \approx 0$) for two pore sizes, $r = 5$ nm and 60 nm: the $(P^{\text{sat}}_{i,\text{curved}}/P^{\text{sat}}_{i,\text{Pl}})$ value is 0.67 and 0.96, respectively, for 5 and 60 nm pore size. The

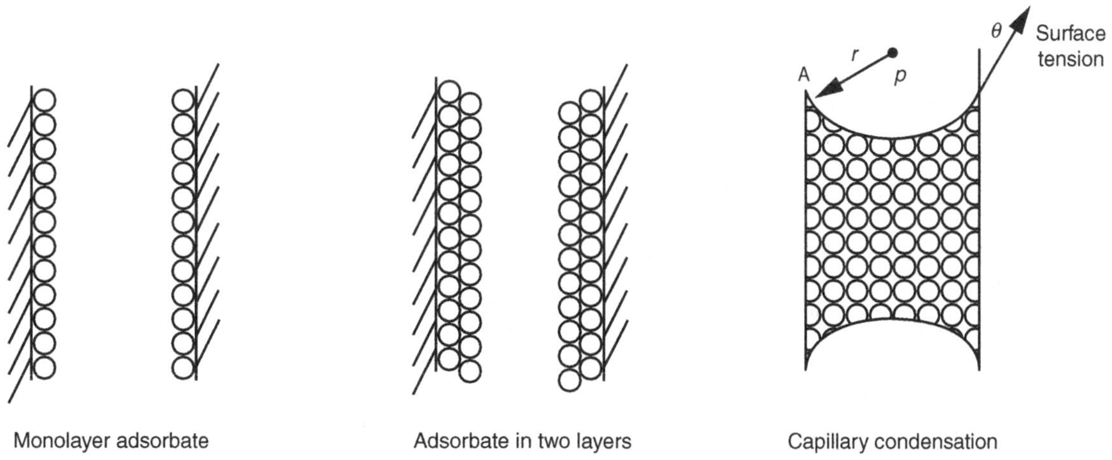

Figure 3.3.8C. Capillary condensation of adsorbate in the adsorbent micropore.

Kelvin effect becomes important for fine capillaries/small pores.

Another important case of a fluid–solid adsorption system is *liquid–solid adsorption*. The general principles of adsorption equilibrium between a vapor and an adsorbent are also applicable to liquid–solid adsorption, except the adsorbate concentrations are often quite high for a liquid phase unlike that in gas–solid adsorption (unless trace concentrations of solute are being adsorbed). Adsorption equilibrium may be viewed here in two different ways. In the more conventional approach, both solute in a solution as well as the solvent molecules get adsorbed on the solid adsorbent. Thus the exchange equilibrium is between a solute molecule and a solvent molecule in both phases. An *i*th species solute molecule in a solution ($j = 2$) displaces a solvent molecule (s) adsorbed on any site in the adsorbent ($j = 1$) and occupies the adsorbent site:

$$i(j=2) + s(j=1) \Leftrightarrow i(j=1) + s(j=2). \qquad (3.3.113a)$$

If the equilibrium constant for this exchange is K, and if the activities of solute i and solvent s molecules in the adsorbed phase are assumed ideal (i.e. $a_{i1} \cong x_{i1}$ and $a_{s1} \cong x_{s1}$), then

$$K = \frac{a_{i1} a_{s2}}{a_{i2} a_{s1}} = \frac{x_{i1} a_{s2}}{a_{i2} x_{s1}}. \qquad (3.3.113b)$$

Further, in the adsorbed phase, $x_{il} + x_{s1} = 1$. Therefore

$$x_{i1} = \left(\frac{K}{a_{s2}}\right) a_{i2} x_{s1} = \left(\frac{K}{a_{s2}}\right) a_{i2} (1 - x_{i1}) \Rightarrow x_{i1} = \frac{\left(\frac{K}{a_{s2}}\right) a_{i2}}{1 + \left(\frac{K}{a_{s2}}\right) a_{i2}} = \theta. \qquad (3.3.113c)$$

Under ideal conditions of almost equal sized solute and solvent molecules and fixed surface area per adsorption site on the adsorbent, $x_{i1}$is equivalent to the fraction θ of the sites occupied by the solute molecules. This isotherm relating x_{i1} to a_{i2} (or x_{i2}) is the equivalent of the Langmuir adsorption isotherm (3.3.112b) for liquid–solid adsorption.

Often this isotherm is also expressed in terms of the molar solute concentration in the liquid phase, C_{i2}, and the moles of solute i adsorbed per unit mass of adsorbent, q_{il}:

$$q_{i1} = q_{i1}|_{\max} \frac{e_i C_{i2}}{1 + e_i C_{i2}}, \qquad (3.3.113d)$$

where $q_{i1}|_{\max}$ is the maximum value of q_{i1} for complete coverage of all sites by the solute species i and e_i is a constant for solute i. Sometimes, the isotherm is similar to the Freundlich isotherm (3.3.112d) and is represented as

$$\theta = e_i \left(C_{i2}\right)^{1/\beta_i}. \qquad (3.3.113e)$$

The graphical method for determining parameters of this isotherm has been described in the text following (3.3.112c).

In an alternative approach, the adsorbent surface is assumed to have a distinct layer of solvent (in a monolayer); the solute i merely partitions between the solvent in solution and the solvent in the monolayer on the adsorbent:

$$i(j=2) \Leftrightarrow i(j=1). \qquad (3.3.114a)$$

If the solute i is such that an infinite dilution standard state is necessary, then, from relation (3.3.79), we get

$$\frac{C_{i1}}{C_{i2}} = \kappa_{i1}.$$

This relation, where C_{i1} and C_{i2} have the units of gmol/cm^3 of adsorbent and gmol/cm^3 of bulk solution, may be expressed in terms of q_{i1}, gmol/mass of adsorbent, and C_{i2} as

$$\frac{q_{i1}}{C_{i2}} = \kappa_{il} \hat{V}_a, \tag{3.3.114b}$$

where $\hat{V}_a$ is the volume of the adsorbed monolayer phase per unit weight of adsorbent (cm^3/mass of adsorbent).

When a *hydrophobic porous adsorbent* is in contact with an aqueous solution which does not wet the adsorbent pores, a hybrid system is formed. Inside the pores containing air, there are solutes in vapor form which are adsorbed on the surfaces in the pore of the adsorbent (Rixey, 1987). Outside the pores, the aqueous solution exists. Only solutes that are volatile are adsorbed on the adsorbent. Solutes that would not be separated by conventional adsorption in the wetted state may be fractionated now. A liquid–solid adsorption system becomes a de facto vapor–solid adsorption system.

3.3.7.7 *Ion exchange systems*

Ion exchangers are porous insoluble solid materials having fixed charges.[27] When immersed in an electrolytic solution, ions having charges opposite to the fixed charges, *counterions*, will enter the porous structure; the ion exchanger as a whole is electrically neutral. If such an ion exchange material in particulate form is immersed in another solution containing a different set or type of counterions, there will be an exchange of the counterions already present in the solution in the ion exchanger's porous structure with those present in the external solution; hence, the characterization of such systems as ion exchange systems. The process is reversible. Porous natural and synthetic insoluble solid materials can act as ion exchangers. These include natural ion exchange minerals (primarily of the alumino-silicate type), synthetic inorganic ion exchangers (e.g. zeolites, zirconium phosphates, etc.), ion exchange coals having weak carboxylic acids and, most importantly, ion exchange resins prepared from organic polymers.

Synthetic ion exchange resin particles consist of a three-dimensional crosslinked porous polymeric structure having a high porosity and pore dimensions exceeding 10 nm. The basic polymeric material in such resins is hydrophobic. Hydrophilic ionic groups (see Figure 3.3.9 for an illustration of such groups) are present at a high density throughout this polymeric network. Although the introduction of such ionic groups, e.g. $-SO_3^-H^+$, makes the polymer soluble in water, crosslinking of different polymer chains in the structure makes them insoluble in water. A common polymer is linear polystyrene; divinyl benzene is employed to prepare a crosslinked polymeric matrix. Resins having fixed negative group, e.g. $-SO_3^-$, $-COO^-$, etc., can exchange cations from an external solution. These

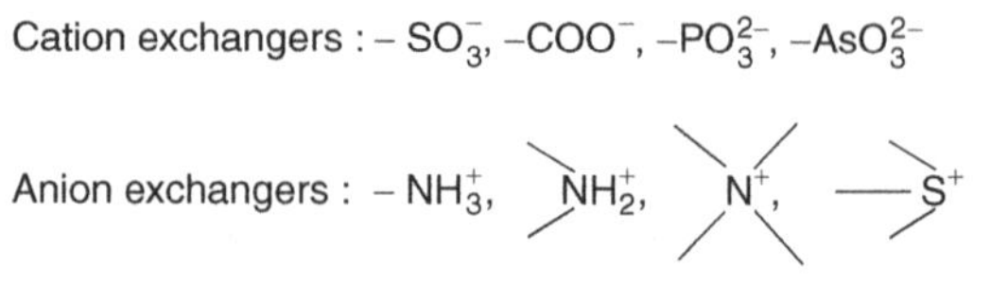

Figure 3.3.9. Ionic groups in the polymeric matrix of ion exchange resins.

are called *cation exchange resins*; they are *acidic*. Resins having fixed positive charges, e.g. $-NH_3^+$, $-N^+(CH_3)_3$, etc., exchange anions from an external solution. They are called *anion exchange resins*; they are *basic* in nature. Consult Helfferich (1995) for a detailed account.

The governing criterion for equilibrium distribution of a solute between an ion exchange resin and an external solution is relation (3.3.8) when the solute is a weak electrolyte or a nonelectrolyte. The nature of the equilibrium is similar to that with nonionic adsorbents. When the solute is a strong electrolyte, criterion (3.3.29) has to be used due to the presence of fixed ionic charges and counterions in the resin.

The distribution equilibrium of a weak electrolyte or a nonelectrolytic solute i between an external solution (w) (generally an aqueous solution) and the ion exchange resin particles ($j = R$) is usually described by a distribution coefficient:

$$\kappa_{iR} = \frac{C_{iR}}{C_{iw}}. \tag{3.3.115}$$

Often, the concentration C_{iR} is described in terms of moles of species i per unit weight of ion exchange resin particle or per unit weight of solvent in the resin particle. In order to make κ_{iR} dimensionless, C_{iw} should have similar units. Sometimes κ_{iR} is constant; for example, acetic acid has a linear isotherm in styrene-type cation exchangers (Helfferich, 1962, p. 126). In general, there are a variety of interactions between the solute and the resin system, leading to a complex behavior.

Distribution equilibria of strong electrolytes between a solution and ion exchange resin will now be considered. First, however, certain unusual aspects of ion exchange systems have to be illustrated. Specifically, the phenomenon of *swelling pressure* in a resin particle is important. When water or a polar solvent enters ion exchange resin particle pores, it has a tendency to dissolve the resin material. The resin swells due to the incorporation of solvent molecules. The resin particle is, however, prevented from dissolution in the solvent due to crosslinks between the neighboring polymer chains. This essentially implies the presence of elastic forces of resin matrix – the whole resin phase including the solvent is now at a higher pressure – higher than that of the surrounding solution. The pore liquid pressure P_R, the pressure in the resin

[27] Liquid ion exchangers are not considered here; see Section 5.2.2.4.

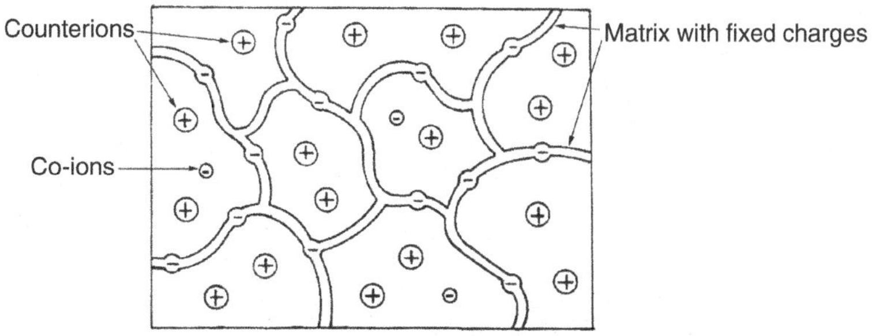

Figure 3.3.10. Structure of an ion exchange resin (schematic). (After Helfferich (1962, 1995).)

phase, can be related to the external solution (say aqueous) pressure P_w using criterion (3.3.16) for equilibrium.

Suppose the external solution is just pure solvent ($i = s$). Then

$$\mu_{sR} = \mu_{sw}, \tag{3.3.116a}$$

where $j = R$ is the resin phase and $j = w$ is the external phase. From expression (3.3.21) for the chemical potential $\mu_{ij}(P_j, T, x_{ij})$, where $j = R$, $i = s$, we have

$$\mu_{sR}(P_R, T, x_{sR}) = \mu_s^0 + \overline{V}_{sR}(P_R - P^0) + RT\ \ell n\, a_{sR}. \tag{3.3.116b}$$

Similarly,

$$\mu_{sw}(P_w, T, x_{sw}) = \mu_s^0 + \overline{V}_{sw}(P_w - P^0) + RT\ \ell n\, a_{sw}. \tag{3.3.116c}$$

Assume standard state pressure P^0 and external solution pressure P_w to be the same. Since there is pure solvent in the external solution, $a_{sw} = 1$. Therefore

$$\mu_{sR}(P_R, T, x_{sR}) = \mu_s^0 + \overline{V}_{sR}(P_R - P^0) + RT\ \ell n\, a_{sR} = \mu_{sw}(P^0, T, 1)$$
$$= \mu_s^0 + \overline{V}_{sw} \times 0 + 0 = \mu_s^0,$$

so that

$$\overline{V}_{sR}(P_R - P^0) = -RT\ \ell n\, a_{sR}, \tag{3.3.117a}$$

where $(P_R - P^0)$ is the swelling pressure in the resin phase and is related to the solvent activity a_{sR} in the resin phase. If $a_{sw} \neq 1$ and $P_w \neq P^0$, then, for $\overline{V}_{sw} = \overline{V}_{sR}$,

$$\overline{V}_{sw}(P_R - P_w) = -RT\ \ell n\,(a_{sR}/a_{sw}). \tag{3.3.117b}$$

It is now possible to develop the distribution equilibrium relation for a strongly electrolytic solute i in an ion exchange system using equilibrium criterion (3.3.29):

$$\mu_{iw} - \mu_{iR} = Z_i \mathcal{F}(\phi_R - \phi_w). \tag{3.3.118a}$$

We know from expression (3.3.21) for μ_{ij},

$$\mu_{iw} = \mu_i^0 + \overline{V}_{iw}(P_w - P^0) + RT\ \ell n\, a_{iw}$$

and

$$\mu_{iR} = \mu_i^0 + \overline{V}_{iR}(P_R - P^0) + RT\ \ell n\, a_{iR}.$$

Substitute these into relation (3.3.118a) to get

$$\phi_R - \phi_w = \frac{1}{Z_i \mathcal{F}}\left[RT\ \ell n\left(\frac{a_{iw}}{a_{iR}}\right) - \overline{V}_i(P_R - P_w)\right], \tag{3.3.118b}$$

using the assumption that the two solute partial molal volumes $\overline{V}_{iw}$ and $\overline{V}_{iR}$ are equal to $\overline{V}_i$. This equilibrium relation is valid for any ionic species i; note that $(P_R - P_w)$ is the swelling pressure of the resin and $(\phi_R - \phi_w)$ is the *Donnan potential*, independent of the species i in a given system.

An important feature of relation (3.3.118b) needs to be reemphasized: it is valid for any ionic species i. The electrolyte or solute i in general is not just one ionic species. Thus the distribution of a strong electrolytic solute AY (which produces A^+ and Y^- ions) between the solution and the resin needs to be known. Let the fixed charge in the resin be negative (Figure 3.3.10). Then A^+ is a counterion. Suppose one mole of electrolyte AY dissociates to give v_A moles of ions A^+ and v_y moles of ions Y^-; we can follow Helfferich's (1962) treatment:

$$Z_A \upsilon_A = -Z_Y \upsilon_Y. \tag{3.3.119a}$$

The partial molar volume of the electrolyte AY is

$$\overline{V}_{AY} = \upsilon_A \overline{V}_A + \upsilon_Y \overline{V}_Y. \tag{3.3.119b}$$

Use relation (3.3.118b) separately for $i = A$ and $i = Y$ and simplify to get

$$RT\ \ell n\left[\left(\frac{a_{Aw}}{a_{AR}}\right)^{\upsilon_A}\left(\frac{a_{Yw}}{a_{YR}}\right)^{\upsilon_Y}\right] = (P_R - P_W)\overline{V}_{AY}. \tag{3.3.119c}$$

Replace the single cation and single anion activities by the mean electrolyte activity $a_\pm$ as follows:

$$a_A^{\upsilon_A} a_A^{\upsilon_Y} = (a_\pm)^{\upsilon},$$

where $\upsilon = \upsilon_A + \upsilon_Y$.

Now use result (3.3.117b) to obtain

$$\left(\frac{a_{\pm R}}{a_{\pm w}}\right)^{\upsilon} = \left(\frac{a_{sR}}{a_{sw}}\right)^{\overline{V}_{AY}/\overline{V}_{sw}}, \tag{3.3.119d}$$

which relates the electrolyte activity in the external solution to that in the resin. Such a relation for the electrolyte activity in the two phases of an ion exchange system can be reduced to an explicit relation between the molalities of the co-ion Y between the two phases.

Consider a cation exchange resin with fixed negative charges - it prefers to exchange only cations from the solution. To demonstrate this preference, rewrite relation (3.3.119d) for the electrolyte AY as

$$a_{AR}^{\upsilon_A}\, a_{YR}^{\upsilon_Y} = a_{Aw}^{\upsilon_A}\, a_{Yw}^{\upsilon_Y}\, (a_{sR}/a_{sw})^{\overline{V}_{AY}/\overline{V}_{sw}}. \quad (3.3.120a)$$

Assume $\upsilon_A = 1, \upsilon_Y = 1$ (for example, NaCl). Rewriting the activities in terms of molalities and activity coefficients, we obtain

$$m_{A,R}\, m_{Y,R} = m_{A,w}\, m_{Y,w}\, (\gamma_{Aw}\, \gamma_{Yw}/\, \gamma_{AR}\, \gamma_{YR})(a_{sR}/a_{sw})^{\overline{V}_{AY}/\overline{V}_{sw}}. \quad (3.3.120b)$$

Now employ electroneutrality relation (3.1.108a) in terms of molalities for the resin phase and the solution phase:

$$\text{solution phase}: \; m_{A,w} = m_{Y,w};$$
$$\text{resin phase}: \; m_{A,R} = m_{Y,R} + m_{F,R}, \quad (3.3.120c)$$

where $m_{F,R}$ is the molality[28] of the fixed negative charges in the resin phase. Relation (3.3.120b) can now be rearranged via the electroneutrality relations to yield

$$m_{Y,R}(m_{Y,R} + m_{F,R}) = m_{Y,w}^2 \underbrace{(\gamma_{Aw}\gamma_{Yw}/\gamma_{AR}\gamma_{YR})\,(a_{sR}/a_{sw})^{\overline{V}_{AY}/\overline{V}_{sw}}}_{b}; \quad (3.3.120d)$$

$$m_{Y,R} = \frac{b\, m_{Y,w}^2}{m_{Y,R} + m_{F,R}} \cong \frac{m_{Y,w}^2}{m_{F,R}} b, \quad (3.3.120e)$$

where we have assumed that the fixed molal charge density in the resin, $m_{F,R}$, is much larger than $m_{Y,R}$; it can be easily in the range of 5 molal. Relation (3.3.120e) suggests that, as long as the external solution is dilute ($m_{Y,w} << 1$), the co-ion concentration $m_{Y,R}$ in the resin will be very small. For example, assuming $b = 1$, if $m_{Y,w} \cong 0.001 - 0.1$ molality, then, for $m_{F,R} = 5$ molality, $m_{Y,R}$ ranges between 0.2×10^{-6} and 2×10^{-3}. Therefore, for dilute external solutions, it is clear that (Helfferich, 1962, 1995), at equilibrium,

$$[\text{molality of Y}]_{\text{resin}} \propto [\text{molality of Y}]^2_{\substack{\text{external}\\ \text{solution}}}; \quad (3.3.120f)$$

i.e. the co-ion concentration in the resin will indeed be very low. This behavior is essential to the success of ion exchange processes with dilute external solutions; correspondingly, the ion exchange process is most often applied to dilute solutions. Dependence of the proportionality factors in relation (3.3.120f) to solution concentration, however, complicates the above relationship.

What is the maximum extent of ion exchange of a counterion with the ion exchange resin? We will briefly

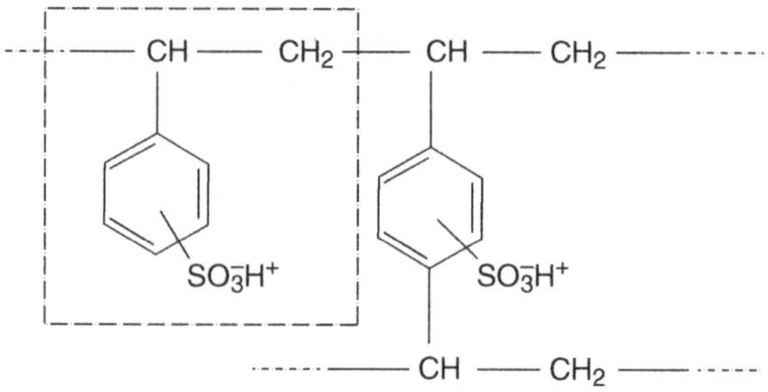

Figure 3.3.11. The basic unit of a cation exchange resin made of crosslinked polystyrene with sulfonic acid groups.

illustrate this aspect following Helfferich (1995). Consider crosslinked polystyrene with sulfonic acid groups, whose basic unit is identified within the dashed lines in Figure 3.3.11. The formula weight of the basic unit $C_8H_8SO_3$ (without the counterion H^+) has the value 184.2; it can carry one counterion of unit charge (here H^+). Therefore the maximum ion exchange capacity of this resin is q_{iR}^m (where subscript R stands for the resin phase), $(1000/184.2) = 5.43$ meq/g. Since the resin is crosslinked using small amounts of divinyl benzene (which is also sulfonated, equivalent weight 210.2) and ethylstyrene (equivalent weight 212.2), the ion exchange capacity (maximum theoretical) will be somewhat less. If such a resin is present in a packed bed, having a void volume fraction of ε, and the water-swollen density of the resin particles is ρ_s, having a fractional water content of ϕ_w, then the volumetric ion exchange capacity of the packed bed, $V_{\text{cap}}^{\text{iex}}$ in equivalents per liter of packed bed volume

$$V_{\text{cap}}^{\text{iex}} = (1-\varepsilon)\rho_s(1-\phi_w)q_{iR}^m. \quad (3.3.120g)$$

Following Helfferich (1995), consider a bed of sulfonated polystyrene resin beads having $\rho_s = 1.25$ g/cm^3, $\phi_w = 0.468$, $\varepsilon = 0.4$. The value of $V_{\text{cap}}^{\text{iex}}$ is

$$V_{\text{cap}}^{\text{iex}} = (1-0.4)1.25 \times (1-0.468)\,5.4\, \frac{\text{eq}}{\text{liter of bed}}$$
$$= 2.2\text{eq/liter of bed}.$$

When the counterion in the ion exchanger is different from the counterion in the external solution, there will be an exchange between the two counterions. For an ion exchanger with negative fixed charges (indicated by R^-, say,), an exchange between cation B^{Z_B} (say) in the external solution and cation A^{Z_A} in the resin (Figure 3.3.12) is indicated by the following ion exchange process:

$$|Z_B|\,(A^{Z_A} + |Z_A|R^-) + |Z_A|B^{Z_B}(w) \Leftrightarrow$$
$$|Z_A|\,(|Z_B|R^- + B^{Z_B}) + |Z_B|A^{Z_A}(w). \quad (3.3.121a)$$

An example of such a *cation exchange* is

[28] Molality means gmol of solute per 1000 g of solvent.

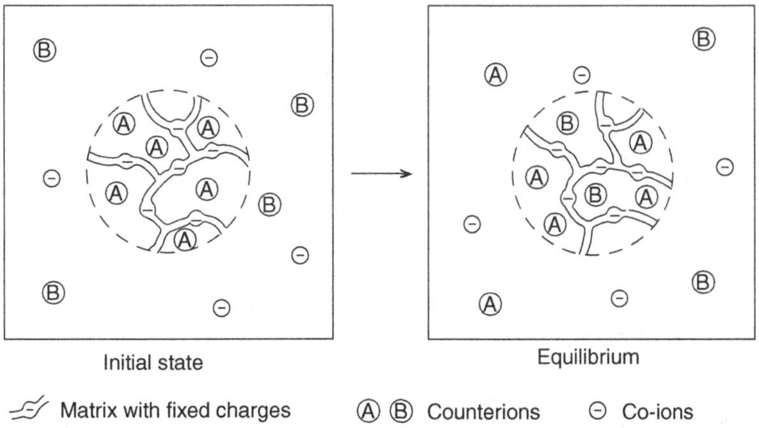

Figure 3.3.12. Ion exchange with a solution (schematic). A cation exchanger containing counterions A is placed in a solution containing counterions B (left). The counterions are redistributed by diffusion until equilibrium is attained (right). (After Helfferich (1962, 1995).)

$$2\,\mathrm{NaR} + \mathrm{CaCl_2(aq)} \Leftrightarrow \mathrm{CaR_2} + 2\,\mathrm{NaCl(aq)}. \qquad (3.3.121b)$$

When instead of $CaCl_2$ in the external solution, there is, say $CaCO_3$, $MgCO_3$, etc., or corresponding sulfate salts, the exchange reactions are called water-softening exchange reactions; such ion exchange processes also remove low levels of iron and manganese sometimes present in the domestic water supply.

For an ion exchanger with positive fixed charges (R^+, say), an exchange between anion B^{Z_B} in the external solution and anion A^{Z_A} in the resin is similarly expressed as

$$|Z_B|\,(\,|Z_A|R^+ + A^{Z_A}) + |Z_A|B^{Z_B} \Leftrightarrow |Z_A|\,(B^{Z_B} + |Z_B|R^+) + |Z_B|A^{Z_A} \qquad (3.3.121c)$$

An example of such an *anion exchange* is

$$2\mathrm{RCl} + \mathrm{Na_2\,SO_4(aq)} \Leftrightarrow \mathrm{R_2SO_4} + 2\mathrm{NaCl(aq)}. \qquad (3.3.121d)$$

The ion exchange equilibrium constant K for such an ion exchange "reaction" may be written using the *law of mass action* as follows:

$$K = \frac{(a_{\mathrm{NaCl}})^2\,(a_{\mathrm{R_2SO_4}})}{(a_{\mathrm{Na_2SO_4}})(a_{\mathrm{RCl}})^2}. \qquad (3.3.121e)$$

We now illustrate the nature of the ion exchange isotherm using the simple cation exchange of Na^+ in solution with the resin in H^+ form in molar concentration units:

$$\mathrm{Na^+(w)} + \mathrm{R^-H^+(resin)} = \mathrm{H^+(w)} + \mathrm{R^-Na^+(resin)}. \qquad (3.3.121f)$$

If the maximum ion exchange capacity of the resin for monovalent ions is C_{FC} (where the subscript FC refers to fixed charge),

$$C_{\mathrm{H^+R}} + C_{\mathrm{Na^+R}} = C_{\mathrm{FC}}. \qquad (3.3.121g)$$

Correspondingly, in the solution,

$$C_{\mathrm{H^+}w} + C_{\mathrm{Na^+}w} = C_{tw}, \qquad (3.3.121h)$$

where C_{tw} is the total molar concentration of H^+ and Na^+ in the aqueous solution. The equilibrium constant for the ion exchange process for ideal solution behavior in both phases is given by

$$K = \frac{C_{\mathrm{Na^+R}}\,C_{\mathrm{H^+}w}}{C_{\mathrm{Na^+}w}\,C_{\mathrm{H^+R}}}. \qquad (3.3.121i)$$

Using the three equations (3.3.121g–i), we can obtain the equilibrium distribution of the counterion Na^+ between the resin and the external solution as

$$C_{\mathrm{Na^+R}} = \frac{K\,C_{FC}\,C_{\mathrm{Na^+}w}}{(C_{tw} + C_{\mathrm{Na^+}w}(K-1))}. \qquad (3.3.121j)$$

To convert this relation to the concentration unit of moles of Na^+/mass of the solid particle (swollen gel), $q_{\mathrm{Na^+R}}$, note that, if ρ_R is the resin mass/volume,

$$C_{\mathrm{Na^+R}} = q_{\mathrm{Na^+}R}\,\rho_R; \quad C_{FC} = q_{\mathrm{max,R}}\,\rho_R; \qquad (3.3.121k)$$

$$q_{\mathrm{Na^+R}} = q_{\mathrm{max},R}\,\frac{K\,C_{\mathrm{Na^+}w}}{(C_{tw} + C_{\mathrm{Na^+}w}(K-1))}, \qquad (3.3.121l)$$

where $q_{\mathrm{max},R}$ is the maximum molar fixed charge density per unit resin mass. Therefore

$$C_{\mathrm{H^+R}} + C_{\mathrm{Na^+R}} = C_{FC} = q_{\mathrm{max},R}\,\rho_R. \qquad (3.3.121m)$$

The equilibrium isotherm relation (3.3.121l) is similar to the Langmuir adsorption isotherm (3.3.112a). In fact, an alternative representation will make them essentially identical (see Figure 3.3.13):

$$\frac{C_{\mathrm{Na^+R}}}{C_{FC}} = \frac{K\,C_{\mathrm{Na^+}w}/C_{tw}}{(1+(C_{\mathrm{Na^+}w}/C_{tw})(K-1))};$$
$$\frac{q_{\mathrm{Na^+R}}}{q_{\max,R}} = \frac{K\,C_{\mathrm{Na^+}w}/C_{tw}}{(1+(C_{\mathrm{Na^+}w}/C_{tw})(K-1))}. \tag{3.3.121n}$$

An equilibrium is achieved in these exchanges after some time. Generally, one of the counterions A^{Z_A} or B^{Z_B} will be preferred by the ion exchanger. This subject of ion exchanger preference will be treated in Section 4.1.6, where expressions for equilibrium separation factor and selectivity between two species A and B will be obtained. Some materials can exchange both cations and anions and are therefore called *amphoteric ion exchangers*. Alternatively a *mixture* of cation exchange resin particles and anion exchange resin particles may be used in exchangeable H^+ and OH^- forms, respectively, where there is no addition to the dissolved solids concentration in the solution. (Otherwise NaCl, for example, will be added via (3.3.121b, d)). Such a strategy is used for water demineralization, e.g. in treating boiler feed water.

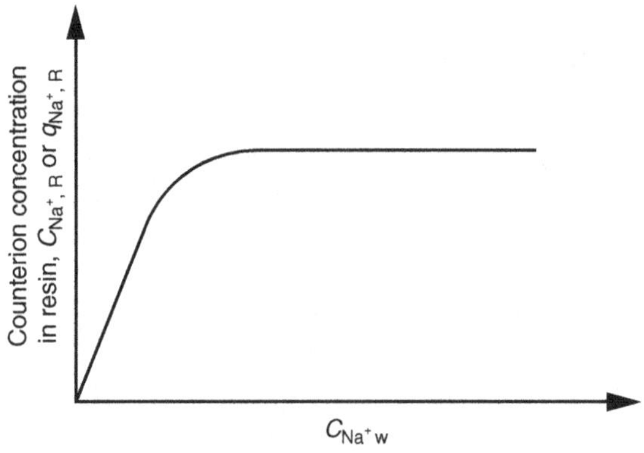

Figure 3.3.13. Langmuir adsorption isotherm type behavior for Na^+ exchange with a cation exchange resin.

Ion exchange processes (3.3.1121a,c) have been written to resemble chemical reactions, but they are *not* chemical reactions; rather, they are redistributions of ions by diffusion within an environment influenced by electrostatic forces. The heat evolved in such ion exchange processes is also very low when compared with chemical reactions or even many adsorption processes (Helfferich, 1962, p. 8).

3.3.7.7.1 Steric mass-action (SMA) ion exchange equilibrium for proteins The treatment of ion exchange equilibrium so far covered, de facto, smaller molecules. Ion exchangers are also frequently used for adsorption/desorption of larger macromolecules, such as proteins, in the presence or absence of various salts (see Section 4.1.6, Figure 4.1.16 and Section 4.1.9.4). Our earlier introduction of adsorption (equations (3.3.112a) and (3.3.113a)) involved an adsorbate molecule with respect to an adsorbent site. A charged macromolecule, e.g. protein, however, has many charged patches over its surface; therefore, adsorption/ion exchange of a protein molecule with the ion exchange resin bead involves multipoint adsorption/binding. However, a large macromolecule, such as a protein, will cover many more adsorbent sites than those dictated by their characteristic charge and the location of the charges (Figure 3.3.14).

Correspondingly, these sites covered by the macromolecule/protein without any charge based interaction will have counterions from the microsolutes in the system, e.g. salts. Such salt counterions bound to the resin particle face steric hindrance and are unavailable for exchange with other macrosolutes/proteins in free solution. Brooks and Cramer (1992) have illustrated a *steric mass-action* (SMA) *ion-exchange equilibrium formalism* to describe the equilibrium ion exchange behavior of a single protein and exchangeable salt counterions with respect to an ion exchange resin. Their model assumes: ideal solution behavior; monovalent salt counterions; the use of concentration C instead of activities (in units of mM) and the law of mass

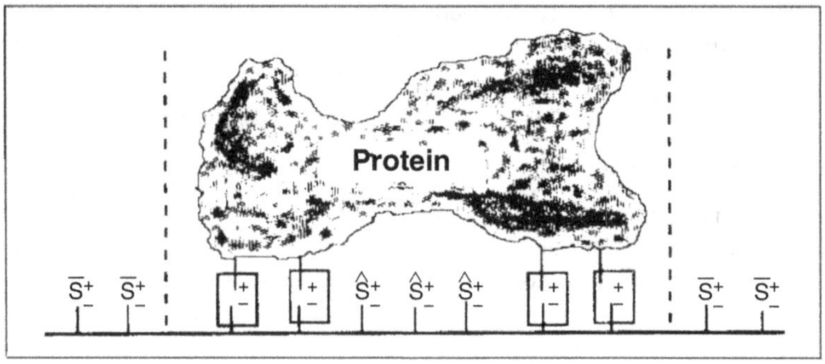

*Figure 3.3.14. Idealized protein binding on an ion exchange bed. $\hat{S}^+$: sterically hindered salt counterions; $\bar{S}^+$: nonsterically hindered salt counterions. Reprinted, with permission, from C. A. Brooks and S. M. Cramer, AIChE J., **38**(12), 1969 (1992). Copyright © [1992] American Institute of Chemical Engineers (AIChE).*

action-based equilibrium; neglect of any effect of co-ions in the ion exchange process; neglect of nonidealities so that equilibrium parameters are constant and independent of macrosolute and salt counterion concentrations.

The stoichiometric exchange of the protein molecule ($i = p$) and the exchangeable salt counterions ($i = s$) may be represented as follows:

$$C_{pw} + \left(\frac{|Z_p|}{|Z_s|}\right) C_{sR} \Leftrightarrow C_{pR} + \left(\frac{|Z_p|}{|Z_s|}\right) C_{sw}. \qquad (3.3.122a)$$

Here $(|Z_p|/|Z_s|)$ represents the ratio of the absolute values of the characteristic charges of the protein and the counterion (n-valent salt); therefore it represents the number of sites with which a protein molecule interacts. Since we have monovalent salt counterions here ($|Z_s| = 1$), Z_p, the characteristic charge of the protein, also provides the number of sites $|Z_p|$ with which it interacts. The equilibrium constant for this binding in the ion exchange process, K_{ps}, may be defined by (since $|Z_s| = 1$)

$$K_{ps} = \left(\frac{C_{pR}}{C_{pw}}\right) \left(\frac{C_{sw}}{C_{sR}}\right)^{|Z_p|}. \qquad (3.3.122b)$$

The total concentration of sterically hindered salt ions unavailable for exchange with protein molecules in free solution is given by

$$\hat{C}_{sp} = \sigma_p \, C_{pR}, \qquad (3.3.122c)$$

where σ_p is a steric factor for the protein. After protein adsorption, the total salt concentration in the stationary (ion exchange) phase, C_{sR}^t is given by

$$C_{sR}^t = C_{sR} + \sigma_p \, C_{pR}. \qquad (3.3.122d)$$

The condition of electroneutrality in the resin phase requires

$$C_{FC} = C_{sR} + \left(|Z_p| + \sigma_p\right) C_{pR}, \qquad (3.3.122e)$$

where C_{FC} represents the concentration of the total fixed charge in the ion exchange resin based on monovalent counterions; it is also the ion exchange capacity (i.e. $m_{F,R}$ in molal units). Substituting equation (3.3.122e) into equation (3.3.122b) and rearranging leads to

$$C_{pw} = \frac{C_{pR}}{K_{ps}} \left(\frac{C_{sw}}{C_{FC} - \left(|Z_p| + \sigma_p\right) C_{pR}}\right)^{|Z_p|}. \qquad (3.3.122f)$$

If the aqueous phase protein concentration C_{pw} is known along with the aqueous phase salt concentration C_{sw}, then this equation defines implicitly the protein concentration, C_{pR}, in the resin phase. Once C_{pR} is known, one can calculate C_{sR} from equation (3.3.122e) given C_{FC}, $|Z_p|$ and σ_p; correspondingly, C_{sR}^t becomes available from (3.3.122d).

Some limiting forms of the isotherm (3.3.122f) for a single protein are useful. Consider linear conditions such that as $C_{pw} \to 0$, $C_{pR} \to 0$. Then we get, from isotherm (3.3.122f),

$$\lim_{C_{pw} \to 0} C_{pR} = K_{ps} (C_{FC}/C_{sw})^{|Z_p|} C_{pw}. \qquad (3.3.122g)$$

The slope of this isotherm decreases drastically as the aqueous phase salt concentration, C_{sw}, increases. As C_{pw} increases to high values, C_{sR} tends to zero and the isotherm approaches the limit from equation (3.3.122e):

$$\lim_{C_{pw} \to \infty} C_{pR} = \frac{C_{FC}}{|Z_p| + \sigma_p} = C_{pR}^{\max}. \qquad (3.3.122h)$$

This value, $C_{pR}^{\max}$, represents the saturation capacity of protein sorption by ion exchange; it appears to be independent of the concentration of the monovalent counterion salt. In reality, such a limit is achieved only at low values of C_{sw}. Unlike other ion exchange systems, where $C_{iR}^{\max}$ for any counterion i is simply obtained from C_{FC} and $|Z_i|$, for proteins/biomolecules, the steric factor σ_p introduces an additional complexity. For illustrations of the isotherm (3.3.122f), see Brooks and Cramer (1992), which gives a number of proteins/biomolecules, e.g. cytochrome-c, α-chymotrypsinogen.

3.3.7.8 Distribution equilibrium representation as a chemical reaction

It is possible to represent the distribution of a species i between phases 2 and 1 by a form usually reserved for chemical reactions (Rony, 1969a):

$$i\,(\text{phase } 2) \Leftrightarrow (i\,(\text{phase } 1). \qquad (3.3.123)$$

Since the condition for chemical equilibrium (not chemical reaction equilibrium) in this case is given by $\mu_{i1} = \mu_{i2}$, the physical reaction (3.3.123) may also be represented by

$$\sum_{j=1} \upsilon_{ij} \, \mu_{ij} = 0, \qquad (3.3.124)$$

where, by convention, $|\upsilon_{ij}| = 1$ and $\upsilon_{i1} = 1$, $\upsilon_{i2} = -1$. The quantities $\upsilon_{ij} - s$ are somewhat analogous to stoichiometric numbers in chemical reaction equilibria. The equilibrium constant for this physical reaction, K, is then given by

$$-RT \,\ell n\, K = \sum \upsilon_{ij} \, \mu_{ij}^0 = \mu_{il}^0 - \mu_{i2}^0 = -RT \,\ell n\, K_i^a,$$

where

$$K_i^a = \frac{a_{i1}}{a_{i2}}. \qquad (3.3.125)$$

Such an approach is consistent with what has been already presented in relations (3.3.76), (3.3.113a), (3.3.114a) and (3.3.121a-d).

3.3.7.9 Equilibria in a system with supercritical fluid

In separation equilibrium involving a *supercritical fluid* (SCF), the two systems commonly used are solid–SCF and liquid–SCF. A solute or solutes are distributed between the two immiscible phases. A supercritical fluid is normally

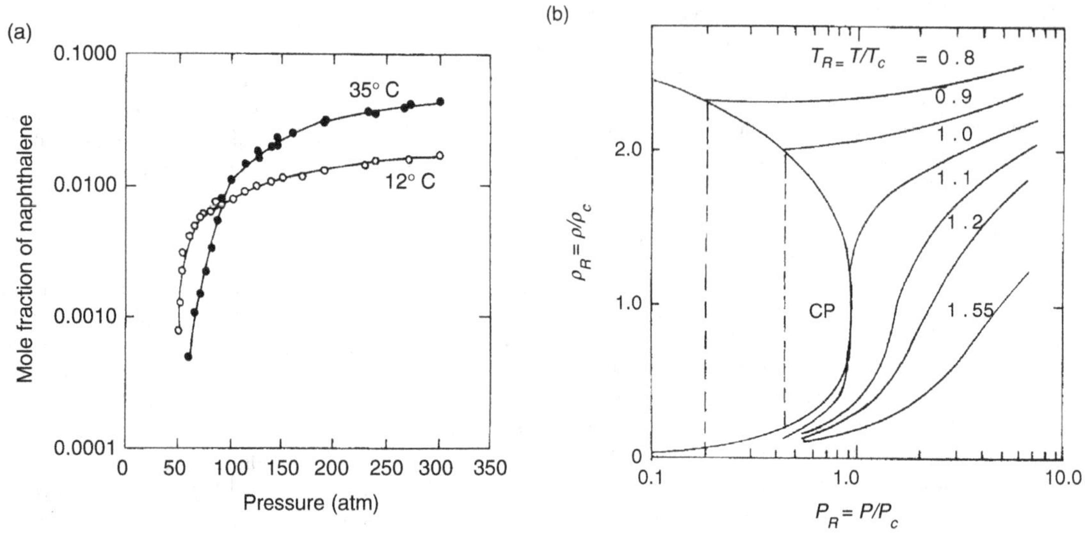

Figure 3.3.15 (a) Solubility behavior of solid naphthalene in supercritical ethylene. (b) Variation of reduced density (ρ_R) of a pure component near its critical point (CP). These figures were published on pp. 5 and 6 of Supercritical Fluid Extraction: Principles and Practice, *M. A. McHugh and V. J. Krukonis, Butterworths-Heinemann (507 pp.). Copyright Elsevier (1986).*

employed to extract a solute from a solid or liquid phase. The solute is recovered later by converting the supercritical fluid into a subcritical fluid (usually a gas), which has much reduced solute solubility. The subcritical fluid is recompressed to the SCF state and recycled for solute extraction. A SCF is therefore often called a *SCF solvent.*

A pure fluid becomes a supercritical fluid when its pressure and temperature exceed its critical pressure (P_c) and critical temperature (T_c). For example, the value of P_c, T_c for the commonly used SCF CO_2 are 72.8 atm and 31.2 °C; the corresponding values for ethylene and propane, sometimes used, are 49.7 atm, 9.4 °C and 41.9 atm, 96.8 °C, respectively. On the other hand, the values of P_c and T_c for water are 217.6 atm and 374 °C; due to such high values of P_c and T_c, water is almost never used as a SCF for separation.

To use a SCF as a solvent for extracting solutes, the solubility of any solute in a SCF must be substantial. It is generally observed that the capacity of a SCF to dissolve a solute is directly related to the density of the SCF. For example, consider the solubility characteristics of solid naphthalene in supercritical ethylene shown in Figure 3.3.15(a) (McHugh and Krukonis, 1986). The solubility of solid naphthalene increases by more than an order of magnitude from a very low value when the ethylene pressure exceeds the $P_c = 49.7$ atm of ethylene (at $T > T_c = 9.4$ °C, of ethylene). The variation of the density of ethylene near its critical point may be seen in Figure 3.3.15(b), a plot of reduced density, $\rho_R = (\rho/\rho_c)$ against reduced pressure $P_R = (P/P_c)$ and reduced temperature $T_R = (T/T_c)$. The 12 °C ($T_R \cong 1.01$) isotherm of ρ_R vs. P_R shows a rapid rise in ρ by more than an order of magnitude around $P_R = 1.0$. The shape of this curve is quite similar to that of the naphthalene solubility curve, suggesting that the solvent power of ethylene for naphthalene is directly related to the SCF solvent density near T_c, P_c.

To develop an estimate of the solute mole fraction x_{il} in the SCF solvent in equilibrium with a solid mixture, the equilibrium relation is

$$\hat{f}_{il} = \hat{f}_{is},$$

where

$$\hat{f}_{il} = x_{il}\,\hat{\Phi}_{il}\,P \quad \text{and} \quad \hat{f}_{is} = x_{is}\,\hat{\Phi}_{is}\,P. \tag{3.3.126}$$

The solid mixture is considered to be an agglomeration of pure species, with each component having its pure component solid fugacity f_{is}, i.e. $\hat{f}_{is} = f_{is}$. At low pressures, this fugacity is almost equal to the sublimation pressure $P_i^{sub}(T)$. However, at the high pressures characteristic of SCF extraction,

$$f_{is} = P_i^{sub}(T)\,\Phi_i^{sub}(T, P_i^{sub})\exp\left[\int_{P_i^{sub}}^{P}\left(\frac{V_{is}}{RT}\right)dP\right], \tag{3.3.127}$$

where Φ_i^{sub} is the fugacity coefficient at T and P_i^{sub}, and the exponential term is the Poynting pressure correction for the fugacity of pure solid having a molar volume V_{is}. Therefore, the mole fraction of solute i in the SCF solvent is

$$x_{il} = \frac{P_i^{\text{sub}}(T)\,\Phi_i^{\text{sub}}(T, P_i^{\text{sub}}) \exp\left[\int_{P_i^{\text{sub}}}^{P} \left(\frac{V_{is}}{RT}\right) dP\right]}{\hat{\Phi}_{il}\,P}. \tag{3.3.128}$$

A review of the literature and thermodynamic calculation procedures for such systems are available in McHugh and Krukonis (1986). The same reference may be studied for thermodynamic equilibrium calculations for the solute distribution between a liquid and a supercritical fluid solvent.

3.3.8 Particle distribution between two immiscible phases

So far, we have been concerned with the distribution of molecules, ions or macromolecules between two immiscible phases. The molecules may have been solutes present in small quantities or major constituents of either or both phases. Classical principles of thermodynamics were used to develop estimates of such solute distributions. When it comes to large particles, such as ore fines, cells or other particulate matter, classical thermodynamics may not appear to be of any use. However, using the phenomenon of wetting based on interfacial thermodynamics, particle separation from one phase is achieved by introducing a second immiscible phase.

There are two types of systems here depending on whether the two immiscible phases are gas-liquid or liquid-liquid. Particle separation in a gas-liquid system is much more common and is called *flotation*. The following three paragraphs will focus on the basic thermodynamic principles in such a system.

Consider two types of mineral particles in an aqueous suspension. If air bubbles can be attached to one type of particle only, the latter will float to the surface of the suspension, due to reduced density, and can be separated from the other type of particles. Normally, mineral particles are wetted completely by water so that air bubbles cannot attach to them. However, if the particle surface can be made sufficiently hydrophobic to prevent wetting, air bubble attachment is possible.

The criterion for air bubble attachment is that the free energy change should be negative:

$$\Delta G = \gamma^{SG} - (\gamma^{SL} + \gamma^{LG}) \tag{3.3.129}$$

where γ^{SG}, γ^{SL} and γ^{LG} are the interfacial tensions for the systems solid-gas, solid-liquid and liquid-gas, respectively. The general relation between these interfacial tensions and a contact angle θ for a gas-liquid-solid system with a flat solid surface (shown in Figure 3.3.16) is given by Young-Dupre' equation:

$$\gamma^{SG} = \gamma^{SL} + \gamma^{LG} \cos\theta \tag{3.3.130}$$

Substituting this into expression (3.3.129), we get

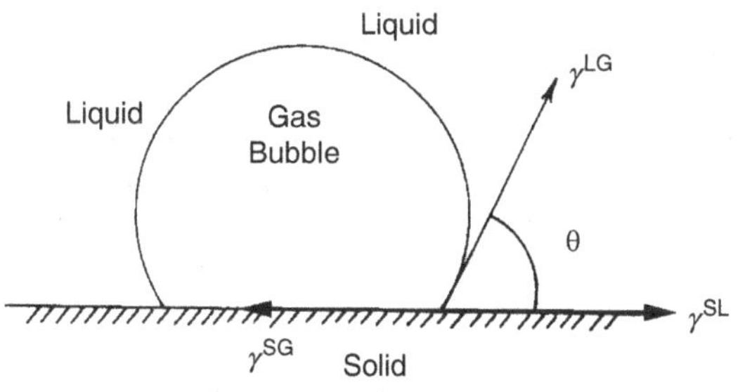

Figure 3.3.16. Interfacial tensions and contact angle for a gas bubble-liquid-solid system with a flat solid surface.

$$\Delta G = \gamma^{LG}(\cos\theta - 1). \tag{3.3.131}$$

If the particle is wetted by the liquid, the contact angle θ should be zero since the liquid spreads over the solid surface completely. Thus, air or a gas bubble cannot become attached to the particle surface. If, however, θ is finite, ΔG is negative; there is now a gas-solid interface, and bubble attachment to the particle is possible.

To create a nonzero contact angle in a gas-liquid-solid particle system where the liquid normally wets the particle (e.g. water and most mineral matter), surface active species called *collectors* are dispersed in the liquid for adsorption on particle surfaces. In aqueous systems, addition of surfactants such as long-chain fatty acids, and their adsorption on particle surfaces, make the surface hydrophobic enough, leading to a nonzero θ. Equation (3.3.40b) may be utilized here to study how the various interfacial tensions are changed by changes in the bulk concentrations of the added surfactants (Fuerstenau and Healy, 1972).

Instead of a gas or air, it is possible to use a second immiscible liquid, usually water or an aqueous solution, to separate particles or cells from the first immiscible liquid. The second immiscible liquid may serve as the collector phase collecting the particles into it (see Raghavan and Fuerstenau (1975)). Alternatively, the solid particles may be pushed to the liquid-liquid interface. Whether the particles are going to be located at the liquid-liquid interface or in the bulk of one of the liquid phases can be determined by relations between the three interfacial tensions $\gamma^{SL}, \gamma^{SL_2}, \gamma^{L_1L_2}$ (see Henry (1984)), where L_1 and L_2 are the two immiscible liquid phases.

3.4 Interphase transport: flux expressions

In Section 3.1.4, we studied how integrated flux expressions can be used to describe mass transport in single-phase systems. In separation processes involving two or more phases, solute and/or solvent is transferred from one phase to another. In processes using membranes, a species is transferred from the feed to the membrane phase and then from the membrane phase to the permeate (or product or receiving) phase; thus three phases are involved in the direction of species transport. It is necessary to have flux expressions, especially integrated ones, for such interphase transport. In general, the approach is to use integrated flux expressions for each phase and couple them together to develop an overall integrated flux expression using interphase equilibrium relations at the phase boundaries. Another objective is to develop a relation between an overall mass-transfer coefficient for the overall interphase transfer process and the mass-transfer coefficient for transfer in each phase.

The following is a partial list of *two-phase* systems encountered in separation processes involving interphase transport: gas-liquid (alternatively vapor-liquid), liquid-liquid, solid-liquid, liquid-ion exchange resin, solid-supercritical fluid, liquid-supercritical fluid, etc. The first four systems are used much more frequently. Note that the two phases in each system are immiscible.

It is possible to have *three-phase* combinations of the above-mentioned two-phase systems: gas-liquid (1)-liquid (2) (or vapor-liquid (1)-liquid (2)), liquid-ion exchange resin (1)-ion exchange resin (2), etc. We are not going to consider here such three-phase systems. The three-phase (or region) systems of interest here involve a *membrane*, the *feed phase* and the *permeate phase*. Some of the common three-phase systems of this type are: liquid-membrane-liquid, gas-membrane-gas, liquid-membrane-gas, gas-membrane-liquid. The two fluid phases on two sides of the membrane are, in general, miscible (exceptions include the liquid-membrane-gas system, etc.). An elementary introduction to species transport through membranes that are either nonporous or porous will be provided here to facilitate development of flux expressions in a few commonly used membrane processes.

3.4.1 Interphase transport in two-phase systems

3.4.1.1 *Gas-liquid systems*

Figure 3.4.1 shows the gas-phase and liquid-phase concentration profiles of species A being transferred from a gas phase into a liquid phase. In the gas phase, the bulk composition is indicated either by the partial pressure p_{Ab} or the gas-phase mole fraction x_{Agb}. This value is reduced to p_{Ai} or x_{Agi} at the gas-liquid interface. In the liquid phase, the corresponding values are C_{Ali} or x_{Ali} at the gas-liquid interface. These values are reduced to the bulk liquid-phase values C_{Alb} or x_{Alb} at some distance away from the gas-liquid interface.

In general,[29] the interfacial compositions in the two phases are related to each other by equilibrium criteria

[29] At high mass-transfer rates and with a contaminated interface, interfacial resistance is possible (see Sherwood *et al.* (1975)).

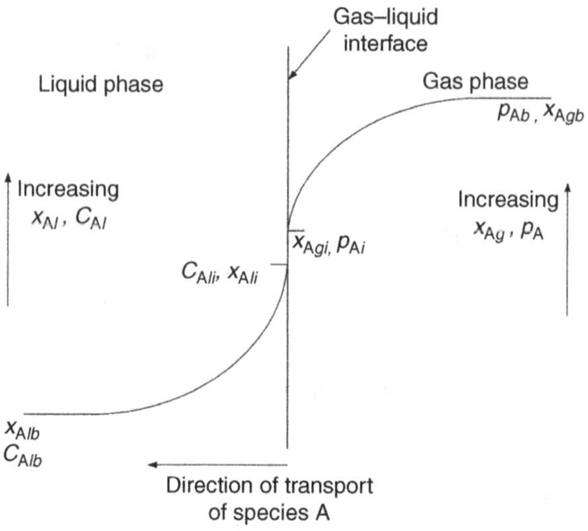

Figure 3.4.1. Concentration profiles of species A being absorbed from a gas into a liquid.

developed in Section 3.3. Thus p_{Ai} or x_{Agi} is related to C_{Ali} or x_{Ali} for an ideal gas-liquid system by Henry's law (relation (3.3.60b)):

$$p_{Ai} = H_A x_{Ali}; \qquad x_{Agi} = (H_A/P)x_{Ali} = H_A^P x_{Ali} \tag{3.4.1a}$$

or

$$p_{Ai} = H_A^C C_{Ali}; \qquad H_A^C = H_A/C_{tl}. \tag{3.4.1b}$$

For nonideal systems, the equilibrium relations

$$p_{Ai} = f_{eq}(x_{Ali}); \qquad x_{Agi} = \bar{f}_{eq}(x_{Ali}) \tag{3.4.2}$$

are to be determined from the equilibrium criterion of fugacities $\hat{f}_{Ag} = \hat{f}_{Al}$ for species A (similarly for all other species).

The flux of species A in both gas and liquid phases can be expressed by either

$$N_{Az} = k_{xl}(x_{Ali} - x_{Alb}) = k_{xg}(x_{Agb} - x_{Agi}), \tag{3.4.3}$$

$$N_{Az} = k_g(p_{Ab} - p_{Ai}) = k_c(C_{Ali} - C_{Alb}), \tag{3.4.4}$$

where k_{xl}, k_{xg}, k_g and k_c are mass-transfer coefficients defined for a particular phase, either gas or liquid. We recall from our earlier discussion in Section 3.1.4 that integrated flux expressions are useful since only concentrations or mole fractions or partial pressures at the two ends of the diffusion/transport path are required instead of their gradients. Here, however, we find that there are quantities like p_{Ai}, x_{Agi}, x_{Ali} corresponding to the gas-liquid interface in single-phase integrated flux expressions; these quantities are very difficult to obtain. On the other hand, the bulk concentrations in each phase, x_{Agb}, C_{Alb}, x_{Alb} and p_{Ab} are much easier to obtain. The answer is provided by an *overall mass-transfer coefficient K at any location.*

The strategy is to define the overall mass-transfer coefficient K with respect to a single phase, either gas or liquid, and known bulk concentrations:

$$N_{Az} = K_{xg}(x_{Agb} - x^*_{Ag}) = K_{xl}(x^*_{Al} - x_{Alb}) \quad (3.4.5)$$

or

$$N_{Az} = K_g(p_{Ab} - p^*_A) = K_c(C^*_{Al} - C_{Alb}). \quad (3.4.6)$$

Here x^*_{Ag} is a hypothetical gas-phase mole fraction which is in equilibrium with x_{Alb}; thus x^*_{Ag} is known. Similarly, x^*_{Al} is a hypothetical liquid-phase mole fraction which is in equilibrium with x_{Agb}; similarly, p^*_A is in equilibrium with C_{Alb} and C^*_{Al} is in equilibrium with p_{Ab}. Now express $x_{Agb} - x^*_{Ag}$ as

$$x_{Agb} - x^*_{Ag} = (x_{Agb} - x_{Agi}) + (x_{Agi} - x^*_{Ag}) \quad (3.4.7)$$

and utilize equilibrium relation (3.4.1a) as well as the definitions of K_{xg}, k_{xg}, k_{xl}:

$$\frac{N_{Az}}{K_{xg}} = \frac{N_{Az}}{k_{xg}} + \frac{H_A}{P}(x_{Ali} - x_{Alb}) = \frac{N_{Az}}{k_{xg}} + \frac{H_A}{P}\frac{N_{Az}}{k_{xl}};$$

$$\frac{1}{K_{xg}} = \frac{1}{k_{xg}} + \frac{H^P_A}{k_{xl}}; \qquad H^P_A = \frac{H_A}{P}. \quad (3.4.8)$$

Therefore, the overall mass-transfer coefficient K_{xg} is known in terms of the individual phase mass-transfer coefficients k_{xg} and k_{xl} and Henry's law constant H^P_A. Further, if the bulk compositions of the two phases are known, the flux N_{Az} of species A from gas to the liquid can be predicted. Note that relation (3.4.8) is also valid if the direction of transport of species A was reversed.

Equation (3.4.8) relates the overall mass-transfer coefficient K_{xg} to the individual mass-transfer coefficients k_{xg} and k_{xl}. Relations between the overall transfer coefficient and the individual phase transfer coefficients for other types of concentration driving gradients are also of considerable use. Consider

$$N_{Az} = K_{xl}(x^*_{Al} - x_{Alb}) = k_{xg}(x_{Agb} - x_{Agi}) = k_{xl}(x_{Ali} - x_{Alb}).$$

But

$$x^*_{Al} - x_{Alb} = (x^*_{Al} - x_{Ali}) + (x_{Ali} - x_{Alb}) = \frac{x_{Agb}P}{H_A} - \frac{x_{Agi}P}{H_A} + \frac{N_{Az}}{k_{xl}}$$

$$\Rightarrow \frac{1}{K_{xl}} = \frac{P}{H_A}\frac{1}{k_{xg}} + \frac{1}{k_{xl}} = \frac{1}{H^P_A k_{xg}} + \frac{1}{k_{xl}}. \quad (3.4.9)$$

Similarly,

$$\frac{1}{K_g} = \frac{1}{k_g} + \frac{H^C_A}{k_c} = \frac{1}{k_g} + \frac{H^C_A}{k_l} \quad (3.4.10)$$

and

$$\frac{1}{K_c} = \frac{1}{k_c} + \frac{1}{H^C_A k_g} = \frac{1}{k_l} + \frac{1}{H^C_A k_g}. \quad (3.4.11)$$

Note that k_l and k_c will be used interchangeably.

An important characteristic of the four relations (3.4.8)–(3.4.11) is that the overall mass-transfer resistance indicated by the $1/K$ term is the sum of the individual gas- and liquid-phase resistances, regardless of the form of the concentration driving gradients. Such linear additive behavior of resistance we will find, again and again, is the rule in interphase transport, although there are a few exceptions.

The individual phase transfer coefficients k_{xl} and k_{xg} are often of the same order of magnitude. However, the value of H^P_A can vary over a wide range depending on the nature of species A and the liquid phase. If a gas species is highly soluble in the liquid (Figure 3.3.3 illustrates such a system), H^P_A has a small value, which makes $(H^P_A/k_{xl}) << (1/k_{xg})$ term in (3.4.8). It is then possible to replace K_{xg} by k_{xg} since

$$\frac{1}{K_{xg}} \cong \frac{1}{k_{xg}}; \quad (3.4.12a)$$

that is, the gas-phase resistance is controlling the species transport. Similarly, if the gas species has a very low solubility in the liquid, H_A has a very high value (e.g. $H_A = 12.5 \times 10^4$ (atm/mole fraction) for helium in water at 20 °C; $H_A = 4.0 \times 10^4$ (atm/mole fraction) for O_2 in water at 20 °C (see Figure 3.3.3). Then

$$\frac{1}{K_{xl}} \cong \frac{1}{k_{xl}}; \quad (3.4.12b)$$

that is, the liquid-phase resistance controls the species transport. Similar relations are easily developed between K_g and k_g for gas-phase control and K_c and k_c for liquid-phase control.

Although the relations given have been developed for any gas–liquid system, they are also valid for any vapor–liquid system. The only difference is that Raoult's law is to be used instead of Henry's law as the equilibrium relation between the phases. For example, consider transfer of species i from vapor phase to liquid phase, x_{igb} to x_{ilb}, with the overall mass-transfer coefficient being expressed in terms of the vapor phase, i.e. K_{xg}:

$$x_{igb} - x^*_{ig} = (x_{igb} - x_{igi}) + (x_{igi} - x^*_{ig}).$$

Here, x^*_{ig} is the vapor-phase mole fraction in equilibrium with the actual liquid-phase bulk composition (in mole fraction) x_{ilb} and x_{igi} is the vapor-phase mole fraction of species i at the vapor–liquid interface. Using definitions of mass-transfer coefficients

$$\frac{N_i}{K_{xg}} = \frac{N_i}{k_{xg}} + K_i x_{ili} - K_i x_{ilb} = \frac{N_i}{k_{xg}} + K_i\frac{N_i}{k_{xl}}$$

$$\Rightarrow \frac{1}{K_{xg}} = \frac{1}{k_{xg}} + \frac{K_i}{k_{xl}}, \quad (3.4.13a)$$

where K_i is the equilibrium ratio (see equilibrium relations (3.3.61) or (3.3.63)). We may use here subscript v for vapor phase instead of g. Similarly, for the overall transfer coefficient expressed in terms of the liquid phase, we have

$$\frac{1}{K_{xl}} = \frac{1}{k_{xl}} + \frac{1}{K_i}\frac{1}{k_{xg}}. \qquad (3.4.13b)$$

The various results considered above for a gas-liquid (or vapor-liquid) system were obtained for absorption of species A from gas (or vapor) into a liquid. If species A is being stripped from the liquid into gas (or vapor) phase, only the direction of transport changes; therefore the flux expressions (3.4.3)–(3.4.6) are changed by a minus sign, as are the concentration profiles in the two phases. However, the relations between the overall transfer coefficient K and the individual phase transfer coefficient k remain unaffected by the direction of transport.

At the beginning of this section, it was indicated that the route of overall mass-transfer coefficients was a convenient way of avoiding the difficulties created by the difficult-to-obtain interfacial concentrations C_{Ali}, p_{Ai}, x_{Agi}, x_{Ali}, etc. We have found it to be so. It is also possible to solve for the interfacial concentration C_{Ali} or x_{Agi}, etc. from either equations (3.4.3) (3.4.4) if k_{xl}, k_{xg} and the equilibrium relation are known for given bulk concentrations. For example, from equation (3.4.3),

$$\frac{k_{xl}}{k_{xg}} = \frac{x_{Agb} - x_{Agi}}{x_{Ali} - x_{Alb}}, \qquad (3.4.14)$$

where, however, $x_{Agi} = H_A^P x_{Ali}$, so that x_{Agi} is the only unknown and can be determined.

3.4.1.2 *Liquid–liquid systems*

Figure 3.4.2 illustrates the concentration profile of solute A being extracted from the aqueous phase ($j = w$) into the organic phase ($j = o$). The molar flux of species A may be expressed by

$$N_{Az} = k_{Aw}(C_{Awb} - C_{Awi}) = k_{Ao}(C_{Aoi} - C_{Aob}), \qquad (3.4.15)$$

where k_{Aw} and k_{Ao} are mass-transfer coefficients for species A in the aqueous and organic phases, respectively. In general, the interfacial concentrations in the two phases C_{Awi} and C_{Aoi} are related to each other by equilibrium (see relation (3.3.79)):

$$\kappa_{Ao} = \frac{C_{Aoi}}{C_{Awi}}. \qquad (3.4.16)$$

As indicated before, the interfacial concentrations are dispensed with by defining overall mass-transfer coefficients:

$$N_{Az} = K_{Aw}(C_{Awb} - C^*_{Aw}) = K_{Ao}(C^*_{Ao} - C_{Aob}). \qquad (3.4.17)$$

Here, C^*_{Aw} is a hypothetical aqueous-phase concentration in equilibrium with the bulk organic-phase concentration C_{Aob}. Similarly, C^*_{Ao} is a hypothetical organic-phase concentration in equilibrium with the bulk aqueous-phase concentration C_{Awb}. If one can assume that the equilibrium

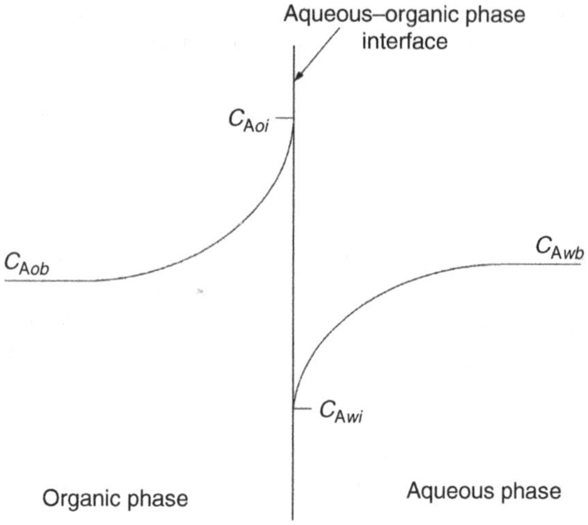

Figure 3.4.2. Concentration profiles of species A being extracted from an aqueous to an organic phase.

distribution coefficient κ_{Ao} is constant over the whole concentration range encountered in this case, then

$$\frac{N_{Az}}{K_{Aw}} = C_{Awb} - C^*_{Aw} = (C_{Awb} - C_{Awi}) + (C_{Awi} - C^*_{Aw})$$

$$= \frac{N_{Az}}{k_{Aw}} + \frac{1}{\kappa_{Ao}}(C_{Aoi} - C_{Aob});$$

$$\frac{N_{Az}}{K_{Aw}} = \frac{N_{Az}}{k_{Aw}} + \frac{N_{Az}}{\kappa_{Ao}k_{Ao}} \Rightarrow \frac{1}{K_{Aw}} = \frac{1}{k_{AW}} + \frac{1}{\kappa_{Ao}k_{Ao}}. \qquad (3.4.18)$$

For the relation between K_{Ao}, k_{Aw} and k_{Ao}, we note that

$$\frac{N_{Az}}{K_{Ao}} = C^*_{Ao} - C_{Aob} = (C^*_{Ao} - C_{Aoi}) + (C_{Aoi} - C_{Aob})$$

$$= \kappa_{Ao}(C_{Awb} - C_{Awi}) + \frac{N_{Az}}{k_{Ao}} \Rightarrow \frac{1}{K_{Ao}} = \frac{\kappa_{Ao}}{k_{Aw}} + \frac{1}{k_{Ao}}. \qquad (3.4.19)$$

Equations (3.4.18) and (3.4.19) provide the desired relations between the individual phase mass-transfer coefficients k_{Aw}, k_{Ao} and the overall mass-transfer coefficients K_{Aw} and K_{Ao}. Although these results are derived for transfer from the aqueous to the organic phase, they are equally valid for transfer in the opposite direction. As before, the overall mass-transfer resistance is the sum of the resistances of the individual phases, the aqueous and the organic. Further, under particular conditions, either the aqueous- or the organic-phase resistance controls. For example, if $\kappa_{Ao} >> 1$ (the solute prefers the organic phase strongly over the aqueous phase), then the aqueous phase resistance is in control:

$$\frac{1}{K_{Aw}} \cong \frac{1}{k_{Aw}}. \qquad (3.4.20)$$

On the other hand, if $\kappa_{Ao} << 1$, then the organic-phase resistance is dominant:

$$\frac{1}{K_{Ao}} \cong \frac{1}{k_{Ao}}. \tag{3.4.21}$$

Both relations are based on the assumption that k_{Aw} and k_{Ao} are of similar orders of magnitude.

Example 3.4.1 Calculate the overall mass-transfer coefficient for the extraction of diethylamine (A) from its dilute solution in water into toluene. The following data on this extraction have been obtained from Treybal (1963, p. 498): κ_{Ao} = 0.735; k_{Aw} = 0.76 lb mol/hr-ft^2-(lb mol/ft^3); k_{Ao} = 0.129 lb mol/hr-ft^2-(lb mol/ft^3). Calculate both K_{AW} and K_{Ao} in units of of cm/s.

Solution Equation (3.4.18) yields

$$\frac{1}{K_{Aw}} = \frac{1}{k_{Aw}} + \frac{1}{\kappa_{Ao} k_{Ao}} = \frac{1}{0.760} + \frac{1}{0.735} \times \frac{1}{0.129} = 1.315 + 10.54;$$

$$K_{Aw} = \frac{1}{11.855} \frac{\text{lb mol}}{\text{hr-ft}^2\text{-}\dfrac{\text{lb mol}}{\text{ft}^3}} = \frac{1}{11.855} \times \frac{1}{3600} \times \frac{30.48}{1} \frac{\text{cm}}{\text{s}}$$

$$= 7.14 \times 10^{-4}\ \text{cm/s};$$

$$\frac{1}{K_{Ao}} = \frac{\kappa_{Ao}}{k_{Aw}} + \frac{1}{k_{Ao}} = \frac{0.735}{0.760} + \frac{1}{0.129} = 0.967 + 7.75;$$

$$K_{Ao} = \frac{1}{8.717} \frac{\text{lb mol}}{\text{hr-ft}^2\text{-}\dfrac{\text{lb mol}}{\text{ft}^3}} = \frac{1}{8.717} \times \frac{1}{3600} \times \frac{30.48}{1} \frac{\text{cm}}{\text{s}}$$

$$= 9.71 \times 10^{-4}\ \text{cm/s}.$$

3.4.1.3 *Solid–liquid system: solution crystallization*

There are two basic rate phenomena of interest in crystallization: (1) the nucleation rate; (2) the growth rate. Let us start with the *nucleation rate*. We have observed via equations (3.3.99) how the addition mechanism leads from kinetic units to a cluster, to an embryo, and then to a nucleus; finally, a critical size r_c is achieved by the nucleus. From the theory of activated process in chemical kinetics, the rate of nucleation in homogeneous nucleation is described by an Arrhenius type expression

$$B^0 = A \exp\left(-\frac{\Delta G_{\text{crit}}}{\text{k}^{\text{B}} T}\right), \tag{3.4.22a}$$

where k^{B} is Boltzman's constant, 1.3806×10^{-23} joules/K $(= \text{R}/\widetilde{\text{N}})$, ΔG_{crit} is the critical free energy change for nucleation and A is a pre-exponential constant. The units of B^0 and A are number of nuclei/cm^3-s; the values of A suggested are of the order 10^{25} (La Mer, 1952).

Introducing the value of ΔG_{crit} from expression (3.3.102), we obtain

$$B^0 = A \exp\left(-\frac{16\pi(\gamma^{12})^3 \overline{V}_{ic}^2}{3\text{k}^{\text{B}}\text{R}^2 T^3 (\ell n\, s)^2}\right) = A \exp\left(-\frac{16\pi(\gamma^{12})^3 \overline{V}_{ic}^2 \widetilde{\text{N}}}{3(\text{R}T)^3 (\ell n\, s)^2}\right) \tag{3.4.22b}$$

for a nonelectrolytic solute, where $\widetilde{\text{N}}$ is Avagadro's number. The nucleation rate therefore increases with an increase in the temperature and the supersaturation ratio, whereas an increase in the interfacial tension decreases it. It has been difficult to verify this equation experimentally since homogeneous nucleation probably never occurs in practice (see McCabe and Smith (1976, p. 867); see, however, references in chap. 2 of Myerson (1993)), whereas heterogeneous nucleation is common. The presence of a solid particle generally reduces the energy barrier ΔG_{crit} needed for nucleation.

When a crystal grows in a supersaturated solution, the *rate of crystal growth* or the rate of solute transfer to the crystal depends on two steps: (1) the rate of diffusion of the solute from the bulk through a laminar layer near the crystal; (2) the rate at which solute molecules are integrated with the surface of the parent crystal.

The second step, known as the *particle integration* rate, involves desolvation of the solute, counterdiffusion of the solvent away from the surface and orientation of the solute into the crystal lattice. Figure 3.4.3(a) illustrates an absorbed layer of solute on the surface of a growing crystal - a sort of a third phase "composed of partially ordered solute, perhaps in a partially desolvated lattice" (Randolph and Larson, 1988). If this surface integration rate is assumed to be linear with the driving force, then the concentration profiles for the diffusional step and the surface integration step are as shown in Figure 3.4.3(b).

The total rate of transfer of solute to the growing crystal at any instant of time can be expressed, respectively, for steps (1) and (2) above by

$$N_{iy} a_p = -k_d a_p (C_{iwb} - C_{iwi}); \tag{3.4.23a}$$

$$N_{iy} a_p = -k_s a_p (C_{iwi} - C_{is}). \tag{3.4.23b}$$

An overall crystal growth coefficient K_o may be defined using an overall concentration difference $(C_{iwb} - C_{is})$:

$$N_{iy} a_p = -K_o a_p (C_{iwb} - C_{is}). \tag{3.4.24}$$

Therefore

$$\frac{1}{K_o} = \frac{1}{k_d} + \frac{1}{k_s} \Rightarrow K_o = \frac{k_d k_s}{k_d + k_s}. \tag{3.4.25}$$

Often it is useful to express the rate of solute transfer to the growing crystal via the rate of growth of a characteristic radius r_p of the particle, such that the particle mass m_p and the surface area a_p may be expressed, respectively, by

$$m_p = \rho_p \psi_v r_p^3 \quad \text{and} \quad a_p = \psi_s r_p^2. \tag{3.4.26}$$

Now, $M_i N_{iy} a_p = (\text{d}m_p/\text{d}t)$, where M_i is the molecular weight of species i. Therefore

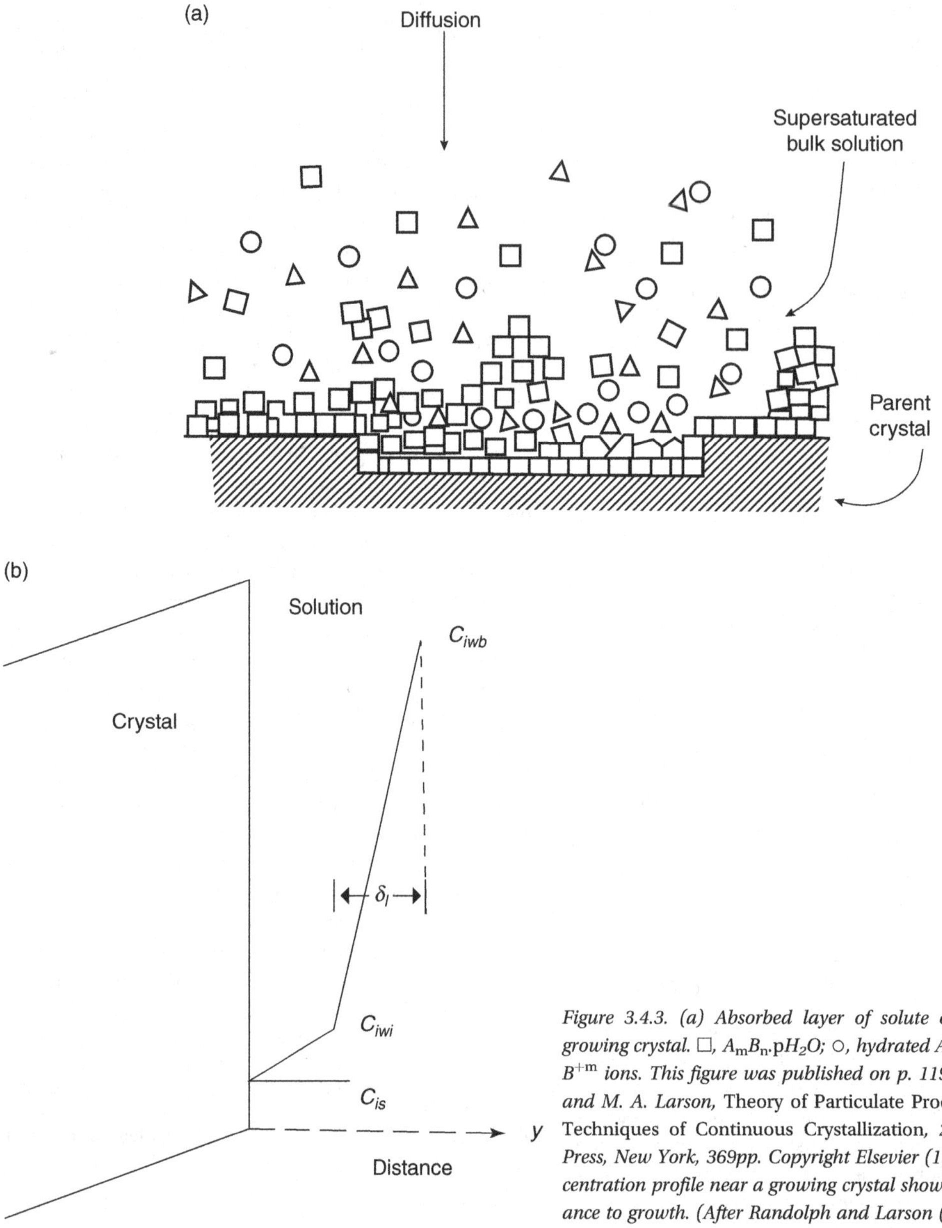

Figure 3.4.3. (a) Absorbed layer of solute on the surface of a growing crystal. □, $A_mB_n.pH_2O$; ○, hydrated A^{+n} ions; Δ, hydrated B^{+m} ions. This figure was published on p. 119 of Randolph, A. D. and M. A. Larson, Theory of Particulate Processes: Analysis and Techniques of Continuous Crystallization, *2nd edn., Academic Press, New York, 369pp. Copyright Elsevier (1988). (b) Solute concentration profile near a growing crystal showing interfacial resistance to growth. (After Randolph and Larson (1988).)*

$$\frac{dm_p}{dt} = \rho_p \psi_v \frac{d(r_p^3)}{dt} = 3\rho_p \psi_v r_p^2 \frac{dr_p}{dt} = K_o M_i \psi_s r_p^2 (C_{iwb} - C_{is});$$
$$\frac{dr_p}{dt} = \frac{K_o M_i \psi_s}{3\rho_p \psi_v}(C_{iwb} - C_s) = G. \tag{3.4.27}$$

The growth rate of the crystal, (dr_p/dt), is said to be diffusion-controlled if an increase in velocity of the supersaturated solution relative to the crystal surface leads to an increase in crystal growth rate. The crystal growth becomes surface integration-controlled if there is no such effect of solution velocity. Often, "G" is used to indicate the growth rate dr_p/dt. Further, L is also used very frequently instead of r_p.

There is no general method of estimating k_s. It has been suggested (McCabe and Smith, 1976, p. 875) that the following growth rate expressions are useful:

$$\text{for hydrated inorganics, } G = A_1 s; \tag{3.4.28a}$$

$$\text{for organic crystals, } G = A_2 \exp\left(-\frac{A_3}{s}\right). \tag{3.4.28b}$$

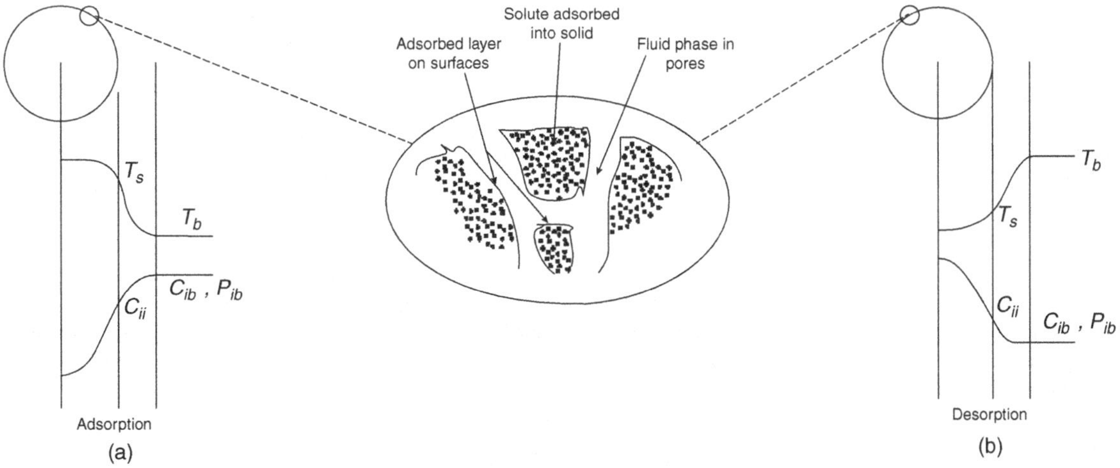

Figure 3.4.4. Concentration profiles of gas or solute during (a) adsorption and (b) desorption in a particle whose structure is shown in the middle. (After Vermeulen et al. *(1973) and Yang (1987).)*

Here, s is the fractional supersaturation and A_1, A_2 and A_3 are constants. The growth rate G of all crystals follows an Arrhenius type of temperature dependence, namely

$$G = A_4 \exp(-\Delta E_s/RT), \tag{3.4.29}$$

where ΔE_s is an activation energy.

Inspection of the growth rates in equations (3.4.28a,b) suggest that, regardless of the crystal size, the growth rate depends only on the supersaturation present in the system/crystallizer; the growth rate G is independent of the crystal size r_p (or L). Therefore

$$\frac{dG}{dr_p} = 0 = \frac{dG}{dL}. \tag{3.4.30}$$

This generalizaton, developed by W.L. Mc Cabe in 1929, is called the ΔL *law of crystal growth*. Correspondingly, over a time period Δt, the total increase in the crystal dimension, ΔL or Δr_p, is given by

$$\Delta L = \Delta r_p = G\Delta t. \tag{3.4.31}$$

3.4.1.4 *Gas-solid and liquid-solid based interfacial adsorption systems*

Separation in interfacial adsorption based gas-solid or liquid-solid systems is generally carried out in stationary beds of solid particles, with the fluid phase (gas or liquid) being mobile. As a result, the concentration profile of any species in each phase is time-dependent. Figure 3.4.4(a) illustrates the concentration profile of a solute species being adsorbed from the fluid phase at any instant of time; the solute concentration $\overline{C}_{ib}$ (or partial pressure $\overline{p}_{ib}$) in the bulk fluid phase decreases to $\overline{C}_{ii}$ (or partial pressure $\overline{p}_{ii}$) at the end of the fluid film at the interface of the porous adsorbent and the fluid. The over bar indicates the average over the flow cross section in the bed. The concentration decreases further inside the pores of the particle. There are three basic and consecutive steps in the adsorption processes for a species present in a mixture around any adsorbent particle.

(1) The species diffuses from the bulk fluid phase to the external surface of the sorbent particle through a so-called fluid film. This film provides what is called the *external film resistance.*
(2) Diffusion and transport of the molecules through the porous structure of the adsorbent particle lead to the *intraparticle transport resistance.*
(3) Actual adsorption or reaction of the molecules occurs at numerous surface sites within the pores of the adsorbents. This step is assumed to be essentially instantaneous compared to the other two resistances identified above.

For desorption processes, the process steps are reversed: starting with desorption at the many sites, continuing with transport/diffusion of the desorbed molecules through the porous structure and finally diffusion from the external surface of the sorbent particle to the bulk fluid phase completes the desorption process. The corresponding species concentration profile is shown in Figure 3.4.4(b). The structure of the porous adsorbent and the different regions in the porous adsorbent (e.g. the fluid phase in the pores, numerous adsorption sites on the adsorbent solid surface, the adsorbed layers on pore surfaces) are schematically illustrated via a magnified section in Figures 3.4.4 (a) and (b).

The molar flux of adsorption of species i from the fluid bulk to the surfaces of the porous adsorbent particle per unit volume of the particle whose surface area per unit volume is a_v is given by

$$N_i a_v = k_i a_v(\overline{C}_{ib} - \overline{C}_{ii}). \tag{3.4.32a}$$

For spherical particles, $a_v = (3/r_p) = (6/d_p)$. In terms of particle surface area a per unit volume of adsorbent particle bed, where $a = a_v(1-\varepsilon)$ (ε being the void volume fraction of the bed),

$$N_i a = k_i a(\overline{C}_{ib} - \overline{C}_{ii}). \tag{3.4.32b}$$

The value of k_i, the species mass-transfer coefficient in the fluid film, may be obtained from a variety of correlations for the gas phase as k_{ig} and for the liquid phase as k_{il}. The experimentally obtained correlations for k_i popularly employed for a gas flowing over a sphere or in a packed bed are:

(1) the *j*-factor correlation (3.1.157);
(2) the Ranz–Marshall type correlation (3.1.154) proposed in Geankoplis (1972);
(3) the Wakao and Funazkri (1978) correlation (3.1.158a). (This equation is recommended when axial dispersion is present in the packed bed. See Section 7.1.1.1 for equations for a packed bed with axial dispersion in the liquid phase; the axial dispersion coefficient is expressed by a correlation in the footnote to Table 3.1.6. The same axial dispersion coefficient may be used for liquid flow over a packed bed also (Ruthven, 1984).)

The external fluid film resistance (the corresponding mass-transfer coefficient k_i from equations (3.4.32a,b)) is in series with the *intraparticle transport resistance*. The flux of a species through a porous/mesoporous/microporous adsorbent particle consists, in general, of simultaneous contributions from the four transport mechanisms described earlier for *gas transport* in Section 3.1.3.2 (for molecular diffusion, where $(D_{AK}/D_{AB}) \gg 1$):

$$N_i = N_i\Big|_{\substack{\text{Knudsen}\\ \text{diffusion}}} + N_i\Big|_{\substack{\text{molecular}\\ \text{diffusion}}} + N_i\Big|_{\substack{\text{surface}\\ \text{diffusion}}} + N_i\Big|_{\substack{\text{Poiseuille/}\\ \text{viscous flow}}}$$

(3.1.115b) (3.1.115d) (3.1.117a) (3.1.116a)

(3.4.33)

Not all of these contributions are simultaneously present/effective. The concentration profile of a gas species inside the particle has to be obtained by solving the mass balance equation within a particle (see equation (7.1.20c)). Such a solution yields a gas-phase concentration profile inside the particle, $C_{is}^p(r)$, which is related to the external resistance via the equality of fluxes at the particle–gas surface $r = r_p$:

$$k_{ig}(\overline{C}_{ib} - \overline{C}_{ii}) = D_{ip}\frac{\partial C_{is}^p}{\partial r}\bigg|_{r=r_p}, \tag{3.4.34}$$

where D_{ip} is the effective diffusion coefficient obtained from the combined processes of equation (3.4.33). The relative contributions of the internal resistance (r_p/D_{ip}) and the external film resistance $(1/k_{ig})$ is determined by the *Biot number*:

$$Bi = \frac{(r_p/D_{ip})}{(1/k_{ig})} = \frac{k_{ig}r_p}{D_{ip}}. \tag{3.4.35}$$

In gas–solid adsorption processes, generally $Bi >> 1$ (Yang, 1987). Therefore the external film resistance may be neglected. The internal diffusion process controls the rate of species adsorption/desorption.

Since intraparticle diffusion/transport controls the species transport rate, a linear transport rate expression is sought to describe it to facilitate solution of the overall adsorption process taking place in a packed bed. The transient rate of adsorption/desorption of a species i may be described by the time rate of change of a particle-averaged species concentration $\overline{C}_{is}^p$ defined for a spherical particle of radius r_p by

$$\left(\frac{4}{3}\pi r_p^3\right)\overline{C}_{is}^p = \int_0^{r_p} C_{is}^p\, 4\pi r^2 \mathrm{d}r. \tag{3.4.36a}$$

Here, C_{is}^p is obtained from the mass balance equation in a spherical particle (see (7.1.19a)):

$$\frac{\partial C_{is}^p}{\partial t} = D_{ip}\left[\frac{1}{r^2}\frac{\partial}{\partial r}\left(r^2\frac{\partial C_{is}^p}{\partial r}\right)\right] = D_{ip}\left[\frac{\partial^2 C_{is}^p}{\partial r^2} + \frac{2}{r}\frac{\partial C_{is}^p}{\partial r}\right]. \tag{3.4.36b}$$

The time rate of change of $\overline{C}_{is}^p$ is described by a linear-driving force based expression:

$$\left(\frac{4}{3}\pi r_p^3\right)\frac{\partial \overline{C}_{is}^p}{\partial t} = \frac{15D_{ip}}{r_p^2}(\overline{C}_{is}^{p^*} - \overline{C}_{is}^p), \tag{3.4.36c}$$

where $\overline{C}_{is}^{p^*}$ is a hypothetical averaged pore-phase concentration in equilibrium with the instantaneous bulk gas-phase concentration at the particle–fluid boundary $(\overline{C}_{ii}, \overline{C}_{ib})$ (both being essentially equal to each other since k_{ig} is, on a relative basis, quite high). This approximation, identified as the *linear driving force* (LDF) *approximation*, was developed first by Glueckauf and Coates (1947), and is widely used in particle–fluid adsorption/desorption processes. For a basis of this approximation, see the development in Yang (1987, p.127) based on a solution of the governing diffusion equation (equation (3.4.36b)) for diffusion of gases within a spherical particle. Sircar and Hufton (2000) have shown for gas–solid systems that any adsorbable species concentration profile describable by

$$C_{is}^p(r,t) = a(t) + b(t)\, F(r), \tag{3.4.37a}$$

where $F(r)$ is any monotonic and continuous function of r in the range $0 \le r \le r_p$ satisfying the boundary condition

$$\frac{\mathrm{d}F(r)}{\mathrm{d}r}\bigg|_{r=0} = 0, \tag{3.4.37b}$$

will satisfy the LDF approximation of (3.4.36c), where $(15D_{ip}/r_p^2)$ may be replaced, in general, by a mass-transfer coefficient k. The restriction for the LDF approximation of Glueckauf and Coates (1947) is $(D_{ip}t/r_p^2) > 0.1$, which often invalidates its use during initial gas sorption periods in activated carbon and zeolites.

The linear driving force approximation (3.4.36c) is also used commonly for separation/transport between a liquid and porous adsorbent particles. In order to account for any deviation between the actual rate of adsorption and the rate obtained by the LDF approximation, a correction factor ψ_p has also been introduced:

$$\left(\frac{4}{3}\pi r_p^3\right)\frac{\partial \overline{C}_{is}^p}{\partial t} = \psi_p \frac{15D_{ip}}{r_p^2}(\overline{C}_{is}^{p^*} - \overline{C}_{is}^p). \tag{3.4.38}$$

On a comparison of the observed value of the rate of uptake of species i with time (as given by the left-hand side of (3.4.36c)) and the rate predicted by the right-hand side, ψ_p may be determined. Some estimates of ψ_p are available in Vermuelen *et al.* (1973).

An alternative approach for estimating the rate of adsorption of species i on the adsorbent employs the method used in chemical kinetics to determine the rate of reaction as an algebraic sum of the forward and backward reaction rates for the second-order reaction (whose reaction equilibrium constant is $K = (= k_f/k_b)$) (Thomas, 1944):

adsorbing species (i)

$$+\,\text{adsorbent site}\,(s) \underset{k_b}{\overset{k_f}{\rightleftharpoons}} \text{adsorbed complex}\,(is). \tag{3.4.39a}$$

If C_{if} is the concentration of adsorbing species i in the pore fluid phase, C_{is} is the concentration of adsorbed species i on the adsorbent phase (s) occupying some of the sites of the adsorbent phase, where C_{ss}^m is the maximum value of the site concentration (s) (first subscript) in the solid phase, then the forward adsorption reaction rate for this second-order reaction is $k_f C_{if}(C_{ss}^m - C_{is})$ and the backward reaction rate is $k_b C_{is}$. The net rate of species i adsorption is given by

$$\frac{\partial C_{is}}{\partial t} = k_f C_{if}(C_{ss}^m - C_{is}) - k_b C_{is} = k_f\left(C_{if}(C_{ss}^m - C_{is}) - \frac{C_{is}}{K}\right). \tag{3.4.39b}$$

The quantity within the brackets on the right-hand side is called the *kinetic driving force*; k_f is the corresponding kinetic coefficient in this reaction-kinetic treatment (Vermeulen *et al.*, 1973). When the rate is zero, the equilibrium concentration, C_{is}^*, of the adsorbent for species i is obtained for C_{if} as follows:

$$C_{if}(C_{ss}^m - C_{is}^*) - \frac{C_{is}^*}{K} = 0 \Rightarrow C_{is}^* = \frac{C_{ss}^m K C_{if}}{1 + KC_{if}}. \tag{3.4.39c}$$

This is a form of Langmuir adsorption isotherm illustrated in Figure 3.3.8B and equations (3.3.112a) and (3.3.113c, d), since (C_{is}^*/C_{ss}^m) is essentially equal to θ, the fraction of the sites occupied.

Often, instead of a molar concentration for the adsorbed species, C_{is} (mole/volume), moles of species i per unit mass of adsorbent, q_{is}, is used. In such a notation, q_{ss}^m is the maximum value of the site concentration. The corresponding form of the rate of adsorption equation (3.4.39b) and the Langmuir adsorption isotherm (3.4.39c) are changed, respectively, to

$$\frac{\partial q_{is}}{\partial t} = k_f\left(C_{if}(q_{ss}^m - q_{is}) - \frac{q_{is}}{K}\right); \tag{3.4.39d}$$

$$q_{is}^* = \frac{q_{ss}^m K C_{if}}{1 + KC_{if}}, \tag{3.4.39e}$$

where q_{is}^* is the equilibrium concentration or saturation capacity of the adsorbent for species i for the fluid phase concentration C_{if}.

A nondimensional form of equation (3.4.39d) is frequently used when q_{is} is nondimensionalized with respect to q_{ss}^m:

$$\frac{\partial(q_{is}/q_{ss}^m)}{\partial t} = \frac{k_f}{K}\left(\frac{\left(KC_{if}\left(\frac{q_{is}^*(1+KC_{if})}{KC_{if}} - q_{is}\right) - q_{is}\right)}{q_{ss}^m}\right)$$

$$= \frac{k_f}{K}(1 + KC_{if})\left(\frac{q_{is}^*}{q_{ss}^m} - \frac{q_{is}}{q_{ss}^m}\right). \tag{3.4.39f}$$

An alternative form of this equation for the rate of adsorption is often used. First, rewrite it as

$$\frac{\partial(q_{is}/q_{ss}^m)}{\partial t} = \frac{k_f}{K}\left\{(1 + KC_{if})\frac{q_{is}^*}{q_{ss}^m} - \frac{q_{is}}{q_{ss}^m}(1 + KC_{if})\right\}. \tag{3.4.40a}$$

Now define $C_{if} = C_{if}^{\text{ref}} x_{if}$ and q_{is}^{ref} corresponding to C_{if}^{ref} in the adsorption system. Then the Langmuir adsorption isotherm (3.4.39e) for C_{if}^{ref} and C_{if} may be written, respectively, as

$$q_{is}^{\text{ref}} = q_{ss}^m \frac{KC_{if}^{\text{ref}}}{1 + KC_{if}^{\text{ref}}} \quad \text{and} \quad q_{is}^* = \frac{q_{ss}^m KC_{if}^{\text{ref}} x_{if}}{1 + KC_{if}^{\text{ref}} x_{if}}, \tag{3.4.40b}$$

leading to

$$q_{is}^*/q_{is}^{\text{ref}} = \frac{1 + KC_{if}^{\text{ref}}}{1 + KC_{if}^{\text{ref}} x_{if}} x_{if}. \tag{3.4.40c}$$

Temporarily assuming that q_{ss}^m equals q_{is}^{ref}, we obtain, for (3.4.40a),

$$\frac{\partial(q_{is}/q_{is}^{\text{ref}})}{\partial t} = \frac{k_f}{K}\left\{(1 + KC_{if}^{\text{ref}})x_{if} - \frac{q_{is}}{q_{is}^{\text{ref}}}(1 + KC_{if}^{\text{ref}} x_{if})\right\}. \tag{3.4.40d}$$

Alternative forms of this equation are used in analyzing the rate of adsorption in adsorbent beds using separation factor relations.

3.4.1.5 *Ion exchange resin-solution systems*

Mass transfer in the ion exchange resin-solution system is briefly considered here for the case where counterion A^{Z_A} in the ion exchanger is exchanged with counterion B^{Z_B} in the external solution. An exchange of this kind may be described by five steps:

(1) the diffusion of ion B^{Z_B} from the bulk of the external solution to the outside surface of the ion exchange resin particle;
(2) the diffusion of ion B^{Z_B} through the solvent-filled pores of the ion exchange matrix to a site containing ion A^{Z_A};
(3) the instantaneous exchange of places between ions B and A;
(4) the diffusion of ion A through the solvent-filled pores of the ion exchange matrix to the outside surface of the resin particle;
(5) the diffusion of ion A from the particle surface to the bulk of the external solution.

Such a scheme was proposed by Boyd *et al.* (1947). Since step (3) is instantaneous, we need only consider steps (1), (2), (4) and (5). However, there are essentially two types of resistances here: steps (1) and (5) account for diffusion through the liquid film on the outside of the resin particle, whereas steps (2) and (4) describe the diffusion of counterions through the pores of the ion exchange resin particle. When steps (1) and (5) control the exchange rate, the ion exchange is said to be *film diffusion controlled.* If steps (2) and (4) are the rate-determining steps, the mechanism is *particle diffusion controlled* (Helfferich, 1962).

The diffusional flux expression for *particle diffusion control* in the exchange of counterion A^{Z_A} in the ion exchanger with the counterion B^{Z_B} in the external solution will now be derived. The Nernst-Plank flux expression for species A in the resin ($j = R$) in the absence of any pressure gradient is obtained from (3.1.106) as

$$\boldsymbol{J}_A^* = -D_{AR}\nabla C_{AR} - D_{AR}\frac{C_{AR}Z_A\mathcal{F}}{RT}\nabla\Phi. \qquad (3.4.41)$$

Since there is no electric current in the system, obtain from the current density expression (3.1.108c)

$$Z_A\boldsymbol{N}_A + Z_B\boldsymbol{N}_B = 0 = Z_A\boldsymbol{J}_A^* + Z_B\boldsymbol{J}_B^* \Rightarrow \boldsymbol{J}_A^* = -\frac{Z_B}{Z_A}\boldsymbol{J}_B^*, \qquad (3.4.42)$$

where $\boldsymbol{N}_i = \boldsymbol{J}_i^*$ due to the absence of any convection in the pores of the resin particle. Further, electroneutrality must be maintained in the resin with a fixed molar charge density of X (see relation (3.3.30b)):

$$Z_A C_{AR} + Z_B C_{BR} = -\omega X.$$

Taking the gradient of this expression yields

$$Z_A\nabla C_{AR} + Z_B\nabla C_{BR} = 0. \qquad (3.4.43)$$

Table 3.4.1. A few diffusion coefficients in ion exchange systems[a]

Ions	Na^+	Cs^+	Ba^{++}	Sr^{++}	Mn^{++}
Infinitely dilute external solution, $D_{il} \times 10^7$ cm^2/s	133	205	84	77.8	70.8
Dowex 50 W-X8 resin,[b] $D_{iR} \times 10^7$ cm^2/s	20.5	30.0	1.16	1.95	2.22

[a] Bajpai *et al.* (1974).
[b] H^+ of sulfonic acid group is the exchangeable cation.

Use these three equations to eliminate $\nabla\Phi$ from $\boldsymbol{J}_A^*$. First, substitute for $\boldsymbol{J}_A^*$ and $\boldsymbol{J}_B^*$ in (3.4.42) using the Nernst-Plank expression (3.4.41) for both A and B and rearrange:

$$(-\nabla\Phi) = \frac{RT}{\mathcal{F}}\frac{Z_A D_{AR}\nabla C_{AR} + Z_B D_{BR}\nabla C_{BR}}{Z_A^2 D_{AR}C_{AR} + Z_B^2 D_{BR}C_{BR}}.$$

Substitute this into expression (3.4.41) for $\boldsymbol{J}_A^*$:

$$\boldsymbol{J}_A^* = -D_{AR}\nabla C_{AR} + \frac{Z_A^2 C_{AR}D_{AR}^2\nabla C_{AR} + Z_A Z_B C_{AR}D_{AR}D_{BR}\nabla C_{BR}}{Z_A^2 D_{AR}C_{AR} + Z_B^2 D_{BR}C_{BR}}.$$

Use here $Z_B\nabla C_{BR} = -Z_A\nabla C_{AR}$ from equation (3.4.43) and simplify:

$$\begin{aligned}\boldsymbol{J}_A^* &= -D_{AR}\nabla C_{AR} + \frac{Z_A^2 C_{AR}D_{AR}^2 - Z_A^2 C_{AR}D_{BR}D_{AR}}{Z_A^2 D_{AR}C_{AR} + Z_B^2 D_{BR}C_{BR}}\nabla C_{AR}\\ &= -\left[\frac{D_{AR}D_{BR}(Z_A^2 C_{AR} + Z_B^2 C_{BR})}{Z_A^2 D_{AR}C_{AR} + Z_B^2 D_{BR}C_{BR}}\right]\nabla C_{AR} = -D_{ABR}\nabla C_{AR}.\end{aligned} \qquad (3.4.44)$$

An unstated assumption in this derivation was that the co-ion was excluded completely from the ion exchanger by Donnan exclusion. Expression (3.4.44) for the flux of counterion A in the particle diffusion controlled exchange of counterions A and B is the equivalent of Fick's first law; however, the diffusion coefficient, D_{ABR} for the exchange between A and B in the resin particle is, in general, not constant and depends on concentrations of counterions A and B. For example, if $C_{AR} << C_{BR}$ with $Z_A \sim Z_B$, then $D_{ABR} \cong D_{AR}$; conversely, if $C_{BR} << C_{AR}$, $D_{ABR} \cong D_{BR}$; i.e. the concentration dependence of D_{ABR} disappears. An idea of the magnitudes of such diffusion coefficients is provided in Table 3.4.1. Note that although ion exchange is usually applied to dilute external solutions where the solution concentration is low (of the order of milliequivalents per liter), the concentration inside the resin is of the order of equivalents per liter.

In *film diffusion controlled* situations, a co-ion is present in the external solution. Consequently, the Nernst-Plank equation for co-ions needs to be considered along with the role of co-ions in the electroneutrality and no electric current conditions (Helfferich, 1962) in deriving the exact flux expressions for film diffusion controlled situations. There is no general solution to this problem. Solutions for some limiting conditions are available (Helfferich, 1962, p. 274).

In a separation process, ion exchange resin particles are generally used in a column. A complex time-dependent differential equation for mass balance in the column has to be combined with the diffusion flux expression for a resin particle, and other appropriate boundary and initial conditions, to determine the extent of separation. It is obvious from the preceding few paragraphs that the diffusion flux expressions are difficult to handle for resin particles. For ion exchange column analysis, practical approaches therefore utilize a linear-driving-force representation of the mass flux to a resin particle (Helfferich, 1962; Vermeulen *et al.*, 1973); this leads to the use of mass-transfer coefficients in resin particle flux expressions.

Before we identify the flux expression in terms of the linear-driving-force (LDF) approximation (see expression (3.4.36c) for fluid–solid adsorption), we will briefly identify the flux expression in ion exchange employing the reaction-kinetic approach already illustrated for fluid–solid adsorption via (3.4.39a) for, say, a cation, Na^+, exchanging with a H^+ ion in a cation exchange resin as a counterion:

$$\begin{array}{cccc} Na^+(w) + (H^+ + R^-) & \underset{k_b}{\overset{k_f}{\rightleftarrows}} & (Na^+ + R^-) + H^+(w) & \\ C_{Aw} \qquad C_{AR}^m - C_{AR} & & C_{AR} \qquad C_{Aw}^t - C_{Aw} & \\ \qquad\quad q_{AR}^m - q_{AR} & & q_{AR} & \end{array} \quad (3.4.45a)$$

In this ion exchange reaction, we have identified the sodium ion concentration in the resin phase on a volumetric basis, C_{AR}, as well as on a per unit resin weight basis, q_{AR}; correspondingly, the H^+ ion concentrations are $C_{AR}^m - C_{AR}$ and $q_{AR}^m - q_{AR}$, where C_{AR}^m and q_{AR}^m represent the corresponding maximum resin ion exchange capacities for Na^+. The net rate of transfer of Na^+ ions to the resin phase may be written for this second-order (pseudo) reaction as (see expression (3.4.39b) or (3.4.39d))

$$\frac{\partial C_{AR}}{\partial t} = k_f \left[C_{Aw}(C_{AR}^m - C_{AR}) - \frac{1}{K} C_{AR}(C_{Aw}^t - C_{Aw}) \right], \quad (3.4.45b)$$

where K is the (pseudo) equilibrium constant ($= k_f/k_b$); alternatively,

$$\frac{\partial q_{AR}}{\partial t} = k_f \left[(q_{AR}^m - q_{AR}) - \frac{1}{K} q_{AR}(C_{Aw}^t - C_{Aw}) \right]. \quad (3.4.45c)$$

However, in reality, the rates are always diffusion controlled (Thomas, 1944) and therefore depend on the particle surface area available:

$$\rho_B \frac{\partial q_{AR}}{\partial t} = k_f a_v \left[C_{Aw}(q_{AR}^m - q_{AR}) - \frac{1}{K} q_{AR}(C_{Aw}^t - C_{Aw}) \right]. \quad (3.4.45d)$$

Here, ρ_B is the bulk density of the resin particle and a_v is the outer particle surface area per unit particle volume. The LDF approximation based expressions are provided next.

Consider the solute concentration profile in any of the two-phase processes considered earlier; specifically, select Figure 3.4.2 with the aqueous phase representing the external solution and the organic phase representing the aqueous solution in the pores of the ion exchange resin particle. The rate of transfer of an ionic species A from the bulk solution ($j = w$) to the interface between the solution and the resin particle is given by

$$N_{Ar} a_v = k_{cw} a_v (C_{Awb} - C_{Awi}), \quad (3.4.45e)$$

where a_v is the outer particle surface area per unit particle volume. For spherical particles of radius r_p,

$$a_v = \frac{3}{r_p}. \quad (3.4.45f)$$

Further, k_{cw} is the external solution-phase mass-transfer coefficient. At the external solution–resin boundary interface, the resin-phase concentration in equilibrium with C_{Awi} is C_{ARi}. The resin-phase mass-transfer expression containing the resin-phase transfer coefficient k_{cR} is

$$N_{Ar} a_v = k_{cR} a_v (C_{ARi} - C_{ARb}). \quad (3.4.46a)$$

Often, concentration units of mole/mass of resin is used by using q_{ARi}, q_{ARb} in the driving force expression given above. It follows that

$$(N_{Ar} a_v/\rho_B) = k_{cR} a_v (q_{ARi} - q_{ARb}), \quad (3.4.46b)$$

where ρ_B is the density of the particle in the mass/volume of the resin.

For film diffusion control, expression (3.4.45e) is used. For particle diffusion control, expression (3.4.45b) is used. **In general, both expressions given above are functions of time when used in an actual ion exchange process.** For mass-transfer conditions where both film diffusion and particle diffusion are important, the two expressions (3.4.45e) and (3.4.45b) may be combined to define an overall mass-transfer coefficient K_c (see Vermeulen *et al.*, 1973, eq. (16–56)):

$$K_c a_v = \left[\frac{1}{\dfrac{1}{k_{cw} a_v} + \dfrac{1}{k_{cR} a_v}} \right]. \quad (3.4.46c)$$

The correlations (3.1.158) by Wakao and Funazkri (1978) and (3.1.159) and (3.1.160) by Wilson and Geankoplis (1966) identified in Table 3.1.6 for flow in a packed bed may be utilized to estimate k_{cw}. For k_{cR}, see equation (3.4.46d) given below.

In Vermeulen *et al.* (1973), the flux expression (3.4.46a) has a correction factor ψ_p to account for the deviation of the actual driving force from the linear driving force approximation used above (originally proposed by Glueckauf and Coates (1947)). The latter had also suggested that

$$k_{cR}a_v = \frac{15D_{AR}}{r_p^2} \qquad (3.4.46d)$$

for resin particles of effective radius r_p and effective diffusion coefficient D_{AR} of species A in resin.

3.4.1.6 *Two-phase systems with a supercritical fluid*

The two-phase systems of interest here are SCF–liquid and SCF–solid. Lahiere *et al.* (1987) have studied the transfer of ethanol or propanol from an aqueous solution into supercritical CO_2. The experimental extraction behavior in a sieve-tray extractor was compared with that predicted from a model used for common liquids. This model for subcritical liquids uses overall mass-transfer coefficients obtained from the individual film coefficients by the conventional sum of the two resistances approach. Lahiere *et al.* (1987) observed reasonable correspondence between experimentally observed values and the performance predicted from conventional models. For SCF–solid systems with pure solids, Debenedetti and Reid (1986) have observed that the mass-transfer coefficient for the SCF phase is very strongly influenced by natural convection. This happens because the SCF has very low viscosity and yet has a high density, leading to a much more important role of natural convection than in normal liquids.

3.4.2 Interphase transport: membranes

There are a variety of membrane structures available. Correspondingly, the variety of transport expressions or transport models is considerable. The nature and magnitude of the driving forces and the frictional resistances can vary, leading to enormous variations in individual species fluxes. It is not possible to cover the whole spectrum of such behavior here. Rather we focus on a few cases of considerable use in practical separation techniques/processes to illustrate integrated flux expressions for membranes.

A membrane separates a feed mixture which can be either liquid or gaseous. In general, the product phase is the same as that of the feed and the concentrate/residue/raffinate/retentate. There are a number of exceptions. For example, a separate sweep, extract or purge stream of a different phase may be introduced on the other side of the membrane. Alternatively, the permeated feed may be in a gaseous/vapor phase from a liquid feed. The membrane can be nonporous or porous from a gross structural point of view. Therefore, we can focus on the following four broad feed phase–membrane type categories: (1) liquid–nonporous; (2) gaseous–nonporous; (3) liquid–porous; (4) gaseous–porous.

Figure 3.4.5 illustrates schematically the four broad feed phase–membrane type categories; it includes some subcategories based on different permeate phases and driving forces excluding electrical potential gradient. An additional category studied is a liquid–porous charged membrane in an electrical field.

We study first liquid separation through practically nonporous membranes and then we move on to porous membranes. Of the known techniques using nonporous membranes, (reverse osmosis, dialysis, liquid membrane permeation and pervaporation), we select the most common, reverse osmosis, to begin our study of integrated flux expression development. Pervaporation is considered next. There is one feature which is, however, common to almost all nonporous membrane processes; i.e. the additional phase, the membrane phase, is stationary in general (except in cases of rapid transient membrane swelling or emulsion liquid membranes). This is in contrast to molecular diffusion processes in a gas or liquid where all species can diffuse (they may or may not).

3.4.2.1 *Liquid permeation through nonporous membranes: reverse osmosis (RO) and pervaporation*

Liquid permeation through practically nonporous membranes can take place under a variety of conditions. These conditions are defined by the nature of the phase on the downstream side of the membrane (liquid or vapor/gas), the nature of the two components in the feed mixture (volatile or nonvolatile) and the level of pressure on the feed side and that on the permeate side (especially including a vacuum). Of the number of combinations possible from this set of variables, the following combinations are of principal interest to us here.

(1) **Reverse osmosis**: feed liquid at high pressure + permeate liquid at lower pressure (with or without a nonvolatile solute species in the feed).
(2) **Pervaporation**: feed liquid at atmospheric pressure + permeate in vapor phase at low pressure (feed liquid components are volatile).
(3) **Liquid permeation**: feed liquid at atmospheric pressure + permeate liquid at atmospheric pressure (feed liquid components volatile/nonvolatile).

Detailed considerations on permeation rates for a volatile feed mixture for a variety of physical states of the permeate with respect to the liquid feed are available in Greenlaw *et al.* (1977) and Shelden and Thompson (1978).

We focus first on *reverse osmosis*. Consider the transport of a salt and water through a homogeneous nonporous membrane (Figure 3.4.6) kept between two *well-stirred* salt solutions, a dilute product solution of salt concentration C_{ip} at a pressure P_p and a more concentrated feed solution of salt concentration C_{if} at a pressure P_f ($>P_p$). If pressure P_f exceeds P_p by an amount larger than the osmotic pressure difference of the two salt solutions, water will be transported from the high-pressure feed solution through the membrane to the low-pressure product

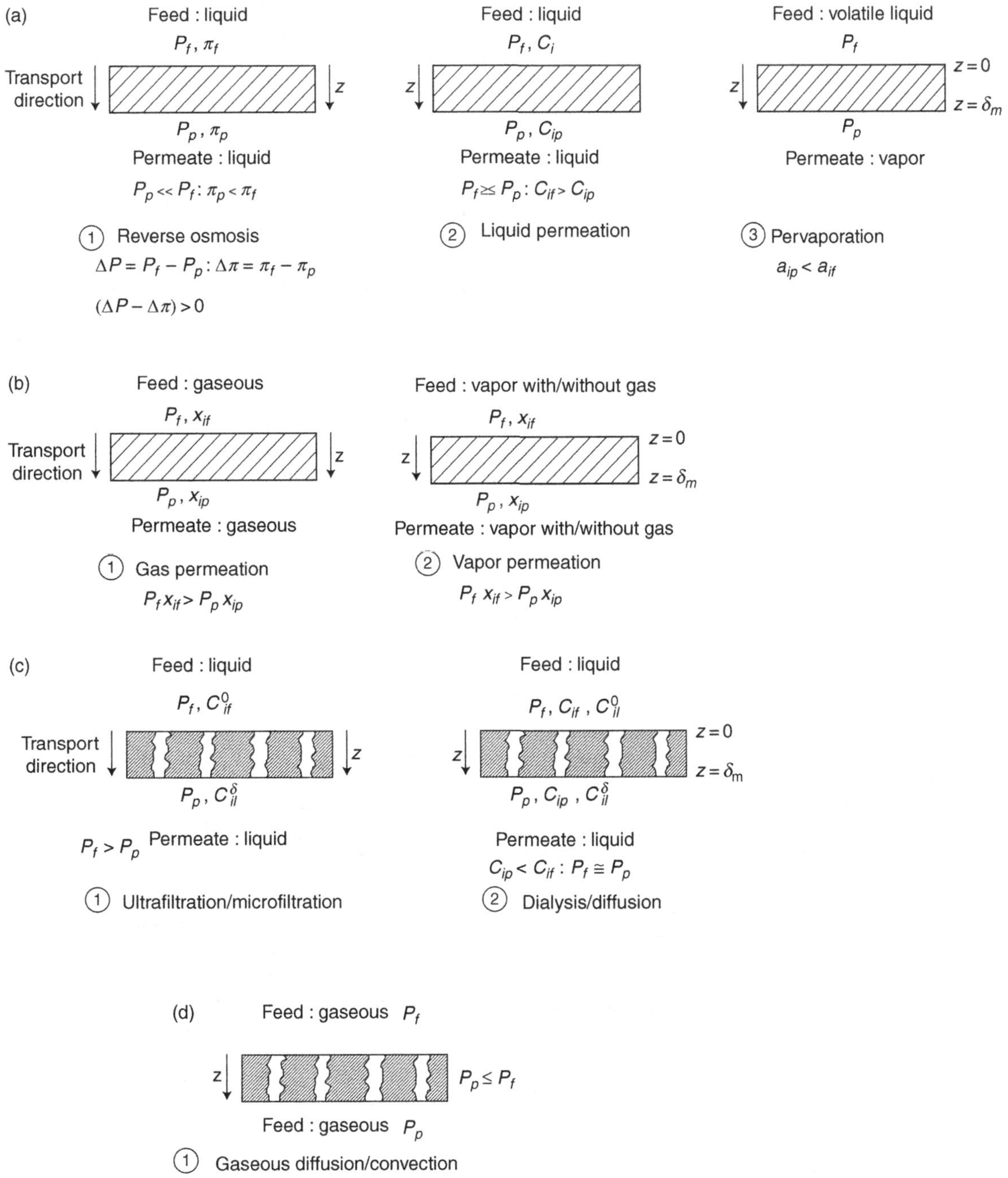

Figure 3.4.5. Schematics of four broad feed phase-membrane type categories including subcategories based on permeate phases and driving forces (electrical potential gradient excluded); membrane based two-phase contacting is not included here. (a) Liquid feed-nonporous membrane; (b) gaseous feed-nonporous membrane; (c) liquid feed-porous membrane; (d) gaseous feed-porous membrane.

solution; the phenomenon is called *reverse osmosis* (RO) (see also Figure 1.5.1). How much salt will also go to the low-pressure product solution depends, amongst other things, on how permeable the membrane is to the salt. In water desalination using a reverse osmosis process, membranes are highly permeable to water and not so much to salt. In this example, one species in the feed (namely salt) is nonvolatile. However, the following treatment is also going to be valid for volatile solutes provided the permeated phase is liquid.

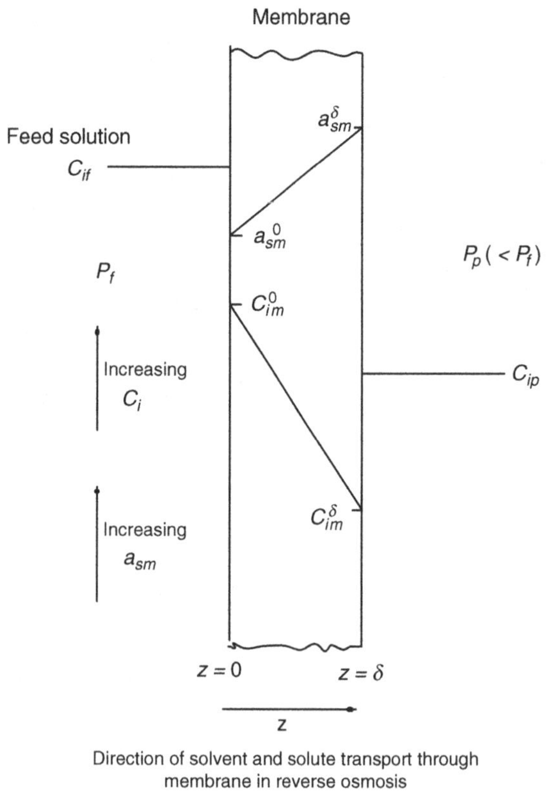

Figure 3.4.6. Solvent and solute transport through a membrane in reverse osmosis with well-mixed feed and permeate solutions.

We ignore the problems, if any, of the transport of solute i and solvent s in the feed solution and the product solution for the time being by assuming them to be well-mixed. In the three-component system of solute–solvent–membrane, the membrane is assumed stationary. Both solvent and solute diffuse through the membrane under the driving force of a chemical potential gradient across the membrane. Such a condition is best studied at first by using the *solution-diffusion theory* (Lonsdale *et al.*, 1965). Both the solvent and the solute are assumed to be dissolved in the membrane at the high-pressure feed solution–membrane interface where the species concentrations in the two phases are related by equilibrium. They then diffuse through the membrane, each under its own chemical potential gradient, and are finally desorbed into the product solution at the low-pressure solution–membrane interface. The system should then be analyzed as if we have two separate binary systems: solvent–membrane and solute–membrane. The membrane thickness is δ_m in the z-direction, the direction of transport. This is a case of diffusion in the membrane. Therefore, we use $\boldsymbol{J}_i^* = \boldsymbol{N}_i - C_i\boldsymbol{v}_t^*$ and equation (3.1.84a) for $\boldsymbol{N}_i$ for both i and s. Note that no external forces are present. But, since ΔP is nonzero, expression (3.1.84d) for $\boldsymbol{U}_i$ has to be used. Inside the membrane phase, $\nabla\mu_i^0 = 0$. Membrane concentrations are indicated by C_{sm} and C_{im} for the solvent and solute, respectively. For the solvent (subscript, s)

$$J_{sz}^* = -\frac{C_{sm}\overline{V}_s}{f_{sm}^d}\left(\frac{\mathrm{d}P}{\mathrm{d}z}\right) - \frac{RTC_{sm}}{f_{sm}^d}\left(\frac{\mathrm{d}\,\ell\mathrm{n}\,a_{sm}}{\mathrm{d}z}\right). \tag{3.4.47}$$

For solute species i,

$$J_{iz}^* = -\frac{C_{im}\overline{V}_s}{f_{im}^d}\left(\frac{\mathrm{d}P}{\mathrm{d}z}\right) - \frac{RTC_{im}}{f_{im}^d}\left(\frac{\mathrm{d}\,\ell\mathrm{n}\,a_{im}}{\mathrm{d}z}\right). \tag{3.4.48}$$

In terms of diffusion coefficients, the frictional coefficients may be replaced by

$$\frac{1}{f_{sm}^d} = \frac{D_{sm}}{RT} \quad \text{and} \quad \frac{RT}{f_{im}^d} = D_{im}. \tag{3.4.49}$$

We could have also obtained the two flux expressions given above by simply considering the uncoupled flux of any solute in the solvent from the irreversible thermodynamic formulation based expression (3.1.208). But the solvent in (3.1.208) is the membrane here. Further, both species being transported, the solute i and the solvent s, are simply solutes:

$$\boldsymbol{J}_i^* = L_{ii}F_i = L_{ii}(-\nabla\mu_i), \tag{3.4.50}$$

where

$$L_{ii} = (D_{im}C_{im}/RT). \tag{3.4.51}$$

Integrate now the flux expression (3.4.47) from $z = 0$ to $z = \delta_m$, assuming $C_{sm}D_{sm}$ and $\overline{V}_s$ to be constants, to obtain at steady state

$$J_{sz}^* = +\frac{C_{sm}D_{sm}\overline{V}_s}{RT\delta_m}(P_f - P_p) - \frac{C_{sm}D_{sm}}{\delta_m}\ell\mathrm{n}\left(\frac{a_{sm}|_{\delta_m}}{a_{sm}|_0}\right). \tag{3.4.52}$$

Now $a_{sm}|_{\delta_m} = a_{sp}$ and $a_{sm}|_0 = a_{sf}$, corresponding to the product and feed solutions. However, relation (3.3.86a) for osmotic equilibrium between two solutions of different solvent activities and osmotic pressures provides

$$\overline{V}_s(\pi_f - \pi_p) = RT\ell\mathrm{n}\left(\frac{a_{sp}}{a_{sf}}\right). \tag{3.4.53}$$

Therefore

$$\begin{aligned} J_{sz}^* &= \frac{C_{sm}D_{sm}\overline{V}_s}{RT\delta_m}[(P_f - P_p) - (\pi_f - \pi_p)] \\ &= \frac{C_{sm}D_{sm}\overline{V}_s}{RT\delta_m}\{\Delta P - \Delta\pi\} = \frac{Q_{sm}}{\delta_m}(\Delta P - \Delta\pi) = A(\Delta P - \Delta\pi), \end{aligned} \tag{3.4.54}$$

where

$$\Delta P = P_f - P_p \quad \text{and} \quad \Delta\pi = \pi_f - \pi_p.$$

This is an **integrated flux expression for the solvent permeating through the reverse osmosis membrane in**

terms of bulk quantities like ΔP and $\Delta\pi$ obtainable outside the membrane with a mass-transfer coefficient-like proportionality constant, $(\mathbf{Q_{sm}}/\boldsymbol{\delta_m})$. Here Q_{sm} is the *permeability coefficient* of the solvent through the membrane of thickness δ_m. In reverse osmosis literature, the quantity (Q_{sm}/δ_m) is often called the *pure water permeability constant A*. It is essentially a mass-transfer coefficient.

Note: For water to permeate through the membrane from feed to the permeate, $\Delta P > \Delta\pi$.

In expression (3.4.48) for the solute flux through the reverse osmosis membrane, the first term is usually unimportant (Merten, 1966). Therefore, the vectorial expression for the solute flux is given by

$$\boldsymbol{J}_i^* = -C_{im}D_{im}\left(\frac{\partial \ln a_{im}}{\partial \ln x_{im}}\right)_{T,P} \nabla \ln x_{im}. \tag{3.4.55}$$

Assume ideal solution in the membrane and a constant total molar density C_{tm} across the membrane. Then

$$\boldsymbol{J}_i^* = -\frac{C_{im}D_{im}}{x_{im}}\nabla x_{im} = -D_{im}\nabla C_{im}. \tag{3.4.56}$$

For one-dimensional steady state transport in the z-direction, on integrating J_{iz}^* one obtains

$$J_{iz}^* = \frac{D_{im}}{\delta_m}(C_{im}^0 - C_{im}^\delta), \tag{3.4.57}$$

where C_{im}^0 and C_{im}^δ are the solute concentrations in the membrane at the two solution–membrane interfaces.

The solute distribution coefficient κ_i may now be used to relate the membrane concentration to the external solution concentration at the two interfaces because of equilibrium:

$$\kappa_{if} = \frac{C_{im}^0}{C_{if}}; \; \kappa_{ip} = \frac{C_{im}^\delta}{C_{ip}}. \tag{3.4.58}$$

If $\kappa_{if} = \kappa_{ip} = \kappa_{im}$, then

$$J_{iz}^* = \frac{D_{im}\kappa_{im}}{\delta_m}(C_{if} - C_{ip}). \tag{3.4.59}$$

The product, $D_{im}\kappa_{im}$, is a *permeability coefficient of solute i* through the membrane of thickness δ_m. The product $(D_{im}\kappa_{im}/\delta_m)$ has been called the *solute transport parameter* in reverse osmosis literature (Sourirajan, 1970). Expression (3.4.59) is an *integrated flux expression for solute transport through a reverse osmosis membrane* in terms of the differences in concentrations of the external solutions and a mass-transfer coefficient-like quantity $(D_{im}\kappa_{im}/\delta_m)$.

A few distinguishing features of transport of any species through a membrane with respect to interphase transport in two-phase systems are as follows.

(1) There are two immiscible phase interfaces: the feed-membrane interface and the membrane-product interface.
(2) There is equilibrium partitioning of species between the solution and the membrane at both interfaces (the two partition coefficients were assumed here to be equal; in general, they are different).
(3) The feed and the product phases here are miscible, unlike conventional interphase transport in two-phase systems (liquid-liquid, gas-liquid, etc.), which are immiscible.
(4) Species transport in both feed and product phases have to be considered in general.

Both J_{sz}^* and J_{iz}^* are diffusive fluxes through the membrane. If there are large pores in the membrane, there will be convection through such pores or defects. The total flux of any species will no longer be completely diffusive and therefore should be expressed in terms of $\boldsymbol{N}_i$ (Soltanieh and Gill, 1981). A simple model for the fluxes through a reverse osmosis membrane having large pores (or defects or imperfections) has been provided by Sherwood *et al.* (1967); it is called the *solution-diffusion-imperfection model*:

$$N_{sz} = J_{sz}^* + A_p\Delta PC_{sf} = A(\Delta P - \Delta\pi) + A_p\Delta PC_{sf}, \tag{3.4.60a}$$

where C_{sf} is the solvent concentration on the high-pressure feed–membrane interface for a *well-mixed* feed. Thus there is simply Darcy or Poiseuille flow through the defects (see Section 3.4.2.3 for these types of solvent flow), and the feed solution at the feed–membrane interface moves unchanged through the defects. Similarly, for the solute,

$$N_{iz} = J_{iz}^* + A_p\Delta PC_{if} = \frac{D_{im}\kappa_{im}}{\delta_m}(C_{if} - C_{ip}) + A_p\Delta PC_{if}, \tag{3.4.60b}$$

where C_{if} is the solute concentration on the high-pressure feed–membrane interface for a *well-mixed* feed. There are *three* transport coefficients in this model, A, $(D_{im}\kappa_{im}/\delta_m)$ and A_p; the solution-diffusion model had only *two* transport coefficients, A and $(D_{im}\kappa_{im}/\delta_m)$.

In practical reverse osmosis, molar fluxes are not usually measured; volume fluxes are. The volume flux (in units of cm^3/cm^2-s), often denoted by J_{vz}, is defined by

$$J_{vz} = J_{sz}^*\overline{V_s} + J_{iz}^*\overline{V_i} \tag{3.4.60c}$$

for conditions when diffusive models hold.[30] When there is some convection through the membrane defects as well, the volume flux measured is to be obtained from the sum of N_{sz} and N_{iz}. In most cases, $N_{iz} << N_{sz}$. Thus N_{sz} may be used to determine the volume flux. In general, if $\overline{V_s}$ is used to indicate the partial molar volume of the permeate, then the volume flux through the reverse osmosis membrane can be expressed by

[30] Here $\overline{V_s}$ and $\overline{V_i}$ are the partial molar volumes of the solvent and the solute, respectively.

$$\text{volume flux}\left(\frac{\text{cm}^3}{\text{cm}^2\text{–s}}\right) \cong (N_{sz}+N_{iz})\overline{V}_s. \quad (3.4.60\text{d})$$

Note that if one were to define a permeation velocity v_z (cm/s) through the membrane, then

$$v_z \cong (N_{sz}+N_{iz})\overline{V}_s \cong (J^*_{sz}+J^*_{iz})\overline{V}_s. \quad (3.4.60\text{e})$$

This relation is to be used only if diffusive fluxes exist. In practical reverse osmosis,

$$v_z \cong N_{sz}\overline{V}_s \cong J^*_{sz}\overline{V}_s. \quad (3.4.60\text{f})$$

For solutions that are *dilute*, the concentration of species i may be represented by

$$C_{ij} = C_t x_{ij}, \quad (3.4.61\text{a})$$

where C_{ij} is the molar concentration of species i in region j ($j = f$ or p), x_{ij} is the corresponding mole fraction and C_t is the molar density of the solution, assumed constant. The osmotic pressure of dilute solutions containing a single solute may also be represented by

$$\pi(C_{ij}) = bC_{ij}, \quad (3.4.61\text{b})$$

where i refers only to the solute species. Using these simplifications, it is now possible to rewrite the solution-diffusion model fluxes for solvent and solute in reverse osmosis (expressions (3.4.54) and (3.4.59)) as

$$J^*_{sz} = A[\Delta P - bC_t(x_{if}-x_{ip})]; \quad (3.4.62\text{a})$$

$$J^*_{iz} = \left(\frac{D_{im}\kappa_{im}}{\delta_m}\right)C_t(x_{if}-x_{ip}). \quad (3.4.62\text{b})$$

Example 3.4.2 Earlier literature on reverse osmosis desalination of seawater at a high $\Delta P = 102$ atm provided the following information for a cellulose acetate membrane: $A = 8.03 \times 10^{-7}$ gmol H_2O/cm^2-s-atm; $\pi(C_{if}) = bC_{if} = 45.7$ atm; $(D_{im}\kappa_{im}/\delta_m)$ for salt $= 1.774 \times 10^{-5}$ cm/s; $x_{if} = 17.7 \times 10^{-3}$; $x_{ip} = 0.4 \times 10^{-3}$. The membrane has a high rejection for the solute i, which is NaCl. Calculate the values of water flux and the salt flux through the membrane. Calculate the salt concentration on two sides of the membrane, assumed to be of the solution-diffusion type. Calculate the salt rejection R. (*Note*: Current membranes are more productive. Therefore ΔP is around 54.4–68 atm (800–1000 psi). One should calculate $\Delta\pi$ based on the final concentration of seawater desired.)

Solution Assume that the osmotic pressure of dilute solutions can be represented by $\pi(C_{ij}) = bC_{ij}$. Therefore $\pi(C_{ip}) = bC_{ip} = C_{ip}\left(\pi(C_{if})/C_{if}\right)$. We will need to calculate (1) C_{if} and (2) C_{ip} to determine $\pi(C_{ip})$.

(1) Now

$$x_{if} = 17.7 \times 10^{-3} = \frac{C_{if}}{C_{if}+C_{sf}} = \frac{C_{if}}{C_{if}+\frac{1}{18.05}},$$

where the partial molar volume of pure water is 18.05 cm^3/gmol ($\overline{V}_s \cong (1/C_{sf})$). We have

$$C_{if}(1-17.7\times10^{-3}) = \frac{17.7}{18.05}\times10^{-3} \Rightarrow C_{if}$$
$$= \frac{17.7\times10^{-3}}{18.05\times0.9823} = \frac{17.7}{18.05\times0.9823}\frac{\text{gmol salt}}{\text{liter}}$$
$$= 58.4 \text{ g salt/liter}.$$

(2) $x_{ip} = 0.4\times10^{-3} = C_{ip}/(C_{ip}+C_{sp})$. This permeate is so dilute that we incur very little error by writing

$$C_{ip} = 0.4\times10^{-3}C_{sp} = 0.4\times10^{-3}\times\frac{1}{18.05}\frac{\text{gmol}}{\text{cm}^3} = \frac{0.4}{18.05}\frac{\text{gmol salt}}{\text{liter}}.$$

Now we can obtain the solvent flux:

$$J^*_{sz} = A\{\Delta P - \pi(C_{if}) + \pi(C_{ip})\}$$
$$= 8.03\times10^{-7}\left\{102-45.7+\frac{C_{ip}}{C_{if}}\pi(C_{if})\right\}\frac{\text{gmol H}_2\text{O}}{\text{cm}^2\text{-s}}$$
$$= 8.03\times10^{-7}\left\{102-45.7\left(1-\frac{0.4\times18.05\times0.9823}{18.05\times17.7}\right)\right\}$$
$$= 8.03\times10^{-7}\{102-45.7\times0.978\}$$
$$= 8.03\times10^{-7}\times(102-44.69)$$
$$= 4.6\times10^{-5}\frac{\text{gmol H}_2\text{O}}{\text{cm}^2\text{-s}} = \frac{4.6\times10^{-5}\times18.05\ \text{cm}^3}{\text{cm}^2\text{-s}}$$
$$\times10^{-3}\frac{\text{liter}}{\text{cm}^3}\times10^4\frac{\text{cm}^2}{\text{m}^2}\times3600\times24\frac{\text{s}}{\text{day}}$$
$$= 4.6\times10^{-5}\times18.05\times10\times3600\times24\frac{\text{liter}}{\text{m}^2\text{-day}}$$
$$= 717\frac{\text{liter H}_2\text{O}}{\text{m}^2\text{-day}};$$

$$J^*_{iz} = \left(\frac{D_{im}\kappa_{im}}{\delta_m}\right)(C_{if}-C_{ip})$$
$$= 1.774\times10^{-5}\left(\frac{17.7\times10^{-3}}{18.05\times0.9823}-\frac{0.4\times10^{-3}}{18.05}\right)$$
$$= \frac{1.774\times10^{-5}\times10^{-3}}{18.05}(18.0-0.4) = \frac{1.774\times10^{-8}}{18.05}\times17.6\frac{\text{gmol}}{\text{cm}^2-\text{s}}$$
$$= \frac{1.774\times17.6}{18.05}\times10^{-8}\times10^4\times3600\times24\frac{\text{gmol}}{\text{m}^2-\text{day}}$$
$$= 14.94\frac{\text{gmol salt}}{\text{m}^2-\text{day}};$$

$$R = 1-\frac{C_{ip}}{C_{if}} = 1-\frac{0.4\times10^{-3}/18.05}{17.7\times10^{-3}/18.05\times0.9823}$$
$$= 1-\frac{0.4\times0.9823}{17.7} = 1-0.022 = 0.978.$$

For a dilute solution having more than one solute (say two solutes, 2 and 3), the osmotic pressures may be represented by

$$\pi_f = b(C_{2f}+C_{3f}) \quad \text{and} \quad \pi_p = b(C_{2p}+C_{3p}). \quad (3.4.62\text{c})$$

Using the approximation (3.4.61a) representation for each solute species 2 and 3 (say), one can develop flux expressions for J^*_{sz}, J^*_{2z} *and* J^*_{3z} as follows:

$$J_{sz}^* = A[\Delta P - bC_t(x_{2f} + x_{3f}) + bC_t(x_{2p} + x_{3p})]; \quad (3.4.63a)$$

$$J_{2z}^* = \left(\frac{D_{2m}\kappa_{2m}}{\delta_m}\right)C_t(x_{2f} - x_{2p}); \quad (3.4.63b)$$

$$J_{3z}^* = \left(\frac{D_{3m}\kappa_{3m}}{\delta_m}\right)C_t(x_{3f} - x_{3p}). \quad (3.4.63c)$$

There are a number of other models of transport of solvent and solute through a reverse osmosis membrane: the Kedem-Katchalsky model, the Spiegler-Kedem model, the frictional model, the finely porous model, the preferential sorption-capillary flow model, etc. Most of these models have been reviewed and compared in great detail by Soltanieh and Gill (1981). We will restrict ourselves in this book to the solution-diffusion and solution-diffusion-imperfection flux expressions for a number of reasons. First, the form of the solution-diffusion equation is most commonly used and is also functionally equivalent to the preferential sorption-capillary flow model. Secondly, the solution-diffusion-imperfection model is functionally representative of a number of more exact three-transport-coefficient models, even though the transport coefficients in this model are concentration-dependent.

So far, the integrated flux expressions for the transport of solvent and solute through a membrane in reverse osmosis assumed *well-mixed feed and permeate solutions*. Therefore any transport resistances in the two liquid phases on two sides of the membrane were eliminated. In practical reverse osmosis, there is, however, significant transport resistance for salt on the feed side. This subject will be briefly treated now.

In practical reverse osmosis with a positive $(\Delta P - \Delta\pi)$, there is considerable flow of solvent from the feed to the permeate. However, the membrane is designed to reject the solute species. Thus, from the feed solution next to the membrane, solvent is continuously withdrawn through the membrane, whereas the solute species is not. This leads to a build-up of solute concentration near the membrane-feed solution interface (Figure 3.4.7) in excess of the bulk feed solute concentration C_{if}. This phenomenon is called *concentration polarization*. The feed-membrane interface is now exposed to a solute concentration C_{il}^0 instead of $C_{if}(\leq C_{il}^0)$. Consequently, the solvent and solute flux expressions in the solution-diffusion model for one solute in a solvent system are changed to

$$J_{sz}^* = A(\Delta P - \{\pi_f^0 - \pi_p\}) = A(\Delta P - bC_{il}^0 + bC_{ip}) = A(\Delta P - bC_t x_{il}^0 + bC_t x_{ip}); \quad (3.4.64a)$$

$$J_{iz}^* = \left(\frac{D_{im}\kappa_{im}}{\delta_m}\right)C_t(x_{il}^0 - x_{ip}) = \frac{D_{im}\kappa_{im}}{\delta_m}(C_{il}^0 - C_{ip}). \quad (3.4.64b)$$

The solvent flux is decreased now and the solute flux is increased. It is desirable to know the fluxes under concentration polarization for which C_{il}^0 has to be determined.

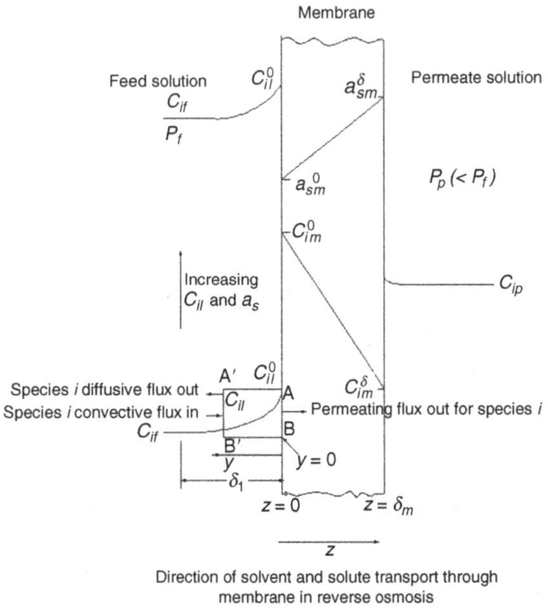

Figure 3.4.7. Transport in reverse osmosis with concentration polarization.

The relationship between C_{il}^0 and C_{if} may be obtained by considering the transport of solute i through the membrane, the diffusion of solute i from the membrane-feed solution interface region to the bulk solution and the volume flux through the membrane. Such a relationship will be derived next. *Note*: If the mixing in the feed solution is vigorous enough to make $C_{il}^0 \cong C_{if}$, then such a relation is no longer needed.

Consider a control volume A′B′BA in the feed solution film region (see Figure 3.4.7). Let the surface A′B′ be located at a distance y from the feed-membrane interface. Then, at steady state, the rate at which solute i enters the control volume via A′B′ must equal the rate at which it leaves the control volume surface AB, which coincides with the feed-membrane interface. There are three sources of solute flux for the control volume:

$$\left.\begin{array}{ll} \text{at } y = y \text{ (surface A}'\text{B}'\text{)} \left\{ \begin{array}{l} \text{diffusive flux out} = -D_{if}\dfrac{dC_{il}}{dy} \\ \text{convective flux in} = |v_y|C_{il} \end{array} \right. \\ \text{at } y = 0 \text{ (surface AB)} \quad \text{permeating flux out} = |v_y|C_{ip} \end{array}\right\}. \quad (3.4.65a)$$

Therefore

$$|v_y|C_{il} - |v_y|C_{ip} = -D_{if}\frac{dC_{il}}{dy}, \quad (3.4.65b)$$

where $|v_y|$ is the magnitude of the permeation velocity through the membrane. We have assumed it to be equal to the velocity normal to the membrane in the liquid film region. If equation (3.4.65b) is rearranged and integrated from $y = 0$ to $y = \delta_l$, the thickness of feed side liquid film, then

$$\int_{C_{il}^0}^{C_{if}} \frac{\mathrm{d}C_{il}}{C_{il} - C_{ip}} = -\int_0^{\delta_l} \frac{|v_y|}{D_{if}} \mathrm{d}y;$$

$$\ell n \left(\frac{C_{il}^0 - C_{ip}}{C_{if} - C_{ip}} \right) = \frac{|v_y| \delta_l}{D_{if}} = \frac{|v_y|}{(D_{if}/\delta_l)} = \frac{|v_y|}{k_{if}},$$

where $k_{if} = (D_{if}/\delta_l)$ is the mass-transfer coefficient in the feed solution. So

$$(C_{il}^0 - C_{ip}) = (C_{if} - C_{ip}) \exp[|v_y|/k_{if}]. \quad (3.4.65c)$$

By relation (3.4.60e), $|v_y|$ is obtained from the two fluxes J_{sz}^* and J_{iz}^* (or N_{sz} *and* N_{iz}) expressed by (3.4.64a) and (3.4.64b). Thus J_{sz}^* and J_{iz}^* may be calculated in terms of the two bulk solute concentrations C_{if} and C_{ip} by solving the three equations simultaneously. If J_{sz}^* and J_{iz}^* are to be expressed only in terms of C_{if}, an additional relation is needed to express C_{ip} in terms of the other quantities.

It is obvious that the flux expressions (3.4.64a) and (3.4.64b) are of an implicit type; both J_{sz}^* and J_{iz}^* depend on C_{il}^0, which in turn depends on J_{sz}^* and J_{iz}^*. It is possible, however, to have explicit flux expressions under the limiting conditions of $C_{ip} \ll C_{il}^0$ and $|v_y| \ll k_{if}$. Then equation (3.4.65c) may be reduced to

$$C_{il}^0 \cong C_{if} \left(1 + \frac{|v_y|}{k_{if}} \right). \quad (3.4.65d)$$

Further, in practical reverse osmosis,

$$v_y \cong N_{sz} \overline{V_s} = J_{sz}^* \overline{V_s}.$$

Therefore,

$$J_{sz}^* \cong A \left(\Delta P - bC_{if} \left\{ 1 + \frac{J_{sz}^* \overline{V_s}}{k_{if}} \right\} \right),$$

leading to

$$J_{sz}^* = \frac{A(\Delta P - bC_{if})}{(1 + AbC_{if} \overline{V_s}/k_{if})}, \quad (3.4.66a)$$

as demonstrated by Rao and Sirkar (1978) for tubular RO desalination and Sirkar *et al.* (1982) for spiral-wound membranes. (See Section 7.2.1.2 for these separation devices.) Similarly,

$$J_{iz}^* \cong \left(\frac{D_{im} \kappa_{im}}{\delta_m} \right) \left\{ C_{if} \left(1 + \frac{J_{sz}^* \overline{V_s}}{k_{if}} \right) \right\}. \quad (3.4.66b)$$

Prasad and Sirkar (1985) may be consulted for details of such approximations for multisolute systems as well as for earlier references providing a quantitative basis for these approximations in a single-solute system.

A major difference between the solute transfer from feed solution to permeate solution through a membrane in reverse osmosis (RO) and the interphase solute transfer shown in Figures 3.4.1 and 3.4.2 is the following: even though there is some solute transfer through the RO membrane, the solute concentration builds up at the feed-membrane interface. It is possible to have such a low value of k_{il} and such a high value of C_{il}^0 that $\Delta P = \Delta \pi$, and the reverse osmosis process stops. Such a situation is unlikely to be encountered in conventional nonmembrane interphase transfer processes.

3.4.2.1.1 Pervaporation and liquid permeation When a volatile liquid mixture is imposed on the feed side of a nonporous membrane and the other side of the membrane has a vapor/gaseous phase, the process is called *pervaporation*, a combination of permeation and evaporation, permeation through the membrane from the feed side and evaporation on the other side (Figure 3.4.5(a)). (When the phase on the other side of the membrane is a liquid, the process is called *liquid permeation.*) The rate of transport of a feed species through a nonporous amorphous polymeric film is apparently a special case of ordinary diffusion through the film, except the species diffusivity is highly concentration-dependent due to the swelling of the polymer by the components of the feed solution (Long, 1965). At steady state, the permeation flux of a species in the z-direction across the film thickness is described by Fick's first law (as long as the pressure difference between the two sides is small):

$$J_{iz}^* = -D_{im} \frac{\mathrm{d}C_{im}}{\mathrm{d}z}. \quad (3.4.67a)$$

The value of D_{im} depends strongly on the species i concentration in the membrane (acting as a solvent swelling the membrane polymer). This concentration dependence has been expressed in a number of ways:

$$D_{im} = D_{im0} \exp(\gamma_i C_{im}); \; D_{im} = D_{im0} C_{im};$$
$$D_{im} = D_{im0}(1 + \gamma_i C_{im}^n). \quad (3.4.67b)$$

In the first and third expressions, the value D_{im0} of the diffusivity of the species i in the membrane is obtained in the limit of zero concentration of species i in the membrane. The quantity γ_i is essentially a plasticization constant (Long, 1965); it illustrates the magnitude of the effect of solvent concentration on the diffusive mobility of species i in the membrane. When γ_i is large, a small value of C_{im} can cause a large change in D_{im} in the case of exponential concentration dependence.

Integrate now equation (3.4.67a) using the exponentially concentration-dependent diffusion coefficient expression from (3.4.67b) between the limits of $z = 0$ (feed mixture) and $z = \delta_m$ (permeate/product, gaseous/vapor phase):

$$\int_0^{\delta_{im}} J_{iz}^* \, dz = -D_{im0} \int_{C_{im}^0}^{C_{im}^\delta} \exp(\gamma_i C_{im}) dC_{im}$$

$$\Rightarrow J_{iz}^* = \frac{D_{im0}}{\gamma_i \delta_m}\left(\exp(\gamma_i C_{im}^0) - \exp(\gamma_i C_{im}^\delta)\right); \quad (3.4.67c)$$

$$J_{iz}^* = \frac{D_{im}}{\delta_m}(C_{im}^0 - C_{im}^\delta), \quad (3.4.67d)$$

where

$$D_{im} = \frac{D_{imo}}{\gamma_i}\frac{\left(\exp(\gamma_i C_{im}^0) - \exp(\gamma_i C_{im}^\delta)\right)}{(C_{im}^0 - C_{im}^\delta)} \quad (3.4.67e)$$

is the effective Fick's law diffusivity under the conditions valid for the system.

This analysis (Long, 1965) is considered quite useful for a single-solvent species (see Example 3.4.3). However, when the feed is a mixture, the presence of both solvent species should be considered. Use now expression (3.4.48) without the pressure gradient term:

$$J_{iz}^* = -D_{im} C_{im} \frac{d \ln a_{im}}{dz}. \quad (3.4.67f)$$

For a binary feed liquid mixture (i =1, 2) and the polymer membrane ($i = m$) constituting a ternary polymeric system, Flory-Huggins theory (Flory, 1953) should be used to express $\ln a_{1m}$ for i =1 (similarly for $i = 2$) as (ϕ denotes a volume fraction):

$$\ln(a_{1m}) = \ln(\phi_{1m}) + (1 - \phi_{1m}) - \phi_{2m}\frac{\overline{V}_1}{\overline{V}_2} - \phi_m \frac{\overline{V}_1}{\overline{V}_m}$$
$$+ (\chi_{12}\phi_{1m} + \chi_{1m}\phi_m) \times (\phi_{2m} + \phi_m) - \chi_{2m}\frac{\overline{V}_1}{\overline{V}_2}\phi_{2m}\phi_m. \quad (3.4.67g)$$

Here, ϕ_{im} is the volume fraction of i in the membrane, ϕ_m is the polymer volume fraction in membrane and χ_{ij} is the Flory interaction parameter between species i and j. Further, D_{im} is a function of ϕ_{1m} and ϕ_{2m}. For detailed analysis, see Neel (1995). If C_{im} in (3.4.67a) is expressed as $(\phi_{im}/\overline{V}_i)$ and D_{im} in (3.4.67b) is expressed as $D_{imo}\exp(\gamma_i \, \phi_{im})$, then expression (3.4.67c)) is reduced to

$$J_{iz}^* = \frac{D_{imo}}{\gamma_i \overline{V}_i \delta_m}(\exp(\gamma_i \phi_{im}^o) - \exp(\gamma_i \phi_{im}^\delta)). \quad (3.4.67h)$$

Example 3.4.3 Schaetzel *et al.* (2004) have studied the pervaporation of a water-ethanol mixture through a polyvinylalcohol based membrane as a function of feed water concentration at 60 °C. They have also measured the equilibrium volume fraction of water and ethanol in the membrane for such feed mixtures. When there is very high vacuum in the permeate side, it is possible to correlate the water flux $J_{H_2O}^*$ (mol/m^2-s) with the feed side water and ethanol volume fractions $\phi_{H_2O,m}|_0$, $\phi_{C_2H_5OH,m}|_0$, respectively, in the membrane in the following fashion:

$$J_{H_2O}^* = 0.002048\left[\exp\left(11.155(\phi_{H_2O,m}|_0 + \phi_{C_2H_5OH,m}|_0\right) - 1\right].$$

However, $\phi_{C_2H_5OH,m}|_0$ may be neglected.

Determine the value of $D_{H_2O\,m0}$ and the plasticization constant for water. The membrane thickness is 2.5 μm; $\overline{V}_{H_2O} = 18.05\ \text{cm}^3/\text{gmol}$.

Solution Flux expression (3.4.67h) may be written in terms of volume fraction of water in the membrane ϕ_{H_2Om} as follows ($i = 1 = H_2O$):

$$J_{1z}^* = \frac{D_{1m0}}{\overline{V}_1 \gamma_1 \delta_m}\left(\exp(\gamma_1 \phi_{1m}^0) - \exp(\gamma_1 \phi_{1m}^\delta)\right).$$

But $\exp(\gamma_1 \phi_{H_2Om}^\delta) = 1$ since $\phi_{H_2Om}^\delta \to 0$ under high vacuum. Therefore

$$J_{H_2Oz}^* = J_{1z}^* = \frac{D_{1m0}}{\overline{V}_1 \gamma_1 \delta_m}\left(\exp(\gamma_1 \phi_{1m}^0) - 1\right).$$

We get $\gamma_1 = 11.155$. Therefore

$$\frac{D_{1m0}}{\overline{V}_1 \delta_{1m} \times 11.155} = 0.002048\,\frac{\text{mol}}{\text{m}^2\text{-s}}$$

$$\Rightarrow D_{1mo} = 0.002048\,\frac{\text{gmol}}{\text{m}^2\text{-s}}\,18.05\,\frac{\text{cm}^3}{\text{gmol}} \times 10^{-6}\,\frac{\text{m}^3}{\text{cm}^3}$$
$$\times 11.155 \times 2.5 \times 10^{-6}\text{m}$$
$$\Rightarrow D_{1m0} = 36.9 \times 10^{-15} \times 11.155 \times 2.5$$
$$= 1.02 \times 10^{-12}\ \text{m}^2/\text{s}.$$

3.4.2.2 *Gas/vapor permeation through nonporous membranes*

Three types of nonporous membranes are relevant here: polymeric, inorganic and liquid. We start this section with polymeric membranes, which are used most often. Later, we briefly treat inorganic and liquid membranes.

We observed in Section 3.3.7.3 that permanent gases dissolve in a nonporous polymeric rubbery membrane according to Henry's law as if the membrane were a liquid. The diffusion of dissolved gases in such a membrane is described by Fick's first law of diffusion, exactly as if the gas were diffusing in a liquid by ordinary diffusion:

$$J_i^* = -D_{im}\nabla C_{im}. \quad (3.4.68)$$

For one-dimensional transport in the positive z-direction (Figure 3.4.5(b)) through a membrane of thickness δ_m exposed to a gas mixture with a partial pressure $p_{if}(= P_f x_{if})$ of species i at $z = 0$ and to a gas mixture having a lower partial pressure $p_{ip}(= P_p x_{ip})$ of species i at $z = \delta_m$,

$$J_{iz}^* = -D_{im}\frac{dC_{im}}{dz}. \quad (3.4.69)$$

Integrating at steady state across the membrane thickness δ_m, one obtains

$$J_{iz}^* = \frac{D_{im}}{\delta_m}(C_{im}^0 - C_{im}^\delta) = \frac{D_{im}}{\delta_m}\Delta C_{im}, \quad (3.4.70)$$

where ΔC_{im} is the concentration difference of species i across the membrane thickness.

Using Henry's law (3.3.80), where S_{im} is constant in the partial pressure range p_{if} to p_{ip} $(p_{if} > p_{ip})$,

$$C^0_{im} = S_{im}p_{if};\ C^{\delta}_{\mathrm{m}} = S_{im}p_{ip}. \qquad (3.4.71)$$

Then the flux J^*_{iz} may be expressed as

$$J^*_{iz} = \frac{D_{im}S_{im}}{\delta_m}(p_{if} - p_{ip}) = \frac{Q_{im}}{\delta_m}(p_{if} - p_{ip}) = \frac{Q_{im}}{\delta_m}\Delta p_i. \quad (3.4.72)$$

Here, Q_{im} is the *permeability coefficient* of species i through the membrane and is a product of D_{im}, the diffusion coefficient, and S_{im}, the solubility coefficient, of species i in the membrane. For permanent gases and rubbery nonporous polymeric membranes, Q_{im} for any species is constant at any temperature. The unit commonly used for Q_{im} is cm^3(STP)-cm/cm^2-s-cm Hg. The unit "barrer," often used for Q_{im}, is 10^{-10} cm^3(STP)-cm/cm^2-s-cm Hg; cm^3(STP) is sometimes written as scc. The units for δ_m, p_i and J^*_{iz} are cm, cm Hg and cm^3(STP)/cm^2-s, respectively. In general, the gas concentration in the membrane at any interface ($z = 0$ or δ_m) is related to the gas partial pressure p_i by

$$C_{im} = S_{im}(C_{im})p_i, \qquad (3.4.73)$$

where the solubility coefficient $S_{im}(C_{im})$ is a function of the concentration C_{im}. For low levels of C_{im}, with a number of permanent gases like H_2, He, Ar, O_2, N_2, CH_4, at $T > T_{ci}$ (critical temperature of gas i), Henry's law is observed (Stern and Frisch, 1981) and

$$\lim_{C_{im}\to 0} S_{im}(C_{im}) = S_{im}, \quad \text{a constant.} \qquad (3.4.74)$$

(For deviations from this behavior, see equations (3.3.81) and (3.3.82a, b).

Similarly, the diffusion coefficient D_{im} of species i in the membrane is also generally dependent on C_{im}, but

$$\lim_{C_{im}\to 0} D_i(C_{im}) = D_{im}, \quad \text{a constant.} \qquad (3.4.75a)$$

When such a limiting behavior does not hold, Q_{im} is defined by

$$Q_{im} = \frac{1}{(p_{if} - p_{ip})}(-)\int_0^{\delta_m} D_{im}\, \mathrm{d}C_{im}. \qquad (3.4.75b)$$

This process of gas transport through a membrane is called *permeation*, and the mechanism has been identified as *solution-diffusion*. Gas species i dissolves at the feed-membrane interface ($z = 0$); by molecular diffusion, the dissolved gas molecules move through the membrane and are finally desorbed into the product gas phase at the product-membrane interface ($z = \delta_m$). Under the simplest of conditions, each species in a mixture diffuses independently of the others according to flux expression (3.4.72). The nature of the dependence of D_{im} on the effective diameter of the gas molecules, the temperature and the polymer for an activated diffusion is illustrated in Section 4.3.3 for an amorphous polymer.

Ideally, it is necessary to use mass-transfer equations in the feed-gas phase and the product-gas phase along with the permeation equation (3.4.72). However, in general, the gas permeation rate through a nonporous membrane is so slow that mass-transfer equations in the feed-gas and product-gas phases are not needed. Note that, for species i transport through the membrane from the feed to the product gas, $p_{if} > p_{ip}$, but P_f need not be greater than P_p, although in practice it generally is. In such a case, flux expression (3.4.72) may also be expressed as

$$J^*_{iz} = \frac{Q_{im}}{\delta_m}(p_{if} - p_{ip}) = \frac{Q_{im}}{\delta_m}(P_f x_{if} - P_p x_{ip}), \qquad (3.4.76)$$

where x_{if} is the mole fraction in the feed mixture and x_{ip} is the mole fraction in the product mixture.

If the gas is *condensable* and/or we have a vapor in the gas mixture, the solubility of the permeating species in the polymeric membrane is higher. The condensable organic/inorganic vapor/gas may plasticize the membrane; the polymeric chains have higher segmental mobility. As a result, the diffusion coefficient is increased drastically. The diffusion coefficient D_{im} can vary with the concentration C_{im} of species in the membrane linearly, or even higher, exponentially:

$$D_{im} = D_{im0}\exp(A_i C_{im}). \qquad (3.4.77a)$$

Assuming the solubility coefficient of species i to be constant, S_{im0}, corresponding to Henry's law, the permeability coefficient Q_{im} may be expressed as

$$Q_{im} = S_{im0}D_{im0}\exp(A_i C_{im}) = Q_{im0}\exp(A_i C_{im}), \quad (3.4.77b)$$

where $Q_{im0} = S_{im0}D_{im0}$. This exponential variation of the permeability coefficient is observed at high partial pressures of condensable gas/vapor species i in rubbery amorphous membranes. Introducing such a variation of D_{im} with C_{im} into Fick's first law, we get

$$J^*_{iz} = D_{im}\frac{\mathrm{d}C_{im}}{\mathrm{d}z} = D_{im0}\exp(A_i C_{im})\frac{\mathrm{d}C_{im}}{\mathrm{d}z}. \qquad (3.4.77c)$$

Integrate it from $z = 0$ to $z = \delta_m$ to obtain

$$\begin{aligned} J^*_{iz} &= \frac{D_{im0}}{A_i\delta_m}\left[\exp(A_i C^0_{im}) - \exp(A_i C^{\delta}_{im})\right] \\ &= \frac{D_{im0}}{A_i\delta_m}\left[\exp(A_i S_{im0} p_{if}) - \exp(A_i S_{im0} p_{ip})\right]. \end{aligned} \qquad (3.4.77d)$$

One could rewrite this as

$$J^*_{iz} = \frac{D_{im0}}{A_i\delta_m}\left[\exp(A_i S_{im0}\Delta p_i) - 1\right]\exp(A_i S_{im0} p_{ip}). \quad (3.4.77e)$$

Taking logarithms of both sides and plotting log (J^*_{iz}) against p_{ip},

$$\log J^*_{iz} = \log\left[\frac{D_{im0}}{A_i\delta_m}\{\exp(A_i S_{im0}\Delta p_i) - 1\}\right] + A_i S_{im0}\, p_{ip}, \qquad (3.4.77f)$$

a straight line results; the slope yields $A_i S_{im0}$ and the intercept can be used to determine D_{im0}. The permeability coefficient Q_{im} will therefore be a strong function of Δp_i. Knowing S_{im0}, the slope yields A_i.

For gas transport through nonporous polymeric membranes of the *glassy type* (i.e. the temperature of permeation, $T < T_g$, the glass transition temperature of the polymer), it has been postulated that the dissolved gas species exists in the membrane in two forms: the Henry's law species and the Langmuir species. The Henry's law species dissolves according to the Henry's law and diffuses in the manner described earlier. The Langmuir species dissolves according to the Langmuir isotherm (see equation (3.3.81)); further, it has a limited mobility at a fraction of that of the Henry's law species (Paul and Koros, 1976). One-dimensional transport of gas molecules i through such a nonporous glassy polymeric membrane according to this *dual sorption-dual transport* model is then described by

$$J_{iz}^* = -D_{iD}\frac{\partial C_{im}^d}{\partial z} - D_{iH}\frac{\partial C_{im}^H}{\partial z} = N_{iz}, \tag{3.4.78}$$

where D_{iD} is the diffusion coefficient of the Henry's law species whose concentration in the membrane is C_{im}^d, D_{iH} is the diffusion coefficient of the Langmuir species of molar concentration C_{im}^H and $(D_{iH}/D_{iD}) \leq 1$.

Paul and Koros (1976) have solved the one-dimensional unsteady state diffusion equation (use equation (3.2.2); alternatively simplify equation (6.2.3a)),

$$\frac{\partial C_{im}}{\partial t} = -\frac{\partial N_{iz}}{\partial z}, \tag{3.4.79}$$

for diffusion of pure species i through the glassy membrane. They assumed that the ratio (D_{iH}/D_{iD}) was constant; further, $C_{im} = C_{im}^d + C_{im}^H$. Additionally, the C_{im}^d and C_{im}^H species are in equilibrium with each other. For the limiting case of $C_{im}^\delta = 0$, i.e. $p_{ip} = 0$, the species profile in the membrane is linear, and the permeability of pure species i, Q_{im}, was found to be

$$Q_{im} = \frac{N_{iz}\delta_m}{p_{if}} = \frac{N_{iz}\delta_m}{P_f} = S_{im}D_{iD}\left(1 + \frac{\left(\frac{D_{iH}}{D_{iD}}\right)\frac{C'_{Hi}b_i}{S_{im}}}{1 + b_i p_{if}}\right). \tag{3.4.80}$$

The quantities C'_{Hi}, b_i and S_{im} are available from the membrane-gas equilibrium behavior (3.3.81).

Note the difference between this permeability expression and that for Q_{im} for rubbery polymeric membranes $(= S_{im}D_{iD}$ with $D_{iD} = D_{im})$. As the feed pressure increases, the permeability coefficient in the glassy membrane decreases. The permeability coefficient has the highest value in the limit of $p_{if}(= P_f) \rightarrow 0$. The integrated flux expression continues to be simply given by

$$N_{iz} = \frac{Q_{im}}{\delta_m}(p_{if} - p_{ip}). \tag{3.4.81a}$$

For the permeation of a binary gas mixture through a glassy polymeric membrane, Koros *et al.* (1981) have developed expressions for the permeability coefficient of each species (their equilibrium behavior is described by (3.3.82a,b)):

$$Q_{1m} = S_{1m}D_{1D}\left[1 + \frac{\left(\frac{D_{1H}}{D_{1D}}\right)\frac{C'_{H1}b_1 p_{1f}/(p_{1f} - p_{1p})}{S_{1m}}}{1 + b_1 p_f + b_2 p_{2f}} - \frac{\left(\frac{D_{1H}}{D_{1D}}\right)\frac{C'_{H1}b_1 p_{1p}/(p_{1f} - p_{1p})}{S_{1m}}}{1 + b_1 p_{1p} + b_2 p_{2p}}\right]. \tag{3.4.81b}$$

The expression for Q_{2m} is

$$Q_{2m} = S_{2m}D_{2D}\left[1 + \frac{\left(\frac{D_{2H}}{D_{2D}}\right)\frac{C'_{H2}b_2 p_{2f}/(p_{2f} - p_{2p})}{S_{2m}}}{1 + b_1 p_{1f} + b_2 p_{2f}} - \frac{\left(\frac{D_{2H}}{D_{2D}}\right)\frac{C'_{H2}b_2 p_{2p}/(p_{2f} - p_{2p})}{S_{2m}}}{1 + b_1 p_{1p} + b_2 p_{2p}}\right]. \tag{3.4.81c}$$

The permeation flux expressions (3.4.76) and (3.4.81a) are valid for membranes whose properties do not vary across the thickness. Most practical gas separation membranes have an asymmetric or composite structure, in which the properties vary across the thickness in particular ways. Asymmetric membranes are made from a given material; therefore the properties varying across δ_m are pore sizes, porosity and pore tortuosity. Composite membranes are made from at least two different materials, each present in a separate layer. Not only does the intrinsic Q_{im} of the material vary from layer to layer, but also the pore sizes, porosity and pore tortuosity vary across δ_m. At least one layer (in composite membranes) or one section of the membrane (in asymmetric membranes) must be nonporous for efficient gas separation by gas permeation. The flux expressions for such structures can be developed only when the transport through porous membranes has been studied.

Flux expressions (3.4.72), (3.4.76) and (3.4.81a) for a diffusing gas species i through a membrane of thickness δ_m describe the observed behavior at steady state achieved after an initial unsteady period. The initial unsteady behavior begins when the gas containing the permeating species i at concentration C_{if} (partial pressure p_{if}) is introduced at time $t = 0$ to the $z = 0$ surface of the membrane, the feed side. The rate of penetration of the membrane by species i is governed by the unsteady state diffusion equation (3.4.79), where $N_{iz} = J_{iz}^*$ is governed by Fick's first law (3.4.68). After some time, the species appears through the membrane into the permeate side, where $z = \delta_m$. For a constant value of D_{im}, the solution of equation (3.4.79) may be obtained for the downstream boundary condition of $p_{ip} \cong 0 (C_{ip} \cong 0)$; from this solution one can calculate the moles of the diffusing species, $m_{iD}|_t$, that have crossed the membrane in time t from $t = 0$ to be (Crank and Park, 1968)

$$\frac{m_{iD}|_t}{\delta_m C_{if}} = \frac{D_{im}t}{\delta_m^2} - \frac{1}{6} - \frac{2}{\pi^2}\sum_1^\infty \frac{(-1)^n}{n^2} \exp\left(-\frac{D_{im}n^2\pi^2 t}{\delta_m^2}\right). \tag{3.4.82a}$$

As time t increases and tends to infinity, a steady state is approached where the exponential term is negligibly small; then, a plot of m_{iD} vs. t becomes a straight line, i.e.

$$m_{iD}|_t = \frac{D_{im}C_{if}}{\delta_m}\left(t - \frac{\delta_m^2}{6D_{im}}\right). \tag{3.4.82b}$$

The intercept of this line on the time axis is $(\delta_m^2/6D_{im})$, called the *time lag* in simple diffusive permeation through a membrane for constant diffusion coefficient in linear diffusion. The value of the time lag provides a good order of magnitude estimate of the time from $t = 0$ after which steady state permeation through the membrane may be assumed. Correspondingly, equations (4.3.24) and (4.3.25) for gas permeation in a device (see also Chapters 6, 7 and 8) are valid only after this initial unsteady state. Experimental measurement of the time lag allows one to measure the value of D_{im}. Knowing Q_{im} from steady state experiments, one can determine S_{im}.

Gas transport through *nonporous inorganic membranes* falls into two categories. It is known that the conventional solution-diffusion permeation mechanism is valid for nonporous membranes of silica, zeolite and inorganic salts. It is no longer so when the membrane is *metallic* in nature (Hwang and Kammermeyer, 1975). Diatomic gases such as O_2, H_2 and N_2 dissolve atomically in the metallic membrane (see (3.3.67)). While a conventional flux expression is valid for atomic species i dissolved in the membrane, i.e.

$$N_{iz} = D_{im}\frac{C_{im}^0 - C_{im}^\delta}{\delta_m}, \tag{3.4.83}$$

the flux expression changes to

$$N_{iz} = \frac{Q_{im}}{\delta_m}\left(p_{if}^{1/2} - p_{ip}^{1/2}\right) \tag{3.4.84}$$

in terms of the species partial pressures in the gas phases, in view of the solubility relation (3.3.70).

As long as there is no chemical reaction of the dissolved gas species with the liquid, the gas species flux expression through a thin liquid layer *acting as* a membrane is identical to that through a rubbery polymeric membrane, as discussed earlier. The major difference comes about in the magnitude of the permeability coefficients. The diffusion coefficient of a gas species in a liquid membrane will, in general, be at least two to three orders of magnitude larger than that through a rubbery polymeric membrane. Furthermore, the magnitude of the solubility coefficient of the gas in the liquid will be larger. However, the liquid layer thickness will, in general, be considerably higher.

3.4.2.3 *Liquid transport through porous membranes*

The nature of solute transport in a porous membrane with solvent flow is complex; the regime is influenced strongly by the ratio of solute diameter to pore diameter at any given transport rate of the solvent. The *solvent transport rate* is, however, commonly described by *Poiseuille flow* through the pores. The nonporous section of the membrane is assumed impermeable. There are two categories of porous membranes: *Ultrafiltration* (UF) membranes, having pore diameters in the range 1–30 nm; *microfiltration* (MF) membranes, having pore diameters between 0.1 and 10 μm (100–10 000 nm). All pores in the membrane may have the same pore size, but generally there is a pore size distribution (Zeman and Zydney, 1996, chap. 4; Cheryan, 1998; Kulkarni *et al.*, 2001). Figure 3.4.5(c) illustrates the basic schematic of such processes. According to IUPAC (the International Union of Pure and Applied Chemistry), pores less than 2 nm in a diameter are called micropores, pores between 2 and 50 nm are called mesopores and larger ones are known as macropores. Correspondingly, microporous membranes should have only pores less than 2 nm, etc. The membrane literature does not follow this practice uniformly.

Consider a membrane where the pores of uniform diameter[31] $2r_p$ occupy a fraction ε_m of the membrane area. Let the membrane thickness be δ_m and the pore tortuosity τ_m. If a pressure ΔP is imposed across the membrane from the feed to the permeate (or the filtrate) side, the volume flux v_z through the pores (in units of cm^3/s-cm^2 of pore area) of a liquid of viscosity μ, and therefore the volume flux $\varepsilon_m v_z$ through the membrane (in units of cm^3/s-cm^2 of membrane area), is given by the Poiseuille law:

$$\varepsilon_m v_z = \frac{\varepsilon_m r_p^2}{8\mu}\frac{\Delta P}{\tau_m \delta_m} = \frac{\varepsilon_m r_p^2}{8\mu}\left[-\frac{1}{\tau_m}\frac{dP}{dz}\right]. \tag{3.4.85}$$

That Poiseuille flow exists through track-etched mica membranes with a pore radius around 5.6 nm has been verified by Anderson and Quinn (1972).

The membrane may have a pore size distribution $f(r_p)$ such that the fraction of pores in the size range r_p to $r_p + dr_p$ is $f(r_p)dr_p$, and they contribute a fraction $d\varepsilon_m$ to the membrane pore area. If $N_{sz}\overline{V_s}$ is the total volume flux through the

[31] Cylindrical pore in a membrane is an idealization widely practiced in membrane literature.

pores of such a membrane, then the contribution of pores of radius r_p to $r_p + dr_p$ to the volume flux is given by

$$(dN_{sz})\overline{V_s} = \frac{(d\varepsilon_m)r_p^2}{8\mu}\frac{\Delta P}{\tau_m\delta_m} = \frac{\varepsilon_m r_p^2 f(r_p)dr_p}{8\mu}\frac{\Delta P}{\tau_m\delta_m}.$$

For a membrane with pores of all radii between r_{max} and r_{min},

$$N_{sz}\overline{V_s} = \int_{\text{membrane}} (dN_{sz})\overline{V_s} = \frac{\varepsilon_m \Delta P}{8\mu\tau_m\delta_m}\int_{r_{min}}^{r_{max}} r_p^2 f(r_p)dr_p = \frac{\varepsilon_m \overline{r_p^2}}{8\mu}\left(\frac{\Delta P}{\tau_m\delta_m}\right). \tag{3.4.86}$$

The quantity $\sqrt{\overline{r_p^2}}$ is the *hydraulic mean pore radius*, defined by

$$(\overline{r_p^2}) = \int_{r_{min}}^{r_{max}} r_p^2 f(r_p)dr_p. \tag{3.4.87}$$

The volume flux (3.4.86) may also be represented by

$$N_{sz}\overline{V_s} = \frac{Q_{sm}}{\mu}\frac{\Delta P}{\delta_m}. \tag{3.4.88}$$

This relation between volume flux, pressure drop, membrane thickness and the fluid viscosity is commonly used for a porous medium and is known as *Darcy's law*. Here, Q_{sm} is called the solvent permeability. Observe that all characteristics of the porous medium that is the membrane, namely ε_m, $\overline{r_p^2}$ and τ_m, are incorporated in Q_{sm}.

Example 3.4.4 The utility of equation (3.4.86) for the determination of the solvent flux through a porous membrane will be briefly illustrated with an example worked out by Cheryan (1987). For an XM100A ultrafiltration (UF) membrane, the mean pore diameter ($\cong 2 \times$ hydraulic mean pore radius) = 17.5 nm; the number of pores/cm^2 of the top membrane surface area (skin) = 3×10^9; δ_m = membrane thickness = 0.2 μm (only of the skin (to be explained later)); viscosity of water (20 °C), μ = 1 cp; $\Delta P = 10^5$ Pa (gauge pressure); $\overline{V_s}$ = 18.05 cm^3/gmol of water. From equation (3.4.86), obtain the following estimate of the solvent volume flux (here, water flux):

$$N_{sz}\overline{V_s}\left(\frac{\text{cm}^3}{\text{cm}^2\text{-s}}\right) = \frac{\varepsilon_m}{8}\frac{(175\times 10^{-8})^2}{4}\text{cm}^2 \times \frac{10^5\text{Pa}\times 10\dfrac{\text{g}}{\text{cm-s}^2\text{-Pa}}}{1\times 10^{-2}\dfrac{\text{g}}{\text{cm-s}}}\frac{1}{0.2\times 10^{-4}\text{cm}},$$

where we have assumed that $\tau_m = 1$ (since no information is available). The membrane porosity

$$\varepsilon_m = \frac{\text{number of pores}}{\text{cm}^2}\times \text{pore area (cm}^2)$$

$$= 3\times 10^9 \times \frac{\pi}{4}(2r_p)^2 = \frac{3\times 10^9\times\pi}{4}\times 175^2 \times 10^{-16}$$

$$= 7.21\times 10^{-3}.$$

So only 0.72% of the membrane surface is covered with pores, i.e. $\varepsilon_m = 0.0072$. We have

$$N_{sz}\overline{V_s} = \frac{7.21\times 10^{-3}\times(175\times 10^{-8})^2\times 10^6}{32\times 0.2\times 10^{-6}}$$

$$= 3.45\times 10^{-3}\text{cm/s};$$

$$N_{sz}\overline{V_s} = 3.45\times 10^{-3}\frac{\text{cm}}{\text{s}}\times 3600\frac{\text{s}}{\text{hour}}\times 1000\frac{\text{liter}}{\text{m}^3}\times\frac{1}{100}\frac{\text{m}}{\text{cm}}$$

$$= 124.3\frac{\text{liter}}{\text{m}^2\text{-hr}}.$$

The experimentally observed value of the volume flux at 20 °C is 80 liter/m^2-hr (lmh).

There are several sources of discrepancy: there is a lack of information about the tortuosity $\tau_m (> 1)$; the deviation of the pores from circularity; the existence of dead-end pores, etc. In the calculation, the membrane thickness of 0.2 μm refers to a thin "skin" on top of the total membrane of substantial thickness (100–300 μm). The rest of the membrane has pores that are orders of magnitude larger (with very little hydraulic resistance), and its porosity is also much higher (as much as 0.3–0.6). Such membranes are called *asymmetric* membranes when the membrane is prepared from one material. Often they are prepared from two different materials to yield a *composite* membrane.

The volume flux given above is for the pure solvent. If there are solute (macrosolute) molecules in the feed solvent, the velocity of the solute molecule in the pore may or may not equal the local pore velocity of the solvent. The latter is especially likely to be true for macrosolutes, macromolecules, proteins, etc. In general, the solute/macrosolute molecule can also diffuse along the pore, down its own concentration gradient. The solute flux N_{iz} through the membrane is related to the solute flux through the pores, N_{iz}^p, as follows:

$$N_{iz} = \varepsilon_m N_{iz}^p, \tag{3.4.89a}$$

where, from flux expression (3.1.113), we can write

$$N_{iz} = \underbrace{C_{im}^p\overline{G_i}v_z\varepsilon_m}_{\text{convective flux}} + \underbrace{J_{iz}^*\varepsilon_m}_{\text{diffusive flux}} = C_{im}^p\overline{G_i}v_z\varepsilon_m - D_{ip}\varepsilon_m\frac{dC_{im}^p}{dz}. \tag{3.4.89b}$$

Here, C_{im}^p is the concentration of solute i averaged over a pore cross section, $\overline{G_i}$ is a convective hindrance factor, accounting for the ratio between the velocity of solute molecules and the averaged pore velocity of the solvent v_z (usually less than 1), and D_{ip} is the diffusion coefficient of solute molecules of species i in the pore (Anderson and Quinn, 1974). We have seen earlier from (3.1.113) that

$$D_{ip} = \frac{D_{il}\overline{G_{Dr}}(r_i,r_p)}{\tau_m}. \qquad (3.4.89c)$$

Depending on the dimensions of the pores and solute molecules, as well as any possible partitioning between the external solution and the pore solution, the nature of the particular terms in the above equation will vary. Integration of the flux expression will be carried out here only for a particular case, namely when the pore concentration of solute C^p_{im} is related to the solution concentration external to the membrane by a partition coefficient or a distribution coefficient,

$$(C^p_{im}/C_{il}) = \kappa_i. \qquad (3.4.90a)$$

Since there is a feed side and a permeate side of the membrane, in general, at $z = 0$,

$$(C^p_{im}/C_{il}) = \kappa_{if} = C^{p0}_{im}/C^0_{il}, \qquad (3.4.90b)$$

and, at $z = \tau_m\delta_m$,

$$(C^p_{im}/C_{il}) = \kappa_{ip} = C^{p\delta}_{im}/C^\delta_{il}. \qquad (3.4.90c)$$

Here, the diffusion length in the z-direction is $\tau_m\delta_m$, where δ_m is the membrane thickness and τ_m is the membrane tortuosity. At steady state, N_{iz} in expression (3.4.89b) is constant, and

$$\int_{C^{p0}_{im}}^{C^{p\delta}_{im}} \frac{dC^p_{im}}{C^p_{im} - \left(N_{iz}/\overline{G_i}v_z\varepsilon_m\right)} = \int_0^{\tau_m\delta_m} \frac{dz}{\left(D_{ip}/\overline{G_i}v_z\right)}. \qquad (3.4.90d)$$

This provides

$$\ell n \left[\frac{C^{p\delta}_{im} - \dfrac{N_{iz}}{\overline{G_i}v_z\varepsilon_m}}{C^{p0}_{im} - \dfrac{N_{iz}}{\overline{G_i}v_z\varepsilon_m}}\right] = \frac{\overline{G_i}v_z\tau_m\delta_m}{D_{ip}}; \qquad (3.4.91a)$$

$$N_{iz} = \frac{\overline{G_i}v_z\varepsilon_m C^{p0}_{im}\left[1 - \dfrac{C^{p\delta}_{im}}{C^{p0}_{im}}\exp\left\{-\dfrac{\overline{G_i}v_z\tau_m\delta_m}{D_{ip}}\right\}\right]}{\left[1 - \exp\left\{-\dfrac{\overline{G_i}v_z\tau_m\delta_m}{D_{ip}}\right\}\right]}. \qquad (3.4.91b)$$

Membrane pore concentrations may be changed to external concentrations using relations (3.4.90b,c):

$$N_{iz} = \frac{\overline{G_i}v_z\varepsilon_m\kappa_{if}C^0_{il}\left[1 - \dfrac{\kappa_{ip}C^\delta_{il}}{\kappa_{if}C^0_{il}}\exp\left\{-\dfrac{\overline{G_i}v_z\tau_m\delta_m}{D_{ip}}\right\}\right]}{\left[1 - \exp\left\{-\dfrac{\overline{G_i}v_z\tau_m\delta_m}{D_{ip}}\right\}\right]}, \qquad (3.4.91c)$$

where C^0_{il} and C^δ_{il} are the solute concentrations in the feed and the permeate (the filtrate) solution, respectively.

The quantity $\{(\overline{G_i}v_z)\tau_m\delta_m/D_{ip}\}$ is called the *pore Péclet number*, Pe^m_i, of solute i since $\overline{G_i}v_z$ is the effective solute velocity in the pore, $\tau_m\delta_m$ is the effective pore length and D_{ip} is the effective solute diffusivity in the pore. When $Pe^m_i \gg 1$, the contribution of solute diffusion to solute flux is unimportant and

$$N_{iz} \cong \overline{G_i}v_z\varepsilon_m\kappa_{if}C^0_{il}. \qquad (3.4.92a)$$

If there is no solute partitioning, $\kappa_{if} = 1$ and $\overline{G_i} = 1$, then

$$N_{iz} = \varepsilon_m v_z C^0_{il}, \qquad (3.4.92b)$$

a result valid for a membrane with pores very large compared to the solute molecules and no solute partitioning.

The integrated flux expression (3.4.91c) for N_{iz} contains the permeate solute concentration C^δ_{il}, which is often related to the solvent velocity v_z in the pore by

$$N_{iz} = \varepsilon_m v_z C^\delta_{il}, \qquad (3.4.93a)$$

since the effective solvent flux through the membrane is $\varepsilon_m v_z$. Substitution of this into the integrated flux expression (3.4.91c) provides the following relation between C^δ_{il} and C^0_{il} (after rearrangement):

$$\frac{C^\delta_{il}}{C^0_{il}} = \frac{\overline{G_i}\kappa_{if}\exp\left(\overline{G}v_z\tau_m\delta_m/D_{ip}\right)}{\overline{G_i}\kappa_{ip} - 1 + \exp\left(\overline{G}v_z\tau_m\delta_m/D_{ip}\right)} = \frac{\overline{G_i}\kappa_{if}\exp\left(Pe^m_i\right)}{\overline{G_i}\kappa_{ip} - 1 + \exp\left(Pe^m_i\right)}. \qquad (3.4.93b)$$

In the absence of any concentration polarization, C^0_{il} and C^δ_{il} are equal to C_{if} and C_{ip}, respectively. The extent of concentration polarization and its effects on the solvent flux and solute transport for porous membranes and macrosolutes/proteins can be quite severe (see Section 6.3.3). This model is often termed the *combined diffusion-viscous flow model* (Merten, 1966), and it can be used in ultrafiltration (see Sections 6.3.3.2 and 7.2.1.3). The relations between this and other models, such as the finely porous model, are considered in Soltanieh and Gill (1981).

The preceding development assumed that solute molecules were present in each pore. The pore size distribution of porous membrane may be such that the solute molecules can enter only pores larger than the solute molecule. This has two effects. First, the development of the solvent flux expression (3.4.86) assumed no effects due to any solute molecules; in reality, the solute molecules increase the solvent viscosity. Second, the solute flux expressed by (3.4.91c) may have to be corrected if all of the membrane pores are not available to solute molecules. Simplified analysis of such a case has been provided by Harriott (1973).

3.4.2.3.1 Solute diffusion through porous liquid-filled membrane We consider here solute diffusion through a porous liquid-filled membrane under the condition of *no convection* (Figure 3.4.5(c)). Examples of separation processes/techniques where such a situation is encountered are: isotonic dialysis, membrane based nondispersive gas absorption/stripping or solvent extraction, supported or contained liquid membrane technique. The governing equation for solute flux N_{iz} of species i per unit membrane cross-sectional area is obtained from Fick's first law applied to a porous membrane having a porosity of ε_m (see equation (3.1.112c)):

$$N_{iz} = -D_{im}\frac{dC^p_{im}}{dz}, \tag{3.4.94a}$$

where D_{im} is the diffusion coefficient of species i in the membrane and C^p_{im} is the pore fluid concentration of species i. If the pore radial dimension, r_p, is much larger than the solute dimension, r_i (i.e. $r_i <<< r_p$) and the tortuosity factor is τ_m, then, in the absence of any specific interaction between the solute and the pore wall,

$$N_{iz} = -\frac{D_{il}\varepsilon_m}{\tau_m}\frac{dC^p_{im}}{dz}. \tag{3.4.94b}$$

Integrate this flux expression across the membrane thickness δ_m to obtain

$$N_{iz} = \frac{D_{il}\varepsilon_m}{\tau_m\delta_m}\left(C^{p0}_{im} - C^{p\delta}_{im}\right). \tag{3.4.94c}$$

Here, D_{il} is the diffusion coefficient of solute i in the liquid present in the pore, and may be obtained from the values provided in Table 3.A.3. To relate the pore concentration C^p_{im} to the external solution concentration C_{il} present on two sides of the membrane ($z = 0$, C^0_{il}; $z = \delta_m, C^\delta_{il}$), consider two cases.

Case (1) The solvent present outside the membrane is identical to the solvent/liquid inside the membrane pores. Since there are no solute-pore interactions and $r_i <<< r_p$ (excludes the geometrical partitioning effect (3.3.88a)), the solute concentration in the solvent in the pores is identical to that immediately outside the pores:

$$C^{p0}_{im} = C^0_{il};\; C^{p\delta}_{im} = C^\delta_{il}. \tag{3.4.95a}$$

Correspondingly,

$$N_{iz} = \frac{D_{il}\varepsilon_m}{\tau_m\delta_m}(C^0_{il} - C^\delta_{il}) = k_{im}(C^0_{il} - C^\delta_{il}) = \frac{Q_{im}}{\delta_m}(C^0_{il} - C^\delta_{il}), \tag{3.4.95b}$$

where the mass-transfer coefficient k_{im} for the membrane is defined by

$$k_{im} = \frac{D_{il}\varepsilon_m}{\tau_m\delta_m}. \tag{3.4.95c}$$

Further, the permeability coefficient Q_{im} is defined by

$$Q_{im} = \frac{D_{il}\varepsilon_m}{\tau_m}. \tag{3.4.95d}$$

Case (2) The solvent present outside the membrane is different from (and immiscible with) the solvent present inside the membrane, the partition coefficient being given by

$$\kappa_{im} = \frac{C^{p0}_{im}}{C^0_{il}} = \frac{C^{p\delta}_{im}}{C^\delta_{il}}. \tag{3.4.96a}$$

The flux expression is given by (see Section 3.4.3.2 for greater details)

$$N_{iz} = \frac{D_{il}\varepsilon_m\kappa_{im}}{\tau_m\delta_m}(C^0_{il} - C^\delta_{il}) = k_{im}(C^{p0}_{im} - C^{p\delta}_{im}) = \frac{Q_{im}}{\delta_m}(C^{p0}_{im} - C^{p\delta}_{im}), \tag{3.4.96b}$$

where Q_{im} and k_{im} are still defined, respectively, by (3.4.95d) and (3.4.95c).

The cases considered above were such that $r_i <<< r_p$ and there was no pore–solute interaction. If the solute dimension is no longer orders of magnitude smaller than the pore dimensions and there are no specific solute–pore wall interactions, we may employ flux expression (3.1.112f) and integrate (Lane and Riggle, 1959) under the conditions of hindered diffusion and geometrical partitioning:

$$N_{iz} = \left(\frac{D_{il}\varepsilon_m\overline{G_{Dr}}(r_i,r_p)}{\tau_m}\right)\kappa_{im}\frac{(C^0_{il} - C^\delta_{il})}{\delta_m}. \tag{3.4.97}$$

We can use Faxen's expression (3.1.112e) for $\overline{G_{Dr}}(r_i,r_p)$ as long as $(r_i/r_p) \leq 0.5$. Further, κ_{im} may be obtained from relation (3.3.88a). The membrane mass-transfer coefficient and the permeability coefficient in the case of hindered diffusion and geometrical partitioning are defined as follows:

$$k_{im} = \frac{D_{il}\varepsilon_m\overline{G_{Dr}}(r_i,r_p)\kappa_{im}}{\delta_m\tau_m};\; Q_{im} = \frac{D_{il}\varepsilon_m\overline{G_{Dr}}(r_i,r_p)\kappa_{im}}{\tau_m}. \tag{3.4.98a}$$

Note that

$$N_{iz} = k_{im}(C^0_{il} - C^\delta_{il}) = \frac{Q_{im}}{\delta_m}(C^0_{il} - C^\delta_{il}). \tag{3.4.98b}$$

In an alternative representation based on the overall membrane phase concentration C_{im}, the flux equation is simply represented as (Figure 3.4.8)

$$N_{iz} = -D_{im}\frac{dC_{im}}{dz} = D_{im}\frac{(C^0_{im} - C^\delta_{im})}{\delta_m}. \tag{3.4.99}$$

Using a partition coefficient κ_{im} between the membrane phase concentration and the external solution concentration (Figure 3.4.8),

$$N_{iz} = D_{im}\kappa_{im}\left(\frac{C^0_{il} - C^\delta_{il}}{\delta_m}\right). \tag{3.4.100}$$

For swollen polymeric gels, an introduction to a free-volume based interpretation of D_{im} through water-filled hydrophilic gels is available in Yasuda *et al.* (1969). Specific applications of such concepts to cellulosic dialysis membranes are provided in Colton *et al.* (1971) and Farrel and

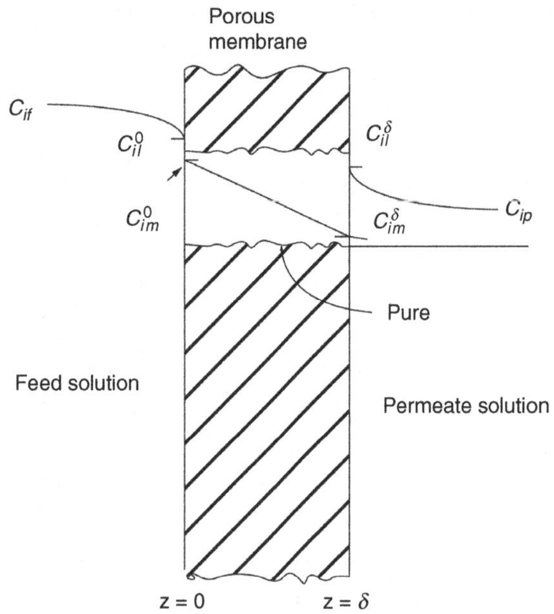

Figure 3.4.8. Solute concentration profile in diffusive transport through a liquid-filled porous membrane from a feed solution to a permeate solution.

Babb (1973). Additional considerations on solute transport, and subsequently selectivity in hemodialysis, are available in Section 4.3.1.

Example 3.4.5 Calculate the membrane mass-transfer coefficients and the permeability coefficients of two solutes, sodium sulfate and sucrose, through a microporous light denitrated cellulose membrane whose properties, along with those of the two solutes, are provided below (Lane and Riggle, 1959). The temperature is 20 °C, and dilute aqueous solutions are under consideration with essentially no convection through the membrane. For sodium sulfate: $M_i = 142$; diffusion coefficient in water $= 7.7 \times 10^{-6}$ cm^2/s; density $\rho_i = 2.698$ g/cm^3. For sucrose: $M_i = 342$; diffusion coefficient in water $=$ 4.5×10^{-6} cm^2/s; density $\rho_i = 1.588$ g/cm^3. Membrane pore diameter $= 3$ nm; membrane swollen thickness $= 9.4 \times 10^{-3}$ cm; membrane porosity $= 0.43$; membrane tortuosity $= 2.6$.

Solution Estimate the solute diameter $2r_i$ from equation (3.3.90b). For Na_2SO_4,

$$2r_i = 1.465 \times 10^{-8}(142/2.698)^{1/3} = 5.479 \times 10^{-8}\text{cm} = 0.5479\,\text{nm}.$$

For sucrose

$$2r_i = 1.465 \times 10^{-8}(342/1.588)^{1/3} = 0.879\,\text{nm}.$$

Employ the definition (3.4.98a) for k_{im} for each solute, where $\overline{G_{Dr}}(r_i, r_p)$ is given by the Faxen relation (3.1.112e) and κ_{im} is given by (3.3.88a). For Na_2SO_4,

$$\begin{aligned} k_{im} &= \left(D_{il}\varepsilon_m \kappa_{im} \overline{G_{Dr}}(r_i, r_p)/\delta_m \tau_m\right) \\ &= \frac{7.7 \times 10^{-6} \times 0.43 \times \left(1 - \dfrac{5.48}{30}\right)^2 \times \left(1 - 2.104(5.48/30) + 2.09(5.48/30)^3 - 0.95(5.48/30)^5 + \cdots\right)}{9.4 \times 10^{-3} \times 2.6} \\ &= \frac{7.7 \times 0.43 \times 10^{-3} \times 0.672}{9.4 \times 2.6}(1 - 0.384 + 0.012 - 0.006) = \frac{7.7 \times 0.43 \times 10^{-3} \times 0.672}{9.4 \times 2.6} \times 0.628 \\ &= 5.69 \times 10^{-5}\ \text{cm/s}. \end{aligned}$$

For sucrose,

$$\begin{aligned} k_{im} &= \frac{4.5 \times 10^{-6} \times 0.43 \times \left(1 - \dfrac{8.79}{30}\right)^2 \times \left(1 - 2.104(8.79/30) + 2.09(8.79/30)^3 - 0.95(8.79/30)^5 + \cdots\right)}{9.4 \times 10^{-3} \times 2.6} \\ &= \frac{4.5 \times 0.43 \times 10^{-3} \times 0.5}{9.4 \times 2.6}(1 - 0.616 + 0.052 - 0.002) = \frac{0.079 \times 10^{-3} \times 0.44}{2} \\ &= 1.74 \times 10^{-5}\ \text{cm/s}. \end{aligned}$$

The larger sucrose molecule partitions into the pore and diffuses more slowly through water and encounters a higher drag from the pore walls, resulting in 3.27 times lower membrane transfer coefficient.

The transport of solute from the feed through the membrane to the receiving solution may be influenced by the solute transport resistance in the feed liquid and the permeate liquid. In the absence of pore convection, the value of the membrane transfer coefficient will be given by expressions (3.4.95c) and (3.4.96b). Referring to Figure 3.4.8, the solute flux at steady state can be expressed as

$$N_{iz} = k_{lf}(C_{if} - C_{il}^0) = \kappa_{im}k_{im}(C_{il}^0 - C_{il}^\delta) = k_{lp}(C_{il}^\delta - C_{ip}), \tag{3.4.101}$$

where k_{lf} and k_{lp} are film transfer coefficients in the feed and permeate for solute i. In the case of hindered diffusion and geometrical partitioning, use instead of $\kappa_{im}\, k_{im}$, k_{im} from (3.4.98a).

An overall solute-transfer coefficient K may be defined by

$$N_{iz} = K(C_{if} - C_{ip}), \tag{3.4.102}$$

so that

$$\frac{1}{K} = \frac{1}{k_{lf}} + \frac{1}{\kappa_{im}k_{im}} + \frac{1}{k_{lp}} \tag{3.4.103}$$

is the relation between K and the individual solute-transfer coefficients.

3.4.2.4 *Gas transport through porous membranes*

The flux expressions for gas transport through porous membranes have been considered in Section 3.1.3.2.4. The steady state Knudsen diffusion flux expression (3.1.115a),

$$N_{iz}^p = -\frac{4r_p}{3}\left(\frac{2RT}{\pi M_i}\right)^{1/2}\frac{dC_i}{dz},$$

of a gas i of molecular weight M_i through straight circular pores of radius r_p and length δ_m based on the pore cross section, can be integrated along the z-direction. Assuming ideal gas behavior, we obtain

$$N_{iz}^p = -\frac{4r_p}{3}\left(\frac{2RT}{\pi M_i}\right)^{1/2}\left(\frac{p_{if} - p_{ip}}{\delta_m}\right)\frac{1}{RT};$$

$$N_{iz}^p = D_{iK}\left(\frac{p_{if} - p_{ip}}{\delta_m}\right)\frac{1}{RT}. \tag{3.4.104}$$

(*Note*: D_{ik} is called the Knudsen diffusivity.) If the membrane has a porosity of ε_m and a tortuosity of τ_m, the integrated membrane flux expression is given by

$$N_{iz} = \frac{4r_p\varepsilon_m}{3\tau_m}\left(\frac{2RT}{\pi M_i}\right)^{1/2}\left(\frac{p_{if} - p_{ip}}{\delta_m}\right)\frac{1}{RT}. \tag{3.4.105}$$

Here,

$$D_{iK} = \frac{4r_p\varepsilon_m}{3\tau_m}\left(\frac{2RT}{\pi M_i}\right)^{1/2}. \tag{3.4.106}$$

For large membrane pores and high gas pressures, leading to a value of (r_p/λ) large enough to ensure Poiseuille or viscous flow, the magnitude of the integrated flux expression for a pure gas i under a positive pressure difference between the feed and the permeate side is given, for a membrane of pore radius r_p and tortuosity τ_m, by

$$N_{iz} = \frac{\varepsilon_m r_p^2}{8\mu_i\tau_m RT}\left(\frac{P_f + P_p}{2}\right)\left(\frac{P_f - P_p}{\delta_m}\right). \tag{3.4.107a}$$

This is obtained simply from the Poiseuille flow relation (3.1.116b) adapted for a tortuous porous medium,

$$\Delta P = \{8\mu_i\delta_m\tau_m v_z/r_p^2\}, \tag{3.4.107b}$$

ideal gas behavior and

$$N_{iz} = v_z C_{\text{mean}}, \tag{3.4.107c}$$

where C_{mean} is the mean density of the gas between the pressures P_f and P_p $[C_{\text{mean}} = (P_f + P_p)/2RT]$. For a gas mixture, a simplistic representation of flux N_{iz} will be

$$N_{iz} = \frac{\varepsilon_m r_p^2}{8\mu\tau_m RT}\left(\frac{P_f^2 - P_p^2}{2}\right)x_{if}, \tag{3.4.108}$$

where x_{if} is the mole fraction of species i in the feed. In reality, the situation is complicated by molecular diffusion of the gas species imposed on the Poiseuille flow.

For slip flow or transitional flow conditions intermediate between Knudsen flow and Poiseuille flow, the molar flux of species A through the pores under such conditions is given by expression (3.1.115e) for a binary gas mixture of species A and B (Scott and Dullien, 1962; Rothfeld, 1963):

$$N_{Az} = -\left[\frac{1}{\dfrac{(1 - x_A/N_R)}{D_{AB}} + \dfrac{1}{D_{AK}}}\right]\frac{P}{RT}\frac{dx_A}{dz}.$$

On integrating between $z = 0$, $x = x_A^0$ and $z = \tau_m\delta_m$, $x_A = x_A^\delta$, we get

$$N_{Az} = \frac{D_{AB}P}{N_R RT\tau_m\delta_m}\ell n\left[\frac{N_R\left(1 + \dfrac{D_{AB}}{D_{AK}}\right) - x_A^\delta}{N_R\left(1 + \dfrac{D_{AB}}{D_{AK}}\right) - x_A^0}\right]. \tag{3.4.109}$$

A similar expression is valid for species B.

We have been introduced to expressions for the gas flux due to *surface diffusion* in a porous medium/membrane: see expression (3.1.117a). Employing relation (3.1.118d) relating the surface excess concentration of the ith species per unit surface area, Γ_{is}^E, and the gas-phase concentration, C'_{ig} moles of i per unit gas phase volume in the pores, we can assume Henry's law to obtain

$$(\mathrm{d}\Gamma_{is}^{E}/\mathrm{d}p_{ig})RT = [H_{i\sigma}^{c}]^{-1}\mathrm{R}T, \qquad (3.2.110a)$$

where

$$p_{ig} = H_{i\sigma}^{c}\Gamma_{is}^{E}, \qquad (3.3.110b)$$

signifying equilibrium between the bulk gas phase being convected through the pores and the surface phase. Here, $H_{i\sigma}^{c}$ is the Henry's law constant. We can now integrate the surface diffusion flux of species i in a mixture to obtain

$$J_{iz} = \left(\frac{D_{is}a_{sp}}{H_{i\sigma}^{c}\mathrm{R}T}\right)\left(\frac{P_f x_{if} - P_p x_{ip}}{\delta_m}\right). \qquad (3.4.111)$$

Here, D_{is} is the surface diffusion coefficient of species i and a_{sp} is the pore surface area of the microporous membrane in cm^2 per unit volume (cm^3) of the membrane defined as a medium of porosity ε_m. Ash *et al.* (1973b) have observed that, as long as the adsorbed films are dilute, each species diffuses independently of the other gas.

Since surface flow rarely exists by itself and is present along with convective and diffusive flow in a pore, it is useful to identify the total flux of gas species i in a pore. For the case of pore diffusion by the Knudsen mechanism, the total flux of i is given by

$$N_{iz} = \left(\frac{D_{iK}}{\mathrm{R}T} + \frac{D_{is}a_{sp}}{H_{i\sigma}^{c}\mathrm{R}T}\right)\left(\frac{P_f x_{if} - P_p x_{ip}}{\delta_m}\right). \qquad (3.4.112)$$

The characteristics of the surface flow are influenced by the gas pressure, the temperature and the condensability of the gas or vapor. The lower the temperature, the more likely it is that the extent of surface flow is higher. Similarly, the higher the condensability of a gas species, the higher the surface flow. The treatments by Ash *et al.* (1967, 1973b), Kammermeyer (1968) and Hwang and Kammermeyer (1975) may be consulted for a detailed study of surface flow through microporous membranes. Under extreme conditions, surface sorption followed by multilayer condensation could block gas-phase flow completely.

Example 3.4.6 Ash *et al.* (1967) have studied the combined phenomena of Knudsen flow and surface flow in graphitized carbon membranes prepared from Graphon and Black Pearl carbon powder compressed to form a plug. The pore surface area measured/volume of these plugs were: $90\,m^2/cm^3$ (Graphon) and $197\,m^2/cm^3$(Black Pearl); the corresponding porosities were 0.42 and 0.43. The temperature of measurement was 303 K. They reported the permeability Q_{im} of argon as the sum of Q_{ig} and Q_{is} (Q_{ig} for Knudsen flow and Q_{is} for surface diffusion), where their gas flux, is given by

$$\left(\frac{Q_{im}}{\delta_m}\right)\Delta C_{ig}{}' = \frac{Q_{im}}{\delta}\left(\frac{\Delta C_{ig}}{\varepsilon_m}\right)\frac{\text{mole}}{\text{cm}^2\text{-s}}$$

and ε_m is the microporous carbon membrane porosity. Graphon membrane, $Q_{ig} \times 10^3 = 4.8\,cm^2/s$, $Q_{is} \times 10^3 = 3.1\,cm^2/s$; Black Pearl membrane, $Q_{ig} \times 10^3 = 1.8\,cm^2/s$, $Q_{is} \times 10^3 = 2.9\,cm^2/s$. The values of Henry's law constant σ_i defined as $(\varepsilon_m \mathrm{d}\Gamma_{is}^{E}/\mathrm{d}C_{ig})$ were, respectively, 3.48×10^{-7} cm and 2.79×10^{-7}cm.

(1) Calculate the values of the surface diffusion coefficient D_{is} for argon through the two different microporous carbon membranes.
(2) Obtain the values of the gas phase permeability Q_{ig} for krypton for the two carbon membranes at the same temperature and pressure.

Solution (1) The flux expression (3.4.112) obtained by combining Knudsen flow and surface diffusion may be rewritten, using equation (3.1.118a), as

$$N_{iz} = -D_{iK}\frac{\mathrm{d}C_{ig}}{\mathrm{d}z} - D_{is}\frac{\mathrm{d}(\Gamma_{is}^{E}a_{sp})}{\mathrm{d}z} = -D_{iK}\varepsilon_m\frac{\mathrm{d}C_{ig}{}'}{\mathrm{d}z} - D_{is}\sigma_i\frac{\mathrm{d}C_{ig}{}'}{\mathrm{d}z}a_{sp}.$$

Integrating across the membrane thickness, we obtain

$$N_{iz} = D_{iK}\varepsilon_m\frac{\Delta C_{ig}{}'}{\delta_m} + D_{is}\sigma_i a_{sp}\frac{\Delta C_{ig}{}'}{\delta_m}.$$

If we rewrite this expression as $N_{iz} = \{(D_{iK}\varepsilon_m) + D_{is}\sigma_i a_{sp}\}(\Delta C_{ig}{}'/\delta_m)$, we observe that $Q_{im} = Q_{ig} + Q_{is} = D_{ik}\varepsilon_m + D_{is}\sigma_i a_{sp}$. Therefore, $D_{is} = (Q_{is}/\sigma_i a_{sp})$. For the Graphon membrane,

$$D_{\mathrm{Ar},s} = \frac{3.1 \times 10^{-3}}{3.48 \times 10^{-7} \times 90 \times 10^4}\frac{\mathrm{cm}^2}{\mathrm{s}} = 989 \times 10^{-5}\frac{\mathrm{cm}^2}{\mathrm{s}};$$

for the Black Pearl membrane

$$D_{\mathrm{Ar},s} = \frac{2.9 \times 10^{-3}}{2.79 \times 10^{-7} \times 197 \times 10^4}\frac{\mathrm{cm}^2}{\mathrm{s}} = 527 \times 10^{-5}\frac{\mathrm{cm}^2}{\mathrm{s}}.$$

(2) Q_{ig} for the Knudsen permeability $= D_{iK}\varepsilon_m$. But $D_{iK}\sqrt{M_i}\varepsilon_m$ is a constant for the porous medium. Therefore

$$D_{\mathrm{Ar}K}\sqrt{M_{\mathrm{Ar}}}\varepsilon_m = D_{\mathrm{Kr}K}\sqrt{M_{\mathrm{Kr}}}\varepsilon_m \Rightarrow D_{\mathrm{Kr}K} = D_{\mathrm{Ar}K}\frac{\sqrt{M_{\mathrm{Ar}}}}{\sqrt{M_K}}.$$

For the Graphon membrane,

$$Q_{\mathrm{Kr},g} = Q_{\mathrm{Ar},g} \times \sqrt{M_{\mathrm{Ar}}/M_{\mathrm{Kr}}}$$
$$= 4.8 \times 10^{-3} \times \sqrt{39.9/83.8} = 3.3 \times 10^{-3}\mathrm{cm}^2/\mathrm{s}.$$

For Black Pearl,

$$Q_{\mathrm{Kr},g} = 1.8 \times 10^{-3} \times 0.69 = 1.24 \times 10^{-3}\mathrm{cm}^2/\mathrm{s}.$$

3.4.2.5 *Transport in a solution with ion exchange membrane*

A number of types of situations are encountered here: (1) transport in the presence of an electrical field and an ion exchange membrane with no convection through it; (2) transport with convection and no external electrical field; (3) transport without any convection and without any electrical field.

We are going to focus on the role of transport without any convection in an ion exchange membrane placed in an electrolytic solution subjected to a constant electrical field. If the aqueous solution of an electrolyte is placed between two electrodes connected outside to two terminals of a battery, the current passing through the solution and the

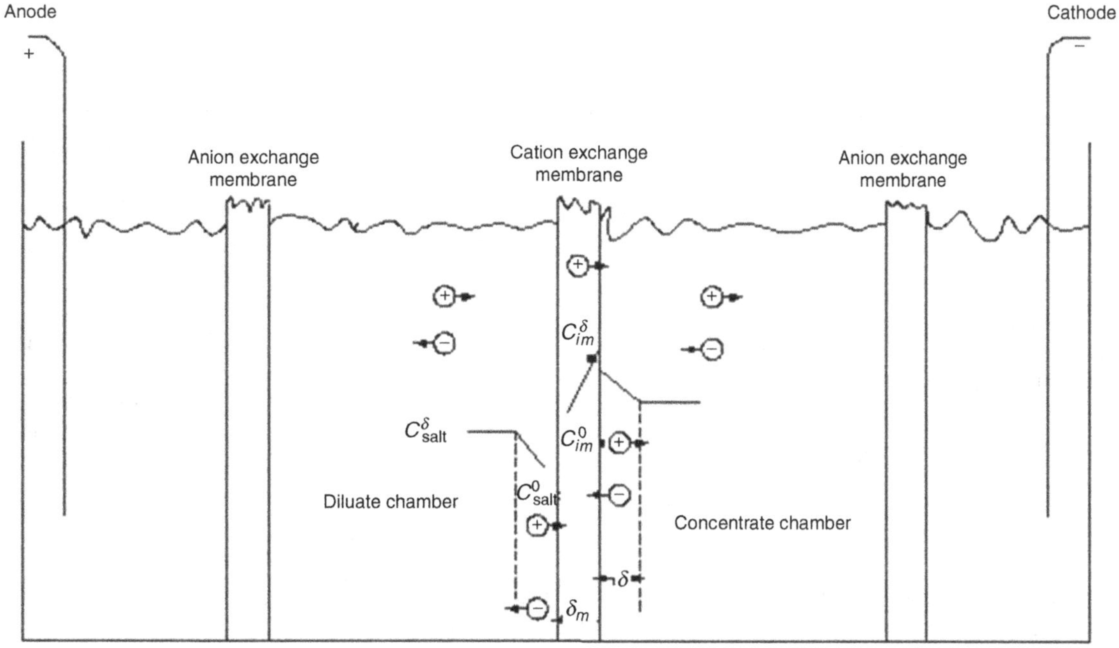

Figure 3.4.9. Concentration profiles in electrodialysis (ED) with an ideal cation exchange membrane (profiles around the anion exchange membrane have not been shown).

membrane will be carried by both cations and anions formed by the dissociation of the electrolyte (Figure 3.4.9). Consider a uni-univalent electrolyte (e.g. NaCl) that is completely dissociated into Na^+ and Cl^-. Identify the different species i where i =1 for Na^+, i = 2 for Cl^- and i = 3 for water. The fractions of the current carried by Na^+ in the solution and in the *cation exchange membrane* are given by its transport numbers t_{1s} in the solution and t_{1m} in the membrane (see definition (3.1.108d)); similarly for the Cl^- ion, the quantities are t_{2s} and t_{2m}. By definition,

$$t_{1s} + t_{2s} = 1 \qquad \text{and} \qquad t_{1m} + t_{2m} = 1. \tag{3.4.113}$$

The cation exchange membrane prefers cations as the counterion. In the limit of an ideal cation exchange membrane, all current in the membrane is carried by cations, i.e. here $t_{1m} = 1$. Therefore $t_{2m} = 0$. Since neither t_{1s} nor t_{2s} is zero in the solution, one observes in Figure 3.4.9 that, at steady state, the rate at which Na^+ ions leave the solution from the left-hand side of the membrane into the membrane is given by $(It_{1m} - It_{1s})/\mathcal{F}$. Due to electroneutrality, the Cl^- ions must go toward the anode from this solution at an equal rate:

$$\frac{It_{1m} - It_{1s}}{\mathcal{F}} = \frac{It_{2s}}{\mathcal{F}}. \tag{3.4.114}$$

As a result, the solution in the left-hand side of the membrane in the so-called *diluate chamber* is depleted of NaCl. On the other hand, the Na^+ ions transported through the membrane arrive in the solution on the right-hand side of the membrane in the so-called *concentrate chamber* at a rate $(It_{1m}/\mathcal{F})$. From this solution, Na^+ ions migrate toward the cathode in the solution at a rate $(It_{1s}/\mathcal{F})$. Since $t_{1s} < t_{1m}$ = 1, the Na^+ ions accumulate in the solution on the right-hand side of the cation exchange membrane at the rate $(It_{1m} - It_{1s})/\mathcal{F}$. To maintain electroneutrality in this solution, Cl^- ions migrate into this solution toward the anode at an equal rate, $It_{2s}/\mathcal{F}$.

The net effect is that the solution on the left-hand side of the cation exchange membrane is depleted of NaCl, whereas that on the right-hand side is concentrated in NaCl. From a Fickian diffusion point of view, solute NaCl should be supplied now to the left of the membrane from the solution to the right of the membrane. Yet the electrical potential does just the reverse. A very brief analysis of these factors is provided below.

The molar rate of transfer of cations by the current through a cation exchange membrane per unit membrane area is $it_{1m}/\mathcal{F}$, where i is the current density (Figure 3.4.9). The direction of this flux is from left to right. A diffusive flux in the opposite direction (from right to left) is simply $Q_{1m}(C^{\delta}_{1m} - C^{0}_{1m})/\delta_m$. The net flux of cations from left to right through the membrane is given by

$$\frac{it_{1m}}{\mathcal{F}} - \frac{Q_{im}}{\delta_m}\left(C^{\delta}_{1m} - C^{0}_{1m}\right). \tag{3.4.115a}$$

At steady state, the solution to the left of the membrane,

depleted of salt to this extent, must be supplied with salt from the bulk solution to the same extent by diffusion and the current to maintain the cation flux, i.e.

$$k_d(C^{\delta}_{\text{salt}} - C^{0}_{\text{salt}}) + \frac{it_{1s}}{\mathcal{F}}. \qquad (3.4.115b)$$

Note that δ_l is the liquid film thickness on the left-hand side of the cation exchange membrane and $k_d = (D_{\text{salt}}/\delta_l)$. Therefore

$$\frac{it_{1m}}{\mathcal{F}} - \frac{Q_{1m}}{\delta_m}(C^{\delta}_{1m} - C^{0}_{1m}) = k_d(C^{\delta}_{\text{salt}} - C^{0}_{\text{salt}}) + \frac{it_{1s}}{\mathcal{F}}. \quad (3.4.116)$$

Most ion exchange membranes are such that the (Q_{1m}/δ_m) values are very low and could be neglected:

$$\begin{aligned} i &= \frac{\mathcal{F}k_d(C^{\delta}_{\text{salt}} - C^{0}_{\text{salt}})}{(t_{1m} - t_{1s})} + \frac{\mathcal{F}\,Q_{1m}(C^{\delta}_{\text{salt}} - C^{0}_{\text{salt}})}{\delta_m(t_{1m} - t_{1s})} \\ i &\cong \frac{\mathcal{F}k_d(C^{\delta}_{\text{salt}} - C^{0}_{\text{salt}})}{(t_{1m} - t_{1s})}. \end{aligned} \qquad (3.4.117)$$

This relation illustrates how the salt mass-transfer coefficient in the region adjacent to the membrane on the left-hand side controls the current through the membrane. Since the solution in the left-hand side is being depleted of salt and is becoming more dilute, it is called the *diluate* (note, k_d). The solution on the right-hand side of the membrane is called the **concentrate** in this electrodialysis (ED) process.

For a given electrolytic solute and values of k_d and C^{δ}_{salt}, the higher the value of *i*, the lower the value of C^{0}_{salt}. In the limit of $C^{0}_{\text{salt}} \cong 0$, the value of *i*, i_{lim}, is called the *limiting current density*; its unit is usually mA/cm^2, and practical values are around 20 mA/cm^2. This condition, identified as *concentration polarization*, specifies the useful limit of *i*. If *i* is increased beyond i_{lim}, the excess current is not useful for cation transport through the cation exchange membrane, and is essentially wasted.

It is useful to explore the steps by which one can arrive at a relation similar to (3.4.117) from Nernst–Plank or other formulations (Helfferich, 1962; Shaffer and Mintz, 1966, 1980). In the electrodialytic separation of a completely dissociated electrolytic solute in water, there are three species: positive ions, negative ions and water. Due to current flow, there will also be heat flow due to any temperature gradient. An irreversible thermodynamic formulation (Hills *et al.*, 1961) of the flux of any one of the three species, due to the chemical potential gradient of each species as well as the temperature gradient, can be provided following the general formulation (3.1.205). Further, there will be four such fluxes (including that of heat), making matters quite complicated. An abstracted treatment has been provided by Shaffer and Mintz (1966, 1980). They have shown, by using order of magnitude estimates for typical ED conditions, that:

(1) the temperature effects across the membrane have a negligible effect;

(2) $\boldsymbol{F}_i$ may be represented as

$$\boldsymbol{F}_i = -Z_i\mathcal{F}\,\nabla\phi - \frac{RT}{C_i}\nabla C_i \qquad (3.4.118)$$

by neglecting a derivative of the activity coefficient of species *i*, where $i = 1$ for a positive ion, $i = 2$ for a negative ion and $i = 3$ for water.

If one assumes that there is no coupling between different fluxes, then, from the general formulation (3.1.205),

$$\boldsymbol{J}^*_i = L_{ii}\boldsymbol{F}_i = L_{ii}\{-Z_i\mathcal{F}\nabla\phi - \frac{RT}{C_i}\nabla C_i\}, \qquad (3.4.119)$$

which has the form of the Nernst–Plank equation (3.1.106) since $(L_{ii}/C_i) \cong (D_{is}/RT)$ from (3.1.210).

Consider now a negatively charged cation-permeable membrane. If it behaves ideally and allows no anions and water to go through, by the Nernst–Plank equation for the cations ($i = 1$), at the *membrane–solution interface*,

$$\boldsymbol{J}^*_1 = -D^N_{1s}\{\nabla C_1 + \frac{C_1 Z_1\mathcal{F}}{RT}\nabla\phi\}. \qquad (3.4.120)$$

Similarly for the anions ($i = 2$),

$$\boldsymbol{J}^*_2 = -D^N_{2s}\{\nabla C_2 + \frac{C_2 Z_2\mathcal{F}}{RT}\nabla\phi\} = 0, \qquad (3.4.121)$$

for this ideal cation-permeable membrane. We know from the definition (3.1.108c) of the current density $\boldsymbol{i}$ that

$$\boldsymbol{i} = \mathcal{F}\sum_{i=1} Z_i\boldsymbol{N}_i = (Z_1\boldsymbol{J}^*_1 + Z_2\boldsymbol{J}^*_2), \qquad (3.4.122)$$

under conditions of no bulk flow. To obtain a simplified result, assume a uni-univalent salt with $v_+ = v_- = 1$. Then $C_1 = C_2 = C_{\text{salt}}$, suggesting that $\nabla C_1 = \nabla C_2 = \nabla C_{\text{salt}}$. Therefore, from $\boldsymbol{J}^*_2 = 0$, we get an expression for ∇C_{salt} in terms of $\nabla\phi$. Substituting this into the expression for $\boldsymbol{J}^*_1$ leads to

$$\begin{aligned} \boldsymbol{J}^*_1 &= -\frac{D^N_{1s}C_{\text{salt}}\mathcal{F}}{RT}\{Z_1 - Z_2\}\nabla\phi \\ &= -D^N_{1s}\{1 - \frac{Z_1}{Z_2}\}\nabla C_{\text{salt}}, \end{aligned} \qquad (3.4.123)$$

where the flux is linearly proportional to the potential gradient.

Therefore,

$$\boldsymbol{i} = -\mathcal{F}Z_1 D^N_{1s}\left\{1 - \frac{Z_1}{Z_2}\right\}\nabla C_{\text{salt}} = -\frac{D^N_{1s}\mathcal{F}^2 Z_1 C_{\text{salt}}}{RT}(Z_1 - Z_2)\nabla\phi \qquad (3.4.124)$$

at the membrane–solution interface of an ideal cation exchange membrane. The first of these relations is essentially very similar to relation (3.4.117).

It is useful to identify some assumptions and considerations used in the above analysis.

(1) There was no convection of water through the ion exchange membrane; the pressures on both sides of

the membrane must be equal. Further, there was no electro-osmotic transport of water in the same direction as the counterion.

(2) The membrane was perfectly selective to cations.

(3) The pore size of the membrane was very small (2–3 nm) so that the diffusive permeability of cation was very low.

(4) The current I or the current density i is determined by the voltage applied between the two electrodes and the ohmic resistances in the current path.

The preceding analysis can be used in analyzing the performance of electrodialytic separations. It is also useful in studying separation in *Donnan dialysis*, where only a single ion exchange membrane (either a cation exchange membrane or an anion exchange membrane) is used without any externally applied electrical field. It is additionally useful in battery separator analysis.

No attempt has been made so far to relate the membrane phase concentration of an ion to its external concentration. A relation of the type (3.3.120e, f) may be derived by using the relation (3.3.118b) for Donnan equilibrium as well as the electroneutrality relations in the membrane and in the external solution. The ionic flux of any ion through an ion exchange membrane may then be obtained from the Nernst-Plank equation (3.1.106). If $\nabla\phi$ is to be eliminated from such an equation, then the following procedure may be adopted. Write down the flux for $\boldsymbol{J}_1^*$ and $\boldsymbol{J}_2^*$. If there is some constraint like $\boldsymbol{J}_2^* = 0$ (as in (3.4.121)), the problem is easily solved. In some cases, $\boldsymbol{J}_1^* = \boldsymbol{J}_2^*$ (for example, in battery separators for alkaline Ni/Zn batteries, KOH is the electrolyte and $\boldsymbol{J}_{K^+} = \boldsymbol{J}_{OH^-}$). This will also allow elimination of $\nabla\phi$. Then one can express $\boldsymbol{J}_i^*$ in terms of the concentrations of two solutions on the two sides of the membrane by integrating across the battery separator membrane at steady state.

3.4.3 Interphase transport in two-phase systems with phase barrier membranes

In Section 3.4.1, simple quantitative representations of mass transport from one phase to another immiscible phase were provided using mass-transfer coefficients. Separation processes where such mass transfer takes place generally utilize the following configurations in industrial practice.

Gas-liquid system:

(1) gas dispersed as bubbles in liquid;

(2) liquid sprayed as drops in gas;

(3) liquid spread as thin films on packings in a continuous gas phase.

Liquid-liquid system: one liquid dispersed as drops in another liquid.

Such dispersion of one phase into another generates a large interfacial area through which mass transfer occurs.

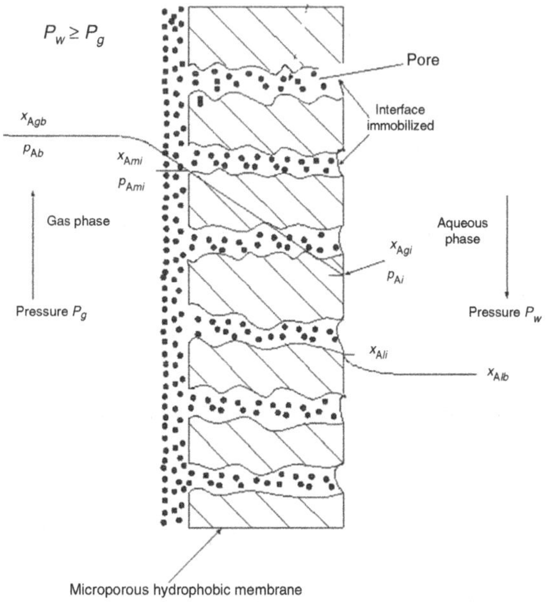

Figure 3.4.10. Immobilized gas-liquid interfaces in a microporous/porous hydrophobic membrane (as in a membrane oxygenator).

The larger the interfacial area per unit volume of the two-phase contacting device, the higher the rate of mass transfer. Generation of interfacial area requires energy input. Further, after mass is transferred, the dispersed phase has to be coalesced. In liquid-liquid systems, coalescence can be a major problem, especially in systems prone to forming stable emulsions. It is possible to have two phases contacted, however, without dispersing one phase into the other as drops or bubbles or thin films by using microporous/porous membranes as phase barrier membranes. Mass transport in such systems is briefly studied below.

3.4.3.1 *Transport in gas-liquid contacting via a microporous/porous membrane*

Membrane oxygenators use a porous/microporous hydrophobic membrane (Figure 3.4.10): on one side of the membrane, the blood from the patient flows, while on the other side air flows. The blood side pressure is maintained slightly above the air pressure to prevent air from bubbling into the blood (Callahan, 1988; Sirkar, 1992). Since the membrane is hydrophobic, most aqueous solutions, including blood, do not wet the pores, which remain filled with air. A gas-liquid interface is created at the mouth of every pore in the porous/microporous hydrophobic membrane; the membrane facilitates the gas-liquid contact. Oxygen is absorbed from the gas into blood and CO_2 is stripped from blood into the gas through such an interface. Unless the liquid pressure exceeds a critical pressure called the *breakthrough pressure*, the gas-liquid interface remains stable.

To determine the rate of interphase transport of any species being absorbed from the gas into the liquid (or desorbed from the liquid into the gas) through such a gas-liquid interface in a microporous/porous hydrophobic membrane, consider the concentration profile of species A shown in Figure 3.4.10. The flux of species A being absorbed at steady state may be written down for the three regions (gas film, membrane pore and liquid film) as follows:

$$N_{Az} = k_{xg}(x_{Agb} - x_{Ami}) = k_{mg}(x_{Ami} - x_{Agi}) = k_{xl}(x_{Ali} - x_{Alb}) \tag{3.4.125}$$

In terms of an overall mass-transfer coefficient, K_{xl} or K_{xg},

$$N_{Az} = K_{xg}(x_{Agb} - x^*_{Ag}) = K_{xl}(x^*_{Al} - x_{Alb})$$

as before (expressions (3.4.5)). Further, from Henry's law relations (3.4.1a, b), we know that $x^*_{Ag} = (H_A/P)\, x_{Alb} = H^P_A x_{Alb}$; $x_{Agb} = H^P_A x^*_{Al}$.

Rearranging these equations and noting that

$$x_{Agb} - x^*_{Ag} = x_{Agb} - x_{Ami} + x_{Ami} - x_{Agi} + x_{Agi} - x^*_{Ag}, \tag{3.4.126}$$

one can write

$$\frac{N_{Az}}{K_{xg}} = \frac{N_{Az}}{k_{xg}} + \frac{N_{Az}}{k_{mg}} + \frac{N_{Az} H^P_A}{k_{xl}} \tag{3.4.127}$$

so that

$$\underbrace{\frac{1}{K_{xg}}}_{\text{overall resistance}} = \underbrace{\frac{1}{k_{xg}} + \frac{1}{k_{mg}}}_{\text{gas-phase resistance}} + \underbrace{\frac{H^P_A}{k_{xl}}}_{\text{liquid-phase resistance}}. \tag{3.4.128}$$

This relation shows that the overall resistance is increased by the membrane resistance, which is now included in the gas-phase resistance. For gas species (e.g. O_2) with low solubility in the patient's blood, it is known from the relation (3.4.12b) that the liquid-phase resistance controls the mass-transfer rate. Thus, the increased gas-phase resistance due to the extra membrane resistance does not influence the rate of oxygenation of the patient's blood at all. If, instead of blood, any aqueous or organic solution is used and oxygen has a low solubility in it, the same conclusion holds, as long as the membrane pores are gas-filled. Obviously, it is true for any other sparingly soluble gas as well. On the other hand, if the gas species is highly soluble in the liquid, it is possible that the gas-phase resistance (which includes that of the membrane) may become important. For more details, see Sirkar (2001).

To determine the liquid-film and the gas-film mass-transfer coefficients k_{xl} and k_{xg}, standard mass transfer correlations valid for the particular geometry, flow regime and conditions may be used (see Tables 3.1.5–3.1.8). The membrane transfer coefficient, k_{mg}, may be estimated at a total gas pressure P from

$$k_{mg} = \frac{D_i \varepsilon_m P}{\tau_m \delta_m RT x_{Bg,lm}}. \tag{3.4.129}$$

Here, D_i is the diffusion coefficient of species i in the gas phase, ε_m is the porosity of the membrane having thickness δ_m and pore tortuosity τ_m and $x_{Bg,lm}$ is the logarithmic mean mole fraction difference of the inert gas species B along the diffusion path (see definition (3.1.132)). Such an expression for k_{mg} is valid for conditions of ordinary diffusion of A through a stagnant gas film of B in the pores of the membrane. If the mean free path conditions are such that Knudsen diffusion or other convection mechanisms are valid, appropriate equations have to be used from Section 3.4.2.4.

From definition (3.4.129), it is clear that solute transport takes place only through the gas-filled pores. Yet the interfacial area to be used for calculating the total rate of mass transfer is the total membrane area; the latter is certainly right for k_{xg} and k_{xl}. For reasonably porous membranes, Keller and Stein (1967) have suggested that the model of the one-dimensional series of resistances (3.4.128) may be valid, even though there is a cross-sectional reduction at the membrane pore. Note that the curvature at the phase interface is usually neglected in the area calculation. Further, there is no transport of gases through the nonporous region.

3.4.3.2 *Solute transport via a microporous membrane*

In Figure 3.4.2, the liquid-liquid interface between an immiscible aqueous-organic system was shown to be planar. In practice, the interface is usually curved, since one phase is dispersed as drops in the other phase. This dispersion can be avoided, and a stable liquid-liquid interface created, by using microporous/porous membranes in a manner somewhat analogous to that shown in Figure 3.4.10. Consider Figure 3.4.11, in which a microporous/porous hydrophobic membrane is shown. An aqueous nonwetting solution containing a solute i to be extracted flows on one side of the membrane. An organic solvent flows on the other side of the membrane. If the organic solvent wets the membrane material, it will spontaneously fill the membrane pores and would have a tendency to disperse as drops in the aqueous phase on the other side. If the aqueous solution pressure is, however, equal to or higher than that of the organic-phase pressure, this dispersion can be prevented and aqueous-organic interface immobilized at each pore mouth (Kiani *et al.*, 1984). Such an interfacial configuration can be easily achieved also if both phases are stagnant.

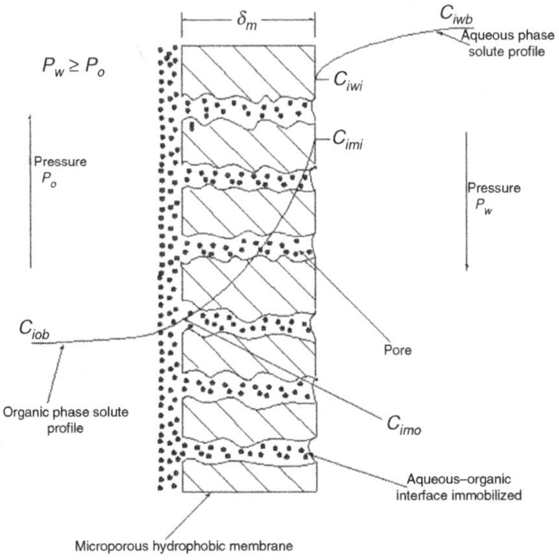

Figure 3.4.11. Schematic of solute concentration profiles in solvent extraction with immobilized aqueous–organic interfaces in a microporous/porous hydrophobic membrane.

The concentration profile of a solute i being extracted from the aqueous solution into the organic solvent is shown in Figure 3.4.11. The flux of solute i may be expressed at steady state in terms of the aqueous film coefficient, the organic film coefficient and a membrane coefficient as follows:

$$N_{iz} = k_w(C_{iwb} - C_{iwi}) = k_{mo}(C_{imi} - C_{imo}) = k_o(C_{imo} - C_{iob}). \tag{3.4.130}$$

The overall mass-transfer coefficient based expressions are

$$N_{iz} = K_o(C_{io}^* - C_{iob}) = K_w(C_{iwb} - C_{iw}^*). \tag{3.4.131}$$

The solute concentrations in the two phases at the phase interface (at the pore mouth) are related by the equilibrium distribution coefficient κ_{io} as follows:

$$\kappa_{io} = \frac{C_{imi}}{C_{iwi}}. \tag{3.4.132}$$

Assuming that κ_{io} is constant over the solute concentration range,

$$\kappa_{io} = \frac{C_{io}^*}{C_{iwb}}; \qquad \kappa_{io} = \frac{C_{iob}}{C_{iw}^*}; \qquad \kappa_{io} = \frac{C_{imi}}{C_{iwi}}. \tag{3.4.133}$$

Now, $C_{io}^* - C_{iob} = \kappa_{io}(C_{iwb} - C_{iwi}) + (C_{imi} - C_{imo}) + (C_{imo} - C_{iob})$, therefore

$$\frac{N_{iz}}{K_o} = \frac{\kappa_{io} N_{iz}}{k_w} + \frac{N_{iz}}{k_{mo}} + \frac{N_{iz}}{k_o},$$

leading to

$$\frac{1}{K_o} = \underbrace{\frac{\kappa_{io}}{k_w}}_{\text{aqueous-phase resistance}} + \underbrace{\frac{1}{k_{mo}} + \frac{1}{k_o}}_{\text{organic-phase resistance}}. \tag{3.4.134}$$

Similarly,

$$C_{iwb} - C_{iw}^* = (C_{iwb} - C_{iwi}) + \left(C_{iwi} - \frac{C_{imo}}{\kappa_{io}}\right) + \left(\frac{C_{imo}}{\kappa_{io}} - C_{iw}^*\right); \tag{3.4.135}$$

$$\frac{N_{iz}}{K_w} = \frac{N_{iz}}{k_w} + \frac{1}{\kappa_{io}} \frac{N_{iz}}{k_{mo}} + \frac{1}{\kappa_{io}} \frac{N_{iz}}{k_o}.$$

Therefore

$$\frac{1}{K_w} = \underbrace{\frac{1}{k_w}}_{\text{aqueous-phase resistance}} + \underbrace{\frac{1}{\kappa_{io}}\left(\frac{1}{k_{mo}} + \frac{1}{k_o}\right)}_{\text{organic-phase resistance}}. \tag{3.4.136}$$

It is useful now to obtain some limiting forms of the two resistances-in-series relations for $\kappa_{io} \gg 1$ and $\kappa_{io} \ll 1$, provided one assumes that k_w, k_o and k_{mo} are of similar orders of magnitude. The condition $\kappa_{io} \gg 1$ means that the solute strongly prefers the organic phase, whereas $\kappa_{io} \ll 1$ means that the solute strongly prefers the aqueous phase:

$$\kappa_{io} \gg 1, \quad \left.\begin{aligned} \frac{1}{K_w} &\cong \frac{1}{k_w} \\ \frac{1}{K_o} &\cong \frac{\kappa_{io}}{k_w} \end{aligned}\right\} \text{aqueous-phase control;} \tag{3.4.137}$$

$$\kappa_{io} \ll 1, \quad \left.\begin{aligned} \frac{1}{K_w} &\cong \frac{1}{\kappa_{io}}\left(\frac{1}{k_{mo}} + \frac{1}{k_o}\right) \\ \frac{1}{K_o} &\cong \left(\frac{1}{k_{mo}} + \frac{1}{k_o}\right) \end{aligned}\right\} \text{organic-phase control.} \tag{3.4.138}$$

For $\kappa_{io} \approx O(1)$, both phases influence the mass transfer. Note that, for $\kappa_{io} \gg 1$, k_{mo} is not important at all; the membrane resistance is irrelevant (Prasad and Sirkar, 1988).

Since for $\kappa_{io} \ll 1$ and for $\kappa_{io} \approx O(1)$ the membrane resistance is important, an estimate of k_{mo} is provided next. The result (3.4.96b) and (3.4.98a) from Section 3.4.2.3.1 is useful here, namely

$$k_{mo} = \frac{D_{io}\varepsilon_m \overline{G_{Dr}}(r_i, r_p)}{\delta_m \tau_m}. \tag{3.4.139a}$$

Here, D_{io} is the diffusion coefficient of solute i in the organic solvent. If there is no hindrance to solute diffusion in the pore due to $r_i \ll r_p$ (at least by two orders of magnitude),

$$k_{mo} = \frac{D_{io}\varepsilon_m}{\delta_m \tau_m}. \quad (3.4.139b)$$

For a detailed introduction to this topic, see Prasad and Sirkar (2001). An alternative configuration employs a porous hydrophilic membrane with the aqueous phase present on one side and in the pores and the organic phase present on the other side at a pressure $P_o \geq P_w$ to stabilize the aqueous–organic interface (Prasad and Sirkar, 1987).

Problems

3.1.1 Consider a pure gas in a tall column of height h in the *gravitational* field. Relate the pressures at the top and the bottom of the column to the height of the column.

3.1.2 Develop an expression for the net vertically downward gravitational force on solute species i present in a solvent s of density ρ_s. The solute density is ρ_i. Assume that the presence of the solute does not influence the force on the solvent. (Ans. $-M_i g(1-(\rho_s/\rho_i))\,\mathbf{k}$.)

3.1.3 In Example 3.1.3, the force calculated for ovalbumin molecules subjected to an electrical field of constant strength $\boldsymbol{E}$ was based on neglecting the effect of an electrical double layer. In this problem, do *not* neglect the effect of the double layer.

(1) Determine an expression for the ovalbumin velocity vector, given that the electrokinetic potential at the shear surface is ζ. (Ans. $\{\varepsilon_d \zeta \boldsymbol{E}/6\pi\mu\}\,\{f(r_0/\lambda)/(1+(r_0/\lambda))\}$.)

(2) For motion in the z-direction, estimate the mobility μ_i^m of ovalbumin defined as the steady ovalbumin velocity in the z-direction per unit value of electrical field strength acting on it in the z-direction. Dielectric constant of the liquid medium $= \varepsilon_d$; ovalbumin molecule may be considered a sphere of radius r_0; the liquid medium vicosity is μ. (Ans. $\{\varepsilon_d \zeta f(r_0/\lambda)/(6\pi\mu\,(1+(r_0/\lambda))\}$.)

(3) Show that the ionic mobility μ_i^m of any small ion i in general in a uniform electrical field $\boldsymbol{E}$ is given by Q_i^{net}/f_i^d, where Q_i^{net} is the net charge on the ion at the plane of motion/shear.

3.1.4 (a) Write down the equation of motion of a charged particle of radius r_p having a charge Q_p in an electrical field of strength $\boldsymbol{E}$ in the manner of equation (3.1.60); include the relevant forces. Employ vectorial form. The fluid phase is gaseous. The drag force is given by Stokes' law.

(b) If the drag force is given by Stokes' law, obtain the steady particle velocity for two cases: (i) gas bulk velocity is $\boldsymbol{v}_t$; (ii) the gas is stagnant. The gas viscosity is μ. The particle velocity in the latter case is the electrical migration velocity.

(c) Obtain the electrical mobility μ_p^m of the charged particle defined as the particle velocity per unit value of the electrical strength when the gas is stagnant.

3.1.5 Consider two flat electrodes at locations z_1 and z_2 with electric potentials ϕ_1 and ϕ_2. For a uniform electrical field $\boldsymbol{E}$ given by $(-\mathrm{d}\phi/\mathrm{d}z)\mathbf{k}$, the electric field at any location z is

$$\boldsymbol{E} = \frac{\phi_1 - \phi_2}{(z_2 - z_1)}\mathbf{k}.$$

Consider a **different electrode configuration**, namely **an annular geometry with the outer electrode at radius r_o having a potential V_o and the inner electrode at radius** r_i having a potential V_i with an electrolytic solution in between. Develop an expression for $\boldsymbol{E}$ in this nonuniform electrical field.

Suggestions: Use Ohm's law for a unit cross-sectional area of solution. Current density i (amp/cm^2) is easily related to potential gradient $(-\mathrm{d}\phi/\mathrm{d}r)$ and conductivity κ mho/cm by $i = k(-\mathrm{d}\phi/\mathrm{d}r)$. For a cross-sectional area $2\pi rh$ at radius r, h being the height of the solution, $i2\pi rh = I$, the total current between the electrodes, which is independent of radial location. Integrate, assuming κ to be constant, to obtain the result

$$\boldsymbol{E} = \frac{(V_i - V_0)}{\ell\text{n}\,(r_0/r_i)}\,\frac{1}{r}\,\mathbf{r}.$$

3.1.6 Consider a single yeast cell in water placed between two cylindrical electrodes with the radius vector $\boldsymbol{r}$ in the vertically upward direction. The density of the yeast cell ρ_p is 1.18 g/cm^3 (that for water, ρ_w, is 1 g/cm^3). If the yeast cell may be considered as a dielectric uncharged particle, and if Brownian motion may be assumed to be negligible, identify the forces acting on this yeast cell. Develop an estimate of the unknown dielectric properties of the cell with respect to the medium (i.e. water) if there is no net force acting on the cell at a given radial location r between the inner and outer cylindrical electrodes (having voltages V_i and V_o, and the corresponding radii are r_i and r_o, respectively).

3.1.7 A *spherical paramagnetic* particle in a *paramagnetic solution* is subjected to a magnetic force due to a local magnetic field intensity H_0^m (in vacuum). Consider the situation where the magnetic permeabilities of the particle and the solution are related to the corresponding susceptibilities χ_p and χ_s via

$$\mu_p^m = \mu_0^m(\chi_p + 1) \text{ and } \mu_s^m = \mu_0^m(\chi_s + 1).$$

Assuming that the only other force in the system is viscous drag from the solution of viscosity μ, obtain the following expression for the terminal velocity of the particle:

$$U_{pzt} = \frac{2}{9}\frac{(\chi_p - \chi_s)\, r_p^2}{\mu}\frac{\mathrm{d}}{\mathrm{d}z}\left(\frac{B_0^2}{2\mu_0^m}\right),$$

under the condition that both susceptibilities χ_p and χ_s are much smaller than unity. *Note*: $H_0^m = (B_0^m/\mu_0^m)$; Stokes' law is valid.

3.1.8 Although in a conventional approach (see the result of Problem 3.1.3 part (3)) the ionic mobility of a small ion is expressed as Q_i^{net}/f_i^d, where f_i^d is the frictional coefficient due to hydrodynamic drag (most often expressed via Stokes' law for a spherical molecule), more advanced theories propose an additional dielectric (charge-induced) frictional drag coefficient,

$$f_i^{dl}\,(\text{Kay}, 1991): \mu_i^m = Q_i^{\text{net}}/(f_i^d + f_i^{dl}). \tag{3.P.1}$$

For a spherical molecule, $f_i^d = 6\pi\mu\, r_i$ and

$$f_i^{dl} = H\left(\tau\frac{\varepsilon_{of} - \varepsilon_{\infty f}}{\varepsilon_{of}^2}\right)\frac{(Z_i\,\mathrm{e})^2}{r_i^3} = \text{constant} \times \frac{(Z_i\mathrm{e})^2}{r_i^3},$$

where Z_i is the algebraic valence of ion i. Although these models assume spherical molecules, in reality other effective structural patterns are common as the ions migrate. It has been found that the volume of the ion may be used in general instead of r_i. Develop an expression for μ_i^m based on (3.P.1), where r_i has been replaced appropriately by V_i, the volume of the ion, assumed spherical.

3.1.9 Consider the particle motion in a fluid where the particle Reynolds number is larger than 1.

(1) Write down in vectorial form the equation of motion of a particle subjected to a total external force of $\boldsymbol{F}_p^{\text{ext}}$. The velocity of the fluid is $\boldsymbol{v}_t$.

(2) Obtain an expression for the steady particle velocity if the fluid is stagnant.

3.1.10 You are required to make some simple calculations here to estimate how long a nuclear winter would last (Sutija and Prausnitz, 1990). The large-scale generation of smoke and dust from urban and forest fires following a nuclear war would lead to a dust cloud in the stratosphere. The dust cloud would block solar radiation and cause a substantial reduction in temperature (as much as 20–40 °C) over large parts of the world (Turco *et al.*, 1990). Assuming that each dust particle diameter is 10 μm and that each particle settles independently of the others, determine the duration of the nuclear winter if the particles have been carried to a height of 50 km by the nuclear blasts and subsequent fires. The particle density is 2 g/cm^3. (Ans. 105 days.)

3.1.11 Determine the force being exerted on two different enzymes diffusing in a slab of an aqueous gel subjected to a uniform electrical field strength of 3 volt/cm. At the pH conditions of the gel, enzyme 1 has a charge of $Z = 4$ while enzyme 2 has $Z = 1$. Calculate the migration velocity of the two enzyme molecules if their diffusion coefficients in the gel are: enzyme 1, 2×10^{-7} cm^2/s; enzyme 2, 4×10^{-7} cm^2/s. Comment on the relative magnitudes of the migration velocities.

3.1.12 Consider a separation system containing no external forces, no pressure gradients, no temperature gradients and for which $[\partial \ln a_i/\partial \ln x_i]_{P,T} = 0$. For such a *ternary system* of solute i, solvent s and membrane m, write down the Maxwell–Stefan equations for solute transport and solvent transport. Identify the diffusion coefficients; note that the membrane velocity is zero.

Write down the equations for solute transport and solvent transport for the same solute-solvent-membrane system using the following irreversible thermodynamics formulation:

$$\boldsymbol{X}_i = \sum_{k=1}^{n} L'_{ik} \boldsymbol{N}_k,$$

where $\boldsymbol{X}_i$ is the force specific to species i, $\boldsymbol{N}_k$ is the flux of the kth species in a stationary frame of reference and L'_{ik} are coefficients are such that $L'_{ik} = L'_{ki}$. For an ideal system satisfying all the assumptions for the Maxwell-Stefan equations, obtain the various relations between the L'_{ik} and $\overline{D}_{ik}$ (from the Maxwell-Stefan formulation).

3.1.13 From the generalized Maxwell-Stefan type equations (3.1.180) and (3.1.181), write down the specific expressions for the following situations in terms of the force $(\boldsymbol{d}_i)$ RTC_t moving species i relative to the solution per unit volume of the solution:

(a) only a uniform electrical field is present;

(b) only a centrifugal force field exists.

3.1.14 Employing equation (3.1.222) to determine an *effective velocity* U for all ions in ambipolar diffusion, develop a relation between the ambipolar diffusion coefficient D for a highly ionized gas and the diffusion coefficients of electrons (D_-) and positively charged ions (D_+) and any other necessary quantities, e.g. ionic mobilities $\mu^m_{+,g}$ and $\mu^m_{-,g}$ of positive ions and electrons, respectively, in the gas phase.

3.1.15 In a dilute solution of a binary electrolyte $A^{Z_+}_{v_+}Y^{Z_-}_{v_-}$, a generalization of equation (3.2.2) for three dimensions and no bulk motion for the solvent $(\boldsymbol{v}^*_t = 0)$ is

$$\frac{\partial C_i}{\partial t} = -\nabla N_i \tag{3.P.2}$$

for each ionic species C_A and C_Y. We are interested in deriving the *general* result (3.1.107),

$$D = D_+ D_- (Z_+ - Z_-) / (Z_+ D_+ - Z_- D_-),$$

using the balance equation for each ionic species and the electrolyte as a whole. Note that the molar concentration of the electrolyte C may be related to those of A and Y by

$$C = \frac{C_+}{v_+} = \frac{C_-}{v_-} \tag{3.P.3}$$

There is no applied potential. Write down the balance equation (3.P.2) for C_+ and C_- using

$$\boldsymbol{N}_i = \boldsymbol{J}^*_i = C_i \boldsymbol{U}_i - D_i \nabla C_i,$$
$$\boldsymbol{U}_i = Z_i D_i \mathcal{F} (-\nabla\phi)/RT \text{ and equation (3.P.3)}.$$

Obtain an expression for $\nabla \cdot (C \nabla\phi)$. Use it in equation (3.P.2) for C_+ along with equation (3.P.3). Compare it with equation (3.P.2) for the electrolyte molar concentration C and obtain the result, assuming that the D_i are concentration-independent in the dilute solution.

3.2.1 (a) Develop an expression for the *migration velocity* of solute species i in a solution rotating in a centrifuge at ω radian/s.

(b) The following information is available: centrifuge cell length = 1 cm; centrifugal force field = 50 000 times the force of gravity; temperature = 75 °F; apparent density of water in solution = 1 g/cm^3; albumin mol. wt., M_i = 45 000; apparent density of albumin in solution = $(M_i/\overline{V}_i)$ = 1.34 g/cm^3; diffusion coefficient of albumin in water = 7×10^{-7} cm^2/s. Calculate the migration velocity of albumin molecules.

(c) How many different species of noninteracting molecules in a solution can be separated in such a centrifuge assuming a linear force profile and albumin molecule properties as characteristic properties to be used in calculation?

3.2.2 It is intended to separate different proteins in an electrical field of constant strength 0.025 volt/cm. The temperature is 75 °F and the diffusion coefficients of all proteins may be considered to be essentially equal to 7×10^{-7} cm^2/s. Focus on three proteins n = 1, 2, 3 and note that the net charge on the protein of each type is $Z_n = n$ = 1, 2, 3 at the given *pH*. Develop an exact expression for the variation of resolution between proteins 1 and 2, 2 and 3 and 1 and 3 in terms of time t and a numerical proportionality constant in each case. You are given $\mathcal{F}$ = 96 490 coulomb/gram equiv.; R = 8.314 joule/mole-K.

3.2.3 In expression (3.2.23) for the resolution between two neighboring peaks being translated in a vessel under uniform potential gradient, it was shown that resolution improved with time. One would like to know, however, the ratio of the standard deviation of a species profile with respect to the distance traveled by the center point of the profile (z_i). Does it depend on the diffusion coefficient as the bandwidth of the profile increases with time? Assume that only external force exists: $\nabla \mu_i^0 = 0$. (Ans.$(\sqrt{2RT}/\sqrt{z_i\, F_{iz}^{ext}})$.)

3.2.4 Consider the concentration profile of a species i being convected down in a uniform continuous potential profile driven external field (e.g. equation (3.2.19)). Obtain the *maximum* value of the concentration C_i at any instant of time. Compare its magnitude with that for a Gaussian profile, expression (2.5.6a). (Ans. $(m_i/l_x l_y)\,(1/\sqrt{2\pi}\,\sigma_i)$.)

3.2.5 A circular tube of length l_z and radius r_0 is filled with a dilute inert gas at a total concentration of C_t. At one end of this tube, called a *drift tube*, a cloud of different types of ions are introduced uniformly across the tube cross section in a delta-function fashion. The drift tube is designed to have a uniform electrical field of strength E_z in the direction of the tube-axis, the z-coordinate. Obtain:

(1) the concentration profile of ionic species i along the drift tube length, z, as a function of time;

(2) the standard deviation of the profile.

You are given that m_i is the number of moles of ionic species i introduced at the inlet.

Suggestion: Employ a drift velocity formalism for the ions. (Ans. $C_i(z,t) = (m_i/\pi r_0^2)((1/\sqrt{4\pi t D_i})$ $\exp\{-(z - U_{iz}t)^2/4 D_i t\}$.)

3.3.1 (1) For a continuous chemical mixture use the expression which is the equivalent of Raoult's law to predict the molecular weight density function $f_v(M)$ for the vapor phase. You are given the molecular weight density function $f_l(M)$ for the liquid phase as well as $P^{sat}(T; M)$. The total pressure P is unknown. Assume an ideal liquid solution.

(2) Let the vapor pressure $P^{sat}(T; M)$ be described via Trouton's rule and the Clausius–Clapeyron equation by

$$P^{sat}(T;M) = P^{+} \exp\left[A\left(1 - \frac{M}{T}\right)\right].$$

Let the liquid-phase function $f_l(M)$ be described by a Gaussian distribution:

$$f_l(M) = \frac{1}{\sqrt{2\pi}\,\sigma_l} \cdot \exp\left[-\frac{(M - \overline{M})^2}{2\sigma_l^2}\right].$$

Obtain an expression for each of P, the total pressure, and $f_v(M)$.

3.3.2 It is known that the *equilibrium solubility* of a solute particle in a solvent increases as the particle diameter is reduced.[32] Assume fugacity ratios are to be replaced by the solubility ratios. Show that the solubilities C_1 and C_2 of two solute particles of radius r_{p1} and r_{p2} are related by

$$\ell n\,(C_1/C_2) = \frac{2\overline{V}_{ij}\,\gamma^{12}}{RT}\left(\frac{1}{r_{p_1}} - \frac{1}{r_{p_2}}\right).$$

Consider separately particles of PbI_2 ($r_p = 0.4$ μm, $\overline{V}_{ij} = 74.8\,cm^3/gmol$, $\gamma^{12} = 130$ dyne/cm) and $BaSO_4$ ($r_p =$ 0.1 μm, $\overline{V}_{ij} = 52.0\,cm^3/gmol$, $\gamma^{12} = 1250$ dyne/cm) in a solvent (see Mullin (1961) for more systems and additional discussion). Calculate the percent increase in solubility of each kind of particle due to their small radius at 25 °C.

3.3.3 The values of the distribution coefficient κ_{i1} of a volatile liquid i between two essentially completely immiscible liquid phases, an organic liquid ($j = 1$) and water ($j = 2$), are provided in Table 3.P.1 for two molar concentrations, C_{i1}, of liquid i in organic liquid phase $j = 1$ at 25 °C.

[32] This process does not continue indefinitely. See Mullin (1961) for a review of the role of possible surface electrical charge and a change of γ^{12} with r_p.

Table 3.P.1.

C_{i1} gmol/l	$\kappa_{i1} = (C_{i1}/C_{i2})$
0.1	29
2.0	35

Determine:

(1) the values of the equilibrium ratio K_{i1} of the volatile liquid i for these two concentrations;

(2) the corresponding ratios of the activity coefficient of liquid i between $j = 2$ and $j = 1$, γ_{i2}/γ_{i1}. You are given: $M_i = 160$; $M_{\text{organic liquid}} = 155$; density $\rho_i = 3\,\text{g/cm}^3$; $\rho_{\text{organic liquid}} = 1.5\,\text{g/cm}^3$; $\rho_{\text{water}} = 1.0\,\text{g/cm}^3$. Assume that if the molar volume of any pure species in liter/gmol is $\overline{V}_i$, then the volume change on mixing two pure species 1 and 2 is zero, so that $C_{1j}\overline{V}_{1j} + C_{2j}\overline{V}_{2j} = 1$. *Note*: $\overline{V}_i$ is normally used for partial molar volume of species i.

3.3.4 The solubility coefficient S_{im} of a gas in a semicrystalline polymer is likely to be linearly proportional to the volume fraction of the amorphous phase since the crystalline region is virtually impenetrable (Section 3.3.7.3). The gas solubility data in Table 3.P.2 were obtained by Michaels and Bixler (1961) in a polyethylene film at 25 °C.

Table 3.P.2. Solubilities of various gases in a semicrystalline polyethylene film

S_{im}, cm^3(STP)/cm^3 of polymer-atm				
He	CO	CO_2	C_2H_6	CH_3Cl
0.008	0.041	0.29	0.84	4.35

In a separate investigation, these authors concluded that the solubility of each of these gases in a completely amorphous polyethylene film having otherwise same properties is as given in Table 3.P.3. Make an estimate of the volume fraction of the amorphous phase in the polyethylene film.

Table 3.P.3. Solubilities of various gases in an amorphous polyethylene film

(25°C)S_{ima}, cm^3(STP)/cm^3 of polymer-atm				
He	CO	CO_2	C_2H_6	CH_3Cl
0.012	0.064	0.451	1.28	6.55

3.3.5 Calculate the *osmotic pressure* of the following dilute solutions using the Van 't Hoff equation: aqueous solution of sucrose of 0.991 molality (ignore density change due to sucrose) at 30 °C; aqueous solution of NaCl, 0.1 Molar, 0.17 Molar and 35 000 ppm. at 25 °C. For solutions of NaCl, calculate also the osmotic pressure if $\pi_i = \upsilon C_i R T$, where $\upsilon = \upsilon_A + \upsilon_B$ for a binary electrolyte AB: υ_A and υ_B are the moles of ion A and ion B, respectively, produced from one mole of electrolyte AB. You may assume NaCl to be completely dissociated.

3.3.6 (1) Derive the expression given below for the geometrical partitioning factor κ_{im} for a spherical rigid solute i between an external solution and the solution in a pore in the shape of a slit:

$$\kappa_{im} = \left(1 - \frac{r_i}{r_p}\right) = \left(1 - \frac{r_i}{b}\right).$$

Here, r_i is the radius of the solute molecule, assumed spherical, and r_p is a characteristic slit pore dimension defined as the half width of the gap between two large flat plates. The slit pore may be imagined as the space bounded by two rectangular plates each of surface area A_{surf} set apart by a distance $2b$.

(2) Derive the expression for the geometrical partitioning factor κ_{im} for a spherical rigid solute of radius r_i between an external solution and a pore in the shape of a rectangle of sides b_1 and b_2. The pore is infinitely long. Rewrite this result in terms of a hydraulic radius r_h of the pore. (Ans. $\kappa_{im} = (1 - (2r_i/b_2))(1 - (2r_i/b_1))$;)

$$\kappa_{im} = \left(1 - \frac{r_i}{r_h(1 + (b_2/b_1))}\right)\left(1 - \frac{r_i(b_2/b_1)}{r_h(1 + (b_2/b_1))}\right).$$

Suggestion: Calculate the partition coefficient of the solute between the pore solution and the external solution as the ratio of two cross-sectional areas, one in which the center of the solute molecule can move and the other as the total pore cross-sectional area. Rewrite the result in terms of a hydraulic radius of the pore.

3.3.7 Determine the solubility of lysozyme in the following aqueous salt solutions:
(1) 1M sodium sulfate solution at $pH = 8$;
(2) 1M sodium chloride solution at $pH = 8$.
You are given (1) for sodium sulfate solution; $\log C_2|_0 = 3.55$, where $C_2|_0$ is in mg/ml; constant $b = 0.712$; (2) for sodium chloride solution: $\log C_2|_0 = 1.18$, where $C_2|_0$ is in mg/ml; constant $b = 0.480$. *Note*: The pI for lysozyme is 10.5. Therefore at $pH = 8$, its solubility is not at its lowest point (without any salt). The data were obtained from a Table quoted in Ladisch (2001).

3.3.8 Employing the Gibbs adsorption isotherm for gas adsorption on the surface of a solid adsorbent and the equivalent of the ideal gas equation of state for the surface adsorbed phase, obtain Henry's law for pure gas adsorption in the form of $(q_{i\sigma}/S_\sigma) = H_i P$, where H_i is the Henry's law constant for species *i*. Henry's law is valid for dilute/low-pressure adsorption systems.

3.3.9 In the adsorption of anthracene from its solution in cyclohexane on alumina adsorbent particles, Thomas (1948) found that the following relation described the equilibrium adsorption behavior:

$$q_{is}^* = \frac{22\,C_{if}}{1 + 375\,C_{if}}.$$

Here, the units of q_{is}^* are millimole of anthracene per gram of the adsorbent and the units of C_{if} are millimole of anthracene per cm^3 of the external cyclohexane solution. Determine the values of the adsorption equilibrium constant K and the maximum possible value of anthracene concentration in the solid phase q_{is}^m for this Langmuirian behavior.

3.3.10 For Langmuir adsorption of pure CO_2 on an adsorbent having an interfacial adsorption surface area S_σ of $4000\,m^2/g$ of adsorbent, the following information on CO_2 is available: surface area covered by one CO_2 molecule is $1.2 \times 10^{-19}\,m^2$.
(1) What is the value of $(q_{i\sigma})_{max}$ for CO_2?
(2) At a pure low CO_2 pressure of 2.5 cm Hg, the equilibrium CO_2 adsorption was observed to be 0.017 gmol of CO_2/g of adsorbent. Obtain the Langmuir adsorption isotherm for CO_2 and the adsorbent.
Note: $(q_{i\sigma})_{max}$ may also be expressed as $q_{i\sigma}^m$ *or* q_{is}^m.

3.3.11 Gas mixtures often have moisture. When such a gas mixture is contacted with a porous inorganic adsorbent material or membrane, water vapor may condense in the pores. If gas mixture adsorption or permeation was the objective, condensed water vapor will be a hindrance. A porous inorganic material has pores of 5 nm radius and the temperature of contacting is 25 °C. Water vapor pressure at a temperature T °F can be determined from the Antoine equation (Henley and Seader, 1981):

$$\ln \frac{P_{H_2O}^{sat}}{P_{H_2O,critical}} = A_1 - \frac{A_2}{T(^\circ F) + A_3}$$

where $A_1 = 6.532$, $A_2 = 7173$, $A_3 = 389$ and $P_{H_2O,\,critical} = 3206$ psia. Determine the relative humidity of the gas mixture $(p_{H_2O,v}/P_{H_2O}^{sat})$ below which one must operate to prevent pore condensation. You are given: $\gamma_{water}^{12} = 72\,dyne/cm = 72\,erg/cm^2$; molar volume of water $= 18\,cm^3/gmol$; contact angle $\theta = 0$ since water wets inorganic adsorbent materials in general.

3.3.12 Rogers *et al.* (1926) have provided data on the decolorization of a petroleum fraction by a clay adsorbent. Their isotherm may be represented in tabular form (see Table 3.P.4).

Table 3.P.4.

$q_{i1}\left(\frac{\text{units of color adsorbed from 100 kg oil}}{\text{kg clay}}\right)$	$C_{i2}\left(\frac{\text{equilibrium color units}}{\text{kg solution}}\right)$
11	200
8.8	85
8.0	60
6.8	38
5.0	13

Determine the parameters of the isotherm if a Freundlich isotherm of the type $C_{i2} = m(q_{i1})^n$ describes the behavior of the data. Relate the parameters here to those of (3.3.113e).

3.3.13 In the calculation of the distribution of the co-ion Y^- from an electrolyte AY between an external aqueous solution and the ion exchanger in A form (i.e. cation exchanger with fixed negative charges) by equation (3.3.120d,e), the importance of two terms which contribute to the term "b" need to be determined. It is known that the term with activity coefficients is important. One would like to know how important is the other term, $(a_{sR}/a_{sw})^{\overline{V}_{AY}/\overline{V}_{sw}}$. Develop an estimate of the value of this term for two cases, namely the swelling pressure is 0 atm and 500 atm, and comment on its influence on the value of $m_{Y,R}$ when $m_{F,R}$ is 5 molal. You are given: $\overline{V}_{AY}$ for most electrolytes varies between 15 and 40 cm^3/gmol; $T = 300$ K.

3.3.14 (1) Develop an *explicit relation* between the external solution phase concentration of an ion and its ion exchange resin phase concentration. Neglect the swelling pressure in the resin.
(2) Use the result (3.3.119d) and other required relations, including the electroneutrality conditions in the resin and the external solution, to arrive at the type of relation which is the basis for the functional dependence suggested in (3.3.120f) for the electrolytic solute AY, where $\nu_A = 2$, $\nu_Y = 1$.

3.3.15 Brooks and Cramer (1992) have determined the following values for various equilibrium parameters of two proteins, α-chymotrypsinogen and cytochrome-c: for α-chymotrypsinogen, $|Z_p| = 4.8$, $\sigma_p = 49.2$, $K_{ps} = 9.22 \times 10^{-3}$; for cytochrome-c, $|Z_p| = 6.0$, $\sigma_p = 53.6$, $K_{ps} = 1.05 \times 10^{-2}$.
The cation exchanger resin has a total ion exchange capacity of 567 mM.
(1) Obtain the SMA based equilibrium adsorption isotherm for each of these proteins.
(2) What protein concentration in the buffer will yield 6 mM protein concentration in the resin for a buffer salt (Na^+ based) concentration of 75 mM for cytochrome-c?

3.3.16 (1) Consider gas–solid equilibrium as encountered in sublimation, in which a species passes from the solid state to the vapor phase without passing through the liquid phase. In the separation of a solid mixture by sublimation, this vapor obtained by vaporizing a solid component may be condensed as such or a carrier gas may be used to entrain the vapor for condensing it later. This is done usually at a low pressure so that ideal gas behavior is obeyed. Obtain an expression for the gas phase mole fraction of species i, assuming the solid phase to be an agglomeration of pure species with each component having its pure component solid fugacity.
(2) Suppose now that gas pressure is much higher. Develop an expression for the gas-phase mole fraction of species i. (Hint: Use equations (3.3.127) and (3.3.128).) Assuming that the gas phase behaves ideally, calculate the mole fraction of naphthalene in ethylene gas at 12 °C, 100 atm pressure, given $P_i^{sub}(T) = 0.0000303$ atm; $V_{is} = 111.9$ cm^3/gmol.
Compare it with the observed mole fraction of around 0.0085 (McHugh and Krukonis, 1986, fig. 1.3) in the extraction of naphthalene by supercritical ethylene.

3.4.1 Determine the values of K_c and K_{xl} for the absorption of O_2 from air into water at 20 °C. Assume $\phi_N \cong 1$ for both the gas film and the liquid film (see expression (3.1.127)). You have been provided with the following information: $H_{O_2} = 4 \times 10^4$ atm/mole fraction; P = 1 atm; k_c' for O_2 in water = 2×10^{-3} cm/s; $D_{O_2 g} = 0.21$ cm^2/s is obtained from D_{AB}, where A and B are O_2 and N_2; δ_g for the gas film is 0.1 cm; to determine C_{tl}, assume that pure water has a molar volume of 18.05 cm^3/gmol. Comment on the mass-transfer control (if any) by one of the two films.

Absorption of O_2 is encountered in the oxygenation of blood. The desorption of O_2 from water is very important in degassing processes employed in ultrapure water production. Comment on which resistance will be important in both processes. (Ans. $K_c = 2 \times 10^{-3}$ cm/s; $K_{xl} = 1.1 \times 10^{-4}$ gmol/cm^2-s-mole fraction.)

3.4.2 Unlike gases like O_2, N_2, etc., which have very low solubility in water, there are **other** gases, such as NH_3, SO_2, etc., which have *high solubility* in water. The following information has been abstracted from an example of absorption of SO_2 from air into water in a packed tower by Flagan and Seinfeld (1988):

$$k_{xg}a = 96.8 \frac{\text{kgmol}}{\text{m}^3\text{-hr-mole fraction}}; \qquad k_{xl}a = 2050 \frac{\text{kgmol}}{\text{m}^3\text{-hr-mole fraction}};$$

the location where these values are useful has $x_{Agi} = 0.07$ and x_{Ali} in equilibrium $= 0.002$. Here, a is the interfacial area per unit total system volume and it is unknown. Determine the value of $K_{xg}a$. If the value of a is $100\,\text{m}^2/\text{m}^3$, determine the value of K_{xg} in units of gmol/cm^2-s-mole fraction.

3.4.3 From a detailed packed tower design by Treybal (1980) for the absorption of benzene vapor present in coal gas by an essentially nonvolatile wash oil, the following information is available:

$$k_{xg} = 1.96 \times 10^{-3} \frac{\text{kgmol}}{\text{m}^2\text{-s-mole fraction}}; \qquad k_{xl} = 4.11 \times 10^{-4} \frac{\text{kgmol}}{\text{m}^2\text{-s-mole fraction}};$$

$K_i = x_{iv}/x_{il} = 0.125$ for benzene at 26 °C. Calculate the values of K_{xg} and K_{xl}. Comment on the relative roles of the two film resistances.

3.4.4 Calculate the overall mass-transfer coefficient for the extraction of solute A between water and an organic phase for the following cases.

(1) An organic compound present in low concentration in water is being extracted into toluene. The distribution coefficient of A between the aqueous phase and toluene is $\kappa_{Ao} = 25$. The aqueous-phase mass-transfer coefficient for solute A is $k_{Aw} = 1.06 \times 10^{-2}$ cm/s. The organic-phase mass-transfer coefficient is of the same order as k_{Aw}. Calculate K_{Aw} and K_{Ao}. (Ans. $K_{Aw} = 1.06 \times 10^{-2}$ cm/s; $K_{Ao} = 4.2 \times 10^{-4}$ cm/s.)

(2) Acetic acid is being extracted from water into methyl isobutyl ketone (MIBK); $\kappa_{Ao} = 0.545$. The values of k_{Aw} and k_{Ao} are, respectively, 0.678 and 2.03 in units of lbmol/(hr-ft^2-ΔC_A) where ΔC_A is measured in lbmol/ft^3. Calculate K_{Aw} in units of cm/s.

(3) Acetone is being extracted from a toluene solution into water: $\kappa_{Ao} = 0.58$; $k_{Ao} = 0.0485$ lbmol/(hr-ft^2-ΔC_A); $k_{Aw} = 0.516$ lbmol/(hr-ft^2-ΔC_A). Calculate K_{Aw} and K_{Ao} in units of cm/s. These data have been obtained from Treybal (1963).

3.4.5 In a reverse osmosis experiment using a flat piece of membrane under the condition of no concentration polarization for a brackish water feed having a salt concentration of 9 kg/m^3, the permeation velocity v_y was found to be 1.512 cm/hour (cm^3/cm^2-hour). The applied pressure difference was 29 atm at a temperature of 25 °C. The observed salt rejection was found to be 0.98. Calculate the values of the pure water permeability constant A and the solute transport parameter $(D_{im}\,\kappa_{im}/\delta_m)$, assuming that the membrane does not have any defects, imperfections, etc. You are given that $b = 0.8$ atm-m^3/kg NaCl. (Ans: $A = 1.06 \times 10^{-6}$ gmol H_2O/cm^2-s-atm; $(D_{im}\,\kappa_{im}/\delta_m) = 8.5 \times 10^{-6}$ cm/s.)

3.4.6 In a particular example of reverse osmosis desalination, the values of A and $(D_{im}\,\kappa_{im}/\delta_m)$ are, respectively, 2×10^{-7} m/(s-atm) and 1.5×10^{-7} m/s. Note the units of A given here refer to the volume flux mode, i.e. $N_{water}\,\overline{V}_w = A(\Delta P - \Delta\pi)$. The feed salt mass concentration is 10 kg/m^3; the temperature is 25 °C; the applied feed pressure is 30 atm. The coefficient b in the osmotic-pressure relation $\pi = b\,C_{ij}M_i$ is provided as 0.77 atm-m^3/kg NaCl; note C_{ij} is in molar units. Calculate the salt rejection. State your assumptions if any. (Ans. $R = 0.967$.)

3.4.7 A rotating hollow cylindrical reverse osmosis (RO) membrane of cellulose acetate was placed in an autoclave containing a salt solution of concentration 0.165 mgmol/cm^3. The pressure of this salt solution with respect to the hollow core of the cylinder around which the RO membrane was wrapped was 605 psi (40.44 atm). At 250 r.p.m. of this membrane-wrapped cylinder, the measured salt flux and water flux into the hollow core were, respectively, 0.00252 and 0.605 mg/s-cm^2; the permeate salt concentration was 0.071 mgmol/cm^3. Calculate the values of the membrane transport parameters A and A_p (equation (3.4.60a)), assuming that the salt diffusion parameter $(D_{im}\,\kappa_{im}/\delta_m)$is negligible. You are given the following: salt diffusion coefficient in feed solution, $D_{if} = 1.48 \times 10^{-5}$ cm^2/s; feed solution density $= 1.006$ g/cm^3; $v_f =$ kinematic viscosity $= 0.0090$ cm^2/s; $\pi = 680\,C_{ij}$ psi, where C_{ij} is in mgmol/cm^3; rotating membrane/cylinder diameter $= 2.54$ cm; membrane surface area $=$ 40.5 cm^2. The values of the J_D factor for the rotating cylinder are as follows: $Re = 10^4$, $J_D = 4.8 \times 10^{-3}$; $Re = 7000$, $J_D = 5.4 \times 10^{-3}$; $Re = 9000$, $J_D = 4.95 \times 10^{-3}$; $Re = 5000$, $J_D = 6 \times 10^{-3}$. The Re value is based on the cylinder diameter and the r.p.m. of the cylinder in saline water.

3.4.8 Develop systematically an explicit solvent flux expression in the reverse osmosis separation of two noninteracting solutes $i = 2, 3$ from water in the manner of equation (3.4.66a). Assume that (3.4.65c) is valid for each solute. Further, $|v_y| << k_{if}$ and $C_{ip} << C_{il}^0$ for $i = 2, 3$. Write down the solute flux expressions as well. Use the solution-diffusion model; assume dilute solution of each solute.

3.4.9 In an experiment on simple diffusive permeation of the permanent gas N_2 through a thin film of the glassy polymer poly(phenolpthalein terephthalate), the time lag determined was 22 seconds. The measured permeability coefficient Q_{N_2m} was found to be 0.7 barrer. The film thickness was 0.8 mil (1 mil $= 25 \times 10^{-4}$ cm $= 25$ μm). Measurements were made at 35 °C. Assume that you have no other information available. Determine approximate values of (1) the diffusion coefficient of N_2 through the film and (2) the solubility coefficient of N_2 in the film. (Ans. (1) $D_{im} = 3 \times 10^{-8}$ cm^2/s; (2) $S_{im} = 0.175$ cm^3 (STP)/cm^3 polymer-atm.)

3.4.10 Permanent gases such as nitrogen, oxygen and carbon dioxide have been found to follow the dual sorption-dual transport model for sorption and diffusion through thin films of the glassy polymer, poly (phenolphthalein terephthalate) (P Pha-tere). Determine: (1) the concentrations of the pure gases CO_2 and N_2 in the polymer film in units of cm^3(STP)/cm^3 of polymer if the pure gas-phase pressure is 5 atm; (2) the permeability coefficient Q_{im} of each of pure CO_2 and N_2 present at a feed gas pressure of 5 atm in the unit of barrer (where 1 barrer $= 1 \times 10^{-10}$ cm^3 (STP)-cm/(cm^2-s-cm Hg).) You are given that the temperature is 35 °C. The data in Table 3.P.5 were obtained by Chern and Brown (1990).

Table 3.P.5.

	S_{im} (cm^3 (STP)/cm^3 polymer-atm)	C'_{Hi} (cm^3 (STP)/cm^3 polymer)	b_i (atm^{-1})	$D_{iD} \times 10^8$ (cm^2/s)	$D_{iH} \times 10^9$ (cm^2/s)
CO_2	1.44	31.2	0.398	10.2	5.85
N_2	0.160	3.89	0.062	3.08	0.755

(Ans. (1) CO_2, N_2: 27.96, 1.72 cm^3(STP)/cm^3 polymer; (2) CO_2, N_2: 22.51, 0.66 barrer.)

3.4.11 Thin films of palladium are of interest because of palladium's extremely high selectivity for H_2 over other gases such as He, N_2, O_2, hydrocarbons, etc. Sievert's law based square root solubility dependency of diatomic gases (3.3.70) leads to the unusual permeation flux vs. partial pressure relation (3.4.84) in such systems where the diffusing species is atomic hydrogen in the palladium membrane.

(1) What are the units of the permeability of Q_{H_2m} if the partial pressure of H_2 is in pascals, the membrane thickness is in meters and the flux units are gmol/m^2-s?

(2) If the reversible dissociative chemisorption of H_2 on the palladium surface influences the hydrogen flux, then hydrogen flux will be proportional to $p^n_{H_2f} - p^n_{H_2p}$, where n values higher than 0.5 are observed. Collins and Way (1993) have correlated their H_2 permeation data successfully with $n = 0.573$ for a 17 μm palladium film at two different temperatures (see Table 3.P.6). Determine an average value of the hydrogen permeability Q_{H_2m} through the palladium film at the two temperatures.

Table 3.P.6

$(p^{0.573}_{H_2f} - p^{0.573}_{H_2p})$ (Pa$^{0.573}$)	873 K N_{H_2m} mole/m^2-s	723 K N_{H_2m} mole/m^2-s
1200	0.5	0.32
1800	0.75	0.5
2400	1.00	0.66

(3) If the temperature dependence of Q_{H_2m} is described by an Arrhenius type of relation,

$$Q_{H_2m} = Q^0_{H_2m} \exp\left(-\frac{E_{H_2}}{RT}\right),$$

where E_{H_2} is the apparent activation energy, determine the values of E_{H_2} and $Q^0_{H_2m}$. (Ans. (1) gmol-m/m^2-s-Pa$^{0.5}$; (2) 873K, 7.08×10^{-9} gmol-m/m^2-s-Pa$^{0.573}$.)

3.4.12 In Example 3.4.5 involving microporous cellulose membranes used in hemodialysis, consider the situation where there are boundary layer resistances on two sides of the membrane. For the transport of a microsolute through the membrane from an aqueous solution, Lane and Riggle (1959) have found that, for their membrane, feed and permeate aqueous solutions, the mass-transfer coefficients on each of the feed and permeate are given by $k_{lf} = k_{lp} = 1000\, D_{il}$ cm/min, where D_{il} is the diffusion coefficient of solute i in water in cm^2/s. Obtain an estimate of the overall mass-transfer coefficient of Na_2SO_4 for such a membrane system with boundary layer resistances. (Ans. $K = 3 \times 10^{-5}$ cm/s.)

3.4.13 In electrodialysis, the analysis in Section 3.4.2.5 assumed a perfect cation exchange membrane and no convection of water through an ion exchange membrane placed in an external electrical field. Consider a pressure difference across this membrane without any applied electric field $\nabla\phi$ such that water flows through the membrane. This water flow will drag the counterions in the membrane downstream in the membrane. Thus the high-pressure interface will be depleted of counterions, whereas the low-pressure interface will have an excess of counterions; this charge separation will create an electrical field. The field created by charge separation will prevent any further counterion flow in the downstream direction. We are interested in finding the ion exchange equilibrium under this condition. Suppose, however, that the membrane is imperfect; a few co-ions enter the membrane (Dresner, 1972). Then a steady flux of counterions can take place as long as there is a flux of co-ions of equal but opposite charge. This is equivalent to, say, a salt flow. This salt flow is determined by the extent of coion flow.

Integrate the extended Nernst–Planck equation for any counterion i in the membrane in the z-direction perpendicular to the membrane surface when the membrane excludes co-ions perfectly. Determine the constant of integration in terms of counterion concentration and resin-phase potential just inside the membrane, $C_{iR}\,(0+)$ and $\phi_R\,(0+)$. Use now the condition of thermodynamic equilibrium for the counterion at the interface (between $C_{iR}(0+)$ and C_{iw} via relation (3.3.118b)) and the assumption that $\phi_w = 0$ to relate $C_{iR}(z)$ to C_{iw}. Obtain $\phi_R(z)$ by using the electroneutrality condition (3.3.30b).

3.4.14 Consider a microporous hydrophilic membrane with an aqueous solution on one side and an organic solvent on the other side. Let the pores of the hydrophilic membrane contain the aqueous solution. If the organic-phase pressure is higher than that of the aqueous phase (but does not exceed a critical pressure difference), the aqueous–organic phase interface will be immobilized at each pore mouth on the organic side of the membrane.

(1) Draw a figure similar to Figure 3.4.11 and show the relevant solute concentration profiles for extraction from the aqueous to the organic phase.
(2) Develop the relations between K_o and the individual coefficients and K_w and the individual coefficients.
(3) Obtain the simplified forms of these two relationships for the limiting conditions $\kappa_{io} << 1$ and $\kappa_{io} >> 1$.

3.4.15 Consider microporous membrane-based solvent extraction using a hydrophilic porous membrane whose pores are filled with the organic solvent used to extract a product, species i, from an aqueous feed solution. The aqueous-phase pressure is maintained equal to or higher than the organic-phase pressure to maintain the aqueous–organic interface for nondispersive solvent extraction conditions.

(1) Identify the interface location and plot the concentration profile of i.
(2) Write down the flux expressions of species i through the various resistances.
(3) Develop a relation between the overall mass-transfer coefficient based on the organic phase and the individual mass-transfer coefficients.
(4) Simplify the above result for conditions where the solute distribution coefficient, κ_{io}, is either very large ($\kappa_{io} >> 1$) or very small ($\kappa_{io} << 1$).

3.4.16 Hydrogen is to be recovered from a gas mixture by permeation through a thin film (thickness δ_m) of palladium (Pd) membrane into an aqueous solution prior to chemical reaction in the aqueous solution. The diatomic gas H_2 is present in the feed gas at a partial pressure p_{H_2gf}. The hydrogen concentration in the bulk water is C_{H_2wb}; the mass-transfer coefficient of H_2 in the aqueous film next to the palladium membrane is k_{H_2w}. The permeability coefficient of H_2 through the Pd membrane via Sievert's law is Q_{H_2m}.

(1) Obtain a relation between an overall H_2 transfer coefficient K_{H_2w} based on the aqueous phase, (Q_{H_2m}/δ_m), k_{H_2w} and any other quantities.
(2) Compare such a relation with a relation like (3.4.11) and comment on the complexities encountered. Suggest a procedure, experimental or otherwise, to overcome the problem. Assume that Henry's law is valid. Use the form $p_{H_2g} = H^c_{H_2}\, C_{H_2w}$. Also assume that the palladium membrane allows only H_2 to pass through.

Appendix

Diffusion coefficients in different systems

Note: Table 3.A.8 includes equivalent ionic conductances of selected ions at infinite dilution.

Table 3.A.1 Binary gaseous mixtures at 1 atm

System A-B	Temperature (°C)	D_{AB} (cm^2/s)
H_2-N_2	0	0.674
He-N_2	20	0.705
Ethanol-H_2	0	0.377
H_2O-air	0	0.220
Ethanol-air	0	0.102
N_2-CO_2	25	0.165
N_2-O_2	0	0.181
Ethanol-CO_2	0	0.0685

Source: Perry *et al.* (1984, table 3-318).

Table 3.A.2. Diffusion coefficients of solute i *in a dilute liquid solution, 1 atm at 25 °C*

System *i*-*s*	Type of solute	$D^0_{is} \times 10^5$ (cm^2/s)
H_2-H_2O	gaseous	5.85
O_2-H_2O	gaseous	2.50
N_2-H_2O	gaseous	1.90
CO_2-H_2O	gaseous	1.96
SO_2-H_2O	gaseous	1.70
H_2S-H_2O	gaseous	1.61
CO_2-ethanol	gaseous	4.0
O_2-glycerol water (106 poise)	gaseous	0.24

Source: Perry *et al.* (1984, table 3-319).

Table 3.A.3. Diffusion coefficients of solute i in a dilute liquid solution, 1 atm at 25 °C

System *i-s*	Type of solute	$D^0_{is} \times 10^5$ (cm^2/s)
Toluene-n-hexane	liquid	4.21
Toluene-dodecane	liquid	1.38
Benzene-CCl_4	liquid	1.53
Acetic acid-toluene	liquid	2.26
Acetic acid-ethylene glycol	liquid	0.13
Acetic acid-H_2O	liquid	1.24
Formic acid-ethylene glycol	liquid	0.094
n-Propanol-H_2O	liquid	1.1
n-Butanol-H_2O	liquid	0.96
Urea-H_2O	liquid	1.37
Sucrose-H_2O	solid	0.56
Glucose-H_2O	solid	0.69

Source: Perry *et al.* (1984, table 3-319).

Table 3.A.4. Diffusion coefficients of selected ions at infinite dilution in water at 25 °C

Ion	Z_i	$D_{iw} \times 10^5$ (cm^2/s)	Ion	Z_i	$D_{iw} \times 10^5$ (cm^2/s)
H^+	1	9.31	OH^-	−1	5.26
Li^+	1	1.03	Cl^-	−1	2.03
Na^+	1	1.33	Br^-	−1	2.08
K^+	1	1.96	I^-	−1	2.04
NH_4^+	1	1.95	NO_3^-	−1	1.90
Mg^{++}	2	0.706	HCO_3^-	−1	1.10
Ca^{++}	2	0.792	SO_4^{2-}	−2	1.065
Cu^{++}	2	0.72	HSO_4^-	−1	1.33
Zn^{++}	2	0.71	$Fe(CN)_6^{3-}$	−3	0.896
$Co(NH_3)_6^{+++}$	3	0.908	$Fe(CN)_6^{4-}$	−4	0.739

Source: Newman (1973, table 75-1).

Table 3.A.5. Diffusion coefficients of proteins and other macromolecules in pure water at 20 °C

Solute *i*	Type of solute	M_i	$D^0_{iw} \times 10^7$(cm^2/s)
Lysozyme	protein	14 400	10.4
Ovalbumin	protein	45 000	7.76
Serum albumin	protein	65 000	5.94
Catalase	protein	250 000	4.1
Urease[a]	protein	480 000	3.46
Myosin[a]	protein	493 000	1.16
Polymethyl methacrylate	macromolecule in acetone	10^6	2.25
DNA	macromolecule	6×10^6	0.13

[a] The conformation, especially deviation from sphericity, will lead to different values of D^0_{iw} for similar M_i. For an interpretation of D^0_{iw}, see Tanford (1961). The macromolecule DNA (deoxyribonucleic acid) is a linear flexible macromolecule, which is essentially a polyelectrolyte and is usually double-stranded in nature (see Russel *et al.* (1989) and Viovy (2000)).
Source: Tanford (1961, tables 21-1, 21-3).

Table 3.A.6. Diffusion coefficient of gases through a polymeric film at 25 °C

Gas i	Material	D_i (cm^2/s)
He	polyvinylchloride	2.3×10^{-7}
O_2	N-methyl dithiocarbamate	6×10^{-8}
Ar		2.3×10^{-9}
N_2		1.7×10^{-9}
Kr		4×10^{-10}

Source: Hwang and Kammermeyer (1975, fig. 11.1).

Table 3.A.7. Diffusion coefficients in miscellaneous solids

Species i	Type of solute	Solid material	Temperature (°C)	D_i (cm^2/s)
He	gas	SiO_2	20	$(2.4\text{–}5.5) \times 10^{-10}$
H_2	gas	Ni	85	1.16×10^{-8}
CO	gas	Ni	950	4×10^{-8}
Cd	metal	Cu	20	2.7×10^{-15}

Source: Geankoplis (1972, table 4.1-1).

Table 3.A.8. Equivalent ionic conductances of selected ions at infinite dilution

Ion	Z_i	Temp. (K)	$\lambda_i^0 \left(\frac{\text{mho-cm}^2}{\text{g equiv.}}\right)$	Ion	Z_i	Temp. (K)	$\lambda_i^0 \left(\frac{\text{mho-cm}^2}{\text{g equiv.}}\right)$
H^+	1	298	350	OH^-	−1	298	198
		308	397				
Li^+	1	298	38.7	Cl^-	−1	298	76.3
						308	92.2
Na^+	1	298	50.1	Br^-	−1	298	78.3
		308	61.5				
K^+	1	298	73.52	I^-	−1	298	76.8
NH_4^+	1	298	73.4	NO_3^-	−1	298	71.44
Mg^{++}	2	298	53.06	HCO_3^-	−1	298	41.5
Ca^{++}	2	298	59.5	SO_4^{2-}	−1	298	80
				HSO_4^-	−1	298	50

Sources: Newman (1973, table 75-1) and Atkinson (1972, pp. 5-249 to 5-263).

4

Separation in a closed vessel

In Section 3.3, we illustrated the thermodynamic relations that govern the conditions of equilibrium distribution of a species between two or more immiscible phases under thermodynamic equilibrium. In Section 4.1, we focus on the value of the separation factor or other separation indices for two or more species present in a variety of two-phase separation systems under thermodynamic equilibrium in a closed vessel. The closed vessels of Figure 1.1.2 are appropriate for such equilibrium separation calculations. There is no bulk or diffusive flow into or out of the system in the closed vessel. The processes achieving such separations are called *equilibrium separation processes*. Separations based on such phenomena in an open vessel with bulk flow in and out are studied in Chapters 6, 7 and 8. No chemical reactions are considered here; however, partitioning between a bulk fluid phase and an individual molecule/macromolecule or collection of molecules for noncovalent solute binding has been touched upon here. The effects of chemical reactions are treated in Chapter 5. Partitioning of one species between two phases is an important aspect ever present in this section.

The criteria for thermodynamic equilibrium in a *single-phase system* in a closed vessel subjected to an *external force field* were also developed in Section 3.3. Based on these criteria, we develop in Section 4.2 estimates of the separation achieved in a single phase in the closed vessel. These estimates are also developed in a closed vessel when an additional property gradient, e.g. density gradient, *pH* gradient, etc., exists across the vessel length. *Focusing* is the term often used to characterize the latter separation techniques. In this section, we cover in addition the extent of separation achieved when a *temperature gradient* is imposed on a single-phase system in a closed vessel not subjected to any external force field.

We consider in Section 4.3 the time-dependent separation in a closed vessel where a *membrane* partitions the closed vessel into two regions. We study the *rate-governed* basis of separation using membranes and identify the need for bulk flow in and out of the vessel to make such processes useful for separation. A single semipermeable ion exchange membrane, however, allows equilibrium separation to be achieved in Donnan dialysis; it has also been covered in Section 4.3.

4.1 Equilibrium separation between two phases or two regions in a closed vessel

The separation achieved between two species distributed between two phases or two regions is considered in this section. In Section 3.3.7, the equilibrium distribution of one species between two phases was determined in many two-phase systems. Those results will be employed here for individual equilibrium separation processes in a closed vessel which becomes an *ideal stage*. There will be additional considerations here on the distribution of one species between two immiscible phases at equilibrium. First, however, a few general results are provided for estimating the separation factor α_{12} between two species 1 and 2 in a two-phase system at equilibrium. We ignore chemical reactions for any separation conditions discussed here.

Consider any species i distributed between two phases $j = 1, 2$ at a *uniform pressure P* and *uniform temperature T*. The chemical potential of species i in phase/region j is given by

$$\mu_{ij}(P, T, x_{ij}) = \mu_{ij}^0(P^0, T) + \overline{V}_{ij}(P_j - P^0) + RT \ \ln \ a_{ij}$$

from relation (3.3.21). Rewrite this as

$$\mu_{ij} = \mu_{ij}^0 + \overline{V}_{ij}(P_j - P^0) + RT \ \ln \ a_{ij}.$$

For species i distributed between phases j =1 and 2, the criterion (3.3.6) for equilibrium, namely $\mu_{i1} = \mu_{i2}$, leads in this case to (for $P_1 = P_2 = P = P^0$)

$$\mu_{i1}^0 + RT\ \ln\ a_{i1} = \mu_{i2}^0 + RT\ \ln\ a_{i2}$$

or

$$\frac{a_{i1}}{a_{i2}} = \exp\left[-\frac{(\mu_{i1}^0 - \mu_{i2}^0)}{RT}\right] = \exp\left[-\frac{\Delta\mu_i^0}{RT}\right] = K_i^a, \qquad (4.1.1)$$

where

$$\Delta\mu_i^0 = \mu_{i1}^0 - \mu_{i2}^0 \qquad (4.1.2)$$

and K_i^a is the equilibrium ratio of species i between two phases 1 and 2 in terms of the activity of species i (instead of mole fraction used in definition (1.4.1)). The equilibrium ratio K_i^a of species i depends on μ_{i1}^0 and μ_{i2}^0, the thermodynamic constants for species i for a given combination of i, $j(=1,\ 2)$, P^0 and T. One can, in principle, calculate μ_{i1}^0 and μ_{i2}^0 from the principles of thermodynamics.

For two species i = 1,2 distributed between phases j = 1,2 at equilibrium, employ (4.1.1) and obtain an activity based separation factor:

$$\frac{a_{11}a_{22}}{a_{12}a_{21}} = \exp\left[-\frac{(\Delta\mu_1^0 - \Delta\mu_2^0)}{RT}\right] = \frac{K_1^a}{K_2^a}. \qquad (4.1.3)$$

Since $a_{ij} = \gamma_{ij}x_{ij}$ by definition (equation (3.1.35)), where γ_{ij} is the activity coefficient of species i in phase j, we obtain

$$\alpha_{12} = \frac{x_{11}x_{22}}{x_{12}x_{21}} = \frac{K_1}{K_2} = \frac{\gamma_{21}\gamma_{12}}{\gamma_{11}\gamma_{22}}\exp\left[-\frac{(\Delta\mu_1^0 - \Delta\mu_2^0)}{RT}\right] = \frac{\gamma_{21}\gamma_{12}}{\gamma_{11}\gamma_{22}}\frac{K_1^a}{K_2^a}. \qquad (4.1.4)$$

For ideal solutions or mixtures, $\gamma_{ij} = 1$ for any i,j combination. A simple relation is obtained for α_{12}:

$$\alpha_{12} = \exp\left[-\frac{(\Delta\mu_1^0 - \Delta\mu_2^0)}{RT}\right] = \frac{K_1}{K_2} \qquad (4.1.5)$$

if ideal solution behavior exists in both phases j = 1,2.

For many systems, the fugacity based equilibrium criterion $\hat{f}_{i1} = \hat{f}_{i2}$ is used instead of the chemical potential based criterion (3.3.6). By expressing $\hat{f}_{ij}$ appropriately in terms of x_{ij} and other thermodynamic quantities, an expression for α_{12} may be developed for every system.

4.1.1 Gas–liquid systems

Consider a gas mixture or vapors in a gas mixture. If this mixture is in contact with a nonvolatile absorbent liquid, which absorbs selectively one of the gases or the vapor species, the process is generally called *absorption* and the absorbent liquid acts as a *mass-separating agent.* In continuous industrial processing studied in Chapter 8, the spent absorbent liquid has to be regenerated. The regeneration is carried out using a stripping gas or a stripping vapor, or by vacuum or heating to remove the absorbed gas species. This process of volatile species removal from a liquid is identified as *stripping.* If a stripping gas or vapor (more commonly steam) is used, both act as a mass-separating agent. If heating is employed to regenerate the absorbent, heat or thermal energy is the energy-separating agent.

The following treatments cover the calculation of separation achieved in a closed vessel after equilibrium has been attained via absorption or stripping.

4.1.1.1 *Gas/vapor absorption*

We consider *gas absorption* first. From Section 3.3.7.1 for gas–liquid equilibrium, since

$$\frac{x_{il}}{x_{ig}} = \frac{\hat{\Phi}_{ig}P}{\gamma_{il}f_{il}^0}$$

we obtain, for gaseous species i = 1,2 and $j = l\ (=1)$ and $j = g\ (=2)$,

$$\alpha_{12} = \frac{x_{1l}x_{2g}}{x_{1g}x_{2l}} = \frac{x_{11}x_{22}}{x_{12}x_{21}} = \frac{\hat{\Phi}_{12}\,f_{21}^0\,\gamma_{21}}{\hat{\Phi}_{22}\,f_{11}^0\,\gamma_{11}}. \qquad (4.1.6)$$

Here the liquid phase is the preferred phase (j =1) for the preferred gas species (i = 1) being absorbed. When Henry's law is applicable for the gas species i = 1,2, the following expression is obtained for the separation factor α_{12} between species 1 and 2:

$$\alpha_{12} = \frac{\hat{\Phi}_{12}\,\gamma_{2l}\,H_2}{\hat{\Phi}_{22}\,\gamma_{1l}\,H_1}, \qquad (4.1.7)$$

where $f_{i1}^0 = H_i$, Henry's constant for species i. If both the liquid phase and the gas phase behave ideally,

$$\alpha_{12} \cong \frac{H_2}{H_1}. \qquad (4.1.8)$$

We can understand this result from the following simplistic explanation based on Henry's law (relation (3.3.60b)):

$$p_{ig} = H_i x_{il}.$$

Typical units of H_i are atmospheres/mole fraction. If $H_2 > H_1$, then, for the same partial pressure ($p_{1g} = p_{2g}$) or mole fraction in the gas phase, the mole fraction of gas species 2 in the liquid phase is less than that of gas species 1. Species 1 is thus more soluble in the liquid phase and therefore may be separated from species 2 by absorption in a suitable liquid. Gas absorption based separation processes utilize this preferential solubility of some gases in selected liquid absorbents. The values of Henry's constant H_i in units of atmospheres/mole fraction for a variety of gases in water are provided in the handbook by Perry and Green (1984). For some species, the values are given at a number of temperatures and values of p_{ig}. The latter does indicate a weak dependence of H_i on p_{ig}.

A common example involves the removal of acid gases, e.g. SO_2, H_2S, CO_2, COS, etc., from a gas stream by absorption in a solvent. Table 4.1.1 identifies the values of H_i for a number of gaseous species in a few absorbents for

Table 4.1.1. Henry's law constants[a] and separation factors for acid gases in absorbents

Solvent (temperature)	H_2S $H_i \times 10^{-4}$	CO_2 $H_i \times 10^{-4}$	CH_4/C_3H_8 $H_i \times 10^{-4}$	$\alpha_{H_2S-CO_2}$	$\alpha_{CO_2-CH_4}$	$\alpha_{CO_2-C_3H_8}$
Water[b] (20 °C)	0.0483	0.142	3.76 (CH_4)	2.94	26.5	-
n-Methyl-pyrrolidone[c] (20 °C)	0.000476	0.00588	0.0829 (CH_4)	12.35	14.11	-
Methanol[d] (7 °C)	0.00405	0.0160	0.115 (CH_4) 0.0245 (C_3H_8)	3.96	7.2	1.53

[a] Units of H_i, atm/mole fraction.
[b] National Research Council (1929).
[c] From Kohl and Riesenfeld (1979, p. 783).
[d] From Astarita *et al.* (1983, table 1.7.2).

removing H_2S and CO_2 from CH_4/C_3H_8. The separation factors have also been calculated to illustrate which species is more selectively absorbed and which absorbent provides a high selectivity. Much more information on a variety of other absorbents is provided in Astarita *et al.* (1983) and Perry and Green (1984).

Table 4.1.1 identifies three pure absorbents, one of which is water. We shall see in Section 5.2.1.2 that a common absorbent for acid gases is water containing either organic ethanolamines or inorganic salts. The value of the Henry's law constant H_i for various gases in such aqueous electrolytic solutions may be related to that in pure water H_i^0 by

$$\log_{10}\left(\frac{H_i}{H_i^0}\right) = hI, \tag{4.1.9a}$$

where I, the ionic strength of the solution, is defined as follows:

$$I = \frac{\sum_n C_{il} Z_i^2}{2}. \tag{4.1.9b}$$

The solution contains n ionic species; species i has a valence of Z_i at a molar concentration of C_{il} gmol/cm^3. Contributions of positive ions, negative ions and dissolved free gas species (h_+, h_- and h_G, respectively) have been tabulated by Danckwerts (1970) for calculating the quantity h in definition (4.1.9a) as

$$h = h_+ + h_- + h_G. \tag{4.1.9c}$$

In general, the solubilities of most gases in an absorbent decrease with an increase in temperature. This implies, from Henry's law, that H_i increases with temperature. In fact, Henry's constant for species i changes with temperature in the manner of

$$\frac{\mathrm{d}\ln H_i}{\mathrm{d}(1/T)} = \frac{\Delta H_i}{\mathrm{R}}, \tag{4.1.10}$$

where T is the absolute temperature, R is the universal gas constant and ΔH_i is the heat of absorption of species i. Thus, if ΔH_i is negative, i.e. heat evolves on absorption of the gas species, H_i increases, reducing the amount of gas dissolved. For many common gases and water, the opposite behavior is observed at temperatures greater than 100 °C and above atmospheric pressure. Equation (4.1.10) may be used to estimate, in general, the variation of α_{12} with temperature.

Absorption of a *condensable vapor species* in a nonvolatile absorbing liquid from a mixture of vapors (or from a mixture of vapors in a gas or a gas mixture) will be briefly treated in the following. Using the expression (3.3.61) for the equilibrium ratio K_i of a vapor species i between a gas and a liquid phase at equilibrium, we may develop the following expression for the separation factor α_{12} between two vapor species $i = 1,2$:

$$\alpha_{12} = \frac{x_{1l}\,x_{2g}}{x_{1g}\,x_{2l}} = \frac{K_2}{K_1} = \frac{\hat{\Phi}_{1g}\gamma_{2l}f_{2l}^0}{\hat{\Phi}_{2g}\gamma_{1l}f_{1l}^0}. \tag{4.1.11}$$

Here, species 1 is preferred for absorption into the preferred phase, the liquid phase. For low to moderate pressures, we know that

$$f_{il}^0 = P_i^{\text{sat}}\Phi_i^{\text{sat}},$$

where P_i^{sat} is the vapor pressure of species i at system temperature T. For ideal gas behavior, $\Phi_i^{\text{sat}} \cong 1$ and $\hat{\Phi}_{ig} \cong 1$. Further, for ideal solution behavior in the liquid phase, $\gamma_{il} \cong 1$. These simplifications (i.e. Raoult's law (3.3.64) holds) lead to:

$$\alpha_{12} \cong \frac{P_2^{\text{sat}}}{P_1^{\text{sat}}}. \tag{4.1.12}$$

This result implies that species 1 will be preferentially absorbed if it has a vapor pressure lower than that of species 2. Thus, heavier species or higher boiling species are more selectively absorbed.

The above treatments primarily consider systems and conditions relevant for the large-scale absorptions of gases/vapors in absorbent liquids employed in the chemical processing industry. These principles are also relevant for *gas–liquid chromatography* (GLC), wherein a mobile gas phase containing sorbable species flows over a nonvolatile stationary liquid phase existing as a thin coating over a porous solid matrix or a capillary tube. The selectivity α_{12} of vapor species 1 over vapor species 2 present in the mobile gas phase (see Section 7.1.5) will be obtained from expression (4.1.11). For low-pressure operations, assume ideal gas behavior $\hat{\Phi}_{ig} \cong 1$ and $\Phi_i^{\text{sat}} \cong 1$. Therefore

$$\alpha_{12} = \frac{x_{1l}}{x_{1g}}\frac{x_{2g}}{x_{2l}} = \frac{\gamma_{2l}P_2^{sat}}{\gamma_{1l}P_1^{sat}}. \quad (4.1.13)$$

Thus, a higher boiling species would prefer the solvent phase, the stationary phase, and, as we shall see later (Section 7.1.5), it will have a higher retention time unless the activity coefficient ratio affects it. The nonvolatile liquid phase, the absorbent, can, however, modify this behavior through the activity coefficients.

4.1.1.2 *Gas/vapor stripping*

The removal of a volatile species, gas or a vapor, from a nonvolatile absorbent or solvent by an inert gas or vapor or by simple heating is called *stripping*. When air is used as the inert gas, the process is called *air stripping*; in the case of steam, the term *steam stripping* is used. The word *desorption* is also employed.

For stripping processes, the preferred species $i = 1$ (which may be a gas or a vapor) should be preferably transferred to the preferred gas phase $j = 1 = g$. When selectivity between *two gas species* $i = 1,2$ being stripped from a nonvolatile liquid by steam or air or just heating is under consideration, we employ the gas–liquid equilibrium relation from Section 3.3.7.1:

$$\hat{f}_{ig} = \hat{f}_{il} = x_{ig}\hat{\Phi}_{ig}P = \hat{f}_{il} = \hat{f}_{i2} = x_{il}\gamma_{il}f_{il}^0 \Rightarrow \frac{x_{ig}}{x_{il}} = \frac{\gamma_{il}f_{il}^0}{\hat{\Phi}_{ig}P}. \quad (4.1.14)$$

Therefore, for $i = 1,2$,

$$\alpha_{12} = \frac{x_{1g}x_{2l}}{x_{1l}x_{2g}} = \frac{\gamma_{1l}\hat{\Phi}_{2g}f_{1l}^0}{\gamma_{2l}\hat{\Phi}_{1g}f_{2l}^0}. \quad (4.1.15a)$$

For gaseous species satisfying Henry's law for the system under consideration, $f_{il}^0 = H_i$. Further, if the liquid phase behaves as an ideal solution and the gas phase behaves as an ideal gas,

$$\alpha_{12} = \frac{x_{1g}x_{2l}}{x_{1l}x_{2g}} \cong \frac{H_1}{H_2}. \quad (4.1.15b)$$

For species 1 and 2, if $H_1 > H_2$, then, for equal liquid-phase mole fractions of 1 and 2 (i.e. $x_{1l} = x_{2l}$), the partial pressure of species 1 will be higher in the gas phase, i.e. species 1 is more easily stripped.

When species 1 and 2 are both vapors at the system temperature and pressure in equilibrium with an appropriate liquid phase, we employ relation (3.3.61) for $i = 1$ and $j = g = 1$ for the preferred species and phase, respectively, to obtain

$$\alpha_{12} = \frac{x_{1g}x_{2l}}{x_{1l}x_{2g}} = \frac{K_1}{K_2} = \frac{\gamma_{1l}P_1^{sat}\Phi_1^{sat}\Phi_{2g}}{\gamma_{2l}P_2^{sat}\Phi_2^{sat}\Phi_{1g}}. \quad (4.1.16)$$

Assuming ideal gas and ideal solution behavior,

$$\alpha_{12} \cong \frac{P_1^{sat}}{P_2^{sat}}. \quad (4.1.17)$$

Species 1, having a higher vapor pressure, is more easily stripped and accumulates preferentially in the gas phase. The example given in the following will show that such a result may not hold if nonideal behavior exists.

A common example of vapor stripping involves removal of volatile organic compounds (VOCs) from water either obtained within a chemical process plant or discharged from it. Such a system rarely displays ideal solution behavior. The selectivity α_{12} then becomes

$$\alpha_{12} \cong \left(\frac{\gamma_{1l}}{\gamma_{2l}}\right)\left(\frac{P_1^{sat}}{P_2^{sat}}\right). \quad (4.1.18)$$

In industrial practice, the vapor–liquid equilibrium of such VOCs employs the ideal gas assumption, and relation (4.1.14) is reformulated as

$$x_{ig} = \frac{\gamma_{il}f_{il}^0}{P}x_{il} = \frac{\gamma_{il}^{\infty}f_{il}^0}{P}x_{il}, \quad (4.1.19a)$$

where γ_{il}^{∞} represents an infinite dilution activity coefficient since x_{il} tends to be very small. A useful form of this relation is

$$Px_{ig} = p_{ig} = H_i^C C_{il} = 18 \times 10^{-6}P \times (f_{il}^0\gamma_{il}^{\infty}/P)C_{il}, \quad (4.1.19b)$$

where P is in atmospheres and C_{il} is in gmol/m^3; the value of $(f_{il}^0\gamma_{il}^{\infty}/P)$, represented as K_i^{∞}, an infinite dilution equilibrium ratio, has been provided for a large number of VOCs by Hwang *et al.* (1992b) at 100 °C (for steam stripping) and 25 °C (for air stripping) along with f_{il}^0 and γ_{il}^{∞} values. Using this notation,

$$\alpha_{12} = \frac{x_{1g}x_{2l}}{x_{1l}x_{2g}} = \frac{K_1^{\infty}}{K_2^{\infty}}, \quad (4.1.19c)$$

where the subscript of K_i^{∞} identifies the species. For example, $K_1^{\infty} = 4.1 \times 10^2$ for toluene (25 °C), whereas the corresponding value for acetone is only 2.1 (25 °C). Thus, toluene appears to be easily strippable by air compared to acetone, even though acetone has a much lower boiling point. Toluene is much more hydrophobic than acetone and has a very large value of γ_{il}^{∞} (~10^4 compared to 17.2 for acetone). The more hydrophobic a compound is, the lower its aqueous solubility and the higher its strippability.

An important aspect of VOC stripping concerns the stripping temperature and therefore the stripping medium. Should one use air stripping (25 °C) or steam stripping (100 °C)? Analysis based on α_{12} of (4.1.19c) is not useful. Comparison of K_i^{∞} values at two different temperatures is more useful (Hwang *et al.*, 1992a).

4.1.2 Vapor–liquid systems

Vapor–liquid systems are encountered frequently in industrial processing. The composition difference between the vapor and the liquid phases made up of volatile species only is the basis for separation by distillation, which involves boiling and condensation. In this section, we first determine the separation factor between the vapor and the liquid phases in a closed vessel at equilibrium under conditions where Raoult's law may or may not be valid. We consider next a variety of vapor–liquid systems encountered in practice. Binary systems are frequently employed here for simplicity and illustration, along with the Gibbs phase rule. Then we study two extreme situations in the closed vessel, when almost all of the material is liquid or almost all of the material is vapor and calculate the unknown phase composition for estimating the separation. Multicomponent systems are of primary interest in such calculations.

Using the equality of fugacity of a species in vapor $(j = v)$ and liquid phases $(j = l)$,

$$\hat{f}_{iv} = x_{iv}\hat{\Phi}_{iv}P = \hat{f}_{il} = x_{il}\gamma_{il}f_{il}^{0},$$

and the relation (3.3.62) for f_{il}^{0} at *low to moderate pressures,* we get:

$$\frac{x_{iv}}{x_{il}} = \frac{\gamma_{il}\Phi_i^{sat}P_i^{sat}}{\hat{\Phi}_{iv}P}. \tag{4.1.20}$$

For two species $i = 1$ and 2,

$$\alpha_{12} = \frac{x_{1v}x_{2l}}{x_{1l}x_{2v}} = \left(\frac{P_1^{sat}}{P_2^{sat}}\right)\left(\frac{\gamma_{1l}}{\gamma_{2l}}\right)\left(\frac{\Phi_1^{sat}}{\hat{\Phi}_{1v}}\right)\left(\frac{\hat{\Phi}_{2v}}{\Phi_2^{sat}}\right), \tag{4.1.21a}$$

where the preferred phase for the preferred (lighter) species $(i = 1)$ is $j = v$, the vapor phase. If the vapor phase is an ideal gas and the liquid phase is an ideal solution, *Raoult's law* (relation (3.3.64)) is valid. The expression for α_{12} is simplified to

$$\alpha_{12} = \frac{P_1^{sat}}{P_2^{sat}} \tag{4.1.21b}$$

at the system temperature T and pressure P. This ratio is also called the *relative volatility,* the value of α_{12} under ideal conditions. If only the vapor phase behaves as an ideal gas, then

$$\alpha_{12} = \left(\frac{P_1^{sat}}{P_2^{sat}}\right)\left(\frac{\gamma_{1l}}{\gamma_{2l}}\right). \tag{4.1.21c}$$

The vapor pressure P_i^{sat} depends on temperature (e.g. Antoine equation).[1] The activity coefficient γ_{il} depends on the composition of the liquid phase and temperature. The distillation separation process utilizing vapor–liquid equilibrium has to achieve changes from the initial composition to the desired composition. A few thermodynamic principles and features of vapor–liquid equilibria are therefore quite useful here. We start with *Gibbs phase rule.*

The Gibbs phase rule for a closed nonreactive system of j phases made up of n species states that, at equilibrium, the number of degrees of freedom F is given by

$$j + F = n + 2. \tag{4.1.22}$$

Once the intensive variables[2] F are specified, the system is completely defined. For example, in a binary system $(n = 2)$ having vapor–liquid equilibrium $(j = 2)$, the degrees of freedom available are 2 $(= F)$. Once the values of any two intensive variables are fixed, the values of all other intensive variables become fixed. If values of pressure and temperature are specified, the compositions of the two phases of the system are determined. Similarly, if the pressure and liquid mole fraction are specified, the composition of the vapor phase and the value of the temperature become fixed. Alternatively, if the temperature and the vapor phase mole fraction are given, the pressure and the liquid phase mole fraction will be fixed automatically from the solution of the governing equations.

Before we present briefly the methodology for such calculations, it is worthwhile illustrating the basic types of vapor–liquid equilibrium (VLE) behavior frequently encountered. The variety in VLE arises primarily from different types of nonideal behavior, as indicated, for example, by the liquid-phase activity coefficients. *Minimum-boiling azeotropes* and *maximum-boiling azeotropes* are two such types of nonideal behavior. To become familiar with such behavior, we start with the ideal solution behavior.

Consider VLE for a binary system containing benzene (species $i = 1$) and toluene $(i = 2)$ at a low pressure, as shown in Figure 4.1.1. Figure 4.1.1(a) (Hougen *et al.*, 1959) illustrates the partial pressure of each species and the total pressure P of the system at 90 °C. Due to ideal gas behavior (low pressure) and ideal liquid solution behavior, Raoult's law (3.3.64) is valid:

$$x_{1v} = \frac{P_1^{sat}}{P}x_{1l}, x_{2v} = \frac{P_2^{sat}}{P}x_{2l}.$$

Since the temperature is constant at 90 °C, the value of the separation factor between species 1 and 2,

$$\alpha_{12} = \frac{P_1^{sat}}{P_2^{sat}} \approx \frac{1000}{400} = 2.5,$$

is also constant with respect to the composition of any phase (see Figure 1.4.1). The total pressure P may be expressed as the sum of the partial pressures p_{1v} and p_{2v} as follows:

---

[1] $\ln P_i^{sat} = A_i - \frac{B_i}{T+C_i}$; A_i, B_i and C_i are constants.

[2] The intensive variables are temperature, pressure and $(n-1)$ mole fractions in each phase present. For a two-phase two-component system, one mole fraction of each phase is enough, since $x_{1j} + x_{2j} = 1$.

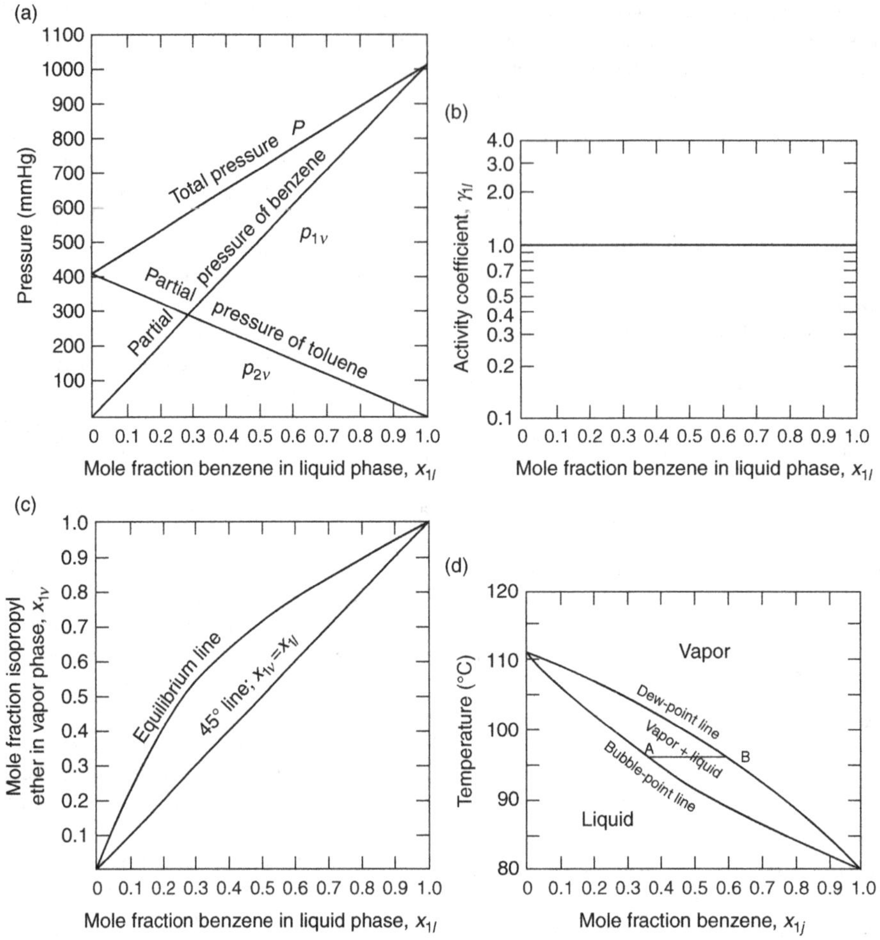

Figure 4.1.1. Ideal solution behavior, benzene-toluene system: (a) partial and total pressures at 90 °C; (b) activity coefficients; (c) vapor-liquid equilibrium; (d) phase diagram at 1 atm pressure. (After Hougen et al. *(1959).)*

$$P = p_{1v} + p_{2v} = x_{1v}P + x_{2v}P$$
$$= P_2^{\text{sat}}(x_{2l} + \alpha_{12}x_{1l}) \Rightarrow P = P_2^{\text{sat}}[1 + x_{1l}(\alpha_{12}-1)]. \quad (4.1.23)$$

This linear dependence of P on x_{1l}, as well as that of p_{1v} on x_{1l} and p_{2v} on x_{2l}, are shown in Figure 4.1.1(a). The ideal liquid solution behavior of $\gamma_{1l} = \gamma_{2l} = 1$ is shown in Figure 4.1.1(b).

Figure 4.1.1(c) illustrates the equilibrium vapor and liquid compositions for the benzene-toluene system at a constant total pressure of 1 atmosphere. The vapor phase is enriched in the lighter component, benzene, at all compositions, except in the case of pure benzene and pure toluene. If such vapor-phase enrichment of benzene did not occur, the straight reference line ($x_{1v} = x_{1l}$) would have represented the VLE. The actual curved equilibrium line relating x_{1v} to x_{1l} may be obtained in the following manner. Consider first Raoult's law expressions for $i = 1$ and $i = 2$:

$$x_{1v} = \frac{P_1^{\text{sat}}}{P}x_{1l}, \; x_{2v} = \frac{P_2^{\text{sat}}}{P}x_{2l}.$$

Divide x_{2v} by x_{1v}, recognize that $x_{2v} = 1 - x_{1v}$ and simplify to obtain (see equation (1.4.76))

$$x_{1v} = \frac{\alpha_{12}x_{1l}}{1 + x_{1l}(\alpha_{12}-1)}. \quad (4.1.24)$$

Note that different values of x_{1l} (and the corresponding x_{1v}) in Figure 4.1.1(c) represent different closed vapor-liquid equilibrium systems at 1 atmosphere but slightly different temperatures. Note further that α_{12} in equation (4.1.24) varies slightly with temperature and is not a constant (since both P_1^{sat} and P_2^{sat} vary with temperature).

The boiling point of a liquid mixture at a given total pressure is shown in Figure 4.1.1(d) as the *bubble-point line* for the benzene-toluene system at 1 atmosphere.

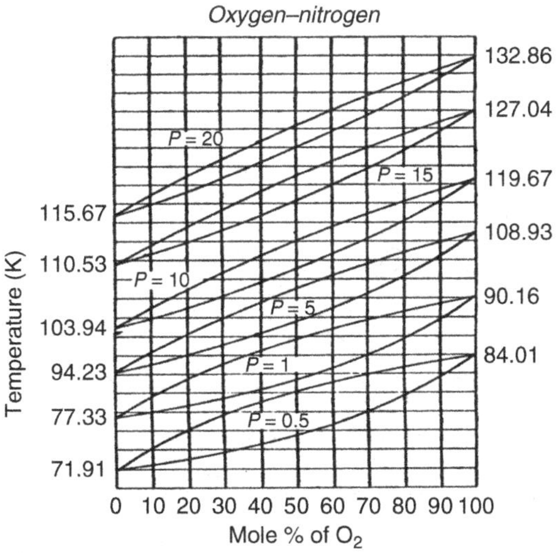

Figure 4.1.2. (T, x_{il}) *curves at 0.5, 1, 5, 10, 15 and 20 atm for an* O_2-N_2 *system (Dodge and Dunbar, 1927). Reprinted, with permission, from* J. Am. Chem. Soc. ***49****, 591 (1927), Figure 6. Copyright (1927) American Chemical Society.*

The composition of the first bubble in equilibrium with the liquid mixture at the boiling point temperature is shown by the upper line. The straight line AB joining these two equilibrium compositions is called a *tie line*. Conversely, when a vapor having any composition is cooled, its temperature comes down to the upper line, the *dew-point line*. The first drop of liquid that condenses from this vapor at this temperature will have its composition given by the bubble-point line at that temperature, the two being connected by another horizontal tie line.

In low-pressure VLE satisfying ideal gas and ideal solution behavior, Raoult's law can be used very effectively to obtain the values of the intensive variables in a closed system at equilibrium. Unfortunately, there are many VLE systems where Raoult's law is not obeyed. Specifically, in many cases of low-pressure VLE, the liquid phase behaves nonideally; γ_{il} varies with composition and is $\neq 1$. The nature of the deviation of γ_{il} from 1 often determines the nature of the VLE. Systems where $\gamma_{il} > 1$ are said to display positive deviations from Raoult's law; when $\gamma_{il} < 1$, the systems show a negative deviation from Raoult's law.

Before we illustrate the VLE system behavior for $\gamma_{il} \neq 1$, note that deviations from Raoult's law occur also due to nonideal behavior in the gas phase. For example, for an O_2-N_2 system at low temperatures, where the gases are liquefied, it has been observed that the partial pressure of N_2 is lower than that predicted by the straight-line behavior shown in Figure 4.1.1(a). Yet the T-x_{il} behavior shown in Figure 4.1.2 for O_2-N_2 appears similar to that in Figure 4.1.1(d); only the N_2 boiling point is lower than that of O_2 (Dodge and Dunbar (1927); Ruhemann (1949)).

Figure 4.1.3 (Hougen *et al.*, 1959) illustrates the VLE of an isopropyl ether-isopropyl alcohol system characterized by $\gamma_{il} > 1$, i.e. positive deviations from Raoult's law. Figure 4.1.3(a) illustrates the variations of the total pressure and the partial pressures as a function of composition at 70 °C. Figure 4.1.3(b) shows the activity coefficients of both species varying with the liquid composition. Since $\gamma_{il} > 1$, the partial pressure of any species is larger than a straight-line behavior exhibited by an ideal solution. Further, the maximum of the total pressure is greater than either P_i^{sat} (recall that there is no such maximum in an ideal solution system). Such behavior leads to the formation of an azeotrope: at a certain composition, both liquid and vapor have the same composition, as shown in Figure 4.1.3(c). Correspondingly, at this composition, the bubble-point line and the dew-point line have a common minimum identifying the minimum boiling point (Figure 4.1.3(d)). There is no separation achieved between the two phases at the azeotrope composition.

Positive deviations from Raoult's law arise when the forces between two molecules of two different species are weaker than those between two molecules of the same species. Negative deviations from Raoult's law happen when the intermolecular forces between two different species are stronger than those between two molecules of the same species. Figure 4.1.4 shows the VLE of a system of this type: that of acetone and chloroform (Hougen *et al.*, 1959).

The negative deviations in the acetone-chloroform system are clearly evident in Figure 4.1.4(a), where the partial pressure of each species is lower than that indicated by the straight-line behavior that would be true in a system exhibiting ideal solution behavior (e.g. benzene-toluene). The escaping tendency of a given species into the vapor phase is reduced by attraction from molecules of other species. The total pressure over the system has a minimum, which leads to a maximum boiling point. The activity coefficient γ_{il} for each species is less than 1 (Figure 4.1.4(b)). Such a system is called a *maximum boiling point azeotrope*. At the azeotropic composition, $x_{iv} = x_{il}$ (Figure 4.1.4(c)); the system exhibits a maximum boiling point at the same composition (Figure 4.1.4(d)).

For systems having substantial positive deviations from Raoult's law, one sometimes encounters limited solubility in the liquid phase. Two liquid phases, substantially immiscible with each other, coexist over a certain composition, whereas, at other compositions, a single liquid-phase exists. An example is provided by the water-n-butanol system (Hougen *et al.*, 1959).

For a binary system like benzene-toluene, exhibiting ideal solution behavior at low pressures, we have already learned how to calculate x_{iv} from x_{il} using Raoult's law (more specifically, using relation (4.1.24)). The calculation procedures for vapor-liquid equilibrium in *multicomponent systems* behaving *nonideally* will now be treated

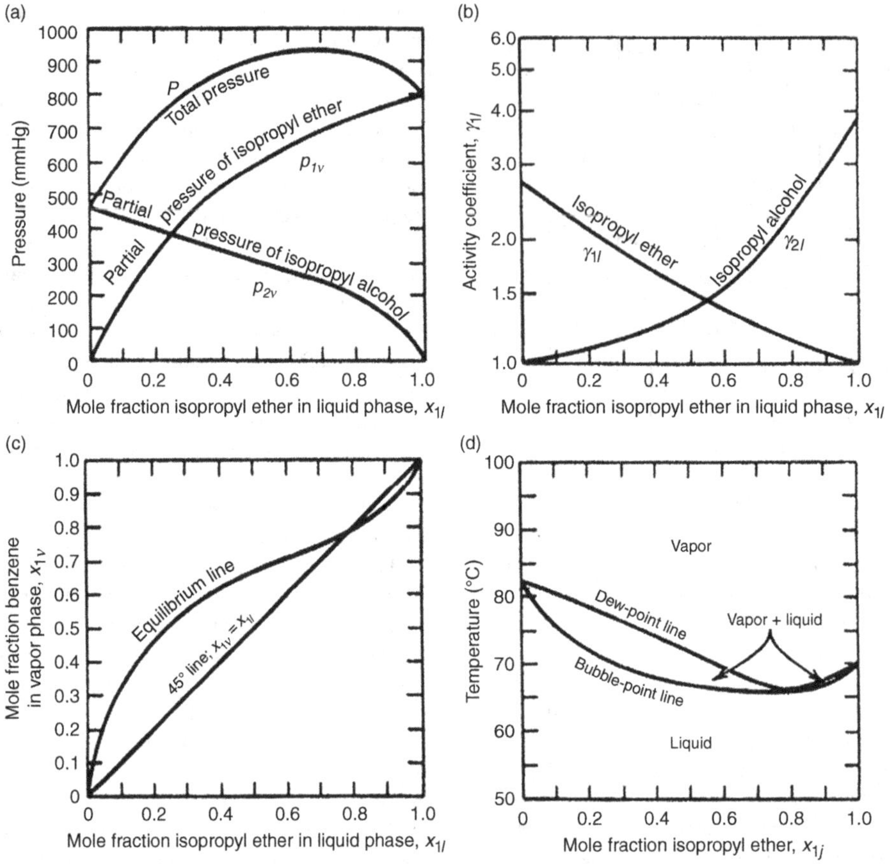

Figure 4.1.3. Minimum-boiling azeotrope, isopropyl ether–isopropyl alcohol system: (a) partial and total pressures at 70 °C; (b) activity coefficients; (c) vapor–liquid equilibria; (d) phase diagram at 1 atm pressure. (After Hougen et al. *(1959).)*

briefly. Detailed calculation procedures are available in standard texts on chemical engineering thermodynamics.

There are four types of VLE calculations for a two-phase n-component system:

(a) The values of T and $x_{1l}, x_{2l}, \ldots, x_{n-1,l}$ are known; the values of P and $x_{1v}, x_{2v}, \ldots, x_{n-1,v}$ need to be determined. This is equivalent to calculating the composition of the first bubble in equilibrium with the specified liquid composition on the *bubble-point line* at the given temperature T.

(b) The values of P and $x_{1l}, x_{2l}, \ldots, x_{n-1,l}$ are known; we need to determine T and $x_{1v}, x_{2v}, \ldots, x_{n-1,v}$.

Both (a) and (b) are identified as *bubble-point calculations*.

(c) The values of P and $x_{1v}, x_{2v}, \ldots, x_{n-1,v}$ are known; we have to calculate T and $x_{1l}, x_{2l}, \ldots, x_{n-1,l}$.

(d) The values of T and $x_{1v}, x_{2v}, \ldots, x_{n-1,v}$ are known; P and $x_{1l}, x_{2l}, \ldots, x_{n-1,l}$ are to be estimated.

Both (c) and (d) are identified as *dew-point calculations*; they involve calculating the composition of the first liquid drop in equilibrium with the given vapor composition on the dew-point line.

Consider the overall composition of a vapor–liquid mixture ($x_{1f}, x_{2f}, \ldots, x_{n-1,f}$) in equilibrium at a particular temperature T and pressure P. Sometimes one would like to know the actual liquid composition ($x_{1l}, x_{2l}, \ldots, x_{n-1,l}$) as well as the actual vapor composition ($x_{1v}, x_{2v}, \ldots, x_{n-1,v}$) along with the relative molar amounts of the liquid phase and the vapor phase under such conditions. The vapor–liquid equilibrium calculation procedure adopted for this fifth case is identified commonly as a *flash calculation*. The flow system equivalent of this calculation is touched upon in Chapter 6, Section 6.3.2.1, for a flash separator.

Each of the five above-mentioned VLE calculations for nonideal multicomponent systems require extensive numerical calculations in an iterative fashion best carried out in a computer. The basic governing equations are:

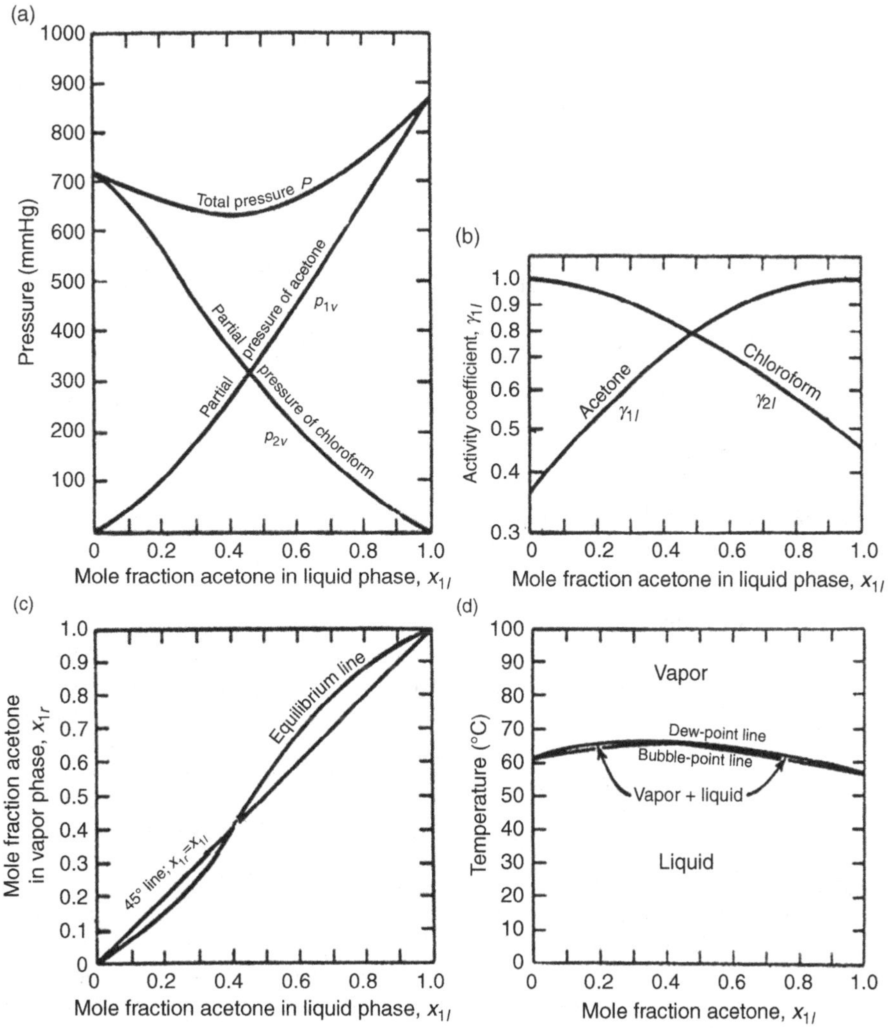

Figure 4.1.4. Maximum-boiling azeotrope, acetone–chloroform system: (a) partial and total pressures at 60 °C; (b) activity coefficients; (c) vapor–liquid equilibria; (d) phase diagram at 1 atm pressure. (After Hougen et al. *(1959).)*

$$\hat{f}_{iv} = \hat{f}_{il}, \qquad i = 1, 2, \ldots, n.$$

Therefore

$$x_{iv}\hat{\Phi}_{iv}P = x_{il}\gamma_{il}f_{il}^0 \qquad i = 1, 2, \ldots, n. \tag{4.1.25}$$

In addition to these equations, we also need

$$\sum_{i=1}^{n} x_{iv} = 1 \quad \text{and} \quad \sum_{i=1}^{n} x_{il} = 1 \tag{4.1.26}$$

for all *bubble-point* and *dew-point calculations.* Equations (4.1.25) are to be solved simultaneously under the constraints of equations (4.1.26). The complexity is most often due to the fact that

(1) $\hat{\Phi}_{iv}$ depends on the x_{iv} terms, P and T;
(2) γ_{il} depends on the x_{il} terms, P and T;
(3) f_{il}^0 depends on P and T.

Further, the functional dependences are complex.

For *flash calculations,* the total number of moles, $\sum_{i=1}^{n} m_{if}$, in the system is known in addition to x_{if} for all $i = 1, \ldots, n$. If m_{iv} and m_{il} represent the number of moles of species i in the vapor phase and the liquid phase, respectively, then the additional equations required to solve the problem of determining the x_{iv} and the relative amounts of the vapor and the liquid phases are:

$$\sum_{i=1}^{n} m_{if} = \sum_{i=1}^{n} m_{iv} + \sum_{i=1}^{n} m_{il}; \tag{4.1.27}$$

$$\left(\sum_{i=1}^{n} m_{if}\right) x_{if} = \left(\sum_{i=1}^{n} m_{iv}\right) x_{iv} + \left(\sum_{i=1}^{n} m_{il}\right) x_{il}. \quad (4.1.28)$$

Simpler expressions for both equations are as follows:

$$m_{tf} = m_{tv} + m_{tl}; \quad (4.1.29a)$$

$$x_{if} m_{tf} = x_{iv} m_{tv} + x_{il} m_{tl}. \quad (4.1.29b)$$

The bubble-point calculations, the dew-point calculations and the flash calculations are not carried out in practice for any set of P, T, x_{iv} and x_{il}. Rather, a subset of conditions which allow certain simplifications are selected and thermodynamic calculations are carried out. They are:

(a) low to moderate pressures;
(b) ideal gas for low to moderate pressures;
(c) high pressures: the Chao–Seader method.

For *low to moderate pressures*, γ_{il} and f_{il}^0 are assumed to be independent of pressure. Following the resulting simplification (3.3.62), equations (4.1.25) are written as

$$x_{iv} \hat{\Phi}_{iv} P = x_{il} \gamma_{il} P_i^{\text{sat}} \Phi_i^{\text{sat}};$$

i.e.

$$x_{iv} = x_{il} \frac{\gamma_{il} P_i^{\text{sat}} \Phi_i^{\text{sat}}}{\hat{\Phi}_{iv} P}. \quad (4.1.30)$$

In an *iterative* computer-based numerical *bubble-point calculation* of T, $x_{1v}, x_{2v}, \ldots, x_{n-1,v}$ when P and $x_{1l}, x_{2l}, \ldots, x_{n-1,l}$ are known, the following approach is often adopted.

(1) Assume T and $\hat{\Phi}_{iv} = 1$.
(2) Calculate the x_{iv} from equations (4.1.30) by calculating the γ_{il}, P_i^{sat} and Φ_i^{sat} using appropriate thermodynamic formulas for each i.
(3) Calculate $\sum_{i=1}^{n} x_{iv}$. For the first iteration, normalize the assumed x_{iv} by dividing each one by $\sum_{i=1}^{n} x_{iv}$. Using these new x_{iv}, calculate new estimates of $\hat{\Phi}_{iv}$ (different from 1) using appropriate thermodynamic formulas.
(4) Estimate new values of x_{iv} from (4.1.30) using the new estimates of $\hat{\Phi}_{iv}$.
(5) Sum all x_{iv} to obtain $\sum_{i=1}^{n} x_{iv}$. If it is different from that obtained earlier, continue the iterative calculation of the x_{iv} using steps (3), (4) and (5). If $\sum_{i=1}^{n} x_{iv}$ changes by less than a small predetermined amount and it is still different from 1, then assume a different T and recalculate the γ_{il}, P_i^{sat}, Φ_i^{sat}; repeat steps (2), (3), etc., till $\sum_{i=1}^{n} x_{iv} = 1$.
(6) For the P and T, when $\sum_{i=1}^{n} x_{iv} = 1$, the results, T and x_{iv} are correct for the problem.

For *low to moderate pressures*, if the gas phase behaves ideally, it is generally observed that $\Phi_i^{\text{sat}} \cong \hat{\Phi}_{iv}$ and/or both are close to unity so that relation (4.1.30) becomes the simpler

$$x_{iv} = x_{il} \frac{\gamma_{il} P_i^{\text{sat}}}{P}. \quad (4.1.31)$$

One can now follow steps (1)-(6) iteratively, except it is much simpler since $\hat{\Phi}_{iv}$ does not have to be calculated. Assume T; calculate the γ_{il} and P_i^{sat} and then the x_{iv}. If $\sum_{i=1}^{n} x_{iv} \neq 1$, assume another T and continue till $\sum_{i=1}^{n} x_{iv} = 1$.

For *higher pressures*, the assumption that γ_{il} and f_{il}^0 are independent of pressure breaks down. A number of approaches have been developed. One of the earliest was proposed by Chao and Seader (1961), wherein equations (4.1.25), written as

$$K_i = \frac{x_{iv}}{x_{il}} = \frac{\gamma_{il} f_{il}^0}{\hat{\Phi}_{iv} P}, \quad (4.1.32a)$$

were simplified to

$$K_i = \frac{\gamma_{il} \Phi_{il}^0}{\hat{\Phi}_{iv}}. \quad (4.1.32b)$$

Thus, the fugacity coefficient Φ_{il}^0 of the pure liquid i at temperature T and pressure P is calculated and substituted for (f_{il}^0/P). Such K_i factors are used frequently in the oil industry; nomographs are available for low molecular weight alkanes and alkenes as a function of temperature and pressure (see Figures 4.1.5 and 4.1.6 (Dadyburjor (1978), adapted from DePriester (1953))). Note that lower molecular weight hydrocarbon species are likely to have K_i greater than 1, whereas higher molecular weight hydrocarbon species are likely to have K_i less than 1. Example 4.1.1 illustrates how K_i values from such plots may be used to carry out dew-point and bubble-point calculations for multicomponent systems at higher pressures. Prausnitz and Chueh (1968) have developed a more complex and exact method of high-pressure VLE calculations.

Example 4.1.1
(a) A vapor mixture of 12% ethane, 32% propane and 56% n-butane is present at 32 °C in a closed vessel. Calculate the dew-point liquid composition and pressure using the charts provided in Figure 4.1.6.
(b) A liquid mixture of 12% ethane, 32% propane and 56% n-butane is present at 32 °C in a closed vessel. Calculate the bubble-point vapor composition and pressure using the charts provided in Figure 4.1.6.

Solution (a) This involves trial-and-error calculations. One has to guess the value of the pressure, draw a line between 32 °C and that pressure and find the values of K_i for different species from the chart. Using these K_i values and the given x_{iv}, one can calculate the x_{il}. If the x_{il} are correct, then $\sum_{i=1}^{n} x_{il} = 1$. If not, a different pressure has to be guessed. If the first guess leads to $\sum_{i=1}^{n} x_{il} > 1$, then a lower pressure is to be employed as the second guess. If $\sum_{i=1}^{n} x_{il} < 1$, then

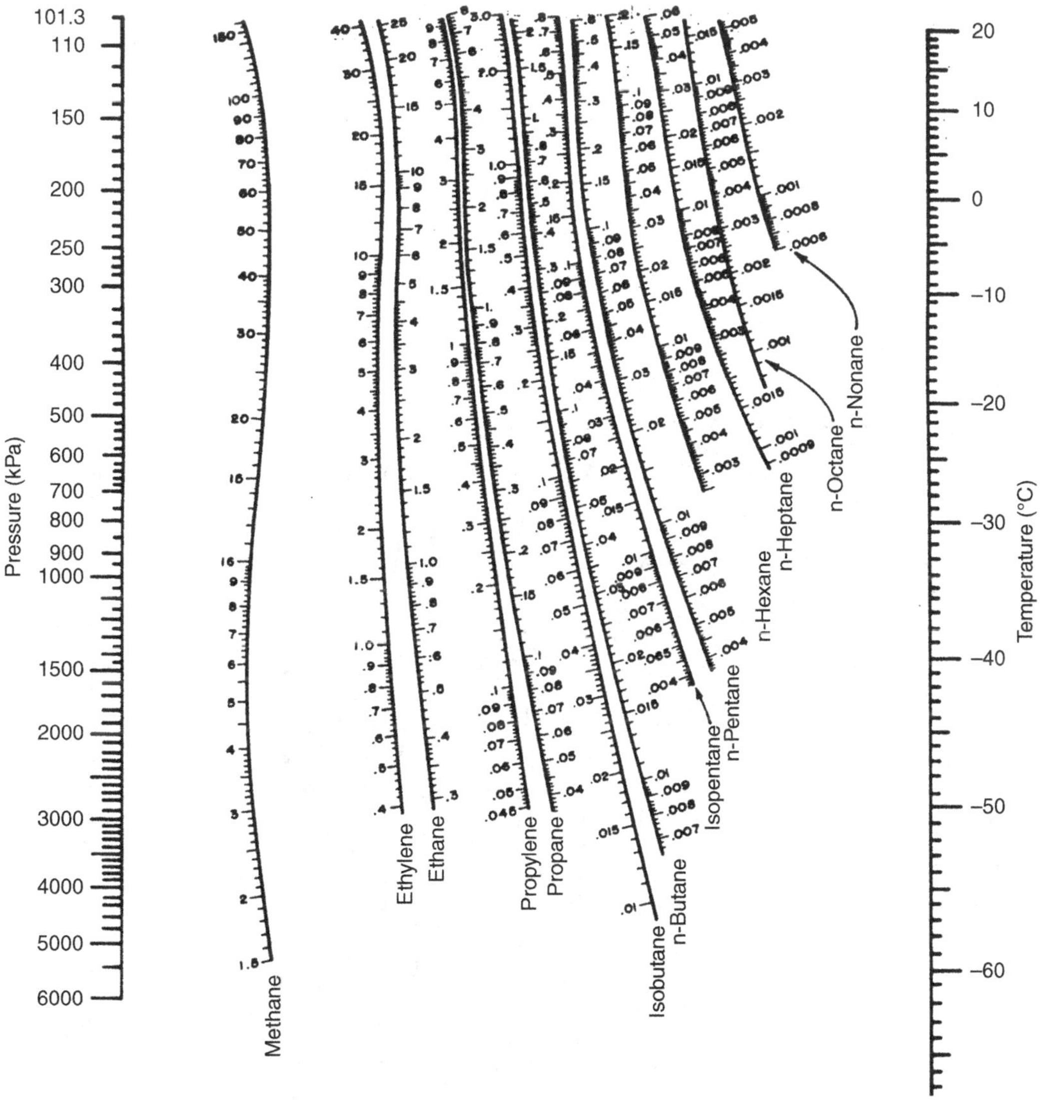

Figure 4.1.5. Low-temperature nomograph for K_i *factors in light hydrocarbon systems. Reprinted, with permission, from D. Dadyburjor,* Chem. Eng. Prog., ***74****(4), 85 (1978). Copyright © [1978] American Institute of Chemical Engineers (AIChE).*

a higher pressure should be used as the second guess. Identify the species ethane, propane and n-butane as $i = 1,2$ and 3. Assume the pressure to be 550 kPa at 32 °C. Obtain the K_i values as shown in Table 4.1.2.

For the next guess, take the pressure to be 500 kPa (see Table 4.1.3). The dew-point liquid composition is 1.85% ethane, 14.5% propane and 84.8% n-butane.

(b) At the bubble-point temperature of 32 °C, assume 690 kPa pressure (see Table 4.1.4). The value of $\sum x_{iv}$ is too high. Increase the pressure to, say, 1050 kPa (Table 4.1.5). This guess of 1050 kPa is just about right.

The procedure for calculating vapor–liquid equilibrium for continuous chemical mixtures will be briefly illustrated here for a *bubble-point calculation*. Let $f_l(M)$ be the known molecular weight density function of the liquid phase. If now the temperature is specified, one would like to know the molecular weight density function $f_v(M)$ of the first vapor bubble formed and the total pressure P of the system. It is known for discontinuous mixtures that the first vapor bubble formed has a composition x_{iv}, which is related to x_{il} by (relation (3.3.61))

$$x_{iv}\hat{\Phi}_{iv}P = x_{il}\gamma_{il}f^0_{il}.$$

At low to moderate pressures, one can easily assume that

$$f^0_{il} = P^{\text{sat}}_i \Phi^{\text{sat}}_i.$$

Table 4.1.2.

550 kPa			
Species	x_{iv}	K_i	$x_{il} = x_{iv}/K_i$
$i = 1$	0.12	6	0.02
$i = 2$	0.32	2	0.16
$i = 3$	0.56	0.6	0.93
			$\sum x_{il} = 1.11$

Table 4.1.3.

500 kPa			
Species	x_{iv}	K_i	$x_{il} = x_{iv}/K_i$
$i = 1$	0.12	6.5	0.0185
$i = 2$	0.32	2.2	0.145
$i = 3$	0.56	0.66	0.848
			$\sum x_{il} = 1.011 \cong 1$

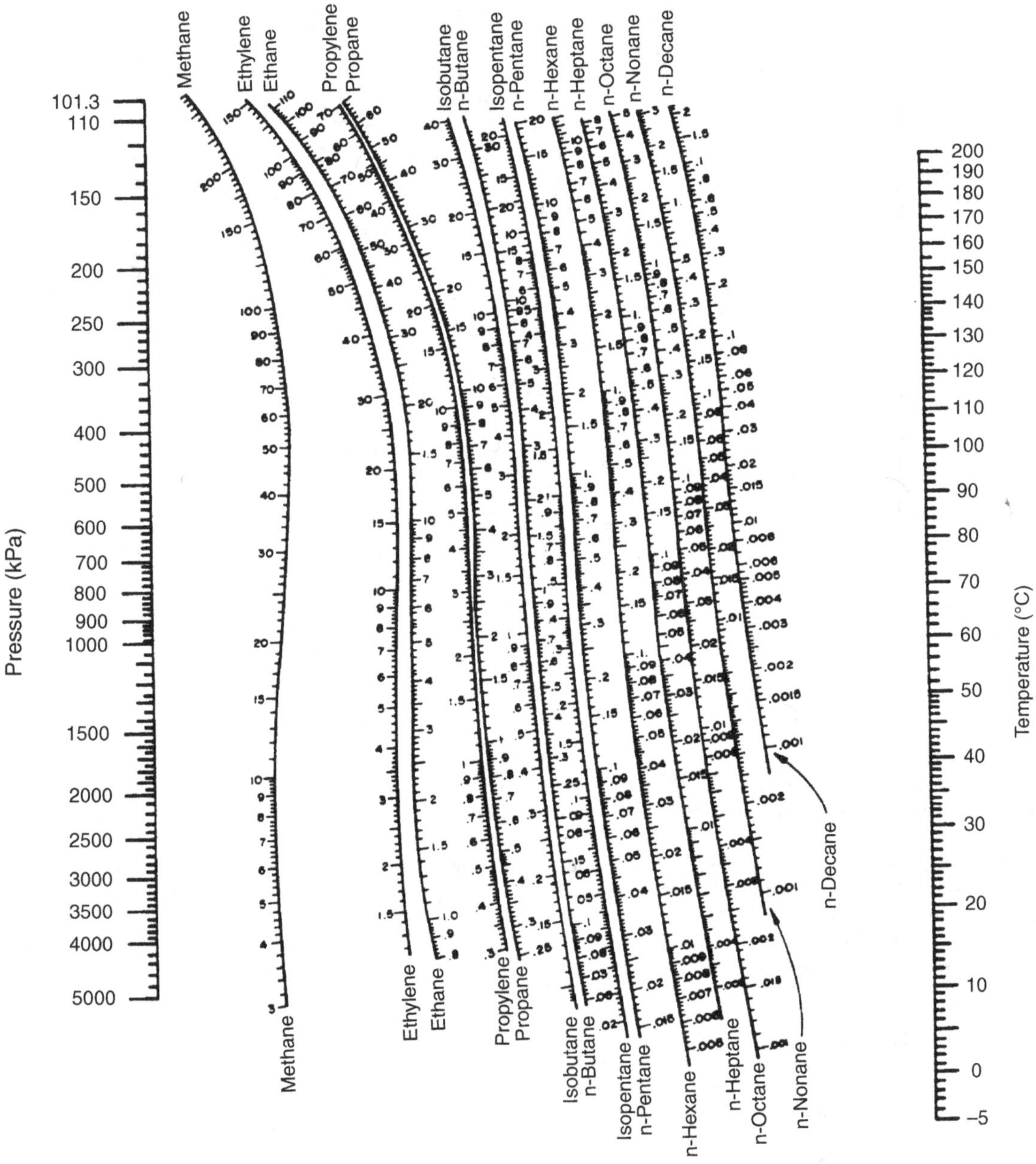

Figure 4.1.6. High-temperature nomograph for K$_i$ *factors in light hydrocarbon systems. Reprinted, with permission, from D. Dadyburjor,* Chem. Eng. Prog., ***74****(4), 85 (1978). Copyright © [1978] American Institute of Chemical Engineers (AIChE).*

Table 4.1.4.

690 kPa			
Species	x_{il}	K_i	$x_{iv} = x_{il}K_i$
$i = 1$	0.12	5	0.6
$i = 2$	0.32	1.6	0.512
$i = 3$	0.56	0.5	0.28
			$\sum x_{iv} = 1.392$ (too high)

Table 4.1.5.

	1050 kPa		
Species	x_{il}	K_i	$x_{iv} = x_{il}K_i$
$i = 1$	0.12	3.4	0.408
$i = 2$	0.32	1.3	0.416
$i = 3$	0.56	0.34	0.19
			$\sum x_{iv} = 1.014$

Therefore,

$$x_{iv}\hat{\Phi}_{iv}\, P = x_{il}\,\gamma_{il}\, P_i^{\text{sat}}\, \Phi_i^{\text{sat}}.$$

The equivalent relation for a continuous chemical mixture is

$$\Big(f_v(M)\text{d}M\Big)P\hat{\Phi}_{Mv} = \Big(f_l(M)\text{d}M\Big)\gamma_{Ml}P_M^{\text{sat}}\Phi_M^{\text{sat}}, \qquad (4.1.33\text{a})$$

where P_M^{sat} is the vapor pressure of the species of molecular weight M at temperature T; other quantities, namely $\hat{\Phi}_{Mv}$, Φ_M^{sat} and γ_{Ml} correspond to the similar quantities for the species of molecular weight M for the given phase at the system condition.

For simplicity, assume an ideal gas phase and an ideal liquid solution: $\hat{\Phi}_{Mv} \sim 1$, $\gamma_{Ml} \sim 1$, $\Phi_M^{\text{sat}} \sim 1$. We obtain, from (3.3.66),

$$Pf_v(M)\text{d}M = P_M^{\text{sat}}f_l(M)\text{d}M, \qquad (4.1.33\text{b})$$

which can be written as

$$f_v(M) = \frac{P_M^{\text{sat}}f_l(M)}{P}, \qquad (4.1.33\text{c})$$

the expression which represents Raoult's law for continuous chemical mixtures. Here P is unknown and can be determined from

$$P = \frac{\int P_M^{\text{sat}}f_l(M)\text{d}M}{\int f_v(M)\text{d}M} = \int P_M^{\text{sat}}f_l(M)\text{d}M \qquad (4.1.33\text{d})$$

over the whole range of molecular weights since $f_l(M)$ is known and P_M^{sat} should be available. For example, Trouton's rule may be used to describe P_M^{sat} for a pure species of molecular weight M as follows:

$$P_M^{\text{sat}} = P^{\text{sat}}(M,T) = P^{+}\exp\left(A\left(1-\frac{M}{T}\right)\right), \qquad (4.1.33\text{e})$$

where P^+ and A are constants. If $f_l(M)$ is known, e.g. it is a Gaussian or Γ distribution, then P can be determined from (4.1.33d) by integration over the molecular weight range of the continuous mixture. From (4.1.33c), the molecular weight density function of the first vapor bubble can now be determined; it is clear from this Raoult's law relation that if $f_l(M)$ is, say, a Γ distribution, then $f_v(M)$ will also be a Γ distribution, with different values of the mean and the standard deviation.

The two molecular weight density functions frequently used in such types of calculations are:

- Γ distribution function

$$f(M) = \frac{(M-\gamma)^{\alpha-1}}{\beta^{\alpha}\Gamma(\alpha)}\exp\left[-\left(\frac{M-\gamma}{\beta}\right)\right]. \qquad (4.1.33\text{f})$$

- For a value of $M = \gamma$, $f(M)$ is zero. The mean of this distribution function is $\alpha\beta + \gamma$ and the standard deviation is $\alpha\beta^2$.
- Gaussian distribution function

$$f(M) = \frac{1}{\sqrt{2\pi}\sigma}\exp\left[-\left(\frac{M-M_{\text{mean}}}{2\sigma^2}\right)\right]. \qquad (4.1.33\text{g})$$

- The mean is M_{mean} and the standard deviation is σ.

4.1.3 Liquid-liquid systems

If a species i is distributed between two immiscible liquid phases $j = 1,2$, the separation achieved between the two liquid phases may be determined by the ratio (a_{i1}/a_{i2}) defined by equation (4.1.1). If separation between two species $i = 1,2$ distributed between two phases $j = 1,2$ is desired, the separation factor is given by equation (4.1.4). The standard state of each solute species $i = 1,2$ has to be specified before these equations can be utilized. When pure solute at the same temperature and pressure of the system under consideration is the standard state, then

$$\mu_{i1}^0 = \mu_{i2}^0. \qquad (4.1.34\text{a})$$

At equilibrium, the distribution of solute i from relation (4.1.1) is given by

$$\frac{a_{i1}}{a_{i2}} = 1 = \frac{\gamma_{i1}x_{i1}}{\gamma_{i2}x_{i2}}; \qquad (4.1.34\text{b})$$

$$\frac{x_{i1}}{x_{i2}} = K_i = \frac{\gamma_{i2}}{\gamma_{i1}}. \qquad (4.1.34\text{c})$$

Nonideal behavior is therefore a prerequisite to separation of solute i between bulk phases 1 and 2.

In the process of solvent extraction of solute i from phase 2 (feed solution) into the solvent, the extract (phase 1), it is important that $K_i > 1$. The activity coefficients γ_{i1} and γ_{i2} are to be estimated from commonly used standard equations (Treybal, 1963, pp. 70–71) e.g. the two-constant based van Laar equation or Margules equation, the three-constant based Redlich–Kister equation. Although the γ_{ij} in these equations are complex functions of concentrations of species i, as well as the concentrations of the two liquids constituting the bulk of the two immiscible liquid phases $j = 1,2$, simple results are obtained for the γ_{ij} under certain limiting conditions (Treybal, 1963, pp. 104–105). The limiting conditions are low concentrations of solute i, the two liquids constituting bulk phases 1 and 2 are very insoluble in each other so that x_{32} and x_{41} (here $i = 3,4$ refer to the species making up the two bulk phases $j = 1,2$, respectively) are essentially zero; the corresponding result is

$$\log_{10} K_i = A_{i2} - A_{i1}, \tag{4.1.34d}$$

where the A_{ij} are appropriate constants in the two-constant or three-constant equations for activity coefficients mentioned earlier (Treybal, 1963, pp. 104–105).

Application of regular solution theory (Prausnitz,1969) leads to the following estimate of γ_{ij} for a solution of i in phase j containing primarily liquid j:

$$\ln \gamma_{ij} = \frac{\overline{V}_i(\delta_i - \delta_j)^2}{RT}, \tag{4.1.34e}$$

where the δ_i and δ_j are the solubility parameters of liquids i and j, respectively. Therefore, from (4.1.34c),

$$\ln K_i = \frac{\overline{V}_i[(\delta_i - \delta_4)^2 - (\delta_i - \delta_3)^2]}{RT}. \tag{4.1.34f}$$

Estimates of solubility parameters for a variety of liquids are available in the literature. Table 4.1.6 illustrates the values of solubility parameters for some common liquids. If the solute prefers the solvent (phase 1, species 3), then δ_i is closer to δ_3 than δ_4, and therefore, $K_i > 1$. (*Note*: Generally, the more polar the liquid, the larger the value of the solubility parameter; water, which is highly polar, has a δ_i of 21, whereas n-pentane, which is extremely nonpolar, has a δ_i of only 7.1. Further, the higher the required energy of vaporization, the higher the solubility parameter.)

Result (4.1.34d) allows rapid estimation of whether K_i is much larger than 1 for solute i being extracted into phase 1 from phase 2. This may be used to select a particular solvent over another solvent. There are many other considerations in the selection of a solvent for extracting a solute from a feed solution. These include: insolubility of the solvent in the feed liquid and vice versa; the solvent should be recoverable easily for reuse; the densities of the solvent and the feed solution when in equilibrium with each other must be different; the interfacial tension of the liquid–liquid system should not be too small to facilitate coalescence of the dispersed phase, since most commercial solvent extraction devices disperse the feed or the solvent phase in the other phase as drops to achieve extractive transfer of solute (membrane solvent extractors are an exception).

Table 4.1.6. Solubility parameters for some common liquids at 25 °C

Liquid	δ_i $(\text{cal/cm}^3)^{1/2}$	Liquid	δ_i $(\text{cal/cm}^3)^{1/2}$
Perfluoroalkanes	6.0	Acetonitrile	11.8
n-Pentane	7.1	Acetic acid	12.4
n-Hexane	7.3	Dimethylsulfoxide	12.8
n-Octane	7.5	Methanol	12.9
Carbon tetrachloride	8.6	Propylene carbonate	13.3
Toluene	8.9	Ethylene glycol	14.7
Benzene	9.2	Formamide	17.9
Ethanol	11.2	Water	21.0

When extractive separation between two solutes $i = 1,2$ present in a feed solution (phase $j = 2$) by an extraction solvent (phase $j = 1$) is desired, the process is called *fractional extraction*. The system now becomes quaternary; however, the separation between solute species $i = 1,2$ can be determined using equation (4.1.4) as if it is a binary system of species 1 and 2. For the standard state condition (4.1.34a), the separation factor α_{12} is given by

$$\alpha_{12} = \frac{x_{11}x_{22}}{x_{12}x_{21}} = \frac{\gamma_{21}\gamma_{12}}{\gamma_{11}\gamma_{22}} = \frac{K_1}{K_2}. \tag{4.1.34g}$$

Treybal (1963, p. 108) has illustrated ways of estimating α_{12} by estimating the different activity coefficients γ_{21}, γ_{12}, γ_{11} and γ_{22}: the system consists of alkyl benzene ($i = 1$), paraffin ($i = 2$), ethylene glycol or furfural (solvent) and n-heptane (feed phase). The solubility-parameter-based approach using relation (4.1.34f) may also be employed.

The estimation of K_i or α_{12} for cases where an infinite dilution standard state for the solute(s) is employed, can be made from equation (4.1.1) as follows:

$$\frac{a_{i1}}{a_{i2}} = \frac{\gamma_{i1}\,x_{i1}}{\gamma_{i2}\,x_{i2}} = \exp\left[-\frac{(\mu_{i1}^0 - \mu_{i2}^0)}{RT}\right], \tag{4.1.34h}$$

where $\mu_{i1}^0 \neq \mu_{i2}^0$. It follows therefore that

$$\frac{x_{i1}}{x_{i2}} = K_i = \frac{\gamma_{i2}}{\gamma_{i1}} \exp\left[-\frac{\Delta\mu_i^0}{RT}\right]. \tag{4.1.34i}$$

For two solutes $i = 1,2$,

$$\frac{x_{11}x_{22}}{x_{12}x_{21}} = \alpha_{12} = \frac{\gamma_{12}\gamma_{21}}{\gamma_{11}\gamma_{22}} \exp\left[-\frac{(\Delta\mu_1^0 - \Delta\mu_2^0)}{RT}\right]. \tag{4.1.34j}$$

If the solutions behave ideally, both γ_{i1} and γ_{i2} tend to 1, resulting in

$$K_i = \exp\left[-\frac{\Delta\mu_i^0}{RT}\right]; \tag{4.1.34k}$$

$$\alpha_{12} = \exp\left[-\frac{(\Delta\mu_1^0 - \Delta\mu_2^0)}{RT}\right]. \tag{4.1.34l}$$

If the value of the Nernst distribution coefficient κ_{i1}^N is known (see (3.3.78)), then, for dilute solutions,

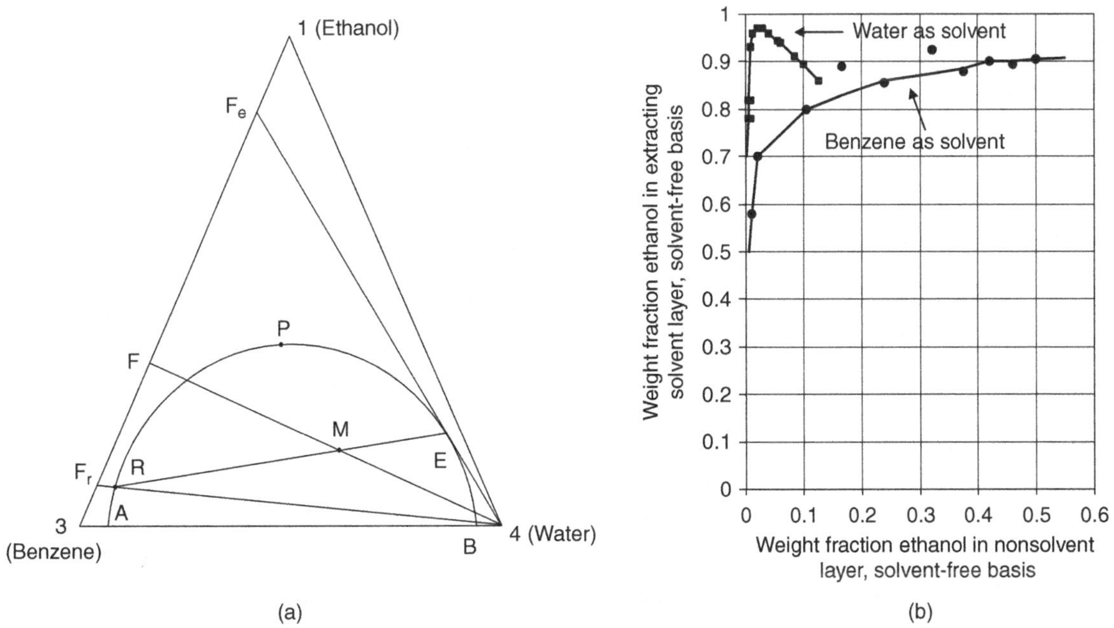

Figure 4.1.7. (a) Selective extraction of ethanol (1) from phase 3 (benzene) by means of a solvent phase 4 (water); (b) selectivity plot on a solvent-free basis for a Type 1 system at 25 °C: ethanol (1)-benzene (3)-water (4); water is the solvent. (After Treybal, 1963, p.124.)

$$\alpha_{12} = \frac{\kappa_{11}^N}{\kappa_{21}^N}. \tag{4.1.34m}$$

When two solutes 1 and 2 are distributed between two immiscible phases $j = 1,2$, we have four species; for the partitioning of one solute between immiscible phases $j = 1,2$, there are three species. To make mass balance and separation calculations, it is useful to recognize that often the two so-called "immiscible" phases may have partial miscibility. For example, consider the ternary system benzene ($i = 3$), water ($i = 4$) and ethanol ($i = 1$, the solute) (see Figure 1.2.2 and replace picric acid by ethanol) whose liquid-liquid equilibrium type is shown in the triangular diagram (see Figure 1.6.1) of Figure 4.1.7(a). As shown, at the given temperature, ethanol and benzene are completely miscible; so are ethanol and water over the whole composition range. However, benzene and water have clearly only limited solubility in each other. So, of the three possible binary pairs, only one pair (benzene-water) has partial miscibility (Type 1 system) (Treybal, 1963).

On the other hand, consider a ternary mixture of n-heptane ($i = 3$), aniline ($i = 4$) and methylcyclohexane ($i = 1$) (the solute) shown schematically in Figure 4.1.8(a). As long as the temperature is less than 59.6 °C, the two bulk liquids n-heptane and aniline will form two immiscible layers with partial solubility (see Example 1.5.2). But aniline and methylcyclohexane also form two immiscible layers having partial solubility of each in the other phase. However, n-heptane and methylcyclohexane are completely miscible with each other over the whole composition range. Such systems having two pairs of partially miscible liquids are called Type 2 systems (Treybal, 1963, p. 15). Obviously, if there is another solute ($i = 2$) beside methylcyclohexane ($i = 1$) in a significant amount, the system will become complex if it is partially miscible with some of the other species in the system.

To understand the separation in solvent extraction in either system (Type 1 or Type 2), focus on a solution of composition F in Figure 4.1.7(a) for a Type 1 system. If the solvent is water ($i = 4$) and it is added to the system, the overall two-phase mixture will be located at M (the object here is to extract ethanol ($i = 1$) from benzene ($i = 3$) into water ($i = 4$)) somewhere along the line joining F and 4; M can be located using the lever rule (see equation (1.6.2)). After sufficient time has been allowed for the two phases (water phase and the benzene phase) to come to equilibrium, the two phases will be located at, say, E and R along a straight line which is called the *tie line*.

The locus of these points E and R for various two-phase mixtures for different feed compositions, ARPEB, provides an envelope in the triangular diagram: the area inside the envelope represents the two-phase region (the water phase and the benzene phase), while the area outside represents a single-phase region. Points A and B represent, respectively, a benzene phase saturated with water and a water phase saturated with benzene without

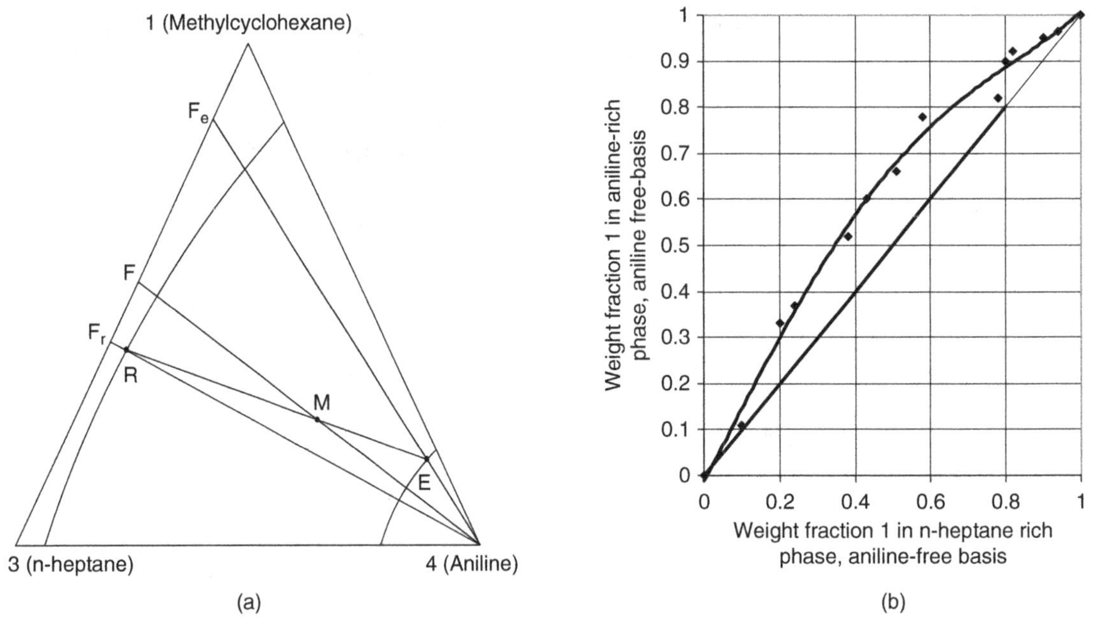

Figure 4.1.8. (a) Selective extraction in Type 2 systems; (b) selectivity plot on a solvent-free basis for a Type 2 system at 25 °C: methylcyclohexane (1)–n-heptane (3)–aniline (4); aniline is the solvent. (After Treybal, 1963, p.125.)

any ethanol. The envelope ARPEB, called the *solubility* or *binodal curve*, represents compositions of individual solutions which are in equilibrium with another immiscible solution on the same curve.

Focus now on the mixture M which splits into two immiscible phases R (raffinate) and E (extract) (Figure 4.1.7(a)); the compositions of R and E are the equilibrium compositions of the two immiscible layers. The water phase representing point E has primarily water, a considerable amount of ethanol and a little bit of benzene. The benzene phase representing point R has primarily benzene, a small amount of ethanol and a very small amount of water. In solvent extraction processes, after the extraction is carried out, the solvent is removed from the extract and the raffinate so that the solvent may be reused. If the solvent (here, water) is removed from the extract (point E), then, by the lever rule, the composition of the water-free organic phase represented by F_e is obtained by extending line 4E to intersect with the line 13. Correspondingly, the water-free raffinate composition is obtained by extending line 4R to intersect with the line 13 at F_r. The net result of the solvent extraction of the benzene–ethanol feed mixture F with water, and the subsequent removal of water from the extract phase and the raffinate phase, are two fractions having composition F_e (ethanol-rich) and F_r (benzene-rich).

Two immiscible phases having different concentrations of the solute to be extracted is a prerequisite to the application of solvent extraction. In Figure 4.1.7(a), the binodal curve ARPEB has two branches: the ARP branch, which is mostly benzene-rich, and the BEP branch, which is mostly water-rich. However, as the curves tend toward P, the concentration of the solute (ethanol) increases and the immiscibility of the two phases disappears at point P, the *Plait point*. It is not necessarily the point in ARPEB which has the highest amount of ethanol.

The quantitative behavior of the distribution of ethanol ($i = 1$) between the two phases ($i = 3$, benzene; $i = 4$, water) can be obtained from the two branches of the binodal curve. A more useful approach would be to plot the ethanol concentration at point F_e against ethanol concentration at point F_r, i.e. on a solvent-free basis after removing water, as if the original feed mixture of benzene–ethanol has been split into two fractions (not phases, since the two fractions are miscible). Figure 4.1.7 (b) illustrates the weight fraction of ethanol in the water layer on a solvent-free basis vs. the weight fraction of ethanol in the benzene layer on a solvent-free basis; here water is the solvent. This figure also has a plot for the same system, with benzene being considered as a solvent to separate ethanol from water.

For a Type 2 system, as shown in Figure 4.1.8(a), the area in the triangle in between the two curves represents the two-phase region. Any feed mixture F of n-heptane and methylcyclohexane when mixed with the solvent aniline will produce a mixture of overall composition M. If we now follow the notation of Figure 4.1.7(a) and the corresponding steps, we will end up with two fractions, F_e and F_r, of n-heptane and methylcyclohexane, free of the solvent.

A plot of the methylcyclohexane weight fraction in the aniline phase vs. that in the n-heptane phase (based on an aniline-free basis) illustrates the separation achieved (Figure 4.1.8(b)).

The partitioning of a solute i between two immiscible liquid phases $j = 1$ and 2 and the selectivity between two solutes in the two-phase system have so far been considered primarily in the context of liquid extraction/solvent extraction in large-scale operations. Many such aspects and some other considerations are also important for the basic equilibria in *liquid–liquid chromatography* (LLC) (see Section 7.1.5). In LLC, a mobile liquid phase ($j = 2$) flows over a porous, finely divided solid phase whose surface has been coated with a liquid, the stationary phase ($j = 1$). Any solute introduced via the mobile phase is partitioned between the two liquid phases. For cases where $\mu_{i1}^0 = \mu_{i2}^0$, the regular solution-theory based expression (4.1.34f) for K_i may be written as

$$\ln K_i = \frac{\overline{V}_i[(\delta_i - \delta_2)^2 - (\delta_i - \delta_1)^2]}{RT}. \qquad (4.1.34n)$$

To achieve a high value of K_i between the two phases, δ_i should tend to δ_1, i.e. the solute i should be similar to the stationary phase $j = 1$. Further, K_{11} should be different from K_{21} for any chromatographic separation between two solutes 1 and 2. Note that the two liquid phases must also be immiscible, for which, generally, $(\delta_1 - \delta_2) \geq 4$ (Karger *et al.*, 1973). There are many combinations of highly polar organic liquids and water where $(\delta_1 - \delta_2) \geq 4$ does not guarantee phase immiscibility. However, the criterion of liquid immiscibility is important for solvent selection.

In liquid–liquid extraction, generally one of the two immiscible liquid phases is aqueous and the other phase is organic, which is nonpolar or mildly polar. There are many examples where both immiscible phases are *primarily organic*: aromatic species such as toluene and benzene are extracted from an essentially nonpolar hydrocarbon feedstock by an immiscible highly polar organic solvent. Large-scale applications utilize highly polar organic solvents, e.g. ethylene glycol, dimethylsulfoxide, n-methylpyrrolidone, etc., with or without a small amount of water (Lo *et al.*, 1983). Figure 4.1.9A illustrates the behavior of many such solvents in terms of their selectivity for aromatics with respect to nonaromatics vs. their capacity for the aromatics. The solvent power or capacity is directly proportional to the equilibrium ratio K_{i1} with the extracting solvent being phase 1. An ideal solvent has a high selectivity as well as high capacity.

Extraction of solute species from one liquid to another immiscible liquid is also carried out when both phases are primarily aqueous. Large-scale purification processes for proteins employ *aqueous two-phase* systems (Albertsson, 1986) containing two water-soluble but incompatible polymers in water, e.g. polyethylene glycol (PEG) and dextran. The PEG-rich layer is at the

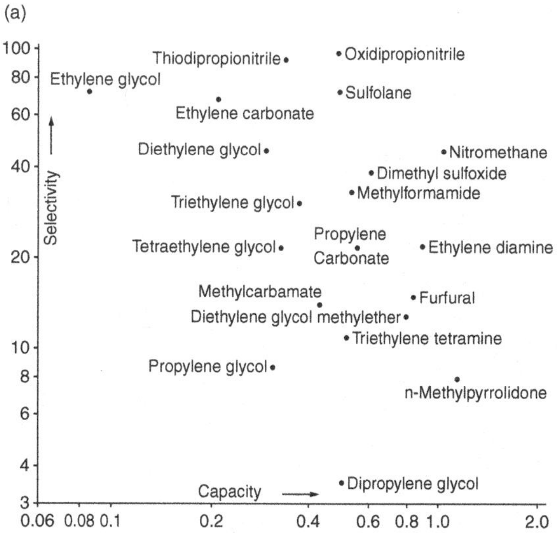

Figure 4.1.9A. (a). Capacity and selectivity of polar organic solvents for aromatics (Lo et al.*, 1983). Reprinted, with permission, from T.C. Lo, M.H.I. Baird and C. Hanson,* Handbook of Solvent Extraction, *Figure 2, p. 525, Wiley-Interscience, 1983. Copyright © 1983, John Wiley & Sons.*

top ($j = 1$); the dextran-rich layer is at the bottom ($j = 2$), with a clear boundary in between. Proteins, cells, etc. are partitioned between the two layers. The partition coefficient, κ_{p1}, decreases from about 2 to 0.2 as the molecular weight increases (Figure 4.1.9B).

The partition coefficient is strongly influenced by the salt (e.g. KCl, K_3PO_4, K_2SO_4, etc.) usually present in such systems. The presence of the salt creates an electrostatic potential gradient across the interface between the two-phase system, which leads to preferential partitioning of the protein into one of the phases, depending on the sign and magnitude of the protein surface charge. Haynes *et al.* (1991) have related the interfacial electrostatic potential difference between the two aqueous phases $j = w_1$ and $j = w_2$, namely $\Delta\phi = (\phi_{w_2} - \phi_{w_1})$,to the salt distribution coefficient κ_{salt}^m in terms of the salt molar concentrations for a 1:1 electrolyte:

$$\Delta\phi = \phi_{w_2} - \phi_{w_1} = \frac{RT}{\mathcal{F}}\ln(\kappa_{\text{salt}}^m) = \frac{RT}{\mathcal{F}}\ln\left(\frac{m_{s,w_1}}{m_{s,w_2}}\right), \qquad (4.1.34o)$$

where $m_{s,j}$ is the salt molality in phase j (mol/kg). For low concentrations of electrolytes and proteins, the protein partition coefficient, κ_{p1}, with lighter phase 1 at the top, has been related to that in the absence of any $\Delta\phi$, κ_{p1}^0, by a model based on virial expansion (King *et al.*, 1988):

$$\begin{aligned}\ln(C_{p1}/C_{p2}) &= \ln\kappa_{p1} = \ln\kappa_{p1}^0 + (Z_p\mathcal{F}/RT)\Delta\phi \\ &= B_{1p}(C_{12} - C_{11}) + B_{2p}(C_{22} - C_{21}) + (Z_p\mathcal{F}/RT)\Delta\phi.\end{aligned} \qquad (4.1.34p)$$

Here B_{1p} and B_{2p} are the second virial coefficients for interaction between polymers 1 and 2, respectively, C_{ij} is

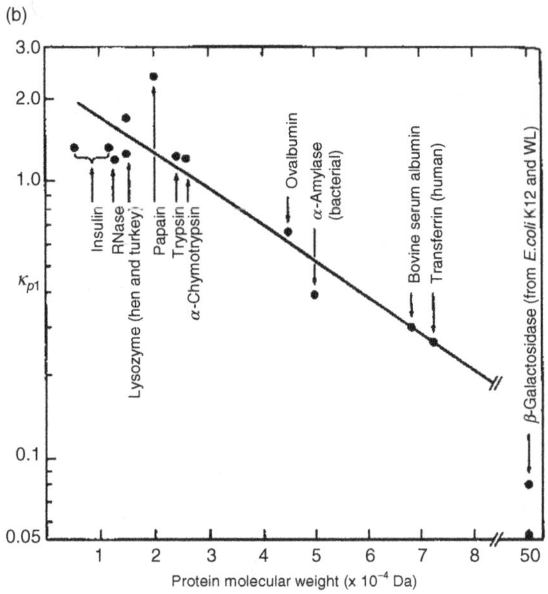

Figure 4.1.9B. Protein molecular weight vs. protein partition coefficient in a PEG 6000-dextran 500 system with pH *at* pI *for all proteins (Sasakawa and Walter, 1972). Reprinted, with permission, from* Biochemistry, ***11****(15), 2760 (1972), Figure 2. Copyright (1972) American Chemical Society.*

the molar concentration of polymer i in region j and Z_p is the algebraic charge number (valence) of the protein. Haynes *et al.* (1993) have developed a model to predict κ_{p1} for a protein in biphasic aqueous systems containing strong electrolytes amongst other things.

4.1.4 Liquid–solid systems

In Section 3.3.7.5, the equilibrium partitioning of a species between a liquid phase and a solid phase was briefly considered for three types of liquid–solid equilibria. The separation between two species i and j for such liquid–solid two-phase systems is briefly considered here. There are systems where three phases can be present; for example, two immiscible solid phases and a saturated solution, as in the case of solid salt, ice and a saturated salt solution. Figure 4.1.10 shows a temperature vs. composition phase diagram where solid phase 1 coexists with solid phase 2 and a saturated liquid solution at the *eutectic point* E. Below the eutectic temperature T_E, immiscible pure solid phase 1 and 2 are present together. For these and more complex systems, the reader should refer to appropriate texts (Darken and Gurry, 1953; DeHoff, 1993). Separation between species i and j in simpler two-phase systems described in Figures 3.3.6A, where the solid phase is a homogeneous solution, will be determined now.

For a solid solution in equilibrium with a liquid, relation (3.3.93) for the equilibrium ratio of species i is

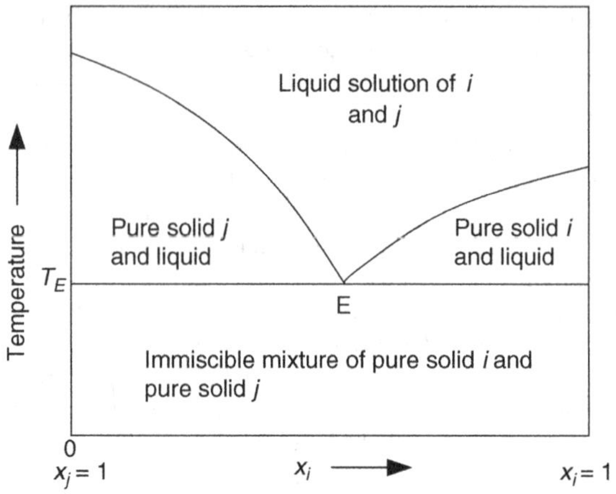

Figure 4.1.10. Temperature vs. composition diagram for two species i *and* j *in solid–liquid systems where the two solid phases are immiscible.*

$$\frac{x_{il}}{x_{is}} = \frac{\gamma_{is}}{\gamma_{il}}\frac{f_{is}^0}{f_{il}^0}.$$

Therefore, for two species i and j,

$$\alpha_{ij} = \frac{x_{il}}{x_{is}}\frac{x_{js}}{x_{jl}} = \frac{\gamma_{is}}{\gamma_{js}}\frac{\gamma_{jl}}{\gamma_{il}}\frac{f_{is}^0}{f_{js}^0}\frac{f_{jl}^0}{f_{il}^0}. \tag{4.1.35}$$

For ideal solutions in both phases, $\gamma_{il} \cong 1 \cong \gamma_{jl}$ and $\gamma_{is} \cong 1 \cong \gamma_{js}$, leading to

$$\alpha_{ij} = \frac{f_{is}^0}{f_{js}^0}\frac{f_{jl}^0}{f_{il}^0}. \tag{4.1.36}$$

At temperature T and pressure P for the system in equilibrium, for any species i,

$$\frac{f_{is}^0(T,P)}{f_{il}^0(T,P)} = \frac{f_{is}^0(T,P)}{f_{is}^0(T_{mi},P)}\cdot\frac{f_{is}^0(T_{mi},P)}{f_{il}^0(T_{mi},P)}\cdot\frac{f_{il}^0(T_{mi},P)}{f_{il}^0(T,P)}. \tag{4.1.37}$$

Here T_{mi} is the melting point of species i. For a pure species i at T_{mi}, $f_{is}^0(T_{mi},P) = f_{il}^0(T_{mi},P)$. Exact expressions have been developed for the product of the two remaining ratios. An expression practically useful and based on particular approximations (Smith *et al.*, 2001, pp. 526–531) is

$$\frac{f_{is}^0(T,P)}{f_{il}^0(T,P)} \cong \exp\left\{\frac{\Delta H_i}{RT_{mi}}\left(\frac{T-T_{mi}}{T}\right)\right\}. \tag{4.1.38}$$

Therefore

$$\alpha_{ij} \cong \exp\left\{\frac{(\Delta H_i - \Delta H_j)\times\left(\frac{T}{T_{mi}}-1\right)}{\left(\frac{T}{T_{mj}}-1\right)}\right\}. \tag{4.1.39}$$

In the case of *leaching* of a solid mixture, using expression (3.3.96), one can develop an appropriate expression for the separation factor, α_{ij}, for species i and j being leached out.

Of the three categories of liquid–solid equilibria considered in Section 3.3.7.5 and briefly considered above, a solid solution in equilibrium with a molten mixture has special importance in the purification of semiconductor materials such as silicon. Here, the bulk solid phase, as well as the molten mixture, consists essentially of silicon; the concentrations of impurities are at a very low level. Therefore the area of focus is very close to either end of the type of phase diagram in Figures 3.3.6A for a given impurity-silicon system. Usually the solution being extremely dilute, the distribution coefficient κ_{is} for the impurity i between the solid phase and the melt (see Example 1.4.3),

$$\kappa_{is} = \frac{C_{is}}{C_{il}}, \tag{4.1.40}$$

is a constant. Illustrative values of the distribution coefficients for a variety of impurities are provided in Table 4.1.7. It appears that, except for oxygen, all impurities will partition more into the molten phase. This is a basis for purification of silicon into silicon single crystals/wafers (see Chapter 6 for the zone refining/melting process). Thus an impure silicon material may have a molten section and a frozen section; this will allow redistribution of impurities between the newly frozen solid phase and the remaining melt in the manner of (4.1.40); the newly frozen solid phase will be purer.

This phenomenon will also allow the creation of a solid phase having a certain desired level of impurity or dopant once a certain amount of impurity level is introduced first into the melt. Doped gallium arsenide is an example of such a material in the semiconductor industry. Table 4.1.7 illustrates some of the dopants and their partition coefficients between the bulk crystal and the melt.

Table 4.1.7. Distribution coefficients for a number of impurities in silicon and dopants in gallium arsenide

Silicon[a]		Gallium arsenide[b]	
Impurity	κ_{is}	Dopant	κ_{is}
Al	0.002	S	0.3–0.5
As	0.3	Se	0.1–0.3
B	0.8	Te	0.059
C	0.07	Be	3
Cu	4×10^{-4}	Mg	0.1
O	1.25	Ge	0.01
P	0.35	C	0.2–0.8
Sb	0.023	Cr	5.7×10^{-4}

[a] From Pearce (1983); [b] from Milnes (1973).

4.1.5 Interfacial adsorption systems

Following Section 3.3.7.4, the separation factor in interfacial adsorption where *both* bulk phases are *fluids* is considered first. For *air–water* systems containing two surface active solutes, $i = 1, 2$ (both being nonelectrolytes), the separation factor α_{12} has been defined as

$$\alpha_{12} = \frac{\Gamma_{1\sigma}^E / C_{11}}{\Gamma_{2\sigma}^E / C_{21}}, \tag{4.1.41}$$

where $j = 1$ corresponds to the bulk phase of water. For a nonelectrolytic surface active solute i, whose presence reduces the interfacial tension linearly in the manner of relation (3.3.105), it is known that

$$\frac{\Gamma_{i\sigma}^E}{C_{i1}} = \frac{b_1^i}{RT}, \tag{4.1.42a}$$

where b_1^i comes from the linear relation between the interfacial tension γ^{12} and the surfactant concentration C_{i1} below the critical micelle concentration (CMC):

$$\gamma^{12} = a_1 - b_1^i C_{i1}. \tag{4.1.42b}$$

Therefore,

$$\alpha_{12} = b_1^1 / b_1^2. \tag{4.1.43}$$

This result indicates that in a closed vessel containing two bulk phases, namely air and water, the selectivity between two nonionic surfactants distributed between the bulk liquid and the gas–liquid interfacial phase will be determined by the ratios of the slopes of the decrease of γ^{12} with the concentration of each individual surfactant. Although this ratio can be easily different from 1, it may not be very large. When, however, the surfactants are ionic, significant selectivity can be achieved. The expressions for this type of system are studied in Section 5.2.5.

Amongst interfacial adsorption systems, where one of the bulk phases is a solid, *gas–solid* systems are important for gas mixture separation. Of the three general approaches to predicting the separation of a gas mixture in gas–solid *adsorption*, namely the solution thermodynamic Gibbs approach, potential-theory methods and the Langmuir-type equation, the first approach will be briefly presented now. Myers and Prausnitz (1965) considered the mixed adsorbate as a solution in equilibrium with the gas mixture and applied solution thermodynamics with the spreading pressure π and the surface area of the sorbent A (more practically, S_σ, the surface area per unit adsorbent mass) replacing, respectively, the pressure P and volume V for the adsorbate. The following three equations, derived by them, allow the prediction of adsorption for a binary mixture:

$$Px_{ig} = P_i^0(\pi)\, \gamma_{i\sigma} x_{i\sigma}$$

(equation (3.3.111a), repeated here for convenience);

$$\frac{1}{\sum_{i=1}^{2} q_{i\sigma}} = \sum_{i=1}^{2} \frac{x_{i\sigma}}{q_{i\sigma}^{p}} + \frac{RT}{S_\sigma} \sum_{i=1}^{2} x_{i\sigma} \left(\frac{\partial \ln \gamma_{i\sigma}}{\partial \pi} \right)_{x_{i\sigma}}; \quad (4.1.44)$$

$$q_{i\sigma} = \left(\sum_{i=1}^{2} q_{i\sigma} \right) x_{i\sigma}. \quad (4.1.45)$$

Here, $q_{i\sigma}^{p}$ is the number of moles of i per unit mass of adsorbent when adsorbed from a pure gas i at the same surface pressure π and temperature T as the mixture.

The determination of the activity coefficients $\gamma_{i\sigma}$ for the adsorbed phase requires accurate experimental data at constant temperature and pressure for the entire range of gas-phase compositions (Myers and Prausnitz, 1965). However, if the *adsorbed solution phase* is considered *ideal*, i.e. $\gamma_{i\sigma} = 1$, then

$$Px_{ig} = P_i^0(\pi) x_{i\sigma}; \quad (4.1.46)$$

$$\frac{1}{\sum_{i=1}^{2} q_{i\sigma}} = \sum_{i=1}^{2} \frac{x_{i\sigma}}{q_{i\sigma}^{p}}; \quad (4.1.47)$$

$$q_{i\sigma} = \left(\sum_{i=1}^{2} q_{i\sigma} \right) x_{i\sigma}. \quad (4.1.48)$$

Note that, from the definition of $P_i^0(\pi)$ (Section 3.3.7.6),

$$\frac{\pi S_\sigma}{RT} = \int_0^{P_1^0} \frac{q_{1\sigma}^{p}}{P} \, dP = \int_0^{P_2^0} \frac{q_{2\sigma}^{p}}{P} \, dP, \quad (4.1.49)$$

where each integral is to be used for *pure gas adsorption* data. For a binary gas mixture, this is the equivalent of a Raoult's law type of situation in vapor-liquid equilibria (see (3.3.64) and (4.1.21b)). The separation factor α_{12} for species 1 and 2 between the adsorbed phase (σ) and the gas phase (g) is

$$\alpha_{12} = \frac{x_{1\sigma} x_{2g}}{x_{1g} x_{2\sigma}} = \frac{P_2^0(\pi)}{P_1^0(\pi)}. \quad (4.1.50)$$

For $\alpha_{12} > 1$, namely the adsorbate phase is enriched in species 1 compared to species 2, we have $P_2^0(\pi) > P_1^0(\pi)$. Obviously,

$$P = P_1^0(\pi) x_{1\sigma} + P_2^0(\pi) x_{2\sigma}. \quad (4.1.51)$$

These results follow directly from *ideal adsorbed solution theory* (Myers and Prausnitz, 1965), whose governing equations are (4.1.46) to (4.1.48). The key quantities are $P_1^0(\pi)$ and $P_2^0(\pi)$. One needs pure component adsorption data for each species to determine $P_1^0(\pi)$ and $P_2^0(\pi)$ and to calculate the separation factor α_{12} for a binary system. The procedure suggested for determining the compositions of the gas phase (x_{ig}) and the adsorbate phase ($x_{i\sigma}$) at any given total pressure P is as follows.

(1) Plot graphically $q_{i\sigma}^{p}/P_i^0$ against P_i^0 at any given T. Calculate the area for different integration limits of P_i^0 (equation (4.1.49)). This will yield a plot of $(\pi S_\sigma/RT)$ against pressure P, correspondingly P_i^0 for each i. If $q_{i\sigma}^{p}$ can be described analytically as a function of P_i^0, the same integration can also be carried out analytically to develop the same plot. See Figure 4.1.11.

(2) To determine relevant quantities for the mixture, select a total pressure P and a value of π for the mixture. In Figure 4.1.11, draw a line parallel to the abscissa for the corresponding π and a vertical line at the selected P. The point of intersection of these two lines, M, defines the mixture location at a certain total pressure P, mixture spreading pressure π and temperature T. Correspondingly, the value of P_i^0 is obtained from the abscissa of the intersection of the line parallel to P-axis (for the corresponding π) and the pure component equilibrium curve. The mixture compositions are obtained via the lever rule (equation (1.6.2)) using equations (4.1.50) and (4.1.51). For example, equation (4.1.51) may be written as

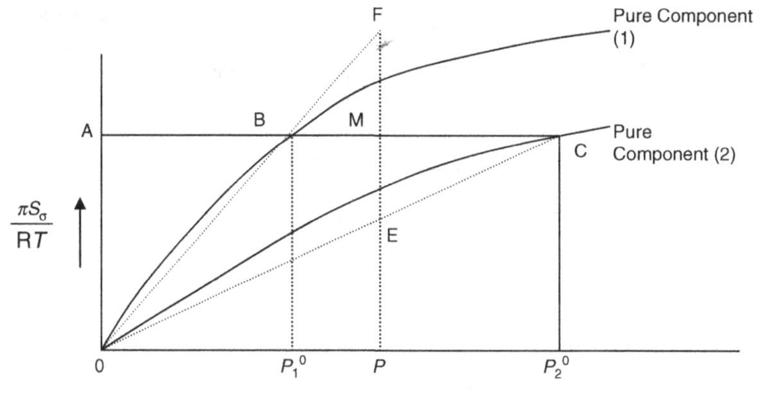

Figure 4.1.11. Graphical calculation of adsorption equilibrium of a gas mixture from pure component spreading pressure information. (After Myers and Prausnitz (1965).)

$$Px_{1\sigma} + Px_{2\sigma} = P_1^0(\pi)x_{1\sigma} + P_2^0(\pi)x_{2\sigma},$$

which may be rearranged to

$$(P - P_1^0(\pi))x_{1\sigma} = (P_2^0(\pi) - P)x_{2\sigma},$$

i.e.

$$\frac{x_{1\sigma}}{x_{2\sigma}} = \frac{P_2^0(\pi) - P}{P - P_1^0(\pi)} = \frac{\text{length of line CM}}{\text{length of line MB}}.$$

Further

$$x_{1\sigma} = \frac{\text{length of line CM}}{\text{length of line CB}} \quad \text{and} \quad x_{2\sigma} = \frac{\text{length of line MB}}{\text{length of line CB}}, \tag{4.1.52a}$$

since $x_{1\sigma} + x_{2\sigma} = 1$ from (4.1.48). Correspondingly,

$$x_{1\sigma} = \frac{P_2^0(\pi) - P}{P_2^0(\pi) - P_1^0(\pi)} \quad \text{and} \quad x_{2\sigma} = \frac{P - P_1^0(\pi)}{P_2^0(\pi) - P_1^0(\pi)}. \tag{4.1.52b}$$

To determine x_{1g} and x_{2g}, we employ (4.1.46) or (4.1.50) along with the above expressions for $x_{1\sigma}$ and $x_{2\sigma}$ in Figure 4.1.11. From (4.1.56),

$$x_{1g} = \frac{P_1^0(\pi)x_{1\sigma}}{P}, \qquad x_{2g} = \frac{P_2^0(\pi)x_{2\sigma}}{P}.$$

Therefore,

$$\begin{aligned}
\frac{x_{1g}}{x_{2g}} &= \frac{P_1^0(\pi)\,x_{1\sigma}}{P_2^0(\pi)\,x_{2\sigma}} = \frac{P_1^0(\pi)\,(P_2^0(\pi) - P)}{P_2^0(\pi)\,(P - P_1^0(\pi))} \\
&= \frac{1 - \dfrac{P}{P_2^0(\pi)}}{\dfrac{P - P_1^0(\pi)}{P_1^0(\pi)}} = \frac{\dfrac{\pi S_\sigma}{RT} - \dfrac{\pi S_\sigma}{RT}\dfrac{P}{P_2^0(\pi)}}{\left(\dfrac{\pi S_\sigma}{RT}\right)\left(\dfrac{P - P_1^0(\pi)}{P_1^0(\pi)}\right)} \\
&= \frac{\text{PM} - \text{PE}}{\text{MF}} = \frac{\text{ME}}{\text{MF}}.
\end{aligned} \tag{4.1.53}$$

So

$$x_{1g} = \frac{\text{ME}}{\text{EF}} \quad \text{and} \quad x_{2g} = \frac{\text{MF}}{\text{EF}}. \tag{4.1.54}$$

Example 4.1.2 Markham and Benton (1931) studied experimentally the adsorption of pure O_2, pure CO as well as CO-O_2 mixtures on 19.6 g of silica at 0°C. The data obtained are provided in Table 4.1.8. The column for the mixtures under "isotherm" contains values obtained from pure component adsorption isotherms via interpolation.

Develop a plot of x_{O_2g} vs. $x_{O_2\sigma}$ for the O_2-CO mixture adsorption on silica at 0 °C at a total pressure of 1 atmosphere using the ideal adsorbed solution theory. Plot the three experimental points for the mixture provided in Tabel 4.1.8 in the same diagram.

Solution To develop a plot of x_{ig} vs. $x_{i\sigma}$, where $i = O_2$, first two plots of $(\pi S_\sigma/RT)$ vs. $P_{O_2}^0$ and $(\pi S_\sigma/RT)$ vs. P_{CO}^0 have to be generated. In order to generate these plots, the integral $\int_0^{P_i^0} (q_{i\sigma}^p/P)\mathrm{d}P$ has to be developed for each species. Employ the data provided in Table 4.1.8 for the adsorption of each pure species; take the pressure as P_i^0 and the volume adsorbed (in cm³) to number of moles of i per gram of adsorbent as $q_{i\sigma}^p$. Now relate $q_{i\sigma}^p/P_i^0$ to P_i^0.

Table 4.1.8.

O_2 at 0 °C		CO at 0 °C	
Pressure (mm Hg)	Volume adsorbed, observed (cm³)	Pressure (mm Hg)	Volume adsorbed, observed (cm³)
83.0	3.32	95.5	7.09
142.4	5.57	127.4	9.48
224.3	8.73	199.7	14.37
329.6	12.68	272.4	19.41
405.1	15.48	367.4	25.54
544.1	20.42	463.7	31.70
602.5	22.48	549.7	37.01
667.5	24.86	647.9	42.65
760.0	28.03	760.0	48.89

CO-O_2 Mixtures at 0 °C

		Volume of oxygen adsorbed (cm³)		Volume of carbon monoxide adsorbed (cm³)	
p_{O_2}	p_{CO}	Isotherm	Obs.	Isotherm	Obs.
230.2	529.8	8.98	8.31	35.90	35.02
391.1	368.9	15.05	14.1	25.82	24.69
585.1	174.9	21.93	21.73	12.8	11.63

Data from Table 4.1.8 for pure oxygen (observed) at 0 °C were used to develop the following relation:

$$\frac{q_{O_2\sigma}^p}{P_{O_2}^0} = -4.228 \times 10^{-6} P_{O_2}^0 + 0.0339 \tag{4.1.55}$$

and $r^2 = 0.967$ (Figure 4.1.12).

Then we have

$$\frac{\pi S_\sigma}{RT} = \int_0^{P_{O_2}^0} \frac{q_{O_2\delta}^p}{P}\,\mathrm{d}P = -\frac{1}{2} \times 4.228 \times 10^{-6} \times (P_{O_2}^0)^2 + 0.0339 \times P_{O_2}^0. \tag{4.1.56}$$

Data from Table 4.1.8 for pure CO (observed) at 0 °C were used to obtain the following relation:

$$\frac{q_{CO\sigma}^p}{P_{CO}^0} = -1.508 \times 10^{-5} P_{CO}^0 + 0.0755 \tag{4.1.57}$$

and $r^2 = 0.987$ (Figure 4.1.13).

Then we have

$$\frac{\pi S_\sigma}{RT} = \int_0^{P_{CO}^0} \frac{q_{CO}^p}{P}\,\mathrm{d}P = \frac{1}{2} \times 1.508 \times 10^{-5} \times (P_{CO}^0)^2 + 0.0755 \times P_{CO}^0. \tag{4.1.58}$$

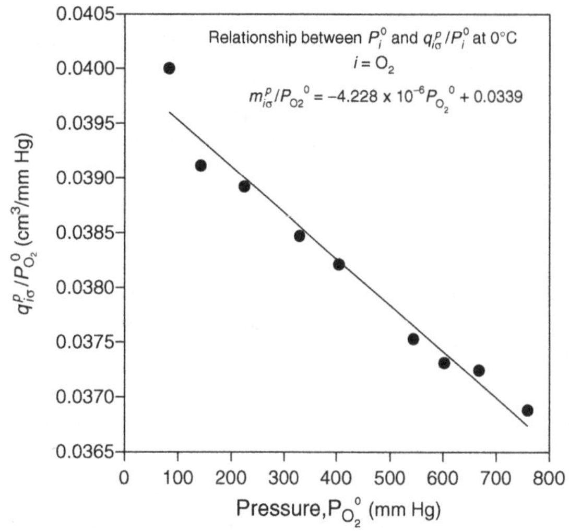

Figure 4.1.12. Plot for oxygen.

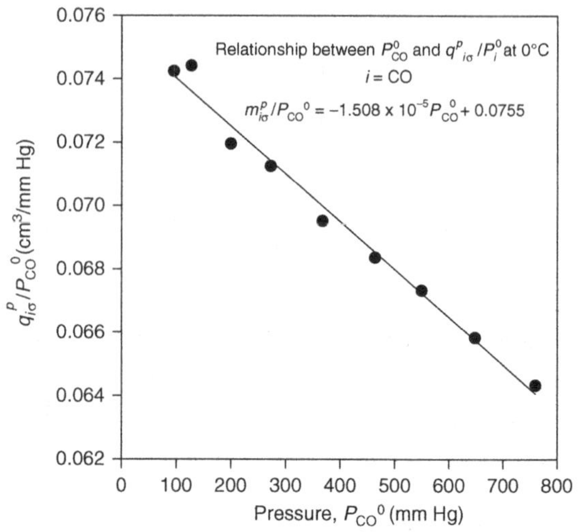

Figure 4.1.13. Plot for carbon monoxide.

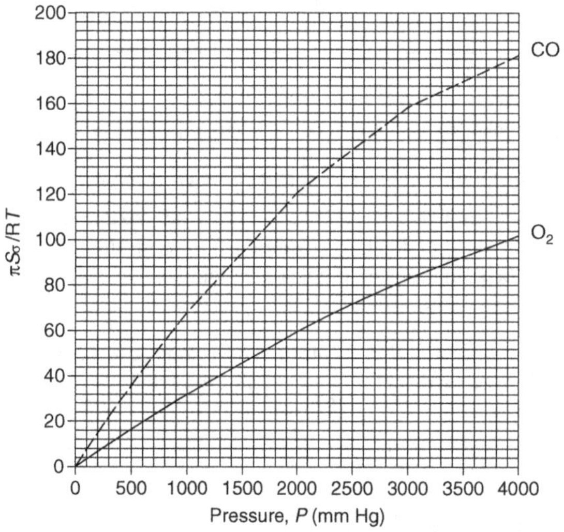

Figure 4.1.14. Calculation for mixture of O_2-CO adsorption equilibia from pure component spreading pressure at 0 °C.

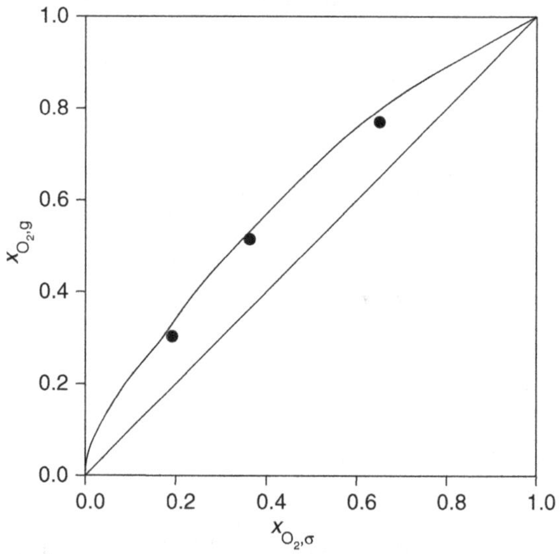

Figure 4.1.15. Prediction of adsorption of O_2-CO mixture on silica at 0 °C, total pressure = 1 atm.

Equations (4.1.56) and (4.1.58) have been plotted in Figure 4.1.14 which is ready to use. Employ equations (4.1.52b) and (4.1.54) to develop estimates of $x_{O_2,\sigma}, x_{CO,\sigma}, x_{O_2,g}$ and $x_{CO,g}$. The predicted behavior of $x_{O_2,g}$ vs. $x_{O_2,\sigma}$ is shown in Figure 4.1.15. The experimental data have also been plotted to indicate how well the ideal adsorbed solution theory predicts the observed behavior.

Another interfacial adsorption system, where one of the bulk phases is a solid, involves adsorption of solutes from a solvent onto the surface of solid adsorbents. Such *liquid-solid adsorption* systems are frequently used in large-as well as small-scale applications. A particular example is the adsorption of organic pollutants from industrial and municipal waste waters onto activated carbon adsorbents. The equilibrium adsorption behavior of a waste water containing multiple solutes may be predicted by the *ideal adsorbed phase* model developed by Radke and Prausnitz (1972) for dilute solutions, provided the single-solute equilibrium adsorption behavior of each solute species is available. The approach is very similar to that of the ideal adsorbed solution theory of Myers and Prausnitz (1965) for gas-solid adsorption.

Radke and Prausnitz (1972) suggested that when the solute species adsorb simultaneously from a dilute solution onto the adsorbent surface at constant temperature and spreading pressure π, the adsorbed phase forms an ideal solution; activity coefficients for all species are unity in the adsorbed phase. The following relations were proposed for a system containing n solutes:

$$\frac{1}{\sum_{i=1}^{n} q_{i\sigma}} = \sum_{i=1}^{n} \frac{x_{i\sigma}}{q_{i\sigma}^{p}} \quad (\text{constant } T, \pi); \tag{4.1.59}$$

$$C_{tl} x_{il} = C_i^0(\pi) x_{i\sigma} \quad (\text{constant } T), \tag{4.1.60}$$

where

$$q_{i\sigma} = x_{i\sigma} \sum_{i=1}^{n} q_{i\sigma}. \tag{4.1.61}$$

Here, C_{tl} is the total concentration of all solutes in the liquid phase, x_{il} is the solvent-free mole fraction of species i in the liquid phase, $x_{i\sigma}$ is the surface phase mole fraction of species i without the solvent, $C_i^0(\pi)$ is the unknown liquid-phase concentration of the pure solute i at spreading pressure π and temperature T, which provides the same surface phase concentration of i as the mixture, and $q_{i\sigma}^p$ is the number of moles of i per unit mass of adsorbent when adsorbed from a pure solution of i at the same π and T as the mixture.

The first relation suggests that the total number of moles of adsorbed solutes is determined by the sum total of the adsorptions of single solutes at the same values of T and π. For the second result, relating the equilibrium mole fractions of a given solute in the two phases, $C_i^0(\pi)$ is unknown. One has to evaluate the spreading pressure π in the manner of equation (4.1.49) from each pure component adsorption data of $q_{i\sigma}^p$ vs. C_i,

$$\frac{\pi S_\sigma}{RT} = \int_0^{C_i^0} \frac{q_{i\sigma}^p(C_i^0)}{C_i^0} \, dC_i^0. \tag{4.1.62}$$

Then, for a given π, C_i^0 can be determined for each species i (just as P_i^0 was determined earlier for gas adsorption), which will yield $x_{i\sigma}$ from relation (4.1.60) for known C_{tl} and x_{il}.

If experimental liquid-solid adsorption equilibrium data are available for each solute species i at a given temperature, then, to obtain the equilibria in a multisolute system, employ relation (4.1.62) for each solute species to obtain a relation between π_i and $f_i(C_i^0)$. For example, for a system of two solutes being present ($i = 1{,}2$), obtain

$$\pi_1 = f_1(C_1^0); \tag{4.1.63a}$$

$$\pi_2 = f_2(C_2^0). \tag{4.1.63b}$$

For any given π for the mixture, C_i^0 is to be defined at the same π; therefore,

$$\pi_1 = \pi = \pi_2. \tag{4.1.63c}$$

For a system of two solutes, relation (4.1.60) provides

$$C_{tl} x_{1l} = C_1^0(\pi) x_{1\sigma}; \tag{4.1.63d}$$

$$C_{tl} x_{2l} = C_2^0(\pi) x_{2\sigma}, \tag{4.1.63e}$$

where $x_{1\sigma} + x_{2\sigma} = 1$. Choose any C_{tl} and x_{1l} (equivalent to choosing the concentration of two solutes, $C_{tl}x_{1l} = C_{1l}$ and $C_{tl}(1-x_{1l}) = C_{tl}x_{2l} = C_{2l}$). Then unknowns C_1^0, C_2^0, π_1, π_2 and $x_{1\sigma}$ (or $x_{2\sigma}$ since $x_{1\sigma} + x_{2\sigma} = 1$) may be determined from equations (4.1.63a–e). One can adopt a graphical procedure of the type suggested by Myers and Prausnitz (1965) and illustrated earlier for gas-solid adsorption. If the single-solute isotherms illustrated by equations (4.1.63a) and (4.1.63b) can be obtained in an integrable analytic form, the set of equations may be solved by computer using the Newton–Raphson iteration scheme (Radke and Prausnitz, 1972). This is especially relevant for a system of solutes numbering more than two.

In some liquid-solid adsorption processes, the number of different components to be adsorbed is very large. One can adopt the formalism of *continuous chemical mixtures* and represent the class of similar adsorbable components by one variable M; a useful choice may be a parameter of the adsorption isotherm, e.g. the slope at infinite dilution (Annesini *et al.*, 1988).

In Section 3.3.7.6, the distribution of a solute between a bulk phase and an interfacially adsorbed phase was considered for a gas-liquid system containing surfactants, gas-solid adsorption systems and liquid-solid adsorption systems. So far, in this section, the separation factor for two solutes has been determined for an air-water system containing surfactants and for gas-solid adsorption. We could do the same also for the liquid-solid adsorption considered in the preceding paragraphs.

In liquid-solid adsorption, if microporous adsorbents are involved, then partitioning of the solute between the external solution and the pore solution has to be considered to obtain accurately the surface adsorption equilibria. If C_{il}^0 is the initial bulk liquid concentration of solute i in the initial bulk solution volume V_l^0, and C_{il} is the bulk liquid concentration of solute i in the final bulk liquid volume V_l, after equilibration of the microporous adsorbent having a specific pore volume (cm^3/g of support) of V_l^p, mass w_s (g) and specific surface area S_σ (m^2/g), then the following solute balance holds:

$$C_{il}^0 V_l^0 = C_{il} V_l + C_{il}^p V_l^p w_s + \Gamma_{i\sigma} w_s S_\sigma, \tag{4.1.64a}$$

where C_{il}^p is the pore liquid concentration of species i and $\Gamma_{i\sigma}$ is the concentration of adsorbed species i per unit surface area of the microporous adsorbent. The different liquid volumes are related by

$$V_l^0 = V_l + V_l^p w_s. \tag{4.1.64b}$$

We can now rewrite (4.1.64a) as

$$C_{il}^0 V_l^0 = C_{il} V_l^0 - C_{il} V_l^p w_s + C_{il}^p V_l^p w_s + \Gamma_{i\sigma} w_s S_\sigma,$$

and rearrange to obtain

$$\frac{(C_{il}^0 - C_{il}) V_l^0}{C_{il} V_l^p} = w_s \left\{ \left(\frac{C_{il}^p}{C_{il}} - 1 \right) + \frac{\Gamma_{i\sigma}}{C_{il}} \frac{S_\sigma}{V_l^p} \right\}. \tag{4.1.65}$$

The quantity (C_{il}^p / C_{il}) is the geometrical partitioning factor κ_{im} defined earlier (see (3.3.88a) and (3.3.89a)) for the partitioning of a solute between a solution and a porous membrane. Here the porous membrane is replaced by the microporous adsorbent. The quantity κ_{im} is less than 1 unless there are specific or nonspecific interactions (electrostatic or van der Waals interactions) between the solute and the pores; it can be quite small if the radius r_i of the solute molecules is of the order of the pore radius r_p. For cylindrical pores (see (3.3.88a)),

$$\kappa_{im} = \left(1 - \frac{r_i}{r_p} \right)^2.$$

In any equilibrium partitioning experiment using a microporous adsorbent and an external solution, one varies C_{il}^0 and w_s for a given adsorbent. From equation (4.1.65), a plot of C_{il}^0 vs. w_s for a given adsorbent will be linear, the slope being equal to the quantity in brackets on the right-hand side of equation (4.1.65). Unless $r_p >>> r_i$, it is clear that $\kappa_{im} = C_{il}^p / C_{il}$ will be less than 1, and the determination of the pore surface adsorption equilibrium relation (between $\Gamma_{i\sigma}$ and C_{il}) will be influenced by κ_{im}. This was demonstrated clearly in the adsorption of aromatic compounds (for example, napthalene) on microporous silica gel adsorbents by Alishusky and Fournier (1990).

4.1.6 Liquid-ion exchanger systems

Ion exchange processes are employed in a variety of situations in process-scale as well as preparative separations. A situation often encountered involves the selectivity of the ion exchanger for counterion A over counterion B. To develop an expression for the selectivity, we employ relation (3.3.118b) describing the equilibrium distribution of ionic species i between an aqueous solution and the ion exchange resin for ionic species A and B:

$$\frac{1}{Z_A \mathcal{F}} \left[RT \ln \left(\frac{a_{Aw}}{a_{AR}} \right) - \overline{V}_A (P_R - P_w) \right] = \phi_R - \phi_w$$

$$= \frac{1}{Z_B \mathcal{F}} \left[RT \ln \left(\frac{a_{Bw}}{a_{BR}} \right) - \overline{V}_B (P_R - P_w) \right].$$

This leads to

$$RT \ln \left[\left(\frac{a_{Aw}}{a_{AR}} \right)^{\frac{1}{Z_A}} \left(\frac{a_{BR}}{a_{Bw}} \right)^{\frac{1}{Z_B}} \right] = \left(\frac{\overline{V}_A}{Z_A} - \frac{\overline{V}_B}{Z_B} \right) (P_R - P_w). \tag{4.1.66}$$

If the swelling pressure, $P_R - P_w$, is negligible, we obtain

$$\ln \left[\left(\frac{a_{Aw}}{a_{AR}} \right)^{\frac{1}{Z_A}} \left(\frac{a_{BR}}{a_{Bw}} \right)^{\frac{1}{Z_B}} \right] = 0, \tag{4.1.67}$$

which leads, after rearrangement, to

$$\left(\frac{a_{Aw}}{a_{AR}} \right)^{Z_B} = \left(\frac{a_{Bw}}{a_{BR}} \right)^{Z_A}. \tag{4.1.68}$$

Suppose $Z_A = 2$, $Z_B = 1$ and $a_{Aw} = a_{Bw}$. Then

$$\left[\frac{a_{AR}}{a_{BR}^2} \right] = \frac{a_{Aw}}{a_{Bw}^2} = \frac{1}{a_{Bw}} \quad \Rightarrow \quad \frac{a_{AR}}{a_{BR}} = \frac{a_{BR}}{a_{Bw}}. \tag{4.1.69}$$

This result indicates that, if, for counterion B, the resin phase is selective over the aqueous phase, i.e. $a_{BR} > a_{Bw}$, then $a_{AR} > a_{BR}$: the resin phase is selective toward counterion A having a higher valence or charge number over counterion B with a lower valence. This result reflects the strong influence of counterion valence on the selectivity: from a mixture of two counterions, A and B, the counterion having a higher valence, A, is always preferred by the ion exchange resin. Correspondingly, the separation factor α_{AB}^a defined using activities, namely

$$\alpha_{AB}^a = \frac{a_{AR}}{a_{Aw}} \frac{a_{Bw}}{a_{BR}}, \tag{4.1.70}$$

will be greater than 1, for example, when $a_{Aw} = a_{Bw}$. For a cation exchange resin, for example, Ca^{++} will be preferred over Na^+.

Helfferich (1962) has provided a detailed analysis of the influence of other factors on ion exchanger selectivity: the ion exchanger tends to prefer (there are exceptions)

(1) the counterion of higher valence,
(2) the counterion having the smaller (solvated) equivalent volume,
(3) the counterion having greater polarizability,
(4) the counterion which interacts more strongly with the fixed ionic groups or the matrix,
(5) the counterion which has a lower tendency of complex formation with the co-ion.

Generally, the order of selectivity among univalent cations is as follows (Helfferich, 1962):

$$Ag^+ > Cs^+ > Rb^+ > K^+ > Na^+ > H^+ > Li^+.$$

For weak-acid resins, the position of H^+ will depend on the acid strength of the fixed anionic group. The corresponding sequence for bivalent cations is:

$$Ba^{2+} > Pb^{2+} > Sr^{2+} > Ca^{2+} > Ni^{2+} > Cd^{2+} > Cu^{2+} > Co^{2+} > Zn^{2+} > Mg^{2+} > UO_2^{2+}.$$

For anions, the selectivity sequence appears to be

$$\text{citrate} > SO_4^{2-} > \text{oxalate} > I^- > NO_3^- > CrO_4^{2-} > Br^- > SCN^- > Cl^- > \text{formate} > \text{acetate} > OH^- > F^-.$$

For weak-base resins, the position of OH^- is variable, but is generally further to the left.

Equation (4.1.70) expresses the selectivity between two particular counterions A and B by a resin. One would also like to know the nature of the corresponding relations that exist between a counterion and a co-ion or two co-ions. Consider the case where there are two counterions, A and B, and a co-ion Y. Using relation (3.3.118b), we obtain

$$\frac{1}{Z_A \mathcal{F}}\left[RT\ln\left(\frac{a_{Aw}}{a_{AR}}\right)-\bar{V}_A(P_R-P_w)\right]$$
$$=\phi_R-\phi_w=\frac{1}{Z_B\ \mathcal{F}}\left[RT\ln\left(\frac{a_{Bw}}{a_{BR}}\right)-\bar{V}_B(P_R-P_w)\right]$$
$$=\frac{1}{Z_Y\ \mathcal{F}}\left[RT\ln\left(\frac{a_{Yw}}{a_{YR}}\right)-\bar{V}_Y(P_R-P_w)\right].$$

A rearrangement leads to

$$RT\ \ln\left[\left(\frac{a_{Aw}}{a_{AR}}\right)^{\frac{1}{Z_A}}\left(\frac{a_{YR}}{a_{Yw}}\right)^{\frac{1}{Z_Y}}\right]=\left(\frac{\bar{V}_A}{Z_A}-\frac{\bar{V}_Y}{Z_Y}\right)(P_R-P_w). \tag{4.1.71}$$

Relation (4.1.66) is another one of these relations. Ignoring the swelling pressure in (4.1.71) leads to

$$\ln\left[\left(\frac{a_{Aw}}{a_{AR}}\right)^{\frac{1}{Z_A}}\left(\frac{a_{YR}}{a_{Yw}}\right)^{\frac{1}{Z_Y}}\right]=0.$$

Therefore,

$$\left(\frac{a_{Aw}}{a_{AR}}\right)^{\frac{1}{Z_A}}=\left(\frac{a_{Yw}}{a_{YR}}\right)^{\frac{1}{Z_Y}}=\left(\frac{a_{Bw}}{a_{BR}}\right)^{\frac{1}{Z_B}}, \tag{4.1.72}$$

where we have employed the result (4.1.67). In fact, this result is valid for all other ions in the solution (Helfferich, 1962).

Consider another co-ion in the system, say X. It follows from the above that

$$\left(\frac{a_{Yw}}{a_{YR}}\right)^{\frac{1}{Z_Y}}=\left(\frac{a_{Xw}}{a_{XR}}\right)^{\frac{1}{Z_X}}. \tag{4.1.73}$$

Let $Z_Y = -1$ and $Z_X = -2$. For the cation exchange resin-based system,

$$\left(\frac{a_{Yw}}{a_{YR}}\right)^{-1}=\left(\frac{a_{Xw}}{a_{XR}}\right)^{-\frac{1}{2}}\Rightarrow\left(\frac{a_{Yw}}{a_{YR}}\right)^{2}=\left(\frac{a_{Xw}}{a_{XR}}\right). \tag{4.1.74a}$$

If $a_{Yw} = a_{Xw}$ and $a_{YR} < a_{Yw}$ for the cation exchange resin, then the separation factor

$$\alpha^a_{YX}=\frac{a_{YR}\,a_{Xw}}{a_{Yw}\,a_{XR}}=\frac{a_{Yw}}{a_{YR}}>1. \tag{4.1.74b}$$

Therefore, a co-ion of lower valence (Y) is more strongly preferred by a cation exchange resin; for example Cl^- will be preferred over SO_4^{2-} by a cation exchanger (correspondingly NaCl over Na_2SO_4). See Problem 4.1.14 for a related exercise.

So far, we have activities of species to arrive at a number of conclusions. In practice, species concentrations are measured. Therefore, let us consider relation (4.1.68) for the selectivity of the ion exchanger for counterion A over counterion B:

$$\left(\frac{a_{Aw}}{a_{AR}}\right)^{Z_B}=\left(\frac{a_{Bw}}{a_{BR}}\right)^{Z_A}\Rightarrow\left(\frac{a_{Aw}}{a_{AR}}\right)=\left(\frac{a_{Bw}}{a_{BR}}\right)^{Z_A/Z_B}$$
$$\Rightarrow\frac{x_{AR}\,\gamma_{AR}}{x_{Aw}\,\gamma_{Aw}}=\left(\frac{x_{BR}\,\gamma_{BR}}{x_{Bw}\,\gamma_{Bw}}\right)^{Z_A/Z_B}\Rightarrow\left(\frac{x_{AR}\,x_{Bw}}{x_{Aw}\,x_{BR}}\right)$$
$$=\alpha_{AB}=\frac{\gamma_{Aw}}{\gamma_{AR}}\left(\frac{\gamma_{BR}}{\gamma_{Bw}}\right)^{Z_A/Z_B}\left(\frac{x_{BR}}{x_{Bw}}\right)^{(Z_A/Z_B)-1};$$
$$\alpha_{AB}=\frac{\gamma_{Aw}}{(\gamma_{Bw})^{Z_A/Z_B}}\frac{(\gamma_{BR})^{Z_A/Z_B}}{\gamma_{AR}}\left(\frac{x_{BR}}{x_{Bw}}\right)^{(Z_A/Z_B)-1}. \tag{4.1.75a}$$

In a dilute solution of counterions A and B in water, one may assume $\gamma_{Aw} \cong 1$, $\gamma_{Bw} \cong 1$. Therefore,

$$\alpha_{AB}=\frac{(\gamma_{BR})^{Z_A/Z_B}}{\gamma_{AR}}\left(\frac{x_{BR}}{x_{Bw}}\right)^{(Z_A/Z_B)-1}.$$

If $Z_A = 2$, $Z_B = 1$, then

$$\alpha_{AB}=\left(\frac{\gamma^2_{BR}}{\gamma_{AR}}\right)\left(\frac{x_{BR}}{x_{Bw}}\right). \tag{4.1.75b}$$

For very low x_{Bw}, α_{AB} will have a large value. For larger x_{Bw}, α_{AB} will be much smaller. Consider the case of A = Ca^{2+} and B = Na^+, as encountered in the removal of hardness (Ca^{2+}) from water. For low values of salt concentration in water (i.e. low x_{Bw}), the selectivity of the ion exchanger for Ca^{2+} over Na^+ will have some value. If, however, x_{Bw} is high, α_{AB} will be much lower, leading to removal of the Ca^{2+} ions from the ion exchanger. At low values of x_{Bw}, the ion exchanger will have many more Ca^{2+} ions, thus removing the cause of hardness from water.

It is useful to consider now another kind of counterion exchange process between the ion exchanger and the counterions in the external solution. The counterions to be considered now are, *macroions*, specifically *proteins* of large molecular weight. Any such protein molecule in solution has positive as well as negative charges distributed on its surface. The net charge of the macromolecular protein due to *pH*-based interactions of various constituent groups is positive if the *pH* is less than the isoelectric point (I.E.P = *pI*); the net charge is negative if the solution *pH* > *pI*. Thus, as long as the solution *pH* is different from the *pI*, the protein surface has some net positive or net negative

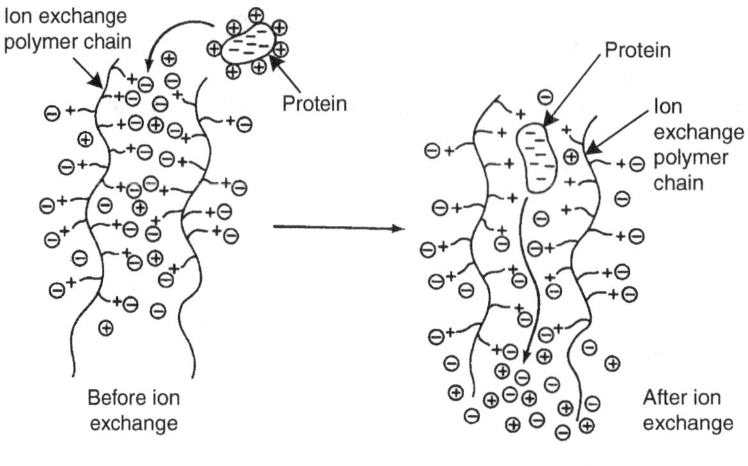

Figure 4.1.16. "Ion exchange" occurring when a negatively charged protein adsorbs to an anion exchanger. Seven positively charged ions (e.g. $HTris^+$) associated with the protein molecule are displaced, together with seven negative ions (Cl^-) from the exchanger. (After Scopes (1987).)

charges. Interactions due to *pH* based chemical reactions in general are considered in Chapter 5.

Due to the principle of electroneutrality, small counterions present in the solution will be distributed around the protein surface to ensure local electroneutrality. Consider now a negatively charged protein molecule (with positively charged microions distributed around it) to be adsorbed on an anion exchanger resin bead (Figure 4.1.16). The fixed positive charges on the anion exchange resin have counterions (e.g. Cl^-) present to start with. The protein molecule has to displace these counterions (Cl^-) near the resin surface in order for it to be ion exchanged. As shown in the figure, after the protein molecule is adsorbed, seven counterions (Cl^-) near the ion exchange resin surface and seven counterions present around the protein in solution (e.g. $HTris^+$)[3] are displaced together into the solution. This addition of Tris-Cl into the solution increases the ionic strength of the solution.

The electrostatic attraction between the negative charges on the protein surface and the positive fixed charges on the anion exchange resin can be considerably weakened by two methods. If the *pH* of the buffered solution is decreased to a lower value, then the electrostatic binding force is considerably reduced. Alternatively, if the ionic strength of the solution is increased, the interaction between the protein and the ion exchanger is reduced and that between the microions (e.g. Cl^-) and the ion exchanger is considerably enhanced. This method is generally preferred if desorption of the adsorbed protein is desired.

[3]Tris stands for tris (hydroxylmethyl) aminomethane.

Another form of ion exchange of interest is that between two different proteins, $i = 1$ and 2, and an appropriate active charged site on the ion exchange resin surface. For any protein i, one can represent this interaction by

$$i(j=2) + \mathrm{S}(j=1) \leftrightarrow i\mathrm{S}(j=1), \tag{4.1.76}$$

where the resin phase is $j = 1$ and S represents the resin site. The equilibrium constant for this binding is usually identified as K_d and represents the processes of adsorption (rate constant is k_a) and the dissociation (rate constant is k_d). At equilibrium, the rates are equal:

$$k_a C_{i2} C_{S1} = k_d C_{iS,1}. \tag{4.1.77a}$$

Now C_{S1} representing the concentration of active charged sites, can be expressed as the difference between the maximum number of charged sites, C_{S1}^m, and the number of sites already occupied by the protein, $C_{iS,1}$:

$$C_{S1} = C_{S1}^m - C_{iS,1}. \tag{4.1.77b}$$

At equilibrium,

$$K_d = \frac{k_d}{k_a} = \frac{C_{i2}(C_{S1}^m - C_{iS,1})}{C_{iS,1}}. \tag{4.1.77c}$$

Correspondingly, at equilibrium, the protein concentration in the resin phase, $C_{iS,1}$ will be related to the protein concentration in the surrounding fluid phase, C_{i2}, by

$$C_{iS,1} = \frac{C_{i2} C_{S1}^m}{K_d + C_{i2}}. \tag{4.1.77d}$$

At any temperature, this relation behaves as a Langmuir adsorption isotherm (see Section 3.3.7.6). The equilibrium constant, K_d, has a low value for proteins which bind strongly to the resin. For any resin, the comparative binding strengths of different proteins will be indicated by the respective values of K_d. Patel and Luo (1998) have indicated that K_d values for many different proteins and a variety of ion exchange resins vary between 10^{-5} and 10^{-7} M. Experimental protein adsorption data plotted as $(1/C_{is,1})$ vs. $(1/C_{i2})$ will yield $(1/C_{S1}^m)$ as intercept and (K_d/C_{S1}^m) as the slope:

$$1/C_{is,1} = (K_d/C_{S1}^m)(1/C_{i2}) + (1/C_{S1}^m). \qquad (4.1.77e)$$

4.1.7 Supercritical fluid–bulk solid/liquid phase

Section 3.3.7.9 describes how a supercritical fluid (SCF), e.g. supercritical CO_2, can extract effectively a solute from a liquid or a solid; the solute is recovered later by reducing the pressure: the SCF becomes a gas having very little solubility of the solute, which precipitates and is recovered. This section briefly considers the systems where two or more solutes are extracted simultaneously. There are two goals here. First, one would like to increase the solubility of a particular species. Second, it would be desirable to enhance the selectivity of the SCF for a particular solute with respect to the second solute.

Generally, it has been found that for a pure SCF (e.g. CO_2), the extent of enhancement of the solubility for a particular solute is very similar for other solutes being extracted over the entire range of the density of the SCF. Thus pure SCFs are not helpful in developing selectivity unless other entrainers (e.g. methanol) are added in small amounts. (*Note*: One of the lures of extraction by SCF is the opportunity to avoid organic extractants, e.g. methanol.) However, Chimowitz and Pennisi (1986) have suggested the use of the crossover region for developing selective SCF separation.

Consider Figure 4.1.17 for the crossover *regime* of a SCF extracting two solid solutes, components 1 and 2, from a solid mixture, where the mole fraction of each solute, x_{il}, in the SCF phase has been plotted against the SCF pressure for two temperatures T_H and T_L ($<T_H$). The crossover point for any solute i is the pressure P_i^* at which $(\partial x_{il}/\partial T)_P = 0$, signifying a change in the temperature dependence of the solubility of species i. Below this pressure, a decrease in the temperature increases the solute solubility in the SCF, whereas above this pressure an increase in the temperature increases the solute solubility.

When there are two solid solutes (extracted into the SCF) each having a different crossover pressure, we have a crossover regime (Figure 4.1.17). At an intermediate pressure between the two crossover pressures P_1^* and P_2^* for solutes 1 and 2, considerable selectivity can be developed between the two solutes by changing the temperature. Suppose the SCF is initially at T_H and is then cooled to a temperature T_L ($<T_H$). For component 2, the process will involve movement from A to B (Figure 4.1.17), i.e. the equilibrium solubility of species 2 is increased. On the other hand, the equilibrium solubility of species 1 decreases as the process involves change from A to B. Thus pure component 1 can be made to drop out of the SCF phase by a change of temperature in the crossover regime. Chimowitz and Pennisi (1986) have provided data for SCF CO_2 based extraction from a solid mixture of 1,10-decanediol and benzoic acid; in the crossover regime, pure decanediol was deposited. For successful use of the resulting selectivity, it is useful to have a wider crossover region (Johnston *et al.*, 1987).

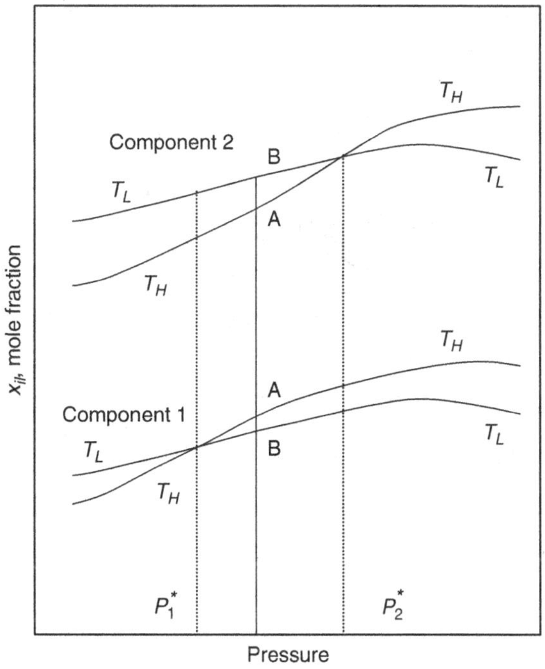

Figure 4.1.17. Separable crossover points with two solutes dissolved in a supercritical solvent. (After Chimowitz and Pennisi (1986).)

4.1.8 Bulk fluid phase – mesophase systems

At the beginning of Section 3.3, bulk phases commonly used for developing separation processes/techniques employing two immiscible phases were identified: gaseous, liquid, solid, supercritical fluid, ion exchange material, membrane. In addition, interfacial phases are capable of developing separation with bulk phases, as has been illustrated in Sections 3.3.5 and 4.1.5. There are other *mesophases*, or *molecular aggregates*, and *associated environments* which can participate in selective partitioning of solute species between itself and another bulk liquid phase. *Micellar systems* provide a particular example of such a mesophase quite useful for separations. Other so-called mesophases useful for partitioning of solutes from a bulk liquid phase are: *reversed micelle, vesicle,*

Surrounding water

Hydrophobic nonpolar hydrocarbon

Locus of solubilization of nonpolar organic compound

Hydrophilic polar head

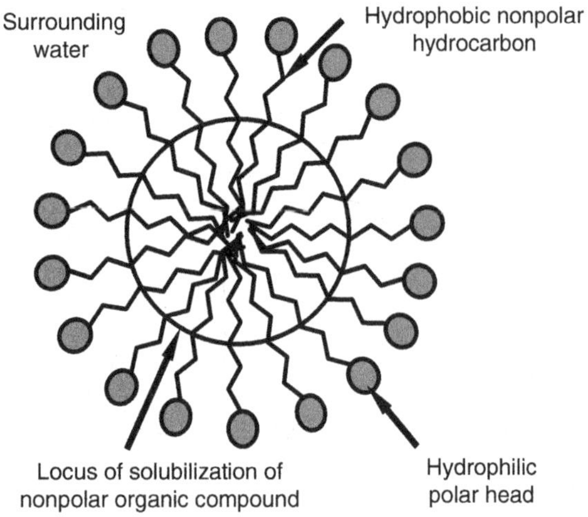

Figure 4.1.18. A spherical micelle can solubilize hydrophobic solutes in the hydrophobic interior.

Table 4.1.9. Solubilization coefficient K_{i1} for a number of organic solutes in different micelles[a]

Surfactant	Organic solute	K_{i1} (M^{-1})
CPC[b]	1-butanol	6
	toluene	121
	p-chlorophenol	495
SDS[c]	phenol	15
	4-tert-butylphenol	365
	hexane	630
Triton X-100[d]	naphthalene	1120
	pyrene	74 000

[a] Data from Christian and Scamehorn (1989) and Jafvert *et al.* (1995).
[b] Cetylpyridinium chloride (CMC = 0.88 mM).
[c] Sodium dodecyl sulfate.
[d] $B_8A_6E_{9.5}$; B_8 is a branched hydrocarbon chain, A_6 is a six-carbon aromatic ring, $E_{9.5}$ is 9.5 repeating –CH$_2$CH$_2$O– group (mol.wt. 628).

liposome, etc. A central feature of these three mesophases is that they are globular dispersions, with water as the inner core into which solute partitioning occurs. Therefore, one can question from a solute partitioning point of view whether we are dealing with a new phase.

Surface-active agents (surfactants) spontaneously form dynamic aggregates called micelles when their concentration in water (or other solvents) increases beyond a certain level, called the *critical micelle concentration* (CMC). These micellar mesophases, or pseudophases, can have several forms, i.e. spherical, cylindrical, disk, etc. When the solvent is water or the solution is aqueous, the micellar aggregate forms with the polar headgroup of the surfactant sticking out on the outside surface and the hydrophobic (or lipophilic) chain on the inside (Figure 4.1.18). For a spherical micelle, the sphere radius will be about 2 nm for a 2 nm long surfactant. The interior of the aggregate is hydrophobic due to the hydrophobic chains of the surfactants.

Such a micelle acts like an oil drop in an aqueous solution and allows solubilization of organic compounds present in the aqueous solution into the hydrophobic interior. If the surfactant is ionic, i.e. either cationic or anionic, then oppositely charged ionic species, namely counterions from the solution, will be adsorbed on the surface of the micelle or bind with the charged micellar surface. Such surface binding will effectively increase the micelle dimension. A unique number of surfactant molecules between 50 and 100 aggregate to form a micelle.

The partitioning of a solute between the micelle (interior for uncharged species, exterior for charged species) has been described in a number of ways, including the equilibrium ratio K_{i}, the distribution coefficient κ_{i}, etc. One method of description for uncharged organic solutes used by a number of investigators defines a solubilization

coefficient (Christian and Scamehorn, 1989; Jafvert *et al.*, 1995) or an equilibrium constant[4] K_{i1} for the solubilization process described by

organic solute + surfactant micelle = organic solute · micelle

as

$$K_{i1} = \frac{\text{solute mole fraction within a micelle}}{\text{concentration of free solute in bulk solution}} = \frac{x_{i1}}{C_{i2}}$$
(4.1.78)

where region $j = 2$ is the bulk solution and region $j = 1$ is the micelle phase. Instead of being dimensionless, the units of this K_{i1} are (mol/liter)$^{-1}$. Values of this solubilization coefficient for a number of organic solutes in micelles prepared from different surfactants are provided in Table 4.1.9. Compounds which are highly hydrophobic and have very low solubilities in water will have high values of K_{i1}.

The solute–micelle equilibrium leads to two basic situations of relevance to separation: the solution surrounding the micelles is stripped of organic solutes; the solutes are transferred to the micellar phase whose dimensions are an order of magnitude larger. These features have been exploited in a number of different separation techniques at large and small scale. Contaminated water may be purified by micelles, which remove the organic pollutants into the micellar core; the purified water may then be removed through an ultrafiltration membrane (see Sections 3.4.2.3, 4.3.4, 5.4.2, 6.3.3.2 and 6.4.2.1), while the pollutant-laden micelles are retained by the membrane (as in micellar-enhanced ultrafiltration (MEUF, see Section 5.4.2)). A micellar solution may be injected subsurface to solubilize and remove hydrophobic pollutants into the

---

[4]It is obviously different from our usual K_{i1}, an equilibrium ratio, defined by equation (1.4.1).

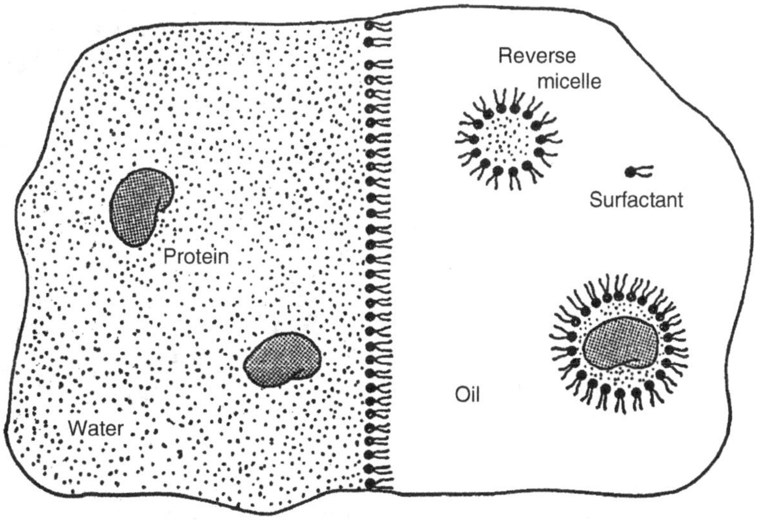

Figure 4.1.19. Schematic representation of the reversed micellar extraction of proteins. (After Göklen and Hatton (1985) and Hatton (1989).)

micellar solution ultimately withdrawn from the ground. In addition, such micellar solutions may aid in chromatographic partitioning by being employed as a mobile phase in chromatographic processes (see Sections 7.1.5 and 8.2).

In the micelle based extraction of organic solutes from an aqueous solution, the mass-separating agent/phase has been developed by the addition of surfactants to the aqueous solution. In most situations, this separating agent needs to be recovered for reuse; a concentrated stream of organic solutes should also be obtained from this micellar phase. When the organic solutes are nonvolatile, chemical reactions can be employed to precipitate the surfactants (Brant *et al.*, 1989). Solvent extraction through a porous membrane has also been attempted (Kitiyanan *et al.*, 1999). When the organic solutes are volatile, they may be removed by stripping or pervaporation and the micellar solution recycled (Abou-Nemeh *et al.*, 1998).

We now consider another kind of pseudophase. In the presence of an organic solvent immiscible with an aqueous phase containing surfactant molecules, new types of surfactant molecular aggregates, called *reversed micelles*, are formed. As shown in Figures 4.1.19, two-tailed surfactants (having a polar water-soluble head group and two hydrocarbon tails insoluble in water) form aggregates in the organic phase that contain an aqueous core and a surfactant shell, where the hydrophobic surfactant tails stretch out into the organic solvent. Figures 4.1.19 illustrates the extraction of a protein molecule from the aqueous phase into the aqueous core of the reversed micelle structure present in the organic solvent phase. A typical surfactant which leads to the development of such structures is sodium di-2-ethylhexyl sulfosuccinate (AOT).

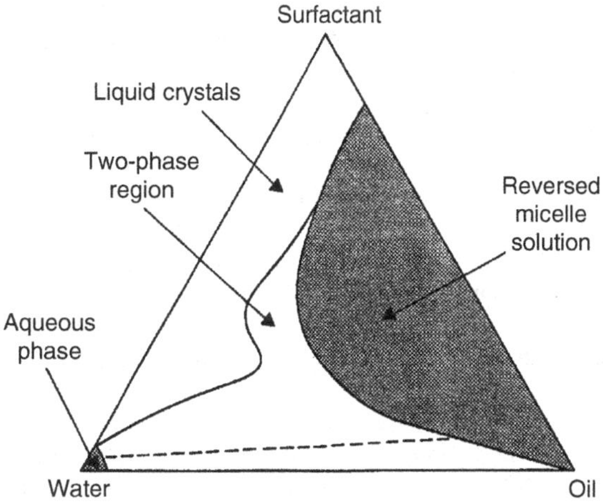

Figure 4.1.20. Typical phase diagram for a water-surfactant-oil system. (After Hatton (1989).)

The reversed micellar structure in an aqueous-organic system is obtained under practical conditions in a certain composition range shown in the phase diagram for water-isooctane-AOT (Figure 4.1.20). Any mixture of water-isooctane-AOT whose composition falls in the two-phase region of this phase diagram will spontaneously split into an aqueous phase and an organic-dominated reverse micellar solution phase via a tie line (the dashed line). The nanometer-sized surfactant aggregate in the organic solvent is stable. A typical example of overall phase concentrations in protein extraction from a protein-containing aqueous buffer solution is: equal volumes

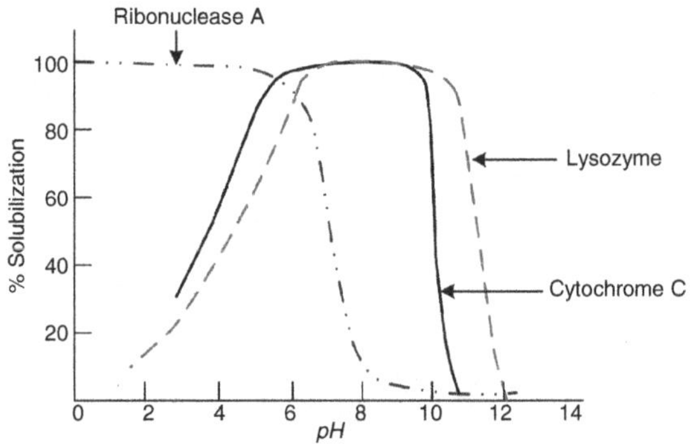

Figure 4.1.21. The effect of pH *on the solubilization of lysozyme, cytochrome C and ribonuclease A in AOT-isooctane solutions. Salt concentration: 0.1 M KCl. (After Göklen and Hatton (1987).)*

of 250 mM AOT in isooctane in equilibrium with 0.1 M KCl in 0.01 M acetate buffer, *pH* 4.0.

The protein so extracted into the reverse micellar (RM) phase may be easily back extracted if the RM phase is contacted with an aqueous buffer solution containing a higher KCl concentration, e.g. 0.1 M, or if isopropyl alcohol is added to the extent of 10–15% to the aqueous phase used for back extraction (Carlson and Nagarajan, 1992). However, the solubilization of the protein in the RM phase or the back extraction of the protein into an aqueous buffer from the RM phase depends on a number of complex interactions of a variety of factors. These include the protein *pI*, the aqueous phase *pH*, the ionic strength (e.g. KCl concentration), the type of salt, the type of surfactant and the interaction between the protein and the charged surfactant headgroup (Hatton, 1989). Figure 4.1.21 illustrates the percent solubilization of three proteins into the RM phase as a function of aqueous phase *pH* (Göklen and Hatton, 1987). Often, solubilization into the RM phase proceeds if the protein $pI > pH$; back extraction is accomplished if $pH > pI$.

A third category of molecular aggregates are called *vesicles, liposomes,* etc. Like reversed micelles, they enclose water inside a spherical shell. However, unlike reversed micelles, the shell is made of a lipid bilayer structure present in biological cell membranes. Specifically, the shell has a bilayer structure with the polar hydrophilic head of two layers of lipids forming the two surfaces of a sandwich of the two layers of lipids, with the long two-tailed nonpolar hydrophobic chains of each lipid in the interior of the sandwich (Figure 4.1.22). The specific lipids have a polar phosphate head and are therefore usually called phospholipids; an example is phosphatidylcholine. The hydrocarbon chains in any such lipid are generally 14 to 18 carbon atoms long, resulting in a typical lipid bilayer thickness of around 3.7 nm. The spherical vesicle dimensions can vary over a wide range, for example 0.03–0.1 μm. Sometimes

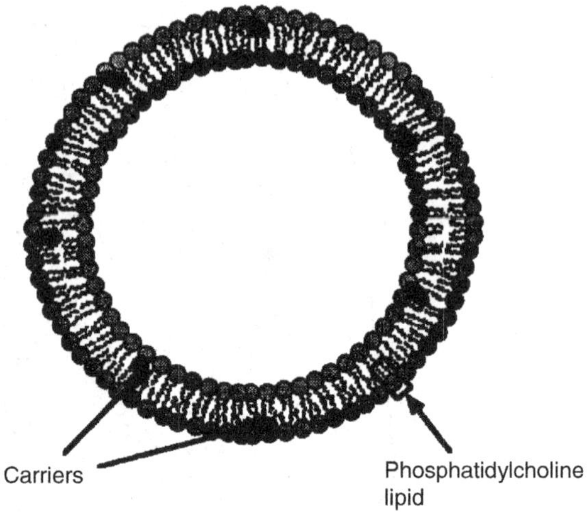

Figure 4.1.22. Vesicle structure.

they can be much bigger, incorporate many layers of the lipid bilayer and are called multilamellar vesicles (MLVs), as opposed to those having a single lipid bilayer (Lasic, 1992). Both structures are also called liposomes.

The 14–18 carbon atoms long chain of the lipid in the lipid bilayer presents a rigid hydrocarbon gel-like structure which is essentially impermeable to charged/polar molecules larger than water. The transfer of polar/charged species from outside to inside of the vesicle requires a carrier; for transferring metal ions, one uses an ionophore as a carrier of the charged/polar species (see Sections 4.1.9.1.3 and 5.4.4) incorporated into the lipid bilayer (Walsh and Monobouquette, 1993). The lipid bilayer is a new material phase that acts as a selective membrane between an external aqueous phase and an aqueous phase enclosed by the lipid bilayer, the vesicle. However, these noncovalent assemblies of lipids are damaged in the presence of detergents, water-soluble alcohols, etc. Covalently closed vesicles can be formed by using polymerizable vesicle-forming surfactants (Shamsai and Monobouquette, 1997).

4.1.9 Partitioning between a bulk fluid phase and an individual molecule/macromolecule or a collection of molecules for noncovalent solute binding

A variety of chemical interactions are possible between two molecules of two different species. Those that result in chemical reactions are considered in Chapter 5 for the purpose of separation of mixtures. There are a number of other, weaker, noncovalent interactions where the bond energy is less than 50 kJ/mol. These include chelation, clathration, hydrogen bonding, hydrophobic interactions, ionic binding, pi bonding, van der Waals interactions, etc. These result in weak chemical complexes, which are reversible in nature. Further reversing the complexes without destroying the complex-forming molecule is possible due to weaker interactions.

King (1980) has enumerated the variety of interactions as originally identified by George K. Keller with respect to the bond energy vs. bond type. These interactions lead to reversible binding of a solute molecule from a bulk fluid phase with a complex-forming individual molecule or collection of individual molecules introduced from outside. The complexing agent may also be a macromolecule of biological (e.g. proteins, etc.) or nonbiological (e.g. conventional polymers) origin. The complex so formed may remain in the original solution phase or may form a second phase, usually a solid. This solid phase may be withdrawn and treated to reverse the complexation.

To avoid the need for separation of the solid phase thus formed, sometimes these molecules/macromolecules may be individually bonded to or deposited on a porous or nonporous bead/particle or the pore surfaces of a membrane at one end of the molecule (for example), while the other ends or parts of the molecule are available to bind the solute species from the solution. Species present in the gas phase may also bind to such molecules on a bead or particle. When the solute molecule is a protein or a large entity, a large number of binding sites are necessary to hold the solute; thus, a number of individual complexing molecules bonded to the porous particle at one end will have their other ends bind at different locations of the solute, which is a protein/large entity. A natural extension of this concept has led to the bonding/coupling of protein A to agarose beads for extremely selective binding of immunoglobulins from a solution containing immunoglobulins and other materials in antibody purification.

Separation generally also requires recovery of the solute species after it has been selectively bound to specific molecules added to the solution or bound to a porous or nonporous bead/particle or membrane pores. When the process takes place without any particles, beads or membranes, the complex produced has to be removed from the solution and decomplexation carried out. The steps usually taken to achieve decomplexation (with or without the use of beads, particles or membranes), i.e. regeneration of the molecules/macromolecules for reuse in binding solute molecules, include: changing the ionic strength of the solution by adding electrolytes; changing the *pH*; raising the temperature; decreasing the pressure, etc.

Although the binding of the solute molecule with the complexing agent is reversible and noncovalent in nature, the mathematical description is generally identical to situations where chemical reactions take place. Thus, a number of such reversible low binding energy based separations are described mathematically in Chapter 5. Here we illustrate the basic nature of the phenomenon of solute partitioning in such processes in a closed system under the following categories:

(1) inclusion compounds;
(2) ionic binding of metals with charged species;
(3) pi bonding/complexation;
(4) hydrophobic interaction.

The phrase "solute partitioning" mentioned above should not suggest that the solute must necessarily be partitioned into a separate phase. There are many examples (to be discussed in the following) where the solute molecule is bound to a different type of molecule in the same phase; such binding facilitates subsequent separation.

4.1.9.1 *Inclusion compounds: adducts, clathrates, crown ethers, cyclodextrins, liquid clathrates*

Different types of molecules acting as mass-separating agents may be added to a liquid mixture in gaseous, liquid or solid form to form preferentially a solid phase of the added external agent molecules and the solvent/one of the solute species in the liquid mixture when cooled. The solid phase of the solvent/solute and the external added agent is generally called an *inclusion compound* (Atwood *et al.*, 1984). We now consider a variety of inclusion compounds: adducts; clathrates; crown ethers; cyclodextrins; liquid clathrates.

4.1.9.1.1 Adducts When the external agent added to the solution forms the bulk of the new crystalline solid phase, the process is called *adductive crystallization* and the compounds so formed are called *adducts*. The solute molecules from the liquid feed will fit into the crystal lattice of this crystalline solid phase, which acts as the "host;" the solute molecules included in the host crystal are the "guest" (an inverse scenario happens in the case of *clathrates*, as we will soon see). The hosts could be either organic or inorganic. The slurry that is produced has to be removed from the raffinate liquid and may be heated to recover the "guest."

An example of this type of process of selective solute partitioning is *urea based adductive crystallization.* A saturated solution of urea in water at 70 °C may be mixed with a mixture of aromatic and paraffinic hydrocarbons present in a solvent at 40 °C (the Edeleanu process). Under conditions of appropriate refrigeration, lumps of urea-n-paraffin adducts appear as crystals (Findlay, 1962; Fuller, 1972). As shown in Figure 4.1.23(a), the host compound, urea, has crystallized into a form having a central tunnel open at both ends; the tunnel accommodates the "guest" paraffinic hydrocarbon molecule and holds it by van der Waals forces with m urea molecules forming the tunnel:

$$m(\text{urea}) + \text{paraffin} \leftrightarrow \text{adduct}. \qquad (4.1.79a)$$

The guest paraffin has significant freedom in the channel, akin to adsorption rather than a chemical reaction. The

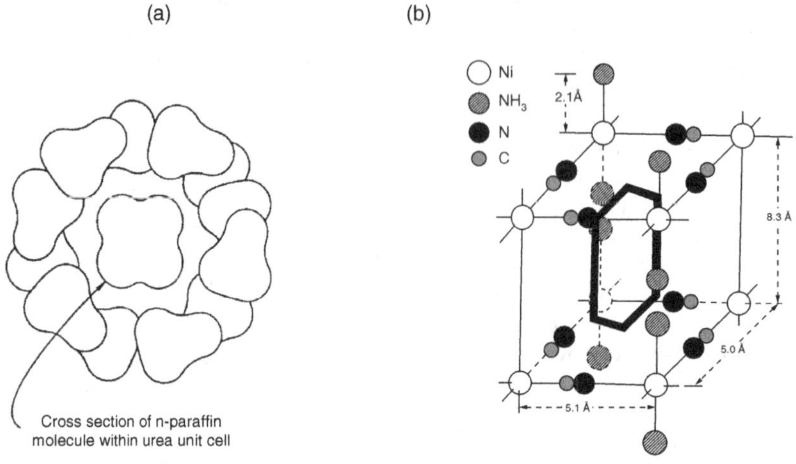

Figure 4.1.23. (a) Representation of a cross section of a urea-n-paraffin adduct showing the urea molecules forming the walls of a tunnel in which the n-paraffin is held by attractive forces. (b) Structure of clathrate formed between monoaminenickel cyanide and benzene, showing the benzene molecule trapped inside a cage formed by the crystal lattice of the former (Findlay, 1962). Reprinted, with permission, from R.A. Findlay, "Adductive crystallization," Chapter 4, Figure 7, p. 283 and Figure 5, p. 275 in Schoen, H.A. (ed.), New Chemical Engineering Separation Techniques, *Interscience Publishers 1962. Copyright © 1962, John Wiley & Sons.*

moles of urea per mole of the pseudoreactant, paraffin, varies, for example, between 6.1 and 23.3: 6.1 for n-heptane and 23.3 for n-detriacontane ($C_{32}H_{66}$). For a given paraffin, m is not fixed; the compounds are therefore nonstoichiometric. Apparently, urea forms adducts with straight-chain paraffinic and olefinic compounds as long as the carbon number is 6 or higher. The heat of this pseudoreaction is approximately 1.6 kcal per carbon atom, which is less than that for heats of adsorption (see Ruthven (1984) for heats of adsorption). Thiourea has also been employed as an adduct-former (Findlay, 1962).

4.1.9.1.2 Clathrates Another class of inclusion compounds are *clathrates*, in which, instead of a channel as in urea adducts, a clathrate or a cage is formed by the "host" molecules, and the "guest" molecule is trapped in this cage (Figure 4.1.23(b)). Generally, the guest molecule is brought into the liquid phase of the host molecules and the clathrate phase crystallizes out as the temperature is lowered. When an aqueous solution is under consideration and the guest is a low molecular weight gas/vapor, e.g. CH_4, SO_2, C_2H_4, H_2S, C_2H_5Cl, Cl_2, Br_2, CO_2, C_2H_4O, etc., the clathrate is called a *gas hydrate*.

The clathrate compound does not have a fixed chemical formula in terms of the number of guest molecules (G_M) which can be accommodated in the cage created by the host molecules; however, there is a maximum number of host molecules. Findlay (1962) has listed the following among others for water as host: for CH_4 guest, $46H_2O{\cdot}8G_M$; for propene guest, $136H_2O{\cdot}8G_M$; for SO_2 guest, $46H_2O{\cdot}8G_M$. In the gas hydrate, water molecules are linked through hydrogen bonding with a cavity diameter inside the cage

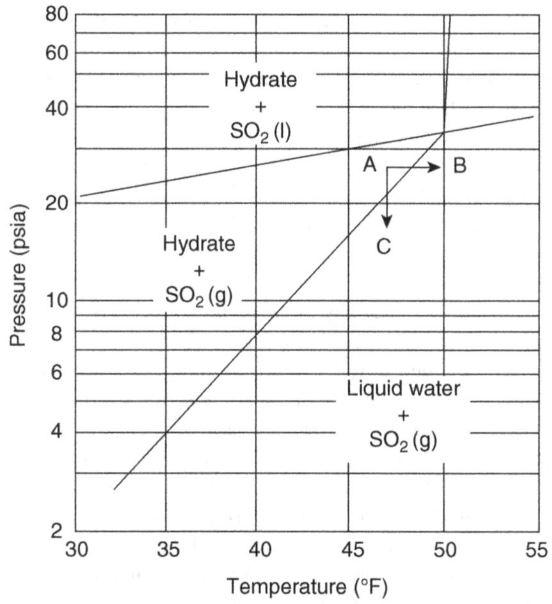

Figure 4.1.24. Phase diagram for the sulfur dioxide-water system. Reprinted, with permission, from G.N. Werezak, in Unusual Methods of Separation, Chem. Eng. Prog. Symp. Ser., No. 91, Vol. 65, AIChE, New York, p. 6 (1969). Copyright © [1969] American Institute of Chemical Engineers (AIChE).

varying between 780 and 920 pm; the guest molecules, which do not interfere with hydrogen bonding and have diameters in the range of 410–580 pm, stabilize the structure under appropriate temperature and pressure (see

Table 4.1.10. Inclusion compounds: types and examples

		Host	Guest (G)	Examples
Multiple molecules create host structure: solid state compounds	organic molecules	urea	n-alkanes, n-alkenes, straight-chain acids, straight-chain esters	adductive crystallization; e.g. 6.1 moles urea per mole of G
		thiourea	cyclopentane, isoparaffins	
		phenol	HCl, HBr, SO_2, CS_2	clathrate compounds; e.g., $12C_6H_5OH \cdot 5G$
		hydroquinone	SO_2, HCl, HBr, HCN, H_2S, CO, etc.	$3C_6H_4(OH)_2 \cdot G$
	inorganic molecules	H_2O	CH_4, C_3H_8, SO_2, Cl_2, H_2S, etc.	$46H_2O \cdot 8G$
		monoaminenickel cyanide	benzene, thiophene, aniline, etc.	
		Ni(4-methyl pyridine)$_4$ $(SCN)_2$	p-xylene	Werner-complex based clathration
Single molecule is host	organic molecules	crown ethers	alkali metal ions, Li^+, K^+, Cs^+, Sr^{2+}, etc.	extraction of such metal salts into an organic phase from aqueous phase/bonded phase in chromatography
		cyclodextrins (CD)	small nonpolar organic molecules; organic isomers; p-xylene	extraction of such species into aqueous phase containing CD or bonded CD phase in chromatography
Multiple molecules create liquid host structure	Organic molecules	benzene	organometallic salt $K[Al_2(CH_3)_6N_3]$	liquid clathrate containing 6 benzene to 1 salt molecules

Englezos (1993) for a review of clathrate hydrates). Such molecules are often called hydrate-formers.

The clathrate hydrates are formed only in given ranges of temperature and pressure. Figure 4.1.24 illustrates the pressure-temperature diagram for an SO_2-water system (Werezak, 1969). Such a diagram will allow determination of the conditions under which the hydrates are formed, or the hydrate crystals decompose, releasing the trapped molecule SO_2 for recycle and reuse. Such clathrate hydrate formation provides a way of recovering almost pure water or concentrating aqueous solutions using an appropriate hydrate-former.

Clathrate compounds are also formed with organic nonaqueous host species, e.g. hydroquinone, phenol, monoaminenickel cyanide, methylnaphthalene, etc. (Findlay, 1962). Clathrate compounds formed with monoaminenickel cyanide as host and benzene/thiophene as guest (Figure 4.1.23 (b)) are of interest in petroleum refining; the selectivity of the host for benzene is much higher.

The nature and types of inclusion compounds are more numerous than the two types, namely adducts and clathrates, considered so far. A broader list is provided in Table 4.1.10. The list is not comprehensive but illustrative of each category. In both adducts and clathrates, the number of molecules needed to develop the host structure is generally larger than 1 and can be as high as 136 (in clathrate hydrates). On the other hand, organic compounds, such as, crown ethers, cryptands, cyclodextrins, etc., incorporate a metallic ion or an organic molecule as guest inside the existing cavity present in the *single host molecule*.

4.1.9.1.3 Crown ethers, cyclodextrins *Crown ethers* are cyclic ethers having repeating $-OCH_2CH_2-$ units; therefore, they are often called macrocyclic polyethers. Figure 4.1.25 identifies a variety of crown ethers, namely 14-crown-4, benzo-15-crown-5, dicyclohexyl-18-crown-6, dibenzo-21-crown-7. The number 14, 15, 18, 21, etc., in this nomenclature identifies the number of atoms in the polyether ring, whereas 4, 5, 6, 7, etc., specify the number of oxygen atoms or repeating units in the ring (Pedersen, 1967). Table 4.1.11 provides an estimate of the diameter of the cavity in the polyether ring in a few crown ethers. This table also provides the diameters of a variety of cations, primarily alkali metal cations (Pedersen, 1988; Steed and Atwood, 2000). As long as the cavity diameter is close to that of an alkali metal cation (M^+), the metal ion is held in the cavity as a guest with several oxygen atoms in the ring

Table 4.1.11. Crown ethers, their properties and cations[a]

Crown ether	Cavity diameter (nm)	Cation diameter (nm)			
		Group I		Group II	
All 14-crown-4[b]	0.12–0.15	Li^+	0.136	Ca^{2+}	0.198
All 15-crown-5	0.17–0.22	Na^+	0.194	Zn^{2+}	0.148
All 18-crown-6	0.26–0.32	K^+	0.266	Sr^{2+}	0.226
		Rb^+	0.294	Cd^{2+}	0.196
All 21-crown 7	0.34–0.43	Cs^+	0.266	Ba^{2+}	0.268
		NH^{4+}	0.286[c]	Ra^{2+}	0.280

[a] Adapted from Pedersen (1988).
[b] "All" refers to various substitutent groups, e.g. dicyclohexyl, dibenzo, etc.
[c] Nonmetallic cations are also amenable to complexation.

Table 4.1.12. Extraction of alkali metal picrates by different crown ethers into methylene chloride[a]

Polyether	Picrate extracted (%)			
	Li^+	Na^+	K^+	Cs^+
Dicyclohexyl-14-crown-4	1.1	0	0	0
Cyclohexyl-15-crown-5	1.6	19.7	8.7	4.0
Dibenzo-18-crown-6	0	1.6	25.2	5.8
Dicyclohexyl-18-crown-6	3.3	25.6	77.8	44.2
Dicyclohexyl-21-crown-7	3.1	22.6	51.3	49.7
Dicyclohexyl-24-crown-8	2.9	8.9	20.1	18.1

[a] From Table 4, p. 538, of "The discovery of crown ethers," Charles J. Pedersen, *Science*, Vol. **241**, July 1988, pp. 536–540. Reprinted with permission from AAAS.

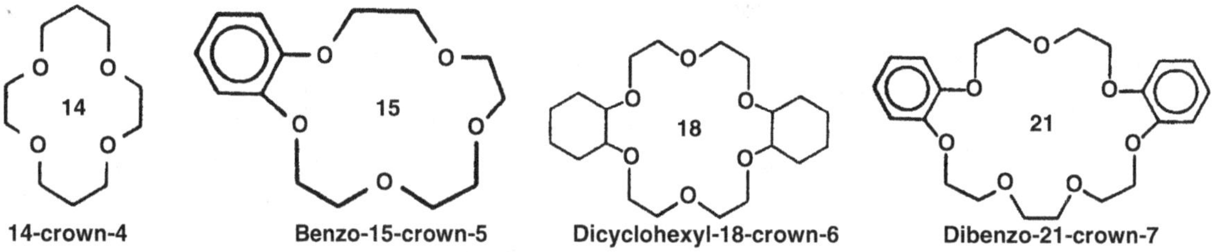

Figure 4.1.25. A few crown ethers.

acting as donors to complex the strongly electropositive alkali metal cations. For example, 18-crown-6 prefers K^+ strongly over Li^+ and Na^+. If the cation is too small or too large, the resulting complexes are not very stable. However, even in such cases, two molecules of crown ethers may form a sandwich to hold one cation.

Crown ethers have low solubility in water but have considerable solubility in organic solvents. Crown ether complexes of alkali metal cations are therefore quite stable in organic phases. The anions of the alkali metal cations present in the aqueous phase, Y^-, are simultaneously extracted into the organic phase; the complex in the organic phase may be represented as $(\text{crown}\cdot M)^+\cdot Y^-$. Due to the likely absence of solvation of such anions in the organic phase, they are likely to be highly reactive. Nevertheless, crown ethers and similar host compounds have been successfully incorporated into organic solvents to extract alkali metal ions/salts from aqueous solutions.

When the organic solvent containing crown ether is used in the form of a liquid membrane between two aqueous solutions (see Section 5.4.4), an alkali metal salt can be selectively transferred through the organic liquid membrane from one aqueous feed solution to an aqueous strip solution. Illustrative treatment of the kinetics and mechanisms of formation and dissociation of the metal complexes with crown ethers is available in Burgess (1988). Crown ethers have been incorporated into numerous polymers; the resulting polymers show the expected order of alkali metal selectivity. Crown ethers have also been bonded to silica. The bonded crown ethers do show selectivity for various metallic cations (Alexandratos and Crick, 1996).

In the case of clathrates and adducts described earlier, the so-called "compounds" formed are separated by crystallization. Although handling of slurry/solid in an industrial context is sometimes not desired, the solid phase conclusively demonstrates the nature of the non-stoichiometric dissociable host–guest compound formed. That macrocyclic ethers form reasonably stable complexes has also been demonstrated by the isolation of the complexes as crystals (Pedersen, 1988). That a particular crown ether having a certain cavity diameter will prefer a certain alkali metal cation having a certain diameter is illustrated in Table 4.1.12 for the extraction of a particular alkali metal picrate salt from water into methylene chloride containing the crown ether. From the table it appears that potassium picrate is most efficiently extracted by 18-crown-6 compared to other crown ethers, since the cavity size of 18-crown-6 is quite close to that of K^+ (see Table 4.1.11).

Cyclodextrins (CDs) are cyclic oligosaccharides having the shape of an asymmetrical doughnut (see Figure 4.1.26 (a)). The three common types of cyclodextrins, α-CD, β-CD

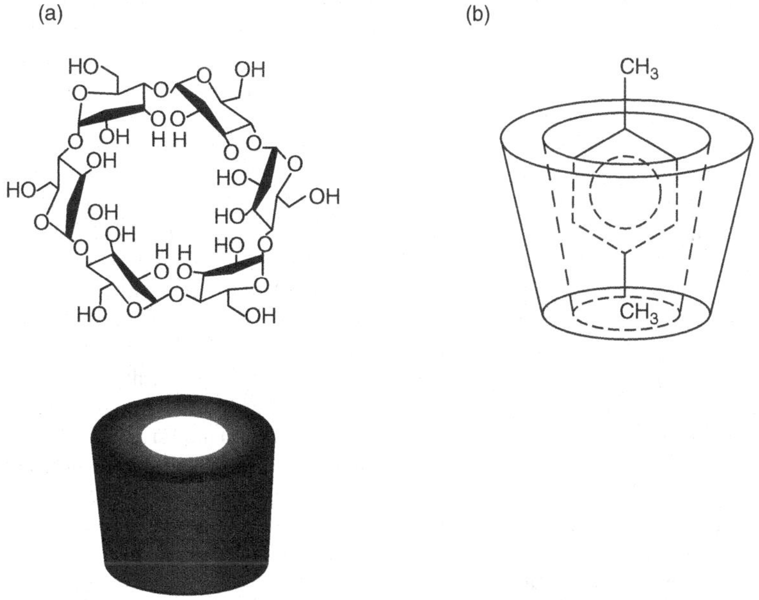

Figure 4.1.26. (a) Schematic of α-cyclodextrin; (b) schematic of p-xylene as a guest in an α-CD host.

and γ-CD, each have a cavity whose diameters are, respectively, 0.57 nm, 0.78 nm and 0.95 nm. The corresponding molecular weights and aqueous solubilities are: 972, 1135, 1297 (molecular weights); 14.5, 1.85, 23.2 (solubilities in g/100 cm^3). (See Bender and Komiyama (1978) for an introduction.) Due to the presence of hydroxyl groups on the outside surface, CDs are soluble in water. However, the interior cavity lined with hydrogen and oxygen atoms is relatively hydrophobic and provides sites for inclusion complex formation with smaller "guest" organic solutes having limited polarity (Figure 4.1.26(b)).

A wide variety of smaller organic guests have been found to form inclusion complexes with a cyclodextrin molecule as host. The stability of the inclusion complex depends on a variety of factors, specifically the geometric fit with the cavity opening and the nature of the interactions. Such differences are sufficient to create different complexation tendencies of different isomers with cyclodextrins. There is a considerable literature on the use of CDs for chiral separations in chromatography; in fact, they are used in a variety of chromatographic columns in analytical chemistry. To this end, the CD is bonded to polysiloxane (Figure 4.1.27), which can be coated onto appropriate silica columns for use in chromatography (see Section 7.1.5) (Armstrong *et al.*, 1993; Jung *et al.*, 1994). The bonding of CD to polysiloxane phase eliminates the need for any crystallization based phase separation. Instead, one has to go through the process of complexation and decomplexation in sequence.

Selective liquid-liquid extraction of xylene isomers (m-, o- and p-) has been demonstrated between organic

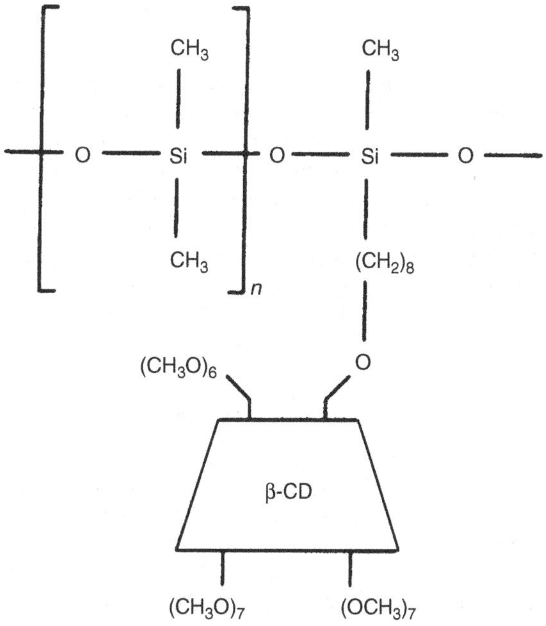

Figure 4.1.27. Structure of polysiloxane-bonded permethylated β-cyclodextrin (Chirasil-Dex) (n ≈ 60). (After Jung et al. *(1994).)*

and aqueous phases using branched α-CDs; the branched CDs have considerably higher aqueous solubility, and p-xylene is strongly preferred (Figure 4.1.26(b)) (Uemasu, 1992). An aqueous solution of β-CD containing urea and NaOH was employed by Armstrong and Jin (1987) and

Mandal *et al.* (1998) as a liquid membrane (see Section 5.4.4) to separate different types of isomeric mixtures effectively.

4.1.9.1.4 Liquid clathrates Unless crown ethers and cyclodextrins are used either in a liquid membrane format, in a bonded state or in solvent extraction–back extraction methodology, the inclusion compound has to be crystallized out. Crystallization, however, has to be practiced in the case of "adducts" and "clathrates." The handling of large-scale solids is difficult in such processes; furthermore, in the case of "adducts," the solute-bonding capacity is limited due to the large load of the host per unit guest molecule. *Liquid clathrates* have therefore been developed so that solvent extraction methodology is available and solids handling can be avoided (Atwood *et al.*, 1984).

Liquid clathrates essentially consist of a liquid made out of guest molecules entrapped in a host species. They are primarily based on low-melting organometallic salts, which have a high solubility in aromatic liquids, as well as a high selectivity for liquid aromatic hydrocarbons. For example,

$$K[Al_2(CH_3)_6N_3](l) + 6C_6H_6(l) \leftrightarrow K[Al_2(CH_3)_6N_3] \cdot 6C_6H_6(l), \tag{4.1.79b}$$

indicating that a liquid of composition 1(salt): 6(aromatic) has been formed; this liquid is immiscible with and heavier than pure benzene phase. The low melting salt $K[Al_2(CH_3)_6N_3]$ has weak interionic interactions, and the benzene molecules surround the anion with the cation outside the cage. There is some order, such that the cations attached to different adjacent cages cannot interact with one another. The stripping/back extraction can be easily implemented, only with particular liquid clathrates, by changing the temperature by only 10 °C.

4.1.9.2 *Ionic binding of metals with charged species*

Metal ions are often present in a variety of aqueous solutions. In hydrometallurgy, the metal ion is to be recovered; in environmental separations, the heavy metal ions[5] have to be removed from water and the water purified; in process streams, it could be either. Weak ionic binding of metal ions to a variety of charged species may be utilized to achieve such goals. The charged species can be polyelectrolytes, ionic surfactants, chelating agents, polyamino acids, etc. The charged species may be an individual molecule in solution or bound to the surface of a bead or a pore in a porous membrane; alternatively, the individual charged molecules may be part of a collection of such molecules, e.g. a surfactant micelle formed from ionic surfactants.

Metals are *most* often present as cations M^{n+} in a solution: Cu^{2+}, Ca^{2+}, Zn^{2+}, Na^+, Pb^{2+}, Cd^{2+}, etc. Some metals exist in anionic forms as well: CrO_4^{2-}, $HCrO_4^-$, etc. Consider now a *polyelectrolyte*, polystyrene sodium sulfonate (PSS), the sodium salt of the polystyrene sulfonic acid (Figure 4.1.28(a)). If such polyelectrolyte macromolecules are added to water containing hardness-causing metal ions, e.g. Ca^{2+}, Mg^{2+}, sodium ions will be replaced by Ca^{2+} or Mg^{2+}, which will remain bound to the polymer via the charged sulfonic acid group. Since a polymer molecule is easily removed by filtration, the undesirable ions remain bound to the anionic polyelectrolyte filtered and concentrated. Each polyelectrolyte molecule (polymer) will have many metallic ions bound to it at every location where there is an oppositely charged ion (Tabatabai *et al.*, 1995).

Such polyelectrolyte macromolecules may also be bound at one end on the surface of a bead or on the surface of a pore in a membrane. The whole polyelectrolyte molecule is essentially available for binding the metal ions in solution. However, since the polyelectrolyte is bound at one end, filtration is no longer necessary to recover it and the bound metal ions. An example of a polyelectrolyte bound at one end on the pore surface of a microfiltration membrane and having various interactions with metal ions is schematically shown in Figure 4.1.28(b). Here, poly (amino acid) molecules (e.g. poly(L-glutamic acid)) are bound to the pore surfaces of porous cellulosic membranes (Hestekin *et al.*, 2001). The COO^- groups created at higher *pH*s bind directly with the metal ion (shown as Me^{2+}) after deprotonation of COOH groups in the poly (amino acid). In addition, the electrostatic potential field created by the neighboring COO^- groups leads to loose retention of metallic counterions (counterion condensation).

The reverse of the metal ion binding process is important in the process of recovering a concentrated solution of metal ions and the polyelectrolytes. The bound heavy metal ions in a homogeneous solution of polyelectrolytes may be released by contacting with a concentrated brine. In the case of PSS in solution used to bind Ca^{2+} and Mg^{2+}, Na^+ ions from brine will replace Ca^{2+} and Mg^{2+}, etc. Such a solution, when filtered, will yield a filtrate concentrated in Ca^{2+}, Mg^{2+}, etc. The concentrate or retentate solution will contain the polyelectrolyte in the Na^+ form and can be recycled for re-use (Tabatabai *et al.*, 1995). An alternative strategy of using high acid/low *pH* will lead to release of the bound metal from the poly(amino acid)s bound to the membrane pore surface (Figure 4.1.28(c)) (Hestekin *et al.*, 2001).

[5]There are a variety of definitions for heavy metals. A few useful definitions are given here: any metal having a specific gravity greater than 5; any metal generally toxic to biological systems; any metal readily precipitated from solution as a sulfide; any metal located in the lower half of the periodic table, etc.

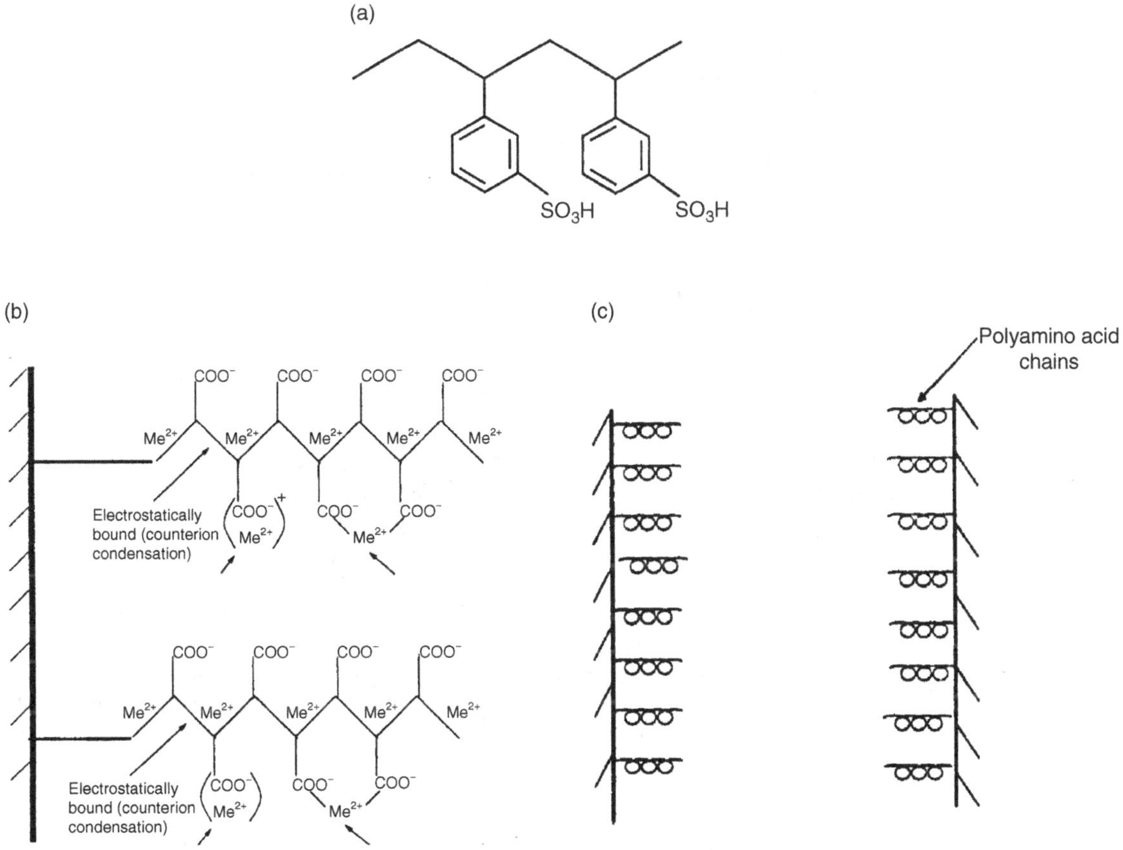

Figure 4.1.28. (a) Polystyrene supported sulfonic acid. (b) Metal binding by two poly(amino acid) chains bound at one end to the pore surface of a membrane (Hestekin et al., *2001). Reprinted, with permission, from* I&EC Research, *40, 2668 (2001), Figure 1(a). Copyright (2001) American Chemical Society. (c) Poly(amino acid) chains bound on the surface of a pore in a membrane.*

Ionic surfactants will form micelles in a solution if their concentrations are above their critical micelle concentration (CMC). The ionic surfactant, for example sodium dodecyl sulfate (SDS), is prepared as the salt in which Na^+ is the counterion and DS^-, representing dodecyl sulfate ($C_{12}SO_4^-$), is the surfactant ion (here an anion). A micelle of such a surfactant will have numerous ionic headgroups, DS^-, around the periphery of individual micelles. If there are other metallic ions present in the system, e.g. Ca^{2+}, Cu^{2+}, Zn^{2+}, Cd^{2+}, etc., there will be an exchange between these ions and Na^+ as the counterions for the DS^- ions. Thus heavy metals, and other metals, present in solution will be bound to the headgroups of the micelle formed from the ionic surfactant. By using an appropriate membrane/filter, one can concentrate the ionic micelles and their bound heavy metallic counterions.

The surfactants may be recovered from the concentrated solution by precipitating the surfactant using high concentrations of a monovalent counterion (e.g. Na^+(aq)). For example, neutral SDS salt may be precipitated via

$$C_{12}SO_4^-(aq) + Na^+ \rightarrow NaC_{12}SO_4(s) \qquad (4.1.80a)$$

(see Brant *et al.* (1989)). A multivalent counterion, such as Ca^{2+}, may also be used to precipitate the surfactant:

$$2C_{12}SO_4^-(aq) + Ca^{2+} \rightarrow Ca(C_{12}SO_4)_2(s). \qquad (4.1.80b)$$

Thus, regeneration of metal ions, surfactants, polyelectrolytes, poly(amino acid)s, etc., participating in the ionic binding of metal ions in solution will require chemical reactions. Chemical reactions are considered in Chapter 5 in general for their effects on separations.

Whether the metal-binding agent is a polyelectrolyte or an ionic surfactant headgroup in a micelle, it is of interest to know the extent of binding of a metal ion to such a species/agent. Based on the work of Oosawa, Scamehorn *et al.* (1989) have suggested equations that relate the fraction of a metal ion that is bound to the micelle to the free (unbound) metal ions. The estimation of these fractions requires additional relations, such as the electroneutrality condition, concentrations of surfactants present as micelles

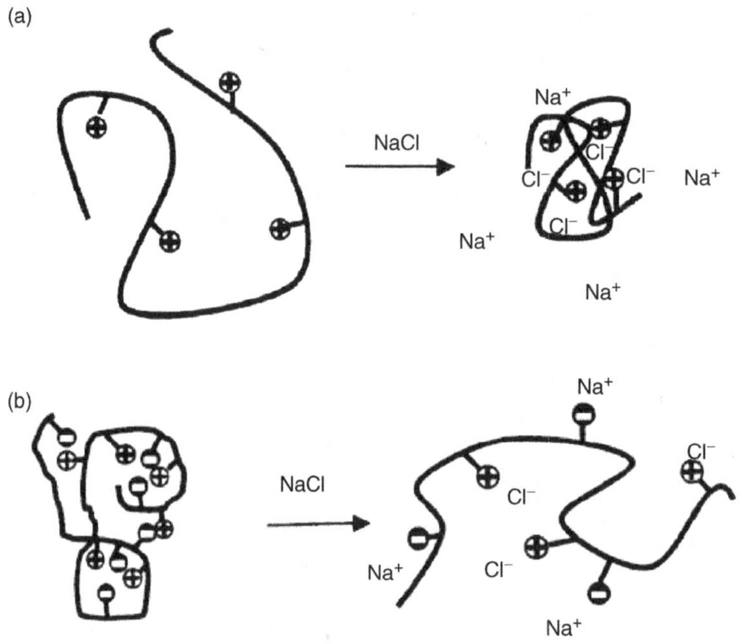

Figure 4.1.29. (a) The polyelectrolyte effect; (b) the anti-polyelectrolyte effect (Lowe and McCormick, 2001). Reprinted, with permission, from Stimuli-Responsive Water Soluble and Amphiphillic Polymers, *ACS Symposium Series 780, 2000, p. 352, C.L. McCormick (ed.), A.B. Lowe and C.L. McCormik (Chapter 1 authors), Figures 1 and 2 on pp. 2 and 3 of Chapter 1. Copyright (2000) American Chemical Society.*

and other parameters involving the CMC value, the charges of various ions, etc.

The binding or presence of metallic ions or their counterions in aqueous solutions of polyelectrolytes considered so far can influence the conformation of free polyelectrolytes in solution. At high concentrations of salt (NaCl), there can be precipitation of polyelectrolytes (equation (4.1.80a)). A range of behavior is possible, having significant effects on separation. Consider a polyelectrolyte, poly (methacrylic acid) (PMMA), which will carry negative charges from carboxylate groups at high *pH*. These will repulse each other, leading to an extended conformation; there will be counterions present near each charged group in the backbone. The end-to-end distance will be large.

When the concentration of the polyelectrolyte (in this case, polysalt) is increased, or the concentration of the salt (NaCl) is increased, the radius of the ionic atmosphere at every charge location is reduced; the polymer chains undergo shrinkage, resulting in a smaller configuration (Figure 4.1.29(a)) often called the *polyelectrolyte effect* (Lowe and McCormick, 2001). On the other hand, if we have a polyzwitterion (i.e. the polymer chain has both anionic and cationic groups) such that the ratio of the anionic to cationic groups is around 1, there will be attractive electrostatic interactions between such groups, leading to a collapsed globular structure. If, however, salt (NaCl) is added, the individual ionic groups in the polyelectrolyte will be shielded by Na^+ and Cl^- ions; the attractive interactions will be reduced, and the polymer will take up a more expanded configuration (the *anti-polyelectrolyte effect*) (Lowe and McCormick, 2001) (Figure 4.1.29(b)).

Such phenomena are the basis of a number of separation techniques. For example, if the polyelectrolyte has a collapsed globular structure, then the diffusional resistance of molecules through such a polyelectrolyte based medium will be drastically increased; larger molecules, e.g. proteins, may not be able to diffuse effectively. On the other hand, a collapsed structure may open up large gaps between the polyelectrolytes and allow easy transport through the overall medium. If a gel-like structure can be created using polyelectrolytes, the swelling or contraction of the gel can open up or shrink transport corridors based on the *pH* of the solvent system. Sometimes, this is carried out in the presence of an uncharged polymer or inside the pores of a rigid membrane. As an extreme case, a separate gel phase can be formed. This *gel phase* (e.g. of crosslinked, partially hydrolyzed polyacrylamide) acts as a size-selective extractive solvent removing water and smaller solutes from an aqueous solution. At low *pH*, the gel shrinks and rejects extra water (soaked in earlier) in a separate vessel (Cussler *et al.*, 1984), just as in solvent extraction and back extraction processes.

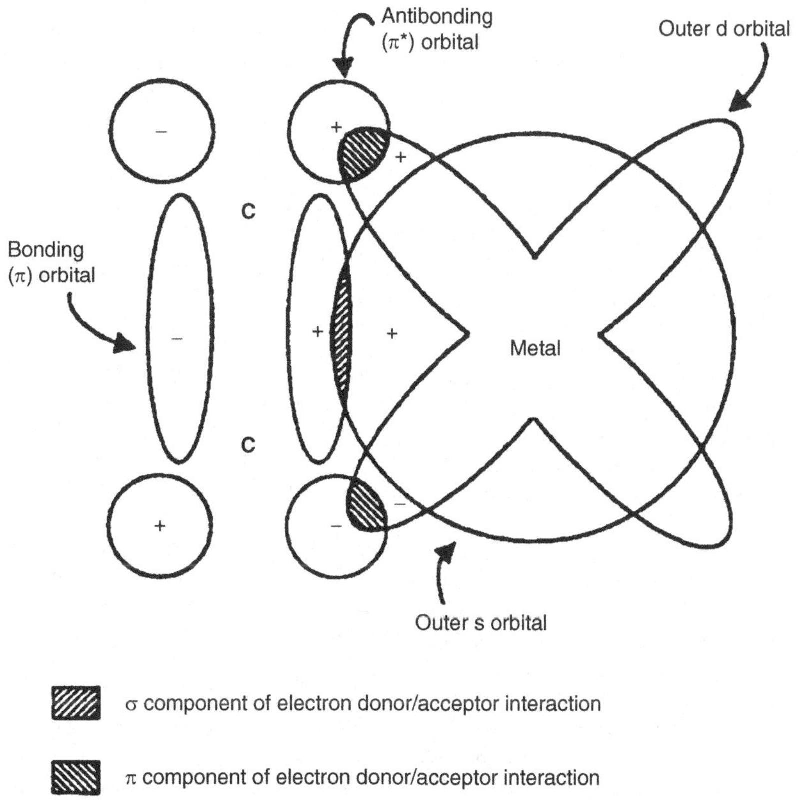

Figure 4.1.30. Dewar-Chatt model of π-complexation (Safarik and Eldridge, 1998). Reprinted, with permission, from Ind. Eng. Chem. Res., ***37****(7), 2572-2581(1998), Figure 1. Copyright (1998) American Chemical Society.*

4.1.9.3 *Pi bonding/complexation*

Mixtures of olefins and paraffins having similar or the same carbon numbers can be successfully separated by what is called *pi bonding* or *π-complexation.* In such a process, the olefin–paraffin mixture in a gaseous or liquid state is contacted with transition metals like Cu, Ag, etc., in a solution or dispersed on a solid substrate having a high surface area. The liquid solution/solid substrate will contain Ag^+ ions; in the case of Cu, cuprous salts are used to make Cu^+ ions available. The π bonds of an olefin molecule interact with the d orbitals of such a metal/ion; also, the metal ion forms σ-bonds with the carbon in the olefin. Collectively this type of metal ion–olefin coordination is identified as π-complexation (Figure 4.1.30). Such complexations are highly selective for olefins; however, they are weak enough to be broken by low pressure or high temperature to release the olefin molecule bound to the transition metal ion.

Olefins have been separated with high selectivity from a mixture with paraffins in gas phase by permeation through a liquid membrane/polymeric membrane containing $AgNO_3$ (aqueous solution, Hughes *et al.* (1986); glycerol solution, Kovvali *et al.* (2002); polyvinyl-alcohol-based polymeric membrane, Ho and Dalrymple (1994); polymeric perfluorosulfonate membrane, Koval *et al.* (1989)). By chemical absorption in a solvent, olefins have been separated successfully from paraffins: aqueous $AgNO_3$ solution for C_2H_4 absorption at 240 psia and 30–40 °C (Keller *et al.*, 1992); cuprous aluminum tetrachloride in an aromatic solvent (commercially used, Gutierrez *et al.* (1978)); $AgNO_3$ solution in water in a microporous membrane based absorber (Davis *et al.*, 1993); cuprous diketonate in α-methylstyrene solvent (Ho *et al.*, 1988). Solid SiO_2 sorbent on which Ag^+ was spread showed considerable selectivity for butene over n-butane at 1 atm and 70 °C (Padin *et al.*, 1999).

A number of different silver salts have been used for complexation with olefins, e.g. $AgNO_3$, $AgBF_4$, etc. The system that has been studied in great detail is $AgNO_3$ in water. The following complexation reactions have been suggested (Herberhold, 1974):

$$Ag^+ + \text{olefin} \leftrightarrow [Ag-\text{olefin}]^+; \qquad (4.1.81a)$$

$$[Ag\,(\text{olefin})]^+ + Ag^+ \leftrightarrow [Ag_2(\text{olefin})]^{2+}; \qquad (4.1.81b)$$

$$[\text{Ag (olefin)}]^+ + \text{olefin} \leftrightarrow [\text{Ag(olefin)}_2]^+. \quad (4.1.81c)$$

For a review of π-complexation for olefin-paraffin separations, see Safarik and Eldridge (1998).

4.1.9.4 *Hydrophobic interaction*

Hydrophobic interaction is utilized generally in the separation of protein mixtures via hydrophobic interaction chromatography. Protein molecules have a distribution of charged hydrophilic regions, as well as hydrophobic regions, on their surface. The hydrophobic regions arise from the side chains of a variety of amino acids making up the protein. The extent of such regions will be dependent on the protein. Such regions will, in aqueous solutions, prefer to interact with other such regions in other protein molecules; alternatively, if there are adsorbent particles having hydrophobic chains, short or long, sticking out, the hydrophobic patches on the protein molecule prefer to interact with such chains. The chains commonly used have a free benzyl group at the end or consist of a $(\text{-}CH_2\text{-})_n$ chain, where n varies from 3 to 10; such chains are bonded to the adsorbent particle via various chemistries. Interactions between such chains on the adsorbent surface and the hydrophobic patches on the protein surface lead to binding of the protein to the hydrophobic adsorbent surface.

If water molecules were to solvate the hydrophobic regions of the protein surface, they have to be highly ordered, requiring a strong decrease in entropy for the process represented as follows (Tanford, 1980):

$$\text{protein} + sH_2O \leftrightarrow \text{protein} \cdot sH_2O, \; \Delta S^0 = -\text{ve}. \quad (4.1.82)$$

This process does happen (so ΔG^0 is negative, correspondingly ΔH^0 is also negative; neglect here any interaction with the charged/hydrophilic regions of the protein). If hydrophobic chains are available on an adsorbent, the same arguments are valid. If the hydrophobic patches on a protein molecule react with the hydrophobic chains of the adsorbent,

$$\text{protein} \cdot rH_2O + \text{adsorbent} \cdot sH_2O \leftrightarrow \text{protein} \cdot \text{adsorbent} + (r+s)H_2O, \; \Delta S^0 = +\text{ve}, \quad (4.1.83)$$

the entropy increases and water molecules are released. If salt ions are available (A^+Y^-) in solution, released water molecules are drawn away to hydrate the salt ions:

$$A^+Y^- + (r+s)H_2O \leftrightarrow A^+ \cdot r_1H_2O + Y^- \cdot s_1H_2O, \quad (4.1.84)$$

where $(r_1 + s_1) = (r + s)$. This will facilitate the forward reaction (4.1.83), enhancing the binding of the protein with the hydrophobic adsorbent. Thus, in hydrophobic interaction chromatography, high concentrations of salt are deliberately maintained to promote reaction (4.1.83).

The effect of a salt on the adsorption of hydrophobic regions of a protein molecule on the hydrophobic patches of an adsorbent has been analyzed by Perkins *et al.* (1997), who have performed a model-independent thermodynamic analysis employing what is known as *the preferential interaction analysis.* A general result of this analysis relates the change in the capacity factor k'_{il} of the protein species $i = 2$ ($j = 1$, adsorbent), k'_{21}, to the variation in the molal salt concentration, $m_{3,w}$, in the solution phase ($j = 2$):

$$\left[\frac{\partial \ln (k'_{21})}{\partial \ln (m_{3,w})}\right]_{T,P,eq.} = \frac{(\Delta\upsilon_+ + \Delta\upsilon_-)}{g} - \frac{n}{m_{1,w}} \frac{\Delta\upsilon_1}{g} m_{3,w}, \quad (4.1.85)$$

where subscripts $i = 1, 2, 3$ correspond to the solvent (e.g. water), the protein and the salt, respectively; further, for the salts which are electrolytes, n is the total number of anions and cations per formula unit, $m_{1,w}$ is the molal concentration of water ($i = 1$) (55.5), $\Delta\upsilon_1$, $\Delta\upsilon_+$ and $\Delta\upsilon_-$ are, respectively, the number of water molecules, the number of cations and the number of anions released during the binding process. The quantity g is defined as $g = (\partial \ln(m_{3,w})/\partial \ln(a_\pm))_{T,P}$. Here $a_\pm$ is the mean ionic activity for electrolytes (see (3.3.119d)) and is given as

$$a_\pm = m_{3,w}\gamma_\pm (n_+^{n_+} n_-^{n_-})^{1/n}, \quad (4.1.86)$$

where $\gamma_\pm$ is the mean ionic activity coefficient and n_+ and n_- are the number of cations and anions per unit of salt. For a few common electrolytes, e.g. NaCl, $(NH_4)_2SO_4$, NaSCN, in hydrophobic interaction based processes, n is, respectively, 2, 3 and 2. The constant g can be calculated for each system and has a range of values (around 1.1–1.9 in Perkins *et al.* (1997)) for a variety of systems. The values of $\Delta\upsilon_1$ can vary over a wide range (5–500) depending on the system.

Perkins *et al.* (1997) have integrated equation (4.1.85) for the system of ovalbumin and ammonium sulfate to obtain

$$\ln (k'_{21}) = a + \frac{(\Delta\upsilon_+ + \Delta\upsilon_-)}{g} \ln(m_{3,w}) - \frac{n \, \Delta\upsilon_1}{m_{1,w} \, g} m_{3,w}. \quad (4.1.87)$$

From the experimental data for k'_{21} obtained against $m_{3,w}$, one can obtain values of a, $(\Delta\upsilon_+ + \Delta\upsilon_-)/g$ and $(n \, \Delta\upsilon_1/m_{1,w} \, g)$.

An alternative scenario is possible in the absence of these hydrophobic chains on the adsorbent/any adsorbent, namely hydrophobic patches in two neighboring protein molecules are bound together; this leads to protein precipitation. This is the *salting-out* phenomenon for protein solutions, wherein a high concentration of a salt, typically ammonium sulfate (1.5–3M), added to the protein solution, precipitates the protein molecules from the solution. Generally, the nature of the cation in the salt is not important. However, the anions are all important. The precipitation ability of the anions follow in decreasing order:

$$PO_4^{3-} > SO_4^{2-} > \text{acetate}^- > Cl^- > NO_3^- > SCN^-.$$

How does one reverse the binding between the hydrophobic patches in a protein with the hydrophobic chains on hydrophobic adsorbents? Detergents, organic solvents (ethylene glycol 50% v/v, i-propanol 30% v/v, acetonitrile), lower temperature, low salt concentration, increased *pH*, chaotropic salts, etc., have been used successfully to desorb the proteins from hydrophobic adsorbents. If hydrophobic adsorption was facilitated by high concentration of a salt, lower salt concentration may be employed. *Chaotropic salts*,[6] such as iodide, lithium bromide, thiocyanate, at high concentrations, are often successful in facilitating desorption since they are less polar and bind water loosely. So water molecules hydrate protein surfaces. On the other hand, a *kosmotropic salt* such as $(NH_4)_2SO_4$, has ions having high polarity, which bind water strongly; therefore they will induce the exclusion of water molecules from the protein surfaces, which will lead to protein precipitation or adsorption on hydrophobic adsorbents. Scopes (1994) has provided a more detailed account of such techniques for eluting proteins in hydrophobic interaction chromatography.

A recent analysis (Marmur, 2000) of the solubility of nonpolar solutes in water suggests that such solutes prefer to aggregate rather than to be dispersed as single molecules. Such aggregation leads to a sufficient loss in entropy, as opposed to a corresponding loss of entropy via ordering of water molecules around hydrophobic patches/chains. This analysis is also valid for the dissolution of water in nonpolar solvents. Therefore, molecular aggregation is due to the more general "solvophobic" effect rather than a "hydrophobic" effect.

4.1.10 Gas–solid particle–liquid system in mineral flotation

In Section 3.3.8, the principle of particle separation via flotation in a gas–liquid system was briefly identified. Those particles whose surfaces are hydrophobic, and to which gas/air bubbles can be attached, will float to the surface of an aqueous suspension due to reduced density. Those particles whose surfaces are hydrophilic and wetted by the liquid (commonly, it is water) cannot become attached to gas/air bubbles and will not therefore float to the surface. Here, following Fuerstenau and Herrera-Urbina (1989), we will briefly illustrate systems where such separations are achieved by converting particular mineral surfaces to a hydrophobic type via the adsorption of surfactants. The system of interest contains two minerals, hematite (Fe_2O_3) and quartz, which can be separated by froth flotation; these two minerals are usual constituents of iron ores.

The primary step involves the use of surface-active collectors, which displace water molecules from the surface of the mineral; if the mineral surface is charged, appropriate long-chain ionic surfactants, which can act as a counterion to the charged surface, are used. The long hydrophobic chains of such surfactants present a hydrophobic surface. Surface-active collectors, which are chemisorbed at the mineral–water interface, can also create a hydrophobic surface. An additional approach employs substances called *depressants*, which inhibit the attachment of gas bubbles to the minerals that should not undergo flotation.

At *pH* 6–7, quartz particles have a negative charge on their surface, whereas hematite particles essentially have no charge. The addition of an alkylamine salt, which is adsorbed on the quartz particle surface, leads to the flotation of quartz particles. Commercially employed alternative separation strategies include the use of starch as a depressant for hematite; simultaneously, calcium ions are used to activate surfaces of quartz particles, which are then floated using sodium oleate at *pH* 11–12 (Fuerstenau and Herrera-Urbina, 1989). Another industrial approach employs flotation of hematite. At *pH* 2–4, the charges on hematite surfaces are positive; an anionic alkyl sulfonate collector is employed to float hematite. Since the quartz surface is negatively charged at these *pH*s, the anionic collector does not become adsorbed on the quartz surface.

The key to any approach is knowing the electrical charge and potential on the surface of the mineral particle in an aqueous suspension. The following four phenomena contribute to the development of the surface charge: specific adsorption of surface-active ions; preferential dissolution of lattice ions; dissociative adsorption of water molecules; isomorphous substitution of ions comprising the mineral lattice (Fuerstenau and Herrera-Urbina, 1989).

4.2 Equilibrium separation in a single phase in an external force field

Consider a liquid solution or a gaseous mixture in a closed vessel subjected to an external force field. We would like to know the change in composition in the solution or in the gaseous mixture in the direction of the force field and calculate the value of the separation achieved between any two locations in the vessel. The external force fields illustrated are centrifugal, electrical and gravitational. This section also explores the separation achieved in the closed vessel subjected to an external force field when a property gradient, e.g. density gradient, *pH* gradient, etc., exists in the single-phase system. Density gradient is important

[6] Chaotropic anions create an increase in entropy resulting from the disruption of water structure around these ions in solution. This facilitates solubilization of hydrophobic proteins by water since many more water molecules are available.

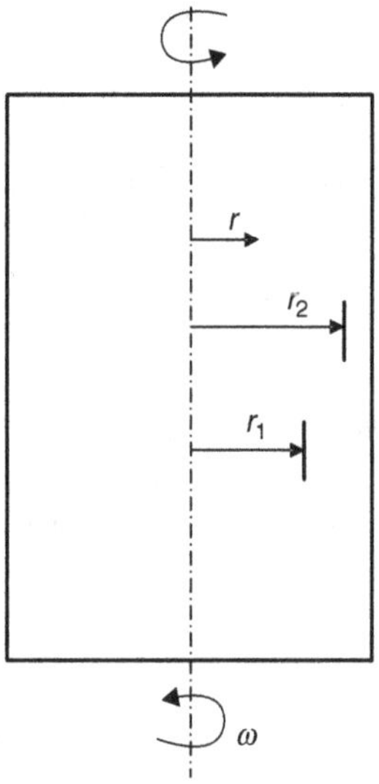

Figure 4.2.1. Closed hollow cylinder rotating at ω radian/s: closed gas centrifuge.

when a centrifugal or gravitational force field is present; a *pH* gradient is useful in the presence of an electrical force field. Finally, the role of a *thermal gradient* in separation is briefly identified in the absence of any external force field.

4.2.1 Centrifugal force field

Three types of mixtures/solutions are considered here: a gaseous mixture, liquid solutions of low molecular weight species and high molecular weight species. We begin with a gas mixture in a centrifuge; gas centrifuges have been used at large scale for the separation of uranium isotopes present in a gaseous mixture of $U^{235}F_6$ and $U^{238}F_6$. Macromolecules and biological particles in solutions are routinely separated and studied in the laboratory using ultracentrifuges. Centrifuges are also extensively used in the chemical process industry to remove particulate matter from a liquid/gas (see Section 7.3.2).

4.2.1.1 *Gas separation*

A simple closed gas centrifuge (Figure 4.2.1) is a rotating hollow cylinder containing a gas or a gas mixture introduced into the stationary hollow cylinder at time $t = 0$. Within a short time,[7] the gas mixture as well as the cylinder are assumed to start rotating at a constant angular velocity ω radian/s. Assume isothermal conditions. To determine the conditions existing at equilibrium, employ the criterion (3.3.25) for equilibrium in the centrifugal field for species *i*:

$$d\mu_i|_T = -d\phi_i^{ext}. \tag{4.2.1}$$

For the centrifugal field, it is known that

$$F_i^{ext}\Big|_{\text{mole of } i} \boldsymbol{r} = M_i\omega^2 r\boldsymbol{r} = -\frac{d\phi_i^{ext}}{dr}\boldsymbol{r}.$$

Further, using relation (3.3.9),

$$d\mu_i\Big|_T = RT\ d\ln\hat{f}_{ij} = M_i\omega^2 r\ dr, \tag{4.2.2}$$

where subscript j refers to the radial location r under consideration. Integrating between two locations, $r = r_1$ ($j = 1$) and $r = r_2$ ($j = 2$), where $r_2 > r_1$, leads to

$$\ln(\hat{f}_{i2}/\hat{f}_{i1}) = (M_i/RT)\frac{\omega^2}{2}(r_2^2 - r_1^2). \tag{4.2.3}$$

If the gas mixture behaves as an ideal gas, then

$$\hat{f}_{i1} = x_{i1}P_1 \quad \text{and} \quad \hat{f}_{i2} = x_{i2}P_2, \tag{4.2.4}$$

where P_1 and P_2 are the total pressures at radial locations $r = r_1$ and $r = r_2$. Result (4.2.3) can now be expressed as

$$\ln\left(\frac{x_{i2}P_2}{x_{i1}P_1}\right) = (M_i/RT)\frac{\omega^2}{2}(r_2^2 - r_1^2). \tag{4.2.5}$$

A general relation between the gas mixture at any location j ($r = r$) and that at $r = r_1$ is

$$\ln\left(\frac{x_{ij}\,P_j|_{r=r}}{x_{i1}P_1}\right) = (M_i/RT)\frac{\omega^2}{2}(r^2 - r_1^2). \tag{4.2.6}$$

The equilibrium separation factor α_{12} between species 1 and 2 in a binary gas mixture between two locations $j = 1$ ($r = r_1$) and $j = 2$ ($r = r_2$) is easily obtained for ideal gas behavior:

$$\frac{x_{11}x_{22}}{x_{12}x_{21}} = \alpha_{12} = \exp\left[\frac{(M_2 - M_1)}{2RT}\omega^2(r_2^2 - r_1^2)\right]. \tag{4.2.7}$$

Two more results are needed to provide a reasonably complete picture of separation in a closed gas centrifuge. The first provides the mole fraction profile $x_{ij}(r)$ of species i at radius r (location j) along the centrifuge radius; the second describes the variation in total pressure. Using relation (4.2.7) for two locations ($r = r$ and $r_1 = 0$), and species $i = 1,2$, one obtains

$$\frac{x_{1j}(r)}{x_{2j}(r)} = \frac{x_{1j}(0)}{x_{2j}(0)}\exp\left[-\frac{(M_2 - M_1)}{2RT}\omega^2 r^2\right]. \tag{4.2.8}$$

Algebraic manipulation leads to

$$x_{1j}(r) = \frac{x_{1j}(0)\exp(Ar^2/2)}{x_{1j}(0)[\exp(Ar^2/2)\ -1] + 1}, \tag{4.2.9}$$

where

$$A = \left((M_1 - M_2)\omega^2/RT\right). \tag{4.2.10}$$

[7] Auvil and Wilkinson (1976) provide estimates of this time.

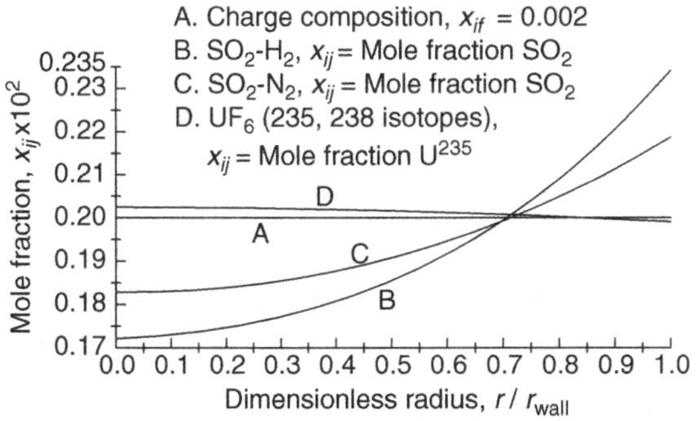

Figure 4.2.2. Mole fraction profiles developed in a simple centrifuge with a peripheral speed of 150 m/s for the gas pairs: sulfur dioxide-hydrogen, sulfur dioxide-nitrogen and UF$_6$ (235, 238 isotopes) at 20 °C. Reprinted, with permission, from Auvil and Wilkinson, AIChE J., **22**, *564 (1976). Copyright © [1976] American Institute of Chemical Engineers (AIChE).*

From relation (4.2.7), we get, for $r_1 = 0$ and $r_2 = r$,

$$\alpha_{12} = \exp\left[-\frac{Ar^2}{2}\right]. \quad (4.2.11)$$

Substituting this into the expression (4.2.9) for $x_{1j}(r)$, we obtain

$$x_{1j}(r) = \frac{x_{1j}(0)}{x_{1j}(0) + \alpha_{12}(1 - x_{1j}(0))}. \quad (4.2.12)$$

By our convention, species 1 is lighter than species 2, making $(M_2 - M_1)$ positive. Therefore α_{12} is greater than 1 as long as $r_2 > r_1$; it also follows that $x_{1j}(r) < x_{1j}(0)$. Lighter species concentrate near the center of the centrifuge and heavier species concentrate near the wall ($r = r_{wall}$). Further, the extent of this composition change is related to the molecular weight difference $(M_2 - M_1)$ since α_{12} increases exponentially with $(M_2 - M_1)$. The extent of separation, ξ, for such a centrifuge is calculated in Problem 1.4.3, and the result has been provided there.

The behavior of the mole fraction profile in a gas centrifuge along the centrifuge radius is illustrated in Figure 4.2.2 for a number of binary gas mixtures (Auvil and Wilkinson, 1976). The simple gas centrifuge rotates with a peripheral speed ($= \omega r_{wall}$) of 150 m/s at 20 °C; the gas initially fed into the centrifuge contained a mole fraction of 0.002 of the species shown. Consider systems B and C, both containing SO_2. The mole fraction of SO_2 ($=1 - x_{1j}(r)$) changes much more rapidly with r in the case of system B since $(M_2 - M_1)$ is much larger for a SO_2-H_2 than for a SO_2-N_2 system (system C). Similarly, the mole fraction variation with radius for $U^{235}F_6$ and $U^{238}F_6$ is much less since $(M_2 - M_1)$ is only 3.

It is also of interest to know the magnitudes of α_{12} achievable in such gas centrifuges. Consider a gas centrifuge of radius 4 cm rotating at an extremely high angular velocity (ω) of 4000 radian/s at 20 °C. The value of α_{12} for a mixture of uranium isotopes ($U^{235}F_6$ and $U^{238}F_6$) will be only 1.0157 in such a centrifuge, whereas that for a SO_2-N_2 mixture is 1.207. However, uranium isotopes are often separated by high-speed centrifuges since other isotope separation methods yield usually a lower α_{12}. A gas centrifuge is never used for SO_2-N_2 separation, however, since other methods provide a far larger separation factor. The mechanical problems encountered at these high speeds are also not inconsiderable.

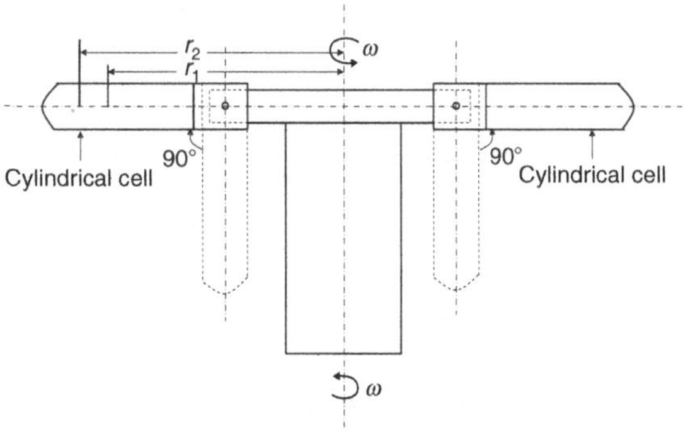

Figure 4.2.3. Swing-bucket rotor with centrifugal tubes which swing 90° upward during centrifugation.

We will now provide the second result, namely the profile of the total gas pressure along the centrifuge radius. The partial pressure of any species is indicated by relation (4.2.6); the total pressure is obtained by summing over all species at any location r (region j):

$$P_j\Big|_{r=r} = P(r) = \sum_{i=1}^{n} P(0)x_{i1}(0)\exp\left(-\frac{M_i\omega^2r^2}{2RT}\right), \quad (4.2.13)$$

where region $j = 1$ corresponds to $r = 0$, there are n number of species in the gas mixture and

$$\sum_{i=1}^{n} x_{i1}P_1 = P_1 = P(r = 0) = P(0). \quad (4.2.14)$$

4.2.1.2 *Liquid separation*

Liquid solutions subjected to a centrifugal field also undergo separation. Such separations, primarily employed in laboratories, are achieved by placing the solution in a small cylindrical cell mounted at the tip of an extremely high-speed rotor (Figure 4.2.3) in an ultracentrifuge. When equilibrium is achieved in the centrifugal force field, criterion (3.3.25) can be applied to species i:

$$d\mu_{ij}\Big|_T = -d\phi_{ij}^{ext} = M_i\omega^2 r\, dr.$$

For the solution, $d\mu_{ij}|_T$ is expressed as

$$d\mu_{ij}\Big|_T = RT\, d\ln a_{ij} + \overline{V}_{ij}\, dP = M_i\,\omega^2\, r\, dr.$$

Using the assumption that the pressure gradient may be determined from the rotation of the solvent only, we have obtained the following result (see relation (3.1.52)):

$$-\overline{V}_{ij}\,dP + M_i\omega^2 r\ dr = M_i\omega^2 r\left[1-\left(\frac{\overline{V}_{ij}}{M_i}\right)\left(\frac{M_s}{\overline{V}_s}\right)\right].$$

Therefore,

$$RT\ d\ln a_{ij} = M_i\left[1-\left(\frac{\overline{V}_{ij}}{M_i}\right)\left(\frac{M_s}{\overline{V}_s}\right)\right]\omega^2 r\ dr.$$

Integrating between two radial locations r_1 and r_2 ($> r_1$), we obtain

$$\frac{a_{i2}}{a_{i1}} = \exp\left\{\frac{M_i}{2RT}\left[1-\left(\frac{\overline{V}_{ij}}{M_i}\right)\left(\frac{M_s}{\overline{V}_s}\right)\right]\omega^2(r_2^2-r_1^2)\right\}. \quad (4.2.15a)$$

For an ideal solution, $a_{ij} = x_{ij}$; therefore $(a_{i2}/a_{i1}) = (x_{i2}/x_{i1})$.

Further, for dilute solutions, $(x_{i2}/x_{i1}) \cong (c_{i2}/c_{i1})$. This last result allows us to relate the concentration of a species at two radial locations at equilibrium:

$$\frac{C_{i2}}{C_{i1}} = \exp\left\{\frac{M_i}{2RT}\left[1-\left(\frac{\overline{V}_{ij}}{M_i}\right)\left(\frac{M_s}{\overline{V}_s}\right)\right]\omega^2(r_2^2-r_1^2)\right\}. \quad (4.2.15b)$$

Consider a dilute aqueous solution of ovalbumin ($M_i \sim 45\,000$) at 24 °C; the density of ovalbumin is 1.34 g/cm^3 ($\cong M_i/\overline{V}_{ij}$). For a dilute solution, $(M_s/\overline{V}_s) \cong 1\text{g/cm}^3$. From the expressions given above for (a_{i2}/a_{i1}) and (C_{i2}/C_{i1}), it is clear that $a_{i2} > a_{i1}$, $C_{i2} > C_{i1}$ and $x_{i2} > x_{i1}$. At equilibrium, there develops a continuous solute concentration profile along the cell length. Solute concentrates at the largest radial location in the cell at the cell wall, and the solution nearer the center becomes depleted in ovalbumin. Such a profile does not change with time at equilibrium. At any radial location, the concentration gradient and the pressure gradient (pressure increases also as the radius increases in the centrifuge) force ovalbumin molecules toward the center ($r = 0$) to balance the centrifugal force driving the molecule away from the center. There is no net radial force on ovalbumin molecules at any radial location once equilibrium separation profile is established.

Ultracentrifuges used in laboratories for solutions employ very high speeds (as much as 60 000 rpm). Measurement of the concentration profile (e.g. (4.2.15b)) is achieved optically; the refractive index gradient is measured (it being proportional to the concentration gradient). For an introduction, see Hsu (1981).

The difference in concentration between $r = 0$ and $r = r_{wall}$ for species like ovalbumin in an ultracentrifuge is substantial. If, however, we had subjected a solution of two low molecular weight liquids, e.g. water and ethanol, to the centrifugal field in an ultracentrifuge, the difference $(C_{i2}\ (r = r_{wall}) - C_{i1}(r = 0))$ would be very small. For the same values of ω, T, r_2, r_1, the change in concentration from one radial location to another is much smaller due to the small value of M_i, since the magnitude of $\{1-(\overline{V}_{ij}/M_i)(M_s/\overline{V}_{sj})\}$ remains comparable. In fact, the higher the molecular weight, the larger the separation. Some of the more common applications of ultracentrifuges, therefore, involve large macromolecules, viruses, cell fragments for preparative scale separations, etc. In centrifuges used for preparative applications, the liquid solution is mounted in the rotor. After centrifugation, a pellet is obtained near the radius $r = r_{wall}$, containing the particles, large macromolecules, cells, etc., separated from the clear supernatant solvent.

We will now focus on the velocity of such types of cells/particles/macromolecules during centrifugal separation, the time required to form the pellet and the relative ease with which different biologically relevant macromolecules/cells, etc., may be pelleted. The analysis of the behavior of larger centrifuges that are open with flow in and out are considered in Section 7.3.2 as well as in Section 6.3.1.3.

For a spherical particle of radius r_p, density ρ_p and mass m_p in a liquid medium of density ρ_t, the net external force acting on the particle (from equation (3.1.59)) in the radial direction (without any gravitational force in that direction) is $m_p r\omega^2\ (1-(\rho_t/\rho_p))\mathbf{r}$ if the angular velocity of the particles is ω radian/s. The drag force experienced by this particle from the liquid resisting this radial particle motion according to Stokes' law is $6\pi\mu r_p U_{pr}$ where U_{pr}, is the radially outward particle velocity. Reformulating the basic equation of motion (3.1.60) for the particle in the r-direction, we have

$$m_p\frac{d^2r}{dt^2}\mathbf{r} = m_p\frac{dU_{pr}}{dt}\mathbf{r} = m_p r\,\omega^2\,(1-(\rho_t/\rho_p))\mathbf{r} - 6\pi\mu r_p U_{pr}\mathbf{r}. \quad (4.2.16a)$$

At steady state the particle radial acceleration is reduced to zero and the terminal radial particle velocity achieved, U_{prt}, is

$$\left.\frac{dr}{dt}\right|_{\text{terminal velocity}} = U_{prt} = \frac{4\pi r_p^3\rho_p r\omega^2(1-(\rho_t/\rho_p))}{3\ 6\pi\mu r_p} = \frac{2}{9}\frac{r_p^2(\rho_p-\rho_t)}{\mu}\omega^2 r;$$

$$U_{prt} = s_p\omega^2 r, \quad (4.2.16b)$$

where s_p is called the *sedimentation coefficient* of the particle of size r_p (correspondingly s_{p1} and s_{p2} for particles of sizes r_{p1} and r_{p2}). If $\omega^2 r$ may be considered as the centrifugal force per unit effective mass[8] (i.e. $m_p(1-(\rho_t/\rho_p)) = 1$) of the particle under a given centrifuge condition, then s_p is an effective indicator of the radial sedimentation particle terminal velocity per unit of centrifugal force. Its units are svedbergs, 1 svedberg having the value 10^{-13} second (Svedberg and Pederson, 1940). The quantity $\omega^2 r$ is often expressed as so many times (e.g. n times) gravity, ng, where g is the acceleration due to gravity.

[8]Magnitude of the centrifugal field strength.

Ivory *et al.* (1995) have collated in svedberg units the sedimentation coefficients of a variety of biologically relevant particles in aqueous systems: yeast and red blood cells ($10^6 > s_p > 10^5$); mitochondria and bacteria ($10^5 > s_p > 10^4$); viruses and phages ($10^3 > s_p > 10^2$); proteins (e.g. serum albumin) ($s_p < 10$). The higher the value of s_p, the easier it is for the particles to settle and the shorter the particle settling time in a given centrifuge for a given angular velocity, ω. Consider now the time required for such a particle to settle in the centrifuge, i.e. reach the centrifuge wall, r_{wall}. Employ equation (4.2.16b); integrate between the starting radius r_{start} anywhere in the centrifuge and r_{wall}:

$$\left.\frac{dr}{dt}\right|_{\text{settling velocity}} = s_p\omega^2 r \Rightarrow \int_{r_{start}}^{r_{wall}} \frac{dr}{r} = s_p\omega^2 \int_{o}^{t_{wall}} dt = s_p\omega^2 t_{wall} = \ln\frac{r_{wall}}{r_{start}};$$

$$t_{wall} = \frac{1}{s_p\omega^2}\ln\frac{r_{wall}}{r_{start}} = \frac{r}{s_p\omega^2 r}\ln\frac{r_{wall}}{r_{start}}. \qquad (4.2.16c)$$

Sometimes, this time is replaced by t_{pellet}, the time needed to form the pellet in an ultracentrifuge.

Example 4.2.1 Harvesting time for a virus in an ultracentrifuge. Lotfian *et al.* (2003) have investigated the centrifugal recovery of a disabled herpes simplex virus type-1 (HSV-1) vector, potentially useful for gene therapy, from a culture medium containing recombinant cells of a suitable kind. The extracellular virus is obtained first in the supernatant of the broth centrifuged at a low value of $\omega^2 r = 1600$ g. This supernatant is then subjected to ultracentrifugation using a swing-bucket rotor and 15 ml volume centrifuge tubes. Consider operation at $\omega^2 r = 26000$g. The radius of the base of the centrifuge tube from the axis of rotation is 13.7 cm.

(a) Calculate the settling velocity of the virus at the following two locations in the ultracentrifuge tube: 15 mm and 50 mm away from the base of the tube.
(b) Estimate the time needed for the virus to deposit if the starting locations are identified in part (a).

You are given the s_p for virus $= 10^2$ svedberg.

Solution (a) The settling velocity of a particle having a sedimentation coefficient s_p at a radial location r is given by

$$U_{prt} = s_p\,\omega^2 r \ \text{cm/s};$$
$$s_p = 10^2 \text{svedberg} = 10^2 \times 10^{-13} = 10^{-11}\text{s};$$
$$\omega^2 r = 26\,000 \text{ g} = 26\,000 \times 980\ \frac{\text{cm}}{\text{s}^2};$$

Therefore,

$$U_{prt} = 10^{-11}\text{s} \times 26\,000 \times 980\ \frac{\text{cm}}{\text{s}^2} = 2.6 \times 0.98 \times 10^{-4}\ \frac{\text{cm}}{\text{s}}$$
$$= \frac{2.6 \times 0.98 \times 10^{-4}}{10^2}\ \frac{\text{m}}{\text{s}}$$
$$\cong 2.55 \times 10^{-6}\ \text{m/s}.$$

If the value of ω is known, then U_{prt} may be calculated for each individual viral location.

(b) To estimate the time needed for the virus to settle, we employ equation (4.2.16c):

$$t_{wall} = \frac{r}{s_p\ \omega^2 r}\ \ln\frac{r_{wall}}{r_{start}}.$$

Since ω is not explicitly given here, we will employ $\omega^2 r = 26\,000$ g and $r = (r_{wall} + r_{start})/2$.

For particles starting at 15 mm from the base, $r_{start} = 137 - 15 = 122$ mm. Therefore $r = 129.5$ mm and

$$t_{wall} = \frac{12.95 \text{ cm}}{10^2 \times 10^{-13}\ \text{s} \times 26\,000 \times 980\ \frac{\text{cm}}{\text{s}^2}} \ln\frac{137}{122}$$
$$= \frac{12.95 \times 10^4}{2.6 \times 0.98} \times 0.1145\ \text{s} = \frac{12.95 \times 1145}{2.6 \times 0.98}\text{s}$$
$$= 5734\ \text{s} = 95.5\ \text{minutes}.$$

For particles starting at 50 mm away from the base of the tube, $r_{start} = 137 - 50 = 87$ mm. Therefore $r = 112$ mm (averaged) and

$$t_{wall} = \frac{11.2 \text{ cm}}{10^{-11}\text{s} \times 26\,000 \times 980\ \frac{\text{cm}}{\text{s}^2}} \ln\frac{137}{87}$$
$$= \frac{11.2 \times 10^4 \times 2.303}{2.6 \times 0.98} \times 0.197\ \text{s}$$
$$= 19941\ \text{s} = 332\ \text{minutes}.$$

Since the sedimentation coefficient s_p for proteins is around 1-10 svedberg, the settling time needed for proteins in Example 4.2.1 would be an order of magnitude larger. Ultracentrifuges are therefore not used for protein separation.

So far we have focused on liquid separation or particle separation where the solvent density was essentially assumed constant throughout the centrifuge. This condition did not pose any problem for separating a given cell or macromolecular species from the solvent; the cells or macromolecular species are obtained as a pellet at the centrifuge tube base. A different strategy has to be pursued if it is necessary to separate different cells or macromolecular species from one another. We focus on that in Section 4.2.1.3.

The mathematical expressions used so far for liquid systems subjected to a centrifugal field have assumed that the solvent density $(M_s/\overline{V}_{sj})$ was essentially constant across the centrifuge. In general, it will vary in the radial direction due to the variation of pressure which influences $\overline{V}_{sj}$. It is particularly true when there is at least one solute species present in the system at a substantial concentration. Under such a condition, we use expression (3.1.58) for centrifugal equilibrium and expression (3.3.17) for $d\mu_{ij}|_{T,P_j}$ at any location. This leads to

$$\left.d\mu_{ij}\right|_{T,P_j} = M_i\left(1 - \rho_{tj}\frac{\overline{V}_{ij}}{M_i}\right)\omega^2 r\ dr = \sum_{k=1}^{n-1}\left(\frac{\partial\mu_{ij}}{\partial x_{kj}}\right)_{T,P_j,x_l,l\neq k} dx_{kj} \qquad (4.2.17)$$

when there are a number of solute species present at substantial concentrations. Consider a system of solvent

($i = s$) and two solutes $i = 1$ and $i = 2$. The particular forms of the above relations are as follows.

$$i=1: (M_1 - \rho_{tj}\overline{V}_{1j})\omega^2 r\mathrm{d}r = \left.\frac{\partial\mu_{1j}}{\partial x_{1j}}\right|_{T,P_j,x_2,x_s}\mathrm{d}x_{1j} + \left.\frac{\partial\mu_{1j}}{\partial x_{2j}}\right|_{T,P_j,x_1,x_s}\mathrm{d}x_{2j};$$

$$i=2: (M_2 - \rho_{tj}\overline{V}_{2j})\omega^2 r\mathrm{d}r = \left.\frac{\partial\mu_{2j}}{\partial x_{1j}}\right|_{T,P_j,x_2,x_s}\mathrm{d}x_{1j} + \left.\frac{\partial\mu_{1j}}{\partial x_{2j}}\right|_{T,P_j,x_1,x_s}\mathrm{d}x_{2j}.$$

(4.2.18)

The use of such equations will be illustrated in the following for a special case.

4.2.1.3 *Isopycnic sedimentation*

Just as a particle denser than a liquid falls through the liquid in the direction of the gravitational force, similarly a particle or species of density ρ_i ($= M_i/\overline{V}_{ij}$) will be thrown radially outward in the direction of the centrifugal force if $\rho_i > \rho_t$, the solution density. This phenomenon is identified in analytical/preparative chemistry as *sedimentation*. When $\rho_t < \rho_i$, the centrifugal force is radially outward. However, if $\rho_t > \rho_i$, the centrifugal force on species i will be radially inward. When $\rho_t = \rho_i$, species i or particle i does not experience any radial centrifugal force (outward or inward); therefore its radial location does not change in the centrifuge with time; it no longer undergoes sedimentation.

These phenomena provide the basis for separating a multicomponent mixture in a liquid centrifuge provided a radial density gradient is created in the centrifuge to start with. Consider Figure 4.2.4A illustrating a density gradient established using a low molecular weight salt such as CsCl in water in the centrifuge. Into such a solution, where ρ_t is a function of r, is introduced a small amount of a mixture of different macromolecular species. Assume that the small amount of sample will not affect the density gradient established in the CsCl solution. Each macromolecular species i will establish its own centrifugal equilibrium and will move to the radial location r, where

$$\rho_{tj}(r) = \frac{M_i}{\overline{V}_{ij}}; \qquad (4.2.19)$$

i.e. the solution density is equal to the species density (which is also the inverse of the partial specific volume of species i). As long as the value of $(M_i/\overline{V}_{ij})$ is different for each ith species, each ith species will be concentrated at different radial locations - provided the range of $\rho_{tj}(r)$ in the centrifuge accommodates the range of the effective macromolecular density $(M_i/\overline{V}_{ij})$.

The separation of different species in a centrifugal force field (the imposed primary force field) having a radial density gradient (secondary physical property based field) is identified as *isopycnic sedimentation* or *density-gradient sedimentation* (Giddings and Dahlgren, 1971). Each species

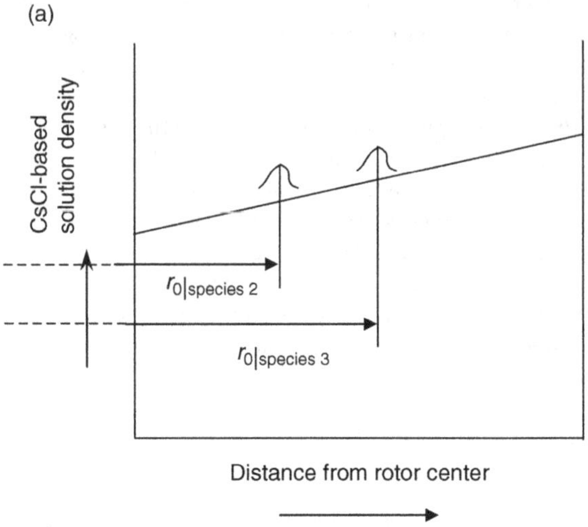

Figure 4.2.4A. Equilibrium density gradient in a centrifuge due to CsCl solution.

forms a small band around a center point where relation (4.2.19) is satisfied. Determinations of the width of such a band and the separation between two neighboring bands are needed to determine the usefulness of isopycnic sedimentation.

Following Meselson *et al.* (1957), we consider the separation of different macromolecules in an aqueous solution of CsCl. For the purpose of analysis, one macromolecular species, $i = 2$, only needs to be considered, where $i = 1$ represents CsCl and $i = s$ is water. To simplify matters, the macromolecular species may be represented as a macromolecular electrolyte P $(\mathrm{Cs})_n$, where P is the macromolecular backbone (e.g. DNA (deoxyribonucleic acid)). We will soon find out the utility of such an assumption.

The centrifugal equilibrium of the macromolecular species $i = 2$ in an aqueous solution of CsCl ($i = 1$) is governed by equation (4.2.18) for $i = 2$:

$$(M_2 - \rho_{tj}\overline{V}_{2j})\omega^2 r\ \mathrm{d}r = \left.\frac{\partial\mu_{2j}}{\partial x_{1j}}\right|_{T,P_j,x_2,x_s}\mathrm{d}x_{1j} + \left.\frac{\partial\mu_{2j}}{\partial x_{2j}}\right|_{T,P_j,x_1,x_s}\mathrm{d}x_{2j}.$$

The radial dependence of the CsCl solution density around the location $r = r_0$, where

$$\rho_{tj}(r_0) = M_2/\overline{V}_{2j}, \qquad (4.2.20)$$

may be described by a Taylor series expression:

$$\rho_{tj}(r) = \rho_{tj}(r_0) + \left.\frac{\mathrm{d}\rho_{tj}}{\mathrm{d}r}\right|_{r_0}(r - r_0). \qquad (4.2.21)$$

There is an implicit assumption here that the macromolecular band has a narrow width around the location $r = r_0$. For ideal solutions of species 1 and 2, we have

$$\left.\frac{\partial \mu_{2j}}{\partial x_{2j}}\right|_{T,P_j,x_1,x_s} \mathrm{d}x_{2j} \cong RT\ \mathrm{d}\ln C_{2j}.$$

Further, we may define a dimensionless factor b, the binding coefficient (Hsu, 1981), to be

$$b = \left(\frac{\partial x_1}{\partial x_2}\right)_{P,T,\mu_1}. \quad (4.2.22)$$

Additionally, by Maxwell's relation for any extensive quantity, e.g. Gibbs free energy G_{tj} (Pitzer and Brewer, 1961),

$$\left.\frac{\partial \mu_{2j}}{\partial x_{1j}}\right|_{T,P_j,x_2,x_s} = \left.\frac{\partial \mu_{1j}}{\partial x_{2j}}\right|_{T,P_j,x_1,x_s} = \left.\frac{\partial G_{tj}}{\partial x_{1j}\partial x_{2j}}\right|_{T,P_j,x_s}, \quad (4.2.23)$$

which leads to

$$\left.\frac{\partial \mu_{2j}}{\partial x_{1j}}\right|_{T,P_j,x_2,x_s} \mathrm{d}x_{1j} = \left.\frac{\partial \mu_{1j}}{\partial x_{2j}}\right|_{T,P_j,x_1,x_s} \mathrm{d}x_{1j} = bRT\ \mathrm{d}\ln C_{1j} \quad (4.2.24)$$

for an ideal solution. Here, b is the effective number of cesium counterions which are to be moved along with the charged polymer molecule $P(Cs)_{n-b}$ to maintain electoneutrality (so far we have gone about this as if we were dealing with nonelectrolytes). The governing equation now becomes, for ideal solutions,

$$\left\{M_2 - \rho_{tj}(r_0)\overline{V}_{2j} - \overline{V}_{2j}\left.\frac{\mathrm{d}\rho_{tj}}{\mathrm{d}r}\right|_{r_0}(r-r_0)\right\}\omega^2 r\ \mathrm{d}r = bRT\ \mathrm{d}\ln C_{1j} + RT\ \mathrm{d}\ln C_{2j}. \quad (4.2.25)$$

Just as the radial dependence of the CsCl solution density around $r = r_0$ was described by a Taylor series expansion (4.2.21), similarly the C_{1j} profile around $r = r_0$ may be described by

$$C_{1j}(r) = C_{1j}(r_0) + \left.\frac{\mathrm{d}C_{1j}}{\mathrm{d}r}\right|_{r_0}(r-r_0). \quad (4.2.26)$$

Introducing this into equation (4.2.25), we get

$$\left[M_2 - \rho_{tj}(r_0)\overline{V}_{2j}\right]\omega^2 r\,\mathrm{d}r - \overline{V}_{2j}\left.\frac{\mathrm{d}\rho_{tj}}{\mathrm{d}r}\right|_{r_0}\omega^2 r(r-r_0)\ \mathrm{d}r$$
$$= bRT\mathrm{d}\ \ln\left[C_{1j}(r_0) + \left.\frac{\mathrm{d}C_{1j}}{\mathrm{d}r}\right|_{r_0}(r-r_0)\right] + RT\mathrm{d}\ \ln C_{2j}.$$

We integrate from r_0 to r to obtain

$$\ln\frac{C_{2j}(r)}{C_{2j}(r_0)} = -b\ \ln\left[1 + \frac{(\mathrm{d}C_{1j}/\mathrm{d}r)_{r_0}}{C_{1j}(r_0)}(r-r_0)\right] + \left[M_2 - \rho_{tj}(r_0)\overline{V}_{2j}\right]\omega^2\frac{(r^2-r_0^2)}{2RT}$$
$$-\overline{V}_{2j}\left.\frac{\mathrm{d}\rho_{tj}}{\mathrm{d}r}\right|_{r_0}\omega^2\left(\frac{r-r_0}{RT}\right)^2\left(\frac{r-r_0}{3}+\frac{r_0}{2}\right). \quad (4.2.27)$$

Since the variation of CsCl concentration over the bandwidth of polymeric species 2 is small,

$$\ln\left[1 + \frac{(\mathrm{d}C_{1j}/\mathrm{d}r)_{r_0}}{C_{1j}(r_0)}(r-r_0)\right] \cong (\mathrm{d}C_{1j}/\mathrm{d}r)_{r_0}\frac{(r-r_0)}{C_{1j}(r_0)}.$$

Further, the bandwidth is small: $|r - r_0| << r_0$.

The concentration profile of species 2 may now be expressed as

$$C_{2j}(r) = C_{2j}(r_0)\exp\left[-\frac{\overline{V}_{2j}\omega^2 r_0}{2RT}\left.\frac{\mathrm{d}\rho_{tj}}{\mathrm{d}r}\right|_{r_0}(r-r_0)^2 - \frac{b(r-r_0)}{C_{1j}(r_0)}\left.\frac{\mathrm{d}C_{1j}}{\mathrm{d}r}\right|_{r_0} + \frac{(r-r_0)}{RT}\left(1-\frac{\rho_{tj}(r_0)\overline{V}_{2j}}{M_2}\right)M_2\omega^2 r_0\right]$$

or

$$C_{2j}(r) = C_{2j}(r_0)\exp\left[-\frac{(r-r_0)^2}{2\sigma_2^2} - \frac{A(r-r_0)}{\sigma_2^2} - \frac{A^2}{2\sigma_2^2} + \frac{A^2}{2\sigma_2^2}\right],$$

where

$$\sigma_2^2 = \frac{RT}{\overline{V}_{2j}\ \omega^2\ r_0\left.\frac{\mathrm{d}\rho_{tj}}{\mathrm{d}r}\right|_{r_0}}, \qquad \overline{V}_{2j} = \frac{M_2}{\rho_{2j}} \quad (4.2.28)$$

and

$$A = \frac{bRT\left.\frac{\mathrm{d}C_{1j}}{\mathrm{d}r}\right|_{r_0}}{\overline{V}_{2j}\left.\frac{\mathrm{d}\rho_{tj}}{\mathrm{d}r}\right|_{r_0}\omega^2 r_0 C_{1j}(r_0)} - \frac{M_2\left(1-\frac{\rho_{tj}(r_0)\overline{V}_{2j}}{M_2}\right)}{\overline{V}_{2j}\left.\frac{\mathrm{d}\rho_{tj}}{\mathrm{d}r}\right|_{r_0}}. \quad (4.2.29)$$

Choosing $A = 0$, we get a simple Gaussian profile of the macromolecular solute 2:

$$C_{2j}(r) = C_{2j}(r_0)\exp\left[-\frac{(r-r_0)^2}{2\sigma_2^2}\right] \quad (4.2.30)$$

in a band (created by diffusion and balanced by centrifugal focusing) centered around r_0 and having a standard deviation σ_2; the standard deviation is inversely proportional to the square root of the molecular weight of the macromolecule.

The choice of $A = 0$ implies

$$\rho_{tj}(r_0) = \frac{M_2}{\overline{V}_{2j}}\left[1 - \frac{bRT\left.\frac{\mathrm{d}C_{1j}}{\mathrm{d}r}\right|_{r_0}}{M_2\omega^2 r_0 C_{1j}(r_0)}\right]. \quad (4.2.31)$$

For the system of DNA ($M_2 \cong 14 \times 10^6$) in an aqueous CsCl solution, it has been suggested that, if the quantity within the brackets is replaced by 1, the error is less than 10% (Meselson *et al.*, 1957); the assumption $A = 0$ therefore does not lead to much error in depicting $C_{2j}(r)$ as a Gaussian profile.

An illustration of such a Gaussian profile is provided in Figure 4.2.4B for a Cs-DNA salt of molecular weight 18×10^6 obtained from the DNA of bacteriophage T4r introduced into the centrifuge rotating at 27 690 rpm in a 7.7 molal CsCl solution at $pH = 8.4$. The mean of the distribution

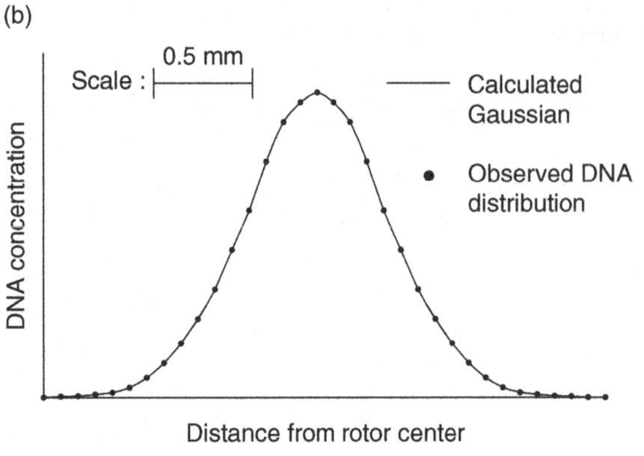

Figure 4.2.4B. The equilibrium distribution of DNA from bacteriophage T4. An aliquot of osmotically shocked T4r containing 3 μg of DNA was centrifuged at 27 690 rpm for 80 hours in 7.7 molal cesium chloride at pH *8.4. Evidence for the attainment of equilibrium was provided by the essential identity of the final band shapes, whether the DNA was initially distributed uniformly in the cell or in an extremely tight band. At equilibrium the observed DNA distribution does not depart appreciably from Gaussian form, indicating essentially uniform molecular weight and effective density. The mean of the distribution corresponds to an effective density of 1.70, and the standard deviation corresponds to a molecular weight for the Cs–DNA salt of 18 × 10⁶. Assuming the base composition for T4 reported by G. R. Wyatt and S. S. Cohen (*Nature, **170**, *1072 (1952)) and the glucose content reported by R. L. Sinsheimer (*Proc. Natl. Acad. Sci., ***42**, *502 (1956)), this corresponds to a molecular weight of 14 × 10⁶ for the sodium deoxyribonucleate. The density gradient is essentially constant over the band and is 0.046 g/cm⁴. The concentration of DNA at the maximum is 20 μg/ml. Figure 2, Meselson* et al. *(1957), reprinted, with permission, from Matthew Meselson, December 5, 2010.*

corresponds to an effective density of 1.7 and is located around 6.6 cm from the rotor center. The density gradient over the band was essentially constant at a value of 0.046 g/cm^4. We can calculate the value of σ_2 from equation (4.2.28) and compare it with the observed $2\sigma_2 \cong 0.5$ mm for $T = 298$ K:

$$\sigma_2^2 = \frac{RT}{\overline{V}_{2j}\,\omega^2 r_0 (\mathrm{d}\rho_{tj}/\mathrm{d}r)_{r_0}}; \quad \overline{V}_{2j} = \frac{M_2}{\rho_{2j}} = \frac{18\times 10^6}{1.70}\ \mathrm{cm^3/gmol};$$

$$\omega = \frac{27\,690 \times 2\pi}{60} = 923\pi\ \mathrm{rad/s};$$

$$r_0 \cong 6.6\mathrm{cm}; \quad RT = 25\times 10^9 \mathrm{erg/gmol},$$

where 1 erg = 1 dyne-cm = 1 g-cm^2/s^2. So we have

$$\sigma_2 = \left(\frac{25\times 10^9 \mathrm{g\text{-}cm^2/s^2\text{-}gmol}}{\frac{18\times 10^6}{1.7}\frac{\mathrm{cm^3}}{\mathrm{gmol}}\times(923\pi)^2\frac{\mathrm{rad^2}}{\mathrm{s^2}}\times 6.6\mathrm{cm}\times 0.046\frac{\mathrm{g}}{\mathrm{cm^4}}}\right)^{1/2}$$

$$\cong 0.03\,\mathrm{cm}$$
$$\Rightarrow 2\sigma_2 = 0.6\,\mathrm{mm},$$

which is reasonably close to the observed value. Analytical ultracentrifuges, where such studies are made, rely on optical systems. The cells have quartz windows (e.g.) and the schlieren system, UV absorption, etc., are used to measure the profiles (Hsu, 1981).

If there is another macromolecular species, $i = 3$, present in the solution in dilute amounts such that $C_{1j}(r)$ is unaffected, a similar analysis can be carried out and a Gaussian distribution of species 3 obtained with the band center located at a radius $r_0|_{\text{species3}}$, where the solution density is given by

$$\rho_{tj}(r_0|_{\text{species 3}}) \cong M_3/\overline{V}_{3j}. \tag{4.2.32}$$

The corresponding standard deviation, σ_3, is obtained from

$$\sigma_3^2 = \frac{RT}{\overline{V}_{3j}\omega^2 r_0\big|_{\text{species 3}} \frac{\mathrm{d}\rho_{tj}}{\mathrm{d}r}\big|_{r_0|_{\text{species 3}}}}. \tag{4.2.33}$$

Figure 4.2.4A provides the profiles for species 2 and 3. Similar results may be developed for other macromolecular species present in small amounts. Multicomponent separation will be achieved as long as the partial specific density of each solute i $(M_i/\overline{V}_{ij})$ is sufficiently different from those of the two solutes with the closest molecular weights.

Following Meselson and Stahl (1958), we can calculate the resolution, R_s, between two different types of DNA in a CsCl solution being centrifuged using the expression of σ developed above and compare it with that measured by Meselson and Stahl. The two different species are isotopically labeled (N^{15}) and unlabeled DNA (N^{14}) of *E. coli*; their density difference is 0.014 g/cm^3.

The molecular weight of the unlabeled DNA in the CsCl solution was estimated to be 9.4 × 10^6. The density of the unlabeled DNA is 1.71 g/cm^3. The value of $\omega^2 r_0$ in the Meselson and Stahl experiment was 140 000 g-cm/s^2 (=140 000 × 980 cm/s^2), corresponding to 44 770 rpm in their ultracentrifuge. The density gradient ($\mathrm{d}\rho_{tj}/\mathrm{d}r$) was estimated to be 0.08 g/cm^4. The resolution between two species located at $r_0|_2$ and $r_0|_3$ is

$$R_s = \frac{r_0|_2 - r_0|_3}{4\sigma}, \tag{4.2.34}$$

where σ is an acceptable average for both bands that can be determined from the expression for one of the species. The resolution expression may be rewritten using

$$r_0|_2 - r_0|_3 = \Delta r_0 = \frac{\rho_2 - \rho_3}{\mathrm{d}\rho_{tj}/\mathrm{d}r} \tag{4.2.35}$$

as

$$R_s = \frac{(\rho_2-\rho_3)}{4}\left(\frac{\overline{V}_{2j}\omega^2 r_0|_2}{RT(\mathrm{d}\rho_{tj}/\mathrm{d}r)}\right)^{1/2} = \frac{0.014}{4}\frac{\mathrm{g}}{\mathrm{cm^3}}$$

$$\left(\frac{(9.4\times 10^6/1.71)\ \frac{\mathrm{cm^3}}{\mathrm{gmol}}\times 140\,000\times 980\ \mathrm{cm/s^2}}{25\times 10^9\ \frac{\mathrm{g\text{-}cm^2}}{\mathrm{gmol\text{-}s^2}}\times 0.08\ \mathrm{g/cm^4}}\right)^{1/2}$$

$$= \frac{0.014}{4}\times 610.8 \cong 2.13.$$

The measured value of R_s reported by Meselson and Stahl (1958) is 1.5.

The osmotic pressure of a concentrated solution of a low molecular weight salt like CsCl (or a solute like sucrose, etc.) employed to create a density gradient is considerable. If isopycnic sedimentation is to be employed for separating a variety of complex physiological cell mixtures in a centrifuge, density gradients created by CsCl or sucrose are unsuitable due to their high osmotic pressure, which may kill or deform the cells. To maintain biological integrity and viability of the cells under consideration, it is necessary that the material used to create the density gradient is isosmotic throughout the gradient. There are many other requirements (e.g. low viscosity, chemically inert, does not dissolve/penetrate biological membranes, etc. (see Van Vlasselaer *et al.* (1998)). Over the years, a number of materials have been used to develop such gradients. These include: Percoll® (colloidal silica covalently coated with polyvinylpyrrolidone); silanized colloidal silica; Ficoll®; dextran; methylcellulose, etc. The reader should refer to Catsimpoolas (1977) and Recktenwald and Radbruch (1998) for more details.

4.2.2 Electrical force field

4.2.2.1 *Isoelectric focusing*

If an electrical field exists in a closed vessel containing an aqueous solution having some charged species, then the charged species will move toward the positively or negatively charged electrode (as the case may be) until the electrode is reached. There is no room for equilibrium separation as such. Suppose, however, there exists a *pH* gradient in the solution and the charge on the species to be separated depends on the local *pH*, as, for example, in proteins. It is known that, at a particular *pH*, the isoelectric *pH*, called *pI*, a particular protein does not have any net charge. (For example, the *pI* for ovalbumin (mol. wt. 45 000) at 25 °C is 4.5, that for bovine serum albumin (mol. wt. 67 000) at 25 °C is 4.9, whereas the *pI* for conalbumin (mol. wt. 86 000) at 25°C is 5.9 (Andrews, 1981)). Such *pI* values vary over a wide range as wide as 1 (pepsin, mol. wt. 33 000) to 10 (lysozyme, mol. wt. 14 000). During the migration of the charged protein through the *pH* gradient, the protein will cross the region where the *pH* is equal to its *pI* (Figure 4.2.5); the net charge on the protein will become zero at this location and so will its migration velocity. Proteins having different *pI*s will be located thereby at different locations having different *pH*s. These locations will become equilibrium positions for different protein species having different *pI*s. A mixture of proteins can therefore be separated. This technique is identified as *isoelectric focusing* (IEF).

The *pH* gradient is established between two electrodes using an aqueous solution of a mixture of ampholytes having molecular weights between 300 and 600. Ampholytes are low molecular weight amphoteric substances (e.g. polyaminopolycarboxylic acids, etc.). They have a net positive charge near the anode (kept in a strong acid) and a net negative charge near the cathode (kept in a strong basic solution). When a voltage is applied to the cell, each ampholyte species migrates to a location where the isoelectric point of the ampholyte is equal to the local *pH*. The *pH* gradient is established by each ampholyte species being at different locations, acidic ones near the anode and basic ones near the cathode. If an ampholyte species is displaced from its IEP, it immediately acquires a charge and is forced back to its position of equilibrium in the electrical field. The *pH* gradient (Figure 4.2.5(a)) is such that the anode or the positive electrode has the most acidic solution and the lowest *pH*, whereas the cathode or the negative electrode has the most basic solution and the highest *pH*. Proteins have a net positive charge at *pH*s below their *pI* and a net negative charge at *pH*s above their *pI* (Figure 4.2.5(c)). If a protein possessing a particular *pI* is introduced at a location where $pH < pI$, it will be positively charged and would therefore migrate toward the cathode. If it is introduced at a location where $pH > pI$, it will be negatively charged and would therefore migrate toward the anode. Thus, regardless of the point of introduction in the *pH* profile, the proteins are forced to move to the regions having their own *pI*s, where they are not subjected to any force and therefore have zero migration velocity.

Although a particular protein species is thus focused to the location $z = z_0$ by the above mechanism, a protein band develops around $z = z_0$ due to molecular diffusion (Figure 4.2.5(d)). Since this band represents the equilibrium profile, there is no net force acting on the particular protein species anywhere in the band. Consider now any location z in the band where $|z-z_0| \geq 0$ for protein species i. Using criterion (3.3.25)

$$\mathrm{d}\mu_{ij}\big|_T = -\mathrm{d}\phi_{ij}^{\mathrm{ext}} = -Z_i\mathcal{F}\,\mathrm{d}\phi.$$

Now,

$$\mathrm{d}\mu_{ij}\big|_T = RT\mathrm{d}\ln a_{ij} = -Z_i\mathcal{F}\frac{\mathrm{d}\phi}{\mathrm{d}z}\mathrm{d}z = Z_i\mathcal{F}E\,\mathrm{d}z, \tag{4.2.36}$$

where E is the strength of the uniform electrical force field. The value of the electrical charge Z_i (+ve or -ve) of a protein molecule of species i depends on its location in the band, i.e. $z - z_0 = z'$, the distance from the band center where Z_i is zero due to $pH = pI$. Since the band is very thin, one can assume that Z_i changes linearly with distance z' from the band center in the following way (Giddings and Dahlgren, 1971):

$$Z_i = \frac{\mathrm{d}Z_i}{\mathrm{d}z}\bigg|_{z=z_0} z' \tag{4.2.37a}$$

$$= \left(\frac{\mathrm{d}Z_i}{\mathrm{d}(pH)}\right)\left(\frac{\mathrm{d}(pH)}{\mathrm{d}z}\bigg|_{z=z_0}\right)z'. \tag{4.2.37b}$$

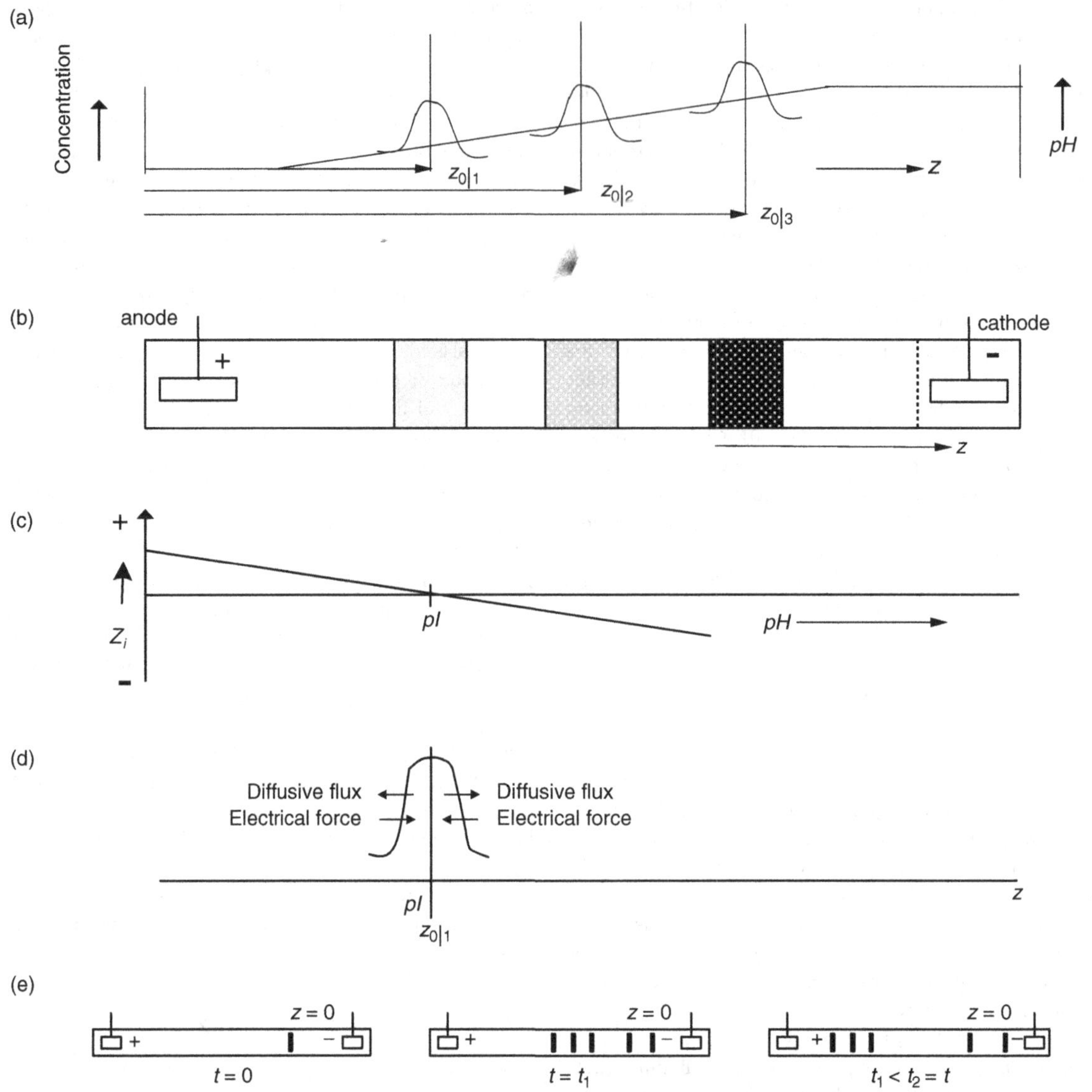

Figure 4.2.5. (a) The pH *gradient along a cell in isoelectric focusing and the concentration profiles of three proteins having three different* pIs. *(b) Locations of the electrodes and the bands in the cell. (c) Variation of the net charge Z_i of a given protein with* pH. *(d) Different fluxes and forces on protein molecules having a particular* pI *at any z around z_0. (e) Zone electrophoresis in a cell; individual solute zone locations at two different times, t_1 and t_2, $t_2 > t_1 > 0$.*

We can assume both variations in the above expression to be linear, independent of z around z_0; this is equivalent to a Taylor's series expansion

$$Z_i = Z_i\big|_{z=0} + \left.\frac{\mathrm{d}Z_i}{\mathrm{d}z}\right|_{z=z_0}(z - z_0) = 0 + \left.\frac{\mathrm{d}Z_i}{\mathrm{d}z}\right|_{z=z_0} z', \tag{4.2.37c}$$

where the first term is zero.

Equation (4.2.36) may now be integrated, after incorporating the above expression for Z_i:

$$\int_{z_0}^{z} \mathrm{d}\ln a_{ij} = \int_0^{z'} \left(\frac{\mathrm{d}Z_i}{\mathrm{d}(pH)}\right)\left(\left.\frac{\mathrm{d}(pH)}{\mathrm{d}z}\right|_{z=z_0}\right)\mathcal{F}\frac{E}{RT} z'\,\mathrm{d}z';$$

$$\frac{a_{ij}|_z}{a_{ij}|_{z_0}} = \exp\left\{-\frac{k_i' z'^2}{2}\right\} \cong \frac{C_{ij}|_z}{C_{ij}|_{z_0}}, \tag{4.2.38a}$$

where

$$k_i' = -\left(\frac{\mathrm{d}Z_i}{\mathrm{d}(pH)}\right)\left(\left.\frac{\mathrm{d}(pH)}{\mathrm{d}z}\right|_{z=z_0}\right)\mathcal{F}\frac{E}{RT}. \tag{4.2.38b}$$

The protein concentration profile (Figure 4.2.5(a)) around $z = z_0$ is Gaussian in nature, having a standard deviation σ_i defined by

$$\sigma_i = (1/k_i')^{1/2}. \quad (4.2.39a)$$

Knowing that the diffusion coefficient of species *i*, D_i, is defined as (RT/f_i^d), we can express σ_i as follows:

$$\sigma_i = (1/k_i')^{1/2} = \left(\frac{D_i f_i^d / \mathcal{F}E}{-\frac{dZ_i}{d(pH)} \frac{d(pH)}{dz}\bigg|_{z=z_0}} \right)^{1/2}. \quad (4.2.39b)$$

This result could have been obtained directly by using the following balance equation at any location *z* in the concentration profile:

$$\begin{pmatrix}\text{diffusive flux of} \\ \text{protein broadening band}\end{pmatrix} = \begin{pmatrix}\text{flux due to electrical force} \\ \text{compressing band toward } z = z_0\end{pmatrix}. \quad (4.2.40)$$

The extent of separation between two contiguous protein bands having centers located at $z_0|_1$ for species 1 and $z_0|_2$ for species 2 may be determined by using the resolution, R_s:

$$R_s = \frac{z_0|_1 - z_0|_2}{2(\sigma_1 + \sigma_2)}. \quad (4.2.41)$$

Since the *pI*s of species 1 and 2 are located, respectively, at $z_0|_1$ and $z_0|_2$, one may approximate $z_0|_1 - z_0|_2$ for two contiguous bands as

$$z_0|_1 - z_0|_2 = \Delta z_0 = \frac{pI|_1 - pI|_2}{\left(d(pH)/dz\right)}. \quad (4.2.42)$$

Further, assuming $\sigma_1 \cong \sigma_2$, one may use σ of any one of the two species given by (4.2.39b) to obtain

$$R_s = \frac{(pI|_1 - pI|_2)}{4} \left(\frac{\mathcal{F}E\left(\frac{-dZ_i}{d(pH)}\right)}{D_i f_i^d \frac{d(pH)}{dz}} \right)^{1/2}. \quad (4.2.43)$$

According to this result, a higher resolution is achieved if D_i is low (bandwidth is smaller), the *pH* gradient is not very steep, the strength of the uniform electrical force field is high and the *pI*s of the contiguous species are substantially different. It is now possible to make an estimate of how far apart the *pI*s of two proteins have to be to achieve a reasonable resolution, say 1: $pI|_1 - pI|_2 \cong 4\sigma'$. We can calculate the value of σ' in *pH* units from expression (4.2.43):

$$\sigma' = \left(\frac{RT \ (d(pH)/dz)}{\mathcal{F}E \ (-dZ_i/d(pH))} \right)^{1/2}. \quad (4.2.44)$$

Following Giddings and Dahlgren (1971), consider ovalbumin, whose $pI = 4.6$. The value of $(-dZ_i/d(pH))$ is 9 at the *pI*. If the *pH* gradient is 0.05 pH/cm and the electrical field strength *E* is 25 volts/cm, the value of $4\sigma'$ at 4 °C is

$$4\sigma' = 4\left\{ \frac{1.987 \times 4.2 \times 277 \left(\frac{\text{joule}}{\text{gmol}}\right) \times 0.05 \ \frac{\text{pH}}{\text{cm}}}{96\,485 \frac{\text{coulomb}}{\text{g}-\text{equiv.}} \times 25 \frac{\text{volt}}{\text{cm}} \times 9 \frac{\text{g}-\text{equiv.}}{\text{pH}-\text{gmol}}} \right\}^{1/2}$$
$$= 4(0.0000052)^{1/2} = 9.12 \times 10^{-3} \cong 10^{-2}\,\text{pH}.$$

Thus, if two neighboring protein *pI*s differ by 10^{-2} pH units, they may be effectively separated by isoelectric focusing.

Isoelectric focusing can be carried out in commercially available columns having as much as 110 ml volume. For analytical applications, they are carried out in capillaries having one dimension as small as 1 mm. To prevent convection currents in larger columns, the solution in the column also has a density gradient developed by having different sucrose concentrations along the column height, amongst other things.

4.2.2.2 *Other electrophoresis techniques*

There are three other electrophoresis techniques carried out in a closed device without any bulk flow in or out: zone electrophoresis (ZE), moving boundary electrophoresis (MBE) and isotachophoresis (ITP). These are not true equilibrium techniques. We will describe very briefly the basics of the zone electrophoresis technique here; all three techniques are suitable for laboratory and preparative applications. For isotachophoresis, see Everaerts *et al.* (1977) and Andrews (1981).

We consider zone electrophoresis first. In Figure 4.2.5(e), we show the electrodes in a closed cell. A thin band of various ions present in a solute mixture is introduced at a location $z = 0$ at time $t = 0$. In the presence of the electrical field, different charged solute species move at different rates; the anionic solutes move toward the positively charged electrode, the anode, while the cationic solutes move toward the negatively charged electrode, the cathode. Neutral species do not move. At another time *t*, different charged species will have moved to different locations and are therefore separated from one another (as shown in Figure 4.2.5(e) for four charged species and one uncharged species). However, unlike that in IEF (isoelectric focusing), the location of each charged solute where a thin solute band is formed will change with time as they move toward the respective electrode. Therefore any separation, purification or identification has to be carried out on a time-dependent basis in cells that are usually several centimeters in length. We have actually studied such a separation in general in Sections 3.2.1 and 3.2.2 without being specific about the driving force: we introduced a thin slab of a mixture of species at one location in a closed vessel and subjected the mixture to an external force in the *z*-direction. When a very thick sample band is utilized containing multiple ions of one particular charge sign, along with an electrolyte having the same ionic charge but a higher mobility than all the ions to be separated, the technique of moving boundary electrophoresis is applied (Everaerts *et al.*, 1977).

4.2.3 Gravitational force field

For separation in a single-phase, the gravitational force field is rather weak compared to the strength of a centrifugal or electrical force field. Still, some amount of separation is achieved in the gravitational field. Employing the criterion (3.3.25) for equilibrium for species i, namely

$$d\mu_i\big|_T = -d\varphi_i^{ext},$$

where $\phi_i^{ext} = M_i gz$ is the gravitational potential of one mole of species i at a height z from the origin, we get

$$d\mu_i\big|_T = -M_i g\ dz$$

for a gaseous system. Using the relationship between μ_i and the fugacity of species $i, \hat{f}_{ij}$, in the region j of the mixture, we get

$$d\mu_i\big|_T = RT\ d\ln \hat{f}_{ij} = -M_i g\ dz. \tag{4.2.45}$$

Integrating between two heights $z = z_1$ ($j = 1$) and $z = z_2$ ($j = 2$), we get

$$\ln(\hat{f}_{i1}/\hat{f}_{i2}) = \frac{M_i g}{RT}(z_2 - z_1), \tag{4.2.46}$$

where $z_2 > z_1$; therefore$\hat{f}_{i1} > \hat{f}_{i2}$. For an ideal gas mixture, this relation leads to

$$\ln(x_{i1}P_1/x_{i2}P_2) = \frac{M_i g}{RT}(z_2 - z_1), \tag{4.2.47a}$$

where P_1 and P_2 are the hydrostatic pressures at heights z_1 and z_2, respectively. For species 2 in a binary mixture,

$$\ln(x_{21}P_1/x_{22}P_2) = \frac{M_2 g}{RT}(z_2 - z_1). \tag{4.2.47b}$$

The separation factor for species $i = 1$ and 2 present in a vertical column of gas mixture is

$$\alpha_{12} = \frac{x_{11}x_{22}}{x_{21}x_{12}} = \exp\left\{\frac{(M_1 - M_2)g(z_2 - z_1)}{RT}\right\}. \tag{4.2.48}$$

If $M_1 > M_2$, heavier species[9] 1 will concentrate more in region 1 i.e. $z = z_1$, since $\alpha_{12} > 1$. (For a pure gas behaving ideally, relation (4.2.46) yields

$$\ln(P_1/P_2) = \frac{M_i g}{RT}(z_2 - z_1), \tag{4.2.49}$$

indicating that P_1 at $z = z_1$ is higher than P_2 at $z = z_2$ ($>z_1$).) This enrichment of the heavier species at $z = z_1$ has been applied to study the isotopic ratios of atmosphere gases trapped in polar ice caps where the heavier isotope are enriched relative to the free atmosphere (Craig *et al.*, 1988; Craig and Wiens, 1996). Sage (1965) has provided a more rigorous procedure for calculating the change of composition in an isothermal vertical column of gas mixture in a gravitational field. Sage (1965) has also considered a liquid mixture of hydrocarbons and illustrated the change in composition over a height of 0–10 000 ft.

[9]Note that this convention where species $i = 1$ is heavier is different from our usual convention.

4.2.3.1 *Batch sedimentation*

Gravitational force is much more often utilized in separating a multiphase mixture, in particular a suspension of particles or drops in a liquid. Consider a stationary vessel containing a suspension of particles in a liquid (Figure 4.2.6(a)) whose top level is at a height z_{HL} from the vessel bottom. Assume that at time $t = 0$ the particles are uniformly distributed throughout the suspension. However, for time $t > 0$, if the particle density ρ_p is larger than that of the liquid ($\rho_f = \rho_t$), the particles will fall toward the bottom of the vessel. As pointed out in Section 3.1.3.1, although there will be particle acceleration initially in the stagnant liquid, very soon the particle will achieve a constant downward velocity, called the terminal velocity, U_{pzt}:

$$U_{pzt} = F_{pz}^{ext}/f_p^d. \tag{4.2.50a}$$

Here, F_{pz}^{ext} is the gravitational force on the particle, taking account of the buoyancy effect, and f_p^d is the frictional coefficient experienced by the particle as it falls through the liquid (here, the suspension).

For a spherical particle of radius r_p, density ρ_p and mass m_p, if the particle motion obeys Stokes' law, then $f_p^d = 6\pi\mu r_p$. From equation (3.1.4), the magnitude of the vertically downward force F_{pz}^{ext} on this particle is $m_p g(1 - (\rho_t/\rho_p))$. Therefore

$$U_{pzt} = \frac{m_p g(1-(\rho_t/\rho_p))}{6\pi\mu r_p} = \frac{2}{9} r_p^2 \rho_p \frac{(1-(\rho_t/\rho_p))}{\mu} g, \tag{4.2.50b}$$

since $m_p = (4/3)\pi r_p^3 \rho_p$. Remember this is the terminal velocity of an isolated spherical particle due to gravitational force. Now we go back to the settling of the suspension of spherical particles in the stationary vessel. (Although the vessel top in the figure is open, it is de facto a closed system.)

What one observes as the particles fall toward the vessel bottom is that a clear layer of liquid, free of particles, emerges near the top of the vessel, whereas near the bottom of the vessel a sludge layer or a dense layer of particles appears (Figures 4.2.6(b), (c)). If sufficient time is allowed for equilibrium to be achieved in the gravitational force field, most of the liquid is clarified and becomes free of particles; all particles accumulate at the bottom of the vessel in a sludge layer of thickness z_{HS}

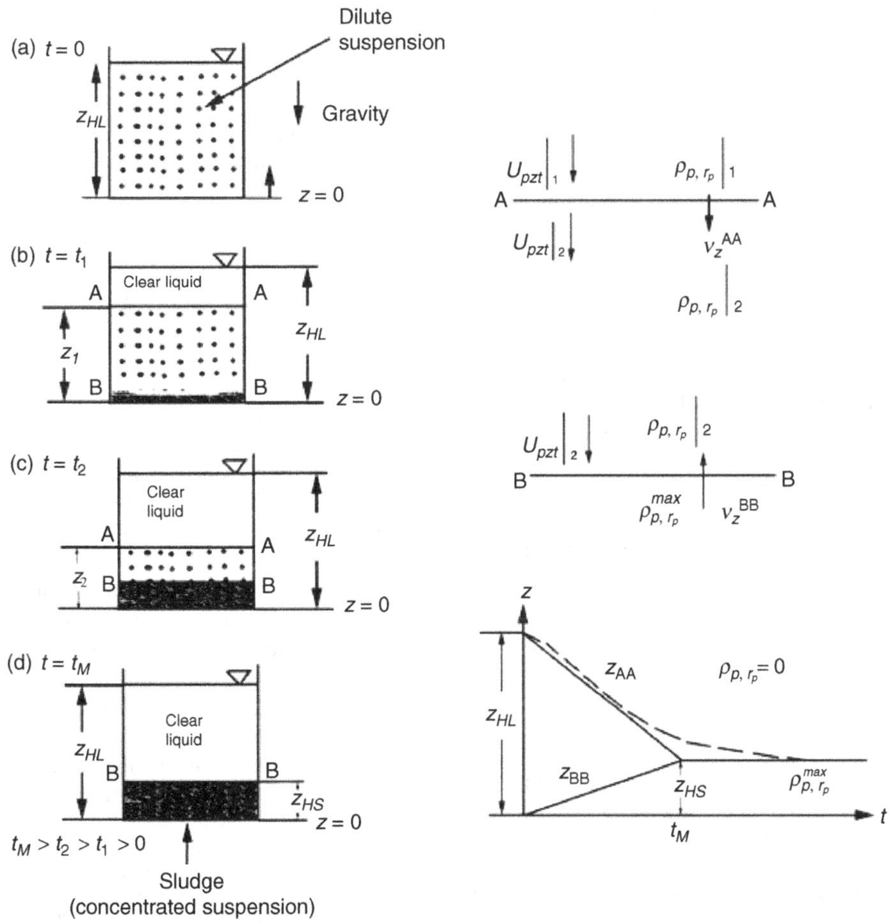

Figure 4.2.6. Batch sedimentation of a dilute suspension of particles under gravitational force: (a) dilute suspension in a vessel at time $t = 0$; (b) condition at time $t_1 > 0$; (c) condition at time $t_2 > t_1 > 0$; (d) final equilibrium state.

(Figure 4.2.6(d)). The questions of interest are as follows. How long (t_M) does it take to achieve this equilibrium in the closed vessel with no flow in or out? What is the height of the sludge layer z_{HS}? Let us assume here that all particles have a density ρ_p and size r_p.

Consider now the different diagrams shown in Figure 4.2.6. In part (b) of this figure, corresponding to a time t_1, where $t_M > t_1 > 0$, a small height of the liquid at the top is clear of any particles. The line AA which separates the clear liquid at the top from the suspension below is a discontinuity in terms of the particle mass concentration in the liquid,

$$\rho_{p,r_p}\left(\frac{\text{grams of particles}}{\text{cm}^3\ \text{of liquid suspension}}\right): \rho_{p,r_p}\big|_1 \text{ above and } \rho_{p,r_p}\big|_2 \text{ below.}$$

If the top liquid is clear, $\rho_{p,r_p|1} = 0$. At any instant of time t_1, the line AA is at location z_1; at another instant of time t_2 ($> t_1$), the line AA (Figure 4.2.6(c)) is at location z_2 ($< z_1$). Thus the line AA is moving downward at a velocity of magnitude $|v_Z^{AA}|$. The net particle flux from the top across the line AA at any time then (diffusion is neglected for larger particles) is $\rho_{p,r_p|1}(|U_{pzt|1}| - |v_Z^{AA}|)$. This value must equal the rate at which particles are removed downward from below the line AA: $\rho_{p,r_p|2}(|U_{pzt|2}| - |v_Z^{AA}|)$. Here $|U_{pzt|2}|$ is the magnitude of the terminal velocity of particles below line AA and $|U_{pzt|1}|$ is that above line AA. For a steady downward movement of this line AA,

$$\rho_{p,r_p|1}(|U_{pzt|1}| - |v_Z^{AA}|) = \rho_{p,r_p|2}(|U_{pzt|2}| - |v_Z^{AA}|). \quad (4.2.51)$$

Rearranging, we get

$$|v_Z^{AA}| = \frac{\rho_{p,r_p|2}|U_{pzt|2}| - \rho_{p,r_p|1}\,|U_{pzt|1}|}{(\rho_{p,r_p|2} - \rho_{p,r_p|1})}. \quad (4.2.52)$$

In the special case of a dilute suspension and the top clarified layer particle density being zero (i.e. $\rho_{p,r_p|1} = 0$), the value of $|v_Z^{AA}|$ is

$$|v_Z^{AA}| = |U_{pzt|_2}| = |U_{pzt|_1}|, \quad (4.2.53)$$

corresponding to the particle settling velocity $|U_{pzt|_1}|$ of a very dilute particle suspension of particle density $\rho_{p,r_p} = \rho_{p,r_p|_2}$. We may assume it to be given by expression (4.2.50b).

On the other hand, the sludge layer builds up with time at the bottom of the vessel. At time t_M its maximum height z_{HS} is reached; above it only clarified liquid exists (Figure 4.2.6(d)). At an earlier time t_2, where $t_M > t_2 > t_1 > 0$, the sludge layer height indicated by the line BB (Figure 4.2.6(c)) is less than z_{HS} (Figure 4.2.6(d)). Let the magnitude of the velocity of the line BB in the upward direction be $|v_Z^{BB}|$. Further, let the particle mass concentration in the sludge layer be ρ_{p,r_p}^{max} since it will be the highest particle mass concentration in the system. For a steady upward movement of line BB, the particle flux into the sludge layer from the top through the line BB for a very dilute suspension, namely $\rho_{p,r_p|_2}(|U_{pzt|_2}| + |v_Z^{BB}|)$ must equal the particle flux built up by the movement of line BB in the upward direction, namely $\rho_{p,r_p}^{max}\ |v_Z^{BB}|$:

$$\rho_{p,r_p}^{max}\ |v_Z^{BB}|\ = \rho_{p,r_p|_2}(|U_{pzt|_2}| + |v_Z^{BB}|); \quad (4.2.54)$$

$$|v_Z^{BB}|\ (\rho_{p,r_p}^{max} - \rho_{p,r_p|_2}) = \rho_{p,r_p|_2}\ |U_{pzt|_2}| = \rho_{p,r_p}\ |U_{pzt|_1}|$$

$$|v_Z^{BB}| = \frac{\rho_{p,r_p}\ |U_{pzt|_1}|}{(\rho_{p,r_p}^{max} - \rho_{p,r_p})}. \quad (4.2.55)$$

When an equilibrium is achieved, line BB will coincide with line AA near the bottom of the vessel (Figure 4.2.6 (d)). If the distance of line AA from the bottom is indicated by z_{AA} and that of line BB by z_{BB}, then at time t_M both must have the same value:

$$z_{AA} = z_{HL} - |U_{pzt|_1}|t_M = z_{BB} = |v_z^{BB}|t_M; \quad (4.2.56)$$

$$t_M = \frac{z_{HL}}{|v_Z^{BB}| + |U_{pzt|_1}|} = \frac{z_{HL}}{|U_{pzt|_1}|}\left\{\frac{\rho_{p,r_p}^{max} - \rho_{p,r_p}}{\rho_{p,r_p}^{max}}\right\}$$

$$= \frac{z_{HL}}{|U_{pzt|_1}|}\left\{1 - \frac{\rho_{p,r_p}}{\rho_{p,r_p}^{max}}\right\}. \quad (4.2.57)$$

The value of z_{HS} is (from (4.2.56) and (4.2.57))

$$z_{HS} = z_{HL} - |U_{pzt|_1}|\ \frac{z_{HL}}{|U_{pzt|_1}|}\left\{1 - \frac{\rho_{p,r_p}}{\rho_{p,r_p}^{max}}\right\}$$

$$= z_{HL}\left\{\frac{\rho_{p,r_p}}{\rho_{p,r_p}^{max}}\right\} \quad (4.2.58)$$

We could have also obtained this result simply by a particle mass balance between the original dilute suspension of height z_{HS} and density ρ_{p,r_p} and the final sludge density of ρ_{p,r_p}^{max}. A very dilute suspension of particles has now been separated by gravity into a clear supernatant liquid and a thin sludge layer of thickness z_{HS} over a time period of t_M.

There are two more aspects of considerable interest.

(1) How do the lines AA and BB move with time?
(2) How are these movements affected by an increased concentration of particles in the starting suspension? This is the subject of *hindered settling*.

The first question has already been answered by expressions (4.2.53) and (4.2.55) for $|v_Z^{AA}|$ and $|v_Z^{BB}|$. If the heights of the interfaces AA and BB are denoted by z_{AA} and z_{BB}, respectively, then

$$|v_Z^{AA}| = |(dz_{AA} / dt)| \quad \text{and} \quad |v_Z^{BB}| = |(dz_{BB} / dt)|. \quad (4.2.59)$$

In Figures 4.2.6(d), the trajectories (solid lines) of z_{AA} and z_{BB} have been plotted as a function of time. They intersect at $t = t_m$ when equilibrium settling has been reached.

The nature of these two trajectories changes if batch sedimentation involves a more concentrated suspension (for an introduction, see the general treatment in Probstein (1989) of the subject). First, the settling velocity of an individual particle is reduced due to the close proximity of many other particles. This is due to an increased viscosity of the liquid, as well as an increased drag force experienced by the particle due to the other particles. The effective settling velocity U_{pzt}^{eff} is less than U_{pzt} for a dilute suspension of the same particle size;

$$U_{pzt}^{eff} = U_{pzt}\ \overline{G}_{D\mu}(\text{particle volume fraction}), \quad (4.2.60)$$

where $\overline{G}_{D\mu}$ is a function of the particle volume fraction in the suspension and has a value less than 1. The functional dependence is often empirically expressed as

$$\overline{G}_{D\mu} = \big(1\text{-(particle volume fraction)}\big)^n, \quad (4.2.61)$$

where the value of n suggested by different workers varies between 4.7 and 5.1. The analysis by Kynch (1952) of hindered settling based on (4.2.60) is a useful first step in this complex subject.

A qualitative discussion of the change in the nature of the interface trajectories shown on the right-hand side of Figures 4.2.6(d) is also in order for a concentrated suspension. The rate at which the interface AA drops, i.e. $|v_z^{AA}|\ (= (dz_{AA}/dt))$ changes with time. As more and more clarified liquid appears at the top, the particle concentration below AA increases. This reduces the value of $|U_{pzt|_2}|$ (equation (4.2.53)), which is no longer equal to $|U_{pzt|_1}|$; further, it decreases with increased t, leading to a more curved surface, shown in Figure 4.2.6(d) by a dashed line. Correspondingly, the speed of the rise of the interface BB decreases with time.

The principle of sedimentation in a closed vessel discussed is the primary basis for the large-scale separation of

Table 4.2.1. Cell densities[a]

Cell type	Density (g/cm^3)
T Cells	1.077
B Cells	1.077
Monocytes	1.064
Granulocytes	1.09
NK Cells	1.06
Erythrocytes	1.115
Osteoblasts	1.005

[a] From Recktenwald and Radbruch (1998). See also Table 7.3.1.

rapidly settling particles in open devices called *thickeners*. Usually, the particle density is significantly higher than that of water, which is usually the liquid. On the other hand, the settling/separation of cells of biological origin by gravitational sedimentation from an aqueous medium/blood, etc., is very difficult: the cell densities are very close to that of water, as shown in Table 4.2.1. From equation (4.2.50b), one can estimate the terminal velocities using appropriate cell dimensions, water viscosity and density (~1.0 g/cm^3). Since these values are quite low, a technology based on *high-density particles* (HDPs) has been developed, and it is described in Section 4.2.3.2. It is implemented in a closed vessel and is useful in biomedical and biotechnology applications.

4.2.3.2 *Sedimentation using high-density particles*

Kenyon *et al.* (1998) have described a technology utilizing high-density particles (HDPs) of nickel approximately 10 μm in diameter and having a density of 9 g/cm^3. Therefore the value of $(\rho_p - \rho_t)$ in the terminal velocity equation (4.2.50b) for such particles will be around 8 g/cm^3, whereas that for the cells in Table 4.2.1 can at the most be 0.1 g/cm^3. The technology involves attaching the cells to the HDPs of nickel so that settling is very rapid. These dense particles have a highly irregular surface, which is easily coated by monoclonal or polyclonal antibodies (Abs) via passive adsorption. When, for example, a sample of whole blood containing certain types of cells whose surface antigens (Ags) interact directly with the Abs is introduced into a suspension of HDPs in a test tube in an appropriate buffer and settling allowed to take place, the particles in the tube settle in about 4 minutes. The attached cells may be removed from the particles by a number of procedures (Kenyon *et al.*, 1998).

This concept is generally useful in a variety of other biologically relevant applications, for example the removal of proteins by dense particles having surface ion exchange groups or other functional groups from a fermentation broth. In the case of HDPs of nickel, to prevent leaching of nickel into the medium, the particles are heated for 3–5 hours at 250 °C; apparently the surface oxidizes, which reduces nickel leaching to levels without any adverse effects. The HDPs with Abs on the surface develop extremely specific interactions with particular cell types so that gravitational settling is accompanied by very high cell type selectivity.

4.2.3.3 *Particle fractionation and density gradient based separation*

In Section 4.2.3.1, we considered the separation of particles of one size and density from the liquid medium via particle settling due to gravity. If two different types of particles having different densities have to be separated by gravity, the liquid medium selected should have a density in between the densities of the two types of particles. The particles, having a higher density than the liquid, will settle as shown in Figure 4.2.6(d); the lighter particles will float to the top of the liquid layer in devices called *sorting classifiers*. The variation of size for particles having a given density is of no consequence in a closed system headed for equilibrium. Although the individual settling velocities will vary due to the particle size variation, this variation is inconsequential since different size particles will ultimately have to settle to the bottom or rise to the top in a closed system slated for equilibrium. For particles of larger sizes employed in such separations, the liquid density may vary from 1.3 upward. Liquids of different densities may be developed by employing varying concentrations of $CaCl_2$ (used for cleaning coal) or fine suspensions of heavy minerals, e.g. magnetite (density ~5.1 g/cm^3), silicon (density ~6.5 g/cm^3), etc., in water.

If there are particles of different densities in a sample, they may be separated by introducing the sample mixture into a vessel (or column) containing a number of liquid layers having different densities; alternatively, the liquid layer density may vary continuously along with column height. Whether one uses a number of discrete densities in a multiple-layer column or a continuous gradient of density, heavier liquids must be lower in the column compared to the lighter liquids to prevent mixing via natural convection (see Section 6.1). Heavier particles will settle to the layer where their densities match the density of the layer. Lighter particles will rise to the liquid layer where their densities match that of the liquid layer, another example of focusing in an external force field.

4.2.3.4 *Inclined settler*

If we look at equation (4.2.58), we can calculate the mass of particles settled per unit bottom surface area of the vessel as $z_{HL}\rho_{p,r_p}$ (g/cm^2). The total mass of particles settled is $z_{HL}\rho_{p,r_p} \times A_{\text{bottom}}$, where A_{bottom} is the cross-sectional area of the vessel; correspondingly, the volume of classified liquid is $z_{HL}A_{\text{bottom}}$. It is clear that the volume of liquid clarified is directly proportional to A_{bottom} for a given z_{HL}.

Consider now Figure 4.2.7, where we show a narrow channel inclined to the vertical by an angle θ. If this channel is filled with a particle suspension, then the surface area over which particles can settle by gravity is equal to $(L \sin\theta + 2b\cos\theta)$ × width of the channel plate (W); i.e. it is given by $W(L\sin\theta + 2b\cos\theta)$. If this channel were vertical ($\theta = 0$), the surface area for settling would have been only $2bW$. Therefore, the ratio of the volume of liquid clarified or the rate of clarification in an inclined settler with respect to a vertical settler is $(L/2b)\sin\theta + \cos\theta$. When $L/2b$ is large (thin lamellae-like channel), this ratio can be very large (when θ is nonzero and has a significant inclination to vertical).

4.2.3.4.1 Inclined electrosettler When the particles are in a colloidal suspension, the particles do not settle due to gravity. An examples is as follows: nonaqueous liquids containing asphaltenes, which adsorb on fine particles present in various liquids encountered in the production of synthetic fuels from coal, oil shale and tar sands; the surface charge carried by asphaltenes leads to a nearly stable colloidal suspension (Gidaspow *et al.*, 1989). An inclined settler whose two walls are turned into two electrodes (one positive, the other negative), with a voltage applied, can produce a clarified liquid at the top and concentrated slurry at the bottom. For example, Gidaspow *et al.* (1989) have illustrated how alumina particles of size 15 μm and having a positive zeta potential (+30 mV) were

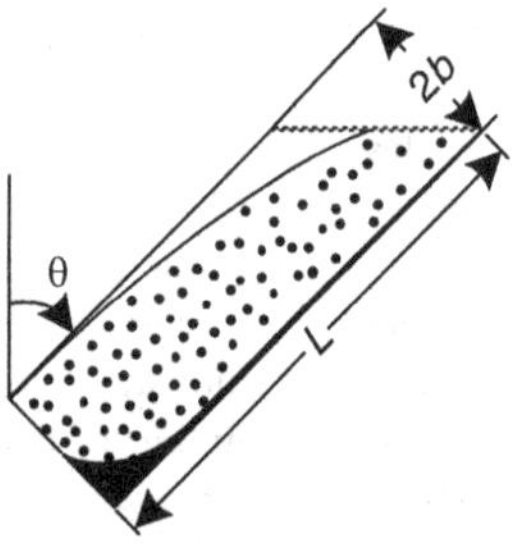

Figure 4.2.7. Batch sedimentation in an inclined settler.

stabilized first in a tetralin suspension by the surfactant Aerosol OT, which stands for dioctyl sodium sulfosuccinate. When an electric field is applied in the inclined settler (Figure 4.2.7), in which the top plate has a positive charge and the bottom plate a negative charge, particles settle onto the bottom plate, having a particle-free liquid at the top and a concentrated suspension over the bottom plate. The higher the voltage gradient applied between the two electrodes, the lower the height of the clear liquid interface from the bottom of the electrosettler.

4.2.4 Particle separation with acoustic forces

The acoustic radiation force F_{acrx} in the x-direction experienced by particles of densities ρ_p suspended in a medium of density ρ_f has already been expressed by equation (3.1.48):

$$F_{acrx} = -\left[\frac{\pi P_0^2 V_p \beta_f}{2\lambda}\right]\left\{\frac{5\rho_p - 2\rho_f}{2\rho_p + \rho_f} - \frac{\beta_p}{\beta_f}\right\}\sin(2kx),$$

The quantity within the second set of brackets is the ϕ-factor:

$$\phi\text{-factor} = \frac{5\rho_p - 2\rho_f}{2\rho_p + \rho_f} - \frac{\beta_p}{\beta_f}.$$

The sign of the *acoustic contrast factor*, or *ϕ-factor*, determines the direction of the force on the particle: a positive ϕ-factor moves those particles to a pressure node, whereas a negative ϕ-factor moves those particles to a pressure antinode (Petersson *et al.*, 2005). This is illustrated in Figures 4.2.8(a) and (b). Consider two types of particles in water in a closed chamber, as shown in Figure 4.2.8(a) without any acoustic force field. The darker particles have positive values of ϕ, whereas the lighter particles have negative values of ϕ. When the acoustic force is switched on, with an acoustic standing wave having a pressure node in the center of the chamber of the closed vessel, the particles are separated: the particles having positive ϕ-values concentrate at the pressure node, whereas those having

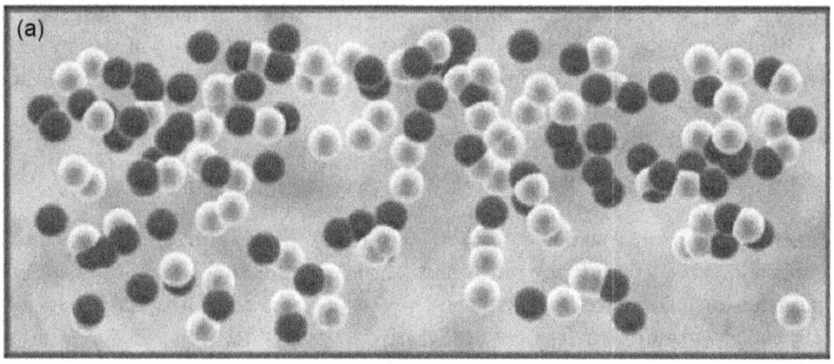

Figure 4.2.8. (a) Schematic illustration of a particle mixture in a closed chamber. No acoustic force field is present in the chamber. The suspended particles have negative (white) and positive (dark) ϕ-factors.

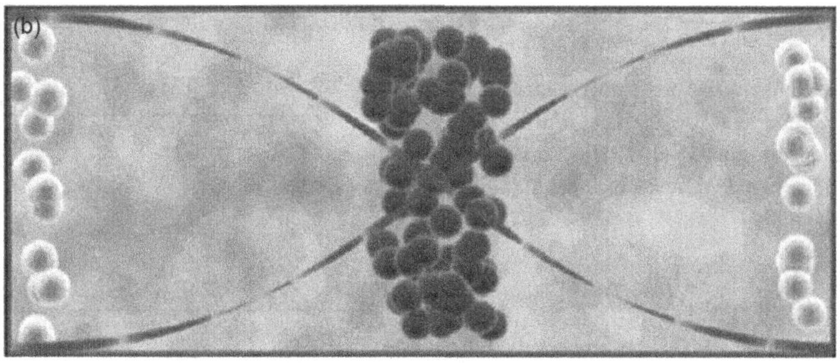

Figure 4.2.8 (cont.) (b) Acoustic standing wave (outlined with the pressure node in the center of the chamber) induced between the walls of the chamber. Under these conditions, the ϕ-factors of the particles will determine whether they move toward the pressure node or the pressure antinodes; particles (white) with negative ϕ-factor are at the wall; dark particles with positive ϕ-factor are at the pressure node, the chamber center. (After Petersson et al. *(2005).)*

negative ϕ-values concentrate at the pressure antinodes near the walls of the closed vessel. An example of such a separation is the separation of lipid particles (emboli) from red blood cells (erythrocytes) for an intraoperative blood wash system.

4.2.4.1 *Gas mixture separation via propagation of acoustic waves*

Consider two reservoirs connected by a narrow duct with the whole volume filled with a 50–50 mixture of He–Ar. If each reservoir is connected to a bellows-sealed piston, and the pistons are driven at, say, a low frequency of 10 Hz with independent phase and amplitude control, the oscillating pressure wave creates a composition difference of as much as 6% between the two chambers at the ends of the narrow connecting duct (Spoor and Swift, 2000). The periodic motion of the pistons produces density fluctuations in the gas; this sound wave creates the composition difference via a complex interaction of thermal diffusion (compression leads to heating), ordinary diffusion, convective motion, etc. (Geller and Swift, 2002a,b).

4.2.5 Externally imposed temperature gradient: thermal diffusion

We have so far considered primarily three externally imposed force fields, centrifugal, electrical and gravitational, and their individual effects in developing separation in a single phase in a closed vessel. Here we consider the effect of an externally imposed temperature gradient on the equilibrium separation in a single phase in a closed vessel. Although it is not an externally imposed force field, the temperature gradient achieves a net separation effect similar to that developed in an external force field.

Consider a binary gas mixture of uniform composition in a vessel shown in Example III of Figures 1.1.3. Initially, the temperature T_2 of the mixture is uniform everywhere. The temperature of one of the chambers is now raised to T_1 ($>T_2$) such that the two bulbs are maintained at two different temperatures. If the thermal diffusion ratio k_T for a given species 1 is negative, then, from definition (3.1.45) and force relation (3.1.44), we can conclude that species 1 moves due to the temperature gradient from the colder bulb to the hotter bulb. This builds up its concentration in the hotter bulb, setting up its back diffusion into the colder bulb. At equilibrium, the temperature-gradient-driven flux of species 1 is balanced by that due to the concentration gradient in the opposite direction, since the two forces in the opposite direction must balance each other. To exploit this dynamic equilibrium, we need to calculate the flux due to the temperature gradient first. Employ expression (3.1.44) for the force on a gmol of species 1 (say) due to the temperature gradient in a mixture of species 1 and 2:

$$\frac{F_{T1}}{f_1^d}C_1 = \text{temperature-gradient-driven molar flux of } 1 = \boldsymbol{J}_1^T$$
$$= -\left(\frac{RT}{D_{12}}\right)\frac{D_1^T}{C_1M_1}(\nabla \ln T)\frac{C_1}{f_1^d} = -\frac{D_1^T}{M_1}(\nabla \ln T), \tag{4.2.62}$$

where f_1^d is the frictional coefficient of species 1 ($= RT/D_{12}$) and there is no bulk velocity. This flux is opposed by the diffusive flux $\boldsymbol{J}_1$ of species 1 from a concentration gradient:

$$\boldsymbol{J}_1^T = -\frac{D_1^T}{M_1}\nabla \ln T = -\boldsymbol{J}_1 = \frac{C_t^2M_1M_2D_{12}}{\rho_tM_1}\nabla x_1. \tag{4.2.63}$$

Use the one-dimensional form of this relation since the species movement is only in one coordinate direction:

$$-\frac{D_1^T\rho_t}{D_{12}C_t^2M_1M_2}\frac{\mathrm{d}T}{T} = \mathrm{d}x_1 = -k_T\frac{\mathrm{d}T}{T}.$$

Often, k_T is expressed as $-k_T'x_1(1-x_1)$, where k_T' is a thermal diffusion constant; this leads to

$$k'_T \frac{dT}{T} = \frac{dx_1}{x_1(1-x_1)}. \qquad (4.2.64)$$

Integrate between region 1 at T_1 and region 2 at T_2. The extent of separation generated by thermal diffusion is rather low; therefore the product x_1 $(1 - x_1)$ is considered to be a constant at an averaged value $\bar{x}_1(\overline{1-x_1})$. Integrating equation (4.2.64) now yields

$$k'_T\bar{x}_1(\overline{1-x_1}) \ln\frac{T_1}{T_2} = x_1|_{T_1} - x_1|_{T_2} = x_{11} - x_{12}.$$

Since x_{11} and x_{12} are quite close, we may substitute $\{\bar{x}_1(\overline{1-x_1})\}$ by $(x_{12})(1 - x_{11})$ (i.e. as a product of two boundary values). Therefore,

$$\frac{x_{11} - x_{12}}{(x_{12})(1-x_{11})} = \frac{(x_{11})(1-x_{12})}{(x_{12})(1-x_{11})} - 1 = \alpha_{12} - 1 = \varepsilon_{12} = k'_T \ln\frac{T_1}{T_2}. \qquad (4.2.65)$$

This relation provides an expression for the enrichment factor ε_{12} for two species 1 and 2 between the two regions in terms of k'_T and the two temperatures. Consider now an equimolar mixture of H_2 and N_2 subjected to a temperature difference of 260 °C in the hot bulb and 15 °C in the cold bulb. In the above result, $(x_{11} - x_{12})$ may be estimated if we can estimate $k'_T(x_{12})(1-x_{11})$, which is equal to $-k_T$; a rough estimate of this value for illustration is 0.066 (Ibbs *et al.*,1939). The value of $(x_{11} - x_{12})$ we obtain is 0.0406, indicating about a 4% difference in mole fraction for the lighter species H_2 between the hot region (T_1) and the cold region (T_2).

4.3 Equilibrium separation between two regions in a closed vessel separated by a membrane

In Sections 4.1 and 4.2, either a two-phase system or a single-phase system was introduced into a vessel which was then kept closed with no mass additions or withdrawals. After allowing the system to reach equilibrium, the separation achieved between the two regions in the closed vessel was calculated. There was always some separation achieved between the two regions. In this section, either we introduce a feed mixture into one region of a membrane device, or we introduce two different mixtures into two regions of the membrane-containing device. We then allow a long time to elapse and calculate the separation achieved at equilibrium between the two regions. This has been studied here for the membrane processes of simple dialysis, Donnan dialysis, gas permeation, reverse osmosis and ultrafiltration.

4.3.1 Separation by dialysis using neutral membranes

Dialysis is a membrane process in which a *microsolute* present in a *feed* solution is removed from the feed to a receiving solution of the same phase called the *dialysate*; the membrane is such that the *macrosolutes*

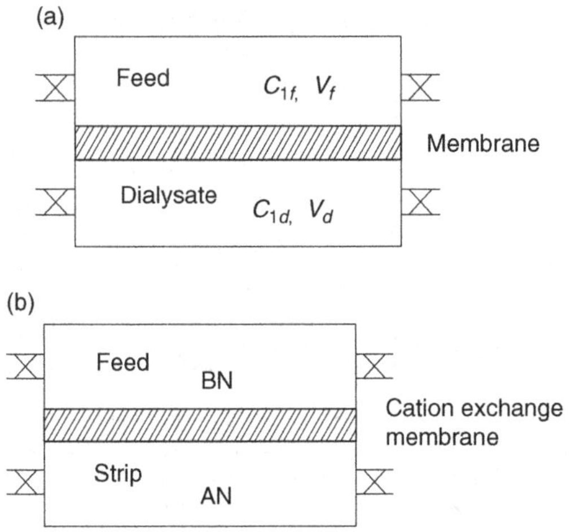

Figure 4.3.1. (a) Dialyzer with closed inlets and outlets. (b) Closed vessel having two aqueous solutions separated by a cation exchange membrane: Donnan dialysis.

present in the feed solution are not allowed to pass through it to the dialysate stream. This process is the basis for blood purification using an artificial kidney (or *hemodialyzer*). The membrane does not have substantial convection of the solvent, namely water. Solutes diffuse through fine pores to the dialysate; metabolic wastes, e.g. urea, uric acid, creatinine, etc., diffuse through the water-filled pores or hydrogels from the blood to the dialysate, but macrosolutes, e.g. blood proteins (albumin, etc.), and larger species are essentially excluded by the membrane. In actual operation, the blood (or feed solution) flows on one side of the membrane and the dialysate flows on the other side into the device and out.

If, in such a device, blood is introduced into the feed side and the dialysate stream is introduced into the receiving side, and the device inlets and exits are closed (Figure 4.3.1(a)), one would like to know what will happen as $t \to \infty$ when equilibrium will be achieved. Would the two regions show separation with respect to the microsolute as $t \to \infty$?

Let the microsolute (e.g. urea) to be removed from feed be species 1, the feed solution region $j = f$ and the dialysate solution region $j = d$. Let the feed and the dialysate concentrations of the solute 1 at time $t = 0$, be C^0_{1f} and $C^0_{1d}(= 0)$. We are interested in knowing the values of C_{1f} and C_{1d} as $t \to \infty$. Assume the two solutions to be well stirred; therefore the two surfaces are exposed to the bulk solution concentrations C_{1f} and C_{1d} at any time. Let the volumes of the feed solution and the dialysate solution be V_f and V_d, respectively. The solute flux expression through the membrane of thickness δ_m and area A_m at any time may be written, following relation (3.4.99), as

$$N_{1z} = k_{1m}(C_{1f} - C_{1d}) = \frac{Q_{1m}}{\delta_m}(C_{1f} - C_{1d}). \tag{4.3.1}$$

A total solute balance on the feed side and the permeate side leads to

$$V_f \frac{\mathrm{d}C_{1f}}{\mathrm{d}t} = -\frac{Q_{1m}}{\delta_m} A_m (C_{1f} - C_{1d}); \tag{4.3.2}$$

$$V_d \frac{\mathrm{d}C_{1d}}{\mathrm{d}t} = \frac{Q_{1m}}{\delta_m} A_m (C_{1f} - C_{1d}). \tag{4.3.3}$$

Laplace transforms of these two coupled ordinary differential equations will convert the t-domain to the s-domain and yield the following two equations:

$$s\overline{C}_{1f} - C_{1f}(t=0) = -\frac{Q_{1m}A_m}{\delta_m V_f}(\overline{C}_{1f} - \overline{C}_{1d});$$

$$s\overline{C}_{1d} - C_{1d}(t=0) = \frac{Q_{1m}A_m}{\delta_m V_d}(\overline{C}_{1f} - \overline{C}_{1d}),$$

where $\overline{C}_{1f}$ and $\overline{C}_{1d}$ are the transformed variables in the s-domain. Introducing the values of C_{1f} and C_{1d} at $t = 0$, we get

$$s\overline{C}_{1f} - C_{1f}^0 = \frac{Q_{1m}A_m}{\delta_m V_f}(\overline{C}_{1d} - \overline{C}_{1f}); \tag{4.3.4}$$

$$s\overline{C}_{1d} = \frac{Q_{1m}A_m}{\delta_m V_d}(\overline{C}_{1f} - \overline{C}_{1d}). \tag{4.3.5}$$

After eliminating $\overline{C}_{1f}$ from these two equations, the following equation is obtained for $\overline{C}_{1d}$:

$$\left[\left(s + \frac{Q_{1m}A_m}{\delta_m V_f}\right)\left(s + \frac{Q_{1m}A_m}{\delta_m V_d}\right)\right]\overline{C}_{1d} - \left(\frac{Q_{1m}A_m}{V_d \delta_m}\right)\left(\frac{Q_{1m}A_m}{V_f \delta_m}\right)\overline{C}_{1d} = \left(\frac{Q_{1m}A_m}{\delta_m V_d}\right)C_{1f}^0; \tag{4.3.6}$$

$$\overline{C}_{1d} = C_{1f}^0\left(\frac{Q_{1m}A_m}{\delta_m V_d}\right)\left[\frac{1}{s^2 + s\left(\frac{Q_{1m}A_m}{\delta_m}\right)\left(\frac{1}{V_f} + \frac{1}{V_d}\right)}\right]. \tag{4.3.7}$$

If we define

$$\left(\frac{Q_{1m}A_m}{\delta_m}\right)\left(\frac{1}{V_f} + \frac{1}{V_d}\right) = a,$$

a constant, then

$$\overline{C}_{1d} = C_{1f}^0\left(\frac{Q_{1m}A_m}{\delta_m V_d}\right)\left(\frac{1}{s^2 + as}\right). \tag{4.3.8}$$

Taking an inverse transform of this, we get

$$C_{1d} = C_{1f}^0\left(\frac{Q_{1m}A_m}{\delta_m V_d}\right)\left[\frac{1}{a}(1 - e^{-at})\right]. \tag{4.3.9a}$$

As $t \to \infty$, this is reduced to

$$C_{1d} = C_{1f}^0\left(\frac{Q_{1m}A_m}{\delta_m V_d}\right)\frac{1}{\left(\frac{Q_{1m}A_m}{\delta_m}\right)\left(\frac{1}{V_f} + \frac{1}{V_d}\right)} = C_{1f}^0 \frac{V_f}{V_f + V_d} \Rightarrow (V_f + V_d)C_{1d} = V_f C_{1f}^0. \tag{4.3.9b}$$

Thus, the total amount of solute $V_f C_{1f}^0$ present initially in the feed is distributed between the feed and the dialysate. As $t \to \infty$, C_{1d} has become a constant, $(\mathrm{d}C_{1d}/\mathrm{d}t)$ in equation (4.3.3) must be zero and $C_{1f} = C_{1d}$.

The process of dialysis has reduced the solute concentration in the feed solution in the closed vessel; at equilibrium, the value of C_{1f} has, however, become equal to that of C_{1d}, the concentration in the dialysate. Thus, if separation of solute 1 between the two regions is the goal, at equilibrium, there is no separation. Further, if separation between two solutes 1 and 2 in the feed is considered and both solutes are permeable through the membrane, then, at equilibrium, each solute will be present in each region of the closed vessel at the same concentration, resulting in no separation.

In practice, separation *is* achieved in dialysis since the device is operated as an open system: fresh dialysate is introduced and then taken out continuously. Thus C_{1d} is always maintained lower than C_{1f}; ultimately, the feed solution concentration of species 1 can be reduced to a very low level. The continuous renewal/removal of the permeate side fluid, in this case the dialysate, is essential to most membrane based processes, which are based on different intrinsic rates of transport of species 1 and 2 through the membrane. **Thus, membrane devices have to be open, in general, to achieve separation.**

A closed device employing a dialysis membrane as described above *can* achieve high purification of the feed solution if special conditions exist. For example, if in the dialysate side there is another species, which reacts chemically with species 1 and produces a product 3 which cannot diffuse through the membrane to the feed side, then it may be possible to purify the feed solution. Obviously, the added reactant in the dialysate side should not be able to diffuse through the membrane to the feed solution if the feed solution has to be purified. One such example has been provided in Section 5.4.3 based on the work carried out by Klein *et al.* (1972, 1973).

The intrinsic separation capability of a dialysis membrane located in a closed vessel can be determined by considering the transport of two solutes $i = 1$ and 2 through the membrane during the initial time period when $t \to 0$. The separation factor between two solutes, α_{12}, between the two regions ($j = 1$, dialysate $\to j = d$; $j = 2$, feed $\to j = f$) for a two-species ($i = 1,2$) system is

$$\alpha_{12} = \frac{x_{11}x_{22}}{x_{12}x_{21}} = \frac{x_{1d}x_{2f}}{x_{1f}x_{2d}} = \frac{\dfrac{C_{1d}}{C_{1d}+C_{2d}}\dfrac{C_{2f}}{C_{1f}+C_{2f}}}{\dfrac{C_{1f}}{C_{1f}+C_{2f}}\dfrac{C_{2d}}{C_{1d}+C_{2d}}}. \tag{4.3.10}$$

Therefore, $\alpha_{12} = C_{1d}C_{2f}/C_{1f}C_{2d}$. To determine its value as $t \to 0$, we assume that, as $t \to 0$,

$$C_{1f} \cong C^0_{1f}, \qquad C_{2f} \cong C^0_{2f}. \tag{4.3.11}$$

Further,

$$C_{1d}|_{t\to 0} = C_{1d}|_{t=0} + \left.\frac{dC_{1d}}{dt}\right|_{t\to 0} \Delta t + \cdots \tag{4.3.12}$$

(Taylor series expansion around $t = 0$). But $C_{1d}|_{t=0} = 0$. From relation (4.3.7), define

$$a_1 = \left(\frac{Q_{1m}A_m}{\delta_m}\right)\left(\frac{1}{V_f}+\frac{1}{V_d}\right) \text{ and}$$

$$a_2 = \left(\frac{Q_{2m}A_m}{\delta_m}\right)\left(\frac{1}{V_f}+\frac{1}{V_d}\right).$$

From relation (4.3.9a),

$$\frac{dC_{1d}}{dt} = C^0_{1f}\left(\frac{Q_{1m}A_m}{\delta_m V_d}\right)e^{-a_1 t} \Rightarrow C_{1d}|_{t=\Delta t}$$
$$= C^0_{1f}\left(\frac{Q_{1m}A_m}{\delta_m V_d}\right)\frac{\Delta t}{(1+a_1\Delta t+\cdots)}.$$

Similarly for $C_{2d}|_{t=\Delta t}$. Therefore

$$\alpha_{12} \cong \frac{\cancel{C^0_{1f}}\left(\dfrac{Q_{1m}A_m}{\delta_m V_d}\right)(1+a_2\Delta t)\Delta t\cancel{C^0_{2f}}}{\cancel{C^0_{1f}}\left(\dfrac{Q_{2m}A_m}{\delta_m V_d}\right)(1+a_1\Delta t)\Delta t\cancel{C^0_{2f}}}.$$

In the limit of $\Delta t \to 0$,

$$\alpha_{12} \cong \frac{Q_{1m}}{Q_{2m}}. \tag{4.3.13}$$

This relation identifies the intrinsic separation capability of the dialysis membrane for two solutes 1 and 2. The Δt value must be larger than the time required for the solutes to diffuse through the membrane and appear in the dialysate.

For porous membranes, one can employ the flux expression (3.4.97) to estimate Q_{im}:

$$Q_{im} = \frac{D_{il}\overline{G}_{Dr}(r_i, r_p)\varepsilon_m\kappa_{im}}{\tau_m}. \tag{4.3.14}$$

Therefore,

$$\alpha_{12}|_{t\to 0} = \frac{D_{1l}\,\overline{G}_{Dr}(r_1, r_p)\,\kappa_{1m}}{D_{2l}\,\overline{G}_{Dr}(r_2, r_p)\,\kappa_{2m}}. \tag{4.3.15}$$

An estimate of $\overline{G}_{Dr}(r_i, r_p)$ may be obtained from the Faxen relation (3.1.112e); the partition coefficient κ_{im} will be determined by the geometric partitioning effect. Such an

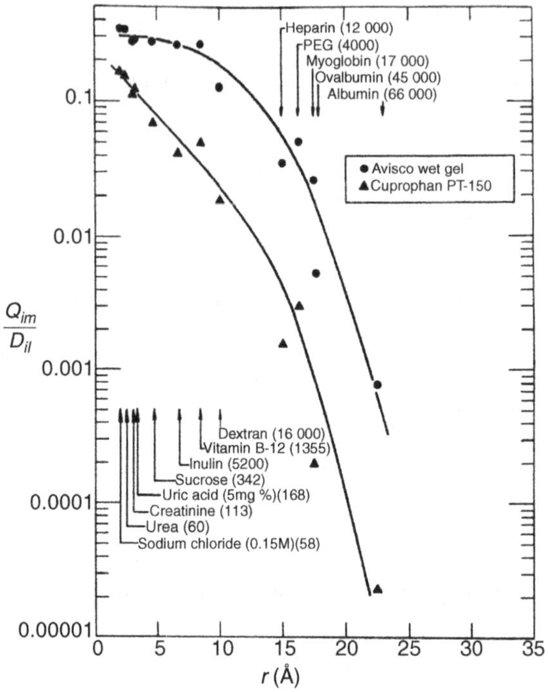

Figure 4.3.2. Permeability reduction as a function of characteristic solute radius (molecular weight in parenthesis) (Colton et al., *1973). Reprinted, with permission, from C.K. Colton, K.A. Smith, E.W. Merrill, P.C. Farrell,* J. Biomed. Mater. Res., *5, 459, (1971), Figure 5, p. 483, © 1971, John Wiley & Sons.*

approach is likely to be quite fruitful, provided the membrane pore size distribution is quite narrow around an average r_p. For many commercially available porous membranes, used, for example, in hemodialyzers, the structure is that of a water-swollen gel, where the transport corridors and channels have varying dimensions. Therefore, models based on cylindrical capillaries of radius r_p may not be accurate enough. However, Klein *et al.* (1979) suggest that their data for Cuprophan 150PM membrane for a variety of solutes may be reasonably modeled by an average pore diameter of 3.5 nm; on the other hand, such an average pore model could not describe the behavior of a number of other membranes with a wider pore size distribution. Estimates of Q_{im} for a variety of solutes and Cuprophan membranes may be obtained from the data of Colton *et al.* (1971) (see Figure 4.3.2) and Farrell and Babb (1973).

4.3.2 Separation between two counterions in two solutions separated by an ion exchange membrane: Donnan dialysis

Consider two electrolytes AN and BN in two different aqueous solutions separated by a cation exchange membrane (as in Figure 4.3.1(b)). Assume that the anion N is impermeable through the membrane. Suppose, at time

$t = 0$, the solution on one side of the membrane (the corresponding quantities are identified by the prime) contains only electrolyte AN, whereas the solution on the other side of the membrane (the corresponding quantities are identified by the double prime) contains only electrolyte BN. After a long time, equilibrium is expected to be established. Therefore the expression for Donnan potential (relation (3.3.118b)),

$$\phi_R - \phi_w = \frac{1}{Z_i \mathcal{F}}\left(RT \ln\left(\frac{a_{iw}}{a_{iR}}\right) - \overline{V}_i(P_R - P_w)\right)$$

for ionic species i, should be valid for both solutions on both sides of the membrane at equilibrium. For cation A and the side of the membrane corresponding to AN,

$$\phi'_R - \phi'_w = \frac{1}{Z_A \mathcal{F}}\left(RT \ln\left(\frac{a'_{Aw}}{a'_{AR}}\right) - \overline{V}_A(P'_R - P_w)\right), \quad (4.3.16a)$$

where ϕ'_w is the feed solution potential, if any. For cation A and the side of the membrane corresponding to BN,

$$\phi''_R - \phi''_w = \frac{1}{Z_A \mathcal{F}}\left(RT \ln\left(\frac{a''_{Aw}}{a''_{AR}}\right) - \overline{V}_A(P''_R - P_w)\right). \quad (4.3.16b)$$

The solution pressure, P_w, is the same on both sides. Assume swelling pressure effects to be such that $P'_R - P''_R \cong 0$. Then

$$\phi'_R - \phi'_w - \phi''_R + \phi''_w = \frac{RT}{\mathcal{F}}\left\{\ln\left[\frac{a'_{Aw}}{a'_{AR}}\right]^{\frac{1}{Z_A}} - \ln\left[\frac{a''_{Aw}}{a''_{AR}}\right]^{\frac{1}{Z_A}}\right\}. \quad (4.3.16c)$$

Now there is no potential difference across the membrane, so $\varphi'_R = \varphi''_R$. Therefore

$$\phi''_w - \phi'_w = \frac{RT}{\mathcal{F}}\left\{\ln\left[\frac{a'_{Aw}}{a''_{Aw}}\right]^{\frac{1}{Z_A}}\right\} \quad (4.3.17a)$$

since $a'_{AR} = a''_{AR}$, both being merely the activity of the counterion A in the membrane. But $\varphi''_w - \varphi'_w$ is independent of any ionic species, which suggests that the relation

$$\phi''_w - \phi'_w = \frac{RT}{\mathcal{F}}\left\{\ln\left[\frac{a'_{Bw}}{a''_{Bw}}\right]^{\frac{1}{Z_B}}\right\} \quad (4.3.17b)$$

is also valid. Consequently (Helfferich, 1962),

$$\left[\frac{a'_{Aw}}{a''_{Aw}}\right]^{\frac{1}{Z_A}} = \left[\frac{a'_{Bw}}{a''_{Bw}}\right]^{\frac{1}{Z_B}} = \text{a constant.} \quad (4.3.18)$$

The implications of this result may be understood if we know the concentrations of anion N in the two chambers. Suppose the chamber for species AN (′) has a high concentration of AN to start with, whereas the chamber for BN (″) has a low concentration of BN to start with. Due to the requirement for electroneutrality and the impermeability of the membrane to anion N, the concentrations of cations (A and B) in the chamber for BN will always be low. Therefore the ratio (a'_{Aw}/a''_{Aw}) will always be large.

Consider now the case where $Z_A = Z_B = 1$. This implies that, at equilibrium, (a'_{Bw}/a''_{Bw}) has a large value. Since the initial concentration of BN in chamber ″ was low to start with, distribution of B between the two chambers will drastically reduce the concentration of B in chamber ″ at equilibrium from its initial concentration. Therefore, if cation B is undesirable in the chamber ″ liquid, the concentration of B in chamber ″ can be drastically reduced via this phenomenon, called *Donnan dialysis*, using an acceptable cation A introduced via chamber ′. Further, by reducing the volume of chamber ′ to a small value, undesirable cation B may now be concentrated to a high level in chamber ′; the solution in chamber ′ is called the *strip solution*, whereas the original solution in chamber ″ containing BN only is called the *feed solution*. After equilibrium is reached, the solution in chamber ″ is called the raffinate (in analogy to liquid–liquid extraction).

Wallace (1967) illustrated this technique by concentrating uranyl ions (UO_2^{++}) from a 0.01 M uranyl nitrate $\left(UO_2(NO_3)_2\right)$ feed with a 2 M nitric acid strip using a cation exchange membrane (see Section 3.4.2.5). The concentration of UO_2^{++} in the raffinate was reduced to 0.67% of the feed solution (~0.01 M), whereas that in the strip was raised to 0.148 M. Wallace (1967) used an open system with feed and the strip solutions flowing on two sides of the membrane. A closed system will also achieve similar partitioning and separation. Another application is water softening. Using a strip solution containing the cation Na^+ at large concentrations, divalent cation Ca^{++} present in dilute concentrations in the feed solution (hard water) on the other side of a cation exchange membrane can be removed substantially and replaced by Na^+.

The difference between the basic result achieved at equilibrium under Donnan dialysis with that achieved in conventional dialysis using a neutral membrane (Section 4.3.1) is due to the assumption of perfect rejection of anion N by the cation exchange membrane in Donnan dialysis. Under such a condition, the requirements of the electroneutrality condition ensure that the two solutions will have radically different cation concentrations at equilibrium for any cation since the anion concentrations of the two chambers are so different.

In practical Donnan dialysis in open systems, with less than perfect rejection of co-ions, the calculation of ion transport rates through the membrane requires a knowledge of the ionic concentrations at the two membrane–solution interfaces on the two sides of the ion exchange membrane. These concentrations are to be determined for the electrolyte AN based on the Donnan potential based equilibrium relation (4.1.73) developed for ion exchange resins:

$$\left(\frac{a'_{Aw}}{a'_{Am}}\right)^{\frac{1}{Z_A}} = \left(\frac{a'_{Nw}}{a'_{Nm}}\right)^{\frac{1}{Z_N}}. \quad (4.3.19)$$

Here, subscript m refers to the membrane, w refers to the solution and the $'$ side of the membrane is under consideration. To determine a'_{Am} or C'_{Am} as a function of known quantities, we need also the electroneutrality relations for the solution and the membrane:

$$\text{solution} \qquad Z_A C'_{Aw} + Z_N C'_{Nw} = 0; \qquad (4.3.20)$$

$$\text{membrane} \qquad Z_A C'_{Am} + Z_N C'_{Nm} - C_m = 0, \qquad (4.3.21)$$

where C_m is the molar concentration of the negatively charged fixed ionic groups in the cation exchange membrane.

Consider a specific electrolyte $CaSO_4$, so $Z_A = 2$ and $Z_N = -2$. It follows from (4.3.19) that

$$(a'_{Aw}\, a'_{Nw}) = (a'_{Am}\, a'_{Nm}). \qquad (4.3.22a)$$

Using molar concentrations and appropriate activity coefficients,

$$(C'_{Aw} C'_{Nw})\,(\gamma^{M'}_{Aw}\gamma^{M'}_{Nw}) = (C'_{Am} C'_{Nm})\,(\gamma^{M'}_{Am}\gamma^{M'}_{Nm}). \qquad (4.3.22b)$$

Assume that $\gamma^{M'}_{Aw}\gamma^{M'}_{Nw} \cong \gamma^{M'}_{Am}\gamma^{M'}_{Nm}$. Then

$$(C'_{Aw} C'_{Nw}) \cong (C'_{Am} C'_{Nm}). \qquad (4.3.22c)$$

From (4.3.20), $C'_{Aw} = C'_{Nw}$, and, from (4.3.21),

$$C'_{Nm} = (-Z_A C'_{Am} + C_m)/Z_N. \qquad (4.3.22d)$$

Therefore,

$$C'^2_{Aw} = -\frac{Z_A}{Z_N} C'^2_{Am} + \frac{C'_{Am} C_m}{Z_N} = +C'^2_{Am} - \frac{C'_{Am} C_m}{2};$$

$$C'^2_{Am} - \frac{C'_{Am} C_m}{2} - C'^2_{Aw} = 0. \qquad (4.3.22e)$$

The acceptable solution of this quadratic is

$$C'_{Am} = \frac{C_m}{4} + \sqrt{\frac{C_m{}^2}{16} + C'^2_{Aw}}. \qquad (4.3.22f)$$

The molar concentration of the anion in the membrane at the interface C'_{Nm} can now be obtained from (4.3.21) since C'_{Am} is known. A similar analysis may be carried out for the $''$ side of the membrane. Note that if the AN electrolyte concentration in the solution at the membrane–solution interface is C'_{sw}, then, for $CaSO_4$, $C'_{Aw} = C'_{sw}$.

4.3.3 Separation of a gas mixture by gas permeation

Gas transport through a nonporous polymeric membrane, called *gas permeation*, has been considered in Section 3.4.2.2. In practical gas separation processes, the feed gas mixture is allowed to flow on one side of the membrane; one of the species preferentially permeates through the membrane to the other side. The permeated gases are enriched in the preferentially permeating species. The residual feed gas, as well as the permeated gas mixture, are removed continuously from the gas permeation membrane device. Consider such a device. At time $t = 0$, a mixture containing m^0_{Af} moles of species A and m^0_{Bf} moles of species B is introduced into the feed side, the permeate side does not have any gas molecules, and both feed and permeate side are kept closed. One would like to know the condition at equilibrium as $t \rightarrow \infty$. The volumes of the feed and permeate sides are V_f and V_p, respectively; the temperature is T, the membrane area is A_m, the membrane thickness is δ_m and ideal gas behavior may be assumed.

Due to ideal gas assumption, the partial pressures of species A in the feed chamber, p_{Af}, and the permeate chamber, p_{Ap}, are related to the corresponding moles of species A, m_{Af} and m_{Ap}, respectively, at any time t by

$$p_{Af} = \frac{m_{Af}RT}{V_f} \quad \text{and} \quad p_{Ap} = \frac{m_{Ap}RT}{V_f}. \qquad (4.3.23)$$

The rates of change in the number of moles of species A in the two chambers are

$$\frac{dm_{Af}}{dt} = -\frac{Q_A A_m}{\delta_m}\left(\frac{m_{Af}RT}{V_f} - \frac{m_{Ap}RT}{V_p}\right); \qquad (4.3.24)$$

$$\frac{dm_{Ap}}{dt} = \frac{Q_A A_m}{\delta_m}\left(\frac{m_{Af}RT}{V_f} - \frac{m_{Ap}RT}{V_p}\right) \qquad (4.3.25)$$

Laplace transforms of these two equations from the t-domain to the s-domain yield

$$s\,\overline{m}_{Af} - m^0_{Af} = \frac{Q_A RTA_m}{V_f\delta_m}\left(-\overline{m}_{Af} + \frac{V_f}{V_p}\overline{m}_{Ap}\right); \qquad (4.3.26)$$

$$s\,\overline{m}_{Ap} = \frac{Q_A RTA_m}{V_f\delta_m}\left(\overline{m}_{Af} - \frac{V_f}{V_p}\overline{m}_{Ap}\right). \qquad (4.3.27)$$

From these two equations, one can easily obtain the following expression for $\overline{m}_{Ap}$:

$$\overline{m}_{Ap} = m^0_{Af}\left(\frac{Q_A RTA_m}{V_f\delta_m}\right)\frac{1}{s^2 + a_1 s} \qquad (4.3.28)$$

where

$$a_1 = \left(\frac{Q_A RTA_m}{V_p\delta_m} + \frac{Q_A RTA_m}{V_f\delta}\right). \qquad (4.3.29)$$

On inverting relation (4.3.28), we get

$$m_{Ap} = m^0_{Af}\left(\frac{Q_A RTA_m}{V_f\delta_m}\right)\left(\frac{1}{a_1}(1 - e^{-a_1 t})\right). \qquad (4.3.30)$$

At time $t \rightarrow \infty$,

$$m_{Ap} = \frac{m^0_{Af}}{a_1}\left(\frac{Q_A RTA_m}{V_f\delta_m}\right) = m^0_{Af}\frac{V_p}{V_f + V_p}. \qquad (4.3.31)$$

Correspondingly, as $t \to \infty$,

$$m_{Af} = m^0_{Af} - m_{Ap}|_{t\to\infty} = m^0_{Af}\frac{V_f}{V_f + V_p}. \tag{4.3.32}$$

Therefore, as $t \to \infty$,

$$p_{Af} = \frac{m^0_{Af}RT}{V_f + V_p} \quad \text{and} \quad p_{Ap} = \frac{m^0_{Af}RT}{V_f + V_p}. \tag{4.3.33}$$

Thus, the partial pressures of species A on the two sides of the membrane are equal at equilibrium. The same result will be obtained for species B present originally in the feed reservoir, i.e.

$$p_{Bf} = \frac{m^0_{Bf}RT}{V_f + V_p} \quad \text{and} \quad p_{Bp} = \frac{m^0_{Bf}RT}{V_f + V_p} \tag{4.3.34}$$

at time $t \to \infty$. Therefore, at equilibrium,

$$(p_{Af}/p_{Bf}) = (m^0_{Af}/m^0_{Bf}) = (p^0_{Af}/p^0_{Bf}), \tag{4.3.35}$$

where p^0_{Af} and p^0_{Bf} are the partial pressures of species A and B in the feed chamber in the feed gas mixture at time $t = 0$. Therefore, no separation has taken place in the gas mixture left on the feed side at equilibrium. Similarly, at equilibrium,

$$(p_{Ap}/p_{Bp}) = (m^0_{Af}/m^0_{Bf}) = (p^0_{Af}/p^0_{Bf}). \tag{4.3.36}$$

Therefore, the permeate side gas mixture at equilibrium has a composition equal to that of the original feed gas mixture at $t = 0$; no separation has taken place in the closed system. This is why gas permeation devices in practice are operated as an open system: any gas mixture produced by permeation is immediately withdrawn from the device. Since the intrinsic membrane transport rates of the two species for unit partial pressure difference between the feed and the permeate are different, the gas mixture produced as a permeate has a composition different from that of the feed. The permeate mixture withdrawn from the device must have a composition different from that of the feed mixture for separation. See Sections 6.3.3.5, 6.4.2.2, 7.2.1.1 and 8.1.9 for open systems. Continuous energy input will yield separation in a closed system - see Figures 8.1.4(b) for a system operated at total reflux.

One can demonstrate such separation capabilities, even in the closed vessel considered so far, if we focus, for example, on the initial time period in the vessel when $t \to 0$. We can calculate the moles of species A and B in the permeate chamber when $t \to 0$. For species A, differentiate the expression (4.3.30) for m_{Ap} with time t to obtain, by Taylor series expansion around $t = 0$,

$$m_{Ap}\Big|_{t\to 0} = m_{Ap}\Big|_{t=0} + \frac{dm_{Ap}}{dt}\Big|_{t\to 0}(\Delta t)$$
$$= 0 + m^0_{Af}\left(\frac{Q_A RTA_m}{V_f\delta_m}\right)e^{-a_1\Delta t}\Delta t. \tag{4.3.37}$$

Since the general expression for m_{Bp} may be written as

$$m_{Bp} = m^0_{Bf}\left(\frac{Q_B RTA_m}{V_f\delta_m}\right)\left(\frac{1}{a_2}(1 - e^{-a_2 t})\right), \tag{4.3.38}$$

where

$$a_2 = \left(\frac{Q_B RTA_m}{V_p\delta_m} + \frac{Q_B RTA_m}{V_f\delta_m}\right),$$

we get, similarly,

$$m_{Bp}|_{t\to 0} = m^0_{Bf}\left(\frac{Q_B RTA_m}{V_f\delta_m}\right)e^{-a_2\Delta t}\Delta t. \tag{4.3.39}$$

The separation factor α_{AB} as $t \to 0$ may be expressed as

$$\alpha_{AB}|_{t\to 0} = \frac{\left(\dfrac{m_{Ap}}{m_{Ap}+m_{Bp}}\right)\left(\dfrac{m_{Bf}}{m_{Af}+m_{Bf}}\right)}{\left(\dfrac{m_{Af}}{m_{Af}+m_{Bf}}\right)\left(\dfrac{m_{Bp}}{m_{Ap}+m_{Bp}}\right)} = \frac{m_{Ap}\, m_{Bf}}{m_{Bp}\, m_{Af}}\Bigg|_{t\to 0}. \tag{4.3.40}$$

For a system where very little of the feed permeates in the short time interval Δt near $t \to 0$, we can assume easily that $(m_{Bf}/m_{Af}) \cong (m^0_{Bf}/m^0_{Af})$. Therefore

$$\alpha_{AB}|_{t\to 0} = \frac{Q_A\, e^{a_2\Delta t}}{Q_B\, e^{a_1\Delta t}} = \left(\frac{Q_A}{Q_B}\right)\left(\frac{1 + a_2\Delta t + \cdots}{1 + a_1\Delta t + \cdots}\right). \tag{4.3.41}$$

This result identifies (Q_A/Q_B) as the initial time separation factor (only after the time lag period is over); as time increases, this factor decreases from this high value. As $t \to \infty$, there is no separation.

In practical gas separation processes in open separators using gas permeation through nonporous membranes, the selectivity (Q_A/Q_B) of gas species A over B is of considerable importance. From the flux expression (3.4.72), we may write

$$\alpha_{AB}|_{t\to 0} = \frac{Q_A}{Q_B} = \left(\frac{D_{AM}}{D_{BM}}\right)\left(\frac{S_{AM}}{S_{BM}}\right). \tag{4.3.42}$$

The selectivity, α_{AB}, is often broken up into a product of two factors, the *diffusivity selectivity* (also known as the *mobility selectivity*), (D_{AM}/D_{BM}), and the *solubility selectivity*, (S_{AM}/S_{BM}). Table 4.3.1 illustrates the contributions of

Table 4.3.1. Transport, solubility and selectivity properties for the gas pair helium (A)-methane (B) at 25 °C for a variety of polymers[a]

Polymer	α_{AB}	D_A/D_B	S_A/S_B
Silicone rubber	0.38	5	0.075
Natural rubber	1.08	24	0.044
Hydropol (hydrogenated polybutadiene)	1.21	28	0.043
Low-density polyethylene, $\rho = 0.9137$ g/cm^3	1.70	35	0.048
High-density polyethylene, $\rho = 0.964$ g/cm^3	2.94	55	0.054
Poly (vinyl acetate), glassy	355.00	5000	0.071

[a] After Paul (1971).

Table 4.3.2. Kinetic[a] sieving diameter of a gas/vapor based on the smallest zeolite window where it can fit

Molecule	d_i, kinetic sieving diameter (nm)
He	0.26
H_2	0.289
NO	0.317
CO_2	0.33
Ar	0.34
O_2	0.346
N_2	0.364
CO	0.376
CH_4	0.38
C_2H_4	0.39
C_3H_8	0.396
n-C_4	0.43
CF_2Cl_2	0.43
C_3H_6	0.44
CF_4	0.45
i-C_4	0.47

[a] Breck (1974).

mobility and solubility selectivity to the overall selectivity α_{AB} for the gas mixture of helium (A)–methane (B) through a variety of polymeric membranes (Paul, 1971).

Due to the smaller size of helium compared to methane (Table 4.3.2), the diffusivity of helium is higher than that of methane through all membranes identified in Table 4.3.1. Therefore, the diffusivity selectivity (D_{AM}/D_{BM}) is greater than 1 in all cases. On the other hand, we know from Section 3.3.7.3, that the more condensable the gas (higher T_c), the higher its solubility coefficient. Since methane is more easily condensable, (S_{AM}/S_{BM}) is less than 1; further, the value of this ratio is independent of the polymer material, whereas (D_{AM}/D_{BM}) depends very much on the structure of the polymer.

Generally, the larger the size or molar volume of the gas/vapor molecule, the more it is impeded in its motion by the polymer chains in a polymer membrane. Thermal and other motions of the chains, creating openings in the membrane through which gas/vapor species can diffuse strongly, influence the value of D_{im} of any species i. For polymers that are glassy at a given temperature T ($<T_g$ of the polymer), the mobility of the gas/vapor molecules, and therefore of D_{im}, is considerably reduced. An increase in the size of the gas/vapor molecule brings about a sharp drop in D_{im}. On the other hand, the polymer chains in a rubbery polymer at a given temperature T ($> T_g$ of the polymer) are very flexible and offer much less hindrance to the permeation of gas/vapor molecules. Therefore, D_{im} values are much larger for any given species in a rubbery polymeric membrane; further, the effect of a change in the size of the gas/vapor molecule is much less. These types of behaviors are illustrated in Figure 4.3.3 for a wide variety of gases and vapors for a rubbery polymer such as natural rubber and a glassy polymer such as polyvinyl chloride. The basis for the wide variation in the mobility selectivity between a rubbery and a glassy polymer in Table 4.3.1 should now be clearer.

Figure 4.3.4 illustrates the solubility coefficient S_{im} (also called the sorption coefficient) of a number of vapors and a few gases in a natural rubber membrane against the molar volume of the vapor/gas species. Generally, the larger the molecule, the more condensible it is. Correspondingly, its solubility in the membrane material is also higher. Therefore, a larger species which has higher condensibility, and therefore higher solubility in the membrane, will have higher solubility selectivity with respect to a smaller species.

The general characteristics of the solubility selectivity and the mobility selectivity, and therefore the membrane selectivity, of gas species A over B through a polymer membrane as described above can be given in a more quantitative basis as follows (Michaels and Bixler, 1968). First, consider the *solubility coefficient* S_{im} of a gas species i in a polymer membrane. Jolley and Hildebrand (1958) have shown that the Henry's law constant for a lot of gases in simple organic liquids at a reference temperature T_0 can be correlated with the Lennard-Jones force constant ε_i for the gas i by an equation of the form

$$\ln S_{im,T_0} = a(\varepsilon_i/\mathrm{k^B}) + \ln b_m; \quad S_{im,T_0} = b_m \exp(a\varepsilon_i/\mathrm{k^B}), \tag{4.3.43a}$$

where the quantity a depends on the temperature, $\mathrm{k^B}$ is the Boltzmann constant and b_m is a constant dependent on the solvent. This equation has been found to describe the gas solubility coefficients S_{im,T_0} in amorphous polymers also, where polymer-gas interactions may be neglected. Further, for the solubility of such gases in amorphous polymers, the van't Hoff equation,

$$\frac{\mathrm{d}\ln S_{im}}{\mathrm{d}(1/T)} = -\frac{\Delta H_s}{\mathrm{R}}, \tag{4.3.43b}$$

has been found to be valid, just as in the case of gas solubilities in organic liquids (see equation (4.1.10)). Here, ΔH_s, the enthalpy of the solution of gas i in polymer m, is related to the Lennard-Jones constant ε_i via

$$\Delta H_s = b'_m - a'_m(\varepsilon_i/\mathrm{k^B}), \tag{4.3.43c}$$

where a'_m and b'_m depend on the polymer. One can now combine (4.3.43a,b,c) to obtain

$$\ln(S_{im}/S_{im,T_0}) = \frac{b'_m - a'_m\,(\varepsilon_i/\mathrm{k^B})}{\mathrm{R}\,T_0}\left(1 - \frac{T_0}{T}\right). \tag{4.3.44a}$$

The solubility selectivity of a polymer for two gases A and B at T_0 may be obtained from (4.3.43a) as follows:

$$\left(\frac{S_{Am,T_0}}{S_{Bm,T_0}}\right) = \exp\left\{a\frac{(\varepsilon_A - \varepsilon_B)}{\mathrm{k^B}}\right\}. \tag{4.3.44b}$$

Thus, the solubility selectivity as a first approximation depends only on the gas pair, and does *not* depend on

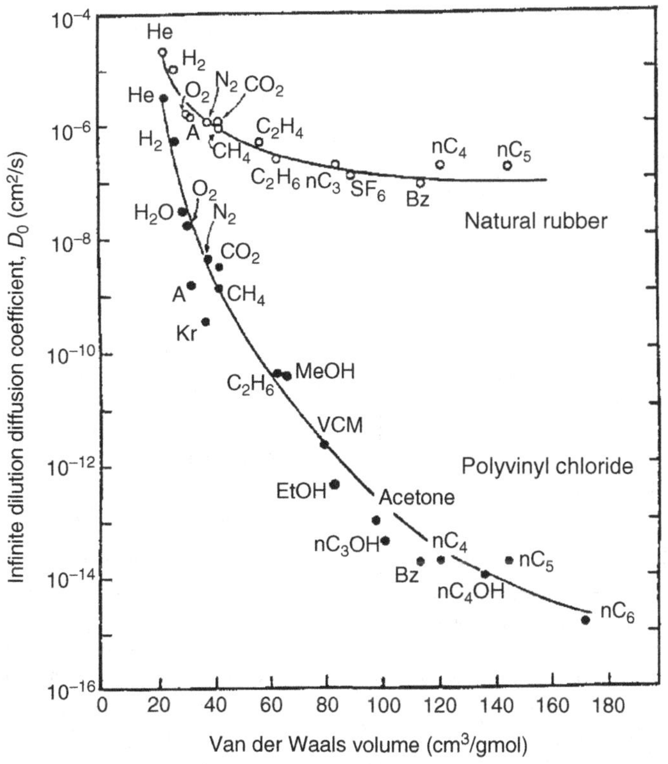

Figure 4.3.3. Dependence of diffusivity on gas/vapor molecular size (Chern et al., *1985). Reprinted, with permission, from* Material Science of Synthetic Membranes, *ACS Symposium Series 269, 1985, p. 492, D. Lloyd (ed.), R.T. Chern, W.J. Koros, H.B. Hopfenberg and V.T. Stannett, (Chapter 2 authors), Figure 1 on p. 28 of Chapter 2. Copyright (1985) American Chemical Society.*

the polymer membrane. However, if we use relation (4.3.44a), we obtain for solubility selectivity, at any temperature T,

$$\left(\frac{S_{Am}}{S_{Bm}}\right) = \left(\frac{S_{AT_0}}{S_{BT_0}}\right) \exp\left\{\frac{a'_m\,(\varepsilon_B - \varepsilon_A)/k^B}{RT_0}\left(1 - \frac{T_0}{T}\right)\right\}. \tag{4.3.45}$$

There is a weak dependence on the membrane.

Consider now the *diffusion coefficient* of a gas species i through a polymer membrane for the purpose of determining the mobility selectivity. The diffusion coefficient of a gas i at a reference temperature T_0 has been found to vary with the effective diameter d_i of the gas species as (Michaels and Bixler, 1968)

$$D_{iT_0} = g_m \exp\,(-f_m d_i^n), \tag{4.3.46a}$$

where g_m and f_m are characteristic of the polymer forming the membrane; n has the value of 2 for rotationally hindered stiff-chain macromolecules such as polypropylene, poly(ethylene) terephthalate, cellulose, polyimide, etc.; n is 1 for flexible-chain polymers such as natural rubber, polyethylene, etc. The estimate of d_i based on the kinetic sieving diameter, which is the smallest zeolite window through which a gas molecule can fit, is considered to be quite appropriate (Table 4.3.2).

As in conventional theories of activated-state-based diffusion in liquids, diffusion of a gas molecule through an amorphous polymer membrane is assumed to be an activated process; it involves the cooperative movements of the gas molecule and the local polymer chain segments around it. Correspondingly, the temperature dependence of D_{im} depends on an Arrhenius relation,

$$\frac{\mathrm{dln}D_{im}}{\mathrm{d}(1/T)} = -\frac{E_{D_i}}{R}, \tag{4.3.46b}$$

which implies

$$D_{im} = D_{iT_0} \exp\left(-\frac{E_{D_i}}{R}\left\{\frac{1}{T} - \frac{1}{T_0}\right\}\right), \tag{4.3.46c}$$

where the activation energy, E_{D_i}, has been related to the effective diameter d_i of the gas species i via

$$E_{D_i} = E_{Dm} + f'_m\, d_i^n. \tag{4.3.46d}$$

Here, E_{Dm} and f'_m are constants, which depend on the polymer. Thus, D_{im} may be expressed as follows:

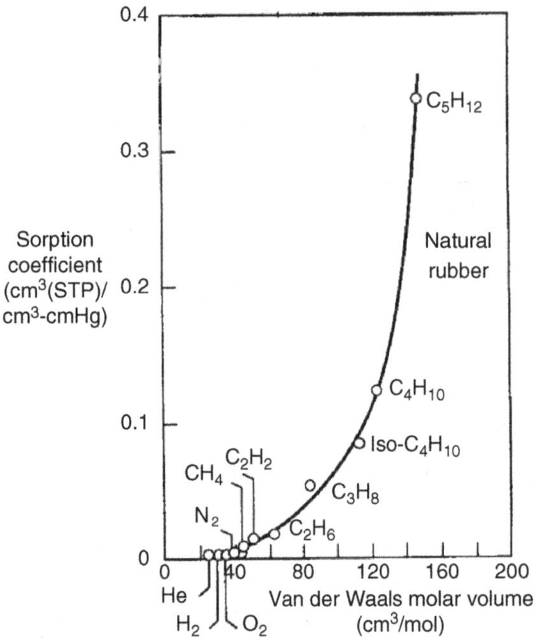

Figure 4.3.4. Henry's law sorption coefficient of vapors and gases as a function of molar volume for natural rubber membranes. From Figure 3, p. 361, in "Membrane Separation of Organic Vapors from Gas Streams'", R.W. Baker and J.G. Wijmans, Chapter 8 in Polymeric Gas Separation Membranes, *D.R. Paul and Y.P. Yampolskii (eds.) 1994, CRC Press; reprinted by permission of the publisher (Taylor & Francis Group, http://www.informaworld.com.)*

$$D_{im} = g_m \exp\left(-f_m d_i^n\right) \exp\left\{\frac{E_{Dm} + f'_m d_i^n}{RT_0}\left(1 - \frac{T_0}{T}\right)\right\}. \tag{4.3.47}$$

Consequently, the mobility selectivity of species A over B through an amorphous polymer membrane is given by

$$\frac{D_{Am}}{D_{Bm}} = \exp\{-f_m(d_A^n - d_B^n)\} \exp\left\{\frac{f'_m(d_A^n - d_B^n)}{RT_0}\left(1 - \frac{T_0}{T}\right)\right\}, \tag{4.3.48}$$

which shows that the mobility selectivity depends almost solely on $(d_A^n - d_B^n)$, i.e. on the relative difference between the nth power of the effective diameters of two types of gas molecules, d_A and d_B, for a particular membrane. Further, it is much larger for stiff-chain polymers ($n = 2$) compared to flexible-chain polymers ($n = 1$).

One can now develop an illustrative expression for the permeability Q_{im} of species i through a membrane by combining expression (4.3.47) for D_{im} and relation (4.3.44a) for S_{im} (Michaels and Bixler, 1968):

$$Q_{im} = D_{im}S_{im} = g_m b_m \exp\left\{\left[a\left(\frac{\varepsilon_i}{k^B}\right) - f_m d_i^n\right] + \left(1 - \frac{T_0}{T}\right)\frac{E_{Dm} + f'_m d_i^n + b'_m - a'_m(\varepsilon_i/k^B)}{RT_0}\right\}. \tag{4.3.49}$$

For a binary gas pair i = A and B, the ideal selectivity $\alpha_{AB}|_{t\to 0} = \alpha^*_{AB}$ displayed by the membrane is therefore

$$\alpha^*_{AB} = \alpha_{AB}|_{t\to 0} = \frac{Q_{Am}}{Q_{Bm}} = \exp\left\{a\,\frac{\varepsilon_A - \varepsilon_B}{k^B} - f_m(d_A^n - d_B^n) + \left(1 - \frac{T_0}{T}\right)\frac{f'_m(d_A^n - d_B^n) - a'_m\dfrac{(\varepsilon_A - \varepsilon_B)}{k^B}}{RT_0}\right\}. \tag{4.3.50}$$

A few comments can now be made about the effect of some important quantities on the ideal selectivity between species A and B, where d_A is, say, smaller than d_B and (ε_A/k^B) is larger than, say, (ε_B/k^B) (Michaels and Bixler, 1968):

(1) For a smaller gas A ($d_A < d_B$), other items remaining favorable, $\alpha^*_{AB} > 1$; i.e. the smaller molecules of species A are more permeable than the molecules of species B. The membrane is A-selective.
(2) For a more condensable gas, which therefore has a higher critical temperature (say, $T_c|_A > T_c|_B$), the selectivity $\alpha^*_{AB} > 1$; i.e. the membrane is more selective to species A. This is so since (ε_i/k^B) is roughly proportional to the critical temperature (e.g. $\varepsilon_i/k^B = 0.77T_c|_i$).
(3) At the reference temperature T_0, the ideal selectivity α^*_{AB} is simply expressed by

$$\alpha^*_{AB}|_{T_0} = \left.\frac{Q_{Am}}{Q_{Bm}}\right|_{T_0} = \exp\left\{a\frac{\varepsilon_A - \varepsilon_B}{k^B} - f_m(d_A^n - d_B^n)\right\}. \tag{4.3.51}$$

At higher temperatures, the ideal selectivity $\alpha^*_{AB}|_T$ will be lower than $\alpha^*_{AB}|_{T_0}$.
(4) Membranes fabricated from stiff-chain polymers, where $n \cong 2$, have much higher selectivity α^*_{AB} than membranes prepared from flexible-chain polymers, where $n \cong 1$.

Since different polymers have been observed to possess, for a gas, permeability coefficients that vary over many orders of magnitude, a question of general interest is: what happens to the corresponding selectivities for a particular gas pair? We adopt here an approach suggested by Freeman (1999) to illustrate how α_{AB} is likely to vary with, say, Q_{Am} where A is the smaller gas molecule. Freeman (1999) may also be consulted for an overview of the subject by the current author (following Michaels and Bixler (1968)), albeit with somewhat different notation and the literature, and also for numerical estimates of various quantities.

First, the ideal selectivity α^*_{AB} may be described via equations (4.3.42) and (4.3.48) as

$$\ln \alpha^*_{AB} = \ln (S_{Am}/S_{Bm}) + \ln (D_{Am}/D_{Bm})$$
$$= \ln (S_{Am}/S_{Bm}) + (d^n_A - d^n_B)\left[-f_m + \frac{f'_m}{RT_0}\left(1 - \frac{T_0}{T}\right)\right]$$
$$= \ln (S_{Am}/S_{Bm}) - \left[\left(\frac{d_B}{d_A}\right)^n - 1\right] d^n_A \left[-f_m + \frac{f'_m}{RT_0}\left(1 - \frac{T_0}{T}\right)\right]. \quad (4.3.52)$$

From expression (4.3.49) for the permeability coefficient Q_{im} and expression (4.3.47) for D_{im}, we get

$$\ln Q_{im} = \ln S_{im} + \ln D_{im}$$
$$= \ln (S_{im} g_m) + \left\{-f_m d^n_i + \frac{E_{Dm} + f_m d^n_i}{RT_0}\left(1 - \frac{T_0}{T}\right)\right\};$$
$$\ln Q_{im} - \ln (S_{im} g_m)$$
$$- \frac{E_{Dm}}{RT_0}\left(1 - \frac{T_0}{T}\right) = d^n_i \left\{-f_m + \frac{f_m}{RT_0}\left(1 - \frac{T_0}{T}\right)\right\}. \quad (4.3.53)$$

Introduce $i = A$ in this relation and substitute it into relation (4.3.52) to obtain

$$\ln \alpha^*_{AB} = -\left[\left(\frac{d_B}{d_A}\right)^n - 1\right]\left\{\ln Q_{Am} - \ln (S_{Am} g_m) - \frac{E_{Dm}}{RT_0}\left(1 - \frac{T_0}{T}\right)\right\}$$
$$+ \ln (S_{Am}/S_{Bm}).$$

Rearrange it to get

$$\ln \alpha^*_{AB} = -\left[\left(\frac{d_B}{d_A}\right)^n - 1\right] \ln Q_{Am}$$
$$+ \left\{\ln (S_{Am}/S_{Bm}) - \left[\left(\frac{d_B}{d_A}\right)^n - 1\right]\right.$$
$$\left.\times\left(-\ln g_m - \frac{E_{Dm}}{RT_0}\left(1 - \frac{T_0}{T}\right) - \ln S_{Am}\right)\right\}. \quad (4.3.54)$$

To understand the implications of this expression for α^*_{AB}, recognize that the solubility selectivity (S_{Am}/S_{Bm}) for the gas pair A-B changes very little with a change in the polymer (see, e.g., Table 4.3.1); further, the variations of the solubility coefficient of gases with different polymers are very limited (thus, g_m, E_{Dm}, etc., may not vary much). On the other hand, it is well known that Q_{im}, as well as α^*_{AB}, varies widely with the polymer. The relation (4.3.54) developed above suggests that a plot of $\ln\alpha^*_{AB}$ against $\ln Q_{Am}$ will decrease linearly with a slope of $-[(d_B/d_A)^n - 1)]$. In fact, the so-called "upper bound" line of Robeson (1991) observed for a variety of gas-pair systems over a wide variety of polymers is well described by this type of suggested behavior (Freeman, 1999). The gas pairs useful in this analysis are: He-H_2; He-CO_2; He-N_2; He-CH_4; H_2-O_2; H_2-N_2; H_2-CO_2; H_2-CH_4; O_2-N_2; CO_2-CH_4. Figure 4.3.5 illustrates the suggested behavior (shown by the solid line) for a H_2-N_2 system for a wide range of polymers.

Table 4.3.3. Permeabilities[a] of carbon dioxide and nitrogen through various polymers at 30 °C

Film	$Q_{CO_2} \times 10^{11}$	$Q_{N_2} \times 10^{11}$	$\alpha_{CO_2\text{-}N_2} = Q_{CO_2}/Q_{N_2}$
Saran	0.29	0.0094	30.9
Mylar	1.53	0.05	30.6
Nylon	1.6	0.10	16.0
Pliofilm NO	1.7	0.08	21.2
Hycar OR 15	746	2.35	31.7
Butyl rubber	518	3.12	17.4
Methyl rubber	75	4.8	15.7
Vulcaprene	186	4.9	37.9
Hycar OR 25	186	6.04	30.9
Pliofilm P4	182	6.2	29.4
Perbunan	309	10.6	29.1
Neoprene	250	11.8	21.1
Polyethylene	352	19	18.5
Buna S	1240	63.5	19.6
Polybutadiene	1380	64.5	21.4
Natural rubber	1310	80.8	16.3

[a] From Stannett (1968).
Units of Q_i are scc-cm/cm^2-s-cm Hg.

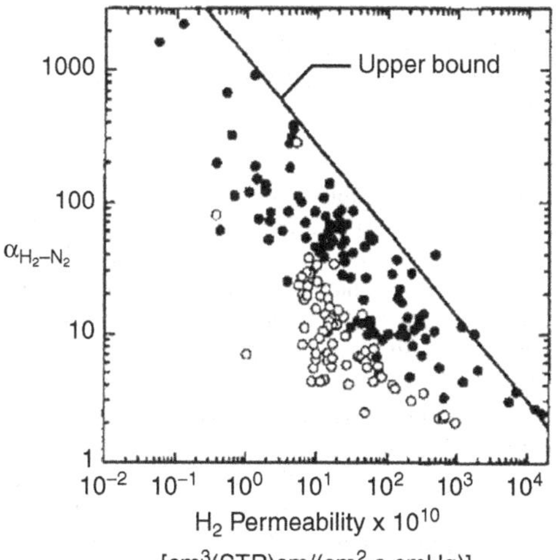

Figure 4.3.5. Relationship between hydrogen permeability and H_2/N_2 selectivity for rubbery (○) and glassy (●) polymers and the empirical upper bound relation of Robeson (1991). (From Freeman (1999).) Reprinted, with permission, from Macromolecules, ***32**, (1999) 375, Figure 1, Copyright (1999) American Chemical Society.*

Over a smaller range of permeability variation of a gas with a series of polymers, the data in Figure 4.3.5 appear to indicate that the separation factor may be almost constant. If we consider flexible polymers ($n = 1$) and equation (4.3.50) at, say, $T = T_0$ (for simplicity), then

$$\alpha_{AB}^{*} = \exp\left\{a\frac{\varepsilon_A - \varepsilon_B}{k^B} - f_m(d_A - d_B)\right\}. \quad (4.3.55)$$

Since, for flexible polymers, the mobility based selectivity plays less of an important role, it is likely that α_{AB}^{*} may depend primarily on the gas-pair A-B. Table 4.3.3 illustrates that, for a number of polymers, the selectivity for the CO_2-N_2 system varies by less than a factor of 2, whereas the permeability coefficient Q_{im} varies by 5000 times (Stannett, 1968).

A simplistic interpretation is provided by the following approach, where the permeability coefficient of species i through a polymeric membrane m, Q_{im}, is assumed to be described as the product of three quantities:

$$Q_{im} = F_m(\text{polymer})\ G_i(\text{gas})\ \gamma(i, m), \quad (4.3.56a)$$

where $\gamma(i, m)$ accounts for the specific interaction between the gas and the polymeric membrane; however, F_m depends only on the polymer and G_i depends only on the gas species i. If $\gamma(i, m) \cong 1$, then

$$\alpha_{AB}^{*} = \frac{Q_{Am}}{Q_{Bm}} = G_A/G_B. \quad (4.3.56b)$$

The limited set of data in Table 4.3.3 lends support to this simplistic explanation since the selectivity for the CO_2-N_2 system does not appear to depend much on the nature of the polymers listed.

Example 4.3.1

(a) Determine the solubility selectivity of helium (A) and methane (B) at T_0, corresponding to 25 °C for a polydimethylsiloxane membrane. Compare with the value quoted in Table 4.3.1. The value of a in equation (4.3.43a) is 0.023 K^{-1} (Freeman, 1999); (ε/k^B) for He and CH_4 are 10.2 and 149 (K), respectively (Freeman, 1999).

(b) Determine the mobility selectivity of a helium (A)-methane (B) pair at 25 °C for a polydimethylsiloxane membrane. Compare with the value quoted in Table 4.3.1. Assume $T = T_0$, corresponding to 25 °C; the value of f_m in equation (4.3.46a) may be obtained from Freeman (1999) as

$$f_m = c\,\frac{(1-a)}{RT} = 250\,\frac{\text{cal}}{\text{mol-Å}^2} \times \frac{(1-0.64)}{RT}.$$

Solution (a) From equation (4.3.44b), the solubility selectivity for the gas pair helium (He)-methane (CH_4) is

$$\frac{S_{He,T_0}}{S_{CH_4,T_0}} = \exp\left(\frac{a(\varepsilon_{He} - \varepsilon_{CH_4})}{k^B}\right) = \exp\left(0.023\ K^{-1}(10.2 - 149)K\right)$$

$$= \exp(-0.023 \times 138.8) = \frac{1}{\exp(3.192)} = \frac{1}{24.23} = 0.041.$$

Therefore, the solubility selectivity of a He-CH_4 pair at $T_0 = T =$ 25 °C is 0.041. Table 4.3.1 quotes a value of 0.075. For other polymers, such as like natural rubber, the values are close to 0.041.

(b) From equation (4.3.48), at $T = T_0$, the mobility selectivity for the helium-methane pair for any membrane is

$$(D_{Hem}/D_{CH_4m}) = \exp\left(-f_m(d_{He}^n - d_{CH_4}^n)\right). \quad (4.3.57)$$

From Table 4.3.2, d_{He} = 2.69 Å (= 0.269 nm) and d_{CH_4} = 3.87 Å(= 0.387 nm); further,

$$f_m = 250\frac{\text{cal}}{\text{mol} - \text{Å}^2}\,\frac{0.36}{1.987\dfrac{\text{cal}}{\text{mol} - \text{K}}\ 298\ \text{K}} = \frac{250 \times 0.36}{1.987 \times 298}\,\frac{1}{\text{Å}^2};$$

$$(D_{Hem}/D_{CH_4m}) = \exp\left(-\frac{250 \times 0.36}{1.987 \times 298}\,\frac{1}{\text{Å}^2}\,(7.23 - 14.9)\ \text{Å}^2\right)$$

$$= \exp\,(1.165) = 3.205.$$

The result quoted in Table 4.3.1 is 5.

4.3.4 Separation of a pressurized liquid solution through a membrane

Consider a liquid solution of a microsolute or a macrosolute under pressure (e.g. by a piston) on one side (side 1) of a membrane; let the other side of the membrane (side 2) be empty at time $t = 0$. As the solvent passes through the membrane (porous/nonporous) under pressure, the other side will fill up, if it has a fixed volume. At this time, and afterwards, what will happen to the solute with respect to the two chambers? If the membrane is semipermeable, i.e. it is impermeable to the solute but permeable to the solvent, then we have seen in Section 3.3.7.4, that, due to osmotic equilibrium, the pressures on the two sides will be related to the osmotic pressure of the feed solution (side 1) on the side of the piston by

$$\overline{V}_s(P_1 - P_2) = \overline{V}_s\pi_1 = RT\ln\frac{a_{s2}}{a_{s1}}. \quad (4.3.58)$$

One side will have pure solvent at activity a_{s2}; the other side will have solvent at activity a_{s1} as well as the solute, and separation will be achieved at equilibrium. *Note*: $P_1 > P_2$ and $a_{s2} > a_{s1}$.

Consider now a porous membrane for the process of ultrafiltration separation of a protein solution (Section 3.4.2.3). Suppose the regions on the two sides of the membrane have become filled with liquid (from the feed region 1 to product region 2). If the membrane does not reject the protein (a macrosolute) completely, the solute protein will slowly diffuse from region 1 to region 2. Ultimately, both sides will have the same protein concentration. The separation achieved initially will be lost. Therefore, in practice, it is necessary to have an open system so that any permeate appearing from the feed side into the permeate side is immediately withdrawn. The same argument is equally valid if the membrane in reverse osmosis (Section 3.4.2.1) has some permeability of the microsolute present in the feed side.

Problems

4.1.1 Dilute waste streams from a chemical plant contain low concentrations of a variety of volatile organic compounds. We would like to know the effectiveness of air stripping of such waste streams at ambient temperature and pressure. Contaminated groundwater at many sites may also have very low concentrations of a variety of organic compounds. The groundwater is pumped to an air stripper and then pumped back into the ground by what is known as "pump-and-treat" processes. We would like to know the effectiveness of such processes for removing volatile organic compounds having similar vapor pressures. Consider two sets of such compounds:

(1) n-hexane and chloroform;

(2) 1-octene and tetrachloroethene.

It is known that, in each set, the first compound is more hydrophobic and will have much higher activity coefficients. The vapor pressures of n-hexane, 1-octene and chloroform at 25 °C are 0.2, 0.029 and 0.258 atm, respectively. The 25 °C activity coefficients for air-stripping conditions are: n-hexane, 5×10^5; 1-octene, 2.3×10^6; chloroform, 7.98×10^2. The following property values are also available: K_i^∞ for chloroform and tetrachloroethene are 2.1×10^2 and 1.5×10^3, respectively. (Data from Hwang *et al.* (1992b).) Obtain the air-stripping separation factors for the two sets of compounds. Comment on the strippability of individual compounds in each set.

4.1.2 The wastewater stream from a pharmaceutical plant on two different days had two different sets of pollutants:

day (1) benezene and n-hexane;

day (2) chloroform and 2-propenal (acrolein).

They are removed by air stripping at 25 °C. For the wastewater of each day:

(a) Which compound on a given day is more easily stripped?

(b) What is the separation factor between the two volatile organic compounds on a given day?

The vapor pressures at 25 °C are: benzene = 0.125 atm; n-hexane = 0.2 atm; chloroform = 0.258 atm; 2-propenal = 0.361 atm. Also K_i^∞ at 25 °C are, for chloroform, 2.1×10^2, and, for benzene, 3×10^2. The activity coefficients at 25 °C under air stripping conditions are, for n-hexane, 5×10^5, and, for 2-propenal, 2.16×10.

4.1.3 One step in the preparation of ultrapure water needed for various applications involves the removal of dissolved gaseous species 1 and volatile organic compound 2. This removal is to be implemented using N_2 and/or vacuum to strip species 1 and 2. Develop an expression for the separation factor between species 1 and 2 in terms of the thermodynamic constants and temperature-dependent physical properties of the two compounds 1 and 2. Assume nonideal behavior in the liquid phase, an infinitely dilute solution of each species and ideal gas behavior.

4.1.4 For the determination of steam/air stripping based separation of sparingly soluble hydrophobic volatile organic compounds (VOCs) $i = 1, 2$ from water, K_i^∞, an infinite dilution vapor-liquid equilibrium ratio for any species i is quite useful. Consider two such species having nearly the same vapor pressure at the stripping temperature. Show that, in such a case, if the two species are liquids at the stripping temperature, $\alpha_{12} \cong \frac{x_{2w/\text{solubility}}}{x_{1w/\text{solubility}}}$.

(*Hint*: To determine γ_{il}^∞, consider an equilibrium between the pure organic phase of species i and water.)

4.1.5 Consider the simple vapor-liquid equilibrium in the following binary systems:

(1) N_2-O_2 at 1 atmosphere and 90 K;

(2) benzene-toluene at 1 atmosphere and 100 °C;

(3) ethylbenzene-styrene at 0.133 atmosphere and 80 °C.

Calculate the mole fraction of each species in the vapor phase and liquid phase, respectively, as well as α_{12}, assuming ideal liquid solution and ideal gas behavior. The following information is available: (1) the vapor pressures of N_2 and O_2 at 90 K are 3.53 atm and 750 mm Hg, respectively (Ruhemann, 1949); (2) the vapor pressures of benzene and toluene at 100 °C are 1.78 atm and 0.73 atm, respectively; (3) the vapor pressures of ethylbenzene and styrene are 0.166 atm and 0.119 atm, respectively. State your assumptions.

4.1.6 A vapor mixture of 7% methane, 8% ethane, 20% propane and 65% n-butane is present at 32 °C in a closed vessel.

(a) Determine the dew-point liquid composition and pressure using the K_i-factor charts.

(b) Determine the bubble-point vapor composition and pressure using the K_i-factor charts.

4.1.7 (a) In a closed vessel, m_{tf} moles of a mixture of n volatile species are present in vapor–liquid equilibrium. If m_{tv} and m_{tl} are the total number of moles present in the vapor and liquid phases, respectively, and if K_i is defined as $K_i = (x_{iv}/x_{i\ell})$, obtain the following relations:

$$x_{iv} = \frac{x_{if} K_i}{1 + \frac{m_{tv}}{m_{tf}}(K_i - 1)}; \quad \sum_{i=1}^{n} \frac{x_{if} K_i}{1 + \frac{m_{tv}}{m_{tf}}(K_i - 1)} = 1.$$

The temperature is given as T °C.
(b) Consider the feed mixture having the following composition in mole %: methane 7%, ethane 8%, propane 20% and n-butane 65%. It is present in a closed vessel at 32 °C and 1000 kPa. What fraction of this mixture is present in the vapor phase? Obtain the vapor phase composition.

4.1.8 An organic feed mixture present at 80 °C and 1 atm consists of 40 mol% acetone, 30 mol % acetonitrile and 30 mol % toluene. Assume that the liquid phase is an ideal solution and that the vapor phase behaves as an ideal gas. Further, the system has both liquid and vapor phases. Caculate the fraction of the feed present in the vapor phase and the compositions of the vapor phase and the liquid phase. The P_i^{sat} values at 80 °C for acetone, acetonitrile and toluene are 196 kPa, 98 kPa and 39 kPa, respectively.

4.1.9 Consider a continuous chemical mixture present in the vapor phase. Its molecular weight density function is known to be $f_v(M)$. Determine the expression of $f_l(M)$ for the first liquid drop formed as the vapor is cooled to temperature T. Make appropriate simplifications for low-pressure gas and ideal liquid solution.

4.1.10 We have observed in Chapter 3 (equation (3.1.25a)) that the reversible work needed to transfer 1 mole of species i from state 1 to state 2 is equal to the difference between the values of the partial molar free energy at the two states: $(\overline{G}_i|_2 - \overline{G}_i|_1) = (\mu_i|_2 - \mu_i|_1)$. Consider the solution of an impurity B in a solvent A. Comment about the possibility of obtaining a pure solvent with respect to the energy required.

4.1.11 Markham and Benton (1931) studied experimentally the adsorption of pure O_2 and pure CO as well as CO–O_2 mixtures on 19.6 g of silica at 100 °C. The data obtained are provided in Table 4.P.1.

Develop a plot of x_{O_2g} vs. $x_{O_2\sigma}$ for the O_2–CO mixture adsorption on silica at 100 °C at a total pressure of 1 atmosphere using the ideal adsorbed solution theory. Plot the three experimental points for the mixture in the same figure.

4.1.12 Consider the following cation exchange process:

$$2\text{NaR} + \text{CaCl}_2(\text{aq}) \Leftrightarrow \text{CaR}_2 + \text{NaCl}(\text{aq}),$$

where R represents the resin phase. Develop an estimate of the activity-based separation factor between the sodium ion and the calcium ion in terms of the activity ratio of the resin phase to that of the water phase for one of the two cations. Neglect swelling effects.

Table 4.P.1.

Pure oxygen at 100 °C		Pure CO at 100 °C		CO–O_2 mixtures at 100 °C					
Volume adsorbed		Volume adsorbed				Volume of oxygen adsorbed		Volume of carbon monoxide adsorbed	
Press. (mm)	Obs. (cc)	Press. (mm)	Obs. (cc)	p_{O_2}	p_{CO}	Isotherm[a]	Observed	Isotherm[a]	Observed
17.9	0.08	26.3	0.26						
88.6	0.54	127.9	1.37	210.3	549.7	1.45	1.56	5.50	5.27
133.9	0.90	224.4	2.36	286.4	473.6	1.95	1.96	4.80	4.40
221.9	1.52	321.0	3.37	335.7	424.3	2.30	2.33	4.34	4.05
329.0	2.19	434.5	4.42						
417.1	2.87	537.2	5.30						
471.4	3.18	639.4	6.33						
570.5	3.92	760.0	7.40						
653.2	4.38								
760.0	5.01								

[a] Contains values obtained from pure component adsorption isotherms via interpolation.

4.1.13 Consider a cation exchange resin in an aqueous solution containing Na_2SO_4. Obtain the selectivity of the resin for the sodium cation over the sulfate anion if the activity based distribution coefficient of the anion between the resin and the external solution is 0.001. (Ans. 31, 630.)

4.1.14 Consider two anions Cl^- and SO_4^{--} and an anion exchange resin. If the activities of both anions in the aqueous solution are equal, determine which anion will be preferred by the resin phase.

4.1.15 The activity based separation factor α^{a}_{AB} between two counterions A and B in an ion exchange resin system has been defined by (4.1.70). Assume an ideal system.

(a) Show that, for an ideal solution in both phases, the separation factor α_{AB} based on mole fractions is equal to the separation factor α^{m}_{AB} based on molalities.

(b) Redefine the expression (3.3.121e) for the law of mass action based equilibrium constant for an "ion exchange reaction" between counterions A and B for ideal solutions as

$$K^{m}_{AB} = \frac{(m_{A,R})^{|Z_B|}\,(m_{B,w})^{|Z_A|}}{(m_{A,w})^{|Z_B|}\,(m_{B,R})^{|Z_A|}}.$$

Show that $K^{m}_{AB} = (\alpha^{m}_{AB})^{|Z_A|}$ when $Z_A = Z_B$.

(c) Show that in ideal solutions, where $\overline{V}_A = \overline{V}_B$, the value of $K^{m}_{AB} = 1$ when $Z_A = Z_B$. The quantity K^{m}_{AB} has been defined as the electroselectivity or *selectivity coefficient* (Helfferich, 1995). Any nonideality and specific preference for an ion by the ion exchanger leads to $\ln K^{m}_{AB} > 0$ when $Z_A = Z_B$.

4.1.16 Define the equivalent ionic fraction $\underline{x}_{ij}$ for $i =$ A, B, a system of two counterions A and B in phase $j = w, R$, by

$$\underline{x}_{ij} = \frac{Z_A\; m_{A,j}}{Z_A\; m_{A,j} + Z_B m_{B,j}}.$$

Develop the following general relation between $\underline{x}_{ij}$ and K^{m}_{ij} defined in part (b) of Problem 4.1.15:

$$\frac{\underline{x}_{AR}{}^{(|Z_B|/|Z_A|)}}{1-\underline{x}_{AR}} = (K^{m}_{AB})^{\frac{1}{|Z_A|}} \left(\frac{\underline{x}_{Aw}{}^{|Z_B|/|Z_A|}}{1-\underline{x}_{Aw}}\right)\left(\frac{Z_A\; m_{A,R} + Z_B\; m_{B,R}}{Z_A\; m_{A,w} + Z_B\; m_{B,w}}\right),$$

where K^{m}_{AB} is the selectivity coefficient between counterions A and B.
Rewrite this result in terms of $m_{F,R}$ and $\sum_i Z_i m_{i,w}$, where $m_{F,R}$ is the molality of fixed charges in the resin phase.

4.1.17 Calculate the equivalent ionic fraction $\underline{x}_{iR}$ of sodium and magnesium ions in a cation exchanger whose selectivity coefficient K^{m}_{AB} is unity due to ideal behavior. (See problem 4.1.16.)
Assume $m_{F,R} = 9 \times 10^3$ mequiv./1000 g H_2O of fixed ionic groups. The total molality of ions in solution is 11 mequiv./1000 g H_2O; the salts NaCl and $MgCl_2$ are present in equal equivalent amounts (assume $\underline{x}_{Mgw} = \underline{x}_{Naw} = 0.5$).
(*Hint*: Employ the results obtained in Problem 4.1.16.) (Ans. $\underline{x}_{MgR} = 0.9756$.)

4.1.18 Chimowitz and Pennisi (1986) have measured the supercritical phase ($j = \ell$) mole fractions of 1,10-decanediol (species $i = d$) and benzoic acid (species $i = b$) in supercritical CO_2 at two temperatures, 318 K and 308 K for a number of pressures. The results from their extensive experiments are summarized in Table 4.P.2.

Table 4.P.2.

Pressure (bar)	Temperature (K)	$x_{d\ell} \times 10^4$ (mole fraction)	$x_{b\ell} \times 10^3$ (mole fraction)
306.8	318	5.335	4.84
	308	3.064	3.874
228.5	318	4.107	3.843
	308	2.542	3.246
163.8	318	3.411	2.755
	308	1.814	2.338
132.2	318	2.03	1.72
	308	1.53	1.79

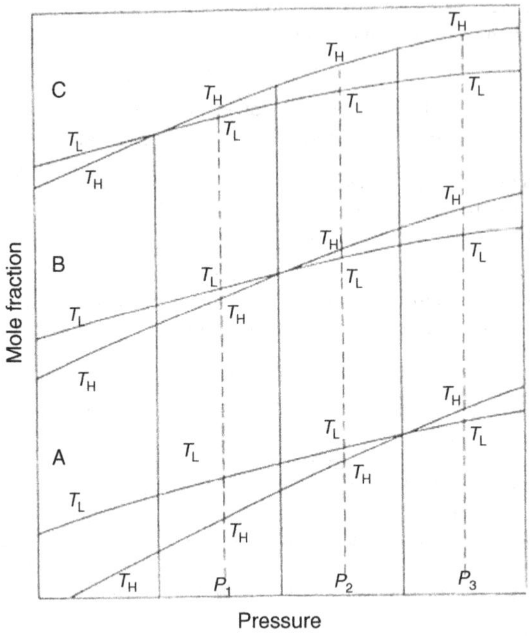

Figure 4.P.1. Crossover behavior for a ternary system.

Determine the solid phase yield (defined as the ratio of moles of solid deposited/mole of CO_2 in feed) of each of the two species, as well as the purity enhancement factor[10] of 1,10-decanediol in the deposited solid phase when the temperature is reduced from 318 K to 308 K at each one of the four pressures identified above.

4.1.19 The separation of a pure solid from a binary solid mixture extracted by supercritical CO_2 via cooling of the mixture in the crossover region has been experimentally demonstrated by Chimowitz and Pennisi (1986). Suppose there is a ternany solid mixture of three species A, B and C. If their solubility behavior in the crossover region is as given in Figure 4.P.1, describe the process sequence you will have to follow to get deposits of pure A, pure B and pure C in different vessels under appropriate conditions.

4.2.1 Calculate the separation factor for each of the following gas mixtures: (1) SO_2-N_2, (2) $U^{235}F_6$-$U^{238}F_6$, present in a gas centrifuge of radius 4 cm rotating at a high angular velocity, $\omega = 4000$ rad/s at 20 °C. The two regions are $r_2 = 4$ cm and $r_1 = 0$ cm. (Ans. $\alpha_{N_2-SO_2} = 1.207$; $\alpha_{U^{235}-U^{238}} = 1.0157$.)

4.2.2 The mole fraction of species 1 in a binary gas mixture of species 1 and 2 introduced into a gas centrifuge is x_{1f}. After the gas centrifuge is rotated at an angular velocity of ω rad/s, a concentration profile is developed for each species in the radial direction. If the centrifuge radius is r_0, indicate a procedure to determine the mole fraction profile, $x_{1f}(r)$.

4.2.3 Calculate the maximum number of macromolecular species whose peaks may be clearly resolved in isopycnic sedimentation if the cell length l_z is 1.5 cm. For the purpose of calculations, use the data provided in the example given in the text to calculate the resolution (R_s) for the isotopically labeled and unlabeled DNA molecules in a CsCl gradient; the density gradient is 0.08 g/cm^4. You may use the values of R_s or σ developed/used in the text calculations for an estimate of the peak width. How does this number change if the R_s value is 1.5? (Ans. $n_{max} = 18.25$; $n_{max} \cong 13$.)

[10] Purity enhancement factor = ratio of mole fraction of decanediol/benzoic acid in the deposited solid phase to that in the feed phase.

4.2.4 A sedimentation cell has a length of 1 cm; it is located 6.0 cm from the rotation axis ($r_1 = 6.0$, $r_2 = 7.0$ cm). A macromolecular solute of molecular weight 10^5 is rotated at 25 °C at an angular speed of ω rad/s. When the rotational speed is 5000 rpm, the concentration of the macromolecule in the solution varies from a maximum to 1% of the maximum over the cell length.

(1) Determine the rotational speed (rpm) which will concentrate the macromolecule to the same extent, over a distance of only 0.01 cm from the tip of the cell ($r = 7.0$ cm).

(2) Determine the ratio of the macromolecular concentration between the two locations $r_2 = 7$ cm and $r_1 =$ 6.5 cm for the macromolecular solute when the rotational speed is 5000 rpm. Assume $R = 8.315 \times 10^7$ erg/gmol-K. (Ans. (1) 48 180 rpm; (2) 2.54.)

4.2.5 The Gaussian concentration profile of protein species i around location z_0 in a cell used for isoelectric focusing, where the pH is equal to the pI of the protein, can also be determined by a balance of the diffusive flux of the protein, leading to band broadening, and the flux, due to the uniform electrical force field $\boldsymbol{E}$ forcing the protein to concentrate at z_0. Obtain the result (4.2.38a), employing appropriate assumptions.

4.2.6 Develop an expression for the maximum number of proteins whose peaks may be clearly resolved via isoelectric focusing with a value of resolution $R_s = 1$ in a cell of length ℓ_z. The values of pH at locations ℓ_z and 0 are pH_ℓ and pH_0, respectively. The variation of Z_i with pH will be incorporated as $(\mathrm{d}Z_i/\mathrm{d}pH)$.

$$\left(\text{Ans. } n_{\max} = \left(\frac{\mathcal{F}E\,(-\mathrm{d}Z_i/\mathrm{d}(pH))\;\ell_z\;(pH_\ell - pH_0)}{16\,RT}\right)^{1/2}.\right)$$

4.2.7 In a natural gas well assumed to consist of methane and butane only, the mole fraction of butane at 3055 m below the surface is 0.16. Calculate the value of the butane weight fraction in the gas well at the surface level, assuming ideal gas behavior and isothermal conditions. (Ans. 0.69.)

4.2.8 Measurements of isotopic ratios or ratios of noble gases (^{84}Kr and ^{36}Ar) in Greenland ice show that (Craig and Wiens, 1996) gravitational separation in the unconsolidated firn layer above the ice is responsible for the observed enrichments relative to atmosphere ratios. Define R to be the ratio of the mole fraction of noble gas 1 over the noble gas 2 at a height z_1, and let R_0 be the corresponding ratio in free atmosphere at the surface z_2 ($>z_1$). Obtain, as a first order of approximation, the following expression:

$$\Delta = \left[\left(\frac{R}{R_0}\right) - 1\right]\;10^3 = \left[\left(\frac{x_{11}}{x_{21}}\Big/\frac{x_{12}}{x_{22}}\right) - 1\right] \times 10^3 = \frac{(M_1 - M_2)\;(z_2 - z_1)}{T} \times 1.18,$$

where region 1 corresponds to z_1 (meters) and region 2 corresponds to z_2; species 1 is heavier than species 2. If $(z_2 - z_1) \sim 70$ m, and species 1 and 2 are ^{84}Kr and ^{36}Ar, respectively, obtain an estimate of Δ when the ice temperature is −20 °C. (Ans. $\Delta(^{84}\mathrm{Kr}/^{36}\mathrm{Ar}) = 15.6$.)

4.3.1 The extent of solubilization of a hydrophobic solute present in water into a micelle is determined by semiequilibrium dialysis (SED) techniques. This technique employs a dialysis cell (see Figure 4.3.1(a)), where the pore size of the dialysis membrane completely rejects the spherical micelle. However, the solute must be able to pass freely through the membrane. The surfactant monomer may also pass freely through the membrane. On addition of surfactant at a concentration of C_{sur} mol/liter to the chamber on one side of the membrane to which C_i^0 mol/liter of solute species i has been added, binding of the solute to the micelles will occur if the surfactant concentration exceeds the CMC for the surfactant.

(a) Identify what will happen after a sufficient time has been allowed to lapse.

(b) After equilibrium has been established, the free solute concentration in the solution chamber on the other side of the membrane has been determined to be C_{ip} mol/liter The volumes of the two chambers are V_p and V_f liter, respectively, where subscripts p and f refer to the permeate side and the feed side (where solute was introduced at concentration C_i^0), respectively. Determine the number of moles of solute i which have been solubilized into the micellar phase.

(c) Assume now that C_{sur} mol/liter >> C_i^0 mol/liter. Further, assume that C_{sur} mol/liter is >> [CMC] and that the moles of surfactant in micelle >> free surfactant moles. Determine the micellar equilibrium ratio, K_{im}, defined as $K_{im} = x_{im}/x_{if}$, where x_{im} and x_{if} are the mole fractions of i in the micellar phase and the feed chamber, respectively. Assume that $V_p = V_f = 0.05$ liter.

4.3.2 Cuprophan membranes made from regenerated cellulose are frequently used in hemodialysis. Model this membrane as one consisting of cylindrical capillaries of radius 18 Å. Determine the separation factors of the dialysis membrane for two solutes, urea and vitamin B_{12}. The characteristic radii of urea and vitamin B_{12} are 2.8 Å and 8.5 Å, respectively. The diffusion coefficients of urea and vitamin B_{12} at infinite dilution in isotonic saline at 37 °C are 1.81×10^{-5} and 0.38×10^{-5} cm^2/s, respectively. Compare the result for the given pore size estimate with that obtained from the data of Colton *et al.* (1971), namely 16 (based on effective diffusion coefficients without any consideration of equilibrium partition coefficients in their Figure 6). (Ans. 23.5.)

4.3.3 Through a microporous denitrated cellulose membrane, the separation factor for two solutes, sodium sulfate and sucrose, has to be determined for the case of simple diffusion with no convection. The properties of the membrane, the operating conditions and the solute properties are available in Example 3.4.5.

4.3.4 In Donnan dialysis based removal and concentration of Cu^{2+} ions from a wastewater into an acidic strip solution through a cation exchange membrane, the equilibrium *pH* of the strip solution is 1 and the equilibrium *pH* of the feed solution is 4. Determine the concentration ratio of the Cu^{2+} ions in the two chambers. The membrane-impermeable anion in both chambers is SO_4^{2-}. (Ans. Strip solution concentration is 10^6 larger than that in the feed.)

4.3.5 In seawater desalination processes by thermal evaporation or a membrane based technique, the hardness of the seawater in terms of calcium carbonate or sulfate (also $MgCO_3$, $MgSO_4$) is a problem, which leads to scaling of evaporator surface or membrane fouling. It has been suggested that one could employ Donnan dialysis using the concentrated seawater from the process as the strip solution to reduce substantially the Ca^{++}/Mg^{++} content of the seawater, which is to undergo desalination by a thermal or a membrane process. If the strip solution Na^+ concentration is twice that of the seawater to be subjected to desalination, identify the type of membrane and the maximum possible extent of reduction of $CaSO_4$ from that present in the feed seawater.

4.3.6 Develop an expression (in terms of known quantities) for the molar concentration of ion Na^+ in a cation exchange membrane at the membrane–solution interface, where the salt Na_2SO_4 is present at a molar concentration of C'_{sw} in the feed solution. The molar fixed charge density in the membrane is C_m. Make appropriate assumptions.

4.3.7 A mixture of CO_2 and CH_4 containing 20 mol% CO_2 has been introduced into one side of a synthetic membrane in a closed vessel; the other side of the membrane is empty to start with. We would like to know the composition of the gas mixture that appears first on the other side of the membrane. The permeability coefficients for CO_2 and CH_4, Q_{CO_2} and Q_{CH_4}, are provided in units of Barrers as $Q_{CO_2} = 15$, $Q_{CH_4} = 0.5$. (Ans. $x_{CO_2,p} = 0.882$.)

4.3.8 In molecular distillation, a liquid mixture of A and B is distilled at a very low pressure of 0.001 mm to prevent heat-sensitive materials from spoiling. At such low pressures, if we are dealing with a pure liquid, the flux of the species escaping the liquid surface is given by equation (3.1.119a):

$$N_A = \frac{P_A^{sat}}{\sqrt{2\pi RTM_A}},$$

where P_A^{sat} is the vapor pressure of pure A of molecular weight M_A. The escaping vapor molecules condense on another surface close by and do not come back to the liquid. For a mixture, the vapor pressure is to be replaced by the corresponding partial pressure. Obtain an expression for the separation factor α_{AB} for species A and B in a liquid mixture subjected to molecular distillation with the product liquid mixture obtained from the condensing surface, assuming ideal behavior.

4.3.9 A feed gas mixture of species A and B at atmospheric pressure is exposed to a microporous membrane of thickness δ_m and pore radius r_p. The permeate side is under vacuum at time $t = 0$. Determine the value of the membrane selectivity α_{AB} as $t \to 0$ in a closed vessel, with the feed gas mixture at total pressure P at time $t = 0$. Assume Knudsen diffusion exists. The membrane porosity is ε_m.

4.3.10 Consider two gases A and B being separated through a microporous carbon membrane. The gas transport may be described as that by Knudsen diffusion through the microporous membrane as well as by surface diffusion of the adsorbed gas layer on the pore surface (3.4.112).

(1) Determine the value of α_{AB} as $t \to 0$ in a closed vessel in the manner of Section 4.3.3.

(2) Speculate what will happen if one of the species (between A and B, for example H_2 and NH_3) is condensable and forms a multilayer condensate on the pore wall, blocking gas-phase transport.

4.3.11 The solubility S_{im} of a gas species i in a polymer has been empirically expressed as

$$\ln S_{im} = M + 0.016\ T_{ci},$$

where M depends on the polymer; the value for M for the polydimethylsiloxane polymer at 30 °C is -8.8. The diffusivity of a gas species i has been correlated as

$$D_{im} = (\tau / V_{ci}^{\eta}),$$

where the value of η for the polydimethylsiloxane (PDMS) polymer is 1.5 and τ has the value of 2.6×10^{-2}. Here, T_{ci} and V_{ci} are the critical temperature and the critical molar volume of gas species i. Calculate the solubility selectivity, mobility selectivity and the ideal selectivity for a PDMS polymer membrane for the two gas pairs O_2–N_2 and CO_2–N_2. Assume T_{ci} (K) = 154.6 (O_2), 126.2 (N_2), 304.2 (CO_2). (Ans. $(S_{O_2m}/S_{N_2m}) = 1.575$; $(S_{O_2m}/S_{N_2m}) = 17.24$; $(D_{O_2m}/D_{N_2m}) = 1.35$;$(D_{CO_2m}/D_{N_2m}) = 0.929$.)

4.3.12 You are in a boat on a lake of brackish water. To recover water from the brackish water (salt concentration ~1000–10 000 ppm) of salt concentration C_{salt} into a concentrated sugar solution C^0_{sugar}, you have a forward osmosis device separating the two solutions by a perfect semipermeable membrane. In *forward osmosis*, water from the salt solution with an osmotic pressure lower than that of the concentrated sugar solution will permeate to the sugar solution. The initial volume of the concentrated sugar solution is V^0_{sugar}. We know that the osmotic pressure of each solution is defined as

$$\begin{aligned} \pi_{sugar} &\cong b_{sugar} C_{sugar}; \\ \pi_{salt} &\cong b_{salt} C_{salt}, \end{aligned}$$

where $b_{salt} > b_{sugar}$. Determine the final volume of the sugar solution, V_{sugar}, that you can achieve and the corresponding sugar concentration, C_{sugar}. Assume that the membrane is perfectly semipermeable.

5

Effect of chemical reactions on separation

Chemical reactions occur in many commonly practiced separation processes. By chemical reactions, we mean those molecular interactions in which a new species results (Prausnitz *et al.*, 1986). In a few processes, there will be hardly any separation without a chemical reaction (e.g. isotope exchange processes). In some other processes, chemical reactions enhance the extent of separation considerably (e.g. scrubbing of acid gases with alkaline absorbent solutions, solvent extraction with complexing agents). In still others, chemical reactions happen whether intended or unintended; estimation of the extent of separation requires consideration of the reaction. For example, in solvent extraction of organic acids, the extent of acid dissociation in the aqueous phase at a given *pH* should be taken into account (Treybal, 1963, pp. 38–41). Chemical equilibrium has a secondary role here, yet sometimes it is crucial to separation.

Familiarity with the role of chemical reactions in separation processes will be helpful in many ways. **This is especially relevant since a few particular types of chemical reactions occur repeatedly in different separation processes/techniques. These include ionization reactions, acid–base reactions and various types of complexation reactions.** The complexation reactions also include the weaker noncovalent low binding energy based bonding/interactions identified in Section 4.1.9. A better understanding and quantitative prediction of separation in a given process is possible, leading to better process and equipment design. In processes where a chemical agent is used, different agents can be evaluated systematically. On occasions, it may facilitate the introduction of reactions to processes for enhancing separation.

The relation between the extent of separation and the extent of reaction is briefly considered in Section 5.1. How chemical reactions alter the separation equilibria in gas–liquid, vapor–liquid, liquid–liquid, solid–liquid, surface adsorption equilibria, etc., is described in Section 5.2. The role of chemical reactions in altering the separation in rate-controlled equilibrium processes is treated in Section 5.3. Section 5.4 illustrates how chemical reactions affect the separation in rate-governed processes using membranes and external field based processes; for example, electrochemical gas separation using membranes in an external electric field.

5.1 Extent of separation in a closed vessel with a chemical reaction

Consider a closed vessel with two regions $j = 1, 2$ and two species $i = 1, 2$. The extent of separation at any time t (or at equilibrium, as the case may be) can be written as follows:

$$\xi(t) = |Y_{11}(t) - Y_{21}(t)| = \left|\frac{m_{11}(t)}{m_1^0} - \frac{m_{21}(t)}{m_2^0}\right|. \quad (5.1.1)$$

Suppose now that region 1 in the separation system has another species 3 with which species 1 reacts to produce a fourth species 4, i.e.

$$1 + 3 \Leftrightarrow 4. \quad (5.1.2)$$

Then species 1 is present in region 1 not only as species 1, but also as species 4. Assume that species 3 and 4 cannot go to region 2 and that species 2 does not react with either species 3 or 4. If, at time t, the total number of moles of species 1 present as free species 1 and species 4 in region 1 as a result of reaction is $m_{11}^r(t)$, the extent of separation with the reaction is

$$\xi^r(t) = |Y_{11}^r(t) - Y_{21}^r(t)| = \left|\frac{m_{11}^r(t)}{m_1^0} - \frac{m_{21}(t)}{m_2^0}\right|, \quad (5.1.3)$$

where we assume that $Y_{21}^r(t) = Y_{21}(t)$ since the distribution of species 2 is unaffected by the reaction. The difference between the two extents of separation is given by

$$\xi^r(t) - \xi(t) = \left|\frac{m_{11}^r(t)}{m_1^0} - \frac{m_{11}(t)}{m_1^0}\right| = \frac{|m_{11}^r(t) - m_{11}(t)|}{m_1^0}. \quad (5.1.4)$$

For the reaction under consideration, the stoichiometric coefficient υ_i for $i = 1$ is −1. The molar extent of reaction is traditionally defined for any ith species participating in a reaction as

$$\frac{\text{the number of moles consumed or produced by the reaction}}{|\upsilon_i|}.$$

We note that the difference in (5.1.4) of the two extents of separation is merely the molar extent of reaction of the ith species normalized with respect to the total number of ith species originally present in the system. Thus, the traditional formalism of reaction engineering or reaction kinetics may be used to estimate directly the change in the extent of separation due to a chemical reaction. Such a straightforward result may not be valid if some of the assumptions made earlier break down.

5.2 Change in separation equilibria due to chemical reactions

The chemical equilibrium in a multiphase system and the corresponding species concentrations in chemical separations are altered when chemical reactions take place. We will illustrate this now for gas-liquid equilibrium (as in gas absorption), vapor-liquid equilibrium (as in distillation), liquid-liquid equilibrium (as in solvent extraction), stationary phase-liquid equilibrium (as in ion exchange, chromatography and crystallization), surface adsorption equilibrium (as in foam fractionation) and Donnan equilibrium.

5.2.1 Gas-liquid and vapor-liquid equilibria

First we consider gas-liquid equilibrium as in gas absorption/stripping.

5.2.1.1 *Gas-liquid equilibrium*

Gas mixtures are commonly separated in industry by absorption in a solution where chemical reactions may take place (the solution is regenerated subsequently in a separate vessel by stripping). The amount of gas absorbed by simple physical absorption is often insufficient; it can be increased considerably by incorporating in the solution chemicals with which the dissolved gas undergoes a chemical reaction. Removal of acid gases, e.g. CO_2, H_2S, SO_2, COS, by alkaline solutions are well-known examples (Danckwerts, 1970; Astarita *et al.*, 1983).

Aqueous absorbent solutions conventionally employed to increase absorption of a species, e.g. CO_2 from a gas, may have a variety of reagents, e.g. K_2CO_3, Na_2CO_3, $NaHCO_3$, NaOH, monoethanolamine, diethanolamine, etc. Even without a reagent, the absorption equilibrium of a species may be affected by reaction. Absorption of SO_2 into water where SO_2 is ionized illustrates such a case.

5.2.1.1.1 Solute ionization in aqueous solution The solubility of a gas in a liquid is often described by Henry's law (see equation (3.3.60b)). Thus the gas-phase partial pressure of SO_2, p_{SO_2}, is related to its aqueous phase mole fraction x_{SO_2} by

$$p_{SO_2} = H_{SO_2} x_{SO_2} \tag{5.2.1}$$

for ideal gas-phase behavior, where H_{SO_2} is the Henry's law constant for SO_2 in water. However, it is valid only for molecular SO_2 in water. Dissolved SO_2 in water ionizes to some extent (Danckwerts, 1970):

$$SO_2 + H_2O \Leftrightarrow HSO_3^- + H^+. \tag{5.2.2}$$

Thus, SO_2 is present in water also as HSO_3^-; the total SO_2 concentration in water is given by

$$C_{SO_2}^t = \left(C_{SO_2} + C_{HSO_3^-}\right) \text{ gmol/liter}. \tag{5.2.3}$$

The ionization equilibrium of (5.2.2) is described by[1]

$$K_d = \frac{C_{H^+}\, C_{HSO_3^-}}{C_{SO_2}}. \tag{5.2.4}$$

The aqueous solution has no net charge anywhere (see the electroneutrality condition (3.1.108a)); therefore,

$$C_{HSO_3^-} = C_{H^+} = (K_d\, C_{SO_2})^{1/2}. \tag{5.2.5}$$

It is convenient to write Henry's law (5.2.1) for free SO_2 as

$$p_{SO_2} = H_{SO_2}^C\, C_{SO_2}. \tag{5.2.6}$$

But the relation between p_{SO_2} and $C_{SO_2}^t$ is of interest here, where

$$C_{SO_2}^t = C_{SO_2} + (K_d C_{SO_2})^{1/2}; \qquad p_{SO_2} = \left[H_{SO_2}^C\right]''\, C_{SO_2}^t. \tag{5.2.7}$$

We find

$$\left[H_{SO_2}^C\right]'' = \frac{1}{\frac{1}{H_{SO_2}^C} + \sqrt{\frac{K_d}{H_{SO_2}^C\, p_{SO_2}}}}, \tag{5.2.8}$$

where $\left[H_{SO_2}^C\right]''$ is the pseudo-Henry's law constant to determine the total SO_2 concentration in the liquid phase for a given p_{SO_2}. As expected, when $K_d = 0$, $\left[H_{SO_2}^C\right]'' = H_{SO_2}^C$. Alternatively, the total molar concentration of SO_2 in water is given by

$$C_{SO_2}^t = \frac{p_{SO_2}}{H_{SO_2}^C} + \left(\frac{K_d\, p_{SO_2}}{H_{SO_2}^C}\right)^{1/2}. \tag{5.2.9}$$

[1] We have generally avoided describing equilibrium constants for any reaction using activities; therefore activity coefficients are absent in (5.2.4) as well as in most relations in this chapter.

Now $H^C_{SO_2} = (1/1.63)$ atm · liter/gmol and $K_d = 1.7 \times 10^{-2}$ gmol/liter at ordinary temperatures (~25 °C) (Danckwerts, 1970). Suppose $p_{SO_2} = 0.01$ atm, then $C_{SO_2} = 0.0163$ gmol/liter but $C^t_{SO_2} = 0.033$ gmol/liter, showing the considerable effect of SO_2 ionization in water on its total aqueous solubility (Prausnitz *et al.*, 1986).

The gas CO_2 also ionizes in water in the manner of (5.2.2), producing HCO_3^- and H^+. However, the equilibrium constant K_d for such a reaction is a couple of orders of magnitude smaller than that of SO_2. Hence, unless alkaline conditions are maintained to shift the reaction to the right, the effect of CO_2 ionization in water has a negligible influence on CO_2 solubility in water. The solubility of an acid gas such as CO_2 or H_2S in water is, however, strongly affected by ionization if a basic gas like NH_3 is present. Ammonia ionizes in water as

$$NH_3 + H_2O \Leftrightarrow NH_4^+ + OH^-, \tag{5.2.10}$$

whereas CO_2 ionizes as

$$CO_2 + H_2O \Leftrightarrow HCO_3^- + H^+. \tag{5.2.11}$$

The H^+ ions and OH^- ions are participants in the ionization

$$H_2O \Leftrightarrow H^+ + OH^-, \tag{5.2.12}$$

where the equilibrium lies far to the left, since water has very little tendency to ionize. Thus both reactions (5.2.10) and (5.2.11) are pushed to the right, resulting in considerable amounts of NH_4^+ and HCO_3^-. This means that the partial pressure of CO_2 (or NH_3) needed for a certain total liquid-phase concentration of CO_2 (or NH_3) is much reduced from that when only either CO_2 or NH_3 is present (Prausnitz *et al.*, 1986). A detailed analysis of the equilibria in a NH_3-CO_2-H_2O system is available in Edwards *et al.* (1978).

Conversely, if one tries to strip NH_3 and CO_2 from the water with steam or air, one finds that the efficiency is quite low compared to the cases when either CO_2 or NH_3 only is present. In fact, sour waste streams in refineries, steel mills and coal conversion wastewaters are typical examples of this difficulty. It has been suggested therefore that one should extract NH_3 by solvent extraction and remove CO_2 (or H_2S) by steam stripping (Cahn *et al.*, 1978; MacKenzie and King, 1985). Such a process has been called *extripping* (simultaneous extraction by a solvent and stripping by steam or air).

5.2.1.1.2 Reaction with an absorbent in solution We now consider enhancement of gas absorption into a liquid by liquid-phase reactions with an absorbent species. One of the common industrial absorbents for removal of CO_2 from gas streams is an aqueous solution of a primary or a secondary amine. A primary amine and a secondary amine may be represented by RNH_2 and R_2NH, respectively. The respective overall reactions with CO_2 in an aqueous amine solution are as follows:

$$\text{primary amine} \quad CO_2 + 2RNH_2 \Leftrightarrow RNH\,COO^- + RNH_3^+; \tag{5.2.13}$$

$$\text{secondary amine} \quad CO_2 + 2R_2NH \Leftrightarrow R_2N\,COO^- + R_2NH_2^+. \tag{5.2.14}$$

Consider an aqueous solution of a primary amine in contact with a gas containing CO_2. The following equilibria are valid:

$$\begin{array}{l} CO_2(g) \\ \uparrow\downarrow \\ CO_2(w) + 2RNH_2(w) \overset{K}{\Leftrightarrow} RNH\,COO^-(w) + RNH_3^+(w). \end{array} \tag{5.2.15}$$

The molar concentration based equilibrium constant K for this aqueous-phase reaction is given by

$$K = \frac{C_{RNHCOO^-} \times C_{RNH_3^+}}{C_{CO_2} \times (C_{RNH_2})^2}. \tag{5.2.16}$$

At 20 °C, its value is 1.1×10^5 gmol/liter (Danckwerts, 1970). The concentration of free CO_2 in water, C_{CO_2}, is very small when there is a substantial amount of free amine RNH_2 present in water. Further, since the concentration of OH^- is very much less than that of $RNHCOO^-$, by the electroneutrality condition, $C_{RNHCOO^-} \cong C_{RNH_3^+}$. The distribution coefficient of CO_2 (species i) between the aqueous ($j = w$) phase and the gas phase ($j = g$) in the absence of the amine is

$$\kappa_{iw} = \frac{C_{iw}}{C_{ig}}. \tag{5.2.17a}$$

In the presence of a large amount of amine and the chemical reaction in water, the effective distribution coefficient κ''_{iw} based on the total CO_2 concentration in water is

$$\kappa''_{iw} = \frac{C_{iw} + C_{RNH\,COO^-}}{C_{ig}} = \frac{C^t_{iw}}{C_{ig}}, \tag{5.2.17b}$$

where C_{iw} is the molar free CO_2 concentration in water. Using the electroneutrality condition and the equilibrium constant relation in the above definition of κ''_{iw}, we find that

$$\kappa''_{iw} = \frac{C_{iw}}{C_{ig}} + \frac{(K\,C_{CO_2})^{1/2}\,C_{RNH_2}}{C_{ig}},$$

where $C_{CO_2} = C_{iw}$. Therefore,

$$\kappa''_{iw} = \kappa_{iw}\left[1 + \frac{K^{1/2}\,C_{RNH_2}}{(C_{iw})^{1/2}}\right]. \tag{5.2.17c}$$

From Danckwerts (1970, p.198), for example, some typical estimates for monoethanolamine at 20 °C are $C_{RNH_2} = 1.75$ gmol/liter, $C_{iw} = 4.1 \times 10^{-7}$ gmol/liter. These values lead to

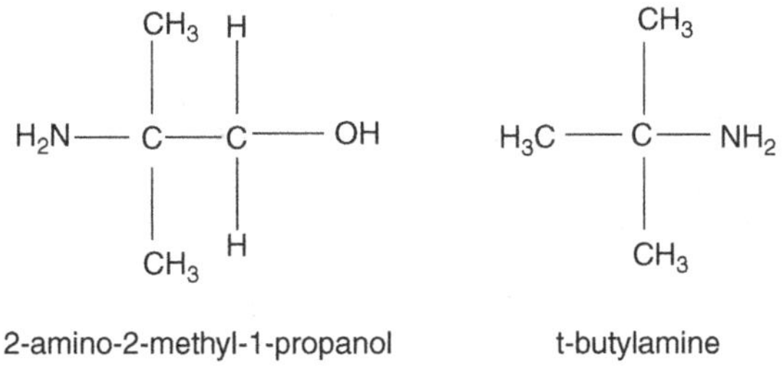

Figure 5.2.1. Examples of hindered amines.

$$\kappa''_{iw} = \kappa_{iw}\left[1 + \frac{10^3}{3} \times \frac{1.75 \times (10^7)^{1/2}}{(4.1)^{1/2}}\right]. \qquad (5.2.17d)$$

This illustrates the extraordinary increase in the effective distribution coefficient of CO_2 due to the chemical reaction with a primary amine added to water. An alternative way to illustrate the same effect would be to use the partial pressure p_{ig} of species i in the gas phase instead of C_{ig}.

An even higher absorption of CO_2 is possible if a *hindered amine* is used. Both reactions (5.2.13) and (5.2.14) indicate that two moles of amine are utilized per mole of CO_2 absorbed and one mole of an amine carbamate $RNHCOO^-$ is produced. This carbamate can hydrolyze as follows:

$$RNHCOO^- + H_2O = RNH_2 + HCO_3^-. \qquad (5.2.18a)$$

An overall reaction in the case of carbamate hydrolysis for CO_2 absorption may be written, using (5.2.13), as follows:

$$CO_2 + RNH_2 + H_2O = RNH_3^+ + HCO_3^-. \qquad (5.2.18b)$$

Thus, for unstable carbamates, only one mole of amine is utilized per mole of CO_2 absorbed; more amine is available for CO_2 absorption.

For conventional amine absorption processes, the amines used (monoethanolamine, diethanolamine, dimethylamine, etc.) are unhindered; a hindered amine has a bulky alkyl group attached to the amino group as in Figure 5.2.1 (Sartori *et al.*, 1987). The carbamates of such hindered amines are much less stable than the carbamates of the corresponding unhindered amines. For example, the equilibrium constant for the hydrolysis reaction (5.2.18a),

$$K = C_{RNH_2}\, C_{HCO_3^-}/C_{RNHCOO^-}, \qquad (5.2.18c)$$

has the value of 0.143 for t-butylamine and 0.06 for the unhindered n-butylamine HO-CH_2-CH_2-NH_2 (see Sartori *et al.*, 1987). Thus, reaction (5.2.18b) will occur if the alkyl group in the amine is bulky and the carbamate is unstable. The net result with a highly hindered amine is that we can approach the theoretical absorption capacity of one mole of CO_2 per mole of amine.

There is an added advantage to having a hindered amine. For gas purification by absorption in an aqueous amine solution in a continuous process with a fixed absorbent, the amine solution has to be regenerated by heating and CO_2 is stripped from the solution. Reactions (5.2.13) and (5.2.18b) now go from right to left. In reaction (5.2.13), all of the absorbed CO_2 cannot be stripped *because* $RNHCOO^-$ for an unhindered amine is stable. On the other hand, there is no such problem in reaction (5.2.18b), leading to much better stripping.

It is useful to consider reactive absorbents in the aqueous solution other than amines, e.g. K_2CO_3. In the *hot potassium carbonate process* widely practiced for CO_2 and H_2S removal, the absorbent is simply K_2CO_3, which ionizes to give K^+ and $CO_3^=$. When CO_2 is present in the gas phase, we have dissolved CO_2. Of course, HCO_3^-, OH^- and H^+ ions are also present:

$$CO_2 + H_2O + CO_3^= = 2HCO_3^-, \qquad (5.2.19a)$$

which follows from the set

$$\begin{bmatrix} CO_3^= + H_2O \Leftrightarrow HCO_3^- + OH^- \\ CO_2 + H_2O \Leftrightarrow HCO_3^- + H^+ \end{bmatrix}.$$

Two moles of HCO_3^- are obtained per mole of CO_2 absorbed. Depending upon the conditions, all or some of the $CO_3^=$ introduced into the system may have reacted. In the amine absorption example, we observed how a chemical reaction increased CO_2 absorption. Here we study how much of the chemical absorbent is being utilized in the reactive absorption; the *fractional consumption* f of the chemical absorbent has been defined as the degree of saturation (Astarita *et al.*, 1983).

The equilibrium constant K for reaction (5.2.19a) is

$$K = \frac{C^2_{HCO_3^-}}{C_{CO_2}\, C_{CO_3^=}}. \qquad (5.2.19b)$$

Since 1 mole of $CO_3^=$ reacts with 1 mole of CO_2, the total molar concentration of $CO_3^=$ introduced as absorbent, $C^t_{CO_3^=}$, includes the $CO_3^=$ concentration present at equilibrium, $C_{CO_3^=}$, and the $CO_3^=$ concentration which has reacted to produce HCO_3^-, $0.5C_{HCO_3^-}$. Therefore,

$$C^t_{CO_3^=} = C_{CO_3^=} + 0.5\, C_{HCO_3^-}. \qquad (5.2.19c)$$

Further, from the definition of f ($f = 1$ means all of the absorbent has been consumed; $f = 0$ means no consumption at all),

$$f C^t_{CO_3^=} = 0.5(C_{HCO_3^-}). \qquad (5.2.19d)$$

We substitute these relations into (5.2.19b) to get

$$C_{CO_2} = \frac{4}{K}\, C^t_{CO_3^=}\, \frac{f^2}{(1-f)}. \qquad (5.2.19e)$$

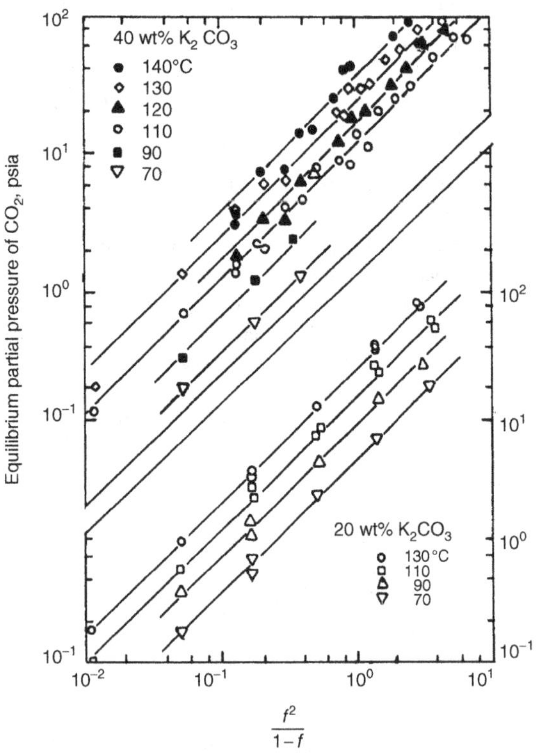

Figure 5.2.2. CO_2 equilibrium vapor pressures over potassium carbonate solutions (Astarita et al.*, 1983.) Reprinted, with permission, from G. Astarita, D.W. Savage, A. Bisio, Gas Treating with Chemical Solvents, Figure 2.4.1, p. 70, John Wiley & Son, 1983. Copyright © 1983, John Wiley & Sons.*

Using Henry's law for free CO_2, $p_{CO_2} = H^C_{CO_2}\, C_{CO_2}$, we obtain

$$p_{CO_2} = \frac{4\, H^C_{CO_2}}{K}\, C^t_{CO_3^=}\, \frac{f^2}{(1-f)}, \qquad (5.2.19f)$$

which relates the partial pressure of CO_2 in the gas at equilibrium with the total molar absorbent concentration and its fractional consumption (Figure 5.2.2). For a given $C^t_{CO_3^=}$, a higher equilibrium CO_2 partial pressure implies a larger value of the fractional absorbent consumption. The temperature dependence of K is also crucial to the partial pressure of CO_2 at equilibrium. Further, the amount of CO_2 absorbed is directly proportional to $C^t_{CO_3^=}$; therefore, K_2CO_3 is preferred to Na_2CO_3, which has a lower solubility. Table 5.2.1 lists common alkaline reagents used for acid gas scrubbing in aqueous or organic solutions.

5.2.1.1.3 Selective absorption in reactive solution Some industrial gas streams contain two volatile species, only one of which needs to be removed. For example, some coal gasification streams have H_2S and CO_2; only H_2S needs to be removed to avoid pollution. Yet there will be absorption of both H_2S and CO_2 in the absorbents traditionally employed. Selective removal of H_2S over CO_2 from gases with a high CO_2/H_2S ratio is highly desirable for several reasons: increased CO_2 absorption increases cost for absorbent regeneration and recirculation; reduction of CO_2 in the acid gas obtained from the stripper-regenerator is useful for sulfur recovery by the Claus process (Astarita *et al.*, 1983; Savage *et al.*, 1986).

Under conditions of absorption equilibrium, the thermodynamic selectivity of an absorbent for H_2S over CO_2 has been defined as (Astarita *et al.*, 1983)

$$\alpha_{H_2S\text{-}CO_2} = \left(\frac{C^t_{H_2S}}{p_{H_2S}}\right)\left(\frac{p_{CO_2}}{C^t_{CO_2}}\right), \qquad (5.2.20)$$

where C^t_i is the total molar concentration of species i in the liquid phase whose equilibrium vapor pressure is p_i. The liquid-phase concentration C^t_i includes the free ith species and the chemically combined form of the ith species.

Most chemical absorbents are aqueous alkaline solutions containing a base, B. The reactions of H_2S and CO_2 are as follows:

$$H_2S + B \overset{K_S}{=} BH^+ + HS^-; \qquad (5.2.21)$$

$$CO_2 + B + H_2O \overset{K^{-1}_{CO_21}}{=} BH^+ + HCO_3^-. \qquad (5.2.22)$$

There is an additional reaction for H_2S, a simple dissociation of HS^-, given by

$$HS^- = H^+ + S^=. \qquad (5.2.23)$$

The equilibrium constant for this reaction is very small (in the range of 10^{-14} to 10^{-12} gion/liter between 20 and 100 °C) compared to that for reaction (5.2.21), which is in the range 0.1 to 10 (Astarita *et al.*, 1983). Therefore we can neglect dissociation of HS^-. We can conveniently consider reactions (5.2.21) and (5.2.22) in terms of

$$HCO_3^- + H_2S \overset{K_{CS}}{=} H_2O + CO_2 + HS^-. \qquad (5.2.24)$$

The equilibrium constant for this reaction, K_{CS}, in terms of the molar concentration based equilibrium constants, K_S and K_{CO_21}, is

$$K_{CS} = \frac{K_S}{K^{-1}_{CO_21}} = \frac{C_{CO_2}\, C_{HS^-}}{C_{HCO_3^-}\, C_{H_2S}}, \qquad (5.2.25)$$

where

$$K_S = \frac{C_{BH^+}\, C_{HS^-}}{C_{H_2S}\, C_B}; \quad K_{CO_21} = \frac{C_{CO_2}\, C_B}{C_{BH^+}\, C_{HCO_3^-}}. \qquad (5.2.26)$$

Table 5.2.1. Common alkaline reagents used in chemical solvents[a]

Monoethanolamine (MEA)	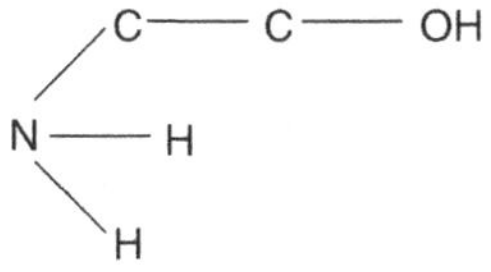
Diethanolamine (DEA)	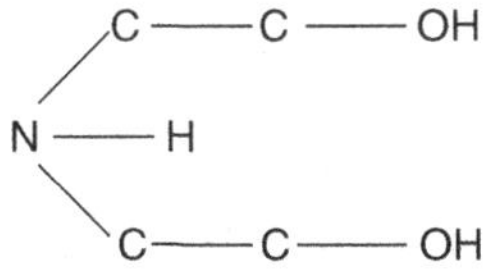
Diisopropanolamine (DIPA)	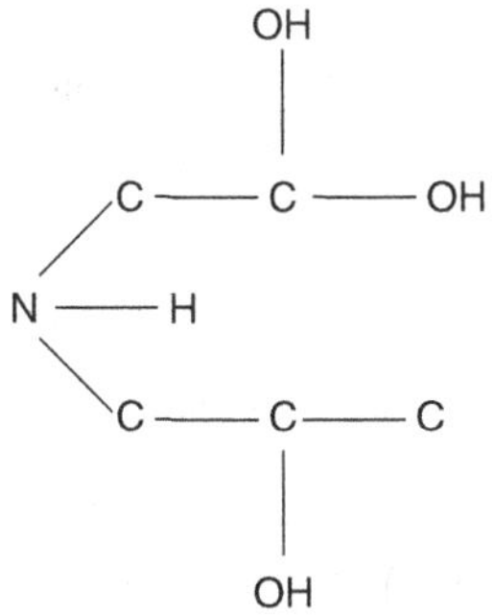
β, β′ Hydroxyaminoethylether (DGA)	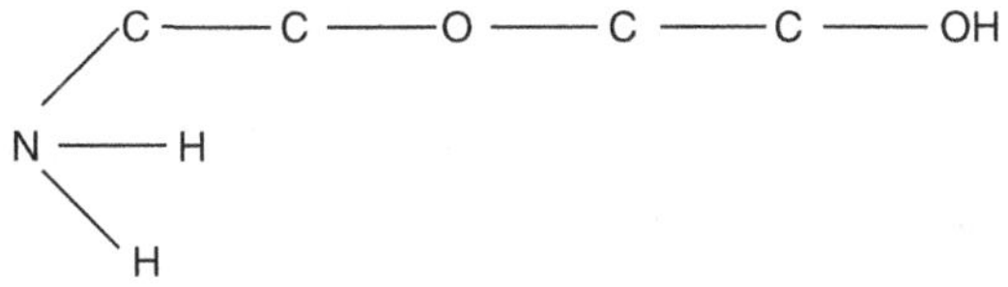
Potassium carbonate (with promoters)	K_2CO_3
Potassium glycinate	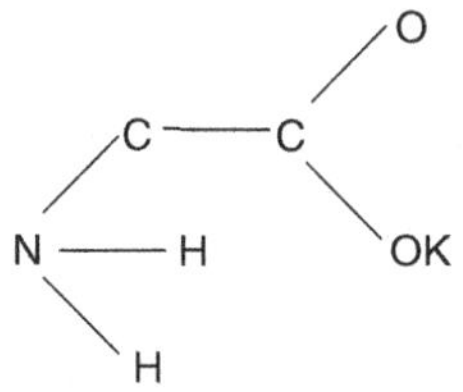
Caustic soda	NaOH

[a] Reprinted with permission from G. Astarita, D.W. Savage, A. Bisio, *Gas Treating with Chemical Solvents*, John Wiley & Sons (1983), Table 1.2.3, p. 9. Copyright © 1983, John Wiley & Sons.

Now, $C^t_{H_2S} = C_{H_2S} + C_{HS^-}$. Assume that $C_{H_2S} << C_{HS^-}$, which is generally valid unless p_{H_2S} is very high. Further,

$$p_{CO_2} = H^C_{CO_2} C_{CO_2} \quad \text{and} \quad p_{H_2S} = H^C_{H_2S}\, C_{H_2S}. \tag{5.2.27}$$

Substitute (5.2.25) and (5.2.27) into the selectivity definition (5.2.20) for $C_{H_2S} << C_{HS^-}$ to obtain

$$\alpha_{H_2S\text{-}CO_2} = \frac{H^C_{CO_2}}{H^C_{H_2S}} K_{CS} \frac{C_{HCO_3^-}}{C_{HS^-}} \frac{C^t_{H_2S}}{C^t_{CO_2}} \cong K_{CS} \frac{H^C_{CO_2}}{H^C_{H_2S}} \frac{C_{HCO_3^-}}{C^t_{CO_2}}. \tag{5.2.28}$$

Now CO_2 exists in the liquid as free CO_2 and an HCO_3^- ion, as well as in other forms, e.g. $CO_3^=$ for an aqueous K_2CO_3 solution, $RNHCOO^-$ for a primary amine solution, etc. Thus, $C_{HCO_3^-} \leq C^t_{CO_2}$, so that

$$\alpha_{H_2S\text{-}CO_2} \leq K_{CS}\left(H^C_{CO_2}/H^C_{H_2S}\right). \tag{5.2.29}$$

Obviously, conditions that lead to $C^t_{CO_2} \cong C_{HCO_3^-}$ provide higher selectivity. This means that the carbamate species $RNHCOO^-$ should be unstable. Savage *et al.* (1986) have justified the use of methyldiethanolamine on such grounds.

The value of K_{CS} is less than 1 for the practical range of temperatures (see Astarita *et al.* (1983, p. 87)). Therefore we obtain

$$\alpha_{H_2S\text{-}CO_2} \leq H^C_{CO_2}/H^C_{H_2S} \tag{5.2.30}$$

with aqueous alkaline absorbents. Note that the ratio $(H^C_{CO_2}/H^C_{H_2S})$ is the selectivity with simple physical absorption (relation (4.1.8)). Therefore the thermodynamic selectivity between H_2S and CO_2 with alkaline aqueous absorbents is less than that achieved without any chemical reaction. For pure water, the limiting value $(H^C_{CO_2}/H^C_{H_2S})$ is around 3.05; it varies slowly with temperature.

In reality, the selectivity achieved in mixed H_2S-CO_2 absorption in alkaline solutions is higher due to the kinetics of the mixed gas absorption process as long as the mass-transfer coefficient of the absorber is high (Danckwerts, 1970). Thus, thermodynamics (or equilibrium behavior) alone is insufficient. See Section 5.3.1.1.

One can also select organic physical solvents with much higher values of $\alpha_{H_2S\text{-}CO_2}$. For example, NMP (n-methyl-pyrrolidone) has a value of around 12 and NMC (N-methyl-ε-caprolactam) has a value of around 12.5; the H_2S solubility in these solvents is also high (Astarita *et al.*, 1983).

5.2.1.2 *Vapor–liquid equilibrium*

The distillation separation factor for an isotopic mixture or a close boiling mixture is very close to 1. Separation by conventional distillation would be highly energy-intensive and capital-intensive. If, however, a reactive but relatively nonvolatile species 3 is added to a close-boiling binary mixture of species 1 and 2, such that the lighter species 1 reacts according to

$$1 + 3 \Leftrightarrow 4 \tag{5.2.31}$$

in the liquid phase while species 2 is nonreactive, a much higher separation can be obtained provided reaction product 4 is relatively nonvolatile. Species 3 is sometimes known as a *reactive entrainer*.

In order to recover 1 and 2 in an unreacted form, it is necessary for reaction (5.2.31) to be reversible. It is also necessary that the volatilities of 3 and 4 should be either more or less than those of both 1 and 2; this ensures that the subsequent separations are easy. Terrill *et al.* (1985) (see also Cleary and Doherty (1985)) have discussed a number of examples of such systems, e.g. the separation of m-xylene and p-xylene using organometallic compounds. The reader should consult Terrill *et al.* (1985) for references on other systems. We follow their treatment below to demonstrate how the vapor–liquid equilibrium is altered by the reaction.

For simplicity, we define the vapor-phase mole fraction, $x_{ij} = x_{i1} = y_i$, where $j = 1$ is the vapor phase and $j = 2$ is the liquid phase. Similarly define $x_{ij} = x_{i2} = x_i$ as the liquid-phase mole fraction. Then relation (1.6.12) becomes

$$y_i = \frac{\alpha_{in}\ x_i}{\sum_{k=1}^{4} \alpha_{kn}\ x_k}, \tag{5.2.32a}$$

where

$$\alpha_{ij} = \frac{y_i\, x_j}{x_i y_j}, \tag{5.2.32b}$$

$$\alpha_{nn} = 1. \tag{5.2.32c}$$

Since species 3 and 4 are assumed nonvolatile relative to components 1 and 2,

$$\alpha_{31} = \alpha_{32} = \alpha_{41} = \alpha_{42} = 0. \tag{5.2.32d}$$

Choose $n = 2$. Also, it is a close separation problem, $\alpha_{12} \cong 1$; obviously $\alpha_{22} = 1$. Therefore relation (5.2.32a) for $i = 1$ and 2 becomes

$$y_1 \cong \frac{x_1}{x_1 + x_2}; \tag{5.2.33}$$

$$y_2 \cong \frac{x_2}{x_1 + x_2}. \tag{5.2.34}$$

The mole fraction based equilibrium constant for reaction (5.2.31) is

$$K^x = \frac{x_4}{x_1\, x_3}. \tag{5.2.35}$$

The liquid-phase mass balance relation is

$$\sum_{i=1}^{4} x_i = 1 \Rightarrow x_1 + x_2 + x_3 + x_4 = 1. \tag{5.2.36}$$

Using relation (5.2.35) for x_4 in (5.2.36) to obtain an expression for $(x_1 + x_2)$ in terms of K^x, x_1 and x_3, we see that, from (5.2.33),

$$y_1 \cong \frac{x_1}{1 - x_3 - (K^x x_3)\ x_1}. \tag{5.2.37}$$

When $x_3 = 0$, $y_1 \cong x_1$, indicating essentially no separation. However, for nonzero x_3, $y_1 > x_1$. Figure 5.2.3 shows plots of y_1 vs. x_1 for $x_3 = 0.1$ and $K^x = 1$, 5 and 10. As K^x increases, the vapor becomes highly enriched in species 1. But, due to the liquid-phase reaction of species 1 with species 3, mole fraction x_1 is considerably reduced from the value of x_1 without a reaction.

Since we are interested in the separation of species 1 and 2, consider y_2 and x_2. From relation (5.2.37) and $y_1 + y_2 = 1.0$ (remember 3 and 4 are nonvolatile),

$$y_2 \cong \frac{x_2}{1 - x_3 - (K^x x_3)\ x_1}. \tag{5.2.38}$$

Again, when $x_3 = 0$, $y_2 \cong x_2$, indicating essentially no separation. However, for the case of a reaction with nonzero x_3, $y_2 > x_2$. Further, since x_1 is small, y_2 can be much larger

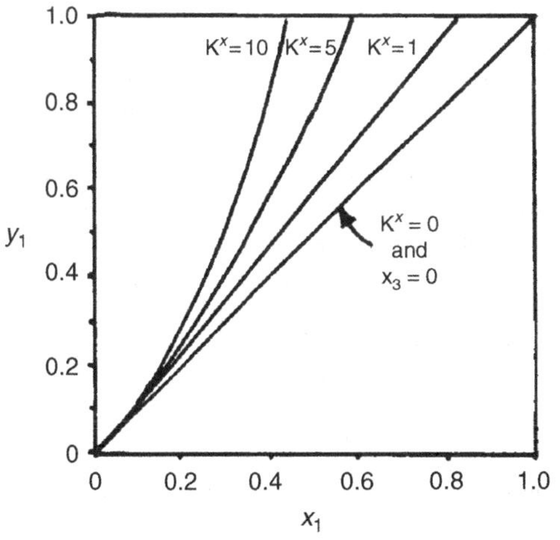

Figure 5.2.3. Effect of reaction equilibrium constant on phase equilibrium (x_3 = 0.1 unless stated otherwise) (Terrill et al., *1985). Reprinted, with permission, from* I & E Chem. Proc. Des. Dev., *24, 1062 (1985). Figure 2. Copyright (1985) American Chemical Society.*

than y_1. For example, suppose $K^x = 10$, $x_3 = 0.1$ and $x_1 = 0.1$. From the above relations, we get $y_1 = 0.125$, $x_2 = 0.7$ and $y_2 = 0.875$. Thus the vapor is highly enriched in species 2 over species 1. Suppose $x_2 = 0.7$ ($x_3 = 0 = x_4$) and there is no chemical reaction. Then $x_1 = 0.3$, $y_1 \cong 0.3$ and $y_2 \cong 0.7$. Obviously, the chemical reaction has led to a much purer vapor in terms of species 2, and therefore to a better separation.

The converse approach is sometimes practiced to improve the conversion from a reaction. By withdrawing selectively one or more products from the reaction mixture, a reversible reaction can be driven to the right, e.g.

$$CH_3COOH + C_2H_5OH \Leftrightarrow CH_3COOC_2H_5 + H_2O. \quad (5.2.39)$$

Removal of water would push the reaction to the right in this esterification process, an example of separation facilitating the main objective, namely a chemical reaction (Suzuki *et al.*, 1971).

5.2.1.2.1 Separation factor in distillation of isotopic mixtures Isotopic mixtures are often separated by distillation. The production of heavy water, D_2O, is a prime example. In the distillation of natural water, which contains variable amounts of deuterium depending on the source (see Table 13.1 in Benedict *et al.* (1981)), deuterium is invariably concentrated in the liquid phase. Water in the tropical oceans contains about 156 ppm deuterium. Multistage distillation (see Section 8.1.3) has to be carried out to obtain a liquid fraction substantially enriched in deuterium. Here we will consider only the equilibrium enrichment achieved in one stage, a closed vessel.

When two isotopic compounds, say XA_n ($i = 1$) and $XA_{n-1}B$ ($i = 2$), are present in a mixture containing two isotopes A and B, the separation factor α_{12} may be easily obtained from (4.1.21b) as $\alpha_{12} = P_1^{sat}/P_2^{sat}$ as long as the gas phase behaves ideally and the liquid phase is an ideal solution. However, in the case of heavy water, there is a third isotopic compound, HDO, present; there is the following reaction in the liquid phase:

$$H_2O + D_2O \overset{K^x}{\Leftrightarrow} 2HDO. \quad (5.2.40)$$

We will now obtain a separation factor α_{12} between H_2O ($i = 1$) and D_2O ($i = 2$) in the presence of HDO ($i = 3$), all three species being distributed between the liquid phase and the vapor phase. Assume (Benedict *et al.*, 1981) that

(1) we have ideal gas behavior and an ideal solution in liquid phase;
(2) the vapor pressure of HDO is a geometric mean of the vapor pressures of H_2O and D_2O, i.e.

$$P_3^{sat} = \sqrt{P_1^{sat}\, P_2^{sat}}; \quad (5.2.41)$$

(3) reaction (5.2.40) is at equilibrium in the liquid phase;
(4) the equilibrium constant K^x of reaction (5.2.40) based on mole fractions is 4.

The separation factor α_{12} based on H_2O ($i = 1$) concentrating in the vapor phase and D_2O ($i = 2$) concentrating in the liquid phase is calculated using the atom fraction of hydrogen and the atom fraction of deuterium, which in turn are calculated from the mole fractions of individual compounds (see (1.6.1d)). For example, the atom fraction of hydrogen in vapor is given by

$$a_{H_2v} = 2x_{H_2O,\,v} + x_{HDO,\,V} \quad (5.2.42a)$$

and the atom fraction of deuterium in vapor is given by

$$a_{D_2v} = x_{HDO,\,v} + 2x_{D_2O,\,v}, \quad (5.2.42b)$$

yielding

$$\alpha_{12} = \left[\frac{a_{H_2v}}{a_{D_2v}}\right]\left[\frac{a_{D_2l}}{a_{H_2l}}\right] = \left[\frac{2x_{H_2O,\,v} + x_{HDO,\,v}}{x_{HDO,\,v} + 2\,x_{D_2O,\,v}}\right]\left[\frac{x_{HDO,\,l} + 2x_{D_2O,\,l}}{2\,x_{H_2O,\,l} + x_{HDO,\,l}}\right]. \quad (5.2.43)$$

From assumption (1),

$$x_{H_2O,\,v} = \frac{P_{H_2O}^{sat}\, x_{H_2O,\,l}}{P}; \qquad x_{D_2O,\,v} = \frac{P_{D_2O}^{sat}\, x_{D_2O,\,l}}{P}. \quad (5.2.44a)$$

From assumptions (2), (3) and (4),

$$P^{sat}_{HDO} = \sqrt{P^{sat}_{H_2O}\, P^{sat}_{D_2O}}; \tag{5.2.44b}$$

$$K^x = 4 = \frac{x^2_{HDO,l}}{x_{H_2O,l}\, x_{D_2O,l}} \Rightarrow x_{HDO,l} = 2\sqrt{x_{H_2O,l}\, x_{D_2O,l}}; \tag{5.2.44c}$$

$$x_{HDO,v} = \frac{P^{sat}_{HDO}\, x_{HDO,l}}{P} = 2\,\frac{\sqrt{P^{sat}_{H_2O}\, P^{sat}_{D_2O}\, x_{H_2O,l}\, x_{D_2O,l}}}{P}. \tag{5.2.44d}$$

Substituting these results into the expression for α_{12}, we get

$$\alpha_{12} = \left[\frac{2\dfrac{P^{sat}_{H_2O}\, x_{H_2O,l}}{P} + \dfrac{2\sqrt{P^{sat}_{H_2O}\, P^{sat}_{D_2O}\, x_{H_2O,l}\, x_{D_2O,l}}}{P}}{2\dfrac{\sqrt{P^{sat}_{H_2O}\, P^{sat}_{D_2O}\, x_{H_2O,l}\, x_{D_2O,l}}}{P} + \dfrac{2P^{sat}_{D_2O}\, x_{D_2O,l}}{P}}\right] \times \left[\frac{2\sqrt{x_{H_2O,l}\, x_{D_2O,l}} + 2x_{D_2O,l}}{2x_{H_2O,l} + 2\sqrt{x_{H_2O,l}\, x_{D_2O,l}}}\right]; \tag{5.2.45}$$

$$\alpha_{12} = \left(\frac{\sqrt{x_{D_2O,l}}}{\sqrt{x_{H_2O,l}}}\right)\left[\frac{\sqrt{P^{sat}_{H_2O}\, x_{H_2O,l}}}{\sqrt{P^{sat}_{D_2O}\, x_{D_2O,l}}}\right] = \sqrt{\frac{P^{sat}_{H_2O}}{P^{sat}_{D_2O}}}. \tag{5.2.46}$$

This value varies between 1.12 at 0 °C to 1.026 at 100 °C (see Table 13.4 of Benedict *et al.* (1981)).

Apparently, using similar assumptions, this type of result has been generalized for a mixture of isotopic compounds. Benedict *et al.* (1981) have suggested that, for a mixture of isotopic compounds, XA_n, $XA_{n-1}B$, $XA_{n-2}B_2$, ..., XB_n, the separation factor α_{AB} for isotopes A and B may be approximately obtained as

$$\alpha_{AB} = \sqrt[n]{\frac{P^{sat}_{XA_n}}{P^{sat}_{XB_n}}}. \tag{5.2.47}$$

For example, in the distillation of ammonia, the separation factor between NH_3 and ND_3 is quite close to the value

$$\alpha_{NH_3,ND_3} = \sqrt[3]{\frac{P^{sat}_{NH_3}}{P^{sat}_{ND_3}}}. \tag{5.2.48}$$

This separation factor was measured to be around 1.042.

5.2.1.2.2 Isotope exchange reactions in a vapor-liquid system Consider the following overall isotope exchange reaction between $HC^{12}N$ vapor and an aqueous $NaC^{13}N$ solution:

$$HC^{12}N(v) + Na^+C^{13}N^-(l) \Leftrightarrow HC^{13}N(v) + Na^+C^{12}N^-(l). \tag{5.2.49}$$

As a result of this exchange between C^{12} and C^{13} isotopes of carbon, the HCN vapor becomes preferentially enriched in the C^{13} isotope. Natural carbon contains 1.11% C^{13} (Benedict *et al.*, 1981). However, the reaction above will lead to a higher percentage of C^{13} in the HCN vapor. If such a behavior could be multiplied in a cascade of many stages (see Chapters 8 and 9), it is possible to produce a highly enriched C^{13} fraction containing as much as 60–70% C^{13} as HCN vapor. One could react this HCN vapor with an aqueous NaCN to obtain a NaCN solution in which 60–70 atom % of C is C^{13}.

Any separation factor describing such an isotope enrichment process has to take into account the vapor-liquid equilibrium for the vapor species HCN. Consider now the isotope exchange reaction (5.2.49) equilibrium along with the vapor-liquid equilibrium of $HC^{12}N$ and $HC^{13}N$. There are a total of six mole fractions, x_{ij}, to deal with:

$HC^{12}N$ (v), $i = 2, j = 1$: x_{21};
$HC^{13}N$ (v), $i = 1, j = 1$: x_{11};
$HC^{12}N$ (l), $i = 2, j = 2$: x_{22};
$HC^{13}N$ (l), $i = 1, j = 2$: x_{12};
$NaC^{13}N$ (l), $i = 3, j = 2$: x_{32};
$NaC^{12}N$ (l), $i = 4, j = 2$: x_{42}.

First, the mole fraction based separation factor in vapor-liquid equilibrium between $HC^{13}N$ and HC^{12} N is given by

$$\alpha^{vl}_{12} = \frac{x_{11}\, x_{22}}{x_{22}\, x_{12}}. \tag{5.2.50}$$

Second, the overall isotope reaction (5.2.49) may be written for the liquid phase as follows:

$$HC^{12}N(l) + Na^+C^{13}N^-(l) \overset{K_l}{\Leftrightarrow} HC^{13}N(l) + Na^+C^{12}N^-(l). \tag{5.2.51}$$

One can now develop the following relation between the various concentrations and the reaction equilibrium constant K_l:

$$K_l = \frac{C_{12}\, C_{42}}{C_{22}\, C_{32}}. \tag{5.2.52a}$$

However, this is also equivalent to

$$K_l = \frac{x_{12}\, x_{42}}{x_{22}\, x_{32}}. \tag{5.2.52b}$$

Define the overall separation factor α^{ov}_{12} between isotopes 1 and 2 and the two phases $j = 1, 2$ as

$$\alpha^{ov}_{12} = \frac{x_{11}}{x_{21}}\frac{(x_{22} + x_{42})}{(x_{12} + x_{32})}. \tag{5.2.53}$$

Pratt (1967) has defined a quantity w as the mole fraction of both isotopic forms of the nonvolatile reactant in the total dissolved reactants as

$$w = \frac{x_{32} + x_{42}}{x_{22} + x_{12} + x_{32} + x_{42}}. \tag{5.2.54}$$

He has then claimed that α^{ov}_{12} is related to K_l, w and α^{vl}_{12} by

$$\alpha^{ov}_{12} = \alpha^{vl}_{12}[(1 - w) + wK_l] \tag{5.2.55}$$

as long as $x_{11} + x_{21} \cong 1.0$ by neglecting the humidity in the vapor phase. For the reaction under consideration, $\alpha_{12}^{vl} = 0.995$ and $K_1 = 1.031$ at 25 °C. It becomes clear that $\alpha_{12}^{ov} > 1$ as long as w is considerable. For example, if w is 0.8, $\alpha_{12}^{ov} = 1.0188$. The experimentally obtained value of α_{12}^{ov} is 1.013 (Pratt, 1967, chap. 6).

Many other isotopes have been enriched by isotope exchange reactions. However, they have primarily employed gas–liquid systems. A few of them are listed below (for a more detailed introduction, see Pratt (1967, chap. 6) and Benedict *et al.* (1981)):

(1) $CO_2^{16}(g) + H_2O^{18}(l) \Leftrightarrow$
$$CO^{16}O^{18}(g) + H_2O^{16}(l), T = 25°C;$$

(2) $N^{15}H_3(g) + N^{14}H_4^+NO_3^-(l) \Leftrightarrow$
$$N^{14}H_3(g) + N^{15}H_4^+NO_3^-(l), T = 25°C;$$

(3) $S^{34}O_2(g) + Na^+HS^{32}O_3^-(l) \Leftrightarrow$
$$S^{32}O_2(g) + Na^+HS^{34}O_3^-(l).T = 25°C. \quad (5.2.56)$$

5.2.2 Liquid–liquid equilibrium

The effective distribution of a solute i between two immiscible liquid phases 1 and 2 is often changed if any one, or some or all of the following happen: species i dissociates in one phase; species i associates in another phase; species i or its dissociated form reacts in one of the phases with another species deliberately added to this phase or deliberately added to the other phase in another form. This may be brought about by using acid–base equilibrium, simple complexation, chelation, ion pairing, solvation, etc. In some cases (for example, in metal extraction), partitioning of the solute species into another phase is generally not possible without some kind of chemical reaction. In the following, we first consider extraction of nonmetallic species in the presence of an acid–base equilibrium, association, dissociation and simple complexation. Metal extraction is treated later.

In many extraction systems in practice, species i can exist not only as i, but also in other forms. The effective distribution (or partition) coefficient κ_{i1}'' will then be based on the total concentration of i in all forms in the two phases. The intrinsic distribution coefficient κ_{i1} of only the free species i between the two phases in the absence of chemical reactions will, in general, be different from κ_{i1}''. The intrinsic distribution coefficient κ_{i1}' of only the free species i between the two phases in the presence of chemical reactions may be different from κ_{i1}. If the presence of other species does not influence the distribution coefficient, then $\kappa_{i1}' = \kappa_{i1}$. Assume the mutual solubility of water and the organic solvent to be unaffected by any of these reactions. We now study how to determine κ_{i1}'' for several different cases.

5.2.2.1 *Dissociation of an organic acid or base in water*

Organic solutes to be extracted from an aqueous phase into an organic solvent phase, or vice versa, are often weak acids or bases. Only unionized solutes are extracted into the organic solvent. But the unionized weak acid solute, HA, will partially dissociate in an aqueous solution[2] as follows:

$$HA \Leftrightarrow H^+ + A^-. \quad (5.2.57)$$

Similarly, an unionized weak base B will partially ionize in the aqueous phase as follows:

$$B + H^+ \Leftrightarrow BH^+. \quad (5.2.58)$$

Thus, the concentration of free HA or free B in water is different from the concentration without ionization.

Consider the distribution equilibrium of the weak acid HA ($i = 1$) between water and a nonionizing organic solvent:

$$\underbrace{HA}_{\substack{\text{solvent phase} \\ (j=1=o)}} \Leftrightarrow \underbrace{HA \Leftrightarrow H^+ + A^-}_{\substack{\text{aqueous phase} \\ (j=2=w)}}. \quad (5.2.59)$$

Now,

$$\kappa_{11}'' = \kappa_{1o}'' = \frac{C_{1o}}{C_{1w} + C_{2w}} = \frac{C_{1o}}{(\text{conc.ofHA} + \text{conc.ofA}^-)_{\text{water}}}, \quad (5.2.60)$$

where ionic species A^- is $i = 2$. Note that the hydrogen ion concentration, C_{H^+w}, here is also equal to C_{2w}. For simplicity, we write $C_{H^+w} = C_{H^+}$. The dissociation or ionization constant K_{d1} of solute 1 is given by

$$K_{d1} = \frac{C_{H^+}C_{2w}}{C_{1w}} \Rightarrow \frac{C_{2w}}{C_{1w}} = \frac{K_{d1}}{C_{H^+}}. \quad (5.2.61a)$$

Further, by definition of κ_{i1}',

$$\kappa_{1o}' = (C_{1o}/C_{1w}). \quad (5.2.61b)$$

Using these relations for K_{d1} and κ_{1o}' in the expression for κ_{1o}'', we get

$$\kappa_{1o}'' = \frac{\kappa_{1o}'C_{1w}}{C_{1w} + \frac{K_{d1}C_{1w}}{C_{H^+}}} = \frac{\kappa_{1o}'}{1 + \frac{K_{d1}}{C_{H^+}}} = \frac{\kappa_{1o}'}{1 + \frac{C_{2w}}{C_{1w}}}. \quad (5.2.62)$$

Thus $\kappa_{1o}'' < \kappa_{1o}'$ in general. Note here that

$$\kappa_{1o} = (C_{1o}/C_{1w}) \quad (5.2.63)$$

in the absence of any ionization; κ_{1o} may equal κ_{1o}'; i.e. the distribution relation of the free acid HA without any

[2]The reaction in water is $HA + H_2O \Leftrightarrow H_3O^+ + A^-$. For dilute solutions, we will avoid writing down water, whose concentration may be assumed to be essentially constant at 55 gmol/liter.

ionization (i.e. no other solute species) may equal the distribution relation of the free acid HA when ionization is present. But $\kappa''_{1o} \neq \kappa'_{1o}$ unless the ionization constant K_{d1} is vanishingly small. On the other hand, in an alkaline solution of a weak acid HA (Treybal, 1963) $C^{+}_{H} \ll K_{d1}$, and

$$\kappa''_{1o} = \kappa'_{1o} C_{H^+}/K_{d1}, \tag{5.2.64}$$

leading to

$$\log \kappa''_{1o} = \log \kappa'_{1o} - pH - \log K_{d1}, \tag{5.2.65a}$$

where $pH = -\log_{10} C^{+}_{H}$. Also, since $\log K_{d1} = -pK_1$, an alternative result is

$$\log \kappa''_{1o} = \log \kappa'_{1o} - pH + pK_1. \tag{5.2.65b}$$

If we now use an alternative form of relation (5.2.61a) using pH and pK_1, we get

$$pH - pK_1 = \log(C_{2w}/C_{1w}) \Rightarrow (C_{2w}/C_{1w}) = 10^{pH-pK_1}. \tag{5.2.65c}$$

Correspondingly, expression (5.2.62) becomes

$$\kappa''_{1o} = \frac{\kappa'_{1o}}{1 + 10^{pH-pK_1}}. \tag{5.2.65d}$$

We now consider the case of a weak organic base B $(i = 1)$ and the ionization reaction (5.2.58) with species BH^+ being $i = 2$:

$$\overbrace{\begin{array}{l} B + H^+ \Leftrightarrow BH^+ \\ \Updownarrow \\ B \end{array}}^{\text{aqueous phase } (j = 2 = w)}_{\underbrace{\qquad\qquad}_{\text{solvent phase } (j = 1 = o)}} \tag{5.2.66}$$

Here,

$$\kappa''_{1o} = \frac{C_{1o}}{C_{1w} + C_{2w}} = \frac{C_{1o}}{(\text{conc.of} B + \text{conc.of} BH^+)_{\text{water}}}. \tag{5.2.67}$$

The dissociation constant[3] K_{d1} of base 1 is

$$\frac{1}{K_{d1}} = \frac{C_{H^+} C_{1w}}{C_{2w}} \Rightarrow C_{H^+} K_{d1} = \frac{C_{2w}}{C_{1w}}. \tag{5.2.68a}$$

An alternative form is

$$\frac{C_{2w}}{C_{1w}} = 10^{-pK_1 - pH}. \tag{5.2.68b}$$

Now,

$$\kappa'_{1o} = C_{1o}/C_{1w}. \tag{5.2.69}$$

Substitute these definitions of K_{d1} and κ'_{1o} into the expression for κ''_{1o} to obtain

[3]Note that this definition of K_{d1} is inverse of that used by Robinson and Cha (1985).

$$\kappa''_{1o} = \kappa'_{1o} \frac{C_{1w}}{C_{1w} + C_{H^+} C_{1w} K_{d1}} = \frac{\kappa'_{1o}}{1 + C_{H^+} K_{d1}}; \tag{5.2.70a}$$

$$\kappa''_{1o} = \frac{\kappa'_{1o}}{1 + 10^{-pK_1 - pH}}. \tag{5.2.70b}$$

The general relations (5.2.62) for a weak organic acid and (5.2.70a) for a weak organic base can also be expressed in the following forms (Robinson and Cha, 1985):

$$\underline{\text{acid}}\ pH - pK_1 = \log\left(\frac{\kappa'_{1o}}{\kappa''_{1o}} - 1\right); \tag{5.2.71}$$

$$\underline{\text{base}}\ -pK_1 - pH = \log\left(\frac{\kappa'_{1o}}{\kappa''_{1o}} - 1\right). \tag{5.2.72}$$

From these relations, we find that, for a weak organic acid, as the pH increases, the value of κ''_{1o} decreases, since most of the acid is ionized and cannot be extracted. This leads to a poor extraction of the unionized acid into the organic phase when compared with the total amount of acid available in the aqueous phase. Conversely, for an organic base, as the pH decreases, κ''_{1o} decreases, reducing the extent of extraction of the unionized base. Figure 5.2.4(a) illustrates these two relations for a weak organic acid (e.g. phenols, formic acid, benzoic acid, etc.) and a weak organic base (Robinson and Cha, 1985). For carboxylic acid extraction, consult Kertes and King (1986). The extraction of pencillins, which are weak acids, is illustrated in Figure 5.2.4(b) (Souders *et al.*, 1970). For the application of solvent extraction to pharmaceutical manufacturing processes for antibiotics and nonantibiotics, see Ridgway and Thrope (1983). The values of pK_1 of a number of biological solutes, e.g. carboxylic acids, amino acids, etc. are provided in Tables 5.2.2 and 5.2.5.

Table 5.2.2. pK values of selected solutes[a]

Solute	pK_1	pK_2	pK_3
Formic acid	3.74		
Acetic acid	4.76		
Propionic acid	4.87		
n-Butyric acid	4.82		
Lactic acid	3.73		
Tartaric acid	3.03	4.37	
Oxalic acid	1.25	3.67	
Citric acid	3.13	4.76	6.40
Ammonium (NH_4^+)	9.25		
Trimethylamine	9.87		
Phosphate H_3PO_4	2.15		
$H_2PO_4^-$		7.22	
HPO_4^{2-}			11.50

[a] Values obtained from Belter *et al.* (1988), Garcia *et al.* (1999) and Table 6.3.1 at 25°C.
See Table 5.2.5 for amino acids.

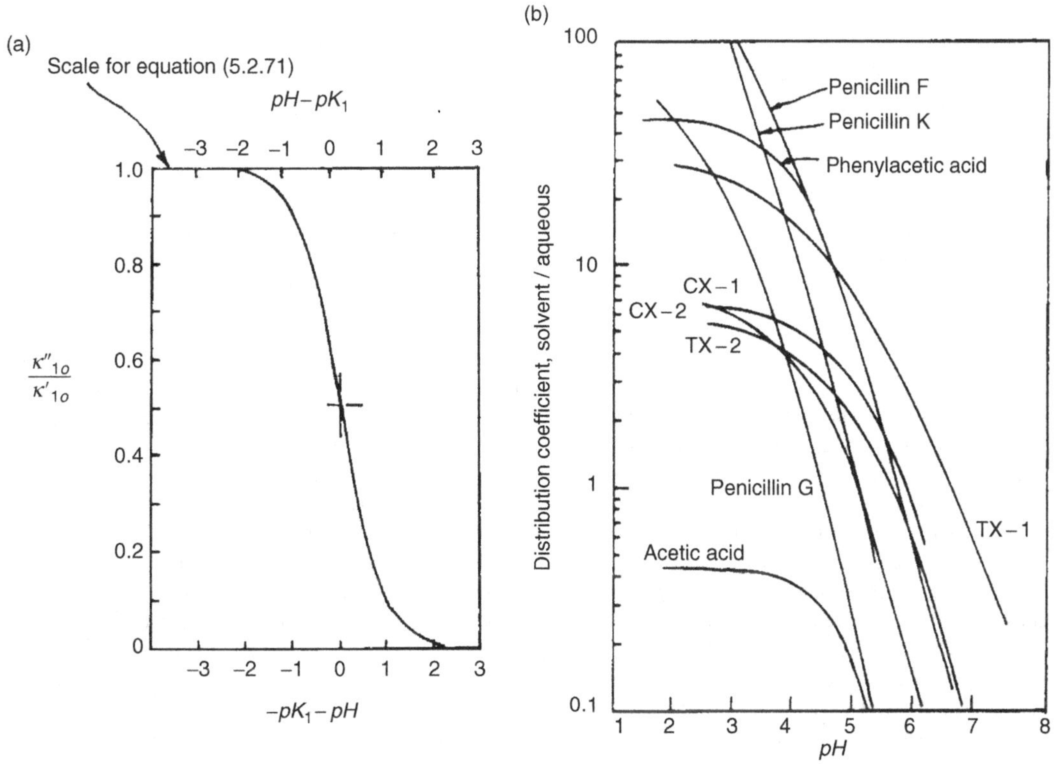

Figure 5.2.4 (a) Solutions of equations (5.2.71) and (5.2.72), respectively. (After Robinson and Cha (1985).) (b) Distribution coefficient data for penicillin broth contents for a solvent. Reprinted, with permission, from Souders et al., Chem. Eng. Prog. Symp. Ser., *66(100), 41 (1970), Figure 2. Copyright © [1970] American Institute of Chemical Engineers (AIChE).*

Some carboxylic acids have two ionizable protons. They are called *dicarboxylic acids*, e.g. oxalic acid, HOOC–COOH. We may represent the unionized acid, species $i = 1$, by H_2A. The first ionization of this acid, represented by

$$\begin{aligned} &H_2A \Rightarrow HA^- + H^+,\\ &(i=1)\quad (i=2) \end{aligned} \tag{5.2.73a}$$

has a dissociation constant K_{d1} corresponding to pK_1 via

$$K_{d1} = \frac{C_{H^+}C_{HA^-}}{C_{H_2A}} \Rightarrow \frac{K_{d1}}{C_{H^+}} = \frac{C_{HA^-}}{C_{H_2A}}, \tag{5.2.73b}$$

which leads to

$$-pK_1 + pH = \log_{10}[C_{HA^-}/C_{H_2A}]. \tag{5.2.73c}$$

The second ionization, represented by

$$\begin{aligned} &HA^- \Rightarrow A^{2-} + H^+,\\ &(i=2)\quad (i=3) \end{aligned} \tag{5.2.74a}$$

has a dissociation constant K_{d2}, corresponding to pK_2 via

$$K_{d2} = \frac{C_{H^+}C_{A^{2-}}}{C_{HA^-}}. \tag{5.2.74b}$$

Rearrangement leads to

$$-pK_2 + pH = \log_{10}[C_{A^{2-}}/C_{HA^-}]. \tag{5.2.74c}$$

We may rewrite this as $-pK_2 + pH = \log_{10}[C_{A^{2-}}/C_{H_2A}] + \log_{10}[C_{H_2A}/C_{HA^-}]$,

$$-pK_2 - pK_1 + 2pH = \log[C_{A^{2-}}/C_{H_2A}], \tag{5.2.74d}$$

by using the result (5.2.73c). We can now define an effective partition coefficient for H_2A between water and an organic solvent in the absence of any complexing extractant for the dicarboyxlic acid H_2A as

$$\kappa''_{1o} = \frac{C_{1o}}{C_{1w} + C_{2w} + C_{3w}} = \frac{\kappa'_{1o}}{1 + \frac{C_{2w}}{C_{1w}} + \frac{C_{3w}}{C_{1w}}}; \tag{5.2.74e}$$

$$\kappa''_{1o} = \frac{\kappa'_{1o}}{1 + 10^{(-pK_1+pH)} + 10^{(2pH-pK_1-pK_2)}}. \tag{5.2.74f}$$

Note: Here the total concentration of the acidic species in water is contributed by three species: H_2A (species 1), HA^-(species 2) and A^{2-} (species 3). In the case of tricarboxylic acids, which may be represented by H_3A, the number of such species will be four: H_3A, H_2A^-, HA^{2-}, A^{3-}.

There are substances that are *amphiprotic*: at low *pH* values they behave as a base, and at high *pH* values they act like organic acids. At intermediate *pH*, the neutral species predominates. An example is 8-quinolinol:

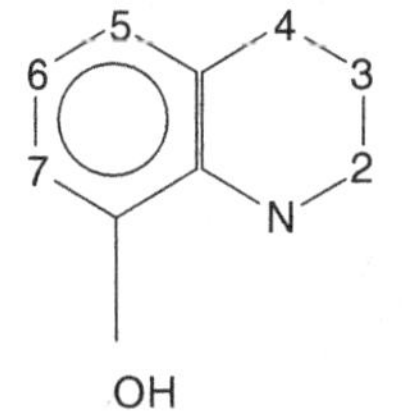

If we represent it as HQ, then, at lower *pH*, H_2Q^+, resulting from the protonation of the nitrogen, will be present in substantial amounts, whereas at high *pH*, Q^- will be preponderant. The neutral species HQ is the major species at intermediate *pH*. Problem 5.2.10 considers the development of a distribution coefficient for such a substance over the whole *pH* range (Karger *et al.*, 1973). A most important example of similar behavior is obtained with amino acids. If we represent an amino acid $NH_2CHRCOOH$ as HA, then, at low *pH*, H_2A^+ dominates; at high *pH*, A^- is preponderant. At intermediate *pH*, the zwitterionic form $HA^{\pm}$ is the major species. See Section 5.2-3.1-2 for an analysis of their ionization behavior.

5.2.2.2 *Two weak organic acids or bases: dissociation extraction*

Many compounds used in practice, especially in the pharmaceutical industry, are weak acids or bases. If two compounds are similar, their distribution coefficients between the aqueous and organic phases would be similar. Since one of the compounds is preferentially sought for its particular properties over the other, conditions for preferential extraction of that compound are required. Select two compounds A and B, both being bases (acids are treated later). Assume for the present that $\kappa'_{Ao} > \kappa'_{Bo}$, i.e. species B is more soluble in the aqueous phase than species A. Remember, however, that κ'_{Ao} is quite close to κ'_{Bo}.

Consider *pH* conditions in the aqueous solution (acidic) so that ionization of A and B has taken place (obviously the bases will hardly ionize at very large *pH*, say $pH > 12$). Then from (5.2.70a),

$$\kappa''_{Ao} = \frac{\kappa'_{Ao}}{1 + (C_{H^+} K_{dA})}; \tag{5.2.75}$$

$$\kappa''_{Bo} = \frac{\kappa'_{Bo}}{1 + (C_{H^+} K_{dB})}. \tag{5.2.76}$$

Now, the ratio $(\kappa''_{Ao}/\kappa''_{Bo}) = [C_{Ao}(C_B + C_{BH^+})_w/(C_A + C_{AH^+})_w C_{Bo}]$. For a dilute solution, we consider the mixture of these two bases on the following basis: organic phase on a solvent-free basis; aqueous solution on a water-free and acid-free basis. Then it is obvious that $(\kappa''_{Ao}/\kappa''_{Bo}) = \alpha''_{AB}$ and $\kappa'_{Ao}/\kappa'_{Bo} = \alpha'_{AB}$. The separation factor with aqueous dissociation, α''_{AB}, and the separation factor α'_{AB} unaffected by the dissociation are then related by

$$\alpha''_{AB} = \frac{\kappa''_{Ao}}{\kappa''_{Bo}} = \alpha'_{AB} \frac{1 + (C_{H^+} K_{dB})}{1 + (C_{H^+} K_{dA})}. \tag{5.2.77}$$

Rewrite this relation as

$$\log\left(\frac{\alpha''_{AB}}{\alpha'_{AB}} - 1\right) = -pK_A - pH + \log\left(\frac{K_{dB}}{K_{dA}} - \frac{\alpha''_{AB}}{\alpha'_{AB}}\right). \tag{5.2.78}$$

Only cases where $K_{dA} \neq K_{dB}$ are useful for our purpose since $\alpha''_{AB} = \alpha'_{AB}$ if $K_{dA} = K_{dB}$. Therefore, we consider cases where the dissociation constants of species A and B are different. For given *pH*, K_{dA} and K_{dB}, one can now solve for the ratio $(\alpha''_{AB}/\alpha'_{AB})$; since α'_{AB} is assumed known, α''_{AB} can be determined. Figure 5.2.5 illustrates how $(\alpha''_{AB}/\alpha'_{AB})$ varies with $-pK_A - pH$ for different positive values of $pK_A - pK_B$ (Robinson and Cha, 1985). Here B is a stronger base than A ($K_{dB} > K_{dA}$); it is also more water soluble. Thus, when the solution becomes more acidic, species B ionizes more and its water solubility increases further. Therefore, κ''_{Bo} will decrease much more rapidly, leading to a high separation factor; species A will be extracted much more into the solvent compared to species B, leading to a higher purity of the extract in species A. Of course, the value of κ''_{Ao} will be decreased due to an increase in *pH*. Consequently, an optimum between the purity of A and its extent of recovery in the solvent has to be struck. The case where $pK_B > pK_A$ ($K_{dA} > K_{dB}$) has not been considered here.

A relationship similar to (5.2.77) can be developed (Treybal, 1963, p. 48) when both species A and B are weak acids:

$$\alpha''_{AB} = \frac{\kappa''_{Ao}}{\kappa''_{Bo}} = \alpha''_{AB} \frac{1 + (K_{dB}/C_{H^+})}{1 + (K_{dA}/C_{H^+})}. \tag{5.2.79}$$

This can be expressed for $K_{dB} >> C_{H^+}$ and $K_{dA} >> C_{H^+}$ (i.e. $pH >> pK_A$ and pK_B) as

$$(\alpha''_{AB}/\alpha'_{AB}) \cong (K_{dB}/K_{dA}), \tag{5.2.80}$$

reflecting the effect of different dissociation constants. Thus, if acid B ionizes more, acid A will be extracted more into the organic solvent at higher *pH* values. For lower *pH* values, if $K_{dB} << C_{H^+}$ and $K_{dA} << C_{H^+}$, then $\alpha''_{AB} = \alpha'_{AB}$, namely the separation is strictly due to preferential extraction of the undissociated species. The separation of two organic acids or two organic bases of similar distribution coefficients by preferential extraction of one of them into a solvent is thus possible only if their dissociation constants are different. By selecting proper *pH* conditions, the separation can be considerably enhanced. This process is known as *dissociation extraction*.

Dissociation extraction can also be utilized to extract preferentially into an aqueous solution a particular organic acid or base from a mixture of organic acids or bases in an organic solvent. For the separation of organic acids, for example, a limited amount of alkali in the aqueous phase is used to create a competition for it by the solutes A and B in the organic phase. The stronger acid is transferred to a large extent to the alkaline aqueous phase, whereas the weaker acid remains essentially in the organic phase.

For the separation of organic bases A and B present in an aqueous solution in completely dissociated forms, an alternative technique of *fractional neutralization* may be adopted. Add a neutralizing alkali corresponding to the weaker base A, which will then revert to A from AH^+ (but B is present as BH^+). This undissociated A can now be extracted by an organic solvent. For additional information on, and analysis of, dissocation extraction, the work by Wise and Williams (1963), the description by Pratt (1967, pp. 327-332) and the papers by Anwar *et al.* (1971) and Wadekar and Sharma (1981) may be consulted.

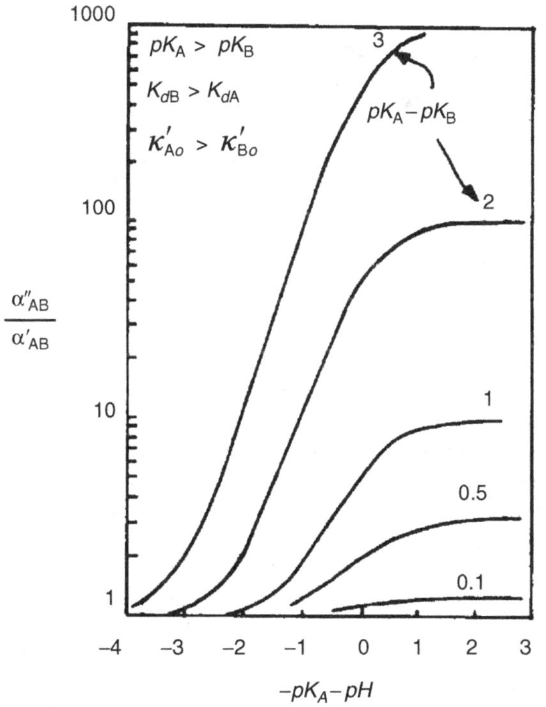

Figure 5.2.5. Solution of equation (5.2.78) for two bases. (After Robinson and Cha (1985).)

5.2.2.3 *Association and/or complexation of a solute in organic solvent*

Solutes to be extracted sometimes associate in the organic solvent. This happens especially when the solute is polar, e.g. organic acids, while the solvent is quite nonpolar, e.g. hexane, benzene, toluene, etc. For example, acetic acid ($i = 1$) is known to dimerize (through hydrogen bonding) significantly in benzene. (Similarly, benzoic acid dimerizes in benzene.) Thus the equilibrium distribution of free acetic acid between the aqueous and organic phase will be different from that of the total amount of acetic acid in the two phases. With a polar solvent, dimerization of solute is likely to disappear. The equilibrium distribution will also be influenced if an agent, present in the organic phase, complexes or reacts with acetic acid. We consider now the distribution of acetic acid, for example, under such conditions. For the time being, we neglect the dissociation of the acid in the aqueous phase. The analysis is useful in general for carboxylic acids.

Conventionally, oil-phase ($j = o$) concentration of acetic acid is not large due to a low value of κ_{1o}. To increase it, organic amines, for example, are added to the oil phase. The basic amine ($i = 3$) complexes with acetic acid to form a complex $i = 4$. In the oil phase, acetic acid is present in three forms, whose concentrations are: free acetic acid C_{1o}, acetic acid dimer C_{2o} and the complex C_{4o}, where $i = 2$ refers to the dimer. The equilibrium constants for the various reactions (shown in Figure 5.2.6) are as follows ($j = w$ for the aqueous phase):

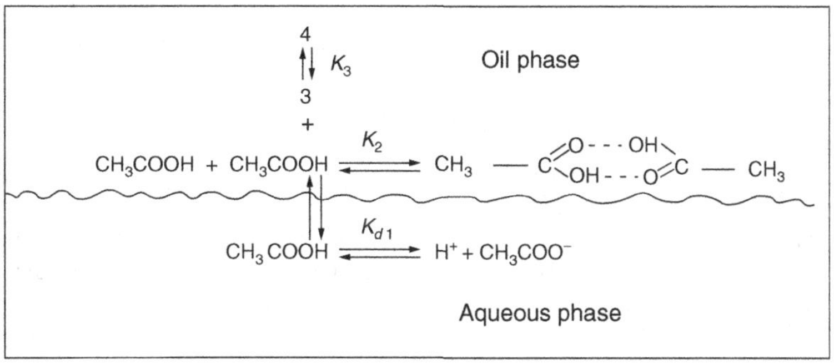

Figure 5.2.6. Distribution of acetic acid between an oil phase and an aqueous phase with attendant reactions.

$$C_{1o} + C_{1o} \overset{K_2}{\Leftrightarrow} C_{2o}; \tag{5.2.81}$$

$$C_{1o} + C_{3o} \overset{K_3}{\Leftrightarrow} C_{4o}. \tag{5.2.82}$$

Assume no acetic acid dissociation in the aqueous phase and that κ'_{1o} is unaffected by these reactions:

$$\kappa_{1o} = \frac{C_{1o}}{C_{1w}} = \kappa'_{1o}. \tag{5.2.83}$$

The effective value of κ_{1o}, κ''_{1o}, however, should be based on C^t_{1o}, the total acetic acid concentration in the oil phase:

$$\kappa''_{1o} = \frac{C^t_{1o}}{C_{1w}} = \frac{C_{1o} + C_{4o} + 2C_{2o}}{C_{1w}}. \tag{5.2.84}$$

From (5.2.82),

$$K_3 = C_{4o}/C_{1o}C_{3o}. \tag{5.2.85}$$

Similarly,

$$K_2 = C_{2o}/C^2_{1o}. \tag{5.2.86}$$

Substituting these relations into (5.2.84), we get

$$\kappa''_{1o} = \frac{C_{1o}}{C_{1w}} + K_3 \frac{C_{1o}}{C_{1w}} C_{3o} + 2K_2 \frac{C^2_{1o}}{C_{1w}}$$
$$= \kappa_{1o} + \kappa_{1o} K_3 C_{3o} + 2\kappa_{1o} K_2 C_{1o}.$$

Using the definition of κ_{1o}, we get a simpler result:

$$\kappa''_{1o} = \kappa_{1o} + \kappa_{1o} K_3 C_{3o} + 2\kappa^2_{1o} K_2 C_{1w}. \tag{5.2.87}$$

The concentration of free amine C_{3o} in the organic phase is the only unknown here (assume that κ_{1o}, K_3 and K_2 are available from the literature and that C_{1w} is known). If C^t_{3o}, the total amine present initially, is known, then

$$C^t_{3o} = C_{3o} + C_{4o} = C_{3o} + K_3 C_{1o} C_{3o}$$
$$= C_{3o}[1 + K_3 \kappa_{1o} C_{1w}].$$

Substitute into (5.2.87) to get

$$\kappa''_{1o} = \kappa_{1o} + 2\kappa^2_{1o} K_2 C_{1w} + \{\kappa_{1o} K_3 C^t_{3o}/(1 + K_3 \kappa_{1o} C_{1w})\}. \tag{5.2.88}$$

Thus, κ''_{1o}, the net distribution coefficient of acetic acid, is considerably increased from κ_{1o}.

Kuo and Gregor (1983) have studied the applicability of such a relation for the solvent decahydronaphthalene ($C_{10}H_{18}$, decalin) and the complexing agent trioctylphosphine oxide (TOPO) under conditions where the extent of dimerization was negligible ($K_2 \cong 0$). Figure 5.2.7 shows the distribution coefficient, κ''_{1o}, of acetic acid with a fixed initial concentration (0.0893 M) as the TOPO concentration was varied from 0 to 0.52 M in decalin. They had earlier observed that a value of $K_3 = 260$ described the value of κ''_{1o} well when $\kappa_{1o} = 0.024$ for a $C^t_{3o} = 0.1295$ M.

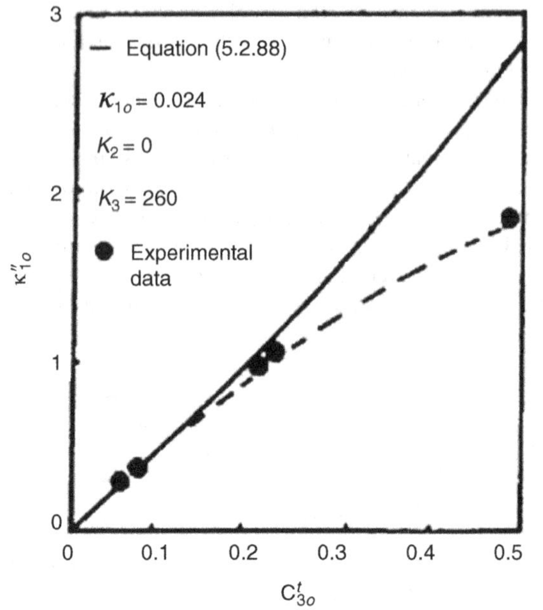

Figure 5.2.7. Effective distribution coefficient of acetic acid as a function of C^t_{3o}. (After Kuo and Gregor (1983).)

This figure shows that, beyond $C^t_{3o} \cong 0.2$ M, relation (5.2.88) no longer holds. This is apparently due to the association or micellization of TOPO at higher concentrations. However, the addition of TOPO has increased the extraction of acetic acid into the solvent phase by an extraordinary amount.

We will briefly identify now the results for extraction of acetic acid into an organic solvent in the absence of any complexing agent in the organic phase; however, the acid can dissociate in the aqueous phase, as shown in Figure 5.2.6. If the dissociation or the ionization constant of acetic acid ($i = 1$) is K_{d1} (see relation (5.2.61a) and Table 5.2.2) and there is dimerization of the acid in the oil phase, then it can be shown that

$$\kappa''_{1o} = \kappa'_{1o} \frac{[1 + 2K_2 \kappa'_{1o} C_{1w}] C_{H^+}}{C_{H^+} + K_{d1}} \tag{5.2.89}$$

in general. If the acid is the only solute, then C_{1w} can be expressed in terms of C_{H^+} using the electroneutrality condition.

5.2.2.4 *Extraction of metals*

Metallic compounds generally ionize in aqueous solution; the metal ion, in addition, is hydrated, i.e. surrounded by basic molecules of water. To extract the ionic metal species into an immiscible organic solvent, an uncharged metal-containing species which is soluble in a water-immiscible organic solvent has to be produced before it is extracted

Table 5.2.3. Structure and properties of various metal extractants[a]

Type	Name	Formula	Mol. wt. of active species	Active species (wt%)	Specific gravity
1a Acidic extractants	di-2-ethylhexyl phosphoric acid	$(C_4H_9CH(C_2H_5)CH_2O)_2POOH$	322	100	0.98
	Versatic 10	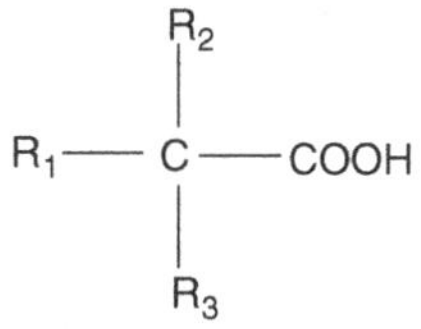	175	99.6	0.91
1b Chelating extractants	LIX 65N aromatic β-hydroxyoximes	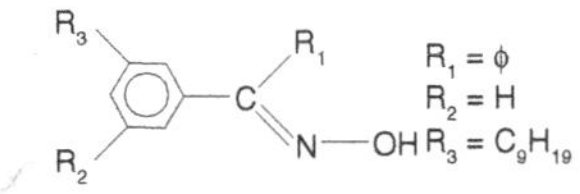	339	39.2	0.88
	Kelex 100 oxime derivative	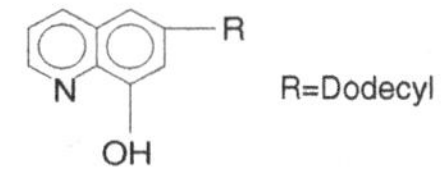	311	74–80	0.99
2 Anion exchangers	tertiary amine, alamine 336	R_3N (R= C_8-C_{10})	~392	95	0.81
	secondary amine, LA-2	R_2NH (R = C_{12}-C_{13})	351-393	100	0.83
	quaternary ammonium compounds, aliquat 336	$(R_3N(CH_3)^+Cl^-)$ (R= C_8-C_{10})	~442	>88	0.88
3 Solvating extractants	phosphorus-oxygen-bonded donors:	$R_1R_2R_3PO$			0.97
	tri-n-butyl phosphate	$R_1 = R_2 = R_3 = C_4H_9O$	226	—	—
	trioctyl phosphine oxide	$R_1 = R_2 = R_3 = C_8H_{17}$	386	—	
	methyl isobutyl ketone	R_1COR_2, $R_1 = CH_3$, $R_2 = (CH_3)_2CHCH_2$	100	—	0.804

[a] Prepared from Tables 2, 3, 4 and 5 in Flett *et al.* (1991).

into the organic solvent. This involves metal charge neutralization as well as replacing some or all waters of hydration by solvent compatible agents. Conventional approaches to achieve this goal are the following (Ritcey and Ashbrook, 1984a, p. 6).

(1) *Compound formation* between the metallic ion and an extractant is achieved by using *acidic extractants* or *chelating extractants*. Acidic extractants have a group like -COOH, $-SO_3H$, whose hydrogen is exchanged for the metal. A chelating extractant has two or more sites to complex with a metal ion into a cyclic compound. A chelating agent dissolved in the organic solvent forms an uncharged metal chelate in the aqueous phase for extraction of the chelate into the organic solvent. The acidic extractant acts similarly.
(2) *Ion-pair formation* is initiated in the aqueous phase between the metallic ion and a large counterion containing a bulky organic group. This large counterion is obtained from a complexing agent added to the organic phase; the agent dissociates in the aqueous phase to provide the counterion. The neutral ion pair is extracted into the organic phase.
(3) The formation of a *loose nonchelating uncharged compound* between the metal ion and the organic solvent itself is achieved with many oxygen-containing solvents like tributyl phosphate. This is identified as *solvation* of the metal ion.

Table 5.2.3 identifies a few extracting agents/extractants/solvents from each of the above categories. In the second category of ion-pair formation, two types of extractants have been identified as *anion exchangers*: long-chain amines (tertiary, secondary or primary) and quaternary ammonium compounds. Some properties of each extractant have been identified in the table.

We study the first of these three approaches now. Consider the *chelation*[4] of metal ions M^{n+} with an organic ion K^- obtained from the chelating agent HK. This agent, dissolved in the organic phase, partitions into the aqueous phase and then dissociates into H^+ and K^-:

$$\underbrace{HK}_{\substack{\text{organic phase}\\(j=1=o)}} \Leftrightarrow \underbrace{HK \Leftrightarrow H^+ + K^-}_{\substack{\text{aqueous phase}\\(j=2=w)}} \qquad (5.2.90)$$

[4] Quantitative representations for acidic or chelating extractants are quite similar. Macrocyclic compounds can complex metal ions in a more stable fashion (Burgess, 1988).

The metal ion M^{n+} in the aqueous phase diffuses to the interface, reacts with the ion K^- and forms MK_n, a neutral species which is extractable into the organic phase:

$$\underbrace{M^{n+} + nK^- \Leftrightarrow MK_n}_{\substack{\text{aqueous phase} \\ (j=2=w)}} \Leftrightarrow \underbrace{MK_n}_{\substack{\text{organic phase} \\ (j=1=o)}} \tag{5.2.91}$$

Generally, the reaction is thought to occur in the aqueous side of the interfacial region; MK_n then partitions into the organic phase. The overall chemical reaction may be represented as

$$M^{n+}(aq) + nHK(o) \Leftrightarrow MK_n(o) + nH^+(aq). \tag{5.2.92}$$

To determine the net distribution coefficient of the metal between the two phases, κ''_{M1}, defined by

$$\kappa''_{M1} = \frac{C_{MK_n,o}}{C_{MK_n,w} + C_{M^{n+},w}} \tag{5.2.93}$$

we proceed as follows. Define the following two distribution coefficients of species MK_n and HK between the organic and the water phase:

$$\kappa'_{MK_n,1} = \frac{C_{MK_n,o}}{C_{MK_n,w}}; \qquad \kappa'_{HK,1} = \frac{C_{HK,o}}{C_{HK,w}}. \tag{5.2.94}$$

The equilibrium constants for the two aqueous-phase reactions (5.2.90) and (5.2.91) are

$$K_{d,HK} = \frac{C_{H^+w}C_{K^-w}}{C_{HK,w}}; \qquad K_{d,MK_n} = \frac{C_{MK_n,w}}{C_{M^{n+},w}(C_{K^-w})^n}. \tag{5.2.95}$$

Using these relations, we can express κ''_{M1} as

$$\kappa''_{M1} = \frac{\kappa'_{MK_n,1}(K_{d,HK})^n K_{d,MK_n}(C_{HK,o})^n}{(C_{H^+,w})^n(\kappa'_{HK,1})^n + K_{d,MK_n}(K_{d,HK})^n(C_{HK,o})^n}. \tag{5.2.96}$$

Thus, as the concentration of the chelating agent HK in the organic phase, $C_{HK,o}$, increases, κ''_{M1} increases. Similarly as the *pH* increases ($C_{H^+,w}$ decreases), κ''_{M1} increases.

Define f_M as

$$f_M = \frac{C_{M^{n+},w}}{C_{M^{n+},w} + C_{MK_n,w}}, \tag{5.2.97}$$

i.e. the fraction of the total metal ion concentration in the aqueous phase present as M^{n+}. Then, using (5.2.94) and (5.2.95), we can rewrite κ''_{M1} as follows:

$$\kappa''_{M1} = \frac{\kappa'_{MK_n,1}(K_{d,HK})^n K_{d,MK_n}(C_{HK,o})^n f_M}{(C_{H^+,w})^n(\kappa'_{HK,1})^n}, \tag{5.2.98}$$

whereby

$$\log\kappa''_{M1} = n(pH) + n\log C_{HK,o} + \log\left[\frac{f_M\kappa'_{MK_n,1}(K_{d,HK})^n K_{d,MK_n}}{(\kappa'_{HK,1})^n}\right]. \tag{5.2.99}$$

Therefore, log κ''_{M1} varies linearly with *pH*, provided the other quantities are unaffected by the *pH* variation. It is obvious that by choosing the right *pH* range, metal extraction into the organic phase can be drastically increased.

If we have to separate two metals M and N having the same valence *n*, at a given *pH* and a chelate concentration $C_{HK,o}$, then

$$\log(\kappa''_{M1}/\kappa''_{N1}) = \log\left[\frac{f_M K_{d,MK_n}\kappa'_{MK_n,1}}{f_N K_{d,NK_n}\kappa'_{NK_n,1}}\right], \tag{5.2.100}$$

which is likely to be unaffected by *pH*, at least over a certain range. To a first approximation, the curves of log κ''_{M1} vs. log *pH* and log κ''_{N1} vs. log *pH* are therefore parallel, with slope *n*. Figure 5.2.8(a) shows such a plot for metals M and N. Recognize that if there is a *pH* at which log κ''_{M1} is 2 and log κ''_{N1} is –2, the two metals are almost completely separated. Here the difference in the two metal dissociation or hydrolysis constants K_{d,MK_n}, K_{d,NK_n}, are crucial. For the extraction of a given metal, depending on the purity and the recovery desired, the *pH* can be chosen. Figure 5.2.8(b) illustrates the extraction of different metals by a particular extractant as a function of *pH* (Cox and Flett, 1991). Note, however, that, as *pH* increases, both log κ''_{M1} and log κ''_{N1} flatten out due to the *pH* variation of f_M and f_N.

If the *pH* of the aqueous solution is increased to a high value, at which the metal hydrolyses, the solvent extractability will drop drastically. The flat behavior we observed in Figures 5.2.8(a) and (b) will change into one where κ''_{M1} will decrease rapidly with *pH*. At very low *pH*, κ''_{M1} also has a very low value since high H^+ concentration prevents ionization of the chelating agent HK (see equation (5.2.90)) and thus hinders formation of the complex MK_n.

Some examples of acidic and chelating agents (Karger *et al.* (1973), Ritcey and Ashbrook (1984a, p. 24) and Lo *et al.* (1983) provide more detailed lists) are provided in Table 5.2.3. Additional examples are given in Figure 5.2.9.

Note than an acidic extractant, such as di-2-ethylhexyl phosphoric acid (DEHPA), may have a more complex chemistry than a chelating agent, HK, due to its dimerization or self-association in the organic phase (Ritcey and Ashbrook, 1984a, p. 28). We have observed such behavior earlier with carboxylic acids in nonpolar organic solvents.

In the extraction of metals by chelation, the neutral product MK_n is a chelated compound. Sometimes, the neutral species is simply a pair of oppositely charged ions which is soluble in the organic phase. Basic extractants, like primary, secondary or tertiary organic amines, are often used in solvent extraction. A tertiary amine R_3N (where R is usually an aliphatic long-chain group) in the presence of an acid HA in the aqueous phase will undergo the following reactions and produce an ion pair $R_3NH^+A^-$:

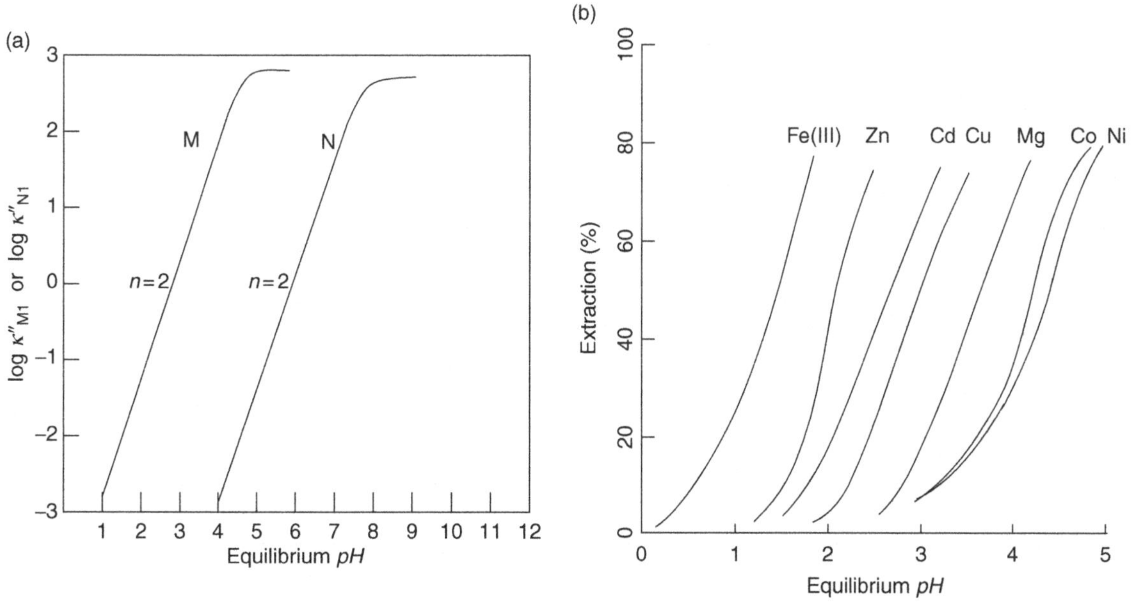

Figure 5.2.8. (a) Equilibrium distribution of a metal species M or N as a function of pH *for M^{n+} or N^{n+}. (b) Extraction of some metals by DEHPA from a sulfate solution (Cox and Flett, 1991). Reprinted, with permission, from M. Cox and D.S. Flett, "Metal extraction chemistry," in T.C. Lo, M.H.I. Baird and C. Hanson,* Handbook of Solvent Extraction, *Wiley-Interscience (1983), Figure 4, p. 59. Copyright © 1983, John Wiley & Sons.*

8-Quinolinol (HOx)

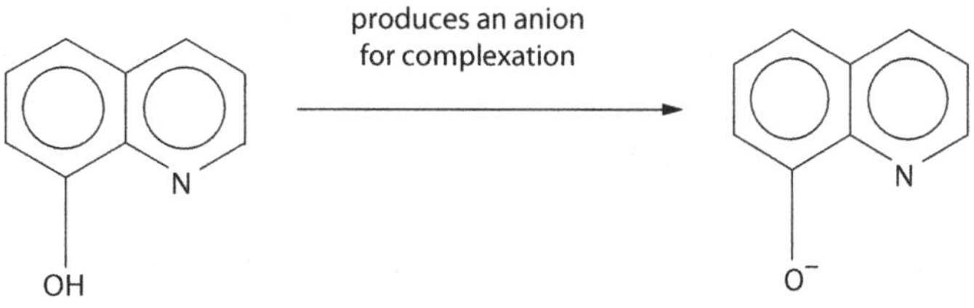

1,3-diketonethenoyltrifluoracetone

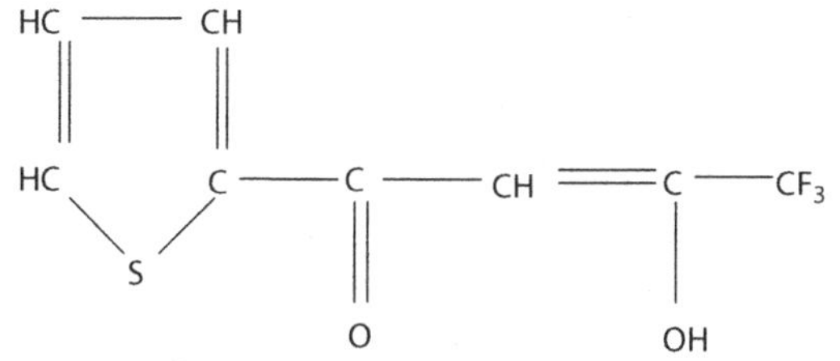

Figure 5.2.9. Examples of acidic and chelating extractants, e.g. both HOx and $M(Ox)_2$ with 8-quinolinol.

$$\begin{array}{lll} R_3N(\text{org}) + HA(\text{aq}) \Leftrightarrow R_3NH^+ \cdot A^-(\text{org}); & & \\ (i=1) \qquad (i=2) & & \\ R_3NH^+ \cdot A^-(\text{aq}) \Leftrightarrow R_3NH^+(\text{aq}) + A^-(\text{aq}). & & \\ (i=3) \qquad\qquad (i=4) \qquad\qquad (i=5) & & \end{array} \tag{5.2.101}$$

In addition, we have the distribution equilibrium of $R_3NH^+ \cdot A^-$ between the two phases:

$$\kappa''_{31} = \frac{C_{3o}}{C_{3w}}. \tag{5.2.102}$$

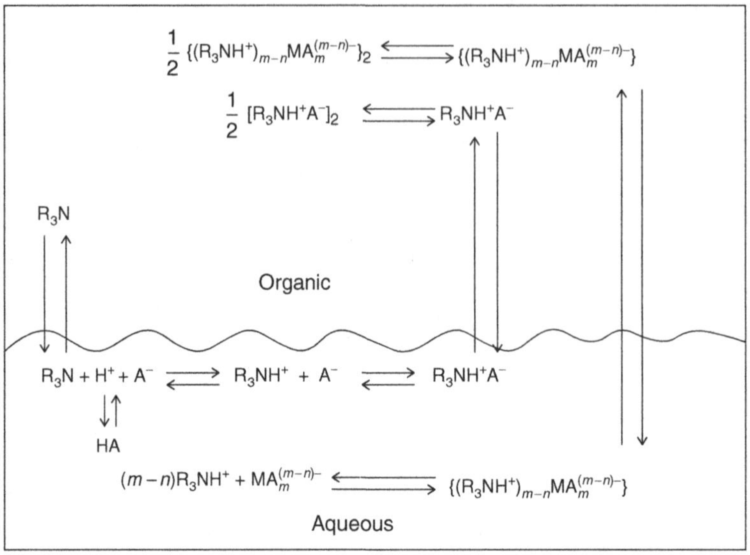

Figure 5.2.10. Ion-pair extraction of a metal ion.

In the aqueous solution, the metal ion M^{n+} in the presence an acid HA exists as various complexes (charged or uncharged) $MA_m^{(m-n)-}$. Typical examples for HA will be HCl, HI, etc. Now, the bulky counterion R_3NH^+ in the aqueous phase will form an *ion pair* with a charged metallic species in aqueous solution, say, $MA_m^{(m-n)-}$ according to

$$\underset{(i=4)}{(m-n)R_3NH^+(aq)} + \underset{(i=7)}{MA_m^{(m-n)-}(aq)} \Leftrightarrow \underset{(i=6)}{(R_3NH^+)_{m-n}\,MA_m^{(m-n)-}(aq)}. \tag{5.2.103}$$

The ion pair $(R_3NH^+)_{m-n}\,MA_m^{(m-n)-}$ with no charge will partition into the organic phase

$$\underset{(i=6)}{(R_3NH^+)_{m-n}\,MA_m^{(m-n)^-}(aq)} \Leftrightarrow \underset{(i=6)}{(R_3NH^+)_{m-n}\,MA_m^{(m-n)^-}(org)} \;\uparrow\downarrow\; \underset{(i=8)}{\frac{1}{2}[(R_3NH^+)_{m-n}\,MA_m^{(m-n)^-}](org)}, \tag{5.2.104}$$

and may even dimerize or polymerize. The various chemical reactions and partition equilibria are schematically illustrated in Figure 5.2.10, where we have included the dimerization of $R_3NH^+A^-$ in the organic phase. Note that the anionic metal complex $MA_m^{(m-n)-}$ is formed with the anion A^- from the acid according to

$$M^{n+}(aq) + mA^-(aq) \Leftrightarrow MA_m^{(m-n)-}(aq). \tag{5.2.105}$$

Further, there can be a number of such complexes at any time, depending on the value of m.

One can also argue that, prior to the ion-pair formation of $(R_3NH^+)_{m-n}$ with $MA_m^{(m-n)-}$ to enable metal extraction, there was an ion pair $R_3NH^+A^-$ whose anion A^- was exchanged with another anion $MA_m^{(m-n)-}$ to effect metal extraction. The tertiary amine compound R_3N had to be protonated before anion exchange is possible. Secondary or primary amines also require protonation before anion exchange. On the other hand, *quaternary ammonium compounds*, used as an extraction agent and representable as $R_4N^+A^-$ (e.g. $R_3N(CH_3)^+Cl^-$, where R may be C_8, C_9 or C_{10}), can readily undergo anion exchange with other anions in the system:

$$\begin{array}{l} R_4N^+A^-(org) \\ \uparrow\downarrow \\ R_4N^+A^-(aq) \Leftrightarrow R_4N^+(aq) + A^-(aq) \end{array} \tag{5.2.106}$$

$$\begin{array}{r} (m-n)\,R_4N^+(aq) + MA_m^{(m-n)-}(aq) \Leftrightarrow (R_4N^+)_{m-n}\,MA_m^{(m-n)-}(aq) \\ \downarrow\uparrow \\ (R_4N^+)_{m-n}MA_m^{(m-n)-}(org). \end{array} \tag{5.2.107}$$

Micellization of $R_4N^+A^-$ is also a problem in the organic phase. Such anion exchange is also a powerful technique used in the solvent extraction of amino acids under appropriate *pH* conditions so that the amino acid (Am) is in a form (Am^-) (see Section 5.2.3.1.2) ready for anion exchange, for example using a quaternary ammonium compound:

$$R_4N^+A^-(org) + Am^-(aq) \Leftrightarrow R_4N^+Am^-(org) + A^-(aq). \tag{5.2.108}$$

The *third* approach to metal extraction uses oxygen-containing solvents like TBP (tributyl phosphate, $(BuO)_3PO$). Widely adopted in the extraction of many radioactive metals in the nuclear industry, this technique may be illustrated for uranium extraction in the presence of nitrate ions in the aqueous phase by

$$UO_2^{2+}(aq) + 2NO_3^-(aq) + 2TBP(org) \Leftrightarrow UO_2(NO_3)_2{\cdot}2TBP(org). \qquad (5.2.109)$$

To shift the equilibrium to the right, the NO_3^- ion concentration is increased in the aqueous phase by adding, say, HNO_3 and the level of free TBP in the organic phase is raised. Under such conditions, nitric acid is also extracted into the organic phase by

$$H^+(aq) + NO_3^-(aq) + TBP(org) \Leftrightarrow HNO_3{\cdot}TBP(org). \qquad (5.2.110)$$

Detailed considerations of equilibria in such systems for the extraction of radioactive metals are available in Benedict *et al.* (1981). Normally water molecules form a shell around the metal atom, creating the hydration sphere. TBP has a highly polar group $\equiv P = O$, which can replace water around the metal. Other extractants, like ethers, alcohols and ketones, have also been used to the same end, except water molecules are not completely excluded. Table 5.2.4 illustrates the variety of solvent extraction agents used commercially for extracting metals in hydrometallurgy.

Extracting agents, such as TPB, are rarely used by themselves. The extracting agent is normally used along with a diluent solvent, e.g. kerosene. The purposes behind the use of a diluent are several. First, the viscosity of the extractant is considerably reduced by the diluent. Second, the inventory of costly extractant is decreased. Diluents, in addition, control the surface-active tendencies of most extractants, leading to improved dispersion-coalescence behavior. A detailed discussion on diluents is available in Ritcey and Ashbrook (1984a, chap. 4). In addition, sometimes modifiers like TBP are added to the organic phase to avoid what is called the third-phase separation (Flett *et al.*, 1991), namely to prevent the splitting of the metal complex-rich organic phase into a metal complex-rich phase at the aqueous interface and a diluent-rich phase above.

5.2.3 Stationary-mobile phase equilibria

Here we consider briefly the effect of chemical reactions on two types of phase equilibrium: ion exchange resin-liquid solution equilibrium; stationary medium-mobile liquid solution equilibrium, as in chromatography (Section 7.1.5). For the first case, the resin phase is generally stationary in a fixed bed; however, the resin phase may also be mobile.

5.2.3.1 *Ion exchange resin-liquid solution equilibrium*

There is an extensive literature in this area. Helfferich (1962, 1995) has provided a comprehensive description of the effect of chemical reactions on ion exchange equilibria. Three types of chemical reactions of interest are: complex formation in solution; complex formation with the ion exchanger; ionization/dissociation in the solution.

5.2.3.1.1 Complex formation The complex or complexes formed in solution or in the ion exchanger can strongly affect ion exchange equilibria. Consider first the equilibria with *cation exchangers* in the presence of a complexing anion (say, Cl^-). Let the two competing cations (counterions) A and B be Zn^{2+} and H^+. In the presence of a free anion (Y), Cl^-, obtained from, say, $ZnCl_2$, and HCl, the cation Zn^{2+} will form a variety of complexes (Helfferich, 1962):

$$\begin{aligned}
Zn^{2+} + Cl^- &\overset{K_1}{\longleftrightarrow} ZnCl^+ : K_1 = \frac{C_{ZnCl^+}}{C_{Zn^{2+}}C_{Cl^-}};\\
Zn^{2+} + 2Cl^- &\overset{K_2}{\longleftrightarrow} ZnCl_2 : K_2 = \frac{C_{ZnCl_2}}{C_{Zn^{2+}}C^2_{Cl^-}};\\
Zn^{2+} + 3Cl^- &\overset{K_3}{\longleftrightarrow} ZnCl_3^- : K_3 = \frac{C_{ZnCl_3^-}}{C_{Zn^{2+}}C^3_{Cl^-}};\\
Zn^{2+} + 4Cl^- &\overset{K_4}{\longleftrightarrow} ZnCl_4^{2-} : K_4 = \frac{C_{ZnCl_4^{2-}}}{C_{Zn^{2+}}C^4_{Cl^-}},
\end{aligned} \qquad (5.2.111)$$

where K_1, K_2, K_3 and K_4 are the stability constants for the complexes formed. Although the overall concentration of zinc in the solution is now given by

$$\begin{aligned}
C^t_{Zn} &= C_{Zn^{2+}} + C_{ZnCl^+} + C_{ZnCl_2} + C_{ZnCl_3^-} + C_{ZnCl_4^{2-}};\\
C^t_{Zn} &= C_{Zn^{2+}}\left[1 + K_1C_{Cl^-} + K_2C^2_{Cl^-} + K_3C^3_{Cl^-} + K_4C^4_{Cl^-}\right],
\end{aligned} \qquad (5.2.112)$$

the concentration of the free cation Zn^{2+} is now substantially lower. The extent of reduction depends on how strong the complexes are. Further, some of the divalent cation Zn^{2+} is now either a monovalent cationic species or even an anionic species; therefore the preference of the cation exchanger for Zn is now substantially reduced. On the other hand, the selectivity of the competing cation B (H^+) is increased substantially in the ion exchange reactions with the cation exchanger (R^-) having fixed negative charges since $ZnCl_3^-$, $ZnCl_4^{2-}$ do not participate in the cation exchange process:

$$\begin{aligned}
Zn^{2+}(aq) + 2HR &\Leftrightarrow ZnR_2 + 2^{H+}(aq),\\
ZnCl^+ + HR &\Leftrightarrow ZnClR + H^+(aq).
\end{aligned} \qquad (5.2.113)$$

The expressions for the selectivity of A over B are available in Helfferich (1962). The net result in this case is that

Table 5.2.4. Solvent extraction reagents for hydrometallurgy[a]

Class	Type	Examples	Manufacturer	Commercial use
Acidic extractants	carboxylic acids	naphthenic acids	Shell Chemical Co.	copper–nickel separation
	alkyl phosphoric acids	di-2-ethylhexyl phosphoric acid (D2EHPA)	Union Carbide Mobile Oil Co.	yttrium recovery, europium extraction, nickel–cobalt separation uranium extraction
		octylphenylphosphoric acid (OPPA)		
	aryl sulfonic acids	SYNEX 1051	King Industries, Inc.	magnesium extraction
Acid chelating extractants	hydroxyoximes	LIX63, LIX64N, LIX65N, LIX 70	Henkel Corporation	copper and nickel extraction
		SME 529	Shell Chemical Co.	copper extraction
		P5000 series	Acorga Ltd.	copper extraction
	oxime derivatives	Kelex 100	Sherex Chemical Co[b].	proposed for copper extraction
	β-diketones	Hostarex DK16	Farbwerke Hoechst AG	proposed for copper extraction from ammoniacal solution
		LIX54	Henkel Corporation	copper extraction from ammoniacal solution
		X151	Henkel Corporation	proposed for cobalt extraction from ammoniacal solution
	alkarylsulfonamide	LIX34	Henkel Corporation	proposed for copper extraction from acidic leach liquors
	polyols		Dow Chemical Co.	boron extraction
Anion exchangers	primary amines	Primene JMT	Rohm and Haas	no known commercial use
	secondary amines	LA-2	Rohm and Haas	zinc and uranium extraction
		Adogen 283	Sherex Chemical Co.[b]	zinc and tungsten extraction
	tertiary amines	various alamines; in particular, Alamine 336	Henkel Corporation	widley used; cobalt, tungsten, vanadium, uranium extractions, etc.
		various adogens; in particular, Adogen 364, Adogen 381, Adogen 382	Sherex Chemical Co.[b]	cobalt, vanadium and uranium extractions
	quaternary amines	Aliquat 336	Henkel Corporation	vanadium extraction; other possible uses are chromium, tungsten and uranium extraction
		Adogen 464	Sherex Chemical Co.[b]	similar to Aliquat 336
Solvating extractants	phosphoric, phosphonic and phosphinic acid esters	tributylphosphate (TBP)	Union Carbide, Albright and Wilson	nuclear fuel reprocessing, U_3O_8 refining, iron extraction, zirconium–hafnium separation, niobium–tantalum separation, rare earth separations, acid extraction
		phosphonic acid esters	Farbwerke Hoechst	no known commercial use
		trioctylphophine oxide (TOPO)	AG, Henkel Corporation, Cyanamid	recovery of uranium from wet process phosphoric acid liquors (with D2EHPA)
	various alcohols, ethers, ketones	butanol-pentanol	various	phosphoric acid extraction
		diisopropyl ether	various	phosphoric acid extraction
		methyl isobutyl ketone (MIBK)	various	niobium–tantalum separation, zirconium–hafnium separation
	alkyl sulfides	Di-n-hexyl sulfide		palladium extraction

[a] Reprinted with permission from D.S. Flett, J. Melling and M. Cox, "Commercial solvent systems for inorganic processes," in T.C. Lo, M.H.I. Baird and C. Hanson, *Handbook of Solvent Extraction*, Table 1, p. 631, Wiley-Interscience (1983). Copyright © 1983, John Wiley & Sons.
[b] Previously Ashland Chemical Co.

Table 5.2.5. A few amino acids: their structures and dissociation constants[a]

Name	R-	pK_1 COOH	pI	$pK_2NH_3^+$	pK_3 R-group
Glycine	H-	2.34	5.97	9.6	–
Alanine	CH_3-	2.31	6.02	9.70	–
Valine	$(CH_3)_2$-CH-	2.33	5.97	9.76	–
Leucine	$(CH_3)_2CHCH_2$-	2.27	5.97	9.57	–
Phenylalanine	$C_6H_5CH_2$-	2.17	5.48	9.11	–
Glutamic acid	$HOOCCH_2CH_2$-	2.18	3.22	9.59	10.68
Lysine	$H_2NCH_2CH_2CH_2CH_2$-	2.19	9.74	9.12	12.48

[a] General formula:

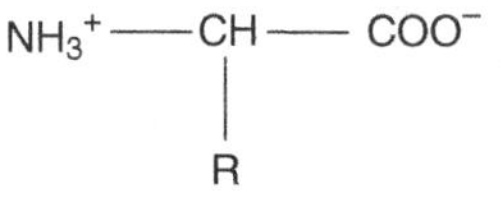

See Bailey (1990); Martell and Smith (1974, 1982); Lehninger (1982); Saunders *et al.* (1989).

uptake of the metal zinc from the solution is reduced by complexation of the metal with a complexing anion (Cl^-) in the solution. Additional examples of excellent complexing agents are anions of weak acids, such as citric acid, ethylenediaminetetraacetic acid (EDTA), etc. (Helfferich, 1962).

The situation regarding cation A uptake by an anion exchange resin is different if there is a complexing anion (e.g. Cl^-) present. For the same cation Zn^{2+}, which would not be preferred by an anion exchanger, we now have anionic complexes $ZnCl_3^-$, $ZnCl_4^{2-}$. These can now successfully participate in the ion exchange process and lead to substantially increased metal uptake when very little would have been possible in the absence of the complexation with the anion Cl^-.

The above two examples considered the role of complexation reactions taking place in the solution on the ion exchange equilibria. The effect of complex-forming cations in the ion exchangers on the partitioning of anionic and nonionic ligands from the solution is also of interest. Examples of complexes in the cation exchange resin (fixed charge group, R^-) formed by cations like Cu^{2+} (and Ni^{2+}, Ag^+, etc.) are $(R^-)_2[Cu(NH_3)_4^{2+}]$, $(R^-)_2Cu(H_2O)_4^{2+}$. Potential ligands from the solution are ammonia, aliphatic amine, polyhydric alcohols, etc. Consider the following ligand exchange between ammonia present in the resin as a ligand with ethylene diamine ($NH_2C_2H_4NH_2$) (EDA):

$$\begin{aligned}&(R^-)_2\left[Cu(NH_3)_4^{2+}\right] + 2NH_2C_2H_4NH_2(aq) \Leftrightarrow\\&(R^-)_2\left[Cu(NH_2C_2H_4NH_2)_2^{2+}\right] + 4NH_3(aq).\end{aligned} \qquad (5.2.114)$$

Ammonia present in the solution had previously formed a complex with the metal ion in the ion exchanger, which acts as a solid carrier for the complexing metal ion. Addition of EDA in the solution allows for an exchange between EDA and ammonia, the two ligands in this case. Since complex formation of a ligand with a metal ion is a highly specific interaction, high selectivities can be obtained. It depends on the coordination valence of the ligands and the solution concentrations (Helfferich, 1962).

5.2.3.1.2 Ionization/dissociation of amino acids Large-scale production of amino acids frequently use ion exchange processes. Amino acids are amphoteric molecules that can exist as anions and cations depending upon the pH of the solution. There are three types of amino acids depending on the nature of the side chain R if we represent the amino acid as $NH_2CHRCOOH$: the R-group is not ionizable; the R-group is negatively charged at pH 7; the R-group is positively charged at pH 7. Consider now the first case with a nonionizable R-group (examples are alanine, leucine, glycine, valine, etc.) (see Table 5.2.5). The following forms of dissociation equilibria exist in solution:

$$\begin{array}{ccc} NH_3^+\text{-}\underset{\underset{\displaystyle (Am^+)}{\displaystyle R}}{\underset{|}{CH}}\text{-}COOH & \overset{K_1}{\Leftrightarrow} & NH_3^+\text{-}\underset{\underset{\displaystyle (Am^\pm)}{\displaystyle R}}{\underset{|}{CH}}\text{-}COO^- + H^+ \end{array} \qquad (5.2.115a)$$

$$\begin{array}{ccc} NH_3^+\text{-}\underset{\underset{\displaystyle (Am^\pm)}{\displaystyle R}}{\underset{|}{CH}}\text{-}COO^- & \overset{K_2}{\Leftrightarrow} & NH_2\text{-}\underset{\underset{\displaystyle (Am^-)}{\displaystyle R}}{\underset{|}{CH}}\text{-}COO^- + H^+ \end{array} \qquad (5.2.115b)$$

Here $Am^\pm$ is the zwitterion form of the amino acid, whereas Am^+ is the cationic form and Am^- is the anionic form of the amino acid.[5] From equation (5.2.115a), the dissociation equilibrium constant K_1 is

[5] The forms should be more appropriately written as $HAm^\pm$ for the zwitterion form, H_2Am^+ for the cationic form and Am^- for the anionic form, with HAm representing the amino acid.

$$K_1 = \frac{C_{\text{Am}^\pm w} C_{\text{H}^+ w}}{C_{\text{Am}^+ w}}. \tag{5.2.116a}$$

Correspondingly,

$$K_2 = \frac{C_{\text{Am}^- w} C_{\text{H}^+ w}}{C_{\text{Am}^\pm w}}. \tag{5.2.116b}$$

If we have a cation exchange resin, we should be interested in the concentration of $C_{\text{Am}^+ w}$, especially as a fraction of the total amino acid concentration, $C^t_{\text{Am}w}$:

$$C^t_{\text{Am}w} = C_{\text{Am}^+ w} + C_{\text{Am}^- w} + C_{\text{Am}^\pm w}. \tag{5.2.117}$$

Employing the two dissociation equilibrium relations (5.2.116a,b), we obtain

$$C^t_{\text{Am}w} = C_{\text{Am}^+ w} + \frac{K_2 C_{\text{Am}^\pm w}}{C_{\text{H}^+ w}} + \frac{K_1 C_{\text{Am}^+ w}}{C_{\text{H}^+ w}}.$$

A repeat application of (5.2.116a) leads to

$$C_{\text{Am}^+ w} = \frac{C^t_{\text{Am}w}}{\left[1 + \frac{K_1}{C_{\text{H}^+ w}} + \frac{K_1 K_2}{C^2_{\text{H}^+ w}}\right]}. \tag{5.2.118a}$$

Correspondingly, the concentration of C_{Am^-} is obtained as a fraction of $C^t_{\text{Am}w}$ as

$$C_{\text{Am}^- w} = \frac{C^t_{\text{Am}w}}{\left[1 + \frac{C_{\text{H}^+ w}}{K_2} + \frac{C^2_{\text{H}^+ w}}{K_1 K_2}\right]}. \tag{5.2.118b}$$

To get an idea of which charged form of the amino acid dominates under what pH condition, consider the amino acid valine, where R = $(CH_3)_2$-CH, $K_1 = 10^{-2.33}$ and $K_2 = 10^{-9.76}$. Further, let the pH be 1, i.e. the solution is highly acidic. Then

$$C_{\text{Am}^+ w} = \frac{C^t_{\text{Am}w}}{\left[1 + \frac{10^{-2.33}}{10^{-1}} + \frac{10^{-(2.33+9.76)}}{10^{-2}}\right]} = \frac{C^t_{\text{Am}w}}{\left[1 + \frac{1}{10^{1.33}} + \frac{1}{10^{10.09}}\right]},$$

whereas

$$C_{\text{Am}^- w} = \frac{C^t_{\text{Am}w}}{\left[1 + \frac{10^{-1}}{10^{-9.76}} + \frac{10^{-2}}{10^{-(12.09)}}\right]} = \frac{C^t_{\text{Am}w}}{[1 + 10^{8.76} + 10^{10.09}]}.$$

Thus, the cationic form Am^+ dominates at low pH and is very close to $C^t_{\text{Am}w}$. At very high pH, the anionic form Am^- dominates. In addition, one can conclude from relation (5.2.116a) that when $pH = pK_1$, $C_{\text{Am}^+ w} = C_{\text{Am}^\pm w}$. Similarly, from (5.2.116b) when $pH = pK_2$, $C_{\text{Am}^- w} = C_{\text{Am}^\pm w}$. Table 5.2.5 provides the isoelectric point of the amino acid, pI, where the concentrations of the positively charged species balance those of the negatively charged amino acids, leading to no charge: $C_{\text{Am}^+ w} = C_{\text{Am}^- w}$.

Although the above conclusions are generally useful, there are other considerations. Consider a low pH where the Am^+ form dominates and is expected to have successful exchange with a cation exchanger; however, there is a competing ion exchange reaction with the H^+ ion:

$$[\text{R}^-][\text{H}^+] + \text{NH}_3^+\text{-}\underset{\substack{| \\ R}}{\text{CH}}\text{-COOH(aq)} \Leftrightarrow [\text{R}^-][\text{NH}_3^+\text{-}\underset{\substack{| \\ R}}{\text{CH}}\text{-COOH}] + \text{H}^+\text{(aq)} \tag{5.2.119}$$

Further, if there are other cations in the system, there will be additional ion exchange reactions. Illustrations of the detailed behaviors of the equilibrium sorption of a number of amino acids by a cation exchange resin are provided in Saunders *et al.* (1989) and Dye *et al.* (1990).

5.2.3.2 *Stationary-mobile phase equilibria in chromatography*[6]

In chromatography, the equilibrium distribution of a species between the mobile phase and a stationary phase is crucial. To separate two species, it is essential that the distribution ratio k'_{i1} for two species i = A and B is significantly different. Note that (from equation (1.4.2))

$$k'_{i1} = \frac{m_{i1}}{m_{i2}} = \frac{C_{i1}\, V_1}{C_{i2}\, V_2} = \kappa_{i1} \frac{V_1}{V_2},$$

where V_1 is the volume of the stationary phase ($j = 1$) and V_2 is the volume of the mobile phase ($j = 2$). For a mobile liquid phase, the stationary phase may be a solid with bonded liquid ion exchange resin or a liquid phase. The chemical reactions may be acid-base equilibrium, complexation, ion pairing, etc. First, we consider acid-base equilibrium for two weak acids HA and HB distributed between the mobile liquid phase and a stationary solid phase, with or without any bonded liquid phase. Assume that their distribution coefficients κ'_{A1} and κ'_{B1} are almost identical but their dissociation constants K_{DA}, K_{DB} are quite different. Assume further that only the form HA or HB partitions; A^- and B^- remain in the mobile phase.

Using relation (5.2.62), write the effective distribution coefficients for A and B as

$$\kappa''_{A1} = \kappa'_{A1} \frac{C_{H^+}}{C_{H^+} + K_{dA}}; \qquad \kappa''_{B1} = \kappa'_{B1} \frac{C_{H^+}}{C_{H^+} + K_{dB}}. \tag{5.2.120}$$

For species A, there are two limits of the distribution ratio k'_{A1} corresponding to the two values of κ''_{A1}, i.e. κ'_{A1} for $K_{dA} \ll C_{H^+}$ and $(\kappa'_{A1} C_{H^+}/K_{dA})$ for $K_{dA} \ll C_{H^+}$ since (V_1/V_2) is fixed for a system. Similarly for species B. It is obvious that if K_{dA} and K_{dB} (therefore pK_A and pK_B) are substantially different, then choosing a pH at an intermediate value anywhere between pK_A and pK_B will lead to substantially different k'_{A1} and k'_{B1}, even though $\kappa'_{A1} \cong \kappa'_{B1}$. This is vitally important for an effective separation of the peaks in chromatography. (*Note*: The detector records both the undissociated and the dissociated species of, say, HA as one species and therefore one peak.)

[6] To be read along with Section 7.1.5.

Next, we briefly examine the role of weak acid (or weak base) dissociation imposed on ion exchange equilibrium between the mobile phase and the stationary ion exchange resin (having, say, fixed positive charge, R^+). The solute of interest is a weak acid HA with a dissociation constant K_{dA}. The mobile phase has a monovalent anion C^-, the counter-ion in the ion exchange resin, which exchanges with A^- obtained from the dissociation of HA:

$$(R^+ + C^-)_{resin} + A^- \Leftrightarrow (R^+ + A^-)_{resin} + C^-. \quad (5.2.121)$$

We assume that the undissociated acid HA does not partition into the resin. The equilibrium constant for this reaction is

$$K_C^A = \frac{C_{A^-R} C_{C^-}}{C_{C^-R} C_{A^-}}, \quad (5.2.122)$$

where C_{A^-R} and C_{C^-R} refer to the ion exchange phase and C_{C^-} and C_{A^-} are for the aqueous eluent phase. The value of the distribution coefficient for species A between the aqueous phase and the ion exchange resin phase ($j = 1$) is

$$\kappa''_{A1} = \frac{C_{A^-R}}{C_{A^-} + C_{HA}} = \left[\frac{K_C^A C_{C^-R}}{C_{C^-}}\right] \frac{C_{A^-}}{C_{A^-} + C_{HA}}. \quad (5.2.123)$$

Now, $K_{dA} = C_{H^+} C_{A^-} / C_{HA}$; use it to rearrange κ''_{A1} to obtain

$$\kappa''_{A1} = \left[\frac{K_C^A C_{C^-R}}{C_{C^-}}\right] \frac{K_{dA}}{K_{dA} + C_{H^+}}. \quad (5.2.124)$$

By choosing the pH properly, κ''_{A1} can be varied over a wide limit. Since $K_{dA} << 1$ for weak acids, if $C_{H^+} >> K_{dA}$, κ''_{A1} is small; therefore k'_{A1} is also small. When, however, $K_{dA} >> C_{H^+}$, the acid is fully dissociated, leading to a much higher κ''_{A1} and k'_{A1}. Note that the value of k'_{A1} for a stationary liquid or solid (uncharged) phase under this condition is the lowest since A^- cannot partition into it.

Complexing agents can be added to the mobile phase to alter the equilibria between the stationary and mobile phases. This is especially useful for separating metal ions using ion exchange resins. For example, suppose the mobile phase has ligands L^- and that a metal ion can complex with it as $M^{n+} + nL^- \Leftrightarrow ML_n$ or as

$$M^{n+} + mL^- \Leftrightarrow ML_m^{(n-m)-} \quad (5.2.125)$$

in general. Now different metals will have different extents of binding with the ligand L. Therefore the partitioning between the mobile phase and the stationary phase will be different for different metals (Karger *et al.*, 1973). On the other hand, when various ligands have to be separated, one can introduce metal ions in the eluent for complexation, as in (5.2.125). Chelating agents like 8-hydroxyquinoline are examples of ligands used for these purposes.

Many ionic solutes are poorly retained by nonpolar stationary phases. However, if a counterion is introduced into an eluent such that it can form a neutral ion pair with the solute, then the ion pair can partition into the nonpolar stationary phase and its retention will be altered. For example, tetramethylammonium chloride ($(CH_3)_4N^+Cl^-$) can be used for pairing with simple anions; sodium dodecyl sulfate ($CH_3(CH_2)_{11}OSO_3^- Na^+$) can be utilized for simple cations.

Still another method is to use a *solute-micelle interaction*:

$$S + M \Leftrightarrow SM, \quad (5.2.126)$$

where S is a solute and M represents a micelle (see Figure 4.1.18), which is a spontaneous aggregate of surfactants. The solute distributes itself between the eluent and the micelle; the latter has a very limited affinity for the stationary phase. Thus, effectively, the retention of the solute is decreased. Different solute-micelle distribution equilibria will then help in separating different solutes. The solutes of interest are ionic or ionizable.

The following aspect of some common mobile phases used in liquid chromatography is important. Polar mobile phases are rarely aqueous; generally they are partially aqueous, e.g. methanol-water or acetonitrile-water, etc. For water, neutral pH is defined as $C_{H^+} = C_{OH^-} = 10^{-7}$ gmol/liter at 25 °C, but what is the neutral pH in a water-methanol system? For water, the ion product or the autopyrolysis constant K_{water} is (using concentrations instead of activities) $K_w = (C_{H^+})\,(C_{OH^-}) = 10^{-14}$. Correspondingly, for methanol, which produces H^+ and CH_3O^-, the autopyrolysis constant is $K_{methanol} = (C_{H^+})(C_{CH_3O^-}) = 10^{-16.7}$ (see Bates (1964)). Thus in methanol-water mixtures, with the autopyrolysis constant varying between 14 and 16.7, neutral pH may be said to vary between 7 and 8.35, depending on the mixture composition. For water-acetonitrile mixtures, the range will be even broader since $K_{acetonitrile}$ is 26.5. The subject is quite complex; refer to Bates (1964) to develop a more correct picture of the state of the protons and anions in such mixed polar solvents.

5.2.4 Crystallization and precipitation equilibrium

We will deliberate first on the role of *pH*, dissociation, etc., on various crystallization/precipitation equilibria. Consider *aqueous solutions* of amino acids. We have seen in Section 5.2.3.1.2 that amphoteric molecules of amino acids can exist in three forms in solution: a cationic form Am^+ at low *pH*; an anionic form Am^- at very high *pH*; and at intermediate *pH*, the zwitterionic form, $Am^\pm$, is important. The relative ratio of Am^+ or Am^- to $Am^\pm$ depends on the *pH* with respect to pK_1 or pK_2. Only the $Am^\pm$ form exists at the *pI* of the amino acid. On the other hand, it is the neutral form $Am^\pm$ which forms the crystal ($j = s$):

$$(NH_3^+RCHCOO^-)_s \overset{\kappa_{iw}}{\Leftrightarrow} (NH_3^+RCHCOO^-)_w, \quad (5.2.127)$$

where κ_{iw} is the equilibrium solubility parameter ($= (NH_3^+RCHCOO^-)_w/(NH_3^+RCHCOO^-)_s$) which depends on the solute, temperature and solute concentration.

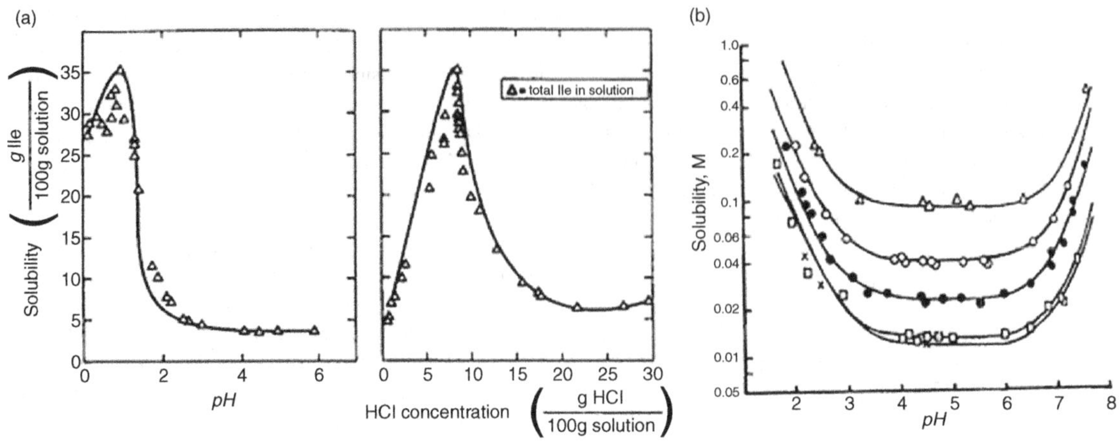

Figure 5.2.11. (a) The solubility of L-isoleucine as a function of pH *and HCl concentration at 25 °C (Zumstein and Rousseau, 1989). Reprinted, with permission, from* Ind. Eng. Chem. Res., ***28**, (1989), 1226, Figure 4. Copyright (1989) American Chemical Society. (b) Solubility-*pH *profiles of amino penicillin at 37 °C. The points are experimental values. The solid curves were generated from equation (5.2.130) and other parameters. Key: Δ, cyclacillin anhydrate; □, ampicillin anhydrate; ●, ampicillin trihydrate; □, amoxicillin trihydrate; and ×, epicillin anhydrate (Tsuji* et al., *1978). Reprinted, with permission, from A. Tsuji, E. Nakashima, S. Hamano, T. Yamana,* J Pharmaceut. Sci., *67(8), 1059 (1978), © 1978, John Wiley & Sons.*

For amino acids there is a band of about 2-3 *pH* units around the *pI* where the net charge is zero. This is identified as the *isoelectric band*; a *pH* value beyond this band on either side increases the amino acid solubility. For example, from equation (5.2.118a), we find that as C_{H^+w} increases, C_{Am^+w} increases; therefore the amino acid solubility increases since C_{Am^+w} increases. Zumstein and Rousseau (1989) have illustrated that, for the amino acid L-isoleucine, the $C_{Am^\pm w}$ species dominates above $pH = 2$ leading to the lowest solubility. As the *pH* decreases, the C_{Am^+w} species forms more and more and the amino acid solubility increases sharply. At $pH = 1$, it reaches a maximum, then it decreases sharply at lower *pH* (Figure 5.2.11a).

As the *pH* is decreased below 2 by the addition of HCl acid, more and more of the acid form of the amino acid is present in solution. However, there is a solubility limit of this form of amino acid which is subject to an equilibrium with the hydrochloric acid salt $(Cl^- \cdot N^+H_3RCOOH \cdot H_2O)_s$ in the solid phase via

$$(Cl^-\cdot N^+H_3RCOOH\cdot H_2O)_s \Leftrightarrow N^+H_3RCOOH(w) + Cl^-(w) + H_2O. \tag{5.2.128}$$

There is a solubility product K_{sp} of the two ions in the aqueous solution:

$$K_{sp} = C_{N^+H_3RCOOH,w} \times C_{Cl^-,w}. \tag{5.2.129}$$

As HCl is added, initially C_{Am^+w} will increase, but so does $C_{Cl^-,w}$. Correspondingly, at some point of acid addition, the increase in $C_{Cl^-,w}$ will reduce C_{Am^+w} when their product has already reached K_{sp}. This decreases C_{Am^+w} and therefore the total amino acid solubility $(= C_{Am^+w} + C_{Am^\pm w})$ from the ionization equilibrium (5.2.116a). One therefore observes a maximum and then a signifcant drop off in the total solubility of the amino acid L-isoleucine on increasing the amount of added HCl and a corresponding decrease in *pH* (Figure 5.2.11(a)).

If the complication of the salt solubility limit being exceeded is not present, then the total concentration of the amino acid in solution in terms of the concentration of the zwitterionic species $C_{Am^+,w}$ may be obtained from equations (5.2.116a, b) as

$$C^t_{Am^\pm w} = C_{Am^\pm w} + C_{Am^+w} + C_{Am^-w}$$

$$= C_{Am^\pm w} + C_{Am^\pm w}\frac{C_{H^+}}{K_1} + C_{Am^\pm w}\frac{K_2}{C_{H^+}}$$

$$= C_{Am^\pm w}\left[1 + \frac{C_{H^+}}{K_1} + \frac{K_2}{C_{H^+}}\right]. \tag{5.2.130}$$

While the crystals are formed from the uncharged $Am^\pm$ species, the solubility of the amino acid in the aqueous solution is obtained from (5.2.130). This solubility decreases at first at a low *pH* as the *pH* is increased, then it has a minimum around the *pI* band since both C_{Am^+w} and C_{Am^-w} are zero at *pI*. Then, as the *pH* is increased further beyond pK_2, the total amino acid solubility increases again.

This type of behavior has been observed also for *proteins*. The solubility of a protein built out of amino acids is minimum when $pH = pI$, the isoelectric point. The minimum is a point not a band. As the pH moves away from the isoelectric point, the protein solubility increases. The presence of a net charge on the protein molecule increases its solubilization due to its interaction with dipolar water molecules (H^+ as well as OH^-). Kirwan and Orella (1993) have illustrated broadly a similar solubility behavior for other biological molecules of smaller molecular weight which produce zwitterions. Observe the solubility variation with pH of amphiprotic solutes such as the *β-lactam antibiotics*, ampicillin, amoxicillin, etc., as determined by Tsuji *et al.* (1978) (Figure 5.2.11(b)). If the species is susceptible to hydrolysis at high pH, then pH effects are observed only at lower pH in the range of $pH < pK_1 \leq pI$.

Optical isomers in solution can be separated by a number of techniques, e.g. enzymatic methods, mechanical resolution, etc. In mechanical resolution of racemic mixtures of optical isomers, a supersaturated solution of a racemic mixture is seeded with the pure crystal of one of the isomers. This crystal grows and one of the isomers is separated from the solution. However, the solution remains supersaturated in the other isomer, which tends to precipitate, resulting in poor separation of the isomers.

If the racemic mixture consists of ionizable species, the addition of acid or base to the solution has been found to stabilize the solution (Asai and Ikegami, 1982). This has been found to be true for ionizable amino acids, e.g. the L and D forms of glutamic acid, the L and D forms of DOPA (3,4-dihydroxy-β-phenylalanine), etc. (Asai, 1985). Here both the free and the salt forms of the acid are racemic mixtures. But the salt forms are much more soluble. The reaction that takes place, for example, in the presence of NaOH is as follows:

$$\mathrm{L-A + D-S \Leftrightarrow L-S + D-A}, \tag{5.2.131}$$

where A refers to the free form of glutamic acid, S refers to the sodium salt of glutamic acid and L and D refer to the two different isomers. When the D–A is seeded, the L–A form of the acid (not seeded) will be pushed to the L–S form since D–A is disappearing by crystallization on the growing crystal. Remember, the L–S form, being more soluble, does not precipitate.

An alternative strategy relies on differing crystal sizes of the two reaction products, which therefore can be successfully separated. For example, in the resolution of DL-DOPA and its hemihydrochloride DL-DOPA · 0.5 HCl, the following reaction occurs:

$$\mathrm{L-DOPA \cdot 0.5\ HCl + D-DOPA \Leftrightarrow}$$

$$\mathrm{L-DOPA + D-DOPA \cdot 0.5\ HCl}. \tag{5.2.132}$$

Not only are the solubilities of the salt form much greater than the free form, but also the crystal sizes of the salt form of the amino acid are much greater than those of the free form. Thus, the mixture of the crystals of L-DOPA and D-DOPA · 0.5 HCL can be easily separated by sieving (Asai, 1985).

A comprehensive account of the effect of chemical reactions on crystallization in *enantiomeric systems* has been provided by Jacques *et al.* (1981). In the resolution of a racemic mixture of acid AH (particular forms are identified as D-AH and L-AH) by an optically active base B, the following dissociations/ionizations, acid–base reactions (leading to the salt, e.g. AHB (undissociated form)) and partitioning equilibria are relevant:

$$\mathrm{AH} \overset{K_d}{\longleftrightarrow} \mathrm{A^- + H^+}; \qquad K_d = \frac{C_{A^-} C_{H^+}}{C_{AH}}. \tag{5.2.133}$$

For both D- and L- forms,

$$K_{dAH} = \frac{C_{DA^-} C_{H^+}}{C_{DAH}}; \qquad K_{dAH} = \frac{C_{LA^-} C_{H^+}}{C_{LAH}}; \tag{5.2.134}$$

$$\mathrm{B + H^+} \overset{K_{dB}}{\longleftrightarrow} \mathrm{BH^+}; \qquad K_{dB} = \frac{C_{BH^+}}{C_B C_{H^+}}; \tag{5.2.135}$$

$$\mathrm{AH + B} \overset{K}{\longleftrightarrow} \mathrm{AHB}; \qquad K = \frac{C_{AHB}}{C_{AH} C_B}, \tag{5.2.136}$$

for both D-AHB (K_D) and L-AHB (K_L). Also,

$$(\mathrm{AHB})_{\text{solid}} \overset{\kappa_{iw}}{\longleftrightarrow} (\mathrm{AHB})_{\text{solution}}. \tag{5.2.137}$$

Here, K_d represents the dissociation constant of the enantiomeric acid, L- or D-form; K_{dB} represents the dissociation constant of the optically active base B, and K represents the acid–base reaction equilibrium constant, K_D for the D-form and K_L for the L-form. Further, κ_{iw} represents the equilibrium solubility parameter of the salt i. However, the process is useful because κ_{iw} varies with the enantiomeric salt species i, i.e. whether it is D-AHB or L-AHB. We identify this via κ_{Dw} and κ_{Lw}.

To start with, we have a certain total concentration of the racemic mixture C^t_{AH}: it consists of C^t_{DAH} ($= 0.5\ C^t_{AH}$) and C^t_{LAH} ($= 0.5\ C^t_{AH}$). However, this total initial concentration after the process is over will consist of, for example,

$$C^t_{DAH} = C_{DAH} + C_{DA^-} + C_{D-AHB} + C_{Dp}, \tag{5.2.138a}$$

where C_{Dp} is the concentration of the precipitate or crystals of the D-form of AHB (C_{Lp} for L-form). Correspondingly,

$$C^t_{LAH} = C_{LAH} + C_{LA^-} + C_{L-AHB} + C_{Lp}. \tag{5.2.138b}$$

From (5.2.134),

$$C_{LAH} + C_{LA^-} = C_{LAH}\left(1 + \frac{K_{dAH}}{C_{H^+}}\right);$$

$$C_{L-AHB} \cong \kappa_{Lw}, \qquad C_{D-AHB} \cong \kappa_{Dw}; \tag{5.2.139}$$

$$C^t_{LAH} - \kappa_{Lw} - C_{Lp} = C_{LAH}\left(1 + \frac{K_{dAH}}{C_{H^+}}\right).$$

Correspondingly,

$$C^t_{\mathrm{DAH}} - \kappa_{\mathrm{D}w} - C_{\mathrm{D}p} = C_{\mathrm{DAH}}\left(1 + \frac{K_{d\mathrm{AH}}}{C_{\mathrm{H}^+}}\right). \quad (5.2.140)$$

If the total initial concentration of the base is C^t_{B}, then

$$C^t_{\mathrm{B}} - \kappa_{\mathrm{L}w} - \kappa_{\mathrm{D}w} - C_{\mathrm{D}p} - C_{\mathrm{L}p} = C_{\mathrm{B}}(1 + K_{d\mathrm{B}}C_{\mathrm{H}^+}). \quad (5.2.141)$$

Now, from (5.2.136), (5.2.140) and (5.2.141), we get, for the L-form,

$$\left(C^t_{\mathrm{LAH}} - \kappa_{\mathrm{L}w} - C_{\mathrm{L}p}\right)\left(C^t_{\mathrm{B}} - \kappa_{\mathrm{L}w} - \kappa_{\mathrm{D}w} - C_{\mathrm{D}p} - C_{\mathrm{L}p}\right) = \frac{\kappa_{\mathrm{L}w}}{K_{\mathrm{L}}}\left(1 + \frac{K_{d\mathrm{AH}}}{C_{\mathrm{H}^+}} + K_{d\mathrm{B}}C^+_{\mathrm{H}} + K_{d\mathrm{AH}}K_{d\mathrm{B}}\right). \quad (5.2.142)$$

Correspondingly, for the D-form, we obtain

$$\left(C^t_{\mathrm{DAH}} - \kappa_{\mathrm{D}w} - C_{\mathrm{D}p}\right)\left(C^t_{\mathrm{B}} - \kappa_{\mathrm{L}w} - \kappa_{\mathrm{D}w} - C_{\mathrm{D}p} - C_{\mathrm{L}p}\right) = \frac{\kappa_{\mathrm{D}w}}{K_{\mathrm{D}}}\left(1 + \frac{K_{d\mathrm{AH}}}{C_{\mathrm{H}^+}} + K_{d\mathrm{B}}C^+_{\mathrm{H}} + K_{d\mathrm{AH}}K_{d\mathrm{B}}\right). \quad (5.2.143)$$

Knowing $C^t_{\mathrm{DAH}} = C^t_{\mathrm{LAH}} = 0.5\, C^t_{\mathrm{AH}}$, we divide the expression for the L-form by that for the D-form to obtain

$$\frac{\left(C^t_{\mathrm{LAH}} - \kappa_{\mathrm{L}w} - C_{\mathrm{L}p}\right)}{\left(C^t_{\mathrm{DAH}} - \kappa_{\mathrm{D}w} - C_{\mathrm{D}p}\right)} = \left(\frac{\kappa_{\mathrm{L}w}K_{\mathrm{D}}}{\kappa_{\mathrm{D}w}K_{\mathrm{L}}}\right),$$

which may be rearranged to give the following:

$$C_{\mathrm{L}p} - C_{\mathrm{D}p} = C_{\mathrm{D}p}\left(\frac{\kappa_{\mathrm{L}w}K_{\mathrm{D}}}{\kappa_{\mathrm{D}w}K_{\mathrm{L}}} - 1\right) - C^t_{\mathrm{DAH}}\left\{\left(\frac{\kappa_{\mathrm{L}w}K_{\mathrm{D}}}{\kappa_{\mathrm{D}w}K_{\mathrm{L}}}\right) - 1\right\} + \kappa_{\mathrm{L}w}\left(\frac{K_{\mathrm{D}}}{K_{\mathrm{L}}} - 1\right). \quad (5.2.144)$$

Here we have arbitrarily assumed that the D-form of the salt is more soluble so that $C_{\mathrm{L}p} - C_{\mathrm{D}p} > 0$. The result above is *pH* independent, even though the individual value of $C_{\mathrm{L}p}$ or $C_{\mathrm{D}p}$ depends on *pH*. Consult Jacques *et al.* (1981, sect. 5.1) for quantitative estimates of various constants for an enantiomeric system. Knowing $\kappa_{\mathrm{L}w}$, $\kappa_{\mathrm{D}w}$, K_{D} and K_{L} for a system, and the value of C^t_{DAH}, one can find out the difference in the amount of crystals of the two enantiomers.

5.2.5 Surface adsorption equilibrium

We have considered the equilibrium distribution of a nonelectrolytic surfactant solute between the bulk liquid phase and the interfacial phase in a gas–liquid system via relation (3.3.106). We have thereby illustrated the application of the Gibbs adsorption isotherm (3.3.40a) to a single nonionic surface-active solute. Chemical reactions can influence such adsorption isotherms in a number of ways. If the surface-active solutes are ionic, the adsorption equilibria are affected. In other cases, the solute to be removed (the colligend) is not surface active but it reacts with or is adsorbed on a solute (collector) which *is* surface active. We treat first the ionic surface-active solute distribution between a bulk liquid phase $j = 1$ and the interfacial phase $j = \sigma$ for a gas–liquid system.

5.2.5.1 *Ionic surface-active solute $R^{\pm}X^{-}$ in solution*

An ionic surface-active solute R^+X^- in a dilute solution may or may not have an excess of another electrolyte present simultaneously in the bulk solution. Further, this electrolyte may or may not have a common ion (R^+Y^-). Select first an ionic surface-active solute R^+X^- without any other electrolyte. From the Gibbs isotherm (3.3.40a),

$$\mathrm{d}\gamma_{12}|_T = -\Gamma^E_{\mathrm{R}^+\sigma}RT\mathrm{dln}C_{\mathrm{R}^+1}|_T - \Gamma^E_{\mathrm{X}^-\sigma}RT\mathrm{dln}C_{\mathrm{X}^-1}|_T \quad (5.2.145a)$$

by considering the R^+ and X^- ions as separate species. If the solute R^+X^- is completely dissociated in the bulk solution ($j = 1$), then

$$C_{\mathrm{R}^+1} = C_{\mathrm{X}^-1} = C_{\mathrm{RX},1}, \quad (5.2.145b)$$

where $C_{\mathrm{RX},1}$ is the bulk concentration of the surfactant solute to start with. The equilibrium relation is now simplified to (since $\Gamma^E_{\mathrm{R}^+\sigma} = \Gamma^E_{\mathrm{X}^-\sigma} = \Gamma_{\mathrm{RX},\sigma}$ by (3.3.42b))

$$\mathrm{d}\gamma_{12}|_T = -2\Gamma^E_{\mathrm{RX},\sigma}RT\mathrm{dln}C_{\mathrm{RX},1};$$

$$\Gamma^E_{\mathrm{RX},\sigma} = -\frac{1}{2RT}\left\{\frac{\mathrm{d}\gamma^{12}}{\mathrm{dln}C_{\mathrm{RX},1}}|_T\right\}. \quad (5.2.146)$$

If $\gamma^{12} = a_1 - b_1C_{\mathrm{RX},1}$, then

$$\Gamma^E_{\mathrm{RX},\sigma} = \frac{b_1C_{\mathrm{RX},1}}{2RT}. \quad (5.2.147)$$

Thus, ionization of the solute has reduced its surface excess by a factor of two (compare relation (3.3.106)) from that for a nonionic solute.

When there is also a large excess of an electrolyte with a common ion (R^+Y^-) present in the liquid, the Gibbs isotherm becomes

$$\mathrm{d}\gamma^{12}|_T = -\Gamma^E_{\mathrm{R}^+\sigma}\mathrm{d}\mu_{\mathrm{R}^+1} - \Gamma^E_{\mathrm{X}^-\sigma}\mathrm{d}\mu_{\mathrm{X}^-1} - \Gamma^E_{\mathrm{Y}^-\sigma}\mathrm{d}\mu_{\mathrm{Y}^-1}. \quad (5.2.148a)$$

Suppose the concentration of the surface active solute R^+X^- is changed in the bulk solution; the concentration of the anion Y^- is essentially unchanged, i.e. $\mathrm{d}\mu_{\mathrm{Y}^-1} = 0$. For a dilute solution, we can then write

$$\mathrm{d}\gamma^{12}|_T = \left[-\Gamma^E_{\mathrm{R}^+\sigma}\mathrm{dln}C_{\mathrm{R}^+1} - \Gamma^E_{\mathrm{X}^-\sigma}\mathrm{dln}C_{\mathrm{X}^-1}\right]RT. \quad (5.2.148b)$$

But, due to any addition of R^+X^-, $\mathrm{d}C_{\mathrm{R}^+1} = \mathrm{d}C_{\mathrm{X}^-1}$. Further, $C_{\mathrm{R}^+1} >> C_{\mathrm{X}^-1}$ due to a large excess of R^+Y^-. Then, assuming $\Gamma^E_{\mathrm{R}^+\sigma}$ and $\Gamma^E_{\mathrm{X}^-\sigma}$ to be of the same order of magnitude,

$$\Gamma^E_{\mathrm{R}^+\sigma}\frac{\mathrm{d}C_{\mathrm{R}^+1}}{C_{\mathrm{R}^+1}} << \Gamma^E_{\mathrm{X}^-\sigma}\frac{\mathrm{d}C_{\mathrm{X}^-1}}{C_{\mathrm{X}^-1}},$$

which leads to

$$d\gamma^{12}|_T = -RT\Gamma^E_{X^-\sigma} d\ln C_{X^-1} = -RT\Gamma^E_{i\sigma} d\ln\, C_{il}. \quad (5.2.149)$$

Thus, the equilibrium relation becomes exactly similar to that for a nonionic surface-active solute.

We can now calculate the separation factor between two surface-active solutes using the above results. We need to assume that the distribution of one does not affect that of the other. For example, for one nonionic solute $i = 1$ and an ionic solute $i = 2$ in the absence of any other electrolyte, the separation factor is

$$\alpha_{12} = \frac{\Gamma^E_{1,\sigma}/C_{1,1}}{\Gamma^E_{2,\sigma}/C_{2,2}} = \frac{2b_1^1}{b_1^2}, \quad (5.2.150)$$

where the superscript for b_1 refers to the solute. If both surface-active solutes are ionic, the distribution equilibria of the two solutes are coupled in general. The requirements of electroneutrality also have to be used. See the treatment for two surface-active ionic solutes Na^+X^- and Na^+Y^-, where $\Gamma^E_{X^-\sigma}$ is a function of $\Gamma^E_{Y^-\sigma}$ and vice versa, in Rubin and Jorne (1969). An extensive treatment of surface adsorption equilibrium for surface-active solutes is available in Davies and Rideal (1963).

5.2.5.2 *Collector-colligend equilibrium*

For solutes that are not surface active, surface adsorption is possible if a surfactant (i.e. the collector) reacts with the solute (the colligend). The product may be an ion pair or a chelate. For an example of the former, consider the following (Grieves, 1982). In a solution of the salt $NaC\ell O_3$, the surfactant added is a quaternary ammonium compound, ethylhexadecyldimethyl-ammonium bromide (EHDA-Br). The following exchange reaction takes place between the colligend $(C\ell O_3^-)_l$ and the collector (EHDA$^+$, the exchanger):

$$(\text{EHDA-Br})_\sigma + \left(C\ell O_3^-\right)_l \Leftrightarrow \left(\text{EHDA}^+\text{-}C\ell O_3^-\right)_\sigma + (\text{Br}^-)_1, \quad (5.2.151)$$

where subscripts σ and l refer to the surface and the bulk liquid phases, respectively. The above represents a surface exchange reaction between ClO_3^- ion in the bulk being exchanged with the Br^- ion in the surfactant molecule adsorbed at the surface.

If there is no surface exchange, ion-pair formation can take place in the bulk solution,

$$(\text{EHDA-Br})_1 + \left(C\ell O_3^-\right)_1 \Leftrightarrow \left(\text{EHDA}^+\text{-}C\ell O_3^-\right)_1 + (\text{Br}^-)_1, \quad (5.2.152)$$

and then surface adsorption of the (EHDA$^+$-$C\ell O_3^-$) species occurs. The selectivity of the process (5.2.151) may be expressed by

$$\alpha_{12} = \frac{\Gamma^E_{1,\sigma}/C_{1,1}}{\Gamma^E_{2,\sigma}/C_{2,1}}, \quad (5.2.153)$$

where species 1 is $C\ell O_3^-$ and species 2 is Br^-.

A different aspect of collector-colligend equilibrium will be considered now. The concentration of the surfactant (collector) w.r.t. the concentration of the colligend in the bulk solution is important if the process is described by (5.2.152) followed by surface adsorption of the complex. Assume the colligend ion to be i and the free surfactant counterion to be k; the collector-colligend complex is i-k. Only species k and i-k can adsorb at the interface. Assume that the adsorption of the complex i-k at the interface is proportional to the adsorption of the surfactant species k (Rubin and Gaden, 1962):

$$\Gamma^E_{i\text{-}k} = b'_{i\text{-}k}\Gamma^E_k. \quad (5.2.154)$$

But $\Gamma^E_k = b'_k C_{k,1}$. Therefore

$$\Gamma^E_{i\text{-}k} = b'_{i\text{-}k} b'_k C_{k,1}. \quad (5.2.155)$$

The effective equilibrium ratio for the colligend is

$$\frac{\Gamma^E_{i\text{-}k}}{C_{i\text{-}k,1} + C_{i,1}} = \frac{b'_{i\text{-}k} b'_k C_{k,1}}{C_{i\text{-}k,1} + C_{i,1}}. \quad (5.2.156)$$

If free surfactant concentration in the bulk solution, $C_{k,1}$, is zero, then there is no separation of the colligend between the bulk phase and the interfacial phase. Such a condition is obtained when the colligend bulk solution concentration is greater than that of the collector. On the other hand, if the surfactant species is present in significant excess in the bulk, there will be no free i, i.e. $C_{i,1} = 0$. Further, the surface excess of the complex is considerable, especially if $C_{k,1} \gg C_{i\text{-}k,1}$.

An altogether different aspect of collector-colligend equilibria arises if the surfactant concentration exceeds the critical micelle concentration (c.m.c.). Beyond this concentration, the surfactant solution becomes colloidal in nature, with individual surfactant molecules coming together and forming aggregates called micelles (Section 4.1.8). Any surfactant added is used for such aggregate formation; the gas-liquid interfacial layer is saturated with surfactant molecules. The adsorption capacity of a collector molecule present in a micelle for a colligend is different from that of a collector molecule present in the gas-liquid interface. Lemlich (1968) has derived an expression for the difference between the colligend bulk concentration below the c.m.c. (C_i^1) and the c.m.c. (C_i^2) in terms of the collector surfactant concentration in excess of the c.m.c. level C_k^c, i.e.

$$\left(C_k^2 - C_k^c\right):\Gamma_i^{E2} = b_i C_i^2; \quad C_i^1 - C_i^2 = \frac{C_k^2 - C_k^c}{E}\frac{\Gamma_i^{E2}}{\Gamma_k^{E2}}. \quad (5.2.157)$$

Here the surfactant molecules in the interfacial layer are E times as effective in absorbing a colligend molecule as a surfactant molecule in a micelle. One possible value of E is 1.

5.2.6 Complexation in Donnan dialysis

We know already from the Donnan equilibrium condition (Section 4.3.2) that two electrolytes AN and BN separated by a cation exchange membrane in two solutions 1 and 2 will distribute themselves such that

$$\left[\frac{C_{A1}}{C_{A2}}\right]^{1/Z_A} = \left[\frac{C_{B1}}{C_{B2}}\right]^{1/Z_B}. \qquad (5.2.158a)$$

(This equation is obtained from equation (4.3.18), where we have replaced the ionic activity by concentration; further, the superscript primes and double primes have been replaced by subscripts 1 and 2, respectively.) Here the membrane has the same charge as the anion N, so that only the cations are transported through the cation exchange membrane. Now, solutions 1 and 2 can contain the same cation complexing agent C or different cation complexing agents. Assuming that the same agent C is present in both solutions, we can write the complexation reaction as

$$A^{Z_{A+}} + C \rightarrow AC, \qquad (5.2.158b)$$

whose equilibrium constant K_A is

$$K_A = \frac{C_{AC}}{C_A C_C} \qquad (5.2.159)$$

for each solution. The total concentration of species A in each solution is the sum of free ions $A^{Z_{A+}}$ and the complex AC:

$$C^t_{A1} = C_{A1} + C_{(AC)1} = C_{A1} + K_A C_{A1} C_{C1} \qquad (5.2.160a)$$

and

$$C^t_{A2} = C_{A2} + C_{(AC)2} = C_{A2} + K_A C_{A2} C_{C2}. \qquad (5.2.160b)$$

The ratio of the total concentrations of species in the two solutions is obtained by using (5.2.159):

$$\frac{C^t_{A1}}{C^t_{A2}} = \frac{C_{A1} + C_{(AC)1}}{C_{A2} + C_{(AC)2}} = \frac{C_{A1}(1 + K_A C_{C1})}{C_{A2}(1 + K_A C_{C2})}. \qquad (5.2.161)$$

For species B, we obtain similarly

$$\frac{C^t_{B1}}{C^t_{B2}} = \frac{C_{B1}(1 + K_B C_{C1})}{C_{B2}(1 + K_B C_{C2})}, \qquad (5.2.162)$$

which assumes that the concentration of the complexing agent C is sufficiently large such that complexation with A still leaves a very large amount of C for complexation with B. In reality, one of the ions should have a high value of K, while that for the other should be low. For example, $K_A >> K_B$, which is true for, say, copper (A) against silver (B) when EDTA is the complexing agent (Wallace, 1967). A selectivity α_{A-B} may be defined by

$$\alpha_{A-B} = \frac{C^t_{A1}\, C^t_{B2}}{C^t_{B1}\, C^t_{A2}} = \frac{C_{A1} C_{B2}\,(1 + K_A C_{C1})\,(1 + K_B C_{C2})}{C_{B1} C_{A2}\,(1 + K_A C_{C2})\,(1 + K_B C_{C1})}. \qquad (5.2.163)$$

Now, if C_{C2} is 0, $K_B << K_A$ and $Z_A = Z_B$ (say), then

$$\alpha_{A-B} \cong (1 + K_A C_{C1}) \frac{C_{A1} C_{B2}}{C_{B1} C_{A2}}. \qquad (5.2.164)$$

It is known that, without the complexation reaction, there is no selectivity between A and B. It is now clear that the complexation provides considerable selectivity in favor of species A. Most of the above analysis is based on the study by Wallace (1967), with one basic underlying assumption, namely that the complexing agent C does not go through the ion exchange membrane.

5.2.7 Enzymatic separation of isomers

Enzymatic reactions can be stereospecific. Thus enzymes can distinguish between two enantiomers and react with only one of the enantiomers (Jones, 1976):

$$\underset{\text{(enantiomer I)}}{\text{X a b c d}} \xrightarrow{\text{enzyme}} \text{X a b c e} + \text{d (fast reaction)}; \qquad (5.2.165)$$

$$\underset{\text{(enantiomer II)}}{\text{X d c b a}} \xrightarrow{\text{enzyme}} \text{very very slow reaction}. \qquad (5.2.166)$$

The enzymatic catalysis has converted the group d to e with enantiomer I, which acted as a good substrate. Enantiomer II is, however, a very poor substrate, so it is essentially unchanged. Such an approach has been used to separate (resolve) racemic mixtures of amino acids. For example, the N-acetyl L-tyrosine ester is available as a racemic mixture of the L-form and the D-form. If the enzyme chymotrypsin (CT) is used to catalyze the hydrolysis of this ester in two forms at $pH = 8$, 25 °C and 18% CH_3OH solution, only the L-form is hydrolyzed to the acid (Jones and Beck, 1976) (see Figure 5.2.12).

We have seen in Section 5.2.4 that the different forms of the racemic species (free and salt forms) have different tendencies, for example for solubility. In the case of the reaction shown in Figure 5.2.12, the L-acid, being highly ionizable compared to the D-ester, can be separated by using an oil membrane (Matson and Quinn, 1986), which allows the ester to pass through it but not the ionized L-acid. Large-scale processes are in operation in Japan where the enzyme aminoacylase is used to separate a racemic mixture of D and L isomers of amino acids containing an acetyl group by selective removal of the acetyl group from the L-isomer.

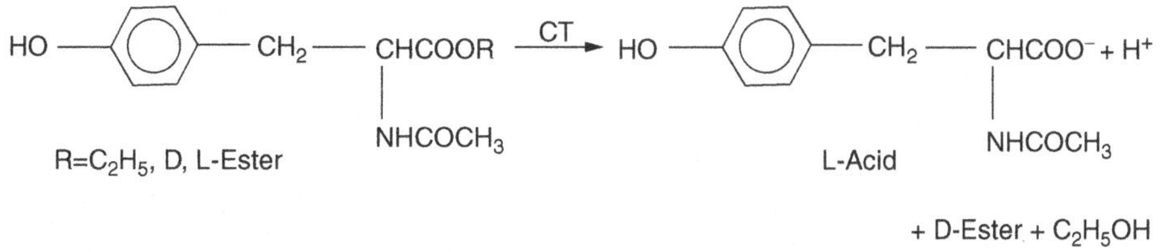

Figure 5.2.12. Enzymatic hydrolysis of the L-form of an ester.

5.3 Rate-controlled equilibrium separation processes: role of chemical reactions

In Section 5.2, we considered the change in separation equilibrium due to chemical reactions. Many such separation processes in practice do not have phases in equilibrium. This may have come about, for example, due to inadequate contact time between the phases in the device being used. The extent of separation achieved will be controlled by the extent of transfer of the species between the phases. Other conditions remaining constant, the higher the rate of transfer, the larger the extent of separation achieved. The species transport rate from one phase to another then controls the actual separation achieved. Chemical reactions can influence this interphase species transport rate. The role of chemical reactions in the separation achieved in rate-controlled equilibrium separation processes is, therefore, the subject of this section. Mass transfer and separation in gas-liquid systems are covered first, followed by liquid-liquid systems.

5.3.1 Absorption of a gas in a reactive liquid

In industrial processes, a particular gas from a gas mixture is absorbed in a liquid under conditions where both phases are flowing, or are agitated in general. During this process, the gas species A (say) has to diffuse through the gas-phase film to the gas-liquid interface, where it is absorbed, and then it diffuses through the liquid film to its bulk or reacts with a species C in the absorbent liquid (Figure 5.3.1(a)). Assume that there is no interfacial resistance to gas absorption and that the interfacial region is planar with no thickness. Assume further that we have steady state so that the concentration at any location does not change with time.

Consider first Figure 5.3.1(a). Gas species A diffuses through the gas film as its partial pressure is reduced from its bulk value p_{Ab} to the value p_{Ai} at the gas-liquid interface. The molar flux and the transfer rate (in an unit of absorber volume) of species A are, respectively,

$$N_A = k_g(p_{Ab} - p_{Ai}) \quad \text{and} \quad W_A = k_g a(p_{Ab} - p_{Ai}), \tag{5.3.1}$$

where a is the gas-liquid interfacial area for absorption per unit absorber volume. At the interface, assume Henry's law,

$$p_{Ai} = H_A^C C_{Ai}, \tag{5.3.2}$$

to be valid, with no interfacial resistance. At steady state, the flux of species A from the gas phase must equal the flux into the liquid phase at the gas-liquid interface (through the liquid film):

$$N_A = k_g(p_{Ab} - p_{Ai}) = k_\ell(C_{Ai} - C_{Ab}), \tag{5.3.3}$$

which can be rearranged to yield

$$N_A = \frac{(p_{Ab} - H_A^C C_{Ab})}{\left(\frac{1}{k_g} + \frac{H_A^C}{k_\ell}\right)}. \tag{5.3.4}$$

If a liquid-phase reaction is taking place between the gas species A being absorbed and a species C in the liquid phase,

$$a_A A + c_C C \rightarrow pP + qQ, \tag{5.3.5}$$

then the first and the *obvious* effect is that the liquid-phase bulk concentration C_{Ab} is reduced. This therefore increases the flux N_A for the absorption of gas species A by increasing the concentration difference driving the mass transfer. Such an increase in flux of A (and therefore separation) is valid for almost all separation processes where a chemical reaction removes free species A from the phase into which species A is transferred (i.e. the receiving phase). If this reaction rate is slow enough not to alter the gradient of species A profile at $y = 0$, the gas-liquid phase interface, the increase in N_A due to a reduction in C_{Ab} is all that can be achieved. In Section 5.4, we will find a similar increase in separation due to chemical reaction in the receiving phase in rate-governed separation processes utilizing membranes.

A chemical reaction of this nature in the receiving phase is helpful from another point of view. The receiving phase volume or flow rate needed to maintain a certain level of C_{Ab} is now reduced. Even though species A is coming to the receiving phase, its concentration cannot build up due to the reaction with species C.

If the chemical reaction in the liquid phase takes place at a substantial rate in the liquid film, it has an additional effect on the species A flux that is fundamentally more complex. Yet it is quite beneficial to separation in general. Chemical

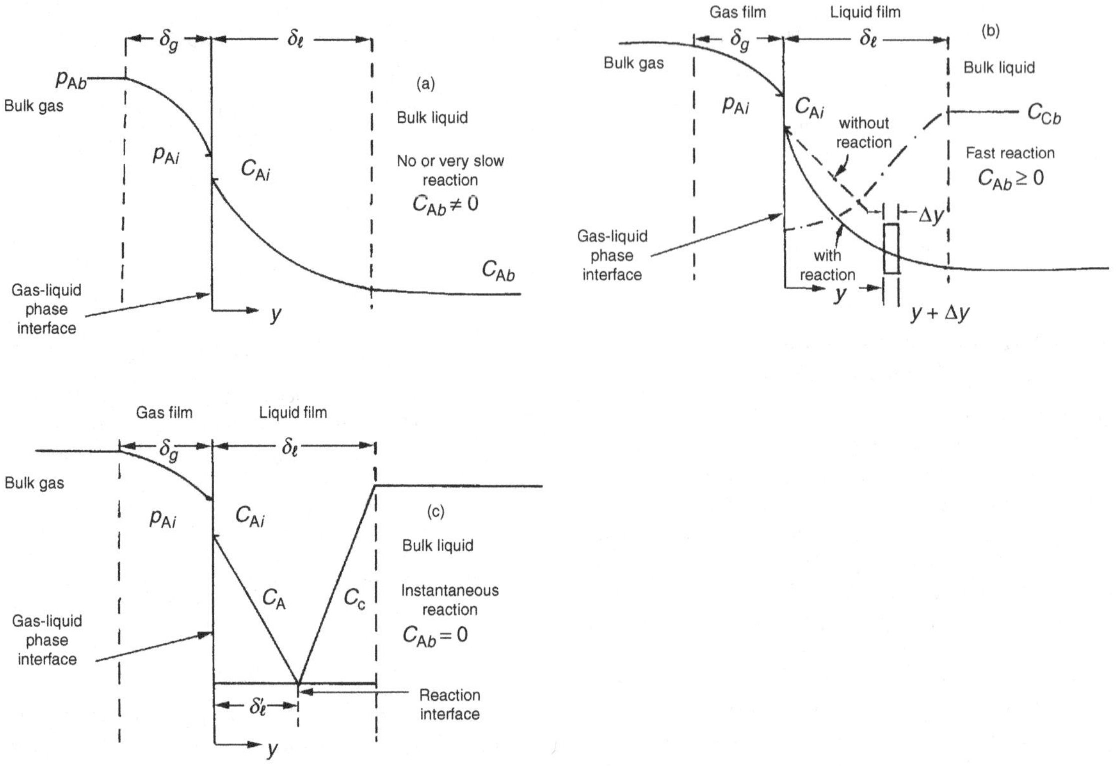

Figure 5.3.1. Different mass-transfer regimes in absorption of gas A in a liquid containing reactant C.

reaction essentially enhances the mass-transfer coefficient k_ℓ of species A in the liquid phase over the value obtained without any reaction other conditions remaining the same. We understand such behavior first by using a system where gas A dissolved in the liquid reacts irreversibly in the liquid film such that the reaction is first order or pseudo-first order (Danckwerts, 1970; Froment and Bischoff, 1979).

At steady state, a mass balance of species A over an element of thickness Δy in the liquid film, with diffusion only in the y-direction, leads to (Figure 5.3.1(b))

$$-D_A \frac{dC_A}{dy}\bigg|_y + D_A \frac{dC_A}{dy}\bigg|_{y+\Delta y} + R_A \Delta y = 0, \tag{5.3.6}$$

where R_A is the molar rate of production of A per unit volume. For a first-order irreversible reaction in A,

$$R_A = -k_1 C_A, \tag{5.3.7}$$

where k_1 is the first-order reaction rate constant. Taking the limit as $\Delta y \rightarrow 0$, we get

$$D_A \frac{d^2 C_A}{dy^2} - k_1 C_A = 0, \tag{5.3.8}$$

where, at

$$y = 0, C_A = C_{Ai}; \qquad y = \delta_\ell, \; C_A = C_{Ab}. \tag{5.3.9}$$

Rearrange this linear second-order homogeneous ordinary differential equation to obtain

$$\frac{d^2 C_A}{dy^2} - \frac{k_1}{D_A} C_A = 0. \tag{5.3.10}$$

The solution satisfying the boundary conditions is (Hatta, 1928, 1932)

$$C_A = \frac{1}{\sinh\sqrt{\gamma_R}}\left[C_{Ab} \sinh y\sqrt{\frac{k_1}{D_A}} + C_{Ai} \sinh(\delta_\ell - y)\sqrt{\frac{k_1}{D_A}}\right], \tag{5.3.11}$$

where $\gamma_R = D_A k_1 / k_\ell^2$. Note that, in film theory descriptions of mass transfer (equations (3.1.141) and (3.1.142)), $k_\ell = (D_A/\delta_\ell)$ when there is no reaction.

The mass flux of species A at the gas–liquid interface ($y = 0$) for this reaction-enhanced gas absorption is obtained by differentiating (5.3.11):

$$N_A^r\big|_{y=0} = -D_A \frac{dC_A}{dy}\bigg|_{y=0} = \frac{D_A\sqrt{\gamma_R}}{\delta_\ell \tanh\sqrt{\gamma_R}}\left[C_{Ai} - \frac{C_{Ab}}{\cosh\sqrt{\gamma_R}}\right], \tag{5.3.12}$$

where the superscript r indicates reaction. If there was no reaction between species A and B in the liquid film (but there will be a reaction in the liquid bulk), then the gas absorption flux is equal to that observed in physical absorption (Figure 5.3.1(a)),

$$N_A|_{y=0} = (D_A/\delta_\ell)(C_{Ai} - C_{Ab}). \tag{5.3.13}$$

Consider now the limiting case of rapid reactions when $\sqrt{\gamma_R} >> 1$. First, $C_{Ai} >> (C_{Ab}/\cosh\sqrt{\gamma_R})$. Often C_{Ab} is assumed to be zero under these conditions. Then assume that $C_{Ab} << C_{Ai}$ for the no reaction case and get

$$\phi = \frac{N_A^r|_{y=0}}{N_A|_{y=0}} = \frac{\sqrt{\gamma_R}}{\tanh\sqrt{\gamma_R}} \tag{5.3.14}$$

(Danckwerts, 1970). For values of $\gamma_R < 1$, the enhancement factor ϕ, by which the chemical reaction enhances the flux of absorption of species A over that in the absence of a reaction, is only marginally greater than 1. For $\gamma_R > 1$, however, the value of ϕ increases sharply and can reach values of the order of 10–1000. Other conditions remaining constant, γ_R increases as the reaction rate constant k_1 increases. In fact, for large values of γ_R, $\tanh\sqrt{\gamma_R} \to 1$ so that $\phi \to \sqrt{\gamma_R}$. Such reactions are characterized as *fast reactions*.

Consider now a special case where $C_{Ab} \cong 0$. Then

$$N_A|_{y=0} = \frac{D_A}{\delta_\ell} C_{Ai} = k_\ell C_{Ai} \tag{5.3.15}$$

and

$$N_A^r|_{y=0} = \frac{D_A}{\delta_\ell}\frac{\sqrt{\gamma_R}}{\tanh\sqrt{\gamma_R}} C_{Ai} = k_\ell^r C_{Ai}, \tag{5.3.16}$$

where k_ℓ^r is the mass-transfer coefficient with reaction. The enhancement factor ϕ can be related to the two mass-transfer coefficients, k_ℓ and k_ℓ^r, by

$$\phi = \frac{N_A^r|_{y=0}}{N_A|_{y=0}} = \frac{k_\ell^r}{k_\ell} = \frac{\sqrt{\gamma_R}}{\tanh\sqrt{\gamma_R}}. \tag{5.3.17}$$

Therefore, the chemical reaction has increased the liquid-phase mass-transfer coefficient to k_ℓ^r from k_ℓ. For high γ_R values, we find

$$N_A^r|_{y=0} \cong \frac{D_A}{\delta_\ell}\sqrt{\gamma_R} C_{Ai} = k_\ell^r C_{Ai}. \tag{5.3.18}$$

This implies

$$k_\ell^r = \sqrt{D_A k_1}; \tag{5.3.19}$$

that is, the mass-transfer coefficient does not depend on the liquid film thickness δ_ℓ for a fast reaction. However, the gas film thickness will still influence the gas-phase film coefficient k_g, which does affect $N_A^r|_{y=0}$. For $C_{Ab} \cong 0$, we can describe this dependence by combining (5.3.12) and (5.3.3) into

$$N_A^r|_{y=0} = \frac{p_{Ab}}{\left[\frac{1}{k_g} + \frac{H_A^C}{k_\ell \phi}\right]}, \tag{5.3.20}$$

where, for high γ_R, $k_\ell\,\phi = k_\ell^r = \sqrt{D_A k_1}$.

An additional feature worthy of note is that, since $N_A^r|_{y=0} > N_A|_{y=0}$, the concentration gradient of species A in the liquid at the gas–liquid interface in the presence of a chemical reaction (see the solid line in Figure 5.3.1(b)) is sharper than that without any chemical reaction (see the dashed line), other quantities like diffusion coefficients, etc., remaining unchanged:

$$|dC_A/dy|_{y=0}^r > |dC_A/dy|_{y=0}.$$

In fact, the faster the reaction, the sharper the profile. Gas molecules of species A dissolved at the gas-liquid interface can diffuse through the film or react. If the reaction rate is very high, molecules of A do not get much of a chance to diffuse far from the interface. Often, molecules of A are completely consumed before the bulk liquid is approached. Such a profile of species A is shown in Figure 5.3.1(c). This behavior is characteristic of *instantaneous reactions*.

The profile of species C with which species A reacts (equation (5.3.5)) is also of interest. Concentration of C decreases from the bulk of the liquid phase to the phase interface to the extent it is consumed by the reaction. In the extreme case of an instantaneous reaction, molecules of A and C cannot coexist. There develops a *reaction plane* in the liquid film at a distance δ'_ℓ from the gas-liquid interface, where the concentrations of both species A and C are zero (Figure 5.3.1(c)). Such a condition is sustained only if the magnitude of the flux of species C toward this reaction interface is enough to react stoichiometrically with that of species A toward the same plane:

$$\frac{c_C}{a_A} N_A^r|_{y=\delta'_\ell} = -N_C^r|_{y=\delta'_\ell}. \tag{5.3.21a}$$

For linear concentration profiles for both species,

$$\frac{c_C}{a_A}\frac{D_A C_{Ai}}{\delta'_\ell} = \frac{D_C C_{Cb}}{\delta_\ell - \delta'_\ell} \tag{5.3.21b}$$

since

$$C_A|_{y=\delta'_\ell} = 0 = C_C|_{y=\delta'_\ell}. \tag{5.3.21c}$$

This provides

$$\delta'_\ell = \frac{\delta_\ell}{\left(1 + \frac{D_C C_{Cb} a_A}{D_A C_{Ai} c_C}\right)}. \tag{5.3.22}$$

Therefore the flux of species A being absorbed under chemical reaction is

$$N_A^r|_{y=0} = D_A \frac{C_{Ai} - 0}{\delta'_\ell} = \frac{D_A C_{Ai}}{\delta_\ell}\left(1 + \frac{D_C C_{Cb} a_A}{D_A C_{Ai} c_C}\right). \tag{5.3.23}$$

The enhancement factor ϕ is given by

$$\frac{N_A^r|_{y=0}}{N_A|_{y=0}} = \left(1 + \frac{D_C C_{Cb} a_A}{D_A C_{Ai} c_C}\right). \tag{5.3.24}$$

Correspondingly, if C_{Ab} is zero for the nonreaction case,

$$N^r_{\rm A}\big|_{y=0} = k^r_\ell C_{{\rm A}i}; \qquad N_{\rm A}\big|_{y=0} = k_\ell C_{{\rm A}i}, \tag{5.3.25}$$

and

$$\phi = \frac{N^r_{\rm A}\big|_{y=0}}{N_{\rm A}\big|_{y=0}} = \frac{k^r_\ell}{k_\ell} = \left(1 + \frac{D_{\rm C} C_{{\rm C}b} a_{\rm A}}{D_{\rm A} C_{{\rm A}i} c_{\rm C}}\right) \tag{5.3.26}$$

yields the enhancement factor for such an instantaneous reaction.

This result indicates that, as C_{Cb} is increased, ϕ increases. From relation (5.3.22), we find that δ'_ℓ decreases as C_{Cb} increases. Although it would appear that, as C_{Cb} is increased, ϕ increases linearly, this is not so. There is an upper limit set by the rate at which gas species A can dissolve into the liquid through the gas film. As C_{Cb} increases, δ'_ℓ decreases with the reaction interface ultimately coming to the gas–liquid interface when

$$-N^r_{\rm C}\big|_{y=0} = +\frac{D_{\rm C} C_{{\rm C}b}}{\delta_\ell} = \frac{c_{\rm C}}{a_{\rm A}} N^r_{\rm A}\big|_{y=0}. \tag{5.3.27}$$

We identify the value of C_{Cb} at which this is possible as

$$C'_{{\rm C}b} = \frac{c_{\rm C}}{a_{\rm A}} \frac{\delta_\ell}{D_{\rm C}} N^r_{\rm A}\big|_{y=0} = \frac{c_{\rm C}}{a_{\rm A}} \frac{D_{\rm A}}{D_{\rm C}} \frac{N^r_{\rm A}\big|_{y=0}}{k_\ell}. \tag{5.3.28}$$

Because the reaction interface is at the gas–liquid interface, C_{Ai} and C_{Ci} are zero. The flux of A through the gas film is

$$N_{\rm A} = k_g(p_{{\rm A}b} - 0) = N^r_{\rm A}\big|_{y=0} = \frac{a_{\rm A}}{c_{\rm C}} \frac{D_{\rm C}}{D_{\rm A}} k_\ell C'_{{\rm C}b}. \tag{5.3.29a}$$

Therefore,

$$C'_{{\rm C}b} = (c_{\rm C}/a_{\rm A})(D_{\rm A}/D_{\rm C})(k_g p_{{\rm A}b}/k_\ell). \tag{5.3.29b}$$

Increasing C_{Cb} to a value higher than this will not lead to any further increase in the flux of species A.

For instantaneous reactions in general, the reaction interface will not always necessarily be at the gas–liquid interface. For such cases, gas film resistance, as well as the liquid film resistance, can be used to describe $N^r_{\rm A}\big|_{y=0}$ in terms of p_{Ab}:

$$N^r_{\rm A}\big|_{y=0} = \frac{p_{{\rm A}b}}{\frac{1}{k_g} + \frac{H^{\rm C}_{\rm A}}{\phi k_\ell}}. \tag{5.3.30a}$$

Alternatively,

$$N^r_{\rm A}\big|_{y=0} = \frac{p_{{\rm A}b} + H^{\rm C}_{\rm A} \frac{D_{\rm C}}{D_{\rm A}} C_{{\rm C}b} \frac{a_{\rm A}}{c_{\rm C}}}{\frac{1}{k_g} + \frac{H^{\rm C}_{\rm A}}{k_\ell}}. \tag{5.3.30b}$$

So far we have considered cases where there is considerable reaction in the liquid film and, as a result, there is enhancement in gas absorption. The role of reaction in the liquid bulk has not been pointed out. For this purpose, observe in Figure 5.3.1(b) the nature of the concentration profile of dissolved gas species A in the liquid film. The concentration gradient is sharper than that without any reaction at the gas–liquid interface. However, the concentration gradient at the liquid film–liquid bulk boundary is smaller:

$$\left|\frac{{\rm d}C_{\rm A}}{{\rm d}z}\right|^r_{y=0} > \left|\frac{{\rm d}C_{\rm A}}{{\rm d}z}\right|_{y=0} > \left|\frac{{\rm d}C_{\rm A}}{{\rm d}z}\right|^r_{y=\delta_\ell}.$$

Therefore, if there is reaction in the film, the flux of gas absorption, $N^r_{\rm A}\big|_{y=0}$, is larger than the flux of A into the bulk liquid, $N^r_{\rm A}\big|_{y=\delta_\ell}$, i.e. $N^r_{\rm A}\big|_{y=0} > N^r_{\rm A}\big|_{y=\delta_\ell}$.

When the reaction is instantaneous, $C_{Ab} = 0$ and $N^r_{\rm A}\big|_{y=\delta_\ell} = 0$. On the other hand, when the reaction is slow, it is possible to have $N^r_{\rm A}\big|_{y=0} = N^r_{\rm A}\big|_{y=\delta_\ell}$; there is no reaction in the liquid film (if there is substantial reaction, it takes place in the liquid bulk). In general, then, for reactions that are not instantaneous, we should allow for a significant amount of flux of a gas species A into the bulk liquid whose concentration of A will change with distance or time as the case may be. More on this will be provided in Chapter 8 when reactive gas absorption is analyzed in a countercurrent contacting device.

The presentation so far has been carried out in the context of a film theory of mass transfer (see Section 3.1.4) and steady state conditions. There is considerable literature on other models of mass transfer, e.g. surface renewal theory. Further, unsteady state analyses exist for a number of cases. Detailed treatments are available in Danckwerts (1970) and Sherwood *et al.* (1975).

5.3.1.1 *Absorption of two gases in a reactive liquid*

There are a number of cases of industrial gas cleanups where two (A and B) or more gases are being simultaneously absorbed in the liquid. The nature of mass transport, and therefore separation, will depend on how fast each species reacts with the nonvolatile reactive species C (often the species C has some volatility, e.g. monoethanolamine). Each of gas species A and B can either react slowly or have a fast or instantaneous reaction with C. Thus, there can be a number of combinations,[7] many of which are treated in Astarita *et al.* (1983). We consider here a particular case of considerable industrial importance, namely simultaneous absorption of H_2S and CO_2 into a reactive aqueous solution containing, say, methyldiethanolamine (MDEA). We want to show how the different rates of absorption lead to a high selectivity of H_2S over CO_2. The specific reactions are (Haimour *et al.*, 1987)

$$H_2S + R_2NCH_3 \rightarrow R_2NH^+CH_3 + HS^-, \tag{5.3.31}$$

$$CO_2 + H_2O + R_2NCH_3 \rightarrow R_2NH^+CH_3 + HCO_3^-. \tag{5.3.32}$$

[7] Fast (A) – fast (B); instantaneous (A) – fast (B); fast (A)–slow (B); instantaneous (A) – slow (B); instantaneous (A) – instantaneous (B); slow (A) – slow (B).

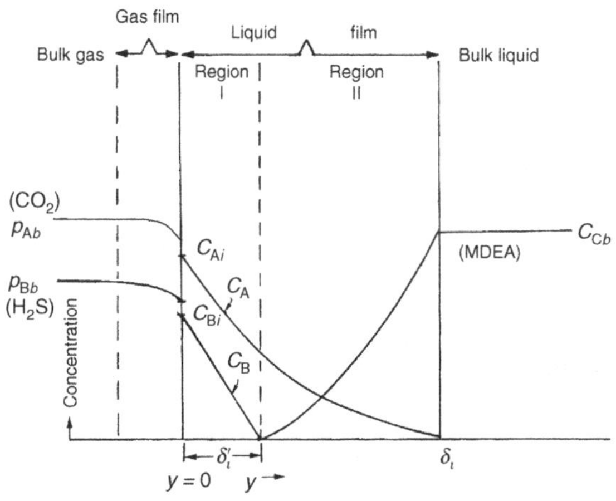

Figure 5.3.2. Concentration profiles for simultaneous absorption of H_2S and CO_2 in an aqueous solution of MDEA. (After Haimour et al. *(1987).)*

The first reaction involves only the transfer of a proton; it is therefore modeled as *instantaneous*. The hydration of CO_2 to HCO_3^- is considered to be a *fast reaction*. The concentration profiles of CO_2, H_2S and MDEA in the liquid film should therefore be as shown in Figure 5.3.2. Identifying species CO_2 as A and H_2S as B, the reactions may be rewritten as follows:

$$\mathrm{A}+\mathrm{C}\overset{k_2}{\rightarrow}\text{products}; \tag{5.3.33a}$$

$$\mathrm{B}+\mathrm{C}\rightarrow\text{products}. \tag{5.3.33b}$$

In Figure 5.3.2, we observe two distinct regions in the liquid film: region I, where there is no C due to instantaneous reaction of H_2S, and region II, where C reacts only with CO_2 species A (species B, H_2S, being nonexistent). The governing diffusion equations in region I will not therefore have the homogeneous reaction term present in equation (5.3.8). On the other hand, the governing diffusion equations for CO_2 and MDEA in region II will have the homogeneous reaction term. For the second-order reaction between CO_2 and MDEA, the molar rate of production of CO_2 can be described as

$$R_\mathrm{A}=-k_2C_\mathrm{C}C_\mathrm{A}. \tag{5.3.34}$$

We now identify the governing equations and the boundary conditions in dimensional form in the two regions. For region I:

$$D_\mathrm{A}\frac{\mathrm{d}^2C_\mathrm{A}}{\mathrm{d}y^2}=0; \tag{5.3.35a}$$

$$D_\mathrm{B}\frac{\mathrm{d}^2C_\mathrm{B}}{\mathrm{d}y^2}=0. \tag{5.3.35b}$$

At $y=0$,

$$C_\mathrm{A}=C_{\mathrm{A}i},\, C_\mathrm{B}=C_{\mathrm{B}i}. \tag{5.3.35c}$$

At $y=\delta'_\ell$,

$$C_\mathrm{A}=C'_{\mathrm{A}\ell},\, C_\mathrm{B}=0. \tag{5.3.35d}$$

For region II:

$$D_\mathrm{A}\frac{\mathrm{d}^2C_\mathrm{A}}{\mathrm{d}y^2}-k_2C_\mathrm{C}C_\mathrm{A}=0; \tag{5.3.35e}$$

$$D_\mathrm{C}\frac{\mathrm{d}^2C_\mathrm{C}}{\mathrm{d}y^2}-k_2C_\mathrm{C}C_\mathrm{A}=0. \tag{5.3.35f}$$

At $y=\delta'_\ell$,

$$C_\mathrm{A}=C'_{\mathrm{A}\ell},\; C_\mathrm{C}=0. \tag{5.3.35g}$$

At $y=\delta_\ell$,

$$C_\mathrm{A}=0,\; C_\mathrm{C}=C_{\mathrm{C}b}. \tag{5.3.35h}$$

The film thickness δ_ℓ is related to $k_{\ell,\mathrm{CO_2}}$ as before by

$$k_{\ell,\mathrm{CO_2}}=(\mathrm{D_A}/\delta_\ell). \tag{5.3.35i}$$

Define the following nondimensional quantities:

$$C_\mathrm{A}^*=C_\mathrm{A}/C_{\mathrm{A}i};\quad C_\mathrm{B}^*=C_\mathrm{B}/C_{\mathrm{B}i};\quad C_\mathrm{C}^*=C_\mathrm{C}/C_{\mathrm{C}b};$$
$$y^*=y/\delta_\ell;\quad y^{*\prime}=\delta'_\ell/\delta_\ell;\quad \gamma_{R_2}=\frac{k_2C_{\mathrm{C}b}D_\mathrm{A}}{k^2_{\ell,\mathrm{CO_2}}}. \tag{5.3.35j}$$

Equations (5.3.35a) to (5.3.35h) can now be nondimensionalized to equations (5.3.36a) to (5.3.36h). For region I:

$$\frac{\mathrm{d}^2C_\mathrm{A}^*}{\mathrm{d}y^{*2}}=0; \tag{5.3.36a}$$

$$\frac{\mathrm{d}^2C_\mathrm{B}^*}{\mathrm{d}y^{*2}}=0. \tag{5.3.36b}$$

At $y^*=0$,

$$C_\mathrm{A}^*=1,\qquad C_\mathrm{B}^*=1. \tag{5.3.36c}$$

At $y^*=y^{*\prime}$,

$$C_\mathrm{A}^*=C'_{\mathrm{A}\ell}/C_{\mathrm{A}i}=C_\mathrm{A}^*(y^{*\prime}),\qquad C_\mathrm{B}^*=0. \tag{5.3.36d}$$

For region II:

$$\frac{\mathrm{d}^2C_\mathrm{A}^*}{\mathrm{d}y^{*2}}=\gamma_{R_2}C_\mathrm{A}^*C_\mathrm{C}^*; \tag{5.3.36e}$$

$$\frac{\mathrm{d}^2C_\mathrm{C}^*}{\mathrm{d}y^{*2}}=\gamma_{R_2}C_\mathrm{A}^*C_\mathrm{C}^*\left(\frac{C_{\mathrm{A}i}}{C_{\mathrm{C}b}}\right)\left(\frac{D_\mathrm{A}}{D_\mathrm{C}}\right). \tag{5.3.36f}$$

At $y^*=y^{*\prime}$,

$$C_\mathrm{A}^*=C_\mathrm{A}^*(y^{*\prime}),\qquad C_\mathrm{C}^*=0. \tag{5.3.36g}$$

At $y^*=1$,

$$C_\mathrm{A}^*=0,\qquad C_\mathrm{C}^*=1. \tag{5.3.36h}$$

The problem is not completely specified unless $y^{*\prime}$ is known. An additional relation needs to be developed at $y^{*\prime}$ based on the instantaneous reaction between H_2S and MDEA (corresponding to equation (5.3.31)):

$$-D_B \frac{dC_B}{dy}\bigg|_{y=\delta'_\ell} = N^r_B\bigg|_{y=\delta'_\ell} = -N^r_C\bigg|_{y=\delta'_\ell} = D_C \frac{dC_C}{dy}\bigg|_{y=\delta'_\ell},$$

which reduces to

$$\frac{dC^*_C}{dy^*}\bigg|_{y^{*\prime}} = -\left(\frac{C_{Bi}}{C_{Cb}}\right)\left(\frac{D_B}{D_C}\right)\frac{dC^*_B}{dy^*}\bigg|_{y^{*\prime}}. \qquad (5.3.36i)$$

Equation (5.3.36a), with the appropriate boundary conditions at $y^* = 0$ and $y^* = y^{*\prime}$, yields

$$C^*_A = \frac{C^*_A(y^{*\prime}) - 1}{y^{*\prime}} y^{*\prime} + 1. \qquad (5.3.36j)$$

Equation (5.3.36b) and the boundary conditions at $y^* = 0$ and $y^* = y^{*\prime}$ generate

$$C^*_B = -\frac{y^*}{y^{*\prime}} + 1. \qquad (5.3.36k)$$

Equation (5.3.36f) and the boundary conditions at $y^* = y^{*\prime}$ and $y^* = 1$ lead to

$$\left(\frac{C_{Ai}}{C_{Cb}}\right)\frac{dC^*_A}{dy^*} - \left(\frac{D_C}{D_A}\right)\frac{dC^*_C}{dy^*} = \frac{\left[\frac{D_C}{D_A} + \frac{C_{Ai}}{C_{Cb}} C^*_A(y^{*\prime})\right]}{1 - y^{*\prime}}. \qquad (5.3.36l)$$

Consider now the location $y^{*\prime}$ only. Obtain (dC^*_A/dy^*) from (5.3.36j). Express (dC^*_C/dy^*) in terms of $(dC^*_B/dy^*)_{y^{*\prime}}$ using (5.3.36i) and then use (5.3.36k). Simplify and obtain the following relation:

$$C^*_A(y^{*\prime}) = 1 + \frac{C_{Bi} D_B}{C_{Ai} D_A} - \left(\frac{\frac{D_B}{D_A} + \frac{C_{Ai}}{C_{Cb}} + \frac{D_B}{D_A}\frac{C_{Bi}}{C_{Ci}}}{C_{Ai}/C_{Cb}}\right) y^{*\prime}. \qquad (5.3.36m)$$

This is the additional relation needed: unknown $y^{*\prime}$ is related to $C^*_A(y^{*\prime})$. By assuming either $y^{*\prime}$ or $C^*_A(y^{*\prime})$, one can now integrate the governing equations numerically till the boundary conditions (5.3.36h) are matched. Goettler and Pigford (1971) developed relation (5.3.36m). Numerical approaches have been studied by Aiken (1982) and Haimour *et al.* (1987).

The enhancement factors ϕ_A and ϕ_B for species A and B (definition (5.3.14)) may be written as

$$\phi_A = \frac{N^r_A\big|_{y=0}}{N_A\big|_{y=0}} = \frac{-D_A \frac{dC_A}{dy}\Big|^r_{y=0}}{-D_A \frac{dC_A}{dy}\Big|_{y=0}} = \frac{-\frac{dC_A}{dy}\Big|^r_{y=0}}{\frac{C_{Ai}}{\delta_\ell}} = -\frac{dC^*_A}{dy^*}\bigg|_{y^*=0}$$

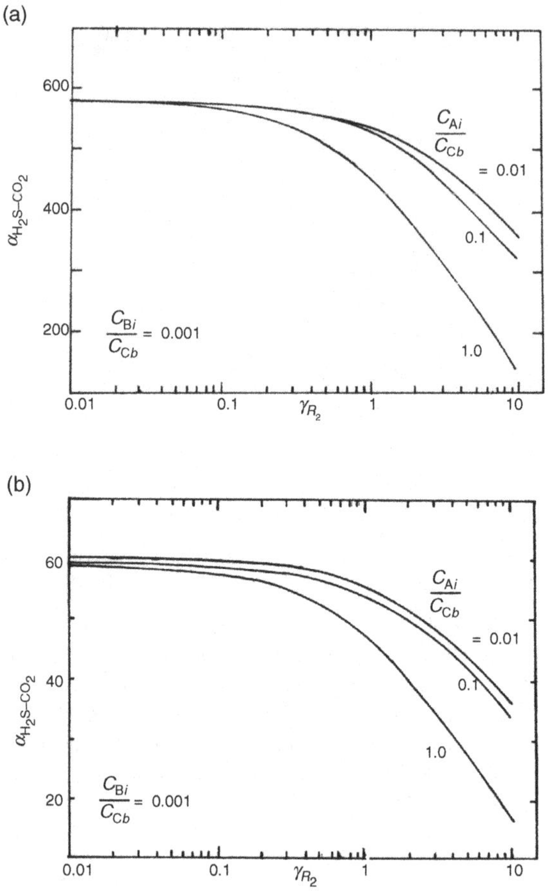

Figure 5.3.3. (a) H_2S-CO_2 selectivity as a function of γ_{R_2} and different concentration ratios. (b) Selectivity as a function of γ_{R_2} and different concentration ratios. (After Haimour et al. *(1987).)*

$$= \frac{1 - C^*_A(y^{*\prime})}{y^{*\prime}}; \qquad (5.3.37a)$$

$$\phi_B = -\frac{dC^*_B}{dy^*}\bigg|_{y^*=0} = \frac{1}{y^{*\prime}}. \qquad (5.3.37b)$$

The selectivity of the absorption process for H_2S over CO_2 can be defined for this rate-controlled process (as opposed to the thermodynamic selectivity of (5.2.20)) as

$$\alpha_{H_2S\text{-}CO_2} = \frac{\phi_B}{\phi_A} = \frac{1}{1 - C^*_A(y^{*\prime})}. \qquad (5.3.37c)$$

Haimour *et al.* (1987) have calculated values of $\alpha_{H_2S-CO_2}$ for different values of the parameters (C_{Bi}/C_{Cb}) and (C_{Ai}/C_{Cb}) against γ_{R2}. A set of such results is shown in Figures 5.3.3(a) and (b); they demonstrate how the differences in the rates of absorption can lead to a very high selectivity in chemically reactive systems even though equilibrium conditions do not provide that high a selectivity (see (5.2.30)).

The results of calculations shown in these figures further indicate that selectivity decreases when either the H_2S content, the CO_2 content or both increase for a given level of MDEA in the solution. In fact, results not shown here indicate that $\alpha_{H_2S-CO_2}$ can easily come to around 3, the thermodynamic limit (see (5.2.30)), for high values of γ_{R2} and high levels of H_2S and CO_2 vis-à-vis MDEA in the solution. High selectivity for H_2S over CO_2 is desirable[8] in industrial gas cleanup processes so that gas obtained from regenerating the spent absorbent solution can be used for producing sulfur by the Clauss process.

5.3.2 Solvent extraction of a species with chemical reaction

We have considered in Section 5.2.2 the nature of equilibrium relationships in solvent extraction of a species in the presence of a chemical reaction in either phase or at the interface.

The rate at which such processes occur depends on a variety of factors. These include: the diffusional rates of the relevant species in the respective phases, the rates of bulk reactions, and the rates of interfacial reactions, if any. Consider the following reaction, where we assume that the reaction takes place on the aqueous side:

$$A_{aq}+B_{org} \Leftrightarrow C_{org}+D_{aq}. \quad (5.3.38)$$

An example would be metal extraction by chelation (see (5.2.92)) with perhaps an interfacial reaction. On the other hand, the reaction

$$A_{aq} + B_{org} \Leftrightarrow C_{org}, \quad (5.3.39)$$

where we assume that the reaction taking place in the organic phase is more typical of complexation reactions of carboxylic acids in the organic phase (see (5.2.82)). In each case, one of the reactants, whose characteristic bulk phase is not the phase where the reaction takes place, has to diffuse to the aqueous–organic interface and partition before reaction can occur.

However, reaction can occur in either phase, contrary to gas absorption processes where reaction almost always takes place only in the liquid phase. Further, in metal extraction processes, interfacial reactions are common and they *may* proceed at a slow rate, unlike instantaneous gas absorption reaction at the gas–liquid interface at high chemical reactant concentrations ($\geq C'_{Cb}$; see (5.3.28)).

[8]Hindered amines (Sartori *et al.*, 1987) are particularly useful here. The reaction rate of CO_2 decreases with an increase in the steric hindrance of amines; k_2 for an unhindered amine, dimethylamine, is 7.9×10^5 (liter/m-s), whereas k_2 for a severely hindered amine is $\sim 1.3 \times 10^2$. But, unlike CO_2, H_2S reacts with an amine at a rate unaffected by steric hindrance since the reaction is a simple protonation.

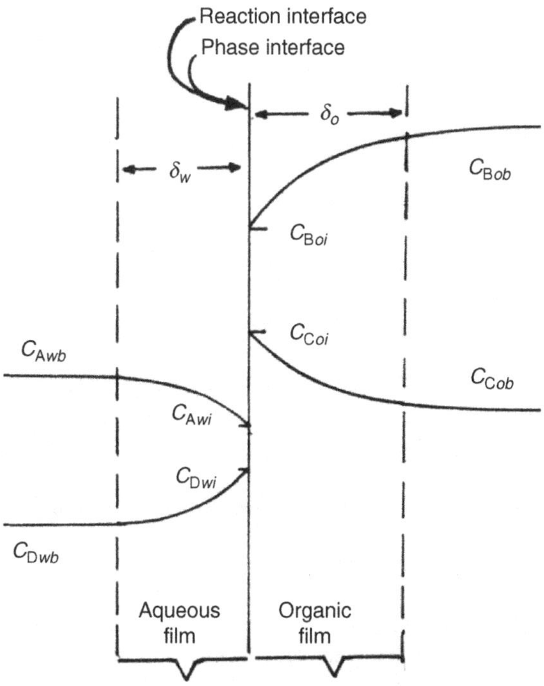

Figure 5.3.4. Concentration profiles in aqueous–organic extraction with an interfacial reaction $A_{aq} + B_{org} \rightleftharpoons C_{org} + D_{aq}$.

We first consider the rates of transport of a solute species being extracted in the presence of a chemical reaction in the interfacial region, the overall reaction being (5.3.38). A schematic is shown in Figure 5.3.4, where the phase interface, shown as a line, actually has a finite thickness. The solute A from aqueous phase bulk (say, $M^{n+}(aq)$) diffuses to the interfacial region, reacts with species B from the organic phase, and the products of the reaction, C_{org} and D_{aq}, diffuse, respectively, into the organic and aqueous phases. At steady state,

$$\begin{aligned} N_A &= k_{Aw}(C_{Awb} - C_{Awi}) = k_{s^+}(C_{Awi}C_{Boi}) - k_{s^-}(C_{Coi}C_{Dwi}) \\ &= k_{Bo}(C_{Bob} - C_{Boi}) = k_{Co}(C_{Coi} - C_{Cob}), \end{aligned} \quad (5.3.40)$$

where the subscripts o and w refer to the organic and aqueous phase, b and i refer to the bulk and the interface, k_{s^+} and k_{s^-} are the rate constants for the forward and backward interfacial reactions, and the diffusional resistance of species D in aqueous phase (say, the H^+ ion) is neglected. For irreversible reactions, the term k_{s^-} (C_{Coi} C_{Dwi}) can also be neglected.

If we ignore the diffusion of reaction products, there are three resistances of importance: the aqueous-phase diffusional resistance of species A, the organic-phase diffusional resistance of species B, and the resistance of the interfacial reaction (Baird, 1980). If the interfacial reaction rate is extremely slow, then $C_{Awb} \cong C_{Awi}$, $C_{Boi} = C_{Bob}$ and

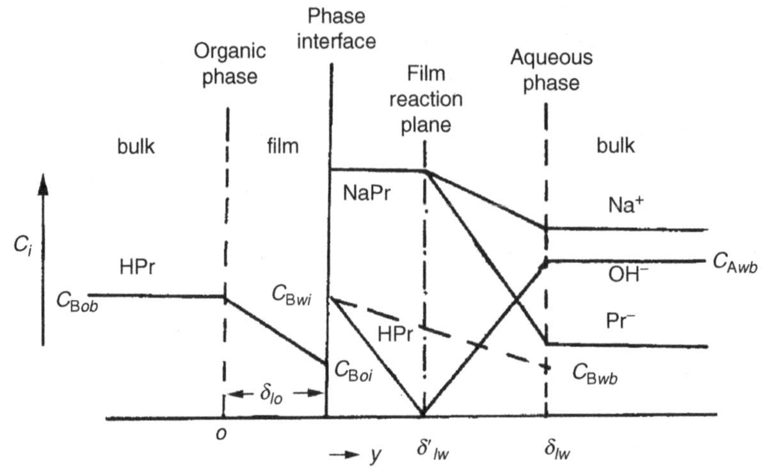

Figure 5.3.5. Extraction and instantaneous reaction of an organic acid B (HPr) in aqueous phase with a base A: film theory representation. (After Grosjeans and Sawistowki (1980).)

$N_A = k_{s+}\ C_{Awb}\ C_{Bob}$; the interfacial reaction controls the extraction rate. If the interfacial reaction is extremely fast, the diffusion rates become controlling. For example, if the diffusional rate of A is slow, then $C_{Awi} = 0$ and $N_A = k_A\ C_{Awb}$ (assuming $k_{Bo}(C_{Bob} - C_{Boi}) > k_{Aw}\ (C_{Awb} - C_{Awi})$). Conversely, if the diffusion of B is quite slow, $C_{Boi} = 0$, leading to $N_A = k_{Bo}\ C_{Bob}$.

We next examine cases where the reaction takes place in a region away from the interface, e.g. in the liquid film or the bulk liquid. For this purpose, the overall reaction can be (5.3.39) with the reaction taking place in the organic phase after species A has been extracted into the organic phase. A number of situations are possible.

(1) No reaction in the organic film; slow reaction in the organic bulk.
(2) Slow or fast reaction in the organic film.
(3) Instantaneous reaction in the organic film, which, under extreme conditions, can push the reaction to the interface.

Each case can be studied in a fashion almost identical to those for gas-liquid absorption (see Section 5.3.1). For further details, consult Astarita (1967), Lo *et al.* (1983), Baird (1980) and Sarkar *et al.* (1980). Similar considerations can be applied if the reaction is

$$A_{aq} + B_{org} \Leftrightarrow C_{aq} + D_{aq} \qquad (5.3.41)$$

and it takes place in the aqueous phase; consider, for example, removal of COS from liquefied C_3-C_4 fractions by treatment with aqueous NaOH or alkanolamine solutions (Dewitt, 1980), or transfer of propionic acid from toluene into aqueous NaOH (Grosjean and Sawistowski, 1980).

The back extraction of propionic acid (C_2H_5COOH, represented here as HPr) present in toluene into an aqueous solution containing NaOH will serve to illustrate liquid-liquid extraction with chemical reaction of the type (5.3.41) with two differences. First,

$$C_2H_5COOH(org) + NaOH(aq) \rightarrow C_2H_5COONa(aq) + H_2O(aq) \qquad (5.3.42)$$

is irreversible; second, the aqueous phase contains, in addition, ions Na^+, OH^-, Pr^-, as shown in Figure 5.3.5. Propionic acid partitions into the aqueous phase and diffuses to the reaction interface at a distance $\delta'_{\ell w}$ from the phase interface, where its concentration is zero due to the instantaneous reaction with the OH^- ion. The latter diffuses from the bulk aqueous phase to the interface. Note, however, that not only in the bulk aqueous phase, but also in the aqueous film region $(\delta_{\ell w} - \delta'_{\ell w})$, electroneutrality has to be maintained between the three ions.

Figure 5.3.5 shows a profile (see the dashed line) of propionic acid (species B) in the aqueous-phase film in the absence of any reaction. The overall mass-transfer coefficient K_o based on the organic phase in such a case is

$$N_B = K_o(C_{Bob} - C_B^*) = k_o(C_{Bob} - C_{Boi}) = k_w(C_{Bwi} - C_{Bwb})$$

and

$$\frac{1}{K_o} = \frac{1}{k_o} + \frac{\kappa''_{Bo}}{k_w}, \qquad (5.3.43)$$

where the distribution coefficient for species B, κ''_{Bo}, is defined by

$$\kappa''_{Bo} = \frac{C_B^*}{C_{Bwb}} = \frac{C_{Boi}}{C_{Bwi}}. \qquad (5.3.44)$$

Note that the film transfer coefficients k_o and k_w are defined by

$$k_o = \frac{D_{Bo}}{\delta_{\ell o}}; \qquad k_w = \frac{D_{Bw}}{\delta_{\ell w}}. \tag{5.3.45}$$

The profile of propionic acid in the case of an instantaneous reaction with caustic soda (species A) is shown by the straight line in the aqueous phase. The reaction plane is located at a distance $\delta'_{\ell w}$ from the phase interface. At this location,

$$N^r_B|_{y=\delta'_{\ell w}} = -N^r_A|_{y=\delta'_{\ell w}}. \tag{5.3.46}$$

For linear concentration profiles in each phase,

$$\frac{D_{BW} C_{Bwi}}{\delta'_{\ell w}} = \frac{D_{Aw} C_{Awb}}{(\delta_{\ell w} - \delta'_{\ell w})}, \tag{5.3.47}$$

since $C_{Bw}|_{y=\delta'_{\ell w}} = 0 = C_{Aw}|_{y=\delta'_{\ell w}}$. This leads to

$$\delta'_{\ell w} = \frac{\delta_{\ell w}}{\left[1 + \frac{D_{Aw}}{D_{Bw}}\frac{C_{Awb}}{C_{Bwi}}\right]}. \tag{5.3.48}$$

Therefore,

$$N^r_B|_{y=0} = N^r_B|_{y=\delta'_{\ell w}} = \frac{D_{Bw}}{\delta'_{\ell w}} C_{Bwi} = \frac{D_{Bw}}{\delta_{\ell w}}\left[C_{Bwi} + \frac{D_{Aw}}{D_{Bw}} C_{Awb}\right]$$

$$= k_w \left[C_{Bwi} + \frac{D_{Aw}}{D_{Bw}} C_{Awb}\right]. \tag{5.3.49a}$$

However, in the organic phase,

$$N^r_B|_{y=0} = k_o[C_{Bob} - C_{Boi}]. \tag{5.3.49b}$$

These two expressions, along with (5.3.43) and (5.3.44) lead to the following flux expression for the propionic acid being extracted:

$$N^r_B|_{y=0} = K_o\left[C_{Bob} + \frac{D_{Aw}}{D_{Bw}} \kappa''_{Bo} C_{Awb}\right], \tag{5.3.50}$$

where K_o is the overall mass-transfer coefficient (5.3.43) for the no reaction case. The extraction enhancement factor ϕ for $C_{Bwb} = 0$ is

$$\phi = \frac{N^r_B|_{y=0}}{N_B|_{y=0}} = \frac{K_o\left[C_{Bob} + \frac{D_{Aw}}{D_{Bw}} \kappa''_{Bo} C_{Awb}\right]}{K_o C_{Bob}}$$

$$= \left[1 + \frac{D_{Aw}}{D_{Bw}} \kappa''_{Bo} \frac{C_{Awb}}{C_{Bob}}\right]. \tag{5.3.51}$$

As C_{Awb} increases, ϕ increases, reaching a critical C_{Awb}, C'_{Awb}; the maximum enhancement is then obtained as the reaction interface coincides with the phase interface:

$$N^r_B|_{y=0} = -N^r_A|_{y=0} = \frac{D_{Aw} C'_{Awb}}{\delta_{\ell w}} = k_o C_{Bob} \tag{5.3.52}$$

since C_{Boi} and C_{Awi} are zero. This critical concentration, C'_{Awb}, is given by

$$C'_{Awb} = \left(\frac{k_o}{k_w}\right)\left(\frac{D_{Bw}}{D_{Aw}}\right) C_{Bob}. \tag{5.3.53}$$

The enhancement factor under this condition is

$$\phi_{max} = \left[1 + \frac{k_o}{k_w} \kappa''_{Bo}\right]. \tag{5.3.54}$$

Grosjean and Sawistowski (1980) have obtained a ϕ_{max} value of up to 3.17 for propionic acid extraction from toluene and its neutralization by NaOH.

It may appear from relation (5.3.51) that increasing the concentration of caustic soda, species A, is uniformly beneficial to the rate of extraction. Consider the extractive hydrolytic removal of COS from liquid hydrocarbons (e.g. propylene at a high pressure and ambient temperature) by contacting with a caustic soda solution (Dewitt, 1980). The rate of hydrolysis of COS by caustic soda in an aqueous solution by the reaction

$$COS + OH^- \rightarrow HCSO_2^- \tag{5.3.55}$$

has been modeled as an irreversible second-order reaction:

$$R_{COS} = -k_2 C_{COS,w} C_{OH^-,w}, \tag{5.3.56}$$

where k_2 is the second-order reaction rate constant. The overall reaction can be written as

$$COS + 4NaOH \rightarrow Na_2CO_3 + Na_2S + 2H_2O. \tag{5.3.57}$$

From the hydrocarbon phase distributed in the aqueous phase as droplets, COS will be partitioned into the aqueous phase, and then a diffusion limited fast reaction will take place in the aqueous boundary layer. Since the rate of hydrolysis of COS in the aqueous phase by caustic soda depends also on the aqueous phase COS concentration, $C_{COS,w}$, increasing $C_{OH^-,w}$ by increasing caustic soda concentration can lead to a reduction in R_{COS} by reducing $C_{COS,w}$. An increase in the electrolyte (NaOH) concentration will reduce the COS solubility according to the Setchenow (1892) relation. Further, COS diffusivity will also be reduced as the caustic soda concentration is increased. Dewitt (1980) has thereby demonstrated experimentally and theoretically that the COS extraction rate is maximum at an intermediate NaOH concentration.

Since the solubility of COS into an aqueous alkaline solution has some importance here, and the reaction takes place only in the aqueous phase, Bhave and Sharma (1983) have introduced species such as tricapryl methyl ammonium chloride (Aliquat 336) into the organic phase. This species acts as a phase-transfer catalyst, transferring OH^- ions into the organic phase near the interface for reaction with COS. Thus COS reacts both in the organic and the aqueous phase, and the rate of COS extraction is increased considerably.

Conventionally, solvent extraction with chemical reaction is implemented in a dispersive system where one liquid phase is dispersed as drops in the other immiscible liquid phase. Such a separation is also implemented using a porous membrane as a phase barrier (see Figure 3.4.11); the immiscible phase interface is immobilized at the pore mouth of, for example, a solvent-resistant hydrophobic microporous/porous membrane. Solvent extraction or back extraction through such an interface is easily implemented without dispersing one phase in the other. Applications of solvent extraction with chemical reaction in such a nondispersive format have been studied for a variety of systems.

Back extraction of phenol from methyl isobutyl ketone into an aqueous caustic solution has been studied using a porous membrane, which can be either hydrophobic (with the organic phase in pores) or hydrophilic (with the aqueous caustic phase in pores) (Basu *et al.*, 1990). Copper extraction by a chelating extractant (HK) in an organic diluent via

$$Cu^{2+}(aq) + 2HK(o) \Leftrightarrow CuK_2(o) + 2H^+(aq) \qquad (5.3.58a)$$

was experimentally studied and modeled by Yun *et al.* (1993). These investigators also demonstrated extraction of Cr^{6+} present as an anion $HCrO_4^-$ in an aqueous solution ($pH = 2.5$) by a long-chain alkyl amine (R_3N) via ion-pair formation through the porous membrane:

$$HCrO_4^-(aq) + H^+(aq) + n[R_3N(o)] \Leftrightarrow \text{complex }(o). \qquad (5.3.58b)$$

As the extraction of a metallic cation (e.g. Cu^{2+}, Ni^{2+}, Zn^{2+}, etc.) proceeds via reactions like (5.3.58a), the protons released into the aqueous solution will decrease the pH. From Figure 5.2.8(b), it is obvious that the distribution coefficient κ''_{mo} for the metallic cation decreases sharply as the pH decreases. Consider the situation encountered in, say, Cr^{6+} extraction via (5.3.58b); as the pH decreases, the extraction ability improves. If, therefore, both processes of (5.3.58a) and (5.3.58b) can be implemented simultaneously in the system, the pH of the aqueous phase may be controlled better. This was implemented in one device in Yang *et al.* (1996a) by using two separate porous hollow fiber membranes containing two different extractants, one containing a chelating extractant for copper and the other containing a long-chain alkylamine for chromium. The aqueous solution contained both copper and chromium; such situations are encountered often in many aqueous waste streams. Two different extract streams were obtained, one containing copper and the other containing chromium. It is possible to have an organic diluent containing both a chelating extractant as well as a long-chain alkyl amine and extract both cation and anion species simultaneously into the diluent (Yang *et al.*, 1996b).

5.4 Rate-governed membrane processes: role of chemical reactions

Separation by preferential transport of one species over another through a membrane can be influenced by chemical reactions. The reaction may take place in any one of the following locations: membrane–feed fluid interface, membrane–product fluid interface, membrane bulk, bulk feed fluid and/or bulk product fluid. Reactions may take place simultaneously at more than one of the above locations. The reaction may be instantaneous, irreversible, reversible, fast, slow, ion pairing, chelating, solute–micelle, etc. We consider various membrane processes in the following order: reverse osmosis, ultrafiltration, dialysis, liquid membrane separation of gases or liquid solution. At the end, we introduce electrochemical cell based reactions at electrodes and selective transport through ion exchange membranes for the purpose of gas separation; we also briefly indicate the role of chemical reactions in separation through a mixed conducting ceramic membrane.

5.4.1 Reverse osmosis: solute ionization

The pH of the aqueous solution will determine the extent of ionization of solutes like phenol, 2-chlorophenol or acetic acid in the solution. If such a solute is ionized, the separation of such a solute from water by reverse osmosis through a membrane will be changed since the ionized solute and the nonionized solute have different solute rejections through the membrane. Consider an ionizable acidic solute species 1; ionization leads to a negatively charged species 2 (see equation (5.2.57)). The total feed solute concentration, C_{1f}^t, is then given by

$$C_{1f}^t = C_{1f} + C_{2f}, \qquad (5.4.1)$$

where the subscript f refers to the feed solution. Similarly, the total solute concentration in the permeate, C_{1p}^t, is given by

$$C_{1p}^t = C_{1p} + C_{2p}, \qquad (5.4.2)$$

with subscript p referring to the permeate.

We assume now that, other conditions remaining constant, the intrinsic solute rejections of the ionized and unionized species remain independent of pH. Let the solute rejection of the unionized solute 1 be R_1 and that of the solute 2 be R_2. Such values for phenol can be determined, for example, by making two measurements with a reverse osmosis membrane (say, the FT-30 membrane manufactured by Filmtec Inc., Minnetonka, MN); at a low pH, phenol is undissociated and so R_1 is obtained, while, at a very high pH, all phenol is dissociated and present as $C_6H_5O^-$ (or, say, as sodium phenolate in caustic solution), yielding R_2.

From the ionization constant K_{d1} of solute 1 (see (5.2.61a))

$$C_{2f} = K_{d1} C_{1f} / C_{H^+f}. \tag{5.4.3}$$

The fraction of this solute present in the feed in the ionized form is given by

$$f_2 = \frac{C_{2f}}{C_{2f} + C_{1f}} = \frac{1}{1 + (C_{H^+f}/K_{d1})}. \tag{5.4.4}$$

Using the definition of $R_2 = (1 - [C_{2p}/C_{2f}])$, we obtain

$$C_{2p} = C_{1f}^t f_2 (1 - R_2) = C_{1f}^t \frac{(1 - R_2)}{1 + \{C_{H^+f}/K_{d1}\}}. \tag{5.4.5}$$

Similarly, the permeate concentration of the unionized solute C_{1p} can be obtained using the definition of intrinsic rejection R_1 of the unionized species 1,

$$R_1 = (1 - [C_{1p}/C_{1f}]), \tag{5.4.6}$$

as

$$C_{1p} = C_{1f}^t (1 - f_2)(1 - R_1) = C_{1f}^t (1 - R_1) \frac{C_{H^+f}/K_{d1}}{1 + \{C_{H^+f}/K_{d1}\}}. \tag{5.4.7}$$

The solute rejection under partially ionized condition,

$$R = 1 - \left(\frac{C_{2p} + C_{1p}}{C_{1f}^t} \right) = 1 - \frac{(1 - R_2)}{1 + \left\{ \frac{C_{H^+f}}{K_{d1}} \right\}} - \frac{(1 - R_1) \frac{C_{H^+f}}{K_{d1}}}{1 + \left\{ \frac{C_{H^+f}}{K_{d1}} \right\}} = \frac{(R_2 - R_1)}{1 + \{C_{H^+f}/K_{d1}\}} + R_1, \tag{5.4.8}$$

where the feed solution $pH = -\log_{10} C_{H^+f}$. When $C_{H^+f} \ll K_{d1}$ at very high feed pH, and all solute is present as ionized species, we see that $R \cong R_2$. When $C_{H^+f} \gg K_{d1}$ at a low pH, and there is virtually no ionization, $R \cong R_1$. Bhattacharyya *et al.* (1987) have provided such an analysis and have demonstrated that expression (5.4.8) describes quite well the observed rejection R at any pH between completely ionized and completely unionized conditions for phenol and various chlorophenols (CP) for an FT-30 membrane used in reverse osmosis (Figure 5.4.1). Obviously, the transition from one regime to the other takes place around $pH = pK_{\text{phenol}}$.

The intrinsic solute rejection value for a particular polymer-solute system can sometimes be qualitatively judged by the solubility parameter δ_{sp}. The solubility parameter has contributions from dispersion forces, dipole-dipole forces and hydrogen bonding forces (Hansen, 1969; Hansen and Beerbower, 1971). For many small molecules that are highly polar, hydrogen bonding forces dominate. Phenol is one such species. Figure 5.4.1 shows that, at low pH, the phenol rejection by the FT-30 membrane is low, around 55% due to hydrogen bonding. In fact, in cellulose acetate membranes, phenol is known to be enriched in the

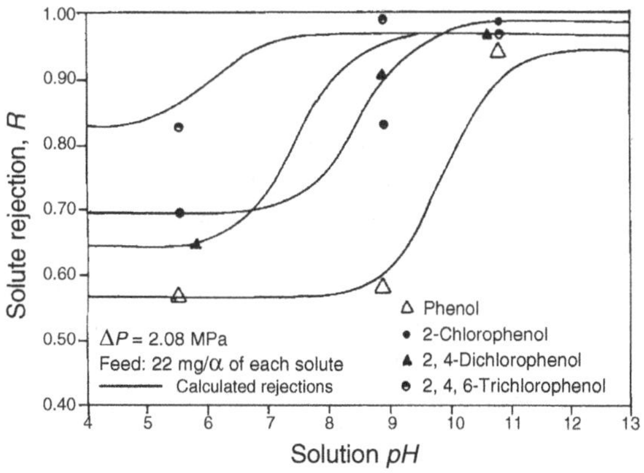

Figure 5.4.1. Comparison of predicted and actual values of solute rejection as a function of pH *for a solution containing 22 mg/liter of chlorophenols. (After Bhattacharyya* et al. *(1987).)*

reverse osmosis permeate (Pusch *et al.*, 1976). However, at high pH (if the membrane can withstand it; the FT-30 membrane can, whereas a cellulose acetate membrane cannot), phenol is ionized, and the reverse osmosis membrane rejection is determined by solute-membrane charge repulsion.

5.4.1.1 *Reverse osmosis: solute complexation*

The rejection behavior of a solute through a reverse osmosis membrane could be significantly altered if the solute species in feed solution forms a chelate or a complex with an added species and the complex dimensions and/or charge are quite different from those of the original solute. Lee *et al.* (1977b) added EDTA (ethylenediaminetetracetic acid) to a feed solution containing $MgCl_2$ and $CaCl_2$ and studied their separation with somewhat open cellulose acetate reverse osmosis membranes. They observed that the rejection of the Mg-EDTA$^=$ complex was around 90%, whereas, without EDTA, the Mg^{2+} rejection was around 50%; this was clearly attributable to a much larger size of the complex. When there was competition for EDTA between magnesium and calcium present in the ratio of 1:1:1, Lee *et al.* (1977b) concluded that almost all of the calcium was present as the complex CaEDTA$^=$, whereas very little of magnesium was complexed ($pH = 6$). Thus the rejection of Mg^{2+} was much lower than that of MgEDTA$^=$ observed without $CaCl_2$ being present.

5.4.2 Ultrafiltration: complexation

Although solute rejection in reverse osmosis can be influenced by the size and shape of the solute, high rejection reverse osmosis depends primarily on membrane-solute interactions in the presence of a preferential solvent interaction with the membrane. On the contrary, solute retention in ultrafiltration (UF) is very largely determined by the dimensions of the solute, especially the macrosolute, with respect to the membrane pore dimensions; sometimes the solute charge vis-à-vis the membrane charge is quite important. However, for a given membrane, if the solute size can be increased, membrane solute retention can be increased. Alternatively, those solutes which are not normally retained by UF membranes can now be held back by the membrane pores if the solutes are attached to other solutes, especially macrosolutes that are retained by the membrane. This phenomenon is illustrated for the following types of systems:

(1) complexation of species to be removed with soluble macroligands rejected by the UF membrane;
(2) complexation of species to be removed with components of a micelle retained by an appropriate UF membrane;
(3) selective binding of a species with soluble proteins retained by the UF membrane.

5.4.2.1 *Complexation/binding with soluble macroligands*

Two different types of systems are briefly illustrated here: complexation of metal ions in solution with soluble macroligands; affinity binding of solutes with water-soluble macroligands. We consider first the example of ultrafiltration of water containing metal ions which have been selectively bound to a water-soluble macromolecular complex. Ordinarily, heavy metals like Cu^{++}, Cd^{++}, Hg^{++} will not be rejected/retained by an UF membrane. But if they are complexed with a macromolecule, the macromolecular complex will have a high retention in the UF membrane (Michaels, 1968a; Strathmann, 1980). The macromolecule should preferably have a high selectivity and a high binding capacity. Such a macromolecule suitable for binding the metal ions is obtained by binding chelating or complexing agents having molecular weights that are not too high with water-soluble high molecular weight polymers with reactive groups. Examples of polymers are polyethylenimine, polyacrylic acid, polyvinyl alcohol, etc.; the chelating agents are illustrated, for example, in Figure 5.4.2.

We will now develop a simple mathematical description of how the rejection of the metal ion M^{n+} through the UF membrane is influenced by complexation with a polymer ligand L^- present as an acid HL form of a polymer. We assume that the free metal ion M^{n+} is not at all rejected by the membrane. Therefore, the concentration of free M^{n+} in the feed solution, $C_{M^{n+},w}$, is the same as in the permeate solution, whereas the effective total concentration of the metal ion (present in free and complexed form) is $C^t_{M^{n+},w}$. Further, the metal complex ML_n is completely rejected by the membrane. Therefore

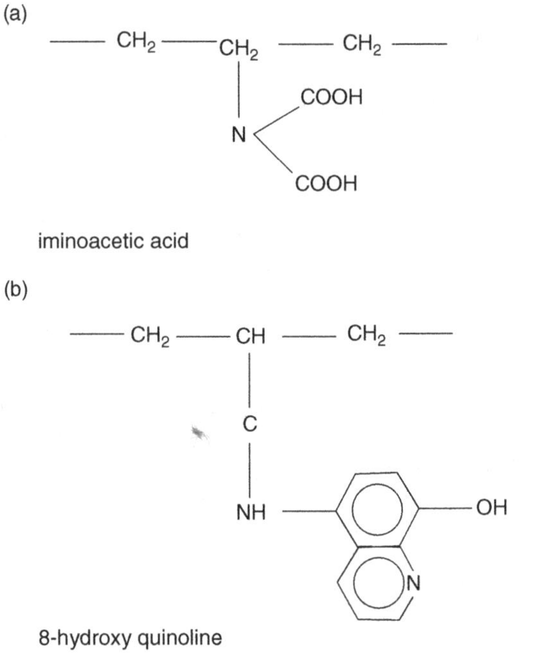

Figure 5.4.2. Examples of complexing agents used to bind metals (shown attached to polymers).

$$R = \text{membrane retention or rejection of metal ion}$$

$$= 1 - \frac{C_{M^{n+},w}}{C^t_{M^{n+},w}} = 1 - \frac{C_{M^{n+}/\text{permeate}}}{C^t_{M^{n+},w}}. \quad (5.4.9)$$

The metal complex ML_n is formed with n negatively charged ligands L^-:

$$M^{n+} + nL^- \Leftrightarrow ML_n. \quad (5.4.10)$$

The aqueous-phase complexation equilibrium constant K_{d,ML_n} for this complexation reaction is

$$K_{d,ML_n} = \frac{C_{ML_n,w}}{C_{M^{n+},w}(C_{L^-,w})^n}. \quad (5.4.11)$$

The total metal ion concentration $C^t_{M^{n+},w}$ present in the feed solution before complexation is

$$C^t_{M^{n+},w} = C_{M^{n+},w} + C_{ML_n,w}. \quad (5.4.12)$$

Using this relation and the definition (5.4.9), we get

$$R = \frac{C_{ML_n,w}}{C^t_{M^{n+},w}}. \quad (5.4.13)$$

From the equilibrium relation (5.4.11), we get

$$K_{d,ML_n} = \frac{RC^t_{M^{n+},w}}{C_{M^{n+},w}(C_{L^-,w})^n} = \frac{R}{(1-R)(C_{L^-,w})^n}. \quad (5.4.14)$$

The ligand L, introduced as an acid HL, is obtained via the following dissociation:

$$HL(w) \Leftrightarrow H^+(w) + L^-(w). \quad (5.4.15)$$

The dissociation constant of this acid, $K_{d,HL}$, is

$$K_{d,\,HL} = \frac{C_{H^+,w} C_{L^-,w}}{C_{HL,w}}. \quad (5.4.16)$$

The sum of the free acid and the dissociated polymer acid is

$$C^T_{HL,w} = C_{HL,w} + C_{L^-,w} = C_{L^-,w}\left[1 + \frac{C_{H^+,w}}{K_{d,HL}}\right]. \quad (5.4.17)$$

However, the total concentration of the polymer acid ligand introduced into the system, $C^t_{HL,w}$, is

$$C^t_{HL,w} = C_{HL,w} + C_{L^-,w} + nC_{ML_n,w}. \quad (5.4.18)$$

Rearrange this to obtain

$$\left(1 - \frac{nC_{ML_n,w}}{C^t_{HL,w}}\right) = \frac{C^T_{HL,w}}{C^t_{HL,w}} = \frac{C_{L^-,w}\left[1 + \frac{C_{H^+,w}}{K_{d,HL}}\right]}{C^t_{HL,w}}. \quad (5.4.19)$$

Introduce from here the expression for $C_{L^-,w}$ into the relation (5.4.14) to obtain

$$K_{d,MLn} = \frac{R\left[1 + \frac{C_{H^+,w}}{K_{d,HL}}\right]^n}{(1-R)\left(C^t_{HL,w}\right)^n\left(1 - \frac{nC_{ML_n,w}}{C^t_{HL,w}}\right)^n}. \quad (5.4.20)$$

Rewrite the expression $\left(1 - nC_{ML_n,w}/C^t_{HL,w}\right)^n$ in the denominator as

$$\left(1 - \frac{n\left(C^t_{M^{n+},w} - C_{M^{n+},w}\right)}{C^t_{HL,w}}\right)^n = \left(1 - \frac{nC^t_{M^{n+},w}}{C^t_{HL,w}} + \frac{nC^t_{M^{n+},w}}{C^t_{HL,w}}\frac{C_{M^{n+},w}}{C^t_{M^{n+},w}}\right)^n$$

$$= \left(1 - \frac{nC^t_{M^{n+},w}}{C^t_{HL,w}}\left(1 - \frac{C_{M^{n+},w}}{C^t_{M^{n+},w}}\right)\right)^n = \left(1 - n\frac{C^t_{M^{n+},w}}{C^t_{HL,w}}R\right)^n.$$

Substitute this result into (5.4.20) to obtain

$$\frac{K_{d,ML_n}}{\left[1 + \frac{C_{H^+,w}}{K_{d,HL}}\right]^n} = \frac{R}{(1-R)}\frac{1}{\left(C^t_{HL,w}\right)^n\left(1 - n\frac{C^t_{M^{n+},w}}{C^t_{HL,w}}R\right)^n}. \quad (5.4.21)$$

Thus, knowing the two equilibrium constants, K_{d,ML_n} and $K_{d,\,HL}$, the *pH* and the total amounts of the metal ion, $C^t_{M^{n+},w}$, and the polymer acid ligand, $C^t_{HL,w}$, added, one can predict the overall effective rejection R of the metal ion by an UF membrane. For a low *pH*, where $C_{H^+,w} >> K_{d,HL}$, $R \to 0$. For a high *pH*, where $C_{H^+,w} << K_{d,HL}$, R tends to a high but constant value. Rumeau *et al.* (1992) have demonstrated the validity of this relation for Ag^{2+}, Cu^{2+}, Ni^{2+} and water-soluble polymeric ligands (50 000 and 100 000 molecular weight) of polyacrylic acid using a 10 000 molecular weight cut off UF membrane of polysulfone. The nature of the dependence of the rejection of the metal ion by the UF membrane on the solution *pH* exhibits some similarities to the solvent extraction of a metal ion by acidic extractants shown in Figure 5.2.8(a). If the values of K_{d,ML_n} are sufficiently different for two metals, selective complexation of one of the metal species at a lower *pH* is also possible.

The preceding analysis was based on the following assumptions: (1) there is only one type of complex ML_n in the solution; (2) there were no metal hydroxides present in the system (it is certainly correct as long as $pH < 7$). In general, however, water-soluble chelating polymers like polyacrylic acid (HL) etc. can form a number of complexes with metal ions. For example, for divalent metal ions such as Cu^{2+}, Ni^{2+}, Co^{2+}, Zn^{2+}, etc., the following types of equilibria have been considered:

$$M^{2+} + HL \overset{K_1}{\longleftrightarrow} LM^+ + H^+; \quad (5.4.22)$$

$$LM^+ + HL \overset{K_2}{\longleftrightarrow} ML_2 + H^+. \quad (5.4.23)$$

The complexation stability constants K_1 and K_2 are

$$K_1 = \frac{C_{LM^+} C_{H^+}}{C_{M^{2+}} C_{HL}}; \qquad K_2 = \frac{C_{ML_2} C_{H^+}}{C_{LM^+} C_{HL}}. \quad (5.4.24)$$

In addition, the following, and other types of, complexes have been found to form (Tomida *et al.*, 2001):

$$ML_2 + n(HL) \overset{K_3}{\longleftrightarrow} L_2M(HL)_n. \quad (5.4.25)$$

If one defines the *coordination number* as the number of ligands L bonding to the central metal atom, then an average coordination number n_{avg} for the three complexation reactions identified above will be (for $n = 2$)

$$n_{avg} = \frac{C_{LM^+} + 2C_{ML_2} + 4C_{L_2M(HL)_2}}{C_{LM^+} + C_{ML_2} + C_{L_2M(HL)_2}}. \quad (5.4.26)$$

Another application of soluble macroligands via complexation is in the application of the removal of smaller biomolecules. For example, Lee (1995) has illustrated how the antibiotic vancomycin (mol. wt. ~1500) binds tightly to a peptide ligand containing D-alanyl-D-alanine: this ligand was coupled to 1-1′-carbonyldiimidazole activated dextran of 500 000 mol. wt., and the complex was rejected by a 20 000 molecular weight cut off (MWCO) UF membrane. The rejection was enhanced when there was substantial concentration polarization (see Section 6.3.3.2) of the dextran on the membrane. Earlier studies illustrated affinity binding of alcohol dehydrogenase (ADH) enzyme to the dye Cibacorn blue, which was bound to starch granules retained by an UF unit having a 500 000 MWCO membrane. This enabled purification of ADH from a crude yeast

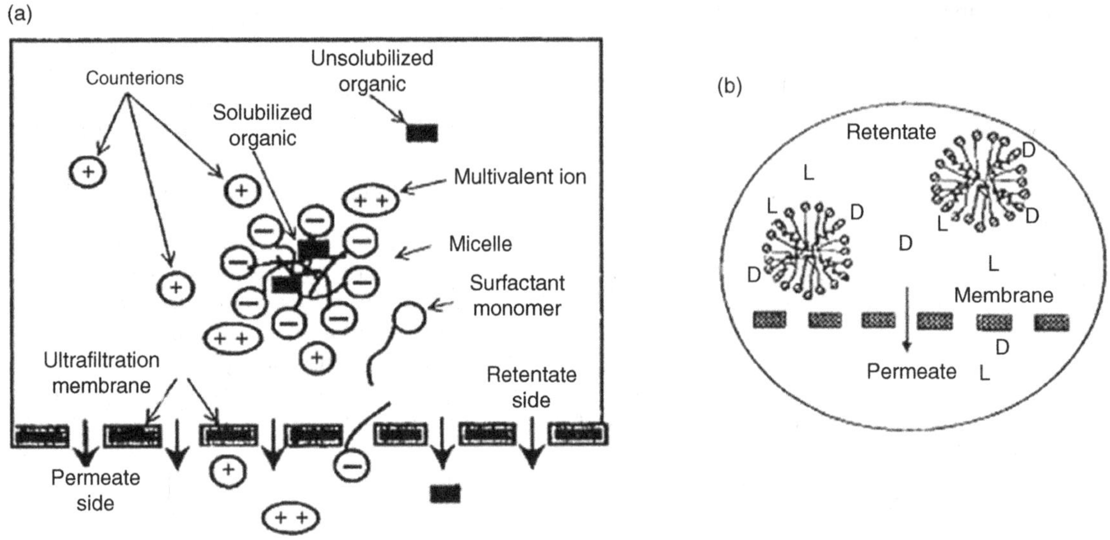

Figure 5.4.3. (a) Micellar-enhanced ultrafiltration (MEUF) for the removal of a target multivalent cationic metal in an aqueous solution which also contains an organic solute that is removed. (After Roberts et al. *(2000).) (b) Enantiomer separation at the UF membrane by an enantioselective micelle. (After Overdevest* et al. *(2000).)*

extract (Mattiasson and Ling, 1986). This example illustrates the purification of an enzyme, a larger molecule, via affinity binding from a solution containing large impurity molecules through a larger-pore UF membrane.

5.4.2.2 *Complexation/binding via a micelle*

Three types of systems will be briefly illustrated here:

(1) binding of multivalent ionic species in water to oppositely charged headgroups of ionic surfactants forming a micelle rejected by an UF membrane;
(2) complexation of metal ions in water with lipophilic complexing agents solubilized in the hydrophobic core of a micelle;
(3) enantioselective micelle preferentially forming a complex with one of the enantiomers.

Micellar systems (introduced in Section 4.1.8) may employ ionic surfactants (Section 4.1.9.2) having an ionic headgroup, for example sodium dodecyl sulfate, where Na^+ is the counterion and there are numerous ionic headgroups DS^- (representing dodecyl sulfate ($C_{12}SO_4^-$)). Multivalent cations in water, e.g. Cu^{2+}, will bind very strongly with these ionic headgroups. These micelles are spheroidal colloidal aggregates containing as many as 50–100 surfactant molecules. One can select an UF membrane which will retain the micelle and, therefore, the multivalent cations which are bound to the charged headgroups of the ionic surfactants. Simultaneously, if there are small organic species dissolved in water, they will become solubilized in the hydrophobic core of the micelles (Leung, 1979). In effect, a very high retention is achieved for multivalent cations (e.g. Cu^{2+}) (Scamehorn *et al.*, 1989; Roberts *et al.*, 2000) (Figure 5.4.3(a)), as well as for small organic molecules in what is called *micellar-enhanced ultrafiltration* (MEUF) (Leung, 1979; Dunn *et al.*, 1985, 1987). Multivalent anions, such as chromate ions (CrO_4^{2-}), can also be removed by using a cationic surfactant, e.g. cetylpyridinium chloride (Christian *et al.* (1988)).

An alternative strategy for removing metal ions by complexation from an aqueous solution involves solubilizing lipophilic metal complexing agents in the hydrophobic core of micelles (Tondre, 2000). A number of different extractants, solubilized in the micelle, for example 8-hydroxyquinoline, Kelex 100, etc. (see Section 5.2.2.4 for these extractants) were used to extract Cu^{2+} successfully. The micelles were retained by an UF membrane. Tondre (2000) has also identified other surfactant based colloidal particles, including vesicles (Section 4.1.8), polymerized micelles, etc., used successfully to the same end.

The two technqiues mentioned above were directed toward removing metallic ions from an aqueous solution by complexation with the constituents of a micelle. The next technique involves resolution of enantiomers using enantioselective micelles, as long as the micelles are retained by an appropriate UF membrane.

For example, Overdevest *et al.* (2000) used a regenerated cellulose membrane (YM3) having a MWCO of 3000 to hold back micelles of the nonionic surfactant, nonylphenylpolyoxyethylene [E10] ether. Chiral selector molecules of cholesteryl-L-glutamate are anchored in the micelle and complex preferentially with the D-form of the enantiomeric mixture of D-phenylalanine and L-phenylalanine present in the feed mixture (schematically illustrated in Figure 5.4.3(b)). It appears that the binding of either enantiomer with the chiral selector could be described by means of a Langmuir iostherm. One can then describe the enantioselectivity between the D and L species via (see equation (3.3.113d)) as follows:

$$\alpha_{DL} = \frac{C_{Dm}/C_{Lm}}{C_{Dw}/C_{Lw}}, \tag{5.4.27}$$

where C_{Dw} and C_{Lw} are the corresponding molar concentrations in the aqueous feed solution, and C_{Dm} and C_{Lm} are the molar concentrations in the micellar phase.

5.4.2.3 *Chiral separation by selective binding with a soluble protein*

Higuchi *et al.* (1993) initated the exploration of the protein bovine serum albumin (BSA) acting as a chirally selective complexing agent for the amino acid tryptophan in an ultrafiltration system. Romero and Zydney (2001) have studied in detail the stereoselective separation of a racemic mixture of D- and L-tryptophan in a solution containing BSA via ultrafiltration. The polyethersulfone UF membrane had a MWCO of 50 000 (sufficient to hold back BSA completely). L-tryptophan preferentially binds with BSA, so less L-tryptophan is present in the filtrate. The observed UF rejection, R_i, of the tryptophan enantiomer i molecule (mol. wt. 240) as such is essentially zero. However, the feed concentration, C_{if}, of the tryptophan enantiomer i is related to total feed concentration, C_{if}^t, by

$$C_{if} = C_{if}^t(1-fr_i), \tag{5.4.28}$$

where fr_i is the fraction of the particular enantiomer i which is bound to BSA and therefore is not available. So the permeate concentration of the particular tryptophan, C_{ip}, is

$$C_{ip} = (1-R_i)C_{if} = (1-R_i)C_{if}^t(1-fr_i). \tag{5.4.29}$$

The selectivity between L-tryptophan ($i = 1$) and D-tryptophan ($i = 2$) will be primarily determined by the relative binding tendencies of L- and D-enantiomers with BSA. For the L-enantiomer, fr_i will be considerably larger than that for the D-enantiomer.

5.4.3 Dialysis: reaction in dialysate

In dialysis (see Section 4.3.1), the rate at which a solute is removed from a feed to the dialyzate through the dialysis membrane is proportional to the concentration difference on two sides of the membrane. If a chemical reaction can be carried out in the dialyzate side between the solute and an added agent such that neither the products nor the agent can diffuse to the feed side, then the solute concentration difference and the solute transfer rate can be increased considerably. For example, phenol from a dilute aqueous waste stream is transferred through a permselective hydrophobic (or oleophilic) membrane to the dialyzate containing caustic soda, where the reaction

$$C_6H_5OH + NaOH \overset{K}{=} C_6H_5O^-Na^+ + H_2O \tag{5.4.30}$$

takes place. The phenolate ion cannot diffuse back to the feed side through the membrane (Klein *et al.*, 1973).

Consider species 1 (e.g. phenol), species 2 (e.g. NaOH) and species 3 (e.g. sodium phenolate) in a batch dialysis system having a feed solution volume V_f and dialyzate solution volume V_d. If the membrane area separating the two solutions is A_m, the molar rate of change of solute 1 in the feed chamber is

$$V_f\frac{dC_{1f}}{dt} = -\frac{Q_{1m}}{\delta_m}A_m(C_{1f} - C_{1d}), \tag{5.4.31}$$

where Q_{1m} is the permeability of solute 1 through the membrane of thickness δ_m. The molar rate of change of solute 1 in the dialyzate chamber is due to permeation from the feed chamber and its disappearance by the reaction (5.4.30):

$$V_d\frac{dC_{1d}}{dt} = \frac{Q_{1m}A_m}{\delta_m}(C_{1f} - C_{1d}) - V_d\frac{dC_{3d}}{dt}. \tag{5.4.32}$$

The membrane is such, that for all $t \geq 0$,

$$C_{2f} = 0 = C_{3f}. \tag{5.4.33a}$$

Further, at $t = 0$, $C_{1f} = C_{1f}^0$,

$$C_{1d} = 0 = C_{3d}. \tag{5.4.33b}$$

Usually, species 2 is present in the dialyzate in large excess; one can assume reaction (5.4.30) to be in equilibrium:

$$K = (C_{3d}/C_{1d}C_{2d}), \tag{5.4.34}$$

with C_{2d} being essentially constant. The two first-order ordinary differential equations can be converted, using the Laplace transform and equation (5.4.34), to

$$s\underline{C_{1f}} - C_{1f}^0 = \frac{Q_{1m}A_m}{\delta_m V_f}\left(\underline{C_{1d}} - \underline{C_{1f}}\right); \tag{5.4.35}$$

$$(1 + KC_{2d})s\underline{C_{1d}} = \frac{Q_{1m}A_m}{\delta_m V_d}\left(\underline{C_{1f}} - \underline{C_{1d}}\right), \tag{5.4.36}$$

where the bar below a quantity indicates a transformed one. Elimination of $\underline{C_{1f}}$ from these two equations yields

$$\underline{C_{1d}} = \frac{\left(C_{1f}^0\frac{A_mQ_{1m}}{V_d\delta_m}\right)}{[1 + KC_{2d}]}\left[\frac{1}{s^2 + Ms}\right], \tag{5.4.37a}$$

where

$$M = \frac{\left[\frac{A_mQ_{1m}}{V_d\delta_m} + \frac{A_mQ_{1m}}{V_f\delta_m}(1 + KC_{2d})\right]}{[1 + KC_{2d}]}. \tag{5.4.37b}$$

Taking the inverse Laplace transform, we obtain

$$C_{1d} = C_{1f}^0 \frac{\left(\frac{A_m Q_{1m}}{V_d \delta_m}\right)}{[1 + KC_{2d}]} \left[\frac{1}{M}\left\{1 - \exp(-Mt)\right\}\right]. \quad (5.4.38)$$

The total amount of phenol transferred from the feed to the dialysate chamber at any time t from $t = 0$, with reaction, is $V_f\left[C_{1f}^0 - C_{1f}(t)\right]^r$. Using equations (5.4.32) and (5.4.34), observe that

$$(1 + KC_{2d})\frac{dC_{1d}}{dt} + \left(\frac{Q_{1m}A_m}{V_d\delta_m}\right)C_{1d} = \left(\frac{Q_{1m}A_m}{V_d\delta_m}\right)C_{1f}.$$

Use (5.4.38) to obtain $C_{1f}(t) = C_{1d}(t) + C_{1f}^0 \exp(-Mt)$. Therefore,

$$V_f\left[C_{1f}^0 - C_{1f}(t)\right]^r = V\left[C_{1f}^0 - C_{1d} - C_{1f}^0\exp(-Mt)\right]$$

$$= V_f C_{1f}^0\left[\frac{1 - \exp(-Mt)}{1 + \frac{V_f}{V_d(1+KC_{2d})}}\right]. \quad (5.4.39)$$

Consider the value of the same quantity, without reaction, $V_f\,[C_{1f}^0 - C_{1f}(t)]$. The ratio of the values with and without reaction, for a very long time, where $\exp(-Mt) \to 0$ is given by

$$\lim_{t\to\infty}\frac{V_f\left[C_{1f}^0 - C_{1f}(t)\right]^r}{V_f\left[C_{1f}^0 - C_{1f}(t)\right]} = \frac{(1+KC_{2d})(V_f+V_d)}{(V_f+V_d\{1+KC_{2d}\})}. \quad (5.4.40)$$

If $V_d << V_f$, this ratio becomes essentially $(1+ KC_{2d})$. For the phenol-caustic system, Klein *et al.* (1973) have calculated K to be around 1.30×10^4. With a caustic solution concentration around, say, 1M, the value of this ratio is large compared to 1. Thus the amount of phenol transferred to the dialyzate in the presence of the reaction is much more than that without any reaction. Klein *et al.* (1973) have recommended several types of membranes made of (for example) silane or ethyl cellulose for this problem. Han *et al.* (2001) have studied such a process using silicone rubber capillary as a polymeric phenol-selective membrane between a complex wastewater and a strongly basic solution. Klein *et al.* (1972) have also illustrated the utility of a similar concept for the removal of a basic solute like aniline from a feed solution by dialysis through a hydrophobic membrane; high concentration of sulfuric acid is used in the dialyzate side to maintain a high value of concentration difference of free aniline between the feed and the dialyzate side.

In hemodialysis, urea and other metabolic wastes are transferred from blood to the dialyzate through the dialysis membrane. By incorporating an enzyme, urease, on the dialysate side, it is possible to convert urea to ammonia. However, it is necessary that the membrane surface facing the dialyzate has a positive charge, so that the toxic NH_4^+ ion cannot diffuse back to the blood.

5.4.4 Chemical reactions in liquid membrane permeation–separation

Figure 5.4.4 illustrates a number of ways by which chemical reactions can influence the transport and separation of a species A by a liquid membrane of thickness δ_m. Figure 5.4.4(a) illustrates the simple permeation of species A through the liquid membrane. The bulk phases on both sides of the liquid membrane could be either gaseous or liquid. Ignore the resistances to transport of A in these two bulk phases for the time being. Of course, the bulk liquid phases should be immiscible with the liquid membrane phase. Figure 5.4.4(b) indicates that if there is a liquid phase on the downstream side, the driving concentration difference between the two sides of the liquid membrane could be increased by reacting permeating species A with an agent E. If the reaction is of the type (5.4.30) and the product AE cannot enter the membrane, then the extent of transport enhancement for a batch permeation process can be estimated from relation (5.4.39).

An example is provided by Cahn and Li (1974), where phenol is removed from a wastewater feed into a receiving phase containing NaOH through an oily liquid membrane. The sodium phenolate is rejected completely by the non-polar oily membrane. When the reaction is instantaneous and the concentration of A is essentially zero at the interface $y = \delta_m$, the flux of species A is simply proportional to the feed concentration of species A, C_{Af}, i.e.

$$N_{Ay} \cong (Q_{Am}/\delta_m)C_{Af} = J_{AZ}^*.$$

Two questions are of interest here. First, phenol (species 1) is the only species permeating through the liquid membrane. Is the concentration of the phenol essentially zero in the caustic-containing receiving phase? Second, if this concentration of phenol is finite (but very small), what is this value so that, when the feed phase has this concentration, permeation will stop? These questions are answered here using calculations available in Cahn and Li (1974).

Phenol is a weak acid and it dissociates in feed water ($j = fw$) or strip water ($j = sw$) as follows:

$$\underset{(i=1)}{C_6H_5OH} = \underset{(i=2)}{C_6H_5O^-} + H^+, \quad (5.4.41a)$$

the dissociation constant being (at 25 °C)

$$K_{d1} = C_2 C_{H^+}/C_1 = 1.28 \times 10^{-10}. \quad (5.4.41b)$$

The ionization product of water, K_w, is

$$K_w = C_{H^+}C_{OH^-} = 10^{-14}. \quad (5.4.41c)$$

Suppose the feed wastewater pH is 7 and the receiving phase contains 10 wt% NaOH (2.773 M).

Electroneutrality in the caustic phase requires

$$C_{Na^+,sw} = C_{OH^-,sw} + C_{2,sw} = 2.773, \quad (5.4.41d)$$

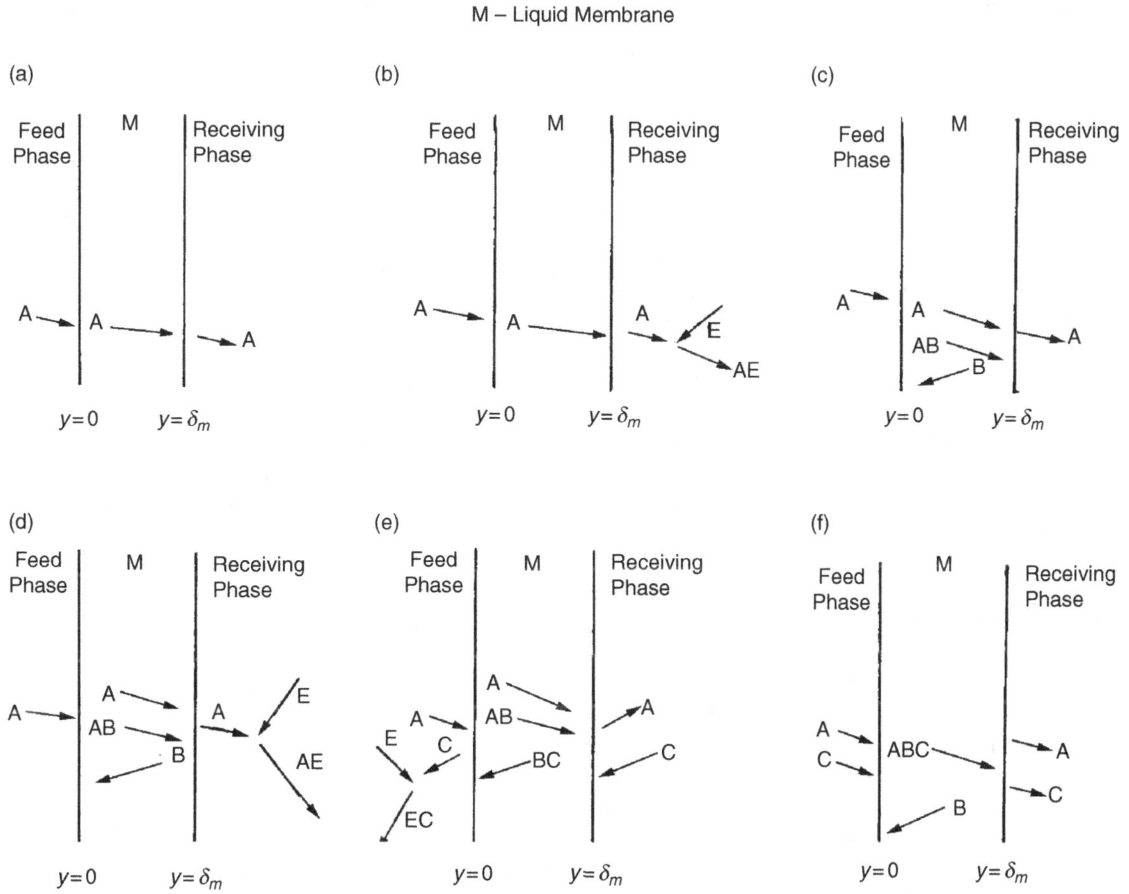

Figure 5.4.4. Various liquid membrane permeation mechanisms. (After Marr and Kopp (1982).) (a) Simple permeation of species A; (b) simple permeation enhanced by reaction of A with an agent E in permeate; (c) facilitated transport with a reversible complexing agent B in the membrane; (d) facilitated transport in the presence of a reactive agent E in permeate; (e) countertransport; (f) cotransport.

the reaction being

$$C_6H_5OH + Na^+ + OH^- = C_6H_5O^-Na^+ + H_2O. \quad (5.4.41e)$$

Therefore

$$C_{OH^-,sw} = 2.773 - C_{2,sw}. \quad (5.4.41f)$$

Substitute this into the ionization product relation (5.4.41c) on the strip water side:

$$C_{H^+,sw} = 10^{-14}/(2.773 - C_{2,sw}). \quad (5.4.41g)$$

To obtain the free phenol concentration in the caustic solution, use (5.4.41b) to obtain

$$C_{1,sw} = \frac{C_{2,sw}C_{H^+,sw}}{K_{d1}} = \frac{C_{2,sw}10^{-4}}{1.28(2.773 - C_{2,sw})}. \quad (5.4.41h)$$

This shows that the total phenol concentration in the caustic phase,

$$C^t_{1,sw} = C_{1,sw} + C_{2,sw} \cong C_{2,sw}, \quad (5.4.41i)$$

is essentially equal to the phenolate ion concentration in the caustic phase; the free phenol concentration is negligible.

In the wastewater phase ($j = fw$), at $pH = 7$ (i.e. $C_{H^+,fw} = 10^{-7}$), obtain from (5.4.41b)

$$C_{2,fw} = 1.28 \times 10^{-10}\frac{C_{1,fw}}{C_{H^+,fw}} = 1.28 \times 10^{-3}C_{1,fw}. \quad (5.4.41j)$$

Therefore, almost all of phenol in the feed phase is present as undissociated phenol:

$$C^t_{1,fw} = C_{1,fw} + C_{2,fw} \cong C_{1,fw}. \quad (5.4.41k)$$

Permeation will stop when

$$C_{1,fw} = C_{1,sw}. \tag{5.4.41l}$$

Use (5.4.41k), (5.4.41h) and (5.4.41i), respectively, to obtain, from (5.4.41l)

$$C^t_{1,fw} - \frac{C^t_{1,sw} \times 10^{-4}}{1.28(2.773 - C^t_{1,sw})}. \tag{5.4.41m}$$

This relation shows that permeation stops when the total phenol concentration in the caustic phase is about 10 000 times that in the feed wastewater. It would be quite useful to calculate how low the phenol concentration in the feed water will become in such a treatment. Cahn and Li (1974) have indicated that a 200 ppm phenol feed will be reduced to 7.4 ppm when 50% of the caustic soda is used up.

5.4.4.1 *Facilitated transport separation*

Figure 5.4.4(c) considers a case where the liquid membrane contains a species B which forms a reversible complex AB with species A in the liquid membrane:

$$A + B \underset{k_b}{\overset{k_f}{\Leftrightarrow}} AB. \tag{5.4.42}$$

Now, not only free species A diffuses through the film (membrane), but also the species AB diffuses from $y = 0$ to $y = \delta_m$. If species AB can exist only within the membrane, then species AB must dissociate at $y = \delta_m$ and release A. Thus the total flux of species A through the membrane is the sum of the flux of free species A and that due to the flux of AB. The flux of species A is *facilitated* by the presence of a complexing agent B (sometimes called the *carrier species*). If other species present in the feed stream do not complex with agent B or AB, their fluxes are not facilitated or increased. This leads to a better separation between species A and other species in the feed stream. *Note*: Species B is chosen such that it does not partition into the feed phase or the receiving phase.

Particular examples of such facilitation are given below. When a gas mixture of an olefin (e.g. ethylene) and paraffin (e.g. ethane) is exposed to an aqueous membrane containing silver nitrate, ethylene complexes with the silver ion (see Section 4.1.9.3):

$$C_2H_4 + Ag^+ + NO_3^- \Leftrightarrow (Ag\text{–}C_2H_4)^+ + NO_3^-, \tag{5.4.43a}$$

while ethane does not. This facilitates the transport of ethylene through the liquid membrane enormously (Hughes *et al.*, 1986). When a mixture of N_2 and CO is exposed to a membrane of an aqueous KCl solution containing cuprous chloride, CO binds reversibly with the ligand complex $Cu(Cl)_3^{-2}$ formed in the KCl solution:

$$CO + Cu(Cl)_3^{-2} \Leftrightarrow CuCO(Cl)_3^{-2}. \tag{5.4.43b}$$

Smith and Quinn (1980) observed that the CO flux was increased by more than 100 times from a 5% CO (rest N_2) feed gas when 0.2M cuprous chloride was present in 1M KCl. Additional reactions were also postulated.

In the above examples, both the feed and the permeate phases were gaseous and aqueous liquid membranes were used. Cahn and Li (1976) have demonstrated facilitated transport separation of the olefin 1-hexene from n-heptane through an aqueous membrane containing cuprous ammonium acetate; the permeate phase was made up of n-octane. Kuo and Gregor (1983) have illustrated facilitated transport of acetic acid from an aqueous solution through an organic membrane of decalin containing trioctylphosphine oxide (TOPO: $(CH_3(CH_2)_7)_3PO$), which complexed with acetic acid.

We will now provide a simplified analysis to determine the rate at which species A is transported through the liquid membrane of thickness δ_m when a complexing agent B present in the membrane facilitates its transport via reaction (5.4.42). The governing differential equation for the species A concentration in the membrane is obtained easily from equation (5.3.6) as

$$D_{Am}\frac{d^2C_{Am}}{dy^2} + R_A = 0, \tag{5.4.44}$$

where R_A, the molar rate of production of A per unit of membrane volume, is

$$R_A = -k_f C_{Am} C_{Bm} + k_b C_{ABm}. \tag{5.4.45}$$

Therefore

$$D_{Am}\frac{d^2C_{Am}}{dy^2} = k_f C_{Am} C_{Bm} - k_b C_{ABm}. \tag{5.4.46}$$

Similarly,

$$D_{Bm}\frac{d^2C_{Bm}}{dy^2} = k_f C_{Am} C_{Bm} - k_b C_{ABm} \tag{5.4.47}$$

and

$$D_{ABm}\frac{d^2C_{ABm}}{dy^2} = -k_f C_{Am} C_{Bm} + k_b C_{ABm}. \tag{5.4.48}$$

The boundary conditions are given by

$$y = 0, \quad C_{Am} = C^0_{Am}, \quad \frac{dC_{ABm}}{dy} = \frac{dC_{Bm}}{dy} = 0; \tag{5.4.49a}$$

$$y = \delta_m, \quad C_{Am} = C^\delta_{Am}, \quad \frac{dC_{ABm}}{dy} = \frac{dC_{Bm}}{dy} = 0. \tag{5.4.49b}$$

The preceding two derivatives imply that neither species B nor the complex AB can leave the system. Further, the total amount of B present in the system is constant at a total concentration of C^t_B:

$$\int_0^{\delta_m} (C_{Bm} + C_{ABm})\mathrm{d}y = C_B^t \delta_m. \tag{5.4.50}$$

Of the four boundary conditions involving gradients, only three are really independent. The total flux of species A at any location in the membrane is given by

$$J_{Ay}^{*t} = -D_{Am}\frac{\mathrm{d}C_{Am}}{\mathrm{d}y} - D_{ABm}\frac{\mathrm{d}C_{ABm}}{\mathrm{d}y} = N_{Ay}^t. \tag{5.4.51}$$

There is no general closed form solution for the concentration profiles of species A, AB and B. Various approximate solutions exist. One limiting solution by Ward (1970) is of interest, namely the *reaction equilibrium limit* achievable either because of *very fast reactions* or *very thick membranes*. Under these conditions, equilibrium exists everywhere in the membrane for reaction (5.4.42):

$$K = (C_{ABm}/C_{Am}C_{Bm}). \tag{5.4.52}$$

Now add equations (5.4.47) and (5.4.48) and obtain, after two integrations, with a_1 and a_2 being integration constants,

$$D_{Bm}C_{Bm} + D_{ABm}C_{ABm} = a_1 y + a_2. \tag{5.4.53}$$

Assume $D_{Bm} = D_{ABm}$ since B is usually a bulky molecule. Then, applying the derivatives' boundary conditions at either $y = 0$ or $y = \delta_m$, we see that $a_1 = 0$. Further substitution in (5.4.50) leads to

$$C_{Bm}^t = \frac{a_2}{D_{Bm}} = C_{Bm} + C_{ABm} = C_B^t. \tag{5.4.54}$$

Therefore,

$$\begin{aligned} C_{ABm} &= KC_{Am}\left[C_{Bm}^t - C_{ABm}\right] \\ \Rightarrow \quad C_{ABm} &= \frac{KC_{Am}C_{Bm}^t}{1 + KC_{Am}}. \end{aligned} \tag{5.4.55}$$

We therefore obtain the total flux of species A through the membrane as

$$N_{Ay}^t = -D_{Am}\frac{\mathrm{d}C_{Am}}{\mathrm{d}y} - \frac{D_{ABm}KC_{Bm}^t}{1}\frac{\mathrm{d}[C_{Am}/(1 + KC_{Am})]}{\mathrm{d}y} \tag{5.4.56}$$

$$N_{Ay}^t = \underbrace{\frac{D_{Am}}{\delta_m}\left(C_{Am}^0 - C_{Am}^\delta\right)}_{\text{flux of free A}} + \underbrace{\frac{D_{ABm}KC_{Bm}^t\left(C_{Am}^0 - C_{Am}^\delta\right)}{\delta_m\left(1 + KC_{Am}^0\right)\left(1 + KC_{Am}^\delta\right)}}_{\text{extra flux due to complexation}}. \tag{5.4.57}$$

Obviously, if species A is more permeable than other species in the feed through the membrane to start with, the addition of species B at a concentration of C_B^t will increase its selectivity, as well as the flux, considerably. The extent of facilitation is expressed by the ratio ψ_{eq} $(= N_{Ay}^t/N_{Ay})$:

$$\psi_{eq} = \left(\frac{N_{Ay}^t}{\frac{D_A}{\delta_m}\left(C_{Am}^0 - C_{Am}^\delta\right)}\right) = 1 + \frac{D_{ABm}KC_B^t}{D_{Am}\left(1 + KC_{Am}^0\right)\left(1 + KC_{Am}^\delta\right)}. \tag{5.4.58}$$

The ideal separation factor between species A and any other species in the feed in the equilibrium limit can also be obtained easily. Equation (5.4.58) is the so-called *thick film* result.

An additional simple approximate analytical solution is available in the so-called "thin film" theory near the diffusion limit when the Damkohler number $\gamma_A = (k_b\delta_m^2/D_A)$ representing the ratio of diffusion time to reaction time approaches zero. Smith *et al.* (1973) have obtained the following estimate of the facilitation ratio by a power series expansion:

$$\psi = 1 + \frac{\gamma_A C_B^* K}{12} + O(\gamma_A^2), \tag{5.4.59a}$$

where

$$C_B^* = \frac{C_{Bm}^t}{1 + K\overline{C}_{Am}}; \tag{5.4.59b}$$

$$\overline{C}_{Am} = \left(C_{Am}^0 + C_{Am}^\delta\right)/2. \tag{5.4.59c}$$

We will now identify some other facilitated transport systems of considerable interest. Separation of CO_2 from inert gases like O_2, N_2, CH_4 has been achieved by using alkaline-concentrated $HCO_3^-/CO_3^=$ solutions:

$$\begin{aligned} CO_2 + H_2O &\Leftrightarrow H^+ + HCO_3^-; \\ CO_2 + OH^- &\Leftrightarrow HCO_3^-. \end{aligned} \tag{5.4.60a}$$

Permeability of CO_2 is enhanced by an additional gradient of HCO_3^- across the film. Simultaneously, permeabilities of O_2, N_2 and CH_4 are reduced by the salting out effect at high concentrations of the added reagent B, say Cs_2CO_3, K_2CO_3, etc. (Ward and Robb, 1967; Bhave and Sirkar, 1987).

Aqueous liquid membranes for gas separation are unstable unless both feed and sweep gas streams are completely humidified. One can, however, use a glycerol based liquid membrane and achieve facilitated transport of CO_2 via reactions (5.4.60a) as long as there is some moisture in the feed stream. Chen *et al.* (1999a) have illustrated considerable facilitation of CO_2 transport and high selectivity for CO_2 over N_2 using different concentrations of Na_2CO_3 in glycerol; the sweep gas flowing was dry; vacuum may be used.

We have seen in Section 5.2.1.1.2 how various amines in water can react with CO_2 and enhance CO_2 absorption. Sodium glycinate has been dissolved in a solution of glycerol to prepare a stable liquid membrane which facilitates the transport of CO_2 via the following reactions in the presence of some moisture in the feed gas stream (Chen *et al.*, 2000):

$$\begin{aligned} CO_2 + RNH_2 &\Leftrightarrow RNHCOO^- + H^+; \\ RNH_2 + H^+ &\Leftrightarrow RNH_3^+. \end{aligned} \tag{5.4.60b}$$

Here, the glycinate ion, designated as RNH_2, is obtained from the dissociation of sodium glycinate:

$$H_2NCH_2COONa \Leftrightarrow H_2NCH_2COO^- + Na^+. \quad (5.4.60c)$$

The transport of H_2S in a K_2CO_3 solution is facilitated by the following reaction (Matson *et al.*, 1977):

$$H_2S + CO_3^= \Leftrightarrow HS^- + HCO_3^-. \quad (5.4.61)$$

In fact, H_2S is selectively transported over CO_2 at low partial pressures of CO_2 and H_2S. The transport of SO_2 is considerably facilitated by an alkaline or neutral solution of sodium salts (NaCl, NaOH, $NaHSO_3$ or Na_2SO_3) (Roberts and Friedlander, 1980a,b).

In all such cases of facilitated transport, the extra flux of species A comes about from the complexation of species A with an added complexing agent B in the solution. Examples will be given now where considerable facilitation takes place because of solute ionization in solvent (without any added reagent B) or solute reaction with the solvent.

5.4.4.1.1 Facilitated transport - solvent as a complexing agent We know from reaction (5.2.2) that SO_2 ionizes significantly in water:

$$SO_2 + H_2O \Leftrightarrow HSO_3^- + H^+.$$

Through a water film, SO_2 diffuses by itself; it is also transferred by a gradient of the HSO_3^- ion in the same direction. Although water may be considered as the complexing agent B here, there is a major difference: we can ignore any gradient of B across the film. For a thick membrane (equivalent to assuming equilibrium everywhere) and using activities in the equilibrium constant K_d, we get

$$K_d = \frac{a_{HSO_3^-} a_{H^+}}{a_{SO_2}(1)} = \frac{\gamma_{HSO_3^-}\gamma_{H^+}}{\gamma_{SO_2}(1)} \frac{C_{HSO_3^-} C_{H^+}}{C_{SO_2}} = \frac{\gamma_{\pm}^2}{\gamma_{SO_2}(\ell)} \frac{C_{HSO_3^-} C_{H^+}}{C_{SO_2}}.$$

The total flux of SO_2 is given by (where the subscript m indicates quantities in the membrane)

$$J_{SO_2}^t = -D_{SO_2,m} \frac{dC_{SO_2,m}}{dy} - D_{HSO_3^-,m} \frac{dC_{HSO_3^-,m}}{dy}. \quad (5.4.62)$$

From local electroneutrality everywhere, we get

$$C_{H^+,m} = C_{HSO_3^-,m} \Rightarrow C_{HSO_3^-,m} = \frac{\sqrt{\gamma_{SO_2} K_d C_{SO_2}}}{\gamma_{\pm}}.$$

Integrate the expression for $J_{SO_2}^t$ at steady state to obtain

$$J_{SO_2}^t = \frac{D_{SO_2,m}}{\delta_m}\left(C_{SO_2}^0 - C_{SO_2}^\delta\right) + \frac{D_{HSO_3^-,m}}{\delta_m \gamma_{\pm}}\left(\sqrt{\gamma_{SO_2} K_d C_{SO_2}^0} - \sqrt{\gamma_{SO_2} K_d C_{SO_2}^\delta}\right). \quad (5.4.63)$$

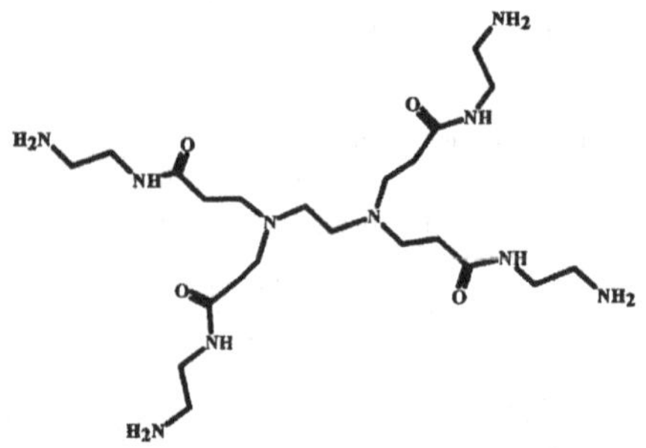

Figure 5.4.5. Polyamidoamine dendrimer of generation 0.

By the electroneutrality condition, the concentration of the two ions HSO_3^- and H^+ should be the same everywhere in the membrane. But the diffusivity of H^+ in water is six times that of HSO_3^- in water. Hence a potential gradient develops, slowing down the mobility of H^+ and increasing that of HSO_3^- (see definition (3.1.107)); the value of the effective diffusivity suggested by Roberts and Freidlander (1980a) for such a case is

$$D_{eff} = D'_{HSO_3^-} = D'_{H^+} = \frac{2D_{H^+} D_{HSO_3^-}}{D_{H^+} + D_{HSO_3^-}}, \quad (5.4.64a)$$

where D_{H^+} and $D_{HSO_3^-}$ are the corresponding true diffusivities in water.

If CO_2 is present in the feed gas containing SO_2 (e.g. flue gas), CO_2 is also transferred through the liquid membrane of water. However, the hydration reaction of CO_2 given by (5.4.60a), $CO_2 + H_2O \rightarrow HCO_3^- + H^+$, is negligible since K_d for CO_2 is a couple of orders of magnitude smaller than that for SO_2. Consequently, the flux of CO_2 is purely diffusive, with essentially no contribution from HCO_3^-. Sengupta *et al.* (1990) have shown theoretically and experimentally that, at 25 °C, the separation factor $\alpha_{SO_2-CO_2}$ through a liquid membrane of pure water is around 90; however, if no dissociation of SO_2 in water is assumed, then the theoretical value is around 30. This dramatically illustrates the role of solute ionization in facilitated transport; conversely, it demonstrates how the solvent itself can act as a complexing agent.

Another example of the solvent acting as the complexing agent is the dendrimer polyamidoamine of generation 0, molecular weight 517, which is a nonvolatile liquid having four terminal primary amine groups and two tertiary amines (see Figure 5.4.5).

Kovvali *et al.* (2000) and Kovvali and Sirkar (2001) have shown that, in the presence of moisture in the feed gas, this pure liquid membrane selectively transports CO_2 extremely efficiently vis-à-vis gases like O_2, N_2, etc., at low CO_2 feed partial pressures. They postulate that reactions (5.4.60b) of

CO_2 with the primary amine groups, and the reaction with the tertiary amine group,

$$CO_2 + R_3N + H_2O \Leftrightarrow HCO_3^- + R_3NH^+, \quad (5.4.64b)$$

are likely to be responsible for the selectivity. The very high charge densities in the pure liquid dendrimer ionized in the presence of moisture will reduce the permeation of N_2, O_2 strongly.

The literature on facilitated transport through liquid membranes is considerable. Interested readers will find additional material and references in Schultz *et al.* (1974), Smith *et al.* (1977), Marr and Kopp (1982), Way *et al.* (1982), Sengupta and Sirkar (1986), Noble and Way (1987), Ho and Li (2001) and Way and Noble (2001).

Figure 5.4.4(d) considers a case of facilitated transport where the receiving phase contains an agent E which reacts with the species A, as illustrated in Figure 5.4.4(b). This will reduce the concentration of A in the receiving phase and therefore lead to a higher flux of species A.

5.4.4.2 *Separation by countertransport*

In this mechanism, solute A from feed solution (see Figure 5.4.4(e)) complexes with complexing agent B in the membrane. The complex AB diffuses down its own concentration gradient to the receiving side where it dissociates. The complexing agent (the carrier) B then complexes with species C from the receiving solution; the complex BC diffuses down its own concentration gradient to the feed side, where it dissociates, releasing B for complexation with species A again. Species C is transferred to the feed solution, whereas species A is transferred to the receiving solution. An example has been provided by Choy *et al.* (1974). A flux of Na^+ (species A) is created from NaOH solution to a NaCl–HCl solution by a flux of protons in the opposite direction through the liquid membrane containing the macrocyclic carrier monesin (B), the complexation reaction being

$$Na^+ + RCOOH \Leftrightarrow RCOO^-Na^+ + H^+. \quad (5.4.65a)$$

In the absence of the carrier, the flux of sodium is orders of magnitude smaller.

Consider the following: complex AB will be transported down its own concentration gradient, even though the receiving phase may have a concentration of A higher than that of the feed phase. Thus species A can be pumped uphill against its own concentration gradient. The conditions are such that complex AB preferentially dissociates and complex BC is preferentially formed at the receiving interface; the reverse happens at the feed interface. Species C concentration in the receiving side is higher than that in the feed side. The transfer of C from the receiving to the feed side provides the energy for transfer of species A against its own concentration gradient. An example from Lee *et al.* (1978) is helpful for the special case where there is no free A in the membrane.

A basic solution of copper ($pH \sim 11$) containing 165 ppm of copper is passed on one side of a liquid membrane M (supported in a porous filter, Figure 5.4.6(a)). On the other side of the membrane, 1N H_2SO_4 solution flows. The organic liquid membrane contains a complexing agent LIX65N,[9] which transfers copper to the acidic side. Figure 5.4.6(a) shows that the copper concentration in the acidic stream goes as high as 700 ppm (from 0), whereas the copper concentration on the basic side drops to about 50 ppm. Further, the *pH* rises on the acidic side and drops on the basic side. The detailed mechanism is also illustrated in Figure 5.4.6(b).

We adopt the simplified analysis of Cussler (1971), wherein everywhere inside the membrane the two complexation reactions

$$A + B \Leftrightarrow AB, \quad (5.4.65b)$$

$$C + B \Leftrightarrow BC, \quad (5.4.65c)$$

are assumed to be in equilibrium. Therefore

$$K_A = \frac{C_{ABm}}{C_{Am}C_{Bm}}; \qquad K_C = \frac{C_{BCm}}{C_{Cm}C_{Bm}}. \quad (5.4.66)$$

If the solutes are ionic, there can be charge gradients in the membrane. The following simplifications are made: there is no charge gradient in the membrane and the diffusion coefficients of all species are equal to D. The governing differential equations for steady state diffusion of the two solutes A and C, the two complexes AB and BC and the complexing agent (the carrier) B are as follows:

$$D\frac{d^2C_{im}}{dy^2} + R_i = 00, \qquad i = A, C; \quad (5.4.67)$$

$$D\frac{d^2C_{iBm}}{dy^2} + R_{iB} = 0, \qquad i = A, C; \quad (5.4.68)$$

$$D\frac{d^2C_{Bm}}{dy^2} - R_{AB} - R_{BC} = 0. \quad (5.4.69)$$

Here R_A, R_C, and R_{AB} and R_{BC} (or R_{CB}), are the molar rates of production of species A, C, and complexes AB and BC, in the liquid membrane per unit volume. The molar concentration C_{Bm}^t of carrier B originally incorporated in the membrane can be accounted for, during separation, in terms of the concentrations of free species B, complex AB and complex BC as follows:

[9] Trade name of extracting agent from Henkel Co. Ltd. Contains 2-hydroxy-5-nonylbenzophenone oxime and 40 volume percent inert aromatic diluent (Ritcey and Ashbrook, 1984a) (Table 5.2.4).

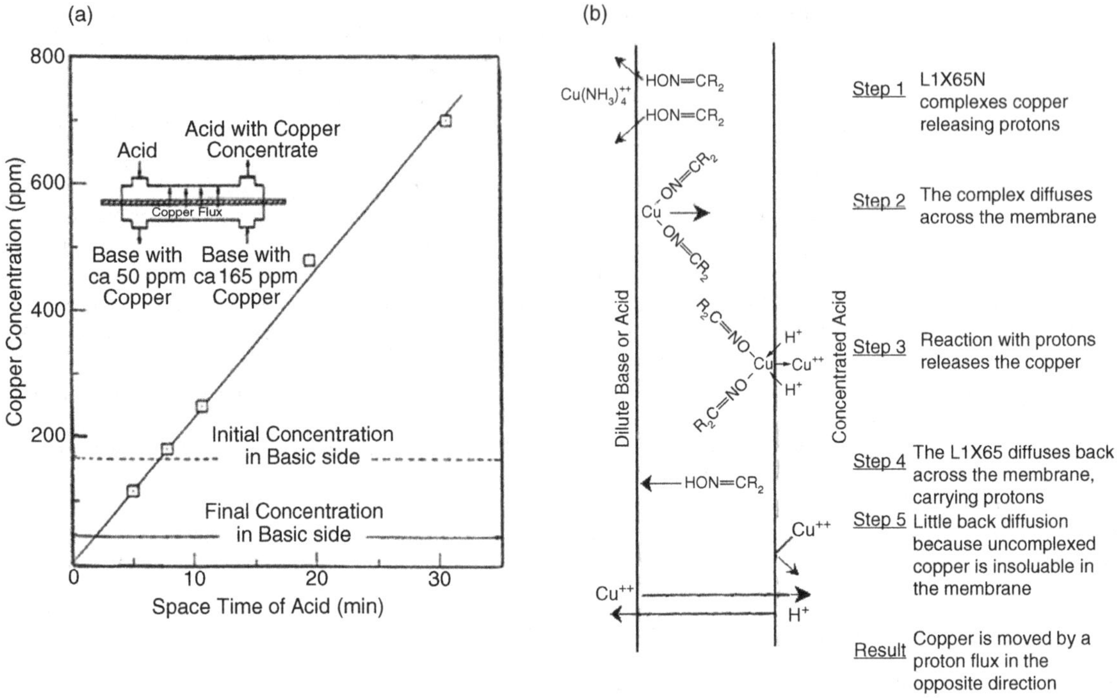

Figure 5.4.6. (a) Copper stripping with a supported liquid membrane. Copper is selectively concentrated by diffusion across a liquid membrane supported by a porous polymer film. (b) How the membranes work. Copper is concentrated by means of the steps shown. (a) and (b) reprinted, with permission, from Lee et al., AIChE J., 24(5), 860 (1978). Copyright © [1978] American Institute of Chemical Engineers (AIChE).

$$\frac{1}{\delta_m}\int_0^{\delta_m}(C_{Bm}+C_{ABm}+C_{BCm})\mathrm{d}y = C^t_{Bm}. \tag{5.4.70}$$

Further, neither species B, nor species AB or BC, can leave the membrane at either end: at

$$y=0,\ y=\delta_m,\ \frac{\mathrm{d}C_{Bm}}{\mathrm{d}y}=0=\frac{\mathrm{d}C_{ABm}}{\mathrm{d}y}=\frac{\mathrm{d}C_{BCm}}{\mathrm{d}y}. \tag{5.4.71}$$

At both ends of the membrane, the membrane concentration of species A and C are related to the feed and receiving phase concentrations by the equilibrium distribution coefficient κ_i':

$$\begin{aligned} y&=0, & \left(C^0_{im}/C_{if}\right)&=\kappa'_{if}, & i&=\mathrm{A,C};\\ y&=\delta_m, & \left(C^{\delta_m}_{im}/C_{ip}\right)&=\kappa'_{ip}, & i&=\mathrm{A,C}. \end{aligned} \tag{5.4.72a}$$

Assume for mathematical simplicity that $\kappa'_{if}=\kappa'_{ip}=\kappa'_i$. (In practice, $\kappa'_{if}\neq\kappa'_{ip}$.) It is now possible to determine the concentration profile of each species in the membrane; the flux of species A and C will follow. The first step is to determine the concentration of free species B. To this end, add equation (5.4.68) for solutes A and C and add it to equation (5.4.69) to obtain

$$D\left\{\frac{\mathrm{d}^2C_{Bm}}{\mathrm{d}y^2}+\frac{\mathrm{d}^2C_{ABm}}{\mathrm{d}y^2}+\frac{\mathrm{d}^2C_{BCm}}{\mathrm{d}y^2}\right\}=0. \tag{5.4.72b}$$

Integrate it once and use equation (5.4.71) to obtain

$$\frac{\mathrm{d}C_{Bm}}{\mathrm{d}y}+\frac{\mathrm{d}C_{ABm}}{\mathrm{d}y}+\frac{\mathrm{d}C_{BCm}}{\mathrm{d}y}=0. \tag{5.4.72c}$$

Integrate once more and put into equation (5.4.70) to yield

$$C^t_{Bm}=C_{Bm}+C_{ABm}+C_{BCm}. \tag{5.4.72d}$$

In this relation, substitute equilibrium relations (5.4.66) to obtain

$$C_{Bm}=\frac{C^t_{Bm}}{1+K_AC_{Am}+K_CC_{Cm}}. \tag{5.4.73}$$

The total flux of solute A can now be obtained as the sum of the fluxes of free species A and the complex AB:

$$\begin{aligned} N^t_{Ay}&=N_{Ay}+N_{ABy}=-D\frac{\mathrm{d}C_{Am}}{\mathrm{d}y}-D\frac{\mathrm{d}C_{ABm}}{\mathrm{d}y}\\ &=-D\frac{\mathrm{d}C_{Am}}{\mathrm{d}y}-D\frac{\mathrm{d}(K_AC_{Am}C_{Bm})}{\mathrm{d}y}. \end{aligned}$$

At steady state, integrate across $y=0$ to $y=\delta_m$:

$$N^t_{Ay} = D\frac{(C^0_{Am} - C^\delta_{Am})}{\delta_m} + \frac{D}{\delta_m}\left[K_A C^0_{Am} C^0_{Bm} - K_A C^{\delta_m}_{Am} C^{\delta_m}_{Bm}\right].$$

Rewrite this using (5.4.72a) and (5.4.73) to obtain

$$N^t_{Ay} = \frac{D\kappa'_A}{\delta_m}(C_{Af} - C_{Ap}) + \frac{D\kappa'_A C^t_{Bm}}{\delta_m}$$
$$\left[\frac{C_{Af}}{1 + K_A\kappa'_A C_{Af} + K_C\kappa'_C C_{Cf}} - \frac{C_{Ap}}{1 + K_A\kappa'_A C_{Ap} + K_C\kappa'_C C_{Cp}}\right].$$

The result can also be expressed as follows (Cussler, 1971):

$$N^t_{Ay} = \frac{D\kappa'_A}{\delta_m}(C_{Af} - C_{Ap})$$
$$+ \frac{D\kappa'_A}{\delta_m} K_A\left[R(1 + \kappa'_C K_C \overline{C}_C)\right](C_{Af} - C_{Ap})$$
$$- \frac{D\kappa'_A}{\delta_m}\left[R\kappa'_C K_C \overline{C}_A\right]\left[C_{Cf} - C_{Cp}\right]. \qquad (5.4.74)$$

Here

$$\overline{C}_i = \frac{C_{if} + C_{ip}}{2}, \qquad i = A, C,$$

and

$$R = \frac{K_A C^t_{Bm}}{(1 + K_A\kappa'_A C_{Af} + K_C\kappa'_C C_{Cf})(1 + K_A\kappa'_A C_{Ap} + K_C\kappa'_C C_{Cp})}. \qquad (5.4.75)$$

If species C does not exist, then the above expression for N^t_{Ay} reduces to expression (5.4.57) for facilitated transport of A in the presence of the complexing agent or carrier B.

Consider now the situation where $C_{Af} = C_{Ap} = \overline{C}_A$, i.e. there is no gradient of free solute A concentration across the membrane. Will there be a flux of A? The above expression for N^t_{Ay} is simplified to

$$N^t_{Ay} = \frac{D\kappa'_A}{\delta_m}\left[R\kappa'_C K_C \overline{C}_A\right]\left[C_{Cp} - C_{Cf}\right]. \qquad (5.4.76)$$

Since $C_{Cp} > C_{Cf}$, species A is indeed being transferred from the feed solution to the receiving solution due to a gradient in concentration of C in the opposite direction. Obviously, the concentration of A in the receiving solution can be built up over that in the feed solution. An estimate of the highest value of C_{Ap} can be obtained by assuming $C_{Cf} = 0$ and $N^t_{Ay} = 0$ for a case where the diffusional rate of the free solute species is negligible (Cussler, 1984).

Most examples of countertransport involve a liquid feed and a liquid receiving solution. Smith *et al.* (1977) have studied countertransport when the feed and receiving phases are gaseous; a gradient of CO_2 in one direction creates transport of H_2S in the other direction.

When one considers separating a metallic species from an aqueous feed solution through an organic liquid membrane into an aqueous receiving solution, it is apparent from Section 5.2.2.4 that there is no free metal species in the organic liquid membrane. The metal A exists only in the complex form AB in the liquid membrane. Species C is usually a H^+ ion. Thus the feed solution is less acidic than the receiving solution. A flux of H^+ ions occurs from the receiving solution to the feed solution, which pumps the metal from the feed to the receiving solution. There is no reaction inside the membrane. Instead the following reactions take place at the feed-membrane and the strip-membrane interfaces, with, for example, an acidic extracting agent HK in the membrane:

$$M^+(aq) + HK(membrane) \underset{receiving}{\overset{K}{\underset{feed}{\Leftrightarrow}}} MK(membrane) + H^+(aq). \qquad (5.4.77)$$

Here K is the equilibrium constant for the reactions at $y = 0$ (feed-membrane interface) and $y = \delta_m$ (strip-membrane interface).

The flux of the metal at the feed-membrane interface could be expressed in terms of the forward and backward interfacial reactions (following (5.3.40)):

$$N_{M^+y} = k^0_{s+} C^0_{M^+i} - k^0_{s-} C^0_{MK_o i}. \qquad (5.4.78)$$

Here k^0_{s+} and k^0_{s-} are pseudo-first-order rate constants (otherwise see (5.3.40)). For very fast interfacial reactions, local equilibrium may be assumed and

$$K = (k^0_{s+}/k^0_{s-}) = \frac{C^0_{MK_o i}}{C^0_{M^+i}} = \kappa''_{Mo}, \qquad (5.4.79)$$

κ''_{Mo} being the effective distribution coefficient for the metal ion. The flux of the metal M across the membrane is merely the flux of the complex MK in the membrane:

$$N_{M^+y} = \frac{D_{MK}}{\delta_m}\left(C^0_{MK_o} - C^\delta_{MK_o}\right). \qquad (5.4.80)$$

In general, there is also a diffusional step in each aqueous boundary layer (neglected here).

When there are two different metals M_1^+ and M_2^+ in the feed solution, the selectivity of the membrane is of interest. For this purpose assume that $C^\delta_{MK_o} \ll C^0_{MK_o}$. Then

$$\frac{N_{M_1^+y}}{N_{M_2^+y}} \cong \frac{D_{M_1K}\, C^0_{M_1K_o}}{D_{M_2K}\, C^0_{M_2K_o}} = \frac{D_{M_1K}\,\kappa''_{M_1o}\, C^0_{M_1^+i}}{D_{M_2K}\,\kappa''_{M_2o}\, C^0_{M_2^+i}}. \qquad (5.4.81)$$

Assume $D_{M_1K} \cong D_{M_2K}$. Further, with the crossflow assumption,[10]

$$\frac{N_{M_1^+y}\, C^0_{M_2^+i}}{N_{M_2^+y}\, C^0_{M_1^+i}} = \frac{C^\delta_{M_1^+i}\, C^0_{M_2^+i}}{C^\delta_{M_2^+i}\, C^0_{M_1^+i}} = \alpha_{M_1-M_2} \cong (\kappa''_{M_1o}/\kappa''_{M_2o}). \qquad (5.4.82)$$

[10] Local concentration ratio at the permeate-membrane interface is given by the flux ratio (equation (7.2.2a)).

The separation of the two metals will be primarily determined by the ratio of the two distribution coefficients, $\kappa''_{M_1 o}/\kappa''_{M_2 o}$, exactly as in solvent extraction. Separation of copper from nickel using LIX65N in the liquid membrane using these principles has been studied by Lee *et al.* (1978). A general review of metal extraction using liquid membranes supported in microporous polymeric membranes is provided by Danesi (1984-85).

Separation of cobalt from nickel using the extractant di (2,4,4-trimethylpentyl)-phosphinic acid, H(DTMPP) or CYANEX 272 in an organic liquid membrane has been studied by Danesi *et al.* (1984). It has been found that H(DTMPP), represented here as HK, exists essentially as a dimer, H_2K_2, in aromatic solvents (e.g. toluene):

$$Co^{2+}(aq) + 2H_2K_2(org) \Leftrightarrow Co(HK_2)_2(org) + 2H^+(aq); \tag{5.4.83a}$$

$$Ni^{2+}(aq) + 3H_2K_2(org) \Leftrightarrow Ni(HK_2)_2(H_2K_2)(org) + 2H^+(aq). \tag{5.4.83b}$$

The equilibrium constant for each reaction can be written as

$$K_{Co} = \frac{C_{Co(HK_2)_2,o}\left(C_{H^+,w}\right)^2}{C_{Co^{2+},w}\left(C_{H_2K_2,o}\right)^2}; \tag{5.4.84a}$$

$$K_{Ni} = \frac{C_{Ni(HK_2)_2(H_2K_2),o}\left(C_{H^+,w}\right)^2}{C_{Ni^{2+},w}\left(C_{H_2K_2,o}\right)^3}. \tag{5.4.84b}$$

The effective distribution coefficient of each metal species at the feed solution-organic membrane interface is

$$\kappa''_{Co,o} = \left(C_{Co(HK_2)_2,o}/C_{Co^{2+},w}\right); \tag{5.4.85a}$$

$$\kappa''_{Ni,o} = \left(C_{Ni(HK_2)_2(H_2K_2),o}/C_{Ni^{2+},w}\right). \tag{5.4.85b}$$

Therefore,

$$\frac{\kappa''_{Co,o}}{\kappa''_{Ni,o}} = \frac{K_{Co}/K_{Ni}}{\left(C_{H_2K_2,o}\right)}. \tag{5.4.86}$$

This suggests that, at low concentrations of free HK in the organic phase (present as H_2K_2), $(C_{H_2K_2,o})$, the value of $(\kappa''_{Co,o}/\kappa''_{Ni,o})$, the separation factor in solvent extraction, can be very high, much higher than (K_{Co}/K_{Ni}), which was found to be 600 from solvent extraction. Danesi *et al.* (1984) have indicated this to be as high as 10^4 when free HK in the organic phase is around 10^{-3} M. From relation (5.4.82), the separation factor for the liquid membrane should also be $(\kappa''_{Co,o}/\kappa''_{Ni,o})$. However, the boundary layer resistances on the two sides of the organic liquid membrane were found to reduce the separation factor from $(\kappa''_{Co,o}/\kappa''_{Ni,o})$ at all except very low concentrations of $C_{H_2K_2,o}$.

By now, it should be clear that the complexing agent HK acts as a chelating agent as studied in Section 5.2.2.4. Further, such an agent in countertransport with the H^+ ion is an acid or is acidic.

5.4.4.3 *Separation by cotransport*

In the case shown in Figure 5.4.4(f), solute A, along with species C in the feed solution, jointly form a complex ABC with the carrier B in the membrane at the feed-membrane interface. Complex ABC diffuses through the membrane and dissociates at the receiving phase-membrane interface. Carrier B, released at $y = \delta_m$, diffuses back to $y = 0$, whereas species A and C are deposited to the receiving solution. In order to transport A, simultaneous transport of C in the same direction is needed. A few examples will be provided in the following.

Babcock *et al.* (1980) have shown that uranium metal in the form of uranyl sulfate anions, $UO_2(SO)_3^{4-}$, is transported through a solvent membrane containing a tertiary amine carrier R_3N by cotransport in the following manner:

$$4R_3N(org) + 4H^+(aq) + UO_2(SO_4)_3^{4-}(aq) \underset{\text{stripside}}{\overset{\text{feedside}}{\Leftrightarrow}} (R_3NH)_4UO_2(SO_4)_3(org) \tag{5.4.87}$$

when the feed solution is highly acidic (e.g. $pH = 1.0$) and the receiving solution is less so ($pH = 4.5$). Thus both hydrogen ions and uranyl sulfate ions are transported in the same direction through the membrane as the complex formed with the carrier amine. Danesi *et al.* (1983) have demonstrated cotransport of Am^{3+} and NO_3^- through an organic membrane containing the complexing carrier CMPO (n-octyl (phenyl)-N, N-diisobutylcarbamoyl methyl phosphine oxide) from a feed containing $LiNO_3$ to a strip solution containing formic acid. Hochhauser and Cussler (1975) have experimentally demonstrated cotransport of chromium present as (say) $HCr_2O_7^-$ and H^+ ions in an acidic feed aqueous solution through an organic liquid membrane containing tridodecylamine, $[CH_3(CH_2)_{11}]_3N$, as a carrier to a basic receiving aqueous solution. The amine freed at the receiving solution-membrane interface shuttled back to the feed side.

A detailed analysis of chromium transport, assuming that $HCr_2O_7^-$ (A) and H^+ (C) forms an ion pair H_2CrO_7 (AC) and then a complex (ABC) with the amine (B) in the organic membrane, is available in Hochhauser and Cussler (1975). The total flux of chromium in the membrane is the sum of the flux of the ion pair $H_2Cr_2O_7$ (AC) and the flux of the complex (ABC):

$$\text{flux}_A = \frac{D_{AC}}{\delta_m}\left[C_{AC}^0 - C_{AC}^\delta\right] + \frac{D_{ABC}}{\delta_m}\left[C_{ABC}^0 - C_{ABC}^\delta\right]. \tag{5.4.88}$$

A general expression for the flux of A is available in Cussler (1984).

Conventionally in cotransport analysis for metal species (A), however, the dominant species is ABC, and only the flux of ABC is analyzed (Danesi, 1984–85) and used for determining the metal transport rate. Note further that, for metal extraction in cotransport, the carrier is neutral or basic (e.g. long-chain alkylamine).

5.4.4.4 *Separation based on dissociation*

Consider a feed aqueous solution containing two weak organic acids A and B on one side of an organic liquid membrane without any carrier species. Then, at the feed-membrane interface, the following distribution equilibria hold (equation (5.2.62)):

$$\kappa''_{\text{Ao}}|_f = \frac{\kappa'_{\text{Ao}}|_f}{1 + (K_{d\text{A}}/C_{\text{H}^+})}; \qquad \kappa''_{\text{Bo}}|_f = \frac{\kappa'_{\text{Bo}}|_f}{1 + (K_{d\text{B}}/C_{\text{H}^+})}. \tag{5.4.89}$$

Suppose $K_{d\text{A}}$ and $K_{d\text{B}}$ are sufficiently apart even if $\kappa'_{\text{Ao}}|_f$ is close to $\kappa'_{\text{Bo}}|_f$. If the *pH* is moderately high, such that the more acidic species B is highly dissociated, but the species A is not so highly dissociated, then

$$\kappa''_{\text{Bo}}|_f \cong \kappa'_{\text{Bo}}|_f (C_{\text{H}^+}/K_{d\text{B}})_f \qquad \text{and} \qquad \kappa''_{\text{Ao}}|_f \cong \kappa'_{\text{Ao}}|_f, \tag{5.4.90}$$

with $K_{d\text{B}} \gg C_{\text{H}^+}$ and $K_{d\text{A}} \ll C_{\text{H}^+}$. Therefore $\kappa''_{\text{Ao}}|_f >> \kappa'_{\text{Bo}}|_f$ and species A is extracted much more at the feed interface. Now, if the strip solution is highly alkaline, it is obvious that species A is going to be pushed to the strip aqueous solution since

$$\kappa''_{\text{Ao}}|_s \cong \kappa'_{\text{Ao}}|_s (C_{\text{H}^+}/K_{d\text{A}})_s \tag{5.4.91}$$

and $K_{d\text{A}} \gg C_{\text{H}^+}$. For lack of specifics, the distribution equilibrium for species B at the organic liquid membrane and the strip aqueous solution interface is

$$\kappa''_{\text{Bo}}|_s = \frac{\kappa'_{\text{Bo}}|_s}{1 + (K_{d\text{B}}/C_{\text{H}^+})_s}. \tag{5.4.92}$$

It is very likely, however, that $\kappa''_{\text{Bo}}|_s \cong \kappa'_{\text{Bo}}|_s (C_{\text{H}^+}/K_{d\text{B}})_s$.

The separation factor α_{AB} for species A and B between the feed aqueous and the strip aqueous solution can be determined in the usual fashion. The flux of each species through the membrane is given by

$$N_{\text{A}y} = \frac{D_{\text{A}m}}{\delta_m}\left(C^0_{\text{A}m} - C^\delta_{\text{A}m}\right); \tag{5.4.93a}$$

$$N_{\text{B}y} = \frac{D_{\text{B}m}}{\delta_m}\left(C^0_{\text{B}m} - C^\delta_{\text{B}m}\right). \tag{5.4.93b}$$

Since $C_{\text{H}+}|_s << K_{d\text{A}}$ and $K_{d\text{B}}$, $C^\delta_{\text{A}m} << C^0_{\text{A}m}$, $C^\delta_{\text{B}m} << C^0_{\text{B}m}$,

$$\frac{N_{\text{A}y}}{N_{\text{B}y}} \cong \frac{D_{\text{A}m}\, C^0_{\text{A}m}}{D_{\text{B}m}\, C^0_{\text{B}m}} \cong \left(\frac{D_{\text{A}m}}{D_{\text{B}m}}\right) \frac{\kappa''_{\text{Ao}}|_f\, C^0_{\text{A}i}}{\kappa''_{\text{Bo}}|_f\, C^0_{\text{B}i}}. \tag{5.4.94}$$

Using the crossflow assumption,

$$\frac{N_{\text{A}y}\, C^0_{\text{B}i}}{N_{\text{B}y}\, C^0_{\text{A}i}} = \frac{C^\delta_{\text{A}i}\, C^0_{\text{B}i}}{C^\delta_{\text{B}i}\, C^0_{\text{A}i}} = \alpha_{\text{AB}} = \frac{D_{\text{A}m}\, \kappa''_{\text{Ao}}|_f}{D_{\text{B}m}\, \kappa''_{\text{Bo}}|_f}.$$

Therefore

$$\alpha_{\text{AB}} \cong \left(\frac{D_{\text{A}m}}{D_{\text{B}m}}\right) \frac{\kappa'_{\text{Ao}}|_f}{\kappa'_{\text{Bo}}|_f} \left(\frac{K_{d\text{B}}}{C_{\text{H}^+}}\right)_f. \tag{5.4.95}$$

This result is based on B being highly dissociated in the feed solution and species A not so highly dissociated. If, however, both species are dissociated significantly, but to different degrees, then

$$\alpha_{\text{AB}} \cong \left(\frac{D_{\text{A}m}}{D_{\text{B}m}}\right) \frac{\kappa'_{\text{Ao}}|_f}{\kappa'_{\text{Bo}}|_f} \left(\frac{K_{d\text{B}}}{K_{d\text{A}}}\right). \tag{5.4.96a}$$

A more general expression would be

$$\alpha_{\text{AB}} \cong \left(\frac{D_{\text{A}m}}{D_{\text{B}m}}\right) \frac{\kappa'_{\text{Ao}}|_f}{\kappa'_{\text{Bo}}|_f} \frac{1 + (K_{d\text{B}}/C_{\text{H}^+})_f}{1 + (K_{d\text{A}}/C_{\text{H}^+})_f}. \tag{5.4.96b}$$

If we discount the role of $(D_{\text{A}m}/D_{\text{B}m})$ in these three expressions for α_{AB}, the separation factors are what would have been achieved in dissociation extraction from an aqueous solution to an organic solvent. If, however, species A were to be reextracted back into an aqueous solution for recovery, then the liquid membrane step achieves the same goal using very little solvent and only one device (instead of an extractor and a back extractor). Such a technique is likely to be highly useful in the pharmaceutical industry.

In this example, the liquid membrane did not have any complexing carrier species. Such a carrier species, incorporated in the liquid membrane, will enhance the flux of species A. There is, however, the distinct possibility also for complexation of the carrier with the limited amount of B that partitions into the membrane.

5.4.4.5 *Diffusional film resistances outside the liquid membrane*

The diffusional film resistances on the two sides of the liquid membrane, the feed side and the receiving (or strip) side, have been ignored in the preceding treatment. They can be quite important when the membrane transport rates and/or the interfacial reaction rates are high. Two examples where external mass-transfer resistances are important are briefly considered below.

5.4.4.5.1 External film resistances in facilitated transport Facilitated transport separation for the reaction

$$\text{A} + \text{B} \underset{k_b}{\overset{k_f}{\Leftrightarrow}} \text{AB}$$

has been discussed in Section 5.4.4.1. If there is a mass-transfer resistance $(1/k_{gf})$ or $(1/k_{lf})$ in the feed phase (gas or liquid) to the transport of species A to the feed-membrane

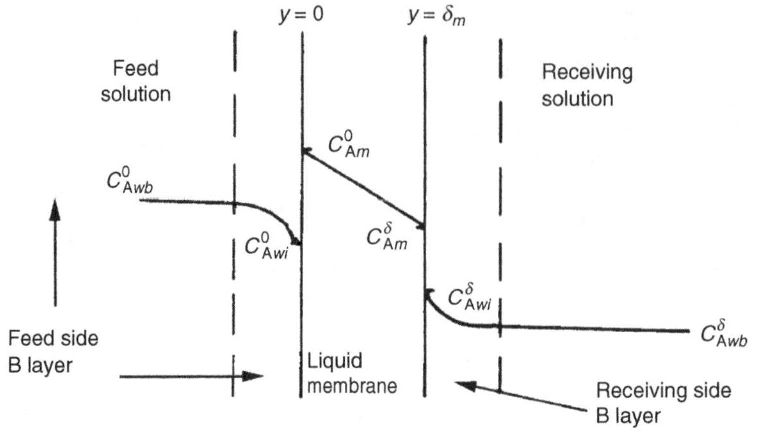

Figure 5.4.7. Concentration profile for species A being transported through liquid membrane with boundary layer resistances.

interface ($y = 0$), then the following boundary condition is needed (Figure 5.4.7):

$$-D_{\mathrm{A}m}\frac{\mathrm{d}C_{\mathrm{A}m}}{\mathrm{d}y}\bigg|_{y=0} = k_{gf}\left(p^0_{\mathrm{A}b} - p^0_{\mathrm{A}i}\right) \qquad (5.4.97\mathrm{a})$$

(gaseous feed) or

$$-D_{\mathrm{A}m}\frac{\mathrm{d}C_{\mathrm{A}m}}{\mathrm{d}y}\bigg|_{y=0} = k_{\ell f}\left(C^0_{\mathrm{A}wb} - C^0_{\mathrm{A}wi}\right). \qquad (5.4.97\mathrm{b})$$

(aqueous feed). Use Henry's law constant H^C_{A} for species A in a gaseous feed or the distribution coefficient $\kappa''_{\mathrm{A}o}$ for an aqueous feed. These relations can be rewritten for gaseous feed as

$$-D_{\mathrm{A}m}\frac{\mathrm{d}C_{\mathrm{A}m}}{\mathrm{d}y}\bigg|_{y=0} = k_{gf}H^C_{\mathrm{A}}\left(\frac{p^0_{\mathrm{A}b}}{H^C_{\mathrm{A}}} - C^0_{\mathrm{A}m}\right) \qquad (5.4.98\mathrm{a})$$

and for aqueous feed with an organic membrane as

$$-D_{\mathrm{A}m}\frac{\mathrm{d}C_{\mathrm{A}m}}{\mathrm{d}y}\bigg|_{y=0} = \left(k_{\ell f}/\kappa''_{\mathrm{A}o}\right)\left(C^{*0}_{\mathrm{A}ob} - C^0_{\mathrm{A}m}\right). \qquad (5.4.98\mathrm{b})$$

Note that $C^0_{\mathrm{A}m}$ is the membrane-phase concentration of A at the feed interface and that $C^{*0}_{\mathrm{A}ob}$ is the hypothetical membrane-phase concentration which would be in equilibrium with the aqueous feed bulk concentration $C^0_{\mathrm{A}wb}$.

A similar boundary condition on the receiving (or strip) side of the membrane will be

$$-D_{\mathrm{A}m}\frac{\mathrm{d}C_{\mathrm{A}m}}{\mathrm{d}y}\bigg|_{y=\delta_m} = k_{gs}\left(p^\delta_{\mathrm{A}i} - p^\delta_{\mathrm{A}b}\right) \qquad (5.4.99\mathrm{a})$$

(gaseous feed) or

$$-D_{\mathrm{A}m}\frac{\mathrm{d}C_{\mathrm{A}m}}{\mathrm{d}y}\bigg|_{y=\delta_m} = k_{\ell s}\left(C^\delta_{\mathrm{A}wi} - C^\delta_{\mathrm{A}wb}\right). \qquad (5.4.99\mathrm{b})$$

These can be rewritten as before using H^C_{A} and $\kappa''_{\mathrm{A}o}$ to relate $p^\delta_{\mathrm{A}i}$ or $C^\delta_{\mathrm{A}wi}$ to $C^\delta_{\mathrm{A}m}$, etc.

Noble *et al.* (1986) have obtained a solution for the facilitated transport of species A through the liquid membrane for such a case when $C_{\mathrm{B}m}$ was constant throughout the film. This is equivalent to assuming a very large excess of the complexing species B in the membrane. Further, there is reaction equilibrium everywhere in the membrane and $C^\delta_{\mathrm{A}wb}$ or $p^\delta_{\mathrm{A}b}$ (as the case may be) is zero. The extent of facilitation ψ_{eq} (see (5.4.58)) was obtained as

$$\psi_{\mathrm{eq}} = \frac{\left(1+\frac{\lambda_1 K^*}{1+K^*}\right)\left(1+\frac{1}{Sh_f}+\frac{1}{Sh_s}\right)}{1+\frac{\lambda_1 K^*}{1+K^*}\frac{\tanh\lambda_2}{\lambda_2}+\left(1+\frac{\lambda_1 K^*}{1+K^*}\right)\left(\frac{1}{Sh_f}+\frac{1}{Sh_s}\right)}, \qquad (5.4.100)$$

where

$$\lambda_l = \frac{D_{\mathrm{AB}}C^t_{\mathrm{B}}H^C_{\mathrm{A}}}{D_{\mathrm{A}}p^0_{\mathrm{A}b}};\quad K^* = \frac{Kp^0_{\mathrm{A}b}}{H^C_{\mathrm{A}}};\quad \lambda_2 = \frac{1}{2}\left(\frac{1+(\lambda_1+1)K^*}{(1+K^*)/\gamma_{\mathrm{A}}}\right)^{1/2};$$

$$Sh_f = k_{gf}H^C_{\mathrm{A}}\delta_m/D_{\mathrm{A}};\quad Sh_s = k_{gs}H^C_{\mathrm{A}}\delta_m/D_{\mathrm{A}};\quad \gamma_{\mathrm{A}} = k_b\delta^2_m/D_{\mathrm{A}}.$$

5.4.4.5.2 External film resistances in countertransport separation with interfacial reaction Consider the following type of interfacial reaction for a metal being transported from an aqueous feed solution through an organic liquid membrane to an aqueous strip solution by countertransport of H^+ ions (as in (5.4.77)):

$$\mathrm{M^+(aq) + HK(org)} \underset{\text{strip}}{\overset{\text{feed}}{\Leftrightarrow}} \mathrm{MK(org) + H^+(aq)},$$

where HK is the complexing carrier in the liquid membrane. At steady state, the metal flux through the aqueous boundary layer on the feed side, the metal flux at the feed-membrane interface, the metal flux through the membrane, the metal flux at the membrane-strip interface and the metal flux through the strip side boundary layer are all

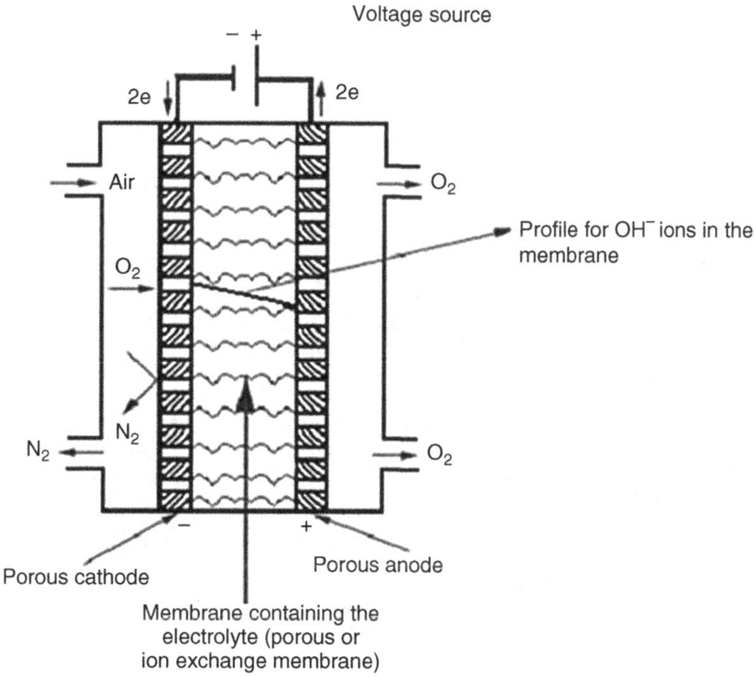

Figure 5.4.8. Electrochemical membrane gas separator for separating oxygen from air.

equal. Represent this flux as N_{M^+z}. Then, for the feed boundary layer,

$$N_{M^+z} = k_{\ell f}\left[C_{M^+b} - C^0_{M^+i}\right]. \tag{5.4.101}$$

Assume psuedo-first-order rate constants $k^0_{s^+1}$ and $k^0_{s^-1}$ to define (see (5.4.78))

$$N_{M^+z} = k^0_{s^+1} C^0_{M^+i} - k^0_{s^-1} C^0_{MK_oi}.$$

Through the membrane (see (5.4.80)),

$$N_{M^+z} = \left(\frac{D_{MK}}{\delta_m}\right)\left(C^0_{MK_oi} - C^\delta_{MK_o}\right).$$

Since strip solutions effectively reduce $C^\delta_{MK_o}$ to essentially a value of zero, we ignore the strip side interfacial reaction as well as the boundary layer equations and write $C^\delta_{MK_o} = 0$. Combining the three equations, we obtain

$$N_{M^+z} = \frac{C_{M^+b}}{\left[\frac{1}{k^0_{s^+1}} + \frac{1}{k_{\ell f}} + \frac{(k^0_{s^-1}/k^0_{s^+1})}{(D_{MK}/\delta_m)}\right]} = \frac{(k^0_{s^+1}/k^0_{s^-1})\,C_{M^+b}}{\left[\left(\frac{\delta_m}{D_{MK}}\right) + \frac{(k^0_{s^+1}/k^0_{s^-1})}{k_{\ell f}} + \frac{1}{k^0_{s^-1}}\right]} \tag{5.4.102}$$

for the general case. For any particular surface reaction, the exact rate equation needs to be considered instead of the pseudo-first-order reaction considered above.

5.4.4.6 *Electrochemical membrane gas separation*

Consider an electrochemical cell having two electrodes, a cathode and an anode, separated by an electrolyte. Let one side of each electrode face the electrolyte while the other side is open to the atmosphere or any gas mixture (Figure 5.4.8). Let the negatively charged cathode be made out of a nickel screen or a highly porous gas diffusion electrode containing a catalyst like nickel dispersed well. If a small voltage of 0.5 volts is applied across the electrodes (the other electrode is similarly prepared) when air is present around the cathode, one finds that virtually pure oxygen is generated/liberated at the positively charged electrode, the anode, in the presence of aqueous KOH solution between the electrodes. At the cathode of this electrochemical gas separator, oxygen from air is reduced at room temperature via

$$\frac{1}{2}O_2 + H_2O + 2e \rightarrow 2OH^-. \tag{5.4.103a}$$

At the anode, the OH^- ions arrive in the presence of the applied voltage difference between the electrodes and are converted to oxygen:

$$2OH^- \rightarrow \frac{1}{2}O_2\uparrow + H_2O + 2e. \tag{5.4.103b}$$

This electrochemical cell has thus produced pure oxygen gas at the anode from oxygen in the air at the cathode

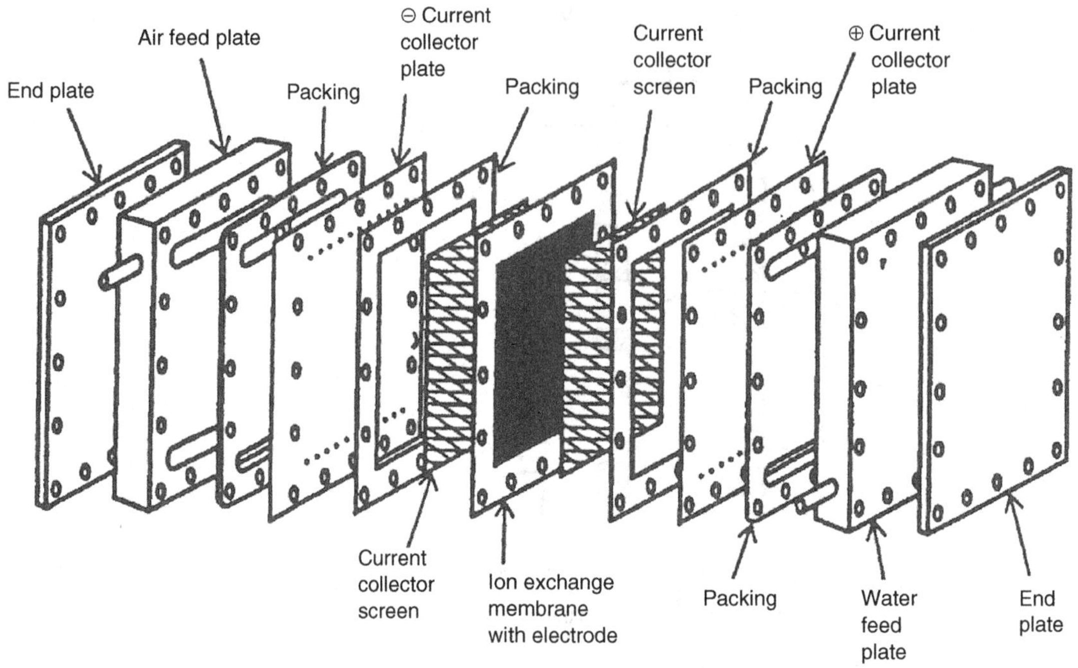

Figure 5.4.9. Exploded view of the electrochemical oxygen separator (100 cm²). Reprinted, with kind permission from Springer Science + Business Media, from the Journal of Applied Electrochemistry*: Y. Fujita, H. Nakamura and T. Muto, "An electrochemical oxygen separator using an ion exchange membrane as the electrolyte,"* J. Appl. Electrochem. *16 935, (1986), Figure 2.*

via electrochemical reactions at the two electrodes (Langer and Haldeman, 1964). Other gases at the cathode do not undergo any such reaction at the applied voltage difference.

The aqueous electrolytic solution can easily leak out or flow into the gas streams at either electrode, creating a number of problems. To avoid this, one can employ a porous membrane or a filter paper soaked in KOH as the electrolytic medium. We have now a *supported liquid membrane* (SLM) through which the OH^- ion is selectively transported; we expect the flux of any dissolved O_2 through this SLM to be extremely small, since O_2 solubility in this concentrated electrolytic solution is likely to be very low. Further, the thickness of the SLM is considerable. However, there is also a possibility of a change in the concentration of the electrolyte due to the evaporation or absorption of water in response to humidity fluctuations in the ambient conditions at the electrodes. A polymeric ion exchange membrane would undergo much more limited change due to environmental humidity fluctuations (Maget, 1970). **This is the genesis of the electrochemical membrane gas separation employing an ion exchange membrane between the electrodes.**

There still remains a problem created by any atmospheric CO_2, which will react with the OH^- to form carbonate:

$$CO_2 + 2OH^- \rightarrow CO_3^{2-} + H_2O. \tag{5.4.104}$$

This will reduce the efficiency of the alkaline membrane with time. One can avoid this problem via an alternative strategy which uses a proton exchange membrane, e.g. Nafion®, and electrodes containing a noble metal platinum as the catalyst. The electrode reactions then are as follows:

$$\begin{aligned} &\text{cathode} \quad \frac{1}{2}O_2 + 2H^+ + 2e \rightarrow H_2O; \\ &\text{anode} \quad H_2O \rightarrow \frac{1}{2}O_2 + 2H^+ + 2e. \end{aligned} \tag{5.4.105}$$

Fujita *et al.* (1986) have demonstrated this by circulating water on the anode side and recovering 98.4% O_2 from the water when 1.4 volts was applied between the two electrodes (Figure 5.4.9).

The general principle illustrated by equations (5.4.103a) and (5.4.103b) has been utilized to separate a number of gas species from their mixtures with other species. For example, Robinson *et al.* (1998) have demonstrated that H_2S can be selectively removed from a sour process gas (coal gas containing H_2S) at the cathode, where H_2S is reduced via

$$H_2S + 2e^- \rightarrow H_2 + S^{2-}, \tag{5.4.106a}$$

whereas, at the anode, the reduced sulfide ion is oxidized to condense as elemental sulfur,

$$S^{2-} \rightarrow S_2 + 2e^-, \qquad (5.4.106b)$$

which is swept away by a hot inert sweep gas (e.g. N_2). This process operates at a high temperature (~650 °C) with a molten carbonate (Li_2CO_3 (62 mol%)-K_2CO_3 (38 mol%)) electrolyte located in the pores of a porous ceramic support membrane. The molten carbonate reacts with H_2S to provide sulfide ions in a molten state:

$$(Li_{0.62}K_{0.38})_2CO_3 + H_2S \longleftrightarrow (Li_{0.62}K_{0.38})_2\,S + CO_2 + H_2O. \qquad (5.4.106c)$$

Complications are encountered since CO_2 and H_2O vapor also participate a little in the cathodic reduction process:

$$CO_2 + H_2O + 2e^- \rightarrow H_2 + CO_3^{2-}. \qquad (5.4.106d)$$

An additional example has been provided by Wauters and Winnick (1998), wherein Br_2 gas is obtained from a waste gas stream containing HBr vapor through a molten salt saturated membrane. At the cathode, HBr vapor is reduced:

$$2HBr + 2e^- \rightarrow H_2 + 2Br^-. \qquad (5.4.107a)$$

At the anode, the Br^- ions are oxidized to Br_2 vapor:

$$2Br^- \rightarrow Br_2 + 2e^-. \qquad (5.4.107b)$$

The electrochemical cell was operated at 300 °C.

In all such processes, the difference in the electrochemical potential (μ_{ij}^{el}, definition (3.3.27)) driving the charged species through the membrane is given by

$$\Delta\mu_{ij}^{el} = \mu_{ij}^{el}\Big|_{\text{cathode}} - \mu_{ij}^{el}\Big|_{\text{anode}}$$

$$= \left(\mu_{ij}\Big|_{\text{cathode}} - \mu_{ij}\Big|_{\text{anode}}\right) + Z_i \mathcal{F} \Delta\phi. \qquad (5.4.108a)$$

Since the membrane essentially has no free species, the process is controlled by the potential difference, $\Delta\phi$, driving the negatively charged species like OH^-, S^{2-}, Br^-, etc. The flux of the ionic species may be obtained from the Nernst–Planck relation (3.1.106) for a membrane of thickness Δy in an integrated form as

$$J_{iy}^* = \frac{C_i D_i Z_i \mathcal{F}}{RT}\frac{\Delta\phi}{\Delta y} + D_i \frac{\Delta C_i}{\Delta y}. \qquad (5.4.108b)$$

The current density is obtained from (3.1.108c) as

$$i = \frac{C_i D_i Z_i^2 \mathcal{F}^2}{RT}\frac{\Delta\phi}{\Delta z} + \mathcal{F} Z_i D_i \frac{\Delta C_i}{\Delta z}, \qquad (5.4.108c)$$

assuming for the time being that no other species contribute to the current and $J_i^* = N_i$. Here $\mathcal{F}$ is Faraday's constant. For considerations on the voltage needed to run such a separation cell, consult Robinson *et al.* (1998).

5.4.5 Separation through solid nonporous membrane

Gaseous mixtures or liquid solutions containing small solute molecules may be separated by transport through a solid nonporous membrane containing reactive groups or sites. This is different from dialysis in the sense that dialysis is generally defined to separate crystalloids from colloids by diffusion through a membrane which is generally microporous. Although Section 5.4.3 considered an example of the transport of small molecules through a nonporous polymeric membrane, dialysis, certainly hemodialysis, employs generally microporous polymeric membranes; no reactions are involved in the membrane. Here we touch very briefly on examples of gas separation through nonporous membranes containing reactive groups/agents.

5.4.5.1 *Fixed carrier membranes*

Solid nonporous polymeric membranes/films have been developed that contain reactive groups or complexing agents. These agents/groups are not mobile, unlike the carrier species in liquid membranes studied in Section 5.4.4. Instead, the permeating solute molecule complexes with such a group/agent molecule and then dissociates/jumps to bind with the next group/agent molecule down the permeation pathway. The permeating solute molecule is assumed to follow two simultaneous pathways. One consists of regular solution diffusion through the polymeric membrane. The second consists of a parallel pathway involving complexation/binding and decomplexation/jumping assisted by the reactive groups/complexing agents dispersed through the polymer. A range of behavior is possible, depending on the nature of the complexation, the density/concentration of the reactive groups/agents and the behavior of the rest of the polymeric matrix. A brief review of the examples and models is available in Way and Noble (2001).

5.4.5.2 *Oxygen separation through thick mixed-conducting solid oxide membrane*

In the examples considered in Section 5.4.4.6, the electrode (cathode) converted one of the gas species, e.g. O_2, from a gas mixture to, say, OH^- in an electrochemical cell in the presence of an applied voltage. The hydroxyl ion so produced was transported through the SLM/ion exchange membrane to the other electrode to regenerate O_2, which was evolved at the electrode as a pure gas; the electron generated in the reaction was supplied to the external electrical circuit. No electron transport was allowed through the SLM/ion exchange membrane.

Mixed-conducting dense (nonporous) solid oxide membranes prepared out of appropriate perovskite ceramic materials are such that both ion conduction and electron transport can occur through the membrane at high temperatures; the material is otherwise impermeable to gases. No voltage is applied across the membrane.

The stoichiometry of a perovskite oxide ideally is ABO_3, where A and B are trivalent cations located at two different crystallographic sites. Consider the B cation, which has a fixed valence. If there is a partial substitution at the A site (typically a rare earth) with a lower valence cation, then an oxygen vacancy ($V_O^{\cdot\cdot}$) is formed to compensate for the reduced charge of the cations. The composition of an oxide with A-site substitution, for example, is $La_{0.6}A_{0.4}Co_{0.8}Fe_{0.2}O_{3-\delta}$ for cases where A = Sr, Ba, Ca; at elevated temperatures and reduced oxygen partial pressures, it becomes highly defective with high oxygen deficiency (δ is the nonstoichiometry of oxygen in the mole formula). In such an oxide film, a partial pressure gradient of oxygen from the feed (p_{O_2f}) to the permeate side (p_{O_2p}) will create an inverse gradient of oxygen vacancies in the film (high concentration of oxygen vacancies on the permeate side, $C_{v,p}$ compared to those on the feed side, $C_{v,f}$ ($< C_{v,p}$)):

$$\text{feed side} \qquad \frac{1}{2}O_2 + V_O^{\cdot\cdot} \longleftrightarrow O_O^x + 2h^{\cdot} \tag{5.4.109}$$

where O_O^x could be considered a lattice oxygen and $h^{\cdot}$ represents an electron hole. Consequently, $V_O^{\cdot\cdot}$, the oxygen vacancy sites, diffuse from the side of p_{O_2p} to the side of p_{O_2f}; correspondingly, electron holes move in the opposite direction, from p_{O_2f} to p_{O_2p}:

$$\text{permeate side} \qquad O_O^x + 2h^{\cdot} \longleftrightarrow V_O^{\cdot\cdot} + \frac{1}{2}O_2. \tag{5.4.110}$$

Thus there are two mobile charged species in the membrane: the oxygen vacancy and the electron hole. The membrane is chargewise neutral everywhere; there is no macroscopic electrical field or potential gradient ($\nabla\phi$) across the membrane between the two membrane surfaces. Therefore there is no current in the system; however, a net transport of O_2 takes place, as was first observed by Teraoka *et al.* (1985).

Quantitative descriptions of these transport processes are available in Heyne (1977) and Gellings and Bouwmeester (1992). The net result of such analysis is an expression for oxygen flux through a membrane of thickness δ_m subjected to a feed side oxygen partial pressure of p_{O_2f} and a permeate side partial pressure of p_{O_2p} ($<p_{O_2f}$):

$$J_{O_2} = \frac{RT\overline{\sigma}_m}{16\mathcal{F}^2\delta_m} \ln\left(\frac{p_{O_2f}}{p_{O_2p}}\right). \tag{5.4.111}$$

Here $\overline{\sigma}_m$ is an average conductivity of the ionic and electronic species in the mixed-conducting ceramic membrane over the oxygen partial pressure range across the membrane. This approach, based on a relatively thick membrane (> 500 μm), has been modified by Lin *et al.* (1994) for thin mixed-conducting oxide membranes by considering, in addition, surface reactions of the adsorbed oxygen species on the two membrane surfaces. Due to the absence of the necessary fundamentals of simultaneous electronic transport and ionic transport in this book, no fundamental analysis leading up to (5.4.111) is provided here.

In common parlance, technologies employing such concepts are often identified as *ion transport membrane* (ITM) technology for oxygen generation/separation/production. Dyer *et al.* (2000) provide details of the various related technologies. For example, air containing CO_2, H_2, etc., at 800–900 °C and 100–300 psia is passed over the ITM membrane. A low-pressure oxygen permeate stream (i.e. pure O_2) is obtained at a fraction of an atmosphere. The ITM is thin and is supported on layers with larger pore dimensions, so that the mechanical load on the membrane is reasonable. The support layer is also made of the same material as the ITM membrane to minimize differential thermal expansions and chemical interaction between the membrane and the support.

Problems

5.2.1 When chlorine is absorbed from an inert gas into water, the following reversible reaction takes place in water: $Cl_2(aq) + H_2O(aq) \longleftrightarrow HOCl(aq) + H^+(aq) + Cl^-(aq)$.
Develop a relation between the chlorine partial pressure in the gas phase and the total molar concentration of chlorine in the liquid phase (the equilibrium constant for the reaction is *K*). Assume ideal behavior.
$\left(\text{Ans. } C^t_{Cl_2} = \frac{p_{Cl_2}}{H^C_{Cl_2}} + \left(K/H^C_{Cl_2}\right)^{1/3}\left(p_{Cl_2}\right)^{1/3}.\right)$

5.2.2 (a) Earlier work in Germany in 1953, quoted in Lightfoot *et al.* (1962), suggested the dissolution of NH_3 from an inert gas into water to be taking place in the following fashion:

$$\begin{array}{l} NH_3(g) \\ \uparrow\downarrow \\ NH_3(aq) + H_2O \overset{K_1}{\longleftrightarrow} NH_4OH(aq) \overset{K_2}{\longleftrightarrow} NH_4^+(aq) + OH^-(aq). \end{array}$$

If Henry's law constant for NH_3 absorption is described via $p_{NH_3} = H^C_{NH_3} C_{NH_3}$, and water concentration is not ignored in the first reaction step (involving K_1), show that the total ammonia concentration in the liquid phase is given by

$$C^t_{NH_3} = \frac{p_{NH_3}}{H^C_{NH_3}}(1 + K_1 C_{H_2O}) + \left(\frac{p_{NH_3}}{H^C_{NH_3}}\right)^{1/2} (K_1 K_2 C_{H_2O})^{1/2}.$$

(b) Modern spectroscopic methods doubt the existence of NH_4OH:

$$\begin{array}{l} NH_3(g) \\ \uparrow\downarrow \\ NH_3(aq) + H_2O \overset{K_b}{\longleftrightarrow} NH_4^+(aq) + OH^-\ (aq), \end{array}$$

where water concentration is ignored and $K_b = 1.86 \times 10^{-5}$ gmol/liter at 25 °C. Show that

$$C^t_{NH_3} = \frac{p_{NH_3}}{H^C_{NH_3}} + \left(\frac{p_{NH_3}}{H^C_{NH_3}}\right)^{1/2} K_b^{1/2}.$$

Given $H^C_{NH_3} = (10^3/2.6)$mmHg-liter/g mol, develop a relation between $C^t_{NH_3}$ and p_{NH_3}.

5.2.3 Consider pure water as an absorbent for SO_2 over CO_2 from an inert gas mixture. Obtain an expression for the separation factor of SO_2 over CO_2 defined by

$$\alpha_{SO_2\text{-}CO_2} = \frac{C^t_{SO_2}\, p_{CO_2}}{p_{SO_2}\, C^t_{CO_2}}$$

at 25 °C in terms of Henry's law constants and any other relevant quantities. Use estimates of thermodynamic quantities like K_{di} to make necessary assumptions and obtain simplified results. You are given $K_{d,SO_2} = 1.7 \times 10^{-2}$ gmol/liter at room temperature and $K_{d,CO_2} = 4.16 \times 10^{-7}$ gion/liter at 20 °C.

$$\left(\text{Ans. } \alpha_{SO_2\text{-}CO_2} = \left\{\frac{H^C_{CO_2}}{H^C_{SO_2}} + \frac{K^{1/2}_{d,SO_2} H^C_{CO_2}}{\left(p_{SO_2} H^C_{SO_2}\right)^{1/2}}\right\}.\right)$$

5.2.4 Develop a relation between the equilibrium partial pressure of H_2S in the gas phase, p_{H_2S}, and the fractional consumption f of a base B which absorbs H_2S in an aqueous solution by the reaction (5.2.21). You are given that C^t_B is the total molar base concentration to start with; K^C_S is the equilibrium constant for the given reaction in terms of molar concentrations; $H^C_{H_2S}$ is Henry's law constant for H_2S in the solution. $\left(\text{Ans. } p_{H_2S} = \frac{H^C_{H_2S}}{K^C_S} C^t_B \frac{f^2}{(1-f)}.\right)$

5.2.5 (a) Develop the equilibrium relation between the mole fraction of CO_2 in the gas phase of an absorber, at a pressure of 2.2 MPa and a temperature of 110 °C, and the fractional consumption, f, of a 3-molar aqueous solution of K_2CO_3. The following information has been summarized by Rousseau and Stanton (1988) for this system:

(1) reaction equilibrium constant: $K = 165 \exp(1312/T)$, where T is in kelvin;

(2) Henry's law constant H for CO_2 at 110 °C in m gmol/liter solution of potassium carbonate is given by $\log_{10}(H/H^0) = 0.125m$, where H^0 is Henry's law constant in water;

(3) the density, ρ, of aqueous K_2CO_3 solution at T K is

$$\rho = 1.1464 - 0.0006T + 1.11 u_{K_2CO_3},$$

where $u_{K_2CO_3}$ is the weight fraction of K_2CO_3 is solution;

(4) the value of H for a 20 wt% solution of K_2CO_3 at 110 °C is 51 MPa-liter/gmol.

(b) What would be the corresponding relation for a stripper being operated at a pressure of 0.2 MPa?
(Answer: (a) $x_{CO_2,g} = 0.0806(f^2/(1-f))$;
(b) $x_{CO_2,g} = 0.887(f^2/(1-f))$.)

5.2.6 The absorption of nitrogen oxides (NO, NO_2, N_2O_4, N_2O_5, N_2O, etc.) into water is employed to control air pollution by nitrogen oxides as well as to produce nitric acid. Of these oxides, NO, NO_2 and N_2O_4 are much more important than the others. In the gas phase, the equilibrium constant K_p for the gas phase reaction

$$2NO_2(g) \overset{K_p}{\longleftrightarrow} N_2O_4(g) \tag{5.P.1}$$

has been described by

$$\log K_p = \frac{2993}{T(K)} - 9.23. \tag{5.P.2}$$

The complex absorption/reaction can be represented by two reactions (Sherwood *et al.*, 1975):

$$2NO_2(N_2O_4)(g) + H_2O(l) \longleftrightarrow HNO_3(l) + HNO_2(l); \tag{5.P.3}$$

$$3HNO_2(l) \longleftrightarrow HNO_3(l) + H_2O(l) + 2NO(g). \tag{5.P.4}$$

These two reactions are equivalent to the following overall reaction, where K_c incorporates the HNO_3 concentration dependence:

$$1.5N_2O_4(g) + H_2O(l) \overset{K_c}{\longleftrightarrow} 2HNO_3(l) + NO(g). \tag{5.P.5}$$

Sherwood *et al.* (1975) have identified HNO_3-concentration-dependent K_c from the literature to be

$$\log K_c = 7.41 - 20.3u_{HNO_3} + 32.5u_{HNO_3}^2 - 30.9u_{HNO_3}^3, \tag{5.P.6}$$

where u_{HNO_3} is the mass fraction of HNO_3 in the liquid phase. For atmospheric pressure absorption at 25 °C from a gas containing NO ($p_{NO} = 0.12$ atm) and combined NO_2 and N_2O_4 ($p_{NO_2+N_2O_4} = 0.27$ atm), determine the equilibrium HNO_3 mass fraction in the aqueous phase. (Ans. 0.61.)

5.2.7 In liquid-phase polycondensation reactions to produce a polymer P_1, say, Nylon 6,6, the following system is separated by distillation (Grosser *et al.*, 1987): nonvolatile adipic acid (A), volatile hexamethylene diamine (B), nonvolatile salt (C) and volatile water (W) (Jacobs and Zimmerman, 1977). The reactions are

$$A + B \longleftrightarrow C \quad (1); \qquad A + B \longleftrightarrow P_1 + W \quad (2).$$

Assume:

(a) the vapor-liquid system to consist only of species A, B, C and W;
(b) only reaction (1) to take place, equilibrium constant is K^x;
(c) the value of α_{WB} is constant and $\alpha_{WB} > 1$.

Express the vapor-phase mole fraction of B, y_B, in terms of the liquid-phase mole fractions of A, B (i.e. x_A, x_B), α_{WB} and K^x. (Ans. $y_B = \{x_B/(x_B + \alpha_{WB}(1 - x_A - x_B - K^x x_A x_B))\}$.)

5.2.8 MacKenzie and King (1985) have studied the extraction of ammonia from sour wastewater into solvents containing acidic cation exchangers. At the *pH* of the wastewaters, ammonia is present primarily as an NH_4^+ ion. An organic acid, e.g. D2EHPA (bis(2-ethylhexyl) phosphate) in a solvent (e.g. toluene) complexes with a cation M^+ as follows:

$$M^+(aq) + HA(org) = M^+A^-(org) + H^-(aq).$$

After the extraction of ammonia, the solvent has to be regenerated by stripping. Stripping is possible when the ion pair M^+A^- in the organic phase dissociates.

(a) Write down the reactions for this dissociation and stripping.
(b) Obtain an expression for the partial pressure of NH_3 in the gas phase in terms of the equilibrium constant K for the reaction in part (a) and various other concentrations.
(c) Suggest conditions that will increase the NH_3 partial pressure in the gas phase.

5.2.9 Biodegradation of organic pollutants in aqueous waste streams is often hindered if the aqueous solution has extreme *pH* and/or high salt concentration. One can, however, extract the pollutant into an organic solvent and then back extract the pollutant into an aqueous stream amenable to biodegradation. p-Nitrophenol (PNP) is one such pollutant. PNP, a weak acid, reacts with aqueous alkalis and forms the phenolate ion, PNP^- (Tompkins *et al.*, 1992):

$$PNP(aq) + OH^-(aq) \longleftrightarrow PNP^-(aq) + H_2O.$$

Usually the ion PNP^- has very low solubility in the organic phases employed for extraction. Show that the effective partition coefficient for PNP in all forms, κ''_{io}, is related to the partition coefficients $\kappa'_{PNP,o}$ for PNP and $\kappa'_{PNP^-,o}$ for PNP^- by

$$\kappa''_{io} = \frac{\kappa'_{PNP,o} + \kappa'_{PNP^-,o}(K_{d1}/C_{H^+})}{1 + (K_{d1}/C_{H^+})},$$

where K_{d1} is the ionization constant of the acid, PNP.

Calculate the value of κ''_{io} as a function of C_{H^+}. You are given $\kappa'_{PNP,o} = 89$; $\kappa'_{PNP^-,o} = 0.14$; pK_1 of PNP = 7.1. (Ans. $\kappa''_{io} = (89 + 0.14(10^{-7.1}/C_{H^+})/[1 + (10^{-7.1}/C_{H^+})]$.)

5.2.10 The species 8-quinolinol (HQ) is amphiprotic. At high *pH*, it preferentially ionizes in the aqueous phase:

$$HQ \overset{K_{d1}}{\longleftrightarrow} HQ^- + H^+.$$

At low *pH* it behaves as a base in the aqueous medium:

$$H_2Q^+ \overset{K_{d3}}{\longleftrightarrow} HQ + H^+.$$

The dissociation constants for the two reactions are K_{d1} and K_{d3}, respectively. Species HQ, Q^- and H_2Q^+ are, respectively, species 1, 2 and 3. Determine the overall distribution coefficient of 8-quinolinol between the aqueous and an organic phase which has only the species HQ. Show that it may be simplified as follows:

$$\kappa''_{1o} = \kappa'_{1o}\frac{C_{H^+}K_{d3}}{K_{d1}K_{d3} + K_{d3}C_{H^+} + C^2_{H^+}},$$

where $\kappa'_{1o} = C_{1o}/C_{1w}$.

5.2.11 Robinson and Cha (1985) have provided the following example for controlled *pH* extraction. Flurbiprofen is a nonsteroidal anti-inflammatory agent. In a synthesis of this compound, a carboxylic acid, a small amount of a '*dimeric acid*' (DA) is formed (Figure 5.P.1). It was found that simple crystallization does not completely eliminate DA from the product. Separation of these two species is possible by high vacuum distillation. It was found to be impractical because of a combination of high boiling and melting points which resulted in sublimation. Although both are carboxylic acids, the measured *pK* values are quite different: 6.3 (flurbiprofen) and 11.4 (DA). It should be noted that the *pK* for DA is a psuedo-*pK* value as it was measured in a 2-phase

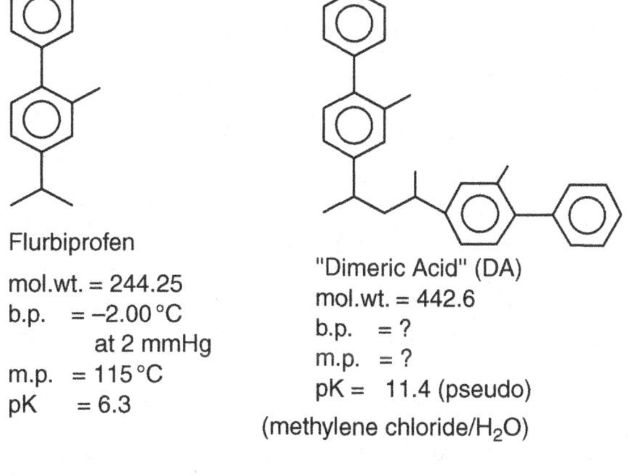

Figure 5.P.1. Properties and structure of Flurbiprofen and "Dimeric Acid" (DA). Reprinted, with permission, from Robinson and Cha, Biotechnol. Progr., *1(1), 18 (1985). Copyright © [1985] American Institute of Chemical Engineers (AIChE).*

(methylene chloride/H_2O) system because of low water solubility. Therefore the extraction of Flurbiprofen from the methylene chloride layer into water should be an efficient way of separating the compounds. Determine the separation factor at $pH = 8.3$. $\left(\text{Ans. } \alpha''_{\text{DA-Flurbiprofen}} = \alpha'_{\text{DA-Flurbiprofen}} \times 101.\right)$

5.2.12 Develop the relation (5.2.89) for the acetic acid distribution coefficient between a nonpolar organic solvent and water. Identify the condition so that κ''_{1o} is independent of the *pH*, given $K_{d1} = 1.5 \times 10^{-5}$. Identify the aqueous acetic acid concentration beyond which κ''_{1o} changes by more than 10% from the value valid for negligible dimerization.

5.2.13 Kuo and Gregor (1983) have measured the distribution coefficient of acetic acid between water and pure decalin at 27 °C at different initial acid concentrations (see Table 5.P.1).

Table 5.P.1

Initial acid concentration (M)	κ''_{1o}
0.4	0.0236
0.133	0.0207
0.0866	0.0264
0.0674	0.0271

If the volumes of the two phases were the same, speculate first about the equilibrium acetic acid concentration in the aqueous phase for each of the four cases. Suggest a way of finding out whether there was any dimerization in the organic phase. Develop an estimate of the dimerization equilibrium constant K_2. (Ans. $K_2 \cong 0$.)

5.2.14 The distribution coefficient of penicillin G between a solvent (isoamyl acetate or methyl isobutyl ketone (MIBK)) and an aqueous solution is shown in Figure 5.2.4(b) as a function of aqueous solution *pH*. In the initial aqueous clarified fermentation broth, there are other impurities from which penicillin has to be separated. Suggest a *pH* of the aqueous broth at which solvent extraction should take place. Suggest also the *pH* of an aqueous solution into which the penicillin is to be back extracted from the solvent extract to purify it further. Assume that the impurity distribution coefficient is independent of *pH*.

5.2.15 Conventionally, penicillin G, a weak monocarboxylic acid, is extracted into polar solvents like n-butyl acetate rapidly at a low temperature to reduce its loss due to instability. Centrifugal extractors are frequently used. Reschke and Schügerl (1984) employed reactive extraction of penicillin G (HP) by a secondary amine, Amberlite LA-2, 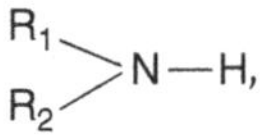

, and suggested the following extraction mechanism:

$$H^+(w) + P^-(w) + A(o) \overset{K}{\longleftrightarrow} AHP(o),$$

where, in the aqueous phase, the penicillin G dissociation takes place via

$$HP(w) = H^+(w) + P^-(w), pK = 2.75.$$

Tamada *et al.* (1990) have suggested a general reaction mechanism of carboxylic acids with amine extractants:

$$mA(o) + nHP(w) \overset{K}{\longleftrightarrow} A_m(HP)_n(o).$$

Spectroscopic studies indicate that, for a secondary amine Amberlite LA-2 present in a nonpolar solvent like kerosene (no extraction of HP as such) and HP in water, the reactive extraction mechanism is

$$2A(o) + 2HP(w) \overset{K}{\longleftrightarrow} A_2(HP)_2(o)$$

(a) Obtain a quartic algebraic equation for the equilibrium concentration of HP in the aqueous phase as a function of *pK*, *pH*, *K* and the total concentrations of penicillin G in water (C^t_{HP}) and amine LA-2 in organic phase (C^t_A) added initially.

(b) How will you obtain the extent of penicillin G extracted?

5.2.16 It is proposed to separate a mixture of monoacidic bases A and B present in an organic solvent at a total concentration of $C_o^t (= C_{Ao} + C_{Bo})$ via dissociation extraction into an immiscible aqueous phase containing a strong extracting agent present at a concentration of C_w^t, which is stoichiometrically deficient to protonate C_o^t. The stronger base, species B, will be protonated more than species A; it will therefore be extracted more into the aqueous phase, provided that κ'_{Ao} is quite close to κ'_{Bo}. Assume, however, that $\kappa'_{Ao} > \kappa'_{Bo}$. Given K_{dA} and K_{dB} for the two bases, and unity for the activity coefficients, develop an expression for $\alpha''_{AB} (= \kappa''_{Ao}/\kappa''_{Bo})$ in terms of κ'_{Ao}, κ'_{Bo}, C_w^t, K_{dA}, K_{dB}, C_{Ao} and C_{Bo}. $\left(\text{Ans. } \alpha''_{AB} = \left(\frac{\kappa'_{Ao}}{\kappa'_{Bo}}\right) \frac{\left\{C_w^t \kappa'_{Bo} + C_{Bo} + \left(\frac{K_{dA}}{K_{dB}}\right)\left(\frac{\kappa'_{Bo}}{\kappa'_{Ao}}\right) C_{Ao}\right\}}{\left\{C_w^t \kappa'_{Bo}\left(\frac{K_{dA}}{K_{dB}}\right) + C_{Bo} + \left(\frac{K_{dA}}{K_{dB}}\right)\left(\frac{\kappa'_{Bo}}{\kappa'_{Ao}}\right) C_{Ao}\right\}}.\right)$

5.2.17 Many amino acids are produced by microbial fermentation processes. Recovery and purification of the amino acid (Am) from the fermentation broth is quite costly. Organic liquid membranes or solvents containing quaternary ammonium chloride ($R_4N^+Cl^-$) have been employed to extract the amino acid in the negatively charged state (Am^-) in exchange for the chloride ion:

$$R_4N^+Cl^-(\text{org}) + Am^-(\text{aq}) \overset{K_e}{\longleftrightarrow} R_4N^+Am^-(\text{org}) + Cl^-(\text{aq}).$$

Knowing the dissociation equilibrium constants K_1 and K_2 for the amino acid in water, i.e.

$$Am^+ \overset{K_1}{\longleftrightarrow} Am^{\pm} + H^+; \qquad Am^{\pm} \overset{K_2}{\longleftrightarrow} Am^- + H^+,$$

and the partition coefficient, $\kappa'_{Cl^-,o}$, of the chloride ion $\left(\kappa'_{Cl^-,o} = C_{R_4N^+Cl^-,o}/C_{Cl^-,w}\right)$ between the aqueous and the organic phases, determine the effective partition coefficient, $\kappa''_{Am,o}$, of the amino acid between the broth and the liquid membrane/organic solvent containing the quaternary compound. You are given that $C_{Am,w}^t$ is the total concentration of the amino acid in the broth in equilibrium with the solvent.

5.2.18 Amino acids are frequently separated by ion exchange based chromatography. Consider a particular amino acid present in a salt solution at a total concentration of C_{Amw}^t.

(a) Identify the equations whose simultaneous solution is needed to determine the concentration of the amino acid in a particular form, for example C_{Am^+w}, C_{Am^-w} as a function of the *pH*, C_{Amw}^t and salt concentration.

(b) Develop an analytical solution for C_{Am^+w}, for example at a very low *pH* (well below pK_1). *Note*: $C_{Cl^-} - C_{Na^+} = C_o$, a quantity used in the expression for the analytical solution.

(Ans. (b) $C_{Am^+w} = 0.5\{(K_1 + C_o + C_{Amw}^t) - \sqrt{(K_1 + C_o + C_{Amw}^t)^2 - 4C_oC_{Amw}^t}\}$.)

5.2.19 In chromatographic separation of two weak acids, the detector records both the dissociated and undissociated species. For an acid species HA, develop an expression for the distribution ratio at any *pH* in terms of the two limiting distribution ratios, one corresponding to fully dissociated acid, $k'_{A1}|_A$, and the other corresponding to no dissociation at all, $k'_{A1}|_{HA}$. Use linear addition for the detector response, with *f* being the fraction present in the dissociated form in the eluent. Assume only that the undissociated species partitions between the mobile and the stationary phase.

5.3.1 Solvent extraction of p-nitrophenol (PNP) from a high-salt aqueous solution into 1-octanol acting as solvent was considered in Problem 5.2.9. After the extraction of the pollutant PNP into 1-octanol, the solvent is regenerated by contacting the solvent containing PNP with another aqueous solution; the solution may be basic or is amenable to biodegradation of PNP (Tompkins *et al.*, 1992). Microporous hydrophobic polypropylene membranes (Figure 3.4.10) were used by these authors to back extract PNP from 1-octanol into the aqueous back extraction phase.

(a) Plot the concentration profile of PNP from the organic phase through the organic-filled membrane pores to the aqueous back extraction phase.

(b) Identify the individual mass transport coefficients or resistances for each region for PNP transport and relate them to an overall PNP mass-transfer coefficient. Specify the role of solute ionization in developing these expressions. Assume that the aqueous-phase mass-transfer coefficients for PNP and PNP^- are essentially similar.

(c) Obtain the limiting expressions for the overall mass transport coefficient for PNP for two conditions in the aqueous back extraction phase: low *pH* and high *pH*.

5.4.1 In pervaporation separation of volatile organics from an aqueous solution through a polymeric membrane (Section 3.4.2.1.1), Böddeker *et al.* (1988) have shown that, for small values of organic concentration in feed aqueous solution, the mass flux of the volatile organic species i (whose feed water concentration is C_{if}) is approximately given by

$$n_i \approx n_{\text{water}} \left[P_i^{\text{sat}} \gamma_{if} C_{if} / P_s \right].$$

Here, P_s is the total pressure on the permeate side, n_{water} is the water mass flux through the membrane, P_i^{sat} is the vapor pressure of species i whose activity coefficient in the feed solution is γ_{if} at a molar concentration C_{if}. Assume that P_i^{sat}, γ_{if} and n_{water} are unaffected by pH levels. For the solute phenol, indicate how n_i will change as the feed water pH is changed. Given $pK_{\text{phenol}} = 10.4$ at 50 °C, show that the fractional reduction in phenol flux with pH is given by $(1/\{1 + 10^{-pK_{\text{phenol}}+pH}\})$.

5.4.2 To utilize poly(acrylic acid) (represented as HL) as a chelating polymer in polymer-assisted ultrafiltration based removal of a divalent heavy metal (M^{n+}, $n = 2$) present in wastewater, it is useful to consider the following polymer ligand-metal complexation equilibria:

$$M^{2+} + HL \overset{K_1}{\longleftrightarrow} LM^+ + H^+; \tag{5.P.7}$$

$$LM^+ + HL \overset{K_2}{\longleftrightarrow} ML_2 + H^+; \tag{5.P.8}$$

$$ML_2 + 2HL \overset{K_3}{\longleftrightarrow} L_2M(HL)_2. \tag{5.P.9}$$

When one carries out batch ultrafiltration, by measuring the concentration of free (uncomplexed) M^{2+}, one can find out q, the amount of metal ions bound to polymer (mol/mol of polymer). One can also determine n_{avg}, the average coordination number. Develop expressions for (a) q and (b) n_{avg} in terms of K_1, K_2, K_3, K_{dHL}, C_{H^+} and $C_{M^{2+}}$ for a divalent metallic cation. You are given that the total poly(acrylic acid) concentration is C_{HL}^t.

5.4.3 In Section 5.4.4, we studied how phenol can be removed very efficiently from wastewater through an oily liquid membrane into a caustic-containing receiving phase.

(a) Suppose the wastewater contains bases like NH_3 and amines ($R\text{-}NH_2$, say). These are quite soluble in the oily liquid membrane. What would be the effect of these species on phenol removal from wastewater?

(b) If the wastewater contains only bases like NH_3 and amines, what should the receiving aqueous phase contain to achieve wastewater purification similar to that for phenol?

(c) If you have any oily membrane and the feed wastewater contains a very strong acid, e.g. HCl, would you use the technique explained in Section 5.4.4?

(d) Wastewater contains phenol and HCl, which are both acidic, and the receiving phase contains caustic soda. Describe the separation that can be achieved.

5.4.4 Consider an oily membrane without any acidic cation exchanger (see Problem 5.2.8). Through such a membrane, which may be a liquid or a polymeric membrane, it is possible to remove ammonia NH_3 ($i = 1$), but not the ammonium ion NH_4^+ ($i = 2$). In a process somewhat analogous to that described for phenol removal in Sections 5.4.3 and 5.4.4, the feed solution is basic ($pH > 10$); small amounts of ammonia in the feed ($j = f$) permeate through the membrane to the other side ($j = p$), where there is a highly acidic solution to react with the ammonia.

(a) Formulate the governing mass balance equations and associated equations for chemical reaction equilibrium for the transfer of NH_3 from a basic solution in the feed vessel (volume V_f) to the permeate/receiving vessel (volume V_p) containing an acid (sulfuric or phosphoric acid) at a high concentration.

(b) Write down the boundary conditions.

(c) Develop a solution for the problem in terms of the total ammonia concentration in the feed solution as a function of time.

(d) Estimate the total ammonia concentration in the feed as $t \to \infty$ for a feed $pH = 10$ and permeate $pH = 2$.

You are given

(i) $NH_3 + H_2O \overset{K_{d1}}{\longleftrightarrow} NH_4^+ + OH^-$; $K_{d1} = 1.86 \times 10^{-5}$ (25 °C) without water concentration;

(ii) $NH_3 + H^+ \overset{K}{\longleftrightarrow} NH_4^+$; $K = 1.86 \times 10^9$ (25 °C).

5.4.5 The facilitated transport separation of CO_2 from a gas mixture through an aqueous liquid membrane of diethanolamine (DEA) occurs via the following reactions:

$$CO_2 + R_2NH \underset{k_{-Am}}{\overset{k_{Am}}{\rightleftarrows}} R_2NCOO^- + H^+ \quad (1); \qquad R_2NH + H^+ \longleftrightarrow R_2NH_2^+. \quad (2)$$

Reaction (2) is considerably faster than reaction (1) (Danckwerts and Sharma, 1966), and is therefore assumed to be in equilibrium (the equilibrium constant for this protonation reaction is K_p, whereas that for (1) is K_{Am}). Assume a negligible H^+ contribution to the electroneutrality relation. Develop two governing ordinary differential equations (one for CO_2 and the other for the free amine R_2NH) describing the facilitated transport of CO_2 and write the boundary conditions. You are given that the total initial amine concentration in the aqueous liquid membrane is $C^t_{R_2NH}$. The governing ordinary differential equations should contain only the concentrations of free CO_2 and free amine and $C^t_{R_2NH}$.

5.4.6 Consider the facilitated transport based separation of species A from C in a feed gas mixture, through a liquid membrane that contains a nonvolatile carrier species B, with which A reacts reversibly via $A + B \longleftrightarrow AB$. The equilibrium constant K for this reaction is 12 (M^{-1}). The molar concentration of species B added to the liquid is 6 M. The feed molar concentration of species A is 0.1 M. Assume $D_{ABM} = D_{AM}$ and a thick membrane. Calculate the ratio of the increased flux of species A with respect to the nonfacilitated flux of species A. The permeate molar concentration of species may be neglected with respect to the feed concentration. State your assumptions.

5.4.7 Platinum chloride anions ($PtCl_6^{2=}$) from an acidic solution are being extracted through a liquid membrane into a basic solution. The organic liquid membrane in which platinum chloride anions are not soluble contains trioctylamine (TOA), with which the following complexation occurs by an interfacial reaction:

$$2R_3N(org) + 2H^+ + PtCl_6^= \longleftrightarrow (R_3NH)_2PtCl_6(org).$$

Assume a fast interfacial reaction. There are boundary layer resistances to $PtCl_6^=$ diffusion on each side of the membrane.

(a) Draw the concentration profiles of various species (H^+, R_3N (org), $PtCl_6^=$ and the complex) in various regions of the system.
(b) Identify which out of facilitated transport, countertransport and cotransport is operative here.
(c) Write down the governing metal flux equations (flux = coefficient × concentration difference) for each region (the boundary layers and the membrane).
(d) What is the maximum value of the metal flux in each region?

5.4.8 Copper is extracted from an acidic solution into an organic phase containing 2-hydroxy 5-nonylbenzophenone oxime (RH) by an interfacial reaction

$$Cu^{2+} + 2\overline{RH} \longleftrightarrow \overline{CuR_2} + 2H^+,$$

where the overbar represents an organic-phase species. The interfacial reaction rate was found by Komasawa and Otake (1983) to be of first order in oxime concentration for concentrations more than 12 gmol/m^3 and of first order in Cu^{2+} concentration. Without neglecting the backward interfacial reaction rate, write down the flux expressions for each of the three resistances leading to the reaction, as well as for the diffusion of products. Assume the diffusion of protons to be extremely rapid. Develop an expression for copper extraction flux in terms of bulk concentrations, mass-transfer coefficients and rate constants. If the oxime concentration is very high, develop the simplified expression for N_A, where A represents the divalent metal cation Cu^{2+}.

6

Open separators: bulk flow parallel to force and continuous stirred tank separators

Chapter 4 described the extent of separation that can be achieved in a closed vessel under three basic categories of separation: phase equilibrium based separations; external force based separations; membrane based separations. Beginning with Chapter 6, we focus on separation achieved in an open vessel: fluid streams and/or solid streams may flow into and/or out of the vessel. Thus, we have bulk flow/s in and/or out of this device. A broad variety of bulk flow patterns can exist in a separation vessel. We will, however, mostly study separations under three general categories of bulk flow configurations defined with respect to the direction of the force which is the source of the basic separation phenomenon. The three general categories of bulk flow-force combinations are:

(a) bulk flow of phase(s) parallel to the force direction;
(b) bulk flow of feed-containing fluid phase/region perpendicular to the force direction;
(c) bulk flow of two fluid phases/fractions/regions perpendicular to the force direction.

In the bulk flow-force combination of (c), there can be cases where, instead of two fluid phases, one can have one fluid phase and another solid phase. Categories (b) and (c) provide a broader and more useful framework than the category of bulk flow perpendicular to the force direction illustrated by Giddings (1991) using a few examples.

Bulk flow(s) in one or more directions in relation to the force direction creates conditions leading to the development of difference in concentrations of species along the directions of bulk flow(s) and/or force. However, in one type of open separator, where the whole vessel may be considered perfectly mixed, there are no spatial gradients in species concentration anywhere in the vessel. Yet the feed is introduced continuously into, and the product stream(s) is (are) withdrawn continuously from, the vessel. Such open separators are analogous to a continuous stirred tank reactor (CSTR) and may be identified as a continuous stirred tank separator (CSTS). None of the three bulk flow-force combinations (a), (b) or (c) can describe the separation in a CSTS; it provides a separate bulk flow-force combination.

In this chapter we illustrate separation achieved in the bulk flow of phase(s) parallel to the force direction. We will also briefly study CSTSs at the end of this chapter. Batch well-stirred tank based separators without any continuous feed in or product out will also be studied.

Chapter 7 will consider separations achieved under the bulk flow-force combination of (b). Separation systems utilizing the configurations of (c) are treated in Chapter 8. (There will be occasional examples of two combinations of bulk flow and force directions.) Chapters 6, 7 and 8 will generally employ one separator vessel. Reactive separations will be treated immediately alongside non-reactive separations as often as possible. Different feed introduction modes will be considered as required in all three configurations, (a), (b) and (c). Multistage separation schemes, widely used in the processes of gas absorption, distillation, solvent extraction, etc., are studied in Chapter 8 when only one vessel is used. When multiple devices are used to form a separation cascade, an introductory treatment is provided in Chapter 9.

Section 6.1 begins with a description of mechanisms/driving forces/sources that cause bulk flow to take place in the separation device. Following a brief description of these sources of flow into and out of the separation device, as well as inside the device, we provide a brief illustration of a variety of ways in which the feed is introduced into the separator. Section 6.2 will identify the general equations of change in an open separator.

First, we consider the variation of concentration of a species in a solution/mixture with space and time inside a separator. An equation that considers such variations in a phase/region will be illustrated. Since such a concentration distribution in time and separator location is influenced strongly by the flow field in the separation device, the equation of continuity and the equations of motion which generally govern the flow field in the separator will be identified. If the flow field is known, it will be introduced into the equation for species concentration change in a given problem to be studied in Chapters 6, 7 and 8. If the flow field is not known, the equation of continuity and the equations of motion have to be solved to determine the flow field (sometimes along with the equation for concentration change). The solutions of such combinations of equations are, in general, quite demanding and will be avoided in general.

Such equations of change to be presented in Section 6.2 for spatio-temporal variations of species concentrations employ the continuum level description. In this approach, the equations of change are developed over a volume element of microscopic dimensions, such that the fluctuations due to individual molecular motions are averaged out yet the volume element is tiny, of the order of micrometers (say). To reduce the level of complexity in single-phase as well as multiphase systems further, volume-averaging techniques are adopted. These techniques allow one to replace the local velocities and concentrations in the separator by values averaged over a unit cell, one of whose dimensions is small compared to the separator dimensions (Lee *et al.*, 1977a). The resulting equations of change of species concentrations are then said to be considered at what is known as the *pseudo-continuum level* (Lee *et al.*, 1977a).

This approach is almost always adopted when dealing with separation in packed beds, or with a porous medium in general. In such an approach, one averages out the species concentration, velocities, etc., in one or two physical dimensions; the concentration, velocity, etc., then remain a function of, say, one remaining physical dimension and time. The usual physical dimension along which variations are retained is the direction of main flow, whereas the directions perpendicular to this main flow are averaged out. For flow in a tubular packed bed, this implies averaging in the radial and circumferential direction of, say, concentrations and velocities, which can now vary only along the direction of mean flow.

Volume-averaging techniques are useful in obtaining the equation of concentration change of a species in a given phase in a multiphase system. At any given cross section of a device, there will be two such equations if there are two phases/regions. These two equations will be coupled through the boundary conditions at the interfaces of the phases/regions. The solution of each equation in each region will provide the concentration of the species in that phase/region. This task can be quite complex, especially if the flow field in the device, packed bed, distillation plate, etc., is complex.

To develop easier routes to the solutions, we will sometimes adopt simpler approaches in the following chapters. The objective is to develop reasonable descriptions of separation achieved rather than a rigorous theory to describe separation development in every case. This will involve, on the one hand, using approximate/known velocity fields. On the other hand, and more often, we will assume, in the case of equilibrium separation processes, the *equilibrium limit*: we will assume that the immiscible phases in contact and in motion in the separator are at thermodynamic equilibrium with respect to each other/one another (for three or more phases). The actual performance may then be determined by means of estimates of stage efficiency, which quantifies the deviation from thermodynamic equilibrium.

The second broad area in Section 6.2 is concerned with particles. For the separation of particles from a fluid or fractionation of particles, one can adopt an *Eulerian approach* to determine the particle concentration variation as observed by an observer located at a fixed coordinate (x,y,z). In such an approach, the fluid velocity is also what is determined by an observer at (x,y,z) as a function of time. However, an alternative approach, the *Lagrangian approach*, is frequently preferred and will be adopted often. The Lagrangian description of particle motion is obtained by an observer who rides on the particle. The geometrical coordinates (x,y,z) of the particle/observer change with time as the particle changes its location in the device in response to fluid motion and other forces, external and/or diffusive, acting on the particle. In such an approach, the coordinates (x,y,z) of a particle are dependent variables whose values as a function of time in the separation device are of interest. These equations, called *trajectory equations*, are also provided in Section 6.2.

In separation processes involving crystallization, hydrosols, aerosols and liquid–liquid dispersions, we encounter a size-distributed population of particles/drops which may be undergoing change in the separation device due to processes of crystal growth, drop/particle breakage, drop/particle coalescence, etc. For such systems, a *population balance equation* in the form of a general dynamic equation for changes in the particle/drop size distribution function is developed in Section 6.2. An integrated form of this equation has also been provided for a CSTS.

Section 6.3 covers three basic categories of separation when the bulk flow is parallel to the direction of the force. Section 6.3.1 describes how separation is achieved when either the gravitational force, or an electrical force, centrifugal force or inertial force, acts on ions/molecules/particles present in a fluid flowing in a direction parallel to the direction of the force. Gravity-driven elutriation, capillary electrophoresis, countercurrent electrophoresis, centrifugal elutriation and inertial impaction are specific techniques considered here. The equilibrium separation processes of flash vaporization, devolatilization, batch distillation for vapor–liquid systems, liquid–liquid extraction in the differential extraction mode, zone melting/normal freezing for solid/melt–liquid systems and drying for solid–vapor systems are studied in Section 6.3.2. Section 6.3.3 illustrates the operation of membrane separation processes of cake filtration/microfiltration, ultrafiltration, reverse osmosis, pervaporation and gas permeation when the bulk fluid flow is parallel to the force acting perpendicular to the membrane. The important role played by the magnitude of the bulk velocity has been highlighted as often as possible.

Section 6.4 covers continuous stirred tank separators. Section 6.4.1 studies equilibrium separation processes; most of this section is devoted to crystallization, with additional coverage of liquid extraction. Membrane separation processes/devices are sometimes modeled as CSTRs. Section 6.4.2 touches upon a few of these examples, encountered, for example, in ultrafiltration and gas permeation. There are brief treatments of batch systems that are well-stirred in Sections 6.4.1 and 6.4.2 for both equilibrium based and membrane separation processes.

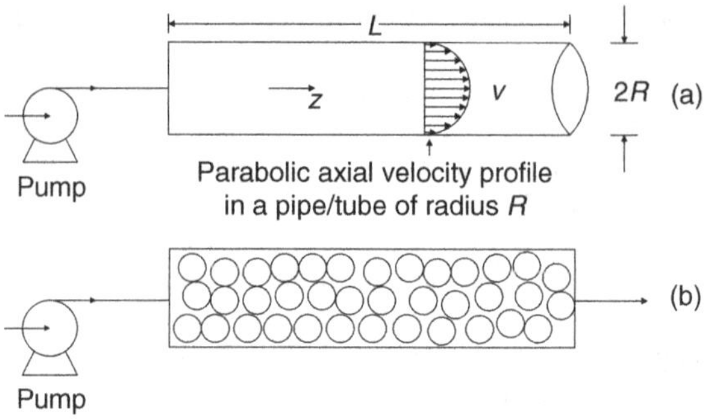

Figure 6.1.1. (a) Pressure-driven laminar flow in a tube/capillary/pipe; (b) pressure-driven flow in a tube/capillary/pipe filled with packings, a packed bed.

6.1 Sources and nature of bulk flow

The bulk flow of fluids, with or without solid particles, can be achieved by a variety of forces: hydrostatic pressure difference, gravity, free convection, capillarity, electrical force causing electroosmotic flow, centrifugal force, surface tension gradient and drag force. For a given force driving the fluid into bulk motion, the nature of the velocity profile will depend on the flow channel geometry and flow obstructions in the channel. Elementary identification of each of the above types of sources of bulk flow will be made in Sections 6.1.1 to 6.1.8, under the assumption that the flow of the liquid/gas is viscous. Bulk flow in separation devices is most often intimately connected with the entry and exit of the feed stream. The various forces identified above have been employed to move feed fluid into and out of separation devices. Such feed fluid flow may be continuous or discontinuous. An elementary illustration of the various types of feed introduction in open systems will be provided in Section 6.1.9.

6.1.1 Hydrostatic pressure induced bulk flow

Consider a separation device, tubular or otherwise, filled with a liquid; assume also that the inlet pipe filled with the liquid is connected to a pump (Figure 6.1.1) at the inlet of the separation device, which is assumed to be horizontal. As the pump runs, it introduces mechanical energy into the liquid, which is driven into the device against whatever flow resistance is offered. As the liquid moves through the device, its hydrostatic pressure is reduced due to frictional losses. The relation between the hydrostatic pressure drop ΔP encountered by the liquid and its volumetric flow rate Q depends on the flow regime, liquid viscosity μ, the flow channel geometry and length L in the mean flow direction (z-coordinate). The following relations are quite useful in a variety of separation devices and processes. Consider first flow in a circular tube of radius R (Figure 6.1.1(a)). Assume steady incompressible flow.

6.1.1.1 *Laminar flow in a straight circular tube of radius* R

- Pressure drop–flow rate relation:

$$\Delta P = \frac{8\mu L}{\pi R^4} Q \,(\text{Hagen–Poiseuille equation}). \tag{6.1.1a}$$

- Parabolic axial velocity profile:[1]

$$v_z(r) = \frac{\Delta P}{4\mu L} R^2 \left[1 - \left(\frac{r}{R}\right)^2\right]. \tag{6.1.1b}$$

- Maximum velocity $v_{z,\max}$ at the pipe center:

$$v_{z,\max} = \frac{\Delta P}{4\mu L} R^2 = 2 v_{z,\text{avg}}, \tag{6.1.1c}$$

where $v_{z,\text{avg}}$ is the average velocity, $\pi R^2 v_{z,\text{avg}} = Q$. These relations are valid for laminar flow of the liquid. Laminar flow continues until around a Reynolds number ($Re = \rho v_{z,\text{avg}} 2R/\mu$) of 2100. For a given Q, ΔP increases linearly with L; it increases drastically as R is reduced.

6.1.1.2 *Laminar flow in a rectangular channel of gap* 2b

The following relations are based on the assumption that the width W of the channel of length L is large compared to the gap between the two parallel walls, $2b$; correspondingly, the viscous effect of the two edges of the channel distance W apart may be neglected.

[1] From now on, the mass average velocity of fluid, v_{tz}, will be replaced by v_z for simplicity.

- Pressure drop–flow rate relation:

$$\Delta P = \frac{3\mu L}{2b^3 W} Q. \tag{6.1.2a}$$

- Axial velocity profile:[2]

$$v_z(y) = \frac{\Delta P b^2}{2\mu L}\left[1 - \left(\frac{y}{b}\right)^2\right]. \tag{6.1.2b}$$

- Maximum velocity $v_{z,\max}$ at the center of the channel:

$$v_{z,\max} = \frac{\Delta P b^2}{2\mu L} = \frac{3}{2} v_{z,\mathrm{avg}}. \tag{6.1.2c}$$

6.1.1.3 *Turbulent flow in a straight tube of radius* R

As the Reynolds number of the flow in a tube or a channel is increased to beyond a certain value (e.g. 2100 for a tube), turbulent flow sets in. For turbulent flow in a horizontal tube of radius R, length L and average axial velocity[3] $\langle v_z \rangle$, a dimensionless quantity f, the Fanning friction factor, is defined as

$$f = \frac{1}{2}\left(\frac{R}{L}\right)\frac{\Delta P}{\rho \langle v_z \rangle^2/2}, \tag{6.1.3a}$$

where ρ is the fluid density. The Fanning friction factor is a function of the tube Reynolds number in the following manner[4]

$$f = \frac{0.0791}{Re^{1/4}} \tag{6.1.3b}$$

over a Re range of $2100 < Re < 10^5$ for fully developed turbulent flow.

6.1.1.4 *Flow in a packed bed*

The flow in a packed bed, where the packing may be spherical, cylindrical, etc., is quite complex (Figure 6.1.1(b)). However, it is often modeled as a collection of cylindrical capillaries of hydraulic radius R_h and length L, which is the packed bed length. Let ε be the fractional void volume of the bed, a_v be the total particle surface area per particle volume, v be the actual interstitial velocity in the void volume between particles and d_p ($= 2r_p$) be the mean particle diameter. Then the superficial velocity v_0 based on the empty cross section of the packed bed is defined as

$$v_0 = v\varepsilon. \tag{6.1.4a}$$

The mean particle diameter d_p is defined as

$$d_p = \frac{6}{a_v}, \tag{6.1.4b}$$

as if the bed consisted of spherical particles of diameter d_p only. The hydraulic radius R_h is defined as

$$R_h = \frac{\text{available flow cross-sectional area}}{\text{perimeter of flow channel in contact with the fluid}}.$$

For a circular capillary of diameter $2R$, $4R_h = 2R$. One can rewrite the hydraulic radius definition as follows:

$$R_h = \frac{\text{void volume available for flow/bed volume}}{\text{particle surface area of bed in contact with fluid/bed volume}} \Rightarrow R_h = \frac{\varepsilon}{a_v(1-\varepsilon)}. \tag{6.1.4c}$$

For *laminar flow* of a fluid of viscosity μ through such cylindrical capillaries of radius $2R_h$, from equations (6.1.1a–c),

$$v_{z,\mathrm{avg}} = \frac{\Delta P}{8\mu L}R^2 = \frac{\Delta P}{8\mu L}4R_h^2 = v = \frac{v_0}{\varepsilon}. \tag{6.1.4d}$$

Substituting (6.1.4b) and (6.1.4c) into the above, one obtains

$$v_0 = \frac{\Delta P}{\mu L}\frac{\varepsilon^3}{(1-\varepsilon)^2}\frac{d_p^2}{72}. \tag{6.1.4e}$$

In reality, the length of the flow path in the hypothetical capillaries in the bed is larger than L due to the path tortuosity. Experimental measurements indicate that the following equation (the *Blake-Kozeny equation*) is instead more accurate as long as $\varepsilon < 0.5$ and $(d_p \rho\, v_0/\mu\,(1-\varepsilon)) < 10$:

$$v_0 = \frac{\Delta P}{\mu L}\frac{\varepsilon^3}{(1-\varepsilon)^2}\frac{d_p^2}{150}. \tag{6.1.4f}$$

Flow in a porous medium and porous membranes is sometimes described by *Darcy's law*, an empirical relation characterized by a hydraulic permeability, Q (see equation (3.4.88)):

$$\text{volume flux} = v_0 = \frac{Q\Delta P}{\mu L}. \tag{6.1.4g}$$

From (6.1.4f), one can express the hydraulic permeability Q as follows:

$$Q = \frac{\varepsilon^3}{(1-\varepsilon)^2}\frac{d_p^2}{150}. \tag{6.1.4h}$$

Such expressions are used also to describe the volumetric flow rate through deposits of particles, macromolecules, proteins, etc., on top of filters, membranes, etc. Often, a tortuosity factor, τ, is used along with L in the definition (6.1.4g) of hydraulic permeability to accommodate

[2] The y-coordinate is normal to the parallel plates and is at the center of the channel gap; thus, the plates have the y-coordinates of $y = \pm b$.

[3] This is the time-averaged axial velocity averaged over the tube radius.

[4] Blasius's law.

empirically the longer effective flow path length (compared to L, the thickness of the deposit).

6.1.1.5 *Compressible flow of gas in a capillary/pore*

Under conventional conditions of bulk flow in separation devices, liquids may be considered incompressible: there is essentially no variation in density even though pressure can vary considerably along the liquid flow direction. In many conventional separation devices, the gas streams may also be considered to be undergoing incompressible flow. However, in the case of some membrane gas separation devices, gas chromatography and flow through porous membranes, variation in gas density, and the consequent change in gas velocity due to a change in the gas pressure along the gas flow path, is considerable. The basic governing equation for compressible flow of gas in a capillary/pore of radius R will be the Hagen–Poiseuille equation[5] applied over a differential length dz in the bulk flow direction:

$$w_{tz} = \frac{\pi R^4}{8\mu}\rho\left(-\frac{dP}{dz}\right) = (Q\rho)_z. \qquad (6.1.5a)$$

Assume also the ideal gas law to be valid. Two forms of ideal gas law are useful. One form relates the gas pressure to the gas density:

$$\left(\frac{P}{\rho}\right)_z = \left(\frac{P_f}{\rho_f}\right), \qquad (6.1.5b)$$

where f refers to the feed gas and z refers to any axial location along the flow direction. The second form relates the gas pressure to the average gas velocity, $v_{z,\text{avg}}$:

$$(P v_{z,\text{avg}})_z = P_f v_{z,\text{avg},f}. \qquad (6.1.5c)$$

Assume isothermal flow and not enough change in pressure along the flow path of length L so that μ can be assumed to be constant. From (6.1.5a) and (6.1.5b), since the total mass flow rate of the gas w_{tz} is constant, integration from $z = 0$ (feed) to $z = L$ yields

$$w_{tz} = \frac{\pi R^4}{16\mu L}\frac{\rho_f}{P_f}(P_f^2 - P_l^2). \qquad (6.1.5d)$$

Here the gas pressure at the capillary exit is P_l. If the average gas density ρ_{avg} can be determined at the average gas pressure in the capillary, $[(P_f + P_l)/2]$, then

$$w_{tz} = \frac{\pi R^4}{8\mu L}\rho_{\text{avg}}(P_f - P_l), \qquad (6.1.5e)$$

which relates the mass flow rate directly to the gas flow pressure drop $\Delta P = P_f - P_l$. To obtain a relation between the gas velocity, gas pressure and the axial location z, we note from (6.1.5c) that

$$\frac{v_{z,\text{avg}}}{v_{z,\text{avg},f}} = \frac{P_f}{P}. \qquad (6.1.5f)$$

From expression (6.1.5d), if, instead of L, any location z is used, we have

$$w_{tz} = \frac{\pi R^4}{16\mu z}\frac{\rho_f}{P_f}(P_f^2 - P^2) = \frac{\pi R^4}{16\mu L}\frac{\rho_f}{P_f}(P_f^2 - P_l^2).$$

Thus

$$P_f^2 - P^2 = \frac{z}{L}(P_f^2 - P_l^2). \qquad (6.1.5g)$$

Therefore

$$\frac{v_{z,\text{avg}}}{v_{z,\text{avg},f}} = \frac{P_f}{\left[P_f^2 - \frac{z}{L}(P_f^2 - P_l^2)\right]^{1/2}}. \qquad (6.1.5h)$$

As (P_f/P_l) increases for any given (z/L), the velocity at z increases substantially over that at the feed location. At $z = L$, for example, if $P_f = 5P_l$, $v_{z,\text{avg}}$ becomes $5v_{z,\text{avg},f}$.

6.1.2 Gravity induced bulk flow

Wetted-wall columns (Figure 6.1.2(a)) and falling-film devices employ gravitational force to allow a thin liquid film to flow down a vertical wall or an inclined wall, often with a gas stream flowing countercurrently upwards. For the coordinate direction of flow being z along a plane surface inclined at an angle θ to the vertical (Figure 6.1.2(a), $\theta = 0$; Figure 6.1.2(b), $\theta \neq 0$), the volumetric flow rate Q of a liquid film of thickness δ, density ρ and viscosity μ is given by

$$Q = \frac{\rho g W \delta^3 \cos\theta}{3\mu}, \qquad (6.1.6)$$

where W is the width of the plate in the x-coordinate perpendicular to the paper. The parabolic velocity profile is given by

$$v_z(y) = \frac{\rho g \delta^2 \cos\theta}{2\mu}\left(1 - \left(\frac{y}{\delta}\right)^2\right). \qquad (6.1.7)$$

The average velocity $v_{z,\text{avg}}$ over the film thickness is

$$v_{z,\,\text{avg}} = \frac{\rho g \delta^2 \cos\theta}{3\mu} = \frac{2}{3}v_{z,\,\text{max}}. \qquad (6.1.8)$$

There are no ripples over the falling-film surface when the Reynolds number $Re = (4\delta v_{z,\text{avg}}\rho/\mu)$ is less than 20 (Bird *et al.*, 2002). The flow becomes turbulent when $Re > 1500$.

6.1.3 Free convection

Bulk motion of a fluid due to a density difference between different fluid elements is called *free convection* or *natural convection*. The density difference may come about due to

[5] Equation (6.1.5a) is also the basis of the result (6.1.1a). However, here ρ varies with z; so does Q.

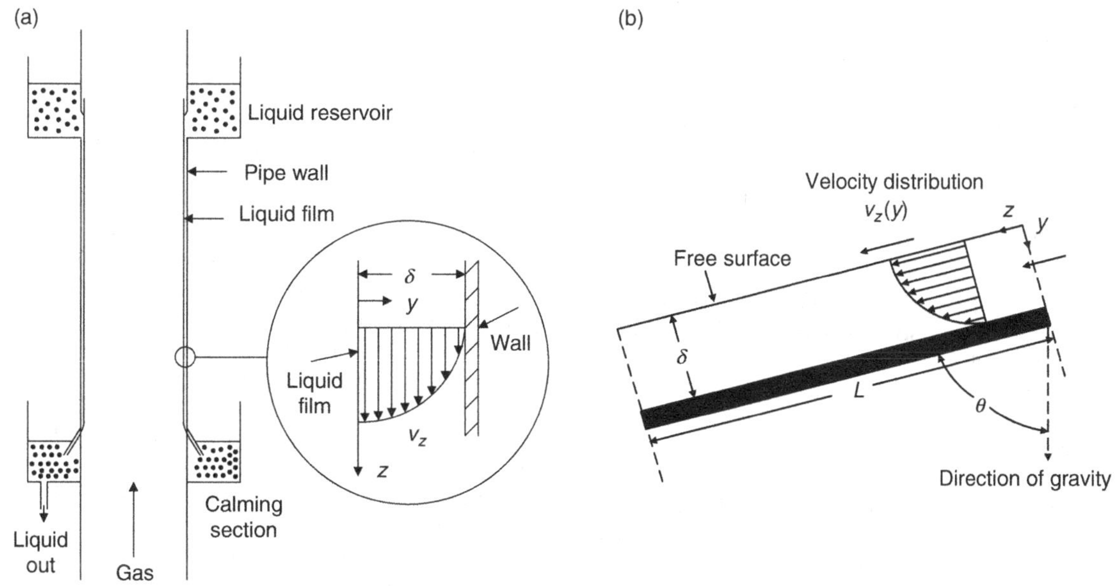

Figure 6.1.2. (a) Wetted-wall column; (b) gravity induced flow in a falling film down an inclined plane.

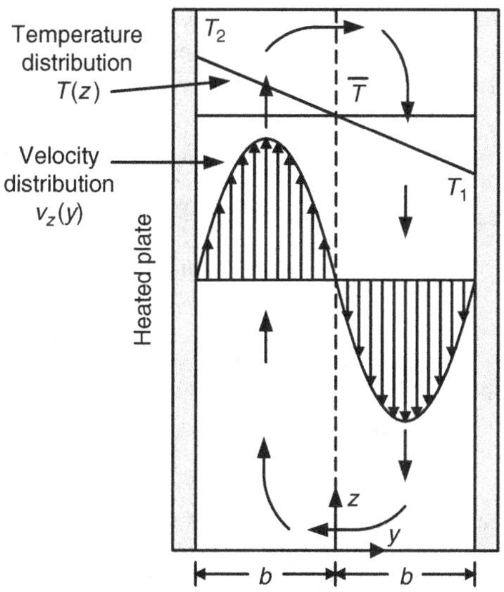

Figure 6.1.3. Bulk flow due to natural convection in a closed rectangular channel with two vertical plates at different temperatures.

temperature or concentration differences. Consider the closed rectangular vessel shown in Figure 6.1.3. The vertical plate on the left is maintained at T_2 and the vertical plate on the right is kept at T_1 ($<T_2$). The fluid (gas or liquid) close to the plate at T_2 is at a temperature higher than the fluid close to the plate at T_1. The fluid close to the plate at T_2 is hotter and lighter, and rises to the top of the closed vessel, where it turns around and flows over to the colder plate; the fluid closer to the colder plate (at T_1) meanwhile descends since it is heavier due to a higher density at a lower temperature. At the bottom of the closed vessel, this colder fluid will turn around and rise up near the plate at T_2 since it becomes lighter.

Let the rectangular channel gap (Figure 6.1.3) between the two vertical plates be $2b$. Assume the fluid viscosity μ to be essentially constant, even though it is dependent on temperature. The buoyancy forces on the lighter fluid elements surrounded by heavier fluid elements (see (3.1.5)) cause the lighter fluid elements to rise and, correspondingly, the heavier fluid elements to sink. The profile of the vertical velocity profile is given by (Bird *et al.*, 2002)

$$v_z = \frac{\overline{\rho}\overline{\beta}\Delta T b^2 g}{12\mu}\left[\left(\frac{y}{b}\right)^3 - \left(\frac{y}{b}\right)\right], \tag{6.1.9}$$

where $\Delta T = T_2 - T_1$, $\overline{\rho}$ is the fluid density at the average temperature $\overline{T} = (T_1 + T_2)/2$ and $\overline{\beta}$ is the coefficient of volume expansion $\left(= (\mathrm{d}\rho/\mathrm{d}T)_{\overline{T}}/\overline{\rho}\right)$ evaluated at $\overline{T}$. The average velocity of the lighter stream flowing upward is given by

$$v_{z,\mathrm{avg}} = \frac{\overline{\rho}g\overline{\beta}\Delta T b^2}{48\mu}. \tag{6.1.10}$$

Such bulk motion is primarily used in separation by thermal diffusion (see Section 8.1). Countercurrent bulk motion of this type has also been employed in a gas centrifuge to separate isotopic mixtures (Bramley, 1940; Martin and Kuhn, 1941). Bulk motion of this type is also developed if the densities of the solutions at the two plates (Figure 6.1.3) are different due to concentration differences, which can be created in the process of *electro-gravitation* (Shaffer and Mintz, 1980) using an electrodialysis cell having an ion exchange membrane and a neutral membrane.

6.1.4 Bulk motion due to capillarity

In Section 3.3.8, we observed that, in a gas–liquid–solid system with a flat solid surface, there is either zero or a finite contact angle θ (see Figure 3.3.16); the liquid attains this contact angle spontaneously without any other external force being present. If the contact angle is zero, the liquid spreads over the solid surface, displacing the gas/air; the solid surface is said to be wetted completely by the wetting liquid. This phenomenon is routinely encountered and exploited when a porous medium, such as a porous membrane or porous bed or paper, comes into contact with a liquid which can wet the material of the porous medium.

The pores in a porous membrane or paper and the channels in a porous bed may be idealized as capillaries of radius R. If we immerse a capillary of radius R made of a particular material into a liquid, we observe either of the following phenomena. If the material has a finite contact angle $\theta < 90°$ (for example, for water and a glass capillary), the liquid rises into the capillary to a finite height h_0 (Figure 6.1.4), given by

$$\rho g h_0 = \frac{2\gamma^{12}\cos\theta}{R}, \tag{6.1.11}$$

where $\rho g h_0$ is the pressure difference ΔP between the atmospheric pressure, P_{atm}, and the pressure in the liquid at the curved gas–liquid interface, P_{liq} (see equation (3.3.48) for a spherical bubble of radius R):

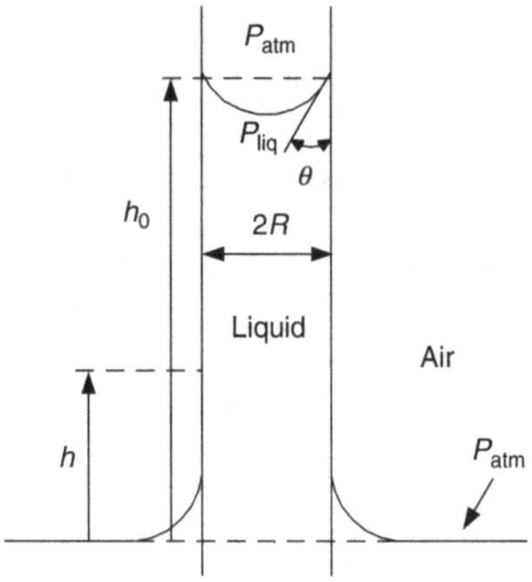

Figure 6.1.4. Rise of a liquid in a vertical capillary wetted by the liquid.

$$\Delta P = \rho g h_0 = P_{\text{atm}} - P_{\text{liq}} = \frac{2\gamma^{12}\cos\theta}{R}. \tag{6.1.12}$$

This quantity is the *capillary pressure*. If the contact angle $\theta > 90°$, then the liquid height is depressed into the tube from the free surface, as is observed, for example, in the case of a glass tube and mercury.

For cases where the contact angle $\theta < 90°$ (especially for completely wetted surfaces), the final magnitude of the liquid height h, namely h_0, is attained after a sufficiently long time. This value of h_0 can be significant for fine capillaries having small values of R. The spontaneous entry of the liquid into the capillary occurs at a certain rate; this bulk motion, regardless of whether the capillary, the membrane or the porous medium/paper is vertical or horizontal, is of interest. The fine dimensions of the capillary/porous medium suggest Poiseuille flow. For a vertical capillary, Levich (1962) has therefore postulated that the rate of rise of the capillary height, dh/dt, is the average velocity in Poiseuille flow (equation (6.1.1c)):

$$\frac{dh}{dt} = v_{z,\text{avg}} = \frac{\Delta P}{8\mu h}R^2, \tag{6.1.13}$$

where ΔP is the pressure drop needed for the liquid to flow. At any given time, when the capillary height is h, if $\theta = 0$, the driving pressure difference, ΔP, is equal to the difference between the capillary pressure (definition (6.1.12)) and $\rho g h$:

$$\frac{dh}{dt} = \frac{R^2}{8\mu h}\left(\frac{2\gamma^{12}}{R} - \rho g h\right), \tag{6.1.14}$$

which leads after integration to

$$t = \frac{8\mu h_0}{R^2 \rho g}\left(\ln\left(\frac{1}{1 - h/h_0}\right) - \frac{h}{h_0}\right). \tag{6.1.15}$$

When $h \to h_0$, $t \to \infty$, as it should be.

If equation (6.1.14) is rewritten as,

$$\frac{dh}{dt} = \frac{\gamma^{12}}{8\mu}\left(\frac{2R}{h} - \frac{\rho g R^2}{\gamma^{12}}\right), \tag{6.1.16}$$

then clearly the rate of rise is fast if the second term is small; conversely, the rate of rise is slow if it is large. This term is defined as the *Bond number* and is the ratio of the gravitational force over the force due to surface tension (γ^{12}/R) if R is a characteristic dimension for the system:

$$\text{Bond number} = Bo = \frac{\rho g R}{\gamma^{12}/R}. \tag{6.1.17}$$

If the capillary is horizontal, then the length, l, of the capillary wetted by the wetting liquid and the average velocity of liquid motion, $v_{z,\text{avg}}$, in the capillary are related by

$$\frac{dl}{dt} = v_{z,\text{avg}} = \frac{\Delta P}{8\mu l}R^2, \tag{6.1.18}$$

where ΔP is given by equation (6.1.12). Therefore,

$$\frac{dl}{dt} = \frac{2\gamma^{12} R \cos\theta}{8\mu l},$$

which, upon integration and using the initial condition of $l = 0$ for $t = 0$, leads to

$$t = \frac{2\mu l^2}{R\gamma^{12}\cos\theta} \qquad (6.1.19)$$

and

$$v_{z,\mathrm{avg}} = \frac{1}{2}\sqrt{\frac{R\gamma^{12}\cos\theta}{2\mu}}\sqrt{\frac{1}{t}}. \qquad (6.1.20)$$

6.1.5 Electroosmotic flow

Electroosmotic flow of a liquid is achieved in the direction of an applied electrical voltage due to the phenomenon of an electrical double layer; the liquid is generally an aqueous electrolytic solution. Consider a capillary of an insulating material, e.g. silica filled with an aqueous electrolytic solution. If the *pH* of the solution is greater than 3, the surface silanol groups (SiOH) get deprotonated and become negatively charged. Cations from the solution (counterions) are attracted to the negative surface charges and form an electrical double layer (in a manner similar to that for a charged particle as shown in Figure 3.1.2D) to achieve electroneutrality on an overall basis (Figure 6.1.5). These counterions, however, have water molecules around them, providing solvation shells.

If now an electrical field of magnitude E is applied along the capillary length, the cations in the electrical double layer will move toward the cathode; the solvent molecules in the solvation shell of the counterion will be dragged along with the cations toward the cathode. This solvent flow is termed the *electroosmotic flow* (EOF); the magnitude of the solvent velocity thus generated is given by (Levich, 1962; Newman, 1973)

$$v_{\mathrm{EOF}} = \frac{\varepsilon_d}{\mu}\zeta E, \qquad (6.1.21)$$

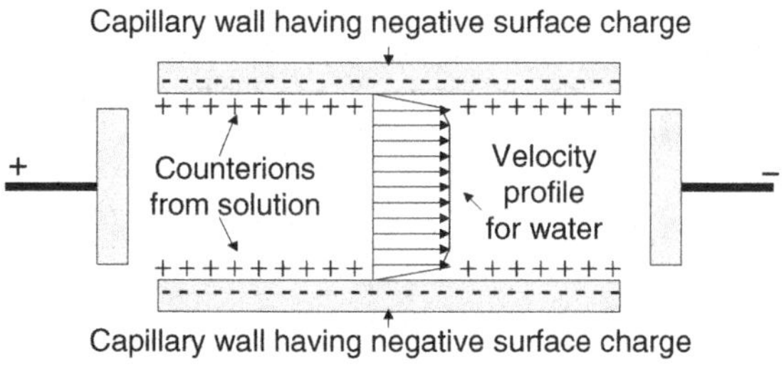

Figure 6.1.5. Electroosmotic flow of water in a capillary with negative surface charge due to an applied electrical potential.

where μ is the fluid viscosity, ζ is the zeta potential of the double layer (see equation (3.1.11a)) and ε_d is the dielectric constant of the solution (= relative dielectric constant of the solution $\times\ \varepsilon_0$, where ε_0 is the permittivity of vacuum). Further, this velocity is essentially constant along the capillary cross section; the velocity at the wall of the insulating capillary will be zero (see Figure 6.1.5). For steady laminar flow, Bird *et al.* (2002) have provided the following expression for the velocity profile in a capillary:

$$v_{\mathrm{EOF}} = \left(\frac{\varepsilon_d \zeta}{\mu}\right) E\left(1 - \exp(-y/\lambda)\right), \qquad (6.1.22)$$

where y is the distance from the wall and λ is the Debye length of the electrolyte solution (definition (3.1.10b)), which is typically around 1 nm for a 0.1 N electrolytic solution. For capillaries having dimensions of 50 μm, one can neglect $\exp(-y/\lambda)$ with respect to 1, essentially throughout the whole capillary, leading to a flat velocity profile.

Newman (1991) has described the fundamental basis for the derivation of the above equation. Newman (1973) has estimated the value of v_{EOF} for an electric field E of 10 volt/cm and a zeta potential ζ of 0.1 volt to be 7.8×10^{-3} cm/s for aqueous systems. Such low values are of limited practical use; electroosmotic flows are therefore generally neglected. However, in the recent applications of capillary electrophoresis (CE), capillary zone electrophoresis (CZE), etc., where silica capillaries of diameter around 30–200 μm and length 25–50 cm are used, with the voltage difference between the two electrodes at the two ends of the capillary being in the range of 9000–20 000 volts, the electric field strength, E, has a much higher value. Values of around 250–400 volts/cm are quite common. The value of v_{EOF} can, under such conditions, be much higher, in the range of 0.2 cm/s. Electroosmotic flow has thus become singularly important in developing the small capillary based techniques of CE and CZE which are being increasingly employed for the separation of proteins and other biomolecules (see Section 6.3.1.2).

Analytical results for the velocity distribution, mass flow rate, pressure gradient, wall shear stress, etc., in mixed electroosmotic/pressure driven flows are available for two-dimensional straight-channel geometry in Dutta and Beskok (2001). The magnitude and direction of the electroosmotic flow inside a microfabricated fluid channel (25 μm high by 100 μm wide) can be controlled by a perpendicular electrical field of 1–5 megavolts/cm generated by a voltage of only 50 volts (Schasfoort *et al.*, 1999).

A somewhat different mechanism of water flow due to the motion of cations having a hydration layer (the solvation shell) has been postulated. It employs porous/ion exchange membranes whose pore diameters are in the range of 1–5 nm. When such a membrane is placed between two electrodes containing an aqueous salt solution, and electrolysis takes place on the application of a voltage, two ions are generated at the electrodes. For example, through a cation exchange membrane such as Nafion®, cations generated at the anode will be driven to the cathode through the membrane pore to maintain electroneutrality. The water molecules dragged by these cations through the membrane contribute to an electrochemically driven flow (EDF) (Norman *et al.*, 2005). The magnitude of such a velocity is small; the flow rates generated are of the order of 1-15 microliter/min.

6.1.6 Centrifugal force driven flow

We have discussed in Section 3.1.2.7 (equation (3.1.51)) that when a solvent/liquid rotates in a centrifugal field with an angular velocity ω (radian/s), the hydrostatic pressure P in the liquid increases radially outward:

$$\frac{\mathrm{d}P}{\mathrm{d}r} = \frac{M_s}{\overline{V}_s}\omega^2 r. \tag{6.1.23}$$

At the radial boundary of the centrifuge (see Figure 4.2.3 or Figure 4.2.1), the liquid pressure will have the highest value (see equation (4.2.13) for the corresponding result for a gas mixture). If the centrifuge boundary is porous or has openings or nozzles, this high liquid pressure will drive the liquid through such openings as if we have hydrostatic pressure induced bulk flow, as described in Section 6.1.1.

This principle is employed in centrifugal filters. Figure 6.1.6 illustrates a basket-type centrifuge. Solid liquid sludge is fed into the device. The particulate phase is radially thrown onto the screen of the basket, through which the liquid flows out, driven by the pressure generated by the centrifugal motion. The solids are removed from the screen by scraping, etc., and discharged from the bottom.

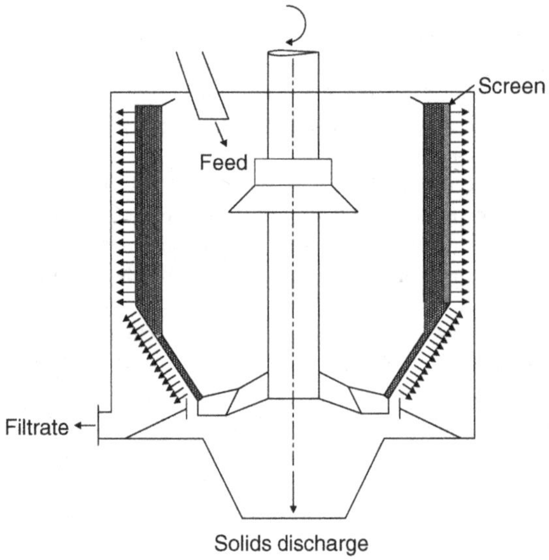

Figure 6.1.6. Centrifugally generated hydrostatic pressure driven flow of liquid through the screen on the basket.

6.1.7 Surface tension gradient based flow

Surface tension on the surface of a liquid at gas-liquid or vapor-liquid interfaces can vary due to a variation in temperature or species concentration. The components of the tangential stress,[6] τ_y, τ_{yx} and τ_{yz}, are related to the corresponding gradients in interfacial tension, γ^{12}, between the liquid phase, $j = 1$, and the gas/vapor phase, $j = 2$ by (Bird *et al.*, 2002):

$$\tau_{yx} = -\frac{\partial \gamma^{12}}{\partial x}; \qquad \tau_{yz} = -\frac{\partial \gamma^{12}}{\partial z}. \tag{6.1.24}$$

Assume that there are no variations of temperature or composition in the x-direction; these vary only in the z-direction. For example,

$$\tau_{yz} = -\left(\frac{\partial \gamma^{12}}{\partial T}\right)\frac{\partial T}{\partial z} \tag{6.1.25}$$

in the case of temperature variation in the z-direction; correspondingly,

$$\tau_{yz} = -\left(\frac{\partial \gamma^{12}}{\partial \Gamma_{i\sigma}}\right)\frac{\partial \Gamma_{i\sigma}}{\partial z} \tag{6.1.26}$$

in the case of variation of surface concentration (mol/cm^2) of species i, $\Gamma_{i\sigma}$ (see definition (3.3.34)) in the z-direction.

Bulk motion created by the variation in surface tension resulting from a variation in liquid temperature is called *thermocapillary flow* (Levich, 1962). In a shallow pan of depth h (shallow in the y-direction but deep in the x-direction perpendicular to the plane of the paper), the two walls of the pan (in the z-direction) are at temperatures T_1 and T_2 ($>T_1$) (Figure 6.1.7). Therefore, the surface tension γ^{12} varies with the distance z as (Figure 6.1.7(a))

$$\gamma^{12} = \gamma^{12}(T_1) + \left(\frac{\partial \gamma^{12}}{\partial T}\right)\frac{T_2 - T_1}{l} z. \tag{6.1.27}$$

Note that surface tension decreases with an increase in temperature. The height of the liquid, h, will depend on the axial location in the pan, i.e. $h(z)$. At the top surface of the liquid, the maximum flow velocity toward the cold surface,

$$v_{z,\max} = \frac{h}{4\mu}\frac{\mathrm{d}\gamma^{12}}{\mathrm{d}z}, \tag{6.1.28}$$

[6] Force per unit area in the tangential direction on a surface whose normal is in the y-direction; the two components of this stress, namely τ_{yx} and τ_{yz}, are relevant here.

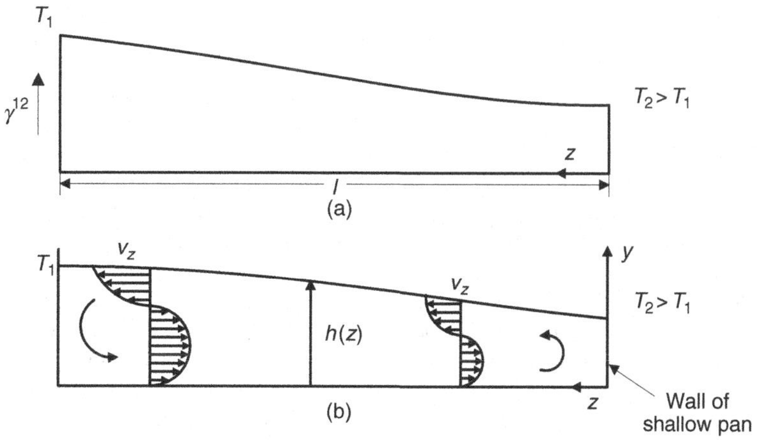

Figure 6.1.7. (a) Surface tension profile in the liquid in a shallow pan whose two ends are at different temperatures, T_1 *and* T_2 *(>*T_1*); (b) velocity profiles in the liquid in the shallow pan.*

is achieved for a liquid of viscosity μ, the velocity profile being given by

$$v_z = \frac{y}{2\mu}\left(\frac{3}{2}\frac{y}{h} - 1\right)\frac{d\gamma^{12}}{dz}. \qquad (6.1.29)$$

The profile is shown in Figure 6.1.7(b). The direction of motion of the z-directional velocity, v_z, is reversed at $y = (2/3)h$, and the flow is toward the hot surface. The basis of this analysis has been provided by Levich (1962) and Yih (1968).

The surface tension gradient driven flow system in Figure 6.1.7(b) has different heights of the liquid film, $h(z)$, at different locations along the z-coordinate. Gravitational force will tend to eliminate it. The relative influence of the two forces is indicated by the Bond number, *Bo*, defined earlier by equation (6.1.17). The characteristic length dimension of the system to be used in the definition of *Bo* is any particular value of $h(z)$:

$$Bo = \frac{\rho g h^2}{\gamma^{12}}. \qquad (6.1.30)$$

A separation process, where the spatial variation of surface tension due to composition is encountered, is distillation. If, along the length direction of flow of a binary liquid mixture, the more volatile species is removed preferentially, one can have two possibilities: a "positive system," where the surface tension increases due to removal of the more volatile species; and a "negative system," where the surface tension decreases due to removal of the more volatile species. Zuiderweg and Harmens (1958) have considered the effect of such motion on the performance of small distillation columns. Such motions are important in general when there are thin liquid layers (compare the shallow pan of Figure 6.1.7 as a basis). The reader may consult Probstein (1989) for a more extended introduction to surface tension gradient based flow. Generally such motions are identified as *Marangoni effects* (see Sternling and Scriven, 1959).

6.1.8 Drag flow

When the viscosity of the feed fluid/material is quite high and/or the material is thermally sensitive, separation of volatiles by heat, or other phase changes like crystallization, often requires mechanical rotors/screws which rotate in the separator. As they rotate, they drag the highly viscous material, which is pushed to one end of the separator due to helical or other screw/rotor geometries. The bulk motion of the fluid/solid is caused by the drag force induced by the moving solid surface, where the fluid/feed may be assumed to have the same velocity as that of the solid rotor surface. Such fluid motion, however, is reduced to zero at the other stationary solid surface, unless there is slip. This is called *drag flow* (Figure 6.1.8). The figure illustrates the velocity profile created when the top plate moves with a velocity v and the bottom plate is stationary. For a fluid having a constant viscosity, μ, the constant shear rate $\dot{\gamma}$ and the velocity profile are defined by

$$\dot{\gamma} = \frac{\partial v_z}{\partial y}, \qquad v_z(y) = \dot{\gamma} y. \qquad (6.1.31)$$

An externally imposed pressure gradient may also be present along with the drag flow, as in devolatilizing screw extruders for polymers.

There is an additional source of bulk flow within a separator vessel, namely a stirrer. It is an integral part of what are considered *well-stirred vessels*. This flow can take place inside the vessel without any bulk flow coming in or going out. The energy for liquid motion is provided by the rotating stirrer.

6.1.9 Feed introduction mode vs. bulk flow

The feed mixture may be introduced into a separation device via a number of forces identified above. What is, however, much more important is the dependence of the feed introduction on the time and spatial coordinates in the separator. We will progressively learn in Sections 6.3, 6.4 and Chapters 7 and 8 that this feed introduction mode is intimately connected with the number of phases/regions

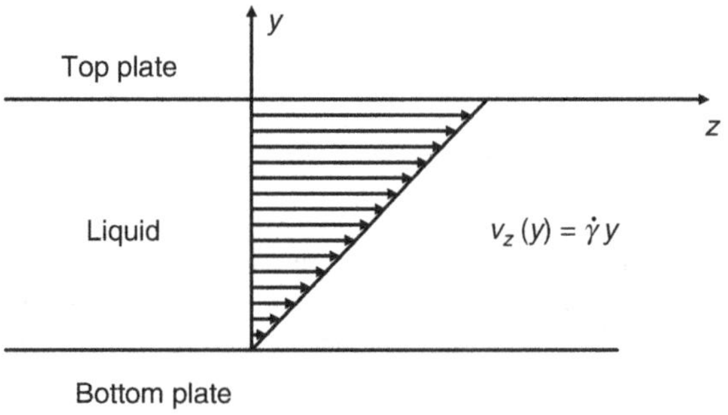

Figure 6.1.8. Simple drag flow of liquid due to the motion of the top plate at a velocity v *while the bottom plate is stationary.*

flowing in the separator and their flow directions in relation to the direction of the force. The feed may be introduced continuously or discontinuously into the separator.

In many separation processes, the feed is introduced continuously at a fixed rate into one end, or in the middle, or other location, of the separator (Figure 6.1.9(a)); the direction, the flow rate and the location of feed introduction are invariant with time. Devices where two phases, regions or fractions flow perpendicular to the direction of the force are of prime interest (Chapter 8) in this mode of feed introduction. Although there is only one feed stream coming in, there will be a total of two/three streams, which will be entering the separator in such a case. The presence of a second mobile stream/phase/fraction within the separator allows species introduced via the feed to be removed continuously as separation is being implemented locally everywhere in the separator.

The same time-invariant mode of feed stream introduction and feed-phase flow perpendicular to the force direction is also utilized in separations where any one stream enters the separator, provided at least two product streams are continuously withdrawn from the separator. Crossflow membrane separators and external force based separators studied in Sections 7.2 and 7.3 are particular examples. Of the at least two product streams, one is the depleted feed stream. The other product stream continuously removes selected species from the feed stream (Figure 6.1.9(b)).

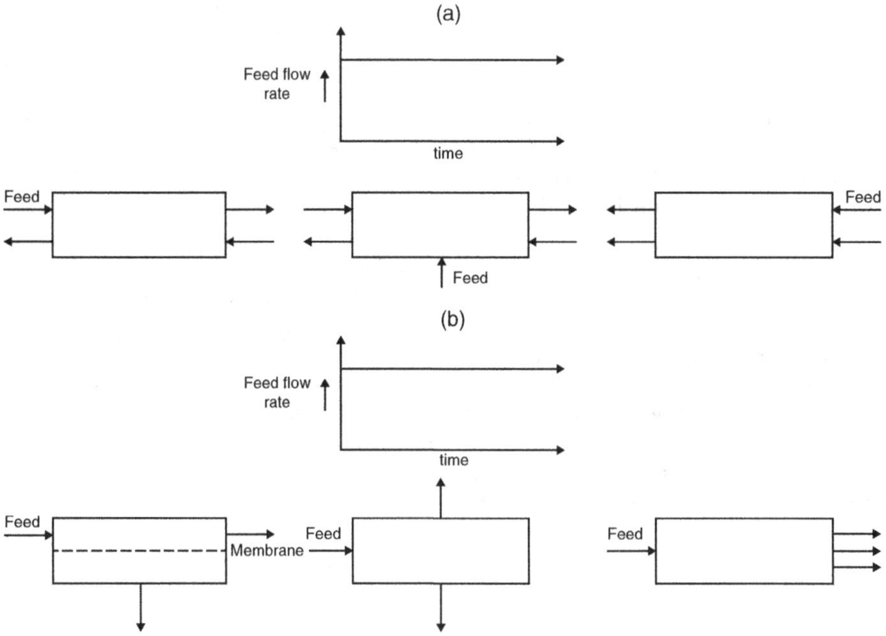

Figure 6.1.9. Constant feed flow rate into a separator. (a) Two/three streams entering/leaving the separator; (b) one stream entering and at least two streams exiting the separator.

Sometimes, in external force based systems, the feed stream is merely a part of one stream continuously entering the separator. In such a case, there may be more than two product streams leaving the separator (see Section 7.3.1.1); further it is difficult to identify a depleted feed stream (Figure 6.1.9(b)). Crossflow fluid–solid adsorption processes employ this mode as well (see Section 8.3.1). A time-invariant mode of feed stream introduction is also practiced with a single-entry separator when the bulk flow is parallel to the force, as in flash based distillation/devolatilization processes having two steady product streams (see Section 6.3.2.1). In all cases covered by Figures 6.1.9(a),(b), a variation of this general mode of feed introduction employs continuous feed introduction with some time-dependent variation of the feed flow rate magnitude. Finally, in this mode of feed introduction (Figure 6.1.9(b)), in continuous stirred tank separators (see Section 6.4), one can have only one product stream, which, however, is a multiphase stream (a configuration not shown here).

A second mode of feed introduction involves a constant flow rate of feed into the separator at a fixed separator location for a fixed period of time, after which feed flow is stopped for a certain amount of time; feed flow is restarted after this pause. The start-stop is carried out cyclically for as many cycles as is feasible/desirable (Figure 6.1.10(a)). Fixed-bed processes in a single-entry separator described in Sections 7.1.1–7.1.3, where one feed stream enters the bed periodically and the purified feed stream is withdrawn in a corresponding manner, are typical examples. The periodic feed is necessitated by the absence of another mobile stream, phase or fraction removing species from the feed in a continuous fashion. During the period of feed stoppage, arrangements such as the introduction of a purge stream, stripping stream or a regenerating solution, are made to remove the species from the feed stream deposited into the separation medium.

A variant of this second mode of feed introduction involves a pulse of feed mixture introduced into the separator at a particular location periodically, as in the chromatographic processes discussed in Section 7.1.5 (Figure 6.1.10(b)). During the rest of the period, another non-feed stream is most likely to be introduced into the separator to continue separation and regenerate the separator. The last mode of feed introduction is simply an amount of fluid/solid introduced into the separator in a batch fashion. The product withdrawal may be batchwise or semi-continuous (stopping when there is no feed).

A third mode of feed introduction employs a constant flow rate of feed into the separator at one end for a fixed period of time; then the same feed flow pattern is implemented at the other end of the separator such that the feed flow direction in the separator is reversed (Figure 6.1.10 (c)). Such a feed flow pattern is observed in continuous

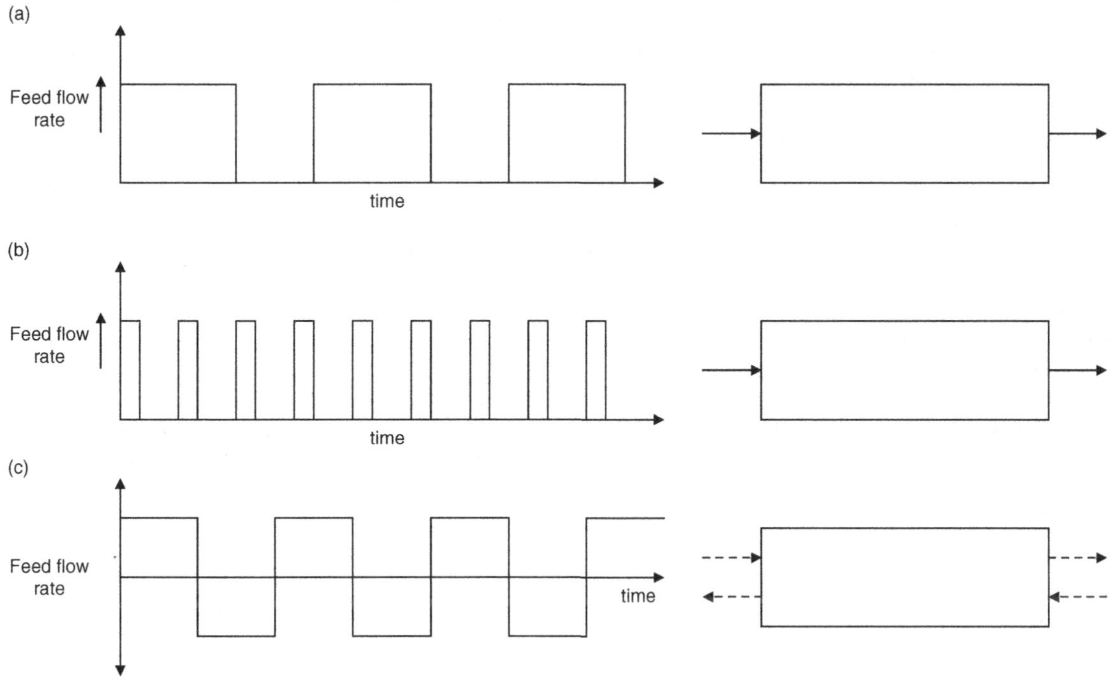

Figurer 6.1.10. Cyclic feed introduction. (a) Constant feed flow rate for part of the cycle in fixed-bed processes; (b) pulse input of feed in a periodic fashion in chromatographic processes; (c) cyclic feed introduction from opposite ends of the separator, as in parametric pumping processes. (After Lee et al.*, 1977a.)*

parametric pumping based processes (see Section 7.1.4). Lee *et al.* (1977a) have provided a comprehensive description of various patterns of feed introduction in relation to the motion/trajectory of solute(s) in separation devices. We note here that product withdrawal modes can also have a variety of patterns, just as in feed introduction.

6.2 Equations of change

The equation of change of concentration of a species i with the spatial coordinates of the separator and time may be obtained by considering the motions of individual molecules present in the fluid mixture. However, this strategy generally leads to extraordinary complexity, which is not useful for us. Rather, we adopt the *continuum hypothesis*. Although the fluid consists of billions of molecules, which leads to tremendous fluctuations in every quantity of interest at any given point, we replace this fluid with a hypothetical continuous phase, such that all physical, chemical and thermodynamic quantities of interest, e.g. density, temperature, composition, energy, are constant over a very small volume, which can be identified as a point in the fluid for our purposes. Although there can and will be variations in various quantities from one such point to the next, one can, at any given point, identify a particular value of any of these quantities mentioned above. Such a point volume is large enough to average all fluctuations due to the individual and random motion of molecules; yet it is small enough to allow for the identification of the actual changes taking place in the fluid during fluid motion and/or separation. After the appropriate differential equations of change are identified, we provide the *volume-averaged equations* employed for separation in *multiphase systems*, where only variations in one direction, the main flow direction, are generally of interest in separator analysis. Subsequently, we provide the equations for a continuous stirred tank separator (CSTS).

To provide an elementary background on separation systems where particles are present in a fluid and particle motion is important for separation, the *equations of motion* of a particle in a fluid are provided next; these are also called *trajectory equations*. These equations have been followed by a *general equation of change* for a particle population, the *population balance equation*. Analysis of a CSTS for particulate systems is considered at the end of this section.

6.2.1 Equations of change for species concentration in a mixture

In Section 3.2.1, a rectangular separator vessel was considered in which there was no bulk velocity of the solvent/fluid in any direction; further, the solute species i was subjected to a force in one coordinate direction (z-coordinate). The balance equation for molar concentration C_i of species i developed there depended only on the z-coordinate and time, t. Here, we will first develop the corresponding equation for C_i, which depends on all three spatial coordinates x, y, z, time t and the fluid molar average velocity vector[7] $\boldsymbol{v}^*$. A similar equation of change for mass concentration ρ_i of species i will also be obtained; it will use the mass average velocity vector $\boldsymbol{v}$. These equations will have a term containing the rate of production of species i by chemical reactions, if any.

Figure 6.2.1 shows a small volume element of dimensions Δx, Δy, Δz in the x, y and z directions in the flow field of the separator at the location (x, y, z). For species i, we will employ the following basic principle of conservation of mass for the small volume element of volume $\Delta x \Delta y \Delta z$:

$$\begin{pmatrix}\text{rate of}\\ \text{accumulation}\\ \text{of species } i\end{pmatrix} = \begin{pmatrix}\text{rate at}\\ \text{which}\\ \text{species } i\\ \text{comes in}\end{pmatrix} - \begin{pmatrix}\text{rate at which}\\ \text{species } i\\ \text{leaves}\end{pmatrix} + \begin{pmatrix}\text{rate of production}\\ \text{of species } i \text{ by}\\ \text{chemical reaction.}\end{pmatrix} \quad (6.2.1)$$

We will now consider each of the above terms in some detail using *molar rates*:

$$\text{rate of accumulation of species } i \text{ in the volume element} \quad \frac{\partial C_i}{\partial t}(\Delta x \Delta y \Delta z); \quad (6.2.2a)$$

$$\text{rate at which species } i \text{ enters the volume through the face perpendicular to the } x\text{-coordinate and having a corner coordinate } (x, y, z) \quad N_{ix}|_x(\Delta y \Delta z); \quad (6.2.2b)$$

$$\text{rate at which species } i \text{ leaves the volume through the face perpendicular to the } x\text{-coordinate and having a corner coordinate } (x+\Delta x, y, z) \quad N_{ix}|_{x+\Delta x}(\Delta y \Delta z); \quad (6.2.2c)$$

$$\text{molar rate of production of species } i \text{ by chemical reaction in the volume element} \quad R_i(\Delta x \Delta y \Delta z). \quad (6.2.2d)$$

Additional terms for the molar rate at which species i enters the volume element and leaves the volume element through the faces perpendicular to the y-axis and the z-axis can be easily developed following (6.2.2b) and (6.2.2c). When all such terms are introduced into the balance equation (6.2.1), we get

$$\begin{aligned}(\Delta x \Delta y \Delta z)\frac{\partial C_i}{\partial t} &= (\Delta y \Delta z)(N_{ix}|_x - N_{ix}|_{x+\Delta x}) \\ &+ (\Delta x \Delta z)(N_{iy}|_y - N_{iy}|_{y+\Delta y}) \\ &+ (\Delta x \Delta y)(N_{iz}|_z - N_{iz}|_{z+\Delta z}) + (\Delta x \Delta y \Delta z)R_i.\end{aligned} \quad (6.2.2e)$$

[7]We drop the subscript t from $\boldsymbol{v}_t^*$ and $\boldsymbol{v}_t$ used in Chapters 2 and 3 from now on.

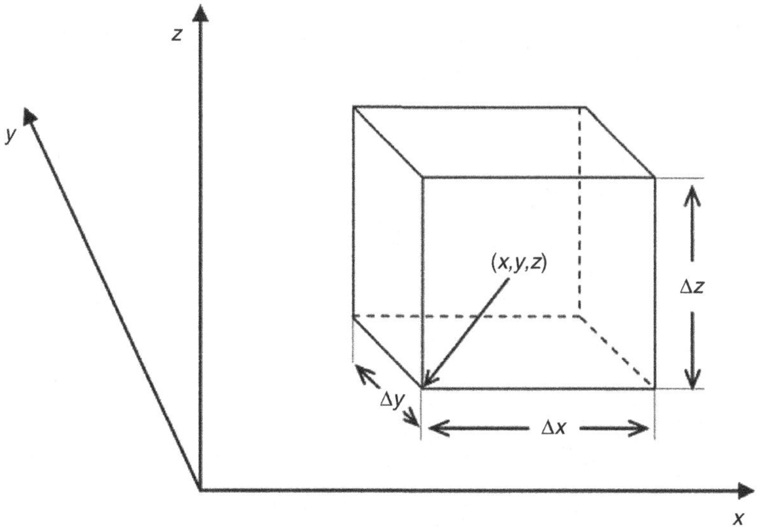

Figure 6.2.1. A small rectangular volume element in the separator having a corner point with coordinates x, y, z.

Dividing by $(\Delta x \Delta y \Delta z)$ and allowing each of Δx, Δy and Δz to go to zero, we obtain in the limit

$$\frac{\partial C_i}{\partial t} = -\frac{\partial N_{ix}}{\partial x} - \frac{\partial N_{iy}}{\partial y} - \frac{\partial N_{iz}}{\partial z} + R_i. \tag{6.2.3a}$$

Here $i = 1, 2, \ldots, n$, if the total number of species in the system is n. This is the *general differential equation of balance* for species i in molar units in a system in the presence of any chemical reactions involving species i. An alternative form in vector notation is

$$\frac{\partial C_i}{\partial t} = -(\nabla \cdot \boldsymbol{N}_i) + R_i. \tag{6.2.3b}$$

The corresponding differential equations in *mass units* for species i are obtained by multiplying the above equations for species i by the molecular weight of species i, M_i:

$$\frac{\partial \rho_i}{\partial t} = -\frac{\partial n_{ix}}{\partial x} - \frac{\partial n_{iy}}{\partial y} - \frac{\partial n_{iz}}{\partial z} + r_i; \tag{6.2.4a}$$

$$\frac{\partial \rho_i}{\partial t} = -(\nabla \cdot \boldsymbol{n}_i) + r_i. \tag{6.2.4b}$$

The nature of the equations of change to be employed for solving a particular separation problem depends on the expressions for molar fluxes $\boldsymbol{N}_i$ or mass fluxes $\boldsymbol{n}_i$. Employing expressions (3.1.84a), (3.1.84b) and (3.1.84d), we may rewrite the general balance equation (6.2.3b) in an essentially equivalent but more detailed form for molar concentration C_i as

$$\frac{\partial C_i}{\partial t} = -\nabla \cdot (C_i \boldsymbol{v}^*) - \nabla \cdot (C_i \boldsymbol{U}_i) + \nabla \cdot \left(C_i \frac{RT}{f_i^d} \nabla \ln a_i \right) + R_i. \tag{6.2.5a}$$

Here the migration velocity $\boldsymbol{U}_i$ of species i is defined by (3.1.84b) or (3.1.84d). Employing expression (3.1.99) for the molar flux $\boldsymbol{J}_i^*$, equation (6.2.5a) becomes

$$\frac{\partial C_i}{\partial t} = -\nabla \cdot (C_i \boldsymbol{v}^*) - \nabla \cdot (\boldsymbol{J}_i^*) + R_i$$

(1) (2) (3) (4) (6.2.5b)

This is the *equation of change for molar concentration of species i*. The different terms in this equation correspond to different aspects of the changing molar concentration at any time and at any point in the separator:

(1) rate of increase of molar concentration of species i per unit volume;
(2) net molar rate of addition of species i by fluid motion per unit volume;
(3) net molar rate of addition of species i by migration and diffusion per unit volume;
(4) net molar rate production of species i by chemical reaction per unit volume.

Equation (6.2.5b) is often rewritten as

$$\frac{\partial C_i}{\partial t} = -C_i \nabla \cdot \boldsymbol{v}^* - \boldsymbol{v}^* \cdot \nabla C_i - \nabla \boldsymbol{J}_i^* + R_i. \tag{6.2.5c}$$

When an appropriate expression of $\boldsymbol{J}_i^*$ is introduced in this equation, it becomes quite useful in analyzing the performance of a separator. Table 6.2.1 illustrates this equation using scalar quantities for different coordinate systems, all of which employ a simplification developed next to express $\nabla \cdot \boldsymbol{v}^*$. Table 6.2.1 also illustrates the same equation in Cartesian coordinates for the case where $\boldsymbol{J}_i^*$ is expressed via equations (3.1.99) and (3.1.89), and C_t and D_{is} are constants.

Table 6.2.1. Equation of change for molar concentration of species i *(equation (6.2.5g))*

Cartesian coordinates (x, y, z)	$\frac{\partial C_i}{\partial t} = -\left[v_x^* \frac{\partial C_i}{\partial x} + v_y^* \frac{\partial C_i}{\partial y} + v_z^* \frac{\partial C_i}{\partial z}\right] - \left[\frac{\partial J_{ix}^*}{\partial x} + \frac{\partial J_{iy}^*}{\partial y} + \frac{\partial J_{iz}^*}{\partial z}\right] + R_i - x_i \sum_{i=1}^{n} R_i$
Cylindrical coordinates (r, θ, z)	$\frac{\partial C_i}{\partial t} = -\left[v_r^* \frac{\partial C_i}{\partial r} + \frac{v_\theta^*}{r} \frac{\partial C_i}{\partial \theta} + v_z^* \frac{\partial C_i}{\partial z}\right] - \left[\frac{1}{r}\frac{\partial}{\partial r}(rJ_{ir}^*) + \frac{1}{r}\frac{\partial J_{i\theta}^*}{\partial \theta} + \frac{\partial J_{iz}^*}{\partial z}\right] + R_i - x_i \sum_{i=1}^{n} R_i$
Spherical coordinates (r, θ, Φ)	$\frac{\partial C_i}{\partial t} = -\left[v_r^* \frac{\partial C_i}{\partial r} + \frac{v_\theta^*}{r} \frac{\partial C_i}{\partial \theta} + \frac{v_\phi^*}{r\sin\theta} \frac{\partial C_i}{\partial \phi}\right] - \left[\frac{1}{r^2}\frac{\partial}{\partial r}(rJ_{ir}^*) + \frac{1}{r\sin\theta}\frac{\partial}{\partial \theta}(J_{i\theta}^* \sin\theta) + \frac{1}{r\sin\theta}\frac{\partial J_{i\phi}^*}{\partial \phi}\right] + R_i - x_i \sum_{i=1}^{n} R_i$
Cartesian coordinates $(x, y, z)^a$	$\frac{\partial C_i}{\partial t} + \left[(v_x^* + U_{ix})\frac{\partial C_i}{\partial x} + (v_y^* + U_{iy})\frac{\partial C_i}{\partial y} + (v_z^* + U_{iz})\frac{\partial C_i}{\partial z}\right] = \left[D_{is}\frac{\partial^2 C_i}{\partial x^2} + D_{is}\frac{\partial^2 C_i}{\partial y^2} + D_{is}\frac{\partial^2 C_i}{\partial z^2}\right] + R_i - x_i \sum_{i=1}^{n} R_i$

C_t is a constant throughout the table; R_i is the molar rate of production of species i by chemical reaction per unit volume.
[a] D_{is} is a constant for this equation only.

If we sum equations (6.2.5c) for all species $i = 1, 2, \ldots, n$ together, we get

$$\frac{\partial C_t}{\partial t} = -C_t \nabla \cdot \boldsymbol{v}^* - \boldsymbol{v}^* \cdot \nabla C_t - \sum_{i=1}^{n} \nabla \cdot \boldsymbol{J}_i^* + \sum_{i=1}^{n} R_i. \tag{6.2.5d}$$

By definition,

$$\boldsymbol{J}_i^* = \boldsymbol{N}_i - C_i \boldsymbol{v}^* \Rightarrow \sum_{i=1}^{n} \boldsymbol{J}_i^* = \sum \boldsymbol{N}_i - \sum C_i \boldsymbol{v}^* = \sum \boldsymbol{N}_i - C_t \boldsymbol{v}^*.$$

From relations (2.1.2) and (3.1.77), $C_t \boldsymbol{v}^* = \sum C_i \boldsymbol{v}_i = \sum \boldsymbol{N}_i$ so that $\sum_{i=1}^{n} \boldsymbol{J}_i^* = 0$. Therefore,

$$\frac{\partial C_t}{\partial t} + C_t \nabla \cdot \boldsymbol{v}^* + \boldsymbol{v}^* \cdot \nabla C_t = \sum_{i=1}^{n} R_i. \tag{6.2.5e}$$

If the total molar density C_t is constant, we obtain

$$C_t \nabla \cdot \boldsymbol{v}^* = \sum_{i=1}^{n} R_i. \tag{6.2.5f}$$

When this result is introduced into equation (6.2.5c), we get

$$\frac{\partial C_i}{\partial t} + \boldsymbol{v}^* \cdot \nabla C_i = -\nabla \cdot \boldsymbol{J}_i^* + R_i - x_i \sum_{i=1}^{n} R_i. \tag{6.2.5g}$$

Note: $\sum R_i$ need not equal zero, since moles are not necessarily conserved in a chemical reaction.

The procedure which led to equation (6.2.5e) from the summation of equation (6.2.5b) for each individual species yields a more important result, the *equation of continuity for the mixture,* if equation (6.2.4b) for the mass density or mass concentration ρ_i of each species i is employed:

$$\sum_i \frac{\partial \rho_i}{\partial t} = \frac{\partial \rho_t}{\partial t} = \sum_{i=1}^{n} -(\nabla \cdot \boldsymbol{n}_i) + \sum_{i=1}^{n} r_i. \tag{6.2.5h}$$

Since mass is conserved in chemical reactions, $\sum_{i=1}^{n} r_i = 0$. Further, by definitions (2.1.1) and (2.1.3),

$$\sum_{i=1}^{n} \nabla \cdot \boldsymbol{n}_i = \nabla \cdot \sum_{i=1}^{n} \boldsymbol{n}_i = \nabla \cdot \rho_t \boldsymbol{v}. \tag{6.2.5i}$$

This allows equation (6.2.5h) to be simplified to

$$\frac{\partial \rho_t}{\partial t} + \nabla \cdot \rho_t \boldsymbol{v} = 0. \tag{6.2.5j}$$

This is called the *equation of continuity for the whole mixture.* If the mixture has constant mass density or mass concentration ρ_t, it is simplified to

$$\nabla \cdot \boldsymbol{v} = 0. \tag{6.2.5k}$$

In the cartesian coordinate system, this equation is written as

$$\frac{\partial v_x}{\partial x} + \frac{\partial v_y}{\partial y} + \frac{\partial v_z}{\partial z} = 0. \tag{6.2.5l}$$

This equation is routinely used in the case of incompressible flow.

Equation (6.2.5j) describes how the total density of the mixture changes with time t and the spatial coordinates, whereas equation (6.2.4b) describes how the mass concentration of species i changes with time t and the spatial coordinates. It is useful to transform the latter equation in the same manner as equation (6.2.3b) for the molar concentration of species i was changed to equation (6.2.5c). To that end, recognize that, from (3.1.84a) and the definition of $\boldsymbol{n}_i$, namely $\boldsymbol{n}_i = \boldsymbol{N}_i M_i$,

$$\begin{aligned}\frac{\partial \rho_i}{\partial t} &= -\left(\nabla \cdot \left\{\rho_i(\boldsymbol{v}^* + \boldsymbol{U}_i) - \rho_i \left(\frac{RT}{f_i^d}\right) \nabla \ln a_i\right\}\right) + r_i \\ &= -\nabla \cdot (\rho_i \boldsymbol{v}^*) - \nabla \cdot (\rho_i \boldsymbol{U}_i) + \nabla \cdot \left\{\frac{\rho_i RT}{f_i^d} \nabla \ln a_i\right\} + r_i.\end{aligned} \tag{6.2.5m}$$

Consider now expression (3.1.99), where, instead of the simplification leading to the ∇x_i term, we use the $\nabla \ln a_i$ term; next use the definition $\boldsymbol{j}_i = \boldsymbol{J}_i M_i$. After a little

Table 6.2.2. Equation of change for mass concentration of species i (equation (6.2.5p))

Cartesian coordinates (*x*, *y*, *z*)	$\frac{\partial \rho_i}{\partial t} = -\left[v_x \frac{\partial \rho_i}{\partial x} + v_y \frac{\partial \rho_i}{\partial y} + v_z \frac{\partial \rho_i}{\partial z}\right] - \left[\frac{\partial j_{ix}}{\partial x} + \frac{\partial j_{iy}}{\partial y} + \frac{\partial j_{iz}}{\partial z}\right] + r_i$
Cylindrical coordinates (*r*, *θ*, *z*)	$\frac{\partial \rho_i}{\partial t} = -\left[v_r \frac{\partial \rho_i}{\partial r} + \frac{v_\theta}{r} \frac{\partial \rho_i}{\partial \theta} + v_z \frac{\partial \rho_i}{\partial z}\right] - \left[\frac{1}{r}\frac{\partial}{\partial r}(r j_{ir}) + \frac{1}{r}\frac{\partial j_{i\theta}}{\partial \theta} + \frac{\partial j_{iz}}{\partial z}\right] + r_i$
Spherical coordinates (*r*, *θ*, *Φ*)	$\frac{\partial \rho_i}{\partial t} = -\left[v_r \frac{\partial \rho_i}{\partial r} + \frac{v_\theta}{r} \frac{\partial \rho_i}{\partial \theta} + \frac{v_\phi}{r \sin\theta} \frac{\partial \rho_i}{\partial \phi}\right] - \left[\frac{1}{r^2}\frac{\partial}{\partial r}(r j_{ir}) + \frac{1}{r\sin\theta}\frac{\partial}{\partial \theta}(j_{i\theta}\sin\theta) + \frac{1}{r\sin\theta}\frac{\partial j_{i\phi}}{\partial \phi}\right] + r_i$

C_t is a constant; r_i is the mass rate of production of species *i* by chemical reaction per unit volume.

Table 6.2.3. Navier-Stokes equation in Cartesian coordinates

x-component	$\rho_t \frac{\partial v_x}{\partial t} + \rho_t\left(v_x \frac{\partial v_x}{\partial x} + v_y \frac{\partial v_x}{\partial y} + v_z \frac{\partial v_x}{\partial z}\right) = -\frac{\partial P}{\partial x} + \mu\left[\frac{\partial^2 v_x}{\partial x^2} + \frac{\partial^2 v_x}{\partial y^2} + \frac{\partial^2 v_x}{\partial z^2}\right] + \rho_t g_x^{\text{ext}}$
y-component	$\rho_t \frac{\partial v_y}{\partial t} + \rho_t\left(v_x \frac{\partial v_y}{\partial x} + v_y \frac{\partial v_y}{\partial y} + v_z \frac{\partial v_y}{\partial z}\right) = -\frac{\partial P}{\partial y} + \mu\left[\frac{\partial^2 v_y}{\partial x^2} + \frac{\partial^2 v_y}{\partial y^2} + \frac{\partial^2 v_y}{\partial z^2}\right] + \rho_t g_y^{\text{ext}}$
z-component	$\rho_t \frac{\partial v_z}{\partial t} + \rho_t\left(v_x \frac{\partial v_z}{\partial x} + v_y \frac{\partial v_z}{\partial y} + v_z \frac{\partial v_z}{\partial z}\right) = -\frac{\partial P}{\partial z} + \mu\left[\frac{\partial^2 v_z}{\partial x^2} + \frac{\partial^2 v_z}{\partial y^2} + \frac{\partial^2 v_z}{\partial z^2}\right] + \rho_t g_z^{\text{ext}}$

Newtonian fluid, constant values of ρ_t and μ.

rearrangement, we will obtain the following equation from equation (6.2.5m):

$$\frac{\partial \rho_i}{\partial t} = -\nabla \cdot (\rho_i \boldsymbol{v}) - \nabla \cdot (\boldsymbol{j}_i) + r_i. \qquad (6.2.5\text{n})$$

This is the *equation of change for the mass concentration of species i*. Each term in this equation represents a contribution similar to that by the corresponding term in equation (6.2.5b), except the molar concentration of species *i* has to be replaced by the mass concentration of species *i*. An alternative form of equation (6.2.5n) is given by

$$\frac{\partial \rho_i}{\partial t} = -\rho_i \nabla \cdot \boldsymbol{v} - \boldsymbol{v} \cdot \nabla \rho_i - \nabla \cdot \boldsymbol{j}_i + r_i. \qquad (6.2.5\text{o})$$

If the mixture has constant total mass density, using (6.2.5k) we get

$$\frac{\partial \rho_i}{\partial t} = -\boldsymbol{v} \cdot \nabla \rho_i - \nabla \cdot \boldsymbol{j}_i + r_i. \qquad (6.2.5\text{p})$$

Table 6.2.2 illustrates this equation using *scalar* quantities.

In all equations of change obtained so far, as well as in the equation of continuity, the fluid velocity appears via $\boldsymbol{v}$ or $\boldsymbol{v}^*$. If the migration velocities of different species do not influence the fluid flow field or the bulk fluid motion, then the fluid velocity field may be obtained from the solution of what is known as the *equation of motion*:[8]

$$\frac{\partial}{\partial t}(\rho_t \boldsymbol{v}) = -[\nabla \cdot (\rho_t \boldsymbol{v}\boldsymbol{v})] - \nabla P - [\nabla \boldsymbol{\tau}] + \rho_t \boldsymbol{g}^{\text{ext}}. \qquad (6.2.6\text{a})$$

Standard textbooks on transport phenomena (Bird *et al.*, 2002) and fluid mechanics provide the derivation of this equation and a detailed interpretation of each term. Certain types of conditions are commonly encountered during bulk motion of fluids in separators; these in turn allow some assumptions to be made that simplify the equation of motion. The three most common assumptions are:

(1) it is a Newtonian fluid;
(2) it has constant fluid density(ρ_t) and viscosity(μ);
(3) fluid acceleration terms are neglected.

Incorporation of the first two assumptions leads to the following simplified form of the equation of motion, called the *Navier-Stokes equation*:

$$\frac{\partial \boldsymbol{v}}{\partial t} + \boldsymbol{v} \cdot \nabla \boldsymbol{v} = -\nabla P + \mu \nabla^2 \boldsymbol{v} + \rho_t \boldsymbol{g}^{\text{ext}}. \qquad (6.2.6\text{b})$$

This vectorial equation has three component equations in three coordinates. Table 6.2.3 provides three equations in the Cartesian coordinates *x*, *y* and *z*. When the viscous terms are neglected (i.e. $\mu \nabla^2 \boldsymbol{v} = 0$) due to $\mu \to 0$, the equations describe motion of an *ideal fluid* or *inviscid fluid*.

The fluid acceleration terms in the Navier-Stokes equation are $(\partial \boldsymbol{v}/\partial t)$ and $(\boldsymbol{v} \cdot \nabla \boldsymbol{v})$. When these two terms are neglected (*usually* for flows that are quite viscous), we obtain

$$-\nabla P + \mu \nabla^2 \boldsymbol{v} + \rho_t \boldsymbol{g}^{\text{ext}} = 0. \qquad (6.2.6\text{c})$$

This differential equation is identified often as the equation for *creeping flow* or *Stokes flow*. The three components of this equation in the Cartesian coordinate system are easily obtained from Table 6.2.3 by equating the left-hand side of each equation to zero.

[8]Here $\rho_t \boldsymbol{g}^{\text{ext}}$ represents all external forces per unit fluid volume acting on the fluid, including gravity, and $\boldsymbol{\tau}$ is the stress *tensor*.

Separation of molecular species i from species j in an open separator having bulk flow requires a knowledge of how C_i and C_j are distributed along the separator length. This knowledge is acquired from a solution of equation (6.2.5g) for each of species i and j. Similarly, solution of equation (6.2.5m) for species i and j will provide the profiles of ρ_i and ρ_j in the separator. Tables 6.2.1 and 6.2.2 illustrate these equations in terms of molar fluxes and mass fluxes, respectively. It is useful, however, to consider such equations in terms of various constituent terms of a flux expression. In Section 6.2.1.1 certain special expressions will be used for the diffusive term in multiphase systems; it is necessary to provide a limited fundamental background here to approaches and treatments that will be routinely employed in that section and in Chapters 7 and 8.

Consider equation (6.2.5g) for species i when there is no chemical reaction (i.e. $R_i = 0$):

$$\frac{\partial C_i}{\partial t} + \boldsymbol{v}^* \cdot \nabla C_i = -\nabla \cdot \boldsymbol{J}_i^*. \tag{6.2.7}$$

Further, focus on a liquid solution of species i in a solvent where the migration velocity $\boldsymbol{U}_i$ is zero; the relevant flux expression for $\boldsymbol{J}_i^*$ is (3.1.101):

$$\boldsymbol{J}_i^* = -C_t D_{is}^0 \left(1 + \frac{\mathrm{d}\,\ell n\,\gamma_i}{\mathrm{d}\,\ell n\,x_i}\right)_{T,P} \nabla x_i \tag{6.2.8}$$
$$\boldsymbol{J}_i^* = D_{is} \nabla C_i,$$

where D_{is} is the molecular diffusivity of species i in the solvent s. On substituting this flux expression in equation of change (6.2.7) for C_i, we get, for an assumed constant D_{is},

$$\frac{\partial C_i}{\partial t} + \boldsymbol{v}^* \cdot \nabla C_i = D_{is} \nabla^2 C_i. \tag{6.2.9}$$

Similarly, consider equation (6.2.5n) for species i in the absence of a chemical reaction (i.e. $r_i = 0$):

$$\frac{\partial \rho_i}{\partial t} + \nabla \cdot (\rho_i \boldsymbol{v}) = -\nabla \cdot \boldsymbol{j}_i. \tag{6.2.10}$$

For an incompressible flow, $\nabla \cdot (\rho_i \boldsymbol{v}) = \boldsymbol{v} \cdot \nabla \rho_i$ Further, from Fick's first law in a binary system of species i and j, we obtain

$$\boldsymbol{j}_i = -\rho_t D_{ij} \nabla u_i. \tag{6.2.11}$$

Substituting these two relations into the equation of change (6.2.10) for ρ_i, we get, for an incompressible flow,

$$\frac{\partial \rho_i}{\partial t} + \boldsymbol{v} \cdot \nabla \rho_i = \rho_t D_{ij} \nabla^2 u_i \tag{6.2.12}$$

for the case of constant D_{ij} and ρ_t. For constant ρ_t in the system, we can also write this as

$$\frac{\partial \rho_i}{\partial t} + \boldsymbol{v} \cdot \nabla \rho_i = D_{ij} \nabla^2 \rho_i. \tag{6.2.13}$$

Dividing by M_i, we get

$$\frac{\partial C_i}{\partial t} + \boldsymbol{v} \cdot \nabla C_i = D_{ij} \nabla^2 C_i, \tag{6.2.14}$$

which becomes identical to equation (6.2.9) for an incompressible system, constant diffusivity and dilute i ($\boldsymbol{v} \cong \boldsymbol{v}^*$). Table 6.2.4 provides detailed expressions for equations (6.2.9) and 6.2.12) in cartesian and cylindrical polar coordinate systems for an incompressible fluid, constant diffusivity D_{ij} and dilute solution of i.

Consider now the laminar flow of a liquid in a straight circular tube of radius R and length L (Figure 6.1.1(a)). Let a solute pulse of species i having m_i moles or $m_i M_i$ grams be introduced across the whole cross-sectional area at the location $z = 0$ of such a straight circular tube shown in Figure 6.2.2(a). As the liquid, having a parabolic velocity profile, flows, the question of interest is as follows. What would this solute pulse look like way down the length of the tube?

We have considered an *apparently* related problem in Section 3.2.1 in a stagnant system. We observed, via solutions of concentration profiles (3.2.15), (3.2.19) and (3.2.20) illustrated in Figure 3.2.3, that the species i concentration pulse introduced at $z = 0$ was convected down the vessel ($z > 0$) at a speed equal to the migration velocity U_{iz} of the

Table 6.2.4. Equations of change for mass concentration[a] and molar concentration[b] of species i *for constant* ρ_t *and* D_{ij} *in cartesian and cylindrical polar coordinates in nonreactive systems*

Mass concentration ρ_i, cartesian coordinates (x,y,z)	$\frac{\partial \rho_i}{\partial t} + v_x \frac{\partial \rho_i}{\partial x} + v_y \frac{\partial \rho_i}{\partial y} + v_z \frac{\partial \rho_i}{\partial z} = D_{ij}\left(\frac{\partial^2 \rho_i}{\partial x^2} + \frac{\partial^2 \rho_i}{\partial y^2} + \frac{\partial^2 \rho_i}{\partial z^2}\right)$
Mass concentration ρ_i, cylindrical polar coordinates (r, θ, z)	$\frac{\partial \rho_i}{\partial t} + v_r \frac{\partial \rho_i}{\partial r} + \frac{v_\theta}{r} \frac{\partial \rho_i}{\partial \theta} + v_z \frac{\partial_i}{\partial z} = D_{ij}\left(\frac{1}{r}\frac{\partial}{\partial r}\left(r \frac{\partial \rho_i}{\partial r}\right) + \frac{1}{r^2}\frac{\partial^2 \rho_i}{\partial \theta^2} + \frac{\partial^2 \rho_i}{\partial z^2}\right)$
Molar concentration C_i, cartesian coordinates (x,y,z)	$\frac{\partial C_i}{\partial t} + v_x^* \frac{\partial C_i}{\partial x} + v_y^* \frac{\partial C_i}{\partial y} + v_z^* \frac{\partial C_i}{\partial z} = D_{ij}\left(\frac{\partial^2 C_i}{\partial x^2} + \frac{\partial^2 C_i}{\partial y^2} + \frac{\partial^2 C_i}{\partial z^2}\right)$
Molar concentration C_i, cylindrical polar coordinates (r,θ,z)	$\frac{\partial C_i}{\partial t} + v_r^* \frac{\partial C_i}{\partial r} + \frac{v_\theta^*}{r} \frac{\partial C_i}{\partial \theta} + v_z^* \frac{\partial C_i}{\partial z} = D_{ij}\left(\frac{1}{r}\frac{\partial}{\partial r}\left(r \frac{\partial C_i}{\partial r}\right) + \frac{1}{r^2}\frac{\partial^2 C_i}{\partial \theta^2} + \frac{\partial^2 C_i}{\partial z^2}\right)$

[a] Equations (6.2.12), (6.2.13).
[b] Equation (6.2.9); $\boldsymbol{U}_i$ is zero.

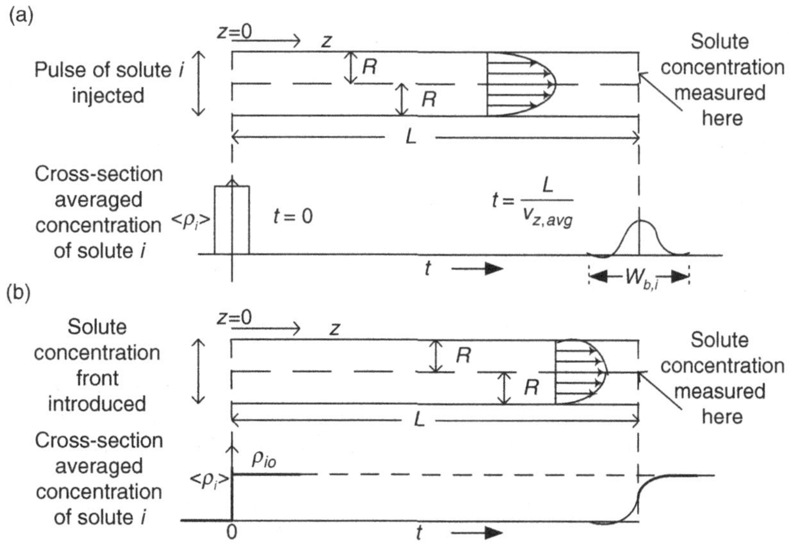

Figure 6.2.2. Axial dispersion of (a) a solute concentration pulse and (b) a solute concentration front introduced at the inlet of a straight circular tube having a laminar liquid flow and a parabolic velocity profile.

solute *i*; furthermore, the narrow solute pulse was becoming a broader band due to molecular diffusion as z increased. This phenomenon was identified as *band broadening* or *dispersion*; note that this phenomenon was strictly due to molecular diffusion.

In the present problem (illustrated in Figure 6.2.2(a)), we will also find that the solute pulse introduced at $z = 0$ will show up (on a radially averaged basis) as a concentration peak with a broadened base as z becomes large. However, this broadening of the solute profile in the z-direction is not due to the molecular diffusion coefficient D_{is} of species i in the solvent. Rather, it arises primarily due to the radially nonuniform axial velocity profile (6.1.1b, c) of flow in a tube. It is identified as an *axial dispersion* or *convective dispersion*. This phenomenon was first studied by Taylor (1953, 1954), and is often described also as *Taylor dispersion*.

What happens is as follows. Let us ignore axial solute diffusion altogether to start with. Since there is a parabolic velocity profile, solute molecules introduced via the pulse further from the center line at $z = 0$ are convected downstream at a lower velocity compared to the solute molecules introduced via the pulse near the center line at $z = 0$. It is very clear that a band will develop around a mean position: sections of the band are at a larger z since they were located at higher velocity regions to start with; conversely, sections of the band are at a smaller z with respect to the mean band position since they were located at lower velocity regions to start with. Radial velocity variation in the tube then creates an axial dispersion of the (radially averaged) solute concentration profile. Radial diffusion, however, will counteract this axial dispersion in the following manner. Regions in the band having a higher concentration at any z vis-à-vis a lower concentration at a larger radial location at the same z location will allow radial diffusion from the higher concentration region to the lower concentration region.

For a cylindrical polar coordinate system needed to describe convective dispersion of the solute pulse in a straight circular tube (Figure 6.2.2(a)), we can discount any θ-dependence due to essential circular symmetry. The governing equation from Table 6.2.4 for species i in mass concentration terms is given by (Bird *et al.*, 2002)

$$\frac{\partial \rho_i}{\partial t} + v_{z,\max}\left[1 - \left(\frac{r}{R}\right)^2\right]\frac{\partial \rho_i}{\partial z} = D_{is}\left(\frac{1}{r}\frac{\partial}{\partial r}\left(r\frac{\partial \rho_i}{\partial r}\right) + \frac{\partial^2 \rho_i}{\partial z^2}\right). \tag{6.2.15}$$

There is no exact analytical solution for this equation. However, if one neglects the axial diffusion term, $D_{is}\,\partial^2\rho_i/\partial z^2$, and solves the equation subject to the following boundary conditions:

$$\text{at } r = 0,\ (\partial \rho_i/\partial r) = 0; \qquad \text{at } r = R,\ (\partial \rho_i/\partial r) = 0, \tag{6.2.16a}$$

one can get a solution for ρ_i. Taylor's solution was obtained by developing $\langle \rho_i \rangle$, which is the value of ρ_i when averaged over the radial cross section of the tube:

$$\langle \rho_i \rangle = \frac{\int_0^{2\pi}\int_0^R \rho_i\, r\,\mathrm{d}r\,\mathrm{d}\theta}{\int_0^{2\pi}\int_0^R r\,\mathrm{d}r\,\mathrm{d}\theta} = \frac{2}{R^2}\int_0^R \rho_i\, r\,\mathrm{d}r \tag{6.2.16b}$$

$$\langle \rho_i \rangle = \frac{m_i M_i}{2\pi R^2 \sqrt{\pi D_{i,\mathrm{eff}}\, t}} \exp\left(\frac{\left(z - v_{z,\mathrm{avg}}\, t\right)^2}{4\, D_{i,\mathrm{eff}}\, t} \right). \qquad (6.2.17)$$

Here, $2v_{z,\mathrm{avg}} = v_{z,\max}$, and $D_{i,\mathrm{eff}}$ is an axial dispersion coefficient such that a solution of

$$\frac{\partial}{\partial t}\langle \rho_i \rangle + v_{z,\mathrm{avg}} \frac{\partial}{\partial z}\langle \rho_i \rangle = D_{i,\mathrm{eff}} \frac{\partial^2}{\partial z^2}\langle \rho_i \rangle \qquad (6.2.18)$$

will yield the profile of the pulse in terms of the radially averaged ρ_i value, $\langle \rho_i \rangle$, as shown in Figure 6.2.2(a). The expression for $D_{i,\mathrm{eff}}$ is

$$D_{i,\mathrm{eff}} = \frac{1}{48} D_{is}\, Pe_i^2, \qquad (6.2.19)$$

where

$$Pe_i = \left(R\, v_{z,\mathrm{avg}}/D_{is}\right). \qquad (6.2.20)$$

Solution (6.2.17), developed by neglecting the axial diffusion term $D_{is}\left(\partial^2\rho_i/\partial z^2\right)$, has been demonstrated to be valid when $Pe_i \geq 70$ and the base width of the pulse ($\sim W_{bi}$) is $\gtrsim 170R$.

Note: If the flow in the tube may be such that it can be hypothesized as a plug flow,[9] there is no axial velocity variation in the radial direction; therefore the value of $D_{i,\mathrm{eff}} \cong 0$. Consequently, the solute pulse will appear down the tube length again as a pulse without any broadening. The value of $Pe_{i,\mathrm{eff}}$, based on $D_{i,\mathrm{eff}}$,

$$Pe_{i,\mathrm{eff}} = R v_z / D_{i,\mathrm{eff}}, \qquad (6.2.21)$$

will then tend toward ∞ as $D_{i,\mathrm{eff}} \to 0$. Aris (1956) has provided a more rigorous analysis, which includes axial diffusion, and has suggested the following expression for $D_{i,\mathrm{eff}}$ for long times:

$$D_{i,\mathrm{eff}} = D_{is} + \frac{R^2 v_{z,\mathrm{avg}}^2}{48\, D_{is}} = D_{is}\left(1 + \frac{Pe_i^2}{48}\right). \qquad (6.2.22)$$

If Pe_i is defined as $(2Rv_{z,\mathrm{avg}}/D_{is})$, the factor 48 in the above equation will be changed to 192. If, instead of a solute pulse, a solute front is introduced at $z = 0$ (Figure 6.2.2 (b)), then the solute concentration front down the end of the tube will be substantially broadened due to the radial variation of the axial velocity in the tube.

The basic conclusions of Taylor dispersion (and Aris's results) have been supported by solutions of the complete equations by a number of investigators. For a brief introduction to these references and literature, consult Froment and Bischoff (1979, p. 621) and Bird *et al.* (2002, p. 646). What is more important from our perspective is that such a model of axial dispersion (namely equation (6.2.18), can be effectively used for complex flow situations in separators. The complexity may arise due to a packed bed of particles (as we will see in modeling adsorbers, chromatographic columns, etc., in Chapter 7), a dispersed gas phase in a continuous liquid phase or vice versa (in absorption/stripping processes, distillation plates, etc., in Chapter 8), a dispersed liquid phase in another liquid phase (in extraction processes in Chapter 8) and other separation processes/techniques where flow situations are complex. Axial dispersion is also of increasing importance in analytical-scale activities involving microfluidics, capillary electrophoresis, electrokinetic flow, etc. (Kirchner and Hasselbrink, 2005). In Section 6.2.1.1, where governing equations are introduced for multiphase systems, the effective axial dispersion coefficient, $D_{i,\mathrm{eff}}$, will be utilized in equations governing species concentration variables averaged over a cross section perpendicular to the mean flow direction. Turbulence, if present, will introduce additional complexity. *Note*: The shortcomings of Taylor dispersion analysis are identified in pp. 225–297 of an article in Marin (2005).

6.2.1.1 *Pseudo-continuum approach for multiphase systems*

Many separation systems have two phases, e.g. gas–liquid, vapor–liquid, liquid–liquid, liquid–solid, gas–solid, vapor–solid, supercritical fluid–solid, supercritical fluid–liquid, etc. Such systems do not possess the geometrical simplicity of, for example, Poiseuille flow in a round tube or the simple drag flow between two flat plates (Figure 6.1.8). Packed beds, multiphase countercurrent or cocurrent contactors, where two immiscible phases flow past each other, etc., have very complex flow patterns. Yet there are two basic requirements: (1) to account for the species transport between the phases at the phase interfaces and (2) to determine the species concentrations and phase flow rate variations along the separator length/bulk flow direction. An appropriate *volume-averaging procedure* is adopted to this end, and the resulting approach is identified as a *pseudo-continuum approach* (Lee *et al.*, 1977a).

"Volume averaging" means determining the average value of a quantity over a certain localized volume; this localized volume should be connected with a point in the flow system. The averaging has to be carried out over three characteristic lengths of the localized volume, l_{loc}, which is much smaller than the characteristic dimension l of the separator,[10] but is much larger than the length, l_{var}, over which the quantity under consideration, say C_i, fluctuates significantly: $l_{\mathrm{var}} << l_{\mathrm{loc}} << l$. However, if we are studying the variations of the species concentrations and phase flow rates along the separator length/bulk flow direction (say l), then the volume element over which the averaging is carried out spans the flow cross-sectional area of the

[9] See Figure 6.1.5 for electroosmotic flow in a capillary.

[10] For an introduction to multiscale analysis, see Marin (2005).

separator and an appropriate dimension over the mean flow direction over which local fluctuations are evened out, and yet the averaged quantities change in the mean flow direction. In what follows, we *utilize* the approaches of Slattery (1972), Whitaker (1973), Gray (1975) and Soo (1989), and *identify the averaged equations* without going through detailed derivations.

Consider k phases or regions present in the separation system. Most often, $k = 1,2$, corresponding to phases 1 and 2 or regions 1 and 2. Correspondingly, the equations for the molar concentration C_{ik} of phase k may be written from (6.2.3b) using the relation

$$(\boldsymbol{N}_i)_k = (\boldsymbol{J}_i)_k + C_{ik}\boldsymbol{v}_k$$

as

$$\frac{\partial C_{ik}}{\partial t} = -\left(\nabla \cdot (\boldsymbol{J}_i)_k\right) - \nabla \cdot (C_{ik}\boldsymbol{v}_k) + \boldsymbol{R}_{ik}. \qquad (6.2.23a)$$

If we extract the species-specific migration velocity $\boldsymbol{U}_{ik}$-containing term out of the expression for $(\boldsymbol{J}_i)_k$ (see expression (3.1.98)), we obtain

$$\frac{\partial C_{ik}}{\partial t} + \nabla \cdot (C_{ik}\boldsymbol{v}_k) + \nabla \cdot (C_{ik}\boldsymbol{U}_{ik}) = -\left(\nabla \cdot (\boldsymbol{J}_i)_k\right) + R_{ik}. \qquad (6.2.23b)$$

Note that the $(\boldsymbol{J}_i)_k$ term no longer has any contribution from $\boldsymbol{U}_{ik}$.

Figure 6.2.3 illustrates two phases $k = 1,2$ in the system. One can define two types of volume averages in such a system. When a quantity is averaged over the volumes of the two phases ($k = 1,2$) in the control volume, we have what is known as the *phase average*. When a quantity is averaged only over a specific phase or region,

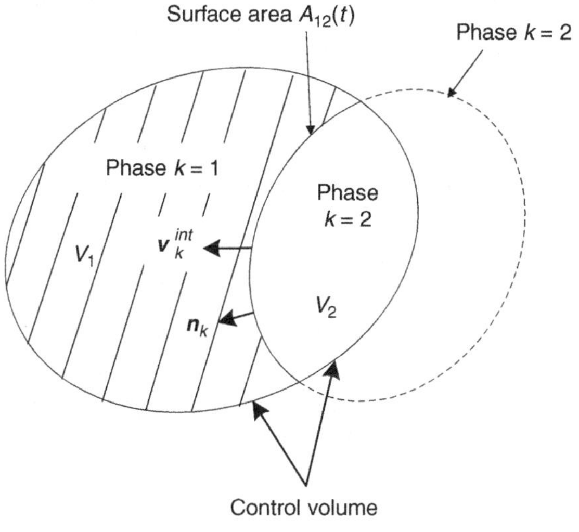

Figure 6.2.3. Configuration of a two-phase system.

it is called an *intrinsic phase average* for that phase or region. The intrinsic phase averages $\rho_{ik,\text{avg}i} = (\overline{\rho}_{ik})$ and $C_{ik,\text{avg}i} = (\overline{C}_{ik})$ of ρ_{ik} and C_{ik}, respectively, over the volume of phase or region k, V_k, are defined by

$$\overline{\rho}_{ik} = \rho_{ik,\text{avg}i} = \frac{1}{V_k}\int_{V_k} \rho_{ik}\text{d}V;$$
$$\overline{C}_{ik} = C_{ik,\text{avg}i} = \frac{1}{V_k}\int_{V_k} C_{ik}\text{d}V. \qquad (6.2.24a)$$

The phase averages $\rho_{ik,\text{avg}}$ and $C_{ik,\text{avg}}$ of ρ_{ik} and C_{ik}, respectively, over the whole volume consisting of all of V_k, corresponding to different values of k, are defined by

$$\overline{\overline{\rho}}_{ik} = \rho_{ik,\text{avg}} = \frac{1}{V_1 + V_2}\int_{V_1+V_2} \rho_{ik}\text{d}V;$$
$$\overline{\overline{C}}_{ik} = C_{ik,\text{avg}} = \frac{1}{V_1 + V_2}\int_{V_1+V_2} C_{ik}\text{d}V, \qquad (6.2.24b)$$

where we have assumed that the system has only two phases or regions, 1 and 2. If ε_k is the volume fraction of phase k in the system, i.e. $\varepsilon_k = \left(V_k/\sum V_k\right)$, then, from the above definitions,

$$\overline{\overline{\rho}}_{ik} = \rho_{ik,\text{avg}} = \varepsilon_k\rho_{ik,\text{avg}i} = \varepsilon_k\overline{\rho}_{ik};$$
$$\overline{\overline{C}}_{ik} = C_{ik,\text{avg}} = \varepsilon_k C_{ik,\text{avg}i} = \varepsilon_k\overline{C}_{ik}. \qquad (6.2.24c)$$

The two types of averages of any quantity ψ are generally identified as follows. The *phase average* of ψ in the k-phase is given by

$$\langle\psi_k\rangle = \frac{1}{\sum_k V_k}\int_{\sum_k V_k} \psi_k\,\text{d}V = \psi_{k,\text{avg}} \qquad (6.2.25a)$$

and the *intrinsic phase average* of ψ in the k-phase is given by

$$\langle\psi_k\rangle^k = \frac{1}{V_k}\int_{V_k} \psi\,\text{d}V = \psi_{k,\text{avg}\,i}. \qquad (6.2.25b)$$

The phase average of, say, the differential equation (6.2.23b) of transport of C_{ik} is given by

$$\left\langle\frac{\partial C_{ik}}{\partial t}\right\rangle + \langle\nabla\cdot(C_{ik}\boldsymbol{v}_k)\rangle + \langle\nabla\cdot(C_{ik}\boldsymbol{U}_{ik})\rangle = \langle -\nabla\cdot(\boldsymbol{J}_i)_k\rangle + \langle R_{ik}\rangle. \qquad (6.2.26a)$$

A number of results from the general transport theorem and the volume-averaging techniques identified in Gray (1975), Soo (1989) and others identified earlier are useful here:

$$\left\langle\frac{\partial\psi_k}{\partial t}\right\rangle = \frac{\partial}{\partial t}\langle\psi_k\rangle - \frac{1}{V}\int_{A_{12}(t)} \psi_k\,\boldsymbol{v}_k^{\text{int}}\cdot\boldsymbol{n}_k\text{d}A; \qquad (6.2.26b)$$

$$\langle\nabla\psi_k\rangle = \nabla\langle\psi_k\rangle + \frac{1}{V}\int_{A_{12}(t)} \psi_k\boldsymbol{n}_k\text{d}A; \qquad (6.2.26c)$$

$$\langle \nabla \cdot \psi_k \rangle = \nabla \cdot \langle \psi_k \rangle + \frac{1}{V} \int_{A_{12}(t)} \psi_k \cdot \boldsymbol{n}_k \mathrm{d}A. \tag{6.2.26d}$$

These relations relate, for example, the average of the derivative to the derivative of the average (e.g. (6.2.26c)). Here, $A_{12}(t)$ is the area of the interface between phases 1 and 2, where $k = 1,2$, $\boldsymbol{v}_k^{\text{int}}$ is the mass average velocity of this interface and $\boldsymbol{n}_k$ is the outwardly directed unit normal on the k-phase surface (Figure 6.2.3). Employing these relations, one can rewrite the phase-averaged equation (6.2.26a) following Gray (1975) as follows:

$$\varepsilon_k \frac{\partial \overline{C}_{ik}}{\partial t} + \overline{C}_{ik} \frac{\partial \varepsilon_k}{\partial t} + \nabla \cdot (\overline{C}_{ik} \langle \boldsymbol{v}_k \rangle) + \nabla \cdot (\hat{C}_{ik} \hat{\boldsymbol{v}}_k) + \nabla \cdot (\overline{C}_{ik} \langle \boldsymbol{U}_{ik} \rangle)$$
$$+ \nabla \cdot (\hat{C}_{ik} \hat{\boldsymbol{U}}_{ik}) + \frac{1}{\sum_k V_k} \int_{A_{12}(t)} C_{ik}(v_k + \boldsymbol{U}_{ik} - \boldsymbol{v}_k^{\text{int}}) \cdot \boldsymbol{n}_k \mathrm{d}A$$
$$= -\nabla \cdot \langle (\boldsymbol{J}_i)_k \rangle - \frac{1}{\sum_k V_k} \int_{A_{12}(t)} (\boldsymbol{J}_i)_k \cdot \boldsymbol{n}_k \mathrm{d}A + \varepsilon_k \langle R_{ik} \rangle^k. \tag{6.2.27}$$

This derivation employed a particular approach of Gray (1975) wherein $\hat{C}_{ik}$, $\hat{\boldsymbol{v}}_{ik}$ and $\hat{\boldsymbol{U}}_{ik}$ are defined via the relations

$$C_{ik} = \overline{C}_{ik} + \hat{C}_{ik}; \qquad \boldsymbol{v}_k = \langle \boldsymbol{v}_k \rangle^k + \hat{\boldsymbol{v}}_k; \qquad \boldsymbol{U}_{ik} = \langle \boldsymbol{U}_{ik} \rangle^k + \hat{\boldsymbol{U}}_{ik}, \tag{6.2.28}$$

such that $\langle \hat{C}_{ik} \rangle^k, \langle \hat{C}_{ik} \rangle, \langle \hat{\boldsymbol{v}}_k \rangle^k, \langle \hat{\boldsymbol{v}}_k \rangle, \langle \hat{\boldsymbol{U}}_{ik} \rangle^k$ and $\langle \hat{\boldsymbol{U}}_{ik} \rangle$ are zero. This assumes that the averages of C_{ik}, $\boldsymbol{v}_k$ and $\boldsymbol{U}_{ik}$ are well-behaved. Gray (1975) had assumed as such for C_{ik}, $\boldsymbol{v}_k$; we have added $\boldsymbol{U}_{ik}$ to the list.

We will now make a number of practical assumptions valid for many two-phase separation systems and simplify equation (6.2.27) by assuming that.

(1) ε_k is constant, i.e. $(\partial \varepsilon_k / \partial t) = 0$;
(2) species i is not being produced by a chemical reaction, i.e. $R_{ik} = 0$;
(3) no phase change occurs, i.e.

$$\int_{A_{12}(t)} C_{ik}(\boldsymbol{v}_k - \boldsymbol{v}_k^{\text{int}}) \cdot \boldsymbol{n}_k \, \mathrm{d}A = 0.$$

These assumptions lead to

$$\varepsilon_k \frac{\partial \overline{C}_{ik}}{\partial t} + \nabla \cdot (\langle \boldsymbol{v}_k \rangle \overline{C}_{ik}) + \nabla \cdot (\langle \boldsymbol{U}_{ik} \rangle \overline{C}_{ik}) + \frac{1}{\sum_k V_k} \int_{A_{12}(t)} C_{ik} (\boldsymbol{U}_{ik}) \cdot \boldsymbol{n}_k \mathrm{d}A$$
$$= -\nabla \cdot \langle (\boldsymbol{J}_i)_k \rangle - \nabla \cdot (\hat{C}_{ik} \hat{\boldsymbol{v}}_k) - \nabla \cdot (\hat{C}_{ik} \hat{\boldsymbol{U}}_{ik}) - \frac{1}{\sum_k V_k} \int_{A_{12}(t)} (\boldsymbol{J}_i)_k \cdot \boldsymbol{n}_k \, \mathrm{d}A \tag{6.2.29}$$

For systems where U_{ik} *is zero* and *molecular diffusion and partitioning between two phases* is the primary mechanism for separation, we can rewrite equation (6.2.29) as follows:

$$\varepsilon_k \frac{\partial \overline{C}_{ik}}{\partial t} + \nabla \cdot (\langle \boldsymbol{v}_k \rangle \overline{C}_{ik}) = -\nabla \cdot \langle (\boldsymbol{J}_i)_k \rangle - \nabla \cdot (\hat{C}_{ik} \hat{\boldsymbol{v}}_k)$$
$$- \frac{1}{\sum_k V_k} \int_{A_{12}(t)} (\boldsymbol{J}_i)_k \cdot \boldsymbol{n}_k \, \mathrm{d}A. \tag{6.2.30}$$

In this equation, the first term on the right-hand side involves molecular diffusion. Usually, the second term on the right-hand side is combined with the first term to represent the complex phenomenon of *solute dispersion*:

$$\langle (J_i)_k \rangle + \hat{C}_{ik} \hat{\boldsymbol{v}}_{ik} = -\varepsilon_k D_{ik} \nabla \overline{C}_{ik} + \hat{C}_{ik} \hat{\boldsymbol{v}}_k = -\varepsilon_k \underline{\underline{D}}_{i,\text{eff},k} \nabla \overline{C}_{ik}, \tag{6.2.31}$$

where $\underline{\underline{D}}_{i,\text{eff},k}$ is the dispersion tensor (Gray, 1975). The second term on the left-hand side, $\nabla \cdot (\langle \boldsymbol{v}_k \rangle \overline{C}_{ik})$, is the contribution of the convective fluid motion to the transport of species i. For incompressible flow, this term is reduced to $\overline{C}_{ik} \nabla \cdot \langle \boldsymbol{v}_k \rangle + \langle \boldsymbol{v}_k \rangle \cdot \nabla \overline{C}_{ik} = \langle \boldsymbol{v}_k \rangle \cdot \nabla \overline{C}_{ik}$, since $\nabla \cdot \langle \boldsymbol{v}_k \rangle$ is zero (see equation (6.2.5k)). The last term on the right-hand side of (6.2.30) represents the rate of interphase transport of species i across the phase interface $A_{12}(t)$; it may be represented by an overall mass-transfer coefficient K_{ik} multiplied by an appropriate surface area and concentration difference: $K_{ik} a (\overline{C}_{ik} - \overline{C}_{ik}^*)$. These simplifications transform equation (6.2.30) as follows:

$$\varepsilon_k \frac{\partial \overline{C}_{ik}}{\partial t} + \langle \boldsymbol{v}_k \rangle \cdot \nabla \overline{C}_{ik} = \varepsilon_k \nabla \cdot (\underline{\underline{D}}_{i,\text{eff},k} \nabla \overline{C}_{ik}) - K_{ik} a (\overline{C}_{ik} - \overline{C}_{ik}^*). \tag{6.2.32}$$

Here, $\overline{C}_{ik}^*$ is a hypothetical concentration of i in the phase k which would be in equilibrium with the concentration of species i in the other phase of the interphase transport system (see equation (3.4.6)); a is the interfacial area of transport per unit control/system volume. In a two-phase system, where $k = 1, 2$, the equation for $\overline{C}_{i1}$ will have a sign for this term which will be opposite to that in the equation for $\overline{C}_{i2}$ since species i comes into one phase while it goes out of the other phase.

When the volume averaging primarily involves an average over the flow cross section with dependent variables and parameters[11] varying only along the mean flow direction along the separator axis, z, we can simplify the equation (6.2.32) as follows:

$$\varepsilon_k \frac{\partial \overline{C}_{ik}}{\partial t} + \langle v_{kz} \rangle \frac{\partial \overline{C}_{ik}}{\partial z} = \varepsilon_k D_{i,\text{eff},k} \frac{\partial^2 \overline{C}_{ik}}{\partial z^2} - K_{ik} a (\overline{C}_{ik} - \overline{C}_{ik}^*). \tag{6.2.33}$$

An overall balance of species i in both phases $k = 1,2$ in this context can be obtained by adding the individual equations for $k = 1$ and 2 (note that $\varepsilon_1 + \varepsilon_2 = 1$):

[11] $D_{i,\text{eff},k}$ is assumed not to vary in the z-direction.

$$\varepsilon_1 \frac{\partial \overline{C}_{i1}}{\partial t} + \varepsilon_2 \frac{\partial \overline{C}_{i2}}{\partial t} + \langle v_{1z} \rangle \frac{\partial \overline{C}_{i1}}{\partial z} + \langle v_{2z} \rangle \frac{\partial \overline{C}_{i2}}{\partial z} = \varepsilon_1 D_{i\text{eff},1} \frac{\partial^2 \overline{C}_{i1}}{\partial z^2} + \varepsilon_2 D_{i\text{eff},2} \frac{\partial^2 \overline{C}_{i2}}{\partial z^2}. \quad (6.2.34)$$

If phase 1 happens to be solid and the stationary phase (as in a fixed-bed process, see Section 7.1), $\langle v_{1z} \rangle = 0$ and $D_{i\text{eff},1} \cong 0$, and we obtain

$$(1-\varepsilon) \frac{\partial \overline{C}_{i1}}{\partial t} + \varepsilon \frac{\partial \overline{C}_{i2}}{\partial t} + \langle v_{2z} \rangle \frac{\partial \overline{C}_{i2}}{\partial z} = \varepsilon D_{i\text{eff},2} \frac{\partial^2 \overline{C}_{i2}}{\partial z^2} \quad (6.2.35)$$

for a fixed-bed process with species i exchanging between the fluid phase 2 and the solid and stationary phase 1, the void volume of the bed being ε. An alternative form of this equation is used more often:

$$\frac{\partial \overline{C}_{i2}}{\partial t} + \frac{(1-\varepsilon)}{\varepsilon} \frac{\partial \overline{C}_{i1}}{\partial t} + \frac{\langle v_{2z} \rangle}{\varepsilon} \frac{\partial \overline{C}_{i2}}{\partial z} = D_{i\text{eff},2} \frac{\partial^2 \overline{C}_{i2}}{\partial z^2}. \quad (6.2.36)$$

Here, $(\langle v_{2z} \rangle / \varepsilon)$ is the interstitial velocity of the fluid phase.

In the case of two fluid phases (gas–liquid, liquid-liquid, etc.) flowing simultaneously in the separation device, we can write down the equations for $k = 1$ and 2 following equation (6.2.33):

$$\varepsilon_1 \frac{\partial \overline{C}_{i1}}{\partial t} + \langle v_{1z} \rangle \frac{\partial \overline{C}_{i1}}{\partial z} = \varepsilon_1 D_{i\text{eff},1} \frac{\partial^2 \overline{C}_{i1}}{\partial z^2} - K_{i1} a (\overline{C}_{i1} - \overline{C}_{i1}^*); \quad (6.2.37\text{a})$$

$$\varepsilon_2 \frac{\partial \overline{C}_{i2}}{\partial t} + \langle v_{2z} \rangle \frac{\partial \overline{C}_{i2}}{\partial z} = \varepsilon_2 D_{i\text{eff},2} \frac{\partial^2 \overline{C}_{i2}}{\partial z^2} - K_{i2} a (\overline{C}_{i2} - \overline{C}_{i2}^*). \quad (6.2.37\text{b})$$

More common forms of these two equations are:

$$\frac{\partial \overline{C}_{i1}}{\partial t} + \frac{\langle v_{1z} \rangle}{\varepsilon_1} \frac{\partial \overline{C}_{i1}}{\partial z} = D_{i\text{eff},1} \frac{\partial^2 \overline{C}_{i1}}{\partial z^2} - K_{i1} \frac{a}{\varepsilon_1} (\overline{C}_{i1} - \overline{C}_{i1}^*); \quad (6.2.37\text{c})$$

$$\frac{\partial \overline{C}_{i2}}{\partial t} + \frac{\langle v_{2z} \rangle}{\varepsilon_2} \frac{\partial \overline{C}_{i2}}{\partial z} = D_{i\text{eff},2} \frac{\partial^2 \overline{C}_{i2}}{\partial z^2} - K_{i2} \frac{a}{\varepsilon_2} (\overline{C}_{i2} - \overline{C}_{i2}^*). \quad (6.2.37\text{d})$$

Generally, when both phases flow in a device, the separation takes place under steady state; therefore the unsteady state terms in these two equations will disappear.

If the species migration velocity $\boldsymbol{U}_{ik}$ is *nonzero*, we will have to go back to equation (6.2.29) and rewrite it as

$$\varepsilon_k \frac{\partial \overline{C}_{ik}}{\partial t} + \nabla \cdot (\langle \boldsymbol{v}_k \rangle \overline{C}_{ik}) + \nabla \cdot (\langle \boldsymbol{U}_{ik} \rangle \overline{C}_{ik}) = -\nabla \cdot \{ \langle (\boldsymbol{J}_i)_k \rangle + (\hat{C}_{ik} \hat{\boldsymbol{v}}_k) + (\hat{C}_{ik} \hat{\boldsymbol{U}}_{ik}) \} - \frac{1}{\sum_k V_k} \int_{A_{12}(t)} \{ C_{ik} (\boldsymbol{U}_{ik}) + (\boldsymbol{J}_i)_k \} \cdot \boldsymbol{n}_k \, dA. \quad (6.2.38)$$

We focus on the first term on the right-hand side of this equation and, in the manner of definition (6.2.31), suggest a dispersion tensor $\underline{\underline{D}}_{i\text{eff},k}$ such that

$$\langle (\boldsymbol{J}_i)_k \rangle + (\hat{C}_{ik} \hat{\boldsymbol{v}}_k) + (\hat{C}_{ik} \hat{\boldsymbol{U}}_{ik}) = -\varepsilon_k D_{ik} \nabla \overline{C}_{ik} + \hat{C}_{ik} \hat{\boldsymbol{v}}_k + \hat{C}_{ik} \hat{\boldsymbol{U}}_{ik} = -\varepsilon_k \underline{\underline{D}}_{i\text{eff},k} \nabla \overline{C}_{ik}. \quad (6.2.39\text{a})$$

The second term on the right-hand side of (6.2.38) represents interphase transport across the phase interface $A_{12}(t)$ due to diffusion and any species-specific migration velocity $\boldsymbol{U}_{ik}$. The second term on the left-hand side of (6.2.38) will be reduced to $\langle \boldsymbol{v}_k \rangle \cdot \nabla \overline{C}_{ik}$ in the case of incompressible flow. The third term on the left-hand side may be expressed as follows:

$$\nabla \cdot (\langle \boldsymbol{U}_{ik} \rangle \overline{C}_{ik}) = \overline{C}_{ik} \nabla \cdot \langle \boldsymbol{U}_{ik} \rangle + \langle \boldsymbol{U}_{ik} \rangle \cdot \nabla \overline{C}_{ik}. \quad (6.2.39\text{b})$$

In the case of some external force, e.g. electrical force in a uniform electrical field, $\nabla \cdot \langle \boldsymbol{U}_{ik} \rangle$ is zero.[12] In other cases, it may not be zero if $\boldsymbol{U}_{ik}$ varies with the location in the separator.

In the context of these considerations, equation (6.2.38) may be written for incompressible flow as follows:

$$\varepsilon_k \frac{\partial \overline{C}_{ik}}{\partial t} + \overline{C}_{ik} \nabla \cdot \langle \boldsymbol{U}_{ik} \rangle + \{ \langle \boldsymbol{v}_{ik} \rangle + \langle \boldsymbol{U}_{ik} \rangle \} \cdot \nabla \overline{C}_{ik} = -\varepsilon_k \underline{\underline{D}}_{i\text{eff},k} \nabla^2 \overline{C}_{ik} - \frac{1}{\sum_k V_k} \int_{A_{12}(t)} \{ C_{ik} \boldsymbol{U}_{ik} + (\boldsymbol{J}_i)_k \} \cdot \boldsymbol{n}_k \, dA. \quad (6.2.40)$$

6.2.1.2 *Continuous stirred tank separator*

In a *continuous stirred tank separator* (CSTS), fluid streams (single or multiphase, with or without solids) enter and leave, and the contents are kept well stirred or well mixed. There is no spatial dependence of species concentrations or any other quantity (such as temperature, pressure, etc.) inside the separator vessel. The species concentrations inside the vessel are at a steady state, sometimes they may change with time; however, the outlet stream concentration is different from the inlet stream concentration, and is equal to that inside the CSTS (Figure 6.2.4). Meanwhile, whatever separation mechanism is employed is operative inside the vessel. (A batch stirred tank separator operates such that there is no spatial gradients of concentration, temperature or pressure inside the vessel. Except when the batch is introduced or withdrawn from the vessel, no fluid/solid streams enter or leave the vessel. The conditions inside the vessel may change with time. Generally there is vigorous bulk motion in the vessel.)

If the volume of a separator, which has, say, one or two feed streams coming in ($j = f$ or f_1 and f_2) and k product streams leaving the separator, is V then, for species i, mass balance in the separator, at any time t, is given by

[12] Here $\boldsymbol{U}_{ik}$ $(= \mu_i^m \boldsymbol{E})$ is independent of the spatial coordinate.

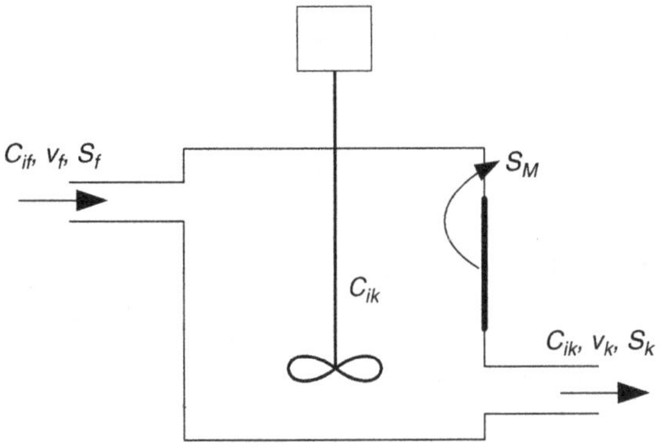

Figure 6.2.4. Continuous stirred tank separator (CSTS) with one incoming stream and one outgoing stream.

$$V\frac{dC_i}{dt} = \int_{S_{f1}} \boldsymbol{N}_{if1}\cdot d\boldsymbol{S}_{f1} + \int_{S_{f2}} \boldsymbol{N}_{if2}\cdot d\boldsymbol{S}_{f2} - \sum_k \boldsymbol{N}_{ik}\cdot d\boldsymbol{S}_k + \int_M \boldsymbol{N}_{iM}\cdot d\boldsymbol{S}_M + R_i V, \tag{6.2.41}$$

where we have assumed that there is a surface ($\boldsymbol{S}_M$) in the vessel (with or without a membrane *M*) through which species *i* is being introduced into the separator vessel in a manner quite different from that by the feed streams; there could be more than one such surface. If, at the separator vessel inlets and outlets, diffusion can be neglected and the averaged stream velocities and concentrations may be utilized, we obtain

$$V\frac{dC_i}{dt} = \langle v_{f_1}\rangle\langle C_{if_1}\rangle S_{f_1} + \langle v_{f_2}\rangle\langle C_{if_2}\rangle S_{f_2} - \sum_k \langle v_k\rangle\langle C_{ik}\rangle S_k + \int_M \boldsymbol{N}_{iM}\cdot d\boldsymbol{S}_M + R_i V. \tag{6.2.42}$$

Generally, a CSTS has one inlet, one outlet and one additional surface that may act like a membrane. Then, by definition of a CSTS, equation (6.2.42) is simplified to

$$V\frac{dC_{i1}}{dt} = \langle v_{f_1}\rangle\langle C_{if_1}\rangle S_{f_1} - \langle v_1\rangle\langle C_{i1}\rangle S_1 + \int_M \boldsymbol{N}_{iM}\cdot d\boldsymbol{S}_M + R_i(C_{i1})\,V. \tag{6.2.43}$$

For steady state nonreactive systems having one membrane surface, the relation is simplified to

$$\langle v_{f_1}\rangle\langle C_{if_1}\rangle S_{f_1} + \int_M \boldsymbol{N}_{iM}\cdot d\boldsymbol{S}_M = \langle v_1\rangle\langle C_{i1}\rangle S_1. \tag{6.2.44}$$

6.2.2 Equation of motion of a particle in a fluid: trajectory equation

The problem of a general equation of motion of a particle suspended in a flowing gas stream or liquid stream is complex. For our purposes, we consider only a relatively dilute suspension with no particle-to-particle interaction. Further, to start with, we employ a Lagrangian description by focusing on a particle, its coordinates as they change with time, its velocity and acceleration and the forces it is subjected to. *Ignoring any random Brownian motion of the particle*, the inertial force $\boldsymbol{F}_p^{\text{iner}}$ on a spherical particle (of radius r_p, mass m_p and volume (m_p/ρ_p)), subjected to an external force $\boldsymbol{F}_{tp}^{\text{ext}}$ and a drag force $\boldsymbol{F}_p^{\text{drag}}$ from the surrounding fluid, satisfies the following basic equation of particle motion, where the particle velocity is $\boldsymbol{U}_p$:

$$\boldsymbol{F}_p^{\text{iner}} = m_p\frac{d\boldsymbol{U}_p}{dt} = \boldsymbol{F}_{tp}^{\text{ext}} - \boldsymbol{F}_p^{\text{drag}}. \tag{6.2.45}$$

Here, $\boldsymbol{F}_{tp}^{\text{ext}}$ may be obtained from expression (3.1.59). Note that we have neglected the following forces:

- the virtual mass force[13];
- the Basset force[14] on the particle due to an unsteady flow field (see Landau and Lifshitz (1959), Friedlander (1977) and Soo (1989) for their expressions and physical basis);
- any force on the particle created by an externally imposed pressure gradient $(-v_p\nabla P)$ (note that expression (3.1.59) includes any ΔP created by centrifugal forces).

Now the particle velocity vector $\boldsymbol{U}_p$ may be represented through the rate of change of its position vector in Cartesian coordinates as

$$\boldsymbol{U}_p = \frac{dx}{dt}\boldsymbol{i} + \frac{dy}{dt}\boldsymbol{j} + \frac{dz}{dt}\boldsymbol{k}. \tag{6.2.46}$$

Therefore the governing equation of particle motion (6.2.45) may be reduced to the following equations for three coordinates, *x*, *y*, *z* of particle motion:

$$m_p\frac{d^2x}{dt^2} = F_{tpx}^{\text{ext}} - F_{px}^{\text{drag}}; \tag{6.2.47a}$$

$$m_p\frac{d^2y}{dt^2} = F_{tpy}^{\text{ext}} - F_{py}^{\text{drag}}; \tag{6.2.47b}$$

$$m_p\frac{d^2z}{dt^2} = F_{tpz}^{\text{ext}} - F_{pz}^{\text{drag}}. \tag{6.2.47c}$$

Corresponding equations may be developed in other coordinate systems. Expressions for the drag force are available in (3.1.63) for particles and flows which obey Stokes' law and in (3.1.64) for larger particles and/or higher velocities.

6.2.3 General equation of change for a particle population

Particle separation devices, aerosol/hydrosol separation in granular media, crystallization and precipitation devices generally involve a size-distributed particle population.

---

[13] $(1/2)m_p\, d(\boldsymbol{v} - \boldsymbol{U}_p)/dt$.

[14] May be neglected for spherical particles of unit density in air (Friedlander, 1977).

The notion of a size-distributed particle population was introduced in Section 2.4 via a particle size density function $f(r_p)$: the quantity $f(r_p)\mathrm{d}r_p$ represents the fraction of particles in the size range of r_p to $r_p + \mathrm{d}r_p$ in a unit fluid volume. It is also the probability of finding a particle having a size in the size range r_p to $r_p + \mathrm{d}r_p$ in a unit fluid volume.

We are interested here in the actual number of particles $\mathrm{d}N(r_p)$ in the size range of r_p to $r_p + \mathrm{d}r_p$ per unit volume of the fluid. This quantity is related to a particle number density function $n(r_p)$ (the *population density function*, see Figure 2.4.3(b) and definition (2.4.2a)) by

$$\mathrm{d}N(r_p) = n(r_p)\,\mathrm{d}r_p; \tag{6.2.48a}$$

$$\int_{r_{min}}^{r_p} n(r_p)\,\mathrm{d}r_p = N(r_{\min}, r_p), \tag{6.2.48b}$$

where $N(r_{\min}, r_p)$ is the number of particles between $r_{\min}$ and r_p in a unit fluid volume. The dimensions of $N(r_p)$ are number/L^3, i.e. numbers per unit volume (L^3); the dimensions of $n(r_p)$ are numbers per length per unit volume, i.e. number/L^4. Instead of numbers of particles, we can also consider the mass of particles in the size range r_p to $r_p + \mathrm{d}r_p$. The relationship between the corresponding probability density functions is illustrated in equation (2.4.2e). When particles coalesce or break up, the total mass, as well as the volume, are usually conserved. In this context, the number of particles $\mathrm{d}N(V_p)$ in the particle volume range of V_p to $V_p + \mathrm{d}V_p$ is related to the corresponding particle number density function $n(V_p)$ by

$$\mathrm{d}N(V_p) = n(V_p)\,\mathrm{d}V_p. \tag{6.2.48c}$$

The changes in the particle number density function $n(r_p)$ of particles of size r_p with the space coordinates (x, y, z) and time t are due to the following: convection in the flowing fluid, having velocity $\boldsymbol{v}$; diffusion in the flowing fluid (if relevant); particle growth from the surrounding medium by molecular level processes; particle birth: an original particle of size r_p undergoes breakage to produce new particles of smaller r_p or coalescence with another particle to create a new particle of larger r_p; particle death via breakage or coalescence, resulting in the disappearance of the original particle of size r_p. When the particles undergo convection, the particle velocity $\boldsymbol{U}_p$ may or may not be different from the fluid velocity $\boldsymbol{v}$ varying with the coordinate location in the separator, namely x, y, z and time t.

There is another kind of particle velocity, $\boldsymbol{U}_p^{\text{int}}$, the internal particle velocity (Randolph and Larson, 1988), which describes the physical growth of the particle due to molecular level processes going on between the particle and the surrounding fluid; for example, crystal growth in a given dimension. If we identify such growth through an internal property coordinate, say crystal length r_p, then the internal particle velocity, $U_{pr_p}^{\text{int}}$, in the r_p direction is $\mathrm{d}r_p/\mathrm{d}t$. We will designate such internal coordinate directions as x^i, y^i, z^i, and the corresponding gradient operator as ∇^i. We will assume from now on that all particles having a given size r_p and located at the same x, y, z will also have the same $U_{pr_p}^{\text{int}}$. This last assumption is equivalent to assuming no growth rate dispersion, although it is known now that crystals may display different growth rates in crystallization even though they have the same size and exist in the same environment (Randolph and Larson, 1988). **For particles/crystals whose growth etc., can be characterized by change in one dimension only, we will use L or r_p interchangeably**; but r_p will be preferred.[15]

We will now derive an equation of change for the particle number density function $n(r_p)$ by developing a *particle population balance*[16] in a small control volume of dimensions Δx, Δy and Δz (Figure 6.2.1):

rate of accumulation of particles of size r_p	=	rate at which particles of size r_p come in	−	rate at which particles of size r_p leave	+	net rate at which particles of size r_p are generated.

(6.2.49)

We will determine the contribution of each of these categories now to the overall balance in the number of particles:

rate of accumulation of particles of size r_p in the volume element

$$\frac{\partial N(r_p)}{\partial t}(\Delta x\Delta y\Delta z) = \frac{\partial n(r_p)}{\partial t}(\Delta x\Delta y\Delta z)\,\mathrm{d}r_p; \tag{6.2.50a}$$

rate at which particles of size r_p enter the volume element through the face perpendicular to the x – coordinate and having a corner coordinate (x, y, z) due to fluid flow

$$\left(U_{px}\mathrm{d}N(r_p)\right)\Big|_x(\Delta y\Delta z) = \left(U_{px}n(r_p)\right)\Big|_x(\Delta y\Delta z)\,\mathrm{d}r_p; \tag{6.2.50b}$$

rate at which particles of size r_p leave the volume element through the face perpendicular to the x – coordinate and having a corner coordinate $(x + \Delta x, y, z)$ due to fluid flow

$$\left(U_{px}n(r_p)\right)\Big|_{x+\Delta x}\Delta y\Delta z\,\mathrm{d}r_p; \tag{6.2.50c}$$

---

[15] If L is used instead of d_p, then $L = 2r_p$.

[16] See Friedlander (1977), Randolph and Larson (1988) and Ramkrishna (2000) and for more extensive treatments of population balance related equations (initiated by Hulbert and Katz (1964)).

rate[17] at which particles of size r_p undergoing physical growth at growth velocity of $U^{\text{int}}_{px^i}$ accumulate in the control volume at an internal particle coordinate of x^i

$$\frac{\partial\left(U^{\text{int}}_{px^i} n(r_p)\right)}{\partial x^i} \Delta x \Delta y \Delta z \, \mathrm{d}r_p; \tag{6.2.50d}$$

rate at which particles of size r_p to $r_p + \mathrm{d}r_p$ enter the volume element face of relation (6.2.50b) by diffusion/Brownian motion

$$-D_p \left.\frac{\partial n(r_p)}{\partial x}\right|_x \Delta y \Delta z \, \mathrm{d}r_p; \tag{6.2.50e}$$

rate at which particles of size r_p to $r_p + \mathrm{d}r_p$ leave the face of relation (6.2.50c) by diffusion/Brownian motion

$$-D_p \left.\frac{\partial n(r_p)}{\partial x}\right|_{x+\Delta x} \Delta y \Delta z \, \mathrm{d}r_p; \tag{6.2.50f}$$

rate at which particles of size r_p to $r_p + \mathrm{d}r_p$ are born per unit volume in the control volume

$$B\,(\Delta x \Delta y \Delta z)\, \mathrm{d}r_p; \tag{6.2.50g}$$

rate at which particles of size r_p to $r_p + \mathrm{d}r_p$ disappear per unit volume in the control volume

$$De\,(\Delta x \Delta y \Delta z)\, \mathrm{d}r_p. \tag{6.2.50h}$$

Here B represents the density function of new particles which are born/created in the size range r_p, $r_p + \mathrm{d}r_p$; De represents the density function of existing particles which disappear in the same size range. Additional terms due to external particle velocities in the y- and z-directions and internal particle velocities in the y^i- and z^i-directions should be considered along with the above terms to obtain from equation (6.2.49) the following equation:

$$\begin{aligned}
\frac{\partial n(r_p)}{\partial t}(\Delta x \Delta y \Delta z) &= \Delta y \Delta z\left[\left(U_{px} n(r_p)\right)_x - \left(U_{px} n(r_p)\right)_{x+\Delta x}\right] \\
&+ \Delta x \Delta z\left[\left(U_{py} n(r_p)\right)_y - \left(U_{py} n(r_p)\right)_{y+\Delta y}\right] \\
&+ \Delta x \Delta y\left[\left(U_{pz} n(r_p)\right)_z - \left(U_{pz} n(r_p)\right)_{z+\Delta z}\right] \\
-\Delta x \Delta y \Delta z &\left[\frac{\partial\left(U^{\text{int}}_{px^i} n(r_p)\right)}{\partial x^i} + \frac{\partial\left(U^{\text{int}}_{py^i} n(r_p)\right)}{\partial y^i} + \frac{\partial\left(U^{\text{int}}_{pz^i} n(r_p)\right)}{\partial z^i}\right] \\
&- \Delta y \Delta z\left[D_p \left.\frac{\partial n(r_p)}{\partial x}\right|_x - D_p \left.\frac{\partial n(r_p)}{\partial x}\right|_{x+\Delta x}\right] \\
&- \Delta x \Delta z\left[D_p \left.\frac{\partial n(r_p)}{\partial y}\right|_y - D_p \left.\frac{\partial n(r_p)}{\partial y}\right|_{y+\Delta y}\right] \\
&- \Delta x \Delta y\left[D_p \left.\frac{\partial n(r_p)}{\partial z}\right|_z - D_p \left.\frac{\partial n(r_p)}{\partial z}\right|_{z+\Delta z}\right] \\
&+ (\Delta x \Delta y \Delta z) B - (\Delta x \Delta y \Delta z) De.
\end{aligned} \tag{6.2.50i}$$

Dividing by $(\Delta x \Delta y \Delta z)$ and allowing the individual dimensions Δx, Δy and Δz to shrink to zero in the limit, we get the *population balance equation* (PBE) in cartesian coordinates:

$$\begin{aligned}
\frac{\partial n(r_p)}{\partial t} = &-\frac{\partial\left(U_{px} n(r_p)\right)}{\partial x} - \frac{\partial\left(U_{py} n(r_p)\right)}{\partial y} - \frac{\partial\left(U_{pz} n(r_p)\right)}{\partial z} \\
&- \frac{\partial\left(U^{\text{int}}_{px^i} n(r_p)\right)}{\partial x^i} - \frac{\partial\left(U^{\text{int}}_{py^i} n(r_p)\right)}{\partial y^i} - \frac{\partial\left(U^{\text{int}}_{pz^i} n(r_p)\right)}{\partial z^i} \\
&+ \frac{\partial}{\partial x}\left(D_p \frac{\partial n(r_p)}{\partial x}\right) + \frac{\partial}{\partial y}\left(D_p \frac{\partial n(r_p)}{\partial y}\right) \\
&+ \frac{\partial}{\partial z}\left(D_p \frac{\partial n(r_p)}{\partial z}\right) + B - De.
\end{aligned} \tag{6.2.51a}$$

In vectorial notation, the population balance equation is

$$\frac{\partial n(r_p)}{\partial t} + \nabla\cdot\left(\boldsymbol{U}_p n(r_p)\right) + \nabla^i\cdot\left(\boldsymbol{U}^{\text{int}}_p n(r_p)\right) = \nabla\cdot\left(D_p \nabla n(r_p)\right) + B - De. \tag{6.2.51b}$$

Here, the ∇^i operator corresponds to the three particle growth coordinate directions x^i, y^i, z^i, even though we are identifying the particle by just one radial dimension, r_p. In most practical situations, one particle/crystal dimension is often sufficient. This equation is a very general equation which describes the change in $n(r_p)$ with x, y, z, t as well as x^i, y^i and z^i.

In processes involving, for example, aerosols in a gas stream, one can replace $\boldsymbol{U}_p$ in the convective term $\nabla\cdot\left(\boldsymbol{U}_p n(r_p)\right)$ by the fluid velocity vector $\boldsymbol{v}$, namely, $\nabla\cdot\left(\boldsymbol{v} n(r_p)\right)$. However, if there are external forces creating

[17] The expression $\partial\left(U^{\text{int}}_{px^i} n(r_p)\right)/\partial x^i$ can be obtained by a process analogous to that carried out in equations (6.2.50b,c), here in a control volume of dimension Δx, Δy, Δz: consider a control volume of internal particle coordinate dimensions Δx^i, Δy^i, Δz^i located in the internal particle coordinate space.

a particle velocity $\boldsymbol{U}_p$, then there will be an additional term, $\nabla\cdot\left(\boldsymbol{U}_p n(r_p)\right)$:

$$\frac{\partial n(r_p)}{\partial t}+\nabla\cdot\left(\boldsymbol{v}n(r_p)\right)+\nabla\cdot\left(\boldsymbol{U}_p n(r_p)\right)+\nabla^i\cdot\left(\boldsymbol{U}_p^{\text{int}} n(r_p)\right)$$
$$=\nabla\cdot\left(D_p\nabla n(r_p)\right)+B-De. \tag{6.2.51c}$$

Such an equation has been identified as a *general dynamic equation* (Friedlander, 1977)) for $n(r_p)$, which, to be exact, should be represented as $n(\boldsymbol{r}_p,\ \boldsymbol{v},\ \boldsymbol{U}_p^{\text{int}},\ t)$; namely, it depends on particle size, fluid velocity, internal particle velocity and time. This equation does not include one term, namely a diffusion term on the right-hand side, $D^i\nabla^{i^2}n(r_p)$, which arises from random fluctuations in crystal growth rate for which the diffusion coefficient is D^i and the coordinate dimensions are x^i, y^i and z^i.

The trajectory equations for the motion of a particle in Section 6.2.2 did not include any random Brownian motion of the particle. Both equations (6.2.51b) and (6.2.51c), on the other hand, include a diffusion coefficient term on the right-hand side for random Brownian motion of the collection of particles. When the particle diameter is less than 1 μm, the effect of random Brownian motion becomes important. In the case of larger particles, the particle motion is generally deterministic and predictable (Friedlander, 1977); equations (6.2.47a-c) should be used. However, turbulent flows may introduce random effects.

6.2.3.1 *Diffusional transport of particles in flowing fluids–convective diffusion*

For particle diameters less than 1 μm, the diffusional transport of particles in a flowing fluid is identified as *convective diffusion* (Friedlander, 1977). The role of Brownian motion is considerable at smaller particle sizes. For cases where there are no birth and death processes ($B = De = 0$) and there is no particle growth/decay, equation (6.2.51c) may be simplified as follows:

$$\frac{\partial n(r_p)}{\partial t}+\nabla\cdot\left(\boldsymbol{v}n(r_p)\right)+\nabla\cdot\left(\boldsymbol{U}_p n(r_p)\right)=D_p\nabla^2 n(r_p). \tag{6.2.52}$$

The second term on the left-hand side may be simplified for incompressible flow via

$$\nabla\cdot\left(\boldsymbol{v}n(r_p)\right)=n(r_p)\nabla\cdot\boldsymbol{v}+\boldsymbol{v}\cdot\nabla\left(n(r_p)\right)=\boldsymbol{v}\cdot\nabla\left(n(r_p)\right). \tag{6.2.53}$$

Therefore, the governing equation for convective diffusion of particle density function $n(r_p)$ of size r_p becomes

$$\frac{\partial n(r_p)}{\partial t}+\boldsymbol{v}\cdot\nabla\left(n(r_p)\right)+\nabla\cdot\left(\boldsymbol{U}_p n(r_p)\right)=D_p\nabla^2 n(r_p). \tag{6.2.54}$$

The fluid velocity field $\boldsymbol{v}$ is quite important in convective diffusion of particles occurring in filters and scrubbers used for gas cleaning. Knowledge of $\boldsymbol{v}$ in the separator is essential in predicting particle separation.

Since there are no birth and death processes here, the only independent variables are x, y, z and t. Thus, for spherical particles, we may multiply each term in (6.2.54) by $\rho_p(4/3)\pi r_p^3\,\mathrm{d}r_p$ to obtain the equation in terms of the dependent variable, $\rho_p(4/3)\pi r_p^3\,\mathrm{d}N(r_p)$, where ρ_p is the mass density of the particle material. However, this quantity is simply the mass density of particles of size r_p to $r_p + \mathrm{d}r_p$ in the flowing phase j; we identify it as $\rho_{p,rp}$ whose governing equation is now given by

$$\frac{\partial\rho_{p,r_p}}{\partial t}+\boldsymbol{v}\cdot\nabla\rho_{p,r_p}+\nabla\cdot\left(\boldsymbol{U}_p\rho_{p,r_p}\right)=D_p\nabla^2\rho_{p,r_p}. \tag{6.2.55}$$

Note that the particle diffusion coefficient, D_p, actually depends on r_p.

6.2.3.2 *Continuous stirred tank separator in particulate systems*

In a continuous stirred tank separator (CSTS) employed in particulate systems, there are no spatial gradients inside the separator: therefore we need to integrate either equation (6.2.51b) or equation (6.2.51c) over the total volume V of the separator. To that end, we may employ the type of relations (6.2.4b-d) utilized in volume-averaging procedures. For example, the second term, $\nabla\cdot\left(\boldsymbol{U}_p n(r_p)\right)$, on the left-hand side of (6.2.51b) will lead to

$$\frac{V}{V}\int_V\nabla\cdot\left(\boldsymbol{U}_p n(r_p)\right)\mathrm{d}V=\nabla\cdot\left[\frac{V}{V}\int_V\left(\boldsymbol{U}_p n(r_p)\right)\mathrm{d}V\right]$$
$$+\frac{V}{V}\int_{A_{12}(t)}\left(\boldsymbol{U}_p n(r_p)\right)\cdot\boldsymbol{n}_k\,\mathrm{d}A_{12}. \tag{6.2.56}$$

The first term in this expression is zero since there are no spatial gradients inside the separator. In the second term, only the particle velocities normal to the surface $A_{12}(t)$ matter; further, in this case, $A_{12}(t)$ represents one or more interfaces at the system boundaries. For the system boundaries with two feed streams (volumetric flow rates Q_{f1}, Q_{f2}) coming in and two product streams (Q_1, Q_2) leaving (the outward-directed normal to $A_{12}(t)$ is positive, therefore Q_{f1}, Q_{f2} are taken negative as coming into V),

$$\int_{A_{12}(t)}\left(\boldsymbol{U}_p n(r_p)\right)\cdot\boldsymbol{n}_k\,\mathrm{d}A_{12}=-\left[Q_{f_1}n_{f_1}(r_p)+Q_{f_2}n_{f_2}(r_p)\right.$$
$$\left.-Q_1 n_1(r_p)-Q_2 n_2(r_p)\right]. \tag{6.2.57}$$

Although we have assumed the existence of two incoming streams and two product streams, there may be only one of each kind, as in a crystallizer. For the system boundary, where there is a free interface which can accumulate particles due to a change in the system volume,

$$\int_{A_{12}(t)} \left(\boldsymbol{U}_p n(r_p)\right) \cdot \boldsymbol{n}_k \,\mathrm{d}A_{12} = \langle n(r_p)\rangle \frac{\mathrm{d}V}{\mathrm{d}t}. \tag{6.2.58}$$

Here we have followed the approach of Randolph and Larson (1988).

The diffusional term (first term on the right-hand side of (6.2.51b)) has generally no contribution at the boundaries of the vessel. The contributions of the terms vis-à-vis the internal coordinates, as well as B and De, may be easily replaced by the corresponding volume-averaged values. Therefore integration of equation (6.2.51b) over the volume V of the CSTS leads to

$$\int_V \left[\frac{\partial n(r_p)}{\partial t} + \nabla\cdot\left(\boldsymbol{U}_p n(r_p)\right) + [\nabla^i\cdot(\boldsymbol{U}_p^{\mathrm{int}})n(r_p)] - D_p\nabla^2 n(r_p) - B + De\right]\mathrm{d}V = 0$$

$$\Rightarrow V\left[\frac{\partial\langle n(r_p)\rangle}{\partial t} - \frac{1}{V}\{Q_{f_1}n_{f_1}(r_p) + Q_{f_2}n_{f_2}(r_p) - Q_1 n_1(r_p) - Q_2 n_2(r_p)\}\right.$$

$$\left. + \frac{\langle n(r_p)\rangle}{V}\frac{\mathrm{d}V}{\mathrm{d}t} + \left(\nabla^i\cdot(\boldsymbol{U}_p^{\mathrm{int}})n(r_p)\right) - B + De\right] = 0. \tag{6.2.59}$$

We rearrange this as follows:

$$\frac{\partial\langle n(r_p)\rangle}{\partial t} + \left(\nabla^i\cdot(\boldsymbol{U}_p^{\mathrm{int}})n(r_p)\right) + \langle n(r_p)\rangle\frac{\mathrm{d}\ln V}{\mathrm{d}t}$$

$$= B - De + \frac{1}{V}\{Q_{f_1}n_{f_1}(r_p) + Q_{f_2}n_{f_2}(r_p) - Q_1 n_1(r_p) - Q_2 n_2(r_p)\}. \tag{6.2.60}$$

This macroscopic particle population balance equation is general and used most often. But the form in which the change with respect to the particle growth coordinate directions is utilized is simpler than $\nabla^i\cdot\left(\boldsymbol{U}_p^{\mathrm{int}}n(r_p)\right)$. Usually, only one particle growth coordinate[18] r_p is employed (instead of the three x^i, y^i, z^i):

$$\nabla^i\cdot(\boldsymbol{U}_p^{\mathrm{int}})n(r_p) = \frac{\partial}{\partial r_p}\left(U_{pr_p}^{\mathrm{int}} n(r_p)\right) = \frac{\partial}{\partial r_p}\left(\frac{\mathrm{d}r_p}{\mathrm{d}t}n(r_p)\right) = \frac{\partial}{\partial r_p}\left(G n(r_p)\right), \tag{6.2.61}$$

where $U_{pr_p}^{\mathrm{int}}$, or G, is the linear crystal growth rate, $\mathrm{d}r_p/\mathrm{d}t$.

6.2.3.3 *Batch stirred tank separator in particulate systems*

For this case, we can start with the general population balance equation (6.2.51b) integrated over the volume of the vessel:

$$\int_V \left[\frac{\partial n(r_p)}{\partial t} + \nabla\cdot\left(\boldsymbol{U}_p n(r_p)\right)\right.$$

$$\left. + [\nabla^i\cdot(\boldsymbol{U}_p^{\mathrm{int}})n(r_p)] - D_p\nabla^2 n(r_p) - B + De\right]\mathrm{d}V = 0. \tag{6.2.62}$$

As we have a batch vessel with no flows in and out, and since the diffusional term has no contributions at the boundaries of the vessel, we end up with (in view of equation (6.2.61))

$$\int_V \left[\frac{\partial n(r_p)}{\partial t} + \frac{\partial}{\partial r_p}\left(G n(r_p)\right) - B + De\right]\mathrm{d}V = 0. \tag{6.2.63}$$

Exchanging the differentiation and integration signs, we get

$$\frac{\partial}{\partial t}\int_V n(r_p)\mathrm{d}V + \frac{\partial}{\partial r_p}\int_V G n(r_p)\mathrm{d}V + \int_V (De - B)dV = 0. \tag{6.2.64}$$

If there is no spatial dependence in the tank of $n(r_p)$ (well-stirred vessel), we get

$$\frac{\partial}{\partial t}(n(r_p)V) + \frac{\partial}{\partial r_p}(n(r_p)GV) + \int_V (De - B)\mathrm{d}V = 0. \tag{6.2.65}$$

6.3 Bulk flow parallel to force direction

We begin now the study of separation concepts/devices/processes/techniques when there is bulk flow of one or more phases. Specifically, in this section, we consider those separation configurations where the bulk flow of one phase is parallel to the direction of the force driving the separation. Generally the magnitude and direction of both the bulk flow as well as the force will be time-invariant: the direction of the bulk flow may or may not be opposite to the direction of the force. In rare cases, the magnitude and direction of the bulk flow and the force may oscillate with time. Our goal in this section is to illustrate important examples of how bulk flow parallel to the direction of the force achieves separation beyond that which can be achieved in a stagnant system, to provide some details about the individual separation technique/process/device and to point out the limits of the magnitudes of the bulk velocity.

In Section 6.3.1, we cover external forces, specifically gravitational, electrical and centrifugal forces; inertial force is also included here. In Section 6.3.2, chemical potential gradient driven equilibrium separation processes involving vapor–liquid, liquid–liquid, solid–melt and solid–vapor systems are considered; the processes are flash vaporization, flash devolatilization, batch distillation, liquid–liquid extraction, zone melting, normal freezing and drying. Section 6.3.3 illustrates a number of membrane separation processes in the so-called dead-end filtration mode achieved when the feed bulk flow is parallel to the

[18] In the crystallization literature, L is used, so that the linear crystal growth rate $G = (\mathrm{d}L/\mathrm{d}t)$; see Randolph and Larson (1988).

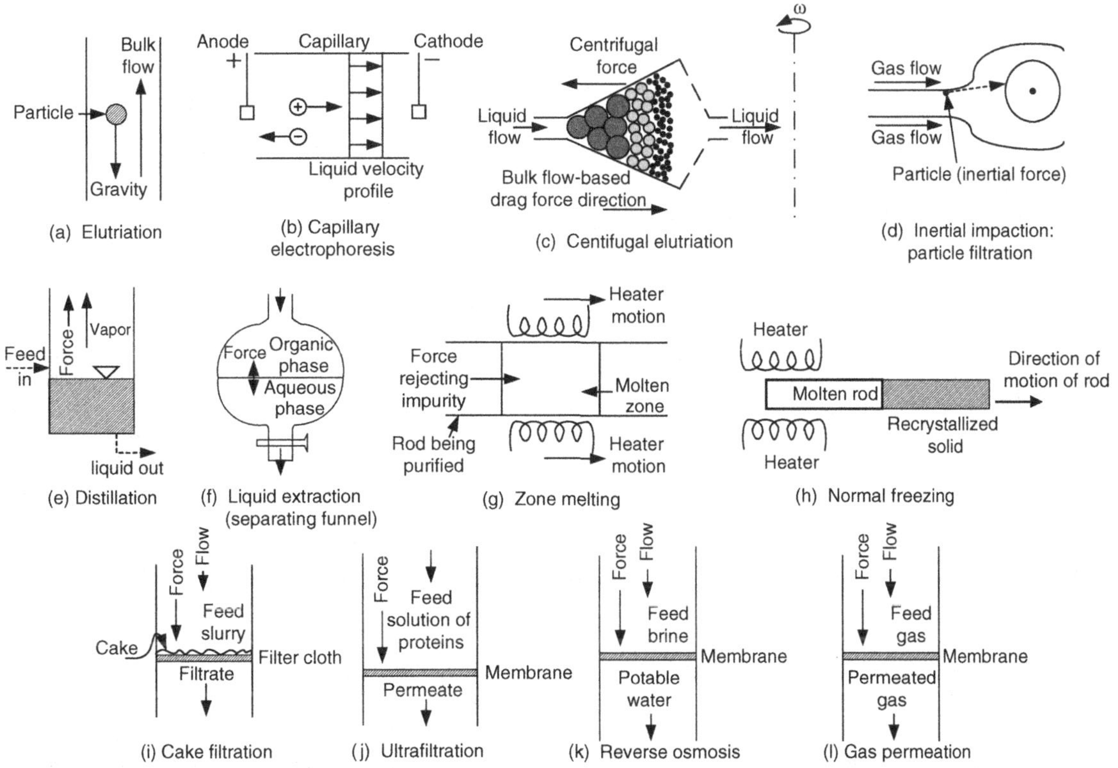

Figure 6.3.1. Separation systems where the bulk flow is parallel to the direction of force: (a) elutriation (particle separation in liquids); (b) capillary electrophoresis; (c) centrifugal elutriation; (d) inertial impaction in particle filtration; (e) distillation (flash/batch); (f) liquid extraction (separating funnel); (g) zone melting; (h) normal freezing; (i) cake filtration; (j) ultrafiltration separation of proteins (dead-end); (k) batch cell reverse osmosis separation of brine; (l) gas permeation.

chemical potential gradient driven force perpendicular to the membrane. The processes considered are: cake filtration/microfiltration, ultrafiltration, reverse osmosis, pervaporation and gas permeation. Figure 6.3.1 summarizes different systems and representative configurations studied in this section where the bulk flow is parallel to the direction of the force.

As mentioned at the end of the first paragraph, the magnitude and direction of the bulk velocity is important in relation to the magnitude and direction of the species/particle velocity due to the force under consideration. For external forces, such as gravitational, electrical and centrifugal forces, we will find that the magnitude of the bulk flow velocity is strongly influenced by the magnitudes of the external force induced individual species/particle velocities if one wants to fractionate a mixture of species/particles. In equilibrium driven dispersive two-phase processes, the bulk flow velocity and the dispersed-phase setting velocity are linked by the need to separate the two phases successfully. In equilibrium driven solid–melt systems, the bulk flow velocity will be limited by the achievable magnitude of the species diffusion velocity in the melt of the impurity rejected at the solid–melt interface. Similarly, in dead-end membrane processes, we will observe that separation is reduced unless the bulk flow velocity (whose direction and magnitude coincide with the force induced velocity through the membrane for incompressible fluids) is controlled to take into account the velocity with which the species rejected at the membrane–feed interface is transferred back to the bulk of the feed fluid.

6.3.1 External forces

We consider separations achieved due to external forces in the following order: gravity; electrical force; centrifugal force; inertial force (considered at the end).

6.3.1.1 *Gravity: elutriation, hydraulic jigging*

Consider the gravitational force acting on a spherical particle of mass m_p, radius r_p, density ρ_p falling vertically downward in a stagnant liquid of density ρ_t $(< \rho_p)$. (All of our considerations are also valid if the medium is gaseous

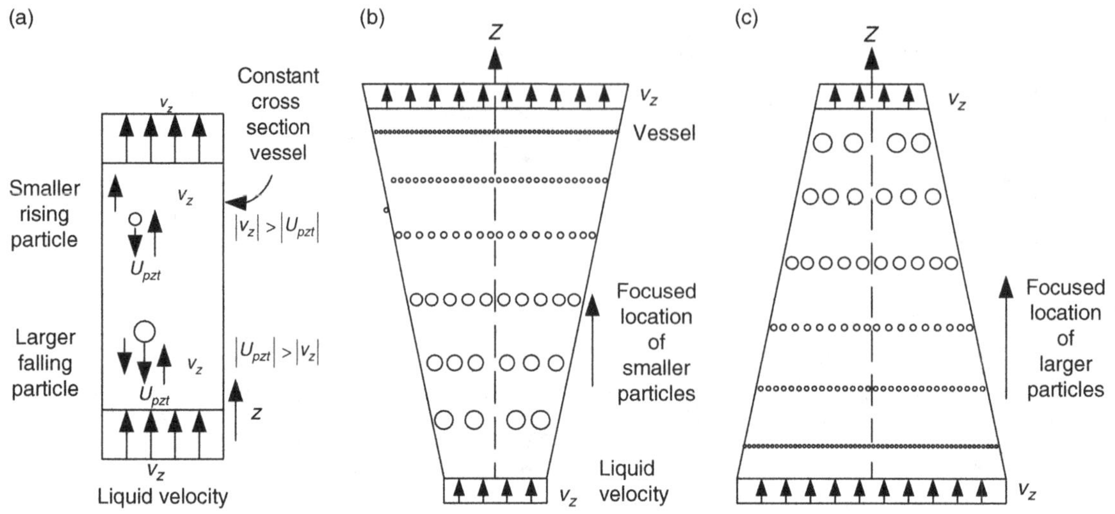

Figure 6.3.2. Gravitational separation of particles in a vertical column of liquid flowing vertically upward: (a) constant column cross section; (b) column cross section increasing upward; (c) column cross section decreasing upward.

instead of water; further, the particle may be replaced by a drop.) We pointed out in Section 3.1.3.1 that the particle velocity becomes constant very soon after a very brief period of initial acceleration; the constant velocity U_{pzt} of the particle is called the *terminal velocity* (equation (3.1.62)):

$$U_{pzt} = F_{pz}^{ext}/f_p^d.$$

If Stokes' law is valid, then $f_p^d = 6\pi\mu r_p$. Since the value of F_{pz}^{ext} for an isolated particle is $m_p g\left(1-(\rho_t/\rho_p)\right)$ (equation (3.1.4)) in a fluid of density ρ_t,

$$U_{pzt} = \frac{m_p g\left(1-(\rho_t/\rho_p)\right)}{6\pi\mu r_p} = \frac{(4/3)\pi r_p^3 \rho_p g\left(1-(\rho_t/\rho_p)\right)}{6\pi\mu r_p} = \frac{2}{9}\frac{r_p^2\rho_p\left(1-(\rho_t/\rho_p)\right)}{\mu} g. \tag{6.3.1}$$

Of the number of particle and fluid properties affecting this terminal velocity, particle radius r_p and density ρ_p are particularly important. Particles will have different terminal velocities if their sizes are different (but densities are same), or their densities are different (but sizes are equal) or both are different.

To explore how such different vertically downward particle terminal velocities may be utilized to separate different particles in a mixture, let there be bulk flow of the liquid vertically upward at a velocity v_z. The vertically upward velocity of the particle will be $(v_z - U_{pzt})$ if $v_z > U_{pzt}$. If U_{pzt} is larger than v_z, the particle will go down against the liquid motion at a velocity (U_{pzt}- v_z). The particle whose terminal velocity U_{pzt} is equal to the upward fluid velocity v_z is said to have the *critical settling velocity* U_{pzt}^c; it will not go up or down. Particles whose velocity U_{pzt} is less than U_{pzt}^c will be carried up by the liquid; particles whose U_{pzt} is greater than U_{pzt}^c will fall vertically downwards against the liquid velocity (Figure 6.3.2(a)). If all particles have the same density ρ_p, then the *critical size* r_{pc} of the particle whose U_{pzt} is U_{pzt}^c is obtained from

$$v_z = U_{pzt}^c = \frac{2}{9}\frac{r_{pc}^2\rho_p\left(1-(\rho_t/\rho_p)\right)g}{\mu}; \qquad r_{pc} = \left(\frac{9 v_z \mu}{2(\rho_p-\rho_t)g}\right)^{1/2} \tag{6.3.2}$$

Particles whose radius is larger than r_{pc} will fall down; smaller particles will be swept upwards by the liquid, generating two particle fractions (Figure 6.3.2(a)). Particles of size r_{pc} will remain stationary at the location of their introduction into the column.

If now the columnar vessel, in which the particle population is being fractionated, does not have a uniform cross section in the vertical direction, then the liquid velocity v_z will change with the vertical location in the vessel. Correspondingly, the value of r_{pc} will now depend on the vertical location in the vessel. Suppose the vessel cross section increases vertically upwards: v_z decreases with increasing z. Particles of a given size r_p will now accumulate in a band around a z-coordinate whose $r_p = r_{pc}$ corresponding to the v_z at that cross section (Figure 6.3.2(b)). Particles having a given size are being *focused* to a given height in the vessel corresponding to a given v_z (equation (6.3.2)); if this particle were located at a lower height, v_z at that location will be larger than U_{pzt} of this particle, which will drag the particle up to its appropriate location. Similarly, if a larger particle were located at a height appropriate for the terminal velocity of a smaller particle, its U_{pzt} will be larger than the local v_z, and therefore it will fall down and ultimately become stabilized around its own v_z (equation (6.3.2)). All situations described in Figure 6.3.2 illustrate the technique of *elutriation*; the column of Figure 6.3.2(a) is called an *elutriator*.

In Section 4.2, we observed the development of focusing in the presence of an external force, e.g. an electrical or centrifugal force, etc. It required the creation of a property gradient such as a *pH* or density gradient, etc., in the closed separator. The development of a liquid velocity gradient parallel to the direction of the external force of gravity in an open separator appears to be equally capable of multicomponent separation. Whereas in Figure 6.3.2(b) the column cross-sectional area increases vertically upwards, leading to focusing of smaller particles at higher locations, an inverse column cross-sectional profile (i.e. the lower section has a larger flow cross-sectional area) will reverse the profile of the focused location of the smaller particles (Figure 6.3.2(c)). The lower sections, having smaller liquid velocities, will stabilize particles of a smaller size.

If, instead of a system of particles of different sizes but having the same density, the system consisted of uniformly sized particles having different densities, a similar separation strategy may be adopted. If the system had variable particle sizes and densities, separation into many fractions is still feasible. However, there is now a possibility that a given terminal velocity may come about due to particular combinations of r_p and ρ_p: thus, at a given column height, particles of different sizes may be located, but they will have different densities. Such particles are called *equal-settling particles*:

$$U_{pzt}\big|_{r_{p_1}} = \frac{2}{9}\frac{r_{p_1}^2 \rho_{p_1} g\left(1-(\rho_t/\rho_{p_1})\right)}{\mu}; \qquad U_{pzt}\big|_{r_{p_2}} = \frac{2}{9}\frac{r_{p_2}^2 \rho_{p_2} g\left(1-(\rho_t/\rho_{p_2})\right)}{\mu}. \tag{6.3.3a}$$

From these we obtain

$$\left(\frac{r_{p_1}}{r_{p_2}}\right) = \left(\frac{\rho_{p_2}-\rho_t}{\rho_{p_1}-\rho_t}\right)^{1/2}, \tag{6.3.3b}$$

a relation that is valid when Stokes' law holds.

For larger particles and/or higher velocities, leading to a value of the particle Reynolds number $Re = (2r_p U_{pzt}\rho_t/\mu) > 0.1$, the drag force on a particle (relation (3.1.64)) is described using a drag coefficient C_D:

$$F_{pz}^{\text{drag}} = C_D \rho (U_{pzt}^2/2) A_p, \tag{6.3.3c}$$

where $A_P = \pi r_p^2$.

Approximate estimates of C_D for two ranges of higher particle Reynolds number are (Bird *et al.*, 1960):

$$2 < Re_p < 500, \textit{intermediate region}, C_D = 18.5/(Re_p)^{0.6};$$
$$500 < Re_p < (2\times10^5), \textit{Newton's law range}, C_D = 0.44. \tag{6.3.4}$$

The corresponding estimates of the particle terminal velocities are obtained as follows.

(a) **Intermediate range**

$$\frac{4}{3}\pi r_p^3 \rho_p g\left(1-(\rho_t/\rho_p)\right) = C_D \rho_t (U_{pzt}^2/2)\pi r_p^2;$$

$$U_{pzt}^2 = \frac{8}{3} r_p \frac{(\rho_p-\rho_t)}{C_D \rho_t} g \Rightarrow U_{pzt} = \frac{0.339 r_p^{1.14}(\rho_p-\rho_t)^{0.71} g^{0.71}}{\mu^{0.42}\rho_t^{0.29}}. \tag{6.3.5a}$$

(b) **Newton's law range**

$$\frac{4}{3}\pi r_p^3 g(\rho_p-\rho_t) = C_D \rho_t (U_{pzt}^2/2)\pi r_p^2;$$

$$U_{pzt}^2 = \frac{8g}{3\times 0.44} r_p \frac{(\rho_p-\rho_t)}{\rho_t} \Rightarrow U_{pzt} = 2.46\left(\frac{r_p g(\rho_p-\rho_t)}{\rho_t}\right). \tag{6.3.5b}$$

Using these expressions for U_{pzt} for the intermediate region and the Newton's law range, one can easily develop expressions for the particle size corresponding to the critical settling velocity, $U_{pzt}^c = v_z$, as well as the case of equal-settling particles. Considerable complexity, however, will be encountered in determining U_{pzt} in *concentrated suspensions*.

There is an upper limit to the value of the upward liquid velocity, beyond which there is no steady state separation in a system containing a distribution of particle sizes and different densities. This value of v_z should equal or exceed the value of U_{pzt} corresponding to $r_{p\max}$ and $\rho_{p\max}$, causing particles to be carried upwards by the liquid stream.

The separation capabilities considered above are significantly reduced in practice due to the actual velocity profiles in the vessels of Figure 6.3.2. Regardless of the flow regime in the vessel, the liquid velocity near the wall is much smaller than in the rest of the cross section at any height. The wall region will therefore accommodate much smaller particles than the rest of the cross section. This will considerably reduce the separation which could have been achieved otherwise.

The general principle of elutriation ought to be valid for a suspension of liquid drops in an immiscible liquid. For a small spherical drop of radius r_p of a liquid of viscosity μ_{dr}, the terminal velocity is obtained from the *Rybczyński-Hadamard formula* (Levich, 1962) as

$$U_{pzt} = \frac{2}{3}\frac{(\rho_{dr}-\rho_t)}{\mu}\frac{\mu+\mu_{dr}}{2\mu+3\mu_{dr}} r_p^2 g. \tag{6.3.6}$$

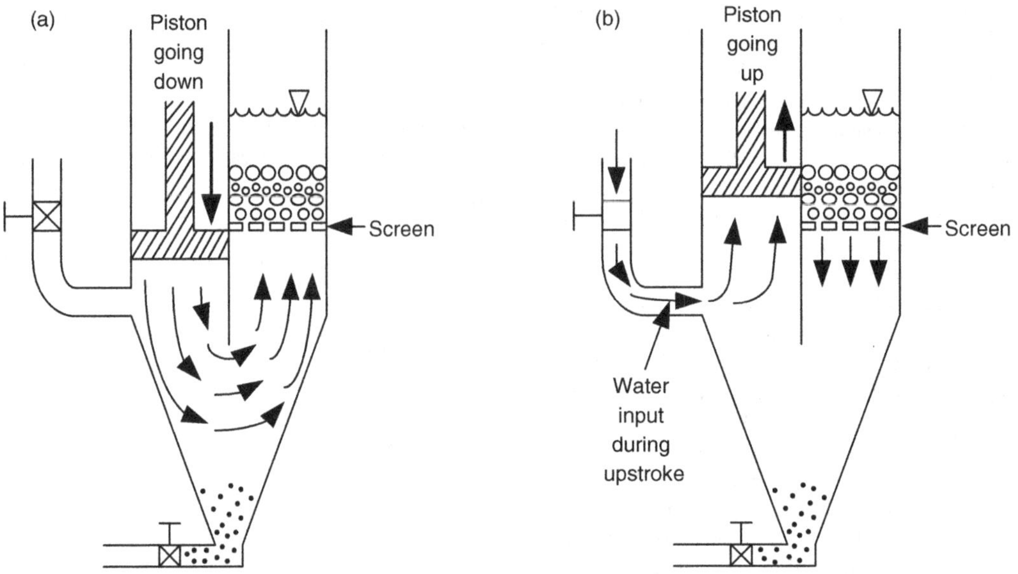

Figure 6.3.3. Hydrualic jig operation schematic: (a) downstroke of piston; (b) upstroke of piston.

When the drop viscosity is very large compared to the surrounding liquid viscosity (i.e. $\mu_{dr} >> \mu$), the above formula is reduced to (6.3.1) for a solid particle. This is valid in the viscous Stokes' law regime. When the drop fluid density $\rho_{dr} < \rho_t$, the drops will rise up to the top and form a separate fluid layer, if the drops can coalesce. In *spray towers*, employed in the separation process of liquid-liquid extraction, the heavier phase is often dispersed as drops; the lighter phase is the continuous phase. The lighter phase moves up and the heavier drop phase falls down through the column. The physical separation involved here is not fractionation of drops of different size. It involves bulk separation of the two phases, the dispersed phase and the continuous phase, as molecular separation occurs between the two phases via solvent extraction.

The bulk liquid flow considered so far has been at steady state. The liquid was flowing either up or down the device at a constant rate. Further, the particles or drops were assumed to be falling with a steady terminal velocity; the initial transient acceleration period was not considered relevant. Correspondingly, the d^2z/dt^2 term in equation (3.1.61) was neglected. In the industrial operation of a *hydraulic jig*, the initial transient is quite important. The principle of operation of such a device (shown in Figure 6.3.3) is as follows.

The device has three basic sections: (1) a chamber open at the bottom on the top left side with a plunger which goes up or down rapidly (this part has attached to it a source of water or a solution of interest which is opened very briefly to introduce water during the upstroke of the piston); (2) a chamber on the top right side containing a screen, through which the smallest particles of the heavy material can pass; (3) a tapered-bottom tank which connects sections 1 and 2. The mixed particulate suspension is introduced into section 2 on top of the screen. During a very rapid downstroke of the piston (Figure 6.3.3(a)), water/solution is rapidly brought up into section 2, which momentarily pushes the whole suspension up on top of the screen (transient bulk velocity in the z-direction). Immediately afterwards, a rapid upstroke of the piston allows the particles to settle in the suspension (Figure 6.3.3(b)) over the screen. The piston cycle time is so small that the particles never have time to attain their terminal velocity. Gravity becomes the primary driving force (equation (3.1.61)):

$$m_p \frac{d^2z}{dt^2} = m_p \frac{dU_{pz}}{dt} \cong \left(\frac{\rho_p - \rho_t}{\rho_p} \right) m_p g. \tag{6.3.7}$$

The heavier particles fall much further than the lighter particles.

Three layers of particles develop on top of the screen. The layer closest to the screen contains the largest particles of the denser material. The next layer has smaller particles of the denser material and the largest particles of the less dense material. The top layer has the smallest particles of the less dense material. The screen aperture is chosen such that the smallest particles of the denser material fall through and are collected at the bottom of the tank (section 3). All four fractions are rapidly removed physically from the tank bottom and from the top of the screen every so often.

The hydraulic jig is used to fractionate mineral/particulate suspensions (especially coal suspensions) larger than 10–20 mesh (0.165–0.083 cm), where the particle/mineral density is substantially larger than that of water. The jig described above is of the fixed-screen type with water forced up through the screen (Perry and Green, 1984); hydraulic jigs can also be of the moving-screen type. The bulk liquid velocity, parallel to the direction of the force, is then introduced into the system by the moving screen.

6.3.1.2 *Electrical force: capillary electrophoresis*

The separation of ions due to a uniform electrical field applied to a liquid solution having a flow parallel to the direction of the electrical field will be studied here. The electrostatic force $\boldsymbol{F}_i$ on 1 gmol of a charged species i due to a uniform electrical field $\boldsymbol{E}$ $(= -\nabla\phi)$ is (see (3.1.8))

$$\boldsymbol{F}_i = Q_i \boldsymbol{E}, \tag{6.3.8a}$$

where Q_i is the charge of 1 gmol of charged species ($Q_i = Z_i \mathcal{F}$, where $\mathcal{F}$ is Faraday's constant of 96 485 coulomb/gram equiv. and Z_i is the algebraic valence of the molecular ion). We may also use this relation for a particular ion instead of 1 gmol of a charged species: Q_i is then the charge on an ion which experiences a force $\boldsymbol{F}_i$. Such an ion will experience fluid friction as it moves toward an electrode having the opposite charge. Using Stokes' law to estimate the drag force experienced by the ion (assumed to be spherical, diameter $2r_i$), the terminal velocity $\boldsymbol{U}_i$ (see (3.1.8) and (3.1.62)) of such an ion is to be obtained from

$$F_i = Q_i \boldsymbol{E} = 6\pi\mu r_i \boldsymbol{U}_i; \tag{6.3.8b}$$

$$\boldsymbol{U}_i = \frac{Q_i \boldsymbol{E}}{6\pi\mu r_i} = \frac{Q_i \boldsymbol{E}}{f_i^d/\tilde{\mathrm{N}}}, \tag{6.3.8c}$$

where $(f_i^d/\tilde{\mathrm{N}})$ is the frictional coefficient experienced by one ion of species i. The quantity $Q_i/(f_i^d/\tilde{\mathrm{N}})$ is identified as the *ionic mobility* μ_i^m (in units of (cm/s)/(volt/cm) or cm^2/s-volt) of ionic species i (see an alternative definition (3.1.108j)):

$$\mu_i^m = \frac{Q_i}{f_i^d/\tilde{\mathrm{N}}} = Z_i \mathcal{F}/f_i^d. \tag{6.3.8d}$$

Therefore

$$\boldsymbol{U}_i = \mu_i^m \boldsymbol{E}. \tag{6.3.8e}$$

For an electrical field of magnitude E applied in the z-direction, the migration velocity U_{iz} of the ith ionic species in the z-direction is

$$U_{iz} = \mu_i^m E. \tag{6.3.8f}$$

If there is a diffuse double layer present around the charged species (see Figure 3.1.2D), the effective charge, $Z_{i,\mathrm{eff}}$, will be less than Z_i, leading to

$$\mu_{i,\mathrm{eff}}^m = Z_{i,\mathrm{eff}}\, \mathcal{F}/f_i^d. \tag{6.3.8g}$$

To separate a mixture of two ions i and j having different ionic mobilities $(\mu_i^m \neq \mu_j^m)$ and moving in one direction (toward an electrode), the solvent may be made to flow at a velocity v_z countercurrent to the direction of movement of the two ions. The magnitude of the solvent velocity, v_z, should, however, be intermediate between the velocities of the two ions. If $\mu_i^m > \mu_j^m$, the slower-moving ion j will be flushed out in the direction of solvent flow, whereas the faster-moving ion i will migrate against the solvent motion toward the electrode. This technique has been identified as *countercurrent electrophoresis* (CCEP) (Wagener *et al.*, 1971):

$$|U_{iz}| = |\mu_i^m E| > |v_z|; \tag{6.3.9a}$$

$$|U_{jz}| = |\mu_j^m E| < |v_z|. \tag{6.3.9b}$$

The exact physical configuration of such a separator has to satisfy a number of constraints. If a solution containing, say, a mixture of cations is to be separated by what is known as the self-stabilizing CCEP, then the feed solution may be introduced at a particular location (A) in separating column 6 (see Figure 6.3.4) as a salt solution. The separating column, as shown, is a trough divided into compartments by porous diaphragms. The flow rate of the electrolyte solution introduced into vessel 1 at a constant rate, the electric field strength applied between the two electrodes (platinum gauzes) at the two ends of column 6 and the solution temperature are preselected. This solution passes through the separating column 6 into a collector vessel 8 through an overflow weir 7 (Wagener *et al.*, 1971).

If, for the different cationic species of interest, $\mu_i^m E > |v_z|$, then these cations will migrate toward the cathode. However, different cations will have different speeds, so the solution near the cathode will now have a cation mixture composition different from that in the feed solution introduced at A. This assumes that we prevent deposition of the cations at the cathode via the use of protective electrolytes. As time passes, the cathode solution will be progressively enriched in the faster cation. The applied voltage level employed is around 1000 volts. Such a process is necessarily a batch process. If, however, a binary cationic mixture separation is desired using $|\mu_i^m E| > |v_z| > |\mu_j^m E|$, then the process can be run continuously (Wagener *et al.*, 1971), with the slower species being flushed out at the opposite end.

There are a number of general issues requiring attention in any such separator. First, the solvent flow velocity, v_z, has to be uniform over the flow channel. However, through each hole in a diaphragm, there will be a parabolic flow profile. Second, the high electrical field required to generate practically useful ionic migration velocities leads to considerable joule heating, which has to be dissipated effectively. Otherwise, the solvent viscosity μ, and therefore

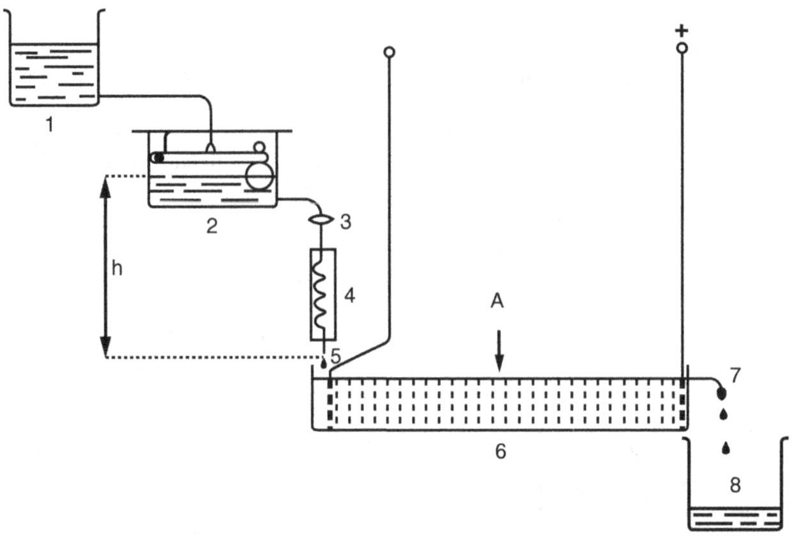

Figure 6.3.4. Construction of an electrolyte separation plant for the self-stabilizing CCEP process (schematic). Explanations are given in the text. From Figure 2, p. 487 of "Countercurrent Electrophoresis," K. Wagener, H.D. Freyer, B.A. Billal, Separ. Sci., ***6**(4), 483 (1971); reprinted by permission of the publisher, Taylor & Francis Group.*

the ionic mobility μ_i^m, will be affected, leading to considerable dispersion and a consequent reduction in the quality of separation.

Both issues have been successfully addressed in the widely used analytical technique of *capillary electrophoresis* (CE) to be considered now. This technique,[19] successfully introduced by Jorgenson and Lukacs (1981), and variations thereof, has been widely adopted for analyzing small samples injected into a glass capillary subjected to an electrical field along the capillary length, generally in the presence of an electroosmotically driven bulk flow along the capillary length. Additional identifications of this technique are *capillary zone electrophoresis* (CZE) and *high-performance capillary electrophoresis* (HPCE).

We will consider this technique below. First, we will describe the basic technique. Next, we will introduce the coupling of the ionic migration velocity and the electroosmotic velocity in the capillary, in the presence of the electrical field. The migration of a sample pulse introduced at one end of the capillary will then be developed in the capillary flow field parallel to the electrical force field. This will lead to expressions for resolution and other quantities for two neighboring peaks. A number of subjects will be considered briefly thereafter: partial ionization of a solute and its effective electrophoretic mobility; the role of a gel in the capillary; enhancement of throughput in CE; separation of uncharged compounds by micellar electrokinetic chromatography. For an introduction to these and other topics, see the articles in Guzman (1993), specifically Karger and Foret (1993).

[19]First attempted in a Teflon® tube by Mikkers *et al.* (1979).

Conventional CE employs a fused-silica *buffer-filled* capillary having an internal diameter between 20 and 200 μm, a wall thickness between 50 and 150 μm and a length in the range 20–200 cm. The two ends of the capillary are immersed in two buffer solution reservoirs containing two platinum electrodes connected to a high-voltage power supply (Figure 6.3.5(a)). The applied voltage can be as high as 25–30 kV; therefore the axial voltage gradient may easily be as high as 400–500 volts/cm. The current level varies between 20 and 100 μA. The sample to be analyzed is injected near the anode into the capillary, either by hydrodynamic flow via pressure or by electromigration. A detector is generally placed at the cathode end to detect the sample peaks traveling toward the cathode. Figure 6.3.5(b) illustrates the separation achieved in one of the earliest investigations.

We have already learned about the presence of an *electroosmotic flow* (EOF) in a silica capillary filled with an electrolytic solution and subjected to a voltage difference at the two ends (Section 6.1.5). This flow is such that the velocity profile, v_{EOF}, is essentially flat, and the liquid motion is toward the cathode (equation (6.1.22)). If the silica capillary tube is open (and not a packed bed of very small particles), and the surface of the capillary has not been coated in any fashion, capillary electrophoresis occurs in the presence of v_{EOF}. Therefore, the net velocity of any ionic species i averaged over the capillary cross section, $\langle v_{iz} \rangle$, is given by

$$\langle v_{iz} \rangle = \langle v_{EOF,z} \rangle + U_{iz}, \qquad (6.3.10a)$$

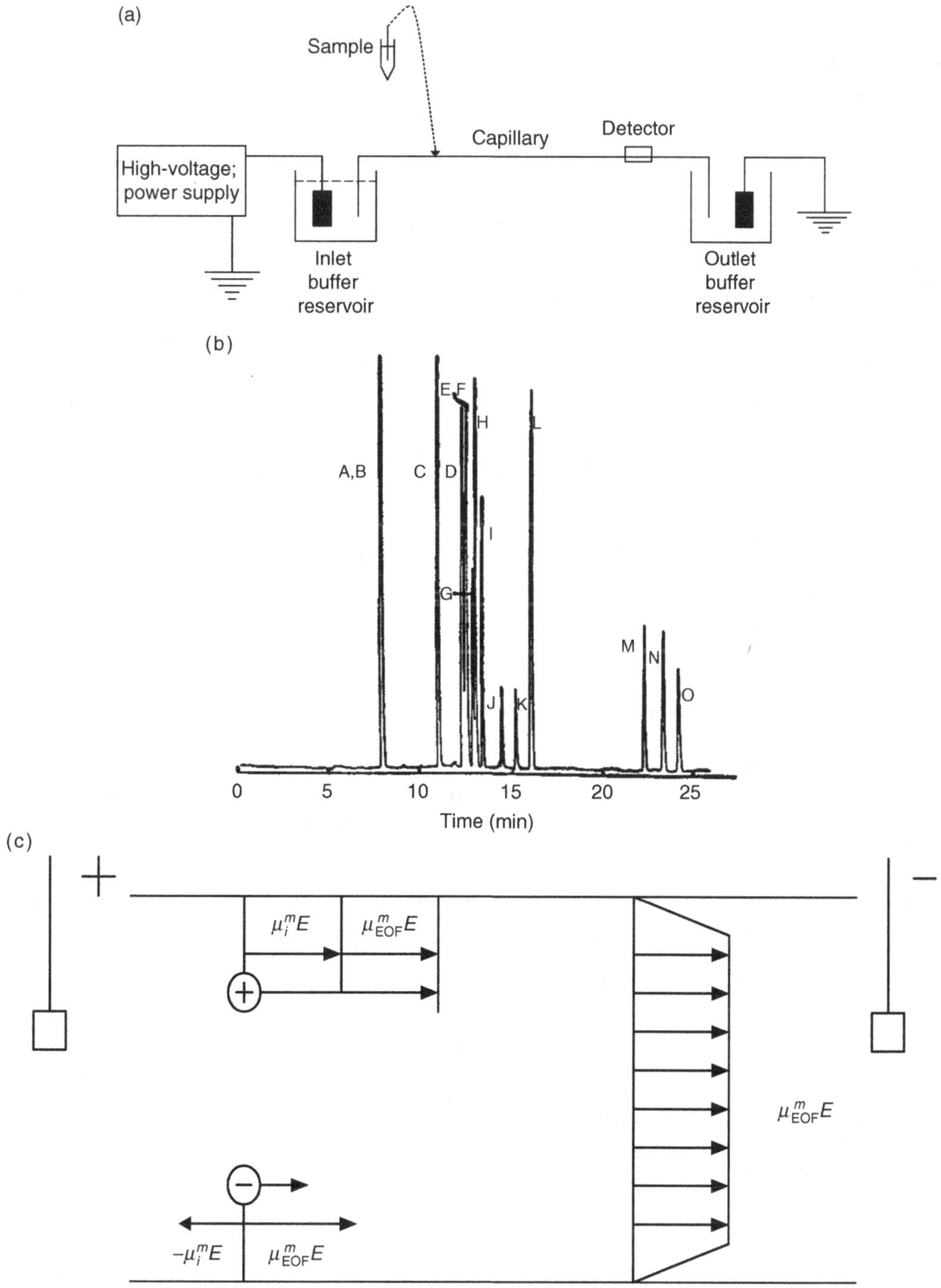

Figure 6.3.5. (a) Basic schematic for a capillary electrophoresis system. (b) Capillary electrophoresis separation of dansyl amino acids: A, unknown impurity; B, ε-labeled lysine; C, dilabeled lysine; D, asparagine; E, isoleucine; F, methionine; G, serine; H, alanine; I, glycine; J and K, unknown impurities; L, dilabeled cysteine; M, glutamic acid; N, aspartic acid; O, cysteic acid. The concentration of each derivative is approximately 5 × 10^{-4} M, dissolved in operating buffer (Jorgenson and Lukacs, 1981). Reprinted, with permission, from Anal. Chem., ***53****, 1298, (1981), Figure 1, p. 1300. Copyright (1981) American Chemical Society. (c) Effective velocities of cationic and anionic species in the presence of electroosmotic flow in a capillary.*

where $\langle v_{\mathrm{EOF},z}\rangle$ is obtained by averaging the expression for v_{EOF} in (6.1.22) over the capillary cross section and U_{iz} is the migration velocity defined by equation (6.3.8f). However, in general, for cations moving toward the cathode, the effective velocity of the cations will be

$$\langle v_{iz}\rangle = \frac{\varepsilon_d \zeta E}{\mu} + \mu_i^m E; \tag{6.3.10b}$$

$$\langle v_{iz}\rangle = \mu_{\mathrm{EOF}}^m E + \mu_i^m E = (\mu_{\mathrm{EOF}}^m + \mu_i^m)E. \tag{6.3.10c}$$

Here, μ_{EOF}^m is often called the *electroosmotic mobility*.

On the other hand, anions having a tendency to move toward the anode will have an effective velocity of

$$\langle v_{iz}\rangle = \mu_{\mathrm{EOF}}^m E - \mu_i^m E = (\mu_{\mathrm{EOF}}^m - \mu_i^m)E. \tag{6.3.10d}$$

As long as $\mu_{\mathrm{EOF}}^m > \mu_i^m$ in (6.3.10d), anionic species will also move toward the cathode. Cationic species will, however, move much faster toward the cathode, at a speed higher than the bulk velocity arising from the electroosmotic flow. These behaviors are schematically illustrated in Figure 6.3.5(c). Obviously, individual cationic species will move at different net velocities, $\langle v_{iz}\rangle$. The detector near the cathode will be detecting the appearance of each such ionic species, cationic or anionic, at different times. If the sample to be analyzed and injected near the anode into the capillary contains different species, one would like to know how well these species will be separated, what would be the value of the resolution, what is the maximum number of species that can be separated, etc.

To provide approximate answers to these questions, recall the analysis of separation development carried out in Section 3.2.1, where a pulse of a solute mixture was introduced into the solvent at time $t = 0$ at one end of the separator liquid, which was stagnant and subjected to an external force field in the z-direction, there being no variations in the x- and y-directions. The only differences here are as follows: (1) there is a convective motion of the buffer solution in the capillary in the z-direction and (2) since we have a cylindrical capillary of radius R, we will assume no variations in the θ- and r-directions (instead of the x- and y-directions). Consider the general equation (6.2.5a) for a molar concentration C_i of species i. Neglect θ- and r-dependencies and assume $\boldsymbol{v}^* = \boldsymbol{v} = \langle v_{\mathrm{EOF},z}\rangle \boldsymbol{k}$ for the electroosmotic flow. Assume $\langle v_{\mathrm{EOF},z}\rangle$ and U_{iz} to be constant. We get

$$\frac{\partial C_i}{\partial t} + \langle v_{\mathrm{EOF},z}\rangle \frac{\partial C_i}{\partial z} + U_{iz}\frac{\partial C_i}{\partial z} = \frac{RT}{f_i^d}\frac{\partial^2 C_i}{\partial z^2}; \tag{6.3.11a}$$

$$\frac{\partial C_i}{\partial t} + (\langle v_{\mathrm{EOF},z}\rangle + U_{iz})\frac{\partial C_i}{\partial z} = \frac{RT}{f_i^d}\frac{\partial^2 C_i}{\partial z^2}. \tag{6.3.11b}$$

We nondimensionalize the independent variables in the following fashion, using the applied voltage V used in the capillary of length $L_T(|E| = V/L_T)$:

$$t_i^+ = \frac{tV}{L_T^2 f_i^d}; \qquad z^+ = z/L_T; \qquad v_{i,\mathrm{eff}}^+ = \frac{(\langle v_{\mathrm{EOF},z}\rangle + U_{iz})}{V} L_T f_i^d, \tag{6.3.12}$$

to obtain

$$\frac{\partial C_i}{\partial t_i^+} + v_{i,\mathrm{eff}}^+ \frac{\partial C_i}{\partial z^+} = \frac{RT}{V}\frac{\partial^2 C_i}{\partial z^{+2}}. \tag{6.3.13}$$

This equation is essentially identical to equation (3.2.10) if we assume the following equalities:

$$v_{i,\mathrm{eff}}^+ = -(\phi_1^+)'; \quad V = (\phi_1^*)_{\max}. \tag{6.3.14a}$$

The solution of equation (6.3.13) using the new set of independent variables

$$\eta = z^+ - v_{i,\mathrm{eff}}^+ t_i^+ \qquad \text{and} \qquad t_i^+, \tag{6.3.14b}$$

corresponding to equation (3.2.11), and a pulse input of m_i moles of species i (3.2.14c) in the sample, can then be obtained from the solution, (3.2.19):

$$C_i(z^+, t_i^+) = \frac{m_i}{(\pi R^2 L_T)}\frac{1}{\sqrt{\frac{4\pi RT t_i^+}{V}}}\exp\left[-\frac{\{z^+ - v_{i,\mathrm{eff}}^+ t_i^+\}^2}{4\{\frac{RT}{V}\} t_i^+}\right]. \tag{6.3.15}$$

The center point of each such concentration profile for any species i corresponds to the following value of z_i^+:

$$z_i^+ = v_{i,\mathrm{eff}}^+ t_i^+ = \frac{(\langle v_{\mathrm{EOF},z}\rangle + U_{iz})}{L_T} t. \tag{6.3.16a}$$

The nondimensional standard deviation σ_{zi}^+ for the Gaussian profile (6.3.15) is, correspondingly,

$$\sigma_{zi}^+ = \{2RTt_i^+/V\}^{1/2} = \left\{2\frac{RT}{f_i^d}\frac{t}{L_T^2}\right\}^{1/2} = \{2D_{i,\mathrm{eff}} t/L_T^2\}^{1/2} = \frac{\{2D_{i,\mathrm{eff}} t\}^{1/2}}{L_T} = \frac{\sigma_{zi}}{L_T}, \tag{6.3.16b}$$

where (RT/f_i^d) may be defined as the effective dispersion/diffusion coefficient, $D_{i,\mathrm{eff}}$, for species i.

The *resolution* R_s between two neighboring peaks for cationic species 1 and 2 is defined by (see definitions (3.2.23) and (2.5.8))

$$R_s = \frac{t_{R_2} - t_{R_1}}{4\sigma_{ti}} = \frac{z_1^+ - z_2^+}{4\sigma_{zi}^+}, \tag{6.3.17}$$

for an averaged value of σ. (Now, the peaks are detected in a CE at a z less than L_T. We ignore this aspect and take $z = L_T$ for the detector location.) Consider the first expression for R_s:

$$t_{R_1} = \left[\frac{z_1^+ L_T}{(\langle v_{\mathrm{EOF},z}\rangle + U_{1z})}\right] = \frac{L_T}{\langle v_{\mathrm{EOF},z}\rangle + U_{1z}}; \qquad t_{R_2} = \frac{L_T}{\langle v_{\mathrm{EOF},z}\rangle + U_{2z}}. \tag{6.3.18a}$$

To determine an averaged value of the profile standard deviations in time units, consider an averaged value of the profile standard deviations in length unit σ_{zi}:

$$\sigma_{zi} = (2D_{i,\text{eff}}t)^{1/2}. \qquad (6.3.18b)$$

Using an averaged velocity $\overline{(\langle v_{\text{EOF},z}\rangle + U_{iz})}$ between species 1 and 2, an averaged

$$\sigma_{ti} = \frac{\sigma_{zi}}{\overline{(\langle v_{\text{EOF},z}\rangle + U_{iz})}}. \qquad (6.3.18c)$$

Therefore

$$\begin{aligned} R_s &= \frac{(U_{1z} - U_{2z})L_T\overline{(\langle v_{\text{EOF},z}\rangle + U_{iz})}}{4(\langle v_{\text{EOF},z}\rangle + U_{1z})(\langle v_{\text{EOF},z}\rangle + U_{2z})(2D_{i,\text{eff}}t)^{1/2}} \\ &= \frac{U_{1z} - U_{2z}}{4\overline{(\langle v_{\text{EOF},z}\rangle + U_{iz})}}\frac{L_T}{(2D_{i,\text{eff}}t)^{1/2}} \\ &= \frac{(\mu_1^m - \mu_2^m)}{4\overline{(\mu_{\text{EOF}}^m + \mu_{iz}^m)}}\frac{L_T}{(2D_{i,\text{eff}}t)^{1/2}}, \end{aligned} \qquad (6.3.19)$$

where we have employed expression (6.3.10c) relating the species velocities to various ionic mobilities, and defined an averaged net species velocity between species 1 and 2, $\overline{(\langle v_{\text{EOF},z}\rangle + U_{iz})}$, as the square root of the product of the individual net species velocities. The above expression for resolution then is related to different ionic mobilities in the applied field, μ_1^m and μ_2^m, electroosmotic mobility μ_{EOF}^m, the capillary length, the effective dispersion coefficient $D_{i,\text{eff}}$ and time (thereby voltage, etc.).

We will now introduce two other quantities frequently used in the literature to analyze this separator (we will see in Section 7.1 that they are also used in analyzing chromatographic and related processes, especially in analytical and preparative chemistry): the *plate height* H_i or H and the *plate number* N_i or N, which are related to the separator length L_T by

$$L_T = H_i N_i = HN. \qquad (6.3.20)$$

To determine the plate height, note that the nondimensional standard deviation σ_{zi}^+ (since z was nondimensionalized using L_T) of definition (6.3.16b) or (3.2.21) may be written in dimensional form, for any time t/separator length L, as follows:

$$\sigma_{zi}^2 = 2D_{i,\text{eff}}t = \frac{2D_{i,\text{eff}}L}{(\langle v_{\text{EOF},z}\rangle + U_{iz})} = \frac{2D_{i,\text{eff}}L}{(\mu_{\text{EOF}}^m + \mu_i^m)E}. \qquad (6.3.21)$$

Now, as L or t increases, σ_{zi}^2 increases:

$$\sigma_{zi}^2 = H_i L; \qquad H_i = \frac{\sigma_{zi}^2}{L} = \frac{2D_{i,\text{eff}}}{(\mu_{\text{EOF}}^m + \mu_i^m)E}. \qquad (6.3.22)$$

The smaller the value of H_i, the lower the extent of the band broadening (see (3.2.23)). Over a certain separator length L or time t from injection of the sample, the smaller the value of H_i, the smaller the standard deviation of the species profile and the narrower the profile base.

The *Péclet number*, Pe_i, for the solute dispersion may also be defined as (see alternative definitions (6.2.21))

$$Pe_i = \frac{2L\langle v_{iz}\rangle}{D_{i,\text{eff}}} = \frac{2L(\langle v_{\text{EOF},z}\rangle + U_{iz})}{D_{i,\text{eff}}}, \qquad (6.3.23a)$$

which leads to

$$\sigma_{zi}^2 = 4L^2/Pe, \qquad (6.3.23b)$$

indicating that the larger the Péclet number, the smaller the bandwidth and the lower the dispersion in the species concentration profile migrating through the capillary. A second quantity, the plate number, N or N_i, can now be defined as follows:

$$N = N_i = \frac{L_T}{H_i} = \frac{L_T^2}{\sigma_{zi}^2} = \frac{(\mu_{\text{EOF}}^m + \mu_i^m)EL_T}{2D_{i,\text{eff}}}. \qquad (6.3.24)$$

To develop an order of magnitude estimate of N_i, consider an electrophoresis situation for ovalbumin: let the total migration distance be 30 cm in one hour. Assume that $D_{i,\text{eff}}$ is essentially D_i (see Table 3.A.5: 7.76×10^{-7} cm^2/s). We can calculate:

$$N_i = \frac{30 \times 30}{2 \times 7.76 \times 10^{-7} \times 3600} = 1.61 \times 10^5;$$
$$\sigma_{zi} \cong 0.075\ \text{cm}; \qquad H_i \cong 1.85 \times 10^{-4}\ \text{cm}.$$

The larger the value of N_i, the better the separation capability of the device. Expression (6.3.19) for the resolution R_s may now be rewritten:

$$R_s = \frac{(\mu_1^m - \mu_2^m)}{4\overline{(\mu_{\text{EOF},z}^m + \mu_{iz}^m)}}N_i^{1/2}; \qquad (6.3.25)$$

indicating that the larger the number of plates, the higher the resolution. The peak capacity $n_{\max}$, (see definition (3.2.32)), for this system is

$$n_{\max} = \frac{L_T}{4\sigma_{zi}} = \frac{L_T/4}{(2D_{i,\text{eff}}t)^{1/2}} = \frac{L_T/4}{\left(2D_{i,\text{eff}}\dfrac{L_T}{(v_{\text{EOF},z} + U_{iz})}\right)^{1/2}}$$
$$= \left(\frac{L_T}{32D_{i,\text{eff}}}\right)^{1/2}(\mu_{\text{EOF}}^m + \mu_i^m)^{1/2}E^{1/2} \qquad (6.3.26a)$$
$$= (V\{\mu_{\text{EOF}}^m + \mu_i^m\})^{1/2}/(32D_{i,\text{eff}})^{1/2}. \qquad (6.3.26b)$$

Note that the peak capacity, $n_{\max}$, may now be related to the plate number N_i by

$$N_i = 16n_{\max}^2. \qquad (6.3.27a)$$

Alternatively, $n_{\max}$ is expressed as

$$n_{\max} = N_i^{1/2}/4. \qquad (6.3.27b)$$

At infinite dilution, a few of these results may be simplified/reexpressed in the absence of any kind of dispersion, except that due to simple molecular diffusion:

$$D_{i,\text{eff}} \cong D_i^0 = \frac{RT}{f_i^d} = \frac{RT\mu_i^m}{Z_i\mathcal{F}}. \tag{6.3.27c}$$

Ignoring electroosmotic flow, from (6.3.24), at 25 °C,

$$N_i = \frac{\mu_i^m E L_T Z_i \mathcal{F}}{2RT\mu_i^m} = \frac{Z_i V \mathcal{F}}{2RT};$$

$$N_i = \frac{Z_i \times 96\,485 \dfrac{\text{coulomb}}{\text{gequiv}} \times V\,\text{volt}}{2 \times 1.987 \dfrac{\text{cal}}{\text{gmol}\cdot\text{K}} \times 4.18 \dfrac{\text{joule}}{\text{cal}} \times 298\,\text{K}} \cong 19.49 Z_i V \tag{6.3.27d}$$

(where, remember, 1 volt = 1 joule/coulomb) (Giddings, 1991). If Z_i is anywhere between 1 and 10, and the applied voltage V in capillary electrophoresis is around 25 000 volt, the plate number can be very high. Further, the peak capacity, n_{max}, can be enormous, as much as $(\sim 20\, Z_i V)^{1/2}/4$.

In the capillary electrophoresis literature, the ratio of the two ionic mobilities is sometimes identified as the separation factor α:

$$\alpha = \frac{\mu_1^m}{\mu_2^m}. \tag{6.3.27e}$$

The expression (6.3.25) for the resolution R_s may be rewritten using α for ionic species 1 and 2 as

$$R_s = \frac{(\alpha - 1)}{4\left(\alpha \pm \frac{\mu_{\text{EOF}}^m}{\mu_2^m}\right)} N_i^{1/2}, \tag{6.3.27f}$$

where we have assumed $(\overline{\mu_{\text{EOF},z}^m + \mu_{iz}^m}) = \mu_{1z}^m \pm \mu_{\text{EOF},z}^m$. The values of α may be close to 1, say 1.1 or 1.01, etc. It is still possible to achieve reasonable resolution between two species having similar ionic mobilities since the value of N_i, and therefore $N_i^{1/2}$, is very high in capillary electrophoresis. A comprehensive theoretical analysis of the performance of a buffer-filled capillary electrophoresis with reference to such relations is available in Datta (1990). The approach employs an electrokinetic dispersion coefficient, which has contributions from the diffusion coefficient of species *i*, and the different bulk velocities (electroosmotic flow and Poiseuille flow if any) present in the system.

Many *solutes* present in the sample to be analyzed may be only partially ionized at the *pH* of the buffered solution. Consider an acid HA undergoing partial ionization (see (5.2.59))

$$HA \rightleftharpoons H^+ + A^-.$$

The neutral solute HA will not have any ionic mobility, $\mu_{HA}^m = 0$. Only the fraction present as the anion A^- will have an ionic mobility, $\mu_{A^-}^m$. The effective mobility of the species related to the acid HA in an electrical field may be defined as

$$\begin{aligned}\mu_{\text{HA,eff}}^m &= \mu_{A^-}^m x_{A^-} + \mu_{HA}^m x_{HA} \\ &= \mu_{A^-}^m x_{A^-}.\end{aligned} \tag{6.3.28a}$$

A more general relation between the effective ionic mobility of a substance A and all other species *i* which are derived from A is (Karger and Foret, 1993)

$$\mu_{\text{A,eff}}^m = \sum_i \mu_i^m x_i. \tag{6.3.28b}$$

From equation (5.2.61a) for the dissociation of the acid,

$$K_{d1} = \frac{C_{H^+} C_{A^- w}}{C_{HAw}};$$

$$\begin{aligned}\log_{10} K_{d1} &= -pK_1 = \log_{10} C_{H^+} + \log_{10} C_{A^- w} - \log_{10} C_{HAw} \\ &= -pH + \log_{10} \frac{C_{A^- w}}{C_{HAw}} \\ &\Rightarrow 10^{pH - pK_1} = \frac{C_{A^- w}}{C_{HAw}};\ 10^{pH - pK_1} + 1 = \frac{C_{A^- w} + C_{HAw}}{C_{HAw}} \\ &\Rightarrow \frac{C_{A^- w}}{C_{HAw} + C_{A^- w}} = \frac{10^{pH - pK_1}}{10^{pH - pK_1} + 1} = x_{A^-} \\ &\Rightarrow x_{A^-} = \frac{10^{-pK_1}}{10^{-pK_1} + 10^{-pH}};\end{aligned} \tag{6.3.29}$$

$$\mu_{\text{HA,eff}}^m = \mu_{A^-}^m \left(\frac{10^{-pK_1}}{10^{-pK_1} + 10^{-pH}} \right). \tag{6.3.30a}$$

Karger and Foret (1993) have illustrated this behavior graphically: we show it in Figure 6.3.6. When $pH < (pK_1 - 2)$, $\mu_{\text{HA,eff}}^m \,\tilde{}\, 0.01\mu_{A^-}^m$: the ionic mobility is essentially zero since the ionization of the acid is negligible. When $pH > (pK_1 + 2)$, $\mu_{\text{HA,eff}}^m \cong \mu_{A^-}^m$ since essentially almost all acid molecules are ionized at a high *pH*. One-half of the acid molecules are ionized when $pH = pK_1$. A not-as-illustrative expression equivalent to (6.3.30a) is

$$\mu_{\text{HA,eff}}^m = \mu_{A^-}^m \frac{(K_{d1}/C_{H^+})}{1 + (K_{d1}/C_{H^+})}. \tag{6.3.30b}$$

The effective mobility of a weak base is considered in Problem 6.3.4. The ionic mobility of a few smaller species and their *pK* values (primarily pK_1) are provided in Table 6.3.1. For a protein, the value of the solution *pH* in relation to its *pI* is important. Since the net charge (Z_i) on a protein is positive at $pH < pI$ (see Figure 4.2.5(c)), the direction of a protein's ionic mobility will correspond to that of a cation. When $pH > pI$, its mobility direction will be reversed.

For compounds which are quite similar (e.g. enantiomers), inclusion complexes formed with cyclodextrin (CD) molecules can lead to selective separation. Cyclodextrin molecules are uncharged and are therefore moved by the electroosmotic flow at a velocity $\langle v_{\text{EOF},z} \rangle$. Any racemic mixture which exists as ions in the solution may be

Table 6.3.1. Ionic mobilities of selected ions[a]

Ions, i	$\mu_i^m \times 10^{-5}$ cm^2/V-s	pK
Ammonium[b]	76.2	-
Hydrogen[b]	362	-
Lithium[b]	40	-
Octadecyl tributylammonium[b]	17.2	-
Potassium[b]	76	-
Sodium[b]	53	-
Triethylammonium[b]	35.2	-
TRIS[+c]	29.5	8.3
Acetate	-42.4	4.74
Chloride	-79.1	-
Hydroxyl	-206	-
Phosphate	-34.1	2.15
	-58.3	7.22
	-71.5	11.50
ACES[d]	-31.3	6.84
BES[e]	-24	7.16
HEPES[f]	-21.8	7.51
MES[g]	-26.8	6.13

[a] Data from Janini and Issaq (1993), Karger and Foret (1993) and Fu and Lucy (1998).
[b] Cations; μ_i^m chosen positive.
[c] tris (hydroxylmethyl) aminomethane.
[d] N-2-acetamide-2-aminoethanesulfonic acid.
[e] N,N-bis (2-hydroxylethyl)-2-aminoethanesulfonic acid.
[f] N-2-hydroxyethylpiperazine-N-2-ethanesulfonic acid.
[g] 2-(N-morpholino) ethanesulfonic acid.

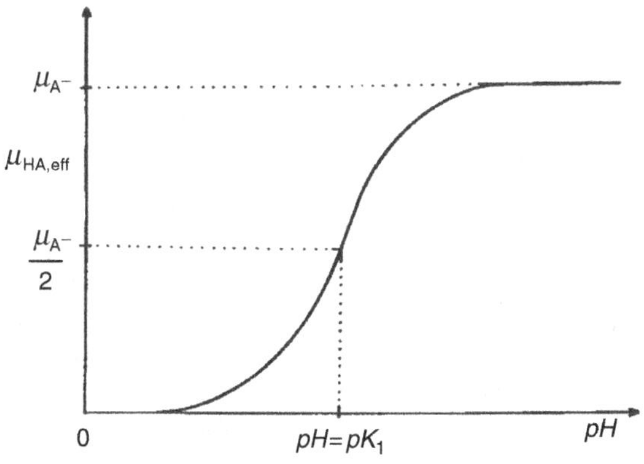

Figure 6.3.6. Dependence of the effective mobility μ of a weak acid on pH. *(After Karger and Foret (1993).)*

separated if one of the compounds preferentially forms an inclusion complex with the CD (see Figure 4.1.26) and has its charge sticking out of the hydrophobic CD cavity. The compound forming the inclusion complex will now move at a velocity different from that of the compound that does not. See Fanali (1993) for details of CD based facilitation of separation in CE.

It has been already pointed out in expression (6.3.8g) that the diffuse double layer around an ion/charged particle will reduce the effective charge on an ion to $Z_{i,\text{eff}}$ from the true charge Z_i. The two are approximately related as follows (Newman, 1973; Wieme, 1975):

$$(Z_{i,\text{eff}}/Z_i) = \frac{(1/h)}{(1/h) + r_i}, \tag{6.3.31a}$$

where $(1/h)$ is a characteristic thickness of the double layer (equivalent to the Debye length λ in equation (6.1.22)) around the ion/charged particle of radius r_i. From Debye–Hückel theory, h is proportional to the square root of the solution ionic strength. Thus, for larger charged particles (larger r_i), at high ionic strengths, $Z_{i,\text{eff}}$ will be related to Z_i by

$$(Z_{i,\text{eff}}/Z_i) \cong (1/h)/r_i. \tag{6.3.31b}$$

Now, Z_i, for a large ion/charged particle, is proportional to the surface area of the particle/ion, $4\pi r_i^2$, if we assume it to be spherical. Therefore its effective mobility, $\mu_{i,\text{eff}}^m$, (Giddings, 1991),

$$\mu_{i,\text{eff}}^m = \frac{Q_{i,\text{eff}}}{f_i^d/\tilde{N}} = \frac{Z_{i,\text{eff}}\mathcal{F}}{\tilde{N}6\pi\mu r_i} = \frac{Z_i\,\mathcal{F}(1/h)}{\tilde{N}6\pi\mu r_i^2} \propto \frac{4\pi r_i^2\mathcal{F}(1/h)}{6\pi\mu r_i^2\tilde{N}} \neq f(r_i), \tag{6.3.32}$$

is no longer a function of the size of the particle. For colloidal particles, ionic moblility is therefore not influenced much by the size of the particle in free solution. Consequently, capillary electrophoresis is no longer useful. The capillary has to have a porous gel/matrix which will allow smaller colloids to pass through with less resistance in the applied electrical field.

The profile development analysis of an injected sample carried out earlier led to the concentration profile (6.3.15) based on a number of assumptions, which included a constant U_{iz} and therefore a constant μ_i^m. Constancy of μ_i^m along the capillary length ensures that the peak dispersion will not have any contribution from variations in μ_i^m. Further, constancy of $v_{\text{EOF},z}$ requires constancy of the zeta potential of the silica surface, ζ, which, in turn, requires constancy of *pH*. Therefore, the *pH* has to be constant and uniform along the capillary, especially for acidic/basic species, suggesting that the solution has to be buffered. The buffer solution usually employed in CE/CZE is 0.01–0.05 M phosphate (0.2 M solution of monobasic sodium phosphate and 0.2 M solution of dibasic sodium phosphate in different ratios). For solutes that are bases, the same buffer is recommended at a low *pH*. For acidic solutes, a borate buffer is recommended at a high *pH*. Sometimes a mixed phosphate–borate buffer may be used.

For the analysis of DNA samples, the capillary is filled with a gel, generally of crosslinked polyacrylamide. Proteins and oligonucleotides have also been separated in such capillaries. The gel provides a sieving matrix, having pores through which smaller biomacromolecules pass much faster. The gel also eliminates electroosmotic flow and substantially decreases the solute diffusion coefficient; thus band broadening is reduced. In fact, for DNA sample analysis, CE employing a gel has become highly successful. DNA sequencing is an area where gel-filled CE has been critically important to develop the speed and selectivity needed to sequence the entire human genome (Zubritsky, 2002). To solve the problem of gel stability, fresh polymer solution was introduced for each sample analysis (Karger and Foret, 1993) and polymerization carried out. After each run, the polymer solution may be blown out and the capillary reloaded with fresh solution. A run may last for a maximum of 80 minutes.

The *sample injection volume* in conventional CE/CZE is around 1–10 nanoliter. The sample amount obtained at the exit in one run is often insufficient for spectral measurement/structure determination. If a semipreparative scale of operation is employed, then the sample amount at the exit would be significant. Since the sample-loading capacity is likely to be proportional to the capillary/column cross-sectional area, use of larger capillaries up to 180 μm diameter (including those of rectangular cross section (Tsuda, 1993)), sometimes packed with octadecyl silica particles (Chen *et al.* (2001) employed capillaries with diameters $\geq$ 550 μm containing 1.5 μm octadecyl silica particles), will allow larger sample injection volumes up to 1 microliter. Alternatively, a bundle of multiple capillaries (up to five) may be introduced into one column to increase the sample-loading capacity (Tsuda, 1993).

This last concept has been extended to what is known as *capillary array electrophoresis* (CAE) (Huang *et al.*, 1992; Kheterpal and Mathies, 1999), wherein there may be as many as 96 capillaries in parallel in instruments currently available (Zubritsky, 2002). This increases the throughput drastically. Such devices were the workhorse of the human genome analysis. Scaling up to 1000 capillaries has also been reported (Kheterpal and Mathies, 1999). Capillary array electrophoresis enables easy and parallel loading of multiple samples as well as rapid and parallel separation and detection using appropriate detection techniques.

An uncharged compound, if present in the injected sample, will move in CE with the bulk liquid at the electroosmotic velocity $v_{EOF,z}$ toward the cathode. Different uncharged compounds would have the same velocity, namely $v_{EOF,z}$. On the other hand, if a charged phase could be created into which different uncharged compounds would partition in a varying fashion, the situation is changed. This is achieved in CE by having an appropriate solution of an ionic surfactant, for example sodium dodecyl sulfate (SDS), as the medium in CE.

If the SDS concentration is higher than the critical micelle concentration (CMC), then the excess surfactants will form spherical micelles (see Section 4.1.8) with negatively charged headgroups (SO_4^{2-}). These micelles will tend to migrate toward the positive electrode in the electric field. However, it is observed that, at $pH > 5$, the value of $v_{EOF,z}$ toward the cathode is higher than the electrophoretic migration velocity of the negatively charged micelles toward the anode, $v_{EPMC,z}$; the net effective direction of movement of the micelles is therefore toward the cathode:

$$v_{MC,z} = v_{EOF,z} - v_{EPMC,z}, \tag{6.3.33}$$

where $v_{MC,z}$ is the net velocity of the negatively charged micelles toward the cathode.

Uncharged compounds in such an environment will partition into the hydrophobic core of the charged micelles. Some will partition much more, the more polar compounds much less, as we have seen in the results illustrated in Table 4.1.9. Therefore, the net velocity of each such compound toward the cathode will be quite different. This is the basis of micellar electrokinetic chromatography (MEKC), first developed by Terabe *et al.* (1984). However, this technique will be covered in Chapter 8, since the partitioning of an uncharged compound from a mobile phase into a micelle, both of which have a bulk motion perpendicular to the force direction, causing the compound partitioning into the micelle, is more appropriate for Section 8.2.2.1.

In CE using silica capillaries having a charged surface, oppositely charged solutes may undergo electrostatic adsorption on the capillary surface. Biopolymers are especially susceptible to adsorption. The capillary surface is therefore often coated with a neutral hydrophilic polymer (e.g. methylcellulose) in a thin layer. A thick neutral coating can eliminate the electroosmotic flow. For these and related issues, the reader should consult a number of relevant chapters in Guzman (1993). Additional details on the experimentally observed effects of wall adsorption of polycations in particular (Towns and Regnier, 1992), and the theoretically predicted consequences of adsorption in general (Schure and Lenhoff, 1993), provide a useful perspective.

6.3.1.3 *Centrifugal elutriation*

In Section 6.3.1.1, we studied elutriation based separation of a mixture of particles in the gravitational field in the presence of a bulk flow of the liquid in the vertical direction. To separate a mixture of different particles (different sizes and/or densities), the liquid flow in the vertical direction occurred in a vessel with changing flow cross-sectional area. Different sized particles (or particles of different densities) found equilibrium positions at different heights where the liquid velocity changing with height equaled the critical terminal velocity of the particle. Gravitational force is not strong enough to fractionate microscopic particles, such as bacterial cells, etc., in the size range of 1–50 μm. However, as we have seen earlier, in Section 4.2.1.2, centrifugal force in the radially outward direction can.

In *centrifugal elutriation*, a mixture of particles is introduced into a rotating chamber through which the liquid

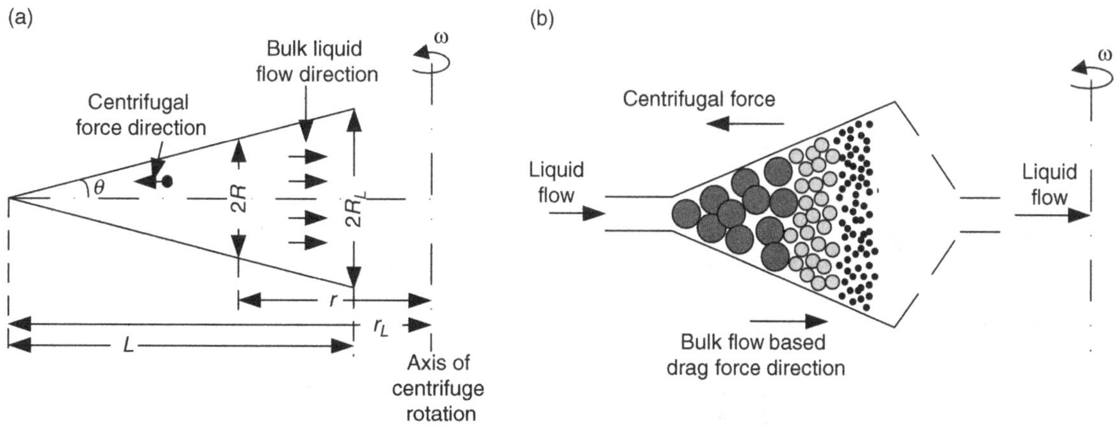

Figure 6.3.7. (a) Schematic of a truncated cone as an elutriation chamber, with bulk flow toward the axis of rotation of the centrifuge. (b) Mixture of cells separated according to their size in centrifugal elutriation. (After Figdor et al. *(1998).)*

flows in a direction opposite to that of the centrifugal force. The rotating chamber is essentially a conical tube with its base near the center of rotation and its apex far away (Figure 6.3.7(a)). If the liquid in which the cells/particles are suspended is made to flow in this conical tube in a direction from the apex to the base, then the liquid velocity increases as the distance r from the center of rotation increases. Correspondingly, at steady state, the drag force on a cell/particle (assumed Stokesian) of radius r_p moving with the liquid velocity v_r at radial location r is $6\pi\mu r_p v_r$; the magnitude of this force increases as r increases. This force is, however, opposed by the radially outward centrifugal force on the particle, $m_p r\omega^2\left(1-(\rho_t/\rho_p)\right)$, which increases linearly with r. Under appropriate conditions, there will be a radial location r where the two forces will balance each other and the cell/particle of specific size/density will concentrate at the radial location:

$$6\pi\mu r_p v_r = m_p r\omega^2\left(1-(\rho_t/\rho_p)\right). \tag{6.3.34a}$$

The inwardly radial medium velocity v_r is related to the medium volume flow rate Q_f and the local cross-sectional area $A(r)$, which depends on the radial location r. Replacing the cell mass m_p by $(4/3)\,\pi r_p^3\rho_p$, we get

$$r_p^2 = \frac{9}{2}\frac{\mu}{\omega^2(\rho_p-\rho_t)}\frac{Q_f}{rA(r)}. \tag{6.3.34b}$$

This equation provides a relation between the radial location and the dimension r_p of the cell population (assuming they are of the same density) which accumulates at r. (Generally, $(\rho_p-\rho_t)$ in equation (6.3.34b) is very small; therefore this technique is not used to separate particles of the same r_p but differing only slightly in ρ_p.) The exact relation will depend on how $A(r)$ varies with r. Consider the conical tube shown in Figure 6.3.7(a), where R_L is the largest radius of the conical tube, L is the distance of this radius from the apex of the conical tube, r_L is the radial distance of the apex of the conical tube from the center of rotation and r is the radial distance from the center of rotation of the cross-sectional area of radius R under consideration. The expression for $A(r)$ is then

$$\pi R^2 = \pi\left(\tan\theta\times(r_L-r)\right)^2 = \pi(R_L/L)^2(r_L-r)^2. \tag{6.3.34c}$$

Correspondingly, equation (6.3.34b) may be reduced to the following expression for r_p:

$$r_p = \frac{3}{\omega(r_L-r)(R_L/L)}\sqrt{\frac{\mu Q_f}{2\pi(\rho_p-\rho_t)r}}. \tag{6.3.34d}$$

From this equation, we can infer that the smaller the particle/cell size r_p, the closer to the center of rotation the equilibrium location of the particles, as illustrated schematically in Figure 6.3.7(b). The withdrawal/separation of each size group of cell population may be accomplished by either a stepwise increasing of the flow velocity enough or a stepwise decreasing of the centrifugal rotational speed, so that the smallest particles in the conical rotating tube come out with the bulk liquid flow. As a practical design, there is a counter-taper at this exit section (closest to the center of rotation), as shown by the dashed-dotted line in Figure 6.3.7(b).

This concept of centrifugal elutriation was developed first by Lindahl (1948, 1956). A recent review of the extensive literature in the biological field on various types of cell separation is available in Figdor *et al.* (1998). Conventionally, the rotating separation chamber volume useful for separation is around 3–4 milliliter. Larger separation chambers with volumes up to 40 milliliter have been

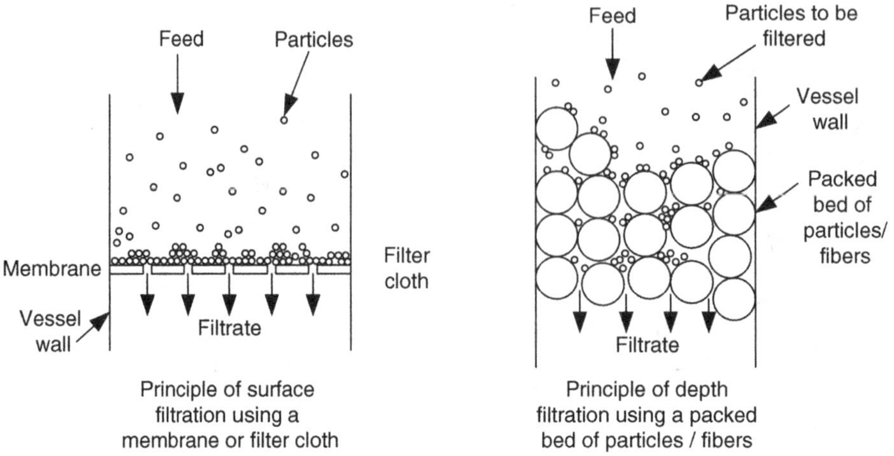

Figure 6.3.8. Principles of surface filtration and depth filtration for removing particles from a fluid.

developed. The highest rotational speed used is 5000 rpm. Each sample fraction that is obtained has a relatively large volume. The volume fraction sizes may vary from 25 to 150 milliliter (Figdor *et al.*, 1998). Pumps of various types are used to generate the flow, whose rate may be around 15–18 milliliter/min.

6.3.1.4 *Inertial deposition of particles on a filter/collector in depth filtration*

Small particles in air or water are conventionally removed by passing the air/gas/water through a filter. Particles in the size range of 0.02 to 10 μm are removed by *microfiltration* as well as *granular filtration*. The latter is also called *depth filtration*. Microfiltration is a general term which describes filtration processes used to remove micron and submicron particles from air/gas/water/solvent. There are two basic mechanisms: *surface filtration* and *depth filtration* (Figure 6.3.8). In surface filtration, the fluid passes through the pores of a relatively thin membrane/filter/cloth, whereas the particles are retained on top of the membrane/filter/cloth whose pores/openings are smaller than the particles. This filtration technique is covered in Section 6.3.3.1.

In *depth filtration*, taking place in a granular/porous/ fibrous medium having a considerably larger thickness in the flow direction, the interstitial openings are usually larger than the particles to be removed. The particles are carried into the interior of the filter by the flowing fluid; the particles are deposited on the surface or collectors of the filter medium via a number of different mechanisms.

For *aerosols*[20] in a gas, the mechanisms are: *inertial impaction, interception, gravitational settling, electrostatic deposition* and *Brownian diffusion*. Of these, only inertial impaction involves bulk flow of the gas parallel to the direction of the force on the particle and will be considered here. The rest of the mechanisms are mentioned in Section 7.2.2 since bulk gas flow will generally be perpendicular to the forces involved in the other mechanisms. The same is true for the capture of *hydrosols*[21] from an aqueous solution in a depth filter/granular filter, except additional forces are involved.

It is useful to illustrate briefly how the mechanism of *inertial impaction* leads to the capture of the particles from a gas stream onto filter elements. Figure 6.3.9A illustrates the streamlines of gas flow in the depth filter around a cylindrical fiber. Far away from the fiber, the gas velocity is uniform: v_∞. For small particles, the velocity is equal to that of the fluid, and the particle flow path follows the streamline. However, near the fiber, the fluid streamline is changed as the fluid goes around the fiber. Due to the inertia of the particle, especially those of diameter larger than 1 μm, the particle trajectory does not coincide with the fluid streamline near the fiber. The particle continues on a path (dashed line in Figure 6.3.9A) which takes it straight to the fiber, and a collision takes place. If the adhesion forces between the particle and the fiber are strong enough for the particle to remain stuck to the fiber surface, the particle is captured.

This situation describes the fate of those particles present in an envelope bounded by distant streamlines called the *limiting streamlines*. The dimension of this envelope is of the order of the fiber diameter and has a width of $2b$; the gas flow streamline at a distance b from the centerline is the limiting streamline, and the particle trajectory that

[20] A dispersion of tiny particles of solid or liquid in a gas.

[21] A dispersion of tiny solid particles or liquid droplets in water.

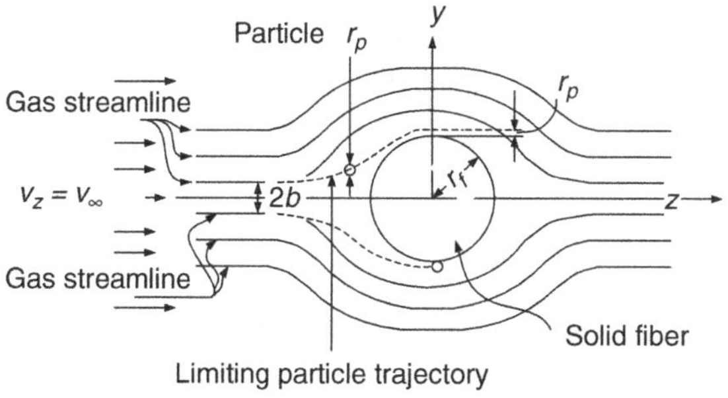

Figure 6.3.9A. Gas flow streamlines around a solid fiber of radius r_f *in a filter bed and the trajectory of a particle of radius* r_p *due to inertial force near the fiber.*

coincides with this streamline far away from the fiber is called the *limiting trajectory*. Far away from the fiber, particles present in the streamlines at distance greater than b from the centerline are not captured.

As the streamline changes near the cylindrical fiber in the filter, the fluid accelerates since both axial and normal fluid velocities change rapidly. Heavier aerosol particles in air, larger than 1 μm in size, are unable to follow the accelerating gas; the effect increases as the free stream velocity v_∞ and the particle mass increases. In the absence of any external forces, equation (6.2.45) may be written as

$$\boldsymbol{F}_p^{\text{iner}} = m_p \frac{\mathrm{d}\boldsymbol{U}_p}{\mathrm{d}t} = -\boldsymbol{F}_p^{\text{drag}}. \qquad (6.3.35)$$

Depending on the Reynolds number, $\text{Re} = 2r_p v_\infty/\nu$ (based on a fixed particle), the drag force may or may not be described by Stokes' law. When Stokes' law (3.1.63) is valid ($Re << 1$)[22],

$$m_p \frac{\mathrm{d}\boldsymbol{U}_p}{\mathrm{d}t} = -6\pi\mu r_p(\boldsymbol{U}_p - v). \qquad (6.3.36)$$

When $Re \geq 1$, the equation for particle motion may be written as (Friedlander, 1977)

$$m_p \frac{\mathrm{d}\boldsymbol{U}_p}{\mathrm{d}t} = -\frac{C_D Re}{24} 6\pi\mu r_p(\boldsymbol{U}_p - \boldsymbol{v}). \qquad (6.3.37)$$

where C_D is the drag coefficient for the particle (see (3.1.64)) and *may be* described by

$$C_D = \frac{24}{Re}(1 + 0.158Re^{2/3}) \qquad (6.3.38)$$

as long as $Re < 1000$. Obviously, C_D is a function of the local Reynolds number of the particle:

$$Re = 2r_p|\boldsymbol{U}_p - \boldsymbol{v}|/\nu. \qquad (6.3.38)$$

[22]The term $6\pi\mu r_p$ is often corrected by a correction factor, for which see equation (6.3.41) and the associated text.

For spherical particles of size r_p, density ρ_p and mass m_p, if we nondimensionalize equation (6.3.36) using the following nondimensional variables:

$$\boldsymbol{U}_p^+ = \frac{\boldsymbol{U}_p}{v_\infty}, \quad t^+ = \frac{t v_\infty}{\ell}, \quad \boldsymbol{v}^+ = \frac{\boldsymbol{v}}{v_\infty}, \qquad (6.3.39)$$

where ℓ is a characteristic length of the filter collector, fiber etc. (a particular value of ℓ for a fiber of diameter $2r_f$ is $r_f + r_p$), we get

$$\left(\frac{2\rho_p v_\infty r_p^2}{9\mu\ell}\right)\frac{\mathrm{d}\boldsymbol{U}_p^+}{\mathrm{d}t^+} = St\frac{\mathrm{d}\boldsymbol{U}_p^+}{\mathrm{d}t^+} = -(\boldsymbol{U}_p^+ - \boldsymbol{v}^+). \qquad (6.3.40)$$

The dimensionless quantity St is called the *Stokes number* and represents the distance such a particle will travel, starting with a velocity v_∞, before stopping if the fluid is stagnant, i.e. $\boldsymbol{v}^+ = 0$. Alternatively, it is the ratio of twice the kinetic energy of a particle moving at a velocity v_∞ and the work done against the drag force experienced by the particle moving at a velocity v_∞ through the fluid over the characteristic distance b (Tien, 1989):

$$St = \frac{m_p v_\infty^2}{(6\pi\mu r_p v_\infty)\ell} = \frac{(4/3)\pi r_p^3 \rho_p v_\infty^2}{(6\pi\mu r_p v_\infty)\ell} = \frac{\rho_p (2r_p)^2 v_\infty}{18\mu\ell}. \qquad (6.3.41)$$

If the particle size is of the order of the mean free path of the gas, there may be slip at the particle surface between the fluid and the particle. To account for this effect, St is multiplied by a quantity C_C, called the *Cunningham correction factor*. For particles of 1 μm diameter at normal temperature and pressure, C_C is around 1.16 (Tien, 1989).

Any solution of equation (6.3.36) or (6.3.37) requires detailed information about the flow field around the filter bed collector/fiber. The flow field may be available via the three velocities v_x, v_y, v_z (or v_r, v_θ, v_z, etc.) or via the stream function ψ, if it can be assumed that the particle motion does not affect the flow field. The solution of such a problem generally requires a numerical solution of the governing equations (e.g. equation (6.2.6b)) for the chosen velocity field around the fiber in the *depth filter*.

The quantity of interest from a utilitarian point of view is the *extent* of removal of particles from the air stream in the depth filter. A predictive approach to this usually involves a number of steps. The first step is to calculate the particle capture efficiency of a single fiber element in a filter via a given mechanism, say E_{IS} due to inertial impaction. In the second step, add up appropriately the corresponding single-fiber capture efficiencies due to the different mechanisms to obtain E_{TS}, the total efficiency for a single fiber, and apply the result to the whole filter bed/depth filter.

An alternative approach calculates the particle capture efficiency of a single fiber element due to the simultaneous action of different capture mechanisms to obtain E_{TS} and then applies such a result to the whole bed. We will consider these approaches in Section 7.2.2 along with the other capture mechanisms primarily for hydrosol removal by granular filters.

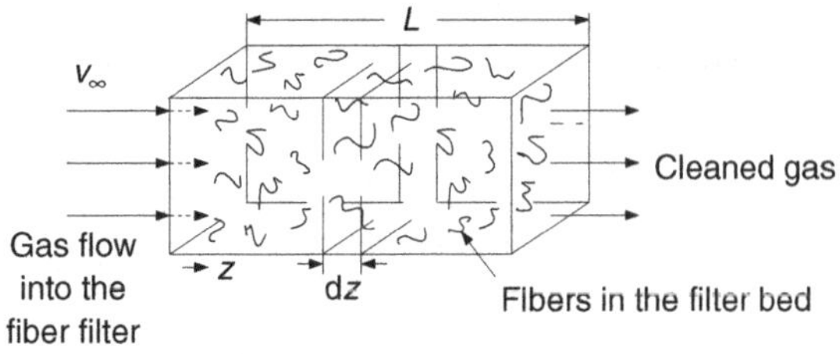

Figure 6.3.9B. Fibrous filter bed of length L *containing a collection of single fibers of radius* r_f *used to clean a flowing gas.*

We will employ here the definition of the inertial impaction based capture efficiency E_{IS} of a single fiber of radius r_f in the context of the definition (2.4.4a) of total efficiency E_T of the depth filter:

$$E_{IS} = \frac{2b}{2r_f}. \tag{6.3.42a}$$

Here, if the fiber of radius[23] r_f is perpendicular to the gas flow, then $2b$ is the width of the region of gas flow (see Figure 6.3.9A) which is cleaned completely of any particles by the single fiber. An essentially identical definition may be employed when the filter bed consists of granular particles of radius r_f; in that case, $2b$ will be the diameter of a cylindrical tube of contaminated gas, which will be cleaned of dust particles by the spherical collector in the filter bed. The value of b is obtained from the solution to the governing equation (6.3.40) and the gas velocity profile. The streamline corresponding to b is the *limiting trajectory*.

Consider a depth filter containing a regular array of cylindrical fibers (of diameter $2r_f$) which occupy a volume fraction $(1-\varepsilon)$ of the fiber bed (Figure 6.3.9B). Generally this volume fraction is of the order of 0.1. Assume the number concentration of particles of diameter $2r_p$ to $2(r_p + \mathrm{d}r_p)$ at a distance z from the inlet of a depth filter of length L to be $\mathrm{d}N(r_p) = n(r_p)\mathrm{d}r_p$. Use the single-fiber capture efficiency E_{IS} defined by (6.3.42a) and focus on a bed length dz (Figure 6.3.9B) in the gas flow direction. The number of fibers per unit width of the bed in this bed length is $(1-\varepsilon)\mathrm{d}z/(\pi r_f^2)$. Since each fiber cleans up a region of width $2b$ of the gas flow (Figure 6.3.9B), the total width of the gas flow region cleared of particles is $[(2b)(1-\varepsilon)\mathrm{d}z]/(\pi r_f^2)$. The number of particles of size r_p removed as a result per unit time is the total width $\times$ fluid approach velocity $\times$ particle number concentration at this location:

$$\frac{2b(1-\varepsilon)\mathrm{d}z}{\pi r_f^2} \times v_\infty \times n(r_p)|_z \mathrm{d}r_p. \tag{6.3.42b}$$

[23] For other collector (fiber) geometries, instead of $2r_f$ use the mass flow rate of particles in the projected cross-sectional area of the collector, whereas, for $2b$, employ the actual mass rate of particles captured.

The extent of change in the particle number concentration, $-\mathrm{d}\big(n(r_p)\big)\mathrm{d}r_p|_z$, as a result of this in the gas flowing into the inlet of the filter at a velocity $v_\infty \cdot \varepsilon \cdot 1$ is given by

$$v_\infty \varepsilon\Big(-\mathrm{d}\big(n(r_p)\big)\mathrm{d}r_p\Big)\Big|_z = \frac{2b(1-\varepsilon)\mathrm{d}z}{\pi r_f^2} v_\infty \times n(r_p)\Big|_z \mathrm{d}r_p. \tag{6.3.42c}$$

However, if we use the single-fiber efficiency E_{IS} due to inertial impaction, then, from definition (6.3.42a), $2b = E_{IS}\, 2r_f$:

$$-\mathrm{d}n(r_p)/n(r_p)|_z = [2E_{IS}(1-\varepsilon)/\pi r_f \varepsilon]\mathrm{d}z. \tag{6.3.42d}$$

Integrate from $z = 0$ to $z = L$ to obtain

$$\ln\frac{n(r_p)|_{z=0}}{n(r_p)|_{z=L}} = \frac{2E_{IS}(1-\varepsilon)}{\pi r_f \varepsilon} L \Rightarrow \frac{n(r_p)|_{z=0}}{n(r_p)|_{z=L}} = \exp\left(\frac{2E_{IS}(1-\varepsilon)}{\pi r_f \varepsilon} L\right).$$

Integrating from $r_{p,\mathrm{min}}$ to $r_{p,\mathrm{max}}$, we get

$$\int_{r_{p,\mathrm{min}}}^{r_{p,\mathrm{max}}} n(r_p)|_{z=0}\mathrm{d}r_p = N(r_{p,\mathrm{min}}, r_{p,\mathrm{max}})\big|_{z=0} = N(r_{p,\mathrm{min}}, r_{p,\mathrm{max}})|_{z=L}\exp\left(\frac{2E_{IS}(1-\varepsilon)}{\pi r_f \varepsilon} L\right). \tag{6.3.43}$$

Therefore,

$$E_{IS} = \frac{\pi r_f \varepsilon}{2(1-\varepsilon)L} \ln\frac{N(r_{p,\mathrm{min}}, r_{p,\mathrm{max}})|_{z=0}}{N(r_{p,\mathrm{min}}, r_{p,\mathrm{max}})|_{z=L}}. \tag{6.3.44}$$

An expression of the overall filter efficiency due to inertial impaction may also be developed as follows:

$$E_T = 1 - \frac{N(r_{p\mathrm{min}}, r_{p\mathrm{max}})|_{z=L}}{N(r_{p\mathrm{min}}, r_{p\mathrm{max}})|_{z=0}} = 1 - \exp\left[-\frac{2E_{IS}(1-\varepsilon)}{\pi r_f \varepsilon} L\right]. \tag{6.3.45}$$

We will very briefly illustrate how E_{IS} may be estimated so that E_T may be calculated. The key to E_{IS} is an estimate of b, the distance of the limiting particle trajectory from the centerline. An exact answer will require a solution of equation (6.3.36) using appropriate initial conditions to determine the particle trajectories. Write equation (6.3.36) in the manner of equations (6.2.47b) and (6.2.47c) in Cartesian coordinates y and z, assuming no variation in the x-direction and that $\boldsymbol{U}_p$ is described by equation (6.2.46):

$$a\frac{\mathrm{d}^2 y}{\mathrm{d}t^2} = v_y - \frac{\mathrm{d}y}{\mathrm{d}t}; \tag{6.3.46a}$$

$$a\frac{\mathrm{d}^2 z}{\mathrm{d}t^2} = v_z - \frac{\mathrm{d}z}{\mathrm{d}t}. \tag{6.3.46b}$$

Here $a = (m_p/6\pi\mu r_p) = \left(\rho_p(2r_p)^2/18\mu\right)$; v_y and v_z are the y- and z-components of the gas velocity field (see Figure 6.3.9A). Particles which are far away from the fiber will essentially follow the gas streamline, namely the v_z-component of the gas velocity ($= v_\infty$) and the v_y-component of the gas velocity ($= 0$). To calculate the number of particles captured by the filter, we need to focus on that particle trajectory coming from far away (where $v_z = v_\infty$ and $v_y = 0$) whose y-value far away from the particle is equal to b such that its trajectory is only at a distance of r_p from the fiber surface at $z = 0$ (fiber centerline). Then all other particle trajectories far away, whose y-values are less than b, will hit the fiber and be captured. So the initial conditions of the particle trajectories far away from the fiber for the solution of equations (6.3.46a,b) are as follows:

$$\text{at } z(t=0), \quad \frac{dx}{dt} = U_{px} = v_\infty; \tag{6.3.46c}$$

$$\text{at } t=0, y=b, \quad \frac{dx}{dt} = U_{py} = 0. \tag{6.3.46d}$$

To obtain a solution which provides a value of the unknown b, one needs to know the gas velocity field, v_y and v_z, as a function of y and z in the filter bed. The filter bed is full of fibers - so the velocity field is complex. The usual approach is to adopt a *cell model* (Happel, 1959), which reduces the problem to finding the gas flow field around a single fiber in a specified volume, such that the ratio of the void volume around the fiber to the total volume of the fiber plus the void volume (i.e. the specified volume) equals the void volume fraction of the fiber bed; there are other conditions imposed on the boundaries of the specified volume which reflect the presence of the surrounding fibers in an appropriate way. Generally, a complete numerical solution is needed for the set of equations and the initial conditions. Using the cell model of Kuwabara (1959) for the velocity field around a fiber, Flagan and Seinfeld (1988) have provided a semianalytical solution to estimate b. The basic approach (Flagan and Seinfeld, 1988) involves developing a solution of the particle trajectory coordinates y and z as a function of t and various parameters for an appropriate upstream distance $-z$ where the fluid velocity $v_z = v_\infty$. Then, for the value of t which brings the fluid streamline to $z = 0$ (namely, the fiber centerline), determine the value of y of the particle trajectory, where $y = r_p + r_f$ so that particle capture occurs. The value of b is obtained from this expression, $y = r_p + r_f$. Figure 6.3.9C illustrates how the value of E_{IS} (defined by equation (6.3.42a)) varies with the Stokes number St for a given value of (r_p/r_f), with the void volume fraction ε as a parameter.

Two items are of importance regarding this analysis and Figure 6.3.9C. First, smaller particles which may follow the gas streamlines, and which therefore come within r_p of the surface of the fiber, will be captured by the mechanism of *interception*. The results shown in Figure 6.3.9C include these particles automatically. Second, results in this figure suggest that, as the Stokes number increases, the particle capture efficiency becomes 1, regardless of how few fibers there are in the bed. Further, the larger the fiber packing fraction in the bed, the higher the capture efficiency for a given St.

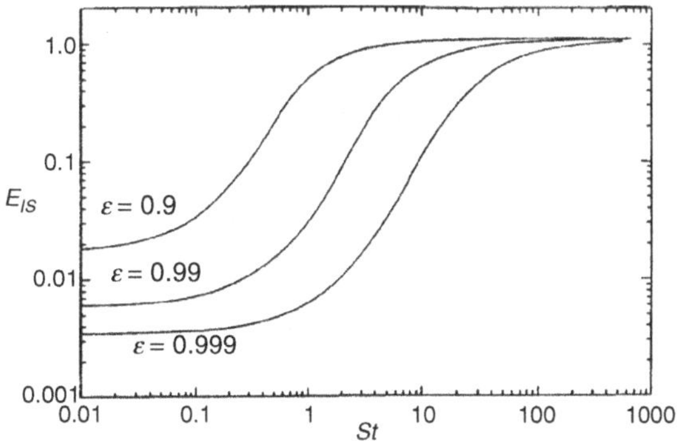

Figure 6.3.9C. Capture efficiency for inertial impaction for a cylindrical fiber of radius r_f placed transverse to the gas flow as a function of Stokes number for $r_p/r_f = 0.01$. (After Flagan and Seinfeld (1988).)

6.3.1.5 *Electrostatic separation of fine particles in a dry state*

We consider here an example where two external forces are acting parallel to the direction of the bulk velocity. Mixtures of very fine particles to be separated are often in a dry state; this is especially true in the processing of various minerals (e.g. coals). When two particles come into contact, electric charges develop via friction; the charge that remains on the particles after "separation of solid-to-solid contacts" is called *triboelectrification* (Inculet, 1984). A fluidized bed of particles is often a convenient method of achieving triboelectrification. One can also predict the polarity of the charge developed, but not necessarily its magnitude.

Consider now a porous horizontal stainless steel plate in a vertical vessel maintained at a high positive voltage (as much as 30 kV) (Figure 6.3.10). Let air flow up through this porous plate and fluidize the bed of particles over this porous charged plate. Let there also be a set of metallic collection troughs at the top of the vertical vessel maintained at ground voltage. Thus there is an electrical field directed vertically upwards. Any particle that develops a positive charge via triboelectrification[24] will be subject to a vertically upward electrostatic force; the particle will also be subjected to gravitational force acting vertically

[24]There will be additional charge development via conductive induction when the particle contacts the stainless steel plate with a positive charge.

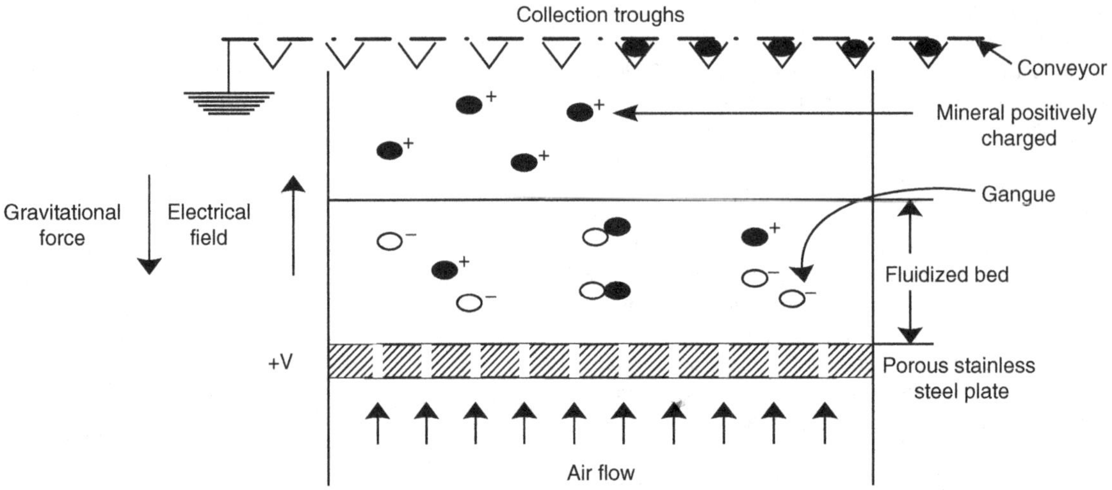

Figure 6.3.10. Separation of minerals from gangue material in an electrical field in the vertical direction after triboelectrification in an air-fluidized bed. (After Inculet (1984).)

downwards, as well as to a drag force. The vertical direction (up or down) of the drag force exerted by the fluidizing air will be dependent on the relative velocity of the particle with respect to the air. If the particles go up, they will be collected in the trough; the electrical field strength may be as high as 200-300 kV/m. The particle sizes treated may vary between 10 and 250 μm. Examples include the separation of Fe_2O_3 from particles of gangue, etc.

6.3.2 Chemical potential gradient driven phase-equilibrium systems

We consider here the role of bulk flow parallel to the direction of the chemical potential gradient based force in phase-equilibrium based open two-phase systems. Vapor-liquid systems of flash vaporization, flash devolatilization and batch distillation are considered first, followed by a liquid-liquid system for extraction. Solid-liquid systems for zone melting and normal freezing are studied thereafter to explore how bulk flow parallel to the force direction is essential to considerable purification of solid systems followed by solid-vapor systems as in drying.

6.3.2.1 *Vapor-liquid systems: flash vaporization, devolatilization and batch distillation*

Distillation based separation of a liquid mixture exploiting the inherent separation achieved in a closed vessel due to vapor-liquid equilibrium (Section 4.1.2) is implemented in an open separator vessel in a variety of ways. The variations are primarily due to the nature and origin of the bulk flows of the liquid and the vapor streams, and to the manner of feed introduction. Here we will consider only those bulk flow configurations where the bulk flows of the vapor and/or the liquid are parallel to the direction of the chemical potential driving force between the vapor and the liquid. Two modes of feed liquid introduction are studied: *continuous liquid feed* and *batch liquid feed.* The case of a continuous liquid feed is identified as *flash vaporization* (Figure 6.3.11(a)). When the liquid is *polymeric* and the vapors generated are *monomers* and volatile solvents used in the polymerization process, the device is called a *flash devolatilizer.* A particular form is shown in Figure 6.3.11b. It is akin to vapor stripping from a liquid.

In flash vaporization, the liquid feed mixture is heated under pressure as it flows through a heater into a pressure-reducing valve, which opens into a flash drum (Figure 6.3.11(a)). The vapor phase formed goes up the flash drum, and the remaining liquid jet and droplets go down into the liquid layer. The vapor phase is withdrawn at a molar flow rate of W_{tv}, and the liquid phase is withdrawn at a molar flow rate of W_{tl} from an incoming feed molar rate of W_{tf}. It is clear that the net velocity of the vapor phase formed is up and that the net velocities of the bulk liquid and the liquid droplets are down. At the surface of the liquid layer in the drum, the direction of the chemical potential driving force is perpendicular to the horizontal liquid surface and parallel to the direction of the net movement of the two bulk phases, the vapor phase and the liquid jet and droplets. Although many of the droplets may have a complex flow pattern, one could similarly argue that, as they fall down, the multiple droplets essentially provide a liquid surface from which molecules escape to the vapor phase vertically upwards.

In reality, the design of the flash drum has to be such that the upward vapor velocity is low enough not to entrain

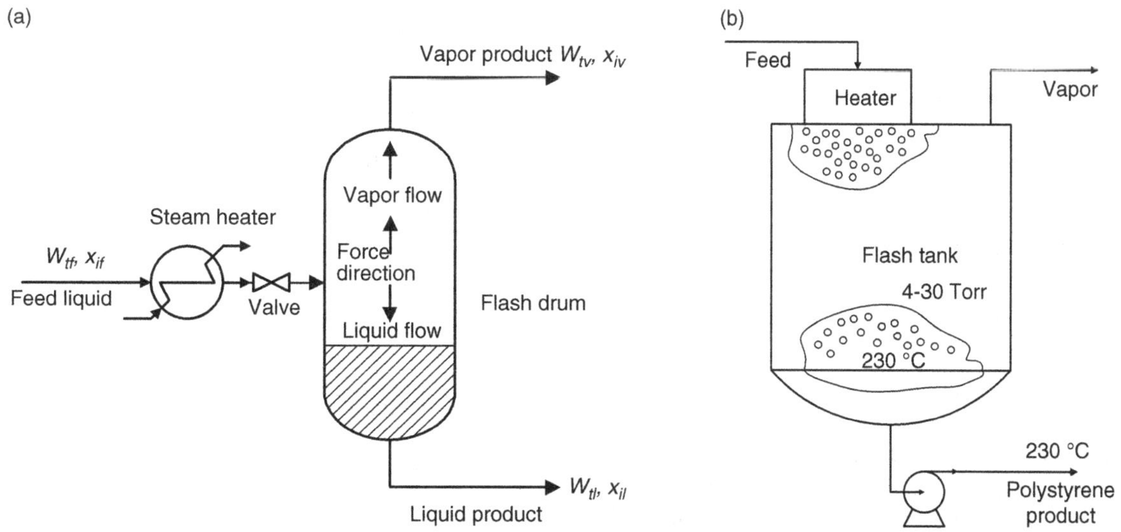

Figure 6.3.11. (a) Continuous flash vaporization. (b) Continuous flash tank devolatilization.

the liquid droplets (see equation (6.3.2)). Therefore, the upward vapor velocity v_{vz} must be lower than the vertically downward terminal velocity U_{pzt} of any liquid droplet:

$$v_{vz} < U_{pzt}. \tag{6.3.47a}$$

The terminal drop velocity may be obtained by balancing the downward gravitational force on the drop of mass m_p, diameter $2r_p$ and density ρ_p, $(m_p g(1 - \rho_v/\rho_p))$, against the drag force $C_D(\pi r_p^2)(\rho_v U_{pzt}^2/2)$ exerted by the vapor of density ρ_v moving up at the terminal velocity, U_{pzt}:

$$m_p g\left(1 - \frac{\rho_v}{\rho_p}\right) = C_D \frac{\pi r_p^2}{2} \rho_v U_{pzt}^2; \tag{6.3.47b}$$

$$U_{pzt} = \sqrt{\frac{8g\left(\rho_p - \rho_v\right) r_p}{3 C_D \rho_v}}. \tag{6.3.48}$$

Here, we have used formula (3.1.61), where the drag force expression used was given in (3.1.64); the value of the projected area A_p of the particle is (πr_p^2) (Bird *et al.*, 1960); m_p, the particle mass, is $(4\pi r_p^3/3)\rho_p$; the drag coefficient C_D (see equation (6.3.4)) may be obtained from Bird *et al.* (1960). In practice, the vertical vapor velocity in the flash drum, v_{vz}, is kept in the range $0.75 U_{pzt} < v_{vz} < U_{pzt}$, where U_{pzt} is the terminal drop velocity. Further, expression (6.3.48) for the terminal drop velocity is written as follows:

$$U_{pzt} = K\sqrt{\frac{\left(\rho_p - \rho_v\right)}{\rho_v}}, \tag{6.3.49}$$

where the value

$$K = \sqrt{\frac{8 g r_p}{3 C_D \rho_v}}$$

is empirically estimated. The units of K determine those of U_{pzt}. Svrcek and Monnery (1993) provide estimates of K for a variety of conditions. The radius of the flash drum, R, should be somewhat larger than that based on the vapor volumetric flow rate Q_v $(= W_{tv}/\rho_v)$ and v_{vz}:

$$R = \left(\frac{Q_v}{\pi v_{vz}}\right)^{1/2}. \tag{6.3.50}$$

The above considerations highlight out the practical limits on the vapor-phase velocity in flash vaporization. Too high a vapor-phase velocity will considerably reduce separation via entrainment of drops. Assuming that there is no drop entrainment, one would like to know the *maximum* separation possible in a flash separator. This will be achieved when the vapor phase and the liquid phase exiting the separator are assumed to be in *equilibrium*. For molar flow rates of feed, vapor stream and the liquid stream being, respectively, W_{tf}, W_{tv} and W_{tl}, a total molar balance and an ith species balance (for a multicomponent system of n species) lead to

$$W_{tf} = W_{tv} + W_{tl}; \tag{6.3.51}$$

$$x_{if} W_{tf} = x_{iv} W_{tv} + x_{il} W_{tl}, \quad i = 1, 2, \ldots, n. \tag{6.3.52}$$

Since the two product streams are in equilibrium,

$$K_i = \frac{x_{iv}}{x_{il}}, \ i = 1, 2, \ldots, n. \tag{6.3.53}$$

Substituting for W_{tv} in terms of W_{tf} and W_{tl} and x_{iv} in terms of $K_i x_{il}$, we get, from (6.3.52), an expression for x_{il}:

$$x_{il} = \frac{x_{if}}{\frac{W_{tl}}{W_{tf}} + K_i\left(1 - \frac{W_{tl}}{W_{tf}}\right)} = \frac{x_{if}}{(1-\theta) + K_i\theta}$$

$$= \frac{x_{if}\left(1 + \frac{W_{tv}}{W_{tl}}\right)}{1 + K_i\frac{W_{tv}}{W_{tl}}}, \qquad i = 1, \ldots, n, \tag{6.3.54}$$

where θ is the stage cut for this single-entry separator if $j = 1 = v$ and $j = 2 = l$ ($\theta = W_{tv}/W_{tf}$). The corresponding expression for x_{iv} in terms of x_{if} is

$$x_{iv} = \frac{x_{if}}{\frac{W_{tv}}{W_{tf}} + \frac{1}{K_i}\left(1 - \frac{W_{tv}}{W_{tf}}\right)} = \frac{x_{if}}{\theta + \frac{1-\theta}{K_i}}$$

$$= \frac{x_{if}\left(1 + \frac{W_{tl}}{W_{tv}}\right)}{1 + \frac{W_{tl}}{K_i W_{tv}}}, \qquad i = 1, 2, \ldots, n. \tag{6.3.55}$$

In a multicomponent system of n components ($i = 1,\ldots, k, \ldots, n$), to obtain the compositions of the vapor phase and the liquid phase leaving the separator under equilibrium, along with the total molar flow rates of the liquid product and the vapor product from the flash drum for a given feed condition, will require the solution of the appropriate governing equations. For a system of n components, there are n equations (6.3.53) describing vapor–liquid equilibrium, n equations (6.3.54) or (6.3.55), describing x_{il} in terms of x_{if} or x_{iv} in terms of x_{if}, and one equation,

$$\sum_{i=1}^{n} x_{iv} = \sum_{i=1}^{n} x_{il} = 1, \tag{6.3.56}$$

describing the relation in any phase between the mole fractions of various components. Thus, there are in total $2n + 1$ equations and $2n + 1$ unknowns (n x_{iv}s, n x_{il}s and (W_{tv}/W_{tf}), or (W_{tl}/W_{tf})), and the solutions can be obtained provided x_{if} is given and P and T are specified. This in turn will allow K_i to be determined as a function of x_{iv} and x_{il}. The required calculations are called *isothermal flash calculations*. In general, it will be an *iterative* process for a *multicomponent* system since K_i depends on x_{iv} and x_{il}; when K_i is almost constant in relation to x_{iv} and x_{il}, calculations become easier.

For a *binary* system of species 1 and 2, where P and T are specified, analytical solutions are possible provided K_i is independent of x_{iv} and x_{il}. Equation (6.3.53) may be written for $i = 1$ and 2 as follows:

$$x_{1v} = K_1 x_{1l}; \qquad x_{2v} = K_2 x_{2l}. \tag{6.3.57}$$

However, $x_{1v} + x_{2v} = 1$ and $x_{1l} + x_{2l} = 1$, leading to

$$x_{1v} = K_1 x_{1l}; \qquad 1 - x_{1v} = K_2(1 - x_{1l}). \tag{6.3.58}$$

Adding these two equations and simplifying, we get

$$x_{1l} = \frac{1 - K_2}{K_1 - K_2}; \qquad x_{2l} = \frac{K_1 - 1}{K_1 - K_2}. \tag{6.3.59}$$

Correspondingly,

$$x_{1v} = \frac{K_1(1 - K_2)}{K_1 - K_2}; \qquad x_{2v} = \frac{K_2(K_1 - 1)}{K_1 - K_2}. \tag{6.3.60}$$

Further, from equations (6.3.55) or (6.3.54), we get

$$x_{1v} = \frac{K_1(1 - K_2)}{K_1 - K_2} = \frac{x_{1f}}{\frac{W_{tv}}{W_{tf}} + \frac{1}{K_1}\left(1 - \frac{W_{tv}}{W_{tf}}\right)}. \tag{6.3.61}$$

We can rearrange this relation to obtain (Lockhart and McHenry, 1958)

$$\frac{W_{tv}}{W_{tf}}\left(\frac{K_1 - 1}{K_1}\right) = \frac{x_{1f}(K_1 - K_2) - 1 + K_2}{K_1(1 - K_2)},$$

which leads to

$$\frac{W_{tv}}{W_{tf}} = \frac{x_{1f}}{1 - K_2} - \frac{x_{2f}}{K_1 - 1}. \tag{6.3.62}$$

Thus, x_{1l}, x_{1v} and (W_{tv}/W_{tf}) can be explicitly determined, since, given P and T, K_1 and K_2 should be known.

For *multicomponent* systems, an iterative procedure for isothermal flash calculation proceeds along the following lines for given P and T. There are two possible cases: (a) K_i is essentially independent of the x_{iv}s and x_{il}s; (b) K_i depends on the x_{iv}s and x_{il}s. When K_i is independent of the x_{iv}s and x_{il}s at a given P and T, then, for given x_{if}, the unknowns are the x_{il}s, the x_{iv}s and (W_{tl}/W_{tf}) or (W_{tv}/W_{tf}). The solution procedure is as follows: determine the K_i values for the given P and T for each i using appropriate formulas, K-factor charts (Figures 4.1.5, 4.1.6), etc. Select a value of (W_{tl}/W_{tf}); calculate the x_{il}s from (6.3.54) *or* the x_{iv}s from (6.3.55). Check whether

$$\sum_{i=1}^{n} x_{il} \cong 1 \;\; or \;\; \sum_{i=1}^{n} x_{iv} \cong 1.$$

If the sum is different from 1, adopt a new guess of (W_{tl}/W_{tf}), calculate the liquid or vapor composition again and check whether the sum of the mole fractions in a given phase equals 1. If and when it does, check whether the sum of mole fractions in the other phase also equals 1. If it does, the guess was correct and the calculation is done.

An alternative, trial-and-error, procedure is to employ

$$\sum_{i=1}^{n} x_{il} \cong 1 \;\; and \;\; \sum_{i=1}^{n} x_{iv} \cong 1$$

in the manner of a function (see Rachford and Rice, 1952a, b),

$$f\left(\frac{W_{tv}}{W_{tf}}\right) = \sum_{i=1}^{n} x_{iv} - \sum_{i=1}^{n} x_{il} = 0, \tag{6.3.63}$$

to determine the value of (W_{tl}/W_{tf}) or (W_{tv}/W_{tf}) which makes the function $f(W_{tv}/W_{tf})$ go to zero. The nature of this function, for ease of calculation, is obtained as follows. From (6.3.55), obtain

$$x_{iv} = \frac{K_i x_{if}}{\frac{W_{tv}}{W_{tf}}(K_i - 1) + 1}. \tag{6.3.64a}$$

From (6.3.54), obtain

$$x_{il} = \frac{x_{if}}{1 - \frac{W_{tv}}{W_{tf}} + K_i \frac{W_{tv}}{W_{tf}}} = \frac{x_{if}}{\frac{W_{tv}}{W_{tf}}(K_i - 1)}. \qquad (6.3.64b)$$

Relation (6.3.63) may now be written as

$$f\left(\frac{W_{tv}}{W_{tf}}\right) = \sum_{i=1}^{n} \frac{K_i x_{if} - x_{if}}{\frac{W_{tv}}{W_{tf}}(K_i - 1) + 1} = 0. \qquad (6.3.65)$$

When the value of the function $f(W_{tv}/W_{tf})$ is essentially zero within a prescribed limit, the correct guess of (W_{tv}/W_{tf}) or (W_{tl}/W_{tf}) has been made. A detailed basis for using such a function of (W_{tv}/W_{tf}) has been discussed in King (1980, pp. 71-87), which also provides different aspects of the calculation procedures for such problems.

One aspect of the calculation procedure needs to be considered at the very beginning of any calculation. The vapor-liquid equilibrium behavior in the flash vaporization problem specification must be such that the system is neither a subcooled liquid nor a superheated vapor. To check that the system has both phases present, calculate $f(W_{tv}/W_{tf})$ for two cases: $(W_{tv}/W_{tf}) = 0$ and $(W_{tv}/W_{tf}) = 1$. For a correctly specified system, $f(W_{tv}/W_{tf})$ will be positive in the first case and negative in the second case (King, 1980, p. 75).

In flash vaporization characterized by the two product phases being in equilibrium, other calculation procedures, if the problem specifications are different from the case of a specified P and T, should be considered. The following *additional* types of problem specifications have been frequently considered (King, 1980, pp. 79-86):

(1) T and the ratio $(x_{iv}W_{tv}/x_{if}W_{tf})$ specified for a particular i;
(2) P and the ratio $(x_{iv}W_{tv}/x_{if}W_{tf})$ specified for a particular i;
(3) P and the ratio (W_{tv}/W_{tf}) specified;
(4) *isenthalpic flash* and P specified. In this case, the sum of the enthalpies of the vapor and the liquid product streams must equal the enthalpy of the feed stream. Illustrations of the solution procedures for a number of different cases have also been provided by King (1980, pp. 75-90).

Example 6.3.1 A liquid mixture containing 12% ethane, 32% propane and 56% n-butane (see Example 4.1.1) is throttled into a flash drum at 32 °C and 700 kPa. The feed composition provided is in mole %. Calculate the fraction of the feed stream which leaves the flash drum as a vapor if vapor-liquid equilibrium may be assumed. Determine the compositions of the vapor stream and the liquid stream. Assume that the K_i values may be obtained from Figures 4.1.5 and 4.1.6.

Solution The governing equations are (6.3.65), (6.3.55) and (6.3.54). First, one must make sure that the problem specifications are such that the flash drum conditions are in the two-phase region of vapor-liquid equilibrium. We calculate the value of $f(W_{tv}/W_{tf})$ from equation (6.3.65) for two values of (W_{tv}/W_{tf}), namely 0 and 1. Consider $(W_{tv}/W_{tf}) = 0$ first. From Figure 4.1.6, we determine K_i for the three species (see Table 6.3.2).

Table 6.3.2.

Species	x_{if}	K_i	$\frac{K_i x_{if} - x_{if}}{\frac{W_{tv}}{W_{tf}}(K_i - 1) + 1}$
Ethane	0.12	4.90	$\frac{4.9\times0.12-0.12}{0\times3.9+1} = \frac{0.588-0.12}{1} = 0.468$
Propane	0.32	1.65	$\frac{1.65\times0.32-0.32}{0\times0.65+1} = \frac{0.528-0.32}{1} = 0.208$
n-Butane	0.56	0.50	$\frac{0.5\times0.56-0.56}{0\times(-0.5)+1} = \frac{-0.28}{1} = -0.28$

From Table 6.3.2 we obtain

$$f\left(\frac{W_{tv}}{W_{tf}}\right) = \sum_{i=1}^{n} \frac{K_i x_{if} - x_{if}}{\frac{W_{tv}}{W_{tf}}(K_i - 1) + 1} = 0.396.$$

Consider now $(W_{tv}/W_{tf}) = 1$ (Table 6.3.3).

Table 6.3.3.

Species	x_{if}	K_i	$\frac{K_i x_{if} - x_{if}}{\frac{W_{tv}}{W_{tf}}(K_i - 1) + 1}$
Ethane	0.12	4.90	$\frac{0.468}{1\times3.9+1} = \frac{0.468}{4.9} = 0.095$
Propane	0.32	1.65	$\frac{0.208}{1\times0.65+1} = \frac{0.208}{1.65} = 0.126$
n-Butane	0.56	0.50	$\frac{-0.28}{1\times(-0.5)+1} = \frac{-0.28}{0.5} = -0.56$

From Table 6.3.3 we obtain

$$f\left(\frac{W_{tv}}{W_{tf}}\right) = \sum_{i=1}^{n} \frac{K_i x_{if} - x_{if}}{\frac{W_{tv}}{W_{tf}}(K_i - 1) + 1} = -0.339.$$

These results indicate that both vapor and liquid are present in the system under the specified conditions of the problem. It is also clear that the correct value of (W_{tv}/W_{tf}) which will make $f(W_{tv}/W_{tf}) = 0$ will be somewhere between $(W_{tv}/W_{tf}) = 0$ and 1, since the function f appears to be almost equally positive and negative at the two limits of 0 and 1. Our *first guess* will be $(W_{tv}/W_{tf}) = 0.6$. If it leads to too negative a value of $f(W_{tv}/W_{tf})$, then we may use $(W_{tv}/W_{tf}) = 0.4$ to find the behavior of the function. See Table 6.3.4.

Table 6.3.4.

	$\frac{K_i x_{if} - x_{if}}{\frac{W_{tv}}{W_{tf}}(K_i - 1) + 1}$	
	$(W_{tv}/W_{tf}) = 0.6$	$(W_{tv}/W_{tf}) = 0.4$
Ethane	$\frac{0.468}{0.6\times3.9+1} = 0.140$	$\frac{0.468}{0.4\times3.9+1} = 0.1828$
Propane	$\frac{0.208}{0.6\times0.65+1} = 0.149$	$\frac{0.208}{0.4\times0.65+1} = 0.165$
n-Butane	$\frac{-0.28}{0.6\times(-0.5)+1} = -0.4$	$\frac{-0.28}{0.4\times(-0.5)+1} = -0.35$
	$f(W_{tv}/W_{tf}) = -0.111$	$f(W_{tv}/W_{tf}) = -0.0022$

The correct guess is around $(W_{tv}/W_{tf}) = 0.4$. Instead of trying to achieve higher accuracy by guessing $(W_{tv}/W_{tf}) = 0.41$ or 0.42 (say), we will use 0.4 and determine the values of x_{iv} and x_{il} by using equations (6.3.55) and (6.3.54).

Table 6.3.5.

	$x_{iv} = \dfrac{x_{if}}{\frac{W_{tv}}{W_{tf}} + \frac{1}{K_i}\left(1 - \frac{W_{tv}}{W_{tf}}\right)}$	$x_{il} = \dfrac{x_{if}}{\frac{W_{tl}}{W_{tf}} + K_i\left(1 - \frac{W_{tl}}{W_{tf}}\right)}$
Ethane	$\frac{0.12}{0.4+0.204\times0.6} = \frac{0.12}{0.522} = 0.23$	$\frac{0.12}{0.6+4.9\times0.4} = \frac{0.12}{2.56} = 0.0468$
Propane	$\frac{0.32}{0.4+0.606\times0.6} = \frac{0.32}{0.763} = 0.419$	$\frac{0.32}{0.6+1.65\times0.4} = \frac{0.32}{1.26} = 0.254$
n-Butane	$\frac{0.56}{0.4+2\times0.6} = \frac{0.56}{1.6} = 0.35$	$\frac{0.56}{0.6+0.5\times0.4} = \frac{0.56}{0.8} = 0.7$
	$\Sigma x_{iv} = 0.999$	$\Sigma x_{il} = 1.0008$

The results (Table 6.3.5) appear to be satisfactory.

The above example illustrates isothermal flash vaporization calculations for a multicomponent chemical mixture. We now consider a continuous chemical mixture (treated earlier in Sections 4.1.2, 3.3.1, 3.3.7.1 and 2.4 (equations (2.4.19) onwards).) A liquid, which is a *continuous chemical mixture*, may also be subjected to an isothermal flash vaporization process. The calculations for such a separation under the condition of equilibrium have been implemented by, among others, Cotterman and Prausnitz (1985) and Kehlen *et al.* (1985), employing a gamma (Γ) distribution and a Gaussian distribution, respectively, as the molecular weight density functions (see equations (4.1.33f, g)). The basic governing relations are given below, employing three molecular weight density functions: feed liquid, $f_f(M)$; vapor fraction from the flash drum, $f_v(M)$; liquid fraction from the flash drum, $f_l(M)$. We focus on species in the molecular weight range M to $M + \mathrm{d}M$.

- Vapor-liquid equilibrium in the flash chamber:

$$\mu_l(M) = \mu_v(M). \tag{6.3.66}$$

- Molar balance

$$W_{tf} f_f(M)\mathrm{d}M = W_{tv} f_v(M)\mathrm{d}M + W_{tl} f_l(M)\mathrm{d}M. \tag{6.3.67}$$

We have seen earlier (e.g. equation (4.1.33c)) how the two density functions of the two phases (vapor and liquid) are related to each other if there is vapor-liquid equilibrium. For example, if the liquid phase has a Gaussian distribution, the vapor phase will also have a Gaussian distribution, with different parameters. In a flash separator, such behavior is still valid under equilibrium conditions. However, the material balance relation (equation (6.3.67)), rewritten as

$$f_f(M) = \theta f_v(M) + (1 - \theta) f_l(M) \tag{6.3.68}$$

(where θ is the fraction of the feed flow rate coming in that is vaporized $(= W_{tv}/W_{tf})$), shows that $f_f(M)$ need *not* have the same type of molecular weight density function as the vapor fraction or the liquid fraction. Since relation (6.3.68) has to be satisfied for all molecular weight species present in the system, and $f_f(M)$ may be given, the problem is quite complex.

Approximate solution formalisms have been developed by Cotterman and Prausnitz (1985) and Kehlen *et al.* (1985) based on the assumption that the same type of density function may describe the molecular weight based distribution in all three streams. Other approximate calculation procedures have also been developed. One such procedure employs the first and second moments of each molecular weight density function and develops appropriate relations from (6.3.68) between the parameters of each density function. Suppose a Γ distribution (see (4.1.33f)) describes the behavior of all three streams. The mean and the standard deviation for each stream are given, respectively, by

$$\begin{aligned} &\text{feed: } \alpha_f\beta_f + \gamma_f;\ \alpha_f\beta_f^2; \\ &\text{vapor: } \alpha_v\beta_v + \gamma_v;\ \alpha_v\beta_v^2; \\ &\text{liquid: } \alpha_l\beta_l + \gamma_l;\ \alpha_l\beta_l^2. \end{aligned}$$

The shift parameter γ (where $f(M)$ is zero) is assumed to be the same for all streams:

$$\gamma_f = \gamma_v = \gamma_l. \tag{6.3.69}$$

Taking the first moment of equation (6.3.68),

$$Mo_f^{(1)} = \int M f_f(M)\mathrm{d}M = \theta\, Mo_v^{(1)} + (1-\theta) Mo_l^{(1)}, \tag{6.3.70}$$

$$\alpha_f\beta_f + \gamma_f = \theta\alpha_v\beta_v + (1-\theta)\alpha_l\beta_l + \theta\gamma_v + (1-\theta)\gamma_l;$$

$$\alpha_f\beta_f = \theta\alpha_v\beta_v + (1-\theta)\alpha_l\beta_l. \tag{6.3.71}$$

Taking the second moment of equation (6.3.68),

$$Mo_f^{(2)} = \int M^2 f_f(M)\mathrm{d}M = \theta\, Mo_v^{(2)} + (1-\theta) Mo_l^{(2)}, \tag{6.3.72}$$

where, for the Γ distribution,

$$Mo^{(2)} = (\alpha\beta + \gamma)^2 + \alpha\beta^2.$$

Therefore, from equation (6.3.72),

$$\begin{aligned}\left(\alpha_f\beta_f + \gamma\right)^2 + \alpha_f\beta_f^2 &= \theta(\alpha_v\beta_v + \gamma)^2 + \theta\alpha_v\beta_v^2 \\ &\quad + (1-\theta)(\alpha_l\beta_l + \gamma)^2 + (1-\theta)\alpha_l\beta_l^2,\end{aligned} \tag{6.3.73}$$

which, after simplification, yields

$$\alpha_f(\alpha_f + 1)\beta_f^2 = \theta\alpha_v(\alpha_v + 1)\beta_v^2 + (1-\theta)\alpha_l(\alpha_l + 1)\beta_l^2. \tag{6.3.74}$$

Equations (6.3.74) and (6.3.71) are the final forms of the two moment based equations. Since feed properties are known, α_f and β_f are known. Unknowns are α_v, β_v, α_l, β_l and θ.

The phase-equilibrium relation (6.3.66) provides additional relations between α_v, β_v and α_l, β_l since both phases satisfy a Γ distribution (see (4.1.33c) for the reason why both phases will have the same distribution). Cotterman and Prausnitz (1985) have provided the following results:

$$\gamma_v = \gamma_l; \qquad \alpha_v = \alpha_l; \qquad \beta_v = \beta_l/\left(1 - C^{(2)}\beta_l\right), \quad (6.3.75)$$

where $C^{(2)}$ is a function of temperature, pressure, etc., but not of M. Use of the last two results (since $\gamma_v = \gamma_l$ was already assumed) in equations (6.3.71) and (6.3.74) leads to two equations for two unknowns α_l, β_l or α_v, β_v for a given degree of vaporization, θ, the stage cut. The results for such a calculation procedure have been illustrated in Cotterman and Prausnitz (1985) for paraffins having a mean molecular weight of 100 and a variance of 800; their calculation procedure was more complicated than that presented above, since they considered a semicontinuous mixture containing 40% CO_2, the rest being a continuous mixture of paraffins. They have also considered multi-ensemble mixtures, where each ensemble is a continuous distribution of certain types of species: for example, they have considered three ensembles, one for paraffinic compounds, one for napthenic compounds and the third for aromatics.

We now briefly consider a *flash devolatilizer* (see Figure 6.3.11(b)) (Biesenberger, 1983; Meister and Platt, 1989) to remove residual monomers and solvents from a polymeric liquid. The polymeric liquid is, however, nonvolatile, whereas the other species are volatile and are to be removed from the polymer. For example, a molten viscous feed from a polystyrene reactor at around 170 °C and containing 85% polymer, the rest being the monomer styrene, ethylbenzene, cumene, styrene dimers and trimers, is pumped into a heater. The polymeric liquid and the associated species are heated up to 230 °C and introduced into the flash tank maintained under vacuum (~4-30 Torr). As the foamy polymeric mass falls to the liquid level (just as in the case of a hydrocarbon liquid in the earlier example), the monomer and the volatile solvent species escape the polymer phase into the gas phase and are withdrawn from the tank at the top. **The net direction of both phases' flow is perpendicular to the liquid surface; the force causing volatilization is also perpendicular to the liquid surface.**

At its best, the volatilizer may be said to operate such that the vapor and liquid phases are in equilibrium (Meister and Platt, 1989), exactly as in Example 6.3.1. This may be achieved if there is enough time for diffusion of the monomer/solvent through the polymer film. In reality, the rate of monomer/solvent diffusion through the foaming mass determines the extent of residual monomer in the polymeric liquid at any given vacuum level. But lack of knowledge of the interfacial area in the foamy mass precludes an exact calculation; further, the values of the diffusion coefficients in such a system are in doubt. See Meister and Platt (1989) for preliminary considerations on mass-transfer control. We will instead briefly consider here the thermodynamic limit where there is equilibrium between the vapor and the polymeric foamy liquid for the volatile solvents and monomers.

To do this, we will first use the definition of the activity, a_{ij}, of a species i in phase j (see (3.3.20b)):

$$a_{ij} = \hat{f}_{ij}/f_{ij}^0.$$

For a volatile species i in a liquid exposed to an ideal gas phase (under vacuum), $f_{ij}^0 = P_i^{sat}\Phi_i^{sat} \cong P_i^{sat}$. Further, under these conditions, $\hat{f}_{il} = \hat{f}_{iv} = p_{iv}$, so that

$$a_{il} = \frac{p_{iv}}{P_i^{sat}}. \qquad (6.3.76)$$

The activity of a volatile species i in a polymeric solution containing volume fractions φ_{pl} and φ_{il} of the polymer and species i, respectively, can be obtained from the Flory-Huggins theory as follows (Flory, 1953):

$$\ln a_{il} = \ln \varphi_{il} + \chi_{ip}\varphi_{pl}^2 + \varphi_{pl}$$
$$\Rightarrow \frac{a_{il}}{\varphi_{il}} = \exp\left(\varphi_{pl}(\chi_{ip}\varphi_{pl} + 1)\right). \qquad (6.3.77)$$

This may be rewritten using (6.3.76) as

$$\frac{p_{iv}}{P_i^{sat}} = \varphi_{il}\exp\left(\varphi_{pl}(\chi_{ip}\varphi_{pl} + 1)\right). \qquad (6.3.78)$$

Here, χ_{ip} is the Flory-Huggins interaction parameter. Under devolatilization conditions, $\varphi_{pl} \sim 1$, so that

$$p_{iv} \cong P_i^{sat}\varphi_{il}\exp(\chi_{ip} + 1) = x_{iv}P. \qquad (6.3.79)$$

If the density of the polymer and the volatile species i are ρ_p and ρ_i, respectively, the volume fraction φ_{il} may be converted into a mass fraction u_{il} to provide

$$u_{il} = \frac{\rho_i}{\rho_p}\frac{p_{iv}}{P_i^{sat}}\frac{1}{\exp\left(\chi_{ip}+1\right)} = \frac{x_{iv}P\rho_i}{P_i^{sat}\rho_p\exp\left(\chi_{ip}+1\right)}, \qquad (6.3.80)$$

where, for a very dilute solution of i in the polymer,

$$\varphi_{il} = \frac{u_{il}/\rho_i}{(u_{il}/\rho_i) + (u_{pl}/\rho_p)} \cong \frac{u_{il}/\rho_i}{(u_{pl}/\rho_p)} \cong \frac{u_{il}\,\rho_p}{\rho_i}, \qquad (6.3.81)$$

since $(u_{il}/\rho_i) << (u_{pl}/\rho_p)$ and $u_{pl} \cong 1$ in the very dilute solution. Analogous to the vapor-liquid equilibrium relation (4.1.31), where the vapor phase behaves ideally, if we use mass fraction instead of mole fraction in the liquid phase, we can rewrite (6.3.80) as

$$u_{il} = \frac{x_{iv}P}{P_i^{sat}}\frac{1}{\gamma_{il,u}^{\infty}}, \qquad (6.3.82)$$

where the infinite dilution activity coefficient on a mass fraction basis is given by

$$\gamma_{il,u}^{\infty} = \frac{\rho_p\exp\left(\chi_{ip}+1\right)}{\rho_i}. \qquad (6.3.83)$$

Based on the work by Vrentas *et al.* (1983) for polystyrene-ethylbenzene, where $\chi_{ip} = 0.35$, and similar systems, $\gamma_{il,u}^{\infty} \sim 5$, resulting in

Table 6.3.6.

Component	Feed composition (wt%)
Polystyrene	84.27
Styrene	11.61
Ethylbenzene	2.68
Cumene	0.626
Others (styrene dimer, etc.)	0.814

$$u_{il} = \frac{x_{iv}P}{5P_i^{\text{sat}}}. \tag{6.3.84}$$

Assuming that essentially all of the volatile monomers and solvents in the feed polymeric liquid are volatilized in the flash devolatilizer, the value of x_{iv} can be determined from the feed composition introduced into the device.

Example 6.3.2 In a commercial flash tank devolatilizer (Meister and Platt, 1989), the output from a polystyrene reactor having the composition (in wt%) shown in Table 6.3.6, is introduced at the rate of 5600 kg/hr at 230 °C from a heater open to the flash tank. The polymer melt is removed from the molten layer at 230 °C at the bottom of the tank by a gear pump. The vapor stream is withdrawn from the tank top by a vacuum pump, which maintains the tank vacuum at 5 Torr. Calculate the residual amount of ethylbenzene and styrene left in the withdrawn polymer melt in ppm (parts per million parts by weight). You are given the vapor pressure of the components at 230 °C in Torr: styrene, 4644; ethylbenzene, 5555; cumene, 4020.

Solution We will employ equation (6.3.84),

$$u_{il} = \frac{x_{iv}P}{5P_i^{\text{sat}}},$$

to calculate the mass fraction of both styrene and ethylbenzene, etc., left in the melt leaving the tank since we do not have data on γ_{il}^{∞} for styrene. Further properties of "others" (styrene dimer, etc.) have not been provided. Therefore we will use a material balance based only on the polymer, styrene, ethylbenzene and cumene (the dimer is supposed to have very low vapor pressure). For the polymer-free basis,

$$\text{wt\% styrene} = \frac{11.61}{11.61 + 2.68 + 0.626} \times 100 = \frac{11.61 \times 100}{14.91} = 77.8\%;$$

$$\text{wt\% ethylbenzene} = \frac{2.68 \times 100}{14.91} = 17.9\%$$

$$\text{wt\% cumene} = 100 - 77.8 - 17.9 = 4.3\%.$$

Define styrene as $i = 1$, ethylbenzene as $i = 2$ and cumene as $i = 3$. Assume now that all of the styrene, ethylbenzene and cumene are in the vapor phase. Therefore, their mole fractions in the vapor phase, x_{iv}, can be obtained from their mole fractions in the feed on a polymer-free basis. The molecular weights of styrene, ethylbenzene and cumene are, respectively, 104,106 and 120. Therefore

$$x_{1v} = \frac{77.8/104}{\frac{77.8}{104} + \frac{17.9}{106} + \frac{4.3}{120}} = \frac{0.748}{0.748 + 0.1689 + 0.0358} = \frac{0.748}{0.9527} = 0.785;$$

$$x_{2v} = \frac{0.1689}{0.9527} = 0.177; \qquad x_{3v} = 0.038.$$

Correspondingly, for styrene,

$$u_{1l} = \frac{0.785 \times 5}{5 \times 4644} = 0.000169 \approx 169\,\text{ppm};$$

for ethylbenzene,

$$u_{2l} = \frac{0.177 \times 5}{5 \times 5555} = 0.000031 \approx 31\,\text{ppm};$$

and for cumene,

$$u_{3l} = \frac{0.038 \times 5}{5 \times 4020} = 0.0000094 \approx 9.4\,\text{ppm}.$$

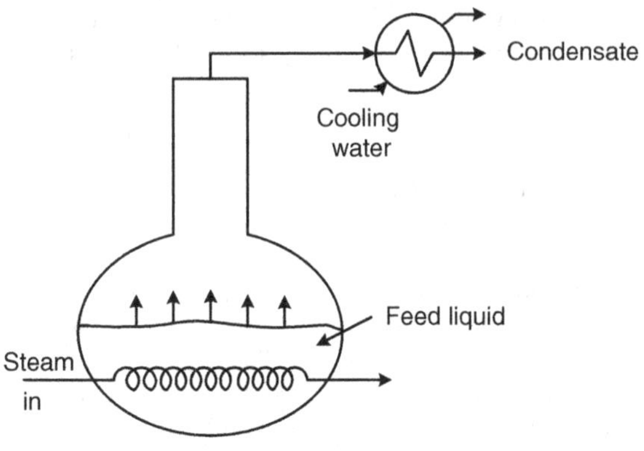

Figure 6.3.12A. Batch distillation with total condensation and no reflux.

We now consider the case of a *batch liquid feed introduced into a distillation vessel, pot or flask* (Figure 6.3.12A). The liquid is heated and vapors continue to leave perpendicularly from the top liquid surface. The escaping vapor flow is parallel to the chemical potential gradient based driving force from the liquid to the vapor phase. Identify the vapor phase as $j = 1$ (or v) and the liquid phase as $j = 2$ (or l), since the more volatile component $i = 1$ will be enriched in the vapor leaving the vessel top at a molar flow rate W_{t1}. Usually there is a condenser at the top of the batch distillation vessel; the total condensate flow rate taken out as the product is also W_{t1}. The total number of moles of liquid mixture charged into the vessel at time $t =$ 0 is $\sum_{i=1}^{n} m_{i2}^0 = m_{t2}^0$. As time progresses, m_{t2} decreases, its composition x_{i2} changes; W_{t1} also changes with time, just as x_{i1} changes with changing x_{i2}. One would like to know how the composition of the liquid changes with the changing amount of the liquid in the heated vessel. (Usually, the total pressure P is held constant; as the liquid composition changes, the temperature will also change if we assume vapor-liquid equilibrium.) Alternatively, one would like to know the condensate composition for a given liquid composition in the vessel.

If in time dt, the change in the total liquid moles for species i is -dm_{i2}, and that in the total number of moles (m_{i1}) which have escaped as vapor is dm_{i1}, then

$$-\mathrm{d}m_{i2} = \mathrm{d}m_{i1} = x_{i1} W_{t1}\,\mathrm{d}t = -\mathrm{d}(x_{i2} m_{t2}) = x_{i1}\,\mathrm{d}m_{t1} = -x_{i2}\,\mathrm{d}m_{t2} - m_{t2}\quad \mathrm{d}x_{i2}. \qquad (6.3.85a)$$

By a total mass balance over any interval dt

$$-\mathrm{d}m_{t2} = \mathrm{d}m_{t1}.$$

Introducing the total mass balance into the ith component balance, we get

$$-x_{i2}\,\mathrm{d}m_{t2} - m_{t2}\,\mathrm{d}x_{i2} = -x_{i1}\,\mathrm{d}m_{t2}$$
$$\Rightarrow m_{t2}\,\mathrm{d}x_{i2} = m_{tl}\,\mathrm{d}x_{il} = (x_{i1} - x_{i2})\mathrm{d}m_{t2} = (x_{iv} - x_{il})\mathrm{d}m_{tl}. \qquad (6.3.85b)$$

Rewrite this as

$$\frac{\mathrm{d}m_{tl}}{m_{tl}} = \frac{\mathrm{d}x_{il}}{x_{iv} - x_{il}}. \qquad (6.3.86)$$

Integrating this from $m^0_{tl} = m^0_{t2}$ to $m_{tl} = m_{t2}$, corresponding to a change in the liquid-phase mole fraction from its initial value x^0_{il} to the value x_{il} at any time t, leads to

$$\int_{m^o_{tl}}^{m_{tl}} \frac{\mathrm{d}m_{tl}}{m_{tl}} = \ln\left(\frac{m_{tl}}{m^0_{tl}}\right) = \int_{x^o_{il}}^{x_{il}} \frac{\mathrm{d}x_{il}}{x_{iv} - x_{il}}. \qquad (6.3.87)$$

This relation is known as the *Rayleigh equation* (this type of relation between the composition of the two fractions, regions, etc., will be encountered later in other separation processes). The right-hand side of this equation may be integrated if the relation between x_{iv} and x_{il} is known.

For a *binary* mixture ($i = 1,2$), if it can be assumed that the vapor leaving the liquid having a composition of x_{iv} is in equilibrium with the liquid of composition x_{il}, then, from (4.1.32a),

$$K_i = \frac{x_{iv}}{x_{il}}.$$

Alternatively,

$$x_{iv} = x_{i1} = \frac{\alpha_{12} x_{il}}{1 + (\alpha_{12} - 1)x_{il}}, \qquad (6.3.88)$$

if α_{12} is the separation factor in vapor–liquid equilibrium (VLE) defined by (4.1.21a) when $i = 1$. Introduction of these two relations between x_{iv} and x_{il} into the Rayleigh equation (6.3.87) and the assumption of constant K_i and α_{12} lead, respectively, to

$$\ln\left(\frac{m_{tl}}{m^0_{tl}}\right) = \frac{1}{K_i - 1}\int_{x^o_{il}}^{x_{il}} \frac{\mathrm{d}x_{il}}{x_{il}} = \frac{1}{K_i - 1}\ln\frac{x_{il}}{x^0_{il}} \qquad (6.3.89)$$

and

$$\ln\left(\frac{m_{tl}}{m^0_{tl}}\right) = \int_{x^o_{il}}^{x_{il}} \frac{\mathrm{d}x_{il}}{\frac{\alpha_{12}x_{il}}{1+(\alpha_{12}-1)x_{il}} - x_{il}} = \frac{1}{\alpha_{12} - 1}\left\{\ln\frac{x_{il}}{x^0_{il}} + \alpha_{12}\ln\frac{1 - x^0_{il}}{1 - x_{il}}\right\}. \qquad (6.3.90)$$

An alternative form of equation (6.3.90) is

$$\ln\left(\frac{m_{tl}}{m^0_{tl}}\right) = \left\{\ln\left(\frac{x_{il}}{x^0_{il}}\right)^{\frac{1}{\alpha_{12}-1}} + \ln\left(\frac{1 - x^0_{il}}{1 - x_{il}}\right)^{\frac{\alpha_{12}}{\alpha_{12}-1}}\right\}$$
$$\Leftrightarrow \frac{m_{tl}}{m^0_{tl}} = \left(\frac{x_{il}}{x^0_{il}}\right)^{\frac{1}{\alpha_{12}-1}}\left(\frac{1 - x^0_{il}}{1 - x_{il}}\right)^{\frac{\alpha_{12}}{\alpha_{12}-1}}. \qquad (6.3.91)$$

Note that $i = 1$ here for the more volatile species. This type of batch distillation is also called *differential distillation* or *Rayleigh distillation*.

Example 6.3.3 A 50% benzene–50% toluene liquid mixture has been introduced as a feed into a batch distillation flask. The distillation is carried out at 120 kPa total pressure. Consider the situation where 50% of the liquid mixture has been collected as the distillate. What is the composition of the distillate mixture? An averaged separation factor α_{12} of 2.40 may be assumed (where $i = 1$ is benzene) over the temperature range under consideration.

Solution We employ equation (6.3.91) and a trial-and-error approach. Here species $i = 1$ is benzene. In this case, $(m_{tl}/m^0_{tl}) = 0.5, x^0_{1l} = 0.5, \alpha_{12} = 2.40$, x_{1l} is unknown. Knowing m_{tl} and x_{1l} means that m_{tv} is known. Therefore, the distillate composition $(x_{1v})_{\text{avg}}$ may be determined from the mass balance over species $i = 1$ (benzene) as follows:

$$m^0_{tl}x^0_{1l} = m_{tl}x_{1l} + m_{tv}(x_{1v})_{\text{avg}} \Rightarrow x^0_{1l} = 0.5 = 0.5x_{1l} + 0.5(x_{1v})_{\text{avg}}.$$

In equation (6.3.91), x_{1l} is unknown. We can solve it numerically, or by straightforward trial-and-error based on an initial guess (see Table 6.3.7).

We introduce a value of $x_{1l} = 0.35$ into the overall mass balance equation:

$$0.5 = 0.5 \times 0.35 + 0.5(x_{1v})_{\text{avg}} \Rightarrow (x_{1v})_{\text{avg}} = (0.5 - 0.175)/0.5 = 0.65.$$

It is clear from Example 6.3.3 for the benzene–toluene system (see Figures 4.1.1 and 1.4.1) that very little separation is achieved in a simple batch distillation. If each

Table 6.3.7.

Guess for x_{1l}	$\left(\frac{x_{1l}}{x^0_{1l}}\right)^{\frac{1}{\alpha_{12}-1}}$	$\left(\frac{1-x^0_{1l}}{1-x_{1l}}\right)^{\frac{\alpha_{12}}{\alpha_{12}-1}}$	(m_{tl}/m^0_{tl})	Comment
0.3	0.695	0.563	0.391	too low
0.4	0.867	0.73	0.633	too high
0.35	0.778	0.636	0.4948	good

individual drop of liquid in the feed liquid mixture were to undergo partial vaporization independently, then the composition of the first vapor bubble in equilibrium with each such drop can be calculated from equation (6.3.88):

$$x_{1v} = \frac{\alpha_{12} x_{1l}}{1 + (\alpha_{12} - 1)x_{1l}} = \frac{2.4 \times 0.50}{1 + (2.4 - 1) \times 0.50} = \frac{1.2}{1 + 0.7} \cong 0.706.$$

As the benzene content of the remaining liquid decreases due to the removal of the first bubble, the benzene content of the second bubble from the remaining liquid drop will be lower than that in the first bubble, and so on. Further, if only part of the liquid lying on top of the liquid pool undergoes vaporization, the rest of the liquid will have a lower benzene content. Thus, the highest amount of vapor having the highest benzene content may be obtained if the whole batch liquid charged into the pot could somehow exist as an infinite number of drops each independently undergoing a vaporization process, with only a small fraction being vaporized.

The flash vaporization process of Figure 6.3.11(a) is a step in that direction only if the extent of vaporization in each drop is quite small so that x_{1v} is close to 0.706. Otherwise, if there is going to be a substantial amount of vaporization in the flash process (as, for example, in Example 6.3.1), then the value of x_{1v} in equilibrium with x_{1l} will be considerably lower. On the other hand, the batch distillation process will start out with values of x_{1v} substantially larger than the x_{1v} value from flash distillation (in the equilibrium limit), since x_{1l} at the beginning corresponds to the feed liquid in batch distillation. On an overall basis, therefore, batch distillation can end up with a higher vapor composition. Also, it is possible that the last drop of liquid in the pot in batch distillation is pure heavy-component liquid (i.e. $x_{2l} \cong 1$); the amount would be infinitesimally small. The flash process alone cannot ever achieve a pure liquid drop or a vapor bubble (King, 1980, p. 141). However, both of these distillation schematics are more productive than a closed separation vessel since the total amount of vapor product in the latter is quite small. **The vessel where the vapor and the liquid are in equilibrium is often identified as the ideal stage.**

A frequent application of batch distillation in pharmaceutical and fine chemical production processes involves *solvent exchange*. The solvent employed in a particular reaction step needs to be replaced by another for the next reaction step. Alternatively, the replacement solvent is needed to carry out crystallization of the final product, which is, in general, nonvolatile. If the final or intermediate product is thermally labile, then distillation is carried out at a lower temperature under partial vacuum. The conventional procedure is as follows. First, the batch is boiled down to remove a lot of the original solvent. Then the replacement solvent is added, and the resulting batch is distilled to reduce further the amount of original solvent present. One could add *additional* replacement solvent next to reduce the original solvent to the level of a low-level impurity (Chung, 1996; Gentilcore, 2002). These references provide illustrative examples of solvent exchange in batch distillation. Of course, the volatile replacement solvent is also lost by distillation.

Gentilcore (2002) has illustrated a different mode of feed introduction for solvent exchange in batch distillation, and has showed that it minimizes the loss of the volatile replacement solvent. This has been named *constant-level batch distillation*. This technique is implemented by adding the replacement solvent continuously to the pot at such a rate that the volume of the solvent mixture in the pot remains constant, hence constant-level batch distillation. We will now develop the relation corresponding to Rayleigh distillation for constant-level batch distillation, where the replacement solvent added to the still is pure species 2 and the original solvent (more volatile) is species 1. We ignore the existence of other organic compounds in the pot from a distillation point of view.

Assume that the molar volumes of the two species are identical. Then, over a small time interval, dt, the total number of moles evaporated (vapor phase $j = 1, j = v$), dm_{t1}, must equal the total number of moles of pure exchange solvent feed added ($j = f$), dm_{tf}:

$$\mathrm{d}m_{t1} = \mathrm{d}m_{tf}. \tag{6.3.92a}$$

Since the total number of moles of liquid, m_{t2}, in the still remains constant, by assumption ($j = 2, j = l$)

$$\mathrm{d}m_{t2} = 0 = \mathrm{d}m_{tl} \tag{6.3.92b}$$

in the time interval dt. Carrying out a molar balance on the more volatile species 1, since the exchange solvent feed is pure species 2, we obtain

$$\begin{aligned} -\mathrm{d}(m_{t2}x_{12}) &= -\mathrm{d}(m_{tl}x_{1l}) = -x_{1l}\,\mathrm{d}m_{tl} - m_{tl}\,\mathrm{d}x_{1l} \\ &= -m_{tl}\,\mathrm{d}x_{1l} = x_{11}\,\mathrm{d}m_{t1}. \end{aligned} \tag{6.3.92c}$$

Rewriting this equality as

$$-\frac{\mathrm{d}m_{t1}}{m_{tl}} = \frac{\mathrm{d}x_{1l}}{x_{11}} = -\frac{\mathrm{d}m_{tf}}{m_{tl}}, \tag{6.3.92d}$$

and integrating between the limits of initial liquid-phase mole fraction x_{1l}^0 and the final mole fraction x_{1l}, when m_{tf} moles of pure exchange solvent have been added since $t = 0$, we get

$$-\int_0^{m_{tf}} \frac{\mathrm{d}m_{tf}}{m_{tl}} = \int_{x_{1l}^0}^{x_{1l}} \frac{\mathrm{d}x_{1l}}{x_{11}} = -\frac{m_{tf}}{m_{tl}} \Rightarrow \frac{m_{tf}}{m_{tl}} = \int_{x_{1l}}^{x_{1l}^0} \frac{\mathrm{d}x_{1l}}{x_{11}} = \int_{x_{1l}}^{x_{1l}^0} \frac{\mathrm{d}x_{1l}}{x_{1v}} \tag{6.3.93}$$

In general, a graphical/numerical solution of the integral on the right-hand side is carried out on the basis of an available vapor–liquid relation $x_{1v} = f(x_{1l})$. If we can assume that the relative volatility/separation factor α_{12}

between the two solvents 1 and 2 is constant and employ relation (4.1.24) in the above integral, we get

$$\frac{m_{tf}}{m_{tl}} = \int_{x_{1l}}^{x_{1l}^0} \frac{1 + x_{1l}(\alpha_{12} - 1)}{\alpha_{12} x_{1l}} \mathrm{d}x_{1l}, \tag{6.3.94}$$

which leads to (Gentilcore, 2002)

$$\frac{m_{tf}}{m_{tl}} = \frac{1}{\alpha_{12}} \ln\left(\frac{x_{1l}^0}{x_{1l}}\right) + \frac{(\alpha_{12} - 1)}{\alpha_{12}}(x_{1l}^0 - x_{1l}). \tag{6.3.95}$$

Gentilcore (2002) has illustrated how the loss of the replacement solvent in the distillate can be considerably reduced by continuously adding the replacement solvent under the condition of a constant-level control of the liquid in the pot (instead of a conventional batch distillation). **This case illustrates how the mode of feed introduction influences the separation achieved, even though the basic separation mechanism and the force vs. flow pattern remain unchanged**. For a general and broad introduction to batch distillation in all forms, consult Diwekar (1996) and Seader and Henley (1998, chap. 13). Section 8.1.3 will focus on other flow vs. force configurations in batch distillation.

6.3.2.1.1 Residue curves in batch distillation The illustrated forms of the Rayleigh equations (6.3.87) and (6.3.91) allow one to relate the change in the total number of moles in the residual liquid phase with the liquid-phase composition change in simple batch distillation with no reflux and total condensation. One would like to know how the liquid-phase composition changes with time. Visualization of this composition change pathway is going to be quite illustrative. For ternary systems, such visualizations are likely to be particularly useful. The topic of *residue curve maps* is briefly introduced here to that end.

From relations (6.3.85a,b) identifying the quantitative changes taking place in simple batch distillation over a time interval dt, namely $m_{tl}\,\mathrm{d}x_{il} + x_{il}\,\mathrm{d}m_{tl} = x_{iv}\,\mathrm{d}m_{tl}$, we obtain

$$(\mathrm{d}(m_{tl}x_{il})/\mathrm{d}t) = x_{iv}(\mathrm{d}m_{tl}/\mathrm{d}t). \tag{6.3.96}$$

This may be rearranged to obtain

$$(\mathrm{d}m_{tl}/\mathrm{d}t)(x_{il} - x_{iv}) = -m_{tl}(\,\mathrm{d}x_{il}/\mathrm{d}t). \tag{6.3.97}$$

Recognizing that $\mathrm{d}t_{mv} = \mathrm{d}m_{t1} = -\mathrm{d}m_{tl} = -\mathrm{d}m_{t2}$, we get

$$(\mathrm{d}x_{il}/\mathrm{d}t) = [(\mathrm{d}m_{tv}/\mathrm{d}t)/m_{tl}](x_{il} - x_{iv}), \tag{6.3.98}$$

which describes how the liquid-phase composition, x_{il}, changes with time. The solution will allow one to follow the composition of the liquid mixture in the batch distillation vessel, *the residue*, with time.

Numerical integration of equation (6.3.98) (see Doherty and Malone (2001) for various approaches) is usually carried out to plot the time series of x_{il}, the liquid-phase mole fraction of the more volatile species. Figure 6.3.12B illustrates how x_{il} changes with time for two different types of systems. For an ideal system (e.g. a benzene–toluene system having a constant $\alpha_{12} = 2.5$ at 90 °C shown in Figure 4.1.1 with $i = 1$ for benzene) of ethanol-isopropanol, the solid lines represent how x_{il} changes with time for a constant total pressure of 1 atm, for three different initial liquid-phase compositions x_{1l}^0. The time coordinate has been appropriately modified (warped time coordinate of Doherty and Malone (2001)) in the plot to vary from 0 to 1. As time progresses, the more volatile species (e.g. ethanol) will be depleted in the liquid solution (whose boiling point will increase) until we are left with only isopropanol, which will ultimately be evaporated. Each line in the figure has an arrow showing the direction of change of the liquid-phase composition of the more volatile species, i.e. ethanol.

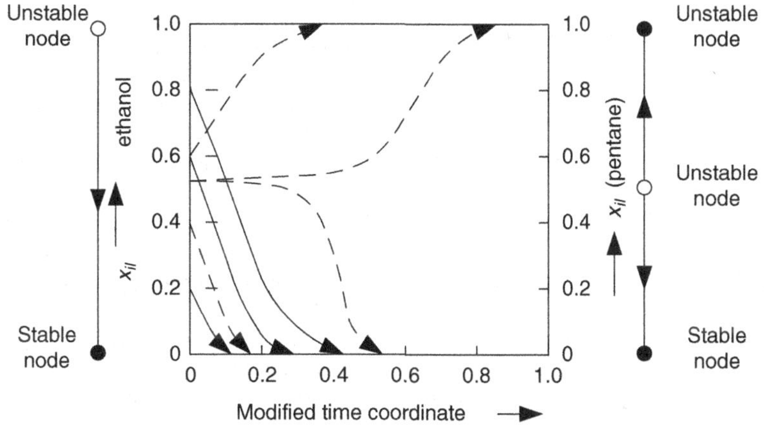

Figure 6.3.12B. Liquid residue composition curves in simple batch distillation for two types of systems: solid lines illustrate residue curves for the ideal system ethanol–isopropanol at 1 atm; dashed lines show the residue curves for the minimum-boiling azeotropic system, pentane–dichloromethane system at 1 atm. (Figure developed from figs. 5.2 and 5.4 of Doherty and Malone (2001).)

The four dashed lines in Figure 6.3.12B correspond to a minimum-boiling azeotrope (see Figure 4.1.3 for an example, isopropyl ether-isopropyl alcohol system) of pentane and dichloromethane at 1 atm total pressure with the azeotropic composition at ~0.52 mole fraction pentane (Doherty and Malone, 2001). For x_{1l}^0 values (for pentane) less than 0.52, where pentane is more volatile, the behavior is similar to that of the ethanol-isopropanol system (solid lines) in that, as time passes, the residual liquid becomes enriched in dichloromethane, the less volatile species. On the other hand, for x_{1l}^0 values higher than 0.52, the dotted lines show a totally different trajectory, with the liquid becoming enriched in pentane as time passes since the vapor-liquid equilibrium behavior has been reversed beyond the azeotropic composition; pentane has become the less volatile species.

For both types of systems the trajectory in time is illustrative. For the ethanol-isopropanol system, the trajectory (solid lines) moves toward $x_{1l} = 0$, which is identified as the *stable node*, whereas $x_{1l} = 1$ is the *unstable node* since the composition of the liquid phase is moving away from it since ethanol ($i = 1$) is the more volatile species. For the pentane-dichloromethane system, we observe two types of trajectories (dashed lines) for the two sides of the azeotropic composition; correspondingly we have two stable nodes and one unstable node.

The directional arrows in the solid and dashed lines of Figure 6.3.12B provide the progression in liquid residue composition with time for two different types of systems. This strategy becomes useful for visualizing the change in liquid residue composition via a triangular diagram for a ternary mixture in the batch distillation setup of Figure 6.3.12A. Figure 6.3.12C illustrates the *residue curve map*

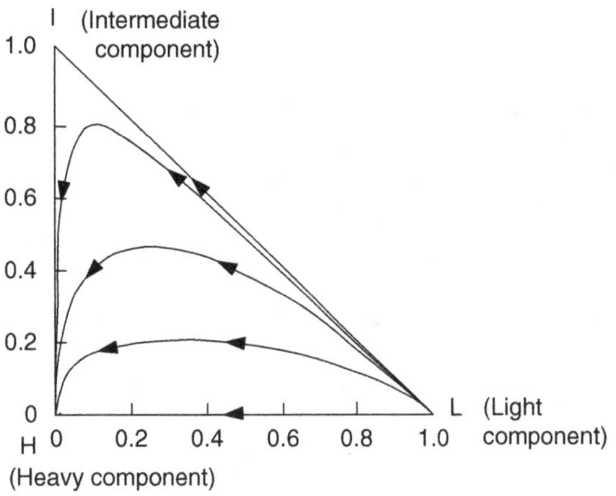

Figure 6.3.12C. Illustrative residue curves of a ternary system without any azeotrope: L represents the lightest component, I represents an intermediate boiling species, and H is the highest-boiling component. Each represents the pure component at each vertex. (After Doherty and Malone (2001).)

for an ideal ternary mixture of a light component (L), intermediate component (I) and heavy component (H) (e.g. methanol, ethanol and n-propanol, respectively, (Doherty and Malone, 2001)) via three solid lines. There is no time coordinate in the triangular diagram; the arrow in each line shows the progression in time. Since the light component, L, is the most volatile of the three species located at three vertices of this diagram, all curves start form this L vertex. As time progresses, the curves first move toward increasing concentration of I, the intermediate boiling species (I is heavier than L), reflecting increasing concentration of I in the liquid residue. As time passes, however, the concentration of H, the heavy component, keeps on increasing, and the curves turn toward the vertex H through a saddle point.

The residue curve maps for ternary mixtures having one binary minimum-boiling azeotrope are much more complicated. They are even more complex when there are multiple azeotropes. See Doherty and Malone (2001) for illustrative figures. Siirola and Barnicki (1997) illustrate residue curves for a variety of systems. Complex properties of the residue maps are analyzed in detail in Doherty and Perkins (1978) and Van Dongen and Doherty (1984).

6.3.2.2 *Liquid-liquid systems*

Bulk flow of one phase parallel to the direction of the force may be visualized for liquid-liquid systems by considering the two arrangements shown in Figure 6.3.13. In arrangement (a), the vessel is initially filled with an aqueous solution of a species *i*. To this solution is introduced, from the bottom of the vessel, an organic extraction solvent in the form of drops. Since the organic solvent is likely to be lighter than the aqueous phase, it will rise to the top and form a layer of the organic extract, from which an *organic extract* could be withdrawn continuously as shown (Treybal, 1963, p. 220). Alternatively, consider a separating funnel containing an organic-phase solution. If an aqueous extracting phase is introduced from the top (as in arrangement (b)) it is likely to be heavier, and therefore drop through the organic phase as it extracts the solute and forms a layer at the bottom of the funnel. A small fraction of it may be continuously withdrawn as the extract from the bottom of the separating funnel. The *organic raffinate* is left in the separating funnel. **In both cases, the effective force for solute partitioning is parallel to the direction of bulk flow of the phase introduced from outside**; specifically, the force is perpendicular to the *bulk interface* between the two phases. However, the transfer from the drops to the sides during their motion introduces a deviation from this model.

This type of contacting two immiscible liquid phases for extracting solutes is termed *differential extraction* (Treybal, 1963, p. 220). There are a number of ways one

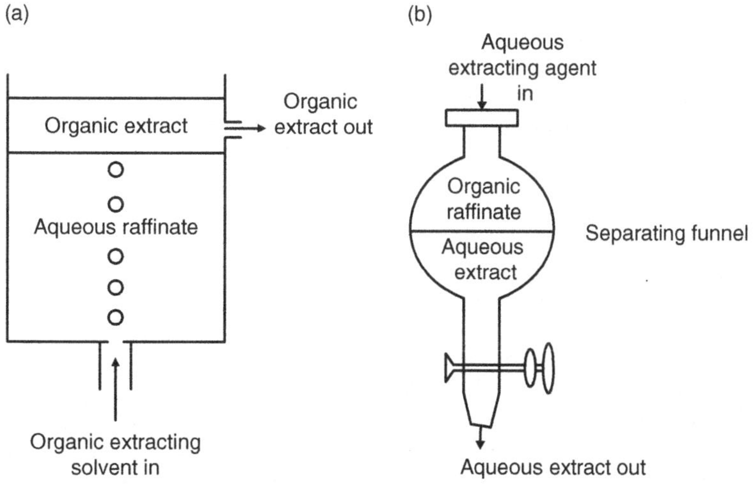

Figure 6.3.13. Two arrangements for bulk flow of one phase in another stationary phase parallel to the direction of the chemical potential driving force in immiscible liquid–liquid systems.

can analyze the separation achieved in the arrangements shown in Figures 6.3.13A and B as long as the rate at which the fresh organic or aqueous phase is introduced is small compared to the volume of the system. First, one can assume that the two phases are *completely immiscible* and *insoluble*; the phase contact merely extracts the solute i from, say, the aqueous phase to the organic phase. Second, we can assume that everywhere in the container, the two phases are in *equilibrium* via efficient mass transport. (If this efficiency is achieved by a well-mixed liquid phase, then the methodology of Section 6.4.1.2 ought to be used.)

Consider the arrangement in Figure 6.3.13A, where an organic solvent having an initial solute i concentration of C^0_{io} is introduced at a low rate into the container containing the aqueous phase. Let the volumes of the organic solvent and the aqueous phase in the vessel be V_o and V_w, respectively. If the solute i concentrations in the two phases are C_{io} and C_{iw}, respectively, then an introduction of dV_o volume of the organic solvent will lead to a small decrease $-dC_{iw}$ in the solute concentration in the aqueous phase:

$$-V_w\, dC_{iw} = \left(C_{io} - C^0_{io}\right) dV_o. \tag{6.3.99}$$

The total solvent volume added, V^t_0, added to the system over the whole process is

$$V^t_o = \int_0^{V^t_o} dV_o = V_w \int_{C_{iw}}^{C^0_{iw}} \frac{dC_{iw}}{\left(C_{io} - C^0_{io}\right)}, \tag{6.3.100}$$

where C^0_{iw} is the initial solute i concentration in the aqueous phase in the vessel and C_{iw} is the final solute i concentration

If the concentrations of solute i in the two phases at equilibrium, namely C_{io} and C_{iw}, are related by a constant distribution coefficient,

$$\kappa_{io} = \frac{C_{io}}{C_{iw}}, \tag{6.3.101a}$$

then we obtain

$$V^t_o = \frac{V_w}{\kappa_{io}} \int_{C_{iw}}^{C^0_{iw}} \frac{dC_{iw}}{\left(C_{iw} - \frac{C^0_{io}}{\kappa_{io}}\right)}. \tag{6.3.101b}$$

Integration leads to

$$\frac{\kappa_{io} V^t_o}{V_w} = \log\left(\frac{C^0_{iw} - C^o_{io}/\kappa_{io}}{C_{iw} - C^o_{io}/\kappa_{io}}\right). \tag{6.3.102a}$$

Alternatively,

$$\frac{C_{iw} - \left(C^0_{io}/\kappa_{io}\right)}{C^0_{iw} - \left(C^0_{io}/\kappa_{io}\right)} = \exp\left(-\frac{\kappa_{io} V^t_o}{V_w}\right). \tag{6.3.102b}$$

Suppose $C_{io}{}^o$ is zero; the solvent added to the vessel is solute-free. **One can never reduce C_{iw} to zero unless an infinite volume of this solute-free organic solvent is passed through the system in the configuration of bulk flow parallel to the force direction.**

Bulk flow via *direct mechanical conveying* can also allow the achievement of bulk flow parallel to the direction of the force. Consider, for example, a liquid phase $j = 2$ of volume V_2 in a vessel containing species $i = 1,2$. Add a certain volume V_1 of an immiscible liquid phase $j = 1$ on top of this liquid phase $j = 2$. This liquid phase will extract preferentially one of the species i. After equilibrium has been achieved, this top liquid phase may be decanted away and fresh liquid phase $j = 1$ of volume V_1 introduced again into the vessel. After equilibration, this top liquid phase may also be withdrawn, and fresh liquid phase j, having the same volume, may be introduced again. Such discrete equilibrium contacts between the two phases may be carried out many times.

In aqueous–organic systems, if the fresh liquid phase j happens to be aqueous, then it will likely become the denser and lower phase after introduction. If the vessel is a separatory funnel, the removal of the water phase through the bottom becomes straightforward. Thus, this introduction of new material and withdrawal can be easily visualized. The direction of phase motion/withdrawal is parallel to the direction of force between the two immiscible phases. One is interested in finding out the amount of separation achieved after, say, n such contacts between the two phases, where all of phase $j = 1$ will be collected after n contacts. We will utilize the extent of separation, ξ, to this end.

After equilibration at the end of the *first contact* between the two phases, we can relate the number of moles of species i in the two phases by (equation (1.4.2))

$$k'_{i1} = \frac{m_{i1}}{m_{i2}} = \kappa_{i1}\frac{V_1}{V_2}.$$

We **assume** that κ_{i1}, V_1 and V_2, and therefore k'_{i1}, are *independent* of the specific number of contacts. Assume the original phase (which becomes the raffinate after the process) containing the solute i is $j = 2$; the total number of moles of solute i originally in this phase is m^0_{i2}. As the first volume V_1 of the extractant is brought in, the segregation fractions after equilibration are given by

$$Y_{i1}(1) = \frac{m_{i1}(1)}{m^0_{i2}}, \quad Y_{i2}(1) = \frac{m_{i2}(1)}{m^0_{i2}};$$

$$Y_{i1}(1) = \frac{m_{i1}(1)}{m_{i1}(1)+m_{i2}(1)} = \frac{k'_{i1}}{k'_{i1}+1}, \quad Y_{i2}(1) = \frac{1}{k'_{i1}+1}. \tag{6.3.103a}$$

For the *second contact*, the total number of moles of i to start with in phase 2 (raffinate) is $m_{i2}(1)$. The number of moles of i in phases 1 and 2 after equilibration with fresh V_1 are $m_{i1}(2)$ and $m_{i2}(2)$, respectively, where $(m_{i1}(2) + m_{i2}(2)) = m_{i2}(1)$. The segregation fractions are:

$$Y_{i1}(2) = \frac{m_{i1}(2)}{m_{i1}(2)+m_{i2}(2)}, \quad Y_{i2}(2) = \frac{m_{i2}(2)}{m_{i1}(2)+m_{i2}(2)};$$

$$Y_{i1}(2) = \frac{k'_{i1}}{k'_{i1}+1}, \quad Y_{i2}(2) = \frac{1}{k'_{i1}+1}. \tag{6.3.103b}$$

If we define the segregation fraction after second contact based on the original amount of solute m^0_{i2} in phase 2 (raffinate), then

$$Y_{i1} = \frac{m_{i1}(2)}{m^0_{i2}} = \frac{m_{i1}(2)}{m_{i2}(1)} \times \frac{m_{i2}(1)}{m^0_{i2}} = \frac{m_{i1}(2)}{m_{i2}(2)}\frac{m_{i2}(2)}{m_{i2}(1)}\frac{m_{i2}(1)}{m^0_{i2}}$$

$$\Rightarrow Y_{i1} = k'_{i1}\left(\frac{m_{i2}(2)}{m_{i1}(2)+m_{i2}(2)}\frac{m_{i2}(1)}{m^0_{i2}}\right) = k'_{i1}Y_{i2}(2)Y_{i2}(1)$$

$$= \frac{k'_{i1}}{k'_{i1}+1}\frac{1}{k'_{i1}+1} = \frac{k'_{i1}}{(k'_{i1}+1)^2}. \tag{6.3.104a}$$

Similarly,

$$Y_{i2} = \frac{m_{i2}(2)}{m^0_{i2}} = \frac{m_{i2}(2)}{m_{i1}(2)+m_{i2}(2)}\frac{m_{i1}(2)+m_{i2}(2)}{m^0_{i2}}$$

$$= Y_{i2}(2)Y_{i2}(1) = \frac{1}{(k'_{i1}+1)^2}. \tag{6.3.104b}$$

If we now carry on a third contact by bringing in fresh V_1 to contact V_2 after second contact (having a total number of moles $m_{i2}(2)$ of species i), the molar distribution now consists of $m_{i1}(3)$, $m_{i2}(3)$, where $(m_{i1}(3) + m_{i2}(3)) = m_{i2}(2)$. Therefore, based on these three contacts, the segregation fractions are

$$Y_{i1} = \frac{m_{i1}(3)}{m^0_{i2}} = \frac{m_{i1}(3)}{m_{i2}(3)}\frac{m_{i2}(3)}{m_{i2}(2)}\frac{m_{i2}(2)}{m_{i2}(1)}\frac{m_{i2}(1)}{m^0_{i2}}$$

$$= k'_{i1}Y_{i2}(3)Y_{i2}(2)Y_{i2}(1) = \frac{k'_{i1}}{(k'_{i1}+1)^3}, \tag{6.3.104c}$$

$$Y_{i2} = \frac{m_{i2}(3)}{m^0_{i2}} = \frac{m_{i2}(3)}{m_{i2}(2)}\frac{m_{i2}(2)}{m_{i2}(1)}\frac{m_{i2}(1)}{m^0_{i2}} = \frac{1}{(k'_{i1}+1)^3}. \tag{6.3.104d}$$

By a process of induction, we may now conclude that, after n contacts, with n volumes of fresh V_1 contacting V_2, the final values of Y_{i1} and Y_{i2} are

$$Y_{i1} = \frac{k'_{i1}}{(k'_{i1}+1)^n} \quad \text{and} \quad Y_{i2} = \frac{1}{(k'_{i1}+1)^n}, \tag{6.3.104e}$$

where Y_{i1} is the fraction of solute which is present in the pooled extract volume nV_1 and Y_{i2} is the fraction of solute left in the raffinate of volume V_2. The extent of separation ξ between two solutes $i = 1,2$ can be calculated after such a process as follows:

$$\xi = \text{abs}|Y_{22} - Y_{12}| = \left|\frac{1}{(k'_{21}+1)^n} - \frac{1}{(k'_{11}+1)^n}\right|. \tag{6.3.105}$$

The number of such contacts, n_{opt}, which yields an optimum separation between the two solutes may be obtained by differentiating ξ with respect to n (see Problem 1.5.1) (Treybal, 1963, p. 290; Rony, 1969):

$$\partial\xi/\partial n = 0, \tag{6.3.106}$$

which results in

$$n_{opt} = \ln\left[\frac{\ln(1+k'_{11})}{\ln(1+k'_{21})}\right] \Big/ \ln\left(\frac{(1+k'_{11})}{(1+k'_{21})}\right). \tag{6.3.107}$$

One could employ a similar strategy to find out the volume ratio (V_1/V_2) which yields the optimum extent of separation:

$$\frac{\partial \xi}{\partial (V_1/V_2)} = 0 = \frac{\partial}{\partial (V_1/V_2)}\left[\frac{1}{\left(\kappa_{21}\frac{V_1}{V_2}+1\right)^n}\right]$$

$$= \frac{\partial}{\partial (V_1/V_2)}\left[\frac{1}{\left(\kappa_{11}\frac{V_1}{V_2}+1\right)^n}\right].$$

The resulting value of the optimum ratio of (V_1/V_2) is

$$(V_1/V_2)_{\text{opt}} = \frac{(\kappa_{11}/\kappa_{21})^{1/(n+1)} - 1}{\kappa_{11} - \kappa_{21}(\kappa_{11}/\kappa_{21})^{1/(n+1)}}, \qquad (6.3.108a)$$

which yields the highest extent of separation after n contacts. This result is also expressed as (Treybal, 1963, p. 290)

$$\kappa_{11}(V_1/V_2)_{\text{opt}} = \frac{(\kappa_{11}/\kappa_{21})\left[(\kappa_{11}/\kappa_{21})^{1/(n+1)} - 1\right]}{(\kappa_{11}/\kappa_{21}) - (\kappa_{11}/\kappa_{21})^{1/(n+1)}}. \qquad (6.3.108b)$$

It is important to note that Treybal (1963, p. 290) has obtained such results for what are called cross current (or cocurrent) multiple contacts. However, the results and analysis are equally valid for the configuration of phase contact selected here.

6.3.2.3 *Solid-liquid systems: zone melting and normal freezing*

In Section 3.3.7.5 we identified three types of liquid-solid equilibrium:

(1) solid solution in equilibrium with a molten mixture;
(2) crystallization equilibrium between a crystal and a mother liquor which is a solution of the crystal material;
(3) leaching of a solute from a solid mixture by a solvent of the solute.

In the first case, there is no external solvent. When the molten mixture freezes, solid crystals are formed. The variety of techniques and processes employed to obtain purified crystals/solids or crystals/solids having the appropriate amount of dopant/impurity are considered to be examples of *fractional solidification* (Zief and Wilcox, 1967). The system may be binary or may have three or more components. On the other hand, crystallization processes always involve a solvent, which has the solute(s) in solution. When one solute crystallizes out of the solution, whereas the other(s) remain(s) in solution, the process is called *fractional crystallization*.

Of the three general kinds of solid-liquid systems described in Section 3.3.7.5, a solid solution in equilibrium with a molten mixture or a solid that melts without decomposition and low vapor pressure, has been separated/purified on a large scale employing bulk flow *parallel* to the direction of the force. The force in this case is due to the chemical potential difference of a species between the melt and the solid/solid solution in contact with it. There are many systems in which an impurity in the solid phase/solution prefers the melt. For example, ice (pure water) separates when seawater is frozen. Table 4.1.7 illustrates the values of the distribution coefficients, κ_{is}, for a variety of impurities in silicon and gallium arsenide, two of the most important semiconductor materials. For almost all such impurities, $\kappa_{is} < 1$. Thus, a melt in contact with the solid will be enriched in the impurity. Conversely, if a melt is frozen, the first solid frozen out will have an impurity concentration $C_{is} = \kappa_{is} C_{il} < C_{il}$, as shown in Figure 6.3.14 (a), which illustrates the temperature-concentration diagram, with the liquidus and solidus lines corresponding to an impurity i whose $\kappa_{is} < 1$.

However, if a large volume of solid and liquid are present together, the equilibrium distribution of an impurity between the two phases is almost impossible to achieve since the diffusional rates in a solid are orders of magnitude smaller. To circumvent this problem and exploit the phenomena of $\kappa_{is} < 1$ for large solid samples and particular impurities, bulk motion is generally introduced in the solid-melt system by two alternative methods: *zone melting* and *normal freezing*. Both are considered important components of a number of techniques of crystallization from a melt, which are known as fractional solidification (Zief and Wilcox, 1967). We consider zone melting first.

In the zone melting method introduced by Pfann (1966), a small section ("zone") of a long sample is melted by a number of heating techniques at any instant of time; the molten zone is slowly made to traverse the whole length L of the sample (Figure 6.3.14(b)). As the molten section refreezes, it rejects the impurity into the adjacent molten section; consequently, the impurities are pushed to the furthest end of the long sample, which is subjected to freezing at the very end (Figure 1.4.3). To prevent contamination by a container, the floating zone technique is used, whereby the zone is held up by surface tension. The zone is kept molten by a radio frequency source (Gandhi, 1983), often in a vertical position; electron beam, induction heating, etc., are also used.

In the zone melting process, at any given time the whole sample/rod is solid except for the small zone, which is molten. In the *normal freezing process*, the whole rod, to start with, is molten. It is slowly pulled out of the heated region, allowing the pulled-out section to freeze and reject the impurity into the adjoining molten section (Figure 6.3.14(c)). The bulk motion of the rod as it is being pulled out of the heater is clearly in a direction parallel to the direction of the chemical potential gradient at the solid-melt interface.

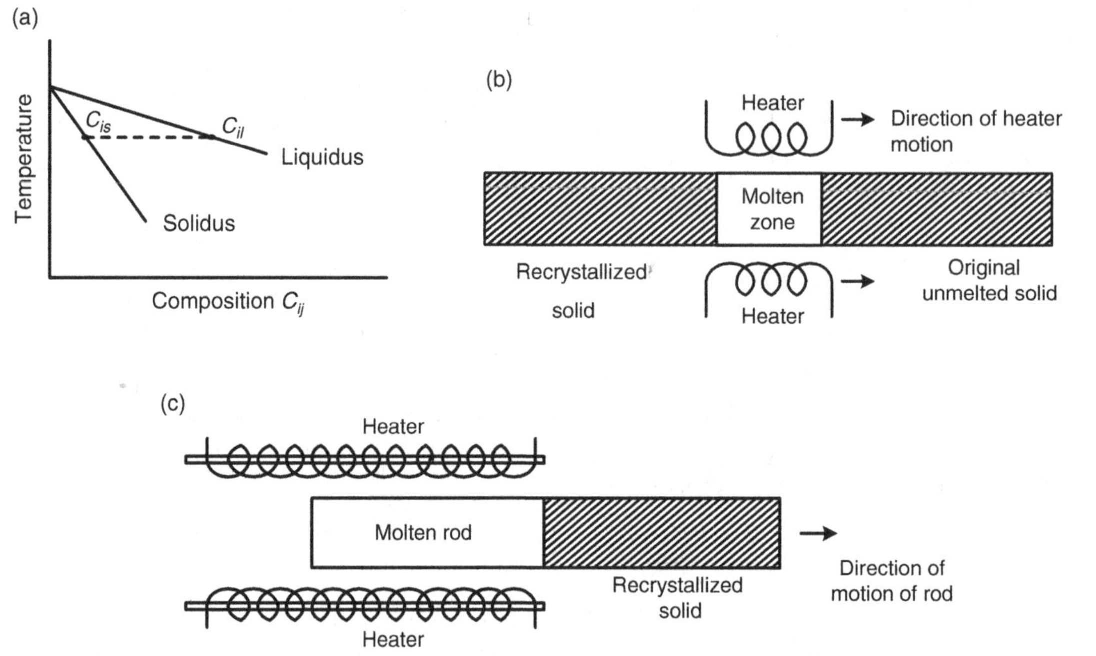

Figure 6.3.14. (a)Temperature-concentration diagram for an impurity whose $\kappa_{is} < 1$: solid-melt system; (b) schematic of zone melting of a solid rod/bar by traversing the molten zone along the rod length by moving the heating band around the rod at a controlled velocity; (c) normal freezing technique for purifying a rod/bar by slowly pulling it out of the heated section.

At first sight, there is no bulk motion of the rod in the zone melting process (Figure 6.3.14(b)), where a heating band around the rod is moved at a given velocity. However, as the heating band moves, so does the small molten zone in the rod along with it; correspondingly, the freezing interface at the back of the molten zone advances in the same direction at the same velocity. Thus, for the zone melting process also, we have a case of bulk motion parallel to the direction of the chemical potential gradient across the two solid-melt interfaces.

In the zone melting process, the molten zone slowly advances along with the heating band along the rod length (Figure 6.3.15(a)). As a small section of length dz from the molten zone freezes at the trailing end of the zone, it rejects the impurities into the nearby molten liquid. If the mixing in the molten zone (of length l) were effective, these impurity molecules would be rapidly mixed with the bulk of the melt, which will then have a uniform impurity concentration all along the length l of the molten zone. Using such a basic assumption, it is possible to develop a theoretical estimate of the impurity concentration profile along the length of the rod after the heater has passed along the rod once, provided the following assumptions are also made: the impurity distribution coefficient is constant; the densities of the melt and the solid are essentially the same; the diffusion coefficient of the impurity in the solid is zero; there is no variation in impurity concentration along the cross section of the rod.

We will work with the mass fractions u_{il} and u_{is} of the impurity in the melt and the solid phase, respectively. The impurity distribution coefficient κ'_{is} is given by

$$\kappa'_{is} = \frac{u_{is}}{u_{il}} = \frac{C_{is}\rho_l}{C_{il}\rho_s} = \kappa_{is}\frac{\rho_l}{\rho_s}. \tag{6.3.109a}$$

Due to the assumption of $\rho_l \cong \rho_s$, we know that $\kappa'_{is} \cong \kappa_{is}$ here. Let the initial impurity mole fraction in the solid be u_{io}. As the molten zone advances (Figure 6.3.15(a)), section ABB′A′ of length dz freezes and a corresponding section CDD′C′ of length dz melts. The total mass of impurity in the molten zone ABCD is $u_{il}\rho_l lA$, where A is the cross-sectional area of the rod. The total amount of impurity in the solid zone CDC′D′ (before melting) is $u_{io}\rho_s A\,\mathrm{d}z$. As the zone advances, the total amount of impurity in the region ABC′D′ ($= u_{il}\rho_l lA + u_{io}\rho_s A\,\mathrm{d}z$) does not change, but is redistributed (there being no impurity diffusion in the solids on the two sides) into the molten zone A′B′C′D′ ($= \{(u_{il} + \mathrm{d}u_{il})\,\rho_l lA\}$) and the refrozen zone ABB′A′ ($= \{(\kappa'_{is}u_{il}\rho_s A\,\mathrm{d}z\}$):

$$u_{il}\rho_l lA + u_{io}\rho_s A\mathrm{d}z = (u_{il} + \mathrm{d}u_{il})\rho_l lA + \kappa'_{is}u_{il}\rho_s A\mathrm{d}z \tag{6.3.109b}$$

$$\Rightarrow (u_{io} - \kappa'_{is}u_{il})\rho_s = \rho_l l\frac{\mathrm{d}u_{il}}{\mathrm{d}z}. \tag{6.3.109c}$$

Rewrite this as

$$-\kappa'_{is}\frac{\rho_s}{\rho_l}\frac{\mathrm{d}z}{l} = \frac{\mathrm{d}(\kappa'_{is}u_{il})}{(\kappa'_{is}u_{il} - u_{io})}, \tag{6.3.109d}$$

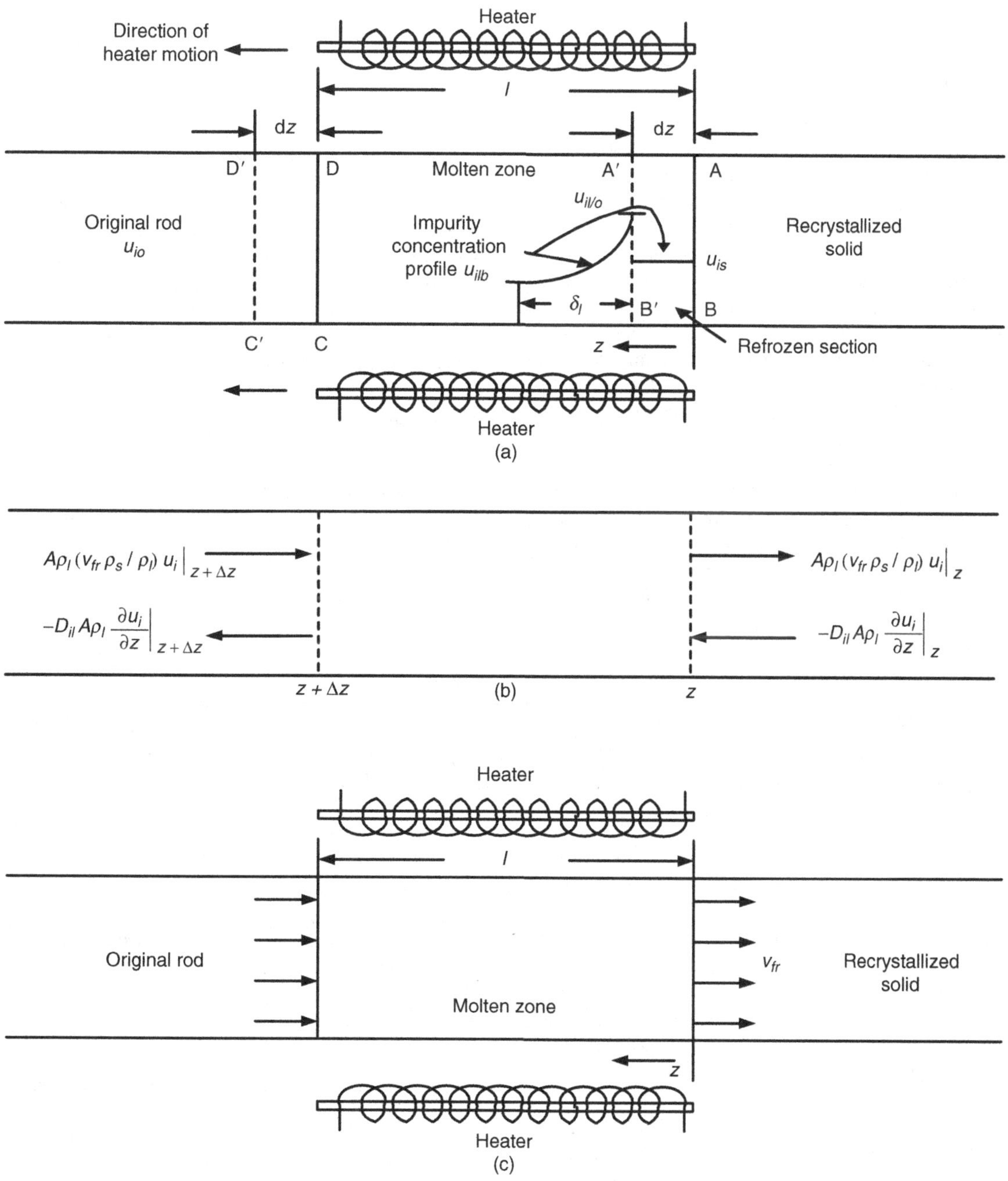

Figure 6.3.15. (a) Zone melting: impurity concentration profile in melt near the freezing interface; (b) a small element in the molten zone illustrating various fluxes (coordinate system based on part (c)); (c) overall configuration in a coordinate system fixed in the heater.

and integrate to obtain

$$-\kappa'_{is}\frac{\rho_s}{\rho_l}\frac{z}{l} = \ln(\kappa'_{is}u_{il} - u_{io}) + \text{constant}. \qquad (6.3.109e)$$

At $z = 0$, when melting has just started, $u_{il} = u_{io}$. Therefore

$$\ln\frac{(\kappa'_{is}u_{il} - u_{io})}{(\kappa'_{is}u_{io} - u_{io})} = -\kappa'_{is}\frac{\rho_s}{\rho_l}\frac{z}{l}. \qquad (6.3.109f)$$

This may be rearranged as follows:

$$\kappa'_{is}\frac{u_{il}}{u_{io}} = 1 - (1 - \kappa'_{is})\exp\left(-\kappa'_{is}\frac{\rho_s}{\rho_l}\frac{z}{l}\right). \qquad (6.3.109g)$$

However, $u_{is} = \kappa'_{is} u_{il}$, so

$$\frac{u_{is}}{u_{io}} = 1 - (1 - \kappa'_{is})\exp\left(-\kappa'_{is}\frac{\rho_s}{\rho_l}\frac{z}{l}\right). \tag{6.3.109h}$$

As long as $\kappa'_{is} \cong \kappa_{is}$, this is reduced to the expression employed in Example 1.4.3, which illustrates also how the impurity mass fraction is distributed along the length of the rod: for $\kappa'_{is} < 1$, the lowest impurity mass fraction is at the starting end and the highest is at the other end.

This derivation was based on the basic assumption that the liquid in the molten zone is well mixed. However, in almost all cases, the mixing effectiveness in the viscous melt is very poor. Essentially, molecular diffusion is the only mechanism available. The rejected impurity molecules develop a concentration profile, which decreases from a high of $u_{il/o}$ (Figure 6.3-15(a)) at the freezing interface to the lower value of u_{ilb} ($<u_{il/o}$) in the bulk of the molten zone at a distance δ_l from the freezing interface. The δ_l is identified as the boundary layer thickness (Burton *et al.*, 1953). As a result, the impurity concentration in the refrozen zone of thickness dz is higher (than what it would have been if the zone were completely mixed) if we assume it to be in equilibrium with the impurity concentration in the melt at the freezing interface. The nature of this impurity concentration profile (Figure 6.3.15(a)) will depend on the impurity partition coefficient, the impurity diffusion coefficient in the melt and the velocity of zone travel, among other things. A solution of the impurity mass-transfer problem in the molten zone will now be developed (Wilcox and Wilke, 1964; Wilcox, 1967).

Consider now a differential element of the molten zone of thickness Δz at location z (Figure 6.3.15(b)). There are no chemical reactions and/or external forces present in the molten zone. Assume a plane interface of area A at both z and $z + \mathrm{d}z$; further, assume that there are no convection and concentration variations in the x- and y-directions at any z. If we assume that the mixture has a constant total mass density ρ_t (valid for a dilute mixture), we get (from Table 6.2.2, equation (6.2.5p), Tables 3.1.3B and 3.1.3C, $\boldsymbol{v}^* = \boldsymbol{v}$, $\boldsymbol{U}_i = 0$ and $C_{tl} =$ constant)

$$\frac{\partial \rho_{il}}{\partial t} + v_z\frac{\partial \rho_{il}}{\partial z} = D_{il}\frac{\partial^2 \rho_{il}}{\partial z^2}. \tag{6.3.110a}$$

Dividing ρ_{il} by ρ_t to get the impurity mass fraction in the melt, u_{il}, we obtain

$$\frac{\partial u_{il}}{\partial t} + v_z\frac{\partial u_{il}}{\partial z} = D_{il}\frac{\partial^2 u_{il}}{\partial z^2}. \tag{6.3.110b}$$

To determine the magnitude and direction of the velocity v_z of the melt, consider the following approach. Fix the xyz coordinate system on the heating band; in this frame of reference, materials flow from left to right in the negative z-direction (Figure 6.3.15(c)). If the freezing rate (therefore the velocity of the freezing interface) is v_{fr} (in cm/s), then the magnitude of the velocity of the melt in the negative z-direction is v_{fr} (ρ_s/ρ_l). This velocity is also utilized in Figure 6.3.15(b). Equation (6.3.110b) may therefore be rewritten as

$$\frac{\partial u_{il}}{\partial t} = v_{fr}\frac{\rho_s}{\rho_l}\frac{\partial u_{il}}{\partial z} + D_{il}\frac{\partial^2 u_{il}}{\partial z^2}. \tag{6.3.110c}$$

This equation may also be developed by considering the rates at which different methods of introduction and withdrawal of impurity i take place across the boundaries of the differential molten element at z and $z + \mathrm{d}z$ (Figure 6.3.15(b)) inside the molten zone:

$$\overbrace{\left[\underbrace{A\rho_t\left(v_{fr}\frac{\rho_s}{\rho_t}\right)u_{il}\Big|_{z+\mathrm{d}z}}_{\text{introduction}} \underbrace{- A\rho_t\left(v_{fr}\frac{\rho_s}{\rho_t}\right)u_{il}\Big|_{z}}_{\text{removal}}\right]}^{\text{convective}} + \overbrace{\left[\underbrace{-D_{il}A\rho_t\frac{\partial u_{il}}{\partial z}\Big|_{z}}_{\text{introduction}} \underbrace{-\left(-D_{il}A\rho_t\frac{\partial u_{il}}{\partial z}\Big|_{z+\mathrm{d}z}\right)}_{\text{removal}}\right]}^{\text{diffusive}} = \underbrace{\rho_t\frac{\partial u_{il}}{\partial t}A\Delta z}_{\text{rate of accumulation}}. \tag{6.3.111a}$$

Dividing this equation by $A\rho_l\Delta z$ and taking the limit $\Delta z \to 0$, we obtain

$$\left(v_{fr}\frac{\rho_s}{\rho_l}\right)\frac{\partial u_{il}}{\partial z} + D_{il}\frac{\partial^2 u_{il}}{\partial z^2} = \frac{\partial u_{il}}{\partial t}. \tag{6.3.111b}$$

The boundary conditions and the initial condition needed for the solution of equation (6.3.110c) will now be developed. Consider the mass-transfer rates at the freezing interface and the melting interface at the two ends of the molten zone in the rod in a coordinate system fixed on the heating band (Figure 6.3.16). At $z = 0$ (the freezing interface),

$$v_{fr}A\rho_s u_{il} = v_{fr}A\rho_s u_{is} - D_{il}A\rho_l\frac{\partial u_{il}}{\partial z}$$

$$\Rightarrow v_{fr}\frac{\rho_s}{\rho_l}(u_{il} - u_{is}) = -D_{il}\frac{\partial u_{il}}{\partial z}. \tag{6.3.112a}$$

Also, at $z = 0$,

$$\frac{u_{is}}{u_{il}} = \kappa'_{is}, \tag{6.3.112b}$$

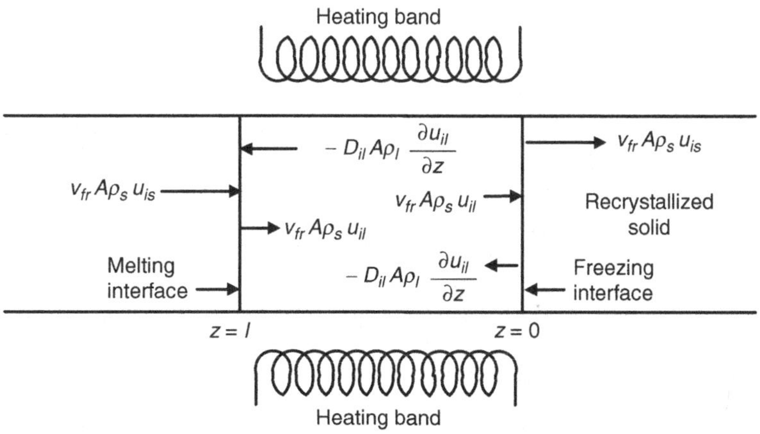

Figure 6.3.16. Different mass fluxes of the impurity at the freezing interface and the melting interface in zone melting in a frame of reference moving with the heating band.

where κ'_{is} is the impurity distribution coefficient. This quantity κ'_{is} is related to that based on molar concentration, namely κ_{is}, by

$$\kappa'_{is} = \frac{\rho_l}{\rho_s}\kappa_{is}. \tag{6.3.112c}$$

At $z = l$ (the melting interface),

$$v_{fr}A\rho_s u_{il} = v_{fr}A\rho_s u_{is} - D_{il}A\rho_l\frac{\partial u_{il}}{\partial z} \Rightarrow v_{fr}\frac{\rho_s}{\rho_l}(u_{il} - u_{is}) = -D_{il}\frac{\partial u_{il}}{\partial z}. \tag{6.3.112d}$$

At $t = 0$,

$$u_{il} = u_{io}, \tag{6.3.112e}$$

where u_{io} is the initial impurity concentration in the rod.

An exact analytical solution of the above set of equations is available for the case of a constant distribution coefficient (Wilcox and Wilke, 1964). It is quite tedious. For the sake of illustration, we will take a simplified approach. If we assume a pseudo steady state condition, equation (6.3.110c) is reduced to

$$v_{fr}\frac{\rho_s}{\rho_l}\frac{\partial u_{il}}{\partial z} + D_{il}\frac{\partial^2 u_{il}}{\partial z^2} = 0. \tag{6.3.113}$$

Following Wilcox and Wilke (1964), we define the following nondimensional quantities:

$$u_{il}^+ = \frac{u_{il}}{u_{io}}; \qquad \eta = \frac{z v_{fr}\rho_s}{D_{il}\rho_l}. \tag{6.3.114}$$

In terms of these nondimensional quantities, equation (6.3.113) is reduced to

$$\frac{\partial u_{il}^+}{\partial \eta} + \frac{\partial^2 u_{il}^+}{\partial \eta^2} = 0 = \frac{\mathrm{d}u_{il}^+}{\mathrm{d}\eta} + \frac{\mathrm{d}^2 u_{il}^+}{\mathrm{d}\eta^2}. \tag{6.3.115}$$

The boundary and initial conditions are changed as follows:

$$\text{at } \eta = 0, \qquad u_{il}^+ - u_{is}^+ = -\frac{\partial u_{il}^+}{\partial \eta} = -\frac{\mathrm{d}u_{il}^+}{\mathrm{d}\eta}; \tag{6.3.116a}$$

$$\text{at } \eta = \frac{l v_{fr}\rho_s}{D_{il}\rho_l}, \qquad u_{il}^+ - \frac{u_{is}}{u_{io}} = -\frac{\partial u_{il}^+}{\partial \eta} = -\frac{\mathrm{d}u_{il}^+}{\mathrm{d}\eta}. \tag{6.3.116b}$$

We can conveniently assume that u_{is} at $z = l$ is equal to u_{io}, the initial impurity concentration. Therefore, the second boundary condition (6.3.116b) is changed to

$$\text{at } \eta = \frac{l v_{fr}\rho_s}{D_{il}\rho_l}, \qquad 1 - u_{il}^+ = \frac{\partial u_{il}^+}{\partial \eta} = \frac{\mathrm{d}u_{il}^+}{\mathrm{d}\eta}. \tag{6.3.116c}$$

Since we have assumed a pseudo steady state, the initial condition (6.3.112e) is no longer needed. A general solution for the ordinary differential equation (6.3.115) is

$$u_{il}^+ = c_1 + c_2\exp(-\eta). \tag{6.3.116d}$$

To determine the constants c_1 and c_2, employ the two boundary conditions at $\eta = 0$ and $\eta = (l v_{fr}\rho_s/D_{il}\rho_l)$:

(1) At $\eta = 0$, from equation (6.3.116a),

$$c_2\exp(-\eta) = c_2 = c_1 + c_2\exp(-0) - \kappa'_{is}(c_1 + c_2\exp(-0)) = c_1 + c_2 - \kappa'_{is}(c_1 + c_2). \tag{6.3.116e}$$

(2) At $\eta = (l v_{fr}\rho_s/D_{il}\rho_l) = \eta_l$, from equation (6.3.116c),

$$1 - c_1 - c_2\exp(-\eta_l) = -c_2\exp(-\eta_l) \Rightarrow c_1 = 1. \tag{6.3.116f}$$

Substituting this value of c_1 into relation (6.3.116e), we get

$$c_2 = (1 + c_2)(1 - \kappa'_{is}) \Rightarrow c_2 = (1 - \kappa'_{is})/\kappa'_{is}. \tag{6.3.116g}$$

The required solution for the impurity mass fraction profile in the molten zone is then given by

$$u_{il}^{+} = \frac{u_{il}}{u_{io}} = 1 + \frac{1-\kappa'_{is}}{\kappa'_{is}} \exp\left(-\frac{z v_{fr} \rho_s}{D_{il} \rho_l}\right). \qquad (6.3.117)$$

Note that at $z = l$ (the melting interface), $u_{il} \neq u_{io}$; however, if $(l v_{fr} \rho_s / D_{il} \rho_l) < 0.1$, then $u_{il} \cong (u_{io}/\kappa'_{is})$, corresponding to an *almost completely mixed* molten zone. In such a case, the velocity of the interface movement due to the heater motion is quite low compared to the rate of diffusive dispersal/mixing of the impurities: a maximum amount of purification takes place in the solid rod. When $(l u_{fr} \rho_s / D_{il} \rho_l) > 5$, u_{il} is essentially equal to u_{io}; no separation/purification takes place. Thus, the above solution provides a perspective on the practical range of interface velocity/heater velocity to achieve separation in zone melting/refining, given the zone length and the mixing rate as provided by D_{il}. **Although bulk motion parallel to the force direction is essential for purifying the solid rod, too rapid a bulk motion compared to the rate at which impurities rejected at the phase interface are transported away from the phase interface is counterproductive. In fact, at higher velocities, separation is lost.**

In the zone melting processes, a more likely achieved condition is illustrated via the impurity concentration profile in the molten zone in Figure 6.3.15(a): the impurity mass fraction u_{il} changes from the value $u_{il}|_o$ at the freezing interface (where it is assumed to be in equilibrium with the solid mass fraction $u_{is}|_o$ in the newly frozen section ABB′A′ via $u_{is}|_o = \kappa'_{is} u_{il}|_o$) to u_{ilb}, the impurity mass fraction in the bulk of the melt at a distance δ_l (usually $<< l$), the *boundary layer thickness*. Ideally, if the complete molten zone is well mixed, then

$$u_{is}|_o = \kappa'_{is} u_{ilb}. \qquad (6.3.118)$$

In the more common cases of *partial liquid mixing* with a boundary layer present near the freezing interface, an *effective partition coefficient*, $\kappa'_{is,\text{eff}}$ is defined as

$$\kappa'_{is,\text{eff}} = \frac{u_{is}|_o}{u_{ilb}} = \frac{\kappa'_{is} u_{il}|_o}{u_{ilb}} = \frac{\kappa'_{is} \dfrac{u_{il}|_o}{u_{io}}}{\dfrac{u_{ilb}}{u_{io}}}$$

$$= \frac{\kappa'_{is}\left(1 + \dfrac{1-\kappa'_{is}}{\kappa'_{is}}\right)}{1 + \dfrac{1-\kappa'_{is}}{\kappa'_{is}} \exp\left(-\dfrac{z|_b v_{fr} \rho_s}{D_{il} \rho_l}\right)}$$

$$\Rightarrow \kappa'_{is,\text{eff}} = \frac{\kappa'_{is}}{\kappa'_{is} + (1-\kappa'_{is}) \exp\left(-\dfrac{\delta_l v_{fr} \rho_s}{D_{il} \rho_l}\right)}, \qquad (6.3.119a)$$

where we have used (6.3.117). This quantity $\kappa'_{is,\text{eff}}$ is always larger than κ'_{is}, as long as $\kappa'_{is} < 1$. Therefore u_{is} is higher in the newly frozen solid due to inadequate mixing in the molten zone near the freezing interface. Correlations for estimating the boundary layer thickness, δ_l, are reviewed in Wilcox (1967). Typical values of parameters and transport and other coefficients are as follows: $v_{fr} \sim 2$ mm/hr - 2 mm/min; $D_{il} \sim 2\text{–}5 \times 10^{-5}$ cm^2/s; $\delta_l \sim 0.02 - 0.1$cm; $l \sim 1\text{–}2$ cm. How the value of $\kappa'_{is,\text{eff}}$ differs from κ'_{is} as a function of the parameter $(\delta_l v_{fr} \rho_s / D_{il} \rho_l)$ is illustrated in Figure 6.3.17 for a variety of values of κ'_{is}, the equilibrium impurity distribution coefficient. The dimensionless ratio $v_{fr}/(D_{il}/z)$, or $v_{fr}/(D_{il}/\delta_l)$, where (D_{il}/δ_l) may be defined as a mass-transfer coefficient, appears to be a key parameter in zone melting, and, we shall see elsewhere in this chapter, in other separation techniques where the bulk flow is parallel to the direction of the force acting on species *i*.

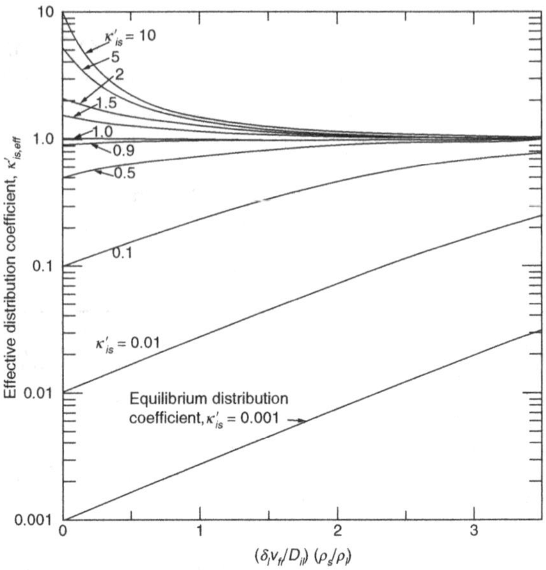

Figure 6.3.17. Effect of parameter $(\delta_l v_{fr}/D_{il})(\rho_s/\rho_l)$ on effective distribution coefficient for partial liquid mixing in molten zone, according to Eq. (6.3.119a.) (After Wilcox (1967).)

In zone melting, if the partial liquid mixing model can be employed, then the impurity mass fraction in the refrozen solid may be obtained from expression (6.3.109h) for the totally mixed case by replacing κ'_{is} by $\kappa'_{is,\text{eff}}$:

$$\frac{u_{is}}{u_{io}} = 1 - (1 - \kappa'_{is,\text{eff}}) \exp\left(-\kappa'_{is,\text{eff}} \frac{\rho_s}{\rho_l} \frac{z}{l}\right). \qquad (6.3.119b)$$

The amount of purification achieved in a single pass of the heating band over the rod is insufficient in many applications. The heating band is generally passed over the rod length a number of times; each pass purifies the bulk of the rod length substantially and concentrates the impurity at the finishing end. Figure 6.3.18 illustrates how the impurity mass fraction, u_{is}, normalized by the starting impurity mass fraction, u_{io}, varies in the solid rod from the starting end ($z = 0$) to the finishing end ($z = L$) for the case of $\kappa'_{is,\text{eff}} \sim 0.1$ as the number of heating passes increases. After an infinite number of heating passes, the *ultimate distribution* of impurity mass fraction, u_{is}^{∞}, is achieved. This distribution cannot be improved any further by additional heating passes: from a practical point of view, this implies the existence of the following relation between N and $N+1$ passes:

$$u_{is}^{N} = u_{is}^{N+1} = \cdots = u_{is}^{\infty}. \tag{6.3.120}$$

An analysis of the ultimate distribution is available in Wilcox (1967):

$$\left(u_{is}^{\infty}/u_{io}\right) = A\exp(Bz). \tag{6.3.121}$$

The constants A and B are obtained from

$$\kappa'_{is,\text{eff}} = Bl/(\exp(Bl) - 1); \tag{6.3.122a}$$

$$A = BL/(\exp(BL) - 1). \tag{6.3.122b}$$

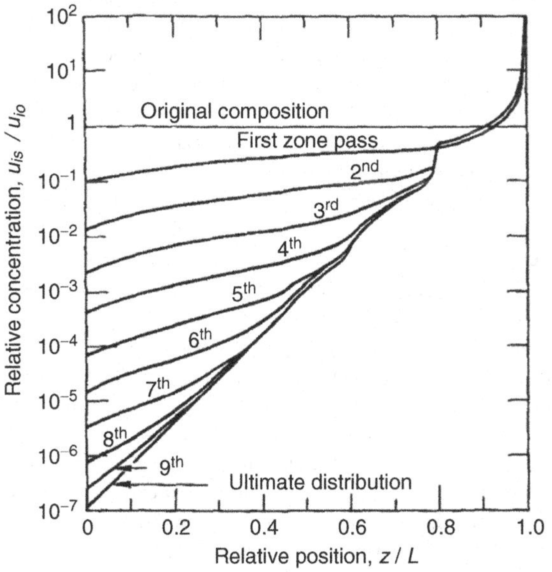

Figure 6.3.18. Calculated concentration profiles for multiple zone passes with $\kappa'_{is,eff} = 0.1$, $l/L = 0.2$. (After Wilcox (1967).)

Results in Figure 6.3.18 highlight the extraordinary levels of purification achieved by multiple passes. In fact, ultrapurity has often been a must in the semiconductor industry; fractional solidification techniques are frequently employed (Zief and Wilcox, 1967).

We will now briefly consider the normal freezing process; it is also sometimes called *controlled progressive freezing*. As illustrated in Figure 6.3.14(c), the molten rod is being pulled out from the heated zone; the section of the rod which has been pulled out is frozen, while the rest remains in a molten state. The velocity of the freezing surface may be assumed to be v_{fr}. In a frame of reference fixed in the refrozen solid being pulled out, the freezing interface has a velocity v_{fr} in the opposite direction, toward the molten part of the rod. Let the total length of the rod be L. If a length z of this rod is out of the heated zone and is solidified at any time t, then, for a rod of uniform cross-sectional area, the weight fraction u_z of the molten rod which is solidified is $u_z L = z$; correspondingly, the weight fraction $1-u_z$ of the rod still remains molten. Employing an impurity mass balance in a section of length dz, which is freezing (Figure 6.3.19A), we get, by equating this gain to the solid phase the loss from the molten phase,

$$u_{is} A \rho_s \,\text{d}z = -\text{d}[u_{ilb} A (L - z)] \rho_l; \tag{6.3.123a}$$

$$u_{is}\rho_s \text{d}z = -\rho_l (L - z)\text{d}u_{ilb} + u_{ilb}\rho_l \,\text{d}z$$

$$\Leftrightarrow (u_{ilb}\rho_l - u_{is}\rho_s)\text{d}z = \rho_l (L - z)\text{d}u_{ilb}; \tag{6.3.123b}$$

$$\int_{u_{is}|_{z=0}}^{u_{is}} \frac{\rho_l \,\text{d}u_{ilb}}{(u_{ilb}\rho_l - u_{is}\rho_s)} = \int_0^z \frac{\text{d}z}{(L - z)} = \int_0^{u_z} \frac{\text{d}u_z}{(1 - u_z)} = -\ln(1 - u_z). \tag{6.3.123c}$$

From studies in zone melting, it is known that models of partial liquid mixing are more realistic, and the effective partition coefficient approach of equation (6.3.119) is quite useful. In this approach,

$$u_{ilb} = \frac{u_{is}}{\kappa'_{is,\text{eff}}}. \tag{6.3.123d}$$

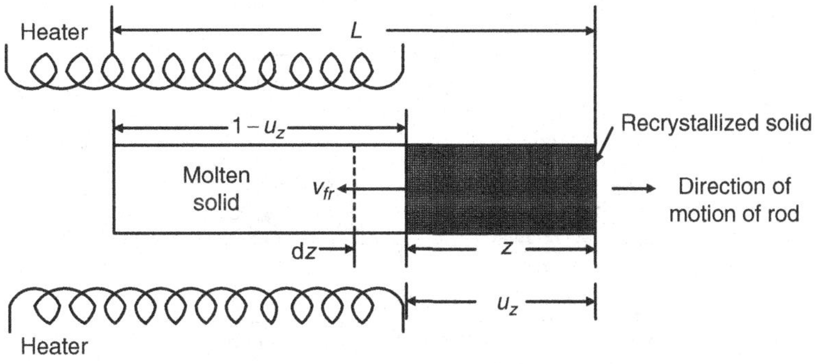

Figure 6.3.19A. Schematic of the normal freezing system for impurity distribution analysis.

Further, the first solid to come out at $z = 0$ will have an impurity mass fraction

$$u_{is}\big|_{z=0} = \kappa'_{is,\text{eff}} u_{io}, \tag{6.3.123e}$$

where u_{io} is the impurity mass fraction in the original rod (it is also the molten rod composition). Assuming further that $\rho_s \cong \rho_l$, we get

$$\int_{u_{is}|_{z=0}}^{u_{is}} \frac{du_{is}}{u_{is}(\kappa'_{is,\text{eff}} - 1)} = \ln(1 - u_z) = \ln\left(1 - \frac{z}{L}\right)$$

$$\Rightarrow \frac{u_{is}}{u_{is}|_{z=0}} = (1 - u_z)^{(\kappa'_{is,\text{eff}}-1)}$$

$$\Rightarrow \frac{u_{is}}{u_{io}} = \kappa'_{is,\text{eff}}(1 - u_z)^{(\kappa'_{is,\text{eff}}-1)}; \tag{6.3.124a}$$

$$\frac{u_{is}}{u_{io}} = \kappa'_{is,\text{eff}}\left(1 - \frac{z}{L}\right)^{(\kappa'_{is,\text{eff}}-1)}. \tag{6.3.124b}$$

An alternative approach of formulating the differential mass balance equation (6.3.123a) would be to use u_z instead of z as the independent variable:

$$u_{is} du_z = -d[u_{ilb}(1 - u_z)]. \tag{6.3.125}$$

This formulation has an advantage in that we do not have to assume a constant cross-sectional area (therefore, $u_z L \neq z$). The result (6.3.124a) is still obtained. Note that both results (6.3.124a) and (6.3.124b) are not valid as $z \rightarrow L$ or $u_z \rightarrow 1$.

Richman *et al.* (1967) have pointed out the following advantages of normal freezing over the zone melting technique:

(1) larger-scale operation is easier;
(2) mechanical stirring of the melt is easier;
(3) the purification achieved in one complete pass is higher than that in zone melting.

Problems 6.3.14 and 6.3.15 illustrate that the purification achieved in a single-pass zone melting process *is* less than that achieved by a normal freezing process.

There is, however, one major disadvantage, namely that the impurities collected at the end of the rod that are last to freeze must be removed before the sample can be subjected to another pass, since, during the remelting of the sample, e.g. rod, the impurities will be homogenized. Zone melting does not suffer from such a problem. Further, in normal freezing, the molten state exists over a large length of the rod for a long time; therefore, evaporation, decomposition, etc., can take place easily. However, normal freezing is often the method of choice for bulk crystal growth, for example of silicon and gallium arsenide: the method is identified as the *Czochralski technique*. A seed crystal is suspended into a melt from a shaft, which is rotated as well as moved vertically upward. The temperature is lowered, the crystal growth initiates and the shaft is moved upward. There is no impurity separation achieved in this process. For an introduction to and brief analysis of how this technique distributes dopant through the crystal, see Lee (1990).

The zone melting/normal freezing processes described so far have been *batch processes*: one solid rod/ingot was used in a device for a period of time needed to achieve the requisite purification. To purify material at a large rate, continuous zone melting may be undertaken, thus avoiding either the necessity of a much larger size batch system or the handling of many batches. The basic structure of a continuous zone melting device is shown schematically in Figure 6.3.19B. Molten feed is introduced near the middle of a column into, for example, a feed hopper, which is connected on one side to the enriching section and on the other side to the stripping section. The feed is a source of the pure material for the enriching section, at the end of which the purified solid is obtained. The purified solid is withdrawn by melting at the end. There is a heating band which traverses the enriching section - it keeps on purifying the material as the band moves toward the feed section. At the feed section, the impurities from the enriching section are rejected. In practice, there may be a number of heating bands to increase the rate of purification; a particular heating band may traverse only part of the length at the enriching section.

The feed flow rate introduced at the feed hopper is split into two parts: the purified product flow rate at the purified product exit and the waste flow rate at the waste exit end of the stripping section. The heating band over the stripping

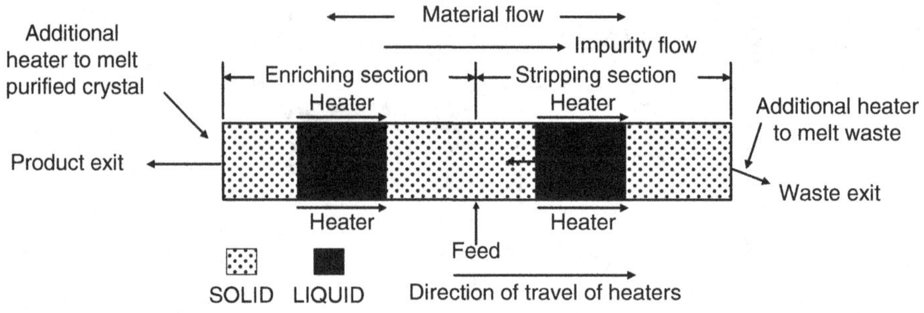

Figure 6.3.19B. Schematic of the principle of a continuous zone refining device.

section moves toward the waste exit and carries the impurities from the feed section toward that end. Thus, the impurities rejected from the enriching section are taken down the column into the stripping section. The actual mechanisms employed to move the heating bands, materials and the purification process, including those developed by W.G. Pfann, are illustrated in Moates and Kennedy (1967).

Fractional solidification processes have been employed to separate/purify a variety of mixtures besides semiconductor materials. These include:

- purification of water, as in desalination by freezing (Orcutt, 1967), since it is well known that ice, which is pure water, is obtained when saline water is frozen;
- purification of organics like naphthalene by the Proabd Refiner (Molinari, 1967), which consists of controlled cooling of the feedstock in a static apparatus, leading to progressive solidification on the cooling surface of the refiner followed by controlled heating which melts the impure fractions the earliest;
- ultrapurification of pharmaceuticals;
- purification of aluminum ingots on a large scale by normal freezing.

Additional details are provided in Zief and Wilcox (1967).

6.3.2.4 *Solid–vapor systems: drying*

We will focus here essentially on the removal of moisture from solids, porous solids, etc., therefore on the drying of solids. "Drying," however, has a broader connotation; the following are treated also as examples of drying: removal of trace amounts of moisture from organic liquids (usually carried out by distillation in a column, Section 8.1.3; also by pervaporation, Section 6.3.3.4); removal of small amounts of moisture from gaseous stream via adsorption (Section 7.1.1.2), membrane gas permeation (Sections 7.2.1.1 and 8.1.8). We restrict ourselves here to a brief treatment of moisture removal from porous solids, only where there is bulk flow parallel to the force direction, namely vacuum drying. A general treatment of drying is available in King (1971), Keey (1972) and Porter *et al.* (1984); additional treatments are available in textbooks on mass transfer and separation; see Treybal (1980), Belter *et al.* (1988) and Geankoplis (2003).

Porous solids, whether they are food materials, nonfood biological materials, pharmaceutical products, etc., often contain a significant amount of moisture. The removal of this moisture is necessary to reduce/eliminate biological activity. Depending on the material, drying can be carried out at higher temperatures or at a low temperature under *freeze-drying* conditions. The process of freeze-drying involves freezing the material first by exposure to a very cold air and then subjecting it to vacuum to sublimate the moisture. Heating of the porous solid may be carried out (1) by circulating hot gas (e.g. air), (2) by conduction from a hot metal plate on which the porous solid is kept, (3) by radiation from hot surfaces as well as from hot gases or (4) by the combination of all three methods. The moisture is removed from the porous solid by evaporation when drying takes place. Circulating hot gas removes the moisture; alternatively, a vacuum removes the evaporated moisture. We will focus here on vacuum driven processes only since the moisture bulk flow is parallel to the direction of the chemical potential gradient driven force, causing evaporation of the moisture. Such a process is called *conduction drying* or *vacuum shelf drying*.

Consider a wet porous solid bed on a tray heated from the bottom by hot water, steam, etc. (Figure 6.3.20(a)) (Belter *et al.*, 1988). From the top surface, where $z = L$, the thickness of the solid bed, moisture is removed by vacuum. At any time, at a particular value of z ($<L$) any water present will be evaporated. Below this value of z ($<L$), the porous solid is wet; at larger values of z, the porous solid is dry. As time progresses this z-value becomes slowly smaller and smaller, and ultimately should become zero.

To determine how z varies with time, and ultimately the time t_{drying} needed to dry the whole bed on the tray, consider the rate of conductive heat transport per unit

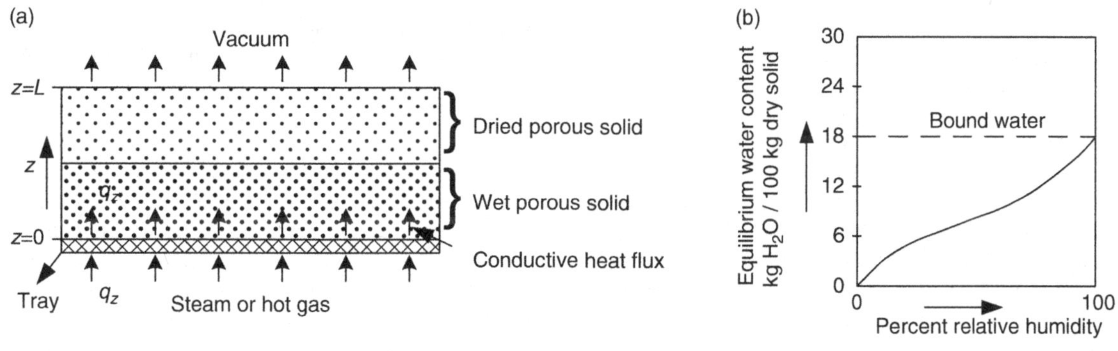

Figure 6.3.20. (a) Conduction drying of porous wet solids on a plate/tray heated from below and subjected to vacuum at the top surface. (b) Typical equilibrium moisture content behavior vs. percent relative humidity.

cross-sectional area, namely the heat flux q_z between the heated plate surface at temperature T_{pan} (where $z = 0$) and the temperature at location z, T_z:

$$q_z A_b = \frac{k_m}{z} A_b (T_{pan} - T_z). \tag{6.3.126}$$

Here k_m is an effective thermal conductivity of the material bed being dried and A_b is the cross-sectional area of the bed. The rate at which the dry–wet interface location coordinate z changes is very low. We may therefore assume a pseudo steady state condition. This heat flux will evaporate water; correspondingly, the dry–wet interface location will change and the volume of the wet bed, zA_b, will decrease. The rate of heat transfer from the plate bottom must equal the rate of evaporation of the local moisture in the bed:

$$q_z A_b = \lambda_{H_2O,bed}\, \rho_{wb} \frac{d}{dt}(zA_b). \tag{6.3.127}$$

Here ρ_{wb} is the mass of water per unit volume of the bed, $\lambda_{H_2O,bed}$ is the enthalpy of vaporization of water per unit mass of water. Combine equations (6.3.126) and (6.3.127) to obtain

$$z\frac{dz}{dt} = \frac{k_m}{\lambda_{H_2O,bed}\, \rho_{wb}}(T_{pan} - T_z) = \frac{1}{2}\frac{dz^2}{dt}. \tag{6.3.128}$$

The initial condition for this drying process corresponds to the whole bed being wet, i.e. the location $z = L$ at $t = 0$. Integrate and use this initial condition to obtain

$$z^2 = L^2 - \frac{2k_m (T_{pan} - T_z)}{\lambda_{H_2O,bed}\, \rho_{wb}} t. \tag{6.3.129}$$

The material in the pan becomes completely dry when $z = 0$ at $t = t_{end}$:

$$t_{end} = \frac{\lambda_{H_2O,bed}\, \rho_{wb}}{2k_m (T_{pan} - T_z)} L^2. \tag{6.3.130}$$

The above analysis ignores any sensible heat contribution as the temperature profile changes in the wet section of the bed and any temperature profile in the dry region. However, one of the more complex items in this analysis is the mass of water per unit volume of the bed (ρ_{wb}).

The mass of water per unit volume of the bed, or alternatively the mass of water per unit mass of dry solid, consists basically of free water (free moisture) and bound water (bound moisture). Free water is ordinary water existing in a porous solid such that it has the same vapor pressure as that of water; therefore it can be evaporated. Further, it is mobile in the porous solid structure, obeys capillarity (described in Section 6.1.4) and therefore could come up to the gas–solid (air–solid) surface. Bound water present in the porous solid, however, has a vapor pressure lower than that of free water; it is not as mobile and has a higher heat of vaporization. Typical examples are: moisture present within cells, biological or otherwise; water in very small-diameter capillaries (the Kelvin effect, see equation (3.3.112e)); hydrated crystals, etc. Biological materials such as cells typically possess significant bound water; wood and food items fall into this category also.

A quantitative description of the mass of water per unit mass of dry solid is provided by plotting it against a quantity called *percent relative humidity*. To understand this quantity, consider air in equilibrium with liquid water. The partial pressure of water vapor in the air, in equilibrium with liquid water at any temperature, is the vapor pressure of water, P^{sat}_{water}. The saturation humidity H^{sat}_{water} is defined as the mass of water vapor per unit mass of air present at a total pressure P:

$$H^{sat}_{water} = \frac{P^{sat}_{water}}{P - P^{sat}_{water}} \times \frac{18}{29}, \tag{6.3.131}$$

where 18 is the molecular weight of water and 29 is the molecular weight of air; the partial pressure ratio provides a ratio of the moles of water to moles of air. If the air is drier than that at the saturation humidity, the humidity H_{water} is defined as

$$H_{water} = \frac{p_{water}}{P - p_{water}} \times \frac{18}{29}, \tag{6.3.132}$$

where $H_{water} < H^{sat}_{water}$. The relative humidity is defined as

$$H^{rel}_{water} = (H_{water}/H^{sat}_{water}), \tag{6.3.133}$$

with percent relative humidity being 100 H^{rel}_{water}.

The equilibrium moisture content of many materials of interest, such as various food items, wood, etc., are described as a function of the percent relative humidity, as illustrated in Figure 6.3.20(b) for a hypothetical system. The level of bound water in this system at 100% relative humidity is 18 kg H_2O/100 kg dry material. If the total level of water in this sample is 26 kg H_2O/100 kg dry material, then the amount of free water at 100% relative humidity is 8 kg H_2O/100 kg dry material. Similarly, one can develop estimates of bound water and free water at other levels of relative humidity.

6.3.3 Filtration and membrane separation processes

We illustrate here those configurations of filtration and important membrane separation processes where the bulk flow of the liquid, solution/suspension or gas mixture is parallel to the direction of the force driving the separation through the membrane/filter. The force consists primarily of that due to the chemical potential gradient; almost invariably there is a pressure gradient across the membrane or the filter. Generally, separation of particles larger than 10 μm by a filter cloth/medium is identified as *filtration*. Particles/colloids of sizes less than 10 μm but larger than 0.02 μm when separated by a membrane are said to undergo *microfiltration*. These two processes are considered first. Next we illustrate *ultrafiltration*, where solutes and macrosolutes having molecular weights between 1000 and 500 000 are separated by an ultrafilter;

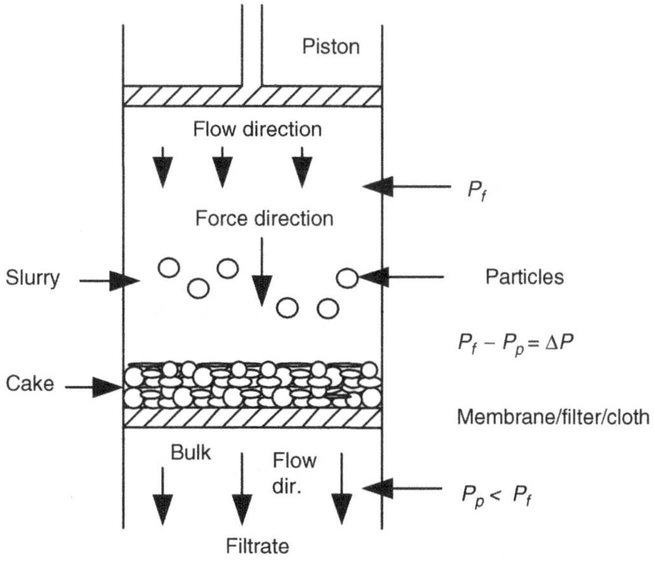

Figure 6.3.21. Deadend filtration: bulk flow parallel to the direction of the force due to applied $\Delta P = P_f - P_p$.

colloids having similar dimensions are also separated by ultrafiltration. The process of *reverse osmosis* separating solutes of molecular weight[25] up to 150–200 is briefly considered next. Separation of volatile solutes from a solution through a dense membrane subjected to a vacuum on the other side of the membrane may be carried out by the *pervaporation* process described subsequently. Finally, the separation of a gas mixture through a dense nonporous membrane is described.

6.3.3.1 *Deadend cake filtration/microfiltration*

In the introduction to Section 6.3.1.4, we introduced the notion of sieving or surface filtration, wherein the fluid under pressure passes through the pores of a relatively thin membrane/filter/cloth whose pores/openings are smaller than the particles. There is an applied pressure difference (ΔP) between the two sides of the filter/membrane, the feed side being at a higher pressure. Particles which cannot pass through the membrane/filter/cloth are deposited on it; the deposited material forms a cake of sorts, a porous bed, a packed bed of particles. When the particle sizes range between 0.02 and 10 μm, the process is called *microfiltration.* For larger particles, as well as particles in this size range, the process is generally called *cake filtration.* When the processes are carried out in the manner shown in Figure 6.3.21, as if a piston is driving the particle suspension in the fluid toward the membrane/filter/cloth of a certain cross-sectional area in a vessel having the same cross-sectional area, we have bulk flow of the particle-containing fluid parallel to the direction of the applied force created by the pressure difference.

Let us focus first on cake filtration and microfiltration for the case where the fluid is a liquid. In the configuration of Figure 6.3.21, the technique is called *deadend filtration.* The same configuration is routinely employed in laboratories with a filter paper on, say, a Büchner funnel and a partial vacuum on the side of the permeate/filtrate: a precipitate/deposit builds up quickly on the filter paper as the slurry is filtered. As time passes, a particle based deposit continues to build up on the filter paper: it is called a *cake.* This cake provides an additional resistance to the flow of the filtrate through the membrane/filter/cloth in deadend filtration. As time passes, deposition of the particles onto/in the cake continues. Therefore the resistance to the flow of the filtrate increases with time. If one wants to maintain a *constant* value of the *filtrate flux,* the applied pressure difference ΔP has to increase with time. Alternatively, for a *constant* applied pressure difference, the flux of the filtrate will decrease with time (Figure 6.3.22).

For both modes of filtration operation in Figure 6.3.22, one would like to know the magnitude of the filtrate flux for a given ΔP, or vice versa for a given slurry-filter system or a suspension-filter system. We will find that the nature of the cake or deposit formed on the membrane/filter/cloth is often crucial to determining the flux level. Sometimes, as in the case of filtration of fermentation broths, the cake formed is highly compressible and filtration rates are quite

[25] Solutes of molecular weight between around 150 and 1000 are usually separated by *nanofiltration*; however, reverse osmosis may be used at the low end of this range, while ultrafiltration may be used at the very high end. There are no sharp boundaries.

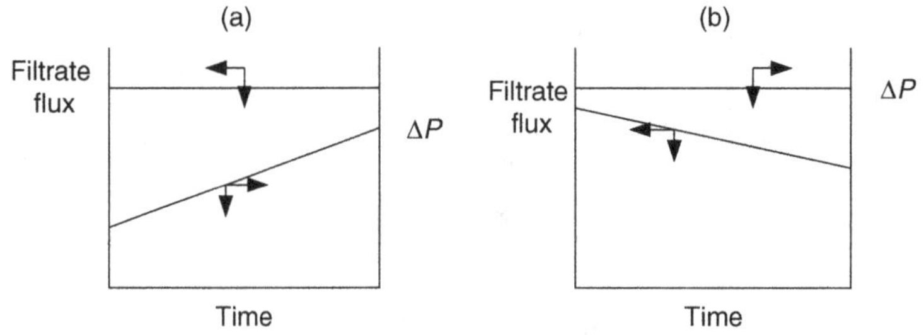

Figure 6.3.22. Two basic models of carrying out deadend filtration: (a) constant flux mode; (b) constant ΔP *mode.*

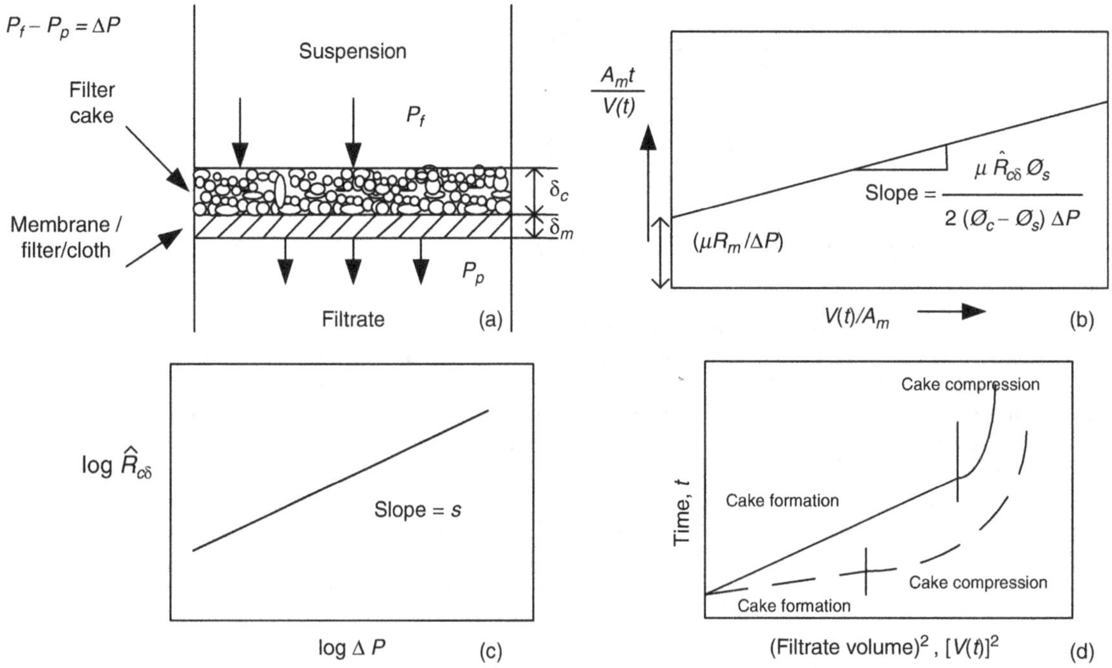

Figure 6.3.23. (a) Cake filtration, bulk flow parallel to force, schematic with a cake of thickness δ_c. *(b) Determination of* R_m *and* $\hat{R}_{c\delta}$ *by plotting* $(A_m t/V(t))$ *against* $V(t)/A_m$ *for an incompressible cake. (c) Determination of* s *for a compressible cake described by equation (6.3.138j). (d) Traditional cake filtration (solid line); nontraditional filtration behavior (dashed line).*

low. To facilitate filtration, *filter aids* are often added to increase the filtration rate by developing a more porous/permeable cake. Filter aids are solid inert porous materials of fine size, such as diatomaceous earths, perlites, asbestos, etc. They increase the cake porosity and also make it much less compressible. They are used especially in the filtration of certain types of fermentation broths.

Consider Figure 6.3.23(a), which shows a membrane/filter/cloth of thickness δ_m, on the top surface of which a cake of thickness δ_c has been formed at time t (starting time $t = 0$). We have observed in Section 3.4.2.3 that the volume flux of the liquid through a porous medium may be described by Darcy's law (equation (3.4.88), repeated here for convenience):

$$v_s = \text{volume flux} = N_s \overline{V}_s = \frac{Q_{sm}}{\mu} \frac{\Delta P}{\delta_m} = \frac{\Delta P}{\mu R_m} \frac{\text{cm}^3}{\text{cm}^2\text{-s}}.$$

Note: The quantity $N_s \overline{V}_s$ is also equal to the solvent velocity v_s (cm/s) through the filter (compare equations (3.4.60d–f) and (3.4.85)). At any instant of time, both the filter and the cake on top of it provide resistances to flow: R_m for the filter and R_c for the cake. If the pressure drops through the membrane/filter and the cake are, respectively, ΔP_m and ΔP_c, then

$$v_s = N_s \overline{V}_s = \frac{Q_{sm} \Delta P_m}{\mu \delta_m} = \frac{Q_{sc}}{\mu} \frac{\Delta P_c}{\delta_c} = \frac{\Delta P_c}{\mu R_c}, \qquad (6.3.134)$$

where Q_{sm} and Q_{sc} and the permeabilities of the membrane/filter and the cake layer, respectively. Since ΔP, the applied pressure difference, is equal to $(\Delta P_m + \Delta P_c)$, where ΔP_m is the pressure drop through the membrane and ΔP_c is that through the cake, we get

$$v_s = \frac{\Delta P}{\mu[R_m + R_c]}; \tag{6.3.135a}$$

$$R_m = (\delta_m/Q_{sm}), \qquad R_c = (\delta_c/Q_{sc}). \tag{6.3.135b}$$

This representation is similar to that in Ohm's law in electricity:

$$\text{current (here velocity)} = \frac{\text{applied potential difference (here } \Delta P)}{\text{resistance}},$$

where the cake resistance and the membrane (filter) resistance are in series. For a membrane/filter having a cylindrical pore size distribution $f(r_p)$ and Poiseuille flow, we obtain, from equation (3.4.86),

$$R_m = \frac{8\,\delta_m}{\varepsilon_m\, r_p^2}. \tag{6.3.135c}$$

If the membrane/filter has N_p pores, each of size r_p, per unit area, and the membrane porosity is ε_m, then

$$\varepsilon_m = N_p \pi r_p^2, \qquad R_m = \frac{8\,\delta_m}{N_p \pi r_p^4}. \tag{6.3.135d}$$

Sometimes, such a cylindrical pore based representation is avoided; instead membrane porosity, ε_m, and pore surface area/membrane volume, s_m, are used. For cylindrical pores of constant size r_p,

$$s_m = \frac{N_p\, 2\pi\, r_p\, \delta_m}{\delta_m} = N_p\, 2\pi\, r_p. \tag{6.3.135e}$$

If either membrane pore volume based, s_{mp}, or the membrane solids volume based, s_{ms}, is used, then

$$s_{mp} = \frac{N_p\, 2\pi\, r_p\, \delta_m}{N_p\, \pi\, r_p^2\, \delta_m} = \frac{2}{r_p}; \tag{6.3.135f}$$

$$s_{ms} = \frac{N_p\, 2\pi\, r_p\, \delta_m}{\delta_m(1-\varepsilon_m)} = \frac{2\pi\, r_p\, N_p}{(1-\varepsilon_m)}. \tag{6.3.135g}$$

(*Note*: $s_{mp} = s$ in equation (3.3.89b).) In terms of s_{ms} and ε_m, R_m may be expressed, from (6.3.135d), as

$$R_m = \frac{2\,(1-\varepsilon_m)^2\, s_{ms}^2\, \delta_m}{\varepsilon_m{}^3}. \tag{6.3.135h}$$

In practice, the constant 2 in this equation may be replaced by other empirically estimated values.

The liquid flow resistance of the cake of deposited particles of diameter d_p on the membrane, R_c, may be estimated from equation (6.1.4e) for a cake porosity of ε_c as

$$R_c = \frac{72\,(1-\varepsilon_c)^2\, \delta_c}{d_p^2\, \varepsilon_c^3}. \tag{6.3.135i}$$

Since the surface area of a particle in the cake per unit particle volume, a_{vc}, is related to d_p by

$$a_{vc} = \frac{4\pi\, d_p^2/4}{(4/3)\pi(d_p/2)^3} = \frac{6}{d_p}, \tag{6.3.135j}$$

$$R_c = \frac{2(1-\varepsilon_c)^2\, a_{vc}^2\, \delta_c}{\varepsilon_c^3} = \frac{a_1\,(1-\varepsilon_c)^2 a_{vc}^2\, \delta_c}{\varepsilon_c^3}, \tag{6.3.135k}$$

a form quite similar to that of expression (6.3.135h) for R_m. From equation (6.1.4f) for flow in a packed bed of particles (the Blake–Kozeny equation), the constant 72 in equation (6.3.135i) should be replaced by 150. Correspondingly, the value of a_1 (= 2) will be increased by a factor of (150/72). Two other quantities are often used instead of R_c : $\hat{R}_{c\delta}$, the specific cake resistance, which is the cake resistance per unit cake thickness, and $\hat{R}_{cw}$, the specific cake resistance per unit mass of the cake,

$$\hat{R}_{c\delta} = \frac{R_c}{\delta_c}; \qquad \hat{R}_{cw} = \frac{R_c}{w_c}, \tag{6.3.135l}$$

where w_c is the mass of cake per unit area of the filter/membrane and is related to δ_c for a mass density ρ_s of the solids by

$$w_c = \rho_s(1-\varepsilon_c)\delta_c. \tag{6.3.135m}$$

Consider now Figure 6.3.23(a). At any instant of time t, if V_p is the total volume of filtrate/permeate collected over the time period 0 to t seconds, then, for a membrane/filter area of A_m, the filtration volume flux v_s is

$$v_s = N_s \overline{V}_s = \frac{1}{A_m}\frac{\mathrm{d}V_p}{\mathrm{d}t} = \frac{\Delta P/\mu}{(R_m + R_c)}. \tag{6.3.136a}$$

Alternatively it may be written as

$$v_s = N_s \overline{V}_s = \frac{1}{A_m}\frac{\mathrm{d}V_p}{\mathrm{d}t} = \frac{\Delta P}{\mu(R_m + \hat{R}_{c\delta}\,\delta_c)}. \tag{6.3.136b}$$

To determine the filtration flux v_s, we need to know δ_c – how it changes with time as the cake layer builds up (Davis and Grant, 2001). To develop an equation for δ_c with time, we need to make a mass balance for the solids in the slurry/suspension being filtered. Define the following quantitites:

ϕ_s – volume fraction of solids in the suspensions being filtered;

ϕ_c – volume fraction of solids in the cake being formed.

The rate of solid deposition in the growing cake per unit membrane area is given by

$$\phi_c \frac{\mathrm{d}\delta_c}{\mathrm{d}t} = \frac{\text{solids volume}}{\text{cake volume}} \times \frac{\text{change in cake volume}}{\text{change in time} \times \text{membrane area}}.$$

The rate of solid deposition in the growing cake per unit membrane area from the production of the filtrate and the cake volume growth is given by

$$\left(v_s + \frac{d\delta_c}{dt}\right)\phi_s = \frac{\text{(filtrate volume + growing cake volume) = total suspension volume}}{\text{membrane area} \times \text{time}} \times \left(\frac{\text{solids volume}}{\text{suspension volume}}\right).$$

Equating these two terms, we obtain (Davis and Grant, 2001)

$$\phi_c \frac{d\delta_c}{dt} = \left(v_s + \frac{d\delta_c}{dt}\right)\phi_s,$$

which may be rearranged to yield

$$\frac{d\delta_c}{dt} = \frac{v_s\,\phi_s}{\phi_c - \phi_s}, \qquad (6.3.136c)$$

where $\delta_c = 0$ at $t = 0$. Solution of equations (6.3.136b) and (6.3.136c) will provide an estimate of how v_s changes with time for a constant ΔP and growing δ_c, or how ΔP changes with time to maintain a constant v_s. Integrating equation (6.3.136c) with respect to time t will lead to the following relation between the volume of filtrate $V_p(t)$ collected in time t through a filter of area A_m:

$$[V_p(t)/A_m] = [(\phi_c - \phi_s)/\phi_s]\,\delta_c(t). \qquad (6.3.136d)$$

Consider first the case of *constant filtration flux* in cake filtration/microfiltration. Since the volume flux v_s is constant,

$$v_s = \frac{1}{A_m}\frac{dV_p}{dt} = \text{constant} = v_{so} \Rightarrow V_p(t) = v_{so}\,A_m\,t. \qquad (6.3.137a)$$

From equation (6.3.136c), it follows that

$$\frac{d\,\delta_c}{dt} = \frac{v_{so}\,\phi_s}{\phi_c - \phi_s} \Rightarrow \delta_c(t) = \frac{v_{so}\,\phi_s}{\phi_c - \phi_s}t, \qquad (6.3.137b)$$

where we have assumed that ϕ_c is constant, i.e. the cake is incompressible. From equation (6.3.136b), the pressure drop ΔP will now increase linearly with time for a constant filtration flux v_s, as shown in Figure 6.3.22(a):

$$\Delta P(t) = \mu v_{so}\left(R_m + \hat{R}_{c\delta}\frac{v_s\,\phi_s}{(\phi_c - \phi_s)}t\right). \qquad (6.3.137c)$$

On the other hand, if the applied pressure drop ΔP is held *constant with time*, we have to recognize that the volume flux, or filtration velocity v_s, will decrease with time as the cake builds up. Introducing expression (6.3.136b) for v_s in (6.3.136c), we obtain

$$\frac{d\delta_c}{dt} = \frac{(\phi_s)}{(\phi_c - \phi_s)} \times \frac{\Delta P}{\mu\,(R_m + \hat{R}_{c\delta}\,\delta_c)}$$

$$(R_m + \hat{R}_{c\delta}\,\delta_c)\frac{d\delta_c}{dt} = \frac{\phi_s}{(\phi_c - \phi_s)}\frac{\Delta P}{\mu}. \qquad (6.3.138a)$$

Integrate this equation for δ_c to obtain

$$\hat{R}_{c\delta}\frac{\delta_c^2}{2} + R_m\delta_c - \frac{\phi_s}{(\phi_c - \phi_s)}\frac{\Delta P}{\mu}t = 0, \qquad (6.3.138b)$$

with $\delta_c = 0$ at $t = 0$.

The solution of the quadratic equation provides an expression for $\delta_c(t)$:

$$\delta_c(t) = \frac{-R_m \pm \sqrt{R_m^2 + (2\hat{R}_{c\delta}\,\phi_s\,\Delta P t/\mu(\phi_c - \phi_s))}}{\hat{R}_{c\delta}}.$$

Take the positive root for a meaningful $\delta_c(t)$:

$$\delta_c(t) = (R_m/\hat{R}_{c\delta})\left[\left\{1 + \frac{2\,\hat{R}_{c\delta}\,\phi_s\,\Delta P\,t}{(\phi_c - \phi_s)\mu\,R_m^2}\right\}^{1/2} - 1\right]. \qquad (6.3.138c)$$

Substitution into equation (6.3.136b) provides an expression for the *time-dependent filtration rate/volume flux* for constant ΔP:

$$v_s = \frac{\Delta P}{\mu\,R_m}\left[1 + \frac{2\,\hat{R}_{c\delta}\,\phi_s\,\Delta P\,t}{(\phi_c - \phi_s)\mu\,R_m^2}\right]^{-1/2}. \qquad (6.3.138d)$$

Figure 6.3.22(a) shows that, for constant ΔP, the filtration flux decreases. Substitution of expression (6.3.138c) for $\delta_c(t)$ in expression (6.3.136d) yields

$$(V(t)/A_m) = \left(\frac{\phi_c - \phi_s}{\phi_s}\right)\frac{R_m}{\hat{R}_{c\delta}}\left[\left\{1 + \frac{2\,\hat{R}_{c\delta}\,\phi_s\,\Delta P\,t}{(\phi_c - \phi_s)\mu\,R_m^2}\right\}^{1/2} - 1\right], \qquad (6.3.138e)$$

which illustrates how the filtrate volume $V(t)$ changes with time for constant ΔP filtration.

The membrane resistance R_m and the specific cake resistance $\hat{R}_{c\delta}$ are of interest here. They are not known in general. To determine them, consider the quadratic equation (6.3.138b) for δ_c and the expression (6.3.136d) for $\delta_c(t)$. Substitution of (6.3.136d) for $\delta_c(t)$ into equation (6.3.138b) results in

$$\frac{\hat{R}_{c\delta}}{2}\frac{V^2(t)\phi_s^2}{A_m^2(\phi_c - \phi_s)^2} + R_m\frac{V(t)\phi_s}{A_m(\phi_c - \phi_s)} = \frac{\phi_s}{(\phi_c - \phi_s)}\frac{\Delta P}{\mu}t.$$

Multiplying both sides by

$$\frac{\mu(\phi_c - \phi_s)}{\Delta P\,\phi_s}\left(\frac{A_m}{V(t)}\right)$$

leads to

$$\left(\frac{A_m}{V(t)}\right)t = \frac{\mu\hat{R}_{c\delta}\,\phi_s}{2(\phi_c - \phi_s)\Delta P}\left(\frac{V(t)}{A_m}\right) + \frac{\mu R_m}{\Delta P}. \qquad (6.3.138f)$$

In any particular case, if one plots $(A_m/V(t))t$ against $(V(t)/A_m)$ (since $V(t)$ at time t is known for a given filter area), the intercept on the ordinate will yield $(\mu R_m/\Delta P)$, and therefore the membrane resistance R_m; the slope will yield $(\mu\hat{R}_{c\delta}\phi_s)/2(\phi_c - \phi_s)\Delta P$, which, for known values of ϕ_s, ϕ_c, ΔP and μ, will yield the specific cake resistance $\hat{R}_{c\delta}$ (Figure 6.3.23(b)).

It is useful to consider some limiting forms of the general flux decline and cake buildup behavior described above. From the expression (6.3.138d) for filtration flux decreasing with time, it appears that, as t becomes large, $v_s \propto (t)^{-1/2}$:

$$v_s \cong \left(\frac{(\phi_c - \phi_s)\Delta P}{2\,\hat{R}_{c\delta}\,\phi_s\,\mu\,t}\right)^{1/2}. \qquad (6.3.138g)$$

Correspondingly, from expression (6.3.138e) for the filtrate volume, $V(t)$ varies with $(t)^{1/2}$ as t becomes large, as does the cake thickness $\delta_c(t)$ (from expression (6.3.138c)):

$$(V(t)/A_m) \cong \left(\frac{2(\phi_c - \phi_s)\Delta P t}{\phi_s\,\mu\,\hat{R}_{c\delta}}\right)^{1/2}; \qquad (6.3.138h)$$

$$\delta_c(t) \cong \left(\frac{2\,\phi_s\,\Delta P t}{(\phi_c - \phi_s)\mu\,\hat{R}_{c\delta}}\right)^{1/2}. \qquad (6.3.138i)$$

The preceding development was based on the assumption that $\hat{R}_{c\delta}$, as well as R_m, were independent of ΔP. It has often been found, however, that $\hat{R}_{c\delta}$ depends on the applied ΔP. This can be true for deposits of inorganic precipitates as well as cells and cell debris encountered in microfiltration or filtration of fermentation broths and lysates. Conventionally, $\hat{R}_{c\delta}$ is then represented as

$$\hat{R}_{c\delta} = \rho_s\phi_c\,\hat{R}_{cw};$$

$$\hat{R}_{c\delta} = \rho_s\phi_c\alpha_0(\Delta P)^s. \qquad (6.3.138j)$$

A logarithmic plot of $\hat{R}_{c\delta}$ against ΔP in a log-log graph will provide s as the slope. Knowing s, one can calculate α_0 from any data point $(\hat{R}_{c\delta}, \Delta P)$ for given ρ_s and ϕ_c (Figure 6.3.23(c)).

One can consider equations (6.3.138h) and (6.3.138j) to be representative of *traditional constant pressure filtration behavior*; the quadratic relation for cake formation between time t and the filtration volume per unit membrane area, $(V(t)/A_m)$, exists for up to 85% of the total time, followed by a short cake compression period (Figure 6.3.23(d), solid line). However, in the filtration of many sludges containing large molecular weight biomacromolecules, sometimes called *supercompactable*, the cake formation period is rather short, whereas the compression phase is very long

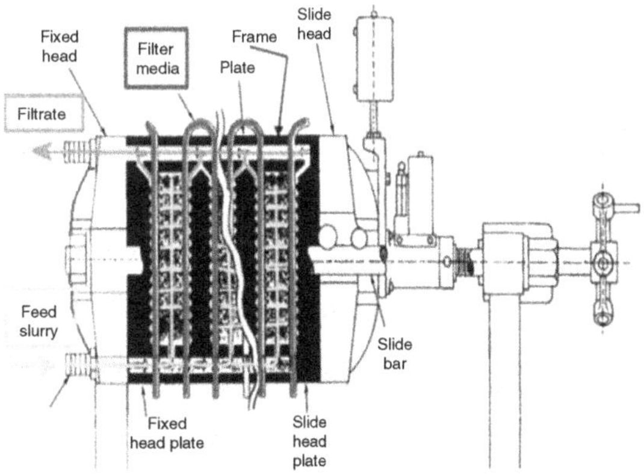

Figure 6.3.24. Plate and frame filter press (http://www.mine-engineer.com).

(Figure 6.3.23(d), dashed line). This behavior is called *nontraditional filtration behavior* under constant pressure conditions. A review of the model and the mathematics of analysis of nontraditional filtration behavior is provided in Strickland *et al.* (2005), along with references to earlier model-building activities.

Deadend filtration with larger particles has often been carried out using the so-called *plate and frame filter press* shown in Figure 6.3.24. It consists, for example, of a set of plates with recessed space between the neighboring plates, creating a chamber for the feed suspension to come in, and a cake, formed on a cloth/canvas on each plate, acting as a filter. The suspension enters each chamber through one corner of the assembled plates; the filtrate passes through the canvas/filter and goes out of the device. Before filtration, the plates are pressed together and sealed via gaskets when pressed by a hydraulic ram/screw. After filtration is over, the press is opened and the cake recovered from the canvas/filter. (Often before such a process, a suspension-free liquid feed is introduced to remove soluble substances from the cake.) Then the filter press is ready for another cycle of operation.

Cake filtration processes are demanding because of the time-dependence of the filtration flux, v_s, and the compressibility of the cake formed. If one could operate under a steady state process in some fashion, it would be quite useful. This essentially means a controlled growth of the cake such that, at any given location of the device, the filtration flux for a given ΔP will be constant with time. The deadend filtration configuration (see Figures 6.3.21 and 6.3.23(a)) of bulk flow of suspension parallel to the force direction has to be changed to develop alternative configurations which limit the growth of the cake. Two such conceptual possibilities are shown in Figures 6.3.25(a) and (b). The first one involves physical motion of the cake plus the filter cloth/membrane in a

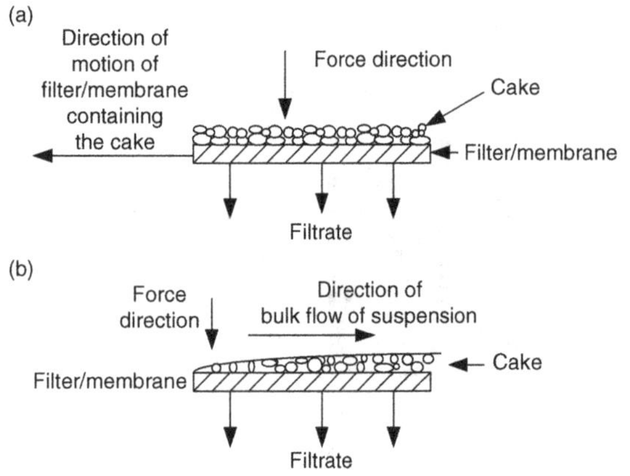

Figure 6.3.25. Steady state filtration/microfiltration of a slurry via bulk flow perpendicular to the force direction. (a) Bulk motion of filter/membrane containing the cake perpendicular to the force direction. (b) Bulk motion of the suspension/slurry perpendicular to the force direction.

direction perpendicular to the force direction, and away from the suspension, so that conditions at any location in the device with respect to the cake thickness remains unchanged (Figure 6.3.25(a)). The cake thickness here will change in the direction of its physical motion. The second one involves a tangential motion of the suspension/slurry over the membrane/filter perpendicular to the force direction (Figure 6.3.25(b)); the shear rate will limit the growth of the cake as the suspension/slurry becomes more concentrated. Both configurations are examples of bulk flow perpendicular to the force direction and are treated in Sections 7.2.1.4 and 7.2.1.5.

Example 6.3.4 Consider expressions (6.3.138c), (6.3.138d) and (6.3.138e) for $\delta_c(t), v_s(t)$ and $V(t)$, respectively. For the limiting case where the cake resistance dominates over the membrane resistance, determine the expressions for $\delta_c(t)$, $v_s(t)$ and $V(t)$.

Solution Consider first expression (6.3.138c) for $\delta_c(t)$:

$$\delta_c(t) = \left(\frac{R_m}{\hat{R}_{c\delta}}\right)\left[\left\{1 + \frac{2\,\hat{R}_{c\delta}\,\phi_s\,\Delta P\,t}{(\phi_c - \phi_s)\mu\,R_m^2}\right\}^{1/2} - 1\right]$$

$$\Rightarrow 1 = \left[\left\{\frac{R_m^2}{\hat{R}_{c\delta}^2\,\delta_c^2} + \frac{2\,\phi_s\,\Delta P\,t}{(\phi_c - \phi_s)\mu\,\hat{R}_{c\delta}\,\delta_c^2}\right\}^{1/2} - \frac{R_m}{\hat{R}_{c\delta}\,\delta_c}\right].$$

If $R_m << \hat{R}_{c\delta}\,\delta_c$, then

$$\delta_c(t) \cong \left[\frac{2\,\phi_s\,\Delta P\,t}{(\phi_c - \phi_s)\mu\,\hat{R}_{c\delta}}\right]^{1/2}.$$

Consider now expression (6.3.138d) for $v_s(t)$:

$$v_s(t) = \frac{\Delta P}{\mu\,R_m}\left[1 + \frac{2\,\hat{R}_{c\delta}\,\phi_s\,\Delta P\,t}{(\phi_c - \phi_s)\mu\,R_m^2}\right]^{-1/2}$$

$$\Rightarrow v_s(t) = \left[\frac{\mu^2 R_m^2}{\Delta P^2} + \frac{2\,\hat{R}_{c\delta}\,\phi_s\,\mu\,t}{(\phi_c - \phi_s)\Delta P}\right]^{-1/2}.$$

By definition,

$$v_s = \frac{\Delta P/\mu}{R_m + R_c} \sim \frac{\Delta P/\mu}{R_c} >> \frac{\Delta P/\mu}{R_m},$$

where $R_c = \hat{R}_{c\delta}\,\delta_c$. Therefore,

$$v_s >> \left(\frac{\Delta P^2}{\mu^2\,R_m^2}\right)^{1/2},$$

which suggests that

$$v_s(t) = \left[\frac{2\,\hat{R}_{c\delta}\,\phi_s\,\mu\,t}{(\phi_c - \phi_s)\Delta P}\right]^{-1/2}.$$

Now consider expression (6.3.138e) for $V(t)$:

$$\left(V(t)/A_m\right) = \left(\frac{\phi_c - \phi_s}{\phi_s}\right)\left(\frac{R_m}{\hat{R}_{c\delta}}\right)\left[\left\{1 + \frac{2\,\hat{R}_{c\delta}\,\phi_s\,\Delta P\,t}{(\phi_c - \phi_s)\mu\,R_m^2}\right\}^{1/2} - 1\right].$$

Write this as

$$\left(V(t)/A_m\,\delta_c\right) = \left(\frac{\phi_c - \phi_s}{\phi_s}\right)\left(\frac{R_m}{\hat{R}_{c\delta}}\right)\left[\left\{1 + \frac{2\,\hat{R}_{c\delta}\,\phi_s\,\Delta P\,t}{(\phi_c - \phi_s)\mu\,R_m^2}\right\}^{1/2} - 1\right]$$

$$= \left(\frac{\phi_c - \phi_s}{\phi_s}\right)\left[\left\{\left(\frac{R_m^2}{R_{c\delta}^2}\right) + \frac{2\,\hat{R}_{c\delta}\,\phi_s\,\Delta P\,t}{(\phi_c - \phi_s)\mu\,R_{c\delta}^2}\right\}^{1/2} - \frac{R_m}{R_{c\delta}}\right].$$

Since $R_m << R_{c\delta}$, we can write this as

$$\left(V(t)/A_m\,\delta_c\right) = \left(\frac{\phi_c - \phi_s}{\phi_s}\right)\left[\left\{\frac{2\,\phi_s\,\Delta P\,t}{(\phi_c - \phi_s)\mu\,\hat{R}_{c\delta}\,\delta_c^2}\right\}^{1/2}\right]$$

$$\Rightarrow V(t) = A_m\left\{\frac{2(\phi_c - \phi_s)\Delta P\,t}{\phi_s\,\mu\,\hat{R}_{c\delta}}\right\}^{1/2}.$$

Example 6.3.5 Calculate the membrane resistance R_m for the following membranes/filters:

(1) $\varepsilon_m = 0.6$, $r_p = 0.1$ μm, $\delta_m = 1$ μm;
(2) $\varepsilon_m = 0.6$, $r_p = 0.05$ μm, $\delta_m = 2$ μm.

Note: The membrane thickness δ_m corresponds to that of a thin layer, having the required pore size and porosity, on top of a thicker highly porous support layer having at least an order of magnitude larger pore size; therefore the support layer resistance may be neglected.

Solution Employ equations (6.3.135c,d) for a model based on uniform straight cylindrical pores:

$$R_m = \frac{8\,\delta_m}{\varepsilon_m\,r_p^2}; \qquad \varepsilon_m = N_p \pi\,r_p^2.$$

Here $\overline{r_p^2} = r_p^2$ for pores of the same size.

Case (1):

$$R_m = \frac{8 \times 10^{-4}\text{cm}}{0.6 \times 0.1 \times 10^{-4} \times 0.1 \times 10^{-4}\text{cm}^2}$$

$$= \frac{8 \times 10^{-4} \times 10^{11}}{6} = 1.5 \times 10^7\text{cm}^{-1}.$$

Case (2):

$$R_m = \frac{8 \times 2 \times 10^{-4}\text{cm}}{0.6 \times 0.05 \times 10^{-4} \times 0.05 \times 10^{-4}\text{cm}^2}$$

$$= \frac{16 \times 10^{-4} \times 10^{13}}{6 \times 5 \times 5} = \frac{16}{150} \times 10^9 = 1.06 \times 10^8\text{cm}^{-1}.$$

Example 6.3.6 Calculate the cake resistance R_c for the following cake properties: void fraction in the cake = 0.35; the particles forming the cake are rigid and have an effective diameter of 10 μm; the cake thickness is 6 cm; $a_1 = 5$.

Solution Employ equation (6.3.135k):

$$d_p = 10 \times 10^{-4}\text{ cm};$$

$$R_c = \frac{a_1(1-\varepsilon_c)^2 a_{vc}^2 \delta_c}{\varepsilon_c^3}\text{cm}^{-1}; \qquad \varepsilon_c = 0.35; \qquad a_{vc} = \frac{6}{d_p}$$

$$\Rightarrow a_{vc} = \frac{6}{10 \times 10^{-4}}\text{cm}^{-1}; \qquad \delta_c = 6\text{ cm}$$

$$\Rightarrow R_c = \frac{5 \times 0.65 \times 0.65 \times 36 \times 6}{0.35 \times 0.35 \times 0.35 \times 10^{-6}}\text{cm}^{-1};$$

$$R_c = \frac{5 \times 0.4225 \times 36 \times 6 \times 10^8}{4.28} = 1.066 \times 10^{10}\text{cm}^{-1}.$$

Example 6.3.7 In the batch filtration of a particular aqueous slurry having a density of 17.5 g of solid per liter of suspension, the vacuum based applied ΔP is 55 cm Hg; the solids volume fraction in the cake is 0.65. The suspension viscosity is 2 cp. The mass density ρ_s of the solid in the suspension is 1.5 g/cm^3. The batch filtration data plotted according to Figure 6.3.23(b) yielded a slope of 10 s/cm^2.

(1) Determine $\hat{\boldsymbol{R}}_{c\delta}$ for an incompressible cake.
(2) Using the approximation for large t, determine the value of R_c at $t = 500$ minutes.

Solution (1) The slope of the plot according to Figure 6.3.23(b) is $\left(\mu \hat{R}_{c\delta}\phi_s/2(\phi_c - \phi_s)\Delta P\right)$. Here $\mu = 2\text{ cp} = 2 \times 10^{-2}$ g/cm-s; $\phi_c = 0.65$;

$$\Delta P = 55\text{ cm Hg} = 550\text{ mm Hg} = 550 \times 1.33 \times 10^3\text{ g/cm-s}^2;$$

$$\phi_s = \text{solids volume fraction in suspension}$$

$$= \frac{17.5\text{ g solid}/1.5\text{ g/cm}^3}{1000\text{ cm}^3} = \frac{11.66}{1000}$$

(here we assume that the effective density of suspension is essentially 1 g/cm^3); $\phi_s = 0.01166$. Therefore, the slope is given by

$$10\text{ s-cm}^2 = \frac{\mu \hat{R}_{c\delta}\phi_s}{2(\phi_c - \phi_s)\Delta P} = \frac{2 \times 10^{-2}\,\frac{\text{g}}{\text{cm-s}} \times \hat{R}_{c\delta} \times 0.01166}{2 \times 0.65 \times 550 \times 1.33 \times 10^3\,\frac{\text{g}}{\text{cm-s}^2}}$$

$$\Rightarrow \hat{R}_{c\delta} = \frac{10 \times 0.65 \times 5.5 \times 1.33 \times 10^{10}}{11.66}\text{cm}^{-2} = 4.07 \times 10^{10}\text{ cm}^{-2}.$$

(2) For large t, we obtain from equation (6.3.138i),

$$\delta_c(t) \cong \left(2\phi_s \Delta P\, t/(\phi_c - \phi_s)\mu \hat{R}_{c\delta}\right)^{1/2}.$$

At $t = 500$ mins,

$$\delta_c(t) = \left(\frac{2 \times 11.66 \times 550 \times 1.33 \times 10^3\,\frac{\text{g}}{\text{cm-s}^2} \times 500 \times 60\text{ s}}{10^3 \times 0.65 \times 2 \times 10^{-2}\,\frac{\text{g}}{\text{cm-s}} \times 4.07 \times 10^{10}\,\frac{1}{\text{cm}^2}}\right)^{1/2}$$

$$= \left(\frac{11.66 \times 5.5 \times 1.33 \times 3 \times 10^9}{0.65 \times 4.07 \times 10^{11}}\text{cm}^2\right)^{1/2}$$

$$= \left(\frac{0.1166 \times 5.5 \times 1.33 \times 3}{0.65 \times 4.07}\right)^{1/2}\text{cm};$$

$$\delta_c(t) = (0.966)^{1/2} = 0.983\text{ cm}$$
$$R_c(t) = \hat{R}_{c\delta}\,\delta_c(t) = 4.07 \times 10^{10} \times 0.983\text{ cm}^{-1};$$
$$R_c(t) = 4 \times 10^{10}\text{ cm}^{-1}.$$

6.3.3.1.1 Membranes and separation characteristics in microfiltration Whereas filter cloths/woven fabrics are useful for filtration of larger particles, microfiltration membranes are widely used in deadend filtration mode to remove effectively suspended matter in the size range 0.1 to 10 μm. The objective can be purification, clarification, sterilization, concentration and analysis. The suspended particle capture/retention efficiency E_T may be described by

$$E_T = \frac{1 - \phi_{\text{out}}}{\phi_{\text{in}}}, \qquad (6.3.139\text{a})$$

where ϕ_{in} and ϕ_{out} are the volume fractions of the particles in the incoming feed stream and the treated stream going through the filter. An alternative index used quite often is the *log reduction value* (LRV) (see Problem 2.4.2(a)):

$$\text{LRV} = \log_{10}\left(\frac{\phi_{\text{in}}}{\phi_{\text{out}}}\right). \qquad (6.3.139\text{b})$$

In many critical applications (see Goel *et al.* (2001)), for example sterilization in the pharmaceutical industry involving water for injection, parenterals, serum and plasma processing, aseptic processing, etc., or clarification and biological sterilization in the beverage industry for absolute microbial removal without heat or chemicals, the microfilters have to be totally retentive of the particulate contaminant/bacteria, etc. For example, sterilizing grade filters are designed to be totally retentive of all bacteria. This is determined by testing the membrane/

microfilter with the bacterium *Pseudomonas diminuta*, which has a diameter of 0.3 μm but an aspect ratio of 1.5 to 1. Sterilizing grade filters which are rated for 0.22 μm have been shown to retain *P. diminuta* completely. The largest pore diameter in the membrane/filter must not allow such a bacteria to enter the membrane. The pore diameter based characterization of the membrane is achieved by the so called *bubble point* technique.

In the "bubble point" technique, the membrane pores are filled with a liquid that spontaneously wets the pores. A gas is brought on one side of the membrane at a pressure. As the pressure is increased, at a certain pressure gas bubbles first appear on the other side of the membrane. The *bubble point pressure*, P_{BP}, corresponds to the lowest gas pressure at which gas bubbles appear on the other side. This value is obtained from the Young–Laplace equation:

$$P_{BP} = \frac{4\gamma \cos\theta}{d_p}, \qquad (6.3.140)$$

were d_p is the pore diameter, γ is the surface tension of the liquid and θ is the contact angle between the membrane surface and the liquid. Often an experimentally determined tortuosity factor, τ_m, is used in the denominator of the expression for P_{BP}; τ_m can vary in the neighborhood of 5 to 3.33. The higher the value of P_{BP}, the lower the size of the largest pore. In practice, when the LRV is greater than 9, i.e. (ϕ_{in}/ϕ_{out}) is greater than 10^9, the filter is considered to be totally retentive (see Goel *et al.* (2001) for a detailed introduction).

Polymers commonly used to make such microfiltration membranes are: polyvinylidene fluoride (PVDF), Nylon 66, polytetrafluoroethylene (PTFE), polysulfone, cellulose, cellulose acetate/cellulose nitrate, polypropylene (PP), polyester, polycarbonate, etc. Ceramic microfiltration membranes are not uncommon. Polymeric membranes may have the following structures.

Homogeneous: porosity and pore size uniform along the membrane thickness.

Asymmetric: the membrane has a thin selective skin region at the top facing the liquid/gas stream to be filtered with a pore size much lower than that of the rest of the membrane.

Composite: thin selective skin on top of the porous membrane is made of a different material.

The treatment so far has focused essentially on microfiltration and cake filtration of a liquid feed. However, microfiltration of process gas streams, as well as gas flowing through vents on tanks, containers, fermentors, etc., are also quite important in the deadend mode. Generally these filters are made of hydrophobic materials like PTFE, PVDF, PP, etc., so that water does not wet the pores and gases can easily flow through the filter pores.

6.3.3.2 *Ultrafiltration*

Ultrafiltration (UF) is a pressure driven membrane separation technique involving solute(s) in a solution, where the solute molecular weight can vary from around 1000 to 500 000. It is used primarily to (1) concentrate a solute in a solution by removing the solvent through the membrane, (2) purify the solution from smaller molecular weight impurities which pass through the membrane along with the solvent and (3) fractionate solute mixtures in a solution. Since the solutes have molecular weights larger than 1000, they are often identified as *macrosolutes*. Specifically, the macrosolutes of relevance are proteins, macromolecules, viruses, etc. Colloidal particles of appropriate dimensions may also be considered as macrosolutes. Comprehensive treatments of ultrafiltration are available in Zeman and Zydney (1996), Cheryan (1998) and Kulkarni *et al.* (2001). At the low end of the solute molecular weight range, this technique overlaps the technique of *nanofiltration/reverse osmosis*; at the high end, it merges with *microfiltration*.

In ultrafiltration, the feed solution containing one or more macrosolutes with or without small molecular weight buffer solutes/impurities is imposed over an ultrafiltration (UF) membrane at above atmospheric pressure (Figure 6.3.1(j)); the other side of the membrane is usually at atmospheric pressure. The solvent, low molecular weight buffer solutes/impurities and the macrosolutes, whose dimensions are smaller than the pore dimensions of the UF membrane, pass through the membrane; larger macrosolutes are retained by the membrane on the feed side. The rates at which the solvent, the microsolutes and/or the macrosolutes pass through the membrane depend on a number of factors. These factors include: applied pressure difference across the membrane (ΔP), the solution properties, the solute dimensions, the membrane structure, the solute–membrane interaction, the bulk feed flow vs. force configuration and the macrosolute transport characteristics of the feed flow field over the membrane. These last two conditions, as well as the nature and magnitude of the bulk flow with respect to the magnitude of the force, are of considerable importance in UF, as will be demonstrated.

A quantitative analysis of separation in UF requires first a knowledge of the transport rates of the solvent and the macrosolutes through the UF membrane. When the macrosolute molecular weights are not high ($\geq$1000), the membrane pores may have dimensions in the range ~1–2 nm; the osmotic pressure of a concentrated solution of such macrosolutes will be significant with respect to the applied pressure difference, ΔP. The molar solvent flux under ideal conditions will be described by the flux expression (3.4.54) (Vilker *et al.*, 1981):

$$N_{sz} = A(\Delta P - \Delta\pi) = \frac{(\Delta P - \Delta\pi)}{R_m + R_c}, \qquad (6.3.141a)$$

where R_m is the resistance of the membrane, R_c is the resistance of any layer deposited over the membrane, π is the osmotic pressure of a solution containing the macrosolute and $\Delta\pi$ is equal to $(\pi_f - \pi_p)$. The subscripts f and p refer, respectively, to the feed side and the permeate (product/filtrate) side. For such cases, the membrane is called a diffusive ultrafilter and the technique is sometimes identified as *diffusive ultrafiltration*. However, instead of J^*_{sz}, N_{sz} is used in UF because of the substantially enhanced role of solvent convection, which will be considered next.

For larger membrane pores, and larger macrosolutes whose solutions have much smaller osmotic pressures, the membrane is sometimes called a *microporous ultrafilter*; the membrane pores are micropores/mesopores/large pores and the solvent volume flux $(N_{sz}\overline{V}_s)$ is described by *Darcy's law*, expression (3.4.88):

$$(N_{sz}\overline{V}_s) = \frac{Q_{sm}}{\mu}\frac{\Delta P}{\delta_m}. \tag{6.3.141b}$$

The molar macrosolute flux, N_{iz}, for macrosolute i may be described in general by equations (3.4.89a,b), which are based on the "combined diffusion-viscous flow model." From the result (3.4.93b), we can obtain an estimate of the macrosolute concentration, C_{ip}, in the permeate as a function of a number of quantities, including the feed macrosolute concentration, C_{if}, and the solvent flux, $N_{sz}\overline{V}_s$, when there is no concentration polarization in the feed solution. Knowing C_{ip} and $N_{sz}\overline{V}_s$, one may calculate the extent of separation achieved in the UF device.

However, in practical UF processes, there is always concentration polarization on the feed side. Figure 6.3.26 (a) illustrates the macrosolute concentration (C_{il}) profile in the feed solution in a batch cell; it increases from the value $C_{ilb}(= C_{if})$ of the bulk solution to the concentration (C^0_{il}) at the membrane-feed solution interface where the macrosolute is rejected by the membrane. This rejection may be complete $(C_{ip} = 0)$ or incomplete $(0 < C_{ip} < C^0_{il})$. The observed value, R_{obs}, of the membrane's *solute rejection* or *solute retention*,

$$R_{obs} = \left(1 - \frac{C_{ip}}{C_{if}}\right), \tag{6.3.141c}$$

is different from the true solute rejection, R_{true}, by the membrane,

$$R_{true} = \left(1 - \frac{C_{ip}}{C^0_{il}}\right). \tag{6.3.141d}$$

Correspondingly, the observed *solute transmission* or *sieving coefficient*, S_{obs}, will be different from the true value, S_{true}, where

$$S_{obs} = \frac{C_{ip}}{C_{if}}; \qquad S_{true} = \frac{C_{ip}}{C^0_{il}} \tag{6.3.141e}$$

and

$$R_{obs} = 1 - S_{obs}; \qquad R_{true} = 1 - S_{true}. \tag{6.3.141f}$$

In the configuration of Figure 6.3.26(a), the rate at which the solvent passes through the membrane, and therefore the solvent velocity through the membrane, is equal to the magnitude of the velocity, v_z, of the solution in the feed side toward the membrane. It is as if there is a piston on the feed solution side generating the feed pressure P_f: the piston velocity toward the membrane can only equal the velocity of the solvent/solution through the membrane for this incompressible liquid (Figure 6.3.26(b)).

Consider a control volume (CV) extending from the membrane surface to a distance z into the feed liquid, where $z \leq \delta_l$, the thickness over which the concentration changes (Figure 6.3.26(a)). Assume a *pseudo steady state*.[26] Then a solute balance over the CV leads to the following:

$$\begin{pmatrix}\text{solute flux into}\\ \text{the CV at } z\\ \text{by convection}\end{pmatrix} - \begin{pmatrix}\text{solute flux out}\\ \text{of the CV at } z=0 \text{ (membrane)}\\ \text{by membrane transport}\end{pmatrix} = \begin{pmatrix}\text{solute flux out of the}\\ \text{CV at } z \text{ into the bulk}\\ \text{liquid by diffusion}\end{pmatrix}$$

$$|v_z|C_{il} \quad - \quad |v_z|C_{ip} \quad = \quad -D_{il}\frac{dC_{il}}{dz}. \tag{6.3.142a}$$

Here, $|v_z|$ is the magnitude of the fluid velocity normal to the membrane. We have assumed that it does not change over $0 \leq z \leq \delta_l$; further, at the wall $(z = 0)$, $|v_z| = N_s\overline{V}_s$, the solvent volume flux through the membrane. Integrate (6.3.142a) over the boundary layer thickness:

[26] In the batch cell, the bulk macrosolute concentration changes with time as the solvent and some of the macrosolutes pass through the membrane.

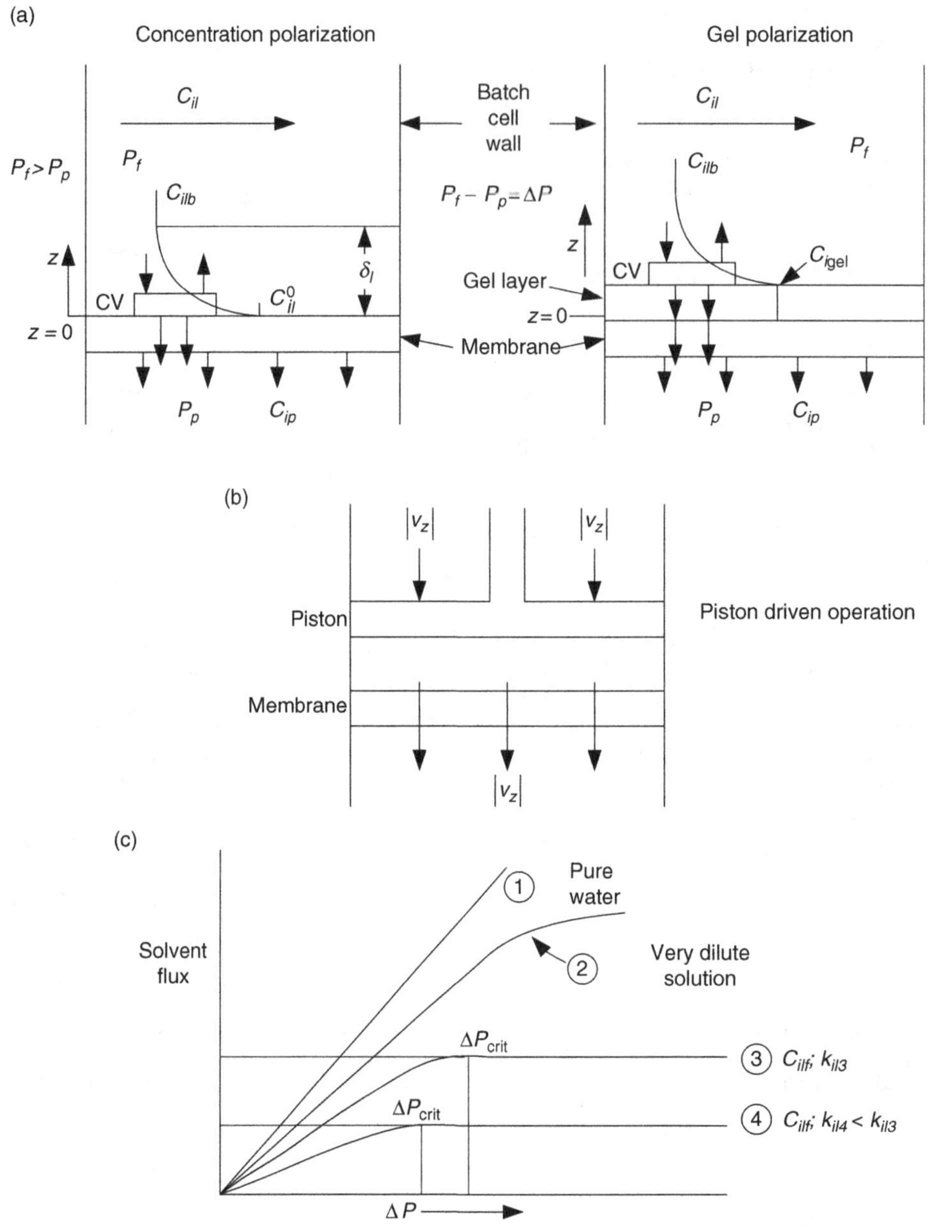

Figure 6.3.26. Ultrafiltration. (a) UF in a batch cell: macrosolute concentration profile in feed side. (b) Piston driven UF in a batch cell: bulk flow parallel to the force. (c) Observed behavior of solvent flux vs. ΔP in macrosolute ultrafiltration. For an explanation of (1)–(4), see the text.

$$\int_{C_{il}^0}^{C_{ilb}} \frac{dC_{il}}{(C_{il} - C_{ip})} = -\frac{|v_z|}{D_{il}} \int_0^{\delta_l} dz \Rightarrow \ln\left(\frac{C_{il}^0 - C_{ip}}{C_{ilb} - C_{ip}}\right) = \frac{|v_z|}{(D_{il}/\delta_l)} = \frac{N_s \overline{V}_s}{k_{il}};$$

$$\frac{C_{il}^0 - C_{ip}}{C_{ilb} - C_{ip}} = \exp\left(\frac{|v_z|}{(D_{il}/\delta_l)}\right) = \exp\left(\frac{|v_z|}{k_{il}}\right). \tag{6.3.142b}$$

For the above result, we have described the solute mass-transfer coefficient in the feed side above the membrane via

$$k_{il} = (D_{il}/\delta_l). \tag{6.3.142c}$$

The result illustrated by (6.3.142b) is singularly important in UF. If in the feed solution, the mass-transfer coefficient k_{il} of the macrosolute i (which is rejected (partially or totally) at the membrane surface) from the interface to the bulk liquid is not sufficiently large relative to the solvent flux through the membrane, the wall concentration,

C_{il}^0, builds up. For *proteins*, a gel layer (Michaels, 1968a) may often be developed on the UF membrane surface when C_{il}^0 is high enough to be C_{igel}, leading to what is called *gel polarization* (Figure 6.3.26(a)):

$$N_s\overline{V_s} = |v_z| = k_{il}\ln\left(\frac{C_{igel} - C_{ip}}{C_{ilb} - C_{ip}}\right). \qquad (6.3.143)$$

Suppose $C_{ip} \cong 0$, i.e. the membrane effectively rejects/retains the macrosolute i. Since C_{igel} is fixed for a particular macrosolute i in a given environment,[27] one can increase the solvent flux only by increasing k_{il}. Although the solvent flux expressions (6.3.141a,b) suggest that an increase in ΔP will increase the solvent flux, no such increase takes place under the condition of gel polarization.

This brings up an important characteristic of UF based separation. If macrosolute molecules are being brought to an interface (in this case, the membrane-feed solution interface), for steady state operation, these macrosolute molecules also have to be removed/evacuated from this interface. Removal of these rejected solute macromolecules from this interface to the feed solution bulk by diffusion is essential to continuing the separation at an acceptable rate in steady state operation (compare the general arguments provided in the two paragraphs around equation (3.2.37)). Here one cannot arbitrarily keep on increasing $|v_z|$, i.e. the solvent flux, by increasing ΔP; the bulk velocity parallel to the force direction (through the membrane) cannot be increased arbitrarily unless the rate of removal of the accumulated macromolecules from the interface back to the bulk is simultaneously increased. The latter rate is inherently low in the bulk flow parallel to the force configuration.

This basic phenomenon in UF has influenced its development in many ways. High concentration of macrosolutes at the interface coupled with a higher solvent flux through the membrane leads to a higher rate of leakage of the macrosolute through the membrane (see equations (3.4.93a,b)); membrane fouling by the macromolecules is also increased. To avoid such conditions, one can operate UF at a lower ΔP and carry on with a lower solvent flux. Figure 6.3.26(c) schematically illustrates the solvent flux behavior in UF as a function of ΔP for a number of conditions involving proteins as macrosolutes: (1) no macrosolute, simple water flow through the membrane, water flux linearly proportional to ΔP; (2) extremely dilute feed solution of macromolecules, leading to a somewhat lower water flux than in (1); (3) higher feed concentration solution of macromolecules at a certain value of k_{il}, displaying gel polarization beyond a critical ΔP; (4) solution concentration as in (3), but operating at a lower k_{il}, displaying gel polarization and a lower water flux level.

[27] There is considerable debate about what this C_{igel} is in equation (6.3.143).

While Figure 6.3.26(c) illustrates how increasing polarization affects the solvent flux, relation (6.3.142b) may be rearranged to quantify how the macrosolute retention/rejection/transmission is affected simultaneously (if the macrosolute is not rejected completely). Rearrange this equation as follows:

$$\frac{\left(1 - \dfrac{C_{ip}}{C_{il}^0}\right)}{\left(1 - \dfrac{C_{ip}}{C_{ilb}}\right)}\left(\frac{C_{il}^0}{C_{ilb}}\right) = \exp\left(\frac{|v_z|}{k_{il}}\right) = \left(\frac{1 - S_{\text{true}}}{1 - S_{\text{obs}}}\right)\left(\frac{S_{\text{obs}}}{S_{\text{true}}}\right)$$

$$\Rightarrow (1 - S_{\text{obs}})\exp\left(\frac{|v_z|}{k_{il}}\right) = \frac{S_{\text{obs}}}{S_{\text{true}}} - S_{\text{obs}}$$

$$\Rightarrow S_{\text{true}} = \frac{S_{\text{obs}}}{(1 - S_{\text{obs}})\exp\left(\frac{|v_z|}{k_{il}}\right) + S_{\text{obs}}} \qquad (6.3.144a)$$

Correspondingly,

$$\exp\left(\frac{|v_z|}{k_{il}}\right) = \left(\frac{R_{\text{true}}}{R_{\text{obs}}}\right)\frac{(1 - R_{\text{obs}})}{(1 - R_{\text{true}})};$$
$$\ln\left(\frac{1 - R_{\text{obs}}}{R_{\text{obs}}}\right) = \ln\left(\frac{1 - R_{\text{true}}}{R_{\text{true}}}\right) + \frac{|v_z|}{k_{il}}. \qquad (6.3.144b)$$

Thus, knowing the solvent volume flux ($|v_z| = N_{sz}\overline{V_s}$) through the membrane and the macrosolute mass-transfer coefficient, k_{il}, in the feed solution, the true macrosolute retention of the membrane, R_{true}, may be determined from the observed retention values, R_{obs}. A plot of $((1 - R_{\text{obs}})/R_{\text{obs}})$ against the observed $|v_z|$ in a semilog plot will yield the value of R_{true} from the intercept $((1 - R_{\text{true}})/R_{\text{true}})$. From the slope of the line, one can also obtain an estimate of k_{il} in the system.

From the solute flux expression (3.4.91c) in the *combined diffusion-viscous flow model*, one can develop an expression for S_{true} or R_{true}. Consider relation (3.4.93b):

$$\frac{C_{il}^\delta}{C_{il}^0} = \frac{\overline{G}_i\kappa_{if}\exp(\overline{G}_i v_z \tau_m \delta_m/D_{ip})}{\overline{G}_i\kappa_{ip} - 1 + \exp(\overline{G}_i v_z \tau_m \delta_m/D_{ip})},$$

where the argument of the exponential function is the pore Péclet number for solute i, Pe_i^m. Therefore

$$\frac{C_{il}^\delta}{C_{il}^0} = \frac{C_{ip}}{C_{il}^0} = S_{\text{true}} = 1 - R_{\text{true}} = \frac{\overline{G}_i\kappa_{if}\exp(Pe_i^m)}{\overline{G}_i\kappa_{ip} - 1 + \exp(Pe_i^m)}. \qquad (6.3.145a)$$

It is also known that, when $Pe_i^m >> 1$, employing relations (3.4.92a) and (3.4.93a), we get

$$\frac{C_{il}^\delta}{C_{il}^0} = \overline{G}_i\kappa_{if} = S_{\text{true}}|_{Pe_i^m \to \infty} = S_\infty = 1 - R_\infty, \qquad (6.3.145b)$$

where S_∞ or R_∞ are asymptotic values achieved at high solvent flux ($Pe_i^m >> 1$). Substitution into relation (6.3.145a), and the assumption that $\kappa_{if} \cong \kappa_{ip}$, leads to

$$S_{\text{true}} = 1 - R_{\text{true}} = \frac{S_\infty \exp(Pe_i^m)}{S_\infty + \exp(Pe_i^m) - 1}. \qquad (6.3.145\text{c})$$

Such an analysis assumes that electrostatic and electrokinetic interactions are not important in macrosolute transport through the membrane. Pujar and Zydney (1994) have considered such effects in protein ultrafiltration through narrow pore membranes.

The above analysis/description of solvent flux and macrosolute rejection/retention/transmission for an ultrafiltration membrane was carried out in the context of a pseudo steady state analysis in a batch cell (Figure 6.3.26 (a)). Back diffusion of the macrosolute from the feed solution–membrane interface to the bulk solution takes place by simple diffusion against the small bulk flow parallel to the force direction. The resulting mass-transfer coefficients for macrosolutes will be quite small; the solvent flux levels achievable will be quite low. For practically useful ultrafiltration rates, the mass-transfer coefficient is increased via different flow configurations with respect to the force.

In a small vessel like that shown in Figure 6.3.26(a), a stirrer is incorporated to increase k_{il}; it may become a well-stirred vessel. A continuous buffer stream may also enter such a well-stirred vessel containing an ultrafilter; however, the conditions in the vessel will change with time, unlike those for a CSTS at steady state (see Section 6.4.2.1). These two configurations are used in laboratory-scale applications. In larger-scale commercial/industrial applications, the bulk flow direction employed is perpendicular to the force direction (see Section 7.2.1.3), a configuration commonly identified as *crossflow*. Such a flow configuration can lead to high values of k_{il}, which can then sustain high solvent flux; further steady state conditions can be maintained at any location in the flow channel bounded by membranes. In the bulk flow parallel to the force configuration of Figures 6.3.26(a) and (b), steady state conditions cannot be maintained. In practice, slow fouling of the membrane leads to a slow reduction in membrane solvent flux.

The treatment discussed so far has not provided a physical picture of how the macrosolute rejection takes place in the UF membrane. Let us imagine this UF membrane (Figure 6.3.27(a)) to consist of a bunch of cylindrical capillaries traversing the thickness of the membrane in an otherwise impervious medium of membrane material (usually polymeric but sometimes ceramic). The diameters of the capillaries may be uniform or variable. The capillaries may be straight or tortuous, with diameters changing along the length; some of the capillaries may be interconnected, others not. A real life UF membrane may have a variety of such features. Consider now a simple case: all

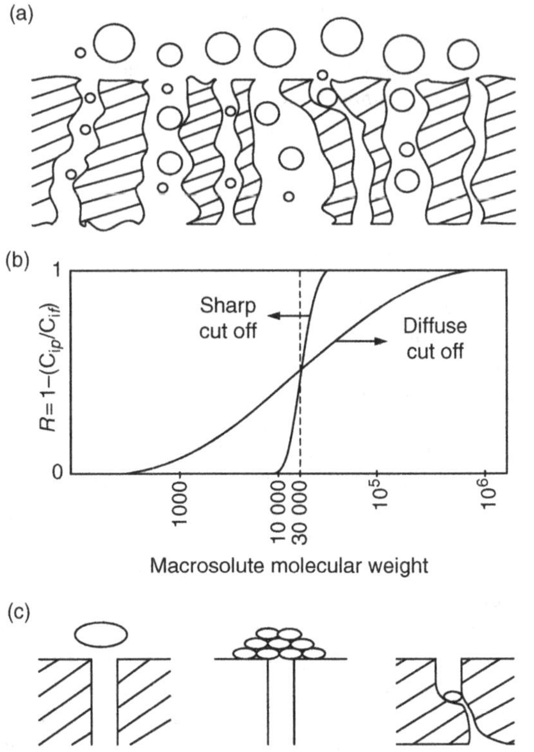

Figure 6.3.27. (a) UF membrane as a bundle of size-distributed tortuous capillaries. (b) Macrosolute retention behavior of two types of UF membranes having a narrow or a broad pore size distribution. (c) Macrosolute retention via smaller membrane pore size, pore mouth adsorption, pore blockage due to pore constriction.

capillaries have the same diameter, and this diameter is constant along the membrane thickness. If we have spherical macromolecules in the feed solution whose diameters are significantly smaller than that of the capillaries, the pores, then these macromolecules will pass through the membrane pores. However, if the macromolecules have diameters larger than that of the pores, they cannot enter the membrane pores and are rejected.

In real life, most membranes have a *distribution* of capillary diameters, what is called a *pore size distribution.* For a given macromolecule, some of the pore diameters in a conventional UF membrane will be larger, some smaller, than the macromolecular dimensions. Therefore, the macromolecules will pass through the larger pores and not through the smaller ones, resulting in a permeate solution concentration of macromolecules less than that of the feed side, i.e. finite solution rejection. A highly simplistic analysis of macrosolute rejection of a microporous ultrafiltration membrane will be carried out now for a membrane having a pore size density function $f(r_p)$, where $r_{\min} < r_p < r_{\max}$ (see the corresponding analysis for solvent flux in Section 3.4.2.3). The treatment follows Michaels' analysis (Michaels, 1968b).

Suppose that the solvated macrosolute radius is r_i. Further, let the macrosolute concentration in the larger pores ($r_p > r_i$) be equal to C_{il}^0, i.e. the macrosolute concentration at the feed solution-membrane interface. On the other hand, the macrosolute concentration in smaller pores ($r_p \leq r_i$) will be zero. The macrosolute flux through the membrane is therefore

$$N_{iz} = C_{il}^0 \times \text{volume flux of solvent through pores whose } r_p > r_i = C_{il}^{\circ}\left[\int_{r_i}^{r_{\max}} \mathrm{d}N_{sz}\right]\overline{V_s}, \tag{6.3.146a}$$

where $(\mathrm{d}N_{sz})\overline{V_s}$ is the contribution of the pores of radius r_p to $r_p + \mathrm{d}r_p$ to the solvent volume flux:

$$(\mathrm{d}N_{sz})\overline{V_s} = \frac{\varepsilon_m r_p^2 f(r_p)\mathrm{d}r_p}{8\mu}\left(\frac{\Delta P}{\tau_m \delta_m}\right). \tag{6.3.146b}$$

The total solvent flux (volume flux) from (3.4.86) is

$$N_{sz}\overline{V_s} = \frac{\varepsilon_m \overline{r_p^2}}{8\mu}\left(\frac{\Delta P}{\tau_m \delta_m}\right).$$

Since the macrosolute concentration, C_{ip}, in the filtrate (permeate) is given by

$$N_{iz} = C_{ip} N_{sz} \overline{V_s}, \tag{6.3.146c}$$

we obtain

$$\frac{C_{ip}}{C_{il}^0} = \frac{N_{iz}}{C_{il}^{\circ} N_{sz}\overline{V_s}} = \frac{C_{il}^0\left[\int_{r_i}^{r_{\max}} \frac{\varepsilon_m r_p^2 f(r_p)}{8\mu}\mathrm{d}r_p\left(\frac{\Delta P}{\tau_m \delta_m}\right)\right]}{C_{il}^0 \frac{\varepsilon_m \overline{r_p^2}}{8\mu}\left(\frac{\Delta P}{\tau_m \delta_m}\right)}$$

$$= \frac{\int_{r_i}^{r_{\max}} r_p^2 f(r_p)\mathrm{d}r_p}{\overline{r_p^2}} = \frac{\int_{r_i}^{r_{\max}} r_p^2 f(r_p)\mathrm{d}r_p}{\int_{r_{\min}}^{r_{\max}} r_p^2 f(r_p)\mathrm{d}r_p}$$

$$= S_{\text{true}} = 1 - R_{\text{true}}. \tag{6.3.146d}$$

Therefore

$$R_{\text{true}} = \frac{\int_{r_{\min}}^{r_i} r_p^2 f(r_p)\mathrm{d}r_p}{\overline{r_p^2}}. \tag{6.3.146e}$$

This interesting, but highly simplified, result suggests that the membrane rejection/retention/transmission of a macrosolute through a microporous ultrafilter depends solely on the characteristics of the pore size distribution of the membrane and the macrosolute dimension. If the membrane has a broad pore size distribution, macromolecules having a wide range of molecular weights and sizes will be able to pass through the membrane. Such a membrane is said to have a *diffuse cut off*. On the other hand, if the membrane has a narrow pore size distribution, the membrane is assumed to have a *sharp cut off*; the membrane is such that the size difference between the macrosolute which is completely retained and the macrosolute which passes through with very little retention is quite small. The macrosolute retention behaviors of these two types of membranes are illustrated in the semilog plot of Figure 6.3.27(b).

It is now useful to explore the factors which control the selectivity of the UF membrane for one macrosolute ($i = 1$) of solvated radius r_1 over another macrosolute ($i = 2$) of solvated radius r_2 ($>r_1$). If we define the *selectivity/separation factor* of the membrane for species 1 over 2 via

$$\alpha_{12} = \frac{x_{1p}x_{2f}}{x_{1f}x_{2p}}, \tag{6.3.147a}$$

then, for dilute solutions, we may rewrite this as

$$\alpha_{12} \cong \frac{C_{1p}C_{2f}}{C_{1f}C_{2p}} = \left(\frac{C_{1p}}{C_{1f}}\right)\Big/\left(\frac{C_{2p}}{C_{2f}}\right). \tag{6.3.147b}$$

From equation (6.3.146d) we obtain, in the absence of concentration polarization,

$$\frac{C_{1p}}{C_{1l}^0} = \frac{C_{1p}}{C_{1f}} = \frac{\int_{r_1}^{r_{\max}} r_p^2 f(r_p)\mathrm{d}r_p}{\overline{r_p^2}}. \tag{6.3.147c}$$

Similarly,

$$\frac{C_{2p}}{C_{2l}^0} = \frac{C_{2p}}{C_{2f}} = \frac{\int_{r_2}^{r_{\max}} r_p^2 f(r_p)\mathrm{d}r_p}{\overline{r_p^2}}. \tag{6.3.147d}$$

Therefore

$$\alpha_{12} \cong \frac{\int_{r_1}^{r_{\max}} r_p^2 f(r_p)\mathrm{d}r_p}{\int_{r_2}^{r_{\max}} r_p^2 f(r_p)\mathrm{d}r_p} = \frac{\int_{r_1}^{r_2} r_p^2 f(r_p)\mathrm{d}r_p + \int_{r_2}^{r_{\max}} r_p^2 f(r_p)\mathrm{d}r_p}{\int_{r_2}^{r_{\max}} r_p^2 f(r_p)\mathrm{d}r_p}$$

$$= 1 + \frac{\int_{r_1}^{r_2} r_p^2 f(r_p)\mathrm{d}r_p}{\int_{r_2}^{r_{\max}} r_p^2 f(r_p)\mathrm{d}r_p}. \tag{6.3.147e}$$

Suppose now that r_2 is sufficiently small compared to r_{max}, i.e. the membrane allows a substantial amount of species 2 to pass through. Therefore, unless r_1 and r_2 are far apart, the integral in the numerator on the right-hand side of (6.3.147e) is small compared to the denominator. The selectivity of a "diffuse cut off" microporous ultrafilter is not very high unless the macrosolutes differ considerably in the values of their solvated radii. Traditionally, therefore, fractionation of macrosolutes like proteins by ultrafiltration is limited to systems where the two macrosolutes differ in size by about seven to ten times (Cherkasov and Polotsky, 1996).

Enhancement of the membrane selectivity for a given macrosolute has been achieved by considering a number of other factors, including increasing/decreasing the effective radius of a charged protein molecule. If the UF membrane has some charge on the surface, then macrosolutes having the same charge will be effectively rejected/repulsed by the membrane. For proteins, this can be achieved by changing the solution *pH*. If the solution *pH* is greater than the *pI* of the protein, the protein will have a net negative charge (see Figure 4.2.5(c)); if the membrane has a negative charge, this protein will be excluded from the membrane pores. If the ionic strength of the solution is, however, increased substantially, then the extent of electrostatic shielding of the charged protein molecule will be substantially increased. The negatively charged membrane, for example, will not be able to reject the protein as much, resulting in decreased selectivity in relation to another protein which may be uncharged (for example, bovine serum albumin, $pI = 4.7$) at the solution *pH* (say 4.7). (See the following references: Saksena and Zydney (1994); van Eijndhoven *et al.* (1995); Nyström *et al.*, (1998).) By stacking three such membranes one over the other, Feins and Sirkar (2004, 2005) were able to obtain one pure protein in the permeate for a binary mixture of proteins whose molecular weight ratio was as low as 1.03 to 2.05. This internal-staging concept has not been treated further.

There are additional factors that influence membrane selectivity or retention. If the macrosolute has tendencies of adsorption on the membrane surface, the adsorbed macromolecules may block the pore entrance or form a layer on top of the membrane changing the membrane selectivity (Figure 6.3.27(c)). If the pore diameter changes along the membrane thickness, a macrosolute which could enter the pore at the feed side may plug the pore, render it useless and change the pore size distribution. Due to lack of convection, the top of this pore surface becomes a stagnant zone. A gel layer formed on top of the membrane from one macrosolute may substantially influence the transport behavior of a smaller macrosolute. A globular macrosolute/protein is likely to have a higher solute rejection than a linear macromolecule of the same molecular weight.

A variety of UF membranes are commercially available. Table 6.3.8 provides an illustration of the properties and performance characteristics of a series of polymeric flat membranes that are used in a batch cell. The data on solute rejection were acquired in a stirred batch UF cell of the type to be considered in Section 6.4.

Example 6.3.8 Saksena and Zydney (1994) have studied the protein transmission characteristics of a 100 000 MWCO ultrafiltration membrane (OMEGA 100K) of polyethersulfone using bovine serum albumin (BSA) as a model protein (mol. wt. 66 430) in a solution of $pH = 7.0$ and ionic strength of 0.15 M NaCl solution. The batch cell operation characteristics and the protein transport properties are as follows:

$k_{i\ell} = 5.2 \times 10^{-6}\ \text{m/s}$; $Pe_i^m = (S_\infty J_v \delta_m / D_{ieff})$; $S_\infty = 0.016$; $D_{ieff} = 1 \times 10^{-13}\ \text{m}^2/\text{s}$; $\delta_m = 0.5\ \mu\text{m}$; $J_v = |v_z| = 10^{-6}\ \text{m/s}$.

Determine the values of the observed and true transmission coefficients, S_{obs} and S_{true}, respectively, of BSA under these conditions. (See Table 6.3.8 for a definition of MWCO.)

Solution We will first employ relation (6.3.145c) between S_{true} and S_∞: $S_{true} = \frac{S_\infty \exp(Pe_i^m)}{S_\infty + \exp(Pe_i^m) - 1}$.

Here

$$S_\infty = 0.016; \qquad Pe_i^m = \frac{S_\infty J_v \delta_m}{D_{ieff}}$$

$$= \frac{0.016 \times 10^{-6}\,\frac{\text{m}}{\text{s}} \times 0.5 \times 10^{-6}\,\text{m}}{1 \times 10^{-13}\,\text{m}^2/\text{s}};$$

$$Pe_i^m = 0.08 \Rightarrow S_{true} = \frac{0.016\exp(0.08)}{0.016 + \exp(0.08) - 1}$$

$$= \frac{0.016 \times 1.0833}{0.0993} \Rightarrow S_{true} = 0.1745.$$

Next we utilize relation (6.3.144a) between S_{obs} and S_{true}:

$$S_{true} = \frac{S_{obs}}{(1 - S_{obs})\exp\left(\frac{|v_z|}{k_{i\ell}}\right) + S_{obs}};$$

$$0.1745 = \frac{S_{obs}}{(1 - S_{obs})\exp\left(\frac{10^{-6}}{5.2 \times 10^{-6}}\right) + S_{obs}}$$

$$= \frac{S_{obs}}{(1 - S_{obs})\exp(0.192) + S_{obs}}$$

$$\Rightarrow 1.212 \times 0.1745(1 - S_{obs}) + 0.1745 S_{obs} = S_{obs}$$

$$S_{obs}(1 + 0.211 - 0.1745) = 0.211 \Rightarrow S_{obs} = \frac{0.211}{1.0355}.$$

For BSA, $S_{obs} = 0.204$.

Table 6.3.8 Properties and performance characteristics of flat Amicon UF membranes[a]

	UM05		UM2	UM10	YM10	YM30	PM10	PM30	XM50	XM100A	XM300
Nominal MWCO[b]	500		1000	10 000	10 000	30 000	10 000	30 000	50 000	100 000	300 000
Average pore diameter (nm)	2.1		2.4	3.0	-	4.0	3.8	4.7	6.6	11	48
Water flux[c] (ml/min/cm^2)	-		-	-	0.10–0.20	0.7–1.1	1.5–3.0	2.0–6.0	1.0–2.5	-	0.5–1.0
Material	polyelectrolyte complex		Polyelectrolyte complex	Polyelectrolyte complex	Regenerated cellulose	Regenerated cellulose	Polysulfone	Polysulfone	PAN-co-PVC[d]	PAN-co-PVC	PAN-co-PVC
Macrosolute (MW)			Solute rejection[e]								
	*pH*5	*pH*10									
D-Alanine (89)	15	80	0	0	-	-	0	0	0	0	0
DL-Phenylalanine (165)	20	90	0	0	-	-	0	0	0	0	0
Tryptophan (204)	20	80	0	0	-	-	0	0	0	0	0
Sucrose (342)	70	80	50	25	-	-	0	0	0	0	0
Raffinose (594)	90		-	50	10	-	0	0	0	0	0
Inulin (5,000)	-		80	60	45	-	-	0	0	0	0
Dextran T10 (10 000)	-		90	90	-	-	5	-	-	-	-
Myoglobin (18 000)	>95		>95	95	80	-	80	35	20	-	-
α-Chymotrypsinogen (24 500)	>95		>98	>95	-	>80	>95	75	85	25	0
Albumin (67 000)	>98		>98	>98	>90	>98	>98	>90	>90	45	10
Aldolase (142 000)	>98		>98	>98	-	-	>98	>98	>95	-	50
IgG (160 000)	>98		>98	>98	-	>98	>98	>98	>98	90	65
Apoferitin (480 000)	>98		>98	>98	-	-	>98	>98	>98	>95	85
IgM (960 000)	>98		>98	>98	-	>98	>98	>98	>98	>98	>98

[a] From Amicon catalogs.
[b] MWCO = molecular weight cutoff; it means $R_i = 0.9$ for a solute of the specified molecular weight.
[c] Measured in a stirred cell after 5 minutes of pressure. All membranes at 55 psi, except XM300 at 10 psi.
[d] Copolymer of acrylonitrile and vinyl chloride.
[e] At 55 psi, except 10 psi for XM100A and XM300

6.3.3.3 *Reverse osmosis*

In Section 3.4.2.1, the phenomenon of reverse osmosis (RO) through a nonporous membrane was introduced. If the hydraulic pressure of a solution containing a microsolute, e.g. common salt, on one side of a nonporous membrane exceeds that of another solution on the other side of the same membrane by an amount more than the difference of the osmotic pressures of the same two solutions, then, according to the solution-diffusion model, the solvent will flow from the solution at higher pressure to the one at a lower pressure (equation (3.4.54)) at the following rate:

$$J^*_{sz} = \frac{C_{sm}\,D_{sm}\,\overline{V}_s}{RT\,\delta_m}((P_{\text{high}} - P_{\text{low}}) - (\pi_{\text{high}} - \pi_{\text{low}})) \cong \frac{v_z}{\overline{V}_s}. \tag{6.3.148a}$$

The solution at a higher pressure P_{high} has a higher solute concentration, therefore a higher osmotic pressure π_{high}, and is the feed solution; the corresponding quantities on the other side of the membrane are P_{low} and π_{low}. For reverse osmosis to take place,

$$\Delta P (= P_{\text{high}} - P_{\text{low}}) > \Delta\pi (= \pi_{\text{high}} - \pi_{\text{low}}). \tag{6.3.148b}$$

The equation describing the solute transport according to the solution-diffusion model is (see equation (3.4.59))

$$J^*_{iz} = \frac{D_{im}\,\kappa_{im}}{\delta_m}(C_{if} - C_{ip}). \tag{6.3.149}$$

Consider Figure 6.3.28(b) for a batch cell containing an RO membrane and a feed solution; as the permeation is going on, we can picture it as if the piston is driving the solution toward the membrane, i.e. the bulk flow of the feed solution is parallel to the direction of the force driving the permeation velocity v_z through the membrane. As the solvent permeates through the membrane, if the RO membrane is effective, the solute (e.g. salt) is rejected and the salt concentration builds up on the feed side. Whether we have a batch cell or not, the salt concentration on the feed side of the membrane will change with time (Nakano *et al.*, 1967). Assume, however, a pseudo steady state for the sake of the following analysis. At any instant of time, *t*, we can assume (Nakano *et al.*, 1967) that we have a bulk solute concentration, $C_{i\ell b}$, which increases to $C^0_{i\ell}$ at the RO membrane surface (Figure 6.3.28(a)). Therefore we have back diffusion of the solute (i.e. salt) from the membrane surface to the bulk solution. From the pseudo steady state analysis of *concentration polarization* carried out for ultrafiltration, resulting in equation (6.3.142b), and the corresponding equation for RO, namely (3.4.65c), we have

$$\frac{(C^0_{i\ell} - C_{ip})}{(C_{i\ell b} - C_{ip})} = \exp\left(\frac{|v_z|}{(D_{i\ell}/\delta_\ell)}\right) = \exp\left(\frac{|v_z|}{k_{i\ell}}\right). \tag{6.3.150}$$

If the value of $k_{i\ell}$ is low and that of $|v_z|$ is high, one can have a situation where the value of $C^0_{i\ell}$ can become large enough so that, for a given ΔP, ΔP can become equal to $(\pi_{\text{wall,feed}} - \pi_{\text{permeate}})$; at this time, there will not be any solvent flux, due to a zero driving force. The membrane is considered polarized. In practice, the $k_{i\ell}$ values are sufficiently high so that the extent of the concentration polarization modulus $((C^0_{i\ell} - C_{ip})/(C_{i\ell b} - C_{ip}))$ is low and not too far from 1. However, in the mode of operation shown in Figures 6.3.28(a) and (b), the value of $k_{i\ell}$ is quite low; in practice, therefore, different configurations of flow vs. force are adopted. However, we will use this configuration and pseudo steady state assumption to illustrate here the extent and nature of separation achieved through the membrane in RO; specifically we will derive expressions for the solute rejection, R_i, and the separation factor, α_{si}, between the solvent *s* and the solute *i* in RO using the solution-diffusion model. We will also point out the inadequacy of the solution-diffusion model, especially, at high ΔP values.

Consider a dilute solution of a solute (specifically, a microsolute such as NaCl) *i* in a solvent *s* (say, water). By definition,

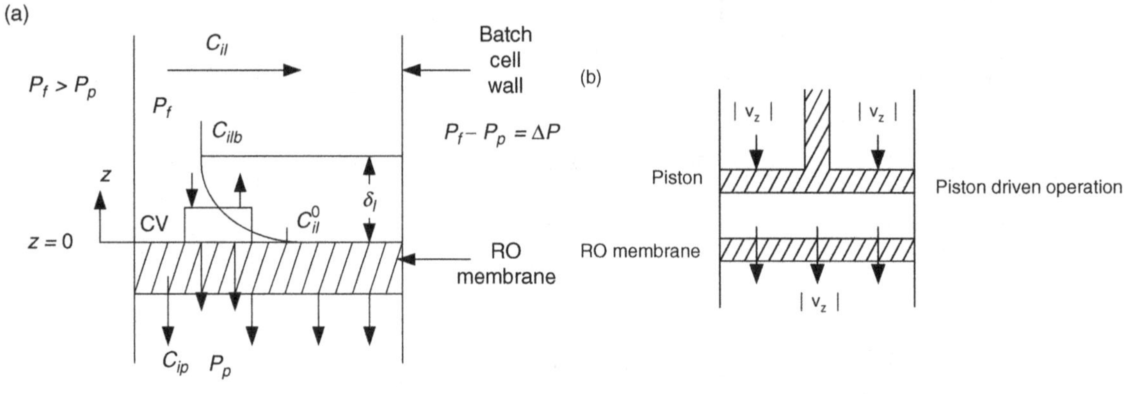

Figure 6.3.28. (a) Reverse osmosis (RO): RO in a batch cell: solute concentration profile in feed side. (b) Piston-driven RO in a batch cell: bulk flow parallel to the force.

$$\alpha_{si} = \frac{x_{sp}(1 - x_{sf})}{(1 - x_{sp})x_{sf}} = \frac{x_{sp}\, x_{if}}{x_{ip}\, x_{sf}}, \tag{6.3.151a}$$

where x_{sp} and x_{sf} are the mole fractions of the solvent in the permeate and the feed stream, respectively. Correspondingly, $(1 - x_{sp})$ and $(1 - x_{sf})$ are the mole fractions of the salt in the permeate and the feed, respectively, in this binary system. The definition for separation factor may be reexpressed as follows:

$$\alpha_{si} = \frac{x_{sp}\, x_{if}}{x_{ip}\, x_{sf}} = \frac{\frac{C_{sp}}{C_{tp}} \frac{C_{if}}{C_{tf}}}{\frac{C_{ip}}{C_{tp}} \frac{C_{sf}}{C_{tf}}} = \frac{C_{sp}\, C_{if}}{C_{ip}\, C_{sf}}, \tag{6.3.151b}$$

where C_{tj} is the total concentration in region j. For a dilute solution, $C_{sp} \cong C_{sf}$; therefore,

$$\alpha_{si} \cong \frac{C_{if}}{C_{ip}}. \tag{6.3.151c}$$

On the other hand, by definition, the solute rejection R_i is (definition (2.2.1b)

$$R_i = 1 - \frac{C_{ip}}{C_{if}}; \tag{6.3.152a}$$

consequently,

$$R_i \cong 1 - (1/\alpha_{si}). \tag{6.3.152b}$$

For the flow vs. force configuration of Figures 6.3.28(a) and (b)

$$\frac{J^*_{iz}}{J^*_{sz}} = \frac{C_{ip}}{C_{sp}} \cong \overline{V}_s\, C_{ip} \tag{6.3.153a}$$

for a dilute salt solution. Substitute here the expressions for J^*_{iz} and J^*_{sz} from expressions (6.3.148a) and (6.3.149) resulting from the solution-diffusion model:

$$\frac{J^*_{iz}}{J^*_{sz}} = \overline{V}_s C_{ip} = \frac{\frac{D_{im}\kappa_{im}}{\delta_m}(C_{if} - C_{ip})}{\frac{C_{sm} D_{sm} \overline{V}_s}{RT\delta_m}(\Delta P - \Delta\pi)} = \frac{RT\, D_{im}\, \kappa_{im}\, (C_{if} - C_{ip})}{C_{sm}\, D_{sm}\, \overline{V}_s\, (\Delta P - \Delta\pi)}. \tag{6.3.153b}$$

Now,

$$\alpha_{si} \cong \frac{C_{if}}{C_{ip}} = \frac{\overline{V}_s\, C_{if}\, C_{sm}\, D_{sm}\, \overline{V}_s(\Delta P - \Delta\pi)}{D_{im}\, \kappa_{im}\, RT(C_{if} - C_{ip})} \tag{6.3.154a}$$

Therefore the solute rejection is given by

$$R_i \cong 1 - \frac{1}{\alpha_{si}} = 1 - \frac{D_{im}\, \kappa_{im}\, RT(C_{if} - C_{ip})}{C_{if}\, C_{sm}\, D_{sm}\, \overline{V}^2_s(\Delta P - \Delta\pi)}. \tag{6.3.154b}$$

This may be rearranged to yield

$$R_i \cong 1 - \frac{D_{im}\, \kappa_{im}\, RT}{C_{sm}\, D_{sm}\, \overline{V}^2_s(\Delta P - \Delta\pi)} \left(1 - \frac{C_{ip}}{C_{if}}\right)$$

$$\Rightarrow R_i = \left[1 + \frac{D_{im}\, \kappa_{im} RT}{C_{sm}\, D_{sm}\, \overline{V}^2_s(\Delta P - \Delta\pi)}\right]^{-1}. \tag{6.3.154c}$$

For membranes showing high rejection, i.e. $C_{ip} << C_{if}$, we obtain, from expression (6.3.154b),

$$R_i \cong 1 - \frac{D_{im}\, \kappa_{im}\, RT}{C_{sm}\, D_{sm}\, \overline{V}^2_s(\Delta P - \Delta\pi)}. \tag{6.3.154d}$$

Correspondingly,

$$\alpha_{si} \cong \frac{C_{sm}\, D_{sm}\, \overline{V}^2_s(\Delta P - \Delta\pi)}{D_{im}\, \kappa_{im}\, RT} \tag{6.3.154e}$$

for $C_{ip} << C_{if}$. Note that expression (6.3.154d) is merely the result of considering only the first term, i.e. replace $(1 + x)^{-1}$ by $1 - x$ (valid when second-order terms, $O(x^2)$, are negligible).

These preceding two expressions for R_i and α_{si} indicate that, as $(\Delta P - \Delta\pi)$ increases, both R_i and α_{si} increase for high rejections. In the limit of ΔP becoming infinitely large, $R_i \rightarrow 1$ and $\alpha_{si} \rightarrow \infty$. In reality, most practical membranes reach a limiting value, $R_{i\infty}$, for solute species i. Solute rejection at high ΔP indicates that RO membranes are not perfect: there is some amount of salt transport, as shown in Figure 6.3.29(a) (Lonsdale, 1966). This is a major limitation of the solution-diffusion model of RO membranes with two parameters, A and $(D_{im}\, \kappa_{im} / \delta_m)$. Models having three parameters can overcome this limitation.

The solution-diffusion-imperfection model based on three parameters and proposed by Sherwood *et al.* (1967) (illustrated in flux expressions (3.4.60a,b)) appears to be able to describe better the observed solute rejections vs. solvent flux behavior in RO membranes (Applegate and Antonson, 1972). Rewrite the flux expressions (3.4.60a,b) for a dilute solution as

$$J_{vz} = K_1(\Delta P - \Delta\pi) + K_3\, \Delta P; \tag{6.3.155a}$$

$$J_{vz} C_{ip} = N_{iz} = K_2(C_{if} - C_{ip}) + K_3\, \Delta P\, C_{if}. \tag{6.3.155b}$$

Then the expression for salt rejection, $R_i = 1 - (C_{ip}/C_{if})$, is given as

$$R_i = \frac{1}{2\pi_f C_{if}} \left\{ C_{if}\Delta P \left(1 + \frac{K_3}{K_1}\right) + \pi_f \left(C_{if} + \frac{K_2}{K_1}\right) - \left[C^2_{if}\, \Delta P^2 \left(1 + \frac{K_3}{K_1}\right) + 2\, C_{if}\, \Delta P\, \pi_f \left(1 + \frac{K_3}{K_1}\right) \times \left(C_{if} + \frac{K_2}{K_1}\right) + \pi^2_f \left(C_{if} + \frac{K_2}{K_1}\right)^2 - 4\pi_f\, \Delta P\, C^2_{if} \right]^{1/2} \right\}, \tag{6.3.156}$$

where π_f is the osmotic pressure of the feed solution. The three parameters in this model, K_1, K_2 and K_3, allow an excellent fit of the experimental data obtained from

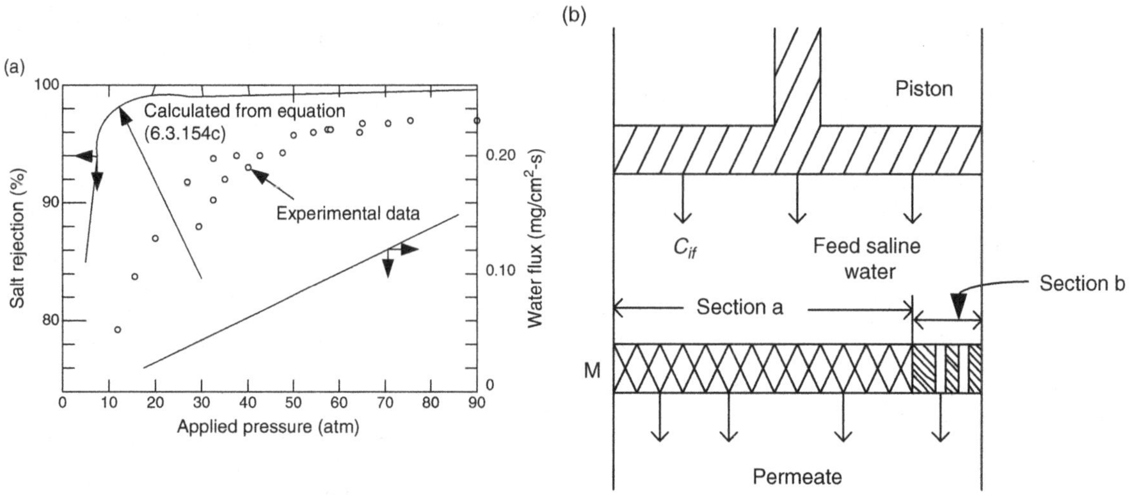

Figure 6.3.29. (a) Water flux and salt rejection vs. pressure for a 40% acetyl cellulose acetate membrane. Conditions: 0.1 M NaCl, 1 μm thick membrane (Lonsdale, 1966). Reprinted, with permission, from Desalination by Reverse Osmosis, *U. Merten (ed)., MIT Press, Cambridge, MA (1966), Figure 49, p. 117. (b) Flow parallel to force through a membrane which has defects (section b) having an area fraction ε_b and nondefective region (section a) of area fraction $(1-\varepsilon_b)$; the corresponding characteristic permeabilities are L_p^b and L_p^a.*

cellulose acetate and polyamide membranes (Applegate and Antonson, 1972). For a detailed consideration of different models in RO, the review by Soltanieh and Gill (1981) should be consulted.

Another three-parameter model of reverse osmosis membrane transport of some importance is the *Spiegler-Kedem* model (Spiegler and Kedem, 1966). In its differential form (similar to equations (3.4.47) and (3.4.48)), the model proposes the following local flux expressions at any location z in the membrane:

$$\text{volume flux} \quad J_{vz} = -Q_{SI}\left(\frac{dP}{dz} - \sigma_i \frac{d\pi}{dz}\right); \tag{6.3.157a}$$

$$\text{solute flux} \quad J_{iz}^* = -\overline{P}\frac{dC_i}{dz} + (1-\sigma_i)C_i J_{vz}. \tag{6.3.157b}$$

(*Note:* Here C_i, π and P correspond to infinitely thin solutions in equilibrium with the local section of the membrane; therefore C_i is the molar concentration of solute i in a solution of osmotic pressure π.) There are three parameters here: Q_{SI} (the intrinsic hydraulic permeability), $\overline{P}$ (the local solute permeability coefficient) and σ_i (the local solute reflection coefficient). When these two equations are integrated across a membrane of thickness δ_m, assuming Q_{SI}, $\overline{P}$ and σ_i to be essentially constant across the membrane thickness, one obtains, for the whole membrane, two equations for the Spiegler-Kedem model (based on the Kedem-Katchalsky model):

$$J_{vz} = L_p(\Delta P - \sigma_i \Delta\pi); \tag{6.3.158a}$$

$$J_{iz}^* = -\omega\,\Delta\pi + (1-\sigma_i)J_{vz}\overline{C}_i. \tag{6.3.158b}$$

The three parameters are: $L_p(=Q_{SI}/\delta_m)$, $\omega(=\overline{P}/2RT\,\delta_m)$ and σ_i; $\overline{C}_i$ is an average of the feed and permeate concentrations. *Note*: Here the osmotic pressure π is related to the solute concentration C_i by $\pi = C_i\,RT$ (van 't Hoff equation (3.3.86l)) for dilute solutions. For electrolytes, $\pi = \upsilon RTC_i$, where υ is the number of ions dissociated from one mole of electrolyte.

One would like to know now what the relation is between the solute rejection, R_i, and these parameters in the Spiegler-Kedem model. Focus on the ratio (J_{iz}^*/J_{sz}^*) as given by the relation (6.3.153a). If, instead, we consider the ratio

$$\frac{J_{iz}^*}{J_{vz}} \cong \frac{C_{ip}\left(\dfrac{\text{moles } i}{\text{cm}^3}\right) \times J_{vz}\left(\dfrac{\text{cm}^3}{\text{cm}^2\text{-s}}\right)}{J_{vz}\left(\dfrac{\text{cm}^3}{\text{cm}^2\text{-s}}\right)} \cong C_{ip}\left(\frac{\text{moles } i}{\text{cm}^3}\right), \tag{6.3.159}$$

then we can rewrite equation (6.3.157b) as

$$\overline{P}\frac{dC_i}{dz} = J_{vz}\{(1-\sigma_i)C_i - C_{ip}\}. \tag{6.3.160a}$$

Integrate across the membrane of thickness δ_m from C_{if} to C_{ip}:

$$\int_{C_{if}}^{C_{ip}} \frac{dC_i}{(1-\sigma_i)C_i - C_{ip}} = \frac{J_{vz}}{\bar{P}} \int_0^{\delta_m} dz; \tag{6.3.160b}$$

$$\frac{1}{(1-\sigma_i)} \ln\left(\frac{(1-\sigma_i)C_{ip} - C_{ip}}{(1-\sigma_i)C_{if} - C_{ip}}\right) = \frac{J_{vz}\,\delta_m}{\bar{P}}; \tag{6.3.160c}$$

$$\exp\left(-\frac{J_{vz}\,\delta_m(1-\sigma_i)}{\bar{P}}\right) = \frac{C_{ip} - (1-\sigma_i)C_{if}}{C_{ip}\,\sigma_i} = \frac{(1-R_i)C_{if} - (1-\sigma_i)C_{if}}{C_{ip}\,\sigma_i}; \tag{6.3.160d}$$

$$\exp\left(-\frac{J_{vz}(1-\sigma_i)\delta_m}{\bar{P}}\right) = \frac{(\sigma_i - R_i)}{\sigma_i(1-R_i)}, \tag{6.3.161}$$

where we have used definition (6.3.152a) for R_i. Rearranging this result leads to the following relation between the solute rejection R_i and the solute reflection coefficient σ_i:

$$R_i = \frac{\sigma_i\left(1 - \exp\left\{-\frac{J_{vz}(1-\sigma_i)\delta_m}{\bar{P}}\right\}\right)}{1 - \sigma_i \exp\left\{-\frac{J_{vz}(1-\sigma_i)\delta_m}{\bar{P}}\right\}}. \tag{6.3.162}$$

Consider the following limit:

$$J_{vz} \to \infty, \text{ when } \Delta P \to \infty; \text{ then } R_i \cong \sigma_i. \tag{6.3.163}$$

Thus, the Spiegler–Kedem model predicts correctly that, at very high ΔP, R_i reaches a limiting value which is less than 1 (σ_i can have a maximum value of 1) for a given solute. Simplistically, a value of $\sigma_i = 1$ means that the solute i is completely rejected by the membrane: it cannot enter the membrane.

A physically clearer interpretation of σ_i has been provided by Spiegler and Kedem (1966). Imagine the membrane to consist of regions (Figure 6.3.29(b)) which are perfectly semipermeable (no solute molecules can pass) and other regions which are "leaky," entirely nonselective, so that feed solution passes through. Let the hydraulic permeability of such perfect regions be identified as L_p^{a} and its fractional cross-sectional area be $(1 - \varepsilon_{\mathrm{b}})$ (note that σ_i for this region is 1); further, let the hydraulic permeability for the defective region be L_p^{b} and its fractional cross-sectional area be ε_{b}. Then the volume flux through the membrane is given by

$$J_{vz} = (1-\varepsilon_{\mathrm{b}})L_p^{\mathrm{a}}(\Delta P - \Delta\pi) + L_p^{\mathrm{b}}\varepsilon_{\mathrm{b}}\,\Delta P, \tag{6.3.164a}$$

since the flux through the defective fraction is governed by Darcy's law (equation (3.4.88)). We can rewrite this volume flux expression as

$$J_{vz} = \left((1-\varepsilon_{\mathrm{b}})L_p^{\mathrm{a}} + \varepsilon_{\mathrm{b}}\,L_p^{\mathrm{b}}\right)\left[\Delta P - \frac{L_p^{\mathrm{a}}(1-\varepsilon_{\mathrm{b}})}{(1-\varepsilon_{\mathrm{b}})L_p^{\mathrm{a}} + \varepsilon_{\mathrm{b}}\,L_p^{\mathrm{b}}}\Delta\pi\right]. \tag{6.3.164b}$$

However, we know from equation (6.3.158a) that

$$J_{vz} = L_p(\Delta P - \sigma_i\,\Delta\pi).$$

Therefore

and
$$L_p = (1-\varepsilon_{\mathrm{b}})L_p^{\mathrm{a}} + \varepsilon_{\mathrm{b}}\,L_p^{\mathrm{b}} \tag{6.3.165a}$$

$$\sigma_i = \frac{L_p^{\mathrm{a}}(1-\varepsilon_b)}{(1-\varepsilon_{\mathrm{b}})L_p^{\mathrm{a}} + \varepsilon_b\,L_p^{\mathrm{b}}}, \tag{6.3.165b}$$

which indicates that σ_i represents the fraction of the volume flux due to the perfect section of the membrane.

Example 6.3.9 Consider the physical picture (Figure 6.3.29(b)) which is the basis for Spiegler-Kedem illustration, namely equation (6.3.164a). We can rewrite it as

$$J_{vz} = K_1(\Delta P - \pi_f) + K_2\,\Delta P. \tag{6.3.166}$$

Here the nondefective membrane section does not allow any solute transport; the defective section allows feed solution to pass through. Therefore the solute flux is given by

$$N_{iz} = K_2\,\Delta P\,C_{if} \tag{6.3.167}$$

Further, the permeate solutions from different sections of the membrane are not mixed because flow is parallel to force and there is no lateral diffusion; therefore in equation (6.3.166), the first term has $(\Delta P - \pi_f)$ instead of $(\Delta P - \Delta\pi)$. Determine the expression for R_i.

Solution From relation (6.3.159), $N_{iz} = C_{ip}J_{vz}$. By definition,

$$R_i = 1 - \frac{C_{ip}}{C_{if}} = 1 - \frac{N_{iz}}{J_{vz}\,C_{if}}$$

$$\Rightarrow R_i = 1 - \frac{K_2 \Delta P\,C_{if}}{C_{if}(K_1(\Delta P - \pi_f) + K_2\,\Delta P)}$$

$$= 1 - \frac{K_2\,\Delta P}{(K_1 + K_2)\Delta P - K_1\,\pi_f}$$

(Applegate and Antonson, 1972).

Example 6.3.10 Expressions (6.3.154c) and (6.3.154d) developed for R_i for a solution-diffusion RO membrane are not truly predictive of R_i since the right-hand side has a $\Delta\pi$ term which contains C_{ip}. Develop an expression for R_i that contains only membrane transport parameters, feed conditions and ΔP; assume that the osmotic pressure–concentration relation is given by $\pi(C_{ij}) = bC_{ij}$ (relation (3.4.61b)).

Solution Assume the following:

$$J_{iz}^* = \frac{D_{im}\,\kappa_{im}}{\delta_m}(C_{if} - C_{ip}) = K_2(C_{if} - C_{ip}); \tag{6.3.168}$$

$$J_{sz}^* = N_{sz} = \frac{C_{sm}\,D_{sm}\,\overline{V}_s}{RT\delta_m}(\Delta P - \Delta\pi) = A(\Delta P - \Delta\pi); \tag{6.3.169}$$

$$J_{sz}^* = A\Delta P - Ab(C_{if} - C_{ip}); \tag{6.3.170}$$

$$\left(J_{iz}^*/J_{sz}^*\right) \cong \overline{V}_s C_{ip} = \frac{K_2\left(C_{if}-C_{ip}\right)}{A\Delta P - Ab\left(C_{if}-C_{ip}\right)} \qquad (6.3.171)$$

$$\Rightarrow A\overline{V}_s \Delta P C_{ip} - Ab\overline{V}_s C_{if} C_{ip} + Ab\overline{V}_s C_{ip}^2 = K_2 C_{if} - K_2 C_{ip}$$

$$\Rightarrow Ab\overline{V}_s C_{ip}^2 + \left(K_2 + A\overline{V}_s\Delta P - Ab\overline{V}_s C_{if}\right)C_{ip} - K_2 C_{if} = 0$$

$$\Rightarrow C_{ip} = \frac{Ab\overline{V}_s C_{if} - K_2 - A\overline{V}_s \Delta P \pm \sqrt{\left(K_2 + A\overline{V}_s\Delta P - Ab\overline{V}_s C_{if}\right)^2 + 4Ab\overline{V}_s K_2 C_{if}}}{2Ab\overline{V}_s};$$

$$R_i = 1 - \frac{C_{ip}}{C_{if}} = \frac{A\overline{V}_s b C_{if} + K_2 + A\overline{V}_s\Delta P - \sqrt{\left(K_2 + A\overline{V}_s\Delta P - A\overline{V}_s b C_{if}\right)^2 + 4A\overline{V}_s K_2 b C_{if}}}{2A\overline{V}_s b C_{if}}.$$

(We have used only the + sign before the square root term in the expression for C_{ip}.)

The treatment so far has been based on a particular feed concentration, C_{il}, in the reverse osmosis cell (Figure 6.3.28 (a)). As time progresses, water from the feed solution will be removed as permeate; therefore the feed concentration of species i, e.g. NaCl, will increase. If we require the process to yield a particular concentration of salt in the permeated water, then the salt rejection required of the membrane, $R_{i,\text{reqd}}$, will have to increase. Further, since the osmotic pressure of the feed solution increases with time, either the solvent flux will go down with time or the driving pressure difference, ΔP, has to go up. To these factors, one has to add the complication of concentration polarization. To illustrate the effect of increasing feed salt concentration with time, we will ignore first the effect of any concentration polarization and then focus on the consequence of different values of *fractional water recovery, re.* For the reverse osmosis cell shown in Figure 6.3.28(a), it is defined as

$$re\big|_{\text{batch}} = \frac{\text{volume of water permeated}}{\text{volume of initial water feed}}. \qquad (6.3.172a)$$

For the flow cells/devices of Section 7.2.1.2s, the corresponding definition is

$$re\big|_{flow} = \frac{\text{water permeation rate}}{\text{water feed flow rate}}. \qquad (6.3.172b)$$

Consider Figure 6.3.28(a) and an initial volume V_{fi} of salt water of salt concentration C_{ili}. Let the volume of salt water remaining after the process is over be V_{fe}; let the corresponding salt concentration be C_{ile}. Let the permeate salt concentration required be C_{ip}. The salt rejection required of the membrane, $R_{i,\text{reqd}}$, may be defined with respect to the final feed salt concentration C_{ile} (an alternative definition may be the average feed concentration, $(C_{ili} + C_{ile})/2$); this, however, will vary with the fractional water recovery, re, defined by (6.3.172a):

$$R_{i,\text{reqd}} = 1 - \frac{C_{ip}}{C_{i\ell e}}. \qquad (6.3.173)$$

Suppose the required permeate salt concentration, C_{ip}, corresponds to 500 ppm. A molar balance of the salt permeation process is

$$\begin{aligned} V_{fi}C_{ili} &= V_{fe}C_{ile} + \left(V_{fi} - V_{fe}\right)C_{ip}; \\ C_{ili} &= \left(V_{fe}/V_{fi}\right)C_{ile} + \left(1 - \left(V_{fe}/V_{fi}\right)\right)C_{ip}; \\ C_{ili} &= (1 - re)C_{ile} + (re)C_{ip}. \end{aligned} \qquad (6.3.174)$$

Use this result in the form $C_{i\ell e} = \left(C_{i\ell i} - (re)C_{ip}\right)/(1 - re)$ in definition (6.3.173):

$$R_{i,\text{reqd}} = 1 - \frac{C_{ip}(1 - re)}{\left(C_{i\ell i} - (re)C_{ip}\right)}. \qquad (6.3.175)$$

We now calculate $R_{i,\text{reqd}}$ for three values of re, namely 0, 0.75, 0.85, for C_{ili} corresponding to 2000 ppm. Our calculations assume very dilute solutions so that ppm values may be used directly without much error:

$$re = 0:\ R_{i,\text{reqd}} = 1 - \frac{500}{2000} = 0.75;$$

$$re = 0.75:\ R_{i,\text{reqd}} = 1 - \frac{500 \times 0.25}{2000 - 0.75 \times 500} = 1 - \frac{125}{1625} = 0.9231;$$

$$re = 0.85:\ R_{i,\text{reqd}} = 1 - \frac{500 \times 0.15}{2000 - 0.85 \times 500} = 1 - \frac{75}{1575} = 0.9524.$$

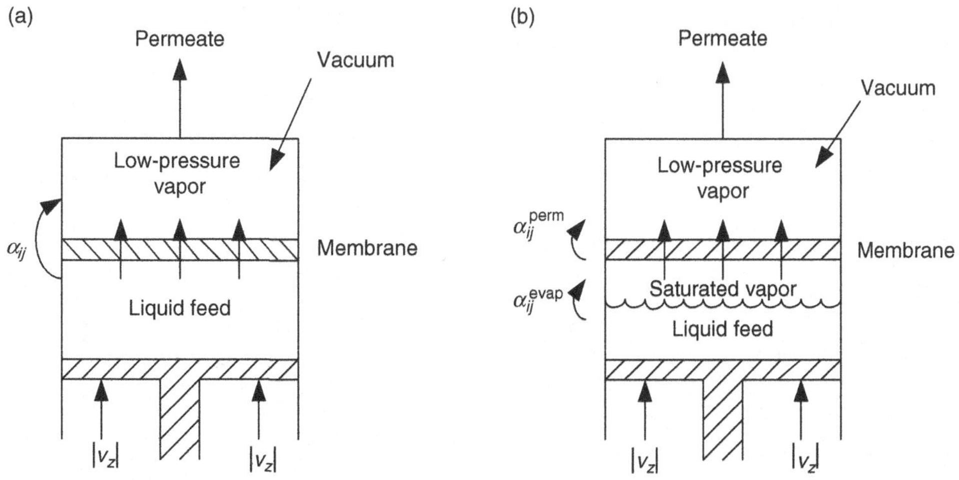

Figure 6.3.30. Vacuum driven pervaporation in a batch cell. (a) Conventional pervaporation with liquid feed imposed on membrane. (b) Thermodynamically equivalent pervaporation configuration with evaporation followed by vapor permeation.

These results show how, as the fractional water recovery changes, the required salt rejection also changes considerably.

6.3.3.4 *Pervaporation*

In Section 3.4.2.1.1, we were introduced to the rate equations for the membrane separation technique of *pervaporation*: a volatile liquid mixture on the feed side of the nonporous membrane at around atmospheric pressure; the other side of the membrane has a vapor/gaseous phase, usually at a lower pressure. The creation and maintenance of a vapor/gaseous phase on the permeate side can be implemented by either passing a sweep gas/vapor on the permeate side or by pulling a vacuum. In the configuration of bulk flow parallel to the direction of the force across a membrane (see Figures 6.3 (i)–(l)), vacuum will provide for permeated vapor flow parallel to the direction of force; sweep gas/vapor will not (in general).

To study separation in this flow vs. force configuration, it is useful to adopt a process model of pervaporation somewhat different from those employed in formulating equations (3.4.67a–h). Consider Figures 6.3.30(a) and (b). In Figure 6.3.30(a), the volatile liquid feed mixture is imposed on the feed membrane: the permeation process may be described by equations in Section 3.4.2.1.1. On the other hand, Figure 6.3.30(b) illustrates a thermodynamically equivalent configuration that is somewhat different (Wijmans and Baker, 1993): the liquid feed is in equilibrium with a vapor phase; this vapor phase is in contact with the feed side of the membrane while there is vacuum on the other side. In this modified configuration, the feed imposed on the membrane is a vapor mixture. Therefore, the transport equation through the membrane for species i may be written as (following equation (3.4.76))

$$J_{iz}^* = (Q_{im}/\delta_m)\left(p_{if} - p_{ip}\right). \tag{6.3.176}$$

However, this hypothetical partial pressure, p_{if}, of species i in the feed side is related to the feed liquid phase mole fraction, x_{if}, by the equilibrium relation (see relations (3.3.61) and (3.3.62) for lower-pressure situations (so that $\phi_i^{sat} \cong 1$))

$$p_{ig} = \hat{f}_{ig} = \hat{f}_{i\ell} = x_{if}\,\gamma_{if} f_{i\ell}^0 = x_{if}\,\gamma_{if} f_{i\ell}$$

$$= x_{if}\,\gamma_{if}\,P_i^{sat}\,\phi_i^{sat} \cong x_{if}\,\gamma_{if}\,P_i^{sat}. \tag{6.3.177}$$

One can therefore write the transport equation for the alternative configuration (Figure 6.3.30(b)) for volatile species i in the feed as

$$J_{iz}^* = (Q_{im}/\delta_m)\left(x_{if}\,\gamma_{if}\,P_i^{sat} - p_{ip}\right). \tag{6.3.178a}$$

For volatile species j,

$$J_{jz}^* = \left(Q_{jm}/\delta_m\right)\left(x_{jf}\,\gamma_{jf}\,P_j^{sat} - p_{jp}\right). \tag{6.3.178b}$$

In the bulk flow parallel to force configuration of Figure 6.3.30(b),

$$\frac{J_{iz}^*}{J_{jz}^*} = \frac{x_{ip}}{x_{jp}} = \left(\frac{Q_{im}}{Q_{jm}}\right)\frac{\left(x_{if}\,\gamma_{if}\,P_i^{sat} - P_p\,x_{ip}\right)}{\left(x_{jf}\,\gamma_{jf}\,P_j^{sat} - P_p\,x_{jp}\right)}, \tag{6.3.179}$$

where we have assumed $p_{ip} = P_p x_{ip}$, etc., for low levels of absolute permeate pressure P_p. Define the following separation factors for the binary system of i and j:

$$\alpha_{ij} = \frac{x_{ip}\, x_{jf}}{x_{jp}\, x_{if}}; \qquad \alpha_{ij}^{\text{perm}*} = (Q_{im}/Q_{jm});$$
$$\alpha_{ij}^{\text{evap}} = \gamma_{if} P_i^{\text{sat}} \Big/ (P_j^{\text{sat}} \gamma_{jf}). \tag{6.3.180}$$

Here, α_{ij} is the separation factor for pervaporation, $\alpha_{ij}^{\text{perm}*}$ is the ideal separation factor for vapor permeation through the membrane in Figure 6.3.30(b) and $\alpha_{ij}^{\text{evap}}$ is the corresponding separation factor in the evaporative equilibrium between the feed liquid and the hypothetical vapor imposed on the membrane. Suppose now that P_p, the permeate side absolute pressure under vacuum, is such that, for both i and j,

$$x_{if}\, \gamma_{if}\, P_i^{\text{sat}} >> P_p\, x_{ip};$$

then equation (6.3.179) may be rearranged to yield

$$\frac{x_{ip}}{x_{jp}} \cong \left(\frac{Q_{im}}{Q_{jm}}\right) \frac{x_{if}\, \gamma_{if}\, P_i^{\text{sat}}}{x_{jf}\, \gamma_{jf}\, P_j^{\text{sat}}} \Rightarrow \frac{x_{ip} x_{jf}}{x_{jp} x_{if}} = \alpha_{ij}$$
$$= \left(\frac{Q_{im}}{Q_{jm}}\right)\left(\frac{\gamma_{if} P_i^{\text{sat}}}{\gamma_{jf} P_j^{\text{sat}}}\right) = (\alpha_{ij}^{\text{perm}*})(\alpha_{ij}^{\text{evap}}) = (\alpha_{ij}^{\text{evap}})(\alpha_{ij}^{\text{perm}*}). \tag{6.3.181}$$

One is interested primarily in finding out the value of x_{ip} vis-à-vis x_{jp} in the pervaporation process. A general expression for x_{ip} can be derived in terms of x_{if} for a binary system. Rewrite relation (6.3.179) as follows:

$$\frac{x_{ip}}{(1 - x_{ip})} = \alpha_{ij}^{\text{perm}*} \frac{\left(x_{if}\, \gamma_{if}\, (P_i^{\text{sat}}/P_p) - x_{ip}\right)}{\left((1 - x_{if})\gamma_{jf} \dfrac{P_j^{\text{sat}}}{P_p} - (1 - x_{ip})\right)}; \tag{6.3.183}$$

$$\alpha_{ij}^{\text{perm}*}\, x_{if}\, \gamma_{if} \left(P_i^{\text{sat}}/P_p\right)(1 - x_{ip}) - \alpha_{ij}^{\text{perm}*}\left(x_{ip} - x_{ip}^2\right)$$
$$= x_{ip}(1 - x_{if})\gamma_{jf}\left(P_j^{\text{sat}}/P_p\right) - \left(x_{ip} - x_{ip}^2\right).$$

Rearrange to obtain

$$x_{ip}^2(\alpha_{ij}^{\text{perm}*} - 1) - x_{ip}\left[(\alpha_{ij}^{\text{perm}*} - 1) + \alpha_{ij}^{\text{perm}*}\left(x_{if}\, \gamma_{if}\, \frac{P_i^{\text{sat}}}{P_p}\right) + (1 - x_{if})\gamma_{jf}(P_j^{\text{sat}}/P_p)\right] + \alpha_{ij}^{\text{perm}*}\, x_{if}\, \gamma_{if}\, \frac{P_i^{\text{sat}}}{P_p} = 0. \tag{6.3.184}$$

This is a quadratic in x_{ip}. The solution is

$$x_{ip} = \frac{\left[(\alpha_{ij}^{\text{perm}*} - 1) + \alpha_{ij}^{\text{perm}*}\, x_{if}\, \gamma_{if}(P_i^{\text{sat}}/P_p) + (1 - x_{if})\gamma_{jf}\, (P_j^{\text{sat}}/P_p)\right]}{2(\alpha_{ij}^{\text{perm}*} - 1)}$$
$$\pm \frac{\{[(\alpha_{ij}^{\text{perm}*} - 1) + \alpha_{ij}^{\text{perm}*}\, x_{if}\, \gamma_{if}(P_i^{\text{sat}}/P_p) + (1 - x_{if})\gamma_{jf}\, (P_j^{\text{sat}}/P_p)]^2 - 4(\alpha_{ij}^{\text{perm}*} - 1)\alpha_{ij}^{\text{perm}*}\, x_{if}\, \gamma_{if}(P_i^{\text{sat}}/P_p)\}^{1/2}}{2(\alpha_{ij}^{\text{perm}*-1})}. \tag{6.3.185}$$

Thus, the selectivity achieved in pervaporation is a product of the selectivity achieved first by distillation/evaporation equilibrium of the feed liquid into vapor ($\alpha_{ij}^{\text{evap}}$) and then the ideal selectivity of the vapor permeation through the membrane $\alpha_{ij}^{\text{perm}*}$. One could write in general:[28]

$$\alpha_{ij} = \frac{x_{ip} x_{jf}}{x_{jp} x_{if}} = \frac{p_{ip}}{p_{jp}} \frac{x_{jf}}{x_{if}} = \frac{p_{ip}}{p_{jp}} \frac{C_{jf}}{C_{if}}$$
$$= \left(\frac{p_{ip}}{p_{jp}} \cdot \frac{p_{jf}}{p_{if}}\right)\left(\frac{p_{if}}{p_{jf}} \cdot \frac{C_{jf}}{C_{if}}\right) = \alpha_{ij}^{\text{perm}}\left(\frac{p_{if}}{p_{jf}} \frac{x_{jf}}{x_{if}}\right)$$
$$= \alpha_{ij}^{\text{perm}} \alpha_{ij}^{\text{evap}}; \quad \alpha_{ij} = \alpha_{ij}^{\text{evap}}\, \alpha_{ij}^{\text{perm}} \tag{6.3.182}$$

[28] All such discussions here assume pseudo-steady state since the feed liquid composition changes with time in this configuration.

The only useful sign in the numerator is the "−" one.

For a given binary system of i and j at a given temperature, the parameters one can vary are the permeate side total pressure, P_p, and the properties of the membrane, and therefore the membrane selectivity, $\alpha_{ij}^{\text{perm}*}$. Any considerations on the limits of selectivity achieved due to the variations of these two quantities should also take into account the vapor pressure ratio $\left(P_i^{\text{sat}}/P_j^{\text{sat}}\right)$, or alternatively $\alpha_{ij}^{\text{evap}}$ given by (6.3.180) (Wijmans and Baker, 1993).

Consider first the case where $\alpha_{ij}^{\text{perm}*}$ is *very large* compared to $\left(P_i^{\text{sat}}/P_j^{\text{sat}}\right)$. It is clear in such a case that, although $p_{ip} \leq p_{if}$,

$$p_{ip} \rightarrow p_{if} \quad \text{if} \quad P_p > p_{if}. \tag{6.3.186a}$$

Further,

$$p_{ip} = P_p \quad \text{if} \quad P_p < p_{if} \tag{6.3.186b}$$

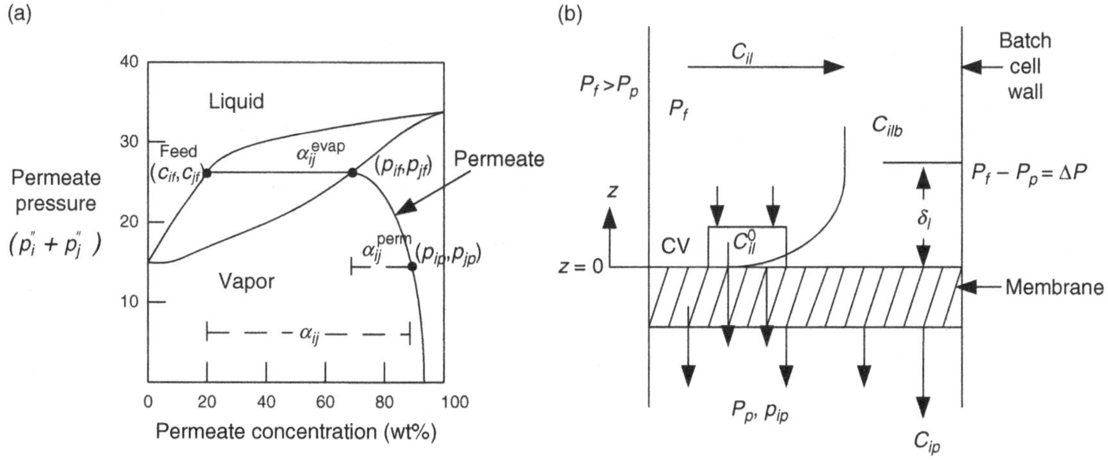

Figure 6.3.31. Consecutive separation corresponding to Figure 6.3.30(b) for ethanol-water azeotropic system. (b) concentration polarization in pervaporation. (After Wijmans and Baker (1993) and Wijmans et al. *(1996).)*

These results become clear if we assume an infinitely large value of α_{ij}^{perm*}. For example, for the second case (equation (6.3.186b)), as long as the permeate total pressure, P_p, is less than p_{if}, one can have pure species i in the permeate due to an infinitely selective vapor permeation membrane. Thus the value of the pressure ratio (P_p/P_f) is important in the case of very large values of α_{ij}^{perm*}. Similarly, consider equation (6.3.186a) and the relation (6.3.182) in the following format for $\alpha_{ij}^{perm*} \rightarrow \infty$:

$$\alpha_{ij} = \left(\frac{p_{ip}}{p_{jp}} \cdot \frac{p_{jf}}{p_{if}}\right) \alpha_{ij}^{evap} \cong \left(\frac{p_{jf}}{p_{jp}}\right) \alpha_{ij}^{evap}. \qquad (6.3.186c)$$

If i is a minor component in the feed,

$$p_{jf} \rightarrow P_f \text{ and } p_{jp} \rightarrow P_p \Rightarrow \alpha_{ij} \cong \left(\frac{P_f}{P_p}\right) \alpha_{ij}^{evap}. \qquad (6.3.186d)$$

In both cases we see that the total pressure ratio (P_p/P_f) is quite important if α_{ij}^{perm*} is *very high*. On the other hand, if $\left(P_i^{sat}/P_j^{sat}\right)$ is large compared to α_{ij}^{perm*}, we obtain a different result. Consider relation (6.3.182) in the following form:

$$\alpha_{ij} = \underbrace{\left(\frac{p_{if}}{p_{jf}} \cdot \frac{C_{jf}}{C_{if}}\right)}_{\alpha_{ij}^{evap}} \left(\frac{p_{ip}}{p_{jp}} \cdot \frac{p_{jf}}{p_{if}}\right); \qquad (6.3.187a)$$

$$\alpha_{ij} = \alpha_{ij}^{evap} \left(\frac{J_{iz}^*}{J_{jz}^*}\right) \left(\frac{p_{jf}}{p_{if}}\right) = \alpha_{ij}^{evap} \left(\frac{Q_{im}}{Q_{jm}}\right) \frac{\left(p_{if} - p_{ip}\right)}{\left(p_{jf} - p_{jp}\right)} \left(\frac{p_{jf}}{p_{if}}\right); \qquad (6.3.187b)$$

$$\alpha_{ij} \cong \alpha_{ij}^{evap} \alpha_{ij}^{perm*}, \qquad (6.3.188)$$

since $p_{ip} << p_{if}$, $p_{jp} << p_{jf}$, if the membrane selectivity is low.

The relationship (6.3.182) between the separation factor α_{ij} for the pervaporation process and the separation factors α_{ij}^{evap} and α_{ij}^{perm} for the hypothetical thermodynamically equivalent process for Figure 6.3.30(b) can be conceptually illustrated via Figure 6.3.31(a). This figure is based on a similar one from Wijmans and Baker (1993) showing vapor-liquid equilibrium of the azeotropic ethanol-water system at 60 °C. The plot illustrates total permeate pressure vs. alcohol concentration in the liquid phase (both feed liquid and permeate vapor). Feed liquid having molar concentrations of C_{if} and C_{jf} (mole fractions x_{if}, x_{jf}) are in equilibrium with a hypothetical vapor phase (Figure 6.3.30(b)) having the partial pressures of p_{if} and p_{jf}. This evaporation process leads to a separation factor of α_{ij}^{evap} (here the permeate pressure equals the feed pressure; therefore there is no membrane selectivity). Permeation of this hypothetical vapor through the membrane leads to a different composition in the permeate vapor phase, where the total permeate pressure is $p_{ip} + p_{jp}$ $(< (p_{if} + p_{jf}))$. The separation factor achieved in this hypothetical process of vapor permeation is α_{ij}^{perm}. It is clear how, in this case, the pervaporation process has created a highly alcohol-enriched permeate vapor phase from a liquid feed of alcohol and water. The overall separation is built on two consecutive separations: evaporation followed by vapor permeation.

Pervaporation processes are carried out naturally with volatile liquid mixtures. The liquid mixtures are of three types:

(1) aqueous solution, containing volatile organic compounds (VOCs);

(2) organic solution, containing water;
(3) organic solution, containing organic species only.

In the liquid mixture of type (1), the VOCs may be present in small amounts from a few parts per million (ppm) to a few percent. Some examples are: wastewater streams containing small amounts of VOCs; process aqueous streams containing small amounts of VOCs (valuable or not); fermentation broth containing volatile bioproducts, e.g. ethanol, butanol, acetone, acetic acid, etc.

In liquid mixtures of type (2), the solutions of primary interest are: azeotropic and other mixtures containing variable amounts of water in organics; dehydration of organic solvents containing very small amounts of water. Removal of water from azeotropic mixtures of ethanol-water, isopropanol-water, etc., is extensively practiced using polymeric membranes (of crosslinked polyvinyl alcohol) that are highly polar and selective for water. On the other hand, the membranes that are used to remove VOCs selectively from aqueous solutions are usually highly nonpolar rubbery polymeric membranes, e.g. dimethyl siloxane (silicone rubber).

For liquid mixtures of type (3), there are a number of different types of mixtures: aromatic-aliphatic separation; sulfur-containing organics from other petroleum constituents; olefins from paraffins, etc.

6.3.3.4.1 Concentration polarization in pervaporation
We have seen in Section 3.4.2.1.1 that the volatile species permeating through the membrane in the pervaporation process is driven by its concentration gradient in the nonporous membrane. The concentration of this species in the membrane at the feed liquid-membrane interface, C_{im}^0, is related to the feed liquid-phase concentration at this interface, C_{il}^0, by a partition coefficient κ_{im}:

$$\kappa_{im} = \frac{C_{im}^0}{C_{il}^0}. \tag{6.3.189}$$

For a steady state pervaporation process to take place, molecules of species i have to diffuse from the bulk liquid (C_{ilb}) to this interface ($C_{il}^0 < C_{ilb}$) (see Figure 6.3.31(b)). Since C_{il}^0 is less than C_{ilb}, C_{im}^0 is less than $\kappa_{im} C_{ilb}$. Therefore, the species i permeation rate through the membrane is reduced. This is exactly the opposite of what happens in ultrafiltration (Figure 6.3.26(a)) or reverse osmosis (Figure 6.3.28(a)), where the solute is being rejected by the membrane and therefore its concentration builds up at the feed liquid-membrane interface since the back diffusion process in the liquid film layer next to the membrane is not fast enough.

To develop a quantitative estimate of this effect, we follow the procedure followed earlier (see equations (6.3.142a) and (3.4.65b)) with due attention to the differences here. Assume a pseudo steady state (for this batch cell in force parallel to bulk flow configuration) for the control volume (CV):

$$\begin{pmatrix}\text{solute flux into}\\ \text{CV by convection}\end{pmatrix} + \begin{pmatrix}\text{solute flux into}\\ \text{CV by diffusion}\end{pmatrix} = \begin{pmatrix}\text{solute flux out of}\\ \text{CV by membrane transport}\end{pmatrix};$$
$$|v_z|\, C_{il} + D_{il}\frac{\mathrm{d}\,C_{il}}{\mathrm{d}z} = |v_z|\, C_{ip}. \tag{6.3.190a}$$

Here $|v_z|$ is the magnitude of the liquid velocity generated in the film layer due to the permeation process in pervaporation. If we have only two species i and j in the feed liquid, then

$$|v_z| = \left(J_{iz}^* + J_{jz}^*\right)/C_{tl}, \tag{6.3.190b}$$

where C_{tl} is the total molar density of the feed liquid. We may now rearrange equation (6.3.190a) as follows:

$$|v_z|\left(C_{il} - C_{ip}\right) = -D_{il}\frac{\mathrm{d}C_{il}}{\mathrm{d}z};$$

$$\int_{C_{il}^0}^{C_{ilb}} \frac{\mathrm{d}C_{il}}{\left(C_{il} - C_{ip}\right)} = -\left(|v_z|/D_{il}\right)\int_0^{\delta_l}\mathrm{d}z \Rightarrow \ln\left(\frac{C_{ilb} - C_{ip}}{C_{il}^0 - C_{ip}}\right)$$

$$= -\frac{|v_z|}{D_{il}/\delta_l} = -\frac{|v_z|}{k_{il}};$$

$$\frac{C_{il}^0 - C_{ip}}{C_{ilb} - C_{ip}} = \exp\left(\frac{|v_z|}{k_{il}}\right) = \frac{C_{ip} - C_{il}^0}{C_{ip} - C_{ilb}}. \tag{6.3.190c}$$

This relation is identical to equation (6.3.142b); however, if the ratio $\left(C_{ip}/C_{ilb}\right)$ is greater than 1, as is true here (which is exactly the reverse of what happens in RO and UF, where the solute is rejected by the membrane), we have severe concentration polarization when $k_{il} << |v_z|$. For example, suppose $(|v_z|/k_{il})\sim 1$, and the membrane is such that $\left(C_{ip}/C_{il}^0\right)$ is, say, 1000. Then, rewriting equation (6.3.190c) as

$$\frac{\left(C_{ip}/C_{il}^0\right) - 1}{\left(C_{ip}/C_{il}^0\right) - \left(C_{ilb}/C_{il}^0\right)} = \exp(1) = e$$

$$= \frac{1000 - 1}{1000 - \left(C_{ilb}/C_{il}^0\right)} = e = 2.718$$

implies that

$$\left(C_{ilb}/C_{il}^0\right) = \frac{2718 - 999}{2.718} = 632.4.$$

Therefore the species i concentration in the liquid imposed on the membrane surface is drastically reduced, leading to a substantial reduction in the species i flux. Consequently, concentration polarization can be severe in pervaporation processes where the membrane enriches the species substantially in the permeate. An example is the selective removal of VOCs from an aqueous solution through a rubbery pervaporation membrane (e.g. silicone polymer based). For a description of such concentration polarization effects in VOC removal from water, consult Wijmans *et al.* (1996).

We will now estimate the effective overall permeance of species $\left(Q_{im}^{ov}/\delta_m\right)$ due to the concentration polarization effect since (Wijmans *et al.* (1996))

$$J_{iz}^* = \left(Q_{im}^{\text{ov}}/\delta_m\right)\left(x_{i\ell b}\,\gamma_{i\ell b}\,P_i^{\text{sat}} - p_{ip}\right); \qquad (6.3.191\text{a})$$

$$J_{iz}^* = k_{i\ell}\,C_{t\ell}\left(x_{i\ell b} - x_{i\ell}^0\right); \qquad (6.3.191\text{b})$$

$$J_{iz}^* = \left(Q_{im}/\delta_m\right)\left(x_{i\ell}^0\,\gamma_{i\ell}^0\,P_i^{\text{sat}} - p_{ip}\right), \qquad (6.3.191\text{c})$$

where

$$C_{i\ell b} = C_{t\ell}\,x_{i\ell b}, \qquad C_{i\ell}^0 = C_{t\ell}\,x_{i\ell}^0. \qquad (6.3.191\text{d})$$

Although γ_{il} depends in general on the liquid-phase composition, γ_{ilb} is not, in general, equal to γ_{il}^0. However, we will assume it to be valid here. Utilize now the concentration polarization relation (6.3.190c) in terms of mole fractions of species i:

$$\exp(|v_z|/k_{il}) = \frac{C_{t\ell}\,x_{ip} - C_{t\ell}\,x_{il}^0}{C_{t\ell}\,x_{ip} - C_{t\ell}\,x_{i\ell b}} = \frac{x_{ip} - x_{i\ell}^0}{x_{ip} - x_{i\ell b}}. \qquad (6.3.191\text{e})$$

Rearrange equations (6.3.191b) and (6.3.191c) as follows:

$$\frac{J_{iz}^*\,\gamma_{i\ell b}\,P_i^{\text{sat}}}{k_{i\ell}\,C_{t\ell}} = \gamma_{i\ell b}\,x_{i\ell b}\,P_i^{\text{sat}} - \gamma_{i\ell b}\,x_{i\ell}^0\,P_i^{\text{sat}};$$

$$\frac{J_{iz}^*}{\left(Q_{im}/\delta_m\right)} = x_{i\ell}^0\,\gamma_{i\ell}^0\,P_i^{\text{sat}} - p_{ip}.$$

Adding these two equations and assuming $\gamma_{i\ell}^0 = \gamma_{i\ell b}$, we get

$$J_{iz}^*\left[\frac{\gamma_{i\ell b}\,P_i^{\text{sat}}}{k_{i\ell}\,C_{t\ell}} + \frac{\delta_m}{Q_{im}}\right] = \left(x_{i\ell b}\,\gamma_{i\ell b}\,P_i^{\text{sat}} - p_{ip}\right),$$

so that, from the definition (6.3.191a) of $\left(Q_{im}^{\text{ov}}/\delta_m\right)$,

$$\frac{\delta_m}{Q_{im}^{\text{ov}}} = \frac{\delta_m}{Q_{im}} + \frac{\gamma_{i\ell b}\,P_i^{\text{sat}}}{k_{i\ell}\,C_{t\ell}}. \qquad (6.3.192)$$

This relation allows one to estimate the effective overall permeance $\left(Q_{im}^{\text{ov}}/\delta_m\right)$ in the presence of concentration polarization.

An alternative expression is often more useful. Rearrange the liquid film transport equation (6.3.191b) and the concentration polarization relation (6.3.191e) as follows:

$$J_{iz}^* = C_{ip}\,|v_z| = k_{i\ell}\,C_{tl}\left(x_{i\ell b} - x_{i\ell}^0\right) \Rightarrow \frac{1}{k_{i\ell}} = \frac{C_{tl}\left(x_{i\ell b} - x_{i\ell}^0\right)}{C_{ip}\,|v_z|};$$

$$\frac{x_{i\ell}^0 - x_{ip}}{x_{i\ell b} - x_{ip}} = \exp(|v_z|/k_{i\ell}) \Rightarrow \frac{x_{i\ell}^0 - x_{i\ell b}}{x_{i\ell b} - x_{ip}} = \exp\left(\frac{|v_z|}{k_{i\ell}}\right) - 1$$

$$\Rightarrow \frac{\left(x_{i\ell}^0/x_{ip}\right) - \left(x_{i\ell b}/x_{ip}\right)}{\left(x_{i\ell b}/x_{ip}\right) - 1} = \left[\exp\left(\frac{|v_z|}{k_{i\ell}}\right) - 1\right]. \qquad (6.3.193\text{a})$$

Defining $\left(x_{ip}/x_{ilb}\right)$ as E_i, the effective enrichment of the species i by the pervaporation process, we get

$$\frac{\left(x_{i\ell b} - x_{i\ell}^0\right)}{x_{ip}} = \left[\exp\left(\frac{|v_z|}{k_{i\ell}}\right) - 1\right]\left[1 - \frac{1}{E_i}\right]. \qquad (6.3.193\text{b})$$

Employ these two results in relation (6.3.192) to replace $(1/k_{il})$:

$$\left(\frac{\delta_m}{Q_{im}^{\text{ov}}}\right) = \left(\frac{\delta_m}{Q_{im}}\right) + \frac{\gamma_{i\ell b}P_i^{\text{sat}}C_{t\ell}x_{ip}\left[\exp\left(\frac{|v_z|}{k_{i\ell}}\right) - 1\right]\left[1 - \frac{1}{E_i}\right]}{C_{t\ell}|v_z|C_{ip}}$$

$$= \left(\frac{\delta_m}{Q_{im}}\right) + \frac{\gamma_{i\ell b}P_i^{\text{sat}}}{|v_z|C_{t\ell}}\left[1 - \frac{1}{E_i}\right]\left[\exp\left(\frac{|v_z|}{k_{i\ell}}\right) - 1\right]. \qquad (6.3.194)$$

This result is more general in the following sense. If a species i does not undergo enrichment ($E_i = 1$), there is no concentration polarization effect. For components which are depleted in the permeate ($E_i < 1$), the permeance is increased over $\left(Q_{im}/\delta_m\right)$, due to an increase in its concentration in the liquid film, and therefore in its transport rate. Further, as the permeation rate increases, $|v_z|$ increases; under such a condition, increasing δ_m will lead to a relative reduction in the liquid film resistance with respect to the membrane resistance (Wijmans *et al.*, 1996).

Example 6.3.11 Membranes made from crosslinked polyvinyl alcohols are known to be highly selective for water over alcohols, ethanol, propanol, etc. Schaetzel *et al.* (2001) have suggested that the diffusion coefficient of each component depends only on the weight fraction of the key component, water, in such a membrane:

$$J_{H_2O}^* = D_{H_2O,0}\exp(\tau_{H_2O}u_{H_2O})\,S_{H_2O} \times\left(x_{H_2O,f}\,\gamma_{H_2O,f}\,P_{H_2O}^{\text{sat}} - x_{H_2O,p}\,P_p\right);$$

$$J_{\text{alc}}^* = D_{\text{alc},0}\exp(\tau_{\text{alc}}u_{H_2O})\,S_{\text{alc}}\left(x_{\text{alc},f}\,\gamma_{\text{alc},f}\,P_{\text{alc}}^{\text{sat}} - x_{\text{alc},p}\,P_p\right).$$

The values of the products $D_{H_2O,0}\,S_{H_2O}$ for water and $D_{\text{alc},0}\,S_{\text{alc}}$ for i-propanol at 45 °C are 8×10^{-4} and 7.4×10^{-7}, respectively, in kg/hr-m^2-mm Hg. If the permeate side partial pressures may be neglected and $\tau_{H_2O} = \tau_{\text{alc}}$, determine the separation factor at the azeotropic composition of around 87% i-propanol by weight at 1 atm.

Solution At the azeotropic composition for any species in vapor–liquid equilibrium at pressure P_{az},

$$x_{ig}\,P_{az} = x_{if}\,\gamma_{if}\,P_i^{sat} \quad \text{and} \quad x_{if} = x_{ig}$$

implies that

$$P_{az} = \gamma_{if}\,P_i^{sat} = \gamma_{H_2O,f}\,P_{H_2O}^{sat} = \gamma_{alc,f}\,P_{alc}^{sat}.$$

The separation factor

$$\alpha_{H_2O-alc} = \frac{x_{H_2O,p}}{x_{alc,p}}\frac{x_{alc,f}}{x_{H_2O,f}} = \frac{J^*_{H_2O}}{J^*_{alc}}\frac{x_{alc,f}}{x_{H_2O,f}}$$

$$= \frac{D_{H_2O,0}S_{H_2O}\exp(\tau_{H_2O}u_{H_2O})\left(x_{H_2O,f}\,\gamma_{H_2O,f}\,P^{sat}_{H_2O}\right)x_{alc,f}}{D_{alc,0}\,S_{alc}\exp(\tau_{alc}u_{H_2O})\left(x_{alc,f}\,\gamma_{alc,f}\,P^{sat}_{alc}\right)x_{H_2O,f}}$$

$$= \left(\frac{D_{H_2O,0}\,S_{H_2O}}{D_{alc,0}\,S_{alc}}\right)\frac{\exp(\tau_{H_2O}u_{H_2O})}{\exp(\tau_{H_2O}u_{H_2O})}\frac{x_{H_2O,f}\,P_{az}}{x_{alc,f}\,P_{az}}\frac{x_{alc,f}}{x_{H_2O,f}}$$

$$= \left(\frac{D_{H_2O,0}\,S_{H_2O}}{D_{alc,0}\,S_{alc}}\right) = \frac{8\times10^{-4}}{7.5\times10^{-7}}\frac{M_{alc}}{M_{H_2O}}$$

$$= \frac{8\times10^{-4}\times60}{7.5\times10^{-7}\times18} = 3555.$$

The separation factor defined in terms of weight fractions will be 1066.

Example 6.3.12 Consider selective pervaporative transport of a VOC species i through a silicone rubber membrane in preference to water. The separation conditions are such that E_i is >> 1, of the order of 1000; further, the permeation rates are quite low with respect to the feed side mass-transfer coefficient (i.e. $|v_z| << k_{il}$). In this pervaporation process, it is known that the membrane resistance is quite low but the boundary layer resistance is quite high for the VOCs. On the other hand, for water, it may be safely assumed that the membrane resistance controls the water transport. Determine an expression for the separation factor, and find the selectivity of the membrane for VOC over water when the permeate pressure is negligible. Assume: that the VOC concentration in water is low for this pervaporation process.

Solution: Consider the general relation (6.3.194):

$$\left(\frac{\delta_m}{Q^{ov}_{im}}\right) = \left(\frac{\delta_m}{Q_{im}}\right) + \frac{\gamma_{ilb}\,P_i^{sat}}{|v_z|C_{tl}}\left[1-\frac{1}{E_i}\right]\left[\exp\left(\frac{|v_z|}{k_{il}}\right)-1\right].$$

For i = VOC,

$$\left[1-\frac{1}{E_i}\right]\sim 1,\ \left[\exp\left(\frac{|v_z|}{k_{il}}\right)-1\right] \cong |v_z|/k_{il}\ \text{for}\ |v_z| << k_{il}.$$

Therefore

$$\left(\frac{\delta_m}{Q^{ov}_{VOC\,m}}\right) = \left(\frac{\delta_m}{Q_{VOC\,m}}\right) + \frac{\gamma_{ilb}\,P_i^{sat}}{C_{tl}\,k_{il}} \cong \frac{\gamma_{ilb}\,P_i^{sat}}{C_{tl}\,k_{il}}(i = \text{VOC}).$$

For water, the corresponding result is

$$\left(\frac{\delta_m}{Q^{ov}_{water\,m}}\right) \cong \left(\frac{\delta_m}{Q_{water\,m}}\right).$$

Now,

$$\alpha_{VOC-water} = \frac{x_{VOC,p}}{x_{H_2O,p}}\cdot\frac{x_{H_2O,f}}{x_{VOC,f}} = \frac{J^*_{VOC,z}}{J^*_{H_2O,z}}\cdot\frac{x_{H_2O,f}}{x_{VOC,f}};$$

$$J^*_{VOC,z} = \frac{Q^{ov}_{VOC,m}}{\delta_m}(\gamma_{i\ell b}\,x_{if}\,P_i^{sat} - x_{ip}\,P_p) \qquad (i = \text{VOC});$$

$$J^*_{VOC,z} = \frac{Q^{ov}_{VOC,m}}{\delta_m}\gamma_{i\ell b}\,x_{if}\,P_i^{sat};$$

$$J^*_{H_2O,z} = \frac{Q^{ov}_{water,m}}{\delta_m}(\gamma_{i\ell b}\,x_{if}\,P_i^{sat} - 0) \qquad (i = \text{water});$$

$$\alpha_{VOC-water} = \frac{Q^{ov}_{VOC,m}/\delta_m}{Q^{ov}_{water,m}/\delta_m}\frac{\gamma_{VOC\ell b}\,x_{VOC,f}\,P^{sat}_{VOC}}{\gamma_{H_2O\ell b}\,x_{H_2O,f}\,P^{sat}_{H_2O}}\cdot\frac{x_{H_2O,f}}{x_{VOC,f}}$$

$$(\gamma_{H_2O\ell b} \cong 1)$$

$$= \frac{C_{t\ell}\,k_{VOC,\ell}}{\gamma_{VOC\ell b}\,P^{sat}_{VOC}}\frac{\delta_m}{Q_{water,m}}\frac{\gamma_{VOC\ell b}\,P^{sat}_{VOC}}{P^{sat}_{H_2O}}$$

$$= \frac{C_{t\ell}\,k_{VOC,\ell}\,\delta_m}{Q_{water,m}\,P^{sat}_{H_2O}}.$$

The VOC–water selectivity, defined as $Q^{ov}_{VOC}/Q^{ov}_{H_2O}$, is given by

$$\frac{C_{t\ell}\,k_{VOC,\ell}\,\delta_m}{\gamma_{VOC\ell b}\,P^{sat}_{VOC}\,Q_{water,m}}$$

6.3.3.5 *Membrane gas permeation*

In Section 4.3.3, we studied the time-dependent potential of separation of a binary gas mixture by permeation through a nonporous membrane in a closed vessel. A certain amount of a binary gas mixture was introduced on one side of the membrane, the other side being empty at $t = 0$ in the closed vessel. Initially there was some separation - but all separation was lost ultimately as an equilibrium was reached between the gases on the two sides of the membrane. Since membrane gas separation relies on intrinsically different permeation rates of different species, we need an open system so that we can continuously remove the separated products in the permeate stream.

Consider now an open system with both feed gas and the permeated gas flowing parallel to the direction of the force (Figure 6.3.32), toward the membrane for the feed and away from the membrane for the permeate; it is as if there is a hypothetical piston driving the feed gas toward the membrane. This is achieved by connecting a gas source at constant pressure to the feed side of the membrane. The gas velocity toward the membrane is determined by the rate at which gases permeate through the membrane and are withdrawn immediately from the permeate side. Note that the cross-sectional area of the membrane through which gas permeation takes place is the same as that of the feed flow channel/piston and the permeate flow channel. Any reduction or increase in the flow cross-sectional area on either side of the membrane will introduce gas velocities in other directions. Such situations are of great

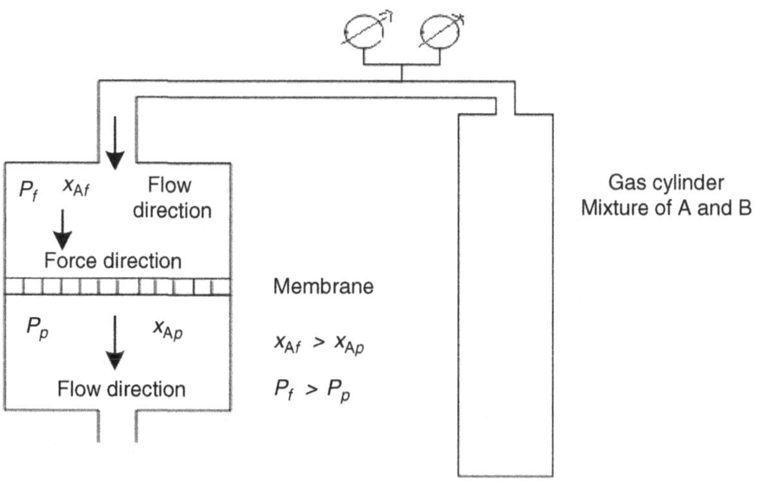

Figure 6.3.32. Gas permeation device, where the bulk gas flow direction is parallel to the direction of force across the membrane for separating a gas mixture; the membrane is selective for species A.

practical use and are considered later in Section 6.4 and Chapters 7 and 8.

For a *binary* feed mixture of gases A and B at feed partial pressures of p_{Af} and p_{Bf} and permeate partial pressures of p_{Ap} and p_{Bp} respectively, the permeation rates per unit membrane area of the two species are, respectively (see equation (3.4.72)),

$$N_{Az} = J^*_{Az} = \frac{Q_{Am}}{\delta_m}(p_{Af} - p_{Ap});$$
$$N_{Bz} = J^*_{Bz} = \frac{Q_{Bm}}{\delta_m}(p_{Bf} - p_{Bp}). \qquad (6.3.195)$$

If the mole fractions of A and B in the feed gas mixture and the gas mixture emerging from the membrane as the permeate are, respectively, x_{Af}, x_{Bf}, x_{Ap}, x_{Bp}, then

$$\frac{N_{Az}}{N_{Bz}} = \frac{J^*_{Az}}{J^*_{Bz}} = \frac{x_{Ap}}{x_{Bp}} = \frac{(Q_{Am}/\delta_m)}{Q_{Bm}/\delta_m)}\frac{\left(p_{Af} - p_{Ap}\right)}{\left(p_{Bf} - p_{Bp}\right)}. \qquad (6.3.196)$$

Rearrange this equation as

$$\frac{x_{Ap}}{x_{Bp}} = \frac{(Q_{Am}/\delta_m)}{(Q_{Bm}/\delta_m)}\frac{(P_f x_{Af} - P_p x_{Ap})}{(P_f x_{Bf} - P_p x_{Bp})},$$

where P_f and P_p are the total gas pressures on the feed and permeate side, respectively. Rearrange further to obtain

$$\alpha_{AB} = \frac{x_{Ap}}{x_{Bp}}\frac{x_{Bf}}{x_{Af}} = \frac{(Q_{Am}/\delta_m)}{Q_{Bm}/\delta_m)}\frac{P_f}{P_f}\frac{\left(1 - \frac{P_p}{P_f}\frac{x_{Ap}}{x_{Af}}\right)}{\left(1 - \frac{P_p}{P_f}\frac{x_{Bp}}{x_{Bf}}\right)}. \qquad (6.3.197)$$

At *any time*, let x_{Af}, x_{Bf} and P_f represent the conditions on the feed side and let x_{Ap}, x_{Bp}, P_p represent those on the permeate side; under such conditions, the expression for the separation factor, α_{AB}, for the membrane between the two gas species A and B is given above. The quantity (P_p/P_f) is identified as γ, the *pressure ratio,* and (Q_{im}/δ_m), is the *permeance* of species *i*.

When $\gamma \to 0$, i.e. the permeate side has a very low pressure compared to that in the feed side, we find

$$\alpha_{AB} = \frac{x_{Ap}}{x_{Bp}}\frac{x_{Bf}}{x_{Af}} \Rightarrow \frac{(Q_{Am}/\delta_m)}{(Q_{Bm}/\delta_m)} = \alpha^*_{AB}, \qquad (6.3.198)$$

where α^*_{AB} is called the *ideal separation factor* achieved in the limit of zero pressure ratio. The condition $\gamma = 0$ leads, for the binary system, to a permeate gas composition x_{Ap} given by

$$\frac{x_{Ap}}{x_{Bp}} = \frac{x_{Ap}}{(1 - x_{Ap})} = \alpha^*_{AB}\frac{x_{Af}}{x_{Bf}} = \alpha^*_{AB}\frac{x_{Af}}{(1 - x_{Af})}.$$

Therefore

$$x_{Ap} = \frac{\alpha^*_{AB}x_{Af}}{1 + x_{Af}(\alpha^*_{AB} - 1)}. \qquad (6.3.199)$$

For nonzero values of γ, the separation factor relation (6.3.197) can now be rewritten as follows:

$$\frac{x_{Ap}}{(1 - x_{Ap})} = \alpha^*_{AB}\frac{(x_{Af} - \gamma x_{Ap})}{(1 - x_{Af}) - \gamma(1 - x_{Ap})}, \qquad (6.3.200)$$

which leads to a quadratic in x_{Ap} in terms of α^*_{AB}, γ and x_{Af}:

$$x^2_{Ap}(\gamma(\alpha^*_{AB} - 1)) - x_{Ap}\left[(\gamma + x_{Af})(x^*_{AB} - 1) + 1\right] + \alpha^*_{AB}x_{Af} = 0.$$

The solution for nonzero values of γ is (Huckins and Kammermeyer, 1953a,b)

$$x_{Ap} = \frac{\left\{(\alpha^*_{AB}-1)\left(\frac{x_{Af}}{\gamma}+1\right)+\frac{1}{\gamma}\right\} \pm \sqrt{\left\{\left(1+\frac{x_{Af}}{\gamma}\right)(\alpha^*_{AB}-1)+\frac{1}{\gamma}\right\}^2 - 4\frac{\alpha^*_{AB}(\alpha^*_{AB}-1)x_{Af}}{\gamma}}}{2(\alpha^*_{AB}-1)} \quad (6.3.201)$$

In this expression for x_{Ap}, only the negative root is meaningful in the numerator since $0 \le x_{Ap} \le 1$ for $0 \le x_{Af} \le 1$.

Given α^*_{AB}, γ and x_{Af}, one can now calculate $x_{Ap}(0 \le x_{Ap} \le 1)$. The permeances of species A and B, (Q_A/δ_m) and (Q_B/δ_m), are likely to be known for the membrane; therefore α^*_{AB} is known. The pressure ratio, γ, is generally specified. For a given feed (known x_{Af}), the permeate mole fraction, x_{Ap} (which will vary between 0 and 1; $x_{Ap} > x_{Af}$ for $\alpha^*_{AB} > 1$), is easily calculated for this mode of operation.

Sometimes one needs to calculate the actual separation factor α_{AB} defined by (6.3.197). Write this expression as

$$\alpha_{AB} = \alpha^*_{AB}\frac{\left(1-\gamma(x_{Ap}/x_{Af})\right)}{1-\gamma(x_{Bp}/x_{Bf})}.$$

Multiply the top and the bottom on the right-hand side by x_{Bf}/x_{Bp}:

$$\alpha_{AB} = \alpha^*_{AB}\left(\frac{x_{Bf}}{x_{Bp}} - \gamma\,\alpha_{AB}\right)\Big/\left(\frac{x_{Bf}}{x_{Bp}} - \gamma\right)$$

$$\Rightarrow \alpha_{AB}\left(\frac{x_{Bf}}{x_{Bp}} - \gamma + \gamma\,\alpha^*_{AB}\right) = \alpha^*_{AB}\,x_{Bf}/x_{Bp}$$

$$\Rightarrow \alpha_{AB} = \alpha^*_{AB}\left(\frac{x_{Bf}/x_{Bp}}{(x_{Bf}/x_{Bp}) + \gamma(\alpha^*_{AB}-1)}\right), \quad (6.3.202a)$$

which expresses α_{AB} in terms of α^*_{AB}, γ, x_{Bf} and x_{Bp}. A more useful expression for α_{AB} eliminates x_{Bp} from the right-hand side to express it as a function of x_{Af}, γ and α^*_{AB} (Stern and Walawender, 1969):

$$\alpha_{AB} = \frac{(\alpha^*_{AB}+1)}{2} - \frac{\gamma(\alpha^*_{AB}-1)}{2} - \frac{1}{2x_{Af}} + \left\{\left(\frac{\alpha^*_{AB}-1}{2}\right)^2 + \frac{(\alpha^*_{AB}-1)-\gamma[\alpha^{*2}_{AB}-1]}{2x_{Af}} + \left[\frac{\gamma(\alpha^*_{AB}-1)+1}{2x_{Af}}\right]^2\right\}^{1/2}. \quad (6.3.202b)$$

In the configuration shown in Figure 6.3.32, the feed gas composition and pressure imposed on the membrane are uniform everywhere along the device, and likewise on the permeate side. Therefore, the total molar permeation rates of species A and B through the membrane are:

$$W_{tAm} = \left(\frac{Q_{Am}}{\delta_m}\right)A_m(p_{Af} - p_{Ap});$$

$$W_{tBm} = \left(\frac{Q_{Bm}}{\delta_m}\right)A_m(p_{Bf} - p_{Bp}). \quad (6.3.203)$$

This mode of operation is such that the rate at which the gas stream enters the device from the gas source equals the rate at which the gas stream permeates through the membrane (θ = stage cut = 1). This ensures a fixed feed pressure P_f. (If the gas stream enters the device from the source at a rate higher than the permeation rate, the feed gas pressure will keep on increasing since it is a compressible fluid, unlike that in the liquid-phase systems considered so far.) However, as permeation takes place, if $\alpha_{AB} > 1$, species A permeates more than species B; the gas mixture left on the feed side has more of species B. Since the feed gas mixture having a constant composition x_{Af} enters the feed side from a gas source, the actual gas composition imposed on the membrane on the feed side will keep on changing with time. This mode of operation cannot therefore maintain a steady state. The permeate composition will also keep on changing with time as the feed side composition has an increasing amount of species B with time.

The permeation rate of gases through nonporous membranes is generally quite low. To achieve a high production rate in larger devices, a large membrane area is needed. In this mode of operation (bulk flow parallel to force), the feed channel cross-sectional area has to be equally large; similarly on the permeate side. This mode of operation is therefore not practical for large-scale operations (where the cross-sectional areas of the incoming feed gas stream and the outgoing permeate gas stream are vanishingly small with respect to the membrane surface area in practical devices). One may use such a mode of operation only for a very small membrane area.

Some of the relations developed earlier, however, are very basic to gas permeation processes: these include (6.3.197) and (6.3.201). For a given x_{Af} and α^*_{AB}, as γ increases, the separation achieved decreases; as γ decreases, the separation factor α_{AB} (definition (6.3.197)) increases, reaching α^*_{AB} in the limit $\gamma \rightarrow 0$. These relations are valid at any point in any gas permeator where x_{Af} is the feed mole fraction of A and x_{Ap} is the corresponding mole fraction in the permeated gas mixture as it emerges from the membrane (before mixing with any other gas mixture in the permeate side of the device). Figure 6.3.33 illustrates the value of the O_2 mole fraction in the permeate side (x_{Ap})

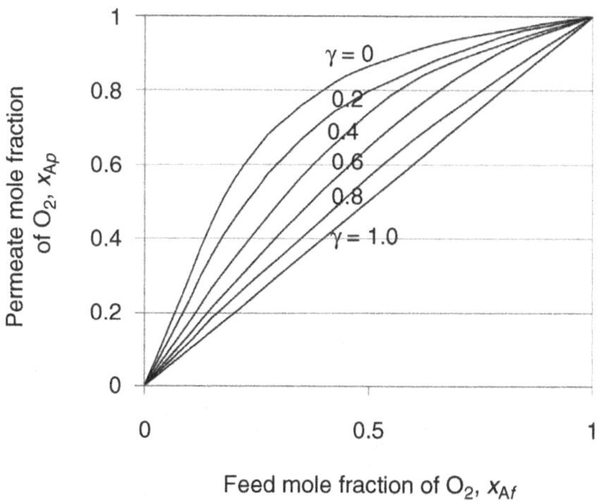

Figure 6.3.33. Effect of pressure ratio, γ, on the permeate mole fraction of oxygen through a cellulose acetate membrane ($\alpha^*_{AB} = 6$) for a feed mixture of O_2-N_2.

for different values of O_2 mole fraction (x_{Af}) on the feed side of a cellulose acetate membrane exposed to various O_2-N_2 gas mixtures; the different curves are for various values of γ. The curve corresponding to $\gamma = 0$ provides the highest O_2 mole fraction in the permeate. This value is easily calculated from relation (6.3.197). All practical gas separation devices having a finite area will yield lower values since $\gamma > 0$, and the feed composition will change along the membrane, as we shall see in the relevant parts of Sections 7.2.1.1 and 8.1.

Example 6.3.13 Consider production of enriched O_2 in a permeate stream from atmospheric air.

Case (1) You have a cellulose acetate membrane having a value of $(Q_{O_2}/Q_{N_2}) = 6.0$. What will be the permeate composition for a pressure ratio, $\gamma = 0$?
Case (2) You would like to produce tonnage oxygen (95% O_2) from atmospheric air. What is the minimum value of $(Q_{O_2}/Q_{N_2}) = 6.0$ required under the condition $\gamma = 0$?

Solution Case (1) We employ equation (6.3.198) valid for $\gamma = 0$, where A $= O_2$, B $= N_2$. We have

$$(x_{Ap}/(1 - x_{Ap})) = \alpha^*_{AB}(x_{Af}/(1 - x_{Af}));$$
$$x_{O_2f} = 0.21; \quad \alpha^*_{AB} = \alpha_{O_2-N_2} = 6.0;$$
$$x_{O_2p}/(1 - x_{O_2p})) = 6(0.21/0.79) = 1.595;$$
$$x_{O_2p} = 1.595 - 1.595\,x_{O_2p} \Rightarrow x_{O_2p} = \frac{1.595}{2.595} = 0.614.$$

Case (2) A $= O_2$, B $= N_2$, so

$$x_{Ap}/(1 - x_{Ap}) = \alpha^*_{AB}(x_{Af}/(1 - x_{Af}));$$
$$x_{Ap} = 0.95; \quad x_{Af} = 0.21 \Rightarrow \frac{0.95}{0.05} = \alpha^*_{AB}\frac{0.21}{0.79}$$
$$\Rightarrow \alpha^*_{AB} = \frac{19}{0.265} \cong 71.7.$$

This is quite a high value. There are almost no polymeric membranes having such an ideal selectivity for O_2 over N_2. There are, however, many different types of polymeric membranes having a selectivity of O_2 over N_2 between 2 (silicone rubber) and 8 (polyestercarbonates) for producing *oxygen-enriched air* (OEA).

In Example 6.3.13, we were calculating/focusing on the permeate composition, i.e the increase in O_2 content over that in the feed. Correspondingly, the N_2 content of the feed gas stream was increasing as O_2 was permeating selectively through the membrane. If a considerable amount of membrane area is provided, the residual gas left on the feed side will be highly concentrated in N_2, producing what is called *nitrogen-enriched air* (NEA); such a gas mixture is quite useful for producing a relatively inert atmosphere, which is required in many applications, e.g. inerting[29] chemical storage tanks, on-board inert gas generation in aircrafts, controlled atmosphere for fruits/vegetables, etc. All such applications, however, utilize bulk flow of feed gas perpendicular to the force direction across the membrane; see the appropriate treatments in Chapters 7 and 8.

Example 6.3.13 and Figure 6.3.33 illustrate separations involving α^* around 6. There are practical membrane gas separation cases where α^* can be as high as 100 and more. Such membranes allow the achievement of a highly purified permeate. For example, there are practical glassy polymeric membranes in use that yield $\alpha^*_{H_2-CH_4}$ of around 100. In the separation of solvent vapors (and volatile organic compounds (VOCs) in general) from N_2/air, selectivities around 50-500 can be achieved using rubbery membranes, leading to a permeate very substantially enriched in VOCs in the permeate.

Considerations on the extent of permeate enrichment in the preferentially permeated component are facilitated if we consider expression (6.3.201) for x_{Ap}. The value of x_{Ap} depends on α^*_{AB}, γ and x_{Af}. For a given x_{Af}, are there regions where α^*_{AB} is controlling or γ is controlling? Consider Figure 6.3.34, where Baker and Wijmans (1991) illustrate the VOC composition in the permeate as $\alpha^*_{VOC-N_2}$ and values of γ are varied for a feed VOC concentration of $x_{VOC,f} = 0.005$. One notices that, in the range $\gamma = 0.1$-1, it essentially does not matter what the ideal membrane selectivity ($\alpha^*_{VOC-N_2}$) is; the permeate composition is essentially controlled by the pressure ratio, γ. On the other hand, in the range $\gamma = 0.01$-0.1 and below 0.01, the permeate composition appears to be determined essentially by $\alpha^*_{VOC-N_2}$, with higher values leading to very high values of x_{Ap}. In this region of γ,

[29] Effort or steps taken to create an inert atmosphere.

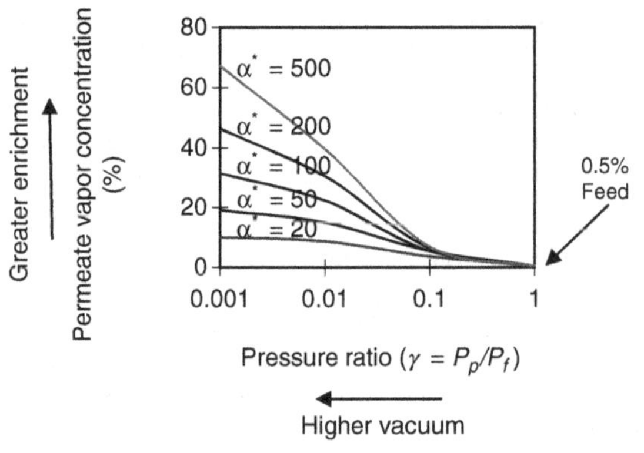

Figure 6.3.34. Permeate vapor concentration for a 0.5% VOC feed as a function of $\alpha^*_{VOC-N_2}$ and pressure ratio γ. (After Baker and Wijmans (1991).)

value of the following ratio (*note*: the mass-transfer coefficient for diffusion in the gas phase was obtained from relation (3.1.132)):

$$\frac{(Q_{im}/\delta_m)}{(D_{ij}/RT\delta_g)}.$$

Assume, for illustration, that at 25 °C we have a silicone rubber membrane which is highly permeable to, say, oxygen:

$$Q_{im} = 5\times10^{-8}\,\text{cm}^3(\text{STP}) - \text{cm}/\left(\text{cm}^2 - \text{s} - \text{cm Hg}\right).$$

Let $\delta_m = 1\,\mu\text{m} = 10^{-4}\,\text{cm}$. From Table 3.A.1, the value of $D_{O_2-N_2}$ may be assumed to be ~0.18 cm^2/s; assume: $\delta_g = 1$ cm; R = 82.05 cm^3-atm/gmol-K; T = 298 K. We have

$$\frac{\dfrac{Q_{im}}{\delta_m}}{\dfrac{D_{ij}}{RT\delta_g}} = \frac{\dfrac{5\times10^{-8}}{10^{-4}}\,\dfrac{\text{cm}^3(\text{STP})-\text{cm}}{\text{cm}^2-\text{s}-\text{cm Hg}}\cdot\dfrac{1}{\text{cm}}}{\dfrac{0.18\,\dfrac{\text{cm}^2}{\text{s}}}{82.05\,\dfrac{\text{cm}^3-\text{atm}}{\text{gmol-K}}\times76\,\dfrac{\text{cm Hg}}{\text{atm}}\times\dfrac{1}{22.4}\,\dfrac{\text{gmol}}{\text{liter}}\times\dfrac{1\text{ liter}}{10^3\text{ cm}^3}\times298\text{ K}\times1\text{ cm}}}$$

$$= \frac{5\times10^{-4}\,\dfrac{\text{cm}^3(\text{STP})}{\text{cm}^2-\text{s}-\text{cm Hg}}}{\dfrac{0.18}{82.05}\times\dfrac{22.4\times10^3}{76\times298\times1}\,\dfrac{\text{cm}^3}{\text{cm}^2-\text{s}-\text{cm Hg}}};$$

$$\frac{(Q_{im}/\delta_m)}{(D_{ij}/RT\delta_g)} = \frac{5\times10^{-4}\times82.05\times76\times298}{0.18\times22.4\times10^3} = \frac{5\times0.82\times0.76\times0.298}{0.18\times22.4} \cong 0.23.$$

then, the permeate composition is controlled by the ideal membrane selectivity. One may conclude the following:

high $\gamma \rightarrow$ pressure ratio controls;
very low $\gamma \rightarrow$ ideal membrane selectivity controls.

There is an important consideration in the separation of a gas mixture through a membrane, as shown in Figure 6.3.32. For a similar configuration, we have seen in Figure 6.3.26 for ultrafiltration that the concentration of the rejected species builds up on the feed side of the membrane: the extent of the buildup is dependent on the solvent flux through the membrane and the rate of back diffusion of the species, specifically the ratio $(|v_z|/k_{i\ell})$. The lower the value of this ratio, the lower the buildup of the rejected species on the feed surface of the membrane. If we employ the same analogy here, then we should determine the

Therefore, for a highly permeable membrane like silicone rubber, if the membrane thickness is smaller than 1 μm, there will be a concentration polarization effect leading to a higher concentration of species B over the membrane (Alpers *et al.*, 1999). Practical membranes of glassy polymers used for many gas separations have lower thicknesses, even though their Q_{im} values are usually much lower. For example, glassy polyimide H_2-selective membranes are reported to have a value of $(Q_{H_2m}/\delta_m) = (100-500)\times10^{-6}\,\text{cm}^3(\text{STP})/\text{cm}^2\text{-s-cm Hg}$ (Zolandz and Fleming, 2001), which is pretty close to the value we used for oxygen through a silicone membrane of 1 μm thickness.

6.3.3.5.1 Composite membranes in gas permeation In practical membrane based gas separation, the membranes as fabricated are often composite in nature: one membrane

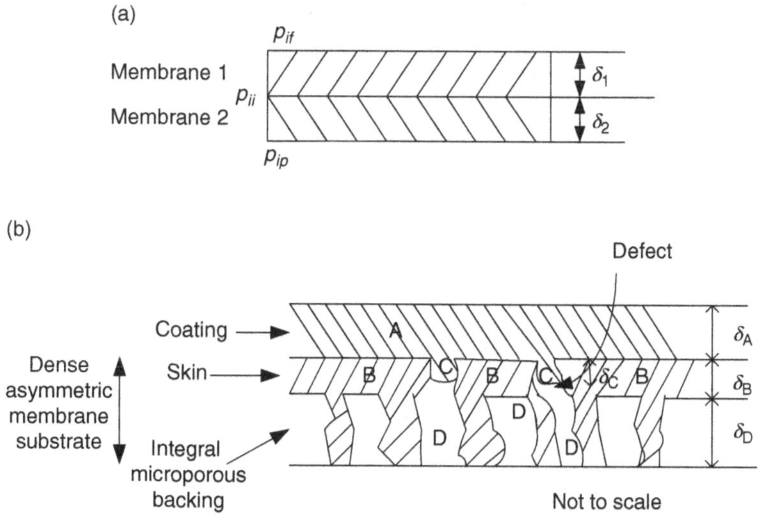

Figure 6.3.35. (a) Schematic of a two-layer composite membrane. (b) Schematic of a composite PRISM-type membrane.

layer of material 1, having a very small thickness δ_1, is supported on another membrane layer of material 2, of thickness δ_2. Let the species i permeance of the first layer of membrane be (Q_{i1}/δ_1) and the species i permeance of the second layer be (Q_{i2}/δ_2). As shown in Figure 6.3.35(a), the feed gas mixture imposed on the membrane in the bulk flow vs. force configuration of flow parallel to force leads to a hypothetical partial pressure profile of gas species i: p_{if} to p_{ii} to p_{ip}, where p_{ii} corresponds to that at the interface of the first and second layers. Therefore, at steady state,

$$N_{iz} = J_{iz}^* = \frac{Q_{i1}}{\delta_1}(p_{if} - p_{ii}); \qquad (6.3.204a)$$

$$N_{iz} = J_{iz}^* = \frac{Q_{i2}}{\delta_2}(p_{ii} - p_{ip}) \qquad (6.3.204b)$$

$$\Rightarrow \left(p_{if} - p_{ii}\right) = N_{iz}/(Q_{i1}/\delta_1); \qquad (6.3.204c)$$

$$\left(p_{ii} - p_{ip}\right) = N_{iz}/(Q_{i2}/\delta_2). \qquad (6.3.204d)$$

Adding the last two equalities results in

$$\left(p_{if} - p_{ii} + p_{ii} - p_{ip}\right) = (p_{if} - p_{ip}) = N_{iz}\left[\frac{1}{Q_{i1}/\delta_1} + \frac{1}{(Q_{i2}/\delta_2)}\right]$$

$$\Rightarrow N_{iz} = \frac{(p_{if} - p_{ip})}{[(\delta_1/Q_{i1}) + (\delta_2/Q_{i2})]} = \left(\frac{Q_{im}}{\delta}\right)(p_{if} - p_{ip}).$$

Therefore, for a composite membrane having two layers 1 and 2, the overall effective membrane permeance (Q_{im}/δ) is related to the permeances of the individual layers by

$$(Q_{im}/\delta) = \frac{1}{[(\delta_1/Q_{i1}) + (\delta_2/Q_{i2})]}. \qquad (6.3.205a)$$

An alternative representation,

$$(\delta/Q_{im}) = R_{im} = [(\delta_1/Q_{i1}) + (\delta_2/Q_{i2})] = R_{i1} + R_{i2}, \qquad (6.3.205b)$$

implies that if the membrane flux of species i, N_{iz}, may be represented in the manner of Ohm's law, namely,

$$\text{current(flux)} = \frac{\text{voltage(partial pressure driving force)}}{\text{resistance}}$$

then the resistance of the two-layer composite membrane, R_{im}, is merely the sum of the resistances of the individual layers in series (as in Ohm's law) (Figure 6.3.35(a)). For n layers in series, relation (6.3.205b) may be generalized as follows:

$$(\delta/Q_{im}) = R_{im} = [(\delta_1/Q_{i1}) + (\delta_2/Q_{i2}) + \cdots + (\delta_n/Q_{in})]. \qquad (6.3.205c)$$

The ideal separation factor, α_{ij}^*, between two species i and j being separated through a composite membrane having two layers, 1 and 2 (Figure 6.3.35(a)), is

$$\alpha_{ij}^* = \frac{(Q_{im}/\delta)}{\left(Q_{jm}/\delta\right)} = \frac{[(\delta_1/Q_{j1}) + (\delta_2/Q_{j2})]}{[(\delta_1/Q_{i1}) + (\delta_2/Q_{i2})]}. \qquad (6.3.206a)$$

Following relation (6.3.197), the actual separation factor, α_{ij}, for a composite membrane having two layers 1 and 2 is

$$\alpha_{ij} = \frac{[(\delta_1/Q_{j1}) + (\delta_2/Q_{j2})]}{[(\delta_1/Q_{i1}) + (\delta_2/Q_{i2})]}\,\frac{\left(1 - \gamma(x_{ip}/x_{if})\right)}{\left(1 - \gamma(x_{jp}/x_{jf})\right)}. \qquad (6.3.206b)$$

Many glassy membranes in commercial use have a structure more complicated than that shown in Figure 6.3.35(a). As shown in Figure 6.3.35(b), the composite membrane (called the "PRISM" membrane (Henis and Tripodi, 1981)) consists of a skin layer, B (of thickness δ_B), of an asymmetric glassy membrane which has some defects. On top of the skin layer, a rubbery coating of material A, having thickness δ_A, has been applied. However, this rubbery material has penetrated the defects of the glassy skin region B; identify these defective regions filled by the rubber material as region C of thickness δ_C. Assume that ε_B is the fraction of the surface area covered by the defects in the glassy skin. Further, let Q_{iA} and Q_{iB} be the permeability coefficients of species i through the materials A and B, respectively. Assume also that the porous substrate of the asymmetric glassy membrane B skin has a thickness δ_D and a species i permeability coefficient Q_{iD}.

It is now useful to postulate that, at distance δ_A from the top, a hypothetical partial pressure p_{iA} exists; similarly, at distance $(\delta_A + \delta_B)$, the hypothetical partial pressure of species i, p_{iB}, exists. Assume now that $\delta_B = \delta_C$. The flux of species i, N_{iz}, may be described as follows:

$$N_{iz} = \frac{Q_{iA}}{\delta_A}(p_{if} - p_{iA}); \qquad (6.3.207a)$$

$$N_{iz} = \frac{Q_{iB}(1-\varepsilon_B)}{\delta_B}(p_{iA} - p_{iB}) + \frac{\varepsilon_B\, Q_{iA}}{\delta_C}(p_{iA} - p_{iB}); \qquad (6.3.207b)$$

$$N_{iz} = \frac{Q_{iD}}{\delta_D}(p_{iB} - p_{ip}). \qquad (6.3.207c)$$

Relation (6.3.207b) indicates that gas species i is diffusing in parallel through region B and region C. Now

$$p_{if} - p_{iA} = N_{iz}/(Q_{iA}/\delta_A);$$

$$p_{iA} - p_{iB} = \frac{N_{iz}}{\left\{\dfrac{Q_{iB}(1-\varepsilon_B)}{\delta_B} + \dfrac{Q_{iA}\,\varepsilon_B}{\delta_C}\right\}};$$

$$p_{iB} - p_{ip} = N_{iz}/(Q_{iD}/\delta_D). \qquad (6.3.207d)$$

Therefore

$$(p_{if} - p_{ip}) = N_{iz}\left[\frac{1}{(Q_{iA}/\delta_A)} + \frac{1}{\left\{\dfrac{Q_{iB}(1-\varepsilon_B)}{\delta_B} + \dfrac{Q_{iA}\,\varepsilon_B}{\delta_B}\right\}} + \frac{1}{\left\{\dfrac{Q_{iD}}{\delta_D}\right\}}\right]$$

since

$$\delta_B = \delta_C \Rightarrow N_{iz} = \frac{Q_{im}}{\delta}(p_{if} - p_{ip}) = \frac{(p_{if} - p_{ip})}{\left[\left(\dfrac{\delta_A}{Q_{iA}}\right) + \dfrac{\delta_B}{(1-\varepsilon_B)Q_{iB} + \varepsilon_B\, Q_{iA}} + \left(\dfrac{\delta_D}{Q_{iD}}\right)\right]}. \qquad (6.3.208)$$

Therefore

$$(\delta/Q_{im}) = R_{im} = \left[R_{iA} + \frac{R_{iB}\,R_{iC}}{R_{iB} + R_{iC}} + R_{iD}\right], \qquad (6.3.209)$$

where

$$R_{iC} = \frac{\delta_C}{\varepsilon_C Q_{iC}} = \frac{\delta_B}{\varepsilon_C Q_{iC}} = \frac{\delta_B}{\varepsilon_C Q_{iA}},$$

$$R_{iA} = (\delta_A/Q_{iA}),\ R_{iB} = \left(\frac{\delta_B}{Q_{iB}(1-\varepsilon_B)}\right),\ R_{iD} = \left(\frac{\delta_D}{Q_{iD}}\right). \qquad (6.3.210)$$

Often the resistance of the porous substrate region, R_{iD}, is neglected in comparison to the others:

$$(\delta/Q_{im}) = R_{im} = \left[R_{iA} + \frac{R_{iB}\,R_{iC}}{R_{iB} + R_{iC}}\right]. \qquad (6.3.211a)$$

The ideal separation factor for two species i and j in the flow vs. force configuration of Figure 6.3.35(b) is given by

$$\alpha^*_{ij} = \frac{(Q_{im}/\delta)}{(Q_{jm}/\delta)} = \frac{R_{jm}}{R_{im}} = \frac{\left[R_{jA} + \dfrac{R_{jB}\,R_{jC}}{R_{jB} + R_{jC}}\right]}{\left[R_{iA} + \dfrac{R_{iB}\,R_{iC}}{R_{iB} + R_{iC}}\right]}. \qquad (6.3.211b)$$

The utility of the structure of Figure 6.3.35(b) was accidentally discovered since it made the glassy polymer skin region B useful by increasing considerably the resistance of permeation through the defect region C by letting the rubbery polymer coating penetrate the defects. Otherwise the high and nondiscriminating permeation through the defects (region C) will drastically reduce the selectivity. Normally the skin region was made up of polysulfone, a glassy polymer, whereas the highly permeable polydimethylsiloxane (silicone rubber) was used to coat the polysulfone material.

Example 6.3.14 Henis and Tripodi (1981) have described the separation performance of a H_2–CO gas mixture through a "PRISM-type" membrane using the following quantitative information for a membrane of glassy polysulfone having a coating of silicone rubber:

$Q_{H_2A} = 5.2 \times 10^{-8}$ scc-cm/cm^2-s-cm Hg; $Q_{COA} = 2.5 \times 10^{-8}$ scc-cm/cm^2-s-cm Hg; $\delta_A = 1 \times 10^{-4}$ cm (~1 μm); $Q_{H_2B} = 1.2 \times 10^{-9}$ scc-cm/cm^2-s-cm Hg; $Q_{COB} = 3 \times 10^{-11}$ scc-cm/cm^2-s-cm Hg; $\delta_B = 1 \times 10^{-5}$ cm (0.1 μm); $\varepsilon = 2 \times 10^{-6}$.

(1) Calculate (Q_{H_2m}/δ) and (Q_{COm}/δ) for this composite membrane.
(2) Calculate the ideal separation factor $\alpha^*_{H_2-CO}$.

Solution (1) We will neglect the resistance of the porous substrate layer D in Figure 6.3.35(b). We now employ equations (6.3.210) and (6.3.211a) to calculate (Q_{H_2m}/δ) and $(Q_{COm}/\ \delta)$:

$$(Q_{H_2m}/\delta) = \cfrac{1}{\cfrac{\delta_A}{Q_{H_2A}} + \cfrac{1}{\cfrac{Q_{H_2B}(1-2\times 10^{-6})}{1\times 10^{-5}} + \cfrac{Q_{H_2A}\, 2\times 10^{-6}}{1\times 10^{-5}}}}$$

$$= \frac{1}{\left[\dfrac{1\times 10^{-4}}{5.2\times 10^{-8}} + \dfrac{1\times 10^{-5}}{1.2\times 10^{-9}(1-2\times 10^{-6}) + 5.2\times 10^{-8}\times 2\times 10^{-6}}\right]}$$

$$= \frac{1}{\left[1923 + \dfrac{1\times 10^{-5}}{1.2\times 10^{-9} + 10.4\times 10^{-14}}\right]} = \frac{1}{[1923+8333]} = 9.75\times 10^{-5}\ \text{scc/cm}^2\text{-s-cm Hg}.$$

Similarly,

$$(Q_{COm}/\delta) = \frac{1}{\left[(\delta_A/Q_{COA}) + \dfrac{1\times 10^{-5}}{Q_{COB}(1-2\times 10^{-6}) + Q_{COA}\times 2\times 10^{-6}}\right]}$$

$$= \frac{1}{\left[(1\times 10^{-4}/2.5\times 10^{-8}) + \dfrac{1\times 10^{-5}}{3\times 10^{-11} + 5\times 10^{-14}}\right]} = \frac{1}{[4000 + 333\,333]}$$

$$= 2.9\times 10^{-6}\ \text{scc/cm}^2\text{-s-cm Hg}.$$

(2) The ideal separation factor,

$$\alpha^*_{H_2-CO} = (Q_{H_2m}/\delta)/(Q_{COm}/\delta)$$

$$= 9.75\times 10^{-5}/2.9\times 10^{-6} = 33.6.$$

Since the ideal separation factor for polysulfone is $(1.2\times 10^{-9}/3\times 10^{-11}) \sim 40$, and that for silicone is 2.08, this composite membrane behaves more like the polysulfone membrane.

6.4 Continuous stirred tank separators

In this section, the flow vs. force configuration of a continuous stirred tank separator (CSTS) will be illustrated with a few examples. The examples cover crystallization, solvent extraction, ultrafiltration and gas permeation.

6.4.1 Well-mixed separators - CSTSs and batch separators

In some separators, the operating conditions may be such that, throughout the separator, pressure, concentration, temperature, etc., are uniform. In separation techniques carried out in the absence of external forces, such a situation creates uniform conditions for separation throughout the separator. A small membrane cell exposed to a feed gas mixture (Figure 6.4.1) is such a separator since very high diffusivity in the gas phase does not allow any concentration gradient to develop due to selective gas permeation through the relatively impermeable membrane. The diffusional rates are considerably slower in liquid solutions; therefore, in some separator vessels, stirring of the liquid is commonly practiced. If the stirring is efficient, one could imagine that the concentration, temperature, etc., are uniform throughout the vessel containing the liquid - making it a well-stirred vessel and a well-mixed separator. When separation requires contacting two immiscible phases, stirring is needed to disperse one phase into the other to generate a high interfacial area and high mass-transfer rates. If such separators have essentially uniform conditions everywhere in the separator, we can also have a well-stirred/well-mixed separator; the term "perfectly mixed" is also used. Note that, in some physical configurations, it will be difficult to create such an environment, for example in highly viscous, high-temperature melts in solid-liquid systems.

If fluid streams with or without solid particles are entering and other streams are exiting the well-stirred separator continuously, we have a *continuous stirred tank separator* (CSTS), *provided that* its properties are uniform throughout the separator. Figure 6.4.1(a) illustrates a CSTS which is a crystallizer. The conditions in such a separator are time- and space-invariant. However, the intensity of mixing conditions in the separator is such that the fresh feed introduced into the separator is mixed in a time interval which is very short compared to the mean residence time of the fluid elements (and solid particles) in the separator. Figure 6.4.1(b) illustrates a *continuous well-stirred extractor*

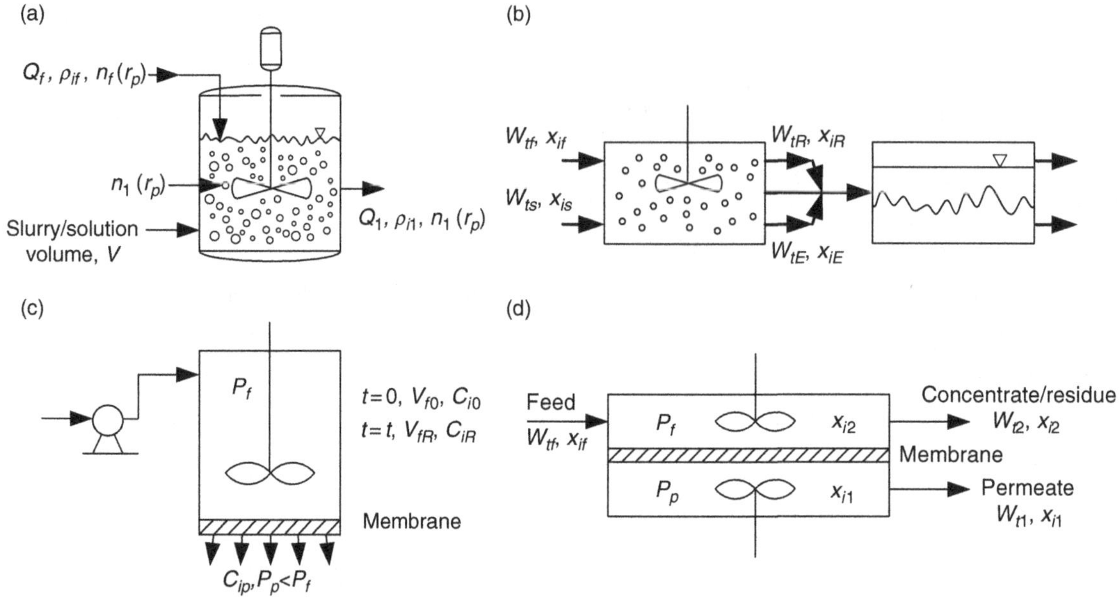

Figure 6.4.1. Well-stirred separations: (a) continuous well-stirred crystallizer (MSMPR); (b) continuous well-stirred solvent extraction device; (c) continuous well-stirred ultrafiltration cell; (d) continuous well-stirred gas separation cell.

based on two immiscible phases. Figure 6.4.1(c) shows a common well-stirred laboratory format for *ultrafiltration,* which was introduced in Section 6.3.3.2. Figure 6.4.1(d) provides the schematic of a continuous well-stirred *membrane gas separator.* This section provides an introduction to these four separation techniques in a CSTS mode. A few other techniques may also be operated in this mode.

The presence of a stirrer in the separator is not a prerequisite for realizing a CSTS. Severe backmixing, etc., can easily create the de facto condition of a well-mixed separator. We will learn in Chapters 8 (and 7) that such conditions are often detrimental to achieving high separation. In CSTSs, there is no specific directional relationship between the bulk flow and the force, unlike those discussed in the rest of Chapter 6 and especially in Chapters 7 and 8. Backmixing can destroy high composition changes achieved along the length of the separator. For phase equilibrium based separation processes, a well-stirred device can merely ensure at its best that the two exiting phases are at equilibrium.

A batch mode of operation in the separation techniques discussed in this section also exists under well-stirred conditions, and these are briefly touched upon.

6.4.1.1 *Solid–liquid systems: solution crystallization, precipitation*

In Section 3.3.7.5, we were introduced to liquid-solid equilibria vis-à-vis crystallization, leaching and precipitation. Here we are going to employ solution crystallization as an example of a solid–liquid system where stirred tanks or well-mixed devices are used routinely for larger-scale/industrial operations. We will focus primarily on a continuous stirred tank crystallizer with one feed stream coming in and one product stream going out from the crystallizer, which is well-mixed. We will consider first various aspects of such a crystallizer with respect to the population density function. Later we will touch upon the progress of crystallization in relation to the cooling (for example) needed for crystallization. Next, we will briefly mention a well-stirred batch/semi-batch crystallizer. We also briefly touch upon precipitation in Section 6.4.1.1.4.

Consider a *continuous crystallizer* of volume V, as shown in Figure 6.4.2(a). A feed stream having a particle (crystal) number density function $n_f(r_p)$ (which is also the population density function), volumetric flow rate Q_f and species i mass concentration ρ_{if} enters the crystallizer continuously. Product stream 1, having a particle (crystal) number density function $n_1(r_p)$, volumetric flow rate Q_1 and species i mass concentration ρ_{i1}, leaves the crystallizer continuously. The particle (crystal) number density function $n(r_p)$ in the well-mixed crystallizer is the same throughout the crystallizer. The macroscopic population balance equation for a stirred tank separator may be written using equations (6.2.60) and (6.2.61) as follows:

$$\frac{\partial n(r_p)}{\partial t}+\frac{\partial}{\partial r_p}(Gn(r_p))=B-De+\frac{1}{V}\{Q_f n_f(r_p)-Q_1 n_1(r_p)\}. \tag{6.4.1a}$$

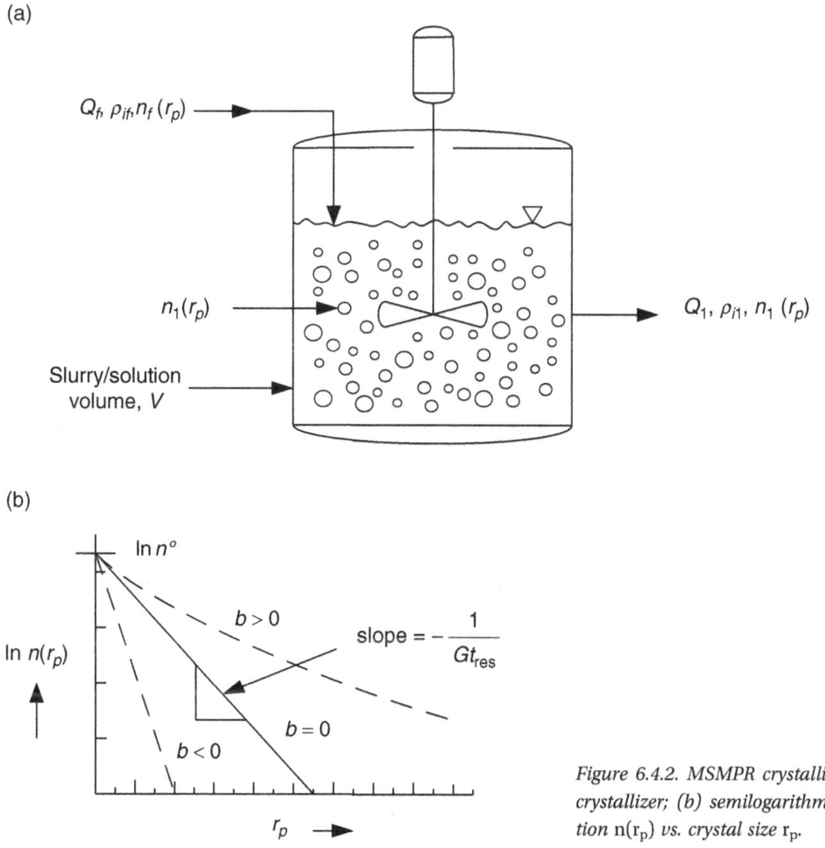

Figure 6.4.2. MSMPR crystallizer. (a) Schematic of a well-mixed crystallizer; (b) semilogarithmic plot of population density function $n(r_p)$ *vs. crystal size* r_p.

If we assume that the processes of crystal breakage, attritional change and agglomeration may be neglected, then $B = 0 = De$. The governing equation for such a continuous stirred tank separator is

$$\frac{\partial n(r_p)}{\partial t} + \frac{\partial}{\partial r_p}(G\,n(r_p)) = \frac{1}{V}\{Q_f\,n_f(r_p) - Q_1\,n_1(r_p)\}, \tag{6.4.1b}$$

except, by definition, the crystal number density function $n_1(r_p)$ in product stream 1 is equal to that present throughout the well-mixed crystallizer, i.e. $n(r_p)$. This implies that there is no crystal size based separation (often called classification) taking place at the crystallizer exit.

We now consider a special case of steady state such that the first term in the above equation is zero. Such a crystallizer is often called a *mixed suspension, mixed product removal* (MSMPR) *crystallizer* (Randolph and Larson, 1988):

$$\frac{d}{dr_p}(G\,n(r_p)) = \frac{1}{V}\{Q_f\,n_f(r_p) - Q_1\,n(r_p)\}. \tag{6.4.2}$$

A special case of this mode of crystallizer operation allows a simplification of this equation: $n_f(r_p) = 0$, corresponding to no crystals in the feed stream, or an *unseeded*[30] system. The corresponding solution for the crystal number density function, $n(r_p)$, is particularly illustrative:

$$\frac{V}{Q_1}\frac{d}{dr_p}(G\,n(r_p)) = -n(r_p). \tag{6.4.3a}$$

The quantity (V/Q_1) is essentially the *mean residence time* of the liquid/particles/crystals in the systems; it is also called the *drawdown time* and will be indicated here by t_{res}. A solution of this equation when the crystal growth rate G $(=dr_p/dt)$ is independent of r_p is quite useful, i.e. $(dG/dr_p) = 0$ (the so-called *size-independent growth*). This condition, encountered earlier as *McCabe's* ΔL *law of crystal growth* (equation (3.4.30) leads to the following form of the crystal population balance equation:

$$\frac{dn(r_p)}{dr_p} = -\frac{n(r_p)}{G\,t_{res}}. \tag{6.4.3b}$$

[30] In some crystallization processes, seed crystals of the same material are introduced with the feed stream.

Define n^0 as the crystal number density function value for crystals of near-zero size, the embryos (see equation (3.3.99)). Integration of this equation for particular values of G and t_{res} yields

$$\int_{n^0}^{n(r_p)} \frac{dn(r_p)}{n(r_p)} = -\int_0^{r_p} \frac{dr_p}{G\,t_{res}} \Rightarrow \ln\left(\frac{n(r_p)}{n^0}\right) = -\frac{r_p}{G\,t_{res}};$$

$$n(r_p) = n^0 \exp\left(-\frac{r_p}{G\,t_{res}}\right); \tag{6.4.4}$$

$$\ln(n(r_p)) = \ln n^0 - (r_p/G\,t_{res}). \tag{6.4.5}$$

A semilogarithmic plot of $n(r_p)$ against the crystal size r_p will yield an intercept of n^0 and a slope of $-(1/G\,t_{res})$ (if the plot is for $\ell og_{10}n(r_p)$ vs. r_p, the slope is $-(1/(2.303\ Gt_{res}))$); since $t_{res}(= (V/Q_1))$ is known, the slope will yield G, the linear crystal growth rate, (dr_p/dt). A semilogarithmic plot of $n(r_p)$ vs. r_p in Figure 6.4.2(b) illustrates this behavior of an MSMPR crystallizer for the case of the crystal-free feed stream $(n_f(r_p) = 0)$ and the ΔL law of crystal growth. Measurement of the crystal number density function then yields two parameters, n^0 and Gt_{res}. Generally, the selected value of t_{res} used to operate the MSMPR crystallizer will influence the value of the growth rate, G, achieved in the crystallizer.

To understand the nature of the other parameter n^0, consider the basic relation between $n(r_p)$ and $N(r_p)$ (relation (6.2.48a)) in the limit of $r_p \to 0$:

$$\lim_{r_p \to 0} n(r_p) = n^0 = \underset{r_p \to 0}{Lt} \frac{dN(r_p)}{dr_p}. \tag{6.4.6}$$

We could express it also as the ratio in the limit

$$n^0 = \lim_{r_p \to 0} \frac{dN(r_p)/dt}{dr_p/dt} = \underset{r_p \to 0}{Lt} \frac{B}{G} = \frac{B^0}{G|_{r_p \to 0}}, \tag{6.4.7}$$

where $dN(r_p)/dt$ in the limit of $r_p \to 0$ is the rate at which the smallest crystals, i.e. nuclei, are being born (i.e. the value of B as $r_p \to 0$, B^0; see (6.2.50g), where the units are

$$\frac{\text{number (of crystals)}}{\text{volume} \times \text{time}} = \frac{\text{number of crystals}}{\text{m}^3\text{-s}},$$

and dr_p/dt in the same limit is the growth rate of the smallest crystals (i.e. G as $r_p \to 0$). The dynamics of the nucleation kinetics in a crystallizing system is captured in this parameter n^0, the nucleation population density parameter, which will depend strongly on the supersaturation level (for B^0, see (3.4.22b); for G, see (3.4.28)).

Equation (6.4.4) provides an expression for the crystal number density function $n(r_p)$ in an MSMPR crystallizer. We now need information about the distribution functions of other properties of MSMPR crystallizers, e.g. the total number of crystals per unit system volume, N_t (equation (2.4.2c)), the total crystal mass per unit system volume, the suspension density M_T (equation (2.4.2f)), etc. The calculation of such quantities appropriately normalized is facilitated by recognizing that equation (6.4.4) may be represented as (Randolph and Larson, 1988)

$$y = \exp(-x), \tag{6.4.8}$$

where

$$y = (n(r_p)/n^0) \text{ and } x = (r_p/G\,t_{res}). \tag{6.4.9}$$

Such calculations are illustrated in the examples below (following Randolph and Larson (1988)).

Example 6.4.1 For the MSMPR crystal number density function (6.4.4), determine the expressions for $N_t, F(r_p), \bar{r}_{p_{1,0}}, A_T$ and M_T. Develop expressions for the cumulative crystal size distribution fraction and the cumulative surface area distribution fraction.

Solution We will employ the simplified representation of (6.4.8) whenever we can. To start with, as per equations (2.4.2c) and (6.4.4), the total number of crystals per unit volume, N_t, is

$$N_t = \int_0^\infty n(r_p)dr_p = \int_0^\infty n^0 \exp\left(-\frac{r_p}{G\,t_{res}}\right) dr_p$$
$$= n^0\,G\,t_{res}\,\frac{\text{number of crystals}}{\text{cm}^3}. \tag{6.4.10}$$

Therefore the crystal size distribution function $F(r_p)$ (defined by (2.4.2d)) is:

$$F(r_p) = \int_0^{r_p} n(r_p)dr_p/N_t = (n^0/N_t)\int_0^x G\,t_{res}\exp(-x)dx$$
$$= (n^0\,G\,t_{res}/n^0\,G\,t_{res})\int_0^x \exp(-x)dx = 1 - \exp(-x)$$
$$= 1 - \exp(-r_p/G\,t_{res}). \tag{6.4.11}$$

This has a value of 1 when $r_p \to \infty$ and a value of 0 when $r_p = 0$.

The number based mean crystal size, $\bar{r}_{p_{1,0}}$, is obtained from definition (2.4.2g) as follows:

$$\bar{r}_{p_{1,0}} = \int_0^\infty \frac{r_p\,n(r_p)dr_p}{N_t} = \int_0^\infty \frac{(G\,t_{res})^2\,n^0\,x\exp(-x)dx}{N_t}$$
$$= \frac{n^0(G\,t_{res})^2}{n^0\,G\,t_{res}}\int_0^\infty x\exp(-x)dx = G\,t_{res}. \tag{6.4.12}$$

A number based cumulative crystal size distribution fraction may be defined by

$$\frac{\int_0^{r_p} r_p \frac{n(r_p)dr_p}{N_t}}{\bar{r}_{p_{1,0}}} = \frac{n^0(G\,t_{res})^2\int_0^x x\exp(-x)dx}{n^0(G\,t_{res})^2}; \tag{6.4.13}$$

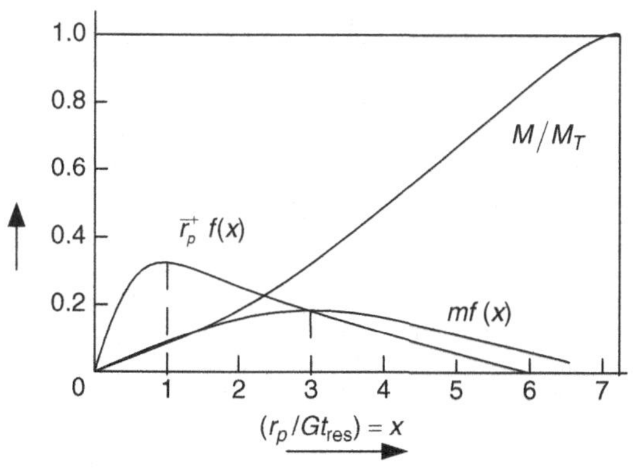

Figure 6.4.3. Mass fraction density function mf(x), *mass distribution function* (M/M_T) *and the density function of* $\overline{r_p}^+ f(x)$ *for an MSMPR crystallizer.*

$$\bar{r}_p^+(x) = \frac{\int_0^{r_p} r_p\, n(r_p)\mathrm{d}r_p}{N_t \bar{r}_{p_{1,0}}} = \int_0^x x\exp(-x)\mathrm{d}x$$
$$= 1-(1+x)\exp(-x). \quad (6.4.14\text{a})$$

The corresponding density function, $\bar{r}_p^+ f(r_p)$, in terms of x is

$$\bar{r}_p^+ f(x) = \mathrm{d}\bar{r}_p^+/\mathrm{d}x = x\exp(-x). \quad (6.4.14\text{b})$$

This is plotted in Figure 6.4.3. When $x\to\infty, \bar{r}_p^+ \to 1$; when $x\to 0, \bar{r}_p^+ \to 0$.

The total crystal surface area per unit liquid volume, A_T, may be defined using a crystal shape factor ψ_s based on the surface area a_p of a particle or a representative radius r_p (definition (3.4.26)):

$$a_p = \psi_s r_p^2; \qquad A_T = \int_0^\infty a_p\, n(r_p)\mathrm{d}r_p = \psi_s \int_0^\infty r_p^2 n(r_p)\mathrm{d}r_p; \quad (6.4.15)$$

$$A_T = \psi_s \int_0^\infty n^0 r_p^2 \exp(-r_p/G\,t_{res})\mathrm{d}r_p = 2\,n^0\,\psi_s\,(G\,t_{res})^3. \quad (6.4.16)$$

A cumulative crystal surface area distribution fraction may be defined as

$$A_p^+ = \int_0^{r_p} a_p n(r_p)\mathrm{d}r_p/A_T = \frac{n^0\,\psi_s \int_0^{r_p} r_p^2 \exp\left(-\frac{r_p}{G\,t_{res}}\right)\mathrm{d}r_p}{2\,n^0\,\psi_s\,(G\,t_{res})^3}$$

$$A_p^+(x) = (1/2)\int_0^x x^2\exp(-x)\mathrm{d}x = 1-\left(1+x+\frac{x^2}{2}\right)\exp(-x). \quad (6.4.17)$$

When $x\to\infty, A_p^+ \to 1$; when $x\to 0, A_p^+ \to 0$.

The suspension density M_T (defined by (2.4.2f)) is

$$M_T = \rho_s\,\psi_v \int_0^\infty r_p^3\, n(r_p)\mathrm{d}r_p = \rho_s\,\psi_v \int_0^\infty n^0 r_p^3 \exp\left(-\frac{r_p}{G\,t_{res}}\right)\mathrm{d}r_p$$
$$M_T = 6\,n^0\,\rho_s\,\psi_v (G\,t_{res})^4\ \mathrm{g/cm^3}. \quad (6.4.18)$$

Example 6.4.2 Define a *dominant crystal size* vis-à-vis a certain property/characteristic of the crystal as that crystal size where the property has a maximum. Determine the dominant crystal size with respect to the crystal mass for an MSMPR crystallizer.

Solution The crystal mass density per unit liquid volume between the crystal dimensions 0 and r_p may be obtained from relation (2.4.2f) as

$$M = \rho_s\,\psi_v \int_0^{r_p} r_p^3\, n(r_p)\mathrm{d}r_p = \rho_s\,\psi_s\, n^0 \int_0^{r_p} r_p^3 \exp\left(-\frac{r_p}{G\,t_{res}}\right)\mathrm{d}r_p.$$

A crystal mass density distribution function may be defined from relation (6.4.18) as

$$\frac{M}{M_T} = \frac{\rho_s\psi_v \int_o^{r_p} r_p^3\, n(r_p)\mathrm{d}r_p}{\rho_s\psi_v \int_o^{\infty} r_p^3\, n(r_p)\mathrm{d}r_p}. \quad (6.4.19)$$

It has a value of 0 at $r_p = 0$ and 1 at $r_p = \infty$ (just like the crystal size distribution function $F(r_p)$ in Example 6.4.1). A density function $mf(r_p)$ of the crystal mass distribution function vis-à-vis the particle size r_p is obtained over a differential size range of r_p to $r_p + \mathrm{d}r_p$:

$$(\mathrm{d}M/M_T) = \mathrm{d}\left(\rho_s\,\psi_v \int_o^{r_p} r_p^3 n(r_p)\mathrm{d}r_p\right)\Big/ M_T$$
$$\Rightarrow (\mathrm{d}M/M_T) = \rho_s\,\psi_v r_p^3 n(r_p)\mathrm{d}r_p/M_T,$$

which implies that the mass fraction density function is

$$mf(r_p) = \frac{\mathrm{d}M/\mathrm{d}r_p}{M_T} = \frac{\rho_s\psi_v r_p^3 n(r_p)}{M_T}. \quad (6.4.20)$$

The dominant crystal size will make the maximum contribution to $mf(r_p)$; therefore, at this dominant size r_{pd},

$$\mathrm{d}(mf(r_p))/\mathrm{d}r_p = 0.$$

For an MSMPR crystallizer,

$$\frac{\mathrm{d}}{\mathrm{d}r_p}\left(\frac{\rho_s\,\psi_v\, r_p^3\, n^0 \exp(-r_p/Gt_{res})}{6\,n^0\,\rho_s\,\psi_v\,(G\,t_{res})^4}\right) = 0;$$
$$\frac{\mathrm{d}}{\mathrm{d}r_p}\left(r_p^3\exp(-r_p/Gt_{res})\right) = 0 \Rightarrow 3\,r_p^2\exp(-r_p/Gt_{res})$$
$$+\,r_p^3(-)\frac{\exp(-r_p/Gt_{res})}{Gt_{res}} = 0 \Rightarrow r_{pd} = 3\,G\,t_{res}. \quad (6.4.21\text{a})$$

The crystal mass density distribution function (M/M_T) for an MSMPR crystallizer from relation (6.4.19) can be written as

$$(M/M_T) = 1 - (\exp(-x))\left(1 + x + \frac{x^2}{2} + \frac{x^3}{6}\right), \quad (6.4.21b)$$

where $x = r_p/G\,t_{res}$. The crystal mass fraction, density function, $mf(x)$, is plotted as

$$d(M/M_t)/dx = (\exp(-x))(x^3/6). \quad (6.4.21c)$$

Figure 6.4.3 illustrates $mf(r_p)$ as a function of x; it has a maximum at $x = 3$, i.e. $r_{pd} = 3Gt_{res}$. This figure also illustrates (M/M_T). The density function of $\bar{r}_p^+(x), \bar{r}_p^+ f(x)$ from Example 6.4.1 is also displayed in this figure.

Example 6.4.3 The crystals obtained from an MSMPR crystallizer were analyzed for their size distribution. A plot of ln $((n(r_p)$ crystals/mm-liter)) against r_p (the abscissa) in mm yielded a slope of $-10\,\text{mm}^{-1}$. The intercept on the ordinate yielded a value of 3×10^8 crystals/ mm-liter. The residence time in the crystallizer was 120 minutes. Determine the growth rate and the nucleation rate.

Solution The slope is given by $(-1/G\ t_{res})$. Since $t_{res} = 120$ minutes,

$$G = \frac{1}{\text{slope}}(-)\frac{1}{t_{res}} \Rightarrow G = (-)\frac{1}{10\,\text{mm}^{-1}}(-)\frac{1}{120\,\text{min}}$$

$$= \frac{1}{1200}\frac{\text{mm}}{\text{min}} = 8.3 \times 10^{-4}\frac{\text{mm}}{\text{min}}.$$

The nucleation rate B as $r_p \to 0$ is given by

$$B^0 = n^0 G = 3 \times 10^8 \frac{\text{crystal}}{\text{mm-liter}} \times 8.3 \times 10^{-4}\frac{\text{mm}}{\text{min}}$$

$$= 2.49 \times 10^5 \frac{\text{crystal}}{\text{liter-min}}.$$

Example 6.4.4 Determine the values of N_t, $\bar{r}_{p_{1,0}}$, r_{pd} and M_T for the MSMPR crystallizer product stream of Example 6.4.3. Assume spherical crystals of density 1.2 g/cm^3.

Solution

$$N_t = n^0 G\,t_{res}$$

$$= 3 \times 10^8 \frac{\text{crystal}}{\text{mm-liter}} \times 8.3 \times 10^{-4}\frac{\text{mm}}{\text{min}} \times 120\,\text{min}$$

$$= 3 \times 8.3 \times 120 \times 10^4 \frac{\text{crystal}}{\text{liter}} = 2.98 \times 10^7 \frac{\text{number}}{\text{liter}}.$$

From expression (6.4.12),

$$\bar{r}_{p_{1,0}} = n^0(G\,t_{res})^2/n^0\,G\,t_{res}$$

$$= G\,t_{res} = 8.3 \times 10^{-4}\frac{\text{mm}}{\text{min}} \times 120\,\text{min}$$

$$= 9.96 \times 10^{-2}\text{mm} = 0.0996\,\text{mm}.$$

On the other hand, from relation (6.4.21a),

$$r_{pd} = 3G\,t_{res} = 3 \times 8.3 \times 10^{-4}\frac{\text{mm}}{\text{min}} \times 120\,\text{min}$$

$$= 2.98 \times 10^{-1}\text{mm} = 0.298\,\text{mm};$$

$$M_T = 6\,n^0\,\rho_s\,\psi_v\,(G\,t_{res})^4$$

$$= 6 \times 3 \times 10^8 \frac{\text{crystal}}{\text{mm-liter}} \times 1.2\frac{\text{g}}{\text{cm}^3} \times \frac{1}{1000}\frac{\text{cm}^3}{\text{mm}^3}$$

$$\times \frac{4}{3} \times (8.3 \times 10^{-4} \times 120\,\text{mm})^4$$

$$= \frac{18 \times 1.2 \times 4 \times 0.984 \times 10^4}{3 \times 1000}\frac{\text{g}}{\text{liter}} = 283\,\text{g/liter}.$$

A brief discussion about the experimental methods used to obtain the crystal number density function, $n(r_p)$, is relevant. One common and classical method is based on *sieving*. The crystal mass is sieved through a vertical stack of sieves of increasingly smaller opening size, the largest being at the top. The mass of crystals retained on a sieve of given size is $\Delta(w^s)$, where w^s is the mass of the crystal material of mass density ρ_s. The particle diameter difference between the two sieves, the one at the top and the one on which $\Delta(w^s)$ was obtained, is identified as Δd_p $(= 2\Delta r_p)$. We know from relation (2.4.2i) that

$$\bar{r}_{p_{4,3}} = \left(\int_0^\infty r_p^4\,n(r_p)\,dr_p\right) \Big/ \left(\int_0^\infty r_p^3\,n(r_p)\,dr_p\right) \quad (6.4.21d)$$

is an average particle radius based on the mass of the particles. If, however, the limits in the integrals above are r_{min} and r_{max}, valid for the crystal fraction obtained for the two sieves under consideration, instead of 0 and ∞, then $\bar{r}_{p_{4,3}}$ will represent the crystal fraction under consideration. By definition (2.4.2f) for the mass density of a population, we can write

$$\Delta w^s = \rho_s\,\psi_v\,r_p^3\,n(r_p)\,\Delta r_p. \quad (6.4.21e)$$

Therefore

$$n(r_p) = \frac{\Delta w^s}{\rho_s\,\psi_v\,r_p^3\,\Delta r_p} = \frac{2\,\Delta w^s}{\rho_s\,\psi_v\,r_p^3\,\Delta d_p}, \quad (6.4.21f)$$

where r_p^3 is now $\bar{r}^3_{p_{4,3}}$. Recall that ψ_v is defined via equation (2.4.2e) based on radius (it is essentially the same as equation (6.4.21e) defined for a finite dr_p), and will be different if equation (2.4.2e) is defined with respect to diameter d_p.

Additional methods to measure crystal size distribution based on particle mass employ light scattering/diffraction methods. Simultaneous measurements of particle size (based on particle volume) and their number are also carried out by Coulter counters. There are a variety of other techniques, based on centrifugal sedimentation, electroacoustic spectroscopy, microelectrophoresis, gravitational sedimentation, scanning electron microscopy, etc. An introduction to these techniques is available in a NIST (National Institute of Standards and Technology) publication by Jillavenkatesa *et al.* (2001).

Although equation (6.4.4) provides an expression for $n(r_p)$ in an MSMPR crystallizer under certain conditions, estimates of additional quantities, such as the *supersaturation*, the *suspension density*, the *growth rate* etc., are also of interest. The supersaturation is of particular interest since the growth rate, as we know from equations (3.4.28) and (3.4.29), depends on the *fractional supersaturation*. Further, the extent of crystallization needs to be determined. These

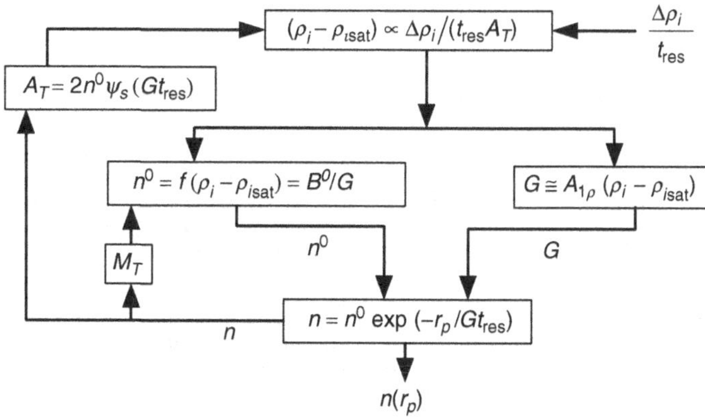

Figure 6.4.4. Interaction between $n(r_p)$, G *and* B^0 *via* n^0. *For an MSMPR crystallizer. (After Randolph and Larson (1988).)*

calculations are facilitated by a *mass balance* on species *i* being crystallized over the MSMPR crystallizer shown in Figure 6.4.2(a):

$$\underbrace{Q_f\rho_{if}}_{\text{mass rate of inflow of species } i} - \underbrace{Q_1\rho_{i1}}_{\text{mass rate of outflow of species } i} = \underbrace{Q_1M_T}_{\text{total mass rate of crystallization.}} \tag{6.4.22}$$

The total mass rate of crystallization, Q_1M_T, may be assumed to be equal to the growth rate of the crystal surfaces:

$$\underbrace{Q_1 \times M_T}_{\frac{\text{flow rate of particle mass}}{\text{time}}} = \underbrace{G}_{\frac{\text{crystal dimension}}{\text{time}}} \times \underbrace{A_T}_{\frac{\text{total crystal surface area}}{\text{solution volume}}} \times \underbrace{V}_{\text{solution volume}} \times \underbrace{\rho_{ip}}_{\text{particle density}} \tag{6.4.23}$$

Two assumptions were made in arriving at the expression on the right-hand side. First, the mass rate of loss of *i* from the solution due to the formation of nuclei only is negligible. Second, since in our treatment the crystal growth rate $G = dr_p/dt$, where r_p is the characteristic radius (and not the diameter), we do not have a factor (1/2) on the right-hand side (unlike that in Randolph and Larson (1988), where $G = \mathrm{d}(2r_p)/\mathrm{d}t$ is based on the growth rate of the diameter; therefore, the value of the shape factor ψ_v in our equations (2.4.2f) and (3.4.26) is different from the shape factor k_v in Randolph and Larson's (1988) eqs. (5.3-2/3)). The mass concentration change $\Delta\rho_i$ achieved in the crystallizer and defined by

$$\Delta\rho_i = M_T = \left(Q_f/Q_1\right)\rho_{if} - \rho_{i1}\ \mathrm{g/cm^3} \tag{6.4.24}$$

can now be utilized to develop an expression for the growth rate *G* from relation (6.4.23):

$$G = \frac{Q_1 M_T}{A_T V \rho_{ip}} = \frac{M_T}{t_{res}\rho_{ip}A_T} = \frac{\Delta\rho_i}{\rho_{ip}t_{res}A_T}; \tag{6.4.25}$$

$$G = \frac{\Delta\rho_i}{\rho_{ip}\,t_{res}\,\psi_s\int_0^\infty r_p^2\,n(r_p)\mathrm{d}r_p}, \tag{6.4.26}$$

where we have employed definition (6.4.15) for A_T. This equation illustrates the inverse dependence of *G* on A_T and therefore on $n(r_p)$, which depends on the growth rate *G* and the nucleation rate $B^0(as\,r_p \to 0)$. Thus there is a feedback loop, as shown in Figure 6.4.4. To understand the role of the inputs $\Delta\rho_i(=M_T)$ and t_{res} in this process, consider the following treatment.

We have observed in an earlier chapter (see equation (3.4.28)) that the crystal growth rate *G* is strongly influenced by the *level of supersaturation.* If we express the growth rate as a linear function of the supesaturation in mass units, i.e. $\rho_i - \rho_{isat}$ (instead of molar supersaturation $C_i - C_{isat}$ or fractional supersaturation *s*, see (3.3.98a, b)),

$$G = A_{1p}(\rho_i - \rho_{isat}) \tag{6.4.27}$$

then

$$(\rho_i - \rho_{isat}) = \text{extent of supersaturation} = \frac{\Delta\rho_i}{A_{1p}\rho_{ip}\,t_{res}\,\psi_s\int_0^\infty n(r_p)r_p^2\,\mathrm{d}r_p}. \tag{6.4.28a}$$

Thus

$$(\rho_i - \rho_{isat}) = \frac{\Delta\rho_i}{A_{1p}\rho_{ip}\,t_{res}\,A_T}. \tag{6.4.28b}$$

From relation (3.4.22b) we know that, in homogeneous nucleation, the nucleation rate

$$B^0 \propto = \exp\left(\cdots \frac{1}{(\ln s)^2}\right), \tag{6.4.29a}$$

where s is the fractional supersaturation in units of molar concentration (easily converted to mass concentration). Therefore, since n^0 from (6.4.7) is given by

$$n^0 \sim B^0/G|_{r_p\to 0}, \tag{6.4.29b}$$

n^0 has a stronger dependence on $(\rho_i - \rho_{isat})$ than G.

We will now follow an illustrative procedure from Randolph and Larson (1988) and Wankat (1990, chap. 4) to find out first the role of the *residence time*, t_{res}, in an MSMPR crystallizer. For the sake of illustration, we will replace the complex dependence of B^0 on supersaturation indicated via (6.4.29a) by (for cases of primarily nucleation only)

$$B^0 = A\,(\rho_i - \rho_{isat})^p, \tag{6.4.30a}$$

where A is a rate constant and the value of p may vary between 2 and 9 for aqueous systems. (The power p may be considered as the *kinetic order* in the expression for the dependence of the nucleation rate on the supersaturation level.) This approach, illustrated in Randolph and Larson (1988), is based on the Miers nucleation model (Miers and Isaac, 1906). Combining it with the linear dependence of the crystal growth rate G on supersaturation (6.4.27), we get, from definition (6.4.7) for n^0,

$$n^0 = A\,(\rho_i - \rho_{isat})^p / A_{1p}(\rho_i - \rho_{isat}) = (A/A_{1p})(\rho_i - \rho_{isat})^{p-1} \\ = (A/A_{1p})(G/A_{1p})^{p-1}. \tag{6.4.30b}$$

Let us now focus on an MSMPR crystallizer having a size-independent growth (McCabe's ΔL law). To determine the effect of t_{res}, we assume two situations where the suspension density M_T is identical but t_{res}, G and n^0 **are different**. From expression (6.4.18) for M_T and for two situations (identified by subscripts a and b) and n^0 from above,

$$M_T = 6\,n_a^0 \rho_s \psi_v (G_a\, t_{res}|_a)^4 = 6\,n_b^0 \rho_s \psi_v (G_b\, t_{res}|_b)^4; \\ (A/A_{1p})(G_a/A_{1p})^{p-1}(G_a)^4\, t_{res}|_a^4 = (A/A_{1p})(G_b/A_{1p})^{p-1} \\ (G_b)^4\,(t_{res}|_b)^4. \tag{6.4.31a}$$

From this we get two results:

$$(G_a/G_b) = (t_{res}|_b/t_{res}|_a)^{4/(3+p)}; \tag{6.4.31b}$$

$$(n_a^0/n_b^0) = (t_{res}|_b/t_{res}|_a)^{4(p-1)/3+p}. \tag{6.4.31c}$$

The exponent p in relation (6.4.30a) for B^0 is clearly quite important. If $p = 1$, by (6.4.31b), the growth rate G depends inversely on the residence time, t_{res}; correspondingly, via relation (6.4.27), the supersaturation also depends inversely on the residence time. Further, $n^0 (= B^0/G)$ is independent of t_{res}, i.e. $n_a^0 = n_b^0$. More importantly, as t_{res} increases, the growth rate decreases. If $t_{res}|_b = 2\,t_{res}|_a$, $G_a = 2G_b$. Also, $(\rho_i - \rho_{isat})_a = 2(\rho_i - \rho_{isat})_b$: as the residence time is doubled, the supersaturation is reduced to half, and so is the growth rate.

On the other hand, for $p > 1$, relations (6.4.31b) and (6.4.31c) indicate, respectively, that, for the same M_T, both n^0 and G increase with t_{res}; therefore, B^0 increases faster (since $B^0 = n^0 G$) than when $p = 1$, where n^0 is independent of t_{res}. A higher nucleation rate (B^0) implies many more crystals; therefore a longer residence time may not necessarily lead to larger crystals since supersaturation is being consumed by many more crystals growing.

We now briefly consider the case where t_{res} is the same but the magma densities are different: $M_{Ta} \neq M_{Tb}$; $t_{res}|_a = t_{res}|_b$. Obviously we are dealing with two different MSMPR crystallizers. From expression (6.4.18) for M_T for the two cases,

$$M_{Ta} = 6\,n_a^0 \rho_s \psi_v (G_a\, t_{res}|_a)^4; \\ M_{Tb} = 6\,n_b^0 \rho_s \psi_v (G_b\, t_{res}|_b)^4; \\ (M_{Ta}/M_{Tb}) = n_a^0\, G_a^4 / n_b^0\, G_b^4. \tag{6.4.32a}$$

Substituting (6.4.30b) for n^0 to obtain

$$(M_{Ta}/M_{Tb}) = G_a^{3+p}/G_b^{3+p} \Rightarrow (G_a/G_b) = (M_{Ta}/M_{Tb})^{1/(3+p)}; \\ (n_a^0/n_b^0) = (G_a/G_b)^{p-1} = (M_{Ta}/M_{Tb})^{(p-1)/(3+p)}. \tag{6.4.32b}$$

If $p = 1$, the nucleation population density function n^0 is independent of the magma densities, but growth rate G increases as magma density increases. If $p > 1$, then n^0 increases as M_T increases; so does G. Similarly, from (6.4.21), the dominant crystal size r_{pd} also increases as M_T increases for constant t_{res} since G increases.

Example 6.4.5 The kinetic order p in the dependence of the nucleation rate B^0 on the supersaturation is an unknown empirical parameter. Suppose you have an option of generating data from an MSMPR for a given system. You can vary t_{res} for a given M_T. Indicate a procedure to determine p.

Solution We know from our earlier analysis that, for a given M_T, if t_{res} is varied, the growth rate G and the nucleation population density function n^0 will vary according to relations (6.4.31b) and (6.4.31c), respectively, for a given p. Therefore a procedure that could be followed is: carry out two different experiments for two values of t_{res} at the same M_T. Obtain, from a plot of $n(r_p)$ vs. r_p, the values of G_a, G_b, n_a^0 and n_b^0. Assume different values of p and check whether the same p can describe both of the following relations:

$$(G_a/G_b) = (t_{res}|_b/t_{res}|_a)^{4/(3+p)}, \tag{6.4.33a}$$

$$(n_a^0/n_b^0) = (t_{res}|_b/t_{res}|_a)^{4(p-1)/(3+p)}, \tag{6.4.33b}$$

for the two residence time values, $t_{res}|_a$ and $t_{res}|_b$. The value of p which satisfies both (6.4.33a) and (6.4.33b) is the desired one.

The decrease in solute mass concentration $\Delta\rho_i\,(= M_T)$ along an MSMPR crystallizer depends on a number of quantities, as one can see from relations (6.4.18) and (6.4.25). If the extent of supersaturation in product stream 1 is not exhausted, the system is identified as a *class I system* with a nonhigh yield. For *class II systems*, the supersaturation is almost exhausted in the product stream, and the system is considered to be a high-yield system.

The basic result (6.4.4) obtained for an MSMPR crystallizer was based on several assumptions.

(1) The crystal growth rate G ($= \mathrm{d}r_p/\mathrm{d}t$) is size-independent.
(2) The product crystals are being withdrawn without any classification, so that $n_1(r_p) = n(r_p)$.
(3) The suspended solids have no effect on the crystal size distribution.
(4) There is no specific relation between the crystal growth rate and the level of supersaturation.
(5) There is no secondary nucleation.
(6) There is no addition of seeds (i.e. $n_f = 0$).
(7) There is no crystal breakage, attritional change or agglomeration in the MSMPR crystallizer.
(8) There is steady state operation.

We now look very briefly at the consequences of relaxing some of these assumptions.

Consider first the case where *crystal growth rate depends on the crystal size* (relaxing assumption (1)). Instead of the simplified equation (6.4.3b), we will have to work with equation (6.4.3a):

$$\frac{\mathrm{d}}{\mathrm{d}r_p}(G\,n(r_p)) = -n(r_p)\frac{Q_1}{V} = -\frac{n(r_p)}{t_{\mathrm{res}}}. \qquad (6.4.34)$$

If empirical relations are available for $G(r_p)$, it may be substituted in the above relation. For example, Abegg *et al.* (1968) have shown that

$$G(r_p) = G_0\left(1+\gamma r_p\right)^b, \qquad b < 1, \qquad (6.4.35)$$

appears to fit most data; here G_0, γ and b are experimentally determined constants. Using such a relation in equation (6.4.34) leads to

$$G_0\,\frac{\mathrm{d}}{\mathrm{d}r_p}\left\{\left(1+\gamma r_p\right)^b n(r_p)\right\} = -\frac{n(r_p)}{t_{\mathrm{res}}}. \qquad (6.4.36)$$

Integration of this equation has led to the following result for size-dependent crystal growth:

$$n(r_p) = \exp[1/G_0\,t_{\mathrm{res}}\,\gamma(1-b)]\,n^0\left(1+\gamma r_p\right)^{-b} \times \exp\left\{\frac{-[(1+\gamma r_p)^{1-b}]}{G_0 t_{\mathrm{res}}\,\gamma(1-b)}\right\}. \qquad (6.4.37)$$

When $b = 0$ and $\gamma = 1$, this relation is simplified to the standard equation (6.4.4) for $G(r_p) =$ constant. The dashed lines in Figure 6.4.2(b) illustrate the nature of these crystal number density functions. Figures 6.4.5(a) and (b) illustrate schematically size-independent and size-dependent crystal growth rates.

Second, we take into account the *effect of suspended solids* (relaxing assumption (3)). Large values of the suspension density M_T means the availability of a large surface area onto which solutes may be deposited from the solution. Potentially this is equivalent to having a higher level of supersaturation, $(\rho_i - \rho_{i\mathrm{sat}})$; alternatively, this means a lower supersaturation is needed to achieve a given M_T. To study its effect, consider expression (6.4.18) for M_T when t_{res} is not changing but M_T is. For two situations a and b, we already observed in results (6.4.32a) that

$$(M_{Ta}/M_{Tb}) = n_a^0\,G_a^4/n_b^0\,G_b^4; \quad (G_a/G_b) = (M_{Ta}/M_{Tb})^{1/(3+p)}; \quad \left(n_a^0/n_b^0\right) = (M_{Ta}/M_{Tb})^{(p-1)/(3+p)}.$$

Correspondingly, the dominant crystal sizes $r_{pd}|_a$ and $r_{pd}|_b$ for the two cases will be related via (6.4.21a) as follows:

$$(r_{pd}|_a/r_{pd}|_b) = G_a/G_b = (M_{Ta}/M_{Tb})^{1/(3+p)}. \qquad (6.4.38)$$

Therefore a larger suspension density will lead to larger crystal sizes. (Remember, it is based on an assumption that G is linearly proportional to $(\rho_i - \rho_{i\mathrm{sat}})$ vis-à-vis assumption (4).)

Third, the effect of secondary nucleation on the population density is explored by considering its effect on the nucleation rate B^0 (relaxing assumption (5)):

$$B^0 = A\,(\rho_i - \rho_{i\mathrm{sat}})^p\,M_T^q. \qquad (6.4.39a)$$

Here we have assumed that M_T^q quantifies the effect of secondary nucleation on B^0, whose rate expression used earlier was (6.4.30a). We can employ the linear dependence of growth rate G on supersaturation given by (6.4.27) to rewrite the above relations as (Randolph and Larson, 1988)

$$B^0 = \left(A/A_{1p}^p\right)G^p\,M_T^q = A_{1p}\,G^p\,M_T^q. \qquad (6.4.39b)$$

Correspondingly, since the nuclei population density $n^0 = B^0/G$, we get

$$n^0 = A_{1p}\,G^{p-1}\,M_T^q. \qquad (6.4.39c)$$

For two situations a and b, we already know that

$$(M_{Ta}/M_{Tb}) = n_a^0\,G_a^4/n_b^0\,G_b^4$$

Therefore, using (6.4.39c), we get

$$\left(n_a^0/n_b^0\right) = \left(\frac{G_a}{G_b}\right)^{p-1}\left(\frac{M_{Ta}}{M_{Tb}}\right)^q = \frac{M_{Ta}}{M_{Tb}}\,\frac{G_b^4}{G_a^4}.$$

On rearrangement, we get

$$\left(n_b^0/n_a^0\right) = \frac{M_{Tb}}{M_{Ta}}\left(\frac{M_{Tb}}{M_{Ta}}\right)^{4(q-1)/(p+3)} = \left(\frac{M_{Tb}}{M_{Ta}}\right)^{(4q+p-1)/(p+3)}. \qquad (6.4.39d)$$

Also

$$(G_b/G_a) = (M_{Tb}/M_{Ta})^{(1-q)/(p+3)}. \qquad (6.4.39e)$$

Often, $q = 1$; we then get

$$(G_b/G_a) = 1; \qquad \left(n_b^0/n_a^0\right) = (M_{Tb}/M_{Ta}). \qquad (6.4.39f)$$

In that case ($q = 1$), the growth rate is unaffected by secondary nucleation; the nuclei population density n^0 changes linearly with M_T.

Fourth, we consider how *seeding* affects the crystallizer performance (relaxing assumption (6)). Seeding can take place along with nucleation. Let us assume, for the sake of simplicity, that nucleation rates are substantially low and may be neglected. Therefore the number of crystals being introduced into the MSMPR crystallizer per unit time will be

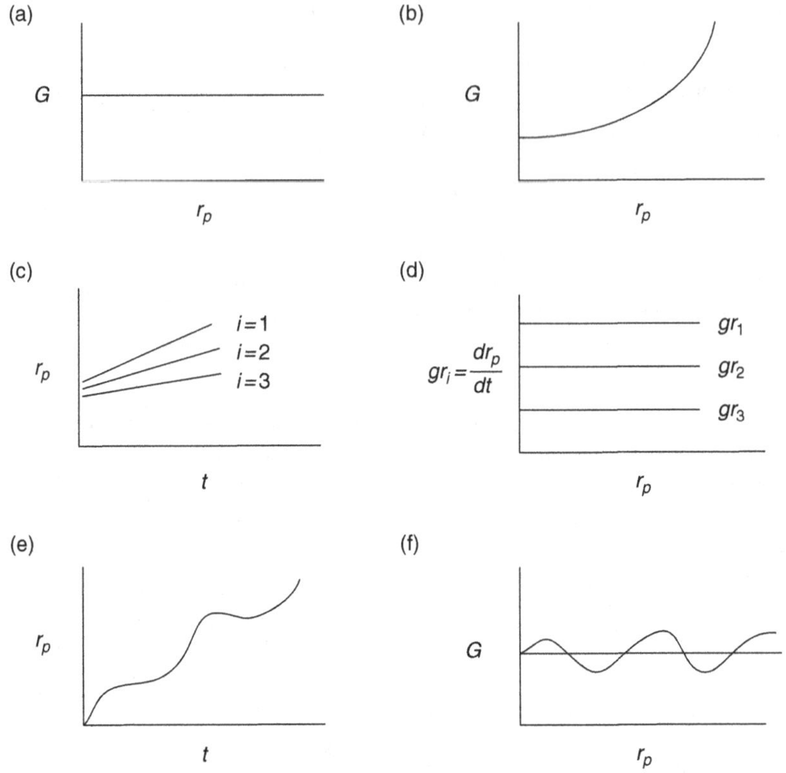

Figure 6.4.5. Crystal growth rates in an MSMPR crystallizer. (a) Constant size-independent growth rate, G. (b) Size-dependent growth rates. (c) Intrinsic growth dispersion: three crystals i = 1, i = 2 and i = 3 have three fixed growth rates gr_1, gr_2 and gr_3, as they grow at different rates, as shown in part (d). (e) Random growth of one crystal with time. (f) Random growth rate of a growing crystal of part (e).

equal to the number of crystals leaving the crystallizer per unit time, except that the sizes of the product crystals will be larger than those in the feed. Let $n_f(r_{ps})$ be the population density function of the seed crystals introduced into the feed; r_{ps} is the size of the seed crystals. Then the seed mass density per unit liquid volume for crystals having dimensions between $r_{ps} = r_{ps\min}$ and $r_{ps} = r_{ps}$, M_{se}, is obtained for crystal volume shape factor ψ_v from relations (2.4.2f) as

$$M_{se} = \rho_s \psi_v \int_{r_{ps\min}}^{r_{ps}} r_{ps}^3 n_f(r_{ps})\, \mathrm{d}r_p. \tag{6.4.40a}$$

Correspondingly, the mass density for crystals between r_{ps} and $r_{ps} + dr_p$ in the seed present in the feed is

$$\mathrm{d}M_{se} = \rho_s \psi_v r_{ps}^3 n_f(r_{ps})\mathrm{d}r_p. \tag{6.4.40b}$$

Let each seed crystal grow by a certain amount. If we assume that the growth rate G is size-independent (assumption (1)), then the net growth in size in the MSMPR crystallizer for all crystals is $\Delta r_p = G t_{\text{res}}$ cm for the given residence time. The mass density per unit liquid volume for crystals in the product stream exiting the crystallizer (population density function $n_1(r_p)$) is

$$M_T = \rho_s \psi_v \int_0^{\infty} r_p^3 n_1(r_p)\mathrm{d}r_p. \tag{6.4.40c}$$

The number of crystals, $\mathrm{d}N_s$, in the seed containing feed stream in the size range r_{ps} to $r_{ps} + dr_p$ is obtained from (6.4.40b) as

$$\frac{\mathrm{d}M_{se}}{\rho_s \psi_v r_{ps}^3} = \mathrm{d}N_s = n_f(r_{ps})\mathrm{d}r_p. \tag{6.4.40d}$$

This number remains unchanged for those specific crystals, but their mass and size have increased in the product stream:

$$\frac{\mathrm{d}M_T}{\rho_s \psi_v (r_{ps} + \Delta r_p)^3} = \mathrm{d}N_s = \frac{\mathrm{d}M_{se}}{\rho_s \psi_v r_{ps}^3}; \tag{6.4.40e}$$

$$\mathrm{d}M_T = \rho_s \psi_v (r_{ps} + \Delta r_p)^3 \mathrm{d}N_s; \tag{6.4.40f}$$

$$\mathrm{d}M_{se} = \rho_s \psi_v r_{ps}^3 \mathrm{d}N_s = \rho_s \psi_v r_{ps}^3 n_f(r_{ps})\mathrm{d}r_p; \tag{6.4.40g}$$

$$M_T = \int_0^{M_T} \mathrm{d}M_T = \int_0^{N_s} \rho_s \psi_v (r_{ps} + \Delta r_p)^3 \mathrm{d}N_s$$
$$= \rho_s \psi_v \int_0^{\infty} (r_{ps} + \Delta r_p)^3 n_f(r_{ps})\mathrm{d}r_p. \tag{6.4.40h}$$

Therefore,

$$r_p^3 n_1(r_p) = (r_{ps} + \Delta r_p)^3 n_f(r_{ps}) = (r_p + \Delta r_p)^3 n_f(r_p);$$

$$\frac{n_1(r_p)}{n_f(r_p)} = (r_p + \Delta r_p)^3 / r_p^3. \tag{6.4.40i}$$

The corresponding relation between M_T and M_{se} is given by

$$(M_T/M_{se}) = \frac{\int_0^\infty r_p^3 n_1(r_p)\mathrm{d}r_p}{\int_0^\infty r_p^3 n_f(r_p)\mathrm{d}r_p} = \frac{\int_0^\infty r_p^3 \frac{(r_p+\Delta r_p)^3}{r_p^3} n_f(r_p)\mathrm{d}r_p}{\int_0^\infty r_p^3 n_f(r_p)\mathrm{d}r_p};$$

$$= \frac{\int_0^\infty (r_p+\Delta r_p)^3 n_f(r_p)\mathrm{d}r_p}{\int_0^\infty r_p^3 n_f(r_p)\mathrm{d}r_p}. \tag{6.4.40j}$$

These two relations allow the prediction of the population density function $n_1(r_p)$ and the suspension density M_T of the product stream from the MSMPR crystallizer given the corresponding quantities of the seed crystals in the feed stream.

6.4.1.1.1 Growth rate dispersion in an MSMPR crystallizer The mechanisms of growth of crystals described very briefly in Section 3.4.1.3 are complex. Crystals may have a growth rate which is size-dependent. However, crystals of the same size may also grow at different rates in an MSMPR crystallizer depending on or influenced by differing environments in the crystallizer or inherent structural difference between crystals. This is called *growth rate dispersion*. We briefly identify first the results of one type of growth rate disperson, namely the *intrinsic growth dispersion*. Here, the growth rate may vary from crystal to crystal, but an individual crystal has an inherent constant growth rate for the whole time, t_{res}, in the crystallizer. Following Randolph and Larson (1988), we identify gr_i as the intrinsic growth rate of the ith crystal (see Figures 6.4.5(c) and (d)). The density function of these growth rates is illustrated by the probability density function $f_g(gr_i)$, such that the total number of crystals per unit volume, N_t, is related to $f_g(gr_i)$ by

$$1 = N_t/N_t = \sum_{i=0}^{\infty} f_g(gr_i) = \int_0^\infty f_g(gr)\,\mathrm{d}(gr), \tag{6.4.41a}$$

where gr is the growth rate variable of the continuous growth rate density function $f_g(gr)$.

Let the population density of crystals of size r_p in the MSMPR crystallizer having a growth rate of gr_i be $n_i(r_p)$; then

$$n_i(r_p) = n_i^0 \exp(-r_p/(gr_i\, t_{res})) \tag{6.4.41b}$$

from equation (6.4.4). Correspondingly, the number density of crystals having a size greater than r_p but possessing a growth rate gr_i is

$$N_{it} - N_i(r_p) = \int_0^\infty n_i^0 \exp\left(\frac{-r_p}{gr_i\, t_{res}}\right)\mathrm{d}r_p$$

$$- \int_0^{r_p} n_i^0 \exp\left(\frac{-r_p}{gr_i\, t_{res}}\right)\mathrm{d}r_p = \int_{r_p}^\infty n_i^0 \exp\left(\frac{-r_p}{gr_i\, t_{res}}\right)\mathrm{d}r_p.$$

Here N_{it} is the total number of crystals per unit volume of all sizes having the growth rate gr_i, and

$$N_{it} - N_i(r_p) = -\frac{n_i^0}{1/(gr_i\, t_{res})} \exp\left(-\frac{r_p}{gr_i\, t_{res}}\right)\Bigg|_{r_p}^{\infty}$$

$$= n_i^0\, gr_i\, t_{res} \exp\left(-\frac{r_p}{gr_i\, t_{res}}\right)$$

$$= B_i^0\, t_{res} \exp(-r_p/(gr_i\, t_{res})). \tag{6.4.41c}$$

The population density $n(r_p)$ of crystals of size r_p having all of the different growth rates is

$$n(r_p) = \sum_{gr_i=0}^{gr_i=\infty} n_i^0 \exp\left(-\frac{r_p}{gr_i\, t_{res}}\right) = \int_0^\infty n_i^0 \exp\left(\frac{-r_p}{gr\, t_{res}}\right)\mathrm{d}(gr). \tag{6.4.41d}$$

We should recall now that N_t, the total number of crystals per unit volume having all sizes and growth rates, has the value $n^0 G\, t_{res}$ $(= B^0 t_{res})$ for an MSMPR crystallizer (see result (6.4.10)). This number includes contributions from different crystal growth rates, gr_i, throughout the crystal population, i.e. $n_i^0\, gr_i\, t_{res} = B_i^0\, t_{res}$, where one can write

$$B_i^0\, t_{res} = B^0\, t_{res} f_g(gr)\mathrm{d}(gr). \tag{6.4.41e}$$

Therefore, the number density of crystals having a size greater than r_p, $(N_t - N(r_p))$, may be obtained by summing over all numbers related to each growth rate gr_i, namely $(N_{it} - N_i(r_p))$:

$$N_t - N(r_p) = \sum_{i=o}^{\infty} (N_{it} - N_i(r_p))$$

$$= \int_0^\infty B^0\, t_{res} \exp(-r_p/gr\, t_{res}) f_g(gr)\mathrm{d}(gr). \tag{6.4.41f}$$

The fraction of crystals having a size greater than r_p is given by

$$1 - F(r_p) = \frac{N_t - N(r_p)}{N_t} = \frac{N_t - N(r_p)}{B^0\, t_{res}}$$

$$= \int_0^\infty \exp\left(-\frac{r_p}{gr\, t_{res}}\right) f_g(gr)\mathrm{d}(gr), \tag{6.4.41g}$$

where $F(r_p)$ is the fraction of crystals having a size between 0 and r_p (definition (2.4.1c)). From the basic relation between $F(r_p)$ and $f(r_p)$ (definition (2.4.1c), namely $f(r_p) = (\mathrm{d}F(r_p)/\mathrm{d}r_p)$), we get

$$0 - \frac{\mathrm{d}F(r_p)}{\mathrm{d}r_p} = \int_0^\infty \frac{\mathrm{d}}{\mathrm{d}r_p}\left(\exp\left(-\frac{r_p}{gr\, t_{res}}\right)\right) f_g(gr)\mathrm{d}(gr);$$

$$f(r_p) = \int_0^\infty \frac{1}{gr\, t_{res}} \exp(-r_p/gr t_{res}) f_g(gr)\mathrm{d}(gr). \tag{6.4.42}$$

Therefore, the crystal size density function $f(r_p)$ of an MSMPR crystallizer can be related to the growth rate density function $f_g(gr)$.

It is useful to develop some relations between the moments of these two density functions. If we multiply both sides by $r_p^j\,\mathrm{d}r_p$ and integrate between the limits of $r_p = 0$ and ∞, we get $Mo_{f_r}^j$, the jth moment of the crystal size density function $f(r_p)$ (see (2.4.1g)):

$$\int_0^\infty r_p^j f(r_p)\mathrm{d}r_p = Mo_{f_r}^j = \int_0^\infty \left(\frac{f_g(gr)\mathrm{d}(gr))}{gr\, t_{\mathrm{res}}}\right) \times \int_0^\infty r_p^j \exp\left(-\frac{r_p}{gr\, t_{\mathrm{res}}}\right)\mathrm{d}r_p; \qquad (6.4.43a)$$

$$\begin{aligned} Mo_{f_r}^j &= \int_0^\infty \left(\frac{f_g(gr)\mathrm{d}(gr)}{gr\, t_{\mathrm{res}}}\right)\int_0^\infty gr^{j+i}\, t_{\mathrm{res}}^{j+1}\, p^j \exp(-p)\mathrm{d}p \\ &= \int_0^\infty \left(\frac{f_g(gr)\mathrm{d}(gr)}{gr\, t_{\mathrm{res}}}\right) gr^{j+i}\, t_{\mathrm{res}}^{j+1} \int_0^\infty p^j \exp(-p)\mathrm{d}p \\ &= t_{\mathrm{res}}^j\, \Gamma(j+1)\int_0^\infty gr^j f_g(gr)\mathrm{d}(gr) \\ &= j!\, t_{\mathrm{res}}^j \int_0^\infty gr^j f_g(gr)\mathrm{d}(gr) \\ &= j!\, t_{\mathrm{res}}^j\, Mo_{gr}^j, \end{aligned} \qquad (6.4.43b)$$

where Mo_{gr}^j is the jth moment of the growth rate density function $f_g(gr)$ (which may be a normal distribution (4.1.33g) or a gamma distribution (4.1.33f), etc.).

Before we use such a relation to determine the nature of the growth rate density function $f_g(gr)$, we should first recognize that the relation between $f(r_p)$ and $n(r_p)$ is

$$\frac{\mathrm{d}N(r_p)}{N_t} = f(r_p)\mathrm{d}r_p = \frac{n(r_p)\mathrm{d}r_p}{N_t}. \qquad (6.4.44a)$$

Therefore

$$\int_0^\infty \frac{r_p^n\, n(r_p)\mathrm{d}r}{N_t} = \int_0^\infty r_p^n f(r_p)\mathrm{d}r_p = Mo_{f_r}^n = \frac{Mo_{f_n}^n}{N_t}, \qquad (6.4.44b)$$

where

$$Mo_{f_n}^n = \int_0^\infty r_p^n n(r_p)\mathrm{d}r_p. \qquad (6.4.44c)$$

Employ now the definition, (2.4.2h), of $\bar{r}_{p_{i+1,i}}$:

$$\bar{r}_{p_{i+1,i}} = \frac{\int_0^\infty r_p^{i+1}\, n(r_p)\mathrm{d}r_p}{\int_0^\infty r_p^i\, n(r_p)\mathrm{d}r_p} = \frac{\int_0^\infty r_p^{i+1} f(r_p)\mathrm{d}r_p}{\int_0^\infty r_p^i f(r_p)\mathrm{d}r_p} = \frac{Mo_{f_r}^{i+1}}{Mo_{f_r}^i} = \frac{Mo_{f_n}^{i+1}}{Mo_{f_n}^i}. \qquad (6.4.44d)$$

Therefore, from relation (6.4.43b) we get

$$Mo_{f_r}^{j+1}/Mo_{f_r}^j = \bar{r}_{p_{j+1,j}} = (j+1)t_{\mathrm{res}}\frac{Mo_{gr}^{(j+1)}}{Mo_{gr}^{(j)}}. \qquad (6.4.44e)$$

Knowing $\bar{r}_{p_{j+1,j}}$ from particle size distribution measurements, one can now make guesses about $f_g(gr)$ and find out which one fits the above relation. Note further that the suspension density M_T, (2.4.2f), is

$$\begin{aligned} M_T &= \rho_s\psi_v\int_0^\infty r_p^3\, n(r_p)\mathrm{d}r_p = \rho_s\psi_v N_t\left(\int_0^\infty r_p^3 f(r_p)\mathrm{d}r_p\right) \\ &= \rho_s\psi_v B^0\, t_{\mathrm{res}}\, Mo_{f_r}^3 = 6\rho_s\psi_v B^0\, t_{\mathrm{res}}^4\, Mo_{gr}^3. \end{aligned} \qquad (6.4.44f)$$

Consult Berglund (1993) and Berglund and de Jong (1990) for further details of such an analysis. They observed that a single growth rate (gr_i) density function $f_g(gr)$ is insufficient to describe the observed population density of crystal size. However, two different growth rate distributions, when combined, could describe the performance of a pilot-scale sugar crystallizer.

There is an additional phenomenon, namely *random fluctuations* in growth rates of the different crystals around a mean value. For example, one crystal will display several growth rates during its growth in the crystallizer for the time period t_{res}; however, there may be a mean, as shown in Figures 6.4.5(e) and (f). The governing macroscopic population balance equation for an MSMPR crystallizer, where $n_f(r_p) = 0$, $B = De = 0$, may be written as

$$\frac{\partial n(r_p)}{\partial t} + \frac{\partial}{\partial r_p}(G\,n(r_p)) = -\frac{Q_1\, n_1(r_p)}{V} + D_{gr}\frac{\partial^2 n(r_p)}{\partial r_p^2}. \qquad (6.4.45)$$

Here we have employed equation (6.4.1a) and added the growth dispersion term on the right-hand side using an empirically determined growth diffusivity, D_{gr}. This was suggested in the paragraph after equation (6.2.51c) (where instead of D_{gr}, D^i was employed). See Randolph and Larson (1988) for further analysis on this topic. Figures 6.4.5 (e) and (f) illustrate the nature of the random variation of a crystal dimension with time, as well as the random variation of the growth rate of this crystal with r_p as it grows.

6.4.1.1.2 Batch crystallization Crystallization from smaller batches of solution is generally carried out in a well-mixed crystallizer into which the batch is charged and either cooling or solvent evaporation or (less often) antisolvent based precipitation is carried out. This method is employed for pharmaceuticals on the one hand and inorganic salts on the other. The solution is kept well-mixed. However, in the cooling crystallizer, for example, the cooling heat exchanger is often externally located and the solution is rapidly circulated between the crystallizer and the heat exchanger. Regardless of the method adopted, the process is inherently unsteady. The equation from Section 6.2.3.2 to be used here now will have an unsteady state term. Therefore $n(r_p)$ will depend both on crystal size r_p and time. Further, in systems where the batch volume is changing (for example, due to evaporation or semibatch operation), the solution volume may not be assumed to be constant. Sometimes, crystal seeds are introduced into the batch at time $t = 0$; this has a strong influence on the final crystal size distribution obtained.

For an unseeded batch crystallizer, equation (6.2.65) may be reduced, for the case where $B = De = 0$, to

$$\frac{\partial(Vn(r_p, t))}{\partial t} + \frac{\partial}{\partial r_p}(GVn(r_p, t)) = 0. \tag{6.4.46a}$$

Here, V may be a function of time. Therefore it is convenient to define a new variable, $\tilde{n}(r_p, t)$, instead of two variables, $V(t)$ and $n(r_p, t)$ (Randolph and Larson, 1988):

$$\tilde{n} = V(t)\, n(r_p, t). \tag{6.4.46b}$$

In terms of this new dependent variable, equation (6.4.46a) is reduced to

$$\frac{\partial \tilde{n}(r_p, t))}{\partial t} + \frac{\partial}{\partial r_p}(G\tilde{n}(r_p, t)) = 0. \tag{6.4.46c}$$

For the two independent variables, r_p and t, boundary/initial conditions are needed to solve the above equation. For the boundary condition with respect to r_p, say $r_p \to 0$, let B^0 be the number of nuclei produced per unit volume; it will, however, depend on time. From definition (6.4.7),

$$n^0 = \frac{B^0}{G|_{r_p \to 0}} \Rightarrow \tilde{n}^0 = n^0 V = \frac{B^0\, V}{G|_{r_p \to 0}}. \tag{6.4.46d}$$

One also needs an estimate of crystal size distribution at $t = 0$, and to know how the operating conditions in the batch crystallizer are changing due to evaporation from the solution or cooling of the solution. An appropriate mass balance on solute species i leads to

$$\frac{d(\rho_i V)}{dt} + \frac{dw_{crys}}{dt} = 0, \tag{6.4.46e}$$

where ρ_i is the mass concentration of species i and w_{crys} is the mass of crystals in suspension, since the sum total of the mass of solute i in two forms, $\rho_i V$ (in solution) and w_{crys} (in suspension), must be constant and invariant with time. In evaporative crystallization, as the solvent evaporates, V decreases and ρ_i increases, leading to crystallization, i.e. w_{crys} increases. In cooling crystallization as temperature decreases, ρ_{isat}, the solubility, decreases (see Figure 3.3.6B) and w_{crys} increases. The procedure for pursuing such calculations is illustrated later in this section.

Considerations of equations (6.4.46c) and (6.4.46e) show that additional equations are needed to express w_{crys} in terms of $\tilde{n}$, etc. These are developed by taking moments of the crystal size distribution, specifically $\tilde{n}$. Consider equation (6.4.46c) along with (6.4.46d) for the zeroth moment for well-behaved functions:

$$\int_0^\infty \frac{\partial \tilde{n}}{\partial t} dr_p + \int_0^\infty \frac{\partial(G\tilde{n})}{\partial r_p} dr_p = 0 = \frac{\partial}{\partial t}\int_0^\infty (\tilde{n}\, dr_p) + \int_0^\infty d(G\tilde{n}) = 0;$$

$$\frac{d\tilde{N}_t}{dt} = G\tilde{n}|_{r_p \to 0} = B^0 V, \tag{6.4.46f}$$

where

$$\tilde{N}_t = \int_0^\infty \tilde{n}\, dr_p. \tag{6.4.46g}$$

In the same fashion, one has to take moments up to order three and then get an expression for w_{crys} in terms of other moment based quantities. The numerical solution of these equations with appropriate boundary conditions would lead to the desired solution. However, it is quite complicated.

To avoid the complexities inherent in the above approach, an alternative approach is often adopted. Many batch crystallization applications involve *cooling crystallization* wherein a hot solution, which is saturated, is cooled at a certain rate. This creates the supersaturation necessary for nucleation and growth of crystal (see Figure 3.3.6B, line AB′). (Other strategies, namely *solvent evaporation*, *antisolvent addition*, etc., may be similarly analyzed.) In this approach, certain assumptions are made to allow the determination of the rate at which the solution temperature is to be cooled. The approach based on earlier studies by Mullin's group has been summarized by Belter *et al.* (1988), Wankat (1990) and Nývlt (1992). It is known as determining the *cooling curve* of a batch crystallizer.

The cooling rate employed is influential in a batch crystallizer. Usually the heat transfer rate Q ($= q\, A_{hex}$, q is the heat flux and A_{hex} is the heat transfer surface area in the cooling crystallizer) is high at the beginning, since the heat transfer rate from the solution into the cooling fluid,

$$Q = qA_{hex} = U_{ov}\, A_{hex}(T_{solu} - T_{cf}), \tag{6.4.47a}$$

is directly proportional to the temperature difference between the solution due to undergo crystallization (T_{solu}) and the temperature of the cooling fluid (T_{cf}). As a result, the solution becomes supersaturated quickly, resulting in the production of a large number of crystal nuclei; consequently, further growth of these crystals becomes limited. Let

w_{s0} = mass of solution charged into the crystallizer;
c_{p0} = specific heat of this solution;
w_{crys} = mass of crystals formed at any time;
ΔH_{crys} = heat of crystallization;
Q_{sub} = heat transfer rate due to subcooling;

Q_{crys} = heat transfer rate due to crystallization;
u_0 = mass fraction of solute in the solution charged;
u_c = mass fraction of solute in the crystallized solution on a solid-free basis;
u_{r0} = kilograms of solute per kilogram of solvent in the solution charged, $u_0/(1-u_0)$;
u_{rc} = kilograms of solute per kilogram of solvent in the crystallized solution, $u_c/(1-u_c)$.

(6.4.47b)

Following Nývlt's (1992) approach, assume that the temperature dependence of the solubility of the solute is given by

$$u_{rc}|_{eq} = \frac{u_c}{1-u_c}\Big|_{eq} = a + bT + cT^2, \qquad (6.4.47c)$$

where $u_{c|eq}$ is the mass fraction u_c of solute in solution at equilibrium with crystals at the solution temperature T_{solu}. Now, any small change in temperature of the solution, dT, over time dt is due to a certain amount of heat transferred due to solution subcooling, -Q_{sub} dt, as well as due to the heat of crystallization, -Q_{crys} dt, where

$$Q = (Q_{sub} + Q_{crys}). \qquad (6.4.47d)$$

But

$$-Q_{sub}\,dt = w_{s0}\,c_{p0}\,dT_{solu} \qquad (6.4.47e)$$

and

$$-Q_{crys}\,dt = -\Delta H_{crys}\,dw_{crys}. \qquad (6.4.47f)$$

Further, the overall heat transfer rate expression (6.4.47d) may be written as

$$Q = \left(\frac{(Q_{sub}\,dt + Q_{crys}\,dt)}{dT_{solu}}\right)\left(\frac{dT_{solu}}{dt}\right); \qquad (6.4.47g)$$

$$\frac{dT_{solu}}{dt} = \frac{U_{ov}\,A_{hex}(T_{solu} - T_{cf})}{\left(\dfrac{(Q_{sub}\,dt + Q_{crys}\,dt)}{dT_{solu}}\right)}. \qquad (6.4.47h)$$

From relations (6.4.47e,f),

$$\left(\frac{(Q_{sub}\,dt + Q_{crys}\,dt)}{dT_{solu}}\right) = -\left(w_{s0}c_{p0} - \Delta H_{crys}\frac{dw_{crys}}{dT_{solu}}\right). \qquad (6.4.47i)$$

Now, from definitions (6.4.47b),

$$w_{crys} = w_{s0}\frac{u_{r0}}{1+u_{r0}} - w_{s0}\frac{u_{rc}}{1+u_{r0}} = w_{s0}\,u_0 - (w_{s0} - w_{crys})u_c$$
$$= w_{s0}\frac{u_{r0} - u_{rc}}{1+u_{r0}}. \qquad (6.4.47j)$$

Therefore

$$\frac{dw_{crys}}{dT_{solu}} = -\frac{w_{s0}}{(1+u_{r0})}\frac{du_{rc}}{dT_{solu}}, \qquad (6.4.47k)$$

since u_{ro} is a constant.

Substituting these results into relation (6.4.47h), we get

$$\frac{dT_{solu}}{dt} = -\frac{U_{ov}\,A_{hex}(T_{solu} - T_{cf})}{w_{s0}\,c_{p0} + \Delta H_{crys}\,w_{s0}(b + 2cT_{solu})/(1+u_{r0})}, \qquad (6.4.48a)$$

where we have employed the temperature dependence (6.4.47c) of the solute solubility. The cooling rate (-dT_{solu}/dt) data may be employed to determine a number of quantities. When there is no crystallization, the cooling rate expression,

$$-\left(\frac{dT_{solu}}{dt}\right) = \frac{U_{ov}\,A_{hex}(T_{solu} - T_{cf})}{w_{s0}\,c_{p0}}, \qquad (6.4.48b)$$

varies linearly with T_{solu}; the slope yields $U_{ov}\,A_{hex}$ since w_{s0} is known and so is c_{p0}. When crystallization begins, the slope of this plot will change. If the two slopes are identified as $\dot{T}_1$ and $\dot{T}_2$ (after crystallization), we obtain, after some algebra, at the temperature, T_{crys}, where crystallization begins

$$\Delta H_{crys} = \frac{\{(\dot{T}_1/\dot{T}_2) - 1\}c_{p0}}{(b + 2c\,T_{cryst})/(1+u_{r0})}. \qquad (6.4.48c)$$

The total amount of heat transferred during the process of cooling crystallization from an initial solution temperature of $T_{initial}$ to the final solution temperature of T_{final} spanning a time period of t_{final} (in seconds) is given by

$$\int_0^{t_{final}} Q\,dt = w_{s0}\left(c_{p0}(T_{initial} - T_{final}) + \Delta H_{crys}\left(\frac{u_0 - u_{final}}{1 - u_{final}}\right)\right), \qquad (6.4.49)$$

where we have used the crystal mass balance expression (6.4.47j). The crystal yield on the same basis is

$$w_{crys} = w_{s0}\frac{(u_{r0} - u_{rfinal})}{(1+u_{r0})} = w_{s0}\frac{(u_0 - u_{final})}{1 - u_{final}}. \qquad (6.4.50)$$

The process described above is somewhat uncontrolled, leading to a high rate of nuclei generation at the beginning. **One can, however, introduce seeds into the batch to minimize/eliminate nucleation**. The rate of cooling in cooling crystallization under such a condition is of interest. The objective would be to carry out crystallization via seeding in the metastable region (see Figure 3.3.6B) and to prevent the uncontrolled rate of nucleation encountered in the supersaturated region. To do this, we focus on a *supersaturation balance* (Belter *et al.*, 1988; Wankat, 1990; Nývlt, 1992), where the supersaturation (ρ_i - ρ_{isat}) in mass units will be used. If we focus on a small time period Δt

when the temperature changed by ΔT, then, per unit solution volume,

$$\begin{pmatrix}\text{change in}\\ \text{super saturation}\end{pmatrix} = \begin{pmatrix}\text{change in}\\ \text{saturation}\\ \text{concentration}\\ \text{due to } \Delta T\end{pmatrix} + \begin{pmatrix}\text{change}\\ \text{due to}\\ \text{crystal}\\ \text{growth}\end{pmatrix} + \begin{pmatrix}\text{change}\\ \text{due to}\\ \text{nuclei}\\ \text{generation}\end{pmatrix}$$

$$\Delta(\rho_i - \rho_{isat}) = \Delta\rho_{isat} + k_{gr}A_T(\rho_i - \rho_{isat})^{p_1}\Delta t + k_{nu}(\rho_i - \rho_{isat})^{p}\Delta t. \quad (6.4.51)$$

Here we have employed relation (6.4.30a) for the dependence of the nucleation rate on the supersaturation level, with k_{nu} being the rate constant; similarly, k_{gr} is a rate constant for growth, where A_T is the crystal surface area per unit liquid volume. In the limiting process of $\Delta t \to 0$, we get

$$\frac{d(\rho_i - \rho_{isat})}{dt} = \frac{d\rho_{isat}}{dt} + k_{gr}A_T(\rho_i - \rho_{isat})^{p_1} + k_{nu}(\rho_i - \rho_{isat})^{p}. \quad (6.4.52)$$

If one can operate in the metastable region (Figure 3.3.6B) where the supersaturation level (ρ_i - ρ_{isat}) remains essentially constant with time and the contribution of nucleation is negligible, we obtain

$$\frac{d\rho_{isat}}{dt} = -k_{gr}A_T(\rho_i - \rho_{isat})^{p_1}. \quad (6.4.53a)$$

However, we can write

$$\frac{d\rho_{isat}}{dt} = \frac{d\rho_{isat}}{dT}\left(\frac{dT}{dt}\right). \quad (6.4.53b)$$

Further, the growth term from the overall crystal growth coefficient K_o in expression (3.4.24) may be used to equate the growth term contribution in (6.4.52):

$$k_{gr}A_T(\rho_i - \rho_{isat})^{p_1} \cong K_o A_T(\rho_i - \rho_{isat}), \quad (6.4.54a)$$

where

$$A_T = \psi_s \int_0^{\infty} r_p^2 n(r_p)dr_p = \psi_s N_t \overline{r}_p^2;$$
$$\overline{r}_p^2 \cong (\overline{r}_{ps} + Gt)^2, \qquad N_t = N_s = \frac{M_{se}}{\rho_s \psi_v \overline{r}_{ps}^3}, \quad (6.4.54b)$$

where N_s is the total number of seed crystals per unit volume, M_{se} is the seed crystal suspension density of size $\overline{r}_{ps}$ (equation (6.4.40a)) and G is the growth rate. Therefore

$$\frac{dT}{dt} = \frac{-K_o A_T(\rho_i - \rho_{isat})}{\left(\frac{d\rho_{isat}}{dT}\right)} = \frac{-K_o \psi_s}{\left(\frac{d\rho_{isat}}{dT}\right)}\left(\frac{M_{se}}{\rho_s \psi_v \overline{r}_{ps}^3}\right) \times (\overline{r}_{ps} + Gt)^2(\rho_i - \rho_{isat}). \quad (6.4.54c)$$

From relation (3.4.27),

$$G = \frac{dr_p}{dt} = \frac{K_o\psi_s}{3\rho_s\psi_v}(M_iC_{iwb} - M_iC_s) = \frac{dr_p}{dt} = \frac{K_o\psi_s}{3\rho_s\psi_v}(\rho_i - \rho_{isat});$$

$$\frac{dT}{dt} = \frac{-3M_{se}G}{(d\rho_{isat}/dT)}\frac{(\overline{r}_{ps} + Gt)^2}{\overline{r}_{ps}^3}. \quad (6.4.54d)$$

Integration of this relation from the crystallization temperature T_{crys} to any temperature T provides (under the assumption $\overline{r_{ps}^3} \cong \overline{r}_{ps}^3$)

$$T = T_{crys} - \left[\frac{M_{se}}{(d\rho_{isat}/dT)}\right]\frac{3Gt}{\overline{r}_{ps}}\left(1 + \frac{Gt}{\overline{r}_{ps}} + \frac{G^2t^2}{3\overline{r}_{ps}^2}\right). \quad (6.4.55)$$

To avoid nucleation, equations (6.4.54c,d) suggest that the cooling rate should be slow at the beginning; then, as the crystal surface area $\left(\alpha\left(\overline{r}_{ps} + Gt\right)^2\right)$ increases, the cooling rate should increase proportionally.

Example 6.4.6 (Adapted from Nývlt (1992).) In a cooling crystallizer charged with a 1000 kg batch of a solution containing a salt at u_{r0} = 0.415 kg/kg of water and 60 °C, the temperature of the solution being cooled was measured at various times; the measured temperatures and the cooling rates are indicated in Table 6.4.1. The value of C_{p0} = 3.25 kJ/kg-K; the cooling water in the heat exchanger is at 13.5 °C. Determine the crystallization temperature from the rate of cooling data. Determine the value of $U_{ov}A_{hex}$ from the data in the unsaturated region after determining the crystallizing

Table 6.4.1.

t (min)	T_{solu} (°C)	-dT_{solu}/dt (K/hr)
0	60	–
10	56.7	19.8
20	53.6	18.6
30	50.7	17.4
40	48.0	16.2
50	45.6	14.4
60	43.3	13.8
70	41.5	10.8
80	39.8	10.2
90	38.2	9.6
100	36.7	9.0
110	35.3	8.4
120	33.9	8.4
130	32.7	7.2
150	30.5	6.0
190	26.8	5.1
250	22.8	3.75
320	19.9	2.10

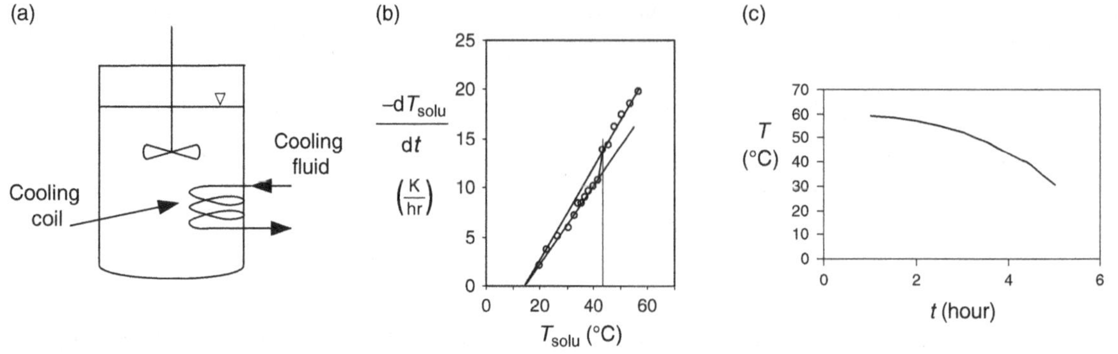

Figure 6.4.6. Batch cooling crystallizer: (a) schematic; (b) cooling rates vs. solution temperature in Example 6.4.6; (c) cooling curve for Example 6.4.7.

temperature. Determine also the enthalpy of crystallization, given $u_{rc}|_{eq} = 0.31 + 0.0017T + 0.000012\,T^2$, where T is in °C.

Solution Figure 6.4.6(a) illustrates a batch crystallizer with an internal heat exchanger. The calculated cooling rates given have been plotted in Figure 6.4.6(b) against the temperature T (°C). It appears that at higher batch solution temperatures, we have a straight line having a particular slope and then, at lower batch solution temperatures, we have another line with a smaller slope. The last data point where the slope starts changing is 43 °C; this must be the temperature at which the crystallization begins.

From the relation for the rate of cooling (6.4.48b),

$$-\frac{dT_{solu}}{dt} = U_{ov}\,A_{hex}\left(T_{solu} - T_{cf}\right)/w_{s0}c_{p0}.$$

Prior to crystallization, for two different solution temperatures $T_{solu}|_1$ and $T_{solu}|_2$,

$$\frac{w_{s0}c_{p0}}{U_{ov}A_{hex}}\left\{\left(-\frac{dT_{solu}}{dt}\right)_1 - \left(-\frac{dT_{solu}}{dt}\right)_2\right\} = T_{solu}|_1 - T_{solu}|_2;$$

$$U_{ov}A_{hex} = w_{s0}\,c_{p0}\left\{\left(-\frac{dT_{solu}}{dt}\right)_1 - \left(-\frac{dT_{solu}}{dt}\right)_2\right\}\Big/\left(T_{solu}|_1 - T_{solu}|_2\right)$$

$$= 1000\,\text{kg} \times 3.25\frac{\text{kJ}}{\text{kg-K}}\left\{\frac{19.8 - 13.8}{56.7 - 43.3}\right\}\text{hr}^{-1} = 1455\frac{\text{kJ}}{\text{hr-K}}.$$

From equation (6.4.48c), the enthalpy of crystallization is given by

$$\Delta H_{crys} = \frac{\{(\dot{T}_1/\dot{T}_2) - 1\}c_{p0}}{\left(\text{b} + 2\text{c}\,T_{crys}\right)/(1 + u_{r0})}.$$

The values of $-\dot{T}_1$ and $-\dot{T}_2$ around the crystallization temperature 43 °C are 13.8 and 10.8. The values of b and c are: b = 0.0017 and c = 0.000012; T_{crys} = 43 °C; u_{r0} = 0.415. Therefore

$$\Delta H_{crys} = \frac{\{\{13.8/10.8\} - 1\}\,3.25(1 + 0.415)}{(0.0017 + 2 \times 0.000012 \times 43\ °\text{C})}$$
$$= \frac{0.277 \times 3.25 \times 1.415}{(0.0017 + 0.00103)} = 466\,\text{kJ/kg}.$$

Example 6.4.7 (Adapted from Nývlt (1992).) A batch cooling crystallizer is to be used to crystallize $CuSO_4$ in a period of 5 hours. The feed solution at 60 °C is to be cooled at 30 °C after seeding the crystallizer with seed crystals of size $\bar{r}_{ps} = 0.075$ mm; these crystals are expected to grow up to 0.65 mm in 5 hours. The solubilities of $CuSO_4$ at the two temperatures are: 50 °C → 0.4 g/cm^3; 20 °C → 0.24 g/cm^3. The seed crystal suspension density is 244 × $10^{\sim 6}$ g/cm^3. How should the crystallizer temperature be reduced to achieve the suggested crystal growth?

Solution We employ equation (6.4.55):

$$T = T_{crys} - \left[\frac{M_{se}}{d(\rho_{isat}/dT)}\right]\frac{3Gt}{\bar{r}_{ps}}\left(1 + \frac{Gt}{\bar{r}_{ps}} + \frac{G^2t^2}{3\bar{r}_{ps}^2}\right);$$

T_{crys} = 60 °C; M_{se} = seed crystal suspension density = 244 × 10^{-6} g/cm^3; growth rate

$$G = \frac{0.65 - 0.075}{10 \times 5 \times 60 \times 60}\frac{\text{cm}}{\text{s}} = \frac{0.575}{5 \times 36} \times 10^{-3}\frac{\text{cm}}{\text{s}}$$
$$= \frac{57.5 \times 10^{-5}}{5 \times 36} = 3.2 \times 10^{-6}\,\text{cm/s};$$

$$T = 60\ °\text{C} - \left[\frac{244 \times 10^{-6}\,\text{g/cm}^3}{\dfrac{(0.4 - 0.24)\text{g/cm}^3}{(60 - 30)\ °\text{C}}}\right]\frac{3 \times 3.2 \times 10^{-6}\,\text{cm/s}}{0.075 \times 10^{-1}\text{cm}}\,t(\text{s})$$
$$\times\left(1 + \frac{3.2 \times 10^{-6}t}{0.075 \times 10^{-1}} + \frac{10.24 \times 10^{-12} \times t^2}{3 \times (0.0075)(0.0075)}\right)$$
$$= 60\ °\text{C} - 30\ °\text{C}\left[\frac{244 \times 10^{-6}}{0.16}\right]$$
$$\times\frac{1.28 \times 10^{-3}}{1}\,t\left(1 + 4.26 \times 10^{-4}t + 6.05 \times 10^{-8}\,t^2\right)$$
$$= 60\ °\text{C} - 30\ °\text{C} \times 15.25 \times 10^{-4} \times 1.28 \times 10^{-3}$$
$$\times t(1 + 4.26 \times 10^{-4}t + 6.05 \times 10^{-8}\,t^2)$$
$$= 60\ °\text{C} - 5.85 \times 10^{-5}t(1 + 4.26 \times 10^{-4}t + 6.05 \times 10^{-8}\,t^2),$$

where t is in seconds. The values of T at a few times are indicated in Table 6.4.2. The cooling curve has been illustrated in Figure 6.4.6(c).

6.4.1.1.3 Series cascade of MSMPR crystallizers In MSMPR crystallizers operated as cooling crystallizers, a better control of temperature can be achieved by having a number of MSMPR crystallizers operated in series, as shown in Figure 6.4.7; this is an example of a series cascade of stages.

Although we have identified the volumetric flow rates from each stage as different from Q_f (i.e. $Q_1, Q_2, Q_3, \ldots, Q_n$ from stages 1, 2, 3,..., n), generally we can assume

$$Q_f = Q_1 = Q_2 = Q_3 = \cdots = Q_n. \qquad (6.4.56)$$

One can also control the residence time in each stage by varying the volumes of each stage. With greater control over temperature in each stage, due to the availability of larger cooling surface area (compared to that in one crystallizer) and also the residence time in each crystallizer, crystal size distribution control may be significantly improved. We will now identify the population balance equations for stages 1, 2 and n following equation (6.4.2) for MSMPR crystallizers, where the growth rate G depends on the stage number (i.e. $G_1, G_2, \ldots, G_n$) and the crystal number density function will obviously vary with each stage (i.e. $n_1(r_p), n_2(r_p), \ldots, n_n(r_p)$):

Stage 1

$$\frac{\mathrm{d}}{\mathrm{d}r_p}(G_1\, n_1(r_p)) = \frac{1}{V_1}\left(Q_f\, n_f(r_p) - Q_1\, n_1(r_p)\right); \qquad (6.4.57a)$$

Table 6.4.2.

t (s)	T (°C)
3600 (1 hour)	59.31
7200 (2 hours)	56.99
10 800 (3 hours)	52.06
14 400 (4 hours)	43.54
16 200 (4.5 hours)	38.63
18 000 (5 hours)	30.34

Stage 2

$$\frac{\mathrm{d}}{\mathrm{d}r_p}(G_2\, n_2(r_p)) = \frac{1}{V_2}(Q_1\, n_1(r_p) - Q_2\, n_2(r_p)); \qquad (6.4.57b)$$

Stage 3

$$\frac{\mathrm{d}}{\mathrm{d}r_p}(G_n\, n_n(r_p)) = \frac{1}{V_n}(Q_{n-1}\, n_{n-1}(r_p) - Q_n\, n_n(r_p)). \qquad (6.4.57c)$$

If we assume that the growth rate G_i in the ith crystallizer is independent of r_p, then we can rewrite these equations in the general form for stage i as follows:

$$\frac{\mathrm{d}n_i(r_p)}{\mathrm{d}r_p} = \frac{Q_{i-1}}{V_i}\frac{n_{i-1}(r_p)}{G_i} - \frac{Q_i}{V_i}\frac{n_i(r_p)}{G_i}. \qquad (6.4.57d)$$

Denoting the residence time in the ith crystallizer as $t_{\mathrm{res},\,i}$, where

$$t_{\mathrm{res},\,i} = (V_i/Q_i), \qquad (6.4.57e)$$

we can rewrite the general form of the ith-stage equation as

$$\frac{\mathrm{d}n_i(r_p)}{\mathrm{d}r_p} = \frac{Q_{i-1}}{Q_i}\frac{n_{i-1}(r_p)}{G_i}\frac{1}{t_{\mathrm{res},\,i}} - \frac{n_i(r_p)}{G_i}\frac{1}{t_{\mathrm{res},\,i}}. \qquad (6.4.58)$$

One should add constraints on the growth rate G_i as well as the nucleation and growth kinetics in each stage to the above equations. See Randolph and Larson (1988) for a detailed treatment. Nývlt (1992) has provided a general solution for equation (6.4.58).

6.4.1.1.4 Precipitation We have been introduced to the phenomenon of precipitation in Sections 3.3.7.5 and 4.1.9.4 via chemical reactions, addition of nonsolvents, salting out via addition of salts like ammonium sulfate and the hydrophobic effect. Precipitation has considerable similarities to crystallization in terms of the devices used and other concepts, except precipitates are amorphous solids, rather than crystals, which have specific geometrical shapes. Further, the particle in a precipitate is often an aggregate of smaller particles and is easily subject to breakage under fluid shear. This is true of inorganic, organic as

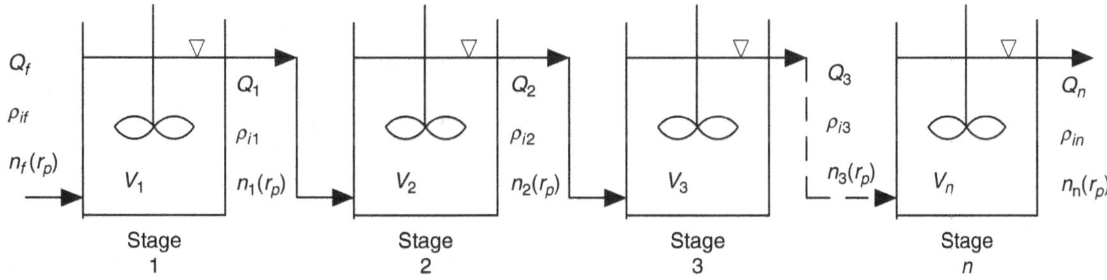

Figure 6.4.7. Series cascade of MSMPR crystallizers.

well as biological (protein) precipitates. More often than not, precipitation is carried out in a batch vessel or in a CSTS (sometimes in tubular devices).

Consider a CSTS being used as a precipitator (Figure 6.4.2(a)). The general equation (6.4.1a) is valid here also. Further, the processes of agglomeration of smaller precipitate particles into a larger aggregates and breakage of the larger particles into smaller ones by shear are quite important (Randolph and Larson, 1988).

The growth rate G of particle aggregates via collision with small primary particles is generally modeled as

$$G(r_p) = (\mathrm{d}r_p/\mathrm{d}t) = K_o r_p, \tag{6.4.59}$$

where $K_o = (\mathrm{A}/4\pi)\dot{\gamma}\phi_1$, where A is a constant, $\dot{\gamma}$ is the shear rate and ϕ_1 is the volume fraction of small primary particles in protein precipitation (Petenate and Glatz, 1983). The death rate, *De*, or the rate of disappearance of larger aggregates into smaller aggregates due to shear in the agitated vessel, has been modeled as

$$De(r_p) = k(\dot{\gamma}) r_p^{\beta} n(r_p), \tag{6.4.60}$$

where β takes care of a higher breakage probability of larger size aggregates (Randolph and Larson, 1988). Specifically, for protein aggregates, Petenate and Glatz (1983) have described

$$De(r_p) = K\mu\dot{\gamma}^2 n(r_p) r_p^3. \tag{6.4.61}$$

The birth rate, $B(r_p)$, is connected by mass balance to the death rate:

$$B(r_p) = f De\left(f^{1/3} r_p\right), \tag{6.4.62}$$

where f is a measure of the number of particles obtained from the breakup of one particle.

In the steady state form of equation (6.4.1a) representing a MSMPR precipitator,

$$\frac{\mathrm{d}}{\mathrm{d}r_p}(Gn(r_p)) = B - De + \frac{1}{V}\{Q_f n_f(r_p) - Q_1 n_1(r_p)\}; \tag{6.4.63}$$

if we introduce these developments, we get

$$\begin{aligned}\frac{\mathrm{d}}{\mathrm{d}r_p}(K_o r_p n(r_p)) &= f k(\dot{\gamma}) f^{\beta/3} r_p^{\beta} n(r_p) - k(\dot{\gamma}) r_p^{\beta} n(r_p)\\ &\quad + \frac{1}{V}\{Q_f n_f(r_p) - Q_1 n_1(r_p)\}\\ &= K_o n(r_p) + K_o r_p \frac{\mathrm{d}n(r_p)}{\mathrm{d}r_p}\end{aligned} \tag{6.4.64}$$

$$\begin{aligned}\Rightarrow \frac{\mathrm{d}n(r_p)}{\mathrm{d}r_p} &= -\frac{n(r_p)}{r_p} + \frac{1}{K_o r_p V}\{Q_f n_f(r_p) - Q_1 n_1(r_p)\}\\ &\quad + \frac{k(\dot{\gamma})}{K_o} r_p^{\beta-1}\{f^{(\beta/3)+1} n\left(f^{1/3} r_p\right) - n(r_p)\}.\end{aligned} \tag{6.4.65}$$

If we have $n_f(r_p) = 0$, then $(V/Q_1) = t_{\text{res}}$ as in (6.4.3b), and we get

$$\begin{aligned}\frac{\mathrm{d}n(r_p)}{\mathrm{d}r_p} &= \frac{k(\dot{\gamma}) r_p^{\beta-1}}{K_o}\left\{f^{(\beta/3)+1} n(f^{1/3} r_p) - n(r_p)\right\}\\ &\quad - \frac{n(r_p)}{r_p}\left(1 + \frac{1}{t_{\text{res}} K_o}\right).\end{aligned} \tag{6.4.66}$$

For values of $f = 2$ and $\beta = 2.3$ found to be reasonable, Glatz *et al.* (1967) have solved this equation numerically for various values of the parameters k and K_o, and compared the results with those obtained from soy-protein precipitation to select those values of $k(\dot{\gamma})$ and K_o which provided agreement. Their model predicted that smaller particles are obtained at larger $\dot{\gamma}$; and further that higher suspension densities led to larger aggregate sizes.

Whereas both birth and death rates are of importance for protein aggregates, in the case of precipitation of sparingly soluble salts, often of an inorganic nature, particle aggregate growth by agglomeration is of primary importance as if the aggregate growth resembles crystal growth. Randolph and Larson (1988) have provided an analytical solution for the population density function $n(r_p)$ in such a case. An additional factor to be considered in the case of protein precipitates is aging. After some time t in a field of shear rate $\dot{\gamma}$, the protein aggregates reach a constant value at high values of the aging parameters $\dot{\gamma}t$; this value is usually slightly smaller than the starting precipitate particle size.

Additional analysis and details of MSMPR precipitators are available in Randolph and Larson (1988). Detailed considerations on the thermodynamics of precipitation for protein-containing systems are available in Ladisch (2001). Models describing precipitate aggregate growth via diffusion and flow in bioseparations have been illustrated in Belter *et al.* (1988). An introductory treatment of precipitation in bioseparations has been provided in Harrison *et al.* (2003, chap. 8). The subject of precipitation, especially with reference to inorganic systems, has been treated by Estrin (1993).

6.4.1.2 *Solvent extraction: mixer-settler*

For contacting two immiscible liquid phases on a laboratory scale, the separating funnel of Figure 6.3.13(b) is used. To achieve rapid mass transfer in such a device with batchwise introductions of two immiscible phases (e.g. aqueous and organic), one may employ a mechanical shaker or many rapid physical inversions. In larger-scale operation, as well as in industrial practice, one uses a *mixer-settler* arrangement: in the *mixer*, there is usually a mechanical stirrer to disperse one phase as drops into the other phase and to agitate the liquid phases vigorously; after the mass transfer/extraction is completed, the two liquids are allowed to settle/separate in a separate *settler*, after which

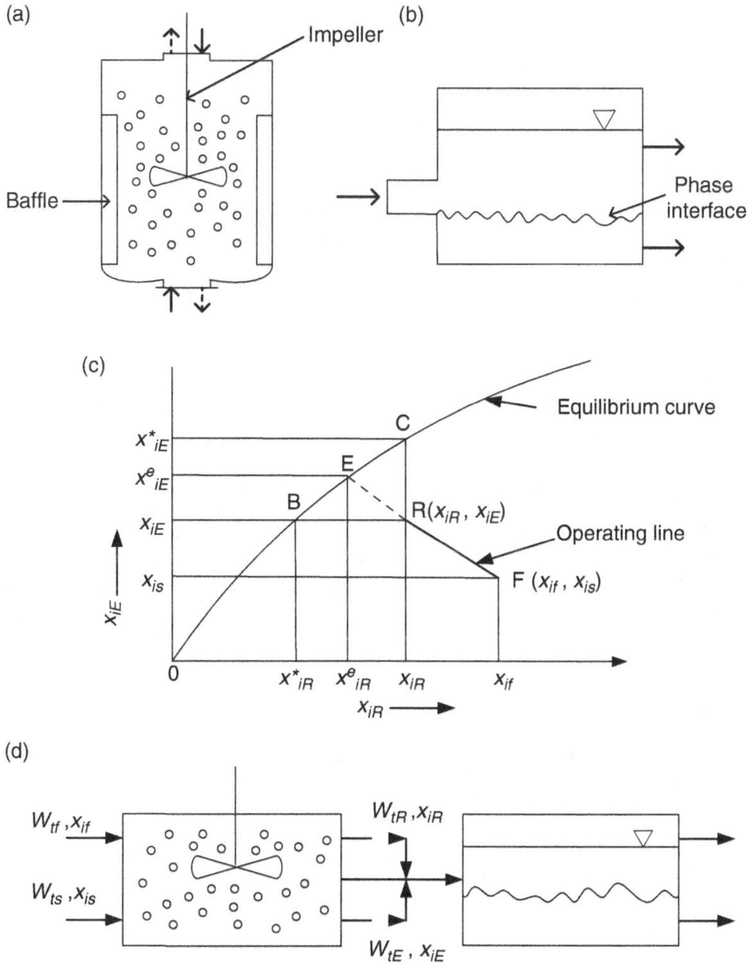

Figure 6.4.8. (a) Mixer - a well-mixed vessel with one phase dispersed in the other. (b) Settler-the two phases have settled into two separate layers. (c) Equilibrium diagram, operating line and stage efficiencies in solvent extraction. (d) Well-mixed continuous mixer-settler.

the two separated phases are withdrawn from the settler. When the two immiscible liquid-phase feeds are introduced in a batch fashion into a vessel for mixing, it is possible to use the same vessel for the settling stage. However, if the two feeds are introduced continuously, settling is carried out in a separate vessel. A simple mixer and a gravity based settler are shown in Figures 6.4.8 (a) and (b). Both liquid phases may be introduced through the vessel bottom, and the dispersion is withdrawn from the top; the latter is introduced into the settler. There are a number of other techniques used in settlers besides gravity, e.g. electrostatic coalescence, centrifugal devices, porous membranes, etc. Treybal (1963, chap. 10) provides a comprehensive introduction to mixer-settlers in solvent extraction and a characterization of their performances. Our treatment is based primarily on Treybal (1963). Perry and Green (1984) provide additional background on these mixer-settler devices.

We consider briefly first the extent of *batch extraction* achieved in a mixer-settler operation carried out in a well-mixed extraction device (Figure 6.4.8(a)). We select a system having the following properties:

(1) dilute solution of species i to be extracted;
(2) extracting solvent (phase $j = s$) (which becomes the extract phase $j = E$ after extraction) and feed/raffinate phases (feed phase $j = f$ and raffinate phase $j = R$, respectively) are immiscible.

At time $t = 0$, m_{tf} moles of feed phase containing x_{if} mole fraction of species i are introduced into the mixer. Simultaneously, m_{ts} moles of extracting solvent are also introduced. We have a dispersion of one phase as drops in the other phase. The dispersion is well-mixed and the composition is assumed to be the same everywhere. As time progresses, the two phases edge closer to equilibrium, as illustrated in Figure 6.4.8(c).

At any time t, if one could separate the raffinate and the extract phases having the mole fraction based compositions of x_{iR} and x_{iE}, respectively, we can write the following molar balance equation for species i:

$$m_{tf}\, x_{if} + m_{ts}\, x_{is} = m_{tR}\, x_{iR} + m_{tE}\, x_{iE}$$
$$\Rightarrow m_{tf}\left(x_{if} - x_{iR}\right) = m_{tE}(x_{iE} - x_{is}), \qquad (6.4.67)$$

where $m_{tf} \cong m_{tR}$ and $m_{ts} \cong m_{tE}$. This progress of the compositions of the feed (raffinate) phase (x_{if}) and the solvent (extract) phase (x_{is}) toward the equilibrium composition of x^e_{iE} and x^e_{iR} is shown in Figure 6.4.8(c). The compositional trajectory of progress toward phase equilibrium is shown by the straight line FR, where point F has coordinates (x_{if}, x_{is}) corresponding to the initial feed and the initial extracting solvent; point R represents the final two phase compositions of the raffinate (x_{iR}) and extract (x_{iE}).

If one rewrites equation (6.4.67) as

$$x_{iE} = -\frac{m_{tf}}{m_{tE}}\, x_{iR} + \frac{m_{tf}}{m_{tE}}\, x_{if} + x_{is}, \qquad (6.4.68)$$

it is a representation of the extract-phase composition, x_{iE}, as a function of the raffinate-phase composition, x_{iR}, and some other quantities defined earlier, namely x_{if} and x_{is}. This straight line, having a slope of $(-m_{tf}/m_{tE})$, is called the *operating line*, since it charts the composition of the two phases as the *extraction operation* progresses toward phase equilibrium, represented by point E (x^e_{iR}, x^e_{iE}), when the two phases are in equilibrium with each other. If, in fact, phase equilibrium were achieved, then the total number of moles of species i transferred would be given by

$$m_i|_{\text{transferred}} = m_{tf}\left(x_{if} - x^e_{iR}\right) = m_{ts}\left(x^e_{iE} - x_{is}\right). \qquad (6.4.69)$$

In practice, when the extraction process is completed in the mixer and the two phases are separated in a settler, the two phases are often not at equilibrium with each other. In such a case, the end of the process is located at a point R (x_{iR}, x_{iE}). The extent of approach to equilibrium achieved by each phase is frequently estimated in the following way. The extract-phase final composition achieved is x_{iE}. However, if the raffinate phase leaving the mixer at a composition x_{iR} were in equilibrium with this extract phase of composition x_{iE}, then this hypothetical extract composition would have a higher value, x^*_{iE} (corresponding to point C on the equilibrium curve in Figure 6.4.8(c)). An estimate of this deficiency is called the *Murphree extract-stage efficiency*, E_{ME}:

$$E_{ME} = \frac{x_{iE} - x_{is}}{x^*_{iE} - x_{is}} = \frac{\text{composition change achieved in extract}}{\text{ideal extract phase composition change}}. \qquad (6.4.70)$$

According to this definition, the value of E_{ME} is generally less than 1.

One can look at the progress toward equilibrium based on the extract phase also. The extract-phase composition exiting the real device is x_{iE}, corresponding to a raffinate composition of x_{iR}. If, however, the extract phase of composition, x_{iE}, leaving the mixer were in equilibrium with the raffinate phase, the hypothetical raffinate composition would have been x^*_{iR}, corresponding to point B on the equilibrium curve. An estimate of this difference is provided by the *Murphree raffinate-stage efficiency*, E_{MR}, which is defined as

$$E_{MR} = \frac{x_{if} - x_{iR}}{x_{if} - x^*_{iR}}. \qquad (6.4.71)$$

On the other hand, one can also define a *stage efficiency E* on the basis of the fractional approach to equilibrium. The feed-phase composition change achieved is $(x_{if} - x_{iR})$. If, however, the phases leaving the mixer were at equilibrium (point E, coordinates (x^e_{iR}, x^e_{iE})), then the feed-phase composition difference achieved would have been $x_{if} - x^e_{iR}$. The *stage efficiency E* is defined as

$$E = \frac{x_{if} - x_{iR}}{x_{if} - x^e_{iR}}. \qquad (6.4.72)$$

On the basis of the extract phase, the corresponding *stage efficiency* is given by

$$E = \frac{x_{iE} - x_{is}}{x^e_{iE} - x_{is}}. \qquad (6.4.73)$$

The two efficiencies are equal, since line FR is part of line FE, and the ratio of their intercepts on the x_{iR} axis and x_{iE} axis is the same:

$$E = \frac{\text{length of FR}}{\text{length of FE}}. \qquad (6.4.74)$$

Consider the case where the equilibrium behavior is linear:

$$x^e_{iE} = \kappa_{iE} x^e_{iR}; \qquad x^*_{iE} = \kappa_{iE} x_{iR}; \qquad x_{iE} = \kappa_{iE}\, x^*_{iR}, \qquad (6.4.75)$$

and the ratio (m_{tE}/m_{tR}) is constant. It has been suggested (Treybal, 1963, table 10.1) that the relations between E, the stage efficiency, and E_{ME} or E_{MR} are as follows, when the distribution coefficient, κ_{iE}, is constant and the solutions of species i in both of the two immiscible phases are dilute:

$$E = \frac{E_{ME}\left(1 + \kappa_{iE}\dfrac{m_{tE}}{m_{tf}}\right)}{1 + \kappa_{iE}\dfrac{m_{tE}}{m_{tf}}\, E_{ME}} = \frac{E_{MR}\left(1 + \kappa_{iE}\dfrac{m_{tE}}{m_{tf}}\right)}{\kappa_{iE}\dfrac{m_{tE}}{m_{tf}} + E_{MR}}. \qquad (6.4.76)$$

The progress of batch extraction in a well-mixed vessel with time can be measured if the overall mass-transfer coefficient and the dispersed phase surface area per unit volume of liquid are known. In Section 3.4.1.2, we were introduced to the overall mass-transfer coefficient of species A, K_{Aw} or K_{Ao}, depending on whether it was based on aqueous phase ($j = w$) or organic phase ($j = o$). These K's were based on concentration differences. Here we will also use K_{xE} and K_{xR}, based on mole fractions and the extract and the raffinate phases, respectively. Let the mole fraction based overall mass-transfer coefficient of the species being extracted with respect to the extract phase be K_{xE} in the vessel, having a liquid volume of V_ℓ and a drop surface area per unit liquid volume of a_v. Then the molar rate of transfer of the species being extracted, at any time, is

$$\frac{d\{m_{ts}\, x_{is}\}}{dt} = m_{ts}\frac{dx_{is}}{dt} = K_{xE}\, a_v\, V_\ell (x_{iE}^* - x_{iE}), \qquad (6.4.77)$$

where x_{iE} is the extract-phase mole fraction of species i being extracted at any time and x_{iE}^* is the hypothetical extract-phase mole fraction of species i which would be in equilibrium with x_{iR} at any time. Further, we are dealing with a dilute solution of i in both phases. Integrating this equation over a time period of t seconds, we get

$$m_{ts}(x_{iE} - x_{is}) = K_{xE}\, a_v\, V_\ell\, (x_{iE}^* - x_{iE})_{\text{avg}}\, t, \qquad (6.4.78)$$

where

$$\int_{x_{is}}^{x_{iE}} (x_{iE}^* - x_{iE})\, dt = (x_{iE}^* - x_{iE})_{\text{avg}}\, t,$$

and $K_{xE}\, a_v$ is assumed to be time-invariant.

In well-mixed extractors operated *continuously* (Figure 6.4.8(d)), the two phases are well-mixed such that essentially each liquid phase has the same composition throughout the vessel; however, each such composition is different from that of the feed stream (or the solvent stream) entering the vessel. For example, the mole fractions of the raffinate and the extract at all locations in the mixer are, respectively, x_{iR} and x_{iE}. The molar rate of extraction of species i per unit extractor volume is

$$N_i\, a_v = K_{cE}\, a_v\, C_{tE}(x_{iE}^* - x_{iE}). \qquad (6.4.79)$$

Here, C_{tE} is the total molar concentration of the extract phase having a species i mole fraction of x_{iE}; also we have used the molar concentration based overall mass-transfer coefficient, K_{cE}, with respect to the extract phase. The total molar rate of extraction of species i in the vessel of volume V_ℓ is given by

$$N_i\, a_v\, V_\ell = K_{cE}\, a_v\, V_\ell\, C_{tE}(x_{iE}^* - x_{iE}). \qquad (6.4.80)$$

The dispersion leaving the extractor (Figure 6.4.8(d)) contains essentially two different streams, the extract and the raffinate.

Equations (6.4.78) and (6.4.80) illustrate how mass-transfer coefficients influence the total number of moles transferred or the total molar rate of transfer taking place in a well-mixed extractor operated batchwise or continuously. One would also like to estimate how difficult this separation is. A quantity called the *number of transfer units* is often employed to quantify this extent of difficulty in separation. To illustrate, consider the well-mixed extractor operated continuously (Figure 6.4.8(d)). The extractor has a liquid-phase volume of V_ℓ and a liquid-phase height H or length L. In a small section of the extractor of length dL or height dH or volume dV_ℓ, the total rate of transfer of species i from the feed liquid into the extract phase is given by

$$W_{tf}\, dx_{iR} = K_{xR}\, a_v(x_{iR} - x_{iR}^*)dV_\ell. \qquad (6.4.81)$$

The number of transfer units based on the raffinate phase, N_{tOR}, is obtained from the above relation as

$$N_{tOR} = \int_{x_{if}}^{x_{iR}} \frac{dx_{iR}}{(x_{iR} - x_{iR}^*)} = \frac{K_{xR}a_v}{W_{tf}}\int_0^{V_\ell} dV_\ell = \frac{K_{xR}\, a_v\, V_\ell}{W_{tf}}; \qquad (6.4.82)$$

$$N_{tOR} = \frac{x_{if} - x_{iR}}{x_{iR} - x_{iR}^*}. \qquad (6.4.83)$$

In this example of a well-mixed continuous extractor, x_{iR} does not change along the extractor length; neither does x_{iE}. Further, x_{iR}^* is the hypothetical composition of the raffinate phase in equilibrium with the exiting extract-phase composition $x_{iE} (= \kappa_{iE}\, x_{iR}^*)$. The composition change achieved in the raffinate phase, i.e. $x_{if} - x_{iR}$, is driven by the mole fraction based concentration difference $x_{iR} - x_{iR}^*$. Thus, N_{tOR} measures the difficulty of separation in this case. If the change $x_{if} - x_{iR}$ desired is large, but the driving mole fraction based concentration difference, $x_{iR} - x_{iR}^*$, is small, then the value of N_{tOR} is large. Correspondingly, for a given $K_{xR}a_v$, V_ℓ (therefore, H or L) has to be large, and the separation is demanding. However, if the change $(x_{if} - x_{iR})$ desired is small and the driving mole fraction difference, $(x_{iR} - x_{iR}^*)$, is large, then the value of N_{tOR} is small, and the separation job is relatively easy; the extractor dimensions are also smaller. One can also demonstrate easily the following relation between the Murphree raffinate-stage efficiency, E_{MR}, and N_{tOR}:

$$N_{tOR} = \frac{x_{if} - x_{iR}}{x_{iR} - x_{iR}^*} = \frac{E_{MR}}{1 - E_{MR}} = \frac{\dfrac{x_{if} - x_{iR}}{x_{if} - x_{iR}^*}}{1 - \dfrac{x_{if} - x_{iR}}{x_{if} - x_{iR}^*}}. \qquad (6.4.84)$$

The volume V_ℓ (or height H or length L) of an extractor needed for the separation may be expressed from equation (6.4.82) as follows:

$$V_\ell = A_{ex}\, H = A_{ex} L = N_{tOR}\left(\frac{W_{tf}}{K_{xR}\, a_v}\right), \qquad (6.4.85)$$

where A_{ex} is the cross-sectional area of the extractor. The term $(W_{tf}/K_{xR}\, a_v\, A_{ex})$ is often called the *height of a transfer unit* (HTU) or *length of a transfer unit* (LTU). This quantity identifies how quickly mass transfer is taking place in a given extractor for a given flow rate. If it is very quick, then the value of HTU or LTU is small for the given separation situation with respect to the composition change; therefore, for a given cross-sectional area, the continuous extractor/separator is short. For a batch extractor, the equivalent situation would lead to a small volume.

Equation (6.4.84) provides a relation between N_{tOR} and one of the Murphree stage efficiencies for a continuous well-mixed extractor. In a *well-mixed batch extractor*, the difficulty or ease of separation will be determined by the *time* needed to achieve the desired change in composition. Thus, the number of transfer units N_{tOE} based on the extract phase in this case may be obtained from equation (6.4.77) as

$$N_{tOE} = \int_{x_{is}}^{x_{iE}} \frac{dx_{is}}{\left(x_{iE}^{*} - x_{iE}\right)} = \frac{(x_{iE} - x_{is})}{\left(x_{iE}^{*} - x_{iE}\right)_{\text{avg}}} = \frac{K_{xE}\, a_v\, V_\ell t}{m_{ts}}. \tag{6.4.86a}$$

A longer time here means a more difficult separation. One may also describe the change needed in terms of the raffinate phase:

$$N_{tOR} = \int_{x_{if}}^{x_{iR}} \frac{dx_{iR}}{\left(x_{iR} - x_{iR}^{*}\right)} = \frac{(x_{if} - x_{iR})}{\left(x_{iR} - x_{iR}^{*}\right)_{\text{avg}}} = \frac{K_{xR}\, a_v\, V_\ell t}{m_{tf}}. \tag{6.4.86b}$$

For such an extraction process in a well-mixed batch extractor, it is also useful to have a relation between the stage efficiency E (definitions (6.4.72) and (6.4.73)) and N_{tOE} or N_{tOR} as defined above. Consider the integral in equation (6.4.86a). Noting that $dx_{is} = dx_{iE}$ and $x_{iE}^{*} = \kappa_{iE}\, x_{iR}$, we rewrite it as

$$N_{tOE} = \int_{x_{is}}^{x_{iE}} \frac{dx_{is}}{\left(x_{iE}^{*} - x_{iE}\right)} = \int_{x_{is}}^{x_{iE}} \frac{dx_{iE}}{\kappa_{iE}\, x_{iR} - x_{iE}} \tag{6.4.86c}$$

However, from the basic mass balance equation (6.4.67), we get

$$x_{iR} = x_{if} - (m_{tE}/m_{tf})(x_{iE} - x_{is}). \tag{6.4.86d}$$

If the two phases were at equilibrium at the end of extraction (x_{iR}^{e}, x_{iE}^{e}), then relation (6.4.67) will become

$$m_{tR}(x_{if} - x_{iR}^{e}) = m_{tE}(x_{iE}^{e} - x_{is}) \Rightarrow \left(x_{if} + \frac{m_{tE}}{m_{tR}} x_{is}\right) = x_{iE}^{e}\left(\frac{m_{tE}}{m_{tR}}\right) + x_{iR}^{e}.$$

Since $x_{iE}^{e} = \kappa_{iE}\, x_{iR}^{e}$, we get

$$\left(x_{if} + \frac{m_{tE}}{m_{tR}} x_{is}\right) = x_{iR}^{e}\left(1 + \frac{\kappa_{iE}\, m_{tE}}{m_{tR}}\right). \tag{6.4.86e}$$

Substitute for x_{iR} from (6.4.86d) and utilize (6.4.86e) in the integral (6.4.86c) to obtain

$$N_{tOE} = \int_{x_{is}}^{x_{iE}} \frac{dx_{iE}}{\kappa_{iE}\, x_{if} - \dfrac{\kappa_{iE}\, m_{tE}}{m_{tf}}(x_{iE} - x_{is}) - x_{iE}}$$

$$= \int_{x_{is}}^{x_{iE}} \frac{dx_{iE}}{-\left(1 + \dfrac{\kappa_{iE}\, m_{tE}}{m_{tf}}\right) x_{iE} + \kappa_{iE}\, x_{iR}^{e}\left(1 + \dfrac{\kappa_{iE}\, m_{tE}}{m_{tf}}\right)}$$

$$= -\frac{1}{\left(1 + \dfrac{\kappa_{iE}\, m_{tE}}{m_{tf}}\right)} \int_{x_{is}}^{x_{iE}} \frac{d\, x_{iE}}{x_{iE} - x_{iE}^{e}}$$

$$= -\frac{1}{\left(1 + \dfrac{\kappa_{iE}\, m_{tE}}{m_{tf}}\right)} \ln\left(\frac{x_{iE} - x_{iE}^{e}}{x_{is} - x_{iE}^{e}}\right); \tag{6.4.86f}$$

$$N_{tOE} = -\frac{1}{\left(1 + \dfrac{\kappa_{iE}\, m_{tE}}{m_{tf}}\right)} \ln(1 - E), \tag{6.4.87a}$$

where for the stage efficiency, E, definition (6.4.73) was used. For a well-mixed batch extractor, the mass-transfer information of relation (6.4.86a) will allow one to determine as a function of time the value of the stage efficiency E achieved in the mixer. The corresponding results, in terms of the raffinate streams, may also be derived in an identical fashion for a well-mixed batch extractor:

$$N_{tOR} = \frac{K_{xR}\, a_v\, V_\ell t}{m_{tf}} = -\frac{\left(\kappa_{iE}\dfrac{m_{tE}}{m_{tf}}\right)}{\left(1 + \dfrac{\kappa_{iE}\, m_{tE}}{m_{tf}}\right)} \ln(1 - E). \tag{6.4.87b}$$

The quantity $(\kappa_{iE}\, m_{tE}/m_{tf})$ encountered earlier in relation (6.4.76) is called the *extraction factor*. For a continuous extractor, the corresponding expression for the extraction factor is $(\kappa_{iE}\, W_{ts}/W_{tf})$ (see Problem 2.2.1).

In a mixer, which is frequently a cylindrical vessel with an agitator (Figure 6.4.8(a)), there are several aspects of importance in solvent extraction. The vessel usually has a dished bottom, with vertical baffles placed along the length of the vessel/extractor and an impeller, propeller or turbine in the center of the vessel (sometimes at an angle to prevent the creation or generation of a vortex). In aqueous–organic systems, one phase is continuous, with the other phase dispersed as drops; the continuous phase must occupy at least 25–30% of the total liquid volume: in general, it is higher. The phase to be dispersed is introduced into the continuous phase already present in the vessel. A minimum impeller speed is needed to eliminate separate layers of two immiscible liquid phases and to ensure that one phase is completely dispersed in the other as drops.

The drop size achieved is important in estimating the mass-transfer surface area developed in the dispersion. In certain situations, the surface area of the dispersion per unit volume of the total liquid phase, a_v, is available from measurements. The average or mean drop size appropriately defined can then be related to a_v. For example, the Sauter mean diameter d_{32} of the drop size number density distribution has been related to a_v via the following relation and the dispersed phase volume fraction ϕ_D:

$$d_{32} = \frac{6\,\phi_D}{a_v}. \qquad (6.4.88)$$

The Sauter mean diameter d_{32} is simply $2\bar{r}_{p_{3,2}}$, and may be obtained from definition (2.4.2h) for $\bar{r}_{p_{i+1,i}}$ with $i = 2$ for a drop size number density function $n(r_p)$ as

$$\bar{r}_{p_{3,2}} = \frac{\int_0^\infty r_p^3 n(r_p)\,dr_p}{\int_0^\infty r_p^2 n(r_p)\,dr_p}. \qquad (6.4.89)$$

Since a_v is the total interfacial area of drops per unit volume of the total liquid phase, and the drop phase volume fraction is ϕ_D, we get

$$a_v = \phi_D \frac{\int_0^\infty 4\pi r_p^2 n(r_p)\,dr_p}{\int_0^\infty \frac{4}{3}\pi r_p^3 n(r_p)\,dr_p} = \frac{3\phi_D \int_0^\infty r_p^2 n(r_p)\,dr_p}{\int_0^\infty r_p^3 n(r_p)\,dr_p} = \frac{3\phi_D}{\bar{r}_{p_{3,2}}} = \frac{6\phi_D}{d_{32}}. \qquad (6.4.90)$$

The mean drop size achieved, or the interfacial area obtained, depends on the inertial forces due to dynamic pressure fluctuations breaking up the drop countered by interfacial tension forces in the drop phase with respect to the continuous phase. This balance is the basis of *Weber number, We*:

$$\begin{aligned} We &= \frac{\text{inertial force in the continuous phase}}{\text{force due to interfacial tension}} \\ &= \frac{\text{pressure fluctuations in the continuous phase}}{\text{countering pressure via interfacial tension}} \\ &= \frac{\rho_c v_c^2}{\gamma / d_p}; \\ We &= \left(\rho_c v_c^2 d_p / \gamma\right). \end{aligned} \qquad (6.4.91)$$

Here, the subscript c refers to the continuous phase, γ is the interfacial tension between the dispersed and the continuous phases, and $\rho_c v_c^2$ represents the kinetic energy of the continuous phase/volume (there is factor ½ missing) whose fluctuations through velocity fluctuations lead to pressure fluctuations, which create drops/drop breakup. Higher-pressure fluctuations leading to a higher Weber number yield smaller size droplets. For an impeller-driven system, with the impeller rotating at N revolutions per minute (rpm), the impeller dimension is important. Therefore

$$We = \left(\rho_c v_c^2 d_{\text{imp}} / \gamma\right) = \left(\rho_c N^2 d_{\text{imp}}^3 / \gamma\right). \qquad (6.4.92)$$

The following correlations are useful for estimating a_v or d_{32} in particular types of agitated devices.

(1) For four-bladed flat-blade paddles and six-bladed turbines,

$$a_v = \frac{100\,\phi_D\,(We)^{0.6}}{C\,d_{\text{imp}}} \qquad (6.4.93)$$

where d_{imp} = impeller diameter range 0.19–0.83 ft; organic liquid volume fraction dispersed in water, ϕ_D = 0–0.20; baffled-vessel diameter range 0.58–1.25 ft; $C = 1 + 3.75\phi_D$ for four-bladed paddles $((d_{\text{imp}}/d_{\text{vessel}}) = 2/3)$; $C = 1 + 9\phi_D$ for six-bladed turbines $((d_{\text{imp}}/d_{\text{vessel}}) = 1/3)$; quoted in Treybal (1963), based on Calderbank's work.

(2)

$$d_{32} = 2\,\bar{r}_{p_{3,2}} = 0.052\,d_{\text{imp}}\,(We)^{-0.6}\,e^{4\phi_D} \text{ for } We < 10\,000; \qquad (6.4.94a)$$

$$d_{32} = 2\,\bar{r}_{p_{3,2}} = 0.39\,d_{\text{imp}}\,(We)^{-0.6} \text{ for } We > 10\,000. \qquad (6.4.94b)$$

These correlations were developed by Gnanasundaram *et al.* (1979).

Additional physical properties and mass-transfer correlations are sometimes needed to determine the performance of the well-mixed baffled device. The viscosity of the two-phase dispersion, μ_{mix}, and its density, ρ_{mix}, are to be determined from (Treybal, 1963)

$$\mu_{\text{mix}} = \frac{\mu_c}{\phi_c}\left(1 + \frac{1.5\,\mu_D\,\phi_D}{\mu_D + \mu_c}\right); \qquad (6.4.95a)$$

$$\rho_{\text{mix}} = \rho_c\,\phi_c + \rho_D\,\phi_D, \qquad (6.4.95b)$$

where μ_c and μ_D are the viscosities of the continuous and the dispersed phases, respectively; the corresponding phase densities are ρ_c and ρ_D. For baffled vessels having an impeller, the mass-transfer coefficient for the continuous phase, k_c, may be obtained from the following correlation (Treybal, 1963):

$$\frac{k_c\,d_{\text{vessel}}}{D_{ic}} = 0.052\,(\text{Re}_{\text{imp}})^{0.833}(\text{Sc}_c)^{0.5}. \qquad (6.4.96)$$

Here d_{vessel} is the diameter of the vessel, D_{ic} is the diffusion coefficient of i in the continuous phase, Re_{imp} is the

Reynolds number of the impeller $\left(= d_{imp}^2 N_{imp} \rho_c/\mu_c\right)$, N_{imp} is the impeller speed in revolutions/time, and the Schmidt number of the continuous phase, Sc_c, is $\mu_c/\rho_c D_{ic}$. The correlations (3.1.167) and (3.1.168) for the dispersed-phase Sherwood number may be utilized to determine the value of the dispersed-phase mass-transfer coefficient k_D. For an aqueous-organic system, if the organic phase is assumed to be the extract phase as well as the dispersed phase, we can follow relations (3.4.18) and (3.4.19), and obtain, in terms of molar concentration differences, the following relations between the overall mass-transfer coefficient based on a particular phase and the individual phase mass-transfer coefficients:

$$\frac{1}{K_{cR}} = \frac{1}{k_{cR}} + \frac{1}{\kappa_{iE}\, k_{cE}}; \tag{6.4.97a}$$

$$\frac{1}{K_{cE}} = \frac{\kappa_{iE}}{k_{cR}} + \frac{1}{k_{cE}}. \tag{6.4.97b}$$

Here $j = w = R$ and $j = o = E$ for the phase identification.

There are a couple of additional factors to be kept in mind. A certain amount of energy has to be supplied through the impeller rotating at a speed beyond the minimum impeller speed. The power consumption, Pr, in such agitated vessels may be obtained from graphical correlations available between the power number, $Po = Pr/(N^3 d_{imp}^5 \rho_{mix})$, and the impeller Reynolds number, Re_{imp} (Treybal, 1963).

In *continuous-flow mixers*, either phase can be made continuous: start the mixing by filling up the vessel with a particular phase, then disperse the other. However, the continuous-phase volume fraction in terms of the flow rate ratio must be greater than 0.25. In *batch mixing*, the situation is more complex.

The length dimensions, or the height of the well-mixed vessel, having liquid extraction in a continuous fashion are quite important. Normally such a system will be analyzed under the condition of two different phases flowing parallel to each other, with the driving force perpendicular to both phases: the phase concentrations will change along the length/height of the device. This is treated in Chapter 8. An extreme case of such a configuration is encountered when there is so much back mixing in both phases that we have a well-mixed tank/vessel. This extreme situation is the only configuration of relevance in this section for the case of continuous flow.

Example 6.4.8 An organic compound present in low concentrations in water is to be extracted into toluene in a well-mixed vessel. The baffled cylindrical vessel, having a six-bladed turbine impeller of 30.48 cm diameter, is 61 cm in height and diameter. The aqueous solution flow rate into this vessel from the bottom is 170 liter/min; the toluene flow rate (from the bottom also) is 34 liter/min. The value of the distribution coefficient of this organic compound, κ_{iE}, is 25. The dispersion is withdrawn continuously from the top into a settler. Determine the values of the continuous-phase based overall mass-transfer coefficient and a_v for this well-mixed extractor if the impeller rotates at 200 rpm. Calculate the value of the Murphree efficiency E_{MR}. The temperature is 20 °C. You are given: density of aqueous solution = 0.985 g/cm^3; density of toluene = 0.866 g/cm^3; viscosity of aqueous solution = 0.96 cp; viscosity of toluene = 0.59 cp; diffusion coefficient of organic compound in water = 1.2×10^{-5} cm^2/s; diffusion coefficient in toluene = 1.68×10^{-5} cm^2/s; interfacial tension $\gamma^{12} = 22$ dyne/cm.

Solution To determine E_{MR} in a continuous-flow system, we need N_{tOR} (equation (6.4.84)). From equation (6.4.85), K_{xR} and a_v are needed to determine N_{tOR} since W_{tf} and V_ℓ are known. So our focus now will be to determine K_{xR} and a_v. Toluene flow rate into the vessel is 20% of the flow rate of the aqueous solution. Therefore, toluene has to be the dispersed phase: $\phi_D = (34/170) = 0.20$; $\phi_c = 0.80$. To calculate k_{cR}, we employ equation (6.4.96) for the six-bladed turbine:

$$(k_{cR}\, d_{vessel}/D_{ic}) = 0.052 \left(d_{imp}^2 N_{imp} \rho_c/\mu_c\right)^{0.833} (\mu_c/\rho_c D_{ic})^{0.5};$$

$$\frac{k_{cR} \times 61}{1.2 \times 10^{-5}} = 0.052 \left(\frac{61^2 \times 200 \times 0.985}{4 \times 60 \times 0.96 \times 10^{-2}}\right)^{0.833} \times \left(\frac{0.96 \times 10^{-2}}{0.985 \times 1.21 \times 10^{-5}}\right)^{0.5};$$

$$k_{cR} = \frac{0.052 \times 10^{-5}}{50.48} (318\,158)^{0.833} \times (805)^{0.5}$$

$$= \frac{0.052 \times 10^{-5} \times 38\,000 \times 28.37}{50.48}$$

$$= 1.06 \times 10^{-2}\ \text{cm/s}.$$

The Weber number is

$$We = \frac{\rho_c v_c^2 d_{imp}}{\gamma^{12}} = \frac{\rho_c N^2 d_{imp}^3}{\gamma^{12}}$$

$$= \frac{0.985 \times 200^2 \times 30.54 \times 932}{60^2 \times 22} = 14\,156.$$

Use equation (6.4.93),

$$a_v = \frac{100 \times \phi_D \times (We)^{0.6}}{C d_{imp}};$$

$C = 1 + 9\,\phi_D$ for a six-bladed turbine (the (d_{imp}/d_{vessel}) ratio needed for this correlation is 1/3), so

$$a_v = \frac{100 \times 0.2 \times (14\,156)^{0.6}}{(1 + 9 \times 0.2) \times 30.48} = \frac{20 \times 309}{2.8 \times 30.48} = 72.55\ \text{cm}^{-1};$$

$$d_p = \frac{6\phi_D}{a_v} = \frac{6 \times 0.2}{72.55} = 0.0165\ \text{cm}.$$

From equation (6.4.97a),

$$\frac{1}{K_{cR}} = \frac{1}{k_{cR}} + \frac{1}{\kappa_{iE}\, k_{cE}} = \frac{1}{k_{cR}} + \frac{1}{25}\frac{1}{K_{cE}} \cong \frac{1}{k_{cR}}$$

(assuming K_{cE} is of the order of K_{cR}). We may as well therefore consider

$$K_{cR} \cong 1.06 \times 10^{-2}\,\text{cm/s}.$$

Further, $k_{xR} = k_{cR}\,C_t \Rightarrow k_{xR} = K_{cR}\,C_t$. From equation (6.4.85), we observe that

$$\frac{K_{xR}\,a_v\,V_\ell}{W_{tf}} = \frac{K_{cR}\,C_t\,a_v\,V_\ell}{W_{tf}} = N_{tOR}; \quad C_t = \frac{1}{18.5}\,\text{gmol/cm}^3;$$

$$N_{tOR} = \frac{1.06 \times 10^{-2} \times 1 \times 72.55 \times \pi\,(d_{\text{vessel}})^2}{18 \times \dfrac{170 \times 10^3}{60} \times 4} \times 61$$

$$= \frac{0.7255 \times 1.06 \times \pi \times 61 \times 61 \times 61 \times 60}{18 \times 170 \times 10^3 \times 4}$$

$$= 2.38\,\text{units}.$$

From equation (6.4.84),

$$N_{tOR} = \frac{E_{MR}}{1 - E_{MR}} \Rightarrow E_{MR} = 2.38 - 2.38\,E_{MR}$$
$$\Rightarrow E_{MR} = (2.38/3.38) = 0.70.$$

If the problem had involved a batch extractor, most of the calculations would be similar, except, instead of equation (6.4.85), one should employ equations (6.4.86a,b) and (6.4.87a,b) to determine the progress of extraction as a function to time.

6.4.2 Well-mixed separators - membrane based devices

In this section, we will briefly describe two membrane processes, ultrafiltration and membrane gas permeation, using configurations where we can assume that the feed mixture region is well-mixed; its composition equals that of the concentrate stream.

6.4.2.1 *Ultrafiltration: well-stirred cell*

We have already encountered general applications of ultrafiltration (UF) at the beginning of Section 6.3.3.2:

(1) the concentration of a macrosolute by removing the solvent through the membrane,
(2) the removal of smaller molecular weight impurities along with the solvent through the membrane via fresh solvent/buffer addition (called diafiltration), and
(3) the fractionation of a mixture of macrosolutes.

In all such applications, if a small volume of solution is to be processed/treated in a laboratory, a well-stirred vessel with a membrane at one end is usually employed. Figure 6.4.9(a) illustrates batch processing of a solution in a vessel that is well-stirred. Flow in this system comes about due to the passage of the solvent and solutes/macrosolutes through the membrane and out of the system. The solution is under pressure; the permeate emerges through the membrane into essentially atmospheric pressure. Figure 6.4.9(b) illustrates the continuous introduction of solvent or buffer solution or feed solution into this well-stirred vessel at a constant pressure.

Consider Figure 6.4.9(a), where, at time $t = 0$, a batch solution of volume, V_{f0}, is introduced as feed to the vessel on top of the membrane; this well-mixed solution has a molar concentration, C_{i0}, of macrosolute species i. If after some time the well-mixed batch feed solution volume is reduced to V_{fR}, the volume of the retentate, by means of ultrafiltration, what is the macrosolute concentration C_{iR} in the retentate? It may be assumed that the observed macrosolute rejection R_i for the species i remains constant during this concentration process (assuming that the macrosolute is substantially rejected). The extent of volume reduction in the well-mixed feed solution is often identified as the

$$\textit{volume concentration ratio}\ (\text{VCR}) = \frac{V_{f0}}{V_{fR}}. \qquad (6.4.98)$$

If, in a differentially small interval of time, dt, a differentially small volume of magnitude $|dV_{fR}|$ permeates through the membrane, and the concentration of the macrosolute in the ultrafiltrate is C_{ip}, whereas the macrosolute concentration in the retentate at this time is C_{iR}, then a simple macrosolute balance leads to

$$-dV_{fR}(C_{iR} - C_{ip}) = V_{fR}\,d\,C_{iR}.$$

(moles retained by the membrane) (moles contributing to increased concentration in retentate) (6.4.99)

By assumption, during this whole process the observed macrosolute rejection R_i is constant:

$$R_i = 1 - \frac{C_{ip}}{C_{iR}}. \qquad (6.4.100)$$

We can now rearrange equation (6.4.99) to yield

$$-dV_{fR}\,C_{iR}\,R_i = V_{fR}\,dC_{iR} \Rightarrow \frac{dC_{iR}}{C_{iR}} = (-)R_i\,\frac{d\,V_{fR}}{V_{fR}}.$$

Integrating from V_{f0} to V_{fR} and C_{i0} to C_{iR}, we obtain

$$\log\frac{C_{iR}}{C_{i0}} = -R_i \log\frac{V_{fR}}{V_{f0}} = \log\left(\frac{V_{f0}}{V_{fR}}\right)^{R_i}; \qquad (6.4.101)$$
$$C_{iR} = C_{i0}\,(V_{f0}/V_{fR})^{R_i} = C_{i0}(VCR)^{R_i}.$$

Therefore, the retentate macrosolute concentration, C_{iR}, at any time t may be estimated by knowing the reduction in the feed volume, as long as the observed macrosolute rejection is assumed to be constant (Cheryan, 1986, 1998). Since in such operations R_i can be less than 1, some of the macrosolute is lost in the permeate. The *yield* of the macrosolute in this batch concentration process is

$$\text{yield} = \text{extent of recovery} = \frac{V_{fR}\,C_{iR}}{V_{f0}\,C_{i0}}; \qquad (6.4.102)$$

$$\text{yield} = \frac{V_{fR}}{V_{f0}}\left(\frac{V_{f0}}{V_{fR}}\right)^{R_i} = \left(\frac{V_{f0}}{V_{fR}}\right)^{R_i - 1} = Y_{iR}, \qquad (6.4.103)$$

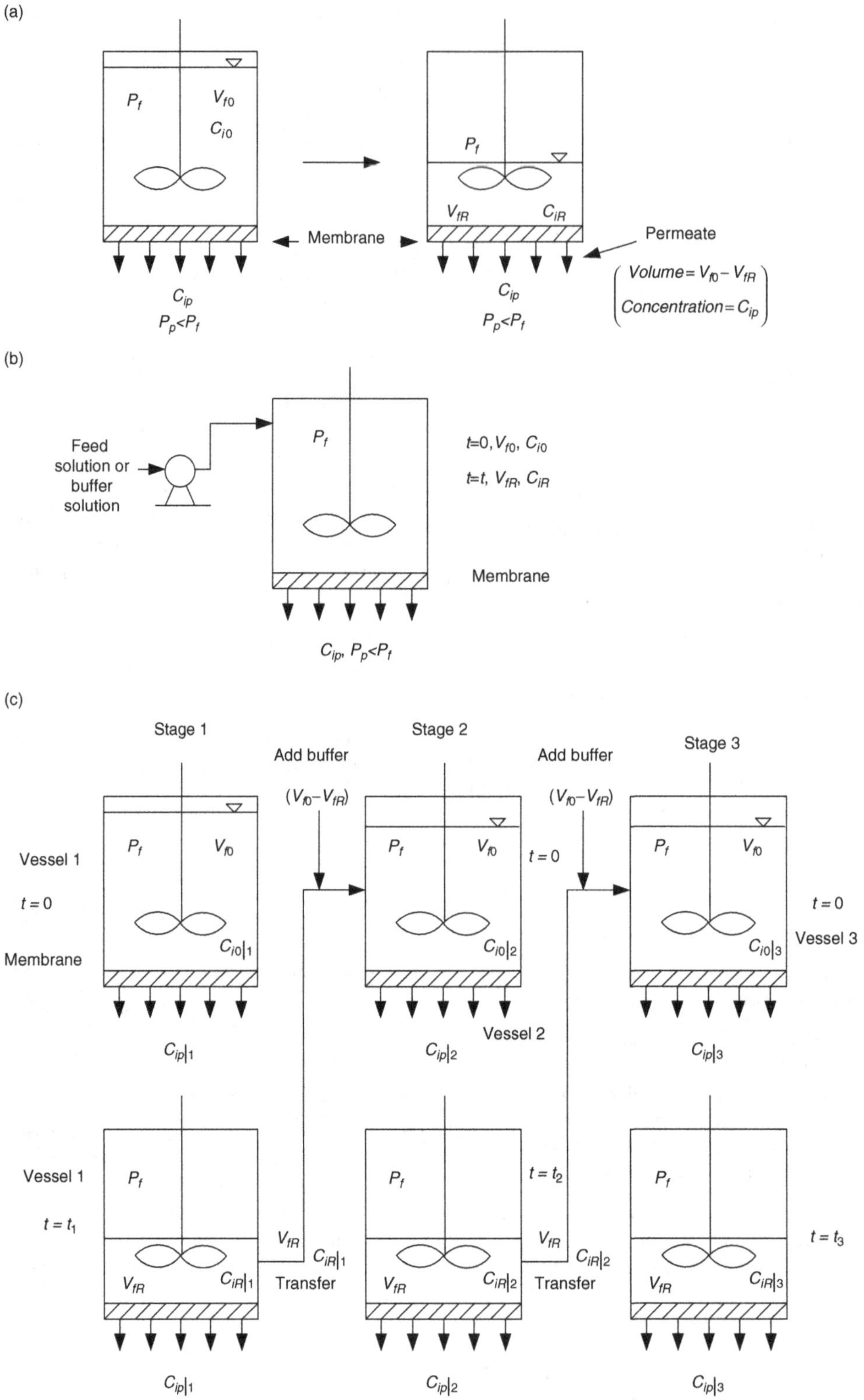

Figure 6.4.9. Ultrafiltration in a well-stirred vessel: (a) batch ultrafiltration; (b) continuous diafiltration/ultrafiltration; (c) discontinuous diafiltration using three stages.

where Y_{iR} is the extent of segregation of species i in the retentate phase (R-phase).

Example 6.4.9 An aqueous buffer solution volume of 10 liter contains by weight 1% inulin (mol. wt. 5000), 1% albumin (mol. wt. 67 000) and 1% apoferitin (mol. wt. 480 000). Batch ultrafiltration is carried out in a well-stirred UF cell having a membrane possessing a molecular weight cut-off value of 30 000 till the solution volume is reduced to 1 liter. Obtain the concentrations of the three proteins in the permeated volume in weight %.

Solution First, select a membrane having a MWCO of 30 000. Table 6.3.8 indicates that membrane PM 30 may be one such membrane. The rejection values (R_i) for different proteins are

inulin 0%, albumin 90%, apoferitin > 98%.

In batch ultrafiltration, the volume concentration ratio (VCR) (definition (6.4.98)) is (10/1) = 10. The retentate concentration of each protein can be determined using $C_{iR} = C_{i0}(VCR)^{R_i}$. Here we assume:

(1) molar concentration of i,

$$C_i = \frac{\text{wt. fraction } i}{\text{mol. wt. } i}\rho_t;$$

(2) ρ_t is essentially constant for all solutions at ρ_t.

For inulin:

$$C_{iR} = C_{i0}(\text{VCR})^{R_i} = \frac{0.01}{M_{\text{inulin}}}\rho_t\,(10)^{0.00} = \frac{0.01}{M_{\text{inulin}}}\rho_t,$$

since $R_i \cong 0.0$; therefore the weight % of inulin in the retentate is 1%.

For albumin:

$$R_i \cong 0.90 \Rightarrow C_{iR} = \frac{0.01}{M_{\text{albumin}}}\rho_t\,(10)^{0.9} = \frac{0.0794}{M_{\text{albumin}}}\rho_t;$$

therefore the weight % albumin in the retentate is 7.94%.

For apoferitin:

$$R_i \cong 0.98 \Rightarrow C_{iR} = \frac{0.01}{M_{\text{apoferitin}}}\rho_t\,(10)^{0.98} = \frac{0.0955}{M_{\text{apoferitin}}}\rho_t;$$

therefore the weight % of apoferitin in the retentate is 9.55%.

If the permeate concentration of a species i is C_{ip}, then

$$9\text{ (liter)}C_{ip} + 1\text{ (liter)}C_{iR} = 10\text{ (liter)}C_{i0}.$$

For inulin:

$$9\times C_{ip} + 1\times\frac{0.01}{M_{\text{inulin}}}\rho_t = 10\times\frac{0.01}{M_{\text{inulin}}}\rho_t$$

$$\Rightarrow C_{ip} = \frac{0.01}{M_{\text{inulin}}}\rho_t \Rightarrow 1.0\,\text{wt.\%}.$$

Inulin, having 5000 mol. wt., with no rejection through the membrane, passes right through. So the permeate and the feed concentrations are identical.

For albumin:

$$9\times C_{ip} + 1\times\frac{0.0794}{M_{\text{albumin}}}\rho_t = 10\times\frac{0.01}{M_{\text{albumin}}}\rho_t$$

$$\Rightarrow C_{ip} = \frac{0.1-0.0794}{9\times M_{\text{albumin}}}\rho_t = \frac{0.0023}{M_{\text{albumin}}}\rho_t,$$

so its weight % in the permeate is 0.23%.

For apoferitin:

$$9\times C_{ip} + \frac{1\times 0.0955}{M_{\text{apoferitin}}}\rho_t = 10\times\frac{0.01}{M_{\text{apoferitin}}}\rho_t$$

$$\Rightarrow C_{ip} = \frac{0.1-0.0955}{9\times M_{\text{apoferitin}}}\rho_t = \frac{0.0005}{M_{\text{apoferitin}}}\rho_t,$$

so its weight % in the permeate is 0.05%.

We will now focus on the process illustrated in Figure 6.4.9 (b), wherein a well-stirred vessel having a volume of solution V_{f0} of a macrosolute i at concentration C_{i0} undergoes ultrafiltration; however, unlike that in Figure 6.4.9(a), a buffer solution having no solute species i is added continuously to this vessel from an external source such that the solution volume remains constant at V_{f0} in the vessel. This mode of operation is called *continuous diafiltration*. There are two goals: if there is a lower molecular weight impurity in the feed solution, it can be eliminated with the ultrafiltrate, since its R_i value is likely to be low; the continuous addition of the solvent via the buffer also prevents an increase in the viscosity of the solution of the macrosolute as ultrafiltration goes on. In the analysis that immediately follows, the subscript i for a solute is valid for the macrosolute as well as for the low molecular weight impurity.

Consider a time interval dt over which a differential volume dV_b of the buffer is added to the well-stirred feed vessel containing a solution volume V_{f0}; to start with, this solution has a solute concentration C_{i0} (the lower molecular weight impurity or the macrosolute of interest). Continuous diafiltration is carried out such that the solution volume in the feed vessel remains constant at V_{f0}. Therefore the volume of ultrafiltrate produced in time dt is dV_b: it has a species i concentration of C_{ip} at time t. A molar balance on species i leads to

$$\underset{\substack{\text{(moles of } i \text{ lost}\\ \text{in permeate)}}}{dV_b\,C_{ip}} = \underset{\substack{\text{(decrease in moles}\\ \text{of } i \text{ in retentate)}}}{V_{f0}(-)dC_{iR}} \qquad (6.4.104)$$

It is further assumed here that, during the process, R_i is constant, even though C_{iR} and C_{ip} are changing. Since $R_i = 1 - C_{ip}/C_{iR}$, we can write $C_{ip} = (1-R_i)C_{iR}$. Introducing this relation into equation (6.4.104), we get

$$dV_b(1-R_i)C_{iR} = (-)\,V_{f0}\,dC_{iR}$$

$$\Rightarrow -\int_{C_{i0}}^{C_{iR}}\frac{dC_{iR}}{C_{iR}} = (1-R_i)\int_0^{V_b}\frac{dV_b}{V_{f0}} = \frac{(1-R_i)}{V_{f0}}V_b$$

$$\Rightarrow -\ln\frac{C_{iR}}{C_{i0}} = (1-R_i)(V_b/V_{f0});$$

$$C_{iR} = C_{i0}\exp\left[-(1-R_i)\,DF\right)], \qquad (6.4.105)$$

where

$$DF = \text{dilution factor} = \frac{V_b}{V_{f0}} = \frac{\text{volume of buffer added}}{\text{feed volume}} = \frac{\text{volume of permeate}}{\text{feed volume}} = \frac{\text{volume of permeate}}{\text{retentate volume}}. \tag{6.4.106}$$

The fraction of solute (macrosolute) left in the feed/retentate (R) liquid is the segregation fraction, given by

$$Y_{iR} = \frac{V_{f0}\, C_{iR}}{V_{f0}\, C_{i0}} = \frac{C_{iR}}{C_{i0}} = \exp[-(1-R_i)DF]. \tag{6.4.107}$$

The purpose of continuous diafiltration in purification of a protein solution from a small molecular weight impurity (e.g. NaCl) can become clear from this result. Suppose the value of DF is 6, i.e. the volume of buffer solution added (= the volume of permeate) is six times that of the feed (here also the retentate) solution. Let R_i for salt (NaCl) = 0.0 and R_i for protein = 0.995. Then, for salt,

$$Y_{iR} = \exp(-6) = \frac{1}{\exp 6} = \frac{1}{403} = 0.002;$$

and, for protein,

$$Y_{iR} = \exp(-0.03) = \frac{1}{\exp(0.03)} = \frac{1}{1.0305} = 0.97.$$

Therefore the impurity (salt) concentration is reduced to 0.2% of the original level at the cost of a loss of 3% of the protein. An ultrafiltration membrane that has a value of $R_i = 1$ for the protein will be better; however, the filtration flux levels would be lower, prompting more membrane area or a higher time for the continuous diafiltration process.

An alternative strategy is often adopted: it is called *discontinuous diafiltration*. Consider Figure 6.4.9(c), which shows three identical well-stirred vessels (vessels 1, 2 and 3) having the same ultrafiltration membrane. At time $t = 0$, a volume V_{f0} of the feed solution containing a macrosolute i and a microsolute (say, salt) j is present in vessel 1. Assume for the time being that $R_i = 1$ and $R_j = 0$. Let the value of the VCR in batch ultrafiltration be 10. Then $C_{iR}|_1 = 10\, C_{i0}$ and $C_{jR}|_2 = C_{j0}$ (from equation (6.4.101)). Now let this small volume of concentrate (volume = $V_{f0}/10$) be transferred to the next well-stirred vessel (vessel 2), where an amount of fresh buffer solution is added to bring the total solution volume to V_{f0}. Therefore the two solute concentrations in this vessel 2 are now:

$$C_{if} = \frac{10\, C_{i0}(V_{f0}/10)}{V_{f0}} = \frac{C_{i0}\, V_{f0}}{V_{f0}} = C_{i0}; \quad \text{and}$$

$$C_{jf} = \frac{C_{j0}(V_{f0}/10)}{V_{f0}} \Rightarrow C_{jf} = (C_{j0}/10).$$

These calculations indicate that before a UF process starts in vessel 2 (stage 2), the macrosolute (i = protein) concentration is identical to C_{i0}, the original concentration we started with in vessel 1 (stage 1); however, the microsolute concentration has been reduced to one-tenth of the original concentration in vessel 1. Therefore, before UF starts in vessel 2 (stage 2), the protein solution is considerably purer since the microsolute impurity concentration has been substantially reduced. If we carry this process further to a third well-stirred vessel (vessel 3, stage 3), we will obtain an impurity (species j) concentration which is one-hundredth of the original concentration (Figure 6.4.9(c)). Multi-vessel discontinuous diafiltration may thus be a substitute for a continuous diafiltration (Figure 6.4.9(b)). The strengths of each approach have been compared by Cheryan (1986). Continuous diafiltration is preferred when retentate viscosity increases greatly and therefore affects the flux of the solvent. Discontinuous diafiltration would be advantageous when a concentrated protein solution is required (amongst other advantages).

A quantitative estimate of the concentration of any solute (microsolute or macrosolute) in the solution remaining in the nth stage after discontinuous diafiltration may be developed as follows. Focus on Figure 6.4.9(c). Consider vessel 1, containing a solute concentration $C_{i0}|_1$ in a solution volume V_{f0}. If we carry out the batch concentration for time t_1 to reduce the solution volume to V_{fR} ($<< V_{f0}$), then the retentate concentration in solution volume V_{fR} is (by equation (6.4.101))

$$C_{iR}|_1 = C_{i0}|_1 \left(V_{f0}/V_{fR}\right)^{R_i}, \tag{6.4.108}$$

provided R_i may be assumed constant during the process. In the discontinuous diafiltration process, this solution is transferred to vessel 2 (stage 2) and fresh buffer is added to increase the volume back to V_{f0} from V_{fR}. Correspondingly, the solute concentration in stage 2 will now be reduced to $C_{i0}|_2$:

$$V_{f0}\, C_{i0}|_2 = V_{fR}\, C_{iR}|_1 \Rightarrow C_{i0}|_2 = C_{i0}|_1 \left(V_{f0}/V_{fR}\right)^{R_i-1}. \tag{6.4.109}$$

Such a solution in vessel 2 (stage 2) now undergoes batch ultrafiltration for time t_2 in a well-mixed vessel with a constant R_i. The retentate concentration $C_{iR}|_2$ is

$$C_{iR}|_2 = C_{i0}|_2 \left(V_{f0}/V_{fR}\right)^{R_i} = C_{i0}|_1 \left(V_{f0}/V_{fR}\right)^{1+2(R_i-1)}. \tag{6.4.110}$$

This solution of volume V_{fR} is now transferred to vessel 3 (stage 3) and fresh buffer is added to bring the solution volume up to V_{f0}. Correspondingly, the concentration $C_{i0}|_3$ will be

$$V_{f0}\, C_{i0}|_3 = V_{fR}\, C_{iR}|_2 \Rightarrow C_{i0}|_3 = C_{i0}|_1 \left(V_{f0}/V_{fR}\right)^{2(R_i-1)}. \tag{6.4.111}$$

This solution of volume V_{f0} now undergoes batch ultrafiltration in vessel 3 for time t_3 to produce a retentate concentration $C_{iR}|_3$:

$$C_{iR}|_3 = C_{i0}|_3 \left(\frac{V_{f0}}{V_{fR}}\right)^{R_i} = C_{i0}|_1 \left(\frac{V_{f0}}{V_{fR}}\right)^{3R_i-2} = C_{i0}|_1 \left(\frac{V_{f0}}{V_{fR}}\right)^{1+3(R_i-1)}. \quad (6.4.112)$$

One can therefore conclude via a process of induction that if there are n vessels (or n stages), the retentate concentration in the nth stage will be

$$C_{iR}|_n = C_{i0}|_1 \left(V_{f0}/V_{fR}\right)^{1+n(R_i-1)}. \quad (6.4.113)$$

This result will allow one to calculate easily the concentration of a small molecular weight impurity after discontinuous diafiltration in n stages. Suppose R_i for the impurity is 0 and $(V_{f0}/V_{fR}) = 10$. Then we see that, for three stages, $C_{iR}|_3$ will be $(C_{i0}|_1/100)$, a considerable degree of purification indeed.

Example 6.4.10 Compare the purification achieved for a protein (species i) in relation to a low molecular weight impurity (species j) for the following two processes: one-stage discontinuous diafiltration vs. continuous diafiltration. The initial volume of solution is 10 liter; the amount of ultrafiltrate produced is 9.5 liter. The UF vessel is well-stirred. The values of R_i and R_j for the membrane used are: $R_i = 1$, $R_j = 0$. Employ a separation factor, α_{ij}, to compare.

Solution The basis of comparison is α_{ij}:

$$\alpha_{ij} = \frac{C_{iR}}{C_{jR}} \frac{C_{j0}}{C_{i0}} = \frac{C_{iR}}{C_{i0}} \frac{C_{j0}}{C_{jR}}. \quad (6.4.114)$$

Calculate α_{ij} for both processes: this means calculate (C_{iR}/C_{i0}) and (C_{jR}/C_{j0}) for both processes.

For continuous diafiltration: use equation (6.4.105), where $DF = 9.5/10 = 0.95$. For the protein,

$$R_i = 1 \Rightarrow C_{iR} = C_{i0} \exp(-(1-1)DF) = C_{i0};$$

for the low molecular weight impurity,

$$C_{jR} = C_{j0} \exp(-(1-0)DF) = C_{j0}\exp(-0.95) = C_{j0}/2.58 \Rightarrow \left(C_{j0}/C_{jR}\right) = 2.58.$$

Therefore, for continuous diafiltration,

$$\alpha_{ij} = \frac{C_{iR}}{C_{i0}} \cdot \frac{C_{j0}}{C_{jR}} = 1 \times 2.58 = 2.58.$$

For one-stage discontinuous diafiltration (or batch ultrafiltration): use equation (6.4.101)

$$C_{iR} = C_{i0}(\text{VCR})^{R_i};$$

$$\text{VCR} = \left(V_{f0}/V_{fR}\right) = \left(\frac{10}{0.5}\right) = 20.$$

For the protein,

$$R_i = 1: \; C_{iR} = C_{i0}\,(20)^1 \Rightarrow (C_{iR}/C_{i0}) = 20;$$

for the low molecular weight impurity,

$$C_{jR} = C_{j0}(20)^0 = C_{j0}.$$

Therefore

$$\alpha_{ij} = \frac{C_{iR}}{C_{i0}} \cdot \frac{C_{j0}}{C_{jR}} = 20 \times 1 = 20.$$

Obviously batch ultrafiltration or one-stage discontinuous diafiltration is a more efficient technique for purifying the protein of a low molecular weight impurity. However, one must ensure that the required solvent flux level is achievable in batch UF when the volume is reduced.

There are two other aspects of importance in practical ultrafiltration: processing time and the membrane area. A brief treatment of processing time with a view to minimizing it will be considered now. The processing time depends on a variety of factors: the total volume to be processed; the concentration of retained solids, especially if it leads to gel polarized operation (equation (6.3.143)); the mode of operation (continuous, batch, etc.) (Cheryan, 1986). One approach suggested by Ng *et al.* (1976) employs the gel-polarized condition and perfect rejection of the protein i for continuous diafiltration operation:

$$N_s\overline{V}_s = |v_z| = k_{il} \ln\left(\frac{C_{igel}}{C_{ilb}}\right). \quad (6.4.115)$$

In our notation used so far, $C_{ilb} = C_{iR}$, where its value changes with time from an initial C_{i0} to C_{iR}. The processing time, t_p, required is

$$t_p = \frac{\text{permeate volume}}{\text{solvent flux}} = \frac{V_b}{|v_z|} = \frac{\text{buffer volume added}}{k_{il} \ln\left(C_{igel}/C_{iR}\right)}. \quad (6.4.116)$$

However, for a given amount of retained proteins, m_i,

$$V_b = (DF)V_{f0} = (DF)\frac{m_i}{C_{iR}}. \quad (6.4.117)$$

Therefore

$$t = \frac{(DF)m_i}{C_{iR}\,k_{i\ell}\ln\left(C_{igel}/C_{iR}\right)}. \quad (6.4.118)$$

If C_{igel} and $k_{i\ell}$ remain unchanged during the continuous diafiltration process, one can determine the optimum processing time, t_{opt}, by determining C_{iR}, which satisfies $(dt/dC_{iR}) = 0$:

$$\frac{dt}{dC_{iR}} = -\frac{(DF)m_i}{C_{iR}^2 k_{i\ell}\ln\left(C_{igel}/C_{iR}\right)} + \frac{(DF)m_i}{C_{iR}^2 k_{i\ell}\left[\ln\left(C_{igel}/C_{iR}\right)\right]^2} = 0$$

$$\Rightarrow \ln\left(C_{igel}/C_{iR}\right) = 1 \Rightarrow C_{iR}|_{\text{optimum}} = \frac{C_{igel}}{\mathrm{e}};$$

$$t_{opt} = \frac{(DF)m_i}{\left(C_{igel}/\mathrm{e}\right)k_{i\ell}\ln(\mathrm{e})}. \quad (6.4.119)$$

A few more aspects about the flux in UF should be remembered. It depends on the temperature as well as other

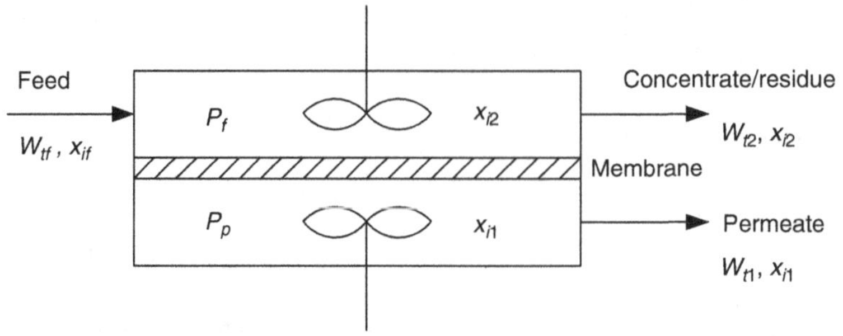

Figure 6.4.10. Gas permeation device where both the feed side and the permeate side are completely mixed.

factors that influence $k_{i\ell}$ besides C_{igel}. Further, continuous UF systems that have a flow pattern over the membrane surface different from that of Figures 6.3.26(b) and 6.4.9(a) and (b) will require additional considerations.

6.4.2.2 *Membrane gas permeation in a CSTS*

In a membrane gas permeator, both the feed side and the permeate side may be well-stirred. Sometimes this is identified as the *complete mixing* case. As shown in Figure 6.4.10, a binary feed gas mixture of species A and B (mole fraction of A, x_{Af}) enters the feed side at a pressure P_f and molar flow rate W_{tf}. The feed side is assumed to be well-mixed, therefore the gas composition of the exiting stream ($j = 2$) x_{A2} (reject mole fraction of A) is present everywhere in the feed side of the permeator. The reject gas stream exits at a molar flow rate W_{t2}. The permeate side may be similarly well-mixed; the permeate molar flow rate is W_{t1}, and the mole fraction of the permeate stream leaving the separator, x_{A1}, is also present throughout the permeate side of the permeator of membrane area A_m. Assume that the pressures on both sides of the permeator are uniform.

An overall molar balance and a species A balance lead to

$$W_{tf} = W_{t1} + W_{t2}; \tag{6.4.120a}$$

$$W_{tf}\, x_{Af} = W_{t1}\, x_{A1} + W_{t2}\, x_{A2}; \tag{6.4.120b}$$

$$x_{Af} = \theta x_{A1} + (1-\theta) x_{A2}, \tag{6.4.120c}$$

where θ is the membrane stage cut (see equation (2.2.10a); $\theta = (W_{t1}/W_{tf})$). In a well-stirred permeator, the gas composition and pressure on the feed side, x_{A2} and P_f, are uniform everywhere along the length of the membrane; similarly on the permeate side for x_{A1} and P_p. The local gas permeation equation over a small membrane area dA_m,

$$x_{A1}\, dW_{t1} = \left(\frac{Q_{Am}}{\delta_m}\right) dA_m \left(P_f x_{A2} - P_p x_{A1}\right), \tag{6.4.121}$$

may therefore be replaced by

$$x_{A1}\, W_{t1} = \left(\frac{Q_{Am}}{\delta_m}\right) A_m \left(P_f x_{A2} - P_p x_{A1}\right). \tag{6.4.122a}$$

The corresponding equation for species B is

$$x_{B1} W_{t1} = (1 - x_{A1}) W_{t1} = \left(\frac{Q_{Bm}}{\delta_m}\right) A_m \left(P_f x_{B2} - P_p x_{B1}\right)$$

$$= \left(\frac{Q_{Bm}}{\delta_m}\right) A_m \left(P_f (1 - x_{A2}) - P_p (1 - x_{A1})\right). \tag{6.4.122b}$$

(*Note*: Using equation (6.4.122a) in mass balance (6.4.120b) for species A, we obtain the equation corresponding to (6.2.44) for a continuous stirred tank separator having one membrane surface.)

In the equations identified above, the following variables appear: (Q_{Am}/δ_m), (Q_{Bm}/δ_m), P_f, P_p, x_{Af}, x_{A1}, x_{A2}, W_{t1}, W_{tf} and A_m. For the solution of any problem in such a system, a few variables are specified. The two common situations encountered are:

(1) (Q_{Am}/δ_m), (Q_{Bm}/δ_m), P_f, P_p, x_{Af}, x_{A2} and W_{tf} are specified; one needs to know x_{A1}, W_{t1} and A_m.
(2) (Q_{Am}/δ_m), (Q_{Bm}/δ_m), P_f, P_p, x_{Af}, W_{tf} and θ are specified; one needs to know x_{A1}, x_{A2} and A_m.

The *first* situation, identified with the Huckins and Kammermeyer (1953a,b) solution, is treated as follows. Divide equation (6.4.122a) by equation (6.4.122b) to obtain

$$\frac{x_{A1}}{(1 - x_{A1})} = \left(\frac{Q_{Am}}{Q_{Bm}}\right) \frac{(x_{A2} - \gamma x_{A1})}{((1 - x_{A2}) - \gamma(1 - x_{A1}))}, \tag{6.4.123}$$

where $(Q_{Am}/Q_{Bm}) = \alpha^*_{AB}$, the ideal separation factor and $\gamma = (P_p/P_f)$, the pressure ratio. We have come across such an equation earlier (equation (6.3.200)). The solution of the quadratic equation in the unknown quantity x_{A1} is obtained as follows:

$$x_{A1} = \frac{\left\{(\alpha^*_{AB}-1)\left(\frac{x_{A2}}{\gamma}+1\right)+\frac{1}{\gamma}\right\} - \sqrt{\left\{\left(1+\frac{x_{A2}}{\gamma}\right)(\alpha^*_{AB}-1)+\frac{1}{\gamma}\right\}^2 - \frac{4\,\alpha^*_{AB}(\alpha^*_{AB}-1)x_{A2}}{\gamma}}}{2(\alpha^*_{AB}-1)}. \quad (6.4.124)$$

From equation (6.4.120c), θ can be easily determined; therefore W_{t2} and W_{t1} are known, since W_{tf} and θ are known (from the definition of θ). The membrane area now becomes available from equation (6.4.122a) since all other quantities are known.

The *second* case was treated by Weller and Steiner (1950a,b) and led to the following set of three algebraic equations (after significant manipulations) for the three unknowns x_{A1}, x_{A2} and A_m:

$$x_{A1} = \left(\alpha^*_{AB} - [(P_f - P_p)/\overline{\beta}]\right)/(\alpha^*_{AB}-1); \quad (6.4.125)$$

$$x_{A2} = \frac{(\overline{\beta}+P_p)}{P_f}x_{A1} = \frac{(\overline{\beta}+P_p)}{P_f}\,\frac{\left(\alpha^*_{AB} - [(P_f-P_p)/\overline{\beta}]\right)}{\alpha^*_{AB}-1}; \quad (6.4.126)$$

$$x_{Af} = \left[\theta + \frac{(1-\theta)(\overline{\beta}+P_p)}{P_f}\right]\frac{\left(\alpha^*_{AB} - [(P_f-P_p)/\overline{\beta}]\right)}{\alpha^*_{AB}-1} = \theta x_{A1} + (1-\theta)x_{A2}, \quad (6.4.127)$$

where

$$\overline{\beta} = \frac{W_{tf}\theta}{(Q_{Am}/\delta)A_m}, \quad \theta = (W_{t1}/W_{tf}), \quad \alpha^*_{AB} = \left(\frac{Q_{Am}/\delta_m}{Q_{Bm}/\delta_m}\right). \quad (6.4.128)$$

From equation (6.4.127), $\overline{\beta}$ can be obtained, since all other quantities are available. From the definition (6.4.128) of $\overline{\beta}$, since W_{tf}, θ and (Q_{Am}/δ) are known, a value of A_m is obtained. Substituting for $\overline{\beta}$ into equation (6.4.125) yields x_{A1}. Substituting x_{A1} now in equation (6.4.126) will yield x_{A2}.

For a multicomponent gas mixture containing species $i = 1, 2, \ldots, n$, a numerical solution is necessary. Consider first the development of the basic governing equations (Sengupta and Sirkar, 1995). Let the permeance of the species i through the membrane of thickness δ_m and a reference species r be, respectively, (Q_{im}/δ_m) and (Q_{rm}/δ_m). Usually, r is one of the species selected out of the n species, $i = 1, 2, \ldots, n$. The ideal selectivity, α^*_{ir}, is defined as

$$\alpha^*_{ir} = (Q_{im}/Q_{rm}). \quad (6.4.129)$$

A molar balance on component i is given by

$$W_{tf}\,x_{if} = W_{t1}\,x_{i1} + W_{t2}\,x_{i2},$$

which leads to

$$W_{tf}\,x_{if} - W_{t2}\,x_{i2} = W_{t1}\,x_{i1} = \left(\frac{Q_i}{\delta_m}\right)A_m\,[P_f\,x_{i2} - P_p\,x_{i1}], \quad i = 1, 2, \ldots, n. \quad (6.4.130)$$

Adding all such equations for n species leads to

$$W_{tf} - W_{t2} = W_{t1} = \left(\frac{Q_{rm}}{\delta_m}\right)A_m P_f\left[\sum_{i=1}^{n}\alpha_{iR}x_{i2} - \gamma\sum_{i=1}^{n}\alpha_{iR}x_{i1}\right], \quad (6.4.131)$$

which can be rearranged to yield

$$1 - (W_{t2}/W_{tf}) = (W_{t1}/W_{tf}) = \theta = \left(\frac{Q_{rm}}{\delta_m}\right)\frac{A_m P_f}{W_{tf}}\left[\sum_{i=1}^{n}\alpha_{iR}\,x_{i2} - \gamma\sum_{i=1}^{n}\alpha_{iR}\,x_{i1}\right], \quad (6.4.132)$$

where

$$\alpha_{iR} = (Q_{im}/Q_{rm}). \quad (6.4.133)$$

One can eliminate W_{t2} from equations (6.4.130) and (6.4.131) to obtain $(n-1)$ equations for $i = 1, 2\ldots, (n-1)$:

$$x_{if} - x_{i2} = \left[\left(\frac{Q_{rm}}{\delta_m}\right)\frac{A_m P_f}{W_{tf}}\right]\left[\alpha_{iR}(x_{i2} - \gamma x_{i1}) - x_{i2}\left\{\sum_{i=1}^{n}\alpha_{iR}\,x_{i2} - \gamma\sum_{i=1}^{n}\alpha_{iR}\,x_{i1}\right\}\right]. \quad (6.4.134)$$

Further, x_{i1} is also obtained from these two equations, (6.4.130) and (6.4.131), as

$$x_{i1} = \alpha_{ir}\left[(x_{i2} - \gamma x_{i1})\Big/\left\{\sum_{i}^{n}\alpha_{iR}\,x_{i2} - \gamma\sum_{i}^{n}\alpha_{iR}\,x_{i1}\right\}\right], \quad i = 1, 2, \ldots, (n-1). \quad (6.4.135)$$

Equations (6.4.132), (6.4.133) and (6.4.134) are $(1 + (n-1) + (n-1) = 2n-1)$ algebraic equations which have to be solved simultaneously to obtain $(n-1)$ x_{i1} variables, $(n-1)$ x_{i2} variables and stage cut θ for known values of $(n-1)$ x_{if} variables, different α_{iR} variables, $[(Q_{rm}/\delta_m)A_m P_f/W_{tf}]$ and γ.

Problems

6.1.1 Calculate the volumetric flow rate of air through a packed bed of spherical particles generated by a pressure drop of 2000 Pa. The air has some other constituents; however, at the temperature of operation, 25 °C and essentially atmospheric pressure, the density and viscosity are as follows: 1.75×10^{-4} g/cm^3 and 1.9×10^{-4} poise (g/cm-s).

The particles in the packed bed have a diameter of 2 mm; the packed bed is 1.5 m long. The fractional void volume of the packed bed is 0.38. (Ans. 125.4 liter/min.)

6.1.2 Determine the magnitude of the electroosmotic velocity, v_{EOF}, generated in a capillary (Figure 6.1.5) under the following conditions: electrical force field $E = 1000$ volt/m (newton/coulomb); zeta potential $\zeta = 0.1$ volt; fluid medium is water, whose dielectric constant $\varepsilon_d = \varepsilon_r$ (relative dielectric constant of water) $\times$ permittivity of vacuum (ε_0); $= 78.3$ ($= \varepsilon_r$) $\times$ 8.854×10^{-12} coulomb/volt-m (ε_0); temperature = 25 °C.
(Ans. 7.8×10^{-3} cm/s.)

6.1.3 A thin film of water at room temperature is present in a shallow rectangular pan. If the two side walls 10 cm apart differ in temperature by 10 K and the value of $(d\gamma^{12}/dT)$ for water is -0.15 (dyne/cm-K), develop an estimate of the maximum velocity for a liquid height of 0.2 mm. Calculate the value of the Bond number and indicate which force dominates. You are given: viscosity of water = 1 centipoise; surface tension of water $\gamma^{12} = 72$ dyne/cm.

6.1.4 To image the surface of a membrane used for separation, scanning electrochemical microscopy (SECM) (Bard and Mirkin, 2001) may be used. The principle consists of moving a tip, which is a conductive disk electrode of radius a, at a given rate over the substrate (here, the membrane) and measuring the variation in the electrochemical response of the tip (the sensor) as it moves over the substrate. The tip is held at a distance d above the substrate as it is moved at a rate of 10 μm/s through the electrolytic solution between the substrate and the tip. The SECM tip of radius a is embedded in an insulating glass sheath of radius $r_{gl} (= 10a)$. Draw a diagram of the setup. What is the velocity profile in the gap between the tip and the substrate if the tip velocity is v_0 cm/s? Make an estimate of the fluid velocity at a distance $d/2$ from the substrate for $d = 17$ μm and $v_0 = 10$ μm/s. Comment on deviations of the velocity profile due to the real-life situation.

6.1.5 You have to develop a means of particle elutriation in microgravity or zero-gravity situations. In extended-duration manned missions in space, there may be need for particle elutriation, and an external force is needed to replace gravity (when there is zero gravity) or augment low gravity (e.g. on the surface of the moon (0.165g) or on Mars (0.37g)). The particles do not/cannot have electrical charge; however, other techniques of particle modification and other external forces may be utilized. Suggest a scheme by which particles may be fractionated; your scheme must be easy to implement. Electrical power is available.

6.3.1 Consider capillary electrophoresis, where there is pressure driven flow of the eluent on top of the electroosmotic flow. This allows greater control of the flow, since controlling the extent of electroosmotic flow requires either changing the capillary surface or the electrolytic solution or both.

(1) Identify the expression for a species velocity in such a configuration. Specify the exact expression for each term. Assume that both bulk flows, Poiseuille flow and electroosmotic flow, are additive.
(2) Pressure driven flow is controllable. However, it will introduce another complication. Identify it and place it in the context of other complexities in capillary electrophoresis.
(3) Write down the equation corresponding to equations (6.3.11a,b) for the present case. Speculate the effect on separation if the pressure driven flow opposes the electroosmotic flow.

6.3.2 Consider two somewhat similar ionic species (e.g. dansyl amino acids, for example serine (1) and alanine (2)) to be separated using capillary electrophoresis. Deliberate on the factors that may lead to increased resolution if $(\mu_1 - \mu_2)$ is small. Comment on the role of the electroosmotic flow in increasing the resolution R_s and affecting the retention time.

6.3.3 Calculate the resolution that may be achieved between Na^+ and K^+ if they are subjected to an electrical field of 300 volt/cm in capillary electrophoresis, where the electroosmotic velocity may be neglected due to an appropriate coating on the capillary surface. The capillary length is 100 cm.

6.3.4 What will be the observed solute mobility, $\mu_{B,eff}$, of a weak base B which will ionize partially in an aqueous solution as

$$B + H^+ \overset{K_{d1}}{\rightleftarrows} BH^+$$

in an applied electrical field in capillary electrophoresis? Develop two expressions, one in terms of C_{H^+} and K_{d1} and the other in terms of pH and pK_1. You are given that the dissociation constant of the base is K_{d1}.

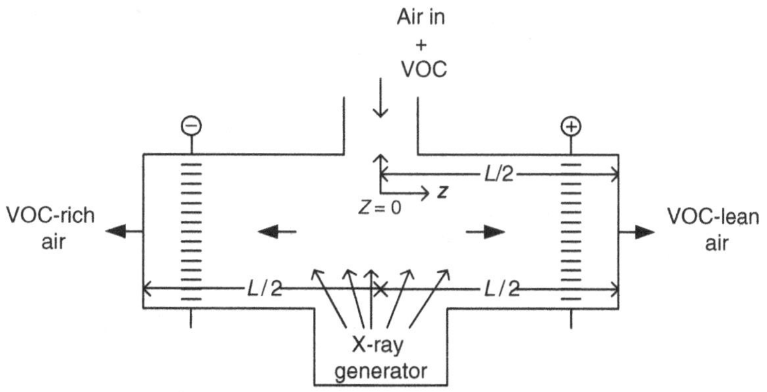

Figure 6.P.1.

6.3.5 Enantiomeric compounds R and S may be separated by capillary electrophoresis by complexing them with a chiral selector compound B. Here the equilibrium binding constants for the R and S enantiomers are, respectively, K_R and K_S (in a molar concentration based expression). Develop expressions for the effective mobility of the R and S species, assuming that the complexes continue to have the same charge as the uncomplexed compounds. Obtain an expression for the mobility difference between the effective mobilities of the R and S species. What contributes to a nonzero value of this mobility difference?

6.3.6 In isoelectric focusing studied in Section 4.2.2.1 for protein separation in a closed vessel with no bulk flow, the focusing point/location in the cell corresponded to no electrical force, since $pH = pI$ at that location. Suppose you have a channel or a tube through which a buffer solution is flowing at velocity v_z. There are two electrodes at the two ends of this channel, creating an electrical field $E(z)$ which varies with the distance z, since we have created an electrical conductivity gradient along z by a selective and continuous decrease in the buffer ion concentration (by some special means, say a dialysis membrane, which allows diffusion of buffer ions out of the solution). If you now inject a sample of two proteins (BSA and hemoglobin, both negatively charged at the buffer pH 8.7) along with the buffer near the anode to the left, as the buffer moves to the right toward the cathode end, is it possible to focus the two proteins to two different locations in the tube? *Write down* the flux expression N_{iz} for a protein i at any location. What would be the value of the protein flux at the location where it is focused? If the total moles of protein injected in the sample is given by m_i, obtain a solution for $C_i(z)$ in terms of $E(z)$ and other relevant quantities. Show why two proteins are focused at two different locations. This technique is part of a family of gradient focusing techniques.

6.3.7 It has been observed that volatile organic compounds (VOCs), such as toluene, ethanol, etc., are selectively ionized to positively charged ions when exposed to soft x-rays and x-rays in the presence of N_2, O_2, etc., constituting air. A VOC separator has been designed, as shown in Figure 6.P.1.

Air-containing VOCs coming in is split into two streams, one moving toward the negative electrode in the gas flow duct and the other moving in the opposite direction, toward the positive electrode. Both streams have the same velocity, $|v_z|$. The first-order reaction rate constant based volumetric rates of formation and disappearance of the VOC ion are, respectively, $k_1 C_i$ and $-k_2 C_{i^+}$, where C_i is the concentration of the VOC species and C_{i^+} is the concentration of the VOC$^+$ ion. The electrical field strength is E volt/cm; the mobility of the VOC ion is $\mu_{i^+}^m$. Write down the equations of change for the molar concentrations of the VOC ions and the neutral VOC molecules in the z-coordinate for the gas stream flowing toward the negative electrode. Assume no radial or circumferential gradients; the gas is very dilute in VOCs. Write down the boundary conditions.

6.3.8 Consider the separation configuration shown in Figure 6.P.2 for separating larger spherical microparticles (~20 μm) from smaller spherical microparticles (~10 μm) flowing in a liquid suspension in a rectangular microchannel. As shown in part (a) of the figure, there are two inlet microchannels through which liquid enters, one (the shaded stream) contains a mixture of particles, the other does not. The two microchannels, each of depth around 50 μm, converge into a straight pinched segment of length 100 μm. Then the outlet from this segment diverges (as shown in part (b)) quickly into a 1 mm broad segment. The widths of these two inlet channels, as well as the converging segment, are 50 μm. The larger particles and the smaller particles in the feed inlet

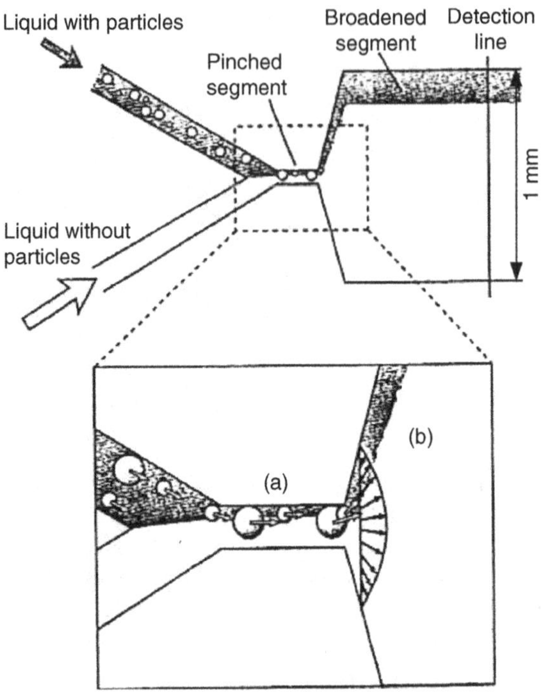

Figure 6.P.2.

channel liquid will follow distinctly different trajectories in the broadened segment. Identify the trajectories of particles of two different sizes. Identify the forces on the particles responsible for separation. Speculate on any other factors influencing such separation.

6.3.9 In flash vaporization using a vertical drum, the terminal drop velocity is important in determining the diameter of the column through which the vapor rises and leaves the column. In petroleum refining, often a three-phase mixture enters a flash drum: a vapor phase, a hydrocarbon liquid and liquid water. The three-phase flash drum separator should allow for the separation of the liquid water from the liquid hydrocarbon separated by an interface, as well as the disengagement of drops from the vapor leaving at the top. The lighter liquid (viscosity μ_ℓ, density ρ_ℓ) leaves via a nozzle below the feed introduction zone; the heavier liquid (viscosity μ_h, density ρ_h) leaves via a nozzle at the bottom of the column.

(1) Calculate the settling velocity (terminal velocity) for the drops in each liquid phase. Simplify the results, assuming that drops behave like solid spherical particles in the Stokes regime.

(2) Determine the settling time for the drops in each phase (heights of each phase in the column are h_h for the heavy phase, h_ℓ for the light phase).

(3) Given the volumetric flow rates of the vapor (Q_v) and the two liquid phases (Q_h and Q_ℓ), determine the criteria for satisfactory separation for all phases. Draw a diagram of the vertical separator.

6.3.10 A liquid mixture containing 60 mole % propane, 10 mole % isobutane and 30 mole % n-butane is throttled into a flash drum at 85 °C and 2500 kPa. Calculate the fraction of the feed stream leaving as a vapor if vapor-liquid equilibrium is assumed. Determine the compositions of the vapor stream and the liquid stream. Obtain K_i values from Figures 4.1.5 and 4.1.6.

6.3.11 Many hydrocarbon mixtures contain a small amount of water. In the processing of such mixtures, three phases are frequently encountered: a vapor phase ($j = v$), a hydrocarbon liquid phase ($j = h$) and a water phase ($j = w$). Consider the case of flash separation of such a hydrocarbon-water mixture. Define

$$\theta_{vf} = (W_{tv}/W_{tf}), \qquad \theta_{hl} = (W_{th}/W_{t\ell}), \qquad W_{t\ell} = W_{th} + W_{tw},$$

where $W_{t\ell}$ is the total molar flow rate of the liquid-phase mixture consisting of the molar flow rates of the hydrocarbon stream (W_{th}) and the water stream (W_{tw}).

(1) Write down the molar balance equations for the total molar flow rates and that for the ith species flow rate. For any species i, show that

$$x_{if} = \theta_{hl}(1-\theta_{vf})x_{ih} + (1-\theta_{h\ell})(1-\theta_{vf})x_{iw} + \theta_{vf}\,x_{iv},$$

where subscript f refers to the feed stream.

(2) Using the following K_i factors, K_{ih} $(= x_{iv}/x_{ih})$ and K_{iw} $(= x_{iv}/x_{iw})$, show that the following relation holds:

$$x_{ih} = x_{if}/\left[\theta_{hl}(1-\theta_{vf}) + \{(1-\theta_{h\ell})(1-\theta_{vf})K_{ih}/K_{iw}\} + \theta_{vf}\,K_{ih}\right].$$

(3) Identify other conditions to be satisfied in any solution procedure to be developed.

6.3.12 A batch distillation pot in a pharmaceutical laboratory can nominally hold up to 110 moles of liquid solvent. It is useful to operate with a lower liquid level to prevent entrainment of drops, around 80 moles. This vessel is agitated to maintain a reasonably uniform liquid composition. For this reason, the lowest liquid level allowed is 20 moles.

(a) Consider the following process. A solvent containing some nonvolatile solute was distilled down to 20 moles from 80 moles solvent (species $i = 1$). Next, another solvent ($i = 2$) was introduced to bring the total solvent level to 80 moles. Distillation was carried out until the final solvent level was reduced to 20 moles. Find out how many moles of solvent $i = 1$ were removed in this last distillation step. Determine how many moles of solvent $i = 2$ were removed at the same time.

(b) Suppose you now follow the Gentilcore (2002) procedure of constant-level batch distillation, when the number of solvent moles to start with in the distillation pot is $m_{t\ell} = m_{1\ell} = 20$ moles; what is the number of moles of solvent $i = 2$ that will be removed if the number of moles of solvent $i = 1$ removed stays the same as in (a)?

Given: $\alpha_{12} = 3$.

6.3.13 Obtain an expression for the molten-phase impurity mass fraction distribution as a function of the molten-phase impurity mass fraction in the bulk melt, employing the partial mixing model in single pass zone melting. State your assumptions.

6.3.14 For the case of a completely mixed melt, compare the purification performance of zone melting of a solid rod with that of normal freezing of the same rod at a distance of $z = 9\ell$ and ℓ, where ℓ is the zone length and the rod length $L = 10\ell$. You are given: $\kappa'_{is} = 0.1$; $\rho_s \cong \rho_\ell$; $v_{fr} = 0.2$ cm/hr; $D_{i\ell} = 10^{-5}$ cm^2/s; $\delta_\ell \cong 0.1$ cm; $\ell = 0.1$ cm. *Note*: This value of κ'_{is} is close to that of carbon in silicon (see Table 4.1.7).

6.3.15 For the case of a partially mixed melt, compare the purification performance of zone melting of a solid rod with that of normal freezing of the same rod at a distance of $z = 8\ell$, where ℓ is the zone length and the rod length $L = 10\ell$. You are given: $\kappa'_{is} = 0.1$; $\rho_s \cong \rho_\ell$; $v_{fr} = 0.2$ cm/hr; $D_{i\ell} = 10^{-5}$ cm^2/s; $\delta_\ell \cong 0.1$ cm; $\ell = 0.1$ cm.

6.3.16 Figure 3.3.6A illustrates the temperature vs. composition phase diagram of the naphthalene-β-naphthol system. In the composition range, where β-naphthol is an impurity, the melt has a concentration of β-naphthol lower than that in the solid. The starting solid mixture contains 5% by weight of β-naphthol. What is the velocity of the freezing interface needed to produce a 2 cm zone composition of overall composition 3% by weight of β-naphthol? Neglect any convection in the molten zone (Wilcox and Wilke, 1964). You are given: $\kappa'_{is} = 1.85$; $D_{i\ell} = 2.9 \times 10^{-5}$ cm^2/s; $\rho_s \cong 1.15$ g/cm^3; $\rho_\ell = 0.97$ g/cm^3. (Ans. $v_{fr} = 1.26 \times 10^{-2}$ cm/hr.)

6.3.17 The permeability characteristics of a filter cake deposited on a highly permeable filter cloth were determined by passing plain water at 20 °C and an applied pressure difference of 0.95 atmosphere. The filter cloth was 5.08 cm in diameter; the water flux was found to be 36 cm^3/min. The filter cake thickness was measured to be 1.5 cm. Determine the value of the filter-cake resistance, R_c, and the value of water permeability, Q_{wm}, through the cake. Neglect the filter cloth resistance. (Ans. $R_c = 0.32 \times 10^{10}$ cm^{-1}; $Q_{wm} = 4.73 \times 10^{-2}$ cm^2-cp/s-atm.)

6.3.18 You have to sterilize an aqueous solution of a heat-labile pharmaceutical compound by a membrane process. *Pseudomonas diminuta* (0.3 μm in diameter for our purpose) is conveniently used as a bacterial model to test the membrane for its integrity. However, you have an opportunity to test and determine only the bubble-point pressure of the membrane with water. Such a test yielded a value of 55 psia. What is the value of the largest pore

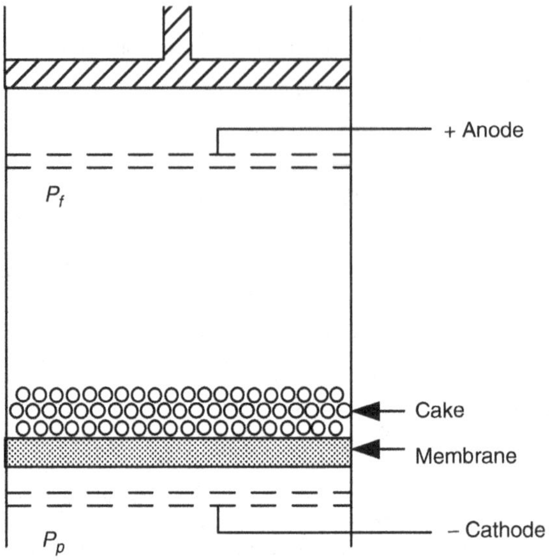

Figure 6.P.3.

diameter of this membrane? Is it okay for sterilization? You are given: surface tension of water = 72 dyne/cm (=g/s^2); conversion factor for pressure, 1 g/cm-s^2 = 1.4504 × 10^{-5} psia; $\tau_m = 4.0$; $\theta = 0$*.

6.3.19 In deadend cake filtration, shown in Figure 6.3.21, suppose there is an arrangement (Figure 6.P.3) in which one can put a cathode below the membrane and an anode up in the region of the piston. (In practice, a piston is not needed; slurry under pressure is what is used.)

(a) Speculate what will happen to the filter cake if the particles being filtered have a net negative charge.

(b) Identify a z-coordinate normal to the membrane/filter cloth pointing away from the membrane toward the anode. Write down the expressions for forces acting on any particle being deposited in the cake. Develop the criterion for removal/no removal of particles from the cake. Deliberate on the various resistances to the filtration flux in the presence of the electrical field.

(c) Would you maintain the voltage between the two electrodes on a steady basis, or apply pulses of voltage?

6.3.20 Equations (6.3.138f/h) in cake filtration indicate a particular type of dependence between the volume of the filtrate, $V(t)$, and the time for filtration, t. We would like to find out a corresponding relation in batch deadend ultrafiltration without any stirring (Figures 6.3.26(a) and (b)). Assume for the time being that $\Delta\pi$ is not important. Recognize that

$$N_{sz}\overline{V}_s = J_v = \text{volume flux } = \frac{1}{A_m}\frac{dV_p}{dt},$$

where V_p is the volume of the permeate collected from time $t = 0$ to $t = t$.

(1) Obtain the overall resistance in terms of the sum of the membrane resistance and a stagnant diffusion layer resistance.

(2) Relate the diffusion layer thickness, δ_ℓ, the bulk concentration of the solute, C_{ilb}, the permeate volume, $V_p(t)$, and the observed solute rejection, R_{obs}. Assume a uniform solute concentration in the diffusion layer, C_{igel}.

(3) Incorporate the result from (2) and (1) above and obtain a dependence of t varying with the square of the permeate volume, i.e. $(V_p(t))^2$. Assume that the value of the stagnant diffusion layer resistance, R_c, is proportional to the diffusion layer thickness, δ_ℓ, and a specific diffusion layer resistance, $\hat{R}_c$.

6.3.21 Saksena and Zydney (1994) have studied the protein transmission characteristics of a 100 000 MWCO ultrafiltration membrane (OMEGA 100K) of polyethersulfone using a binary mixture of bovine serum albumin (BSA) (mol.wt., 67 000) and immunoglobulin (IgG) (mol.wt., 155 000) in a solution of $pH = 7$ and an ionic strength of 0.15 M NaCl solution. The batch cell operation characteristics and BSA transport properties have already been identified in Example 6.3.8. The transport properties of IgG are as follows: $S_\infty = 0.0026$; $D_{ieff} = 4 \times 10^{-15}$ m^2/s; $k_{il} = 3.3 \times 10^{-6}$ m/s.

(1) Calculate the values of the observed and true transmission coefficients, S_{obs} and S_{true}, respectively, of IgG.

(2) Determine the selectivity (~ the separation factor) of BSA over IgG for this membrane.

Assume that the presence of one protein in a dilute concentration does not interfere with the ultrafiltration of the other protein also present in a dilute concentration. (Ans. (1) $S_{\text{obs}} = 0.0119$; $S_{\text{true}} = 0.009$; (2) $(S_{\text{obs}}|\text{BSA}/S_{\text{obs}}|\text{IgG}) = 17.14$.)

6.3.22 In a stirred ultrafiltration cell using a flat UF membrane, an aqueous solution of the polymer Dextran 20 was ultrafiltered. Data were gathered at different values of the water flux, and the solute rejection was measured. Dextran 20 is a linear polymer, and, as the solvent flux was increased, the rejection observed for Dextran 20 decreased. A plot of the solvent flux against the quantity $(1-R_{\text{obs}})/R_{\text{obs}}$ in a semilogarithmic plot (logarithmic on the abscissa for $(1-R_{\text{obs}})/R_{\text{obs}}$) yielded a straight line with a positive slope and an intercept of 0.05 on the abscissa.

(1) Provide an estimate of the true sieving coefficient for this Dextran 20 macromolecule of mean molecular weight $\overline{M}_w$ 20 000 and the membrane used. (Ans. $S_{\text{true}} = 0.0476$.)

(2) For the same Dextran polymer and a stirred cell, water flux data were gathered at 1 atmospheric pressure gauge and 25 °C for different feed concentrations C_{ilb} of Dextran; simultaneously, the permeate Dextran concentrations, C_{ip}, were also measured:

water flux (gallons/ft^2-day)(GFD)	24	14.5	7.5
$(C_{ilb} - C_{ip})\overline{M}_w$ (wt%)	0.6	2.4	6.5

Plot these data appropriately and determine the value of C_{igel} corresponding to zero flux. Speculate about the relation between C_{igel} and zero flux.

6.3.23 Consider ultrafiltration based separation of a protein in the configuration of Figures 6.3.26(a) and (b). Suppose now another membrane, identical to the first membrane, is located below the first membrane at some distance from it. Permeate generated from the first membrane having a protein concentration of C_{ip_1} becomes the feed to the second UF membrane. The protein concentration in the permeate from the second membrane is C_{ip_2}. The protein concentration in the feed to the first membrane is C_{if_1}. Assume a pseudosteady state; the macrosolute observed rejection value R_1 for the first membrane may be assumed to be valid for the second membrane as well, in relation to its feed and permeate.

(a) Determine the value of the permeate concentration, C_{ip_2}, as a function of R_1 and C_{if_1}.

(b) What is the overall solute rejection, R_{ov}, for the permeate C_{ip_2} with respect to C_{if_1}?

(c) If instead of two membranes, we have three membranes in an identical configuration, express R_{ov} in terms of R_1. Assume $R_1 = R_2 = R_3$.

6.3.24 Employ the principles illustrated in Problem 6.3.23 to calculate the value of the rejection of the protein myoglobin (mol. wt., 17 566) at a pH of 4.35 through a regenerated cellulose membrane having a molecular weight cut off of 30 000. The pI of myoglobin is 7.3. At a pH of 4.35, the myoglobin molecules are charged, have larger hydrodynamic radius and therefore undergo considerable rejection in the regenerated cellulose membrane ($R \cong 0.96$) (Feins, 2004). Calculate the value of R_{ov}, the overall myoglobin rejection, if two regenerated cellulose membranes are stacked together. Obtain the corresponding value for a stack of three such membranes. (Ans. two membranes $\Rightarrow R_{\text{ov}} = 0.9984$; three membranes $\Rightarrow R_{\text{ov}} = 0.999936$.)

6.3.25 For reverse osmosis desalination of seawater at a high $\Delta P = 102$ atm using a cellulose acetate membrane, the following information is available for a solution–diffusion model based analysis (employ Example 6.3.10): $A = 8.03 \times 10^{-7}$ gmol H_2O/cm^2-s-atm; $\pi(C_{if}) = bC_{if} = 45.7$ atm; $(D_{im}\kappa_{im}/\delta_m)_{\text{salt}} = K_2 = 1.774 \times 10^{-5}$ cm/s; $\overline{V}_s = 18.05$ cm^3/gmol. Determine the value of the salt rejection, R_{salt}. (Ans. $R_{\text{salt}} = 0.979$.)

6.3.26 The physical picture of a reverse osmosis membrane illustrated in Figure 6.3.29b is sometimes identified as the *sieve transport model*: one section of the membrane (fractional area ε_b) has pores that are completely open, i.e. nonrejecting, while the rest of the membrane (fractional area $(1 - \varepsilon_b)$) has complete rejection of the solute. However, variations in this model arise depending on the extent of solute mixing downstream of the membrane from different sections and the driving force for solvent transport. Consider the following case: the driving force for solvent transport through both regions of membrane is $(\Delta P - \Delta\pi)$. Employ the notation of Example 6.3.9. *Write down* the expression for J_{vz} and N_{iz}. *Obtain* an expression for the solute rejection R_i. *What* would be the value of σ_i in such a model?

6.3.27 To meet the requirement of a particular salt concentration in the permeate, the quantity $R_{i,\text{reqd}}$ was defined by (6.3.173) based on the final salt concentration, C_{ile}. Define the $R_{i,\text{reqd}}$ instead based on an average feed salt concentration,

$$C_{il\text{avg}} = (C_{ili} + C_{ile})/2, \text{ as } R_{i,\text{reqd}} = 1 - (C_{ip}/C_{il\text{avg}}).$$

(1) Develop an expression for $R_{i,\text{reqd}}$ (corresponding to (6.3.175)) in terms of the fractional water recovery, *re*.
(2) Calculate the values of $R_{i,\text{reqd}}$ for a brackish water feed containing 2000 ppm salt for the following values of *re*: 0, 0.75, 0.85, 0.95. The required C_{ip} corresponds to 500 ppm.
(3) Calculate the values of $R_{i,\text{reqd}}$ for a seawater feed containing 35 000 ppm salt for the following values of *re*: 0, 0.25, 0.3 and 0.5. The required C_{ip} corresponds to 500 ppm salt. Comment on the requirement for pressure as the fractional water recovery is increased.
(4) Calculate the final feed salt concentration, C_{ile}, in ppm, corresponding to the recoveries specified in (3) above.

6.3.28 (1) To remove CO_2 from flue gases being released from a coal-burning power plant, a membrane process is being explored. What should be the minimum selectivity of the polymeric membrane for CO_2 over N_2 if the flue gas may be assumed to be 10% CO_2, the rest being essentially N_2? The permeate must have at least 90% CO_2. (Ans. 81.)
(2) Consider an undesirable scenario where the atmosphere has 400 ppmv (parts per million by volume) of CO_2 and the rest may be assumed to be N_2 from the perspective of membrane separation. How high should the ideal membrane selectivity be to recover a 90% CO_2 stream in the permeate for the purpose of sequestering the CO_2? (Ans. 22 500.)

6.3.29 In a small cell of the type shown in Figure 6.3.32, feed gas mixture containing 50% helium and 50% methane (mole %) is introduced at 17 atm. The helium permeability coefficient through the nonporous membrane has the value 240 (units) and that for methane is 3 (units). The permeate pressure is quite low. Calculate the composition of the permeate side gas emerging at the beginning of the process. (Ans. $x_{\text{He}p} = 0.987$.)

6.3.30 Consider the composite "PRISM-type" membrane illustrated in Example 6.3.14. For such a polysulfone membrane having a thin coating of silicone rubber, the values of the permeability coefficients through layers A and B for a H_2–CO mixture are:

$$Q_{\text{H}_2\text{A}} = 5.2 \times 10^{-8} \text{ scc-cm/cm}^2\text{-s-cm Hg};$$
$$Q_{\text{H}_2\text{B}} = 1.2 \times 10^{-9} \text{ scc-cm/cm}^2\text{-s-cm Hg};$$
$$Q_{\text{COA}} = 2.5 \times 10^{-8} \text{ scc-cm-cm}^2\text{-s-cm Hg};$$
$$Q_{\text{COB}} = 3 \times 10^{-11} \text{ scc-cm/cm}^2\text{-s-cm Hg};$$
$$\varepsilon = 2 \times 10^{-6}.$$

Calculate the values of $\left(Q_{H_2m}/\delta\right)$ and $\alpha^*_{\text{H}_2-\text{CO}}$ for the following cases:
(a) $\delta_A = 0.1$ μm; $\delta_B = 0.05$ μm;
(b) $\delta_A = 0.1$ μm; $\delta_B = 0.25$ μm;
(c) $\delta_A = 5$ μm; $\delta_B = 0.05$ μm;
(d) $\delta_A = 5$ μm; $\delta_B = 0.25$ μm.
Comment on the effect of the thickness, δ_A, of the coating as well as the substrate skin thickness, δ_B. (Ans. (a) 230×10^{-6}; 38. (b) 48×10^{-6}; 39. (c) 70×10^{-6}; 13. (d) 33×10^{-6}; 28.)

6.4.1 Define the crystal mass density distribution function per unit solution volume between the crystal dimensions r_p and ∞ to be $M(r_p, \infty) = \rho_s \psi_v \int_{r_p}^{\infty} r_p^3 n(r_p) dr_p$.
(a) Obtain an expression for $M(r_p, \infty)/M_T$ for an MSMPR crystallizer whose $n(r_p)$ is given by expression (6.4.5).
(b) Obtain an expression for the density function of $M(r_p, \infty)/M_T$.
(c) Locate the value of r_p where this density function has a maximum.

6.4.2 In Example 6.4.1, analytical expressions for the distribution functions $\overline{r}_p^+(x)$ and $A_p^+(x)$ were derived for an MSMPR crystallizer. Develop the density functions $\overline{r}_p^+ f(x)$ and $A_p^+ f(x)$ of these two distribution functions in terms of x $(= r_p/G\ t_{\text{res}})$. Determine the values of x where these density functions may have a maximum. *Suggestion*: Follow the approach of Example 6.4.2.

6.4.3 For a particular system, the dependence of the nucleation rate B^{o} on the supersaturation has a kinetic order p. The dependence of the growth rate on the supersaturation has a kinetic order of growth of h (instead of 1). From an MSMPR crystallizer, you have an option of generating data such that you can vary t_{res} for a fixed M_T. Express the ratio of the two growth rates G_a and G_b for two values of $t_{\text{res}}|_a$ and $t_{\text{res}}|_b$. Similarly, obtain a value of the ratio $(n_a^{\text{o}}/n_b^{\text{o}})$ in terms of $t_{\text{res}}|_a$ and $t_{\text{res}}|_b$. What would be the slope of a plot of log n^{o} vs. log G?

6.4.4 A batch cooling crystallizer is to be used to crystallize a salt. The feed solution at 60 °C is to be cooled after seeding the crystallizer with seed crystals of size $\overline{r}_{ps} = 0.075$ mm; these crystals are expected to grow up to 0.65 mm. The solubility change with respect to temperature in the temperature range of interest is 0.0053 g/cm^3-K. The seed crystal suspension density is 244×10^{-6} g/cm^3. The solid density, ρ_s, is 3.6 g/cm^3. The supersaturation is around 0.01 g/cm^3. The diffusion-controlled crystal growth rate mass-transfer coefficient has a value of 1.15×10^{-3} cm/s in the agitated crystallizer. Determine the expression for the cooling curve for this problem. The crystals may be assumed spherical for the purpose of your calculations; in reality, they have a different shape.

6.4.5 Consider an unseeded batch crystallizer where all crystals are generated from nuclei via growth at a constant growth rate, G. The governing equation is (6.4.51), where we assume that the supersaturation level is constant with time. The size of the nuclei is $\overline{r}_{pn}$, the volume based shape factor for all nuclei/crystal size is ψ_v and B^{o} is the nuclei generation rate per unit volume.

(1) Obtain the following expressions for A_T:

$$A_T = \int_o^t \frac{k_{nu}(\rho_i - \rho_{isat})^p\, \psi_s\, \overline{r}_p^2\, \mathrm{d}t}{\psi_v\, \rho_s\, \overline{r}_{pn}^3},$$

(where $\overline{r}_p^2 = \overline{r_p^2}, \overline{r}_{pn}^3 = \overline{r_{pn}^3}$ and ψ_s is the constant surface-area based shape factor). Show that this is equivalent to

$$A_T = \frac{k_{nu}(\rho_i - \rho_{\text{isat}})^p\, \psi_s}{3\,\psi_v\, \rho_s\, G}\left(\frac{\overline{r}_p^3}{\overline{r}_{pn}^3} - 1\right).$$

(2) Using the expression for growth in equation (6.4.51) for one crystal, show that

$$G = \frac{k_{gr}\, \psi_s(\rho_i - \rho_{\text{isat}})^{p_1}}{3\,\rho_s\, \psi_v}.$$

(3) Employing these results in equation (6.4.51) for a constant supersaturation level, obtain the following expression for the cooling curve:

$$T_0 - T(t) = \frac{k_{nu}(\rho_i - \rho_{\text{isat}})^p\, \overline{r}_{pn}}{4(d\rho_{\text{isat}}/\mathrm{d}T)G}\left[\left\{1 + \left(\frac{Gt}{\overline{r}_{pn}}\right)\right\}^4 - 1\right],$$

where the temperature of the solution at $t = 0$ is T_0 and that at time t is $T(t)$.

6.4.6 A mixer for solvent extraction is 61 cm in diameter and height. It has a six-bladed turbine impeller of 30.48 cm diameter. The mixer is a regular cylindrical baffled vessel. Batch extraction of an organic pharmaceutical compound is to be carried out from its dilute aqueous solution into p-xylene. The ratio of its molar concentration in p-xylene to that in water is 20. The ratio of the organic-phase volume to the aqueous-phase volume in the charge to the vessel is 0.2. The impeller rotates at 200 rpm. The temperature is 20 °C.

(1) Calculate the time needed to achieve 90% of the equilibrium extraction.

(2) If the impeller was rotated at 100 rpm without changing the phase being dispersed, what would be the time needed to achieve 90% extraction?

You are given: aqueous solution density = 0.985 g/cm^3; density of p-xylene = 0.861 g/cm^3; viscosity of aqueous solution = 0.96 cp; viscosity of p-xylene = 0.60 cp; diffusion coefficient of organic compound in

Table 6.P.1.

		R_i - rejection %	
Membrane	Inulin (5000)	Albumin (67000)	Apoferitin (480000)
UM20 (MWCO 20000)	5	95	99
XM 100A (MWCO 100000)	1	35	98

water $= 1.2 \times 10^{-5}$ cm^2/s; diffusion coefficient of organic compound in p-xylene $= 1.6 \times 10^{-5}$ cm^2/s; interfacial tension $\gamma^{12} = 21$ dyne/cm. (Ans. (1) 0.96 s; (2) 3.96 s.)

6.4.7 Batch ultrafiltration is carried out in a well-stirred membrane cell using 10 liter of an aqueous buffer solution containing 1 wt% inulin, 1 wt% albumin and 1 wt% apoferitin until the retentate volume is 1 liter. The membrane is UM-20, having a MWCO of 20 000.

(1) Obtain the concentrations of the three proteins in the permeate and the retentate in wt%.

(2) Now add 9 liter of buffer to the retentate and carry out batch ultrafiltration using the UF membrane XM-100A, having a MWCO of 100 000. The retentate volume obtained after UF is 1 liter. Obtain the concentration of the three proteins in the retentate in wt%. (Ans. (1) retentate: inulin, 1.123%; albumin, 8.91%, apoferitin, 9.97%; permeate: inulin 0.996%; albumin, 0.12%; apoferitin, 0.025%. (2) inulin, 0.115%; albumin, 1.09%; apoferitin, 9.32%.)

You are given the data in Table 6.P.1.

6.4.8 The problem involves purification of a monoclonal antibody species 1 present in 100 liter of a solution which contains another protein species 2 as well as a small molecular weight impurity, 3. Species 1, a monoclonal antibody, has a molecular weight of 450,000; protein 2 is albumin of molecular weight 67,000; impurity 3 has a molecular weight of 250 dalton. The initial feed solution concentrations are: Protein 1, C_{10} g/liter; Protein 2, C_{20} g/liter; low molecular weight impurity 3, C_{30} g/liter. To reduce the small molecular weight impurity level as well as the level of albumin substantially from the solution of the monoclonal antibody, the following two-step process has to be implemented:

(1) Continuous diafiltration where the buffer volume added is 500 liter.
Determine the concentrations of species 1 and species 2 and the low molecular weight impurity in the retentate in terms of their original concentrations for PM 30 UF membrane (MWCO, 30,000). Given: $R_1 = 1.0$; $R_2 = 0.92$; $R_3 = 0.0$.

(2) We have to now reduce substantially the concentration of albumin from the solution resulting from step (1) above with the monoclonal antibody being our product. Employ continuous diafiltration again using XM100A membrane with the buffer volume added being 500 liter. Determine the concentrations of protein 1 and protein 2 and the low molecular weight impurity 3 in the final retentate in terms of their original concentrations C_{10}, C_{20}. and C_{30} before step (1). Given: MWCO of XM100A membrane: 100,000; $R_1 = 0.96$; $R_2 = 0.45$; $R_3 = 0.0$.

7

Separation in bulk flow of feed-containing phase perpendicular to the direction of the force

Separations in which the feed phase or feed-containing phase has a bulk motion parallel to the direction of the force causing separation have been studied in Chapter 6. In many separation techniques/processes/operations, the feed phase or feed-containing phase has a bulk motion perpendicular to the direction of the driving force. In a number of situations, the feed is introduced in small amounts in a carrier fluid whose bulk motion is perpendicular to the force direction. Such separations will be studied in this chapter.

For separations based on distribution of species between two phases in equilibrium, we study in Section 7.1 primarily those two-phase systems where the second phase (e.g. adsorbent particles in a packed adsorbent bed) is stationary; the first phase, which is more often the feed solution/mixture, moves perpendicular to the direction of chemical potential driving force between the two immiscible phases (Figure 7.0.1(a)). This first phase (the mobile phase) bulk motion is generally in one direction. The benefits of such bulk motion in terms of extreme purification achievable in two-component systems and multicomponent separation capability will be illustrated. In the cyclic processes studied next, the direction of motion of the mobile phase is periodically reversed; the direction of force is also reversed, except it remains perpendicular to the bulk-phase flow direction (Figures 7.0.1(b) and (c)). The mobile phase is sometimes generated by the separation operation, as in the case of the blowdown phase of pressure swing adsorption (PSA). The elution chromatographic process considered next involves injection of a sample to be separated into a carrier fluid flowing perpendicular to the direction of the force between the fluid and the stationary adsorbent phase (Figure 7.0.1(d)); this and other related chromatographic processes are also studied in Section 7.1.5. The imposition of an electrical force parallel to the direction of flow in a packed chromatographic column is studied in Section 7.1.6 in what is called counteracting chromatographic processes.

Many of the membrane separation processes considered in the preceding six chapters produce a permeate phase or fraction from the feed stream. The permeate phase/fraction is generally miscible with the feed phase/fraction from which it remains separated by the membrane; the residual fraction of the feed stream is either called the *reject* or the *concentrate*. Such processes are considered in Section 7.2.1 under conditions where the permeate stream generated by the force present in the system flows parallel to the direction of the force, while the feed/concentrate/reject bulk flow direction is perpendicular to that of the force. Such a flow pattern is commonly termed *crossflow*. The particular membrane techniques studied in order are: gas permeation (Figure 7.0.1(e)); reverse osmosis (Figure 7.0.1(f)); ultrafiltration (Figure 7.0.1(g)); microfiltration (Figure 7.0.1(g)); rotary vacuum filtration (7.0.1(h)). In the last example, the membrane/filter moves instead of the liquid feed. In Section 7.2.2, we consider briefly the process of granular filtration, in which bulk flow of the feed liquid takes place through a loosely packed bed of granular filtration media, where most of the forces, except that causing inertial impaction (see Section 6.3.1.4), are operating perpendicular to the direction of the bulk flow and lead to particle capture by the granular media.

Section 7.3 covers separation operations/processes/techniques in which an external force field is applied perpendicular to the direction of bulk flow of a single-phase solution or a dispersed/particulate multiphase mixture (Figures 7.0.1(i)–(p)). The external force fields considered are: electrical, centrifugal, gravitational and magnetic. Specific processes treated include, among others, electrophoresis (Figure 7.0.1(i)), dielectrophoresis,

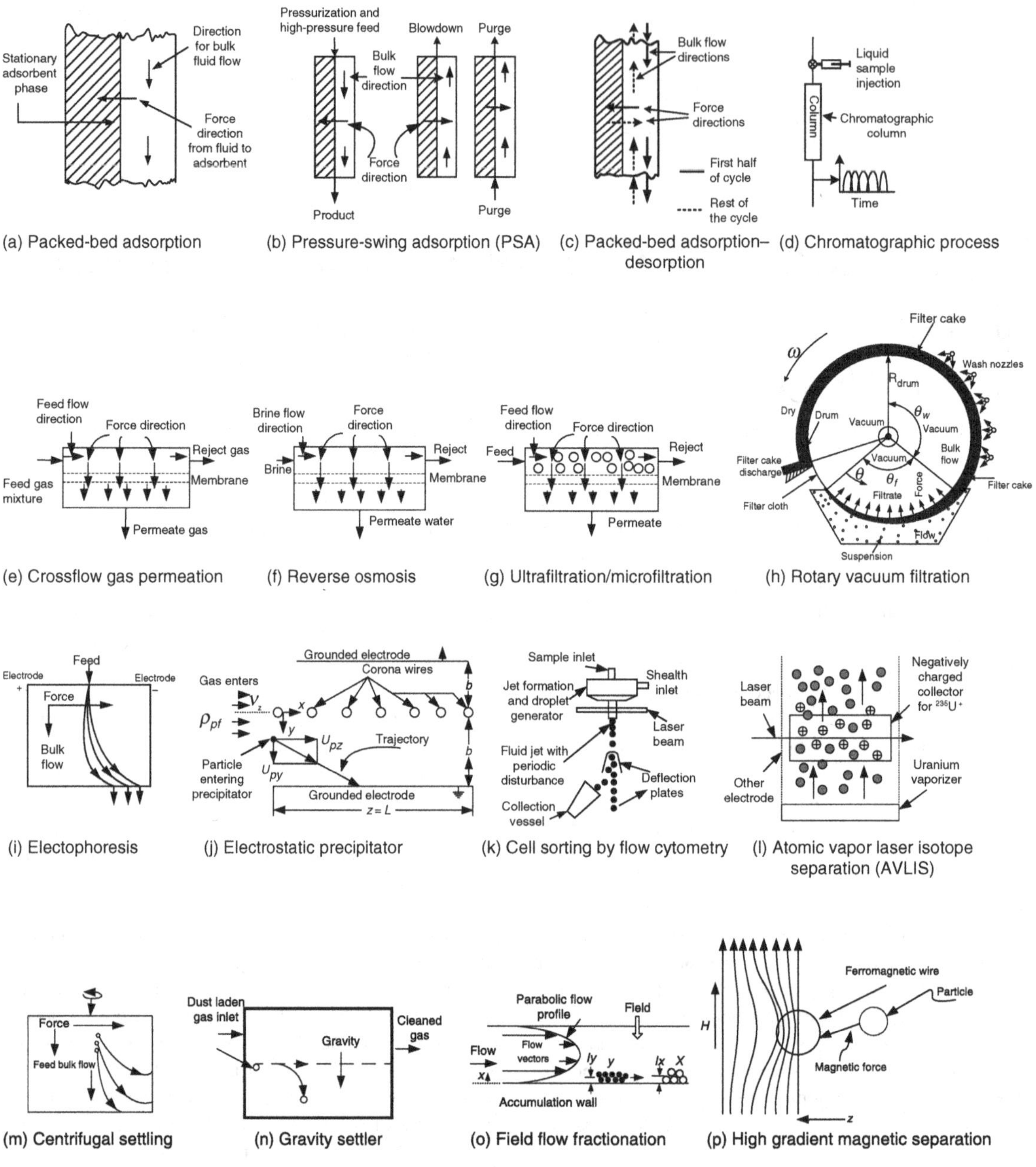

Figure 7.0.1. Bulk flow of feed-containing phase/region perpendicular to the direction of force: examples.

electrostatic precipitation (Figure 7.0.1(j)), flow cytometry (Figure 7.0.1(k)), laser isotope separation (Figure 7.0.1(l)), particle separation in centrifuges/cyclones, etc. (Figure 7.0.1(m)), gas separation by a separation nozzle process, gravity settling (Figure 7.0.1(n)), high-gradient magnetic separation (7.0.1(p)). When any one of the external force fields is coupled with the velocity field of feed bulk flow perpendicular to the external force field, unique separation capabilities are achieved, as in field flow fractionation (Figure 7.0.1(o)). Column 7 in each of Tables 1–7 in the introductory chapter illustrates the large number of separation operations/processes/techniques studied here via a few distinguishing characteristics under three broad categories: phase equilibrium based separations; membrane based separations; separations in external and other force fields.

There is one basic phenomenon in the three types of separations identified above and described in detail in this chapter. It arises from the force direction being perpendicular to the direction of the bulk flow of the feed-containing phase (phases in one stream in the case of particle separation). Focus on Figures 7.0.1(i) and 7.0.1(m). In a given period of time, a feed species or a protein molecule, or a cell or a particle, moves a certain distance away from the axial line of its introduction, as defined by its trajectory. The latter is shaped by the magnitude of its species-specific/particle-specific velocity perpendicular to the bulk flow and the magnitude of its velocity toward the separator outlet, determined by the convective bulk fluid velocity. The higher the species-specific/particle-specific velocity perpendicular to the bulk flow, the further the end-point of the species/particle trajectory from the point of introduction. This is a successful recipe for multicomponent separation into fractions for both molecular/macromolecular species/ions as well as particles in external force based systems.

For crossflow membrane processes (Figure 7.0.1(e)-(g)), the species which encounters the least resistance from the membrane (when subjected to its chemical potential based driving force across the membrane) will have the highest velocity of movement through the membrane. Correspondingly, its rate of disappearance from the feed stream will be highest, and its concentration in the bulk flow of the reject stream will be lowest. Current practice in membrane separation processes does not exploit the differences in individual species velocities through the membrane for multicomponent separation.

In Figures 7.0.1(e)-(g), 7.0.1(i)-(l) and 7.0.1(m)-(p), solutes/macromolecules/particles moving in the direction of the force are removed continuously from the separator by removing the mobile fluid phase associated with them. (In Figure 7.0.1(h), the membrane/filter phase containing the particles is removed.) However, in the fixed-bed processes/techniques of Figures 7.0.1 (a)-(d), when solute molecules move into a particle in a packed bed, the stationary phase, these molecules cannot be continuously moved out of the separator via a fluid stream as in Figures 7.0.1(e)-(p). The higher the driving force on the solute molecules toward the adsorbent particle in the packed bed and the higher their affinity, the greater is their extent of disappearance from the feed stream. It is as if such molecules penetrate deeper into the stationary phase of particles due to their higher species-specific migration velocity perpendicular to the bulk fluid velocity. Correspondingly, when a desorption process is initiated, the slower will be the emergence of these molecules from the stationary phase into the moving fluid used to desorb. Multicomponent separation capability is an adjunct to such a phenomenon, as we will see in Section 7.1.5.

7.1 Chemical potential gradient based force in phase equilibrium: fixed-bed processes

All separations which involve the distribution equilibrium of a solute or solutes between a fluid phase and a solid phase are implemented *commonly* with a fixed bed of particles in a vessel. Although there are separation processes where the solid particles are moved through the separation device, they have somewhat limited use due to the difficulty of moving solid particles. The mobile fluid phase is either liquid or gaseous or supercritical. The solid phase may provide surface adsorption/desorption, ion exchange, or partitioning (e.g. between a liquid-phase coating on solid particles and the mobile phase). In leaching processes, the solid particles are mixtures of different species, one of which is extracted by the leaching solvent. The mobile phase can be a feed liquid or gaseous stream of binary or multicomponent nature. Alternatively, it can be a strip/sweep/eluent stream to strip/desorb/extract species from the solid phase. The mobile phase can be obtained also by stripping of the adsorbed species from the solid surface during bed regeneration (in gaseous feed systems).

The driving force for species i transfer from the mobile phase to the surface of the solid phase or from the solid phase surface to the mobile phase in adsorption or leaching processes is simply $-\nabla\mu_i$. Since the total pressures of both mobile and solid phases are equal (except in ion exchange resins), only $-\nabla\mu_i^0$ and the concentration gradient component of $-\nabla\mu_i$ are important. For a mobile gas phase, the partial pressure gradient is the relevant quantity. In ion exchange systems, the driving force is $-\nabla\mu_i^{el}$.

The systems considered here have, in the carrier fluid, at least one component, which is to be preferentially adsorbed onto or desorbed from particles in simple fixed-bed processes. The number of components in leaching is at least three: leaching solvent and two components in solid phase, one of which is to be preferentially extracted. All ion exchange systems have more than two species if we identify each ion as a separate species. Finally, chromatographic processes are concerned with true multicomponent systems to separate a number of species in the feed fluid sample from one another in the product stream, the carrier fluid.

7.1.1 Fixed-bed adsorption/desorption processes

A simplified analysis of either adsorption onto or desorption from adsorbent particles in a fixed bed is carried out here. The mobile phase may be liquid or gaseous. The separation objective may be purification of the carrier fluid or recovery of the species (one or more) present in the carrier fluid. The basis of separation is preferential adsorption/desorption or partitioning between the carrier fluid and the adsorbent surface phase. (A list of common adsorbents and their uses is given in Table 7.1.1.) The bulk

Table 7.1.1. Common adsorbents and their uses in fixed-bed processes[a]

Adsorbent	Applications	Comments
Activated alumina	drying warm gases or air	not very common
Activated carbon	solvents or odors from air, indoor air pollutants, organics from vent streams, gasoline vapor recovery in automobiles, decolorizing solutions, refining oils, organics from water in water purification	widely used
Activated clays	decolorizing petroleum products	
Bauxite	decolorizing petroleum products, drying gases	
Carbon molecular sieve (CMS) (or molecular sieve carbon, MSC)	production of N_2 from air by faster O_2 penetration through micropore	relatively recent
Silica gel	drying of air and gases (desiccant), preferential adsorption of polar compounds over nonpolar ones such as saturated hydrocarbons	common
Starch, cellulose, biomass solvents	dehydration of ethanol and other organics	relatively recent and limited use
Zeolites: a variety of crystalline silicates with pore diameters between 0.3 and 1 nm	dehydration of gases, dehydration of liquids (e.g. alcohol), n-paraffins from iso-paraffins, CO_2 removal from natural gas, desulfurization of natural gas and other streams, fructose/glucose separation, sulfur compounds from organics, N_2/O_2 separation	very widely used

[a] Yang (1987, 2003).

flow direction is perpendicular to the direction of the chemical potential driving force between the fluid and the particle in the packed bed.

7.1.1.1 *Fixed-bed adsorption: mobile feed liquid in axial flow*

The first separation operation considered in this mode is the adsorption separation of solutes between a mobile feed liquid and a stationary bed of adsorbent particles. The use of such an operation is widespread. Packed beds of powdered activated carbon (PAC) or granular activated carbon (GAC) are commonly used to reduce the toxicity of effluents obtained from biological treatment processes for wastewater.[1] Toxic organics are removed by adsorption on GAC or PAC. The packed bed of carbon is then regenerated thermally (by steam, for example) or by using a wet air oxidation process to oxidize the organics. Packed-bed adsorption from liquid feeds is also used on a large scale to decolorize aqueous sugar solutions and petroleum fractions, to remove moisture from gasoline and alcohol, and to fractionate aromatics from paraffinic hydrocarbons. A number of new adsorbents, e.g. carbon nanotubes, π-complexation sorbents (such as $AgNO_3/SiO_2$, $AgNO_3$/clays), Fe-Mn-Ti oxides, etc., are being developed for challenging applications (Yang, 2003).

Consider a solution of species i in a liquid feed (C_{i2}^0) entering a packed bed (Figure 7.1.1(a)) containing adsorbent particles suitable for adsorbing solute species i. If the adsorption of the solute species imparts a color to the adsorbent particles, then what we observe with time is as follows. The color near the liquid entrance of the bed has changed after some time. A color front is slowly moving toward the liquid exit with time. As more time passes, the color of an increasing length of the bed changes. Finally, after an additional amount of time, the color of the whole bed is changed. Suppose we had an opportunity to measure the concentration of species i in the effluent stream from the bed with time. We would find that for a long time from the beginning there is no trace of solute species i in the effluent; however, once the whole bed color has changed, there is suddenly quite a concentration of species i in the effluent. Obviously, at this time, the adsorbent bed is no longer useful. What is of interest from a process operation point of view is the determination of the time at which the solute species i broke through the adsorbent bed. Alternatively, one can determine the velocity of the color front in the bed; knowing the bed length, we can then predict the time for breakthrough of the solute in the effluent.

The solution of the problem of fixed-bed adsorption of a strongly adsorbed species i present in an inert mobile feed liquid is at hand when the concentration C_{i2} of the desired species in the liquid phase (moles of species i per unit volume of the liquid phase) is obtained as a function of time t and spatial coordinates x, y, z. Complexity of the fluid dynamics in a packed bed and the presence of convective dispersion require, however, considerable reduction in this goal for practical purposes. In the "pseudo-continuum" approach (Lee *et al.* (1977a); see Section 6.2.1.1), only the axial (mean flow direction, z-coordinate) variation is retained, i.e. we look only for a

[1] Remember, these adsorbent particles are highly porous, and thus adsorption takes place primarily on the inside surface of the pores (see Sections 3.3.7.6 and 3.4.1.4).

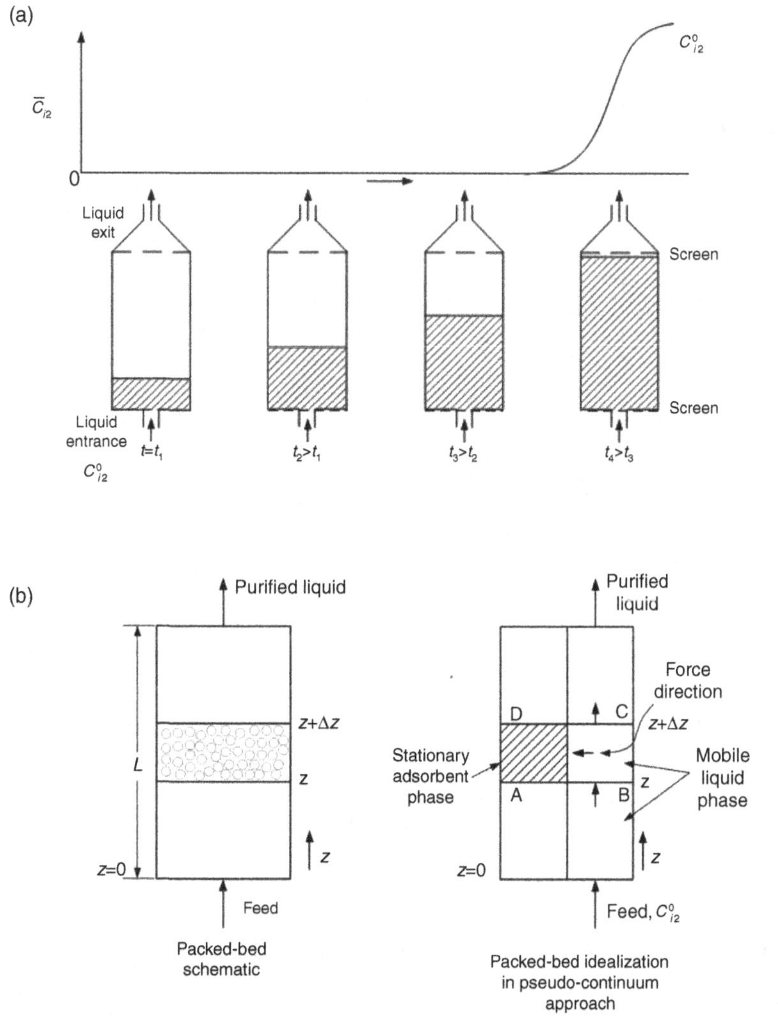

Figure 7.1.1. (a) Movement of the color front in the packed adsorbent bed with time and the corresponding concentration of the coloring species at the bed outlet; (b) packed-bed schematic and its idealization in pseudo-continuum approach for adsorption from a liquid. (After Lightfoot et al.*, 1962.)*

solution $\overline{C}_{i2}(z, t)$. This may be achieved by averaging over the packed-bed cross section in the x- and y-directions. However, to allow for species transport from the mobile phase to the stationary (fixed-bed particles) phase and vice versa, it is necessary to consider each phase separately without any x and y dependence, as shown in Figure 7.1.1(b) (Lightfoot *et al.*, 1962). This figure shows the cross section of only a small length, Δz, of the packed bed. *Note*: $\overline{C}_{i1}(z,t)$ represents the species i concentration per unit volume of the solid phase.

A mass balance on species i over the shell ABCD which spans the whole cross section of the packed bed of length Δz may now be carried out. The total bed cross-sectional area is S_c; the cross-sectional area for flow of fluid is εS_c. Here, ε is the void volume fraction of the packed bed[2] and the interstitial fluid velocity[3] in z-direction is v_z. The effective diffusion coefficient of solute i in the liquid in the z-direction is $D_{i,\text{eff},z}$. The particle surface area per unit particle volume is a_v. The particles may be porous, with a porosity ε_p. We will consider it later; for now, $\varepsilon_p = 0$. Referring to Figure 7.1.1(b), a species i mass balance may be written in the absence of any chemical reaction as

[2] In a capillary bundle model of the packed bed.

[3] From now on, we will use v_z instead of v_{tz} for the average fluid velocity.

$$\begin{pmatrix}\text{rate of accumulation}\\ \text{of species } i\\ \text{in volume ABCD}\end{pmatrix} = \begin{pmatrix}\text{rate of inflow}\\ \text{of species } i \text{ into}\\ \text{volume ABCD}\end{pmatrix} - \begin{pmatrix}\text{rate of outflow}\\ \text{of species } i \text{ out of}\\ \text{volume ABCD}\end{pmatrix}; \quad (7.1.1)$$

$$S_c\Delta z\left[\varepsilon\frac{\partial\overline{C}_{i2}}{\partial t}+(1-\varepsilon)\frac{\partial\overline{C}_{i1}}{\partial t}\right] = \varepsilon S_c\left[v_z\overline{C}_{i2}-D_{i,\text{eff},z}\frac{\partial\overline{C}_{i2}}{\partial z}\right]_z - \varepsilon S_c\left[v_z\overline{C}_{i2}-D_{i,\text{eff},z}\frac{\partial\overline{C}_{i2}}{\partial z}\right]_{z+\Delta z}. \quad (7.1.2)$$

The mobile phase has been identified here by $j = 2$ and the stationary phase by $j = 1$. The assumptions are: (1) species i enters the volume ABCD by convection and diffusion at location z; (2) species i leaves by similar mechanisms at location $z + \Delta z$; (3) the particle phase ($j = 1$) occupies a volume fraction $(1-\varepsilon)$ of the packed-bed volume $S_c\Delta z$. We now implement a limiting process where, as $\Delta z \to 0$, the differential equation

$$\frac{\partial\overline{C}_{i2}}{\partial t}+\frac{(1-\varepsilon)}{\varepsilon}\frac{\partial\overline{C}_{i1}}{\partial t}+\frac{\partial(v_z\overline{C}_{i2})}{\partial z}=\frac{\partial}{\partial z}\left[D_{i,\text{eff},z}\frac{\partial\overline{C}_{i2}}{\partial z}\right] \quad (7.1.3)$$

is obtained for $\overline{C}_{i2}$. *Note*: $\overline{C}_{i1}$ is the molar concentration of species i based on particle volume only. Conventionally, for liquid-phase systems, v_z is independent of z and so is $D_{i,\text{eff},z}$. Therefore

$$\frac{\partial\overline{C}_{i2}}{\partial t}+\frac{(1-\varepsilon)}{\varepsilon}\frac{\partial\overline{C}_{i1}}{\partial t}+v_z\frac{\partial\overline{C}_{i2}}{\partial z}=D_{i,\text{eff},z}\frac{\partial^2\overline{C}_{i2}}{\partial z^2}. \quad (7.1.4)$$

This is the starting differential equation for almost all liquid-phase based fixed-bed processes for a single adsorbable solute in an inert liquid under isothermal conditions with particles, where $\varepsilon_p = 0$. (*Note*: We could have obtained this equation directly from equation (6.2.35).) However, this equation does not take into account the transfer of species i from liquid phase 2 to adsorbent particle phase 1; such a relation between $\overline{C}_{i1}$ and $\overline{C}_{i2}$ is needed before a solution of $\overline{C}_{i2}(z,t)$ may be obtained. In terms of an overall mass-transfer coefficient K_{il} based on the liquid-phase concentration and the surface area of the particle (where a is the particle surface area per unit bed volume and a_v is the specific surface area of the particle, i.e. particle surface area per unit particle volume so that $a = a_v(1-\varepsilon)$), the mass-transfer rate per unit bed volume is

$$(1-\varepsilon)\frac{\partial\overline{C}_{i1}}{\partial t}=K_{il}a(\overline{C}_{i2}-\overline{C}^*_{i2}), \quad (7.1.5\text{a})$$

where $\overline{C}^*_{i2}$ is a hypothetical mobile-phase concentration in equilibrium with $\overline{C}_{i1}$. (We could have obtained this equation also from equation (6.2.32) for the stationary phase.) If the equilibrium relation is known, $\overline{C}^*_{i2}$ is easily expressed in terms of $\overline{C}_{i1}$. Thus, the solution for $\overline{C}_{i2}(z,t)$ can be obtained *in principle* for appropriate initial and boundary conditions imposed on the fixed bed of adsorbents. Extensive heat effects from adsorption will introduce additional complexity.

Species i is transferred from the mobile liquid phase to an adsorbed state on the particle surface via a number of steps: diffusion in the fluid phase around the particle, diffusion across the fluid–particle interface, diffusion in liquid in the pores of the particle, adsorption on available sites on the particle pore surface and surface diffusion, if any (see Sections 3.1.3.2.3/4 and 3.4.2.3/4). A fundamental approach would be to develop a species balance for such a particle phase and couple it with equation (7.1.4) via an additional mass-transfer relation for diffusion in the fluid phase around the particle. Such an approach, with some simplifications by Rosen (1952, 1954), has been illustrated at the very end of this section.

There are a number of other solutions primarily based on various assumed mechanisms of mass transfer. One such approach replaces the mass balance equation for diffusion within the particle by simplifying assumptions. In what is known as the *linear driving force* assumption, Glueckauf (1955b) suggested that, for $(D_{ip}t/r_p^2) > 0.1$,

$$\frac{\partial\overline{C}_{i1}}{\partial t}=\frac{15D_{ip}}{r_p^2}(\overline{C}^*_{i1}-\overline{C}_{i1}), \quad (7.1.5\text{b})$$

where $\overline{C}^*_{i1}$ is in equilibrium with the bulk liquid concentration $\overline{C}_{i2}$, and D_{ip} is the effective diffusion coefficient of species i in the pores of particle of radius r_p. A nonlinear driving force approximation has been suggested by Vermuelen (1953). In both models the liquid-phase resistance around the particle is neglected.

Before focusing on some of the simpler solutions of such a system to develop an understanding of the single solute separation capabilities of a fixed-bed adsorption process, we will provide a few more definitions for a packed adsorbent bed in general. The density of the solid material of the adsorbent particles is ρ_s. The bulk density of the adsorbent bed ρ_b is related to the void volume ε of the packed bed, solid material density ρ_s and the porosity of the particle ε_p via

$$\rho_b=\rho_s(1-\varepsilon_p)(1-\varepsilon). \quad (7.1.5\text{c})$$

This definition assumes the pores inside the particles to be empty. If the bulk density is defined with respect to particles containing a fluid phase $j = 2$ inside the pores, then

$$\rho_b=\rho_s(1-\varepsilon_p)(1-\varepsilon)+\varepsilon_p\rho_2, \quad (7.1.5\text{d})$$

where ρ_2 is the fluid-phase density. Unless specified, we will assume (7.1.5c) to be the valid relation for ρ_b.

We can now write down the differential equation for solute i in the manner of equation (7.1.3) for porous adsorbent particles whose pores contain the external solution (therefore $\overline{C}_{i2}$):

$$\varepsilon\frac{\partial\overline{C}_{i2}}{\partial t}+\varepsilon_p(1-\varepsilon)\frac{\partial\overline{C}_{i2}}{\partial t}+(1-\varepsilon)\frac{\partial\overline{C}_{i1}}{\partial t}+\varepsilon\frac{\partial(v_z\overline{C}_{i2})}{\partial z}=\varepsilon\frac{\partial}{\partial z}\left[D_{i,\text{eff},z}\frac{\partial\overline{C}_{i2}}{\partial z}\right]. \quad (7.1.5e)$$

This equation is based on a number of assumptions, one being that partitioning of solute i between the external solution and that in the pores of the adsorbent is such that the partitioning coefficient is 1. This is true when the solute size is small in relation to the adsorbent pore size (by about two orders of magnitude). An alternative form of this equation is as follows:

$$\left[1+\varepsilon_p\frac{(1-\varepsilon)}{\varepsilon}\right]\frac{\partial\overline{C}_{i2}}{\partial t}+\frac{(1-\varepsilon)}{\varepsilon}\frac{\partial\overline{C}_{i1}}{\partial t}+\frac{\partial(v_z\overline{C}_{i2})}{\partial z}=\frac{\partial}{\partial z}\left[D_{i,\text{eff},z}\frac{\partial\overline{C}_{i2}}{\partial z}\right]. \quad (7.1.5f)$$

When the solute size is larger (e.g. proteins) or the pore sizes are smaller, the partitioning of the solute between the mobile phase and the pore phase liquid (i.e. $\kappa_{im}=(C_{im}^p/C_{i2})$ as in (3.3.89a)) has to be taken into account:

$$\left[1+\frac{(1-\varepsilon)}{\varepsilon}\varepsilon_p\,\kappa_{im}\right]\frac{\partial\overline{C}_{i2}}{\partial t}+\frac{(1-\varepsilon)}{\varepsilon}\frac{\partial\overline{C}_{i1}}{\partial t}+\frac{\partial(v_z\overline{C}_{i2})}{\partial z}=\frac{\partial}{\partial z}\left[D_{i,\text{eff},z}\frac{\partial\overline{C}_{i2}}{\partial z}\right]. \quad (7.1.5g)$$

The simplest solution of a fixed-bed adsorption problem is provided by **isothermal equilibrium nondispersive operation** of the fixed bed. For single solute adsorption from the liquid phase to the adsorbent particles (where $\varepsilon_p=0$) under isothermal conditions,

(1) assume that the liquid-phase concentration $C_{i2}(z,t)$ in the column everywhere is locally in equilibrium with the solid-phase concentration $C_{i1}(z,t)$, and
(2) neglect the contribution of the axial diffusion and dispersion term, $D_{i,\text{eff},z}(\partial^2\overline{C}_{i2}/\partial z^2)$, in equation (7.1.4) (the plug flow assumption).

Define $\overline{q}_{i1}$ to be moles of species i in solid phase 1 per unit mass of solid phase. Then $\overline{C}_{i1}$ may be replaced by $\overline{q}_{i1}$ using the following relation:

$$\overline{C}_{i1}=\overline{q}_{i1}\rho_b/(1-\varepsilon);\qquad (1-\varepsilon)\frac{\partial\overline{C}_{i1}}{\partial t}=\rho_b\frac{\partial\overline{q}_{i1}}{\partial t}, \quad (7.1.6)$$

where ρ_b is the bulk density of the packed bed. Describe the equilibrium relation of the first assumption now by

$$\overline{q}_{i1}=q_i(\overline{C}_{i2}). \quad (7.1.7)$$

Using this and the second assumption in equation (7.1.4), we get

$$\left[1+\frac{\rho_b q_i'(\overline{C}_{i2})}{\varepsilon}\right]\frac{\partial\overline{C}_{i2}}{\partial t}+v_z\frac{\partial\overline{C}_{i2}}{\partial z}=0, \quad (7.1.8)$$

where $q_i'(\overline{C}_{i2})=\partial q_i(\overline{C}_{i2})/\partial\overline{C}_{i2}$. Note that relation (7.1.7) serves the purpose of relating $\overline{C}_{i2}$ to $\overline{C}_{i1}$, i.e. $\overline{q}_{i1}$, and we now have only one equation, (7.1.8), to solve to obtain $\overline{C}_{i2}(z,t)$. This equation is sometimes called the *De Vault equation* (De Vault, 1943). An alternative form of equation (7.1.8) is more convenient:

$$\frac{\partial\overline{C}_{i2}}{\partial t}+\frac{v_z}{\left[1+\dfrac{\rho_b q_i'(\overline{C}_{i2})}{\varepsilon}\right]}\frac{\partial\overline{C}_{i2}}{\partial z}=0. \quad (7.1.9)$$

A key question in fixed-bed adsorption separation is: what is the time needed for the fixed bed to become saturated if it is fed continuously and steadily with a feed of constant concentration C_{i2}^0? The feed initially introduced into the column (Figure 7.1.1(a)) displaces the liquid already present in the column. If the column particles did not have any species i to start with, the liquid displaced from the column and appearing at the outlet will be free of species i. Meanwhile, species i from the feed liquid will be adsorbed near the feed entry and, soon after, feed liquid free of species i will appear at the column outlet. As this process continues, the particle surfaces will become saturated with species i. Ultimately, all particles will lose their capacity of adsorbing species i; the feed solution of concentration C_{i2}^0 will appear at the column outlet. At this time, the adsorber is taken off the feed line and subjected to regeneration treatment so it may be used again for adsorption.

Obviously, the velocity with which the concentration C_{i2}^0 moves down the column (the concentration wave velocity) is less than that of the interstitial liquid. If one could ride with the wave having a concentration C_{i2}^0, one will only witness C_{i2}^0 around oneself; therefore, for any wave having a fixed value of specific concentration $\overline{C}_{i2}$, $d\overline{C}_{i2}=0$. Now, $\overline{C}_{i2}$ is a function of the independent variables z and t; therefore, for any small change in z and t, the change in $\overline{C}_{i2}$ is given by

$$d\overline{C}_{i2}=\left(\frac{\partial\overline{C}_{i2}}{\partial t}\right)_z dt+\left(\frac{\partial\overline{C}_{i2}}{\partial z}\right)_t dz. \quad (7.1.10)$$

For any particular concentration $\overline{C}_{i2}$, the concentration wave velocity must satisfy $d\overline{C}_{i2}=0$. Compare a reformulated (7.1.10) under such a condition,

$$\left(\frac{\partial\overline{C}_{i2}}{\partial t}\right)_z+\frac{dz}{dt}\left(\frac{\partial\overline{C}_{i2}}{\partial z}\right)_t=0, \quad (7.1.11)$$

with equation (7.1.9) to obtain (De Vault, 1943)

$$v_{Ci}^*=\frac{dz}{dt}=\frac{v_z}{\left[1+\dfrac{\rho_b q_i'(\overline{C}_{i2})}{\varepsilon}\right]}. \quad (7.1.12a)$$

Here, v_{Ci}^* is the velocity with which a wave of concentration $\overline{C}_{i2}$ travels along the column (the z-direction). For a given $\overline{C}_{i2}$, the above expression provides a unique relation between the location z along the column and time t since

$q_i'(\overline{C}_{i2})$ is fixed. The velocity v_{Ci}^* is variously called the *concentration wave velocity of species i*, the *migration rate of species i* or the *concentration front propagation velocity of species i* (Sherwood *et al.*, 1975). A more formal method of arriving at expression (7.1.12a) is given below.

This formal procedure is based on the method of characteristics (Aris and Amundson, 1973). If a solution for $\overline{C}_{i2}$ were available, we may write any change in $\overline{C}_{i2}$ as

$$d\overline{C}_{i2} = \left(\frac{\partial \overline{C}_{i2}}{\partial t}\right)_z dt + \left[\frac{\partial \overline{C}_{i2}}{\partial z}\right]_t dz = d\overline{C}_{i2}. \tag{7.1.12b}$$

Now one can solve for the two derivatives of $\overline{C}_{i2}(z,t)$ using equation (7.1.9):

$$\left(\frac{\partial \overline{C}_{i2}}{\partial t}\right)_z + \left(\frac{\partial \overline{C}_{i2}}{\partial z}\right)_t \frac{v_z}{\left[1+\dfrac{\rho_b q_i'(\overline{C}_{i2})}{\varepsilon}\right]} = 0.$$

The two expressions that result are

$$\left(\frac{\partial \overline{C}_{i2}}{\partial t}\right)_z = \frac{\begin{vmatrix} 0 & \dfrac{v_z}{\left[1+\dfrac{\rho_b q_i'(\overline{C}_{i2})}{\varepsilon}\right]} \\ d\overline{C}_{i2} & dz \end{vmatrix}}{\begin{vmatrix} 1 & \dfrac{v_z}{\left[1+\dfrac{\rho_b q_i'(\overline{C}_{i2})}{\varepsilon}\right]} \\ dt & dz \end{vmatrix}}; \tag{7.1.12c}$$

$$\left(\frac{\partial \overline{C}_{i2}}{\partial z}\right)_t = \frac{\begin{vmatrix} 0 & 1 \\ d\overline{C}_{i2} & dt \end{vmatrix}}{\begin{vmatrix} 1 & \dfrac{v_z}{\left[1+\dfrac{\rho_b q_i'(\overline{C}_{i2})}{\varepsilon}\right]} \\ dt & dz \end{vmatrix}}. \tag{7.1.12d}$$

The denominator (i.e. the coefficient determinant) in both expressions is zero under certain conditions. One such condition is

$$dz = \frac{v_z}{\left[1+\dfrac{\rho_b q_i'(\overline{C}_{i2})}{\varepsilon}\right]} dt \Rightarrow \frac{dz}{dt} = \frac{v_z}{\left[1+\dfrac{\rho_b q_i'(\overline{C}_{i2})}{\varepsilon}\right]}. \tag{7.1.12e}$$

If the denominator is zero, for the derivatives to be finite the numerator also has to be zero. From relation (7.1.12c), this implies

$$d\overline{C}_{i2} = 0. \tag{7.1.12f}$$

Result (7.1.12e) is valid if $\overline{C}_{i2}$ is constant along the characteristic line traced by (7.1.12e), which is the same as (7.1.12a).

It is clear from relation (7.1.12a) that the value of v_{Ci}^* for a given $\overline{C}_{i2}$ will depend on the value of $q_i'(\overline{C}_{i2})$, i.e. on the nature of the equilibrium relation $\overline{q}_{i1} = q_i(\overline{C}_{i2})$. Figure 7.1.2(a) shows two linear adsorption equilibrium isotherms for two different species $i = 1$ and $i = 2$. As shown, species $i = 1$ is more strongly adsorbed than species $i = 2$ and for $\overline{C}_{12} = \overline{C}_{22}$:

$$q'_1(\overline{C}_{12}) > q'_2(\overline{C}_{22}). \tag{7.1.13a}$$

Therefore, by relation (7.1.12a)

$$v_{C1}^* < v_{C2}^* \tag{7.1.13b}$$

in equilibrium nondispersive operation; the species which is more strongly adsorbed moves through the column more slowly and will take longer to appear at the column exit.

There is another way to visualize the migration velocity of a solute species down the column. When some extra solute is added to the packed-bed section of length Δz

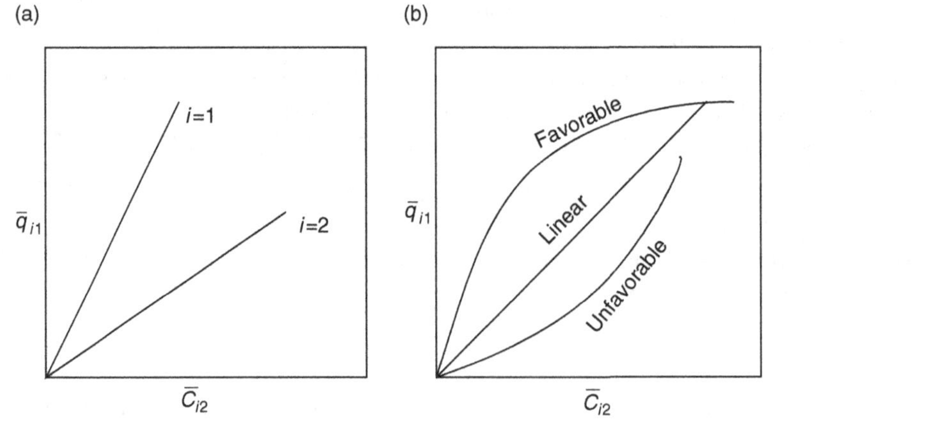

Figure 7.1.2. Adsorption isotherms: (a) two linear isotherms for two species i =1, 2; (b) three isotherms, favorable, linear and unfavorable.

which may already contain some solute (Figure 7.1.1(b)), the following will happen. Due to a higher solute concentration in the mobile phase, more solute will be adsorbed now on the adsorbent. Therefore the fraction of the extra solute added that remains in the mobile phase will be less than 1, reducing the mobile-phase concentration. Suppose the small increase in mobile-phase concentration and stationary-phase concentration of species i are, respectively, $\Delta\overline{C}_{i2}$ and $\Delta\overline{C}_{i1}$, then

$$\begin{pmatrix}\text{fraction of solute}\\ \text{added that remains}\\ \text{in the mobile phase}\end{pmatrix} = \frac{\varepsilon S_c \Delta\overline{C}_{i2}}{\varepsilon S_c \Delta\overline{C}_{i2} + (1-\varepsilon) S_c \Delta\overline{C}_{i1}} = \frac{1}{1+\left(\frac{1-\varepsilon}{\varepsilon}\right)\frac{\Delta\overline{C}_{i1}}{\Delta\overline{C}_{i2}}}.$$

From relation (7.1.6), this ratio is given by

$$\lim_{\Delta\overline{C}_i \to 0} = \frac{1}{1+\frac{\rho_b}{\varepsilon} q_i'(\overline{C}_{i2})}. \qquad (7.1.13c)$$

We now focus on beds with no dispersion and with a constant interstitial fluid velocity v_z (used to derive equation (7.1.12a)). Due to adsorption, the probability of any solute molecule being in the mobile phase after introduction into the control volume is reduced from 1 by the ratio given above. Thus the velocity with which these molecules migrate down the column is reduced from the fluid velocity v_z by the same fraction:

$$v_{Ci}^* = \frac{v_z}{1+\frac{\rho_b q_i'(\overline{C}_{i2})}{\varepsilon}}. \qquad (7.1.13d)$$

The mechanistic basis of the result (7.1.13b) is clearer from such an approach.

The result (7.1.13d) was based on ε being the void volume fraction occupied by the mobile phase in the packed bed. If the adsorbent particles are also porous with a void volume fraction of ε_p, and ρ_s is the actual density of the solid adsorbent particle material, then, using arguments as before, one can show that

$$v_{Ci}^* = \frac{v_z}{1+\frac{(1-\varepsilon)}{\varepsilon}\varepsilon_p + \frac{(1-\varepsilon)}{\varepsilon}(1-\varepsilon_p)\rho_s q'_{i1}(\overline{C}_{i2})}. \qquad (7.1.13e)$$

The denominator in (7.1.13e) has (unlike that in expression (7.1.13c)) three sources that contribute: the fraction of total solutes that remain in the mobile phase, the fraction of total solutes present in the liquid in the pores of the adsorbent and the fraction of total solutes adsorbed on the surfaces of the pores of the adsorbent and other outside surfaces. Note that $\rho_s(1-\varepsilon_p)(1-\varepsilon)$ is the bulk density, ρ_b, of the bed used in earlier expressions. The above expression is valid when the solute size is small with respect to the adsorbent pore size so that the solute concentration in the adsorbent pore liquid phase is the same as that in the mobile phase. For larger solute sizes, the partitioning of the solute between the mobile-phase and the pore-phase liquid has to be taken into account. The corresponding result will be (see equation (7.1.5g))

$$v_{Ci}^* = \frac{v_z}{1+\frac{(1-\varepsilon)}{\varepsilon}\varepsilon_p \kappa_{im} + \frac{(1-\varepsilon)}{\varepsilon}(1-\varepsilon_p)\rho_s q'_{i1}(\overline{C}_{i2})}. \qquad (7.1.13f)$$

This result indicates that larger solutes (larger with respect to the pore size) will have larger values of v_{Ci}^*.

The solute molecules of a particular species staying in the mobile phase move down the column with the fluid velocity v_z, whereas those that are adsorbed cannot move down the column. Thus, a given species concentration moves down the column at an effective velocity v_{Ci}^* lower than v_z. For a given species concentration, the higher the fraction of the species in the mobile phase, the higher its speed through the column.

Figure 7.1.2(b) illustrates a few more adsorption equilibrium isotherms. For the isotherm identified as *favorable*, it is obvious that if two concentrations $\overline{C}_{i2}|_1$ and $\overline{C}_{i2}|_2$ are significantly apart and $\overline{C}_{i2}|_2 > \overline{C}_{i2}|_1$, then

$$q'_{i2}(\overline{C}_{i2}|_2) < q'_{i2}(\overline{C}_{i2}|_1). \qquad (7.1.14a)$$

From (7.1.12a) and (7.1.13f),

$$v_C^*|_2 > v_C^*|_1. \qquad (7.1.14b)$$

Therefore, a higher solute concentration exits the column faster for a favorable isotherm in equilibrium nondispersive operation. For the isotherm identified as *unfavorable*, it is easily shown that an exactly reverse behavior holds.

These features help one to visualize the movement of trajectories of constant $\overline{C}_{i2}$ in the (z, t)-plane shown in Figure 7.1.3(a). These lines (called *characteristics*) which start at $z = 0$ represent different concentrations entering the column ($z = 0$) at different times. If a feed of constant concentration C_{i2}^0 enters the column, then all lines starting at $z = 0$ and $t \geq 0$ will be parallel; the slope of each line is equal to v_{Ci}^* for $\overline{C}_{i2} = C_{i2}^0$. The interstitial liquid at the column inlet is represented by the trajectory AB, whose slope is v_z, the liquid velocity. To the left of it lie the trajectories of liquid originally present in the column at different locations in the column (different z, $t = 0$). On the right side of line AB are characteristics of the feed liquid which have $\overline{C}_{i2} = 0$. The wave velocity of this zero concentration v_{Ci}^*(for $\overline{C}_{i2} = 0$) should provide the slope of these characteristics. If the adsorption isotherm is favorable (Figure 7.1.2(b)), $v_{Ci}^*|_{\overline{C}_{i2}=0} < v_{Ci}^*|_{\overline{C}_{i2}\neq 0}$. These intersect the characteristics for C_{i2}^0 along the line AS. At any point on line AS, characteristics having two different values $\overline{C}_{i2} = 0$ and $\overline{C}_{i2} = C_{i2}^0$ simultaneously exist. Line AS is like a *shock wave* having two different concentrations on two sides of a line, a discontinuity (Sherwood *et al.*, 1975). The point of intersection of this line of discontinuity AS with $z = L$, the column length, is crucial. The value of time t

corresponding to the coordinates (L, t) of this point defines the time when the bed is completely saturated. As shown in Figure 7.1.3(b), the exit concentration in the liquid suddenly jumps from $\overline{C}_{i2} = 0$ to $\overline{C}_{i2} = C^0_{i2}$ at this time; the feed solution breaks through the bed at this time, and the adsorption operation must stop. The sudden jump part of the square concentration wave and its subsequent constant value at C^0_{i2} is identified as the *column breakthrough curve.*

If the isotherm is linear instead of favorable, the characteristics for $C_{i2} = 0$ in between AB and AS in Figure 7.1.3(a) will be parallel to the characteristics for C^0_{i2}. Even then, when the characteristics for C^0_{i2} first hit $z = L$, there will be a concentration discontinuity from $C_{i2} = 0$ to $\overline{C}_{i2} = C^0_{i2}$ at the column outlet. For more details, see Wankat (1986, vol. I, pp 16–22).

To determine the value of time $t\,(=\bar{t})$ when the feed solution breaks through a column of length L, a solute mass balance is carried out over the column from time $t = 0$ to the time t of breakthrough. For the sake of generality, the mass balance is carried out in a column which may have some solute present in the column at $t = 0$, i.e. $\overline{C}_{i2}(z, 0)$ and $\overline{C}_{i1}(z, 0)$ are nonzero. Further, $\varepsilon_p \neq 0$. Before the feed concentration breaks through the column end in the liquid effluent, the liquid phase everywhere in the column would have C^0_{i2}; thus the total number of moles of solute in the column at time t of breakthrough is given by

$$LS_c\left[\varepsilon C^0_{i2} + (1-\varepsilon)\,\varepsilon_p\, C^0_{i2}\,\kappa_{im} + (1-\varepsilon)C^0_{i1}\right]$$
$$= LS_c\left[\varepsilon C^0_{i2} + (1-\varepsilon)\,\varepsilon_p\, C^0_{i2}\,\kappa_{im} + \rho_b q_{i1}(C^0_{i2})\right].$$

This must equal the number of moles introduced into the column by the steady inflow of feed from time $t = 0$ to time t for breakthrough, i.e. $\varepsilon v_z S_c t C^0_{i2}$ plus the moles of solute originally present in the column,

$$S_c\int_0^L \left\{\varepsilon\overline{C}_{i2}(z,0) + (1-\varepsilon)\,\varepsilon_p\,\overline{C}_{i2}(z,0)\,\kappa_{im} + (1-\varepsilon)\,\overline{C}_{i1}(z,0)\right\} \mathrm{d}z.$$

The solute balance relation is therefore

$$\varepsilon v_z S_c t C^0_{i2} + S_c\int_0^L \left\{\varepsilon\overline{C}_{i2}(z,0) + (1-\varepsilon)\,\varepsilon_p\,\overline{C}_{i2}(z,0)\,\kappa_{im} + (1-\varepsilon)\overline{C}_{i1}(z,0)\right\} \mathrm{d}z \qquad (7.1.15\mathrm{a})$$
$$= LS_c\left[\varepsilon C^0_{i2} + (1-\varepsilon)\,\varepsilon_p\, C^0_{i2}(z,0)\,\kappa_{im} + \rho_b q_i(C^0_{i2})\right].$$

For the special case studied so far, $\overline{C}_{i2}(z, 0) = 0 = \overline{C}_{i1}(z, 0)$; in such a case,

$$t = \frac{L}{v_z}\left[1 + \frac{(1-\varepsilon)}{\varepsilon}\,\varepsilon_p\,\kappa_{im} + \frac{\rho_b q_i(C^0_{i2})}{\varepsilon C^0_{i2}}\right]. \qquad (7.1.15\mathrm{b})$$

Alternatively, the direct relation between the column length and the time for breakthrough for a liquid feed of constant concentration C^0_{i2} fed at a constant interstitial velocity v_z is

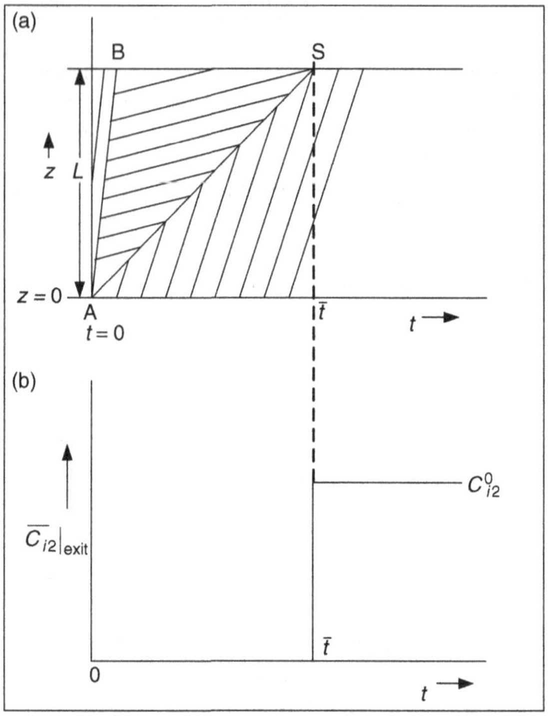

Figure 7.1.3. (a) Characteristics for feed concentration C^0_{i2} for adsorption in the (z, t)-plane for the equilibrium nondispersive model for a favorable isotherm. (b) Breakthrough curve for equilibrium nondispersive model with constant feed concentration in an initially empty column.

$$\frac{L}{t} = \frac{v_z}{\left[1 + \dfrac{(1-\varepsilon)}{\varepsilon}\,\varepsilon_p\,\kappa_{im} + \dfrac{\rho_b\, q_i(C^0_{i2})}{\varepsilon\quad C^0_{i2}}\right]}$$
$$= \frac{v_z}{\left[1 + \dfrac{(1-\varepsilon)}{\varepsilon}\,\varepsilon_p\,\kappa_{im} + \dfrac{(1-\varepsilon)}{\varepsilon}\dfrac{C^0_{i1}}{C^0_{i2}}\right]} = \frac{L}{\bar{t}} = v^*_{Ci}|_{\mathrm{shock}}. \qquad (7.1.15\mathrm{c})$$

This also happens to characterize the square wave front of concentration C^0_{i2}. For time $t < t$ (breakthrough) $= \bar{t}$, the location of the front will be inside the column where $z < L$; for such times, relation (7.1.15c) should be expressed as (z/t) instead of $(L/\bar{t})$. A more general expression for $\bar{t}$ or $v^*_{Ci}|_{\mathrm{shock}}$ may be derived for the case where $C^0_{i2}(z,0) = C^i_{i2}$ and $C^0_{i1}(z,0) = C^i_{i1}$ (in equilibrium with C^i_{i2}). The relation equivalent to (7.1.15a) is

$$\varepsilon\, v_z\, S_c t C^0_{i2} + L\,S_c\left[\varepsilon\, C^i_{i2} + (1-\varepsilon)\,\varepsilon_p\, C^i_{i2}\,\kappa_{im} + (1-\varepsilon)\, C^i_{i1}\right]$$
$$= L\,S_c\left[\varepsilon\, C^0_{i2} + (1-\varepsilon)\,\varepsilon_p\, C^0_{i2}\,\kappa_{im} + \rho_b\, q_i(C^0_{i2})\right]. \qquad (7.1.15\mathrm{d})$$

On rearranging, one can obtain

$$\frac{L}{t} = \frac{L}{\bar{t}} = v^*_{Ci}|_{\text{shock}}$$
$$= \frac{v_z}{\left[\left(1-\frac{C^i_{i2}}{C^0_{i2}}\right)\left(1+\frac{(1-\varepsilon)}{\varepsilon}\varepsilon_p\kappa_{im}\right)+\frac{(1-\varepsilon)}{\varepsilon}\frac{C^0_{i1}-C^i_{i1}}{C^0_{i2}}\right]}. \tag{7.1.15e}$$

Is this operation of fixed-bed adsorption with bulk liquid phase flow perpendicular to the direction of the chemical potential gradient force from the liquid to the solid beneficial for separation when compared with just plain equilibration between the feed liquid and the adsorbent without any bulk flow in a given direction?

There are a number of ways of looking at this issue. *First*, the bulk liquid-phase flow perpendicular to force produces a very substantial volume of absolutely solute-free feed liquid at the column exit. Simple equilibration in a batch process in a vessel will never yield solute-free liquid. For example, feed solution of concentration C^0_{i2} will be reduced in plain batch equilibration to $\overline{C}_{i2}$, determined by (for $\varepsilon_p = 0$)

$$\varepsilon S_c v_z t C^0_{i2} = \varepsilon S_c v_z t \overline{C}_{i2} + (1-\varepsilon) S_c v_z t \overline{C}_{i1}, \tag{7.1.16a}$$

which leads to

$$\overline{C}_{i2} = \frac{C^0_{i2}}{1+\frac{(1-\varepsilon)}{\varepsilon}\kappa_{i1}}. \tag{7.1.16b}$$

Here, κ_{i1} is the distribution coefficient of solute i between phases 1 and 2 at equilibrium, i.e. $\kappa_{i1} = \overline{C}_{i1}/\overline{C}_{i2}$. The fraction of the total solute i present in the liquid phase is obtained from the ratio R'_i,

$$R'_i = \frac{\varepsilon S_c v_z t \overline{C}_{i2}}{\varepsilon S_c v_z t \overline{C}_{i2} + (1-\varepsilon) S_c v_z t \overline{C}_{i1}} = \frac{\varepsilon \overline{C}_{i2}}{\varepsilon \overline{C}_{i2} + (1-\varepsilon)\overline{C}_{i1}}$$
$$= \frac{1}{1+\frac{(1-\varepsilon)}{\varepsilon}\frac{\overline{C}_{i1}}{\overline{C}_{i2}}} = \frac{1}{1+(m_{i1}/m_{i2})} = \frac{1}{1+k'_{i1}}. \tag{7.1.16c}$$

Thus, the purification capability of this pattern of bulk flow vs. force is excellent for plain batch equilibration.

Second, one can compare the total number of moles transferred to the adsorbent phase in rival modes of operation. If C^0_{i1} is the solid-phase concentration in equilibrium with the influent concentration C^0_{i2}, then the moles of species i transferred to the solid adsorbent phase per unit volume of adsorbent is $(C^0_{i1} - 0)$ for an initially solute-free adsorbent bed when the bulk flow is perpendicular to the force. A batch operation without bulk flow will transfer according to relation (7.1.16a) only $(\overline{C}_{i1} - 0)$ moles per unit volume of adsorbent, where $\overline{C}_{i1} < C^0_{i1}$. Thus, bulk flow of feed along the bed length perpendicular to the force direction achieves a better utilization of the intrinsic adsorption capacity of the adsorbents.

Third, the longer the bed, the larger the volume of feed liquid that can be purified almost completely. A batch adsorption process lacks any such feature.

The volume $\overline{V}$ of feed solution needed to be passed through the column of length L so as to saturate it with C^0_{i2} under equilibrium nondispersive mode of operation is:

$$\overline{V}C^0_{i2} = S_c L\left[\varepsilon C^0_{i2} + (1-\varepsilon)\varepsilon_p C^0_{i2}\kappa_{im} + (1-\varepsilon)C^0_{i1}\right]$$
$$= S_c L\left[\varepsilon C^0_{i2} + (1-\varepsilon)\varepsilon_p C^0_{i2}\kappa_{im} + \rho_b q_i(C^0_{i2})\right];$$

$$\overline{V} = S_c L\left[\varepsilon + (1-\varepsilon)\varepsilon_p\kappa_{im} + \frac{\rho_b q_i(C^0_{i2})}{C^0_{i2}}\right]. \tag{7.1.16d}$$

It will be found later that $\overline{V}$ is a useful quantity in the study of the actual separation behavior of real columns which may not satisfy equilibrium nondispersive conditions. The value of t corresponding to $\overline{V}$ is $\bar{t}$, and it is obtained from the expression (7.1.15c) for t.

The concentration wave front moving out of a column in equilibrium nondispersive operation need not be a square wave front as shown in Figure 7.1.3(b). The square wave front resulted from an initially solute-free column and a constant influent concentration C^0_{i2}. Suppose the column has in its entry region a linear distribution of solute concentration from C^0_{i1} to 0, from column inlet to some distance into the column. The liquid in immediate contact is in equilibrium, and its concentration then varies from C^0_{i2} to 0, as shown in Figure 7.1.4. As the feed liquid enters the column, concentration waves having values from C^0_{i2} to 0 will be moving down the column due to this initial column loading. The concentration wave velocity of each concentration will depend on the value of $q'_i(\overline{C}_{i2})$ for the particular $\overline{C}_{i2}$, which in turn will depend on the nature of the adsorption isotherm.

Consider first the "favorable isotherm" of Figure 7.1.2(b). This type of isotherm is characterized by $\left(\mathrm{d}^2 q_i/\mathrm{d}\overline{C}^2_{i2}\right) < 0$ for all values of $\overline{C}_{i2}$. For such an isotherm, if the values of $\overline{C}_{i2}$ at two locations z_1 and z_2 ($> z_1$) are such that $\overline{C}_{i2}|_{z_1} > \overline{C}_{i2}|_{z_2}$, then $q'_i(\overline{C}_{i2})|_{z_1} < q'_i(\overline{C}_{i2})|_{z_2}$ and $v^*_{Ci}|_{z_2} < v^*_{Ci}|_{z_1}$. Therefore, the velocity v^*_{Ci} of the higher concentration nearer the column inlet is higher than that of the lower concentration further from the column inlet. After some time t (Figure 7.1.4), an entirely different concentration profile will be observed along the column. The concentration profile has become quite sharp, almost like a square wave, since the higher concentrations have moved much faster and have caught up with the much slower moving lower concentrations further down. This is identified as the *self-sharpening wave front*. When C^0_{i2} is zero, that is pure solvent (called sometimes eluent) comes in, the desorption behavior of the solute is often called *elution*.

We would now like to focus on calculating such *elution behavior* from a preloaded column (Figure 7.1.4). The nondispersive isothermal equilibrium operation of the packed adsorption bed described by equation (7.1.9) may be written using the following new independent variable s (instead of time, t),

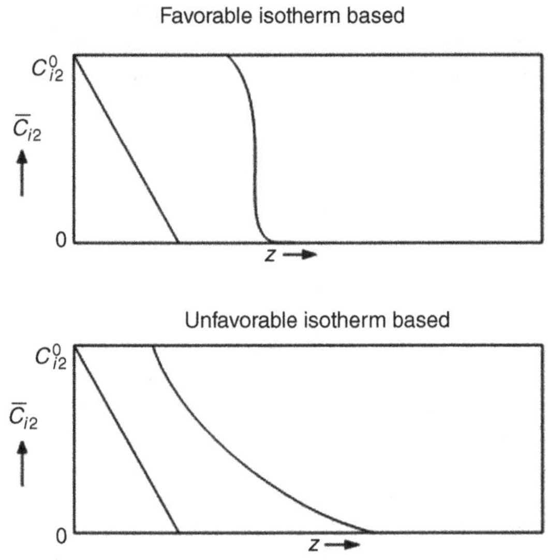

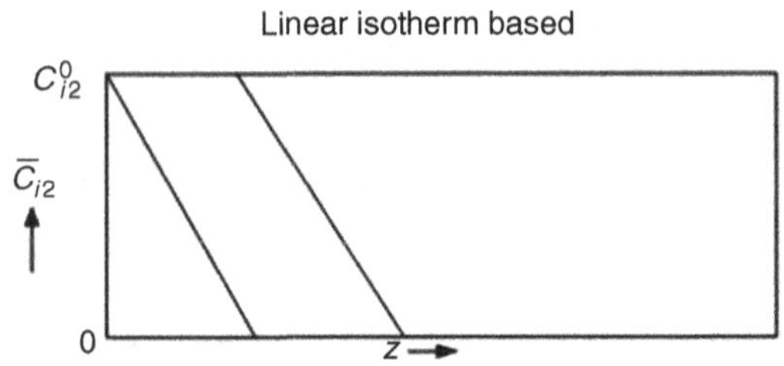

Figure 7.1.4. Movement of concentration profile along the column for three different types of isotherms when the initial column section is loaded linearly with solute from $\overline{C}_{i2} = C^0_{i2}$ at z = 0 to $\overline{C}_{i2} = 0$ some distance down.

$$(\mathrm{d}s/\mathrm{d}t) = \varepsilon v_z \tag{7.1.17a}$$

(where s is the volume of solution fed to the column per unit empty cross section of the bed (Lightfoot *et al.*, 1962)) as (for $\varepsilon_p = 0$)

$$\frac{\partial \overline{C}_{i2}}{\partial z} + \left[\varepsilon + \rho_b q_i'(\overline{C}_{i2})\right] \frac{\partial \overline{C}_{i2}}{\partial s} = 0, \tag{7.1.17b}$$

since $s = \int_0^t \varepsilon v_z \,\mathrm{d}t = \varepsilon v_z t$ (for a constant feed flow rate).

In general, we may write

$$\mathrm{d}\overline{C}_{i2} = (\partial \overline{C}_{i2}/\partial z)\mathrm{d}z + (\partial \overline{C}_{i2}/\partial s)\mathrm{d}s. \tag{7.1.17c}$$

Following the procedure used in equations (7.1.12b) and (7.1.12e), we get

$$\left.\frac{\partial s}{\partial z}\right|_{\overline{C}_{i2}} = \left[\varepsilon + \rho_b q_i'(\overline{C}_{i2})\right]. \tag{7.1.17d}$$

(*Note*: Equation (7.1.12e) for v^*_{Ci} continues to be valid for any $\overline{C}_{i2}$.)

For a constant feed concentration, $\overline{C}_{i2}(0, t)$, we can integrate this equation to obtain

$$\left\{s/\left[\varepsilon + \rho_b q_i'(\overline{C}_{i2})\right]\right\} = z - z_0(\overline{C}_{i2}), \tag{7.1.17e}$$

where $z_0(\overline{C}_{i2})$ is the distance corresponding to any $\overline{C}_{i2}$ at the start of the operation. We may rewrite this relation as

$$\frac{s}{z - z_0(\overline{C}_{i2})} = \left[\varepsilon + \rho_b q_i'(\overline{C}_{i2})\right]. \tag{7.1.17f}$$

This relation may be interpreted as follows. Consider a column of cross-sectional area S_c. The volume of solution being passed, V, is given by sS_c. Let the volume of solution passed before elution is started be V^o. Then the elution solution volume passed is $V - V^o$. If $z_0(\overline{C}_{i2}) = 0$, we obtain from relation (7.1.17f)

$$V - V^0 = S_c(z - z_0(\overline{C}_{i2}))\left[\varepsilon + \rho_b q_i'(\overline{C}_{i2})\right]$$
$$= S_c z\left[\varepsilon + \rho_b q_i'(\overline{C}_{i2})\right] = S_c z\left[\varepsilon + (1-\varepsilon)\frac{\mathrm{d}\overline{C}_{i1}}{\mathrm{d}\overline{C}_{i2}}\right]. \tag{7.1.17g}$$

This expression allows us to calculate the eluent volume V that has to be passed corresponding to a given $\overline{C}_{i2}$ at the column outlet.

If the isotherm is a simple linear one in Figure 7.1.2(b) (therefore $\left(\mathrm{d}^2 q_i/\mathrm{d}\overline{C}^2_{i2}\right) = 0$ for all $\overline{C}_{i2}$), then all concentrations at all locations in the initial part of the column (Figure 7.1.4), have the same wave velocity. As time progresses, the same linear concentration profile is transported down the column without any change in shape. On the other hand, if the isotherm is unfavorable $\left(\left(\mathrm{d}^2 q_i/\mathrm{d}\overline{C}^2_{i2}\right) > 0\right)$ in Figure 7.1.2(b)), lower concentrations further down from the inlet of the column will have higher v^*_{Ci} than higher concentrations near the beginning of the column inlet. This will lead to a more spread out profile along the column as time increases and concentrations move down the column. Such a condition is identified as a *dispersive* (or *diffusive*) *wave front*.

So far, the adsorption based purification of a liquid feed flowing down a bed of adsorbent particles has been analyzed using the equilibrium nondispersive approximation. For a liquid feed of constant concentration C^0_{i2} and a bed initially free of any solute i, such an approximation suggests a square concentration wave exiting the column end at time $\overline{t}$ defined by result (7.1.15c). Alternatively, a volume $\overline{V}$ (given by relation (7.1.16d)) of feed liquid has to pass through the column before the feed solution concentration C^0_{i2} suddenly breaks through. In reality, the column effluent concentration has more of an S-shape (Figure 7.1.5(a)) than a square wave. This requires

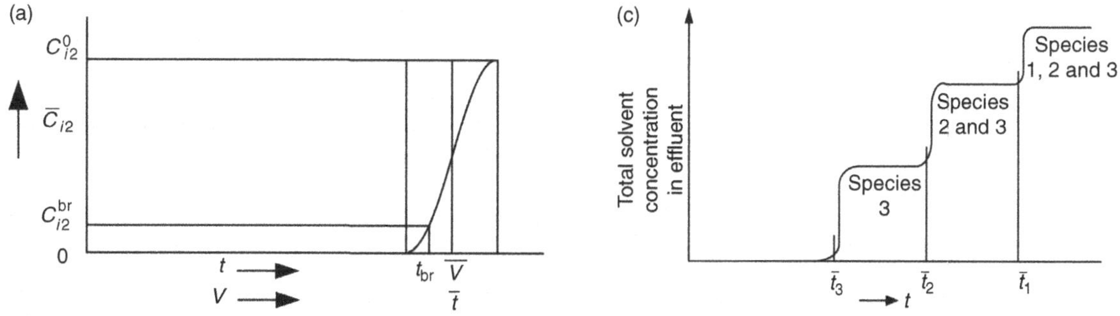

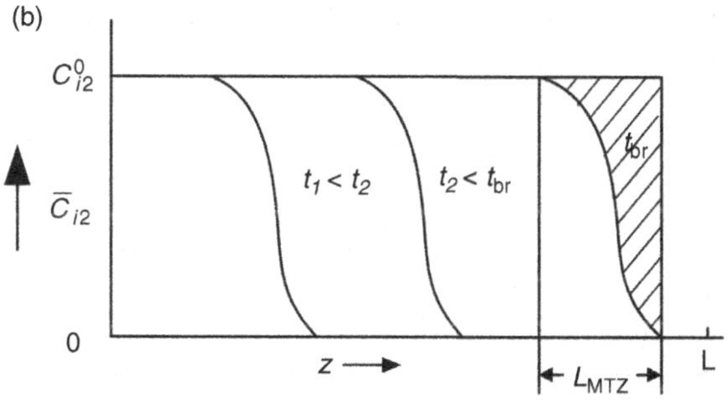

Figure 7.1.5. (a) Effluent concentration profile from a fixed bed (breakthrough curve) for one solute i in feed. (b) Column concentration profile at three different times t_1, t_2 (>t_1) and t_{br} for one solute i in feed. (c) Breakthrough curve from a fixed bed for a feed containing three solutes i = 1, 2, 3: species 3 is least strongly adsorbed; species 1 is most strongly adsorbed.

termination of the adsorption operation at time t_{br} less than that ($\overline{t}$) defined by (7.1.15c); the volume V_{br} of purified liquid obtained would be less than $\overline{V}$. The *breakthrough concentration* of the solute C^{br}_{i2} at t_{br} is determined by the nature of the operation. This value is usually around 0.05–0.1 of C^0_{i2}; when the liquid is to be purified, its value is much lower. Such a diffuse breakthrough of solute instead of a sharp front may be caused by a lack of equilibrium between the two phases or by the presence of axial diffusion and dispersion, or both. The lack of equilibrium between the two phases may come about due to a finite mass-transfer resistance in the liquid film surrounding the particle or due to diffusional resistances in the pores of the particle, or both.

If one were to determine the solute concentration profile in the liquid phase in the packed bed at any time prior to breakthrough, one will find a shape inverse to that shown in Figure 7.1.5(a). Figure 7.1.5(b) illustrates two bed solute profiles at two instants of time t_1 and $t_2(> t_1)$ that are far apart. In both profiles, the initial section of the profile has a constant concentration C^0_{i2} corresponding to the feed liquid coming in; the bed is saturated here and no longer has the capacity to adsorb any more solute. The final sections of the profiles in each case have a value of $\overline{C}_{i2} = 0$; they are without any solute and retain their intrinsic capacity for solute adsorption. In between, each profile changes from $\overline{C}_{i2} = C^0_{i2}$ to 0; this section of the bed has some solute adsorption capacity left. It is obvious that in the first section, where $\overline{C}_{i2} = C^0_{i2}$, solute transfer is no longer going on, although some occurred at an earlier time; in the final section, with $\overline{C}_{i2} = 0$, no mass transfer has started yet. Thus, mass transfer occurs only in the intermediate region: this region of the bed is called the *adsorption zone* or the *mass-transfer zone* (MTZ). As time increases, this adsorption zone travels through the bed; finally, it shows up in the bed effluent as concentration increasing with time (Figure 7.1.5(a)). The length of this zone is identified as L_{MTZ}.

A number of analyses have been developed to account for such features in real fixed-bed adsorption processes. Two such analyses and their major results will be briefly identified below. A *linear equilibrium model with dispersion* is considered first. Lapidus and Amundson (1952) solved equation (7.1.4) which included the axial dispersion term under the following assumptions:

(1) local equilibrium exists everywhere between the liquid and solid phase and it is linear, e.g.

$$\overline{C}_{i1} = \kappa_{i1}\overline{C}_{i2}, \qquad (7.1.18a)$$

where κ_{i1} is a constant;

(2) the column is infinitely long.

The conditions used for the solution are: a steady liquid feed of constant inlet concentration C^0_{i2}, uniform solute concentration of C^i_{i2} and C^i_{i1} (at equilibrium) throughout the bed for time $t \leq 0$. Define a new dependent variable

$$\theta_i(z,t) = (\overline{C}_{i2}(z,t) - C^i_{i2})/(C^0_{i2} - C^i_{i2}). \qquad (7.1.18b)$$

Under the assumption of local equilibrium, we can rewrite equation (7.1.4) as

$$\left[1+\frac{(1-\varepsilon)}{\varepsilon}\frac{C^0_{i1}}{C^0_{i2}}\right]\frac{\partial\theta_i(z,t)}{\partial t} + v_z\frac{\partial\theta_i(z,t)}{\partial z} = D_{i,\text{eff},z}\frac{\partial^2\theta_i(z,t)}{\partial z^2}. \qquad (7.1.18c)$$

The boundary and initial conditions are as follows:

$$z=0, \theta_i = 1 \text{ for } t>0; \qquad z=\infty, \theta_i = 0 \text{ for } t>0;$$
$$t=0,\ \theta_i = 0 \text{ for } z>0. \qquad (7.1.18d)$$

The solution for equation (7.1.18c) under the boundary and initial conditions is

$$\theta_i = \frac{1}{2}\left[1+\text{erf}\left\{\sqrt{\frac{v_z^2 t}{4D_{i,\text{eff},z}\gamma}} - z\sqrt{\frac{\gamma\varepsilon}{4t\varepsilon D_{i,\text{eff},z}}}\right\} + \left\{\exp\frac{v_z z}{D_{i,\text{eff},z}}\right\} \times \text{erfc}\left\{\sqrt{\frac{v_z^2 t}{4D_{i,\text{eff},z}\gamma}} + z\sqrt{\frac{\gamma\varepsilon}{4t\varepsilon D_{i,\text{eff},z}}}\right\}\right], \qquad (7.1.18e)$$

where

$$\gamma = (1 + ((1-\varepsilon)/\varepsilon)\, C^0_{i1}/C^0_{i2}). \qquad (7.1.18f)$$

The erfc quantity is a small number, and under conditions where $(v_z z/D_{i,\text{eff},z})$ is not too large it may be neglected. Then we get

$$\theta_i = \frac{1}{2}\left[1+\text{erf}\left\{\frac{v_z t}{\sqrt{4D_{i,\text{eff},z}t\gamma}} - z\sqrt{\frac{\gamma}{4t\,D_{i,\text{eff},z}}}\right\}\right], \qquad (7.1.18g)$$

which may be rearranged to yield

$$\theta_i = \frac{\overline{C}_{i2}(z,t) - C^i_{i2}}{C^0_{i2} - C^i_{i2}} = \frac{1}{2}\left\{1+\text{erf}\left[\frac{Pe^{1/2}_{z,\text{eff}}(V-\overline{V})}{2(V\overline{V})^{1/2}}\right]\right\}, \qquad (7.1.18h)$$

where $V = \varepsilon v_z S_c t$ is the volume of feed solution which has passed through the column of length z from $t = 0$ to time t,

$$\overline{V} = S_c z[\varepsilon + (1-\varepsilon)\kappa_{i1}] = S_c z\left[\varepsilon + (1-\varepsilon)\frac{C^0_{i1}}{C^0_{i2}}\right] = \varepsilon v_z S_c \overline{t} \qquad (7.1.18i)$$

is the volume of solution required to saturate a column of length z in the absence of dispersion, and $Pe_{z,\text{eff}} = (z v_z/D_{i,\text{eff},z})$ is a Péclet number based on the effective axial diffusion coefficient.

Equation (7.1.18e) provides the liquid-phase solute concentration in an infinitely long column at location z from the inlet. Although the solution was developed for an infinitely long column, the value of the concentration profile at any z will also provide the output concentration profile for a column of length z. The truncated form[4] (7.1.18h) of the solution (7.1.18e) represents the complete solution accurately, provided $Pe_{z,\text{eff}}$ is not too large (Lapidus and Amundson, 1952). Sometimes it is desirable to have the solution (7.1.18h) expressed in terms of t and $\overline{t}$ (remember $\text{erf}(-x) = -\text{erf}(x)$):

$$\frac{\overline{C}_{i2}(z,t) - C^i_{i2}}{C^0_{i2} - C^i_{i2}} = \frac{1}{2}\left\{1-\text{erf}\left[\frac{1-(t/\overline{t})}{2\left(D_{i,\text{eff},z}t/v_z z\overline{t}\right)^{1/2}}\right]\right\}. \qquad (7.1.18j)$$

Lapidus and Amundson (1952) have shown that, as the value of v_z increases or $D_{i,\text{eff},z}$ decreases (increasing $Pe_{z,\text{eff}}$), the S-shaped breakthrough curve approaches the square wave form of nondispersive operation. On the other hand, if we consider the extent of dispersion around the center of the profile $\overline{t}$ and $\overline{V}$, it varies as $t^{1/2}$ or $V^{1/2}$: the extent of dispersion increases with time or volume. We will employ equation (7.1.18h) sometimes to describe the breakthrough.

There are a number of models which do not assume the existence of equilibrium between the liquid and the solid phases. However, they do not incorporate the effect of axial dispersion. In such models of *nonequilibrium nondispersive operation* of the column, the mass-transfer rate between the liquid and the surface of the adsorbent (primarily in the pores) is not infinitely fast; rather it is finite. Further diffusion in the porous adsorbent particle is quite important. One of the earlier models of this type is by Rosen (1952, 1954).

Description of the mass-transfer rate between the liquid and the adsorbent particle is facilitated by a specification of the particle geometry. Relatively large particles are often used in industrial adsorbents to reduce pressure drops. Rosen considered spherical particles; the mass balance equation for the concentration, C^p_{i2}, of species i in the pore liquid of such an adsorbent particle (where the effective diffusion coefficient of species i is D_{ip}) is

$$\frac{\partial C^p_{i2}}{\partial t} = D_{ip}\left(\frac{1}{r^2}\frac{\partial}{\partial r}\left(r\frac{\partial C^p_{i2}}{\partial r}\right)\right) \qquad (7.1.19a)$$

if spherical symmetry is assumed. (See Table 6.2.1; $v = 0$; $J^*_{i\theta} = J^*_{i\phi} = 0$.) In adsorption, the rate of transfer of species i from the liquid to this particle can be described by means of a mass-transfer coefficient k_c and a particle-average concentration $\overline{C}^p_{i2}$ of species i for particles of radius r_p:

$$\overline{C}^p_{i2}\left(\frac{4}{3}\pi r_p^3\right) = \int_0^{r_p} C^p_{i2}4\pi r^2\,dr \quad\Rightarrow\quad \overline{C}^p_{i2} = \frac{3}{r_p^3}\int_0^{r_p} C^p_{i2}r^2\,dr; \qquad (7.1.19b)$$

[4]This solution is often the basis for what is called *linear chromatography* based on *linear adsorption equilibrium* (7.1.18a).

$$\left(\frac{4}{3}\pi r_p^3\right)\frac{\partial \overline{C}_{i2}^p}{\partial t} = k_c 4\pi r_p^2\left(C_{i2} - C_{i2}^p|_{r=r_p}\right)$$
$$\Rightarrow \frac{\partial \overline{C}_{i2}^p}{\partial t} = \frac{3k_c}{r_p}\left(\overline{C}_{i2} - C_{i2}^p|_{r=r_p}\right). \quad (7.1.19c)$$

Rosen had assumed linear adsorption equilibrium between the liquid-phase concentration in the pore, C_{i2}^p, and the solid-phase concentration in the pore, C_{i1}^p:

$$C_{i1}^p = \kappa_{i1} C_{i2}^p.$$

A particle-average concentration $\overline{C}_{i1}$ based on the particle volume is then obtained simply from definitions (7.1.19b) as

$$\overline{C}_{i1} = \frac{3}{r_p^3}\int_0^{r_p} \kappa_{i1} C_{i2}^p r^2\, dr. \quad (7.1.19d)$$

These four equations, along with the form of equation (7.1.4) without any axial dispersion, namely

$$\frac{\partial \overline{C}_{i2}}{\partial t} + \frac{(1-\varepsilon)}{\varepsilon}\frac{\partial \overline{C}_{i1}}{\partial t} + v_z \frac{\partial \overline{C}_{i2}}{\partial z} = 0, \quad (7.1.19e)$$

have been solved by Rosen for the following conditions:

$$\text{at } t = 0, C_{i2}^p = 0 \text{ for } z > 0 \text{ and } 0 \le r \le r_p;$$
$$\text{for } t \ge 0, \text{at } z = 0, \overline{C}_{i2} = \overline{C}_{i2}^0. \quad (7.1.19f)$$

The solution is obtained as an integral:

$$\frac{\overline{C}_{i2}}{C_{i2}^0} = \frac{1}{2} + \frac{2}{\pi}\int_0^\infty \exp\{-\eta H_1(\lambda,\upsilon)\}\sin\{\tau\lambda^2 - \eta H_2(\lambda,\upsilon)\}\frac{d\lambda}{\lambda}, \quad (7.1.20a)$$

where

$$\eta = \frac{3D_{ip}\kappa_{i1}z}{v_z r_p^2}\left(\frac{1-\varepsilon}{\varepsilon}\right); \quad \upsilon = \frac{D_{ip}\kappa_{i1}}{k_c r_p}; \quad \tau = \frac{2D_{ip}(t - \{z/v_z\})}{r_p^2};$$

$$H_1(\lambda,\upsilon) = \frac{H_{D1} + \upsilon(H_{D1}^2 + H_{D2}^2)}{(1+\upsilon H_{D1})^2 + (\upsilon H_{D2})^2};$$

$$H_2(\lambda,\upsilon) = \frac{H_{D2}}{(1+\upsilon H_{D1})^2 + (\upsilon H_{D2})^2};$$

$$H_{D1} = \frac{\lambda[\sinh 2\lambda + \sin 2\lambda]}{[\cosh 2\lambda - \cos 2\lambda]} - 1; \quad H_{D2} = \frac{\lambda[\sinh 2\lambda - \sin 2\lambda]}{[\cosh 2\lambda - \cos 2\lambda]}, \quad (7.1.20b)$$

and λ is the variable of integration. This exact solution of the solute breakthrough curve is numerically determined due to the nature of the integral in solution (7.1.20a). If the adsorbent bed length L is very large (specifically, $\eta \ge 50$ and $\upsilon \le 0.01$), an asymptotic solution of (7.1.20a) is

$$\frac{\overline{C}_{i2}}{C_{i2}^0} = \frac{1}{2}\left[1 + \text{erf}\left(\frac{(3/2)\tau - \eta}{2(\eta/5)^{1/2}}\right)\right]. \quad (7.1.20c)$$

Due to the restriction of $\upsilon \le 0.01$, this solution also represents negligible external mass-transfer film resistance if the bed is long. Note that this solution is symmetrical around $(\overline{C}_{i2}/C_{i2}^0) = 1/2$. Since these solutions are expressed in dimensionless quantities, results from one particular column could be used to predict the breakthrough curve for other columns. The value of k_c needed to make calculations in such systems may be obtained from (Dwivedi and Upadhyay, 1977)

$$j_D = \left(\frac{k_c(Sc)^{2/3}}{v_z}\right) = \frac{0.458}{\varepsilon}\left(\frac{2r_pG}{\mu}\right)^{-0.407}, \quad (7.1.20d)$$

where G is the superficial mass velocity based on empty column cross-sectional area and the Reynolds number $(2r_pG/\mu) > 10$.

There are a number of other analytical solutions for $\overline{C}_{i2}(z,t)$ available in the literature for linear isotherms. These take into account axial dispersion, model particles using macropores and micropores, etc.; they have been summarized by Ruthven (1984) in his Table 8.1. Widespread use of powerful computers and sophisticated numerical methods have reduced the importance of such analytical solutions for breakthrough curves.

For a constant liquid feed of concentration C_{i2}^0 fed into the bed, an immediate result of the diffuse breakthrough of the feed solute through the column end is that the total solute adsorption capacity of the mass-transfer zone of length L_{MTZ} cannot be fully utilized. At the breakthrough time t_{br} for the breakthrough solute concentration C_{i2}^{br} at the column end $z = L$, the solute concentration profile in the solution in the mass-transfer zone is shown in Figure 7.1.5(b). This profile, assumed to be symmetrical, shows that the hatched area representing the integral

$$S_c\rho_b\int_0^{L_{MTZ}} [q_i(C_{i2}^0) - q_i(\overline{C}_{i2})]\, dL \quad (7.1.21a)$$

represents the additional moles of solute i which could have been adsorbed if there were no diffuse breakthrough. Due to the symmetric concentration profile, one can argue that essentially the total number of solute moles adsorbed in this region,

$$S_c\rho_b\int_0^{L_{MTZ}} q_i(\overline{C}_{i2})\, dL, \quad (7.1.21b)$$

is approximately equal to the number of moles not adsorbed,

$$S_c\rho_b\int_0^{L_{MTZ}} q_i(\overline{C}_{i2})\, dL = S_c\rho_b\int_0^{L_{MTZ}} [q_i(C_{i2}^0) - q_i(\overline{C}_{i2})]\, dL,$$

unless the sorption process is highly nonlinear. Therefore,

$$S_c\rho_b\int_0^{L_{MTZ}} q_i(\overline{C}_{i2})\,dL=\frac{1}{2}\left[S_c\rho_b\int_0^{L_{MTZ}}\left[q_i(C_{i2}^0)\right]dL\right]$$
$$-S_c\rho_b\int_0^{L_{MTZ}}\left[q_i(C_{i2}^0)-q_i(\overline{C}_{i2})\right]dL, \qquad (7.1.21c)$$

where the second expression in brackets represents the total solute adsorption capacity of L_{MTZ} at t_{br} for the feed solution concentration C_{i2}^0. The total adsorption capacity of the bed of length L for the same feed solution concentration C_{i2}^0 under the nondispersive equilibrium condition is

$$S_c\rho_b\int_0^{L} q_i(C_{i2}^0)\,dL=S_c\rho_b q_i(C_{i2}^0)L. \qquad (7.1.21d)$$

Correspondingly, the lost adsorption capacity of the bed from (7.1.12c) is

$$S_c\rho_b\int_0^{L_{MTZ}}\left[q_i(C_{i2}^0)-q_i(\overline{C}_{i2})\right]dL=\frac{1}{2}\left[S_c\rho_b\int_0^{L_{MTZ}}q_i(C_{i2}^0)\,dL\right]$$
$$=\frac{1}{2}S_c\rho_b q_i(C_{i2}^0)L_{MTZ}. \qquad (7.1.21e)$$

An estimate of the fractional loss of the bed adsorption capacity due to the diffuse breakthrough is therefore given by

$$\frac{\frac{1}{2}S_c\rho_b q_i(C_{i2}^0)L_{MTZ}}{S_c\rho_b q_i(C_{i2}^0)L}\cong\frac{1}{2}\frac{L_{MTZ}}{L}. \qquad (7.1.21f)$$

The quantity the *length of the unused bed* (*LUB*), is often used to describe this loss:

$$LUB=\left(\frac{1}{2}\frac{L_{MTZ}}{L}\right)L. \qquad (7.1.21g)$$

The smaller the length of the MTZ, the lower the loss of bed adsorption capacity due to the diffuse breakthrough. Methods for calculation of *LUB* are available in Treybal (1980) and Wankat (1990, pp. 366–375).

It is useful to recall the basic liquid adsorption system considered so far. There is one solute i to be adsorbed from an inert solvent flowing through a column of adsorbent particles under isothermal conditions. There can be many systems, however, with more than one adsorbable solute species in a solvent which may or may not be inert. For an overview, the reader is referred to Ruthven (1984); original studies of considerable importance in this area are by Glueckauf (1949), Helfferich and Klein (1970) and Rhee *et al.* (1970a).

The adsorption behavior and the breakthrough curve for systems containing a number of adsorbable solutes in a solvent, as illustrated in the above-mentioned references, are generally quite complex. It is useful, however, to consider an elementary (but highly inexact) analysis of breakthrough in such a system containing more than one solute. Assume that there are, say, three solutes $i = 1, 2, 3$ present in the feed solution at concentrations of C_{12}^0, C_{22}^0 and C_{32}^0, respectively. Assume further that each solute adsorbs independently of the other two and that the adsorption isotherm for each is linear in the following fashion (Figure 7.1.2(a)):

$$\frac{q_{11}(\overline{C}_{12})}{\overline{C}_{12}}>\frac{q_{21}(\overline{C}_{22})}{\overline{C}_{22}}>\frac{q_{31}(\overline{C}_{32})}{\overline{C}_{32}}. \qquad (7.1.22a)$$

Therefore species 3 is least adsorbed, whereas species 1 is most strongly adsorbed, with species 2 being in between.

In the context of a nondispersive equilibrium adsorption model, the breakthrough times $\overline{t}_i$ (7.1.15c) for the three species are related by

$$\overline{t}_1=\frac{L}{v_z}\left[1+\frac{\rho_b}{\varepsilon}\frac{q_1(C_{12}^0)}{C_{12}^0}\right]>\overline{t}_2=\frac{L}{v_z}\left[1+\frac{\rho_b}{\varepsilon}\frac{q_2(C_{22}^0)}{C_{22}^0}\right]>$$
$$\overline{t}_3=\frac{L}{v_z}\left[1+\frac{\rho_b}{\varepsilon}\frac{q_3(C_{32}^0)}{C_{32}^0}\right]. \qquad (7.1.22b)$$

The least strongly adsorbed species 3 will come out of the adsorber first (Figure 7.1.5(c)). Correspondingly there will be a pure solution of species 3 at the adsorber outlet during the time period $\overline{t}_2-\overline{t}_3$. At time $\overline{t}_2$, species 2 will break through and will be present along with solute 3; Figure 7.1.5(c) plots the total concentration of all solutes at the column outlet as a function of time. We observe a staircase behavior, with all three solutes present at the outlet after $\overline{t}_1$. Figure 7.1.5(c) displays this behavior in the context of a dispersive model. Such an outlet breakthrough profile development is called a *frontal development*.

Frontal development is employed in large-scale purification applications, for example in decolorization of sugar, corn syrup, for removal of oxidation products from waxes, used oils, etc. Adsorbents, such as activated charcoal, are used to adsorb the strongly adsorbing color-causing impurity, etc. One pure species, the least adsorbing one, is obtained. However, the feed concentration of the least adsorbed species can be substantial, making this technique industrially useful.

Example 7.1.1 Breakthrough calculations for equilibrium nondispersive operation of a fixed bed with a liquid feed.

(a) Lightfoot *et al.* (1962) have illustrated the breakthrough behavior of a solution of lauric acid in petroleum ether flowing through a packed bed of activated carbon adsorbent, 3 cm long, having $\varepsilon = 0.4$ and a cross-sectional area of 1.32 cm^2. The concentration of lauric acid is 0.035M. The equilibrium adsorption isotherm (Freundlich type, see (3.3.112c)) is given as $C_{i1} = 2.26\, C_{i2}^{0.324}$. Determine the breakthrough volume of liquid when the solute feed concentration will appear as a shock wave at the column end for a similar column 6 cm long.

(b) Baker and Pigford (1971) have studied the adsorption of acetic acid on activated carbon from a dilute aqueous solution at 60 °C. The sorption behavior of the solute was found to be

$$q_i \frac{\text{gmol}}{\text{kg dry carbon}} = 3.02\, C_{i2}^{0.41},$$

for C_{i2} in gmol/liter. The activated carbon bed characteristics are as follows: $\varepsilon = 0.43$; $\varepsilon_p = 0.57$; $\rho_s = 1.82\,\text{g/cm}^3$; $L = 100$ cm; $S_c = 3\,\text{cm}^2$. The feed solution at a concentration of 0.1 gmol/liter is introduced at 15 cm^3/min. Determine the time for solute breakthrough and the volume of solution that will have passed through the bed at the time of breakthrough.

(c) An enzyme present in a dilute solution is to be adsorbed on cellulosic adsorbent particles in a 100 cm long packed bed 4 cm in diameter, having a bed porosity of 0.4. The enzyme feed concentration is 1 mg/liter; the linear adsorption isotherm is as follows:

$$C_{i1}\,(\text{mg/cm}^3) = 40\, C_{i2}\,(\text{mg/cm}^3).$$

Determine the breakthrough solution volume.

Solutions (a) The breakthrough volume $\overline{V}$ of the solution to be passed through the column for the feed concentration to break through in equilibrium nondispersive operation is given by relation (7.1.16d):

$$\overline{V} = S_c L \left[\varepsilon + \frac{\rho_b q_i(C_{i2}^0)}{C_{i2}^0}\right].$$

Employing definition (7.1.6), we can rewrite this as

$$\overline{V} = S_c L\left[\varepsilon + (1-\varepsilon)\frac{C_{i1}^0}{C_{i2}^0}\right] = 1.32 \times 6\left[0.4 + 0.6\frac{2.26(C_{i2}^0)^{0.324}}{C_{i2}^0}\right];$$

$$C_{i2}^0 = 0.035\,\text{M} \Rightarrow \overline{V} = 7.92\left[0.4 + 1.356\frac{1}{(0.035)^{0.676}}\right]$$

$$\Rightarrow \overline{V} = 7.92\,[0.4 + 13.1] = 7.92 \times 13.5 = 107\,\text{cm}^3.$$

(b) Since the adsorbent particle porosity has been provided, we will employ equation (7.1.15c) to determine the time $\bar{t}$ for breakthrough. For acetic acid, assume $\kappa_{im} = 1$. Here the interstitial fluid velocity

$$v_z = \frac{\text{volumetric flow rate}}{\text{cross-sectional area} \times \varepsilon} = \frac{15\,(\text{cm}^3/\text{min})}{3 \times 0.43\,\text{cm}^2} = 11.6\,\text{cm/min}.$$

Now

$$\frac{L}{\bar{t}} = \frac{v_z}{\left[1 + \dfrac{(1-\varepsilon)}{\varepsilon}\varepsilon_p\,\kappa_{im} + \dfrac{\rho_b q_i(C_{i2}^0)}{\varepsilon\, C_{i2}^0}\right]},$$

So

$$\frac{100}{\bar{t}} = \frac{11.6}{1 + \dfrac{0.57}{0.43} \times 0.57 \times 1 + \dfrac{\rho_s(1-\varepsilon_p)(1-\varepsilon)}{\varepsilon}\dfrac{q_i(C_{i2}^0)}{C_{i2}^0}}$$

$$= \frac{11.6}{1.755 + \dfrac{1.82 \times 0.43 \times 0.57}{0.43}\dfrac{3.02}{(C_{i2}^0)^{0.59}}};$$

$$\bar{t} = \frac{100}{11.6}\left[1.755 + \frac{1.037 \times 3.02}{(0.1)^{0.59}}\right]$$

$$= 8.62\left[1.755 + \frac{3.13}{0.257}\right] = 8.62 \times 13.925.$$

Now, $\bar{t} = 120$ min. From equation (7.1.16d), the breakthrough volume

$$\overline{V} = S_c L\varepsilon\left[1 + \frac{(1-\varepsilon)}{\varepsilon}\varepsilon_p\,\kappa_{im} + \frac{\rho_b q_i C_{i2}^0}{\varepsilon\, C_{i2}^0}\right]$$

$$= 3 \times 100 \times 0.43 \times 13.925 = 1796\,\text{cm}^3.$$

(c)

$$\overline{V} = S_c L\left[\varepsilon + (1-\varepsilon)\frac{C_{i1}^0}{C_{i2}^0}\right]$$

$$= \frac{\pi}{4}(4)^2 \times 100[0.4 + 0.6 \times 40]\,\text{cm}^3$$

$$= \pi \times 400[24.4] = 30.66\,\text{liter}.$$

Note: Use of equations (7.1.13d, e) is not recommended for determining $\bar{t} = (L/v_{Ci}^*)$ for adsorption examples with nonlinear equilibrium behavior since the characteristic lines for different concentrations can overlap, a physically impossible situation (Sherwood *et al.*, 1975; Wankat, 1986).

Example 7.1.2 Elution calculation for equilibrium nondispersive operation of a fixed bed with a liquid feed. In Example 7.1.1(a), 150 cm^3 of the solution of lauric acid was passed. The bed was completely saturated and the exiting solution had feed concentration 0.035 M. Now pure petroleum ether flow is initiated to elute the adsorbed lauric acid. Calculate the volume of pure petroleum ether passed corresponding to the following concentrations of lauric acid at the bed outlet: 0.03 M, 0.02 M, 0.01 M and 0.005 M.

Solution From equation (7.1.17g), we can write

$$V - V^0 = S_c(z - z_0(\overline{C}_{i2}))\,\left[\varepsilon + (1-\varepsilon)\,(\mathrm{d}\overline{C}_{i1}/\mathrm{d}\overline{C}_{i2})\right].$$

Here

$$z_0(\overline{C}_{i2}) = 0; \qquad (\mathrm{d}\overline{C}_{i1}/\mathrm{d}\overline{C}_{i2}) = 2.26 \times 0.324 \times \overline{C}_{i2}^{(-0.676)}.$$

Now, $V^0 = 150\,\text{cm}^3$. Therefore

$$V = 150 + 1.32\,z\left[0.4 + 0.6 \times 2.26 \times 0.324 \times \left(1/\overline{C}_{i2}^{0.676}\right)\right]$$

$$\Rightarrow V = 150 + 3.168 + \left[3.47/\overline{C}_{i2}^{-0.676}\right]$$

(since $z = 6$ cm for the outlet concentration). The volume of pure petroleum ether passed is

$$V - 150 = 3.168 + \left[3.47/\overline{C}_{i2}^{-0.676}\right]$$

for the $\overline{C}_{i2}$ values of interest (see Table 7.1.2).

Table 7.1.2.

Outlet concentration, $\overline{C}_{i2}$ (M)	$(V - 150)$ cm^3
0.03	~34
0.02	~52
0.01	~82
0.005	~120

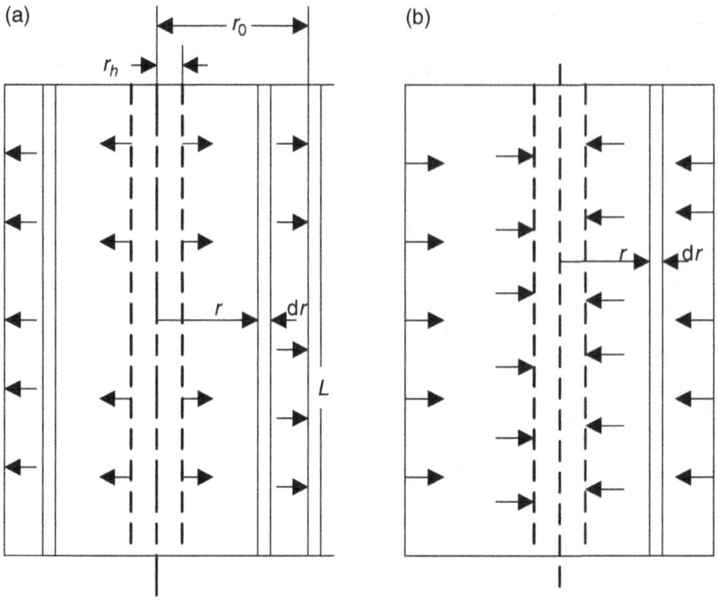

Figure 7.1.6 Radial flow fixed bed (adsorbents not shown): (a) flow direction radially outward; (b) flow direction radially inward. Arrows show flow direction.

7.1.1.1.1 Fixed-bed adsorption - radial flow Fixed-bed adsorption from a solution is sometimes carried out in a packed bed where the liquid flows radially instead of axially along the column length. Typically there is a central duct or hole in the packed column. The feed solution may be added to the central hole and it flows radially outward (Lapidus and Amundson, 1950); alternatively the feed solution may be forced to enter the packed bed from the periphery, move through the bed radially into the central duct of radius r_h (Huang *et al.*, 1988a, b). These two configurations are shown in Figures 7.1.6(a) and 7.1.6(b), respectively. Note that in each case the radial bulk flow is perpendicular to the force direction, which is vertically upward or downward depending on the location of the flow and the particle.

A differential equation for the concentration $\overline{C}_{i2}$ of species i in the mobile phase (averaged in the z- and θ-directions) may be developed in the manner of equation (7.1.2). Consider a cylindrical layer of adsorbent, of thickness Δr and bed length L, at radial location r for the flow configuration of Figure 7.1.6(a) with a radially outward flow. A species i balance, in the absence of any chemical reaction, is

(rate of accumulation) = (rate of inflow at r)
 − (rate of outflow at $r + \Delta r$);

$$2\pi r \Delta r L\left\{\varepsilon\frac{\partial \overline{C}_{i2}}{\partial t} + (1-\varepsilon)\frac{\partial \overline{C}_{i1}}{\partial t}\right\}_{\overline{r}} = \left(\dot{V}\,\overline{C}_{i2}\right)_r - 2\pi L\left(rD_{i,\text{eff},r}\,\varepsilon\frac{\partial \overline{C}_{i2}}{\partial r}\right)_r - \left(\dot{V}\,\overline{C}_{i2}\right)_{r+\Delta r} + 2\pi L\left(rD_{i,\text{eff},r}\,\varepsilon\frac{\partial \overline{C}_{i2}}{\partial r}\right)_{r+\Delta r}, \quad (7.1.23\text{a})$$

where $\dot{V}$ is the radial volume rate of flow of solution from the inside hole to the periphery of the packed bed. In the limit $\Delta r \to 0$, one obtains

$$2\pi r L\left\{\varepsilon\frac{\partial \overline{C}_{i2}}{\partial t} + (1-\varepsilon)\frac{\partial \overline{C}_{i1}}{\partial t}\right\}_{\overline{r}} + \dot{V}\frac{\partial \overline{C}_{i2}}{\partial r} = 2\pi L\varepsilon D_{i,\text{eff},r}\frac{\partial}{\partial r}\left(r\frac{\partial \overline{C}_{i2}}{\partial r}\right), \quad (7.1.23\text{b})$$

where $\dot{V}$ and $D_{i,\text{eff},r}$ are assumed constants. Further, $r < \overline{r} < r + \Delta r$, and r varies from r_h to r_0. An alternative form is given by (this could also be obtained from (6.2.32) in the manner of (6.2.35), with the z-coordinate replaced by r)

$$\varepsilon\frac{\partial \overline{C}_{i2}}{\partial t} + (1-\varepsilon)\frac{\partial \overline{C}_{i1}}{\partial t} + \frac{\dot{V}}{2\pi r L}\frac{\partial \overline{C}_{i2}}{\partial r} = \varepsilon\frac{D_{i,\text{eff},r}}{r}\frac{\partial}{\partial r}\left(r\frac{\partial \overline{C}_{i2}}{\partial r}\right)_{\overline{r}}. \quad (7.1.23\text{c})$$

If diffusion and dispersion in the radial direction are neglected, this equation is reduced to

$$\varepsilon\frac{\partial \overline{C}_{i2}}{\partial t} + (1-\varepsilon)\frac{\partial \overline{C}_{i1}}{\partial t} + \frac{\dot{V}}{2\pi r L}\frac{\partial \overline{C}_{i2}}{\partial r} = 0. \quad (7.1.23\text{d})$$

A solution of this nondispersive packed-bed adsorption with radial flow is available (Lapidus and Amundson, 1950) under the condition of elution chromatography (we will soon learn about this) for two cases: (1) equilibrium theory with linear adsorption isotherm; (2) nonequilibrium theory based on a suitable rate equation. The governing equation analogous to (7.1.23b) for the radial flow packed-bed adsorption where the effluent flows into the central duct from the bed periphery is

$$-2\pi rL\left\{\varepsilon\frac{\partial\overline{C}_{i2}}{\partial t}+(1-\varepsilon)\frac{\partial\overline{C}_{i1}}{\partial t}\right\}_{\bar{r}}+\dot{V}\frac{\partial\overline{C}_{i2}}{\partial r}$$
$$=2\pi L\varepsilon D_{i,\mathrm{eff},r}\frac{\partial}{\partial r}\left(r\frac{\partial\overline{C}_{i2}}{\partial r}\right). \qquad (7.1.24\mathrm{a})$$

If the diffusion term is neglected, this equation is simplified as follows:

$$\varepsilon\frac{\partial\overline{C}_{i2}}{\partial t}+(1-\varepsilon)\frac{\partial\overline{C}_{i1}}{\partial t}-\frac{\dot{V}}{2\pi rL}\frac{\partial\overline{C}_{i2}}{\partial r}=0. \qquad (7.1.24\mathrm{b})$$

The recovery of bioactive products from their dilute solutions is generally carried out using a fixed bed of adsorbents. Since such solutions are dilute, often large solution volumes have to be processed; also, the solutions may be viscous. In a conventional fixed bed (Figure 7.1.1) of considerable length, this will lead to large pressure drops. A radial flow fixed bed will, however, have a much lower pressure drop. Results of the purification of an enzyme and the removal of proteases from human plasma using radial flow cartridges are available in Huang *et al.* (1988b).

7.1.1.2 *Fixed-bed adsorption: mobile feed gas*

Preferential adsorption of a gas species or a vapor species from a flowing gas mixture by solid adsorbent particles in a packed bed is frequently practiced in industry (see Table 7.1.1). A large number of gas purifications (solvents or odors from air, sulfur compounds from natural gas, vent streams, etc., CO_2 from natural gas, H_2O from a variety of gas streams) and bulk separations (O_2-N_2, n-paraffins from iso-paraffins, mixtures of aromatics, etc.) are achieved by adsorption in a packed bed of adsorbent such as activated carbon, zeolites, carbon molecular sieves, silica gel, etc. (Yang, 1987). When the adsorbent particles are saturated with adsorbates, the bed is regenerated, by heating, by a purge gas or by lowering the pressure. The latter process, known as *pressure swing adsorption* (PSA), is quite important and is treated in Section 7.1.2.

An analysis of fixed-bed adsorption with mobile feed gas (instead of feed liquid) should start with the mass balance equation (7.1.3) for species i. Unlike liquid feeds, the gas velocity v_z is a function of z unless some trace compounds are being removed. (For *trace compound removal*, the volume $\overline{V}$ of feed gas needed to saturate the column is given by (7.1.16d); the shock velocity is obtained from (7.1.15c)). Therefore, instead of (7.1.4), the general mass balance equation for species i (phase $j=1$, solid; phase $j=2$, gas) is

$$\frac{\partial\overline{C}_{i2}}{\partial t}+\frac{(1-\varepsilon)}{\varepsilon}\frac{\partial\overline{C}_{i1}}{\partial t}+v_z\frac{\partial\overline{C}_{i2}}{\partial z}+\overline{C}_{i2}\frac{\partial v_z}{\partial z}=D_{i,\mathrm{eff},z}\frac{\partial^2\overline{C}_{i2}}{\partial z^2}, \qquad (7.1.25)$$

where $\overline{C}_{i2}$ is the gas-phase molar concentration and $\overline{C}_{i1}$ is the solid-phase molar concentration. Analyses of adsorber performance for a gaseous feed often assume nondispersive operation. Models of nondispersive operation either postulate equilibrium between the two phases or describe the mass-transfer rate of species in particular ways. Analysis of the adsorber using an *equilibrium nondispersive* mode of operation is considered next.

It is customary to use the mole fraction $\overline{x}_{i2}$ of the gas phase instead of $\overline{C}_{i2}$, the two being related by $\overline{C}_{i2}=\overline{x}_{i2}C_t$, where C_t is the total molar concentration. Here, $\overline{x}_{i2}$ is the value of x_{i2} space-averaged over the x- and y-coordinates in the same manner as $\overline{C}_{i2}$ is developed from C_{i2}. If the gas pressure drop along the packed bed is very small, C_t may be considered constant along the bed since C_t is proportional to the total gas pressure. The mass balance equation (7.1.25) may now be written in terms of x_{i2} as (Ruthven, 1984)

$$\frac{\partial\overline{x}_{i2}}{\partial t}+\frac{(1-\varepsilon)}{C_t\varepsilon}\frac{\partial\overline{C}_{i1}}{\partial t}+v_z\frac{\partial\overline{x}_{i2}}{\partial z}+\overline{x}_{i2}\frac{\partial v_z}{\partial z}=0 \qquad (7.1.26)$$

under nondispersive condition. The variation of the gas velocity v_z with the axial coordinate z is due to adsorption of species onto the adsorbent particles. For a binary gas mixture, either both species may be adsorbed or only one will. Assuming only one species is adsorbed, a mass balance for species i being adsorbed is

$$(1-\varepsilon)\frac{\partial\overline{C}_{i1}}{\partial t}=-\varepsilon C_t\frac{\partial v_z}{\partial z}, \qquad (7.1.27\mathrm{a})$$

where, following relations (7.1.6) and (7.1.7), we can write

$$\frac{\partial\overline{C}_{i1}}{\partial t}=\frac{\rho_b}{(1-\varepsilon)}\frac{\partial\overline{q}_{i1}}{\partial t}=\frac{\rho_b}{(1-\varepsilon)}q'_{i1}\left(\overline{C}_{i2}\right)\frac{\partial\overline{C}_{i2}}{\partial t}$$
$$=\frac{\rho_b q'_{i1}\left(\overline{C}_{i2}\right)C_t}{(1-\varepsilon)}\frac{\partial\overline{x}_{i2}}{\partial t}. \qquad (7.1.27\mathrm{b})$$

Note that if the species adsorbed is in trace amounts, the change in v_z along z due to adsorption would be negligible. Thus, for nontrace systems, using these two relations, relation (7.1.26) can be simplified to

$$\left[1+\frac{\rho_b q'_{i1}\left(\overline{C}_{i2}\right)}{\varepsilon}(1-\overline{x}_{i2})\right]\frac{\partial\overline{C}_{i2}}{\partial t}+v_z\frac{\partial\overline{C}_{i2}}{\partial z}=0. \qquad (7.1.28)$$

Following (7.1.9) and (7.1.11), we can obtain for the wave velocity v^*_{Ci} of concentration $\overline{C}_{i2}$ for species i in a gaseous feed of one adsorbed species i (inert carrier)

$$v_{Ci}^* = \frac{dz}{dt} = \frac{v_z}{\left[1 + \frac{\rho_b q'_{i1}(\overline{C}_{i2})}{\varepsilon}(1 - \overline{x}_{i2})\right]}. \qquad (7.1.29)$$

Note that v_z changes with bed distance z and mole fraction $\overline{x}_{i2}$; it would therefore be convenient to replace it by terms containing $\overline{x}_{i2}$. Assume the isotherm to be linear so that $q'_{i1}(\overline{C}_{i2}) =$ constant. Then, combining relations (7.1.27a) and (7.1.27l), the following expression is obtained:

$$\frac{\partial v_z}{\partial z} = -\frac{\rho_b q'_{i1}}{\varepsilon}\frac{\partial \overline{x}_{i2}}{\partial t}. \qquad (7.1.30)$$

Using now relation (7.1.29) in the form of dz/dt in the above equation, we get

$$\frac{\partial v_z}{\partial \overline{x}_{i2}} = -\frac{\rho_b q'_{i1}}{\varepsilon}\frac{v_z}{\left[1 + \frac{\rho_b q'_{i1}}{\varepsilon}(1 - \overline{x}_{i2})\right]}. \qquad (7.1.31)$$

This can be integrated between the limits of v_z^0, x_{i2}^0 (at the adsorber inlet) and v_z, $\overline{x}_{i2}$ anywhere else in the adsorber to yield

$$\frac{v_z}{v_z^0} = \frac{1 + \frac{\rho_b q'_{i1}}{\varepsilon}(1 - x_{i2}^0)}{1 + \frac{\rho_b q'_{i1}}{\varepsilon}(1 - \overline{x}_{i2})}; \qquad (7.1.32)$$

$$v_{Ci}^* = \frac{dz}{dt} = \frac{v_z^0\left[1 + \frac{\rho_b q'_{i1}}{\varepsilon}(1 - x_{i2}^0)\right]}{\left[1 + \frac{\rho_b q'_{i1}}{\varepsilon}(1 - \overline{x}_{i2})\right]^2}. \qquad (7.1.33)$$

This result implies that, as $\overline{x}_{i2}$ increases, v_{Ci}^* increases. In liquid systems with a linear isotherm, equation (7.1.12a), however, indicated no such concentration dependence of v_{Ci}^*. The effect of concentration on v_{Ci}^* in a nontrace gaseous system is, then, similar to that observed with a favorable isotherm (see Figure 7.1.4). A spread out profile along the column is compressed to a front sharper than would have been possible otherwise.

This type of behavior aids in creating what is known as *constant-pattern behavior*. Consider the mass-transfer zone (MTZ) in Figure 7.1.5(b). Near the entrance to the column, such a profile is created spontaneously by the effects of axial dispersion and mass-transfer effects. Further down the column, as the gas velocity decreases to v_z from the inlet value v_z^0 due to substantial adsorption, the concentration front tends to become compressed: this compression acts counter to the tendency of axial dispersion and mass transfer to expand the front (as in an unfavorable isotherm of Figure 7.1.4). Often these two opposing tendencies balance each other, so that the MTZ (Figure 7.1.5 (b)) travels through a long column without any change. Such constant-pattern behavior may be achieved also in systems where the adsorbed species is present in trace systems, provided the species displays a favorable isotherm. The adsorption of moisture from air onto silica gel adsorbent is a case in point. Although dispersion and mass-transfer effects, if any, tend to broaden the front,

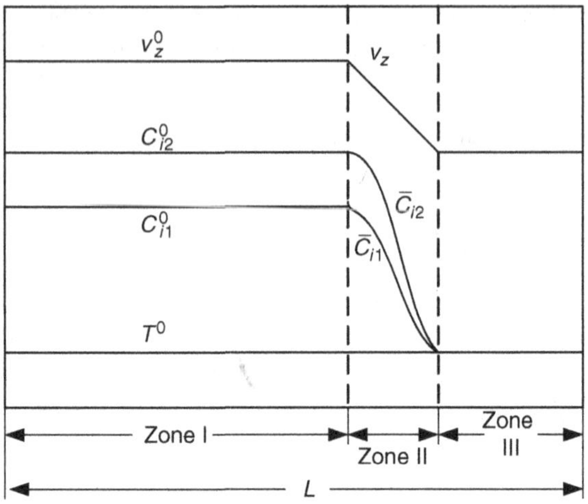

Figure 7.1.7. Constant-pattern behavior in isothermal single-component adsorption for a gas mixture. (After Sircar and Kumar (1983).)

the favorable isotherm tends to sharpen the front, leading often to a constant-pattern behavior.

The nature of the constant-pattern behavior for isothermal adsorption of a single gas species i in the presence of an inert gas is shown in Figure 7.1.7, where the profiles of gas velocity v_z, mobile-phase concentration $\overline{C}_{i2}$ of species i, adsorbent-phase concentration $\overline{C}_{i1}$ and temperature T are shown along the column. There are three zones in the column: Zone I is saturated with species i, therefore the values of v_z, C_{i2} and C_{i1} are constant; Zone II is the MTZ, where the values of these quantities vary from saturation to values characteristic of the bed where no adsorption has taken place yet, the latter being Zone III. Using the constant-pattern model, Sircar and Kumar (1983) have provided analytical expressions relating the time difference $(t_2 - t_1)$ in the breakthrough curve corresponding to two arbitrary composition levels $\overline{C}_{i2}|_1$ and $\overline{C}_{i2}|_2$ for a number of cases. They have also provided analytical expressions for the length of the MTZ. The cases studied include bulk single-component adsorption satisfying Langmuir adsorption isotherm under either gas film control or solid diffusion control.

7.1.1.3 *Fixed adsorbent bed regeneration*

After the passage of a feed gas mixture or liquid solution through a fixed adsorbent bed for some time, the adsorbent particles are saturated with the solute. For reuse, the bed has to be regenerated; the particles in the bed acting as mass-separating agents need to be restored to their original state so that they may be useful again as mass-separating agents. The process of desorption needed for the regeneration of adsorbents may be understood from

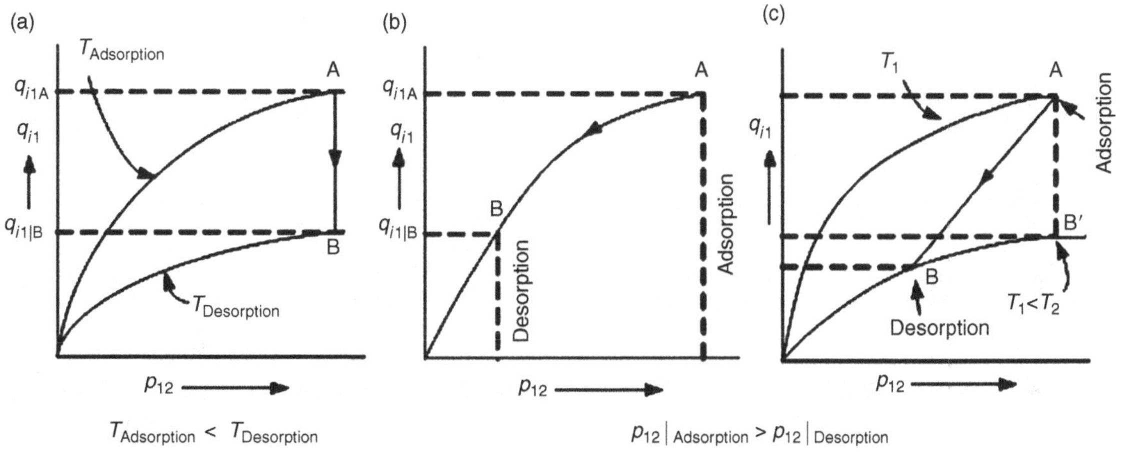

Figure 7.1.8. Adsorption isotherms for a gas mixture and conditions for adsorption–desorption: (a) thermal swing; (b) purge gas or pressure swing; (c) combined thermal swing and purge gas.

the adsorption isotherms shown in Figure 7.1.8 for gas separation, where p_{i2} is the partial pressure of species i in the mobile gas phase; p_{i2} is related to C_{i2} for ideal gas behavior by $p_{i2} = C_{i2}RT$.

If, at the end of adsorption, the adsorbate concentration on adsorbent particles is represented by point A (Figure 7.1.8(a)) and the temperature is, say, $T_{\text{Adsorption}}$, for desorption the temperature is increased to $\text{T}_{\text{Desorption}}$, which is higher. The adsorption isotherm now is considerably lower; q_{i1} is reduced to a lower value, even if p_{i2} remains the same. The temperature of the bed is raised by supplying either a hot gas or steam. The hot gas may be obtained using the feed gas itself or some other gas, e.g. air. When the partial pressure of species i in the hot gas is lower than that at point A, there is additional desorption since $q_{i1}|_{\text{B}} < q_{i1}|_{\text{B}'}$ (Figure 7.1.8(c)). After desorption, the bed is ready for adsorption again at a lower temperature, in a cyclic fashion, in processes characterized as *thermal-swing adsorption* (TSA).

An alternative procedure to desorbing by a hot gas or vapor is to reduce the pressure of the bed. As shown in Figure 7.1.8(b), although the temperatures of A and B are the same (they are on the same isotherm), the partial pressure of species i in gas phase at B is much lower; therefore, the adsorbate concentration is much lower. The pressure reduction in the bed is achieved either by lowering the pressure of the bed from the higher pressure, in a process usually called *blowdown*, or by pulling a vacuum. Both strategies lower the value of p_{i2}, which can also be lowered by passing a purge gas. The process whereby the absolute pressure level of the adsorber is reduced essentially at constant temperature, with or without a purge gas, is called *pressure-swing adsorption* (PSA) (see Section 7.1.2). Such processes are cyclic in nature; after desorption, the bed is ready for adsorption again.

An elementary basis for quantifying the movement of concentration of the adsorbed species in the column during desorption will now be developed. If we consider the adsorption/desorption of a trace amount of gaseous species from/into a carrier gas vis-à-vis an adsorbent, then equation (7.1.25) is reduced to equation (7.1.4), which is normally used for liquid-phase feeds. If now nondispersive operation is assumed, the governing species balance equation in the column is (7.1.8). Thus, the relevant equation continues to be equation (7.1.9),

$$\frac{\partial \overline{C}_{i2}}{\partial t} + \frac{v_z}{\left[1 + \frac{\rho_b q_i'(\overline{C}_{i2})}{\varepsilon}\right]} \frac{\partial \overline{C}_{i2}}{\partial z} = 0,$$

whereby the concentration wave velocity v_{Ci}^* is obtained as (equation (7.1.12a))

$$v_{Ci}^* = \frac{\mathrm{d}z}{\mathrm{d}t} = \frac{v_z}{\left[1 + \frac{\rho_b q_i'(\overline{C}_{i2})}{\varepsilon}\right]}.$$

Let the fluid-phase concentration of a uniformly saturated bed be C_{i2}^0. This fluid, of concentration C_{i2}^0, will be pushed out by the desorbing mobile phase entering the column at a velocity v_z. The fluid at $z = 0$ and $t = 0$ (having a concentration C_{i2}^0) will exit from the column of length $z = L$ at time $t = (L/v_z)$. Fluid elements at $z > 0$ and $t = 0$ having the initial saturation concentration C_{i2}^0 will exit earlier (Figure 7.1.9(a)). Let the desorbing mobile-phase concentration be $C_{i2} = 0$ (it should be $< C_{i2}^0$). The value of v_{Ci}^* corresponding to any $\overline{C}_{i2}$ is obtained from the expression (7.1.12a)

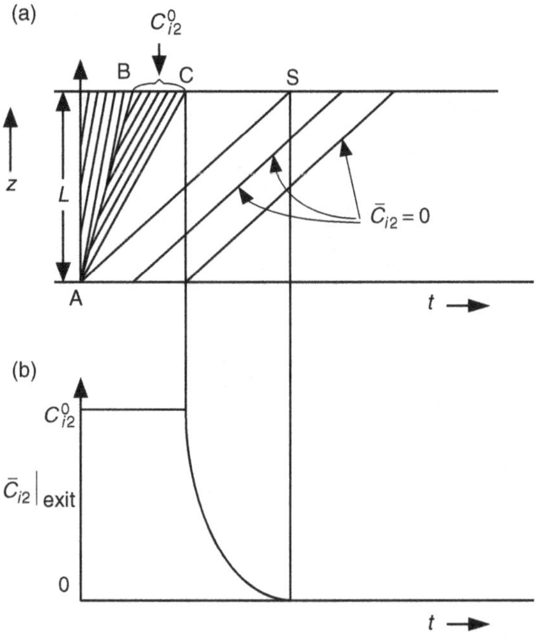

Figure 7.1.9. (a) Characteristics for desorption of a bed saturated with C_{i2}^0 with an inert purge: equilibrium nondispersive model, favorable isotherm. (b) Column exit concentration for desorption of column in (a).

given earlier. Consider any favorable adsorption isotherm (Figure 7.1.2(b)). Obviously,

$$q'_{i1}(\overline{C}_{i2})\,|_{C_{i2}=0} > q'_{i1}(\overline{C}_{i2})\,|_{C_{i2}^0}. \tag{7.1.34a}$$

Therefore

$$v_{Ci}^*|_{C_{i2}=0} < v_{Ci}^*|_{C_{i2}^0} < v_z. \tag{7.1.34b}$$

The characteristics lines for $C_{i2} = 0$ have been shown in Figure 7.1.9(a). When such a line, AS, starting at $t = 0$ intersects $z = L$, the desorbing fluid breaks through the column, i.e. the column desorption is complete.

Between this line AS and the line AB (whose slope is that of the fluid velocity, v_z) lie two regions. The first region is between lines AB and AC, where the desorbing fluid has arrived but its concentration is still C_{i2}^0; such lines have origins on AB at different column locations and they exit the column with velocity $v_{Ci}^*|_{C_{i2}^0}$. Such a line originating at $z = 0$ and $t = 0$ is shown as AC. To the right of such a line, the mobile-phase column concentrations are changing. Further, the fluid exiting the column now has concentrations between C_{i2}^0 and $C_{i2} = 0$. The column exit concentrations as a function of time are shown in Figure 7.1.9(b). For a favorable isotherm, this implies that the slopes of lines originally at $t = 0$, $z = 0$ now decrease. Ultimately, the line AS is reached when the bed is completely desorbed. Note that the column outlet concentration, $\overline{C}_{i2}(z = L)$, changes with t (Figure 7.1.9(b)).

To obtain such concentrations, i.e. $\overline{C}_{i2}(z = L)$ as a function of time, use the definition for v_{Ci}^*, namely

$$\frac{\mathrm{d}z}{\mathrm{d}t} = \frac{v_z}{\left[1 + \frac{\rho_b q'_i(\overline{C}_{i2})}{\varepsilon}\right]}, \tag{7.1.35}$$

for any given favorable isotherm, e.g. Langmuir isotherm (equation (3.3.112b)),

$$\theta_i = \frac{b_{i1} C_{i2}}{1 + b_{i1} C_{i2}} = \frac{q_{i1}}{q_{i1}^S}, \tag{7.1.36a}$$

where q_{i1}^S is the moles of species i in solid phase 1 at saturation per unit solid-phase mass. Therefore

$$q'_i(C_{i2}) = (\mathrm{d}q_{i1}/\mathrm{d}C_{i2}) = \frac{b_{i1} q_{i1}^S}{(1 + b_{i1} C_{i2})^2}. \tag{7.1.36b}$$

Substitute this into (7.1.35) using $\overline{C}_{i2}$ instead of C_{i2} and integrate between $z = 0$ and $t = 0$ to z and t for any given $\overline{C}_{i2}$:

$$z = \int_0^z \mathrm{d}z = \frac{v_z}{\left[1 + \frac{\rho_b b_{i1} q_{i1}^S}{\varepsilon\left(1 + b_{i1}\overline{C}_{i2}\right)^2}\right]} \int_0^t \mathrm{d}t = \frac{v_z t}{\left[1 + \frac{\rho_b b_{i1} q_{i1}^S}{\varepsilon\left(1 + b_{i1}\overline{C}_{i2}\right)^2}\right]}. \tag{7.1.37a}$$

Rearrange this as

$$\frac{\overline{C}_{i2}}{C_{i2}^0} = \frac{1}{b_i C_{i2}^0}\left\{\left[\frac{z\rho_b b_{i1} q_{i1}^S}{\varepsilon(v_z t - z)}\right]^{1/2} - 1\right\}. \tag{7.1.37b}$$

For $z = L$, the bed length, this expression (Walter, 1945) provides the breakthrough curve for desorption at the column outlet, a relation between $\overline{C}_{i2}|_{\mathrm{exit}}$ and t, as shown in Figure 7.1.9(b). Further, the whole bed is completely desorbed when $\overline{C}_{i2} = 0$ at $z = L$; this happens when

$$\frac{L}{t} = \frac{v_z}{\left[1 + \frac{\rho_b b_{i1} q_{i1}^S}{\varepsilon}\right]}. \tag{7.1.37c}$$

Recall that these two results are valid for nondispersive equilibrium desorption of a bed, initially saturated throughout at a level of C_{i2}^0 by a mobile phase without any solute species, when Langmuir adsorption isotherm characterizes the adsorption equilibrium between the solute in the mobile phase and the adsorbent.

The breakthrough curve developed above for desorption was based on a pure mobile phase, i.e. $C_{i2} = 0$, sent in to desorb. Often such a phase is at a higher temperature, e.g. a hot purge gas. In such a case, the species balance equation (e.g. (7.1.9)) is coupled with an overall heat balance equation. A solution of such a system is schematically illustrated in Figure 7.1.10(a) for nonequilibrium, nonisothermal desorption of a single adsorbed species

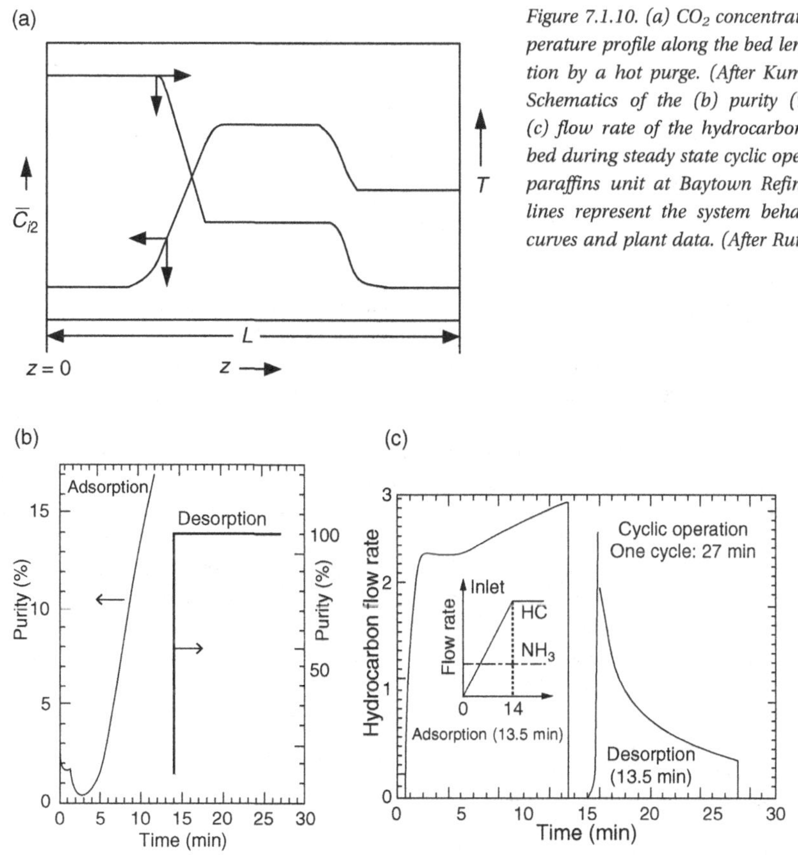

Figure 7.1.10. (a) CO_2 concentration profile ($\overline{C}_{i2}$) and temperature profile along the bed length at any time in desorption by a hot purge. (After Kumar and Dissinger (1986).) Schematics of the (b) purity (% normal paraffins) and (c) flow rate of the hydrocarbon stream leaving the sieve bed during steady state cyclic operation of the Ensorb linear paraffins unit at Baytown Refinery (arbitrary units). The lines represent the system behavior based on theoretical curves and plant data. (After Ruthven (1984).)

(e.g. CO_2) from an adsorbent (0.5 nm molecular sieve) by a hot purge of N_2 (Kumar and Dissinger, 1986). This figure shows the temperature and CO_2 concentration profile along the bed length at any instant of time. The concentration of CO_2 in the saturated bed is shown near the column end (near $z = L$), where no desorption process has been initiated yet. In the initial sections of the column near $z = 0$, where desorption is complete, the very low value of $\overline{C}_{i2}$ reflects the purge concentration of CO_2, if any, just as the high purge gas temperature is present. Then there is a MTZ, where CO_2 concentration in the bed rises and the temperature falls. Finally, there is an additional MTZ further down the column which pulls down the higher temperature and gas-phase concentration to the values existing in the saturated column prior to purge introduction.

There have been a number of investigations into the behavior of a column subjected to desorption. Although these include studies on isothermal desorption (Zwiebel *et al.*, 1972; Garg and Ruthven, 1973), nonisothermal desorption based studies are often more realistic. The reader interested in detailed understanding should refer to papers by Rhee *at al.* (1970b, 1972), Basmadjian *et al.* (1975a,b), Kumar and Sircar (1984) and Kumar and Dissinger (1986). A comprehensive introduction to bed regeneration is available in the treatment of cyclic batch processes by Ruthven (1984).

There are additional methods for regenerating an adsorbent bed beside TSA and PSA. In the case of gas/vapor mixtures, the introduction of a *displacement purge gas* (as opposed to an inert purge gas) into the bed serves first the same function as in PSA, namely the reduction of the partial pressure, p_{i2}, which facilitates desorption of the adsorbed gas species *i*. Second, there is a separate and more important function: the displacement purge gas adsorbs strongly and competes with the adsorbed feed species to be separated. Further, if the temperature of the displacement purge gas/vapor is higher, as in the case of steam stripping of activated carbon to strip organic solvents or volatile organic compounds, then we have the combined effects of TSA, the reduction of p_{i2} and desorption via displacement.

A major requirement for a successful displacement purge gas/vapor, the *displacer*, however, is that it should not be too strongly adsorbed compared to the feed species to be desorbed. If it is, then it would be difficult to desorb it during the adsorption part of the next cycle. In fact, ideally, it should have an almost identical affinity for adsorption on the adsorbent as the feed species to be desorbed. Then, during the beginning of the next cycle, the species in the feed mixture which is preferentially adsorbed will displace the displacer adsorbed during the regeneration part of the cycle. Even under optimal conditions, this displacer will contaminate the preferentially nonadsorbed (*raffinate*) species/stream, leaving the adsorber during the adsorption part of the cycle. It is desirable therefore to use a displacer which is easily separated from the raffinate stream.

An example of such a displacer is ammonia in the large-scale separation of linear paraffins of medium molecular weight (C_{10}–C_{16}) from branched chain and cyclic isomers. The adsorbent is a 0.5 nm molecular sieve, which, at 550–600 °F and slightly above atmospheric pressure, adsorbs the linear paraffins strongly from the feed mixture (Figures 7.1.10(b) and (c)). Such a high feed temperature precludes a TSA process for desorption lest it should lead to cracking/coking. Ammonia is, therefore, used to desorb these paraffins. A higher ammonia flow rate is required since it is adsorbed somewhat less strongly than the paraffins (Ruthven, 1984), but it is easily separated from the paraffins due to its high volatility after cooling the mixture and flashing it out via distillation. Another displacer for the system is n-hexane, which also possesses a high volatility with respect to the other species in the system. A brief introduction to these and related processes has been provided by Ruthven (1984). An introduction to the problems of modeling in such multicomponent systems is provided in Wankat (1990, pp. 394–400).

Displacer based desorption is practiced more often in liquid-phase processes, particularly in ion-exchange resin based processes. Conventional adsorption based liquid-phase processes are not particularly suitable for displacer based processes. Further, displacer based processes are invariably used in multicomponent separations (Rhee and Amundson, 1982) and are properly considered under chromatographic processes (see Section 7.1.5) (Frenz and Horvath, 1985).

7.1.1.4 *Ion exchange beds*

To remove a particular ion from an aqueous solution, a column containing ion exchange resin beads is often used, especially if the ion is present at low concentrations. The ion exchange bed/column may also be used to replace one particular ion in solution by another ion. The resin beads are typically introduced in a vertical cylindrical column and are supported from the bottom; the top section is free so that, as the resins expand, there is room for expansion; this is operationally important since resin beads swell in water. Figure 7.1.11(a) illustrates a column filled with vertical resin beads.

Consider the following process scenario. The ion exchange resin beads are to participate in an ion exchange process between two counterions A and B, with Y being the co-ion; the ion exchange resin beads in the column to start with are in B form and the solution being continuously fed from the column top contains essentially AY. As the solution keeps coming down the column, the beads at the top of the column are converted to the A form. Just as we observed in Figure 7.1.5(b) how the column concentration profile developed in an adsorbent column fed continuously with a solution of concentration C_{i2}^0, similarly here the top of the column will be converted to the A form. There will also be a mass-transfer zone (MTZ, as in Figure 7.1.5(b)) where, at one end, the column is in A form and, at the other end, it is in B form, with a continuous transition going on in between (Figure 7.1.11(b)).

What would be of interest here is to find out when the feed ionic species A is going to break through from the column end, as shown in Figure 7.1.11(c). As in Figure 7.1.5(a), the actual breakthrough curve is going to be diffuse, shown by the dashed line in Figure 7.1.11(c). However, we will calculate the time corresponding to a step jump in the column outlet concentration of ionic species A (solid line in Figure 7.1.11(c)) in the manner of Figure 7.1.3(b). The governing equation for the concentrations of ionic species A in the two different regions, mobile phase ($j = 2$) and the ion exchange resin phase ($j = 1$), namely C_{A2} and C_{A1}, respectively, will continue to be equation (7.1.4)):

$$\frac{\partial \overline{C}_{A2}}{\partial t} + \frac{(1-\varepsilon)}{\varepsilon}\frac{\partial \overline{C}_{A1}}{\partial t} + v_z\frac{\partial \overline{C}_{A2}}{\partial z} = D_{A,\mathrm{eff},z}\frac{\partial^2 \overline{C}_{A2}}{\partial z^2}.$$

An item to be noted here is that, even though the resin is swollen and highly porous ($\varepsilon_p \neq 0$), the ions inside the ion exchange resin beads are essentially next to the fixed charges in the ion exchange resin beads. Therefore, the concept employed to develop equation (7.1.13e) is not valid here; we therefore employ, de facto, $\varepsilon_p = 0$.

The analysis carried out earlier to arrive at the concentration wave velocity expression (7.1.12a) for v_{Ci}^* is also valid here. Therefore, for equilibrium nondispersive operation of an ion exchange column, we can write, for ionic species A,

$$v_{CA}^* = \frac{v_z}{\left[1 + \frac{\rho_b q'_A(\overline{C}_{A2})}{\varepsilon}\right]}. \tag{7.1.38}$$

What we will do now is to deliberate on the nature of the ion exchange equilibrium behavior of a given ionic species, say A. However, note what the quantity $\rho_b(\partial q_A(\overline{C}_{A2})/\partial \overline{C}_{A2}) = \rho_b q'_A(\overline{C}_{A2})$ used for adsorption of solute A stands for here (see notation in equations (3.3.121f-n)):

$$\rho_b\frac{\partial q_A(\overline{C}_{A2})}{\partial \overline{C}_{A2}} = \rho_R(1-\varepsilon)\frac{\partial q(\overline{C}_{A2})}{\partial \overline{C}_{A2}} = (1-\varepsilon)\frac{\partial\{\rho_R\, q(\overline{C}_{A2})\}}{\partial \overline{C}_{A2}}$$
$$= (1-\varepsilon)\frac{\partial \overline{C}_{AR}}{\partial \overline{C}_{A2}}. \tag{7.1.39a}$$

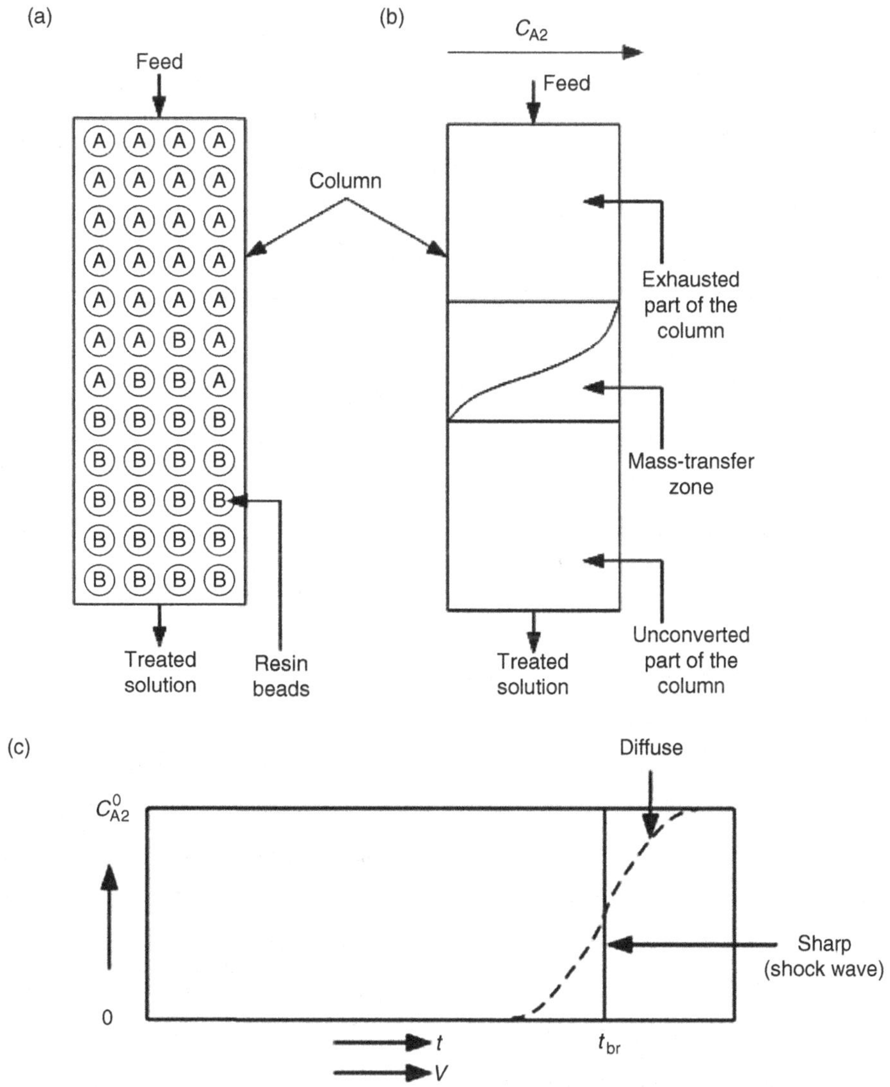

Figure 7.1.11. Mass-transfer/concentration profiles in an ion exchange column: (a) sections of the column where it is in A form, B form or in mixed AB form; (b) concentration profile of the A form of the resin in the column at any time; (c) breakthrough behavior at the column bottom, sharp breakthrough and diffuse breakthrough.

However in view of relations (3.3.121g) for the resin phase and (3.3.121h) for the external solution phase, we can express the resin phase and the external solution phase concentrations in terms of mole fractions $\overline{x}_{AR}$ and $\overline{x}_{Aw}$ in the resin and water phase, respectively:[5]

$$\overline{x}_{AR} = \frac{\overline{C}_{AR}}{\overline{C}_{AR} + \overline{C}_{BR}} = \frac{\overline{C}_{AR}}{\overline{C}_{FC}}; \qquad \overline{x}_{Aw} = \frac{\overline{C}_{Aw}}{\overline{C}_{Aw} + \overline{C}_{Bw}} = \frac{\overline{C}_{Aw}}{\overline{C}_{tw}}. \tag{7.1.39b}$$

[5]The bar "‾" on top of any quantity refers to a quantity averaged over the column cross section.

Here we define the system in terms of the binary mixture of counterions A and B and ignore the presence of any co-ion Y on other ions. Therefore, the *concentration wave velocity* v^*_{CA} of ionic species A through the column is given by (for a given concentration, say $\overline{C}_{Aw}$ or mole fraction $\overline{x}_{Aw}$)

$$v^*_{CA} = \frac{v_z}{\left[1 + \frac{(1-\varepsilon)}{\varepsilon}\frac{\overline{C}_{FC}}{\overline{C}_{tw}}\frac{\partial \overline{x}_{AR}}{\partial \overline{x}_{Aw}}\right]}. \tag{7.1.40a}$$

The corresponding expression for $v^*_{CA}|_{shock}$ may be obtained from a mass balance (as in equation (7.1.15c)):

$$v_{CA}^{*} = \frac{v_z}{\left[1 + \frac{(1-\varepsilon)}{\varepsilon} \frac{\overline{C}_{FC}}{\overline{C}_{tw}} \frac{\Delta \overline{x}_{AR}}{\Delta \overline{x}_{Aw}}\right]}$$

$$= \frac{v_z}{\left[1 + \frac{(1-\varepsilon)}{\varepsilon} \frac{\overline{C}_{FC}}{\overline{C}_{tw}} \frac{\overline{x}_{AR}|_{\text{shock}} - \overline{x}_{AR}|_{\text{initial}}}{\overline{x}_{Aw}|_{\text{shock}} - \overline{x}_{Aw}|_{\text{initial}}}\right]}. \qquad (7.1.40b)$$

For an initially empty column, $\overline{x}_{AR}|_{\text{initial}} = \overline{x}_{Aw}|_{\text{initial}} = 0$. (*Note*: In determining $(\mathrm{d}x_{AR}/\mathrm{d}x_{Aw})$, the overbar has no relevance since the equilibrium relation is independent of column cross section. However, there may be variations along the cross section of x_{Aw}, therefore $\overline{x}_{Aw}$ is relevant.)

In such a context, the Langmuir ion exchange isotherm (3.3.121n) for counterion A may be written as

$$\frac{q_{AR}}{q_{\max R}} = \frac{C_{AR}}{C_{FC}} = x_{AR} = \frac{K_{AB}\, C_{Aw}/C_{tw}}{(1 + (C_{Aw}/C_{tw})(K_{AB} - 1))}. \qquad (7.1.41)$$

Note that here the counterionic species under consideration along with counterion A is counterion B, so that (compare (3.3.121g,h))

$$C_{AR} + C_{BR} = C_{FC}; \qquad C_{Aw} + C_{Bw} = C_{tw}; \qquad (7.1.42a)$$

$$x_{AR} + x_{BR} = 1; \qquad x_{Aw} + x_{Bw} = 1, \qquad (7.1.42b)$$

and the equilibrium constant K_{AB} for the ion exchange process between counterions A and B is

$$K_{AB} = \frac{C_{AR}\, C_{Bw}}{C_{Aw}\, C_{BR}}. \qquad (7.1.42c)$$

From isotherm (7.1.41), one can calculate $(\partial x_{AR}/\partial x_{Aw})$ for the Langmuir isotherm as

$$\frac{\mathrm{d}x_{AR}}{\mathrm{d}x_{Aw}} = \frac{K_{AB}}{(1 + x_{Aw}(K_{AB} - 1))^2}. \qquad (7.1.42d)$$

Following the column mass balance based development of (7.1.15c), one can determine the concentration wave velocity of the shock wave at time $\overline{t}$ when the exit concentration jumps from $C_{i2} = 0$ to C_{i2}^0, the incoming concentration:

$$\frac{L}{\overline{t}} = \overline{v}_{Ci}^{*} = \overline{v}_{Ci}^{*}|_{\text{shock}} = \frac{v_z}{\left[1 + \frac{(1-\varepsilon)}{\varepsilon} \frac{\overline{C}_{i1}^{0}}{\overline{C}_{i2}^{0}}\right]}. \qquad (7.1.43a)$$

The corresponding quantities in the ion exchange process for the counterion A are

$$\frac{L}{\overline{t}} = \overline{v}_{Ci}^{*}|_{\text{shock}} = \frac{v_z}{\left[1 + \frac{(1-\varepsilon)}{\varepsilon} \frac{\overline{C}_{AR}}{\overline{C}_{Aw}}\right]} \qquad (7.1.43b)$$

for an initially empty column. Therefore

$$\frac{L}{\overline{t}} = \overline{v}_{Ci}^{*}|_{\text{shock}} = \frac{v_z}{\left[1 + \frac{(1-\varepsilon)}{\varepsilon} \frac{\overline{C}_{FC}}{\overline{C}_{tw}} \frac{\overline{x}_{AR}}{\overline{x}_{Aw}}\right]}. \qquad (7.1.43c)$$

Example 7.1.3 Using the equations developed above, we will now illustrate an ion exchange system worked out by Wankat (1990, pp. 465–470) for removal of K^+ from a solution using an ion exchange bed in the form of Na^+. Langmuir isotherm describes this ion exchange with $K_{AB} = 1.54$, where $A = K^+$ and $B = Na^+$. The mole fraction of potassium in the 0.2 N feed solution $\overline{x}_{Kw} = 0.7$. Determine the concentration wave velocity of K^+ and determine when this shock wave exits the column of length 30 cm and diameter 2 cm if the volumetric flow rate of the solution is 20 cm^3/min.

Next, regeneration of the bed is implemented by passing a solution of 0.2N NaCl. Calculate the K^+ concentration in the effluent as a function to time (eluent volume). You are given: $\varepsilon = 0.4$, $\overline{C}_{FC} = 3.96$ equivalent/liter of resin.

Solution To employ equation (7.1.43c) for the shock wave velocity $v_{CA}^{*}|_{\text{shock}}$, we need v_z, the interstitial velocity:

$$20\frac{\text{cm}^3}{\text{min}} = \frac{\pi}{4} d^2 \times v_z \times \varepsilon = \frac{\pi}{4} \times 4\,\text{cm}^2 \times v_z \times 0.4 \Rightarrow v_z = \frac{20}{\pi \times 0.4};$$

$$v_z = 15.9\,\text{cm/min} \Rightarrow v_{CA}^{*}|_{\text{shock}} = \frac{15.9\,\text{cm/min}}{\left[1 + \left(\frac{0.6}{0.4}\right)\left(\frac{3.96}{0.2}\right)\frac{\overline{x}_{AR}}{\overline{x}_{Aw}}\right]}.$$

For the Langmuir isotherm, (7.1.41),

$$x_{AR} = \frac{K_{AB} x_{Aw}}{1 + x_{Aw}(K_{AB} - 1)} = \frac{1.54 \times 0.7}{1 + 0.7(0.54)} = \frac{1.078}{1.378} \Rightarrow x_{AR} = 0.782.$$

Therefore

$$v_{CA}^{*}|_{\text{shock}} = \frac{15.9\,\text{cm/min}}{[1 + 1.5 \times 18 \times (0.782/0.7)]}$$

$$= \frac{15.9\,\text{cm/min}}{34.18} = 0.465\,\text{cm/min} = v_{CK}^{*}|_{\text{shock}}.$$

Since the column length is 20 cm, this shock wave will appear at the column end at 43 $(= 20/v_{CK}^{*}|_{\text{shock}})$ minutes. In a process to remove K^+ from the solution, the feed solution flow will be stopped somewhat earlier so that the effluent solution from the column is K^+-free.

We will now focus on the regeneration of the bed by the 0.2 N NaCl stream at $t = 43$ minute. If we employ equation (7.1.17g) for $z_0(\overline{C}_{i2}) = 0$ for $\overline{C}_{K2}$ (at the column inlet), then

$$V - V^0 = S_c z \varepsilon \left[1 + \frac{(1-\varepsilon)}{\varepsilon} \frac{\mathrm{d}\overline{C}_{K1}}{\mathrm{d}\overline{C}_{K2}}\right]$$

$$= S_c z \varepsilon \left[1 + \frac{(1-\varepsilon)}{\varepsilon} \frac{\overline{C}_{FC}}{\overline{C}_{tw}} \frac{\mathrm{d}\overline{x}_{AR}}{\mathrm{d}\overline{x}_{Aw}}\right].$$

We know that $V^0 = 43\,\text{minute} \times 20\,\text{cm}^3/\text{min} = 860\,\text{cm}^3$. Now

$$(\mathrm{d}\overline{x}_{AR}/\mathrm{d}\overline{x}_{Aw}) = \frac{K_{AB}}{[1 + x_{Aw}(K_{AB} - 1)]^2}$$

$$= \frac{K_{KNa}}{[1 + x_{Kw}(K_{KNa} - 1)]^2} = \frac{1.54}{[1 + x_{Kw} \times 0.54]^2}.$$

The results in Table 7.1.3 illustrate the diffuse nature of the K^+-containing stream exiting the column. The book by Wankat (1990, pp. 463–494) should be read for greater details of such equilibrium nondispersive calculation procedures for ion exchange columns. See Sherwood *et al.* (1975) for models involving mass-transfer resistance.

Table 7.1.3.

x_{Kw}	$(d\overline{x}_{AR}/d\overline{x}_{Aw})$	$S_c\, z\, \varepsilon \left(1 + \frac{(1-\varepsilon)}{\varepsilon} \frac{\overline{C}_{FC}}{\overline{C}_{tw}} \frac{d\overline{x}_{AR}}{d\overline{x}_{Aw}}\right) = (V - V^0)\,\text{cm}^3$
0.7	0.81	630
0.5	0.95	734
0.3	1.14	850
0.1	1.4	1045
0.0	1.54	1174

7.1.2 Pressure-swing adsorption process for gas separation

The fixed-bed processes studied earlier use adsorbents which are mass-separating agents. These mass-separating agents have to be regenerated for reuse. An introduction to the process of fixed-bed adsorbent regeneration was provided in Section 7.1.1.3. The bed may be regenerated by a temperature increase, a pressure decrease, a purge stream, a displacer stream or various combinations of these. **In an actual fixed-bed separation process, the adsorption step and the desorption step are combined in a cyclic fashion.** The manner in which these cycles are operated has led to a few distinct classes of fixed-bed processes. The nature of the introduction of the feed (semicontinuous, pulse, temperature programmed, etc.) is integral to the development of such distinct processes. However, the direction of the force between the fluid and the fixed bed in the following processes always remains perpendicular to the bulk fluid flow direction:

(1) thermal-swing adsorption;
(2) pressure-swing adsorption (PSA);
(3) inert purge;
(4) potential-swing adsorption;
(5) parametric pumping;
(6) cycling zone adsorption;
(7) chromatographic processes.

We focus first on pressure-swing adsorption processes. Potential-swing adsorption is mentioned in passing. In the three sections that follow, we consider parametric pumping, cycling zone adsorption and chromatographic processes. Thermal-swing adsorption and inert purge have already been briefly considered.

The basic mode of operation in *pressure-swing adsorption* will be illustrated first using the Skarstrom cycle (Skarstrom, 1960, 1975) employed for air drying (Figure 7.1.12). The process uses two identical beds of a granular solid adsorbent which preferentially adsorbs moisture from air. Wet air at 40 psig enters bed A through the connection shown by a solid line in a four-way solenoid valve. The air leaving the other end of bed A is essentially dry air at 40 psig. About half of the wet air entering is taken off as dry air product; the other half, also produced as dry air, is throttled down to essentially atmospheric pressure and introduced into bed B from the opposite end (i.e. the exit end of bed A). This low-pressure dried air acts as a purge to desorb the moisture from bed B, which has already been saturated with H_2O during a previous cycle. The resulting wet air is taken out through the front end of bed B and discharged to the atmosphere.

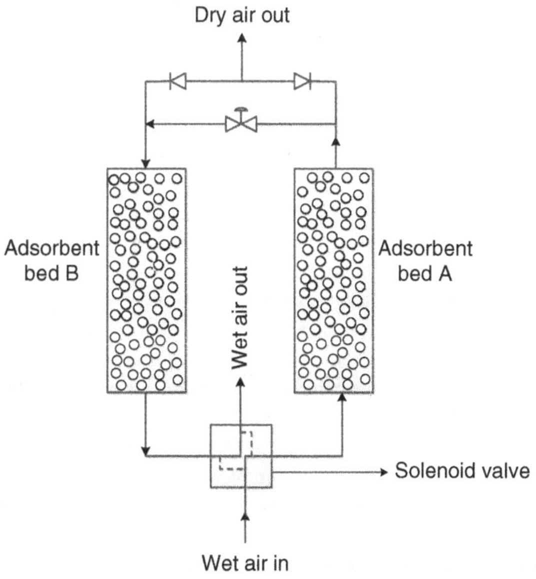

Figure 7.1.12. Continuous adsorption drying system using pressure-swing adsorption. (After Skarstrom (1960, 1975).)

This process is continued for 3 minutes, by the end of which bed A is saturated with moisture and bed B is essentially dry and regenerated. Then the solenoid valve connection is changed. Wet air is introduced at 40 psig into bed B by a connection in the solenoid valve, shown in Figure 7.1.12 as dashed lines. This wet air leaves bed B at the other end dry. Half of this exiting dried air is taken out as dry product at 40 psig; the other half is throttled down to near atmospheric pressure and sent to the previous product end of bed A, whose pressure has now been quickly reduced. This dry air, acting as a purge, flows through bed A in the opposite direction and exits at the other end as wet air through the connection, indicated by the dashed lines of the four-way solenoid valve. It is released to the atmosphere. This process continues for another 3 minutes, when bed B becomes saturated and bed A is regenerated. At this time, the cycle is complete, and wet air is sent to bed A ready for adsorption as before, while bed B has to be regenerated. Meanwhile, dry air at 40 psig is obtained continuously as the product.

Such a process was originally known as *heatless fractionation* or *heatless adsorption* since no heat was used to regenerate the adsorbents. Currently the term pressure-swing adsorption (PSA) is used instead. Skarstrom (1975) demonstrated that air containing 3800 volume ppm of H_2O (11 mm Hg H_2O partial pressure) at 40 psig was dried to 1 ppm H_2O at the same pressure. However, it took some time (5 days) to achieve a steady state in the moisture concentration in the dried air output. In modeling PSA processes, therefore, steady state achievement is usually indicated by $n \rightarrow \infty$, where n is the number of cycles.

Since the original patent of Skarstrom was issued in 1960 (Skarstrom, 1960), there has been considerable progress in the research and development of PSA processes. A large number of commercial plants based on PSA are used to purify or fractionate a variety of gas mixtures and vapor mixtures. These include: drying of air and other gases; air separation by zeolites (N_2 is more strongly adsorbed than O_2 at equilibrium) and by carbon molecular sieves (O_2 more rapidly adsorbed, therefore the high-pressure product is a highly purified N_2 stream); H_2 purification to very high levels (~99.9999%) from various refinery streams; adsorption removal of n-paraffins from mixtures containing branched-chain isomers and cyclic hydrocarbons, etc. (vacuum desorption is used for n-paraffins with carbon numbers less than 10). The adsorbent particle diameter varies between 1/16″ pellets to as low as 60 mesh. The number of beds used simultaneously is often more than two, going all the way up to ten. The cycle time can go as low as a few seconds, especially if a single-bed based rapid PSA is used (Keller, 1983). For feed gases slightly above atmospheric pressure, vacuum-swing adsorption has been used to desorb via vacuum (Sircar and Zondlo, 1977). An excellent treatment of PSA based processes is available in Yang (1987), which also provides a critical account of PSA models and experiments.

Before we consider PSA models, one should recognize that separation by adsorption in PSA can take place in two general ways: equilibrium separation and kinetic separation. In equilibrium separation, the adsorbent preferentially adsorbs one or more species in preference to others on an equilibrium basis. In zeolite adsorbents, as well as molecular sieves, molecules having an appropriately small dimension can enter the zeolite pores and are adsorbed, whereas others cannot enter the pores due to steric effects and are excluded. Such adsorption processes are considered under equilibrium separation processes (Yang, 2003). On the other hand, adsorption separation by kinetic effects involves adsorbents through whose pores one of the species diffuses much more readily than others (e.g. O_2 diffuses 30 times faster than N_2 through carbon molecular sieves having the right pore opening dimensions). Such separations come under a kinetic separation mechanism in PSA processes.

7.1.2.1 *An equilibrium nondispersive PSA model*

Consider the two columns or two beds used in Skarstrom's scheme for air drying. A simple PSA cycle for this process involves four steps, shown schematically in Figure 7.1.13(a), where two beds, bed 1 and bed 2, are identified by 1 and 2, respectively. We provide first the cycle description according to Chan *et al.* (1981).

A high-pressure feed mixture is continuously supplied to the top of the bed 1 in step 1 and the adsorbable species are taken up by the bed. The effluent from the end of the bed is sufficiently purified and still at a high pressure. A part of this effluent becomes the desired product; the pressure of the remaining effluent is reduced without any change in mole fraction to the low-pressure level of bed 2, where it is introduced to the column top as a low-pressure purge to desorb species adsorbed by bed 2 during an earlier cycle. Figure 7.1.13(b) shows the essentially constant bed 1 pressure and steady flow rates of high-pressure feed and product and low-pressure purge during step 1 (Weaver and Hamrin, 1974).

After some time, bed 1 approaches breakthrough conditions; step 2 is therefore initiated, wherein bed 1 is depressurized and there is removal of the adsorbed species in the blowdown stream from the end of the bed where the high-pressure feed was introduced earlier. Simultaneously, bed 2, ready for adsorption, is pressurized by the high-pressure feed gas mixture. Step 2 is of short duration, and is followed by step 3, in which bed 2 is fed with high-pressure feed gas continuously. In step 3, bed 2 behaves as bed 1 in step 1, just as bed 1 behaves in step 3 as bed 2 did in step 1. Similarly, in step 4, bed 1 is subjected to pressurization with high-pressure feed (as bed 2 was in step 2); while bed 2 undergoes blowdown prior to complete regeneration by a low-pressure product purge (as in step 1)

A model developed by Chan *et al.* (1981) for the purification of a **binary gas mixture of species A and B by such a PSA cycle when A is present at a trace level** will be presented below. Focus on bed 1's operation. The assumptions employed in this model are:

(1) isothermal operation;
(2) negligible pressure drop along the bed during high-pressure feed flow or low-pressure purge flow;
(3) gas–solid equilibrium exists always and everywhere;
(4) linear adsorption isotherms are valid, with species A being preferentially adsorbed:

$$\kappa_{A1} > \kappa_{B1} : \; C_{A1} = \kappa_{A1} C_{A2} \tag{7.1.44a}$$

$$C_{B1} = \kappa_{B1} C_{B2}; \tag{7.1.44b}$$

(5) no axial dispersion;
(6) ideal gas law is valid;
(7) the interstitial gas velocity v_z is constant during constant-pressure steps 1 and 3.

Any model for PSA should provide ways of calculating the fractional recovery of the less adsorbed species, its purity, the purity of the strongly adsorbed species dominated fraction, the minimum bed length, etc. We follow to these ends the treatment by Chan *et al.* (1981) and focus on steady state operation obtained after initial transients. The governing

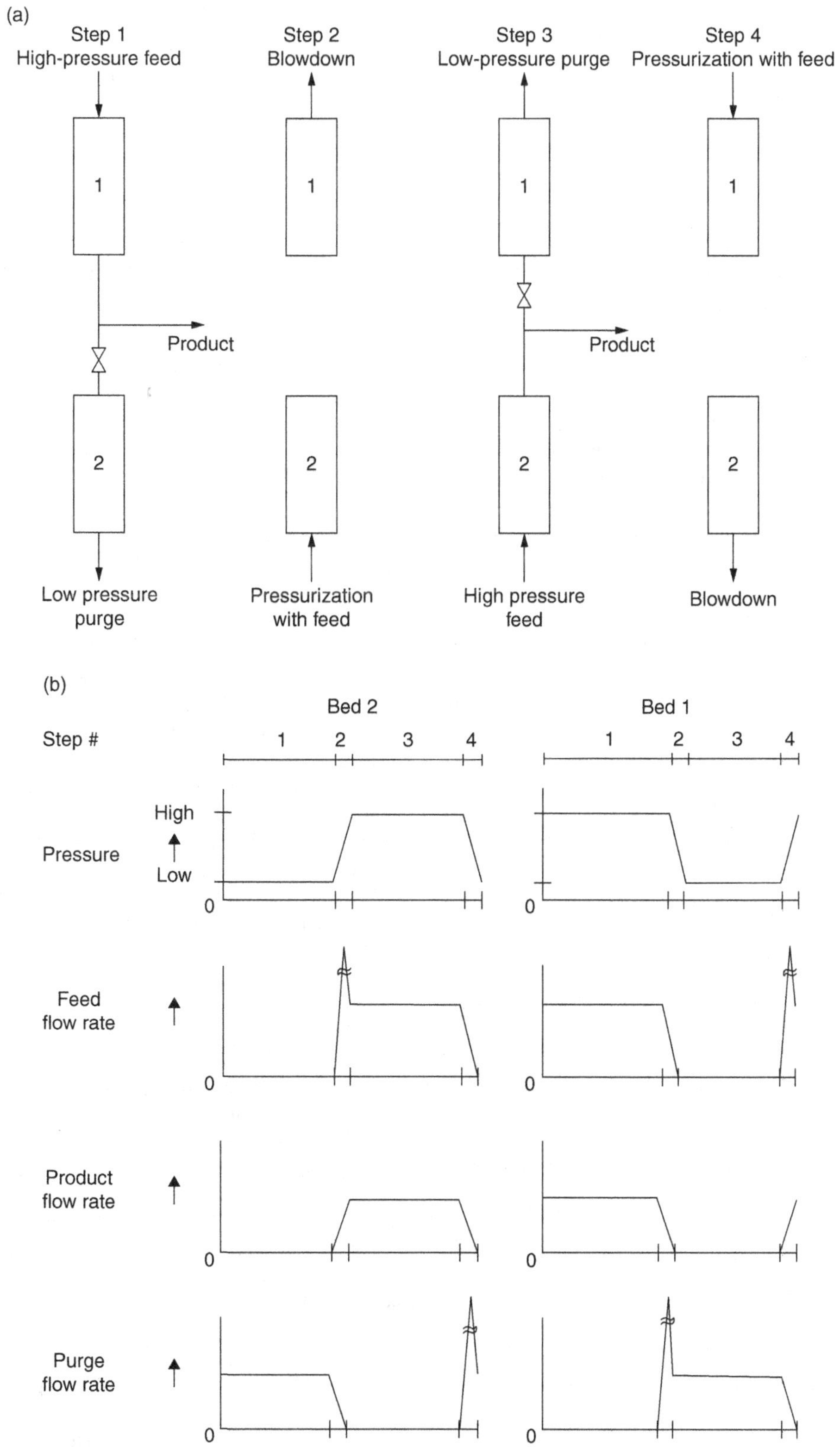

Figure 7.1.13. (a) Four steps in a PSA cycle. (b) Diagrams of pressure and flow rate changes for beds 1 and 2 during one cycle. (After Weaver and Hamrin (1974).)

balance equations for species A and B are obtained from equation (7.1.3) under nondispersive condition as

$$\varepsilon\left[\frac{\partial \overline{C}_{A2}}{\partial t}+\frac{\partial(v_z\overline{C}_{A2})}{\partial z}\right]+(1-\varepsilon)\frac{\partial \overline{C}_{A1}}{\partial t}=0; \tag{7.1.45a}$$

$$\varepsilon\left[\frac{\partial \overline{C}_{B2}}{\partial t}+\frac{\partial(v_z\overline{C}_{B2})}{\partial z}\right]+(1-\varepsilon)\frac{\partial \overline{C}_{B1}}{\partial t}=0. \tag{7.1.45b}$$

Recall that $\overline{C}_{A2}$ and $\overline{C}_{B2}$ refer to (z, t)-dependent molar concentrations of species A and B in the gas phase, while $\overline{C}_{A1}$ and $\overline{C}_{B1}$ refer to those in the solid adsorbent phase, as before. Replace κ_{ig} by y_i here. Due to ideal gas behavior,

$$\overline{C}_{A2}=p_A/RT=(Py_A/RT); \qquad \overline{C}_{B2}=p_B/RT=(Py_B/RT), \tag{7.1.46}$$

where P is the total pressure and y_A and y_B are the local (z, t)-dependent mole fractions of species A and B in the gas phase. Since species A is present in very dilute concentration, we may assume that $y_B \cong 1$ and represent y_A henceforth only by y. Substituting these into equations (7.1.45a) and (7.1.45b) (where we now use the equilibrium relations (7.1.44a) and (7.1.44b)) leads to

$$\varepsilon\left[\frac{\partial(Py)}{\partial t}+\frac{\partial(v_z Py)}{\partial z}\right]+(1-\varepsilon)\kappa_{A1}\frac{\partial(Py)}{\partial t}=0; \tag{7.1.47a}$$

$$\varepsilon\left[\frac{\partial P}{\partial t}+\frac{\partial(v_z P)}{\partial z}\right]+(1-\varepsilon)\kappa_{B1}\frac{\partial(P)}{\partial t}=0. \tag{7.1.47b}$$

Since $(\partial P/\partial z)$ is zero by assumption in either bed/column at any time, we get

$$\varepsilon\left[y\frac{\partial P}{\partial t}+P\frac{\partial y}{\partial t}+v_z P\frac{\partial y}{\partial z}+Py\frac{\partial v_z}{\partial z}\right]+(1-\varepsilon)\kappa_{A1}\left[y\frac{\partial P}{\partial t}+P\frac{\partial y}{\partial t}\right]=0; \tag{7.1.48a}$$

$$\varepsilon\left[\frac{\partial P}{\partial t}+P\frac{\partial v_z}{\partial z}\right]+(1-\varepsilon)\kappa_{B1}\frac{\partial P}{\partial t}=0. \tag{7.1.48b}$$

Now multiply equation (7.1.48b) by y and subtract it from equation (7.1.48a) to obtain

$$[\varepsilon+(1-\varepsilon)\kappa_{A1}]\frac{\partial y}{\partial t}+\varepsilon v_z\frac{\partial y}{\partial z}=-(1-\varepsilon)(\kappa_{A1}-\kappa_{B1})y\frac{\partial \ln P}{\partial t}. \tag{7.1.49a}$$

To use the method of characteristics (see equations (7.1.12b) to (7.1.12f)), we can write any solution of y as

$$\left(\frac{\partial y}{\partial t}\right)dt+\left(\frac{\partial y}{\partial z}\right)dz=dy \tag{7.1.49b}$$

to obtain

$$\left(\frac{\partial y}{\partial t}\right)=\frac{\begin{vmatrix} dy & dz \\ -(1-\varepsilon)(\kappa_{A1}-\kappa_{B1})y\dfrac{d\ln P}{dt} & \varepsilon v_z \end{vmatrix}}{\begin{vmatrix} dt & dz \\ \varepsilon+(1-\varepsilon)\kappa_{A1} & \varepsilon v_z \end{vmatrix}} \tag{7.1.50}$$

If the determinant in the denominator is zero, we get

$$\frac{dz}{dt}=\frac{\varepsilon v_z}{[\varepsilon+(1-\varepsilon)\kappa_{A1}]}=\frac{v_z}{\left[1+\frac{(1-\varepsilon)}{\varepsilon}\kappa_{A1}\right]}. \tag{7.1.51a}$$

For $(\partial y/\partial t)$ to be finite, the numerator has to be zero, which leads to

$$\varepsilon v_z\frac{d\ln y}{dz}=-(1-\varepsilon)(\kappa_{A1}-\kappa_{B1})\frac{d\ln P}{dt}.$$

Using (7.1.51a), this may be rearranged to yield

$$\frac{d\ln y}{dt}=-\frac{(1-\varepsilon)(\kappa_{A1}-\kappa_{B1})}{[\varepsilon+(1-\varepsilon)\kappa_{A1}]}\frac{d\ln P}{dt}\Rightarrow\frac{d\ln y}{dt}=\left(\frac{\beta_A}{\beta_B}-1\right)\frac{d\ln P}{dt}, \tag{7.1.51b}$$

where

$$\beta_A=\frac{\varepsilon}{\varepsilon+(1-\varepsilon)\kappa_{A1}},\quad \beta_B=\frac{\varepsilon}{\varepsilon+(1-\varepsilon)\kappa_{B1}},\quad \beta_A<\beta_B. \tag{7.1.52a}$$

Note that, if the species are strongly adsorbed, i.e. $(1-\varepsilon)\kappa_{A1}>>\varepsilon$ and $(1-\varepsilon)\kappa_{B1}>>\varepsilon$, then the separation factor α_{AB} for species A and B is

$$\alpha_{AB}\cong\frac{\kappa_{A1}}{\kappa_{B1}}=\frac{\beta_B}{\beta_A}=\frac{1}{\beta}. \tag{7.1.52b}$$

Since the interstitial velocity v_z is constant during steps 1 and 3 by assumption (7), the characteristic velocity of any y value, (dz/dt), is also constant from (7.1.51a) since κ_{A1} is constant. On the other hand, v_z varies during steps 2 and 4; the pressure also changes, just as mole fraction y also changes. The interstitial velocity during steps 1 and 3 may be obtained by dividing the volumetric flow rate by εS_C, where S_C is the column cross-sectional area.

We will now calculate the changes in the position of the characteristics for these two kinds of changes. For steps 1 and 3, integrate (7.1.51a) and obtain, respectively,

$$\Delta z_H=\beta_A v_{zH}\Delta t; \tag{7.1.53a}$$

$$\Delta z_L=\beta_A v_{zL}\Delta t. \tag{7.1.53b}$$

Here Δz_H and Δz_L, called the penetration distances, are the net displacements of the concentration wave front in the high-pressure feed step 1 (subscript H), having a steady interstitial gas velocity v_{zH}, and the low-pressure purge step 3 (subscript L), having a steady interstitial gas velocity v_{zL}, respectively.

For the blowdown (step 2) and pressurization with feed (step 4), consider equation (7.1.51b) first. This allows the gas mole fraction y to be related to pressure P in the column in two cases: as the high pressure, P_H, is reduced very quickly to a low pressure, P_L, during blowdown step 2, y is increased from y_H to y_L; as the low pressure, P_L, is increased to P_H during the pressurization with feed in step 3, y is decreased from y_L to y_H. Integration of equation (7.1.51b) leads to estimates of such changes:

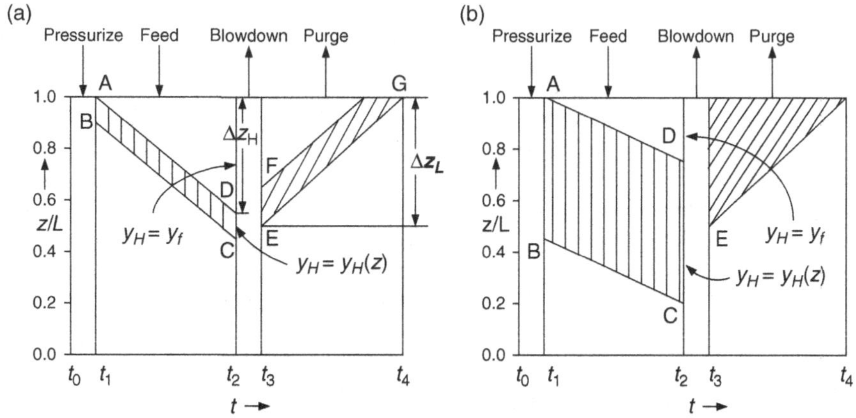

Figure 7.1.14. Characteristics movement during a PSA cycle for the adsorption of a binary mixture of A and B very dilute in A; $G = G_{crit}$, $(L/\Delta z_L) = 2$, $(P_H/P_L) = 10$. *(a)* $\beta = 0.05$, $\left[1-[P_H/P_L]^{-\beta}\right]L-\Delta z_H < 0$; *(b)* $\beta = 0.4$, $\left[1-[P_H/P_L]^{-\beta}\right]L-\Delta z_H \geq 0$. *Reprinted from* Chem. Eng. *Sci.,* ***36****, (1981), p. 243, Y.N.I Chan, F.B. Hill, Y.W. Wong, "Equilibrium theory of a pressure swing adsorption process," copyright (1981), with permission from Elsevier.*

$$(y_H/y_L) = (P_H/P_L)^{\beta-1}, \qquad (7.1.54a)$$

where

$$\beta = (\beta_A/\beta_B) < 1. \qquad (7.1.54b)$$

It is necessary to calculate how the characteristics position will change as the transition takes place rapidly from P_H to P_L (blowdown) or from P_L to P_H (pressurization with feed). For this purpose, we use equation (7.1.48b) in a rearranged form,

$$\frac{\partial v_z}{\partial z} = -\left\{\frac{\varepsilon + (1-\varepsilon)\kappa_{B1}}{\varepsilon}\right\}\frac{\partial \ln P}{\partial t},$$

and integrate with respect to z (since the right-hand side is not a function of z, remember $(\partial P/\partial z) = 0$) to obtain

$$v_z = -\frac{1}{\beta_B}\left(\frac{\partial \ln P}{\partial t}\right)z + \text{constant}. \qquad (7.1.55a)$$

If the coordinate system of the column is such that the closed end of the column of length L (Figure 7.1.14) is identified as $z = 0$, then, during the blowdown or repressurization steps, $v_z = 0$ at $z = 0$.

That the constant in equation (7.1.55a) is zero implies that

$$v_z = -\frac{1}{\beta_B}\left(\frac{\partial \ln P}{\partial t}\right)z. \qquad (7.1.55b)$$

Substitute this expression for v_z into the basic equation (7.1.51a) for characteristics movement along the column to get

$$\frac{dz}{dt} = -\left(\frac{\partial \ln P}{\partial t}\right)\frac{z\beta_A}{\beta_B}. \qquad (7.1.55c)$$

Rearrange it as

$$\frac{d\ln z}{dt} = -\beta\frac{d\ln P}{dt}. \qquad (7.1.55d)$$

Integrate to obtain

$$\ln\frac{z_H}{z_L} = \ln\left[P_H^{-\beta}\Big/P_L^{-\beta}\right] \Rightarrow (z_H/z_L) = (P_H/P_L)^{-\beta}, \qquad (7.1.55e)$$

where z_H and z_L are the z-coordinates of the characteristics when the pressures are P_H and P_L, respectively. Obviously, the absolute value of z at the end of any such process will depend on the z-coordinate at the beginning of the process, which will be influenced by prior steps.

It is useful now to calculate the net displacement of a concentration front during a complete cycle of four steps in PSA. Assume the column/bed length to be L and that step 1 is as shown in Figure 7.1.13(a) with the high-pressure feed continuously entering at the top end ($z = L$). Prior to this, step 4 occurred wherein the gas (mole fraction y_f, feed mole fraction) entered the column ($z = L$) at a low pressure, P_L. Gas entering the column a short time later enters at a slightly higher column pressure, and the final gas in this step enters at pressure P_H. The characteristics location of the initial entering gas is obtained from (7.1.55e) as

$$z_H = L(P_H/P_L)^{-\beta}, \qquad (7.1.56a)$$

where z_L is assumed to be L. Based on this assumption, the z_H location is now at B. The specific characteristic path is difficult to plot for this step. This gas concentration front (Figure 7.1.13(a)) (now at pressure P_H) moves down the column by Δz_H during step 1 (high-pressure feed flow) so that the lowest z-coordinate at the end of step 1 (point C in Figure 7.1.14(a)) has the value

$$z_H - \Delta z_H = L(P_H/P_L)^{-\beta} - \beta_A v_{zH} \Delta t. \quad (7.1.56b)$$

The triangular region above AD has only a feed-gas composition. The gas composition changes between C and D, it being highest at D ($= y_f$). The blowdown step that follows (step 2) pushes the characteristics back. The z_H and z_L for this step are still related by (7.1.55e). Since z_H in this relation is valid at the start of this step, it must be equal to that of the location described by (7.1.56b):

$$z_L(P_H/P_L)^{-\beta} = L(P_H/P_L)^{-\beta} - \beta_A v_{zH} \Delta t$$
$$\Rightarrow z_L = L - \beta_A v_{zH} \Delta t (P_H/P_L)^{\beta}, \quad (7.1.56c)$$

corresponding to point E, just as point F corresponds to starting point D. Now comes the low-pressure purge step, which pushes the characteristics still further toward $z = L$ by Δz_L (equation (7.1.53b)); thus, the final z-coordinate (location G) at the end of step 3 is

$$L - \beta_A v_{zH} \Delta t (P_H/P_L)^{-\beta} + \beta_A v_{zL} \Delta t. \quad (7.1.56d)$$

At the start of step 4, the characteristics front was at $z = L$ (the feed gas entered the column at $P = P_H$). Therefore, in the whole cycle, the net displacement of the characteristics front from $z = L$ has been

$$\Delta z|_{\text{overall}} = -\beta_A v_{zH} \Delta t (P_H/P_L)^{-\beta} + \beta_A v_{zL} \Delta t$$
$$= \beta_A \Delta t \left[v_{zL} - v_{zH} (P_H/P_L)^{-\beta} \right]. \quad (7.1.57a)$$

Figure 7.1.14(a) is drawn such that $\Delta z|_{\text{overall}} = 0$.

The ratio of v_{zL}, the velocity during the low-pressure purge step, and v_{zH}, the velocity during the high-pressure feed step, is an important parameter, γ, in PSA:

$$\gamma = (v_{zL}/v_{zH}). \quad (7.1.57b)$$

If $\Delta z|_{\text{overall}}$ for a cycle is zero, then a critical value of (v_{zL}/v_{zH}) is reached:

$$\gamma_{\text{crit}} = \left(\frac{v_{zL}}{v_{zH}} \right)_{\text{crit}} = \left(\frac{P_H}{P_L} \right)^{\beta}. \quad (7.1.57c)$$

This definition is valid only when Δt for the low-pressure purge step 3 is equal to that for the high-pressure feed step 1.

Consider now step 4 and then step 1. Assuming that the characteristics start at $z_L = L$ (top of the column where fresh feed enters) with $P = P_L$ and $y = y_f$ in step 4, z_H by equation (7.1.55e) for any P is

$$z_H(P) = L(P_L/P)^{\beta}, \qquad P_L \leq P \leq P_H. \quad (7.1.58a)$$

By equation (7.1.54a), the y value at any P and $z_H(P)$ is then related to y_f by

$$y_H(P) = (P_L/P)^{1-\beta} y_f, \qquad P_L \leq P \leq P_H \quad (7.1.58b)$$

so that

$$y_H(z) = y_f (z/L)^{((1/\beta)-1)}. \quad (7.1.58c)$$

This expression provides the value of the gas mole fraction of species A in the advancing front in the pressurization step 4 (Figure 7.1.5(b) illustrates how the composition decreases in the front from the feed composition to the exiting composition). When $P = P_H = P_f$, step 1 begins, and the characteristics are pushed further into the column toward $z = 0$ by an amount Δz_H. As long as $z_H(P_H)$ from (7.1.58a) is larger than Δz_H, the feed-gas composition never appears at $z = 0$ and therefore never breaks through into the high-pressure product leaving the column at $z = 0$. This condition is identified by

$$z_H(P_H) = L(P_L/P_H)^{\beta} \geq \Delta z_H = \beta_A v_{zH} \Delta t. \quad (7.1.58d)$$

If, in addition, $(v_{zL}/v_{zH}) \geq \gamma_{\text{crit}}$, then Chan *et al.* (1981) have indicated that, at steady state, complete removal of trace component A from the high-pressure product stream is obtained. The same result was theoretically demonstrated earlier by Shendalman and Mitchell (1972) for the same system, except that species B was inert and was not adsorbed at all.

A more general index to identify the onset of complete purification is G_{crit}, the value of the fraction of the feed introduced during steps 1 and 3, just required as purge in order to obtain complete purification:

$$G_{\text{crit}} = \frac{v_{zL}}{v_{zH}} \frac{P_L}{P_H} = \frac{\text{amount of species B in purge}}{\text{amount of species B in feed}}. \quad (7.1.58e)$$

On the basis of (7.1.57c),

$$G_{\text{crit}} = (P_L/P_H)^{\beta - 1}. \quad (7.1.58f)$$

A plot of G_{crit} against (P_H/P_L) for various values of β is provided in Figure 7.1.15. The fraction of feed required as purge decreases as (1) the ratio of the high feed pressure to the low purge pressure increases and (2) the value of β decreases, i.e. the separation factor increases. However, too high a ratio of (P_H/P_L) does not reduce G_{crit} by much.

The fraction of the major component (species B) fed to the process which is recovered in pure form in the high-pressure product stream at steady state (identified by subscript ∞) is

$$f_{\infty} = \frac{m_{\text{B,feed}} - m_{\text{B,purge}}}{m_{\text{B,feed}} + m_{\text{B,blowdown}}}, \quad (7.1.59a)$$

where m_{B} represents the number of moles of species B introduced to or leaving the process per half-cycle. Assuming the ideal gas law to be valid, for any column/bed having a total volume V, cross-sectional area S_c ($= V/L$) and a void volume fraction ε,

$$m_{\text{B,feed}} = \varepsilon (V/L) v_{zH} \Delta t (P_H/RT); \quad (7.1.59b)$$

$$m_{\text{B,purge}} = \varepsilon (V/L) v_{zL} \Delta t (P_L/RT), \quad (7.1.59c)$$

since the feed gas occupies the void volume at pressure P_H, whereas the purge occupies the void volume at pressure

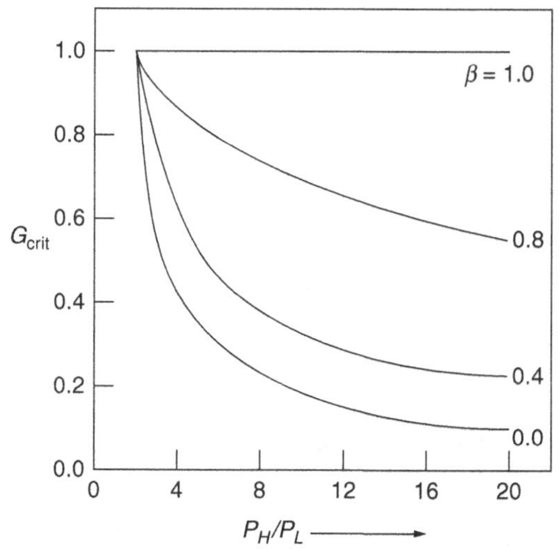

Figure 7.1.15. Critical purge-to-feed ratio. (After Chan et al. *(1981).)*

P_L. The number of moles removed during blowdown (when pressure is reduced from P_H to P_L) is

$$m_{B,\text{blowdown}} = [\varepsilon + (1-\varepsilon)\kappa_{B1}]V[P_H - P_L]/RT = \frac{\varepsilon V[P_H - P_L]}{\beta_B RT}, \tag{7.1.59d}$$

which includes the gas leaving the void volume of the bed over a pressure range (P_H–P_L) and the adsorbed gas in the pressure range (P_H–P_L). Therefore an expression for f_∞ is

$$f_\infty = \frac{\varepsilon(V/L)\frac{\Delta t}{RT}(v_{zH}P_H - v_{zL}P_L)}{\varepsilon(V/L)\frac{\Delta t}{RT}v_{zH}P_H + \frac{\varepsilon V[P_H - P_L]}{\beta_B RT}} = \frac{[P_H/P_L]^{1-\beta} - 1}{[P_H/P_L]^{1-\beta} + \left[\frac{P_H}{P_L} - 1\right]\frac{L\beta}{\Delta z_L}}, \tag{7.1.59e}$$

where we have used the definition for γ_{crit} from (7.1.57c) to replace (v_{zH}/v_{zL}).

If the preferentially adsorbed species A has to be recovered in a more enriched form, the blowdown and the purge stream would be the source; species A mole fraction y_{BDPG} in this combined stream may be obtained from

$$\frac{y_{BDPG}}{y_f} = \frac{\left(\frac{\beta L}{\Delta z_L}\right)\left(\frac{P_H}{P_L} - 1\right) + \left(\frac{P_H}{P_L}\right)^{1-\beta}}{\left(\frac{\beta L}{\Delta z_L}\right)\left(\frac{P_H}{P_L} - 1\right) + 1} = \frac{m_{B,\text{feed}} + m_{B,\text{blowdown}}}{m_{B,\text{purge}} + m_{B,\text{blowdown}}}. \tag{7.1.59f}$$

This ratio is identified as the enrichment of species A by Chan *et al.* (1981). When $\beta \to 0$, this enrichment increases linearly with an increase in (P_H/P_L).

The above expressions provide theoretical estimates of the productivity and process stream quality from a PSA process for a trace component removal by an equilibrium based theory. The column/bed length was given. An alternative design oriented goal would be to calculate the shortest length of the column, L_{min}, i.e. the least amount of adsorbent required for complete removal of species A. This is achieved when $G = G_{crit}$ (equation (7.1.58f)) and $L = \Delta z_L$. These conditions ensure that component A is just completely purged from a column at the end of purge step 3; full adsorption capacity of all adsorbents in the column is available in step 4 onwards. An expression for L_{min} is obtained by replacing v_{zH} in (7.1.59b) using (7.1.57c) and (7.1.53b):

$$L_{min} = \frac{\beta_A m_{B,\text{feed}}(P_H/P_L)^\beta}{\varepsilon S_c(P_H/RT)}. \tag{7.1.60}$$

This equilibrium based analytical theory of a conventional PSA process predicts that the mole fraction of species A in the high-pressure product gas, y, will continue to decrease with time (Shendalman and Mitchell, 1972; Chan *et al.*, 1981). This is contrary to experimental observations. More exact theories based on axial diffusion and mass-transfer effects are needed to predict the observed behavior.

A number of other analyses for PSA systems have been carried out. Shendalman and Mitchell (1972) studied analytically and experimentally the PSA separation of trace amounts of CO_2 from helium, assuming that helium is completely inert and a linear isotherm is valid for CO_2 with instantaneous equilibrium. Fernandez and Kenney (1983) studied air separation in a single-column PSA using a linear equilibrium adsorption model without any axial diffusion. Knaebel and Hill (1985) studied two-column equilibrium PSA for a general binary system using the same framework. Raghavan *et al.* (1985) numerically simulated the CO_2–He system of Shendalman and Mitchell (1972) using axial dispersion and finite gas-to-solid mass-transfer resistance. Bulk separation of a ternary gas mixture, $H_2/CH_4/CO_2$, has been experimentally and numerically studied by Doong and Yang (1986). Variations in the simple PSA cycle, especially in the pressurization and blowdown steps, have been studied. Suh and Wankat (1987) studied a combined cocurrent-countercurrent blowdown cycle; Liow and Kenney (1990) investigated the backfill part of the cycle involving pressurization of the adsorption bed with the product in a direction opposite to that of the feed mixture. The design of PSA systems has been treated by White and Barkley (1989).

Example 7.1.4 A refinery gas stream at 10.13×10^5 Pa contains primarily H_2 besides some impurities. Consider this as a binary mixture where the impurity species A adsorbs at the gas temperature onto the porous adsorbent used according to q_{A1} (gmol/g adsorbent) $= 40C_{A2}$ (gmol/cm^3). This H_2 stream has to be purified by a two-bed PSA process using the above adsorbent ($\rho_s = 2$ g/cm^3); the porosity of the adsorbent is 0.3. The adsorbent bed porosity is $\varepsilon = 0.4$. The bed diameter in a small-scale study to be conducted is 4 cm; the high-pressure feed-gas flow rate is 0.52 liter/s. The low pressure, P_L, is 1.013×10^5 Pa. The durations of different steps in the four-step (Figure 7.1.13(a)) process are: pressurization, 3 s; high-pressure feed flow, 80 s; blowdown, 3 s; low-pressure purge, 80 s. Determine the length of the bed needed for essentially complete purification of H_2. Determine the values of γ_{crit} and G_{crit}. Calculate the enrichment of the impurity in relation to its feed concentration in the blowdown-purge. Employ the equilibrium nondispersive PSA model of Chan *et al.* (1981).

Solution The equilibrium model of Chan *et al.* was formulated and solved using an equilibrium relation of the type for impurity A: $C_{A1} = \kappa_{A1} C_{A2}$. We will now convert $q_{A1} = 40 C_{A2}$ to this form using

$$q_{A1}(\text{gmols A/g adsorbent})\,\rho_s\left(\frac{\text{g adsorbent}}{\text{cm}^3\text{of adsorbent material}}\right)(1-\varepsilon_p)\left(\frac{\text{cm}^3\text{ of adsorbent}}{\text{cm}^3\text{ of porous adsorbent}}\right)$$

$$= C_{A1}(\text{mols A/cm}^3\text{ of porous adsorbent})$$
$$= 40 \times \rho_s(1-\varepsilon_p)C_{A2} = 40 \times 2 \times 0.7 C_{A2} = 56\, C_{A2}.$$

We will adopt an approximate approach. If L is assumed to be equal to Δz_L, then, by equation (7.1.60),

$$L_{min} = \beta_A m_{B,feed}(P_H/P_L)^\beta / \{\varepsilon S_c (P_H/RT)\} = \Delta z_L.$$

But from (7.1.59b),

$$m_{B,feed} = \varepsilon (V/L)\, v_{zH} \Delta t (P_H/RT).$$

Therefore

$$L_{min} = \frac{\beta_A \varepsilon (V/L) v_{zH} \Delta t (P_H/RT)(P_H/P_L)^\beta}{\varepsilon\, S_c (P_H/RT)}$$

$$\Rightarrow L_{min} = \beta_A v_{zH} \Delta t (P_H/P_L)^\beta.$$

But $\beta = (\beta_A/\beta_B) = \beta_A$ since H_2 is assumed not to be getting adsorbed, and

$$\beta_A = \frac{\varepsilon}{\varepsilon + (1-\varepsilon)\kappa_{A1}} = \frac{0.4}{0.4 + 0.6 \times 56} = \frac{0.4}{0.4 + 33.6} = 0.01176.$$

Further, $\Delta t = 80$ s, so

$$v_{zH} = \frac{520\,\text{cm}^3/\text{s}}{\frac{\pi}{4} 4^2 \times 0.4\,\text{cm}^2} = \frac{520}{1.6\pi}\,\frac{\text{cm}}{\text{s}}$$

$$\Rightarrow L_{min} = 0.01176 \times \frac{520}{1.6\pi} \times 80 \times (10)^{0.01176}$$

$$\Rightarrow L_{min} = \frac{0.01176 \times 520 \times 80 \times 1.027}{1.6\pi}\,\text{cm}$$

$$= 99.95\,\text{cm}.$$

We could have also obtained this length by employing equation (7.1.58d) for the high-pressure feed step. However, there is a pressurization step ahead of this step where, by equation (7.1.55e),

$$(z_H/z_L) = (P_H/P_L)^{-\beta} \Rightarrow (z_H/z_L) = (10)^{-0.0117} = 0.973.$$

To account for the short length of the bed (near $z = L$) which is consumed as the pressure rises from P_L to P_H in 3 s, we employ the above result:

$$99.95\,\text{cm} + (z_L - z_H) = z_L.$$

But

$$(z_L - z_H) = 0.027 z_L \Rightarrow 99.95 + 0.027 z_L = z_L \Rightarrow z_L = \frac{99.95}{0.973}.$$

The required bed length

$$z_L = 102.7\text{cm} = L.$$

The critical value of the (purge/feed) ratio, γ_{crit}, is obtained from (7.1.57c):

$$\gamma_{crit} = (v_{zL}/v_{zH})_{crit} = (P_H/P_L)^\beta = (10)^{0.0117}$$
$$\Rightarrow \gamma_{crit} = 1.027.$$

From result (7.1.58e),

$$G_{crit} = (P_H/P_L)^{\beta-1} = (10)^{-0.9883} = \frac{1}{9.735} = 0.102.$$

To determine the enrichment of the impurity species in the blowdown-purge stream in relation to the feed stream, we use (7.1.59f):

$$\frac{y_{BDPG}}{y_f} = \frac{\left(\frac{\beta L}{\Delta z_L}\right)\left(\frac{P_H}{P_L} - 1\right) + \left(\frac{P_H}{P_L}\right)^{1-\beta}}{\left(\frac{\beta L}{\Delta z_L}\right)\left(\frac{P_H}{P_L} - 1\right) + 1}$$

$$= \frac{\left(\frac{0.01176 \times 102.7}{99.95}\right)(10-1) + (10)^{0.9883}}{\frac{0.01176 \times 102.7}{99.95}(10-1) + 1}$$

$$= \frac{0.012 \times 9 + 9.73}{0.012 \times 9 + 1};$$

$$(y_{BDPG}/y_f) = \frac{0.1087 + 9.73}{1.1087} = 8.87.$$

This analysis ignores the second term, $((1-\varepsilon)\varepsilon_p/\varepsilon)$, in the denominator of relation (7.1.13e). To include this term, terms in the analysis of Chan *et al.* (1981) have to change.

Except for the study by Fernandez and Kenney (1983), all the studies mentioned above were based on multicolumn arrangements. Turnock and Kadlec (1971) and Kowler and Kadlec (1972) initiated studies of PSA processes using only a single bed. There are basically two steps in such a process, adsorption and desorption; the adsorption step consists of rapid pressurization and then high-pressure feed flow, while desorption is achieved by depressurization and adsorbed species removal from the feed introduction end. Short cycle times (≤ 20 seconds) are used, and product can be continuously withdrawn from the opposite end in a variety of ways using a surge tank and pressure-reducing valve. Hill (1980) used a single-column PSA to recover a weakly adsorbed impurity. A somewhat different single-column PSA process for oxygen production has been described by Keller (1983).

The PSA processes mentioned so far generally produce a high-pressure product stream purified of the species strongly adsorbed. Further, the models employed assume selectivity between the different species generated by different equilibrium relationships (see (7.1.52a) and (7.1.52b)). There are PSA processes already developed or being developed that are based on differing rates of diffusion of gases through molecular-sieve carbons (MSCs). For example, oxygen diffuses very rapidly through MSCs compared to nitrogen, although their equilibrium uptakes are very similar. This property is being exploited in a few PSA processes; see Yang (1987) for an introduction. Schork *et al.* (1993) have provided a model for such a process.

We have in this section used porous solid adsorbents to adsorb selectively specific gas species from the feed gas stream and then go through a cycle of desorption followed by adsorption in PSA processes. Absorption–desorption of specific gas species from a feed gas stream vis-à-vis an absorbent liquid has been carried out also using a porous hollow-fiber membrane based rapid pressure-swing absorption (RAPSAB) process (Bhaumik *et al.*, 1996). In this cyclic separation process, a well-packed microporous hydrophobic hollow-fiber module was used to achieve non-dispersive gas absorption (see Sections 3.4.3.1 and 8.1.2.2.1) from a high-pressure feed gas into a stationary absorbent liquid on the shell side of the module during a certain part of the cycle, followed by desorption of absorbed gases from the liquid in the rest of the cycle. The total cycle time was varied from 20 s upwards. Separation of mixtures of N_2 and CO_2 (around 10%), where CO_2 is the impurity to be removed, was studied using an absorbent liquid such as pure water and a 19.5% aqueous solution of diethanolamine (DEA). Three RAPSAB cycles studied differ in the absorption part. Virtually pure N_2 streams were obtained with DEA as absorbent, demonstrating the capability of bulk separation to very high levels of purification.

7.1.3 Potential-swing adsorption

Sometimes, the amount of species adsorbed from a solution onto an adsorbent, e.g. activated carbon, depends on the electrical potential applied to the adsorbent: the amount adsorbed *at equilibrium* depends on the potential applied. This equilibrium phenomenon of potential dependent adsorption could provide a basis for potential-swing adsorption. Eisinger and Keller (1990) have utilized such a phenomenon (see Zabasajja and Savinell (1989) for electrosorption of alcohols on graphite surfaces and references to earlier studies) to develop a potential-swing adsorption process. They have experimentally studied

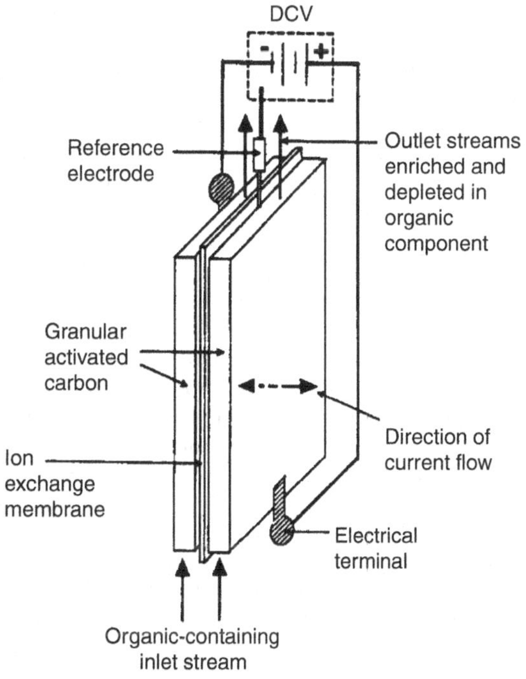

Figure 7.1.16. A suggested design of a single cell of two layers of carbon particles separated by an ion exchange membrane for potential-swing adsorption. Reprinted, with permission, from Figure 5 of R.S. Eisinger and G.E. Keller, "Electrosorption: a case study on removal of dilute organics from water," Env. Progr., ***9****(4), Nov., 235 (1990). Copyright © [1990] American Institute of Chemical Engineers (AIChE).*

adsorption of ethylenediamine (EDA) on activated carbon from an aqueous alkaline brine at negative potentials of upto −1.0 volt; the potential was then changed to 0.0 volt or positive to desorb the EDA into a suitable aqueous solution such that the solution will be highly concentrated in EDA.

An engineering assessment of this concept, also called *electrosorption*, was carried out by Eisinger and Keller (1990) using flat cells: a single cell consisted of two layers of granular activated carbon separated by an ion exchange membrane (Figure 7.1.16). Organic-containing inlet streams flow through each carbon layer. The carbon on one side of the membrane is held at the adsorbing potential (−1.0 volt) while the other side is at the desorbing potential (0 volt). When the carbon on the adsorbing side approaches saturation with EDA, and therefore breakthrough is imminent, the potentials are switched: the adsorbing side becomes the desorbing side, and vice versa. At the same time, the outlet stream switching valves would be activated. Note that the electrical potential in the carbon particles is generated by a current flow transverse to the carbon layer thickness via electrodes on the two sides. The electrical field here merely generates the requisite electrical potential on the adsorbents. The economic advantages and technical uncertainties of this concept have been considered. Scale-up problems have also been identified. Desalination of water by a similar technique, called *capacitive deionization* (CDI), is being studied (Farmer, 1995; Farmer *et al.*, 1996). Here two carbon electrodes kept at 1.3 volts apart are used during ion collection.

7.1.4 Parametric pumping

Parametric pumping is a cyclic fixed-bed process in which the axial fluid flow direction and the driving force direction (perpendicular to the axial flow direction) are changed periodically and synchronously throughout the bed in the presence of reflux[6] of the fluid at one or both ends of the bed (Sweed, 1972). The most common driving force is $-\nabla\mu_i$ created by a difference in temperature between the mobile phase and the stationary phase. Other driving forces used include a *pH* difference or a partial pressure difference. Some of the earliest descriptions of parametric pumping appear in Wilhelm *et al.* (1966, 1968).

Consider Figures 7.1.17(a) and (b), illustrating the batch operation of a thermal parametric pump in the direct mode. To start with, the packed bed is filled with the feed liquid containing an adsorbable species *i*. The bottom reservoir is also filled with the same feed fluid, whereas the top reservoir is essentially empty. The cyclic process was initiated earlier by suddenly heating the fluid and the solid in the packed bed to T_{hot} via the jacket and simultaneously forcing the fluid to flow upward into the top reservoir in a synchronous fashion (Figure 7.1.17(c)). As this figure shows, the velocity and temperature at any point in the bed remain constant for a time equal to half the cycle if a square wave is used (Pigford *et al.*, 1969a). Generally, the amount of species *i* retained by the adsorbent is reduced as the temperature is increased. Further, the bottom reservoir has the same concentration as that in the packed bed. The fluid flowing into the top reservoir, however, is enriched or concentrated in species *i*, compared to conditions where the bed is colder.

During the next half of the square wave cycle, the fluid from the top reservoir is pushed downward into the packed bed, which is suddenly cooled to T_{cold}; the colder fluid is now pushed into the bottom reservoir with the same speed. Since the adsorbent particles are cold, their adsorption capacity is much higher, thus purifying the fluid of solute *i*. The fluid coming into the bottom reservoir is substantially purified of species *i*. This cycle is ended when the volume of fluid introduced into the bottom reservoir during the second half of the cycle is equal to that removed from the same reservoir during the first half of the cycle. A new cycle is then initiated. Note that the top reservoir now has a solute concentration more than that in the initial feed, and the bottom reservoir has a solute concentration less than that in the initial feed. More cycles lead to a higher solute enrichment in the top reservoir and greater solute depletion in the bottom reservoir.

A very high degree of separation can be achieved after multiple cycles, limited only by axial dispersion and finite mass-transfer rate between the mobile and fixed phases.

[6]Reflux is considered in general in Chapter 8.

7.1.4.1 *A nondispersive equilibrium theory: batch operation*

The earliest nondispersive linear equilibrium based theory of parameteric pumping was developed by Pigford *et al.* (1969a). A number of others have since appeared (Aris, 1969; Gregory and Sweed, 1970; Rhee and Amundson, 1970; Chen and Hill, 1971). We illustrate here the model of parametric pumping developed by Pigford *et al.* (1969a) for a *batch system*. The model has the following assumptions.

(1) There is perfect local equilibrium between the solid and the fluid phases.
(2) A linear equilibrium relationship exists that depends on bed temperature.
(3) No axial dispersion or diffusion occurs, and there is no variation in the radial direction.
(4) There is an instantaneous change in bed temperature and velocity direction in a square-wave pattern, as shown in Figure 7.1.17(c).
(5) During each half-cycle, the interstitial velocity and temperature have the same value throughout the length of the bed.

Employing assumptions (3) and (5) in equation (7.1.3), the mass balance equation for species *i* is obtained as ($j = 1$, solid phase, $j = 2$, mobile phase):

$$\varepsilon\frac{\partial\overline{C}_{i2}}{\partial t}+\varepsilon v_z\frac{\partial\overline{C}_{i2}}{\partial z}+(1-\varepsilon)\frac{\partial\overline{C}_{i1}}{\partial t}=0. \tag{7.1.61}$$

If ρ_s is the density of the solid particles (mass/volume) and q_{i1} represents the moles of species *i* per unit mass of solid particles, then

$$\overline{C}_{i1}=\overline{q}_{i1}\rho_s. \tag{7.1.62a}$$

If y_i (instead of κ_{i2} for phase 2) represents the moles of species *i* per mole of fluid, and ρ_{fm} stands for the number of moles per unit fluid volume, then

$$\overline{C}_{i2}=y_i\rho_{fm}. \tag{7.1.62b}$$

The linear equilibrium assumption (2) is used to relate $\overline{q}_{i1}$ to y_i by

$$\overline{q}_{i1}=\mathrm{a}_i(T)\,y_i, \tag{7.1.63}$$

where $\mathrm{a}_i(T)$ is the local temperature-dependent species *i* equilibrium constant. Introducing definitions (7.1.62a) and (7.1.62b) into (7.1.4.1) leads to

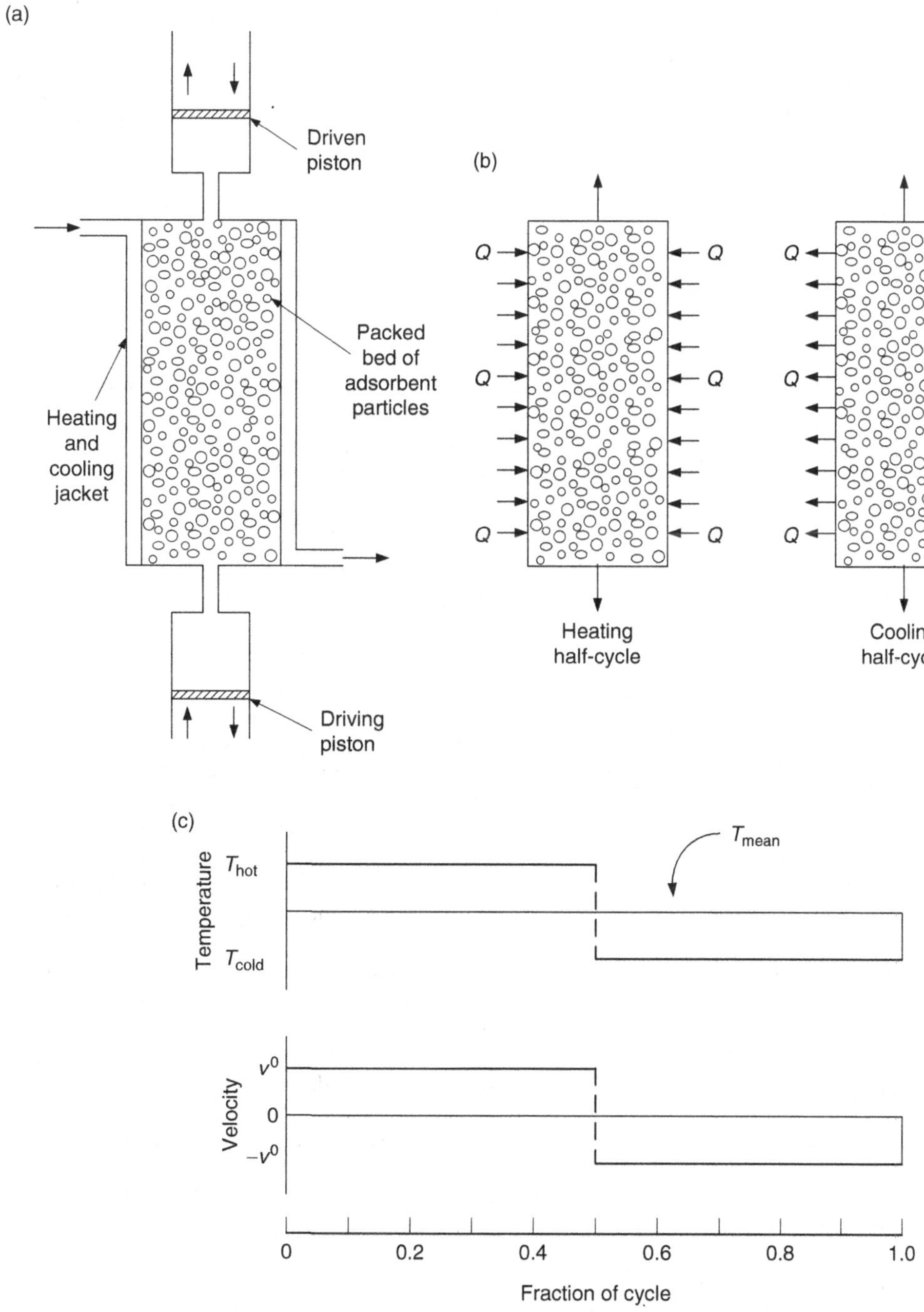

Figure 7.1.17. (a) Diagram of column for parametric pumping in direct mode. (b) Heat flow and fluid flow direction during two half-cycles (Wilhelm et al., *1968). Reprinted, with permission, from I & EC Fund.,* **7***(3), 337 (1968), Figure 3. Copyright (1968) American Chemical Society. (c) Velocity and temperature at a point in the bed as a function of time.*

$$\varepsilon\rho_{fm}\frac{\partial y_i}{\partial t}+\varepsilon v_z\rho_{fm}\frac{\partial y_i}{\partial z}+(1-\varepsilon)\,\rho_s\frac{\partial \overline{q}_{i1}}{\partial t}=0. \quad (7.1.64)$$

Introduce now the equilibrium relation (7.1.63) into the above equation. This results in

$$(1+\mathrm{A}_i(T))\frac{\partial y_i}{\partial t}+v_z\frac{\partial y_i}{\partial z}=-\left(\frac{\mathrm{dA}_i(T)}{\mathrm{d}T}\right)\left(\frac{\partial T}{\partial t}\right)y_i, \quad (7.1.65)$$

where

$$\mathrm{A}_i(T)=\frac{(1-\varepsilon)\,\rho_s \mathrm{a}_i(T)}{\varepsilon\rho_{fm}} \quad (7.1.66)$$

is a modified equilibrium constant.

To use the method of characteristics (equations (7.1.12b)–(7.1.12f)), write

$$\left(\frac{\partial y_i}{\partial t}\right)\mathrm{d}t+\left(\frac{\partial y_i}{\partial z}\right)\mathrm{d}z=\mathrm{d}y_i \quad (7.1.67)$$

to obtain

$$\left(\frac{\partial y_i}{\partial t}\right)=\frac{\begin{vmatrix}\mathrm{d}y_i & \mathrm{d}z\\ -y_i\left(\frac{\partial T}{\partial t}\right)\left(\frac{\mathrm{dA}_i(T)}{\mathrm{d}T}\right) & v_z\end{vmatrix}}{\begin{vmatrix}\mathrm{d}t & \mathrm{d}z\\ 1+\mathrm{A}_i(T) & v_z\end{vmatrix}}. \quad (7.1.68)$$

If the determinant in the denominator is zero, for $(\partial y_i/\partial t)$ to be finite the determinant in the numerator has to be zero. These two determinants with a zero value lead, respectively, to

$$\frac{\mathrm{d}z}{\mathrm{d}t}=\frac{v_z}{1+\mathrm{A}_i(T)} \quad (7.1.69)$$

and

$$v_z\frac{\mathrm{d}\ln y_i}{\mathrm{d}z}=-\left(\frac{\partial T}{\partial t}\right)\frac{\mathrm{dA}_i(T)}{\mathrm{d}T}. \quad (7.1.70)$$

Use of equation (7.1.69) in (7.1.70) yields

$$\frac{\mathrm{d}\ln y_i}{\mathrm{d}t}=-\left(\frac{\partial T}{\partial t}\right)\frac{1}{1+\mathrm{A}_i(T)}\frac{\mathrm{dA}_i(T)}{\mathrm{d}T}=-\frac{\mathrm{d}\ln(1+\mathrm{A}_i(T))}{\mathrm{d}t}. \quad (7.1.71)$$

Equations (7.1.69) and (7.1.71) represent, respectively, the characteristics and y_i variation along the characteristics.

The square-wave mode of variation of fluid velocity and bed temperature (and therefore equilibrium constant $\mathrm{A}_i(t)$) may be represented (via assumption (4)), respectively, by (Figure 7.1.17(c))

$$v_z=v_z(\omega t)=v_z^0\,\mathrm{sq}(\omega t)=v^0\mathrm{sq}(\omega t) \quad (7.1.72a)$$

and

$$\mathrm{A}_i(t)=\mathrm{A}_i^0-\hat{\mathrm{a}}_i\,\mathrm{sq}(\omega t). \quad (7.1.72b)$$

When the bed is heated, the fluid flows upward through the bed into the top reservoir. Due to the higher temperature, q_{i1} decreases and y_i increases because of reduced adsorption capacity; hence $\mathrm{A}_i(t)$ is reduced from A_i^0, the value at the mean temperature (T_{mean}), leading to a negative sign before $\hat{\mathrm{a}}_i\,\mathrm{sq}(\omega t)$. Introduction of these square-wave variations into the two governing equations (7.1.69) and (7.1.70) gives (here $v_z^0=v^0$)

$$\frac{\mathrm{d}z}{\mathrm{d}t}=\frac{v_z^0\,\mathrm{sq}(\omega t)}{1+\mathrm{A}_i^0-\hat{\mathrm{a}}_i\,\mathrm{sq}(\omega t)}=\frac{v^0\,\mathrm{sq}(\omega t)}{1-\mathrm{b}_i\,\mathrm{sq}(\omega t)} \quad (7.1.73)$$

and

$$\frac{\mathrm{d}\ln y_i}{\mathrm{d}t}=-\frac{\mathrm{d}\ln(1+\mathrm{A}_i(T))}{\mathrm{d}t}=-\frac{\mathrm{d}\ln[1-\mathrm{b}_i\,\mathrm{sq}(\omega t)]}{\mathrm{d}t}. \quad (7.1.74)$$

Equation (7.1.73) provides the velocity of propagation of any y_i through the bed and represents the characteristics. During the half-cycle when the bed is hot ($T=T_{\mathrm{hot}}$), the concentration wave velocity is given by

$$T=T_{\mathrm{hot}},\qquad \frac{\mathrm{d}z}{\mathrm{d}t}=\left(\frac{v^0}{1-\mathrm{b}_i}\right). \quad (7.1.75a)$$

During the half-cycle when the bed is colder ($T=T_{\mathrm{cold}}$), the concentration wave velocity is given by

$$T=T_{\mathrm{cold}},\qquad \frac{\mathrm{d}z}{\mathrm{d}t}=-\left(\frac{v^0}{1+\mathrm{b}_i}\right). \quad (7.1.75b)$$

Focus now on Figure 7.1.18, which illustrates the characteristics lines for the process initiated by upflow of the feed fluid in the column to the top reservoir while the bed temperature is instantaneously raised to T_{hot}. The feed fluid species concentration in equilibrium with the bed adsorbents at this temperature is C_{i2}^0 and the corresponding y_i value is y_i^0. Since this fluid is in equilibrium with the adsorbents, as this concentration moves along the z–t characteristics, it remains unchanged during the half-cycle $0<\omega t<\pi$. Those characteristic lines which cross $z=L$, the column length, discharge a hot fluid of y_i equal to y_i^0 into the top reservoir. Denoting by $\langle y_{iT}\rangle_n$ the mean composition of the fluid entering the top reservoir from the column during the nth cycle, it would appear from the figure that only the characteristics in region (A) of Figure 7.1.18 enter the top reservoir during the first hot period ($n=1$) $0<\omega t<\pi$. All characteristics in the column, however, have $y_i=y_i^0$. Therefore

$$\langle y_{iT}\rangle_1=y_i^0. \quad (7.1.76)$$

Consider the next half-period $\pi<\omega t<2\pi$, when the bed temperature and the fluid temperature have been lowered to T_{cold}. The fluid entering the top of the column from the top reservoir will now have the composition y_i^0, and their characteristics with changed slopes lie in region (C) of the z–t diagram. These characteristics do not cross $z=0$, the column bottom. The characteristics in region (D) do. These characteristics originated earlier from $z=0$ when $T=T_{\mathrm{hot}}$ during $0<\omega t<\pi$ and therefore had a value of

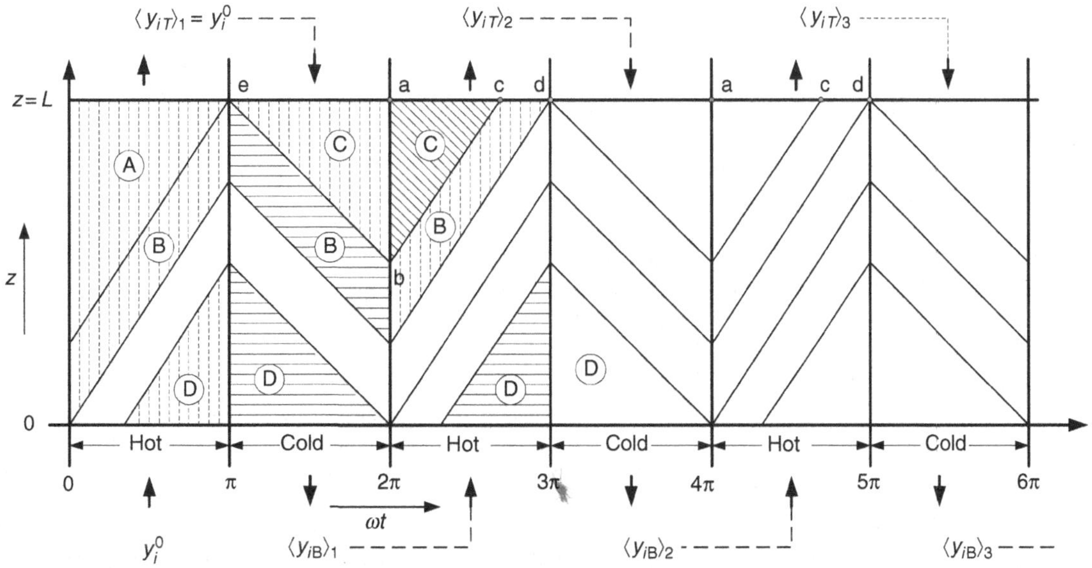

Figure 7.1.18. Characteristic lines used for solution by equilibrium theory: upflow and heating in phase (Pigford et al., *1969a). Reprinted, with permission, from I & EC Fund.,* **8***(1), 144 (1969), Figure 3. Copyright (1969) American Chemical Society.*

$y_i(0,t) = y_i^0$. This composition will be changed in region (D) of $\pi < \omega t < 2\pi$, where the characteristic slope is changed to that of (7.1.75b) when $\omega t \geq \pi$. Further, from the second governing equation (7.1.74), we get

$$y_i(1 - \mathrm{b}_i\,\mathrm{sq}(\omega t)) = \text{constant},$$

i.e.

$$\{y_i(1 - \mathrm{b}_i\ \mathrm{sq}(\omega t))\}_{\text{cold}} = \{y_i(1 - \mathrm{b}_i\ \mathrm{sq}(\omega t))\}_{\text{hot}}. \quad (7.1.77\text{a})$$

Therefore

$$\frac{y_i|_{\text{cold}}}{y_i|_{\text{hot}}} = \frac{1 - \mathrm{b}_i}{1 + \mathrm{b}_i}. \quad (7.1.77\text{b})$$

Since, for $\omega t \geq \pi$, the characteristics for $y_i|_{\text{hot}}$ change to those for $y_i|_{\text{cold}}$,

$$y_i|_{\text{cold}} = y_i|_{\text{hot}}\left(\frac{1 - \mathrm{b}_i}{1 + \mathrm{b}_i}\right) = y_i^0\left(\frac{1 - \mathrm{b}_i}{1 + \mathrm{b}_i}\right). \quad (7.1.78)$$

The characteristics crossing $z = 0$ will enter the bottom reservoir. For the period $\pi \leq \omega t \leq 2\pi$, only the characteristics in region (D) enter $z = 0$. Therefore, the composition of this fluid entering the bottom reservoir, represented by $\langle y_{iB}\rangle_1$, is

$$\langle y_{iB}\rangle_1 = y_i^0\left(\frac{1 - \mathrm{b}_i}{1 + \mathrm{b}_i}\right). \quad (7.1.79)$$

In the second cycle, beginning $2\pi \leq \omega t \leq 3\pi$, this very same fluid with $\langle y_{iB}\rangle_1$ will enter the column bottom, which is now at $T = T_{\text{hot}}$ from the bottom reservoir, and the characteristic line slope is changed. However, we are more interested in those characteristics which cross $z = L$ and therefore enter the top reservoir. There are two regions in $2\pi \leq \omega t \leq 3\pi$, region (B) and region (C), through which such characteristics pass. The characteristics in region (C) have originated from those in region (C) of $\pi \leq \omega t \leq 2\pi$, with ultimate origin from the top reservoir having a composition of y_i^0. This composition, remaining constant up to $\omega t = 2\pi$, is changed in the $\omega t \geq 2\pi$ region since the temperature is changed to T_{hot} from T_{cold}:

$$\text{region (C)}, \omega t \geq 2\pi : y_i|_{\text{hot}} = y_i|_{\text{cold}}\left(\frac{1 + \mathrm{b}_i}{1 - \mathrm{b}_i}\right) = y_i^0\left(\frac{1 + \mathrm{b}_i}{1 - \mathrm{b}_i}\right). \quad (7.1.80)$$

These characteristics thus yield a higher species i composition in the fluid entering the top reservoir. But the top reservoir also receives characteristics in region (B), $2\pi \leq \omega t \leq 3\pi$. These characteristics originated in region (B) for $\pi \leq \omega t \leq 2\pi$, where they originated from region (B) of $0 \leq \omega t \leq \pi$. The composition of these last characteristics is simply y_i^0; as these characteristics cross $\omega t = \pi$, from T_{hot} to T_{cold}, the composition is lowered to $y_i^0(1 - \mathrm{b}_i)/(1 + \mathrm{b}_i)$. This last composition is increased to y_i^0 as these characteristics cross $\omega t = 2\pi$ and change from T_{cold} to T_{hot}.

We now know the compositions of the characteristics in regions (B) and (C) for $2\pi \leq \omega t \leq 3\pi$, where the fluid in upflow enters the top reservoir at $z = L$. To determine the mean composition $\langle y_{iT}\rangle_2$ of these characteristics entering the top reservoir via regions (B) and (C), the weighted contribution of each region is needed since each region occupies a different time interval:

$$\langle y_{iT}\rangle_2 = \left\{ y_i^0 \left(\frac{1+\mathrm{b}_i}{1-\mathrm{b}_i}\right) \times \left(\frac{ac}{ad}\right) + y_i^0 \left(\frac{cd}{ad}\right) \right\}, \quad (7.1.81)$$

where ac, cd and ad are identified in Figure 7.1.18 for $2\pi \le \omega t \le 3\pi$. To determine ratios (ac/ad) and (cd/ad), focus on the characteristics lines in both regions $\pi \le \omega t \le 2\pi$ and $2\pi \le \omega t \le 3\pi$. Specifically, in region (C) of time period $\pi \le \omega t \le 2\pi$, characteristic line eb has a slope of magnitude $(v^0/(1+\mathrm{b}_i))$. Therefore, the distance ab is $\pi(v^0/(1+\mathrm{b}_i))$ (ideally the length of time t used should be π/ω, but ω cancels out later anyway). From the slope of the characteristics in region (C) for $2\pi \le \omega t \le 3\pi$,

$$\frac{ab}{ac} = \left(\frac{v^0}{1-\mathrm{b}_i}\right)\frac{1}{\omega} \Rightarrow ac = \frac{(ab)\omega}{(v^0/(1-\mathrm{b}_i))} = \frac{\omega\pi(1-\mathrm{b}_i)}{(1+\mathrm{b}_i)}.$$

Since ad is $\omega\pi$, we get

$$(ac/ad) = (1-\mathrm{b}_i)/(1+\mathrm{b}_i). \quad (7.1.82a)$$

The ratio (cd/ad) is obtained as

$$\frac{cd}{ad} = 1 - \frac{ac}{ad} = \frac{2\mathrm{b}_i}{1+\mathrm{b}_i}. \quad (7.1.82b)$$

Thus

$$\langle y_{iT}\rangle_2 = y_i^0 \left\{ \left(\frac{1+\mathrm{b}_i}{1-\mathrm{b}_i}\right)\frac{(1-\mathrm{b}_i)}{(1+\mathrm{b}_i)} + \frac{2\mathrm{b}_i}{1+\mathrm{b}_i} \right\} = y_i^0 \left\{ 1 + \frac{2\mathrm{b}_i}{1+\mathrm{b}_i} \right\}. \quad (7.1.83)$$

This result demonstrates that the hot stream entering the top reservoir at the beginning of the second cycle $2\pi \le \omega t \le 3\pi$ is considerably enriched beyond y_i^0 (the value for $\langle y_{iT}\rangle_1$).

Although fluid of composition $\langle y_{iT}\rangle_2$ enters the column top from the top reservoir for the next half-cycle, $3\pi \le \omega t \le 4\pi$, the bottom region of the column is more important for determining $\langle y_{iB}\rangle_2$, the stream composition entering the bottom reservoir. Characteristics in region (D) here originated in region (D) for $2\pi \le \omega t \le 3\pi$. The composition of the characteristics in this later region is $\langle y_{iB}\rangle_1$; this composition is changed when $\omega t \ge 3\pi$. The changed composition $y_i|_{\mathrm{cold}}$ is related to $\langle y_{iB}\rangle_1 = y_i|_{\mathrm{hot}}$ by (7.1.80); thus, for $3\pi \le \omega t \le 4\pi$ region (D),

$$\langle y_{iB}\rangle_2 = y_i|_{\mathrm{cold}} = y_i|_{\mathrm{hot}}\left(\frac{1-\mathrm{b}_i}{1+\mathrm{b}_i}\right) = y_i^0\left(\frac{1-\mathrm{b}_i}{1+\mathrm{b}_i}\right)^2. \quad (7.1.84)$$

Thus the bottom reservoir solution is depleted in species i after the second cycle, while the top reservoir is enriched in species i.

Pigford *et al.* (1969a) have provided general expressions for $\langle y_{iT}\rangle_n$ and $\langle y_{iB}\rangle_n$ after n cycles ($n > 0$) in terms of those immediately proceeding:

$$\langle y_{iT}\rangle_n = \langle y_{iT}\rangle_{n-1}\left(\frac{1+\mathrm{b}_i}{1-\mathrm{b}_i}\right)\left(\frac{1-\mathrm{b}_i}{1+\mathrm{b}_i}\right) + \langle y_{iB}\rangle_{n-3}\left(\frac{2\mathrm{b}_i}{1+\mathrm{b}_i}\right); \quad (7.1.85a)$$

$$\langle y_{iB}\rangle_n = \langle y_{iB}\rangle_{n-1}\left(\frac{1-\mathrm{b}_i}{1+\mathrm{b}_i}\right). \quad (7.1.85b)$$

This preceding relation leads to

$$\langle y_{iB}\rangle_n = y_i^0\left(\frac{1-\mathrm{b}_i}{1+\mathrm{b}_i}\right)^n. \quad (7.1.86a)$$

Use of both relations (7.1.85a) and (7.1.85b) yields

$$\langle y_{iT}\rangle_n = y_i^0\left\{1 + \frac{2\mathrm{b}_i}{1+\mathrm{b}_i} + \left[1 - \left(\frac{1-\mathrm{b}_i}{1+\mathrm{b}_i}\right)^{n-2}\right]\right\}. \quad (7.1.86b)$$

The composition ratio between the top reservoir and the bottom reservoir is

$$\frac{\langle y_{iT}\rangle_n}{\langle y_{iB}\rangle_n} = \left(2 + \frac{2\mathrm{b}_i}{1+\mathrm{b}_i}\right)\left(\frac{1+\mathrm{b}_i}{1-\mathrm{b}_i}\right)^n - \left(\frac{1+\mathrm{b}_i}{1-\mathrm{b}_i}\right)^2. \quad (7.1.87)$$

These results indicate that, as n increases, $\langle y_{iB}\rangle_n$, the bottom product composition, tends to zero (relation (7.1.86a)). Although as n increases $\langle y_{iT}\rangle_n$ increases, relation (7.1.85a) suggests that $\langle y_{iT}\rangle_n = \langle y_{iT}\rangle_{n-1}$. In a real system, the ratio $\langle y_{iT}\rangle_n/\langle y_{iB}\rangle_n$ does not tend to infinity, as suggested by (7.1.87) for large n; instead, mass-transfer rate limitations between the phases and axial diffusion limit the possible enrichment (Pigford *et al.*, 1969a). Figure 7.1.19 compares the predictions from this equilibrium theory with the experimental data on an n-heptane–toluene separation

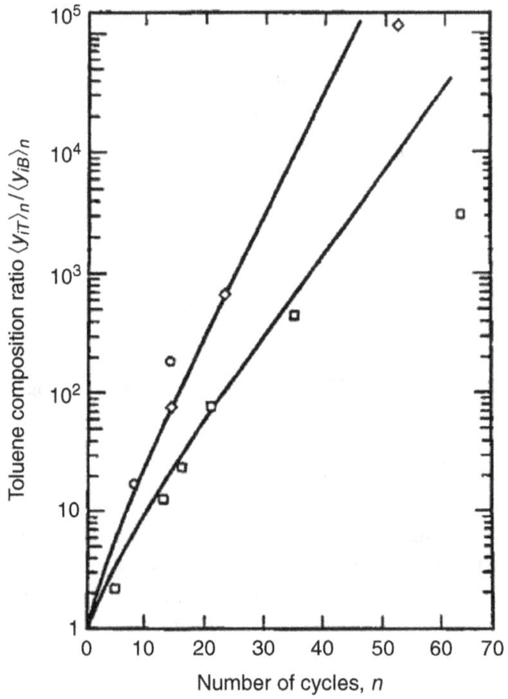

Figure 7.1.19. Toluene composition ratio according to equilibrium theory and the data of Wilhelm et al. *(1968) for the system n-heptane-toluene in a silica gel column (Pigford* et al., *1969a). Reprinted, with permission, from I & EC Fund.,* ***8****(1), 144, (1969), Figure 6. Copyright (1969) American Chemical Society.*

in a silica gel column (Wilhelm and Sweed, 1968; Wilhelm *et al.*, 1968).

That separation achieved in batch parametric pumping can be demonstrated more easily by what is known as the *Tinkertoy model* developed by Wilhelm *et al.* (1968). This model is considered next.

7.1.4.2 *Tinkertoy model of parametric pumping*

Consider Figure 7.0.1(c) from the very beginning of this chapter. The hatched section represents the stationary phase or the packings in the manner of Figure 7.1.1, i.e. a pseudo-continuum approach. Focus on a section of the column of differential length dz. As before, subscript $j = 1$ represents solid packing phase and $j = 2$ represents the mobile fluid phase. The mass balance equation (7.1.61) for species i at any axial column location z may be rearranged to yield

$$v_z(\omega t)\frac{\partial \overline{C}_{i2}}{\partial z} = -\left[\frac{\partial \overline{C}_{i2}}{\partial t} + \frac{(1-\varepsilon)}{\varepsilon}\frac{\partial \overline{C}_{i1}}{\partial t}\right] \tag{7.1.88}$$

for nondispersive column operation. In parametric pumping, the flow velocity $v_z(\omega t)$ is alternating in direction (a particular form is illustrated by equation (7.1.72a)); the force direction, and therefore the mass flux direction, are simultaneously alternating in direction. During fluid upflow, if the bed is heated, solute is released by adsorbents into the fluid (shown by the dashed lines in Figure 7.0.1(c)); during fluid downflow, the bed is cooled and the solute is adsorbed by adsorbents from the fluid (shown by solid lines in Figure 7.0.1(c)).

It can now be argued (Wilhelm *et al.*, 1968; Sweed, 1971, pp. 175–180) that if the force is oscillating with a frequency ω (the same as the velocity), the concentrations $\overline{C}_{i2}$ and $\overline{C}_{i1}$ will also be periodic in frequency ω. Therefore, $(\partial\overline{C}_{i2}/\partial t)$ and $(\partial\overline{C}_{i1}/\partial t)$ will also be periodic in ω. Equation (7.1.88) may be expressed now as

$$v_z(\omega t)\frac{\partial \overline{C}_{i2}}{\partial z} = f_c(\omega t) \tag{7.1.89a}$$

$$\Rightarrow \quad \frac{\partial \overline{C}_{i2}}{\partial z} = \frac{f_c(\omega t)}{v_z(\omega t)}. \tag{7.1.89b}$$

The change in mobile-phase composition $\overline{C}_{i2}$ over the column of differential length dz is provided by the above relation *at any instant of time t*. We are more interested in knowing what this change is over one cycle ($t = 2\pi/\omega$), and ultimately over many cycles.

Before determining a time average of $(\partial\overline{C}_{i2}/\partial z)$ over one cycle, let it be recognized that, in general, $f_c(\omega t)$ and $v_z(\omega t)$ are displaced by a phase angle ε_a. Therefore, an averaged behavior over a time period $0 \le t \le (2\pi/\omega)$ is obtained from

$$\frac{\omega}{2\pi}\int_0^{\frac{2\pi}{\omega}}\left(\frac{\partial \overline{C}_{i2}}{\partial z}\right)\mathrm{d}t = \overline{\frac{\partial \overline{C}_{i2}}{\partial z}} = \frac{\omega}{2\pi}\int_0^{\frac{2\pi}{\omega}}\frac{f_c(\omega t+\varepsilon_a)}{v_z(\omega t)}\,\mathrm{d}t. \tag{7.1.90a}$$

It is assumed that the integral on the right-hand side of (7.1.90a) exists. To calculate its value, assume specific forms for $f_c(\omega t + \varepsilon_a)$ and $v_z(\omega t)$:

$$v_z(\omega t) = \mathrm{A}_1\cos\omega t; \quad f_c(\omega t+\varepsilon_a) = \mathrm{A}_2\cos(\omega t+\varepsilon_a), \tag{7.1.90b}$$

where A_1 and A_2 are arbitrary constants. Then

$$\overline{\frac{\partial \overline{C}_{i2}}{\partial z}} = \frac{\mathrm{A}_2}{\mathrm{A}_1}\cos\varepsilon_a, \tag{7.1.91}$$

a result indicating that the time-averaged local axial concentration gradient in $\overline{C}_{i2}$ can be nonzero provided ε_a is anything but $(\pi/2)$ or $(3\pi/2)$, i.e. there will be separation provided the oscillations in velocity and concentration are not out of phase by $(\pi/2)$ or $(3\pi/2)$.

7.1.4.3 *Continuous parametric pumping*

Batch parametric pumping in a closed system was studied in the context of an equilibrium nondispersive model in Section 7.1.4.1. Continuous open parametric pumps have been modeled by a number of investigators (Gregory and Sweed, 1970; Chen and Hill, 1971). A number of experimental systems have been studied. These include *pH* based operations (Sabadell and Sweed, 1970; Chen *et al.*, 1979, 1980) and pressure based gas separation operations (Keller and Kuo, 1982).

7.1.4.4 *Cycling zone adsorption*

It is useful to recapitulate briefly the basics of the two separation techniques described earlier, namely PSA and parametric pumping. Both are cyclic processes. In PSA, one gaseous species is preferentially adsorbed from the feed gas flowing into the bed of adsorbents at a high pressure, allowing the remaining gas species to be purified (for a binary mixture). The preferentially adsorbed species is then desorbed from the adsorbent bed at a much lower pressure, usually with a low-pressure purge flowing in the opposite direction (see Figure 7.1.13(a)). For liquid-phase adsorption–desorption processes, pressure has a very limited effect; therefore, temperature swing is useful for liquid-phase adsorption. In parametric pumping, the liquid flow up or down the adsorbent bed is synchronized with heating or cooling of the liquid and the bed, leading to the reservoirs at the top and bottom being enriched or depleted in the solute (see Figures 7.1.17–7.1.19). The reversal of the direction of liquid flow through the adsorbent bed (practiced in parametric pumping) is avoided in the technique called *cycling zone adsorption*, which was proposed first by Pigford *et al.* (1969b).

Focus on Figure 7.1.20(a), which shows an adsorbent column being heated or cooled in the direct mode (via a jacket). Unlike that in Figure 7.1.17(a), there are no pistons/reservoirs at the top and bottom of the column as such. Liquid flow in the bed is unidirectional; liquid feed is

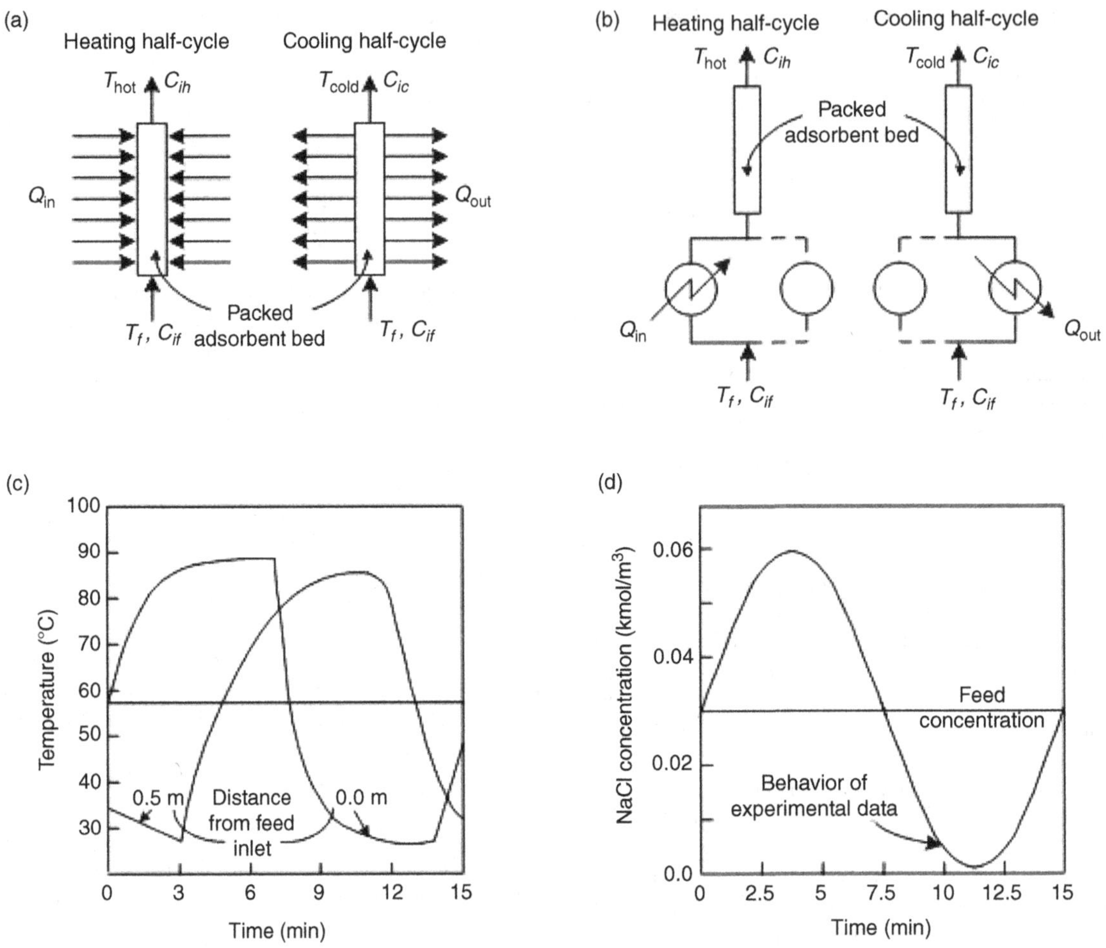

Figure 7.1.20. Cycling zone adsorption. (a) Direct mode of operation with heat supplied/removed directly into the bed through the wall. (b) Recuperative mode of operation with heat supplied or removed directly into the feed. (c) Temperature variation with time at two distances from feed inlet. (After Knaebel and Pigford (1983).) (d) NaCl concentration of feed outlet as a function of time. (After Knaebel and Pigford (1983).)

continuously flowing up the column of adsorbents (or down the column of adsorbents!). However, the temperature of the column is oscillating, for example, as in Figure 7.1.17(c). If the liquid feed is continuously flowing up the column, then, during the first half of the cycle when the bed is heated up (the temperature is T_{hot}), in an empty bed at the start, there will be very little adsorption. However, in the next half-cycle, as the temperature is reduced to T_{cold}, the liquid feed undergoes purification since solutes are adsorbed much more at the low temperature. The liquid exiting the bed at the top has much less solute and is purified.

At the beginning of the next cycle, the bed is heated up; the adsorbed solutes are desorbed and the feed liquid entering the bed becomes highly enriched in the solute as it leaves the column at the top. This half of the cycle, in succeeding cycles also, will keep on producing a solute-enriched liquid stream. Therefore, by cycling the zone temperature, we are going through a cycling of the zone adsorption behavior. (*Note*: Here we have assumed that the liquid feed temperature is essentially the same as that of the bed in the direct mode of operation.) One can have an alternative arrangement where the liquid feed will be heated or cooled outside in a heater/cooler and introduced into the bed in the so-called recuperative mode (useful in parametric pumping as well) of operation (Figure 7.1.20 (b)). Correspondingly, the outlet liquid in the "hot" part of the cycle is enriched in the solute, whereas in the "cold" part of the cycle the outlet liquid is purified of the solute (in both modes of operation).

A variety of systems have been investigated using the cycling zone adsorption technique. See Wankat *et al.* (1975) and Knaebel and Pigford (1983) for a list of useful references. Figures 7.1.20(c) and (d) illustrate schematically the recuperative mode of cycling of temperature in a bed

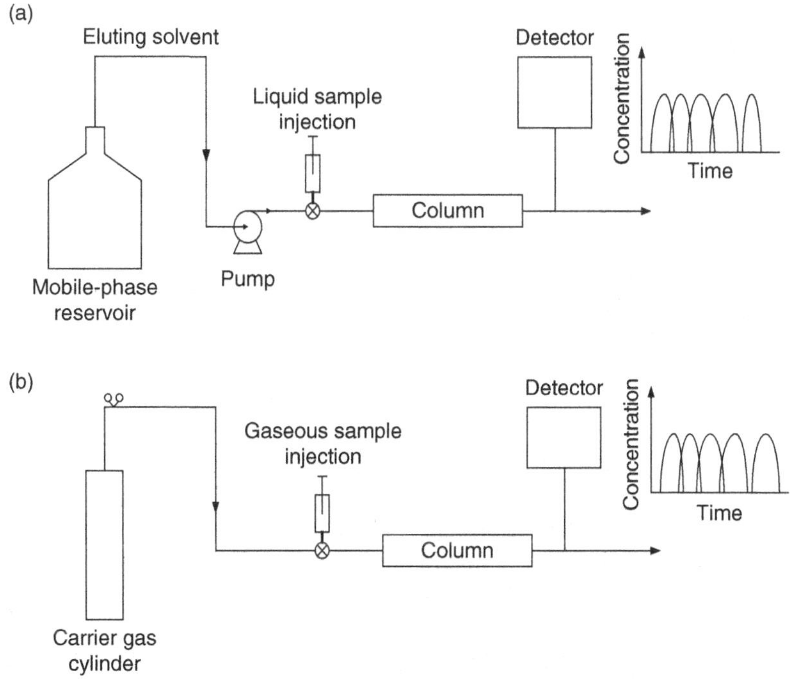

Figure 7.1.21. Schematics for a basic chromatographic system for (a) liquid sample or (b) a gaseous sample.

(Figure 7.1.20(b)) and the corresponding bed outlet liquid concentration profile for purification of a brackish water (feed NaCl concentration, 0.030 kmol/m^3) using thermally regenerable ion exchange resins Amberlite XD. Two basic models of the cycling zone adsorption are available in Baker and Pigford (1971) and Gupta and Sweed (1971).

7.1.5 Chromatographic processes

The fixed-bed processes studied so far are generally useful for separating one solute or one ionic species from the solvent; in some cases, more than one solute or one ionic species are also separated from the solvent. Separation of each individual species present in a multicomponent gas mixture or liquid solution can be achieved in a fixed bed of adsorbent particles if the mode of operation, e.g. the method of feed introduction, is changed from that used so far. A number of methods are commonly used to this end. They are elution chromatography, displacement chromatography and frontal chromatography (see Figure 7.1.5(c)).

7.1.5.1 *Elution chromatography*

Consider first the process known as *elution chromatography* used to separate components from one another present in a multicomponent gas mixture or liquid solution. In such a process, a small volume, V^0, of the feed solution or feed gas mixture is first introduced into the column at the column top. Different species present in this feed sample are adsorbed in a small band of stationary adsorbent particles in the region of the column top. (The nature of the adsorbent in relation to the nature of the feed fluid, gas or liquid will be discussed in detail in Section 7.1.5.1.2.) Next the solvent used in the feed solution, another solvent or an inert carrier gas is introduced into the column top at a steady rate. If one monitors[7] the effluent concentration at the column bottom or end, one observes, after some time, a series of concentration pulses of different species appearing one after another (Figure 7.1.21). By collecting effluent volumes between appropriate times (see Section 2.5), it is possible to have each particular solute species in the feed essentially segregated into separate and different effluent volumes; thus, all species present in the multicomponent feed have been separated from one another. However, they are obtained as a solution in the mobile-phase solvent or carrier gas. The solvent or carrier gas is identified as the eluent, and this process of multicomponent separation using a fixed bed of adsorbents is called elution chromatography.

[7]Generally, two types of detectors are used to monitor gas compositions: a thermal conductivity detector (TCD) for common permanent gases like O_2, N_2, CO_2, He, Ar, etc., or a flame ionization detector (FID) for common combustible hydrocarbon gases/vapors such as C_nH_{2n+2}, $C_n\,H_{2n}$ and all sorts of volatile organic compounds. Ultraviolet spectrophotometric detectors, refractive index detectors and conductivity monitors are used to monitor liquid-phase compositions.

The mechanism of separation may be described as follows. As the solute-free solvent or carrier gas is introduced into the column, it desorbs the solutes adsorbed on the column-top particles from the initial feed sample. Those solutes that are adsorbed less strongly desorb more easily and are quickly carried downstream by the eluent. All solute species are desorbed and carried downstream, where they are readsorbed on fresh adsorbents immediately downstream. As fresh eluent appears and contacts these particles, the solute molecules go through cycles of repeated desorption–adsorption along the column length. This process continues till they appear in the effluent at column exit, with the least strongly adsorbed solute appearing first and the most strongly adsorbed appearing last.

7.1.5.1.1 Liquid–solid adsorption based elution chromatography First, we consider systems where the eluent is a liquid; further, the mechanism of solute partitioning onto the solid particles is simply adsorption, e.g. adsorption on silica or alumina particles. We assume further that the eluent pressure drop along the bed does not influence the adsorption process. There are a number of ways by which elution chromatography in such a system may be described mathematically. The differential equation describing the solute adsorption–desorption process for a dilute liquid-phase system in a fixed bed of adsorbent particles continues to be equation (7.1.4):

$$\frac{\partial \overline{C}_{i2}}{\partial t} + \frac{(1-\varepsilon)}{\varepsilon}\frac{\partial \overline{C}_{i1}}{\partial t} + v_z \frac{\partial \overline{C}_{i2}}{\partial z} = D_{i,\text{eff},z}\frac{\partial^2 \overline{C}_{i2}}{\partial z^2}.$$

We will develop a solution for elution chromatography in two ways. In the first method, a direct solution of this equation is developed for the initial and boundary conditions appropriate for elution chromatography using linear local equilibrium assumption everywhere. In the second method, the solution (7.1.18h) developed earlier for a fixed-bed process with a continuous liquid feed stream is differentiated to approximate the conditions for elution chromatography. We consider the first method now.

Rewrite the governing equation using the local equilibrium assumption and relations (7.1.6) and (7.1.7) as

$$\frac{\partial \overline{C}_{i2}}{\partial t} + \frac{v_z}{\left[1+\frac{\rho_b}{\varepsilon}q_i'(\overline{C}_{i2})\right]}\frac{\partial \overline{C}_{i2}}{\partial z} = \frac{D_{i,\text{eff},z}}{\left[1+\frac{\rho_b}{\varepsilon}q_i'(\overline{C}_{i2})\right]}\frac{\partial^2 \overline{C}_{i2}}{\partial z^2}. \tag{7.1.92}$$

For *linear adsorption equilibrium*, encountered in dilute systems,

$$q_i'(\overline{C}_{i2}) = \frac{(1-\varepsilon)}{\rho_b}\kappa_{i1} \tag{7.1.93a}$$

is a reasonable assumption, with κ_{i1} constant. For porous adsorbent particles, employing equation (7.1.13f), we get the following form of equation (7.1.92):

$$\frac{\partial \overline{C}_{i2}}{\partial t} + \frac{v_z}{\left[1+\frac{(1-\varepsilon)}{\varepsilon}\varepsilon_p\kappa_{im} + \frac{(1-\varepsilon)(1-\varepsilon_p)\rho_s}{\varepsilon}q_i'(\overline{C}_{i2})\right]}\frac{\partial \overline{C}_{i2}}{\partial z} = \frac{D_{i,\text{eff},z}}{\left[1+\frac{(1-\varepsilon)}{\varepsilon}\varepsilon_p\kappa_{im} + \frac{(1-\varepsilon)(1-\varepsilon_p)\rho_s}{\varepsilon}q_i'(\overline{C}_{i2})\right]}\frac{\partial^2 \overline{C}_{i2}}{\partial z^2}. \tag{7.1.93b}$$

Additionally, the total local molar concentration $\overline{C}_{it}$ of species i per unit volume (which incorporates both mobile- and stationary-phase solute contributions) is given, for nonporous particles, by

$$\overline{C}_{it} = \varepsilon\overline{C}_{i2} + (1-\varepsilon)\overline{C}_{i1} = \varepsilon\overline{C}_{i2} + (1-\varepsilon)\frac{\overline{C}_{i1}}{\overline{C}_{i2}}\overline{C}_{i2}$$
$$= \left(\varepsilon + (1-\varepsilon)\frac{\overline{C}_{i1}}{\overline{C}_{i2}}\right)\overline{C}_{i2} = [\varepsilon + (1-\varepsilon)\kappa_{i1}]\,\overline{C}_{i2}, \tag{7.1.94}$$

where $[\varepsilon + (1-\varepsilon)\kappa_{i1}]$ is now a constant. Equation (7.1.92) may be reformulated now with $\overline{C}_{it}(z,t)$ as the dependent variable:

$$\frac{\partial \overline{C}_{it}}{\partial t} + \frac{v_z}{\left[1+\frac{(1-\varepsilon)}{\varepsilon}\kappa_{i1}\right]}\frac{\partial \overline{C}_{it}}{\partial z} = \frac{D_{i,\text{eff},z}}{\left[1+\frac{(1-\varepsilon)}{\varepsilon}\kappa_{i1}\right]}\frac{\partial^2 \overline{C}_{it}}{\partial z^2}. \tag{7.1.95a}$$

This is a *linear* equation with *constant coefficients* and is essentially equivalent to equation (3.2.10). Only the coefficients are different. For porous adsorbent particles, we need to change equation (7.1.94) and obtain instead the following:

$$\overline{C}_{it} = \overline{C}_{i2}\,(\varepsilon + (1-\varepsilon)\varepsilon_p\kappa_{im} + (1-\varepsilon)\,\kappa_{i1}]. \tag{7.1.95b}$$

Equation (7.1.93b) may now be reformulated with $\overline{C}_{it}(z,t)$ as the dependent variable:

$$\frac{\partial \overline{C}_{it}}{\partial t} + \frac{v_z}{\left[1+\frac{(1-\varepsilon)}{\varepsilon}\varepsilon_p\kappa_{im} + \frac{(1-\varepsilon)}{\varepsilon}\kappa_{i1}\right]}\frac{\partial \overline{C}_{it}}{\partial z} = \frac{D_{i,\text{eff},z}}{\left[1+\frac{(1-\varepsilon)}{\varepsilon}\varepsilon_p\kappa_{im} + \frac{(1-\varepsilon)}{\varepsilon}\kappa_{i1}\right]}\frac{\partial^2 \overline{C}_{it}}{\partial z^2}. \tag{7.1.95c}$$

Since the initial solute band on top of the column (very near $z = 0$) is a very thin one, we will assume that all of the solute species i (m_i moles) introduced via the feed sample are essentially located at $z = 0$ at the start of solvent elution operation ($t = 0$); therefore, at $t = 0$,

$$\overline{C}_{it} = (m_i/S_c)\,\delta(z). \tag{7.1.96}$$

Although the length of the column is finite, we will assume that the solution for $z \rightarrow \infty$ is usable for our purpose (exactly as in the boundary condition (3.2.14a)). For all t, then, at $z = \infty$,

$$\overline{C}_{it} = 0 \quad \text{and} \quad \frac{\partial \overline{C}_{it}}{\partial z} = 0. \tag{7.1.97}$$

The solution for $\overline{C}_{it}$ from equation (7.1.95a) may be obtained following equation (3.2.19) as the Gaussian concentration profile,

$$\overline{C}_{it}(z,t) = \frac{m_i}{S_c}\frac{1}{\sqrt{\dfrac{4\pi D_{i,\text{eff},z}t}{\left[1+\dfrac{(1-\varepsilon)}{\varepsilon}\kappa_{i1}\right]}}}\exp\left[-\frac{\left\{z-\dfrac{v_z t}{\left[1+\dfrac{(1-\varepsilon)}{\varepsilon}\kappa_{i1}\right]}\right\}^2}{4\left\{\dfrac{D_{i,\text{eff},z}}{\left[1+\dfrac{(1-\varepsilon)}{\varepsilon}\kappa_{i1}\right]}\right\}t}\right]. \tag{7.1.98a}$$

Correspondingly, the solution for the profile $\overline{C}_{i2}$ of species i in the mobile phase is the Gaussian concentration profile,

$$\overline{C}_{i2}(z,t) = \frac{m_i}{S_c}\frac{[\varepsilon+(1-\varepsilon)\kappa_{i1}]^{-1}}{\sqrt{\dfrac{4\pi D_{i,\text{eff},z}t}{\left[1+\dfrac{(1-\varepsilon)}{\varepsilon}\kappa_{i1}\right]}}}\exp\left[-\frac{\left\{z-\dfrac{v_z t}{\left[1+\dfrac{(1-\varepsilon)}{\varepsilon}\kappa_{i1}\right]}\right\}^2}{4\left\{\dfrac{D_{i,\text{eff},z}}{\left[1+\dfrac{(1-\varepsilon)}{\varepsilon}\kappa_{i1}\right]}\right\}t}\right]. \tag{7.1.98b}$$

This concentration profile may be written in a compact form as follows:

$$\overline{C}_{iz}(z,t) = \overline{C}_{iz}(z,t)\Big|_{\max}\cdot\exp\left\{\frac{(z-R_i'v_z t)^2}{2\sigma_{ti}^2}\right\}. \tag{7.1.98c}$$

The quantities R_i' and σ_{ti} are defined below as we characterize this profile, with $\overline{C}_{iz}(z,t)|_{\max}$ being the pre-exponential factor in (7.1.98b) as well as the value of $\overline{C}_{i2}(z,t)$ when $z = R_i' v_z t$.

The center point of the profile of the total concentration $\overline{C}_{it}$ of species i (also of $\overline{C}_{i2}$) at any time is located at

$$z = \frac{v_z t}{\left[1+\dfrac{(1-\varepsilon)}{\varepsilon}\kappa_{i1}\right]} = R_i' v_z t = z_{i0}. \tag{7.1.99a}$$

Since the quantity

$$\left(v_z\Big/\left[1+\frac{(1-\varepsilon)}{\varepsilon}\kappa_{i1}\right]\right)$$

is a constant, the z-directional velocity of the center point of the $\overline{C}_{i2}(z,t)$, as well as the $\overline{C}_{it}(z,t)$, profiles is

$$\left(\frac{dz}{dt}\right)_{\text{center point of } i} = \frac{v_z}{\left[1+\dfrac{(1-\varepsilon)}{\varepsilon}\kappa_{i1}\right]} = R_i' v_z, \tag{7.1.99b}$$

where R'_i, the fraction of the total solute i present in the mobile liquid phase, was defined earlier by relation (7.1.16c). The standard deviation of this profile is given by

$$\sigma_{ti} = \left\{\frac{2D_{i,\text{eff},z}\,t}{\left[1+\dfrac{(1-\varepsilon)}{\varepsilon}\kappa_{i1}\right]}\right\}^{1/2}. \tag{7.1.99c}$$

The retention time, t_{R_i}, of any species i is obtained from (7.1.99a) when $z = L$, the column length:

$$t_{R_i} = (L/R_i' v_z). \tag{7.1.99d}$$

If we can assume that the movement of each species (present in the feed sample introduced in the column prior to elution with the solvent) is independent of those of all other species (generally valid for dilute systems), then it is obvious from relation (7.1.99b) that the migration velocity of each species i down the column will be different as long as each κ_{i1} is different.

If the migration velocity of each species i down the column is different from those of others, the concentration peaks monitored in the detector at the column end (Figure 7.1.21) at different times will represent different species. Generally, this is indicated by stating that each species i has a different *retention time* t_{R_i} (see (2.5.3)) or *retention volume* V_{R_i} after which it appears at the column exit. The retention volume, V_{R_i}, is the volume of elution solvent that passes through the column after which species i appears at the exit. Since the mobile-phase velocity v_z is constant, this volume V_{R_i} is obtained from

$$\begin{aligned} V_{R_i} &= (t_{R_i}) \times (\text{volumetric flow rate of mobile phase}) \\ &= t_{R_i} \times \varepsilon S_c v_z. \end{aligned} \tag{7.1.99e}$$

If the column length is z, then, for any species i, t_{R_i} and z are related by (7.1.99a). Substituting that relation into (7.1.99e), we get

$$V_{R_i} = z\varepsilon S_c\left[1+\frac{(1-\varepsilon)}{\varepsilon}\kappa_{i1}\right], \tag{7.1.99f}$$

where, however,

$$z\varepsilon S_c = V_M \tag{7.1.99g}$$

is the mobile-phase volume within the column of length z. Rewriting (7.1.99f) as

$$V_{R_i} = V_M + V_M\left[\frac{(1-\varepsilon)}{\varepsilon}\right]\kappa_{i1} = V_M/R'_i = V_M + V_S\kappa_{i1}, \tag{7.1.99h}$$

where V_S is the stationary adsorbent-phase volume in the column, the contributions of the quantities V_M, V_S and κ_{i1} to V_{R_i} become clear.

Therefore, as long as κ_{i1} for each species i between the mobile fluid phase and the stationary adsorbent particles is different, each V_{R_i} will be different. Correspondingly, each t_{R_i} will be different. Further, both V_{R_i} and t_{R_i} are increased by an increase in V_M, V_S and κ_{i1}. Conversely, a reduction of V_{R_i} and t_{R_i} is achieved for any given system by a reduction in V_M and V_S. The resolution, R_S, between peaks of two neighboring species i and j is

$$\begin{aligned} R_S &= \frac{t_{R_i}-t_{R_j}}{2\sigma_{ti}+2\sigma_{tj}} = \frac{L}{2v_z}\frac{\left[\frac{1}{R_i'}-\frac{1}{R_j'}\right]}{(\sigma_{ti}+\sigma_{tj})} \\ &= \frac{L\left[\left\{1+\dfrac{(1-\varepsilon)}{\varepsilon}\kappa_{i1}\right\}-\left\{1+\dfrac{(1-\varepsilon)}{\varepsilon}\kappa_{j1}\right\}\right]}{2v_z\left[\left\{\dfrac{2D_{i,\text{eff},z}t_{R_i}}{1+\dfrac{(1-\varepsilon)}{\varepsilon}\kappa_{i1}}\right\}^{1/2}+\left\{\dfrac{2D_{j,\text{eff},z}t_{R_j}}{1+\dfrac{(1-\varepsilon)}{\varepsilon}\kappa_{j1}}\right\}^{1/2}\right]}. \end{aligned} \tag{7.1.99i}$$

The resolution is proportional to $(t_{R_i} - t_{R_j})$ and thus to $(V_{R_i} - V_{R_j})$, i.e. to $V_S(\kappa_{i1} - \kappa_{j1})$. Therefore, one often uses the net retention volume, V_{N_i}, for species i:

$$V_{N_i} = V_{R_i} - V_M = V_S \kappa_{i1}. \tag{7.1.99j}$$

Reflect now on what is being achieved in elution chromatography. By having the bulk flow of the mobile phase (solvent) perpendicular to the direction of force (for adsorption from the mobile phase to the stationary particle phase; for desorption from the stationary particle phase to the mobile phase), a multicomponent mixture of solutes is separated, as long as R_S values are reasonable. If the feed mixtures were simply equilibrated with the adsorbent particle phase without any mobile-phase flow perpendicular to the force direction, no such separation would have been achieved.

Consider, in addition, the nature of the force being used in separation. Since there is adsorption and/or desorption of solutes between the mobile fluid and the stationary solid phases, the potential profile under consideration for each solute is discontinuous, a simple step function (see Figure 3.2.2); there are no external forces. According to Section 3.2, such a system in a closed vessel without any flow does not have any multicomponent separation capability. Multicomponent separation capability is, however, achieved in elution chromatography by having bulk flow perpendicular to the direction of the discontinuous chemical potential profile. The velocity v_z here functions exactly like the quantity $(-\phi_1^+)'$ in equation (3.2.37).

To illustrate this, ignore the diffusion coefficient based term in (7.1.95a) to obtain

$$\frac{\partial \overline{C}_{it}}{\partial t} = -\frac{v_z}{\left[1 + \frac{(1+\varepsilon)}{\varepsilon}\kappa_{i1}\right]}\frac{\partial \overline{C}_{it}}{\partial z}. \tag{7.1.100}$$

In the trailing edge of the profile (along the column length z), $(\partial \overline{C}_{it}/\partial z)$ is positive and $(\partial \overline{C}_{it}/\partial t)$ is negative. Thus, solute i is being picked up from the trailing edge region and carried forward to increasing z to make room for the slower-moving species j to come and get adsorbed in the stationary adsorbent particles. Correspondingly, in the leading edge of the $\overline{C}_{it}$ profile, $\overline{C}_{it}$ is decreasing with z, therefore $\overline{C}_{it}$ is increasing with time due to solute i brought over from the trailing edge region by mobile-phase convection. This is how separate profiles of multiple species are developed along the column length, and ultimately the chromatographic detector output shown in Figure 7.1.21 is obtained as a function of time.

The results of elution chromatography were illustrated above using the solution of the equation for the local value of the total solute concentration $\overline{C}_{it}$. In reality, a series of profiles of mobile-phase concentration $\overline{C}_{i2}$ show up at the column end at different times in elution chromatography. Such a result may be obtained by using the *second method* mentioned earlier.

Consider solution (7.1.18h) for a fixed-bed process having a liquid feed of constant inlet concentration C_{i2}^0 into an initially solute-free column (i.e. $C_{i2}^i = 0$). The solution may be expressed compactly as

$$\overline{C}_{i2}(z,t) = C_{i2}^0\,\theta_i(z,V), \tag{7.1.101a}$$

where $\theta_i(z,V)$ is the solution (7.1.18h) for the linear dispersive equilibrium condition. Suppose that a volume V^0 of solution of concentration C_{i2}^0 is passed through the column from time $t = 0$ to $t = t^0$. Now, for time $t > t^0$, suppose that the interstitial liquid velocity remains the same, but that the incoming solution concentration is changed to C_{i2}^{00}. One is interested in knowing what happens at $t > t^0$, especially at very large times. Since the system (including the governing equation, the boundary and initial conditions) is linear, the sum of two solutions may be used to develop a solution for this changed condition. For $V \leq V^0$,

$$\overline{C}_{i2} = C_{i2}^0\,\theta_i(z,V) \qquad V \leq V^0. \tag{7.1.101b}$$

For $V > V^0$, one can assume that an inlet concentration of $(C_{i2}^{00} - C_{i2}^0)$ has been introduced into the column for a volume flow equal to $(V - V^0)$; however, the effect of C_{i2}^0 must continue. Therefore for $V^0 < V < \infty$, the solution is additive:[8]

$$\overline{C}_{i2} = C_{i2}^0\theta_i(z,V) + (C_{i2}^{00} - C_{i2}^0)\,\theta_i(z,V-V^0) \qquad V^0 < V < \infty. \tag{7.1.101c}$$

The second term reflects the contribution of the extra solute concentration $(C_{i2}^{00} - C_{i2}^0)$ as if the volume passed is 0 and the feed concentration is $(C_{i2}^{00} - C_{i2}^0)$. When $C_{i2}^{00} = 0$, we find

$$\overline{C}_{i2} = C_{i2}^0\{\theta_i(z,V) - \theta_i(z,V-V^0)\}. \tag{7.1.101d}$$

If V^0 is small, as in elution chromatography, a limiting process can be initiated:

$$\frac{\overline{C}_{i2}}{C_{i2}^0} = V^0\left[\frac{\theta_i(z,V) - \theta_i(z,V-V^0)}{V^0}\right]; \tag{7.1.101e}$$

$$\frac{\overline{C}_{i2}}{C_{i2}^0} = V^0\,\frac{\partial \theta_i(z,V)}{\partial V}. \tag{7.1.101f}$$

The approximation leading to the derivative is acceptable only in the limit of $V^0 \to 0$. Using the solution (7.1.18h) for $\theta_i(z,V)$,

[8]This is the *principle of superposition*. Since equation (7.1.18c) or equation (7.1.95a) are linear equations due to constant κ_{i1}, the superposition of two simpler solutions by simple addition leads to the solution of the problem under consideration. This is the advantage of *linear chromatography models*.

$$\frac{\overline{C}_{i2}}{C^0_{i2}} = \frac{V^0}{2}\frac{\partial}{\partial V}\left[1+\text{erf}\left\{\frac{Pe^{1/2}_{z,\text{eff}}(V-\overline{V})}{2(V\overline{V})^{1/2}}\right\}\right]$$

$$= V^0\frac{Pe^{1/2}_{z,\text{eff}}}{2\sqrt{\pi}(V\overline{V})^{1/2}}\left\{1+\frac{1}{2}\left(\frac{V-\overline{V}}{V}\right)\right\}\exp\left\{-\frac{(V-\overline{V})^2}{\frac{4V\overline{V}}{Pe_{z,\text{eff}}}}\right\}. \quad (7.1.101\text{g})$$

Since $(V-\overline{V})/2V$ is, in general, small compared to 1 in the breakthrough region, we can rewrite the above as

$$\overline{C}_{i2} \cong (V^0 C^0_{i2})\frac{Pe^{1/2}_{z,\text{eff}}}{2\sqrt{\pi}(V\overline{V})^{1/2}}\exp\left\{-\frac{(V-\overline{V})^2}{\frac{4V\overline{V}}{Pe_{z,\text{eff}}}}\right\}, \quad (7.1.102\text{a})$$

where $V^0 C^0_{i2} = m_i$ is the number of moles of solute i injected. This is a Gaussian profile around $\overline{V}$, with a standard deviation σ_{Vi} given by

$$\sigma_{Vi} = \left(\frac{2V\overline{V}}{Pe_{z,\text{eff}}}\right)^{1/2}. \quad (7.1.102\text{b})$$

For such a profile, the following approximations are valid around[9] $V = \overline{V} = \overline{V}_i$:

$$(V\overline{V})^{1/2} \cong \overline{V} \quad \text{and} \quad \sigma_{Vi} \cong \overline{V}(2/Pe_{z,\text{eff}})^{1/2}. \quad (7.1.102\text{c})$$

Employing the definition of $\overline{V}$ and $Pe_{z,\text{eff}}$ in (7.1.18i),

$$\sigma^2_{Vi} = \frac{2S^2_C z^2[\varepsilon + (1-\varepsilon)\kappa_{i1}]^2 D_{i,\text{eff},z}}{z v_z} \propto \frac{2D_{i,\text{eff},z}}{v_z}z, \quad (7.1.102\text{d})$$

where z is the column length. For illustrative examples of the above procedure, consult Mayer and Tompkins (1947) and Lightfoot *et al.* (1962). Expressions for similar profiles may be developed for other species present in the small feed sample of volume V^0 introduced initially at the top of the column. To determine κ_{i1} and $Pe_{z,\text{eff}}$ from $V = \overline{V}$, see Problem 7.1.13, in which column information and flow rates are provided.

A simpler representation of the solute i elution profile (7.1.102a) is

$$\overline{C}_{i2} = \overline{C}_{i2}\big|_{\max}\cdot\exp\left\{-\frac{(V-\overline{V})^2}{2\sigma^2_{Vi}}\right\}, \quad (7.1.102\text{e})$$

where

$$\overline{C}_{i2}\big|_{\max} = \left[(V^0 C^0_{i2})\,Pe^{1/2}_{z,\text{eff}}/\{2\sqrt{\pi}(V\overline{V})^{1/2}\}\right] \quad (7.1.102\text{f})$$

is the maximum value of $\overline{C}_{i2}$ achieved when $V = \overline{V}$ (see equation (2.5.6b)).

The volume of eluent collected during the elution of a solute band is ultimately the source of recovering the solute. The presence of multiple eluting solutes with overlapping bands forces us to make a decision on the eluent volume acting as the cut point t_c in time (see Figure 2.5.1(b)). Correspondingly, one is interested in knowing what is the fractional recovery of a particular solute for a given amount of solute input moles $(= m_i = V^0 C^0_{i2} = m^0_i)$. For the species 1 type of profile shown in Figure 2.5.1(b), the fractional recovery of solute i from time $t = 0$ (alternatively $t = -\infty$) to time $t = t_c$, the cut point, is given by

$$\text{fractional solute recovery} = \frac{m_{ii}}{m^0_i}$$

$$= \frac{\int_{-\infty}^{t} Q_f\,\overline{C}_{i2}(t)\,dt}{m^0_i} = \frac{\int_{-\infty}^{t_c} \varepsilon S_c v_z\,\overline{C}_{i2}(t)\,dt}{\int_{-\infty}^{\infty} \varepsilon S_c v_z\,\overline{C}_{i2}(t)\,dt}. \quad (7.1.102\text{g})$$

However, from equations (2.5.11)–(2.5.13), we already know that this ratio for Gaussian profiles is given by

$$\frac{m_{ii}}{m^0_i} = \frac{1}{2} + \frac{1}{2}\text{erf}\left(\frac{t_c - t_{R_i}}{\sqrt{2}\,\sigma_{ti}}\right) = \frac{1}{2} + \frac{1}{2}\text{erf}\left(\frac{V_c - V_{R_i}}{\sqrt{2}\,\sigma_{Vi}}\right), \quad (7.1.102\text{h})$$

where the final expression has been written by analogy, since the volumes V_c (the volume at the cut point), V_{R_i} and σ_{Vi} in the elution process are linearly proportional to t_c, t_{R_i} and $\sigma_{ti}(V_c = \varepsilon\, v_z\, S\, t_c$, etc.). Similarly, for the species 2 type of profile shown in Figure 2.5.1(b),

$$\text{fractional solute recovery} = \frac{m_{ii}}{m^0_i}$$

$$= \frac{\int_{t_c}^{\infty} Q_f\,\overline{C}_{i2}(t)\,dt}{m^0_i} = \frac{\int_{t_c}^{\infty} \varepsilon S_c v_z\,\overline{C}_{i2}(t)\,dt}{\int_{-\infty}^{\infty} \varepsilon S_c v_z\,\overline{C}_{i2}(t)\,dt}. \quad (7.1.102\text{i})$$

From equations (2.5.14)–(2.5.16), we already know that this ratio for Gaussian profiles is given by

$$\text{fractional solute recovery} = \frac{m_{ii}}{m^0_i}$$

$$= \frac{1}{2} + \frac{1}{2}\text{erf}\left(\frac{t_{R_i} - t_c}{\sqrt{2}\,\sigma_{ti}}\right) = \frac{1}{2} + \frac{1}{2}\text{erf}\left(\frac{V_{R_i} - V_c}{\sqrt{2}\sigma_{Vi}}\right). \quad (7.1.102\text{j})$$

Example 7.1.5 In the purification of an enzyme by an appropriate column of adsorbents, the elution process provided the information given in Table 7.1.4 about the enzyme concentration (in suitable units) in the eluent leaving the column (volume 2 liter) and the total volume of eluent passed.

---

[9] $\overline{V}$ here refers to expression (7.1.16d) for solute i and therefore should more correctly be represented as $\overline{V}_i$, which is not to be confused with the partial molar volume of species i.

Table 7.1.4.

$\overline{C}_{i2}$	V (liter)
6.5	1.8
15.0	2.0[a]

[a] It is known that the $\overline{C}_{i2}$ value of 15.0 was the maximum; therefore $\overline{V} = 2$ liter.

Determine the value of σ_{Vi} (liter) and the fractional recovery of the enzyme (assuming that the concentration profile is Gaussian) when the eluent volume passed is 2.1 liters.

Solution Consider the Gaussian elution profile (7.1.102e):

$$\overline{C}_{i2} = \overline{C}_{i2}\big|_{\max} \cdot \exp\left\{-\frac{(V-\overline{V})^2}{2\sigma_{Vi}^2}\right\}.$$

Apply the profile to the two data points provided and then divide one by the other:

$$\frac{\overline{C}_{i2}|_1}{\overline{C}_{i2}|_2} = \frac{\exp\left\{-\dfrac{(V_1-\overline{V})^2}{2\sigma_{Vi}^2}\right\}}{\exp\left\{-\dfrac{(V_2-\overline{V})^2}{2\sigma_{Vi}^2}\right\}} \Rightarrow \frac{0.65}{1.5}$$

$$= \frac{\exp\left(-\dfrac{(1.8-2.0)^2}{2\sigma_{Vi}^2}\right)}{\exp\left(-\dfrac{(2-2)^2}{2\sigma_{Vi}^2}\right)} \Rightarrow 0.43$$

$$= \exp\left(-\frac{(0.2)^2}{2\sigma_{Vi}^2}\right) \Rightarrow \exp\left(+\frac{0.02}{2\sigma_{Vi}^2}\right) = 2.325$$

$$\Rightarrow (0.02/\sigma_{Vi}^2) = 0.844 \Rightarrow \sigma_{Vi} = 0.154\,\text{liter}.$$

Since the eluent volume passed, 2.1 liters, is larger than $\overline{V} = 2$ liter, the cut point $t_c > t_{R_i}$. So the profile corresponding to species 1 in Figure 2.5.1(b), is relevant. The formula for fractional recovery is given by (7.1.102h):

$$\text{fractional recovery} = \frac{1}{2} + \frac{1}{2}\,\text{erf}\left(\frac{V_c - V_{R_i}}{\sqrt{2}\,\sigma_{Vi}}\right)$$

$$= \frac{1}{2} + \frac{1}{2}\,\text{erf}\left(\frac{2.1-2.0}{1.414 \times 0.154}\right) = 0.5 + 0.5\,\text{erf}(0.459)$$

$$= 0.5 + 0.5 \times 0.484 = 0.742.$$

Elution chromatography based on liquid–solid adsorption for small molecules typically employs adsorbents such as alumina, charcoal, silica, hydrophobic silica, etc. Surface area, water content and the chemical nature of the adsorbent (e.g. polar or nonpolar) distinguish one adsorbent from another. For larger molecules/macromolecules/charged species, other adsorbents needed are briefly touched upon in Sections 7.1.5.1.6–7.1.5.1.8.

7.1.5.1.2 Types of elution chromatography The type of elution chromatography described earlier was liquid–solid adsorption chromatography. The mobile phase was liquid, the stationary phase consisted of porous particles, and the local mechanism of separation of various species between the mobile phase and the stationary phase was simply adsorption. There can be a number of other types of elution chromatography depending on the combination of mobile and stationary phases and the separation mechanism locally operative. The more well-known and frequently used of these techniques are summarized in Table 7.1.5.

There are variations in a given technique identified in Table 7.1.5. For example, in liquid–liquid chromatography (LLC), normally the stationary-phase liquid is polar and the mobile phase is nonplolar, both being immiscible with each other. In *reversed-phase LLC*, the stationary-phase liquid is nonpolar, whereas the mobile phase is polar; the polar phase can even be water. In many cases, the stationary liquid phase may be a monomolecular layer chemically bonded to the surface groups of the porous support material. This is often achieved with a nonpolar hydrocarbon liquid phase.

A very common column configuration in elution chromatography is simply a tubular column packed with porous particles, the packings, with or without a bonded liquid phase on the particle surfaces. Other column configurations include capillary columns or open tubular columns, in which a thin liquid film of adsorbents has been applied (or bonded) to the internal surface of the capillaries. A potential variation of this is the microporous hollow fiber membrane based column, wherein the stationary phase is held in the pores of fiber wall and the eluent is passed through the bore of the fiber (Ding *et al.*, 1989).

Analytical-scale column chromatography, used to identify the components in a sample or to determine the feed sample composition, uses small columns; for example, in gas–liquid chromatography, a typical column may be 6 ft long. In preparative chromatography, column diameters of 1 to 2 cm are used, and the feed sample volumes are considerably larger. While this is used to prepare larger amounts of purified materials or compounds for laboratory use, industrial-scale column chromatography, producing 100 metric ton/year, is being practiced (*Chem. Eng.*, 1980, 1981). The same organization (Elf Aquitane) has demonstrated a variety of separations in gas chromatography using industrial-scale columns of diameter 40 cm, length 1.5 m, containing 60–80 mesh packing operating at 180 °C for injection times between 5 and 50 s and sample sizes between 10 and 50 g/s (Bonmati *et al.* 1980).

In column or capillary chromatography, the gas or liquid eluent is driven to flow by the application of a pressure gradient along the column length. In *paper* or *planar chromatography*, the adsorbent material is in a thin granular form (thin-layer chromatography) or in fibrous form (as in paper chromatography); no such pressure gradient can be independently applied. Instead, the liquid eluent is driven along the layer by capillary action (i.e. the capillary pressure, see Section 6.1.4). The rate of

Table 7.1.5. Phase combinations and local retention mechanisms in chromatography

Mobile phase	Stationary phase	Local mechanism of separation	Name
Liquid	porous solid adsorbent	adsorption	liquid–solid adsorption chromatography (LSC)
Liquid	porous ion exchange resin particles	ion exchange	ion exchange chromatography
Liquid	liquid stationary phase coated on a porous solid support	partitioning (as in solvent extraction between two immiscible liquid phases)	liquid–liquid chromatography (LLC)
Liquid	porous gels	molecular size exclusion via gel pore size	gel permeation chromatography (GPC) **or** size exclusion chromatography (SEC)
Liquid	immobilized affinant on an insoluble support having strong affinity for macromolecules	reversible binding between solute and affinant	affinity chromatography
Gas	liquid stationary phase with low vapor pressure coated on a porous solid support	differential solubility in the stationary phase and differing solute vapor pressure	gas–liquid chromatography (GLC)
Gas	porous solid adsorbent	adsorption	gas–solid chromatography
Supercritical fluid	porous solid adsorbent **or** liquid stationary phase coated on a porous solid support	adsorption **or** differential solubility and differing solute vapor pressures	supercritical fluid chromatography

spreading of the liquid cannot be precisely controlled. Further, the movement of the liquid is two-dimensional, unlike in column chromatography, in which the movement is in one dimension only, along the column length.

7.1.5.1.3 Elution chromatography with mobile gas phase Elution chromatography with a mobile gas phase (gas chromatography, GC) can be carried out with either porous solid adsorbent or a liquid absorbent coated on porous solid support particles (see Table 7.1.5). The former is identified as gas–solid chromatography (GSC), whereas the latter is called gas–liquid chromatography (GLC). Sometimes in GLC, the liquid absorbent exists as a coating on the inside surface of a capillary tube. The liquid absorbent in GLC must have low enough vapor pressure to be considered essentially nonvolatile.

The analysis of elution chromatography with a liquid eluent that was presented earlier was based on a constant eluent velocity, v_z. Unlike simple fixed-bed adsorption processes with low pressure drops, analytical-scale chromatographic techniques for a mobile gas phase employ long packed columns, where the gas undergoes a considerable pressure drop. As a result, the gas velocity changes with column location. But the gas velocity is proportional to the molar gas volume at every location. Therefore, by Boyle's law, the gas pressure P and velocity v_z at any location are related to those at the inlet (subscript, in) and outlet (subscript, out) by

$$P v_z = P_{\text{in}} v_{z,\text{in}} = P_{\text{out}} v_{z,\text{out}}. \tag{7.1.103a}$$

The gas velocity v_z and the gas pressure gradient $(\mathrm{d}P/\mathrm{d}z)$ at any location are related by Darcy's law for the packed bed acting as a porous medium (see the solvent flux expressions (3.4.85), (3.4.88) and (6.1.4g)):

$$v_z = -\frac{Q_g}{\varepsilon\mu}\frac{\mathrm{d}P}{\mathrm{d}z}. \tag{7.1.103b}$$

Remember: the Darcy permeability Q_g is proportional to r_p^2, where r_p is the pore radius, and therefore to d_p^2, where d_p is the packing diameter (since it is essentially proportional to the effective diameter of the interpacking opening).

Substituting for v_z in terms of P from (7.1.103a) into the above expression, then integrating and rearranging, we get

$$\frac{v_z(z)}{v_{z,\text{in}}} = \left\{\frac{(P_{\text{in}}/P_{\text{out}})^2}{(P_{\text{in}}/P_{\text{out}})^2 - \left[\left\{(P_{\text{in}}/P_{\text{out}})^2 - 1\right\}\frac{z}{L}\right]}\right\}^{1/2}, \tag{7.1.103c}$$

where μ is assumed to be constant and independent of gas pressure. The retention time t_{R_i} may now be obtained from relation (7.1.99b) if v_z is considered to be a function of the z-coordinate:

$$\int_0^{t_{R_i}} \mathrm{d}t = \int_0^L \frac{\mathrm{d}z}{R_i' v_z(z)} = t_{R_i}. \tag{7.1.103d}$$

Substituting into this an expression for $v_z(z)$ from (7.1.103c) and integrating, one obtains

$$t_{R_i} = \frac{L}{R_i'}\frac{1}{v_{z,\text{avg}}}, \tag{7.1.103e}$$

where

$$\frac{v_{z,\text{avg}}}{v_{z,\text{out}}} = \frac{3}{2}\left[\frac{(P_{\text{in}}/P_{\text{out}})^2 - 1}{(P_{\text{in}}/P_{\text{out}})^3 - 1}\right], \tag{7.1.103f}$$

as demonstrated first by James and Martin (1952). This last ratio allows the conversion of the known gas velocity at column outlet, $v_{z,out}$, to an average gas velocity, $v_{z,avg}$, in the column for calculating the retention time t_{R_i} of species i. Correspondingly, the retention volume of gas needs to be corrected due to this gas compressibility. In addition, if the column temperature T is different from the exit temperature T_{out} at which the gas flow rate is measured, an additional correction has to be made.

Given the flow conditions and column dimensions, the retention time or volume of a species is known, if κ_{i1} is known. Since the variation in V_{R_i} or t_{R_i} between two neighboring peaks or species depends on values of κ_{i1} for $i = 1, 2$ (say), it is useful to enquire what properties of the gas–stationary phase combination influence κ_{i1}.

For gas–liquid chromatography (GLC) with liquid coating on packings, assume equilibrium to exist between the two phases for any species i. Then, from Section 3.3.7.1, we get $\hat{f}_{ig} = \hat{f}_{il}$, $\hat{f}_{ig} = x_{ig}\phi_{ig}P$ and $\hat{f}_{il} = \gamma_{il}x_{il}f^0_{il}$. Since the gas phase may be assumed to behave ideally at the low-pressure characteristics of GLC, $\phi_{ig} \cong 1$. For condensible solute species (vapors) at low pressures, $f^0_{il} \cong P^{sat}_i$ from equation (3.3.62). Further, the solute concentration in the stationary liquid phase on packing is very low, so that $\gamma_{il} \cong \gamma^\infty_{il}$, corresponding to infinite dilution. Therefore

$$\frac{x_{il}}{x_{ig}} = \frac{P}{\gamma^\infty_{il}P^{sat}_i} \quad \text{and} \quad \alpha_{12} = \frac{x_{1l}x_{2g}}{x_{1g}x_{2l}} = \frac{\gamma^\infty_{2l}P^{sat}_2}{\gamma^\infty_{1l}P^{sat}_1}, \tag{7.1.104a}$$

where species 2 elutes earlier from the column than species 1. Generally, $P^{sat}_2 > P^{sat}_1$. However, the ratio of the activity coefficients is quite important, especially in the separation of two species of essentially equal vapor pressure. The selectivity is obtained by different molecular interactions between the solute species to be separated and the stationary liquid phase. For a brief introduction to the contribution of the activity ratio to the selectivity, see Karger *et al.* (1973).

The net retention volume, V_{N_i} (definition (7.1.99j)), for species i can also be expressed in terms of P^{sat}_i and γ^∞_i. Consider the distribution coefficient κ_{i1} first and assume ideal gas behavior:

$$\kappa_{i1} = \frac{\overline{C}_{i1}}{\overline{C}_{i2}} = \frac{\overline{C}_{i1}}{\overline{C}_{ig}} = \frac{\overline{C}_{i1}RT}{\overline{x}_{ig}P} = \frac{m_{il}RT}{V_{Sl}\,\overline{x}_{ig}P}$$
$$= \frac{m_{il}RT}{m_{sl}\,\overline{x}_{ig}P}\left(\frac{m_{sl}}{V_{Sl}}\right) \cong \left(\frac{\overline{x}_{il}}{\overline{x}_{ig}}\right)\frac{RT}{P}\left(\frac{m_{sl}}{V_{Sl}}\right),$$

where m_{il} denotes the moles of solute i in a stationary liquid phase of volume V_{Sl} and solvent moles m_{sl}. Therefore,

$$\kappa_{il} = \frac{RT}{\gamma^\infty_{il}P^{sat}_i}\left(\frac{m_{sl}}{V_{Sl}}\right). \tag{7.1.104b}$$

To get this result, we have used relation (7.1.104a) for (x_{il}/x_{ig}) and assumed $m_{il} << m_{sl}$, the subscript s representing the solvent species in the stationary coated liquid phase. The net retention volume, by definition (7.1.99), is

$$V_{N_i} = V_{Sl}\kappa_{i1} = \frac{RTm_{sl}}{\gamma^\infty_{il}P^{sat}_i}. \tag{7.1.104c}$$

Since the value of the quantity m_{sl} of the coating solvent is particular to a given column and packing, an alternative quantity, the specific retention volume (V_{N_i}/w^l), is sometimes used:

$$\frac{V_{N_i}}{w^l} = \frac{RT}{\gamma^\infty_{il}P^{sat}_iM_{Sl}}, \tag{7.1.104d}$$

where w^l is the mass of the stationary liquid phase used as a coating and M_{sl} is the molecular weight of this coating liquid. The specific retention volume of species i decreases as its vapor pressure increases; however, the coating liquid also influences this value through γ^∞_{il}.

A large number of organic liquids with maximum operating temperatures varying between 100 and 350 °C and varying degrees of polarity have been used as stationary phases in GLC. Nonstandard liquid phases used include: liquid crystals; addition of nonvolatile complexing agents in the stationary liquid phase which preferentially complex with particular solutes (e.g. $AgNO_3$ with olefins). Porous, chemically inert solid supports often used to hold the liquid phase include diatomaceous earth (called Chromosorb), fluorine-containing polymers that are quite inert, glass microbeads, etc.

Whereas GLC is more often used to separate a variety of thermally stable volatile nonelectrolytic species with molecular weight up to 300, gas–solid chromatography (GSC) is primarily used to separate permanent gases, gaseous hydrocarbons, etc. Porous adsorbents used in GSC include graphitized thermal carbon black, silica, porous glass, molecular sieves, porous polymer beads, etc. Gas–solid adsorption equilibrium provides the local separation mechanism. Complications arise from chemisorption, degradation of samples due to higher column temperatures in analytical separations and large retention volumes. Since the surfaces of porous adsorbents are not well-defined, there are difficulties in developing effective models for species retention volumes.

Gas chromatographic techniques are primarily used for analytical separations. However, they have begun to be used in industrial-scale separations. A schematic of an industrial-scale gas chromatographic unit (*Chem. Eng.*, 1981) is illustrated in Figure 7.1.22. As described by Bonmati *et al.* (1980), liquid feed from a reservoir is fed by a pump to a vaporizer. The vaporizer is connected to the injector via a valve, whose opening and closing is controlled by an electronic programmer. In the injector, the vaporized feed is mixed with the purified carrier gas, which is generally compressed hydrogen. This vaporized feed in the carrier gas is introduced into the chromatographic column at particular

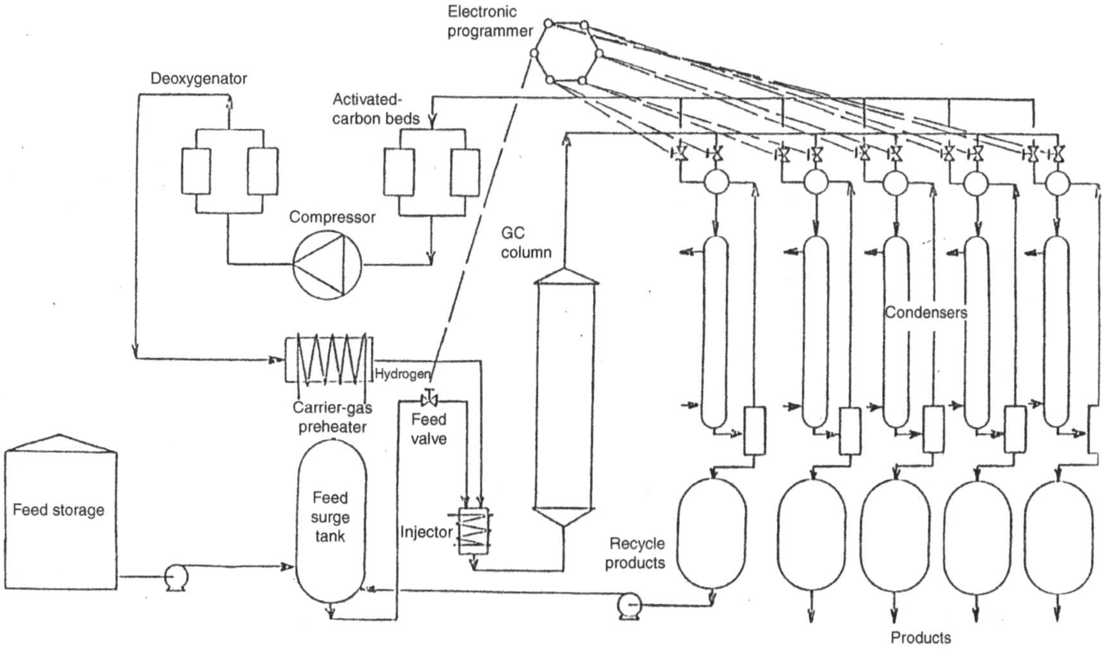

Figure 7.1.22. Basic flow diagram of a large-scale gas chromatographic process. (After Bonmati et al. *(1980) and* Chem. Eng. *(1980).)*

intervals. Typical injection times are between 5 and 50 s, and the sample size varies between 10 and 50 g/s. The column diameter can be as much as 40 cm, and the column length is 1.5 m. It is uniformly packed with Chromosorb of size 60–80 mesh and coated with a suitable liquid.

The end of the column is connected by a valving sequence to a set of condensers and product collectors. Different species move through the column at different speeds and exit at different times. A given fraction of the exiting stream from the column containing a particular species is routed to a particular condenser and product collector via a particular valve opened at that time by the electronic programmer. The overhead of each condenser produces a stream of the carrier gas (e.g. hydrogen), which is recycled to the injector after passing it through a carrier-gas cleaner, compressor, catalytic deoxygenator and a preheater. The bottoms from each condenser comprise the particular product fraction. A portion of the liquid output is recycled to the feed storage or feed vaporizer. The residence time in the column is of the order of a minute or two; thus, thermal degradation is avoided at column temperatures as high as 180 °C.

7.1.5.1.4 Number of theoretical plates and plate height in elution chromatography In an ideal equilibrium stage of distillation, the vapor and liquid mixtures are in equilibrium, and each phase has the same composition everywhere. Such an equilibrium stage is often identified as a *theoretical plate* (the ideal stage in Section 6.3.2.1). Similarly, in elution chromatography, using a column packed with adsorbent particles, one often encounters the notion of a number N of theoretical plates. On each plate, the adsorbent particles are assumed to be always in equilibrium with the eluent vis-à-vis any solute species. The eluent passes continuously and without mixing through these plates from one plate to the next (Figure 7.1.23). Different expressions have been developed to calculate the solute concentration along the column of N plates as a function of the eluent volume passed through the column (Martin and Synge, 1941; Mayer and Tompkins, 1947; Said, 1956). The method of analysis followed here is somewhat related to that of Said (1956). The earlier derivations by Martin and Synge (1941) and Mayer and Tompkins (1947) are limiting cases.

The following assumptions are made in the context of linear chromatography:

(1) The chromatographic column length L essentially consists of N theoretical plates, on each of which the adsorbent particles are always in equilibrium with the eluent vis-à-vis any solute species. Further, each plate contains $V_S/N\,\text{cm}^3$ of adsorbent, where V_S is the total volume of adsorbent; similarly, each plate contains V_M/N volume of mobile phase, where $L\varepsilon S_c = V_M$. The distance between two consecutive plates is $L\varepsilon/N$, where ε is the void volume in the column occupied by the eluent.

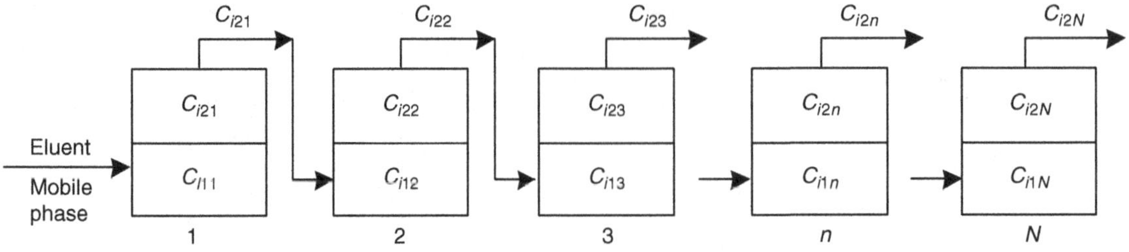

Figure 7.1.23. Theoretical plate model of a chromatographic column with N *well-mixed plates; both phases in each plate are in equilibrium, and the eluent passes continuously and without mixing through these plates.*

(2) The eluent passes continuously and without mixing through these plates.
(3) The eluent phase concentration of species i on the nth plate, C_{i2n}, is linearly related to the adsorbent-phase concentration C_{i1n} by

$$(C_{i1n}/C_{i2n}) = \kappa_{i1}, \tag{7.1.105a}$$

where κ_{i1} is a constant.

Figure 7.1.23 shows a schematic of the column with N plates. At any time t during the elution process, the adsorbent-phase concentrations on plates 1, 2, 3..., n,..., N of species i are C_{i11}, C_{i12}, C_{i13},..., C_{i1n},..., C_{i1N}, respectively. (At the beginning of the process ($t = 0$), the values of these quantities were $C^0_{i11}, C^0_{i12}, C^0_{i13}, \ldots, C^0_{i1n}, \ldots C^0_{i1N}$.) The corresponding mobile-phase concentrations on each plate at any time t are C_{i21}, C_{i22}, C_{i23},..., C_{i2n},..., C_{i2N}.

Consider plate 1 in Figure 7.1.23. If the total volume of eluent that has crossed plate 1 at any time is V, then a differential material balance (for a short time interval dt) on species 1 around plate 1 provides

$$-(C_{i11}/\kappa_{i1})\mathrm{d}V = \frac{V_S}{N}\,\mathrm{d}C_{i11} + \frac{V_M}{N}\,\mathrm{d}C_{i21}; \tag{7.1.105b}$$

$$-(C_{i11}/\kappa_{i1})\mathrm{d}V = \frac{V_S}{N}\,\mathrm{d}C_{i11} + \frac{V_M}{N\kappa_{i1}}\,\mathrm{d}C_{i11};$$

$$\frac{\mathrm{d}C_{i11}}{C_{i11}} = -\frac{\mathrm{d}V}{\kappa_{i1}\frac{V_S}{N} + \frac{V_M}{N}} \Rightarrow \log C_{i11} = -\frac{V}{\frac{V_M}{N} + \kappa_{i1}\frac{V_S}{N}} + \text{constant}_1.$$

When $V = 0$, $C_{i11} = C^0_{i11}$. Therefore

$$(C_{i11}/C^0_{i11}) = \exp\left[-\frac{V}{\frac{V_M}{N} + \kappa_{i1}\frac{V_S}{N}}\right]. \tag{7.1.105c}$$

Therefore, the eluent concentration leaving plate 1 and entering plate 2 is given by

$$C_{i21} = \frac{C_{i11}}{\kappa_{i1}} = \frac{C^0_{i11}}{\kappa_{i1}}\exp\left(-\frac{V}{\frac{V_M}{N} + \kappa_{i1}\frac{V_S}{N}}\right). \tag{7.1.105d}$$

A differential mass balance around plate 2 for species i in the manner of equation (7.1.105b) yields

$$\left(C_{i21} - \frac{C_{i12}}{\kappa_{i1}}\right)\mathrm{d}V = \frac{V_S}{N}\,\mathrm{d}C_{i12} + \frac{V_M}{N}\,\mathrm{d}C_{i22}.$$

Substituting for C_{i21} from (7.1.105d) and rearranging, we get

$$\frac{\mathrm{d}C_{i12}}{\mathrm{d}\left(\frac{V\kappa_{i1}}{\kappa_{i1}\frac{V_S}{N} + \frac{V_M}{N}}\right)} + \frac{C_{i12}}{\kappa_{i1}} = \frac{C^0_{i11}}{\kappa_{i1}}\exp\left[-\frac{V}{\frac{V_M}{N} + \kappa_{i1}\frac{V_S}{N}}\right].$$

Its general solution is

$$C_{i12} = \left[\exp\left(-\frac{V}{\frac{V_M}{N} + \kappa_{i1}\frac{V_S}{N}}\right)\right]\left[\left(\frac{C^0_{i11}V}{\frac{V_M}{N} + \kappa_{i1}\frac{V_S}{N}} + \text{constant}_2\right)\right].$$

When $V = 0$, $C_{i12} = C^0_{i12}$. Therefore

$$C_{i12} = \left(\frac{C^0_{i11}V}{\frac{V_M}{N} + \kappa_{i1}\frac{V_S}{N}} + C^0_{i12}\right)\left[\exp\left(-\frac{V}{\frac{V_M}{N} + \kappa_{i1}\frac{V_S}{N}}\right)\right]. \tag{7.1.105e}$$

Similarly, a differential material balance around plate 3 leads to

$$\left(C_{i22} - \frac{C_{i13}}{\kappa_{i1}}\right)\mathrm{d}V = \frac{V_S}{N}\,\mathrm{d}\,C_{i13} + \frac{V_M}{N}\,\mathrm{d}C_{i23}.$$

Employing the solution for C_{i12} from (7.1.105e), we obtain

$$\frac{\mathrm{d}C_{i13}}{\mathrm{d}\left(\frac{V\kappa_{i1}}{\frac{V_M}{N} + \kappa_{i1}\frac{V_S}{N}}\right)} + \frac{C_{i13}}{\kappa_{i1}} = \frac{1}{\kappa_{i1}}\left[\exp\left(-\frac{V}{\frac{V_M}{N} + \kappa_{i1}\frac{V_S}{N}}\right)\right]$$
$$\left[C^0_{i12} + \frac{C^0_{i11}V}{\left(\frac{V_M}{N} + \kappa_{i1}\frac{V_S}{N}\right)}\right]. \tag{7.1.105f}$$

A solution of this for the initial condition of $C_{i13} = C^0_{i13}$, when $V = 0$, is

$$C_{i13} = \left[\exp\left(-\frac{V}{\frac{V_M}{N} + \kappa_{i1}\frac{V_S}{N}}\right)\right]\left[\frac{C^0_{i11}V^2}{2\left(\frac{V_M}{N} + \kappa_{i1}\frac{V_S}{N}\right)^2}\right.$$
$$\left. + \frac{C^0_{i12}V}{\left(\frac{V_M}{N} + \kappa_{i1}\frac{V_S}{N}\right)^2} + C^0_{i13}\right]. \tag{7.1.105g}$$

For plate 4, the corresponding expression is

$$C_{i14}=\left[\exp\left(-\frac{V}{\frac{V_M}{N}+\kappa_{i1}\frac{V_S}{N}}\right)\right]\left[\frac{C^0_{i11}V^3}{2\times 3\left(\frac{V_M}{N}+\kappa_{i1}\frac{V_S}{N}\right)^3}+\frac{C^0_{i12}V^2}{2\left(\frac{V_M}{N}+\kappa_{i1}\frac{V_S}{N}\right)^2}+\frac{C^0_{i13}V}{\left(\frac{V_M}{N}+\kappa_{i1}\frac{V_S}{N}\right)}+C^0_{i13}\right]. \quad (7.1.105h)$$

By induction, the expression for plate n is

$$C_{i1n}=\left[\exp\left(-\frac{V}{\frac{V_M}{N}+\kappa_{i1}\frac{V_S}{N}}\right)\right]\left[\frac{C^0_{i11}V^{n-1}}{(n-1)!\left(\frac{V_M}{N}+\kappa_{i1}\frac{V_S}{N}\right)^{n-1}}+\frac{C^0_{i12}V^{n-2}}{(n-2)!\left(\frac{V_M}{N}+\kappa_{i1}\frac{V_S}{N}\right)^{n-2}}+\cdots+C^0_{i1n}\right], \quad (7.1.105i)$$

which may be written in a compact fashion as follows:

$$C_{i1n}=\left[\sum_{r=1}^{n}C^0_{i1r}\frac{V^{n-r}}{(n-r)!\left(\frac{V_M}{N}+\kappa_{i1}\frac{V_S}{N}\right)^{n-r}}\right]\left[\exp\left(-\frac{V}{\frac{V_M}{N}+\kappa_{i1}\frac{V_S}{N}}\right)\right]. \quad (7.1.105j)$$

Special case: Only the first plate is loaded with solute before the beginning of the elution process, i.e. $C^0_{i12}=C^0_{i13}=\cdots=C^0_{i1n}=0$; therefore

$$C_{i1n}=C^0_{i11}\frac{V^{n-1}}{(n-1)!\left(\frac{V_M}{N}+\kappa_{i1}\frac{V_S}{N}\right)^{n-1}}\left[\exp\left(-\frac{V}{\left(\frac{V_M}{N}+\kappa_{i1}\frac{V_S}{N}\right)}\right)\right], \quad (7.1.105k)$$

which is a Poisson distribution function. If m_i moles of solute i were introduced in the first plate, then

$$m_i=C^0_{i21}\left(\kappa_{i1}\frac{V_S}{N}+\frac{V_M}{N}\right)=\frac{C^0_{i11}}{\kappa_{i1}}\left(\kappa_{i1}\frac{V_S}{N}+\frac{V_M}{N}\right), \quad (7.1.105l)$$

leading to

$$C_{i2n}=\frac{m_i}{\left(\frac{V_M}{N}+\kappa_{i1}\frac{V_S}{N}\right)}\frac{1}{(n-1)!}\frac{V^{n-1}}{\left(\frac{V_M}{N}+\kappa_{i1}\frac{V_S}{N}\right)^{n-1}}\left[\exp\left(-\frac{V}{\left(\frac{V_M}{N}+\kappa_{i1}\frac{V_S}{N}\right)}\right)\right]. \quad (7.1.105m)$$

When the value of n is large, the Poisson distribution becomes a Gaussian distribution. Giddings (1965) has indicated that this discontinuous Poisson distribution may be approximated by a Gaussian distribution when the number of plates exceeds 100.

The Gaussian distribution on plate j, where j is a large number, may be described by

$$C_{i2j}=\frac{m_i}{((V_M/N)+\kappa_{i1}(V_S/N))}\frac{1}{\sqrt{2\pi j}}\exp\left[-\frac{\left(v^{*+}_{Ci}-j\right)^2}{2j}\right], \quad (7.1.105n)$$

where the nondimensional concentration wave velocity for species i, v^{*+}_{Ci}, is defined by

$$v^{*+}_{Ci}=\frac{V}{((V_M/N)+\kappa_{i1}(V_S/N))}=\left[\frac{\varepsilon S_c v_z t}{\frac{L}{N}S_c[\varepsilon+\kappa_{i1}(1-\varepsilon)]}\right]=\frac{v_z t}{(L/N)\left[1+\frac{(1-\varepsilon)}{\varepsilon}\kappa_{i1}\right]}=\frac{v^*_{Ci}t}{H}. \quad (7.1.105o)$$

Here H is the height of a theoretical plate if there are N plates in a column of length L:

$$H=\frac{L}{N}, \quad (7.1.105p)$$

and v^*_{Ci} is given by (7.1.15c) and (7.1.12a). The value of C_{i2j} at the column outlet, where $j=N$, will represent the peak maximum when $v^{*+}_{Ci}=N$:

$$C_{i2N}=C_{i2N}|_{\max}\exp\left(-\frac{\left(v^{*+}_{Ci}-N\right)^2}{2N}\right), \quad (7.1.105q)$$

where

$$C_{i2N}|_{\max}=\frac{m_i}{((V_M/N)+\kappa_{i1}(V_S/N))}\frac{1}{\sqrt{2\pi N}}. \quad (7.1.105r)$$

Further, the value of the dimensionless standard deviation σ^+_{Vi} is given as

$$\sigma^+_{Vi}=\sqrt{N}=\sqrt{L/H}. \quad (7.1.105s)$$

Rewriting equation (7.1.105n) with v^{*+}_{Ci} replaced by $v^*_{Ci}t/H$ and $j=N$, we get

$$C_{i2N}=C_{i2N}|_{\max}\exp\left(-\frac{\left(\frac{v^*_{Ci}}{H}t-\frac{L}{H}\right)^2}{2(L/H)}\right). \quad (7.1.105t)$$

This expression may be written in two alternative forms to illustrate a Gaussian profile in terms of time t or distance z:

$$C_{i2N}=C_{i2N}|_{\max}\exp\left(-\frac{\left(t-t_{R_i}\right)^2}{2H\,t^2_{R_i}/L}\right), \quad (7.1.105u)$$

$$\sigma_{ti}=t_{R_i}/\sqrt{N}; \quad (7.1.105v)$$

$$C_{i2N}=C_{i2N}|_{\max}\exp\left(-\frac{(z_{i0}-L)^2}{2(LH)}\right), \quad (7.1.105w)$$

$$\sigma_{zi}=\sqrt{LH}. \quad (7.1.105x)$$

Two conclusions from these results are: σ_{zi} increases with $L^{1/2}$; σ_{ti} decreases as N increases in the manner of $N^{-1/2}$. Also, N can be determined from any data on t_{R_i} and σ_{ti}; therefore, H can be determined from (L/N).

The strength of the plate model is that, for a discrete number of stages, it can predict useful quantities, e.g. which plate will have the highest solute concentration at any time (volume of mobile phase) (Said, 1956; Giddings, 1965; King, 1980, pp. 379–387; Wankat, 1990, pp. 313–316). It is also useful for comparing two chromatographic columns in terms of the number N of plates/stages available. If the column is of length L and contains N theoretical plates, then the two are related through a plate height H (sometimes called the *height of an equivalent theoretical plate*, i.e. HETP) via (see the corresponding definition (6.3.24) in capillary electrophoresis)

$$N = \frac{L}{H}. \qquad (7.1.106a)$$

For a fixed L, the larger the value of N, the more efficient the column, and correspondingly, the smaller the value of the plate height. However, in the plate model of a chromatographic column, the number of plates, N, is unknown; therefore H is unknown. In fact, H has to be predicted independently via alternative models (Giddings, 1965).

Giddings (1965) has defined the plate height H for a uniform column free from concentration and velocity gradients as

$$H = \sigma_{zi}^2/L, \qquad (7.1.106b)$$

where σ_{zi} is the standard deviation of the eluting concentration profile of species i. For a given column length L, the extent of band broadening (equation (7.1.99c) as well as the text around equations (3.2.21a, b) and (3.2.23)) is directly related to H. As σ_{zi}^2 increases, H increases. Correspondingly, the number of plates, N, is reduced; therefore, the peak capacity, $n_{\max}$, is reduced. From expression (7.1.99c), since σ_{zi}^2 is proportional to $D_{i,\text{eff},z}$, H increases linearly with $D_{i,\text{eff},z}$. A more general definition of H for nonuniform columns (Giddings, 1965) is

$$H = (\mathrm{d}\sigma_{zi}^2/\mathrm{d}z), \qquad (7.1.106c)$$

i.e. "the plate height is the increment in the variance σ_i^2 per unit length of migration," the migration taking place in the z-direction.

An expression for N in terms of L and σ_{zi},

$$N = (L/\sigma_{zi})^2 \qquad (7.1.106d)$$

(obtained from definitions (7.1.106a) and (7.1.106b)), indicates that the smaller the bandwidth or dispersion and the higher the column length, the larger the number of theoretical plates. The σ_i used in the definitions given above is in length units. On the other hand, the σ_i used in definition (7.1.102b) is in volume units, σ_{Vi}. The number of plates N is then defined by

$$N = \left(\frac{V_{Ri}}{\sigma_{Vi}}\right)^2, \qquad (7.1.106e)$$

where σ_{Vi} is used along with the retention volume V_{Ri} for species i. From relation (7.1.102c) for σ_{Vi}, observe that

$$N = \left[Pe_{z,\text{eff}}/2\right]. \qquad (7.1.106f)$$

Therefore, the plate number increases linearly with the axial column Péclet number and the plate height decreases as N increases. Also N increases as $D_{i,\text{eff},z}$ decreases, and as v_z and the column length increases.

It is also useful to develop a relation in elution chromatography between the resolution R_S between peaks of two species i and j and the plate number in the column (among other quantities). We have already developed such a relation (e.g. (6.3.25)) in capillary electrophoresis). Consider for this purpose the expression (7.1.99i) for resolution between species i and j:

$$R_S = \frac{t_{R_i} - t_{R_j}}{2\sigma_{ti} + 2\sigma_{tj}} = \frac{t_{R_i} - t_{R_j}}{2(\sigma_{ti} + \sigma_{tj})}, \qquad (7.1.106g)$$

where the standard deviations are in time units. However, in relation (7.1.106d), σ_i is in length units, σ_{zi}. We may write

$$(\sigma_{ti} + \sigma_{tj}) = 2\overline{\sigma}_t = 2\overline{\sigma}_z/(v_{Ci}^* v_{Cj}^*)^{1/2}, \qquad (7.1.106h)$$

where $\overline{\sigma}_t$ is an averaged standard deviation in time units, $\overline{\sigma}_z$ is an averaged standard deviation in length units and $(v_{Ci}^* v_{Cj}^*)^{1/2}$ is a geometrical average of the center-point velocities of species i and j. Further, we have written $(\mathrm{d}z/\mathrm{d}t)_{\text{center point of } i}$ and v_{Ci}^* in equation (7.1.99b). Now, from relation (7.1.106d) in terms of length units, we may write

$$\sqrt{N} = \frac{L}{\sigma_z} \cong \frac{L}{\overline{\sigma}_z}, \qquad (7.1.106i)$$

where

$$\overline{\sigma}_z = (1/2)\,(\sigma_{zi} + \sigma_{zj}) \qquad (7.1.106j)$$

is an averaged standard deviation of the two profiles in length units. We note that, from (7.1.99d),

$$v_{Ci}^* = R'_i\, v_z \qquad (7.1.106k)$$

and

$$t_{R_i} = (L/R'_i\, v_z).$$

Therefore, from (7.1.106g) and (7.1.106h),

$$R_S = \frac{t_{R_i} - t_{R_j}}{2(\sigma_{ti} + \sigma_{tj})} = \frac{L}{4v_z}\frac{\left(\dfrac{1}{R'_i} - \dfrac{1}{R'_j}\right)}{\overline{\sigma}_t} = \frac{L(v_{Ci}^* v_{Cj}^*)^{1/2}\,(R'_j - R'_i)}{4v_z\overline{\sigma}_z\, R'_i R'_j}$$
$$= \frac{L}{4}\frac{v_z^2(v_{Cj}^* - v_{Ci}^*)(v_{Ci}^* v_{Cj}^*)^{1/2}}{v_z^2\,\overline{\sigma}_z\,(v_{Ci}^* v_{Cj}^*)} = \frac{L}{4\overline{\sigma}_z}\frac{(v_{Cj}^* - v_{Ci}^*)}{(v_{Ci}^* v_{Cj}^*)^{1/2}}. \qquad (7.1.106l)$$

Employing relation (7.1.106i) for N and defining $(\overline{v_{Cj}^* + v_{Cj}^*})$ as the geometrical average of v_{Ci}^* and v_{Cj}^* (i.e. $(v_{Ci}^* v_{Cj}^*)^{1/2}$), we get

$$R_S = \frac{1}{4}\frac{(v_{Cj}^* - v_{Ci}^*)}{(\overline{v_{Ci}^* + v_{Cj}^*})}\sqrt{N}. \qquad (7.1.106m)$$

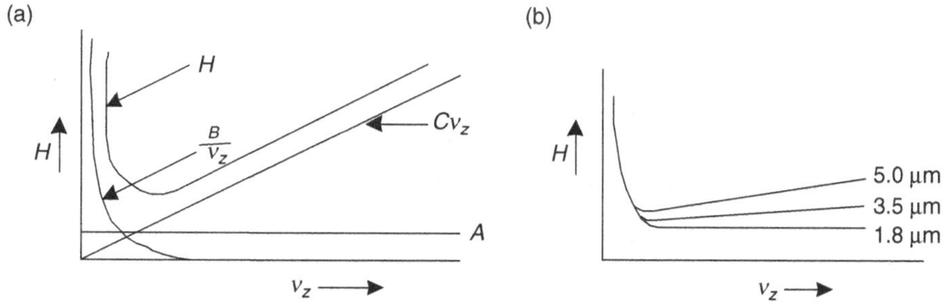

Figure 7.1.24. (a) Dependence of plate height H *and its various components on* v_z *from the Van Deemter equation (7.1.107a). (b) Dependence of plate height* H *on* v_z *for smaller particle sizes in HPLC. (After Majors, 2005.)*

Therefore as the plate number increases, the resolution between two species increases. We will see in Section 7.1.5.1.5 that as the particle size is reduced, leading to a larger N and lower H, the value of R_S increases.

A more detailed description of the plate height H is needed for a more exact description of chromatographic separation in a column of length L, which will allow also a knowledge of N. Van Deemter *et al.* (1956) developed such a description by comparing the Gaussian distribution equivalent of the concentration at the outlet of the column (plate N) given by (7.1.105m) with the concentration profile for linear equilibrium, a long column and a differential solute feed pulse (equations (7.1.101f), (7.1.101g), $V^0 \to 0$). This concentration profile was a simplified form of a solution developed by Lapidus and Amundson (1952) based on mass-transfer effects as well as dispersion. The result of Van Deemter *et al.* (1956) is usually described by the following equation, called the Van Deemter equation:

$$H = A + \frac{B}{v_z} + C\,v_z. \tag{7.1.107a}$$

Here, the A term is due to eddy dispersion and flow contribution to plate height and is independent of v_z; it is a function of the particle size and the packing efficiency. The next term includes B, which depends on the molecular diffusion coefficient in the longitudinal direction, i.e. the mobile-phase diffusivity. The third term includes C, which results from mass transfer between the mobile and stationary phases and has contributions from (1) diffusion in the film around the particle in the column, (2) diffusion in the liquid phase that is stagnant in the pores and (3) diffusion in the liquid-phase coating on particles.

The dependences of each of the three terms on v_z, the mobile-phase interstitial velocity, are shown in Figure 7.1.24(a). The nature of their variation with v_z leads to an overall dependence of H with v_z, which has a minimum at a low value of v_z. The minimum is obtained from

$$(\mathrm{d}H/\mathrm{d}v_z) = 0 \Rightarrow v_z|_{\mathrm{min}} = (B/C)^{1/2}. \tag{7.1.107b}$$

Therefore

$$H_{\mathrm{min}} = A + 2\sqrt{BC}. \tag{7.1.107c}$$

The optimum velocity in gas chromatography is around 10 cm/s (Giddings, 1965) or 10–30 cm/s (Mittlefehldt, 2002). The optimum velocity in high-performance liquid chromatography (HPLC) using small particles (~2 μm) is around 0.1 cm/s (Majors, 2005) or 10^{-3} cm/s (Giddings, 1965). To the extent H may be represented by the dependence of σ_{zi}^2 on $D_{i,\mathrm{eff},z}$, values of H will be drastically different for gas chromatography and liquid chromatography since diffusion coefficients for gaseous systems are larger than those for liquid systems by factors of 10^4–10^5. For example, the values of H in HPLC vary between 5×10^{-4} to 30×10^{-4} cm (for ~2 μm particles (Majors, 2005)). On the other hand, the values of H in gas chromatography may vary, say, between 5×10^{-2} cm to 3×10^{-1} cm (Mittlefehldt, 2002). The range of variation observed in gas chromatography for the constants A, B and C in equation (7.1.107a) are: A = 0–1 mm; B = 10 mm^2/s; C = 0.001 - 0.1s; H_{min} = 0.5 - 2 mm (Moody, 1982). *Note*: Most commercial chromatographic columns are run at a value of v_z larger than $v_z|_{\mathrm{min}}$. There is another reason for it. As the particle sizes become smaller (5 μm, 3.5 μm, 1.8μm), the Van Deemter curve shown in Figure 7.1.24(b) flattens out; therefore, if operation is at $v_z > v_z|_{\mathrm{min}}$, the increase in HETP is very little (Majors, 2005). This same reference provides an estimate of the number of plates in a 15 cm column as the particle size is varied: 100 μm, 200; 50 μm, 1000; 10 μm, 6000; 5 μm, 12 000; 2.5 μm, 25 000; 1.8 μm, 32 500.

Additional developments have led to a different form of the Van Deemter equation, where the plate height H may be represented in the following fashion (Giddings, 1965, eq. (2.11.2), 1991; Horvath and Lin, 1978):

$$H = \frac{B}{v_z} + C\,v_z + \sum \frac{1}{(1/A) + (1/C_m v_z)}. \tag{7.1.107d}$$

The following alternative representation is also employed (Jennings, 1987; Karger *et al.*, 1993):

$$H = H_D + H_M + H_{SM} + H_S, \tag{7.1.107e}$$

where

H_D is due to the longitudinal molecular diffusion,
H_M is due to the mobile-phase contributions in a packed bed,
H_{SM} is due to the stagnant mobile phase in particle pores,
H_S is duc to the stationary-phase sorption–desorption.

Expressions for the individual plate height contributions are given below:

$$H_D = \frac{2D_{il}\gamma_1}{v_z}, \quad (7.1.107f)$$

where D_{il} is the diffusion coefficient of species i in the mobile-phase liquid and γ_1 is the obstruction factor (<1.0) due to particles in a packed bed, which increases the diffusion path length;

$$H_M = \frac{1}{[1/\gamma_2 d_p] + [D_{il}/\gamma_3 d_p^2 v_z]}, \quad (7.1.107g)$$

where γ_2 and γ_3 are functions of the packing structure in the packed bed;

$$H_{SM} = (\text{constant}) \frac{(1-\phi_m + k_{i1})^2 d_p^2 v_z}{(1-\phi_m)(1+k_{i1})^2 \tau_p D_{il}}, \quad (7.1.107h)$$

where ϕ_m denotes the fraction of mobile phase in particle pores of tortuosity factor τ_p and k_{i1} is the distribution ratio of species i between the stationary and mobile phases;

$$H_S = \frac{\gamma_4 k_{i1} d_p^2 v_z}{(1+\kappa_{i1})^2 D_{is}}, \quad (7.1.107i)$$

where γ_4 is a configuration factor dependent on the shape of the liquid dispersed on packing (as in LLC, see Table 7.1.5) and D_{is} is the diffusion coefficient of species i in the stationary phase.

Example 7.1.6 In conventional gas chromatography, the typical quantity measured for species i is the retention time t_{R_i}, under given conditions. To develop an estimate of the number of plates N in the chromatographic column of length L, one also needs to determine the standard deviation σ_i in the column. Show that

$$N = 16\left(\frac{t_{R_i}}{W_{bi}}\right)^2$$

will provide such an estimate for peaks that are assumed Gaussian. Here W_{bi} is the base width of the chromatographic peak.

Solution The chromatographic peaks appear in the time domain. The base width W_{bi} of a Gaussian peak is essentially equal to $4\sigma_{ti}$ (see Section 2.5), where we have added a subscript t to indicate time units. In equation (7.1.106d), $N = (L/\sigma_{zi})^2$, where σ_{zi} is in length units. The effective migration velocity of a peak of species i is $R_i' v_z$ (from relation (7.1.99a)); this velocity also characterizes the time rate of migration of the base width of a chromatographic peak. Therefore

Table 7.1.6.

He flow rate (cm^3/min)	Retention time (cm)	Peak base width (cm)	H (cm)
7.1	14	3.3	0.529
34.4	5.3	0.95	0.306
78.9	3.28	0.66	0.386

$$N = \left(\frac{L}{\sigma_{zi}}\right)^2 = \left(\frac{L/R_i' v_z}{\sigma_{zi}/R_i' v_z}\right)^2 = \left(\frac{t_{R_i}}{\sigma_{zi}/R_i' v_z}\right)^2$$
$$= \left(\frac{t_{R_i}}{\sigma_{ti}}\right)^2 = \left(\frac{t_{R_i}}{W_{bi}/4}\right)^2 = 16\left(\frac{t_{R_i}}{W_{bi}}\right)^2.$$

Example 7.1.7 One would like to determine the values of the constants A, B and C in the Van Deemter equation (7.1.107a) for plate height H in a gas chromatographic column by conducting a few experiments. A minimum of three sets of data are needed to determine the three constants A, B and C. The data in Table 7.1.6 were obtained by Moody (1982) from a 152.4 cm (5 ft) long 1/8″ column operated at 82 °C with a particular packing for a C_4-compound, and helium as the carrier gas. *Note*: The retention time and the peak base width have been converted to length units. Determine the values of A, B and C. What is the value of $v_z|_{min}$?

Solution Substitute the values of the helium flow rate and plate height into the Van Deemter equation for the three cases as follows:

$$0.529 = A + \frac{B}{7.1} + 7.1C;$$
$$0.306 = A + \frac{B}{34.4} + 34.4C;$$
$$0.386 = A + \frac{B}{78.9} + 78.9C.$$

Note: In the absence of the exact column cross section for gas flow, we are using the gas flow rate instead of the actual gas velocity in the Van Deemter equation. Therefore the units of B and C will be different from those if v_z were used instead of the gas volumetric flow rate. Simultaneous solutions of these three equations yield the following values of the constants:

$$A = 0.1329\,\text{cm}; \quad B = 2.665\,\text{cm}^4/\text{min}; \quad C = 0.00278\,\text{min/cm}^2.$$

If v_z (cm/min, say) were used, the units of B would be cm^2/min and those for C would be min. These units will become self evident if the conventional expressions for A, B and C in terms of fundamental quantities of the chromatographic column and system are considered:

$$A = \lambda d_p,$$

where λ is a dimensionless factor related to eddy diffusion;

$$B = 2\gamma D_{i\ell},$$

where γ is a dimensionless factor related to the tortuosity of the flow path in the packed column; C has a number of contributions, as shown in equations (7.1.107g–i).

7.1.5.1.5 Scale up or scale down packed columns for adsorption/chromatography (elution) If one considers the plate height expressions provided in expressions (7.1.107a) and (7.1.107f–i), one observes that reducing the particle diameter d_p in the column in general leads to lower values of the plate height, and therefore a larger number of plates. Further, the peak capacity is increased due to a reduction in σ_i. This has been the driving force for a continuous decrease in particle sizes in various chromatographic techniques from values of 100 μm in the 1950s to as low as 2 μm in 2003 (Majors, 2005). It is useful to enquire about the effects of a change in the particle size in a packed column for adsorption and chromatographic processes.

We have studied in earlier subsections how separation takes place as a fluid containing various species to be separated moves through a packed bed of adsorbents or chromatographic medium. As the fluid flows, there is a significant pressure drop as a certain level of separation is achieved. The question of interest here is: what happens if the particle size is reduced? This is part of a broader question: how can one improve separation in a packed column based process?

The pressure drop experienced by a fluid as it flows through a packed bed of particles of diameter d_p may be described by the Blake–Kozeny equation (6.1.4f), among others:

$$\Delta P = \mu\, v_0 \frac{(1-\varepsilon)^2 150}{\varepsilon^3\, d_p^2} L. \tag{7.1.108a}$$

Since v_0, the superficial fluid velocity based on the empty cross section, is related to the actual interstitial velocity v_z by $v_0 = v_z\,\varepsilon$, we can rewrite the above relation as

$$\Delta P = \frac{\mu\, v_z\,\varepsilon}{d_p^2} L \left(\frac{150\,(1-\varepsilon)^2}{\varepsilon^3}\right). \tag{7.1.108b}$$

Generally, for rigid spherical particles, the bed porosity or fractional void volume ε is independent of particle size. Unless the bed length and the bed diameter are vanishingly small (of the order of d_p), ε is also independent of the bed length L and the bed diameter. Consider now two different packed beds:

bed 1: length L_1, particle diameter d_{p1}, column radius R_1, fluid volumetric flow rate Q_1, velocity v_{z1}, pressure drop ΔP_1;

bed 2: length L_2, particle diameter d_{p2}, velocity v_{z2}, column radius R_2, fluid volumetric flow rate Q_2, pressure drop ΔP_2.

The ratio of the pressure drops in the two beds is as follows (from relation (7.1.108b)):

$$\frac{\Delta P_1}{\Delta P_2} = \left(\frac{v_{z1}}{v_{z2}}\right)\left(\frac{L_1}{L_2}\right)\left(\frac{d_{p2}^2}{d_{p1}^2}\right). \tag{7.1.108c}$$

However, the ratio of the interstitial velocities is related to the two volumetric flow rates and the two column radii by

$$\frac{v_{z1}}{v_{z2}} = \frac{\dfrac{Q_1}{\pi R_1^2\,\varepsilon}}{\dfrac{Q_2}{\pi R_2^2\,\varepsilon}} = \left(\frac{Q_1}{Q_2}\right)\frac{R_2^2}{R_1^2}. \tag{7.1.108d}$$

Substituting this relation into the ratio (7.1.108c), we get

$$\frac{\Delta P_1}{\Delta P_2} = \left(\frac{Q_1}{Q_2}\right)\left(\frac{R_2^2}{R_1^2}\right)\left(\frac{L_1}{L_2}\right)\left(\frac{d_{p2}^2}{d_{p1}^2}\right). \tag{7.1.108e}$$

If the two packed beds employed two different particle sizes but had the same Q and R, then, to maintain the same ΔP, we must ensure

$$(L_1/L_2) = (d_{p1}^2/d_{p2}^2). \tag{7.1.108f}$$

If d_{p2} is smaller than d_{p1}, L_2 is going to be much smaller than L_1. Correspondingly, from relation (7.1.106c) for the breakthrough time $\bar{t}$, for a fixed v_z (i.e. fixed Q and R), $\bar{t}$ will decrease with d_p. Therefore, for the adsorption–desorption cycle, the cycle time will have to be reduced considerably as the particle size is reduced:

$$(L_1/\bar{t}_1) = (L_2/\bar{t}_2) \Rightarrow (\bar{t}_1/\bar{t}_2) = (L_1/L_2) = (d_{p1}^2/d_{p2}^2). \tag{7.1.108g}$$

Consequently, if $d_{p2} < d_{p1}$, $\bar{t}_2 << \bar{t}_1$ (Wankat, 1987). Smaller particles will dictate a much faster breakthrough time and a much shorter cycle time unless the pressure drop is allowed to rise considerably in the following fashion:

$$(\Delta P_1/\Delta P_2) = (d_{p2}^2/d_{p1}^2) \tag{7.1.108h}$$

for constant Q, R and L. In fact, in high-performance liquid chromatography applications in practice, the particle sizes can be as low as ~2 μm, leading to very high values of pressure drops, of the order of 100–1000 atm in columns which are around 5 cm long (Majors, 2005). On the other hand, in larger-scale process applications, the adsorbent particle sizes are ~1 mm. The pressure drops are much more moderate, the column diameter and length are larger.

We have not discussed the nature of the separation being achieved as the scale of operation is changed from one column having one particle size to another with a smaller or larger particle size. One could wish to maintain as constant the fractional column length lost in the adsorption column, i.e.

$$(L_{\mathrm{MTZ}}/L)_1 = (L_{\mathrm{MTZ}}/L)_2, \tag{7.1.108i}$$

between the two scales of operation, where MTZ stands for mass-transfer zone.

Alternatively, from a chromatographic operation point of view, it may be desirable to maintain the same number of theoretical plates, N, between the two scales of operation:

$$N_1 = N_2. \qquad (7.1.108j)$$

In practical chromatographic separations, on the other hand, usually the driving force for smaller particles is higher and higher N, since we have already pointed out that smaller d_p leads to lower H and therefore higher N for a fixed L.

From a quantitative point of view vis-à-vis the chromatographic operation, if the plate height is controlled by pore diffusion (in the stagnant mobile phase, equation (7.1.107h)) and /or stationary-phase sorption-desorption, the plate height $H \propto d_p^2 v_z$. Since $N = (L/H)$, for a fixed L,

$$L = N_1 H_1 = N_2 H_2.$$

Therefore the ratio of the number of plates for the two columns of same length is

$$(N_1/N_2) = (H_2/H_1) = (d_{p2}^2/d_{p1}^2), \qquad (7.1.108k)$$

assuming v_z is constant. However, the ΔP will be quite different for the two columns by (7.1.108c). If, we cannot assume pore diffusion control, but can represent the HETP via

$$H = a\, v_z^n\, d_p^{n+1}, \qquad (7.1.108l)$$

then Wankat (1987) has obtained the following result in a straightforward fashion:

$$\text{constant } Q\text{: } (N_1/N_2) = (d_{p2}/d_{p1})^{n+1}\,(L_1/L_2)\,(R_2/R_1)^{-2n}. \qquad (7.1.108m)$$

Additional results for a number of ratios, (cycle time)$_2$/(cycle time)$_1$, (adsorbent volume)$_2$/(adsorbent volume)$_1$, etc., may be developed. (*Note*: $n = 1$ represents pore diffusion control.) Similar considerations may be employed for the estimation of the ratio (L_{MTZ}/L) in relation (7.1.108i) provided an expression for L_{MTZ} is employed in terms of the mass-transfer coefficient, the particle surface area, etc. (Wankat, 1986, 1987). Wankat (1987) has provided the following result:

$$\frac{(L/L_{\mathrm{MTZ}})_1}{(L/L_{\mathrm{MTZ}})_2} = \left(\frac{d_{p2}}{d_{p1}}\right)^{n+1}\left(\frac{L_1}{L_2}\right)\left(\frac{R_1}{R_2}\right)^{2n} \qquad (7.1.108n)$$

for the same volumetric flow rate through the two columns.

7.1.5.1.6 Elution chromatography with a mobile liquid phase - ion exchange chromatography The treatment of elution chromatography with a mobile liquid phase considered so far involved liquid-solid adsorption chromatography (LSC) (see Sections 7.1.5.1.1, 7.1.5.1.2, 7.1.5.1.4 and 7.1.5.1.5). There are three other forms of stationary solid phase commonly used: *ion exchange resin beads, porous hydrogel beads* (generally crosslinked) and *affinity beads*, having functional groups with which the solute has an extremely strong affinity to the exclusion of other solutes. Although all such chromatographic media may be used for the separation of microsolute species, all three techniques are used extensively for the separation of mixtures of proteins/biomacromolecules. On the other hand, when the stationary phase is a liquid coated on solid particles/beads, we have liquid-liquid chromatography (LLC); this technique is used for smaller molecules. We will now provide an extremely brief introduction to each of these elution techniques. A reasonably comprehensive introduction to analytical-scale chromatography using all four techniques is available in Karger *et al.* (1973). A comprehensive introduction to the first three techniques for bioseparations, including process-scale operations, is available in Ladisch (2001).

The general characteristics of liquid-solid adsorption based elution chromatography developed in Sections 7.1.5.1.1 and 7.1.5.1.4 are also valid here. For example, the retention volume V_{R_i} for species i in a column having a stationary phase volume V_S is related to the mobile-phase volume within the column, V_M, and the solute partition coefficient κ_{i1} by relation (7.1.99h):

$$V_{R_i} = V_M + V_S\,\kappa_{i1},$$

where

$$\kappa_{i1} = (C_{i1}/C_{i2}) = (C_{iR}/C_{iw}), \qquad (7.1.109a)$$

with subscripts R and w representing the resin phase (the stationary phase) and the mobile phase (aqueous phase $j = w$), respectively. This partition coefficient has now to be related to the concentration/activity of the salt (for example) species in the eluent. We have seen earlier (in Section 4.1.9.4) that salts such as ammonium sulfate are used in hydrophobic interaction chromatography to displace proteins from ion exchange resins. In ion exchange chromatography, NaCl plays a similar role in modifying the ion exchange behavior of proteins. In the case of ion exchange chromatography of small ions, a variety of different species can act as the eluent species, e.g. KCl, HCl, K_3PO_4, M_2O_3 (where M stands for a metal), etc. We identify such an eluent component/salt (a counterion) by s and assume that it is monovalent, $s^{\pm}$, i.e. $Z_s = \pm 1$.

We first consider the separation of two charged/ionic species A and B, where they are represented as A^{Z_A} and B^{Z_B}. The counterion eluent component/salt s initially present in the ion exchange resin will exchange with each of A and B via the corresponding ion exchange (see (3.3.121c) for guidance):

$$|Z_A|\,(|Z_s|R^{\pm} + s^{Z_s}) + |Z_s|A^{Z_A} \Leftrightarrow |Z_s|\,(|Z_A|R^{\pm} + A^{Z_A}) + |Z_A|s^{Z_s}; \qquad (7.1.109b)$$

$$|Z_B|\,(|Z_s|R^{\pm} + s^{Z_s}) + |Z_s|B^{Z_B} \Leftrightarrow |Z_s|\,(|Z_B|R^{\pm} + B^{Z_B}) + |Z_B|s^{Z_s}. \qquad (7.1.109c)$$

The equilibrium constants K_{As} and K_{Bs} for these two ion exchange "reactions" are (in activity units; see (3.3.121e) and (3.3.122b) for activity and molar concentration units)

$$K_{As} = \frac{(a_{sw})^{|Z_A|}(a_{AR})^{|Z_s|}}{(a_{sR})^{|Z_A|}(a_{Aw})^{|Z_s|}}; \qquad K_{Bs} = \frac{(a_{sw})^{|Z_B|}(a_{BR})^{|Z_s|}}{(a_{sR})^{|Z_B|}(a_{Bw})^{|Z_s|}}. \tag{7.1.109d}$$

We need to recognize two realities in elution based ion exchange chromatography: (1) the concentrations of A and B are quite small, therefore the eluent counterion s, whose concentration is much larger, occupies virtually all of the resin sites; (2) the ion exchange equilibrium constant K_{As}, or K_{Bs}, is essentially a constant vis-à-vis the concentration levels.

We are interested in κ_{i1} (see (7.1.109a)) so that we can predict V_{R_i}, i.e. V_{R_A} and V_{R_B}. If for species A we write

$$\kappa_{A1} = \left(\frac{C_{AR}}{C_{Aw}}\right) \cong \left(\frac{a_{AR}}{a_{Aw}}\right), \tag{7.1.109e}$$

then, if $|Z_s| = 1$ (say, the salt is NaCl),

$$\kappa_{A1} = K_{As}\,(a_{sR})^{|Z_A|}/(a_{sw})^{|Z_A|}. \tag{7.1.109f}$$

Correspondingly,

$$\kappa_{B1} = K_{Bs}\,(a_{sR})^{|Z_B|}/(a_{sw})^{|Z_B|}. \tag{7.1.109g}$$

Using definition (7.1.99j) of the net retention volume of species i, V_{N_i}, we get

$$\log V_{N_i} = \log\,(V_{R_i} - V_M) = \log V_S\,\kappa_{i1} = \log V_S + \log \kappa_{i1}; \tag{7.1.109h}$$

$$\begin{aligned}\log V_{N_A} &= \log V_S + \log\ \kappa_{A1}\\ &= \log V_S + \log K_{As} + |Z_A|\,(\log a_{sR} - \log a_{sw}).\end{aligned} \tag{7.1.109i}$$

Further,

$$\begin{aligned}\log\ \kappa_{A1} &= \log K_{As} + |Z_A|\,(\log a_{sR} - \log a_{sw});\\ \log\ \kappa_{i1} &= \log\ K_{is} + |Z_i|\,(\log a_{sR} - \log a_{sw}).\end{aligned} \tag{7.1.109j}$$

Therefore plots of $\log V_{N_A}$ (or $\log V_{N_i}$) and $\log \kappa_{A1}$ (or $\log \kappa_{i1}$) against $\log a_{sw}$ will be straight lines with negative slopes of $-|Z_i|$, as shown in Figure 7.1.25(a). For two species A and B whose V_{N_i} differ because κ_{A1} and κ_{B1} are different, the difference between V_{N_A} and V_{N_B} becomes substantial as a_{sw} decreases; at low concentrations of the salt (say), the centers of the two elution peaks A and B are substantially far apart. On the other hand, if $|Z_A| = |Z_B|$, the plots of $\log \kappa_{A1}$ and $\log \kappa_{B1}$ vs. $\log a_{sw}$ will be parallel to each other.

In chromatographic literature, the quantity k'_{i1}, the distribution ratio or the capacity factor, is frequently used:

$$k'_{i1} = \frac{V_S}{V_M}\,\kappa_{i1} = \frac{V_{R_i} - V_M}{V_M} = \frac{(1-\varepsilon)}{\varepsilon}\,\kappa_{i1}, \tag{7.1.109k}$$

which leads to

$$\log k'_{i1} = \log \frac{(1-\varepsilon)}{\varepsilon}\,K_{is} + |Z_i|\,(\log a_{sR} - \log a_{sw}). \tag{7.1.109l}$$

Consequently, plots of the capacity factor will have a dependence on a_{sw} similar to that of κ_{i1} or V_{N_i}.

The behavior of $\log V_{N_i}$ at low concentrations of the eluent ion, C_{sw} (or a_{sw}) (Figure 7.1.25(a)) indicates that the difference between, say, V_{N_A} and V_{N_B}, is quite large. Therefore, the retention volumes for ions A and B will be very far apart, i.e. the peaks will emerge from the column at very different times, leading to considerable wastage of the mobile phase. The conventional method of reducing this mobile-phase loss and causing the peaks to come out closer in time is *gradient elution*. This is achieved by changing the eluent ion concentration C_{sw} (or a_{sw}) with time to a larger value so that V_{N_B} is much less compared to what it would have been at a lower C_{sw}. (Constant eluent composition, temperature and flow rate is identified as an *isocratic mode of operation*.)

We may explain this using the retention time t_{R_i} for a given species i. By definition (7.1.99d), $t_{R_i} = (L/R'_i v_z)$. However, for a species which passes through without interacting with the ion exchanger or the stationary-phase material, $R'_i = 1$, since its $v^*_{Ci} = v_z$. Identify this as the retention time for the mobile phase, t_{R_M}, the time taken by the mobile phase introduced with the sample to come to the column exit. Therefore,

$$t_{R_M} = (L/v_z). \tag{7.1.109m}$$

We may now write, for any species i which undergoes adsorption, partitioning, ion exchange, etc., with the stationary phase,

$$\begin{aligned}k'_{i1} &= \frac{V_{R_i} - V_M}{V_M} = \frac{(V_M/R'_i) - V_M}{V_M} = \frac{\frac{1}{R'_i} - 1}{1}\\ &= \frac{\frac{L}{R'_i v_z} - \frac{L}{v_z}}{L/v_z} = \frac{t_{R_i} - t_{R_M}}{t_{R_M}}.\end{aligned} \tag{7.1.109n}$$

From relation (7.1.109l), for two eluent ion concentrations $a_{sw}|_1 < a_{sw}|_2$, we get (see Figure 7.1.25(a))

$$k'_{i1}|_1 > k'_{i1}|_2 \quad \text{and} \quad t_{R_i}|_1 > t_{R_i}|_2. \tag{7.1.109o}$$

In the separation of a number of proteins by elution based ion exchange chromatography (see Ladisch (2001) for much greater detail), first the sample is injected. Different proteins undergo ion exchange based binding with the resin (see Figure 4.1.16). Soon after the sample injection, the salt concentration in the mobile phase is increased with time, say, in a linear or stepwise fashion. Therefore, the proteins which are bound more strongly, and therefore have a larger value of V_{R_i} or t_{R_i} at a given a_{sw}, will come out much earlier, since a_{sw} has been increased. Figures 7.1.25(b) and (c) illustrate this behavior. Figure 7.1.25(b) illustrates the elution profiles for proteins A and B, where the salt concentration in the eluent is constant at a low value of $a_{sw}|_1$; as a result, t_{R_A} and t_{R_B} are far apart. Figure 7.1.25(c) shows that gradient elution is started close to the time when protein A elutes by a step increase in salt

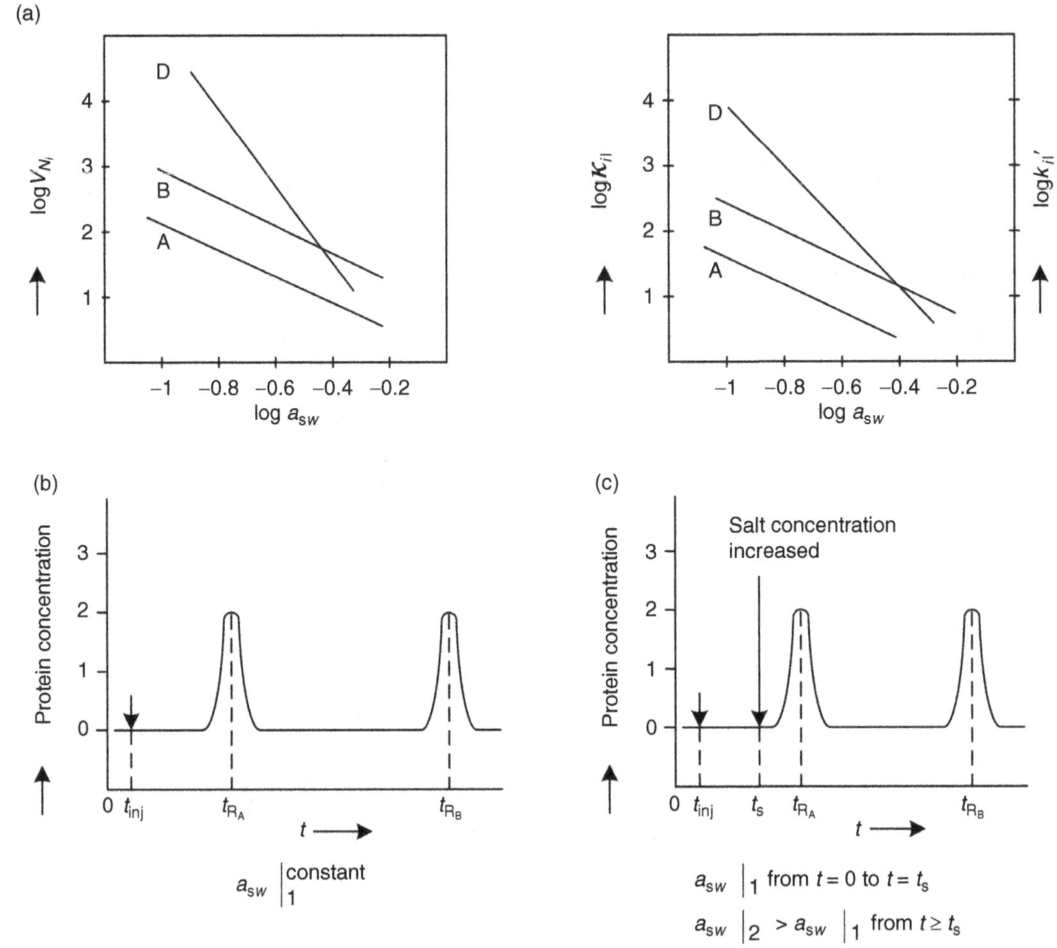

Figure 7.1.25. Ion exchange chromatography. (a) Variations of net retention volume V_{N_i} and partition coefficient κ_{i1} as well as the capacity factor (k'_{i1}) with eluent ion activity. Gradient elution is depicted in (b) for constant eluent ion concentration and in (c) for stepwise variation of eluent ion concentration.

concentration. The result is that the protein B peak comes out much earlier. Instead of a step change in salt concentration, the salt concentration in the eluent may be varied continuously, for example in a linear fashion. The typical concentration variation of NaCl used as a result is from 0.01 M to 0.1 M (up to 0.3 M).

The discussion of elution of two protein/ionic species has so far employed species A and B identified in Figures 7.1.25(a)–(c); both species had identical charges. Species D in Figure 7.1.25(a), however, has a different charge, in fact a higher charge $|Z_D| > |Z_A| = |Z_B|$. Therefore, the negative slope of log V_{N_D} or log κ_{D1} vs. log a_{sw} is larger than that for species A or B. The difference in the two slopes leads to the following: at lower salt concentrations, D will elute much later; however, at the salt concentration where the two lines intersect, two species will elute at the same time (no separation). Beyond the intersection, at higher salt concentrations, species D will elute faster than species B (or A if there is an intersection). Such behavior leads to complications; alternatively, it may be an opportunity for improved separation.

7.1.5.1.7 Elution chromatography with a mobile liquid phase – size exclusion chromatography We will now briefly consider *size exclusion chromatography* (SEC), which is also called *gel permeation chromatography* (GPC), *gel filtration* or *gel chromatography*. In Section 3.3.7.4, we came across porous crosslinked polymers called *hydrogels* and the partitioning behavior of macromolecules/proteins between their solutions and the hydrogels (e.g. κ_{im} given by (3.3.90i)). Generally, the lower the molecular weight and the smaller the effective radius (e.g.

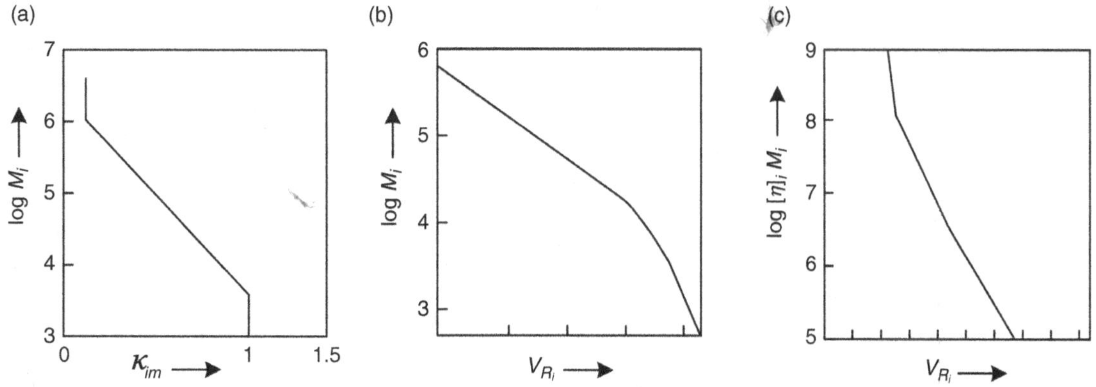

Figure 7.1.26. Characteristics of size exclusion chromatography/gel permeation chromatography. (a) Partition coefficient of macromolecular species i *between solution and gel particle:* M_i *is the molecular weight of macromolecular species. (b) Elution volume of macromolecular species* i. *(c) Universal calibration plot of many polymers in tetrahydrofuran (THF): log* $[\eta]_i$ M_i *vs. elution volume* V_{R_i}. *(After Grubisic* et al. *(1967).)*

the radius of gyration, r_g; hydrodynamic radius, r_h), the higher the value of κ_{im} (Figure 7.1.26(a)). A larger protein/macromolecule has a lower κ_{im}; consequently, it will spend less time in the porous hydrogel phase and will be eluted faster in a chromatographic column containing hydrogel beads as the packing/sorbent phase. Let there be no other type of interaction between the macromolecular solute and the porous hydrogel except geometrical partitioning between the pore liquid and the external liquid (e.g. (3.3.88a) and (3.3.89a)); this interaction is dependent on the pore size, solute size, pore shape and solute shape. Here the value of $\kappa_{im} \to 1$ when the solute size tends to zero. In such a situation, there is no solute adsorption as such, so that $q'_{i1}(\overline{C}_{i2})$ in equation (7.1.13f) is zero and

$$v^*_{Ci} = \frac{v_z}{1 + \frac{(1-\varepsilon)}{\varepsilon}\,\varepsilon_p\,\kappa_{im}}. \tag{7.1.110a}$$

Therefore, a large macromolecule/protein which cannot enter the pores of the hydrogel beads will have a concentration velocity $v^*_{Ci} = v_z$, the interstitial fluid velocity. On the other hand, a small solute such as salt will probably have $\kappa_{im} = 1$ and its $v^*_{Ci} << v^*_{Ci}|_{\text{protein}}$.

We have also obtained earlier, for the case of solute adsorption, the following relation for species i between the retention volume V_{R_i} and the solute adsorption/partition coefficient κ_{i1} (equation (7.1.99h)):

$$V_{R_i} = V_M + V_S\,\kappa_{i1} = V_M + V_M\left[\frac{(1-\varepsilon)}{\varepsilon}\right]\kappa_{i1}.$$

This was obtained from the differential equation (7.1.95a) for elution chromatography. If instead we worked with the differential equation (7.1.95c), where $\kappa_{i1} = 0$, we would have obtained

$$V_{R_i} = V_M + V_M\left[\frac{(1-\varepsilon)}{\varepsilon}\right]\varepsilon_p\,\kappa_{im}; \tag{7.1.110b}$$

$$V_{R_i} = V_M + V_S\,\varepsilon_p\,\kappa_{im} = V_M + V_{Sp}\,\kappa_{im}, \tag{7.1.110c}$$

where V_{Sp} is the volume of the mobile phase present inside the porous hydrogel particle. Therefore the retention volume of a species, V_{R_i}, varies between the value of $V_M + V_{Sp}$ for a very small molecule, such as salt, to V_M for a protein/macromolecule completely excluded from the pores of the gel particle. Employing one of the relations (3.3.90i) (e.g. $\kappa_{im} = \mathrm{a} + \mathrm{b}\log M_i$), the following relation is obtained between V_{R_i} and the molecular weight M_i of the macromolecular species:

$$V_{R_i} = V_M + V_S\,\varepsilon_p\mathrm{a} + V_S\,\varepsilon_p\mathrm{b}\log M_i, \tag{7.1.110d}$$

which may be rearranged as follows:

$$V_{R_i} = p_1 + p_2\log M_i. \tag{7.1.110e}$$

As shown in Figure 7.1.26(b), the retention volume in size exclusion chromatography/gel chromatography appears to increase almost linearly with a decrease in M_i in a semi-logarithmic plot, except at the low end of M_i.

Important separation applications of this behavior appear in *desalting protein solutions* and *exchanging buffer solutions*. Salts like ammonium sulfate (used in the precipitation of selected proteins (see Section 3.3.7.4 and Ladisch (2001)) or NaCl (used in ion exchange chromatography) may be easily removed from the solution of the protein/s by passing the salt-containing solution through a gel particle filled column. Molecules used to develop buffer solutions are usually much smaller than proteins; gel chromatography will also allow the purification of a protein solution vis-à-vis the constituents of a buffer solution so that other buffer constituents may be added to the purified protein solution.

Such size exclusion based separations are carried out using hydrogel particles, which have a variety of commercial names, e.g. Sephadex™, Biogel™, Fractogel™, Sepharose™, Ultragel™, etc. Sephadex™, the trade name for the earliest gel particle, was developed by the reaction of dextran and epichlorohydrin (Porath and Flodin, 1959). The range of molecular weights of molecules/biomolecules/proteins separated by the list of commercialized gel particles identified above vary between 2×10^2 and 1×10^8. Although these gel particles do have a size distribution, typical diameters are around 40–165 μm. Further, these particles are somewhat soft and compressible; therefore the columns are operated at quite low pressure drops. For example, typical column pressure drops allowed vary between 10 and 300 cm H_2O. For details on laboratory-scale operation, see Scopes (1994).

Gel permeation chromatography has another very important application in the separation of macromolecules which are not biomacromolecules. Many organic polymers are soluble in organic solvents. A number of gel particles are obtained from crosslinked styrene-divinylbenzene, polyvinylacetate, porous silica, etc. (e.g. Styragel™, Bio-Beads™, Porasil™, etc.), which can be used with a variety of organic solvents, nonpolar as well as polar. A solution of an organic polymer in an organic solvent will generally have a molecular weight distribution (see Figure 2.4.8). Such a sample solution when injected into an appropriate gel column will elute various monodisperse fractions (the species in such a fraction have molecular weights in a very narrow band), whose average molecular weights will decrease with increasing time. Therefore, GPC will allow fractionation of a polymer sample into narrow molecular weight fractions. If an appropriate calibration curve is available which correlates the retention volume of a specific eluted fraction with its molecular weight, then one can find out the molecular weight distribution of any sample injected into the column.

This goal is achieved in the following fashion. Polystyrene (PS) is available commercially in narrow molecular weight fractions. One first determines the elution volumes of individual PS fractions in a solvent, say tetrahydrofuran (THF). It has been found that a semilog plot of the product $[\eta](M)$ of different polystyrene fractions against the elution volumes of different PS fractions in THF provides a universal calibration curve in GPC; here $[\eta]$ is the intrinsic viscosity of the PS fraction of molecular weight M (see relation (3.3.90f)) (for PS, standard samples of defined molecular weight are available commercially). Many different polymers have been found to obey the same behavior developed for PS fractions (Figure 7.1.26(c)). Therefore, if such a universal calibration curve is available, the elution volumes of different polymer fractions of a sample of any other polymer i coming out in solvent THF will have its

$$\log\,[\eta]_i M_i = \log\,([\eta]M)|_{\mathrm{PS}} \qquad (7.1.110\mathrm{f})$$

at the same elution volumes (Grubisic *et al.*, 1967). Since the Mark–Houwink equation relates the intrinsic viscosity $[\eta]$ of a polymer fraction to its molecular weight via

$$[\eta]_i = K_i[M_i]^{a_i}; \quad [\eta]_{\mathrm{PS}} = K_{\mathrm{PS}}[M_{\mathrm{PS}}]^{a_{\mathrm{PS}}}, \qquad (7.1.110\mathrm{g})$$

we obtain, for a given elution volume, the molecular weight M_i of the polymer i fraction as

$$\log M_i = \frac{1}{(1+a_i)}\log\frac{K_{\mathrm{PS}}}{K_i} + \frac{(1+a_{\mathrm{PS}})}{(1+a_i)}\log M_{\mathrm{PS}}. \quad (7.1.110\mathrm{h})$$

Therefore, knowing a_i, a_{PS}, K_{PS}, K_i and M_{PS} for the given elution volume, M_i may be determined corresponding to that elution volume.

One needs, however, the intrinsic viscosity relation parameters for both PS as well as the polymer i under consideration for the given solvent, say THF, and the gel particle under consideration. This procedure works because of the following reason. We have observed the following in relations (3.3.90e–g):

(1) the effective dimension of the polymer molecule in the form of a polymer coil given by r_g, the radius of gyration, is proportional to the hydrodynamic viscosity based radius, r_h;
(2) the product $[\eta]_i\, M_i \propto V_h \propto r_h^3$, where V_h is the hydrodynamic volume of the polymer fraction;
(3) r_h (or r_g) controls the partitioning behavior of the macromoleculer vis-à-vis the gel, and therefore the macromolecular separation behavior.

There are two other important aspects of macromolecular fractionation in organic solvents. For crosslinked polystyrene based gel beads (e.g. Styragel), only certain solvents such as THF, benzene, xylene, etc., may be used so that the pores in the gel remain open, since PS is soluble in these solvents. Porous glass beads are therefore often used, since a variety of solvents, including water and polar organic solvents, are easily used without any pore closing problems. Pore diameters can vary between 50 and 10 000 nm. Column pressures may go up to 1000–4000 psi (68–272 atm). For an introduction to GPC using organic solvents, see Allcock *et al.* (2003).

Example 7.1.8 In a small laboratory column for SEC for proteins, it is known that the biomacromolecule blue dextran (mol. wt. 2×10^6, having a blue dye attached to it), is completely excluded. The total column volume is 20.5 milliliter (ml). An injected sample of blue dextran comes out with the center of its peak around 7.2 ml. An injected sample of salt (NaCl) comes out with the center of its peak around 15.1 ml. An introduced unknown protein sample has an elution volume of 10 ml.

(1) Determine the void volume of this gel column.
(2) Calculate the fractional void volume of the gel particles.

(3) Estimate the value of κ_{im} for the unknown protein.
(4) Estimate the radius of this unknown protein if the gel pores are assumed to be cylindrical with $r_p = 10$ nm.

Solution

(1) Since blue dextran is completely excluded, $\kappa_{im} = 0$ and its $V_{R_i} = V_M = 7.2$ ml. Since the total volume of the column is 20.5 ml, the void volume of the column

$$\varepsilon = (7.2/20.5) = 0.351.$$

(2) Since salt will most certainly have a value of $\kappa_{im} = 1.0$ in the gel particles, we have

$$V_{R_i} = V_M + V_{Sp}\,\kappa_{im} = 7.2 + V_{Sp} \times 1.0 = 15.1\ \text{ml},$$

where V_{Sp} is the volume of the mobile phase present inside the porous particles. Therefore

$$V_{Sp} = 7.9\ \text{ml}.$$

However, the volume of the stationary phase V_S in the column is (20.5 – 7.2), i.e. 13.3 ml. The fractional void volume of the gel particles, is given by

$$\varepsilon_p = \frac{V_{Sp}}{\text{column volume} - \text{mobile phase volume}}$$

$$= \frac{V_{Sp}}{\text{stationary phase volume}} = \frac{7.9}{13.3} \Rightarrow \varepsilon_p = 0.594.$$

(3) $$V_{R_i} = V_M + V_{Sp}\,\kappa_{im} \Rightarrow 10 \Rightarrow 7.2 + 7.9\,\kappa_{im}$$

$$\Rightarrow \kappa_{im} = \frac{2.8}{7.9} = 0.35.$$

(4) Employ equations (3.3.88a) and (3.3.89l):

$$\frac{C_{im}^p}{C_{i1}} = \kappa_{im} = \left(1 - \frac{r_i}{r_p}\right)^2;$$

$$0.35 = \left(1 - \frac{r_i}{10\,\text{nm}}\right)^2 \Rightarrow 0.595 = \left(1 - \frac{r_i}{10\,\text{nm}}\right)$$

$$\Rightarrow r_i = 10 \times 0.405 \quad \Rightarrow r_i = 4.05\,\text{nm}.$$

7.1.5.1.8 Elution chromatography with a mobile liquid phase – affinity chromatography, membrane chromatography Two chromatographic methods described earlier, size exclusion chromatography (SEC) and ion exchange chromatography, are frequently used for the separation of biomacromolecules. However, both processes have limited selectivity. For example, in the process of SEC for proteins, whose molecular weights/sizes are not too far apart, one cannot achieve high selectivity for one protein with respect to another (however, desalting and buffer exchange can be implemented with ease). Similarly, in ion exchange chromatography, specificity between two proteins is often quite limited. On the other hand, in *affinity chromatography*, specific ligands exist on the adsorbent/bead surface which interact only with a specific protein. An example of such an interaction is that between an antigen and an antibody. Further, this binding is reversible, using an appropriate *pH* or ionic strength, or a species which acts as a displacer or modifier releasing the protein. However, the protein will have to be purified from the displacer/modifier. *Note*: In chromatography, competing species adsorb/bind to the beads/substrate to different extents. However, in affinity chromatography, under ideal/expected conditions, only the target protein is adsorbed/bound. Therefore *affinity adsorption* would also be an appropriate name.

A simplistic pictorial representation of such a process is provided in Figure 7.1.27(a). An insoluble hydrophilic stationary phase of polyacrylamide, polysaccharide, crosslinked dextran, etc., provides the matrix, or support generally, in the form of a bead. On the surface of the pores in such a matrix, ligands are attached by specific chemistries. In the figure, we show that protein A, has been attached to the surface of a hydrogel bead pore. This protein (mol. wt. 41 000) is obtained from the outer coat of the bacteria *Staphylococcus aureus*. Antibodies of various kinds are of great importance in modern therapeutic approaches. The antibodies have the Y-shaped structure of immunoglobulins (IgGs). It is known that protein A interacts with a great deal of specificity with the so-called Fc region of the IgG molecule. A variety of other proteins, impurities, etc., present in the feed/sample solution do not get attached to the surface-bound protein A. Once the affinity adsorption/attachment process is over, elution is initiated by reducing the *pH*/ionic strength. Generally, at a selected *pH*, the attached IgG is desorbed and eluted out. For a comprehensive introduction to such separations, consult Boyle *et al.* (1993), Scopes (1994) and Ladisch (2001).

It is useful to note a few other characteristics of this separation technique.

(1) The *pH* used to elute the adsorbed IgG should be selected to ensure that the bonding of the affinity ligand (sometimes called an affinant) to the matrix is unaffected.
(2) In spite of such steps, such affinity ligands are leached out of the matrix at very low levels; they have to be removed from the eluate.
(3) Although the targeted biomacromolecule (e.g. IgG) is adsorbed with high specificity, other proteins from the feed/sample may also be adsorbed somewhat via ion exchange/hydrophobic interactions, etc., between these proteins and the substrate ligands.

This technique provides, in principle, a way of selectively removing a target protein, enzyme, antibody, etc., from a mixture via highly specific molecule-to-molecule interactions. Such an activity requires highly selective ligands, which are currently being developed via specialized techniques.

These ligands are quite costly. It is desirable that the ligands present inside a substrate matrix, e.g. crosslinked

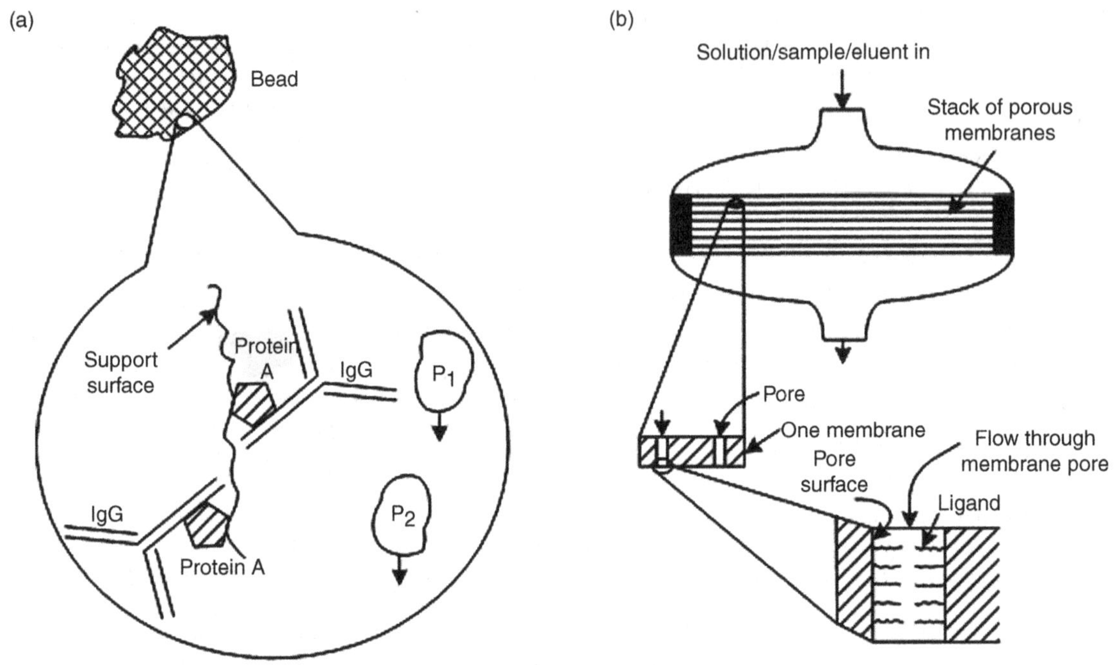

Figure 7.1.27. Affinity chromatography. (a) Protein A based affinity chromatographic removal of an immunoglobulin by a bead in a packed bed. (b) Flat porous membrane based affinity chromatography.

agarose, etc., are utilized for binding to the fullest extent. Although the beads are highly porous gels, diffusional rates of the large proteins into the pores are necessarily low. Therefore fractional ligand utilization is low, resulting in a costly inventory, especially if the mobile phase has a higher velocity. These limitations are overcome by a technique called *membrane chromatography*.

We have already observed in Figures 4.1.28(b) and (c) how polymeric chains anchored on the surfaces of pores of a membrane can bind metals, etc., present in the solution flowing through the membrane pores. This concept essentially applied the principle of membrane chromatography developed first with affinity adsorption of proteins to solve the problem of removal of metals from a solution (Brandt *et al.*, 1988). Affinity ligands attached to the membrane pore surfaces pick up proteins/antibodies/enzymes from the solution in a highly selective fashion. The membranes may be in the form of flat films (or hollow fibers, see Section 7.2). However, the amount of pore surface area available in one flat membrane (which is usually quite thin) is not substantial. Therefore a stack of membranes is used (Figure 7.1.27(b)).

Usually highly porous microfiltration membranes (Sections 6.3.3.1.1 and 7.2.1.4) are employed. The convective velocity along the pore length brings the species to be adsorbed to each ligand very quickly. Ligand utilization is very high. As a result, the ligand inventory is quite low compared to that in porous beads.

Modeling of the performance of a stack of flat membranes, each having ligands in the pores (Figure 7.1.27(b)) as a protein-containing solution flows through, has been carried out by Suen and Etzel (1992) and Gebauer *et al.* (1997). The governing equation for adsorption of a protein from a solution as it flows through the membrane pore is still equation (7.1.4). However, with proteins, local equilibrium cannot be assumed; attention has to be focused on the protein-binding kinetics. Suen and Etzel (1992) have employed $P + S \underset{k_d}{\overset{k_a}{\leftrightarrow}} PS$ to describe the protein–ligand interaction (see equation (4.1.76)) kinetics. Their detailed model should be studied to determine the effects of axial diffusion, flow velocity, protein–ligand association kinetics, membrane porosity and the variation in membrane thickness.

A few details are, however, useful. The value of the ligand capacity in one membrane is quite low. Therefore, breakthrough can happen quickly unless the flow velocity is kept low. When a number of membranes are stacked together (Figure 7.1.27(b)), the breakthrough can be substantially delayed; alternatively a higher flow velocity may be utilized. Further, the breakthrough can be delayed if the ligand capacity on the membrane pore surfaces is large compared to the protein concentration in the feed solution. The pressure drop can, however, be controlled to a lower level compared to that in a packed bed of affinity beads where intra-bead diffusion is quite slow and therefore ligand utilization will be low.

We will now very briefly describe another mobile-liquid-phase chromatography technique having "affinity" in the name: *immobilized metal affinity chromatography* (IMAC). It is also used in bioseparations, specifically in protein separations. In Section 5.4.2.1, we observed macromolecules having specific ligands bind metal ions in solution. In particular, polymers having iminoacetic acid ligands are chelating agents for divalent cations such as Cu^{++}, Zn^{++}, Ni^{++}, etc. If an adsorbent is formed by attaching to its surface a molecule having, say, an iminoacetic acid chelator group hanging out, then such a group will bind a divalent cation, e.g. Cu^{++}, present in solution in the mobile phase. Although this divalent cation is tightly bound now to the adsorbent particle via chelation, not all of its coordination spheres (d-orbitals) are occupied. It still has complexation ability. It has been found that basic groups in proteins, histidine residues (for example), will bind with the partially complexed divalent metal ion (which is part of the adsorbent particle). Other electron-donating side chains in proteins will also complex with the adsorbent-bound divalent metal. Trivalent metal ions of iron and aluminum have also been used to this end.

In practice (Scopes, 1994), columns are filled with the adsorbent particles having the ligands. Then a dilute salt solution of the metal, e.g. $CuSO_4$, etc., is passed through the column till it is saturated with the metal ion. Excess metal ions are washed out using an appropriate weakly metal-complexing compound such as glycine to ensure that there are no free metal ions floating around. Only then can proteins be separated by chromatography, namely IMAC, through such a column. The actual operation is carried out at a high ionic strength, e.g. 1 M NaCl. The technique was introduced by Porath *et al.* (1975); an extensive earlier review is available (Porath, 1992).

7.1.5.1.9 Elution chromatography with a mobile liquid phase: liquid–liquid chromatography In traditional liquid–liquid chromatography (LLC), the mobile liquid phase flows through a column packed with porous solid beads/particles, whose surface has been coated with a liquid, which acts as the stationary phase. Partitioning of the solute takes places between the mobile-phase liquid ($j = 2$) and the stationary-phase ($j = 1$) liquid coating. Partitioning of the solute i between the mobile- and stationary-phase liquids may be determined, for example, from a knowledge of the solubility parameters in the same manner as in liquid–liquid extraction (e.g. equation (4.1.34n)), repeated here for convenience:

$$\ell n K_i = \frac{\overline{V}_i\left[(\delta_i - \delta_2)^2 - (\delta_i - \delta_1)^2\right]}{RT}.$$

The equilibrium ratio K_i increases as δ_i becomes closer to δ_1, the solubility parameter for the stationary phase.

In normal LLC, the stationary-phase liquid is usually more polar than the mobile phase. The liquid immiscibility criterion, however, requires $(\delta_1 - \delta_2) \geq 4$ (Karger *et al.*, 1973). Karger *et al.* (1973) have demonstrated that, for hydrocarbon compounds which are relatively nonpolar, $\delta_2 - \delta_i$ cannot be made much less than 0, which with $(\delta_1 - \delta_2) \geq 4$ leads to $K_i \approx 0$. This had led to reverse-phase chromatography, where the stationary-phase liquid is nonpolar and the mobile phase is polar. Nonpolar hydrocarbon compounds then have high K_i values.

The problem with the coated liquid phase is that it may be lost by solubilization with time in the mobile phase. To avoid it, the stationary-phase liquid may be chemically bonded to the stationary-phase particle material. For a review of the preparation of nonpolar bonded stationary phases having hydrocarbon-like chains grafted onto the particle support surface, see Snyder and Kirkland (1979).

7.1.5.2 *Frontal chromatography and displacement chromatography*

We have already encountered *frontal development* in Figure 7.1.5(c). The breakthrough times for different fronts were provided in equation (7.1.22b) on the basis of the nondispersive equilibrium adsorption model. Here the mobile phase moving through the column is the feed solution to be separated; a pure output is obtained only for the species that is least strongly adsorbed. Subsequent components of the feed appear as a step contaminated by the species which have come out earlier.

Displacement chromatography involves initial feeding of a large feed sample pulse into a column free of anything except pure mobile phase, say liquid flowing through the column. Elution of the adsorbed species is next achieved by passing a "displacer," generally a strongly adsorbed species, which displaces/replaces the species already adsorbed from the sample. For a gaseous mobile phase, we have already seen an example of NH_3 as a displacer for the linear paraffines of medium molecular weight adsorbed in molecular sieve adsorbents (Figures 7.1.10 (b) and (c)). In general, controlled removal of the adsorbed species from the adsorbent is achieved by a more strongly adsorbed substance, the displacer. The column has to be regenerated afterwards vis-à-vis the displacer, for example by heating before the next feed pulse is introduced.

For multiple solutes present in the feed sample pulse, there will be adsorption of each solute in the column. Usually the solutes are present at a high concentration. When the strongly adsorbed eluting solute displaces the adsorbed solutes, each adsorbed solute goes down the column and appears as a band at the column outlet; the multiple solutes appear one after another, with the bands overlapping. The displacer peak/band appears at the very end.

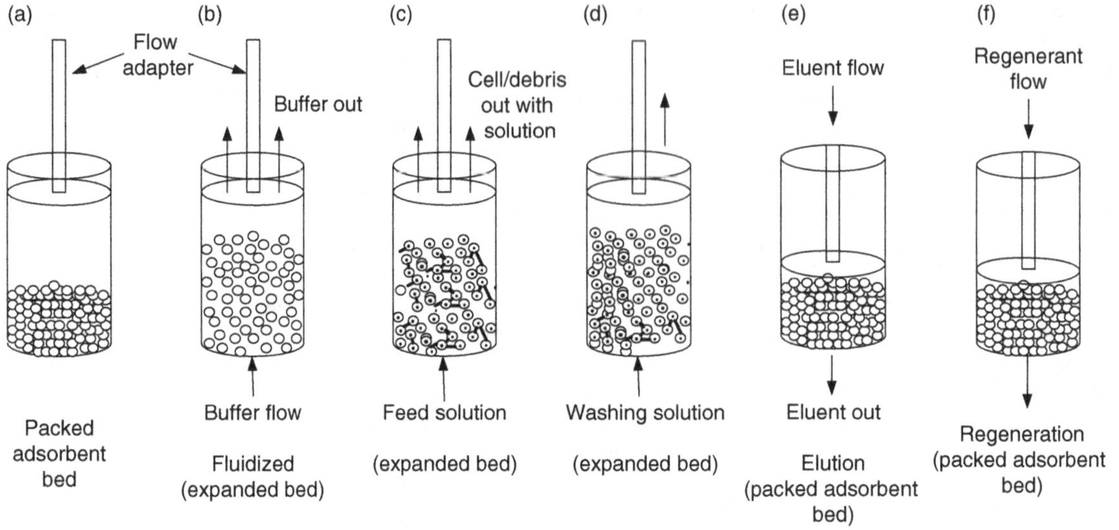

Figure 7.1.28. Different steps in a cyclic expanded-bed adsorption process for protein recovery/purification from a fermentation broth/lysate.

7.1.6 Expanded bed adsorption (EBA) from a broth/lysate aided by gravitational force

Products obtained via biotechnology are either extracellular or intracellular. Product recovery is implemented correspondingly from a fermentation broth with whole cells or from a lysate containing ruptured bacterial cells in a suspension and product in solution. Conventionally, the cellular material is often removed first via filtration/centrifugation; the product in the solution is then recovered/purified by a variety of steps including adsorption/chromatography. (These processes are carried out in different devices generally with a holding tank in between. Xu *et al.* (2005) have developed one hollow fiber membrane device whose shell side is filled with chromatographic resin particles to achieve cyclical filtration, adsorption and elution of different protein peaks (see Figure 11.1.3). Here we focus on an adsorbent bed where the functions of filtration, adsorption and elution are carried out cyclically.)

If such a feed is introduced from the bottom into a fixed adsorbent bed, the bed will immediately become clogged by the whole cells or the lysate particulates/cell fragments, etc. (Figure 7.1.28(a)). The fixed bed will act as a granular filter (see Section 7.2.2). However, if the bed is expanded somehow, for example by fluidization, then the cells and cellular debris can pass in between the adsorbent particles and out from the top of the bed (Figure 7.1.28(b)). Obviously, before the start of the fluidization process, there has to be a significant distance between a flow adapter (also called a column adapter) at the top of the bed and the actual top of the fixed bed (as shown in Figure 7.1.28(a)).

We will now describe the cyclic process followed in the expanded-bed adsorption process. The cyclic process is implemented to start with using the buffer solution introduced from the bottom. Then the feed solution containing cells/cell debris is introduced into the expanded bed (Figure 7.1.28(c)). The cells/cell debris are carried through the space between the adsorbent particles, since the fluid velocity exceeds the terminal velocity of the cells and cell debris. Due to the same reason, the density of the adsorbent particles is increased by having a quartz core, or even a metallic alloy core, in the particles; these heavier adsorbent particles should not be swept away, but should undergo mild/incipient fluidization. Further, often a size-distributed adsorbent particle population (50–400 μm) is employed, with smaller particles moving around near the top and the larger particles near the bottom of the expanded bed (see Figure 6.3.3(b)). Meanwhile, proteins/biomacromolecules from the unclarified feed are adsorbed on the surface of these adsorbent particles.

Once the adsorption process is over, the flow of a washing solution is initiated to wash away undesirable material (Figure 7.1.28(d)). Then the liquid flow is stopped and the particles settle down. Next, a flow of the buffer is initiated in the opposite direction, leading to densification of the expanded bed into a packed bed on the bottom support screen. Finally, the eluent is passed (Figure 7.1.28(e)) from the top to the bottom. Now the bed is packed, so there is chromatographic elution, different peaks of different protein species emerging at different times in the eluent, which leaves through the bottom of the column. Usually, such a chromatographic elution produces only partial purification of the proteins since the whole bed is loaded with the protein in varying amounts. Additional chromatographic purification steps are needed. After elution through the bed, the bed is washed using a buffer solution flowing downward to remove undesirable proteins and regenerate the bed to be used for the next cycle (Figure 7.1.28(f)).

In this cyclic process of adsorption/elution, the forces causing adsorption/elution are perpendicular to the fluid flow direction (upward or downward). However, the force of gravity acting on the particles is parallel to the direction of bulk fluid flow. Two different bulk flow vs. force directions are utilized to achieve two different goals: (1) adsorption/elution of proteins vis-à-vis adsorbent particles; (2) expansion of the bed of adsorbent particles to allow cells/cell debris to pass through the particles undergoing incipient/mild fluidization without being swept out of the column.

A few of the characteristics of the expanded-bed device and process are provided as follows: adsorbent particle size, 50–400 μm; particle density ~1.2–1.3 g/cm^3; ratio of maximum bed height to packed bed height, 2–3; flow velocity range in expanded beds, 200–500 cm/hr; cycle time, ~140 min. More details are provided on the website (http://www.biotech.pharmacia.se). Recovery and purification of proteins from bacterial cultures, mammalian cell cultures, yeasts, milk as well as blood have been studied (Batt *et al.*, 1995; Chang and Chase, 1996; Frej *et al.*, 1997). The types of adsorbent resin beads used include cation exchangers, anion exchangers, protein A for affinity adsorption and chelating resins for immobilized metal affinity chromatography.

7.1.7 Counteracting chromatographic electrophoresis and electrochromatography

We will study here the role of two different forces that are active in a bulk flow field perpendicular to one of the forces but parallel to the other. In a fixed-bed/packed-bed column, we have seen that a solute/macrosolute is convected down the column with a velocity ($v^*_{Ci} = R'_i v_z$ for species i) given by expressions (7.1.99b) and (7.1.12a). This velocity is different from the liquid-phase velocity v_z $(= v_0/\varepsilon$, where v_0 is the superficial velocity of the liquid based on the empty column cross-sectional area), and is determined by the interactions between the species in the mobile-phase and the stationary-phase adsorbent as a result of the chemical potential gradient based force acting perpendicular to the bulk liquid flow. Now, if an electrical field is imposed in the direction of the bulk liquid flow and the solutes/macrosolutes are charged (e.g. proteins), we will have an additional z-directional imposed velocity of U_{iz} on the solute/macrosolute (Figure 7.1.29(a)) due to the electrical field.

Assume now that the direction of this electrophoretic velocity is opposite to that of v_z, and therefore that it opposes the solute velocity in the chromatographic column, $R'_i v_z = v^*_{Ciz}$. Therefore the net velocity of species i down the column is

$$v_{iz,\text{net}} = v^*_{Ciz} - |U_{iz}|_{\text{net}}, \qquad (7.1.111)$$

where $|U_{iz}|_{\text{net}}$ is less than $|U_{iz}|$ due to the partitioning going on between the mobile and the stationary phases. Assume now that the chromatographic/adsorbent column consists of two distinct regions $j = 1$ and $j = 2$ so that v^*_{Ciz} has two different values in the two regions, v^*_{Ci1} and v^*_{Ci2}. Consequently, solute/macrosolute i will have a net z-direction velocity which will be different in the two regions:

$$v_{iz,\text{net}}|_{j=1} = v^*_{Ci1} - |U_{iz}|_{\text{net},1}; \qquad (7.1.112a)$$

$$v_{iz,\text{net}}|_{j=2} = v^*_{Ci2} - |U_{iz}|_{\text{net},2}. \qquad (7.1.112b)$$

Consider the case where the electrophoretic velocities of solute i in the two zones are assumed to be identical in magnitude (they have the same direction), i.e.

$$|U_{iz}|_{\text{net},1} = |U_{iz}|_{\text{net},2} \qquad (7.1.112c)$$

It is now possible to create a situation where the magnitude of $v_{iz,\text{net}}|_{j=1}$ and $v_{iz,\text{net}}|_{j=2}$ are identical but their directions are different:

$$|v_{iz,\text{net}}|_{j=1}| = |v_{iz,\text{net}}|_{j=2}|; \qquad (7.1.112d)$$

$$v_{iz,\text{net}}|_{j=1} + v_{iz,\text{net}}|_{j=2} = 0. \qquad (7.1.112e)$$

Therefore at the interfacial region between the two regions $j = 1$ and $j = 2$ in the column, species i will be focused and will concentrate.

On the other hand, other solute/macrosolute species/proteins will move through this interfacial region since the magnitudes of this solute's $v_{iz,\text{net}}|_{j=1}$ and $v_{iz,\text{net}}|_{j=2}$ are not identical. If, however, there is a third zone in the column, $j = 3$, next to $j = 2$ such that, for this solute,

$$|v_{iz,\text{net}}|_{j=2}| = |v_{iz,\text{net}}|_{j=3}|, \qquad (7.1.112f)$$

$$v_{iz,\text{net}}|_{j=2} + v_{iz,\text{net}}|_{j=3} = 0, \qquad (7.1.112g)$$

then this second solute species will accumulate in the interfacial region between regions 2 and 3. This basic principle is called *counteracting chromatographic electrophoresis* (CACE) and was first proposed by O'Farrell (1985). Additional treatments are available in McCoy (1986), Hunter (1988) and Ivory and Gobie (1990).

We will now develop a quantitative illustration of the results (7.1.112d,e) for any solute i and then provide an example from O'Farrell (1985) in terms of protein separation. **Later we will deal with electrochromatography in general as applied to a chromatographic column having one stationary phase with an imposed electrical field in the direction of the bulk liquid flow perpendicular to the force direction between this mobile phase and the stationary phase.**

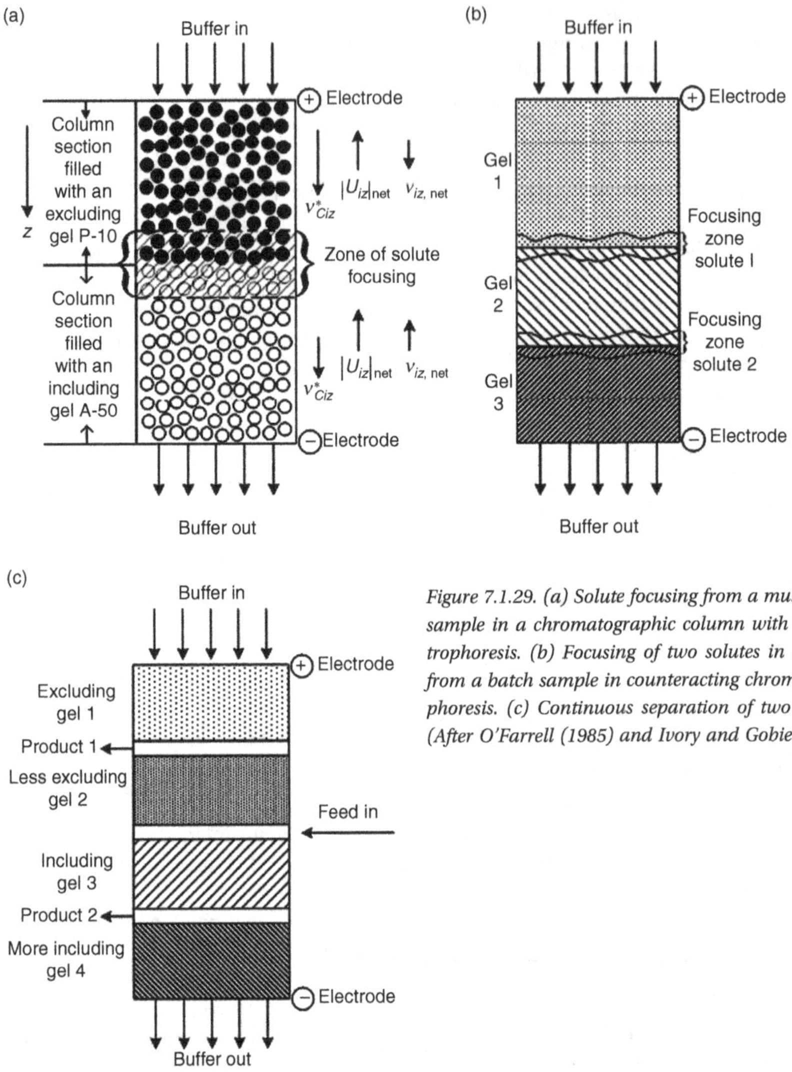

Figure 7.1.29. (a) Solute focusing from a multicomponent batch sample in a chromatographic column with counteracting electrophoresis. (b) Focusing of two solutes in a three-gel column from a batch sample in counteracting chromatographic electrophoresis. (c) Continuous separation of two proteins in CACE. (After O'Farrell (1985) and Ivory and Gobie (1990).)

Consider equation (7.1.2) for a charged solute species i in a packed bed or a chromatographic column. If we have a voltage applied in the z-direction, creating an electrophoretic velocity U_{iz} of charged species i, then equation (7.1.2) may be reformulated as

$$S_c\Delta Z\left[\varepsilon\frac{\partial \overline{C}_{i2}}{\partial t}+(1-\varepsilon)\frac{\partial \overline{C}_{i1}}{\partial t}\right]=\varepsilon S_c\left[(v_z+U_{iz})\overline{C}_{i2}-D_{i,\text{eff},z}\frac{\partial \overline{C}_{i2}}{\partial z}\right]_z -\varepsilon S_c\left[(v_z+U_{iz})\overline{C}_{i2}-D_{i,\text{eff},z}\frac{\partial \overline{C}_{i2}}{\partial z}\right]_{z+\Delta z}. \tag{7.1.113a}$$

Applying the limiting process of $\Delta z\rightarrow 0$ after dividing by $S_c\Delta z$, we get

$$\frac{\partial \overline{C}_{i2}}{\partial t}+\frac{(1-\varepsilon)}{\varepsilon}\frac{\partial \overline{C}_{i1}}{\partial t}+\frac{\partial\left[(v_z+U_{iz})\overline{C}_{i2}\right]}{\partial z}=\frac{\partial}{\partial z}\left[D_{i,\text{eff},z}\frac{\partial \overline{C}_{i2}}{\partial z}\right]. \tag{7.1.113b}$$

Conventionally, v_z and U_{iz} may be assumed to be independent of the z-coordinate.[10] Therefore

$$\frac{\partial \overline{C}_{i2}}{\partial t}+\frac{(1-\varepsilon)}{\varepsilon}\frac{\partial \overline{C}_{i1}}{\partial t}+(v_z+U_{iz})\frac{\partial \overline{C}_{i2}}{\partial z}=\frac{\partial}{\partial z}\left[D_{i,\text{eff},z}\frac{\partial \overline{C}_{i2}}{\partial z}\right]. \tag{7.1.113c}$$

[10]There may be variation of U_{iz} along the column due to concentration effects.

If we assume isothermal equilibrium nondispersive operation and adopt the methodology of equations (7.1.6)–(7.1.8), we can rewrite this equation as

$$\left[1+\rho_b\frac{q_i'(\overline{C}_{i2})}{\varepsilon}\right]\frac{\partial\overline{C}_{i2}}{\partial z}+(v_z+U_{iz})\frac{\partial\overline{C}_{i2}}{\partial z}=0. \quad (7.1.113d)$$

Using the methodology adopted to obtain the migration rate of species i, v^*_{Ci}, via equations/relations (7.1.10), (7.1.11) and (7.1.12a), we can write, for any particular section of the column,

$$v_{iz,\text{net}}\big|_{j=1}=\frac{v_z+U_{iz}}{\left[1+\frac{\rho_b q_i'(\overline{C}_{i2})}{\varepsilon}\right]}$$

$$\Rightarrow v_{iz,\text{net}}\big|_{j=1}=\frac{v_z}{\left[1+\frac{\rho_b q_i'(\overline{C}_{i2})}{\varepsilon}\right]}+\frac{U_{iz}}{\left[1+\frac{\rho_b q_i'(\overline{C}_{i2})}{\varepsilon}\right]}. \quad (7.1.114a)$$

At the interface between column regions 1 and 2 (Figure 7.1.29(a)), if the species i is to accumulate there, we should have

$$\left|\left.\frac{v_z+U_{iz}}{\left[1+\frac{\rho_b q_i'(\overline{C}_{i2})}{\varepsilon}\right]}\right|_{\text{column region1}}\right|=\left|\left.\frac{v_z+U_{iz}}{\left[1+\frac{\rho_b q_i'(\overline{C}_{i2})}{\varepsilon}\right]}\right|_{\text{column region2}}\right|. \quad (7.1.114b)$$

There are a number of assumptions implicit here since we are dealing with charged species; further, if we are dealing with charged macrosolutes, there are additional assumptions: we neglect the concentration polarization of proteins on the surface of gel matrix in the presence of the electrical field; we have not considered the issue of intraparticle porosity; the nature of the partitioning equilibrium (7.1.6) assumed has to be relevant; there is no electroosmotic velocity.

We will now briefly illustrate the example originally provided by O'Farrell (1985) for CACE using beads employed in gel chromatography. Region 1, at the top of the column, contains a gel that tends to exclude the protein (O'Farrell employed the protein ferritin); therefore the value of v^*_{Ciz} is larger and very close to v_z without the electrical field. The electrically induced velocity U_{iz} in the opposite direction reduces this downward velocity in region 1. But the net velocity in region 1 is downward. On the other hand, region 2 has a rather open gel with a solute partition coefficient around 1. Therefore the downward value of v^*_{Ciz} in region 2 is much smaller and the electrically induced velocity in the opposite direction is larger. This leads to focusing of the protein into the intermediate region between the two regions having two different types of gel particles. The protein ferritin is negatively charged at the *pH* employed; accumulation of this protein in the intermediate region leads to a buildup of counterions, which increases the electrical conductivity of this zone, and therefore decreases the voltage gradient E (= dV/dz) in the z-direction. This reduces the electrophoretic velocity of the protein sufficiently so that the net protein velocity becomes zero in this zone in between region 1 and 2. An advantage of this method is that focusing occurs in the native electrolyte as opposed to the ampholytes needed in isolectric focusing (IEF; see Section 4.2.2.1).

It should be noted that all of the above treatment took place in a *batch solute introduction mode* with a *continuous buffer flow*. For example, the experimental arrangement used by O'Farrell (1985) was as follows. A 0.7 cm diameter, 50 cm long glass column was packed with Biogel A-50 m beads upto a height of 25 cm from the bottom; the top of the column was filled with BioGel P-10 beads, which excludes all of the proteins under consideration. Two electrode reservoirs (positively charged at the top and negatively charged at the bottom) were connected to the glass column through a large-bore tubing filled with polyacrylamide gel. Another port connected to the column top allowed a flow from a peristaltic pump of the carrier solvent (10 mM tris acetate, *pH* 7.4) to come in at 0.17 cm^3/min. The voltage applied between the electrodes was 600 volts. Ferritin was introduced in a batch sample and was found to concentrate in a narrow band 2 mm wide around the interface between the two sections of resins (Figure 7.1.29(a)).

By having successively different layers of gel matrices (beyond two), one can have different proteins focused in different intermediate regions from a batch sample. By having three gel matrices, one can have two intermediate regions, each providing a focusing region for a separate protein, thus separating a binary mixture (Figure 7.1.27(b)). By having four gel matrices, one can have three focusing regions for three different proteins (note, only for a batch sample).

In the *batch mode of sample introduction* described above, both sample loading and product extraction are considered cumbersome. Ivory and Gobie (1990) have demonstrated a *continuous mode* of CACE operation in a column, as illustrated in Figure 7.1.29(c). The buffer solution is introduced at the top. The feed solution mixture is introduced into the middle of the column containing a number of layers of different gel particles. Two concentrated product streams are withdrawn from an appropriate intermediate position at the top half and bottom half of the column; this is done through a port, which, however, contains no gel particles. For the exact configuration employed, see Ivory and Gobie (1990). Figure 7.1.29(c) illustrates a column having five ports: one for feed sample introduction, two for withdrawal of two products and two for buffer introduction and withdrawal, all on a continuous basis. There are four different gels stacked in the column to achieve continuous fractionation of two proteins. Multicomponent separation is not possible in this configuration on a continuous basis.

This last point needs some elaboration. Consider in Figure 7.1.29(c), any one of the two ports used to withdraw two different protein products continuously. If we have three proteins present in the feed stream, then let the liquid coming to one of the ports be pure in one protein; it is clear that the liquid passing over the other port will have two other proteins: therefore the liquid stream withdrawn from this port cannot be pure in one protein. This is an inherent limitation of processes where the bulk flow is parallel to the direction of the force. Here, even though we have one force perpendicular to the bulk flow, in the presence of the electrical force parallel to bulk flow, the system in continuous operation is just like the countercurrent electrophoresis of Figure 6.3.4 and equations (6.3.9a,b). However, in a batch mode, one can have multicomponent separation since each species will be focused to its own zone and therefore can be withdrawn at a later time. Multicomponent separation in electrochromatography needs a different flow vs. force configuration, to be discussed further in Section 8.2.

7.1.7.1 *Electrochromatography*

The CACE technique described above is a specialized offshoot of the general technique of electrochromatography, where in a column containing chromatographic beads as the buffer solution flows, an electrical field is simultaneously applied through two electrodes mounted at the top and bottom of the column. The sample is generally introduced in a batch fashion (an exception is shown in Figure 7.1.29(c)). Only one kind of chromatographic medium is present in the column, unlike that in CACE; the medium will depend on the species to be separated. It could be gel permeation beads for a mixture of proteins (Rudge and Ladisch, 1988), as studied first by Nerenberg and Pogojeff (1969) for separation of blood plasma proteins; alternatively, for smaller molecules there is a vast literature (Tsuda, 1995). One could also employ a capillary column, as in capillary electrophoresis (Section 6.3.1.2); however, the capillary is packed with a suitable chromatographic medium. Alternatively, there may be a coating on the wall of an open capillary column. Electroosmotic flow is possible with an appropriate silica-type tubular capillary wall in the applied electrical field. When carried out in a capillary column, this combination of capillary chromatography with capillary electrophoresis is generally called *capillary electrochromatography* (CEC).

We will now provide an elementary approach to calculating *selectivity* or *resolution* in CEC for a batch sample. Suppose, in the capillary system under consideration, there is an electroosmotic velocity $v_{\text{EOF},z}$, a pressure driven velocity v_z and a solute migration velocity U_{iz} due to the electrophoretic driving force, due to a constant potential gradient E in the z-direction. Therefore the effective velocity of an ionic species i is (see equation (6.3.10c))

$$\langle v_{iz}\rangle = \langle v_z\rangle + \mu_{\text{EOF}}^m E + \mu_i^m E, \tag{7.1.115}$$

where μ_i^m is the ionic mobility of the charged species i, μ_{EOF}^m is the electroosmotic mobility of electroosmotic flow (EOF) (see (6.3.10c)) and $\langle v_z\rangle$ is the cross-sectional averaged value of the pressure driven capillary flow. We may now follow the standard procedure for conventional packed-bed/chromatographic columns (see relations (7.1.99b) and (7.1.99h)) that the centerline of the peak of species i moves down at a velocity

$$v_{iz,\text{net}} = \frac{\langle v_{iz}\rangle}{\left[1+\dfrac{1-\varepsilon}{\varepsilon}\kappa_{i1}\right]} = \frac{\langle v_{iz}\rangle}{\left[1+\rho_b\dfrac{q_i'(\overline{C}_{i2})}{\varepsilon}\right]} = \frac{\langle v_{iz}\rangle}{V_{R_i}/V_M}; \tag{7.1.116}$$

$$v_{iz,\text{net}} = \frac{\langle v_z\rangle + \mu_{\text{EOF}}^m E + \mu_i^m E}{\left[1+\dfrac{(1-\varepsilon)}{\varepsilon}\kappa_{i1}\right]}. \tag{7.1.117}$$

The selectivity α_{ij} between two solute species i may now be defined in a number of ways. For example,

$$\alpha_{ij} = \frac{v_{iz,\text{net}}}{v_{jz,\text{net}}} = \frac{(\langle v_z\rangle + \mu_{\text{EOF}}^m E + \mu_i^m E)}{(\langle v_z\rangle + \mu_{\text{EOF}}^m E + \mu_j^m E)}\,\frac{\left[1+\dfrac{(1-\varepsilon)}{\varepsilon}\kappa_{j1}\right]}{\left[1+\dfrac{(1-\varepsilon)}{\varepsilon}\kappa_{i1}\right]}. \tag{7.1.118}$$

A major advantage of CEC over capillary electrophoresis (CE) (Section 6.3.1.2) lies in its capacity to separate uncharged molecules. Whereas in CE uncharged molecules will be convected at the bulk velocity, v_{EOF}, of the electroosmotic flow in an unselective manner, the incorporation of a chromatographic medium in CEC will allow the separation of uncharged compounds as well. For charged molecules/macromolecules, the chromatographic medium (for example, gel beads for proteins) provides additional selectivity to that achieved by the electrophoretic mobility differences (see (6.3.19)).

A simplistic analysis, provided below, can allow the quantification of which mechanism contributes to what extent to the overall selectivity in, say, CEC. To this end, we define a selectivity between species i and j via

$$\alpha_{ij}\big|_{t_{R_i}} = \frac{t_{R_i} - t_{R_j}}{t_{R_i}}, \tag{7.1.119a}$$

where the retention time for species i in the column of length L, t_{R_i}, is obtained from

$$t_{R_i} = (L/v_{iz,\text{net}}). \tag{7.1.119b}$$

This yields

$$\alpha_{ij}\big|_{t_{R_i}} = \frac{\dfrac{\left[1+\frac{(1-\varepsilon)}{\varepsilon}\kappa_{i1}\right]}{\langle v_z\rangle + \mu_{\mathrm{EOF}}^m E + \mu_i^m E} - \dfrac{\left[1+\frac{(1-\varepsilon)}{\varepsilon}\kappa_{j1}\right]}{\langle v_z\rangle + \mu_{\mathrm{EOF}}^m E + \mu_j^m E}}{\dfrac{\left[1+\frac{(1-\varepsilon)}{\varepsilon}\kappa_{i1}\right]}{\langle v_z\rangle + \mu_{\mathrm{EOF}}^m E + \mu_i^m E}}, \tag{7.1.119c}$$

leading to

$$\alpha_{ij}\big|_{t_{R_i}} = \frac{\left(\mu_j^m - \mu_i^m\right)E + \left(\langle v_z\rangle + \mu_{\mathrm{EOF}}^m E\right)\left(\frac{1-\varepsilon}{\varepsilon}\right)\left(\kappa_{i1} - \kappa_{j1}\right) + \left(\frac{1-\varepsilon}{\varepsilon}\right)E\left(\kappa_{i1}\mu_j^m - \kappa_{j1}\mu_i^m\right)}{\left[1+\frac{(1-\varepsilon)}{\varepsilon}\kappa_{i1}\right]\left[\langle v_z\rangle + \mu_{\mathrm{EOF}}^m E + \mu_j^m E\right]}. \tag{7.1.120}$$

The three terms in the numerator identify three different contributions to $\alpha_{ij}|_{t_{R_i}}$: the first term, $(\mu_j^m - \mu_i^m)E$, is the contribution by the differences in the electrophoretic mobility of the two species; the second term relates to the difference in the partitioning characteristics and the capacity of the two phases for the two species i and j; the third term refers to the interaction between chromatographic partitioning capacity and electrophoretic mobility (Deng *et al.*, 1998).

Columns in CEC have discontinuities of the electric field strength and flow velocity at the interface of the packed and open sections of the column. Rathore and Horvath (1998) have provided a methodology for treating such a configuration.

7.2 Crossflow membrane separations, crossflow filtration and granular filtration

This section will describe a number of membrane separation processes as well as filtration processes where the force(s) causing separation is (are) acting perpendicular to the direction of flow of the feed phase. In Section 7.2.1, we will treat the following membrane processes which rely on crossflow: crossflow gas permeation, crossflow reverse osmosis (RO), crossflow ultrafiltration and crossflow microfiltration. We will encounter how the configuration of bulk feed flow parallel to the membrane surface but perpendicular to the force direction between the feed fluid and the membrane can lead to steady state operation (see Figures 7.0.1(e)–(h); a relatively slow process of membrane fouling can introduce unsteadiness in an otherwise steady process. We observed earlier, in Section 6.3.3, that, for all such membrane/filtration processes, the dead-end mode of operation with bulk feed flow parallel to the direction of the force through the membrane led to unsteady state operation. We also noted (e.g. in Section 6.3.3.5) that it is often impractical to employ a large membrane surface area in the dead-end configuration. Membrane separation and filtration processes having crossflow and studied in this chapter do not suffer from these limitations. However, when the permeate phase collected from different sections of the membrane cannot be removed by simple crossflow and starts having a bulk motion parallel to the membrane on the permeate side, the nature of separation achieved may be affected; this configuration is treated in Sections 8.1.9 and 8.2.4.2. Both the mode of feed introduction and the mode of product withdrawal vis-à-vis the force direction are important in separation devices.

At the end of Section 7.2.1, the process of rotary vacuum filtration is described. Here the membrane containing the particulate cake/deposit (see Figure 6.3.25(a)) is moved perpendicular to the pressure induced crossflow causing filtration through the filter. If we imagine the coordinate system to be fixed to the membrane, then the system configuration is similar to the crossflow microfiltration briefly considered earlier (Figure 6.3.25(b)) where the liquid slurry/suspension moves parallel to the membrane but perpendicular to the direction of the force.

In Section 7.2.2, we consider fixed-bed granular filtration for clarifying dilute suspensions of particles from a liquid. We will identify the similarities between this process and the fixed-bed adsorption/desorption processes studied in Section 7.1.1: both processes employ a low velocity bulk liquid flow through a bed of fixed particles and solute molecules or particles are removed from the liquid by forces acting perpendicular to the fixed particles. Both processes achieve high levels of purification. From a black-box type of perspective, granular filtration may also look very similar to the dead-end filtration studied in Section 6.3.3.1, but there are significant differences.

7.2.1 Crossflow membrane separations, crossflow filtration

The membrane/filtration processes have been considered in the following order: crossflow gas permeation, crossflow reverse osmosis, crossflow ultrafiltration, crossflow microfiltration and rotary vacuum filtration.

7.2.1.1 *Crossflow gas permeation*

In Sections 6.3.3.5 and 6.4.2.2, we described gas separation by permeation through a membrane under the conditions of bulk flow parallel to the force and having a well-mixed flow on both sides of the membrane. We learned that the configuration of bulk gas flow parallel to the force direction was not very useful: it could not be used in practice with large membrane surface area, and it led to unsteady state operation. Since the permeation rates of gases through membranes are quite slow, large membrane surface areas are needed in industrial practice; also, steady state operation is highly desirable. The disadvantage of having a well-mixed flow is simply stated: it leads to lower permeate enrichment and lower retentate purification. If we consider the membrane as a concentration amplifier of the more permeable species, then we achieve the highest permeate concentration at the permeator location where the feed has the highest concentration of the more permeable species. In the well-mixed mode of operation, this feature is lost, since both the permeate and the feed side are well-mixed. Further, in most practical gas permeation devices used at larger scales, the feed is unlikely to be well-mixed; the same is true of the permeate side.

In this section, we therefore consider the *crossflow configuration* introduced in Figure 7.0.1(e). The gas flows on the feed side of the membrane under the following conditions:

(1) there is no concentration polarization in the feed gas phase normal to the membrane surface;
(2) there is no longitudinal diffusion/dispersion along the feed flow direction[11] (no-mixing case);
(3) the feed gas and the permeate gas streams undergo negligible flow pressure drops.

The permeated gas is withdrawn in a direction perpendicular to the direction of flow of the feed gas and parallel to the direction of the force causing separation, the partial pressure gradient component of the chemical potential gradient of the permeating species. Figure 7.2.1(a) illustrates this separation configuration schematically.

An essential feature of the existing analyses of this crossflow configuration carried out by Weller and Steiner (1950a, b) is that the permeate gas stream emerging from the membrane (at any axial location) does not mix with any other permeates emerging from the membrane at other axial locations. Therefore the membrane permeation flux N_{im} at any permeator location is given by

$$N_{im} = \left(\frac{Q_{im}}{\delta_m}\right)(P_f x_{if} - P_p x_{ip}), \tag{7.2.1}$$

where the permeate side mole fraction x_{ip} of species i is determined by the permeation fluxes of different species. For a binary system of two species i and j,

$$\frac{N_{im}}{N_{jm}} = \frac{(Q_{im}/\delta_m)}{(Q_{jm}/\delta_m)}\frac{(P_f\,x_{if} - P_f\,x_{ip})}{(P_f\,x_{jf} - P_f\,x_{jp})} = \frac{x_{ip}}{x_{jp}}. \tag{7.2.2a}$$

An asymmetric or a composite membrane having a porous substrate through which the permeate emerges satisfies this postulate (Figure 7.2.1(b)).

[11] Chen *et al.* (1986) have studied the role of axial diffusion in gas permeator analysis.

At this time, it is useful to consider such an asymmetric or composite structure (a more complex structure is shown in Figure 6.3.35(b)) in detail and explore what membrane device configuration can allow the achievement of such operating assumptions. The complex structure of, for example, an asymmetric membrane having a nonporous skin may be modeled as a composite having a thin homogeneous skin on a porous backing. Conventionally, the permeate emerging from the skin into the pores beneath will have a composition indicated by x'_{ip} and x'_{jp} and a total pressure of $P'_p\,(> P_p)$:

$$\frac{N_{im}}{N_{jm}} = \frac{(Q_{im}/\delta_m)}{(Q_{jm}/\delta_m)}\frac{(P_f\,x_{if} - P'_p\,x'_{ip})}{(P_f\,x_{jf} - P'_p\,x'_{jp})}. \tag{7.2.2b}$$

If there is essentially no pressure drop in the transport of these permeated gases through the porous substrate, then $P'_p \cong P_p$, which is the pressure in the bulk gas on the permeate side. Further, if this gas mixture does not mix with any other gas stream emerging from other membrane locations, then only $x'_{ip} = x_{ip}$ and $x'_{jp} = x_{jp}$, leading to the crossflow relation (7.2.2a).

These conditions may be achieved in a hollow fiber permeator where the gas is fed through the hollow fiber bore; the permeate is radially withdrawn from the shell side and there is a low pressure drop in the shell side in the radial direction (Figure 7.2.1(c)). Two other commonly used configurations, a shell-side fed hollow fiber module and a spiral-wound device shown in Figures 7.2.1(d) and (e), will bring permeates emerging from different feed concentrations together at every pore mouth. The hollow fibers in these membrane processes may have internal diameters varying between 50 and 1000 μm and wall thicknesses varying between 5 and 300 μm. The analysis of such configurations is best considered in Chapter 8 since the compositions and pressure drops in both streams flowing perpendicular to the force direction are important. However, the crossflow analysis is still useful because the analytical results allow a quick estimation of the separation performance.

We will now develop the governing equations for such a crossflow permeator. For the permeator configuration of Figure 7.2.1(a) and the assumptions of (1) no concentration polarization and (2) no longitudinal diffusion/dispersion in the flow direction (z-coordinate), a control volume analysis (in the manner of Figure 6.2.1 but without any x-coordinate dependence) leads to (under conditions of no chemical reaction ($R_i = 0$) and steady state ($(\partial C_i/\partial t) = 0$))

$$0 = \Delta x \Delta y\,(N_{iz}|_z - N_{iz}|_{z+\Delta z}) - \Delta x \Delta z\, N_{im}. \tag{7.2.3}$$

Here, the control volume dimensions Δx and Δy are along the membrane flow channel of width W and height h, respectively. Further, the expression for N_{iz} in view of assumption (2) is simply

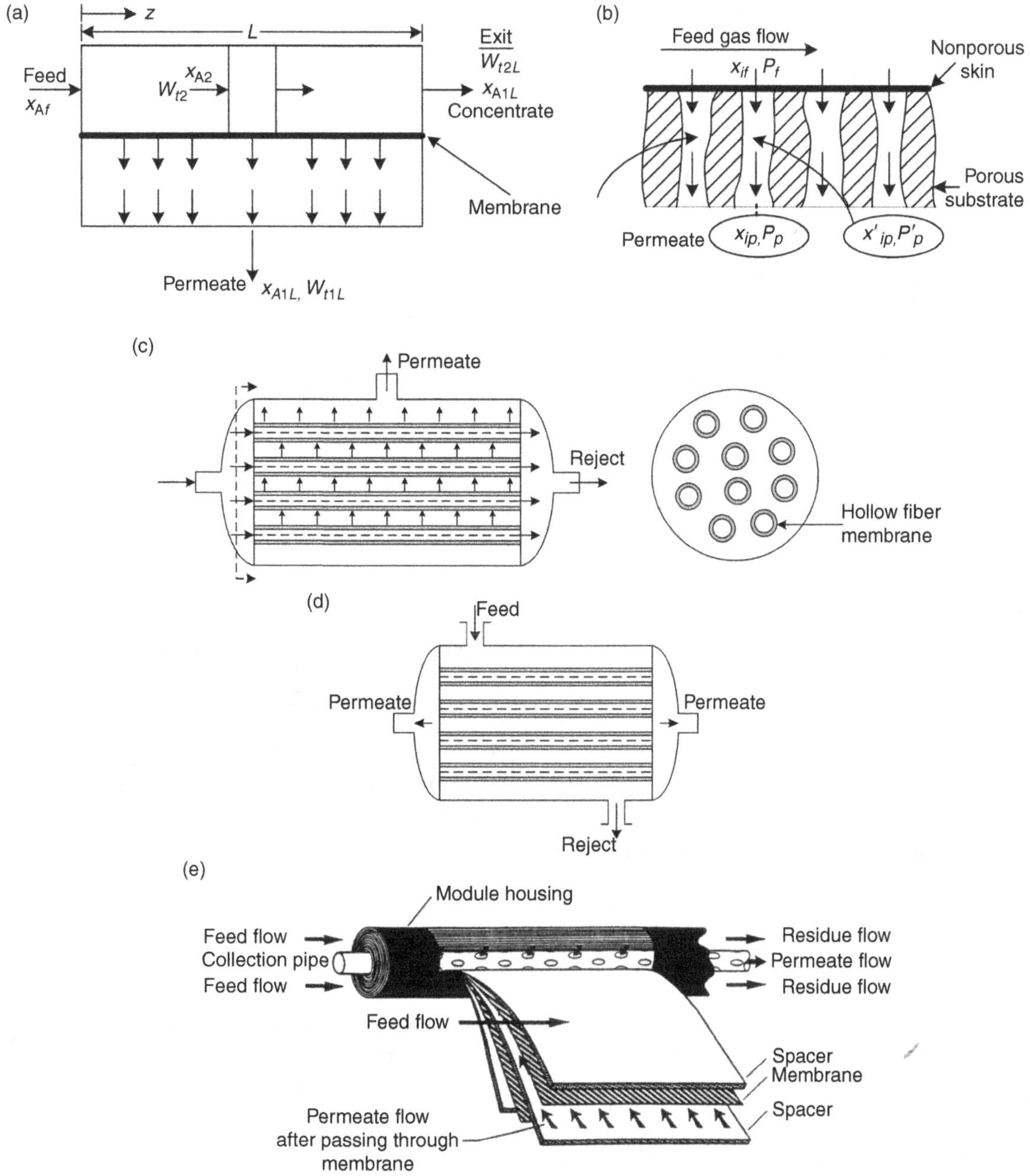

Figure 7.2.1. (a) Crossflow membrane module for gas permeation; (b) model for crossflow gas permeation in an asymmetric or composite membrane; (c) tube-side feed crossflow hollow fiber module; (d) shell-side feed hollow fiber module; (e) spiral-wound module.

$$N_{iz} = C_{i2}v^* = (W_{t2}\, x_{i2})/Wh, \tag{7.2.4}$$

where the flow stream on the feed side is $j = 2$. For the permeate $j = 1, p$. At $z = 0$, $N_{iz} = (W_{tf}\, x_{if})/Wh$, corresponding to the feed location. We are interested in finding out what would be the value of W_{t2} and x_{i2} at $z = L$, the end of the permeator on the feed side. Rewriting equation (7.2.3) as

$$0 = Wh(N_{iz}|_z - N_{iz}|_{z+\Delta z}) - W\Delta z\, N_{im},$$

and taking the limit of $\Delta z \to 0$, we get

$$\frac{\mathrm{d}(N_{iz}Wh)}{\mathrm{d}z} = -WN_{im} \Rightarrow \frac{\mathrm{d}(W_{t2}x_{i2})}{W\,\mathrm{d}z} = -N_{im}$$

$$\Rightarrow \frac{\mathrm{d}(W_{t2}x_{i2})}{\mathrm{d}A_m} = -N_{im} = -\frac{Q_{im}}{\delta_m}(P_f x_{i2} - P_p x_{i1}), \tag{7.2.5}$$

where

$$\mathrm{d}A_m = W\,\mathrm{d}z \quad \text{and} \quad A_m^t = WL, \tag{7.2.6}$$

L being the permeator length in the z-direction for a total membrane area A_m^t. This is the *governing equation for the feed side*. The governing equation for the permeate side is primarily dictated by the crossflow relation (7.2.2a). An alternative form of this relation for a multicomponent system is

$$\frac{N_{im}}{\sum_{j=1}^{n} N_{jm}} = \frac{x_{ip}}{\sum_{j=1}^{n} x_{jp}} = x_{ip}. \tag{7.2.7}$$

However,

$$\sum_{i=1}^{n} \mathrm{d}(W_{t2}x_{i2}) = \mathrm{d}W_{t2} = -\sum_{i=1}^{n} N_{im}\,\mathrm{d}A_m = -\sum_{j=1}^{n} N_{jm}\,\mathrm{d}A_m.$$

Therefore,

$$\frac{N_{im}\,\mathrm{d}A_m}{\sum_{j=1}^{n} N_{jm}\,\mathrm{d}A_m} = \frac{\mathrm{d}(W_{t2}x_{i2})}{\mathrm{d}W_{t2}} = x_{ip} = x_{i1} \tag{7.2.8}$$

is the required relation for the *permeate side composition*. In addition, we have to satisfy the following two relations:

$$\sum_{i=1}^{n} x_{i2} = 1, \qquad \sum_{i=1}^{n} x_{i1} = 1. \tag{7.2.9}$$

One can now follow the solution procedure of Pan and Habgood (1978a) instead of the Weller–Steiner (1950a, b) approach, which is valid for a binary system only.

We will now develop an analytical solution for a binary feed gas mixture of species A and B by assuming that the local separation factor α_{AB} is independent of the membrane separator length/location. This approach (Stern and Walawender, 1969) is an adaptation of the Naylor and Backer (1955) approach employed for gaseous diffusion with porous barriers. Rewrite equation (7.2.8) for $i = \mathrm{A}$ as

$$x_{\mathrm{A}2}\,\mathrm{d}W_{t2} + W_{t2}\,\mathrm{d}x_{\mathrm{A}2} = x_{\mathrm{A}1}\,\mathrm{d}W_{t2};$$

$$\frac{\mathrm{d}W_{t2}}{W_{t2}} = \frac{\mathrm{d}x_{\mathrm{A}2}}{(x_{\mathrm{A}1} - x_{\mathrm{A}2})} = \left\{\frac{1 + x_{\mathrm{A}2}(\alpha_{\mathrm{AB}} - 1)}{x_{\mathrm{A}2}(\alpha_{\mathrm{AB}} - 1)(1 - x_{\mathrm{A}2})}\right\}\mathrm{d}x_{\mathrm{A}2}, \tag{7.2.10}$$

where we have used definitions and relations such as

$$\alpha_{\mathrm{AB}} = \frac{x_{\mathrm{A}1}(1 - x_{\mathrm{A}2})}{(1 - x_{\mathrm{A}1})x_{\mathrm{A}2}}, \qquad (\alpha_{\mathrm{AB}} - 1) = \frac{x_{\mathrm{A}1} - x_{\mathrm{A}2}}{(1 - x_{\mathrm{A}1})x_{\mathrm{A}2}},$$

$$x_{\mathrm{A}1} = \frac{\alpha_{\mathrm{AB}}\,x_{\mathrm{A}2}}{1 + x_{\mathrm{A}2}(\alpha_{\mathrm{AB}} - 1)}, \qquad (1 - x_{\mathrm{A}1}) = \frac{(1 - x_{\mathrm{A}2})}{1 + x_{\mathrm{A}2}(\alpha_{\mathrm{AB}} - 1)}. \tag{7.2.11}$$

Integrating this relation between the limits of W_{t2} and $x_{\mathrm{A}2}$ along the permeator length from $z = 0$, $W_{t2} = W_{tf}$, $x_{\mathrm{A}2} = x_{\mathrm{A}f}$ to $z = L$, $W_{t2} = W_{t2L}$, $x_{\mathrm{A}2} = x_{\mathrm{A}2L}$, $W_{t1L} = W_{tp}$, $x_{\mathrm{A}1L} = x_{\mathrm{A}p}$, we get

$$\ell n\left(\frac{W_{t2L}}{W_{tf}}\right) = \frac{1}{(\alpha_{\mathrm{AB}} - 1)}\int_{x_{\mathrm{A}f}}^{x_{\mathrm{A}2L}} \frac{\mathrm{d}x_{\mathrm{A}2}}{x_{\mathrm{A}2}} + \frac{\alpha_{\mathrm{AB}}}{(\alpha_{\mathrm{AB}} - 1)}\int_{x_{\mathrm{A}f}}^{x_{\mathrm{A}2L}} \frac{\mathrm{d}x_{\mathrm{A}2}}{(1 - x_{\mathrm{A}2})}$$

$$= \ell n\left(\frac{x_{\mathrm{A}2L}}{x_{\mathrm{A}f}}\right)^{\frac{1}{(\alpha_{\mathrm{AB}}-1)}} + \ell n\left(\frac{1 - x_{\mathrm{A}f}}{1 - x_{\mathrm{A}2L}}\right)^{\frac{\alpha_{\mathrm{AB}}}{(\alpha_{\mathrm{AB}}-1)}};$$

$$\frac{W_{t2L}}{W_{tf}} = \frac{W_{tf} - W_{t1L}}{W_{tf}} = 1 - \theta$$

$$= \left(\frac{x_{\mathrm{A}2L}}{x_{\mathrm{A}f}}\right)^{\frac{1}{(\alpha_{\mathrm{AB}}-1)}}\left(\frac{1 - x_{\mathrm{A}f}}{1 - x_{\mathrm{A}2L}}\right)^{\frac{\alpha_{\mathrm{AB}}}{(\alpha_{\mathrm{AB}}-1)}}. \tag{7.2.12}$$

Therefore, knowing α_{AB} and $x_{\mathrm{A}f}$, we have a relation between the stage cut, θ, and the high-pressure outlet composition (concentrate/reject) $x_{\mathrm{A}2L}$. If either θ or $x_{\mathrm{A}2L}$ is specified, the other quantity is obtained from the above equation. Note that if the integration is not carried out to the permeator end, this relates $x_{\mathrm{A}2}$ at any location to the corresponding W_{t2}. It is also important to determine the value of $x_{\mathrm{A}p}$, the mole fraction of species A in the permeate stream:

$$x_{\mathrm{A}p} = \frac{\int_{x_{\mathrm{A}f}}^{x_{\mathrm{A}2L}} x_{\mathrm{A}1}\,\mathrm{d}W_{t2}}{\int_{x_{\mathrm{A}f}}^{x_{\mathrm{A}2L}} \mathrm{d}W_{t2}}. \tag{7.2.13}$$

The integral in the numerator may be rearranged by using expression (7.2.11) for $x_{\mathrm{A}1}$ in terms of $x_{\mathrm{A}2}$, expression (7.2.10) for $\mathrm{d}W_{t2}$ in terms of $x_{\mathrm{A}2}$ and expression (7.2.12) for W_{t2} in terms of $x_{\mathrm{A}2}$:

$$x_{\mathrm{A}p} = -\frac{\int_{x_{\mathrm{A}f}}^{x_{\mathrm{A}2L}} \left\{\frac{\alpha_{\mathrm{AB}}x_{\mathrm{A}2}}{1 + x_{\mathrm{A}2}(\alpha_{\mathrm{AB}} - 1)}\right\}\left\{\frac{1 + x_{\mathrm{A}2}(\alpha_{\mathrm{AB}} - 1)}{x_{\mathrm{A}2}(\alpha_{\mathrm{AB}} - 1)(1 - x_{\mathrm{A}2})}\right\} W_{t2}\,\mathrm{d}x_{\mathrm{A}2}}{(W_{tf} - W_{t2L})}$$

$$= -\frac{\int_{x_{\mathrm{A}f}}^{x_{\mathrm{A}2L}} \frac{\alpha_{\mathrm{AB}}(1-\theta)}{(\alpha_{\mathrm{AB}}-1)(1-x_{\mathrm{A}2})} W_{tf}\left(\frac{x_{\mathrm{A}2}}{x_{\mathrm{A}f}}\right)^{\frac{1}{(\alpha_{\mathrm{AB}}-1)}}\left(\frac{1-x_{\mathrm{A}f}}{1-x_{\mathrm{A}2}}\right)^{\frac{\alpha_{\mathrm{AB}}}{(\alpha_{\mathrm{AB}}-1)}} \mathrm{d}x_{\mathrm{A}2}}{\theta W_{tf}}$$

$$= -\frac{\alpha_{\mathrm{AB}}(1-\theta)}{(\alpha_{\mathrm{AB}}-1)\theta}\frac{(1-x_{\mathrm{A}f})^{\frac{\alpha_{\mathrm{AB}}}{(\alpha_{\mathrm{AB}}-1)}}}{(x_{\mathrm{A}f})^{\frac{1}{(\alpha_{\mathrm{AB}}-1)}}}\int_{x_{\mathrm{A}f}}^{x_{\mathrm{A}2L}} \frac{(x_{\mathrm{A}2})^{\frac{1}{(\alpha_{\mathrm{AB}}-1)}}}{(1-x_{\mathrm{A}2})^{\frac{2\alpha_{\mathrm{AB}}-1}{(\alpha_{\mathrm{AB}}-1)}}}\mathrm{d}x_{\mathrm{A}2}$$

$$= \frac{(1-\theta)}{\theta}\frac{1}{(x_{A2L})^{\frac{1}{(\alpha_{AB}-1)}}}\left[(1-x_{A2L})^{\left(\frac{\alpha_{AB}}{\alpha_{AB}-1}\right)}\left(\frac{x_{Af}}{1-x_{Af}}\right)^{\left(\frac{\alpha_{AB}}{\alpha_{AB}-1}\right)} - (x_{A2L})^{\frac{\alpha_{AB}}{\alpha_{AB}-1}}\right]. \tag{7.2.14}$$

This expression relates the permeate mole fraction x_{Ap} to the residue composition x_{A2L} as a function of x_{Af}, θ and α_{AB}. Since x_{Ap}, x_{A2L}, x_{Af} and θ are related via

$$x_{Af} = \theta x_{Ap} + (1-\theta)x_{A2L}, \tag{7.2.15}$$

we can rewrite expression (7.2.14) for x_{Ap} as

$$x_{Ap} = \frac{(1-\theta)}{\theta}\frac{1}{(x_{A2L})^{\frac{1}{\alpha_{AB}-1}}}\left[(1-x_{A2L})^{\frac{\alpha_{AB}}{\alpha_{AB}-1}}\left\{\frac{(1-\theta)x_{A2L}+\theta x_{Ap}}{1-[(1-\theta)x_{A2L}+\theta x_{Ap}]}\right\}^{\frac{\alpha_{AB}}{\alpha_{AB}-1}} - (x_{A2L})^{\frac{\alpha_{AB}}{\alpha_{AB}-1}}\right]. \tag{7.2.16}$$

This is an implicit expression for x_{Ap} in terms of x_{A2L}, θ and α_{AB}. The actual separation factor, α_{AB}, is to be determined from the ideal separation factor α^*_{AB}, the pressure ratio $\gamma(=P_p/P_f)$ and the exit composition for the concentrate x_{A2L} (since α_{AB} is assumed to be independent of the local variation of x_{A2} from x_{Af} to x_{A2L} at the exit) via relation (6.3.202b), where x_{A2} can be used instead of x_{Af}.

The total membrane area $A^t_m(=WL)$ required for this separation can be obtained by combining equations (7.2.5) and (7.2.8):

$$dA_m = -\frac{d(W_{t2}x_{A2})}{\left(\frac{Q_{im}}{\delta_m}\right)(P_f x_{A2} - P_p x_{A1})} = -\frac{x_{A1}\, dW_{t2}}{\left(\frac{Q_{im}}{\delta_m}\right)(P_f x_{A2} - P_p x_{A1})};$$

$$A^t_m = \int_{x_{A2L}}^{x_{Af}} \frac{x_{A1}\, dW_{t2}}{\left(\frac{Q_{im}}{\delta_m}\right)(P_f x_{A2} - P_p x_{A1})}. \tag{7.2.17}$$

Now employ equation (7.2.10) for dW_{t2}, equation (7.2.12) for W_{t2} and equation (7.2.11) expressing x_{A1} in terms of x_{A2} and α_{AB} in the above integral for A_m, and numerically evaluate it.

Pan and Habgood (1978a) have provided a solution to this crossflow permeator problem **without assuming a constant value of the selectivity α_{AB} for a binary mixture** characteristic of the Naylor–Backer approach described above. We use their notation here: species 1, base component (less permeable); x_i, mole fraction of species i in feed side; y_i, mole fraction of species i in permeate side; α_i for $i \neq 1$ is the selectivity of species i with respect to species 1, where $\alpha_1 = 1$; $\beta_i = 1-(1/\alpha_i)$; $Y_i = \beta_i\, y_i$; γ = pressure ratio (= permeate pressure/feed pressure);

$$a = (\gamma\alpha_2\beta_2+1)/(\alpha_2\beta_2(1-\gamma)); \quad b = (\gamma\alpha_2\beta_2-\alpha_2)/(\alpha_2\beta_2(1-\gamma)). \tag{7.2.18}$$

Their results are as follows:

$$1-\theta = \left(\frac{Y_2}{Y_{2f}}\right)^a\left(\frac{\beta_2-Y_2}{\beta_2-Y_{2f}}\right)^b\left(\frac{1-Y_2}{1-Y_{2f}}\right), \tag{7.2.19}$$

where Y_{2f} is the value of Y_2 at feed location;

$$Y_2 = \frac{\left\{1+(\alpha_2-1)(\gamma+x_2)-\sqrt{[1+(\alpha_2-1)(\gamma+x_2)]^2-4\gamma\alpha_2(\alpha_2-1)x_2}\right\}}{2\gamma\alpha_2}; \tag{7.2.20}$$

$$S = (1-Y_2)\theta + \int_{Y_{2f}}^{Y_2}\theta\, dY_2; \tag{7.2.21}$$

$$S = (Q_1/d)(P/L_f)(1-\gamma)s, \tag{7.2.22}$$

where S is the nondimensional membrane area, (Q_1/d) is the permeance (= permeability coefficient/membrane thickness) of species 1, P is the feed side pressure, L_f is the molar feed gas flow rate, s is the actual membrane area (= WL in our notation). For the special case of $\alpha_2>>1$ (i.e. $\beta_2 \cong 1$, $b \cong -1$). These results are reduced to

$$\theta = 1 - (Y_2/Y_{2f})^a; \tag{7.2.23}$$

$$S = 1 - \frac{a}{a+1}Y_{2f} - \left(\frac{Y_2}{Y_{2f}}\right)^a\left(1 - \frac{a}{a+1}Y_2\right). \tag{7.2.24}$$

Example 7.2.1 Consider the production of nitrogen-enriched air from atmospheric air using an asymmetric cellulose acetate membrane having an ideal selectivity of $(Q_{O_2m}/Q_{N_2m}) = 6.0$. If the permeate side is maintained at a considerable vacuum, and if you can assume crossflow and no permeate side pressure drop in the module (spiral-wound or hollow fiber), then determine the stage cut and the permeate composition for the following values of oxygen mole fraction in the concentrate: $x_{A2L} = 0.1, 0.05, 0.02, 0.01$, where species A is oxygen. Plot the oxygen mole fraction in the permeate and concentrate as a function of the stage cut; include values for $\theta = 0$ and 1.

Solution There is no specification of the permeate side pressure, except we have considerable vacuum. Since the pressure ratio γ = (permeate pressure/feed pressure) is very low, we may assume that $\alpha_{AB} \cong \alpha^*_{AB}$; therefore, $\alpha_{AB} = \alpha^*_{O_2-N_2} = 6.0$. Since α_{AB} is constant throughout the module (due to negligible pressure drop along the permeate side and very low γ), we can use the result (7.2.12) based on the Naylor–Backer method to calculate θ from

$$1-\theta = (x_{A2L}/x_{Af})^{\frac{1}{(\alpha^*_{AB}-1)}}((1-x_{Af})/(1-x_{A2L}))^{\frac{\alpha^*_{AB}}{(\alpha^*_{AB}-1)}}$$

given $x_{Af} = 0.21$, $\alpha^*_{AB} = 6$ and x_{A2L} (different values; see Table 7.2.1).

Table 7.2.1.

x_{A2L}	$(x_{A2L}/x_{Af})^{\frac{1}{(\alpha^*_{AB}-1)}}((1-x_{Af})/(1-x_{A2L}))^{\frac{\alpha^*_{AB}}{(\alpha^*_{AB}-1)}}$	θ	x_{Ap}
0.1	$(0.1/0.21)^{0.2}(0.79/0.9)^{1.2} = 0.73$	0.27	0.507
0.05	$(0.05/0.21)^{0.2}(0.79/0.95)^{1.2} = 0.600$	0.40	0.45
0.02	$(0.02/0.21)^{0.2}(0.79/0.98)^{1.2} = 0.481$	0.519	0.386
0.01	$(0.01/0/21)^{0.2}(0.79/0.99)^{1.2} = 0.3297$	0.6703	0.308
0.21	1	0	0.61[a]

x_{Ap} is determined from (7.2.15): $x_{Af} = \theta\, x_{Ap} + (1-\theta)\, x_{A2L}$.
[a] In this case, x_{Ap} has to be determined from the relations (6.3.199) for $\gamma \cong 0$ and (6.3.201) for the nonzero γ.

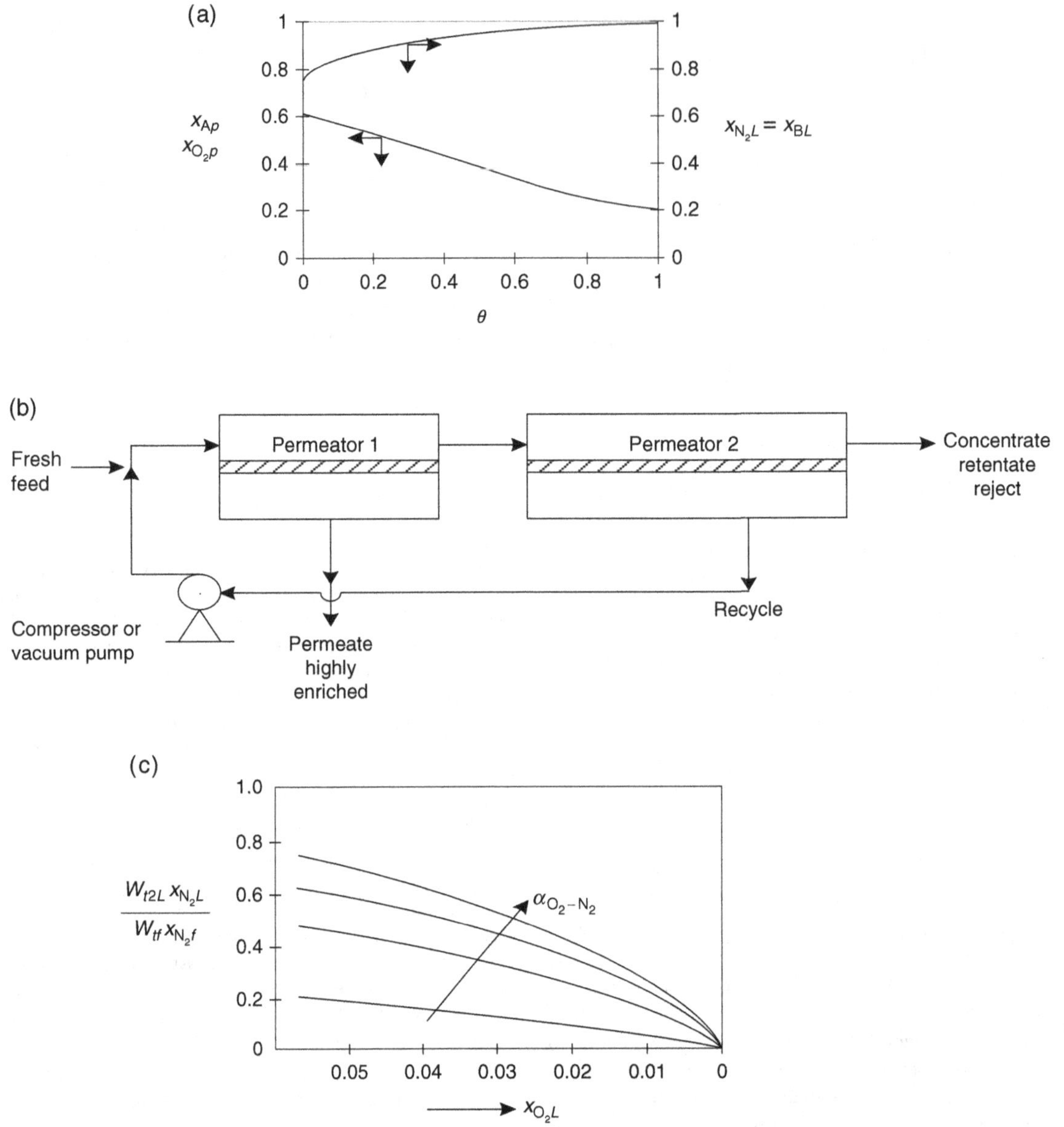

Figure 7.2.2. (a) Oxygen mole fraction in permeate vs. stage cut for a membrane having $\alpha_{O_2\text{-}N_2} = 6.0$. (b) Two permeators in series operation with recycle of permeate from the second permeator to the feed. (c) Fractional nitrogen recovery vs. residual oxygen mole fraction in NEA for different membrane selectivities. (part(c) after Baker (2004).)

Since $\gamma \cong 0$ and $\alpha_{AB} \cong \alpha^*_{AB}$ here,

$$x_{Ap} = \frac{\alpha^*_{AB}\, x_{Af}}{1 + x_{Af}(\alpha^*_{AB} - 1)} = \frac{6 \times 0.21}{1 + 0.21 \times 5} = \frac{1.26}{2.05} = 0.61,$$

the maximum value achievable anywhere in the permeator. The plot of x_{Ap} vs. the stage cut, θ, is illustrated in Figure 7.2.2(a).

For $\theta = 1$, $x_{Ap} = x_{Af} = 0.21$, since all of the feed gas appears as the permeate. If a value of permeate side pressure was provided, then γ would be known; α_{AB} may then be determined from (6.3.202b) knowing α^*_{AB}, γ and x_{Af}.

What is clear from this example is that, as the value of the stage cut, θ, increases, the permeate mole fraction of O_2 in the oxygen-enriched air (OEA) decreases and the mole fraction of N_2 in the retentate (the nitrogen-enriched air

(NEA)) increases. However, the loss of nitrogen in the permeate is also increased. If the goal is to obtain oxygen-enriched air (OEA) as well as nitrogen-enriched air (NEA), the behavior of x_{Ap} vs. θ shown in Figure 7.2.2(a) provides guidance. One can take the permeate up to a stage cut of, say, 0.1–0.2; the corresponding permeate oxygen mole fraction will be around 0.55. On the other hand, the concentrate mole fraction of nitrogen for a stage cut of 0.6703 will be 0.99, a reasonable quality nitrogen-enriched air. These two product streams are shown in Figure 7.2.2(b). However, in order to show what to do with the *rest of the permeate*, Figure 7.2.2(b) shows that the gas permeator has been split into two permeators: the first permeator produces OEA as the permeate; the concentrate from this permeator goes to a second permeator having a high stage cut and producing the NEA as the concentrate. The composition of the permeate from this second permeator is of interest.

Let us make a rough calculation concerning the composition of this second permeate, identified as $x_{Ap|_2}$(the first one being $x_{Ap|_1}$) for x_{A2L} leaving the permeators being 0.01 (99% N_2). For 1 mole of feed air (21% O_2) entering the system per unit time, and $W_{t1|_1}$ and $W_{t1|_2}$ being the permeate flow rates from the two permeators,

$$1 \times 0.21 = W_{t1}\Big|_1 \times x_{Ap}\Big|_1 + W_{t1}\Big|_2 \times x_{Ap}\Big|_2 + W_{t2}\, x_{A2L}.$$

Let the overall stage cut $= 0.67 \rightarrow W_{t2} = 0.33$; $W_{t1|_1} + W_{t1|_2} = 0.67$. Let $W_{t1|_1} = 0.1$, $x_{Ap|_1} = 0.55$, $x_{A2L} = 0.01$. Then

$$0.21 = 0.1 \times 0.55 + 0.57 x_{Ap|_2} + 0.33 \times 0.01 \rightarrow x_{Ap|_2} = 0.266,$$

a somewhat O_2-enriched permeate. This stream, if recycled to the fresh feed air, will increase the O_2 concentration of the net feed entering the permeator 1. Therefore the oxygen concentration in the permeate from permeator 1 will be even higher (as discussed in Section 2.2.2). On the other hand, if we operate permeator 2 as a new permeator with a N_2-enriched feed stream coming as the reject from permeator 1, then we may have a different situation if we operate it at a high θ. The oxygen concentration in this feed may be around 0.15. If we operate the second permeator with a very high stage cut (say $\theta = 0.67$), it is entirely likely that $x_{Ap|_2} < 0.21$. Such a stream, if recycled to the fresh feed air, will increase N_2 concentration in the feed gas. This will lead to a better NEA from permeator 2, but a poorer OEA from permeator 1.

If the production of NEA is the goal, then the variation of the production rate as a function of nitrogen concentration is also of interest. The higher the production rate of this NEA stream, the higher will be the fractional recovery of N_2 from the feed air in this product stream. However, as the purity of N_2 increases in this product, the fractional N_2 recovery $(W_{t2L} x_{N_2 2L} / W_{tf} x_{N_2 f})$, will decrease. Figure 7.2.2 (c) illustrates this behavior qualitatively for membranes having different O_2-N_2 selectivities (varying between 2 and 8) (after Baker (2004)). Obviously, membranes having higher oxygen selectivity will lead to a lower loss of N_2 in the permeate and therefore a higher fractional N_2 recovery in the NEA. For a more detailed introduction to different membranes used in O_2/N_2 separation, see Zolandz and Fleming (2001) and Baker (2004).

Other gas separation systems of interest are as follows.

(1) Separation of H_2 from a variety of gas streams: H_2 from N_2 in purge streams in ammonia plants; recovery of H_2 from various streams in petroleum refineries and petrochemical plants; H_2/CO ratio adjustment in synthesis gas for oxo-alcohol plants. The membrane materials used are polysulfone, cellulose acetate, polyimides and polyaramides. Selectivities of H_2 over N_2, CH_4, CO in these applications vary between 30 and 200.
(2) Separation of CO_2 from hydrocarbons: CO_2 separated from light hydrocarbons in natural gas streams at high pressures; CO_2 separated from biogas (containing CO_2 and methane); CO_2 separated from breathing gas mixture.
(3) Dehydration of air.
(4) Recovery of C_{3+} condensable hydrocarbons from natural gas using rubbery polymeric membranes.
(5) Recovery of helium from natural gas.

A quantity of considerable practical utility is the membrane area required to treat a gas stream of given flow rate to achieve a certain concentrate mole fraction or permeate mole fraction under given conditions; these conditions involve specification of the value of (Q_{im}/δ_m) for one of the species, namely species *i*, α^*_{AB} for a binary mixture of gases A and B, feed pressure and permeate pressure, alternatively, one of the stream pressures and a pressure ratio. In general, there is a pressure drop in the feed side and the permeate side. However, the analysis provided so far in this section is based on zero pressure drops on either side. Quite often, such an assumption does provide reasonable estimates unless the pressure drop is such that the γ value changes considerably along the permeator length.

Example 7.2.2 We wish to estimate the membrane area required to produce nitrogen-enriched air having $x_{N_2 2L} = 0.95$ using a silicone rubber-coated hollow fiber membrane, with $(Q_{N_2 m}/\delta_m)$ at 25 °C $= 4.6 \times 10^{-10}$ gmol/s-cm^2-cm Hg and $\alpha^*_{O_2-N_2} = 2.1$. The feed air is introduced at 10 bar; the permeate is at 1 atm. In the microporous hollow fiber membrane with a silicone coating on the fiber internal diameter, the feed flows through the fiber bore. The permeate is withdrawn in crossflow on the shell side all along the module length (Figure 7.2.1(c)). Neglect any pressure drop along the module length. The feed air flow rate at 25 °C is 900 kgmol/hr.

Solution Since there is no pressure drop in either stream along the module length, and we have pure crossflow, we can use the Pan and Habgood (1978a) analysis for non-constant α_{AB}, i.e. α_{AB} can change with composition.

Table 7.2.2.

$x_2(=x_{O_2 2})$	Y_2	θ
0.21	0.178	0
0.18	0.156	0.219
0.14	0.124	0.47
0.10	0.092	0.66
0.05	0.0476	0.858

The equations of relevance are (7.2.18) for α_{O_2}, β_{O_2}, a and b, (7.2.19) for θ, (7.2.20) for $Y_{O_2 f}$ and Y_{O_2}, (7.2.22) for the definition of nondimensional membrane area S and (7.2.21) for the calculation of this S to obtain the actual membrane area s. In this approach

$$\alpha_{N_2}=1; \qquad \alpha_{O_2}=2.1;$$

$$\beta_{O_2}=1-\left(\frac{1}{\alpha_{O_2}}\right)=0.524; \qquad \gamma=0.1;$$

$$a=\left[(\gamma\alpha_{O_2}\beta_{O_2}+1)/(\alpha_{O_2}\beta_{O_2}(1-\gamma)\right]$$

$$=\frac{0.1\times 2.1\times 0.524+1}{2.1\times 0.524\times 0.9}=1.12;$$

$$b=\left[(\gamma\alpha_{O_2}\beta_{O_2}-\alpha_{O_2})/(\alpha_{O_2}\beta_{O_2}(1-\gamma)\right]$$

$$=\frac{0.1\times 2.1\times 0.524-2.1}{2.1\times 0.524\times 0.9}=-2.01.$$

To determine S from equation (7.2.21), we need to have the value of θ for a few values of Y_2 and integrate. Determine $Y_{2f}=0.178$ (from (7.2.20)) for $x_2=x_{2f}=x_{O_2 f}=0.21$. Now determine Y_2 for different values of $x_2(=x_{O_2 2})$ from (7.2.20) and then determine the value of the stage cut θ from (7.2.19). The results of these calculations are summarized in Table 7.2.2.

The value of the integral $\int_{Y_{2f}}^{Y_2}\theta\,dY_2$ in (7.2.21) is 0.0698. Therefore

$$S=(1-Y_2)\theta+0.0698=(1-0.0476)0.858+0.0698=0.887.$$

So,

$$S=\left(\frac{Q_{N_2 m}}{d}\right)\left(\frac{P}{L_f}\right)(1-\gamma)s$$

$$=\frac{4.6\times 10^{-10}\text{gmol}}{\text{s-cm}^2\text{-cm Hg}}\;\frac{10\,\text{atm}\times 76\,\frac{\text{cm Hg}}{\text{atm}}\times 3600\,\frac{\text{s}}{\text{hr}}}{900\,\frac{\text{kgmol}}{\text{hr}}\times\frac{1000\text{gmol}}{\text{kgmol}}}\times 0.9\times s;$$

$$s=\frac{0.887\times 900\times 1000\times 10^{10}}{4.6\times 10\times 76\times 3.6\times 10^3\times 0.9}\text{cm}^2\,\frac{1}{10^4}\,\frac{\text{m}^2}{\text{cm}^2}$$

$$=7.05\times 10^4\,\text{m}^2.$$

7.2.1.2 *Reverse osmosis*

As we have seen in Section 6.3.3.3, the dead-end mode of operation of reverse osmosis creates major problems: steady state is not possible; it is difficult to generate a high value of the solute mass-transfer coefficient $k_{i\ell}$; further, as in Section 6.3.3.5 for gas permeation, it is often easier to pack a lot of membrane surface area when the bulk liquid flow is not parallel to the direction of the force. The configuration studied here is crossflow (see Figure 7.0.1(f)): feed solution (e.g. brine in desalination) flows along the membrane module length parallel to the membrane length direction while the permeate flows perpendicular to the feed solution flow direction, i.e. in the direction of the driving force. In many devices, the permeated liquid collected along the membrane length flows in a particular direction. This second bulk flow (of the permeate) may have a significant effect on separation; this case of two bulk flows, where both are perpendicular to the direction of the force across the membrane, is studied in general in Chapter 8. Here we study those cases where the second bulk flow is of no consequence in so far as separation is concerned. Thus crossflow as in Figure 7.0.1(f) is sufficient for analysis.

The physical configurations studied are: (1) spiral-wound module; (2) tubular module. Let us consider first a highly simplified, but reasonably useful, model for a spiral-wound reverse osmosis module. The basic structure of a single-leaf spiral-wound membrane module is shown in Figure 7.2.1(e). An actual module consists of a number of leaves. A single-leaf module as shown consists of a rectangular membrane packet (sandwich) spirally wound around a central collection pipe. Each membrane packet (sandwich) is separated from the membrane in the next and nearest membrane packet by a spacer screen. There are two membranes in each membrane packet, with a product water side backing material in between (Figure 7.2.3(a)). Three ends of this packet are glued and sealed, and the fourth end empties into the collection pipe through a glued joint. As brine (or any other aqueous feed) flows along the module length over the spacers in between the membrane packets, permeation occurs and permeate flows along the spirally wound permeate channel in the membrane packet through the product water side backing toward the central collection pipe (see Figure 7.2.1 (e)). By having a number of leaves in parallel, the length of the membrane permeate side is reduced; correspondingly, the permeate pressure drop is reduced.

If we open a spirally wound membrane channel and lay it flat, then, as shown in Figure 7.2.3(b), brine flows at a high pressure in the mean flow direction (the residue flow direction in Figure 7.2.1(e)), designated here as the z-coordinate; the membrane length in this direction is L. The permeate flows inside the membrane packet in the x-direction, the so-called permeate channel toward the central collection tube; the width of the membrane packet is W (in the x-direction).

Our basic assumptions in the following analysis are as follows.

(1) The pressure difference ΔP changes very little along the z-direction from the value ΔP_f at the feed entrance location ($z=0$).
(2) The permeate channel pressure drop along the x-direction is negligible.

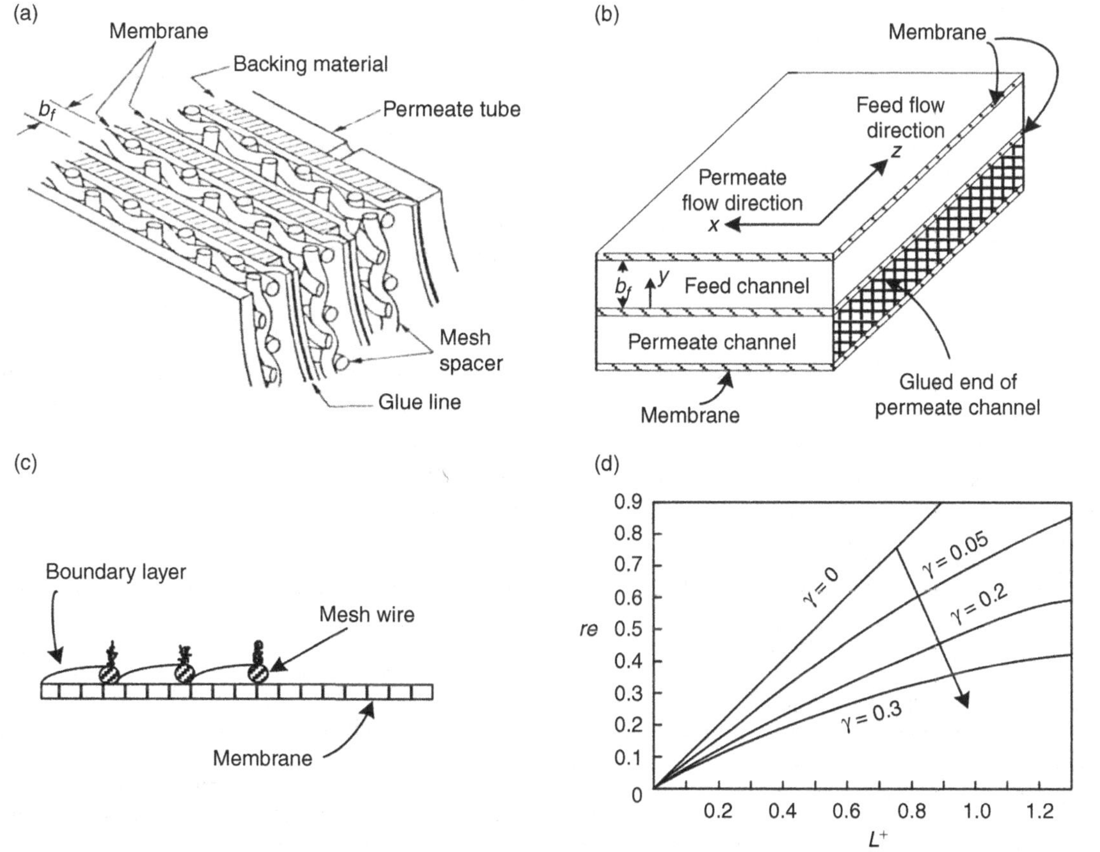

Figure 7.2.3. (a) Details of a spiral-wound module. (After Ohya and Taniguchi (1975).) (b) Channels and flow configurations in an unwrapped spiral-wound module. (c) How mesh wires in the brine side spacer screen disturb the brine flow and create new boundary layers (schematically). (d) Fractional water recovery re *as a function of normalized channel length* L^+ *for various parametric values of* γ, *indicating the feed osmotic pressure level with respect to* ΔP_f.

(3) The membrane has a very high solute rejection (salt rejection $\sim$ 99%+, i.e. $R_{salt} \cong 0.99 \rightarrow 1.0$).
(4) The solute mass-transfer coefficient $k_{i\ell}$ in the membrane feed channel has a high value and can be considered to be essentially constant along the z-coordinate.

The species conservation equation for solute i (here, salt) in the feed flow channnel may be written using the species conservation equation (6.2.5b) for steady state, no external force based species velocities ($\boldsymbol{U}_i = 0$) and no chemical reactions as

$$\nabla \cdot (C_{is}\, \boldsymbol{v}^*) = -\nabla \cdot (\boldsymbol{J}_i^*) = \nabla \cdot (D_{is} \nabla C_{is}). \qquad (7.2.25)$$

The flow conditions in the thin feed brine channel with feed brine spacers in between is complex (see Schwinge *et al.* (2002, 2003) for the unsteady flow, vortex shedding and mass-transfer enhancement in spacer-filled channels). The complexity is captured by using an empirically derived mass-transfer coefficient $k_{i\ell}$ reflective of the flow conditions (Schock and Miquel, 1987). One could represent the flow in a simplistic way as a chopped laminar flow (Solan *et al.*, 1971) with a boundary layer beginning at each cross wire in the spacer (illustrated schematically in Figure 7.2.3(c)). However, the mass-transfer coefficient $k_{i\ell}$ in the very thin boundary layer is high even if the Reynolds number is quite low (essentially laminar flow at this low Reynolds number, even though there is a lot of mixing). We will represent this velocity field as two-dimensional, v_y (normal to the membrane) and v_z (along the membrane length, z-direction). Equation (7.2.25) may be represented in this context in the (y, z)-coordinate system as

$$\frac{\partial (C_{is}\, v_y^*)}{\partial y} + \frac{\partial (C_{is}\, v_z^*)}{\partial z} = \frac{\partial}{\partial y}\left(D_{is}\frac{\partial C_{is}}{\partial y}\right) + \frac{\partial}{\partial z}\left(D_{is}\frac{\partial C_{is}}{\partial z}\right). \qquad (7.2.26)$$

Since the axial concentration gradient of the solute, C_{is}, is very small,[12] the axial diffusion term may be neglected, leading to

$$\frac{\partial(C_{is}\,v_y^*)}{\partial y}+\frac{\partial(C_{is}\,v_z^*)}{\partial z}=\frac{\partial}{\partial y}\left(D_{is}\frac{\partial C_{is}}{\partial y}\right); \tag{7.2.27}$$

$$\frac{\partial(C_{is}\,v_z^*)}{\partial z}=-\frac{\partial}{\partial y}\left(+C_{is}\,v_y^*-D_{is}\frac{\partial C_{is}}{\partial y}\right). \tag{7.2.28}$$

Integrating this equation along the membrane feed channel height b_f (as if it is a flow between two flat plates with a gap b_f in between), we get

$$\frac{\mathrm{d}}{\mathrm{d}z}\int_0^{b_f} C_{is}\,v_z^*\,\mathrm{d}y=-\int_0^{b_f}\mathrm{d}\left(C_{is}\,v_y^*-D_{is}\frac{\partial C_{is}}{\partial y}\right). \tag{7.2.29}$$

For dilute solutions, $v_z^*\cong v_z$; therefore, for one membrane channel of width W, height b_f and a brine flow rate of Q, corresponding to an averaged axial brine velocity of $v_{z,\mathrm{avg}}$, we get

$$\frac{\mathrm{d}}{\mathrm{d}z}(v_{z,\mathrm{avg}}\,C_{isb})=-\frac{1}{b_f}\left[\left(C_{is}\,v_y-D_{is}\frac{\mathrm{d}C_{is}}{\mathrm{d}y}\right)_{b_f}-\left(C_{is}\,v_y+D_{is}\frac{\mathrm{d}C_{is}}{\mathrm{d}y}\right)_0\right], \tag{7.2.30}$$

where C_{isb} is the bulk solution concentration. Note that if we write the magnitude of the solute flux through either of the membranes on two sides of the feed brine channel as $|N_{iy}|$, then the above expression is reduced to

$$\frac{\mathrm{d}}{\mathrm{d}z}(v_{z,\mathrm{avg}}\,C_{isb})=-\frac{1}{b_f}\left[|N_{iy}|+|N_{iy}|\right]=-\frac{2}{b_f}|N_{iy}|. \tag{7.2.31}$$

An exactly similar analysis may be carried out for the solvent (water), yielding

$$\frac{\mathrm{d}}{\mathrm{d}z}(v_{z,\mathrm{avg}}\,C_{wsb})=-\frac{2}{b_f}|N_{wy}|, \tag{7.2.32}$$

where $i=w$. By assumption (3), the solute flux $|N_{iy}|$ through the membrane may be neglected for the time being:

$$\frac{\mathrm{d}}{\mathrm{d}z}(v_{z,\mathrm{avg}}\,C_{isb})\cong 0\Rightarrow v_{z,\mathrm{avg}}|_f\,C_{isb}|_f=v_{z,\mathrm{avg}}\,C_{isb} \tag{7.2.33}$$

at any axial location (any z). Now, if we are dealing with a very dilute solution, $C_{wsb}(z)\sim$ constant $=(1/\overline{V}_w)$. Combine now equations (7.2.31) and (7.2.33):

$$\frac{\mathrm{d}}{\mathrm{d}z}\left(\frac{v_{z,\mathrm{avg}}|_f\,C_{isb}|_f}{C_{isb}\,\overline{V}_w}\right)=-\frac{2}{b_f}|N_{wy}|;$$

$$\frac{v_{z,\mathrm{avg}}|_f\,C_{isb}|_f}{(C_{isb})^2}\frac{\mathrm{d}C_{isb}}{\mathrm{d}z}=+\frac{2}{b_f}|N_{wy}|\,\overline{V}_w. \tag{7.2.34}$$

[12] The permeation rate of water through the membrane is quite low. Therefore the axial concentration change, $(\partial C_{is}/\partial z)$, is small. No wonder we need a lot of membrane surface area!

If we use the solution-diffusion model to describe the solvent flux, then, from equations (3.4.60c–f), $N_{sy}=N_{wy}=J_{wy}^*$:

$$|N_{wy}|=|J_{wy}^*|=A(\Delta P-\Delta\pi)=A(\Delta P-bC_{il}^0+bC_{ip}), \tag{7.2.35}$$

where (see relation (3.4.64a)) C_{il}^0 is the wall concentration of the solute and C_{ip} is the solute concentration in the permeate. By assumption (3), we can neglect C_{ip} here. Further, using the film model analysis and assumption (4) with $|N_{wy}|\,\overline{V}_w/k_{il}<<1$, we have already obtained the following result (see (3.4.66a)):

$$|J_{wy}^*|=\frac{A(\Delta P-b\,C_{isb})}{\left(1+\frac{A\,b\,C_{isb}\,\overline{V}_w}{k_{il}}\right)}=|N_{wy}|. \tag{7.2.36}$$

Remember this C_{is} is really $C_{isb}(z)$. We will next integrate equation (7.2.34) along the z-coordinate, the membrane length direction:

$$\frac{2}{b_f}\int_0^L\frac{\mathrm{d}z}{v_{z,\mathrm{avg}}|_f}=\frac{2L}{b_f\,v_{z,\mathrm{avg}}|_f}=C_{isb}|_f\int_{C_{isb}|_f}^{C_{isbL}}\frac{\mathrm{d}C_{isb}}{|N_{wy}|\,\overline{V}_w(C_{isb})^2}. \tag{7.2.37}$$

The development of this equation did not involve any assumption, other than an extremely low solute flux through the membrane (assumption (3)). Therefore, any general expression for $|N_{wy}|=|J_{wy}^*|$, the membrane solvent flux (which is the water flux, for example, in desalination), as a function of $C_{isb}(z)$ may be used here with the limitation of assumption (3), namely that $C_{ip}\cong 0$. An option is to use the simplified water flux expression (7.2.36) after defining a nondimensional membrane length L^+ as follows:

$$L^+=\frac{2v_{ywf}L}{v_{z,\mathrm{avg}}|_f\,b_f}=v_{ywf}\,C_{isb}|_f\int_{C_{isb}|_f}^{C_{isbL}}\frac{\left[1+\frac{AbC_{isb}(z)\overline{V}_w}{k_{i\ell}}\right]\mathrm{d}C_{isb}}{A_f(\Delta P_f-bC_{isb}(z))(C_{isb}(z))^2\overline{V}_w}, \tag{7.2.38}$$

where

$$v_{ywf}=A_f\,\Delta P_f\,\overline{V}_w \tag{7.2.39}$$

is the volume flux or permeation velocity through the membrane at the feed entrance location; further, by assumption (2), the transmembrane pressure drop ΔP everywhere is assumed to be constant at the value ΔP_f corresponding to the feed entrance location. In reality, ΔP as well as A will decrease somewhat along the membrane length; so will $k_{i\ell}$. The result of integration provides the following analytical expression for L^+ if we rewrite expression (7.2.38) as

$$L^+=\Delta P_f\,C_{isb}|_f\int_{C_{isb}|_f}^{C_{isbL}}\frac{\left[1+\frac{A\,b\,C_{isb}(z)\,\overline{V}_w}{k_{i\ell}}\right]\mathrm{d}C_{isb}}{(\Delta P_f-bC_{isb}(z))(C_{isb}(z))^2}, \tag{7.2.40}$$

leading to (Sirkar *et al.*, 1982)

$$L^+ = \left(1 - \frac{C_{isb}|_f}{C_{isbL}}\right) + \left(\frac{b\,C_{isb}|_f}{\Delta P_f} + \frac{A_f b\, C_{isb}|_f \overline{V}_w}{k_{il}}\right) \times \ln\left[\frac{C_{isbL}}{C_{isb}|_f}\frac{(\Delta P_f - b\,C_{isb}|_f)}{(\Delta P_f - b\,C_{isbL})}\right]. \tag{7.2.41}$$

If we define the fractional water recovery as *re* (definition (6.3.172b)), then, from the solute balance relation (7.2.33), valid for a membrane having a very high solute rejection,

$$re = 1 - \frac{v_{z,\text{avg}}|_L}{v_{z,\text{avg}}|_f} = 1 - \frac{C_{isb}|_f}{C_{isbL}}. \tag{7.2.42}$$

Correspondingly,

$$L^+ = re + \left(\frac{bC_{isb}|_f}{\Delta P_f} + \frac{A_f bC_{isb}|_f \overline{V}_w}{k_{il}}\right) \times \ln\left[\frac{1}{1-re}\frac{\left(1-\frac{bC_{isb}|_f}{\Delta P_f}\right)}{\left(1-\frac{bC_{isbL}}{\Delta P_f}\right)}\right]. \tag{7.2.43}$$

For feed solutions whose osmotic pressure is negligibly small, i.e. $bC_{isb}|_f << \Delta P_f$, this result is reduced to (Ohya and Taniguchi, 1975)

$$L^+ \cong re; \qquad L = (re)\left(\frac{v_{z,\text{avg}}|_f b_f}{2\,A_f\,\Delta P_f\,\overline{V}_w}\right). \tag{7.2.44}$$

The quantity $(b_f/2)$ needs to be elaborated on. Although b_f is the gap between the membranes in the feed channel, one can also define $(2/b_f)$ as the total active membrane surface area per unit volume of the membrane feed channel:

$$2/b_f = (2WL/WLb_f). \tag{7.2.45}$$

An additional feature of a spiral-wound module is of interest; i.e. the number of feed brine channels in a module. There can be as many as four feed brine channels around one central collection tube; two membrane packets next to each other and separated by a feed side spacer create one membrane channel. The membrane surface area per unit device volume achieved in spiral-wound revere osmosis devices is around 800 m^2/m^3 (245 ft^2/ft^3) (Bhattacharyya *et al.*, 1992). This reference provides additional references and results for more detailed spiral-wound module design models developed by Prasad and Sirkar (1985) and Evangelista and Jonsson (1988). The mass-transfer coefficient $k_{i\ell}$ in the membrane channel may be determined from the Schock and Miquel (1987) correlation (3.1.170), provided in Table 3.1.8.

Our analysis has so far provided only an estimate of the solvent recovery, *re*. Since the membrane is not perfect, it is also necessary to determine the solute content (i.e. the salt concentration in desalination) of the permeate produced. Earlier we employed the solution-diffusion model (e.g. equation (7.2.35)). The solute flux in the same model is given by equation (3.4.64b):

$$J_{iy}^* = \frac{D_{im}\,\kappa_{im}}{\delta_m}\,(C_{i\ell}^0 - C_{ip}). \tag{7.2.46}$$

In the current problem, by assumption (3), $C_{ip} << C_{il}^0$. Further, the approximation leading to equation (7.2.36) will also lead to (see equation (3.4.65d))

$$C_{il}^0(z) \cong C_{isb}(z)\left(1 + \frac{|N_{wy}|\overline{V}_w}{k_{i\ell}}\right). \tag{7.2.47}$$

Combining equation (7.2.36) with the above equation leads to

$$J_{iy}^* = \frac{D_{im}\,\kappa_{im}}{\delta_m}\left[\frac{\Delta P}{b} - \frac{|N_{wy}|}{Ab}\right]. \tag{7.2.48}$$

The total molar solute permeation rate through the two membranes lining one feed brine channel may now be employed to determine $C_{ip}|_{\text{avg}}$, the average concentration of solute *i* in the permeate whose volumetric production rate through the two membranes is Q_{permeate}:

$$Q_{\text{permeate}}\,C_{ip}|_{\text{avg}} = \int_0^L J_{iy}^*\,2W\,dz. \tag{7.2.49}$$

Introduce in the right-hand side integral expression (7.2.48) for J_{iy}^* and the expression for dz in terms of dC_{isb} from relation (7.2.34) to obtain

$$C_{ip}|_{\text{avg}} = \frac{Wb_f\,v_{z,\text{avg}}|_f\,C_{isb}|_f}{Q_{\text{permeate}}}\left(\frac{D_{im}\,\kappa_{im}}{\delta_m}\right)\int_{C_{isb}|_f}^{C_{isbL}}\left[\frac{\Delta P}{b} - \frac{|N_{wy}|}{Ab}\right]\frac{dC_{isb}}{|N_{wy}|\overline{V}_w(C_{isb})^2}. \tag{7.2.50}$$

As before, along the module length ΔP, A and k_{il} are assumed to be constants. On integration, we get (Sirkar *et al.*, 1982)

$$C_{ip}|_{\text{avg}} = \frac{Wb_f\,v_{z,\text{avg}}|_f\,C_{isb}|_f}{Q_{\text{permeate}}}\left(\frac{D_{im}\,\kappa_{im}}{\delta_m}\right)\left[\frac{1}{Ab}\int_{C_{isb}|_f}^{C_{isbL}}\frac{A\,\Delta P C_{isb}|_f}{|N_{wy}|\overline{V}_w(C_{isb})^2} - \frac{1}{Ab\overline{V}_w}\left\{1 - \frac{C_{isb}|_f}{C_{isbL}}\right\}\right]. \tag{7.2.51}$$

To be noted here is that

$$re = \frac{Q_{\text{permeate}}}{v_{z,\text{avg}}|_f\,Wb_f}. \tag{7.2.52}$$

Recognizing the result (7.2.42) for high rejection membranes as well as the expression for L^+ from (7.2.37)–(7.2.39), we obtain the following result:

$$C_{ip}|_{\text{avg}} = \frac{(D_{im}\,\kappa_{im}/\delta_m)}{re\,A\,b\,\overline{V}_w}\,(L^+ - re), \tag{7.2.53}$$

where no assumption has been made regarding $|N_{wy}|$ except a very low solute permeation rate (leading to the solute mass balance result (7.2.33)).

We will briefly illustrate graphically the productivity trend in a spiral-wound module. Figure 7.2.3(d) shows a plot of the behavioral trend exhibited by equation (7.2.43). The objective is to predict the fractional water recovery *re* as a function of the nondimensional membrane channel length L^+ for various parametric values of $(b\,C_{isb}|_f / \Delta P_f) = \gamma$ for a given value of the ratio $(k_{il}/A_f\,\Delta P_f\,\overline{V}_w)$. Figure 7.2.3(d) does not provide exact values corresponding to equation (7.2.43), but it provides the trends as illustrated by Ohya and Taniguchi (1975).

Example 7.2.3 Brackish water containing NaCl at the level of 2600 ppm is passed at 34 atm (gauge) through a spiral-wound module. The gap between the membranes lining the brine feed channel is 1.1 mm. The membrane length in the brine flow direction is 70 cm. The length of the unwrapped membrane in the permeate flow direction is 150 cm. The pure water permeability constant for the membrane exposed to the feed brine at 20 °C is 66.15×10^{-6} gmol/cm^2-min-atm. The osmotic pressure of the brine solution is provided via $\pi_f = 0.0115\,(\text{ppm})_f$, where π_f is in psi and the salt concentration is in ppm. The average brine velocity in the brine channel is 320 cm/min. Estimate the fractional water recovery by assuming a highly rejecting membrane. You are given that $k_{il} = 0.51$ cm/s.

Solution We will employ equation (7.2.43) under the assumptions of a highly rejecting membrane, constant ΔP, k_{il}, A, corresponding to feed location and $(|N_{wy}|\overline{V}_w/k_{il}) << 1$:

$$L^+ = re + \left(\frac{b\,C_{isb}|_f}{\Delta P_f} + \frac{b\,C_{isb}|_f}{\Delta P_f}(A_f\,\Delta P_f \overline{V}_w)\frac{1}{k_{il}}\right) \times \ln\left[\frac{1}{1-re}\,\frac{\left(1-\frac{b\,C_{isb}|_f}{\Delta P_f}\right)}{\left(1-\frac{b\,C_{isb}|_f}{\Delta P_f}\left(\frac{1}{1-re}\right)\right)}\right].$$

The only unknown here is the fractional water recovery *re*, which has to be determined by trial and error. Now

$$v_{ywf} = A_f \Delta P_f \overline{V}_w$$
$$= 66.15 \times 10^{-6}\,\frac{\text{gmol}}{\text{cm}^2\text{-min-atm}}\,34.0\,\text{atm}\,\frac{18.05\,\text{cm}^3}{\text{gmol}}$$
$$= 66.15 \times 10^{-6} \times 34.0 \times 18.05\,\frac{\text{cm}}{\text{min}} = 4.06 \times 10^{-2}\,\frac{\text{cm}}{\text{min}};$$

$\pi_f =$ osmotic pressure of feed solution $= 0.0115 \times 2600$ psi.

But

$$\pi_f = b\,C_{isb}|_f \Rightarrow \frac{b\,C_{isb}|_f}{\Delta P_f} = \frac{1.15 \times 26}{34 \times 14.7} = 0.0598.$$

$$L^+ = \frac{2L v_{ywf}}{v_{z,\text{avg}}|_f b_f} = \frac{2 \times 70\,\text{cm} \times 4.06 \times 10^{-2}(\text{cm/min})}{320\,(\text{cm/min}) \times 0.11\,\text{cm}} = 0.16.$$

Substitute these values into the expression for L^+ above:

$$0.16 = re + \left(0.0598 + \frac{0.0598 \times 4.06 \times 10^{-2}}{0.51}\right) \times \ln\left(\frac{1}{1-re}\,\frac{(1-0.0598)}{(1-\frac{0.0598}{1-re})}\right).$$

Guess 1: $re = 0.14$ leads right-hand side to be 0.15. Therefore our *re* guess is raised to 0.15.
Guess 2: $re = 0.15$ leads right-hand side to be 0.161, which is close to the left-hand side value of 0.16.

(Recall that in the limit of $b\,C_{isb}|_f << \Delta P_f$, $L^+ \rightarrow re$ (see result (7.2.44)); this provides a basis for our guess of *re*, but it should also be remembered that $re < L^+$.)

Example 7.2.4 If in Example 7.2.3 the value of $(D_{im}\,\kappa_{im}/\delta_m)$ is known to be 2.4×10^{-5} cm/s, develop a rough estimate of the salt concentration in the permeate.

Solution We employ equation (7.2.53) for a rough estimate of the average salt concentration in the permeate:

$$C_{ip}|_{\text{avg}} = \frac{(D_{im}\,\kappa_{im}/\delta_m)}{re\,A\,b\,\overline{V}_w}(L^+ - re) = \frac{(D_{im}\,\kappa_{im}/\delta_m)\,(L^+ - re)}{re\left(\frac{A\,\Delta P_f\,\overline{V}_w}{C_{isb}|_f}\right)\left(\frac{bC_{isb}|_f}{\Delta P_f}\right)}$$

$$\Rightarrow \frac{C_{ip}|_{\text{avg}}}{C_{isb}|_f} = \frac{(D_{im}\,\kappa_{im}/\delta_m)\,(L^+ - re)}{re\,(A\,\Delta P_f\,\overline{V}_w)\left(\frac{bC_{isb}|_f}{\Delta P_f}\right)}.$$

we have $(D_{im}\,\kappa_{im}/\delta_m) = 2.4 \times 10^{-5}$ cm/s, $L^+ = 0.16$ and $re = 0.15$ so

$$A\,\Delta P_f\,\overline{V}_w = 4.06 \times 10^{-2}\,\frac{\text{cm}}{\text{min}}$$
$$= \frac{4.06 \times 10^{-2}}{60}\,\frac{\text{cm}}{\text{s}};\quad \frac{b\,C_{isb}|_f}{\Delta P_f} = 0.0598;$$

$$\left(C_{ip}|_{\text{avg}}/C_{isb}|_f\right) = \frac{2.4 \times 10^{-5}\,(0.16 - 0.15)}{0.15 \times \frac{4.06 \times 10^{-2}}{60} \times 0.0598}$$
$$= \frac{2.4 \times 60 \times 10^{-3}}{0.15 \times 4.06 \times 5.98};$$

$$C_{ip}|_{\text{avg}} = C_{isb}|_f \times 0.0395 \cong 0.0395 \times 2600 = 102\,\text{ppm}.$$

The actual experimental value observed by Ohya and Taniguchi (1975) was higher, at around 200 ppm. *Note:* Although the units of $C_{ip}|_{\text{avg}}$ and $C_{isb}|_f$ are mol/cm^3, here we have used ppm units.

In the above examples, the fractional water recovery is around 0.15. If one increases ΔP_f, the fractional water recovery will increase, and so will the energy cost via the cost of pumping. On the other hand, usually a higher ΔP_f is used with a higher feed salt concentration; therefore water recovery may not increase. To increase the water recovery, a number of spiral-wound modules are connected in series (as shown in Figure 7.2.4(a)) inside the pressure vessel. There is a brine seal between the module and the pressure vessel so that the brine is forced to go through the channels of the module. The permeate tubes are connected in series. Concentrated brine from one module enters the next module as feed and so on. The fractional water recovery is ultimately limited by the difference between the concentrated feed pressure and the osmotic pressure of the concentrate and the level of acceptable flux.

To characterize the total performance of such a configuration of a number of modules in series, it is useful to

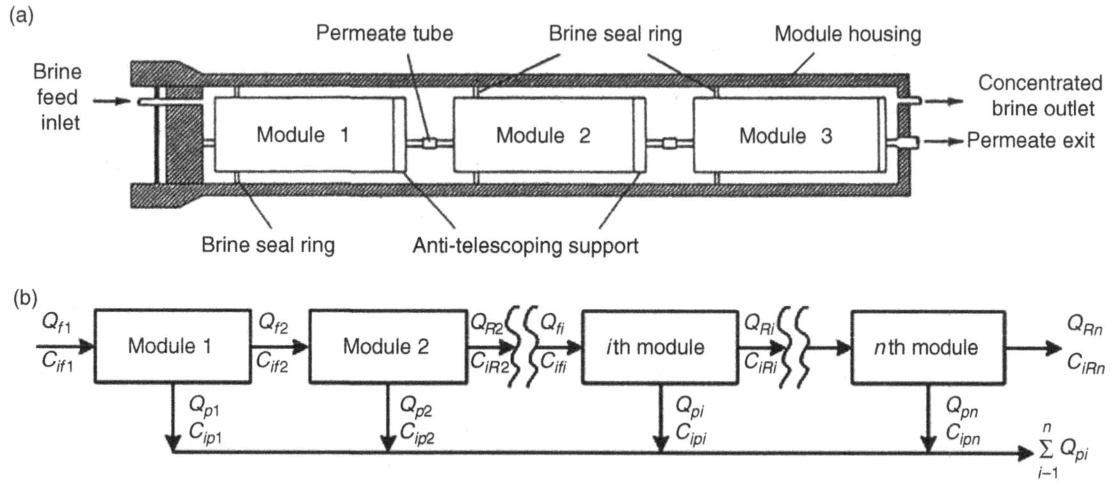

Figure 7.2.4. (a) Schematic of a number of spiral-wound modules in series in a module housing which is a pressure vessel. (b) Configuration for multiple modules in series or multi-tube system for reverse osmosis operation.

consider the configuration of a multi-tube system based desalination, as shown in Figure 7.2.4(b). Although such multi-tube based systems are rarely used at this time, the configuration is useful for spiral-wound modules of Figure 7.2.4(a). Let the feed and permeate parameters for any given ith module/tube be as follows:

feed: $Q_{fi}\,\text{cm}^3/\text{s}$, $C_{ifi}\,\text{mol/cm}^3$, P_{fi};
permeate: $Q_{pi}\,\text{cm}^3/\text{s}$, $C_{ip}|_{\text{avg}}\,\text{mol/cm}^3$.

The retentate quantities needed are $Q_{Ri}\,\text{cm}^3/\text{s}$, $C_{iRi}\text{mol/cm}^3$ and P_{Ri}. Note that, for the ith module/membrane tube,

$$P_{Ri} = P_{f(i+1)}; \qquad Q_{Ri} = Q_{f(i+1)}; \qquad C_{iRi} = C_{if(i+1)}\,\text{mol/cm}^3.$$

Further,

$$P_{fi} - P_{Ri} = \Delta P_i,$$

the pressure drop in brine in the ith tube. The cumulative permeate volumetric flow rate entering the permeate side of the $(i+1)$th module is given by: $\sum_{i=1}^{i} Q_{pi}$, and therefore permeate concentration entering the permeate side

$$\text{of}\,(i+1)\text{th module} = \frac{\sum_{i=1}^{i} Q_{pi}\,C_{ipi}}{\sum_{i=1}^{i} Q_{pi}}. \tag{7.2.54}$$

Since in the analysis of the performance of a spiral-wound module fed with Q_{fi}, C_{ifi} and P_{fi}, we can obtain Q_{pi}, $C_{ip}|_{\text{avg}}$, and all other quantities in (7.2.54) may be obtained without difficulty.

7.2.1.2.1 More detailed model for reverse osmosis in a module The simplified model already presented for reverse osmosis desalination is quite useful, not just for illustrative reasons, but also for preliminary design purposes. We repeat here our earlier assumptions (not all of them explicitly).

(1) Axial diffusion term in equation (7.2.26) neglected.
(2) Film theory employed for concentration polarization estimate.
(3) Effect of wall permeation on the axial and normal velocity profiles in the flow channels neglected.
(4) Constant physical properties.
(5) Salt permeation for solvent flux determination neglected.
(6) Gravitational considerations neglected.

The flow becomes highly complex in a spiral-wound module containing a feed-side spacer screen. Numerical solutions of the governing equations incorporating most of these complexities have been/are being implemented (Wiley and Fletcher, 2003) using computational fluid dynamics models (see Schwinge *et al.* (2003) for the complex flow patterns in a spacer-filled channel).

In these models, the equations governing the flow and species balance in the feed channel are standard and are coupled together via the variation of viscosity with concentration and wall permeation. On the other hand, the wall permeation conditions of solvent and solute permeation need to be identified. If the membrane rejection of the solute is identified via a salt rejection coefficient, R_i, then the following conditions hold at the membrane wall ($y = 0$) (Figure 7.2.3(b)) for the solute transport:

$$\text{feed channel side}: D_{il}\frac{\mathrm{d}\,C_{is}}{\mathrm{d}y} = |v_{yw}|C_{isw}\,R_i;$$

$$\text{permeate channel side}: D_{il}\frac{\mathrm{d}C_{ip}}{\mathrm{d}y} + |v_{yw}|C_{ip} = |v_{yw}|C_{isw}(1-R_i), \tag{7.2.55}$$

where

$$R_i = \left(1 - \frac{C_{ipw}}{C_{isw}}\right). \tag{7.2.56}$$

The solvent permeation condition at the wall employs, along with the no-slip condition (i.e. $v_z = 0$ at $y = 0$), the following type of boundary condition (Brian, 1965):

$$v_{yw} = v_{ywf}\left(1-\beta\left(\frac{C_{isw}}{C_{isf}}-1\right)\right) = v_{ywf}\left(1-\beta\left(\frac{\pi_w}{\pi_f}-1\right)\right), \tag{7.2.57}$$

where β is defined as

$$\beta = \frac{R_i\,\pi_f}{\Delta P - R_i\pi_f} \tag{7.2.58}$$

and we have assumed that $\pi_w = b\,C_{isw}, \pi_f = b\,C_{isf}$. This may be written as follows:

$$\frac{v_{yw}}{v_{ywf}} = \left(1 - \frac{R_i\,\pi_f}{\Delta P - R_i\pi_f}\left(\frac{\pi_w}{\pi_f} - 1\right)\right) = \frac{\Delta P - R_i\,\pi_w}{\Delta P - R_i\,\pi_f}. \tag{7.2.59}$$

This is a straightforward solution–diffusion model, since

$$\frac{\Delta P - R_i\,\pi_w}{\Delta P - R_i\,\pi_f} = \frac{\Delta P - \left(1 - \frac{C_{ipw}}{C_{isw}}\right)\pi_w}{\Delta P - \left(1 - \frac{C_{ipw}}{C_{isf}}\right)\pi_f} = \frac{\Delta P - (\pi_w - \pi_p)}{\Delta P - (\pi_f - \pi_p)}, \tag{7.2.60}$$

where, for the feed solution driving force, R_i has been suitably defined with respect to the feed concentration.

7.2.1.3 *Ultrafiltration*

In Sections 6.3.3.2 and 6.4.2.1, we studied ultrafiltration (UF) in the configurations of a batch unstirred cell and in a well-stirred cell (and in a CSTS), respectively. In the batch unstirred cell, operation with bulk flow parallel to the force direction led to the dead-end mode. The bulk protein concentration changed with time, leading to unsteady state operation. The analyses of solvent flux and solute rejection were carried out there, however, based on a *pseudosteady state assumption*. It was also pointed out (following equation (6.3.145c)) that bulk flow perpendicular to the direction of the force can lead to steady state operation. In this section, one such common configuration, *crossflow ultrafiltration*, is briefly studied; it is also sometimes called *high-performance tangential flow selective filtration.*

In this configuration (Figure 7.2.5(a)), the feed solution is convected along the membrane length at a pressure in the range of 0.2–2 atm gauge; the permeate is withdrawn perpendicular to the membrane, and the permeate pressure is assumed to be the same all along the membrane length. In practice, the permeate channel may be that for a hollow fiber module (as in Figure 7.2.1(c)) or a spiral-wound flat membrane module (as in Figure 7.2.1(e)). The flow regime may be laminar or turbulent. Conditions at any location are essentially steady with respect to time, except for any slow change due to membrane fouling. There are three general questions of importance here.

(1) How does one predict the solvent flux in such a configuration?
(2) What would be the average solvent flux if there is considerable change in bulk concentration during UF?
(3) How does one treat the problem of purification of one protein over another?

The general species conservation equation (6.2.5b) for macromolecular species i in solvent s may be written under the conditions of steady state, no external force based species velocities ($\mathbf{U}_i = 0$) and no chemical reactions as

$$\nabla\cdot(C_{is}\,\mathbf{v}^*) = -\nabla\cdot(\mathbf{J}_i^*) = +\nabla\cdot(D_{is}\,\nabla\,C_{is}). \tag{7.2.61}$$

For the two-dimensional flow conditions encountered in the membrane device of Figure 7.2.5(a), the two coordinates relevant are y (normal to the membrane) and z (parallel to the length of the membrane) in the Cartesian system. Therefore

$$\frac{\partial(C_{is}\,v_y^*)}{\partial y} + \frac{\partial(C_{is}\,v_z^*)}{\partial z} = \frac{\partial}{\partial y}\left(D_{is}\frac{\partial C_{is}}{\partial y}\right) + \frac{\partial}{\partial z}\left(D_{is}\frac{\partial C_{is}}{\partial z}\right). \tag{7.2.62}$$

Generally, the axial diffusion term (z-direction) is quite small compared to the axial convection term. Therefore

$$\frac{\partial(C_{is}\,v_y^*)}{\partial y} + \frac{\partial(C_{is}\,v_z^*)}{\partial z} = \frac{\partial}{\partial y}\left(D_{is}\frac{\partial C_{is}}{\partial y}\right). \tag{7.2.63}$$

The solution of such an equation for an actual membrane device for ultrafiltration is difficult to obtain (see Zeman and Zydney (1996) for background information). One therefore usually falls back on the stagnant film model for determining the relation between the solvent flux and the concentration profile (see result (6.3.142b)). To use this result, we need to estimate the mass-transfer coefficient $k_{i\ell}(= D_{i\ell}/\delta_\ell)$, for the protein/macromolecule. One can focus on the entrance region of the concentration boundary layer, assume D_{is} to be constant for a dilute solution, $\mathbf{v}^* \cong \mathbf{v}$, $\mathbf{v}_y \cong 0$ in the thin boundary layer, $\mathbf{v}_z = \dot{\gamma}_w y$ (where $\dot{\gamma}_w$ is the wall shear rate of magnitude $|\mathrm{d}v_z/\mathrm{d}y|$) and obtain the result known as the Leveque solution at any location z in terms of the Sherwood number:

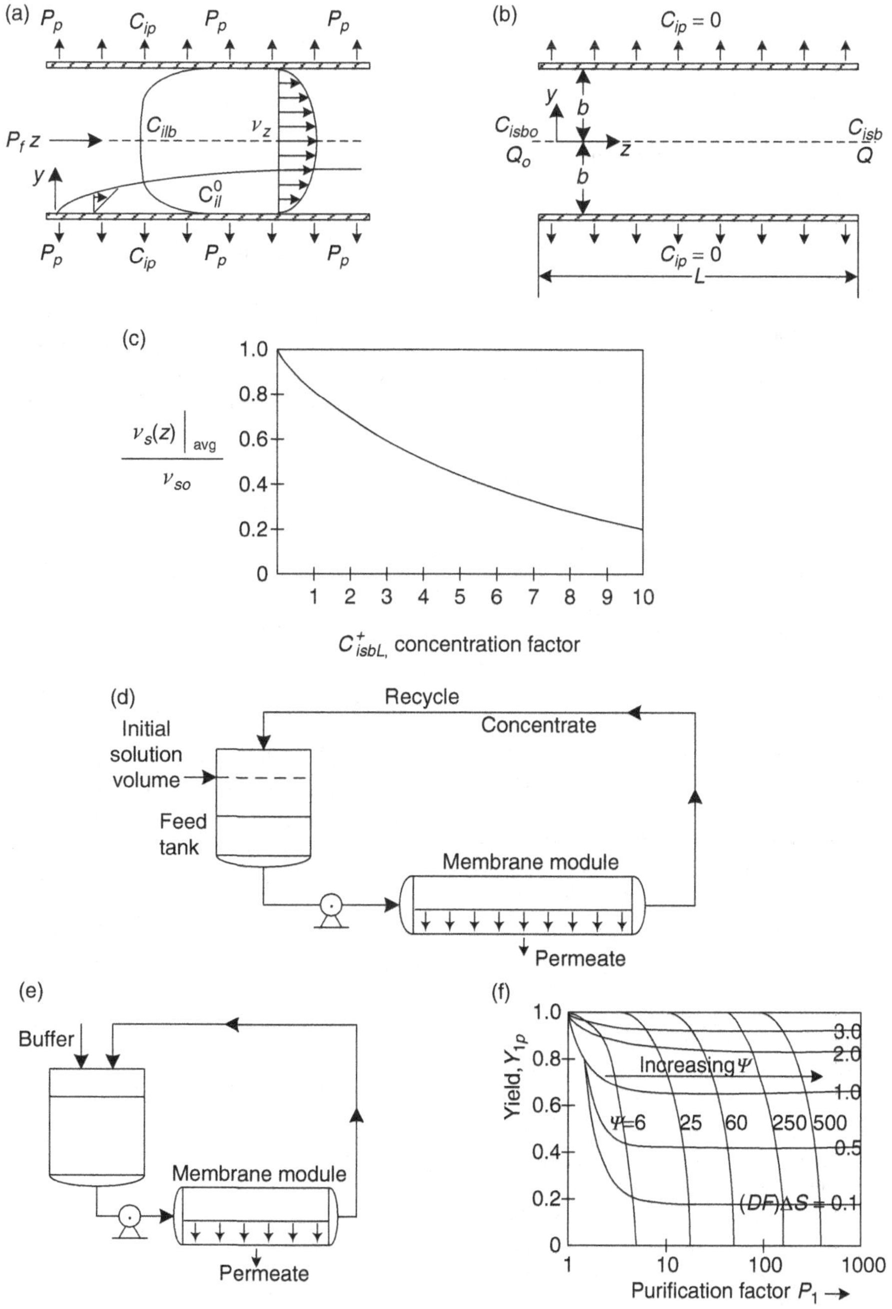

Figure 7.2.5. (a) Concentration profile of species i *and axial velocity profile in a crossflow UF membrane channel. (b) Parallel-plate crossflow UF membrane channel. (c) Reduction of averaged membrane module filtration flux with the extent of solute concentration. (d) Process schematic for batch UF with a crossflow membrane module. (e) Process schematic for membrane diafiltration. (f) Variation of the yield of purified product species 1 in the filtrate with its purification factor for different values of the parameters* ψ, (DF)ΔS. *(After Van Reis and Saksena (1997).)*

$$Sh_z = \frac{k_{i\ell}\, 2R}{D_{i\ell}} = \frac{2R}{0.893}\left(\frac{\dot{\gamma}_w}{9\, D_{i\ell} z}\right)^{1/3}, \qquad (7.2.64a)$$

where we have used a tube of diameter $2R$ as the membrane flow channel of interest. In practice, experimentally obtained correlations for mass-transfer coefficients (illustrated in Tables 3.1.8 and 3.1.5) are utilized in the concentration polarization relation (6.3.142b)):[13]

$$N_s \overline{V}_s = |v_y| = k_{i\ell}\, \ell n\left(\frac{C_{i\ell}^0 - C_{ip}}{C_{ilb} - C_{ip}}\right). \qquad (7.2.64b)$$

The solvent flux expression is[14]

$$N_s = A\,(\Delta P - \Delta\pi). \qquad (7.2.64c)$$

The solute rejection expression (from (6.3.145c)) is

$$R_{\text{true}} = 1 - \frac{C_{ip}}{C_{i\ell}^0} = 1 - \frac{S_\infty \exp\,(Pe_i^m)}{S_\infty + \exp\,(Pe_i^m) - 1}. \qquad (7.2.64d)$$

There are three unknowns here: $|v_y|$ (the solvent volume flux), $C_{i\ell}^0$ (the wall solute concentration) and C_{ip} for given $C_{i\ell b}$, ΔP, $k_{i\ell}$, S_∞, $\overline{G}_i$, τ_m, δ_m, D_{ip}. One can solve the problem by simultaneous numerical solution of the three equations. For systems having perfect rejection of the macrosolute, $C_{ip} = 0$, $R_{\text{true}} = 1$; only two equations ((7.2.64b) and (7.2.64c)) have to be solved. The complexities that have to be taken into account in this procedure include the concentration dependences of the transport properties like $D_{i\ell}$ and μ.

As ultrafiltration progresses along the module length, the increased macrosolute concentration is likely to reduce the solvent flux through the membrane. We will therefore consider now the problem of *predicting the average solvent flux* obtained in a crossflow ultrafiltration membrane module having a given membrane length or membrane area when the following three conditions are satisfied at a minimum: (1) membrane rejection of the macrosolute is perfect; (2) gel polarization condition exists; (3) longitudinal diffusion is negligible compared to longitudinal convective transport. A variety of module configurations exist. We may ignore these by assuming that the following characterization may suffice: in the bulk flow direction of the z-coordinate, the module length is L; the relevant transverse dimension (e.g. the y-coordinate) is R_h, the hydraulic radius, so that, for cylindrical tubes (or hollow fibers), $4R_h = 2R$, where R is the radius of the tube or hollow fiber, and for rectangular channels of gap $2b$, $4R_h = 2b$. In channels having a spacer, there are added complexities.

For the macrosolute species i, consider equation (7.2.63). For dilute solutions, we can replace v_z^* by v_z. Integrate this equation across the membrane channel transverse dimension y (from the center line $y = 0$ to $y = b$ for one wall of a symmetric channel having a gap $2b$, Figure 7.2.5(b))

$$\int_0^b \frac{\mathrm{d}(C_{is}\, v_z^*)}{\mathrm{d}z}\,\mathrm{d}y + \int_0^b \mathrm{d}(C_{is}\, v_y^*) = \int_0^b \mathrm{d}\left(D_{is}\frac{\mathrm{d}C_{is}}{\mathrm{d}y}\right). \qquad (7.2.65a)$$

But

$$\frac{\mathrm{d}}{\mathrm{d}z}\int_0^b C_{is}\, v_z^*\,\mathrm{d}y = \frac{1}{2W}\frac{\mathrm{d}}{\mathrm{d}z}(QC_{isb}), \qquad (7.2.65b)$$

where Q is the volumetric flow rate of solution through the channel and C_{isb} is the bulk concentration of the solute i in solution; further, W is the width of channel, therefore $2W$ is the membrane surface area per unit length. Equation (7.2.65a) becomes

$$\frac{1}{2W}\frac{\mathrm{d}}{\mathrm{d}z}(QC_{isb}) + C_{is}\, v_y^*|_b - C_{is}\, v_y^*|_0 = D_{is}\frac{\mathrm{d}C_{is}}{\mathrm{d}y}\bigg|_b - D_{is}\frac{\mathrm{d}C_{is}}{\mathrm{d}y}\bigg|_{y=0}. \qquad (7.2.66)$$

Due to symmetry, we can write $v_y^*|_{y=0} = 0$, $(\mathrm{d}C_{is}/\mathrm{d}y)|_{y=0} = 0$. We get

$$\frac{\mathrm{d}}{\mathrm{d}z}(Q\, C_{isb}) = \left(-C_{is}\, v_y^*|_b + D_{is}\frac{\mathrm{d}C_{is}}{\mathrm{d}y}\bigg|_b\right)2W. \qquad (7.2.67)$$

The magnitude of the right-hand side in parentheses is the wall flux of species i, the flux through the membrane. Since the membrane rejects the macrosolute completely, we get

$$\frac{\mathrm{d}}{\mathrm{d}z}(Q\, C_{isb}) = 0 = \frac{Q\,\mathrm{d}C_{isb}}{\mathrm{d}z} + C_{isb}\frac{\mathrm{d}Q}{\mathrm{d}z}. \qquad (7.2.68)$$

For the solvent, usually water in ultrafiltration of biological solutions, the equation corresponding to (7.2.63) is

$$\frac{\partial(C_{ws}\, v_y^*)}{\partial y} + \frac{\partial(C_{ws}\, v_z^*)}{\partial z} = \frac{\partial}{\partial y}\left(D_{ws}\frac{\partial C_{ws}}{\partial y}\right). \qquad (7.2.69)$$

Carrying out the integration in the same manner, we get

$$\frac{1}{2W}\frac{\mathrm{d}}{\mathrm{d}z}(QC_{wsb}) = -C_{ws}\, v_y^*|_{y=b} + D_{ws}\frac{\mathrm{d}C_{ws}}{\mathrm{d}y}\bigg|_{y=b} = -(N_{sy}|_{y=b});$$

$$\frac{\mathrm{d}}{\mathrm{d}z}(QC_{wsb}) = \left(-N_{sy}|_{y=b}\right)2W = \left(-N_{sy}|_{y=b}\right)a_{ms\ell}, \qquad (7.2.70)$$

where $a_{ms\ell}$ is the surface area of membrane per unit channel length and $\left(-N_{sy}|_{y=b}\right)$ is the solvent flux through the membrane at the walls of the channel; recall that the channel walls are identical and have the same membrane. Since, in most ultrafiltration applications, the solutions are quite dilute in proteins[15] or otherwise, we can rewrite this equation as

$$\frac{\mathrm{d}Q}{\mathrm{d}z} = -\left(N_{sy}|_{y=b}\right)a_{ms\ell}/C_{wsb} = -v_s\, a_{ms\ell}, \qquad (7.2.71)$$

[13] Note the change in the coordinate : we have $|v_y|$ here.
[14] $\Delta\pi$ is to be replaced by $-\sigma_i\Delta\pi$ (see (6.3.158a)).

[15] C_{wsb} is essentially constant.

where v_s is the volume flux of the permeate (units of cm^3/cm^2-s). Equations (7.2.68) and (7.2.71) are the basic balance equations for determining how Q and C_{isb} vary with distance z along the membrane module (Figure 7.2.5(b)). However, the solvent permeation flux $v_s(= N_s\overline{V}_s)$ depends on, among other things, the bulk solute concentration C_{isb} (as well as the local ΔP across the membrane). We have assumed earlier (assumption (2)) that a gel polarization condition exists. From result (6.3.143), we know that

$$v_s = k_{i\ell}\,\ell n\left(\frac{C_{igel} - C_{ip}}{C_{i\ell b} - C_{ip}}\right) = k_{i\ell}\,\ell n\left(\frac{C_{igel}}{C_{isb}}\right), \qquad (7.2.72)$$

since $C_{ip} = 0$ by assumption and $\ell = s$. Therefore v_s changes along the membrane length as Q and C_{isb} change. Note that $k_{i\ell}$, the mass-transfer coefficient of the macrosolute species i, depends, among other things, on the solution velocity, the solution kinematic viscosity as well as the solute diffusivity D_{is}. A common way of representing these dependences is a correlation of the following type (see Tables 3.1.5 and 3.1.8):

$$\frac{k_{i\ell}\,2b}{D_{i\ell}} = a_1\left(\frac{v_z b}{v}\right)^{a_2}\left(\frac{v}{D_{i\ell}}\right)^{a_3}. \qquad (7.2.73)$$

Note that v_z, v and $D_{i\ell}$ are all local quantities and vary with z in the flow direction. We will now follow the procedure of Lopez-Leiva (1980) to develop an expression for the local permeate volume flux $v_s(z)$ as well as a length-averaged value $v_s(z)|_{avg}$. Define C_{isb0}, Q_0, v_{s0}, v_0, $D_{i\ell 0}$, $k_{i\ell 0}$, to be the values of C_{isb}, Q, v_s, v, $D_{i\ell}$ and $k_{i\ell}$ at the inlet of the membrane channel (Figure 7.2.5(b)). Define further the following nondimensional quantities:

$$C^+_{isb} = \frac{C_{isb}}{C_{isb0}}, \qquad Q^+ = \frac{Q}{Q_0}, \qquad z^+ = \frac{z}{L}. \qquad (7.2.74)$$

From equations (7.2.68) and (7.2.71), we get

$$\frac{Q}{C_{isb}}\frac{\mathrm{d}C_{isb}}{\mathrm{d}z} = v_s a_{ms\ell}. \qquad (7.2.75)$$

Further, from equation (7.2.68),

$$QC_{isb} = \text{constant} = Q_0\,C_{isb0}. \qquad (7.2.76)$$

We can rewrite equation (7.2.75) as

$$\frac{Q_0\,C_{isb0}}{C^2_{isb}}\frac{\mathrm{d}C_{isb}}{\mathrm{d}z} = \frac{Q_0}{C^{+2}_{isb}L}\frac{\mathrm{d}C^+_{isb}}{\mathrm{d}z^+} = v_s\,a_{ms\ell}. \qquad (7.2.77)$$

Integrating along the membrane length, we obtain

$$v_s(z)|_{avg} = \int_0^1 v_s \mathrm{d}z^+ = \frac{Q_0}{a_{ms\ell}L}\int_1^{C^+_{isbL}}\frac{\mathrm{d}C^+_{isb}}{C^{+2}_{isb}} = \frac{Q_0}{a_{ms\ell}L}\left[1 - \frac{1}{C^+_{isbL}}\right], \qquad (7.2.78)$$

where C^+_{isbL} is the value of C^+_{isb} at $z^+ = 1$. This expression is essentially a solvent balance expression for the membrane device of length L: the average solvent flux over the length L is obtained as a function of $Q_0, a_{ms\ell}, L$ and the extent of increase in concentration from C_{isb0} to C_{isbL}.

Suppose the flux of the solvent is available for the inlet/starting concentration, v_{s0} corresponding to C_{isb0}. One would like to know how v_s is changing with concentration C_{isb} or z. Correspondingly, how does $v_s(z)|_{avg}$ change vis-à-vis v_{s0} as C_{isb} changes from C_{isb0} to C_{isbL}. We will still employ the basic differential equation (7.2.77) as well as (7.2.76). However, we will need to employ assumption (2), namely that a gel polarization condition exists, i.e. equation (7.2.72) is valid. Therefore

$$(v_s(z)/v_{s0}) = (k_{i\ell}(z)/k_{i\ell 0})\left[\ell n\left(\frac{C_{igel}}{C_{isb}}\right)\Big/\ell n\left(\frac{C_{igel}}{C_{isb0}}\right)\right]$$

$$= (k_{i\ell}(z)/k_{i\ell 0})\left[1 - \frac{\ell n(C_{isb}/C_{isb0})}{\ell n(C_{igel}/C_{isb0})}\right]. \qquad (7.2.79)$$

From relation (7.2.73), we get

$$\left(\frac{k_{i\ell}(z)}{k_{i\ell 0}}\right) = \left(\frac{Q(z)}{Q_0}\right)^{a_2}\left(\frac{v_0}{v(z)}\right)^{a_2}\left(\frac{v(z)}{v_0}\right)^{a_3}\left(\frac{D_{i\ell}}{D_{i\ell 0}}\right)^{1-a_3}$$

$$= (Q^+(z))^{a_2}\left(\frac{D_{i\ell}}{D_{i\ell 0}}\right)^{1-a_3}\left(\frac{v(z)}{v_0}\right)^{a_3-a_2}. \qquad (7.2.80)$$

The variation of $D_{i\ell}$ with viscosity μ, concentration C_{isb} and other quantities may be obtained for dilute solutions of species i from relations (3.1.89) and (3.1.91a). We can thereby obtain the following relation for $(D_{i\ell}/D_{i\ell 0})$ if we assume that the solution is ideal enough for $(d\,\ell n\,\gamma/d\,\ell n\,x_i)_{T,P} \cong 0$ and the solution density ρ does not change much with the concentration change, i.e. $v(z) = (\mu(z)/\rho_0)$, $v_0 = (\mu_0/\rho_0)$:

$$(D_{i\ell}(z)/D_{i\ell 0}) = v_0/v(z). \qquad (7.2.81)$$

This leads to

$$(k_{i\ell}(z)/k_{i\ell 0}) = (Q^+(z))^{a_2}\left(\frac{v(z)}{v_0}\right)^{(2a_3-a_2-1)}. \qquad (7.2.82)$$

Most often $a_3 = 1/3$ (see Tables 3.1.5 and 3.1.8). Further, employing relation (7.2.76),

$$(k_{i\ell}(z)/k_{i\ell 0}) = \left(\frac{C_{isb0}}{C_{isb}}\right)^{a_2}\left(\frac{v(z)}{v_0}\right)^{-(a_2+1/3)}. \qquad (7.2.83)$$

Since one can obtain a relation for v as a function of C_{isb}, we may rewrite this relation as

$$\frac{k_{i\ell}(z)}{k_{i\ell 0}} = (k_{i\ell}(C_{isb})/k_{i\ell 0}) = \left(\frac{1}{C^+_{isb}}\right)^{a_2}\left(\frac{v(C^+_{isb})}{v_0}\right)^{-(a_2+1/3)}. \qquad (7.2.84)$$

We now go back to the basic differential equation (7.2.77), introduce for $v_s(z)$ expression (7.2.79), wherein for $(k_{i\ell}(z)/k_{i\ell 0})$, we introduce expression (7.2.84) given above:

$$\frac{Q_0}{(C^+_{isb})^2 a_{ms\ell} L}\frac{dC^+_{isb}}{dz^+} = v_{s0}\left[1-\frac{\ell n C^+_{isb}}{\ell n C^+_{igel}}\right]\left[\frac{1}{C^+_{isb}}\right]^{a_2}\left[\frac{v(C^+_{isb})}{v_0}\right]^{-(a_2+1/3)}. \tag{7.2.85}$$

Introducing expression (7.2.78) for $v_s(z)|_{\text{avg}}$ and integrating from $z^+ = 0$ to 1 and $C^+_{isb} = 1$ to C^+_{isbL}, we get

$$\frac{v_s(z)|_{\text{avg}}}{v_{s0}} = \left[1-\frac{1}{C^+_{isbL}}\right]\frac{1}{\displaystyle\int_1^{C^+_{isbL}} \frac{dC^+_{isb}}{(C^+_{isb})^{2-a_2}\left[\frac{v(C^+_{isb})}{v_0}\right]^{-(a_2+1/3)}\left[1-\frac{\ell n C^+_{isb}}{\ell n C^+_{igel}}\right]}}$$

$$= \frac{1-\frac{1}{C^+_{isbL}}}{I(C^+_{isbL})}. \tag{7.2.86}$$

Numerical integration to obtain the value of the integral $I(C^+_{isbL})$ for given dependences of v (as well as $D_{i\ell}$) on C_{isb} will yield values of the relative reduction in the solvent flux with increase in concentration along the membrane under the condition of gel polarization. Figure 7.2.5(c) illustrates how the ratio $(v_s(z)|_{\text{avg}}/v_{s0})$ decreases with an increase in the value of C^+_{isbL}, which is sometimes called the *concentration factor* (Lopez-Leiva, 1980).

In Figure 7.2.5(c), the values of the concentration factor can be as high as 5–10. Although such an increase in concentration can be achieved in long modules, a more likely scenario is crossflow ultrafiltration of a feed solution from a batch vessel with the retentate recycled to the feed tank (Figure 7.2.5(d)). At any given time, for the given feed concentration entering the module, the module performance analysis may be implemented using equation (7.2.86). However, with time, this feed concentration of the rejected protein keeps on increasing substantially, resulting in high concentration factors. The disadvantages of such a scheme are the potentially long time taken and the increase in viscosity, leading to a significant reduction in flux and potential bacterial growth in many biological systems.

A common alternative used where possible is the *diafiltration mode* with a crossflow UF membrane unit and concentrate recycle (Figure 7.2.5(e)). Here the solution concentration and viscosity are not allowed to increase due to the continuous addition of buffer replacing the permeate volume lost. Equations developed in Section 6.4.2.1 for well-stirred UF cells having *continuous diafiltration* may be used here with appropriate care since we can treat the crossflow UF device as a blackbox for the purpose of an overall process mass balance and solute selectivity analysis. Similarly, the equations developed in Section 6.4.2.1 for a *batch concentration process* may be utilized here to determine various quantities, such as the yield of macrosolute, retentate concentration, etc.

In many ultrafiltration applications, purification of one protein from another protein is often required from a batch solution of volume V_{f0}. Both proteins are permeable through the membrane; however, one of the solutes ($i = 1$) may be more selectively permeable than the other ($i = 2$). Thus one would like to have a high yield of $i = 1$ in the permeate; the *yield* of this solute in the permeate (segregation fraction) is defined for a cumulative permeate volume V_p as

$$Y_{ip} = \frac{\int_0^{V_p} C_{ip}\, dV_p}{V_{f0}\, C_{i0}} = \frac{V_p \overline{C}_{ip}}{V_{f0}\, C_{i0}} = \frac{V_p \overline{C}_{1p}}{V_{f0}\, C_{10}} = Y_{1p}. \tag{7.2.87}$$

Correspondingly, the other protein ($i = 2$) should be retained as much as possible in the retentate. Therefore the yield of solute 2 in the retentate (in the continuous diafiltration mode) may be defined as

$$Y_{iR} = \frac{\int C_{iR}\, dV_f}{V_{f0}\, C_{i0}} = \frac{V_{f0}\, C_{iR}}{V_{f0}\, C_{i0}} = \frac{C_{2R}}{C_{20}} = Y_{2R}. \tag{7.2.88}$$

We have observed earlier (in Section 6.4.2.1) that, for continuous diafiltration (result (6.4.107)),

$$Y_{iR} = Y_{2R} = \exp(-(1-R_2)DF). \tag{7.2.89}$$

Correspondingly, we may obtain

$$Y_{ip} = Y_{1p} = \frac{V_p \overline{C}_{1p}}{V_{f0}\, C_{10}} = \frac{V_{f0}\, C_{10} - V_{f0} C_{1R}}{V_{f0} C_{10}} = 1 - \frac{V_{f0} C_{1R}}{V_{f0} C_{10}}$$

$$= 1 - Y_{1R} = 1 - \exp(-(1-R_1)DF). \tag{7.2.90}$$

Both results may be expressed in terms of the sieving coefficients S_1 ($= 1 - R_1$) and S_2 ($= 1 - R_2$) of the two solutes and the dilution factor DF (also called the *number of diavolumes*) as

$$Y_{1p} = 1 - \exp(-S_1(DF)); \tag{7.2.91a}$$

$$Y_{2R} = \exp(-S_2(DF)), \tag{7.2.91b}$$

where

$$DF = \text{dilution factor} = V_b/V_{f0} = V_p/V_{f0}$$
$$= \text{number of diavolumes} \tag{7.2.91c}$$

and V_b is the buffer volume added ($=V_p$).

It is not enough to have a high yield of species 1 in the permeate; we want the permeate to be as pure as possible in species 1. The selectivity of the UF process for one protein (more permeable species 1) over the less permeable protein (species 2) has been described by a number of indices (see Van Reis and Saksena (1997)):

$$\psi = \frac{S_1}{S_2} = \frac{1-R_1}{1-R_2} = \frac{C_{1p}/C_{1f}}{C_{2p}/C_{2f}} = \alpha_{12}, \tag{7.2.92}$$

where α_{12} was defined in (6.3.147b) for dilute solutions. Another selectivity index employed is

$$\Delta S = S_1 - S_2. \tag{7.2.93}$$

To combine the properties of quantitative yield (Y_{ip}, Y_{iR}) and the quality of the permeated product ($\psi = S_1/S_2$), Van Reis and Saksena (1997) have proposed an additional parameter called $N\,\Delta S$, which in our notation is

$$(DF)\,\Delta S = \frac{V_p}{V_{f0}}(S_1 - S_2) \tag{7.2.94}$$

(N being the number of diavolumes, which is equal to our dilution factor DF (6.4.106)). Noting that S (= S_{obs}) is C_{ip}/C_{if}, which is C_{ip}/C_{iR}, we see that

$$(DF)\,\Delta S = \frac{V_p C_{1p}}{V_{f0} C_{1f}} - \frac{V_p C_{2p}}{V_{f0}\, C_{2f}} = \frac{V_p C_{1p}}{V_{f0}\, C_{1R}} - \frac{V_p C_{2p}}{V_{f0} C_{2R}}, \tag{7.2.95}$$

a quantity similar to the extent of separation, ξ, defined in Chapters 1 and 2. It is exactly equal to ξ at the initial stages of UF when $C_{1R} \cong C_{1f}$ and $C_{2R} \cong C_{2f}$.

Van Reis and Saksena (1997) have identified the product yield and the product purification factor as the two dimensions of selective protein separation with, for example, ψ or $(DF)\,\Delta S$ as parameters. The *product purification factor* P_1 for solute 1 in the permeate was defined as

$$P_1 = \left(\frac{V_p \overline{C}_{1p}}{V_{f0} C_{10}}\right) \Big/ \left(\frac{V_p \overline{C}_{2p}}{V_{f0} C_{20}}\right) = \alpha_{12}^{hf}. \tag{7.2.96}$$

Using expression (7.2.90) for the yield Y_{1p} and applying it to solute 2 as well (namely Y_{2p}), we get

$$P_1 = \frac{Y_{1p}}{Y_{2p}} = \frac{1 - \exp\left(-S_1(DF)\right)}{1 - \exp\left(-S_2(DF)\right)}. \tag{7.2.97}$$

For solute 2, the purification factor P_2 in the retentate is defined as

$$P_2 = \frac{V_{f0} C_{2R}}{V_{f0} C_{20}} \Big/ \frac{V_{f0} C_{1R}}{V_{f0} C_{10}} = \alpha_{12}^{ft}. \tag{7.2.98}$$

Using expression (7.2.91b) for the yield Y_{2R}, we get

$$P_2 = \frac{Y_{2R}}{Y_{1R}} = \frac{\exp\left(-S_2(DF)\right)}{\exp\left(-S_1(DF)\right)} = \exp\left((DF)(S_1 - S_2)\right). \tag{7.2.99}$$

One can now plot the yield vs. the purification factor for two cases: the desired protein is in the permeate; the desired protein is in the retentate. Commercialized UF processes target both high product yield and high selectivity in biopharmaceutical industries when it comes to biomacromolecules/proteins: yields in the range of 75% -100% with tenfold or higher product purification factors (Van Reis and Saksena,1997). Before we illustrate such a plot, we need to recognize the relation between P_i and Y_i (Y_{ip} or Y_{iR}) and their dependence on the important parameters for this process DF, S_1 and S_2. We have indicated earlier that often the following two combinations, $\psi = (S_1/S_2)$ and (DF) $S_1 - (DF)S_2 = (DF)\Delta S$, of these parameters are of interest. For example, if one combines the selectivity definition (7.2.92) with expression (7.2.91a) for Y_{1p} and expression (7.2.97) for the purification factor P_1 for the desired protein species 1 in the filtrate, we obtain the following relation between P_1 and Y_{1p}:

$$P_1 = = \frac{1 - \exp\left(-S_1(DF)\right)}{1 - \exp\left(-S_2(DF)\right)} = \frac{Y_{1p}}{1 - (1 - Y_{1p})^{1/\psi}}, \tag{7.2.100}$$

since, from (7.2.91a),

$$\begin{aligned}(1 - Y_{1p}) &= \exp(-S_1(DF)) \Rightarrow \ell n(1 - Y_{1p}) = -(DF)S_2\psi \\ &\Rightarrow \ell n(\exp - S_2(DF)) = \frac{1}{\psi}\,\ell n\,(1 - Y_{1p}) \\ &\Rightarrow \exp(-S_2(DF) = (1 - Y_{1p})^{1/\psi}).\end{aligned}$$

The above expression relates the product purification factor P_1 for species 1 in the filtrate to its yield in the filtrate Y_{1p} through the selectivity ψ. The corresponding relation in terms of the parameter $(DF)\Delta S$ is obtained from equations (7.2.91a) and (7.2.97) as

$$P_1 = \frac{Y_{1p}}{1 + (Y_{1p} - 1)\exp\left((DF)\,\Delta S\right)}, \tag{7.2.101}$$

since, from (7.2.91a),

$$(Y_{1p} - 1) = -\exp(-S_1(DF))$$

and

$$-\{\exp\left(-S_1(DF)\right)\}\exp\left((DF)\Delta S\right) = -\exp\left(-S_2(DF)\right)).$$

If the protein species 2 in the retentate is of interest, the purification factor for species 2, P_2, defined by (7.2.99) is related to the yield of species 2 in the retentate Y_{2R} by

$$P_2 = (Y_{2R})^{1-\psi} \tag{7.2.102}$$

(from (7.2.99),

$$\begin{aligned}\log P_2 &= (DF)(S_1 - S_2) = -S_2(DF)(1 - \psi) \\ &\Rightarrow P_2 = (\exp\left(-S_2 \mathrm{DF}\right))^{1-\psi} = (Y_{2R})^{1-\psi}),\end{aligned}$$

with ψ as the parameter. The corresponding relation between P_2 and Y_{2R} with $(DF)\Delta S$ as the parameter is

$$P_2 = \exp\left((DF)\,\Delta S\right). \tag{7.2.103}$$

It is useful now to illustrate graphically the relation between, for example, P_1 and Y_{1p} with either ψ or $(DF)\Delta S$ as a parameter. Figure 7.2.5(f) illustrates the behavior. There are two types of curves here (Van Reis and Saksena, 1997); one type of curve depends on the selectivity parameter ψ, while the other type is based on the parameter $(DF)\Delta S (= (DF)(S_1 - S_2))$. The key features of these two types of curves are as follows.

(1) When purification increases, yield decreases, regardless of the parameter used. Yield is highest when the purification is lowest and vice and versa.

(2) The highest value of the purification factor P_1 for a constant selectivity ψ is ψ:

$$\lim_{Y_{1p}\to 0} P_1 = \lim_{Y_{1p}\to 0}\left[\frac{Y_{1p}}{1-(1-Y_{1p})^{1/\psi}}\right] \cong \lim_{Y_{1p}\to 0}\left[\frac{Y_{1p}}{1-1+\frac{Y_{1p}}{\psi}}\right] = \psi. \quad (7.2.104)$$

(3) The lowest value of the yield Y_{1p} for a given a value of the parameter $(DF)\Delta S$ is of interest; then, from (7.2.101), we get

$$Y_{1p} = \frac{P_1\,[1 - \exp\,(\Delta S(DF))]}{1 - P_1 \exp\,(\Delta S(DF))},$$

Which, in the limit of $P_1 \to \infty$, is reduced to

$$\lim_{P_1\to\infty} Y_{1p} = -\,[1 - \exp(\Delta S(DF))]\exp(-\Delta S(DF))$$
$$= 1 - \exp(-\,\Delta S(DF)). \quad (7.2.105)$$

If one knows what the selectivity ψ is during a particular purification process, then one can move along a constant ψ and find out how Y_{1p} will change with P_{1p} from figures of the type shown in Figure 7.2.5(f).

Example 7.2.5 This example will illustrate the level of solvent flux reduction in a hollow fiber UF unit as a protein solution is substantially concentrated. Consider a feed solution of BSA (bovine serum albumin) containing 2 g/100 cm^3. This is to be concentrated to 15 g/100 cm^3. Kozinski and Lightfoot (1972) obtained the following correlation for the viscosity of a BSA solution:

$$\mu = \exp\,(0.00244(\rho_{\text{solu}} \times 100)^2),$$

where ρ_{solu} is the density of the BSA solution in g/cm^3. It is known that the value of a_2 in the correlation (7.2.73) of the Sherwood number dependence on the Reynolds number is 1/3. It is further known from Kozinski and Lightfoot (1972) that the gel concentration for the BSA solution is 58.5 g/100 cm^3. Determine the fractional reduction in solvent flux as the BSA solution is concentrated along the module length (z) (or time) from 2g/100 cm^3 to 15 g/100 cm^3.

Solution In the gel polarization model, we know from results (6.3.143 and 7.2.92) that, between locations $z = 0$ and z,

$$[v_s(z)/v_s(0)] = (k_{il}(z)/k_{il}(0))\left[\ln\left(\frac{C_{igel}}{C_{isb}(z)}\right)\Big/\ln\left(\frac{C_{igel}}{C_{isb}(0)}\right)\right]$$

$$\Rightarrow [v_s(z)/v_s(0)] = (k_{il}(z)/k_{il}(0))\left[\ln\left(\frac{58.5/M_{\text{BSA}}\times 100}{15/M_{\text{BSA}}\times 100}\right)\Big/\ln\left(\frac{58.5/M_{\text{BSA}}\times 100}{2/M_{\text{BSA}}\times 100}\right)\right]$$

$$\Rightarrow [v_s(z)/v_s(0)] = (k_{il}(z)/k_{il}(0)]\left[\ln\left(\frac{58.5}{15}\right)\Big/\ln\left(\frac{58.5}{2}\right)\right]$$

$$\Rightarrow [v_s(z)/v_s(0)] = [\log 3.9/\log 29.25][k_{il}(z)/k_{il}(0)]$$

$$= 0.41\left[\frac{k_{il}(z)}{k_{il}(0)}\right].$$

We will now calculate the mass-transfer coefficient ratio using result (7.2.84):

$$(k_{il}(z)/k_{il}(0)) = \left(\frac{1}{C^+_{isb}}\right)^{a_2}\left(\frac{v(C^+_{isb})}{v_0}\right)^{-(a_2+1/3)}; \quad a_2 = 1/3;$$

$$C^+_{isb} = C_{isb}(z)/C_{isb0} = \left(\frac{15}{100M_i}\Big/\frac{2}{100M_i}\right) = (15/2) = 7.5.$$

Assuming that the density change in the solution for the concentration levels is not substantial,

$$\frac{v(C^+_{isb})}{v_0} \cong \frac{\mu(C^+_{isb})}{\mu_0} = \frac{\exp\,(0.00244(\rho_{\text{solu}}(z) \times 100)^2)}{\exp\,(0.00244(\rho_{\text{solu}}(0) \times 100)^2)}$$

$$= \exp\left(0.00244\,\{15^2 - 2^2\}\right) = \exp\,(0.54).$$

Therefore

$$(k_{il}(z)/k_{il0}) = (1/7.5)^{1/3}\left(\frac{1}{\exp\,(0.54)}\right)^{2/3}$$
$$= 0.515 \times 0.723 = 0.4145.$$

Therefore

$$[v_s(z)/v_s(0)] = 0.41 \times 0.4145 = 0.170,$$

indicating a very significant drop in flux with length or time as the case may be.

Example 7.2.6 Calculate the value of $v_s(0)$ in Example 7.2.5 using the following additional information. Hollow fiber: I.D., 750 μm; length, 50 cm. Feed solution of BSA: density ~1 g/cm^3; viscosity, 0.9 cp. Diffusivity of BSA: $D_{il0} = D_{il\infty} = 5.94 \times 10^{-7}$ cm^2/s (Table 3.A.5). Velocity of feed solution through fiber bore, 80 cm/s. For hollow fibers, the Leveque solution (equation (3.1.145)) should be employed to determine the mass-transfer coefficient for the hollow fiber UF membrane.

Solution Hollow fiber: I.D., $d_i = 750$ μm $= 750 \times 10^{-4}$ cm $=$ 0.075 cm; length, $L = 50$ cm. Fluid velocity, $v_z = 80$ cm/s. Fluid density $\rho_0 \cong 1$ g/cm^3. Fluid viscosity, $\mu_0 = 0.9$ cp $=$ 0.009 g/cm-s. *Leveque solution* (equation (3.1.145)):

$$Sh_{\text{avg}} = \frac{k_{il0}\,d_i}{D_{il}} = 1.615\left(\frac{d_i}{L}\right)^{1/3}(ReSc)^{1/3};$$

$$Re = \frac{d_i v_z \rho}{\mu} = \frac{0.075 \times 80 \times 1}{0.009} = 667;$$

$$Sc = \frac{\mu}{\rho D_{il}} = \frac{0.009}{1 \times 5.94 \times 10^{-7}} = 1.51 \times 10^4;$$

$$Sh_{\text{avg}} = 1.615 \times \left(\frac{0.075}{50}\right)^{1/3} \times (667 \times 1.51 \times 10^4)^{1/3}$$
$$= 1.615 \times (0.0015)^{1/3} \times (1010 \times 10^4)^{1/3} = 40 = \frac{k_{il0}\,d_i}{D_{il}};$$

$$k_{il0} = \frac{40 \times 5.94 \times 10^{-7}}{0.075} = 3.17 \times 10^{-4}\,\text{cm/s}.$$

From the gel polarization model result:

$$v_{s0} = k_{il0}\,\ell n(C_{igel}/C_{isb0});$$

$$v_{s0} = 3.17 \times 10^{-4} \times 2.303 \times \log(58.5/2)$$

$$\Rightarrow v_{s0} = 1.07 \times 10^{-3}\ \mathrm{cm/s}.$$

In industrial practice, a different unit, liter/m^2-hr (lmh) is employed; $v_{s0} = 38.5$ liter/m^2-hr.

7.2.1.4 *Crossflow microfiltration*

In Figure 6.3.25(b), we illustrated how the unsteady behavior in dead-end cake filtration/microfiltration may be avoided (among other methods) by having a bulk motion of the suspension/slurry perpendicular to the direction of the force, causing, for example, microfiltration. At any axial location, the time-dependent buildup of a cake layer is avoided; instead, the thickness of the boundary layer of particle suspension over the membrane keeps on increasing in the bulk flow direction. This increase accommodates the particles rejected at any axial location, as shown in Figure 7.2.6(a). However, there is a thin cake layer or an immobile layer of particles over essentially the whole length of the microfiltration membrane unless the shear rate due to the tangential flow is high enough to drag this layer and prevent the formation of a cake layer. Therefore, in general, one can envisage a steady state operation delivering a filtrate/permeate free of particles at a steady rate and a concentrated slurry at the exit of the separator. In reality, most systems display a decline in permeation flux with time, due to what is called *fouling* of the membrane. Membrane fouling results from *internal membrane fouling* and *external cake fouling* (Davis, 2001). Deposition or attachment of materials within the porous structure of the membrane by adsorption, adhesion, pore plugging (Figure 6.3.27(c)) and precipitation leads to reduced solvent permeability of the membrane: this is called internal membrane fouling. External cake fouling results from the formation of a stagnant cake layer on the membrane surface aided by surface adsorption, pore mouth blockage, etc. (Figure 6.3.27(c)). We will describe later strategies adopted to overcome such deficiencies. However, we will first briefly describe a model that yields a realistic estimate of solvent flux in crossflow microfiltration, sometimes identified as *tangential-pass microfiltration* (Romero and Davis, 1988; Davis and Sherwood, 1990) under the assumption of no fouling.

As shown in Figure 7.2.6(a), consider a microfiltration membrane with the tangential flow of the suspension over it in the direction of the membrane length coordinate z. The suspension pressure is likely to be 1 to 2 atmospheres above the pressure on the other side of the membrane. The coordinate normal to the membrane into the suspension is y. A suspension having a solids volume fraction ϕ_s contacts the membrane at $z = 0$ and flows along z. As filtrate goes through the membrane, rejected solid particles build up near the membrane a higher volume fraction of particles than ϕ_s. The higher particle concentration near the membrane leads to diffusion of the particles to the bulk of the suspension. There develops a boundary layer of higher particle concentration and therefore particle volume fraction ϕ from ϕ_w at the membrane surface ($y = 0$) to ϕ_s ($<\phi_w$) at $y > \delta_p$, the boundary layer thickness. Note that the value of this δ_p is different from δ_v, the thickness of the velocity boundary layer. The particle concentration boundary layer thickness, δ_p, depends on z and increases with z, as shown in Figure 7.2.6(a). However, δ_p is very much influenced by the velocity field, amongst other things.

There are other important physical features in this flow configuration. At the inlet region of the membrane, $z \geq 0$, the filtration flux v_s is determined only by the resistance of the membrane, R_m (from equation (6.3.136a)), ΔP and μ:

$$v_{s0} = (\Delta P/\mu_0 R_m), \qquad (7.2.106)$$

where we have specifically identified this flux value as v_{s0} (μ_0 being the solvent viscosity). However, a short distance (z_{cr}) downstream, a thin cake layer forms on the membrane, and it grows with the downstream distance coordinate z. Correspondingly, the filtration flux is reduced from v_{s0} to a v_s which decreases as z increases beyond z_{cr}:

$$v_s = v_{s0}/(1 + (\hat{R}_{c\delta}\,\delta_c/R_m)). \qquad (7.2.107)$$

Here we have employed the filtration volume flux expression (6.3.136b) and the expression (7.2.106) for v_{s0}. For an estimated value of the specific cake resistance $\hat{R}_{c\delta}$, variation of δ_c with z will yield the dependence of v_s on z; one can then integrate v_s along z and obtain an expression for the length-averaged filtration flux $v_{s,\mathrm{avg}}$ for the membrane length. This is a major goal for any model.

Solution of the particle concentration profile in the particle concentration boundary layer from ϕ_s in the feed suspension liquid to the concentration ϕ_w on top of the cake (and equal to the concentration in the cake) requires consideration of the particle transport equation in the boundary layer. We will proceed as follows. We will first identify the basic governing differential equations and appropriate boundary conditions (Davis and Sherwood, 1990) and then identify the required equations for an integral model and list the desired solutions from Romero and Davis (1988). However, we will first simplify the population balance equation (6.2.51c) for particles under conditions of steady state $(\partial n(r_p)/\partial t = 0)$, no birth and death processes ($B = 0 = De$), no particle growth ($\boldsymbol{U}_p^{\mathrm{int}} = 0$) and no particle velocity due to external forces ($\boldsymbol{U}_p = 0$), namely

$$\nabla \cdot (\boldsymbol{v} n(r_p)) = \nabla \cdot (D_p \nabla n(r_p)) \qquad (7.2.108)$$

to an equation in terms of particle volume fraction ϕ in the flowing liquid. Since $n(r_p)\mathrm{d}r_p$ represents the number of particles between the size r_p to $r_p + \mathrm{d}r_p$ per unit total

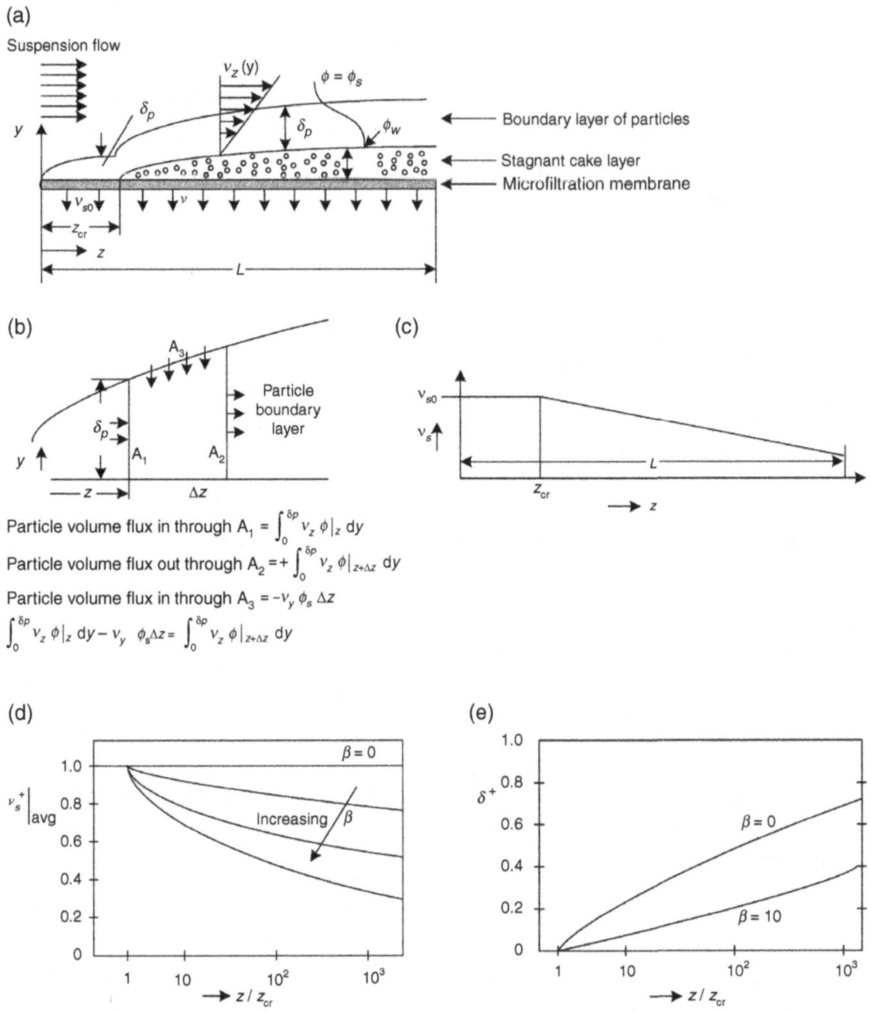

Figure 7.2.6. (a) Boundary layer of particles and a growing stagnant cake layer over a crossflow microfiltration membrane. (After Romero and Davis (1988).) (b) Control volume balance for obtaining equation (7.2.121). (c) Local filtration flux profile along membrane length. (d) Nondimensional length-averaged filtration flux for different β. (After Davis (1992).) (e) Nondimensional cake thickness profile. (After Davis (1992).)

volume, the volume fraction of such particles per unit total volume is $n(r_p)\ ((4/3)\pi r_p^3)\mathrm{d}r_p$. If the particles are of only one size, r_p, then

$$n(r_p)\left(\frac{4}{3}\pi r_p^3\right)\mathrm{d}r_p = \phi,$$

where ϕ is the volume fraction of particles in the liquid, and equation (7.2.108) becomes

$$\nabla\cdot(\boldsymbol{v}\phi) = \nabla\cdot(D_p\nabla\phi). \qquad (7.2.109)$$

The particle concentration boundary layer on top of the membrane is essentially two-dimensional: equation (7.2.109) is simplified to

$$\frac{\partial(v_y\phi)}{\partial y} + \frac{\partial(v_z\phi)}{\partial z} = \frac{\partial}{\partial y}\left(D_p\frac{\partial\phi}{\partial y}\right). \qquad (7.2.110)$$

The equation needed along with it is the two-dimensional equation of continuity for an incompressible liquid (6.2.51):

$$\frac{\partial v_y}{\partial y} + \frac{\partial v_z}{\partial z} = 0. \qquad (7.2.111)$$

While these two equations govern the particle concentration boundary layer for a given liquid velocity field, their solution has to be carried out with the applicable boundary conditions at $y = 0$ (surface of the membrane for $z \le z_{cr}$ and surface of the cake for $z \ge z_{cr}$) and $y = \delta_p(z)$ (the surface of the growing particle concentration boundary layer). If we consider the general expression for particle number flux in a given coordinate direction (equation (3.1.70), the expression for the y-direction is

$$n_{py} = -D_p \frac{\partial N_t}{\partial y} + U_{pyt} N_t, \tag{7.2.112}$$

where N_t is the total number of particles per unit volume of the liquid. Since U_{pyt} is equal to the liquid velocity v_y (by our earlier assumption) and $N_t(4/3)\pi r_p^3 = \phi$, the particle volume fraction in the liquid, the volume flux of particles in the y-direction is given by

$$\left(\frac{4}{3}\pi r_p^3\right) n_{py} = -D_p \frac{\partial \phi}{\partial y} + v_y \phi. \tag{7.2.113}$$

At the surface of the membrane ($z \leq z_{cr}$) and on the surface of the cake on the membrane ($z \geq z_{cr}$), the particle volume flux is zero, ϕ hits ϕ_w and the liquid velocity magnitude in the y-direction is equal to the local filtration volume flux v_s:

$$\text{at } y=0,\ -D_p \frac{\partial \phi}{\partial y} + v_y \phi = 0; \tag{7.2.114}$$

$$\text{at } y=0, \quad -v_y = v_s, \quad v_z = 0, \quad \phi = \phi_w; \tag{7.2.115}$$

$$\text{at } y > \delta_p, \phi = \phi_s. \tag{7.2.116}$$

When we neglect the axial convection in equation (7.2.110) and use boundary condition (7.2.114) in the integrated form of the particle concentration boundary layer equation (7.2.110), we obtain

$$\int_0^y d(v_y \phi) = \int_0^y d\left(D_p \frac{\partial \phi}{\partial y}\right); \qquad v_y\phi - v_y\phi|_0$$
$$= D_p \frac{\partial \phi}{\partial y} - D_p \frac{\partial \phi}{\partial y}\bigg|_0; \qquad v_y \phi = D_p \frac{\partial \phi}{\partial y} \tag{7.2.117}$$

in general, with the restriction that it is not valid at $y = \delta_p$. It is obviously always valid at $y = 0$.

A completely general solution of the governing convective diffusion equation (7.2.110), and equation (7.2.111), subject to the boundary conditions (7.2.114), (7.2.115) and (7.2.116) is not available. There are two types of solutions, similarity solutions and integral boundary layer solutions (apart from complete numerical solutions). Common to both of these solutions is the assumption that the particle concentration boundary layer is very thin compared to the membrane channel dimension normal to the axial flow; further, the shear stress due to the axial velocity gradient in the particle concentration boundary layer is equal to that at the wall, namely τ_w:

$$\mu\,(\partial v_z/\partial y) = \tau_w \tag{7.2.118}$$

for all $y \leq \delta_p$. We should note that the values of τ_w for laminar flow in tubes (of radius R) and channels (channel half height b, channel width W) are, respectively (Bird *et al.*, 2002),

$$\text{tubes: } \tau_w = (4\mu Q/\pi R^3); \tag{7.2.119a}$$

$$\text{channels: } \tau_w = (3\mu Q/2\,b^2\,W). \tag{7.2.119b}$$

These two membrane channel configurations are often encountered in crossflow microfiltration.

We will now briefly illustrate how the governing equation for the "integral model" for crossflow microfiltration developed by Romero and Davis (1988) is obtained. Then we will provide the results from such an analysis. Consider equation (7.2.110) and integrate it along the boundary layer thickness $y = 0$ to δ_p:

$$\int_0^{\delta_p} \frac{\partial(v_z\phi)}{\partial z} dy + \int_0^{\delta_p} \frac{\partial(v_y\phi)}{\partial y} dy = \int_0^{\delta_p} \frac{\partial}{\partial y}\left(D_p \frac{\partial \phi}{\partial y}\right) dy. \tag{7.2.120a}$$

Exchanging the integration and differentiation processes for the first term, we get

$$\frac{\partial}{\partial z}\int_0^{\delta_p} v_z\phi\, dy + \int_0^{\delta_p} d(v_y\phi) = \int_0^{\delta_p} d\left(D_p \frac{\partial \phi}{\partial y}\right)$$
$$= D_p \frac{\partial \phi}{\partial y}\bigg|_{\delta_p} - D_p \frac{\partial \phi}{\partial y}\bigg|_{y=0}. \tag{7.2.120b}$$

Now, at $y = \delta_p$, $(\partial\phi/\partial y) = 0$. Further, from the continuity equation (7.2.111),

$$\int_0^y dv_y = -\int_0^y \frac{\partial v_z}{\partial z} dy = v_y - v_y\big|_{y=o} = v_y + v_s. \tag{7.2.120c}$$

Substituting these into equation (7.2.120b), we get

$$\frac{\partial}{\partial z}\int_0^{\delta_p} v_z\phi\, dy + \int_0^{\delta_p} d\left\{\left(-v_s - \int_0^y \frac{\partial v_z}{\partial z} dy\right)\phi\right\} = -D_p \frac{\partial \phi}{\partial y}\bigg|_{y=0};$$

$$\frac{\partial}{\partial z}\int_0^{\delta_p} v_z\phi\, dy + \left(-v_s - \int_0^y \frac{\partial v_z}{\partial z} dy\right)\phi_s + v_s\phi\big|_{y=0} = -D_p \frac{\partial \phi}{\partial y}\bigg|_{y=0};$$

$$\frac{\partial}{\partial z}\int_0^{\delta_p} v_z(\phi - \phi_s)dy = +v_s\phi_s + \left(-v_s\phi\big|_{y=0} - D_p \frac{\partial \phi}{\partial y}\bigg|_{y=0}\right). \tag{7.2.120d}$$

The quantity in the parentheses on the right-hand side is the particle flux at $y = 0$; by the boundary conditions (7.2.114) and (7.2.115), it is zero. We are left with

$$\frac{\partial}{\partial z}\int_0^{\delta_p} v_z(\phi - \phi_s)dy = v_s\,\phi_s, \tag{7.2.121}$$

which is the governing equation for the integral model. If we take a thin control volume spanning the boundary layer thickness, having a thickness Δz in the axial direction (Figure 7.2.6(b)), then we can derive the above equation easily by a balance on the particle flux: the left-hand side identifies the net increase in particle flux across the boundary layer over thickness Δz, whereas the right-hand side identifies the particle flux coming in from the top surface of length Δz (remember, there is no flux at the bottom surface of the control volume, where all particles are rejected by the membrane-cake on top of it (Figure 7.2.6(b))).

We can derive result (7.2.121) from considerations of flux through these three surface areas of the control volume of length Δz in Figure 7.2.6(b):

particle volume flux in through surface A_1 of unit width

$$\text{perpendicular to the paper at } z = \left\{\int_0^{\delta} v_z\,\phi\,\mathrm{d}y\right\}_z. \tag{7.2.122a}$$

The corresponding particle volume flux in at surface A_2 at $z+\Delta z = \left\{\int_0^{\delta} v_z\,\phi\,\mathrm{d}y\right\}_{z+\Delta z}$.

Particle volume flux in through top boundary layer

$$\text{surface } A_3 \text{ is} = \phi_s(-v_y)|_{\delta_p}\,\Delta z. \tag{7.2.122b}$$

At steady state,

$$\left\{\int_0^{\delta} v_z\,\phi\,\mathrm{d}y\right\}_z - \left\{\int_0^{\delta} v_z\,\phi\,\mathrm{d}y\right\}_{z+\Delta z} + \phi_s(-v_y)|_{\delta_p}\Delta z = 0. \tag{7.2.122c}$$

In the limit of $\Delta z \rightarrow 0$, we get, first by dividing by Δz throughout and then taking the limit,

$$\frac{\partial}{\partial z}\int_0^{\delta_p} v_z\,\phi\,\mathrm{d}y + \phi_s\,(v_y)|_{\delta_p} = 0. \tag{7.2.122d}$$

From the continuity equation and result (7.2.120c),

$$v_y = -v_s - \int_0^{y}\frac{\partial v_z}{\partial z}\mathrm{d}y \Rightarrow (v_y)|_{\delta_p} = -v_s - \int_0^{\delta_p}\frac{\partial v_z}{\partial z}\mathrm{d}y. \tag{7.2.122e}$$

Substituting this into equation (7.2.122d), we get

$$\frac{\partial}{\partial z}\int_0^{\delta_p} v_z(\phi - \phi_s)\mathrm{d}y = v_s\,\phi_s, \tag{7.2.122f}$$

which is identical to equation (7.2.121).

Solution of equation (7.2.121) requires a knowledge of the particle concentration profile $\phi(y)$ and the velocity profile $v_z(y)$. This is achieved by solving equations (7.2.117) and (7.2.118) for $v_z(y)$ and $\phi(y)$ (Davis and Leighton, 1987). These profiles have been employed by Romero and Davis (1988) to calculate a quantity called the *excess particle flux*,

$$Q_p = \int_0^{\delta_p(z)} v_z(\phi - \phi_s)\mathrm{d}y = \int_0^{z} v_s\,\phi_s\,\mathrm{d}z. \tag{7.2.123}$$

The quantity on the right-hand side identifies the rate at which particles have been carried into the boundary layer over the membrane length 0 to z through the top boundary layer surface due to the filtration from the bulk suspension; the left-hand side reflects this via the excess particle volume fraction $(\phi - \phi_s)$ (over that in the feed suspension) in the boundary layer being convected axially. In Romero and Davis's (1988) analysis, a nondimensional form of Q_p is utilized:

$$Q_p^+ = \frac{Q_p\,\mu_0^3\,v_s^2}{\tau_w^3\,r_p^4}, \tag{7.2.124}$$

where r_p is the particle radius and τ_w is the wall shear stress.

To develop an expression for Q_p or Q_p^+, Romero and Davis (1988) proceeded as follows. Using equation (7.2.117) between the limits of ϕ_w and ϕ, and assuming $v_y = -v_s$,

$$\int_{\phi}^{\phi_w}\frac{D_p\,\mathrm{d}\phi}{\phi} = +\int_y^0 v_y\,\mathrm{d}y = -\int_y^0 v_s\,\mathrm{d}y = v_s y. \tag{7.2.125}$$

However, the shear-induced particle diffusivity $D_p(\phi)$ may be described via relation (3.1.74) and a nondimensional particle diffusivity D_p^+ as

$$D_p^+(\phi) = \frac{D_p(\phi)}{r_p^2\,\dot{\gamma}} = \frac{D_p(\phi)}{r_p^2|(\mathrm{d}v_z/\mathrm{d}y)|}. \tag{7.2.126}$$

Further, from the shear stress relation (7.2.118),

$$|\mathrm{d}v_z/\mathrm{d}y| = \tau_w/\mu(\phi) = \frac{\tau_w}{\mu_0\,\mu^+(\phi)},$$

where

$$\mu^+(\phi) = (\mu(\phi)/\mu_0) \tag{7.2.127}$$

is a nondimensional viscosity. Substituting these two relations into (7.2.125), we obtain

$$v_s y = \int_{\phi}^{\phi_w}\frac{D_p^+(\phi)}{\phi}r_p^2\left|\frac{\mathrm{d}v_z}{\mathrm{d}y}\right|\mathrm{d}\phi = \left(\int_{\phi}^{\phi_w}\frac{D_p^+(\phi)\,\mathrm{d}\phi}{\phi\,\mu^+(\phi)}\right)\frac{\tau_w}{\mu_0}r_p^2;$$

$$\frac{v_s\,\mu_0 y}{r_p^2\,\tau_w} = \int_{\phi}^{\phi_w}\frac{D_p^+(\phi)\mathrm{d}\phi}{\phi\,\mu^+(\phi)} = \left(\int_{\phi}^{\phi_w}\frac{D_p^+(\phi')\,\mathrm{d}\phi'}{\phi'\,\mu^+(\phi')}\right), \tag{7.2.128}$$

an implicit equation for the particle volume fraction ϕ as a function of y, the distance from the wall. The value of ϕ_w is, however, unknown; its maximum value, $\phi_{\max}$, is known, and is equal to the maximum packing volume fraction of, say, spheres, i.e. ~0.6 (Davis and Leighton, 1987). We need also an expression for v_z to develop an expression for Q_p^+ via (7.2.123) and (7.2.124). Use the wall shear stress relation (7.2.118) to obtain

$$v_z = \int_0^{y}\frac{\tau_w}{\mu}\mathrm{d}y = \frac{\tau_w}{\mu_0}\int_0^{y}\frac{\mathrm{d}y}{\mu^+(\phi)} \tag{7.2.129}$$

(since $v_z = 0$ at $y = 0$). Substitute this result, along with the differential form of (7.2.128) in terms of $\mathrm{d}y$, into (7.2.123) and (7.2.124):

$$Q_p^+ = \frac{\mu_0^3 v_s^2}{\tau_w^3 r_p^4}\int_0^{\delta_p(z)} \frac{\tau_w}{\mu_0}\left[\int_0^y \frac{1}{\mu^+(\phi)}\frac{r_p^2\tau_w}{v_s\mu_0}\frac{D_p^+(\phi)\,\mathrm{d}\phi}{\phi\,\mu^+(\phi)}\right](\phi-\phi_s)\frac{r_p^2\tau_w}{v_s\mu_0}\frac{D_p^+}{\phi\mu^+(\phi)}\,\mathrm{d}\phi$$

$$= \int_{\phi_s}^{\phi_w}\left[\int_{\phi}^{\phi_w}\frac{D_p^+(\phi')\,\mathrm{d}\phi'}{\phi'\mu^{+2}(\phi')}\right]\frac{(\phi-\phi_s)\,D_p^+(\phi)}{\phi\,\mu^+(\phi)}\,\mathrm{d}\phi$$

$$= \left(\int_0^z v_s\,\phi_s\,\mathrm{d}z\right)\frac{\mu_0^3 v_s^2}{\tau_w^3 r_p^4}. \qquad (7.2.130)$$

This is an implicit double integral based equation for the unknown ϕ_w. For given values of ϕ_s and Q_p^+, one can determine the value of ϕ_w at the wall via numerical integration (Romero and Davis, 1988), provided the functional dependences of $D_p(\phi)$ and $\mu(\phi)$ on ϕ are known. These relations are:

$$D_p^+(\phi) = 0.33\,\phi^2(1+0.5\,\exp\,(8.8\phi)) \qquad (7.2.131a)$$

(Leighton and Acrivos, 1987; Romero and Davis, 1988);

$$\mu^+(\phi) = ((0.58-0.13\phi)/(0.58-\phi))^2 \qquad (7.2.131b)$$

(Davis and Leighton, 1987).

Let us now focus on the right-hand side of equations (7.2.130) or (7.2.123). At $z=0$, the particle boundary layer starts forming (Figure 7.2.6(a)). The filtration flux value or the permeation velocity magnitude v_{s0} is constant for a certain length of the membrane. Therefore, on integration we obtain

$$Q_p^+ = \frac{\mu_0^3 v_{s0}^3\,\phi_s}{\tau_w^3 r_p^4}\,z. \qquad (7.2.132)$$

However, the wall volume fraction of particles, ϕ_w, keeps on increasing from ϕ_s at $z=0$. When it reaches the maximum value ϕ_{max} at z_{cr}, a cake layer starts forming for $z \geq z_{cr}$. The cake layer thickness δ_c keeps on increasing beyond z_{cr}:

$$z_{cr} = \frac{\tau_w^3 r_p^4}{\mu_0^3 v_{s0}^3\,\phi_s}\,Q_p^+|_{cr}. \qquad (7.2.133)$$

As a result, the filtration flux, v_s, defined by (7.2.107) keeps on decreasing with increasing δ_c, with $\hat{R}_{c\delta}$ assumed constant for an incompressible cake. The range of values of $Q_{p,cr}^+$ in crossflow microfiltration can vary between 10^{-7} and 10^3 (Davis and Leighton, 1987). These authors have also suggested that, for very dilute suspensions, the value of Q_p^+ should exceed 1×10^{-4} for $z \geq z_{cr}$. However, Romero and Davis (1988) have suggested[16] that, for $z \geq z_{cr}$ and a given ϕ_s, Q_p^+ is constant at the value given by (7.2.132),

$$Q_p^+(\phi_s) = Q_p^+(\phi_s)|_{cr} = \frac{\mu_0^3 v_{s0}^3\,\phi_s}{\tau_w^3 r_p^4}\,z_{cr}. \qquad (7.2.134)$$

If we now introduce the expression (7.2.107) for the filtration flux v_s influenced by the developing cake layer into the right-hand side integral of expression (7.2.130) for Q_p^+, we get

$$Q_p^+ = Q_p^+(\phi_s)|_{cr} = \frac{\mu_0^3 v_{s0}^2\,\phi_s \displaystyle\int_0^z \frac{v_{s0}\,\mathrm{d}z}{(1+(\hat{R}_{c\delta}\,\delta_c/R_m))}}{\tau_w^3 r_p^4(1+(\hat{R}_{c\delta}\,\delta_c/R_m))^2};$$

$$Q_p^+(\phi_s)|_{cr} = \frac{\mu_0^3 v_{s0}^3\,\phi_s}{\tau_w^3 r_p^4}\frac{1}{[1+(\hat{R}_{c\delta}\,\delta_c/R_m)]^2}\int_0^z \frac{\mathrm{d}z}{(1+(\hat{R}_{c\delta}\,\delta_c/R_m))}. \qquad (7.2.135)$$

Although one would like to consider this expression and determine the unknown δ_c as a function of z, the growing cake layer thickness in the axial direction, so that v_s may be determined from (7.2.107) as a function of z, we have to consider the effect of a growing δ_c on the shear stress τ_w. If we consider flow in a *membrane tube* (or *hollow fiber*) of radius R, then, for a constant bulk liquid flow rate Q, we may obtain τ_{wc}, the wall shear stress in the cake region, from (7.2.119a) as

$$\frac{\tau_w\,\pi R^3}{4\mu} = \frac{\tau_{wc}\,\pi(R-\delta_c)^3}{4\mu} \Rightarrow \tau_{wc} = \tau_w R^3/(R-\delta_c)^3$$

$$= \tau_w\Big/\left(1-\frac{\delta_c}{R}\right)^3. \qquad (7.2.136a)$$

Similarly for a *channel configuration*, from (7.2.119b) we obtain

$$\tau_{wc} = \tau_w\frac{b^2}{(b-\delta_c)^2} = \tau_w\Big/\left(1-\frac{\delta_c}{b}\right)^2. \qquad (7.2.136b)$$

The quantity τ_w in equation (7.2.135) should now be replaced by τ_{wc}. On substituting these two results into equation (7.2.135), we get,

membrane tube :

$$Q_p^+(\phi_s)|_{cr} = \frac{\mu_0^3 v_{s0}^3\,\phi_s\,(1-\frac{\delta_c}{R})^9}{\tau_w^3 r_p^4\,[1+(\hat{R}_{c\delta}\,\delta_c/R_m)]^2}\int_0^z \frac{\mathrm{d}z}{(1+(\hat{R}_{c\delta}\,\delta_c/R_m))}; \qquad (7.2.137a)$$

membrane channel :

$$Q_p^+(\phi_s)|_{cr} = \frac{\mu_0^3 v_{s0}^3\,\phi_s\,(1-\frac{\delta_c}{b})^6}{\tau_w^3 r_p^4\,[1+(\hat{R}_{c\delta}\,\delta_c/R_m)]^2}\int_0^z \frac{\mathrm{d}z}{(1+(\hat{R}_{c\delta}\,\delta_c/R_m))}, \qquad (7.2.137b)$$

where τ_w corresponds to the membrane tube/channel without any cake layer. Define the following nondimensional quantities (Davis, 1992):

[16] Davis and Leighton (1987) have shown that, for a given ϕ_s, the value of the double integral representing $Q_p^+(\phi_s)|_{cr}$ becomes constant after $\phi_w \Rightarrow \phi_{max} \sim 0.6$.

$$z^+ = z/z_{cr}; \quad v_s^+ = v_s/v_{s0} \Rightarrow \delta^+|_{\text{tube}} = \delta_c/R; \quad \beta|_{\text{tube}} = \hat{R}_{c\delta}\, R/R_m;$$

$$\delta^+|_{\text{channel}} = \frac{\delta_c}{b}; \quad \beta|_{\text{channel}} = \hat{R}_{c\delta} b/R_m. \tag{7.2.138}$$

Equation (7.2.137a) for a membrane tube is reduced to

$$\frac{(1-\delta^+)^9}{(1+\beta\,\delta^+)^2}\left[\int_0^1 \frac{dz^+}{1} + \int_1^{z^+} \frac{dz^+}{1+\beta\delta^+}\right] = 1. \tag{7.2.139}$$

Equation (7.2.137b) for a membrane channel is similarly reduced to

$$\frac{(1-\delta^+)^6}{(1+\beta\,\delta^+)^2}\left[\int_0^1 \frac{dz^+}{1} + \int_1^{z^+} \frac{dz^+}{1+\beta\delta^+}\right] = 1. \tag{7.2.140}$$

For given values of the parameter β, one can solve this equation numerically for either a membrane tube or a membrane channel and describe how δ^+ varies with z^+. Since a nondimensional expression for the filtration flux may be developed from (7.2.107) and (7.2.138) as

$$v_s^+ = \frac{1}{(1+\beta\delta^+)}, \tag{7.2.141}$$

the length-averaged filtration flux, $v_s^+|_{\text{avg}}$, is obtained as

$$\frac{1}{z^+}\int_0^{z^+} \frac{v_s^+\, dz^+}{1} = v_s^+|_{\text{avg}}(z^+). \tag{7.2.142}$$

Analytical expressions may be developed for the following limiting cases (Davis, 1992):

Membrane resistance controls

$$v_s^+=1; \;\; \delta^+|=1-(z^+)^{-1/9}; \;\; \delta^+|=1-(z^+)^{-1/6} \; \left(\beta<<1, z^+\geq 1\right).$$
$(\beta<<1)$ tube channel

(7.2.143a)

Filtration flux is constant at v_{s0} along the whole length (Figure 7.2.6(d)). For large values of z ($z^+ \geq 1$), δ_c is large and becomes a significant part of the membrane tube radius or channel height (Figure 7.2.6(e)).

Cake resistance controls ($\beta >> 1$, $z^+ \geq 1$)

$$\delta^+ = \frac{1}{\beta}\left[\left(\frac{3z^+}{2} - \frac{1}{2}\right)^{1/3} - 1\right]; \quad v_s^+ = \left(\frac{3z^+}{2} - \frac{1}{2}\right)^{-1/3}. \tag{7.2.143b}$$

Filtration flux decreases significantly as z^+ increases; δ_c increases with z^+, but slowly, and its magnitude remains small. The average flux over a length L of the membrane is obtained from (7.2.142) by using (7.2.143a, b). For $\beta >> 1$, and $z^+ = (L/z_{cr}) >> 1$ with the cake resistance controlling, we can obtain

$$v_s^+|_{\text{avg}} \cong \frac{1}{(L/z_{cr})}\int_0^{(L/z_{cr})} \frac{dz^+}{\left(\frac{3}{2}z^+\right)^{1/3}} \cong \left(\frac{2}{3}\right)^{1/3}\frac{3}{2}\frac{1}{L^+}(L^+)^{2/3}, \tag{7.2.144a}$$

where $L^+ = L/z_{cr}$;

$$v_s^+|_{\text{avg}} \cong \left(\frac{3}{2}\right)^{2/3}\left(\frac{z_{cr}}{L}\right)^{1/3} \Rightarrow v_s|_{\text{avg}} = 1.31\, v_{so}\, z_{cr}^{1/3}/L^{1/3} \tag{7.2.144b}$$

$$\Rightarrow v_s|_{\text{avg}} = 1.31\, v_{so}\left(\frac{\tau_w^3 r_p^4\, Q_p^+(\phi_s)|_{cr}}{L\mu_0^3\, v_{so}^3\, \phi_s}\right)^{1/3}$$
$$= 1.31\frac{\tau_w}{\mu_0}\left(\frac{r_p^4 Q_p^+(\phi_s)|_{cr}}{\phi_s L}\right)^{1/3};$$

$$v_s|_{\text{avg}} = 1.31\,\dot{\gamma}\,\mu^+(\phi_s)\left(\frac{r_p^4\, Q_p^+(\phi_s)|_{cr}}{\phi_s L}\right)^{1/3}, \tag{7.2.145}$$

where we have employed relations (7.2.133), (7.2.118) and (7.2.127); $\dot{\gamma}$ is the value of the shear rate in the boundary layer. The averaged filtration flux is proportional to the shear rate in the boundary layer; it is independent of the pressure drop applied. However, in the case of membrane resistance control,

$$v_s = v_{s0} \propto \Delta P, \tag{7.2.146}$$

i.e. the averaged filtration flux is proportional to the applied ΔP.

We will now illustrate graphically the behavior of the length-averaged flux and the cake layer thickness with distance along the membrane length. In Figure 7.2.6(c), the filtration flux profile along the membrane length has been shown for a given set of conditions. The flux is constant till $z = z_{cr}$, after which it decreases steadily. In Figure 7.2.6(d), the nondimensional filtration flux profile along the nondimensional membrane length has been illustrated parametrically for β as a parameter; it shows how the filtration flux decreases as the cake resistance increases, relative to the membrane resistance. This decrease is due to increasing cake thickness along the membrane length for nonzero values of β, i.e. $\hat{R}_{c\delta}$, as shown in Figure 7.2.6(e).

We observed earlier in dead-end cake filtration (equation (6.3.135k)) that $\hat{R}_{c\delta}$ varies inversely with the square of the particle radius; therefore, in effect, the filtration flux v_s varies with the square of the particle radius for cake dominated filtration. The larger the particle radius, the higher the filtration flux. As shown in equation (7.2.145), in cross-flow microfiltration also the averaged filtration flux increases with particle radius, here as $(r_p)^{4/3}$. Romero and Davis (1988) have shown via calculations based on their integral model that, for suspensions having particles larger than 4.3 μm radius, the flux of solvent will correspond to that with zero cake resistance, i.e. the flux will correspond to that of a clean membrane. However, the shear rate is also important. Unless the shear rate is around 660 s^{-1}, the flux for a suspension of particles of 5 μm radius was shown by Romero and Davis (1988) to be less than that of a clean filter.[17] An additional quantity to be taken into account is the deformability of the particles, especially valid for cellular suspensions, e.g. red blood cells in blood; the cake resistance will be much higher and the cake will be more compact.

One would now like to determine the length-averaged flux for a given membrane undergoing microfiltration under given conditions. Although we have the nondimensional expressions (7.2.143a) for the solvent flux for the membrane dominated case and (7.2.143b) for the cake resistance controlled situation, determination of the length-averaged flux, as well as the local flux for the latter, is complicated due to two quantities, $Q_p^+(\phi_s)|_{cr}$ and z_{cr}. Knowledge of one will yield the other from relation (7.2.133). The quantity $Q_p^+(\phi_s)|_{cr}$ has to be obtained from the double integral (7.2.130) for the limit of $\phi_w = \phi_{max}$ for appropriate dependences of $D_p^+(\phi)$ and $\mu^+(\phi)$ on ϕ. Davis (1992) and Romero and Davis (1988) have provided results (1) and (2), and (3), respectively:

$$(1)\ \lim_{\phi_s\to 0} Q_p^+(\phi_s)|_{cr} = 9.79\times 10^{-5}; \qquad (7.2.147a)$$

$$(2)\ Q_p^+(\phi_s)|_{cr} = 9.79\times 10^{-5}(1-4.38\,\phi_s), \text{for}\ \phi_s \le 0.1; \qquad (7.2.147b)$$

$$(3)\ \left[\frac{Q_p^+(\phi_s)|_{cr}}{\lim_{\phi_s\to 0} Q_p^+(\phi_s)|_{cr}}\right] = 1-3.8\,\phi_s, \text{for}\ \phi_s \le 0.2. \qquad (7.2.147c)$$

One can now estimate the required quantities, at least for $\phi_s \le 0.2$, which is a high enough particle volume fraction for suspensions.

Example 7.2.7 In a ceramic tubular microfilter of diameter 0.30 cm and length 40 cm, a suspension having particles of radius 0.35 μm flows; the suspension particle volume fraction is 0.005. The value of the viscosity, μ_0, is 10^{-2} g/cm-s (1 centipoise) and the wall shear stress is 10 g/cm-s^2. The observed flux for the microfilter without any particles is 3.5 × 10^{-3} cm/s for an applied pressure difference of 10^5 kPa. The conditions are such that one can assume a cake resistance controlled operation. Make an estimate of what the average filtration flux will be at 25 °C. Estimate the same if the particle radius were 4.3 μm.

Solution Since we can assume cake resistance control, we will employ equation (7.2.145) for the average filtration flux:

$$v_s|_{avg} = 1.31\,\dot{\gamma}\,\mu^+(\phi_s)\left(\frac{r_p^4\,Q_p^+(\phi_s)|_{cr}}{\phi_s\,L}\right)^{1/3};$$

$$\dot{\gamma}\,\mu^+(\phi_s) = \frac{\tau_w}{\mu_0} = \frac{10\,\text{g/cm-s}^2}{10^{-2}\,\text{g/cm-s}} = 10^3\ \text{s}^{-1};$$

$r_p = 0.35\times 10^{-4}$cm; $\phi_s = 0.005$; $L = 40$ cm. From equation (7.2.147b),

$$Q_p^+(\phi_s)|_{cr} = 9.79\times 10^{-5}(1-4.38\times 0.005) = 9.79\times 10^{-5}\times 0.979;$$

from equation (7.2.131b),

$$\mu^+(\phi_s) = \left(\frac{(0.58-0.13\times 0.005)}{(0.58-0.005)}\right)^2 \cong 1.01;$$

$$v_s|_{avg} = 1.31\times 10^3\times 1.01\left(\frac{3.5^4\times 10^{-20}\times 9.79\times 0.979\times 10^{-5}}{0.005\times 40}\right)^{1/3}$$

$$= 1310\left(\frac{150\times 9.58\times 10^{-25}}{0.2}\right)^{1/3} = 11867\times 10^{-8} = 0.00011\,\text{cm/s}.$$

We will now calculate the value of $v_s|_{avg}$ by assuming the particle radius to be 4.3 μm. We should not assume either membrane control or cake resistance control. However, that will require an extensive numerical effort. We will use the same formula as used earlier to estimate the effect of a particle size increase:

$$v_s|_{avg} = 1.18\times 10^{-4}\left(\frac{4.3}{0.35}\right)^{4/3}\frac{\text{cm}}{\text{s}} = 1.18\times 10^{-4}(12.28)^{4/3}$$

$$= 12.28\times 2.308\times 1.18\times 10^{-4}$$

$$\Rightarrow v_s|_{avg} = 3.344\times 10^{-3}\,\text{cm/s},$$

which is close to the value of the clean filter flux.

The above example illustrates clearly how large a flux reduction can take place in microfiltration compared to the flux in a clean filter. In general, when crossflow microfiltration is initiated, one observes immediately a sharp flux decline with time. This immediate flux decline is due to the increase in the filtration resistance due to the buildup of the cake layer on the membrane; this buildup occurs within minutes to an hour from startup and can easily reduce the value of $v_s|_{avg}$ to one-tenth of the clean filter flux value, or an even greater reduction. A brief introduction to modeling such a transient flux behavior on its way to steady state has been provided by Davis (1992). There is an additional flux decay over a much longer period of time, e.g. days, which takes place due to slow processes of membrane fouling, compaction of the cake under the applied ΔP and membrane compaction.

Such losses of membrane solvent flux is avoided now a days in practice by what is known as *backflushing*. The cake layer on the membrane is built up by the applied ΔP over a period of time as the liquid flows perpendicular to this

[17] As the particle size becomes larger, the phenomenon of *inertial lift* (Green and Belfort, 1980) becomes important.

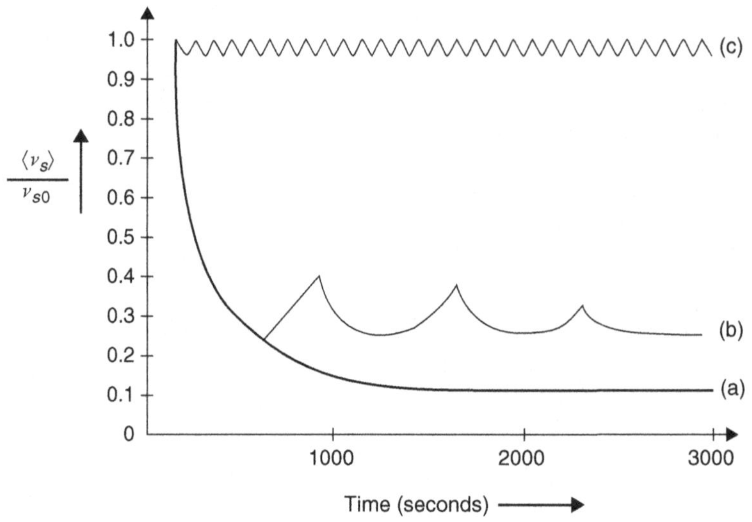

Figure 7.2.7. Filtration flux decay with time after starting and the effect of backflushing: (a) no backflushing; (b) slow backflushing introduced after some time (low frequency); (c) rapid backflushing with high frequency initiated immediately after starting. (After Bhave (1991).)

force direction. If the force direction can be reversed, the cake layer may be completely lifted off the membrane and swept away by the tangential flow. The backflushing mode is based on this strategy. Every so often, the permeate side pressure is suddenly increased to above the feed pressure. A small amount of earlier permeated filtrate is forced back into the feed side (and is therefore lost from the permeate): as it comes, it lifts off the cake and cleans the membrane pores as well if there is some membrane fouling. The tangential flow on the feed side sweeps off the dislodged debris from the destruction of the cake layer. Figure 7.2.7 illustrates this behavior. Curve (a) illustrates the rapid flux decay during the first 30 minutes to 1 hour of conventional crossflow microfiltration. Curve (b) illustrates the flux behavior for low values of backflush frequency started after some time. Curve (c) illustrates the flux for a high backflush frequency initiated after a very short interval, as shown in the figure. Experience indicates that it is necessary to introduce rapid backflushing if one wants to obtain high flux, and further that it must be introduced immediately after the flow is introduced (Bhave, 1991).

Recent research into the reduction of filtration flux that occurs as soon as the microfiltration process is started (Figure 7.2.7(a)) indicates that it is very much dependent on the level of the flux and therefore the applied ΔP. Beyond a *critical flux*, $v_{s,\mathrm{cr}}$, deposits form on the membrane since the forces dragging the colloidal particles in suspension toward the membrane are larger than the forces causing the particles to move away from the membrane. If this *critical flux value* is reached at the end of the membrane channel, where the boundary layer is the thickest, then the membrane is unlikely to experience fouling. Thus the value of applied ΔP should be reduced and operation continued at a *subcritical flux level* to avoid membrane fouling (Field *et al.*, 1995; Howell, 1995).

7.2.1.4.1 crossflow microfiltration-operational configurations In Section 6.3, we observed in many examples that the magnitude of the bulk flow velocity which was parallel to the direction of the force could not be arbitrarily increased, otherwise the separation achieved could be severely damaged or the separation process halted. Further, in many processes, there was unsteadiness due to this bulk flow vs. force configuration. Clearly, in crossflow microfiltration we have overcome this unsteadiness. Further, in order to maintain a thin caker layer, we maintain a high wall shear rate in the tangential flow field. In most crossflow microfiltration devices[18] this means that the bulk flow velocity in the tangential flow direction (z) is quite high; on the other hand, the filtration velocity continues to be limited by the membrane, the cake or both. Thus, the ratio (v_z/v_s) in crossflow microfiltration is high, unlike that in dead-end microfiltration, where its value is 1. In addition, in the latter, v_s is likely to be significantly smaller than that in crossflow microfiltration due to a larger cake thickness; also, in the dead-end mode, with time, the flux decreases at constant ΔP.

---

[18] In those devices where the tangential shear rate can be generated by means other than bulk motion, namely Dean vortices or Taylor vortices, the correlation between shear rate and the bulk velocity may be weak or zero.

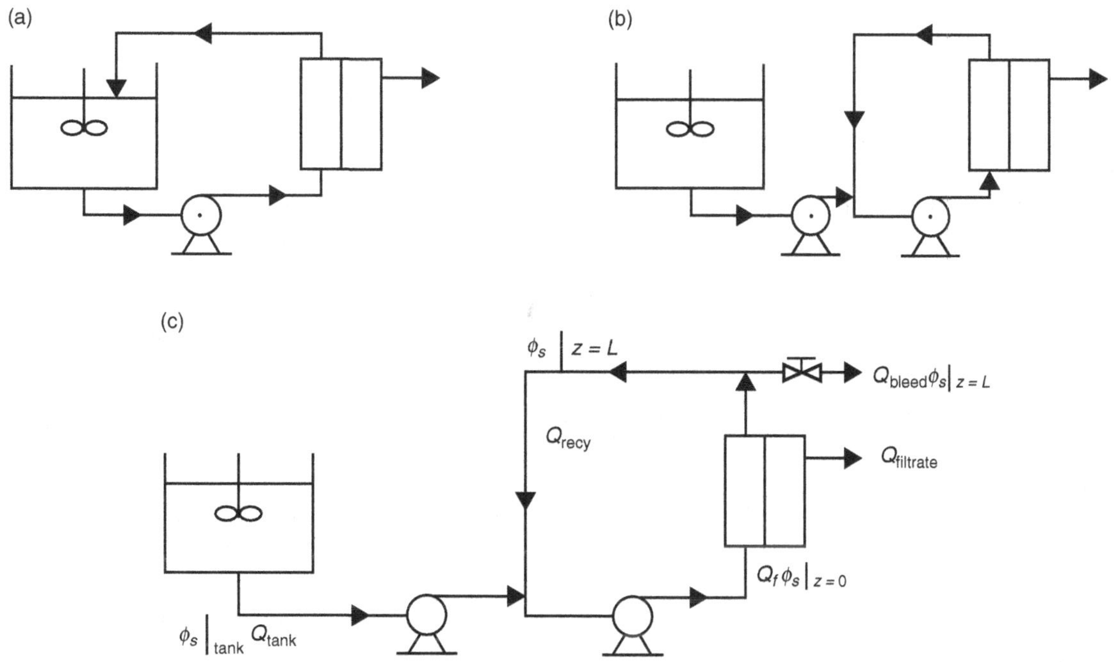

Figure 7.2.8. Operational configurations of microfiltration: (a) batch concentration - open system; (b) batch concentration - closed system; (c) feed and bleed operation in the continuous mode.

A result of this high bulk flow velocity and high shear rate in crossflow microfiltration is that the suspension circulation rate is high when compared with the microfiltration rate through the membrane. Either one provides a very long membrane device or one keeps on circulating the solution from a reservoir/tank over a much smaller length of the membrane for an extended length of time. The latter is the most common mode of crossflow microfilter operation and is shown in Figure 7.2.8. This mode of batch concentration can be carried out in two ways: open system and closed system (Bhave, 1991). In an *open system*, the concentrate from the module is sent back to the reservoir for recirculation through a single pump; in a *closed system*, there are two pumps, and the energy loss associated with bringing the liquid suspension from the tank to full pressure is avoided. Mass balance considerations for a suspension volume of $V_s(t)$ having a particle volume fraction $\phi_s(t)$ being microfiltered through a membrane area A_m leads to

$$V_s(t)|_{t=0} - V_s(t)|_t = A_m \int_0^t v_s|_{avg}\, dt, \qquad (7.2.148)$$

where $v_s|_{avg}$ for the membrane of area A_m changes with time since the suspension particle volume fraction $\phi_s(t)$ is time-dependent, as the suspension volume is reduced with time. The corresponding increase in the volume fraction of particles in suspension from $\phi_s(t)|_{t=0}$ to $\phi_s(t)$ at any time t is related to the suspension volume reduction by

$$V_s(t)|_{t=0}\ \phi_s(t)|_{t=0} = V_s(t)\ \phi_s(t), \qquad (7.2.149)$$

since all solid particles are assumed to be rejected by the membrane.

One feature of the closed system described above is a rapid increase in ϕ_s in the recirculation loop. One could avoid it by continually bleeding off a small portion of the retentate/concentrate, as shown in Figure 7.2.8(c), in a method called *feed and bleed*. This bleed volume is small since the increase in concentration of ϕ_s per pass through the filter is also small. If we indicate by $\phi_s|_{z=0}$ and $\phi_s|_{z=L}$ the values of ϕ_s at the inlet and outlet of the microfilter, respectively, then, for a totally rejecting filter,

$$Q_f\ \phi_s|_{z=0} = Q_2\ \phi_s|_{z=L}, \qquad (7.2.150)$$

where Q_f and Q_2 are the volumetric suspension flow rates at the filter inlet and the concentrate end, respectively. The flow rate ratio Q_f/Q_2 (>1) is close to 1 since the recirculation rates are high; therefore $\phi_s|_{z=L}\ /\ \phi_s|_{z=0}$ is also close to 1 (but >1). However, this flow rate Q_f into the filter consists of the sum of the flow rate Q_{tank} out of the tank and the flow being recirculated Q_{recy} (= Q_2), the latter being R times the bleed flow rate Q_{bleed}:

$$Q_f = Q_{tank} + Q_{recy} = Q_{tank} + R\,Q_{bleed};$$
$$Q_f\ \phi_s|_{z=0} = Q_{tank}\ \phi_s|_{tank} + Q_{recy}\ \phi_s|_{z=L}. \qquad (7.2.151)$$

From the second equation, one can obtain

$$\phi_s|_{z=0} = (Q_{\text{tank}}/Q_f)\phi_s|_{\text{tank}} + (RQ_{\text{bleed}}/Q_f)\phi_s|_{z=L}. \quad (7.2.152)$$

However, to obtain steady state operation, the increase in solids volume flow rate due to filtration $(Q_f - Q_2)\,\phi_s|_{z=0}$ must not affect steady state operation:

$$\text{volume flow rate balance}: Q_f = Q_{\text{recy}} + Q_{\text{bleed}} + Q_{\text{filtrate}}; \quad (7.2.153)$$

The particle volume fraction balance around the filter is given by

$$\begin{aligned} Q_f\,\phi_s|_{z=0} &= Q_{\text{recy}}\,\phi_s|_{z=0} + Q_{\text{bleed}}\,\phi_s|_{z=0} + Q_{\text{filtrate}}\,\phi_s|_{z=0} \\ &= Q_{\text{recy}}\,\phi_s|_{z=L} + Q_{\text{bleed}}\,\phi_s|_{z=L}. \end{aligned} \quad (7.2.154)$$

Rearrange the last equation to obtain

$$\begin{aligned} Q_{\text{filtrate}}\,\phi_s|_{z=0} &= (Q_f - Q_2)\,\phi_s|_{z=0} \\ &= (Q_{\text{bleed}} + Q_{\text{recy}})\,(\phi_s|_{z=L} - \phi_s|_{z=0}); \end{aligned}$$

$$\frac{Q_f - Q_2}{Q_{\text{bleed}} + Q_{\text{recy}}} = \frac{\phi_s|_{z=L}}{\phi_s|_{z=0}} - 1 = \frac{Q_{\text{filtrate}}}{Q_2};$$

$$\frac{\phi_s|_{z=L}}{\phi_s|_{z=0}} = 1 + \frac{Q_{\text{filtrate}}}{Q_2} = \frac{Q_f}{Q_2}, \quad (7.2.155)$$

where the ratio on the left-hand side is also called the volume concentration factor (VCF). When the desired VCF value is higher, then the bleed stream from the first stage could be used as if it were the feed stream from the tank for the second stage in a multistage arrangement.

7.2.1.5 *Rotary vacuum filtration*

Consider Figure 7.2.9 which illustrates schematically a drum rotating around its axis. The drum length (not shown) is L. As the drum rotates at an angular velocity of ω radian/s, part of the outside of the drum remains submerged in the suspension to be filtered. The drum surface has a filter medium exposed to the suspension at atmospheric pressure. Inside the porous drum surface, a partial vacuum is maintained so that the filter medium has a ΔP across it from the suspension. As filtration takes place, particles are deposited on the outside of the filter medium surface. As a particular section of the drum/filter medium rotates through the suspension bath, there is continuing filtration; so there is a cake buildup on the filter medium as it goes into the suspension and then comes out of the suspension after some time. During this movement through the suspension, the partial vacuum applied brings out filtrate into the inside of the drum.

As this section of the drum comes out of the suspension, nozzles on the outside spray clean liquid onto the drum filter medium surface and wash out any solution remaining in the cake built up on the filter. This removes any valuable product left in the cake via filtration into the filtrate inside the drum operating under partial vacuum. Only a section of the drum circumference is exposed to this cake washing. The next environment encountered by the cake on the rotating drum surface (Figure 7.2.9) is essentially dry air. The vacuum inside the drum continues to remove any remaining liquid in the cake on the filter medium. The fourth and final step in this cyclic process involves removal of the filter cake formed on the filter medium via a knife blade.

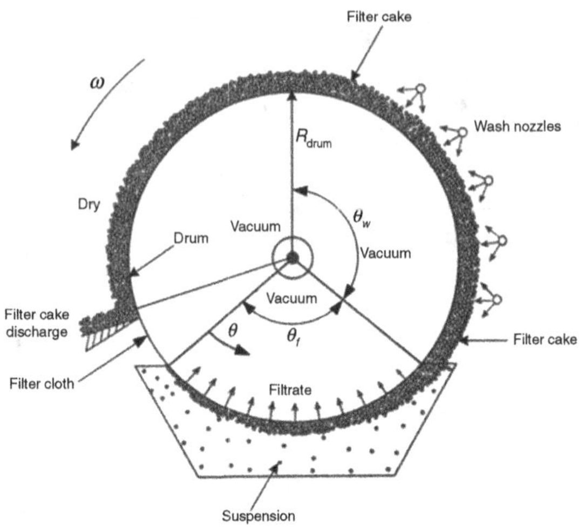

Figure 7.2.9. Schematic of a rotary vacuum filter.

In conventional application, no coating is applied to the filter medium. In the filtration of fermentation broths, however, a thin but controlled layer of a filter aid, e.g. diatomaceous earth, is applied onto the filter medium before the drum enters the suspension as a "precoat." The filtration of the fermentation broth leads to a buildup of the cake over it. When the knife blade is applied at the end of one cycle, most often only a particular fraction of this "precoat" of filter aid is removed. The porous precoat facilitates extended treatment of the broth, which tends to create a complex compressible cake. If the filtrate contains the desired and valuable material, the cake discharged by the knife blade is waste; filter aid material added as a precoat contaminates the cake and increases the volume of this waste. It also changes the nature of the waste, which will now contain, in addition to cellular debris, cells, etc., siliceous material.

We have already pointed out in the introduction to Section 7.2 that the separation configuration of this rotary vacuum filtration technique corresponding to Figure 6.3.25(a) may be conceived of as a crossflow filtration technique (Figure 6.3.25(b)) if we fix the coordinate system to the filter medium/drum surface. In this configuration, the suspension/slurry will be moving with the linear velocity of the drum surface where the filter medium is located. However, this velocity, and the corresponding tangential shear stress magnitude, are rather low and are incapable of affecting the cake thickness buildup (unlike that in crossflow microfiltration studied in Section 7.2.1.4) (Davis and Grant, 1992). Therefore, one should consider the cake buildup with rotation (as angle θ in Figure 7.2.9 increases) as if we have dead-end filtration (Section 6.3.3.1), where (θ/ω) may replace time, t. Correspondingly, one can employ expressions developed earlier in dead-end cake filtration for the time-dependent cake thickness $\delta_c(t)$ (6.3.138c) and the filtration rate/volume flux for constant ΔP, $v_s(t)$, (6.3.138d) and write here (Davis and Grant, 1992)

$$\delta_c(\theta) = (R_m/\hat{R}_{c\delta})\left[\left\{1+\frac{2\hat{R}_{c\delta}\,\phi_s\,\Delta P\,\theta}{(\phi_c-\phi_s)\,\mu\,R_m^2\,\omega}\right\}^{1/2}-1\right]; \quad (7.2.156)$$

$$v_s(\theta) = \frac{\Delta P}{\mu R_m}\left\{1+\frac{2\hat{R}_{c\delta}\,\phi_s\,\Delta P\,\theta}{(\phi_c-\phi_s)\,\mu\,R_m^2\,\omega}\right\}^{-1/2}. \quad (7.2.157)$$

Note that here R_m includes the resistance due to any precoat applied to the membrane.

The question of interest is: what is the volumetric rate of filtration in this rotating drum (of total surface area $2\pi R_{\text{drum}}L$) due to the submerged fraction of its surface? That will yield the device filtration rate (unless the suspension properties change with time - they will for a batch operation!) since, as long as the drum rotates, part of it is always carrying out filtration (fractional surface area submerged, $\theta_f/2\pi$; total surface area submerged at any time, $LR_{\text{drum}}\,\theta_f$). This quantity may be obtained by integrating expression (7.2.157) for the filtration flux over the range of angles $\theta = 0$ to θ_f, the angle of contact with the suspension (over the time period the surface area of the drum is in contact with the suspension):

$$Q_{\text{filtrate}} = R_{\text{drum}}\,L\int_0^{\theta_f} v_s(\theta)\,d\theta$$

$$= R_{\text{drum}}\,L\int_0^{\theta_f}\frac{\Delta P}{\mu R_m}\left[1+\frac{2\hat{R}_{c\delta}\,\phi_s\,\Delta P\,\theta}{(\phi_c-\phi_s)\,\mu\,R_m^2\,\omega}\right]^{-1/2} d\theta; \quad (7.2.158)$$

$$Q_{\text{filtrate}} = \frac{R_{\text{drum}}L(\phi_c-\phi_s)R_m\omega}{\hat{R}_{c\delta}\,\phi_s}\left\{\left[1+\frac{2\hat{R}_{c\delta}\,\phi_s\,\Delta P\,\theta_f}{(\phi_c-\phi_s)\,\mu\,R_m^2\,\omega}\right]^{-1/2}-1\right\}. \quad (7.2.159)$$

If the cake resistance is much larger than the membrane (+ precoat) resistance, then $\hat{R}_{c\delta}\delta >> R_m$; correspondingly, we have already observed in Example 6.3.4 that

$$v_s(t) = \left\{\frac{2\hat{R}_{c\delta}\,\phi_s\,\mu\,t}{(\phi_c-\phi_s)\,\Delta P}\right\}^{-1/2}.$$

We follow now the same procedure we followed earlier to obtain expression (7.2.159) for Q_{filtrate} from expression (7.2.157) for $v_s(\theta)$:

$$v_s(\theta) = \left\{\frac{2\hat{R}_{c\delta}\,\phi_s\,\mu\,\theta}{(\phi_c-\phi_s)\,\Delta P\,\omega}\right\}^{-1/2}; \quad (7.2.160)$$

$$Q_{\text{filtrate}} = R_{\text{drum}}\,L\int_0^{\theta_f} v_s(\theta)\,d\theta$$

$$= R_{\text{drum}}\,L\left\{\frac{2\,(\phi_c-\phi_s)\omega\,\Delta P\,\theta_f}{\hat{R}_{c\delta}\,\phi_s\,\mu}\right\}^{1/2}. \quad (7.2.161)$$

However, since the cakes developed in rotary vacuum filtration are most often compressible, we employ expression (6.3.138j) for $\hat{R}_{c\delta}$ for a compressible cake to obtain

$$Q_{\text{filtrate}} = R_{\text{drum}}\,L\,\theta_f\left\{\frac{2\,(\phi_c-\phi_s)\,(\Delta P)^{1-s}\,\omega}{\rho_s\,\phi_s\,\phi_c\,\alpha_0\,\mu\,\theta_f}\right\}^{1/2}. \quad (7.2.162)$$

If t_{cycle} is the time taken by the drum to go through one rotation, then the fraction of the time the drum is in contact with the suspension is given by

$$\frac{t_{\text{contact}}}{t_{\text{cycle}}} = \frac{\theta_f}{2\pi}; \qquad t_{\text{contact}} = \frac{\theta_f}{2\pi}\,t_{\text{cycle}} = \frac{\theta_f}{\omega}. \quad (7.2.163)$$

Correspondingly, the drum surface area in contact with the suspension is $R_{\text{drum}}\;L\;\theta_f$. Therefore, from (7.2.162), the filtrate production rate, Q_{filtrate}, is linearly proportional to $R_{\text{drum}}\;L\;\theta_f$ as well as to $(t_{\text{contact}})^{-1/2}$. Further, in expression (7.2.162) for dilute suspensions ($\phi_s << 1$), $\{\phi_c - \phi_s)/\phi_c\}$ is ~1 and $\rho_s\;\phi_s$ is the mass concentration of particles in suspension (g/cm^3 or g/liter of suspension).

We will briefly consider now the step of cake washing. This step is undertaken to recover the solutes left in the solution trapped in the cake mass. As the wash liquid displaces this trapped solution, the latter goes through the filter medium and is recovered in the filtrate inside the drum. Therefore, washing enhances the recovery of soluble fermentation products left in the liquid in the cake mass in the case of a fermentation broth. Since the wash liquid is suspension-free as well as solute-free, the cake filtration characteristics are unlikely to be altered. Hence, the filtration rate during washing should correspond to the filtration rate existing at the location where the drum came out of the suspension:

$$v_s(\theta_f) = \left[\frac{2\hat{R}_{c\delta}\,\phi_s\,\mu\,\theta_f}{(\phi_c-\phi_s)\,\Delta P\,\omega}\right]^{-1/2}. \quad (7.2.164)$$

Correspondingly, the overall filtration rate during washing is

$$Q_{\text{filtrate/washing}} = R_{\text{drum}}L\int_0^{\theta_w}\left[\frac{2\hat{R}_{c\delta}\phi_s\mu\theta_f}{(\phi_c-\phi_s)\Delta P\omega}\right]^{-1/2} d\theta$$

$$= R_{\text{drum}}L\theta_w\left[\frac{2\hat{R}_{c\delta}\phi_s\mu\theta_f}{(\phi_c-\phi_s)\Delta P\omega}\right]^{-1/2}. \quad (7.2.165)$$

Introducing the expression (6.3.138j) for $\hat{R}_{c\delta}$, we get

$$Q_{\text{filtrate/washing}} = R_{\text{drum}}\, L\, \theta_w \left[\frac{(\phi_c - \phi_s)\,(\Delta P)^{1-s}\,\omega}{2\,\rho_s\,\phi_c\,\phi_s\,\alpha_0\,\mu\,\theta_f}\right]^{1/2}. \tag{7.2.166}$$

The volume of the wash liquid employed has been found to be a factor in the extent of recovery of soluble materials. The fraction of solute remaining after the wash has been related to the washing efficiency, η_{wash}, by Choudhury and Dahlstrom (1957) via the following expression:

$$\text{solute fraction remaining} = (1 - \eta_{\text{wash}})^{vr}, \tag{7.2.167}$$

where vr = volume of wash liquid employed divided by the volume of liquid left in the unwashed cake (i.e. a volume ratio), and η_{wash} = washing efficiency. When $vr = 1$, the solute fraction remaining is (1 - washing efficiency); the value of vr in practice can vary up to 5. The value of η_{wash} has been found to go up to 0.86 by Choudhury and Dahlstrom (1957). A semilog plot of the solute fraction remaining against vr yields a slope of log(1- η_{wash}). Belter *et al.* (1988) have illustrated data for the extraction of the antibiotic lincomycin from a cake of bacteria at two values of *pH* which appears to influence the extraction efficiency significantly.

There exists a simple relation between Q_{filtrate} and $Q_{\text{filtrate/washing}}$ and θ_f and θ_w (alternatively t_{contact} (7.2.163) and the time for washing t_{wash}). Relations (7.2.162) and (7.2.166) yield directly

$$\frac{Q_{\text{filtrate}}}{Q_{\text{filtrate/washing}}} = 2\frac{\theta_f}{\theta_w} = 2\frac{t_{\text{contact}}}{t_{\text{wash}}}$$
$$\Rightarrow \frac{\theta_w}{\theta_f} = 2\frac{Q_{\text{filtrate/washing}}}{Q_{\text{filtrate}}} = 2\frac{V_{\text{filtrate/washing}}}{V_{\text{filtrate}}}$$
$$\Rightarrow \frac{\theta_w}{\theta_f} = 2\frac{V_{\text{filtrate/washing}}}{V_{\text{retained}}}\frac{V_{\text{retained}}}{V_{\text{filtrate}}} = \frac{t_{\text{wash}}}{t_{\text{contact}}}; \tag{7.2.168}$$

$$\frac{t_{\text{wash}}}{t_{\text{contact}}} = 2(vr)\,(V_{\text{retained}}/V_{\text{filtrate}}). \tag{7.2.169}$$

For fixed t_{contact}, the variation of t_{wash} plotted against (vr) will yield a straight line whose slope is $2(V_{\text{retained}}/V_{\text{filtrate}})$.

7.2.2 Granular filtration of hydrosols (and aerosols)

Sand filters are used extensively in large-scale water treatment and wastewater treatment. In such a filtration process, the water to be treated flows very slowly down a filter bed, which is generally 1 m deep and is made out of sand grains of diameter varying between 0.6 and 2 mm. The pressure drop encountered by purified water leaving the bottom of the bed varies between 2.5 and 10 meters of H_2O. Particulate contaminants/impurities are collected in the granular filtration media in the manner shown in Figure 6.3.8 for depth filtration. The word *clarification* is used for liquid feeds containing suspended materials/particulates, namely hydrosols. Just like the adsorbent bed in Section 7.1.1.1, other conditions remaining constant, the purification rate decreases with time. Further, the pressure drop across the filter increases with time for a given filtration rate. As in Section 7.1.1.1, the granular sand bed/media has to be regenerated after some time to reverse filter clogging. The regeneration is carried out by *backwash*: liquid/water is brought in the bottom of the bed at a high rate and is made to flow up while the bed is expanded substantially.

We have already been introduced to aerosol removal from gas/air streams via depth filtration in a fiber-filled bed in Section 6.3.1.4 (see Figures 6.3.8, 6.3.9A and 6.3.9B). The general technique is identified as *fibrous filtration*, where surface filtration is also possible. For cases where surface filtration does not exist in fibrous filtration, the process is very close to granular filtration of hydrosols. The only differences are as follows: the fibrous bed is highly porous compared to granular filtration beds; the bed pressure drop is lower; the medium consists of fibers/cylinders as opposed to granules/spheres used in granular filtration; the fibrous bed loaded with captured aerosol materials cannot be regenerated easily (unlike that in granular filtration) (Tien, 1989). We will focus primarily in this section on the granular filtration of hydrosols.

We will briefly describe the filtration behavior of hydrosols in a granular filter bed, but first, it is useful to conceptualize the structure of the granular medium. The bed is assumed to consist of a large number of unit bed elements (UBEs) in series (Tien, 1989). Each UBE has a certain type of flow channel, and the granular medium surface acts as a particle collector, collecting particles from the fluid flowing in the channel. The porous medium in each UBE may be represented in a number of ways.

(1) The porous medium may be assumed to consist of a bundle of straight capillaries of equal size (*capillaric model*). A particle follows a certain trajectory and is captured when it hits the wall.
(2) If each granule is assumed to be a sphere, the porous medium is a collection of spherical collectors collecting the particles as the liquid flows by (*spherical collector model*).
(3) As the liquid flows through the granular medium, the flow may be described as the combination of flow through the large void spaces followed by flow through the constricted region in the area of contact between neighboring granules (*constricted tube model*). This model appears to be capable of describing filter clogging more efficiently (Tien, 1989).

Very detailed models of granular filtration have been developed using such idealizations of the porous medium (Tien, 1989). A simple aspect of this filtration is that, as the liquid flows down the porous medium, particles are transported perpendicular to the flow by electrostatic force, Brownian diffusion, various surface interaction forces and interception. Gravitational force is also present: its

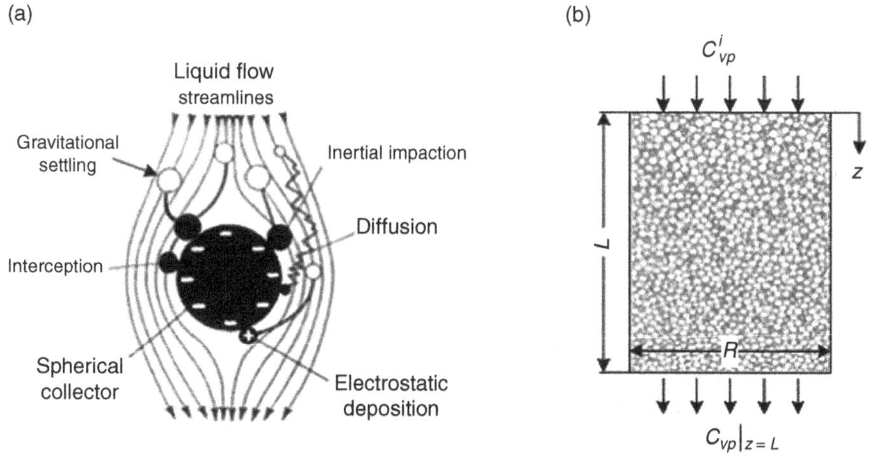

Figure 7.2.10. Fixed-bed granular filtration: (a) various mechanisms of particle capture by a spherical collector granule (after Davis (2001)); (b) granular packed-bed schematic for filtration.

direction is same as that of overall bulk flow; but, as the particle follows a curved streamline, it falls vertically downward and therefore at an angle to the local liquid flow. We have already observed in Figures 6.3.9A and 6.3.9B how inertial impaction can bring particles to be captured for the case of flow parallel to force in gas flow through a fibrous filter. In granular filtration, inertial impaction exists as discussed in Section 6.3.1.4; but so do a number of other forces/mechanisms just mentioned (see Figure 7.2.10(a)).

We will now provide a brief introduction to the modeling of the dynamic behavior of the fixed bed of granules as water containing a small concentration of particles (in the range of 100 parts per million (ppm) to 1% by weight) flows down the granular bed. Our simplified treatment will be guided by the comprehensive account to granular filtration of hydrosols and (aerosols) by Tien (1989).

Consider a granular filter bed (of radius R, length L and porosity ε) through which a fluid (here, water) flows, as shown in Figure 7.2.10(b). The superficial velocity of the fluid (through the cross section of the empty bed) is v_0. The actual interstitial velocities of the fluid in the three coordinate directions (r, θ, z) are v_r, v_θ and v_z, respectively. Due to radial symmetry, $v_\theta = 0$. As in Section 7.1.1.1, we can assume that there are no radial variations in this filter bed where $R >>> d_{gr}$, the average diameter of the grains in the bed. Although locally around a granule there may be a v_r component of velocity, it is not considered here in the lumped model of the overall bed performance. We will now use the general dynamic equation (6.2.51c) in the form of equation (6.2.55) for transport of particles in flowing fluids-convective diffusion; however, we will retain the term De in equation (6.2.51c), while $B = 0$. Here De will be used to indicate the disappearance of particles from the fluid. Since each term in (6.2.51c) is multiplied by $\rho_p(4/3)\pi r_p^3\, dr_p$ to arrive at equation (6.2.55), the De term will become $(De)\rho_p\,(4/3)\pi r_p^3\, dr_p$. From the description of the De-containing term in (6.2.50h), we conclude that $(De)\rho_p\,(4/3)\pi r_p^3\, dr_p$ will represent the mass rate of disappearance of particles of size r_p per unit filter volume; we will represent it as $(De)_{mv}$. Further, for incompressible flow, if we assume that all particles are of the same size r_p, then we can rewrite equation (6.2.55) after a control volume analysis in the manner of equation (6.2.50i) as follows (with D_p replaced by $\underline{\underline{D}}_{p,\text{eff}}$ reflecting dispersion effects and where ρ_{p,r_p} identifies the particle mass concentration in the fluid as opposed to the total control volume used earlier):

$$\frac{\partial(\varepsilon\rho_{p,r_p})}{\partial t} + \varepsilon\boldsymbol{v}\cdot\nabla\left(\rho_{p,r_p}\right) + \varepsilon\nabla\cdot\left(\boldsymbol{U}_p\rho_{p,r_p}\right) = \varepsilon\,\underline{\underline{D}}_{p,\text{eff}}\nabla^2\left(\rho_{p,r_p}\right) - (De)_{mv}, \qquad (7.2.170\text{a})$$

where ε is the porosity of the bed at any location. We need to ignore the third term on the left-hand side since any particle-specific velocity leading to particle capture or otherwise is lumped into the $(De)_{mv}$ term. In addition, in general for aerosols/hydrosols, particle velocity and fluid velocity may be assumed to be identical for the overall filtration behavior of interest here. (When an actual particle trajectory model is considered, this assumption is invalid.) We can now rewrite equation (7.2.170a) as

$$\frac{\partial(\varepsilon\rho_{p,r_p})}{\partial t} + \varepsilon\mathbf{v}\cdot\nabla\left(\rho_{p,r_p}\right) = \varepsilon\,\underline{\underline{D}}_{p,\text{eff}}\nabla^2\rho_{p,r_p} - (De)_{mv}. \qquad (7.2.170\text{b})$$

If we ignore any dispersion effects and employ only the interstitial velocity in the z-direction, v_z (v_r, v_θ not being relevant), we get the time-dependent z-directional equation for particle mass concentration, ρ_{p,r_p}, in the liquid:

$$\frac{\partial(\varepsilon \rho_{p,r_p})}{\partial t} + v_z \varepsilon \frac{\partial \rho_{p,r_p}}{\partial z} + (De)_{mv} = 0, \tag{7.2.171}$$

where $v_z \varepsilon = v_0$, the superficial velocity through the bed, and ρ_{p,r_p} is the particle mass concentration in the flowing liquid.

As depth filtration continues, particles are deposited on the filter-bed granules. The filter-bed porosity ε decreases from the porosity ε_0 of the clean filter (the initial porosity at time $t=0$); also, the pressure drop for fluid flow through the bed increases, due to a reduction in flow cross-sectional area as well as increased form drag from the changes in the geometrical shape and dimensions of the granules. The pressure gradient in the axial direction increases from the initial value $(\partial P/\partial z)_0$ to $(\partial P/\partial z)$. As time increases, ε keeps on decreasing and $(\partial P/\partial z)$ keeps on increasing. Granular bed filtration is therefore an inherently unsteady state operation. To facilitate the development of a solution of the problem, some relations have to be developed between the time-dependent decrease of ε from ε_0, the time-dependent increase of $(\partial P/\partial z)$ and the term $(De)_{mv}$.

We identify σ_v as the volume of particles deposited per unit filter volume, a specific volume based deposit. Then

$$(De)_{mv} = \rho_p \frac{\partial \sigma_v}{\partial t}, \tag{7.2.172}$$

where ρ_p is the actual density of the particle material. Further, the parameters and variables that influence this may be described as

$$(De)_{mv} = \rho_p \frac{\partial \sigma_v}{\partial t} = \rho_p f(\alpha_f, \rho_{p,r_p}, \sigma_v), \tag{7.2.173}$$

where α_f is a parameter. Similarly, σ_v will also influence the increase in the pressure gradient with time:

$$\left[(\partial P/\partial z)/(\partial P/\partial z)_0\right] = \mathrm{Gr}(\beta_f, \sigma_v), \tag{7.2.174}$$

where β_f is a parameter.

Introducing the definition (7.2.172) into the particle mass concentration balance equation (7.2.171), we get

$$v_z \varepsilon \frac{\partial \rho_{p,r_p}}{\partial z} + \frac{\partial(\varepsilon \rho_{p,r_p})}{\partial t} + \rho_p \frac{\partial \sigma_v}{\partial t} = 0. \tag{7.2.175}$$

We will now introduce a new dependent variable, C_{vp}, the volume concentration of particles, i.e. the volume of particles per unit volume of the fluid:

$$C_{vp} = \rho_{p,r_p}/\rho_p. \tag{7.2.176}$$

Further, if ε_d is the porosity of the particle deposit on the filter media, then the porosity ε of the filter medium at any time is related to σ_v and the initial porosity ε_0 of the clean filter by

$$\varepsilon = \varepsilon_0 - (\sigma_v/(1-\varepsilon_d)). \tag{7.2.177}$$

We can now rewrite the balance equation (7.2.175) as

$$v_z \varepsilon \frac{\partial C_{vp}}{\partial z} + \frac{\partial(\varepsilon C_{vp})}{\partial t} + \frac{\partial \sigma_v}{\partial t} = 0. \tag{7.2.178}$$

The two independent variables here are z and t. If we introduce the transformation

$$\theta = t - \int_0^z \frac{dz}{v_z} \tag{7.2.179a}$$

and change the independent variables to z and θ, then, since

$$dC_{vp} = \left(\frac{\partial C_{vp}}{\partial z}\right)_\theta dz + \left(\frac{\partial C_{vp}}{\partial \theta}\right)_z d\theta, \tag{7.2.179b}$$

we get

$$\left(\frac{\partial C_{vp}}{\partial z}\right)_t = \left(\frac{\partial C_{vp}}{\partial z}\right)_\theta + \left(\frac{\partial C_{vp}}{\partial \theta}\right)_z \left(\frac{\partial \theta}{\partial z}\right)_t$$

$$= \left(\frac{\partial C_{vp}}{\partial z}\right)_\theta + \left(\frac{\partial C_{vp}}{\partial \theta}\right)_z (-)\frac{1}{v_z}. \tag{7.2.179c}$$

Further,

$$\left(\frac{\partial(\varepsilon C_{vp})}{\partial t}\right)_z = \left(\frac{\partial(\varepsilon C_{vp})}{\partial \theta}\right)_z; \quad \left(\frac{\partial \sigma_v}{\partial t}\right)_z = \left(\frac{\partial \sigma_v}{\partial \theta}\right)_z. \tag{7.2.179d}$$

Therefore the balance equation (7.2.178) is reduced to

$$v_z \varepsilon \left(\frac{\partial C_{vp}}{\partial z}\right)_\theta - \varepsilon \left(\frac{\partial C_{vp}}{\partial \theta}\right)_z + \varepsilon \left(\frac{\partial C_{vp}}{\partial \theta}\right)_z + C_{vp}\left(\frac{\partial \varepsilon}{\partial \theta}\right)_z + \frac{\partial \sigma_v}{\partial \theta} = 0, \tag{7.2.180}$$

where, if we introduce expression (7.2.177) for ε, we get

$$v_z \varepsilon \left(\frac{\partial C_{vp}}{\partial z}\right) + \left(1 - \frac{C_{vp}}{1-\varepsilon_d}\right) \frac{\partial \sigma_v}{\partial \theta} = 0. \tag{7.2.181}$$

If $C_{vp} << 1$, this equation is reduced to

$$v_z \varepsilon \frac{\partial C_{vp}}{\partial z} + \frac{\partial \sigma_v}{\partial \theta} = 0. \tag{7.2.182}$$

This equation, along with equation (7.2.177) for ε, relation (7.2.174) for the increase in pressure gradient and expression (7.2.173) for the time-dependent increase of σ_v, govern the granular filtration process subject to the following initial and boundary conditions:

$$\text{for } \theta \le 0, z \ge 0, \quad C_{vp} = 0, \sigma_v = 0; \tag{7.2.183a}$$

$$\text{for } \theta > 0, \text{ at } z=0, \quad C_{vp} = C^i_{vp}, \tag{7.2.183b}$$

where C^i_{vp} is the volume concentration of the particles in the liquid stream entering the filter.

Solution of these equations requires knowledge of two unknown functions, $f(\alpha_f, \rho_{p,r_p}, \sigma_v)$, (relation (7.2.173)) and $\mathrm{Gr}(\beta_f, \sigma_v)$ (relation (7.2.174)). A variety of models have been developed; these have been summarized in Tien (1989). The solution has to be carried out numerically (in general).

However, since we have the $(\partial\sigma_v/\partial\theta)$ term in equation (7.2.182), which via postulate (7.2.173), is related to $f(\alpha_f, \rho_{p,r_p}, \sigma_v)$ or $f_1(\alpha_f, C_{vp}, \sigma_v)$ (obtained via substitution of ρ_{p,r_p} into $f(\alpha_f, \rho_{p,r_p}, \sigma_v)$ by C_{vp} via (7.2.176)), we need to have this functional dependence. We will now briefly introduce a few expressions for $f_1(\alpha_f, C_{vp}, \sigma_v)$.

Experimental observations during the initial filtration period with a clean filter indicate that the volume concentration of particles in the liquid follow a logarithmic law, i.e. the particle concentration decreases exponentially with z:

$$\left(C_{vp}/C^i_{vp}\right) = \exp(-\lambda z), \tag{7.2.184}$$

which leads to

$$(\partial C_{vp}/\partial z) = -\lambda C_{vp}. \tag{7.2.185}$$

From equation (7.2.182), we get

$$(\partial\sigma_v/\partial\theta) = v_z\,\varepsilon\,\lambda C_{vp}, \tag{7.2.186}$$

which suggests that the time rate of filtration per unit filter volume is proportional to the volume concentration of particles in the liquid, a first-order behavior. Here λ, known as the *filter coefficient*, has the dimension of (length)$^{-1}$, as the dimensional consideration of equation (7.2.186) indicates:

$$\frac{\sigma_v\left(\dfrac{\text{volume}}{\text{volume}}\right)}{\theta\ (\text{time})} = \frac{v_z\left(\dfrac{\text{length}}{\text{time}}\right)\varepsilon\left(\dfrac{\text{volume}}{\text{volume}}\right)\lambda\, C_{vp}\left(\dfrac{\text{volume}}{\text{volume}}\right)}{1}.$$

However, it is also known that the filter coefficient λ varies with time and may be expressed as

$$\lambda = \lambda_0 f_\lambda(\alpha_f, \sigma_v), \tag{7.2.187}$$

where $f_\lambda(\alpha_f, \sigma_v) = 1$ when $\sigma_v = 0$. Therefore the filtration rate expression (7.2.186) becomes

$$\partial\sigma_v/\partial\theta = v_z\varepsilon\lambda_0 f_\lambda(\alpha_f, \sigma_v)\, C_{vp}. \tag{7.2.188}$$

Models that have been developed employ different functional dependences of $f_\lambda(\alpha_f, \sigma_v)$ on σ_v and other (adjustable) filter parameters. Herzig *et al.* (1970) have in addition, developed the following results from an extended analysis of particle conservation:

$$C_{vp}/C^i_{vp} = \sigma_v/\sigma^i_v; \tag{7.2.189}$$

$$(\partial\sigma_v/\partial z)_\theta = -\lambda_0 f_\lambda(\alpha_f, \sigma_v)\sigma_v, \tag{7.2.190}$$

where σ^i_v is the value of σ_v at $z = 0$ (filter inlet) (it depends on θ, time); the corresponding value of C_{vp} is C^i_{vp}.

One can now postulate specific functional forms of $f_\lambda(\alpha_f, \sigma_v)$ and develop an expression for C_{vp}, the particle volume concentration in the liquid phase. For example, if particle deposition results primarily in filter clogging, then Ornatski *et al.* (1955) (as quoted in Tien (1989)) postulate that

$$f_\lambda(\alpha_f, \sigma_v) = 1 - \text{a}\,\sigma_v, \tag{7.2.191}$$

where a is a positive constant. Substituting this form of f_λ into the filtration rate expression (7.2.188) applied to location $z = 0$ leads to

$$\left(\text{d}\sigma^i_v/(1-\text{a}\sigma^i_v)\right) = v_z\varepsilon\lambda_0\, C^i_{vp}\,\text{d}\,\theta. \tag{7.2.192}$$

On integration, using the value $\sigma^i_v = 0$ at $\theta = 0$, we get

$$(1-\text{a}\sigma^i_v) = \exp(-v_z\varepsilon\lambda_0\, C^i_{vp}\,\text{a}\,\theta). \tag{7.2.193}$$

From equation (7.2.190), we can now write

$$(\text{d}\sigma_v/\sigma_v(1-\text{a}\sigma_v)) = -\lambda_0\,\text{d}z.$$

Integrating and using the value of σ_v at $z = 0$, namely σ^i_v, we get

$$\frac{\sigma_v(1-\text{a}\sigma^i_v)}{\sigma^i_v(1-\text{a}\sigma_v)} = \exp(-\lambda_0 z). \tag{7.2.194}$$

From the result (7.2.189) developed by Herzig *et al.* (1970), we now obtain

$$\left(\frac{C_{vp}}{C^i_{vp}}\right)\frac{(1-\text{a}\sigma^i_v)}{\left(1-\text{a}\sigma^i_v\left(C_{vp}/C^i_{vp}\right)\right)} = \exp(-\lambda_0 z). \tag{7.2.195}$$

We can now solve for $\left(C_{vp}/C^i_{vp}\right)$ from this equation:

$$\left(\frac{C_{vp}}{C^i_{vp}}\right) = \frac{\exp(-\lambda_0 z)}{1-\text{a}\sigma^i_v+\text{a}\sigma^i_v\exp(-\lambda_0 z)} = \frac{1}{\text{a}\sigma^i_v+(1-\text{a}\sigma^i_v)\text{e}^{\lambda_0 z}}. \tag{7.2.196}$$

If we substitute into this relation expression (7.2.193) for $\text{a}\sigma^i_v$ in terms of C^i_{vp} and θ, we get

$$\frac{C_{vp}}{C^i_{vp}} = \frac{1}{1-\exp(-v_z\varepsilon\lambda_0\,C^i_{vp}\text{a}\theta)+\exp(-\lambda_0 z)\exp(-v_z\varepsilon\lambda_0\,C^i_{vp}\text{a}\theta)}.$$

Rearranging,

$$\frac{C_{vp}}{C^i_{vp}} = \frac{\exp(v_z\,\varepsilon\,\lambda_0\,C^i_{vp}\,\text{a}\theta)}{\exp(\lambda_0 z)+\exp(v_z\,\varepsilon\,\lambda_0\,C^i_{vp}\,\text{a}\theta)-1}. \tag{7.2.197}$$

This expression of the volume concentration of particles in the liquid phase as a function of time variable, θ, the location in the bed, z, and the incoming fluid-phase particle concentration, C^i_{vp}, is a useful result illustrative of the modeling efforts. *In general*, numerical solutions are carried out. However, a few parameters are needed; these parameters are estimated from experimental data by a complex procedure which has been summarized in Tien (1989). The following details from an example illustrated in Tien (1989) are illustrative. It is based on the following assumed functional dependences of f_λ (defined by (7.2.187)) and Gr (defined by (7.2.174)):

$$f_\lambda = \frac{1}{(1+\text{a}_1\sigma_v)^2}; \tag{7.2.198a}$$

$$\text{Gr} = 1 + \text{b}_1. \tag{7.2.198b}$$

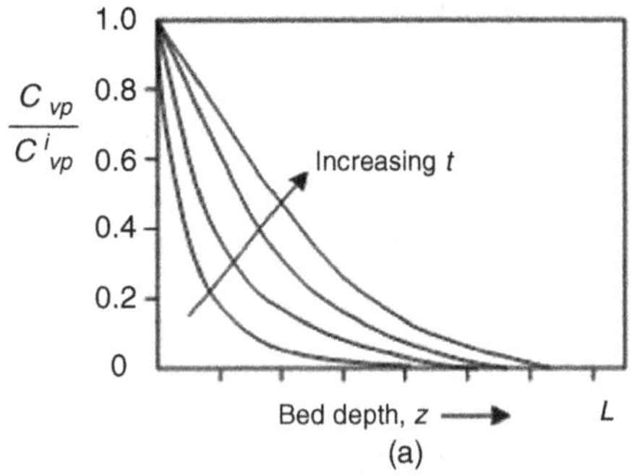

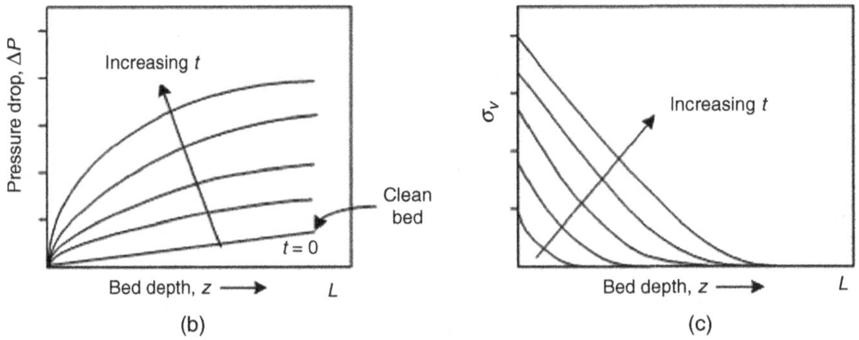

Figure 7.2.11. (a) suspension concentration profile along the bed as a function of time; (b) pressure drop behavior; (c) specific particle deposit growth with time along the bed.

The filter bed consists of granular activated carbon (GAC) of the "DARCO" 20 × 40 brand with a mean grain diameter (d_{gr}) of 0.594 mm. The filter bed length (L) is 1.3 m and the diameter is 7.6 cm. The clean filter void volume fraction ε_0 is 0.49. Suspended clay particles of diameter 4–40 μm were introduced at a concentration of $C_{vp}^{i} = 88 \times 10^{-6}\,\text{cm}^3/\text{cm}^3$. The superficial velocity of water, $v_0 (= \varepsilon v_z)$, is 1.38 mm/s. A parameter estimation method led to the following values: $\lambda_0 = 1.02 \times 10^{-2}\,\text{mm}^{-1}$; $b_1 = 464$; $a_1 = 30.46$. The value of $(\partial P/\partial z)_0$ corresponding to time $t = 0$ is 3.36×10^{-2} atm/m. The suspension concentration distribution in terms of C_{vp} as a function of the bed depth was found to agree well with measurements for time periods varying between 1 hr and 16 hrs. The behavior is shown schematically in Figure 7.2.11(a). The increase in pressure drop along the bed length (from the bed inlet to a specific location in the bed) with time is illustrated in Figure 7.2.11(b), where the straight line at the bottom for $t = 0$ represents the pressure drop through a clean filter. One can see that the pressure drop is much larger in the initial sections of the filter, which collect much more of the particulate deposit. These types of behavior have been experimentally verified (Tien, 1989). Figure 7.2.11(c) shows the growth in σ_v, the specific particle volume based deposit with time across the filter-bed length, as obtained from the solution of the equations considered so far.

It is useful to summarize the common operating conditions in granular filtration in drinking water and wastewater treatment from Tien (1989) (we have identified some of these at the beginning of this section); see Table 7.2.3.

Example 7.2.8 Consider a filter bed operating under the conditions described in the example from Tien (1989) given previously. Specifically, the conditions are as follows: $C_{vp}^{i} = 88 \times 10^{-6}\,\text{cm}^3/\text{cm}^3$; $\varepsilon_0 = 0.49$; $\varepsilon_0 v_z = v_0 = 0.138$ cm/s; $\lambda_0 = 1.02 \times 10^{-2}\,\text{mm}^{-1} = 1.02 \times 10^{-1}\,\text{cm}^{-1}$. However, the function f_λ may be described for lower levels of deposit via the Ornatski expression $f_\lambda(\alpha_f, \sigma_v) = 1 - a\,\sigma_v$, where the value of a has been found to be 42. Determine the value of (C_{vp}/C_{vp}^{i}) for $t = 2$ hours at bed locations of $z = 20$ cm and 100 cm. Compare with the values 0.24 and 0.05 obtained experimentally.

Solution We will employ expression (7.2.197) for C_{vp}/C_{vp}^{i}:

$$\frac{C_{vp}}{C_{vp}^{i}} = \frac{\exp\,(v_z\,\varepsilon\,\lambda_0\,C_{vp}^{i}\,a\theta)}{\exp\,(\lambda_0 z) + \exp\,(v_z\,\varepsilon\,\lambda_0\,C_{vp}^{i}\,a\theta)}.$$

From definition (7.2.179a),

$$\theta = t - \int_0^z \frac{dz}{v_z} = t - \frac{\varepsilon_0}{v_0}\,z,$$

where we have assumed that, for the short time period, $v_z = (v_0/\varepsilon)$ may be approximated by v_0/ε_0. For time $t = 2$ hours $= 7200$ s, the values of θ corresponding to $z = 20$ and 100 cm are, respectively,

$$\theta|_{z=20} = 7200 - \frac{0.49}{0.138\,(\text{cm/s})} \times 20\,\text{cm} = (7200 - 71)\,\text{s} = 7129\,\text{s};$$

Table 7.2.3.

Variables under consideration	Drinking water treatment	Wastewater (tertiary) treatment
Filter-bed depth	1 m	1 m
Filter-grain diameter	0.06 cm	0.2 cm
Filtration rate	5 m/hr	7.5 m/hr
Concentration of particles in feed water to be treated	50 mg/liter	30 mg/liter
Treated water quality	0.1 mg/liter	5 mg/liter
Pressure drop	2.5–10 m of water	2.5–10 m of water
Running time	10–100 hr	20 hr
Backwashing water requirement	3% of throughput	3% of throughput

A minimum rate of 15 gal/ft^2/ min over a 15 minute period is required. A rate of 20 gal/ft^2-min to provide 50% bed expansion is recommended.

$$\theta|_{z=100} = 7200 - \frac{0.49}{0.138\,(\text{cm/s})} \times 100\,\text{cm}$$
$$= (7200 - 355)\,\text{s} = 6945\,\text{s}.$$

For $z = 20$ cm,

$$\frac{C_{vp}}{C^i_{vp}} = \frac{\exp\left(0.138\dfrac{\text{cm}}{\text{s}} \times 1.02 \times 10^{-1}\dfrac{1}{\text{cm}} \times 88 \times 10^{-6} \times 42 \times 7129\,\text{s}\right)}{\exp(1.02 \times 10^{-1} \times 20) + \exp(0.138 \times 1.02 \times 88 \times 10^{-7} \times 42 \times 7129)}$$
$$= \frac{\exp(0.37)}{\exp(2.04) + \exp(0.37)} = \frac{1.447}{7.69 + 1.447} = 0.158.$$

For $z = 100$ cm,

$$\frac{C_{vp}}{C^i_{vp}} = \frac{\exp\left(0.37 \times \dfrac{6945}{7129}\right)}{\exp(10.2) + \exp\left(0.37 \times \dfrac{6945}{7129}\right)}$$
$$= \frac{\exp(0.36)}{24\,000 + \exp(0.36)} = \frac{1.433}{24\,000 + 1.433} \cong 0.$$

In the original method of solution, $f_\lambda(\alpha_f, \sigma_v) = (1/(1 + a_1\sigma_v)^2)$. Replacing this by the linear function of the Ornatski expression, $(1 - a\sigma_v)$, leads to an element of underprediction. Further, the value of "a" provided could be improved.

7.2.2.1 *Brief overview of mechanistic models in granular filtration*

The granular bed filtration model presented so far employed expressions for f_λ, Gr, etc. (7.2.198a,b), which are empirical in nature. A more fundamental approach, based on the operative mechanism, is desirable. Refer to Tien (1989) for a detailed treatment. Here we are going to provide a brief introduction to a more fundamental approach. As indicated in the introduction to Section 7.2.2, the granular filter bed, which is a porous medium, may be thought of as consisting of a large number of UBEs in series. Three types of UBEs were identified earlier as alternative representations of the porous medium that is the granular filter bed:

(1) bundle of straight capillaries as particle collectors (Figure 7.2.12(a)) (capillaric model);
(2) collection of spherical collectors (Figure 7.2.12(b)) (spherical collector model);
(3) constricted tube collectors (Figure 7.2.12(c)) (constricted tube model).

The granular filter bed contains n UBEs in series with $n \to \infty$. For a filter bed of length L, the length l of a UBE is therefore

$$n = L/\ell. \tag{7.2.199}$$

The overall filter efficiency E_T (see definition (6.3.45))

$$E_T = 1 - \left(C_{vp}|_{z=L}/C^i_{vp}\right), \tag{7.2.200a}$$

where $C_{vp}|_{z=L}$ denotes the volume concentration of particles at the filter exit for an incoming concentration of C^i_{vp}. Consider now the ith UBE: the influent particle concentration (from the $(i-1)$th UBE above it), $C_{vp}|_{i-1}$, enters this ith UBE at a distance z_i from the top; the effluent concentration leaving this ith UBE is $C_{vp}|_i$. The particle collection efficiency E_i of the ith UBE is given by

$$E_i = 1 - \left(C_{vp}|_i/C_{vp}|_{i-1}\right). \tag{7.2.200b}$$

Therefore

$$E_T = 1 - \prod_{i=1}^{n}(1 - E_i), \tag{7.2.201}$$

relating the particle collection efficiency E_i of UBEs to the overall filter efficiency E_T.

We know from an empirical understanding of clean filter behavior (see (7.2.185)) that

$$\frac{\partial C_{vp}}{\partial z} = -\lambda\, C_{vp},$$

where λ is the time-dependent filter coefficient. If we focus on this behavior (and equation), obtain a general solution

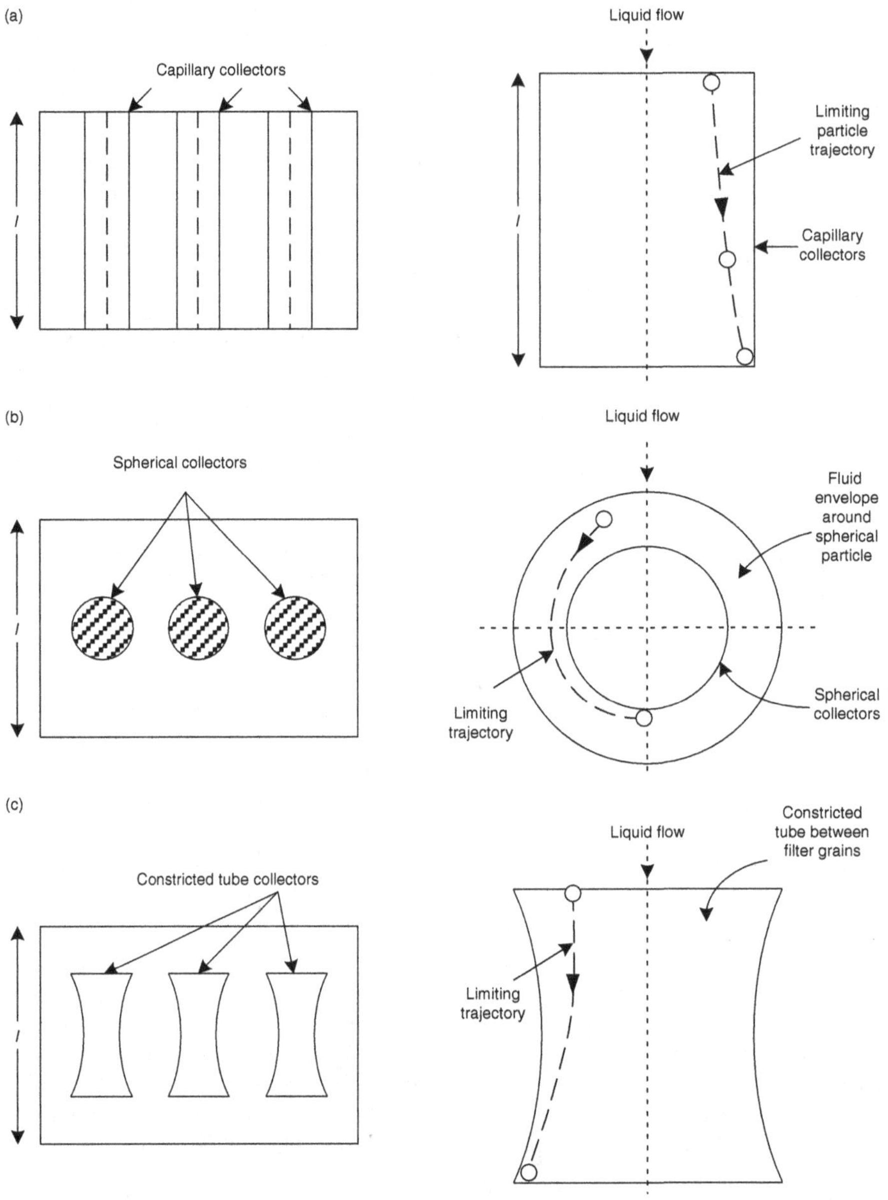

Figure 7.2.12. Three models of granular media filtration mechanism. (a) UBEs, of capillary tube collectors and limiting particle trajectory of a particle in a capillary. (b) UBEs of spherical collectors and limiting particle trajectory around a spherical collector. (c) UBEs of constricted tube collectors and limiting particle trajectory in a constricted tube. (After Tien (1989).)

$$C_{vp} = C_{vp}|_{z=0} \exp(-\lambda z) \tag{7.2.202}$$

and write it for the two concentrations of relevance around the *i*th UBE, $C_{vp}|_i$ and $C_{vp}|_{i-1}$,

$$C_{vp}|_{i-1} = C_{vp}|_{z=0} \exp(-\lambda z_{i-1});$$
$$C_{vp}|_i = C_{vp}|_{z=0} \exp(-\lambda z_i), \tag{7.2.203}$$

we get

$$C_{vp}|_{i-1}/C_{vp}|_i = \exp(-\lambda(z_i - z_{i-1})) \Rightarrow \ell n\,(C_{vp}|_{i-1}/C_{vp}|_i) = \lambda \ell, \tag{7.2.204}$$

where $(z_i - z_{i-1}) = \ell$. Therefore, using relation (7.2.200b),

$$\lambda = \ell n(1/(1-E_i))/\ell, \tag{7.2.205}$$

the required relation providing an estimate of λ in terms of E_i of the UBE and ℓ. In the limit of small values of E_i ($E_i << 1$),

$$\lambda \cong (E_i/\ell). \tag{7.2.206}$$

We will now provide an estimate of ℓ in terms of the filter-bed granule dimensions and the bed porosity ε. Assume that the filter-bed granules are uniformly spherical particles of diameter d_{gr}. Assume further that the filter may be considered to be a cube, the dimension of each side of which is $n\ell$, where there are n UBEs in series, each of length ℓ (note that $n \rightarrow \infty$). Therefore $(1 - \varepsilon)\ (n^3\ell^3) =$ volume of granules in the filter. Assuming one granule per one cubic UBE of side ℓ, there are n^3 granules in total in the bed, so that

$$(1-\varepsilon)\,(n^3\ell^3) = n^3 \times \frac{4}{3}\pi\frac{d_{\text{gr}}^3}{8} \Rightarrow \ell = \left(\frac{\pi}{6(1-\varepsilon)}\right)^{1/3} d_{\text{gr}}. \tag{7.2.207}$$

In general, granule diameters show a significant variation; therefore an average grain diameter should be used. The only unknown that is now required to estimate λ, the filter coefficient, is E_i, the particle collection efficiency of the ith UBE.

In the development of an expression for E_i, a more detailed model of the porous medium is needed. Of the three types of models of a granular filter as a porous medium, the capillaric model is not preferred. For the sake of simplicity, we will consider one of the other two, namely the *spherical collector model*. In this model, the filter grain is assumed to be a sphere. There are a number of alternative approaches based on a spherical collector. We will illustrate the approach by Happel (1958). In Happel's model, the granular porous medium is assumed to consist of a large collection of identical cells, where each cell consists of a spherical particle of radius $(\langle d_{\text{gr}}\rangle/2)$ (i.e. half of the average grain diameter) surrounded by a liquid envelope of radius b, such that the void volume of this cell is identical to the void volume of the porous medium:

$$\left[\frac{4}{3}\pi\frac{\langle d_{\text{gr}}\rangle^3}{8} \Big/ \frac{4}{3}\pi b^3\right] = (1-\varepsilon). \tag{7.2.208}$$

Consequently,

$$(\langle d_{\text{gr}}\rangle/2b) = (1-\varepsilon)^{1/3}. \tag{7.2.209a}$$

Further, in Happel's cell model, there is no interaction between the contiguous cells through the cell surface boundaries. The number of spherical collectors in a UBE, n_c, is obtained as follows:

$$\ell(1-\varepsilon) = n_c\frac{4}{3}\pi\left(\frac{\langle d_{\text{gr}}\rangle}{2}\right)^3;$$

$$n_c = \frac{6(1-\varepsilon)}{\pi\langle d_{\text{gr}}\rangle^3}\ell = \left[\frac{6(1-\varepsilon)}{\pi}\right]^{2/3}\langle d_{\text{gr}}\rangle^{-2}, \tag{7.2.209b}$$

where we have employed expression (7.2.207) for ℓ.

Particles in suspension flowing through the granular filter media are collected by individual granules; the collection efficiency of an individual collector/granule (to be identified by E_{MechS}, where the subscript "Mech" refers to the particular mechanism and subscript "s" refers to the single collector) has to be related to the UBE collection efficiency E_i defined by (7.2.200b). The incoming liquid, flowing with a superficial velocity v_0 (into any unit cell of Happel's model, say) and particle mass concentration of ρ_{p,r_p} for monosized particles (of radius r_p), brings in particles at a rate of $\pi b^2 v_o \rho_{p,r_p}$ over the projected cross-sectional area πb^2 of Happel's unit cell of radius b. If only a fraction E_{MechS} of these particles is captured by the granule, then the rates of particle capture by a single granule and by n_c collectors are, respectively, $\pi b^2 v_0 \rho_{p,r_p} E_{\text{Mech S}}$ *and* $n_c\pi b^2 v_0 \rho_{p,r_p} E_{\text{Mech S}}$. The total rate of particle introduction into the UBE of unit cross-sectional area is $v_0\rho_{p,r_p}$. Therefore the ith UBE collection efficiency, E_i, is given by

$$E_i = \frac{n_c\pi b^2 v_0 \rho_{p,r_p} E_{\text{Mech S}}}{v_0\rho_{p,r_p}}; \tag{7.2.209c}$$

$$E_i = n_c\pi b^2 E_{\text{Mech S}} = \left[\frac{6(1-\varepsilon)}{\pi}\right]^{2/3}\frac{\pi\langle d_{\text{gr}}\rangle^2 E_{\text{Mech S}}}{\langle d_{\text{gr}}\rangle^2 4(1-\varepsilon)^{2/3}}$$
$$= [6]^{2/3}E_{\text{Mech S}}\frac{\pi^{1/3}}{4} = 1.209\,E_{\text{Mech S}}. \tag{7.2.209d}$$

We will now briefly illustrate how to calculate $E_{\text{Mech S}}$ for a given capture mechanism.

As we have seen in Section 6.3.1.4 on the removal of particles from air by a fibrous bed via the mechanism of inertial deposition, if one can locate the limiting trajectory (dimension b in Figure 6.3.9A), the particle capture efficiency can be determined (e.g. definition (6.3.42a)). Determination of the limiting trajectory is achieved via *particle trajectory analysis* in the porous medium, i.e. the granular filter medium. The governing equation for particle motion in the inter-particle space is equation (6.2.45):

$$\boldsymbol{F}_p^{\text{iner}} = m_p\frac{\mathrm{d}\boldsymbol{U}_p}{\mathrm{d}t} = \boldsymbol{F}_{tp}^{\text{ext}} - \boldsymbol{F}_p^{\text{drag}}, \tag{7.2.210}$$

where $\boldsymbol{U}_p$ depends on the x-, y- and z- coordinates of the particle.

For hydrosols in granular filtration, the external force $\boldsymbol{F}_{tp}^{\text{ext}}$ consists of gravitational force, particle–collector surface interaction forces, such as the unretarded London attraction force (defined in (3.1.16)) and electrokinetic force (3.1.17) in the double layer, and electrostatic forces, if any, such as coulombic attraction/repulsion forces (3.1.15) (usually important in aerosol-removal processes unless the collector particles are deliberately charged). In addition, one has to consider inertial impaction and interception (see Section 6.3.1.4) if it is aerosol filtration, since particle velocities in hydrosol filtration are very low and inertial impaction (but not interception) may be neglected. Further, one has to take into account the role, if any, of Brownian diffusion. One next solves such an equation ((7.2.210)) to determine the trajectory of the particle in terms of the particle coordinates (x, y, z) as a function of time in the geometrical configuration of the porous media model adopted; it may be the capillaric model, the spherical collector model or the constricted tube model. Tien (1989) has provided detailed treatments for the latter two models in view of their greater success in realistically describing the particle deposition process in a granular filter. However, these treatments are quite complex; we will therefore avoid describing them here. Instead we will provide illustrative estimates of $E_{\text{Mech}\,s}$ via simple analysis for a single force like gravity or Brownian motion (equation (3.1.43)).

Consider first Figure 7.2.10(a), where the liquid is coming down vertically in the negative z-direction against the particle/grain of diameter $\langle d_{gr}\rangle$. The particle of radius r_p and actual density ρ_p settling in the liquid of density ρ_t will have a terminal velocity whose magnitude is given by (see equation (6.3.1))

$$|U_{pzt}| = \frac{2}{9}\frac{\left\langle \frac{d_{gr}}{2}\right\rangle^2 \rho_p\left(1-\frac{\rho_t}{\rho_p}\right)g}{\mu}. \tag{7.2.211}$$

Consider now Happel's cell model (Happel, 1958) in the spherical collector model. The total rate at which the particles enter one cell in Happel's model is given by the following expression:

$$\text{superficial liquid velocity} \times \text{cell incoming cross-sectional area} \times \text{particle mass concentration in incoming liquid} = (v_0)\times(\pi b^2)\times(\rho_{p,r_p}). \tag{7.2.212a}$$

The rate at which particles settle on the spherical collector of radius $(\langle d_{gr}\rangle/2)$ due to gravitational terminal velocity $|U_{pzt}|$ is given by

$$|U_{pzt}|\times\pi\left(\frac{\langle d_{gr}\rangle}{2}\right)^2\times(\rho_{p,r_p}). \tag{7.2.212b}$$

The particle collection efficiency due to gravitational force E_{GrS} is

$$E_{\text{GrS}} = \frac{|U_{pzt}|\times\pi\left(\frac{\langle d_{gr}\rangle}{2}\right)^2\times(\rho_{p,r_p})}{v_0\times\pi b^2\times\rho_{p,r_p}}. \tag{7.2.213}$$

If we employ result (7.2.209a) from Happel's model, we get

$$E_{\text{GrS}} = \frac{9}{2}\frac{\left(\frac{\langle d_{gr}\rangle}{2}\right)^2\rho_p\left(1-\frac{\rho_t}{\rho_p}\right)g\times\pi\left(\frac{\langle d_{gr}\rangle}{2}\right)^2\times(\rho_{p,r_p})}{\mu\times v_0\times\pi\left(\frac{\langle d_{gr}\rangle}{2}\right)^2(1-\varepsilon)^{-2/3}\times\rho_{p,r_p}}$$
$$= \frac{2\left(\frac{\langle d_{gr}\rangle}{2}\right)^2\rho_p\left(1-\frac{\rho_t}{\rho_p}\right)g}{9\mu v_0}(1-\varepsilon)^{2/3}. \tag{7.2.214}$$

If gravity were the only particle capture mechanism, then, for the ith UBE, the particle collection efficiency is given by

$$E_i = 1.209\,E_{\text{GrS}}. \tag{7.2.215}$$

We consider next the particle collection efficiency due to Brownian motion for very small particles (submicron dimensions). Brownian motion or Brownian diffusion may be incorporated by considering it simply as a diffusion process, where the diffusion coefficient for particles of radius r_p is

$$D_{\text{BR}} = \frac{k^B T}{6\pi\mu r_p}C_c \tag{7.2.216}$$

(see (3.3.90c)); here C_c is the Cunningham correction factor (the slip correction factor, (3.1.215)). The particle flux due to Brownian motion may be multiplied by the surface area of the spherical filter grain, $\pi\langle d_{gr}\rangle^2$, to determine the rate of particle capture:

$$\text{rate of particle capture} = (\text{particle flux})\times(\pi\langle d_{gr}\rangle^2). \tag{7.2.217a}$$

If the particle flux is represented via the traditional mass flux expression based on a mass-transfer coefficient (here, k_{pp}), then

$$\text{particle mass flux} = n_p = k_{pp}\left(\rho_{p,r_p} - \rho_{p,r_p}|_{\text{particle surface}}\right). \tag{7.2.217b}$$

We can assume $\left(\rho_{p,r_p}|_{\text{particle surface}}\right)$ to be zero.

The mass-transfer coefficient, k_{pp}, may be estimated from correlations of the Sherwood number, Sh, in a packed bed (where the diameter based definition is used):

$$Sh = Sh_p = \frac{k_{pp}(\langle d_{gr}\rangle)}{D_{\text{BR}}}. \tag{7.2.218a}$$

Therefore, the particle mass flux to the spherical collector, n_p, is

$$n_p = \frac{Sh\,D_{\text{BR}}}{\langle d_{gr}\rangle}\rho_{p,r_p} = \frac{Sh_p\,D_{\text{BR}}}{\langle d_{gr}\rangle}\rho_{p,r_p}. \tag{7.2.218b}$$

Since the rate at which particles enter the cell containing the spherical collector is given by (7.2.212a), the single-particle collector efficiency due to Brownian diffusion, E_{BRS}, is given by

$$E_{BRS} = \frac{\text{rate of particle collection}}{\text{rate of particle entry}} = \frac{n_p \times \pi\left(\langle d_{gr}\rangle^2\right)}{v_0 \times \pi b^2 \times \rho_{p,r_p}}; \tag{7.2.219}$$

$$\begin{aligned} E_{BRS} &= \frac{Sh_p\, D_{BR}\, \rho_{p,r_p}\, \pi\left(\langle d_{gr}\rangle^2\right)}{\langle d_{gr}\rangle v_0 \pi\, b^2 \rho_{p,r_p}} \\ &= \frac{Sh_p \langle d_{gr}\rangle D_{BR}}{v_0\, b^2}. \end{aligned} \tag{7.2.220}$$

If we use Happel's model, specifically result (7.2.209a), we get

$$E_{BRS} = \frac{4\, Sh_p (1-\varepsilon)^{2/3} D_{BR}}{v_0 \langle d_{gr}\rangle}. \tag{7.2.221}$$

Often, the mass-transfer correlation of Pfeffer (1964) based on Happel's (1958) cell model is used:

$$Sh_p = \left\{\frac{2\left[1-(1-\varepsilon)^{5/3}\right]}{2-3(1-\varepsilon)^{1/3}+3(1-\varepsilon)^{5/3}-2(1-\varepsilon)^2}\right\}^{1/3} \left(\frac{v_0 \langle d_{gr}\rangle}{D_{BR}}\right)^{1/3}, \tag{7.2.222a}$$

where the Péclet number Pe_p, defined by

$$Pe_p = \frac{v_0 \langle d_{gr}\rangle}{D_{BR}}, \tag{7.2.222b}$$

is employed to yield

$$Sh_p = (f(\varepsilon))^{1/3}\, Pe_p^{1/3}. \tag{7.2.222c}$$

Here, $f(\varepsilon)$ represents the complicated expression in Sh_p dependent on ε. This leads to

$$E_{BRS} = 4\,(f(\varepsilon))^{1/3}\,(1-\varepsilon)^{2/3}\,(Pe_p)^{-2/3}. \tag{7.2.223}$$

This estimate assumes that there are no surface interactive forces, e.g. repulsive or attractive double-layer forces between the filter grain and the particle to be captured (equation (3.1.17)). The double-layer force is attractive or repulsive depending on whether the zeta potentials of the particle and the collector grain are of opposite or the same sign. This happens because any particle or colloid (or a protein/macromolecule) in an aqueous environment has developed a double layer and a zeta potential (see Figure 3.1.2D and equations (3.1.11a–c)). For different expressions of E_{BRS}, where surface interaction forces are involved, see Tien (1989, sect. 4.5-2.3) for expressions developed by Spielman and Friedlander, Chiang and Tien and Rajagopalan and Karis.

In the absence of any external forces, particles can also be deposited on a collector granule by *interception* if during its motion it comes to within the particle radius from the spherical collector's surface. The result based on Happel's model is

$$E_{IS} = E_{InterceptionS} = 1.5(f(\varepsilon))^{1/3}\,(1-\varepsilon)^{2/3}\,(2r_p/\langle d_{gr}\rangle)^2. \tag{7.2.224}$$

Expressions for the efficiency of particle collection by *electrostatic forces* are available in Tien (1989) for four types of electrical forces:

(1) a Coulombic force, when both particles and collectors are charged;
(2) a charged collector inducing opposite charge on the particle;
(3) a charged particle inducing opposite charge on the collector;
(4) particles having similar charges producing a repulsive force among particles.

Many of these mechanisms will be operative simultaneously in the granular filter, in which case the efficiencies due to individual mechanisms have to be added up. For example, if gravitational force, Brownian motion and interception are operative simultaneously,

$$\sum_{Mech} E_{MechS} = E_{GrS} + E_{BRS} + E_{IS} \tag{7.2.225}$$

and

$$E_i = 1.209\left(\sum E_{MechS}\right) \tag{7.2.226}$$

from (7.2.209d) for Happel's cell model for spherical collectors representing the packed bed.

The same approach has been used successfully for aerosol filtration in a fibrous filter bed. Rubow (1981) has developed the following expression for the single-fiber efficiency for fibers of radius r_f when Brownian diffusion and interception operate simultaneously:

$$\sum E_{MechS} = E_{BRS} + E_{IS}, \tag{7.2.227a}$$

where

$$\begin{aligned} E_{BRS} = 2.86 &\left(\frac{1-\varepsilon}{f_K}\right)^{1/3} \left(\frac{v_0\, 2r_f}{(1-\varepsilon)D_{BR}}\right)^{-2/3} \\ \times &\left[1 + 0.389\left(\frac{(1-\varepsilon)}{f_K}\left(\frac{v_0\, 2r_f}{(1-\varepsilon)D_{BR}}\right)^{1/3}\right)\right]; \end{aligned} \tag{7.2.227b}$$

$$E_{IS} = \frac{(1-\varepsilon)}{f_K}\,\frac{(r_p/r_f)^2}{1+(r_p/r_f)}\left(1+\frac{2}{(r_p/r_f)}\right), \tag{7.2.227c}$$

where the Kuwabara hydrodynamic factor

$$f_K = -0.75 - 0.5\,\ell n\,\varepsilon + \varepsilon - 0.25\varepsilon^2 + fr_s(-0.5 - \ell n\,\varepsilon + 0.5\varepsilon^2);$$

fr_s = constant describing fraction of molecules reflected diffusely from the surface = 1.14 for air; v_0 = superficial face velocity. The overall fibrous filter efficiency, E_T, defined by (6.3.45) may be determined with this estimate of single-fiber efficiency $\sum E_{MechS}$.

7.3 External force field based separation: bulk flow perpendicular to force

This section covers separation by an electrical force field (Section 7.3.1), a centrifugal force field (Section 7.3.2), a gravitational force field (Section 7.3.3) and a magnetic force field (Section 7.3.5); Section 7.3.4 covers field flow fractionation.

7.3.1 Electrical force field

Charged bodies move when exposed to an electrical field. The charged bodies may be molecules, macromolecules, cells or particles in a liquid phase, or they may be ionized gas molecules or charged particles in the gas phase or in a plasma. The charge on the body may be free or natural for the system to be separated, or it may be induced into the bodies by external sources including a nonuniform electrical field.

Electrophoresis is used to separate naturally charged macromolecules, particles and proteins in a liquid phase via a uniform or nonuniform electrical field perpendicular to the direction of the bulk liquid motion (Figure 7.0.1(i)). If a neutrally charged particle or cell becomes polarized in a nonuniform electrical field and moves as a result in the liquid phase partly in the direction of the applied force, the phenomenon is called *dielectrophoresis* (see equation (3.1.13) for an expression for the force). Neutral molecules in a gas or vapor may be ionized by a laser beam via loss of electrons; the positively charged species are then attracted toward and collected by a negatively charged electrode, as in *laser isotope separation*. *Electronegative impurities* in a gas may have electrons attached to them if the gas is passed through a *corona-discharge reactor*. Such impurities may then be collected by the positive electrode.

Particles in a gas having no charge or very slight charge are allowed to acquire substantial negative charge in a corona generated by an intense electrical field around a wire; the particles then move in the electrical field to a grounded collecting electrode in *electrostatic precipitation* (see Figure 7.0.1(j)). When different particles (especially of plastics) acquire different charges via *triboelectrification* and are exposed to an electrical field, different particles with different charge to mass ratios fall to different locations at the end of the *electrostatic separator* and are separated. In the process of *flow cytometry* used for *cell sorting* (separation of different cells), droplets containing a particular type of cell have different charges imparted to them and are collected in different bins (containers) when passing through two vertical (or inclined) electrodes applying a voltage perpendicular to the vertically downward bulk motion of the drops. In all such cases treated in this section, the bulk gas or liquid phase or a droplet flows perpendicular to the direction of the electrical force exerted on the charged or polarized or ionized particle or molecule or drop. This is in contrast to the bulk flow of liquid parallel to the direction of the uniform electrical force field employed in capillary electrophoresis (Section 6.3.1.2).

7.3.1.1 *Electrophoresis*

A brief analysis of continuous free-flow electrophoresis (CFE) of charged macromolecular species is provided below. (For a brief introduction to electrophoresis, see Bier's chapter in Karger *et al.* (1973). For a more detailed introduction, see Andrews (1981). For capillary electrophoresis, see Section 6.3.1.2; for isoelectric focusing, see Section 4.2.2.1; for time-dependent separation of a batch sample by electrophoresis in a closed vessel, see Section 4.2.2.2.) The conditions are such that the effective bulk motion of the conducting aqueous buffer solution into which a feed mixture is introduced is perpendicular to the direction of the electrical force field. A number of apparatus geometries have been studied. One of the most common devices employs *thin-film continuous-flow electrophoresis* in which the buffer solution flows in a narrow channel of width $2b$ between two flat plates with cooling jackets; two oppositely charged electrode plates provide the other two surfaces of this rectangular liquid flow duct (Figure 7.3.1). *Note*: The feed mixture is introduced over a limited area of buffer flow.

The electrical force field between these two electrode plates is perpendicular to the bulk buffer flow direction. Positively charged species move toward the cathode in the positive x-direction. The axial buffer velocity (v_z) profile in the laminar channel flow is parabolic. At the channel inlet ($z = 0$), the feed solution is introduced continuously in a narrow band at (y, x_0), whereas the buffer solution flows into the channel through the entire channel cross section. One would like to know the locations (x, y) at $z = L$, the channel length, where each individual charged species present in the feed will appear. Note that only a *uniform electrical field* is assumed to exist between the electrodes.

The general governing equation for concentration C_{il} of species i in a nonreactive liquid system is (equation (6.2.3b))

$$\frac{\partial C_{il}}{\partial t} = -(\nabla \cdot \boldsymbol{N}_i).$$

The flux $\boldsymbol{N}_i$ of species i in a fixed coordinate frame of reference for a dilute noninteracting multicomponent system at constant temperature and pressure is (see equation (3.1.84a))

$$\boldsymbol{N}_i = C_{il}\left(\boldsymbol{v}_t^* + \boldsymbol{U}_i\right) - D_{is} \nabla C_{il},$$

where $\boldsymbol{U}_i$ is the migration velocity vector of solute species i defined by (3.1.84b). Incorporation of this expression of $\boldsymbol{N}_i$ into the conservation equation for species i and the one-dimensional (x-direction) migration of species i due to the

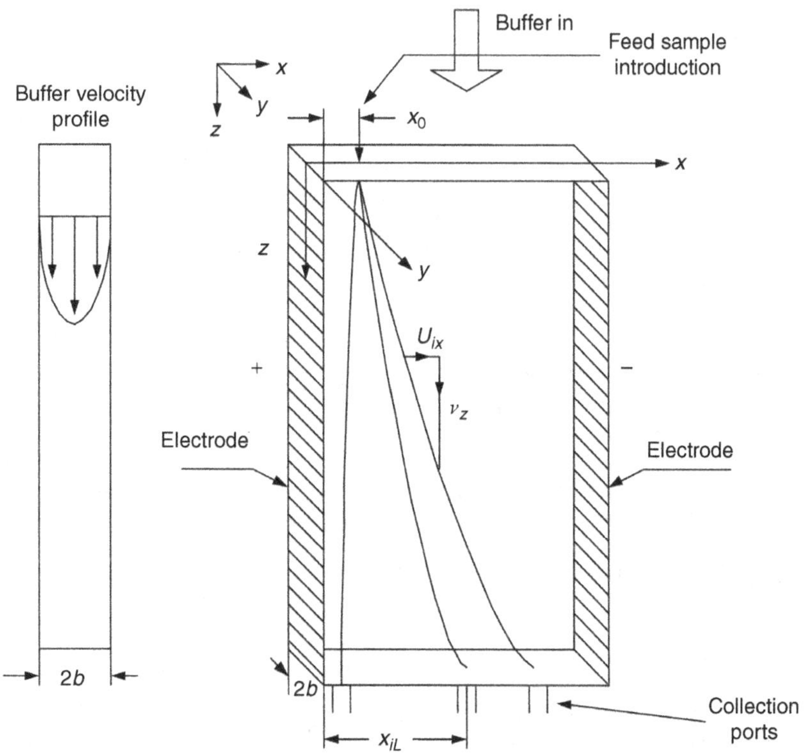

Figure 7.3.1. Thin-film continuous-flow electrophoresis separator and idealized solute trajectory. (After Gobie et al. *(1985).)*

electrical field (Figure 7.3.1), leads to the following equation in rectangular coordinates (see Table 6.2.1):

$$\frac{\partial C_{il}}{\partial t} + v_z \frac{\partial C_{il}}{\partial z} + U_{ix} \frac{\partial C_{il}}{\partial x} = D_{is}\left(\frac{\partial^2 C_{il}}{\partial x^2} + \frac{\partial^2 C_{il}}{\partial y^2} + \frac{\partial^2 C_{il}}{\partial z^2}\right), \tag{7.3.1}$$

where, since the system is dilute, we have assumed $v_z \cong v_z^*$. Further, v_y and v_x are assumed to be zero.

A *highly simplified* approach to the above problem is adopted first by making the following assumptions:

(1) there is no solute diffusion in any of the three axial directions;
(2) the axial buffer moves in plug flow in the x-direction;
(3) there is no electroosmotic flow velocity induced by the electrical field in the z-direction;
(4) the effect of Joule heating in the system due to the current is negligible;
(5) the system is of constant physical properties, e.g. density, viscosity and diffusion coefficients are constant.

Some of the complexities introduced by relaxation of these assumptions are treated later.

Using the *nondiffusive limit* of assumption (1), one finds at steady state from equation (7.3.1)

$$v_z \frac{\partial C_{il}}{\partial z} + U_{ix} \frac{\partial C_{il}}{\partial x} = 0. \tag{7.3.2}$$

Therefore, a constant macromolecular solute concentration C_{il} introduced at $z = 0$ at the inlet end of the rectangular duct at a particular $x = x_0$ will have an (x, z) trajectory defined by

$$\frac{\mathrm{d}x}{\mathrm{d}z} = \frac{U_{ix}}{v_z}. \tag{7.3.3}$$

This relation is obtained by using the methodology employed earlier in deriving (7.1.12e). For constant U_{ix} and v_z, integration leads to

$$x_L - x_0 = \frac{U_{ix} L}{v_z}. \tag{7.3.4}$$

Note that (L/v_z) is the solute (as well as buffer) residence time in the rectangular device of length L. Thus, the straight-line trajectory of charged solute i arrives at $z = L$ displaced a distance $(x_L - x_0)$ from its original x-location due to the electrical field. The y-coordinate of the solute is assumed to remain unaffected between x_0 and x. There is neither convection nor diffusion in the y-direction (by assumptions (1) and (2)).

The migration velocity U_{ix} of charged species i in a uniform electrical field $\boldsymbol{E}$ in the x-direction is known to be simply (from equations (6.3.8e) and (3.1.108m))

$$U_{ix} = E\mu_i^m, \tag{7.3.5}$$

where μ_i^m is the electrophoretic mobility of the ith ionic species (in units of $\text{cm}^2/\text{s-volt}$). Identify by x_{iL} the x-coordinate location of species i at $z = L$; then,

$$x_{iL} = \left((E\mu_i^m)/(v_z/L)\right) + x_0, \tag{7.3.6}$$

suggesting that different species introduced into the device via the feed solution at $z = 0$, $x = x_0$ and y will appear at different x-coordinate locations at $z = L$ as long as each μ_i^m is different. Since the buffer solution and the feed mixture are flowing in continuously, this allows continuous separation of a multicomponent charged species mixture if we collect separately and continuously the buffer solution at each x_{iL} location (Figure 7.3.1). The difference between the x_{iL} values for species i and j is

$$x_{iL} - x_{jL} = E(L/v_z)\left(\mu_i^m - \mu_j^m\right). \tag{7.3.7}$$

Obviously species having a particular charge (here positive) will have x_{iL} values greater than x_0; others having the opposite charge will arrive at $z = L$ with x_{iL} values less than x_0.

There are a number of real-life features which result in significantly lower separation in actual practice. Consider the role of solute diffusion (assumption (1)). Diffusional spreading cannot be prevented. Restricting diffusional broadening for the time being to only the x-direction and ignoring the z-direction convection, the following equation is obtained from equation (7.3.1):

$$\frac{\partial C_{il}}{\partial t} + U_{ix}\frac{\partial C_{il}}{\partial x} = D_{is}\frac{\partial^2 C_{il}}{\partial x^2} \tag{7.3.8a}$$

instead of equation (7.3.2). The unsteady state term $(\partial C_{il}/\partial t)$ is needed here since, in the absence of a z-direction convection term, the time t in effect defines the position along the x-axis where the C_{il} dependence on z is to be determined (see Reis *et al.* (1974) for an exact procedure, where a solution for C_{il} is obtained as $C_{il} = C_{il}^1(z,y,t) \times C_{il}^2(x,t)$, where $C_{il}^2(x,t)$ satisfies (7.3.8a) and the requisite boundary conditions given below). For simplicity, instead of a constant solute feed, we consider a pulse of solute at $z = 0$ (i.e. $t = 0$). Any solution for C_{il} obtained from (7.3.8a) should then satisfy the following conditions:

$$\text{at } x = \pm\infty, C_{il} = 0; \qquad \text{at } t = 0,\ C_{il} = \frac{m_i}{l_z l_y}\delta(x), \tag{7.3.8b}$$

where we have conveniently assumed that $x_0 = 0$. The second condition implies that originally m_i moles of all solute species i are contained in a volume of dimensions l_z, l_y and $\delta(x)$ (see Section 3.2). The solution for this is (see equation (3.2.19))

$$C_{il}(x,t) = \left(\frac{m_i}{l_z l_y}\right)\frac{1}{(4\pi D_{is}t)^{1/2}}\exp\left(-\frac{(x - U_{ix}t)^2}{4D_{is}t}\right). \tag{7.3.9}$$

The concentration profile $C_{il}(x,t)$ may be calculated at $t = (L/v_z)$, corresponding to the device exit where various fractions are to be collected:

$$C_{il}(x,L/v_z) = \frac{m_i}{(l_z l_y)}\frac{1}{2\left(\sqrt{\frac{\pi D_{is}L}{v_z}}\right)}\exp\left[-\frac{\left(x - \frac{U_{ix}L}{v_z}\right)^2}{\frac{4D_{is}L}{v_z}}\right]. \tag{7.3.10}$$

This result, which is based on a number of assumptions, including $x_0 = 0$, indicates a Gaussian concentration profile of species i around a center point of

$$x = \frac{U_{ix}L}{v_z} = \frac{E\mu_i^m L}{v_z} \tag{7.3.11}$$

and a standard deviation σ_{ix} of

$$\sigma_{ix} = (2D_{is}L/v_z)^{1/2}. \tag{7.3.12a}$$

The width of the collection port for each species i at $z = L$ should accommodate such a profile. The base width of such a profile is $4\sigma_i$ (see Figure 2.5.2) to accommodate 95% of species i introduced into the device. Therefore,

$$\text{collection port width} \cong 4(2D_{is}L/v_z)^{1/2}. \tag{7.3.12b}$$

Philpot (1940) obtained such a result in terms of $t = L/v_z$. The number of different charged species which can therefore be separated in a free-flow electrophoresis device is limited by solute dispersion in the x-direction, amongst other things. The peak-to-peak distance between two species may be estimated from equation (7.3.7). A more exact solution, which includes the effect of dispersion in the z-direction, has been provided by Reis *et al.* (1974) for a parabolic velocity profile $v_z(y)$ in the y-direction; effectively, the Philpot (1940) model underestimates x-directional band broadening substantially. The base width of the profile described by equation (7.3.9) depends on t; at z values less than L, the base width will be smaller. Note, however, the separation of Figure 7.3.1 is essentially at steady state. The time coordinate used here allows a specification of position along the z-axis.

The feed solution is normally introduced into the mobile buffer solution at $z = 0$ of a continuous free-flow electrophoresis (CFE) apparatus in a thin band of width 2δ in the y-direction (Figure 7.3.2(a)). If assumption (2) of plug flow of buffer in the z-direction were truly valid, the preceding analyses (e.g. result ((7.3.6)) would hold also for a solute band of width 2δ, i.e. the band will appear undeformed but displaced by $(E\mu_i^m L/v_z)$ in the x-direction.

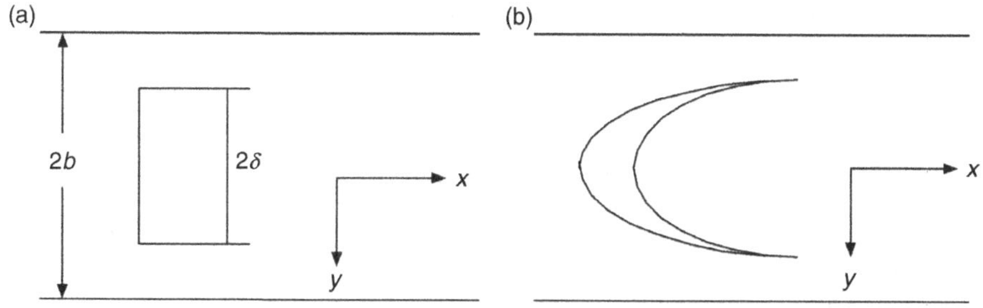

Figure 7.3.2. (a) Solute band at inlet of device shown in (x,y)-plane; (b) solute band deformed into a crescent shape at the exit of the device.

In conventional CFE equipment, the buffer flow, however, has a parabolic velocity profile:

$$v_z(y) = v_{\max}\left(1 - \frac{y^2}{b^2}\right). \qquad (7.3.13)$$

Therefore, solute molecules at different y-coordinate locations within the band of width 2δ experience different values of z-direction convection velocity v_z. Consequently, the residence time for solute species in the CFE device varies, depending on its y-location. For example, solute species in the band closer to the wall have a lower $v_z(y)$ and therefore a longer residence time $L/v_z(y)$; it will be displaced further along the x-axis than those molecules which are introduced near the channel center. Obviously $(x_{iL} - x_0)$, given by equation (7.3.6), will now depend on the y-location. Figure 7.3.2(b) illustrates the shape of the thin band in the (x, y)-plane.

In reality, additional complications are introduced by three factors. First, diffusional spreading has to be superimposed on this band distortion. Second, there is an electroosmotic velocity v_{EOF} in the x-direction in addition to the electrophoretic velocity U_{ix}; therefore the solute velocity in the x-direction is (Ivory, 1980):

$$v_{tx} = \left\{v_{\text{EOF}} - \frac{3}{2}v_{\text{EOF}}\left(1 - \frac{y^2}{b^2}\right) + U_{ix}\right\}. \qquad (7.3.14)$$

Displacement of solute molecules in the x-direction at $z = L$ compared to their x-coordinate at $z = 0$ now depends on the y-coordinate in a much more complex fashion:

$$x_{iL}(y) - x_0(y) = \frac{\left\{v_{\text{EOF}} - \frac{3}{2}v_{\text{EOF}}\left(1 - \frac{y^2}{b^2}\right) + U_{ix}\right\}}{v_{\max}\left(1 - \frac{y^2}{b^2}\right)/L}. \qquad (7.3.15)$$

Ivory (1980) has described the deformation of the rectangular solute band into a "crescent" shape. Figure 7.3.3 illustrates the solute output behavior as a result of these phenomena. To overcome such problems, Gobie *et al.* (1985) have suggested employing recycling of the effluent fractions from the CFE apparatus and introducing them into appropriate locations at the device inlet. Sharnez and Sammons (1992) have designed a new type of CFE chamber comprising a series of constrictions along the x-axis of separation to increase the peak-to-peak distance substantially.

The nonidealities considered so far fall into two categories: those due to diffusion and those due to other effects, e.g. parabolic velocity profile, electroosmotic flow. A measure of the role of diffusion is provided by the axial Péclet number ($Pe_z = v_z b/D_{is}$) obtained by nondimensionalizing x, y, z and t in equation (7.3.1):

$$\frac{\partial C_{il}}{\partial t^*} + Pe_z\frac{\partial C_{il}}{\partial z^*} + \frac{bU_{ix}}{D_{is}}\frac{\partial C_{il}}{\partial x^*} = \frac{\partial^2 C_{il}}{\partial x^{*2}} + \frac{\partial^2 C_{il}}{\partial y^{*2}} + \frac{\partial^2 C_{il}}{\partial z^{*2}}, \qquad (7.3.16)$$

where $x^* = (x/b)$, $y^* = (y/b)$, $z^* = (z/b)$ and $t^* = (tD_{is}/b^2)$.

In a typical CFE apparatus, Gobie *et al.* (1985) have suggested that the dimensionless Péclet number, Pe_z, is of the order of 10 000 or greater. (An example would be as follows: L = 16 cm, b = 0.187 cm, $v_z \approx 0.7$ cm/s, $D_{is} = 5 \times 10^{-7}$cm/s for a solute similar to albumin, so $Pe_z = 2.6 \times 10^5$.) Therefore diffusional effects may be neglected. The dimensionless number (bU_{ix}/D_{is}) is called the *electrophoretic Péclet number*; it is also expressed as $(bE\mu_i^m/D_{is})$.

Example 7.3.1 Consider a CFE device such that the two flat electrodes may be treated as if they were the walls of a flat slit of two infinite parallel plates spaced a distance b apart. The axial velocity v_z in such a configuration in laminar flow is given[19] by

$$v_z = \mathrm{d}z/\mathrm{d}t = 6v_{z,\text{avg}}\left((y/b) - (y/b)^2\right),$$

where y is the transverse distance to a point in the channel from one of these plates; $y = b$ at the other plate. For a uniform electrical potential profile between the two electrodes, develop an equation for the trajectory of a

[19] See Happel and Brenner (1965); equation (6.1.2b) is for a slit flow where the channel gap << width of the channel < length of the channel in the mean flow direction (Bird *et al.*, 2002).

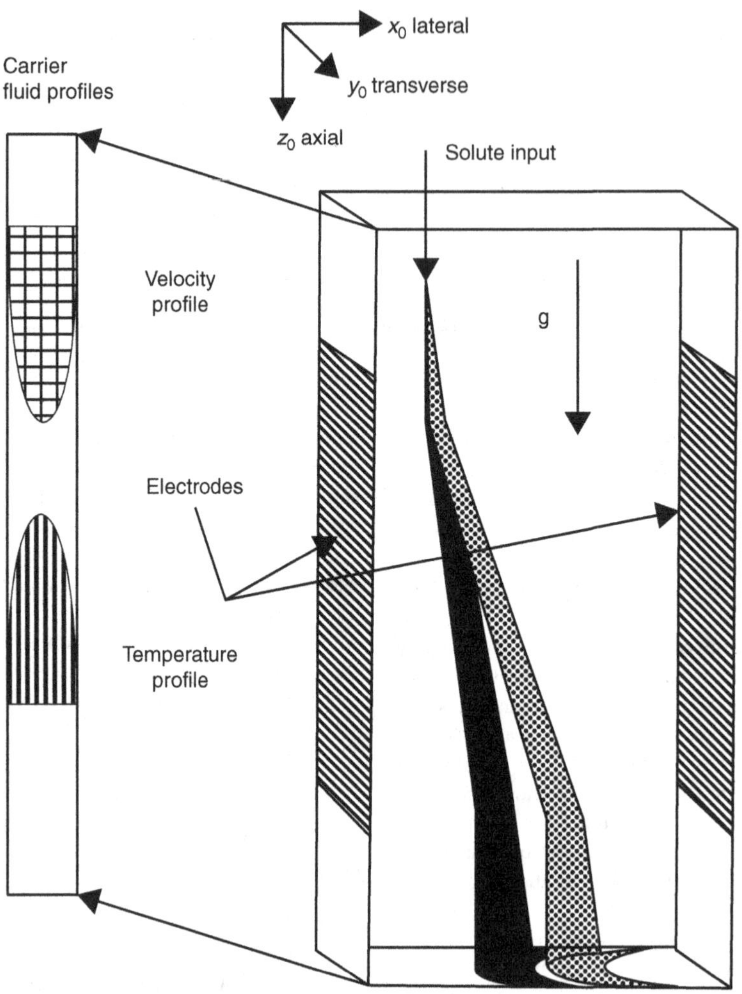

Figure 7.3.3. Classical vertical CFE with buffer flowing cocurrent to gravity; observe the crescent shape in the solute output. Reprinted, with permission, from W.A. Gobie, J.B. Beckwith, C.F. Ivory, Biotechnol. Progr., *1(1), 60 (1985). Copyright © [1985] American Institute of Chemical Engineers (AIChE).*

charged protein relating its z-coordinate to its y-coordinate, in the absence of any diffusive band broadening.

Solution Let the y-coordinate be normal to one of the flat electrodes; its y-value will be zero and that for the other electrode will be b. The velocity of any protein in the axial (z) direction will be that of the buffer:

$$v_z = (dz/dt) = 6\,v_{z,\text{avg}}\left\{(y/b) - (y/b)^2\right\}.$$

Its y-direction velocity will be

$$v_y = dy/dt = \mu_i^m E.$$

The trajectory equation is obtained following (7.3.3) as

$$(dz/dy) = 6v_{z,\text{avg}}\left\{(y/b) - (y/b)^2\right\}/\mu_i^m E$$

(no diffusive effects). Integrating implies that

$$z - 0 = \frac{6v_{z,\text{avg}}}{\mu_i^m E}\left(\frac{y^2}{2b} - \frac{y^3}{3b^2}\right),$$

which relates z to y, assuming the starting point is $z = 0$, $y = 0$ (one of the electrode locations). (In reality, y will have a finite value at $z = 0$, say y_0.) If one defines the following nondimensional variables:

$$z^+ = \frac{z}{b}\,\frac{\mu_i^m E}{v_{z,\text{avg}}}; \qquad y^+ = \frac{y}{b},$$

then

$$z^+ - z_0^+ = 6\left\{\frac{y^{+2}}{2} - \frac{y^{+3}}{3}\right\} - 6\left\{\frac{y_0^{+2}}{2} - \frac{y_0^{+3}}{3}\right\},$$

where the coordinates of solute injection are z_0, y_0, correspondingly

$$z_0^+ = z_0 \mu_i^m E / b v_{z,\text{avg}}; \qquad y_0^+ = y_0/b.$$

Therefore all solute trajectories may be described by the nondimensional result just derived. For a given solute of μ_i^m and $z = L$, z^+ will have a given value; correspondingly y^+ will have a particular value.

An effect not yet touched upon is that covered by assumptions (4) and (5). Due to the passage of the electric current, a considerable amount of heat (identified in the literature as Joule heating) is generated in the buffer solution, leading to natural convection, which mixes different regions of the buffer and reduces separation. In smaller devices that have a considerable amount of cooling surface area available for a given device volume (Hannig, 1961), such an effect is reduced. The problem becomes magnified for larger-scale operation. As a result, a scale up of continuous FFE is problematic. Vermuelen *et al.* (1971) used small spherical particles as a bed packing through which the buffer flowed and which ensured uniform flow. Mattock *et al.* (1980) developed an annular device in which the outer cylinder is rotated and the inner cylinder is at rest. An axial flow imposed on this rotational flow allows free-flow electrophoresis to be carried out since the laminar axial flow is stabilized by the radial angular velocity gradient in the buffer against any natural convection due to Joule heating.

Figure 7.3.4 illustrates such a continuous free-flow electrophoretic separator. The feed sample to be separated (identified as the migrant mixture input) is introduced through a circumferential slit in the stationary inner cylinder (the stator, 8 cm diameter) near the bottom of the device, from where the different species flow up. The buffer solution (identified as the carrier) is introduced at the bottom of the stator, it flows up and between the stator and the outer rotating cylinder (the rotor, 9 cm diameter). Sections of each of the inner and outer cylinders act as the two electrodes across which the voltage is applied. Individual species migrate to different radial distances from the inner cylinder as they are moved up by axial flow; the latter is divided into various fractions and collected simultaneously through respective openings in the stator near the top of the 1 m long annulus. As many as 20 fractions can be collected simultaneously by a collection of stacked discs in the stator above the electrode section; see Mattock *et al.* (1980) for details of electrode design and other aspects. A sample throughput of up to 20 ml/min can be achieved; the temperature rise is limited to 20–25 °C during the short residence time (30–60 seconds) in the buffer flowing at a rate of $\approx$ 500 ml/min. This minimizes the loss of biological activity of proteins. The extent of temperature rise in such a device has been modeled by Noble (1985).

Continuous free-flow electrophoresis in a thin rectangular channel of Figure 7.3.1 involved a uniform electrical field. The electrical field in the rotationally stabilized CFE separator (Figure 7.3.4) is strictly *nonuniform*. The electrical field varies inversely with the radius (see Problem 3.1.5); the electrophoretic radial migration velocity therefore decreases as the radial coordinate of the charged solute molecules increases. Due to the very small thickness of the annular region in the rotationally stabilized CFE separator, the electrical field is, however, almost uniform. Rolchigo and Graves (1988) have modeled continuous electrophoresis processes in general when subjected to a nonuniform electrical field.

Example 7.3.2 If one solute is injected at $z = 0$ of the CFE separator of Figure 7.3.4 at z_0^+ (see Example 7.3.1 for the notation as applied to Figure 7.3.4, where the annulus is very thin, i.e. $((r_o - r_i)/r_o) << 1$ so that the velocity profile of a thin flat slit of Example 7.3.1 is applicable) and another solute is injected at the same location ($z = 0$) such that the values of z_0^+ of the two differ by Δz_0^+, how would the center lines of these two solute trajectories differ from each other in the limit of $\Delta z_0^+ \to 0$? Assume $y_0^+ = 0$.

Solution In the limit of $((r_o - r_i)/r_o) << 1$, the axial flow profile corresponds to that between infinite parallel plates. Therefore the trajectory equation developed in Example 7.3.1 for solute 1 is usable here. Therefore

$$z^+ - z_0^+ = 6\left(\frac{y_1^{+2}}{2} - \frac{y_1^{+3}}{3}\right)$$

Since $y_0^+ = 0$. Let the trajectory for a second solute, whose z_0^+ value is $z_0^+ + \Delta z_0^+$, be given by

$$z^+ - (z_0^+ + \Delta z_0^+) = 6\left(\frac{y_2^{+2}}{2} - \frac{y_2^{+2}}{3}\right).$$

(*Note*: $y_1^+ = y_1/b; y_2^+ = y_2/b$.) Therefore

$$\Delta z_0^+ = 6\left\{\left(\frac{y_1^{+2}}{2} - \frac{y_1^{+3}}{3}\right) - \left(\frac{y_2^{+2}}{2} - \frac{y_2^{+3}}{3}\right)\right\}.$$

Let

$$y_2^+ = y_1^+ + \Delta y_1^+ \Rightarrow \Delta z_0^+ = 6\left\{\left(\frac{y_1^{+2}}{2} - \frac{y_1^{+3}}{3}\right) - \left(\frac{y_1^{+2}}{2} + y_1^+\Delta y_1^+ + \frac{\Delta y_1^{+2}}{2} - \frac{1}{3}(y_1^{+3} + 3y_1^{+2}\Delta y_1^+ + 3y_1^+\Delta y_1^{+2} + \Delta y_1^{+3})\right)\right\}$$

$$= 6\left\{(-\Delta y_1^+)(y_1^+ - y_1^{+2}) + O\left\{(\Delta y_1^+)^2\right\}\right\}.$$

Therefore

$$\Delta z_0^+ = 6(y_1^+ - y_1^{+2})(-\Delta y_1^+) + 6O\left\{(\Delta y_1^+)^2\right\}.$$

Differentiating, we obtain

$$\lim_{\Delta z_0^+ \to 0} \frac{d(\Delta y_1^+)}{d(\Delta z_0^+)} = (-)\frac{1}{6(y_1^+ - y_1^{+2})}.$$

This implies that, at the wall location, the gap between the center lines of the two solute trajectories will be very large (since $y_1^+ \to 1$). Therefore it is desirable to withdraw the product bands somewhere in between. These results were obtained by Beckwith and Ivory (1987).

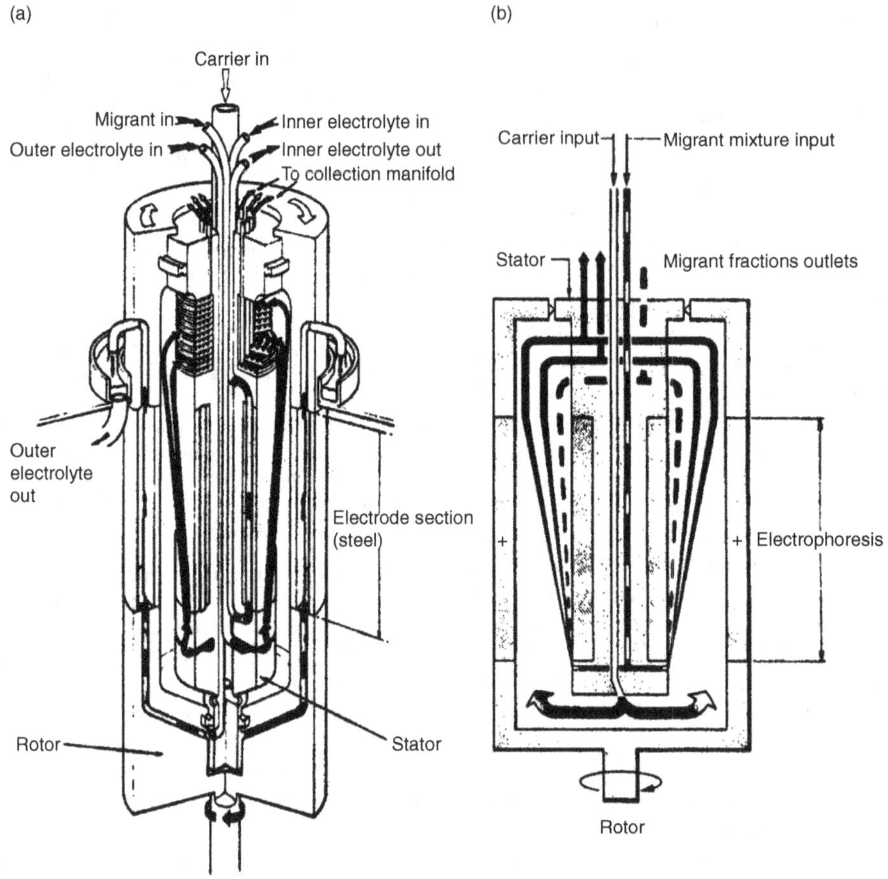

Figure 7.3.4. Diagrammatic representation of the continuous-flow electrophoretic separator. (a) Cutaway drawing of the system showing the construction of the rotor, stator and main flow channels. (b) Separation of three components as they are carried up through the annulus. From Figure 2, p. 7 of "Velocity gradient stabilized continuous free flow electrophoresis. A review," P. Mattock, G.F. Aitchison, A.R. Thomson, Separ. Purif. Meth., ***9**(1), 1 (1980); reprinted by permission of the publisher (Taylor and Francis Group, http://www.informaworld.com).*

7.3.1.1.1 Continuous dielectrophoresis In expression (3.1.13) for the dielectrophoretic force on an uncharged particle in a nonuniform electrical field, the Clausius-Mossoti function includes ε_p and ε_d, which are complex quantities. However, only the real part is useful:

$$\mathbf{F} = 2\pi r_p^3 \varepsilon_d \,\mathrm{Re}\left(\frac{\varepsilon_p - \varepsilon_d}{\varepsilon_p + 2\,\varepsilon_d}\right)\nabla \mathbf{E}^2. \tag{7.3.17a}$$

The sign of the quantity $(\varepsilon_p - \varepsilon_d)$ is important. If ε_p is greater than ε_d, then the direction of the force is toward the region of high electrical field strength; it is called *positive dielectrophoresis*. On the other hand, if ε_p is smaller than ε_d, the direction of the force is reversed, and one has what is called *negative dielectrophoresis*: the particle is forced in the direction of lower field strength (Pohl and Kaler, 1979). (Of course, the particle may experience zero force as well.) Note further that the dielectrophoretic force is proportional to the gradient of the square of the electrical field strength. Whereas in electrophoresis, the sign of the charge (+ve or -ve) on a particle or a macromolecule or a molecule and the direction of the electrical field are quite important, here in dielectrophoresis the direction of the applied electrical field is of no consequence; therefore dielectrophoresis can work with alternating current generating fields.

Continuous dielectrophoresis of whole cells using a "stream-centered" introduction of sample cell suspension is illustrated in Figure 7.3.5 (Pohl, 1977). A stream of cells or other particles to be separated is fed into the center of a cocurrent liquid stream (the carrier liquid) very similar in composition to that carrying the cells or particles. A sample to carrier flow rate ratio of 1:5 to 1:100 is employed to keep the sample stream as close to the center of the flow channel (as far from the walls and the electrodes) as possible.

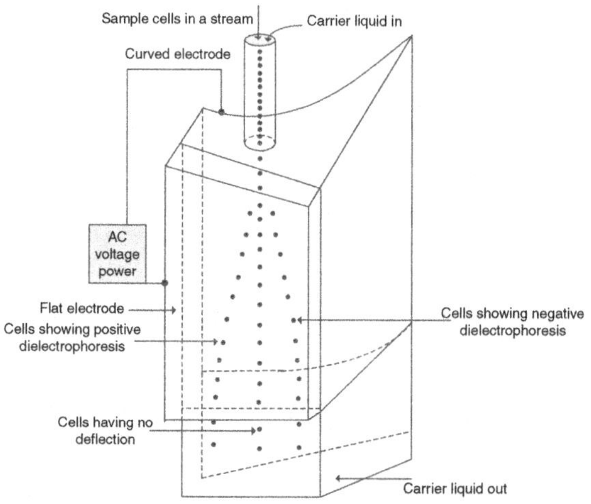

Figure 7.3.5. Schematic diagram of continuous dielectrophoresis using a "stream-centered" introduction of the sample cell suspension. (After Pohl (1977).)

Cells undergoing positive dielectrophoresis move in the radial direction of the merging of the curved rear electrode and the flat front electrode, whereas cells experiencing negative dielectrophoresis move in the opposite direction. Meanwhile, the bulk carrier flow perpendicular to this electrical force carries the cells forward toward the carrier exit. The cells or particles unaffected by the nonuniform electrical field continue to move along the line of original introduction. Thus different cells trace out different trajectories and therefore can be withdrawn at different locations from the stream near the liquid outlet.

Typical deflections experienced by cells in such a device amount to several millimeters for axial flow path lengths of 100 mm for an applied potential field of 4–8 volts rms (root mean square) in a medium of appropriate conductivity (Pohl, 1977). The particular electrode configuration employed is called the *isomotive electrode* configuration; it creates a constant dielectrophoretic force on a particle or cell over a wide region, i.e. over a large range of radial locations (Pohl and Kaler, 1979). The magnitude of this force is

$$F = ar^0, \tag{7.3.17b}$$

where a is a constant for a particular type of particle or cell and the zeroth power for r indicates r-independence of F. This ensures that particles or cells of a certain type and size will experience the same external force independent of their radial location. This, along with the "stream-centered" introduction of sample stream containing cells, avoids the placement of living cells close to the wall, thus avoiding higher residence time and possible damage to the cells.

We will now provide from Kralj *et al.* (2006) a brief example of continuous dielectrophoretic size based particle sorting in laminar flow (in the z-direction) in a rectangular parallel-plate type of microfluidic channel (cross-sectional view of channel in Figures 7.3.6(a) and (b), (x, y)-plane) containing an array of slanted planar electrodes on a glass channel floor going from one side of the channel to the other in one of the parallel plates (Figures 7.3.6(a) and (b), (x, z)-plane); here the main liquid flow is in the z-direction. The particle mixture to be separated is introduced through one inlet in one of the side walls, next to which there is another inlet used to introduce a particle-free solution to get the particles away from the sidewall; the main channel inlet introduces the bulk liquid flowing in the z-direction. Kralj *et al.* (2006) have approximated the electrical field due to this slanted array of electrodes in the (x, z)-plane via

$$|E^2| = AV^2 \sin\left(\frac{2\pi z}{\lambda_z} - \frac{2\pi x}{\lambda_x}\right) + AV^2, \tag{7.3.18}$$

where A is the amplitude of the wave, V is the voltage between the electrodes, λ_z and λ_x are the electrode spacings in the z- and x-directions, respectively (determined by the slant angle and the normal distance between the electrodes). Since the dielectrophoretic force is proportional to $\nabla \mathbf{E}^2$, the two components of interest are

$$\frac{\partial |E^2|}{\partial x} = \frac{2\pi}{\lambda_x} AV^2 \cos\left(\frac{2\pi z}{\lambda_z} - \frac{2\pi x}{\lambda_x}\right) \tag{7.3.19a}$$

and

$$\frac{\partial |E^2|}{\partial z} = -\frac{2\pi}{\lambda_z} AV^2 \cos\left(\frac{2\pi z}{\lambda_z} - \frac{2\pi x}{\lambda_x}\right). \tag{7.3.19b}$$

Figure 7.3.6(c) illustrates the variation of $|E^2|$ with the x-coordinate at a small distance from the top plate.

Kralj *et al.* (2006) studied the separation of polystyrene (PS) particles of diameters 2–6 μm in a dilute aqueous sucrose solution (having a density such that gravitational forces were zero since $\rho_p = \rho_t$). At the low particle Reynolds numbers employed (<0.01), the drag force will be described by Stokes' law. At steady state, the drag force will balance the dielectrophoretic force (equations (6.2.47)):

$$2\pi r_p^3 \varepsilon_d \,\mathrm{Re}\left(\frac{\varepsilon_p - \varepsilon_d}{\varepsilon_p + 2\varepsilon_d}\right) \nabla \boldsymbol{E}^2 = 6\pi\mu r_p \boldsymbol{v}. \tag{7.3.20}$$

From (3.1.14), we know that $\nabla \boldsymbol{E}^2 = \nabla(\boldsymbol{E}\cdot\boldsymbol{E}) = \nabla|E^2|$. Therefore, the two components of this force balance lead to two components of the particle velocity:

$$\text{particle velocity}|_{x\text{-dir}} = \frac{\mathrm{d}x}{\mathrm{d}t} = \frac{2\pi r_p^3 \varepsilon_d}{6\pi\mu r_p} \mathrm{Re}\left(\frac{\varepsilon_p - \varepsilon_d}{\varepsilon_p + 2\varepsilon_d}\right) \frac{\partial |E^2|}{\partial x}; \tag{7.3.21a}$$

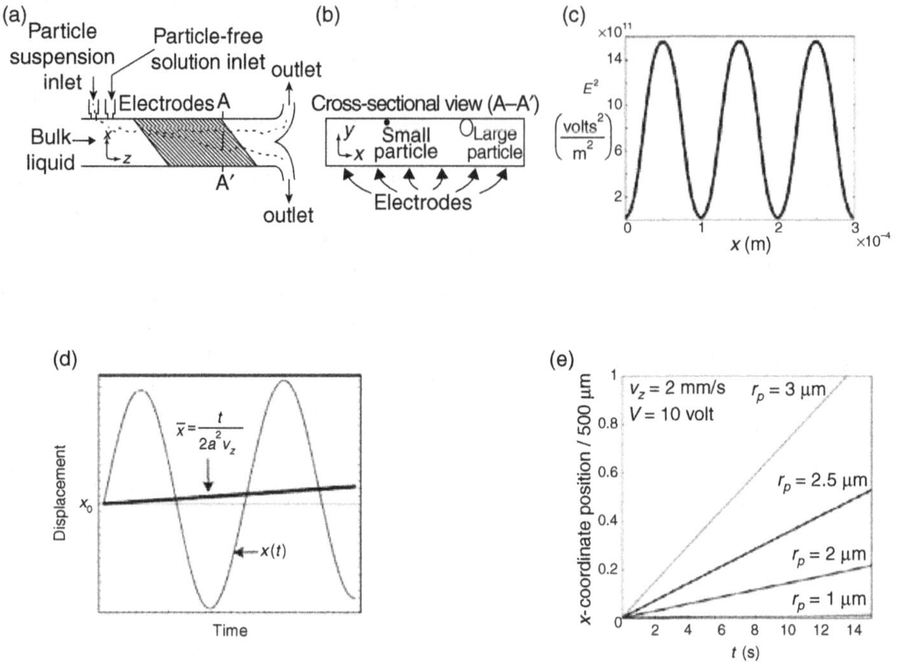

Figure 7.3.6. Continuous dielectrophoretic separation of a particle mixture in microfluidic channel flow. (After Kralj et al. *(2006); channel height 28 μm, channel width 500 μm.) (a) Cutaway view of the channel in the (x,z)-plane showing the inlets of particle suspension, particle-free solution, bulk liquid flow, two liquid outlets for two particle fractions and slanted electrodes. (b) Cross-sectional view at the A-A′ plane of (a), showing electrodes on the channel floor, larger and smaller particles separated in the* x*-direction; (c) Value of* E^2 *of the voltage variation in the periodic array of electrodes at a distance 3 μm from the channel top. (d) Behavior of* x(t) *per equation (7.3.24) and a plot of* $\bar{x}$ *against* t *per equation (7.3.27). (e)* x*-coordinate position of particles of different* r_p *normalized with channel width of 500 μm against time* t.

$$\text{particle velocity}|_{z\text{-dir}} = \frac{dz}{dt} = v_z + \frac{2\pi r_p^3 \varepsilon_d}{6\pi\mu r_p} \operatorname{Re}\left(\frac{\varepsilon_p - \varepsilon_d}{\varepsilon_p + 2\varepsilon_d}\right) \frac{\partial |E^2|}{\partial z}, \tag{7.3.21b}$$

where v_z is the liquid medium velocity in the z-direction, the direction of mean flow. *Note*: (1) The real part of the dielectric constant for the PS particle, ε_p, is 2.5, much less than that of the solution (~70); therefore we have negative dielectrophoresis. (2) The magnitude of v_z is much larger than that of the second term on the right-hand side of equation (7.3.21b). Therefore

$$(dz/dt) = v_z \Rightarrow z = v_z t. \tag{7.3.22}$$

We may now use this result in the particle trajectory equation resulting from combining equations (7.3.19a) and (7.3.21a):

$$\frac{dx}{dt} = \frac{2}{3}\frac{\pi r_p^2}{\mu \lambda_x} \varepsilon_d A V^2 \operatorname{Re}\left(\frac{\varepsilon_p - \varepsilon_d}{\varepsilon_p + 2\varepsilon_d}\right) \cos\left(\frac{2\pi v_z t}{\lambda_z} - \frac{2\pi x}{\lambda_x}\right). \tag{7.3.23}$$

Thus the extent of lateral particle displacement, x, due to the negative dielectrophoretic force may be determined as a function of time t (which is linearly related to the axial displacement of the particle). Kralj *et al.* (2006) have provided the following analytical solution of the above particle trajectory equation when $\lambda_z = \lambda_x = \lambda$ for an electrode slant angle of 45°:

$$x(t) = v_z t - \frac{2}{w_1} \arctan\left(\frac{\sqrt{a^2 v_z^2 - 1}}{a v_z + 1} \tan\left(\frac{w_1 \sqrt{a^2 v_z^2 - 1}}{2a} t\right)\right), \tag{7.3.24}$$

where

$$(1/a) = (2\pi r_p^2 / 3\mu\lambda_x)\, \varepsilon_d \operatorname{Re}\left(\frac{\varepsilon_p - \varepsilon_d}{\varepsilon_p + 2\varepsilon_d}\right) A V^2, \qquad w_1 = \frac{2\pi}{\lambda}. \tag{7.3.25}$$

This time-dependent oscillating x-directional displacement of the particle (shown in Figure 7.3.6(d)) may be characterized by means of an averaged x-directional displacement $\bar{x}$ at a time t defined by the half-period of the function

$$t = \frac{(2N+1)\pi}{w_1 \sqrt{v_z^2 - (1/a^2)}}, \qquad N \geq 0, \tag{7.3.26}$$

which is shown by the thick straight line in Figure 7.3.6(d). The slope of the thick straight line is given by

$$\left(\frac{\overline{x}}{t}\right) = v_z - \sqrt{v_z^2 - \frac{1}{a^2}} = \frac{1}{2a^2 v_z} + O\left(\frac{1}{a^4}\right), \quad (7.3.27)$$

which is acceptable since $(1/a)$ is usually less than $0.02v_z$. One can now calculate the x-directional displacement of the particle at the end of the channel (where $z = L$ and $t = (L/v_z)$):

$$x|_{z=L} = \left(\frac{\overline{x}}{t}\right) t = \frac{1}{2a^2 v_z}\left(\frac{L}{v_z}\right)$$

$$= \frac{\left(\frac{2\pi}{3\mu\lambda_x}\varepsilon_d \operatorname{Re}\left(\frac{\varepsilon_p - \varepsilon_d}{\varepsilon_p - 2\varepsilon_d}\right)A\right)^2 L}{2}\left(\frac{r_p^4 V^4}{v_z^2}\right). \quad (7.3.28)$$

This result shows that the displacement in the x-direction is proportional to the fourth power of the particle radius and the voltage applied and inversely proportional to the square of the flow velocity in the z-direction. Figure 7.3.6(e) illustrates the x-direction displacement of particles of different radius normalized with respect to the channel width of 500 μm. This figure shows that the largest particle (radius, 3 μm) has a very large displacement. Kralj *et al.* (2006) have experimentally verified the dependence of the x-displacement vis-à-vis $r_p^4 V^4/v_z^2$, as indicated by the result (7.3.28).

7.3.1.2 *Separation via ionization in gas/vapor phase: laser excitation of isotopes, electron attachment in a corona-discharge reactor*

In liquid-phase separations, charged proteins, macromolecules, cells or particles have different mobilities in an applied electrical field. The charge may be free or excess charge, as in electrophoresis, or induced by polarization in a neutral particle, as in dielectrophoresis. In a gas or vapor phase of neutral species, if one of the species can be selectively ionized in preference to the others, then the ionized species with the charge may be selectively removed by applying an electrical field perpendicular to the direction of the bulk flow of the gas or vapor. This principle has been applied to the following types of systems: separation of uranium isotopes in the atomic vapor laser isotope separation (AVLIS) technique; electron attachment to electronegative impurities in a corona-discharge reactor.

In the AVLIS technique, raw uranium metal containing 0.7% ^{235}U and 99.3% ^{238}U is vaporized at >4000 °F. The vapor flows up; as it flows up, a laser beam (from a copper-vapor laser) crosses this beam at a right angle (Figure 7.3.7 (a)). The design is such that the laser beam traverses a maximum amount of the vapor. This vapor goes up between two long plates acting as electrodes. The laser excitation is such that U^{235} is ionized to $^{235}U^+$ by losing an electron. However, ^{238}U is not ionized. The narrow-band laser source selectively excites only one isotope from its ground state to the ionized state. The negatively biased electrode collects the positively charged $^{235}U^+$ ions; the electrical field between the two plates is perpendicular to the upward vapor flow, which consists essentially of ^{238}U (Grossman and Shepp, 1991), and the tails stream. In one such device, the ^{235}U concentration is increased to 3.2%, which is enough for a reactor grade uranium. The performance achieved is somewhat reduced by charge-transfer collisions in the neutral background vapor:

$$^{235}U^+ + {}^{238}U \Leftrightarrow {}^{235}U + {}^{238}U^+. \quad (7.3.29)$$

Removal of vapor-phase impurities from gas streams may be implemented by exposing the gas stream to a *corona-discharge reactor*, where the low-energy electrons produced collide with gas molecules. Electronegative impurities, such as sulfur compounds and halogen compounds, have a much higher probability of becoming negative ions by *electron attachment*:

$$A + e \Leftrightarrow B^- + C,$$

for example

$$SF_6 + e \Leftrightarrow SF_5^- + F. \quad (7.3.30a)$$

Some of the other species which can be removed from air via the mechanism of (7.3.30a) are SO_2, CS_2, H_2S, COS, etc. For the second type of reaction mechanism, illustrated by (7.3.30b), the species which may be removed are I_2, O_2 as well as SF_6:

$$A + e \Leftrightarrow A^-. \quad (7.3.30b)$$

Tamon *et al.* (1995) have illustrated the devices and principles of gas purification using such ionization mechanisms. Consider a cylindrical reactor where a high voltage (≈ 3–15 kV) is applied between an inner piano wire acting as the cathode and an outer cylinder, which is the anode (Figure 7.3.7(b)). Electrons produced by the corona discharge drift toward the anode in the nonuniform electrical field. As they drift toward the anode, some of them collide with the gas molecules and generate negative ions by electron attachment. The negatively charged ions will also drift toward the anode. Removal of the negatively charged ions at the anode may be carried out in two ways.

In the *deposition-type reactor*, the negatively charged ions are attracted to the anode and are deposited there; the species adhere to the anode surface after they lose the electrons. This arrangement is not suitable if the electronegative impurities are not deposited by the negative ions at the anode. An alternative arrangement uses the *sweep-out type reactor*, where the anode is a sintered metal based porous pipe (of brass, for example). The negatively charged ions will come to the anode; many of the ions will

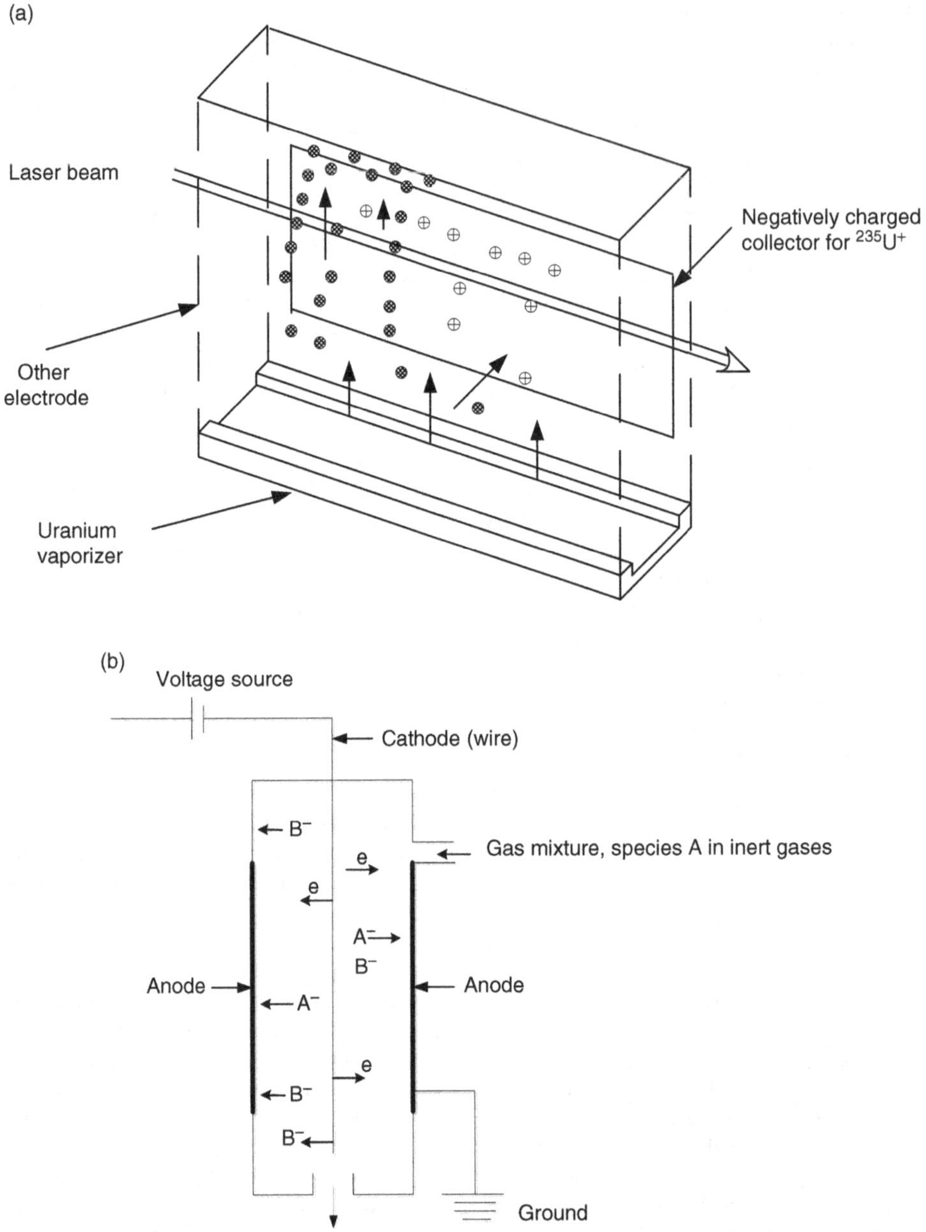

Figure 7.3.7. (a) Exploded view of the AVLIS device for uranium isotope separation. (b) Electron attachment based gas purification in a corona-discharge reactor (see equations (7.3.30a, b)). (After Tamon et al. *(1995).)*

lose their electrons and become neutral molecules. The enriched neutral molecules will be swept out through the porous anode (Tamon *et al.*, 1995). Although there is considerable back diffusion of the enriched neutral molecules toward the reactor center, an adequate sweep-out rate will substantially reduce the back diffusion rate. Consult Tamon *et al.* (1995) for earlier literature, especially on the electron attachment phenomenon as well as on gas-discharge technology for ultrahigh purification. Bulk flow of the gas perpendicular to the electrical field can be employed also to analyze/separate/capture different ions in the gas produced from different chemicals. This is the basis of *ion mobility spectrometry*, employed for explosives detection; see Problem 7.3.6.

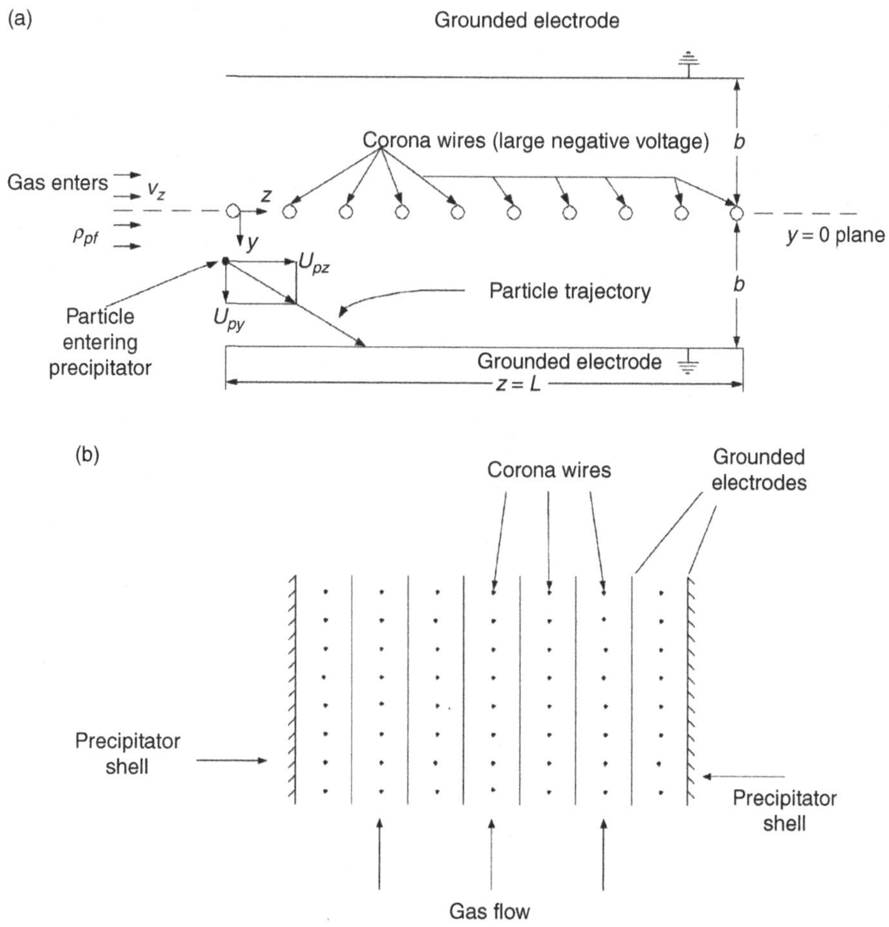

Figure 7.3.8. (a) Particle trajectory in a rectangular precipitator duct showing the configuration of corona wires and the collecting electrodes; (b) multi-duct arrangement in a precipitator shell. (After White (1963).)

7.3.1.3 *Electrostatic precipitation of particles*

Suspended particles are routinely separated from gas streams in large-scale industrial processes by electrostatic precipitators, which were first introduced in 1906. The process of separation consists essentially of three steps (White, 1963).

(1) Suspended particles become electrically charged.
(2) The charged particles move in an electrical field and are collected at an electrode.
(3) The electrode plates are cleaned by rapping and dropping by gravity of collected dust particles, which are removed from the precipitator (into a hopper).

Step (3) is followed in dry electrostatic precipitator systems only. In wet electrostatic precipitation systems, a liquid (usually water) film is used to flush the collecting electrode surface of particles. This may be done in a variety of ways: spray irrigation, film irrigation, intermittently flushed irrigation, self-irrigation, etc. (Jaasund, 1987).

Most particles in industrial gas streams have some electrical charge; however, the magnitude of this charge is very small. In practice, suspended particles are imparted substantial negative charge by passing the gas through a device containing a fine wire at a large negative voltage and a large particle-collecting electrode which is grounded (Figure 7.3.8(a)). There exists a strong electrical field near the wire with a corona glow due to the large number of ions generated. Although both positive and negative ions are generated near the wire, the bulk of the gas space contains essentially negative ions generated by collisions with electrons from the corona region; therefore most suspended particles acquire a negative charge (White, 1963) and move toward the grounded large electrode, which acts as the particle collector. (The general process of particles acquiring charge via collision with ions generated in an electrical field is called *field charging*. Particles, especially smaller ones, may also acquire charge via ion-particle collisions in the absence of an electrical field; this is called *diffusion charging*. See Friedlander (1977) and Flagan and Seinfeld (1988) for a comparison between the two processes.)

The movement of the negatively charged suspended particle toward the large grounded electrode takes place due to a large electrical field. This large electrical field may be due to that between two large oppositely charged plates through which the gas containing charged particles flows. Or the electrical field may simply be the corona between the negatively charged wire and the grounded electrode (Figure 7.3.8(a)); this is identified as the single-stage or Cottrell precipitator. The former arrangement is usually preceded by a first or corona stage to charge the particles; hence it is called a two-stage precipitator. The voltage used in the two-stage precipitator varies between 10 and 15 kV, whereas Cottrell precipitators employ voltages as high as 30–100 kV.

In an electrostatic precipitator, the path or trajectory of a suspended particle is determined by the nature of the convective motion of the gas, the electrical force on the particle and, in a limited way, by the particle size. Consider first *laminar flow* of a gas containing charged particles in a horizontal rectangular duct, as shown in Figure 7.3-8(a), which illustrates gas flowing at a constant velocity v_z in the z-direction parallel to the large electrodes. Assuming no slip between a particle and the gas, we observe that the migration velocity U_{py} of the particle toward the large collecting electrode is perpendicular to the bulk velocity U_{pz} of the particle due to the axial gas motion. (U_{pz} is also essentially equal to gas velocity v_z.) For a spherical particle of radius r_p and charge Q_p in an electrical field of constant strength E, the migration velocity U_{py} may be written, according to equations (3.1.62) and (3.1.214a), as

$$U_{py} = \frac{Q_p E C_c}{6\pi\mu r_p} = \frac{Q_p E}{f_p^d} \tag{7.3.31a}$$

if Stokes' law is assumed to be valid in the gas of viscosity μ with the slip correction factor C_c included as submicron particles are also involved. Such a particle will be captured by the large grounded electrode (of length L in the z-direction) if it traverses a distance b (between the high-voltage corona generating wires and the grounded collecting electrode) in the y-direction in time t needed to traverse axially the precipitator length L:

$$t = (L/v_z) \geq (b/U_{py}) = \frac{6\pi\mu\, r_p b}{Q_p E C_c}. \tag{7.3.31b}$$

For a gas stream with a particle size distribution (r_{min} and r_{max}) and charge distribution, all particles will be collected if

$$\frac{C_c L E}{6\pi\mu b v_z} \geq \frac{r_{min}}{Q_{p,min}} \tag{7.3.31c}$$

for a precipitator with a maximum dimension of b perpendicular to the gas flow. (Here $Q_{p,\,min}$ is the charge on the smaller particle size r_{min}.) The net vector of any particle path in such a case is linear (Figure 7.3.8(a)).

Generally the charge on a particle of radius r_p developed in an electrical field of strength E_c via the phenomenon of field charging is given as (Flagan and Seinfeld, 1988)

$$Q_p = 4\pi\varepsilon_0 \left(\frac{3\,\varepsilon_p}{\varepsilon_p + 2}\right) E_c\, r_p^2, \tag{7.3.32a}$$

where ε_0 is the electrical permittivity of vacuum (8.854×10^{-14} coulomb/volt-cm (see text near equation (3.1.10c)) and ε_p is the relative dielectric constant of the particle.

For a particle whose $\varepsilon_p >> 1$ (conducting), the value of Q_p is

$$Q_p \cong 12\pi\,\varepsilon_0\, E_c r_p^2. \tag{7.3.32b}$$

Flagan and Seinfeld (1988) have illustrated the charge of a particle of radius 0.5 μm in an electrical field strength of 500 kV/m to be 4.17×10^{-17} coulombs:

$$\begin{aligned} Q_p &= 12 \times \pi \times 8.854 \times 10^{-14}\,\frac{\text{coulomb}}{\text{volt-cm}} \\ &\quad \times \frac{500 \times 10^3}{10^2}\,\frac{\text{volt}}{\text{cm}} \times 0.5^2 \times 10^{-8}\text{cm}^2 \\ &= 12 \times \pi \times 8.854 \times 5 \times 0.25 \times 10^{-19}\,\text{coulomb} \\ &= 4.17 \times 10^{-17}\text{coulomb}. \end{aligned}$$

One can now write, from (7.3.31a),

$$U_{py} = 2\,\varepsilon_0 E_c E r_p C_c/\mu. \tag{7.3.32c}$$

If E_c and E are provided in units of volt/meter, r_p, the particle radius (in μm) and μ, the gas viscosity (in kg/m-hr), then U_{py} (in m/s) is provided by the following expression:

$$\begin{aligned} U_{py} &= 2 \times 8.854 \times 10^{-14}\,\frac{\text{coulomb}}{\text{volt-cm}} \times 10^2\,\frac{\text{cm}}{\text{m}} \\ &\quad \times E_c\,E\,\frac{\text{volt}^2}{\text{m}^2}\,r_p(\mu\text{m})\,\frac{\text{m}\,10^{-6}}{\mu\text{m}}\,\frac{C_c}{\mu\frac{\text{kg}}{\text{m-hr}}}\,3600\frac{\text{s}}{\text{hr}}. \end{aligned}$$

Knowing that 1 volt-coulomb = 1 newton-meter and 1 newton = 1 (kg-m)/s², we get

$$U_{py}(\text{m/s}) = \frac{6.37 \times 10^{-14}\,r_p(\mu\text{m})\,E_c(\text{volt/m})\,E(\text{volt/m})\,C_c}{\mu(\text{kg/m-hr})}, \tag{7.3.32d}$$

which is close to that recommended by Wark and Warner (1976, chaps. 4, 5).

Instead of a single set of corona wires in the duct between two collecting grounded electrodes

(Figure 7.3.8(a)), industrial practice often employs a precipitator shell containing a large number of parallel ducts, with a set of corona wires in each duct and horizontal gas flow (Figure 7.3.8(b)). Optimization of the duct spacing in such a configuration has been analyzed in White (1963). Such optimization must consider the possibility of bridging of the collected dust layers within a gas flow channel. The dust buildup on the wall near the bottom of vertical precipitators needs to be considered in any design of precipitators, especially in the multiduct design, to prevent short circuits.

In the analysis of dust precipitation in a duct, attention has so far been paid to a single particle only. In a precipitator, as the gas enters and flows horizontally, particles exist at all locations of the cross section. One is then interested in finding out the density of particles as a function of two coordinates: z, in the direction of bulk gas flow, and y, normal to it. To that end, for particles of size r_p, a more general analysis may be carried out to determine the fraction of particles collected. Consider laminar gas flow using the particle diffusion equation (6.2.55) in the gas in the presence of an external force, no particle generation or agglomeration, dilute suspension and a quasi-stationary state:

$$\mathrm{v}\cdot\nabla\rho_{pg} + \frac{m_p}{f_p^d}\nabla\cdot\left(\rho_{pg}\hat{\boldsymbol{F}}_p\right) = \nabla\cdot\left(D_p\nabla\rho_{pg}\right), \qquad (7.3.33)$$

where ρ_{pg} is the particle mass concentration in the gas. For one-dimensional plug flow of the gas in the z-direction, uniform electrical field of constant strength E in the y-direction, and a constant particle diffusivity D_p, the above equation is simplified for particles of mass m_p and radius r_p to (Soo and Rodgers, 1971)

$$v_z\frac{\partial\rho_{pg}}{\partial z} + \left(\frac{Q_pE}{f_p^d}\right)\frac{\partial\rho_{pg}}{\partial y} = D_p\frac{\partial^2\rho_{pg}}{\partial y^2}. \qquad (7.3.34a)$$

The boundary conditions for two locations, namely the wires ($y = 0$) and the collector plates ($y = \pm b$) are

$$y = 0: \quad \left(Q_pE/f_p^d\right)\rho_{pg} = D_p\left(\partial\rho_{pg}/\partial y\right); \qquad (7.3.34b)$$

$$y = \pm b: \quad (1-\sigma)\left(Q_pE/f_p^d\right)\rho_{pg} = D_p\left(\partial\rho_{pg}/\partial y\right), \qquad (7.3.34c)$$

where σ is the sticking probability of the particle on the collector plate. (The basis for the collector plate condition is provided in the footnote.)[20] The initial condition for the suspension entering the device is $\rho_{pg} = \rho_{pf}$. When $\sigma = 1$, a particle arriving at the plate cannot reenter the gas phase.

Soo (1989) had used the nondimensional variables

$$z^* = (z/b), y^* = (y/b), Pe = \left(bv_z/D_p\right), \beta_e = \left(Q_pEb/f_p^dD_p\right) \qquad (7.3.34d)$$

to obtain the following nondimensional form of equation (7.3.34a) and the boundary conditions:

$$Pe\frac{\partial\rho_{pg}}{\partial z^*} + \beta_e\frac{\partial\rho_{pg}}{\partial y^*} = \frac{\partial^2\rho_{pg}}{\partial y^{*2}}. \qquad (7.3.35a)$$

$$\text{At } y^* = 0; \quad \frac{\partial\rho_{pg}}{\partial y^*} = \beta_e\rho_{pg}; \qquad (7.3.35b)$$

$$\text{at } y^* = 1; \quad \frac{\partial\rho_{pg}}{\partial y^*} = (1-\sigma)\beta_e\rho_{pg}. \qquad (7.3.35c)$$

An exact solution by separation of variables has been provided by Soo and Rodgers (1971). The solution for small values of β_e and σ (Soo, 1989) is

$$\left(\frac{\rho_{pg}}{\rho_{pf}}\right) = [1 + \beta_e y^*]\exp[-\sigma\beta_e z^*/Pe]. \qquad (7.3.36)$$

The fraction $\eta|_z$ of particles of size r_p collected in an electrostatic precipitator of length z is (it is also equal to the grade efficiency function $G_r|_z$)

$$\eta|_z = 1 - \frac{\left(\int_0^b \rho_{pg}v_z\,dy\right)_z}{\left(\int_0^b \rho_{pg}v_z\,dy\right)_{z=0}} = 1 - \frac{\left(\int_0^1 \rho_{pg}v_z\,dy^*\right)_{z^*}}{\left(\int_0^1 \rho_{pg}v_z\,dy^*\right)_{z^*=0}} = G_r|_z. \qquad (7.3.37)$$

Substituting the solution for ρ_{pg}, we get the following expression for $z = L$ (i.e. $z^* = L/b$):

[20] Integrate equation (7.3.34a) from $y = 0$ to $y = +b$ (for example) to obtain

$$\frac{\partial}{\partial z}\int_0^b v_z\rho_{pg}\,dy + \left(\frac{Q_pE}{f_p^d}\right)\left\{\rho_{pg}|_{y=b} - \rho_{pg}|_{y=0}\right\}$$
$$= D_p\left(\partial\rho_{pg}/\partial y\right)|_{y=b} - D_p\left(\partial\rho_{pg}/\partial y\right)|_{y=0}.$$

Since $v_z \cong U_{pz}$, we can make a total particle balance based on the net rate of particle deposition at the wall:

$$\frac{\partial}{\partial z}\int_0^b U_{pz}\rho_{pg}dy = \lim_{\Delta z\to 0}\frac{\int_0^b U_{pz}\rho_{pg}dy|_{z+\Delta z} - \int_0^b U_{pz}\rho_{pg}dy|_z}{\Delta z}$$
$$= -ve\,\text{quantity} = -(\text{rate of particle loss to wall})$$
$$= -\sigma\left(\frac{Q_pE}{f_p^d}\right)\rho_{pg}|_{y=b} \cong \frac{\partial}{\partial z}\int_0^b v_z\rho_{pg}dy.$$

Using boundary condition (7.3.34b) at $y = 0$ and this result, we get

$$(1-\sigma)\left(Q_pE/f_p^d\right)\rho_p|_{y=b} = D_p\left(\partial\rho_{pg}/\partial y\right)|_{y=b}.$$

$$\eta|_L = 1 - \frac{\left\{\int_0^1 (1+\beta_e y^*)\mathrm{d}y^*\right\}\exp[-\sigma\beta_e L/bPe]}{\left\{\int_0^1 (1+\beta_e y^*)\mathrm{d}y^*\right\}} = G_r|_L;$$

$$G_r|_L = \eta|_L = 1 - \exp[-\sigma\beta_e L/bPe]. \tag{7.3.38}$$

Soo (1989) has provided the solutions for large values of β_e as well as other types of conditions. Soo and Rodgers (1971) provide a general expression for $\eta|_L$ as well as for $\rho_{pg}(y, z)$.

The previous analysis was based on laminar gas flow in the precipitator ducts. In practice, turbulent flow exists. The complications introduced by turbulent flow for small dust particles may be appreciated if we consider *first*, among others, the expression (7.3.32c) for the particle migration velocity:

$$U_{py} = 2\,\varepsilon_0\,E_c\,E\,r_p\,C_c/\mu. \tag{7.3.39}$$

The magnitude of the particle migration velocity decreases as the particle becomes smaller (see, however, the discussion following equation (7.3.42)). For smaller particles ($\leq 1\,\mu$m), $U_{py} < 1$ ft/s, whereas axial duct gas velocities v_z are generally of the order of 10 ft/s (White, 1963), between 1 and 20 ft/s (Wark and Warner, 1976).

Secondly, the gas flow, and therefore the particle flow or vector in turbulent flow, is highly irregular due to three-dimesional flow fluctuations. Further, the magnitude of the instantaneous velocity fluctuations in the $\pm y$- and $\pm x$-directions are easily 10–30% of the fluctuating axial gas velocity v_z. Such velocity fluctuations, especially in the $\pm y$-direction mask the small y-directional migration velocity due to the electrical field. Only if such turbulent-flow fluctuations bring the small particle close to the collector plates does the electrical field take over and the particle is captured, since the gas flow velocities are much smaller in the wall region.

A simple model has been developed by Deutsch (1922) to predict the extent of dust particle capture and consequent gas cleanup in *turbulent flow* in a duct using a number of simplifying assumptions.

(1) Due to turbulent mixing, the particle concentration is uniform across the duct cross section A_c at any axial location.
(2) The gas moves in plug flow with a velocity v_z along the duct; however, close to the duct walls, the gas velocity is significantly reduced.
(3) The particle migration velocity U_{py} due to the electrical field in the region close to the duct wall is constant independent of particle size, etc. All particles entering this wall region of thickness δ_g will be captured.

Let the particle capture described in assumption (3) take place in time Δt. Therefore

$$\delta_g = U_{py}\Delta t. \tag{7.3.40}$$

The change in particle mass concentration, $\Delta\rho_{pg}$, as a result of this capture from the value of ρ_{pg} may be described by

$$\frac{\Delta\rho_{pg}}{\rho_{pg}} = -\frac{B_p\delta_g}{A_c} = -\frac{B_p U_{py}\Delta t}{A_c},$$

where B_p is the perimeter of the duct since $(B_p\delta_g/A_c)$ is really the fractional cross section of the duct occupying the slow-moving wall region of the gas. In this time interval Δt, the bulk gas moves an axial distance Δz, where $\Delta z = v_z\Delta t$. Therefore

$$\frac{\Delta\rho_{pg}}{\rho_{pg}} = -\frac{B_p U_{py}}{A_c v_z}\Delta z.$$

In the limits of $\Delta\rho_{pg} \to 0$ and $\Delta z \to 0$, we integrate to obtain

$$\ln\rho_{pg}|_{\rho_{pf}}^{\rho_{pg}} = -\frac{B_p U_{py}}{A_c v_z}L \Rightarrow \rho_{pg} = \rho_{pf}\exp\left\{-\frac{B_p U_{py}}{A_c v_z}L\right\}. \tag{7.3.41}$$

The fraction of particles collected in a precipitator of length L can be obtained from above as (it is also equal to the total efficiency E_T)

$$E_T = \eta|_L = \frac{\rho_{pf} - \rho_{pg}}{\rho_{pf}} = 1 - \exp\left\{-\frac{B_p U_{py}}{A_c v_z}L\right\}. \tag{7.3.42}$$

This result, known as the *Deutsch equation*, is identical to equation (7.3.38) if the sticking probability $\sigma = 1$ and $(A_c/B_p) = b$. Both equations point to certain characteristics of the precipitator behavior. Increases in particle migration velocity and precipitator length increase the extent of particle capture, whereas an increase in gas velocity decreases $\eta|_L$. Since U_{py} is proportional to the product of the electrical fields EE_c, an increase in the electrical field E and the particle charging field E_c will increase the particle charging efficiency. By the same argument, larger particles will have higher collection efficiency. Such a conclusion, however, has to be tempered by the observation of a minimum in $\eta|_L$ with respect to $r_p \cong 0.5\,\mu$m, possibly due to transition from diffusion to field charging (Friedlander, 1977). This value can be high, which is one reason why such devices are quite useful for smaller particles unlike cyclones.

The quantity (B_pL/A_cv_z) in the Deutsch equation carries additional significance. It is essentially the ratio of the collecting electrode surface area over the gas flow rate. Thus, for identical values of U_{py}, similar particle collection efficiencies will be obtained in two different size precipitators if the value of (B_pL/A_cv_z) is identical. This provides a basis for precipitator scaleup. A survey of the volumes of electrostatic precipitators needed for required gas flow rates is given in Figure 7.3.9 (Soo and Rodgers, 1971).

The Deutsch equation is based on a highly simplified situation where particle capture is dependent on its entering the wall region in a probabilistic sense. It does not include effects of particle reentrainment, particle agglomeration and other effects related to the corona. Additional important effects are due to particle size distribution, time-varying voltages applied to generate the electrical field, displacements in corona wire locations, etc. To the extent, the Deutsch equation describes a situation where $\sigma = 1$ (no particle reentrainment), to that extent it can describe wet precipitators, which are usually more efficient than dry precipitators (Jaasund, 1987).

We consider now briefly one of the effects not considered in the derivation of the Deutsch equation (7.3.42), namely the presence of a particle size distribution. Let $f_f(r_p)$ be the particle size density function and let $F_f(r_p)$ be the corresponding distribution function of the dust cloud entering the precipitator. Then

$$\mathrm{d}F_f(r_p) = f_f(r_p)\mathrm{d}r_p,$$

where $f_f(r_p)\mathrm{d}r_p$ represents the fraction of particles in the size range r_p to $r_p + \mathrm{d}r_p$ entering the precipitator. Applying the form of the Deutsch equation (7.3.41) only to particles of size r_p having a migration velocity of $U_{py}(r_p)$, we may write, from equation (7.3.41),

$$f_1(r_p) = f_f(r_p)\ \exp\left\{-\frac{B_p U_{py}(r_p)}{A_c v_z} L\right\} \tag{7.3.43}$$

for the fraction of particles between size r_p and $r_p + \mathrm{d}r_p$ in the exiting gas. Integrating over all particle sizes between 0 and r_p, we get

$$F_1(r_p) = \int_0^{r_p} f_1(r_p)\,\mathrm{d}r_p = \int_0^{r_p} f_f(r_p)\ \exp\left\{-\frac{B_p U_{py}(r_p)}{A_c v_z} L\right\}\mathrm{d}r_p, \tag{7.3.44}$$

where $F_1(r_p)$ is the fraction of particles in the exiting gas having sizes between 0 and r_p. When r_p tends to ∞, equation (7.3.44) yields the fraction of all particles which entered the precipitator escaping the device of length L. Therefore the fraction of particles collected in the precipitator of length L is (it is also equal to the total efficiency E_T)

$$E_T = \eta|_L = 1 - F_1(r_p)|_0^\infty = 1 - \int_0^\infty f_f(r_p)\ \exp\left\{-\frac{B_p U_{py}(r_p)}{A_c v_z} L\right\}\mathrm{d}r_p. \tag{7.3.45}$$

More realistically, the limits of the integral are r_{min} and r_{max} for the minimum and maximum particle sizes.

The integral expression (7.3.45) has an exponential term containing $U_{py}(r_p)$. One could, in principle, employ the expression (7.3.39) for U_{py} and rewrite $\eta|_L$ as

$$\eta|_L = 1 - \left[\int_0^\infty f_f(r_p)\ \exp\left\{-\left(\frac{2B_p L\varepsilon_0 E_c E C_c}{\mu A_c v_z}\right) r_p\right\}\mathrm{d}r_p\right] = E_T. \tag{7.3.46}$$

This allows the calculation of $\eta|_L$ for any given $f_f(r_p)$. Such equations, however, have limited utility in actual design calculations.

The fraction $\eta|_L$ of particles collected is sometimes reported in terms of percent collection efficiency (after multiplication by 100). Values of such efficiencies in a few operating plants are indicated in Figure 7.3.9, which identifies only a few major types of installations. The use of electrostatic precipitators is widespread; additional examples of where they are used are cement plants, in the steel industry, in smelting operations, in the chemical industry, the petroleum industry, in carbon black factories, foundries, etc. In such industrial applications, industrial practice suggests techniques other than electrostatic precipitation for particles in the range 0.1–1 μm since the value of $\eta|_L$ is <0.99 and decreases with a decrease in r_p; (Donovan, 1985; Eggerstedt *et al.*, 1993).

For turbulent gas flow in an electrostatic precipitator, an alternative model may be developed by considering the flux of particles toward the wall due to eddy diffusion, Brownian diffusion and electrical migration:

$$n_p(z) = -\left(D_p^{Br} + \varepsilon_p\right)\frac{\partial \bar{\rho}_p}{\partial y} - U_{py}\bar{\rho}_p, \tag{7.3.47}$$

where $\overline{\rho_p}$ is the time-averaged particle density at any (y, z)-location.

7.3.1.4 *Electrostatic separation of a particulate mixture of different plastics/minerals*

In the electrostatic precipitator studied in Section 7.3.1.3, the objective was to separate all particles present in the gas stream; this is achieved by creating an electrical field perpendicular to the main gas flow and ensuring that the particles become negatively charged for removal in the electrical field. In electrostatic separation of a mixture of plastics (or mineral) particles which are of relatively large dimensions, the particles first undergo triboelectric charging in a charger (particle-charging device) where the particles contact one another; as a result, electrons are transferred from one particle to another in contact (a sort of charge exchange between different material particles). For example, triboelectric charging can change the sign of the charge on polyethylene (PE) particles in a mixture of particles of the plastics polyvinylchloride (PVC), PE and polystyrene (PS). When different particles having different charges are exposed to an electrical field via a high voltage (30–60 kV), particles of different plastics (or minerals) can have different trajectories; therefore different particles of different plastics arrive at different locations in the separator and can potentially be separated (similarly for minerals). The roles of the particle-charging device (e.g. cyclones, rotating drum, fluidized bed, vibratory feeder) and method, including humidity, pretreatment

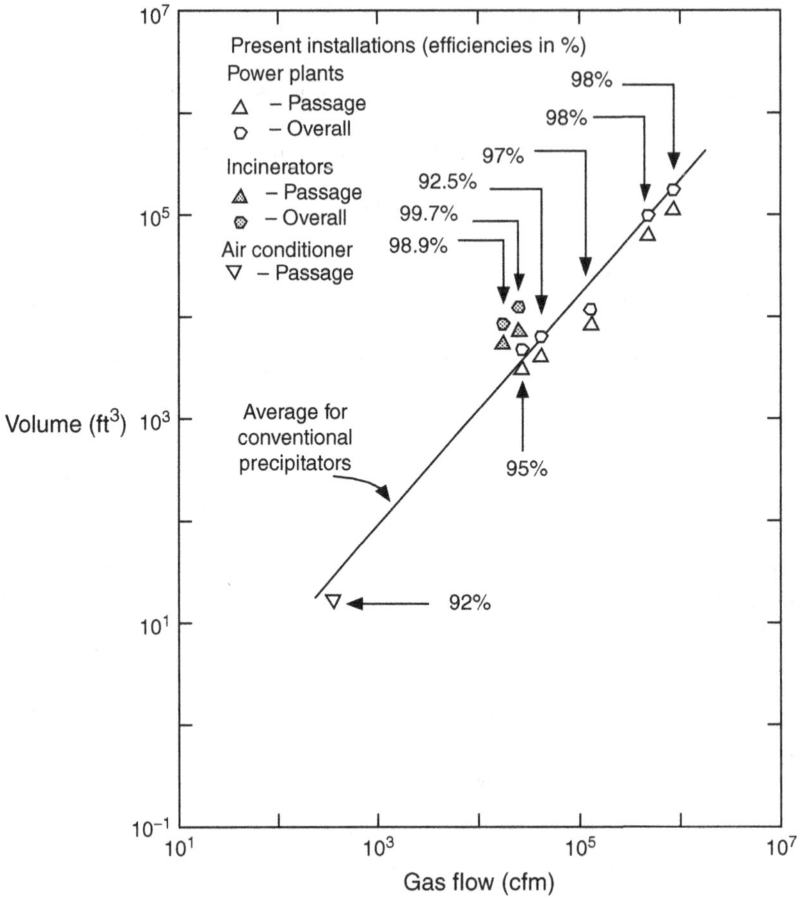

Figure 7.3.9. Survey of size of electrostatic precipitators vs. flow capacity. Reprinted from Powder Technol., 5, 43 (1971), S.L. Soo, L.W. Rogers, "Further studies on the electro-aerodynamic precipitator." Copyright (1971), with permission from Elsevier.

using surfactants, etc., are crucial in introducing an appropriate feed mixture to the *electrostatic separator*. Such separators are very useful in separating mixtures of different minerals (plastics for recycling), especially when their densities are close. Refer to Wei and Realff (2003) for appropriate references in the area of plastics separation. For mineral separation, see Inculet (1984).

There are different types of electrostatic separators, e.g. rotating drum, free-fall, etc. We will focus on the *free-fall electrostatic separator* here. Figures 7.3.10(a) and (b) illustrate, respectively, two different configurations: parallel vertical plates and diverging vertical plates (inclined at an angle θ to the vertical direction). The plates have length L and the gap between the plates at the feed location is c for both the vertical plate as well as the diverging plate configuration. The origin of the (y, z)-coordinates is at the bottom of the plates. There are three bins for collecting particles: the left side bin (for which $y < b_1$,) the right side bin (for which $y > b_2$) and the middle bin ($b_1 \le y \le b_2$). The left side vertical plate is at a voltage $+V$ volts; the right side vertical plate voltage is at $-V$ volts. Therefore the electrical field strength (-voltage gradient) in the y-direction for the parallel vertical plate configuration is

$$E_y = -\frac{\partial \Phi}{\partial y} = -\frac{-V-(V)}{c} = +\frac{2V}{c}, \qquad (7.3.48)$$

which is a constant. However, E_y for the diverging vertical plate configuration will vary with the y-coordinate. We will focus first on *particle trajectories* in these two configurations. We will then very briefly consider *particle recoveries* at different locations of the separator exit influenced by particle charge-to-mass distributions at the separator inlet.

Consider the configuration of *parallel vertical plates* to start with, where E_y is a *constant*. The properties that

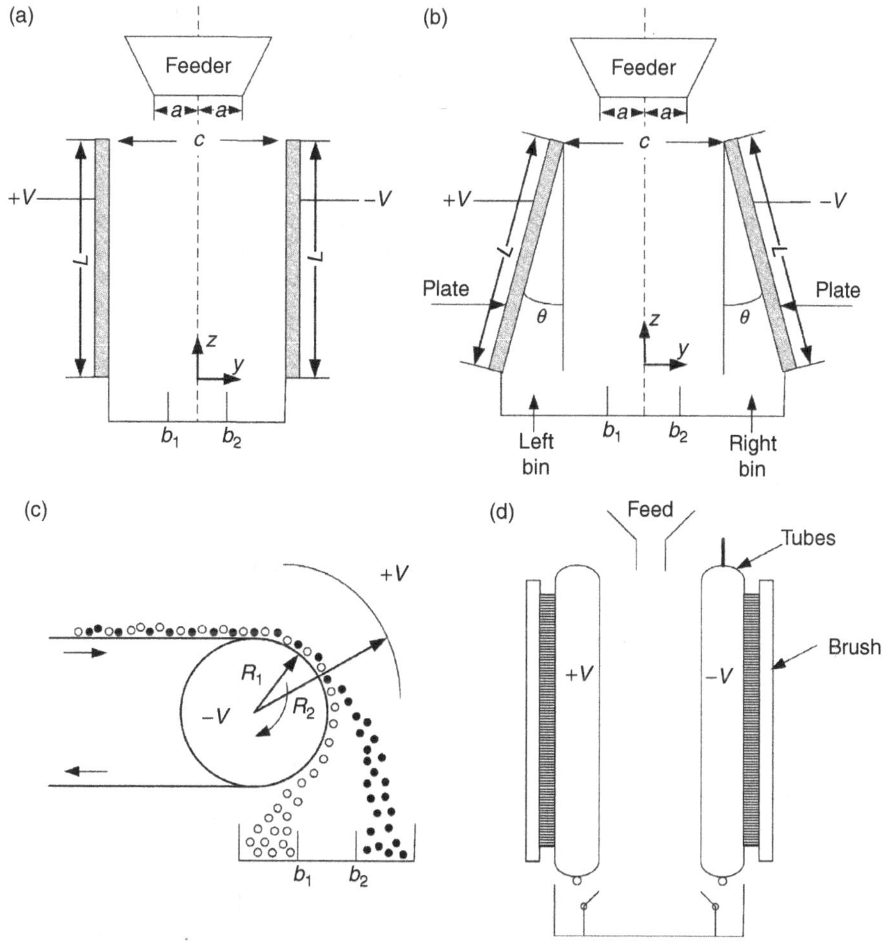

Figure 7.3.10. (a) Free-fall electrostatic separator with vertical plates; (b) free-fall electrostatic separator with inclined vertical plates. (After Wei and Realff (2003).) (c) Drum-type electrostatic separator; (d) tubular free-fall separator. (After Beier and Stahl (1997).)

vary from particle to particle are charge, mass, shape and size. It has been found that the effect of the drag force due to air through which the particles are falling is very small and may be neglected. Therefore the particle shape and size are unimportant. This, and a number of other assumptions (described below), are integral to the particle trajectory analysis that follows. The Coulombic force of interaction between two particles is neglected compared to the electrostatic force on a particle. The effects of interparticle collisions are negligible. The walls of the vertical plates are inelastic: any particle colliding with a vertical wall drops into the local bin. The plate length/vertical particle fall height (L for parallel-plate configuration and $L\cos\theta$ for the diverging-plate configuration) is much larger than the gap between the plates, c. The electrical field E_y is unaffected by the charge Q_p on the particles.

Let a particle start falling at $t = 0$ from coordinates $y = y_0$ and $L - z = 0$ at the entrance of the separator. The two governing equations for particle trajectory coordinates (y, L-z) are (Figure 7.3.10(a)) (from equations (6.2.47c,b))

$$m_p \frac{d^2(L-z)}{dt^2} = m_p g; \qquad (7.3.49a)$$

$$m_p \frac{d^2 y}{dt^2} = E_y Q_p, \qquad (7.3.49b)$$

since drag forces are negligible. From the first equation, we get, after integration,

$$\frac{d(L-z)}{dt} = gt + a_1, \qquad L - z = \frac{gt^2}{2} + a_1 t + a_2.$$

At time $t = 0$, $L - z = 0 \Rightarrow a_2 = 0$. At time $t = 0$, the vertical velocity of the particle is zero; therefore $a_1 = 0$. We get

$$L - z = \frac{gt^2}{2}. \qquad (7.3.50)$$

From the second equation, (7.3.49b), we get

$$\frac{d^2y}{dt^2} = E_y Q_p/m_p = \left(\left(\frac{2V}{c}\right)Q_p\right)/m_p.$$

Integrating, we get

$$\frac{dy}{dt} = \left\{\left(\left(\frac{2V}{c}\right)Q_p\right)/m_p\right\}t + a_3,$$

$$y = \left\{\left(\left(\frac{2V}{c}\right)Q_p\right)/m_p\right\}\frac{t^2}{2} + a_3 t + a_4.$$

At $t = 0$, $y = y_0$, therefore[21] $a_4 = y_0$. Further, at $t = 0$, $dy/dt = 0$, leading to $a_3 = 0$. Therefore, we get

$$y = y_0 + \left\{\left(\left(\frac{2V}{c}\right)Q_p\right)/m_p\right\}\frac{t^2}{2}. \qquad (7.3.51)$$

It is of interest to know what is the particle trajectory, i.e. the relation between y and $(L - z)$ as t increases. From equation (7.3.50), we get t in terms of $(L - z)$; therefore,

$$y = y_0 + \left\{\left(\left(\frac{2V}{c}\right)Q_p\right)/m_p\right\}\frac{(L-z)}{g}. \qquad (7.3.52)$$

If we nondimensionalize the two Cartesian coordinates (y, z) via

$$y^+ = y/L \quad \text{and} \quad z^+ = z/L, \qquad (7.3.53)$$

then

$$y^+ = y_0^+ + \left\{\left(\left(\frac{2V}{c}\right)Q_p\right)/m_p g\right\}(1 - z^+), \qquad (7.3.54)$$

which is the desired trajectory equation relating y with $(L - z)$, the two coordinates characterizing the particle position at any time t; the corresponding nondimensional coordinates are y^+ and $(1 - z^+)$. At the bottom of the two plates $(z = 0, z^+ = 0)$,

$$y^+ = y_0^+ + \left\{((2V/c)Q_p)/m_p g\right\}. \qquad (7.3.55)$$

Consider now the *second configuration* (Figure 7.3.10(b)), where we have two diverging plates of length L at an angle θ to the vertical. The first item to note here is that the electrical field strength in the y-direction is no longer a constant:

$$E_y = -\frac{\partial \phi}{\partial y} = \text{not a constant}. \qquad (7.3.56)$$

[21] Here y_0 is a random variable. A particle can be anywhere at $t = 0$ along the y-coordinate and $z = L$.

One has to determine this quantity first. This is done, in general, by solving Poisson's equation (Newman, 1973, 1991),

$$\nabla \cdot (\varepsilon_d \boldsymbol{E}) = -\nabla \cdot (\varepsilon_d \nabla \phi) = \rho_e, \qquad (7.3.57)$$

where ρ_e is the electric charge density per unit volume and ε_d is the dielectric constant of the medium. Since the dielectric constant of the medium is constant and, by our assumption, E_y is unaffected by the charge on the particles, the above equation is reduced to the Laplace equation in (y, z)-directions:

$$-\nabla \cdot (\nabla \phi) = \frac{\rho_e}{\varepsilon_d} = 0 \Rightarrow \nabla^2 \phi = 0;$$

$$\frac{\partial^2 \phi}{\partial y^2} + \frac{\partial^2 \phi}{\partial z^2} = 0. \qquad (7.3.58)$$

Since the plate height (in the z-direction) is much larger than the gap between the plates, the second term may be neglected. Therefore

$$\frac{\partial^2 \phi}{\partial y^2} = 0. \qquad (7.3.59a)$$

The boundary conditions are as follows:

$$\text{at } y = -\frac{c}{2} - (L - z)\sin\theta: \quad \phi = +V;$$
$$\text{at } y = +\frac{c}{2} + (L - z)\sin\theta: \quad \phi = -V. \qquad (7.3.59b)$$

Introducing the nondimensional variables

$$\phi^+ = (\phi/V), y^+ = (y/L\cos\theta), z^+ = (z/L\cos\theta) \text{ and } c^+ = (c/L\cos\theta) \qquad (7.3.59c)$$

into equations (7.3.59a,b), we get

$$\frac{\partial^2 \phi^+}{\partial y^{+2}} = 0. \qquad (7.3.60a)$$

$$\text{At } y^+ = -(c^+/2) - (1 - z^+)\tan\theta: \phi^+ = 1;$$
$$\text{at } y^+ = (c^+/2) + (1 - z^+)\tan\theta: \phi^+ = -1. \qquad (7.3.60b)$$

The solution of equation (7.3.60a) is

$$\phi^+ = a_5 y^+ + a_6.$$

Applying the two boundary conditions, we get, at any z^+,

$$a_6 = 0 \quad \text{and} \quad a_5 = -\frac{2}{(c^+ + 2(1 - z^+)\tan\theta)}.$$

Therefore

$$\phi^+ = -\frac{y^+}{(c^+/2) + (1 - z^+)\tan\theta} \qquad (7.3.61)$$

and

$$E_y = -\frac{\partial\phi}{\partial y} = -\frac{\partial\phi}{\partial\phi^+}\frac{\partial\phi^+}{\partial y^+}\frac{\partial y^+}{\partial y}$$
$$= (-)\,V\,(-)\frac{1}{(c^+/2)+(1-z^+)\tan\theta}\frac{1}{L\cos\theta}$$
$$= \frac{V}{L\cos\theta}\frac{1}{(c^+/2)+(1-z^+)\tan\theta}. \quad (7.3.62)$$

Since we have an expression for E_y, we can now substitute it into equation (7.3.49b):

$$\frac{d^2y}{dt^2} = \frac{V(Q_p/m_p)}{L\cos\theta}\frac{1}{(c^+/2)+(1-z^+)\tan\theta}.$$

Introducing a nondimensional time t^+ via

$$t^+ = t/(2L\cos\theta/g)^{1/2}, \quad (7.3.63)$$

we get

$$\frac{d^2y^+}{dt^{+2}}\left(\frac{dt^+}{dt}\right)^2 L\cos\theta = \frac{V(Q_p/m_p)}{L\cos\theta}\frac{1}{(c^+/2)+(1-z^+)\tan\theta};$$

$$\frac{d^2y^+}{dt^{+2}} = \frac{2VQ_p}{cm_pg}\frac{2}{1+(1-z^+)\tan\theta\,\dfrac{2L\cos\theta}{d}} = \frac{2a_7}{1+a_8(1-z^+)}, \quad (7.3.64)$$

where

$$a_7 = \{(2V/c)Q_p/(m_pg)\}, \qquad a_8 = \frac{2L\sin\theta}{c}.$$

We now need to solve the other equation, (7.3.49a). However, the particle trajectory coordinates here are (y, $L\cos\theta - z$) (from Figure 7.3.10(b)). Therefore, the corresponding equation is

$$m_p\frac{d^2(L\cos\theta - z)}{dt^2} = m_pg. \quad (7.3.65a)$$

The corresponding solution is

$$L\cos\theta - z = gt^2/2. \quad (7.3.65b)$$

When nondimensionalized, this expression is reduced to

$$1 - z^+ = t^{+2}, \quad (7.3.66)$$

a result which allows us to rewrite equation (7.3.64) as

$$\frac{d^2y^+}{dt^{+2}} = \frac{2a_7}{1+a_8t^{+2}}, \quad (7.3.67)$$

which is an ordinary differential equation of y^+ in terms of t^+. Since t^+ is directly related to z^+ via (7.3.66), a solution of this equation will yield the required relation between y^+ and z^+, the particle trajectory coordinates. The two conditions needed to solve this equation are:

$$\text{at } t=0, y=y_0; \qquad y^+ = (y_0/L\cos\theta) = y_0^+;$$

also

$$\text{at } t=0; \quad (dy/dt)=0, (dy^+/dt^+)=0. \quad (7.3.68)$$

The solution of equation (7.3.67) subject to these two conditions (Wei and Realff, 2003) is

$$y^+ = y_0^+ + (2a_7/a_8)\left[\sqrt{a_8}\,t^+\arctan(\sqrt{a_8}\,t^+) - \ell n(1+a_8t^{+2})\right]. \quad (7.3.69)$$

The particle trajectory coordinates y^+ and z^+ are therefore related via

$$y^+ = y_0^+ + (2a_7/a_8)\Big[\sqrt{a_8(1-z^+)}\,\arctan\left(\sqrt{a_8(1-z^+)}\right) - \ell n\left\{\sqrt{1+a_8(1-z^+)}\right\}\Big]. \quad (7.3.70)$$

The y-coordinate of the particle at the bottom of the separator is of greatest interest. For this location, $z=0$, $z^+=0$:

$$y^+ = y_0^+ + (2a_7/a_8)\left[\sqrt{a_8}\arctan(\sqrt{a_8}) - \ell n\sqrt{1+a_8}\right]. \quad (7.3.71)$$

Note here that the constant a_8 has only the geometric dimensions of the separator, but that the constant a_7 has dimension (Q_p/m_p), the particle charge per unit particle mass. The particle charge-to-mass ratio is a random variable; the initial particle coordinate at the start of the fall, y_0, is also a random variable. The situation is identical in the case of the separator with vertical plates; see equation (7.3.55).

We will not develop the details of a particle recovery model here beyond introducing the basic approach (for details, see Wei and Realff (2003)). For the free-fall electrostatic separators shown in Figures 7.3.10(a) and (b), one would like to determine the probability that the final particle position is either $y < b_1$ (corresponding to the left bin) or $y > b_2$ (corresponding to the right bin), since the probability of going to the central bin is obtained by subtracting the two above-mentioned probabilities from 1. For the left bin ($y < b_1$) at the separator exit, from result (7.3.71) we can write, for particles being treated in the separator of Figure 7.3.10(b),

$$b_1 > y\left\{= y_0 + \frac{\{2VQ_p/(m_pg)\}}{\tan\theta}\left[\sqrt{a_8}\arctan\sqrt{a_8} - \ell n\sqrt{1+a_8}\right]\right\}. \quad (7.3.72)$$

One needs to know the following probability to determine the fraction collected in the left bin:

$$\Pr\left\{y_0 + \frac{2VQ_p}{m_pg\tan\theta}\left[\sqrt{a_8}\arctan\sqrt{a_8} - \ell n\sqrt{1+a_8}\right] < b_1\right\}. \quad (7.3.73)$$

For the separator of Figure 7.3.10(a), one needs to know, correspondingly, the probability of (from result (7.3.55))

$$\Pr\left\{ y_0 + \frac{2VQ_pL}{c\,m_p g} < b_1 \right\}. \tag{7.3.74}$$

For both cases ($\theta > 0$, $\theta = 0$), one can express the probability (alternatively, the fractional recovery, *re*) as

$$\Pr\left\{ (Q_p/m_p) < (b_1 - y_0)\, a_9 \right\}, \tag{7.3.75}$$

where, for $\theta > 0$,

$$a_9 = \frac{g \tan\theta}{2V\left[\sqrt{a_8}\arctan\sqrt{a_8} - \ell n\sqrt{1+a_8}\right]}$$

and, for $\theta = 0$,

$$a_9 = \frac{cg}{2VL}. \tag{7.3.76}$$

Now for the right bin ($y > b_2$) for $\theta \geq 0$, we can similarly obtain the probability as

$$\Pr\left\{ y\left\{ = y_0 + \frac{\left\{2VQ_p/(m_p g)\right\}}{\tan\theta} \left[\sqrt{a_8}\arctan\sqrt{a_8} - \ell n\sqrt{1+a_8}\right] \right\} > b_2 \right\}, \tag{7.3.77}$$

which is reduced to

$$\Pr\left\{ (Q_p/m_p) > (b_2 - y_0)\, a_9 \right\}. \tag{7.3.78}$$

Now consider the case of the fractional recovery of the left bin ($y < b_1$). The probability of a particle coming into this bin has been expressed via (7.3.75), where

$$(Q_p/m_p) < (b_1 - y_0)\, a_9.$$

We have two density functions (or probability functions) to deal with. First, the initial (starting coordinate) particle coordinate y_0, which can be anywhere between $-a$ and $+a$; therefore the quantity (b_1 - y_0) can vary between b_1 - a and $b_1 + a$. Further, this probability is uniform: any starting location is as probable as any other starting location. Consequently, the probability density function for the initial particle location is

$$f_{y0} = \frac{1}{b_1 + a - (b_1 - a)} = \frac{1}{2a}. \tag{7.3.79}$$

Second, the other important random variable is the particle charge per unit particle mass, Q_p/m_p, which can be assumed to vary as a Gaussian distribution:

$$f_{Qm} = \frac{1}{\sqrt{2\pi}\sigma_{Qm}} \exp\left\{ - \frac{\left((Q_p/m_p) - (\overline{Q_p/m_p})\right)^2}{2\sigma_{Qm}^2} \right\}, \tag{7.3.80}$$

where $(\overline{Q_p/m_p})$ is the mean of the distribution and σ_{Qm}^2 is the variance. The joint probability of an initial particle location y_0 and a particle charge-to-mass ratio (Q_p/m_p) will be the product of the two independent probabilities,

$$f_{y0} f_{Qm} = \frac{1}{2a} \times \frac{1}{\sqrt{2\pi}\,\sigma_{Qm}} \exp\left\{ - \frac{\left((Q_p/m_p) - (\overline{Q_p/m_p})\right)^2}{2\sigma_{Qm}^2} \right\}. \tag{7.3.81}$$

The probability of finding all particles in the left bin (therefore the fractional particle recovery, *re*) can be obtained by integrating this joint probability density function over the required limits:

$$re = \Pr\left\{ \begin{matrix} (Q_p/m_p) < (b_1 - y_0)a_9 \\ b_1 - a < y_0 < b_1 + a \end{matrix} \right\}$$
$$= \int_{b_1-a}^{b_1+a} \left[\frac{1}{2a} \int_{-\infty}^{(b_1-y_0)a_9} \frac{1}{\sqrt{2\pi}\sigma_{Qm}} \exp - \left\{ \frac{\left[\left(\frac{Q_p}{m_p}\right) - \left(\frac{\overline{Q_p}}{m_p}\right)\right]^2}{2\sigma_{Qm}^2} \right\} d\left(\frac{Q_p}{m_p}\right) \right] dy_0$$
$$= \int_{b_1-a}^{b_1+a} \frac{1}{2a} \left[\int_{-\infty}^{0} \mathrm{I}\,d\left(\frac{Q_p}{m_p}\right) + \int_0^{(b_1-y_0)a_9} \mathrm{I}\,d\left(\frac{Q_p}{m_p}\right) \right] dy_0, \tag{7.3.82}$$

where I is the integrand describing the Gaussian distribution. Using the variable transformation

$$\frac{(Q_p/m_p) - (\overline{Q_p/m_p})}{\sqrt{2}\sigma_{Qm}} = -\eta,$$

the integral between the limits $-\infty$ and 0 could be changed to the limits of 0 to ∞ (see equation (2.5.13) for erf(x)):

$$re = \int_{b_1-a}^{b_1+a} \frac{1}{2a}\left[\frac{1}{2} + \int_0^{(b_1-y_0)a_9} \mathrm{I}\,d\left(\frac{Q_p}{m_p}\right) \right] dy_0.$$

This integral reduces to (Wei and Realff, 2003)

$$re = \frac{1}{2} + \frac{\psi(g_2) - \psi(g_1)}{2(g_2 - g_1)}, \tag{7.3.83}$$

where

$$\psi(\eta) = \eta\,\mathrm{erf}(\eta) + \frac{\exp(-\eta^2)}{\sqrt{\pi}} \tag{7.3.84}$$

and

$$g_1 = \frac{(b_1 - a)a_9 - (\overline{Q_p/m_p})}{\sqrt{2}\sigma_{Qm}}, \qquad g_2 = \frac{(b_1 + a)a_9 - (\overline{Q_p/m_p})}{\sqrt{2}\sigma_{Qm}}. \tag{7.3.85}$$

Related expressions for the fractional recovery in the right and the middle bins are available in Wei and Realff (2003).

A few details about the process and the device (Wei and Realff, 2003) are useful. The voltage V applied between the plates can be as high as 50–80 kV. The electrical field strength for the plate gaps used is of the order of 4×10^5 volts/m. A typical estimate of Q_p/m_p is 3×10^{-6} coulomb/

kg. The plate angle θ varies between 0 and 15°. The plate lengths are in the range of 1-2 meters, while the plate widths (perpendicular to the plane of the paper in Figure 7.3.10(a) and (b)) are ~1 meter. The gap between the plates is $\leq(L/3)$.

An important issue is the charge developed in various components of a plastic mixture. Polyethylene (PE) and polyvinylchloride (PVC) are very common plastics. In a copper-lined cyclone based tribocharging device, Yanar and Kwetkus (1995) showed that PE particles develop positive charges and that PVC particles develop negative charges; they then demonstrated successful separation: PE fraction contained >90% PE at a recovery of >60%. Wei and Realff (2003) have discussed two-stage operations with or without recycle to improve purity and considered several design options. They have also illustrated the charging tendencies of different plastics via a triboelectric series of plastics.

There is another useful geometrical configuration of an electrostatic separator for separation of plastic particles: the drum-type separator shown in Figure 7.3.10(c). There is a rotating drum charged with a high voltage ($-V$) and an outer plate also charged with a high voltage ($+V$). As shown, the outer plate is curved, but it can be straight as well. Particles coming over the belt are subjected to three forces as they approach the angular location of the outer plate: the electrostatic force due to the electrostatic potential gradient between the two electrodes, a centrifugal force and the gravitational force. If the net outwardly radial force is positive for a particle, it will become detached from the belt. The location and force of detachment will determine the particle trajectory and therefore the collection bin where it lands.

The use of electrostatic separators for mineral particle separation is common. It is primarily used for separating mineral ores, ore tailing, pulverized coals containing clay particles (Inculet, 1984), coal ash, etc. For example, triboelectrification of coal ash from power plant exhaust leads to two fractions: carbon-rich material, which becomes positively charged, and the rest of the ash, which is negatively charged (Holusha, 1993). Similarly, high mineral matter containing coal will be separated via triboelectrification into a positively charged coal-rich fraction and a negatively charged mineral-rich fraction.

Very-large-scale separation of raw salts by electrostatic separators is practiced at the level of 900 tons/hr (Beier and Stahl, 1997). In these processes, used for the separation of the potash ore for example, the ore is first ground to free the minerals of individual crystals, then the salt crystals undergo a conditioning treatment, where they are coated with a reagent (or a combination of the reagents), followed by warming and adjustment of the relative humidity, which allows charge exchange to take place. The chemical conditioning and humidity treatment are generally carried out in a fluidized bed using an air stream having the required relative humidity. The charged salt mixture is then separated by letting the mixture fall through a vertical charged separator. Multiple stages are used to process various fractions into increasingly pure fractions. The conditioning agent's role is quite important. Beier and Stahl (1997) have shown that aliphatic monocarboxylic acids allow one to recover almost all of KCl in a mixture of KCl, NaCl and $MgSO_4 \cdot H_2O$ in the positive electrode, with very limited recoveries of NaCl and $MgSO_4 \cdot H_2O$. The addition of aromatic monocarboxylic acid or NH_4 salts of aliphatic monocarboxylic acids leads to different patterns of recovery.

In the above examples, for the successful separation of minerals at the electrodes, a different arrangement is used. A major problem is the adherence of dust/particles on the electrodes in the configurations of Figures 7.3.10(a) and (b) which reduces the field strength. Instead, a number of vertical tubes standing side by side constitute the electrode; however, these tubes slowly rotate around their axis, and any adhering dust/minerals on their surfaces are scraped away by brushes at the back (Figure 7.3.10(d)).

7.3.1.5 *Separation of cells using flow cytometry*

Flow cytometry is a successful technique for "cytometry" (i.e. cell measurements) when a suspension of cells in a liquid stream "flows" through the device and the cells are separated one by one. Cells of interest introduced via a sample injected into the core liquid stream surrounded by a much thicker liquid flowing stream (called the *sheath flow*) in laminar flow are analyzed via their optical signals, usually by a laser beam (Figure 7.3.11). Prior to the sample injection, fluorescent dyes are introduced into the mixtures of cells; particular cells are tagged with particular fluorochromes. The laser beam "interrogates" individual cells as they flow downward. The cells emitting the desired signals are identified extremely rapidly and then tagged individually in the following fashion. The liquid jet/stream beyond the laser beam sensing zone is broken down into tiny droplets (Figure 7.3.11). Knowing which droplet contains a particular cell, the system introduces a charge on such a droplet; then, by electrostatic forces generated by charged deflection plates (as in Section 7.3.1.4), the droplet containing a particular cell is directed to the right or left by a certain distance and collected in several containers. The central container is for the liquid waste stream consisting of droplets (without or with cells) having no charge on them since they are not of interest. For a comprehensive introduction to flow cytometry, see Shapiro (1995). For shorter accounts, consult Hoffman and Houck (1998) and Robinson (2004).

A few more details about the technique will provide a better prespective. The cell population has different subgroups having (potentially, for example) different antigens on the cell surface. The sample to be analyzed is treated

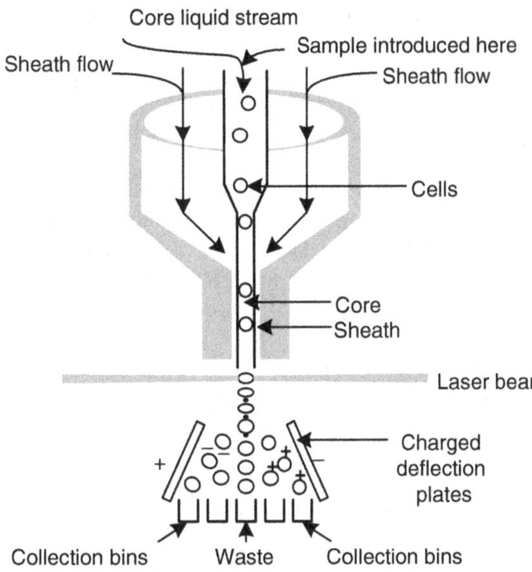

Figure 7.3.11. Schematic view of a flow cytometer based on fluorescence-activated cell sorting via droplet sorting.

(stained) with, say, a number of different antibodies (Abs) which bind with specific antigens (Ags) on the surface of the cells. The antibodies are labeled with (e.g.) specific fluorochromes, each of which provides a distinct color and therefore a distinct identity to the laser probe, of which there may be more than two in some systems. The analysis of the spectral emission from the cell surface is carried out extremely rapidly, at rates as high as 10 000–25 000 cells per second. To ensure that the fluid flow regime is correct, laminar flow is maintained. The fluid jet diameter varies between 50 and 400 μm. The frequency of droplet formation varies between 2000/s and 100 000/s. Vibration at frequencies of 10 000–300 000 Hz via a piezoelectric crystal oscillator is most often used to produce drops. Drops of interest are imparted positive or negative charges. The voltage applied between the two plates can be as high as 5000 volts to separate various drops. The technique is often identified in general as *flow sorting* via droplet sorters, sorting out one type of cell from another via different droplets. The overall technique is also known as ***fluorescence-activated cell sorting*** **(FACS)**.

The efficiency of cell sorting or cell purification has two aspects: (1) How pure is the cell sorted in a particular droplet? (2) What is the rate at which this sorting is taking place? For the technique to be successful, the rate of cell sorting has to be reasonably high; unduly slow techniques are not acceptable. High rates of cell sorting can lead to situations where a cell that is undesirable appears in the same drop with a cell that is desirable; this happens because of the presence of an undesirable cell with a desirable cell in the liquid in the laser interrogation zone. The presence of the desirable cell triggers droplet formation, which can enclose an undesirable cell. Hoffman and Houck (1998) have summarized the different modes of sorting, not only for droplet sorters, but also for other enclosed sorting methods, e.g. catcher-tube sorting, fluidic-switching sorting (here droplets are not generated from the sample stream; rather when there are desired cells, a catcher tube catches it, etc.). These modes are:

(a) single-cell mode: reject if there is more than one cell even though both may be desirable;
(b) enrichment mode: collect as long as there is a desirable cell, even though there may also be an undesirable cell;
(c) exclusion mode: reject when there is an undesirable cell; collect even if there are two desirable cells.

It would be useful to illustrate quantitative measures of purity of the sorted cells with respect to the rate, *ra*, at which cells appear in the sensing zone, the time period, *T*, during which a cell can be sorted for selection or rejection and sub_{fr}, the fraction of the subpopulation of desirable cells to be sorted. The results, illustrated in Hoffman and Hauk (1998), based on analyses available in literature are as follows. The efficiency, *E*, of sorting cells is defined as the fraction of desired cells passing through the sensing zone that are captured/sorted since sometimes the desired cells may not be sorted due to the presence of undesired cells.

(i) Single-cell mode:

$$E = \exp(-(ra)T). \tag{7.3.86}$$

(ii) Exclusion mode:

$$E = \exp(-(1 - sub_{fr})(ra)T). \tag{7.3.87}$$

Illustration of the values of the parameters is useful. The time period, *T*, may vary from 25 to 200 μs. As *T* increases, *E* decreases substantially since unwanted cells appear more often. As the rate of arrival, *ra*, of cells to the sensing zone increases (*ra* varying between, say, 300 cells/s and 10 000 cells/s or more), the efficiency decreases. As the value of the subpopulation fraction, sub_{fr}, decreases, the efficiency decreases, reaching a limit when $sub_{fr} \rightarrow 0.01$. The purity of the droplets sorted out depends on the probability that no other type of cell appears along with the sorted cell in the droplet sorter or other sorters. Expressions (7.3.86) and (7.3.87) provided above for the efficiencies are also used to estimate the purity of the cells in the sorted droplet population.

7.3.2 Centrifugal force field

Large-scale devices in industry employ centrifugal forces to separate particles from a fluid or droplets from an immiscible continuous phase. Usually, these large devices are single-entry separators: a multiphase feed stream enters the separator and two multiphase product streams, an overflow and an underflow, leave the equipment. Centrifuges of much smaller dimensions are used for preparative work or in analytical studies. The centrifugal force is imparted to the multiphase fluid by rotating the whole device around an axis or by imparting a swirl to the feed, introduced tangentially to a stationary cylindrical device, as in a cyclone or hydrocyclone separator. For a review, see Svarovsky (1977, 1981) and Hsu (1981).

In all such separators, the multiphase fluid has a tangential motion as it rotates around a central axis. However, there exists an axial movement, which is superimposed on the primary circulatory motion, which generates the centrifugal force. This axial movement is essential to continuous removal through an outlet of the entering fluid from the centrifugal separating equipment. More important, from a separation point of view, is that this bulk motion is perpendicular to the direction of the centrifugal force generated by the rotation of the multiphase fluid around the main axis. This enhances the multicomponent separation capacity considerably for larger-scale operations. However, this capability is generally unutilized since industrial processing focuses more on particle separation from a fluid or phase separation from a feed which is a multiphase dispersion. Note that, even without the axial bulk flow, the primary circulatory flow is perpendicular to the direction of the centrifugal force.

In this section, particle separation from a liquid having one component of bulk motion along the axis of a *tubular bowl centrifuge* is studied first. Following the particle separation, the separation of two immiscible liquids of different densities is briefly considered. The performance of a *disk centrifuge* is treated next; here the bulk liquid motion toward the outlet is not perpendicular, but is at an angle to the centrifugal force. Separation of particles from a gas in a *cyclone separator* is then considered, followed by brief treatments of the separation nozzle process for gas separation and hydrocyclones.

7.3.2.1 *Tubular bowl centrifuge*

Consider first particle separation from a liquid suspension in a simple tubular bowl centrifuge; a schematic of this centrifuge is shown in Figure 7.3.12. It is essentially a cylindrical bowl rotating around its axis at a high rpm (revolutions per minute). Smaller centrifuges may have rpms as high as 15 000. A dilute feed suspension is introduced through a central nozzle at the bottom. The feed suspension may be assumed to attain the angular velocity of the bowl very quickly as it spreads out from the feed nozzle to the bottom of the bowl at $z = 0$. However, the feed liquid moves axially upward from $z = 0$ to $z = L$, the length of the centrifuge. The radial location of the free surface of this liquid, r_f, is set essentially by the liquid outlet weirs at the top of the bowl.

The feed liquid is assumed to be uniformly distributed at the bottom of the centrifuge (i.e. $z = 0$) from $r = r_f$ to r_0. Therefore, particles of all sizes found in the feed suspension are present at all radial locations (between $r = r_f$ to $r = r_0$) at the centrifuge bottom ($z = 0$). However, as soon as the particles are at various r-values at $z = 0$, they are immediately subjected to sedimentation. The particles are thrown toward the bowl wall by the centrifugal force as they move vertically upward ($z > 0$) due to the axial vertical motion of the liquid. (The circumferential liquid velocity due to the rotation is not important in itself except for the calculation of the centrifugal force on the particles.)

Since each particle has a velocity in the z-direction due to the z-directional liquid motion, as well as a centrifugally generated radially outward velocity, it follows a particular path. These paths are identified in Figure 7.3.12 as *effective particle trajectories*. If the trajectory of a particle hits the walls of the centrifuge (i.e. $r = r_0$) before $z = L$, then the particle is separated from the liquid and is deposited along the wall. When the deposit becomes thick, the centrifuge is stopped to remove the particle deposits, since a thick deposit reduces the cross-sectional area for liquid flow. This particle deposit *may be* considered as the *underflow* in a continuous single-entry particle fractionators, even though, on an extended-time basis, the operation is intermittent.

A particle trajectory is a series of (r, z)-coordinates occupied by a particle as it moves in the centrifuge. A particle trajectory, whose r value is less than r_0 but $z = L$, will not hit the centrifuge wall at $r = r_0$; this particle will leave the centrifuge with the liquid overflow and is not captured by the device. A separation analysis should be able to predict the particle size density function, $f_0(r_p)$, in the liquid overflow for a given feed liquid particle size density function $f_f(r_p)$. Alternatively, the separation analysis should provide an expression for the grade efficiency function, $G_r(r_p)$ of the device.

Separation analysis in a centrifuge for particles begins with a particle trajectory analysis. This is conveniently initiated by following the z-coordinate and the r-coordinate of the particle with time. Assuming that there is no slip between the liquid and the particle, the z-direction particle velocity (U_{pz}) is equal to that of the liquid, $v_z(z, r)$. Therefore,

$$U_{pz} = (\mathrm{d}z/\mathrm{d}t) = v_z(z,r). \tag{7.3.88}$$

If Q_f is the volumetric flow rate of the liquid, and the liquid is assumed to have plug flow between radial locations r_0 and r_f, then

$$U_{pz} = (\mathrm{d}z/\mathrm{d}t) = v_z(z,r) = v_z = Q_f/\pi\left(r_0^2 - r_f^2\right). \tag{7.3.89}$$

For the r-direction motion of a particle of radius r_p, the general equation of particle motion (equation (6.2.45)) is

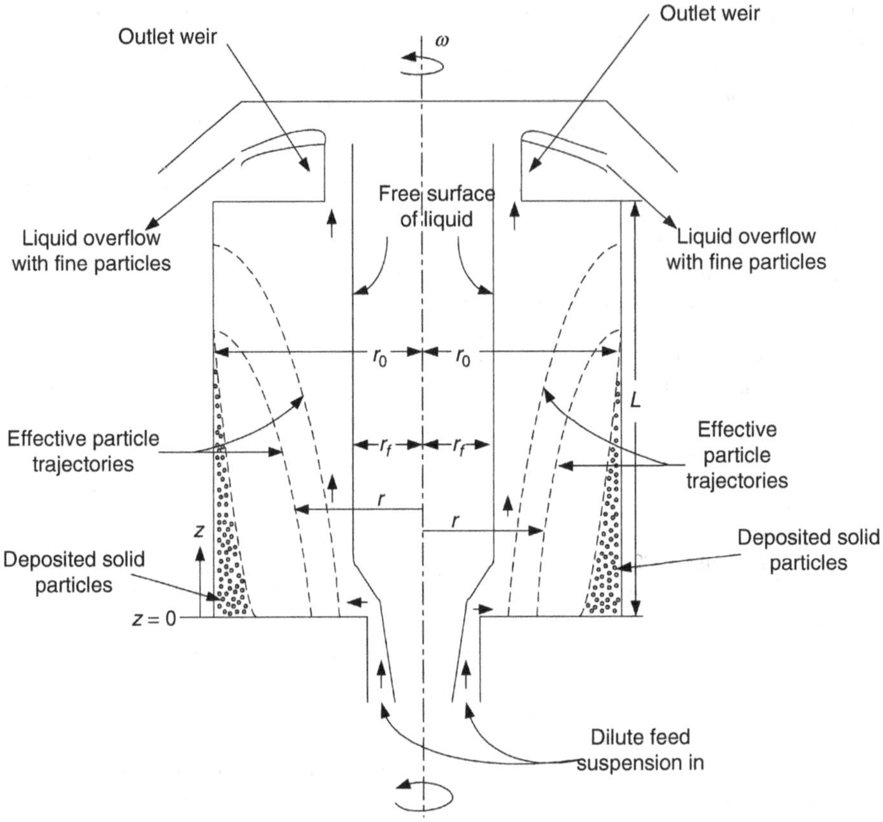

Figure 7.3.12. Schematic for particle separation from a liquid in a tubular bowl centrifuge.

$$m_p \frac{dU_{pr}}{dt} = m_p \frac{d^2 r}{dt^2} = F_{pr}^{ext} - F_{pr}^{drag}. \quad (7.3.90a)$$

Here centrifugal force is the external force along with the centrifugal buoyancy term (see equation (3.1.59)) and its magnitude is given by $m_p r\omega^2 \left(1 - (\rho_t/\rho_p)\right)$. If the size of the particle (assumed spherical) is small enough for Stokes' law to be valid, the radial resistive drag force magnitude is $6\pi\mu\, r_p v_r$. If we further assume that the particles settle always with their terminal radial velocity (i.e. $(dU_{pr}/dt) = (d^2r/dt^2) = 0$ is achieved very quickly), we obtain

$$F_{pr}^{ext} = m_p r\,\omega^2 \left(1 - \frac{\rho_t}{\rho_p}\right) = 6\pi\mu\, r_p U_{pr} = F_{pr}^{drag}. \quad (7.3.90b)$$

This leads to an expression for the radial particle terminal velocity:

$$U_{pr} = U_{prt} = \frac{dr}{dt} = \frac{2r_p^2 \omega^2 r}{9\mu}\left(\rho_p - \rho_t\right). \quad (7.3.91)$$

(*Note*: If gravity is the external force, U_{pz} is obtained by replacing $\omega^2 r$ by g; $U_{pr}|_{centrifuge} = U_{pz}|_{gravity}(\omega^2 r/g)$, where the ratio $(\omega^2 r/g)$ is called the *centrifuge effect*.) Combining this equation with equation (7.3.89), one gets

$$\frac{dz}{dr} = \frac{9Q_f\mu}{2\pi\, r_p^2\, \omega^2 r\left(r_0^2 - r_f^2\right)\left(\rho_p - \rho_t\right)}. \quad (7.3.92)$$

Rearrange it and integrate between $(r, z = 0)$ and $(r = r_0, z = L)$ to obtain

$$\int_r^{r_0} \frac{dr}{r} = \ln\left(\frac{r_0}{r}\right) = \frac{2\pi\, r_p^2\, \omega^2\left(\rho_p - \rho_t\right)\left(r_0^2 - r_f^2\right)}{9Q_f\mu} L. \quad (7.3.93)$$

Since r and z are the coordinates of a particle of radius r_p and density ρ_p as it moves along its trajectory, equation (7.3.93) defines the smallest value of the radial location r of the particle at $z = 0$, where it must be located if it is to hit the wall $(r = r_0)$ and be captured when $z = L$. If the particle is located at a value of r greater than that defined by equation (7.3.93) at $z = 0$, then it will surely hit the wall $(r = r_0)$ before $z = L$. However, if the particle at $z = 0$ is located at an r smaller than that defined by equation (7.3.93), it will not hit the wall $(r = r_0)$ by the time $z = L$; then the particle escapes with the overflow. **A critical radius r is thus identified with each particle size r_p via equation** (7.3.93).

An alternative form of equation (7.3.93) is also quite useful. Suppose the particle starts out at $r = r_f$ when $z = 0$. Then, if its $r = r_0$ when $z = L$,

$$\ln\left(\frac{r_0}{r_f}\right) = \frac{2\pi r_{\max}^2 \omega^2 \left(\rho_p - \rho_t\right)\left(r_0^2 - r_f^2\right)}{9Q_f\mu} L, \tag{7.3.94}$$

where r_p is identified as $r_{\max}$. This is the largest particle size with any nonzero probability of not being captured. Any particle with a smaller size ($<r_{\max}$) and starting at $r = r_f$ and $z = 0$ will not be captured by the time $z = L$, and therefore must appear in the overflow stream. Any particle of size $r \geq r_{\max}$ must settle in the centrifuge.

There is another way of considering the general question of particle settling in the centrifuge. Since the particle velocity in the z-direction is assumed to be equal to the z-direction liquid velocity, the maximum time available for settling is equal to the liquid residence time

$$t_{\text{res}} = L/U_{pz} = L\pi\left(r_0^2 - r_f^2\right)/Q_f = V_{\text{eff}}/Q_f, \tag{7.3.95}$$

where V_{eff} is the liquid-filled volume of the centrifuge. Knowing the radial distance to be traversed by a particle starting out at any r and $z = 0$, the particle settles if t_{res} is greater than or equal to the time required by the radial particle terminal velocity to reach $r = r_0$.

We calculate now the grade efficiency function $G_r(r_p)$ of the tubular bowl centrifuge. Since the z-directional liquid motion is assumed to be of the plug flow type, the following equality holds:

$$\left[\frac{\text{liquid flow rate through an annulus between } r_0 \text{ and } r}{\text{liquid flow rate through an annulus between } r_0 \text{ and } r_f}\right] = \frac{\pi\left(r_0^2 - r^2\right)}{\pi\left(r_0^2 - r_f^2\right)}. \tag{7.3.96}$$

For a particle of size r_p (uniformly present in the entering liquid at all radii between r_f and r_0) to settle by the time $z = L$, it must be at a radius larger than or equal to the r given by equation (7.3.93). Thus particles of size r_p which are in the liquid flow cross-sectional area $\pi\left(r_0^2 - r^2\right)$ to start with ($z = 0$) will settle. Let the total volume fraction of particles in the feed liquid be ϕ_f and the particle size distribution in the feed liquid be $f_f(r_p)$ (based on mass fraction). Then the mass flow rate of particles of size r_p entering the centrifuge between r and r_0 is

$$Q_f\left[\frac{\pi\left(r_0^2 - r^2\right)}{\pi\left(r_0^2 - r_f^2\right)}\right]\phi_f \rho_p f_f(r_p).$$

Remember these particles settle.

The corresponding mass flow rate of particles of size r_p between r_f and r_0 at $z = 0$ is $Q_f\phi_f\rho_p f_f(r_p)$, which is the total incoming rate of particles of size r_p. Therefore, the value of $G_r(r_p)$ (by equation (2.4.4b)) is

$$G_r(r_p) = Q_f\left[\frac{\left(r_0^2 - r^2\right)}{\left(r_0^2 - r_f^2\right)}\right]\phi_f \rho_p f_f(r_p) \Big/ Q_f \phi_f \rho_p f_f(r_p) = \frac{r_0^2 - r^2}{r_0^2 - r_f^2}, \tag{7.3.97}$$

where r for any r_p is to be obtained from the relation (7.3.93) as

$$r = r_0 \exp\left(-\frac{2\pi\omega^2\left(\rho_p - \rho_t\right)\left(r_0^2 - r_f^2\right)L}{9Q_f\mu} r_p^2\right). \tag{7.3.98}$$

From equation (7.3.94), the value of $G_r(r_p) = 1$ when $r_p \geq r_{\max}$. Using (7.3.98), we get, from relation (7.3.97) for $G_r(r_p)$,

$$\left(r_0^2 - r_f^2\right)G_r(r_{\max}) = \left(r_0^2 - r_f^2\right) = r_0^2\left[1 - \exp\left\{-\frac{4\pi\omega^2\left(\rho_p - \rho_t\right)\left(r_0^2 - r_f^2\right)L}{9Q_f\mu} r_{\max}^2\right\}\right].$$

Therefore

$$G_r(r_p) = \frac{\left[1 - \exp\left\{-\dfrac{4\pi\omega^2\left(\rho_p - \rho_t\right)\left(r_0^2 - r_f^2\right)L}{9Q_f\mu} r_p^2\right\}\right]}{\left[1 - \exp\left\{-\dfrac{4\pi\omega^2\left(\rho_p - \rho_t\right)\left(r_0^2 - r_f^2\right)L}{9Q_f\mu} r_{\max}^2\right\}\right]}; \tag{7.3.99a}$$

$$G_r(r_p) = \frac{r_0^2}{\left(r_0^2 - r_f^2\right)}\left[1 - \exp\left\{-\frac{4\pi\omega^2(\rho_p - \rho_t)(r_0^2 - r_f^2)L}{9Q_f\mu} r_p^2\right\}\right] \tag{7.3.99b}$$

for $0 \leq r_p \leq r_{\max}$ (Svarovsky, 1977, p.130).

The above analysis was based on the assumption of plug flow of liquid in the z-direction. Schachman (1948) analyzed the same problem using a velocity profile for annular axial flow of the liquid between r_f and r_0. The drag force experienced by the spherical particle was obtained above from Stokes' law. For larger particle sizes or higher radial velocities, the resistive radial drag force may be expressed by (see equation (3.1.64))

$$F_{pr}^{\text{drag}} = C_D \frac{\rho_t U_{pr}^2}{2} A_p. \tag{7.3.100}$$

This changes the expression (7.3.91) for U_{prt} to

$$U_{prt} = \frac{\mathrm{d}r}{\mathrm{d}t} = \left\{\frac{8r_p\omega^2 r\left[\left(\rho_p/\rho_t\right) - 1\right]}{3C_D}\right\}^{1/2}, \tag{7.3.101}$$

where, for a spherical particle, the value of $A_p = \pi r_p^2$. One can now develop an expression for $(\mathrm{d}z/\mathrm{d}r)$ for the particle trajectory similar to (7.3.92). Expressions for C_D as a function of the Reynolds number are provided by relations

(6.3.4) (see Bird *et al.* (1960, 2002), where the symbol f for the friction factor is used instead of C_D, the drag coefficient). Hsu (1981) has considered various flow regimes specifically for spherical particles in a centrifuge.

Although the grade efficiency function enables one to determine easily the particle size distribution in the overflow and the underflow if the feed particle size distribution is known, a quantity, Σ, called *sigma*, is frequently used in practice to compare different centrifuges (Ambler, 1952). It is defined by the relation

$$Q_f = 2U_{pz}|_{\text{gravity}}\Sigma, \tag{7.3.102}$$

where $U_{pz}|_{\text{gravity}}$ represents the gravitational settling velocity of a particle which obeys Stokes' law and has the equiprobable size, i.e. $r_{p,50}$ (see Section 2.4). It represents the cross-sectional area of a gravity based separator which will process the same feed rate Q_f with somewhat similar particle separation characteristics (via $r_{p,50}$). The gravitational settling velocity is defined as

$$U_{pz}|_{\text{gravity}} = \frac{2\, r_{p,50}^2\, g\left(\rho_p - \rho_t\right)}{9\mu}. \tag{7.3.103}$$

The equiprobable size, $r_{p,50}$, is obtained by equating the expression for $G_r(r_p)$, (7.3.99b) to 0.50:

$$0.50 = \frac{r_0^2\left[1 - \exp\left\{-\frac{4\pi\omega^2\left(\rho_p-\rho_t\right)\left(r_0^2 - r_f^2\right)L}{9Q_f\mu} r_{p,50}^2\right\}\right]}{\left(r_0^2 - r_f^2\right)},$$

which yields

$$r_{p,50}^2 = \frac{Q_f}{2\pi\left(r_0^2 - r_f^2\right)L}\left(\frac{9\mu}{2\omega^2\left(\rho_p - \rho_t\right)}\right)\ln\left[\frac{2r_0^2}{r_0^2 + r_f^2}\right]. \tag{7.3.104}$$

An expression for sigma is now obtained for a tubular bowl centrifuge:

$$\Sigma = \frac{Q_f}{2U_{pz}|_{\text{gravity}}} = \frac{\omega^2}{g}\frac{\pi\left(r_0^2 - r_f^2\right)L}{\ln\left(\frac{2r_0^2}{r_0^2 + r_f^2}\right)}. \tag{7.3.105}$$

This expression for Σ shows that it depends on the physical dimensions of a centrifuge operating with a given liquid and the thickness of the liquid in the bowl rotating at a given condition. It has been observed in practice that particle separation characteristics of centrifuges of different sizes are quite similar if the following equality holds between different centrifuges:

$$\left.\frac{Q_f}{\Sigma}\right|_{\text{centrifuge 1}} = \left.\frac{Q_f}{\Sigma}\right|_{\text{centrifuge 2}} = \left.\frac{Q_f}{\Sigma}\right|_{\text{centrifuge 3}}. \tag{7.3.106}$$

Obviously such a relation is quite useful for scaleup.

Table 7.3.1. Dimensions and/or density of some cells

Cell/particle	Density (g/cm^3)	Dimensions (μm)
Acetobacter oxydans		0.5 × 1
Candida mycoderma		3
Chinese hamster ovary (CHO) cells	1.06	
Chlorella	1.03	~8–10
Coliform bacteria		0.5
Escherichia coli (*E. coli*)	1.09	1
Red blood cell		2.4 × 8.4
Single-cell protein (Candida)		4–7
Yeast (*Saccharomyces*)		
Baker's	~1.1	7–10
Brewer's	~1.1	5–8
S. bayanus		3

Data obtained from: Belter *et al.* (1988); Goel *et al.* (2001); Harrison *et al.* (2003).

Tubular bowl centrifuges are routinely used in smaller-scale separations of suspended material from fermentation beer. The objective may be to recover the cells which are to be disrupted to recover intracellular bioproducts. Alternatively, for extracellular bioproducts, the beer has to be clarified (be free of cells, cellular debris, etc.). Table 7.3.1 provides the dimensions and/or densities of a few types of microorganisms/cells/particles. Two special features of the separation of the suspensions in fermentation beer as feed are their very low density difference $\left(\rho_p - \rho_t\right)$ and smaller particle sizes. In the separation of suspended materials from the fluid in nonbiological systems, $\left(\rho_p - \rho_t\right)$ is considerably larger. Further, the particle dimensions have wide variations.

Example 7.3.3 This example illustrates the procedure by which one can estimate the volumetric flow rate of a suspension which can be effectively treated by a tubular bowl centrifuge having given dimensions. The centrifuge inside radius (r_0) is 10 cm; the centrifuge design creates a free surface radius r_f of 4 cm. The bowl length is 90 cm. The centrifuge rotates at 5000 revolutions per minute (rpm). Determine the volumetric capacity of operation of the centrifuge for the following suspensions and the desired separations.

(a) *E. coli* cells of diameter 1 μm, density 1.09 g/cm^3, suspension viscosity 1.5 cp: complete recovery of the cells is desired for subsequent intracellular product recovery by cell rupturing. The suspension is quite dilute.
(b) Solid particles of density 2 g/cm^3, diameter 10 μm, suspension viscosity 3 cp and density 1.3 g/cm^3: complete clarification of the liquid is desired along with recovery of the solids.

Solution (a) We will employ equation (7.3.94) to determine the value of Q_f for the given $r_{\max}$. Here the cell dimension of *E. coli* will be taken as $r_{\max}$. The value of Q_f determined will be recommended as an upper limit. Here $r_{\max} = 0.5$ μm $= 0.5 \times 10^{-4}$ cm; $r_f = 4$ cm; $r_0 = 10$ cm; $\rho_p = 1.09$ g/cm^3; $\rho_t = 1.00$ g/cm^3 (very dilute suspension);

$$\omega = \frac{5000 \times 2\pi\,\text{radian}}{\text{min}} = \frac{5000 \times 2\pi\,\text{radian}}{60\quad\text{s}};$$

$\mu = 1.5$ cp $= 1.5 \times 0.01$ g/cm-s; L = 90 cm. Introducing these values into equation (7.3.94), we get

$$\ln\left(\frac{10}{4}\right) = \frac{2\pi(0.5\times10^{-4})^2 5000^2 \times 4\pi^2 \times (1.09-1.00)(100-16)90}{9Q_f(1.5\times10^{-2})60^2},$$

So (on rearranging)

$$Q_f = \frac{8\pi^3 \times 0.25 \times 10^{-8} \times 25 \times 10^6 \times 0.09 \times 84 \times 90\,\text{cm}^3}{2.302\log 2.5 \times 9 \times 1.5 \times 10^{-2} \times 36 \times 100\quad \text{s}}$$

$$= \frac{8 \times 31 \times 0.25 \times 25 \times 0.09 \times 84 \times 90 \times 60 \times 10^{-3}\,\text{liter}}{2.302 \times 0.3979 \times 13.5 \times 36 \times 100\quad \text{min}}$$

$= 1.406$ liter/min.

For Part (b), $r_{max} = 5$ μm $= 5 \times 10^{-4}$ cm; $r_0 = 10$ cm; $\rho_p = 2\text{g/cm}^3$; $\rho_t = 1.3\text{g/cm}^3$; $\mu = 3$ cp $= 3 \times 0.01$g/cm-s;

$$Q_f = \frac{8\pi^3 \times (5\times10^{-4})^2 \times 5000^2 \times (2-1.3)(100-16)90\,\text{cm}^3}{2.302\log 2.5 \times 9 \times 3 \times 10^{-2} \times 60^2\quad \text{s}}$$

$$= \frac{8 \times 31 \times 25 \times 10^{-8} \times 25 \times 10^6 \times 0.7 \times 84 \times 90 \times 60\,\text{liter}}{2.302 \times 0.3979 \times 27 \times 36 \times 1000\quad \text{min}}$$

$= 552$ liter/min.

Larger particle size and higher density enables a drastic increase in the centrifuge processing capacity.

Example 7.3.4 Determine the expression for the grade efficiency function $G_r(r_p)$ for the centrifuge under the conditions of Example 7.3.3(a). Calculate the value of the grade efficiency function for cells of two sizes, $r_p = 0.25$ μm and $r_p = 0.5$ μm.

Solution Equation (7.3.99b) provides the expression for the grade efficiency function $G_r(r_p)$ of the cylindrical centrifuge:

$$G_r(r_p) = \frac{r_0^2}{(r_0^2 - r_f^2)}\left[1 - \exp\left\{-\frac{4\pi\omega^2(\rho_p-\rho_t)(r_0^2-r_f^2)L}{9Q_f\mu}r_p^2\right\}\right].$$

Given that $r_0 = 10$ cm; $r_f = 4$ cm; $(\rho_p - \rho_t) = 0.09\,\text{g/cm}^3$; $\mu = 1.5$ cp $= 1.5 \times 10^{-2}$ g/cm-s; L = 90 cm;

$$\omega = \frac{5000 \times 2\pi}{60}\ \frac{\text{radian}}{\text{s}};\ Q_f = \frac{1406}{60}\ \frac{\text{cm}^3}{\text{s}},$$

we have

$$G_r(r_p) = \frac{100}{(100-16)}\left[1 - \exp\left\{-\frac{4\pi(5000)^2 4\pi^2 \times 0.09 \times (100-16) \times 90\,r_p^2}{60^2 \times 9 \times \frac{1406}{60} \times 1.5 \times 10^{-2}}\right\}\right]$$

$$= \frac{100}{84}\left[1 - \exp\left\{-\frac{16\pi^3 \times 25 \times 10^6 \times 8.1 \times 84\,r_p^2}{540 \times 14.06 \times 1.5}\right\}\right] = 1.19\left[1 - \exp\left\{-\frac{16 \times 31 \times 25 \times 10^6 \times 680.4}{11388}r_p^2\right\}\right].$$

So

$$G_r(r_p) = 1.19\left[1 - \exp\left\{-0.7404 \times 10^9 \times r_p^2\right\}\right],$$

where r_p is in cm, and

$$G_r(r_p) = 1.19\left[1 - \exp\left\{-0.7404 \times 10\,r_p^2\right\}\right],$$

where r_p is μm. For

$$r_p = 0.25\,\mu\text{m} \Rightarrow G_r(r_p) = 1.19[1 - \exp\{-0.7404 \times 10 \times 0.0625\}],$$

$$G_r(r_p) = 1.19[1 - \exp\{-0.462\}] = 1.19\left\{1 - \frac{1}{\exp(0.462)}\right\}$$

$$= 1.19\left[1 - \frac{1}{1.588}\right] = 1.19\{1 - 0.629\} = 0.4415.$$

For

$$r_p = 0.5\,\mu\text{m} \Rightarrow G_r(r_p) = 1.19[1 - \exp\{-0.7404 \times 10 \times 0.25\}],$$

$$G_r(r_p) = 1.19[1 - \exp\{-1.851\}] = 1.19\left\{1 - \frac{1}{\exp(1.851)}\right\}$$

$$= 1.19\left[1 - \frac{1}{6.36}\right] = 1.19[1 - 0.157] \cong 1.000.$$

Therefore, cells of 0.5 μm radius will be captured completely in this device according to the theory. (*Note*: The theory is not exact, so allow a safety factor and operate at a lower flow rate than 1.4 liter/min.)

7.3.2.1.1 Continuous separation of two immiscible liquids of different densities Consider a dispersion of one liquid phase as drops in another immiscible liquid phase. If such a dispersion is allowed to stand in a vessel, the heavier phase will settle to the bottom and the lighter phase will rise to the top to form two distinct layers, unless the dispersion is stabilized. However, the rate at which the phase separation will take place is generally quite slow and often incomplete. If, however, centrifugal force is employed, the phase separation can be implemented much more rapidly than is possible with gravity.

A tubular bowl centrifuge can be used to separate continuously two immiscible liquids having different densities. Near the top of the rotating bowl in such a centrifuge (Figure 7.3.13), two liquid outlets are provided at different radii (r_{1i}, r_{2o}) and different elevations via two weirs (see McCabe and Smith (1976, p. 966) Hsu (1981) and Perry and Green (1984, pp. 21-65)). The feed liquid mixture enters through a bowl opening in the bottom through a stationary nozzle. Two layers of liquid are formed in the bowl: the outer one is for the heavier liquid and the inner one is for the lighter liquid. Both leave through separate outlets at the top of the device. There is a liquid–liquid interface of radius r_{int} between the two liquid layers: $r_{int} > r_{1i}$, $r_{int} > r_{2o}$. At r_{int}, the radial location of the interface, the pressure, P_{int} is the same for both liquids. Equation (3.1.51) for centrifugal equilibrium in any pure liquid integrated between two radial locations will yield for the light liquid ($\rho = \rho_1$) and the heavy liquid ($\rho = \rho_2 > \rho_1$), respectively,

$$P_{int} - P_{1i} = \left(\frac{M_s}{\overline{V}_s}\right)_1 \omega^2 \left(\frac{r_{int}^2 - r_{1i}^2}{2}\right) = \rho_1 \omega^2 \left(\frac{r_{int}^2 - r_{1i}^2}{2}\right);$$

$$P_{int} - P_{2o} = \left(\frac{M_s}{\overline{V}_s}\right)_2 \omega^2 \left(\frac{r_{int}^2 - r_{2o}^2}{2}\right) = \rho_2 \omega^2 \left(\frac{r_{int}^2 - r_{2o}^2}{2}\right). \tag{7.3.107}$$

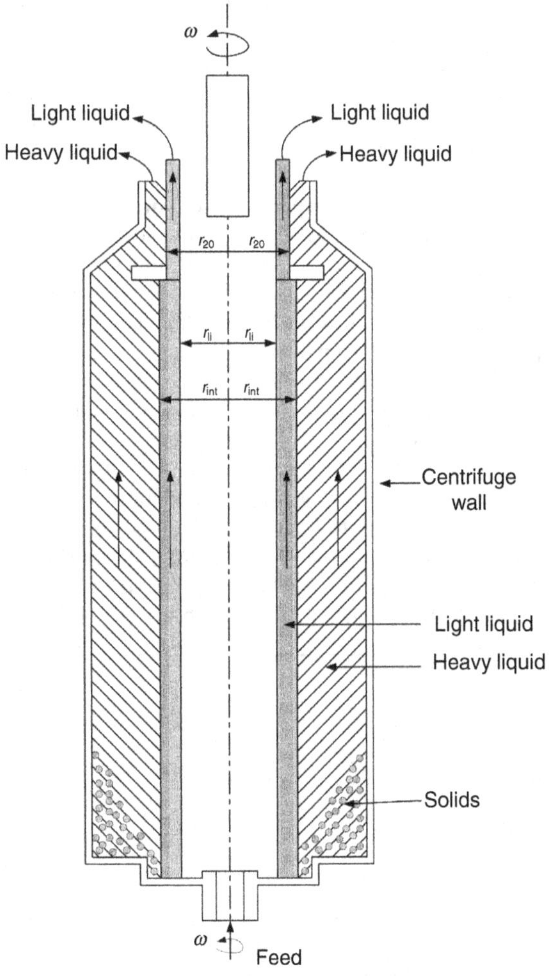

Figure 7.3.13. Schematic of a tubular centrifuge for separating two immiscible liquids.

The two liquid outlet radii (r_{1i}, r_{2o}) are known (via mechanical adjustments); the unknown radial interface location is obtained by equating the hydraulic pressure drops in the two liquid layers reflecting mechanical equilibrium with atmospheric pressure being exerted at both r_{1i} and r_{2o}. Therefore

$$\rho_1 \omega^2 \left(\frac{r_{int}^2 - r_{1i}^2}{2}\right) = \rho_2 \omega^2 \left(\frac{r_{int}^2 - r_{2o}^2}{2}\right),$$

which leads to

$$r_{int}^2 = \left(r_{2o}^2 - \frac{\rho_1}{\rho_2} r_{1i}^2\right) \Big/ \left(1 - \frac{\rho_1}{\rho_2}\right). \tag{7.3.108}$$

When $\rho_1 \rightarrow \rho_2$, the location of the interface becomes unstable. The above relation also shows that if r_{2o} is increased, r_{int} is also increased, pushing the liquid–liquid interface closer to the wall of the bowl. Meanwhile, if there are any solid particles in the fluids, they are thrown to the bowl wall and can be collected from the bottom. Such liquid–liquid centrifuges have a narrower and taller bowl than those used for particle separations (Figure 7.3.12). Note that the bulk flow of each liquid phase is perpendicular to the direction of the external force causing the separation.

7.3.2.2 *Disk centrifuge*

This liquid–particle separation device provides an example in this chapter where the bulk liquid motion toward the device outlet is at an angle to the centrifugal force. The rotating bowl in this device has a flat bottom and a conical top rotating at an angular speed ω rad/s around the vertical axis (Figure 7.3.14). A stack of closely spaced conical disks is set at an angle θ to the vertical axis between the conical top and the bottom. The gap between the consecutive disks is $2b$; the disks rotate with the bowl at the same angular velocity. The feed suspension introduced at the top of the device through a pipe flows through the bottom of the device between the disk stack and the device flat bottom to the periphery of the bowl. From this region, the liquid suspension goes up, enters the gap between two consecutive disks and flows toward the central device axis. Particles in the suspension having a density ρ_p greater than the fluid density ρ_t are radially thrown outward and hit the bottom surface of the top disk in every channel. The collected particles slide outside and, at the end of each top disk surface, get thrown outward to the bowl wall. The clarified liquid leaves the device top through an annulus between the end of the disk stack and the feed inlet pipe.

Figure 7.3.14 illustrates the trajectory of a particle in between two consecutive disks (dashed lines). Define two coordinate axes: the positive z-axis parallel to the disks, and in the direction of main liquid flow toward the main vertical device axis, and the positive r-axis radially

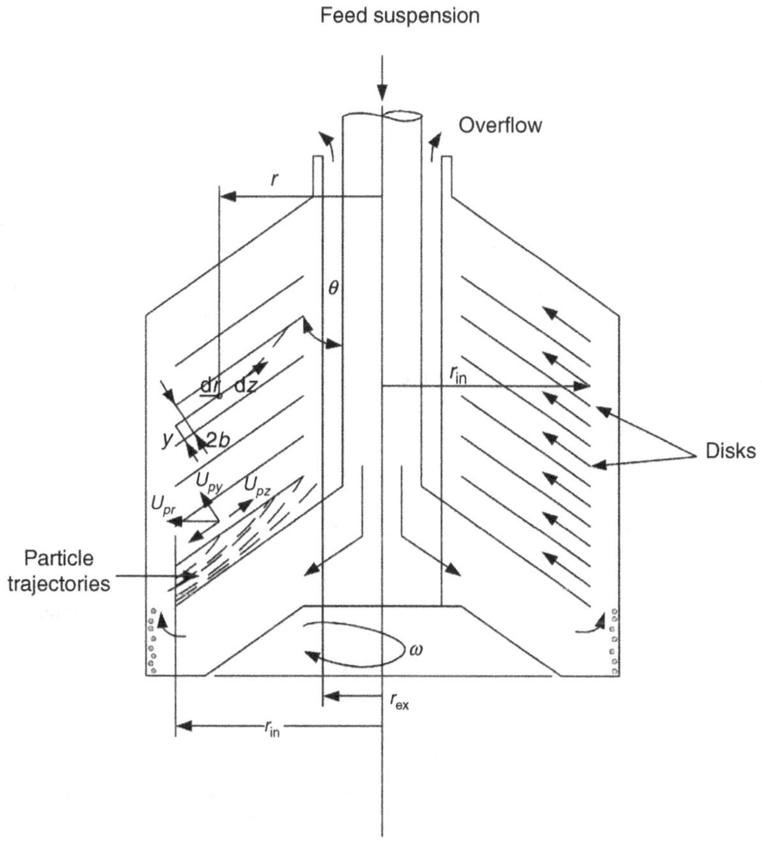

Figure 7.3.14. Schematic of a disk centrifuge.

outward. A particle entering the channel between two consecutive disks moves along the positive z-axis carried by the main liquid flow at a velocity U_{pz}; simultaneously, the centrifugal force imparts a radially outward velocity U_{pr} $(= \mathrm{d}r/\mathrm{d}t)$ to the particle.

This radially outward particle velocity, U_{pr}, can be broken down into two component velocities, U_{py} $(= \mathrm{d}y/\mathrm{d}t)$ normal to the disk wall, or the z-direction, and another component in the negative z-direction reducing the positive z-direction velocity, U_{pz}. The device operation will be affected if U_{pz} is negative or zero. The y-component of the radially outward velocity causes the particle to hit the bottom surface of the top disk in every channel as it moves in the positive z-direction. Those particles which are small enough not to hit the top disk by the time the channel end is reached escape with the liquid and are not captured by the centrifuge.

We assume first that the positive z-component of the particle velocity U_{pz} $(= \mathrm{d}z/\mathrm{d}t)$ is essentially equal to the positive z-component of the liquid velocity v_z. We further assume that this liquid is in *plug flow*. To calculate this velocity v_z, the flow cross-sectional area has to be determined. Since the gap $2b$ between the two plates is small compared to the value of r at any radial location, this flow area is $\approx 2\pi r(2b)$ at any radial location r. So

$$U_{pz} = (\mathrm{d}z/\mathrm{d}t) \cong v_z \cong \frac{Q_f}{4\pi\, r\, \mathrm{b}\, n_c}, \tag{7.3.109}$$

where n_c is the total number of channels through which the liquid slurry having a volumetric flow rate Q_f flows. The total number of disks is therefore $n_c + 1$. Since the two coordinates of interest in the particle trajectory before the particle hits the bottom of the top plate in any channel are y and z, we need next U_{py} $(= \mathrm{d}y/\mathrm{d}t)$. But

$$U_{py} = U_{pr} \cos\theta = (\mathrm{d}y/\mathrm{d}t). \tag{7.3.110}$$

To determine U_{pr}, we assume terminal settling velocity conditions and applicability of Stokes' law (equation (7.3.91)) (but ignore the contribution $(-\, v_z \sin\theta)$):

$$U_{pr} = \frac{\mathrm{d}r}{\mathrm{d}t} = \frac{2r_p^2\omega^2 r}{9\mu}(\rho_p - \rho_t).$$

Therefore

$$\frac{dy}{dt} = \frac{2r_p^2\omega^2 r}{9\mu}(\rho_p - \rho_t)\cos\,\theta. \tag{7.3.111}$$

One can now obtain a relation between y and z for the particle trajectory,

$$\frac{dy}{dz} = \frac{8\pi\, b\, n_c r_p^2 \omega^2 r^2}{9\mu Q_f}(\rho_p - \rho_t)\cos\,\theta, \tag{7.3.112}$$

and integrate between the limits of y and z from the entrance to the flow channel between any pair of disks to the exit. It is, however, more useful to express the particle location in terms of the radial coordinate r instead of the z-coordinate. Define r_{ex} and r_{in} to be the radial coordinates, respectively, for the inner location (where liquid exits the channel) and the outer location (where liquid enters the channel). Any differential positive change dz in the z-coordinate of a particle along the channel involves a differential change in r which is negative:

$$dr = -\, dz \sin\,\theta. \tag{7.3.113}$$

Therefore

$$\frac{dy}{dr} = -\frac{8\pi\, b\, n_c(\rho_p - \rho_t)}{9\mu Q_f} r_p^2\omega^2 r^2 \cot\,\theta. \tag{7.3.114}$$

Integration of this relation between the limits (r_{in}, y_{in}) and (r_{ex}, y_{ex}) yields

$$y_{ex} - y_{in} = \frac{8\pi\, b\, n_c(\rho_p - \rho_t)\cot\,\theta}{27\mu Q_f} r_p^2\omega^2\left(r_{in}^3 - r_{ex}^3\right). \tag{7.3.115}$$

For a particle to settle on the bottom surface of the top plate at $r = r_{ex}$, y_{ex} should equal $2b$, the channel gap. At the inlet end ($r = r_{in}$), the particle can enter the channel with the liquid at any value of $y = y_{in}$. Thus, for any particle of size r_p entering at any y_{in} to settle by the time the liquid exits the channel,

$$2b - y_{in} = \frac{8\pi\, b\, n_c(\rho_p - \rho_t)\cot\,\theta}{27\mu\, Q_f} r_p^2\omega^2\left(r_{in}^3 - r_{ex}^3\right). \tag{7.3.116}$$

We can now assume, as we did for the tubular bowl centrifuge, that particles of size r_p are present uniformly at all y-values at $r = r_{in}$. Then the grade efficiency $G_r(r_p)$ is given by

$$G_r(r_p) = \frac{2b - y_{in}}{2b}, \tag{7.3.117}$$

since those particles of size r_p which have $y < y_{int}$ at $r = r_{in}$ do not settle. Now, for $r_p \geq r_{max}$, $G_r(r_p) = 1$ only if $y_{in} = 0$, i.e. all particles entering the channel settle before $r = r_{ex}$. Therefore

$$2b = \frac{8\pi b n_c(\rho_p - \rho_t)\cot\,\theta}{27\mu Q_f} r_{max}^2\omega^2\left(r_{in}^3 - r_{ex}^3\right), \tag{7.3.118}$$

leading to

$$G_r(r_p) = \left(r_p^2/r_{max}^2\right) \text{ for } r_p \leq r_{max}. \tag{7.3.119}$$

Further, the largest particle size with a nonzero probability of not being captured is obtained from equation (7.3.118) as (Ambler, 1952)

$$r_{max}^2 = \frac{27\mu Q_f}{4\pi n_c(\rho_p - \rho_t)\cot\,\theta\left(r_{in}^3 - r_{ex}^3\right)\omega^2}. \tag{7.3.120}$$

All particles of size $r_p > r_{max}$ will be captured in the device. Thus the grade efficiency function for a disk-stack centrifuge does not depend on the channel height $2b$ but instead depends on the flow rate per channel. Also, the larger the value of ω, the smaller the value of r_{max}. Note that the particle capture efficiency E for particles of a particular size is equal to the grade efficiency $G_r(r_p)$ for that particle size.

An important assumption in the derivation of equation (7.3.120) was that the value of the particle z-velocity component (dz/dt) was equal to the liquid velocity v_z. By this assumption, the component of the velocity in the negative z-direction due to the centrifugal force was neglected. If we do not neglect it, then the radial velocity of the particle is (using equation (7.3.91))

$$\frac{dr}{dt} = \frac{2r_p^2\omega^2 r}{9\mu}(\rho_p - \rho_T) - v_z \sin\,\theta. \tag{7.3.121}$$

Employ now expression (7.3.111) for (dy/dt) and expression (7.3.109) for v_z to obtain

$$\frac{dy}{dr} = \frac{r^2\cos\,\theta}{r^2 - b_1^2}, \tag{7.3.122}$$

where

$$b_1 = \left(\frac{9Q_f\,\mu\sin\,\theta}{8\pi\, b\, n_c\omega^2 r_p^2(\rho_p - \rho_t)}\right)^{1/2}. \tag{7.3.123}$$

Integration has to be carried out between the limits (r_{in}, y_{in}) and (r_{ex}, y_{ex}), where $y_{ex} = 2b$:

$$\int_{y_{in}}^{y_{ex}} dy = \cos\,\theta \int_{r_{in}}^{r_{ex}} \frac{r^2\, dr}{r^2 - b_1^2}. \tag{7.3.124}$$

This analysis was developed by Jury and Locke (1957); they have also provided an analytical solution to the integral in terms of the grade efficiency function:

$$G_r(r_p) = \frac{2b - y_{in}}{2b} = \frac{\cos\,\theta}{2b}\left\{\frac{b_1}{2}\ln\left[\left(\frac{b_1 + r_{in}}{b_1 - r_{in}}\right)\right.\right. \times \left.\left.\left(\frac{b_1 - r_{ex}}{b_1 + r_{ex}}\right)\right] - (r_{in} - r_{ex})\right\}. \tag{7.3.125}$$

Jury and Locke (1957) have provided, in addition, an expression for the thickness of the sludge film flowing down the cone wall.

Typical values and ranges for the operating conditions and dimensions of a disk centrifuge are provided now for perspective: $35° \leq \theta \leq 50°$; $n_c \approx 23$; $2b \approx 0.17$ cm; bowl diameter 15 cm–100 cm; rotational speeds up to 12 000 rpm ($\omega = 1257$ rad/s); Q_f up to 16 m^3/min. The disk centrifuges generally employed are larger than tubular bowl centrifuges and can be run continuously.

7.3.2.3 *Cyclone dust separator*

To separate particles from gaseous streams, cyclones are frequently used in large-scale practices. In fact, *cyclone based dust collectors* are one of the most widely used devices for removing larger-sized particles. Virtually all cyclones used industrially are reverse-flow cyclones. There are two other types of cyclones: rotary-flow (Ciliberti and Lancaster, 1976a, b) and uniflow (Ter Linden, 1949). Only the *reverse-flow cyclone* will be considered here.

A reverse-flow cyclone (Figure 7.3.15(a)) is a simple hollow structure consisting of two parts: a cylindrical cyclone barrel having an annular *vortex finder* or *exit pipe*; and a truncated cone at the bottom joined at the top to the cyclone barrel. The dusty gas is introduced tangentially into the cyclone barrel through the inlet. The gas, having a swirling motion, goes around the cyclone as it moves down toward the *dust exit*; however, the swirling gas soon reverses its gross axial movement, rises up and exits through the top of the vortex finder. The rotational motion of the gas generates the radial centrifugal force on dust particles, which makes them hit the wall of the cyclone as shown in the idealized flow pattern in an ideal cyclone (Flagan and Seinfeld, 1988) shown in Figure 7.3.15(b). The centrifugal force on the particle may be as much as 5-2500 times larger than the gravitational force (Perry and Green, 1984, p. 20–83). If there is no reentrainment, the dust particles settle and are removed through the dust exit at the bottom. Figure 7.3.15(b) shows only the circular gas motion in an idealized flow. The gas and the particles have an axial (z-directional) motion, which is not shown. An analysis of the ideal flow cyclone for particle separation is available in Flagan and Seinfeld (1988).

The flow pattern of the vortex motion of the gas in reverse-flow cyclone is quite complex. First, it is three-dimensional; second, the flow is turbulent. An exact analysis is therefore difficult. Soo (1989) has summarized a fundamental analysis of velocity profiles and pressure drops in such a cyclone. He has also analyzed the governing particle diffusion equation in the presence of electrostatic, gravitational and centrifugal forces. He has then provided an analytical expression for particle collection efficiency under a number of limiting conditions. We will, however, opt here for a much simpler model of particle separation in a cyclone developed by Clift *et al.* (1991). This approach is based on a modification of the original model by Leith and Licht (1972). The model will be presented in the framework of a three-region model of cyclone by Dietz (1981) to introduce the inherent complexities. The treatment begins with the Dietz (1981) model.

The three regions of a cyclone particle separator in the model of Dietz (1981) consist of the entrance region (region 1), the downflow region (region 2) and the upflow region (region 3) (see Figure 7.3.15(a)). Regions 2 and 3 are sometimes called the *annular region* and the *core region*, respectively. In these cyclone models, the cyclone of Figure 7.3.15(a) is replaced by a right circular cylinder of radius r_c and length L below the exit tube, equal to the length of the actual cyclone below the exit pipe; the exit pipe is of radius r_t (Figures 7.3.15(b), (c)).

Regions 1 and 2 have the swirling gas flowing downward at a volumetric flow rate of Q_{v0} and $Q_v(z)$, respectively, where Q_{v0} is the total volumetric gas flow rate entering the cyclone. In Dietz's model, $Q_v(z)$ is related to Q_{v0} by

$$Q_v(z) = Q_{v0}\{1 - (z/L)\}. \qquad (7.3.126)$$

The linear decrease in $Q_v(z)$ with increasing z is due to the radial gas velocity, $-v_r(z)$, from the annular to the core region (the upflow region):

$$|v_r(z)| = v_{r0} = \frac{Q_{v0}}{2\pi r_t L}. \qquad (7.3.127)$$

Dietz (1981) assumed (i) $|v_r(z)|$ to be a constant, v_{r0}, and (ii) that, due to turbulent mixing, the particle mass concentration profile in each region was radially uniform. However, each such profile,[22] $\rho_{p1}(z)$, $\rho_{p2}(z)$ and $\rho_{p3}(z)$ for regions 1, 2 and 3, respectively, changes with the z-coordinate due to particle deposition at the cyclone wall and/or particle exchange between regions 2 and 3:

$$\text{region 1}: \frac{\mathrm{d}}{\mathrm{d}z}\left(\rho_{p1}(z)\,Q_{v0}\right) = -2\pi\, r_c \Gamma_w(z); \qquad (7.3.128)$$

$$\text{region 2}: \frac{\mathrm{d}}{\mathrm{d}z}\left(\rho_{p2}(z)\,Q_v(z)\right) = -2\pi\, r_c \Gamma_w(z) - 2\pi\, r_t \Gamma_v(z); \qquad (7.3.129)$$

$$\text{region 3}: -\frac{\mathrm{d}}{\mathrm{d}z}\left(\rho_{p3}(z) Q_v(z)\right) = 2\pi r_c \Gamma_v(z). \qquad (7.3.130)$$

Here the quantity $\Gamma_w(z)$ is the particle mass flux at the wall at axial coordinate z. For region 1, $\Gamma_w(z)$ is the product of the particle mass concentration, $\rho_{p1}(z)$, and the radial particle velocity at the wall, U_{prw}:

$$\text{region 1}: \Gamma_w(z) = \rho_{p1}(z) U_{prw}. \qquad (7.3.131)$$

For regions 2 and 3, $\rho_{p1}(z)$ has to be replaced by the corresponding particle mass concentrations, $\rho_{p2}(z)$ and $\rho_{p3}(z)$, respectively. Further, U_{prw} is a function of the z-coordinate:

[22] Instead of Dietz's number density profiles, $n_1(z)$, $n_2(z)$ and $n_3(z)$, the particle density profiles, are used here.

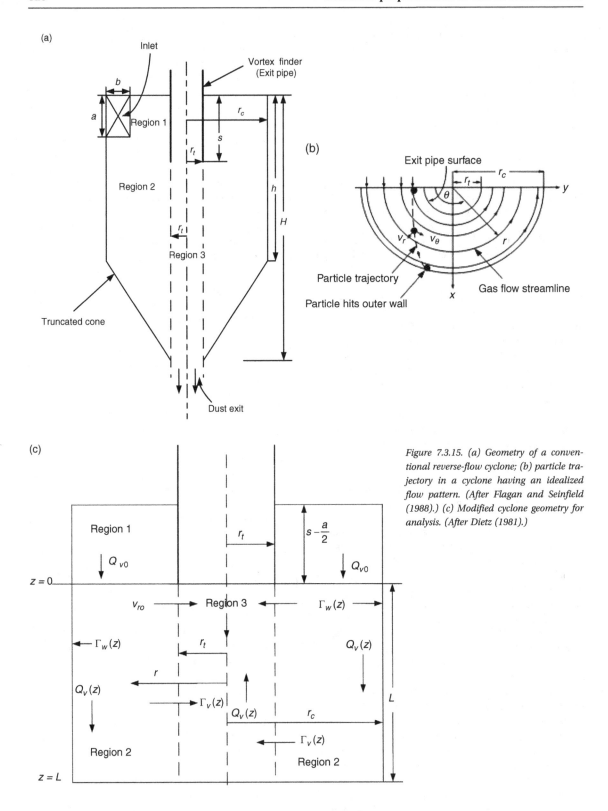

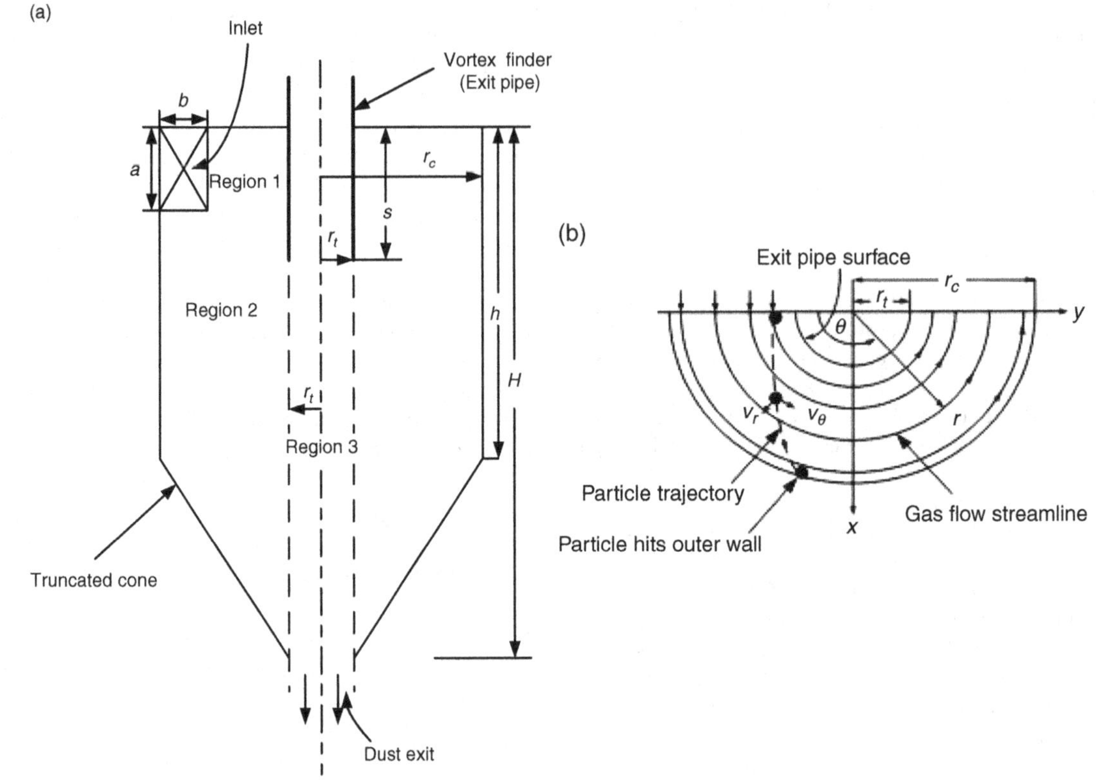

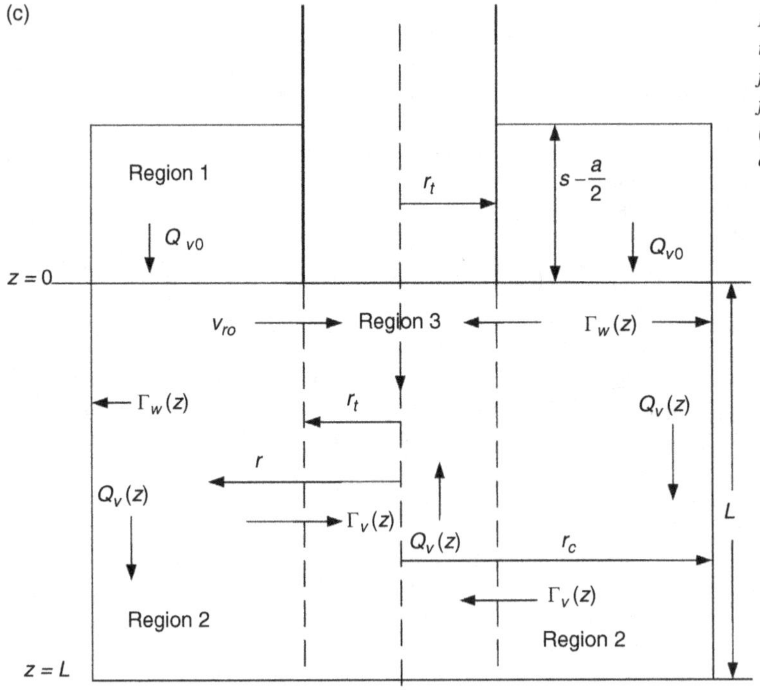

Figure 7.3.15. (a) Geometry of a conventional reverse-flow cyclone; (b) particle trajectory in a cyclone having an idealized flow pattern. (After Flagan and Seinfield (1988).) (c) Modified cyclone geometry for analysis. (After Dietz (1981).)

$$\text{Region 2}: \ \Gamma_w(z) = \rho_{p2}(z)\, U_{prw}(z). \qquad (7.3.132)$$

Note: In each region the radial particle flux due to the radial centrifugal force is perpendicular to the bulk flow of the fluid and the particles in the main flow (z) direction.

We need to know what U_{prw} and U_{prw} (z) are; further, an expression for $\Gamma_v(z)$ has to be found if the three equations (7.3.128) - (7.3.130) are to be solved. We assume that

- Stokes' law is valid;
- the particle is hitting the wall at the terminal velocity without any radial acceleration.

From equation (7.3.90b), we obtain at the wall ($r = r_c$)

$$m_p r_c \omega^2 \left(1 - (\rho_t/\rho_p)\right) = 6\pi\mu r_p U_{prw}. \qquad (7.3.133)$$

However, $r_c\omega^2$ is simply (U^2_{ptw}/r_c), where U_{ptw} is the tangential particle velocity $(= U_{p\theta})$ at the wall $(= U_{p\theta w})$. Now, if one assumes no slip between the particle and the gas in the tangential direction (θ-direction), the tangential particle velocity at the wall, U_{ptw}, is also equal to the corresponding tangential fluid velocity, v_{tw}, at the wall. At the wall, the fluid velocity v_{tw} is zero. Note that, in the boundary layer close to the wall, it is nonzero. Further, the tangential gas velocity $v_t(r)$ in the vortex may be related to the radial location by an equation valid for a free vortex in an ideal fluid:

$$v_t(r)\, r^m = \text{constant}; \qquad v_t(r) = v_{tw}\left(\frac{r_c}{r}\right)^m. \qquad (7.3.134)$$

There is little error in replacing $v_t(r)$ in the wall region by v_{tw} for $m = 1$ (ideal vortex); in general for cyclones, the vortex index m is between 0.5 and 1. We can therefore write

$$U_{prw} = \frac{2\, r_p^2 \rho_p v_{tw}^2}{9\mu r_c} \qquad (7.3.135)$$

from equation (7.3.133) if $m_p = 4\pi r_p^3 \rho_p/3$ and $\rho_t <<< \rho_p$ for particles of density ρ_p.

Since U_{prw} and $U_{prw}(z)$ are now replaced by v_{tw} and $v_{tw}(z)$, we have to estimate the characteristic gas velocity in the wall region, v_{tw}, as a function of z. There are three regions in the cyclone. In region 1, v_{tw} is constant since there is no change in gas flow velocities; the gas flow rate and region dimensions are constant throughout. A simple procedure for estimating v_{tw} in region 1 provided by Leith and Licht (1972) is to equate it to the average gas velocity at the cyclone duct inlet of cross-sectional area ab (Figure 7.3.15(a)):

$$v_{tw} = Q_{v0}/ab. \qquad (7.3.136)$$

In region 2 also, v_{tw} may be assumed constant and equal to the above. Thus U_{prw} and U_{prw} (z) for regions 1 and 2 are known.

The only unknown now (except $\rho_{p1}, \rho_{p2}, \rho_{p3}$) is $\Gamma_v(z)$. Dietz (1981) described it by

$$\Gamma_v(z) = \rho_{p2}(z)\,|v_r(z)| - \rho_{p3}(z)\, U_{prv}(z) \qquad (7.3.137a)$$

as a balance between the particle flux entering the core via the radial gas velocity $|v_r(z)|$ from outside to inside and the particle flux going radially outward due to the centrifugal force on particles in the gas vortex in the core. The quantity $U_{prv}(z)$, i.e. the radial particle velocity at the region 2/region 3 boundary, may be estimated in the manner of equation (7.3.135)

$$U_{prv}(z) = 2\, r_p^2 \rho_p v_{tv}^2/(9\mu r_t). \qquad (7.3.137b)$$

The tangential velocity v_{tv} at $r = r_t$ may be estimated from equation (7.3.134) for a free vortex. An alternative, and more realistic, description of $\Gamma_v(z)$ based on a particle diffusivity between regions 2 and 3 has been developed by Mothes and Löffler (1984).

By applying appropriate boundary conditions, Dietz (1981) obtained solutions for equations (7.3.128)-(7.3.130). For example, region 1 extends from $z = -s + a/2$ to $z = 0$. Equation (7.3.128) may be written via equation (7.3.131) as

$$\frac{d\rho_{p1}(z)}{\rho_{p1}(z)} = d\ln \rho_{p1}(z) = -\frac{2\pi\, r_c U_{prw}}{Q_{v0}}\, dz. \qquad (7.3.138a)$$

Integrating between $-s + a/2$ and $z = z$, we get

$$\ln\left(\frac{\rho_{p1}(z)}{\rho_{p0}}\right) = -\frac{2\pi\, r_c U_{prw}}{Q_{v0}}\left(z + s - \frac{a}{2}\right);$$

$$\rho_{p1}(z) = \rho_{p0} \exp\left(-\frac{2\pi\, r_c U_{prw}}{Q_{v0}}\left(z + s - \frac{a}{2}\right)\right), \qquad (7.3.138b)$$

where ρ_{p0} is the particle density at the inlet, i.e. $z = -s + a/2$.

Using the solutions for $\rho_{p1}(z)$, $\rho_{p2}(z)$ and $\rho_{p3}(z)$ for a given inlet condition, namely

$$\rho_{p1}\left(z = -s + \frac{a}{2}\right) = \rho_{p0},$$

the particle collection efficiency E_T of the cyclone is obtained as Follows (Dietz, 1981):

$$E_T = 1 - \frac{\rho_{p3}(z = 0)}{\rho_{p0}}$$

$$= 1 - \left[a_0 - \left\{a_1^2 + a_2\right\}^{1/2}\right] \exp\left[\frac{-2\pi r_c U_{prw}\left(s - \frac{a}{2}\right)}{Q_{v0}}\right], \qquad (7.3.139a)$$

where

$$a_0 = \frac{r_c U_{prw} + r_t(v_{r0} + U_{prv})}{2 r_t U_{prv}} \qquad (7.3.139b)$$

$$a_1 = \frac{r_t U_{prv} - r_t v_{r0} - r_c U_{prw}}{2 r_t U_{prv}}, \qquad (7.3.139c)$$

$$a_2 = \frac{r_c\, U_{prw}}{r_t\, U_{prv}} = \left(\frac{r_t}{r_c}\right)^{2m}. \qquad (7.3.139d)$$

To obtain E_T for a given cyclone-particle separation system, a value of m needs to be assumed. Typical values assumed vary from 0.5–0.7 (Leith and Licht, 1972; Dietz, 1981).

We now consider the modification of the Leith and Licht (1972) model by Clift *et al.* (1991); the result is a relatively simple procedure for estimating E_T for a cyclone. It assumes that the particle mass concentration profile, ρ_{pg}, is uniform radially, but varies axially; it is represented by $\rho_{pg}(z)$ for the whole cyclone. Consider now a length dz of the cylindrical part of the cyclone. The rate at which particle masses are being deposited from this volume to the cyclone wall at $r = r_c$ is $(2\pi r_c \, dz)\rho_{pg}(z)U_{prw}$. However, this must equal the rate of change of particle mass in that volume:

$$\frac{d}{dt}\left(\pi r_c^2 \rho_{pg}(z)\, dz\right) = -2\pi r_c \rho_{pg}(z)\, U_{prw}\, dz. \tag{7.3.140}$$

Simplifying

$$\frac{d\left(\ln \rho_{pg}(z)\right)}{dt} = -\frac{2U_{prw}}{r_c}. \tag{7.3.141}$$

Substituting for U_{prw} from equation (7.3.135), we get

$$\frac{d\left(\ln \rho_{pg}(z)\right)}{dt} = -\frac{4r_p^2 \rho_p v_{tw}^2}{9\mu r_c^2}, \tag{7.3.142}$$

which, when integrated from $t = 0$ to $t = t_{res}$, the mean residence time of the gas in the cyclone, yields

$$\ln\left(\rho_{pe}/\rho_{p0}\right) = -\frac{4\, r_p^2 \rho_p v_{tw}^2}{9\mu r_c^2} t_{res}, \tag{7.3.143}$$

where ρ_{pe} is the particle mass concentration of the gas exiting the cyclone. *Note*: The result is meant for particles of radius r_p only.

An alternative expression for particles of radius r_p in terms of the grade efficiency function or the particle collection efficiency for size r_p, $E_T(r_p)$,

$$G_r(r_p) = E_T(r_p) = 1 - \frac{\rho_{pe}}{\rho_{p0}} = 1 - \exp\left[-\frac{\rho_p}{9\mu}\left(\frac{2r_p v_{tw}}{r_c}\right)^2 t_{res}\right], \tag{7.3.144}$$

directly illustrates the functional dependences of the collection efficiency on important quantities. Clift *et al.* (1991) suggest that the residence time t_{res} of the gas in the cyclone may be calculated as follows:

$$t_{res} = \frac{\text{cyclone volume}}{\text{gas flow rate } (= Q_{v0})}, \tag{7.3.145}$$

where the cyclone volume corresponds to that from the inlet to the vortex finder (or exit pipe). They have also suggested that the volume of the inlet region is given by $\pi a(r_c^2 - r_t^2)$. One approximate procedure is to estimate the time taken by the gas to complete one complete turn from the inlet:

$$t_{turn|inlet} = \frac{\text{inlet region volume}}{\text{gas flow rate}} = \frac{\pi a(r_c^2 - r_t^2)}{Q_{v0}}. \tag{7.3.146a}$$

Then, if one knows empirically the number of turns, n_t the gas takes in the cyclone before finding the exit pipe, the residence time t_{res} of the gas may be estimated by

$$t_{res} = n_t t_{turn|inlet} = \frac{n_t \pi a(r_c^2 - r_t^2)}{Q_{v0}}. \tag{7.3.146b}$$

Clift *et al.* (1991) have compared the particle collection efficiency predictions by various models plotted against the quantity $\left(U_{prv}/v_{r0}\right)^{1/2}$; note that the latter is proportional to the particle diameter $2r_p$. That particle collection efficiency by a reverse-flow cyclone rises very rapidly with $2r_p$ is shown clearly in Figure 7.3.16. This figure is important from a particle separation point of view. Cyclones are, in general, quite efficient in removing coarse particles $\geq$ 5–10 μm. For finer particles, other separation techniques, such as electrostatic precipitation (see Section 7.3.1.3), filters, etc., should be used. However, an *electrocyclone* concept has been developed wherein an electrode at a high voltage is introduced into the cyclone through the center of the vortex finder where the cyclone wall is grounded. At low gas flow rates, it has a much higher efficiency than conventional cyclones (Chen, 2001) for smaller particles. For submicron particles, filters of different kinds are very efficient (Eggerstedt *et al.*, 1993).

The cyclone models illustrated above are meant merely to introduce the reader to the complex subject of particle separation from gases by cyclones. More interested readers should pursue the original references, as well as the brief overview on devices, in Perry and Green (1984, pp. 20-83–20-89). A model of particle separation in a down-exhaust cyclone separator, of uniflow type, is available in Chen *et al.* (1999b). These separators have a very low gas pressure drop.

Example 7.3.5 A cyclone is used to remove as many particles as possible coming out of a cement kiln. The inlet gas velocity entering the cyclone is 1524 cm/s. The particle density and the gas viscosity are 2.9 g/cm^3; and 2×10^{-4} g/cm-s, respectively, The dimension b of the gas inlet to the cyclone is 76 cm. The radius of the cyclone (r_c) is 152 cm. The number of turns, n_t, the gas goes through in the cyclone before entering the exit pipe is 5 (Theodore, 2005). Determine the value of the particle radius $r_{p,50}$ which yields a value of $G_r(r_p) = 0.5 = E_T(r_p)$ (this is the equiprobable size, or the cut diameter, $(2r_{p,50})$). Calculate also the value of $E_T(r_p)$ for the following values of r_p: 0.5 μm; 2.5 μm; 10 μm; 20 μm; 30 μm.

Solution We will employ equation (7.3.144) to calculate $E_T(r_p)$ for a given r_p. To employ $E_T(r_p)$, we need to know, amongst other things, t_{res}. From equation (7.3.146b),

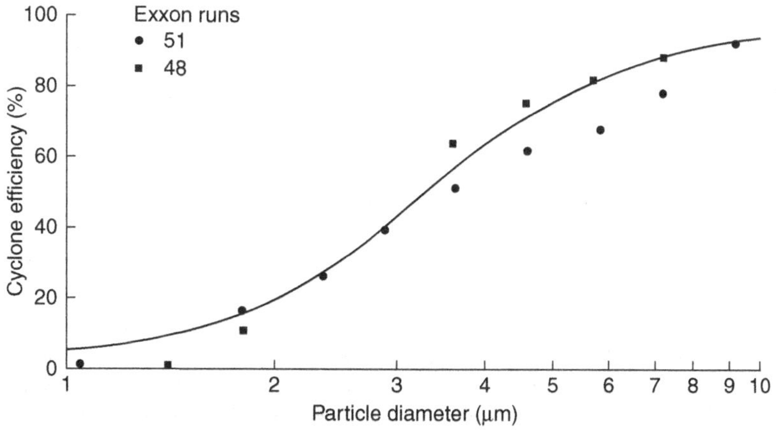

Figure 7.3.16. Comparison between theory and experiment for the second-stage cyclone at Exxon's miniplant (assumed m = 0.7). Reprinted, with permission, from P.W. Dietz, "Collection efficiency of cyclones," AIChE J., 27(6), 888 (1981). Copyright © [1981] American Institute of Chemical Engineers (AIChE).

$$t_{\text{res}} = n_t \frac{\pi a \left(r_c^2 - r_t^2\right)}{Q_{v0}}.$$

Now

$$v_{tw} = 1524 \frac{\text{cm}}{\text{s}} = \frac{Q_{v0}}{ab}.$$

Therefore

$$Q_{v0} = v_{tw}\, ab = 1524 \times a \times 76.$$

We do not know r_t. We will neglect r_t^2 $\left(<< r_c^2\right)$. Therefore;

$$t_{\text{res}} = \frac{5 \times \pi \times (152^2 - 0) \times a}{1524 \times a \times 76}\,\text{s}$$

$$= \frac{5 \times \pi \times 152 \times 152}{1524 \times 76} \cong \pi\,\text{s}.$$

Now calculate $r_{p,50}$:

$$E_T(r_{p,50}) = 0.5 = G_r(r_{p,50}) = 1 - \exp\left[-\frac{\rho_p 4 r_{p,50}^2 v_{tw}^2}{9 \mu r_c^2} t_{\text{res}}\right]$$

$$\Rightarrow \exp\left[-\frac{2.9 \frac{\text{g}}{\text{cm}^3} \times 4 \times r_{p,50}^2\,\text{cm}^2 \times 1524 \times 1524 \frac{\text{cm}^2}{\text{s}^2} \times \pi\,\text{s}}{9 \times 2 \times 10^{-4} \frac{\text{g}}{\text{cm-s}} \times 152 \times 152}\right] = 0.5$$

$$\Rightarrow 0.5 = \exp\left[-\frac{2.9 \times 4 \times 1524 \times 1524 \times \pi}{18 \times 152 \times 10^{-4} \times 152} r_{p,50}^2 (\text{cm})^2\right]$$

$$\Rightarrow 0.5 = \exp\left[-203 \times 10^{+4} \times 10^{-8} r_{p,50}^2 (\mu\text{m})^2\right]$$

$$\Rightarrow 0.0203\, r_{p,50}^2 = 0.7$$

$$\Rightarrow r_{p,50}^2 = 34.48 \Rightarrow r_{p,50} = 5.87\,\mu\text{m}$$

$$\Rightarrow \text{cut diameter} = 11.75\,\mu\text{m} (= 2 r_{p,50}).$$

We will now calculate the values of E_T (r_p) for different values of r_p. We will also list the values of E_T from Theodore (2005) based on Lapple's curve (Lapple, 1951). The value of $2r_{p,50}$ calculated by Theodore (2005) was ~10 μm. Theodore (2005) has recommended the following equation for calculating $E_T(r_p)$:

$$E_T(r_p) = \frac{1}{1 + \left\{2r_{p,50}/(2r_p)^2\right\}}.$$

The results are shown in Table 7.3.2.

The analysis based on Clift *et al.* (1991) and equantions (7.3.140)-(7.3.144) led to result (7.3.144), based on a particular size. If one is interested in the total particle collection efficiency E_T, we have to integrate over all sizes using the incoming particle number density function $n_f(r_p)$ as follows:

$$E_T = 1 - \frac{\int_0^{r_{\max}} \rho_p n_f(r_p) \exp\left[-\frac{\rho_p}{9\mu}\left(\frac{2 r_p v_{tw}}{r_c}\right)^2 t_{\text{res}}\right] \mathrm{d}r_p}{\int_0^{r_{\max}} \rho_p n_f(r_p)\, \mathrm{d}r_p}. \quad (7.3.147)$$

7.3.2.3.1 Hydrocyclones The success of gas-solid cyclones for dust collection has led to hydrocyclones for solid-liquid and liquid-liquid separations. The latter involves oil-water emulsions. A number of extended treatments are available on particle separation in hydrocyclones (Bradley, 1965; Svarovsky, 1982), which are extensively used in a variety of industries, including the corn milling industry. As in cyclones used for gas cleaning, the vortex tube at the top produces an overflow, which is the clarified liquid containing fines, whereas the underflow coming down the conical wall is concentrated slurry of larger particles. Frequently, multiple cyclones are used in series, with the overflow from one cyclone being the feed to the next one to improve the final clarification of the suspension. An introduction to various aspects of particle separation in hydrocyclones,

Table 7.3.2.

r_p (μm)	$\exp\left[-203\times10^{-4}\, r_p^2(\mu m)\right]$	$E_T(r_p) = 1-\exp\left[-203\times10^{-4} r_p^2(\mu m)\right]$	E_T (from Theodore, 2005)
0.5	exp [−0.00507]	~0	0
2.5	exp [−0.1268]	~0.12	0.20
5.0	exp [−0.5075]	~0.40	0.50
10	exp [−2.028]	0.869	0.80
20	exp [−8.112]	~1	0.93
30	exp [−18.252]	~1	0.98

including computational fluid dynamics (CFD) based modeling, has been provided by Salcudean *et al.* (2003).

Liquid-liquid dispersions/emulsions are also increasingly separated by hydrocyclones. Analyses based on droplet trajectories are being developed (Wolbert *et al.*, 1995). Rietema (1969) developed a hydrocyclone design analogous to cyclones for gas separation: the continuous phase leaves the cyclone at the top through the vortex finder as the overflow; the oil droplets in a concentrated emulsion appear in the underflow outlet at the bottom. Colman and Thew (1988) have developed an alternative hydrocyclone design wherein the clarified liquid leaves through the bottom outlet and the concentrated emulsion leaves through the top vortex-finder tube. A factor of importance here is the density of the dispersed phase in relation to the continuous phase. The heavier dispersed phase will move toward the wall; the lighter dispersed phase will move toward the hydrocyclone core.

7.3.2.4 *Separation nozzle process for gas separation*

So far, the description of centrifugal force based separation has been limited to the separation of particles from a fluid (gas or liquid) and the separation of a dispersion of two immiscible liquids. In each case, the bulk flow velocity was perpendicular to the direction of the centrifugal force created by a rotational bulk flow of the gas-particle/liquid-particle/liquid-liquid in the device. This rotational bulk flow was developed either by rotation of the device itself (as in a tubular bowl centrifuge, etc.) or by introducing the particle-loaded gas stream into a cyclone via a swirling motion. Here we will focus on separating a gas mixture via centrifugal force (see Section 4.2.1.1), where the bulk motion direction itself is the source of the centrifugal force and is perpendicular to the centrifugal force. It is called the *separation nozzle process*; for a comprehensive introduction, see Benedict *et al.* (1981).

Consider Figure 7.3.17, which shows a convergent-divergent slit opening into a curved nozzle of radius r_o. The gas mixture entering through the throat of dimension t diverges, is turned around 180° by the curved nozzle and comes up to a thin wall acting as a gas flow divider at a radius r_c. Two gas fractions are collected from two sides of this wall; the fraction at a smaller radius ($r < r_c$) is the light fraction enriched in the lighter species and the fraction at a larger radius ($r > r_c$) is the heavy fraction enriched in the heavier species. This enrichment is possible because the bulk gas is undergoing a rotational motion in the nozzle. The magnitude of the centrifugal force is substantially enhanced by the magnitude of the tangential velocity $r\omega$, where ω is the angular velocity in the semicircular groove of radius r_0: this gas velocity is in the range of 100–1000 m/s. The process is carried out at subatmospheric pressures for the separation of uranium isotopes $U^{235}F_6$ from $U^{238}F_6$. The feed pressure, for example, is 290 Torr, whereas the product fractions are at a pressure of around 138 Torr and at a temperature T_p less than that of the feed (T_f) (Benedict *et al.*, 1981).

We will now calculate, following Benedict *et al.* (1981), the separation factor and other quantities for this process, primarily explored for isotope separation. The fugacity profile in the radial direction for a component i in a mixture undergoing centrifugal rotation at an angular velocity of ω radian/s is given by expression (4.2.3) for centrifugal equilibrium. We can rewrite this expression as a ratio of mass densities of the component i at the two locations, $r = r_1$ and r_2:

$$(\hat{f}_{i2}/\hat{f}_{i1}) = (\rho_{i2}/\rho_{i1}) = \exp(M_i\omega^2\,(r_2^2 - r_1^2)/2RT_p). \tag{7.3.148}$$

If we define temporarily the mass density of species i at $r = 0$ as $\rho_i(0)$, knowing that the species mass density ρ_{ij} varies with the radial location, we can also rewrite the above as

$$\rho_{ij}(r) = \rho_i(0)\exp\,(M_i\omega^2 r^2/2RT_p). \tag{7.3.149}$$

Define now the mass flow rate of species i between the radial locations $r_1 = 0$ and $r_2 = r_c$ as w_{i1} for the product region $j = 1$, the light fraction. Then

$$w_{i1} = \int_{r_1=0}^{r_2=r_c} \omega r\rho_{ij}(r)\mathrm{d}r = \int_{r_1=0}^{r_2=r_c} \omega r\rho_i(0)\exp\,(M_i\omega^2 r^2/2RT_p)\mathrm{d}r$$
$$= (RT_p\rho_i(0)/M_i\omega)\left[\exp(M_i\omega^2 r_c^2/2RT_p) - 1\right]. \tag{7.3.150}$$

Correspondingly, the mass flow rate of species i between the radial locations $r_1 = r_c$ and $r_2 = r_0$ representing the product region $j = 2$ is given by

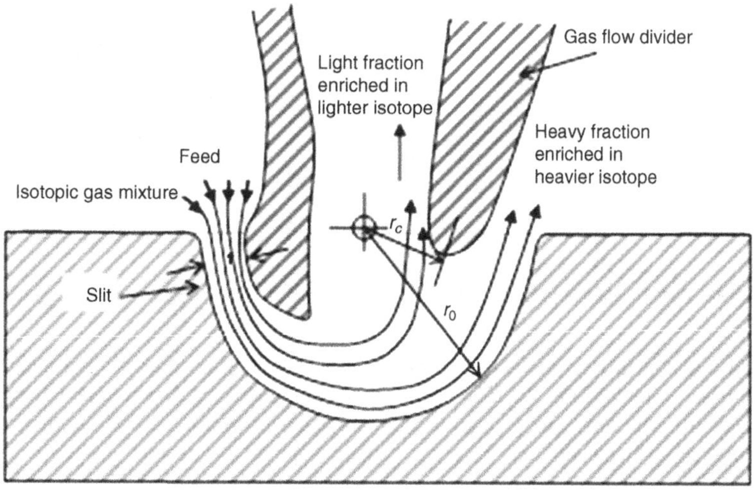

Figure 7.3.17. Cross section of the slit and curved nozzle in the separation nozzle process. (After Benedict et al. *(1981).)*

$$w_{i2} = \int_{r_1=r_c}^{r_2=r_0} \omega r \rho_{ij}(r) dr = \int_{r_1=r_c}^{r_2=r_0} \omega r \rho_i(0) \exp(M_i \omega^2 r^2/2RT_p) dr$$
$$= (RT_p \rho_i(0)/M_i \omega)\left[\exp(M_i\omega^2 r_0^2/2RT_p) - \exp(M_i \omega^2 r_c^2/2RT_p)\right]. \quad (7.3.151)$$

Define now a constant

$$A_{20} = M_2 \omega^2 r_0^2/2RT_p. \quad (7.3.152)$$

We may now define a separation factor α_{12} between two species $i = 1$ and 2 between the two product regions, region $j = 1$ between $r = 0$ and $r = r_c$ and region $j = 2$ between $r = r_c$, and $r = r_0$ via the following species mass flow rate based definition (see definition (2.2.9a)):

$$\alpha_{12} = \frac{w_{11}}{w_{12}} \frac{w_{22}}{w_{21}} = \frac{\left[\exp(M_1\omega^2 r_c^2/2RT_p) - 1\right]}{\left[\exp(M_2\omega^2 r_c^2/2RT_p) - 1\right]} \frac{\left[\exp(M_2\omega^2 r_0^2/2RT_p) - \exp(M_2\omega^2 r_c^2/2RT_p)\right]}{\left[\exp(M_1\omega^2 r_0^2/2RT_p) - \exp(M_1\omega^2 r_c^2/2RT_p)\right]}; \quad (7.3.153a)$$

$$\alpha_{12} = \frac{\left[\exp(A_{20}(M_1/M_2)(r_c^2/r_0^2)) - 1\right]}{\left[\exp(A_{20}(r_c^2/r_0^2)) - 1\right]} \frac{\left[\exp(A_{20}) - \exp(A_{20}(r_c^2/r_0^2))\right]}{\left[\exp(A_{20}(M_1/M_2)) - \exp(A_{20}(M_1/M_2)(r_c^2/r_0^2))\right]}. \quad (7.3.153b)$$

One can now develop an estimate of the separation factor α_{12} between $U^{235}F_6$ ($i = 1$) and $U^{238}F_6$ ($i = 2$) at $T_p = 300$ K, when $r_0 = 0.1$ mm, $r_c = 0.05$ mm and $\omega r_0 = 300$ m/s, based on the value of $A_{20} = 6.350$ (Benedict *et al.*, 1981, figs. 14.27 and 14.22). Further, $M_1 = 349$, $M_2 = 352$, and $R = 8.31 \times 10^7$ g-cm^2/s^2-gmol-K. First, we check the following:

$$A_{20} = (M_2 \omega^2 r_0^2/2RT_p)$$
$$= \frac{352(\text{g/gmol}) \times (300 \times 10^2)^2 (\text{cm}^2/\text{s}^2)}{2 \times 10^7 \times 8.3(\text{g-cm}^2/\text{s}^2\text{-gmol-K}) \times 300\text{K}} \Rightarrow A_{20}$$
$$= 6.353;$$

this is, okay. So, for $(r_c/r_0)^2 = 0.25$; $(M_2/M_1) = 1.008596$,

$$\alpha_{12} = \frac{\exp\left(\dfrac{6.353 \times 0.25}{1.008596}\right) - 1}{\exp(6.353 \times 0.25) - 1} \times \frac{\exp(6.353) - \exp(6.353 \times 0.25)}{\exp(6.353/1.008596) - \exp\left(\dfrac{6.353 \times 0.25}{1.008596}\right)}$$
$$= \frac{(\exp(1.5747) - 1)}{(\exp(1.588) - 1)} \times \frac{[\exp(6.353) - \exp(1.588)]}{[\exp(6.2988) - \exp(1.5747)]}$$
$$= \frac{(4.830 - 1)}{(4.895 - 1)} \times \frac{(574 - 4.895)}{(544 - 4.830)} = \frac{3.830 \times 569.105}{3.895 \times 539.17} = 1.037.$$

This value is considerably larger than values achieved in other isotope separation processes. However, the range of values of α_{12} is 1.01–1.04, more commonly around 1.015. One important characteristic of this process has not been mentioned yet. The isotopic gas mixture is diluted using a low molecular weight gas, e.g. H_2. This increases the peripheral velocity ωr_0 which is, at the sonic level, increasing α_{12}; also, the diffusion coefficients of UF_6 species are increased, which permits operation at a higher gas pressure and higher gas throughput without affecting separation (Benedict *et al.*, 1981). Note that a major difference

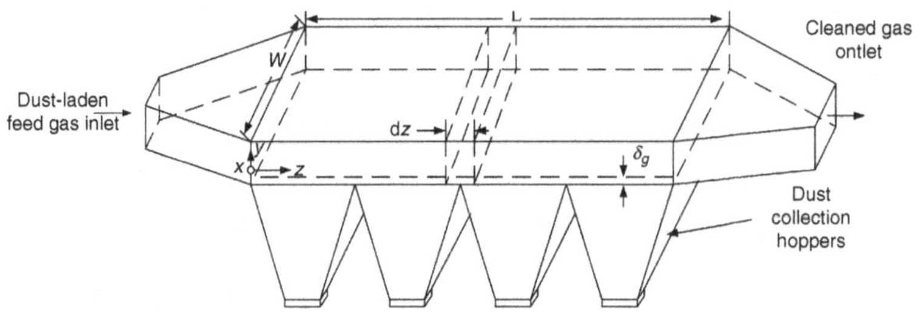

Figure 7.3.18. Gravity based dust settling chamber.

between this method and the first turn of the particle-loaded gas in a cyclone separator (Figures 7.3.15(a)–(c)) is the use of a convergent–divergent nozzle to generate sonic speeds and, correspondingly, a high centrifugal force.

7.3.3 Gravitational force field

The gravitational force field as the sole driver for the separation of particles shows up in two stages of removal of particles from a gaseous stream. At the end of all cleanup processes, the waste gas is released to the atmosphere through a stack rising high above the ground. The particles in this waste gas plume will be deposited on the ground as the gas is dispersed into the atmosphere over a substantial length of the surrounding landscape. Gravitational force based settling of dust particles from a gas stream is also employed as a preliminary cleanup step for removing larger particles prior to a more thorough cleanup. We will first consider this settling of larger particles in a *gravity based dust settling chamber* (Figure 7.3.18). We will now briefly describe the settling of particles from a gaseous plume leaving a stack. Gravity is also utilized to settle particles from a liquid stream. Inclined settlers are used to separate cells in a cell culture broth. Such devices will be briefly described at the end of the section.

In a gravity based dust settling chamber (Figure 7.3.18), the gas stream is allowed to flow into a large horizontal rectangular chamber which reduces the gas velocity. The particles, especially larger ones, settle toward the bottom surface of the duct, which are essentially the top of multiple dust-collecting hoppers. The gravitational force based settling velocity of particles is perpendicular to the mean gas flow direction. The gas velocity inside the device is considerably smaller than that in the duct at the gas inlet. One could have laminar flow of the gas or turbulent flow in the chamber. Turbulent flow is characterized by a high degree of mixing across the flow cross section. The observed behavior of settling in such chambers suggests a significant degree of vertical mixing. We will briefly illustrate here the extent of particle removal under conditions of turbulent mixing. We follow the treatment of Flagan and Seinfeld (1988). The model is very similar to that of Deutsch (1922) for an electrostatic precipitator.

Let the rectangular settling chamber (Figure 7.3.18) have the following dimensions: L (length in the main flow direction, z-coordinate) $\times$ $2b$ (height perpendicular to the main flow direction, y-coordinate; gravity is in the negative y direction) $\times$ W (width of the flow cross section, x-coordinate). We adopt the following simplifying assumptions.

(1) Due to turbulent mixing, particle concentration is uniform across the duct cross-sectional area $2bW$.
(2) The gas moves in plug flow with an axial velocity of v_z along the duct length; the spherical particles have the same axial velocity. However, near the bottom surface of the chamber, the wall region, the gas flow velocity is significantly reduced. The wall region thickness is δ_g.
(3) The particles in the wall region have a gravitationally induced vertically downward velocity corresponding to the terminal velocity U_{pyt} of a spherical Stokesian particle (see expression (6.3.1)) of radius r_p:

$$U_{pyt} = \frac{2}{9}\frac{r_p^2 \rho_p g}{\mu}. \tag{7.3.154}$$

Here we have neglected the density of air vis-à-vis the density of the particle.

Consider a control volume of length dz spanning the cross-sectional area $2bW$ of the rectangular duct. The number of particles in the size range r_p to $r_p + \mathrm{d}r_p$ per unit volume is indicated by $n(r_p)\mathrm{d}r_p$. Any change in the total number of particles in this size range entering and leaving this control volume per unit volume per unit time,

$$2bWv_z\big(n(r_p)\,|_z - n(r_p)\,|_{z+\mathrm{d}z}\big)\mathrm{d}r_p, \tag{7.3.155}$$

is due to the deposition of particles on the bottom chamber wall. For a particle of radius r_p to settle on the bottom wall with a terminal velocity U_{pyt}, it must traverse vertically a distance δ_g during the time $\mathrm{d}t$ needed to traverse axially a distance dz of the control volume:

$$\mathrm{d}t = (\mathrm{d}z/v_z); \quad U_{pyt}\mathrm{d}t = \delta_g; \quad \delta_g/U_{pyt} = \mathrm{d}z/v_z. \tag{7.3.156}$$

Since only the particles within the wall layer δ_g can settle, the fraction of particles entering the control volume that can settle is $\delta_g/2b$. Since the total rate of particle entry into the control volume is given by $2bWv_z n(r_p)|_z \, dr_p$, we now obtain, at steady state,

$$2bWv_z\left[n(r_p)|_z - n(r_p)_{z+dz}\right]dr_p = \left(\frac{\delta_g}{2b}\right)2bWv_z n(r_p)|_z dr_p = \left(\frac{dz\,U_{pyt}}{2bv_z}\right)2bWv_z n(r_p)|_z dr_p; \tag{7.3.157}$$

$$\frac{n(r_p)|_{z+dz} - n(r_p)|_z}{dz} = -\frac{U_{pyt}}{2bv_z}n(r_p)\bigg|_z.$$

As $dz \to 0$, we obtain

$$\frac{dn(r_p)}{dz} = -\frac{U_{pyt}}{2bv_z}n(r_p). \tag{7.3.158}$$

If, at the chamber entrance ($z = 0$), the value of $n(r_p)$ is $n_f(r_p)$, integration leads to, at location z,

$$n(r_p)|_z = n_f(r_p)\exp\left(-\frac{U_{pyt}\,z}{2bv_z}\right). \tag{7.3.159}$$

If we define the dust collection efficiency for particles of radius r_p as $E(r_p)$, then, for a chamber of length L,

$$E(r_p) = G_r = 1 - \frac{n(r_p)|_L}{n_f(r_p)} = 1 - \exp\left(-\frac{U_{pyt}L}{2bv_z}\right), \tag{7.3.160}$$

where G_r is the grade efficiency function. For a Stokesian particle,

$$G_r = E(r_p) = 1 - \exp\left(-\frac{2r_p^2\rho_p g L}{9\mu(2b)v_z}\right) = 1 - \exp\left(-\frac{2r_p^2\rho_p g L W}{9\mu Q_f}\right), \tag{7.3.161}$$

where Q_f is the volumetric flow rate of the dust-laden air. As $r_p \to \infty$, $E(r_p) \to 1$. Obviously, a larger value of length L of the chamber, and a larger cross section to reduce the gas velocity v_z, will increase the particle collection efficiency. In practice, limitations in the physical dimensions of the gravitational settling chambers do not allow these devices to capture particles smaller than 50 μm (Flagan and Seinfeld, 1988). Correspondingly, particles having settling velocities higher than 13 cm/s may be removed in a gravitational settling chamber (Wark and Warner, 1976). However, the observed gas cleaning behaviors of such settling chambers are approximately described by relation (7.3.161).

One can also have gravitational settling of dust particles when the dust-laden gas moves in laminar flow between the top and the bottom plates in the chamber. Unlike that in turbulent flow, the axial velocity has the following parabolic profile (6.1.2b, c):

$$v_z(y) = \frac{3}{2}v_{z,avg}\left[1 - \left(\frac{y}{b}\right)^2\right].$$

Flagan and Seinfeld (1988) have described the particle trajectory in such a flow field and calculated the particle collection efficiency of particles of radius r_p as

$$G_r = E(r_p) = \frac{LU_{pyt}}{v_{z,avg}(2b)}. \tag{7.3.162}$$

Since the gas residence time in the chamber is $(L/v_{z,avg})$, one concludes that, as long as $(2b/U_{pyt})$ is smaller than or equal to $(L/v_{z,avg})$ for any particle, those particles will settle.

It is useful also to calculate the total efficiency of particle collection over all particle sizes. From equation (7.3.159), we obtain the total number concentration of particles of sizes from $r_p = 0$ to $r_p = r_p$ as

$$\int_0^{r_p} n(r_p)|_z dr_p = N(r_p)|_z = \int_0^{r_p} n_f(r_p)\exp\left(-\frac{U_{pyt}z}{2bv_z}\right)dr_p. \tag{7.3.163}$$

The total efficiency is given by

$$E_T = 1 - \frac{N(r_{p_{max}})|_L}{N_f(r_{p_{max}})} = 1 - \frac{\int_0^{r_{max}} n_f(r_p)\exp\left(-\frac{U_{pyt}z}{2bv_z}\right)dr_p}{\int_0^{r_{max}} n_f(r_p)dr_p}. \tag{7.3.164}$$

Example 7.3.6 Consider gravitational settling of dust particles in a chamber 8 meters long and 1 meter high, where air may be assumed to move in laminar flow at an average velocity of 25 cm/s. The viscosity of air at the temperature of operation is 1.80×10^{-4} g/cm-s. The dust particle density is 2 g/cm^3. Develop an expression for the minimum size of particles which will settle in this chamber. What is the value of this minimum size for the conditions provided?

Solution The particle that enters at the top of the chamber has the greatest chance to escape. If we assume that the residence time of this particle corresponds to the average gas velocity (a safer assumption than the local velocity near the top), then the particle residence time at this axial velocity is given by

$$t_{res} = \frac{L}{v_{z,avg}}. \tag{7.3.165}$$

The particle which enters at the center of the channel will have half of this residence time ($= L/v_{z,max} = L/2v_{z,avg} = t_{res}/2$); however, we will see soon that is not a problem. Assuming the particle settles with a settling velocity given by (7.3.154), then the time required for the particle entering at the top to settle is

$$t_{settling} = \frac{2b}{U_{pyt}}. \tag{7.3.166}$$

For this particle to settle and be captured in this device,

$$t_{\text{settling}} \le t_{\text{res}} \Rightarrow \text{assume } t_{\text{settling}} = t_{\text{res}}$$

$$\Rightarrow 2b/U_{pyt} = L/v_{z,\text{avg}} \qquad (7.3.167)$$

$$\Rightarrow U_{pyt} = \frac{2b\, v_{z,\text{avg}}}{L} = \frac{2}{9} r_p^2 \frac{\rho_p g}{\mu}$$

(from relation (7.3.154)

$$\Rightarrow r_p = \left(\frac{9b\, v_{z,avg}\, \mu}{\rho_p g L}\right)^{1/2}.$$

(*Note*: In (7.3.167), 2b multiplies $v_{z,\text{avg}}$; if the particle is at centerline (entrance point) b will multiply $v_{z,\text{max}}$. The product will still be the same.) The minimum size of particle which will settle is

$$r_p|_{\text{min}} = \left(\frac{18}{2}\frac{2b}{\rho_p}\frac{v_{z,\text{avg}}}{g}\frac{\mu}{L}\right)^{1/2} \Rightarrow r_p|_{\text{min}}$$

$$= \left(\frac{9 \times 100\,\text{cm} \times 25\,\frac{\text{cm}}{\text{s}} \times 1.80 \times 10^{-4}\,\frac{g}{\text{cm-s}}}{2\frac{g}{\text{cm}^3} \times 980\,\frac{\text{cm}}{\text{s}^2} \times 800\,\text{cm}}\right)^{1/2};$$

$$r_p|_{\text{min}} = \left(\frac{9 \times 25 \times 1.80 \times 10^{-4}}{16 \times 980}\right)^{1/2}$$

$$= 16 \times 10^{-4}\text{cm} = 16\,\mu\text{m}.$$

We will now very briefly illustrate the procedure followed to determine the *rate of deposition* of particles on the ground from stack emissions by gravity. Recognize first that it is quite a complex problem. A stack or a chimney rises a distance h above the ground (in the y-direction) (Figure 7.3.19). However, before such a gaseous plume coming out of the stack turns around and follows the wind, it (being usually hotter than the surroundings) rises further, then turns around at a height H ($>h$) above the ground and follows the wind direction (z-coordinate). As it goes with the wind in the z-direction, the gaseous pollutants diffuse and disperse in all three directions, x, y and z (the ultimate objective of stack emission of pollutants). Since the axial velocity in the z-direction is substantial, dispersion in this direction may be neglected when compared with convection. If one can assume steady state, the equation describing the dispersion of a gaseous pollutants species i may be written from equation (6.2.9) (as described in Table 6.2.4 for Cartesian coordinates)

$$v_x \frac{\partial C_i}{\partial x} + v_y \frac{\partial C_i}{\partial y} + v_z \frac{\partial C_i}{\partial z} = D_{ij}\left(\frac{\partial^2 C_i}{\partial x^2} + \frac{\partial^2 C_i}{\partial y^2} + \frac{\partial^2 C_i}{\partial z^2}\right) \qquad (7.3.168)$$

where we have assumed $\mathbf{v}^* \cong \mathbf{v}$. Since the z-direction is the mean direction of the gas (also wind) motion, we can neglect the convective terms in the x-and y-directions and incorporate their fluctuating contributions through dispersion terms (see Section 6.2.1.1):

$$v_z \frac{\partial C_i}{\partial z} = D_{i,\text{eff},x}\frac{\partial^2 C_i}{\partial x^2} + D_{i,\text{eff},y}\frac{\partial^2 C_i}{\partial y^2}. \qquad (7.3.169)$$

(For an introduction to this approach, see Wark and Warner (1976), chap. 4.)

An approximate solution of this equation has been obtained subject to an emission rate $Q_e C_{if}$ from the stack as a point source (at ground level, i.e. $x = 0$, $y = 0$, $z = 0$) as well as additional boundary conditions:

$$\int_{-\infty}^{\infty}\int_0^{\infty} v_z\, C_i(x,y,z)\, \mathrm{d}x\, \mathrm{d}y = Q_e C_{if},\; z > 0. \qquad (7.3.170)$$

$$\text{As} \quad z \to 0, \quad C_i \to \infty \quad (\text{point source, delta function}); \qquad (7.3.170\text{b})$$

$$C_i \to 0, \quad \text{when } x, y, z \to \infty\, (\text{zero concentration, far away}); \qquad (7.3.170\text{c})$$

$$\text{as } y \to 0,\, D_{i,\text{eff},y}\frac{\partial C_i}{\partial y} \to 0\; (\text{no diffusion to the ground surface}). \qquad (7.3.170\text{d})$$

Here, Q_e is the volumetric flow of the gas releasing pollutants at a concentration C_{if}. The solution is (see Wark and Warner (1976), pp. 141–2)

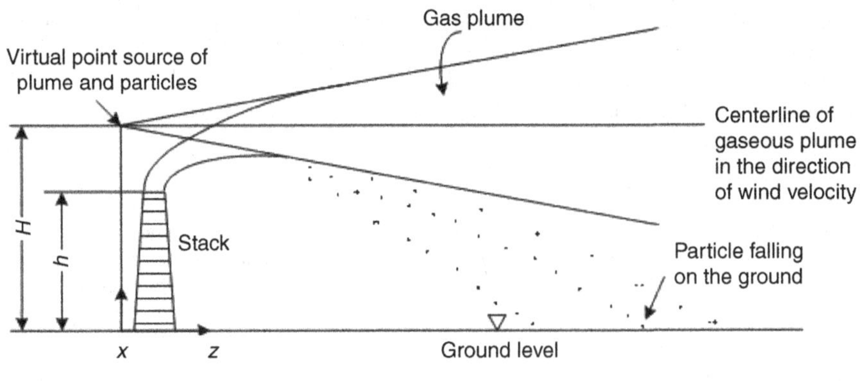

Figure 7.3.19. Gaseous plume rising from a stack and dispersing in the atmosphere as particles fall to the ground by gravity. (After Wark and Warner (1976).)

$$C_i(x,y,z) = \frac{Q_e C_{if}}{2\pi z (D_{i,\text{eff},x} D_{i,\text{eff},y})^{1/2}} \exp\left[- \left(\frac{x^2}{D_{i,\text{eff},x}} + \frac{y^2}{D_{i,\text{eff},y}} \right) \frac{v_z}{4z} \right]. \tag{7.3.171}$$

If you consider dispersion in either the x-or y-direction and replace the z-coordinate by $t = z/v_z$, this problem and the solution follows in a manner similar to equations (3.2.12) and (3.2.15), except that the limits for x and y here are $(-\infty, \infty)$ and that for z is 0 to ∞. In reality, the pollutant source coordinates are $x = 0$, $y = H$ and $z = 0$. Therefore the solution above will be modified to

$$C_i(x,y,z,H) = \frac{Q_e C_{if}}{2\pi z (D_{i,\text{eff},x} D_{i,\text{eff},y})^{1/2}} \exp\left\{ - \left[\left(\frac{x^2}{D_{i,\text{eff},x}} + \frac{(y-H)^2}{D_{i,\text{eff},y}} \right) \right] \frac{v_z}{4z} \right\}. \tag{7.3.172}$$

This solution has the form of a Gaussian distribution if it is written as follows:

$$C_i(x,y,z,H) = \frac{Q_e C_{if}}{2\pi v_z \sigma_x \sigma_y} \exp\left\{ - \left[\left(\frac{x^2}{2\sigma_x^2} + \frac{(y-H)^2}{2\sigma_y^2} \right) \right] \right\},$$

if

$$\sigma_x^2 = \frac{2 D_{i,\text{eff},x}\, z}{v_z}, \qquad \sigma_y^2 = \frac{2 D_{i,\text{eff},y}\, z}{v_z}. \tag{7.3.173}$$

Note that $(z/v_z) = t$; therefore, as time increases, the standard deviation of the profile from the centerline ($x = 0$, $y = H$, z) increases (Figure 7.3.19) in both the x- and y-directions.

We are interested in the rate at which particles come out of this gaseous plume and hit the ground via gravity. For particles of a given size/density having a particle settling velocity U_{pyt}, the particle having the same coordinates x, z as the pollutants will, however, have fallen toward the ground by a distance $U_{pyt}\, z/v_z$, where (z/v_z) is the time t over which the plume has traveled an axial distance z. If the particle mass concentration at the source is ρ_{p0}, we can rewrite the solution for particle mass concentration $\rho_p(x, y, z, H)$ from equation (7.3.173) in an analogous fashion as

$$\rho_p(x,y,z,H) = \frac{Q_e \rho_{p0}}{2\pi v_z \sigma_x \sigma_y} \exp\left\{ - \left[\frac{x^2}{2\sigma_x^2} + \frac{(y-H)^2}{2\sigma_y^2} \right] \right\}. \tag{7.3.174}$$

At this time, we can introduce a correction for the vertical drop by the particle from $y = H$ to $y = H - (U_{pty} z/v_z)$ in time t:

$$\rho_p(x,y,z,H) = \frac{Q_e \rho_{p0}}{2\pi v_z \sigma_x \sigma_y} \exp\left\{ - \left[\frac{x^2}{2\sigma_x^2} \right] - \frac{(y - (H - (U_{pyt} z/v_z)))^2}{2\sigma_y^2} \right\}. \tag{7.3.175}$$

Therefore the centerline of the particle plume slopes downward (toward $y = 0$) as z and t increase (Wark and Warner, 1976).

We now need to calculate the rate at which particles are hitting the ground ($y = 0$) at $x = 0$, the centerline of the plume. The corresponding particle concentration is given by

$$\rho_p(0,0,z,H) = \frac{Q_e \rho_{p0}}{2\pi v_z \sigma_x \sigma_y} \exp\left\{ - \frac{(H - (U_{pyt} z/v_z))^2}{2\sigma_y^2} \right\}. \tag{7.3.176}$$

The mass rate of transport of these particles into the ground per unit area is the mass flux of particles at location z, $\Gamma_{\text{ground},r_p}(z)$ (mass/time-area):

$$\Gamma_{\text{ground},r_p}(z) = U_{pyt}\, \rho_p(0,0,z,H); \tag{7.3.177}$$

$$\Gamma_{\text{ground},r_p}(z) = \frac{Q_e \rho_{p0} U_{pyt}}{2\pi v_z \sigma_x \sigma_y} \exp\left\{ - \frac{(H - (z U_{pyt}/v_z))^2}{2\sigma_y^2} \right\}. \tag{7.3.178}$$

Since U_{pyt} depends on the particle size and density, the above result is valid for particular values of U_{pyt}, i.e. particles of a particular size and density having a particular U_{pyt}. To determine the total particle deposition rate, we will have to sum it over all particle sizes/densities, or alternatively all particle terminal velocities.

Unlike separation devices generally encountered in the laboratory or used for industrial operations, in this case nature and natural phenomena provide the environment for the separation of dust particles from the gaseous plume released into the atmosphere from the stack. The key unknowns in the equation (7.3.178) for $\Gamma_{\text{ground},r_p}(z)$ are σ_x and σ_y, although there is some empiricism involved in estimating U_{pyt} also. Turner (1969) provides charts to estimate values of σ_x (his σ_y), σ_y (his σ_z) as a function of the axial distance z (his x) from the source for six types of atmospheric conditions.

Gravity is also the primary force for separation of droplets from a gas/vapor phase into a liquid layer in what are called *knockout drums* in many chemical and petrochemical industries. Pressure relieving equipment generally contains a large cylindrical drum into which the vapor/gas and liquid mixture are introduced at one end of the drum at the highest point. The droplets formed are subjected to downward motion by gravity, an upward buoyancy force and a drag force by the gas/vapor escaping through the top exit at the other end of the drum. Droplets follow a downward trajectory, ultimately falling into the liquid layer occupying about half the height of the drum. Those that do not will escape with the gas/vapor. Knockout drum design therefore often incorporates conditions such that droplets 100 μm and larger are captured by coming down and touching the liquid; to ensure this, the vapor velocity should be low enough to provide enough residence time.

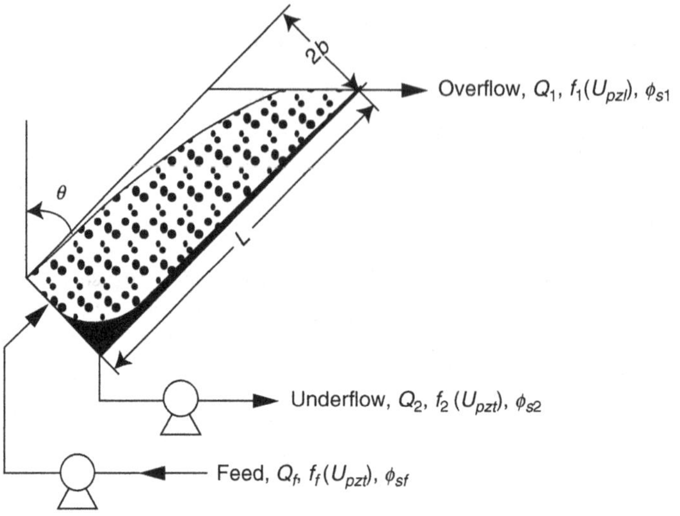

Figure 7.3.20. Inclined lamella settler with continuous flow for particle classification at steady state.

7.3.3.1 *Inclined settlers*

We have seen in Section 4.2.3.4 that inclined settlers provide a much larger vessel bottom surface area for particles to settle by gravity than simple vertical vessels; therefore, inclined setters can clarify a much larger volume of liquid by simple gravity. If such an inclined settler, in the form of a lamella settler (Figure 4.2.7) has, the suspension flow up the channel slowly, the particles will slowly settle onto the bottom channel plate, slide down the channel wall and form an underflow stream. Larger particles will settle quickly and are going to show up in the underflow stream. The overflow stream leaving at the top of the channel will contain the finer particles (Figure 7.3.20). In this configuration, only a component of the external gravitational force, namely, g sin θ is perpendicular to the bulk flow direction (similar to that in a disk centrifuge, Figure 7.3.14).

When a suspension is introduced into the inclined lamella settler, the feed suspension may be characterized by means of its solids volume fraction ϕ_{sf} and its particle size density function $f_f(r_p)$. The corresponding quantities for the overflow and underflow streams are: ϕ_{s1}, $f_1(r_p)$; ϕ_{s2}, $f_2(r_p)$. Often such problems are analyzed instead using the solids volume fraction ϕ_s and the particle settling (terminal) velocity density function $f(U_{pzt})$, where the particle settling velocity U_{pzt} in the Stokes' law range is related to the particle radius r_p by relation (6.3.1):

$$U_{pzt} = (2/9)\, r_p^2(\rho_p - \rho_t)g/\mu.$$

Using such a formalism (Davis *et al.*, 1989), we can obtain the following mass balance relations for the total volumetric flow rate, total solids flow rate and the flow rate of solids having a particular settling velocity U_{pzt} (subscript $j = 1$, overflow; subscript $j = 2$, underflow):

$$\text{total vol. flow rate}: \; Q_f = Q_1 + Q_2; \tag{7.3.179}$$

$$\text{total solids vol. flow rate}: Q_f\phi_{sf} = Q_1\phi_{s1} + Q_2\phi_{s2}. \tag{7.3.180}$$

The total solids volumetric flow rate for particles having U_{pzt} is given by

$$Q_f\phi_{sf}f_f(U_{pzt}) = Q_1\phi_{s1}f_1(U_{pzt}) + Q_2\phi_{s2}f_2(U_{pzt}). \tag{7.3.181}$$

Earlier theories (e.g. PNK theory), briefly identified in Davis *et al.* (1989), indicate that if one can assume that all particles settle with the same settling (terminal) velocity U_{pzt}, then the volumetric rate of production of the clarified liquid (Q_c) will be

$$Q_c = U_{pzt}W(L\sin\theta + 2b\cos\theta), \tag{7.3.182}$$

since we have already identified in Section 4.2.3.4 the surface area for settling to be $W(L \sin\theta + 2b \cos\theta)$ for a lamella settler having a channel plate width of W, channel plate length L and channel plate gap of $2b$. Now, this clarified liquid flow rate, Q_c, will combine with the unsettled feed suspension flow rate, namely (Q_1 - Q_c), to make up the overflow volume flow rate Q_1. For a balance on particle volume fraction, this implies

$$Q_1\,\phi_{s1} = (Q_1 - Q_c)\,\phi_{sf}. \tag{7.3.183}$$

(This analysis assumes that the particle concentration in the unsettled suspension entering the overflow is the same as that in the feed suspension (Davis *et al.*, 1989).) If there is a distribution of particle settling velocities, then, for particles having a particular settling velocity U_{pzt}, the corresponding balance is

$$Q_1\,\phi_{s1}f_1(U_{pzt}) = (Q_1 - Q_c(U_{pzt}))\,\phi_{sf}f_f(U_{pzt}). \tag{7.3.184}$$

If the clarified liquid production rate for particles having a settling velocity U_{pzt} is $Q_c(U_{pzt})$ and all particles have the same U_{pzt}, then

$$Q_c(U_{pzt}) \geq Q_1 \quad \text{and} \quad f_1(U_{pzt}) = 0. \tag{7.3.185}$$

However, when there is a distribution in U_{pzt}, the analysis is somewhat complicated. Let $Q_c(U_{pzt})$ in relation (7.3.184) now denote the volumetric rate at which the suspension containing particles with settling velocities less than U_{pzt} is produced due to settling of particles having settling velocities greater than U_{pzt}. One can now integrate relation (7.3.184) over all particle settling velocities such that the overflow production rate Q_1 is now defined as

$$Q_1 = Q_c(U_{pztc}), \tag{7.3.186}$$

where U_{pztc} is the cutoff settling (terminal) velocity such that particles having $U_{pzt} > U_{pztc}$ settle to the bottom plate before reaching the overflow:

$$\int_0^\infty Q_1\phi_{s1} f_1(U_{pzt})\mathrm{d}U_{pzt} = Q_1\phi_{s1} = \phi_{sf}\int_0^\infty (Q_1 - Q_c(U_{pzt}))f_f(U_{pzt})\,\mathrm{d}U_{pzt}$$

$$\Rightarrow \phi_{s1} = \phi_{sf}\int_0^{U_{pztc}} \frac{Q_1 - Q_c(U_{pzt})}{Q_1} f_f(U_{pzt})\mathrm{d}U_{pzt} + \int_{U_{pztc}}^\infty \frac{Q_1 - Q_c(U_{pzt})}{Q_1} f_f(U_{pzt})\mathrm{d}U_{pzt}.$$

The second integral on the right-hand side is zero from definition (7.3.186), leading to

$$\phi_{s1} = \phi_{sf}\int_0^{U_{pztc}} \frac{Q_1 - Q_c(U_{pzt})}{Q_1} f_f(U_{pzt})\,\mathrm{d}U_{pzt}. \tag{7.3.187}$$

The particle volume fraction in the underflow, ϕ_{s2}, is now obtained from the total particle balance (7.3.180) as

$$\phi_{s2} = \frac{\phi_{sf}\left\{Q_f - \int_0^{U_{pztc}}\{Q_1 - Q_c(U_{pzt})\}f_f(U_{pzt})\,\mathrm{d}U_{pzt}\right\}}{Q_f - Q_1}. \tag{7.3.188}$$

One can obtain the probability density function $f_1(U_{pzt})$ of the overflow stream by substituting (7.3.187) in to relation (7.3.184):

$$f_1(U_{pzt}) = \frac{(Q_1 - Q_c(U_{pzt}))f_f(U_{pzt})}{\int_0^{U_{pztc}} Q_1 - Q_c(U_{pzt})f_f(U_{pzt})\,\mathrm{d}U_{pzt}}, \tag{7.3.189}$$

where

$$U_{pzt} < U_{pztc} \tag{7.3.190}$$

and

$$f_1(U_{pzt}) = 0 \quad \text{for} \quad U_{pzt} \geq U_{pztc}. \tag{7.3.191}$$

Correspondingly, the probability density function $f_2(U_{pzt})$ of the underflow stream may be obtained by substituting for ϕ_{s1}, ϕ_{s2} and $f_1(U_{pzt})$ in equation (7.3.181):

$$f_2(U_{pzt}) = \frac{Q_f\phi_{sf}f_f(U_{pzt}) - \phi_{sf}f_1(U_{pzt})\int^{U_{pztc}}(Q_1 - Q_c(U_{pzt}))f_f(U_{pzt})\,\mathrm{d}U_{pzt}}{(Q_f - Q_1)\phi_{sf}\dfrac{\left\{Q_f - \int_0^{U_{pztc}}\{Q_1 - Q_c(U_{pzt})\}f_f(U_{pzt})\,\mathrm{d}U_{pzt}\right\}}{(Q_f - Q_1)}} \tag{7.3.192}$$

$$\Rightarrow f_2(U_{pzt}) = \frac{Q_f f_f(U_{pzt}) - f_1(U_{pzt})\left[\int_0^{U_{pztc}}(Q_1 - Q_c(U_{pzt}))f_f(U_{pzt})\,\mathrm{d}U_{pzt}\right]}{Q_f - \left[\int_0^{U_{pztc}}(Q_1 - Q_c(U_{pzt}))f_f(U_{pzt})\,\mathrm{d}U_{pzt}\right]}, \tag{7.3.193}$$

where we have employed equations (7.3.179) and (7.3.187). Substituting for $f_1(U_{pzt})$ from equation (7.3.189), we get

$$f_2(U_{pzt}) = \frac{(Q_f - (Q_1 - Q_c(U_{pzt})))f_f(U_{pzt})}{Q_f - \left[\int_0^{U_{pztc}}(Q_1 - Q_c(U_{pzt}))f_f(U_{pzt})\,\mathrm{d}U_{pzt}\right]}. \tag{7.3.194}$$

Note: We have generally employed particle size density functions $f_f(r_p)$, $f_1(r_p)$ and $f_2(r_p)$, where the probability density function depends on the random variable r_p, the particle radius. In the analysis considered here for inclined settlers, we are dealing with density functions $f_f(U_{pzt})$, $f_1(U_{pzt})$ and $f_2(U_{pzt})$. Since, by relation (6.3.1), the relation between U_{pzt} and r_p is (if Stokes' law is valid)

$$U_{pzt} = (2/9)r_p^2(\rho_p - \rho_t)g/\mu,$$

we have

$$f(U_{pzt}) = f(r_p)/|(\mathrm{d}U_{pzt}/\mathrm{d}r_p)|; \tag{7.3.195a}$$

$$f(U_{pzt}) = 9\mu f(r_p)/(4(\rho_p - \rho_t)g r_p). \tag{7.3.195b}$$

7.3.4 Field-flow fractionation for colloids, macromolecules and particles

This technique is primarily utilized in a relatively small scale of operation. It involves a somewhat different interaction of the bulk flow with the force operating perpendicular to the bulk flow direction compared to what we have seen so far. So far, whenever we have employed the bulk flow, limited attention was paid to the fact that there is a velocity profile in the bulk flow taking place in the channel/device. Such velocity profiles can often be damaging to the separation achievable (for example, see, the comments in the paragraph preceding equation (6.3.6)) or they can introduce complications/distortions in the separation achieved (see Figures 7.3.2(b) and 7.3-3 in thin-film continuous-flow electrophoresis). In the *field-flow fractionation* (FFF) technique, however, the velocity profile of bulk flow in the channel is crucial to the separation achieved. Before we discuss this feature, let us describe the effect of a force field applied perpendicular to the bulk flow direction in a channel where a suspension of colloids, macromolecules or particles is flowing. The force field can be electrical, centrifugal, gravitational, thermal diffusion or crossflow from a pressure gradient across the channel wall.

Let the dimensions of such materials in suspension vary between the wide range of 1 nm and around 100 μm (Giddings, 1993). As shown in Figure 7.3.21(a) , a force field applied in the negative y-direction in the channel creates a flux of macromolecules/colloids/particles toward the wall. For macromolecular species i, we may write the following expression[23] (See 3.1.88):

$$N_{iy} = C_{is}(v^*_{ty} + U_{iy}) - D_{is}\frac{dC_{is}}{dy}. \qquad (7.3.196)$$

If the flow in the channel is laminar and fully developed, then $v^*_{ty} = 0$ and

$$N_{iy} = C_{is}U_{iy} - D_{is}\frac{dC_{is}}{dy}, \qquad (7.3.197)$$

where U_{iy} is the force field induced migration velocity of species i in the y-direction. For particles, an expression for the particle number flux n_{py} in the y-direction in terms of the total number density N_t of particles may be written from (3.1.70) as

$$n_{py} = N_t\, U_{py} - D_p\frac{dN_t}{dy}. \qquad (7.3.198)$$

This field induced macromolecular migration velocity U_{iy} normal to the channel wall and toward the wall at $y = 0$ may be created by a variety of fields (to be discussed soon). At steady state, there should be no net flux of species i in the y-direction as the flux due to the field is counteracted by back diffusion of species i away from the wall:

$$N_{iy} = 0 = C_{is}U_{iy} - D_{is}\frac{dC_{is}}{dy}; \qquad (7.3.199a)$$

$$D_{is}\frac{dC_{is}}{dy} = C_{is}U_{iy}. \qquad (7.3.199b)$$

A solution of this equation is obtained from

$$\frac{dC_{is}}{C_{is}} = \frac{U_{iy}}{D_{is}}\,dy \Rightarrow \ell n\, C_{is} = \frac{U_{iy}}{D_{is}}y + a \Rightarrow C_{is} = C_{i0}\exp\left[\frac{U_{iy}}{D_{is}}y\right], \qquad (7.3.200)$$

where C_{i0} is the concentration of species i at the wall, $y = 0$. The direction of the force field applied is such that U_{iy} is a negative quantity. We can therefore write the above concentration profile as

$$C_{is} = C_{i0}\exp\left(-\frac{|U_{iy}|}{D_{is}}y\right), \qquad (7.3.201)$$

where $|U_{iy}|$ is the magnitude of the force field induced velocity of species in the y-direction. One could define a characteristic[24] thickness δ_i of this concentration profile by defining

$$\frac{D_{is}}{|U_{iy}|} = \delta_i. \qquad (7.3.202)$$

Rewrite the concentration profile (7.3.201) now as

$$C_{is} = C_{i0}\exp\left(-\frac{y}{\delta_i}\right). \qquad (7.3.203)$$

If there are two macromolecular species a and b, the characteristic thicknesses of their profiles extending from the wall out are δ_a and δ_b , respectively. Species subjected to a lower $|U_{iy}|$ and possessing a higher D_{is} will have a larger δ_i. Therefore the larger species, whose diffusion coefficient at infinite dilution D^0_{is} may be described by the Stokes-Einstein equation (3.3.90c)

$$D^0_{is} = \frac{k^B\, T}{6\pi\, r_i\, \mu}, \qquad (7.3.204)$$

will have a smaller D^0_{is}, and therefore a smaller D_{is}. For two macromolecular species a and b having similar U_{iy} (say), if $r_a > r_b$, $D_{ay} < D_{by}$; therefore $\delta_a < \delta_b$. Thus the smaller macromolecular species will have a profile whose average extends out further from the wall (Figure 7.3.21(a)).

[23] There is also an axial flux N_{iz} since $C_{is}(y,z)$ is a function of y and z. Here we assume that $C_{is}(y,z)$ may be represented as $C_{is}(y)\, f(z)$ and focus on $C_{is}(y)$.

[24] The characteristic thickness here corresponds to a distance where the value of C_{is} has been reduced to 36% of its value at the wall; therefore the bulk of the molecules are contained in the region $y = 0$ to $y = \delta_i$.

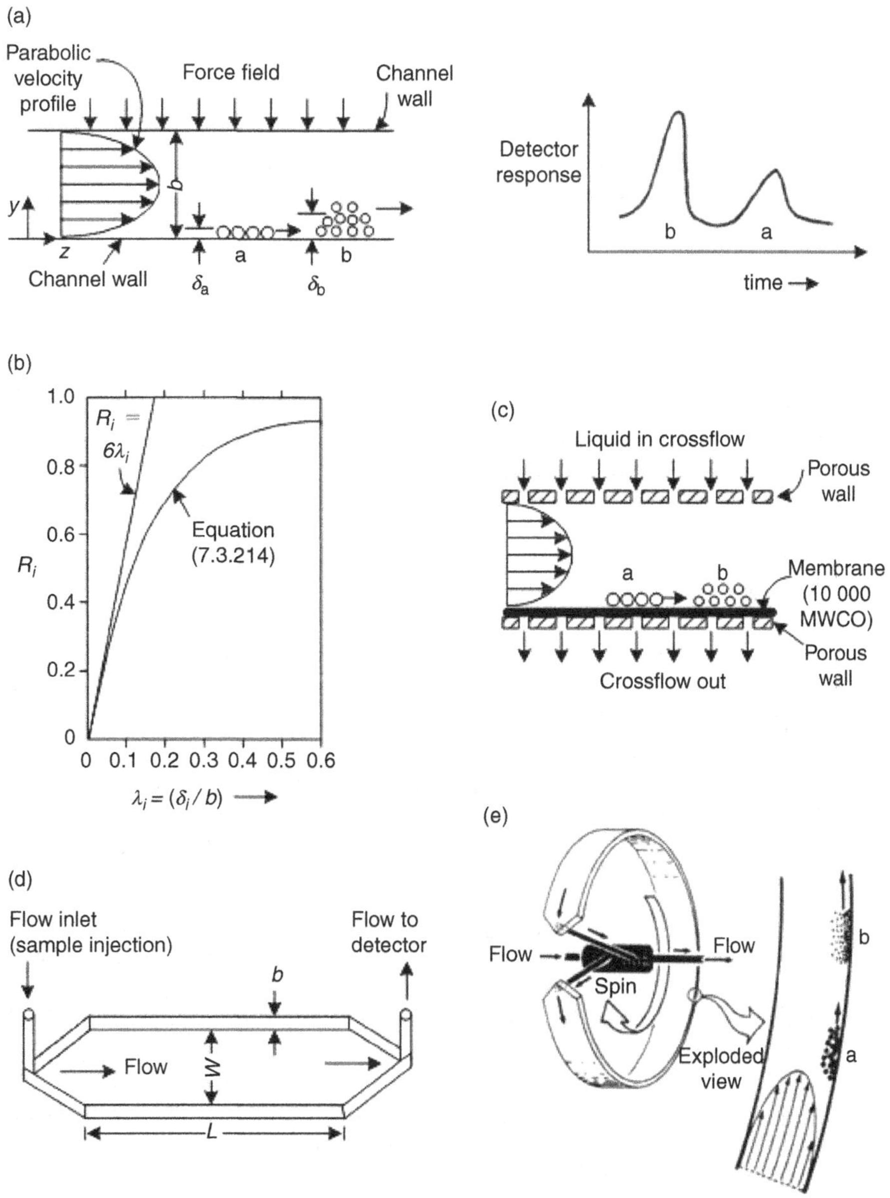

Figure 7.3.21. Field-flow fractionation (FFF). (a) Basic configuration of field-flow fractionation device and the detector response to a sample. (b) Retention ratio of species i vs. retention parameter λ_i. (c) Schematic for flow field-flow fractionation with a crossflow. (d) Physical configuration/dimensions of a FFF channel with inlet/outlet. (e) Configuration of the channel in sedimentation FFF. (After Giddings (1993).)

Field-flow fractionation exploits this difference in the distance of the mean of the species profile from the channel wall by coupling it with the velocity profile in the channel. In laminar channel flow with a parabolic velocity profile (equation (6.1.2b)), the larger molecules are then concentrated in the slower axial velocity zones closer to the wall, whereas the smaller molecules reside primarily in higher axial velocity zones further away from the wall. Therefore the smaller molecules will show up (via a detector) at the channel exit faster than the larger molecules if there is an injection of a sample containing different macromolecular species upstream in the channel

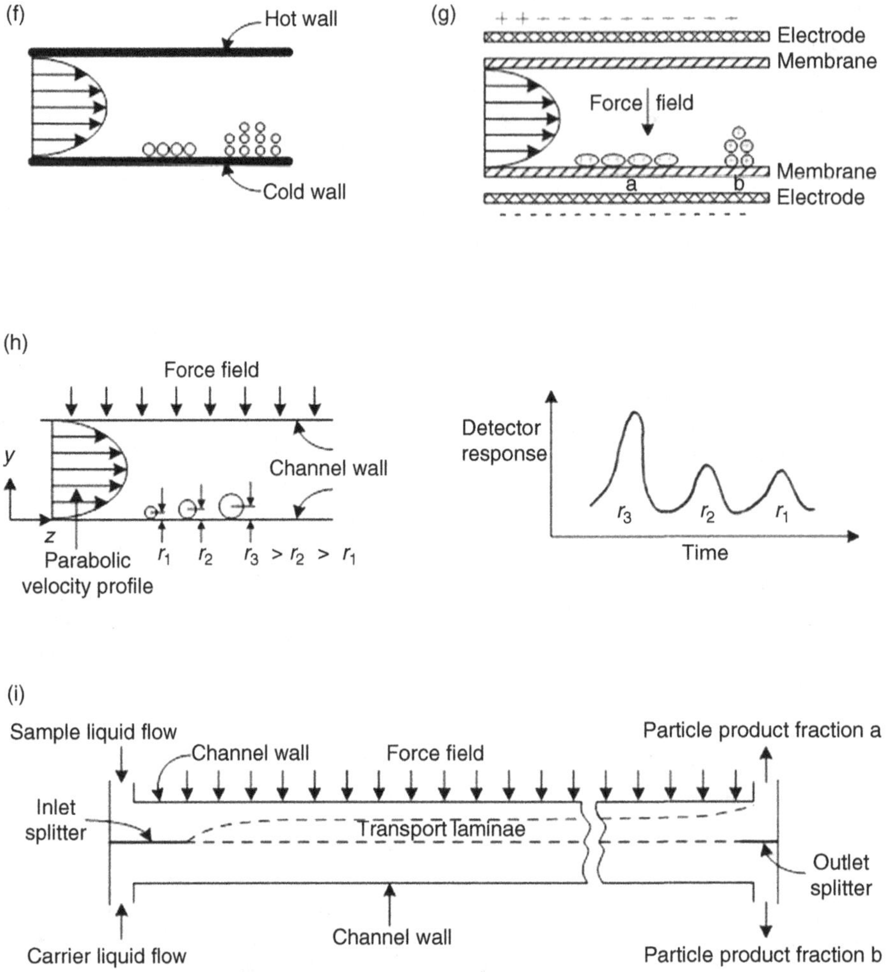

Figure 7.3.21. (cont.) (f) Thermal FFF. (g) Electrical FFF. (h) Steric FFF and the detector response to a sample injection. (i) SPLITT fractionation.

(Figure 7.3.21(a)). It is as if fluid lamellae (laminae) of each species located at different distances from the wall ($y = 0$) travel at different velocities; therefore they arrive at different times at the channel end. (For a comparison, in elution chromatography (Section 7.1.5.1) different species also arrive at different times at the end of the column; however, each species exists throughout the column cross section as the species peak travels. Similarly, in capillary electrophoresis (Section 6.3.1.2) species arriving at the end of the capillary at different times exist throughout the capillary cross section.) This technique was first proposed by Giddings (1966).

Quantitative analysis of the retention time, t_{R_i}, of a macrosolute species in the channel after sample injection may be carried out as follows (Giddings, 1991). The axial velocity profile, $v_z(y)$, of a liquid of viscosity μ in a thin rectangular channel formed between two infinite parallel plates (Figure 7.3.21(a)) spaced a distance b apart is given by[25] (Happel and Brenner, 1965; Grushka *et al.*, 1973; Giddings, 1991)

$$v_z(y) = 6v_{z,\text{avg}}\left(\frac{y}{b} - \frac{y^2}{b^2}\right), \tag{7.3.205}$$

where

$$v_{z,\text{avg}} = \frac{\Delta P b^2}{12\mu L}; \tag{7.3.206}$$

L is the channel length and ΔP is the axial liquid pressure drop. If a macrosolute which is not affected by the force field is injected into the liquid upstream, then t_{R_0}, its

[25] Equation (6.1.2b) provided earlier is for a slit flow where the channel gap << width of the channel plates < length of the channel in the mean flow direction (Bird *et al.*, 2002).

retention time (i.e. the time when it shows up at the exit) will be essentially determined by the average liquid velocity, $v_{z,\text{avg}}$:

$$t_{R_0} = L/v_{z,\text{avg}}. \tag{7.3.207}$$

The retention time t_{R_i} of any macrosolute species i affected by the force field can be determined if we have the knowledge of an average velocity of the liquid zone carrying this species i, v_{ci}^*:

$$v_{ci}^* = \frac{\int_o^b C_{is}(y)\, v_z(y)\mathrm{d}y}{\int_o^b C_{is}(y)\mathrm{d}y}. \tag{7.3.208}$$

Substituting expressions (7.3.203) and (7.3.205) into the above relation, we get

$$v_{ci}^* = \frac{6v_{z,\text{avg}}\int_o^b [\exp(-y/\delta_i)]\left(\frac{y}{b} - \frac{y^2}{b^2}\right)\mathrm{d}y}{\int_o^b \exp(-y/\delta_i)\mathrm{d}y}. \tag{7.3.209}$$

Defining t_{R_i} as the retention time of species i in the chromatographic column (here the channel) as

$$t_{R_i} = L/v_{ci}^*, \tag{7.3.210}$$

one can determine a retention ratio for species i, R_i, defined as

$$R_i = \frac{v_{ci}^*}{v_{z,\text{avg}}} = \frac{t_{R_0}}{t_{R_i}}. \tag{7.3.211}$$

From expression (7.3.209) for v_{ci}^*, we get

$$R_i = \frac{\frac{6}{b}\int_o^b y\exp(-y/\delta_i)\mathrm{d}y - \frac{6}{b^2}\int_o^b y^2\exp(-y/\delta_i)\mathrm{d}y}{\int_o^b \exp(-y/\delta_i)\mathrm{d}y}$$

$$= \frac{-6\delta_i\left[\exp(-\lambda_i^{-1})(\lambda_i+1)-\lambda_i\right] + 6\,\delta_i\left[\exp(-\lambda_i^{-1}) + 2\lambda_i\exp(-\lambda_i^{-1})(\lambda_i+1) - 2\lambda_i^2\right]}{-\delta_i\left[\exp(-\lambda_i^{-1})-1\right]}, \tag{7.3.212}$$

where λ_i is a retention parameter for species i defined as

$$\lambda_i = (\delta_i/b). \tag{7.3.213}$$

Hovingh *et al.* (1970) and Giddings (1991) have shown that this result for R_i yields, after rearrangement,

$$R_i = 6\lambda_i\left[\coth\left(\frac{1}{2\lambda_i}\right) - 2\lambda_i\right]. \tag{7.3.214}$$

Figure 7.3.21(b) (after Grushka *et al.* (1973)) schematically illustrates the retention ratio R_i as a function of the retention parameter λ_i. As the characteristic thickness of the concentration profile δ_i decreases, the retention ratio R_i decreases linearly. In the limit as $(\delta_i/b) \to 0$ (i.e. $\lambda_i \to 0$)

$$\lim_{\lambda_i \to 0} R_i = 6(\delta_i/b) = \frac{6\,D_{is}}{|U_{iy}|b} = 6\lambda_i; \tag{7.3.215}$$

$$\lim_{\lambda_i \to 0} R_i = 6\,\mathrm{R}T/F_{tiy}^{\text{ext}}\,b, \tag{7.3.216}$$

where we have employed definition (3.1.82) for the migration velocity U_{iy} of species i in the y-direction and $D_{is} = D_{is}^0 = \mathrm{R}T/f_i^d$ (relation (3.1.86)). To be able to detect two different species A and B at the outlet of the channel, we need to have a reasonable value of the difference, $t_{R_a} - t_{R_b} = \Delta t_R$, which is given by

$$\Delta t_R = t_{R_0}\left(\frac{1}{R_a} - \frac{1}{R_b}\right) = \frac{t_{R_0}\,b}{6\mathrm{R}T}\left(F_{tay}^{\text{ext}} - F_{tby}^{\text{ext}}\right). \tag{7.3.217}$$

To that end Δt_R should be such that the ratio, $\Delta t_R/t_{R_0}$, should be of order 1:

$$\frac{\Delta t_R}{t_{R_0}} = \frac{b\left(F_{tay}^{\text{ext}} - F_{tby}^{\text{ext}}\right)\Big|_{\text{gmole}}}{6\mathrm{R}T} = \frac{b\left(F_{tay}^{\text{ext}} - F_{tby}^{\text{ext}}\right)\Big|_{\text{macromolecule}}}{6\mathrm{k}^{\mathrm{B}}T}, \tag{7.3.218}$$

since $\mathrm{R} = \mathrm{k}^{\mathrm{B}}\tilde{N}$, where k^{B} is Boltzmann's constant. The channel gap in FFF techniques is around 100-300 μm (typical length, 30 cm, 2 cm width). Assume b = 300 μm. For $T = 298$ K, if $(\Delta t_R/t_{R_0})$ is going to be around 1, then the force difference $\Delta F_t^{\text{ext}}|_{\text{molecule}}$ required is

$$\Delta F_t^{\text{ext}}|_{\text{macromolecule}} = \left(F_{tay}^{\text{ext}} - F_{tby}^{\text{ext}}\right)\Big|_{\text{macromolecule}} \cong \frac{6\mathrm{k}^{\mathrm{B}}T}{b}$$

$$= \frac{6 \times 1.3807 \times 10^{-23}\ \text{newton-m} \times 298\,\text{K}}{300 \times 10^{-6}\text{m} \qquad \text{K}}$$

$$= 8.22 \times 10^{-17}\text{newton} \cong 10^{-16}\text{newton},$$

which is not much (Giddings, 1993).

Most of the above analysis was carried out for macromolecules. One could also carry out an almost identical analysis for particles using the particle flux expression (7.3.198), as long as the particle dimensions are less than 0.1 μm. The following selectivity definitions have been used in FFF techniques for macromolecules (molecular weight M) and particles (particle diameter d_p):

$$\text{macromolecules}: \quad S_M = \left|\frac{\mathrm{d}\log t_R}{\mathrm{d}\log M}\right|; \tag{7.3.219a}$$

$$\text{particles}: \quad S_{d_p} = \left|\frac{\mathrm{d}\log t_R}{\mathrm{d}\log d_p}\right|. \tag{7.3.219b}$$

Alternative definitions of selectivity, employing the retention ratio R_i instead of t_{R_i}, are:

$$S_M = \left| \frac{d \log R_i}{d \log M} \right|, \tag{7.3.220a}$$

$$S_{d_p} = \left| \frac{d \log R_i}{d \log d_p} \right|. \tag{7.3.220b}$$

It is useful now to explore the force fields typically employed and the corresponding FFF techniques. These are:

(1) flow FFF (viscous drag force in crossflow);
(2) sedimentation FFF (centrifugal force);
(3) thermal FFF (thermal diffusion due to a temperature gradient);
(4) electrical FFF (electrical potential gradient).

In *flow FFF* (Giddings *et al.*, 1976), the channel walls are porous (e.g. frits are used), as shown in Figure 7.3.21(c). Liquid crossflow is imposed at the porous top wall and it goes out through the porous bottom wall of the channel. This flow in the negative y-direction is superimposed on the main channel flow in the z-direction; this crossflow drags macromolecules/particles to the bottom porous wall (frit). To prevent macromolecules/particles from leaving through the porous channel wall, a membrane that retains all macromolecules/particles to be separated is employed on top of the bottom porous wall; the liquid goes through the membrane and the porous channel wall at a velocity $|U_y|$. The macromolecules (or particles) experience a drag force in the negative y-direction due to this negative y-directional fluid motion, whose magnitude is given by (if we assume Stokes' law)

$$|F_{tiy}|_{\text{molecule}} = f_i^d|_{\text{molecule}} |U_y| = 3\pi \mu d_i |U_y| = \frac{k^B T |U_y|}{D_{is}}. \tag{7.3.221}$$

(*Note*:

$$f_i^d|_{\text{gmol}} = \tilde{N} f_i^d|_{\text{molecule}}; \quad D_{is} = \frac{RT}{f_i^d|_{\text{gmol}}} = \frac{k^B \tilde{N} T}{\tilde{N} f_i^d|_{\text{molecule}}} = \frac{k^B T}{3\pi\mu d_i};$$

see relations (3.1.86), (3.1.91d) and (3.3.90c)). One can clearly see now from definition (7.3.202) that, for a given crossflow velocity U_y,

$$\delta_i = \frac{D_{is}}{|U_{iy}|} = \frac{k^B T}{3\pi \mu d_i |U_y|}, \tag{7.3.222}$$

leading to different values of δ_i for macromolecules experiencing different drag forces due to their different sizes. Larger molecules will have smaller δ_i closer to the wall. The same analysis is also valid for microsized particles.

This technique, often identified as FLFFF (flow field-flow fractionation), is generally the most useful of the different FFF techniques (Kirkland and Dilks, 1992). It has been used to separate polymers when the molecular weights (MWs) of polymers are in the range 10^4–10^7 and similarly for particles (<1 μm) of size down to 1 nm (Giddings, 1993). Consider a channel (Figure 7.3.21(d)) where the channel volume is V_{channel}. Let the volumetric flow rate of the crossflow be $Q_{\text{crossflow}}$. Then, from relation (7.3.222),

$$\lambda_i = \frac{\delta_i}{b} = \frac{D_{is}}{|U_{iy}| b} = \frac{D_{is}}{\dfrac{Q_{\text{crossflow}}\, b}{WL}} = \frac{D_{is}}{b^2} \frac{(WLb)}{Q_{\text{crossflow}}} = \frac{D_{is}}{b^2} \frac{V_{\text{channel}}}{Q_{\text{crossflow}}}, \tag{7.3.223}$$

where W is the width of the channel, L is its length and b is the gap between the channel plates. Experimentally, one can determine R_i from (7.3.211) from measurements of t_{R_i} and t_{R_0}. If one then employs the result (7.3.214), one can determine λ_i. Substitution of this λ_i into the result (7.3.223) derived above will yield D_{is} for that fraction/species/the particle for known experimental values, of b, V_{channel} and $Q_{\text{crossflow}}$. Then the Stokes–Einstein relation (3.3.90c) will yield r_i for the macromolecules. If relations like (3.3.90d) are available, the molecular weight will be known for the fraction. Thus, a prediction of which macromolecule will appear at what time (t_{R_i}) is possible in this technique. It is a similar case for particles; however, when particles are larger than a few hundred nanometers, there are wall effects due to the particle–wall interactions, for example, by geometrical exclusion (see relations (3.3.88a,b)). This will be considered later.

For *sedimentation FFF*, consider the thin channel going around a centrifuge basket like a thin belt (Figure 7.3.21(e)) with appropriate arrangements for incoming flow and outgoing flow. The radial force component on a particle of density ρ_p, volume V_p, radius r_p in a liquid of density ρ_t rotating at an angular velocity ω at radial location r is obtained, from expression (3.1.59), as

$$F_{tpr}^{\text{ext}} = \frac{4}{3}\pi r_p^3 \rho_p \left(1 - \frac{\rho_t}{\rho_p}\right) r\,\omega^2. \tag{7.3.224}$$

This may be rearranged in terms of the magnitude of the force $|F_{tpr}^{\text{ext}}|$ as follows:

$$|F_{tpr}^{\text{ext}}| = \frac{\pi d_p^3}{6} |\rho_p - \rho_t|\, r\,\omega^2. \tag{7.3.225}$$

Similarly, the magnitude of the radial force component on one gmol of species i may be obtained from (3.1.52) as

$$|F_{tir}^{\text{ext}}|_{\text{gmol}} = M_i \left[1 - \left(\frac{\overline{V}_i}{M_i}\right)\left(\frac{M_s}{\overline{V}_s}\right)\right] r\omega^2. \tag{7.3.226}$$

Correspondingly, the force on one macromolecule will be

$$|F_{tir}^{\text{ext}}|_{\text{molecule}} = \frac{M_i}{\tilde{N}} \left[1 - \left(\frac{\overline{V}_i}{M_i}\right)\left(\frac{M_s}{\overline{V}_s}\right)\right] r\omega^2. \tag{7.3.227}$$

Now, the radius of this thin channel inside the centrifuge is large; so the radial variation at different radial locations in the channel for a given radial direction may be neglected.

Therefore, regardless of the radial location where the solute/particle band is (in Figure 7.3.21(e)), the centrifugal acceleration, $r\omega^2$, may be assumed to be a constant, equal to G_c.

We may now estimate δ_i for a given macromolecular species i from definition (7.3.202) as

$$\delta_i = \frac{D_{is}}{|U_{ir}|} = \frac{D_{is}}{|F_{tir}^{\text{ext}}|_{\text{gmol}}/f_i^d} = \frac{RT}{|F_{tir}^{\text{ext}}|_{\text{gmol}}} = \frac{RT}{M_i\left[1-\left(\frac{\overline{V}_i}{M_i}\right)\left(\frac{M_s}{\overline{V}_s}\right)\right]G_c}.$$

Correspondingly,

$$\lambda_i = \frac{RT}{M_i\left[1-\left(\frac{\overline{V}_i}{M_i}\right)\left(\frac{M_s}{\overline{V}_s}\right)\right]G_c\, b}. \tag{7.3.228}$$

The equivalent result for particles of radius r_p, diameter d_p and density ρ_p is

$$\delta_i(r_p) = \frac{D_p}{|U_{pr}|} = \frac{D_p f_p^d}{|F_{tpr}^{\text{ext}}|} = \frac{D_p f_p^d}{\frac{\pi d_p^3}{6}|\rho_p-\rho_t|G_c}$$
$$= \frac{k^B T\, 6\pi\mu r_p}{6\pi\mu r_p \frac{\pi d_p^3}{6}|\rho_p-\rho_t|G_c} = \frac{6k^B T}{\pi d_p^3|\rho_p-\rho_t|G_c}, \tag{7.3.229}$$

where we have used the Stokes-Einstein equation (3.3.90c) for the particle diffusivity D_p and $f_p^d = 6\pi\mu r_p$ from Stokes' law. One can now determine the retention ratio R_i from equation (7.3.214) for any macromolecular species or particle. For particle separation, it is clear from expression (7.3.229) that $\delta_i(r_p)$ will differ significantly from particle to particle due to the d_p^{-3} dependency; a small difference in r_p will be magnified substantially. However, to separate smaller particles and macromolecules, high values of ω are required, which leads to problems of sealing liquid inflow/outflow. Therefore sedimentation FFF is applied to particles of size larger than 10–30 nm and to polymers having a molecular weight $>10^6$–10^7 dalton.

In *thermal FFF*, one channel wall (Figure 7.3.21(f)) is kept hotter than the other to create a temperature gradient $(\mathrm{d}T/\mathrm{d}y)$ of sufficient magnitude; in practice, values ~10^4 K/cm are maintained. For a channel gap b of, say, 300 μm, a temperature difference ΔT of 300×10^{-4} cm $\times\ 10^4$ K/cm $= 300$ K is required. Polymers and particles are supposed to be driven toward the colder wall (as in thermal diffusion), and this motion is opposed by diffusion. (See relation (4.2.63) for gas species; the approach to be used here is similar.) Although there are many complications, Grushka *et al.* (1973) have employed the following expression for the net molar flux for a polymer species (which becomes zero at steady state):

$$J_{iy}^*|_{\text{total}} = \underbrace{-D_{is}C_t\frac{\mathrm{d}x_{is}}{\mathrm{d}y}}_{\text{diffusive flux}(J_{iy}^{*D})} \underbrace{-D_{is}^T C_t x_{is}(1-x_{is})\frac{\mathrm{d}T}{\mathrm{d}y}}_{\text{flux due to thermal diffusion }(J_{iy}^{*T})} = 0,$$

where D_{is}^T is the thermal diffusion coefficient for species i whose mole fraction in the dilute solution is x_{is} (therefore $(1 - x_{is}) \sim 1$).

Due to the very high temperature gradient, the solution density, and therefore C_t, will vary with distance y. Since the macrosolute concentration $C_{is} = C_t\, x_{is}$, we may write

$$\frac{\mathrm{d}C_{is}}{\mathrm{d}y} = C_t\frac{\mathrm{d}x_{is}}{\mathrm{d}y} + x_{is}\frac{\mathrm{d}C_t}{\mathrm{d}y}; \tag{7.3.230}$$

$$\frac{\mathrm{d}C_{is}}{\mathrm{d}y} = C_t\frac{\mathrm{d}x_{is}}{\mathrm{d}y} + x_{is}\left(\frac{\mathrm{d}C_t}{\mathrm{d}T}\right)\left(\frac{\mathrm{d}T}{\mathrm{d}y}\right). \tag{7.3.231}$$

Therefore, at steady state,

$$J_{iy}^*|_{\text{total}} = 0 = -D_{is}\left\{\left(\frac{\mathrm{d}C_{is}}{\mathrm{d}y} + \frac{D_{is}^T}{D_{is}}C_{is}\frac{\mathrm{d}T}{\mathrm{d}y}\right) - \frac{C_{is}}{C_t}\left(\frac{\mathrm{d}C_t}{\mathrm{d}T}\right)\left(\frac{\mathrm{d}T}{\mathrm{d}y}\right)\right\}; \tag{7.3.232}$$

$$\frac{1}{C_{is}}\frac{\mathrm{d}C_{is}}{\mathrm{d}y} = \frac{\mathrm{d}\ln C_{is}}{\mathrm{d}y} = -\left(\frac{D_{is}^T}{D_{is}} - \frac{1}{C_t}\left(\frac{\mathrm{d}C_t}{\mathrm{d}T}\right)\right)\frac{\mathrm{d}T}{\mathrm{d}y}. \tag{7.3.233}$$

The quantity $(-C_t^{-1}(\mathrm{d}C_t/\mathrm{d}T))$ is a coefficient of thermal expansion analogous to $\overline{\beta}$ (see equation (6.1.10)) and will be indicated here as $\overline{\beta}_c$. Therefore

$$\frac{\mathrm{d}\ln C_{is}}{\mathrm{d}y} = -\left(\frac{D_{is}^T}{D_{is}} + \overline{\beta}_c\right)\frac{\mathrm{d}T}{\mathrm{d}y}. \tag{7.3.234}$$

As the temperature decreases in the negative y-direction (T increasing with y), C_{is}, however, decreases with increasing y from the bottom wall. From the form of equation (7.3.199b), we see that

$$\mathrm{d}\ln C_{is} = -\left(\left(\frac{D_{is}^T}{D_{is}} + \overline{\beta}_c\right)\frac{\mathrm{d}T}{\mathrm{d}y}\right)\mathrm{d}y.$$

Therefore,

$$\left|\frac{U_{iy}}{D_{is}}\right| = \left|\left(\frac{D_{is}^T}{D_{is}} + \overline{\beta}_c\right)\frac{\mathrm{d}T}{\mathrm{d}y}\right| = \frac{1}{\delta_i}. \tag{7.3.235}$$

Since $\overline{\beta}_c$ is negligible under the conditions employed in thermal FFF, we get

$$\delta_i \cong \frac{D_{is}}{D_{is}^T(\mathrm{d}T/\mathrm{d}y)}. \tag{7.3.236}$$

For a linear temperature gradient,

$$\delta_i \cong \frac{b\,D_{is}}{D_{is}^T\,\Delta T}. \tag{7.3.237}$$

Therefore (δ_i/b) should be linearly proportional to ΔT^{-1}; this behavior has been verified experimentally. See, for example, the behavior of (δ_i/b) for the macromolecule polystyrene of molecular weight 51 000 in the solvent ethylbenzene in Grushka *et al.* (1973). If we can assume a certain type of dependence of D_{is} on the molecular weight of the polymer (e.g. $r_i \propto M_i^\beta$ (see relations 3.3.90d–f); therefore, from the Stokes-Einstein relation (3.3.90c)

$D_{is}^0 \propto M_i^{-\beta}$), it is clear from relations (7.3.211), (7.3.214) and (7.3.215) that, as $\lambda_i \to 0$, $R_i = 6\lambda_i$ and

$$\frac{1}{R_i} = \frac{t_{R_i}}{t_{R_0}} = \frac{1}{6\lambda_i} = \frac{1}{6}\frac{D_{is}^T \Delta T}{D_{is}} \propto D_{is}^T \Delta T M_i^\beta, \quad (7.3.238)$$

allowing one to fractionate the polymers of different molecular weights, provided the variation of D_{is}^T with M_i does not interfere. The basis of variations of D_{is}^T is, however, quite complex.

In *electrical FFF*, the channel walls are bounded by semipermeable membranes (having a molecular weight cutoff of 10 000 daltons), with electrodes located on the outside of the channel walls. As the electrical field is applied perpendicular to the channel flow, small ions in the buffer solution move toward the electrodes through the semipermeable membranes. However, charged macromolecules/proteins/colloids cannot go through the membranes and are pressed to the channel walls. The migration velocity of species i toward the wall, U_{iy}, in an electrical field of magnitude E_y, is $\mu_i^m E_y$ (see relation (6.3.8f)). So

$$\delta_i = \frac{D_{is}}{\mu_i^m E_y}, \quad (7.3.239)$$

where μ_i^m is the ionic mobility of species i. As long as the ratio D_{is}/μ_i^m of the macromolecular species i varies significantly with i under otherwise constant conditions, one can separate different charged macromolecules/proteins/charged colloids by electrical FFF (Grushka *et al.*, 1973).

Caldwell and Gao (1993) have studied this subject in some detail. For colloidal particles, they have pointed out that, for smaller ζ-potentials (see equation (3.1.11a)), the electrophoretic mobility μ_i^m can be described by

$$\mu_i^m = (2\,\varepsilon_d k^B\, T/3\mu e) f(r_i/\lambda), \quad (7.3.240)$$

where ε_d is the dielectric constant of the medium, k^B is Botlzmann's constant, μ is the liquid viscosity, λ is the Debye length and $f(r_i/\lambda)$, the Henry function, increases monotonically from 1 to 1.5 as r_i/λ ranges from zero to infinity. Therefore μ_i^m is relatively insensitive to particle/colloid/macromolecule size, varying by a maximum factor of 1.5 at a given ionic strength in a given liquid. Consequently, the variation of δ_i appears primarily due to variation in D_{is}, which varies essentially inversely with the r_i of the particle/colloid/charged macromolecule.

Caldwell and Gao (1993) applied voltages $\leq$2 volts across the channel walls (the channel thickness ~178 μm) and were able to separate a variety of polystyrene latex particles with sizes ranging from 60 nm to much larger values. Grushka *et al.* (1973) have illustrated the separation of protein mixtures such as albumin, hemoglobin and γ-globulin. In both of these, and other related, studies, the voltage gradient applied was orders of magnitude smaller than that in capillary electrophoresis (see Section 6.3.1.2). Thermal effects were found to be minimal.

At this time, it is necessary to point out two very important characteristics of the separation techniques being studied. *First*: So far, the technique is a chromatographic technique (Section 7.1.5) with samples being injected upstream and different macromolecular species/particles/proteins detected/separated at different retention times. *Second*: No wall effects were considered (see discussion after equation (7.3.223)). This second issue is quite important, and it ends up influencing the first issue as well.

When the particles are larger than a few hundred nanometers, a few other factors not considered so far become important. *First*: The variation of particle diffusion coefficient is no longer important. *Second*: The magnitudes of the forces become much larger. *Third*: Most importantly, a region of the wall up to a thickness of r_i from the wall is no longer accessible to the center of the particle due to geometrical exclusion or steric exclusion (Figure 3.3.5A). Since in normal laminar flow there is almost no back diffusion for larger particles, the force acting on the particle pushes it to the bottom wall ($y = 0$) and the particle essentially stays there. Thus, the centerline of the particle, which is at $y = r_i$, is also equal to δ_i, a characteristic thickness of the layer of particles of radius r_i. These particles will move with the fluid velocity characteristic of the region $y = r_i$. Larger particles will elute much more quickly (and have smaller retention times) since their centers are at a larger y and are exposed to a higher fluid velocity (Figure 7.3.21(h)). This is called the *Steric elution order.* These techniques (depending on the force field) are useful for larger particles (0.5–200 μm) and are generically identified as *steric FFF*. On the other hand, all the other techniques studied earlier (for smaller particles and macromolecules), where geometrical exclusion effects are negligible and D_{is} is important, are called *normal-mode FFF*, and they display *normal elution order*, where smaller particles elute faster/earlier.

From Figure 7.3.21(h), it is clear that the value of the retention parameter λ_i of particles of radius r_i should be equal to (r_i/b) and that for $\lambda_i \to 0$, the retention ratio for particle i, $R_i = 6\lambda_i$ from equations (7.3.213) and (7.3.214). In general, however, there exist hydrodynamic effects which tend to increase δ_i beyond r_i. This is accommodated via the following expression for R_i (Ratanathanawongs and Giddings, 1992):

$$R_i = 6\gamma_i\, r_i/b, \quad (7.3.241)$$

where γ_i is a dimensionless steric correction factor. In steric FFF, $\gamma_i \cong 1$. However, there often exist hydrodynamic lift forces which raise the particle band higher up from the wall, and $\gamma_i \geq 2$. Such operating modes are called *lift hyperlayer FFF*. *Note*: The shear-induced diffusion coefficient of particles is proportional to r_p^2, therefore r_i^2 (here); so larger particles have higher shear-induced diffusivity. However, regardless of the general mode of operation used in FFF, normal or steric, it is desirable that $\delta_i < 0.2b$ (Ratanathathanawongs and Giddings, 1992).

One of the important characteristics of steric FFF for particle fractionation is that it can take place continuously; i.e. the feed particle mixture can be fed continuously at the channel inlet, and one can obtain continuously several product fractions at the channel outlet (Figure 7.3.21(i)). Usually, the outlet flow is split into two fractions by an outlet splitter (Fuh *et al.*, 1992), although multisplit outlets have also been realized (Giddings, 1985) to produce several different fractions simultaneously. In both of these configurations, there is also an inlet splitter, with the sample being introduced from the top and the carrier liquid from the bottom. However, unlike normal-mode FFF, which can develop a complete particle size distribution or molecular weight distribution via the elution mode of operation, steric FFF produces a finite number of fractions regardless of the mode of operation, elution or continuous.

In steric FFF with an inlet splitter and an outlet splitter (the technique is called SPLITT fractionation (SF) (Giddings, 1985; Fuh *et al.*, 1992), the sample-containing liquid enters through the top while the sample-free carrier liquid is introduced from the bottom, with the inlet splitter in between (Figure 7.3.21(i)). Usually the carrier liquid flow rate exceeds that of the sample-containing stream and compresses it to a thin laminae. The y-location of the outlet splitter is at a lower level; the in-between region is called the *transport laminae*. A particle must cross the thickness of this transport laminae to appear in the outlet splitter stream b.

If the velocity U_{py} may be described for a sedimentation process (gravity and/or centrifugal forces, see equation (4.2.16b) for example) via

$$U_{py} = s_p\,\omega^2 r = s_p\,G_c, \tag{7.3.242}$$

then a particle having a diameter larger than a critical diameter $d_{p,\mathrm{cr}}$ will appear in the outlet splitter stream b (bottom), whereas those that are smaller will appear in the upper outlet splitter stream a. The value of $d_{p,\mathrm{cr}}$ is given by (Fuh *et al.*, 1992)

$$d_{p,\mathrm{cr}} = \left\{\frac{18\mu(\dot{V}(a) - \dot{V}(a'))}{W L \omega^2 r(\rho_p - \rho_t)}\right\}^{1/2}, \tag{7.3.243}$$

where $\dot{V}(a)$ is the volume flow rate through the upper outlet splitter stream a, $\dot{V}(a')$ is that coming in through the feed inlet splitter, W is the width of the channel of length L, $\omega^2 r$ (or g for a gravitational field) is the centrifugal field strength applied normal to the flow and s_p is the sedimentation coefficient from (4.2.16b) defined here as $((\rho_p - \rho_t)d_p^2/18\mu)$. Thus larger and heavier particles emerge through the bottom stream at the outlet.

There is a complication in steric FFF. As we have seen earlier, when exposed to a high shear field near the wall due to a higher flow rate, particles tend to get lifted off, and the shear-induced particle diffusivity becomes substantial (see, e.g., relations (3.1.74) and (7.2.131a)). Larger particles will then have their centers located at a y larger than $(d_p/2)$.

The existence of diffusion becomes useful in another separation context. If a sample stream comes in through the feed inlet splitter (Figure 7.3.21(i)) and it has larger and smaller particles/species, what can happen in the absence of any force field is as follows. The smaller particles/molecules can diffuse rapidly into the lower stream generated by the carrier liquid. Therefore the liquid leaving through the upper outlet stream a is now substantially purified of smaller molecules/particles (Williams *et al.*, 1992). The only force that exists here perpendicular to the feed flow direction is that due to diffusion (which we do not identify as a force in the sense of other forces; see the explanation following equation (3.2.22b)).

In all the FFF techniques considered so far, the flow took place in the environment of a channel between two wide flat plates. The channel plate widths are orders of magnitude larger than the gap between the plates (e.g. 2 cm × 200–300 μm). However, there is always an edge effect at the two ends of the width of the plate. Much more important, however, is the requirement in flow FFF and electrical FFF that there be a membrane lining the channel to allow crossflow permeation in flow FFF and buffer ion transport in electrical FFF.

If the flow channel is provided by the lumen of a porous-wall hollow fiber membrane, then both aspects are taken care of. Edge effects are eliminated due to the cylindrical geometry of the hollow fiber. Flow entering the hollow fiber at one end is automatically divided into two parts: one part that permeates through the wall creating the needed crossflow, and the other part that flows straight through with different particles/macromolecules having different retention times (Lee *et al.*, 1974). In the case of electrical FFF, the hollow fiber is placed in the wide gap between two planar electrodes (Lightfoot *et al.*, 1981). This technique has been called *electropolarization chromatography*.

Example 7.3.7 Determine the dependence of δ_i on the radius r_p or r_i of the particle or macromolecule in (1) normal FFF techniques: sedimentation FFF; flow FFF and (2) steric FFF. Comment on how the selectivity S_{dp} will vary with r_p. You may exchange r_i for r_p or d_p for d_i anywhere. (3) Are there any implications vis-à-vis the separation time?

Solution (1) Consider *sedimentation FFF*. From relation (7.3.229),

$$\delta_i(r_p) \propto \frac{1}{d_p^3} \propto \frac{1}{r_p^3}.$$

From the definition of selectivity (7.3.220b),

$$S_{d_p} = \left|\frac{\mathrm{d}\log R_i}{\mathrm{d}\log d_p}\right|.$$

Assume now that $\lambda_i \to 0$; then $R_i = 6\lambda_i = 6\delta_i/b$. Since, from (7.3.229),

$$R_i = \frac{6k^B T}{\pi d_p^3 |\rho_p - \rho_t| G_c b} \Rightarrow S_{d_p} = \left|\frac{d\log d_p^{-3}}{d\log d_p}\right| = |-3| = 3.$$

Next consider flow FFF. From (7.3.222), $R_i = 6\lambda_i = 6\delta_i/b$; so

$$R_i = \frac{6k^B T}{3\pi d_i |U_y| b} \Rightarrow S_{d_p} = \left|\frac{d\log d_i^{-1}}{d\log d_i}\right| = |-1| = 1.$$

(2) Consider *steric FFF*. We have

$$R_i = 6\gamma_i r_i/b \Rightarrow S_{d_p} = \left|\frac{d\log R_i}{d\log d_p}\right| = \left|\frac{d\log r_i}{d\log d_p}\right| = 1$$

(assuming γ_i is independent of r_i).

(3) From relation (7.3.211), we obtain for two particles/species a and b ($r_a > r_b$),

$$\frac{t_{R_a}}{t_{R_b}} = \frac{R_b}{R_a} \cong \frac{6\lambda_b}{6\lambda_a} = \frac{6\delta_b}{6\delta_a}.$$

Now, in sedimentation FFF from (7.3.229), other things being equal,

$$\frac{t_{R_a}}{t_{R_b}} = \frac{\delta_b}{\delta_a} = \frac{r_a^3}{r_b^3}.$$

So if $r_a = 10 r_b$, $t_{R_a} = 1000\, t_{R_b}$. Correspondingly, in *flow FFF*

$$\frac{t_{R_a}}{t_{R_b}} = \frac{\delta_b}{\delta_a} = \frac{r_a}{r_b} \Rightarrow t_{R_a} = 10 t_{R_b}.$$

In steric FFF, from (7.3.241) (assuming $\gamma_a = \gamma_b$),

$$\frac{t_{R_a}}{t_{R_b}} = \frac{R_b}{R_a} = \frac{r_b}{r_a}; \qquad t_{R_b} = 10 t_{R_a}.$$

These calculations indicate that the time needed to achieve resolution between the largest and smallest particles (if they vary by a ratio of 10) can be very high in sedimentation FFF, whereas that in flow FFF/steric FFF is much less.

Example 7.3.8 Grushka *et al.* (1973) report obtaining good values of the retention ratio for the two virus particles T2 ($M_i = 49 \times 10^6$, $\rho_i = 1.57$) and T7 ($M_i = 240 \times 10^6$, $\rho_i = 1.57$) when using sedimentation FFF at a moderate angular velocity of 2000 rpm. Assume that the centrifuge radius is 15 cm and the channel gap is 300 μm at 25 °C. Calculate the values of the retention parameter and retention ratio for both virus particles.

Solution: We will calculate first the value of λ_i. From the result (7.3.228),

$$\lambda_i = \frac{RT}{M_i\left(1 - \left(\frac{\overline{V}_i}{M_i}\right)\left(\frac{M_s}{\overline{V}_s}\right)\right) G_c b},$$

where

$$R = 8.317 \text{ joule/gmol-K} = 8.317 \times 10^7 \frac{\text{erg}}{\text{gmol-K}} = 8.317 \times 10^7 \frac{\text{g-cm}^2}{\text{s}^2\text{-gmol-K}}$$

and $T = 298$ K. Consider first the *T2 virus particle* of $M_i = 49 \times 10^6$g/gmol. Also we have that

$$G_c = r\omega^2 = 15 \text{ cm}\left(\frac{2000 \text{ rpm}}{60\text{s/min}} \times 2\pi \frac{\text{rad}}{\text{rev}}\right)^2 = 15\left(\frac{200\pi}{3}\right)^2 \frac{\text{cm}}{\text{s}^2}; \qquad b = 300 \times 10^{-4}\text{cm},$$

so we have

$$\lambda_i = \frac{8.317 \times 10^7 \times 298 \dfrac{\text{g-cm}^2}{\text{s}^2\text{-gmol}}}{49 \times 10^6 \dfrac{\text{g}}{\text{gmol}} \times \dfrac{15 \times 4 \times 10^4 \times \pi^2}{9} \dfrac{\text{cm}}{\text{s}^2} \times 300 \times 10^{-4}\text{ cm} \times \left(1 - \dfrac{1}{1.57}\right)}$$

$$= \frac{8.317 \times 10^7 \times 298 \times 9 \times 10^4}{49 \times 10^6 \times 60 \times 10^4 \times \pi^2 \times 300 \times 0.363} = 0.0705.$$

We can safely assume that for such low λ_i, $R_i \cong 6\lambda_i = 0.423$.

For the *T7 virus particle* of $M_i = 240 \times 10^6$, and having the same density 1.57 as before,

$$\lambda_i = 0.0705 \times \frac{49}{240} = 0.0144; \qquad R_i \cong 6\lambda_i = 0.0863,$$

a substantially different retention parameter and ratio (almost inversely proportional to M_i).

7.3.5 Magnetic force field

A magnetic force field has been employed to separate particles having differing magnetic properties, primarily from a liquid, via the technique of *high-gradient magnetic separation* (HGMS). This technique requires the presence of ferromagnetic cylinders (steel wools, etc.), to develop strong variations in the magnetic force field on the particles in the local regions around the ferromagnetic cylinders, leading to the capture of particles on those cylinders. The technique of *free-flow magnetophoresis*, on the other hand, is simpler: it does not employ additional ferromagnetic cylinders or spheres, and is somewhat similar to free-flow electrophoresis (Section 7.3.1). We will describe initially the free-flow magnetophoresis based particle separation. Then we will provide a brief treatment of HGMS based particle separation from a fluid. However, a few basic features of particle motion in a magnetic field will be considered first.

As pointed out in Section 3.1.2.4, there are three classes of magnetic materials: ferromagnetic materials (Fe, Co, Ni), which are strongly magnetic; paramagnetic materials (which are far less magnetized compared to ferromagnetic materials); and diamagnetic materials. Expression (3.1.24) for the magnetic force on a nonferromagnetic spherical particle is

$$\boldsymbol{F}_p^{\text{mag}} = 2\pi\,\mu_s^m\,r_p^3\left(\frac{\mu_p^m-\mu_s^m}{\mu_p^m+2\mu_s^m}\right)\nabla \boldsymbol{H}_0^{m^2}. \qquad (7.3.244)$$

If the magnetic permeabilities μ_p^m and μ_s^m of the particle and solution may be described, respectively, via

$$\mu_p^m = \mu_0^m\,(1+\chi_p), \qquad \mu_s^m = \mu_0^m(1+\chi_s), \qquad (7.3.245)$$

where μ_0^m is the magnetic permeability of free space and χ_p and χ_s are, respectively, the magnetic susceptibilities of the particle and the solution, then

$$\boldsymbol{F}_p^{\text{mag}} = 2\pi\,\mu_s^m\,r_p^3\left(\frac{\chi_p-\chi_s}{\chi_p+2\chi_s+3}\right)\nabla\left(\frac{\boldsymbol{B}_0}{\mu_0^m}\right)^2. \qquad (7.3.246)$$

Here the magnetic field strength $\boldsymbol{H}_0^m$ is related to the magnetic flux density $\boldsymbol{B}_0$ by $(\boldsymbol{B}_0/\mu_0^m)$. This relation may be rewritten as follows:

$$\boldsymbol{F}_p^{\text{mag}} = 2\mu_0^m\,(1+\chi_s)\frac{3V_p}{4}\left(\frac{\chi_p-\chi_s}{\chi_p+2\chi_s+3}\right)\frac{\nabla(\boldsymbol{B}_0)^2}{(\mu_0^m)^2}$$

$$= 3(1+\chi_s)V_p\left(\frac{\chi_p-\chi_s}{\chi_p+2\chi_s+3}\right)\frac{\nabla(\boldsymbol{B}_0)^2}{2\mu_0^m}. \qquad (7.3.247)$$

When χ_p and χ_s are much smaller than unity, we obtain

$$\boldsymbol{F}_p^{\text{mag}} = \frac{V_p(\chi_p-\chi_s)}{2\mu_0^m}\nabla(\boldsymbol{B}_0)^2, \qquad (7.3.248\text{a})$$

which is essentially expression (3.1.22), sometimes written as

$$\boldsymbol{F}_p^{\text{mag}} = V_p(\chi_p-\chi_s)(\boldsymbol{H}_0^m\cdot\nabla)\boldsymbol{B}_0. \qquad (7.3.248\text{b})$$

In one-dimensional form (say the z-direction), this force expression is given by

$$F_{pz}^{\text{mag}} = V_p\,(\chi_p-\chi_s)\,\frac{\mathrm{d}}{\mathrm{d}z}\left(\frac{\boldsymbol{B}_0^2}{\mu_0^m}\right) = V_p(\chi_p-\chi_s)H_0^m\,\frac{\mathrm{d}B_0}{\mathrm{d}z}. \qquad (7.3.248\text{c})$$

For paramagnetic particles, $(\chi_p-\chi_s)$ is positive; for diamagnetic particles, it is negative.

A nonferromagnetic particle moving in the direction of the nonuniform magnetic force field would encounter a drag force from the fluid and very soon achieve a terminal velocity, usually identified as the *magnetic migration velocity* U_{pzt}^{mag}, where, if Stokes' law is valid,

$$F_{pz}^{\text{drag}} = 6\pi r_p\mu U_{pzt}^{\text{mag}} = V_p\,(\chi_p-\chi_s)\frac{\mathrm{d}}{\mathrm{d}z}\nabla\left(\frac{B_0^2}{2\mu_0^m}\right); \qquad (7.3.249)$$

$$U_{pzt}^{\text{mag}} = \frac{2}{9}r_p^2\left(\frac{\chi_p-\chi_s}{\mu}\right)\frac{\mathrm{d}}{\mathrm{d}z}\left(\frac{B_0^2}{2\mu_0^m}\right). \qquad (7.3.250)$$

This magnetic migration velocity U_{pzt}^{mag} (Watson, 1973), also called the *magnetic velocity*, may be expressed in terms of a *magnetophoretic mobility* mo_p^{mag} and the *magnetic field force strength* S_{mag} (Moore *et al.*, 2004) by

$$U_{pzt}^{\text{mag}} = mo_p^{\text{mag}}\,S_{\text{mag}}; \qquad mo_p^{\text{mag}} = \frac{2}{9}r_p^2\frac{(\chi_p-\chi_s)}{\mu};$$

$$S_{\text{mag}} = \frac{\mathrm{d}}{\mathrm{d}z}\left(\frac{B_0^2}{2\mu_0^m}\right). \qquad (7.3.251)$$

7.3.5.1 *Free-flow magnetophoresis*

We now briefly illustrate *free-flow magnetophoresis* in analogy to free-flow electrophoresis (Section 7.3.1.1). Consider Figure 7.3.22, in which a rectangular flat separation chamber is shown. Here the vertical coordinate y is normal to the (x,z)-plane. There are a number of inlet and corresponding outlet channels at the two ends of the flat separation chamber. Buffer liquid comes in through all inlet channels except one at one end, where a particle mixture is introduced along with the liquid flow. A nonuniform magnetic field is applied perpendicular to this laminar liquid flow. Those particles which are nonmagnetic, e.g. polystyrene microspheres (Pamme and Manz, 2004), go straight through without any deflection in the direction of the magnetic field gradient to the corresponding outlet channels in the z-direction. Paramagnetic particles, however, follow a trajectory deflected in the direction of the magnetic field (x-coordinate), the extent of deflection depending on the magnetic susceptibility χ_p and the particle radius r_p since

$$U_{pxt}^{\text{mag}} \propto r_p^2(\chi_p-\chi_s). \qquad (7.3.252)$$

The particle trajectory equations are (see equations (7.3.88) and (7.3.91)):

$$\frac{\mathrm{d}x}{\mathrm{d}t} = U_{pxt}^{\text{mag}} = \frac{2}{9}r_p^2\frac{(\chi_p-\chi_s)}{\mu}\frac{\mathrm{d}}{\mathrm{d}x}\left(\frac{B_0^2}{2\mu_0^m}\right); \qquad (7.3.253\text{a})$$

$$\frac{\mathrm{d}z}{\mathrm{d}t} = U_{pzt} = v_z, \qquad (7.3.253\text{b})$$

where v_z is the fluid velocity in the mean flow direction (z-direction). Typical magnitudes of these flow velocities in microfluidic environments employed by Pamme and Manz (2004) were 0.1-0.4 mm/s. As shown in Figure 7.3.22, paramagnetic particles are deflected in the x-direction; different particles follow different trajectories, arriving at different outlet channels, and are therefore separated from one another. The larger the particle volume (or size, r_p) and the higher the magnetic susceptibility χ_p of the particle material, the higher the deflection in the x-direction since U_{pxt}^{mag} is higher. The particle trajectory is obtained from equations (7.3.253a,b) as

$$\frac{\mathrm{d}x}{\mathrm{d}z} = \frac{2}{9v_z}r_p^2\frac{(\chi_p-\chi_s)}{\mu}\frac{\mathrm{d}}{\mathrm{d}x}\left(\frac{B_0^2}{2\mu_0^m}\right); \qquad (7.3.254\text{a})$$

$$\frac{\mathrm{d}x}{\mathrm{d}z} = \frac{2\mu_0^m}{9v_z}r_p^2\frac{(\chi_p-\chi_s)}{\mu}H_0^m\frac{\mathrm{d}H_0^m}{\mathrm{d}x}. \qquad (7.3.254\text{b})$$

To identify the exact extent of deflection of the particle in the x-direction, one has to know the manner in

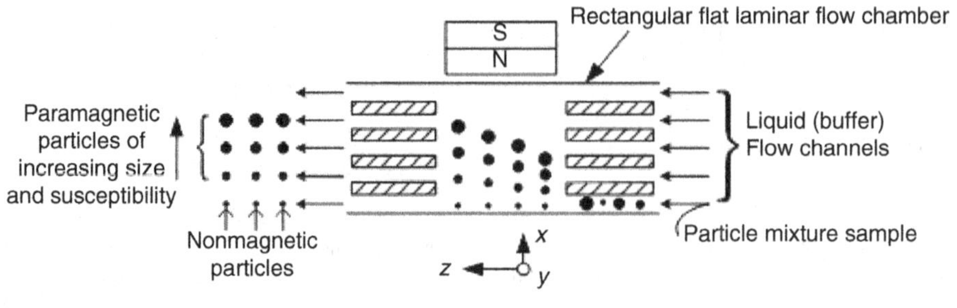

Figure 7.3.22. Free-flow magnetophoresis in a horizontal channel. (After Pamme and Manz (2004).)

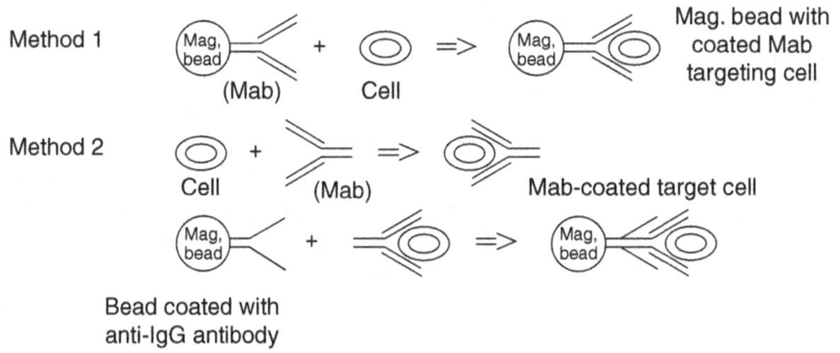

Figure 7.3.23. Two methods of cell separation using magnetic beads and antibody–antigen binding. (After Gee (1998).)

which the nonuniform magnetic field is varying in the x-direction.

Pamme and Manz (2004) have provided estimates of the magnetic force being exerted on small superparamagnetic particles.[26] Particles of diameter 4.5 μm were found to traverse a distance 2 mm in the x-direction when the value of the distance tranversed in the z-direction was 6 mm; the axial fluid velocity was 0.3 mm/s. From equation (7.3.249), we obtain for a liquid consisting essentially of water (viscosity $\mu = 1$ cp)

$$F_{px}^{mag} = 6\pi r_p \mu U_{pxt}^{mag} = 6\pi \times 2.25\ \mu\text{m} \times 1\text{cp} \times \frac{2}{6} \times 0.3\frac{\text{mm}}{\text{s}}$$

$$= 6\pi \times 2.25 \times 10^{-6}\ \text{m} \times 1 \times 10^{-3}\frac{\text{kg}}{\text{m-s}} \times 0.1 \times \frac{10^{-3}\text{m}}{\text{s}}$$

$$= 6\pi \times 2.25 \times 0.1 \times 10^{-12}\frac{\text{kg-m}}{\text{s}^2}$$

$$= 4.24 \times 10^{-12}\frac{\text{kg-m}}{\text{s}^2} = 4.24\ \text{pN (piconewton)}.$$

The force on the smaller particles of similar χ_p will be correspondingly smaller by the volume ratio (equation (7.3.248c)) of the particles.

[26]Dynabeads, Dynal Biotech, Oslo, Norway.

The manner in which such free-flow magnetophoresis (FFM) is applied in practice is related to the needs of life sciences vis-à-vis separation of native biological particles. These native biological particles by themselves do not lend to practical separation by FFM. However, they can be fixed to magnetic or magnetizable microspheres having specific antibodies (Abs) on their surfaces via antigen (Ag)–Ab bridges from antigens on the surfaces of the biological particles. Two examples of this strategy are illustrated in Figure 7.3.23 (Gee, 1998). A common form of the magnetic or magnetizable microspheres consists of superparamagnetic, polystyrene based particles 2.0–4.5 μm in diameter having iron oxide homogeneously distributed throughout the particle (Hausmann *et al.*, 1998); these are called Dynabeads (as mentioned earlier). For the separation of cells and organelles, the FFM method has been identified as *Continuous Immunogenic Sorting.*

7.3.5.2 *High-gradient magnetic separation*

A high-gradient magnetic separator contains a loosely packed bed of ferromagnetic wool of stainless steel; there is one inlet at the bottom for feed fluid to come in and one outlet at the top for treated feed fluid to go out (Figure 7.3.24). A liquid feed containing paramagnetic particles to be removed enters the device from the bottom. A uniform magnetic field is applied in the same direction

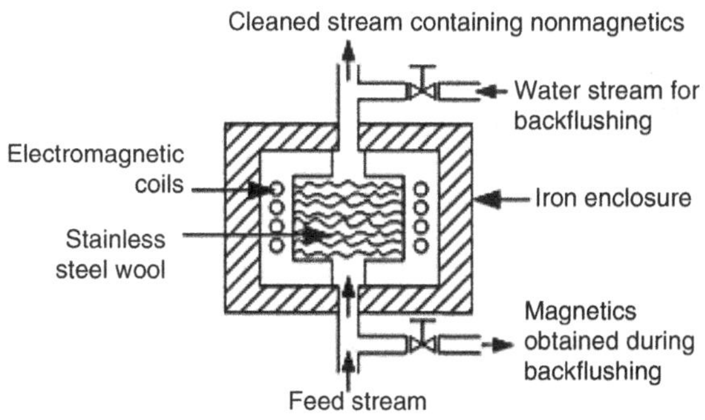

Figure 7.3.24. Schematic of a HGMS unit for cyclic operation.

as that of the fluid flow. Around each strand of the stainless steel wool, the uniform magnetic field becomes highly distorted, creating a nonuniform magnetic field whose gradient can be of the order of 10^5 tesla/m. Particles in the size range of microns are forced to the stainless steel strands and retained there. The nonmagnetic particles pass through with the fluid.

The matrix of stainless steel wool will soon be covered with paramagnetic particulates captured from the fluid, at which time the liquid feed flow is stopped. A wash water stream is passed. Then the magnet is turned off and a water stream is passed in the direction opposite to that of the feed stream; this stream carries away the paramagnetic particles, which fall off the strands of stainless steel wool. After these particles are removed, the magnet is turned on and the feed liquid starts flowing in. The process thus operates cyclically; its operation is very similar to the operation of the packed-bed adsorber in pressure-swing adsorption (Figure 7.1.13(a)). The only difference here is that, instead of the adsorption process, we have magnetic force based particle capture in a direction perpendicular to the main flow direction. Shutting off the magnetic force field leads to release, as in desorption processes.

This HGMS technique is widely employed in industry for the separation of paramagnetic particles from a stream with or without nonmagnetic particles. This technique can also be employed to remove paramagnetic particles from a gas stream (Gooding and Felder, 1981). The stainless steel wool matrix is very open, with a porosity around 0.95. However, it is important that the matrix openings are not very large and that the wire dimensions are quite small since the magnetic field variation around a wire is strongest over a range of distances very close to the wire (Birss and Parker, 1981) and of the order of the particle diameter. We will now provide a very constricted glimpse into the modeling of such a separation device.

The particle trajectory equation (see equation (6.2.45) in such a device is as follows:

$$\underset{\text{inertial force}}{m_p \frac{d\boldsymbol{v}}{dt}} = \underset{\text{magnetic force}}{\boldsymbol{F}^{\text{mag}}} + \underset{\text{gravitational force}}{\boldsymbol{F}^{\text{gr}}} - \underset{\text{drag force}}{\boldsymbol{F}^{\text{drag}}}. \tag{7.3.255}$$

The strands in the stainless steel wool may be modeled as a thin cylinder of radius r_w. The random configurations of the wires are ignored at this time; they may be considered as the sum total of interactions involving three arrangements: longitudinal (uniform magnetic field $\boldsymbol{H}_0^m$ perpendicular to the wire axis and particle velocity parallel but opposed); transverse (H_0^m perpendicular to the wire axis and perpendicular to the particle velocity, which is perpendicular to the wire axis); axial (H_0^m perpendicular to the wire axis as in the longitudinal arrangement, but particle velocity parallel to the wire axis). One could represent a general configuration of the particle-to-wire interaction as shown in Figure 7.3.25 (a) (Oak, 1977; Birss and Parker, 1981; Liu and Oak, 1983). (The physical configuration is similar to that of a particle flowing with air around a solid fiber of radius r_f in depth filtration; See Figure 6.3.9A.) Based on such a configuration and the analyses of Birss and Parker (1981) and Oak (1977), we can provide the following expressions for the r-component and the θ-component of the individual force terms on a spherical particle of radius r_p and volume V_p in equation (7.3.255), after neglecting the gravitational and inertial force components vis-à-vis the other forces. In the following expressions, M_s stands for the saturation magnetization of the wire, i.e. the value of M_w (for example, in equation (3.1.23) when it is saturated). For the: *r-component:*

$$F_{pr}^{\text{mag}} = \left[-\frac{1}{2}\mu_0^m(\chi_p - \chi_s)V_p\left(\frac{4M_s r_w^3}{2r_w r^3}\right)\left(H_0^m \cos 2\theta + \frac{M_s r_w^2}{2r^2}\right)\right]; \tag{7.3.256a}$$

$$F_{pr}^{\text{drag}} = \left[6\pi\mu r_p\left\{\frac{dr}{dt} - v_\infty \cos(\theta - \gamma)(1 - (r_w/r)^2)\right\}\right]. \tag{7.3.256b}$$

As long as $(\chi_p - \chi_s)$ is positive, the radial magnetic force F_{pr}^{mag} on the particle is negative; i.e. the particle is attracted to the ferromagnetic wire as long as $(H_0^m \cos 2\theta + (M_s r_w^2/2r^2)) > 0$; the force magnitude is largest for $\theta = 0$ (for $\gamma = 0$). The configuration becomes that of flow parallel to force (Section 6.3.1.4). There are regions around the wire where the particle is repulsed by the wire (Birss and Parker, 1981). For diamagnetic particles ($(\chi_p - \chi_s)$ is negative), the attractive force is largest for $\theta = \pi/2$.

For the *θ-component:*

$$F_{p\theta}^{\text{mag}} = -\frac{1}{2}\mu_0^m(\chi_p - \chi_s)V_p\left(\frac{4M_s r_w^3 H_0^m}{2r_w r^3}\right)\sin 2\theta; \tag{7.3.257a}$$

$$F_{p\theta}^{\text{drag}} = \left[6\pi\mu r_p\left\{\frac{d\theta}{dt}r + v_\infty \sin(\theta - \gamma)\left(1 + \left(\frac{r_w}{r}\right)^2\right)\right\}\right]. \tag{7.3.257b}$$

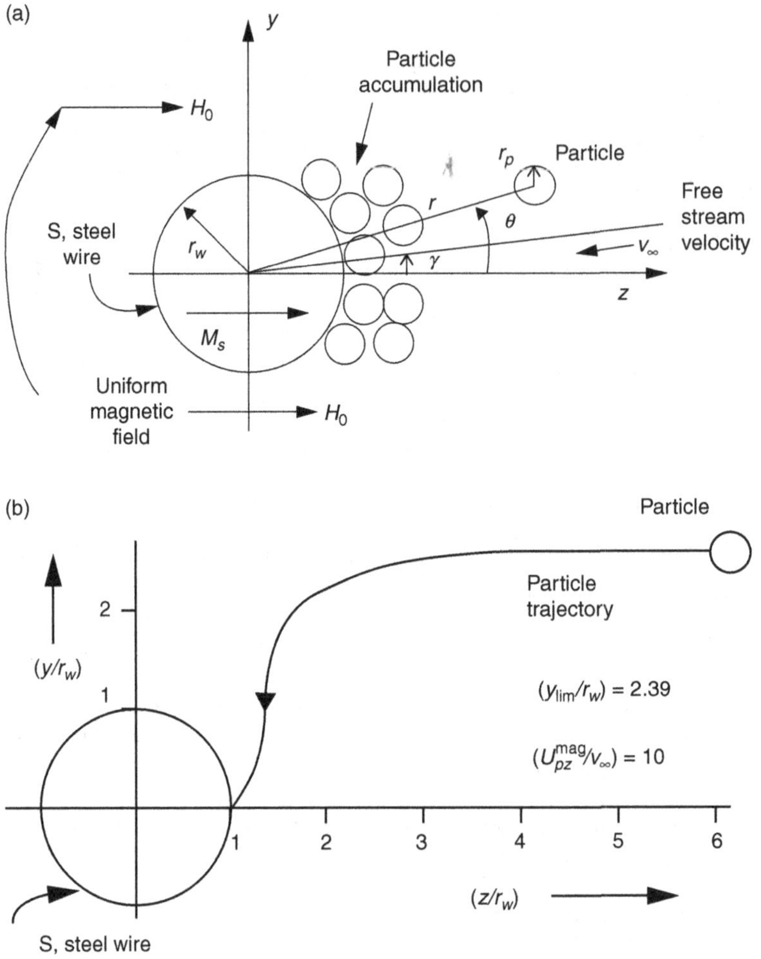

Figure 7.3.25. (a) Configuration of the paramagnetic particle, magnetized stainless steel wire and a particle; coordinate system for analysis (Birss and Parker, 1981; Liu and Oak, 1983). (b) The trajectory of a particle being captured. (After Watson (1973).)

Therefore the governing equations of particle motion in the r-direction and the θ-direction are as follows. For the *r-component:*

$$\left[-\frac{1}{2}\mu_0^m(\chi_p-\chi_s)V_p\left(\frac{4M_s r_w^3}{2r_w r^3}\right)\left(H_0^m\cos 2\theta+\frac{M_s r_w^2}{2r^2}\right)\right]$$
$$-\left[6\pi\mu r_p\left\{\frac{dr}{dt}-v_\infty\cos(\theta-\gamma)(1-(r_w/r)^2)\right\}\right]=0. \tag{7.3.258a}$$

This leads to the following for the r-component equation from equation (7.3.255):

$$\frac{dr}{dt}=v_\infty\cos(\theta-\gamma)\left(1-\left(\frac{r_w}{r}\right)^2\right)$$
$$-\frac{\mu_0^m(\chi_p-\chi_s)V_p}{12\pi\mu r_p}\left(\frac{4M_s r_w^3}{2r_w r^3}\right)\left(H_0^m\cos 2\theta+\frac{M_s r_w^2}{2r^2}\right). \tag{7.3.258b}$$

Similarly for the *θ-component:*

$$-\frac{1}{2}\mu_0^m(\chi_p-\chi_s)V_p\left(\frac{4M_s r_w^3 H_0^m}{2r_w r^3}\right)\sin 2\theta$$
$$-\left[6\pi\mu r_p\left\{\frac{d\theta}{dt}r+v_\infty\sin(\theta-\gamma)\left(1+\left(\frac{r_w}{r}\right)^2\right)\right\}\right]=0, \tag{7.3.259a}$$

leading to

$$r\frac{d\theta}{dt}=-v_\infty\sin(\theta-\gamma)\left(1+\left(\frac{r_w}{r}\right)^2\right)$$
$$-\frac{\mu_0^m(\chi_p-\chi_s)}{12\pi\mu r_p}V_p\left(\frac{4M_s r_w^3 H_0^m}{2r_w r^3}\right)\sin 2\theta. \tag{7.3.259b}$$

Dividing equation (7.3.258b) by equation (7.3.259b), we get the particle trajectory equation:

$$\frac{1}{r}\frac{\mathrm{d}r}{\mathrm{d}\theta} = \frac{\cos(\theta-\gamma)\left(1-\left(\frac{r_w}{r}\right)^2\right) - \frac{\mu_0^m(\chi_p-\chi_s)V_p}{12\pi\mu r_p v_\infty}\left(\frac{4M_s r_w^3}{2r_w r^3}\right)\left(H_0^m\cos 2\theta + \frac{M_s r_w^2}{2r^2}\right)}{-\sin(\theta-\gamma)\left(1+\left(\frac{r_w}{r}\right)^2\right) - \frac{\mu_0^m(\chi_p-\chi_s)}{12\pi\mu r_p v_\infty}V_p\left(\frac{4M_s r_w^3 H_0^m}{2r_w r^3}\right)\sin 2\theta}; \tag{7.3.260a}$$

$$\frac{1}{r}\frac{\mathrm{d}r}{\mathrm{d}\theta} = \frac{\left(1-\left(\frac{r_w}{r}\right)^2\right)\cos(\theta-\gamma) - \frac{2\mu_0^m(\chi_p-\chi_s)r_p^2 M_s H_0^m}{9r_w\mu v_\infty}\left(\frac{r_w}{r}\right)^3\left(\cos 2\theta + \frac{M_s}{2H_0^m}\left(\frac{r_w}{r}\right)^2\right)}{-\left(1+\left(\frac{r_w}{r}\right)^2\right)\sin(\theta-\gamma) - \frac{2\mu_0^m(\chi_p-\chi_s)r_p^2 M_s H_0^m}{9r_w v_\infty\mu}\left(\frac{r_w}{r}\right)^3\sin 2\theta}. \tag{7.3.260b}$$

From definition (7.3.260) for the magnetic migration velocity U_{pzt}^{mag}, we have

$$U_{pzt}^{\mathrm{mag}} = \frac{2}{9}r_p^2\frac{(\chi_p-\chi_s)}{\mu}\frac{\mathrm{d}}{\mathrm{d}z}\left(\frac{H_0^{m^2}\mu_0^m}{2}\right) = \frac{2}{9}r_p^2\frac{\mu_0^m H_0^m}{\mu}(\chi_p-\chi_s)\frac{\mathrm{d}H_0^m}{\mathrm{d}z}; \tag{7.3.261a}$$

$$U_{pzt}^{\mathrm{mag}} = \frac{2}{9}\frac{\mu_0^m r_p^2(\chi_p-\chi_s)H_0^m}{\mu}\frac{M_s}{r_w} = \frac{2\mu_0^m r_p^2(\chi_p-\chi_s)H_0^m M_s}{9\mu r_w}. \tag{7.3.261b}$$

Using this definition, we can rewrite the expression (7.3.260b) as follows:

$$\frac{1}{r}\frac{\mathrm{d}r}{\mathrm{d}\theta} = \frac{\left(1-\left(\frac{r_w}{r}\right)^2\right)\cos(\theta-\gamma) - \frac{U_{pzt}^{\mathrm{mag}}}{v_\infty}\left(\frac{r_w}{r}\right)^3\left(\cos 2\theta + \frac{M_s}{2H_o^m}\left(\frac{r_w}{r}\right)^2\right)}{-\left(1+\left(\frac{r_w}{r}\right)^2\right)\sin(\theta-\gamma) - \frac{U_{pzt}^{\mathrm{mag}}}{v_\infty}\left(\frac{r_w}{r}\right)^3\sin 2\theta}. \tag{7.3.262a}$$

Defining a nondimensional radial coordinate $r^+ = (r/r_w)$, we get

$$\frac{1}{r^+}\frac{\mathrm{d}r^+}{\mathrm{d}\theta} = \frac{(1-(1/r^+)^2)\cos(\theta-\gamma) - \left(U_{pzt}^{\mathrm{mag}}/v_\infty\right)\left(\frac{1}{r^+}\right)^3\left(\cos 2\theta + \frac{M_s}{2H_0^m r^{+2}}\right)}{-(1+(1/r^+)^2)\sin(\theta-\gamma) - \left(U_{pzt}^{\mathrm{mag}}/v_\infty\right)\left(\frac{1}{r^+}\right)^3\sin 2\theta}. \tag{7.3.262b}$$

The solution of this trajectory equation can be developed numerically. What is of interest, however, is the y-coordinate of the limiting trajectory of a particle which will be captured: if the particle starts out at a y-value in the range $(y_{\mathrm{lim}}/r_w) \geq (y/r_w) \geq -(y_{\ell\mathrm{im}}/r_w)$, where the magnitude of the y-value of the limiting trajectory is $|y_{\ell\mathrm{im}}|$ (at $z = \infty$), then its trajectory will surely hit the stainless steel wire ($r = r_w$) at differing values of θ. Luborsky and Drummond (1975), as well as Liu and Oak (1983), have provided a few simplified expressions for $y_{\ell\mathrm{im}}$:

$$(y_{\ell\mathrm{im}}/r_w) = (3\sqrt{3}/4)\left(U_{pzt}^{\mathrm{mag}}/v_\infty\right)^{1/3} \text{ if } \left(U_{pzt}^{\mathrm{mag}}/v_\infty\right) > 1; \tag{7.3.263a}$$

$$(y_{\ell\mathrm{im}}/r_w) = \left(U_{pzt}^{\mathrm{mag}}/2v_\infty\right) \text{ if } \left(U_{pzt}^{\mathrm{mag}}/v_\infty\right) \leq \sqrt{2}. \tag{7.3.263b}$$

Watson (1973) also provided the last result. This limiting trajectory coordinate $y_{\ell\mathrm{im}}$ far away from the stainless steel wire is similar to the coordinate b in Figure 6.3.9A except the particle trajectory does not follow the streamline at all near the wire (Figure 7.3.25(b)).

A few general features are to be noted here. In Figure 7.3.25(a), the particle may come in and hit the wire at any angle between $\theta = 0$ and 90°. When $\theta = 0$ and $\gamma = 0$, it may appear as a case of bulk flow parallel to force (as in Section 6.3.1.4). However, if $\theta \neq 0$, there is always a component of force perpendicular to the bulk flow (as long as $\gamma = 0$; it is true even for $\gamma \neq 0$ as long as $\theta \neq \gamma$). But the overall configuration continues to be particles taken out of the bulk flow in the perpendicular direction and deposited in the woolen medium.

7.3.5.2.1 Performance of a HGMS filter We will develop now an expression for the fractional capture of the paramagnetic particles as the liquid flows through the HGMS filter bed of ferromagnetic wool. We assume that the trajectory model has yielded a value of the y-coordinate of the limiting particle trajectory, $y_{\ell\mathrm{im}}$, equal to b (see Figures 7.3.25(b) and 6.3-9A). If the void volume fraction of the bed is ε, then $(1 - \varepsilon)$ is the bed volume fraction occupied by strands of the ferromagnetic wool. For a length dz in the feed flow direction of the filter bed of unit cross-sectional area perpendicular to fluid flow, the length of the ferromagnetic wire in the wool is $((1-\varepsilon)\mathrm{d}z/\pi r_w^2)$. As in Figures 7.3.25(b), and 6.3.9A and 6.3.9B, a total width of $2y_{\ell\mathrm{im}} (= 2b)$ in the y-direction and unit length of the wire provides a fluid flow cross-sectional area of $2y_{\ell\mathrm{im}}$ per unit wire length: all particles coming in through this capture cross section are ultimately captured by the stainless steel wires. Therefore the total capture cross section for the filter bed of thickness dz is $(2(1-\varepsilon)y_{\ell\mathrm{im}}\,\mathrm{d}z/\pi r_w^2)$ (assuming that there are no overlaps between the capture cross sections for different location of the wire).

For a velocity v_∞ of the fluid entering the bed and a feed fluid particle number concentration of $n(r_p)\mathrm{d}r_p$ (the total number of particles of size r_p to $r_p + \mathrm{d}r_p$ per unit fluid volume), the number of particles removed per unit time per unit flow cross-sectional area is given by

$$\frac{v_\infty \times n(r_p)|_z\,\mathrm{d}r_p \times 2(1-\varepsilon)\,y_{\ell\mathrm{im}}\,\mathrm{d}z}{\pi r_w^2} = v_\infty\,\varepsilon\,(-\mathrm{d}(n(r_p))\mathrm{d}r_p|_z, \tag{7.3.264a}$$

where the right-hand side provides the corresponding total change in particle flow rate over the distance dz (see the development for equation (6.3.43)). We can rearrange this relation for particles of size r_p as follows:

$$-\mathrm{d}n(r_p)/n(r_p)|_z = \left[2\,y_{\ell im}(1-\varepsilon)/\varepsilon\pi\,r_w^2\right]\mathrm{d}z. \quad (7.3.264b)$$

Integrate from $z = 0$ to $z = L$, the bed length, to obtain

$$\ell n\left[n(r_p)|_{z=0}/n(r_p)|_{z=L}\right] = \left[2y_{\ell\mathrm{im}}(1-\varepsilon)/\varepsilon\pi\,r_w^2\right] L; \quad (7.3.264c)$$

$$\left(n(r_p)|_{z=0}/n(r_p)|_{z=L}\right) = \exp\{\left[2y_{\ell\mathrm{im}}(1-\varepsilon)/\varepsilon\pi\,r_w^2\right] L\}. \quad (7.3.264d)$$

When all particles have the same size r_p,

$$\int_{r_{\min}}^{r_{\max}} n(r_p)|_{z=0}\,\mathrm{d}r_p = \left[\int_{r_{\min}}^{r_{\max}} n(r_p)|_{z=L}\,\mathrm{d}r_p\right] \exp\{\left[2y_{\ell\mathrm{im}}(1-\varepsilon)/\varepsilon\pi r_w^2\right] L\} \quad (7.3.264e)$$

$$\Rightarrow N(r_p)|_{z=0} = N(r_p)|_{z=L}\,\exp\{\left[2y_{\ell\mathrm{im}}(1-\varepsilon)/\varepsilon\pi r_w^2\right] L\} \quad (7.3.264f)$$

$$\Rightarrow N(r_p)|_{z=L} = N(r_p)|_{z=0}\,\exp\{-\left[2y_{\ell\mathrm{im}}(1-\varepsilon)/\varepsilon\pi r_w^2\right] L\}, \quad (7.3.265)$$

where $N(r_p)$ represents the total concentration of particles, all of which have the same size, r_p. If there is a particle size distribution, then we will obtain, instead of (7.3.264e),

$$N(r_{\min}, r_{\max})|_{z=L} = \int_{r_{\min}}^{r_{\max}} n(r_p)|_{z=L}\,\mathrm{d}r_p$$

$$= \int_{r_{\min}}^{r_{\max}} n(r_p)|_{z=0}\exp\{-\left[2y_{\ell\mathrm{im}}(1-\varepsilon)/\varepsilon\pi r_w^2\right] L\}\,\mathrm{d}r_p, \quad (7.3.266)$$

since $y_{\ell\mathrm{im}}$ depends on the particle size r_p.

Some comments are needed to qualify the results developed above. As shown in Figure 7.3.25(a), the magnetic field $\boldsymbol{H}_0$ is perpendicular to the stainless steel wire. However, about one-third of the strand length is parallel to $\boldsymbol{H}_0$. Therefore the argument of the exponential in equation (7.3.265) should be multiplied by 2/3 (Watson, 1973). Further, any change in the bed void volume fraction ε due to particle buildup on the wires is neglected. Ignoring additional assumptions inherent in the above analysis, let us make an estimate of the particle concentration reduction

$$\left(N(r_p)|_{z=L} = N(r_p)|_{z=0}\right)$$

achieved in a bed of length $L = 51$ cm (say) following Liu and Oak (1983). The bed void volume fraction ε is around 0.95; therefore $(1 - \varepsilon) \sim 0.05$. An average wire radius is around 45 μm ($= r_w$). The value of $y_{\ell\mathrm{im}}$ depends very strongly on the ratio $(U_{pzt}^{\mathrm{mag}}/v_\infty)$; the value of this last parameter in practice is generally greater than 0.5 (the minimum value needed for particle buildup (Liu and Oak, 1983)) and can go to quite high values (as much as 100–400). For values of the parameter less than $\sqrt{2}$, we know that $(y_{\ell\mathrm{im}}/r_w) = 0.5(U_{pzt}^{\mathrm{mag}}/v_\infty)$ (equation (7.3.263b)). For higher values of the parameter, $(y_{\ell\mathrm{im}}/r_w)$ keeps on increasing and can be as high as 7 for $(U_{pzt}^{\mathrm{mag}}/v_\infty) = 100$. Let us, for the sake of illustration, take $(U_{pzt}^{\mathrm{mag}}/v_\infty)$ to be 1. Then

$$\left(N(r_p)|_{z=L}/N(r_p)|_{z=0}\right)$$

$$= \exp\left\{-\frac{2 \times 0.5 \times 1 \times r_w \times 0.5 \times 51}{0.95 \times \pi \times r_w^2}\right\}$$

$$= \exp\left\{-\frac{0.05 \times 51}{0.95 \times \pi \times 45 \times 10^{-4}}\right\}$$

$$= \exp\left\{-\frac{5 \times 51 \times 10^2}{0.95 \times \pi \times 45}\right\} = \exp\{-189\}.$$

This result shows that the exiting fluid stream will have very few of the particles left; almost all of them will be captured by the bed. Performances of HGMS units have generally been found to support such conclusions. The influence of the captured particle buildup on the stainless steel wires is important; see Liu and Oak (1983) for a detailed analysis.

7.3.6 Radiation pressure - optical force

It is known (equation (3.1.47)) that radiation pressure from continuous wavelength (cw) visible light is known to apply a force on and accelerate freely suspended particles in the direction of the light. This optical force (radiation pressure force) induced migration of particles, especially colloidal particles, is called *photophoresis* (PP). Such forces have very little impact on larger particles. However, micron and submicron particles can encounter substantial acceleration. Different refractive indices of transparent particles vis-à-vis the surrounding liquid, as well as light absorption, are the bases of this force. Helmbrecht *et al.* (2007) have demonstrated that a simple crossflow setup, in which a laser beam illuminated perpendicular to the mean flow direction of the fluid containing suspended microparticles causes a lateral shift in different particles. Different particles follow different trajectories, as in other cases of bulk flow perpendicular to the direction of the force.

To describe such trajectories, consider the *net photophoretic force* F_{rad} on a particle of radius r_p by a light beam (whose waist size wa is much larger, i.e. $r_p << wa$); here the refractive index of the particle (n_{par}) is larger than that of the surrounding medium (n_{med}) so that the "particle acts like a spherical lens" (Helmbrecht *et al.*, 2007). The net force F_{rad} resulting from the refraction of incident light (a laser beam) which transfers momentum to the particle has been described by Helmbrecht *et al.* (2007) as

$$F_{\text{rad}} = 2\frac{n_{\text{med}}Pr}{\text{c}}\left(\frac{r_p}{wa}\right)^2\eta, \qquad (7.3.267)$$

where Pr is the total power of the laser light, c is the velocity of light, η is a measure of the conversion of incident momentum into particle movement (ranging from 0, no conversion, to 2 for total reflection). This expression is more detailed than (3.1.47).

Any such particle motion will immediately encounter a drag force, which, from Stokes' law, may be written as

$$F_{\text{drag}} = 6\pi\mu r_p U_{px}, \qquad (7.3.268)$$

where the direction coordinate of the photophoretic motion is x, whereas the main bulk fluid is in the z-direction (see, for example, Figure 7.3.22). Soon, a terminal migration velocity U_{pxt} is achieved:

$$F_{\text{drag}} = 6\pi\mu r_p U_{pxt} = F_{\text{rad}} = \frac{2\,n_{\text{med}}\,\text{Pr}}{\text{c}}\left(\frac{r_p}{wa}\right)^2\eta, \qquad (7.3.269)$$

where

$$U_{pxt} = \frac{1}{3}\,\frac{n_{\text{med}}\,Pr}{\pi\,\text{c}\,\mu}\,\frac{r_p}{(wa)^2}\,\eta. \qquad (7.3.270)$$

The magnitudes of such terminal velocities identified as photophoretic velocities are in the range 1 to 80 μm/s for particles having diameters in the range of 1 to 10 μm. However, for particles of comparable diameter, the velocity of silica particles (which were transparent to the laser light used) was six times less than that of melamine particles. Therefore, separation of particles of the same size but having different optical properties appears feasible (Helmbrecht *et al.*, 2007).

There are unresolved complexities in such a continuous separation technique for particles due to the uncertainties in determining the effective power ($= 2Pr\,(r_p/wa)^2$). Also, the incident light has two components, an axial and a radial component, each leading to a separate PP velocity. See Helmbrecht *et al.* (2007) for additional references on the subject.

Problems

7.1.1 You have an *adsorber* packed with an *adsorbent* such that $\varepsilon = 0.40$. However, the adsorbent particles are also porous, with a void volume fraction of $\varepsilon_p = 0.30$, and ρ_s ($= 2.0$ g/cm^3) is the actual density of the material of the solid adsorbent particles.

(1) If the interstitial velocity v_z of the liquid in the packed bed is 53 cm/min, calculate the migration velocity (concentration wave velocity) of species 1 whose equilibrium relation between the solid-phase concentration of solute C_{11} and the mobile/liquid-phase concentration C_{12} is given by $C_{11} = 4C_{12}$. Compare it with the situation where $\varepsilon_p = 0$. (Ans. 7.11 cm/min; 7.57 cm/min.)

(2) Neglect now any ε_p ($= 0$). The adsorber is 12 cm long. If it takes 20 cm^3 of the mobile phase to saturate the column completely, determine the diameter of the column. (Ans. diameter = 0.867 cm.)

(3) Calculate the breakthrough time for solute 1 if the bulk liquid flow rate is 12 cm^3/min under the conditions of part (2). (Ans. 1.65 min.)

(4) Calculate the moles of species i adsorbed from a feed solution of concentration C_{if} from 20 cm^3 of mobile phase needed to saturate the column as in (2). Calculate now the volume of adsorbents needed to remove 99% of the amount adsorbed above in a closed vessel with no flow of mobile phase. Assume $\varepsilon_p = 0$, $C_{i1} = 4C_{i2}$ and $C_{i2} = 0.01C_{if}$; the units of C_{if} are gmol/liter. Compare it with the adsorbent volume in part (2).

7.1.2 (a) An adsorber containing nonporous adsorbent particles is removing a C_{18} compound from a feed solution of heptane. The adsorber is 15 cm long. The interstitial velocity is 30 cm/min. The C_{18}-compound concentration in the feed solution is 0.001 gmol/liter. The corresponding equilibrium concentration in the adsorbent phase is 0.009 gmol/liter. Calculate the breakthrough time for the C_{18} compound assuming nondispersive equilibrium adsorption behavior. If the feed solution comes in at a higher temperature and the equilibrium adsorbent phase concentration at that temperature is 0.006 gmol/liter, calculate the breakthrough time. You are given that the bed void volume fraction = 0.45.

(b) The adsorbent column diameter in (a) is 1 cm. Calculate the volume of feed solution which will completely saturate the column in the absence of any dispersion at the lower temperature of operation.

(c) Determine the solute migration velocity through such a column if the adsorbent particles are porous with a porosity of 0.30. Use the lower temperature of operation.

7.1.3 (a) A low molecular weight pharmaceutical compound is in an aqueous solution containing a low molecular weight impurity which is to be removed by adsorption on activated carbon. Experiments were done in the laboratory with this aqueous solution at 20 °C in a 45 cm long column of 2 cm diameter. The adsorbent properties are: $\varepsilon = 0.45$; $\varepsilon_p = 0.6$; $\rho_f = 1$ g/cm^3; $\rho_s = 2$ g/cm^3; $\kappa_{im} = 1$. After passing 6900 cm^3 of the aqueous solution in the laboratory over a time of 90 minutes, the low molecular weight impurity broke through and appeared at the column outlet. In the pilot plant you want to have a fatter and much longer column (150 cm long) fed with the superficial velocity used in the laboratory column. The adsorbent properties are: $\varepsilon = 0.45$; $\varepsilon_p = 0.5$; $\rho_f = 1\,\text{g/cm}^3$; $\rho_s = 2\,\text{g/cm}^3$; $\kappa_{im} = 1$. Assume that the equilibrium nondispersive adsorption model is valid for the laboratory as well as the pilot plant adsorbers. Further assume a linear adsorption isotherm. Calculate the breakthrough time from the equilibrium nondispersive operation model for the pilot plant column.

(b) Solve the above problem if $\varepsilon_p = 0.6$ in the laboratory column as well as in the pilot plant.

7.1.4 A Pd-containing organo-metallic catalyst soluble in an organic reaction mixture is to be removed by adsorption on QuadraPure TU resin beads packed in a column of length 20 cm, bulk packing density $0.5\,\text{g/cm}^3$, cross-sectional area $1.69\,\text{cm}^2$, bed void volume fraction 0.4 and interstitial liquid velocity of 10.2 cm/hr. The liquid feed solution density is $1\,\text{g/cm}^3$. For a molar feed solution Pd concentration of $570\,\text{ppm}/M_{\text{Pd}}$, where M_{Pd} is the molecular weight of Pd and ppm means (parts of metal in grams per gram of liquid) $\times\, 10^6$, the equilibrium solid adsorbent-phase molar Pd concentration is $1.8 \times 10^4\,\text{ppm}/M_{\text{Pd}}$. where ppm means (parts of metal in grams per gram of solid adsorbent material) $\times\, 10^6$.

(1) Calculate the time for breakthrough of the Pd compound.

(2) Calculate the volume of solution that will pass through before the breakthrough happens. Use the equilibrium nondispersive adsorption model. Assume a linear adsorption equilibrium model for the system.

7.1.5 (a) A particular protein follows the Langmuir adsorption isotherm on an ion exchange resin (equation (4.1.77d)): $K_d = 0.12$ g/liter; $C_{s1}^m = 0.11$ g / liter. *Adsorption* and *elution* of the protein using this resin is to be carried out in a resin bed ($\varepsilon = 0.40$) 15 cm long having a cross-sectional area of $3\,\text{cm}^2$. The feed flow rate is $1.5\,\text{cm}^3$/min. Calculate the breakthrough volume $\overline{V}$, the time $\bar{t}$ for breakthrough and the shock velocity $v_{Ci}^*|_{shock}$ for adsorption using two different feed solutions:

(1) feed protein concentration, 1.5 g/liter;

(2) feed protein concentration, 0.15 g/liter.

Compare the two results and comment.

(b) Elution of the protein adsorbed on the resin particles described in part (a) is to be carried out using an eluent for which the Langmuir isotherm holds with the same parameters. The fresh eluent has zero protein concentration. Elution is initiated when the bed is completely saturated. Using nondispersive equilibrium model derived expressions, calculate the volume of solution that you have to pass, including the volume needed to saturate the column at the following protein concentrations at the outlet of the column for a feed concentration of 0.15 g/liter: 0.13, 0.1, 0.07, 0.04, 0.01, 0.0 g/liter. The column dimensions, eluent flow rate and packed-bed properties are identical to those in part (a). Calculate also the corresponding time for each concentration at the column outlet. (*Note*: Eluent flow rate = feed flow rate.)

7.1.6 A solution of praseodymium chloride at $pH = 3.0$ citrate buffer is being passed continuously through a Dowex 50 resin column at a constant inlet concentration. The column was originally empty. Lightfoot *et al.* (1962) have shown that equation (7.1.18h) describes the breakthrough behavior of the effluent concentration of this solute effectively. Calculate the *nondimensional solute concentration profile* θ_i for this example as a function of an appropriate nondimensional V if the following information is available: $Pe_{z,\text{eff}} = 409$; the value of a nondimensional $\overline{V}$, $\overline{V}^+$ $(= \overline{V}/\varepsilon L S_c)$, is 113. Assume further that you have a long column so that the variation of V around $\overline{V}$ is small enough for you to assume $(V\,\overline{V})^{1/2} \cong \overline{V}$.

7.1.7 This problem involves the determination of the length of an adsorbent bed of silica gel being used at 25 °C and 2 atm pressure to adsorb benzene from air. The following data are taken from an example in Flagan and Seinfeld (1988). The feed air benzene mass fraction is 0.025 kg benzene/kg air. The silica gel bed bulk density is $625\,\text{kg/m}^3$. The superficial velocity of the benzene-containing air is 1 m/s. The bed cycle time is 180 min: 90 minutes for adsorption, 90 minutes for regeneration. The nonlinear adsorption isotherm for benzene is represented as

$$u_{\text{benzene, air}}\left(\frac{\text{kg benzene}}{\text{kg air}}\right) = 0.167\left(u_{\text{benzene, gel}}\left(\frac{\text{kg benzene}}{\text{kg gel}}\right)\right)^{1.5}.$$

What is the length of the silica gel bed needed if benzene breaks through as a square wave at 90 minutes? Employ an equilibrium nondispersive mode of operation of the bed being saturated by a trace gas. If the length of the mass-transfer zone is known from Flagan and Seinfeld (1988) to be 0.42 m, what would be the required bed length for a diffuse breakthrough?

7.1.8 A refinery gas stream at 10.13×10^5 Pa contains primarily H_2 and another trace impurity species A. The impurity A adsorbs onto the porous adsorbent used according to q_{A1} (mol/g adsorbent) $= 41C_{A2}$ (mol/cm^3) at the gas stream temperature. To provide highly purified H_2 at a pressure close to 10.13×10^5 Pa, *a two-bed PSA process* is proposed using the above adsorbent in a bed 105 cm long. The low pressure, P_L, reached during blowdown is 1.013×10^5 Pa. The durations of different steps in the Four-step PSA process (See Figure 7.1.13(a)) are: pressurization, 3 s; high-pressure feed flow, 80 s; blowdown, 3 s; low-pressure purge, 80 s. The diameter of each bed is 8 cm; the high-pressure refinery gas stream flow rate obtained via a slip stream is 2.0 liter/s. Determine whether this process can provide an essentially pure H_2 product at a pressure very close to 10.13×10^5 Pa employing an equilibrium nondispersive model. You are given that $\rho_s = 2.1\ \text{g/cm}^3; \varepsilon_p = 0.31; \varepsilon = 0.41$.

7.1.9 (a) For the process of *pressure-swing adsorption of a nontrace binary gas mixture*, Rege and Yang (2001) have defined a dimensionless sorbent selection parameter S based on the Langmuir isotherm parameters $(q_{A\sigma})_{max} b_{A\ell}$, $(q_{B\sigma})_{max}\, b_{B\ell}$ for species A and B, respectively (see equation (3.3.112 b)):

$$S = \left(\frac{(q_{A\sigma})_{max} b_{A\ell}}{(q_{B\sigma})_{max} b_{B\ell}}\right) \frac{\Delta q_{A\sigma}}{\Delta q_{B\sigma}}.$$

Assume linear isotherm behavior corresponding to the lower-pressure region of the Langmuir isotherm. Show that this sorbent selection parameter S, based on the product of the selectivity ratio $((q_{A\sigma})_{max} b_{A\ell}/\ (q_{B\sigma})_{max}\, b_{B\ell})$ and the ratio of the working capacity of the two species A and B (working capacity is the difference between the adsorbed amounts at adsorption (high) pressure P_H and at desorption (low) pressure P_L, preferably with the gas mixture) ($(\Delta q_{A\sigma})/(\Delta q_{B\sigma})$, may be described as follows:

$$S = \left[\frac{(q_{A\sigma})_{max} b_{A\ell}}{(q_{B\sigma})_{max} b_{B\ell}}\right]^2 \frac{y_{AH}}{y_{BH}} \left[\frac{1 - (P_L/P_H)^{(q_{A\sigma})_{max} b_{A\ell}/(q_{B\sigma})_{max} b_{B\ell}}}{1 - (P_L/P_H)^{\frac{(q_{B\sigma})_{max} b_{B\ell}}{(q_{A\sigma})_{max} b_{A\ell}}}}\right] \quad \text{for } P_L \ll P_H,$$

provided relation (7.1.54a), $(y_H/y_L) = (P_H/P_L)^{\beta-1}$, may be assumed valid for each species with appropriate changes for β. (*Note*: Relation (7.1.54a) was developed for a trace system; therefore it is a drastic assumption).

(b) Rege and Yang (2001) have provided the data given in Table 7.P.1 for choosing between two different zeolite adsorbents for PSA separation of N_2 and O_2 from air. Calculate the value of the parameter S for each of these two zeolites from the expression developed in (a); however, neglect the term in brackets containing (P_L/P_H). The feed gas contained 22% O_2, 78% N_2. The strongly adsorbed species A was N_2. Compare with the values of the parameter S obtained by Rege and Yang (2001) from a detailed computational effort over the range (P_H/P_L) from 2 to 10: LiX, 201 to 255; NaX, 110 to 132. In this problem, feed air contained Ar as an impurity, which is incorporated in the O_2 volume percent.

7.1.10 Consider the *equilibrium isothermal nondispersive PSA, separation* of a *binary gas mixture* of A and B employing the cycle shown in Figure 7.1.13(a). Let the binary feed have an arbitrary composition of species A denoted by y, unlike the very dilute mixture of A in B treated in Chan *et al.* (1981). Knaebel and Hill (1982) have developed an analysis similar to that of Chan *et al.* (1981) when $k_A > k_B$. Show that the two key equations corresponding to equations (7.1.51a) and (7.1.51b) are, in this case,

$$\frac{dz}{dt} = \frac{\beta_A\, v_z}{1 + (\beta - 1)y}$$

Table 7.P.1.

	N_2		O_2	
	$(q_{A\sigma})_{max}$ (mmol/g)	$b_{A\ell}$ (atm^{-1})	$(q_{B\sigma})_{max}$ (mmol/g)	$b_{B\ell}$ (atm^{-1})
LiX(Si/Al = 1)	2.653	0.946	2.544	0.086
NaX	0.982	0.901	0.276	0.624

Table 7.P.2.

$\overline{C}_{i2}$, enzyme concentration in eluent exiting the column (in suitable units)	Volume of eluent passed (in liter), V
15	2.0
6.5	2.2

and

$$\frac{dy}{dP} = \frac{(\beta - 1)(1 - y)y}{[1 + (\beta - 1)y]\,P}$$

where $\beta = (\beta_A/\beta_B)$; β_A and β_B are defined in equation (7.1.52a).

(*Suggestion*: Work with the mass balance equations for species A and total moles (i.e. A plus B); then obtain the following equation:

$$\frac{\partial y}{\partial t} + \frac{\beta_A\, v_z}{1 + (\beta - 1)y}\frac{\partial y}{\partial z} = \frac{(\beta - 1)(1 - y)y}{1 + (\beta - 1)y}\,\frac{1}{P}\,\frac{dP}{dt}$$

and employ the method of characteristics.)

7.1.11 For *liquid-solid adsorption based chromatography*, two separate expressions were developed for the chromatographic output profile $\overline{C}_{i2}(z,t)$ by two methods: See (7.1.98b) and (7.1.102a). Show that these two expressions are identical.

7.1.12 *Elution based purification* of an enzyme in an adsorbent column yielded significant data, some of which are listed in Table 7.P.2.

It is known that the $\overline{C}_{i2}$ value of 15.0 was the maximum. Determine the value of σ_{iV} (liter) and the percent recovery of enzyme when the eluent volume passed is 2.2 liter. The elution profile may be assumed to be Gaussian.

7.1.13 The *elution profile* of praseodymium chloride ($PrCl_3$) by a citrate buffer at $pH = 3$ from a Dowex 50 resin column (originally studied by Mayer and Tompkins (1947)) can be described by equation (7.1.102a) (Lightfoot *et al.* (1962) have described it in detail) in the presence of cerium ($CeCl_3$); the sorption equilibria are linear and noninterfering.

(1) From an elution profile of praseodymium, show how you can determine the value κ_{i1} and $Pe_{z,\mathrm{eff}}$ for praseodymium for the given device and system.

(2) Calculate the values of κ_{i1} and $Pe_{z,\mathrm{eff}}$ for praseodymium when you have the following experimental values available: the maximum effluent concentration of $C_{i2} = 0.0505\,(C_{i2}^0\, V^0/\varepsilon L\, S_c)$ is reached at an elution volume $V = 113\varepsilon\, L\, S_c$; $\varepsilon = 0.4$.

(3) Calculate the plate number.

(Ans. (2) $\kappa_{i1} = 74.6$; $Pe_{z,\mathrm{eff}} = 409$; (3) $N = 204.5$.)

7.1.14 In the recovery of a product from a clarified fermentation broth, the adsorption time needed is 80 minutes. In a given cycle, the individual steps of washing, elution and regeneration (see Section 7.1.6) require the same amount of time. The data were acquired in a small laboratory column having a flow rate of $x\,\mathrm{cm}^3/\mathrm{min}$. **This process has to be scaled up for ten times higher production rate of the bioproduct** using the same overall cycle time and individual step times. Consider three cases:

(1) same pressure drop in both columns;

(2) same σ_{iV} in both columns during elution;

(3) same plate number.

Determine for each case a relation between the lengths and diameters of the two columns.

Assume that the adsorbent particle diameter remains unchanged. The value of $D_{i,\mathrm{eff}}$ may be assumed to be the same in the two columns.

7.1.15 Dai *et al.* (2003) obtained the following linear isotherm behavior for the two proteins myoglobin and α-lactalbumin (α-LA) with the ion exchange resin DEAE-Sepharose fast flow, at low concentrations in 20 mM Tris-HCL buffer solution ($pH = 8.5$):

$$\begin{aligned} q_{i1}(\mathrm{mg/mliter}) &= K_i C_{i2}(\mathrm{mg/mliter}); \\ K_i(\mathrm{myoglobin}) &= 43.3;\ K_i(\alpha\text{-lactalbumin}) = 31\,099. \end{aligned}$$

Myoglobin elution was easily accomplished using 0.05 M NaCl, whereas α-LA elution needed 0.5 M NaCl solution; α-LA was quite tightly bound. which elution procedure, gradient or stepwise, would you recommend if it is known that small amounts of α-LA start coming out at salt concentrations less than 0.5 M?

7.1.16 *Size exclusion chromatographic separation* of a protein mixture from salt, as well as the separation of two proteins present, in the mixture, are being carried out in a column of diameter 3 cm and length 30 cm. The column void volume fraction is 0.39. The gel particle void volume fraction is 0.6. The values of the partition coefficient κ_{im} of the two proteins 1 and 2 are $\kappa_{im}|_1 = 0.8$, $\kappa_{im}|_2 = 0.1$, with protein 2 being much larger than protein 1.

(1) Estimate the values of the elution volumes for the salt, protein 1 and protein 2.
(2) What would be the corresponding value for blue dextran if it cannot access the pores in the gel particle at all?
(3) If the pore diameter of the gel is estimated to be around 20 nm, estimate the diameters of protein 1 and protein 2, assuming that they are spherical and the pores are cylindrical.

7.1.17 Many organic synthesis reactions employ metal-catalyzed processes. As a result, the solution containing the reaction products or waste streams from the synthesis process will contain *residual metals* in solution. Such residual metals have to be removed. In one organic synthesis process, the organic solution/aqueous waste stream contained residual metals such as Cu, Al and Ni. It is proposed to remove these metals via functionalized macroporous polystyrene resin beads in a packed-bed format. The beads are crosslinked enough to undergo only limited swelling in organic solvents. Suggest appropriate functional groups in the resin to remove such metals in solutions.

7.1.18 Consider *capillary electrochromatography based separation* of two enantiomers identified as Q_1 and Q_2 respectively. A chiral complexing agent (C) is bonded to the stationary phase in the capillary. The enantiomers Q_1 and Q_2 in the mobile phase (subscript m) react differently with the stationary phase complexing agent (C with subscripts):

$$Q_{1m} + C_s \overset{K_{f1}}{\leftrightarrow} [Q_1C]_s, \qquad K_{f1} = \frac{[Q_1C]_s}{Q_{1m}C_s};$$

$$Q_{2m} + C_s \overset{K_{f2}}{\leftrightarrow} [Q_2C]_s, \qquad K_{f2} = \frac{[Q_2C]_s}{Q_{2m}C_s}.$$

The enantiomers also partition between the mobile phase and the stationary phase in the free form:

$$Q_{1m} \overset{K_1}{\leftrightarrow} Q_{1s}; \qquad Q_{2m} \overset{K_2}{\leftrightarrow} Q_{2s}.$$

Assume that $K_1 = K_2$, $\mu_1^m = \mu_2^m$ for the enantiomeric pair 1 and 2. Develop an expression for $\alpha_{12}|_{t_{R_1}}$, the separation factor based on the retention time t_{R_1}. It is known that one enantiomer forms the complex $[QC]_s$ much more readily than the other one, so that K_{f1} and K_{f2} are far apart. Further, the mobile-phase volume and stationary-phase volume in the column are V_m and V_s respectively. Assume that the C_s concentration is essentially very high and unaffected by the reactions. (Ans.

$$\alpha_{12}|_{t_{R_1}} = (V_M/V_s)(K_{f1} - K_{f2}) \Big/ \left[1 + \frac{V_m}{V_s}(K_{f1} + K_1)\right].)$$

7.1.19 In problem 7.1.18, there was a stationary phase containing a chiral complexing agent C which reacted differently with the different enantiomers. Suppose in a column in which the buffer solution flowing vertically has a chiral selector molecule C (sulfated β-cyclodextrin chiral selector) added as the two enantiomers flow with the buffer. The chiral selector binds to form enantiomer–ligand complexes with each enantiomer. However, these complexes have different electrophoretic mobilities. If there are two electrodes at the two ends of the column as shown in Figure 7.P.1, then, by selecting an appropriate hydrodynamic velocity in the counterflow at the column bottom, one could separate the fast enantiomer-ligand complex (mobility μ_{iF}^m) from the slow enantiomer–ligand complex (mobility μ_{iS}^m).

(1) Write down the electrophoretic velocities of the two complexes as functions of their ionic mobilities.
(2) Obtain the hydrodynamic velocities in the four sections of the column shown. The column cross-sectional area is S_c. The various volumetric flow rates entering or leaving the column are shown in Figure 7.P.1.

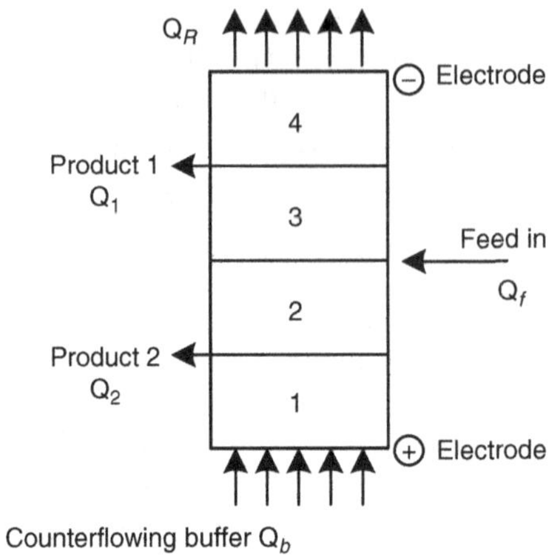

Figure 7.P.1.

(3) If both enentiomer–ligand complexes have negative charges, identify which of the two product streams will have the faster complex. Identify the criteria vis-à-vis the hydrodynamic velocities in a given section. Neglect competitive binding aspects. Assume uniform electrical field gradient.

7.2.1 Air is being separated into nitrogen-enriched air (NEA) and oxygen-enriched air (OEA) in a *crossflow membrane separator* (See Figure 7.2.1(a)). The permeate side total pressure is very low so that the pressure ratio $\gamma \rightarrow 0$. The nitrogen mole fraction in the NEA is 0.96. Calculate the highest and the lowest oxygen mole fractions achieved in the permeate at different locations in such a permeator for two membranes, one from cellulose acetate ($\alpha^*_{O_2-N_2} = 6$) and the other from an unknown glassy polymer ($\alpha^*_{O_2-N_2} = 9$).

7.2.2 To remove the acid gas CO_2 from natural gas at 500 psia, a *polymeric membrane module* is being used. The permeate side pressure may be assumed to be quite low. The feed gas has 10% CO_2. The purified natural gas should have only 2% CO_2. The permeability values (in barrer) for CO_2 and the dominant consitutent of natural gas CH_4, are 200 and 5 units, respectively. Calculate the highest and lowest values of the CO_2 mole fraction in the permeate side if you can assume pure crossflow in the module. Identify the locations. What will be the values of CH_4 in the permeate at these locations? Assume: a binary CO_2–CH_4 system.

7.2.3 In *crossflow gas permeation*, we have identified in Figure 7.2.1(c) a tube-side feed crossflow hollow fiber module. Unlike flat membrane systems, the radial cross-sectional area for permeation through the hollow fiber wall (inner radius r_1, outer radius r_2) increases as the radius increases. Develop the following expression for the molar permeation rate of species i through a hollow fiber of length Δz:

$$(\Delta z)\pi d_{\ell m}\left(\frac{Q_{im}}{\delta_m}\right)(p_{i1} - p_{i2}),$$

where

$$d_{\ell m} = \frac{(2r_2 - 2r_1)}{\ln\left(\frac{2r_2}{2r_1}\right)}, \qquad Q_{im} = D_{im}\,S_{im}, \qquad \delta_m = (r_2 - r_1).$$

You are given: that $C_{im}|_{r_1} = S_{im}|_{r_1}\,p_{i1}$; $C_{im}|_{r_2} = S_{im}|_{r_2}p_{i2}$; $S_{im} = S_{im}|_{r_1} = S_{im}|_{r_2}$. Assume ideal gaseous mixture behavior. Employ $N_{ir} = J^*_{ir}$ from Table 3.1.3C; neglect U_i; the partial pressure of species i in the feed gas at radius r_1 is p_{i1}; the partial pressure of species i in the permeate gas at radius r_2 in crossflow is p_{i2}; $d_{\ell m}$ is the logarithmic mean diameter of the hollow fiber membrane.

7.2.4 This problem is concerned with removing small amounts of solvent vapor (volatile organic compounds, VOCs) from an air stream through a hollow fiber membrane which is *highly selective for the VOC* over air. The feed air stream moves through the bore of the hollow fiber, as in Figure 7.2.1(c). The permeate side in crossflow is maintained under high vacuum such that $p_{if} >> p_{ip}$, where subscripts f and p refer to the feed and the permeate, respectively. The change in the total feed gas flow rate with the permeator length z may be neglected. Employing the notation of Figure 7.2.1(a), develop an analytical relation between the length L of the permeator needed to reduce the VOC mole fraction in air from x_{if} (feed) to x_{i2} (concentrate, reject). Assume a negligible feed gas pressure drop along the length of the hollow fiber module. Employ the result of Problem 7.2.3.

You are given that (Q_{im}/δ_m) for VOC, $i = \text{a}\exp(\text{b}P_f x_{if})$, i.e. the VOC permeability varies exponentially with the feed-gas mole fraction of VOC.

7.2.5 For a *crossflow gas permeator* and a binary system, adopt the Naylor-Backer approach and the starting equation (7.2.10). Define W_{t2} and x_{A2} as the molar gas flow rate and mole fraction of species A at any location in the feed side of the permeator; let the values of W_{t2} and x_{A2} at the permeator exit be W_{t2}^o and x_{A2}^o. Develop a relation somewhat similar to the result (7.2.12) relating W_{t2} to W_{t2}^o as a function of x_{A2}, x_{A2}^o and α_{AB}, where the more permeable species is A in the binary mixture of species A and B.

7.2.6 Consider the *spiral-wound module* for *reverse osmosis desalination* described in Example 7.2.3. Determine the fractional water recovery if the feed brine has 10 000 ppm salt. All other conditions are as in the Example 7.2.3. What will be the fractional recovery if the feed brine has 20 000 ppm salt?

7.2.7 Calculate the value of the average permeate salt concentration for the two cases of spiral-wound RO desalination given in Problem 7.2.6: feed brine has 10 000 ppm salt; feed brine has 20 000 ppm salt. Make comments about the possible errors in calculation. You are given that the salt permeation parameter $(D_{im}\,\kappa_{im}/\,\delta_m) = 2.4 \times 10^{-5}$ cm/s.

7.2.8 Consider reverse osmosis desalination in a spiral-wound module for a feed having negligible osmotic pressure. If the fractional water recovery, *re*, is such that the osmotic pressure of the concentration from the reverse osmosis process still has a negligible osmotic pressure vis-à-vis the liquid pressure, derive the result (7.2.44), i.e. $L^+ = re$, from simple mass balance considerations. Assume that the membrane has a very high rejection of salt.

7.2.9 The Colorado River has become brackish at the level of 600 ppm. This water is passed at 34 atm (gauge) through a spiral-wound module. The gap between the membranes lining the brine feed channel is 1.1 mm. The membrane length in the brine flow direction is 70 cm. The length of the unwrapped membrane in the permeate flow direction is 150 cm. The pure water permeability constant for the membrane exposed to the feed brine at 20° C is 66.15×10^{-6} gmol/cm^2-min-atm. The osmotic pressure of the brine solution is provided via $\pi_f = 0.0115$ $(\text{ppm})_f$, where π_f is in psi and the salt concentration is in ppm. The average brine velocity in the brine channel is 320 cm/min. Estimate the fractional water recovery by assuming a highly rejecting membrane and any other appropriate assumption. (*Note*: You do not have any method to estimate the value of $k_{i\ell}$ for this spiral-wound module.)

7.2.10 Milk is to be concentrated by *ultrafiltration* in a hollow fiber module having the following dimensions and characteristics: number of hollow fibers, 500; fiber internal diameter, 800 μm; fiber length, 60 cm. The milk will flow on the tube side at an average flow velocity of 90 cm/s. The relevant properties of milk are: viscosity, 0.8 cp; density, 1.02 g/cm^3; diffusion coefficient of proteins in milk, 6.5×10^{-7} cm^2/s; milk protein contents, 2.8% w/v; gel concentration C_{igel}, 25% w/v.

(1) Determine the value of the water flux through this module using the length-averaged Leveque solution. (Ans. 25.3 liter/m^2-hr).

(2) Determine the fractional feed water removal through this module. Justify the use of the form of Leveque solution recommended for use. (Ans. 0.0235.)

7.2.11 Consider a protein solution having the characteristics of the feed solution in Examples 7.2.5 and 7.2.6. This solution has to be *desalted* via *continuous diafiltration* (Section 6.4.2.1) in the hollow fiber unit of Example 7.2.6. The diafiltration configuration is illustrated in Figure 7.2.5(e). For 2000 liters of feed protein solution to be desalted, determine the crossflow membrane area required if the following levels of salt removal are desired in 2 hours: (a) 99% of the salt removed; (b) 99.9% of the salt removed; (c) 99.99% of the salt removed. It is known that salt has zero rejection. Specify the volume of buffer solution required in each case.

7.2.12 Consider Example 7.2.5 dealing with the *concentration* of a *dilute solution of BSA*. The diffusion coefficient of BSA has been found to vary as

$$D_{i\ell} = D_{i\ell}^0\,\frac{\tanh(0.159\,(\rho_{\text{solu}}\,\times\,100))}{0.159\,(\rho_{\text{solu}}\,\times\,100)}.$$

It is known further that the Leveque solution (see Example 7.2.6) is valid here. If all other conditions in the problem are similar to those of Example 7.2.6, determine the fractional reduction in solvent flux as the BSA solution is concentrated from 2 g/100 cm^3 to 15 g/100 cm^3.

7.2.13 For the UF based protein concentration example provided in Example 7.2.5, develop a graphical illustration of $(v_z(z)|_{\text{avg}}/v_{s0})$ against C^+_{isbL} as per equation (7.2.86).

7.2.14 Consider the *batch ultrafiltration configuration* of Figure 7.2.5(d). We would like to determine the time required to go from an initial volume V_{f0} of the solution to the final volume V_{fe} ($<V_{f0}$) for the following cases. The feed concentration of the protein C_{if} in the batch solution varies from the initial value C^0_{if} to the final value C^e_{if}.

(1) The value of C_{if} is small enough and the protein molecular weight large enough to assume that the osmotic pressure of the feed solution is negligible with respect to ΔP; further, there is essentially no concentration polarization in the membrane device. Determine an expression for the time required. Assume $C_{ip} = 0$.

(2) Unlike that in (1), the osmotic pressure is not negligible; however, the concentration polarization *is* still negligible. Determine an expression for the time required. Assume that $C_{ip} = 0$ and that van 't Hoff's law (3.3.86b) holds for osmotic pressure.

(3) Gel polarization exists in the membrane module. Deliberate on how you would approach the problem. The solution for $V(t)$ as a function of time is not needed.

You are given that the membrane area = A_m.

7.2.15 *Ultrafiltration of latex solutions,* (e.g., PVC (polyvinyl chloride) latex) may be carried out in a tubular device containing membrane tubes of 2.54 cm I.D. The solvent flux v_{s0} was found in a few experiments to be proportional to $Q_f^{b_1}$, where Q_f is the volumetric latex flow rate through the tube and $b_1 > 1.0$. Assume that the Sherwood number $(k_{i\ell o}\, d_i/D_{i\ell})$ for the mass-transfer process is such that $Sh = (k_{i\ell o} d_i/D_{i\ell}) \propto (Re)^{b_1}$.

(1) Show that, if gel polarization exists,

$$v_{s0} \propto v_{z,\text{avg}}^{b_1}\, d_i^{b_1 - 1},$$

where $v_{z,\text{avg}}$ is the bulk average velocity and d_i is the tube internal diameter.

(2) Define e_p, a measure of power consumption, as

$$e_p = \frac{\text{power dissipated in circulating fluid}}{\text{volume rate of permeate production}} = \frac{\Delta P_{\text{loss}}\, Q_f}{v_{s0}\, A_m}.$$

Employ the Blasius relation (6.1.3a,b) for the pressure drop in turbulent flow through a smooth pipe and assume A_m is the tubular membrane surface area. If we require equal flux performance from tubes of diameters d_1 and d_2, show that

$$(e_{p_1}/e_{p_2}) = (d_1/d_2)^{(-3b_1+2.75)/b_1}.$$

7.2.16 It has been observed that particle flux expressions developed based on shear-induced particle diffusivity describe the observed solvent flux through a *microfiltration membrane* much better than those based on Brownian diffusivity of a particle. For the following system properties, determine the ratio of the solvent fluxes based on shear-induced particle diffusivity and Brownian diffusivity. You should employ the particle volume fraction based solvent flux expression based on the gel polarization model used in ultrafiltration (equation (7.2.72)):

$$v_s = (D_p/\delta_p)\,\ell n\,(\phi_w/\phi_s).$$

You are given the following: $\phi_s = \phi_p = 0.1$, $r_p = 10^{-4}$ cm, $\dot{\gamma} = 10^3\,\text{s}^{-1}$, $\mu = 10^{-2}$ g/cm-s, $k^B = 1.380 \times 10^{-16}$g-cm^2/s^2-K, temperature = 25 °C.

7.2.17 In a *ceramic tubular microfilter* of diameter 0.2 cm and length 30 cm, a suspension having a particle volume fraction of 0.1 ($= \phi_s$) and 1 μm particle radius is flowing at 25 °C. The suspension viscosity μ_0 is 1 cp ($= 10^{-2}$ g/cm-s); the wall shear stress is 10 g/cm-s^2. The clean membrane resistance is $R_m = 10^9$/cm. The applied pressure difference is 10 psi (7×10^5 g/cm-s^2). The filter cake porosity $\varepsilon_c = 0.4$.

(1) Determine whether the cake layer resistance controls or the membrane resistance controls.

(2) Obtain an estimate of z_{crit} and determine the ratio (L/z_{crit}).

7.2.18 In a ceramic tubular *microfilter* of radius 0.13 cm and length 37 cm, a suspension having particles of radius 0.32 μm flows; the value of $\phi_s = 0.00015$. The wall shear rate at inlet $\dot{\gamma}_0 = 2400\,s^{-1}$. The observed flux in the microfilter at the inlet, v_{s0}, is 0.0034 cm/s. It is known that $\beta >> 1$. Calculate the value of $v_s|_{avg}$ for this microfilter. The values given above are valid for 25 °C.

7.2.19 Many aqueous solutions used as a feed for separations contain a mixture of particulate matter such as cells, cell fragments, macromolecules and low molecular weight solutes. These are found in the food industry, the biopharmaceutical industry, etc. **One could use a combination of reverse osmosis (RO) membrane devices, ultrafiltration (UF) units and microfiltration (MF) membrane modules**. What order of feed solution processing should you follow: RO → UF → MF or MF → UF → RO? Why?

7.2.20 Employing the correct **sequence of** *crossflow membrane processes* from Problem 7.2.19, suggest a solution for the following processing problems.

(1) Consider the fractionation of a mixture of the following proteins and other solutes: aldolase (MW, 142 000), ovalbumin (MW, 45 000), cytochrome C (MW, 12 400), cyanocobalamin (MW, 1355). Draw an appropriate sequence of crossflow membrane processes. For ultrafiltration based processes, identify an appropriate membrane for the particular separation from Table 6.3.8.

(2) Animal blood from slaughterhouses can be fractionated to recover constituent components: red blood cells (0.8 μm), bacteria (> 0.2 μm), virus (MW, 2–3 × 10^6), serum albumins (MW, 60 000 to 100 000). Draw an appropriate sequence of crossflow membrane processes. Identify the membrane pore size or MWCO selected in each process. Specify which species is concentrated in the reject stream and which species is primarily in the permeate.

7.2.21 *Rotary vacuum filtration* of a fermentation broth is to be carried out at the rate of 5000 liter/hr. The properties of the dilute suspension, e.g. ϕ_s, ρ_s, μ and the cake properties, ϕ_c,$\hat{R}_{c\delta}$ are provided in Example 6.3.7. The vacuum based applied ΔP is 55 cm Hg. Assume an incompressible cake whose resistance dominates filtration resistance. The filter cycle time is 80 s; the filtration time duration is 20 s. Determine the total filter area A_m of the rotary filter needed for this operation. The temperature of operation is the same as that in Example 6.3.7 so that the physical properties are identical.

7.2.22 Determine the value of the normalized volume concentration of particles $\left(C_{vp}/C_{vp}^i\right)$ at 3 hours from the start at bed locations at a distance $z = 30$ and 70 cm, respectively, from the inlet of a *granular filter*. You have been provided the following information about the bed characteristics and operating conditions: $\varepsilon_0 = 0.50$; $C_{vp}^i = 100 \times 10^{-6}\,cm^3/cm^3$; $\lambda_0 = 0.1\,cm^{-1}$; $v_z = 0.28\,cm/s$; $a = 50$ (Ornatski parameter).

7.2.23 In *deep-bed filtration*, very fine particles of uniform radius r_p in suspension in water are deposited on the surface of the filter bed material as the water flows down. The porous medium of filter bed may be modeled as a bundle of straight capillaries of radius r_c and length L through which water flows vertically downward at a velocity $v_z(r)$, considered parabolic. Figure 7.P.2 shows the limiting trajectory of a particle which enters the capillary ($z = 0$) at a radius r_c^{cr} such that it is deposited on the capillary wall at $z = L$. Particles entering at a smaller r are not captured.

(1) Obtain an integral expression for the value of G_r, the grade efficiency of this capillary, and E_T, the overall efficiency, assuming that the particle concentration entering the capillary is uniform across the capillary radius.

(2) If the rate of particle deposition per unit volume of the bed is first order with respect to the volume concentration of particles C_{vp} in water ($= k_1 C_{vp}$, where k_1 is the rate constant), develop a simple first-order differential equation for C_{vp} in the capillary by mass balance in a differential control volume at location z. Use $v_z|_{avg}$.

(3) Solve this equation for C_{vp} and related C_{vp} at $z = L$ to C_{vp}^i at $z = 0$. Using these two values of C_{vp}, develop an estimate of G_r in terms of L, $v_z|_{avg}$ and k_1.

7.3.1 For typical proteins such as albumin being separated by *continuous free-flow electrophoresis* in a rectangular device (Figure 7.3.1) the following information is available: $2b = 4$ mm; $L = 30$ cm; $v_{max} = 1$ cm/s; $E = 30$ volt/cm; $\mu_i^m = 1.6 \times 10^{-4}\,cm^2/s-volt$; $D_{is} = 6 \times 10^{-7}\,cm^2/s$.

Calculate $(x_{iL} - x_0)$, σ_i, collection port width and $x_{iL} - x_{jL}$ (if $\mu_j^m = 1.1 \times \mu_i^m$ and $D_{js} \cong D_{is}$). Comment on whether it is possible to separate species i from species j.

Notes: (1) Average axial velocity is equal to $(2/3)v_{max}$ for parabolic velocity profile; (2) assume plug flow and neglect electroosmotic velocity.

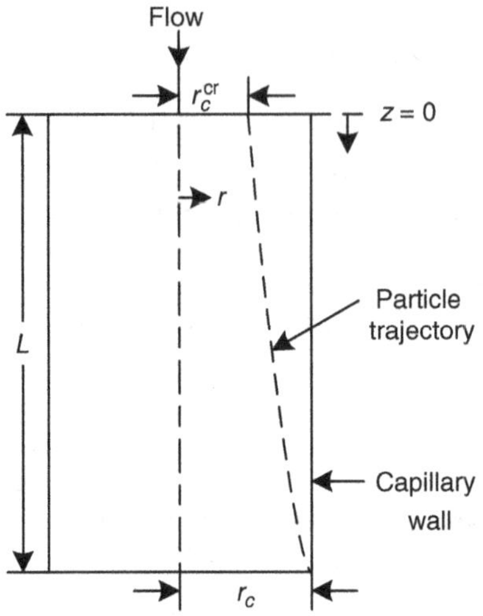

Figure 7.P.2.

7.3.2 A *cylindrical electrophoresis column* with electrodes at the center radius (r_b) and at the outer circumference (radius r_a) with lengthwise flow (z-direction) through inert packings can be used for continuous fractionation of ionic solutes on a scale approaching kilograms per hour. A feed mixture is fed in a ring of radius r_F at the upstream end, and solute fractions are withdrawn from eight concentric rings at the downstream end at a distance L from the inlet. A background electrolyte ("elutant") is fed downward continuously and uniformly across the bed cross section with a linear velocity of v_z.

(1) It is desired to determine the radial location of the ring at the product withdrawal end for any given ith species.

(2) Obtain a relation between the radial product withdrawal locations r_i and r_j of two ionic species (i and j) with the closest ionic mobilities. (Ans.

$$(r_i^2 - r_j^2) = \frac{2\,L\,E_{ab}\,(\mu_i^m - \mu_j^m)}{v_z\,\ell n(r_a/r_b)}.\Big)$$

You are given the following: ionic mobility of solute $i = \mu_i^m \text{cm}^2/\text{s-volt}$; E_{ab} is the voltage difference between the two electrodes; the voltage gradient varies inversely with the radius (see problem 3.1.5); the solution conductivity is uniform throughout.

7.3.3 The *rotationally stabilized continuous free-flow electrophoretic (CFE) separator* of Figure 7.3.4, sometimes known as the Philpot–Harwell electrophoretic separator, is of interest for determining the trajectory of a protein introduced at $z = z_0$ and $r = r_i$, the radius of the inner cylinder, which is stationary and acts as the cathode. The outer rotating cylinder of radius $r = r_o$ acts as the anode. Neglect solute diffusion, electroosmotic flow velocity, Joule heating and any change in physical properties. Determine the trajectory of the protein for this annular flow configuration if the outer electrode has a voltage V_o while the inner electrode has a voltage V_i ($< V_o$). You are given the following information:

(1) the electrical force field is given by (Problem 3.1.5)

$$\mathbf{E} = \frac{(V_i - V_o)}{\ell n(r_o/r_i)}\,\frac{1}{r}\,\mathbf{r};$$

(2) from Bird *et al.* (2002), the average axial velocity, $\langle v_z \rangle$, is given by

$$\langle v_z \rangle = \frac{(P_0 - P_L)\,r_o^2}{8\mu L}\left[\frac{(1-(r_i/r_o)^4)}{(1-(r_i/r_o)^2)} - \frac{(1-(r_i/r_o)^2)}{\ln(r_o/r_i)}\right];$$

the axial velocity profile in the annulus is given by

$$\langle v_z \rangle = \frac{(P_0 - P_L)\, r_o^2}{4\mu L}\left[1 - \left(\frac{r}{r_o}\right)^2 - \frac{(1-(r_i/r_o)^2)}{\ell n(r_o/r_i)}\,\ell n\left(\frac{r_o}{r}\right)\right];$$

(3) the length of the cylinders is L;
(4) the protein charge makes it move radially outward;
(5) the total pressure drop (includes gravity) $(P_0 - P_L)$ is not known.

7.3.4 Consider continuous free-flow electrophoresis in the annular separator of Figure 7.3.4. Let the negatively charged stator outside diameter be R_o and the rotor inner diameter be R_i. Assume steady state operation in the nondiffusive limit.

(1) Obtain the equations corresponding to equations (7.3.2) and (7.3.3) for C_i of any species i.
(2) Using the steps and results of Problem 3.1.5, show that the radial location r_i where negatively charged species i will appear at a height L from the point of feed introduction at the bottom is related to various relevant quantities by

$$\mu_i^m I = \sigma \int_{R_i}^{r_i} 2\pi r\, v_z(r)\, dr,$$

where I is the total current between the electrodes, σ is the average buffer electrical conductivity, μ_i^m is the electrophoretic mobility of species i and $v_z(r)$ is the axial velocity profile in the annulus. Note that the integral on the right-hand side of the above result is the total axial (z-axis) volumetric flow rate of buffer between the wall of the stator ($r = R_i$) and the radial location r_i at ($z = L$) where species i exits and is collected.

(3) The radial dependence of v_z may be obtained from standard solutions of flow through an annulus (Bird *et al.*, 1960):

$$v_z(r) = \frac{2Q\left[1 - (r/R_o)^2 + \left\{\left(1 - \frac{R_i^2}{R_o^2}\right)\Big/\ell n\left(\frac{R_o}{R_i}\right)\right\}\ell n\,(r/R_o)\right]}{\pi R_o^2\left[1 - (R_i/R_o)^4 - \left(1 - \frac{R_i^2}{R_o^2}\right)^2\Big/\ell n\left(\frac{R_o}{R_i}\right)\right]}.$$

Obtain an algebraic equation for r_i in terms of μ_i^m, I, R_o, R_i, Q and σ.

7.3.5 *Capillary electrophoresis* (Section 6.3.1.2) of live and dead bacteria show identical retention times; their electrophoretic mobilities (μ_i^m) are indistinguishable (Armstrong and He, 2001). However, the electrical conductivity of the cell membrane of a live bacteria is around 10^{-4} μS/mm, whereas that of the cell's interior is much higher, around 10^3 μS/mm due to many ions. When a bacterium dies, the conductivity of the cell membrane (which becomes permeable) becomes much higher (around 1μS/mm). If dead bacteria and live bacteria are suspended in deionized water having a conductivity of 2.2 μS/mm, and if *dielectrophoresis* is employed, describe what kind of *dielectrophoretic separation* can be achieved. What would be the observed separation behavior of a plastic particle of radius 0.5 μm if its conductivity is 18.5 μS/mm?

You are given that the Clausius–Mossoti function in equation (3.1.13) contains dielectric constants which are complex quantities. However, their real parts at low frequencies are electrical conductivities, σ (see (3.1.108p)); one can work with the real part only. Further, assume that the cell membrane conductivity is controlling.

7.3.6 In *ion mobility spectrometry* (IMS) used for explosive detection purposes, ions are generated from injected sample species, e.g. nitroglycerine, RDX, trinitrotoluene, etc., in the reaction region (includes other ion sources); the ions so generated are injected into a drift tube/channel where separations occur in an electrical field perpendicular to the bulk gas (air) flow containing the ions. For (E/N) values (see (3.1.219) and (3.1.220)) below 10 Td (Eiceman *et al.*, 2004), $\mu_{\text{ion},g}^m$ is independent of the electrical field. For higher (E/N) values employed in differential mobility spectrometry (DMS), $\mu_{\text{ion},g}^m$ may be characterized as

$$\mu_{\text{ion},g}^m = \mu_{\text{ion},g}^m(0)\,[1 + \alpha_{\text{ion}}(E/N)].$$

The values of $\alpha_{\text{ion}}(E/N)$ at a value of $(E/N) = 100$ Td for nitroglycerine, dinitrobenzene and TNT were found to be -0.260×10^{-2}, 2.63×10^{-2} and 1.24×10^{-2}, respectively. Plot qualitatively the trajectories of each of these species, whose mass over charge as well as the sign of the charge are provided here: nitroglycerine, $m/Z = 289$

(M^-); TNT, $m/Z = 227$ (M^-); dinitrobenzene, $m/Z = 168$ (M^-). Develop the trajectory equations for such species.

Note: In DMS methodology (Eiceman *et al.*, 2004), the electric field applied across the two channel plates fluctuates rapidly between 20 and 100 Td (for E/N), the separation between two ions depends more on $\left(\mu^m_{\text{ion},g}\right)_1 - \left(\mu^m_{\text{ion},g}\right)_2$ rather than on $\left(\mu^m_{\text{ion},g}\right)_1$ or $\left(\mu^m_{\text{ion},g}\right)_2$.

7.3.7 In an *electrostatic precipitator* of the type shown in Figure 7.3.8, the rectangular collector plates (the grounded electrodes) are 30 ft long with a gap of 25 cm in between them. The dust-laden hot gas velocity is 2 m/s. The height of the rectangular plates is 30 ft, and the applied voltage difference between the corona wires and the plate is 40 kV.

(1) What is the value of the particle migration velocity for a dust particle of radius 0.25 μm?

(2) Assume the gas to be in laminar flow. What would be the duct length needed for complete settling of the dust particle?

(3) Calculate the value of the total efficiency E_T of particle collection for such particles.

You are given that the hot gas viscosity may be assumed to be 0.085 kg/m-hr; $C_c \sim 1.3$; the hot gas density is 0.60 kg/m^3.

7.3.8 Develop an expression for the fraction of particles collected in a *dry electrostatic precipitator*, $\eta|_L$, if the particle size has a Gaussian distribution with a mean value of r_{mean} and a standard deviation σ; the Deutsch equation may be used to describe the precipitator performance. It is preferable to employ expression (7.3.39) for U_{py} in terms of r_p for this purpose. Calculate the value of $\eta|_L$ (White, 1963) for $r_{\text{mean}} = 10$ μm, $\sigma = 2$ μm, $(B_pL\ E_cE/2\pi\mu A_c v_z) = 3 \times 10^3\ \text{cm}^{-1}$.

7.3.9 Consider a particle falling vertically downward due to gravity. If there is an electrical field of uniform strength E_x in the horizontal direction, and the particle charge is Q_p, we model the particle motion in the horizontal direction using a terminal settling velocity (an example is equation (7.3.31)).

(1) Write down the equation governing the particle motion in the horizontal direction (as it falls through air vertically) due to the electrical field.

(2) Develop a solution for the horizontal component of the particle velocity as a function of time.

(3) Indicate when the terminal settling velocity in the horizontal direction is achieved. (You are given that the x-coordinate is in the horizontal direction.) To calculate the terminal settling velocity achieved in a given time for the third part, the following information is provided: particle size, 0.5 mm radius; particle density, 1100 kg/m^3; viscosity of air 1.8×10^{-5} Pa-s (kg/m-s).

7.3.10 Consider an inclined plate electrode connected to the ground; let there be a tubular high-voltage electrode of ellipsoidal cross section at a distance, as shown in Figure 7.P.3.

The particulate mixture to be separated is introduced to the surface of the inclined plate. Insulating particles slide down the plate electrode without being affected by the electric field. The conducting particles, on the other hand, become charged by induction in contact with the inclined plate electrode (grounded); these particles are attracted by the high-voltage electrode. The two different types of particles will have two different trajectories and can be collected by two separate bins (1 and 2).

(1) Identify the different forces that act on a conducting particle of radius r_p as it slides down the inclined plate; particle density $= \rho_p$.

(2) Specify the criterion based on different forces when the particle will be lifted from the plate. Make assumptions that are realistic.

(3) Identify the value of the minimum electric field strength needed to lift the conducting particle from the inclined electrode/plate.

You are given that the maximum charge $Q_{\max}$ acquired by the spherical particle in a locally uniform electrical field of strength E is $Q_{\max} = (2\pi^3/3)\varepsilon_a r_p^2 E$, where ε_a is the permittivity of air. The frictional coefficient between the sphere and the inclined plate is μ_{fr}.

7.3.11 You have a *tubular bowl centrifuge* available for collecting 0.5 μm radius bacterial cells from a cell culture broth. The centrifuge dimensions are: inside radius, 5 cm; free surface radius, 2 cm; length, 100 cm. The centrifuge rotates at 5000 rpm. You have a fermentation broth of 100 liter. How long will you take to process the broth so that you can have a completely clarified broth? Assume the following: bacterial cells are spherical; the cell density is 1.10 g/cm^3; the broth is dilute; the broth viscosity is equal to that of water (~1 cp).

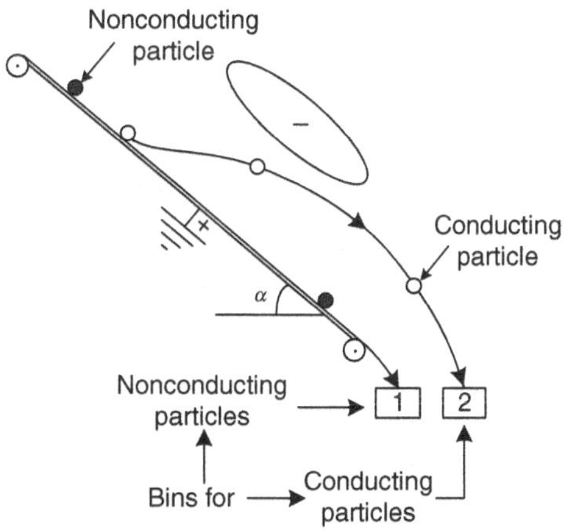

Figure 7.P.3.

7.3.12 You have a *tubular bowl centrifuge* available for collecting bacterial cells from a cell culture broth. The centrifuge dimensions are inside radius, 5 cm; free surface radius, 2 cm; length, 100 cm. The centrifuge rotates at 5000 rpm. You have a fermentation broth of 80 liter; it has to be processed in two hours. What should be the minimum dimension of the bacterial cells so that you can have a completely clarified broth? Assume the following: bacterial cells are spherical; the cell density is 1.10 g/cm^3; the broth is quite dilute; the broth viscosity is that of water (~1 cp).

7.3.13 You have a number of *tubular bowl centrifuges* available for collecting 1 μm radius bacterial cells from a cell culture broth. The centrifuge dimensions are: inside radius, 5 cm; free surface radius, 2 cm; length, 100 cm. The centrifuge rotates at 5000 rpm. You have a fermentation broth of 900 liters. You may not take more than 2 hours to process the broth so that you can have a completely clarified broth. How many centrifuges do you need? Assume the following: bacterial cells are spherical; the cell density is 1.10 g/cm^3; the broth is quite dilute; the broth viscosity is that of water (~1 cp).

7.3.14 Svarovsky (1977, p. 134) has provided an experimental *grade efficiency curve* for a *tubular centrifuge*. The following information is available about the centrifuge: $r_0 = 2.2$ cm; $r_f = 1.1$ cm; $L = 19$ cm. The operating conditions are: $Q_f = 7.5$ cm^3/s; $\omega = 8000$ rpm. The feed is a very dilute solution (0.5% by volume) of a fine mineral pigment having a density of 4 g/cm^3; the feed solution viscosity is 1 cp at the operating temperature. Determine an expression for the grade efficiency function $G_r(r_p)$ as a function of r_p only. Calculate the value of the grade efficiency function for $r_p = 0.3$ μm which was found experimentally to be the cut point, $r_{p,50}$.

7.3.15 A *disk centrifuge* is operated with 52 disks on a viscous solution containing crystals whose smallest diameter is 0.9 μm. The centrifuge rotates at 5800 rpm, where the various geometrical dimensions and angles are: $\theta = 45°$, $r_{in} = 6''$, $r_{ex} = 2''$. The liquid density is 0.9 g/cm^3 and the liquid viscosity is 80 cp. Determine the liquid flow rate that will allow capture of all crystals in the slurry.

7.3.16 To overcome the drawbacks of conventional mist eliminators (e.g. wire-mesh devices, etc.) in high-pressure applications, an *axial-flow cyclone mist eliminator* has been proposed. For the time being, imagine it as if you are analyzing the cyclone gas separator by Clift *et al.* (1991) (See equations (7.3.140) and (7.3.144)) except for the following: the drop density is ρ_ℓ; the hydrodynamic drag is described via Stokes' law for radial droplet motion; gas and droplet tangential velocities are identical; the tangential gas velocity v_θ and the radial coordinate r are related by the free vortex relation (7.3.134) $v_\theta r^n = v_{\theta,0} r_0^n =$ constant (see Figure 7.3.15(b), but replace particle by droplet); there is complete radial mixing (as in Clift *et al.* (1991) and used in (7.3.140)).

(1) Determine an expression for the radial settling velocity of the droplets in terms of the reference values, $v_{\theta,0}$ and r_0.

(2) Develop now a mass balance over a sector with thickness dz in the axial direction and angle dθ as shown in Figure 7.P.4, where N is the droplet concentration and No is the total number of droplets in the control volume:

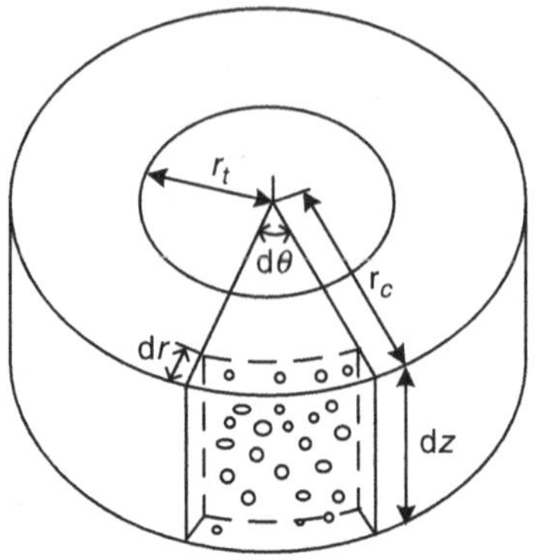

Figure 7.P.4.

$$\frac{dNo}{No} = \frac{2\,r_c\,dr}{(r_c^2 - r_t^2)},$$

where r_c is the radius of the cyclone and r_t is the exit pipe radius. This analysis is based on the assumption that droplets contained near the wall in the region of radial thickness dr and height dz will be deposited on the wall.

(3) Introduce now an expression for the droplet radial location r as a function of t from part (1) in the result from Part (2), integrate over the residence time t_{res} of the gas/droplets (diameter, d_d) in the cyclone and obtain the following expression for droplet capture efficiency:

$$\eta = 1 - \exp\left[-\frac{2}{(r_c^2 - r_t^2)} \times \frac{1}{(r_c^2)^{-(2n+1)/(2n+2)}} \left\{ \frac{2(n+1)d_d^2(\rho_\ell - \rho_g)}{18\mu_g} v_{\theta,c}^2\, t_{res} \right\}^{1/(2n+2)} \right],$$

where $v_{\theta,c}\, r_c^n = v_{\theta,0}\, r_0^n$. The expression for t_{res}, the residence time in the cyclone, may be obtained for the cyclone length L for a gas flow rate of Q_{v0} and the two cyclone radii, r_c and r_t.

7.3.17 Mass (1979) has provided the following suggested dimensions for laminar flow settling of dust particles from a hot flue gas at 300 °F, 1 atm pressure in a *gravitational settling chamber*: length 38.6 ft, height 6 ft, width 2.78 ft. The particle density is 125 lb/ft^3; the gas viscosity is 0.0238 cp; the gas density is 0.055 lb/ft^3; the actual gas flow rate is 10, 000 acfm (actual cubic feet per minute). The smallest particle diameter is 0.0005 inch; the largest particle diameter is 0.004 inch. Calculate the grade efficiency function for the smallest and the largest particles using the cgs system and assuming validity of Stokes' law. Check whether Stokes' law is valid for the largest particle.

7.3.18 In chemical and petrochemical plants, a horizontal knockout drum[27] is commonly employed to separate larger-size droplets (~300–400 μm) from a vapor stream. The vapor-droplet stream enters through the top of the drum of radius $2R$ at the left-hand end. The vapor leaves through the top of the drum at the right-hand end at a horizontal distance L from the inlet location. The liquid level in the drum is maintained at, say, 50% of the total drum volume. The droplets are supposed to settle by gravity into the liquid layer before the vapor stream escapes via the right-hand end exit at the top. Expression (6.3.3c) is valid for the drag force on the liquid droplets.

Develop a criterion for designing the value of L with respect to the drum radius, vapor flow rate Q_{vap}, the vertical settling velocity U_{pyt} of the drops after expressing U_{pyt} in terms of the droplet radius r_p, vapor density ρ_v, liquid density ρ_ℓ and other relevant quantities.

7.3.19 In an experimental study of particle separation via *electrical field-flow fractionation* (EFFF), Caldwell and Gao (1993) estimated the following values of λ_i for polystyrene (PS) latex spheres of diameters 272 nm and 496 nm, respectively: 0.0273, 0.0184. In an experiment using EFFF for separating a mixture of such latex spheres, the

[27] The basic objective is to prevent a large liquid slug to go into the vapor stream which is sent to a flare.

channel dimensions were 64 cm long 2 cm wide, 178 μm channel gap. However, the measured void volume of the channel was 2.7 ml. For a carrier liquid flow rate of 0.5 cm^3/min at a fixed conductivity of the aqueous NaCl carrier solution, estimate the retention times of the latex spheres. The applied potential was 1.41 V and the potential gradient was ~78 V/cm. The conductivity of the aqueous NaCl was 10.0 μS/cm.

7.3.20 Determine the selectivity for the following FFF techniques: sedimentation FFF; flow FFF; steric FFF. Employ the following definition of selectivity:

$$S_{r_i} = \left| \frac{\mathrm{d} \log t_R}{\mathrm{d} \log r_i} \right| .$$

7.3.21 Consider the combination of *crossflow FFF* and gravity in the channel used for FFF by keeping it horizontal. Ordinarily, to avoid gravitational force, the channel shown in Figure 7.3.21(c) is kept vertical so that crossflow is the only external force perpendicular to the bulk flow field in flow FFF.

(a) Write down the force experienced by a particle of radius r_p in the presence of a crossflow velocity of magnitude $|U_y|$ in a horizontal channel.

(b) Comment on the effect of gravity being present on the selectivity or the retention ratio for the technique between two different particles having the same density.

(c) Comment on the particle size range where this could be useful. Is there any other effect one should be concerned about?

7.3.22 The silica-alumina zeolite catalyst particles in the FCC (fluid catalytic cracking) process used to produce gasoline from gasoil are subjected to a number of deposits such as coke and heavy metals, e.g. Ni, Va, Fe, as well as nickel oxides. It is known that the catalyst particles as such are diamagnetic. In order to continue operation of the FCC process continuously, it is desired to remove selectively the heavy metal contaminated catalyst particles. Suggest a process and provide a schematic for such a goal of removing heavy metal contaminated/spent catalyst particles; similarly, speculate about the removal of catalyst particles having a significant amount of coke.

8

Bulk flow of two phases/regions perpendicular to the direction(s) of the force(s)

In this chapter, we cover those separation processes/techniques where two bulk phases or regions flow through the separation device such that the force(s) driving the separation act perpendicular to the directions of flow of both phases/regions. Section 8.1 describes those separation devices where the two phases or regions have bulk motions countercurrent to each other in the separation device. This section also covers those larger multistage separation devices where the overall pattern of flow vs. force corresponds to countercurrent flow, even through the local flow configurations in a stage may utilize other flow vs. force arrangements for the two flowing phases. Invariably, these separations will be studied under continuous flow of two phases/regions and steady state conditions. The focus will be primarily on phase equilibrium based processes driven by chemical potential gradient such as gas absorption, distillation, solvent extraction, melt crystallization and adsorption. Limited attention has been paid to membrane based processes such as dialysis, electrodialysis, liquid membrane processes, gas permeation and external force driven processes of gas centrifuge (thermal diffusion and mass diffusion are also considered here). Section 8.2 will be concerned with separations where the two phases or regions flow in a cocurrent fashion. Continuous flow of two phases/regions as well as discontinuous flow of one of the phases will be covered. Section 8.3 will focus on those configurations where the two bulk phases are in crossflow in the device. The force(s) will continue to act perpendicular to the flow directions of the two bulk phases in both Sections 8.2 and 8.3.

At the beginning of each of Sections 8.1, 8.2 and 8.3, a brief description of the overall flow patterns of the two phases/regions in various separation technologies/processes will be provided. This description will be followed by a brief consideration of the multicomponent separation capability of the particular bulk flow vs. force(s) configuration. Then detailed treatments of individual separation processes/techniques will follow. There will be an occasional example of particle separation in such bulk flow vs. force configurations.

8.1 Countercurrent bulk flow of two phases or regions perpendicular to the direction(s) of the force(s) driving species

We begin here with a brief description, analysis and elementary design of a variety of separation devices, processes and techniques where two bulk phases or regions move countercurrent to each other as the force(s) driving the species from one phase/region to the other phase/region act perpendicular to the directions of bulk motion of the two phases/regions. First, in Section 8.1.1, we will provide a simplified description of four basic types of separation systems employing countercurrent bulk flow of two phases/regions. How separation develops in such configurations is of intrinsic interest. An introduction to the governing equations for such systems will also be provided (Sections 8.1.1.1 and 8.1.1.2). This will be followed by Section 8.1.1.3, where an elementary illustration will be provided of why a device, having a steady countercurrent flow of two phases/regions, can separate a binary mixture only. Next, we will present extended treatments of various separation processes and techniques. Sections 8.1.2 to 8.1.6 will cover chemical potential gradient driven two-phase systems; Sections 8.1.7 to 8.1.9 will treat membrane separation systems; Sections 8.1.10 and 8.1.11 will cover external force based and single-phase based systems.

8.1.1 Development of separation in countercurrent flow systems

In the flow configuration of a countercurrent bulk flow of two phases or regions in one device or column, with the force acting perpendicular to the directions of bulk flow of both phases or regions, four basic types of separation systems are encountered.

System type (1) The same phase flows countercurrently in two regions of the device in the presence of an external force or temperature gradient, or concentration gradient or partial pressure gradient.

System type (2) Two different pre-existing immiscible phases introduced into the device/column move countercurrent to each other and leave the device/column. Examples include gas–liquid absorption/stripping; solvent extraction/back extraction; moving-bed adsorption; supercritical extraction.

System type (3) Two different immiscible phases flow countercurrently in the device/column; different immiscible phases are generated from one feed input stream into the column by the application/withdrawal of thermal energy. Examples include distillation (vapor–liquid system), melt crystallization (solid–liquid system).

System type (4) Two miscible phases flow countercurrently in two regions of the device separated by a membrane. One of the phases may be generated from one feed phase by the application of pressure energy. Examples include reverse osmosis, ultrafiltration, microfiltration, gas permeation, pervaporation. Examples where the other phase is introduced from outside are electrodialysis, dialysis, sweep vapor/liquid based system.

An additional feature which distinguishes different separation systems is whether there occurs any recycle or reflux of a part or whole of one or more of the flowing streams at the exit ends of the device back into the device.

We will now illustrate these four categories of separation systems in countercurrent flow. An example from system type (1) will be considered first to illustrate some basic separation results achieved in such a countercurrent flow configuration. Then we will consider examples from each of the system types (2), (3) and (4) sequentially.

Consider a vertical channel consisting of two parallel plates (Figure 8.1.1). The bottom and top of this channel are closed, and so are the other two sides. At time $t = 0$, this enclosed channel contains a binary gas mixture of (50 mole % of species 1)–(50 mole % of species 2) at a temperature T_{low} (Figure 8.1.1(a)). For time $t > 0$, the right-hand plate temperature is raised to a high value (e.g. see Example III, 50% H_2–50% N_2 separation by thermal diffusion in Figure 1.1.3; the gas mixture in the hot bulb becomes enriched in the lighter species, H_2, whereas the cold bulb becomes enriched in the heavier gas, N_2 (see calculations after equation (4.2.65)); the other plate is maintained at the initial low temperature. If species 1 is the lighter species, e.g. H_2, then the gas mixture next to the hot plate will become enriched in H_2, whereas the gas mixture next to the cold plate will become enriched in the heavy species 2, e.g. N_2. In Figure 8.1.1(b), we illustrate the changed composition in terms of light gas (e.g. H_2) by assuming a 4% change in thermal diffusion based equilibrium composition: the gas mixture next to the hot plate is 52% H_2, whereas that next to the cold plate is 48% H_2.

The enclosed channel with the particular temperature profile will lead spontaneously to natural convective motion of the gaseous regions (see Figure 6.1.3): the gaseous mixture next to the heated plate will go up, whereas that next to the cold plate will come down. As the gaseous regions move up and down slightly, we will have somewhat changed situations near the top and bottom of the column. We illustrate this by dividing the vertical channel into eight identical sections at different vertical locations. As Grew and Ibbs (1952) first illustrated (since described also by Powers (1962, pp. 1-98, 36-37)), the top section will have both regions of the channel containing a gas mixture of 52% H_2, whereas the bottom section will contain 48% H_2 (Figure 8.1.1(c)). However, due to the temperature difference between the two plates, there will be thermal diffusion based separation, creating a changed situation, as shown in Figure 8.1.1(d), in these two sections. Now the bottom section will have 50% species 1 near the hot plate and 46% near the cold plate, whereas, in the top section, the region next to the hot plate will have 54% species 1 and there will be 50% species 1 near the cold plate. This separation is achieved after thermal diffusion based equilibriation is achieved following the natural convective motion. This way of describing the separation development assumes consecutive processes: Figure 8.1.1(a) – time $t = 0$; Figure 8.1.1(b) – first thermal diffusion takes place; Figure 8.1.1(c) – first natural convection occurs; Figure 8.1.1(d) – second thermal diffusion based equilibriation takes place. In reality, these steps/processes go on simultaneously.

Let us continue with this process. As the natural convection continues (the second natural convection, Figure 8.1.1(e)), we have gas compositions in the top two sections and bottom two sections that are different from those based on local thermal diffusion equilibrium. Establishment of local thermal diffusion based equilibrium will lead to the compositions shown in Figure 8.1.1(f) (third diffusion), which shows that the top section now has 55% species 1 near the hot plate and 51% species 1 near the cold plate; further, the bottom section now has 45% species 1 near the cold plate and 49% species 1 near the hot plate. What is clear from this description is that the top section near the hot plate now has 55% species 1, whereas the bottom section near the cold plate has 45% species 1. Figure 8.1.1(g) shows the further enhancement in this composition difference between the top and the bottom column locations after the third natural convection and the fourth thermal diffusion based equilibriation steps have taken place.

There are two basic differences that exist here with respect to the separation achieved in any one section

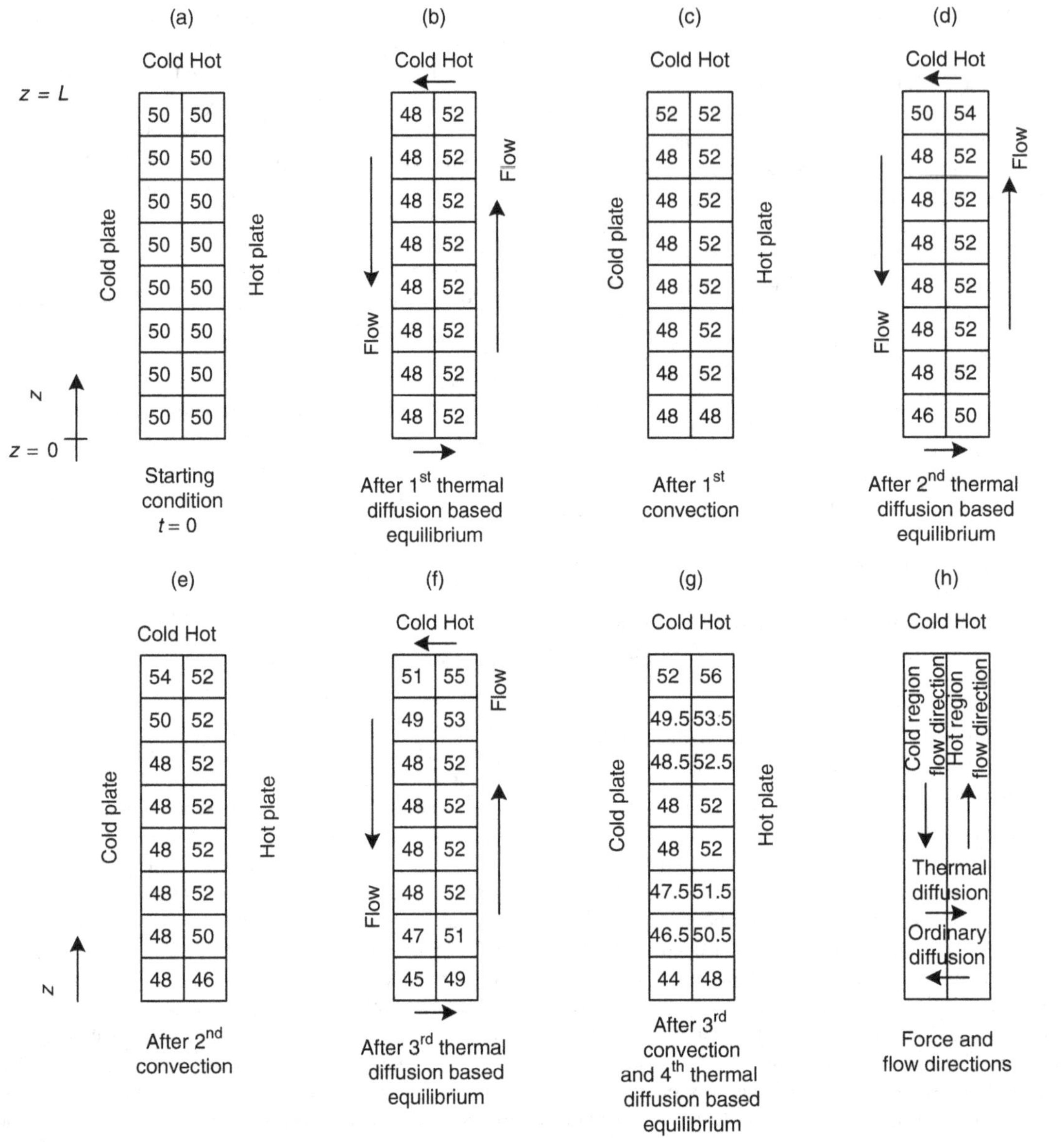

Figure 8.1.1. Type (1) system. Development of concentration profile for a binary gaseous system along the height of a vertical column of parallel plates having closed ends and subjected to thermal diffusion equilibrium and countercurrent flow of two regions by natural convection. Numbers in boxes in each figure represent mole % of lighter species (e.g. H_2). (After Grew and Ibbs (1952) and Powers (1962).)

of this channel (or column), the latter simulating the condition in Figure 1.1.3. First, there is countercurrent flow of two regions perpendicular to the directions of the two forces operating here: those due to thermal diffusion and ordinary diffusion (not really a force as such) (Figure 8.1.1(h)). Second, due to the top end of the column being closed, the hot fluid enriched in species 1 is turned around (recycled) in this case, or refluxed in the case of two-phase systems, to the cold part of the channel (total recycle or total reflux). This allows even further enhancement of the hot side concentration of species 1 due to the thermal diffusion phenomenon, since the cold side has higher species 1 concentration than before. Similarly the cold fluid is turned around in the bottom part of the channel to the hot part, leading to further enrichment on the hot side. (If we did not have this recycle or reflux, and the top and bottom sections were open, we would just have 52% species 1 in the hot stream exiting the top and 48% species 1 in the cold stream exiting the bottom.) By making the column taller, one can achieve a very large change in concentration between the top right and the bottom left section; one can achieve almost pure streams at the top and at the bottom for an exceedingly tall column. However, in practice, one should recognize that one may not achieve the separation corresponding to equilibration in each section of the vertical column (called the Clusius–Dickel column).

Let us reflect on the net directional movement of species 1 and species 2 in this countercurrent column. It is clear that the lighter species 1, H_2, moves up the column and the heavier species 2, N_2, moves down the countercurrent column. One could argue that, in effect, the concentration wave velocity, v^*_{C1}, of species 1 is positive vis-à-vis the z-coordinate in Figure 8.1.1, whereas that for species 2, N_2, is negative. We illustrate the quantitative criterion for this type of species movement in a countercurrent system of two immiscible phases in Section 8.1.1.3, where we illustrate also that only two species may be separated in such a countercurrent column having steady state flow.

We will now consider a system of type (2). Figure 8.1.2(a) illustrates a vertical channel as in Figure 8.1.1(a). However, the top and the bottom ends of this vertical separator are open. Let a liquid absorbent stream enter at the top on the left-hand side and flow down the column. Let a gas to be cleaned up containing at least two species, 1 and 2, enter on the right-hand side at the column bottom and flow up. (Conventionally the two phases are contacted in a dispersive mode; in Section 8.1.2 on gas absorption and stripping, we will discuss the actual physical configuration of the column.)

Consider what happens as species 2 in the gas stream is absorbed in the liquid and species 1 is not absorbed. As we assumed in our description of the type (1) system, the gas and the liquid may be assumed to be in equilibrium everywhere in the column, even as they move countercurrently. If the liquid absorbent entering at the top does not have any species 2 at all, and the liquid, as well as the gas stream, are assumed to be in equilibrium, then the gas stream exiting the column top through section 1 will be essentially purified of species 2. Correspondingly, the liquid stream leaving at the column bottom through section n will be in equilibrium with the gas stream entering at the bottom of the column; so the liquid absorbent leaving will be saturated with species 2 corresponding to the feed-gas composition. Therefore, in this type of bulk flow vs. force schematic, the extent of purification of the gas stream leaving the column top is dependent on the purity of the liquid stream entering at the top of the column. For simplicity, we have not divided the vertical column into n sections as we have done in Figure 8.1.1 (where $n = 8$). *Note*: We can argue that the unabsorbed gas species 1 moves up the column in the positive z-direction, whereas species 2, absorbed by the liquid absorbent, moves in the negative z-direction. A common example

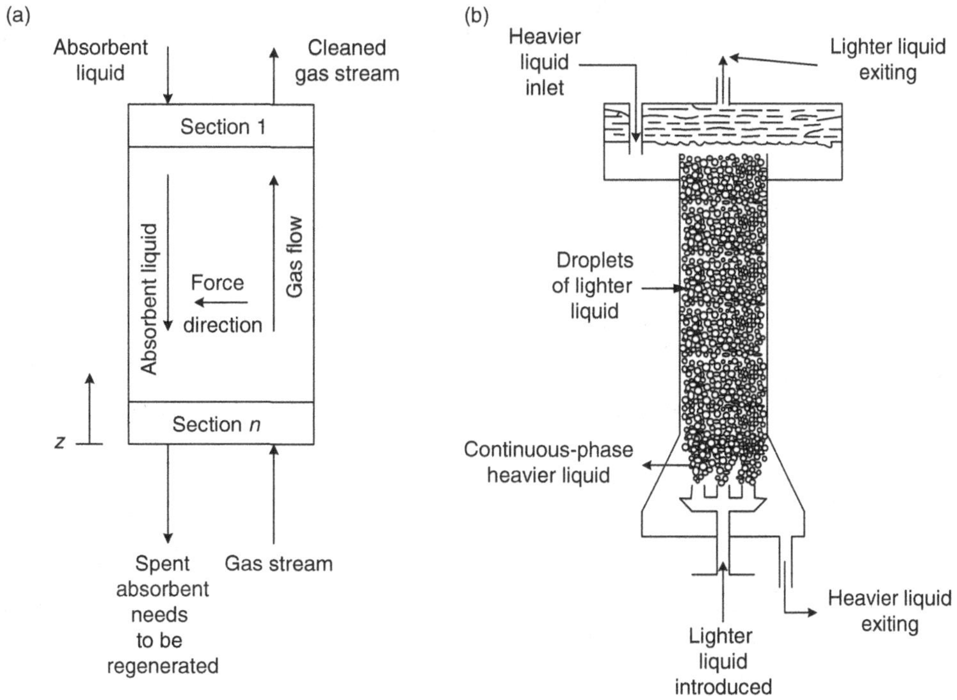

Figure 8.1.2. (a) Type (2) systems. Countercurrent flow of a gas stream and an absorbent liquid in a vertical column for absorption of a species from the gas; countercurrent separation system without recycle or reflux. (b) Countercurrent solvent extraction column: lighter liquid introduced at column bottom by dispersing it as droplets as it rises through a continuous heavier liquid, which flows downward through the column. In aqueous-organic extraction systems, usually the aqueous phase is heavier.

would be as follows: species 2 is CO_2 being absorbed in the absorbent liquid, whereas species 1, say N_2 or O_2, etc., is absorbed very little.

If, instead of absorption, we are engaged in stripping a volatile species from the absorbent liquid entering at the top of the column by means of a stripping gas stream (e.g. air) or a stripping vapor stream (e.g. steam) flowing countercurrently, similar considerations will hold, but in reverse. The extent of stripping based purification of the liquid stream leaving the column bottom will depend on how pure the stripping gas/steam is as it enters the column at the bottom. An important distinction between this mode of operation for gas absorption or stripping and the mode of operation illustrated in Figures 8.1.1(a)-(h) is that there is recycle/reflux (in two-phase systems) of the streams in the latter arrangement. Usually there is no such recyle or reflux in Figure 8.1.2(a). A comparable separation system of type (2) is routinely encountered in liquid-liquid systems for solvent extraction or back extraction processes. Figure 8.1.2 (b) illustrates a dispersive countercurrent extraction column where the heavy liquid flows down the column as the immiscible lighter liquid rises up the column as droplets.

A system of type (3) will now be discussed. Consider a tall column (Figure 8.1.3(a)), in which a binary volatile liquid mixture containing a more volatile species 1 and less volatile species 2 is flowing down the column, and a binary vapor mixture of the same species moves up the column. Details of the actual flow arrangements and column internals will be considered later. At the top of the column, the vapor stream goes out, is condensed in a condenser by an external coolant, and the condensed liquid is refluxed back to the column as the liquid stream going down the column. At the bottom of the column, the liquid is taken out to a reboiler, vaporized completely by an external heat source, and introduced into the bottom of the column as the vapor stream going up the column. Witness, however, the basic similarity in the flow pattern and the arrangements of recycle (here reflux) of both streams at the top of the column and the bottom of the column between Figures 8.1.1(a)-(h) and the present arrangement (Figure 8.1.3(a)). We may assume that, everywhere in the column, the liquid and the vapor streams are in vapor-liquid equilibrium. Just as in Figures 8.1.1(a)-(h), one can develop a very large difference in composition between the top section, containing a stream highly enriched in the lighter species 1, and the bottom section, containing a stream highly enriched in the heavier species 2. This arrangement is known as *total reflux* in the process of column distillation; in the case of a tall column, it can provide compositions at the top and bottom that are quite close to pure species.

During the actual operation of such a distillation column for continuous separation of a feed mixture to obtain two product streams, the feed stream (shown by the dashed lines in Figure 8.1.3(a)) is introduced somewhat near the middle of the column. From the total condenser at the top, a fraction of the total condensate stream is continuously withdrawn as the top product stream; similarly, a fraction of the liquid to be introduced into the reboiler at

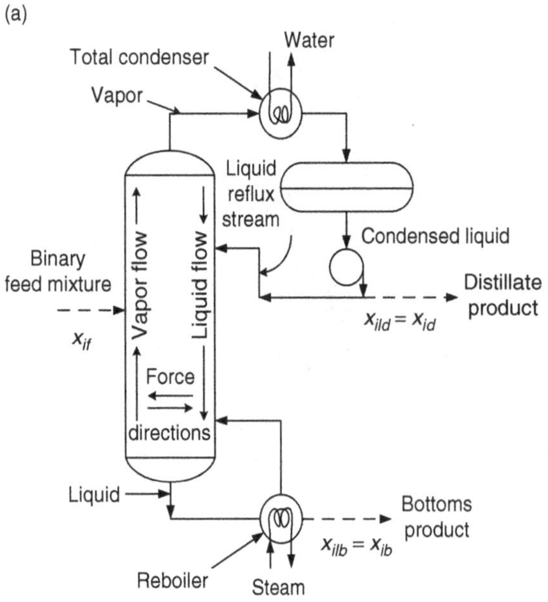

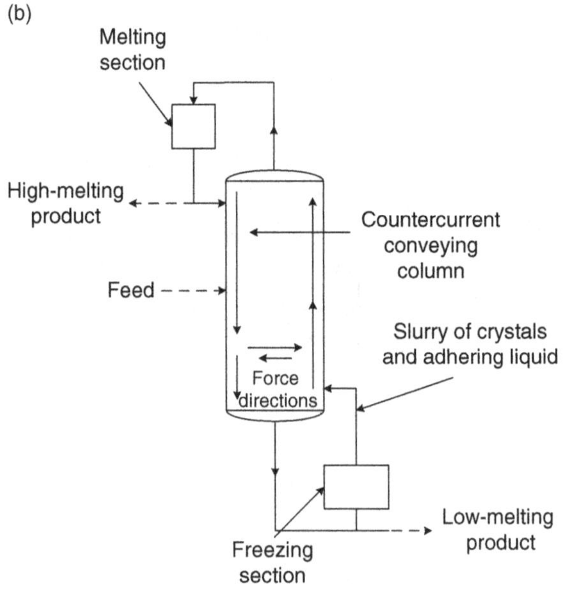

Figure 8.1.3. (a) Type (3) systems. Countercurrent flow of a vapor stream and a liquid stream in a distillation column operated at total reflux, no feed, no bottom product. (b) Countercurrent melt crystallizer (column crystallizer) for contacting of a slurry of crystals and adhering liquid being conveyed countercurrent to a liquid reflux obtained by melting crystals.

the bottom is continuously withdrawn as the bottom product stream (also shown by a dashed line in Figure 8.1.3(a)). The same strategy is employed in the column of Figures 8.1.1(a)–(h) for continuous feed introduction and product withdrawals, except the two streams are of identical phase. One must note here that this countercurrent column is such that the lighter species 1 has a net velocity in the positive z-direction (considered vertically upward) and the heavier species 2 has a net velocity in the negative z-direction.

An analogous separation system of type (3) is illustrated in Figure 8.1.3(b) for a solid–liquid system as in *melt crystallization* (an alternative title might be *column crystallization)*. The solid solution of Figure 3.3.6A would be one such system; eutectic systems are additional examples. Crystals and the liquid adhering to crystals are carried by means of a spiral conveying arrangement from one end of the column (the freezing section end) to the other end: this flow is countercurrent to the flow of the free liquid. During this process, impurities from the crystals and the liquid adhering to the crystals are transferred to the counterflowing free liquid stream. Once the crystals and the liquid adhering to the crystals reach the column end, they are taken to a melting section, and the molten liquid is sent (refluxed) back to the column for countercurrent flow (after withdrawing some amount of product which melts at a higher temperature). Conversely, the free liquid exiting the other end of the column is introduced to a freezing section to produce some low-melting product, while a slurry of crystals and adhering liquid is introduced to the conveying arrangement (Albertins *et al.*, 1967).

We will now briefly consider the membrane gas permeation process to illustrate a separation system of type (4). A binary gas mixture at a high pressure flows along the device length on one side of the membrane. The gas mixture permeating through the membrane and emerging on the other side enriched in the more permeable species flows countercurrent to the feed gas stream. No separate gas stream is introduced in the permeate side. Additional variations of this type include: (1) a sweep gas/vapor stream introduced on the permeate side in countercurrent flow (shown by a dashed line in Figure 8.1.4(a)); in this case one can have the same total pressure on both sides of the membrane; (2) a fraction of the permeated gas stream (shown by a dashed line) is mixed with the feed gas stream (recycled), compressed and introduced back into the column top. An additional countercurrent configuration appears somewhat analogous to the distillation column of Figure 8.1.3(a); the permeated low-pressure gas is compressed and introduced (recycled) back into the column top (after withdrawal of a product stream), with the feed gas introduced somewhere in the middle of the column (with a resemblance to the top section of the distillation column (Figure 8.1.4(b)). This latter configuration is called the *continuous-membrane column*.

Figure 8.1.4(c) introduces through the example of dialysis (see Section 4.3.1) a countercurrent flow configuration of a membrane device where two separate feed streams are entering the separator as in the separation system type (2). In the electrodialysis (Section 3.4.2.5) process of selective transport of ions through an ion exchange membrane, the liquid solutions on two sides of any ion exchange membrane are sometimes in countercurrent flow.

The countercurrent flow system for two phases/regions with the force responsible for separation acting perpendicular to both phases/regions is very effective in developing separation, especially with high levels of purifications, as we will see later in this section. It is convenient to implement this physically for many two-phase/two-region separation systems. However, for a few systems, either (1) other modes of operation of two phases/regions are much more convenient or (2) there are inherent problems to achieving direct countercurrent flow. In both cases, separation schemes/arrangements have been developed to achieve de facto countercurrency, even though locally countercurrent flow is not present. We will now illustrate each case.

Consider situation (1) identified above. In the separation processes using a vapor–liquid system (distillation) or a gas–liquid system (absorption/stripping), a most common and convenient mode of phase contacting employs crossflow on a horizontal plate (see Figure 2.1.2(b)): vapor/gas coming from the bottom and bubbling through a liquid layer flowing perependicular to this vapor/gas flow in crossflow. However, many such plates are located in a vertical column (Figure 8.1.5) such that one can achieve, on an overall basis, a countercurrent flow of the vapor and the liquid or the gas and the liquid. Each plate acts as a separator or a stage; but the vertical column integrates the vapor or gas and the liquid flow as if we have countercurrent flow of gas and liquid phases (Figure 8.1.5). The vertical column in such a case becomes a multistaged countercurrent separation device instead of a device where the two phases/regions are in continuous countercurrent flow/contact throughout the device/column length.

We now illustrate examples corresponding to situation (2). In some two-phase configurations based on a fluid–solid system, with the solid phase consisting of, say, solid adsorbent particles, the flow of the solid particles in a vertical, or any other, direction is difficult to implement without encountering substantial attritional losses, eroding the device wall, etc. To avoid such losses, solid particles remain in a fixed bed, as in Sections 7.1.1 and 7.1.5; however, the arrangement of the fluid inlets and exits is manipulated as if there is countercurrency via the movement of the fluid past the particles. In Figure 8.1.6 one such arrangement (*simulated moving bed* (SMB)) is shown. A vertical packed-bed column may be divided into four sections, with the liquid flowing up from the bottom to the

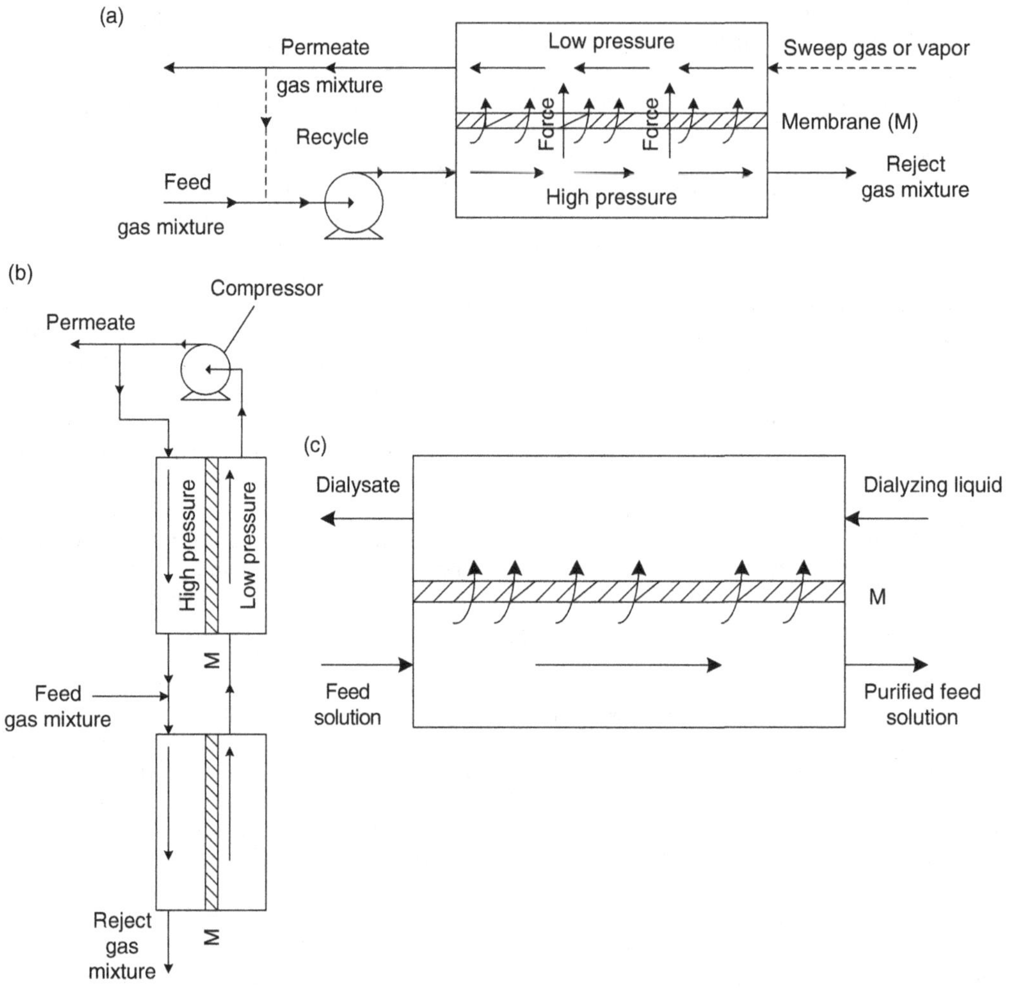

Figure 8.1.4. (a) Type (4) systems. Countercurrent flow of feed gas mixture and permeated gas mixture in a membrane device. (b) Continuous membrane column method of gas mixture separation. (c) Countercurrent dialyzer with the feed solution and the dialyzing liquid entering the device countercurrently on two sides of the membrane.

top section of the column. Liquid streams are introduced or withdrawn at different locations, shown as an eluent (desorbent), extract, feed, raffinate, etc., as if we had true countercurrent flow with the adsorbent particles flowing vertically downward, as shown in Figure 8.1.6. However, in a SMB, adsorbent particles do not come down. Therefore the upward relative velocity of the liquid vis-à-vis the particles has to be increased. This is achieved as follows: at selected time intervals, the two liquid inlet points and the two liquid exit points are advanced upward in the column by means of an external rotary valve in a programmed sequence (e.g. the Sorbex® process of UOP Inc.) as if we have countercurrent flow.

Figure 8.1.6 has been drawn for *true* countercurrent flow of two phases, solid adsorbent and the liquid, for the separation of two species 1 and 2. Species 1 is preferentially adsorbed by the adsorbent. Therefore in sections 2 and 3, it comes down with the adsorbent if the adsorbent particles were in countercurrent flow. In section 4, desorbent liquid or eluent introduced into the column bottom strips both species 1 and 2 from the adsorbent and moves them up the column in the direction of liquid flow. Part of this stream is withdrawn as the product/extract in between sections 3 and 4. This stream is primarily enriched in the strongly adsorbed species 1, since the species 2 content of the adsorbent entering section 4 is quite low. In section 3, the upflowing desorbent/eluent liquid strips the adsorbent stream of species 2. The adsorbent had picked up species 1 and 2 from the feed liquid stream introduced in between sections 2 and 3. However, the conditions are such that species 1 moves down and species 2 moves up. Species 2 is removed by the stream labeled "Raffinate" between

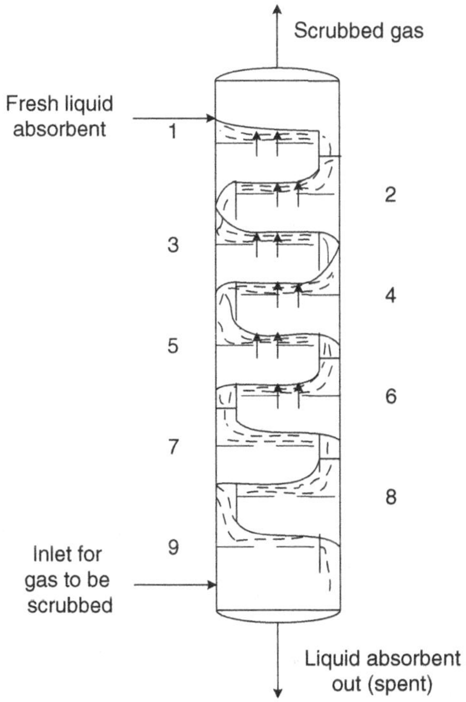

Figure 8.1.5. Nine-plate (nine-stage) gas–liquid scrubber with the absorbent liquid in crossflow over each perforated plate, through which the gas bubbles move up; the overall pattern of flow of the gas and liquid phases is countercurrent.

sections 2 and 1. In Section 8.1.1.3, the concentration wave velocity of a species has been developed for countercurrent flow. We will employ this information to provide an elementary analysis of the SMB in Section 8.1.6.

In the following part of this section, we provide simple mathematical descriptions of a few common features of two-phase/two-region countercurrent devices, specifically some general considerations on equations of change, operating lines and multicomponent separation capability. Sections 8.1.2, 8.1.3, 8.1.4, 8.1.5 and 8.1.6 cover two-phase systems of gas–liquid absorption, distillation, solvent extraction, melt crystallization and adsorption/SMB. Sections 8.1.7, 8.1.8 and 8.1.9 consider the countercurrent membrane processes of dialysis (and electrodialysis), liquid membrane separation and gas permeation. The subsequent sections cover very briefly the processes in gas centrifuge and thermal diffusion.

In the following developments, we rely on the results of Section 6.2.1.1 and identify the equations of change of concentration of a species i in a countercurrent two-region/two-phase system; we focus on two-phase systems. Next we consider the equations for operating lines in such devices. The multicomponent separation capability of such systems is treated next in the context of a two-phase system.

8.1.1.1 *Introduction to equations of change of concentration in a countercurrent device*

Now that we have an understanding of the variety of separation systems where two phases/regions are in continuous countercurrent flow along the device length, it would be useful to specify the equations of change of concentration of species i in the two phases/regions along the direction of the mean flow coordinate, z (positive for one stream, negative for the other). For systems where U_{ij} is zero and molecular diffusion and partitioning between two phases driven by $-\nabla\mu_i$ is the primary driving force, equations (6.2.30) and (6.2.31) are needed. For systems in incompressible flow, where U_{ij} is nonzero, the governing equation (6.2.40) is relevant.

In practice, equations employing quantities averaged over the cross section of the bulk flow streams are employed. For two immiscible phase based separation systems, the corresponding equations for phases $j = 1$ and $j = 2$ may be obtained from equation (6.2.33) as

$$\varepsilon_1 \frac{\partial \overline{C}_{i1}}{\partial t} + \langle v_{1z}\rangle \frac{\partial \overline{C}_{i1}}{\partial z} = \varepsilon_1 D_{i,\text{eff},1} \frac{\partial^2 \overline{C}_{i1}}{\partial z^2} - K_{i1c} a\left(\overline{C}_{i1} - \overline{C}^*_{i1}\right); \tag{8.1.1a}$$

$$\varepsilon_2 \frac{\partial \overline{C}_{i2}}{\partial t} + \langle v_{2z}\rangle \frac{\partial \overline{C}_{i2}}{\partial z} = \varepsilon_2 D_{i,\text{eff},2} \frac{\partial^2 \overline{C}_{i2}}{\partial z^2} + K_{i2c} a\left(\overline{C}^*_{i2} - \overline{C}_{i2}\right) \tag{8.1.1b}$$

if species i is being transferred from phase 1 to phase 2, with a being the interfacial area between phases 1 and 2 per unit total volume. Here $\overline{C}^*_{i1}$ is the hypothetical concentration of species i in phase 1, which will be in equilibrium with $\overline{C}_{i2}$, the bulk concentration of species i in phase 2 averaged over the flow cross section. Correspondingly, $\overline{C}^*_{i2}$ is the hypothetical concentration of species i in phase 2, which will be in equilibrium with $\overline{C}_{i1}$, the bulk concentration of species i in phase 1. Further, K_{i1c} and K_{i2c} are the corresponding overall mass-transfer coefficients for molar concentrations of species i based on phases 1 and 2, respectively. *Note*: One of the two velocities, $\langle v_{1z}\rangle$ and $\langle v_{2z}\rangle$, is positive in the z-direction, whereas the other is negative. Note further that, in equations (8.1.1a,b),

$$K_{i1c}\left(\overline{C}_{i1} - \overline{C}^*_{i1}\right) = K_{i2c}\left(\overline{C}^*_{i2} - \overline{C}_{i2}\right). \tag{8.1.1c}$$

We will briefly develop equation (8.1.1a) now by developing a mass balance over the cross section of the countercurrent column with respect to phase 1 (one can do it similarly for phase 2 in the manner of equation (7.1.4)). Figure 8.1.7 illustrates a countercurrent flow based column of length L, in which phase 1 flows upward in a positive z-direction and phase 2 flows downward; phase 1 is losing species i and phase 2 is gaining species i. Focus on the column cross-sectional area bounded by axial locations z and $z + dz$: phase 1 is imagined to flow through the region ABCD and phase 2 is imagined to flow through the region EFHG, wherein the interface between phases 1 and 2 is

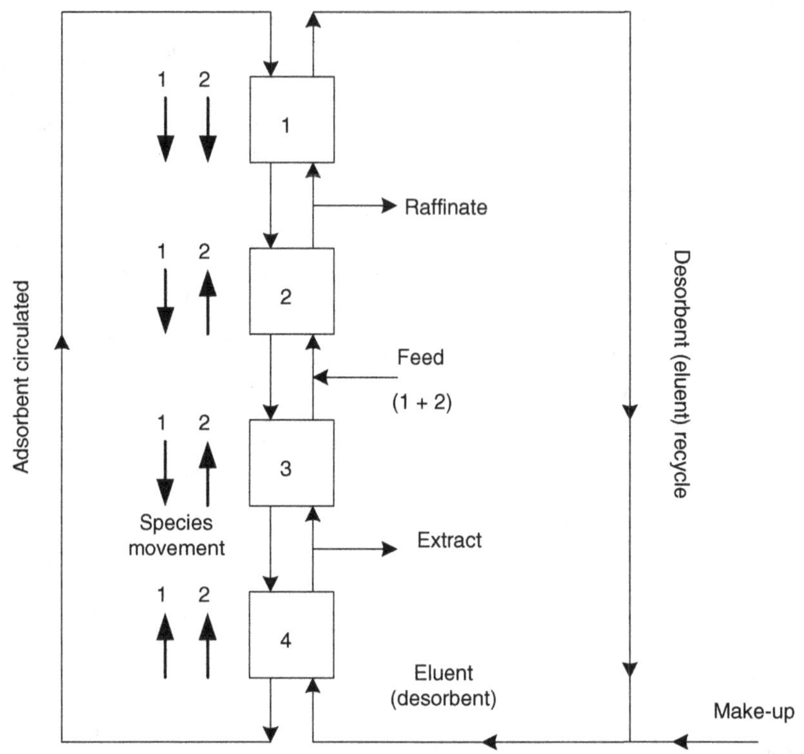

Figure 8.1.6. Various sections, species movement and phase velocity directions in a true vertical countercurrent system for solid adsorbent and liquid phase for separation of two species 1 and 2.

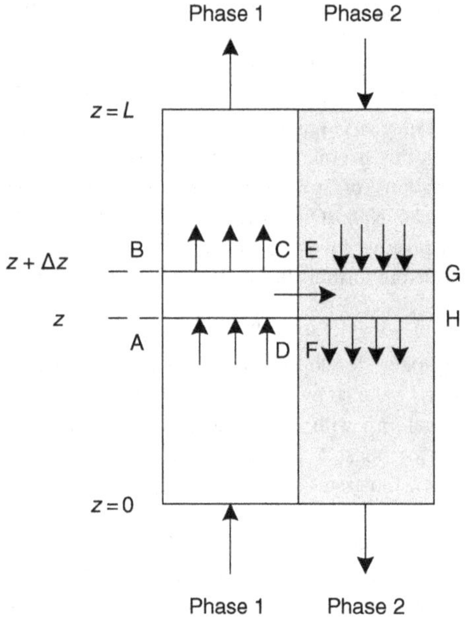

Figure 8.1.7. Countercurrent flow of two phases in a column: schematic and control volume ABCDEFGH.

located between lines CD and EF (lines CD and EF are identical). In reality, the phases are most often dispersed in each other and therefore flow over the whole cross-section. The column cross-sectional area is S_c (see Figure 7.1.1); the cross-sectional area occupied by flowing phase 1 is $\varepsilon_1 S_c$ and that by phase 2 is $\varepsilon_2 S_c$. We now adopt the procedure employed to develop equations (7.1.1) and (7.1.3), except we focus only on phase 1 (adopting the pseudocontinuum approach). (*Note*: v_{1z} is the actual velocity of phase 1 based on the phase 1 cross section; therefore $\varepsilon_1 v_{1z}$ is the superficial velocity of phase 1 through the column, $\langle v_{1z} \rangle$.) We have

$$\begin{pmatrix}\text{rate of}\\ \text{accumulation}\\ \text{of species } i\\ \text{in volume}\\ \text{ABCD}\end{pmatrix} = \begin{pmatrix}\text{rate of convective and}\\ \text{diffusive/dispersive}\\ \text{inflow of species } i \text{ into}\\ \text{volume ABCD through AD}\end{pmatrix}_{\text{at } z} - \begin{pmatrix}\text{rate of convective}\\ \text{and diffusive/dispersive}\\ \text{outflow of species } i \text{ out}\\ \text{of volume ABCD}\\ \text{through BC}\end{pmatrix}_{\text{at } z+\Delta z} - \begin{pmatrix}\text{mass-transfer}\\ \text{rate out through}\\ \text{interface CDEF}\end{pmatrix}; \qquad (8.1.1\text{d})$$

$$\varepsilon_1 S_c \Delta z \frac{\partial \overline{C}_{i1}}{\partial t} = \varepsilon_1 S_c \left[\underset{\text{convective}}{v_{1z}\overline{C}_{i1}} - \underset{\text{diffusive/dispersive}}{D_{i,\text{eff},1}} \frac{\partial \overline{C}_{i1}}{\partial z} \right]_z$$

$$- \varepsilon_1 S_c \left[\underset{\text{convective}}{v_{1z}\overline{C}_{i1}} - \underset{\text{diffusive/dispersive}}{D_{i,\text{eff},1}} \frac{\partial \overline{C}_{i1}}{\partial z} \right]_{z+\Delta z}$$

$$- K_{i1c}(aS_c\Delta z)\left(\overline{C}_{i1} - C_{i1}^*\right), \quad (8.1.1e)$$

where a is the interfacial area between phases 1 and 2 at CDEF per unit of total volume in ABCDEFHG and $S_c\Delta z$ is the total volume of ABCDEFHG. Dividing all terms by $S_c\Delta z$ and taking the limit of $\Delta z \to 0$, we get

$$\varepsilon_1 \frac{\partial \overline{C}_{i1}}{\partial t} = -\varepsilon_1 \frac{\partial (v_{1z}\overline{C}_{i1})}{\partial z} + \underset{\text{diffusive/dispersive}}{\varepsilon_1 D_{i,\text{eff},1}} \frac{\partial^2 \overline{C}_{i1}}{\partial z^2}$$

$$- K_{i1c}\, a\left(\overline{C}_{i1} - \overline{C}_{i1}^*\right). \quad (8.1.1f)$$

If v_{1z} is constant along z, then we can write

$$\varepsilon_1 \frac{\partial \overline{C}_{i1}}{\partial t} + \langle v_{1z}\rangle \frac{\partial \overline{C}_{i1}}{\partial z} = \underset{\text{diffusive/dispersive}}{\varepsilon_1 D_{i,\text{eff},1}} \frac{\partial^2 \overline{C}_{i1}}{\partial z^2} - K_{i1c}\, a\left(\overline{C}_{i1} - \overline{C}_{i1}^*\right). \quad (8.1.1g)$$

To be noted here is that $D_{i,\text{eff},1}$ is developed based on the whole cross section, but, for the phase 1 equation, ε_1 is needed.

Generally, countercurrent flow of two immiscible phases contacting each other in one continuous contacting device (Figures 8.1.2 and 8.1.3) is carried out under steady state conditions. Therefore equations (8.1.1a) and (8.1.1b) are reduced to

$$\langle v_{1z}\rangle \frac{\partial \overline{C}_{i1}}{\partial z} = \varepsilon_1 D_{i,\text{eff},1} \frac{\partial^2 \overline{C}_{i1}}{\partial z^2} - K_{i1c}\, a\left(\overline{C}_{i1} - \overline{C}_{i1}^*\right); \quad (8.1.2a)$$

$$\langle v_{2z}\rangle \frac{\partial \overline{C}_{i2}}{\partial z} = \varepsilon_2 D_{i,\text{eff},2} \frac{\partial^2 \overline{C}_{i2}}{\partial z^2} + K_{i2c}\, a\left(\overline{C}_{i2}^* - \overline{C}_{i2}\right). \quad (8.1.2b)$$

For a given two-phase based countercurrent flow system, there will be appropriate boundary/initial conditions. If we neglect any axial/longitudinal dispersion/diffusion in the flowing system, these two steady state equations are reduced to

$$\langle v_{1z}\rangle \frac{\partial \overline{C}_{i1}}{\partial z} = -K_{i1c}\, a\left(\overline{C}_{i1} - \overline{C}_{i1}^*\right); \quad (8.1.3a)$$

$$\langle v_{2z}\rangle \frac{\partial \overline{C}_{i2}}{\partial z} = +K_{i2c}\, a\left(\overline{C}_{i2}^* - \overline{C}_{i2}\right). \quad (8.1.3b)$$

Correspondingly, an overall balance of species i for both phases $j = 1, 2$ is available from equation (6.2.34) as (or by adding equations (8.1.3a) and (8.1.3b))

$$\langle v_{1z}\rangle \frac{\partial \overline{C}_{i1}}{\partial z} + \langle v_{2z}\rangle \frac{\partial \overline{C}_{i2}}{\partial z} = 0. \quad (8.1.4a)$$

If we multiply this equation by the device flow cross-sectional area S_c, we get

$$\langle v_{1z}\rangle S_c \frac{\partial \overline{C}_{i1}}{\partial z} + \langle v_{2z}\rangle S_c \frac{\partial \overline{C}_{i2}}{\partial z} = 0. \quad (8.1.4b)$$

If the molar concentrations of the individual species may be expressed in terms of the corresponding mole fractions $\overline{x}_{i1}$ and $\overline{x}_{i2}$, then

$$\overline{C}_{i1} = C_{t1}\,\overline{x}_{i1}; \qquad \overline{C}_{i2} = C_{t2}\,\overline{x}_{i2}, \quad (8.1.4c)$$

where C_{t1} and C_{t2} are the total molar concentrations of phases 1 and 2, respectively. If the total molar flow rates of phases 1 and 2 are W_{t1} and W_{t2}, respectively, then, if W_{t1} and W_{t2} do not change along the z-coordinate, we get

$$\langle v_{1z}\rangle S_c\, C_{t1} \frac{\partial \overline{x}_{i1}}{\partial z} + \langle v_{2z}\rangle S_c\, C_{t2} \frac{\partial \overline{x}_{i2}}{\partial z} = 0; \quad (8.1.4d)$$

$$W_{t1} \frac{\partial \overline{x}_{i1}}{\partial z} + W_{t2} \frac{\partial \overline{x}_{i2}}{\partial z} = 0, \quad (8.1.5)$$

provided C_{t1} and C_{t2} are *essentially constant along the z-coordinate* or change very little. This is valid under the following two conditions:

(1) species i being transferred from one phase to another is present in a very dilute solution;
(2) the species transport between the two phases is taking place such that the total molar flow rate in a given phase does not change in the direction of mean flow (z-direction). This situation is realized approximately under the condition of equimolar counterdiffusion (see Figure 3.1.4); this condition is also characterized by the assumption of *constant molar overflow*[1] in column distillation. In the absence of axial diffusion/dispersion, equation (8.1.5) is the governing equation for *transfers in dilute solutions* or under conditions of *constant molar overflow*.

There are separation systems where the molar flow rate in each phase/region ($j = 1, 2$) changes very substantially along the z-direction (Figures 8.1.4(a), (b)). Then, instead of equation (6.2.33), we will obtain the following equation for region j from the general equation (6.2.30) if $U_{ik} = 0$:

$$\varepsilon_j \frac{\partial \overline{C}_{ij}}{\partial t} + \frac{\partial (\langle v_{jz}\rangle \overline{C}_{ij})}{\partial z} = \varepsilon_j D_{i,\text{eff},j} \frac{\partial^2 \overline{C}_{ij}}{\partial z^2} - K_{ijc} a\left(\overline{C}_{ij} - \overline{C}_{ij}^*\right). \quad (8.1.6)$$

The final term, describing transport of species i via a mass-transfer coefficient, may be replaced by a membrane transport rate expression for a membrane process. For steady state, if we neglect the contribution of longitudinal

[1] Requires the latent heat of the two species (e.g.) to be identical and the dependences of the enthalpy of both the vapor and the liquid mixture to vary linearly with the mole fraction based composition (see Treybal (1980, pp. 402–403) and Doherty and Malone (2001, pp. 507–512)).

diffusion/dispersion, we get, for species i balance in region/phase j,

$$\frac{\partial(\langle v_{jz}\rangle \overline{C}_{ij})}{\partial z} = -K_{ijc}a\left(\overline{C}_{ij} - \overline{C}_{ij}^{*}\right). \tag{8.1.7a}$$

Multiplying by the device flow cross-sectional area S_c, we get

$$\frac{\partial(S_c\langle v_{jz}\rangle \overline{C}_{ij})}{\partial z} = -K_{ijc}aS_c\left(\overline{C}_{ij} - \overline{C}_{ij}^{*}\right) = \frac{\partial(x_{ij}W_{tj})}{\partial z}. \tag{8.1.7b}$$

Summing such a relation for two phases/regions ($j = 1, 2$), we get

$$\frac{\partial(\overline{x}_{i1}W_{t1})}{\partial z} + \frac{\partial(\overline{x}_{i2}W_{t2})}{\partial z} = 0, \tag{8.1.8}$$

a very useful result for a number of membrane systems where W_{t1} or W_{t2} is likely to vary considerably along the device length. This equation is also needed for those phase equilibrium based systems where the flow rate changes substantially due to substantial changes in concentration of the species transferred along the column length. We have not considered here any specific one-dimensional z-directional balance equations for external force field based countercurrent separation systems. These will be considered at the end of Section 8.1.

8.1.1.2 *Equation for the operating line in a countercurrent device*

To develop a better understanding of how the stream compositions change with respect to each other along the countercurrent device length in the absence of dispersion and axial diffusion, equations (8.1.5) or (8.1.8) are frequently used. However, instead of a differential equation, algebraic equations relating $\overline{C}_{il}$ to $\overline{C}_{i2}$ or $\overline{x}_{i1}$ to $\overline{x}_{i2}$ are developed. The lines represented by such equations are usually called *operating lines*; sometimes the relationship is linear.

Consider equation (8.1.5) and a countercurrent device for a *separation system of type (2)*, as shown in Figure 8.1.8(a). We may rewrite this equation as

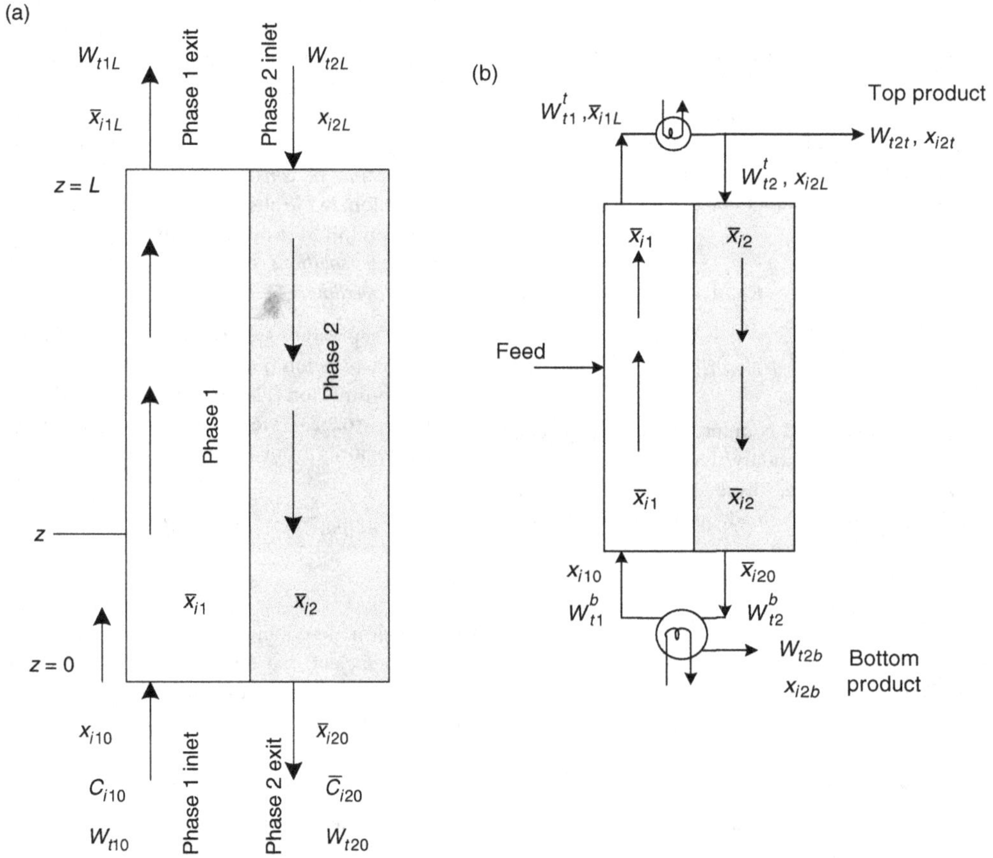

Figure 8.1.8. (a) Schematic of the countercurrent flow device for a separation system of type (2). (b) Schematic of the countercurrent flow device for a separation system of type (3).

$$\langle v_{1z}\rangle S_c\, C_{t1}\, \mathrm{d}\overline{x}_{i1} = -\langle v_{2z}\rangle S_c\, C_{t2}\, \mathrm{d}\overline{x}_{i2}; \tag{8.1.9}$$

$$\frac{\mathrm{d}\overline{x}_{i2}}{\mathrm{d}\overline{x}_{i1}} = \frac{\langle v_{1z}\rangle S_c\, C_{t1}}{-\langle v_{2z}\rangle S_c\, C_{t2}} = \frac{|W_{t1}|}{|W_{t2}|}, \tag{8.1.10}$$

since phase 2 flows in the negative z-direction and $|W_{t1}|$ and $|W_{t2}|$ are the magnitudes of the total molar flow rates in phases/regions 1 and 2. For dilute systems or systems satisfying the condition of constant molar overflow, the ratio $|W_{t1}|/|W_{t2}|$ is essentially constant along the z-coordinate. Integrating from $z = 0$ (phase 1 inlet, x_{i10}, C_{i10}; phase 2 exit, $\overline{x}_{i20}$, $\overline{C}_{i20}$) to any z ($\overline{x}_{i1}$, $\overline{C}_{i1}$; $\overline{x}_{i2}$, $\overline{C}_{i2}$), we get

$$\overline{x}_{i2} = \frac{|W_{t1}|}{|W_{t2}|}(\overline{x}_{i1} - \overline{x}_{i10}) + \overline{x}_{i20} \tag{8.1.11}$$

as the equation for the operating line for separation systems of type (2) relating the two species i mole fractions, $\overline{x}_{i2}$ and $\overline{x}_{i1}$, in the two phases/regions at any value of z along the device for dilute systems or systems having constant molar overflow. An alternative representation of the operating line may be developed by integrating equation (8.1.10) between any location z and $z = L$ (where $\overline{x}_{i1} = \overline{x}_{i1L}$ and $\overline{x}_{i2} = \overline{x}_{i2L}$):

$$\overline{x}_{i2} = \frac{|W_{t1}|}{|W_{t2}|}(\overline{x}_{i1} - \overline{x}_{i1L}) + \overline{x}_{i2L}. \tag{8.1.12}$$

We now develop the equation for an operating line for a separation system of type (3), illustrated in Figure 8.1.8(b), with a reflux at the top and at the bottom. Under the condition of constant molar overflow, we use equation (8.1.10) again between any axial location z and the top of the column $z = L$. After integration, we get

$$|W_{t2}^t|\, x_{i2L} - |W_{t2}^t|\, \overline{x}_{i2} = |W_{t1}^t|\, \overline{x}_{i1L} - |W_{t1}^t|\, \overline{x}_{i1}, \tag{8.1.13}$$

where the superscript t on the molar flow rate indicates the top half of the column; subscripts t and b imply top product and bottom product, respectively. However, a molar balance on species i between stream 1 leaving the column top, stream 2 entering the column top and the top product (x_{i2t}) flowing out at the molar flow rate, W_{t2t}, yields

$$|W_{t1}^t|\, \overline{x}_{i1L} = |W_{t2}^t|\, x_{i2L} + W_{t2t}\, x_{i2t}. \tag{8.1.14}$$

Introducing this species i molar balance into relation (8.1.13), we get

$$|W_{t2}^t|\, \overline{x}_{i2} = |W_{t1}^t|\, \overline{x}_{i1} - W_{t2t}\, x_{i2t};$$

$$\overline{x}_{i2} = \frac{|W_{t1}^t|}{|W_{t2}^t|}\, \overline{x}_{i1} - \frac{W_{t2t}}{|W_{t2}^t|}\, x_{i2t}. \tag{8.1.15}$$

An alternative result is

$$\overline{x}_{i1} = \frac{|W_{t2}^t|}{|W_{t1}^t|}\, \overline{x}_{i2} + \frac{W_{t2t}}{|W_{t1}^t|}\, x_{i2t}, \tag{8.1.16}$$

relating the phase 1/region 1 composition $\overline{x}_{i1}$ with that ($\overline{x}_{i2}$) for phase 2/region 2 at location z in the column and the top product flow rate, W_{t2t}, and composition x_{i2t}. These represent the equations for the operating line of the top half of the column, the so-called *enriching section,* since the top product stream becomes enriched in the more volatile species in, for example, distillation.

Separation systems of type (3), with the feed entering near the middle of the column and two product streams, one at the top and the other at the bottom with reflux/recycle of the products at the top and the bottom, have two sets of operating lines. The operating line equation given by either (8.1.15) or (8.1.16) relates the two local compositions $\overline{x}_{i1}$ and $\overline{x}_{i2}$ in the top part of the column above the feed entry location (often called the enriching section) to the top product stream (W_{t2t}, x_{i2t}). Now we follow a similar procedure and develop a molar balance relation between $\overline{x}_{i1}$ and $\overline{x}_{i2}$ in the bottom half of the column and the bottom product stream (W_{t2b}, x_{i2b}). First, we require a relation between $\overline{x}_{i1}$, $\overline{x}_{i2}$ and $\overline{x}_{i10}$, $\overline{x}_{i20}$:

$$|W_{t2}^b|\, \overline{x}_{i2} = |W_{t1}^b|\, \overline{x}_{i1} - |W_{t1}^b|\, \overline{x}_{i10} + |W_{t2}^b|\, \overline{x}_{i20}, \tag{8.1.17}$$

where the superscripts b on W_{t1} and W_{t2} reflect the bottom half of the column. However, a molar balance of species i between stream 1 entering the column bottom, stream 2 leaving the column bottom and the bottom product stream (W_{t2b}, x_{i2b}) yields

$$|W_{t2}^b|\, \overline{x}_{i20} = W_{t2b}\, x_{i2b} + |W_{t1}^b|\, \overline{x}_{i10}. \tag{8.1.18}$$

Introducing this relation into (8.1.17), we get

$$|W_{t2}^b|\, \overline{x}_{i2} = |W_{t1}^b|\, \overline{x}_{i1} + W_{t2b}\, x_{i2b}, \tag{8.1.19}$$

which may be rewritten as either

$$\overline{x}_{i2} = \frac{|W_{t1}^b|}{|W_{t2}^b|}\overline{x}_{i1} + \frac{W_{t2b}}{|W_{t2}^b|}\, x_{i2b} \tag{8.1.20}$$

or

$$\overline{x}_{i1} = \frac{|W_{t2}^b|}{|W_{t1}^b|}\overline{x}_{i2} - \frac{W_{t2b}}{|W_{t1}^b|}\, x_{i2b}. \tag{8.1.21}$$

These are considered to be equations for the operating line of the bottom half of the column, the so-called *stripping section,* since the bottom product stream gets stripped of the more volatile species in, for example, distillation.

As mentioned earlier, in Section 8.1.1, in separation systems of both type (2) and type (3) (illustrated in Figures 8.1.2 and 8.1.3), the local flow conditions of the two phases/regions vis-à-vis each other may not be countercurrent; often, it is crossflow, even though the overall arrangement of these two flow streams is in countercurrent flow. We consider here an arrangement characteristic of separation systems of type (2), where we have many local two-phase contacting stages in crossflow, with the connection between the stages in countercurrent flow (Figure 8.1.9). The stage characterizing number increases toward the inlet of phase 1 from $n = 1$ at the top of the column to $n = N$ at

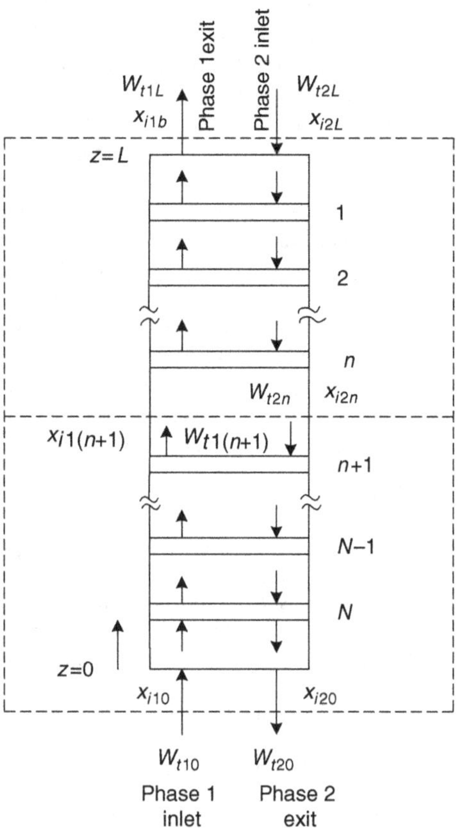

Figure 8.1.9. Multistage countercurrent flow schematic for a separation system of type (2) with individual stages having, say, crossflow.

the bottom. At $z = 0$, the molar flow rates and compositions are indicated by: phase 1, W_{t10}, x_{i10}; phase 2, W_{t20}, x_{i20}. At $z =$ L, the top of the column, the corresponding quantities are W_{t1L}, x_{i1L}; W_{t2L}, x_{i2L}. Stream 2 entering a stage, say $n+1$, from stage n above it will be characterized as W_{t2n}, x_{i2n}. Stream 1 entering a stage, say n, from stage $n+1$ below it will be characterized as $W_{t1(n+1)}$, $x_{i1(n+1)}$. We will now consider molar balances on species i, first around plate n and then from plate n to the top of the column ($z =$ L). Consider the balance around plate n as follows:

$$W_{t1(n+1)}\, x_{i1(n+1)} + W_{t2(n-1)}\, x_{i2(n-1)} = W_{t1n}\, x_{i1n} + W_{t2n}\, x_{i2n}. \tag{8.1.22}$$

If we now consider a balance over the envelope covering the top exit of the column and a location between stages n and $n+1$, we get

$$W_{t1(n+1)}\, x_{i1(n+1)} + W_{t2L}\, x_{i2L} = W_{t1L}\, x_{i1L} + W_{t2n}\, x_{i2n}. \tag{8.1.23}$$

It is now possible to develop a relation between $x_{i1(n+1)}$ and x_{i2n}, the compositions of the two phases/streams at any location of the column to those at the column exit/inlet at the top:

$$x_{i1(n+1)} = \frac{W_{t2n}}{W_{t1(n+1)}}\, x_{i2n} + \frac{W_{t1L}\, x_{i1L}}{W_{t1(n+1)}} - \frac{W_{t2L}\, x_{i2L}}{W_{t1(n+1)}}. \tag{8.1.24}$$

In the case of dilute streams 1 and 2 or a constant molar overflow assumption, this relation will be simplified to

$$x_{i1(n+1)} = \frac{W_{t2}}{W_{t1}}\, x_{i2n} + x_{i1L} - \frac{W_{t2}}{W_{t1}} x_{i2L}$$
$$= \frac{W_{t2}}{W_{t1}} (x_{i2n} - x_{i1L}) + x_{i1L}. \tag{8.1.25}$$

We can also develop an ith species balance over an envelope covering the bottom exit of the column and a location between the stages n and $n+1$:

$$W_{t1(n+1)}\, x_{i1(n+1)} + W_{t20}\, x_{i20} = W_{t10}\, x_{i10} + W_{t2n}\, x_{i2n}. \tag{8.1.26}$$

Correspondingly, an operating line equation relating $x_{i1(n+1)}$ and x_{i2n} with respect to the quantities at the bottom exit/inlet of the column has the following form:

$$x_{i1(n+1)} = \frac{W_{t2n}}{W_{t1(n+1)}}\, x_{i2n} + \frac{W_{t10}\, x_{i10}}{W_{t1(n+1)}} - \frac{W_{t20}\, x_{i20}}{W_{t1(n+1)}}. \tag{8.1.27}$$

8.1.1.3 *Multicomponent separation capability in a device with a countercurrent flow system*

We now focus on the multicomponent separation capability of a countercurrent flow configuration in a two-phase system. For two phases $j = 1, 2$ moving with superficial velocities of $\langle v_{1z}\rangle$ and $\langle v_{2z}\rangle$ in the z-direction, an overall balance of species i in both phases $j = 1, 2$ is obtained from equation (6.2.34) as

$$\varepsilon_1 \frac{\partial \overline{C}_{i1}}{\partial t} + \varepsilon_2 \frac{\partial \overline{C}_{i2}}{\partial t} + \langle v_{1z}\rangle \frac{\partial \overline{C}_{i1}}{\partial z} + \langle v_{2z}\rangle \frac{\partial \overline{C}_{i2}}{\partial z}$$
$$= \varepsilon_1\, D_{i,\mathrm{eff},1} \frac{\partial^2 \overline{C}_{i1}}{\partial z^2} + \varepsilon_2\, D_{i,\mathrm{eff},2} \frac{\partial^2 \overline{C}_{i2}}{\partial z^2}. \tag{8.1.28}$$

Note that with the z-coordinate being vertically upward and, say, the phase $j = 2$ moving vertically upward, $\langle v_{2z}\rangle$ is a positive quantity; phase $j = 1$ will be then moving vertically downward, and therefore $\langle v_{1z}\rangle$ is a negative quantity. We will represent it as $-|\langle v_{1z}\rangle|$, where $|\langle v_{1z}\rangle|$ is the magnitude of the velocity $\langle v_{1z}\rangle$. (To allow comparison with the fixed-bed processes of Section 7.1.1, we assume here that phase $j = 2$ moves up along the positive z-coordinate.) We now make the following three assumptions: that the operation is isothermal and nondispersive, and that the phases are locally in equilibrium. (These are exactly the same ones made to develop the de Vault equation (7.1.8) for fixed-bed adsorption.) The assumption that the two phases $j = 1$ and 2 are everywhere in equilibrium in the vertical column with respect to species i may be illustrated by the following equilibrium relation:

$$\overline{C}_{i1} = \kappa_{i1}\overline{C}_{i2}. \tag{8.1.29}$$

Due to the assumptions of isothermal nondispersive operation, equation (8.1.28) is simplified to

$$\varepsilon_1 \frac{\partial \overline{C}_{i1}}{\partial t} + \varepsilon_2 \frac{\partial \overline{C}_{i2}}{\partial t} - |\langle v_{1z}\rangle| \frac{\partial \overline{C}_{i1}}{\partial z} + \langle v_{2z}\rangle \frac{\partial \overline{C}_{i2}}{\partial z} = 0. \tag{8.1.30}$$

If the magnitudes of the actual/interstitial velocities of the two phases through the column are $|v_{1z}|$ and v_{2z} for phases $j = 1$ and 2, respectively, then the superficial velocities and the interstitial velocities are related as follows:

$$|\langle v_{1z}\rangle| = \varepsilon_1 |v_{1z}|, \qquad \langle v_{2z}\rangle = \varepsilon_2\, v_{2z}. \tag{8.1.31}$$

Employing these two relations, as well as the linear equilibrium relation (8.1.29), the overall balance equation (8.1.30) is simplified to

$$(\varepsilon_1 \kappa_{i1} + \varepsilon_2)\frac{\partial \overline{C}_{i2}}{\partial t} + \langle v_{2z}\rangle \left(1 - \frac{\kappa_{i1}|\langle v_{1z}\rangle|}{\langle v_{2z}\rangle}\right)\frac{\partial \overline{C}_{i2}}{\partial z} = 0. \tag{8.1.32}$$

Using now the relations (8.1.31), we get

$$\left(1 + \frac{\varepsilon_1}{\varepsilon_2}\kappa_{i1}\right)\frac{\partial \overline{C}_{i2}}{\partial t} + v_{2z}\left(1 - \frac{\kappa_{i1}\,\varepsilon_1\,|v_{1z}|}{\varepsilon_2\, v_{2z}}\right)\frac{\partial \overline{C}_{i2}}{\partial z} = 0, \tag{8.1.33}$$

which may be rewritten as

$$\frac{\partial \overline{C}_{i2}}{\partial t} + v_{2z}\left(\frac{1 - \kappa_{i1}\dfrac{\varepsilon_1\,|v_{1z}|}{\varepsilon_2\, v_{2z}}}{\left(1 + \dfrac{\varepsilon_1}{\varepsilon_2}\kappa_{i1}\right)}\right)\frac{\partial \overline{C}_{i2}}{\partial z} = 0. \tag{8.1.34}$$

This equation has an exactly similar form to equation (7.1.9) (the de Vault equation). Therefore, following a procedure similar to that used to obtain equations (7.1.12a) and (7.1.12e), we can obtain the following expression for the concentration wave velocity, v^*_{Ci}, of species i (Fish *et al.*, 1989):

$$v^*_{Ci} = v_{2z}\frac{\left(1 - \kappa_{i1}\dfrac{\varepsilon_1\,|v_{1z}|}{\varepsilon_2\, v_{2z}}\right)}{\left(1 + \dfrac{\varepsilon_1}{\varepsilon_2}\kappa_{i1}\right)}. \tag{8.1.35}$$

When $\kappa_{i1}\,(\varepsilon_1|v_{1z}|)/(\varepsilon_2 v_{2z}) < 1$, v^*_{Ci} is positive, and species i moves up the column along the positive z-direction, the direction of movement of phase 2. When $\kappa_{i1}\,(\varepsilon_1|v_{1z}|)/(\varepsilon_2 v_{2z}) > 1$, v^*_{Ci} is negative, and species i moves down the column along the negative z-direction, the direction of movement of phase 1. Therefore, if there is a binary mixture of species A and B (i = A, B), and we want species A to go up the column and species B to go down the column, the following relation has to be satisfied:

$$\kappa_{A1}\frac{\varepsilon_1\,|v_{1z}|}{\varepsilon_2\, v_{2z}} < 1 < \kappa_{B1}\frac{\varepsilon_1\,|v_{1z}|}{\varepsilon_2\, v_{2z}}. \tag{8.1.36}$$

Alternatively,

$$\kappa_{A1} < \frac{\varepsilon_2\, v_{2z}}{\varepsilon_1\,|v_{1z}|} < \kappa_{B1}; \tag{8.1.37}$$

$$\kappa_{A1} < \frac{\langle v_{2z}\rangle}{|\langle v_{1z}\rangle|} < \kappa_{B1} \tag{8.1.38}$$

for species A to go up and for species B to go down the column (Barker, 1971; Fish *et al.*, 1989). For a fixed bed, $|v_{1z}| = 0$ in relation (8.1.35). Therefore all species move in one direction, that of the fluid phase $j = 2$; however, they move with different values of v^*_{Ci}. For elution operation in the transient mode with a fixed adsorbent bed and a small sample injected, elution chromatography (Section 7.1.5.1) allows separation of multiple species.

Continuous countercurrent flow of two phases in a column is, however, normally implemented in a steady state fashion. Therefore an isothermal nondispersive equilibrium operation of a column will lead to the following balance equation for any species i, from equation (8.1.33):

$$v_{2z}\left(1 - \frac{\kappa_{i1}\,\varepsilon_1\,|v_{1z}|}{\varepsilon_2\, v_{2z}}\right)\frac{\partial \overline{C}_{i2}}{\partial z} = 0. \tag{8.1.39}$$

For all species moving in a given direction in a column, the solution of this equation will indicate that, at that column exit, all such species will appear/exist; therefore multicomponent separation is not possible. Only a binary separation is possible with one species moving in the opposite direction in the column and therefore available as a pure species. This is the primary reason why we will see that a countercurrent column used for steady state processes such as distillation, absorption, extraction, crystallization, etc., separates a binary mixture only. For ternany mixture separation, two columns are needed. Three columns are employed to separate a four-component mixture (see Chapter 9 for various schematics). However, if a feed sample injection is made, as in elution chromatography, into a mobile phase in countercurrent flow vis-à-vis another mobile phase, transient multicomponent separation *would* appear to be feasible. If pulse injection of one phase containing feed is introduced countercurrent to the other phase, it may be possible to achieve a multicomponent separation capability (as is true for cocurrent flow, considered in Section 8.2).

8.1.2 Gas (vapor) absorption/stripping

In Section 3.3.7.1, we were introduced to gas–liquid equilibrium relations describing how one can quantify the distribution of a volatile solute between a gas phase and a liquid phase characterized as the absorbent phase. The solute i was a gas or a vapor under the operating conditions. In Section 4.1.1, we developed expressions for the selective absorption of one gas species 1 over the other gas species 2 in a liquid; similarly, we developed the separation factor expression for absorption between two condensable vapor species. In Section 4.1.1.2, we considered selective gas/vapor stripping from a liquid absorbent by air, steam or vacuum. In Section 5.2.1.1, we were introduced to the role of chemical reactions in enhancing gas absorption (or inhibiting stripping). All these treatments were carried out in the context of an assumed equilibrium existing between the two phases. In actual gas-liquid contacting devices, equilibrium may not exist, and rates of gas absorption or stripping will control the extent of separation. Section 3.4.1.1 treated such rate processes for gas absorption without any chemical reaction. Section 5.3.1 considered in some detail the role of chemical reaction in enhancing gas absorption, as well as in developing selectivity between two gas species such as H_2S and CO_2 to levels far higher than what is achievable under equilibrium conditions.

Given this background, we are ready to consider now the gas absorption performances of countercurrent devices, or de facto countercurrent multistage devices, referred to in Sections 8.1.1 and 8.1.1.2. We will first pay attention to the nature of the physical devices where the gas and liquid phases in continuous contact flow countercurrently. We next analyze the simplest possible case, namely physical absorption of a gas/vapor present in a very dilute concentration. Then we treat the case of physical absorption of a species present in the gas stream at higher concentrations; the device schematic for both cases is represented in Figure 8.1.2(a), corresponding to a separation system of type (2). The role of axial diffusion/dispersion in the mean flow direction in a continuous countercurrent device is briefly touched upon next. Since absorption of acid gases, for example, is frequently carried in the reactive absorption mode, a simplified analysis of a chemical absorber is provided in Section 8.1.2.5. This is followed by the case of gas/vapor absorption in a multistaged device in one column.

8.1.2.1 *Devices used in countercurrent gas-liquid separation and flooding correlations for dispersive devices*

There are two broad classes of countercurrent gas-liquid separation devices: *dispersion based* and *nondispersive*. Dispersion based devices are ubiquitous and used throughout industry, and we consider these first. We will discuss the two common types of these devices for gas-liquid separation: packed towers and plate towers.

A packed tower usually consists of a vertical column, a tower, which is filled with packings (solid elements having a large surface area) in a random fashion, resulting in *random packings*. The gas flows up the column; the liquid distributed into thin films over the surfaces of random packings flows down the column, creating a large gas-liquid interfacial area through which mass can be transferred (Figure 8.1.10(a)). Instead of random tower packings (Figure 8.1.10(b)), one could use regular packings or stacked packings. Here the packed column is filled with an ordered arrangement of packings of various types. Often these packings are prepared from corrugated metal gauze and sheet metal cobbled together very carefully so that the fluid flow pressure drops are quite low due to reduced presence of form drag. Consequently they are quite useful in vacuum applications. They also fit the column diameter and are routinely called *structured packings* (SPs) (Figure 8.1.10(c)).

When considering these devices for countercurrent gas-liquid separation, a few topics are of great practical importance. What are the practical ranges of gas and liquid flow rates in such columns? What is the pressure drop encountered by the gas phase? A brief discussion on the flow conditions in a packed column is useful. When the liquid flow rate is quite low, the liquid will just flow through a few random channels: there will not be enough liquid flow to cover the available packing surfaces, resulting in what is called *channeling*. The gas will be flowing up through sections of the column where there will be no liquid, so no gas-liquid mass transfer will occur in these locations.

As the liquid flow rate is increased, the surfaces of all packings will have liquid flowing over them. If the gas flow rate is kept fixed, increasing the liquid flow rate means increasing the liquid film thickness over the packing surfaces and decreasing the flow cross-sectional area for gas. Therefore the gas flow pressure drop starts increasing as the liquid holdup (defined as the volume of liquid in the column divided by the volume of the packed column) in the column increases. This liquid holdup increases rapidly with the gas flow rate, leading to a condition called *loading*. Operation under loading conditions ensures complete utilization of the packing surface area for gas-liquid transport. The minimum liquid velocities recommended to ensure complete wetting of different packings are (Seader and Henley, 1998, p. 327): ceramic, 0.00049 ft/s; plastic, 0.00394 ft/s; bright metal, 0.0029 ft/s.

As one increases the gas flow rate or liquid flow rate further, one encounters what is generally known as *flooding* of the column. Flooding describes a variety of conditions: the appearance of a layer of liquid at the top of the packings through which gas appears as bubbles; liquid fills the tower completely and gas appears dispersed as bubbles; slugs of foam rise through the column (Treybal, 1980). The gas pressure drop is increased considerably. It is not advisable to run the packed tower under such conditions.

It is useful to know the conditions under which flooding may be encountered so as to avoid it during gas-liquid absorption or stripping; it is also useful to estimate the gas-phase pressure drop as a function of the two phase flow rates in packed beds containing random or structured packings. Such estimations for random packings are generally made from the Sherwood-Eckert generalized pressure drop correlation chart, illustrated graphically in

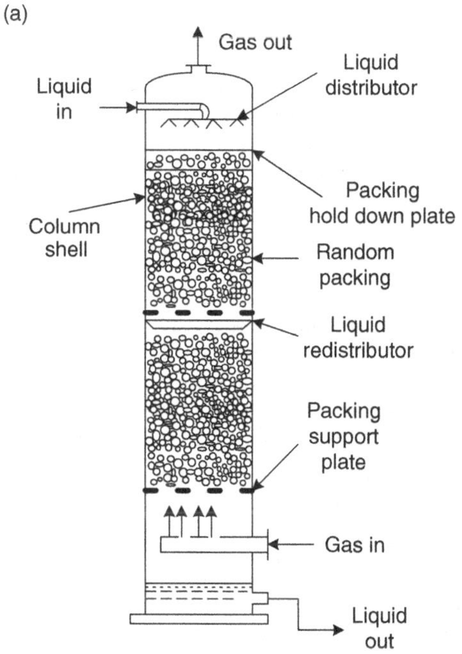

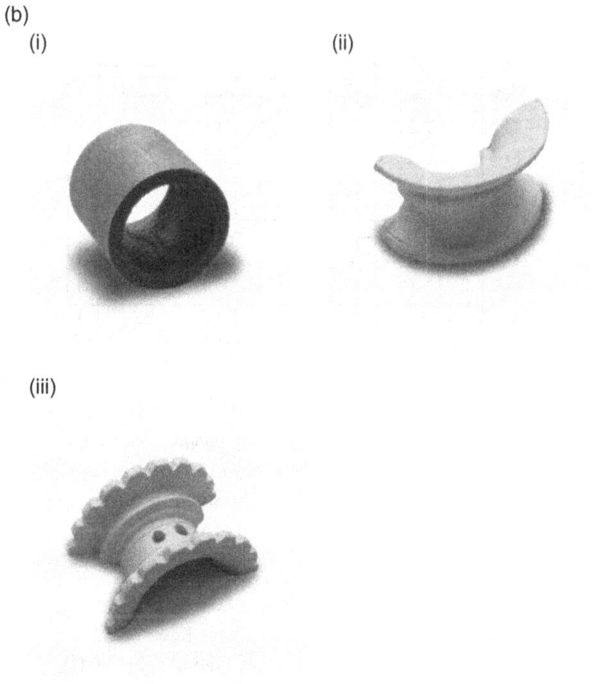

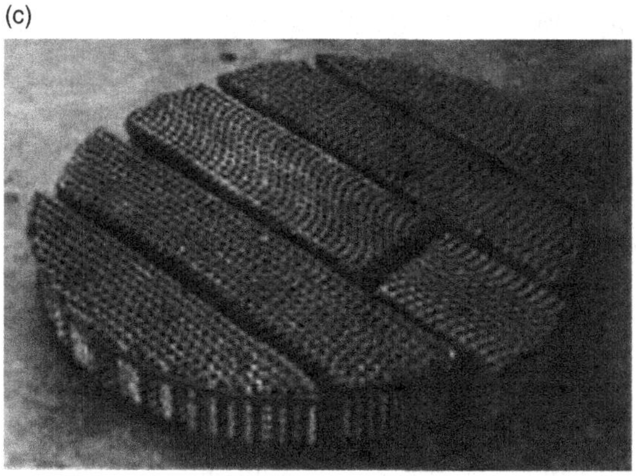

Figure 8.1.10. (a) Randomly packed column for gas absorption/stripping. (b) Some random column packings: (i) Saint-Gobain NorPro Carbon Raschig ring; (ii) Saint-Gobain NorPro Norton™ Saddle; (iii)Saint-Gobain NorPro Norton™ Supper Saddle. Copyright 2012, Saint-Gobain NorPro, Stow, Ohio, used with permission. (c) Mellapak 250.Y with the elements separated (Fitz et al., *1999). Reprinted, with permission, from* I & E Chem. Res., *38, 512 (1999), Figure 1. Copyright (1999) American Chemical Society.*

Figure 8.1.11. The ordinate of this log-log plot (Eckert, 1970), $\left(G_g^2 F_p \psi \mu_\ell^{0.2} / \rho_g \rho_\ell g_c\right)$, often called the *capacity factor*, C_F, contains the following quantities, specified with their units: G_g is the superficial mass average gas velocity based on empty column cross section $v_{0g}\rho_g$ (lbmass/ft^2-s); F_p is a packing factor (unit of 1/ft); ψ is the ratio of $\rho_{\text{water}}/\rho_\ell$, where ρ_ℓ is the density of the liquid in lbmass/ft^3; μ_ℓ is the liquid viscosity in centipoise; ρ_g is the gas density in lbmass/ft^3; $g_c = 32.2$. This correlation is such that the specified units for each quantity in the ordinate has to be used. The quantity F_p (unit, 1/ft): is approximately inversely proportional to the packing size (Wankat, 2007): $F_p \propto$ (characteristic packing dimension, in inches)$^{-1.1}$. The abscissa of this plot (often called the *flow parameter*), $(v_{0\ell}\rho_\ell / v_{0g}\rho_g)(\rho_g/\rho_\ell)^{0.5}$, is a product of the ratios of two quantities: the ratio of the superficial mass average velocity of the liquid $v_{0\ell}\rho_\ell$ ($= G_\ell$) and the superficial mass average velocity of the gas $v_{0g}\rho_g$ ($= G_g$) (recall that v_0 is the superficial velocity based on empty cross section of the column (equation (6.1.4a)); and the ratio of the gas density to the liquid density. There are a number of curves in this plot depending on the value of the gas pressure drop in inches of water per foot of packed column. This parameter varies between the values of 0.05 and 1.50, beyond which lies the flooding limit; specifically, the following ranges have been suggested (Cussler, 1997; Wankat, 2007): 0.2–0.6 for absorbers, 0.1–0.4 for vacuum operations, 0.4–0.8 for higher-pressure operations.

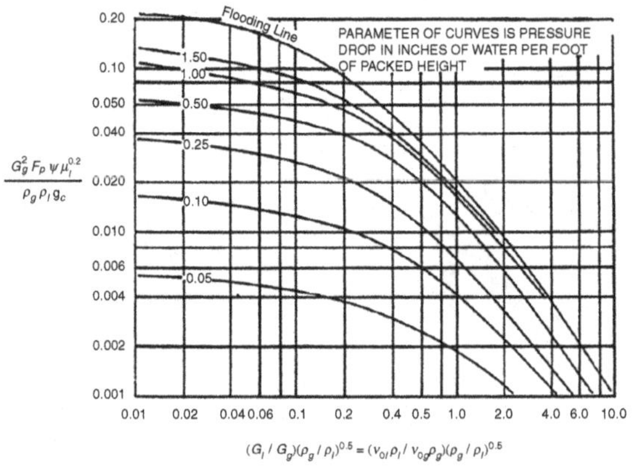

Figure 8.1.11. Generalized pressure drop and flooding correlation. Reprinted, with permission, from Eckert, Chem. Eng. Progr., 66(3), 39 (1970). Copyright © [1970] American Institute of Chemical Engineers (AIChE).

An interpretation of the ordinate of Figure 8.1.11 is possible following Kister *et al.* (2007), who have provided instead a semilog chart (Strigle's chart, (see Strigle 1994)), where the ordinate containing the quantities G_g, ρ_g, ρ_ℓ is essentially a square root of the ordinate of Figure 8.1.11:

$$G_g/(\rho_g \rho_\ell)^{0.5} = \rho_g v_{0g}/\sqrt{\rho_g \rho_\ell} = v_{0g}(\rho_g/\rho_\ell)^{0.5} \cong v_{0g}(\rho_g/(\rho_\ell - \rho_g)). \quad (8.1.40)$$

Note that this may be interpreted as the upward gas velocity divided by the terminal velocity of a vertically falling liquid droplet before being entrained by the upward gas velocity (see equations (6.3.48) and (6.3.49) for the terminal velocity of a droplet falling in an upflowing gas/vapor phase). The abscissa in Figure 8.1.11 reflects the relative flow rates of the liquid and the gas phase, alternatively a square root of the ratio of the kinetic energy of the liquid phase to that of the gas phase.

The pressure drop at the flooding point has been correlated with the packing factor, F_p (Kister and Gill, 1991; Strigle, 1994):

$$\Delta P_{\text{flooding point}} = 0.12 F_p^{0.7}. \quad (8.1.41)$$

Kessler and Wankat (1988) have correlated the ordinate with the abscissa of Figure 8.1.11 for the flooding curve as follows:

$$\log_{10}(\text{ordinate}) = -1.6678 - 1.085\log_{10}(\text{abscissa}) - 0.29655(\log_{10}(\text{abscissa}))^2. \quad (8.1.42)$$

A quantity of importance in the ordinate for Figure 8.1.11 is the packing factor, F_p. Table 8.1.1 provides values of the packing factor F_p for some common random packings (Cussler, 1997); Wankat (2007, Tables 10-3, 10-4) provides additional values of F_p for random and structured packings). Table 8.1.1 also provides limited information about F_p values of some structured packings (Kister *et al.*, 2007; Wankat, 2007).

Example 8.1.1 Calculation of diameter of and pressure drop in a column filled with random packings

A synthesis gas may be assumed to consist of 20% CO_2, 2% CO, 50% H_2, 28% N_2. It is at 250 psia and 30 °C. It is to be scrubbed with an aqueous solution of monoethanolamine (MEA) in a packed tower having 1.5 inch ceramic Raschig rings. The absorbent solution flow rate is 16.5 kg/s, whereas the gas flow rate is 9.7 kg/s. The liquid absorbent viscosity is 2 centipoise. The column should be operated at 50% of flooding velocity. Using the Sherwood–Eckert generalized plot, develop an estimate of the diameter of the packed column. Ignore the change in the gas flow rate as CO_2 is absorbed. Assume liquid density to be 63 lb/ft³. Use the ideal gas law to estimate ρ_g. What is the gas-phase pressure drop?

Solution We will determine the value of the abscissa in the Sherwood–Eckert plot first:

$$(G_\ell/G_g)(\rho_g/\rho_\ell)^{0.5} \Rightarrow (G_\ell/G_g) = (16.5/9.7) = 1.7.$$

The liquid density $\rho_\ell = 63\,\text{lb/ft}^3$; the molecular weight of the gas mixture $= 0.2 \times 44 + 0.02 \times 28 + 0.5 \times 2 + 0.28 \times 28 = 18.2$ lb/lb mole. The gas density, assuming ideal gas behavior, is given by

$$\rho_g = PM/RT = \left(\frac{250}{14.7}\right)\text{atm} \times 18.2\frac{\text{lb}}{\text{lb mol}} \Big/ \left(1.314\frac{\text{atm ft}^3}{\text{lb mol K}} \times 303\text{K}\right); = 0.77\,\text{lb/ft}^3.$$

So

$$(G_\ell/G_g)(\rho_g/\rho_\ell)^{0.5} = 1.7(0.77/63)^{0.5} = 0.109.$$

The ordinate from the plot for the flooding limit is 0.13; 50% of this value is 0.065. (*Note*: If we use correlation (8.1.42), we will get 0.1265 as the value of the ordinate.) Therefore

$$(G_g^2 F_P \psi \mu_\ell^{0.2}/(\rho_g \rho_\ell g_c)) = 0.065.$$

For 1.5 inch caramic Rashcig rings, Table 8.1.1 provides $F_p = 93\,\text{ft}^{-1}$; $\psi \cong (63/63) = 1$; $\rho_g = 0.77\,\text{lb/ft}^3$; $\rho_\ell = 63\,\text{lb/ft}^3$; $g_c = 32.2$; $\mu_\ell = 2$ cp. Therefore,

$$0.065 = \left(G_g^2 \times 93 \times 1 \times (2)^{0.2}/((0.77) \times (63) \times 32.2)\right)$$
$$= G_g^2 \times 93 \times 1.16/0.77 \times 63 \times 32.2$$
$$\Rightarrow G_g^2 = 0.065 \times 0.77 \times 63 \times 32.2/(93 \times 1.16)$$
$$= 0.94\,\text{lbmass/ft}^2\text{-s}$$

Since the gas flow rate is 9.7 kg/s, the diameter of the column d_i is obtained as

$$d_i = \left(\frac{9.7 \times 2.2 \times 4}{0.94 \times \pi}\right)^{0.5} = 5.37\,\text{ft} = 1.63\,\text{m}.$$

The value of the pressure drop in inches of water per foot of the packed column for the values of 0.109 and 0.065 for the abscissa and ordinate, respectively, obtained from the generalized Sherwood–Eckert plot is 0.9 inches of water per foot

Table 8.1.1. Packing factor[a] F_p *and surface area per packed volume*[b] *(per foot),* a, *for random packings and structured packings*[c]

Random packings	Nominal packing size (inches)								
	1/4	3/8	1/2	5/8	3/4	1	1 1/2	2	3
Raschig rings (ceramic)	1600	1000	580(111)	380(100)	255(80)	179(58)	93(38)	65(28)	37(19)
Raschig rings (1/32 inch metal)	700	390	300(128)	170	155(84)	115(63)			
Raschig rings (1/16 inch metal)			410(118)	300	220(72)	144(57)	83(41)	57(31)	32(21)
Berl saddles (ceramic)	900		240(142)		170(82)	110(76)	65(44)	45(32)	
Pall rings (metal)				81(104)		56(63)	40(39)	27(31)	18
Pall rings (plastic)				95(104)		55(63)	40(39)	26(31)	17(26)
Intalox saddles (ceramic)	725	330	200(190)		145(102)	92(78)	52(60)	40(36)	22
Hy-Pak rings (metal)						45(69)	29(42)	26(33)	16(31)
Hy-Pak rings (plastic)						25		12	

Structured packings	Flexipac			Gempak		Intalox		Mellapak Plus	Munters		Sulzer	
	2	4	HC2Y	2A	4A	2T	3T	252Y	12060	19060	CY	BX
	22(68)	6	13	16(67)	32(138)	17	13	12	27	15	70(213)	21(150)

[a] Packing factors obtained from Cussler (1997), Geankoplis (2003), Kister *et al.* (2007, pp. 28–38) and Wankat (2007, pp. 336–338).
[b] Values of *a*, the surface area of packing per unit packed volume, are provided in parentheses after the value of F_p from Cussler (1997) and Geankoplis (2003).
[c] Structured packing values are from Geankoplis (2003), Kister *et al.* (2007) and Wankat (2007). Additional information about other random packings are available in Seader and Henley (1998, table 6.8).

of the column length. If we employ the correlation $\Delta P_{\text{flood}} = 0.12 F_p^{\,0.7}$, we obtain

$$\Delta P_{\text{flood}} \text{ inches of water/ft of the column} = 0.12(93)^{0.7} = 0.12 \times 23.9 = 2.868,$$

indicating that ΔP at 50% flooding should be around 1.43 inches of water/ft. of the column length.

We will now briefly touch upon *plate towers* for dispersive gas–liquid absorption/stripping. The treatment of the plate tower dynamics will be given in greater detail in Section 8.1.3.5 on vapor–liquid distillation. A schematic of the plate tower with a sieve plate for countercurrent gas–liquid absorption/stripping has already been shown in Figure 8.1.5.

Vertical packed-bed columns and plate towers dominate the landscape of dispersive gas–liquid countercurrent separation devices. However, other devices have been/are being explored. Chief among them is the rotating-packed-bed (RPB) concept called HIGEE (high gravity) (Ramshaw and Mallinson, 1981; Mallinson and Ramshaw, 1982). Figure 8.1.12 illustrates the basic schematic: it consists of a rotor and a stationary casing. The rotor houses an appropriate packing. The liquid introduced at the center through a stationary liquid distributor flows radially outward due to centrifugal force. At the end of the packed bed, it is sprayed onto the stationary housing casing and collected from the bottom. The gas is introduced through the periphery of this casing and flows countercurrent to the liquid through the packing and is withdrawn through the center, around the liquid absorbent inlet.

The centrifugal field generated by rotation of the packed bed imposed on the density difference between the liquid and the gas $(\rho_\ell - \rho_g)$ develops a high gas–liquid interfacial shear stress, leading to a very high value of $k_\ell a$ (as much as 3 to 7 s^{-1}); usually, a high surface area packing, having values in the range 1000–5000 m^{-1} (= m^2/m^3), is used. The net result of these developments is a highly compact device (leading to the notion of *process intensification*, where the volume and weight of the device carrying out the process are drastically reduced), wherein the values of HTU_{og} can be as low as 14–21 cm compared to the high values of 1–4 m in conventional randomly packed columns (Jassim *et al.*, 2007) (see Section 8.1.2.2 for the definition of *HTU*).

We will now briefly consider *nondispersive gas–liquid absorption* and *stripping devices*. In Section 3.4.3.1, we illustrated the notion that a gas phase and an aqueous liquid phase can contact each other at the mouth of the pores of a porous hydrophobic membrane whose pores are filled with the gas phase, as long as the nonwetting aqueous solution pressure did not exceed the gas pressure by the value of the breakthrough pressure. The gas–liquid interface where the gas absorption or gas stripping takes place is located on the liquid-facing side of the membrane (see Figure 3.4.10). When the membrane is in the form of a hollow fine fiber having dimensions in the range 150–400 μm I.D. and 200–500 μm O.D. (see Figures 7.2.1(c) and (d) for hollow fibers) and the device is well packed, the membrane surface area per unit packed volume becomes very high, as much as 30–40 cm^{-1} (3000–4000 m^{-1}). Therefore the *HTU* values become quite small (see Section 8.1.2.2). The hollow fiber membranes behave as a specialized packing, where, however, the gas or the liquid is not

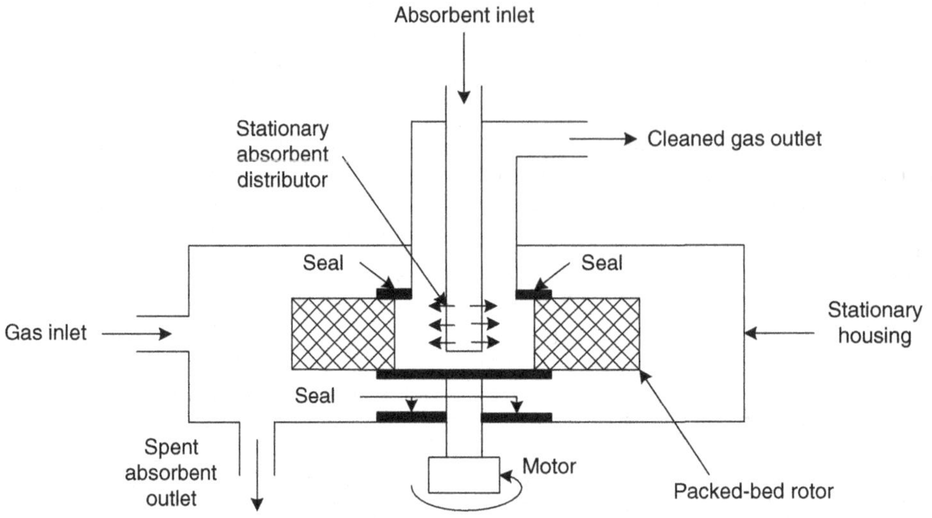

Figure 8.1.12. Rotating-packed-bed gas absorber.

dispersed in the other phase as bubbles or drops/droplets. The contacting gas and liquid phases can be at very different pressures with no effect on mass transfer as such. Therefore the problem of flooding is removed. The problem of channeling is easily eliminated. Such devices are frequently characterized as *membrane contactors.* The hollow fibers are primarily polymeric in nature. Any solution (even nonaqueous) may be used as long as the hydrophobic membrane is not spontaneously wetted.

Figure 8.1.13(a) illustrates a commercially used device which looks like a baffled shell-and-tube heat exchanger with each porous hollow fiber of polypropylene acting like a tube. One of the fluids, usually the liquid, flows through a central tube with perforations all along its periphery, which allows the liquid to flow radially out in crossflow across the hollow fibers, through the bores of which gas flows. Figure 8.1.13(b) provides a view of a commercial system for degassing an aqueous solution. This system has 48 hollow fiber membrane modules, each module having a membrane surface area of 220 m^2 (Sirkar, 2008). Such modules are frequently used for gas absorption as well as gas stripping. A common application is removing dissolved O_2 from water in ultrapure water production, which is required in semiconductor industries to the level of ~1 ppb by applying vacuum and sweep N_2 on the gas side. For an introduction to such hollow fiber membrane based devices, see Sirkar (1992) and Reed *et al.* (1995).

An additional nondispersive method of contacting a gas and a liquid stream involves what is called the Fiber-FILM™ technology, specific applications of which are identified as THIOLEX^{SM}, CHLOREX^{SM}, AQUAFINING^{SM} and AMINEX^{SM}. The device basically consists of a cylinder packed with very fine proprietary metal fibers. The liquid phase (especially caustic-containing phases) flows along the length of each fiber, preferentially wetting it. The gas phase flows through the cylinder parallel to the fibers in between the fibers. The ultrathin films of the liquid around each metal fiber and the gas stream contact each other efficiently. Introduced by Merichem Company, this technique is primarily employed in hydrocarbon processing. Laboratory studies of gas absorption in a wetted-wire column have been illustrated in Migita *et al.* (2005).

8.1.2.2 *Countercurrent continuous contact gas–liquid absorber or stripper for a dilute species in a gas stream*

We have two goals here. First, we will derive the equation for the operating line for an absorber relating $\overline{x}_{i1}$ $(= \overline{x}_{ig})$ to $\overline{x}_{i2}$ $(= \overline{x}_{i\ell})$ (gas phase $j = 1 = g$; absorbent phase $j = 2 = \ell$); the stripper will be treated next. Next, we estimate the height L of the countercurrent column shown in Figure 8.1.14(a) required to achieve the change in species i concentration from the gas inlet at $z = 0$ (i.e. $C_{ig0} = C_{igf}$, $x_{ig0} = x_{igf}$) to that at the gas outlet ($z = L$) (i.e. C_{igL}; $\overline{x}_{igL}$); correspondingly, the liquid absorbent composition of species i changes from $C_{i\ell L}$ $(= C_{i\ell f}, x_{i\ell f})$ at $z = L$ to $\overline{C}_{i\ell 0}$ (also $\overline{x}_{i\ell 0}$) at $z = 0$. Since species i is present as a dilute mixture in the gas flowing up the column, we assume that the total molar gas flow rate up the column, W_{tg}, is essentially constant. Correspondingly, the molar liquid absorbent flow rate down the column is assumed to be effectively constant at the value $W_{t\ell}$. In the absence of any longitudinal diffusion/axial dispersion in the z-direction (i.e. plug flow exists), the governing equation is obtained from relation (8.1.10) as

$$\frac{d\overline{x}_{i\ell}}{d\overline{x}_{ig}} = \frac{|W_{tg}|}{|W_{t\ell}|}. \tag{8.1.43a}$$

(a)

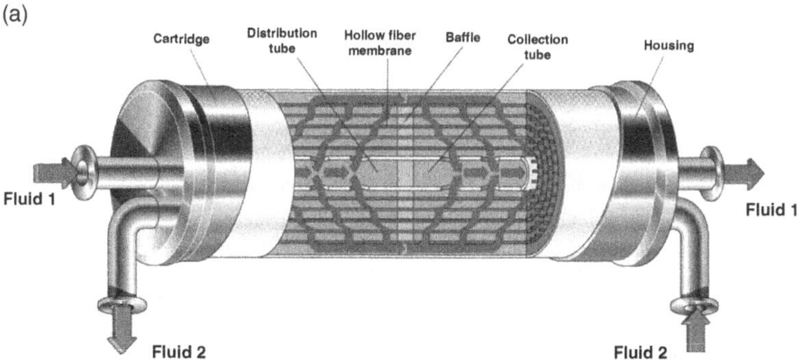

(b)

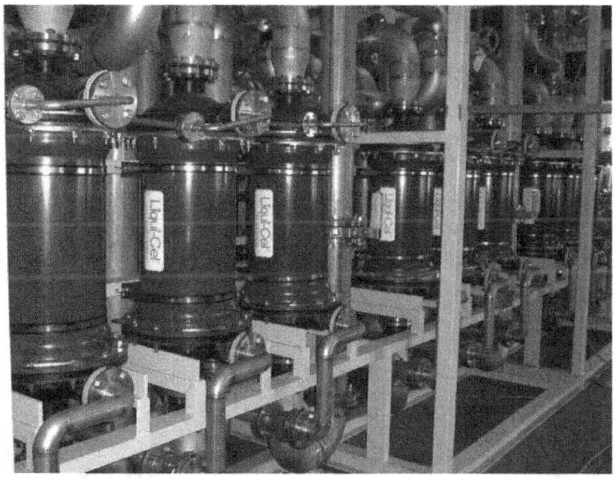

Figure 8.1.13. (a) Commonly used porous hollow fiber based membrane contactor. (Liqui-Cel® is a registered trademark of Membrane-Charlotte, a Division of Celgard, LLC Copyright © 2011. All rights reserved.) (b) Commercial water degassing system with 48 modules, each having a surface area of 220 m². (Courtesy of Membrana, Charlotte, NC.)

For the gas stream entering the bottom ($z = 0$) with x_{igf} ($= x_{ig0}$) and the absorbent stream leaving the bottom with $\bar{x}_{il0}$, we get the following equation for the *operating line* after integrating the above equation:

$$\bar{x}_{ig} = -\frac{|W_{t\ell}|}{|W_{tg}|}(\bar{x}_{i\ell 0} - \bar{x}_{i\ell}) + x_{igf}, \tag{8.1.43b}$$

since $\bar{x}_{ig}$ is less than x_{igf}. This equation relates the cross-sectional averaged compositions of the two phases, $\bar{x}_{ig}$ and $\bar{x}_{i\ell}$, at any height z of the column. The slope of the operating line is $|W_{t\ell}|/|W_{tg}|$; it is essentially constant since we have absorption of a dilute species in a gas stream and $|W_{t\ell}|$ and $|W_{tg}|$ are essentially constants (Figure 8.1.14(b)). *Note*: The mass balance equation for species i, (8.1.43a), could also have been derived by focusing on the mass balance over a small element of thickness dz across the packed tower in Figure 8.1.14(a):

$$|W_{tg}|(\bar{x}_{ig}|_z - \bar{x}_{ig}|_{z+\Delta z}) + |W_{t\ell}|(\bar{x}_{i\ell}|_{z+\Delta z} - x_{i\ell}|_z)$$
$$= 0 \Rightarrow \frac{d\bar{x}_{i\ell}}{d\bar{x}_{ig}} = \frac{|W_{tg}|}{|W_{t\ell}|}. \tag{8.1.43c}$$

An alternative form for relation (8.1.43b) is

$$\bar{x}_{i\ell} = \frac{|W_{tg}|}{|W_{t\ell}|}(\bar{x}_{ig} - x_{igf}) + x_{i\ell 0}. \tag{8.1.43d}$$

The slope of the operating line for a given gas flow rate will depend on the liquid absorbent flow rate. Correspondingly, for the given gas flow rate, as the gas composition of species i being absorbed changes from the feed composition x_{igf} ($= x_{ig0}$) to the specified outlet composition $\bar{x}_{igL}$, the liquid absorbent outlet composition rises to $\bar{x}_{i\ell 0}$ from the inlet value $x_{i\ell L}$. If the absorbent flow rate is reduced, the value of $\bar{x}_{i\ell 0}$ will rise; the diffusional driving concentration difference will decrease and a taller column will be needed. There is, however, a limit to how low this liquid absorbent flow rate can be. As the slope of the operating line $|W_{t\ell}|/|W_{tg}|$ decreases (Figure 8.1.14(b)), at a certain value of $|W_{t\ell}|$, $|W_{t\ell}|_{\min}$ the operating line touches the gas absorption equilibrium curve at point M. At this location, the driving concentration difference between the gas and the liquid absorbent disappears; therefore an infinitely tall

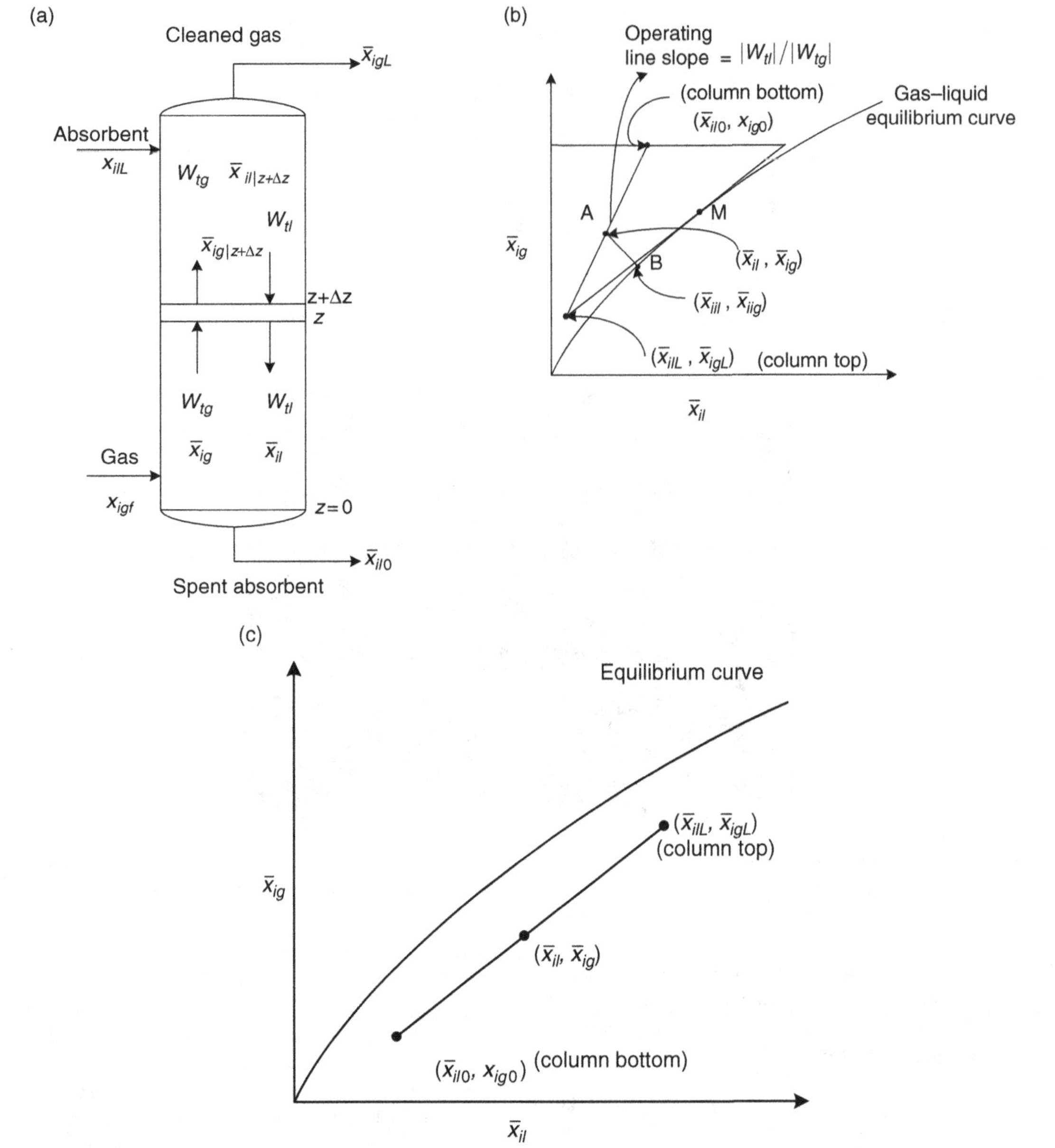

Figure 8.1.14. (a) Schematic of a countercurrent continuous contact gas-liquid absorber. (b) Operating line and equilibrium curve for a countercurrent gas-liquid absorber. (c) Operating line and equilibrium curve for a countercurrent gas-liquid stripper.

column will be needed to scrub the gas for the desired concentration change. Obviously, a practical gas absorption process is carried out with a higher absorbent flow rate.

The operating line equation for a *stripper* may be derived by considering the required composition changes identified in Figure 8.1.14(c). The governing mass balance equation at any height of the column is still (8.1.43a). This is also true of the equation for the operating line, (8.1.43b) which may, however, be rewritten as

$$\bar{x}_{ig} - x_{igf} = \frac{|W_{t\ell}|}{|W_{tg}|}(\bar{x}_{i\ell} - \bar{x}_{i\ell 0}) \tag{8.1.44}$$

since the stripping gas inlet composition x_{igf} ($= x_{ig0}$) is less than $\bar{x}_{ig}$ (unlike that in a gas absorber); further, the absorbent outlet composition $\bar{x}_{i\ell 0}$ is less than $\bar{x}_{i\ell}$ since the absorbent is being regenerated (its composition is being reduced from $x_{i\ell L}$ to $\overline{x}_{i\ell 0}$).

Let us now determine the height of the absorber/stripper. It is clear that the driving concentration difference between the operating line and the equilibrium curve is going to be important. The slope of the gas–liquid equilibrium curve and the slope of the operating line become important through a quantity called the *absorption factor A*:

$$A = \frac{|W_{t\ell}|}{|W_{tg}|}\frac{1}{H_i^P}, \tag{8.1.45}$$

where H_i^P is a Henry's law type constant for the gas–liquid system (see relation (8.1.49)).

We will now estimate the column length L required to change the gas composition from x_{igf} $(= x_{ig0})$ to $\overline{x}_{igL}$. Assuming no longitudinal dispersion and axial diffusion, the governing equation will be (8.1.3a) multiplied by S_c, the column cross-sectional area:

$$\langle v_{gz}\rangle S_c \frac{d\overline{C}_{ig}}{dz} = |W_{tg}|\frac{d\overline{x}_{ig}}{dz} = -K_{igc} a S_c(\overline{C}_{ig} - C_{ig}^*), \tag{8.1.46}$$

where K_{igc} is the overall gas-phase based mass-transfer coefficient for species i being absorbed, $\overline{C}_{ig}$ is the molar concentration of species i in the bulk gas averaged over the column cross section and C_{ig}^* is the hypothetical gas-phase composition in equilibrium with $\overline{C}_{i\ell}$. For a dilute gas stream,

$$\overline{C}_{ig} = C_{tg}\,\overline{x}_{ig}, \qquad C_{ig}^* = C_{tg}\,x_{ig}^*, \tag{8.1.47}$$

where C_{tg} is assumed essentially constant along z. Now rewrite equation (8.1.46) as[2]

$$\left|\frac{W_{tg}}{S_c}\right|\frac{d\overline{x}_{ig}}{dz} = -K_{igc}\,a\,C_{tg}(\overline{x}_{ig} - x_{ig}^*) = -K_{igx}a(\overline{x}_{ig} - x_{ig}^*). \tag{8.1.48}$$

Before integration, we assume Henry's law[3] to be valid for this dilute gas stream:

$$x_{ig}^* = H_i^P\,\overline{x}_{i\ell}. \tag{8.1.49}$$

Here, $\overline{x}_{i\ell}$ is the bulk liquid concentration averaged over the column cross section. Employing the operating line equation (8.1.43), we integrate the differential equation (8.1.48) from $z = 0$ to $z = L$:

$$\int_0^L dz = -\int_{x_{ig0}}^{\overline{x}_{igL}} \frac{|W_{tg}|/S_c}{K_{igx}\,a}\frac{d\overline{x}_{ig}}{(\overline{x}_{ig} - x_{ig}^*)}$$
$$= -\frac{|W_{tg}|/S_c}{K_{igx}\,a}\int_{x_{ig0}}^{\overline{x}_{igL}} \frac{d\overline{x}_{ig}}{(\overline{x}_{ig} - H_i^P\overline{x}_{i\ell})} = L; \tag{8.1.50}$$

$$L = -\frac{|W_{tg}|/S_c}{K_{igx}\,a}\int_{x_{ig0}}^{\overline{x}_{igL}} \frac{d\overline{x}_{ig}}{\left\{\overline{x}_{ig}\left(1 - H_i^P\frac{|W_{tg}|}{|W_{t\ell}|}\right) + H_i^P\frac{|W_{tg}|}{|W_{t\ell}|}\overline{x}_{ig0} - H_i^P\,\overline{x}_{i\ell 0}\right\}}; \tag{8.1.51}$$

$$L = \frac{(|W_{tg}|/S_c)}{K_{igx}\,a}\frac{1}{\left(1 - H_i^P\frac{|W_{tg}|}{|W_{t\ell}|}\right)}\,\ell n\left[\frac{x_{ig0}\left(1 - H_i^P\frac{|W_{tg}|}{|W_{t\ell}|}\right) + H_i^P\frac{|W_{tg}|}{|W_{t\ell}|}x_{ig0} - H_i^P\overline{x}_{i\ell 0}}{\overline{x}_{igL}\left(1 - H_i^P\frac{|W_{tg}|}{|W_{t\ell}|}\right) + H_i^P\frac{|W_{tg}|}{|W_{t\ell}|}x_{igo} - H_i^P\overline{x}_{i\ell o}}\right]; \tag{8.1.52}$$

$$L = \left[\frac{|W_{tg}|}{K_{igx}\,a\,S_c}\right]\left\{\frac{1}{\left(1 - H_i^P\frac{|W_{tg}|}{|W_{t\ell}|}\right)}\,\ell n\left[\frac{x_{ig0} - H_i^P\overline{x}_{i\ell 0}}{\overline{x}_{igL} - H_i^P x_{i\ell L}}\right]\right\}, \tag{8.1.53a}$$

where, to obtain the denominator of the logarithmic term, we have employed equation (8.1.43b) at location $z = L$. Here the units of K_{igx} are gmol/cm^2-s-mole fraction, a is in cm^2/cm^3 and $(|W_{tg}|/S_c)$ is given in gmol/cm^2-s. Therefore the unit of $((|W_{tg}|/S_c)/K_{igx}a)$ is cm, that of length or height, since mole fraction has no units. An alternative form of the above results for the absorber length, first derived by Colburn (1939), is

$$L = \left[\frac{|W_{tg}|}{K_{igx}\,a\,S_c}\right]\frac{1}{\left(1 - H_i^P\frac{|W_{tg}|}{|W_{t\ell}|}\right)}\,\ell n\left[\left(1 - \frac{H_i^P|W_{tg}|}{|W_{t\ell}|}\right)\left(\frac{x_{ig0} - H_i^P x_{i\ell L}}{\overline{x}_{igL} - H_i^P x_{i\ell L}}\right) + \left(\frac{H_i^P|W_{tg}|}{|W_{t\ell}|}\right)\right]. \tag{8.1.53b}$$

Often the results (8.1.53a) and (8.1.53b) for the length or height of the countercurrent column are expressed as

$$L = (HTU_{og}) \times (NTU_{og}), \tag{8.1.54a}$$

where

HTU_{og} = height of a transfer unit based on overall gas phase $= ((|W_{tg}|/S_c)/K_{igx}a)$,

NTU_{og} = number of transfer units (overall gas-phase based). (8.1.54b)

The latter is defined by

$$NTU_{og} = \int_{\overline{x}_{igL}}^{x_{ig0}} \frac{dx_{ig}}{(\overline{x}_{ig} - x_{ig}^*)}. \tag{8.1.54c}$$

[2] Note that $K_{igx} = K_{igc}C_{tg}$ (see Table 3.1.4B).

[3] Relation (3.3.60b) ⇒ $p_{ig} = H_i\,x_{i\ell}$ ⇒ $Px_{ig} = H_i\,x_{i\ell}$ ⇒ $x_{ig} = (H_i/P)\,x_{il} = H_i^P\,x_{il}$.

These definitions follow from equation (8.1.50). The quantity NTU_{og} provides an estimate of how difficult the separation is as we change the gas composition from x_{ig0} (inlet) to $\bar{x}_{igL}$ (outlet) at the column top. If $(\bar{x}_{ig} - x^*_{ig})$ may be assumed to be a reflection of the composition difference between the two phases driving the transport of species i from the gas to the liquid phase, then the larger this value is to achieve a given change in $\bar{x}_{ig}$, the smaller is going to be the value of *NTU*, and the easier will be the separation. We see this interpretation reflected in the following result if $(\bar{x}_{ig} - x^*_{ig})$ is assumed to be constant over the desired gas-phase concentration change from x_{ig0} to $\bar{x}_{igL}$ in definitions (8.1.54b, c):

$$NTU_{og} = \frac{x_{ig0} - \bar{x}_{igL}}{\left(\bar{x}_{ig} - x^*_{ig}\right)}. \tag{8.1.54d}$$

An expression for NTU_{og} may also be written down for the Henry's law case from equation (8.1.53b) as

$$NTU_{og} = \left\{\frac{1}{\left(1 - H_i^P \frac{|W_{tg}|}{|W_{t\ell}|}\right)}\right\} \ell n\left[\left(1 - \frac{H_i^P |W_{tg}|}{|W_{t\ell}|}\right)\left(\frac{x_{ig0} - H_i^P x_{i\ell L}}{\bar{x}_{igL} - H_i^P x_{i\ell L}}\right) + \left(\frac{H_i^P |W_{tg}|}{|W_{t\ell}|}\right)\right]. \tag{8.1.54e}$$

Note the following points.

(1) We have used $x_{i\ell 0}$ in many equations here. Usually in gas absorption $x_{i\ell L}$ is known from the entering liquid composition; $x_{i\ell 0}$ is unknown. It has to be calculated by mass balance, knowing x_{ig0} and $\bar{x}_{igL}$.
(2) These derivations were carried out for species i absorption from the gas phase to the liquid phase. However, they are equally valid for the stripping of species i from the liquid to the gas phase (or vapor phase if steam or some other vapor is used).
(3) All derivations were carried out using gas-phase concentration change. We will now provide a few corresponding results using liquid-phase concentration changes.
(4) When species i is a vapor, use Raoult's law (3.3.64):

$$x^*_{ig} = \left(P_i^{sat}/P\right) x_{i\ell}. \tag{8.1.54f}$$

The equation corresponding to equation (8.1.48) for a liquid-phase based calculation is

$$\langle v_{\ell z}\rangle \frac{d\bar{C}_{i\ell}}{dz} = \frac{|W_{t\ell}|}{S_c}\frac{d\bar{x}_{i\ell}}{dz} = -K_{i\ell x} a\left(x^*_{i\ell} - \bar{x}_{i\ell}\right). \tag{8.1.55a}$$

If we assume Henry's law to be valid for the dilute gas stream, then

$$\left(\bar{x}_{ig}/H_i^P\right) = x^*_{i\ell}. \tag{8.1.55b}$$

Therefore

$$\frac{|W_{t\ell}|}{S_c}\frac{d\bar{x}_{i\ell}}{dz} = -K_{i\ell x} a\left(\frac{\bar{x}_{ig}}{H_i^P} - \bar{x}_{i\ell}\right); \tag{8.1.55c}$$

$$\int_0^L dz = L = -\int_{\bar{x}_{i\ell o}}^{x_{i\ell L}} \frac{|W_{t\ell}|/S_c}{K_{i\ell x} a}\frac{d\bar{x}_{i\ell}}{\left(\frac{\bar{x}_{ig}}{H_i^P} - \bar{x}_{i\ell}\right)} = \frac{|W_{t\ell}|}{S_c K_{i\ell x}\, a}\int_{x_{i\ell L}}^{\bar{x}_{i\ell o}} \frac{d\bar{x}_{i\ell}}{\left(\frac{\bar{x}_{ig}}{H_i^P} - \bar{x}_{i\ell}\right)}. \tag{8.1.55d}$$

Introducing the operating line equation (8.1.43b) for $\bar{x}_{ig}$ in terms of $\bar{x}_{i\ell}$, we get

$$L = \frac{|W_{t\ell}|}{S_c K_{i\ell x}\, a}\int_{x_{i\ell L}}^{\bar{x}_{i\ell 0}} \frac{d\bar{x}_{i\ell}}{\bar{x}_{i\ell}\left[\frac{|W_{t\ell}|}{|W_{tg}|}\frac{1}{H_i^P} - 1\right] - \frac{|W_{t\ell}|}{|W_{tg}|}\frac{\bar{x}_{i\ell 0}}{H_i^P} + \frac{x_{ig0}}{H_i^P}}$$

$$= \left[\frac{(|W_{t\ell}|)}{S_c K_{i\ell x}\, a}\right]\left\{\frac{1}{\left(\frac{|W_{t\ell}|}{|W_{tg}|}\frac{1}{H_i^P} - 1\right)}\right\} \ell n\left[\frac{\bar{x}_{i\ell 0}\left(\frac{|W_{t\ell}|}{|W_{tg}|}\frac{1}{H_i^P} - 1\right) - \frac{|W_{t\ell}|}{|W_{tg}|}\frac{\bar{x}_{i\ell 0}}{H_i^P} + \frac{x_{ig0}}{H_i^P}}{x_{i\ell L}\left(\frac{|W_{t\ell}|}{|W_{tg}|}\frac{1}{H_i^P} - 1\right) - \frac{|W_{t\ell}|}{|W_{tg}|}\frac{\bar{x}_{i\ell 0}}{H_i^P} + \frac{x_{ig0}}{H_i^P}}\right]; \tag{8.1.56a}$$

$$L = \left[\frac{|W_{t\ell}|}{S_c K_{i\ell x}\, a}\right]\left\{\frac{1}{\left(1 - \frac{|W_{t\ell}|}{|W_{tg}|}\frac{1}{H_i^P}\right)} \ell n\left[\frac{\frac{\bar{x}_{igL}}{H_i^P} - x_{i\ell L}}{\frac{x_{ig0}}{H_i^P} - \bar{x}_{i\ell 0}}\right]\right\}; \tag{8.1.56b}$$

$$L = [HTU]_{o\ell}\, [NTU]_{o\ell} \tag{8.1.57a}$$

where

$$HTU_{o\ell} = \frac{|W_{t\ell}|}{S_c K_{i\ell x}\, a} \tag{8.1.57b}$$

and

$$NTU_{o\ell} = \left\{\frac{1}{1 - \frac{|W_{t\ell}|}{|W_{tg}|}\frac{1}{H_i^P}}\right\} \ell n\left[\left(1 - \frac{|W_{t\ell}|}{H_i^P |W_{tg}|}\right)\left(\frac{x_{i\ell L} - \frac{x_{ig0}}{H_i^P}}{\bar{x}_{i\ell o} - \frac{x_{ig0}}{H_i^P}}\right) + \frac{|W_{t\ell}|}{|W_{tg}| H_i^P}\right]; \tag{8.1.57c}$$

$$NTU_{o\ell} = \left\{\frac{1}{\left(1 - \frac{|W_{t\ell}|}{|W_{tg}|}\frac{1}{H_i^P}\right)}\right\} \ell n\left[\frac{x_{i\ell L} - \frac{\bar{x}_{igL}}{H_i^P}}{\bar{x}_{i\ell 0} - \frac{x_{ig0}}{H_i^P}}\right]. \tag{8.1.57d}$$

Here, $HTU_{o\ell}$ stands for the height of the transfer unit based on the overall liquid phase; correspondingly, $NTU_{o\ell}$ represents the number of transfer units based on the overall liquid phase.

Example 8.1.2 Air at atmospheric pressure containing acetone vapor at the level of 0.01 mole fraction is to be scrubbed by pure water at 15 °C. The value of H_i^P for acetone and water under these conditions is 1.2 (Sherwood *et al.*, 1975). To scrub the air entering the column bottom at 7 kgmol/hr, the pure water flow rate into the packed column is maintained at 23 kgmol/hr. The value of $K_{igx}a$ is known to be 20 gmol/m^3-s-mole fraction under these flow conditions for acetone in a column of diameter 33 cm. Determine the height of the column needed to scrub 90% of the acetone from air.

Solution

$$H_i^P = 1.2; S_c = \frac{\pi}{4}d^2 = \frac{\pi}{4}\times(0.33)^2\,\text{m}^2 = 0.085\,\text{m}^2;$$

$$|W_{tg}| = 7\,\text{kgmol/hr} = 7000\,\text{gmol}/3600\,\text{s} = 1.944\,\text{gmol/s};$$

$$|W_{t\ell}| = 23\frac{\text{kgmol}}{\text{hr}} = 23\,000\,\text{gmol}/3600\,\text{s} = 6.389\,\text{gmol/s}.$$

We employ equation (8.1.53b) to calculate the column height. (*Note*: $x_{i\ell L} = 0$ since pure water is being used.) The column height is given by

$$L = \left[\frac{|W_{tg}|}{K_{igx}\,a\,S_c}\right]\left\{\frac{1}{\left(1 - H_i^P\dfrac{|W_{tg}|}{|W_{t\ell}|}\right)}\right\}$$

$$\ell n\left[\left(1 - \frac{H_i^P|W_{tg}|}{|W_{t\ell}|}\right)\left(\frac{x_{ig0} - H_i^P x_{i\ell L}}{x_{igL} - H_i^P x_{i\ell L}}\right) + \left(\frac{H_i^P|W_{tg}|}{|W_{t\ell}|}\right)\right]$$

$$= \frac{1.944\,\dfrac{\text{gmol}}{\text{s}}}{20\,\dfrac{\text{gmol}\times 0.085\text{m}^2}{\text{m}^3\text{-s-mol fraction}}}\left\{\frac{1}{1 - 1.2\dfrac{1.944}{6.389}}\right\}$$

$$2.303\log\left[\left(1 - 1.2\frac{1.944}{6.389}\right)\left(\frac{0.01}{0.001}\right) + \frac{1.2\times 1.944}{6.389}\right].$$

So

$$L = \frac{1.944\,\text{m}}{20\times 0.085}\left\{\frac{2.303}{1 - 0.365}\right\}\log[(0.635)\times 10 + 0.365]$$
$$= 1.143\times 3.626\times\log[6.715] = 4.144\times 0.827 = 3.42\,\text{m}.$$

The previous results expressing the length of the absorber/stripper in terms of *HTU* and *NTU* for the absorption of species *i* present in dilute concentrations in the feed were developed in terms of an overall mass-transfer coefficient K_{igx} (or K_{igc}) and $K_{i\ell x}$ for species *i*. It is also useful to develop such results in terms of individual phase based mass-transfer coefficients, $k_{i\ell x}$, k_{igx}, k_{igc}, $k_{i\ell c}$. (When species *i* is not specifically identified, these quantities will be represented as $k_{\ell x}$, k_{gx}, k_{gc}, $k_{\ell c}$; see Section 3.4.1.1.) To develop expressions based on individual phase based mass-transfer coefficients, we need to define/locate the gas-liquid interface composition at any location in the absorber in Figure 8.1.14(b), namely $(\overline{x}_{ii\ell}, \overline{x}_{iig})$, representing point B on the gas-liquid equilibrium curve, corresponding to bulk gas-liquid compositions $(\overline{x}_{i\ell}, \overline{x}_{ig})$, represented by point A on the operating line.

The flux of species *i*, N_i, across the gas-liquid interface at this location may be expressed as (Figure 8.1.14(b))

$$N_i = k_{igx}(\overline{x}_{ig} - \overline{x}_{iig}) = k_{i\ell x}(\overline{x}_{ii\ell} - \overline{x}_{i\ell}) \tag{8.1.58}$$

for species *i* transfer from the gas-phase bulk to the gas-liquid interface and then to the liquid bulk.[4] Therefore

$$(\overline{x}_{ig} - \overline{x}_{iig})/(\overline{x}_{i\ell} - \overline{x}_{ii\ell}) = -\frac{k_{i\ell x}}{k_{igx}}, \tag{8.1.59}$$

representing the negative slope of the line AB connecting the two bulk-phase compositions to the gas-liquid interface composition. We can now rewrite equation (8.1.46) as follows:

$$\frac{|W_{tg}|}{S_c}\frac{d\overline{x}_{ig}}{dz} = -k_{igx}a\,(\overline{x}_{ig} - \overline{x}_{iig}) = -k_{i\ell x}a\,(\overline{x}_{ii\ell} - \overline{x}_{i\ell}), \tag{8.1.60}$$

which leads to

$$\int_0^L dz = L = -\int_{x_{ig0}}^{\overline{x}_{igL}}\frac{|W_{tg}|/S_c}{k_{igx}a}\frac{d\overline{x}_{ig}}{(\overline{x}_{ig} - \overline{x}_{iig})}$$
$$= -\int_{x_{i\ell 0}}^{x_{i\ell L}}\frac{|W_{t\ell}|/S_c}{k_{i\ell x}a}\frac{d\overline{x}_{i\ell}}{(\overline{x}_{ii\ell} - \overline{x}_{i\ell})}, \tag{8.1.61}$$

where, for the last integral, we used equation (8.1.43a).

We observed earlier (in Section 3.1.4, equations (3.1.127), repeated below for convenience) that there are basically two types of molecular diffusion processes in a binary gaseous mixture: equimolar counterdiffusion and nonequimolar counterdiffusion. In general, we may express the mass-transfer coefficient for a general case in terms of that for equimolar counterdiffusion via

$$k_{ix} = \frac{k'_{ix}}{\phi_N}; \qquad k_{ic} = \frac{k'_{ic}}{\phi_N}; \qquad k_{ig} = \frac{k'_{ig}}{\phi_N},$$

where ϕ_N is the bulk flow correction factor. Expression (8.1.61) for the gas absorber column height *L* may also be expressed in terms of the individual phase based mass-transfer coefficient for equimolar counterdiffusion as

[4] The quantity $(\overline{x}_{ig} - \overline{x}_{iig})$ may be considered as the cross-sectional averaged $(x_{ig} - x_{iig})$.

$$L = -\int_{\overline{x}_{ig0}}^{\overline{x}_{igL}} \frac{(|W_{tg}|/S_c)}{(k'_{igx}a/\phi_N)} \frac{\mathrm{d}\overline{x}_{ig}}{(\overline{x}_{ig} - \overline{x}_{iig})}$$

$$= \left[\frac{(|W_{tg}|/S_c)}{k'_{igx}a}\right] \int_{\overline{x}_{igL}}^{x_{ig0}} \frac{\phi_N \,\mathrm{d}\overline{x}_{ig}}{(\overline{x}_{ig} - \overline{x}_{iig})}. \quad (8.1.62a)$$

For

$$L = HTU_g \times NTU_g,$$

$$HTU_g = (|W_{tg}|/S_c)/k'_{igx}a; \qquad NTU_g = \int_{\overline{x}_{igL}}^{x_{ig0}} \frac{\phi_N \,\mathrm{d}\overline{x}_{ig}}{(\overline{x}_{ig} - \overline{x}_{iig})}, \quad (8.1.62b)$$

so

$$L = -\int_{\overline{x}_{i\ell 0}}^{\overline{x}_{i\ell L}} \frac{(|W_{t\ell}|/S_c)}{k_{i\ell x}a} \frac{\mathrm{d}\overline{x}_{i\ell}}{(\overline{x}_{ii\ell} - \overline{x}_{i\ell})} = \left[\frac{(|W_{t\ell}|/S_c)}{k'_{i\ell x}a}\right] \int_{\overline{x}_{igL}}^{\overline{x}_{i\ell 0}} \frac{\phi_N \,\mathrm{d}\overline{x}_{i\ell}}{(\overline{x}_{ii\ell} - \overline{x}_{i\ell})}; \quad (8.1.62c)$$

for

$$L = HTU_\ell \times NTU_\ell,$$

$$HTU_\ell = (|W_{t\ell}|/S_c)/k'_{i\ell x}a; \qquad NTU_\ell = \int_{x_{i\ell L}}^{\overline{x}_{i\ell 0}} \frac{\phi_N \,\mathrm{d}\overline{x}_{i\ell}}{(\overline{x}_{ii\ell} - \overline{x}_{i\ell})}. \quad (8.1.62d)$$

In equations (8.1.62a,b), k'_{igx} corresponds to equimolar counterdiffusion and k_{igx} may be assumed to represent the case of species i diffusing through a stagnant gas consisting of species other than i (see equations (3.1.127)). In such a case for a binary system, we note from (3.1.130) that, for $N_R = 1$,

$$\phi_N = \frac{\left(1 - \overline{x}_{ig}^{\delta}\right) - \left(1 - \overline{x}_{ig}^{0}\right)}{\ell\mathrm{n}\left[\left(1 - \overline{x}_{ig}^{\delta}\right) / \left(1 - \overline{x}_{ig}^{0}\right)\right]} = \frac{(1 - \overline{x}_{iig}) - (1 - \overline{x}_{ig})}{\ell\mathrm{n}[(1 - \overline{x}_{iig}) / (1 - \overline{x}_{ig})]}$$

$$= \left(1 - \overline{x}_{ig}\right)_{i\ell m}, \quad (8.1.63)$$

where $\left(1 - \overline{x}_{ig}\right)_{i\ell m}$ represents a logarithmic mean of the bulk and the interface mole fractions of stagnant gas species other than species i. For a dilute system vis-à-vis species i, ϕ_N is close to 1 and can often be neglected. When it cannot be neglected, and yet may be considered as very weakly dependent on composition in the range under consideration, we may simplify equation (8.1.62a) as follows:

$$L = \frac{(|W_{tg}|/S_c)}{k'_{igx}a} \left(1 - \overline{x}_{ig}\right)_{i\ell m} \int_{\overline{x}_{igL}}^{x_{ig0}} \frac{\mathrm{d}\overline{x}_{ig}}{(\overline{x}_{ig} - \overline{x}_{iig})}. \quad (8.1.64a)$$

If we can assume now that the equilibrium curve shown in Figure 8.1.14(b) is approximately a straight line (the operating line being essentially straight for the dilute gas mixture), we may express the difference $(\overline{x}_{ig} - \overline{x}_{iig})$ as varying linearly with $\overline{x}_{ig}$:

$$(\overline{x}_{ig} - \overline{x}_{iig}) = b_1 \overline{x}_{ig} + b_2. \quad (8.1.64b)$$

Therefore,

$$L = \frac{(|W_{tg}|/S_c)}{k'_{igx}a} (1 - \overline{x}_{ig})_{i\ell m} \int_{\overline{x}_{igL}}^{x_{ig0}} \frac{\mathrm{d}\overline{x}_{ig}}{b_1 \overline{x}_{ig} + b_2}$$

$$= \frac{(|W_{tg}|/S_c)}{k'_{igx}a} (1 - \overline{x}_{ig})_{i\ell m} \left[\left(\frac{1}{b_1}\right) \ell\mathrm{n} \frac{(b_1 x_{ig0} + b_2)}{(b_1 \overline{x}_{igL} + b_2)}\right]; \quad (8.1.64c)$$

$$L = \frac{(|W_{tg}|/S_c)}{k'_{igx}a} \left(1 - \overline{x}_{ig}\right)_{i\ell m} \frac{(x_{ig0} - x_{igL})}{(x_{ig0} - x_{iig})_{\ell m}}, \quad (8.1.64d)$$

where

$$\left(x_{ig0} - \overline{x}_{iig}\right)_{\ell m} = \frac{(x_{ig0} - \overline{x}_{iig0}) - (\overline{x}_{igL} - \overline{x}_{iigL})}{\ell\mathrm{n}\left[(x_{ig0} - \overline{x}_{iig0}) / (\overline{x}_{igL} - \overline{x}_{iigL})\right]}. \quad (8.1.64e)$$

The corresponding expression in terms of $k'_{i\ell x}$ is

$$L = \frac{(|W_{tg}|/S_c)}{k'_{i\ell x}a} (1 - \overline{x}_{i\ell})_{i\ell m} \frac{(\overline{x}_{i\ell 0} - x_{i\ell L})}{(\overline{x}_{iiL} - \overline{x}_{i\ell 0})_{\ell m}}. \quad (8.1.64f)$$

So far, we have obtained the height of the gas–liquid absorber, L, expressed in terms of the HTU_g and NTU_g based on the individual gas film (equation (8.1.62b) *or* in terms of HTU_ℓ or NTU_ℓ based on the individual liquid film (equation (8.1.62d)) *or* in terms of the overall gas film (equations (8.1.54a–c)) *or* in terms of the overall liquid film (equations (8.1.57a–d)). It is very useful to have these quantities based on individual film coefficients related to those based on the overall film coefficients. To develop such relations, recall the basic relation (3.4.8) between the individual film transfer coefficients and the overall gas-phase based mass-transfer coefficient in a gas–liquid system:

$$\frac{1}{K_{xg}} = \frac{1}{k_{xg}} + \frac{H_A^p}{k_{x\ell}}; \qquad x_{Agi} = H_A^p \, x_{A\ell i}.$$

Here we are using the notation $k_{i\ell x}$, where i refers to the species i, ℓ to the liquid phase (second subscript is usually for the phase) and x for mole fraction based calculations; similarly for k_{igx} and K_{igx}. Therefore we rewrite the above relation here (species i for A) as

$$\frac{1}{K_{igx}} = \frac{1}{k_{igx}} + \frac{H_i^p}{k_{i\ell x}}. \quad (8.1.65a)$$

We can rewrite this relation as

$$\frac{1}{K'_{igx}} = \frac{1}{k'_{igx}} + \frac{H_i^p}{k'_{i\ell x}} \quad (8.1.65b)$$

for a dilute system, where $\phi_N \cong 1$. It follows therefore that

$$\frac{|W_{tg}|}{S_c K'_{igx}\, a} = \frac{|W_{tg}|}{S_c k'_{igx}\, a} + \frac{|W_{t\ell}|}{S_c k'_{i\ell x}\, a} \frac{H_i^P |W_{tg}|}{|W_{t\ell}|}. \tag{8.1.65c}$$

Define

$$\lambda = H_i^P \left(|W_{tg}|\right)/\left(|W_{t\ell}|\right) \tag{8.1.65d}$$

to get

$$HTU_{og} = HTU_g + HTU_\ell \cdot \lambda. \tag{8.1.65e}$$

If we now employ the basic relations between the column height L and the HTU and NTU values based on individual films and the overall transfer coefficients, equation (8.1.65e) is reduced to

$$\frac{L}{NTU_{og}} = \frac{L}{NTU_g} + \frac{\lambda L}{NTU_\ell}, \tag{8.1.66a}$$

i.e.

$$\frac{1}{NTU_{og}} = \frac{1}{NTU_g} + \frac{\lambda}{NTU_\ell}. \tag{8.1.66b}$$

The corresponding relations between an overall liquid and the individual film coefficients are based on relation (3.4.9):

$$\frac{1}{K_{x\ell}} = \frac{1}{k_{x\ell}} + \frac{1}{H_i^P k_{xg}}.$$

Rewrite this as

$$\frac{1}{K'_{i\ell x}} = \frac{1}{k'_{i\ell x}} + \frac{1}{k'_{igx} H_i^P}, \tag{8.1.67a}$$

leading to (for $\phi_N \cong 1$ in a dilute system)

$$\frac{|W_{t\ell}|}{S_c K'_{i\ell x}\, a} = \frac{|W_{t\ell}|}{S_c k'_{i\ell x}\, a} + \frac{|W_{tg}|}{S_c k'_{igx}\, a} \frac{|W_{t\ell}|}{|W_{tg}|} \frac{1}{H_i^P}; \tag{8.1.67b}$$

$$HTU_{o\ell} = HTU_\ell + \frac{HTU_g}{\lambda}, \tag{8.2.67c}$$

$$\frac{1}{NTU_{o\ell}} = \frac{1}{NTU_\ell} + \frac{1}{\lambda (NTU_g)}. \tag{8.2.67d}$$

8.1.2.2.1 Porous membrane contactor-stripper for deoxygenation of ultrapure water We will provide a simplified mathematical description of a gas stripper using a membrane contactor, as illustrated in Figure 8.1.13(a). The specific example is deoxygenation of ultrapure water by applying a vacuum through the bore of the fibers as an aqueous solution flows on the outside of the hollow fibers. The aqueous solution (fluid 1) enters through a central tube, having many openings on its surface, through which the fluid flows out radially in crossflow across the hollow fibers. As shown in the simplified schematic of Figure 8.1.15, the water flows radially outward in the first half of the device, goes to the outermost radius, turns around by flowing over the baffle to the second half of the device, where it flows radially inward, and finally enters the other half of the central tube and flows out. The baffle completely separates the two sides of the device including the central tube. Fluid 2 in Figure 8.1.13(a) is the gaseous stream taken out through the bore of the hollow fibers via vacuum. The overall pattern of flow between the two streams is countercurrent; however, locally in each half the flow pattern is crossflow.

The analysis illustrated below is based on Sengupta *et al.* (1998). The basic assumptions are as follows.

(1) The partial pressure of the gas species being stripped into the fiber bore is negligible compared to the equilibrium partial pressure of the species over the liquid.
(2) The dissolved gas concentration in the liquid phase in any half of the device changes only radially; it has no axial variation.

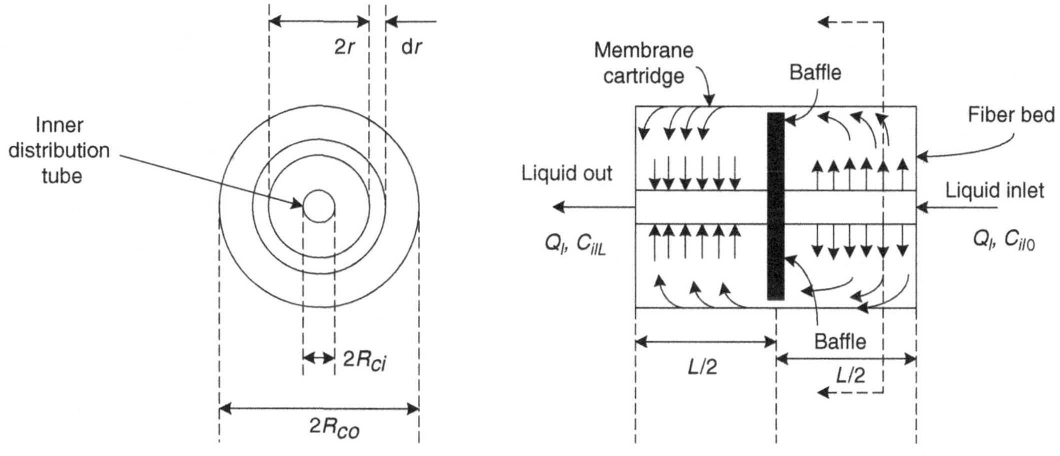

Figure 8.1.15. Simplified configuration of hollow fiber membrane contactor device for stripping of a gas from a liquid. (After Sengupta et al. *(1998).)*

Consider now the first half of the device where the liquid is flowing radially out. Take a thin cylinder of radial thickness dr spanning the length $L/2$ of the device. As the liquid having an averaged radial velocity $\langle v_{\ell r}\rangle$ flows out through a radial cross-sectional area of S_{cr}, oxygen (species i) is being transferred through a hollow fiber membrane surface area dA_s via the liquid-phase based overall transfer coefficient $K_{i\ell}$ (following (8.1.46)):

$$\langle v_{\ell r}\rangle S_{cr}\frac{d\overline{C}_{i\ell}}{dr} = -K_{i\ell}\,dA_s\left(\overline{C}_{i\ell} - C^*_{i\ell}\right). \tag{8.1.68}$$

By assumption (1), $C^*_{i\ell} << \overline{C}_{i\ell}$. If Q_ℓ is the volumetric radial flow rate of the water in the device, then $Q_\ell = \langle v_{\ell r}\rangle S_{cr}$. If the total number of hollow fibers of outside diameter d_o and length L in the device is n^t_{fib}, then the number of hollow fibers in the thin cylinder of radius r and thickness dr is given by

$$dn^t_{\text{fib}} = \frac{\varepsilon_{\text{fib}}\,2\pi r\,dr}{(\pi/4)\,d_o^2}, \tag{8.1.69a}$$

where ε_{fib} is the volume fraction of the hollow fibers in the device. Correspondingly, the hollow fiber outside surface area dA_s is obtained as

$$dA_s = (L/2)\,(\pi\,d_o)\,dn^t_{\text{fib}} = (4\pi\varepsilon_{\text{fib}}L/d_o)r\,dr. \tag{8.1.69b}$$

Substituting this into equation (8.1.68), we get

$$\frac{d\overline{C}_{i\ell}}{\overline{C}_{i\ell}} = -\frac{4\pi\,\varepsilon_{\text{fib}}L}{Q_\ell\,d_o}K_{i\ell}r\,dr. \tag{8.1.70}$$

From (3.4.12b), for a sparingly soluble gas species such as O_2 in water, $K_{i\ell} \sim k_{i\ell}$. Further, from Table 3.1.8 and the functional relations of the type (3.1.143d) between the Sherwood number and the Reynolds number, we may express, for the hollow fiber crossflow system,

$$k_{i\ell} = (\text{constant}) \times \langle v_{\ell r}\rangle^{b_1} \cong (\text{constant}) \times \langle v_{\ell r}\rangle^{0.42}, \tag{8.1.71}$$

where $b_1 \cong 0.42$ (range 0.38–0.46, see Sengupta *et al.* (1998)) for the hollow fiber devices under consideration. It is useful to indicate the complete form of the mass-transfer correlation adopted:

$$\frac{k_{i\ell}\,d_o}{D_{i\ell}} = a_1\left(\frac{d_o\langle v_{\ell r}\rangle\rho_\ell}{\mu_\ell}\right)^{b_1}\left(\frac{\mu_p}{\rho_\ell D_{i\ell}}\right)^{c_1}. \tag{8.1.72}$$

This suggests that the value of (constant) in (8.1.71) is as follows:

$$(\text{constant}) = \frac{a_1}{d_o^{(1-b_1)}}\left(\frac{\rho_\ell}{\mu_\ell}\right)^{(b_1-c_1)}(D_{i\ell})^{(1-c_1)}. \tag{8.1.73}$$

The averaged radially outward liquid velocity $\langle v_{\ell r}\rangle$ at any radial location in the first half of the membrane device and the volumetric radial water flow rate Q_ℓ are related via the following relation:

$$Q_\ell = \langle v_{\ell r}\rangle \times (2\pi r(L/2) \times \varepsilon_{Ar}, \tag{8.1.74}$$

where ε_{Ar} is the fraction of the radial area $(2\pi r)(L/2)$ which is open to fluid flow between the fibers. One can now express the overall transfer coefficient $K_{i\ell}$, and therefore $k_{i\ell}$, as follows:

$$K_{i\ell} = k_{i\ell} = (\text{constant}) \times (Q_\ell)^{b_1}/(\pi\,rL\,\varepsilon_{Ar})^{b_1}. \tag{8.1.75}$$

We now utilize this result in the basic mass balance equation (8.1.70) to obtain

$$\int_{C_{i\ell 0}}^{\overline{C}_{i\ell m}}\frac{d\overline{C}_{i\ell}}{\overline{C}_{i\ell}} = -\int_{R_{ci}}^{R_{co}}\frac{4\pi\,\varepsilon_{\text{fib}}(L)^{(1-b_1)} \times (\text{constant}) \times (Q_\ell)^{b_1-1}}{d_o\,(\pi\,\varepsilon_{Ar})^{b_1}}r^{(1-b_1)}\,dr. \tag{8.1.76}$$

This equation is being integrated between the two limits of r, namely R_{ci} and R_{co}, the inside radius of the fiber bundle and the outside radius of the fiber bundle in the device, respectively; the corresponding limits of species i concentration in the device are $C_{i\ell 0}$ and $\overline{C}_{i\ell m}$, where the subscript m indicates the midway point between the first and second halves of the porous membrane based gas absorption/stripping device:

$$-\ell n\left[\frac{\overline{C}_{i\ell m}}{C_{i\ell 0}}\right] = \frac{4\pi^{(1-b_1)}\,\varepsilon_{\text{fib}}(L)^{(1-b_1)} \times (\text{constant})}{(2-b_1)\,d_o\varepsilon_{Ar}^{b_1}(Q_\ell)^{(1-b_1)}}\left(R_{co}^{(2-b_1)} - R_{ci}^{(2-b_1)}\right). \tag{8.1.77a}$$

We now focus on the second half of the device where the liquid flows radially inward from the outside radius of the fiber bundle, R_{co}, with the species i concentration $\overline{C}_{i\ell m}$ coming out of the first half. The governing equation continues to be the same, except the limits of solution concentration are $\overline{C}_{i\ell m}$ at R_{co} and $\overline{C}_{i\ell L}$ at R_{ci}. Therefore the result for the second half of the device is:

$$-\ell n\left[\frac{\overline{C}_{i\ell L}}{\overline{C}_{i\ell m}}\right] = \frac{4\pi^{(1-b_1)}\,\varepsilon_{\text{fib}}(L)^{(1-b_1)} \times (\text{constant})}{(2-b_1)\,d_o\varepsilon_{Ar}^{b_1}(Q_\ell)^{(1-b_1)}}\left(R_{co}^{(2-b_1)} - R_{ci}^{(2-b_1)}\right). \tag{8.1.77b}$$

Adding equations (8.1.77a) and (8.1.77b), we get

$$-\ell n\left[\frac{\overline{C}_{i\ell L}}{C_{i\ell 0}}\right] = \frac{8\pi^{(1-b_1)}\,\varepsilon_{\text{fib}}(L)^{(1-b_1)} \times (\text{constant})}{(2-b_1)\,d_o\varepsilon_{Ar}^{b_1}(Q_\ell)^{(1-b_1)}}\left(R_{co}^{(2-b_1)} - R_{ci}^{(2-b_1)}\right), \tag{8.1.77c}$$

where the quantity (constant) is defined by equation (8.1.73). If the goal of separation is complete stripping of species i from the solution, the stripping device efficiency E_i may be defined as

$$E_i = \frac{C_{i\ell 0} - \overline{C}_{i\ell L}}{C_{i\ell 0} - 0} = 1 - \frac{\overline{C}_{i\ell L}}{C_{i\ell 0}}, \tag{8.1.78a}$$

so that

$$\frac{-\ell n(1-E_i)}{L^{(1-b_1)}\left[R_{co}^{(2-b_1)}-R_{ci}^{(2-b_1)}\right]}=\frac{8\,\pi^{(1-b_1)}\,\varepsilon_{\text{fib}}\times(\text{constant})}{(2-b_1)\,d_o\varepsilon_{Ar}^{b_1}(Q_\ell)^{(1-b_1)}}. \tag{8.1.78b}$$

Log-log plots of $\ell n(1-E_i)$ against the liquid flow rate Q_ℓ will allow the determination of b_1 from the slope of the line, $-(1-b_1)$, if b_1 is unknown.

It may not be out of place to mention that porous hydrophobic hollow fiber membrane devices having various designs are used extensively as membrane oxygenators or blood oxygenators. Air as a source of oxygen is supplied through the bore of the hollow fibers while blood flows on the outside of the fibers; the blood is supplied with O_2 while CO_2 in the blood is stripped into the air stream.

8.1.2.3 *Countercurrent gas absorption from a concentrated gas stream*

If species i to be absorbed is present in considerable amounts in the feed gas to be scrubbed by the absorbent liquid, the gas flow rate will change substantially along the column length. Therefore the assumption of $|W_{tg}|$ and $|W_{t\ell}|$ being essentially constant along the z-coordinate is no longer valid. The governing equation in the absence of longitudinal dispersion/axial diffusion is now equation (8.1.8):

$$\frac{\partial\,(\overline{x}_{ig}\,W_{tg})}{\partial z}+\frac{\partial(\overline{x}_{i\ell}\,W_{t\ell})}{\partial z}=0. \tag{8.1.79}$$

Since only species i is being absorbed, we assume that the molar flow rate of inerts, $W_{tg}{}^{\text{in}}$, is constant along the column length; further, it is related to W_{tg} via

$$W_{tg}\,(1-\overline{x}_{ig})=W_{tg}^{\text{in}}=\left|W_{tg}^{\text{in}}\right|. \tag{8.1.80}$$

Correspondingly, the variable molar absorbent liquid flow rate $W_{t\ell}$ in the opposite direction ($-z$) is related to the molar flow rate of nonvolatile liquid $W_{t\ell}^{\text{in}}$ via

$$W_{t\ell}\,(1-\overline{x}_{i\ell})=W_{t\ell}^{\text{in}}=-\left|W_{t\ell}^{\text{in}}\right| \tag{8.1.81}$$

since the liquid flows in the negative z-direction. Introduce these two relations in (8.1.79):

$$\left|W_{tg}^{\text{in}}\right|\frac{\partial(\overline{x}_{ig}/(1-\overline{x}_{ig}))}{\partial z}-\left|W_{t\ell}^{\text{in}}\right|\frac{\partial(\overline{x}_{i\ell}/(1-\overline{x}_{i\ell}))}{\partial z}=0;$$
$$d(\overline{x}_{ig}/(1-\overline{x}_{ig}))=+\left(\left|W_{t\ell}^{\text{in}}\right|/\left|W_{tg}^{\text{in}}\right|\right)d(\overline{x}_{i\ell}/(1-\overline{x}_{i\ell})). \tag{8.1.82}$$

Integrate this between locations $z=0$ and $z=z$ to obtain

$$[\overline{x}_{ig}/(1-\overline{x}_{ig})]-[\overline{x}_{ig0}/(1-\overline{x}_{ig0})]=\left(\left|W_{t\ell}^{\text{in}}\right|/\left|W_{tg}^{\text{in}}\right|\right)\left(\frac{\overline{x}_{i\ell}}{(1-\overline{x}_{i\ell})}-\frac{\overline{x}_{i\ell0}}{(1-\overline{x}_{i\ell0})}\right). \tag{8.1.83}$$

A rearrangement of this result yields (Cussler, 1997)

$$\overline{x}_{ig}=\frac{\dfrac{\overline{x}_{ig0}}{(1-\overline{x}_{ig0})}+\left(\left|W_{t\ell}^{\text{in}}\right|/\left|W_{tg}^{\text{in}}\right|\right)\left\{\dfrac{\overline{x}_{i\ell}}{(1-\overline{x}_{i\ell})}-\dfrac{\overline{x}_{i\ell0}}{(1-\overline{x}_{i\ell0})}\right\}}{1+\dfrac{\overline{x}_{ig0}}{(1-\overline{x}_{ig0})}+\left(\left|W_{t\ell}^{\text{in}}\right|/\left|W_{tg}^{\text{in}}\right|\right)\left\{\dfrac{\overline{x}_{i\ell}}{(1-\overline{x}_{i\ell})}-\dfrac{\overline{x}_{i\ell0}}{(1-\overline{x}_{i\ell0})}\right\}}. \tag{8.1.84}$$

This is the equation for the operating line relating $\overline{x}_{ig}$ to $\overline{x}_{i\ell}$ at any location in the countercurrent gas absorption column. Obviously this line is not linear, unlike that for the scrubbing of a dilute gas mixture.

We now determine the height L of the column required to reduce the gas-phase composition of species i from C_{ig0} or (x_{ig0}) to $\overline{C}_{igL}$ (or $\overline{x}_{igL}$) for the concentrated feed gas stream. The relevant starting equation is (8.1.7b):

$$d(\overline{x}_{ig}\,W_{tg})/dz=-K_{igc}a\,S_c\left(\overline{C}_{ig}-\overline{C}_{ig}^{*}\right)$$
$$=W_{tg}^{\text{in}}\,(d(\overline{x}_{ig}/(1-\overline{x}_{ig}))/dz), \tag{8.1.85}$$

where we have introduced relation (8.1.80) on the right-hand side. Therefore

$$W_{tg}^{\text{in}}\frac{d\overline{x}_{ig}}{(1-\overline{x}_{ig})^2}=-K_{igc}a\,S_c\,C_{tg}\left(\overline{x}_{ig}-\overline{x}_{ig}^{*}\right)dz, \tag{8.1.86}$$

leading to

$$\int_0^L dz=L=\int_{\overline{x}_{igL}}^{x_{ig0}}\frac{\left(W_{tg}^{\text{in}}/S_c\right)}{K_{igx}a}\frac{d\overline{x}_{ig}}{(1-\overline{x}_{ig})^2\left(\overline{x}_{ig}-\overline{x}_{ig}^{*}\right)}. \tag{8.1.87}$$

As shown in equations (8.1.54), we can express this column length L as the following product, provided we can assume an average value of $\left(W_{tg}^{\text{in}}/S_c\right)/K_{igx}a$ without much error:

$$L=HTU_{og}\times NTU_{og};\qquad HTU=\left(\left(W_{tg}^{\text{in}}/S_c\right)/K_{igx}a\right); \tag{8.1.88}$$

$$NTU=\int_{\overline{x}_{igL}}^{x_{ig0}}\frac{d\overline{x}_{ig}}{(1-\overline{x}_{ig})^2\left(\overline{x}_{ig}-\overline{x}_{ig}^{*}\right)}. \tag{8.1.89}$$

As the flow rate of the gas phase changes through the column, the quantity $K_{igx}\,a$ will change; however, since its dependence on flow rate has a power less than one, the change in K_{igx} with length will be less than that of the gas flow rate. Therefore taking it out of the integral in (8.1.87) will involve somewhat less error, especially if preliminary estimates of L are being developed.

One could also develop an analysis based on individual film coefficients, as we have done in equations (8.1.62(a)–d). Further, if the film coefficient/overall coefficient available is based on equimolar counterdiffusion, then a correction using the bulk flow correction factor ϕ_N has to be used for gas absorption/stripping; therefore the expression for NTU will contain additional terms via ϕ_N.

Example 8.1.3 A packed tower containing ceramic rings each of radius 2.54 cm is being used to absorb SO_2 from air at 20 °C using water as the absorbent in countercurrent flow. The air stream flow rate at the column bottom is 6 kgmol/hr. The SO_2 mole fraction in this entering stream is 0.20; it has to be reduced to 0.01 at the column top using water entering at a rate of 400 kgmol/hr. The Henry's law constant H_i^P over this concentration range may be approximated as 28 (i.e. $(x_{ig}^* = 28\bar{x}_{i\ell})$). The column diameter is 0.5 m.

(a) Determine the SO_2 concentration in the liquid exiting the column; obtain the equation for the operating line.
(b) Based on the data of Whitney and Vivian (1949), Geankoplis (2003, p. 681) has provided the following correlations for individual film coefficients for 2.54 cm ceramic rings:

$$k_g' a = 0.0594 G_g^{0.7} G_\ell^{0.25}; \qquad k_\ell' a = 0.152 G_\ell^{0.82},$$

where the mass-average velocities have the following units. For G_g, total gas flow rate (kg/s) per m^2 of tower cross section; for G_l, total liquid flow rate (kg/s) per m^2 of tower cross section; i.e. kg/s-m^2. The units of $k_g' a$ and $k_\ell' a$ are kgmol/s-m^3-mol fraction. Calculate the value of HTU_{og} using an averaged value of the mass-averaged velocities and overall mass-transfer coefficient $K_{igx}\,a$. What is the height of the column?

Solution (a)

$$|W_{tg}|_{z=0} = 6\,\text{kgmol/hr}; \qquad |W_{t\ell}|_{z=L} = 400\,\text{kgmol/hr};$$
$$x_{ig0} = 0.20; \qquad \bar{x}_{igL} = 0.01; \qquad x_{i\ell L} = 0; \bar{x}_{i\ell 0} = ?$$

Employ (8.1.80) and (8.1.81) to determine $|W_{tg}^{\text{in}}|$ and $|W_{t\ell}^{\text{in}}|$:

$$|W_{tg}^{\text{in}}| = W_{tg}|_{z=o}\left(1 - x_{ig0}\right) = 6 \times (1 - 0.2) = 6 \times 0.8$$
$$= 4.8\,\text{kgmol/hr};$$
$$|W_{t\ell}^{\text{in}}| = |W_{t\ell}|_{z=L}\left(1 - x_{i\ell L}\right) = 400 \times (1 - 0) = 400\,\text{kgmol/hr}.$$

Now use these values in equation (8.1.83) at $z = L$ to determine $\bar{x}_{i\ell 0}$ since $x_{i\ell L}$ is known:

$$\left[\bar{x}_{igL}/(1 - \bar{x}_{igL})\right] - \left[x_{ig0}/(1 - (x_{ig0})\right]$$
$$= \left(|W_{t\ell}^{\text{in}}|/|W_{tg}^{\text{in}}|\right)\left(\frac{x_{i\ell L}}{(1 - x_{i\ell L})} - \frac{\bar{x}_{i\ell 0}}{(1 - \bar{x}_{i\ell 0})}\right);$$

$$\left[\frac{0.01}{0.99}\right] - \left[\frac{0.2}{0.8}\right] = \left(\frac{400}{4.8}\right)\left[\frac{0}{1} - \frac{\bar{x}_{i\ell 0}}{(1 - \bar{x}_{i\ell 0})}\right]$$

$$\Rightarrow 0.0101 - 0.25 = 83.34 \times (-)\frac{\bar{x}_{i\ell 0}}{(1 - \bar{x}_{i\ell 0})};$$

$$\frac{0.2399}{83.34} = \frac{\bar{x}_{i\ell 0}}{(1 - \bar{x}_{i\ell 0})} \Rightarrow \bar{x}_{i\ell 0} = 0.00288.$$

From equation (8.1.84), the operating line equation is

$$\bar{x}_{ig} = \frac{0.25 + 83.34\{[\bar{x}_{i\ell}/(1 - \bar{x}_{i\ell})] - 0.00288\}}{1.25 + 83.34\{[\bar{x}_{i\ell}/(1 - \bar{x}_{i\ell})] - 0.00288\}}.$$

(b)

Tower cross-sectional area $= (\pi/4) \times 0.25\,\text{m}^2 = 0.196\,\text{m}^2$.

We will calculate first the mass-averaged velocity at the entrance and exit locations of the column. At the gas entrance, $z = 0$:

$$G_g = \frac{6\dfrac{\text{kgmol}}{\text{hr}} \times [29 \times 0.8 + 64.1 \times 0.2]}{3600\,\text{s/hr} \times 0.196\,\text{m}^2}$$
$$= \frac{6 \times 5.102}{3600} \times 36.02 = 8.5 \times 10^{-3} \times 36.02$$
$$= 0.307\,\text{kg/s-m}^2.$$

At the exit $z = L$; at $z \to L$ (or $z = L^-$):

$$\frac{d\bar{C}_{ig}}{dz} = 0;\ G_g = \frac{4.8 \times 5.102}{3600} \times 29.3 = 0.199\,\text{kg/s-m}^2.$$

Therefore

$$G_g\Big|_{\text{avg}} = (0.307 + 0.199)/2 = 0.253\,\text{kg/s-m}^2.$$

The variation of liquid flow rate is limited since $\bar{x}_{i\ell 0} = 0.00288$. Therefore,

$$G_\ell\Big|_{\text{avg}} = \frac{400\dfrac{\text{kgmol}}{\text{hr}} \times [18 \times 0.9986 + 64.1 \times 0.0014]}{3600\,\text{s/hr} \times 0.196\,\text{m}^2}$$
$$= \frac{400 \times 18.06}{3600 \times 0.196} = 10.23\,\text{kg/s-m}^2.$$

From equation (3.4.8)

$$\frac{1}{K_{igx}a} = \frac{1}{k_{igx}a} + \frac{H_i^P}{k_{i\ell x}a} \cong \frac{1}{0.0594(0.253)^{0.7}(10.23)^{0.25}}$$
$$+ \frac{28}{0.152(10.23)^{0.82}} = \frac{1}{0.0594 \times 0.382 \times 1.78}$$
$$+ \frac{28}{0.152 \times 6.72} = 24.75 + 27.41 = 52.16;$$

$$HTU_{og} = \frac{|W_{tg}^{\text{in}}|/S_c}{K_{igx}a} = \frac{\dfrac{4.8}{3600}\dfrac{\text{kgmol}}{\text{s}} \times \dfrac{1}{0.196\,\text{m}^2}}{\dfrac{1}{52.16}\dfrac{\text{kgmol}}{\text{s-m}^3\text{-mol frac}}}$$
$$= \frac{4.8 \times 52.16}{3600 \times 0.196}\,\text{m} = 0.354\,\text{m}.$$

The integral (8.1.87) may be determined with the quantity $(W_{tg}^{\text{in}}/S_c)/K_{igx}a$ inside the integral. As the flow rate of the gas varies substantially through the column, $K_{igx}a$ will also vary. Here, however, the variation being somewhat limited, we assume an average value of $K_{igx}a$ for the column, take it outside the integral and then determine the value of the integral:

$$I = \int_{0.01}^{0.20} \frac{d\bar{x}_{ig}}{(1 - \bar{x}_{ig})^2\left(\bar{x}_{ig} - \bar{x}_{ig}^*\right)}.$$

To determine the value of this integral, we need to determine $\bar{x}_{ig}^* = H_i^P \bar{x}_{i\ell}$. Using the operating line equation obtained in part (a), we obtain the following values for the corresponding pairs: $\bar{x}_{ig} = 0.05$, $\bar{x}_{i\ell} = 0.005$; $\bar{x}_{ig} = 0.1$, $\bar{x}_{i\ell} = 0.00121$; $\bar{x}_{ig} = 0.15$, $\bar{x}_{i\ell} = 0.00198$; $\bar{x}_{ig} = 0.20$, $\bar{x}_{i\ell} = 0.00288$; $\bar{x}_{ig} = 0.01$, $\bar{x}_{i\ell} \cong 0.00$. The ordinate $(1/(1-\bar{x}_{ig})^2(\bar{x}_{ig}-\bar{x}^*_{ig}))$ is calculated and given in Table 8.1.2.

The value of the integral I was found to be 5.024. Therefore the height of the column

$$L = HTU_{og} \times NTU_{og} = 0.354 \times 5.024 = 1.78\text{m}.$$

8.1.2.4 *Countercurrent gas absorption/stripping in a column with axial diffusion/dispersion*

The analysis of countercurrent gas absorption/stripping carried out so far in Sections 8.1.2.2 and 8.1.2.3 assumed plug flow, leading to an absence of axial dispersion, as well as no longitudinal diffusion in the device. Further, we utilized the pseudo-continuum level of modeling with a fluid velocity averaged over the column cross section (Section 6.2.1.1); correspondingly, we worked on the axial variation of species concentrations averaged over the column cross section. Figure 8.1.16(a) plots the column compositions of the two phases via the operating line and the equilibrium curve in such a context (they are identical to those in Figures 8.1.14(b) and (c). If we now plot each phase composition as a function of the axial distance z along the column (Miyauchi and Vermeulen, 1963), we

Table 8.1.2.

$\bar{x}_{ig}$	$\bar{x}_{i\ell}$	$\bar{x}^*_{ig}$	$\left[\left(1/(1-\bar{x}_{ig})^2\left(\bar{x}_{ig}-\bar{x}^*_{ig}\right)\right]\right.$
0.20	0.00288	0.0806	13.08
0.15	0.00198	0.0554	14.63
0.1	0.00121	0.0338	18.65
0.05	0.0005	0.014	30.79
0.01	0.00	0.0	102.04

will get two dashed lines (AB and CD in Figure 8.1.16(b)) for the case of gas absorption and two dashed lines (EF and GH) for stripping of a gas from a liquid. For gas absorption in this figure, the gas-phase mole fraction of species i decreases from $x_{igf}(= x_{ig0})$ to $\bar{x}_{igL}$ as z goes from 0 to L. The absorbent phase mole fraction of species i increases from the inlet value of $x_{i\ell L}$ to $\bar{x}_{i\ell 0}$ as z decreases from L to 0. We have also indicated the corresponding quantities in terms of molar concentrations, C_{ig} and $C_{i\ell}$. The dashed lines drawn may also be thought of as curves of $\overline{C}_{ig}$ vs. z and $\overline{C}_{i\ell}$ vs. z. For desorption, the corresponding quantities have been specified.

The essential features of such modeling in Sections 8.1.2.2. and 8.1.2.3 were as follows.

(1) At any axial distance along the column, each phase may be characterized by a composition averaged over the column cross section.
(2) Similarly, the velocity of each phase at any axial distance is obtained as an average over the column cross section, i.e. we have plug flow of each phase.
(3) There is no longitudinal diffusion/axial dispersion.

In practice, there is no way to eliminate longitudinal (or axial) diffusion due to concentration gradients in any phase. For the gas phase, which has a decreasing concentration in the flow direction in gas absorption, longitudinal diffusion will introduce an even sharper concentration decrease along the z-coordinate; correspondingly, in the absorbent phase, there will be a lesser increase in concentration in the $-z$-direction from $z = L$ to $z = 0$. The axial dispersion due to a nonuniform velocity profile (e.g. Figure 6.2.2) will be superimposed on this. In addition, there *may be* additional possibilities of backmixing at the locations of

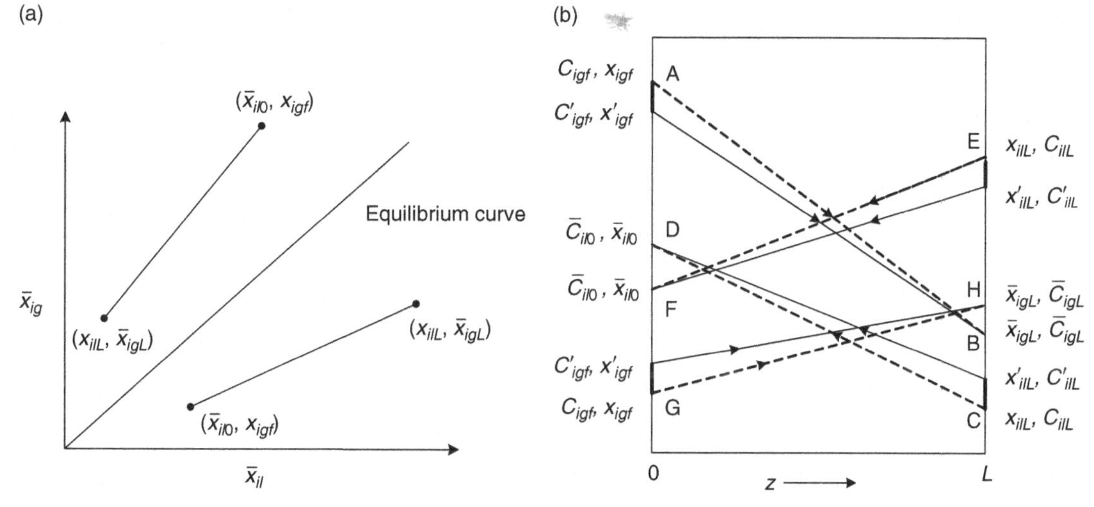

Figure 8.1.16. Countercurrent gas absorption/stripping column with axial dispersion/diffusion. (a) Operating lines and equilibrium curve. (b) Concentration profiles of gas and liquid phases along the tower length for gas absorption/stripping with (solid lines) or without (dashed lines) axial dispersion/diffusion.

the inlets of the two streams, via local circulations, turbulence, etc. There are two net effects of such phenomena, as shown by the two solid lines (for gas absorption and, correspondingly, for gas stripping) in Figure 8.1.16(b) representing the actual changes in species i mole fractions in the two phases along the column length: the concentration difference or mole fraction difference between the two streams at any column location will be reduced,[5] indicating that the species i flux from the gas to the absorbent phase will be reduced at every column location; at the inlet location of each phase there is a sharp change in concentration from the incoming values For the incoming gas, x_{igf} is reduced to x'_{igf}; for the incoming absorbent, $x_{i\ell L}$ is raised to $x'_{i\ell L}$. We can similarly describe the situation for stripping.

The modeling of such behavior is generally carried out by retaining the longitudinal diffusion/axial dispersion term in the steady state form of equations (8.1.1a,b) for both gas and absorbent phases, namely equations (8.1.2a,b):

$$\langle v_{gz}\rangle \frac{d\overline{C}_{ig}}{dz} - \varepsilon_g D_{i,\text{eff},g} \frac{d^2\overline{C}_{ig}}{dz^2} + K_{igc}a\left(\overline{C}_{ig} - C^*_{ig}\right) = 0; \quad (8.1.90\text{a})$$

$$+|\langle v_{\ell z}\rangle| \frac{d\overline{C}_{i\ell}}{dz} + \varepsilon_\ell D_{i,\text{eff},\ell} \frac{d^2\overline{C}_{i\ell}}{dz^2} + K_{i\ell c}a\left(\overline{C}^*_{i\ell} - \overline{C}_{i\ell}\right) = 0. \quad (8.1.90\text{b})$$

Here, $\langle v_{gz}\rangle$ is a positive quantity with respect to the z-coordinate, whereas $\langle v_{\ell z}\rangle$ is a negative quantity with respect to the z-coordinate; we use the magnitude of this velocity $|\langle v_{\ell z}\rangle|$, a positive quantity. Further, instead of $\varepsilon_g D_{i,\text{eff},g}$ and $\varepsilon_\ell D_{i,\text{eff},\ell}$, we will use $\langle D_{i,\text{eff},g}\rangle$ and $\langle D_{i,\text{eff},\ell}\rangle$, reflecting quantities based on the column cross section. These equations are usually solved using the so-called "Danckwerts boundary conditions" at the entrance and exit locations of the two streams (Danckwerts, 1953). At $z \to 0$ (or $z = 0^+$):

$$\langle v_{gz}\rangle C_{igf} - \langle v_{gz}\rangle C'_{igf} = -\langle D_{i,\text{eff},g}\rangle \frac{d\overline{C}_{ig}}{dz}; \quad (8.1.91\text{a})$$

$$\frac{d\overline{C}_{i\ell}}{dz} = 0. \quad (8.1.91\text{b})$$

At $z \to L$ (or $z = L^-$):

$$\frac{d\overline{C}_{ig}}{dz} = 0; \quad (8.1.91\text{c})$$

$$|v_{\ell z}|\, C'_{i\ell L} - |v_{\ell z}|\, C_{i\ell L} = -\langle D_{i,\text{eff},\ell}\rangle \frac{d\overline{C}_{i\ell}}{dz}. \quad (8.1.91\text{d})$$

[5]This implies a reduction in the $(\overline{x}_{ig} - x^*_{ig})$ term in the denominator of the *NTU* expressions (8.1.54c) and (8.1.55d); therefore, an increase in *NTU* happens and a more demanding mass-transfer situation develops.

If we nondimensionalize the axial coordinate z and employ the following dimensionless numbers and quantities,

$$z^+ = z/L; \quad Pe_{zg} = \frac{L\langle v_{gz}\rangle}{\langle D_{i,\text{eff},g}\rangle}; \quad Pe_{z\ell} = \frac{L\langle v_{\ell z}\rangle}{\langle D_{i,\text{eff},\ell}\rangle};$$

$$N_{og} = \frac{K_{igc}aL}{\langle v_{gz}\rangle}; \quad N_{o\ell} = \frac{K_{i\ell c}aL}{\langle v_{\ell z}\rangle}, \quad (8.1.92)$$

then equations (8.1.90a,b) are converted to

$$\frac{d^2\overline{C}_{ig}}{dz^{+2}} - Pe_{zg}\frac{d\overline{C}_{ig}}{dz^+} + Pe_{zg}N_{og}\left(\overline{C}_{ig} - \overline{C}^*_{ig}\right) = 0; \quad (8.1.93\text{a})$$

$$\frac{d^2\overline{C}_{i\ell}}{dz^{+2}} + Pe_{z\ell}\frac{d\overline{C}_{i\ell}}{dz^+} + Pe_{z\ell}N_{o\ell}\left(\overline{C}_{i\ell} - \overline{C}^*_{i\ell}\right) = 0. \quad (8.1.93\text{b})$$

The boundary conditions (8.1.91a–d) will be changed correspondingly. Solutions of such equations have been obtained (Sleicher, 1959a; Miyauchi and Vermeulen, 1963; Hartland and Mecklenburgh, 1966) in general for two counterflowing phases, for linear two-phase equilibrium behavior.

In gas–liquid absorption–desorption processes, the dispersion in the gas phase is often not very important. For linear desorption equilibrium described by

$$C^*_{i\ell} = \kappa_{i\ell}\, C_{ig} + q_i, \quad (8.1.94)$$

an analytical solution has been developed to describe the effect of axial dispersion in the liquid phase on the desorption/separation achieved in the column (Miyauchi and Vermeulen, 1963; Sherwood *et al.*, 1975):

$$\left\{\frac{\overline{C}_{i\ell 0} - (\kappa_{i\ell}\, C_{igf} + q_i)}{C_{i\ell L} - (\kappa_{i\ell}\, C_{igf} + q_i)}\right\} = \frac{4b\exp\left(Pe_{z\ell}/2\right)}{(1+b)^2\exp\left(Pe_{z\ell}\, b/2\right) - (1-b)^2\exp\left(-Pe_{z\ell}\, b/2\right)} = \exp(-N_{o\ell P}). \quad (8.1.95)$$

Here

$$b = \left[1 + 4\left(N_{o\ell}/Pe_{z\ell}\right)\right]^{1/2}; \quad (8.1.96\text{a})$$

$N_{o\ell} = L/H_{o\ell}$, the true number of transfer units under dispersion; (8.1.96b)

$H_{o\ell}$ = true height of a transfer unit based on dispersion $= \dfrac{\langle v_{\ell z}\rangle}{K_{i\ell c}a}$; (8.1.96c)

$N_{o\ell P} = L/H_{o\ell P}$, the number of transfer units under conditions of plug flow, (8.1.96d)

$H_{o\ell P}$ = height of a transfer unit under plug flow. (8.1.96e)

Further, exp(–(–$N_{o\ell P}$)) represents the fractional unattained approach of the liquid-phase concentration with respect to the gaseous feed (Sherwood *et al.*, 1975).

Note the following points.

(1) From Henry's law, relation (3.3.60b), $\overline{p}_{ig} = H_i \overline{x}_{i\ell}$, we obtain

$$C^*_{i\ell} = \frac{RT\,\overline{C}_{t\ell}}{H_i}\overline{C}_{ig} = \kappa_{i\ell}\overline{C}_{ig}. \tag{8.1.97}$$

(2) Estimates of axial dispersion coefficients $\langle D_{i,\text{eff},\ell}\rangle$ and $\langle D_{i,\text{eff},g}\rangle$ are available in literature for both the gas and liquid phases for a number of packings; more recent correlations are described in Macias-Salinas and Fair (2002) for both random and structured packings. Examples of earlier experimental data from countercurrent flow of air and water through a few random packings are available in Sherwood *et al.* (1975, fig 11.10) as a plot of the random packing diameter based Péclet number against the gas- or liquid-phase mass velocity (G_g, G_ℓ).

(3) As the superficial velocity of the phase from which solute is being transferred increases, with other values (e.g. column height and the axial dispersion coefficient) remaining constant in the Péclet number, the plug flow case increasingly approximates the axial dispersion case.

Example 8.1.4 Sherwood and Holloway (1940) determined the mass-transfer behavior in O_2 desorption from water from a series of Raschig rings of sizes 0.5″, 1.0″, 1.5″ and 2″ in a column of diameter 20″ at 25 °C. Sherwood *et al.* (1975), have considered a particular example from this study for 2″ Raschig rings, wherein, at superficial liquid and gas velocities of $\langle v_{\ell z}\rangle = 0.542$ cm/s, $\langle v_{\ell z}\rangle = 26.3$ cm/s and a packed height of 15.5 cm, the values of $H_{o\ell P}$ and $N_{o\ell P}$ were found to be 29.26 cm and 0.529, respectively. We employ this worked example from Sherwood *et al.* (1975) to illustrate the application of the axial dispersion analysis just provided. Determine the true values of $H_{o\ell}$ and $N_{o\ell}$ for this case, given the Henry's law constant for O_2 and water at 20 °C and 30 °C to be 4.01×10^4 and 4.75×10^4 atm/mole fraction, respectively. You are given that $\langle D_{i,\text{eff},\ell}\rangle = 0.0148\ \text{ft}^2/\text{s}$.

Solution We will employ equations (8.1.95) and (8.1.96) after assuming that dispersion in the gas phase is negligible. The basis for this assumption is that liquid-phase resistance must be the controlling factor due to the very low solubility of O_2 in water (equation (3.4.12b)). Correspondingly, the absorption factor A (definition (8.1.45)) is given by

$$A = \frac{|W_{t\ell}|}{|W_{tg}|}\frac{1}{H_i^P} = \frac{P}{H_i}\frac{\langle v_{\ell z}\rangle}{\langle v_{gz}\rangle}\frac{C_{t\ell}}{C_{tg}} = \frac{\langle v_{\ell z}\rangle \times RTC_{t\ell}}{H_i\langle v_{gz}\rangle}.$$

Assume, at 25 °C, $H_i = ((4.01 + 4.75)/2) \times 10^4 = 4.38 \times 10^4$ atm/mol frac; $R = 82.057\ \text{cm}^3$-atm/gmol-K; $T = 298\ K$; $C_{t\ell} = (1/18\ \text{cm}^3/\text{gmol})$

$$\Rightarrow A = \frac{0.542\ \text{cm/s} \times 82.05\ \text{cm}^3\text{-atm/gmol-K} \times 298\ \text{K} \times 1\ \text{gmol}}{4.38 \times 10^4\ \dfrac{\text{atm}}{\text{mol frac}} \times 26.3\ \text{cm/s} \times 18\ \text{cm}^3}$$

$$= \frac{0.542\ \text{cm/s} \times 82.05 \times 298}{4.38 \times 26.3 \times 18 \times 10^4} = 63.9 \times 10^{-5},$$

which is negligibly small. Now,

$$C^*_{i\ell} = \kappa_{i\ell}C_{ig}$$
$$\Rightarrow x^*_{i\ell}C_{t\ell} = \kappa_{i\ell}p_{ig}/RT \Rightarrow x^*_{i\ell} = (\kappa_{i\ell}/RT\,C_{t\ell})\,p_{ig}$$
$$\Rightarrow p_{ig} = \frac{RT\,C_{t\ell}}{\kappa_{i\ell}}x^*_{i\ell} \Rightarrow H_i = \frac{RT\,C_{t\ell}}{\kappa_{i\ell}}$$
$$\Rightarrow \kappa_{i\ell} = \frac{RT\,C_{t\ell}}{H_i} = 82.05\,\frac{\text{cm}^3\text{-atm}}{\text{gmol-K}}\,\frac{298\ \text{K mol frac}}{4.38 \times 10^4\text{atm}}\,\frac{\text{gmol}}{18\ \text{cm}^3}$$
$$= \frac{82.05 \times 298}{4.38 \times 10^4 \times 18} = 0.0031.$$

Employ equation (8.1.95) to determine $N_{o\ell}$ and therefore $H_{o\ell} = (L/N_{o\ell})$, since other quantities are known or can be determined. First, determine $Pe_{z\ell}$:

$$Pe_{z\ell} = \frac{L\langle v_{\ell z}\rangle}{\langle D_{i,\text{eff},\ell}\rangle} = \frac{15.5 \times 0.542}{0.0148 \times (30.48)^2} = 0.61.$$

Then

$$b = [1 + 4\,(N_{o\ell}/0.61)]^{0.5} = [1 + 6.55\,N_{o\ell}]^{0.5};$$
$$\exp(-N_{o\ell P}) = \exp(-0.529) = 0.59$$
$$= \frac{4b\exp(0.305)}{(1+b)^2\exp(0.305b) - (1-b)^2\exp(-0.305b)}.$$

The value of $N_{o\ell}$, and therefore b, can be determined by trial and error. The first guess should be $N_{o\ell}$=0.53. The right-hand side (rhs) yields 0.64. Since $N_{o\ell}$ is larger than $N_{o\ell P}$, the next guess should be higher than 0.53, say $N_{o\ell} = 0.61$. We get the rhs to be 0.577. The average of these two guesses would yield on the rhs (0.61 + 0.577)/2 = 0.593, which is pretty close to the left-hand side. So a good guess for $N_{o\ell}$ would be the average of (0.61 + 0.69)/2 = 0.65 = $N_{o\ell}$. Therefore $H_{o\ell} = 15.5/0.65\ \text{cm} = 23.84\ \text{cm}$.

Example 8.1.5 We continue here with the example from Sherwood *et al.* (1975) on the desorption of O_2 from water in a tower, as described in Example 8.1.4, except now the tower has a different height. The packings and the two phase flow rates are identical. The quantity $C_{i\ell L} - (\kappa_{i\ell}\,C_{igf} + q_i)$ in the denominator on the left-hand side (lhs) of equation (8.1.95) is $((C_{i\ell L} - C^*_{i\ell}|_{C_{igf}})$, the composition change in the liquid phase needed to bring the feed liquid to equilibrium with the feed gas, C_{igf}. If the actual change achieved in the liquid, $(C_{i\ell L} - \overline{C}_{i\ell o})$, is $0.98(C_{i\ell L} - C^*_{i\ell}|_{C_{igf}})$, then determine the value of $N_{o\ell}$ and the height of the column needed if $H_{o\ell}$ is the same as in Example 8.1.4, i.e. 23.84 cm. Determine the column height if axial diffusion were neglected.

Solution The lhs of equation (8.1.95) may be written as

$$\left(\overline{C}_{i\ell o} - C^*_{i\ell}|_{C_{igf}}\right)/\left(C_{i\ell L} - C^*_{i\ell}|_{C_{igf}}\right) = ratio;$$

then

$$1 - ratio = \frac{C_{i\ell L} - \overline{C}_{i\ell 0}}{\left(C_{i\ell L} - C^*_{i\ell}|_{C_{igf}}\right)} = 0.98.$$

Therefore,

$$ratio = 0.02.$$

We may now focus on various terms in equation (8.1.95):

$$Pe_{z\ell} = \frac{L \times 0.542}{0.0148 \times (30.48)^2} = 0.039L.$$

Now, $N_{o\ell} H_{o\ell} = L$

$\Rightarrow L = N_{o\ell} \times 23.84 \Rightarrow Pe_{z\ell} = 0.039 \times 23.84 N_{o\ell} - 0.929 N_{o\ell}$.

So

$$b = \left[1 + 4(N_{o\ell}/Pe_{z\ell})\right]^{0.5} = \left[1 + (4/0.929)\right]^{0.5} = \left[1 + 4.305\right]^{0.5} = \left[5.305\right]^{0.5} = 2.303.$$

Using these results in equation (8.1.95), we get

$$0.02 = \frac{4 \times 2.303 \exp(0.4645 N_{o\ell})}{10.91 \exp\left(\dfrac{0.929 \times 2.303}{2} N_{o\ell}\right) - 1.698 \exp(-1.069 N_{o\ell})}$$

$$\Rightarrow 0.02 = \frac{9.212 \exp(0.4645 N_{o\ell})}{10.91 \exp(1.069 N_{o\ell}) - 1.698 \exp(-1.069 N_{o\ell})}.$$

By trial and error, we find $N_{o\ell}$=6.19. Therefore the height of the column is given by

$$L = 6.19 \times 23.84 \text{ cm} = 147.5 \text{ cm}.$$

If axial dispersion were neglected, then, from equation (8.1.95),

$$\exp(-N_{o\ell P}) = 0.02 \Rightarrow N_{o\ell P} = 3.913$$
$$\Rightarrow L = N_{o\ell P} \times H_{o\ell P} = 3.913 \times 29.26 = 114.5 \text{ cm},$$

much lower than the actual value needed, 147.5 cm.

Readers should consult the extensive study by Macias-Salinas and Fair (2002) on axial mixing effects in packed gas–liquid contactors. Various types of situations have been considered: neglect of gas-phase dispersion (desorption of O_2 from water); dispersion in both phases (SO_2 absorption by water); neglect of liquid-phase dispersion (water cooling with air). They have employed the rigorous analytical solution of Hartland and Mecklenburgh (1966) developed for countercurrent liquid extraction and applied it to gas absorption with gas as the raffinate phase and the liquid as the extract phase by a trial-and-error algorithm.

8.1.2.5 *Countercurrent gas absorption in a chemical absorber*

This is a subject of considerable complexity. For a comprehensive treatment, see Astarita *et al.* (1983). First we will introduce a method for analyzing such a problem numerically (following Froment and Bischoff (1979)). Consider a packed column with the gas flowing up and the chemically reactive liquid flowing down. There are two major distinctions to be kept in mind: the nature of the flow (axial dispersion/diffusion present or not); whether the reaction takes place completely in the liquid film or in the bulk liquid as well.

Consider now the case of no axial dispersion/diffusion. If we carry out a balance of gas species A and the chemical absorbent C from the column top ($z = L$) to any location z in the column for the following reaction:

$$a_A\,\text{A} + c_C\,\text{C} \rightarrow pP + qQ, \qquad (8.1.98)$$

we obtain

$$W_{tg}\left[\overline{x}_{Ag} - \overline{x}_{AgL}\right] = \frac{a_A}{c_C} Q_\ell \left[C_{C\ell L} - \overline{C}_{C\ell}\right] + Q_\ell \overline{C}_{A\ell}, \qquad (8.1.99)$$

where a_A, c_C are the stoichiometric coefficients for the reaction, Q_ℓ is the volumetric flow rate of the downflowing absorbent liquid, $\overline{C}_{C\ell}$ is the bulk concentration of chemical absorbent C in the liquid at location z and $\overline{C}_{A\ell}$ is the bulk concentration of gas species A in the absorbent liquid (at z) in case there is bulk reaction; $C_{C\ell L}$ is the molar concentration of the chemical absorbent C entering the column at the top. Now, instead of having equation (8.1.46) as the governing equation for $\overline{C}_{Ag}$ or $\overline{x}_{Ag}$ variation with z, we will have the following equation:

$$\langle v_{gz}\rangle S_c \frac{d\overline{C}_{Ag}}{dz} = -N_A^r\Big|_{y=0} a\, S_c = \langle v_{gz}\rangle S_c C_{tg} \frac{d\overline{x}_{Ag}}{dz}, \qquad (8.1.100)$$

where a is the gas–liquid interfacial area per unit absorber volume and $N_A^r|_{y=0}$ is the mass flux of species A at the gas–liquid interface for the reaction-enhanced gas absorption; expressions for it are available, for example, in equations (5.3.12), (5.3.16), (5.3.20), (5.3.30b), etc., for a few different types of reactions. In the two equations given above, there are three unknowns: $\overline{x}_{Ag}$, $\overline{C}_{C\ell}$ and $\overline{C}_{A\ell}$. The third unknown $\overline{C}_{A\ell}$ appears because of the possibility of reaction in the bulk; if the reaction were instantaneous or fast enough to be completed essentially in the film, such a quantity would not appear. Any change in this bulk concentration of the gaseous reactant A, $\overline{C}_{A\ell}$, in the bulk liquid phase is due to a nonzero flux $N_A^r|_{y=\delta_\ell}$ (see Figures 5.3.1(a) and (b)) at $y = \delta_\ell$, the liquid film–bulk liquid boundary and the reactive consumption of A going on in the bulk liquid:

$$N_A^r\Big|_{y=\delta_\ell} a S_c = (-R_A)(1 - \delta_\ell a)(1 - \varepsilon_g) S_c + Q_\ell \frac{d\overline{C}_{A\ell}}{dz}. \qquad (8.1.101)$$

Here, R_A is the molar rate of production of species A per unit bulk liquid volume by reaction, a is the gas–liquid interfacial area per unit liquid volume, $(1-\delta_\ell a)$ is the fractional liquid volume that constitutes the bulk liquid, δ_ℓ is the liquid film thickness, ε_g is the void volume of the gas phase, and S_c dz is the total volume of the absorber column of length dz.

Simultaneous solution of the three equations (8.1.99), (8.1.100) and (8.1.101) has to be carried out numerically. One can start the numerical solution at $z = L$, the column top, where $C_{C\ell L}$ is usually specified for the absorbent concentration at the liquid inlet. Most often, $\overline{x}_{AgL}$ or $\overline{C}_{AgL}$ is also specified to indicate the degree of cleaning/scrubbing of the gas required from the incoming valve C_{Ag0}. Further, $C_{A\ell L}$ should be known; for example, it can be negligibly small if the spent absorbent has been regenerated well in a stripper and introduced at the column top. For a Δz step, starting from $z = L$, where $C_{C\ell} = C_{C\ell L}$,

$\overline{C}_{Ag} = \overline{C}_{AgL}$, $C_{A\ell} = C_{A\ell L}$ are known, we can calculate the values of $C_{C\ell}$, $\overline{C}_{Ag}$, $C_{A\ell}$ at $L - \Delta z$. In this fashion, by marching to $z = 0$ (inlet of gas stream), one can find out $\overline{C}_{C\ell 0}$, $\overline{C}_{Ag0}$ and $C_{A\ell 0}$. In reality, we do not know L. The value of $\Sigma\Delta z$ where one matches the calculated value with the given $\overline{C}_{Ag0}$ and the corresponding $\overline{C}_{C\ell 0}$ becomes the column inlet, $L = \Sigma\Delta z$.

If we have an *instantaneous reaction*, equation (8.1.101) is no longer needed; further, $\overline{C}_{A\ell}$ in equation (8.1.99) is zero. From equation (5.3.30b),

$$N^r_A\big|_{y=0} = \frac{p_{Ab} + H^C_A \dfrac{D_{C\ell}}{D_{A\ell}} \overline{C}_{C\ell} \dfrac{a_A}{c_C}}{(1/k_g) + (H^C_A/k_\ell)}, \tag{8.1.102}$$

where p_{Ab} is related to $\overline{x}_{Ag}$ via $p_{Ab} = P\overline{x}_{Ag} = p_{Ag}$, $\overline{C}_{C\ell}$ can be expressed in terms of $\overline{x}_{Ag}$ or $\overline{C}_{Ag}$ from equation (8.1.99) with $\overline{C}_{A\ell} = 0$ and $W_{tg} = \langle v_{gz}\rangle S_c C_{tg}$. Therefore

$$\langle v_{gz}\rangle S_c C_{tg} \frac{d\overline{x}_{Ag}}{dz} = -aS_c \frac{P\overline{x}_{Ag} + H^C_A \dfrac{D_{C\ell}}{D_{A\ell}} \dfrac{a_A}{c_C}\left\{C_{C\ell L} - \dfrac{c_C}{a_A}\dfrac{W_{tg}}{Q_\ell} \times (\overline{x}_{Ag} - \overline{x}_{AgL})\right\}}{(1/k_g) + (H^C_A/k_\ell)}. \tag{8.1.103}$$

This first-order ordinary differential equation can be easily integrated to develop an analytical solution of $\overline{x}_{Ag}$ or $\overline{C}_{Ag}$ or p_{Ab} against L. If the limits of $\overline{x}_{Ag}$ are specified, L can be determined, and vice versa:

$$\int_0^L dz = L = -\frac{\left\{\left(\dfrac{1}{k_g}\right) + (H^C_A/k_\ell)\right\}}{a} \int_{x_{Ag0}}^{x_{AgL}} \frac{\langle v_{gz}\rangle C_{tg}\, d\overline{x}_{Ag}}{\left[P\overline{x}_{Ag} + H^C_A \dfrac{D_{C\ell}}{D_{A\ell}} \dfrac{a_A}{c_C}\left\{C_{C\ell L} - \dfrac{c_C}{a_A}\dfrac{W_{tg}}{Q_\ell}(\overline{x}_{Ag} - \overline{x}_{AgL})\right\}\right]}. \tag{8.1.104}$$

Noting that $W_{tg} = \langle v_{gz}\rangle C_{tg} S_c$ and $LS_c a$ = total interfacial area in the device, we get

$$\frac{(LS_c a) K_g P}{W_{tg}} = -\int_{p_{Ag0}}^{p_{AgL}} \frac{d\overline{p}_{Ag}}{\overline{p}_{Ag}\left[1 - H^C_A \frac{D_{C\ell}}{D_{A\ell}} \frac{W_{tg}}{Q_\ell P}\right] + H^C_A \frac{D_{C\ell}}{D_{A\ell}} \frac{W_{tg}}{Q_\ell} \frac{\overline{p}_{AgL}}{P} + H^C_A \frac{D_{C\ell}}{D_{A\ell}} C_{C\ell L} \frac{a_A}{c_C}}. \tag{8.1.105}$$

We have seen earlier (in Section 5.3.1) that, as the concentration of the reactant C in the absorbent phase is increased, the instantaneous reaction interface is pushed from inside the liquid film to the gas-liquid interface. Correspondingly, from equation (5.3.29a),

$$N^r_A\big|_{y=0} = k_g(p_{Ab}) = \frac{a_A}{c_C}\frac{D_{C\ell}}{D_{A\ell}} k_\ell C'_{Cb}. \tag{8.1.106}$$

Substituting this expression for $N^r_A\big|_{y=0}$ into equation (8.1.100), we get

$$S_c\langle v_{gz}\rangle C_{tg} \frac{d\overline{x}_{Ag}}{dz} = -N^r_A\big|_{y=0} a S_c = -a S_c k_g p_{Ab} = \frac{\langle v_{gz}\rangle S_c}{P} C_{tg} \frac{dp_{Ab}}{dz}; \tag{8.1.107}$$

$$\frac{a S_c P k_g}{W_{tg}} \int_0^L dz = \frac{(aS_c L) P k_g}{W_{tg}} = -\int_{p_{Ab}|_{z=0}}^{p_{Ab}|_{z=L}} \frac{dp_{Ab}}{p_{Ab}} = \ell n\left(\frac{p_{Ab}|_{z=0}}{p_{Ab}|_{z=L}}\right), \tag{8.1.108}$$

a relatively simple result (Froment and Bishcoff, 1979). For a reactive absorption system involving an instantaneous reaction, these two results, (8.1.105) and (8.1.108), suggest the following approach for a countercurrent gas absorption tower. The calculation has to be carried out in two steps: from the column top, where $C_{Cb} > C'_{Cb}$ at $z = L$, to $z = z'$, where $C_{Cb} = C'_{Cb}$ using equation (8.1.108); then, for the column section from $z = z'$ to $z = 0$, where $C_{Cb} \leq C'_{Cb}$ using equation (8.1.105). Correspondingly, the relation to be used to determine the value of $\overline{p}_{Ag}$ or $\overline{x}_{Ag}$ where the transition takes place (i.e. location z') from $C_{Cb} \geq C'_{Cb}$ to $C_{Cb} \leq C'_{Cb}$ is obtained from (8.1.99) as

$$W_{tg}\left[\overline{x}_{Ag} - \overline{x}_{AgL}\right] = \frac{a_A}{c_C} Q_\ell \left[C_{C\ell L} - C'_{Cb}\right], \tag{8.1.109}$$

where we have introduced $\overline{C}_{C\ell} = C'_{Cb}$. A design example of scrubbing of ammonia from an air stream by an aqueous solution of sulfuric acid in a countercurrent packed tower has been illustrated by Froment and Bischoff (1979). We illustrate this example here for different flow conditions.

Example 8.1.6 A solution of sulfuric acid having a sulfuric acid concentration of 0.6 kgmol/m^3 is used to absorb NH_3 from an air stream at 25 °C and essentially atmospheric pressure containing 4 mole % NH_3 in a countercurrent packed tower via the instantaneous reaction

$$2NH_3(A) + H_2SO_4\,(C) \rightarrow (NH_4)_2\,SO_4(Q).$$

The airflow rate coming in is 1900 ft^3/min. The scrubbing solution flow rate is 0.5 m^3/min. The gas-phase mass-transfer coefficient for NH_3 is $k_g = 0.35$ kgmol/m^2-hr-atm; the corresponding liquid-phase mass-transfer coefficient $k_\ell = 0.5$ cm/hr. Henry's law constant (H^C_A) is given as 0.0133 m^3-atm/kgmol. Assume $D_{A\ell} = D_{C\ell}$ in the liquid film. Determine the total gas-liquid interfacial area required to remove 90% of NH_3 in the column.

Solution We will first determine the value of C'_{Cb} by using equation (5.3.29a):

$$k_g\, p_{Ab} = \frac{a_A}{c_C}\frac{D_{C\ell}}{D_{A\ell}} k_\ell C'_{Cb} = \frac{2}{1} \times 1 \times 0.5\,\frac{\text{cm}}{\text{hr}} \times \frac{1}{100}\frac{\text{m}}{\text{cm}}$$

$$C'_{Cb} = 0.35\,\frac{\text{kg mol}}{\text{m}^2\text{-hr-atm}}\, p_{Ab} \Rightarrow C'_{Cb} = 35 p_{Ab}.$$

At the column bottom, $p_{Ab} = 0.04$ atm; at the column top, $p_{Ab} = 0.04 \times 0.1 = 0.004$ atm. Therefore the values of C'_{Cb} at the two column ends are as follows.

$$\text{Column top}: C'_{Cb} = 0.35\ \frac{\text{kgmol}}{\text{m}^2\text{-hr-atm}} \times 0.004\,\text{atm} \times 100\ \frac{\text{hr}}{\text{m}} = 0.14\ \frac{\text{kgmol}}{\text{m}^3};$$

$$\text{column bottom}: C'_{Cb} = 0.35\ \frac{\text{kgmol}}{\text{m}^2\text{-hr-atm}} \times 0.04\,\text{atm} \times 100\ \frac{\text{hr}}{\text{m}} = 1.4\ \frac{\text{kgmol}}{\text{m}^3}.$$

This means that, at the column top, the feed H_2SO_4 concentration of 0.6 kgmol/m^3 is certainly much higher than the C'_{Cb} value of 0.14 kgmol/m^3; therefore, to start with, the reaction interface will coincide with the gas–liquid interface. To find out the condition as we approach the column bottom, we need to carry out a molar balance due to the reaction over the column length to see whether C_{Cb} at the column bottom ($= C_{C\ell o}$) is higher than $C'_{Cb} = 1.4$ kgmol/m^3 using the mass balance equation (8.1.109):

$$W_{tg}\left[\bar{x}_{Ag0} - \bar{x}_{AgL}\right] = \frac{a_A}{c_C}\, Q_\ell\left[C_{C\ell L} - C_{C\ell o}\right].$$

Determine W_{tg} first. Calculate ρ_g via the molecular weight of the gas mixture: $(0.04 \times 17 + 0.96 \times 29 = 28.52$ kg/kgmol). Assume an ideal gas mixture:

$$\rho_g = \frac{PM}{RT} = \frac{1\,\text{atm} \times 28.52\,\dfrac{\text{lb mass}}{\text{lb mol}}}{1.314\,\dfrac{\text{atm-ft}^3}{\text{lb mol-K}} \times 298\,\text{K}} = \frac{1 \times 28.52}{1.314 \times 298}\ \frac{\text{lb mass}}{\text{ft}^3} = 0.072\ \frac{\text{lb mass}}{\text{ft}^3}.$$

The molar gas flow rate entering the column is given by

$$W_{tg} = 1900\ \frac{\text{ft}^3}{\text{min}} \times 0.072\ \frac{\text{lb mass}}{\text{ft}^3} \times 60\ \frac{\text{min}}{\text{hr}} \times \frac{1}{2.2}\ \frac{\text{kg}}{\text{lb mass}} \times \frac{1}{28.52}\ \frac{\text{kgmol}}{\text{kg}}$$

$$= \frac{1900 \times 0.072 \times 60}{2.2 \times 28.52}\ \frac{\text{kgmol}}{\text{hr}} = 130.8\,\text{kgmol/hr}.$$

So,

$$130.8\,[0.04 - 0.004]\ \frac{\text{kgmol}}{\text{hr}} = \frac{2}{1} \times 0.5\ \frac{\text{m}^3}{\text{min}} \times \frac{60\,\text{min}}{\text{hr}} \left[0.6\ \frac{\text{kgmol}}{\text{m}^3} - C_{Cl0}\right];$$

$$130.8 \times 0.036 = 60\,[0.6 - C_{C\ell 0}] \Rightarrow C_{C\ell 0} = 0.52\,\text{kgmol/m}^3 < C'_{Cb}.$$

Therefore a section at the bottom of the column will operate with the reaction interface away from the gas–liquid interface (for which one needs 1.4 kgmol/m^3). We need to find the dividing line between these two regions. We use equation (8.1.109) from the gas inlet to the location z, where $C_{Cb} = C_{C\ell} = C'_{Cb}$:

$$W_{tg}\left[\bar{x}_{Ag0} - \bar{x}_{Ag}\right] = \frac{a_A}{c_C}\, Q_\ell\left[C'_{Cb} - C_{C\ell 0}\right];$$

$$130.8\ \frac{\text{kgmol}}{\text{hr}} \times \left[0.04 - \bar{x}_{Ag}\right] = \frac{2}{1} \times 0.5\ \frac{\text{m}^3}{\text{min}} \times \frac{60\,\text{min}}{\text{hr}} \left[C'_{Cb} - 0.52\ \frac{\text{kgmol}}{\text{m}^3}\right];$$

$$130.8 \times 0.04\ \frac{\text{kgmol}}{\text{hr}} - \frac{130.8\,p_{Ag}}{P}\ \frac{\text{kgmol}}{\text{hr}} = 60 C'_{Cb} - 31.2$$

$$\Rightarrow 5.232 - 130.8\,\frac{p_{Ag}}{P} = 60 \times 35\,\frac{p_{Ag}}{P} - 31.2$$

$$\Rightarrow \left[p_{Ag}/P\right] = (36.43/2230.0) = 0.0163 = \bar{x}_{Ag}.$$

Correspondingly,

$$C'_{Cb} = 35 \times 0.0163 = 0.5705\,\text{kgmol/m}^3.$$

We will now calculate the interfacial area required in each section. We start with the top section, where the reaction interface coincides with the gas–liquid interface (relation (8.1.108)):

$$\frac{(a\,S_c(L - z'))\,Pk_g}{W_{tg}} = \ell n\left(\frac{p_{Ab}|_{z=z'}}{p_{Ab}|_{z=L}}\right)$$

(for this section we integrated from $z = L$ at the top to $z = z'$, where $C'_{Cb} = 0.5705$ kgmol/m^3 and $p_{Ab} = 0.0163P$). So,

$$\frac{\{a\,S_c\,(L - z')\} \times 1\,\text{atm} \times 0.35\,\dfrac{\text{kgmol}}{\text{m}^2\text{-hr-atm}}}{130.8\,\text{kgmol/hr}} = \ell n\left(\frac{0.0163 \times 1}{0.004}\right);$$

$$\{a\,S_c\,(L - z')\} = \frac{130.8}{0.35} \times 2.303 \times \log 4.075 = 525\,\text{m}^2.$$

For the bottom part of the column, use equation (8.1.105), where instead of L use z' going from $z' = 0$:

$$\frac{(z'\ S_c\ a)\,K_g P}{W_{tg}} = -\int_{p_{Ago}}^{\bar{p}_{Agz'}} \frac{\mathrm{d}\,\bar{p}_{Ag}}{\bar{p}_{Ag}\left[1 - H_A^C \frac{D_{C\ell}}{D_{A\ell}} \frac{W_{tg}}{Q_\ell P}\right] + H_A^C \frac{D_{C\ell}}{D_{A\ell}} \frac{W_{tg}}{Q_\ell} \frac{\bar{p}_{Agz'}}{P} + H_A^C \frac{D_{C\ell}}{D_{A\ell}} \frac{a_A}{c_C} C_{C\ell z'}}.$$

Now

$$\frac{1}{K_g} = \frac{1}{k_g} + \frac{H_A^C}{k_\ell} = \frac{1}{0.35\ \dfrac{\text{kgmol}}{\text{m}^2\text{-hr-atm}}} + \frac{0.0133\,\text{m}^3\text{-atm}}{\text{kgmol} \times 0.005\ \dfrac{\text{m}}{\text{hr}}} = 2.85 + 2.66 = 5.51\ \frac{\text{m}^2\text{-hr-atm}}{\text{kgmol}};$$

$$K_g = 0.18\ \frac{\text{kgmol}}{\text{m}^2\text{-hr-atm}};$$

$W_{tg} = 130.8$ kgmol/hr; $\quad H_A^C = 0.0133$ m^3-atm/kgmol;

$D_{C\ell} \cong D_{A\ell}$; $\quad Q_\ell = 30$ m^3/hr;

$$\frac{(z'\,S_c\,a)\,0.18\,\dfrac{\text{kgmol}\times 1\,\text{atm}}{\text{m}^2\text{-hr-atm}}}{130.8\,\dfrac{\text{kgmol}}{\text{hr}}} = -\int_{\overline{p}_{\text{Ago}}}^{\overline{p}_{\text{Agz}'}} \frac{\text{d}\,\overline{p}_{\text{Ag}}}{\overline{p}_{\text{Ag}}\left[1-\dfrac{0.0133\times 130.8}{30\times 1}\right]+\dfrac{0.0133\times 130.8}{30\times 1}\times\overline{p}_{\text{Agz}'}+0.0133\times 0.5705};$$

$$(z'\,S_c\,a) = (-)\frac{130.8}{0.18}\int_{\overline{p}_{\text{Ago}}}^{\overline{p}_{\text{Agz}'}} \frac{\text{d}\overline{p}_{\text{Ag}}}{0.942\overline{p}_{\text{Ag}}+0.0579\overline{p}_{\text{Agz}'}+0.00758}$$

$$= \frac{726.67\times 2.303}{0.942}\log\frac{0.942\,\overline{p}_{\text{Ag0}}+0.0579\,\overline{p}_{\text{Agz}'}+0.00758}{0.942\,\overline{p}_{\text{Agz}'}+0.0579\,\overline{p}_{\text{Agz}'}+0.00758}.$$

Now $\overline{p}_{\text{Ag0}} = 0.04$ atm, $\overline{p}_{\text{Agz}'} = 0.0163$ atm. Therefore,

$$(z'\,S_c\,a) = 1776.5\log\frac{0.0376+0.00094+0.00758}{0.0153+0.00094+0.00758}$$

$$= 1776.5\log\left(\frac{0.0461}{0.0238}\right) = 1776.5\log(1.935)$$

$$= 1776.5\times 0.29 = 515.1\text{m}^2.$$

Therefore

$$L\,S_C a = \{a\,S_c\,(L-z')\} + \{a\,S_c z'\} = 525 + 515.1 = 1040\ \text{m}^2.$$

The height of the tower will depend on the type of packing selected, the tower diameter and the extent of pressure drop allowed; see Example 8.1.1.

8.1.2.5.1 Operating lines and equilibrium curves in a chemical absorber In general, the analysis of a countercurrent chemical absorber is complex. The question addressed here is: Could one represent it graphically via operating lines and equilibrium curves in a manner analogous to physical absorption for binary gas mixtures? We illustrate the procedure here following Rousseau and Staton (1988) for the reactive absorption of species A by reactant C:

$$a_{\text{A}}\,\text{A} + C_{\text{C}}\,\text{C} \rightarrow p\,\text{P}, \tag{8.1.110}$$

representing, for example, reaction (5.2.19a) in the hot potassium carbonate process for CO_2 absorption,

$$\text{CO}_2\,(\text{A}) + \text{CO}_3^{=}\,(\text{C}) + \text{H}_2\text{O} \rightarrow 2\text{HCO}_3^{-}\,(\text{P}), \tag{8.1.111}$$

where the concentration of water is not relevant as such. A number of results developed in Section 5.2.1.1.2 are relevant here, namely

$$p_{\text{CO}_2} = \frac{4\,H^{\text{C}}_{\text{CO}_2}}{K}\,C^t_{\text{CO}_3^=,\ell}\,\frac{f^2}{(1-f)} \Rightarrow p_{\text{A}} = \frac{4\,H^{\text{C}}_{\text{A}}}{K}\,C^t_{\text{C}\ell}\,\frac{f^2}{(1-f)}; \tag{8.1.112}$$

$$f C^t_{\text{CO}_3^=,\ell} = 0.5 C_{\text{HCO}_3^-,\ell} \Rightarrow f C^t_{\text{C}\ell} = 0.5 C_{\text{P}\ell} \Rightarrow f C^t_{\text{C}\ell} = \frac{a_{\text{A}}}{p}\,C_{\text{P}\ell}, \tag{8.1.113}$$

where f is the fractional consumption of the potassium carbonate reactant. Assume now that the physical solubility of the gaseous solute (i.e. CO_2) may be neglected compared to its chemical solubility:

$$C^t_{\text{CO}_2,\ell} = C_{\text{CO}_2,\ell} + 0.5\,C_{\text{HCO}_3^-,\ell}$$

$$\Rightarrow C^t_{\text{A}\ell} = C_{\text{A}\ell} + 0.5\,C_{\text{P}\ell} \Rightarrow C^t_{\text{A}\ell} \cong \frac{a_{\text{A}}}{p}\,C_{\text{P}\ell}. \tag{8.1.114}$$

We will now implement a species A balance from the column inlet to the column exit using the following assumption, namely that the carrier gas (inerts) is insoluble in the solvent, implying constancy of the carrier gas flow rate, i.e. from (8.1.80),

$$W_{tg}(1-\overline{x}_{\text{Ag}}) = W^{\text{in}}_{tg} = |W^{\text{in}}_{tg}|. \tag{8.1.115}$$

Species A balance over the whole column (from $z = 0$ to $z = L$) is given by

$$W_{tg}\big|_{z=0} x_{\text{Ag0}} + |W_{t\ell}|_{z=L} C^t_{\text{A}\ell L} = W_{tg}\big|_{z=L}\overline{x}_{\text{Ag}L} + |W_{t\ell}|_{z=0} C^t_{\text{A}\ell 0}. \tag{8.1.116}$$

Employing relations (8.1.114) and (8.1.115), we get

$$W^{\text{in}}_{tg}\frac{x_{\text{Ag0}}}{(1-x_{\text{Ag0}})} + |W_{t\ell}|\left(\frac{a_{\text{A}}}{p}\right)C_{\text{P}\ell L}$$

$$= W^{\text{in}}_{tg}\frac{\overline{x}_{\text{Ag}L}}{(1-\overline{x}_{\text{Ag}L})} + |W_{t\ell}|\left(\frac{a_{\text{A}}}{p}\right)C_{\text{P}\ell 0}. \tag{8.1.117}$$

Introducing (8.1.113) here, we get

$$\frac{x_{\text{Ag0}}}{(1-x_{\text{Ag0}})} = \frac{\overline{x}_{\text{Ag}L}}{(1-\overline{x}_{\text{Ag}L})} + \frac{|W_{t\ell}|}{W^{\text{in}}_{tg}}\,C^t_{\text{C}\ell}(f_0-f_L), \tag{8.1.118}$$

where f_0 and f_L correspond to $z = 0$ and $z = L$, respectively. If the species A balance is carried out from the gas inlet ($z = 0$) to any location in the column (z), we will get, correspondingly,

$$\frac{\overline{x}_{\text{Ag}}}{(1-\overline{x}_{\text{Ag}})} = \frac{x_{\text{Ag0}}}{(1-x_{\text{Ag0}})} + \frac{|W_{t\ell}|}{W^{\text{in}}_{tg}}\,C^t_{\text{C}\ell}(f-f_0). \tag{8.1.119}$$

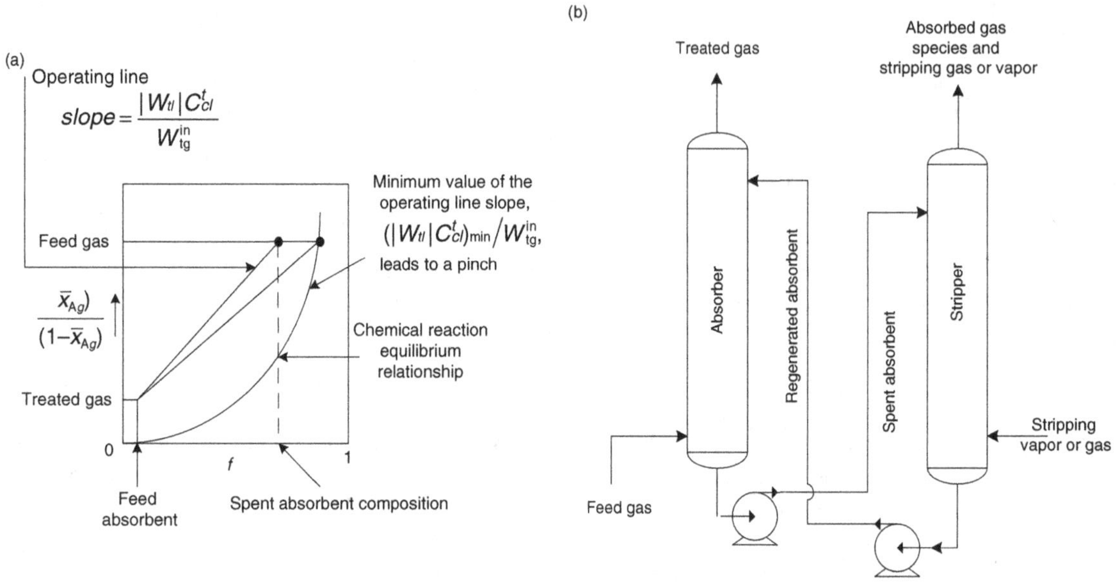

Figure 8.1.17. (a) Feed gas composition vs. fractional chemical absorbent consumption: operating line and equilibrium behavior. (After Rousseau and Staton (1988).) (b) Schematic for absorption stripping process using a stripping vapor or gas.

This is the equation for the operating line, providing a linear relationship between the ordinate $\bar{x}_{Ag}/(1-\bar{x}_{Ag})$ and the abscissa f, representing the relation between the fractional consumption of the chemical absorbent C and the gas phase of species A at any location z in the column. This line has been plotted in Figure 8.1.17(a) along with the equilibrium curve (8.1.120) represented as

$$\frac{\bar{x}_{Ag}}{(1-\bar{x}_{Ag})} = \frac{4H_A^C}{KP(1-\bar{x}_{Ag})} C_{C\ell}^t \frac{f^2}{(1-f)}. \qquad (8.1.120)$$

The slope of the operating line is $\left(|W_{t\ell}|\, C_{C\ell}^t / W_{tg}^{in}\right)$. The minimum slope of the operating line will correspond to the minimum value of the product, $|W_{t\ell}|\, C_{C\ell}^t$, leading to a pinch, an intersection of the operating line and the equilibrium curve at the bottom of the column where the feed gas enters. One may also have a pinch point at the top of the column. For regular operation of the column in a stagewise fashion, one can draw steps between the equilibrium curve and the operating line, starting, for example, at the outlet gas location.

Detailed modeling of CO_2 absorption by the hot carbonate process is demanding. One such analysis, along with references to earlier literature, is available in Sanyal *et al.* (1988). In actual operation, the absorbent, e.g. the aqueous K_2CO_3 solution, loaded with CO_2 is regenerated in a separate column, a stripper, by a CO_2-free vapor, say steam at a lower pressure, e.g. 0.14 MPa, when the absorber operates at a much higher pressure, say 2.2 MPa.

In general, stripping of the loaded absorbent to regenerate the absorbent may be carried out by heating it using steam which flows countercurrently. Alternatively, in some systems air may be used to strip, or vacuum may be employed. A general scheme for absorption stripping is illustrated in Figure 8.1.17(b). Such schemes are also valid for physical solvents where there is no reaction. In general, the spent absorbent is often heated up (using a heater with an external heat source or heat exchanged with the heated regenerated absorbent) before being introduced into the stripper. Correspondingly, the regenerated absorbent is often cooled before entering the absorber. Note, however, that regeneration in reactive absorbents is feasible only with reversible reactions.

8.1.2.6 *Gas absorption/stripping in multistage devices having an overall countercurrent flow pattern*

We have introduced in Figures 8.1.5 and 8.1.9 schematics for multistage devices having an overall countercurrent flow pattern. The performance of such devices for equilibrium separation processes, such as gas absorption/stripping, distillation (Section 8.1.3), solvent extraction (Section 8.1.4), is conventionally studied in two steps. First, one assumes that in each stage the two exiting streams, e.g. the gas stream to be scrubbed and the absorbent liquid for scrubbing the gas, are in equilibrium; each stage is an ideal equilibrium stage. In the second step, any deviation from equilibrium is taken care of by using a quantity called *stage efficiency* or *tray efficiency* (for devices having multiple trays). We will treat first the problem of determining the number of ideal stages required for a specified change in gas composition in a countercurrent multistage device where the two exiting streams from any ideal stage are in equilibrium. The resulting analytical expression for the number of stages N is called the Kremser equation (Kremser, 1930). This development will be followed by a commonly adopted graphical procedure.

Assume first that we are dealing with absorption/stripping from a dilute gas stream; specifically, we assume constant flow rates of phases 1 (= g) and 2 (= ℓ) along the length of the device and constant temperature and pressure of operation. The equation for the operating line (8.1.24) becomes equation (8.1.25) under the following conditions:

$$x_{ig(n+1)} = x_{igL} - \frac{|W_{t\ell}|}{|W_{tg}|} x_{i\ell L} + \frac{|W_{t\ell}|}{|W_{tg}|} x_{i\ell n}. \qquad (8.1.121)$$

This equation provides a relation between $x_{ig(n+1)}$, the gas composition leaving plate $(n+1)$, and $x_{i\ell n}$, the liquid composition leaving plate n (above plate $(n+1)$). They are at the same vertical level of the column. If we assume, in addition, that the gas-liquid absorption equilibrium relation is linear,

$$x^*_{ig} = m_i x_{i\ell} + b_i, \qquad (8.1.122a)$$

then we can express $x_{i\ell L}$ and $x_{i\ell n}$ via

$$x^*_{igL} = m_i x_{i\ell L} + b_i; \qquad x^*_{ign} = m_i x_{i\ell n} + b_i = x_{ign}, \qquad (8.1.122b)$$

where $x^*_{ign} = x_{ign}$ since x_{ign} is in equilibrium with $x_{i\ell n}$ by assumption for stage n; further, x^*_{igL} is the composition of the hypothetical gas stream in equilibrium with $x_{i\ell L}$, the entering liquid. Using these relations, rewrite (8.1.121) as

$$x_{ig(n+1)} = x_{igL} - \frac{|W_{t\ell}|}{|W_{tg}|}\left(\frac{x^*_{igL} - b_i}{m_i}\right) + \frac{|W_{t\ell}|}{|W_{tg}|}\left(\frac{x_{ign} - b_i}{m_i}\right). \qquad (8.1.123)$$

Rearranging this equation, we get a linear first-order finite difference equation in x_{ign}:

$$x_{ig(n+1)} - \frac{|W_{t\ell}|}{|W_{tg}|m_i} x_{ign} = x_{igL} - \frac{|W_{t\ell}|}{|W_{tg}|m_i} x^*_{igL}. \qquad (8.1.124)$$

It can also be written in the following form:

$$\left(D - \frac{|W_{t\ell}|}{|W_{tg}|m_i}\right) x_{ign} = x_{igL} - \frac{|W_{t\ell}|}{|W_{tg}|m_i} x^*_{igL}, \qquad (8.1.125)$$

where D is a finite difference operator. A solution of this difference equation for any n is

$$x_{ig(n+1)} = \left(\frac{1 - A^{n+1}}{1 - A}\right) x_{igL} - \left(\frac{A(1 - A^n)}{1 - A}\right) x^*_{igL}, \qquad (8.1.126)$$

where A is the absorption factor $\left(|W_{t\ell}| / |W_{tg}|m_i\right)$ defined earlier by relation (8.1.45) when Henry's law is valid. Therefore, for a column having a total of N stages,

$$x_{ig(N+1)} = \left(\frac{1 - A^{N+1}}{1 - A}\right) x_{igL} - \left(\frac{A(1 - A^N)}{1 - A}\right) x^*_{igL} = x_{ig0}. \qquad (8.1.127)$$

It is one form of the Kremser equation.

Rewrite equation (8.1.124) in the following form for $n = N$:

$$\frac{x_{ig(N+1)} - x_{igL}}{x_{igN} - x^*_{igL}} = \frac{|W_{t\ell}|}{|W_{tg}|}\frac{1}{m_i}. \qquad (8.1.128)$$

Further, we obtain for stage $N-1$ from equation (8.1.126)

$$x_{igN} = \left(\frac{1 - A^N}{1 - A}\right) x_{igL} - \left(\frac{A(1 - A^{N-1})}{1 - A}\right) x^*_{igL}, \qquad (8.1.129)$$

where x_{igN} is in equilibrium with $x_{i\ell 0}(= x_{i\ell N})$, i.e. $x_{igN} = m_i x_{i\ell N} + b_i = x^*_{igN}$. From equations (8.1.127) and (8.1.29), we get

$$x_{ig(N+1)} - x_{igN} = x_{igL}\left(\frac{A^N - A^{N+1}}{1 - A}\right) - x^*_{igL}\left(\frac{A^N - A^{N+1}}{1 - A}\right), \qquad (8.1.130)$$

resulting in

$$\frac{x_{ig(N+1)} - x_{igN}}{x_{igL} - x^*_{igL}} = A^N \Rightarrow N = \frac{\left[\ell n \dfrac{x_{ig(N+1)} - x_{igN}}{x_{igL} - x^*_{igL}}\right]}{\ell n A}. \qquad (8.1.131)$$

Alternatively, using (8.1.128), we also get

$$N = \frac{\left[\ell n \dfrac{x_{ig(N+1)} - x_{igN}}{x_{igL} - x^*_{igL}}\right]}{\left[\ell n \dfrac{x_{ig(N+1)} - x_{igL}}{x_{igN} - x^*_{igL}}\right]}. \qquad (8.1.132)$$

By algebraic manipulation of (8.1.131), we can have an additional result,

$$N = \frac{\ell n\left[\left(1 - \dfrac{1}{A}\right)\left(\dfrac{x_{ig(N+1)} - x^*_{igL}}{x_{igL} - x^*_{igL}}\right) + \dfrac{1}{A}\right]}{\ell n A}. \qquad (8.1.133)$$

Relations (8.1.131)-(8.1.133) are alternative forms of the *Kremser equation*. This equation was developed with a number of restrictions, one of them being constant gas and liquid flow rates. Even if we have dilute gas absorption, the flow rates of the two phases (particularly the gas phase) will vary somewhat along the column. To determine the compositions of the gas/liquid streams, the actual flow rate of each stream at the column ends has

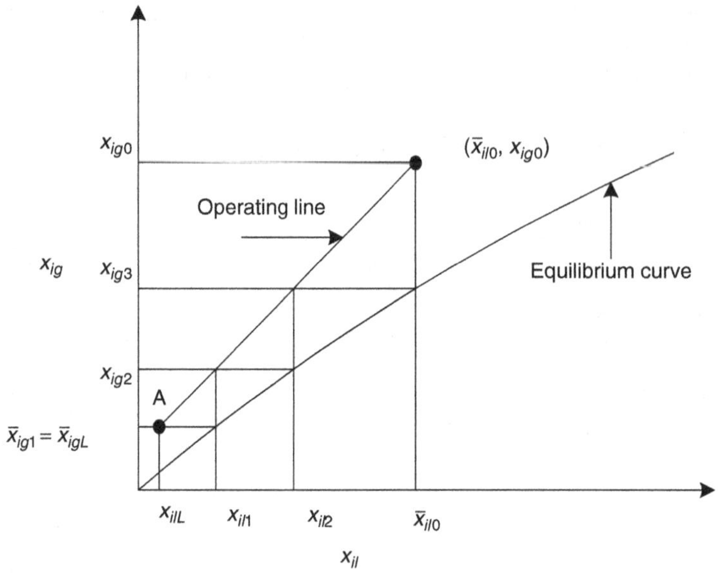

Figure 8.1.18. Graphical determination of the number of ideal stages required for a gas absorption problem.

to be used. However, in using the Kremser equation, one may use an average value of the flow rate of each stream to calculate A.

The Kremser equation has a different form when applied to a stripper. The form of the Kremser equation (Kremser, 1930) for a stripper in which the stripping gas or vapor having a mole fraction $x_{ig(n+1)}$ enters at the column bottom and the liquid to be stripped containing species i mole fraction $x_{i\ell L}$ enters the column at the top is (Treybal, 1980, p. 291)

$$N = \ell n \left[\frac{x_{i\ell L} - (x_{ig(N+1)}/m_i)}{x_{i\ell 0} - (x_{ig(N+1)}/m_i)} \left(1 - \frac{1}{S}\right) + \frac{1}{S} \right] \Big/ \ell n\, S. \tag{8.1.134}$$

Here, $x_{i\ell 0}$ is the species i mole fraction in the stripped liquid leaving the column bottom, and the stripping factor is given by

$$S = m_i |W_{tg}| / |W_{t\ell}|. \tag{8.1.135}$$

Example 8.1.7 Consider the example of acetone scrubbing from air by a pure water stream at 15 °C, as described in Example 8.1.2. If this case of dilute gas absorption is carried out in a multistage countercurrent absorber containing N plates, determine the number of ideal plates required by the Kremser equation.

Solution

$$H_i^{\rm P} = 1.2 = m_i \text{ at } 15\,^\circ\text{C}; \qquad x_{ig0} = x_{ig(N+1)} = 0.01;$$
$$x_{igL} = (0.1) \times 0.01 = 0.001; \qquad x_{i\ell L} = 0 \Rightarrow x_{igL}^* = 1.2 \times x_{i\ell L} = 0;$$
$$x_{igN} = H_i^{\rm P} x_{i\ell N} = H_i^{\rm P} x_{i\ell 0} = 1.2 \times x_{i\ell 0}.$$

Determine $x_{i\ell 0}$ via an overall species i molar balance. First, determine the exit flow rate of the gas stream:

$$|W_{tg}|_{\rm inert} = |W_{tg}|_{z=0} \times 0.99 = 7 \times 0.99 = 6.93\,\text{kgmol/hr};$$

$$|W_{tg}|_{z=L} \times 0.999 = |W_{tg}|_{\rm inert} = 6.93\,\text{kgmol/hr} \Rightarrow |W_{tg}|_{z=L} = 6.937\,\text{kgmol/hr}.$$

The variation in absorbent flow rate is neglected.

For species i balance:

$$x_{i\ell L}|W_{t\ell}|_{z=L} + x_{ig0}\,|W_{tg}|_{z=0} = x_{igL}\,|W_{tg}|_{z=L} + x_{i\ell 0}|W_{t\ell}|_{z=0}$$
$$\Rightarrow 0 \times |W_{t\ell}|_{z=L} + 0.01 \times 7 = 0.001 \times 6.937 + x_{i\ell 0} \times 23$$
$$\Rightarrow x_{i\ell 0} = \frac{0.06307}{23} = 0.00274.$$

Therefore

$$x_{igN} = 1.2 \times 0.00274 = 0.00329.$$

Determine N from

$$N = \ell n \left[\frac{x_{ig(N+1)} - x_{igN}}{x_{igL} - x_{igL}^*} \right] \Big/ \ell n\, A$$
$$= \ell n \left[\frac{0.01 - 0.00329}{0.001 - 0} \right] \Big/ \ell n \left[|W_{t\ell}|_{\rm avg} / |W_{tg}|_{\rm avg} m_i \right]$$
$$= \ell n \left[\frac{0.00671}{0.001} \right] \Big/ \ell n [23/(6.968 \times 1.2)]$$
$$= \ell n[6.71]/\ell n[2.75] = \frac{0.8267}{0.4393} = 1.88.$$

Therefore two ideal stages are required.

The equilibrium behavior in gas-liquid absorption may not always be described by a linear relation (8.1.122a). Further, the operating line equation may not be linear since $|W_{t\ell}|/|W_{tg}|$ may vary significantly along the column. A graphical procedure may be conveniently adopted to determine the number of ideal stages required. It is based on the McCabe–Thiele method employed in distillation, where the operating lines are generally linear (Section 8.1.3).

Consider Figure 8.1.18, where the operating line and the equilibrium relation (x_{ig} vs. x_{il}) have been plotted for a gas absorption tower. At the tower top in plate 1 or stage 1, the gas leaving has a composition $\overline{x}_{igL}$ and the liquid coming in has a composition $x_{i\ell L}$. These two coordinates define point A on the operating line. Draw a horizontal line from this point to the equilibrium curve; the point of intersection has an abscissa of $x_{i\ell 1}$, which is in equilibrium with x_{ig1} ($= x_{igL}$) since it is the equilibrium curve. Now go up vertically and intersect the operating line at an ordinate of x_{ig2}, which is the composition of the gas stream coming from stage/plate 2 (one stage down from plate 1 or stage 1). The triangle formed in the process is identified as stage 1.

Next draw a horizontal line from the x_{ig2} value on the operating line till it intersects the equilibrium line at ($x_{i\ell 2}$, x_{ig2}). Then move vertically from this point of intersection to the operating line and obtain the stage 2/plate 2. Follow this procedure till one hits ($x_{i\ell 0}$, x_{ig0}) in the operating line at the column bottom. Essentially we are developing a staircase structure between the operating line and the equilibrium curve. We could have started this construction from the top, provided $x_{i\ell 0}$ was known. In general, for gas absorption problems, x_{ig0} and $x_{i\ell L}$ are given, and the fractional gas purification is specified, i.e. x_{igL} is known. Consequently, the bottom starting point ($x_{i\ell L}$, x_{igL}) corresponds to the top of the column exiting stage 1 at the top of the column. The reader may practice such a graphical procedure with the data provided in the worked example.

8.1.3 Distillation

The equilibrium distribution of a mixture of volatile liquids between a vapor phase and a liquid phase in a closed vessel was introduced in Sections 3.3.7.1 and 4.1.2 as the basis for the separation process of distillation. The preferential enrichment of the vapor phase with the more volatile species and the liquid phase with the less volatile species was illustrated in Section 4.1.2 for a variety of systems, along with the procedures for calculating the composition of each phase in a closed system. How chemical reactions in the liquid phase affect such vapor–liquid equilibrium was demonstrated in Section 5.2.1.2. In Section 6.3.2.1, open systems of flash vaporization and batch distillation in the context of bulk flow of the vapor and liquid phases parallel to the direction of the force were studied, and the separation achieved was quantified. The most common configuration of separation based on vapor–liquid equilibrium employs, however, a vertical column in which the vapor stream flows up and the liquid stream flows down. How the vapor and the liquid phases may contact each other was illustrated, for example, in Figure 2.1.2(b) for a sieve plate in a column. The chemical potential gradients of various volatile species between these two phases drive the separation; these forces act essentially perpendicular to the vertical directions of the bulk flow of both phases. We consider here the separation attained in such a configuration in a vertical column. Note, however, that the column has a refluxing arrangement at the column top and column bottom, as shown in Figure 8.1.3(a), a Type (3) system. All such considerations hold for a distillation column, whether it is of a multiplate or a packed-bed type.

Column distillation usually involves a multistage (multiplate) arrangement of vapor–liquid contacting in the manner of Figures 8.1.5 and 8.1.9, with important differences. (Column distillation also employs continuous vapor–liquid contact in packed columns.) As shown in Figure 8.1.19(a), the vapor stream leaving the top is sent to a condenser, where cooling water is used to condense the vapor. Part of the condensed liquid is withdrawn as the distillate product and the rest is sent back to the column as the liquid reflux, which flows down the column countercurrent to the rising vapor. Somewhere near the middle of the column, the feed mixture is introduced: the feed may be liquid, a mixture of liquid and vapor or vapor. At the bottom of the column, the liquid stream is withdrawn: part of it is withdrawn as the bottom product and the rest is vaporized in a reboiler (using, say, steam) and introduced back into the column to become the vapor stream going up the column.

The vapor stream going up the column bubbles through the liquid layer on each plate in Figure 8.1.19(a), which shows a total of 11 plates. As the vapor stream goes up the column, it undergoes a certain amount of pressure drop over the column length depending on the number of plates, the vapor–liquid contacting arrangement, the liquid/foam height on each plate and the spacing between each plate. For an introductory analysis, one can assume that the pressure has essentially the same value throughout the length of the column; however, the temperature is decreasing from the bottom of the column (plate N, having the highest temperature) to the top of the column (plate 1, having the lowest temperature) (Figure 8.1.19(b)). The changes in vapor/liquid composition from the bottom to the column top, or vice versa, are usually studied in a diagram called the $y - x$ diagram, here the $x_{ig} - x_{i\ell}$ or $x_{iv} - x_{i\ell}$ diagram. (See Figures 1.4.1, 4.1.1(c), 4.1.3(c), 4.1.4(c) in earlier chapters and Figure 8.1.20 here.) A point on the equilibrium curve, shown in Figure 8.1.20, represents the vapor–liquid equilibrium composition distribution at a given temperature, corresponding to a given column location, at a pressure assumed essentially constant throughout the column. *Note*: Here, species i in all plots is the more volatile species in a binary mixture. Further, unless otherwise mentioned, *assume a binary system*.

In such a vertical column, the vapor stream coming from a plate below comes up and bubbles through the flowing liquid layer on a plate. During this process, some

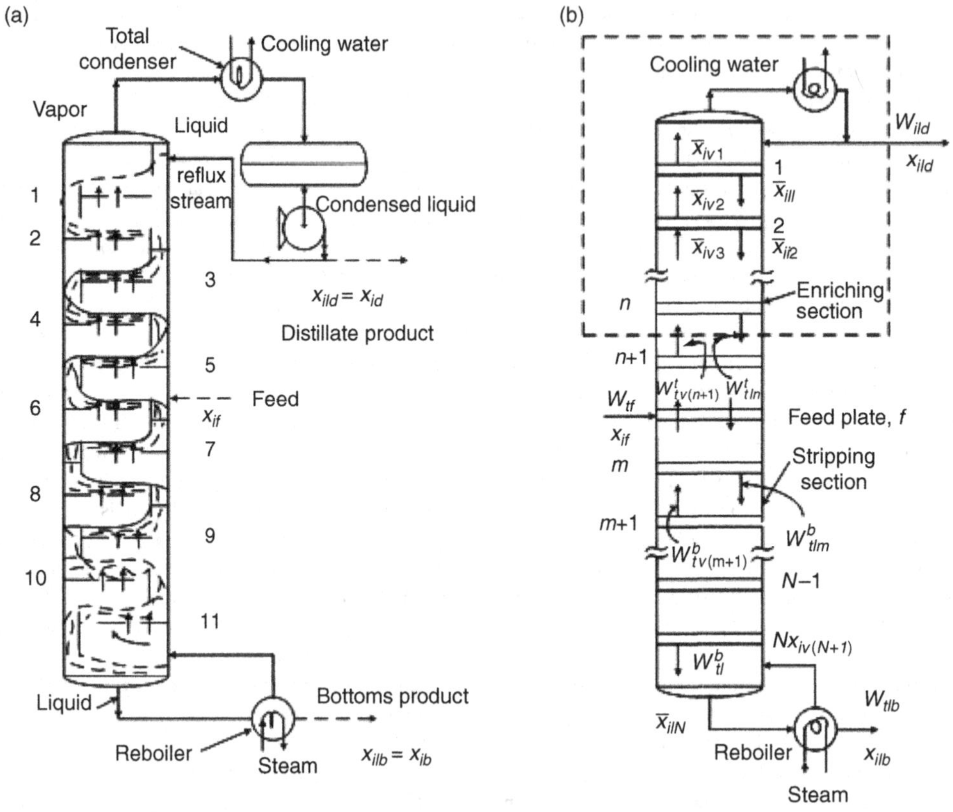

Figure 8.1.19. (a) Eleven-plate sieve tray distillation column; (b) a multi-plate distillation column.

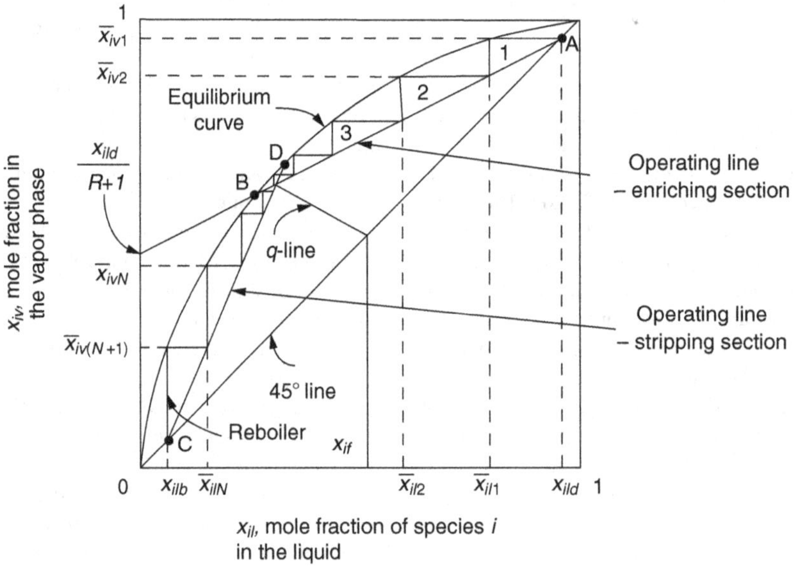

Figure 8.1.20. Vapor–liquid equilibrium curve, operating lines for enriching section and stripping section and construction of ideal stages and q line for a feed consisting of vapor and liquid in a multistage distillation column having ideal equilibrium stages: the McCabe–Thiele method.

less volatile species in the vapor are condensed and some more volatile species in the liquid are evaporated. The temperatures of the two streams are somewhat different, the vapor temperature being somewhat higher than that of the downcoming liquid flowing horizontally on the sieve tray. Therefore, there is heat exchange involving latent heat of vaporization/condensation as well as sensible heats of the two streams. Analysis of the changes in the enthalpy of each phase as the composition changes coupled with the equilibrium curve of Figure 8.1.20 is one way of studying the separation performance of the distillation column. It is known as the *Ponchon–Savarit method.* We will, however, utilize a slightly less exact method here, the *McCabe–Thiele method,* in which the role of enthalpy balance is drastically reduced; it is almost nonexistent. A basic assumption in the following analysis is that in each stage/plate in the column, the exiting vapor stream and the liquid stream are in equilibrium. Deviations from such a model are treated in Section 8.1.3.4.

8.1.3.1 *McCabe–Thiele method for binary mixtures*

We have already been introduced in Section 8.1.1.2 to the equations for the operating lines in the enriching section and the stripping section for a separation system of type (3). However, it did not take into account a multistaged configuration. A multiplate configuration's operating line equation was developed in equations (8.1.22) and (8.1.27) for a type (2) system. Here we develop it for distillation, a type (3) system. Consider first a species i balance over an envelope of the enriching section covering the distillate product and a location between the stages n and $n+1$ (Figure 8.1.19(b)):

$$W^t_{tv(n+1)}\,\overline{x}_{iv(n+1)} = W^t_{t\ell n}\,\overline{x}_{i\ell n} + W_{t\ell d}\,x_{i\ell d}. \quad (8.1.136a)$$

The operating line equation is then

$$\overline{x}_{iv(n+1)} = \frac{W^t_{t\ell n}}{W^t_{tv(n+1)}}\,\overline{x}_{i\ell n} + \frac{W_{t\ell d}}{W^t_{tv(n+1)}}\,x_{i\ell d}, \quad (8.1.136b)$$

where $W_{t\ell d}$ is the molar distillate product flow rate having a mole fraction $x_{i\ell d}$ of species i.

We will now provide a few relations based on overall material balance over the column (Figure 8.1.19(b)) as well as an overall balance of the more volatile species i over the column. For the overall molar balance,

$$W_{tf} = W_{t\ell d} + W_{t\ell b}; \quad (8.1.136c)$$

For the overall species i balance,

$$x_{if}\,W_{tf} = x_{i\ell d}\,W_{t\ell d} + x_{i\ell b}\,W_{t\ell b}. \quad (8.1.136d)$$

Employing equation (8.1.136c) to eliminate either $W_{t\ell b}$ or $W_{t\ell d}$ in equation (8.1.136d), we get the following two useful results:

$$\left(W_{t\ell d}/W_{tf}\right) = \frac{x_{if} - x_{i\ell b}}{x_{i\ell d} - x_{i\ell b}}; \quad (8.1.136e)$$

$$\left(W_{t\ell b}/W_{tf}\right) = \frac{x_{i\ell d} - x_{if}}{x_{i\ell d} - x_{i\ell b}}. \quad (8.1.136f)$$

We could have employed the "lever rule" (equation (1.6.2)) to obtain these two results, even though we have flow rates here as opposed to two fractions in a closed vessel.

We will now consider in the following section *a binary system* as we illustrate the utility of the McCabe–Thiele method. Focus now on Figure 8.1.19(b). In general, $W^t_{t\ell n}$ and $W^t_{tv(n+1)}$ can vary throughout the enriching section (the section of the column between the feed plate and the top of plate 1 at the top) with any n. However, we utilize the *constant molar overflow assumption,* namely that both $W^t_{t\ell n}$ and $W^t_{tv(n+1)}$ are independent of n and have constant values of $W^t_{t\ell}$ and W^t_{tv}, respectively, throughout the enriching section (see the paragraph prior to equation (8.1.6) for the basic conditions needed to justify this assumption) in the absence of any heat loss to the surroundings from the column. Using the following definition for the *external reflux ratio, R,*

$$R = \left(W^t_{t\ell}/W_{t\ell d}\right), \quad (8.1.137)$$

and an overall molar balance over the envelope, shown by dashed lines in Figure 8.1.19(b),

$$W^t_{t\ell} + W_{t\ell d} = W^t_{tv}, \quad (8.1.138a)$$

the *operating line equation* (8.1.16) and (8.1.136b) for the *enriching section* may be rewritten as

$$\overline{x}_{iv(n+1)} = \frac{R}{R+1}\,\overline{x}_{i\ell n} + \frac{1}{R+1}\,x_{i\ell d}. \quad (8.1.138b)$$

This it is the equation of a straight line relating $\overline{x}_{iv(n+1)}$ to $\overline{x}_{i\ell n}$ at any level/height in the multiplate distillation column, since the value of R is constant under any given conditions of operation.

In Figure 8.1.20, we have plotted this enriching section operating line between the equilibrium curve and the 45° line, $x_{iv} = x_{i\ell}$. The line starts at A on the 45° line, where the liquid distillate product composition $x_{i\ell d}$ is equal to $\overline{x}_{iv1}$, the composition of the vapor leaving the top plate ($n = 1$) and becoming totally condensed in the condenser without any subcooling. The latter condition is identified as refluxing the condensate at the bubble point. The enrichment section operating line intersects the x_{iv}-coordinate at the x_{iv} intercept value of ($x_{i\ell d}/$ $(R+1)$); however, only the section AB (with point B on the equilibrium curve) between the equilibrium curve and the 45° line is relevant. As we have illustrated in Figure 8.1.18, we will now try to determine the number of ideal vapor–liquid equilibrium stages in the enrichment section.

From the point of intersection of a horizontal line from the point ($x_{i\ell d}$, $\overline{x}_{iv1}$) on the 45° line with the equilibrium curve, draw a vertical line to the operating line where it intersects the point $((\overline{x}_{i\ell 1}, \overline{x}_{iv2})$. Note that this composition corresponds to the condition between plate 1 and plate 2, liquid from plate 1 $(\overline{x}_{i\ell 1})$ coming down and vapor from plate 2 $(\overline{x}_{iv2})$ coming up. The staircase represents plate 1. We can now start drawing a horizontal line from this point on the operating line $(\overline{x}_{i\ell 1}, \overline{x}_{iv2})$ to the equilibrium curve where it intersects at a point whose coordinates are $(\overline{x}_{i\ell 2}, \overline{x}_{iv2})$. Draw a vertical line from this point to the operating line: the point of intersection has the coordinates $(\overline{x}_{i\ell 2}, \overline{x}_{iv3})$, corresponding to the condition between plates 2 and 3. The staircase here represents plate 2. In this fashion, one can go all the way down to the feed plate. However, the process cannot be completed since the condition of the feed needs to be known; further, one has to find out how the stripping section merges into the feed section and therefore the enriching section.

We now focus on the *operating line for the stripping section* by considering an envelope over the stripping section covering the bottoms product and between plates m and $(m + 1)$ (Figure 8.1.19(b)). A molar balance of species i over this envelope (using a notation having b as the superscript for the molar flow rates in this bottom part, see equation (8.1.17)) leads to

$$W^b_{tv(m+1)}\,\overline{x}_{iv(m+1)} + W_{t\ell b}\,x_{i\ell b} = W^b_{t\ell m}\,\overline{x}_{i\ell m}. \tag{8.1.139a}$$

The total molar balance over the same envelope is

$$W^b_{tv(m+1)} + W_{t\ell b} = W^b_{t\ell m}. \tag{8.1.139b}$$

The equation for the operating line is obtained from (8.1.139a) as

$$\overline{x}_{iv(m+1)} = \frac{W^b_{t\ell m}}{W^b_{tv(m+1)}}\,\overline{x}_{i\ell m} - \frac{W_{t\ell b}}{W^b_{tv(m+1)}}\,x_{i\ell b}. \tag{8.1.140}$$

Employing relation (8.1.139b), we rewrite the above as

$$\overline{x}_{iv(m+1)} = \frac{W^b_{t\ell m}}{W^b_{t\ell m} - W_{t\ell b}}\,\overline{x}_{i\ell m} - \frac{W_{t\ell b}}{W^b_{t\ell m} - W_{t\ell b}}\,x_{i\ell b}. \tag{8.1.141}$$

The operating line equations (8.1.140) and (8.1.141) represent a straight line of slope $(W^b_{t\ell}/W^b_{tv})$ since, by the constant molar overflow assumption,

$$W^b_{t\ell m} = W^b_{t\ell}; \qquad W^\ell_{tv(m+1)} = W^b_{tv}. \tag{8.1.142}$$

We therefore rewrite equations (8.1.140) and (8.1.141) as

$$\overline{x}_{iv(m+1)} = \frac{W^b_{t\ell}}{W^b_{tv}}\,\overline{x}_{i\ell m} - \frac{W_{t\ell b}}{W^b_{tv}}\,x_{i\ell b}; \tag{8.1.143a}$$

$$\overline{x}_{iv(m+1)} = \frac{W^b_{t\ell}}{W^b_{t\ell} - W_{t\ell b}}\,\overline{x}_{i\ell m} - \frac{W_{t\ell b}}{W^b_{t\ell} - W_{t\ell b}}\,x_{i\ell b}. \tag{8.1.143b}$$

We can draw such an operating line CD starting at $x_{i\ell b} = x_{ib}$ (point C) on the 45° line with a slope of $(W^b_{t\ell}/W^b_{tv})$ in between the equilibrium curve and the 45° line (Figure 8.1.20).

As in the enriching section, we now start making a construction of staircases identifying various stages. However, we need to focus first on the mode of operating with the reboiler. In the reboiler, the fraction $((W^b_{t\ell} - W_{t\ell b})/W^b_{t\ell})$ of the downcoming liquid flow rate $W^b_{t\ell}$ is evaporated: this fraction is called the external *reboil ratio* (or *boilup ratio*); the remaining fraction $(W_{t\ell b}/W^b_{t\ell})$ becomes the bottom liquid product.

There are two ways of carrying this out. As shown in Figure 8.1.19(b), the whole liquid stream exiting the column bottom is introduced into the reboiler to generate vapor stream. Part of the liquid in the reboiler is withdrawn at a rate $W_{t\ell b}$ to provide the bottoms product stream. In such a configuration the vapor stream generated (flow rate, $W^b_{t\ell} - W_{t\ell b}$) is in equilibrium with the bottoms product stream (flow rate, $W_{t\ell b}$). This configuration, which is commonly utilized, is called a *partial reboiler* (Figure 8.1.19 (b)). On the other hand, if the fraction $(W^b_{t\ell} - W_{t\ell b}/W^b_{t\ell})$ of the liquid leaving the Nth plate and the column bottom is taken out, introduced into the reboiler and completely evaporated, then the vapor stream generated (flow rate, $W^b_{t\ell} - W_{t\ell b}$) will have the same composition as the liquid leaving the Nth plate. The rest of the liquid (flow rate, $W_{t\ell b}$) does not enter the reboiler. The configuration is called the *total reboiler*.

Complete evaporation of the fraction sent back to the N-plate column via the partial reboiler produces a vapor composition $\overline{x}_{iv(m+1)} = \overline{x}_{iv(N+1)}$ in an N-plate column which is in equilibrium with the bottoms product from the reboiler, $x_{i\ell b} = x_{ib}$. Therefore we go up from $x_{i\ell b}$ on the 45° line to $\overline{x}_{iv} = \overline{x}_{iv(N+1)}$ on the equilibrium curve. Next we draw a horizontal line from this point $(x_{i\ell b}, \overline{x}_{iv(N+1)})$ to intersect the operating line at $\overline{x}_{i\ell N}$. This staircase corresponds to the ideal equilibrium stage which is the partial reboiler. From the point $(\overline{x}_{i\ell N}, \overline{x}_{iv(N+1)})$ on the operating line, we go up to $\overline{x}_{ivN}$ on the equilibrium curve. Then we go horizontal and intersect the operating line at $(\overline{x}_{i\ell(N-1)}, \overline{x}_{ivN})$, corresponding to the column location between stages N and $(N - 1)$. In this fashion, one goes up now from the staircase representing stage N to stage $(N - 1)$ and so on in the stripping section.

Consider now the operating lines AB and CD for the enriching section and the stripping section, respectively (Figure 8.1.20). In the middle region of the column where the feed is introduced, the vapor and the liquid flow rates change from those above the feed location to those below it. This transition is of great importance. To locate this transition, focus on the feed plate identified as f in Figure 8.1.19(b). Figure 8.1.21(a) illustrates the flow rate, composition and enthalpy of the feed stream being introduced from outside onto the feed plate: W_{tf} is the molar feed flow rate having a composition of x_{if}; W^v_{tf} is the vapor fraction of the molar feed flow rate; W^ℓ_{tf} is the liquid fraction of the molar feed flow rate; x^v_{if} is the composition of the vapor fraction of the feed; x^ℓ_{if} is the composition of the liquid

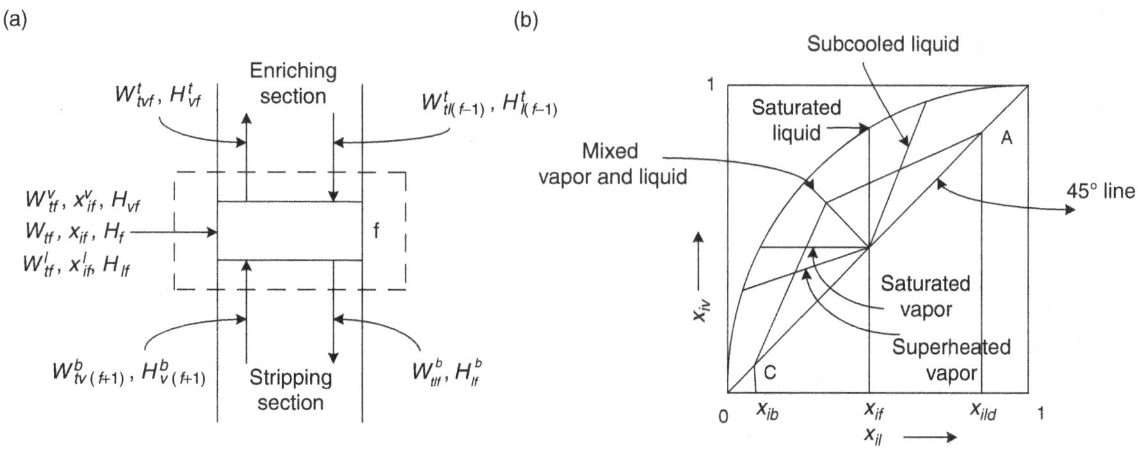

Figure 8.1.21. (a) Conditions around the feed plate; (b) slope of q*-line for different feed conditions:* q = 1*, saturated liquid;* q = 0*, saturated vapor;* 0 < q < 1*, mixture of vapor and liquid;* q > 1*, subcooled liquid;* q < 0*, superheated vapor.*

fraction of the feed; $H_{\ell f}$ is the molar enthalpy of the liquid fraction of the feed; H_{vf} is the molar enthalpy of the vapor fraction of the feed; H_f is the molar enthalpy of the feed. Figure 8.1.21(a) also illustrates the molar flow rates and enthalpies of the two streams, vapor as well as liquid, on both sides of the feed plate: enriching section, $W^t_{t\ell(f-1)}$, $H^t_{\ell(f-1)}$, W^t_{tvf}, H^t_{vf}; stripping section, $W^b_{t\ell f}$, $H^b_{\ell f}$, $W^b_{tv(f+1)}$, $H^b_{v(f+1)}$. Now carry out a total molar balance and an enthalpy balance around the feed plate as follows.

Molar flow rate balance around feed plate:

$$W_{tf} + W^t_{t\ell(f-1)} + W^b_{tv(f+1)} = W^t_{tvf} + W^b_{t\ell f}. \tag{8.1.144}$$

Enthalpy balance around feed plate:

$$W_{tf}\,H_f + W^t_{t\ell(f-1)}\,H^t_{\ell(f-1)} + W^b_{tv(f+1)}\,H^b_{v(f+1)} = W^t_{tvf}\,H^t_{vf} + W^b_{t\ell f}\,H^b_{\ell f}. \tag{8.1.145}$$

Simplification of these equations is carried out by recognizing that the variation in the molar enthalpy of the saturated liquid, as well as the saturated vapor phase, do not vary much over one plate. Therefore,

$$H^t_{\ell(f-1)} \cong H^b_{\ell f} \cong H_{\ell f}; \qquad H^b_{v(f+1)} \cong H^t_{vf} \cong H_{vf}, \tag{8.1.146}$$

which simplifies equation (8.1.145) to

$$W_{tf}\,H_f + W^t_{t\ell(f-1)}\,H_{\ell f} + W^b_{tv(f+1)}\,H_{vf} = W^t_{tvf}\,H_{vf} + W^b_{t\ell f}\,H_{\ell f} \tag{8.1.147}$$

$$\Rightarrow \left(W^b_{t\ell f} - W^t_{t\ell(f-1)}\right) H_{\ell f} = W_{tf}\,H_f + \left(W^b_{tv(f+1)} - W^t_{tvf}\right) H_{vf}. \tag{8.1.148a}$$

Substitute now from relation (8.1.144),

$$W^b_{tv(f+1)} - W^t_{tvf} = W^b_{t\ell f} - W^t_{t\ell(f-1)} - W_{tf}, \tag{8.1.148b}$$

into (8.1.148a), to obtain

$$\left(W^b_{t\ell f} - W^t_{t\ell(f-1)}\right) H_{\ell f} = W_{tf}\,H_f + \left(W^b_{t\ell f} - W^t_{t\ell(f-1)}\right) H_{vf} - W_{tf}\,H_{vf}; \tag{8.1.149}$$

$$\frac{\left(W^b_{t\ell f} - W^t_{t\ell(f-1)}\right)}{W_{tf}} = \frac{H_{vf} - H_f}{H_{vf} - H_{\ell f}} = q. \tag{8.1.150}$$

The quantity q defined above identifies the moles of liquid added to the liquid stream in the column per mole to total feed introduced.

Since $(H_{vf} - H_{\ell f})$ represents the amount of heat needed to vaporize one mole of saturated liquid under feed plate conditions, q also represents the fraction of that heat required to vaporize completely the feed as it is introduced into the column. If the feed introduced into the column is a saturated liquid at feed plate conditions, i.e. $H_f = H_{\ell f}$, $q = 1$. If the feed is introduced as saturated vapor at feed plate condition, then $H_f = H_{vf}$, $q = 0$. If the feed introduced is a mixture of liquid and vapor, then $0 < q < 1$. If the feed is introduced as a subcooled liquid, then $H_f < H_{\ell f}$, the molar enthalpy of the saturated liquid at the temperature and pressure of the feed plate; correspondingly, $q > 1$. If the feed is introduced as a superheated vapor vis-à-vis the feed plate conditions, then $H_f > H_{vf}$ and $q < 0$. Note at this time the overall material and component i balance for the column:

$$W_{tf} = W_{t\ell d} + W_{t\ell b}; \qquad W_{tf}\,x_{if} = W_{t\ell d}\,x_{i\ell d} + W_{t\ell b}\,x_{i\ell b}. \tag{8.1.151}$$

We will now determine the locus of the intersection of the operating lines of the enriching section and the stripping section as a function of the condition of the feed introduced into the column. Recognize that at the locus of the intersection the vapor phase and liquid phase compositions of both sections are identical, i.e. $\overline{x}_{iv(n+1)}$ and $\overline{x}_{i\ell n}$ for the enriching section operating line equations (8.1.136b) and (8.1.138b) are identical to $\overline{x}_{iv(m+1)}$ and $\overline{x}_{i\ell m}$, respectively, for the stripping section operating line equation (8.1.140). Therefore, we may rewrite the enriching section operating line at this location as

$$W_{tv}^{t}\,\overline{x}_{ivf} = W_{t\ell}^{t}\,\overline{x}_{i\ell f} + W_{t\ell d}\,x_{i\ell d}. \tag{8.1.152}$$

Here $\overline{x}_{ivf}$ and $\overline{x}_{i\ell f}$ refer to the composition vis-à-vis the feed plate. The corresponding form of the stripping section operating line is

$$W_{tv}^{b}\,\overline{x}_{ivf} = W_{t\ell}^{b}\,\overline{x}_{i\ell f} - W_{t\ell b}\,x_{i\ell b}. \tag{8.1.153}$$

Subtract equation (8.1.152) from the above equation to obtain

$$\left(W_{tv}^{b} - W_{tv}^{t}\right)\overline{x}_{ivf} = \left(W_{t\ell}^{b} - W_{t\ell}^{t}\right)\overline{x}_{i\ell f} - W_{t\ell b}\,x_{i\ell b} - W_{t\ell d}\,x_{i\ell d}. \tag{8.1.154}$$

From equation (8.1.151), we may rewrite this as

$$\left(W_{tv}^{b} - W_{tv}^{t}\right)\overline{x}_{ivf} = \left(W_{t\ell}^{b} - W_{t\ell}^{t}\right)\overline{x}_{i\ell f} - W_{tf}\,x_{if}. \tag{8.1.155}$$

Recognize now that

$$\left(W_{t\ell f}^{b} - W_{t\ell(f-1)}^{t}\right) = \left(W_{t\ell}^{b} - W_{t\ell}^{t}\right). \tag{8.1.156a}$$

Therefore, from (8.1.150),

$$\left(W_{t\ell}^{b} - W_{t\ell}^{t}\right) = q\,W_{tf}. \tag{8.1.156b}$$

Further, from (8.1.148b) and the above two simplifications,

$$W_{tv}^{b} - W_{tv}^{t} = q\,W_{tf} - W_{tf} = W_{tf}(q-1). \tag{8.1.156c}$$

Use results (8.1.156b) and (8.1.156c) in equation (8.1.155):

$$\overline{x}_{ivf} = \frac{q}{(q-1)}\overline{x}_{i\ell f} - \frac{1}{(q-1)}x_{if}. \tag{8.1.157}$$

This is the locus of the intersection of the operating lines of the enriching section and the stripping section of the multiplate distillation column; it is often called the *q-line*. It is a straight line whose slope is $q/(q-1)$. The point $\overline{x}_{ivf} = \overline{x}_{i\ell f} = x_{if}$ also satisfies the equation; it is located on the 45° line. Figure 8.1.20 has a q-line corresponding to a feed containing both vapor as well as liquid, i.e. $0 < q < 1$. Figure 8.1.21(b) illustrates the q-line for a variety of feed conditions representing different values of q: $q = 1$, vertical line, saturated liquid; $q = 0$, horizontal line, saturated vapor; $0 < q < 1$, mixed vapor and liquid; $q > 1$, subcooled liquid; $q < 0$, superheated vapor.

Example 8.1.8 Figure 4.1.1(c) illustrates the vapor-liquid equilibrium diagram for a benzene-toluene system at 1 atm. A distillation column fed with 3 kgmol/min of a 50-50 benzene-toluene mixture is producing a distillate having 95% benzene-5% toluene and a bottoms product containing 5% benzene-95% toluene. The feed contains both vapor and liquid. The latent heat of vaporization of the feed at the feed temperature is 32 000 J/gmol; further, the difference between the feed enthalpy and the saturated liquid enthalpy at the feed plate temperature is 18 000 J/gmol. The reflux ratio is 2.

(1) Determine the molar flow rates of the distillate and the bottoms product.
(2) Determine the equations of the operating lines and plot them.
(3) Using the McCabe-Thiele method, obtain the number of ideal stages needed to achieve this separation.

Solution (1) Let $i = 1$ stand for benzene. Then $x_{1f} = 0.50$; $x_{1\ell d} = 0.95$; $x_{1\ell b} = 0.05$. From equations (8.1.136e,f) involving total and species molar balances, we get

$$\left(W_{t\ell d}/W_{tf}\right) = \frac{x_{1f} - x_{1\ell b}}{x_{1\ell d} - x_{1\ell b}} = \frac{0.5 - 0.05}{0.95 - 0.05} = \frac{0.45}{0.9} = 0.5.$$

Therefore

$$W_{t\ell d} = 0.5W_{tf} = 0.5 \times 3\,\text{kgmol/min} = 1.5\,\text{kgmol/min}.$$

Further,

$$W_{t\ell b} = W_{tf} - W_{t\ell d} = (3 - 1.5)\text{kgmol/min} = 1.5\,\text{kgmol/min}.$$

Part (2) Given that the reflux ratio R equals 2, equation (8.1.138b) for the operating line for the enriching section is

$$\overline{x}_{1v(n+1)} = 0.667\,\overline{x}_{1\ell n} + 0.333\,x_{1\ell d}.$$

The slope of this operating line is 0.667; it intersects the ordinate at $(x_{1\ell d}/(R+1))$, namely $(0.95/3) = 0.317$. We will now calculate the slope of the q-line and obtain its intersection with the equation for the operating line of the enriching section. From equation (8.1.150),

$$q = \frac{H_{vf} - H_f}{H_{vf} - H_{\ell f}}.$$

However,

$$H_{vf} - H_{\ell f} = \text{latent heat of vaporization of the feed at feed temperature} = 32\,000\,\text{J/gmol}.$$

Further,

$$H_{vf} - H_f = H_{vf} - H_{\ell f} + H_{\ell f} - H_f = \left(H_{vf} - H_{\ell f}\right) - \left(H_f - H_{\ell f}\right) = 32\,000\ \text{J/gmol} - 18\,000\ \text{J/gmol} = 14\,000\ \text{J/gmol}.$$

Therefore

$$q = \frac{14\,000}{32\,000} = 0.4375.$$

The slope of the q-line (equation (8.1.156c)) is

$$(q/(q-1)) = (0.4375/-(0.5625)) = -0.777.$$

It starts at $x_{1f} = 0.5$ on the 45° line; as shown in Figure 8.1.22, draw it with the calculated slope of -0.777 to intersect the enriching section operating line. Draw a line between this point of intersection and $x_{1\ell b} = 0.05$ on the 45° line; that is the operating line of the stripping section.

The slope of this stripping section operating line is $(W_{t\ell}^{b}/W_{tv}^{b})$; the determination of the values of $W_{t\ell}{}^{b} =$

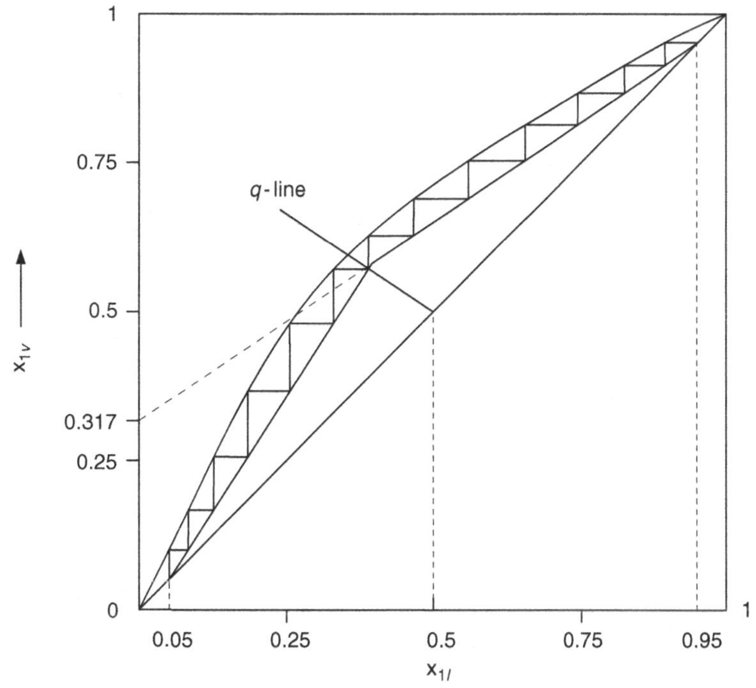

Figure 8.1.22. McCabe-Thiele diagram for Example 8.1.8.

4.312 kgmol/min and $W_{t\ell}^{b}$ = 2.812 kgmol/min may be carried out af follows. We know from relation (8.1.156c) that

$$W_{tv}^{b} - W_{tv}^{t} = W_{tf}(q-1) = \frac{3\,\text{kgmol}}{\text{min}}(0.4375-1).$$

Since $R = 2$,

$$R = \left(W_{t\ell}^{t}/W_{t\ell d}\right) \Rightarrow W_{t\ell}^{t} = 2\,W_{t\ell d}.$$

From (8.1.138a),

$$W_{t\ell}^{t} + W_{t\ell d} = W_{tv}^{t} = 3\,W_{t\ell d}.$$

Therefore

$$W_{tv}^{b} = W_{tv}^{t} - 3 \times 0.5625 \Rightarrow W_{tv}^{b} = 3\,W_{t\ell d} - 1.6875$$
$$= 4.5 - 1.6875 = 2.812\,\text{kgmol/min};$$
$$W_{t\ell}^{b} = W_{tv}^{b} + W_{t\ell b} = 2.812 + 1.5 = 4.312\,\text{kgmol/min}.$$

(3) To find out the number of equilibrium stages, start drawing steps from the distillate end, at the intersection of the 45° line and $x_{1\ell}$= 0.95. Keep drawing the steps, go over the intersection of the q-line with the operating lines and go down to the value $x_{1\ell}$=0.05 on the 45° line. We find approximately 13 stages in total, 6 stages in the enriching section and the rest in the stripping section (Figure 8.1.22).

8.1.3.2 *Two limits of reflux ratio: total reflux and minimum reflux*

The reflux ratio R, defined by equation (8.1.137) as ($W_{t\ell}^{t}$/ $W_{t\ell d}$), has considerable influence on the separation achieved in a distillation column. We will focus here on two limits of column operation: (a) the reflux ratio R achieved when $W_{t\ell d}$=0; (b) the reflux ratio for the highest possible value of $W_{t\ell d}$. When $W_{t\ell d}$=0, the reflux ratio R becomes infinite, no distillate product is withdrawn and $W_{t\ell}^{t}$=W_{tv}^{t} (see equation (8.1.138a)). This is achieved by condensing the overhead vapor stream completely and returning it into the column with no product withdrawal. At the same time, the bottoms product withdrawal rate $W_{t\ell b}$=0 for total reboil, i.e. the liquid stream leaving the column bottom, is completely evaporated in the reboiler and the vapor is returned to the column. For steady state operation, the feed flow rate W_{tf} into the column has to be zero. These conditions define *total reflux*:

$$W_{t\ell d} = 0; \qquad W_{t\ell b} = 0; \qquad W_{tf} = 0;$$
$$W_{t\ell}^{t} = W_{tv}^{t}; \qquad W_{t\ell}^{b} = W_{tv}^{b}, \qquad (8.1.158)$$

where the final two equalities follow from the bottom product withdrawal rate being zero. Correspondingly, the slope of the enriching section operating line, $W_{t\ell}^{t}/W_{tv}^{t}$, and that for the stripping section operating line, $W_{t\ell}^{b}/W_{tv}^{b}$, become equal to 1. Both operating line equations coincide with the 45° line where x_{iv}=$x_{i\ell}$ (Figure 8.1.23(a)).

If now one were to draw the steps between the equilibrium curve and the operating line(s), we would have the least numbers of steps, and therefore the equilibrium stages needed to change the composition from the column bottom composition $\overline{x}_{i\ell N} = \overline{x}_{iv(N+1)} = x_{i\ell b}$ (where $W_{t\ell b}$=0

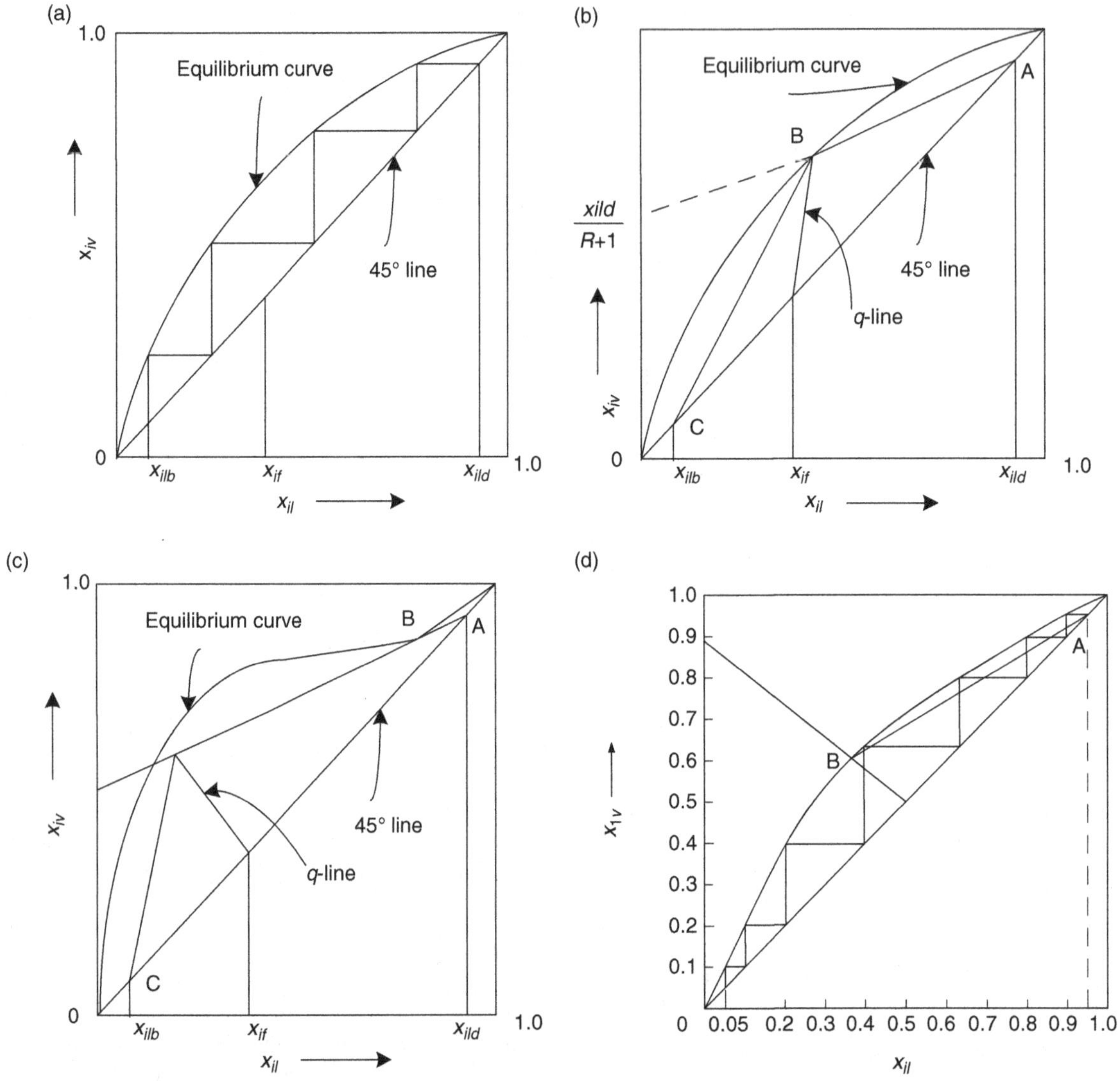

Figure 8.1.23. (a) McCabe-Thiele diagram for total reflux distillation and the minimum number of stages needed; (b) minimum reflux ratio and the operating lines; (c) minimum reflux ratio and the enriching section operating line being tangent to the equilibrium curve; (d) graphical determination of minimum reflux ratio and the number of ideal stages at total reflux for benzene-toluene distillation in Example 8.1.8.

for total reboil) to the column top composition $\overline{x}_{iv1} = \overline{x}_{i\ell d}$ (where $W_{t\ell d}$=0) in a column of N plates. Essentially, the composition change being achieved on any stage or plate is the largest possible value. Another way to look at "total reflux" would involve considering the composition change achievable in a distillation column having a certain number of plates or ideal stages. The reflux would achieve the highest possible change in composition in the distillation column containing a certain number of plates. Figure 8.1.23(a) illustrates the few steps (ideal stages) between the particular equilibrium curve and the total reflux operating line, which is the 45° line.

Fenske (1932) developed an equation to estimate the number of ideal stages needed under total reflux to go from the bottoms composition $\overline{x}_{i\ell b}$ to the top distillation composition $\overline{x}_{i\ell d}$. We will start from the reboiler and then the bottommost plate N. The total reboil mode of operation without any bottoms liquid product means that the following relation is valid for the reboiler:

$$\left[\overline{x}_{iv(N+1)}/\left(1-\overline{x}_{iv(N+1)}\right)\right] = \left[\overline{x}_{i\ell b}/\left(1-\overline{x}_{i\ell b}\right)\right] \qquad (8.1.159)$$

since $\left(\overline{x}_{i\ell b} = \overline{x}_{iv(N+1)}\right)$. Due to the total reflux condition, we have already observed that $\overline{x}_{iv(N+1)} = \overline{x}_{i\ell N}$. Therefore

$$\bar{x}_{i\ell N}/(1-\bar{x}_{i\ell N}) = [\bar{x}_{i\ell b}/(1-\bar{x}_{i\ell b})]. \quad (8.1.160)$$

Focus now on the bottommost plate N, where, under total reflux,

$$[\bar{x}_{ivN}/(1-\bar{x}_{ivN})] = \alpha_{ijN}[\bar{x}_{i\ell N}/(1-\bar{x}_{i\ell N})]. \quad (8.1.161)$$

Introduce relation (8.1.160) now in this result:

$$[\bar{x}_{ivN}/(1-\bar{x}_{ivN})] = \alpha_{ijN}[\bar{x}_{i\ell b}/(1-\bar{x}_{i\ell b})]. \quad (8.1.162)$$

Focus now on the next plate up, $N-1$, where

$$\left[\bar{x}_{iv(N-1)}/\left(1-\bar{x}_{iv(N-1)}\right)\right] = \alpha_{ij(N-1)}\left[\bar{x}_{i\ell(N-1)}/\left(1-\bar{x}_{i\ell(N-1)}\right)\right]; \quad (8.1.163)$$

but, under total reflux, $\bar{x}_{i\ell(N-1)} = \bar{x}_{ivN}$. Therefore

$$\left[\bar{x}_{iv(N-1)}/\left(1-\bar{x}_{iv(N-1)}\right)\right] = \alpha_{ij(N-1)}\,\alpha_{ijN}[\bar{x}_{i\ell b}/(1-\bar{x}_{i\ell b})]. \quad (8.1.164)$$

In this fashion, we can go all the way up to the column top:

$$[\bar{x}_{iv1}/(1-\bar{x}_{iv1})] = \alpha_{ij1}\,\alpha_{ij2}\cdots\alpha_{ijN}[\bar{x}_{i\ell b}/(1-\bar{x}_{i\ell b})]. \quad (8.1.165)$$

However, with total reflux, $\bar{x}_{iv1} = \bar{x}_{i\ell d}$. Therefore

$$[\bar{x}_{i\ell d}/(1-\bar{x}_{i\ell d})] = \alpha_{ij1}\,\alpha_{ij2}\cdots\alpha_{ijN}[\bar{x}_{i\ell b}/(1-\bar{x}_{i\ell b})]. \quad (8.1.166)$$

Using an average relative volatility $\bar{\alpha}_{ij}$,

$$[\bar{x}_{i\ell d}/(1-\bar{x}_{i\ell d})] = \left(\bar{\alpha}_{ij}\right)^N[\bar{x}_{i\ell b}/(1-\bar{x}_{i\ell b})], \quad (8.1.167)$$

where

$$\left(\bar{\alpha}_{ij}\right)^N = \alpha_{ij1}\,\alpha_{ij2}\cdots\alpha_{ijN}. \quad (8.1.168)$$

This result, the Fenske equation, is commonly illustrated as follows:

$$N = N_{\min} = \frac{\log\left[\dfrac{\bar{x}_{i\ell d}}{(1-\bar{x}_{i\ell d})}\,\dfrac{(1-\bar{x}_{i\ell b})}{\bar{x}_{i\ell b}}\right]}{\log\bar{\alpha}_{ij}}, \quad (8.1.169)$$

where $N_{\min}$ is the minimum number of plates or stages required under total reflux to achieve a distillate composition of $\bar{x}_{i\ell d}$ and a reboiler bottoms composition $\bar{x}_{i\ell b}$ under total reboil. Often $\bar{\alpha}_{ij}$ is used as a geometric mean of α_{ij1} and α_{ijN}, i.e. $\sqrt{\alpha_{ij1}\,\alpha_{ijN}}$, provided the variation along the column is minimal.

The total reflux condition is usually employed during distillation column startup; then slowly one can start withdrawing products from the column top and the column bottom. Normal column operation involves product withdrawls at both the column top and the column bottom. The column performance in terms of composition changes achieved will depend on the reflux ratio employed. If the reflux ratio $(W_{t\ell}^t/W_{t\ell d})$ is decreased by increasing $W_{t\ell d}$, the distillate product withdrawl rate, the slope of the enriching section operating line, will decrease. As one keeps on decreasing R, the number of plates required to achieve a certain extent of composition change increases. Ultimately, location, B, where the enriching section operating line AB and the q-line intersect (the feed plate region), hits the equilibrium curve and we reach the limit of *minimum reflux ratio*, $R_{\min}$ (Figure 8.1.23(b)). Point B on the equilibrium curve may be reached only via an infinite number of ever-decreasing smaller steps, effectively an infinite number of stages. Point B is identified as a *pinch point*, and leads to the other limit of the reflux ratio, the limit of minimum reflux,

$$R_{\min} = \left(W_{t\ell}^t/W_{t\ell d}\right)_{\min}. \quad (8.1.170)$$

Practical operating conditions in a column utilize $R > R_{\min}$; usually, it varies between $1.1R_{\min}$ and $1.5R_{\min}$, depending on the separation factor α_{12}. For higher values of α_{12}, higher values of R are utilized; for lower α_{12}, lower values are used. It is based on an optimization of the capital costs of the distillation unit and the operating costs of running the column. As the reflux ratio increases, the number of stages required decreases, which reduces the capital cost. Simultaneously, the operating costs increase since the amount of heat utilized in the reboiler/condenser increases (see equation (10.1.40)). Therefore an optimum cost is usually obtained at $1.1R_{\min} < R < 1.5R_{\min}$.

The equilibrium curve shown in Figures 8.1.23(a) and (b) is that of a fairly well-behaved system exhibiting ideal solution behavior. Often the systems behave nonideally. Extreme nonideal behaviors have been illustrated in Figures 4.1.3 and 4.1.4. We are not considering here such systems forming azeotropes. However, there are systems where, near the top of the column, the enriching section operating line may become a tangent to the equilibrium curve; Figure 8.1.23(c) shows such a system with the contact point B between the operating line and the equilibrium curve; contact point B is also considered as a pinch point. These behaviors occur when the relative volatility α_{12} is not constant over the whole composition range.

It is possible to obtain an analytical expression for $R_{\min}$, provided the equilibrium curve is concave downward over the whole composition range, as shown in Figure 8.1.23(b) and we have straight operating lines. At location B on the enriching section operating line and equilibrium curve intersection, the q-line also intersects. Underwood (1932) utilized the fact that the intersection of the operating line for the enriching section (equation (8.1.138b)) and the q-line (equation (8.1.157)) lie on the equilibrium curve point B, a pinch point. Denoting this location via the composition coordinates $(\bar{x}_{i\ell B}, \bar{x}_{ivB})$, we can rewrite equations (8.1.138b) and (8.1.157) as

$$\bar{x}_{ivB} = \frac{R_{\min}}{R_{\min}+1}\,\bar{x}_{i\ell B} + \frac{1}{R_{\min}+1}\,x_{i\ell d}; \quad (8.1.171)$$

$$\bar{x}_{ivB} = \frac{q}{q-1}\,\bar{x}_{i\ell B} - \frac{1}{q-1}\,x_{if}. \quad (8.1.172)$$

Simultaneous solution of these two equations yields

$$\overline{x}_{i\ell B}=\frac{(R_{\min}+1)x_{if}+(q-1)x_{i\ell d}}{R_{\min}+q};\qquad \overline{x}_{ivB}=\frac{R_{\min}x_{if}+x_{i\ell d}\,q}{R_{\min}+q}. \tag{8.1.173}$$

Since this point lies on the equilibrium curve, the two coordinates are related by the separation factor or relative volatility relation (4.1.24) between two species i and j:

$$\alpha_{ij}=\frac{\overline{x}_{ivB}\,(1-\overline{x}_{i\ell B})}{(1-\overline{x}_{ivB})\,\overline{x}_{i\ell B}}. \tag{8.1.174}$$

Substituting relations (8.1.173) for $\overline{x}_{i\ell B}$ and $\overline{x}_{ivB}$ in definition (8.1.174), we obtain the following general relation:

$$\frac{R_{\min}x_{if}+q\,x_{i\ell d}}{R_{\min}(1-x_{if})+q(1-x_{i\ell d})}=\frac{\alpha_{ij}\{x_{i\ell d}(q-1)+x_{if}(R_{\min}+1)\}}{(R_{\min}+1)(1-x_{if})+(q-1)(1-x_{i\ell d})}. \tag{8.1.175}$$

For specific cases, e.g. feed is a saturated liquid so that $q=1$, we can simplify the above to obtain

$$R_{\min}=\frac{1}{(\alpha_{ij}-1)}\left[\frac{x_{i\ell d}}{x_{if}}-\alpha_{ij}\frac{(1-x_{i\ell d})}{(1-x_{if})}\right], \tag{8.1.176}$$

a result often identified as the *Underwood equation.* When $x_{i\ell d}\rightarrow 1$ for high-purity distillate products, it is simplified to

$$R_{\min}=\frac{1}{(\alpha_{ij}-1)}\,\frac{1}{x_{if}};\qquad W^t_{t\ell}=\frac{1}{(\alpha_{ij}-1)}\,\frac{W_{t\ell d}}{x_{if}}=\frac{W_{tf}}{(\alpha_{ij}-1)}. \tag{8.1.177}$$

When the feed is introduced as a saturated vapor, i.e. $q=0$, the corresponding result is

$$R_{\min}=\frac{1}{(\alpha_{ij}-1)}\left(\frac{\alpha_{ij}\,x_{i\ell d}}{x_{if}}-\frac{(1-x_{i\ell d})}{(1-x_{if})}\right)-1. \tag{8.1.178}$$

Example 8.1.9 Consider Example 8.1.8 for the separation of a 50-50 benzene-toluene mixture producing a distillate having 95% benzene and a bottoms product having 95% toluene. All conditions of operation, including that of the feed, remain the same as in Example 8.1.8. Determine the following:

(a) the minimum reflux ratio given the separation factor $\alpha_{\text{benzene-toluene}}=2.5$ near the feed plate;
(b) the total number of ideal equilibrium stages if the column is operated at total reflux ($\alpha_{\text{benzene-toluene}}=2.5$).

Employ both graphical and analytical approaches.

Solution (a) The minimum reflux ratio may be determined analytically as well as graphically. For an *analytical* solution, employ the general relation (8.1.175) for the unkown $R_{\min}$; here, $x_{1f}=0.5$, $x_{1\ell d}=0.95$ and $q=0.4375$ for species $i=1=$ benzene:

$$\frac{R_{\min}\times 0.5+0.4375\times 0.95}{R_{\min}\times 0.5+0.4375\times 0.05}$$
$$=\frac{2.5\{0.95(-0.5625)+0.5\,(R_{\min}+1)\}}{(R_{\min}+1)(0.5)+(-0.5625)(0.05)}$$
$$\Rightarrow 0.375\,R^2_{\min}-0.4596\,R_{\min}-0.1978\Rightarrow R_{\min}=1.56.$$

For the graphical solution, draw the q-line in Figure 8.1.23(d) with a slope of $(q/(q-1))=-0.777$ from $x_{1f}=0.5$ on the 45° line. It intersects the equilibrium curve at point B. Draw a line from this point to point A on the 45° line, where $x_{1\ell d}=0.95$. The slope of this line is $(R_{\min}/(R_{\min}+1))$; from the graph, we find the value to be 0.6, leading to $R_{\min}=1.5$. The difference between the two values is due to inaccuracies in graphical measurement of the slope.

(b) We will first calculate the total number of ideal equilibrium stages needed under total reflux via the Fenske equation (8.1.169):

$$x_{1\ell d}=0.95;\qquad x_{1\ell b}=0.05,\qquad \overline{\alpha}_{ij}=2.5;$$
$$N_{\min}=\log\left[\frac{0.95}{0.05}\times\frac{0.95}{0.05}\right]\Big/\log(2.5);$$
$$N_{\min}=\log[361]/0.3979=2.557/0.397=6.427.$$

Graphically in Figure 8.1.23(d), we have drawn the steps from $x_{1\ell d}=0.95$ on the 45° line to the equilibrium line and down to the 45° line in the manner of Figure 8.1.23(a). The number of stages so found is ~7 ($=N_{\min}$).

8.1.3.3 *Additional modes of distillation column operation*

There are a number of additional modes of operating a distillation column. These include: partial condenser; open steam introduced at the column bottom without a reboiler; enriching distillation column; stripping distillation column; operation with a side stream, etc. We will briefly describe each one of these in the context of equilibrium stages/plates and constant molal overflow.

8.1.3.3.1 Partial condenser Sometimes an overhead condenser is operated such that all of the vapor stream leaving the top plate in the column is not condensed, and the noncondensed vapor is withdrawn as the product instead of the condensed liquid as the product. Figure 8.1.24(a) illustrates this mode of operation. The overall and more volatile species molar balance equations for this condition over the dashed envelope crossing the column between the nth and $(n+1)$th plate are:

$$W^t_{tv}=W^t_{t\ell}+W_{tvd}; \tag{8.1.179a}$$

$$W^t_{tv}\,\overline{x}_{iv(n+1)}=W^t_{t\ell}\,\overline{x}_{i\ell n}+W_{tvd}\,x_{ivd}. \tag{8.1.179b}$$

The operating line equation of the enriching section may be obtained from equation (8.1.179b) as

$$\overline{x}_{iv(n+1)}=\frac{W^t_{t\ell}}{W^t_{tv}}\,\overline{x}_{i\ell n}+\frac{W_{tvd}}{W^t_{tv}}\,x_{ivd}. \tag{8.1.180a}$$

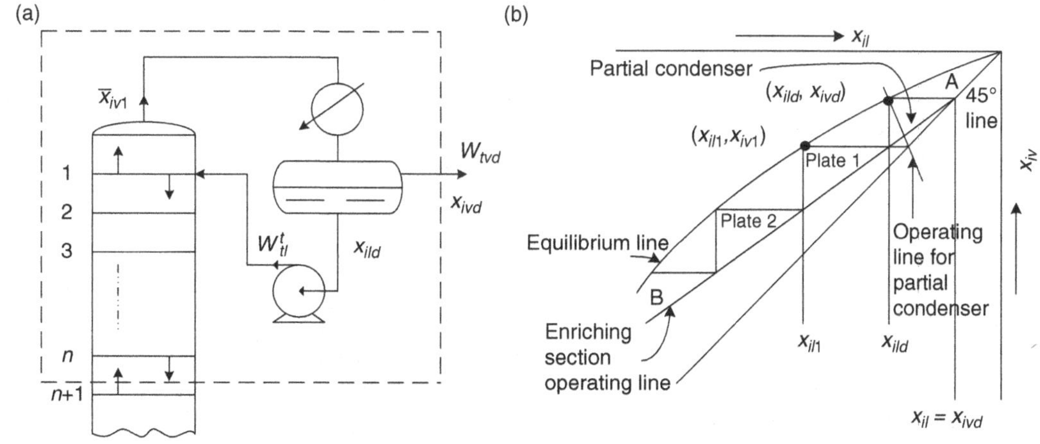

Figure 8.1.24. (a) Schematic for a partial condenser in a distillation column (feed and stripping section not shown). (b) McCabe–Thiele diagram for the partial condenser and the enriching section operating line.

Using the overall molar balance equation, this is simplified to

$$\bar{x}_{iv(n+1)} = \frac{W^t_{t\ell}}{W^t_{tv}} \bar{x}_{i\ell n} + \left(1 - \frac{W^t_{t\ell}}{W^t_{tv}}\right) x_{ivd}, \tag{8.1.180b}$$

a line with a slope of $(W^t_{t\ell}/W^t_{tv})$. If the partial condenser is operated such that $x_{i\ell d}$ and x_{ivd} are in equilibrium, then one can write for the partial condenser also

$$W^t_{tv} = W^t_{t\ell} + W_{tvd}, \tag{8.1.181a}$$

$$W^t_{tv} \bar{x}_{iv1} = W^t_{t\ell} x_{i\ell d} + W_{tvd} x_{ivd}, \tag{8.1.181b}$$

leading to

$$x_{ivd} = -\frac{W^t_{t\ell}}{W_{tvd}} x_{i\ell d} + \frac{W^t_{tv}}{W_{tvd}} \bar{x}_{iv1}, \tag{8.1.182}$$

which relates $x_{i\ell d}$ and x_{ivd}, the compositions of the two product streams from the condenser; this straight line has a slope of $(-W^t_{t\ell}/W_{tvd})$. Figure 8.1.24(b) shows a McCabe–Thiele diagram having two operating lines, one for the enriching section (slope, $W^t_{t\ell}/W^t_{tv}$), the other for the partial condenser (slope, $(-W^t_{t\ell}/W_{tvd})$. The latter operating line intersects the 45° line at $x_{i\ell} = x_{iv1} = x_{i\ell d}$.

8.1.3.3.2 Open steam introduced at the column bottom without a reboiler If the bottoms product in a distillation column is essentially water, then one could avoid having a reboiler and introduce steam directly at the bottom of the column as if it is coming from the reboiler. The molar steam introduction flow rate is $W_{tv}{}^b$ and the corresponding mole fraction for the more volatile species, $x_{iv(N+1)}$, is equal to zero (Figure 8.1.25(a)). The enriching section remains unchanged in this mode of operation: if the reflux ratio R is specified, the operating line (8.1.138b) may be plotted knowing $x_{i\ell d}$. However, in this mode of operation, the stripping line equation is somewhat changed. (The q-line equation is unchanged; see Figure 8.1.25(b).) To understand the column behavior, consider the following molar balances. For the *overall column*:

$$W_{tf} + W^b_{tv} = W_{t\ell d} + W^b_{t\ell}; \tag{8.1.183}$$

for the overall column, more volatile species:

$$W_{tf}\, x_{if} = W_{t\ell d}\, x_{i\ell d} + W^b_{t\ell} \bar{x}_{i\ell N}. \tag{8.1.184}$$

However,

$$\bar{x}_{i\ell N} = x_{i\ell b}, \qquad W^b_{t\ell} = W_{t\ell b}. \tag{8.1.185}$$

So, for the stripping section operating line we have

$$W^b_{tv(m+1)} \bar{x}_{iv(m+1)} + W_{t\ell b}\, x_{i\ell b} = W^b_{tv} x_{iv(N+1)} + W^b_{t\ell} \bar{x}_{i\ell m}. \tag{8.1.186a}$$

(Compare this with equation (8.1.139a).)

Equation (8.1.186a) may be rearranged to yield

$$W^b_{tv(m+1)} \bar{x}_{iv(m+1)} + W_{t\ell b}\, x_{i\ell b} = W^b_{t\ell} \bar{x}_{i\ell m}, \tag{8.1.186b}$$

leading to

$$\bar{x}_{iv(m+1)} = \frac{W^b_{t\ell}}{W^b_{tv}} (\bar{x}_{i\ell m} - x_{i\ell b}). \tag{8.1.187}$$

This stripping section operating line has a slope of $(W^b_{t\ell}/W^b_{tv})$, and it intersects the x_{iv}-axis at $-(W^b_{t\ell}/W^b_{tv}) x_{i\ell b}$ (when $\bar{x}_{i\ell m} = 0$); further, it intersects the $x_{i\ell}$-axis at $x_{i\ell b}$ (when $\bar{x}_{iv(N+1)} = \bar{x}_{iv(m+1)} = 0$). Therefore this operating line may be drawn as a line joining $(x_{i\ell b}, \bar{x}_{iv(N+1)} = (0))$ at the $x_{i\ell}$-axis with the point of intersection of the q-line with the enriching section operating line (Figure 8.1.25(b)). As a result, stage construction at the column bottom begins with the point $(x_{i\ell} = x_{i\ell b} = \bar{x}_{i\ell N}, \bar{x}_{iv(N+1)} = 0)$, goes up, hits the equilibrium diagram, a horizontal line is drawn, and so on.

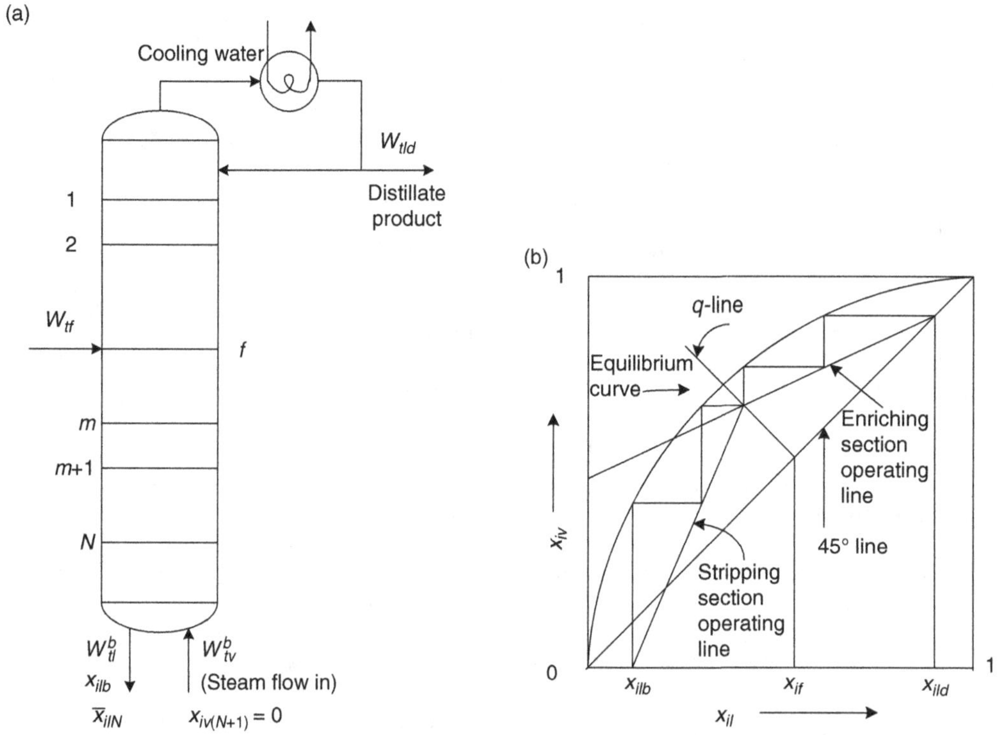

Figure 8.1.25. (a) Distillation column with open steam injection at the bottom instead of a reboiler; (b) the corresponding McCabe-Thiele plot.

8.1.3.3.3 Stage requirements for a distillation column having only an enriching section or a stripping section – Kremser equations Consider what a distillation column effectively achieves. The feed has a certain level of the more volatile species. In the enriching section of the column, the countercurrent exchange ends up developing a distillate product stream highly enriched in the more volatile species. The downflowing liquid from the feed plate has a considerable amount of the more volatile species; the stripping section strips this more volatile species from the downcoming liquid so that the bottoms product can become highly enriched in the less volatile species and is stripped of the more volatile species. These column separation capabilities suggest the following possibilities.

From a feed that comprises only vapor that is introduced at the bottom of a column, one could expect to achieve a top distillate stream highly purified in the more volatile species if there is a condenser at the top and there is reflux of the condensate at the top. Therefore such a distillation column has only an enriching section (Figure 8.1.26(a)) which can produce a highly purified distillate. However, there will be considerable loss of the more volatile species in the liquid leaving the column bottom where the vapor feed is introduced since there is no stripping section in the column. Correspondingly if a liquid feed is introduced at the top of a column, acting as a stripping section only with a reboiler at the bottom, one can achieve a highly purified bottoms product with very little contamination by the more volatile species (Figure 8.1.26(b)). However, the vapor leaving the top will have a composition close to that of the feed liquid.

If the vapor-liquid equilibrium behavior can be described as linear in either of the two configurations discussed above and the operating lines are linear, we can use the Kremser equations (8.1.131)–(8.1.133) to determine the number of stages required to achieve the desired compostion change. The equilibrium relation is assumed to be

$$x_{iv} = m_i x_{i\ell}. \qquad (8.1.188)$$

Note: For the conditions shown in Figures 8.1.26(a) and (b), the mode of operation is very similar to that of a type (2) separation system (Figure 8.1.2(a)), for example a gas absorber or a stripper. The condenser at the top of Figure 8.1.26(a) provides essentially a downflowing liquid stream which acts as the absorbent for the less volatile species. The reboiler at the bottom of Figure 8.1.26(b) provides an upflowing vapor stream which strips the downflowing liquid of the more volatile species as in a stripper.

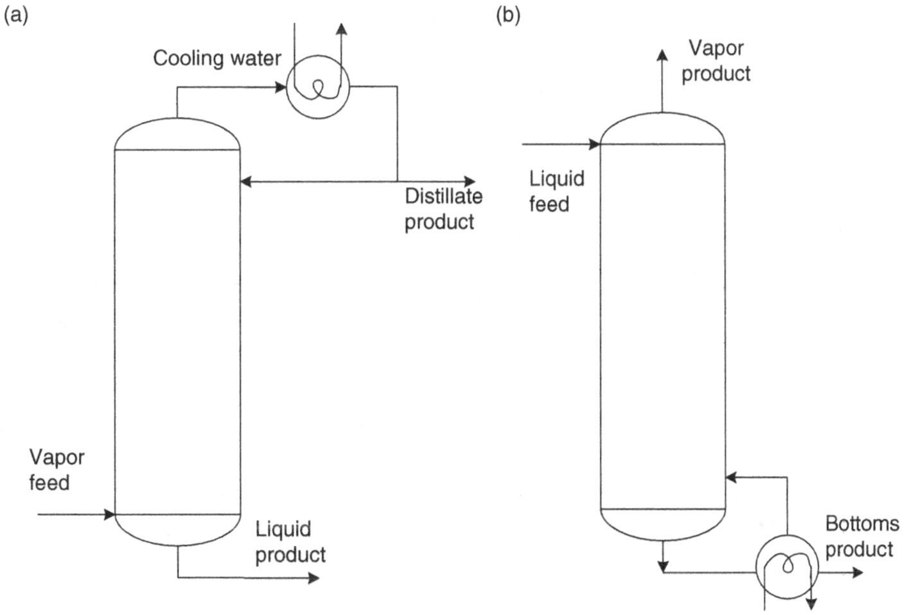

Figure 8.1.26. (a) Enriching distillation column; (b) stripping distillation column.

Therefore we will use different forms of the Kremser equation to estimate the number of ideal stages for the two different configurations. For Figure 8.1.26(a), an enricher, the result (8.1.133) is based on the more volatile species i in the gas stream going up from plate $N+1$ (feed stream) into the column of N plates. Here the feed vapor stream is introduced onto the last plate (N), which becomes the feed plate and goes up from plate N (composition x_{ivN}); therefore result (8.1.133) based on the more volatile species in the binary system is changed to

$$N-1=\frac{\ell n\left[\left(\frac{(1-x_{ivN})-(1-x_{i\ell L})m_i}{(1-x_{iv1})-(1-x_{i\ell L})m_i}\right)\left(1-\frac{1}{A}\right)+\frac{1}{A}\right]}{\ell n A}, \tag{8.1.189a}$$

since it is based on the less volatile species. The vapor leaving the top plate $n=1$ has a composition x_{iv1} and the liquid composition entering the top from the condenser is $x_{i\ell L}$. The quantity A is defined as $(m_i\,|W_{t\ell}|/|W_{tv}|)$. Correspondingly, the number of plates required for the configuration of Figure 8.1.26(b), a stripping distillation column, may be obtained as follows (Treybal, 1980, p. 422):

$$N+1=\ell n\left[\frac{x_{i\ell L}-\left(x_{iv(N+1)}/m_i\right)}{x_{i\ell 0}-\left(x_{iv(N+1)}/m_i\right)}\left(1-\frac{1}{S}\right)+\frac{1}{S}\right]\Big/\ell n\, S,$$

where

$$S=m_i|W_{tv}|/|W_{t\ell}|. \tag{8.1.189b}$$

The basis is that Figure 8.1.26(b) is acting as a stripper of the more volatile species present in the liquid feed; therefore equation (8.1.134) is applicable for the more volatile species. Further, the reboiler acts as one more ideal stage, therefore $N+1$.

8.1.3.3.4 Column operation with a side stream/two feed streams In the simplest form of a distillation column operation, there is one overhead liquid product stream, one liquid reflux stream and one feed stream, one bottoms product stream and one vapor stream from the reboiler, in addition to a liquid stream going out at the bottom to the reboiler and one overhead vapor stream going out to the condenser. There are additional configurations in terms of the number of streams entering or leaving the column. When there is a need for additional product stream(s) beyond the two product streams (one from the top and the other from the bottom), a side stream or two may be withdrawn from the column. Figure 8.1.27(a) illustrates two such side streams: the first one, in between the feed and the liquid reflux at the top, is taken out as a liquid product of intermediate composition; the second one, in between the feed and the vapor input from the reboiler, is withdrawn as a vapor stream.

The stage requirements for a column may be determined by considering the McCabe–Thiele plot for the case of only one liquid side stream between the feed and the top liquid reflux. As shown in Figure 8.1.27(b), the operating line CD for the stripping section remains unchanged since operation up to the feed introduction point from the bottom is unchanged. Similarly, the q-line for the feed plate region and its intersection with line CD is unchanged.

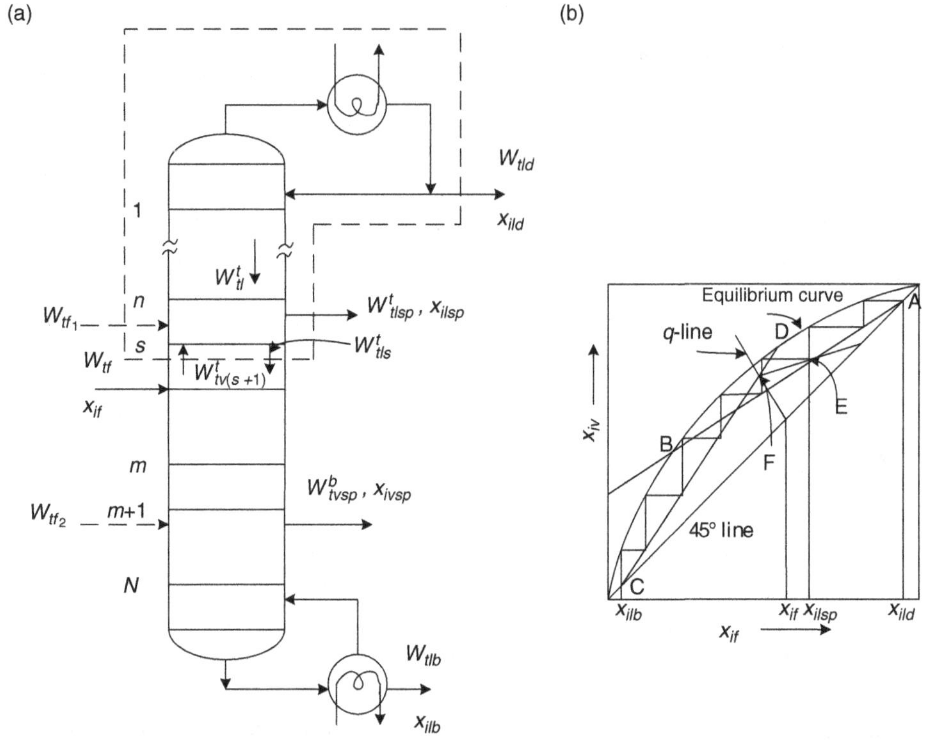

Figure 8.1.27. Column operation with additional product streams. (a) Schematic with two additional product streams, one in the enriching section, one in the stripping section; (b) McCabe–Thiele diagram for the column having one additional product stream in the enriching section.

The slope and the origin of the enriching section operating line AB is the same, except it ceases to be valid below the point of side stream withdrawal; at this location, due to the liquid side stream product withdrawal, the liquid flow rate up to the feed point changes and we have a new operating line EF, F being the point of its intersection with the q-line. If point E is fixed by $x_{i\ell sp}$, the liquid-phase composition of the liquid product being withdrawn and its intersection with the enriching section operating line AB, then knowing the slope of this intermediate operating line, one can draw it till it intersects the q-line and the stripping section operating line.

For the envelope shown in Figure 8.1.27(a) around plate s somewhere above which the liquid product side stream is withdrawn at the rate $W^t_{t\ell sp}$ and the overhead product stream rate is $W_{t\ell d}$, a total molar balance is given by

$$W^t_{tv(s+1)} = W^t_{t\ell s} + W_{t\ell d} + W^t_{t\ell sp}. \tag{8.1.190}$$

Now, $W^t_{t\ell n} = W^t_{t\ell}$ for the enriching section plates above the side product withdrawal plate, whereas $W^t_{t\ell s}$ is valid for plates below it. Therefore

$$W^t_{t\ell} = W^t_{t\ell s} + W^t_{t\ell sp}. \tag{8.1.191}$$

However, the vapor flow rates remain unaffected, i.e.

$$W^t_{tv(s+1)} = W^t_{tv(n+1)} = W^t_{tv}, \tag{8.1.192}$$

where the plate number s is obviously larger than the enriching section plate number n. A balance on more volatile species i for the balance equation (8.1.190) yields

$$W^t_{tv(s+1)}\,\overline{x}_{iv(s+1)} = W^t_{t\ell s}\,\overline{x}_{i\ell s} + W_{t\ell d}\,x_{i\ell d} + W^t_{t\ell sp}\,x_{i\ell sp}; \tag{8.1.193}$$

$$\overline{x}_{iv(s+1)} = \frac{W^t_{t\ell s}}{W^t_{tv(s+1)}}\,\overline{x}_{i\ell s} + \frac{W_{t\ell d}\,x_{i\ell d} + W^t_{t\ell sp}\,x_{i\ell sp}}{W^t_{tv(s+1)}}. \tag{8.1.194}$$

The slope of this intermediate operating line, $W^t_{t\ell s}/W^t_{tv(s+1)}\ (= W^t_{t\ell s}/W^t_{tv})$, is less than that of the top section operating line AB, namely $W^t_{t\ell}/W^t_{tv}$. The vertical step down for the plate where the side product is being withdrawn must intersect the two top operating lines, AB and EF, E being the point where they intersect.

Figure 8.1.27(a) shows via dashed lines two feed streams having molar flow rates of W_{tf1} and W_{tf2}, respectively (see Problem 10.2.4 for an example). The condition of each feed stream dictates where it is going to be introduced into the column. For example, if one of the feed streams is a vapor stream, it should be introduced in the bottom half of the column, where its composition is likely to match the local vapor stream composition. Similarly, if one of the feed streams is a liquid stream, it should be introduced at an appropriate location in the top half of the column.

8.1.3.4 *Stage efficiency, tray efficiency, plate efficiency*

Before we initiated the illustration of the McCabe–Thiele method of analyzing the distillation column (Section 8.1.3.1), we specified a basic assumption: on each stage/plate/tray in the column, the exiting vapor and the liquid streams are in equilibrium with each other; therefore the stages/plates/trays are ideal stages. In reality, the two exiting streams from a given stage/plate are generally not in equilibrium. Consequently, the number of stages/plates N needed to achieve the required change in composition from the column top to the column bottom is larger than N_{ideal}, the ideal number of stages predicted by the assumption of the exiting vapor and liquid streams for any stage being in equilibrium. This difference is usually described by an overall efficiency E_o:

$$E_o = N_{\text{ideal}}/N. \tag{8.1.195}$$

The plate number N refers to the *actual* plates/trays in a distillation column.

The correlation by O'Connell (1946) allows one to estimate the overall efficiency (given as a percentage) when the relative volatility of the key components and the feed viscosity are known (Figure 8.1.28). Several empirical relations provide different approximations of O'Connell's (1946) correlation.

(1) Doherty and Malone (2001):

$$\frac{E_o - a}{1 - a} = \exp\left(-\sqrt{\alpha_{12}(\mu/\mu_0)}\right), \tag{8.1.196}$$

where $a = 0.24$, $\mu_o = 10^{-3}$ Pa-s (1 cp), α_{12} is the relative volatility of the key components 1 and 2, μ is the viscosity of the liquid mixture at the feed composition (see Section 2.4.3 for definitions of the key components).

(2) Seader and Henley (1998):

$$E_o = 50.3\,(\alpha_{12}\,\mu)^{-0.226}, \tag{8.1.197a}$$

valid for a range of 0.1 to 10 cp for the product $\alpha_{12}\mu$, where μ is in centipoise. Here E_o is in percent.

(3) Wankat (2007):

$$E_o = 0.52782 - 0.27511\,\log_{10}(\alpha_{12}\mu) + 0.044923\,(\log_{10}(\alpha_{12}\,\mu))^2, \tag{8.1.197b}$$

where μ is in centipoise.

The overall efficiency is ultimately a reflection of the performance being achieved on each plate/tray/stage. The performance of any tray (say the nth) is usually judged by the Murphree vapor efficiency E_{MV} or the Murphree liquid

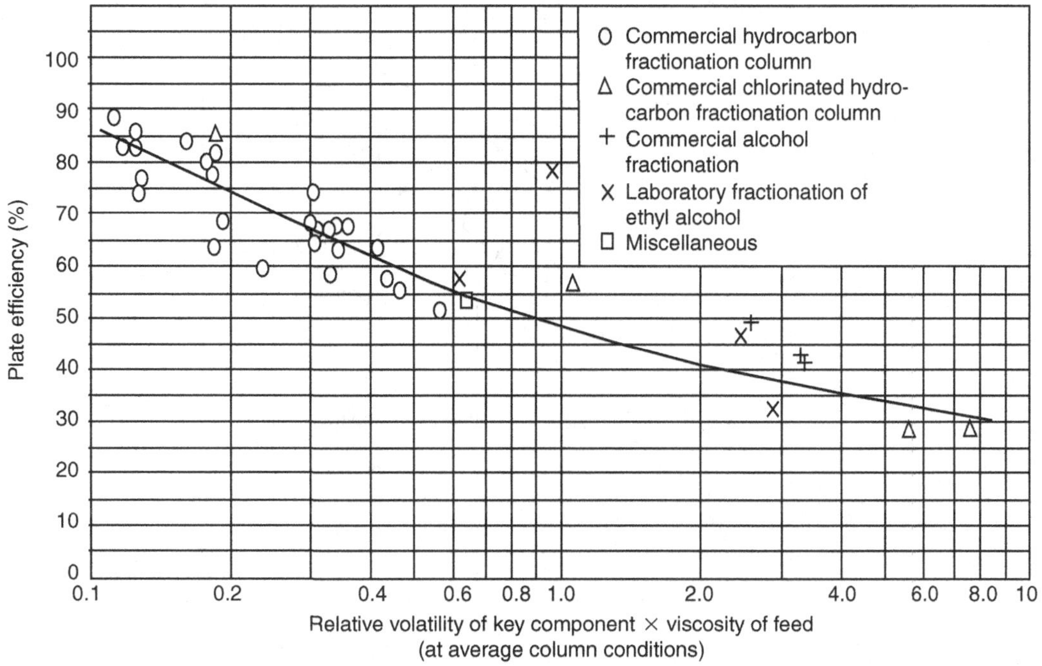

Figure 8.1.28. Effect of relative volatility and viscosity on plate efficiency of fractionating columns. Reprinted, with permission, from O'Connell, Trans. AIChE, *42, 741 (1946). Copyright © [1946] American Institute of Chemical Engineers (AIChE).*

efficiency E_{ML} for the more volatile species i in a binary mixture:

$$E_{MV} = \frac{\overline{x}_{ivn} - \overline{x}_{iv(n+1)}}{x^*_{ivn} - \overline{x}_{iv(n+1)}}; \qquad E_{ML} = \frac{\overline{x}_{i\ell n} - \overline{x}_{i\ell(n-1)}}{x^*_{i\ell n} - \overline{x}_{i\ell(n-1)}}. \quad (8.1.198)$$

Here $\overline{x}_{ivn}$ and $\overline{x}_{i\ell n}$ are the actual vapor and liquid compositions, respectively, leaving the nth tray, while x^*_{ivn} and $x^*_{i\ell n}$ are hypothetical vapor and liquid compositions, respectively, leaving the nth tray that are in equilibrium with the actual liquid composition, $\overline{x}_{i\ell n}$, and the actual vapor composition, $\overline{x}_{ivn}$, respectively, leaving the nth tray; correspondingly,

$$x^*_{ivn} = f_{\text{eq}}(\overline{x}_{i\ell n}); \qquad x^*_{i\ell n} = g_{\text{eq}}(\overline{x}_{ivn}), \qquad (8.1.199)$$

where functions f_{eq} and g_{eq} represent vapor-liquid equilibrium relations. Correspondingly, $(\overline{x}_{ivn} - \overline{x}_{iv(n+1)})$ is the actual vapor-phase enrichment achieved in plate n, whereas $(x^*_{ivn} - \overline{x}_{iv(n+1)})$ would have been the vapor-phase enrichment had the vapor phase been in equilibrium with the liquid phase leaving the nth tray. Similarly, $(\overline{x}_{i\ell n} - \overline{x}_{i\ell(n-1)})$ is the actual liquid-phase composition change achieved in plate n compared to the change $(x^*_{i\ell n} - \overline{x}_{i\ell(n+1)})$ which would have been achieved had the liquid phase leaving tray n been in equilibrium with $\overline{x}_{ivn}$. The notion of a Murphree stage efficiency was introduced in Section 6.4.1.2 for solvent extraction in a well-mixed extractor. However, the overall flow pattern on a distillation plate/tray is quite different and involves the crossflow of two different immiscible phases with an overall direction of force being perpendicular to both flowing phases (Section 8.3.2).

If the value of E_{MV} is available for a given stage/plate, then knowing $(x^*_{ivn} - \overline{x}_{iv(n+1)})$ one can determine the actual change achieved $(\overline{x}_{ivn} - \overline{x}_{iv(n+1)})$. Such a procedure provides a basis for the graphical determination of the total number of actual plates in the McCabe-Thiele method. Figure 8.1.29 illustrates this approach, with $x_{i\ell b}$ providing the start of the vertical line. Focus on the bottom of the column to the last plate and on the horizontal line intersecting the operating line at $(\overline{x}_{i\ell N}, \overline{x}_{iv(N+1)})$, say point G. When we go up from this point, we intersect the equilibrium curve at $(\overline{x}_{i\ell N}, \overline{x}_{ivN})$, point H. The distance GH, i.e. $(\overline{x}_{ivN} - \overline{x}_{iv(N+1)})$, is the maximum vapor-phase composition change achievable in plate N, since $\overline{x}_{ivN}$ is the equilibrium composition valid for an ideal stage. Multiply this vertical distance by E_{MV} (<1) and go up this distance (= E_{MV} × GH) from point G to point K, which represents the actual exit vapor-phase composition. Draw a horizontal line from this point K to the operating line point M: the resulting step now represents that actual stage. One can now carry out the same procedure from point M up and create the real staircase. The locus of the new points like point K (represented by a dashed line) represents a curve that behaves

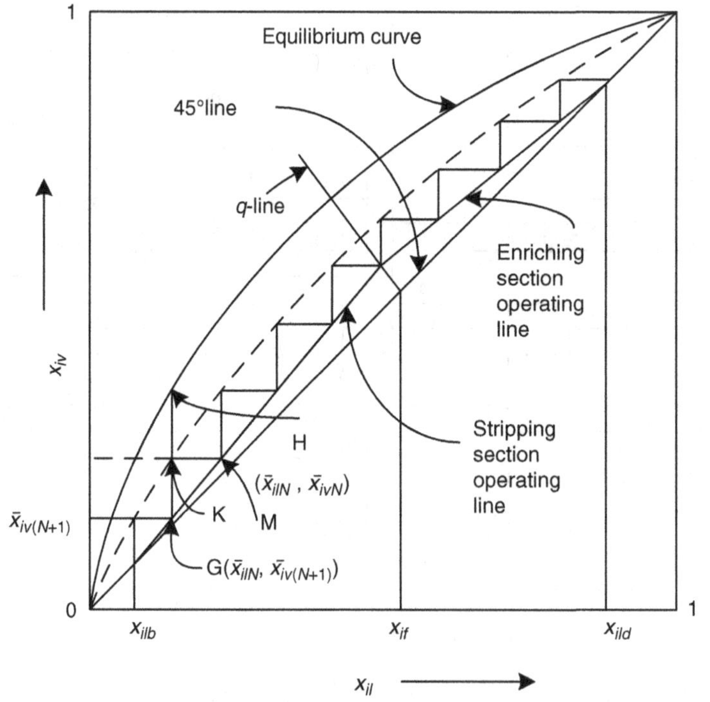

Figure 8.1.29. Graphical method of determining the actual number of plates using the Murphree vapor efficiency, E_{MV}.

like the equilibrium curve in the McCabe–Thiele method. Obviously the total number of actual plates required to traverse the composition change required from $x_{i\ell b}$ to $x_{i\ell d}$ is increased in practice from the number required if the plates were ideal stages.

It is possible to develop an estimate of the total number of actual plates required by theoretically estimating E_o given E_{MV}, provided the equilibrium curve as well as the operating lines are straight but not necessarily parallel. If

$$x^*_{iv} = m_i x_{i\ell} + b_i, \tag{8.1.200}$$

then Lewis (1936) showed that

$$E_o = \frac{\log\left[1 + E_{MV}\left(\left(m_i |W_{tg}|/|W_{t\ell}|\right) - 1\right)\right]}{\log\left(m_i\, |W_{tg}|/\,|W_{t\ell}|\right)} \tag{8.1.201}$$

for constant m_i and $|W_{tg}|/|W_{t\ell}|$. When, however,

$$m_i \Rightarrow |W_{t\ell}|/|W_{tg}|, \qquad E_o \to E_{MV}. \tag{8.1.202}$$

In a distillation column, such a relation may be used for approximate calculations in a particular section of the column where the equilibrium curve may be considered straight with the operating line being straight. Specifically, if the E_{MV} value is available, E_o may be determined so that the number of actual trays for that section may be determined. One way to determine the local value of m_i is to use the separation factor relation (Equation (4.1.24)):

$$x^*_{iv} = \frac{\alpha_{ij}\, x_{i\ell}}{1 + (\alpha_{ij} - 1)x_{i\ell}} \Rightarrow \frac{dx^*_{iv}}{dx_{i\ell}} = m_i \frac{\alpha_{ij}}{\left[1 + (\alpha_{ij} - 1)x_{i\ell}\right]^2}. \tag{8.1.203}$$

Therefore the value of the local m_i may be determined if the local α_{ij} is known for a given $x_{i\ell}$. *Note*: When $b_i = 0$, $m_i = K_i$.

Example 8.1.10 Consider the benzene-toulene distillation from Example 8.1.8. Calculate the overall efficiency E_o using the correlations by Doherty and Malone (2001), Seader and Henley (1998) and Wankat (2007). You are given the following information: feed mixture (50-50 benzene-toluene) viscosity, 0.278 cp; separation factor, $\alpha_{\text{benzene-toluene}} = 2.5$ near the feed composition.

Solution Start with the Doherty and Malone (2001) approximation (8.1.196):

$$\frac{E_o - a}{1 - a} = \exp\left(-\sqrt{\alpha_{12}(\mu/\mu_0)}\right); \qquad a = 0.24; \mu_0 = 1\text{ cp};$$

$$\frac{E_o - 0.24}{1 - 0.24} = \frac{E_o - 0.24}{0.76} = \exp\left(-\sqrt{2.5 \times (0.278/1)}\right);$$

$$E_o - 0.24 = 0.76\left\{\exp\left(-\sqrt{0.695}\right)\right\}$$
$$= 0.76 \times 0.43 \Rightarrow E_o = 0.57.$$

Seader and Henley (1998):

$$E_o = 50.3(2.5 \times 0.278)^{-0.226} = \frac{50.3}{0.92}\%;$$

$$E_o = 54.6\% = 0.546\text{ (fractional value)}.$$

Wankat (2007):

$$E_o = 0.52782 - 0.27511 \log_{10}(\alpha_{12}\mu) + 0.044923(\log_{10}(\alpha_{12}\mu))^2.$$

Now

$$\alpha_{12}\mu = 2.5 \times 0.278 = 0.695,$$

So

$$E_o = 0.52782 - 0.27511 \times (-)0.158 + 0.044923 \times (0.158)^2$$
$$= 0.52782 - 0.0434 + 0.00112 = 0.57234.$$

Example 8.1.11 In the benzene-toluene distillation of Example 8.1.8, the Murphree vapor efficiency is estimated to be around 0.6. Make an estimate of the value of E_o for the enriching section using the Lewis relation (8.1.201). Employ the following values of $x_{i\ell}$: 0.58 and 0.78 (i = benzene, ℓ), and assume that the local equilibrium curve may be assumed to be linear, with $\alpha_{\text{benzene-toluene}} = \alpha_{\text{b-t}} = 2.5$.

Solution Determine the value of the slope $(dx_{iv}/dx_{i\ell})$ of the VLE curve at the two values of $x_{i\ell}$. Given

$$(dx_{iv}/dx_{i\ell}) = \frac{\alpha_{\text{b-t}}}{\left[1 + (\alpha_{\text{b-t}} - 1)\, x_{i\ell}\right]^2}$$

we have, at $x_{i\ell} = 0.58$,

$$(dx_{iv}/dx_{i\ell}) = \frac{2.5}{[1 + 1.5 \times 0.58]^2} = \frac{2.5}{[1.87]^2} = \frac{2.5}{3.49} = 0.71.$$

For $x_{i\ell} = 0.78$,

$$[dx_{iv}/dx_{i\ell}] = \frac{2.5}{[1 + 1.5 \times 0.78]^2} = \frac{2.5}{(2.17)^2} = 0.532.$$

Note that $|W_{tg}|$ in the Lewis relation is here $|W^t_{tv}|$ and $|W_{t\ell}|$ in the Lewis relation is $|W^t_{t\ell}|$. For $R = 2$,

$$|W^t_{t\ell}| = 2W_{t\ell d} = 2 \times 1.5 = 3\,\frac{\text{kgmol}}{\text{min}};$$

$$|W^t_{tv}| = |W^t_{t\ell}| + W_{t\ell d} = 3W_{t\ell d} = 4.5\text{ kgmol/min}.$$

From the Lewis relation,

$$E_o = \frac{\log\left[1 + E_{MV}\left(\left(m_i|W_{tg}|/|W_{t\ell}|\right) - 1\right)\right]}{\log\left(m_i|W_{tg}|/|W_{t\ell}|\right)};$$

$$x_{i\ell} = 0.58 \Rightarrow E_o = \frac{\log\left[1 + 0.6\,(0.71 \times 1.5 - 1)\right]}{\log\,(0.71 \times 1.5)}$$
$$= \frac{\log\left[1 + 0.6 \times 0.065\right]}{\log\,(1.065)} = \frac{\log\,(1.039)}{\log\,(1.065)}$$
$$= \frac{0.0166}{0.027} = 0.61;$$

$$x_{i\ell} = 0.78 \Rightarrow E_o = \frac{\log\left[1 + 0.6\,(0.532 \times 1.5 - 1)\right]}{\log\,(0.532 \times 1.5)}$$
$$= \frac{\log\left[1 + 0.6 \times (0.798 - 1)\right]}{\log\,(0.798)} = \frac{\log(1 - 0.121)}{\log(0.798)}$$
$$= \frac{\log\,[0.879]}{\log\,[0.798]} = \frac{0.944 - 1}{0.902 - 1} = \frac{0.056}{0.098} = 0.57.$$

8.1.3.5 *Vapor–liquid contacting plate or tray in a distillation column*

Each stage or plate in a multiplate/multistage distillation column is designed to provide good contacting between the vapor stream and the liquid stream. The most common configuration involves a vertical cylindrical column of metal, glass-lined metal, plastic, etc., in which there are a number of horizontal trays or plates supported appropriately. Each plate, as shown in Figures 8.1.30(a) and (b), has many perforations/openings through which vapor from the plate below comes up and is distributed as vapor bubbles through the liquid layer flowing horizontally. After the bubbles have contacted the crossflowing liquid, often through a frothy/foamy condition, the vapor stream goes up to the next plate in order to contact the liquid layer in crossflow on this plate.

The intimate vapor–liquid contacting on a plate is implemented in a variety of ways. Two methods are common: *sieve plate* and *bubble-cap plate*. The plate, usually of sheet metal, has holes punched or drilled into it to make a sieve plate (Figure 8.1.30(c)); through this hole, vapor goes up and bubbles through the flowing liquid layer. Sieve plates (also called sieve trays or perforated trays) are inexpensive and quite common. Very high vapor velocity through such holes leads to "entrainment," the carryover of liquid by the vapor bubbles to the next higher tray, reducing rhe separation achieved. Very low vapor

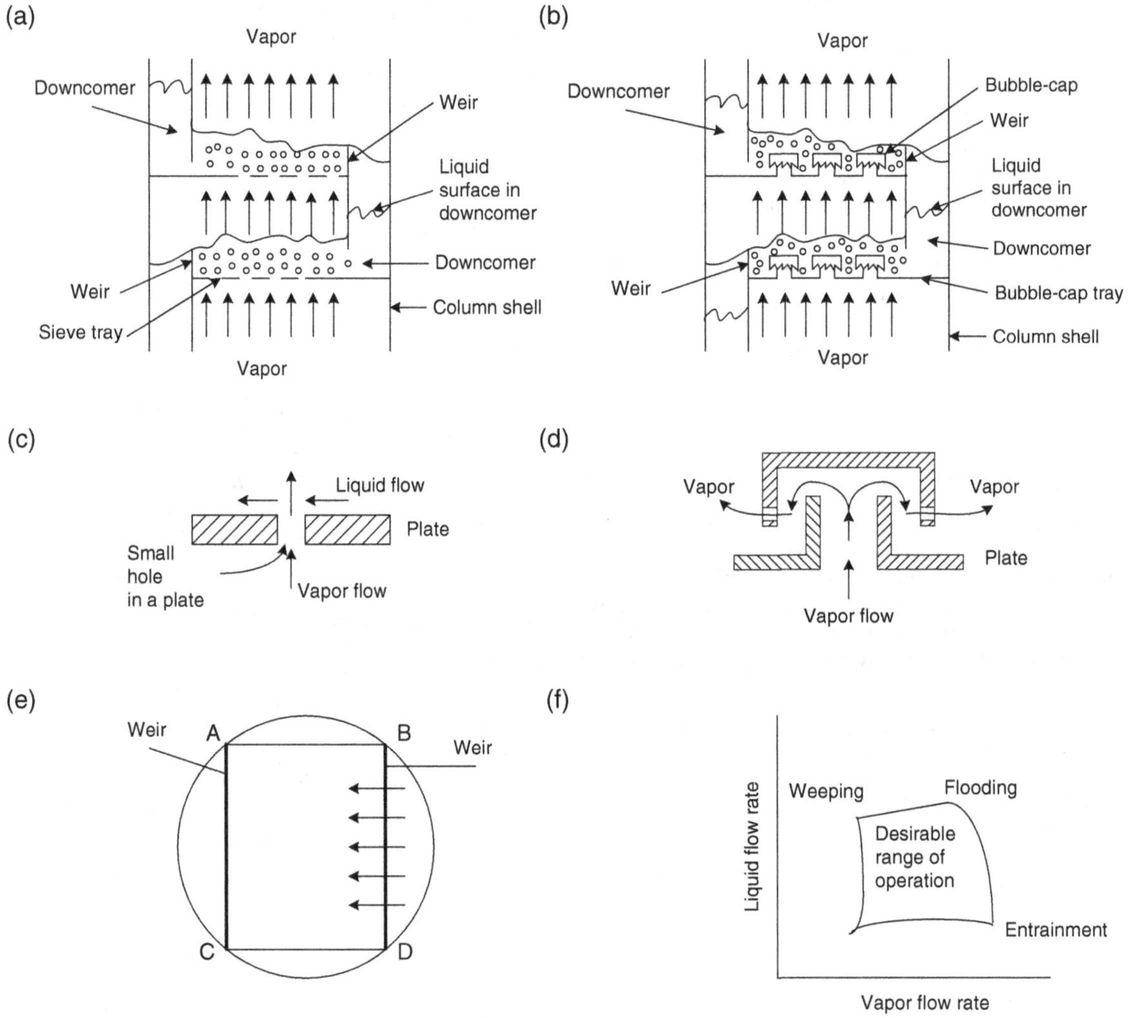

Figure 8.1.30. (a) Sieve trays in a distillation column; (b) bubble-cap trays in a column; (c) sieve plate structure; (d) bubble-cap on a tray; (e) gross liquid flow pattern in a plate from one downcomer to the next; (f) desirable range of operating vapor and liquid flow rate conditions on a plate in a distillation column.

velocity leads to "weeping," with liquid falling down to the next plate through the hole. Both are undesirable.

In a bubble-cap, the vapor enters through the central opening in the plate under each bubble-cap, is turned around by the cap and then bubbles through the liquid through slots cut in the bottom section of the periphery of the cap (Figure 8.1.30(d)). There are additional designs, e.g. valve-cap trays. Here the top cover can move up or down depending on how high or low the vapor velocity is. At low vapor flow rates, it prevents liquids from coming into the valve-cap. Valve-cap trays, although costlier, have a broader range of operating velocities of the two phases.

On any plate, the liquid flows horizontally, goes over the weir and comes down the downcomer (downspout) and then flows horizontally along the next plate down in a direction opposite to that on the plate above (Figure 8.1.30(e)). As it flows, it contacts the vapor bubbles coming up through the perforations in the sieve plate, bubble-caps or valve-caps. When the vapor flow rate is excessive, the vapor pressure drop increases, and the entrainment of liquid from the tray below increases the liquid flow rate on the plate above beyond the gravity driven rates through the downcomer, leading to increased liquid height in the downcomer and ultimately flooding, with the liquid level reaching the tray above it. Eventually the interplate spacing becomes filled with the liquid, a condition identified as *flooding*. Note, however, that the chord sections of AB and CD contain almost stagnant liquid layers, since the horizontal liquid flow from one downcomer to the next weir bypasses the regions in these sections.

The desirable range of operating conditions in a plate tower is illustrated in Figure 8.1.30(f). In this figure, the vapor flow rate (or gas flow rate in absorption/stripping processes) and the liquid flow are plotted on the abscissa and ordinate, respectively, for a sieve plate column (it is generally valid for other plate columns) (Treybal, 1980, p. 163; Fair *et al.*, 1984). For a given liquid flow rate, a high vapor flow rate leads to excessive entrainment. For a given vapor flow rate, a very high liquid flow rate may lead to the following two conditions: at low vapor flow rates, weeping takes place; at high vapor flow rates, the vapor pressure drop increases substantially and flooding takes place. The vapor pressure drop usually encountered is in the range of 0.05–0.15 psi per tray for normal operation. One contribution to this drop in pressure comes from the pressure drop arising from the flow through the holes in the trays; this is required to overcome liquid surface tension and the liquid height on the plate.

The spacing between consecutive trays is determined primarily by the requirements of cleaning, reducing entrainment from the plate below and controlling the height of the column. The smallest tray spacing used is 6 inches; it can go up to as much as 36 inches. Usually the tray spacing is around 12–20 inches. Tray spacing is an important parameter in calculating flooding conditions in a plate column.

8.1.3.5.1 Estimating the column diameter We have seen how to calculate the diameter of a flash drum where continuous flash vaporization is carried out (Section 6.3.2.1); the basis was to operate at a vapor velocity less than that needed to entrain the falling liquid droplets; i.e. the vapor velocity should be lower than the downward liquid droplet settling velocity. In a multiplate distillation column, the same physical constraint is valid, except, instead of falling liquid droplets, we need to prevent entrainment of liquid droplets from the liquid in the plate below to the plate above to prevent flooding. Relation (6.3.49) is valid here also, except the quantity K depends on a number of other factors and we use v_{vz} instead of U_{pzt} and ρ_ℓ for ρ_p:

$$v_{vz}|_{\text{flood}} = C_F \left(\gamma^{12}/20\right)^{0.2} F_f \{(\rho_\ell - \rho_v)/\rho_v\}^{0.5}. \qquad (8.1.204)$$

Here, γ^{12} is the surface tension in mN/m (dyne/cm), expressed as a ratio with the reference surface tension of 20 dyne/cm, F_f is the foaming factor ($= 1$ for nonfoaming systems; otherwise $F_f < 1$) and C_F is a capacity factor analogous to that in Figure 8.1.11 for packed beds. Correspondingly, the variation of C_F with the flow parameter of Figure 8.1.11, namely $(v_{o\ell}\rho_\ell / v_{og}\rho_g)(\rho_g/\rho_\ell)^{0.5}$ $(= F\ell_p)$, is also employed here, with the tray spacing as the parameter (varying between 6 and 36 inches).

Fair (1961) developed a widely used graphical correlation between C_F and the flow parameter ($F\ell_p$). Doherty and Malone (2001) have suggested that the following relation between C_F and $F\ell_p$ represents Fair's (1961) correlation well (checked against Fair *et al.* (1984)):

$$C_F = \frac{C_{F0}}{1 + C_{F1}\left(F\ell_p\right)^{C_{F2}}}, \qquad (8.1.205)$$

where the units of C_F and C_{F0} are m/hr. The values of the constants C_{F0}, C_{F1} and C_{F2} for three tray spacings are given in Table 8.1.3. Additional analytical expressions for C_F are available in Wankat (2007) for tray spacings between 6 and 36 inches.

In practice, the vapor velocity used is less than that at the flooding velocity $v_{vz}|_{\text{flood}}$ by a factor (<1) chosen based on best practice. In addition, one has to account for a reduction in the actual vapor flow cross-sectional area from the column diameter d_i based area $((\pi/4)d_i^2)$ due to the area of the downcomer, $area|_{dc}$. All such considerations are included in the following expression for the given total vapor flow rate in a column:

Table 8.1.3.

Tray spacing (inches)(m)	C_{F0}(m/hr)	C_{F1}	C_{F2}
12 (0.31)	252	2	1
18 (0.46)	329	2.3	1.1
24 (0.61)	439	2.5	1.2

$$W_{tv}\frac{\text{mol}}{\text{hr}} = (factor)\left[C_F\left(\frac{\gamma^{12}}{20}\right)^{0.2} F_f\left(\frac{\rho_\ell - \rho_v}{\rho_v}\right)^{0.5}\right]$$
$$\left[\frac{\pi}{4}d_i^2 - area|_{dc}\right]\frac{\rho_v}{M_v} \qquad (8.1.206)$$

The net fractional area of vapor flow after deducting the downcomer area from the column area is often 0.8; the quantity (*factor*) is usually around 0.6. One calculates the value of d_i for W_{tv}^t as well as W_{tv}^b; the larger value is used for the whole column. As indicated in Section 8.1.3.8, packed towers are used for towers whose diameter turn out to be less than 2 ft.

Example 8.1.12 Consider benzene–toluene distillation in a tray column, as in Example 8.1.8.

(1) Determine the tower cross-sectional area needed if your design is based on the conditions existing at the top plate (temperature ~82 °C).
(2) Do you expect considerable change if it is based on the bottom plate (temperature ~107 °C)?

The liquid densities may be assumed to be: benzene ~0.879 g/cm^3; toluene ~0.866 g/cm^3. The mixture surface tension may be assumed to be ~20 dyne/cm in both cases. The tray spacing is 12 inches.

Solution (1)

$$R = 2 \Rightarrow W_{t\ell}^t = 2W_{t\ell d}; \qquad W_{t\ell}^t + W_{t\ell d} = W_{tv}^t$$
$$\Rightarrow W_{tv}^t = 3W_{t\ell d} = 3 \times 1.5\,\text{kgmol/min} = 4.5.$$

Now employ equation (8.1.206):

$$W_{tv}^t = 0.6\left[C_F\left(\frac{\gamma^{12}}{20}\right)^{0.2} F_f\left(\frac{\rho_\ell - \rho_v}{\rho_v}\right)^{0.5}\right]\left[\frac{\pi}{4}d_i^2 - area|_{dc}\right]\frac{\rho_v}{M_v}.$$

Assume $F_f = 1$; $area|_{dc} = 0.2\,(\pi/4)d_i^2$; $\gamma^{12} = 20$ dyne/cm. To determine C_F, employ equation (8.1.205); therefore, we calculate the flow parameter as follows:

$$F\ell_p = (v_{o\ell}\rho_\ell / v_{og}\rho_g)(\rho_g/\rho_\ell)^{0.5} = (W_{t\ell}^t/W_{tv}^t)\,(\rho_v/\rho_\ell)^{0.5};$$
$$(W_{t\ell}^t/W_{tv}^t) = (2/3) = 0.667;$$

$$\rho_\ell\,(\text{top plate}) = 0.879 \times 0.95 + 0.866 \times 0.05 = 0.878\,\text{g/cm}^3.$$

To calculate $\rho_v(=\rho_g)$ at the top plate, calculate the mol. wt. of the vapor first:

$$\text{mol. wt.} = 0.95\,(78) + 0.05\,(92) = 78.7\,\text{kg/kmol}.$$

Therefore, at the top plate temperature of ~82 °C,

$$\rho_v = \frac{78.7}{22.4\,\text{liter}} \times \frac{273}{273+82} = 2.7\,\text{kg/m}^3.$$

We can now calculate

$$F\ell_p = 0.667\left(\frac{2.7}{878}\right)^{0.5} = 0.667 \times 0.0554 = 0.037.$$

From equation (8.1.205) for 12 inch tray spacing,

$$C_F = \frac{252(\text{m/hr})}{1 + 2(0.037)^1} = \frac{252}{1.074} = 234\,\text{m/hr};$$

$$W_{tv}^t = \frac{4.5 \times 60}{1}\frac{\text{kgmol}}{\text{hr}}$$
$$= 0.6\left[234\,\frac{\text{m}}{\text{hr}}\left(\frac{20}{20}\right)^{0.2} 1\left(\frac{878 - 2.7}{2.7}\right)^{0.5}\right]\frac{0.8\pi}{4}\,d_i^2\,\frac{2.7}{78.7};$$

$$d_i^2(\text{m}^2) = \frac{4.5 \times 4 \times 78.7 \times 60}{0.6 \times 234 \times 1 \times 1 \times 18 \times 0.8 \times \pi \times 2.7} = 4.95$$
$$\Rightarrow d_i = 2.22\,\text{m} = 7.28\,\text{ft}.$$

Such a diameter suggests a larger tray spacing, e.g. 2 ft.

(2) The primary change will be in W_{tv}; from Example 8.1.11, $W_{tv}^b = 2.812$ kgmol/min; ρ_v and $F\ell_p$ will be changed slightly due to a higher temperature. Otherwise there will be limited change. A smaller diameter will be obtained.

8.1.3.6 *Rate based approach for modeling distillation*

The standard approach identified earlier in terms of the Murphree efficiency has been found to be quite useful for binary systems which behave as ideal, close-boiling mixtures and for multicomponent systems behaving as ideal/near-ideal mixtures. However, for multicomponent systems in general, the Murphree efficiency can have wide-ranging values, including negative values and values much larger than 1. Theoretical studies on diffusion in ternary systems show complex results, such as the fact that the presence of a concentration gradient of a species does not lead to its diffusion, a species diffuses without any concentration gradient being present, etc. (Toor, 1957; Krishna *et al.*, 1977); the latter study experimentally verified such unusual predictions, leading to values of E_{MV} from negative to much larger than 1.

To resolve such problems, rigorous mass-transfer theory has been applied to a distillation stage in combination with the required heat transfer models (Krishna and Standart, 1979; Taylor and Krishna, 1993). Based on such theories, numerical models have been developed wherein correlations of mass-transfer and heat-transfer coefficients for the distillation device, of packed or plate type, are incorporated (Krishnamurthy and Taylor, 1985; Taylor *et al.*, 1994). For multicomponent systems, Maxwell–Stefan formalism (Section 3.1.5.1) provided a structural framework for such models. Such theories are known as a *rate based approach for modeling distillation* where equilibrium between phases is nonexistent except at the vapor–liquid interface.

It should be stated that the number of equations in this approach is very large, and specialized computer programs have been developed for their solution. Our objective here is very limited, namely to provide an idea of the basis of these equations. The basic set of equations in such rate based or nonequilibrium models of a distillation plate/stage is sometimes referred to as the MERSHQ equations (Taylor *et al.*, 2003), with each letter representing equations for a particular aspect of the problem for any plate n of the n-component system.

- M stands for the *material balance* for each species i: one balance equation for liquid phase and one for gas phase. These balance equations are lumped mass balance equations instead of an ordinary differential equation of species i along the column length for each phase on the plate. A total of $n + 1$ equations for each phase (e.g. equation (8.1.3a) for vapor phase and equation (8.1.3b) for the liquid phase will be replaced for the nth plate by, for example,

$$-W^t_{tv(n+1)}\, x_{iv(n+1)} + W^t_{tvn}\, \overline{x}_{ivn} = N^t_{ivn}\, A_n \qquad (8.1.207)$$

for the vapor phase, where A_n is the total vapor-liquid interfacial area on plate n, N^t_{ivn} is the species i flux into the vapor phase on the nth plate in the enriching section). Allowance has to be made for feed introduction/product withdrawal.
- E stands for the *energy balance* equations for each phase on any plate: one equation for the liquid phase and one for the vapor phase, with an additional equation on each plate at the phase interface representing no interfacial accumulation of heat.
- R represents the *equation for transport* of each species i into each phase through the interfacial flux of that species ((n - 1) equations for each phase).
- S refers to the *mole fraction summation equation* for different species i (first subscript) for each phase applied to the vapor-liquid interface (the second subscript i refers to the interface region):

$$\sum x_{ii\ell} = 1; \qquad \sum x_{iiv} = 1. \qquad (8.1.208)$$

- H stands for an equation describing the *pressure drop* on each plate (one per stage).
- Q refers to the *equilibrium* existing on plate n (say) between the two phases at the phase interface for each species i via, for example, a $K_{i,n}$ factor (n equations):

$$x_{iiv,n} = K_{i,n}\, x_{ii\ell,n}. \qquad (8.1.209)$$

The $K_{i,n}$ values for each species i and the enthalpies used in the energy balance equations for any stage n are obtained from conventional approaches used in multistage distillation analysis. However, the species flux N^t_{ivn} is expressed in terms of the sum of a convective component and a diffusive component. The diffusive component is modeled using the Maxwell-Stefan approach (Section 3.1.5.1) for this complex multicomponent system in a matrix framework. For an illustrative introduction, see Seader and Henley (1998).

8.1.3.7 *Separation of a multicomponent mixture in a distillation column*

We have seen in Section 8.1.1.3 that in a two-phase countercurrent flow system, where an equilibrium separation process is going on, the separation system lacks multicomponent separation capability. If one needs to separate a three-component system, one needs two distillation columns. For n components, one needs n - 1 columns, since it is usually difficult to separate more than one component with high purity in one column in a multicomponent system. Here we will briefly consider separation achieved with a multicomponent mixture in one distillation column. A brief discussion of multicomponent separation in a multicolumn cascade is provided in Section 9.2.

We have been introduced to multicomponent vapor-liquid equilibrium calculations for a closed vessel in Section 4.1.2. Section 6.3.2.1 illustrated briefly the calculation methods for multicomponent systems in a flash separator. The notion of key components (e.g. heavy key (HK) and light key (LK)) was introduced in Section 2.4.3. The definition of a separation factor and the relations between x_{iv} and $x_{i\ell}$ were introduced through phases $j = 1$ and $j = 2$ for a multicomponent system in Section 1.6, equations (1.6.12) and (1.6.13). Here, our objectives are limited. First, we would like to find out the minimum number of stages/plates required to achieve a given extent of separation between two species for a multicomponent system at total reflux: the resulting equation, known as the *Fenske equation*, is often used with the heavy key component and the light key component. Next we go to the other limit, namely minimum reflux, and illustrate briefly the *Underwood equations* used to obtain this limiting value. The empirical *Gilliland correlation*, described next, is often used to determine the equilibrium plate number for any other reflux ratio.

8.1.3.7.1 Fenske equation for the number of total plates at total reflux In Section 8.1.3.2, the number of ideal equilibrium stages needed to separate a binary mixture under total reflux conditions was obtained via the Fenske equation (8.1.169). A similar procedure will be followed here, except we focus on two different species i, k among n species present in the multicomponent system. We start with the bottommost plate and the reboiler. Under total reflux condition with total reboil, we know that, for stage N at the column bottom,

$$\overline{x}_{iv(N+1)} = \overline{x}_{i\ell N} = \overline{x}_{i\ell b}. \qquad (8.1.210)$$

For the reboiler, we therefore obtain, for species i and k,

$$\left(\overline{x}_{iv(N+1)}/\,\overline{x}_{kv(N+1)}\right) = \left(\overline{x}_{i\ell b}/\,\overline{x}_{k\ell b}\right). \qquad (8.1.211)$$

Utilizing relation (8.1.210) again in (8.1.211), we get

$$\left(\overline{x}_{i\ell N}/\,\overline{x}_{k\ell N}\right) = \left(\overline{x}_{i\ell b}/\,\overline{x}_{k\ell b}\right). \qquad (8.1.212)$$

Now, for the bottommost column plate N, under total reflux with the two streams leaving at equilibrium,

$$\left(\overline{x}_{ivN}/\,\overline{x}_{kvN}\right) = \alpha_{ikN}\left(\overline{x}_{i\ell N}/\,\overline{x}_{k\ell N}\right). \qquad (8.1.213)$$

Combining the last two results, we get

$$(\overline{x}_{ivN}/\overline{x}_{kvN}) = \alpha_{ikN}(\overline{x}_{i\ell b}/\overline{x}_{k\ell b}). \tag{8.1.214}$$

Move up to the next plate up, $N-1$, where

$$\left(\overline{x}_{iv(N-1)}/\overline{x}_{kv(N-1)}\right) = \alpha_{ik(N-1)}\left(\overline{x}_{i\ell(N-1)}/\overline{x}_{k\ell(N-1)}\right). \tag{8.1.215}$$

Under total reflux, $\overline{x}_{i\ell(N-1)} = \overline{x}_{ivN}$, $\overline{x}_{k\ell(N-1)} = x_{kvN}$. Therefore,

$$\left(\overline{x}_{iv(N-1)}/\overline{x}_{kv(N-1)}\right) = \alpha_{ik(N-1)}\,\alpha_{ikN}\left(\overline{x}_{i\ell b}/\overline{x}_{k\ell b}\right). \tag{8.1.216}$$

We carry out this process to the column top and obtain

$$(\overline{x}_{iv1}/\overline{x}_{kv1}) = \alpha_{ik1}\,\alpha_{ik2}\cdots\alpha_{ikN}\left(\overline{x}_{i\ell b}/\overline{x}_{k\ell b}\right). \tag{8.1.217}$$

Using an average relative volatility $\overline{\alpha}_{ik}$ defined via

$$(\overline{\alpha}_{ik})^N = (\alpha_{ik1}\,\alpha_{ik2}\cdots\alpha_{ikN}), \tag{8.1.218}$$

we get

$$(\overline{x}_{iv1}/\overline{x}_{kv1}) = (\overline{\alpha}_{ik})^N\left(\overline{x}_{i\ell b}/\overline{x}_{k\ell b}\right). \tag{8.1.219}$$

However, at total reflux, $\overline{x}_{iv1} = \overline{x}_{i\ell d}$, $\overline{x}_{kv1} = \overline{x}_{k\ell d}$; therefore,

$$(\overline{x}_{i\ell d}/\overline{x}_{k\ell d}) = (\overline{\alpha}_{ik})^N(\overline{x}_{i\ell b}/\overline{x}_{k\ell b}). \tag{8.1.220}$$

Correspondingly,

$$N = N_{\min} = \frac{\log\left[\dfrac{\overline{x}_{i\ell d}}{\overline{x}_{k\ell d}}\cdot\dfrac{\overline{x}_{k\ell b}}{\overline{x}_{i\ell b}}\right]}{\log\overline{\alpha}_{ik}}, \tag{8.1.221}$$

where $N_{\min}$ is the minimum number of plates in the column required at total reflux with a distillate composition of $\overline{x}_{i\ell d}$, $\overline{x}_{k\ell d}$ and a reboiler bottoms composition of $\overline{x}_{i\ell b}$, $\overline{x}_{k\ell b}$. This is the Fenske equation (Fenske, 1932) for a multicomponent system.

There is an alternative form for this result. The quantity $(\overline{x}_{i\ell d}/\overline{x}_{k\ell d})$ may be written as

$$(\overline{x}_{i\ell d}/\overline{x}_{k\ell d}) = (W_{t\ell d}\,\overline{x}_{i\ell d}/W_{t\ell d}\,\overline{x}_{k\ell d}). \tag{8.1.222}$$

Similarly,

$$(\overline{x}_{i\ell b}/\overline{x}_{k\ell b}) = (W_{t\ell b}\,\overline{x}_{i\ell b}/W_{t\ell b}\,\overline{x}_{k\ell b}), \tag{8.1.223}$$

even though $W_{t\ell d}$ and $W_{t\ell b}$ are zero in this total reflux mode. In terms of a feed flow rate W_{tf} and feed composition x_{if}, x_{kf}, we may define the following component cuts θ_{id} and θ_{kd} for species i and k in the distillate for the single-entry separator the distillation column is (see equation (2.2.10b) for the component cut definition):

$$\theta_{id} = \left(W_{t\ell d}\,\overline{x}_{i\ell d}/W_{tf}\,x_{if}\right); \qquad \theta_{kd} = \left(W_{t\ell d}\,\overline{x}_{k\ell d}/W_{tf}\,x_{kf}\right). \tag{8.1.224}$$

Correspondingly,

$$(1-\theta_{id}) = \left(W_{t\ell b}\,\overline{x}_{i\ell b}/W_{tf}\,x_{if}\right)$$

and

$$(1-\theta_{kd}) = \left(W_{t\ell b}\,\overline{x}_{k\ell b}/W_{tf}\,x_{kf}\right). \tag{8.1.225}$$

Therefore

$$N = N_{\min} = \frac{\log\left[\dfrac{\theta_{id}}{\theta_{kd}}\cdot\dfrac{(1-\theta_{kd})}{(1-\theta_{id})}\right]}{\log\overline{\alpha}_{ik}}. \tag{8.1.226}$$

If species k represents the heavy key (HK) component and species i represents the light key (LK) component, and we intend to use them, then one can use the Fenske equation. The only issue is the uncertainty in the value of $\overline{\alpha}_{ik}$ to be used as it may be influenced by the presence of other species.

The Fenske equation in the form of (8.1.226) may be used to estimate how other components in the multicomponent system are present at the top and bottom of the column, provided $N_{\min}$ has been determined for two key components, i and k (light key and heavy key). Convert equation (8.1.226) to the following form:

$$\theta_{id}/(1-\theta_{id}) = \left((\theta_{kd}/(1-\theta_{kd}))(\alpha_{i,k})^{N_{\min}}\right), \tag{8.1.227}$$

where component i is not the light key component. Employing the definition (8.1.224) of θ_{id}, we get from the above equation

$$W_{t\ell d}\,\overline{x}_{i\ell d} = W_{t\ell b}\,\overline{x}_{i\ell b}\left(\theta_{kd}/(1-\theta_{kd})\right)(\alpha_{i,k})^{N_{\min}}. \tag{8.1.228}$$

Since

$$W_{tf}\,x_{if} = W_{t\ell d}\,\overline{x}_{i\ell d} + W_{t\ell b}\,\overline{x}_{i\ell b}, \tag{8.1.229}$$

we get

$$W_{t\ell b}\,\overline{x}_{i\ell b} = \frac{W_{tf}\,x_{if}}{\left[1+(\theta_{kd}/(1-\theta_{kd}))\,(\alpha_{i,k})^{N_{\min}}\right]}; \tag{8.1.230a}$$

$$W_{t\ell d}\,\overline{x}_{i\ell d} = \frac{W_{tf}\,x_{if}\,(\theta_{kd}/(1-\theta_{kd}))(\alpha_{i,k})^{N_{\min}}}{\left[1+(\theta_{kd}/(1-\theta_{kd}))\,(\alpha_{i,k})^{N_{\min}}\right]}, \tag{8.1.230b}$$

allowing us to determine the distribution of species i vis-à-vis the heavy key species k.

8.1.3.7.2 Underwood equations for minimum reflux The Underwood equations (Underwood, 1948) provide a shortcut method for determining the minimum reflux ratio, $R_{\min}$, in multicomponent distillation under the following assumptions: constant relative volatilities and constant molal overflows in the stripping section as well as in the enriching section. The minimum reflux ratio, $R_{\min}$, is obtained from a solution of the following two equations for n components:

$$\sum_{i=1}^{n}\frac{\alpha_{ik}\,W_{tf}\,x_{if}}{\alpha_{ik}-\phi} = W_{tf}(1-q) = W_{tv}^{t} - W_{tv}^{b}; \tag{8.1.231a}$$

$$\sum_{i=1}^{n}\frac{\alpha_{ik}\,x_{i\ell d}}{\alpha_{ik}-\phi} = R_{\min}+1 = \frac{W_{tv}^{t}}{W_{t\ell d}}. \tag{8.1.231b}$$

Here α_{ik} refers to the relative volatility/separation factor of species i vis-à-vis the reference species, namely k.

The exact calculation procedure is dependent on how the feed species distribute between the distillate and the bottoms product. For multicomponent systems of class 1 type, all species in the feed are present in the distillate and the bottoms product. In such a system, we have a pinch point of the type represented by point B around the feed point in Figure 8.1.23(b) meant for a binary system. For multicomponent systems of class 2 type, either the distillate or the bottoms or both will not have one or more species present in the feed. One encounters two pinch points in these types of systems, one in the rectifying (enriching) section and the other in the stripping section due to the presence of the so-called nondistributing species which do not show up either in the distillate or the bottoms or both (Shiras *et al.*, 1950).

To obtain the value of R_{min} from equations (8.1.231a, b), one has to solve equation (8.1.231a) for the different ϕ values which satisfy the equation for n feed components if they are distributed, i.e. present in the distillate and the bottoms. Use these ϕ values and solve for the value of R_{min} that satisfies these equations. For illustrative examples involving a variety of multicomponent systems, see Treybal (1980, pp. 521–526), McCabe *et al.* (1993, pp. 657–674), Seader and Henley (1998, chap. 9), Doherty and Malone (2001, chap. 4) and Wankat (2007, chaps. 5–7).

8.1.3.7.3 Gilliland correlation and Fenske–Underwood–Gilliland method for the number of stages Gilliland (1940) developed an empirical shortcut method to determine the number of stages N needed in a distillation column for a multicomponent system for a reflux ratio R larger than the minimum value, R_{min}. Figure 8.1.31 illustrates a plot of $(N - N_{min})/(N + 1)$ vs. $(R - R_{min})/(R + 1)$, where N_{min} refers to the minimum number of stages needed at total reflux (e.g. that obtained from the Fenske equation (8.1.221)) and R_{min} is the minimum reflux ratio from the Underwood methodology. One determines the values of N_{min} and R_{min} for the problem first. Then, for a given value of the reflux ratio, R, the abscissa is known; the corresponding value of the ordinate from the graph will yield an estimate of N for the given R. Modifications to this correlation have been provided by Liddle (1968) as well as by Erbar and Maddox (1961). Eduljee (1975) has provided the following equation to describe the Gilliland correlation:

$$\{(N - N_{min}/(N+1)\} = 0.75\left\{1 - \left(\frac{R + R_{min}}{R+1}\right)^{0.5688}\right\}. \tag{8.1.232}$$

8.1.3.8 Distillation in a packed tower: continuous vapor–liquid contact in countercurrent flow

In Sections 8.1.2.2 and 8.1.2.3, we studied a countercurrent continuous contact gas–liquid absorber or stripper. Here we study a somewhat similar physical configuration, except we have a vapor stream rather than a gas stream going up. However, there are some major differences. This is a type (3) separation system, just like a multiplate distillation system with the reflux of a liquid stream at the column top obtained from the condensation of the overhead vapor

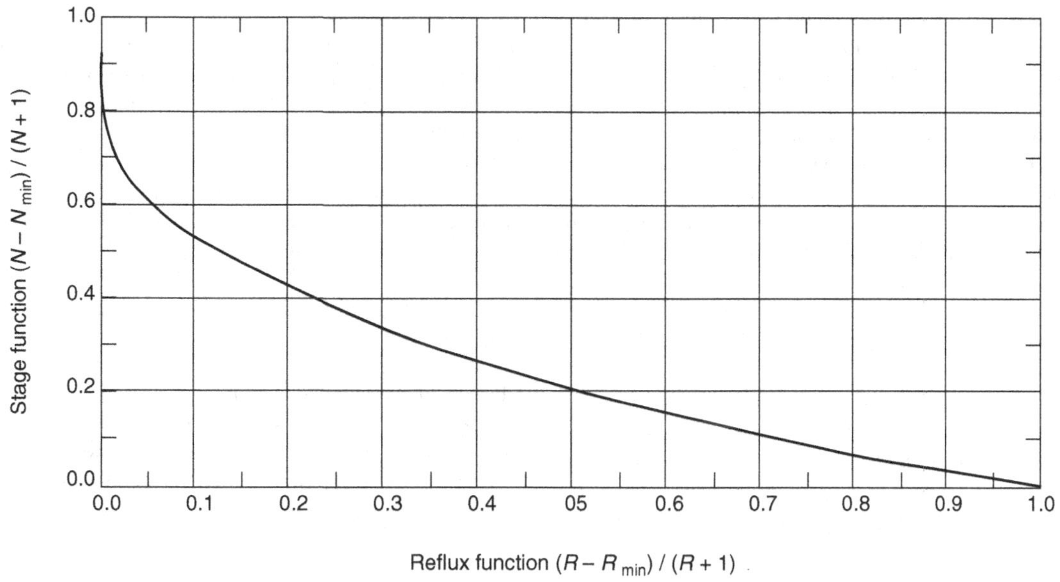

Figure 8.1.31. Gilliland (1940) correlation relating the function of stage numbers and the function of reflux ratios. Reprinted with permission from Ind. Eng. Chem., *32, p. 1220 (1940), Figure 4. Copyright (1940) American Chemical Society.*

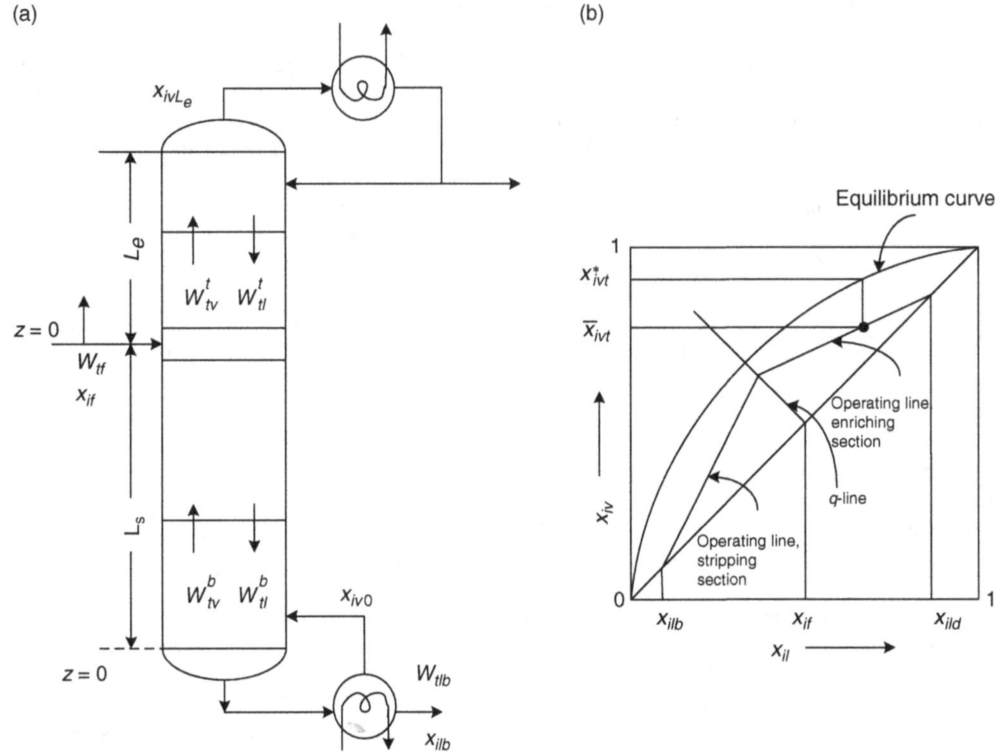

Figure 8.1.32. (a) Packed tower continuous contact distillation column. (b) McCabe–Thiele plot of operating lines in a packed tower distillation column.

stream and the reboiler based introduction of a vapor stream at the bottom. The only difference is that, instead of a feed plate, there is an empty section near the middle for an appropriate distributed introduction of the feed liquid (Figure 8.1.32(a)). Consequently, the operating lines of the enriching section and the stripping section are obtained from equations (8.1.136b) and (8.1.140), respectively, except the discrete stage number dependent composition variables $\overline{x}_{iv(n+1)}$, $\overline{x}_{i\ell n}$, $\overline{x}_{iv(m+1)}$, $\overline{x}_{i\ell m}$ become continuous variables, i.e. $\overline{x}_{iv}$ and $\overline{x}_{i\ell}$. Note that, in the two sections, the flow rates continue to be different and also become continuous variables leading to the following two equations for the two operating lines. For the *enriching section*:

$$\overline{x}_{iv} = \frac{W^t_{t\ell}}{W^t_{tv}}\,\overline{x}_{i\ell} + \frac{W_{t\ell d}}{W^t_{tv}}\,x_{i\ell d}; \tag{8.1.233}$$

for the *stripping section*:

$$\overline{x}_{iv} = \frac{W^b_{t\ell}}{W^b_{tv}}\,\overline{x}_{i\ell} - \frac{W_{t\ell b}}{W^b_{tv}}\,\overline{x}_{i\ell b}. \tag{8.1.234}$$

Assuming constant molar overflow throughout the column, it is clear that the slopes $(W^t_{t\ell}/W^t_{tv})$ and $(W^b_{t\ell}/W^b_{tv})$ are constant in each section. Further, the operating line equation is unchanged. Therefore one can draw the McCabe–Thiele diagram as before (Figure 8.1.20) with the two operating lines and the q-line (Figure 8.1.32(a) and (b)). However, determination of the column height from knowledge of the plate number and the plate gap is not useful here. We will employ a procedure similar to that which led to equations (8.1.51) and (8.1.87).

The governing equation for transfer of the more volatile vapor species i may be written at any location z of the packed tower from equation (8.1.3a) as

$$\langle v^t_{vz}\rangle\, S_c\,\frac{\mathrm{d}\overline{C}_{ivt}}{\mathrm{d}z} = |W^t_{tv}|\,\frac{\mathrm{d}\overline{x}_{ivt}}{\mathrm{d}z} = K_{ivc}\,a\,S_c\left(C^*_{ivt} - \overline{C}_{ivt}\right), \tag{8.1.235}$$

where we have used $j = 1 = v$ for the vapor phase; note that the more volatile species is being transferred from the liquid to the vapor phase both in the enriching section and the stripping section. One could also develop a similar equation for the liquid phase following equation (8.1.3b):

$$\langle v^t_{\ell z}\rangle\, S_c\,\frac{\mathrm{d}\overline{C}_{i\ell t}}{\mathrm{d}z} = |W^t_{t\ell}|\,\frac{\mathrm{d}x_{i\ell t}}{\mathrm{d}z} = K_{i\ell c}\,a\,S_c\left(\overline{C}_{i\ell t} - C^*_{i\ell t}\right). \tag{8.1.236}$$

Focus now on equation (8.1.235) for the vapor phase; we may rewrite it as

$$\frac{|W_{tv}^t|}{S_c}\frac{d\overline{x}_{ivt}}{dz} = K_{ivc}\,a\,C_{tv}^t\left(x_{ivt}^* - \overline{x}_{ivt}\right). \tag{8.1.237}$$

For the enriching column section of length L_e, we integrate this expression from $z = 0$ (corresponding to the feed section) to $z = L_e$ (top of column) corresponding to the vapor-phase composition change from $\overline{x}_{ivf}$ to $\overline{x}_{ivL_e}$ at the top, and obtain for L_e

$$\int_0^{L_e} dz = \int_{\overline{x}_{ivf}}^{\overline{x}_{ivLe}} \frac{|W_{tv}^t|}{K_{ivx}\,a\,S_c}\frac{d\overline{x}_{ivt}}{\left(x_{ivt}^* - \overline{x}_{ivt}\right)} = L_e, \tag{8.1.238}$$

where $K_{ivx} = K_{ivc}C_{tv}^t$. Under conditions of constant molar overflow in each section of the column, $|W_{tv}^t|$ is constant along the z-direction. Further, assuming equimolar counterdiffusion, we observe $K_{ivx} = K'_{ivx}$; if it is assumed to be essentially constant over L_e, then

$$L_e = \frac{W_{tv}^t}{S_c\,K'_{ivx}\,a}\int_{x_{ivf}}^{\overline{x}_{ivLe}} \frac{d\overline{x}_{ivt}}{\left(x_{ivt}^* - \overline{x}_{ivt}\right)}. \tag{8.1.239}$$

In Figure 8.1.32(b), the denominator $\left(x_{ivt}^* - \overline{x}_{ivt}\right)$ is the vertical height between the equilibrium curve and the operating line for the vapor-phase composition $\overline{x}_{ivt}$ on the operating line in the enriching section. Therefore a plot of $(1/\left(x_{ivt}^* - \overline{x}_{ivt}\right))$ against $\overline{x}_{ivt}$ over the length of the enriching section will allow the determination of the integral in equation (8.1.239) and therefore the length L_e provided K'_{ivx} may be assumed constant. We will deliberate on this soon.

As practiced in gas absorption/stripping analyses (equations (8.1.54a–c), we may define for the enriching section the height of a transfer unit based on the overall vapor phase, HTU_{ove}, and the number of transfer units based on overall vapor performance, NTU_{ove}, as follows:

$$HTU_{ove} = \frac{|W_{tv}^t|}{S_c\,K'_{ivx}\,a}; \tag{8.1.240}$$

$$NTU_{ove} = \int_{x_{ivf}}^{\overline{x}_{ivLe}} \frac{d\overline{x}_{ivt}}{\left(x_{ivt}^* - \overline{x}_{ivt}\right)}. \tag{8.1.241}$$

The enriching section length L_e is related to these via

$$L_e = HTU_{ove} \times NTU_{ove} \tag{8.1.242}$$

provided K'_{ivx} is constant. We will soon see that the approach adopted here is not correct.

Similarly, we can develop an estimate of the height of the stripping section L_s as

$$L_s = HTU_{ovs} \times NTU_{ovs}, \tag{8.1.243}$$

where

$$HTU_{ovs} = \frac{|W_{tv}^b|}{S_c\,K'_{ivx}\,a}; \tag{8.1.244a}$$

$$NTU_{ovs} = \int_{x_{iv0}}^{\overline{x}_{ivf}} \frac{d\overline{x}_{ivb}}{\left(x_{ivb}^* - \overline{x}_{ivb}\right)}. \tag{8.1.244b}$$

Note that the value of K'_{ivx} in the stripping section will be different from that in the enriching section. The assumption of K'_{ivx} being constant is not correct here.

The problem with the above approaches of equations (8.1.241), (8.1.242) and (8.1.244a,b) is as follows. Consider relation (8.1.65b), where the gas phase g is replaced with v, the vapor phase, and the Henry's law constant H_i^P is replaced by the slope m_i of the VLE diagram at any location:

$$\frac{1}{K'_{ivx}} = \frac{1}{k'_{ivx}} + \frac{m_i}{k'_{ilx}}. \tag{8.1.245a}$$

Here, m_i is the slope of the vapor–liquid equilibrium curve, $m_i = dx_{ivi}/dx_{i\ell i}$, and it varies very strongly with $x_{i\ell i}$. An alternative relation equivalent to (8.1.65e) is

$$HTU_{ov} = HTU_v + (HTU_\ell)\lambda, \tag{8.1.245b}$$

where

$$\lambda = m_i\,|W_{tv}|/|W_{t\ell}|. \tag{8.1.245c}$$

Therefore $HTU_{ov\ell}$ or HTU_{ovs} will depend strongly on the composition since λ depends strongly on the composition via m_i; however, HTU_v and HTU_ℓ are essentially independent of composition. Therefore L_e and L_s have to be determined as follows:

$$L_e = \int_{\overline{x}_{ivf}}^{\overline{x}_{ivLe}} HTU_{ove}\frac{d\overline{x}_{ivt}}{\left(x_{ivt}^* - \overline{x}_{ivt}\right)}; \tag{8.1.246a}$$

$$L_s = \int_{\overline{x}_{iv0}}^{\overline{x}_{ivf}} HTU_{ovs}\frac{d\overline{x}_{ivb}}{\left(x_{ivb}^* - \overline{x}_{ivb}\right)}. \tag{8.1.246b}$$

In the evaluation of the integrands now, the local values of HTU_{ove} and HTU_{ovs} have to be utilized:

$$HTU_{ove} = HTU_{ve} + (HTU_{\ell e})\,\lambda_e; \tag{8.1.247a}$$

$$\lambda_e = m_i\left(|W_{tv}^t|\right)/\left(|W_{t\ell}^t|\right); \tag{8.1.247b}$$

$$HTU_{ovs} = HTU_{vs} + (HTU_{\ell s})\,\lambda_s; \tag{8.1.247c}$$

$$\lambda_s = m_i\left(|W_{tv}^b|\right)/\left(|W_{t\ell}^b|\right), \tag{8.1.247d}$$

where m_i will be a very strong function of the composition in the VLE diagram.

Note: The expression (8.1.203) for $m_i = \left(dx_{iv}^*/dx_{i\ell}\right)$ should be used, i.e.

$$m_i = \frac{\alpha_{ij}}{\left[1 + \left(\alpha_{ij} - 1\right)x_{i\ell}\right]^2}. \tag{8.1.247e}$$

Example 8.1.13 Consider the problem of benzene-toluene distillation illustrated in Example 8.1.8. It is to be carried out in a packed tower at 1 atm with a reflux ratio of 2 using a 50-50 benzene-toluene feed mixture coming in at 3 kgmol/min; $x_{1\ell d} = 0.95$, $x_{1\ell d} = 0.05$, where $i = 1$ is for benzene. The HTU values in these flow rate ranges are given in Table 8.1.4 (Seader and Henley, 1998).

Determine the height of the enriching section of the packed column. Example 8.1.8 did not provide the data to construct the VLE diagram. Construct the VLE diagram first, assuming α_{12} = constant. You are given that at 90 °C, $P_1^{sat} = 1016$ mm Hg, $P_2^{sat} = 405$ mm Hg.

Solution From the vapor pressure data,

$$\alpha_{12} = \left(P_1^{sat}/P_2^{sat}\right) = \alpha_{\text{benzene-toluene}} = 1016/405 = 2.508;$$

we will work with 2.5, as identified earlier in Example 8.1.9. Employ

$$x_{1v}^* = \frac{\alpha_{12}\, x_{1\ell}}{1 + (\alpha_{12} - 1)\, x_{1\ell}}$$

to determine x_{1v}^* vs.$x_{1\ell}$ (see Table 8.1.5).

Plot the VLE diagram (Figure 8.1.33(a)). Draw the enriching section operating line, the q-line and the stripping section operating line per Example 8.1.8. Note the value of $x_{i\ell t} = 0.375$ at the intersection of the two operating lines. From Example 8.1.8, $|W_{tv}^t| = 4.5$ kgmol/min and $|W_{t\ell}^t| = 3$ kgmol/min. Therefore $\lambda_e = m_i\,(4.5/3) = 1.5m_i$. We will now calculate m_i, λ_e, HTU_{ove} and the value of the integrand $HTU_{ove}/\left(x_{ivt}^* - \overline{x}_{ivt}\right)$ for different values of $x_{i\ell t}$. For HTU_{ove}, employ relation (8.1.247a). See Table 8.1.6.

Plot $HTU_{ove}/\left(x_{ivt}^* - \overline{x}_{ivt}\right)$ against x_{ivt} (Figure 8.1.33(b)), compute the area; the unit of HTU_{ove} is feet. The area calculated yielded 10.1 ft, which is the height of the enriching section of the packed tower. The height of the stripping section is calculated in Problem 8.1.15.

Distillation in a packed tower is preferred over a plate tower for small diameter towers having a diameter less than 2 ft. For corrosive liquids, ceramic/plastic packings are preferred along with a packed tower over plate towers. For low pressure drops or vacuum distillation applications, structured packings are preferred. The details provided in Section 8.1.2.1 for gas absorption/stripping processes are equally valid for distillation processes. Calculation methods for flooding based design illustrated there may also be used here.

Table 8.1.4.

	HTU_v (ft)	HTU_ℓ (ft)
Enriching section	1.6	0.48
Stripping section	0.9	0.53

Table 8.1.5.

$x_{1\ell}$	x_{1v}^*	$x_{1\ell}$	x_{1v}^*
0.05	0.116	0.5	0.714
0.10	0.217	0.6	0.789
0.20	0.384	0.75	0.882
0.30	0.517	0.90	0.957
0.40	0.625	0.95	0.979

Table 8.1.6.

$x_{i\ell t}$	$\overline{x}_{ivt}$	m_i	λ_e	HTU_{ove}	$\left(x_{ivt}^*-\overline{x}_{ivt}\right)$	$HTU_{ove}/\left(x_{ivt}^*-\overline{x}_{ivt}\right)$
0.375	0.575	1.0245	1.536	1.897	0.03	63.2
0.5	0.655	0.816	1.224	1.747	0.105	16.63
0.6	0.728	0.692	1.038	1.658	0.087	19.05
0.75	0.83	0.553	0.83	1.558	0.055	28.32
0.85	0.9	0.483	0.7245	1.508	0.032	47.12
0.95	0.95	0.425	0.6375	1.466	0.03	48.87

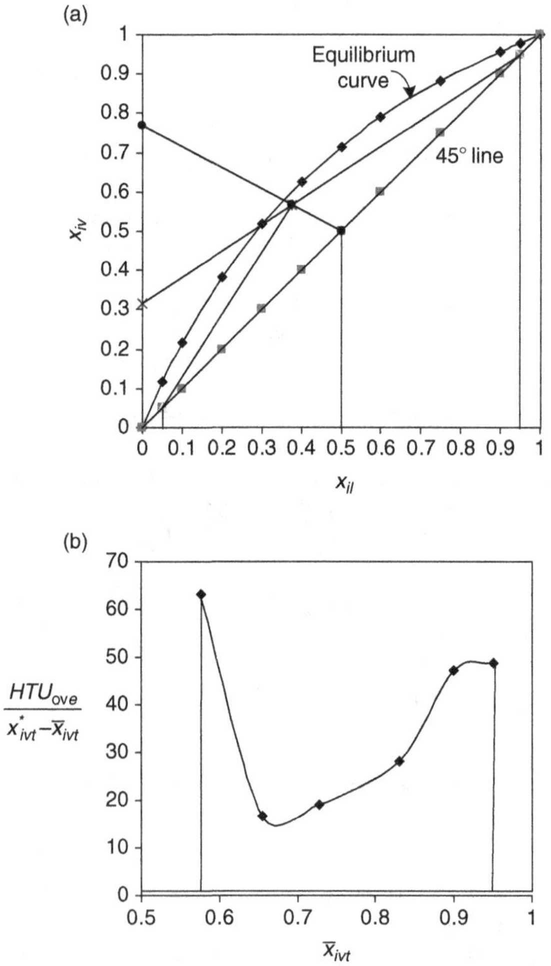

Figure 8.1.33. (a) VLE diagram for benzene-toluene system. (b) Plot for integration in Example 8.1.13.

8.1.3.9 *Batch distillation with reflux in a multistage column*

In Section 6.3.2.1, we studied batch distillation in a simple vessel, a pot/reboiler, with total condensation and no reflux; we had a case of bulk flow of the vapor phase parallel to the direction of the chemical potential gradient based driving force between the liquid in the pot and the vapor flowing up. The highest separation achieved corresponded to that for one ideal equilibrium stage. If from the total condenser at the top, part of the condensed distillate is recycled back to the column, the flow vs. force configuration will involve two phases flowing countercurrent to each other, with the $-\nabla\mu_i$ driving force being perpendicular to the flow of both phases. Such a configuration is usually utilized in a multiplate (or a packed) column configuration (Figure 8.1.34(a)) on top of the reboiler/pot at the bottom. Therefore the distillate product composition achieved is considerably higher than that achieved in the simple one-stage batch distillation of Figure 6.3.12A. However, in both cases, we have a time-dependent behavior. *Note*: The pot/reboiler will continue to be one plate in this multiplate configuration.

The time-dependent behavior in terms of the distillate composition $x_{i\ell d}$ vs. $x_{i\ell b}$ is illustrated in Figure 8.1.34(b); it does not show instantaneous composition changes. The composition $x_{i\ell}^0$ of the original liquid charge into the pot is related to the highest distillate composition $x_{i\ell d}^0$ for a given reflux ratio R through the enriching section operating line AB. As time progresses, the pot composition becomes more enriched in the heavier species, i.e. $x_{i\ell b} < x_{i\ell}^0$; correspondingly, $x_{i\ell d}$ becomes reduced, i.e. $x_{i\ell d} < x_{i\ell d}^0$.

Figure 8.1.34(b) shows three enriching section operating lines corresponding to three different values of $x_{i\ell d}$ at three different times, $t = 0$, $t = t_1$ and $t = t_2$, for a fixed value of the reflux ratio R. Each operating line represents a pseudo steady state based situation. It is clear that $x_{i\ell d}^0 > x_{i\ell d}(t_1) > x_{i\ell d}(t_2)$ for $t_2 > t_1 > 0$; the operating lines for times t_1 and t_2 are A_1 B_1 and A_2 B_2, respectively; for the pot liquid composition, the values at different times are: $x_{i\ell b}(t_2) < x_{i\ell b}(t_1) < x_{i\ell}^0$. The enriching section operating line equation for any one of these lines is (equation (8.1.138b))

$$\overline{x}_{iv(n+1)}(t) = \frac{R}{R+1}\, x_{i\ell n}(t) + \frac{1}{R+1}\, x_{i\ell d}(t), \tag{8.1.248}$$

where R is constant.

If the rate of evaporation from the pot is known, the time needed to achieve a time-averaged distillate product composition can be determined for constant reflux ratio by an equation developed by Block (1961) under the assumption that the holdup in the column is negligible. Using a complex control system, a multiplate batch distillation column could also be operated such that $x_{i\ell d}(t_1) = x_{i\ell d}(t_2)$, with the reflux varying with time. Calculation methods for such a method of operation have been illustrated in Bogart (1937) and Ellerbe (1973). However, there comes a time when it is no longer possible to have a distillate product due to the ever-increasing liquid refluxing requirement.

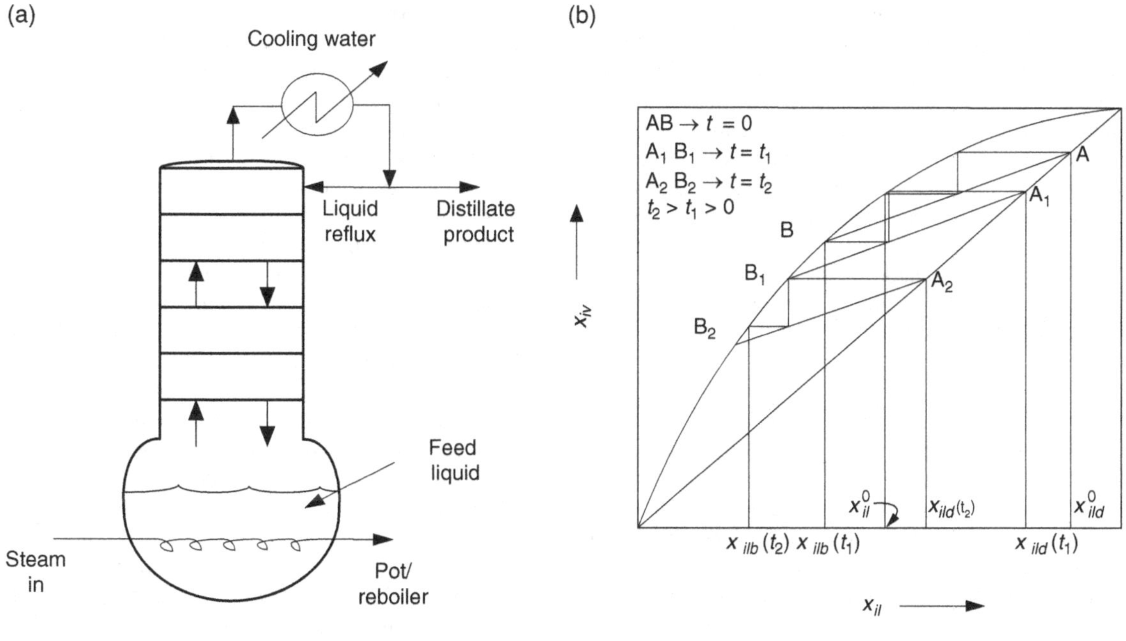

Figure 8.1.34. Batch distillation with reflux in a multiplate column. (a) Schematic of the device with total condensation, liquid reflux and distillate product. (b) McCabe–Thiele diagram for constant reflux ratio with enriching section operating lines for three times, t = 0, t = t_1, t = t_2.

An additional mode of operation of a batch still with a multiplate distillation column and total condensation involves no distillate product, total condensate reflux and a total reboiler, with fractions of bottom liquid taken out at different times. Various other combinations are possible (Diwekar, 1996; Seader and Henley, 1998; Wankat, 2007); these references also provide a window onto various calculation methods, especially for multicomponent systems.

8.1.4 Countercurrent solvent extraction

We have been introduced to various aspects of solvent extraction in the following sections: Section 3.3.7.2, liquid–liquid equilibria in aqueous–organic, organic–organic, aqueous–aqueous systems; Section 3.4.1.2, flux expressions in liquid–liquid systems; Section 3.4.3.2, solute transport in phase barrier membranes; Section 4.1.3, separation achieved in a closed vessel; Section 5.2.2, role of chemical reaction in liquid extraction; Section 5.3.2, rate controlled aspects of chemical reaction in liquid–liquid systems; Section 6.3.2.2, bulk flow parallel to force; Section 6.4.1.2, mixer–settler, CSTS system.

In this section, we will focus first on solvent extraction devices having countercurrent flow of the two immiscible phases. The devices are primarily dispersive, with one liquid phase dispersed generally as drops in the other immiscible liquid phase; however, the emerging nondispersive porous membrane based solvent extraction devices will also be described briefly. Next we will consider a countercurrent arrangement of a mixer–settler type of devices which achieve an overall countercurrent configuration of each stage operating as a CSTS. Such countercurrent cascades, as well as their continuous contact varieties, are used extensively in the following industrial applications: antibiotics extraction; hydrometallurgical applications, including those in the nuclear industry; use in the chemical and petrochemical industries; enzyme/protein separation in two-aqueous-phase systems. (See Treybal (1963, 1980, chap. 10); Benedict *et al.* (1981, chap. 4); Belter *et al.* (1988); Harrison *et al.* (2003, chap. 6); Seader and Henley (1998).)

8.1.4.1 *Continuous countercurrent extraction devices*

Almost all countercurrent extraction devices utilize dispersion of one immiscible phase as drops in another immiscible phase; we will provide a brief introduction here. At the end, we will introduce porous hollow fiber membrane based nondispersive countercurrent solvent extraction devices. The dispersive devices may involve continuous agitation or no agitation at all. Dispersive devices without any agitation as such are of three types: spray towers; packed towers; perforated plate towers. Spray towers were illustrated in Figure 8.1.2(b).

A packed tower for extraction is similar to that for towers for other two-phase systems, e.g. a gas–liquid absorber, except for the heavy liquid introduction and lighter liquid withdrawal (Figure 8.1.35(a)). However, the packing wetting properties require attention. It is necessary that the continuous phase wets the packing material so that the dispersed phase can move as droplets through the continuous phase; further, the dispersed-phase distributor should be located inside the packing. Generally organic liquids preferentially wet plastic packings, whereas aqueous liquids spontaneously wet ceramic packings. Normally the heavy liquid is the continuous phase and the dispersed phase is the lighter liquid rising up as droplets and coalescing at the top into a continuous phase. The possibility of flooding in these devices without agitation is high, especially since the driving force for motion of the droplets relies on the density difference between the two phases, which is at most around 0.08 g/cm^3.

Perforated-plate towers are similar to sieve-plate towers used in vapor–liquid/gas–liquid contacting (Figure 8.1.30(a)) except for the weir at the liquid outlet on a tray (Figure 8.1.35(b)). There are, however, important differences. The heavier liquid flows along a plate, comes down via a downspout to the next plate and so on. The lighter liquid is introduced as a continuous phase at the bottom of the last plate, it rises as drops through the plate perforations and then coalesces into a continuous phase below the plate above and so on. For systems with low interfacial tension, the plate tower works well without any mechanical agitation. The sieve-plate perforation diameters vary between 3 and 8 mm. Treybal (1980, pp. 521–526) provides considerable details regarding the drop diameters, the drop terminal velocities, the dispersed phase holdup, the sieve tray mass-transfer during drop formation, the drop rise, the drop coalescence and associated topics.

Figure 8.1.35(c) illustrates a mechanically agitated countercurrent extractor of the earliest version of a Scheibel extractor. Basically there is a central shaft with different kinds of impellers spaced at equal intervals with wire meshes, rings, etc., in between to provide, for example, drop coalescence and settling. These intermediate regions between impeller agitated regions are like the settler after the mixer (see Figure 6.4.8(d)). The dispersed liquid phase must wet the wire mesh material. An alternative configuration is provided by a Karr column, which is essentially a number of perforated plates mounted on a central shaft which goes up and down, say, twice per second. The reciprocating motion is carried out using plates that are 50%–60% open and fit loosely within the shell of the extractor column (Figure 8.1.35(d)).

We have described very briefly two general types of extractors: those requiring mechanical agitation and those without it. For systems where the liquid viscosities or the surface tensions are high, or the liquid-phase density differences are small, or all three conditions are present, it is

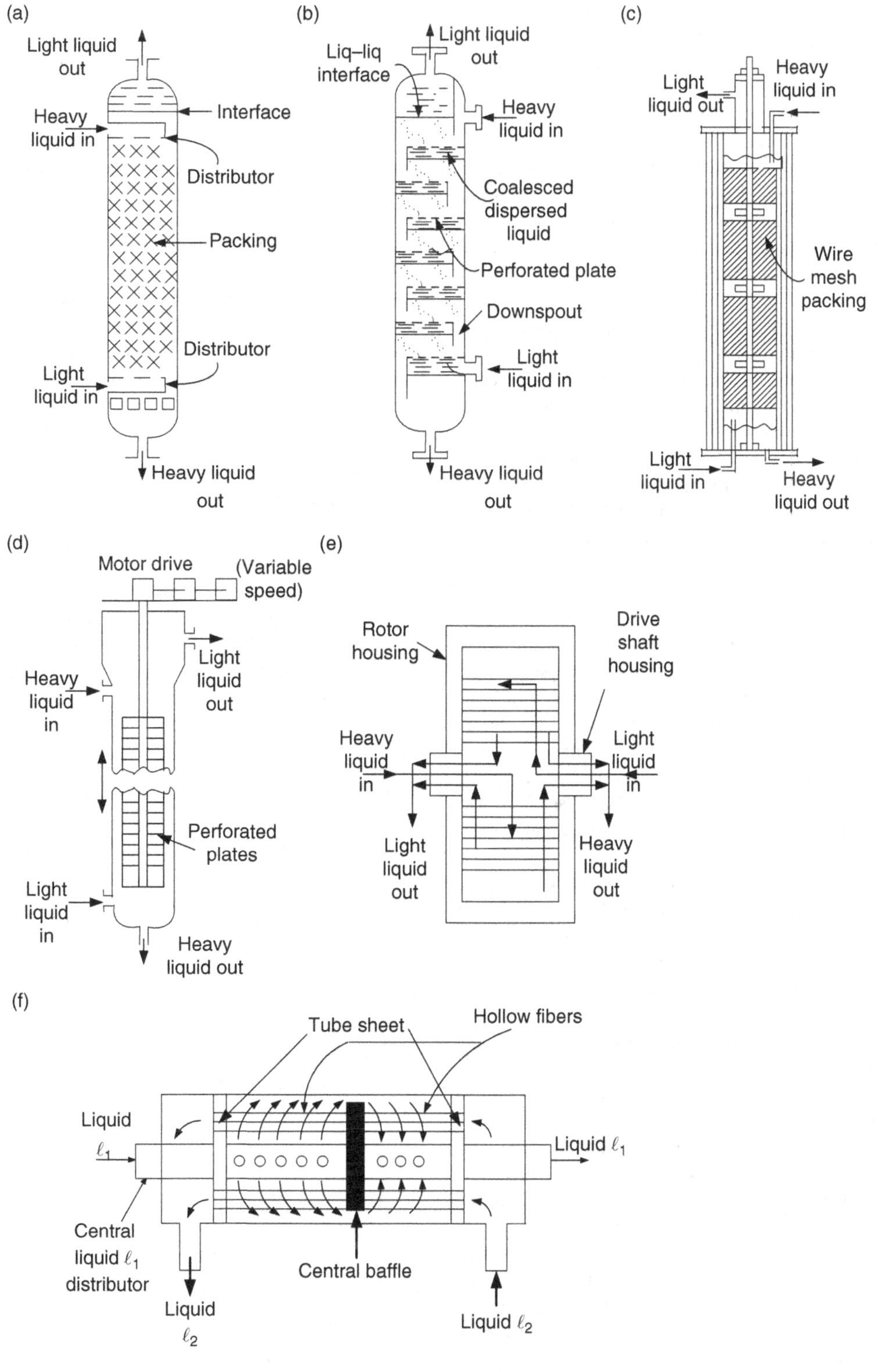

Figure 8.1.35. Schematic of various large-scale liquid–liquid extraction devices. (a) Packed tower for solvent extraction; (b) sieve-plate extraction column; (c) an early Scheibel column extraction design; (d) Karr column, in which the plates have reciprocating motions; (e) centrifugal extractor; (f) porous hollow fiber membrane solvent extraction device (see Figure 8.1.13(a) for a detailed design).

necessary to employ extractors having mechanical agitation to achieve appreciable dispersion. In all such devices, the possibility of significant backmixing exists, and the design and operating conditions should seek to reduce it. Attention should also be paid to the location of the liquid–liquid interface in nonagitated devices. As shown in Figure 8.1.35(a), the interface is at the top of the column with a continuous heavy liquid phase and dispersed light liquid phase. This is achieved by maintaining the heavy phase pressure at the column bottom at a certain level. A lower pressure level will lead to the lighter phase becoming the continuous phase.

There are liquid–liquid extraction systems where the residence time in the device should be very small (~10 s) to reduce degradation, for example in penicillin extraction carried out at a low temperature. Centrifugal devices (e.g. Podbielniak) are utilized for such systems. As illustrated schematically in Figure 8.1.35(e), this device contains a number of concentric perforated shells around a central shaft rotating rapidly around a central axis. The two liquids are introduced from opposite ends into different tubular channels in the shaft: the heavy liquid is introduced near the shaft, whereas the light liquid is introduced near the periphery.

Figure 8.1.35(f) illustrates a porous membrane based solvent extraction device. Figure 8.1.13(a) illustrated the basic design of such devices in greater detail. The basic principle of nondispersive contact of two immiscible liquid phases at the mouth of the pore of a membrane has been described earlier (see Figure 3.4.11, Section 3.4.3.2) (Kiani *et al.*, 1984; Prasad and Sirkar, 1988). Large-scale devices built based on such a principle and related patents (Sirkar, 1991, 1995) are being used. One of the liquids (liquid ℓ_1) is brought in through a central liquid distributor with circumferential perforations, which allows the liquid to flow out radially through the hollow fiber bundle. The device has a central baffle, which turns around the shell-side liquid ℓ_1, to the second half of the device, where this liquid flows radially inward in the fiber bundle to the central liquid distributor tube, which is blocked in the middle to insulate inlet liquid ℓ_1 from exiting liquid ℓ_1.

Liquid ℓ_2 is introduced through the bore of the porous hollow fibers from one end and flows out to the other side, in effective countercurrent flow to liquid ℓ_1. However, locally on both sides of the baffle, liquid ℓ_1 is in crossflow around each hollow fiber and therefore around liquid ℓ_2. Since crossflow introduces very efficient liquid-phase mass transfer at a low Reynolds number in such systems, it is preferable to introduce that liquid phase to the shell side whose resistance is likely to control the mass-transfer/solvent extraction rate. Further, the liquid phase wetting the pores of the membrane should prefereably be the lower resistance phase. These considerations are described in detail in Prasad and Sirkar (2001) for both porous hydrophobic as well as porous hydrophilic membranes.

A most important additional aspect of such devices is that, as long as the phase interfaces are immobilized via appropriate pressure/wetting conditions, one can have a very wide range of flow rate ratios between the two phases. There is no need for any density difference between the phases. The issue of flooding does not arise, emulsification is unlikely to arise, and the need for coalescence is absent. However, surfactant impurities, if present, could interfere with interface immobilization. Further, the solvents must not swell the membrane very much. Therefore the compatibility of the membrane with the solvents to be used should be checked. Smaller pore membranes will lead to a broader range of pressure difference between the two phases for nondispersive operation. The value of $K_{\ell}a$ for such devices can be larger than conventional devices by 5–50 times.

8.1.4.2 *Extraction of a dilute solute between immiscible solvents in a continuous device*

We will now consider a dilute solution of solute i in a countercurrent continuous-contact solvent extraction system consisting of feed and extract phases that are essentially insoluble in each other (Figure 8.1.36). We can now ignore the variation in the flow rate of either phase along the length of the counterflowing extraction column. The situation is similar to that of a steady state gas absorption system for a dilute species in a gas stream. The general

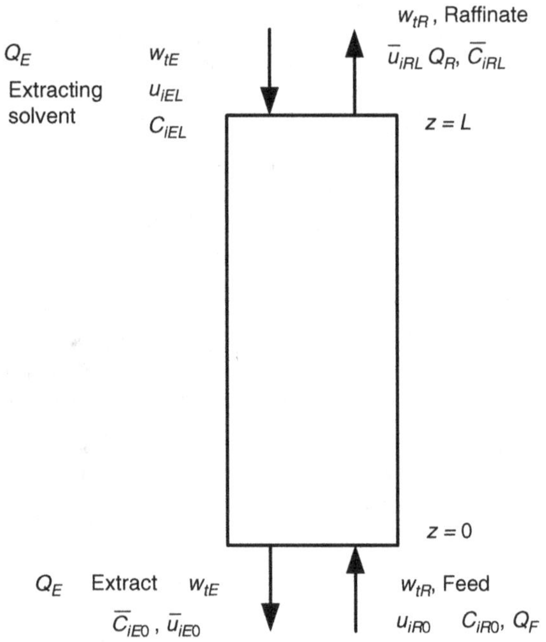

Figure 8.1.36. Countercurrent solvent extraction in a column with continuous contact.

equations for such dilute systems are (8.1.2a,b) for any two immiscible phases and (8.1.90a,b) for gas absorption systems. Here we first describe them in terms of molar concentrations for phase $j = R$ (feed, raffinate) and $j = E$ (extract, solvent) at steady state as (phase $j = R$ moving in the positive z-direction):

$$\varepsilon_R D_{i,\mathrm{eff},R} \frac{\mathrm{d}^2 \overline{C}_{iR}}{\mathrm{d}z^2} - \langle v_{Rz} \rangle \frac{\mathrm{d}\overline{C}_{iR}}{\mathrm{d}z} - K_{iRc} a \left(\overline{C}_{iR} - \overline{C}^*_{iR} \right) = 0; \tag{8.1.249}$$

$$\varepsilon_E D_{i,\mathrm{eff},E} \frac{\mathrm{d}^2 \overline{C}_{iE}}{\mathrm{d}z^2} + |\langle v_{Ez} \rangle| \frac{\mathrm{d}\overline{C}_{iE}}{\mathrm{d}z} + K_{iEc} a \left(\overline{C}^*_{iE} - \overline{C}_{iE} \right) = 0, \tag{8.1.250}$$

where $\overline{C}^*_{iR}$ is the solute concentration in raffinate in equilibrium with that in the bulk extract. Similarly, $\overline{C}^*_{iE}$ is the solute concentration in the extract in equilibrium with the bulk raffinate. (*Note*: In general, an organic liquid is lighter and would rise up vertically. The schematic in Figure 8.1.36 therefore involves an organic feed and an aqueous extract phase.) To start with, we will ignore axial dispersion for a simplified analysis. We obtain

$$\langle v_{Rz} \rangle \frac{\mathrm{d}\overline{C}_{iR}}{\mathrm{d}z} = -K_{iRc} a \left(\overline{C}_{iR} - \overline{C}^*_{iR} \right); \tag{8.1.251a}$$

$$|\langle v_{Ez} \rangle| \frac{\mathrm{d}\overline{C}_{iE}}{\mathrm{d}z} = -K_{iEc} a \left(\overline{C}^*_{iE} - \overline{C}_{iE} \right). \tag{8.1.251b}$$

In mass concentration units, these are reduced to

$$\langle v_{Rz} \rangle \frac{\mathrm{d}\overline{\rho}_{iR}}{\mathrm{d}z} = -K_{iR\rho} a \left(\overline{\rho}_{iR} - \overline{\rho}^*_{iR} \right); \tag{8.1.252a}$$

$$|\langle v_{Ez} \rangle| \frac{\mathrm{d}\overline{\rho}_{iE}}{\mathrm{d}z} = -K_{iE\rho} a \left(\overline{\rho}^*_{iE} - \overline{\rho}_{iE} \right). \tag{8.1.252b}$$

Composition in solvent extraction processes is normally described by mass fractions instead of mole fractions. Assuming that the total mass concentrations ρ_{tR} and ρ_{tE} are essentially constant throughout the respective phases, we can express the above relations in terms of solute mass fractions $\overline{u}_{iR}$ and $\overline{u}_{iE}$ of the raffinate and the extract phases, respectively:

$$\rho_{tR} \langle v_{Rz} \rangle \frac{\mathrm{d}\overline{u}_{iR}}{\mathrm{d}z} = -K_{iR\rho} \rho_{tR} a \left(\overline{u}_{iR} - \overline{u}^*_{iR} \right) = -K_{iRu} a \left(\overline{u}_{iR} - \overline{u}^*_{iR} \right); \tag{8.1.253a}$$

$$\rho_{tE} |\langle v_{Ez} \rangle| \frac{\mathrm{d}\overline{u}_{iE}}{\mathrm{d}z} = -K_{iE\rho} \rho_{tE} a \left(\overline{u}^*_{iE} - \overline{u}_{iE} \right) = -K_{iEu} a \left(\overline{u}^*_{iE} - \overline{u}_{iE} \right). \tag{8.1.253b}$$

Here

$$\overline{\rho}_{iR} = \rho_{tR} \overline{u}_{iR}, \quad \overline{\rho}_{iE} = \rho_{tE} \overline{u}_{iE}, \quad K_{iRu} = K_{iR\rho} \rho_{tR}; \tag{8.1.254a}$$

$$K_{iEu} = K_{iE\rho} \rho_{tE}; \quad \overline{u}_{iE} = m_i \overline{u}^*_{iR}, \quad \overline{u}^*_{iE} = m_i \overline{u}_{iR}, \tag{8.1.254b}$$

m_i being the distribution coefficient between the extract phase and the raffinate phase for composition in mass fraction units. The raffinate phase undergoes the following composition changes:

$$z = 0, \ \overline{u}_{iR} = u_{iR0}; \qquad z = L, \ \overline{u}_{iR} = \overline{u}_{iRL}. \tag{8.1.255}$$

Integrate now equation (8.1.253a) from $z = 0$ to $z = L$:

$$\int_0^L \mathrm{d}z = L = -\frac{\rho_{tR} \langle v_{Rz} \rangle}{K_{iRu} a} \int_{\overline{u}_{iR0}}^{\overline{u}_{iRL}} \frac{\mathrm{d}\,\overline{u}_{iR}}{\left(\overline{u}_{iR} - \overline{u}^*_{iR} \right)}. \tag{8.1.256}$$

Integration of the right-hand side requires expressing $\overline{u}^*_{iR}$, therefore $\overline{u}_{iE}$, in terms of $\overline{u}_{iR}$. That can be implemented via the equation of an operating line. Equations (8.1.252a) and (8.1.252b) lead to

$$\langle v_{Rz} \rangle \frac{\mathrm{d}\overline{\rho}_{iR}}{\mathrm{d}z} = |\langle v_{Ez} \rangle| \frac{\mathrm{d}\overline{\rho}_{iE}}{\mathrm{d}z} \tag{8.1.257}$$

since the rate of solute loss from the feed/raffinate is equal to that gained by the extract. In terms of mass fractions, we get

$$\langle v_{Rz} \rangle \rho_{tR} \frac{\mathrm{d}\overline{u}_{iR}}{\mathrm{d}z} = |\langle v_{Ez} \rangle| \rho_{tE} \frac{\mathrm{d}\overline{u}_{iE}}{\mathrm{d}z}. \tag{8.1.258}$$

Multiplication by A_c, the column cross-sectional area, yields

$$\langle v_{Rz} \rangle \rho_{tR} A_c \frac{\mathrm{d}\overline{u}_{iR}}{\mathrm{d}z} = w_{tR} \frac{\mathrm{d}\overline{u}_{iR}}{\mathrm{d}z} = |\langle v_{Ez} \rangle| \rho_{tE} A_c \frac{\mathrm{d}\overline{u}_{iE}}{\mathrm{d}z} = |w_{tE}| \frac{\mathrm{d}\overline{u}_{iE}}{\mathrm{d}z}. \tag{8.1.259}$$

Here, w_{tR} and $|w_{tE}|$ are the mass flow rates of the raffinate and the extract phase, respectively. Integrating this equation leads to

$$\overline{u}_{iE} = \frac{w_{tR}}{|w_{tE}|} \overline{u}_{iR} + \text{constant}. \tag{8.1.260}$$

At $z = L$, $\overline{u}_{iE} = u_{iEL}$, $\overline{u}_{iR} = \overline{u}_{iRL}$

$$\Rightarrow \text{constant} = u_{iEL} - \frac{w_{tR}}{|w_{tE}|} = \overline{u}_{iRL}. \tag{8.1.261}$$

Therefore,

$$\overline{u}_{iE} = u_{iEL} + \frac{w_{tR}}{|w_{tE}|} \left(\overline{u}_{iR} - \overline{u}_{iRL} \right), \tag{8.1.262}$$

which is the equation for the operating line. Alternatively,

$$\overline{u}^*_{iR} = \left\{ u_{iEL} + \frac{w_{tR}}{|w_{tE}|} \left(\overline{u}_{iR} - \overline{u}_{iRL} \right) \right\} \Big/ m_i. \tag{8.1.263}$$

It is now possible to integrate the right-hand side of equation (8.1.256) using the above expression for $\overline{u}^*_{iR}$ in terms of $\overline{u}_{iR}$:

$$L = -\frac{\langle v_{Rz} \rangle \rho_{tR}}{K_{iRu} a} \int_{\overline{u}_{iR0}}^{\overline{u}_{iRL}} \frac{\mathrm{d}\,\overline{u}_{iR}}{\left(\overline{u}_{iR} - \frac{1}{m_i} \left\{ u_{iEL} + \frac{w_{tR}}{|w_{tE}|} \left(\overline{u}_{iR} - \overline{u}_{iRL} \right) \right\} \right)}; \tag{8.1.264}$$

$$L = -\frac{\langle v_{Rz}\rangle \rho_{tR}}{K_{iRu}a} \int_{\overline{u}_{iR0}}^{\overline{u}_{iRL}} \frac{d\overline{u}_{iR}}{\overline{u}_{iR}\left(1 - \frac{w_{tR}}{|w_{tE}|}\frac{1}{m_i}\right) - \frac{1}{m_i}\left\{u_{iEL} - \frac{w_{tR}}{|w_{tE}|}\overline{u}_{iRL}\right\}}; \tag{8.1.265}$$

$$L = \left[\frac{w_{tR}}{K_{iRu}aS_c}\right] \frac{1}{\left(1 - \frac{w_{tR}}{|w_{tE}|}\frac{1}{m_i}\right)} \ell n \left[\frac{\overline{u}_{iR0}\left(1 - \frac{w_{tR}}{|w_{tE}|}\frac{1}{m_i}\right) - \frac{1}{m_i}\left\{u_{iEL} - \frac{w_{tR}}{|w_{tE}|}\overline{u}_{iRL}\right\}}{\overline{u}_{iRL}\left(1 - \frac{w_{tR}}{|w_{tE}|}\frac{1}{m_i}\right) - \frac{1}{m_i}\left\{u_{iEL} - \frac{w_{tR}}{|w_{tE}|}\overline{u}_{iRL}\right\}}\right]. \tag{8.1.266}$$

Using relation (8.1.262), at $z = 0$ we can simplify the above expression to

$$L = \underbrace{\left[\frac{w_{tR}}{K_{iRu}aS_c}\right]}_{H_{oR}} \underbrace{\left[\frac{1}{\left(1 - \frac{w_{tR}}{|w_{tE}|}\frac{1}{m_i}\right)}\right] \ell n \left[\frac{u_{iR0} - (\overline{u}_{iE0}/m_i)}{\overline{u}_{iRL} - (u_{iEL}/m_i)}\right]}_{N_{oR}} \tag{8.1.267}$$

where

$H_{oR} = w_{tR}/(K_{iRu}\,a\,S_c)$
= height of overall transfer unit based on raffinate phase

and

$$N_{oR} = \int_{\overline{u}_{iR0}}^{\overline{u}_{iRL}} \frac{\overline{u}_{iRL}\, d\overline{u}_{iR}}{(\overline{u}_{iR} - \overline{u}_{iR}^*)} = \left[\frac{1}{\left(1 - \frac{w_{tR}}{|w_{tE}|}\frac{1}{m_i}\right)}\right] \ell n \left[\frac{u_{iR0} - (\overline{u}_{iE0}/m_i)}{\overline{u}_{iRL} - (\overline{u}_{iEL}/m_i)}\right]$$

= number of overall transfer units based on raffinate phase. (8.1.268)

The quantity $(|w_{tE}|\, m_i/w_{tR})$ is called the *extraction factor, E.*

Example 8.1.14 A pharmaceutically active compound present in a low concentration in an aqueous solution is to be extracted into an organic solvent. The m_i value is known to be 12. The extraction is to be carried out from the feed aqueous solution in a packed tower of diameter 11 cm. Earlier experiments in the packed tower indicated that the $K_{iru}a$ value for this compound is $4 \times 10^{-4}\ s^{-1}$. The feed aqueous solution flow rate is 7000 cm^3/hr. The solvent flow rate to achieve the given mass-transfer conditions is 3500 cm^3/hr. Determine the height of the column needed to achieve 95% recovery of the compound. The organic solvent enters solute-free.

Solution We will employ equation (8.1.267) to determine L. We are given $w_{tR} = 7000\,\text{cm}^3/\text{hr} = 1.945\,\text{cm}^3/\text{s}$; $K_{iRu}a = 4 \times 10^{-4}\ s^{-1}$; $|w_{tE}| = 3500\,\text{cm}^3/\text{hr} = 0.972\,\text{cm}^3/\text{s}$; $m_i = 12$. However, values of $\overline{u}_{iR0}$, $\overline{u}_{iRL}$, $\overline{u}_{iE0}$ and u_{iEL} are not known, except $u_{iEL} = 0$. We can, however, express the unknown $\overline{u}_{iE0}$ via a mass balance in terms of the fractional recovery of the compound:

$$\begin{aligned} |w_{tE}|\overline{u}_{iE0} &= w_{tR}\,(u_{iR0} - \overline{u}_{iRL}) \\ &\Rightarrow (\overline{u}_{iE0}/u_{iR0}) = (w_{tR}/|w_{tE}|)(1 - (\overline{u}_{iRL}/u_{iRo})) \\ &\Rightarrow (\overline{u}_{iE0}/u_{iR0}) = (1.945/0.972)\,(1 - 0.05) = 1.90. \end{aligned}$$

The value of S_c, the column cross-sectional area, is $(\pi/4)(11)^2 = 95\,\text{cm}^2$. Put these numbers into equation (8.1.267):

$$\begin{aligned} L &= \left[\frac{1.945\,\text{cm}^3/\text{s}}{4 \times 10^{-4}(\text{s}^{-1}) \times 95\,\text{cm}^2}\right] \times \left[\frac{2.303}{1 - \frac{1.945}{0.972} \times \frac{1}{12}} \log\left\{\frac{1 - \left(\frac{\overline{u}_{iE0}}{m_i u_{iR0}}\right)}{\frac{\overline{u}_{iRL}}{u_{iR0}} - 0}\right\}\right] \\ &= \left[\frac{1.945 \times 10^4}{4 \times 95}\right]\left[\frac{2.303}{1 - 0.166}\log\left\{\frac{1 - \frac{1.90}{12}}{0.05}\right\}\right] \\ &= 51.18 \times \frac{2.303}{0.834}\log\left\{\frac{0.842}{0.05}\right\} \\ &= 130.7\log(16.84) = 130.7 \times 1.226 = 160.2\,\text{cm}. \end{aligned}$$

In countercurrent continuous solvent extraction in a column, availability of $K_{iRu}a$ is crucial. There are two parts to this quantity: the mass-transfer coefficient and the interfacial area per unit device volume. Once both are available, determining the column height is relatively easy. However, as pointed out earlier, determination of $K_{iRu}a$ in dispersive solvent extraction is highly demanding. We will now illustrate a *simplified determination* of $K_{iRc}\,a$ in a nondispersive membrane solvent extraction device. The device uses porous hydrophobic hollow fibers (Figure 3.4.11) in a simple countercurrent flow format - one fluid flowing through the bores of hollow fibers while the other fluid flows countercurrently on the shell side. There are no baffles, unlike that shown in Figure 8.1.35(f). The local value of the overall organic-phase based mass-transfer coefficient K_{iEo} for solvent extraction from an aqueous phase may be described via (3.4.134) as

$$\frac{1}{K_{iEo}} = \frac{\kappa_{io}}{k_{iRw}} + \frac{1}{k_{imo}} + \frac{1}{k_{iEo}} \tag{8.1.269}$$

For an elementary analysis over the length of the membrane extractor, we may use a length-averaged form:

$$\frac{1}{\overline{K}_{iEo}} = \frac{\kappa_{io}}{\overline{k}_{iRw}} + \frac{1}{\overline{k}_{imo}} + \frac{1}{\overline{k}_{iEo}} \tag{8.1.270}$$

The overall length-averaged mass-transfer coefficient $\overline{K}_{iEo}$ is defined via a logarithmic concentration difference

$$\Delta C_i\big|_{\ell m} = \frac{\Delta C_{i1} - \Delta C_{i2}}{\ell n(\Delta C_{i1}/\Delta C_{i2})} \tag{8.1.271}$$

(see Figure 8.1.36), where

$$\Delta C_{i1} = \kappa_{i1}\, C_{iR0} - \overline{C}_{iE0}; \qquad (8.1.272)$$

$$\Delta C_{i2} = \kappa_{i1}\, \overline{C}_{iRL} - C_{iEL}. \qquad (8.1.273)$$

The membrane mass-transfer coefficient defined by (3.4.139b) for a hollow fiber of inside and outside diameter of d_i and d_o, respectively, is

$$k_{imo} = \frac{D_{io}\,\varepsilon_m}{\tau_m (d_o - d_i)/2}, \qquad (8.1.274)$$

with the organic phase having solute diffusivity of D_{io} occupying the pores. Let the organic extract phase flow on the shell side and the aqueous feed phase flow on the tube side. The shell-side film mass-transfer coefficient $\overline{k}_{iEo}$ in a nonbaffled countercurrent extractor may be described by (Prasad and Sirkar, 1988)

$$\overline{k}_{iEo} = \frac{D_{io}}{d_h}\beta\,[d_h\,(1-\phi)/L]\, Re_o^{0.6}\, Sc_o^{0.33}, \qquad (8.1.275)$$

where d_h is the hydraulic diameter on the shell side (= 4 × flow cross-sectional area/wetted perimeter), ϕ is the fiber packing fraction on the shell side, Re_o is the shell-side Reynolds number, $d_h v_o/\nu_o$, where v_o is the actual velocity of the organic phase of kinematic viscosity ν_o and Sc_o is ν_o/D_{io}. The fiber bore side mass-transfer coefficient $\overline{k}_{iRw}$ may be described by the Leveque solution (3.1.145) (subject to $Gz = (\pi d_i\, Re_i\, Sc_i/4L) \geq 400$):

$$\overline{k}_{iRw} = \frac{1.615\, D_{iw}}{d_i}\,[(d_i/L)\, Re_i\, Sc_i]^{0.33}. \qquad (8.1.276)$$

Here

$$Re_i = d_i\, v_w/\nu_w;\; Sc_i = \nu_w/D_{iw}. \qquad (8.1.277)$$

For $Gz < 400$, see Prasad and Sirkar (1988).

8.1.4.3 *Extraction of a dilute solute between immiscible solvents in a continuous countercurrent multistage device/cascade*

Often solvent extraction is carried out continuously in a countercurrent multistage device/cascade. The sieve-plate tower is an example of one multistage device, whereas Figure 8.1.37(a) illustrates a multistage arrangement of N mixer-settler devices (one such device is studied in Section 6.4.1.2). First, we analyze a dilute solution of solute i in a feed-extract phase system assumed to be essentially insoluble in each other (Cussler, 1997). Then we will analyze extraction systems with some mutual solubility of the feed and the extraction solvent.

Consider the cascade of Figure 8.1.37(a) and the solute i satisfying the linear equilibrium relation (8.1.254b), for example,

$$u_{iE} = m_i\, u_{iR}, \qquad (8.1.278)$$

employing solute mass fractions in the extract and the raffinate phases. The exiting streams from each stage are

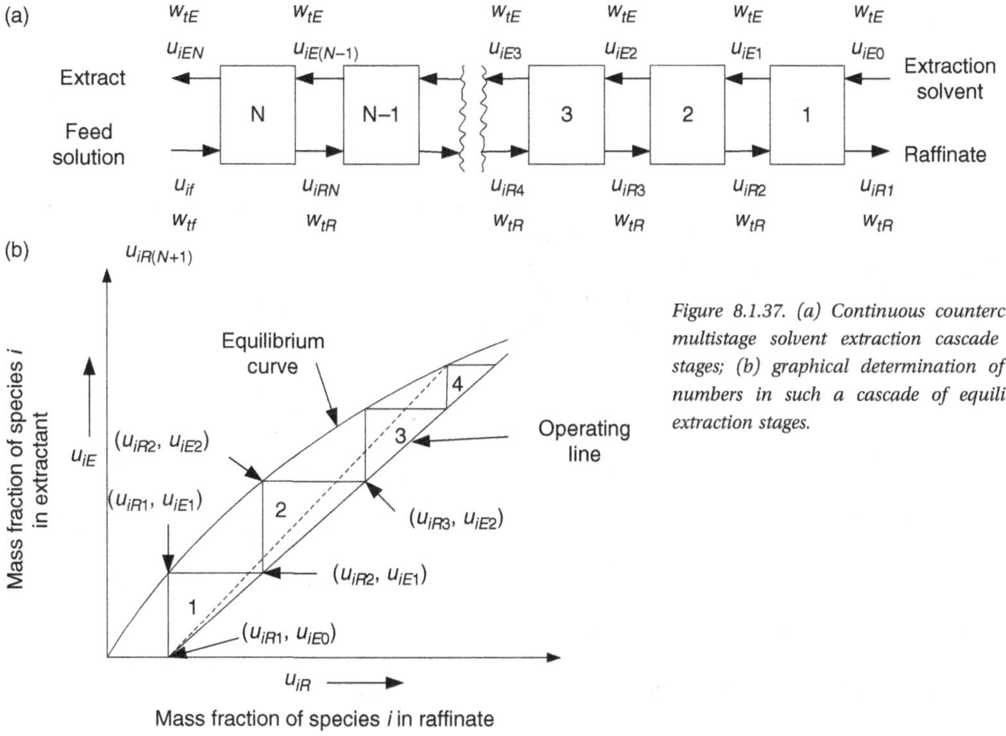

Figure 8.1.37. (a) Continuous countercurrent multistage solvent extraction cascade of N *stages; (b) graphical determination of stage numbers in such a cascade of equilibrium extraction stages.*

in equilibrium. Our objective is to relate the raffinate composition u_{iR1} to feed composition u_{if} or $u_{iR(N+1)}$. Consider species i balance over stage 1:

$$w_{tR}\, u_{iR2} + w_{tE}\, u_{iE0} = w_{tR}\, u_{iR1} + w_{tE}\, u_{iE1}. \tag{8.1.279}$$

A similar balance over stage 2 yields

$$w_{tR}\, u_{iR3} + w_{tE}\, u_{iE1} = w_{tR}\, u_{iR2} + w_{tE}\, u_{iE2}. \tag{8.1.280}$$

Substitute for u_{iE1} in (8.1.279) from relation (8.1.278) and $E = (m_i\, w_{tE}/w_{tR})$ to obtain

$$u_{iR2} = u_{iR1}\,(1+E) - \frac{w_{tE}}{w_{tR}}\, u_{iE0} = u_{iR1}\,(1+E) - \frac{E}{m_i}\, u_{iE0}. \tag{8.1.281}$$

Utilize this relation now in the mass balance relation (8.1.280) for stage 2 to obtain

$$u_{iR3} = u_{iR1}\left(1+E+E^2\right) - \frac{E}{m_i}\, u_{iE0}\,(1+E). \tag{8.1.282}$$

A mass balance relation for stage 3 is

$$w_{tR}\, u_{iR4} + w_{tE}\, u_{iE2} = w_{tR}\, u_{iR3} + w_{tE}\, u_{iE3}, \tag{8.1.283}$$

which yields

$$u_{iR4} = u_{iR1}\left(1+E+E^2+E^3\right) - \frac{E}{m_i}\, u_{iE0}\left(1+E+E^2\right). \tag{8.1.284}$$

On employing the process of induction, we can write for stage N, and therefore for the feed mass fraction $u_{if} = u_{iR(N+1)}$,

$$u_{iR(N+1)} = u_{if} = \left(1+E+E^2+\cdots+E^N\right) u_{iR1} - \frac{E\, u_{iE0}}{m_i}\left(1+E+E^2+\cdots+E^{N-1}\right); \tag{8.1.285}$$

$$u_{iR(N+1)} = \left[\frac{E^{N+1}-1}{E-1}\right] u_{iR1} - \left[\frac{E^N-1}{E-1}\right]\frac{E\, u_{iE0}}{m_i}. \tag{8.1.286}$$

This result is analogous to expression (8.1.127) developed for countercurrent multistage gas absorption, known as the Kremser equation; it is also valid for distillation. If the extraction solvent entering stage 1 is solute-free ($u_{iE0} = 0$), we have a simpler result:

$$u_{iR(N+1)} = \left[\frac{E^{N+1}-1}{E-1}\right] u_{iR1}. \tag{8.1.287a}$$

Therefore the fractional extraction of solute i from the feed in the N-stage cascade is

$$\left(1 - \frac{u_{iR1}}{u_{iR(N+1)}}\right) = \frac{E^{N+1}-E}{E^{N+1}-1}. \tag{8.1.287b}$$

In the purification of antibiotics, for example, first the antibiotic is extracted into a solvent from the aqueous feed at a low pH via a cascade, as illustrated in Figure 8.1.37(a). Then it is back extracted into an aqueous solution at a higher pH. Figure 8.1.37(a) can also represent a countercurrent arrangement where the feed is an organic solvent and the extract phase is an aqueous solution. Results (8.1.287a,b) are still valid; however, in the definition of the extraction factor, namely $|w_{tE}|m_i/w_{tR}$, the raffinate and extract phase flow rates are, respectively,

$$w_{tR} = w_{to},\ |w_{tE}| = w_{tw}; \tag{8.1.288a}$$

further

$$m_i = (\text{aqueous phase wt. fraction}/\text{organic phase wt. fraction}) = (u_{iw}/u_{io}), \tag{8.1.288b}$$

and

$$u_{iR(N+1)} = u_{if}\big|_{\text{org feed}} = \left[\frac{E^{N+1}-1}{E-1}\right] u_{iR1}, \tag{8.1.289}$$

where $u_{iR1} = u_{io}\big|_{\text{stripped org}}$.

An important goal in many solvent extraction applications is the purification of the desired solute i vis-à-vis another solute j present in the feed. One could employ various separation indices to determine the extent of purification achieved. If we employ the definition of separation factor α_{ij} between the extract phase ($j = E$) and the feed phase ($j = f$), then

$$\alpha_{ij} = \frac{x_{iE}\, x_{jf}}{x_{if}\, x_{jE}} = \frac{(x_{iE}/x_{if})}{(x_{jE}/x_{jf})}; \tag{8.1.290}$$

$$\alpha_{ij} \cong \frac{(u_{iE}/u_{if})}{u_{jE}/u_{jf}} = \frac{(w_{tE}\, u_{iE}/w_{tf}\, u_{if})}{(w_{tE}\, u_{jE}/w_{tf}\, u_{jf})}. \tag{8.1.291}$$

From relation (8.1.287b), we can write

$$\left(1 - \frac{w_{tR}\, u_{iR1}}{w_{tR}\, u_{iR(N+1)}}\right) = \left(1 - \frac{w_{tf}\, u_{iR1}}{w_{tf}\, u_{iR(N+1)}}\right) = \frac{E_i^{N+1}-E_i}{E_i^{N+1}-1} = \frac{w_{tE}\, u_{iEN}}{w_{tf}\, u_{iR(N+1)}}, \tag{8.1.292}$$

where $E_i = |w_{tE}|m_i/w_{tR}$ and similarly for E_j. Therefore, for solutes i and j,

$$\alpha_{ij} = \frac{(E_i^{N+1}-E_i)/(E_j^{N+1}-E_j)}{(E_i^{N+1}-1)/(E_j^{N+1}-1)} = \text{separation factor}. \tag{8.1.293}$$

Example 8.1.15

(a) Solvent extraction of weakly acidic penicillin type antibiotics is illustrated in Figure 5.2.4(b); at a $pH = 3$, the partition coefficient κ_{i1} of penicillin G antibiotic is 25. From a filtered fermentation broth containing the antibiotic at the level of 0.3 g/liter and flowing at 300 liter/hr, the antibiotic is to be extracted into the solvent of Figure 5.2.4(b) flowing at 30 liter/hr. Determine the number of ideal stages required to achieve 98% recovery of this antibiotic in a countercurrent extraction cascade via analytical means. Assume that the extracting solvent has no antibiotic in it.

(b) Determine the concentration of the antibiotic in the extract phase from the cascade of part (a). This antibiotic is to be now back extracted from this solvent extract into water at $pH = 5$, where $\kappa_{i1} = 0.2$ (Figure 5.2.4(b)). The water flow rate is 8 liter/hr; 90% recovery of the antibiotic into the water is needed. How many equilibrium stages are needed in a countercurrent extraction cascade? Use analytical methods. What is the concentration of the antibiotic in the aqueous back extract?

Solution (a) Focusing on Figure 8.1.37(a), we need to know the number of stages required to go from u_{if} (feed, $u_{iR(N+1)}$) to u_{iR1} (raffinate) to achieve 98% recovery. Therefore

$$\left(u_{iR1}/u_{iR(N+1)}\right) = 1 - 0.98 = 0.02.$$

We should now determine the value of the extraction factor E, $(((|w_{tE}|)m_i)/w_{tR})$, where m_i has been defined by (8.1.254b). Determine first the value of m_i from the given value of $\kappa_{i1} = 25$:

$$\kappa_{i1} = \frac{C_{i1}}{C_{i2}} = \frac{C_{i\text{organic}}}{C_{i\text{water}}} = \frac{C_{io}}{C_{iw}} = \frac{C_{iE}}{C_{iR}} = \frac{\rho_{iE}}{\rho_{iR}} \cong \frac{u_{iE}}{u_{iR}} = \frac{u_{io}}{u_{iw}} = m_i.$$

This result is an assumption since we do not know the solvent and its molecular weight. Since $w_{tR} = 300$ liter/hr (feed is filtered broth) and $w_{tE} = 30$ liter/hr (solvent),

$$E \cong \frac{30 \times 25}{300} = 2.5.$$

(Here also we are not exact since the volumetric flow rate ratio is not equivalent to the mass flow rate ratio due to the aqueous phase being in general somewhat heavier than the organic extract phase.)

Employ now relation (8.1.287a):

$$\frac{u_{iR(N+1)}}{u_{iR1}} = \left[\frac{E^{N+1} - 1}{E - 1}\right] = \frac{1}{0.02} = 50 = \left[\frac{(2.5)^{N+1} - 1}{2.5 - 1}\right] \Rightarrow 76$$
$$= (2.5)^{N+1} \Rightarrow N + 1 = 4.72 \Rightarrow N = 3.74\text{~}4.$$

Therefore four ideal extraction stages are required.

(b) We employ equation (8.1.289):

$$u_{if} = \left[\frac{E^{N+1} - 1}{E - 1}\right] u_{iR1}; \qquad \frac{u_{iR1}}{u_{if}} = \frac{u_{io}|_{\text{stripped feed}}}{u_{io}|_{\text{feed}}} = 0.1;$$

$$E = \frac{|w_{tw}|}{w_{to}}\frac{(u_{iw})}{(u_{io})}; \qquad \kappa_{i1} \cong \frac{(u_{io})}{(u_{iw})} \cong 0.2;$$

$$(w_{to}/w_{tw}) = \frac{30}{8} = 3.75; \qquad E = \frac{5.0}{3.75} = 1.334;$$

$$\frac{u_{if}}{u_{iR1}} = \frac{1}{0.1} = 10 = \left[\frac{(1.334)^{N+1} - 1}{1.334 - 1}\right] = \frac{(1.334)^{N+1} - 1}{0.334}$$

$$\Rightarrow (1.334)^{N+1} = 4.34 \Rightarrow N = 4.1,$$

suggesting that five equilibrium stages are needed for back extraction. The concentration of antibiotic in the organic feed from part (a) is

$$\text{concentration}|_{\text{organic feed}} = \frac{300\,\text{liter/hr} \times 0.3\,\text{g/liter} \times 0.98}{30\,\text{liter/hr}}$$
$$= 2.94\,\text{g/liter};$$

$$\text{concentration}|_{\text{aqueous extract}} = \frac{2.94\,\text{g/liter} \times 30\,\text{liter/hr} \times 0.90}{8\,\text{liter/hr}}$$
$$= 9.92\,\text{g/liter}.$$

Example 8.1.16 Obtain an analytical expression for the number of equilibrium stages N required in a countercurrent solvent extraction cascade in terms of the extraction factor E and the fractional solute recovery $\left(1 - \left(u_{iR1}/u_{iR(N+1)}\right)\right)$. Assume that the extracting solvent has zero solute concentration, the equilibrium distribution is linear and the two phases are immiscible.

Solution For the conditions specified, the equation under consideration is

$$u_{iR(N+1)} = \left[\frac{E^{N+1} - 1}{E - 1}\right] u_{iR1} \Rightarrow E^{N+1} = \frac{u_{iR(N+1)}}{u_{iR1}}(E - 1) + 1$$
$$\Rightarrow (N + 1)\,\ell n\, E = \ell n\left(\frac{u_{iR(N+1)}}{u_{iR1}}(E - 1) + 1\right);$$

$$N = \frac{\ell n\left(\frac{u_{iR(N+1)}}{u_{iR1}}(E - 1) + 1\right)}{\ell n E} - 1.$$

The fractional solute recovery is

$$\left(1 - \frac{u_{iR1}}{u_{iR(N+1)}}\right) = fr.$$

Therefore

$$N = \frac{\ell n\left(\frac{1}{1 - fr}(E - 1) + 1\right)}{\ell n E} - 1.$$

The analysis and examples illustrated above involve a linear equilibrium relation (8.1.278). Often the solute partitioning relation is nonlinear; this is especially true when chemical complexation (Section 5.2.2) takes place. Graphical methods are frequently utilized in such situations to solve the problem of determining the number N of ideal stages. We briefly illustrate the graphical method used via Figure 8.1.37(b). Plot the equilibrium relation first in terms of the solute i mass fraction in extract (u_{iE}) vs. that in the raffinate u_{iR}. Next we follow the procedure adopted earlier for gas–liquid absorption (Figure 8.1.18) and distillation (Figure 8.1.20) via the McCabe–Thiele method. Obtain the equation for the operating line by a solute i mass balance over an envelope covering the $N - 1$ stage to stage 1:

$$w_{tR}\, u_{iRN} + w_{tE}\, u_{iE0} = w_{tR}\, u_{iR1} + w_{tE}\, u_{iE(N-1)}. \qquad (8.1.294)$$

Rearrange it as follows:

$$u_{iE(N-1)} = \frac{w_{tR}}{w_{tE}} u_{iRN} + u_{iE0} - \frac{w_{tR}}{w_{tE}} u_{iR1}. \qquad (8.1.295)$$

This operating line equation plots the two immiscible phase compositions at any level of the cascade, u_{iRN} and $u_{iE(N-1)}$, as a straight line with a slope of (w_{tR}/w_{tE}). Start plotting at (u_{iR1}, u_{iE0}), corresponding to the raffinate composition u_{iR1} and the extract composition u_{iE0}; go up and intersect the equilibrium curve at (u_{iR1}, u_{iE1}), corresponding to the two stream compositions at equilibrium leaving stage 1. Then draw a horizontal line to the operating line, intersecting at (u_{iR2}, u_{iE1}). Then go up to (u_{iR2}, u_{iE2}) and go horizontal, complete stage 2 construction and so on.

From equation (8.1.295), we can obtain the following ratio of the organic extraction solvent flow rate to that of the aqueous feed flow rate:

$$(w_{tE}/w_{tR}) = (u_{iRN} - u_{iR1})/(u_{iE(N-1)} - u_{iE0}). \qquad (8.1.296)$$

Had we implemented a solute mass balance using an envelope covering stage N instead of stage $(N - 1)$, we would have obtained

$$(w_{tE}/w_{tR}) = (u_{iR(N+1)} - u_{iR1})/(u_{iEN} - u_{iE0}) = \frac{u_{if} - u_{iR1}}{u_{iEN} - u_{iE0}}. \qquad (8.1.297)$$

When we reduce w_{tE} relative to w_{tR}, the slope of the operating line equation, (w_{tR}/w_{tE}), increases, and ultimately it crosses the equilibrium line in Figure 8.1.37(b) (shown by the dashed line). The point of intersection is $(u_{iR(N+1)}, u_{iEN})$ or (u_{if}, u_{iEN}), where

$$u_{iEN} = m_i\, u_{if} = u^*_{iEN}. \qquad (8.1.298)$$

It will require an infinite number of stages to reach this composition when the extract flow rate is so reduced. The flow rate ratio under such a condition is called the *minimum flow ratio,*

$$(w_{tE}/w_{tR})_{\min} : (w_{tE}/w_{tR})_{\min} = \frac{u_{if} - u_{iR1}}{m_i\, u_{if} - u_{iE0}}. \qquad (8.1.299)$$

8.1.4.4 *Separation of two solutes by fractional extraction in a countercurrent extraction-scrubbing cascade*

If there are two solutes, 1 and 2, to be separated efficiently, the countercurrent cascade of Figure 8.1.37(a) can yield highly purified fractions at the two ends of the cascade provided α_{12} (definition (8.1.293)) is large, which is achieved when (m_1/m_2) has a large value. However, a different type of cascade is recommended in general to this end. In this technique, called *fractional extraction*, the feed mixture of solutes 1 and 2 present in a phase ($j = f$, flow rate w_{tf}) completely miscible with one of the two phases ($j = R$) in the cascade (Figure 8.1.38(a)) is introduced into that phase near the middle of the cascade, between stages n and k. This phase ($j = R$) introduced into the cascade at the left end at stage 1 (flow rate w_{tR}) flows in one direction.

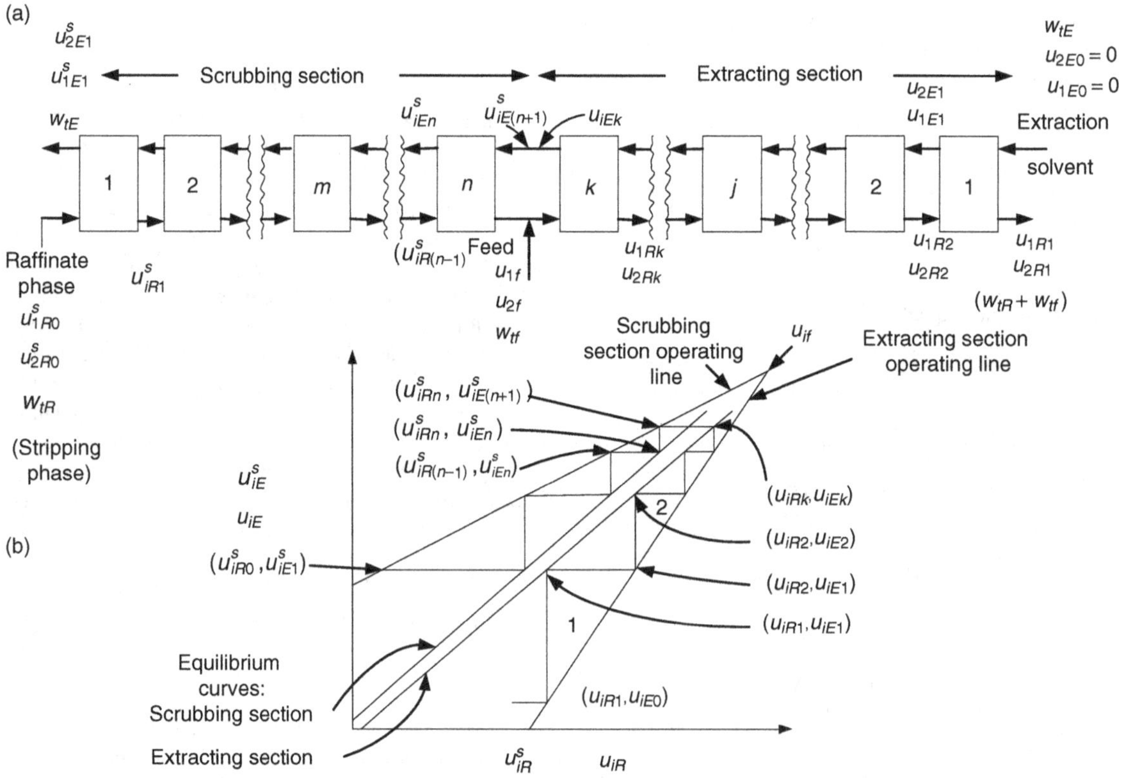

Figure 8.1.38. Countercurrent extraction-scrubbing cascade for fractional extraction of two solutes: (a) cascade diagram; (b) equilibrium curves, operating lines and equilibrium stages. (After Benedict et al. *(1981, pp. 181–186).)*

The extracting solvent phase ($j = E$) flows countercurrently in the opposite direction from the other end of the cascade (flow rate, w_{tE}). Both phases entering the cascade do not have solutes 1 and 2. If the extract phase $j = E$ prefers solute 1 over solute 2, then the following occurs. As the phase R carries both solutes 1 and 2 to the right from stage k, phase E will preferentially extract solute 1 from this phase, so that the phase R stream leaving the cascade on the right will have primarily solute 2 and will have very low contamination of solute 1. Similarly, as phase E moves to the left of the feed location, phase R will strip any solute 2 in phase E, so that phase E moving out to the left will have primarily solute 1 with a very low contamination of solute 2.

For solute 1, the cascade to the left of the feed stage is called the *scrubbing section* or the *stripping section* since solute 2 is stripped/scrubbed away by phase R from the extract phase product at the left end. The cascade to the right of the feed section is called the *extraction section* since solute 1 is extracted from phase R into phase E in this section. Employing some simplifying assumptions, we can determine the extent of separation achieved in this cascade. First, we assume that phases E and R are completely immiscible. Second, we are dealing with dilute solutions and constant partition coefficients for both solutes. Third, both the extract and the raffinate phases have no solute ($i = 1,2$), as they are introduced into the cascade.

Consider now the cascade section from stage k to stage 1. We can conveniently analyze its performance as if we are dealing with the cascade of Figure 8.1.37(a) since $u_{1E0} = 0$, $u_{2E0} = 0$ and the feed concentration entering cascade N in Figure 8.1.37(a) will be equivalent to the net feed concentration entering stage k in Figure 8.1.38(a). To this end, carry out the following mass balances for any solute i. The overall solute i balance over the cascade:

$$w_{tE}\, u^s_{iE1} + \left(w_{tR} + w_{tf}\right) u_{iR1} = w_{tf}\, u_{if}; \tag{8.1.300}$$

solute i balance for feed to stage k:

$$w_{tR}\, u^s_{iRn} + w_{tf}\, u_{if} = \left(w_{tR} + w_{tf}\right) u_{iR(k+1)}. \tag{8.1.301}$$

Apply now equation (8.1.287a) to this cascade:

$$u_{iR(k+1)} = \left[\frac{E_e^{k+1} - 1}{E_e - 1}\right] u_{iR1}. \tag{8.1.302}$$

Here the extraction factor E_e for the extracting section is defined by

$$E_e = |w_{tE}| m_i / \left(w_{tR} + w_{tf}\right). \tag{8.1.303}$$

Similarly, we apply equation (8.1.289) to the cascade from stage n to stage 1 to the left of the feed section:

$$u^s_{iE(n+1)} = \left[\frac{E_s^{n+1} - 1}{E_s - 1}\right] u^s_{iE1} = u_{iEk}, \tag{8.1.304}$$

where the extraction factor E_s for the scrubbing section is defined as $E_s = |w_{tR}| m^s_i / w_{tE}$. Note for any stage 1 to n here,

$$m^s_i = \frac{u^s_{iRn}}{u^s_{iEn}}, \tag{8.1.305a}$$

whereas

$$m_i = \left(u_{iEk} / u_{iRk}\right) \tag{8.1.305b}$$

for stages k to 1. If the partition coefficient is constant along the whole cascade, then

$$m^s_i = (1/m_i). \tag{8.1.306}$$

We can then develop analytical expressions for different quantities in the cascade in terms of other specified quantities, as illustrated below.

Consider the extracting section of the cascade of stages k to 1. The fractional recovery f of species i in this cascade is defined as

$$f = \frac{w_{tE}\, u_{iEk}}{\left(w_{tR} + w_{tf}\right) u_{iR(k+1)}} = 1 - \frac{\left(w_{tR} + w_{tf}\right) u_{iR1}}{\left(w_{tR} + w_{tf}\right) u_{iR(k+1)}} = 1 - \frac{u_{iR1}}{u_{iR(k+1)}}. \tag{8.1.307}$$

By equation (8.1.302),

$$f = 1 - \frac{E_e - 1}{E_e^{k+1} - 1} = \frac{E_e^{k+1} - E_e}{E_e^{k+1} - 1}. \tag{8.1.308}$$

Therefore

$$\frac{u_{iEk}}{u_{iR(k+1)}} = \frac{\left(w_{tR} + w_{tf}\right)}{w_{tE}} \frac{E_e\left(E_e^k - 1\right)}{\left(E_e^{k+1} - 1\right)} = \frac{m_i\left(E_e^k - 1\right)}{\left(E_e^{k+1} - 1\right)}. \tag{8.1.309}$$

Reapplying equation (8.1.302),

$$u_{iEk} = \frac{m_i\left(E_e^k - 1\right)}{\left(E_e^{k+1} - 1\right)} \frac{\left(E_e^{k+1} - 1\right)}{\left(E_e - 1\right)} u_{iR1}, \tag{8.1.310}$$

which yields

$$u_{iEk} = m_i \frac{\left(E_e^k - 1\right)}{E_e - 1} u_{iR1}. \tag{8.1.311}$$

However, from relation (8.1.304) for the scrubbing section of the cascade of stages 1 to n,

$$u^s_{iE(n+1)} = \frac{\left(E_s^{n+1} - 1\right)}{E_s - 1} u^s_{iE1} = u_{iEk}. \tag{8.1.312}$$

Equating the last two results, we get

$$m_i \frac{\left(E_e^k - 1\right)}{E_e - 1} u_{iR1} = \frac{\left(E_s^{n+1} - 1\right)}{E_s - 1} u^s_{iE1}. \tag{8.1.313}$$

Quantities u_{iR1} and u^s_{iE1} are the species i compositions of the two product streams. They are also related by the overall solute i balance (8.1.300), rewritten as follows:

$$u_{iR1} = -\frac{w_{tE}}{(w_{tR}+w_{tf})}u^s_{iE1} + \frac{w_{tf}}{w_{tR}+w_{tf}}u_{if}. \quad (8.1.314)$$

Equations (8.1.313) and (8.1.314) can be simultaneously solved for two product stream compositions u_{iR1} and u^s_{iE1} in terms of known quantities, such as m_i, and two flow rate ratios, such as (w_{tR}/w_{tE}), (w_{tf}/w_{tE}). This approach is illustrated in Example 8.1.17 below. Alternatively, if u_{iR1} and u^s_{iE1} are specified along with m_i, E_e and E_s, we can determine the number of stages required in each section of the cascade, namely k and n. The procedure for the latter method is as follows. Consider equation (8.1.313) for both species $i = 1,2$ to be separated:

$$m_1\frac{(E^k_{e1}-1)}{E_{e1}-1}u_{1R1} = \frac{(E^{n+1}_{s1}-1)}{E_{s1}-1}u^s_{iE1}; \quad (8.1.315)$$

$$m_2\frac{(E^k_{e2}-1)}{E_{e2}-1}u_{2R1} = \frac{(E^{n+1}_{s2}-1)}{E_{s2}-1}u^s_{2E1}. \quad (8.1.316)$$

Here the extraction factors for the individual species are:

$$E_{e1} = \frac{|w_{tE}|m_1}{(w_{tR}+w_{tf})}; \qquad E_{e2} = \frac{|w_{tE}|m_2}{(w_{tR}+w_{tf})};$$

$$E_{s1} = \frac{|w_{tR}|m^s_1}{w_{tE}}; \qquad E_{s2} = \frac{|w_{tR}|m^s_2}{w_{tE}};$$

$$m^s_1 = (1/m_1); \qquad m^s_2 = (1/m_2). \quad (8.1.317a)$$

For a more exhaustive and illustrative treatment of fractional extraction based separation of metals, see Benedict *et al.* (1981, p. 181).

Example 8.1.17 A countercurrent extraction-stripping cascade for fractional extraction is being utilized to purify zirconium from hafnium from a 3.5 M $NaNO_3$/3 M HNO_3 feed solution containing 0.123 M Zr and 0.00246 M Hf flowing in at 48 liter/min. The extracting solvent is 2.25 M tributylphosphate (TBP) containing 1.6 M HNO_3 flowing at a rate of 100 liter/min; assume no Zr or Hf in this solvent. The stripping phase is a solution of 3.5 M $NaNO_3$/3 M HNO_3 without any Zr or Hf - flowing in at a rate of 48 liter/min. Determine the molar concentrations of Zr in the organic extract and the aqueous raffinate for such a cascade where $k = 12.9$ and $n = 2.21$ according to Benedict *et al.* (1981, sect. 6.5). Assume that the densities of the two solutions for mass balance calculation purposes are identical; further, the molar concentration of each solution is sufficiently high so that the mass fraction of Zr or Hf is proportional to the mole fraction of Zr or Hf, the solutions being highly dilute in these two species. The m_i for Zr is 1.20.

Solution We assume the density of the solutions used to be ρ_s g/cm^3. Therefore $w_{tE} = 100\rho_s$ kg/min, $w_{tf} = 48\rho_s$ kg/min, $w_{tR} = 48\rho_s$ kg/min. So

$$E_e = \frac{|w_{tE}|m_i}{(w_{tR}+w_{tf})} = \frac{100\times1.20}{48+48} = 1.04\times1.20 = 1.248;$$

$$E_s = \frac{|w_{tR}|m^s_i}{|w_{tE}|} = \frac{48}{100}\times\frac{1}{1.2} = 0.4.$$

Equation (8.1.313) may now be employed:

$$m_i\frac{(E^k_e-1)}{(E_e-1)}u_{iR1} = \frac{(E^{n+1}_s-1)}{(E_s-1)}u^s_{iE1}; \quad (8.1.317b)$$

$$u_{iR1}\frac{1.2((1.248)^{12.9}-1)}{(1.248-1)} = \frac{((0.4)^{3.21}-1)}{(0.4-1)}u^s_{iE1}$$

$$\Rightarrow u_{iR1}\frac{1.2(17.4-1)}{0.248} = \frac{(0.053-1)}{-0.6}u^s_{iE1} = 0.0198\,u^s_{iE1}. \quad (8.1.318)$$

Now use the overall solute mass balance equation (8.1.300):

$$u_{iR1} = -\frac{w_{tE}}{(w_{tR}+w_{tf})}u^s_{iE1} + \frac{w_{tf}}{w_{tR}+w_{tf}}u_{if}. \quad (8.1.319)$$

Assume that $u_{if}|_{Zr} = 0.123a_1$, where a_1 is a constant independent of the concentration of Zr at the low levels used. Therefore

$$u_{iR1} = -\frac{100}{96}u^s_{iE1} + \frac{48}{96}0.123a_1.$$

Substitute result (8.1.318) here:

$$0.0198u^s_{iE1} = -1.04\,u^s_{iE1} + 0.0651a_1 \Rightarrow 1.0598\,u^s_{iE1} = 0.0615a_1 \Rightarrow u^s_{iE1} = 0.0580a_1.$$

Therefore the molar concentration of Zr in the extract stream is 0.0580 M. Correspondingly,

$$u_{iR1} = -1.04\times0.0580a_1 + 0.0615a_1 = 0.0012a_1.$$

So the molar concentration of Zr in the raffinate stream leaving the cascade is 0.0012 M.

For the extraction-stripping (extraction-scrubbing) cascade, one can plot the operating lines of each section of the cascade following Figure 8.1.37(b). However, there are considerable differences. In Example 8.1.17 and the above analysis, we assumed a constant distribution coefficient throughout the cascade. In general, the constant distribution coefficient will be different in the two sections; further, there will be concentration dependence (Benedict *et al.*, 1981, sect. 6.5). Two such equilibrium curves are shown in Figure 8.1.38(b): one for the extraction section, the other for the stripping section for one of the species being extracted. We will now indicate the equations for the operating lines of each section of the cascade.

For the extracting section, consider a species i balance covering stage j (less than k but more than 2) to the last stage 1:

$$w_{tE}\,u_{iE0} + (w_{tf}+w_{tR})\,u_{iR(j+1)} = w_{tE}\,u_{iEj} + (w_{tR}+w_{tf})\,u_{iR1}; \quad (8.1.320)$$

$$u_{iEj} - u_{iE0} = \frac{(w_{tf}+w_{tR})}{w_{tE}}(u_{iR(j+1)} - u_{iR1}). \quad (8.1.321)$$

The operating line for the extracting section represented by this equation is shown in Figure 8.1.38(b): it has a slope of $((w_{tf}+w_{tR})/w_{tE})$ and passes through the point (u_{iR1}, u_{iE0}), where u_{iE0} is not zero for generality. For the scrubbing section, consider a species i balance covering stage m to stage 1:

$$w_{tR}\, u_{iR0}^{s} + w_{tE}\, u_{iE(m+1)}^{s} = w_{tE}\, u_{iE1}^{s} + w_{tR}\, u_{iR(m+1)}^{s}; \tag{8.1.322}$$

$$u_{iE(m+1)}^{s} - u_{iE1}^{s} = \frac{w_{tR}}{w_{tE}}\left(u_{iR(m+1)}^{s} - u_{iR0}^{s}\right). \tag{8.1.323}$$

The scrubbing section operating line represented by this equation has a slope of (w_{tR}/w_{tE}) and it passes through the point $(u_{iR0}^{s}, u_{iE1}^{s})$.

It may be shown that the operating lines (8.1.321) and (8.1.323) intersect at the point u_{if}, the feed composition (Benedict *et al.*, 1981, sect. 6.5). Therefore, if (u_{iR1}, u_{iE0}) is specified along with u_{if}, and the different flow rates are known, the extraction section operating line can be drawn; knowing its intersection with u_{if} and the slope of the scrubbing section operating line, the scrubbing operating line may be drawn. The stage construction in the extraction section begins with the point (u_{iR1}, u_{iE0}) on the operating line. Go up and hit the equilibrium curve at (u_{iR1}, u_{iE1}). Then draw a horizontal line to hit the operating line at u_{iR2} to complete the stage. Now go up from (u_{iR2}, u_{iE1}) to the equilibrium curve at (u_{iR2}, u_{iE2}), and then move horizontally to complete the stage. Keep going up in this fashion until you hit the equilibrium line (u_{iRk}, u_{iEk}). Since

$$u_{iEk} = u_{iE(n+1)}^{s}, \tag{8.1.324}$$

we draw a horizontal line from (u_{iRk}, u_{iEk}) in the equilibrium line to the scrubbing section operating line where it hits $(u_{iRn}^{s}, u_{iE(n+1)}^{s})$. Now go vertically down and hit the equilibrium curve of the scrubbing section at $(u_{iRn}^{s}, u_{iEn}^{s})$ - this completes the first equilibrium stage n in the scrubbing section. Draw a horizontal line to the operating line on the left at $(u_{iR(n-1)}^{s}, u_{iEn}^{s})$. From this point go vertically down to hit the scrubbing section equilibrium line at $(u_{iR(n-1)}^{s}, u_{iE(n-1)}^{s})$ to complete the second equilibrium stage $(n - 1)$ in the scrubbing section. In this fashion go down to the stage number 1 - operating line coordinates $(u_{iR0}^{s}, u_{iE1}^{s})$ and equilibrium curve coordinates $(u_{iR1}^{s}, u_{iE1}^{s})$.

8.1.4.5 *Solvent extraction in systems showing partial solubility of the bulk phases*

In the preceding sections, the feed phase and the extract phase were assumed to be insoluble over the range of solute concentrations. Therefore the process analysis and design of solvent extraction was simpler. However, in many systems, the feed phase (primary constituent A, for example water) and the extract phase (primary constituent is solvent B) have some degree of solubility in each other as the solute species C is transferred from the feed phase to the extract phase. Therefore both phases are ternary mixtures with significant concentrations of species A and B. Analysis of such ternary systems of ideal stages is conveniently done using triangular diagrams (see Figures 1.6.1, 4.1.7(a) and 4.1.8(a)). The composition of a species i in such a diagram is indicated in weight fraction, u_{if} for feed phase f or u_{iR} for the raffinate phase R, and u_{iS} for solvent phase S or u_{iE} for the extract phase E.

Consider Figure 8.1.39(a) for a continuous countercurrent multistage solvent extraction cascade of N equilibrium stages where, unlike Figure 8.1.37(a), the mass flow rate in each phase is changing with the stage number: w_{tEs}, the mass flow rate of the extract phase out of stage s, its composition being u_{iEs} for species i; w_{tRs}, the mass flow rate of the raffinate phase out of stage s, its composition being u_{iRs}. We follow the treatment of Treybal (1980, pp. 497–506). Consider first a total mass balance of the two phases over the whole cascade and then a balance on solute species C:

$$w_{tf} + w_{ts} = w_{tRN} + w_{tE1} = w_{tM}. \tag{8.1.325}$$

In the triangular diagram (Figure 8.1.39(b)) containing the binodal curve $R_N R_1 E_1 E_N$, the feed composition (u_{if}) is represented by point F, the solvent composition is represented by point S (u_{iS}) and the mixture of these two phases by point M (u_{iM}). Also M lies on a straight line joining the points E_1 and R_N since it represents that mixture as well. (An extension of the line between S and R_N hits the AC line on R'_N representing R_N on solvent-free coordinates.) Correspondingly, a balance on species C being extracted is:

$$w_{tf}\, u_{Cf} + w_{ts}\, u_{Cs} = w_{tRN}\, u_{CRN} + w_{tE1}\, u_{CE1} = w_{tM}\, u_{CM}, \tag{8.1.326}$$

where

$$u_{CM} = \frac{w_{tf}\, u_{Cf} + w_{ts}\, u_{Cs}}{w_{tf} + w_{ts}} = \frac{w_{tRN}\, u_{CRN} + w_{tE1}\, u_{CE1}}{w_{tRN} + w_{tE1}}, \tag{8.1.327}$$

which indicates that point M also represents the two product phases w_{tRN} and w_{tE1} leaving stage N and stage 1 (where feed is entering), represented in Figure 8.1.39(b) as points R_N and E_1, respectively; the value of N in this figure is 5. Point R'_N has not been shown.

The overall mass balance may also be described as

$$w_{tRN} - w_{ts} = w_{tf} - w_{tE1} = \Delta_R, \tag{8.1.328}$$

where Δ_R represents the net mass flow out of the system at the last stage N as well as at the first stage 1. Both equalities in (8.1.328) represent straight lines intersecting at Δ_R located on the left-hand side of the triangular diagram. Under some conditions (generally high solvent flow rate), the point Δ_R may be located on the right-hand side of the triangle. If we make a total mass balance using an envelope between any stage n and the last stage N, we get

$$w_{tR(n-1)} + w_{ts} = w_{tRN} + w_{tEn} \tag{8.1.329}$$

which yields also

$$w_{tR(n-1)} - w_{tEn} = w_{tRN} - w_{ts} = \Delta_R. \tag{8.1.330}$$

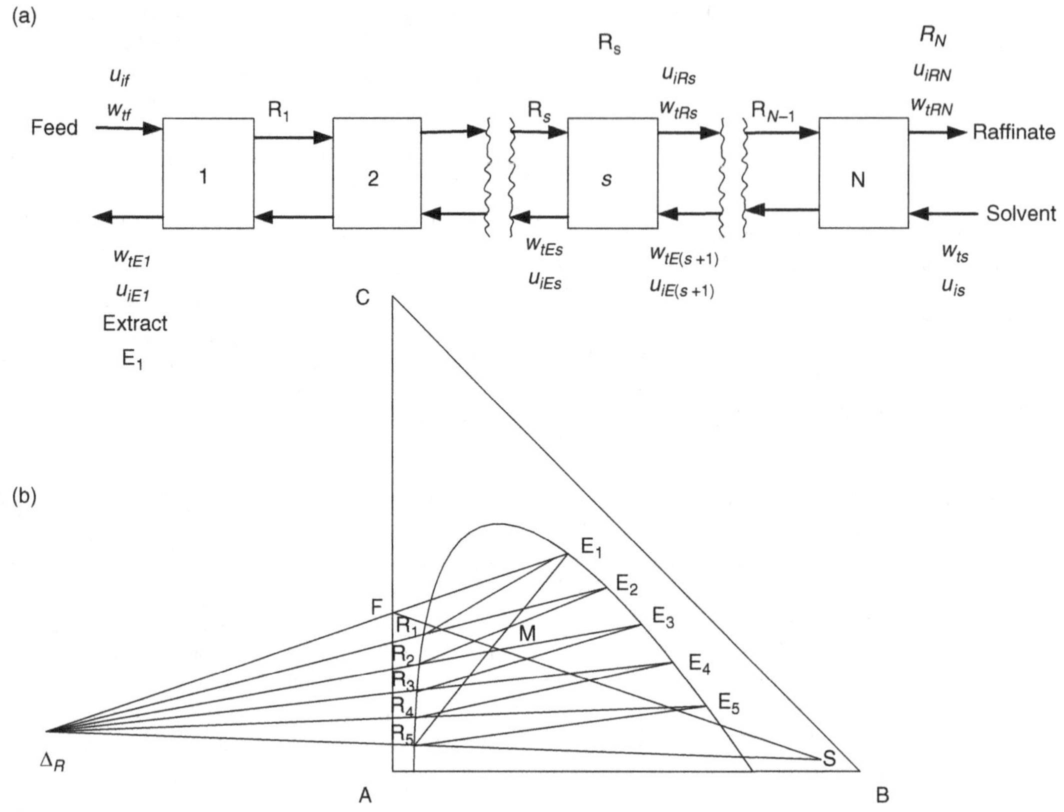

Figure 8.1.39. (a) Countercurrent multistage extraction system for a partially miscible system. (b) Right triangular diagram for countercurrent multistage extraction, solute C being extracted from feed phase containing primarily A by solvent B (here N = *5).*

Therefore Δ_R also represents the excess mass flow rate of the raffinate stream from the cascade and the extract stream at all levels of the countercurrent cascade. As with equation (8.1.328), points E_n and R_{n-1} when jointed together intersect at Δ_R.

We can now initiate the drawing of the individual stages in the triangular diagram. Recognize now, for the equilibrium stage 1, streams R_1 and E_1 are at equilibrium; a tie line for E_1 will therefore hit R_1 on the binodal curve (see Figure 4.1.7(a) for the tie line). Since E_2 and R_1 in between stages 1 and 2 are also related by (8.1.330) (for $n = 2$), joining Δ_R and R_1 and extending it to hit the other side of the binodal curve will lead to point E_2. We can continue this process to complete the multistage construction as shown in Figure 8.1.39(b) to end with E_N and R_N corresponding to the N stages needed to achieve the desired change.

As with gas absorption/distillation processes, the flow rate of the solvent for a given feed flow rate will control the number of stages required to achieve a given composition change. When the solvent flow rate is increased, point M is shifted toward S; correspondingly, for a given R_N, E_1 comes closer to S, and therefore the number of stages required is reduced. Also, the point Δ_R is pushed further to the left of the triangle. As the solvent flow rate is increased even further, E_1 comes very close to S, and point Δ_R may shift to the right of the triangle. On the other hand, when the solvent flow rate is reduced, it is possible to have a line for Δ_R merge with a tie line in the mixed-phase system under the binodal curve: under such a condition, one requires an infinite number of stages and the solvent to feed ratio is considered the minimum possible. Additional approaches may be taken to analyze the performance of such a countercurrent cascade of ideal solvent extraction stages. It is based on the solvent-free coordinate system illustrated in Figures 4.1.7(b) and 4.1.8(b). For illustrative treatments, consult Treybal (1963, chap. 6, 1980, pp. 497–507).

Example 8.1.18 Equilibrium data of the ternary system water (A)–isopropyl ether (B)–acetic acid (C) have been provided in Table 8.1.7. The source of this data acquired at 20 °C is Treybal (1980). Using these data, solve the following problem. Pure isopropyl ether is being used in a countercurrent multistage cascade to extract acetic acid from a feed aqueous solution of 37 wt% acetic acid. The final raffinate product on a solvent-free basis should be 2 wt% acetic acid. The feed solution flow rate is 2000 kg/hr.

(a) Determine the minimum solvent flow rate needed to achieve the goal.
(b) If the solvent flow rate is three times the minimum value, calculate the number of ideal stages needed to achieve the goal.

Solution Figure 8.1.40(a) illustrates the triangular diagram, where the ternary equilibrium data of water-isopropyl ether-acetic acid from Table 8.1.7 have been plotted.

Table 8.1.7. Equilibrium data for Water-isopropyl ether-acetic acid (from Treybal (1980)); all data in wt%

Water layer			Isopropyl ether layer		
Acetic acid	Isopropyl ether	Water	Acetic acid	Isopropyl ether	Water
0.69	1.2	98.1	0.18	99.3	0.5
1.41	1.5	97.1	0.37	98.9	0.7
2.89	1.6	95.5	0.79	98.4	0.8
6.42	1.9	91.7	1.93	97.1	1.0
13.3	2.3	84.4	4.82	93.3	1.9
25.5	3.4	71.1	11.4	84.7	3.9
36.7	4.4	58.9	21.6	71.5	6.9
44.3	10.6	45.1	31.1	58.1	10.8
46.4	16.5	37.1	36.2	48.7	15.1

Table 8.1.8.

0.0	0.01	0.02	0.03	0.047	0.058	0.075	$0.085 = u_{iE1}$
0.02	0.075	0.125	0.175	0.225	0.275	0.325	$0.375 = u_{if}$

Part (a) The feed containing 37.5% acetic acid is shown as point F on the AC arm of the triangle. Since the solvent is pure isopropyl ether, point S concides with the vertex point B, pure solvent. Join this point with 0.02 on the AC coordinate, point R'_N, corresponding to the solvent-free raffinate product; it intersects the u_{CR} vs. u_B curve at R_N. Join R_N to B and extend it. Then have the tie line, which, when extended to the left, hits point F on AC extended to the right; this line will intersect the $R_N B$ line at Δ_{Rm}. This tie line will provide the minimum value of the extract phase flow rate E_{1m}. Line $E_{1m}R_N$ intersects line FB at point M_m, where $u_{CE} = 0.168$. Represent this point as u_{CM}. Then equation (8.1.327) provides (since $u_{Cs} = 0$)

$$u_{CM} = 0.168 = \frac{2000 \times 0.375 + (w_{ts})_{\min} \times 0}{2000 + (w_{ts})_{\min}}$$

$$\Rightarrow (w_{ts})_{\min} = 2464 \text{ kg/hr},$$

which is the minimum solvent flow rate.

Part (b)

$$w_{ts} = 3 \times 2464 = 7392 \text{ kg/hr};$$

$$u_{CM} = \frac{2000 \times 0.375 + 4928 \times 0}{2000 + 7392} = \frac{750}{9392} = 0.0798.$$

Locate this point M on the line FB. Extend now the line $R_N M$ to intersect the u_{CE} vs. u_B line at E_1 at $u_{CE} = 0.082$. Focus now on line FE_1: extend it to intersect the extended $R_N B$ line at Δ_R. Having located Δ_R, we will now draw a number of lines from Δ_R to intersect u_{CR} vs. u_B and u_{CE} vs. u_B to obtain the values of $u_{iE(s+1)}$ and u_{iRs} (Table 8.1.8) for drawing the operating line in Figure 8.1.40(b). Plot

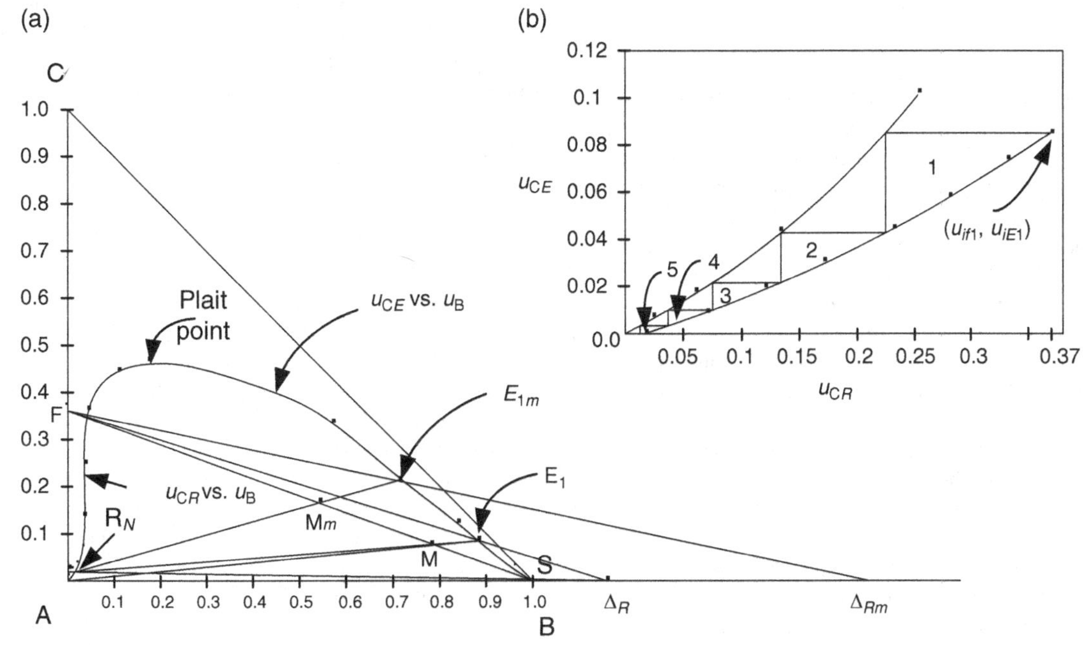

Figure 8.1.40. (a) Triangular diagram for Example 8.1.18; (b) McCabe–Thiele type plot for Example 8.1.18.

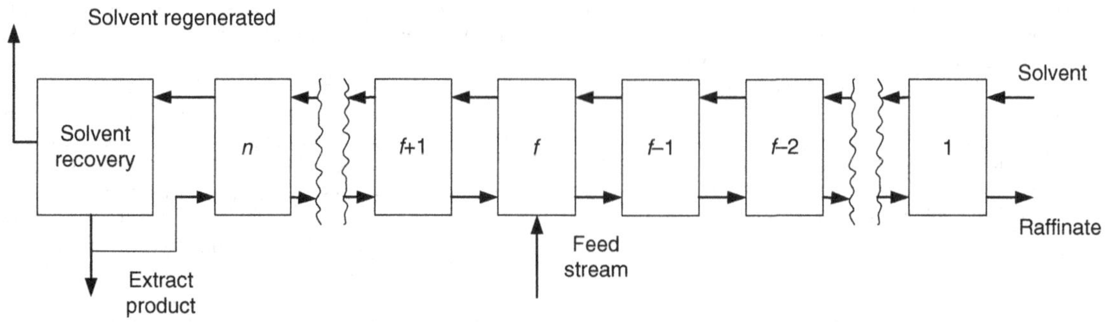

Figure 8.1.41. Countercurrent multistaged solvent extraction with reflux.

these data in Figure 8.1.40(b) as the operating line (a curved line); also plot the equilibrium data. Then start from the stage 1 end of the cascade (u_{if}, u_{iE1}), draw a horizontal line, hit the equilibrium curve, draw a vertical line to the operating line to complete stage 1; then continue the process and end up with five ideal stages.

8.1.4.6 *Role of reflux in a continuous countercurrent extraction system*

In a solvent extraction system, the extract solvent phase is highly enriched in the solutes preferentially extracted. From this extract, the solvent stream has to be recovered for reuse at the other end of the cascade; simultaneously, a highly solute-enriched stream is obtained as the product after separation of the solvent. The solute concentration of this latter stream is (much) higher than that of the feed stream (see Figures 8.1.36 and 8.1.37(a)). This is obviously one of the ultimate product streams from the overall extraction process.

If a fraction of this product stream is introduced back into a countercurrent solvent extraction cascade, as shown in Figure 8.1.41, the countercurrent solvent extract stream is contacting a highly solute-enriched stream. Therefore the solute concentration in the countercurrent extract stream at the top of the cascade (stage n, Figure 8.1.41) is considerably enhanced. In such an arrangement, the feed stream miscible with the ultimate refluxed product stream is introduced further down (stage f), where its concentration matches that in the cascade. The part of the cascade below the feed section is like a traditional extraction cascade, namely like that in Figures 8.1.36 and 8.1.37(a). Note that the top section of this cascade (Figure 8.1.41), where the solvent is separated from the extract, and part of the separated and highly solute-enriched phase is recycled back to the column/cascade, is like that on top of a distillation column with a condenser and condensate reflux (see Figure 8.1.19). The solvent extraction system now behaves like a type (3) system instead of a type (2) system.

8.1.4.7 *Reverse micellar and aqueous two-phase extraction processes*

The previous sections considered countercurrent solvent extraction primarily in the context of aqueous-organic systems. Section 4.1.3 illustrated solvent extraction in polar organic-organic systems (see Figure 4.1.9(a)) and aqueous two-phase systems (Figure 4.1.9(b)) as well. An additional two-phase system for solvent extraction of proteins is the reverse micelle-aqueous phase system shown in Figures 4.1.19–4.1.21.

Centrifugal contacting devices have been employed to demonstrate reverse micellar extraction of enzymes on an industrial scale (Dekker *et al.*, 1991). Spray columns have been employed in a number of studies to study mass transfer in such systems. The mass-transfer rate of protein lysozyme being extracted was satisfactorily described in terms of the mass-transfer coefficients of the reverse micelle (solvent) phase and the aqueous phase (Lye *et al.*, 1996). Spray extraction columns have also been used to study the extraction of enzymes in aqueous two-phase systems (Arsalani *et al.*, 2005).

8.1.4.8 *Backmixing in solvent extraction columns*

Equations (8.1.249) and (8.1.250) for a continuous countercurrent extractor contain an axial dispersion term in the raffinate phase and the extractive phase, respectively. Virtually all continuous countercurrent solvent extraction devices encounter backmixing/axial dispersion. Spray towers exhibit the highest amount of backmixing. There is considerable literature on analyzing/modeling axial dispersion in countercurrent continuous extraction devices. Earlier modeling studies yielding analytical solutions of the governing equations include those by Sleicher (1959), Miyauchi and Vermeulen (1963) and Hartland and Mecklenburgh (1966). Staged columns (Figures 8.1.35 (c) and (d)) also encounter substantial backmixing. Sleicher (1960) developed a "backflow model" for the compartmental-type columns to that end. The complex analytical solutions of such studies were based on the linear equilibrium relation between the two phases. We have illustrated one such result for gas-liquid absorption in Section 8.1.2.4.

Solvent extraction systems often display nonlinear equilibrium behavior. Pratt (1983) and Pratt and Stevens (1991) have illustrated methods in which results of sufficient accuracy are obtained by representing the equilibrium curve by means of two or three straight segments for both continuous as well as multicompartment columns. Additional simple and shortcut procedures have been described by Von Stockar and Lu (1991). The extent of backmixing in the continuous phase in Karr reciprocating-plate extraction columns (Figure 8.1.35(d)), widely used in the petroleum, pharmaceutical and hydrometallurgical industries, has been studied experimentally by Stella *et al.* (2006).

It would be of interest to find out the role (if any) of backmixing in nondispersive porous membrane based solvent extraction devices. Here, only the shell-side liquid phase can encounter backmixing, or, more correctly, bypassing and channeling. Simple countercurrent devices often display such flow behavior. However, the baffled construction with radial flow illustrated in Figure 8.1.35(f) is thought to be generally free of such flow distortions.

8.1.4.9 *Stage efficiency*

So far, whenever multistage solvent extraction cascades have been considered, ideal stages were assumed with both exiting streams considered to be at equilibrium. As we have seen in Section 6.4.1.2, the performances of the stages were likely to be less than ideal. To that end, the notion of the Murphree extract-stage efficiency E_{ME} and the Murphree raffinate stage efficiency E_{MR} was introduced via definitions (6.4.70) and (6.4.71) using mole fractions. One can replace mole fractions by the mass fractions extensively used here and employ the same concepts. Further, one can employ the procedure employed in Section 8.1.3.4 and illustrated in Figure 8.1.29 for distillation: develop a hypothetical equilibrium curve shown by the dashed line and then draw the steps between this curve and the operating line(s) to determine the number of actual stages required.

8.1.5 Countercurrent melt crystallization in a column

We have studied crystallization from a solution in a CSTS in Section 6.4.1.1. Crystallization from a melt was studied in Section 6.3.2.3 under conditions where there was bulk flow parallel to the force. Further, the systems dealt with were solid solutions, as shown in Figures 6.3.14(a) and 3.3.6A, and extremely high purifications were involved. Here we consider countercurrent movement of crystals and molten liquid in a column. The scale of operation is much larger. Chemicals such as naphthalene, p-xylene, acrylic acid, bisphenol A, monochloroacetic acid, dichlorobenzene, etc., are being routinely purified via such a technique (Wynn, 1992).

We have already provided a schematic for such a countercurrent crystal-melt system in Figure 8.1.3(b), and we have identified this technique as a type (3) system, like a vapor-liquid distillation column with a liquid reflux at the top and vapor reboil at the bottom. A few features, however, make this technique stand apart from distillation. The temperature level of operation is much lower in crystal-melt systems; the latent heat of melting is much smaller (~3-5 times) than the latent heat of vaporization of organic compounds separated by distillation; and the equipment volume is much smaller, since we do not have to deal with the large volumes of vapor (Ulrich, 1993). Because of much lower temperatures, degradation reactions are much less likely in crystal-melt systems. In addition, a most important characteristic of crystal-melt system is the high degree of purification possible: in eutectic systems (see Figure 4.1.10), one equilibrium stage can lead to pure product from a conceptual point of view, although the fractional product recovery is low (Ulrich, 1993) (One such system comprises xylene isomers; the freezing points are: o-xylene, 248 K; m-xylene, 225.4 K; p-xylene 286.6 K. These values are sufficiently far apart. On the other hand, the relative volatility of these isomers varies between 1.02 and 1.16, pointing out how costly distillation is likely to be.) However, for solid solution systems, multiple equilibrium stages are required in crystal-melt systems.

Focus now on the center-fed column concept shown in Figure 8.1.3(b). A more detailed schematic representation of the column containing a spiral-type conveyor is provided in Figure 8.1.42(a). This spiral-type conveyor is located in between two concentric tubes; the spiral conveyor is rotated. It is needed to push the crystals in one direction while the melt is moving in the other direction, since the densities of the melt and the crystals are close. The spirals scrap the crystals formed by cooling on the surface of the tubes and push it toward the melting section. The molten liquid moves countercurrently to the freezing section.

A crystal in such a column has a certain amount of molten liquid adhering to it (Figure 8.1.42(b)). The impurities in the crystal and the "adhering liquid" have to be transferred to the free liquid flowing in the column (Albertins *et al.*, 1967). The countercurrent mass transfer of the impurity from the crystal phase

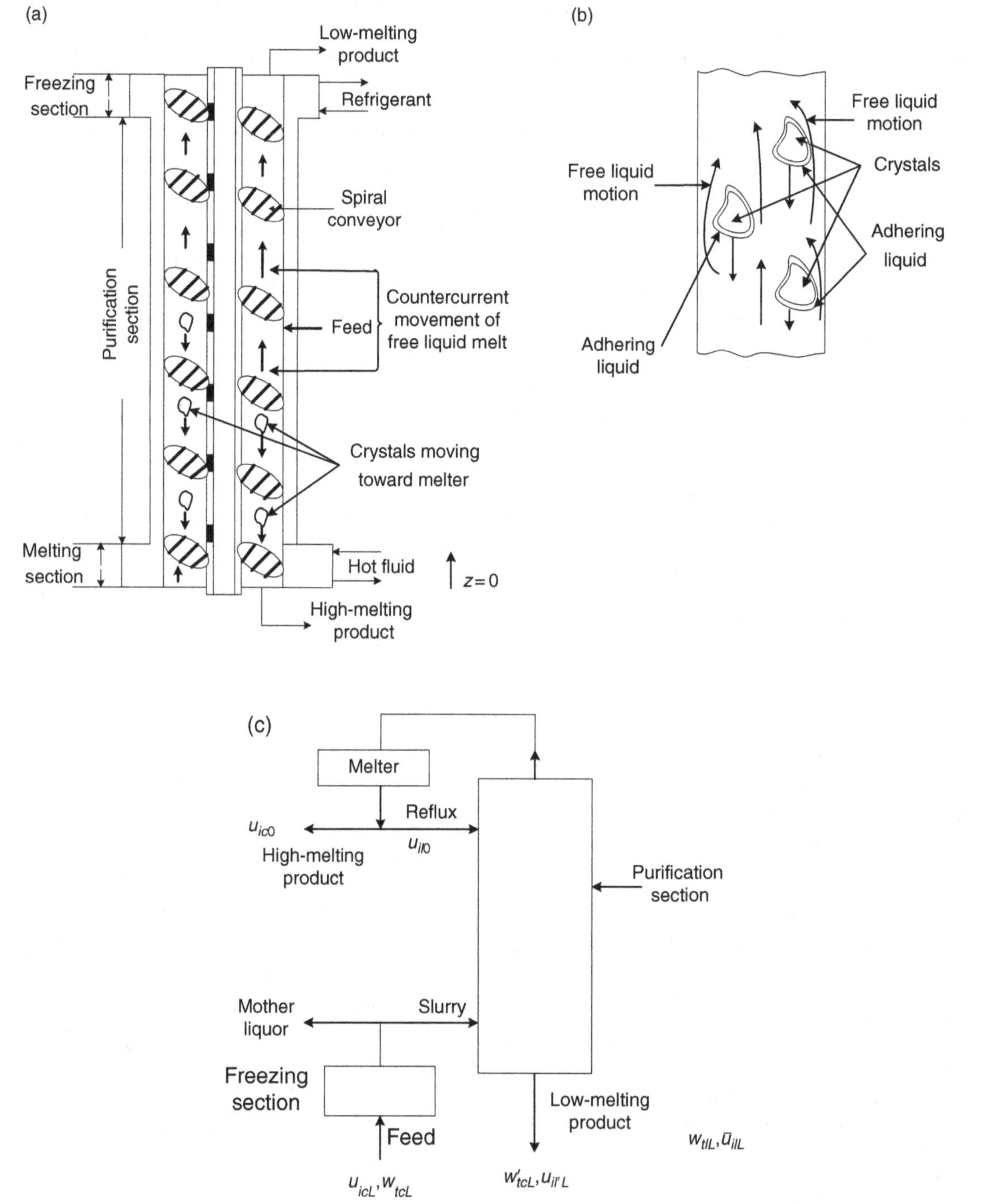

Figure 8.1.42. Column crystallizer with countercurrent flow. (a) Spiral conveyor in a centrally fed arrangement; (b) adhering liquid around a crystal with countercurrent free liquid motion; (c) end-feed arrangement. (After Albertins et al. *(1967).)*

(which includes the liquid/melt tightly adhering to the crystal surface) to the flowing melt liquid phase takes place primarily between the adhering liquid and the surrounding melt flowing countercurrently. The column shown in Figure 8.1.42(a) is centrally fed. The crystals pushed toward the melting section by the spiral generate a melt (part of which is the product with the higher melting point, e.g. 99%+ p-xylene in the case of xylene isomer separation) at that column end; thereby, the impurities in the crystal phase are transferred to the liquid, which moves countercurrently up the column and is equivalent to the reboiler vapor in a distillation column. Correspondingly, the liquid melt pushed toward the freezing section at the other end undergoes some amount of freezing which generates the solid-phase reflux pushed by the spiral to the opposite end of the column. The purification section is insulated. At both ends of the column, only liquid products are taken out; the top end liquid product is the lower melting product.

Another type of melt crystallization column used is the end-feed column shown in Figure 8.1.42(c). In this design, the feed is provided to the freezing section; the low-melting product is also withdrawn at this end of the column through a filter so that the crystals may be fed back to the column. Although the figure shows the freezing section to be separate from the column, it could also be an integral part of the column as in Figure 8.1.42(a).

Our discussion has so far assumed that all of the impurity is present in the adhering liquid around a crystal. It is possible that the crystals have trapped impurities inside. The only way to get the impurities out would be to melt the crystals. In such a case, the crystals move in the direction of increasing temperature and therefore undergo varying degrees of melting. Simultaneously the melt moving in the other direction is subjected to freezing, which leads to purification also.

Two types of systems are of interest: eutectic and solid solution. Since eutectic systems produce relatively pure crystals, the melt crystallizer purifies the adhering liquid. We will now briefly provide for an eutectic system without any solid of eutectic composition an elementary mass-transfer based description of transport taking place in such a countercurrent melt crystallization column (Figure 8.1.42(a)). (For an introduction to such an analysis for a solid solution system, see Albertins *et al.* (1967).) Identifying the melt going up as phase $j = \ell$ and the adhering liquid phase on the crystal ($j = c$) as phase $j = \ell'$, we can write (following equation (8.1.3b)) for the steady state non-dispersive mass transport of impurity species i from the adhering liquid flowing down into the upflowing melt phase:

$$\langle v_{\ell z}\rangle \frac{d\overline{C}_{i\ell}}{dz} = K_{i\ell c}a\left(\overline{C}_{i\ell'} - \overline{C}_{i\ell}\right), \qquad (8.1.331)$$

where $\overline{C}'_{i\ell}$ is the concentration of impurity i in the adhering liquid phase;

$$\langle v_{\ell z}\rangle C_{t\ell} \frac{d\overline{x}_{i\ell}}{dz} = K_{i\ell c}C_{t\ell}a(\overline{x}_{i\ell'} - \overline{x}_{i\ell}), \qquad (8.1.332)$$

$$\frac{|W_{t\ell}|}{A_c}\frac{d\overline{x}_{i\ell}}{dz} = K_{i\ell x}a(\overline{x}_{i\ell'} - \overline{x}_{i\ell}). \qquad (8.1.333)$$

Generally, mass fraction based concentration units are used in melt crystallization. The corresponding equation is as follows:

$$\frac{|w_{t\ell}|}{A_c}\frac{d\overline{u}_{i\ell}}{dz} = K_{i\ell u}a(\overline{u}_{i\ell'} - \overline{u}_{i\ell}), \qquad (8.1.334)$$

where $w_{t\ell}$ is the total mass flow rate, A_c is the cross-sectional area for flow and $K_{i\ell u}$ is an overall mass-transfer coefficient for species i between the two liquid phases based on mass fractions having the units of g/cm^2-s. Over a certain length L of the column, the melt composition changes from $u_{i\ell 0}$ (at $z = 0$) to $\overline{u}_{i\ell L}$ (at $z = L$):

$$\int_0^L dz = L = \frac{|w_{t\ell}|}{A_c K_{i\ell u}\, a}\int_{u_{i\ell 0}}^{\overline{u}_{i\ell L}} \frac{d\overline{u}_{i\ell}}{(\overline{u}_{i\ell'} - \overline{u}_{i\ell})}, \qquad (8.1.335)$$

$$L = HTU_{o\ell} \times NTU_{o\ell}, \qquad (8.1.336)$$

$$HTU_{o\ell} = (|w_{t\ell}|/A_c K_{i\ell u}\, a); \qquad NTU_{o\ell} = \int_{u_{i\ell 0}}^{\overline{u}_{i\ell L}} \frac{d\overline{u}_{i\ell}}{(\overline{u}_{i\ell'} - \overline{u}_{i\ell})}. \qquad (8.1.337)$$

For an end-feed column with the feed having a crystal flow rate of w_{tcL}, impurity mass fraction u_{icL}, a low-melting product flow rate of $w_{t\ell L}$, impurity mass fraction of $\overline{u}_{i\ell L}$, an adhering liquid flow rate of w'_{tcL} and an impurity mass fraction of $u_{i\ell' L}$, the value of $NTU_{o\ell}$ has been calculated as (Wankat, 1990, p. 182)

$$NTU_{o\ell} = \frac{1}{\left[\dfrac{w_{t\ell L}}{w'_{tcL}} - 1\right]} \ln\left[u_{i\ell' L} - \left\{1 - \frac{w_{t\ell L}}{w'_{tcL}}\right\}\overline{u}_{i\ell} - \frac{w_{t\ell L}}{w'_{tcL}}\overline{u}_{i\ell L}\right]_{u_{i\ell 0}}^{\overline{u}_{i\ell L}}, \qquad (8.1.338)$$

where the flow rates $w_{t\ell L}$, w'_{tcL} and w_{tcL} have been assumed to be constant along the column length. Note the lower limit $u_{i\ell 0}$ is related to the crystal product u_{ic0} since it is obtained by refluxing the melted product,

$$u_{ic0} = u_{i\ell 0}. \qquad (8.1.339)$$

Further, u_{ic0} is usually specified in terms of the purity of the crystal product. An additional impurity balance at the melter end of the column yields the melted product composition u_{ic0} via

$$\left(w_{tcL} + w'_{tcL}\right) u_{ic0} = w'_{tcL}\, u_{i\ell' 0}, \qquad (8.1.340)$$

where we have assumed that crystals at this end for a eutectic are pure, i.e. $u_{icL} \cong 0$. Knowing $u_{i\ell' 0}$, we can now have a mass balance around the crystallizer to determine $\overline{u}_{i\ell L}$, the other limit in the expression (8.1.338) for $NTU_{o\ell}$:

$$w_{t\ell L}\,\overline{u}_{i\ell L} = w'_{tcL}\, u_{i\ell' L} + w_{t\ell 0}\, u_{i\ell 0} - w'_{tcL}\, u_{i\ell 0}. \qquad (8.1.341)$$

Since $w_{t\ell L} = w_{t\ell 0}$, we get

$$\overline{u}_{i\ell L} = \frac{w'_{tcL}}{w_{t\ell L}}\left(u_{i\ell' L} - u_{i\ell 0}\right) + u_{i\ell 0}. \tag{8.1.342}$$

Knowing $u_{i\ell 0}$, $u_{i\ell' L}$ and the two flow rates, $\overline{u}_{i\ell L}$ can be determined. The unknowns in this analysis from a parametric point of view are the fraction of adhering melt, w'_{tcL}/w_{tcL}, and the value of $HTU_{o\ell}$. Albertins *et al.* (1967) obtained a value of $HTU_{o\ell} = 12.3$ cm for a particular eutectic system (purifying azobenzene).

For an earlier introduction to column crystallization vis-à-vis the variety of equipment and patents, see Albertins *et al.* (1967). Ulrich (1993) provides a more modern version of melt crystallization devices, plants and processes. For mathematical modeling and optimization of multistage crystallization processes, see Gilbert (1991). A more involved model of continuous countercurrent contacting with axial dispersion in melt crystallization has been illustrated in Henry and Moyers (1984).

8.1.6 Countercurrent adsorption and simulated moving bed system

In the introductory Section 8.1, we briefly mentioned problems related to having a countercurrent flow of solid particles and another fluid phase, such as attritional losses, erosion of device walls, etc. An additional problem in the circulation of the solid adsorbent particles arises due to the considerable mixing of the solid particles as well as that of the fluid phase. For example, an adsorbent-coated moving belt has been employed with a countercurrent flowing liquid. It was found that the belt motion induced recirculation of the liquid effectively reduced the performance of the system to that of a single equilibrium stage; on the other hand, the same group demonstrated the achievement of five to ten equilibrium stages of separation for countercurrent vapor-phase adsorption with a moving adsorbent-coated belt (Phelps and Ruthven, 2001). Continuous countercurrent ion exchange has, however, been carried out with an endless flexible belt incorporating an ion exchange resin moving continuously over horizontal rollers and vertically, with many passes, through which the solution to be treated must travel (Karlson and Edkins, 1975; Treybal, 1980, p. 613).

The problems posed by the motion of solid particles or a moving bed of particles in countercurrent adsorption processes have led to a number of different operational configurations. In a *gas-fluidized multistage countercurrent adsorber*, there are *multiple sieve-tray plates* in a vertical column. The adsorbent particles are fluidized by the gas stream to be cleaned entering through the perforations in the sieve tray. The fluidization level is not very vigorous. Particles are not swept away by the gas. The particles on a plate are transferred to the plate below (since they behave like a liquid) via a downcomer (Figure 8.1.43(a)). Such an arrangement has been used to dry air with silica gel particles. The wet air enters the column somewhere in the middle of the column. As the silica gel particles come down further, there is a partition in the column and they are introduced into the next section, where they are regenerated by exposing them to dry hot gas coming from below in the bottom section of the tower which strips the adsorbent moisture. It is like the stripping section of a distillation column. The dried silica gel particles are then lifted by compressed air to the top of the adsorber where it comes down into the column at the top sieve plate. Such a countercurrent scheme is in contrast to the PSA process for drying air (see Figure 7.1.12) employing a fixed bed of silica gel particles and a pressure-swing based cyclic process.

We will now describe another scheme utilized to carry out solid–liquid contacting in a countercurrent fashion. It is an intermittent/cyclic process. Consider Figure 7.1.28, which illustrates a cyclic expanded bed adsorption process using resin beads to adsorb bioproducts directly from a fermentation broth containing cells, cell debris, etc. Suppose now that there are a number of such stages arranged vertically. Instead of eluent/regenerated liquid flowing in steps (e) and (f) of Figure 7.1.28 through the contracted bed of particles, let the liquid briefly flow in the same direction (as in Figure 7.1.28) and push the solid adsorbents to the next lower plate/stage. The cycle of fluidization and adsorption can now begin on each plate/stage. This device is called a Cloethe–Streat contactor and is used in ion exchange based metal separation.

Another successful strategy used to achieve the performance of a countercurrent solid–fluid adsorber is to employ a simulated moving bed system. We have briefly described it in Figure 8.1.6 in Section 8.1.1. We will provide an illustrative treatment of it later in this section.

8.1.6.1 *Separation in a countercurrent fluid–solid adsorber*

We will consider here two basic aspects of separation in a countercurrent fluid–solid adsorber of the type illustrated in Figure 8.1.43(b). If the feed fluid has two species $i = 1$ (or A) and 2 (or B) in the bulk fluid phase, which species will go up to the top of the column and where will the other species appear? The next item of interest is: what is the height of the column needed for the desired extent of separation for any given species? Let us consider the first question.

Let the feed fluid stream introduced into the column contain species A and B in, say, a carrier fluid. In a countercurrent flow system containing two phases $j = 1,2$ moving

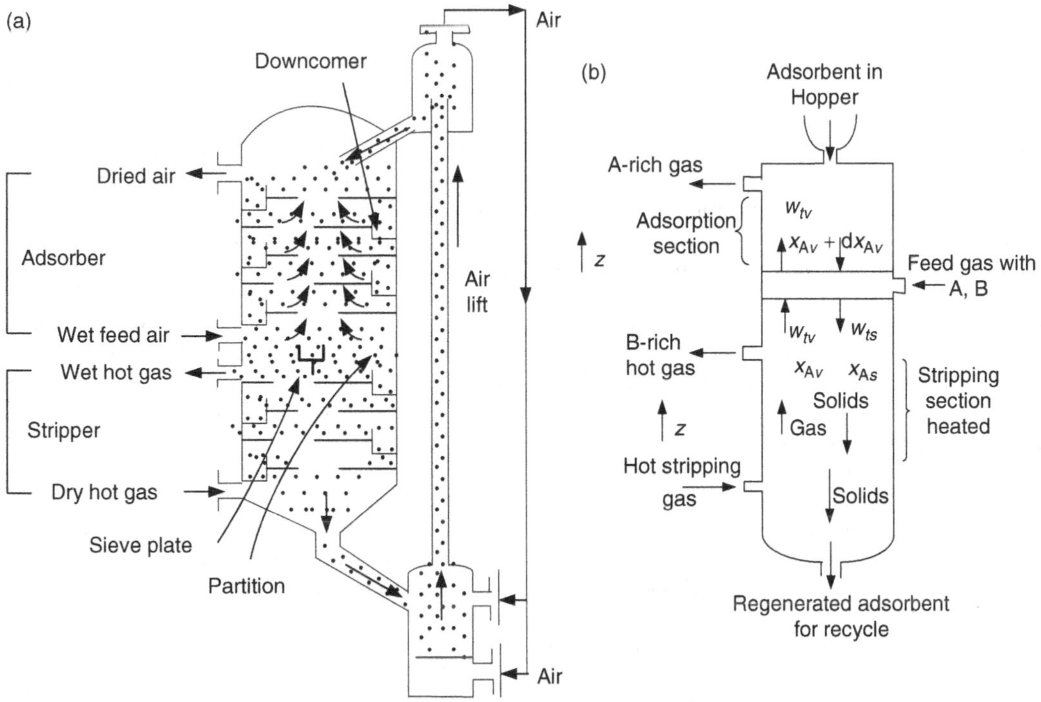

Figure 8.1.43. (a) Gas-fluidized multistage countercurrent adsorber with sieve plates and adsorbent regeneration with hot gas. (b) Countercurrent continuous adsorber/stripper separates species A and B present in feed gas.

in opposite directions in the column, the concentration wave velocity of species i from equation (8.1.35) (repeated here for convenience) is

$$v^*_{Ci} = v_{2z}\frac{\left(1-\kappa_{i1}\dfrac{\varepsilon_1}{\varepsilon_2}\dfrac{|v_{1z}|}{v_{2z}}\right)}{\left(1+\dfrac{\varepsilon_1}{\varepsilon_2}\kappa_{i1}\right)}.$$

We have already learnt via equation (8.1.36) that, for $j = 1$ (s, solid) being the solid/adsorbent phase moving down and $j = 2$ (say, v for vapor) being the fluid phase moving up, species A moves up the column and species B moves down the column if

$$\kappa_{A1}\frac{\varepsilon_1}{\varepsilon_2}\frac{|v_{1z}|}{v_{2z}} < 1 < \kappa_{B1}\frac{\varepsilon_1}{\varepsilon_2}\frac{|v_{1z}|}{v_{2z}}.$$

Alternatively, via equation (8.1.37),

$$\kappa_{A1} < \frac{\varepsilon_2}{\varepsilon_1}\frac{v_{2z}}{|v_{1z}|} < \kappa_{B1},$$

and, via equation (8.1.38),

$$\kappa_{A1} < \frac{\langle v_{2z}\rangle}{\langle|v_{1z}|\rangle} < \kappa_{B1}.$$

This result implies that species B, which is adsorbed more strongly by the adsorbent, moves *down* the column and species A moves *up* the column; note that

$$\kappa_{i1} = C_{i1}/C_{i2} = C_{i,\mathrm{solid}}/C_{i,\mathrm{fluid}}. \tag{8.1.343}$$

One could also develop this result by a simple species balance in the feed introduction zone of the column. Let us assume that species A concentration increases upwards in the column. Under the assumption of a dilute feed stream so that W_{tv} is essentially constant along the column height (similarly for W_{ts}), a species A balance around a differential length of the column around the feed point (Figure 8.1.43(b)) leads to

$$W_{tf}\,x_{Af} + W_{tv}\,x_{Av} + W_{ts}(x_{As} + \mathrm{d}x_{As}) = W_{tv}(x_{Av} + \mathrm{d}x_{Av}) + W_{ts}x_{As} \tag{8.1.344a}$$

$$\Rightarrow W_{tv}\,\mathrm{d}x_{Av} - W_{ts}\,\mathrm{d}x_{As} = W_{tf}x_{Af} \tag{8.1.344b}$$

$$\Rightarrow W_{tv}\,\mathrm{d}x_{Av} > W_{ts}\,\mathrm{d}x_{As} \tag{8.1.345a}$$

$$\Rightarrow W_{tv}/W_{ts} > \frac{\mathrm{d}x_{As}}{\mathrm{d}x_{Av}} \tag{8.1.345b}$$

$$\Rightarrow (\langle v_{vz}\rangle\, C_{tv}\, S_c/\,\langle|v_s|\rangle\, C_{ts}\, S_c) > \frac{\mathrm{d}x_{As}}{\mathrm{d}x_{Av}} \tag{8.1.345c}$$

$$\Rightarrow (\langle v_{vz}\rangle/\langle|v_{sz}|\rangle) > \frac{\mathrm{d}C_{As}}{\mathrm{d}C_{Av}} = \kappa_{As}. \tag{8.1.346a}$$

Correspondingly, we can show that

$$\kappa_{Bs} = \frac{\mathrm{d}C_{Bs}}{\mathrm{d}C_{Bv}} > \frac{\langle v_{vz}\rangle}{\langle|v_{sz}|\rangle} \tag{8.1.346b}$$

will lead to the concentration of species B increasing down the column. The overall relation is given by

$$\kappa_{Bs} > \frac{\langle v_{vz} \rangle}{\langle |v_{sz}| \rangle} > \kappa_{As}. \tag{8.1.347}$$

However, as shown in Figures 8.1.43(a) and (b), in the lower half of the column, where hot gas is introduced to strip the adsorbed species, the gas/vapor flow rate is larger than W_{tv} in the top half of the column; further, κ_{Bs} is reduced due to the higher temperature. Therefore the condition that exists is $(\langle v_{vz} \rangle / \langle |v_{sz}| \rangle) > \kappa_{Bs}$, leading to stripping of species B from the downward moving adsorbent and its enrichment in the upward moving hot gas which exits the column near the middle of the column.

The next topics of interest are the operating line for the countercurrent adsorption device and the height of the device or the number of stages needed for the device to achieve a given separation. The device may be of continuous contact type or stagewise contact type. For continuous contact type devices, the approach will follow that in Section 8.1.2.3 employed for gas absorption from a concentrated gas stream. We focus on one particular species i (= A) which is being adsorbed, say, from a gas stream $j = g$; here the inert (solute-free) gas stream mass flow rate w_{tg}^{in} is related to the gas-phase mass flow rate w_{tg} (going up the column in the positive z-direction) by

$$w_{tg} = w_{tg}^{\text{in}} (1 + ur_{ig}), \tag{8.1.348}$$

where the solute mass concentration ratio with respect to the inerts, ur_{ig}, is defined as

$$ur_{ig} = \frac{\text{mass of solute } i \text{ in gas}}{\text{mass of solute-free material in gas}}. \tag{8.1.349}$$

Correspondingly, the inert solute-free solid adsorbent flow rate is w_{ts}^{in} and the solute mass concentration ratio in the adsorbent is

$$ur_{is} = \frac{\text{mass of solute in adsorbent}}{\text{mass of solute-free adsorbent}}. \tag{8.1.350}$$

The two are related to the total solid mass flow rate w_{ts} via

$$w_{ts} = w_{ts}^{\text{in}} (1 + ur_{is}). \tag{8.1.351}$$

In Figure 8.1.44(a), a countercurrent adsorption device has an upflowing gas stream and a downflowing adsorbent stream in continuous contact. Adopting the approach of Section 8.1.1.2, the following solute i balance may be developed between any location z and the column top ($z = L$):

$$w_{ts}^{\text{in}} ur_{isL} + w_{tg}^{\text{in}} \overline{ur}_{ig} = w_{ts}^{\text{in}} \overline{ur}_{is} + w_{tg}^{\text{in}} \overline{ur}_{igL}, \tag{8.1.352}$$

where we have introduced an overbar above each composition to indicate a column cross section averaged quantity unless it is entering from outside (e.g. ur_{isL}). Rearranging, we get the equation for the operating line in terms of the composition variables $\overline{ur}_{ig}$ and $\overline{ur}_{is}$:

$$\overline{ur}_{ig} = \frac{w_{ts}^{\text{in}}}{w_{tg}^{\text{in}}} (\overline{ur}_{is} - ur_{isL}) + \overline{ur}_{igL}. \tag{8.1.353}$$

This line is shown in Figure 8.1.44(b) along with the equilibrium adsorption curve $\overline{ur}_{ig}^{*}$ against $\overline{ur}_{is}$; the column end compositions are also identified. Such a diagram may be utilized to determine the minimum adsorbent to gas flow rate ratio, $\left(w_{ts}^{\text{in}}/w_{tg}^{\text{in}}\right)_{\text{min}}$, when the operating line and the equilibrium curve intersect. Note that the above treatment is based on the assumption of a constant temperature throughout the column; it breaks down when the heat of adsorption is substantial. The analysis for a countercurrent stripper will be similar, only the operating line will be on the other side of the equilibrium curve.

We will now determine the height of the countercurrent adsorber removing a gas/vapor species i = A from the upflowing gas phase. Following equations (8.1.3a) and (8.1.334), we may write the following species i transfer rate over a column length dz at axial location z:

$$w_{ts}^{\text{in}} \frac{d\overline{ur}_{is}}{dz} = -K_{igur} a \left(\overline{ur}_{ig} - ur_{ig}^{*}\right) = w_{tg}^{\text{in}} \frac{d\overline{ur}_{ig}}{dz}. \tag{8.1.354}$$

Here ur_{ig}^{*} is the gas-phase composition in equilibrium with the value of $\overline{ur}_{is}$ valid for the local value of the axial location; the quantity $(\overline{ur}_{ig} - ur_{ig}^{*})$ is the concentration difference driving the adsorption of species i from the gas phase to the adsorbent; the quantity K_{igur} is an overall coefficient based on the gas phase and ur as the concentration variable; one can write similar relations using individual mass-transfer coefficients k_{igur} and k_{isur} for the gas phase and the solid phase, respectively. Over the length L of the column, the gas-phase composition changes from $\overline{ur}_{ig0}$ at the inlet $z = 0$ to $\overline{ur}_{igL}$ at the outlet $z = L$:

$$\int_0^L dz = L = \int_{\overline{ur}_{igL}}^{ur_{ig0}} \frac{w_{tg}^{\text{in}}}{K_{igur}\, a} \frac{d\,\overline{ur}_{ig}}{\left(\overline{ur}_{ig} - \overline{ur}_{ig}^{*}\right)}. \tag{8.1.355}$$

If $K_{igur}\, a$ may be assumed constant along the column height, we can rewrite the above relation as follows:

$$L = \left(\frac{w_{tg}^{\text{in}}}{K_{igur}\, a}\right) \int_{\overline{ur}_{igL}}^{ur_{ig0}} \frac{d\,\overline{ur}_{ig}}{\left(\overline{ur}_{ig} - \overline{ur}_{ig}^{*}\right)}, \tag{8.1.356}$$

$$L = HTU_{og} \times NTU_{og}, \tag{8.1.357a}$$

where

$$HTU_{og} = \left(w_{tg}^{\text{in}}/K_{igur}\, a\right) \tag{8.1.357b}$$

and

$$NTU_{og} = \int_{\overline{ur}_{igL}}^{ur_{ig0}} \frac{d\,\overline{ur}_{ig}}{\left(\overline{ur}_{ig} - \overline{ur}_{ig}^{*}\right)}. \tag{8.1.357c}$$

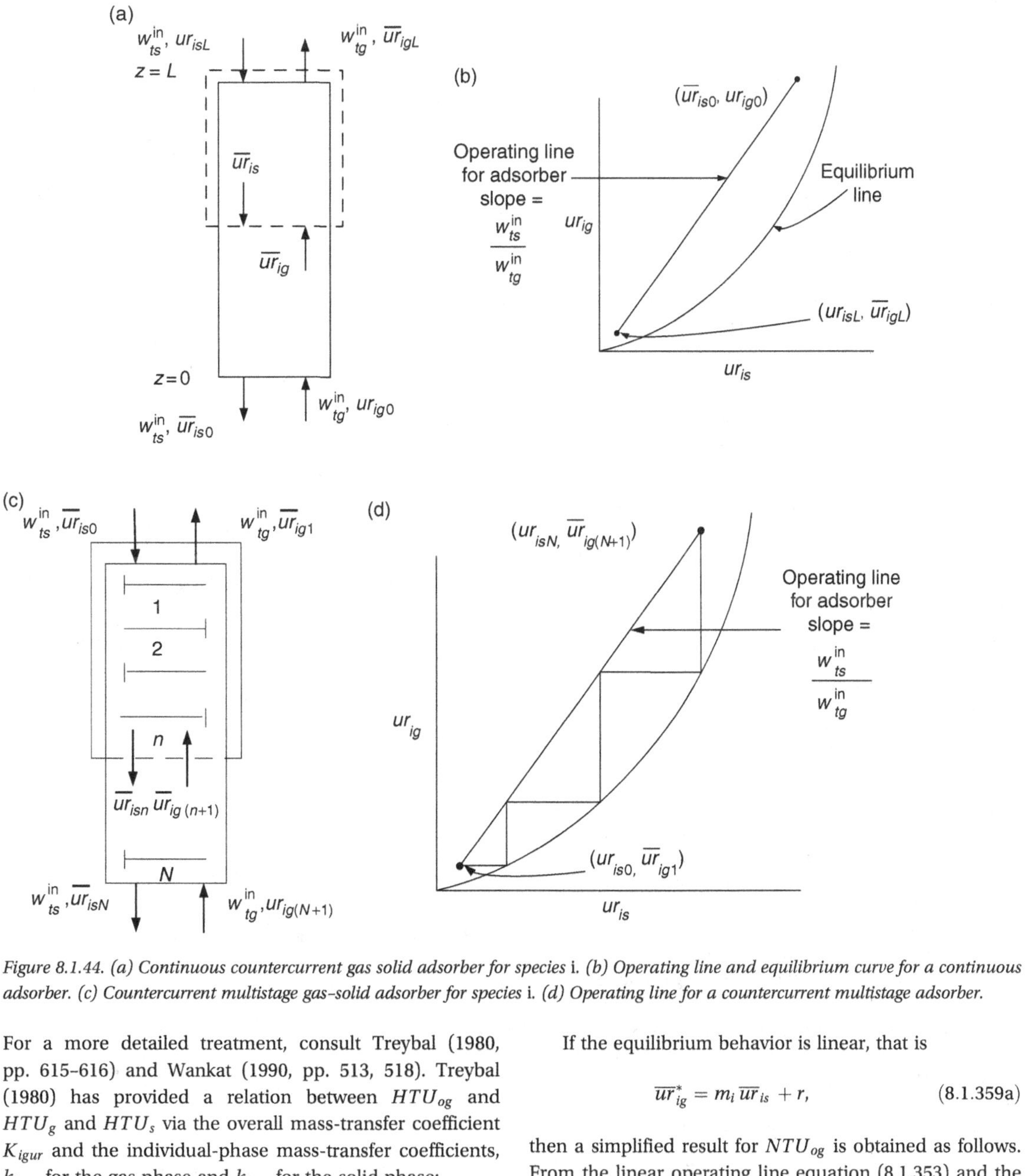

Figure 8.1.44. (a) Continuous countercurrent gas solid adsorber for species i. (b) Operating line and equilibrium curve for a continuous adsorber. (c) Countercurrent multistage gas–solid adsorber for species i. (d) Operating line for a countercurrent multistage adsorber.

For a more detailed treatment, consult Treybal (1980, pp. 615–616) and Wankat (1990, pp. 513, 518). Treybal (1980) has provided a relation between HTU_{og} and HTU_g and HTU_s via the overall mass-transfer coefficient K_{igur} and the individual-phase mass-transfer coefficients, k_{igur} for the gas phase and k_{isur} for the solid phase:

$$\frac{w_{tg}^{\text{in}}}{K_{igur}\,a} = \frac{w_{tg}^{\text{in}}}{k_{igur}\,a} + \frac{m_i\,w_{tg}^{\text{in}}}{w_{ts}^{\text{in}}}\,\frac{w_{ts}^{\text{in}}}{k_{isur}\,a}, \qquad (8.1.358\text{a})$$

$$HTU_{og} = HTU_g + \frac{m_i\,w_{tg}^{\text{in}}}{w_{ts}^{\text{in}}}\,HTU_s. \qquad (8.1.358\text{b})$$

Here

$$m_i = \mathrm{d}ur_{ig}/\mathrm{d}ur_{is}|_{\text{equilibrium}} \qquad (8.1.358\text{c})$$

is the slope of the equilibrium curve (Figure 8.1.44(b)).

If the equilibrium behavior is linear, that is

$$\overline{ur}_{ig}^{*} = m_i\,\overline{ur}_{is} + r, \qquad (8.1.359\text{a})$$

then a simplified result for NTU_{og} is obtained as follows. From the linear operating line equation (8.1.353) and the equilibrium relation (8.1.359a), we get

$$\overline{ur}_{ig} - ur_{ig}^{*} = \overline{ur}_{ig} - m_i\,\overline{ur}_{is} - r; \qquad (8.1.359\text{b})$$

$$\overline{ur}_{ig} - ur_{ig}^{*} = \overline{ur}_{ig} - m_i\left\{\left(\overline{ur}_{ig} - \overline{ur}_{igL}\right)\frac{w_{tg}^{\text{in}}}{w_{ts}^{\text{in}}} + ur_{isL}\right\} - r$$

$$= \left(1 - m_i\,\frac{w_{tg}^{\text{in}}}{w_{ts}^{\text{in}}}\right)\overline{ur}_{ig} + m_i\,\frac{w_{tg}^{\text{in}}}{w_{ts}^{\text{in}}}\,\overline{ur}_{igL} - m_i ur_{isL} - r$$

$$= a\overline{ur}_{ig} + b. \qquad (8.1.359\text{c})$$

Therefore

$$\int_{\overline{ur}_{igL}}^{ur_{ig0}} \frac{d\,\overline{ur}_{ig}}{\left(\overline{ur}_{ig} - \overline{ur}^{*}_{ig}\right)} = \int_{\overline{ur}_{igL}}^{ur_{ig0}} \frac{d\,\overline{ur}_{ig}}{a\overline{ur}_{ig} + b} \tag{8.1.359d}$$

$$= NTU_{og} = \frac{1}{\left(1 - m_i \frac{w_{tg}^{\text{in}}}{w_{ts}^{\text{in}}}\right)} \ell\text{n}\left[\frac{a\,\overline{ur}_{ig0} + b}{a\,\overline{ur}_{igL} + b}\right]; \tag{8.1.359e}$$

$$NTU_{og} = \frac{1}{\left(1 - m_i \frac{w_{tg}^{\text{in}}}{w_{ts}^{\text{in}}}\right)} \ell\text{n}\left[\frac{\left(\overline{ur}_{ig} - ur^{*}_{ig}\right)_{z=0}}{\left(\overline{ur}_{ig} - ur^{*}_{ig}\right)_{z=L}}\right]; \tag{8.1.360}$$

$$NTU_{og} = \frac{\dfrac{\overline{ur}_{ig}|_{z=0} - \overline{ur}_{ig}|_{z=L}}{\left(\overline{ur}_{ig} - ur^{*}_{ig}\right)_{z=0} - \left(\overline{ur}_{ig} - ur^{*}_{ig}\right)_{z=L}}}{\ell\text{n}\left[\left(\overline{ur}_{ig} - ur^{*}_{ig}\right)_{z=0} / \left(\overline{ur}_{ig} - ur^{*}_{ig}\right)_{z=L}\right]} \tag{8.1.361}$$

since

$$\left(\overline{ur}_{ig} - ur^{*}_{ig}\right)_{z=0} - \left(\overline{ur}_{ig} - ur^{*}_{ig}\right)_{z=L} = \left(1 - m_i \frac{w_{tg}^{\text{in}}}{w_{ts}^{\text{in}}}\right)\left(\overline{ur}_{ig}|_{z=0} - \overline{ur}_{ig}|_{z=L}\right). \tag{8.1.362}$$

Example 8.1.19 A countercurrent continuous contact silica gel adsorber is being used to dry humid air entering with $ur_{H_2O,g} = 0.006\text{g}\, H_2O/\text{g}$ dry air to the level of 0.0002 g H_2O/g dry air. The silica gel may be assumed to be essentially dry as it enters the top of the adsorber. Treybal (1980, p. 616) has employed data from Eagleton and Bliss to determine the values for HTU_g and HTU_s for silica gel particles of 1.727 mm diameter for a certain range of gas and solid flow rates. Assume on that basis HTU_g and HTU_s to be 0.03 m and 0.69 m for an inert gas flow rate of $w_{tg}^{\text{in}} = 0.4$ kg/s and dry silica gel flow rate of $w_{ts}^{\text{in}} = 0.225$ kg/s in the column of diameter 56 cm. The temperature of the air coming in is 25 °C. The linear equilibrium relation is $ur^{*}_{H_2O,g} = 0.02\, ur_{H_2O,\text{silicagel}}$. Determine the height of the column.

Solution To determine the column height L, determine HTU_{og} and NTU_{og} (equation (8.1.357a)). From equation (8.1.358b)

$$HTU_{og} = HTU_g + \frac{m_i\, w_{tg}^{\text{in}}}{w_{ts}^{\text{in}}} HTU_s,$$

where $m_i = 0.02$; $w_{tg}^{\text{in}} = 0.4$ kg/s; $w_{ts}^{\text{in}} = 0.225$ kg/s. So

$$HTU_{og} = \left(0.03 + \frac{0.02 \times 0.4}{0.225} \times 0.69\right)\text{m} = (0.03 + 0.0245)$$

$$\Rightarrow HTU_{og} = 0.0545\,\text{m}.$$

To determine NTU_{og}, focus on equation (8.1.360). We need to determine $(\overline{ur}_{ig} - ur^{*}_{ig})$ at the two ends of the adsorption column. Since the silica gel is entering essentially dry,

$$ur^{*}_{H_2O,g}|_{z=L} = 0.02\, ur^{*}_{H_2O,\,\text{silica gel}} = 0.02 \times 0 = 0.$$

We now need to determine $ur^{*}_{H_2O,g}\big|_{z=0} \left(= 0.02\, ur_{H_2O,\,\text{silica gel}}\right)\big|_{z=0}$ by determining $ur_{H_2O,\,\text{silica gel}}$ at $z = 0$. Do an overall column balance:

$$w_{tg}^{\text{in}}\, ur_{H_2O,g}\big|_{z=0} + w_{ts}^{\text{in}}\, ur_{H_2O,\,\text{silica gel}}\big|_{z=L} = w_{tg}^{\text{in}}\, ur_{H_2O,g}\big|_{z=L} + w_{ts}^{\text{in}}\, ur_{H_2O,\,\text{silica gel}}\big|_{z=0}$$

$$\Rightarrow 0.4 \times 0.006 + 0.225 \times 0 = 0.4 \times 0.0002 + 0.225 ur_{H_2O,\,\text{silica gel}}\big|_{z=0}$$

$$\Rightarrow ur_{H_2O,\,\text{silica gel}}\big|_{z=0} = \frac{0.4 \times 0.0058}{0.225} = 0.0103$$

$$\Rightarrow ur^{*}_{H_2O,g}\big|_{z=0} = 0.02 \times 0.0103 = 0.000206;$$

$$NTU_{og} = \frac{1}{\left(1 - m_i \frac{w_{tg}^{\text{in}}}{w_{ts}^{\text{in}}}\right)} \ell\text{n}\left[\frac{\left(\overline{ur}_{ig} - ur^{*}_{ig}\right)_{z=0}}{\left(\overline{ur}_{ig} - ur^{*}_{ig}\right)_{z=L}}\right]$$

$$= \frac{1}{\left(1 - 0.02\,\frac{0.4}{0.225}\right)} \ell\text{n}\left[\frac{0.006 - 0.000206}{0.0002 - 0}\right]$$

$$= \frac{1}{(1 - 0.0355)} \ell\text{n}\left[\frac{0.00574}{0.0002}\right]$$

$$= \frac{1}{0.9645}\, 2.303 \times 1.462 = 3.49$$

$$\Rightarrow L = 0.0545 \times 3.49 = 0.19\,\text{m}.$$

Often if countercurrent adsorption is carried out, it may be of the multistaged type like that in Figure 8.1.43(a), or the Cloethe–Streat type or other configurations. The operating line for such operation may be developed following Section 8.1.2.6 for gas absorption. As shown in Figure 8.1.44(c), for a gas–solid system, for example, a mass balance for species i between the top and the nth plate leads for a dilute system to a linear relation between $\overline{ur}_{ig(n+1)}$ and $\overline{ur}_{isn}$ at the bottom of plate n:

$$w_{tg}^{\text{in}}\, \overline{ur}_{ig(n+1)} + w_{ts}^{\text{in}}\, ur_{is0} = w_{tg}^{\text{in}}\, \overline{ur}_{ig1} + w_{ts}^{\text{in}}\, \overline{ur}_{isn}; \tag{8.1.363a}$$

$$\overline{ur}_{ig(n+1)} = \frac{w_{ts}^{\text{in}}}{w_{tg}^{\text{in}}}\, \overline{ur}_{isn} + \left(\overline{ur}_{ig1} - \frac{w_{ts}^{\text{in}}}{w_{tg}^{\text{in}}}\, \overline{ur}_{is0}\right). \tag{8.1.363b}$$

In Figure 8.1.44(d), we can now develop a McCabe–Thiele type diagram for the adsorber starting from the point $(ur_{is0}, \overline{ur}_{ig1})$ on the operating line (8.1.363a,b). Draw a horizontal line to the equilibrium curve; from the point where this line intersects the equilibrium curve, draw a vertical line to the operating line; from the point where this vertical line intersects the operating line, draw a horizontal line to the equilibrium curve, and so on till you hit the value $\overline{ur}_{ig(N+1)}$ on the operating line corresponding to the feed gas stream; the figure shows a total of three stages. Similarly, one can develop a McCabe–Thiele type of diagram for the regenerator or the stripper. When the equilibrium curve is also a straight line, one could employ the Kremser equation in the manner of Section 8.1.2.6.

8.1.6.2 *Analytical solution for countercurrent continuous adsorption of one species in a dilute mixture with dispersion in fluid phase*

The analysis we carried out earlier resulting in solutions, e.g. equation (8.1.360), was carried out for plug flow of both phases. However, there is usually some axial dispersion at least in the fluid phase (e.g. a liquid phase). The result of existing analyses as described by Ruthven (1984) will be provided here for downward plug flow of the solid phase $j = 1$ and upward ($z > 0$) axially dispersed flow of fluid phase $j = 2$. The governing balance equation from equation (6.2.34) is therefore

$$\varepsilon_1 \frac{\partial \overline{C}_{i1}}{\partial t} + \varepsilon_2 \frac{\partial \overline{C}_{i2}}{\partial t} + \langle v_{1z} \rangle \frac{\partial \overline{C}_{i1}}{\partial z} + \langle v_{2z} \rangle \frac{\partial^2 \overline{C}_{i2}}{\partial z} = \varepsilon_2 D_{i,\text{eff},2} \frac{\partial \overline{C}_{i2}}{\partial z^2}; \tag{8.1.364}$$

$$\frac{\varepsilon_1}{\varepsilon_2} \frac{\partial \overline{C}_{i1}}{\partial t} + \frac{\partial \overline{C}_{i2}}{\partial t} = D_{i,\text{eff},2} \frac{\partial^2 \overline{C}_{i2}}{\partial z^2} - v_{2z} \frac{\partial \overline{C}_{i2}}{\partial z} + \frac{\varepsilon_1}{\varepsilon_2} |v_{1z}| \frac{\partial \overline{C}_{i1}}{\partial z}. \tag{8.1.365}$$

Note that since there are only two phases, $\varepsilon_1 = (1 - \varepsilon_2)$. Consider steady state: the terms on the left-hand side disappear. We now employ the linear driving force assumption (7.1.5b) along with linear equilibrium relation (7.1.18a), $\overline{C}_{i1}/\overline{C}_{i2} = \kappa_{i1}$ and the relation $|v_{1z}| = -(z/t)$ relating the solid-phase velocity to time and axial coordinates:

$$\frac{\partial \overline{C}_{i1}}{\partial t} = \frac{15 D_{ip}}{r_p^2} \left(\overline{C}_{i1}^* - \overline{C}_{i1} \right) = k_{ip} \left(\overline{C}_{i1}^* - \overline{C}_{i1} \right)$$
$$= k_{ip} \left(\kappa_{i1} \overline{C}_{i2} - \overline{C}_{i1} \right) = -|v_{1z}| \frac{\partial \overline{C}_{i1}}{\partial z}. \tag{8.1.366}$$

These assumptions allow us to express equation (8.1.365) as

$$D_{i,\text{eff},2} \frac{\partial \overline{C}_{i2}}{\partial z^2} - v_{2z} \frac{\partial \overline{C}_{i2}}{\partial z} - \frac{(1 - \varepsilon_2)}{\varepsilon_2} k_{ip} \left(\kappa_{i1} \overline{C}_{i2} - \overline{C}_{i1} \right) = 0. \tag{8.1.367}$$

The overall species i balance over the adsorber column length L for a dilute mixture of species i to be adsorbed may be carried out by assuming that v_{2z} and $|v_{1z}|$ remain essentially constant along the column:

$$\varepsilon_2 v_{2z} \left(C_{i20} - \overline{C}_{i2L} \right) = -|v_{1z}| (1 - \varepsilon_2) \left(C_{i1L} - \overline{C}_{i10} \right). \tag{8.1.368}$$

The boundary conditions similar to those for a gas absorber (8.1.91a–d) are in this case as follows. For $z \to 0$ (or $z = 0^+$),

$$\varepsilon_2 v_{2z} C_{i20} - \varepsilon_2 v_{2z} C'_{i20} = -\varepsilon_2 D_{i,\text{eff},2} \frac{d \overline{C}_{i2}}{dz}; \tag{8.1.369a}$$

$$\frac{d \overline{C}_{i1}}{dz} = 0; \tag{8.1.369b}$$

For $z \to L$ (or $z = L^-$),

$$\frac{d \overline{C}_{i2}}{dz} = 0. \tag{8.1.369c}$$

Using the following dimensionless quantities:

$$\phi = \overline{C}_{i2}/C_{i20}; \quad z^+ = z/L; \quad Pe = \frac{v_{2z} L}{D_{i,\text{eff},2}};$$
$$St = \frac{k_{ip} L}{v_{2z}}; \quad \gamma = \frac{(1 - \varepsilon_2) \kappa_{i1} |v_{1z}|}{\varepsilon_2 v_{2z}}, \tag{8.1.370}$$

Ruthven (1984) has expressed the solution of the differential equation (8.1.367) and boundary conditions (8.1.369a–c) in a dimensionless form,

$$\frac{\phi(1 - (1/\gamma)) + (1/\gamma) - (C_{i10}/\kappa_{i1} C_{i20})}{1 - (C_{i10}/\kappa_{i1} C_{i20})}$$
$$= \frac{m_1 \exp(m_1 + m_2 z^+) - m_2 \exp(m_2 + m_1 z^+)}{m_1 (\exp m_1)(1 - (m_2/Pe)) - m_2 (\exp m_2)(1 - (m_1/Pe))}. \tag{8.1.371}$$

See equations (8.1.373) and (8.1.374) below for m_1 and m_2.

For an adsorption process where $\gamma > 1$, $C_{i1L} = 0$, the overall mass balance relation (8.1.368) yields

$$\frac{\overline{C}_{i10}}{\kappa_{i1} C_{i20}} = \frac{1 - \left(\overline{C}_{i2L}/C_{i20} \right)}{\gamma}, \tag{8.1.372}$$

which reduces the general solution (8.1.371) (Ruthven, 1984) to

$$\frac{\phi(\gamma - 1) + \phi_L}{\gamma - 1 + \phi_L}$$
$$= \frac{m_1 \exp(m_1 + m_2 z^+) - m_1 \exp(m_2 + m_1 z^+)}{m_1 (\exp(m_1))(1 - (m_2/Pe)) - m_2 (\exp(m_2))(1 - (m_1/Pe))}, \tag{8.1.373}$$

where m_1 (+ve sign) and m_2 (-ve sign) are the roots of the following equation:

$$m = \frac{1}{2} \left\{ (Pe + St) \pm \left[(Pe + St)^2 + 4 Pe St(\gamma - 1) \right]^{1/2} \right\}, \tag{8.1.374}$$

leading to

$$\phi_L = \left(\overline{C}_{i2L}/C_{i20} \right) = \frac{\gamma - 1}{(\gamma/A) - 1}, \tag{8.1.375}$$

where

$$A = \frac{m_1 - m_2}{m_1 (\exp(-m_2))(1 - (m_2/Pe)) - m_2 (\exp(-m_1))(1 - (m_1/Pe))}. \tag{8.1.376}$$

8.1.6.3 *Simulated moving bed (SMB) process*

We have already introduced the basic notion of the SMB process in the material accompanying Figure 8.1.6. Here we elaborate on it. Consider Figure 8.1.45(a), which shows four packed adsorbent beds each of length L (they could be different lengths) located in one column (they do not have to be located in the same column; one can have four separate packed-bed columns connected to one another externally (Wankat, 1990, p. 527)). The solid lines with arrows in Figure 8.1.45(a) indicate the lines along which the mobile phase having a particular composition is moving. The dashed vertical line on the left of this figure identifies the hypothetical recirculation of the adsorbent particles from the column bottom to the top for the hypothetical countercurrent system.

Consider now how SMB operates. As indicated earlier, if the adsorbent particles were actually coming down in any one of the four beds, the relative velocity between the upward mobile phase and the downward adsorbent particles will be significantly higher than the actual upward mobile-phase velocity. Under such a condition, the mobile phase would have been in contact with particles higher up in the bed than it is in actual contact within a fixed bed. Therefore, after a defined time period, t^*, called the switching time, the mobile phase is introduced into a bed port at a higher up location in the packed-bed system. This arrangement is simultaneously implemented in four locations in between the four columns, as shown in Figure 8.1.45(b). Further, this switching is continued, as shown in Figures 8.1.45(c) and (d) till we are back to the original scheme in Figure 8.1.45(a); then the cycle is repeated.

To understand the mechanics better, focus on the separation of species A (glucose) and B (fructose) from a feed solution using ion exchange resin in the Ca^{2+} form which adsorbs fructose strongly with respect to glucose. In bed 1 (Figure 8.1.45(a)), the goal is to produce purified desorbent for recycle to column 4 at the bottom; any glucose remaining after withdrawal of the raffinate product (enriched in glucose) should not be allowed to contaminate the desorbent product via breakthrough. One can calculate the time $\bar{t}$ needed for glucose breakthrough using the concentration wave velocity expression (7.1.13e) for v^*_{CA} and the bed length L for an interstitial velocity v_z

$$\frac{L}{\bar{t}} = v^*_{CA,1} \Rightarrow \bar{t} = \frac{L}{v^*_{CA,1}} \tag{8.1.377}$$

(in $v^*_{CA,1}$, the subscript 1 refers to column 1). Therefore, after a time $\bar{t} > t^*$ where

$$t^* = L/v_{port}, \qquad v^*_{CA,1} = \alpha_{1A}\, v_{port}, \qquad \alpha_{1A} \leq 1, \tag{8.1.378}$$

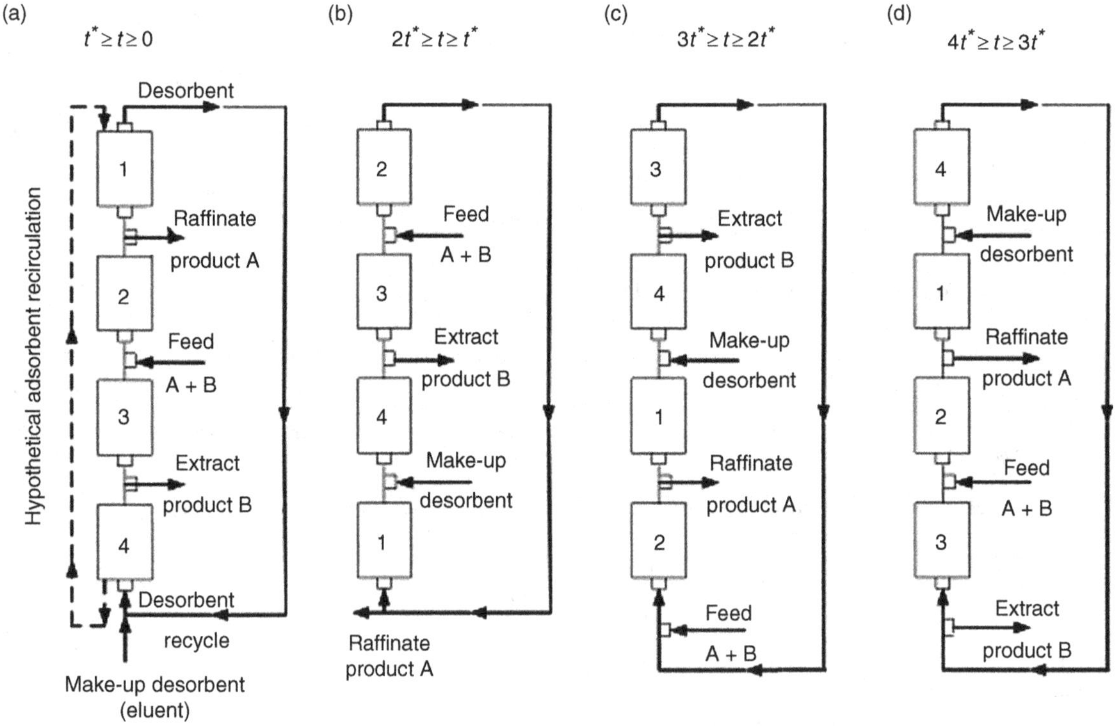

Figure 8.1.45. Simulated moving bed operation with four beds and four equal time steps for separation of a binary mixture of glucose (A) and fructose (B).

v_{port} being defined as the average port velocity (Wankat, 1990, p. 529, 2007, pp. 649–654), the liquid introduction at the top column 1 is now from the feed solution (Figure 8.1.45(b)) and the top column now behaves as if it were column 2 of Figure 8.1.45(a). Correspondingly, column 4 of Figure 8.1.45(a) now (in Figure 8.1.45(b)) behaves as if it were column 1 of Figure 8.1.45(a). Similarly, the other liquid introduction ports now have different liquid streams being introduced (see Figure 8.1.45(b))).

In bed 2 of Figure 8.1.45(a), species A (glucose) moves up toward column 1 and species B (fructose), which is more strongly adsorbed, moves down. If the raffinate product stream from bed 2 (Figure 8.1.45(a)) is to be a glucose-enriched stream, then glucose will break through bed 2 during 0 to t^*, i.e. before port switching. This suggests that the concentration wave velocity of glucose in bed 2, $v^*_{CA,2}$, is

$$v^*_{CA,2} = \alpha_{2A}\, v_{port} \qquad (\alpha_{2A} \geq 1). \tag{8.1.379}$$

Further, we do not want fructose to break through into the raffinate product stream. Therefore

$$v^*_{CB,2} = \alpha_{2B}\, v_{port} \qquad (\alpha_{2B} \leq 1). \tag{8.1.380}$$

In bed 3 of Figure 8.1.45(a), we want similar behavior in terms of movement of glucose and fructose: glucose should move up and break through, whereas fructose should have a net downward movement and not break through into the stream exiting bed 3:

$$v^*_{CA,3} = \alpha_{3A} v_{port} \qquad (\alpha_{3A} \geq 1); \tag{8.1.381}$$

$$v^*_{CB,3} = \alpha_{3B} v_{port} \qquad (\alpha_{3B} \leq 1). \tag{8.1.382}$$

On the other hand, in bed 4 of Figure 8.1.45(a), we want fructose to break through into the top exiting stream from which the extract product enriched in fructose is withdrawn before port switching occurs. Therefore

$$v^*_{CB,4} = \alpha_{4B} v_{port} \qquad (\alpha_{4B} \geq 1). \tag{8.1.383}$$

An important item to be noted here is that the interstitial velocity v_z is different in different beds because of the extract product withdrawal between beds 3 and 4, feed introduction between beds 3 and 2 and raffinate product withdrawal between beds 2 and 1 (Figure 8.1.45(a)). Therefore

$$v_z\big|_{bed\,4} - v_z\big|_{bed\,3} = v_z\big|_{extract\,product}, \tag{8.1.384}$$

assuming the same cross section for beds 3 and 4 and appropriate allowance for extract product mass balance. Similarly,

$$v_z\big|_{bed\,2} - v_z\big|_{bed\,1} = v_z\big|_{raffinate\,product}. \tag{8.1.385}$$

Equations (8.1.378)–(8.1.383) have to be solved simultaneously with additional constraints of equations such as (8.1.384) and (8.1.385). See Wankat (1990, 2007) for numerical illustrations. Earlier analyses of such a process developed at UOP are available in Ching and Ruthven (1984, 1985) and Ching *et al.* (1986).

8.1.7 Membrane processes of dialysis and electrodialysis

We consider briefly the processes of countercurrent dialysis and electrodialysis.

8.1.7.1 *Countercurrent dialysis, buffer exchange, hemodialyzer*

The process of dialysis, in which small solutes diffuse from a feed solution through a microporous membrane into the dialyzing solution, was illustrated in Section 4.3.1 via a membrane kept in a closed vessel. To achieve separation in a continuous fashion, we realized there that the dialyzer had to be open. Such a device is schematically shown in Figure 8.1.46(a). A small hollow fiber membrane based dialyzer used as an artificial kidney is illustrated in Figure 8.1.46(b). Both utilize countercurrent flow of feed solution (for example, blood) and the dialyzing solution on two sides of the membrane.

In hemodialysis, blood flows through the bore of the hollow fiber, and the dialyzing solution flows countercurrently on the shell side. Small molecular weight metabolic waste products, like urea, uric acid, creatinine, etc., are transferred from the blood by diffusion through the water-filled membrane pores to the aqueous dialysate solution (Figure 8.1.46(c)). Traditionally, the pore radii of dialysis membranes are >1.5 nm (see the discussion following equation (4.3.15)) large enough for metabolic waste products to diffuse through rapidly, but not large enough to allow blood plasma proteins, for example serum albumin, to pass through. Many modern hemodialysis membranes have much larger pores, with pore radii in the range of 7.5 to 10 nm to facilitate removal of solutes of "middle" molecular weight; however, as soon as these membranes are exposed to blood, substantial protein adsorption occurs, eliminating the possibility of protein loss through membrane pores, whose open diameters become substantially reduced. The fiber wall thickness varies between 10 and 50 μm. The hollow fiber internal diameters are generally around 200 μm. For a comprehensive introduction to these topics, see Kessler and Klein (2001).

A major nonhemodialysis application of countercurrent dialysis (CCD) is the buffer exchange of proteins in the biopharmaceutical industry. In multistep biopharmaceutical production processes, the buffer solution constituents often have to be exchanged for protein purification steps. The original buffer solution, for example, may contain, say, 0.5 M ammonium sulfate and 0.05 M sodium phosphate (plus protein and other constituents); the CCD

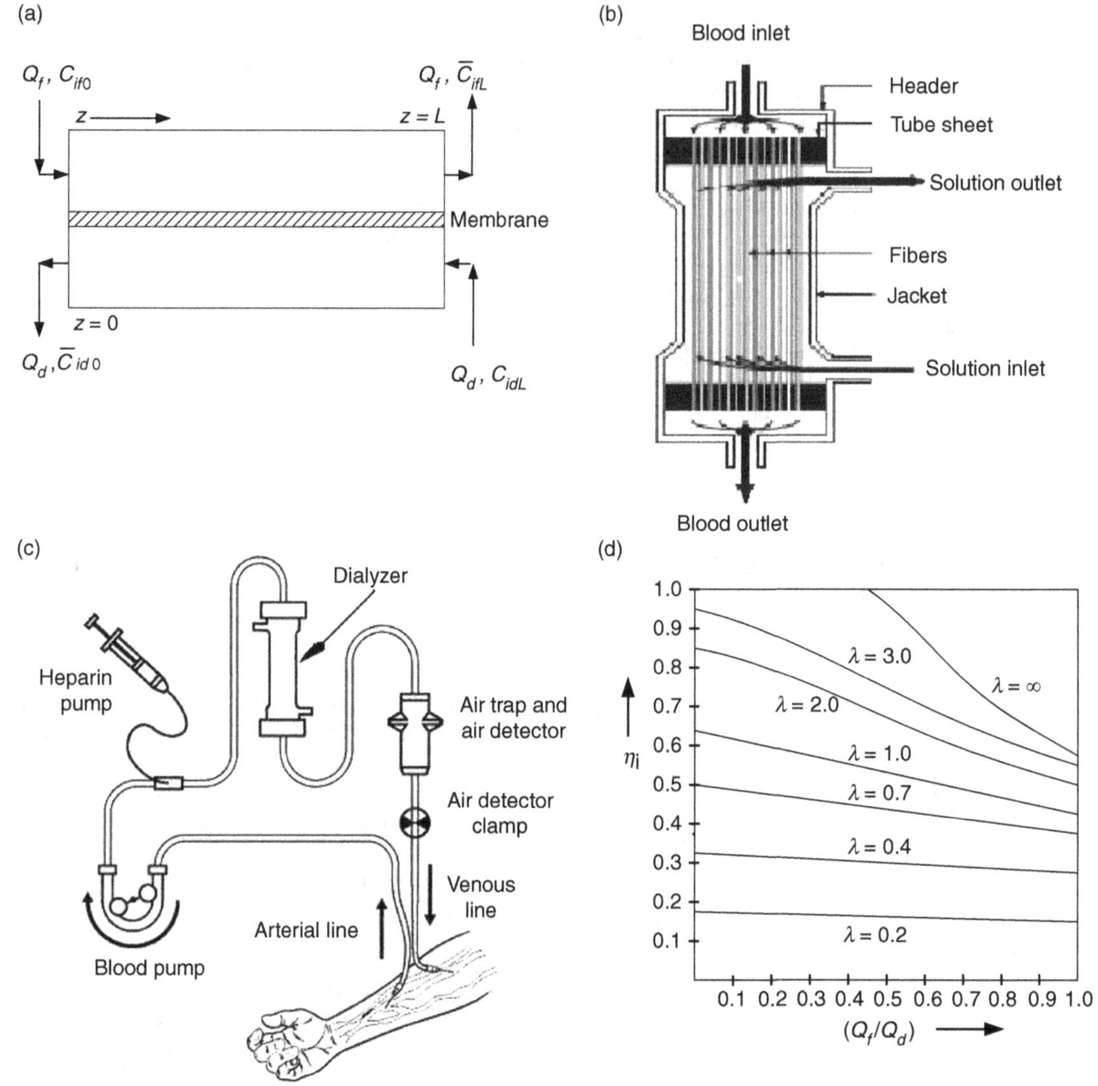

Figure 8.1.46. Countercurrent dialyzer: (a) flow and process schematic; (b) hemodialyzer, device schematic; (c) hemodialysis, blood flow loop; (d) hemodialyzer efficiency η_i *plotted against* (Q_f/Q_d) *for various values of parameter* λ. *(Part (d) after Michaels (1966).)*

process should remove, say, 99% of both ammonium sulfate and sodium phosphate into an appropriate aqueous dialyzing solution (Kurnik *et al.*, 1995). These authors have demonstrated that CCD is more efficient than size exclusion chromatography (see Section 7.1.5.1.7) for buffer exchange. Correspondingly, desalting of protein solutions can also be efficiently implemented using CCD. There have been many other industrial applications of dialysis, the earliest being the recovery of NaOH from cellulose-steeping liquors; the dialysis membrane blocked cellulosic polymers, but allowed NaOH molecules to diffuse from a 20% solution to an aqueous phase leading to a dilute NaOH solution (~4%) as dialysate (Kessler and Klein, 2001). This reference also illustrates various types of dialysis membrane devices developed and used over the years.

We will now employ a lumped model of a countercurrent dialyzer to illustrate its performance in the context of hemodialysis (Figures 8.1.46(a) and (b)). We will use a device-averaged overall solute-transfer coefficient $\overline{K}_{ic}$ instead of employing a *z*-coordinate dependent analysis of the type illustrated in (8.1.46) for countercurrent phase equilibrium based devices. In Figure 8.1.46(a), a feed solution having a molar solute concentration of C_{if0} (for example, blood) enters at $z = 0$ with a volumetric flow rate Q_f, and treated solution (Q_f, $\overline{C}_{ifL}$) exits at $z = L$. The membrane area is A_m. The dialyzing solution (Q_d, C_{idL}) enters at $z = L$ and leaves at $z = 0$ as (Q_d, $\overline{C}_{id0}$). The overall molar balance relations for the transfer of solute i using a device-averaged $\overline{K}_{ic}$ are:

$$Q_f\,(C_{if0} - \overline{C}_{ifL}) = Q_d(\overline{C}_{id0} - C_{idL}) = \overline{K}_{ic}A_m\,(\Delta\,C_i)_{\ell m}, \tag{8.1.386}$$

where

$$(\Delta\,C_i)_{\ell m} = \frac{[(C_{if0} - \overline{C}_{id0}) - (\overline{C}_{ifL} - C_{idL})]}{\ell n[(C_{if0} - \overline{C}_{id0}) / (\overline{C}_{ifL} - C_{idL})]} \tag{8.1.387}$$

is the overall logarighmic-mean concentration difference between the blood side and the dialysate driving the solute transport.

Hemodialyzer performance is analyzed using a number of indices: solute transfer efficiency η_i, clearance C, and dialysance D_B. The solute transfer efficiency η_i (Michaels, 1966),

$$\eta_i = (C_{if0} - \overline{C}_{ifL}) / (C_{if0} - C_{idL}), \tag{8.1.388}$$

represents the feed solution purification achieved as a fraction of the highest possible concentration change $(C_{if0} - C_{idL})$ in the bloodstream. Dialysance D_B of the bloodstream is defined for a countercurrent dialyzer as (Sweeney and Galletti, 1964; Michaels, 1966)

$$D_B = \left(Q_f(C_{if0} - \overline{C}_{ifL}) / (C_{if0} - C_{idL})\right). \tag{8.1.389}$$

Therefore D_B is merely $Q_f\,\eta_i$. The clearance C of a solute i is essentially blood dialysance D_B, where $C_{idL} = 0$:

$$C = Q_f(C_{if0} - \overline{C}_{ifL})\,/C_{if0}. \tag{8.1.390}$$

We will now develop an expression for η_i in terms of known parameters and variables. From relation (8.1.386), we obtain

$$\left(1 - \left(Q_f/Q_d\right)\right) = 1 - \frac{(\overline{C}_{id0} - C_{idL})}{(C_{if0} - \overline{C}_{ifL})} = \frac{(C_{if0} - \overline{C}_{id0}) - (\overline{C}_{ifL} - C_{idL})}{(C_{if0} - \overline{C}_{ifL})}. \tag{8.1.391}$$

Employing this result, we can rewrite relation (8.1.386) as

$$\overline{K}_{ic}A_m(\Delta\,C_i)_{\ell m} = \overline{K}_{ic}\,A_m\,\frac{(C_{if0} - \overline{C}_{ifL})\left(1 - \left(Q_f/Q_d\right)\right)}{\ell n\,[(C_{if0} - \overline{C}_{id0})/(\overline{C}_{ifL} - C_{idL})]} = Q_f(C_{if0} - \overline{C}_{ifL}), \tag{8.1.392}$$

resulting in

$$\ell n\left[\frac{C_{if0} - \overline{C}_{id0}}{\overline{C}_{ifL} - C_{idL}}\right] = \frac{\overline{K}_{ic}\,A_m\left(1 - \left(Q_f/Q_d\right)\right)}{Q_f}. \tag{8.1.393}$$

Define

$$\lambda = \overline{K}_{ic}\,A_m\,/Q_f. \tag{8.1.394}$$

Therefore

$$\frac{C_{if0} - \overline{C}_{id0}}{\overline{C}_{ifL} - C_{idL}} = \exp\left(\lambda\left(1 - \left(Q_f/Q_d\right)\right)\right). \tag{8.1.395}$$

Now obtain the following two results:

$$1 - \exp\left(\lambda\left(1 - \left(Q_f/Q_d\right)\right)\right) = \frac{\overline{C}_{ifL} - C_{idL} - C_{if0} + \overline{C}_{id0}}{\overline{C}_{ifL} - C_{idL}}; \tag{8.1.396}$$

$$\left(Q_f/Q_d\right) - \exp\left(\lambda\left(1 - \left(Q_f/Q_d\right)\right)\right) = \frac{\overline{C}_{id0} - C_{idL}}{C_{if0} - \overline{C}_{ifL}} - \frac{C_{if0} - \overline{C}_{id0}}{\overline{C}_{ifL} - C_{idL}}. \tag{8.1.397}$$

A little algebra leads to the following result, providing an expression for η_i:

$$\frac{1 - \exp\left(\lambda\left(1 - \left(Q_f/Q_d\right)\right)\right)}{\left(Q_f/Q_d\right) - \exp\left(\lambda\left(1 - \left(Q_f/Q_d\right)\right)\right)} = \frac{C_{if0} - \overline{C}_{ifL}}{C_{if0} - C_{idL}} = \eta_i. \tag{8.1.398}$$

Figure 8.1.46(d) illustrates a few curves from a more detailed plot by Michaels (1966) with λ as a parameter. As Q_d is decreased for a given blood flow rate Q_f, η_i decreases, since C_{id} increases, leading to a reduced solute transfer rate. At any given Q_f/Q_d, η_i increases as λ increases, since the solute transfer rate is increased. There is an additional interpretation of the parameter λ. From an analysis of the packed countercurrent column design equation (8.1.54a), we know

$$L = \frac{|W_{tg}|}{K_{ixg}\,a\,S_c}\,NTU_{og}.$$

However, $K_{ixg}\,a\,S_c\,L = K_{ixg}\,A$, where A is the total interfacial area in the transfer device of total volume $S_c\,L$, the interfacial area/volume being a. In the dialyzer, A will be A_m; therefore,

$$NTU_{og} = \frac{K_{ixg}\,a\,S_c\,L}{|W_{tg}|} \cong \frac{\overline{K}_{ic}\,A_m}{Q_f} = \lambda \tag{8.1.399}$$

for the dialyzer. Since NTU_{og} is an estimate of how difficult the separation is, it provides a guide to how large the dialyzer should be (Kessler and Klein, 2001). These authors have plotted a quantity called the extraction ratio E, which is equivalent to η_i when $C_{idL} = 0$, against λ, with $\left(Q_f/Q_d\right)$ as a parameter.

There are a number of other important issues in hemodialysis. So far, for solute transport through the membrane, we have assumed only diffusion; but the device design and pressure conditions employed may lead to ultrafiltration via convective motion through the pores. For an introduction to this topic, see Kessler and Klein (2001) for the considerable contribution of convection to the accelerated removal of larger molecules whose diffusive transport rates are quite low. Another important issue is the role of the boundary layer mass-transfer resistances. The local value of the overall solute-transfer coefficient K_{ic} in such a configuration, illustrated by relation (3.4.103), may be described in terms of the local values of the feed side and dialysate side mass-transfer coefficients k_{ifc} and k_{idc}:

$$\frac{1}{K_{ic}} = \frac{1}{k_{ifc}} + \frac{1}{\kappa_{im}}\frac{1}{k_{im}} + \frac{1}{k_{idc}}. \tag{8.1.400}$$

Here the membrane mass-transfer coefficient k_{im} can be described by (3.4.95c,d). It is known that each of the local boundary layer transfer coefficients, k_{ifc} and k_{idc}, is a function of the distance from the liquid channel inlet z in the manner of $(1/z)^{0.33}$. This z-dependence is influenced by the nature of the flow channel geometry. Consult Kessler and Klein (2001) to develop a better understanding of the important role of the boundary-layer transfer coefficients.

Example 8.1.20 Countercurrent dialysis may be explored to eliminate substantially 0.05M sodium phosphate from a 1 mg/ml solution of bovine serum albumin (BSA) and 25 mM Tris buffer at a *pH* of 7 (via NaOH). The dialyzing buffer composition is 25 mM Tris and the *pH* is 7 (via HCl). The dialysate flow rate is 100 ml/min. The membrane module has 0.2 m^2 surface area. The value of $\overline{K}_{ic}$ for the phosphate salt for one membrane module is 0.020 cm/min and that for two modules in series is 0.015 cm/min. Determine the values of $(C_{if0}/\overline{C}_{ifL})$ for sodium phosphate using the expression developed in Problem 8.1.22 for the following cases: (a) feed flow rate, 10 cm^3/min; (b) feed flow rate, 20 cm^3/min; (c) feed flow rate, 10 cm^3/min and two membrane modules in series. Comment on the effects of feed flow rate and putting membrane modules in series.

Solution *Part (a)* The expression for $(C_{if0}/\overline{C}_{ifL})$ is:

$$\frac{C_{if0}}{\overline{C}_{ifL}} = \frac{Q_d \exp\left(\lambda\left(1-\left(Q_f/Q_d\right)\right)\right)}{\left(Q_d - Q_f\right)}$$

(here $C_{idL} = 0$);

$$(C_{if0}/\overline{C}_{ifL}) = 100 \exp\left(\frac{0.02\times 0.2\times 10^4}{10}\left(1-\frac{10}{100}\right)\right)\Big/(100-10)$$

$$= \frac{100}{90} \times \exp(3.6) = 40.67.$$

Part (b)

$$(C_{if0}/\overline{C}_{ifL}) = Q_d \exp\left(\lambda\left(1-(Q_f/Q_d)\right)\right)/\left(Q_d - Q_f\right)$$

$$= \frac{100}{80}\exp\left(\frac{0.02\times 0.2\times 10^4}{20}\times 0.8\right) = 6.18.$$

Part (c)

$$(C_{if0}/\overline{C}_{ifL}) = \frac{100}{90}\exp\left(\frac{0.015\times 0.4\times 10^4}{10}\times 0.9\right) = 246.$$

A lower feed flow rate leads to a much higher level of purification. Increasing the membrane area does lower the mass-transfer coefficient; however, the purification is increased drastically.

8.1.7.2 *Countercurrent electrodialysis*

We have been introduced to the ion transport processes taking place in an electrolyte-containing solution adjacent to a cation exchange membrane in the diluate chamber of an electrodialysis (ED) device in Section 3.4.2.5. Figure 3.4.9 illustrated there shows two anion exchange membranes (AEMs) and one cation exchange membrane (CEM) in the ED device. An ED device in practice has many AEMs and CEMs between the two electrodes; the assembly is sometimes known as a membrane stack. One configuration employed shows (Figure 8.1.47(a)) the brine feed to be desalted introduced into one chamber and the same brine feed introduced countercurrently as a waste stream into the adjacent chamber. However, as shown, the brine feed is introduced at one end of the cascade while the brine feed acting as the wash water is introduced at the other end of the cascade. The region containing one AEM, the diluate solution in between this membrane and the adjacent CEM on the right, the CEM and concentrate solution on the right of this CEM all the way upto the next AEM constitutes a *cell pair.*

In Figure 8.1.47(a), the brine feed to be desalted is introduced on the left-hand side into chamber 1 between a CEM and an AEM denoted as C and A, respectively. It then is passed on to chambers 2, 3, 4 and 5, respectively. As it moves, Na^+ ions from this solution are moved through CEMs to the right and Cl^- ions move through the AEMs to the left. So it becomes progressively desalted as it moves further down. On the other hand, the brine waste feed introduced into chamber 9 becomes concentrated in NaCl as Na^+ ions come into it from the CEMs on the left and Cl^- ions come from the AEMs to the right; it finally leaves the cascade of flat cells as a concentrate. There are two electrode chambers into which brine/wash water will be introduced. The following reactions take place at the anode:

$$\begin{aligned}(1/2)H_2O &\rightarrow H^+ + (1/4)O_2(g) + e^-;\\ Cl^- &\rightarrow (1/2)Cl_2(g) + e^-,\end{aligned} \tag{8.1.401a}$$

leading to evolution of $Cl_2(g)$ and $O_2(g)$. Various steps are taken to reduce the evolution of $Cl_2(g)$ and isolate this water (Shaffer and Mintz, 1980) exiting the anode chamber. At the *cathode*, the reaction

$$H_2O + e^- \rightarrow OH^- + (1/2)H_2(g) \tag{8.1.401b}$$

occurs, leading to the formation of NaOH from the Na^+ ions coming in and evolution of $H_2(g)$.

Two alternative configurations introduce feed brine into each cell pair in a countercurrent fashion in the same cell pair (Figure 8.1.47(b)) or in a cocurrent fashion

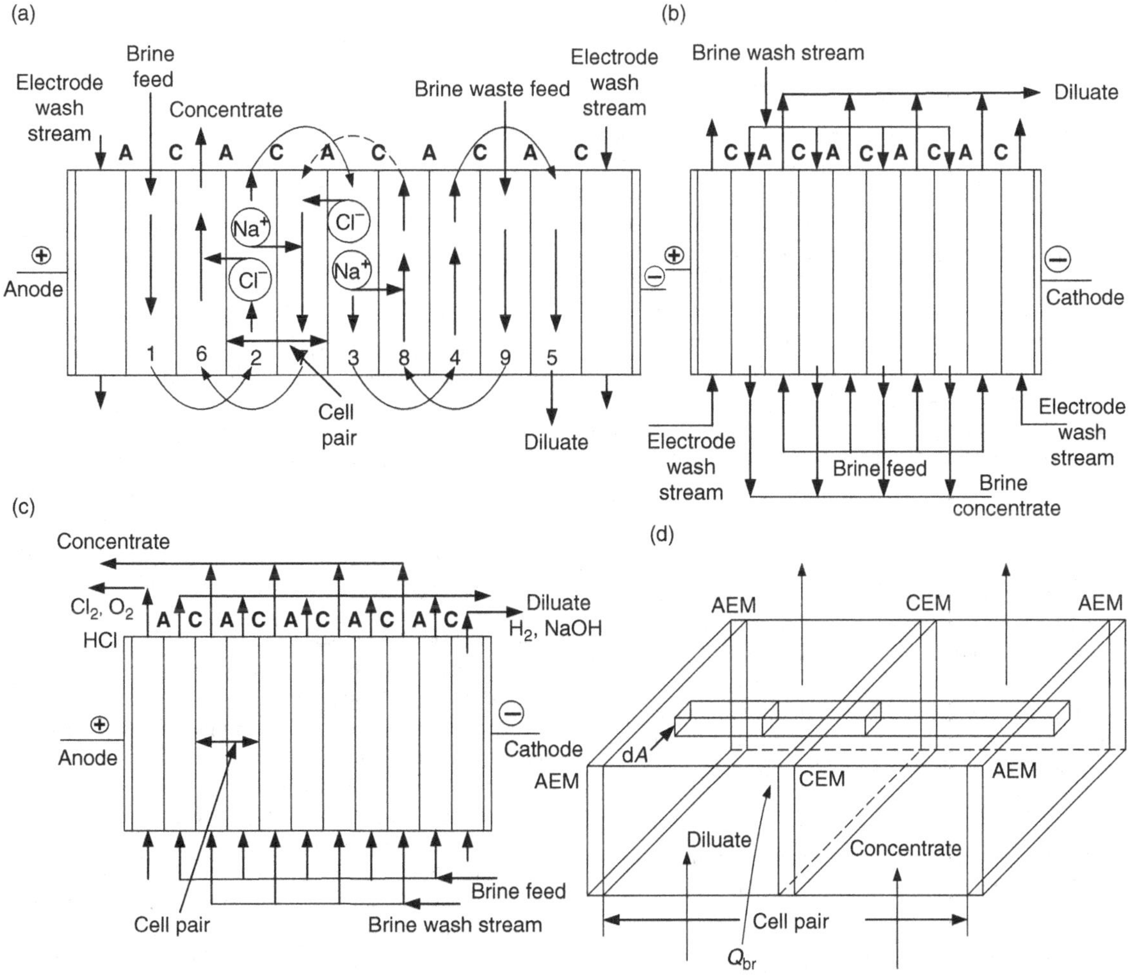

Figure 8.1.47. Schematic of various electrodialysis cell configurations using CEMs and AEMs: (a) internally staged countercurrent configuration; (b) simple countercurrent configuration; (c) simple cocurrent configuration; (d) process/transport analysis in a cell pair.

(Figure 8.1.47(c)). A fourth configuration employs a number of such membrane stacks with the partly desalted brine product from one stack introduced into the next stack for additional desalting.

We will now utilize the result (3.4.114) along with Ohm's law and the dependences of various resistances on salt concentration to develop an expression for the extent of desalting in the diluate chamber. From equation (3.4.114),

$$\frac{I\,t_{1m}}{\mathcal{F}} = \frac{I\,(t_{1s}+t_{2s})}{\mathcal{F}} = I/\mathcal{F}. \qquad (8.1.402)$$

Assume the CEM to be perfectly selective so that $t_{1m} = 1$. So the rate at which Na^+ ions are being transported through the membrane, and therefore the rate at which the solution is being desalted, is $I/\mathcal{F}$ gm-eq/s. We follow now the treatment of Shaffer and Mintz (1980). Consider a distance dz along the brine flow direction in the diluate chamber, where the brine volumetric flow rate is Q_{br} liter/s and the entering brine normality at location z in the diluate is N_{dil} (Figure 8.1.47(d)); the magnitude of the decrease in brine normality due to ion transport is $-dN_{dil}$, leading to the following expression for the total rate of salt transport over the length dz, $-Q_{br}\,dN_{dil}$ gm-equiv/s.

Let the membrane area (of each of the CEMs and the AEMs) in the diluate chamber over the brine flow path length dz be dA_m. The current density i amp/cm^2 in this section of the cell pair (Figure 8.1.47(d)) for a total current dI is

$$i = \frac{dI}{dA_m} \Rightarrow dI = \frac{\phi_p}{R_p}\,dA_m, \qquad (8.1.403)$$

where ϕ_p is the voltage drop over the cell pair and R_p is the resistance of 1 cm^2 cross section of the cell pair at location z. In most cases, resistance R_p (in units of ohm-cm^{-2}) may be described as

$$R_p = \frac{a_1}{N_{\text{dil}}} \text{ ohm-cm}^{-2}. \tag{8.1.404}$$

The cell pair resistance is the sum of the resistances of the AEM, the diluate chamber solution, the CEM and the concentrate chamber solution. Usually the resistance of the two membranes is quite low (around 3–20 ohm-cm^{-2}) because of the high density of ions and ionic fixed charges in the membrane. However, the diluate chamber will have considerably fewer electrolytes than the concentrate chamber, indicating that the normality of the diluate chamber N_{dil} will most likely control the solution resistance. For example, the solution resistance R for a NaCl solution may be represented in terms of the solution resistivity (ohm-cm), solution thickness δ (cm) and cross-sectional area A (cm^2) via

$$R = (\text{resistivity}) \frac{\delta}{A} \text{ ohm}; \tag{8.1.405}$$

the unit of resistivity is ohm-cm: a 0.5 gm-equiv/liter solution of NaCl at 25 °C has a resistivity value of 25 ohm-cm, whereas a 0.001 gm-equiv/liter solution's resistivity is 8000 ohm-cm.

In general, a small fraction of the current is not utilized in salt transport (it is lost, for example, in the brine connection channels at the end of the membranes); the fraction η_{IF} (pretty close to 1) utilized is called the current utilization factor. Introduce this factor in equation (8.1.403) and equate it to the gram-equivalents of salt transferred per unit time over 1 cm^2 area at location z (Figure 8.1.47(d)):

$$\frac{\text{d}I}{\mathcal{F}} \eta_{IF} = \frac{\phi_p}{R_p} \text{d}A_m \frac{\eta_{IF}}{\mathcal{F}} = -\, Q_{\text{br}} \, \text{d}N_{\text{di}\ell}. \tag{8.1.406}$$

Introduce here expression (8.1.404) for the cell-pair resistance:

$$-\, Q_{\text{br}} \, \text{d}N_{\text{di}\ell} = \frac{\phi_p}{a_1} \frac{\eta_{IF}}{\mathcal{F}} N_{\text{dil}} \, \text{d}A_m. \tag{8.1.407}$$

Integrate over the membrane area 0 to $A_{m,cp}$ of the cell pair along which the diluate solution concentration decreases from $N_{\text{dil},0}$ to $N_{\text{dil},L}$:

$$-\int_{N_{\text{dil},o}}^{N_{\text{dil},L}} \frac{\text{d}N_{\text{dil}}}{N_{\text{dil}}} = \frac{\phi_p \, \eta_{IF}}{Q_{\text{br}} \, \mathcal{F} \, a_1} \int_0^{A_{m,cp}} \text{d}A_m; \tag{8.1.408}$$

$$\ell\text{n}\left(\frac{N_{\text{dil},0}}{N_{\text{dil},L}}\right) = \frac{\phi_p \, \eta_{IF}}{Q_{\text{br}} \, \mathcal{F} \, a_1} A_{m,cp}. \tag{8.1.409}$$

If $A_{m,cp} = W \times L$ where L is the brine path length and W is the membrane width, we get

$$\ell\text{n}\,(N_{\text{dil},0}/N_{\text{di}\ell,L}) = \frac{\phi_p \, \eta_{IF}}{\mathcal{F} \, a_1} \left(\frac{W}{Q_{\text{br}}}\right) L. \tag{8.1.410}$$

Utilizing (8.1.403) and (8.1.404), one can express the current density i as

$$\frac{i}{N_{\text{dil}}} = \frac{\phi_p}{R_p N_{\text{dil}}} = \frac{\phi_p}{a_1}, \tag{8.1.411}$$

leading to

$$\ell\text{n}\,(N_{\text{dil},0}/N_{\text{di}\ell,L}) = \frac{\eta_{IF}}{\mathcal{F}} \left(\frac{i}{N_{\text{dil}}}\right) \left(\frac{W}{Q_{\text{br}}}\right) L = \ell\text{n}\,(D_r). \tag{8.1.412}$$

The quantity (i/N_{dil}) is an important one. An estimate of the limiting value of (i/N_{dil}) is around 480 (milliamp-liter/cm^2-gm-equiv) when concentration polarization sets in (Shaffer and Mintz, 1980). *Note*: The result (8.1.412) includes the desalination ratio D_r defined by (2.2.1a) achieved in the diluate chamber.

The analysis carried out and the result (8.1.412) achieved do not indicate any influence due to the overall cocurrent or countercurrent flow pattern; there is no apparent influence of the concentrate chamber concentration. Actually there is; see equation (3.4.115a). But the value of (Q_{im}/δ_m) is so small that there is hardly any transfer of cations from the concentrate chamber to the diluate chamber through the CEM, and similarly for anions through the AEM. This is an example where an external force trumps the generally assumed influence of flow patterns in two fluid-phase systems.

There is another important issue, however, which decides whether countercurrent or cocurrent flow pattern is to be used: pressure drop. The channel dimension across the flow between an AEM and an adjacent CEM is very small, ~1 mm; the membranes are thin to keep membrane resistances low. Yet the diluate stream has to flow over long distances (~150–300 cm) at a reasonable velocity to reduce polarization and increase the limiting current density i_{lim} (see equation (3.4.117)). This leads to a considerable pressure drop along the length of the diluate channel as well as the concentrate channel. If the cocurrent flow is utilized, then the pressure difference across any membrane is reduced to a low level. In countercurrent flow, the pressure difference across the membrane is increased considerably. This is the major reason for adopting the cocurrent flow pattern (Figure 8.1.47(c)) in general in ED. However, Figure 8.1.47(a) with countercurrent flow and internal staging has one advantage – one can reduce the electrolyte concentration substantially. Configurations shown in Figures 8.1.47(b) and (c) require the diluate from one stack of membranes to be fed to additional stacks of membranes to achieve considerable demineralization.

A few more details about ED devices will be useful. There may be as many as 150–300 cell pairs in an ED stack between the anode and the cathode. The value of ϕ_p for a cell pair varies between 0.5 and 3 volts, whereas the stack voltage may vary between 100 and

300 volts. Liquid velocities in a compartment in a cell pair may vary around 6 cm/s.

Example 8.1.21 Determine the extent of desalination achieved in terms of the desalination ratio D_r in an ED stack in which the diluate chamber liquid velocity is 10 cm/s, the diluate chamber cell thickness is 0.1 cm, $\eta_{IF} = 0.9$, the membrane length along the diluate flow path is 250 cm and (i/N_{dil}) has the value of 1000 milliamp-liter/cm^2-gm-equiv. The feed brine concentration is 3000 ppm.

Solution We will employ equation (8.1.412):

$$\ell n\,(D_r) = \frac{\eta_{IF}}{\mathcal{F}}\left(\frac{i}{N_{\text{dil}}}\right)\left(\frac{W}{Q_{\text{br}}}\right)L,$$

where $L = 250$ cm; $\eta_{IF} = 0.9$; $\mathcal{F} = 96\,500$ C/gm-equiv. So

$$\left(\frac{i}{N_{\text{dil}}}\right) = 1000\frac{\text{milliamp-liter}}{\text{cm}^2\text{-gm-equiv}}$$

$$= \frac{1000\times 10^{-3}\dfrac{\text{amp}}{\cancel{\text{milliamp}}}\cdot\cancel{\text{milliamp}}\times 1000\dfrac{\text{cm}^3}{\cancel{\text{liter}}}\cancel{\text{liter}}}{\text{cm}^2\text{-gm-equiv}};$$

$$Q_{\text{br}} = v_{\text{br}}\times W\times\delta \Rightarrow (W/Q_{\text{br}}) = \frac{1}{v_{\text{br}}\times\delta} = \frac{1}{10\dfrac{\text{cm}}{\text{s}}\times 0.1\,\text{cm}};$$

$$\ell n(D_r) = \frac{0.9}{96\,500\,\cancel{C}/\cancel{\text{gm-equiv}}}\,1000\frac{\cancel{\text{amp-cm}}^{\,3}}{\cancel{\text{cm}}^{\,2}\text{-}\cancel{\text{gm-equiv}}}$$

$$\times\frac{1\cancel{s}}{10\times 0.1\cancel{\text{cm}}^{\,2}}\times 250\cancel{\text{cm}}$$

$$= \frac{0.9\times 10\cancel{0}\cancel{0}\times 250}{965\cancel{0}\cancel{0}\times 1} = 2.3315 \Rightarrow \log_{10}(D_r) = 1.011$$

$$\Rightarrow D_r = \text{desalination ratio} = 10.3.$$

A broader introduction to stage analysis in an ED cell pair has been provided by Mason and Kirkham (1959). Examples of applications of ion exchange membranes in ED from earlier literature is available in Mason and Juda (1959). Strathmann (2001) has provided a comprehensive overview of ED, including considerable details on ion exchange membranes.

8.1.8 Countercurrent liquid membrane separation

In Section 5.4.4, we studied a variety of chemical reaction facilitated separation where the reaction was taking place in a thin liquid layer acting as the liquid membrane; Figure 5.4.4 illustrated a variety of liquid membrane permeation mechnisms. Here we will identify first the structural configuration of the liquid membranes as they are used in separators with countercurrent flow pattern (as well as for the cocurrent flow pattern). There are three general classes of liquid membrane structures: *emulsion liquid membrane* (ELM); *supported liquid membrane* (SLM) or *immobilized liquid membrane* (ILM); *hollow fiber contained liquid membrane* (HFCLM). Each will be described very briefly.

A most comprehensive description of the emulsion liquid membrane is available in Ho and Sirkar (2001, chaps. 36–40); our treatment relies on these chapters. Figure 8.1.48(a) describes an emulsion liquid membrane via a large emulsion globule. The large globule diameter may vary between 100 and 2000 μm. Inside each large globule, there are many tiny droplets of diameter 1–3 μm. Although this figure does not show it, the tiny droplets are tightly packed together. The large globule is dispersed in an immiscible continuous liquid phase, in this case a feed aqueous phase. Specifically, the tiny internal droplets consist of an aqueous phase, in this case a solution of NaOH in water. These tiny droplets are inside an organic solvent phase making up the rest of the large emulsion globule. This organic solvent phase contains surfactants and additives in a solvent and acts like a membrane between the external aqueous solution and the internal tiny aqueous phase droplets. From the external solution, phenol present in the continous aqueous phase partitions into the surface of the emulsion globule, the organic membrane phase. It then diffuses through the organic liquid membrane and hits one of the tiny aqueous phase droplets, the receiving phase, the permeate phase here. This particular separation with reaction example was briefly considered at the beginning of Section 5.4.4. This liquid membrane structure has also been used with an aqueous liquid membrane phase surrounding tiny organic phase droplets in an emulsion globule which is dispersed in an external organic phase, a continuous phase.

Figure 8.1.48(b) illustrates how the *ELM based separation* works. First, the internal droplet phase is dispersed into the membrane – the process identified as emulsification. Next this emulsion is dispersed into the external continuous feed phase, creating the large emulsion globules. For the separation system of Figure 8.1.48(a), the emulsion is oil/water (O/W) since the external continuous phase is oil; however, for the system shown in Figure 8.1.48 (b) the external continuous phase is water, therefore it is considered a W/O system. The overall system is a double emulsion W/O/W. An alternative double emulsion configuration would be O/W/O, where the feed phase and the permeate phase would be oils with the liquid membrane being water. Figure 8.1.48(b) illustrates a well-mixed stage where the permeation process goes on. After the permeation is over, the original O/W phase settles out of the feed aqueous phase in a settler, followed by withdrawal of the spent aqueous phase, the raffinate; simultaneously, the O/W emulsion is separated and the oil phase (membrane phase) is recycled while the aqueous phase becomes the extract.

In actual use, a number of such stages are cobbled together in a countercurrent configuration, as shown in Figure 8.1.48(c). The countercurrent permeation column shown has the feed aqueous solution (a waste water) containing zinc being introduced at the column top. The permeate phase contains a strong solution of H_2SO_4; it is dispersed as tiny droplets in the organic liquid membrane

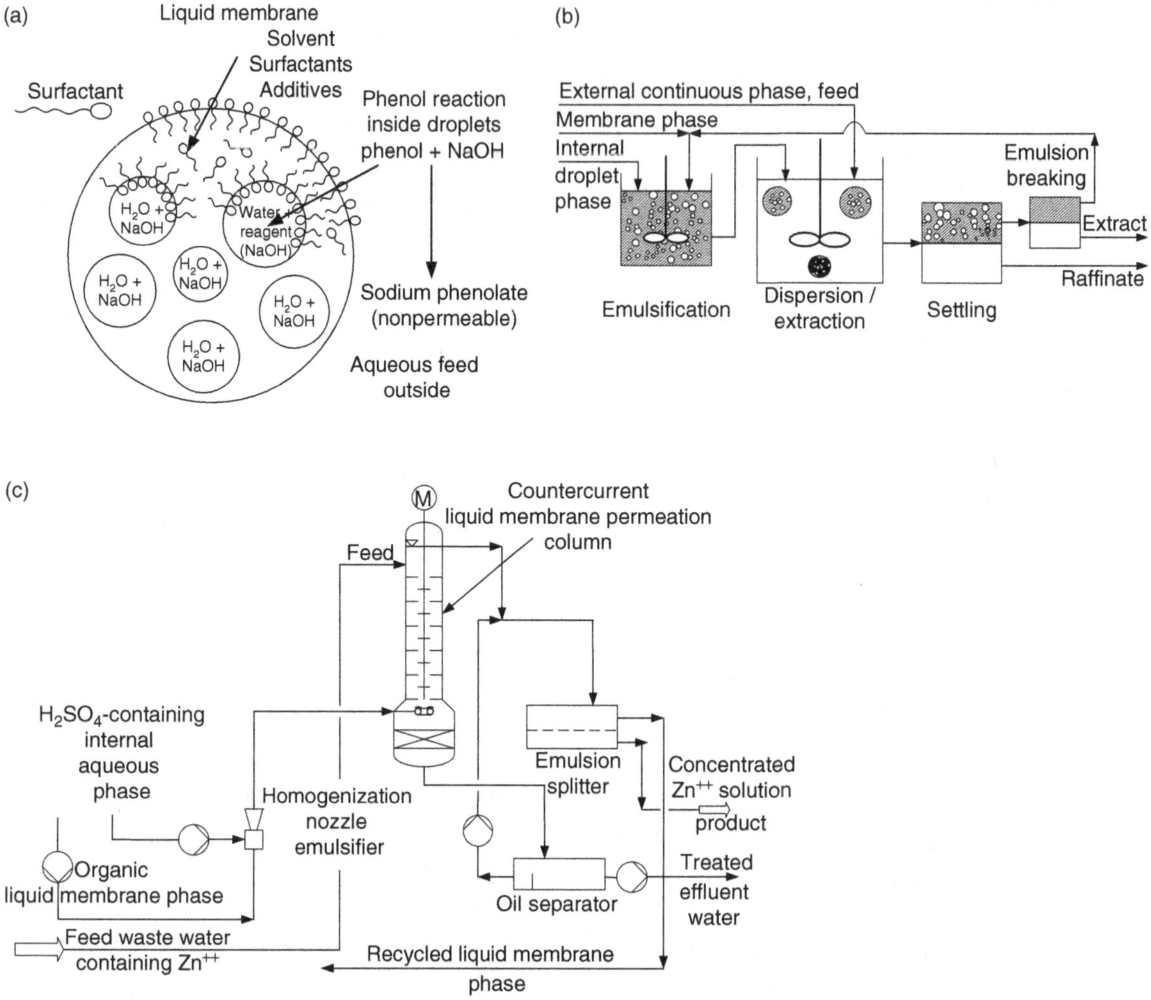

Figure 8.1.48. Emulsion liquid membrane: (a) a large emulsion globule containing many tiny droplets; (b) continuous ELM process schematic; (c) countercurrent ELM column for removing Zn^{++} from wastewater. (After Ho and Sirkar (2001), chaps. 36, 39.)

phase containing the extractant in a diluent (see the description following equation (5.2.110)) along with surfactants needed to form a stable emulsion. This emulsion is introduced at the column bottom, and it rises up the column, allowing Zn^{++} to be extracted into the internal H_2SO_4-containing aqueous phase via reactions of the type shown in (5.4.77) and (5.4.83a) in a countertransport reaction scheme. As in Figure 8.1.48(b), the various exiting streams are separated out into the effluent treated aqueous stream (the raffinate), the aqueous H_2SO_4 solution contaning the extract and concentrated Zn^{++} product and the organic liquid membrane, which is recycled for membrane making.

In this particular example, the extractant used was a sulfur-containing derivative of di-2-ethylhexylphosphoric acid (see Table 5.2.3) at the level of 5% in an aliphatic diluent with a high flash point such as Escaid 120 containing 3% of a long-chain polyamine surfactant. The Zn concentration in the feed waste water varied between 150 and 500 mg/liter, and the raffinate Zn concentration was reduced to around 1 mg/liter. The permeate phase H_2SO_4 concentration was as much as 250 mg/liter. The liquid membrane was achieving first solvent extraction of Zn^{++} into the liquid membrane phase. Then Zn^{++} was back extracted and concentrated into the aqueous H_2SO_4-containing permeate phase, from which H^+ ions were transported to the feed waste water being treated during the countertransport process.

Due to the presence of a strong concentration of H_2SO_4 in the internal droplet phase, its osmotic pressure is far

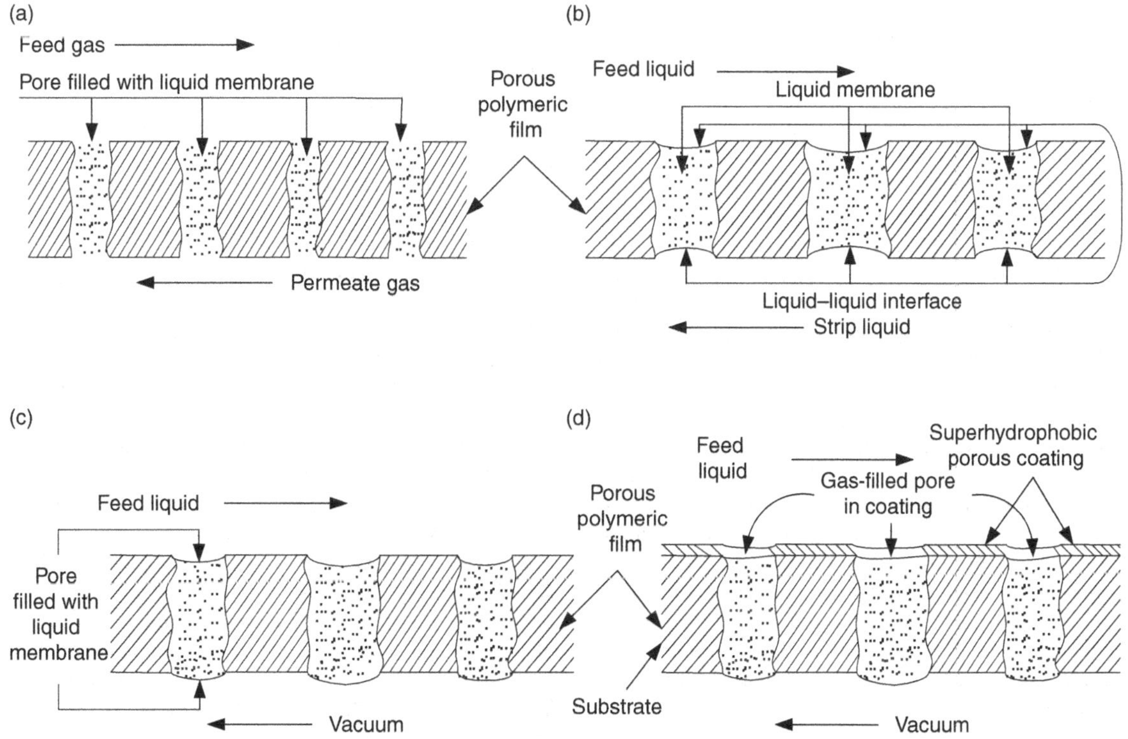

Figure 8.1.49. (a) ILM for gas separation; (b) SLM for liquid separation; (c) SLM for pervaporation with a porous membrane; (d) SLM for pervaporation with a composite membrane having a porous superhydrophobic coating on the feed liquid side.

higher than that of the feed aqueous solution. Consequently, water molecules are transported through the liquid membrane into the internal droplet phase, leading to its swelling (entrainment of the external aqueous phase also contributes). Swelling of the internal droplet phase reduces the driving force in countertransport, reduces the permeate solute concentration level and makes the liquid membrane thinner, potentially introducing instability. Another demanding aspect of the ELM process is the demulsification step (Figure 8.1.48(b), emulsion breaking) needed to recycle the liquid membrane. Demulsification is successfully implemented by using an electrostatic field, which leads to the coalescence of dispersed aqueous droplets from nonconducting oil (Ho and Sirkar, 2001, chap. 38).

The countercurrent emulsion liquid membrane extractor shown in Figure 8.1.48(c) has been modeled by Ho and Li (1984b). Two types of modeling were implemented: a multistage mixer-settler and a mechanically agitated column. A multistage countercurrent mixer-settler cascade of n stages operating continuously was shown to be more effective than a cocurrent cascade of similar mixer-settlers. It was found to be true for mechanically agitated columns as well.

The two phases on two sides of the liquid membrane are liquid phases that are immiscible with the liquid membrane phase in the ELM technique. There is very little information on an ELM being employed for separation from a gas phase into another gas phase. The next liquid membrane technique to be considered, namely the SLM or ILM technique, however, allows operation with both situations: (1) feed-gas phase, permeate-gas phase; (2) feed-liquid phase, sweep-liquid phase (both immiscible with the liquid membrane phase).

The membrane structure in the SLM/ILM technique is as follows. Consider a porous hydrophilic film. If it is contacted with an aqueous solution, more than likely it will be spontaneously wetted. The pores will be filled with the aqueous solution, which will be held by capillary force (see equation (6.1.12)). Similarly, if we have a porous hydrophobic film and it is contacted with an organic solution, more than likely it will be wetted by the solution and the pores will be filled with the organic liquid. Consider now a feed-gas phase on one side of this film (Figure 8.1.49 (a)). If the solid polymeric region of the film is considered relatively impervious, then species in the feed-gas mixture will be absorbed in the liquid in the pores, diffuse through the liquid in the pores and be desorbed at the other side of the liquid film in the pore into the gas phase on the other side of the porous membranes, provided there is a positive partial pressure difference for this gas species across the film. The liquid film stabilized by capillary forces in the pore can act as a membrane. However, if the liquid is volatile, it will be evaporated into the gas phase on either side, ultimately destroying the liquid membrane. Nonvolatile immobilized liquid membranes have been used successfully for CO_2 separation (Kovvali *et al.*, 2000; Kovvali and Sirkar, 2001; Kouketsu *et al.*, 2007).

One could use the liquid immobilized in the pores for separation of liquid mixtures as well, provided the feed liquid and the permeate (strip) liquid are immiscible with the liquid membrane in the pores (Figure 8.1.49(b)). Such membranes are called *supported liquid membranes* (SLMs). However, at the pore mouths, the two immiscible phases contact each other. The liquid membrane phase has some solubility, however small, in the feed/strip liquid phase. Therefore the life of the liquid membrane is limited (see Kovvali and Sirkar (2003) for a brief review), and periodically it needs to be regenerated (Yang and Kocherginsky, 2006). The liquid membrane is most likely to have extractants used with the emulsion liquid membranes or solvent extraction in the case of chemically complexing extractants.

An intermediate configuration of a liquid membrane employs it for pervaporation: feed liquid contacting the liquid membrane supported in a porous membrane with vacuum or sweep gas on the other side (Qin *et al.*, 2003). Such a configuration still cannot eliminate loss of the liquid membranes to the feed liquid (Figure 8.1.49(c)). One technique solves this problem: introduce a thin porous hydrophobic coating on the porous substrate membrane on the feed side such that neither the feed solution nor the liquid membrane can spontaneously wet the pores of the porous hydrophobic coating (Figure 8.1.49(d)). Volatile species are evaporated from the feed solution into the gas-filled pores of the coating and are then absorbed into the liquid membrane in the pores of the porous substrate as they are moved to the permeate side where the species are continuously removed by the permeate side vacuum, condensed and recovered (Thongsukmak and Sirkar, 2007, 2009).

There are a few points to be kept in mind. The liquid membrane must be nonvolatile since it is subjected to vacuum on the permeate side and can evaporate on the feed side and be lost to the feed solution (Figure 8.1.49(d)). Further, wetting of the porous membrane structure spontaneously by the membrane liquid present outside requires that the surface tension of the membrane liquid should be equal to or lower than the critical surface tension γ_{cr} of the polymeric or ceramic substrate being employed. For example, γ_{cr} for polypropylene (PP) is ~33 dyne/cm. Many organic liquids will spontaneously wet it. However, γ_{cr} values for various fluoropolymers (FPs) are usually <25 dyne/cm. Therefore an organic liquid with a γ value in between 33 and 25 dyne/cm will wet spontaneously a porous PP substrate but not a FP based porous coating on its surface (Thongsukmak and Sirkar, 2007). Here we have an asymmetric porous membrane in terms of wettability via the composite structure. We have briefly described other asymmetric or composite membrane structures in Figures 7.2.1(b), and 6.3.35(b) and the description related to Example 3.4.4.

The third liquid-membrane technique is the *hollow fiber contained liquid membrane* (HFCLM). In the SLM/ILM technique, the liquid membrane is in contact with the feed liquid/feed gas and the strip liquid/permeate gas. The membrane liquid may be lost by solubilization/volatilization; in addition, there may be irreversible reactions with extractants, complexing agents inside the liquid membrane, which could reduce the performance over time. The HFCLM structure can take care of such problems. In this structure (Figure 8.1.50), the membrane module (cylindrical or otherwise) has two sets of porous hollow fiber membranes. The shell-side interstitial space between the fibers is filled with a liquid acting as the membrane. If this membrane liquid wets the membrane pores and is therefore present in the pores, the pressure conditions in the feed liquid and the sweep/strip fluid should be such (higher) that the membrane liquid is not dispersed in the feed/sweep/strip fluids. If the membrane liquid does not wet the membrane pores, but the feed/sweep fluids do, then the membrane liquid pressure should be higher than the feed/sweep fluid pressures.

The conditions described above ensure that the feed-liquid membrane interface and the liquid membrane-sweep/strip interface will be stable and mass transport will continue. Any loss or change in the liquid membrane is continuously made up by a pressurized membrane liquid reservoir connected to the shell-side membrane fluid. Majumdar *et al.* (2001) provide a concise description of this technique. The basic technique has been illustrated in Majumdar *et al.* (1988) and Guha *et al.* (1992) for gas separation and in Sengupta *et al.* (1988) for liquid separation. For gas separation, Guha *et al.* (1992) have demonstrated that feed-gas separation may be implemented with conventional gas permeation, permeate side under vacuum, sweep gas and sweep liquid. In both techniques the overall resistance to transfer may be described as a sum of five resistances in series:

$$\begin{aligned}\text{total resistance} = {} & \text{feed-side film resistance} \\ & + \text{feed-side substrate resistance} \\ & + \text{contained liquid membrane resistance} \\ & + \text{permeate-side substrate resistance} \\ & + \text{permeate-side film resistance.}\end{aligned} \tag{8.1.413}$$

In systems where interfacial reactions are taking place, as, for example, in metal transport (e.g. equation (5.4.77)),

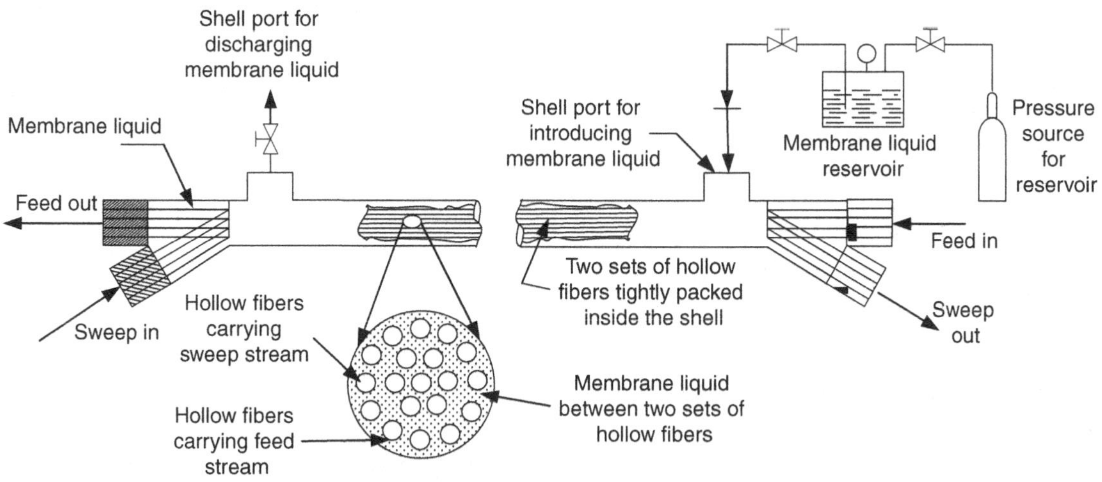

Figure 8.1.50. Basic hollow fiber contained liquid membrane (HFCLM) permeator structure.

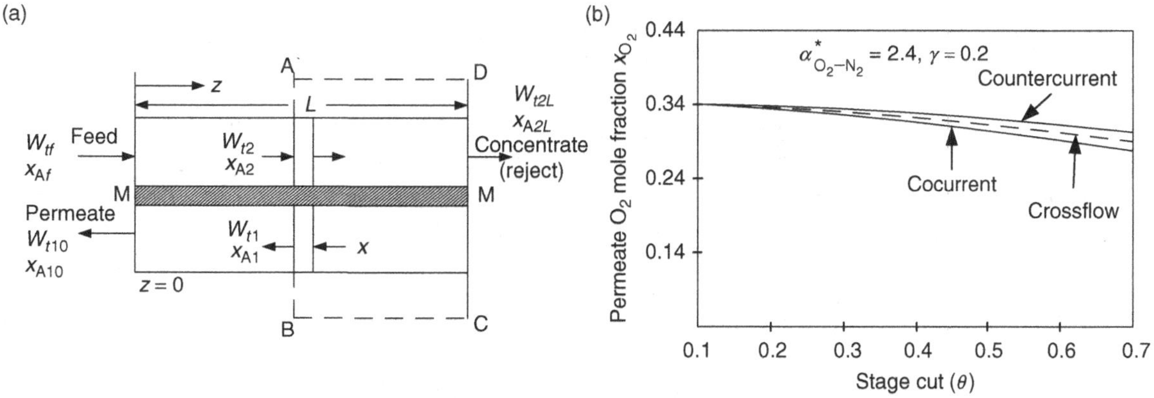

Figure 8.1.51. (a) Schematic for a countercurrent gas permeator; (b) permeate O_2 composition vs. stage cut for a silicone copolymer membrane having $\alpha^*_{O_2-N_2} = 2.4$, $\gamma = 0.2$. The plots are qualitative and indicate the relative performance of a countercurrent permeator, vis-à-vis a crossflow permeator (Figure 7.2.1(a)) and a cocurrent permeator. (After Walawender and Stern (1972).)

additional resistance terms at the two interfaces between the liquid membrane and the external phases have to be taken into account. *Note*: These two interfacial resistances exist in the ELM system as well as in the SLM system. In the SLM system there is only one substrate resistance, which is equivalent to the liquid membrane resistance:

$$\begin{aligned}\text{total resistance} = {} & \text{feed-side film resistance} \\ & + \text{liquid membrane resistance in the substrate} \\ & + \text{strip-side film resistance.}\end{aligned} \qquad (8.1.414)$$

In the ELM system there is no substrate as such; we have simply the liquid membrane resistance beside the two film resistances, unless interfacial resistances exist. The presence of surfactants at the ELM interfaces is relevant.

8.1.9 Countercurrent gas permeation

In earlier chapters we have studied gas separation via gas permeation in two permeator configurations: in a completely mixed membrane gas permeation cell (Section 6.4.2.2); in crossflow membrane permeators (Section 7.2.1.1). In the shell-side fed hollow fiber gas permeator of Figure 7.2.1(d), if the permeate is withdrawn from the fiber bore at the feed end only (the other end is closed), then we have countercurrent flow between the feed side and the permeate side. We consider such a gas permeator configuration here via Figure 8.1.51(a); this figure does not show a hollow fiber. However, one could easily model a hollow fiber permeator with the permeate side being the fiber bore (I.D., d_i; O.D. d_o). We will now develop the governing balance equations for the countercurrent gas permeator for a binary gas mixture of species A and B. *Note*: There are two general types of problems.

(1) Design problem. The concentrate outlet composition x_{A2L} is provided. We have to determine the total membrane area A_m^t and other necessary quantities. There can be other types of specifications instead of x_{A2L}.
(2) Rating problem. A permeator of area A_m^t is given. We need to determine x_{A2L}, x_{A10}, etc.

To start with we adopt the formalism of equations (7.2.3) and (7.2.5) under the assumptions of no concentration polarization and no longitudinal diffusion/dispersion in the bulk flow direction (z-coordinate), and obtain for species A permeation through membrane area dA_m over a length dz:

$$\frac{d(W_{t2}\,x_{A2})}{dA_m} = -N_{Am} = -\frac{Q_{Am}}{\delta_m}\,(P_f x_{A2} - P_p x_{A1}). \quad (8.1.415)$$

For a rectangular feed-gas flow channel of width W and length L,

$$dA_m = W\,dz, \qquad A_m^t = WL. \quad (8.1.416)$$

For a hollow fiber device with a *symmetrical hollow fiber membrane of diameters d_i and d_o* and total number of fibers being n,

$$dA_m = n\pi d_{\ell m}\,dz, \qquad d_{\ell m} = \frac{d_o - d_i}{\ell n(d_o/d_i)}, \quad (8.1.417)$$

where $d_{\ell m}$ is the logarithmic mean of d_i and d_o. For species B, an equation similar to (8.1.415) is

$$\frac{d(W_{t2}\,(1-x_{A2}))}{dA_m} = -N_{Bm} = -\frac{Q_{Bm}}{\delta_m}\,(P_f(1-x_{A2}) - P_p(1-x_{A1})). \quad (8.1.418)$$

If we focus on the permeate side, we find also

$$\frac{d(|W_{t1}|\,x_{A1})}{dA_m} = -N_{Am} = -\frac{Q_{Am}}{\delta_m}\,(P_f\,x_{A2} - P_p\,x_{A1}), \quad (8.1.419)$$

and similarly for species B, leading to

$$\frac{d(W_{t2}\,x_{A2})}{dA_m} + \frac{d(W_{t2}\,(1-x_{A2}))}{dA_m} = \frac{dW_{t2}}{dA_m} = \frac{d(|W_{t1}|\,x_{A1})}{dA_m} + \frac{d(|W_{t1}|\,(1-x_{A1}))}{dA_m} = \frac{d|W_{t1}|}{dA_m} = -N_{Am} - N_{Bm}. \quad (8.1.420)$$

We will now develop equations for the axial variation of species A composition on the feed and the permeate sides. From equation (8.1.420) and equations (8.1.415) and (8.1.418), we get

$$\frac{dW_{t2}}{dA_m} = -\frac{Q_{Am}}{\delta_m}\,(P_f x_{A2} - P_p x_{A1}) - \frac{Q_{Bm}}{\delta_m}\,(P_f(1-x_{A2}) - P_p(1-x_{A1})). \quad (8.1.421)$$

Also, we can express the left-hand side of equation (8.1.415) as

$$\frac{d(W_{t2}\,x_{A2})}{dA_m} = W_{t2}\frac{dx_{A2}}{dA_m} + x_{A2}\frac{dW_{t2}}{dA_m}. \quad (8.1.422)$$

Therefore,

$$\frac{dx_{A2}}{dA_m} = \frac{1}{W_{t2}}\left[\frac{d(W_{t2}\,x_{A2})}{dA_m} - x_{A2}\frac{dW_{t2}}{dA_m}\right]; \quad (8.1.423)$$

$$\frac{dx_{A2}}{dA_m} = \frac{1}{W_{t2}}\left[-\frac{Q_{Am}}{\delta_m}\,(P_f x_{A2} - P_p x_{A1}) - x_{A2}\left\{-\frac{Q_{Am}}{\delta_m}\,(P_f x_{A2} - P_p x_{A1}) - \frac{Q_{Bm}}{\delta_m}\,(P_f(1-x_{A2}) - P_p(1-x_{A1}))\right\}\right]; \quad (8.1.424)$$

$$\frac{dx_{A2}}{dA_m} = -\frac{1}{W_{t2}}\left[(1-x_{A2})\frac{Q_{Am}}{\delta_m}\,(P_f x_{A2} - P_p x_{A1}) - x_{A2}\frac{Q_{Bm}}{\delta_m}\,(P_f(1-x_{A2}) - P_p(1-x_{A1}))\right]. \quad (8.1.425)$$

Define the following nondimensional quantities:

$$\alpha^*_{AB} = (Q_{Am}/Q_{Bm}); \qquad A_m^+ = ((Q_{Bm}/\delta_m)P_f A_m/W_{tf});$$

$$A_m^+ = \frac{Q_{Bm}}{\delta_m}\frac{P_f\,n\pi\,d_{\ell m}z}{W_{tf}};$$

$$A_m^{t+} = \frac{Q_{Bm}}{\delta_m}\frac{P_f\,n\pi\,d_{\ell m}L}{W_{tf}} = A_m^+\frac{L}{z} = \frac{Q_{Bm}}{\delta_m}\frac{P_f\,A_m^t}{W_{tf}};$$

$$W_{t2}^+ = (W_{t2}/W_{tf}); \qquad \gamma = (P_p/P_f). \quad (8.1.426)$$

Equation (8.1.425) may now be written as

$$\frac{dx_{A2}}{dA_m^+} = -\frac{1}{W_{t2}^+}\left[\alpha^*_{AB}(1-x_{A2})\,(x_{A2} - \gamma x_{A1}) - x_{A2}\{\,(1-x_{A2}) - \gamma(1-x_{A1})\}\right]. \quad (8.1.427)$$

Solution of this equation will yield the variation of x_{A2} with the dimensionless membrane area A_m^+. The dependent variables here are x_{A2}, W_{t2}^+ and x_{A1} (if γ is assumed constant with A_m^+). Solution of this problem needs at least two more equations (unless we encounter another dependent variable) so that we may know how x_{A2} is varying from the feed location ($z = 0$, $W_{t2}^+ = 1$, $x_{A2} = x_{Af}$) to the exit location where $z = L$ and $A_m = A_m^t$. Equation (8.1.421) when nondimensionalized provides the second required equation:

$$\frac{dW_{t2}^+}{dA_m^+} = -\alpha^*_{AB}\,(x_{A2} - \gamma x_{A1}) - ((1-x_{A2}) - \gamma(1-x_{A1})). \quad (8.1.428)$$

For the third equation, focus on the governing equations (8.1.419) and (8.1.420) for the permeate side; from the latter we get

$$\frac{\mathrm{d}|W_{t1}|}{\mathrm{d}A_m} = -\frac{Q_{\mathrm{A}m}}{\delta_m}\left(P_f x_{\mathrm{A2}} - P_p x_{\mathrm{A1}}\right) - \frac{Q_{\mathrm{B}m}}{\delta_m}\left(P_f(1-x_{\mathrm{A2}}) - P_p(1-x_{\mathrm{A1}})\right). \tag{8.1.429}$$

From (8.1.419), we can obtain

$$\begin{aligned}\frac{\mathrm{d}(|W_{t1}|x_{\mathrm{A1}})}{\mathrm{d}A_m} &= |W_{t1}|\frac{\mathrm{d}x_{\mathrm{A1}}}{\mathrm{d}A_m} + x_{\mathrm{A1}}\frac{\mathrm{d}|W_{t1}|}{\mathrm{d}A_m}\\ &= |W_{t1}|\frac{\mathrm{d}x_{\mathrm{A1}}}{\mathrm{d}A_m} - x_{\mathrm{A1}}\frac{Q_{\mathrm{A}m}}{\delta_m}\left(P_f x_{\mathrm{A2}} - P_p x_{\mathrm{A1}}\right)\\ &\quad - x_{\mathrm{A1}}\frac{Q_{\mathrm{B}m}}{\delta_m}\left(P_f(1-x_{\mathrm{A2}}) - P_p(1-x_{\mathrm{A1}})\right)\\ &= -\frac{Q_{\mathrm{A}m}}{\delta_m}\left(P_f x_{\mathrm{A2}} - P_p x_{\mathrm{A1}}\right),\end{aligned} \tag{8.1.430}$$

leading to

$$|W_{t1}|\frac{\mathrm{d}x_{\mathrm{A1}}}{\mathrm{d}A_m} = -(1-x_{\mathrm{A1}})\frac{Q_{\mathrm{A}m}}{\delta_m}\left(P_f x_{\mathrm{A2}} - P_p x_{\mathrm{A1}}\right) + x_{\mathrm{A1}}\frac{Q_{\mathrm{B}m}}{\delta_m}(P_f(1-x_{\mathrm{A2}}) - P_p(1-x_{\mathrm{A1}})). \tag{8.1.431}$$

Nondimensionalize this equation to obtain

$$\frac{\mathrm{d}x_{\mathrm{A1}}}{\mathrm{d}A_m^+} = -\frac{1}{|W_{t1}^+|}\left[\alpha_{\mathrm{AB}}^*(1-x_{\mathrm{A1}})(x_{\mathrm{A2}} - \gamma x_{\mathrm{A1}}) - x_{\mathrm{A1}}((1-x_{\mathrm{A2}}) - \gamma(1-x_{\mathrm{A1}}))\right]. \tag{8.1.432}$$

Due to the presence of the variable $|W_{t1}^+|$ in this equation, we need another equation. Convert equation (8.1.429) to

$$\frac{\mathrm{d}|W_{t1}^+|}{\mathrm{d}A_m^+} = -\alpha_{\mathrm{AB}}^*(x_{\mathrm{A2}} - \gamma x_{\mathrm{A1}}) - ((1-x_{\mathrm{A2}}) - \gamma(1-x_{\mathrm{A1}})). \tag{8.1.433}$$

Equations (8.1.427), (8.1.428), (8.1.432) and (8.1.433) provide a set of four coupled nonlinear ordinary differential equations for four variables x_{A2}, W_{t2}^+, x_{A1} and $|W_{t1}^+|$ as a function of the nondimensional membrane area variable A_m^+ subject to the feed inlet conditions and other specified values of P_f, P_p, $x_{\mathrm{A}f}$, W_{tf}, $(Q_{\mathrm{A}m}/\delta_m)$, $(Q_{\mathrm{B}m}/\delta_m)$, A_m^t (total membrane area corresponding to $z = L$) or $x_{\mathrm{A2}L}$. We should be able to solve these four equations numerically in a computer utilizing a few overall balance equations to be described soon; there is one problem, however. Equation (8.1.432) is indeterminate at $A_m = A_m^t$ $(z = L)$, where $|W_{t1}^+| = 0$. Employ L'Hospital's rule here:

$$\left.\frac{\mathrm{d}x_{\mathrm{A1}}}{\mathrm{d}A_m^+}\right|_{z=L} = -\frac{\frac{\mathrm{d}}{\mathrm{d}A_m^+}\left[\alpha_{\mathrm{AB}}^*(1-x_{\mathrm{A1}})(x_{\mathrm{A2}} - \gamma x_{\mathrm{A1}}) - x_{\mathrm{A1}}((1-x_{\mathrm{A2}}) - \gamma(1-x_{\mathrm{A1}}))\right]_{z=L}}{\left[\mathrm{d}|W_{t1}^+|/\mathrm{d}A_m^+\right]_{z=L}}; \tag{8.1.434}$$

$$\left.\frac{\mathrm{d}x_{A1}}{\mathrm{d}A_m^+}\right|_{z=L} = -\frac{-\alpha_{\mathrm{AB}}^*(x_{\mathrm{A2}} - \gamma x_{\mathrm{A1}})\left.\frac{\mathrm{d}x_{\mathrm{A1}}}{\mathrm{d}A_m^+}\right|_{z=L} + \alpha_{\mathrm{AB}}^*(1-x_{\mathrm{A1}})\left(\left.\frac{\mathrm{d}x_{\mathrm{A2}}}{\mathrm{d}A_m^+}\right|_{z=L} - \gamma\left.\frac{\mathrm{d}x_{\mathrm{A1}}}{\mathrm{d}A_m^+}\right|_{z=L}\right) - ((1-x_{\mathrm{A2}}) - \gamma(1-x_{\mathrm{A1}}))\left.\frac{\mathrm{d}x_{\mathrm{A1}}}{\mathrm{d}A_m^+}\right|_{z=L} + x_{\mathrm{A1}}\left.\frac{\mathrm{d}x_{\mathrm{A2}}}{\mathrm{d}A_m^+}\right|_{z=L} - \gamma x_{\mathrm{A1}}\left.\frac{\mathrm{d}x_{\mathrm{A1}}}{\mathrm{d}A_m^+}\right|_{z=L}}{-\alpha_{\mathrm{AB}}^*(x_{\mathrm{A2}} - \gamma x_{\mathrm{A1}}) - ((1-x_{\mathrm{A2}}) - \gamma(1-x_{\mathrm{A1}}))} \tag{8.1.435}$$

Rearranging, we get

$$\left.\frac{\mathrm{d}x_{A1}}{\mathrm{d}A_m^+}\right|_{\substack{z=L\\ A_m = A_m^t}} = -\frac{\{\alpha_{\mathrm{AB}}^*(1-x_{\mathrm{A1}}) + x_{\mathrm{A1}}\}\left.\frac{\mathrm{d}x_{\mathrm{A2}}}{\mathrm{d}A_m^+}\right|_{z=L}}{2\{\alpha_{\mathrm{AB}}^*(x_{\mathrm{A2}} - \gamma x_{\mathrm{A1}}) + (1-x_{\mathrm{A2}}) - \gamma(1-x_{\mathrm{A1}})\} + \gamma\{\alpha_{\mathrm{AB}}^* - (\alpha_{\mathrm{AB}}^* - 1)x_{\mathrm{A1}}\}|_{z=L}}, \tag{8.1.436}$$

where the values of x_{A1}, x_{A2} are to be calculated at $A_m = A_m^t$ $(z = L)$.

For both cases of a design problem or a rating problem for a countercurrent gas permeator, if we assume that γ is constant throughout the permeator (i.e. P_f and P_p are constant along the permeator), the set of four equations (8.1.427), (8.1.428), (8.1.432) and (8.1.433) may now be solved using (8.1.436). However, x_{A1} is unknown at $A_m = A_m^t$. It is determined using the crossflow equations (6.3.200) and (6.3.201) using known $x_{\mathrm{A2}L}$ at $A_m = A_m^t$ for the design problem and for the rating problem as if it is known (assume some $x_{\mathrm{A2}L}$):

$$x_{A1L} = \frac{\left\{(\alpha^*_{AB}-1)\left(\frac{x_{A2L}}{\gamma}+1\right)+\frac{1}{\gamma}\right\}-\sqrt{\left\{\left(1+\frac{x_{A2L}}{\gamma}\right)(\alpha^*_{AB}-1)+\frac{1}{\gamma}\right\}^2-\frac{4\alpha^*_{AB}(\alpha^*_{AB}-1)x_{A2L}}{\gamma}}}{2(\alpha^*_{AB}-1)}. \tag{8.1.437}$$

A few additional relations based on overall mass balance are needed.

Total mass balance over the permeator:

$$W_{tf} = W_{t2L} + |W_{t10}|; \tag{8.1.438a}$$

species A balance over the permeator:

$$W_{tf}\, x_{Af} = W_{t2L}\, x_{A2L} + |W_{t10}|\, x_{A10}. \tag{8.1.438b}$$

Divide both equations by W_{tf} to obtain the following relation in terms of the stage cut:

$$\theta = \frac{|W_{t10}|}{W_{tf}}; \tag{8.1.438c}$$

$$x_{Af} = (1-\theta)x_{A2L} + \theta\, x_{A10} \Rightarrow x_{A10} = \frac{x_{Af}-(1-\theta)\,x_{A2L}}{\theta}. \tag{8.1.438d}$$

We will now identify the numerical procedure to be followed via a computer to solve particular types of problems. First, we consider *design problems*. Either x_{A2L} (concentrate outlet) or the stage cut θ is given (W_{t2L} and $|W_{t10}|$ are known) (an additional problem: x_{A10} is given). Suppose x_{A2L} is given. Assume a particular value of θ; then W_{t2L} and $|W_{t10}|$ are obtained from (8.1.438a) and (8.1.438c)). Calculate x_{A1L} at the exit end from (8.1.437) for specified x_{A2L}. Now calculate (dx_{A2}/dA^+_m) at the exit end for a small negative increment of area dA^+_m from A^{t+}_m (unknown) using equation (8.1.427) since all quantities on the right-hand side of (8.1.427) are known. We can now calculate $(dx_{A1}/dA^+_m)_{z=L}$ from equation (8.1.436). We are now in a position to use equations (8.1.427), (8.1.432), (8.1.428) and (8.1.433) to calculate the values of x_{A2}, x_{A1}, W^+_{t2} and W^+_{t1} for $A^{t+}_m - dA^+_m$. Since A^{t+}_m is unknown (we may assume it to be zero and obtain its value when the summation of $-dA^+_m$ leads to the correct compositions, etc.), the sum total of $-dA^+_m$ will yield $-A^{t+}_m$, where A^{t+}_m would yield A^t_m from definitions (8.1.417) and (8.1.426). *Note*: $|W^+_{t1}|$ needs to be calculated at $A^t_m - dA^+_m$ first from (8.1.433) using x_{A1}, x_{A2} for A^{t+}_m since equation (8.1.432) needs it. Now we march toward the feed end stepwise with increments of $-dA^+_m$ and stop when $x_{A2} = x_{Af}$. Correspondingly, we will obtain at that axial location a value of x_{A1}. Check from equation (8.1.438d) whether, for the specified x_{A2L} and assumed θ, the value x_{A10} calculated coincides with the value of x_{A1} obtained numerically. If it does, the negative area A^{t+}_m calculated is correct. Otherwise assume another value of θ for the specified x_{A2L} and march toward the feed inlet. These are examples of what are called *boundary-value problems*.

An alternative procedure suggested by Walawender and Stern (1972) and Pan and Habgood (1974) reduces the number of equations to be integrated for the problem. Consider equations (8.1.432) and (8.1.427). Obtain from them the following equation:

$$\frac{dx_{A1}}{dx_{A2}} = \frac{W^+_{t2}}{|W^+_{t1}|}\frac{[\alpha^*_{AB}(1-x_{A1})(x_{A2}-\gamma x_{A1}) - x_{A1}\{(1-x_{A2})-\gamma(1-x_{A1})\}]}{[\alpha^*_{AB}(1-x_{A2})(x_{A2}-\gamma x_{A1}) - x_{A2}\{(1-x_{A2})-\gamma(1-x_{A1})\}]}. \tag{8.1.439}$$

The ratio of $(W^+_{t2}/|W^+_{t1}|)$ is expressed in terms of compositions via the following equations. Consider the envelope ABCD in Figure 8.1.51(a). Mass balances for this envelope lead to

$$W_{t2} = W_{t2L} + |W_{t1}|, \tag{8.1.440a}$$

$$W_{t2}\, x_{A2} = W_{t2L}\, x_{A2L} + |W_{t1}|\, x_{A1}. \tag{8.1.440b}$$

First eliminate $|W_{t1}|$ to obtain

$$(W_{t2}/W_{t2L}) = (x_{A2L}-x_{A1})/(x_{A2}-x_{A1}). \tag{8.1.441a}$$

Then eliminate W_{t2} to obtain

$$(|W_{t1}|/W_{t2L}) = (x_{A2L}-x_{A2})/(x_{A2}-x_{A1}), \tag{8.1.441b}$$

leading to

$$(W^+_{t2}/|W^+_{t1}|) = (x_{A1}-x_{A2L})/(x_{A2}-x_{A2L}). \tag{8.1.442}$$

Introduction of this into equation (8.1.439) yields

$$\frac{dx_{A1}}{dx_{A2}} = \frac{(x_{A1}-x_{A2L})}{(x_{A2}-x_{A2L})}\frac{[\alpha^*_{AB}(1-x_{A1})(x_{A2}-\gamma x_{A1}) - x_{A1}\{(1-x_{A2})-\gamma(1-x_{A1})\}]}{[\alpha^*_{AB}(1-x_{A2})(x_{A2}-\gamma x_{A1}) - x_{A2}\{(1-x_{A2})-\gamma(1-x_{A1})\}]}. \tag{8.1.443}$$

However, the ratio $(x_{A1}-x_{A2L})/(x_{A2}-x_{A2L})$ appearing first on the right-hand side of this equation makes the ratio (dx_{A1}/dx_{A2}) indeterminate at $z = L$. One can determine it from the ratio $(dx_{A1}/dA^+_m)_{z=L}/(dx_{A2}/dA^+_m)_{z=L}$ from equations (8.1.436) and (8.1.427), respectively. In the design problem solution procedure illustrated earlier, this equation may then be used as follows. Calculate as before for specified x_{A2L}, x_{A1L}. Then, for a given increment of $-dA^+_m$, calculate dx_{A2} from (8.1.427) and calculate dx_{A1} from

(8.1.443). In this fashion go to the feed section as before in the trial-and-error procedure to determine A_m^t.

Consider now the rating problem where the membrane area is known; the two product flow rates and compositions have to be determined. Here also assume a value of x_{A2L} at the exit end. Then follow the procedure described earlier for the design problem and arrive at the feed end where A_m^+ now would be zero since A_m^t is known at $z = L$ and therefore $A_m^t - dA_m$ is still positive at $z = L - dL$ and so on till $z = 0$. Check now whether the value of x_{A10} and θ obtained numerically satisfy the relation (8.1.438d) for the assumed x_{A2L}. If they do not, make an additional guess of x_{A2L} and repeat the procedure.

Figure 8.1.51(b) illustrates qualitatively the permeate composition in terms of oxygen as a function of stage cut for a silicone-copolymer membrane which has $\alpha_{O_2-N_2} = 2.4$ and $\gamma = 0.2$ and the feed-gas mixture is air $(x_{O_2f} = 0.209; x_{N_2f} = 0.791)$. The countercurrent results are compared with those from a cocurrent permeator and a crossflow permeator (Figure 7.2.1(a)). The countercurrent permeator performs the best and the cocurrent permeator the worst. However, the performance differences are not that much. Note that the value of x_{A1} at $\theta = 0$ is ~0.34; it can be obtained from (8.1.437) for $\alpha^*_{AB} = 2.4$, $\gamma = 0.2$, $x_{A2} = 0.209 = x_{O_2f}$.

As pointed out at the beginning of this section, in the case of hollow fibers, the permeate will flow through the bore of the fibers for a shell-side feed. **Under such conditions, it is quite likely that the permeate will undergo a significant pressure drop; P_p will not remain constant along** z. We can determine the pressure drop at any location using the Hagen–Poiseuille equation (6.1.1a) for a local gas flow rate per fiber of (Q/n):

$$\frac{\Delta P_p}{L} = \frac{8\mu Q}{n\pi R^4} = -\frac{dP_p}{dz} = \frac{128\,\mu\,Q}{n\,\pi\,d_i^4}, \tag{8.1.444}$$

where the hollow fiber internal diameter $d_i = 2R$ and the volumetric gas flow rate is Q/n. Assuming that the gas satisfies the ideal gas law (n_T, total number of moles having volume V), write

$$V = n_T\,\frac{RT}{P_p} \Rightarrow Q = |W_{t1}|\,\frac{RT}{P_p}. \tag{8.1.445}$$

Therefore

$$-\frac{dP_p}{dz} = \frac{128\,\mu\,|W_{t1}|\,RT}{n\,\pi\,d_i^4 P_p} \Rightarrow \frac{dP_p^2}{dz} = -\frac{256\,\mu\,RT}{n\,\pi\,d_i^4}\,|W_{t1}|. \tag{8.1.446}$$

An alternative equation will be

$$-\frac{d\gamma}{dz} = \frac{128\,\mu\,|W_{t1}|\,RT}{n\,\pi\,d_i^4\gamma}. \tag{8.1.447}$$

Now γ, the pressure ratio, becomes an additional variable and equation (8.1.447) has to be solved at every location. See Pan and Habgood (1978b) for an introduction.

There are a number of other aspects of significant importance in membrane gas permeator analysis and design.

(1) There is a tube sheet at the end of the hollow fiber permeator for potting every single hollow fiber. The permeated gas in the fiber bore will undergo some pressure drop in this section; however, there is no gas permeation.
(2) The feed gas may flow through the fiber bore and the permeated gas on the shell side may be in countercurrent flow. An equation of the type (8.1.447) has to be used using P_f/P_{fi} as the new variable, whereas γ, redefined as P_p/P_{fi}, will become constant; here P_{fi} is the value of P_f at the feed-gas inlet.
(3) Asymmetric hollow fiber membranes will have species flux expressions defined via Figure 7.2.1(b) and equation (7.2.2b) for membrane skin on the fiber O.D.
(4) The analysis of spirally wound gas permeation devices shown in Figure 7.2.1(e) will not follow the traditional countercurrent analysis formalism. See Pan (1983, 1986).

A chapter by Sengupta and Sirkar (1995) provides greater details on each one of the above aspects. As shown in Figure 7.2.1(b), the feed gas is imposed on the skin of the hollow fiber O.D. In air-separation the compressed feed air is often brought to the I.D. of such a fiber with the skin on the O.D. Sidhoum *et al.* (1988) have studied this configuration.

8.1.10 Countercurrent gas centrifuge

In Section 4.2.1.1, we studied how separation is achieved in a closed gas centrifuge. As shown in Figure 4.2.2, the heavier gas species (mol. wt. M_2) concentrates on the periphery and the lighter gas species (mol. wt. $M_1 < M_2$) concentrates near the center of the cylindrical gas centrifuge. Also, relation (4.2.13) indicates that the gas pressure at the cylinder periphery is considerably larger than that at the center ($r = 0$). Further, the separation achieved between the peripheral region ($r = r_2$) and any other region near the center (say, $r = r_1$) is dependent on the term $\exp\left[(M_2 - M_1)\,\omega^2\left(r_2^2 - r_1^2\right)/2RT\right]$ for species 1 and 2. We observed in Figure 4.2.2 that the separation achieved for a gas mixture such as SO_2–N_2 was not attractive enough vis-à-vis other techniques. On the other hand, the separation achieved with isotopic mixtures is considerably better, $\alpha_{12} \sim 1.03$ to 1.1, compared to that of gaseous diffusion, 1.004. In fact, separation of a mixture of uranium isotopes, $U^{235}F_6$ and $U^{238}F_6$, is carried out in large scale around the world using gas centrifuges; however, this gas centrifuge is an open

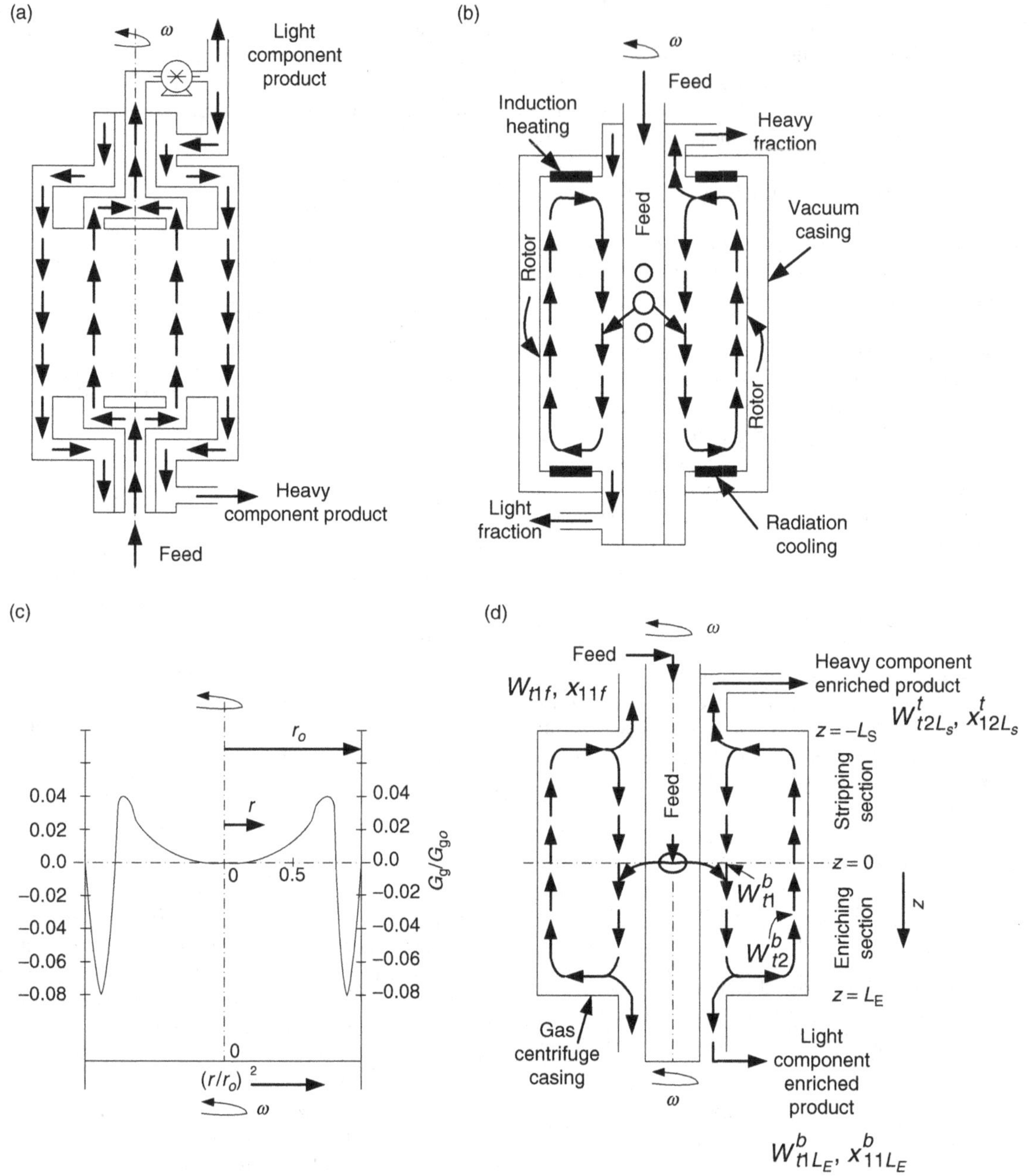

Figure 8.1.52. Countercurrent gas centrifuge schematics. (a) Externally generated flow by pump; (b) Groth design employing induction heating at top and radiation cooling at bottom; (c) mass velocity profile across centrifuge radius for flow pattern in (a); (d) CGC analysis schematic details. (After Olander (1972) and Benedict et al. *(1981, chap. 14).)*

device which uses countercurrent flow. Due to security reasons, the exact description of countercurrent gas centrifuge technology is not available in the public domain. We will describe below the basic concept available in the open literature.

Our brief treatment of a countercurrent gas centrifuge (CGC) relies on the following sources: Cohen (1951); Pratt (1967, pp. 333–346); Olander (1972); Benedict *et al.* (1981, pp. 847–875). We will first illustrate the overall gas flow pattern in a CGC. In Figure 8.1.52(a) we show two different flow patterns in the countercurrent format. Figure 8.1.52(a) shows a cylindrical centrifuge where the lighter stream enriched (or to be enriched) in the gas species of lower molecular weight is pumped from an external source and it moves up near the center of the CGC; the gas stream enriched (or to be enriched) in the heavier species is pumped down from the same external source to move down the CGC near its external periphery.

Figure 8.1.52(b) illustrates an alternative arrangement to generate the vertical flow pattern (the Groth design). At the top, there is an electromagnetically driven induction heating arrangement to circulate the gas, the heavier stream on the periphery going up and the lighter stream near the center going down; correspondingly, at the bottom there is a radiation cooling arrangement to cool the bottom section of the CGC, and thereby the gas, and turn the lighter gas stream around. The temperature difference is around 20 °C (Olander, 1972). An additional arrangement (Zippe's machine) employs scoops at the top and the bottom to withdraw appropriate product streams. Generally the feed-gas mixture is introduced near the center of the cylindrical gas centrifuge into the lighter gas stream in both designs. The centrifuge rotor is kept in a vacuum casing.

The overall arrangement in such a scheme appears to be quite similar to the distillation scheme shown in Figure 8.1.3(a), a type (3) system: the feed is introduced at the middle of the column, one stream going up, one stream coming down, with the centrifugal separation force, as well as the pressure gradient force, acting perpendicular to the two flow streams. Further, after an appropriate fraction of the upflowing stream is withdrawn as product at the top, the rest of the stream starts coming down as if that is the reflux stream coming from the condenser in Figure 8.1.3(a). The only difference is that the counterflowing streams are of the same phase, i.e. they are miscible, unlike those in Figure 8.1.3(a); it is therefore a type (1) system. A similar flow arrangement exists at the bottom of the column, where an appropriate fraction of the downflowing stream is withdrawn as product and the rest gets turned around as if we have a reboiler at the column bottom. Note one difference in Figure 8.1.52(b). The lighter component appears enriched in the bottom product stream, which is part of the stream coming down the center in the Groth/Zippe design; the heavier product stream is withdrawn at the top from the stream flowing upward near the periphery of the gas centrifuge.

There exist, however, some peculiarities in this technique of CGC in terms of the pressure conditions varying across the radius as well as the flow profile. At any axial coordinate z, the value of $P(r)$ in a closed centrifuge is given by equation (4.2.13). For the isotopic gas mixture of $U^{235}F_6$ and $U^{238}F_6$ at $T = 300$ K, Benedict *et al.* (1981) have illustrated the value of the ratio of the total gas pressure at any radius $P(r)$ to that at the periphery of the cylindrical centrifuge $P(r_o)$ (where $r = r_o$, the radius of the cylindrical gas centrifuge), $P(r)/P(r_o)$; when $(r/r_o) = 0.9$ the value of the ratio $P(r)/P(r_o)$ is 1.4×10^{-3} for a peripheral speed of 700 m/s, whereas when $(r/r_o) = 0.5$, the ratio has the value of 5×10^{-12}. Essentially there is very little gas near the center; most of the gas is located in a thin region close to the wall. This is also reflected in the profile of the mass velocity G_g $(= \rho_g v_z)$, where v_z is the gas velocity in the z-direction (vertical) and ρ_g is the local gas density. Figure 8.1.52(c) illustrates approximately (for the case where the stream near the center moves up) the radial variation of the axial mass velocity profile (G_g/G_{go}) where G_{go} is proportional to the gas circulation rate (following Benedict *et al.* (1981)); note that the values near the center are uncertain even though they are very small. Further, around $(r^2/r_o^2) = 0.88$, the flow turns around. At $r = r_o$ (i.e. $(r/r_o) = 1$), the mass velocity is zero due to the no slip boundary condition. As shown, the lighter fraction is moving up and the heavier fraction is moving down as per Figure 8.1.52(a). It could easily have been the other way around, as in Figure 8.1.52(b). The radius location $r = r_1$ will be used to indicate where the flow direction turns around.

8.1.10.1 *Axial composition variation in a CGC*

We will now provide a very brief introduction to the method adopted to determine the variation of gas composition with axial distance z along the column length. Our treatment follows that of Olander (1972) and Benedict *et al.* (1981). It involves the following core assumptions in the context of the equations of change (6.2.5b) and (6.2.5g) in cylindrical polar coordinates (see Table 6.2.1).

(a) The whole gas is rotating with an angular velocity of ω rad/s as if it is a solid body rotation with ω rad/s, the rotational velocity of the centrifuge.
(b) The above also implies $v_r^* = 0$, an assumption that is not valid at the top and bottom of the centrifuge where product streams are being withdrawn.
(c) Analyze the system in a frame of reference moving with an angular velocity ω rad/s. Therefore $v_\theta^* = 0$.
(d) Assume $J_{i\theta}^* = 0$.
(e) Neglect temperature and pressure dependence of quantities like $C_t D_{AB}$ (equation (3.1.91b) indicates that $PD_{AB}|_T \sim$ constant; $P = C_t RT$). Note that the temperature variation from top to bottom of the centrifuge is not much.
(f) Assume $\boldsymbol{v}^* \cong \boldsymbol{v}$.
(g) Assume steady state operation.
(h) Due to the extreme variation of pressure P, we cannot assume C_t to be constant throughout the centrifuge.
(i) The variation of composition in the radial direction, although significant and the fundamental basis for enrichment of the isotopes, is negligible compared to that achieved axially. Therefore for an axial

composition variation calculation one could take an average across the cross section of the concentration or mole fraction and determine its z-variation. For illustration: $r_o = 9.145$ cm; $L = 66 - 335$ cm.

We will now employ these assumptions and focus on the equation of change of concentration of the lighter isotope species using molar concentration C_i and mole fraction x_i. The CGC under consideration is shown in Figure 8.1.52(d), with the z-coordinate vertically downwards having a value of $z = 0$ at the location of feed introduction. Since C_t is not constant, we should focus first on equation (6.2.5b) and later to its C_t - constant version, equation (6.2.5g) (see Table 6.2.1), in cyclindrical polar coordinates (for $R_i = 0$):

$$\nabla \cdot (C_i \boldsymbol{v}) = -\nabla \cdot (J_i^*). \qquad (8.1.448)$$

Since $v_\theta = 0$, $v_r = 0$ via assumptions (b) and (c), and

$$\frac{\partial}{\partial z}(C_i v_z) = -\frac{1}{r}\frac{\partial}{\partial r}(rJ_{ir}^*) - \frac{\partial J_{iz}^*}{\partial z} \qquad (8.1.449)$$

in cylindrical polar coordinates, where we have also employed assumption (d). Here we note that the total axial flow rate ($C_t\, v_z \times$ cross-sectional area) does not change with z; what changes is its composition, x_i. Therefore

$$\frac{\partial}{\partial z}(C_i v_z) = C_t v_z \frac{\partial x_i}{\partial z}, \qquad (8.1.450)$$

which leads to

$$C_t v_z \frac{\partial x_i}{\partial z} = -\frac{1}{r}\frac{\partial}{\partial r}(rJ_{ir}^*) - \frac{\partial J_{iz}^*}{\partial z}. \qquad (8.1.451)$$

Since there is no pressure gradient driven diffusion in the axial direction, we have only molecular diffusion

$$J_{iz}^* = -D_{AB}\, C_t \frac{dx_i}{dz}. \qquad (8.1.452)$$

For J_{ir}^*, the following approach is normally adopted. In a closed gas centrifuge with no axial motion, there is an equilibrium radial composition profile developed via a balance between the radially inward pressure gradient (fugacity gradient) and the radially outward centrifugal force; it is illustrated by relations (4.2.8) and (4.2.9) for a binary system of species 1 and 2:

$$\frac{x_{1j}(r)}{(1-x_{1j}(r))}\frac{(1-x_{1j}(0))}{x_{1j}(0)} = \exp-\left(\frac{(M_2-M_1)\omega^2 r^2}{2RT}\right). \qquad (8.1.453)$$

To obtain a radial gradient of the profile, we differentiate this expression in the following fashion:

$$d\left(\ln\left[x_{1j}(r)/(1-x_{1j}(r)]\right)\right)/dr = -\left(\frac{(M_2-M_1)\omega^2 r}{2RT}\right); \qquad (8.1.454)$$

$$\frac{d\left[x_{1j}(r)/(1-x_{1j}(r))\right]/dr}{x_{1j}(r)/(1-x_{1j}(r))} = -\frac{(M_2-M_1)\omega^2 r}{2RT}$$

$$\Rightarrow \left.\frac{dx_{ij}(r)}{dr}\right|_{\text{equilibrium}} = -\frac{(M_2-M_1)\omega^2 r}{2RT} x_{1j}(r)(1-x_{1j}(r)). \qquad (8.1.455)$$

Although this profile exists under centrifugal equilibrium, we may assume that the actual radial profile will be somewhat different. In the countercurrent gas centrifuge, we can expect that the radial diffusive flux $-C_t D_{AB}\, dx_{1j}(r)/dr$ will be superimposed on that due to the equilibrium profile, leading to the following expression for J_{1r}^*:

$$J_{1r}^* = -C_t D_{AB}\left[\frac{dx_{1j}(r)}{dr} - \left.\frac{dx_{1j}(r)}{dr}\right|_{\text{equilibrium}}\right] \qquad (8.1.456)$$

$$= -C_t D_{AB}\left[\frac{dx_{1j}(r)}{dr} + \frac{(M_2-M_1)\omega^2 r}{2RT} x_{1j}(r)\,(1-x_{1j}(r))\right]. \qquad (8.1.457)$$

We will replace $x_{1j}(r)$ by $x_1(r)$ from now on. Substituting this expression, as well as that for J_{1z}^* from (8.1.452), we get for species $i = 1$ from equation (8.1.451)

$$C_t v_z \frac{\partial x_1(r)}{\partial z} = \frac{C_t D_{AB}}{r} \times \frac{\partial}{\partial r}\left[r\frac{dx_1(r)}{dr} + \frac{(M_2-M_1)\omega^2 r^2}{2RT} x_1(r)\,(1-x_1(r))\right] + (D_{AB}C_t)\frac{d^2 x_1}{dz^2}. \qquad (8.1.458)$$

There is no exact solution available for this equation. However, a radial-averaging procedure is implemented along with the following assumption: the axial diffusion term on the right-hand side of the equation may be neglected compared with the axial convection term. Such a procedure is implemented in both the enriching section and the stripping section of this CGC column (Figure 8.1.52(d)) on the following simpler equation:

$$C_t v_z \frac{\partial x_1(r)}{\partial z} = \frac{C_t D_{AB}}{r} \times \frac{\partial}{\partial r}\left[r\frac{dx_1(r)}{dr} + \frac{(M_2-M_1)\omega^2 r^2}{2RT} x_1(r)\,(1-x_1(r))\right]. \qquad (8.1.459)$$

Note that at $r = 0$ or $r = r_o$, the bracketed quantity will be zero. Integrate this equation from $r = 0$ to r, after multiplying it by r, to get

$$r\frac{dx_1(r)}{dr} = -\frac{(M_2-M_1)\omega^2 r^2}{RT} x_1(r)\,(1-x_1(r)) + \frac{1}{C_t D_{AB}}\frac{\partial x_1}{\partial z}\int_0^r C_t v_z\, r\, dr, \qquad (8.1.460)$$

where we have assumed (1) $(\partial x_1/\partial z)$ to be independent of r and (2) that $(x_1(r)\,(1-x_1(r))$ is a constant.

Now consider the CGC column of Figure 8.1.52(d), where the light component enriched bottom product flow rate of $W^b_{t1L_E}$ at $z = L_E$ has a composition x_{11L_E}. The molar flow rate of the downflowing stream at any level of the column in the enriching section is W^b_{t1}; the upflowing stream flow rate at that axial location is W^b_{t2} (upflowing and downflowing reversed with respect to distillation) vis-à-vis the light component composition. The feed coordinate location is $z = 0$. The heavy component enriched top product flow rate of $W^t_{t2L_S}$ at $z = -L_S$ has a composition of x_{12L_S}. We can have a total mass flow balance in the enriching section with an envelope crossing the light-component enriched bottom product flow out $(W^b_{t1L_E})$:

$$W^b_{t1L_E} = W^b_{t1} - W^b_{t2} = 2\pi \int_0^{r_o} C_t v_z r \, dr. \tag{8.1.461}$$

However, the light component balance over this envelope will involve not only convective, but also diffusive contributions:

$$W^b_{t1L_E} x_{11L_E} = 2\pi \int_0^{r_o} r N_{1z} \, dr = 2\pi \int_0^{r_o} x_{11b} C_t v_z r \, dr - 2\pi D_{AB} C_t \int_0^{r_o} \frac{\partial x_{11b}}{\partial z} r \, dr, \tag{8.1.462}$$

where

$$N_{1z} = x_{11b} C_t v_z - D_{AB} C_t \frac{dx_{11b}}{dz}. \tag{8.1.463}$$

After detailed steps involving integration of the first term by parts, the following result has been obtained (see Olander (1972) and Benedict *et al.* (1981)) for the *composition variation of the light fraction with height for the enriching section*:

$$\frac{d\overline{x}_{11b}}{dz} = \frac{a_1}{a_5} \overline{x}_{11b} (1 - \overline{x}_{11b}) - \frac{W^b_{t1L_E} (\overline{x}_{11L_E} - \overline{x}_{11b})}{a_5}, \tag{8.1.464}$$

where

$$a_1 = \frac{(M_2 - M_1)\omega^2}{RT} \int_0^{r_o} F(r) r \, dr, \tag{8.1.465a}$$

$$a_2 = \pi D_{AB} C_T r_o^2, \tag{8.1.465b}$$

$$a_3 = \frac{1}{2\pi D_{AB} C_T} \int_0^{r_o} [F(r)]^2 \frac{dr}{r}, \tag{8.1.465c}$$

$$a_5 = a_2 + a_3, \tag{8.1.465d}$$

$$F(r) = 2\pi \int_0^{r} C_t v_z r \, dr. \tag{8.1.465e}$$

Note: $\overline{x}_{11b}$ is an averaged quantity over the cross section of the column.

The equation corresponding to equation (8.1.464), for the *stripping section*), where the light component mole fraction is denoted by $\overline{x}_{11t}$, is

$$\frac{d\overline{x}_{11t}}{dz} = \frac{a_1^*}{a_5^*} \overline{x}_{11t} (1 - \overline{x}_{11t}) - \frac{W^t_{t2L_s}(x_{11t} - \overline{x}_{11L_s})}{a_5^*}. \tag{8.1.466}$$

Note: $\overline{x}_{11t}$ is an averaged quantity over the column cross section. For this equation, constants a_1^*, a_2^*, a_3^*, a_5^* have exactly the same form as a_1, a_2, a_3, a_5, except the quantity $F(r)$, called the *flow function*, is to be evaluated using the v_z-profile in the stripping section. This profile is only slightly different from that in the enricher.

Both equations (8.1.464) and (8.1.466) may be written as (we use the enriching section)

$$h \frac{d\overline{x}_{11b}}{dz} = (\alpha_{12} - 1)\overline{x}_{11b}(1 - \overline{x}_{11b}) - \frac{W^b_{t1L_E}(x_{11L_E} - \overline{x}_{11b})}{N}, \tag{8.1.467}$$

where

$$a_5 = hN, \tag{8.1.468a}$$

$$a_1/a_5 = (\alpha_{12} - 1)/h, \tag{8.1.468b}$$

$$N = 2\pi \int_0^{r_1} C_t v_z r \, dr, \tag{8.1.468c}$$

$$N = 2\pi \int_{r_1}^{r_o} C_t v_z r \, dr. \tag{8.1.468d}$$

Here, r_1 is the radius where the vertical flow turns around and N is the molar flow rate of the lighter stream going down or the heavier stream going up, respectively. The unit of h is cm (or m); it is similar to the height of transfer unit, e.g. $HTU_g = (|W_{tg}|/S'_c)/k'^a_{igx}$, equation (8.1.62b).

In general, the quantity N (equation (8.1.468d)) for the heavy component enriched stream flow rate and the corresponding quantity for the light component enriched stream flow rate (equation (8.1.468c)) vary with the height of the column. Note that v_z has different signs for $0 \le r \le r_1$ and $r_1 \le r \le r_o$. If we take the magnitudes of v_z and evaluate the *magnitude* of the *total internal flow rate* as

$$N^t = 2\pi \int_0^{r_o} C_t |v_z| r \, dr, \tag{8.1.469}$$

and assume it to be essentially constant along with the radial velocity profile, the parameters a_1 and a_5 in equation (8.1.464) in the enriching section and a_1^* and a_5^* in equation (8.1.466) for the stripping section are independent of z. Although a_1 and a_5 are slightly different from a_1^* and a_5^*, respectively, we assume them to be the same here (this is true at total reflux). Then, for the case of low concentration or low enrichment of the light component (i.e. $\overline{x}_{11b} << 1$), we can rewrite equation (8.1.464) as

$$a_5 \frac{\mathrm{d}\overline{x}_{11b}}{\mathrm{d}z} = \left(a_1 + W^b_{t1L_E}\right) \overline{x}_{11b} + W^b_{t1L_E}\, \overline{x}_{11L_E} \quad (8.1.470)$$

for the enriching section. Similarly, we can rewrite equation (8.1.466) for the stripping section as (for $(\overline{x}_{11t} << 1)$)

$$a_5 \frac{\mathrm{d}\overline{x}_{11t}}{\mathrm{d}z} = \left(a_1 - W^t_{t2L_S}\right) \overline{x}_{11t} + W^t_{t2L_S}\, \overline{x}_{11L_S}. \quad (8.1.471)$$

Now it is possible to integrate each of these two equations to obtain the two profiles as a function of the column height. For the *enriching section profile* we go from $z = 0$, where $\overline{x}_{11b} = x_{11f}$, to $z = L_E$, where $\overline{x}_{11b} = x_{11L_E}$. Equation (8.1.470) when integrated,

$$\int\limits_{x_{11f}}^{\overline{x}_{11L_E}} \frac{\mathrm{d}\overline{x}_{11t}}{\left(a_1 + W^t_{t1L_E}\right)\overline{x}_{11t} - W^t_{t1L_E}\, x_{11L_E}} = \frac{1}{a_5}\int\limits_0^{L_E} \mathrm{d}z, \quad (8.1.472)$$

yields, after appropriate rearrangement,

$$\frac{\overline{x}_{11L_E}}{\overline{x}_{11f}} = \frac{a_1 + W^b_{t1L_E}}{W^b_{t1L_E} + a_1 \exp\left[-\left(a_1 + W^b_{t1L_E}\right) L_E/a_5\right]}. \quad (8.1.473)$$

Similarly, the *profile for the stripping section* is obtained by integrating equation (8.1.471) from $z = 0$, where $\overline{x}_{11t} = x_{11f}$, to $z = -L_S$, where $\overline{x}_{11t} = \overline{x}_{11L_S}$, under the assumption that the feed gas is introduced at the location $(z = 0)$ where $\overline{x}_{11b} = x_{11t} = x_{11f}$:

$$\int\limits_{x_{11f}}^{\overline{x}_{11L_S}} \frac{\mathrm{d}\overline{x}_{11t}}{\left(a_1 - W^b_{t2L_S}\right)\overline{x}_{11t} + W^b_{t2L_S} x_{11L_S}} = \frac{1}{a_5}\int\limits_0^{-L_S} \mathrm{d}z = -(L_S/a_5); \quad (8.1.474)$$

$$\frac{\overline{x}_{11L_S}}{\overline{x}_{11f}} = \frac{W^b_{t2L_S} - a_1}{W^b_{t2L_S} - a_1 \exp\left[-\left(W^b_{t2L_S} - a_1\right) L_S/a_5\right]}. \quad (8.1.475)$$

We will now introduce a few parameters in CGC such as the *flow pattern efficiency* E_f and the *flow number* N^f. Then we will focus on the separation achieved in a *CGC under total reflux*. The flow pattern efficiency E_f is defined as

$$E_f = \frac{4\left[\int_0^1 F(r^+)\, r^+\, \mathrm{d}r^+\right]^2}{\int_0^1 \{[F(r^+)]^2\, \mathrm{d}r^+/r^+\}}, \quad (8.1.476a)$$

where

$$F(r^+) = F(r) = 2\pi r_o^2 \int\limits_0^{r^+} C_t\, v_z\, r^+ \mathrm{d}r^+, \qquad r^+ = r/r_o, \quad (8.1.476b)$$

from equation (8.1.465e). The flow number is defined as

$$N^f = \frac{N^t}{4\left\{\int_0^1 [F(r^+)]^2\, \mathrm{d}r^+/r^+\right\}^{1/2}}. \quad (8.1.477)$$

In essence, scaling factors have been introduced in the denominators of E_f and N^f making them independent of the magnitude of the flow rates of the feed and the internal flow (circulation). In terms of these parameters, the coefficients a_1 and a_5 become

$$a_1 = \frac{(M_2 - M_1)\, \omega^2 r_o^2}{2RT} \left(\sqrt{E_f}/4N^f\right) N^t \quad (8.1.478)$$

and

$$a_5 = \left\{\frac{1}{2}\left(2\pi D_{\mathrm{AB}}\, C_T\, r_o^2\right) + \frac{(N^t)^2}{16\,(N^f)^2}\, \frac{1}{2\pi\, D_{\mathrm{AB}} C_t}\right\}. \quad (8.1.479)$$

We can now focus on the separation achieved at total reflux ($W^b_{t1L_E} = 0$; $W^t_{t2L_S} = 0$) in a CGC for low enrichment cases (i.e. $\overline{x}_{11b} << 1, \overline{x}_{11t} << 1$). In this case, the governing equations are obtained from equations (8.1.470) and (8.1.471) as

$$a_5 \frac{\mathrm{d}\overline{x}_{11b}}{\mathrm{d}z} = a_1\, \overline{x}_{11b}, \quad (8.1.480a)$$

$$a_5 \frac{\mathrm{d}\overline{x}_{11t}}{\mathrm{d}z} = a_1\, \overline{x}_{11t}. \quad (8.1.480b)$$

(*Note*: Due to total reflux, $a_1 \cong a_1^*$, $a_5 \cong a_5^*$; Olander (1972).) Integrate now from $z = 0$ (where $\overline{x}_{11b} = x_{11f} = \overline{x}_{11t}$) to $z = L_E$ for equation (8.1.480a) and $z = -L_S$ to $z = 0$ for equation (8.1.480b):

$$\left(\frac{\overline{x}_{11L_E}}{x_{11f}}\right) = \exp\left\{\frac{a_1}{a_5} L_E\right\}, \quad (8.1.481a)$$

$$\left(\frac{x_{11f}}{\overline{x}_{11L_S}}\right) = \exp\left\{\frac{a_1}{a_5} L_S\right\}. \quad (8.1.481b)$$

(*Note*: There is no feed introduction at total reflux (see Section 8.1.3.2); x_{11f} merely identifies the composition at the feed location, $z = 0$.) Multiplying these two equations together, we get an expression for the separation factor between the light fraction end ($\overline{x}_{11L_E}$) and the heavy fraction end ($\overline{x}_{11L_S}$) at total reflux:

$$\left(\frac{\overline{x}_{11L_E}}{\overline{x}_{11L_S}}\right) \cong \frac{\overline{x}_{11L_E}(1 - \overline{x}_{11L_S})}{(1 - \overline{x}_{11L_E})\,\overline{x}_{11L_S}} = \alpha_{12}\big|_{\text{total reflux}}$$

$$= \exp\left\{\frac{a_1}{a_5}(L_E + L_S)\right\}. \quad (8.1.482)$$

The separation factor over the whole cylindrical gas centrifuge of length $(L_E + L_S)$ at total reflux is determined by (a_1/a_5), which depends on N^t, the total internal flow rate.

When $N^t \to 0$, $a_1 \to 0$, yielding $\alpha_{12}|_{\text{total reflux}} \to 1$, since, without internal circulation, compositions at different locations would be levelized by axial diffusion. When $N^t \to \infty$, axial circulation mixes up regions of different concentrations, substantially reducing the axial enrichment. At intermediate values of N^t, we have an optimum, a maximum value of $\alpha_{12}|_{\text{total reflux}}$. To obtain the optimum value of α_{12} at total reflux, obtain the value of N^t which maximizes (a_1/a_5):

$$\frac{d(a_1/a_5)}{dN^t} = 0 = \frac{d}{dN^t}\left\{\frac{\frac{(M_2-M_1)\omega^2 r_o^2}{2RT}\left(\sqrt{E_f}/4N^f\right)N^t}{\frac{1}{2}\left(2\pi D_{AB}\, C_t\, r_o^2\right) + \frac{(N^t)^2}{16(N^f)^2}\frac{1}{2\pi D_{AB}\, C_t}}\right\}. \quad (8.1.483)$$

Define (a_1/a_5) as $\left(AN^t/B + N^{t^2}C\right)$. Then the optimum N^t is given by

$$N^t = (B/C)^{1/2} = 2\sqrt{2}(2\pi D_{AB}\, C_t\, r_o)\, N^f. \quad (8.1.484)$$

Correspondingly,

$$\alpha_{12}\Big|_{\text{total reflux}}^{\text{opt}} = \exp\left[\frac{a_1}{a_5}(L_E + L_S)\right]$$

$$= \exp\left\{(L_E+L_S)\frac{\frac{(M_2-M_1)\omega^2 r_o^2}{2RT}\sqrt{E_f}\frac{2\sqrt{2}(2\pi C_t D_{AB}\, r_o)N^f}{4N^f}}{\frac{1}{2}(2\pi D_{AB}\, C_t\, r_o^2) + \frac{8(2\pi C_t D_{AB}\, r_o)^2 (N^f)^2}{16(N^f)^2\, 2\pi D_{AB}\, C_t}}\right\}; \quad (8.1.485)$$

$$\alpha_{12}\Big|_{\text{total reflux}}^{\text{opt}} = \exp\left\{\left(1/\sqrt{2}\right)\frac{(M_2-M_1)\omega^2 r_o^2}{2RT}\sqrt{E_f}\frac{(L_E+L_S)}{r_o}\right\}. \quad (8.1.486)$$

Example 8.1.22 below compares this optimal separation factor at total reflux at $z = L_E$ and the heavy fraction $z = -L_S$ with the radial equilibrium separation factor (4.2.7) at any axial location in the column.

Example 8.1.22 Consider a countercurrent gas centrifuge separating a $U^{235}F_6$ and $U^{238}F_6$ mixture at 27 °C. The radius of the centrifuge is 9 cm. The flow pattern efficiency E_f = 0.8 (Olander, 1972). The centrifuge is rotating at 32 000 rpm. Calculate the values of the optimum separation factor of the centrifuge at total reflux for the following values of $(L_E + L_S)$: 66 cm; 132 cm. Compare them with the separation factor achieved in the radial direction at any axial location.

Solution We will first calculate the value of the separation factor in the radial direction via equation (4.2.7) for $r = r_o =$ 9 cm: $(M_2 - M_1)$ = 3 g/gmol; $\omega = (32\,000/60) \times 2\pi = 3351$ rad/s; $T = 300$ K; R = 8.31×10^7 g-cm^2/gmol-K-s^2; so

$$\alpha_{12}|_{\text{radial}} = \exp\left[+\frac{3\frac{\cancel{g}}{\cancel{gmol}} \times 3351^2 \frac{\text{radian}^2}{\cancel{s}^2} \times 81\,\cancel{cm}^2}{2 \times 8.31 \times 10^7 \times 300\cancel{K} \times \frac{\cancel{g}\text{-}\cancel{cm}^2}{\cancel{g}\text{mol-}\cancel{K}\text{-}\cancel{s}^2}}\right]$$

$$= \exp\left[\frac{3 \times 11.23 \times 10^6 \times 81}{16.62 \times 10^7 \times 300}\right] = \exp(0.054)$$

$$\Rightarrow \alpha_{12}|_{\text{radial}} = 1.055.$$

We will now calculate the optimum column separation factor at total reflux using equation (8.1.486). Let $(L_E + L_S) = 66$ cm. Then

$$\alpha_{12}\Big|_{\text{total reflux}}^{\text{opt}} = \exp\left\{\frac{1}{1.414}\frac{3 \times 3351^2 \times 81}{2 \times 8.31 \times 10^7 \times 300}\sqrt{0.8}\frac{66}{9}\right\}$$

$$= \exp\left\{\frac{3 \times 11.23 \times 10^6 \times 81 \times 0.894 \times 66}{1.414 \times 2 \times 8.31 \times 10^7 \times 300 \times 9}\right\}$$

$$\Rightarrow \alpha_{12}\Big|_{\text{total reflux}}^{\text{opt}} = \exp(0.254) = 1.289.$$

When $(L_E + L_S) = 132$ cm,

$$\alpha_{12}|_{\text{total reflux}}^{\text{opt}} = \exp(0.508) = 1.661.$$

The calculations in Example 8.1.22 show how the separation factor is considerably enhanced by the countercurrent gas flow compared to what is achieved at any axial location between the center and the wall. However, these much higher values are achieved at total reflux. When products are withdrawn at both ends, the enrichment achieved will be lower. See Olander (1972) and Benedict *et al.* (1981) for further details.

8.1.11 Thermal diffusion and mass diffusion

We have observed that countercurrent separation devices achieve considerable separation under driving forces such as chemical potential gradient and the external force of a centrifugal force field. As illustrated in equations (3.1.44) and (3.1.50), there is another type of force, the thermal diffusion force. The separation achieved thereby in a closed two-bulb cell has been illustrated in Section 4.2.5.1. We have already illustrated conceptually how thermal diffusion can achieve separation in a countercurrent column via Figures 8.1.1(a)–(h). For UF^6 isotope separation, however, the radial separation factor in a two-bulb cell is much smaller than for other isotope separation processes. In a countercurrent column, this value is reduced by about 50%. As a result, thermal diffusion columns are not used at all for any practical/large-scale separation. More details on thermal diffusion columns are available in Pratt (1967, chap. viii) and Benedict *et al.* (1981, pp. 906–915), where one can find information on the primary references.

The mass diffusion process relies on the following phenomenon. If we have a gas mixture of a heavier gas (say, N_2) and a lighter gas (say, H_2 or He) exposed to a layer of a sweep vapor (say, steam), the lighter gas will diffuse much faster through this layer than the heavier gas; the sweep vapor will become enriched in the lighter gas (H_2 or He). A practical way of implementing it would involve flow of the two gas streams: the feed-gas mixture of, say, N_2 and H_2 and the sweep vapor of steam countercurrently on two sides of a screen containing small holes, preferably below 10 μm in diameter; the hole dimensions are chosen to reduce the bulk flow of the feed-gas mixture and the sweep vapor through the screen holes for a given pressure difference between the two sides. The lower the pressure difference between the two sides, the lower the extent of bulk mixing. The net result is that the sweep vapor stream (of steam) is enriched in the lighter gas species, whereas the feed-gas stream is enriched in the heavier species. When steam is condensed out of both the streams, we will get two separate gas streams, the light fraction and the heavy fraction.

Modeling of such devices and processes using the Maxwell–Stefan equations (3.1.178) (see equations (3.1.193) and (3.1.196)) has been illustrated briefly in Pratt (1967, chap. VIII). More recent efforts on this process, termed "FricDiff," have been described by Peters *et al.* (2008). The variation they have studied involves the separation of a vapor mixture, say of water and alcohol, with the sweep stream being a gas, say CO_2. The condensation of both streams at the exits of the device will yield two liquid streams having considerably different compositions.

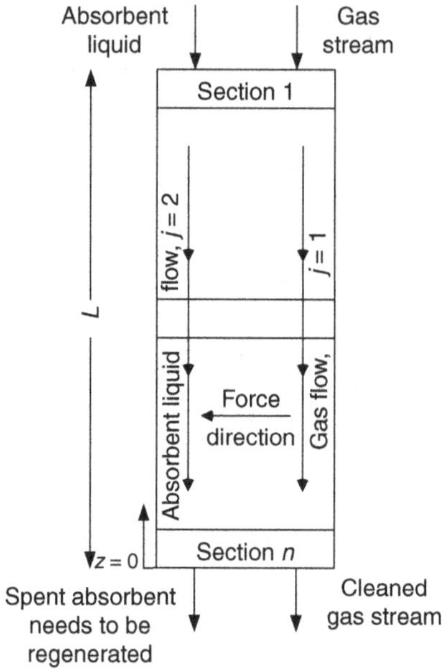

Figure 8.2.1. Cocurrent flow of a gas stream and an absorbent liquid in a vertical column for absorption of a species from the gas.

8.2 Cocurrent bulk motion of two phases or regions perpendicular to the direction(s) of the force(s) driving species/particles

We provide here a brief introduction to the separation achieved/achievable when two different immiscible phases or regions flow in a device in the same direction. Consider Figure 8.2.1(a), where phase $j = 1$ (say, the gas phase) and $j = 2$ (say, the liquid phase) move along the negative z-coordinate as they are in contact with each other. Divide the device length into multiple segments (say, $n = 8$). As the two phases contact each other, species i will transfer from one phase to the other in each segment. We may assume that none of the earlier segments are in equilibrium; however, the later segments may/will achieve two-phase equilibrium distribution of any species i. So the ultimate distribution of species i achieved between the two phases just corresponds to that achieved when two feed phases are contacted in a static vessel after a long time (Section 4.1). If we assume that the earlier segments in the cocurrent device are in equilibrium, then the rest of the segments in the flow direction do not have any particular function. This is in direct contrast to the countercurrent operational mode where the longer the countercurrent column, the higher the extent of separation.

Why would one utilize cocurrent contacting of two immiscible phases at all? We have observed in Sections 8.1.2.1 and 8.1.3.5 for gas absorption stripping and distillation processes, respectively, that there are strict limitations in the flow rates of the two counterflowing phases to avoid loading and ultimately flooding of the columns (unless porous membrane based nondispersive contacting is implemented, see Section 8.1.2.2.1). If the two phase flow rates needed are high and a more vigorous contacting is desired, cocurrent flow may be employed provided no more separation than that achieved in one ideal stage is desired. In fact, one can have countercurrent multistage separation in which each stage has cocurrent contact and therefore potentially a high mass-transfer coefficient due to much higher flow rates and dispersion possible in cocurrent flow. Such a configuration of cocurrent flow stages in a countercurrent cascade may lead to a smaller overall device at the expense of a more complex arrangement (King, 1980, pp. 642–643). Reiss (1967) has illustrated cocurrent gas–liquid contacting in packed columns and its advantages over countercurrent flow.

Classical cocurrent separators in two-phase systems generally use dispersive contacting with one phase dispersed as drops (or bubbles). Porous membrane based systems can avoid such dispersive contacting, as we have observed in Section 8.1.2.2.1. Porous/microporous membranes are being used also to contact two miscible phases, such as blood and aqueous dialysate solution, in dialysis processes, as studied in Sections 8.1.7.1 and 8.2.4.1. Microfluidic devices have been developed where laminar, parallel and cocurrent flow of two or three streams take place - the streams are of miscible phases. It is known from stability analysis that the interface between two viscous fluid layers in a two-dimensional horizontal channel will be stable at very low Reynolds numbers, as long as the surface tension has a nonzero value and the viscosities are close (Pozrikidis, 1997). Microfluidic designs and devices having dimensions in the range of 10–300 μm allow adjacent fluid layers to flow in parallel without any physical mixing (Larsen and Shapley, 2007). However, a diffusive gradient of solutes from one miscible phase to another develops, achieving separation. Cocurrent separators for miscible gaseous phases with a membrane in between are also relevant for this section.

We will first provide a very brief illustration of the governing equations for mass transport and the operating line for a two-phase continuous cocurrent separation system in a conventional chemical engineering context. This will be followed by a brief treatment of the multicomponent separation capability of such a system. Cocurrent chromatographic separation in a two-phase system, where both phases are mobile and in cocurrent flow, will be introduced next. The systems of interest are: micellar electrokinetic chromatography (MEKC); chromatography with two mobile phases, a gas phase and a liquid phase; capillary electrochromatography, with mobile nanoparticles in the mobile liquid phase. Continuous separation of particles from a gas phase to a cocurrent liquid phase in a scrubber will then be illustrated. Finally, cocurrent membrane separators will be introduced.

8.2.1 Cocurrent two-phase flow devices - general considerations

We treat three aspects here: governing equations of change, operating lines and multicomponent separation capability. The equations of change are considered first.

8.2.1.1 *Equations of change*

For two immiscible phase based separation systems, employing equation (6.2.33), we have already obtained equations (8.1.1a,b) for phases 1 and 2 if species i is being transferred from, say, phase 1 to phase 2 for constant phase velocities/flow rates. For steady state operation without any dispersion term, the governing equations are reduced to (Figure 8.2.1)

$$\langle v_{1z}\rangle \frac{\partial \overline{C}_{i1}}{\partial z} = -K_{i1c}a\left(\overline{C}_{i1} - \overline{C}_{i1}^{*}\right); \tag{8.2.1a}$$

$$\langle v_{2z}\rangle \frac{\partial \overline{C}_{i2}}{\partial z} = K_{i2c}a\left(\overline{C}_{i2}^{*} - \overline{C}_{i2}\right). \tag{8.2.1b}$$

Since we have superficial velocities of each phase in the above equations, employing the device cross-sectional area S_c, we obtain, in the manner of equations (8.1.4b–d) and (8.1.5),

$$W_{t1}\frac{\partial \overline{x}_{i1}}{\partial z} + W_{t2}\frac{\partial \overline{x}_{i2}}{\partial z} = 0, \tag{8.2.2}$$

for essentially constant C_{t1} and C_{t2} along the z-coordinate. For conditions where the superficial phase velocities vary along the separator length, equation (8.1.6) is to be used. In such a case for nondispersive steady state operation, we find for the jth phase, from which species i is being transferred out,

$$\frac{\partial\left(\langle v_{jz}\rangle \overline{C}_{ij}\right)}{\partial z} = -K_{ijc}\,a\left(\overline{C}_{ij} - \overline{C}_{ij}^{*}\right). \tag{8.2.3}$$

The governing equations and solutions for concentration profiles in two cocurrent phases for unsteady state operation have been provided in Tan (1994).

8.2.1.2 *Equation for the operating line*

For dilute systems with constant phase velocities, equation (8.2.2) may be written in the manner of equation (8.1.9) as

$$\langle v_{1z}\rangle S_c C_{t1}\, d\overline{x}_{i1} = -\langle v_{2z}\rangle S_c C_{t2}\, d\overline{x}_{i2}; \tag{8.2.4a}$$

$$(d\overline{x}_{i2}/d\overline{x}_{i1}) = \{\langle v_{1z}\rangle S_c C_{t1}/(-\langle v_{2z}\rangle S_c C_{t2})\}; \tag{8.2.4b}$$

$$(d\overline{x}_{i2}/d\overline{x}_{i1}) = -\{|W_{t1}|/|W_{t2}|\}, \tag{8.2.4c}$$

since both phases flow in the negative z-direction and $|W_{t1}|$ and $|W_{t2}|$ are the magnitudes of the total molar flow rates in phases 1 and 2, respectively. For dilute systems (as in gas absorption/stripping) or systems with constant molar overflow (as in distillation), the ratio $(|W_{t1}|/|W_{t2}|)$ may be assumed to be constant along the cocurrent separator length. Integrating from $z = L$, where the inlet conditions are known (i.e. $\overline{C}_{i1L}$, $\overline{C}_{i2L}$, $\overline{x}_{i1L}$, $\overline{x}_{i2L}$) to any z, we get

$$\overline{x}_{i1} = -\frac{|W_{t2}|}{|W_{t1}|}(\overline{x}_{i2} - \overline{x}_{i2L}) + \overline{x}_{i1L}, \tag{8.2.5a}$$

an operating line with a negative slope $(-|W_{t2}|/|W_{t1}|)$ starting at $(\overline{x}_{i2L}, \overline{x}_{i1L})$, as shown in Figure 8.2.2, and

$$\frac{(\overline{x}_{i1L} - \overline{x}_{i1})}{(\overline{x}_{i2} - \overline{x}_{i2L})} = \frac{|W_{t2}|}{|W_{t1}|}. \tag{8.2.5b}$$

If the two phases achieve equilibrium as they leave the device at $z = 0$, then $\overline{x}_{i1}$ and $\overline{x}_{i2}$ at $z = 0$ are in equilibrium, i.e.

$$\overline{x}_{i10} = f_{eq}(\overline{x}_{i20}). \tag{8.2.5c}$$

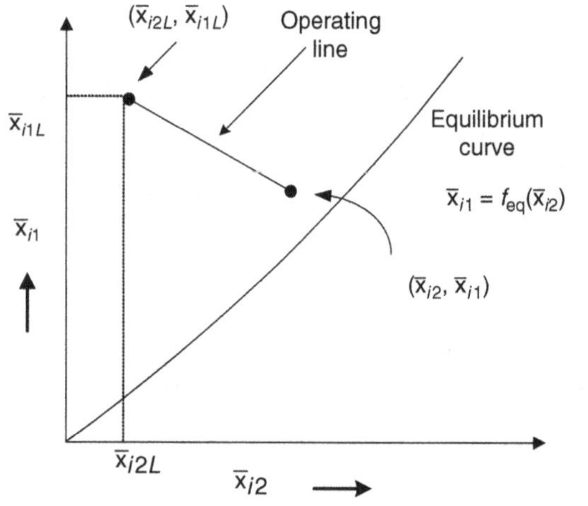

Figure 8.2.2. Operating line and equilibrium curve for a cocurrent gas–liquid absorber.

If, for example, we have simple linear equilibrium $\overline{x}^*_{i10} = K_i \overline{x}_{i20}$, then

$$\frac{(\overline{x}_{i1L} - \overline{x}^*_{i10})}{((\overline{x}^*_{i10}/K_i) - \overline{x}_{i2L})} = \frac{|W_{t2}|}{|W_{t1}|}. \tag{8.2.6}$$

The two-phase systems of interest are generally gas–liquid and vapor–liquid systems with molecular transport from one phase to the other. Cocurrent liquid–liquid systems are additional candidates in the same category. Another separation category includes gas–liquid systems where the gas phase has particles which are transferred into the liquid phase and provide very efficient gas cleaning; a gas/vapor species or two may be simultaneously absorbed into the cocurrent liquid flow.

8.2.1.3 *Multicomponent separation capability in a cocurrent flow separator*

If two phases $j = 1, 2$ move with superficial velocities of $\langle v_{1z}\rangle$ and $\langle v_{2z}\rangle$ in the z-direction, an overall balance of species i in both phases $j = 1$ and 2 is provided by equation (6.2.34) as

$$\varepsilon_1 \frac{\partial \overline{C}_{i1}}{\partial t} + \varepsilon_2 \frac{\partial \overline{C}_{i2}}{\partial t} + \langle v_{1z}\rangle \frac{\partial \overline{C}_{i1}}{\partial z} + \langle v_{2z}\rangle \frac{\partial \overline{C}_{i2}}{\partial z}$$
$$= \varepsilon_1 D_{i,\text{eff},1} \frac{\partial^2 \overline{C}_{i1}}{\partial z^2} + \varepsilon_2 D_{i,\text{eff},2} \frac{\partial^2 \overline{C}_{i2}}{\partial z^2}. \tag{8.2.7}$$

Here, if the z-axis is vertically upward, both phases $j = 1$ and 2 move either vertically upward or vertically downward. If both phases move upward, both $\langle v_{1z}\rangle$ and $\langle v_{2z}\rangle$ are positive quantities. If both phases move downward, both $\langle v_{1z}\rangle$ and $\langle v_{2z}\rangle$ are negative quantities and should be described in the above equation as $-|\langle v_{1z}\rangle|$ and $-|\langle v_{2z}\rangle|$; $|\langle v_{1z}\rangle|$ and $|\langle v_{2z}\rangle|$ are the magnitudes of the velocities $\langle v_{1z}\rangle$ and $\langle v_{2z}\rangle$.

We make three assumptions now: we have isothermal, nondispersive and equilibrium operation (these are exactly the same assumptions used to develop the De Vault equation (7.1.8) for fixed-bed adsorption processes). The assumption of equilibrium everywhere between phases $j = 1$ and $j = 2$ suggests

$$\overline{C}_{i1} = \kappa_{i1} \overline{C}_{i2}. \tag{8.2.8}$$

Correspondingly, the assumptions of isothermal nondispersive operation imposed onto equation (8.2.7) lead to for flow in the positive z-direction:

$$\varepsilon_1 \frac{\partial \overline{C}_{i1}}{\partial t} + \varepsilon_2 \frac{\partial \overline{C}_{i2}}{\partial t} + |\langle v_{1z}\rangle| \frac{\partial \overline{C}_{i1}}{\partial z} + |\langle v_{2z}\rangle| \frac{\partial \overline{C}_{i2}}{\partial z} = 0. \tag{8.2.9}$$

If both phases flow in the negative z-direction,

$$\varepsilon_1 \frac{\partial \overline{C}_{i1}}{\partial t} + \varepsilon_2 \frac{\partial \overline{C}_{i2}}{\partial t} - |\langle v_{1z}\rangle| \frac{\partial \overline{C}_{i1}}{\partial z} - |\langle v_{2z}\rangle| \frac{\partial \overline{C}_{i2}}{\partial z} = 0. \tag{8.2.10}$$

Using the equilibrium relation (8.2.8), these two equations are converted, respectively, to

$$(\varepsilon_1 \kappa_{i1} + \varepsilon_2) \frac{\partial \overline{C}_{i2}}{\partial t} + |\langle v_{2z}\rangle| \left(1 + \kappa_{i1} \frac{|\langle v_{1z}\rangle|}{|\langle v_{2z}\rangle|}\right) \frac{\partial \overline{C}_{i2}}{\partial z} = 0 \tag{8.2.11a}$$

and

$$(\varepsilon_1 \kappa_{i1} + \varepsilon_2) \frac{\partial \overline{C}_{i2}}{\partial t} - |\langle v_{2z}\rangle| \left(1 + \kappa_{i1} \frac{|\langle v_{1z}\rangle|}{|\langle v_{2z}\rangle|}\right) \frac{\partial \overline{C}_{i2}}{\partial z} = 0. \tag{8.2.11b}$$

Consider now the case where both phases move vertically upward with actual/interstitial velocities of v_{1z} and v_{2z}. Then the superficial velocities and interstitial velocities are related by

$$\langle v_{1z}\rangle = \varepsilon_1 v_{1z}, \qquad \langle v_{2z}\rangle = \varepsilon_2 v_{2z}. \tag{8.2.12}$$

We can now rewrite equation (8.2.11a) as

$$\frac{\partial \overline{C}_{i2}}{\partial t} + \frac{\langle v_{2z}\rangle \left(1 + \kappa_{i1} \dfrac{|\langle v_{1z}\rangle|}{|\langle v_{2z}\rangle|}\right)}{(\varepsilon_1 \kappa_{i1} + \varepsilon_2)} \frac{\partial \overline{C}_{i2}}{\partial z} = 0, \tag{8.2.13}$$

which may be rewritten as

$$\frac{\partial \overline{C}_{i2}}{\partial t} + v_{2z} \frac{\left(1 + \kappa_{i1} \dfrac{\varepsilon_1}{\varepsilon_2} \dfrac{v_{1z}}{v_{2z}}\right)}{\left(1 + \dfrac{\varepsilon_1}{\varepsilon_2} \kappa_{i1}\right)} \frac{\partial \overline{C}_{i2}}{\partial z} = 0. \tag{8.2.14}$$

If we recall the development of equation (8.1.35), we obtain the following expression for the concentration wave velocity v^*_{Ci} for species i:

$$v_{Ci}^* = v_{2z}\frac{\left(1+\kappa_{i1}\dfrac{\varepsilon_1}{\varepsilon_2}\dfrac{v_{1z}}{v_{2z}}\right)}{\left(1+\dfrac{\varepsilon_1}{\varepsilon_2}\kappa_{i1}\right)}. \tag{8.2.15}$$

Therefore, as long as the κ_{i1} values for different species i are different, the v_{Ci}^* for different species i will be different. However, different species i move in the same direction. As we have learnt in Section 7.1 (Figure 7.1.5(c)), if we have a continuous input of different species i in the entering stream, then, after some time, the most rapidly moving species i will appear first at the end of the cocurrent two-phase separation device ($z = L$); after some time, the next most rapidly moving species i will appear, and so on for other species (in the manner of Figure 7.1.5(c)). Therefore it is not possible to have more than one pure species as a product (for a brief period) in such a flow vs. force configuration when both phases are flowing on a steady state basis.

If we have one of two other conditions, then it is possible to have separation of multiple species at the device outlet.

(1) If the feed mixture is injected as a sample into one phase or both phases at $z = 0$, then we have a chromatographic situation, and it will be possible to have different pulses (peaks) of different species appearing in the flowing two-phase system at the device outlet $z = L$.
(2) Suppose the two-phase cocurrent flow is not at steady state. Instead, one of the phases is introduced as a pulse or in an interrupted fashion so that we have pulses of the second phase moving in the positive z-direction as the first phase moves continuously. If this pulsed second phase is the source of the solute mixture, or if we have pulses of mixture introduced, then we have the possibility that multicomponent separation may be achieved.

8.2.2 Chromatographic separations in cocurrent two-phase flow devices

Three types of cocurrent two-phase systems will be considered here for chromatographic separations: a micellar phase in cocurrent flow in a bulk mobile liquid phase; a dispersed gas phase in cocurrent flow in a bulk mobile liquid phase; mobile nanoparticles in the mobile liquid phase.

8.2.2.1 *Micellar electrokinetic chromatography*

Capillary electrophoresis (Section 6.3.1.2) can provide transient separation from an injected sample of various charged species in solution as they migrate at different speeds toward the negatively charged electrode. However, uncharged species cannot be separated by this technique. To overcome this shortcoming, the technique of micellar electrokinetic chromatography (MEKC) has been developed. This technique involves the cocurrent flow of two phases, an aqueous solution and a micellar phase containing ionic surfactants in the aqueous solution.

Consider the electroosmotic flow of an ionic surfactant containing aqueous solution in a capillary of the type used in capillary electrophoresis (see Section 6.3.1.2, Figures 6.3.5(a) and (c)). As shown in Figure 8.2.3, the electroosmotic flow (EOF) is driving the aqueous micellar solution toward the negative electrode. If this ionic surfactant happens to be sodium dodecyl sulfate (SDS), the micelles

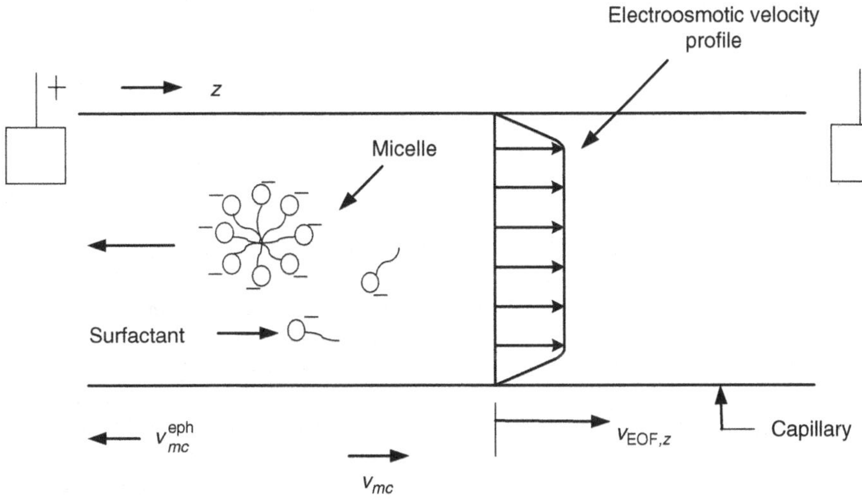

Figure 8.2.3. Micellar electrokinetic chromatography (MEKC) in a capillary with electroosmotic flow of the aqueous solution with micelles in cocurrent flow.

will be negatively charged and would tend to migrate via electrophorectic flow in the opposite direction, toward the positive electrode. As long as the pH of the solution is above 5, the EOF velocity $v_{\text{EOF},z}$ toward the negative electrode is larger than the electrophoretic migration velocity of the SDS micelles, v_{mc}^{eph}, toward the positive electrode, causing the micelles to move in the direction of the negative electrode with a velocity of v_{mc}:

$$v_{mc} = v_{\text{EOF},z} - v_{mc}^{\text{eph}}. \qquad (8.2.16)$$

Focus now on uncharged solutes in the sample injected into the micellar solution. Such a solute, if present outside the micelles, will move with a velocity of $v_{\text{EOF},z}$. If it has partitioned into the micelle interior, those molecules will move toward the negative electrode with the velocity v_{mc} via definition (8.2.16). The extent of partitioning of a solute between the micelle interior and the external solution (Section 4.1.8) will determine its retention time in the capillary. To calculate the retention time for any uncharged solute, we follow the analysis by Terabe (1993) of the MEKC technique developed first by Terabe *et al.* (1984).

In a given volume of the micellar solution of the solute species i, the micellar phase capacity factor k'_{imc} (see definition (1.4.1), also called the distribution ratio) may be defined as

$$k'_{imc} = \frac{n_{imc}}{n_{iw}}, \qquad (8.2.17)$$

where n_{imc} are moles of species i in the interior of the micelles (the micelle phase $j = mc$) and n_{iw} are moles of species i in the external solution ($j = w$). Then the fraction of solute species in the interior of the micelles is given by

$$\frac{n_{imc}}{n_{imc} + n_{iw}} = \frac{k'_{imc}}{1 + k'_{imc}}. \qquad (8.2.18a)$$

The fraction of the solute species in the external solution is

$$\frac{n_{iw}}{n_{iw} + n_{imc}} = \frac{1}{1 + k'_{imc}}. \qquad (8.2.18b)$$

Since the external solution and the charged micelles move with velocities $v_{\text{EOF},z}$ and v_{mc}, respectively, in the positive z-direction, the effective z-directional migration velocity of solute species i will be

$$v_{iz} = v_{\text{EOF},z}\left(\frac{1}{1 + k'_{imc}}\right) + v_{mc}\left(\frac{k'_{imc}}{1 + k'_{imc}}\right). \qquad (8.2.19)$$

As we have noted in equations (6.3.10a–c), it is more correct to replace the velocities by their capillary cross section averaged values:

$$\langle v_{iz}\rangle = \langle v_{\text{EOF},z}\rangle\left(\frac{1}{1 + k'_{imc}}\right) + \langle v_{mc}\rangle\left(\frac{k'_{imc}}{1 + k'_{imc}}\right). \qquad (8.2.20)$$

Define t_{R_i} as the retention time for species i; similarly, let t_{EOF} and t_{mc} be the residence times of the aqueous phase and the charged micellar phase, respectively, in the capillary of length L. Then

$$\langle v_{iz}\rangle = \frac{L}{t_{R_i}} = \left(\frac{L}{t_{\text{EOF}}}\right)\frac{1}{1 + k'_{imc}} + \left(\frac{L}{t_{mc}}\right)\frac{k'_{imc}}{1 + k'_{imc}}, \qquad (8.2.21)$$

leading to

$$t_{R_i} = \frac{1 + k'_{imc}}{1 + k'_{imc}\,(t_{\text{EOF}}/t_{mc})}\,t_{\text{EOF}}, \qquad (8.2.22)$$

which shows that the retention time t_{R_i} for an uncharged solute i in the capillary will depend on its micellar distribution ratio or capacity factor, k'_{imc}, for a given micellar electrokinetic system. Uncharged solutes having different k'_{imc} values will therefore have different retention times, as long as $t_{\text{EOF}} \neq t_{mc}$.

Now, following Ghowsi *et al.* (1990), we will briefly treat the migration velocity of the uncharged solute species i in terms of the electroosmotic mobility μ_{EOF}^m of the electroosmotic flow and the ionic mobility μ_i^m due to electrophoresis as in definition (6.3.10c), namely

$$\langle v_{iz}\rangle = (\mu_{\text{EOF}}^m + \mu_i^m)E.$$

Here the solute is not an ionic species; however, we can determine its electrophoresis based ionic mobility μ_i^m based on the electrophoretic mobility of the ionic micelle, namely μ_{mc}^m, and the fraction of the solute present in the micelle, $k'_{imc}/(1 + k'_{imc})$:

$$\mu_i^m = \left(\frac{k'_{imc}}{1 + k'_{imc}}\right)\mu_{mc}^m, \qquad (8.2.23a)$$

where

$$\langle v_{mc}^{\text{eph}}\rangle = \mu_{mc}^m\,E. \qquad (8.2.23b)$$

Therefore the effective solute velocity is

$$\langle v_{iz,\text{eff}}\rangle = \left(\mu_{\text{EOF}}^m + \left(\frac{k'_{imc}}{1 + k'_{imc}}\right)\mu_{mc}^m\right)E; \qquad (8.2.23c)$$

$$\langle v_{iz,\text{eff}}\rangle = \mu_{i,\text{eff}}^m E. \qquad (8.2.24)$$

In the absence of the micelle, $\mu_{mc}^m = 0$, and the effective veleocity of the neutral (uncharged) solute will be that due to the electroosmotic flow only. Correspondingly, now neutral solutes having different values of k'_{imc} can be separated by the technique of MEKC. More detailed results on resolution for such a system are illustrated in Ghowsi *et al.* (1990). Note that for negatively charged micelles (e.g. SDS based micelles), μ_{mc}^m is negative.

The resolution between two solutes 1 and 2 subjected to CZE has been illustrated in expressions (6.3.19) and (6.3.25). We can express relation (6.3.25) here for two nonionized solutes in MEKC as

$$R_s = \frac{\mu^m_{1,\text{eff}} - \mu^m_{2,\text{eff}}}{(0.5)\left[\mu^m_{1,\text{eff}} + \mu^m_{2,\text{eff}}\right]}\left(\frac{N^{1/2}}{4}\right), \tag{8.2.25}$$

where the denominator is merely an average of the two effective mobilities of nonionic solutes, 1 and 2.

8.2.2.2 *Chromatography with two immiscible mobile fluid phases in cocurrent flow*

Consider a long capillary column of the type used in Section 7.1.5 except there are no packings/beads and the capillary wall surface has no coating. Therefore conventional chromatography with a stationary phase is not possible since there is no stationary phase. However, let there be cocurrent flow of two mobile immiscible phases, one liquid and the other gaseous, simultaneously pumped into the capillary. An example would be methanol and CO_2 at 61 bar and 60 °C, with methanol content varying between 5 and 55% CO_2 in the feed (Wang *et al.*, 2006); the liquid phase in this study was primarily methanol and the gaseous phase was primarily CO_2. Any injected sample would partition between the two flowing phases. For example, neon gas hardly partitioned into the liquid phase and therefore had a residence time equal to that of the gas phase (CO_2 phase). On the other hand, benzene injected into the column was partitioned into the liquid as well as the gas phase and therefore had a higher residence time. There is no formal stationary phase in this cocurrent configuration. Consequently there is no problem of gradual deterioration of the stationary phase as is true in all conventional chromatographic columns.

In such a configuration, the notion of residence time of a given flowing phase in the column is useful. The residence time of the gas phase, $t_{\text{res},g}$, may be determined from

$$t_{\text{res},g} = \frac{n_{ig}}{W_{tg}x_{ig}}, \tag{8.2.26a}$$

where W_{tg} is the total molar flow rate of the gas phase, n_{ig} represents the total number of moles of species i present in the column in the gas phase and x_{ig} is the mole fraction of species i in the gas phase. Correspondingly, the residence time of the liquid phase, $t_{\text{res},\ell}$, may be expressed as

$$t_{\text{res},\ell} = \frac{n_{i\ell}}{W_{t\ell}\,x_{i\ell}}. \tag{8.2.26b}$$

If the feed flow rate and feed mole fraction of species i for a continuous feed into the column are W_{tf} and x_{if}, respectively, then the residence time $t_{\text{res},i}$ of the species in the column is

$$t_{\text{res},i} = \frac{n_{ig} + n_{i\ell}}{W_{tf}x_{if}}. \tag{8.2.27}$$

Utilizing these three equations, one can write

$$\frac{n_{ig} + n_{i\ell}}{t_{\text{res},i}} = \frac{n_{ig}}{t_{\text{res},g}} + \frac{n_{i\ell}}{t_{\text{res},\ell}}, \tag{8.2.28}$$

since

$$W_{tf}x_{if} = W_{tg}x_{ig} + W_{t\ell}x_{i\ell}. \tag{8.2.29}$$

If we define a distribution ratio (or capacity factor) (see equation (1.4.1)) of species i between the gas phase and the liquid phase as

$$k'_{i\ell} = \frac{n_{i\ell}}{n_{ig}}, \tag{8.2.30}$$

then

$$\frac{n_{ig}}{n_{ig} + n_{i\ell}} = \frac{1}{1 + k'_{i\ell}}; \qquad \frac{n_{i\ell}}{n_{ig} + n_{i\ell}} = \frac{k'_{i\ell}}{1 + k'_{i\ell}}. \tag{8.2.31}$$

We can recast equation (8.2.28) as

$$\frac{1}{t_{\text{res},i}} = \left(\frac{1}{1 + k'_{i\ell}}\right)\frac{1}{t_{\text{res},g}} + \left(\frac{k'_{i\ell}}{1 + k'_{i\ell}}\right)\frac{1}{t_{\text{res},\ell}}. \tag{8.2.32}$$

If we define v^*_{Ci} as the average velocity of component i in the column of length L,

$$v^*_{Ci} = L/t_{\text{res},i}, \tag{8.2.33}$$

then the following result is obtained:

$$v^*_{Ci} = \left(\frac{1}{1 + k'_{i\ell}}\right) v_{g,\text{avg}} + \left(\frac{k'_{i\ell}}{1 + k'_{i\ell}}\right) v_{\ell,\text{avg}}, \tag{8.2.34}$$

where $v_{g,\text{avg}}$ and $v_{\ell,\text{avg}}$ are average velocities of the gas phase and liquid phase, respectively, in the column of length L; i.e.

$$v_{g,\text{avg}} = (L/t_{\text{res},g}); \qquad v_{\ell,\text{avg}} = (L/t_{\text{res},\ell}). \tag{8.2.35}$$

It is illustrative to show that the above result reduces to the classical result in gas chromatography with a stationary phase, (7.1.99h), if we assume a stationary liquid phase (i.e. $v_{\ell,\text{avg}} = 0$). Employ (8.2.32) for $t_{\text{res},\ell} = \infty$:

$$\begin{aligned} t_{\text{res},i} &= t_{\text{res},g}(1 + k'_{i\ell}) \Rightarrow \varepsilon S_c\, v_{g,\text{avg}}\, t_{\text{res},i} \\ &= \varepsilon S_c\, v_{g,\text{avg}}\, t_{\text{res},g}\,(1 + k'_{i\ell}) \\ \Rightarrow V_{R,i} &= V_M\left(1 + \frac{n_{i\ell}}{n_{ig}}\right) = V_M\left(1 + \frac{(1-\varepsilon)}{\varepsilon}\kappa_{i\ell}\right), \end{aligned} \tag{8.2.36}$$

where $V_{R,i}(= \varepsilon S_c\, v_{g,\text{avg}}\, t_{\text{res},i})$ is the retention volume of species i in the column, $V_M\ (= \varepsilon S_c\, v_{g,\text{avg}}\, t_{\text{res},g} = \varepsilon S_c L)$ is the mobile-phase volume in the column, ε is the column fractional volume occupied by the gas, S_c is the column cross-sectional area and $\kappa_{i\ell}\ (= C_{i\ell}/C_{ig})$ is the partition coefficient of species i between the mobile gas phase and the stationary liquid phase.

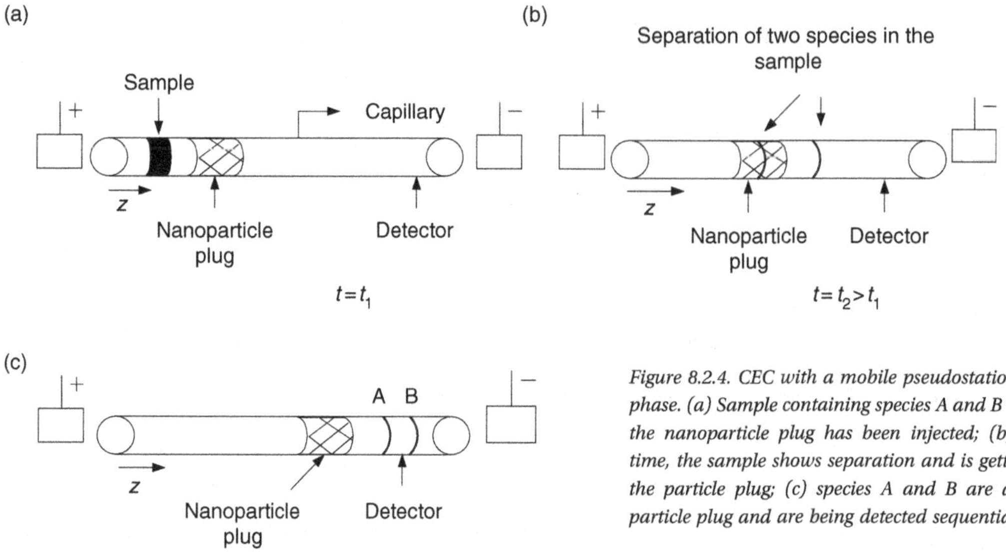

Figure 8.2.4. CEC with a mobile pseudostationary particle phase. (a) Sample containing species A and B injected after the nanoparticle plug has been injected; (b) after some time, the sample shows separation and is getting ahead of the particle plug; (c) species A and B are ahead of the particle plug and are being detected sequentially.

8.2.2.3 *Capillary electrochromatography with a mobile pseudostationary particle phase*

We have been introduced very briefly to capillary electrochromatography (CEC) in Section 7.1.7.1. The commonly employed fused-silica capillaries are often packed with a chromatographic medium. Such systems have a number of shortcomings, including a demanding packing procedure, the need for a frit at the column end to retain the particulate medium, leading to band broadening, etc. One can avoid such problems by injecting a small plug of nanoparticles into the capillary prior to injecting the sample; the nanoparticles act as a stationary phase which is moving via electroosmotic flow toward the negative electrode (Amini *et al.*, 1999; Schweitz *et al.*, 2000). We have therefore two phases in cocurrent flow, albeit that they are in an unsteady state. The positively charged species in the sample move ahead rapidly and are eluted ahead of the nanoparticles (see Figure 8.2.4) since their z-directional velocities are greater than those of the nanoparticles. The nanoparticles generally move with the electroosmotic velocity unless they are positively charged. Therefore the positively charged solutes move faster and are detected earlier than the exit of the nanoparticles in solution. Amongst other advantages, this technique allows the selection of a new nanoparticle based stationary phase for a given separation.

8.2.3 Particle separation in cocurrent gas–liquid flow–Venturi scrubber

We have observed in Section 6.3.1.4 that particles from flowing air streams can be collected by stationary fibers in a filter bed by inertial impaction. There are additional particle collection mechanisms by fiber collectors and granular media, such as Brownian diffusion, interception, electrostatic deposition, gravitational settling, etc. (see Figure 7.2.10(a)). Venturi scrubbers and related wet scrubber devices effectively generate a cocurrent spray of water droplets in the moving gas stream containing particles to capture and remove them by inertial impaction onto the droplets from all sides. When a particle impacts a droplet, it is very likely to be collected by the droplet. These droplets are then collected separately, and thereby the particles removed from the gas streams are in suspension in this liquid stream. The droplet size may vary quite a bit from less than 10 μm to 100–200 μm+. Particles larger than 1–5 μm are collected into the liquid droplets with very high efficiency; particles in the submicron range are also collected with high efficiency, especially when the gas phase undergoes a high pressure drop.

Venturi scrubber configurations vary. Generally the particle-laden gas stream enters a converging flow section in the vertically downward direction toward a throat (Figure 8.2.5), where water is injected through fine nozzles into the gas streams moving at velocities of the order of 60–180 m/s. The atomized droplets of water are accelerated by the gas, during which stage particles in the gas impact the accelerating drops and are collected via inertial impaction. The chances of inertial impaction are substantially reduced when the relative velocity of the droplets vis-à-vis the gas is reduced further downstream of the throat. Liquid flow rate into the scrubber is in the range of 1–3 m^3 per 1000 m^3 of gas.

Flagan and Seinfeld (1988) have illustrated a general procedure for developing an expression for the differential change in $N(r_{p,\min}, r_{p,\max})$, the total number density of

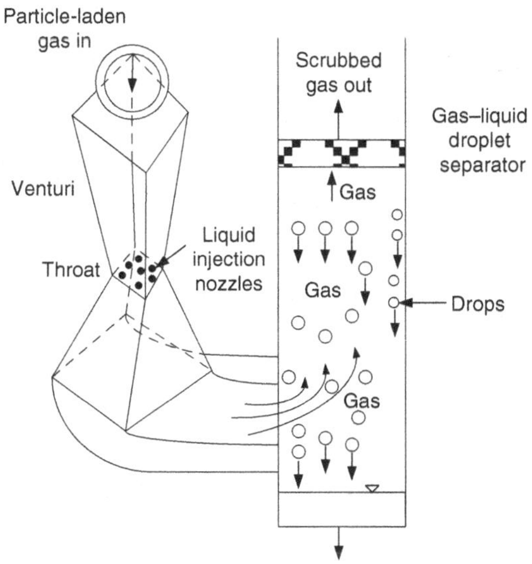

Figure 8.2.5. Venturi scrubber with liquid injection and a subsequent gas–liquid droplet separator.

particles in the gas stream, with distance z along a spray chamber, and suggested the following for a cocurrent scrubber:

$$\frac{dN}{dz} = -\left[\frac{3}{2}\eta\left(\frac{Q_\ell}{Q_g}\right)\left(\frac{v_g - U_{dr}}{U_{dr}}\right)\frac{1}{d_{dr}}\right]. \tag{8.2.37}$$

Here, Q_ℓ and Q_g are the volumetric liquid and gas flow rates, respectively, in m^3/s; d_{dr} and U_{dr} are the diameter and velocity of the drop falling in a gas moving with a velocity v_g, so that $(v_g - U_{dr})$ is the relative velocity between the drop and the gas; η is the particle collection efficiency of an individual droplet, somewhat analogous to the single fiber efficiency E_{IS} due to inertial impaction, (6.3.42a). Calvert (1984) has obtained the following expression for the overall Venturi scrubber efficiency E_T after incorporating an appropriate expression for the individual drop collection efficiency η:

$$E_T = 1 - \exp\left[\frac{1}{55}\left(\frac{Q_\ell}{Q_g}\right)\frac{v_g \rho_\ell d_{dr}}{\mu_g} F(K_p f)\right]. \tag{8.2.38}$$

Here

$$F(K_p f) = \frac{1}{K_p}\left[-0.7 - K_p f + 1.4\,\ell n\left(\frac{K_p f + 0.7}{0.7}\right) + \frac{0.49}{0.7 + K_p f}\right]; \tag{8.2.39}$$

$$K_p = 2St; \qquad St = \frac{C_c \rho_\ell d_p^2 (U_{dr} - v_g)}{18 \mu_g d_{dr}}, \tag{8.2.40}$$

where St is the Stokes number (see (6.3.41)) of the falling drop with the Cunningham correction factor C_c, particle diameter d_p, and $(U_{dr} - v_g)$, the drop velocity magnitude relative to the gas velocity. Further, the empirical parameter f has the value of 0.5 for large scrubbing units and 0.25 for smaller units and hydrophobic particles.

As shown in Figure 8.2.5, the cross-sectional area of the Venturi scrubber expands beyond the throat; finally the gas–liquid droplet mixture is turned around in a vertical structure which allows the gas to go out through the top and the liquid droplets to collect at the bottom. The vertical structure often employs cyclone or other configurations to separate the droplets; a demister is used to ensure no drop carryover by the scrubbed gas. However, at the Venturi scrubber throat area, there is considerable pressure drop in the gas; an estimate in the Venturi is provided by the following formula (Wark and Warner, 1976, pp. 200–205):

$$\Delta P_{gas} = 1.03 \times 10^{-6}\, v^2_{g,\,throat}(Q_\ell/Q_g)\ \text{cm}\ H_2O. \tag{8.2.41}$$

Here $v_{g,\,throat}$ is the gas velocity at the throat in cm/s and (Q_ℓ/Q_g) has the units of liter/m^3. An earlier result by Calvert *et al.* has been illustrated by Wark and Warner (1976) for total efficiency estimation if ΔP_{gas} is known:

$$E_T = 1 - \exp\left(-\frac{6.1 \times 10^{-3} \times \rho_\ell \times \rho_p \times C_c \times d_p^2 \times f^2 \times \Delta P_{gas}}{\mu_g^2}\right). \tag{8.2.42}$$

Here ΔP_{gas} is the gas pressure drop across the Venturi in cm H_2O, ρ_p is the particle density in g/cm^3, μ_g is the gas viscosity in poise, ρ_ℓ is the liquid density in g/cm^3, d_p is the particle diameter in micron, C_c is the Cunningham correction factor and f is an experimental coefficient varying between 0.1 and 0.4.

8.2.4 Cocurrent membrane separators

We briefly describe cocurrent dialysis and gas permeation processes.

8.2.4.1 *Cocurrent hemodialyzer performance*

In Section 4.3.1, we were introduced to a hemodialyzer with blood on one side of the membrane and the dialyzing solution on the other side. Solutes (metabolic waste products) from blood diffused through the liquid filled pores of the membrane to the dialysate side. Using a simple lumped analysis based on the overall solute mass-transfer coefficient $\overline{K}_{ic}$, we will develop an expression for the solute removal efficiency of a hemodialyzer in which blood as well as the dialyzing solution are in steady cocurrent flow (Section 8.1.7 treated countercurrent dialyzers). The analysis is valid for any other system, not just hemodialysis.

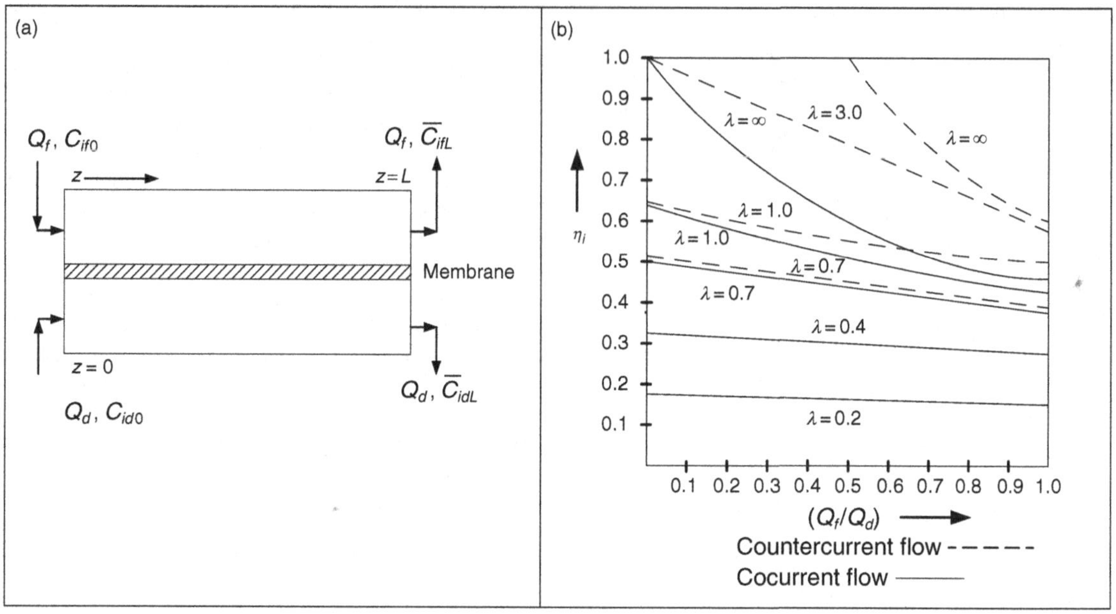

Figure 8.2.6. Cocurrent dialyzer. (a) Schematic; (b) plots of dialyzer efficiency η_i against (Q_f/Q_d) with λ as a parameter for cocurrent flow along with a few for countercurrent flow. (Part (b) after Michaels (1966).)

Figure 8.2.6(a) illustrates the schematic of a cocurrent dialyzer with the feed solution (e.g. blood side) (Q_f, C_{if0}) entering at $z = 0$ and exiting at $z = L$, $(Q_f, \overline{C}_{ifL})$. The dialyzing solution (Q_d, C_{id0}) is entering at $z = 0$ also and exiting at $z = L$, $(Q_d, \overline{C}_{idL})$. The overall molar solute i balance relations for the whole dialyzer based on a device averaged overall solute mass-transfer coefficient $\overline{K}_{ic}$ are:

$$Q_f(C_{if0} - \overline{C}_{ifL}) = Q_d(\overline{C}_{idL} - C_{id0}) = \overline{K}_{ic}A_m(\Delta C_i)_{\ell m}, \tag{8.2.43}$$

where

$$(\Delta C_i)_{\ell m} = \frac{[(C_{if0} - C_{id0}) - (\overline{C}_{ifL} - \overline{C}_{idL})]}{\ell n\,[(C_{if0} - C_{id0})/(\overline{C}_{ifL} - \overline{C}_{idL})]} \tag{8.2.44}$$

is the overall logarithmic mean concentration difference between the two solutions driving the solute transport. We are interested in obtaining an expression for the solute transfer efficiency

$$\eta_i = \frac{C_{if0} - \overline{C}_{ifL}}{C_{if0} - C_{id0}} \tag{8.2.45}$$

which represents the actual concentration change achieved in the feed solution (for example, blood) as a fraction of the maximum achievable concentration change, $C_{if0} - C_{id0}$.

To that end, we carry out the following manipulations. From the mass balance relations (8.2.43), we get

$$\frac{C_{if0} - \overline{C}_{ifL}}{\overline{C}_{idL} - C_{id0}} = \frac{Q_d}{Q_f},$$
$$\left(1 + \frac{Q_d}{Q_f}\right) = \frac{(C_{if0} - C_{id0}) - (\overline{C}_{ifL} - \overline{C}_{idL})}{(\overline{C}_{idL} - C_{id0})} \tag{8.2.46}$$

and

$$1 + \frac{Q_f}{Q_d} = \frac{(C_{if0} - C_{id0}) - (\overline{C}_{ifL} - \overline{C}_{idL})}{(C_{if0} - \overline{C}_{ifL})}. \tag{8.2.47}$$

Therefore

$$Q_f(C_{if0} - \overline{C}_{ifL}) = \overline{K}_{ic}A_m \frac{\left(1 + (Q_d/Q_f)\right)(\overline{C}_{idL} - C_{id0})}{\ell n((C_{if0} - C_{id0})/(\overline{C}_{ifL} - \overline{C}_{idL}))}$$
$$= \overline{K}_{ic}A_m \frac{\left(1 + (Q_f/Q_d)\right)(C_{if0} - \overline{C}_{ifL})}{\ell n\,[(C_{if0} - C_{id0})/(\overline{C}_{ifL} - \overline{C}_{idL})]}. \tag{8.2.48}$$

This result leads to

$$\ell n\left[\frac{(C_{if0} - C_{id0})}{(\overline{C}_{ifL} - \overline{C}_{idL})}\right] = \frac{\overline{K}_{ic}A_m}{Q_f}\left(1 + \frac{Q_f}{Q_d}\right); \tag{8.2.49}$$

$$\frac{(\overline{C}_{ifL} - \overline{C}_{idL})}{(C_{if0} - C_{id0})} = \exp\left[-\lambda\left(1 + \frac{Q_f}{Q_d}\right)\right], \tag{8.2.50}$$

where

$$\lambda = \left(\overline{K}_{ic}A_m/Q_f\right). \tag{8.2.51}$$

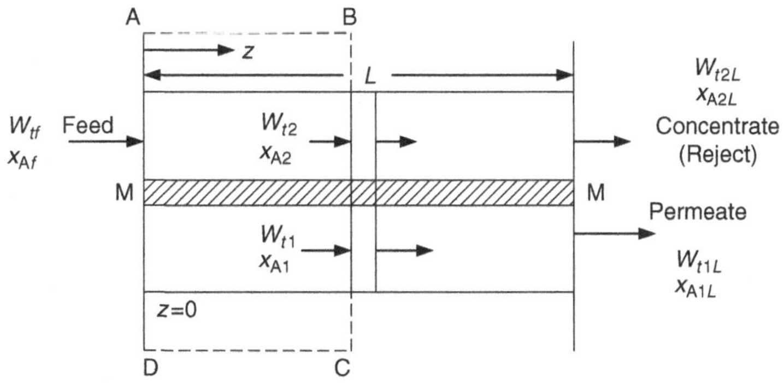

Figure 8.2.7. Schematic for a cocurrent gas permeator.

However,

$$\left(1+\frac{Q_f}{Q_d}\right)\left(\frac{C_{if0}-\overline{C}_{ifL}}{C_{if0}-C_{id0}}\right)=1-\frac{\overline{C}_{ifL}-\overline{C}_{idL}}{C_{if0}-C_{id0}}=\left(1+\frac{Q_f}{Q_d}\right)\eta_i \tag{8.2.52}$$

from (8.2.47) and (8.2.45). Therefore

$$\eta_i=\frac{1}{\left[1+\left(Q_f/Q_d\right)\right]}\left[1-\exp\left(-\lambda\left(1+\left(Q_f/Q_d\right)\right)\right)\right]. \tag{8.2.53}$$

Michaels (1966) provided this analysis based on an earlier analysis by Leonard and Bleumle (1959) and plotted the efficiency η_i as a function of (Q_f/Q_d) with λ as a parameter. As we have observed in Section 8.1.7, the efficiency η_i decreases as (Q_f/Q_d) increases (see Figure 8.2.6(b)). If Q_d decreases vis-à-vis Q_f, C_{id} increases, which reduces the rate of solute transport to the dialyzing liquid, resulting in lower η_i. Correspondingly, when Q_d increases vis-a-vis Q_f, C_{id} decreases, which increases the rate of solute transport and increases η_i. At any value of (Q_f/Q_d), increasing the parameter λ leads to higher η_i since the solute transfer rate is increased. However, this effect is much more pronounced at low values of (Q_f/Q_d); at higher values of (Q_f/Q_d) a change in λ introduces much less change in η_i.

A brief comparison between a cocurrent flow dialyzer and a countercurrent one is in order. In Figure 8.2.6(b), we have also shown the behavior of a countercurrent dialyzer for a few values of λ (dashed lines). It is clear that the countercurrent dialyzer performs significantly better only if λ is high, which essentially means that $\overline{K}_{ic}$ has to be high (A_m and Q_f remaining constant). At lower λ values, the performances are similar. This is in sharp contrast to phase equilibrium based separation devices, where cocurrent flow can achieve very limited separation, unlike those with countercurrent flow.

8.2.4.2 *Cocurrent gas permeation*

We will follow here the treatment of countercurrent gas permeation for a binary mixture of species A and B as in Section 8.1.9. As shown in Figure 8.2.7 for a cocurrent gas permeator, one can obtain, for species A permeation through membrane area dA_m over a length dz,

$$\frac{d(W_{t2}\,x_{A2})}{dA_m}=-N_{Am}=-\frac{Q_{Am}}{\delta_m}\left(P_f x_{A2}-P_p x_{A1}\right). \tag{8.2.54}$$

For a rectangular feed-gas flow channel of width W and length L,

$$dA_m=W\,dz,\qquad A_m^t=WL. \tag{8.2.55}$$

For a hollow fiber device with a *symmetrical hollow fiber membrane of diameters d_i and d_o* and total number of fibers being n

$$dA_m=n\pi d_{\ell m}\,dz,\qquad d_{\ell m}=\frac{d_o-d_i}{\ln(d_o/d_i)}, \tag{8.2.56}$$

where $d_{\ell m}$ is the logarithmic mean of d_i and d_o. For species B, an equation similar to (8.2.54) is

$$\frac{d(W_{t2}\,(1-x_{A2}))}{dA_m}=-N_{Bm}=-\frac{Q_{Bm}}{\delta_m}\left(P_f(1-x_{A2})-P_p(1-x_{A1})\right). \tag{8.2.57}$$

If we focus on the permeate side, we find, unlike equation (8.1.419),

$$\frac{d(W_{t1}\,x_{A1})}{dA_m}=+N_{Am}=+\frac{Q_{Am}}{\delta_m}\left(P_f\,x_{A2}-P_p\,x_{A1}\right), \tag{8.2.58}$$

and similarly for species B, leading to

$$\frac{d(W_{t2}\,x_{A2})}{dA_m}+\frac{d(W_{t2}\,(1-x_{A2}))}{dA_m}$$

$$=\frac{dW_{t2}}{dA_m}=-\frac{d(W_{t1}\,x_{A1})}{dA_m}-\frac{d(W_{t1}\,(1-x_{A1}))}{dA_m}$$

$$\Rightarrow\frac{dW_{t1}}{dA_m}=+N_{Am}+N_{Bm}. \tag{8.2.59}$$

Correspondingly, the equations for the axial variation of species A composition on the feed and the permeate side are (from equation (8.1.425)) as follows:

$$\frac{\mathrm{d}x_{A2}}{\mathrm{d}A_m} = -\frac{1}{W_{t2}}\left[(1-x_{A2})\frac{Q_{Am}}{\delta_m}(P_f x_{A2} - P_p x_{A1}) - x_{A2}\frac{Q_{Bm}}{\delta_m}(P_f(1-x_{A2}) - P_p(1-x_{A1}))\right]. \quad (8.2.60)$$

Define the following nondimensional quantities:

$$\alpha^*_{AB} = (Q_{Am}/Q_{Bm});$$

$$A^+_m = ((Q_{Bm}/\delta_m)P_f A_m/W_{tf});$$

$$A^+_m = \frac{Q_{Bm}}{\delta_m}\frac{P_f\, n\pi\, d_{\ell m} z}{W_{tf}};$$

$$A^{t+}_m = \frac{Q_{Bm}}{\delta_m}\frac{P_f\, n\pi\, d_{\ell m} L}{W_{tf}} = A^+_m\frac{L}{z} = \frac{Q_{Bm}}{\delta_m}\frac{P_f A^t_m}{W_{tf}};$$

$$W^+_{t2} = (W_{t2}/W_{tf}); \qquad \gamma = (P_p/P_f). \quad (8.2.61)$$

Equation (8.2.60) may be written now as

$$\frac{\mathrm{d}x_{A2}}{\mathrm{d}A^+_m} = -\frac{1}{W^+_{t2}}\left[\alpha^*_{AB}(1-x_{A2})(x_{A2}-\gamma x_{A1}) - x_{A2}\{(1-x_{A2}) - \gamma(1-x_{A1})\}\right]. \quad (8.2.62)$$

Also, we get

$$\frac{\mathrm{d}W^+_{t2}}{\mathrm{d}A^+_m} = -\alpha^*_{AB}(x_{A2}-\gamma x_{A1}) - ((1-x_{A2}) - \gamma(1-x_{A1})). \quad (8.2.63)$$

For the third permeate-side equation, focus on the governing equations (8.2.58) and (8.2.59) for the permeate side; from the last one, we get

$$\frac{\mathrm{d}W_{t1}}{\mathrm{d}A_m} = +\frac{Q_{Am}}{\delta_m}(P_f x_{A2} - P_p x_{A1}) + \frac{Q_{Bm}}{\delta_m}(P_f(1-x_{A2}) - P_p(1-x_{A1})). \quad (8.2.64a)$$

From (8.2.58), we obtain

$$\begin{aligned}\frac{\mathrm{d}(W_{t1}x_{A1})}{\mathrm{d}A_m} &= W_{t1}\frac{\mathrm{d}x_{A1}}{\mathrm{d}A_m} + x_{A1}\frac{\mathrm{d}W_{t1}}{\mathrm{d}A_m}\\ &= W_{t1}\frac{\mathrm{d}x_{A1}}{\mathrm{d}A_m} + x_{A1}\frac{Q_{Am}}{\delta_m}(P_f x_{A2} - P_p x_{A1}) + x_{A1}\frac{Q_{Bm}}{\delta_m}(P_f(1-x_{A2}) - P_p(1-x_{A1}))\\ &= +\frac{Q_{Am}}{\delta_m}(P_f x_{A2} - P_p x_{A1}),\end{aligned} \quad (8.2.64b)$$

leading to

$$W_{t1}\frac{\mathrm{d}x_{A1}}{\mathrm{d}A_m} = +(1-x_{A1})\frac{Q_{Am}}{\delta_m}(P_f x_{A2} - P_p x_{A1}) - x_{A1}\frac{Q_{Bm}}{\delta_m}(P_f(1-x_{A2}) - P_p(1-x_{A1})). \quad (8.2.65)$$

Nondimensionalize this equation to obtain:

$$\frac{\mathrm{d}x_{A1}}{\mathrm{d}A^+_m} = +\frac{1}{W^+_{t1}}\left[\alpha^*_{AB}(1-x_{A1})(x_{A2}-\gamma x_{A1}) - x_{A1}((1-x_{A2}) - \gamma(1-x_{A1}))\right]. \quad (8.2.66)$$

Due to the presence of the variable W^+_{t1} in this equation, we need another equation. Convert equation (8.2.64a) to

$$\frac{\mathrm{d}W^+_{t1}}{\mathrm{d}A^+_m} = +\alpha^*_{AB}(x_{A2}-\gamma x_{A1}) + ((1-x_{A2}) - \gamma(1-x_{A1})). \quad (8.2.67)$$

Equations (8.2.62), (8.2.63), (8.2.66) and (8.2.67) provide a set of four coupled nonlinear ordinary differential equations for four variables x_{A2}, W^+_{t2}, x_{A1} and W^+_{t1} as a function of the nondimensional membrane area variable A^+_m subject to the feed inlet conditions and other specified values of P_f, P_p, x_{Af}, W_{tf}, (Q_{Am}/δ_m), (Q_{Bm}/δ_m), A^t_m (total membrane area corresponding to $z = L$) or x_{A2L}. We should be able to solve these four equations numerically in a computer, utilizing a few overall balance equations to be described soon; there is one problem, however. Equation (8.2.66) is (as employed to obtain (8.1.436)) indeterminate at $A_m = 0$ ($z = 0$), where $W^+_{t1} = 0$. Employ L'Hospital's rule:

$$\left.\frac{\mathrm{d}x_{A1}}{\mathrm{d}A^+_m}\right|_{z=0} = \frac{\frac{\mathrm{d}}{\mathrm{d}A^+_m}\left[\alpha^*_{AB}(1-x_{A1})(x_{A2}-\gamma x_{A1}) - x_{A1}((1-x_{A2}) - \gamma(1-x_{A1}))\right]_{z=0}}{\left[\mathrm{d}W^+_{t1}/\mathrm{d}A^+_m\right]_{z=0}}; \quad (8.2.68)$$

$$\left.\frac{\mathrm{d}x_{A1}}{\mathrm{d}A^+_m}\right|_{z=0} = \frac{-\alpha^*_{AB}(x_{A2}-\gamma x_{A1})\left.\frac{\mathrm{d}x_{A1}}{\mathrm{d}A^+_m}\right|_{z=0} + \alpha^*_{AB}(1-x_{A1})\left(\left.\frac{\mathrm{d}x_{A2}}{\mathrm{d}A^+_m}\right|_{z=0} - \gamma\left.\frac{\mathrm{d}x_{A1}}{\mathrm{d}A^+_m}\right|_{z=0}\right) - ((1-x_{A2})-\gamma(1-x_{A1}))\left.\frac{\mathrm{d}x_{A1}}{\mathrm{d}A^+_m}\right|_{z=0} + x_{A1}\left.\frac{\mathrm{d}x_{A2}}{\mathrm{d}A^+_m}\right|_{z=0} - \gamma x_{A1}\left.\frac{\mathrm{d}x_{A1}}{\mathrm{d}A^+_m}\right|_{z=0}}{+\alpha^*_{AB}(x_{A2}-\gamma x_{A1}) + ((1-x_{A2})-\gamma(1-x_{A1}))} \quad (8.2.69)$$

Rearranging, we get

$$\left.\frac{\mathrm{d}x_{A1}}{\mathrm{d}A^+_m}\right|_{\substack{z=0\\ A_m=0}} = -\frac{\{\alpha^*_{AB}(1-x_{A1}) + x_{A1}\}\left.\frac{\mathrm{d}x_{A2}}{\mathrm{d}A^+_m}\right|_{z=0}}{2\{\alpha^*_{AB}(x_{A2}-\gamma x_{A1}) + (1-x_{A2}) - \gamma(1-x_{A1})\} + \gamma\{\alpha^*_{AB} - (\alpha^*_{AB}-1)x_{A1}\}|_{z=0}} \quad (8.2.70)$$

where the values of x_{A1}, x_{A2} are to be calculated at $A_m = 0$ ($z = 0$).

For both cases of a design problem or a rating problem for a cocurrent gas permeator, if we assume that γ is constant throughout the permeator (i.e. P_f and P_p are constant along the permeator), the set of four equations (8.2.62), (8.2.63), (8.2.66) and (8.2.67) may now be solved using (8.2.70). However, x_{A1} is unknown at $A_m = 0$. It is determined using the crossflow equations (6.3.200) and (6.3.201) using known x_{Af} at $A_m = 0$ for the design problem and for the rating problem:

$$x_{A1}\big|_{z=0} = \frac{\left\{(\alpha^*_{AB}-1)\left(\frac{x_{Af}}{\gamma}+1\right)+\frac{1}{\gamma}\right\} - \sqrt{\left\{\left(1+\frac{x_{Af}}{\gamma}\right)(\alpha^*_{AB}-1)+\frac{1}{\gamma}\right\}^2 - \frac{4\alpha^*_{AB}(\alpha^*_{AB}-1)x_{Af}}{\gamma}}}{2(\alpha^*_{AB}-1)} = x_{A10}. \quad (8.2.71)$$

A few additional relations based on overall mass balance are needed.

$$\text{Total mass balance over the permeator: } W_{tf} = W_{t2L} + W_{t1L}; \quad (8.2.72a)$$

$$\text{species A balance over the permeator: } W_{tf}\,x_{Af} = W_{t2L}\,x_{A2L} + W_{t1L}\,x_{A1L}. \quad (8.2.72b)$$

Divide both equations by W_{tf} to obtain the following relation in terms of the stage cut:

$$\theta = \frac{W_{t1L}}{W_{tf}}; \quad (8.2.72c)$$

$$x_{Af} = (1-\theta)\,x_{A2L} + \theta\,x_{A1L} \Rightarrow x_{A1L} = \frac{x_{Af} - (1-\theta)\,x_{A2L}}{\theta}. \quad (8.2.72d)$$

We will now identify the numerical procedure to be followed via a computer to solve particular types of problems. First, we consider *design problems*. Either x_{A2L} (concentrate outlet) or the stage cut θ is given (W_{t2L} and W_{t1L} are known) (an additional problem: x_{A1L} is given). Suppose x_{A2L} is given. Assume a particular value of θ, then W_{t2L} and W_{t1L} are obtained from (8.2.72a,c). Calculate x_{A10} at the inlet end from (8.2.71) for given x_{Af}. Now calculate (dx_{A2}/dA^+_m) at the inlet end for a small increment of area dA^+_m using equation (8.2.62) since all quantities on the right-hand side are known. We can now calculate $(dx_{A1}/dA^+_m)_{z=0}$ from equation (8.2.70). We are now in a position to use equations (8.2.62), (8.2.63), (8.2.66) and (8.2.67) to calculate the values of x_{A2}, x_{A1}, W^+_{t2} and W^+_{t1} for dA^+_m. *Note*: W^+_{t1} needs to be calculated at dA^+_m first from equation (8.2.67) using x_{A10}, x_{Af} since equation (8.2.66) needs it. Now we march toward the product end stepwise with increments of $+dA^+_m$ and stop when $x_{A2} = x_{A2L}$. Correspondingly, we will obtain at that axial location a value of x_{A1L}. Check from equation (8.2.72d) whether, for the specified x_{A2L} and assumed θ, the value x_{A1L} calculated coincides with the value of x_{A1L} obtained numerically. If it does, the area A^{t+}_m calculated is correct. Otherwise assume another value of θ for the specified x_{A2L} and march toward the reject location.

An alternative procedure suggested by Walawender and Stern (1972) and Pan and Habgood (1974) reduces the number of equations to be integrated for the problem. Consider equations (8.2.66) and (8.2.62). Obtain from them the following equation:

$$\frac{dx_{A1}}{dx_{A2}} = -\frac{W^+_{t2}}{W^+_{t1}} \frac{[\alpha^*_{AB}(1-x_{A1})(x_{A2}-\gamma x_{A1}) - x_{A1}\{(1-x_{A2})-\gamma(1-x_{A1})\}]}{[\alpha^*_{AB}(1-x_{A2})(x_{A2}-\gamma x_{A1}) - x_{A2}\{(1-x_{A2})-\gamma(1-x_{A1})\}]}. \quad (8.2.73)$$

The ratio of (W^+_{t2}/W^+_{t1}) is expressed in terms of compositions via the following equations. Consider the envelope ABCD in Figure 8.2.7. Mass balances for this envelope lead to

$$W_{t1} + W_{t2} = W_{tf}, \quad (8.2.74a)$$

$$W_{t1}\,x_{A1} + W_{t2}\,x_{A2} = W_{tf}\,x_{Af}. \quad (8.2.74b)$$

First eliminate W_{tf} to obtain

$$(W^+_{t2}/W^+_{t1}) = (x_{Af} - x_{A1})/(x_{A2} - x_{Af}). \quad (8.2.75)$$

Introduction of this into equation (8.2.73) yields

$$\frac{dx_{A1}}{dx_{A2}} = \frac{(x_{A1}-x_{Af})}{(x_{A2}-x_{Af})} \times \frac{[\alpha^*_{AB}(1-x_{A1})(x_{A2}-\gamma x_{A1}) - x_{A1}\{(1-x_{A2})-\gamma(1-x_{A1})\}]}{[\alpha^*_{AB}(1-x_{A2})(x_{A2}-\gamma x_{A1}) - x_{A2}\{(1-x_{A2})-\gamma(1-x_{A1})\}]}. \quad (8.2.76)$$

However, the ratio $(x_{A1}-x_{Af})/(x_{A2}-x_{Af})$ appearing first on the right-hand side of this equation makes the ratio (dx_{A1}/dx_{A2}) indeterminate at $z = 0$. One can determine it from the ratio $(dx_{A1}/dA^+_m)_{z=0}/(dx_{A2}/dA^+_m)_{z=0}$ from equations (8.2.70) and (8.2.62), respectively. In the design problem solution procedure illustrated earlier, this equation may be used as follows. Calculate as before for specified x_{A2L} and assumed θ. Then, for a given increment of dA^+_m, calculate dx_{A2} from (8.2.62) and calculate dx_{A1} from (8.2.76). In this fashion go to the reject section as before in the trial-and-error procedure to determine A^t_m.

Figure 8.1.51(b) illustrates qualitatively the permeate composition in terms of oxygen as a function of stage cut for a silicone-copolymer membrane which has $\alpha_{O_2-N_2} = 2.4$ and $\gamma = 0.2$ and the feed gas mixture is air $((x_{O_2f} = 0.209; x_{N_2f} = 0.791)$. The cocurrent results are compared with those from a countercurrent permeator and a crossflow permeator (Figure 7.2.1(a)). The performance differences are not that much for this symmetric membrane.

8.3 Crossflow of two bulk phases moving perpendicular to the direction(s) of the driving force(s)

There are a number of phase equilibrium driven separation processes where the separation devices are such that crossflow of two bulk phases exists. Crossflow is utilized to enable continuous contacting between two immiscible phases, vapor and liquid, in an efficient fashion, as in a plate located in a distillation column. In chromatographic processes, crossflow of the solid adsorbent particles and the mobile fluid phase (liquid or gas) can lead to continuous separation of a multicomponent feed mixture introduced at one location of the mobile fluid (eluent) phase. We will illustrate first how crossflow of adsorbent particles or the adsorbent bed and the mobile fluid phase overcomes the batch nature of multicomponent separation in elution chromatography. Then we will focus on the crossflow plate in a distillation column.

8.3.1 Continuous chromatographic separation

We learned in Section 7.1 that adsorption processes carried out in fixed adsorbent beds are unsteady state processes. When injected with a multicomponent sample and fed continuously with an eluent or a carrier gas, the chromatographic process (Section 7.1.5) separates various species; however, it is a batch process and separates the sample injected. If one could carry out continuously the separation of a multicomponent feed mixture via adsorption in one vessel, it would be quite useful. Continuous separation via adsorption must satisfy the following criteria. (1) The adsorbent has to be regenerated. (2) The adsorbent must not suffer attritional losses encountered if the bed of particles moves. (3) Each constituent of the feed must appear as a separate product stream; such a goal has been achieved in devices having a crossflow of the adsorbent phase and the mobile/carrier/feed/eluent phase.

It is useful to recall how multicomponent separation is achieved, for example in electrophoresis, before we focus on continuous chromatographic separation. As illustrated in Figure 7.3.1 in a thin-film continuous-flow electrophoresis separator, the feed mixture of proteins introduced at the feed point location ($x = x_0$) moves with the buffer in the z-direction at buffer velocity v_z. However, due to the electrical force in the perpendicular direction (x-coordinate), different protein species move in the perpendicular direction with different migration velocities U_{ix}. As a result, the net trajectory of each protein species is different, so each species arrives at different values of x at $z = L$. *Note*: For this multicomponent separation to be achieved in one vessel, the feed solution is introduced continuously over a very small area of entry for the buffer solution. Had the feed been introduced over the whole buffer cross section, multicomponent separation would not have been possible.

8.3.1.1 *Continuous annular chromatograph*

We will now try to understand different device configurations for continuous chromatographic separation. Consider a regular packed column with a sample of a multicomponent feed injected at $z = 0$, the mobile phase moving with an interstitial fluid velocity v_z and equilibrium nondispersive operation. It is known that the velocity with which species i moves down the bed along the z-coordinate is given by equation (7.1.99a) for nonporous resin beads ($\varepsilon_p = 0$)

$$\frac{z}{t} = \frac{v_z}{\left(1 + \frac{(1-\varepsilon)}{\varepsilon}\kappa_{i1}\right)} = v^*_{Ci}. \tag{8.3.1}$$

If, instead of injecting a sample, the feed solution is continuously injected at $z = 0$, then, from equation (7.1.15c), we can obtain a similar result (for $\varepsilon_p = 0$) (till the bed is completely saturated):

$$\frac{L}{t} = \frac{v_z}{\left(1 + \frac{(1-\varepsilon)}{\varepsilon}\kappa_{i1}\right)}. \tag{8.3.2}$$

Every species i is moving in the z-direction at a different velocity, v^*_{Ci}. If we now introduce a constant velocity in a perpendicular direction, we can create a separate trajectory for each species (exactly as in Figure 7.3.1) provided we introduce the feed solution over a small area at $z = 0$.

Figures 8.3.1(a) and (b) illustrate a possible configuration to that end. Let the adsorbent particles be secured in an annulus. The figure shows one layer of such particles in the annulus. The eluent is flowing vertically down. The forces of adsorption/desorption are perpendicular to the eluent as well as the particles. Therefore the velocity with which a species is moving vertically down is given by relations (8.3.1) and (8.3.2). If the annulus containing the adsorbent particles is rotated at an angular velocity ω as a solid body around its center, the particles have a tangential velocity $r\omega$ which is perpendicular to the forces of adsorption/desorption between the downflowing liquid and the particles (Figure 8.3.1(b)). This tangential motion will introduce a curved trajectory to the net motion of a species i profile; each species will have a different trajectory, each a separate helical band.

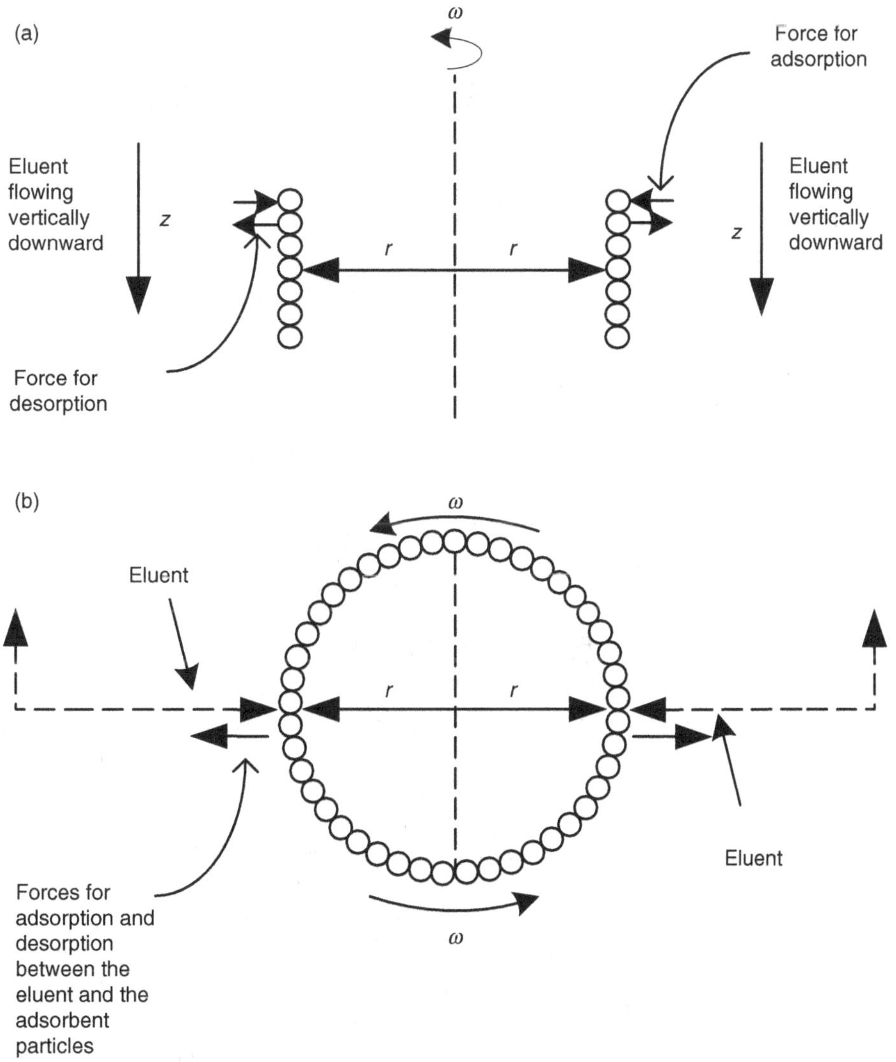

Figure 8.3.1. Configuration of bulk flow directions and the adsorption-desorption force directions in the crossflow arrangement of continuous liquid-phase chromatography with eluent flowing vertically downward and adsorbent particles in an annulus rotating against a vertical axis at an angular velocity of ω *rad/s. (a) Cross-sectional view of one vertical layer of adsorbent rotating at an angular velocity* ω *at radius* r*: eluent flowing vertically downward; forces for adsorption/desorption perpendicular to eluent flow direction. (b) Plan view of a layer of adsorbent rotating at an angular velocity* ω *at radius* r*; forces for adsorption/desorption perpendicular to adsorbent rotational motion in the tangential direction.*

A device that achieves this goal for liquid solutions is shown in Figure 8.3.2 (Begovich and Sisson, 1984). First developed by Scott *et al.* (1976), the device, called a continuous annular chromatograph (CAC), consists of an annular bed of adsorbent particles between two concentric open cylinders of diameters 27.9 cm and 30.5 cm; the radial bed thickness is ~1.27 cm and the adsorbent particles are Dowex 50W-X8 ion exchange resin in Ca^{+2} form, 50–60 μm in diameter. The column is 60 cm long and the resin particles are packed in the annulus. The inner cylinder is closed at the top (Howard *et al.*, 1987), with a layer of 0.18 mm glass beads packed on top of the adsorbents to enable efficient distribution of feed and eluent at the top. As shown in Figure 8.3.2, the feed solution is continuously introduced at the top at a specific location by a stationary nozzle. Also, there are a number of stationary nozzles (usually four) around the bed circumference at the top through which eluent solution is continuously introduced

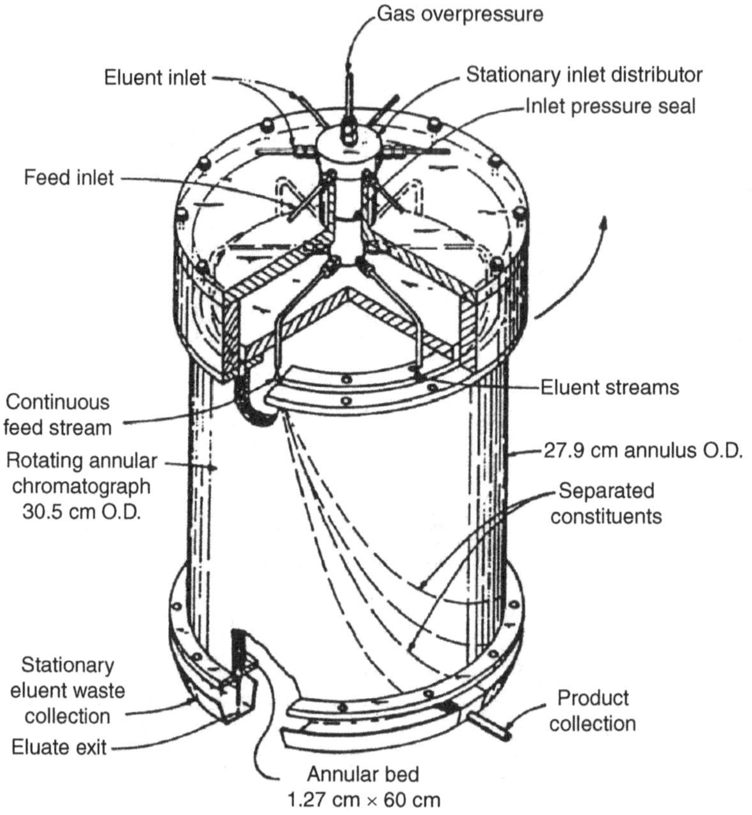

Figure 8.3.2. Schematic of the 279 nm diameter. continuous annular chromatograph (CAC-II). Reprinted, with permission, from Begovich and Sisson, AIChE J., *30(5), 705 (1984). Copyright © [1984] American Institute of Chemical Engineers (AIChE).*

into the bed through the glass beads at the top. Eluent nozzle locations are set away from the feed nozzle location. The column system is slowly rotated at, say, 15 to 90 degrees/hr (Begovich and Sisson, 1984). The flow rate of the eluent is controlled by gas pressure at the top of the column and influenced by the annular gap width. The parts of the bed not receiving feed at a given time will undergo elution or regeneration. There are stationary product collectors at different locations at the bottom of the column collecting different species at the end of different trajectories.

A highly simplified model of separation in this rotating annular chromatograph will be summarized here following Wankat (1977) and Begovich and Sisson (1984). The general differential equation for liquid-phase fixed-bed processes for constant v_z in the z-direction is equation (7.1.4):

$$\frac{\partial \overline{C_{i2}}}{\partial t} + \frac{(1-\varepsilon)}{\varepsilon}\frac{\partial \overline{C_{i1}}}{\partial t} + v_z \frac{\partial \overline{C_{i2}}}{\partial z} = D_{i,\text{eff},z}\frac{\partial^2 \overline{C_{i2}}}{\partial z^2}. \tag{8.3.3}$$

Since we have variation in the θ-direction, the corresponding equation in cylindrical polar coordinates is obtained from Table 6.2.4 for negligible radial variation and $v_\theta \cong 0$ as

$$\frac{\partial \overline{C_{i2}}}{\partial t} + \frac{(1-\varepsilon)}{\varepsilon}\frac{\partial \overline{C_{i1}}}{\partial t} + v_z \frac{\partial \overline{C_{i2}}}{\partial z} = D_{i,\text{eff},z}\frac{\partial^2 \overline{C_{i2}}}{\partial z^2} + D_{i,\text{eff},\theta}\frac{\partial^2 \overline{C_{i2}}}{\partial \theta^2}. \tag{8.3.4}$$

However, the time coordinate t may be related to the angular coordinate θ via

$$\theta = \omega t \tag{8.3.5}$$

for an angular bed velocity of ω. Equation (8.3.4) may now be written also as

$$\omega\frac{\partial \overline{C_{i2}}}{\partial \theta} + \omega\frac{(1-\varepsilon)}{\varepsilon}\frac{\partial \overline{C_{i1}}}{\partial \theta} + v_z \frac{\partial \overline{C_{i2}}}{\partial z} = D_{i,\text{eff},z}\frac{\partial^2 \overline{C_{i2}}}{\partial z^2} + D_{i,\text{eff},\theta}\frac{\partial^2 \overline{C_{i2}}}{\partial \theta^2}. \tag{8.3.6}$$

Taking the equilibrium nondispersive limit of both equations, we obtain the following:

$$\frac{\partial \overline{C_{i2}}}{\partial t} + \frac{(1-\varepsilon)}{\varepsilon}\frac{\partial \overline{C_{i1}}}{\partial t} + v_z \frac{\partial \overline{C_{i2}}}{\partial z} = 0; \tag{8.3.7}$$

$$\omega\frac{\partial \overline{C_{i2}}}{\partial \theta} + \omega\frac{(1-\varepsilon)}{\varepsilon}\frac{\partial \overline{C_{i1}}}{\partial \theta} + v_z \frac{\partial \overline{C_{i2}}}{\partial z} = 0. \tag{8.3.8}$$

We know from expressions (7.1.12a) and (7.1.100a) that the migration velocity of species i satisfying equation (8.3.7) may be expressed for $\varepsilon_p = 0$ as

$$\frac{\mathrm{d}z}{\mathrm{d}t} = v_{Ci}^{*} = \frac{v_z}{\left[1 + \frac{\rho_b q_i'(C_{i2})}{\varepsilon}\right]} = \frac{v_z}{\left[1 + \frac{(1-\varepsilon)}{\varepsilon}\kappa_{i1}\right]}. \quad (8.3.9)$$

Employing relation (8.3.5), we can change the above to

$$\omega\frac{\mathrm{d}z}{\mathrm{d}\theta} = \frac{v_z}{\left[1 + \frac{\rho_b q_i'(C_{i2})}{\varepsilon}\right]} = \frac{v_z}{\left(1 + \frac{(1-\varepsilon)}{\varepsilon}\kappa_{i1}\right)}. \quad (8.3.10)$$

For a linear equilibrium relation (q_i', κ_{i1} constants), we can obtain a relation between the average θ_i value (i.e. $\overline{\theta}_i$) of the species i peak and its axial location (z) by integrating (8.3.10),

$$\overline{\theta}_i = \frac{\omega z}{v_z}\left[1 + \frac{\rho_b q_i'(C_{i2})}{\varepsilon}\right] = \frac{\omega z}{v_z}\left[1 + \frac{(1-\varepsilon)}{\varepsilon}\kappa_{i1}\right], \quad (8.3.11)$$

since $\overline{\theta}_i = 0$ when $z = 0$. The average $\overline{\theta}_i$ value arises since there is some dispersion in the θ-direction for each species i. At $z = L$, i.e. at the end of the column, $\overline{\theta}_i$ for each species will be different as long as κ_{il} for each species is different. Correspondingly, the resolution R_s between two species 1 and 2 is described by (Begovich and Sisson, 1984)

$$R_s = \frac{2(\overline{\theta}_2 - \overline{\theta}_1)}{W_{b2} + W_{b1}}. \quad (8.3.12)$$

In such a continuous annular chromatograph, the pressure of the gas in the headspace above the annulus drives the eluent flow through the packed adsorbent bed in the annulus. So far, the highest pressures used have been 274 kPa for plexiglass columns and 1135 kPa for metallic columns. Smaller adsorbent particles (which lead to smaller plate heights and larger plate numbers; see the description around equations (7.1.107c,d)) require a much higher pressure drop than has been demonstrated so far in such CAC devices for liquid-phase chromatographic processes.

8.3.1.2 *Continuous-surface chromatograph*

Continuous gas chromatography using a continuous-surface chromatograph was demonstrated somewhat earlier (Sussman and Huang, 1967; Sussman *et al.*, 1974) than CAC based liquid chromatography. In this separation scheme, shown in Figures 8.3.3(a) and (b), two slowly rotating flat glass disks, with their surfaces facing each other and coated with a chromatographic solvent, are held apart by plastic spacers forming a channel for passing gas (Sussman *et al.*, 1972). The center of the top glass disk has an orifice, into which, through a stationary nozzle located on the orifice periphery, feed-gas mixture is introduced. Throughout the periphery of this orifice, except near the feed nozzle location, carrier gas is introduced; this carrier gas flows out radially through the channel between the two parallel disks and is insoluble in the chromatographic solvent. However, the carrier gas carries forward components of the feed species radially outward through the process of sorption and desorption in the chromatographic medium coated on the two glass surfaces. Therefore there are two bulk phases whose bulk motions are perpendicular to the direction of force: the chromatographic phase moves tangentially (as in Figure 8.3.1(b)) and the carrier-gas phase moves radially outwards. The two bulk phase motions are perpendicular to each other.

Figure 8.3.3(b) illustrates the trajectories of two hypothetical gas species A and B introduced via the feed nozzle into the carrier gas as the disks are rotating at a slow angular velocity of ω rad/s (Sussman *et al.*, 1972). Each such species is carried radially outward by the carrier gas. However, during the sorption–desorption process undergone by each species between the carrier gas and the chromatographic medium on the disk surface, each gas species will be displaced by a certain angle from the original introduction point along the circumference as it leaves the outer radius of the disks. This concept has been demonstrated by Sussman's group as capable of efficient and continuous separation of a multicomponent feed-gas mixture. There is some dispersion in the peak of each species as it comes out. The issue with such a concept is that the feed-gas mixture flow rate is low, say ~1 cm^3/min. Computer simulations suggest higher flow rates of 50 cm^3/min in 12 inch diameter disks in this technique, sometimes dubbed as continuous-surface chromatography (CSC) (Sussman *et al.*, 1972). Larger systems containing multiple disks will increase capacity; however, mechanical difficulties with rotation remain (Siegell *et al.*, 1986). This system could also utilize very small particulate medium on the surface of the rotating disks. This and the CAC concept avoid/bypass the attritional problem if the particles themselves are moving in crossflow configuration.

8.3.1.3 *Crossflow magnetically stabilized bed chromatography*

Movement of adsorbent particles is generally not practiced because of attritional losses as well as larger-scale particle handling problems. Section 8.1.6 illustrated the simulated moving bed concept, which bypassed such problems. Magnetically stabilized fluidized bed (MSB) can, however, be operated as if the stably fluidized solids almost resemble a liquid (Siegell *et al.*, 1986). This happens to a fluidizable bed of magnetizable solids when a magnetic field is applied. Such a fluidized bed operates without bubbles and pulsations. In the following, we illustrate what occurs in the context of continuous chromatography.

Figure 8.3.4 illustrates the principle of such a continuous chromatographic separation operation. In the device

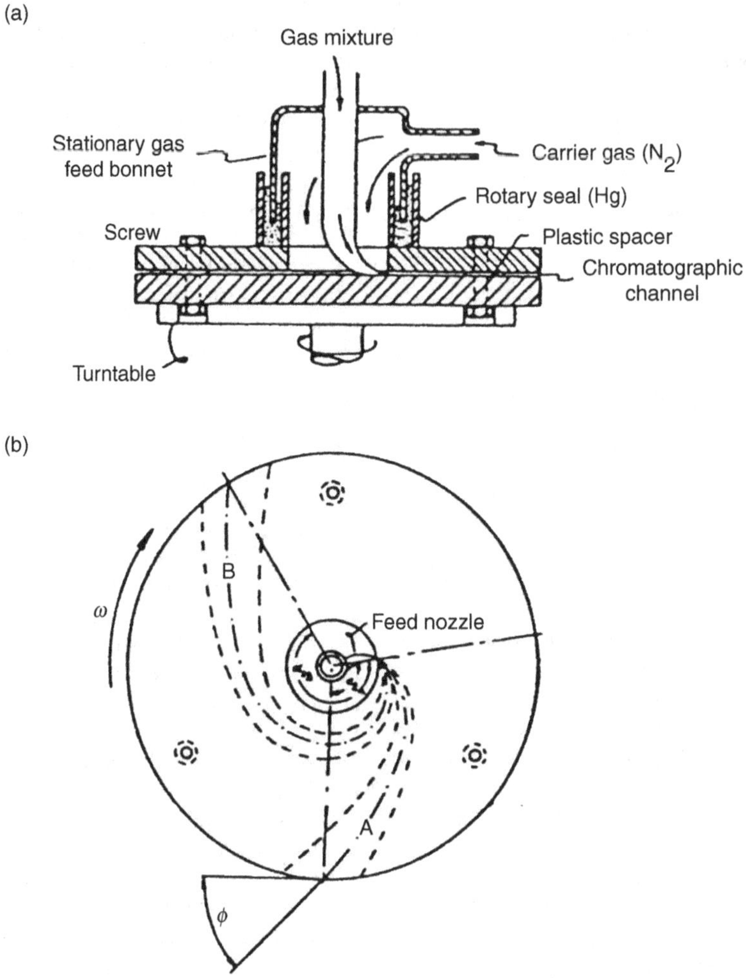

Figure 8.3.3. Continuous disk chromatograph. (a) Vertical section: the chromatograph channel is formed between solvent-coated surfaces of flat glass disks separated by three small plastic spacers. (b) Bottom: disks showing central supply orifice and paths of hypothetical mixture components A and B. The disks rotate in the direction of the outer arrow. From M.V. Sussman and C.C. Huang, "Continuous gas chromatography," Science ***156****, 974 (1967). Reprinted, with permission, from AAAS.*

shown (Siegell *et al.*, 1986), the solids in the magnetically stabilized bed flow horizontally (in the z-direction) in essentially plug flow without any gradients of the axial z-directional velocity in the x- and y-directions. The carrier gas is introduced along the length of the bottom surface to fluidize the particulate bed, except at the beginning, where the feed-gas mixture is introduced. Any species in the feed-gas mixture is subjected to two different velocities perpendicular to each other. The carrier gas generates a migration velocity, $v^*_{Ci,y}$, of species i in the y-direction; simultaneously, there is a crossflow velocity in the z-direction of each species due to the axial movement of the adsorbent particles. The resulting trajectory of each species i ends at different z-locations at the top of the bed depending on the value of $v^*_{Ci,y}$. For example, a species poorly adsorbed will have a high $v^*_{Ci,y}$ and appear on the top surface at a small value of z. On the other hand, a species strongly adsorbed will travel further downstream before showing up at the top surface. As long as each species has a separate κ_{i1}, its trajectory will be different from that of another species. However, we cannot avoid a dispersive band of each species around its center point.

It is useful to provide some details of such a system that has been used successfully. Siegell *et al.* (1986) provide the following data: vessel dimensions, z-direction, 69 cm, 7.6 cm wide; composite adsorbent of 30 wt% alumina, 70 wt % stainless steel; solids density, 1.7 g/cm^3; bed depth, 14 cm (y-direction); fluidizing air velocity (y-direction)

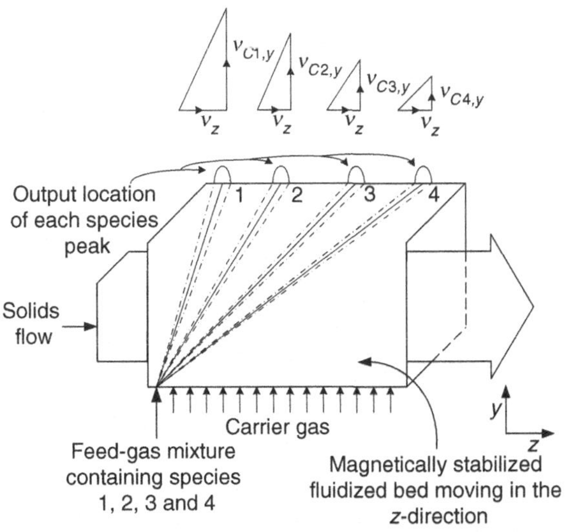

Figure 8.3.4. Continuous chromatographic separation of a four-component feed-gas mixture in a magnetically stabilized fluidized bed moving in the z-direction with carrier gas and feed-gas mixture introduced via crossflow in the y*-direction. (After Siegell* et al. *(1986).)*

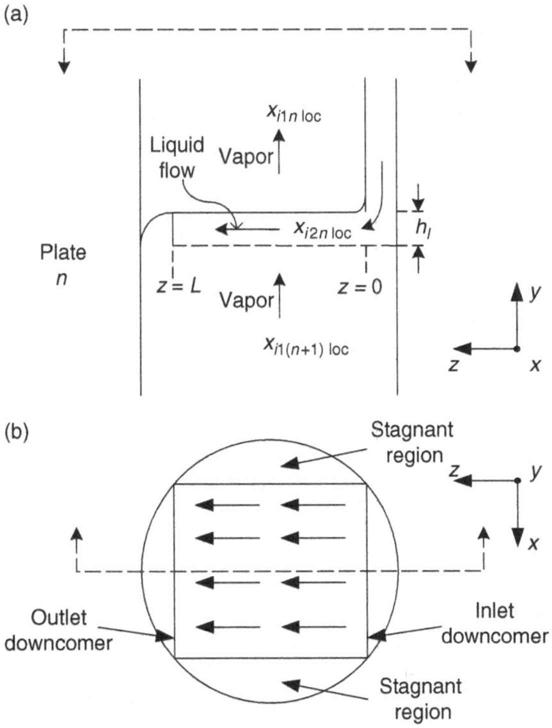

Figure 8.3.5. Crossflow plate in a distillation column: (a) gross flow pattern in a cross-sectional view; (b) gross flow pattern in a plan view.

30 cm/s (superficial velocity); z-direction solid particle velocity 0.56 cm/s; applied magnetic field, 190 oersted. Increasing bed height from 14 to 22 cm leads to better separation with less peak overlap. Such a concept may also be used for the continuous chromatographic separation of a liquid feed mixture.

A major feature of the three techniques of continuous chromatography in the crossflow configuration is that the feed stream is introduced only over a small section of the flow cross section of the carrier fluid. Had the feed stream been introduced throughout the flow cross section in the carrier fluid flow direction, multicomponent separation would not have been possible: feed introduction mode is important. We see in Section 8.3.2 that introduction of the feed fluid across the whole flow cross section for this fluid in a crossflow system is a common feature of separation in a plate in a distillation column which can produce only a binary separation.

8.3.2 Crossflow plate in a distillation column

We have seen in Section 8.1.3 that multistage distillation is generally carried out in a vertical multiplate column in which each plate has crossflow. Vapor bubbles rise vertically through a liquid layer flowing horizontally in crossflow over the plate (Section 8.1.3.5); the same two-phase flow scheme is employed in a plate for gas–liquid absorption stripping. We were introduced to the notion of stage efficiency, plate efficiency, tray efficiency, etc., in Section 8.1.3.4. We will introduce here the models used to determine the stage efficiency, specifically the Murphree vapor efficiency E_{MV}, among others. *Note*: Such a countercurrent multiplate column with crossflow on each plate can only produce two pure species in one column.

Unless a rate based approach for modeling distillation (Section 8.1.3.6) is adopted, the usual method of designing a column or understanding its separation behavior requires an estimate of the stage efficiency, usually the Murphree vapor efficiency, E_{MV} (see definition (8.1.198)). A number of models have been developed to estimate the value of E_{MV} for a crossflow plate in a distillation column (Figures 8.3.5(a) and (b)). Although the composition of the vapor coming to any plate n from the plate below, $(n + 1)$, varies with the location, most models assume a uniform entering vapor composition across the length and cross section of the plate. Most models also make the constant molar overflow assumption. In addition, the liquid is generally assumed to be well-mixed in the vertical direction, the direction of flow of the vapor bubbles. Therefore the symbol $\overline{C}_{i2}$ for species i in the liquid phase ($j = 2$) means averaged over the vertical coordinate y (Figure 8.3.5(a)).

As is clear from the plan view of the plate (Figure 8.3.5(b)), the liquid coming down from the plate above flows in the z-direction over a very large section of the circular plate. However, two sections at the two edges of the plate are stagnant - the liquid coming down at the inlet and going over the liquid outlet weir avoid these two sections. If these stagnant regions are not directly modeled, $\overline{C}_{i2}$ could also include an average in the x-coordinate direction. (Although these two regions have no net z-directional liquid-phase velocities, there are local transient circulating velocities.) Similarly, $\overline{C}_{i1}$ for species i in the vapor phase ($j = 1$) means averaged over the x-coordinate direction.

An additional point is important vis-à-vis relating E_{MV} to the mass-transfer process between the vapor and the liquid phases. It is carried out via a point efficiency E_{oG} at any location on the plate:

$$E_{oG} = \frac{x_{i1n\,\mathrm{loc}} - x_{i1(n+1)\mathrm{loc}}}{x^*_{i1n\,\mathrm{loc}} - x_{i1(n+1)\mathrm{loc}}}. \tag{8.3.13}$$

This quantity illustrates the local vapor-phase composition change achieved between the vapor coming in from the bottom plate (n +1) and the vapor going out from plate n at that location, with respect to the change that would have been achieved if the vapor phase leaving plate n at that location were in equilibrium with the local liquid-phase composition $x_{i2n\mathrm{loc}}$ (i.e. $x^*_{i1n\mathrm{loc}} = f_{\mathrm{eq}}(x_{i2n\mathrm{loc}})$). We will see later how E_{oG} is related to the local mass-transfer process.

We will now write down the local species i balance equations for phase 1 flowing up in the y-direction and for phase 2 flowing axially in the z-direction on plate n in the manner of equation (6.2.33); however, for the convective transport term, we will have to use the particular coordinate component of the $\nabla \cdot (\langle v_k \rangle \overline{C}_{ik})$ term from equation (6.2.30) since the $\langle v_k \rangle$ values are changing in both the y- and z-directions due to mass transport.

$$\text{Phase 1: (gas or vapor)}\ \varepsilon_1 \frac{\partial \overline{C}_{i1}}{\partial t} + \frac{\partial(\langle v_1 \rangle \overline{C}_{i1})}{\partial y} = \varepsilon_1 D_{i,\mathrm{eff},1,y} \frac{\partial^2 \overline{C}_{i1}}{\partial y^2} - K_{ilc}a\left(\overline{C}_{i1} - \overline{C}^*_{i1}\right). \tag{8.3.14}$$

Here we have neglected any dispersion/diffusion in the x- and z-directions for phase 1 in upflow.

$$\text{Phase 2: (liquid)}\ \varepsilon_2 \frac{\partial \overline{C}_{i2}}{\partial t} + \frac{\partial(\langle v_2 \rangle \overline{C}_{i2})}{\partial z} = \varepsilon_2 D_{i,\mathrm{eff},2,z} \frac{\partial^2 \overline{C}_{i2}}{\partial z^2} + \varepsilon_2 D_{i,\mathrm{eff},2,z} \frac{\partial^2 \overline{C}_{i2}}{\partial x^2} - K_{i2c}a\left(\overline{C}_{i2} - \overline{C}^*_{i2}\right). \tag{8.3.15}$$

Here we have neglected any dispersion/diffusion in the y-direction for phase 2 in crossflow along the plate length. In most analysis of the crossflow plate/stage/device used for distillation, the liquid is assumed to be well-mixed vertically at any axial z-coordinate location. Therefore when one obtains an overall species i balance equation at any location by combining the above two equations,

$$\varepsilon_1 \frac{\partial \overline{C}_{i1}}{\partial t} + \varepsilon_2 \frac{\partial \overline{C}_{i2}}{\partial t} + \frac{\partial(\langle v_1 \rangle \overline{C}_{i1})}{\partial y} + \frac{\partial(\langle v_2 \rangle \overline{C}_{i2})}{\partial z} = \varepsilon_1 D_{i,\mathrm{eff},1,y} \frac{\partial^2 \overline{C}_{i1}}{\partial y^2} + \varepsilon_1 D_{i,\mathrm{eff},2,z} \frac{\partial^2 \overline{C}_{i2}}{\partial z^2} + \varepsilon_1 D_{i,\mathrm{eff},2,x} \frac{\partial^2 \overline{C}_{i2}}{\partial x^2}, \tag{8.3.16}$$

and integrates the following form of this equation,

$$\varepsilon_1 \frac{\partial \overline{C}_{i1}}{\partial t} + \varepsilon_2 \frac{\partial \overline{C}_{i2}}{\partial t} + \frac{\partial(\langle v_1 \rangle C_{t1} \overline{x}_{i1})}{\partial y} + \frac{\partial(\langle v_2 \rangle C_{t2} \overline{x}_{i2})}{\partial z} = \varepsilon_2 D_{i,\mathrm{eff},2,z} C_{t2} \frac{\partial^2 \overline{x}_{i2}}{\partial z^2} + \varepsilon_2 D_{i,\mathrm{eff},2,x} C_{t2} \frac{\partial^2 \overline{x}_{i2}}{\partial x^2} + \varepsilon_1 D_{i,\mathrm{eff},2,y} C_{t1} \frac{\partial^2 \overline{x}_{i1}}{\partial y^2}, \tag{8.3.17}$$

over the liquid-film height h_ℓ in the y-direction (from $y = 0$ to $y = h_\ell$) for plate n, we obtain the following for steady state and no dispersive term in the gas phase ($j = 1$):

$$-\langle v_1 \rangle_{n+1} C_{t1} \overline{x}_{i1(n+1)} + \langle v_1 \rangle_n C_{t1} \overline{x}_{i1n} + \langle v_2 \rangle h_\ell C_{t2} \frac{\partial \overline{x}_{i2}}{\partial z} + \overline{x}_{i2} \frac{\partial(\langle v_2 \rangle h_\ell C_{t2})}{\partial z} = \varepsilon_2 h_\ell C_{t2} D_{i,\mathrm{eff},2,z} \frac{\partial^2 \overline{x}_{i2}}{\partial z^2} + \varepsilon_2 h_\ell C_{t2} D_{i,\mathrm{eff},2,x} \frac{\partial^2 \overline{x}_{i2}}{\partial x^2}. \tag{8.3.18}$$

In the derivation of this equation, we assumed that the total molar density C_{t2} of the liquid phase is essentially constant and that the vapor entering plate n from plate (n + 1) below has a uniform composition $x_{i1(n+1)}$. We will now obtain simpler forms of equation (8.3.18) for a variety of situations by making additional assumptions.

Case (1) Neglect transverse mixing (x-direction) in the liquid phase ($j = 2$):

$$\overline{D}_{i,\mathrm{eff},2,z} h_\ell C_{t2} \frac{d^2 \overline{x}_{i2}}{dz^2} - \langle v_2 \rangle h_\ell C_{t2} \frac{d\overline{x}_{i2}}{dz} - \frac{\overline{x}_{i2}\, d(\langle v_2 \rangle h_\ell C_{t2})}{\partial z} + \langle v_1 \rangle_{n+1} C_{t1} \overline{x}_{i1(n+1)} - \langle v_1 \rangle_n C_{t1} \overline{x}_{i1n} = 0. \tag{8.3.19}$$

Here $\varepsilon_2 D_{i,\mathrm{eff},2,z} = \overline{D}_{i,\mathrm{eff},2,z}$. This equation is the basis of the AIChE model (AIChE, 1958) for tray efficiency in distillation.

Case (2) Neglect convective liquid movement in the z-direction:

$$-\langle v_1 \rangle_{n+1} C_{t1} \overline{x}_{i1(n+1)} + \langle v_1 \rangle_n C_{t1} \overline{x}_{i1n} = \varepsilon_2 h_\ell C_{t2} D_{i,\mathrm{eff},2,z} \frac{\partial^2 \overline{x}_{i2}}{\partial z^2} + \varepsilon_2 h_\ell C_{t2} D_{i,\mathrm{eff},2,x} \frac{\partial^2 \overline{x}_{i2}}{\partial x^2}. \tag{8.3.20}$$

On a large plate, there are two stagnant regions (shown in Figure 8.3.5(b)) where there is no liquid replenishment. Ideally therefore, the crossflow plate in a distillation column may be analyzed via the simultaneous solution of equations (8.3.18) and (8.3.20) coupled via appropriate boundary conditions (Porter *et al.*, 1972).

Case (3) Neglect transverse and axial mixing in liquid phase (plug flow of liquid) and assume constant molal overflow (i.e. $\langle v_1\rangle = \langle v_1\rangle_{n+1} = \langle v_1\rangle_n$; $(\partial(\langle v_2\rangle h_\ell C_{t2})/\partial z) = 0$):

$$\left(\overline{x}_{i1n} - \overline{x}_{i1(n+1)}\right) = -\frac{\langle v_2\rangle h_\ell C_{t2}}{\langle v_1\rangle C_{t1}}\frac{d\overline{x}_{i2}}{dz}. \tag{8.3.21}$$

Case (4) Liquid on the plate is completely mixed in the direction of flow and is at steady state. Therefore, the local vapor-liquid mass-transfer efficiency controls the stage efficiency; use equation (8.3.14). We will treat this case first; however, before we do, we look at local mass transfer.

The equations given above focus on the whole plate in a distillation column. However, it is quite useful to focus on a particular axial location on a plate vis-à-vis the liquid height. Consider equation (8.3.14) for vertical movement of gas/vapor bubbles through the liquid layer of height h_ℓ. Assume that the vertical liquid column is completely mixed and that there is negligible axial mixing in the vapor phase. Equation (8.3.14) is then reduced, under steady state, to

$$\frac{d(\langle v_1\rangle C_{i1})}{dy} = \frac{d(\langle v_1\rangle C_{t1}x_{i1})}{dy} = +K_{i1c}\,a\left(C^*_{i1} - C_{i1}\right)$$
$$= K_{i1c}a\,C_{t1}\left(x^*_{i1} - x_{i1}\right). \tag{8.3.22}$$

For constant molar overflow, we may rewrite it as

$$\langle v_1\rangle C_{t1}\frac{dx_{i1}}{dy} = K_{i1c}\,aC_{t1}\left(x^*_{i1} - x_{i1}\right)$$
$$\Rightarrow \int_0^{h_\ell} dy = \frac{\langle v_1\rangle}{K_{i1c}a}\int_{x_{i1(n+1)\mathrm{loc}}}^{x_{i1n\mathrm{loc}}}\frac{dx_{i1}}{x^*_{i1} - x_{i1}}. \tag{8.3.23}$$

If we assume a linear equilibrium relation, namely

$$x^*_{i1} = m_i\overline{x}_{i2} + b_i, \tag{8.3.24}$$

where $\overline{x}_{i2}$ is fixed for this vertical liquid column, the value of x^*_{i1} is fixed. Integration leads to

$$h_\ell = \frac{\langle v_1\rangle}{K_{ilc}\,a}\,\ell n\left(\frac{x^*_{i1} - x_{i1(n+1)\mathrm{loc}}}{x^*_{i1} - x_{i1n\,\mathrm{loc}}}\right). \tag{8.3.25}$$

At location z, the liquid height h_ℓ may be expressed as

$$h_\ell = NTU_{og} \times HTU_{og} = NTU_{og} \times \frac{\langle v_1\rangle}{K_{i1c}\,a}, \tag{8.3.26}$$

as if it is a gas absorber (see equation (8.1.54a)); therefore

$$NTU_{og} = \ell n\left[\frac{x^*_{i1} - x_{i1(n+1)\mathrm{loc}}}{x^*_{i1} - x_{i1n\mathrm{loc}}}\right]; \tag{8.3.27a}$$

$$\frac{x^*_{i1} - x_{i1n\,\mathrm{loc}}}{x^*_{i1} - x_{i1(n+1)\mathrm{loc}}} = \exp\left(-NTU_{og}\right) = \exp\left(-h_\ell/HTU_{og}\right); \tag{8.3.27b}$$

$$1 - \frac{x^*_{i1} - x_{i1n\mathrm{loc}}}{x^*_{i1} - x_{i1(n+1)\mathrm{loc}}} = \frac{x_{i1n\,\mathrm{loc}} - x_{i1(n+1)\mathrm{loc}}}{x^*_{i1} - x_{i1(n+1)\mathrm{loc}}}$$
$$= 1 - \exp\left(-NTU_{og}\right) = E_{oG}, \tag{8.3.28}$$

which is the Murphree point efficiency. Here x^*_{i1} refers to the vapor in equilibrium with the local $\overline{x}_{i2}$, i.e. $\overline{x}_{i2\mathrm{loc}}$.

We consider now case (4), namely that the liquid on plate n is completely mixed and has the value $\overline{x}_{i2n}$ corresponding to the exiting liquid composition. Therefore the vapor leaving different locations on plate n will have the same composition, namely x_{i1n} and $x_{i1n\,\mathrm{loc}}$. We have already assumed that the vapor entering plate n has a uniform composition $x_{i1(n+1)}$. Therefore E_{MV} defined by (8.1.198) becomes equal to E_{oG}:

$$E_{MV} = \frac{x_{i1n} - x_{i1(n+1)}}{x^*_{i1n} - x_{i1(n+1)}} = \frac{x_{i1n\,\mathrm{loc}} - x_{i1(n+1)\mathrm{loc}}}{x^*_{i1n} - x_{i1(n+1)\mathrm{loc}}}$$
$$= 1 - \exp\left(-NTU_{og}\right). \tag{8.3.29}$$

We will now focus on case (1), namely the solution of equation (8.3.19) to develop the expression for Murphree vapor efficiency E_{MV} for plate n under the condition of constant molar overflow; this is the AIChE model. Equation (8.3.19) may be written as

$$\frac{\overline{D}_{i,\mathrm{eff},2z}h_\ell C_{t2}}{W'_{t\ell}}\frac{d^2\overline{x}_{i2}}{dz^2} - \frac{d\overline{x}_{i2}}{dz} + \frac{W''_{tg}}{W'_{t\ell}}\left(\overline{x}_{i1(n+1)} - \overline{x}_{i1n}\right) = 0, \tag{8.3.30}$$

where $W'_{t\ell} = \langle v_2\rangle h_\ell C_{t2}$, and the molar liquid flow rate per unit width (= $W_{t\ell}$/weir width = $W_{t\ell}/W$) is given by

$$W''_{tg} = \langle v_1\rangle C_{t1}$$
$$= \left(W_{tg}/(\text{weir width} \times \text{plate length in } z\text{-direction})\right). \tag{8.3.31}$$

Nondimensionalize z using plate length along z, L:

$$z^+ = z/L \tag{8.3.32}$$

to yield

$$\frac{\overline{D}_{i,\mathrm{eff},2z}h_\ell C_{t2}}{W'_{t\ell}L}\frac{d^2\overline{x}_{i2}}{dz^{+2}} - \frac{d\overline{x}_{i2}}{dz^+} + \frac{W''_{tg}L}{W'_{t\ell}}\left(\overline{x}_{i1(n+1)} - \overline{x}_{i1n}\right) = 0. \tag{8.3.33}$$

Define a liquid-phase Péclet number as

$$Pe_\ell = \frac{\langle v_2\rangle L}{\overline{D}_{i,\mathrm{eff},2z}}. \tag{8.3.34}$$

Equation (8.3.33) may now be written as

$$\frac{1}{Pe_\ell}\frac{d^2\overline{x}_{i2}}{dz^{+2}} - \frac{d\overline{x}_{i2}}{dz^+} + \frac{W_{tg}}{W_{t\ell}}\left(\overline{x}_{i1(n+1)} - \overline{x}_{i1n}\right) = 0, \tag{8.3.35}$$

where

$$W_{tg} = W''_{tg}\,W\,L. \tag{8.3.36}$$

This equation for the dependent liquid-phase composition variable of interest, namely $\overline{x}_{i2}$, has additional unknowns in the third term. We remove them by utilizing the Murphree point efficiency definition (8.3.29) and the linear equilibrium relation (8.3.24):

$$E_{oG} = \frac{x_{i1n\,\mathrm{loc}} - x_{i1(n+1)\mathrm{loc}}}{x_{i1}^{*} - x_{i1(n+1)\mathrm{loc}}} = \frac{x_{i1n\,\mathrm{loc}} - x_{i1(n+1)\mathrm{loc}}}{m_i\,\overline{x}_{i2} + b_i - m_i\,x_{i2}^{*} - b_i}$$

$$= \frac{x_{i1n\mathrm{loc}} - x_{i1(n+1)\mathrm{loc}}}{m_i\left(\overline{x}_{i2} - x_{i2}^{*}\right)} \qquad (8.3.37a)$$

$$\Rightarrow x_{i1n\,\mathrm{loc}} - x_{i1(n+1)\mathrm{loc}} = \overline{x}_{i1n} - \overline{x}_{i1(n+1)} = E_{oG}\,m_i\left(\overline{x}_{i2} - \overline{x}_{i2}^{*}\right), \qquad (8.3.37b)$$

where we are assuming that there are no variations along z of the vapor coming from plate $(n+1)$ and going out of plate n. Further, x_{i2}^{*} is the liquid-phase mole fraction in equilibrium with the entering vapor-phase mole fraction $x_{i1(n+1)\mathrm{loc}} = x_{i1(n+1)}$. Equation (8.3.35) is now simplified to

$$\frac{1}{Pe_{\ell}}\frac{\mathrm{d}^2\overline{x}_{i2}}{\mathrm{d}z^{+2}} - \frac{\mathrm{d}\overline{x}_{i2}}{\mathrm{d}z^{+}} - \left(\frac{W_{tg}\,m_i}{W_{t\ell}}\right)E_{oG}\left(\overline{x}_{i2} - x_{i2}^{*}\right) = 0. \qquad (8.3.38)$$

The quantity $\left(m_i W_{tg}/W_{t\ell}\right)$ is usually denoted by the parameter λ. Two boundary conditions used to solve this equation are that, at the tray liquid outlet,

$$z = L, \qquad z^{+} = 1, \qquad \overline{x}_{i2} = \overline{x}_{i2n} = \overline{x}_{i2n,\mathrm{out}} \quad \text{and} \quad \frac{\mathrm{d}\overline{x}_{i2}}{\mathrm{d}z^{+}} = 0. \qquad (8.3.39)$$

Gerster *et al.* (1958) provided the following analytical solution for this equation and the boundary conditions:

$$\frac{\overline{x}_{i2} - x_{i2}^{*}}{\overline{x}_{i2n} - x_{i2}^{*}} = \frac{\exp\left[(\eta + Pe_{\ell})(z^{+} - 1)\right]}{1 + (\eta + Pe_{\ell})/\eta} + \frac{\exp\left(\eta\,(1 - z^{+})\right)}{1 + \eta/(\eta + Pe_{\ell})}, \qquad (8.3.40)$$

where

$$\eta = \frac{Pe_{\ell}}{2}\left[\left(1 + \frac{4\lambda\,E_{oG}}{Pe_{\ell}}\right)^{0.5} - 1\right]. \qquad (8.3.41)$$

Since the goal is to determine an expression for E_{MV}/E_{oG} for an assumed constant E_{oG} along the plate, we will first determine $x_{ivn} - x_{iv(n+1)} = x_{i1n} - x_{i1(n+1)}$ in the definition (8.1.198) of E_{MV}. The quantity x_{i1n} is an average of all the vapors leaving plate n:

$$x_{i1n} - x_{i1(n+1)} = \int_0^1 \left(x_{i1n\,\mathrm{loc}} - x_{i1(n+1)\mathrm{loc}}\right)\mathrm{d}z^{+}$$

$$= m_i\,E_{oG}\int_0^1 \left(\overline{x}_{i2} - x_{i2}^{*}\right)\mathrm{d}z^{+}, \qquad (8.3.42)$$

where we have utilized definitions (8.3.37a,b) for E_{oG}. In the definition (8.1.198) of E_{MV}, the denominator $\left(x_{ivn}^{*} - x_{iv(n+1)}\right) = \left(x_{i1n}^{*} - x_{i1(n+1)}\right)$ may be expressed via the following equilibrium relations:

$$x_{i1n}^{*} = m_i\,\overline{x}_{i2n} + b_i, \qquad (8.3.43a)$$

$$x_{i1(n+1)} = m_i\,x_{i2}^{*} + b_i, \qquad (8.3.43b)$$

as

$$x_{i1n}^{*} - x_{i1(n+1)} = m_i\left(\overline{x}_{i2n} - x_{i2}^{*}\right). \qquad (8.3.43c)$$

Therefore

$$E_{MV} = \frac{x_{i1n} - x_{i1(n+1)}}{x_{i1n}^{*} - x_{i1(n+1)}} \Rightarrow \left(E_{MV}/E_{oG}\right) = \int_0^1 \frac{\overline{x}_{i2} - x_{i2}^{*}}{\overline{x}_{i2n} - x_{i2}^{*}}\,\mathrm{d}z^{+}. \qquad (8.3.44)$$

Introducing expression (8.3.40) for the integrand leads to the following result known as the "AIChE manual recommended expression for partially mixed trays" (AIChE, 1958):

$$\frac{E_{MV}}{E_{oG}} = \frac{1 - \exp\left[-(\eta + Pe_{\ell})\right]}{(\eta + Pe_{\ell})\left[1 + ((\eta + Pe_{\ell})/\eta)\right]} + \frac{\exp(\eta) - 1}{\eta\left[1 + (\eta/(\eta + Pe_{\ell}))\right]}. \qquad (8.3.45)$$

Figure 8.3.6 illustrates the value of (E_{MV}/E_{oG}) as a function of λE_{oG} for various parametric values of Pe_{ℓ}. This figure shows that, when the plate is completely mixed, $Pe_{\ell} \to 0$, $E_{MV} \to E_{oG}$. On the other hand, when $Pe_{\ell} \to \infty$, we have no axial mixing, and the situation should correspond to case (3), a case of plug flow of the liquid on the plate:

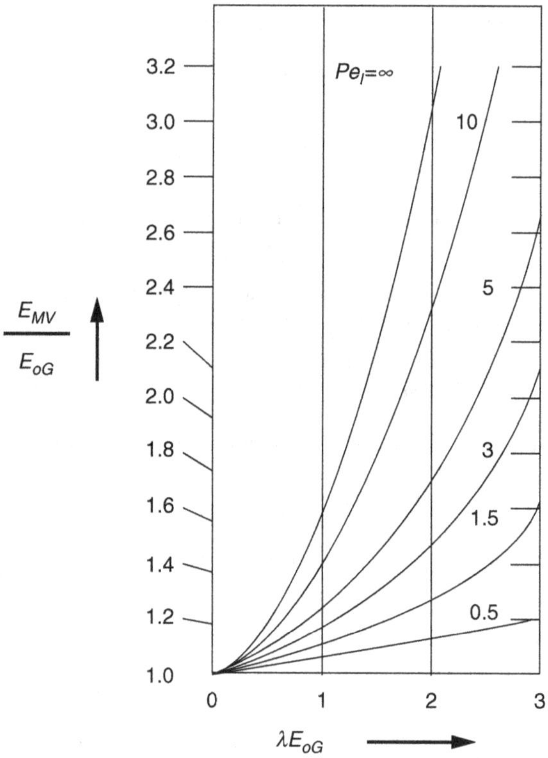

Figure 8.3.6. Plot of equation (8.3.45) in terms of E_{MV}/E_{oG} *against* λE_{oG} *for various values of the parameter,* Pe_{ℓ}*, the liquid-phase Péclet number. (After AIChE (1958).)*

$$\left(\overline{x}_{i1n\,\text{loc}} - \overline{x}_{i1(n+1)}\right)\langle v_1\rangle C_{t1}\,dz = -\langle v_2\rangle h_\ell\, C_{t2}\, d\overline{x}_{i2}, \quad (8.3.46)$$

where we have identified $\overline{x}_{i1n\,\text{loc}}$ as the local mole fraction in the vapor phase. Using equilibrium relation (8.3.24), we obtain

$$m_i\left(\overline{x}_{i1n\,\text{loc}} - \overline{x}_{i1(n+1)}\right)\frac{\langle v_1\rangle\, C_{t1}\,dz}{\langle v_2\rangle\, h_\ell\, C_{t2}} = -dx^*_{i1}, \quad (8.3.47)$$

where x^*_{i1} is in equilibrium with the local liquid mole fraction $\overline{x}_{i2}$. Further, using the point efficiency definition (8.3.28), we get

$$\frac{x_{i1n\,\text{loc}} - x_{i1(n+1)\text{loc}}}{x^*_{i1} - x_{i1(n+1)\text{loc}}} = E_{oG} \Rightarrow dx_{i1n\,\text{loc}} = E_{oG}\, dx^*_{i1}. \quad (8.3.48)$$

Noting that we may use x-coordinate-averaged $d\overline{x}_{i1n\,\text{loc}}$ instead of $dx_{i1n\,\text{loc}}$, we get

$$\frac{m_i\langle v_1\rangle\, C_{t1}\, E_{oG}}{\langle v_2\rangle\, h_\ell\, C_{t2}}\,dz = -\frac{dx_{i1n\,\text{loc}}}{\overline{x}_{i1n\,\text{loc}} - \overline{x}_{i1(n+1)}} = -\frac{d\overline{x}_{i1n\,\text{loc}}}{\overline{x}_{i1n\,\text{loc}} - \overline{x}_{i1(n+1)}}. \quad (8.3.49)$$

We may rewrite this as follows:

$$E_{oG}\frac{m_i\langle v_1\rangle\, C_{t1}\, WL}{\langle v_2\rangle\, h_\ell\, C_{t2}\, W}\int_L^z \frac{\langle v_1\rangle\, C_{t1}\, W\,dz}{\langle v_1\rangle\, C_{t1}\, WL}$$

$$= E_{oG}\frac{m_i\, W_{tg}}{W_{t\ell}}\int_L^z df = \int_{\overline{x}_{i1n,\,\text{out}}}^{\overline{x}_{i1n}} \frac{d\overline{x}_{i1n\,\text{loc}}}{\overline{x}_{i1n\,\text{loc}} - \overline{x}_{i1(n+1)}}; \quad (8.3.50)$$

$$E_{oG}\,\lambda\int_L^z df = E_{oG}\,\lambda f = \ell n\,\frac{\overline{x}_{i1n\,\text{loc}} - \overline{x}_{i1(n+1)}}{\overline{x}_{i1n,\,\text{out}} - \overline{x}_{i1(n+1)}}, \quad (8.3.51)$$

where f = fraction of the vapor flow rate to the left of point z, so that, when $z = L, f = 0$, and when $z = 0, f = 1$. The averaged vapor composition leaving plate n is obtained by averaging the product of the fractional vapor flow rate and composition over the plate length L:

$$\int_0^1 \overline{x}_{i1n\,\text{loc}}\,df = \overline{x}_{i1n,\text{avg}}$$

$$= \int_0^1\left\{\overline{x}_{i1(n+1)} + \left(\overline{x}_{i1n,\text{out}} - \overline{x}_{i1(n+1)}\right)\exp(E_{oG}\lambda f)\right\}df; \quad (8.3.52)$$

$$\overline{x}_{i1n,\text{avg}} = \overline{x}_{i1(n+1)} + \left(\overline{x}_{i1n,\text{out}} - \overline{x}_{i1(n+1)}\right)\frac{\exp(E_{oG}\lambda) - 1}{\lambda\, E_{oG}}. \quad (8.3.53)$$

However, from the point efficiency definition (8.3.28) at $z = L$,

$$\left(\overline{x}_{i1n,\text{out}} - \overline{x}_{i1(n+1)}\right) = E_{oG}\left(x^*_{i1n,\text{out}} - \overline{x}_{i1(n+1)}\right). \quad (8.3.54)$$

Therefore

$$\frac{\overline{x}_{i1n,\text{avg}} - \overline{x}_{i1(n+1)}}{x^*_{i1n,\text{out}} - \overline{x}_{i1(n+1)}} = E_{MV} = \frac{\exp(E_{oG}\lambda) - 1}{\lambda}. \quad (8.3.55)$$

This expression corresponds essentially to the limit of $Pe_\ell \to \infty$ shown in Figure 8.3.6. A comprehensive background on stage efficiency for a plate in a distillation column, its relation to point efficiency, as well as a brief description of the "AIChE tray-efficiency prediction method" has been provided by King (1980, pp. 621–622).

A few comments about E_{MV} in general, and these results in particular, are useful. First, the models for different cases (except case (2) which was not analyzed) predict that $E_{MV} \geq E_{oG}$. Since E_{oG} is determined by the mass-transfer conditions existing locally, as indicated by $(HTU)_{og}$ or $(NTU)_{og}$ for a given liquid height h_ℓ, it appears that crossflow of the liquid on the plate without any mixing or with partial mixing leads to a higher value of E_{MV}. A basic reason for this is as follows. The liquid coming down from plate $(n-1)$ is richer in the light component $i = 1$. Therefore the vapor emerging from this liquid is richer than that emerging from the liquid leaving plate n, namely $\overline{x}_{i2n} = \overline{x}_{i2n,\text{out}}$. This increases E_{MV} over E_{oG}. Second, for a binary mixture, the E_{MV} values are identical for $i = 1, 2$:

$$E_{MV}\Big|_1 = \frac{\overline{x}_{11n,\text{avg}} - \overline{x}_{11(n+1)}}{x^*_{11n,\text{out}} - \overline{x}_{11(n+1)}} = \frac{1 - \overline{x}_{21n,\text{avg}} - 1 + \overline{x}_{21(n+1)}}{1 - x^*_{21n,\text{out}} - 1 + \overline{x}_{21(n+1)}}$$

$$= \frac{\overline{x}_{21n,\text{avg}} - \overline{x}_{21(n+1)}}{x^*_{21n,\text{out}} - \overline{x}_{21(n+1)}} = E_{MV}\Big|_2. \quad (8.3.56)$$

However, for multicomponent systems, the values of E_{MV} are component-dependent. The number of such E_{MV} values is less than the total number of components by 1, since $\sum x_{i1n} = 1$ when summed over all species.

Third, the value of the axial dispersion coefficient $\overline{D}_{i,\text{eff},\ell z}$ in the liquid phase on a few different types of trays has been correlated (King, 1980, p. 619):

$$\left(\overline{D}_{i,\text{eff},\ell z}\right)^{0.5} = (0.0124 + 0.0171\langle v_1\rangle + 0.00250Q_\ell + 0.0150 \times \text{weir height})\ \text{ft}^2/\text{s}, \quad (8.3.57)$$

where the units of $\langle v_1\rangle$ are ft^3/s of vapor flow divided by the bubbling area in ft^2, Q_ℓ is the liquid volumetric flow rate per unit plate width in gallon/min-ft and the weir height is in inches. Typical values vary between 0.02 and 0.2 ft^2/s (Seader and Henley, 1998). Fourth, a plate in a small-diameter column has a much higher probability of having a completely mixed liquid ($E_{MV} \cong E_{oG}$). As the crossflow plate becomes longer in larger columns, partially mixed models are more likely to be valid, making $E_{MV} > E_{oG}$. Here the effect of transverse mixing in the x-direction (Figure 8.3.5) becomes important.

The subject of plate efficiency of a crossflow distillation plate is complex and has a wide literature. See Lockett (1986), AIChE (1987) and Kister (1990, 1992, 2008). Many factors influence it, including flow path length, fractional hole area and hole dimension, weir height, entrainment, vapor–liquid equilibrium data, etc. Larger-diameter columns employ a split-flow design with two downcomers or a crossflow cascade with a series of weirs on one plate.

A better estimate of the stage efficiency is a practical way to realize a cheaper distillation column.

Example 8.3.1 Laboratory sieve-plate crossflow distillation columns, called Oldershaw columns, having diameters around 2.54 cm to 5.08 cm, have been found to yield plate efficiencies which generally agree with those in columns larger in diameter in the range of 45–120 cm (Fair *et al.*, 1983). The plates in Oldershaw columns are generally well-mixed. Consider now the benezene-toluene distillation illustrated in Examples 8.1.11 and 8.1.12, where $\alpha_{\text{benzene-toluene}} = 2.5$. The value of E_{MV} for this system in an Oldershaw column is 0.6.

(1) Determine the values of HTU_{og} and NTU_{og} for a larger column, where $h_\ell = 55$ mm.
(2) Obtain the value of E_{MV} for a sieve plate in Example 8.1.12 assuming plug flow; the value of $x_{\text{benzene},\ell}$ on this plate is around 0.8.
(3) Obtain the value of E_{MV} for a sieve plate in Example 8.1.11.

Solution: Part (1) Oldershaw columns have well-mixed sieve trays; therefore $E_{MV} = E_{oG}$ for such a tray. This value of E_{oG} is valid for larger trays as well: we therefore have, from equation (8.3.28),

$$E_{oG} = 0.6 = 1 - \exp(-h_\ell/HTU_{og}). \qquad (8.3.58)$$

Since $h_\ell = 55$ mm in the larger tray,

$$0.4 = \exp(-5.5/HTU_{og}) \Rightarrow 0.92 = \frac{5.5}{HTU_{og}}$$

$$\Rightarrow HTU_{og} = 5.98\,\text{cm}; \qquad NTU_{og} = 0.92.$$

Part (2) We need to determine the value of $\lambda = m_i W^t_{tv}/W^t_{t\ell}$. To determine m_i we will use $\alpha_{12} = 2.5$ and $x_{1\ell} = 0.7$ in the basic separation factor relation, namely equation (4.1.24):

$$x_{1v} = \frac{\alpha_{12}\, x_{1\ell}}{1 + x_{1\ell}(\alpha_{12} - 1)}.$$

Differentiate this relation with respect to x_{1L}:

$$(\mathrm{d}x_{1v}/\mathrm{d}x_{1\ell}) = m_i = \frac{\alpha_{12}(1 + x_{1\ell}(\alpha_{12}-1) - \alpha_{12}^2 x_{1\ell} + \alpha_{12}x_{1\ell}}{(1 + x_{1\ell}(\alpha_{12}-1))^2}$$

$$= \frac{\alpha_{12}}{(1 + x_{1\ell}(\alpha_{12}-1))^2} \Rightarrow m_i = \frac{2.5}{(1 + 0.8 \times 1.5)^2}$$

$$= \frac{2.5}{(2.2)^2} = 0.516.$$

From Example 8.1.12, $W^t_{tv} = 4.5$ kgmol/min, $W^t_{t\ell} = 3$ kgmol/min. Therefore $\lambda = 0.516 \times 4.5/3 = 0.774$. For the plug flow model, use expression (8.3.55) for E_{MV}:

$$E_{MV} = \frac{\exp(E_{oG}\lambda) - 1}{\lambda} \Rightarrow E_{MV} = \frac{\exp(0.6 \times 0.774) - 1}{0.774}$$

$$= \frac{1.59 - 1}{0.774} = 0.76.$$

Part (3) $m_i = 2.0$, $W^b_{tv} = 2.812$ kgmol/min, $W^b_{t\ell} = 4.312$ kgmol/min, $\lambda = 2 \times 2.812/4.312 = 1.30$, so

$$E_{MV} = \frac{\exp(0.6 \times 1.3) - 1}{1.3} = \frac{2.18 - 1}{1.30} = 0.91.$$

Problems

8.1.1 Water is being used to scrub 4 mole% of NH_3 in an air stream flowing at 2000 ft^3/min, 30 °C and 1 atm through an absorption tower packed with 1.5 inch ceramic Raschig rings. The water flow rate is 600 lb/min. The viscosity and density of water are 0.80 cp and 0.995 g/cm^3. Develop an estimate of the diameter of the packed column if the column is to be operated at 50% of the flooding velocity. Calculate the gas-phase pressure drop. Use a Sherwood-Eckert generalized plot. (Ans. 3.115 ft; 0.85 inch water/foot of packed bed.)

8.1.2 A waste air stream leaving a manufacturing plant is contaminated with acetone vapor; the acetone mole fraction in this stream is 0.08. This stream must be cleaned so that the acetone level in the air stream to be discharged is only 0.01 mole fraction. This stream at 25 °C is proposed to be scrubbed in a countercurrent packed tower with water at 25 °C containing acetone at the level of 0.0001 mole fraction. The tower diameter is 40 cm. The waste gas flow rate is 10.3 kgmol/hr. The scrubbing water flow rate is 40 kgmol/hr. Available information about this packed tower suggests a value of $K_{igx}a$ for acetone to be 30 gmol/m^3-s-mol fraction. Sherwood *et al.* (1975) have indicated that, in the scrubbing concentration range, H_i^P for acetone is 3.2. Develop an estimate of the packed tower height assuming dilute solutions. (Ans. 3.53 m.)

8.1.3 If Problem 8.1.2 has to be solved by contacting the gas stream and the liquid scrubbing solution of water containing acetone at the level of 0.0001 mole fraction in a countercurrent multistage device containing sieve plates, determine the number of ideal stages/plates needed to solve the problem by assuming dilute solutions. In determining the compositions at the two end locations in the device, include the changes in the flow rates of the two streams and use average values for determining A. (Ans. 4.)

8.1.4 The SO_2 present in a process air stream is being scrubbed out in a 6 m long packed tower using pure water entering at the top. The feed air flow rate is 0.1 kgmol/s; its SO_2 mole fraction is 0.015. Ninety-five percent of the

SO_2 is being scrubbed out. The temperature is 25 °C. The water flow rate is 3 kgmol/s. Benitez (1993) has indicated that the linear absorption equilibrium relation may be described by $x_{ig}^* = 10x_{il}$. Determine the value of $K_{igx}a$ for this operating condition if the packed tower diameter is 1.25 m. Determine also the value of NTU_{og}. (Ans. $K_{igx}a = 0.053$ kgmol/m^3-s-mole fraction; $NTU_{og} = 3.9$.)

8.1.5 Benzene vapor present in an air stream at 26 °C is being scrubbed with wash oil in a countercurrent packed tower 50 ft tall and 1.5 ft in diameter. The air flow rate is 1.3 lb-mol/min; the wash oil flow rate is 0.3 lb-mol/min. The benzene mole fraction in the entering air is 0.01; the scrubbed air leaving at the top has a benzene mole fraction of 0.0008. The entering wash oil has a benzene mole fraction of 0.0015. The vapor pressure of benzene at 26 °C is 100 mm Hg. The entering air, as well as the exiting air, may be assumed to be at 1 atm. Raoult's law may be assumed to be valid for the benzene-wash oil system. Develop an estimate of $K_{igx}a$ for benzene under the operating conditions in the units of lb-mol/hr-ft^2. The wash oil may be assumed to be nonvolatile. (Ans. $K_{igx}a = 3.15$ lb-mol/hr-ft^3 mole fraction.)

8.1.6 In the scrubbing of highly soluble/reactive gas/vapors (e.g. HCl vapor by water), the value of H_i^P may be at least an order of magnitude smaller than the ratio (W_{tg}/W_{tl}). If the scrubbing liquid does not have any solute species as it enters a countercurrent scrubber, obtain a simplified expression for NTU_{og} based only on the inlet gas-phase composition x_{ig0} and the outlet gas-phase composition x_{igL}. Assume that (1) the gas is very dilute in the species to be scrubbed, (2) the scrubbing is very efficient and the gas is highly purified.

8.1.7 Consider the hollow fiber membrane contactor device of Figure 8.1.15 for deoxygenation of water. Sengupta *et al.* (1998), among others, have studied a 10 × 28 contactor containing 224 640 hollow fibers of internal diameter 240 μm and outside diameter 300 μm. The fiber length is 61 cm; the outside diameter of the cartridge, $2R_{co}$, is 24.5 cm; the inside diameter of the cartridge, $2R_{ci}$, is 11.4 cm. The water flow rate is 79.2 gallon/min (5000×10^{-6} m^3/s). At this flow rate, data from Sengupta *et al.* (1998, fig. 10) for deoxygenation indicate that, at 20 °C, the value of the right-hand side of equation (8.1.78b) is 0.01. Determine the value of the stripping device efficiency E_i for oxygen at the specified flow rate for $b_1 = 0.38$. (Ans. $E_i = 0.9947$.)

8.1.8 Steam introduced at the column bottom is used to strip a volatile impurity i from an aqueous solution introduced at the column top. Develop the following simplified form of the Kremser equation for the steam stripper:

$$\frac{x_{il0}}{x_{ilL}} = \frac{S-1}{S^{N+1}-1}.$$

8.1.9 Hwang *et al.* (1992a) carried out experiments on steam stripping of high-boiling pollutants from water in a countercurrent multistage column. Specifically, the nitrobenzene concentration in the feed water was 1000 ppm. It is to be reduced to 28 ppm in the treated water leaving the column bottom. The steam-to-water molar flow rate ratio is given as 0.056 in a particular run. The K_i^∞ value for nitrobenzene is 28 (see equations (4.1.19a–c)) at the operating temperature. The Kremser equation for steam stripping is provided in Problem 8.1.8, where S is the stripping factor (see Problem 2.2.1) $K_i^\infty |W_{tg}|/|W_{tl}| = m_i |W_{tg}|/|W_{tl}|$. Determine the number N of ideal equilibrium stages needed for the specified stripping goal. Identify the overall stripper efficiency if ten actual stages were needed. (Ans. six ideal equilibrium stages; 57.9%.)

8.1.10 In the steam stripping of organic pollutants from water considered in Problem 8.1.9, the steam-to-water molar flow rate ratio used was 0.056. In reality, quite a significant part of the steam is used up by the subcooled water. When this is coupled with heat losses to the surroundings, the effective steam-to-water molar ratio is around 0.02. Consider the Kremser equation given in Problems 8.1.8 and 8.1.9. Develop a guideline in terms of values of K_i^∞ (or m_i) for the following performance levels: (a) the stripper removes more than 99.9% of species i; (b) the stripper removes less than 20%; (c) the stripper removes in between 20% and 99.9%. These guidelines will identify which compounds are easily strippable and which are not. Assume $N = 10$ ideal equilibrium stages. (Ans. (a) 100; (b) 10; (c) $10 < K_i^\infty < 100$.)

8.1.11 Perry and Green (1984, table 13-1) provide the vapor-liquid equilibrium data in Table 8.P.1 for an ethanol-water system at 101.3 kPa.

The distillation column used to separate this binary mixture which forms an azeotrope (azeotrope composition is 0.8943) is operating at 1 atm. The feed flow rate is 2 kgmol/min; the feed, introduced as saturated liquid, has 25 mole% ethanol. The column is operating with a total condenser, a distillate alcohol mole fraction of 0.85 and a reflux ratio of 1.5; the column has a partial reboiler and a bottoms composition of 0.01 alcohol mole fraction.

Table 8.P.1.

Temperature (°C)	95.5	89	86.7	85.3	84.1	82.3	80.7	79.7	78.74	78.41	78.15
Mole fraction of ethanol											
Vapor	0.17	0.3891	0.4375	0.4704	0.5089	0.5580	0.6122	0.6599	0.7385	0.7815	0.8943
Liquid	0.019	0.0721	0.0966	0.1238	0.1661	0.2608	0.3965	0.5198	0.6763	0.7472	0.8943

(1) Determine the molar flow rates of the distillate and the bottoms products.
(2) Obtain the equations for the two operating lines and plot them in the $x_{iv} - x_{il}$ plot of the alcohol-water system.
(3) Graphically determine the number of ideal stages needed to achieve the desired separation.
(Ans. (1) $W_{tld} = 0.606$ kg mol/min; $W_{tlb} = 1.394$ kgmol/min; (3) 14.)

8.1.12 In Example 8.1.9, the minimum reflux ratio was estimated by two methods, one analytical, employing equation (8.1.175), and the other based on drawing a line between the intersection point B and point A on the 45° line. There is another simpler graphical method available which does not require estimation of the slope of the line AB.
(1) Employ such a method to determine the value of R_{min} for Example 8.1.9.
(2) Next assume that the actual reflux ratio is $1.4R_{\text{min}}$. Determine the number of stages in the enrichment section for this separation, knowing that there is an uncertainty to the tune of a stage at the intersection point between the two operating lines.
(Ans. (1) $R_{\text{min}} = 1.5$; (2) 7.)

8.1.13 Consider the case where the equilibrium curve is concave downward over the whole composition range and the operating lines are straight. Obtain the following expression for the minimum reflux ratio R_{min} when the feed stream into a distillation column is saturated vapor:

$$R_{\text{min}} = \frac{1}{(\alpha_{ij} - 1)}\left(\frac{\alpha_{ij}\, x_{ild}}{x_{if}} - \frac{(1 - x_{ild})}{(1 - x_{if})}\right) - 1.$$

8.1.14 Consider benzene-toluene separation by distillation in a column at 1 atm. A feed containing 40-60 mole% benzene-toluene is introduced at 4 kgmol/min. The distillate is 98 mole% benzene; the bottoms product contains 98 mole% toluene. The properties of the feed are as follows: the latent heat of vaporization of the feed at the feed temperature is 32 000 J/gmol; the difference between the enthalpy of the saturated vapor at the feed plate condition and the feed enthalpy at the feed plate temperature is 16 000 J/gmol. The reflux ratio is $1.4R_{\text{min}}$.
(1) Determine the molar flow rates of the distillate and the bottoms product. (Ans. $W_{tld} = 1.58$ kgmol/min; $W_{tlb} = 2.41$ kgmol/min.)
(2) Determine the equations of the operating lines and plot them. (Ans. ($\overline{x}_{1v(n+1)} = 0.765\,\overline{x}_{1\ell n} + 0.235\,x_{1\ell d}$.)
(3) Using the McCabe–Thiele method, obtain the number of ideal stages needed for this separation. (Ans. 14.)
(4) Determine the actual number of trays for this column if the feed mixture viscosity is 0.28 cp and $\alpha_{\text{benzene-toluene}} = 2.5$. The VLE data for the system are available in Example 8.1.13. (Ans. 25.)

8.1.15 In the packed tower based distillation of benzene-toluene illustrated in Example 8.1.13, only the height of the enriching section was calculated. Calculate the height of the stripping section of the packed tower. (Ans. 11.5 ft.)

8.1.16 A small molecular weight drug present in a clarified fermentation broth at the level of 2 g/liter is being extracted into an organic solvent in a 150 cm tall packed column of diameter 8 cm. The feed flow rate is 2 cm^3/s and the solvent rate is 1 cm^3/s. It was observed that 95% of the drug was being extracted into the organic solvent. Obtain an estimate of the mass-transfer parameter $K_{iRu}a$ for this system, given $m_i = 10$. (Ans. $K_{iRu}a = 9.2 \times 10^{-4}\ \text{s}^{-1}$.)

8.1.17 (a) A fermentation broth containing 3 mg/liter of an antibiotic is being processed at 400 liter/hr. The antibiotic is to be extracted using the solvent amyl acetate whose flow rate will be 40 liter/hr. The partition coefficient m_i (= 50) of the antibiotic is defined as $u_{iE} = m_i u_{iR}$, where u_{iR} is the mass fraction of the antibiotic in the aqueous fermentation broth and u_{iE} is the corresponding value in the solvent, amyl acetate. If you need to extract 90% of the antibiotic into the solvent, determine the number of stages required in an idealized staged countercurrent extraction. (Ans. 2.)

(b) After you have extracted the antibiotic into amyl acetate (as in part (a) flowing at 40 liter/hr, you have to strip the antibiotic into water at *pH* 6; under such conditions, the value of m_i (= ratio of organic- over aqueous-phase concentrations) is 0.15. You have three stages available in an idealized staged counter-current extraction cascade. You must achieve 95% antibiotic recovery. What should be the water flow rate? (Ans. 13.8 liter/hr.)

8.1.18 Zirconium (Zr) and zirconium based alloys are preferred materials for cladding and structural elements in certain types of nuclear power reactors due to their very low neutron absorption cross section. However, in nature zirconium is invariably contaminated by hafnium (Hf), which has a very high neutron absorption cross section (Benedict *et al.*, 1981, p. 318). Zirconium is purified from hafnium present in its aqueous solution in 3.5 M $NaNO_3$ and 3 M HNO_3 by extracting it in 2.25 M tributyl phosphate solution (which also contains 1.6 M HNO_3). The feed aqueous solution contains 0.123 M Zr, 0.00246 M Hf. The values of κ_{i1} for Zr between the organic phase ($j = 1$) and the aqueous phase is 1.2. Calculate the number of ideal equilibrium stages needed in a countercurrent extraction cascade if the organic solvent flow rate to the aqueous feed flow rate ratio is (a) 1.2, (b) 1 for 98% fractional recovery of zirconium. (Ans. (a) 8; (b) 13.)

8.1.19 An important goal of the zirconium extraction example given in Problem 8.1.18 is to obtain a solvent extract stream containing zirconium substantially purified of hafnium. It is known that the value of κ_{i1} of Hf is 0.11. Determine the value of the separation factor for zirconium vis-à-vis hafnium in the extract stream for case (b) of Problem 8.1.18. This separation factor has also been called the decontamination factor (Benedict *et al.*, 1981, p. 185). (Ans. $\alpha_{Zr\text{-}Hf} = 8.9$.)

8.1.20 Example 8.1.17 dealt with fractional extraction in an extraction–stripping cascade for the Zr-Hf feed mixture, but determined only the values for Zr in the extract and the raffinate. Determine the concentration of the Hf in the extract and in the raffinate for the same problem if the value of $m_i|_{Hf} = 0.12$. (Ans. Molar concentration of Hf in the extract, 0.0000059 M; in the raffinate, 0.00121 M.)

8.1.21 A commom problem in gas separation is to separate a ternary gas/vapor mixture. One can use a continuous countercurrent adsorption technique using two columns such that, for each column, the principle is applied for a binary separation. Consider two countercurrent moving bed columns for a ternary separation of (1 : 1 : 1) mixture of compounds 1, 2 and 3 (Figure 8.P.1).

The individual species flow rates in the feed are w_{1f}, w_{2f} and w_{3f}. It is known that compound 1 is most strongly adsorbed in the liquid coating acting as the adsorbent and compound 3 is least adsorbed. On the basis of the flow rates on the desorbent gas phase and the adsorbent, determine suitable criteria, and on that basis indicate the nature of the three products, A, B and C, from the two-column system. Identify the desorbent gas flow rates and adsorbent liquid flow rates in various columns and sections of the columns beyond those identified. The density of any species *i* in the gas phase of any section of the two-column systems is indicated by ρ_{igA}, ρ_{igB} and ρ_{igC}; those for the density of any species *i* in the solvent coating (the adsorbent) are $\rho_{i\ell A}$, $\rho_{i\ell B}$ and $\rho_{i\ell C}$ for sections A, B and C, respectively. The volumetric flow rates of the gas in the two top sections are Q_{gA} and Q_{gB}; the liquid coating flow rates in columns A and B are $Q_{\ell B}$ and $Q_{\ell A}$, respectively.

8.1.22 In countercurrent dialysis, a particular salt species (*i*) is to be removed from the feed solution of a protein into the dialyzing solution, which does not have this salt to start with. Develop an expression for $C_{if0}/\overline{C}_{ifL}$ in terms of λ, Q_f and Q_d. (Ans. $(C_{if0}/\overline{C}_{ifL}) = [Q_d \exp(\lambda(1 - (Q_f/Q_d))) - Q_f]/[Q_d - Q_f]$.)

8.1.23 To implement buffer exchange, coutercurrent dialysis has been explored to eliminate 0.05 M sodium phosphate from a 1 mg/ml solution of bovine serum albumin (BSA) and 25 mM Tris buffer at a *pH* of 7 (via NaOH). The dialyzing buffer composition is 25 mM Tris and the *pH* is 7 (via HCl). The dialysate flow rate is 100 ml/min. Two membrane modules are in series, each having a surface area of 0.2 m^2. The value of $(C_{if0}/\overline{C}_{ifL})$ for sodium phosphate at a feed flow rate of 10 cm^3/min was found to be 246. Determine the value of the overall mass-transfer coefficient $\overline{K}_o$. (Ans. 1.5×10^{-2} cm/min.)

8.1.24 Countercurrent dialysis is being utilized to remove ammonium sulfate from a feed solution of bovine serum albumin, 1 mg/ml, containing 25 mM Tris at $pH = 7$. This sulfate ion is to be removed into a 25 mM Tris buffer solution at the same *pH*. The dialysate flow rate is 100 ml/min. The membrane module has a membrane surface area of 0.2 m^2. Determine the value of $(C_{if0}/\overline{C}_{ifL})$ for two situations:

(a) four membrane modules are in series, the overall mass-transfer coefficient for the sulfate ion being 0.012 cm/min;

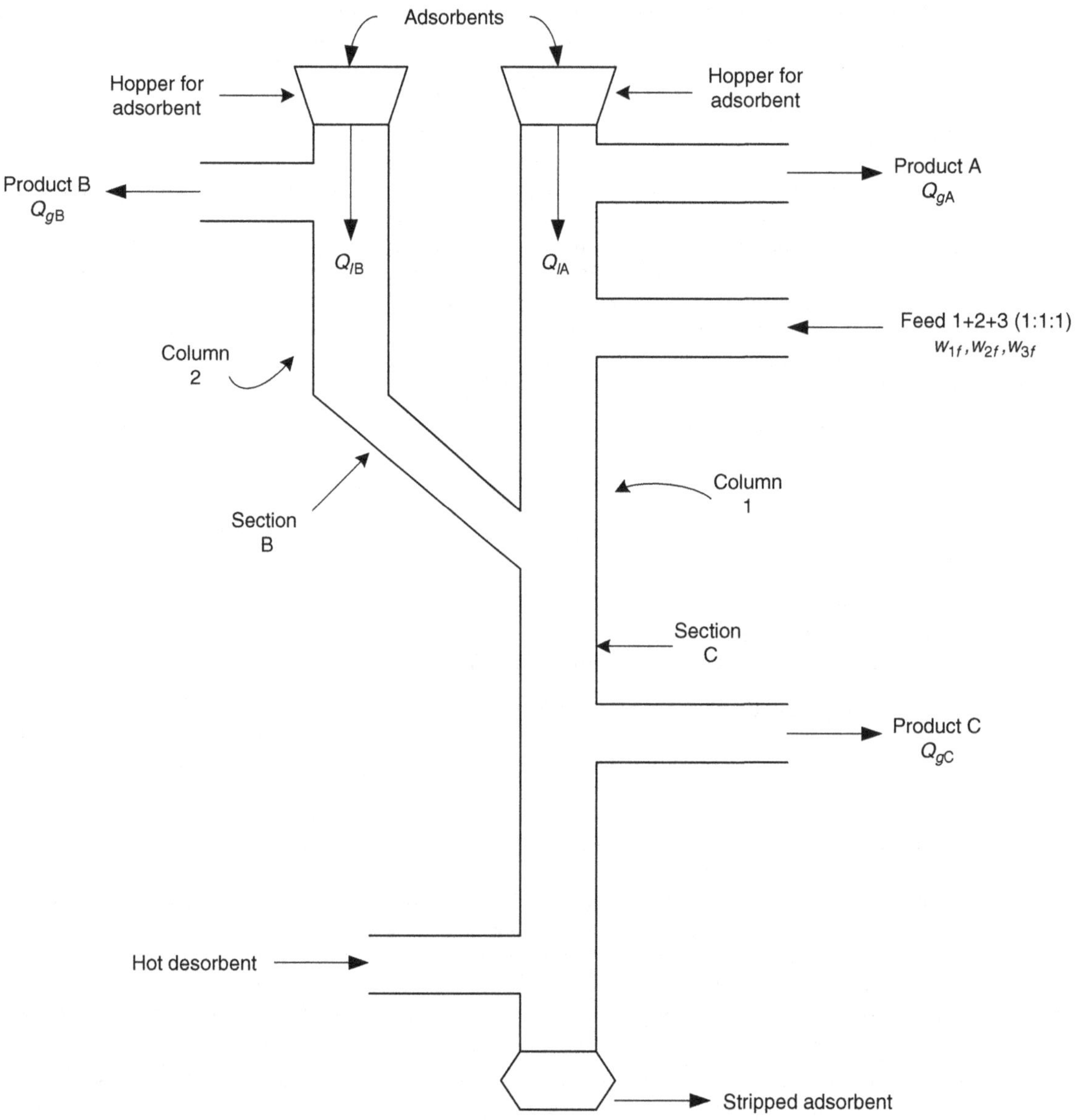

Figure 8.P.1.

(b) one membrane module only, the overall mass-transfer coefficient being 0.026 cm/min. You are given that the feed flow rate is 10 ml/min.
(Ans. (a) 6210; (b) 119.5.)

8.1.25 Consider a dialyzer for a solute i where the dialysate side is well-mixed. Show that the following expression for mass-transfer efficiency η_i is valid for the case of well-mixed dialysate flow:

$$\eta_i = \frac{1 - \exp(-\lambda)}{1 + \dfrac{Q_f}{Q_d}(1 - \exp(-\lambda))},$$

where $\lambda = \overline{K}_{ic} A_m / Q_f$.

8.1.26 Salt is to be removed from a solution containing 100 g/liter of salt and 200 g/liter of raffinose by dialysis in a hollow fiber dialyzer operating countercurrently. The overall mass-transfer coefficient for the salt was determined in the dialyzer with 200 cm^3/min feed flow rate and 500 cm^3/min pure water dialysate flow rate to be 0.0415 cm/min. If 90% of the salt is to be removed for the same feed and dialysate flow rates, what is the hollow fiber membrane area required? Assume that the overall transport coefficient remains unchanged due to any changed conditions. (Ans. 14 860 cm^2.)

8.1.27 An aqueous solution has to be desalted by electrodialysis (ED). The cell pairs in the ED stack of membranes are connected in countercurrent arrangement (see Figure 8.1.47(b)). The aqueous feed solution flow rate per cell pair is 108 cm^3/min. The ED stack is being operated at 80% of the limiting current density such that $(i_{\ell\text{im}}/N_{\text{di}\ell}) = 0.25$, where i_{lim} is in amp/cm^2 and $N_{\text{di}\ell}$ is the normality of the aqueous solution entering the diluate chamber. Calculate the fractional demineralization $f\,(= (N_{\text{di}\ell,0} - N_{\text{di}\ell,L})/\,N_{\text{di}\ell,0})$ being achieved; here $N_{\text{di}\ell,0}$ and $N_{\text{di}\ell,L}$ are the normality of the salt solution in the diluate at the cell-pair inlet and exit, respectively. The cell-pair membrane area is 600 cm^2; $\eta_{IF} \cong 1$. Would the result be different if you employ the scheme of Figure 8.1.47 (c) showing parallel flow? State your assumptions. (Ans. $f = 0.497$.)

8.1.28 In an ED stack containing five cell pairs, the brine feed flow rate to the diluate chambers is 0.60 liter/min. The value of $((i_{\text{lim}}/N_{\text{dil}})$ is 0.20, where i_{lim} is in amp/cm^2 and N_{dil} is the diluate normality. The current utilization factor $\eta_{IF} = 0.9$. Determine the membrane area in the cell pair if the value of fractional demineralization, f, is 0.395, where $f = (N_{\text{dil},0} - N_{\text{dil},L})/N_{\text{dil},0}$. (Ans. 538 cm^2.)

8.1.29 Consider the demineralization of a 3000 ppm NaCl solution to 400 ppm. This is being implemented using a couple of ED membrane stacks in series. The diluate coming out of the ith membrane stack, $(N_{\text{dil},L})_i$, becomes the feed brine to the next $(i + 1)$th membrane stack $(N_{\text{dil},0})_{i+1}$. Each membrane stack is operating in cocurrent mode between the diluate and the concentrate streams. Suppose there are four membrane stacks in series and that the fractional demineralization f_i for the ith ED membrane stack is constant and equal for each stack $(= f)$. Determine the value of f being achieved in any membrane stack. (Ans. 0.395.)

8.1.30 Consider a countercurrent gas permeator as well as a cocurrent gas permeator for the separation of a particular binary gas mixture at feed pressure P_f and permeate pressure P_p. Assume that there are no pressure drops on either the feed side or the permeate side. If the pressure ratio $\gamma \to 0$, show that the separation achieved would be independent of the flow pattern. Assume a symmetric membrane with constant values of permeances. Will the situation be any different if we used the crossflow permeator of Section 7.2.1.1?

8.1.31 Consider equation (8.1.427) describing the variation of x_{A2} with membrane area in countercurrent membrane gas separation in a hollow fiber device with no longitudinal diffusion/dispersion and symmetrical hollow fiber membrane. The feed-gas composition varies from x_{Af} to x_{A2L} as per Figure 8.1.51(a).

(1) Develop a relation for the length L of the hollow fibers as a product of two quantities, the length of a membrane transfer unit (*LMU*) and the number of membrane transfer units (*NMU*) in the manner of equations (8.1.50) and (8.1.54a). The gas flow rate changes quite a bit along the membrane length. Assume the pressure on each side to be constant.

(2) Show that if the pressure ratio γ is very small,

$$NMU = \left(1/\left(\alpha^*_{AB} - 1\right)\right) \ell n \left[x_{Af}/x_{A2L}\right)\left((1 - x_{A2L})/(1 - x_{Af})\right)].$$

(3) Assume that the variation of W_{t2} with z may be represented over the locations z_a to z_b as follows:

$$\int_{z_a}^{z_b} \frac{dz}{W_{t2}} = \frac{z_b - z_a}{W_{t2}|_a - W_{t2}|_b} \left[\ell n\, W_{t2}|_a - \ell n W_{t2}|_b\right].$$

Obtain a simplified expression for *LMU*.

8.1.32 Figure 8.1.4(b) illustrates two membrane gas permeation devices coupled together with a compressor at the top. Suppose the whole system is filled with a binary gas mixture and the two sides of the membrane are at two different pressures as shown. No permeate product is being withdrawn, however, so there is no reject gas mixture. Further, there is no feed. It is a closed system, except for the energy input by the compressor at the top. This is similar to the operation of a distillation column at total reflux. The gas flowing on the high-pressure side must completely permeate to the low-pressure side; the overall stage cut is 1. However, local compositions on two sides become identical, but permeation continues since the partial pressures are different. Based on the results of part (1) of Problem 8.1.31, develop the following expression for *NMU* at total reflux:

$$NMU = \frac{1}{(1-\gamma)(\alpha^*_{AB}-1)}\,\ell n\left\{\left(\frac{x_{A2T}}{x_{A2L}}\right)\left(\frac{1-x_{A2L}}{1-x_{A2T}}\right)\right],$$

where x_{A2T} and x_{A2L} are the mole fractions of species A at the compressor outlet and the reject gas mixture location on the high-pressure side.

8.1.33 Reverse osmosis based concentration of a dilute solution is to be implemented in a hollow fiber module with the high-pressure feed flowing through the fiber bore. To develop an optimum design, modeling of pure water permeation is to be implemented. You are given the following information:

(1) Poiseuille flow exists inside the fiber bore (for pressure drop calculation purposes);
(2) the active length of the hollow fiber is l; the inactive length at the tube sheet is l_s;
(3) the feed enters the fiber bore at pressure P_{1i} and volumetric flow rate Q_f per fiber;
(4) the shell-side pressure may be assumed to be constant at P_3;
(5) the shell-side fluid is in countercurrent flow;
(6) the high-pressure concentrate/reject leaves the fiber bore at $P_3 + \Delta P_{sp}$;
(7) P_3 is constant throughout the shell length;
(8) fiber dimensions are r_i, r_o;
(9) the pure water permeability through the hollow fiber membrane is A; water viscosity is μ_w;
(10) ignore tube sheet length l_s for pressure drop calculations.

Write down the governing equations for (a) pure water permeation; (b) pressure inside the fiber bore. Identify the boundary conditions in terms of P_1. Solve the problem of obtaining the expressions for water permeation velocity as a function of fiber length and the pressure inside the fiber bore.

8.1.34 The radius of a countercurrent gas centrifuge separating a $U^{235}F_6$ and $U^{238}F_6$ mixture is 6 cm and its total length is 48 cm. At 27 °C, the radial separation factor at any axial location was found to be 1.038. Determine the optimum separation factor achievable at total reflux along the length of the CGC. You are given that the flow pattern efficiency $E_f = 0.8$. (Ans. 1.213.)

8.2.1 Consider micellar electrokinetic chromatography.

(1) Specify the condition under which the micelles act as a stationary phase and chromatographic separations take place. Obtain an expression for the retention time of any solute under such a condition. Assume the micelle to be built out of sodium dodecyl sulfate.
(2) Suppose electroosmotic flow is completely suppressed. Using an appropriate criterion to characterize this condition, obtain an expression for the retention time of any solute i. If sodium dodecyl sulfate is employed, describe how the migration velocity of one analyte will vary vis-à-vis another one having a smaller k'_{imc}.

8.3.1 In a methanol–water distillation column operating at atmospheric pressure with the feed introduced at 58 °C, the liquid and vapor flow rates at the top and bottom sections are close to those used in Treybal (1980, p. 414): $W^t_{tv} = 162$ kgmol/hr; $W^t_{t\ell} = 79$ kgmol/hr; $W^b_{tv} = 171$ kgmol/hr; $W^b_{t\ell} = 304$ kgmol/hr. The slope of the equilibrium curve at $x_{\text{methanol},\ell} = 0.8$ is 0.48; the slope of the equilibrium curve at $x_{\text{methanol},\ell} = 0.01$ is 1.785. In an Oldershaw column, the value of E_{MV} determined at $x_{\text{methanol},\ell} = 0.8$ in the enriching section was 0.6; the corresponding value determined at the bottom of the stripping section at $x_{\text{methanol},\ell} = 0.015$ was 0.5.

(1) Calculate the values of E_{MV} in the large distillation column in the enriching and stripping sections, assuming plug flow on the plate.
(2) Determine the values of appropriate mass-transfer parameters of the system on the large plate if the liquid height is 5.5 cm.

8.3.2 Consider the conversion of equation (8.3.18) for a crossflow plate in a distillation column to one for a crossflow gas permeation stage (Section 7.2.1.1). Replace both phases 1 and 2 by gas phases; further, assume that phase 2 is the feed-gas phase flowing at pressure P_f on the feed side of the gas permeation stage (see Figure 7.2.1(a)) and that phase 1 is the gas phase generated by permeation on the other side of the membrane. Therefore there is no input of gas phase 1 from the plate below $(n + 1)$, this stage being the only one under consideration. Obtain equation (7.2.5) for a membrane of width W and length L in the crossflow stage. Neglect dispersion in both the x- and z-directions.

9

Cascades

We have studied in Section 8.1 how a number of double-entry separation stages can be connected together in a countercurrent fashion. We have also learnt that such an arrangement achieves a far greater extent of separation in separation processes such as distillation, extraction, absorption and adsorption compared to what can be achieved in a single stage. Such multistage configurations are generally identified as "cascades," and in the case of configurations of an overall countercurrent flow as "countercurrent cascades." Such countercurrent cascades can have significant variations, for example ideal cascades, squared off cascades, tapered cascades, etc. Further, there can be novel countercurrent cascades with single-entry separation stages. In addition, there can be other types of connections between a number of single stages, resulting in, for example, crosscurrent cascades, two-dimensional cascades, etc.

We will briefly focus in Section 9.1 of this chapter on many such configurations of cascades, beginning with countercurrent cascades. Special attention will be paid to single-entry separation stages in, for example, the configuration of an ideal cascade. These cascades are generally employed to enhance substantially the separation achieved in a single stage for a binary mixture. Such cascades can also reduce the amount of energy or mass-separating agent needed for the separation of a binary mixture. In the separation of a multicomponent mixture by distillation, a number of such cascades may be joined together in particular ways - Section 9.2 will provide a brief introduction to this topic. Section 9.3 briefly considers cascades for multicomponent mixture separation involving other separation processes.

9.1 Types of cascades

Of the variety of cascades identified we will consider first countercurrent cascades of various types.

9.1.1 Countercurrent cascades, ideal cascade

We have already studied simple countercurrent cascades built out of double-entry separation stages via Figures 8.1.5 (gas–liquid scrubber), 8.1.19 (distillation column), 8.1.35 (solvent extraction), 8.1.37 (solvent extraction), 8.1.43, 8.1.44 (adsorption) and 8.1.48 (emulsion liquid membrane). One such cascade is shown in Figure 9.1.1(a), without reference to any particular separation process, with the condenser and the reboiler representing refluxing devices. Note that the numbering of the stages here is different from that in Chapter 8. If, however, a single-entry separation stage is employed, the countercurrent cascade may be configured as shown in Figure 9.1.1(b). There are two major sections on two sides of the feed introduction point - the enriching section at the top and the stripping section below. *Note*: The stages are numbered 1 to N, with N being the top most stage in the enriching section, similarly they are numbered 1 to M in the stripping section.

In any single-entry separator stage n in the enriching section of the cascade of Figure 9.1.1(b), the feed stream has a composition x_{ifn} and the stage produces two product streams: the light fraction (heads), having a composition of x_{i1n} leaving the stage of the top; the heavy fraction (tails) of composition x_{i2n} leaving the stage from the side. If species i is the lighter species and if α_{ij}^{ht} is greater than 1 (equation (2.2.2e)), then $x_{i1n} > x_{i2n}$. Obviously the product stream from the cascade top is highly enriched in species i since the species i enriched stream from any stage n becomes the feed to an upper stage and so on. On the other hand, the heavy fraction (tails) stream ($\sim x_{i2n}$) from stage n depleted in species i and enriched in species j should be introduced lower into the multistage cascade, which produces at the bottom a heavy product highly enriched in the heavier species j. The question to be answered now is: In which locations of the cascade does one introduce these two product streams from stage n?

In one of the most common configurations (Figures 9.1.1(b) and (c)), the light fraction of composition x_{i1n} from

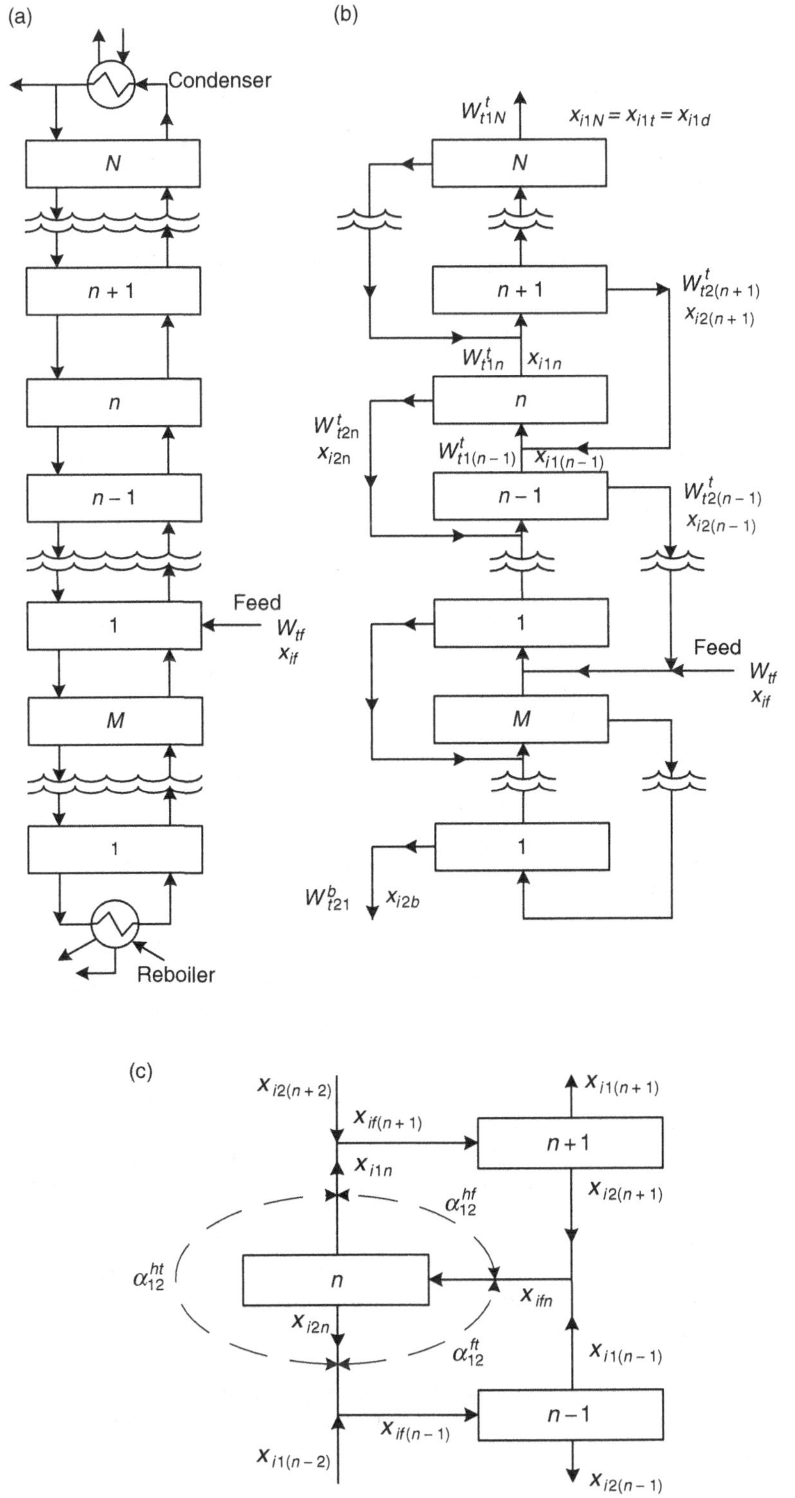

Figure 9.1.1. (a) Countercurrent cascade of double-entry separating elements as in distillation (e.g.). (b) Countercurrent cascade of single-entry separating elements. (c) Compositions of various streams for stages (n − 1), n *and* (n + 1) *and separation factors.*

stage n is introduced as the feed $x_{if(n+1)}$ to the next higher stage $(n + 1)$. The heavy fraction of composition x_{i2n} from stage n is mixed with the light fraction leaving the lower $(n - 2)$ stage, $x_{i1(n-2)}$, and the combined stream enters stage $(n - 1)$ as the feed, $x_{if(n-1)}$ (Figures 9.1.1(b) and (c)). In the configuration of the *ideal cascade*, the mixing of these two streams is implemented under the condition of

$$x_{i1(n-2)} = x_{if(n-1)} = x_{i2n}, \tag{9.1.1}$$

so that the compositions of both streams being mixed to form the feed for stage $(n - 1)$ are identical to start with. When we consider the feed to stage n, this relation is changed to

$$x_{i1(n-1)} = x_{ifn} = x_{i2(n+1)}; \tag{9.1.2}$$

the feed to stage n comes from the blending of the heavy fraction from stage $(n + 1)$ above and the light fraction from stage $(n - 1)$ below, both having the same composition (Figure 9.1.1(c)).

We will learn in the next chapter that separation of any mixture requires an energy input, resulting in two fractions having different compositions, both being different from that of the feed. If we mix these two fractions, or any two other fractions having different compositions as such, we lose the energy input used to create these two fractions of different compositions. Therefore, in the ideal cascade, when two streams are mixed together to create the feed stream for any stage n, both streams must have the same composition to prevent loss any of separative work spent earlier. This condition, namely relations (9.1.1) and (9.1.2), is called the "no-mixing condition." There are additional assumptions, namely that the two streams being mixed together are of the same phase and at the same pressure and temperature. Therefore perhaps a more general criterion for "no-mixing condition" should be (see Figure 9.1.1(c))

$$\mu_{i1(n-1)} = \mu_{ifn} = \mu_{i2(n+1)}, \tag{9.1.3}$$

based on the equality of the chemical potential of each species in the two streams being mixed to develop the feed to any stage n. (It is conceivable that two streams having different temperatures, pressures and compositions can have the same chemical potential for i.) For a broad introduction to ideal cascades, see Pratt (1967, chap. II) and Benedict *et al.* (1981, pp. 658–701).

Single-entry separation stages are commonly used in most membrane separation processes (exceptions include dialysis, electrodialysis, Donnan dialysis, most liquid membrane processes, sweep stream based gas permeation and pervaporation processes and membrane based contacting processes); particle separation processes also employ, in general, single-entry separation stages (an exception is the Venturi scrubber, etc.). In the case of pressure driven membrane separation processes, the three stages $(n - 2)$, $(n - 1)$ and n participating in producing the feed for the $(n - 1)$ stage in an ideal cascade will appear as shown in Figure 9.1.2. The no-mixing condition in this case is (9.1.1). Further, the pressure of the permeate stream from each stage has to be raised to the level of the feed stream for the stage where it is being introduced after mixing with the other stream. The earliest and largest application of such cascades involved separation of uranium isotopes by gaseous diffusion process.

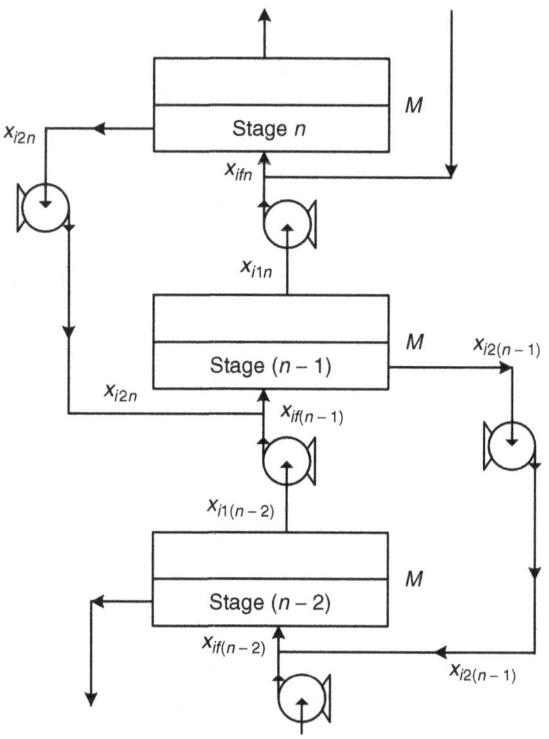

Figure 9.1.2. Configuration of three contiguous single-entry membrane separation stages for an ideal cascade.

The preceding discussion presented for an ideal cascade presumed that we are dealing with single-entry separation stages. However, from an overall system perspective, a distillation column with reflux and reboiler is a single-entry separator with one feed stream and two product streams. Therefore one could have many stages in the ideal cascade of Figure 9.1.1(b) in which each stage consists of a countercurrent distillation column containing many stages, each stage being a double-entry separator in the manner of the individual stages in Figure 8.1.19. Such a situation is rarely encountered in processes involving distillation (except in heavy-water distillation where the separation factor between H_2O and D_2O is very low). However, there are other examples in isotope separation where an individual stage in Figure 9.1.1(b) is not like a membrane unit or a particle separation unit.

Consider Figure 8.1.52, which shows countercurrent gas centrifuges where, at each location, we have two different gas streams in countercurrent flow, just like that in the continuous contact packed tower distillation column of Figure 8.1.32. Yet on an overall basis it is a single-entry separator with a feed in and two product streams going out at two ends. We learnt via Example 8.1.22 that the overall separation factor achieved between the two product streams is going to be at the most 1.66. However, in many isotope separation problems we need very high enrichment, requiring many such countercurrent gas centrifuges connected as individual stages in an ideal cascade.

Example 9.1.1 In the cascades of Figure 9.1.1(b) and 9.1.2 built out of single-entry separation stages, the highest concentration/purification of the lighter species or more selectively permeable species is achieved at the top of the cascade. Consider now a cascade of single-entry membrane separation stages where the stage number increases upward, as in Figures 9.1.1(b) and 9.1.2. However, the highest concentration/purification of the more permeable species is achieved at the bottom of the cascade. Draw such a cascade and obtain the criteria similar to (9.1.1) and (9.1.2) describing the "no mixing condition."

Solution Figure 9.1.3 illustrates the cascade desired, in which the permeate concentrated in the more permeable species is taken downward in the cascade and becomes the feed to a lower stage. Therefore if the stage number increases upward, unlike that in Figures 9.1.1(b) and 9.1.2, where the most permeable species or the lightest species is concentrated or purified at the top of the cascade, here the bottom most stage yields in the permeate the highest concentration of the more permeable species. In terms of the requirement of the "no mixing condition" in an ideal cascade, consider the feed to stage n in Figure 9.1.3:

$$x_{ifn} = x_{i1(n+1)} = x_{i2(n-1)}. \tag{9.1.4}$$

Alternatively

$$x_{i1(n+1)} = x_{ifn} = x_{i2(n-1)}, \tag{9.1.5}$$

i.e. the permeate from stage $(n+1)$, $x_{i1(n+1)}$, is sent down by one stage to become the feed to stage n; this means that stages lower in the cascade operate with feed streams more enriched in the more permeable species. Correspondingly, the reject from stage $(n - 1)$, lower in the cascade, $x_{i2(n-1)}$, is brought up as the feed for stage n. In this context, the "no mixing condition" for stage $(n+1)$ is $x_{i1(n+2)} = x_{if(n+1)} = x_{i2n}$. See Lightfoot (2005) for a membrane cascade to this end.

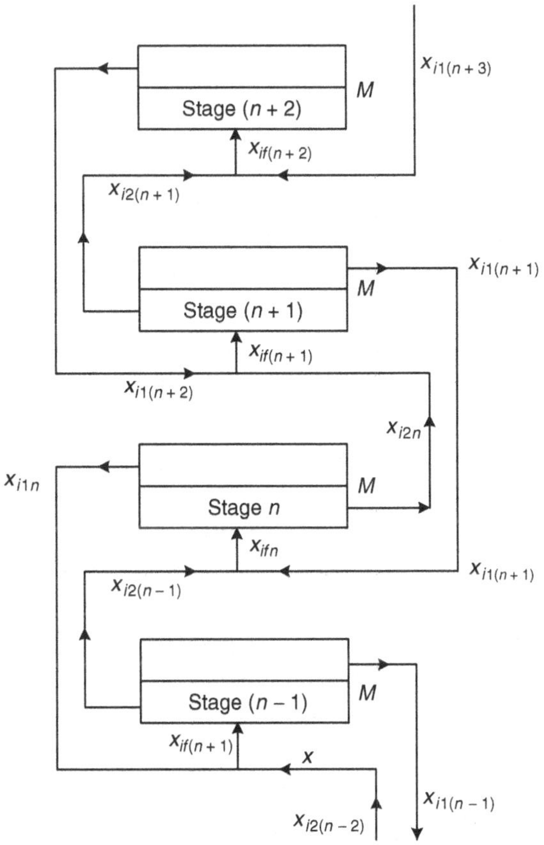

Figure. 9.1.3. Ideal cascade of single-entry membrane separation stages where the highest concentration/purification of the more permeable species is achieved at the lowest point in the cascade.

9.1.1.1 *Number of stages required in an ideal cascade*

Consider Figure 9.1.1(b) in greater detail. There are two major sections: an enriching section, containing N stages above the feed entry point, and a stripping section, containing M stages below the feed entry point, just like other countercurrent cascades of double-entry separation elements (see Figures 8.1.19(b) and 9.1.1(a)). Focus now on Figure 9.1.1(c) containing any three stages $(n - 1)$, n and $(n + 1)$ in the enrichment section of total N stages. From definitions (2.2.2a), (2.2.3) and (2.2.4) for the stage separation factor α_{12}^{ht}, the heads separation factor α_{12}^{hf} and the tails separation factor α_{12}^{ft}, respectively, we obtain, for stage n for a binary system of species 1 and 2,

$$\alpha_{12}^{ht} = \frac{x_{11n}\, x_{22n}}{x_{12n}\, x_{21n}} = \frac{x_{11n}\, x_{22n}}{x_{21n}\, x_{12n}} = \frac{X_{11n}}{X_{12n}}, \qquad X_{11n} = \alpha_{12}^{ht}\, X_{12n}, \tag{9.1.6}$$

where X_{11n} and X_{12n} are the corresponding mole ratios (see definitions (1.4.6) and (1.4.7a)). *Note*: This is a linear relation for constant α_{12}^{ht}. Similarly, we can express definitions for α_{12}^{hf} and α_{12}^{ft} as

$$\alpha_{12}^{hf} = \frac{x_{11n}\, x_{2fn}}{x_{1fn}\, x_{21n}} = \frac{x_{11n}\, x_{2fn}}{x_{21n}\, x_{1fn}} = \frac{X_{11n}}{X_{1fn}}; \tag{9.1.7a}$$

$$\alpha_{12}^{ft} = \frac{x_{1fn}\, x_{22n}}{x_{12n}\, x_{2fn}} = \frac{x_{1fn}\, x_{22n}}{x_{2fn}\, x_{12n}} = \frac{X_{1fn}}{X_{12n}}; \tag{9.1.7b}$$

$$X_{11n} = \alpha_{12}^{hf}\, X_{1fn}; \qquad X_{1fn} = \alpha_{12}^{ft}\, X_{12n}. \tag{9.1.8}$$

(These are additional linear relations.) From relations (9.1.6) and (9.1.8), it is clear that

$$\alpha_{12}^{ht} = \alpha_{12}^{hf}\, \alpha_{12}^{ft}. \tag{9.1.9}$$

This result is valid, in general, from the definitions of α_{12}^{ht}, α_{12}^{hf} and α_{12}^{ft} (see relation (2.2.5)).

Consider now stage n in an ideal cascade. From relation (9.1.2), we obtain, for a binary system,

$$\frac{x_{11(n-1)}}{\left(1 - x_{11(n-1)}\right)} = \frac{x_{1fn}}{\left(1 - x_{1fn}\right)} = \frac{x_{12(n+1)}}{\left(1 - x_{12(n+1)}\right)};$$

$$X_{11(n-1)} = X_{1fn} = X_{12(n+1)}. \tag{9.1.10}$$

Apply this result to relation (9.1.8) to obtain

$$X_{11n} = \alpha_{12}^{hf} X_{1fn} = \alpha_{12}^{hf} X_{12(n+1)}. \tag{9.1.11}$$

Assume now that α_{12}^{ht}, α_{12}^{hf} and α_{12}^{ft} are independent of n. Considering stage $(n + 1)$, we find for its feed in an ideal cascade

$$X_{11n} = X_{1f(n+1)} = \alpha_{12}^{ft}\, x_{12(n+1)}. \tag{9.1.12}$$

From these two relations, we see that

$$\alpha_{12}^{hf} = \alpha_{12}^{ft}, \tag{9.1.13}$$

which, when substituted in to the basic result (9.1.9) for a single-entry separator, yields

$$\alpha_{12}^{hf} = \alpha_{12}^{ft} = \left(\alpha_{12}^{ht}\right)^{1/2} \tag{9.1.14}$$

for an ideal cascade.

These basic results allow us to determine the number of stages needed in the stripping section and the enriching section to achieve a certain difference in the composition from the top product x_{11d} to the bottoms product x_{12b} in the ideal cascade having a feed of composition x_{1f}. For the enriching section, we can start from *stage 1*:

$$X_{111} = \alpha_{12}^{hf} X_{1f1} = \alpha_{12}^{hf}\, X_{1f} = \alpha_{12}^{hf}\, X_{122}. \tag{9.1.15a}$$

Similarly for *stage 2*:

$$X_{112} = \alpha_{12}^{hf} X_{1f2} = \alpha_{12}^{hf}\, X_{111} = \left(\alpha_{12}^{hf}\right)^2 X_{1f}. \tag{9.1.15b}$$

For *stage 3*:

$$X_{113} = \alpha_{12}^{hf} X_{112} = \left(\alpha_{12}^{hf}\right)^3 X_{1f}, \tag{9.1.15c}$$

and for *stage N*:

$$X_{11N} = \alpha_{12}^{hf} X_{11(N-1)} = \left(\alpha_{12}^{hf}\right)^N X_{1f}. \tag{9.1.15d}$$

Equation (9.1.15d) may be written as

$$N = \frac{\ln(X_{11N}/X_{1f})}{\ln \alpha_{12}^{hf}} = \frac{\ln(X_{11d}/X_{1f})}{\ln\left(\alpha_{12}^{hf}\right)}, \tag{9.1.16}$$

providing an estimate of the number of stages N in the enriching section of an ideal cascade of single-entry separation stages with a feed x_{1f} and producing a top product $x_{11N} = x_{11d}$.

We can follow a similar procedure for the number of stages M in the stripping section. For *stage 1*:

$$X_{1f1} = \alpha_{12}^{ft} X_{12b} = X_{122}; \qquad X_{111} = \alpha_{12}^{hf}\, X_{1f1}; \tag{9.1.17a}$$

stage 2:

$$X_{112} = \alpha_{12}^{hf} X_{1f2} = \alpha_{12}^{hf}\, X_{111} = \left(\alpha_{12}^{hf}\right)^2$$
$$X_{1f1} = \left(\alpha_{12}^{hf}\right)^2 \left(\alpha_{12}^{ft}\right) X_{12b} = X_{1f3}; \tag{9.1.17b}$$

stage 3:

$$\begin{aligned} X_{113} &= \alpha_{12}^{hf} X_{1f3} = \alpha_{12}^{hf} \left(\alpha_{12}^{hf}\right)^2 \left(\alpha_{12}^{ft}\right) X_{12b} \\ &= \left(\alpha_{12}^{hf}\right)^3 \left(\alpha_{12}^{hf}\right) X_{12b}; \end{aligned} \tag{9.1.17c}$$

stage M:

$$X_{11M} = \left(\alpha_{12}^{hf}\right)^M \left(\alpha_{12}^{hf}\right) X_{12b}. \tag{9.1.17d}$$

However,

$$X_{11M} = X_{1f}, \tag{9.1.17e}$$

therefore

$$M + 1 = \ln(X_{1f}/X_{12b})/\ln \alpha_{12}^{hf}. \tag{9.1.18}$$

The total number of stages in the cascade producing a top product X_{11N} ($\simeq X_{11d}$) and a bottom product X_{12b} from the feed X_{1f} is $N + M$:

$$N + M = \left[\ln(X_{11d}/X_{12b})/\ln \alpha_{12}^{hf}\right] - 1; \tag{9.1.19a}$$

$$N + M = 2\left[\ln(X_{11d}/X_{12b})/\ln \alpha_{12}^{ht}\right] - 1. \tag{9.1.19b}$$

One could utilize the arguments employed to derive equation (9.1.16) to develop the following relation between X_{11N} and X_{11n} for any stage n in the enriching section:

$$X_{11N} = \left(\alpha_{12}^{hf}\right)^{N-n} X_{11n}. \tag{9.1.20a}$$

Correspondingly,

$$X_{11N} = \left(\alpha_{12}^{hf}\right)^{N-n+1} X_{11(n-1)}; \tag{9.1.20b}$$

$$\begin{aligned} \frac{x_{11N}}{1 - x_{11N}} &= \left(\alpha_{12}^{hf}\right)^{N-n+1} \frac{x_{11(n-1)}}{1 - x_{11(n-1)}} = \frac{x_{11t}}{1 - x_{11t}}; \\ x_{11(n-1)} &= \frac{x_{11t}}{x_{11t} + \left(\alpha_{12}^{hf}\right)^{N-n+1} (1 - x_{11t})}. \end{aligned} \tag{9.1.21}$$

Such a relation will be useful for developing expressions for the variations in various flow rates as a function of n.

9.1.1.2 *Flow rates between stages, tapered cascades*

Our goal in this section is to determine the changes in the flow rates of various streams as we move from the location of feed introduction between enriching stage 1 and stripping stage M to the top of the enriching cascade (stage N); one could similarly go to the bottom of the stripping cascade (Figure 9.1.1(b)). Focus now on the enriching section. Consider an envelope in the enriching section covering stages N to n. The total molar balance is given by

$$W^t_{t1(n-1)} = W^t_{t2n} + W^t_{t1N}; \qquad (9.1.22a)$$

the species i balance is

$$W^t_{t1(n-1)} x_{i1(n-1)} = W^t_{t2n}\, x_{i2n} + W^t_{t1N}\, x_{i1t}. \qquad (9.1.22b)$$

Eliminating $W^t_{t1(n-1)}$ from (9.1.22b) via (9.1.22a) yields

$$(x_{i1(n-1)} - x_{i2n}) = +\frac{W^t_{t1N}}{W^t_{t2n}}(x_{i1t} - x_{i1(n-1)}). \qquad (9.1.23)$$

We can rewrite this result as follows:

$$\frac{W^t_{t2n}}{W^t_{t1N}} = \frac{x_{i1t} - x_{i1(n-1)}}{x_{i1(n-1)} - x_{i2n}}. \qquad (9.1.24)$$

The flow rate ratio (W^t_{t2n}/W^t_{t1N}) is called the *internal reflux ratio*, as opposed to external reflux ratio of definition (8.1.137) since no formal reflux stream is generated in Figure 9.1.1(b). This flow rate ratio provides an estimate of the heavy fraction flow rate downward in the cascade below stage n in the enriching section. We are interested in its variation with the stage number n.

To that end, first note that x_{i2n} in relation (9.1.24) is equal to $x_{if(n-1)}$ in an ideal cascade. Further, $x_{if(n-1)}$ is related to $x_{i1(n-1)}$ by the separation factor relation (9.1.8), rewritten for the binary system as

$$\frac{x_{11(n-1)}}{(1 - x_{11(n-1)})} = \alpha^{hf}_{12}\,\frac{x_{1f(n-1)}}{(1 - x_{1f(n-1)})}, \qquad (9.1.25a)$$

resulting in

$$x_{1f(n-1)} = \frac{x_{11(n-1)}}{\alpha^{hf}_{12} - x_{11(n-1)}(\alpha^{hf}_{12} - 1)}. \qquad (9.1.25b)$$

We can now rewrite expression (9.1.24) for the binary system of species $i = 1, 2$ as

$$W^t_{t2n} = W^t_{t1N}\,\frac{x_{11t} - x_{11(n-1)}}{x_{11(n-1)} - \dfrac{x_{11(n-1)}}{\alpha^{hf}_{12} - x_{11(n-1)}(\alpha^{hf}_{12} - 1)}}, \qquad (9.1.26)$$

since $x_{if(n-1)} = x_{i2n}$. Therefore,

$$W^t_{t2n} = W^t_{t1N}\,\frac{(x_{11t} - x_{11(n-1)})(\alpha^{hf}_{12} - x_{11(n-1)}(\alpha^{hf}_{12} - 1))}{x_{11(n-1)}(\alpha^{hf}_{12} - 1)(1 - x_{11(n-1)})}. \qquad (9.1.27)$$

Now insert relation (9.1.21), expressing $x_{11(n-1)}$ in terms of x_{11t}, α^{hf}_{12}, N and $(n - 1)$, in to the above relation to obtain, after rearrangement,

$$W^t_{t2n} = \frac{W^t_{t1N}\left\{\left(\alpha^{hf}_{12}\right)^{N-n+1} - 1\right\}\left\{x_{11t} + \alpha^{hf}_{12}\left(\alpha^{hf}_{12}\right)^{N-n+1}(1 - x_{11t})\right\}}{(\alpha^{hf}_{12} - 1)\left(\alpha^{hf}_{12}\right)^{N-n+1}}, \qquad (9.1.28)$$

expressing the heavy fraction flow rate W^t_{t2n} downward in the cascade below stage n of the enriching section in terms of the top product flow rate W^t_{t1N}, top product composition x_{11t}, α^{hf}_{12}, N and stage n. One can also obtain the value of the light fraction flow rate, W^t_{t1n}, leaving stage n and flowing upward in the cascade via the following overall mass balance from the top to stage $(n + 1)$ in the manner of (9.1.22a):

$$W^t_{t1n} = W^t_{t2(n+1)} + W^t_{t1N}. \qquad (9.1.29)$$

Expressions for W^t_{t1n} as well as W^t_{tfn} are available in Pratt (1967) and Benedict *et al.* (1981). The light fraction flow rate W^t_{t1n} leaving stage n and going upward is obtained as

$$\frac{W^t_{t1n}}{W^t_{t1N}} = \frac{1}{(\alpha^{hf}_{12} - 1)}\left\{x_{11t}\,\alpha^{hf}_{12}\left(1 - \frac{1}{(\alpha^{hf}_{12})^{N-n+1}}\right) + (1 - x_{11t}) \times \left((\alpha^{hf}_{12})^{N-n+1} - 1\right)\right\}. \qquad (9.1.30)$$

Further, the feed flow rate to stage n can be calculated from

$$\frac{W^t_{tfn}}{W^t_{t1N}} = \frac{W^t_{t1n}}{W^t_{t1N}} + \frac{W^t_{t2n}}{W^t_{t1N}}. \qquad (9.1.31)$$

Example 9.1.2 Consider an ideal cascade built out of the countercurrent gas centrifuges (CGCs) shown in Figure 8.1.52, where such a CGC (or a collection of CGCs connected in parallel) is a stage in the ideal cascade of Figure 9.1.1(b). This cascade is to be used to separate feed natural uranium containing $U^{235}F_6$ at the level of $x_{if} = 0.0072$ (the rest is $U^{238}F_6$) to an overhead light fraction containing 90% $U^{235}F_6$, $x_{11t} = x_{11d} = x^t_{11N} = 0.9$ and a bottoms product containing $U^{235}F_6$, $x_{11b} = 0.003$. Determine the number of stages in the enriching section and the stripping section if the stage separation factor $\alpha^{ht}_{12} = 1.4$ and $\alpha^{hf}_{12} = 1.18$. Assume x_{ijn} may be used for this isotope separation collection.

Solution From relation (9.1.16), we obtain the number of stages in the enriching section:

$$N = \frac{\ln(X_{11d}/X_{1f})}{\ln \alpha^{hf}_{12}} = \frac{\log\left(\dfrac{x_{11d}(1 - x_{1f})}{(1 - x_{11d})x_{1f}}\right)}{\log \alpha^{hf}_{12}}$$

$$= \frac{\log \dfrac{0.9 \times 0.9928}{0.1 \times 0.0072}}{\log 1.18} = \frac{\log 1241}{\log 1.18} = 43.0.$$

From relation (9.1.18), we obtain the number of stages in the stripping section:

$$M+1=\frac{\ln(X_{1f}/X_{12b})}{\ln\alpha_{12}^{hf}}=\frac{\log\left(\dfrac{x_{1f}\,(1-x_{12b})}{(1-x_{1f})x_{12b}}\right)}{\log 1.18}$$

$$=\frac{\log\left(\dfrac{0.0072\times 0.997}{0.9928\times 0.003}\right)}{0.0719}=\log 2.410/0.0719$$

$$=0.3820/0.0719=5.313$$

$$\Rightarrow M\ =4.3\ \cong 5.$$

Example 9.1.3 Calculate the values of (W_{t1n}^t/W_{t1N}^t) and (W_{t2n}^t/W_{t1N}^t) for the values of $n = 30$, 20 and 10 in the enriching section of the ideal cascade of Example 9.1.2 for uranium isotope separation using a countercurrent cascade of gas centrifuges.

Solution First we employ expression (9.1.30):

$$\frac{W_{t1n}^t}{W_{t1N}^t}=\frac{1}{(\alpha_{12}^{hf}-1)}\left\{x_{11t}\alpha_{12}^{hf}\left(1-\frac{1}{\left(\alpha_{12}^{hf}\right)^{N-n+1}}\right)+(1-x_{11t})\times\left(\left(\alpha_{12}^{hf}\right)^{N-n+1}-1\right)\right\}.$$

For $n = 30$, $\left(\alpha_{12}^{hf}\right)^{N-n+1}=(1.18)^{43-30+1}=(1.18)^{14}=10.14$; for $n = 20$, $\left(\alpha_{12}^{hf}\right)^{N-n+1}=(1.18)^{24}=53.05$; for $n=10$, $\left(\alpha_{12}^{hf}\right)^{N-n+1}=(1.18)^{34}=264.2$. So, since $x_{11t}=x_{11d}=0.9$

$$n=30\Rightarrow\frac{W_{t1n}^t}{W_{t1N}^t}=\frac{1}{1.18}\left\{0.9\times 1.18\left(1-\frac{1}{10.14}\right)+0.1(10.14-1)\right\}$$

$$\Rightarrow (W_{t1n}^t/W_{t1N}^t)=1.585;$$

$$n=20\Rightarrow\ (W_{t1n}^t/W_{t1N}^t)=5.29;$$

$$n=10\Rightarrow\ (W_{t1n}^t/W_{t1N}^t)=23.19;$$

$$n=1\Rightarrow\ (W_{t1n}^t/W_{t1N}^t)=100.1.$$

Next we employ expression (9.1.28) for W_{t2n}^t/W_{t1N}^t:

$$\frac{W_{t2n}^t}{W_{t1N}^t}=\frac{\left\{\left(\alpha_{12}^{hf}\right)^{N-n+1}-1\right\}\left\{x_{11t}+\alpha_{12}^{hf}\left(\alpha_{12}^{hf}\right)^{N-n+1}(1-x_{11t})\right\}}{(\alpha_{12}^{hf}-1)\left(\alpha_{12}^{hf}\right)^{N-n+1}}.$$

This leads to

$$n=30\Rightarrow\ (W_{t2n}^t/W_{t1N}^t)=\frac{\{(1.18)^{14}-1\}\{0.9\times 1.18\,(1.18)^{14}\times 0.1\}}{0.18\times(1.18)^{14}}=10.49\ ;$$

$$n=20\Rightarrow (W_{t2n}^t/W_{t1N}^t)=39.02;$$

$$n=10\Rightarrow (W_{t2n}^t/W_{t1N}^t)=177.5.$$

The results of the calculations in Example 9.1.3 are of significant interest. They show that the light fraction flow rate W_{t1n}^t coming out of stage n in the enrichment section of the cascade varies substantially with stage number n. At the top, where $n = N$, the ratio (W_{t1n}^t/W_{t1N}^t) is equal to 1. This ratio increases substantially as n decreases. For example, whereas at stage $n = 30$ the ratio, 1.5855, is close to 1 existing at stage $n = N = 43$, at stage $n = 30$ the ratio is 5.29, and at stage $n = 10$ the ratio is 23.19. In fact, this ratio keeps on increasing to a value of 100.1 at $n = 1$ where the feed is introduced. Figure 9.1.4(a) illustrates this behavior in general. The variation of W_{t2n}^t/W_{t1N}^t with n is much larger. Therefore the variation of W_{tfn}^t/W_{t1N}^t will be even larger via relation (9.1.31). The nature of this variation is dictated by the very low concentration of light component $i = 1$ in the feed, the very low separation factors encountered in isotope separation problems and the high value of $x_{11N}^t = x_{11d}$ needed for high purification.

The processing capacity of a separator invariably depends on its dimensions, especially its cross section perpendicular to the mean flow direction of the major streams. We saw in Example 9.1.3 how the rates of upflow, downflow or feed flow vary with stage number n. Correspondingly, the flow cross-sectional area will vary strongly with the stage number n. Since, in many separation devices, the cross-sectional area of an individual separator is limited and cannot be varied continuously, a number of separators are employed in parallel for any given stage. The number of such separators in parallel will depend strongly on the stage number n. This is illustrated in Figure 9.1.4(b). It is clear that the ideal cascade cross-sectional area decreases strongly with an increase in N, resulting in what is called a "tapered cascade." When it is inconvenient to vary the stage cross-sectional area by small amounts as n changes by, say, 1 or 2, the notion of "squared off" cascades is adopted. Here the stage cross-sectional area remains constant for a few stages, i.e. a few values of n, and then the cross-sectional area is again changed for a few more stages, and so on (Figure 9.1.4(c)).

9.1.2 Other cascade configurations

We will illustrate briefly a number of other types of cascades:

(1) tapered cascade of crossflow membrane modules;
(2) crosscurrent cascades;
(3) two-dimensional cascades;
(4) series cascades.

9.1.2.1 *Tapered cascade of crossflow membrane modules*

We have studied crossflow reverse osmosis, ultafiltration and microfiltration in Sections 7.2.2–7.2.4. In any such process, if a very substantial fraction of the solvent

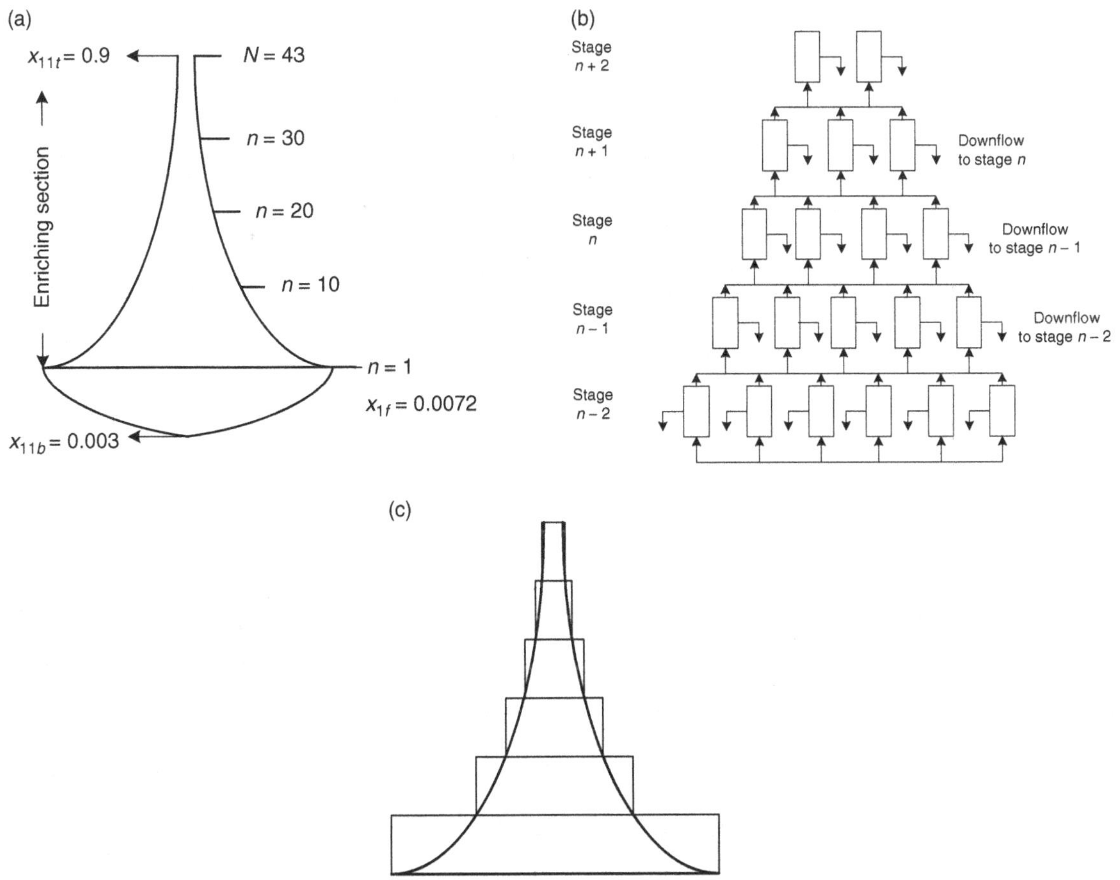

Figure 9.1.4. (a) Variation of light fraction flow rate W^t_{t1n} with stage number n in the enriching section of the ideal cascade of Example 9.1.3. (b) Tapered cascade: hypothetical variation of numbers of separators in parallel in the enriching section of an ideal cascade. (c) Squared off cascade in the enriching section replacing the actual stage variation.

(i.e. water in most cases) has to be removed from the feed, a considerable amount of membrane area is needed. In, for example, the process of reverse osmosis, it means putting quite a few spiral-wound membrane modules in series (see Figure 7.2.4(a)). Consider now the case of high water recovery (fractional water recovery, 0.8 to 0.9) from a brackish water feed containing salt, say, at the level of 2000 ppm. The feed flow rate will be reduced near the concentrate outlet to as low as one-eighth or one-ninth of the feed flow rate. If the same number of spiral-wound membrane modules are used, the feed velocity will be reduced to one-eighth or one-ninth of that at the feed location; however, the salt concentration has been increased eight to nine times, leading to high concentration polarization.

The solution adopted is a tapered cascade of crossflow membrane modules, as shown in Figure 9.1.5. In this cascade the number of membrane modules in parallel, and therefore the feed flow cross-sectional area, is progressively reduced as the feed becomes more concentrated. This ensures an adequate liquid velocity at every location in the cascade and therefore control over concentration polarization. This method is also adopted for ultrafiltration and microfiltration processes with crossflow membrane modules when operating in large scale, especially as part of pretreatment prior to reverse osmosis (see Section 11.2).

9.1.2.2 *Crosscurrent cascades*

In the tapered cascade of crossflow membrane modules in Figure 9.1.5, the flow cross-sectional area varies in a stepwise fashion. Therefore the taper is not continuous; it varies also in a stepwise fashion. In each step, crossflow is generated from the feed stream itself in each membrane module and collected together. In the so-called *crosscurrent cascades* employed in solvent extraction (for example), fresh extraction solvent is introduced into each

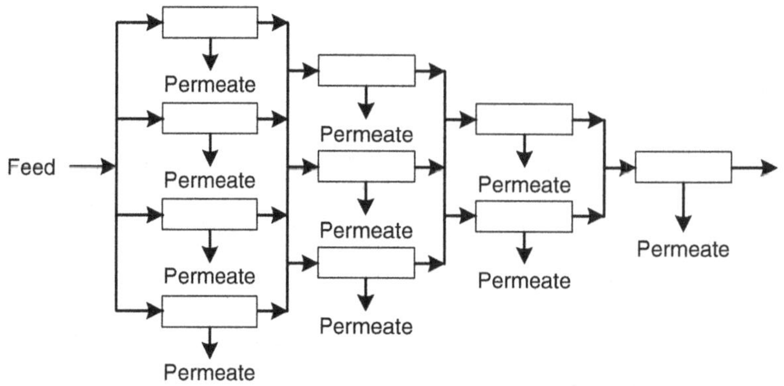

Figure 9.1.5. Tapered cascade of crossflow membrane modules.

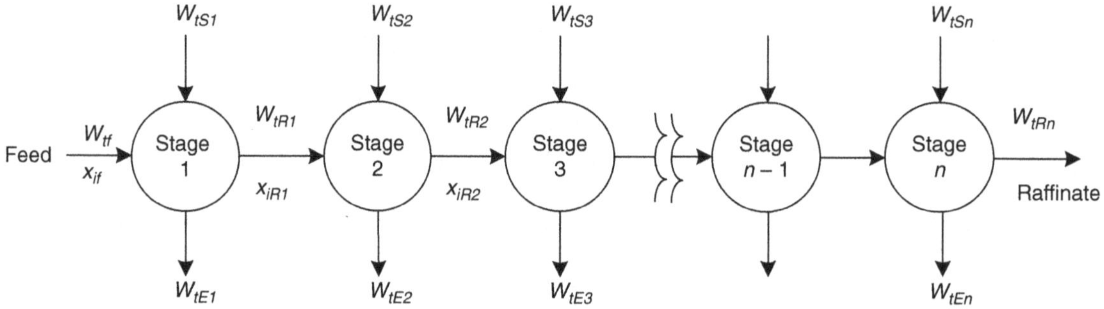

Figure 9.1.6. Crosscurrent cascade in solvent extraction.

stage in the cascade and corresponding extract is withdrawn as the immiscible feed solution is moved from the first stage to the second stage and so on (Figure 9.1.6). There is an actual crossflow of an immiscible phase with respect to the feed/raffinate phase at all stages in this cascade on the basis of an overall flow configuration. However, in each cascade the manner of contacting/mixing is variable; it may be well-mixed (see Section 6.4.1.2), cocurrent or countercurrent or batch. Treybal (1963, pp. 201–219) has provided a detailed analysis of solvent extraction in such a cascade. A few results from this reference are provided here for linear equilibrium behavior and immiscible solvents. Let

u_{ris} = weight of solute i per unit weight of fresh solvent,
u_{riE} = weight of solute i per unit weight of solvent extract,
ur_{iEn} = value of ur_{iE} leaving stage n,
ur_{if} = weight of solute i per unit weight of feed,
ur_{iR} = weight of solute i per unit weight of raffinate,
ur_{iRn} = weight of solute i per unit weight of raffinate leaving stage n. (9.1.32)

The linear equilibrium behavior is represented by

$$ur_{iEn} = m_i ur_{iRn}. \tag{9.1.33}$$

The relation between the number of extraction stages n, ur_{if}, ur_{iRn} and ur_{is} for the crosscurrent cascade is

$$n = \frac{\log\left(\dfrac{ur_{if} - (ur_{is}/m_i)}{ur_{iRn} - (ur_{is}/m_i)}\right)}{\log\,(E+1)}, \tag{9.1.34}$$

where

$$E = m_i(w_{tE}/w_{tR}), \tag{9.1.35}$$

the extraction factor.

Note: The notation w_{tE}, w_{tR} indicates mass flow rates. However, each stage is operated in a batch fashion. Therefore w_{tE} and w_{tR} should be considered here as the individual masses of phase E and R, respectively, over the time needed to operate each stage. The total amount of solvent added to n stages is $w_{tE}^t = nw_{tE}$. Further, if the solvent extract from each stage is joined together, the composition of the combined extract u_{iRE}^t will be

$$u_{iRE}^t = \frac{w_{tR}\,(u_{rif} - u_{riRn})}{n\,w_{tE}} + ur_{is}. \tag{9.1.36}$$

If there are two solutes $i = 1, 2$ present in the feed, the stage number where the maximum amount of separation is achieved between them can also be obtained (Treybal, 1963, eqn. 7.28).

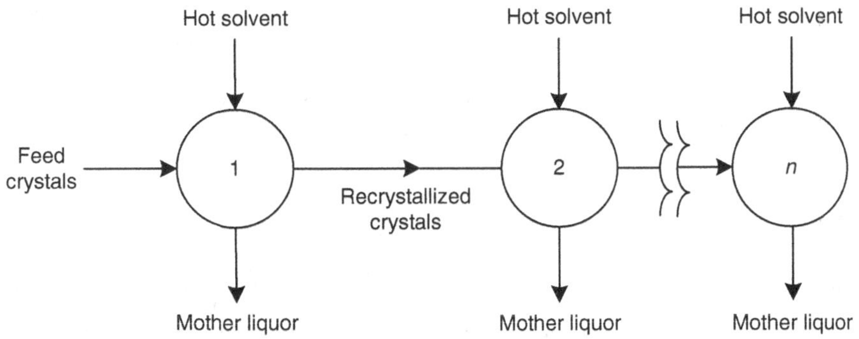

Figure 9.1.7. Crosscurrent cascade for recrystallization via cooling crystallization.

The simple crosscurrent cascade of Figure 9.1.6 has wider utility, e.g. in recrystallization. In conventional crystallization from solution, the crystals obtained may contain a certain amount of impurity. If melt crystallization can be implemented to purify the crystals, as in Section 8.1.5, then countercurrent melt crystallization can be implemented. However, if solution crystallization has to be used for purification via, for example, *cooling crystallization*, then a *crosscurrent cascade* may be employed with hot solvent, as in Figure 9.1.7. Here the crystals to be purified are introduced in stage 1 along with pure hot solvent in a cooling crystallizer. The crystals are redissolved and then recrystallized via cooling, leading to purer crystals. Of the two products from stage 1, the purer crystals become feed to the next stage 2, where again fresh hot solvent is introduced to redissolve and recrystallize after cooling. The mother liquor from stage 1 becomes enriched in the impurity; similarly, for the mother liquor from stage 2. In this fashion, one can have a number of crosscurrent stages ultimately to produce highly purified crystals from multiple crystallizations, as shown in Figure 9.1.7. The mother liquors remaining from each stage, however, contain a significant amount of the desired solute along with the impurities.

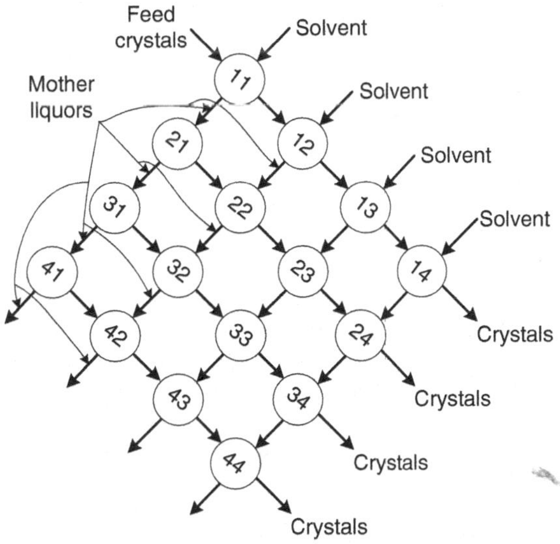

Figure 9.1.8. Two-dimensional cascade useful for improving purity in solution crystallization.

9.1.2.3 *Two-dimensional cascades*

The shortcoming of the relatively simple cascade of Figure 9.1.7 is a significant loss of the desired material with the mother liquor from each stage. The two-dimensional cascade shown in Figure 9.1.8 overcomes this deficiency by reusing the mother liquors rejected from each of stages 1, 2, 3, etc. In Figure 9.1.8, the feed crystals are introduced into stage 11 along with hot solvent. The recrystallized crystals obtained are fed to the next stage 12, where again fresh hot solvent is introduced, as in Figure 9.1.7, and so on to stages 13 and 14. However, the mother liquor from stage 11 introduced into stage 21 becomes a feed solution, from which further cooling crystallization leads to additional crystals, which are then fed to stage 22, with the resulting mother liquor fed to stage 31. In stage 22, the new crystals introduced from stage 21 are mixed with the mother liquor from stage 12 - the process of redissolving by raising the temperature and then cooling crystallization will lead to a purer batch of crystals, which is sent to stage 23, the residual mother liquor is sent to stage 32, and so on.

The utility of this cascade may be appreciated if we think of the feed containing two species 1 and 2, where species 1 is the primary result of cooling crystallization, with species 2 being an impurity or a solute that does not crystallize as well due to, say, higher solubility. In the cascade, stages 14, 24, 34 and 44 will obviously produce crystals that are highly enriched in species 1.

However, as we go down and stay left in the cascade, the presence of species 2 increases substantially, with stages 41 and 42 having high concentrations of species 2 in the mother liquors produced. Such types of cascades have also been studied for solvent extraction, with the feed solution containing the solute(s) to be extracted and solvents introduced crosswise (Treybal, 1963, pp. 363–373).

9.1.2.4 *Series cascades*

We have already discussed simple series cascades in Figure 6.4.7 for crystallization and Figure 7.1.23 for the eluent passing through n plates. In Figure 6.4.7, multiple cooling crystallizers in series provided greater control of temperature and residence time in each stage, leading to better control of the crystal size distribution. Figure 7.1.23 shows how such a bed of adsorbent in a series of stages could lead to separation of a multicomponent mixture introduced via a sample feed in stage 1 as the eluent flows through the stages in series.

9.2 Cascades for multicomponent mixture separation via distillation

We have seen in Section 8.1.1.3 that continuous operation of a distillation column with countercurrent flow of vapor and liquid can separate a binary mixture only. If we have a multicomponent nonazeotropic mixture as feed, then only *one* of the two product streams can have one species with sufficient purity; the other product stream will contain all other species in quantities reflecting their feed concentration. This stream has to be fed to another distillation column that can produce two product streams, where each product stream will have sufficient purity with respect to one of the two remaining species for a ternary feed to the first distillation column. Similarly, if the feed to the first column has four components, in general, we will need three columns to obtain four product streams, each product stream being sufficiently purified in one of the species. In general, to separate n species in the feed by distillation in simple distillation columns of the type shown in Figure 8.1.19(b), we will need $(n - 1)$ distillation columns.

The problem, however, has considerable complexity, since $(n - 1)$ distillation columns may be operated/connected in a number of ways. Consider first a ternary nonazeotropic mixture of three species A, B and C, with C having the highest boiling point (lowest volatility) and A having the lowest boiling point (highest volatility). In the configuration shown in Figure 9.2.1(a), often called the *direct sequence*, the mixture fed to column 1 yields a distillate product containing primarily species A; the bottoms product contains essentially a mixture of B and C. The bottoms product is then fed to column 2, yielding B as the top product and C as the bottoms product. An alternative configuration/column sequence known as the *indirect sequence* shown in Figure 9.2.1(b) produces from the first column a bottoms product which is essentially C; therefore the top product contains both A and B. This top distillate product is now fed to the second column, yielding a top distillate product containing essentially A and a bottoms product containing B.

Using simple column configurations, these are the only two alternative column sequences that are possible for a nonazeotropic ternary mixture. If we use a more complex column operational mode, for example with two feed streams and/or a side product stream, a number of additional column configurations (as many as six or seven) are possible using two distillation columns for a ternary mixture. King (1980, pp. 710–717) and Doherty and Malone (2001, pp. 296–300) have illustrated such configurations. Considerations on capital costs, as well as the energy requirements, are important considerations when deciding which alternative column sequences to use (see Chapter 10).

To separate a four-component nonazeotropic feed mixture containing species D in addition to species A, B and C, with D being even less volatile than species C, the number of possible column sequences using conventional column operational mode is significantly larger, five for example. Figure 9.2.2 illustrates these different configurations. The configuration shown in Figure 9.2.2(a) represents an example of a direct sequence: column 1 takes out the lightest component A in the overhead; column 2 takes out species B in the distillate; column 3 takes out species C in the distillate product and species D in the bottoms product. Four other configurations, shown in Figures 9.2.2 (b)–(e), follow alternative separation schemes. King (1980, p. 713) has briefly indicated results of earlier research and has suggested that the number of different column configurations for separating n components is given by the following relation:

$$\text{no. of different configurations} = \frac{\{2(n-1)!\}}{n!(n-1)!}. \qquad (9.2.1)$$

Using this equation, the number of different configurations for $n = 2$, 3, 4 and 5 are, respectively, 1, 2, 5 and 14. As n increases, the number of different configurations increases at an astounding rate.

Considerable research has been carried out to facilitate the selection of desirable configurations. A few simple heuristics are generally employed to guide the selection of a few sequences, which may be subjected to detailed analysis (King, 1980; Doherty and Malone, 2001). Consideration of the energy required for separation is very important in these deliberations. (An elementary introduction to the energy required for separation in distillation is provided in Sections 10.1.3, 10.1.4.2, 10.2.2.1 and 10.2.2.2.) We list a few simple sequencing heuristics below; the rest

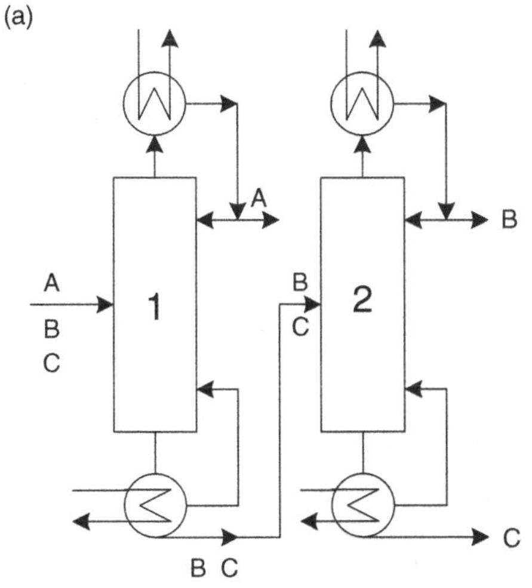

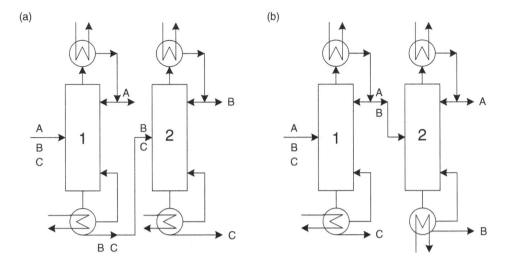

Figure 9.2.1. Number of possible column sequences in multicomponent distillation of nonazeotropic mixture using simple distillation columns. Ternary mixture of species A, B and C: (a) direct sequence; (b) indirect sequence.

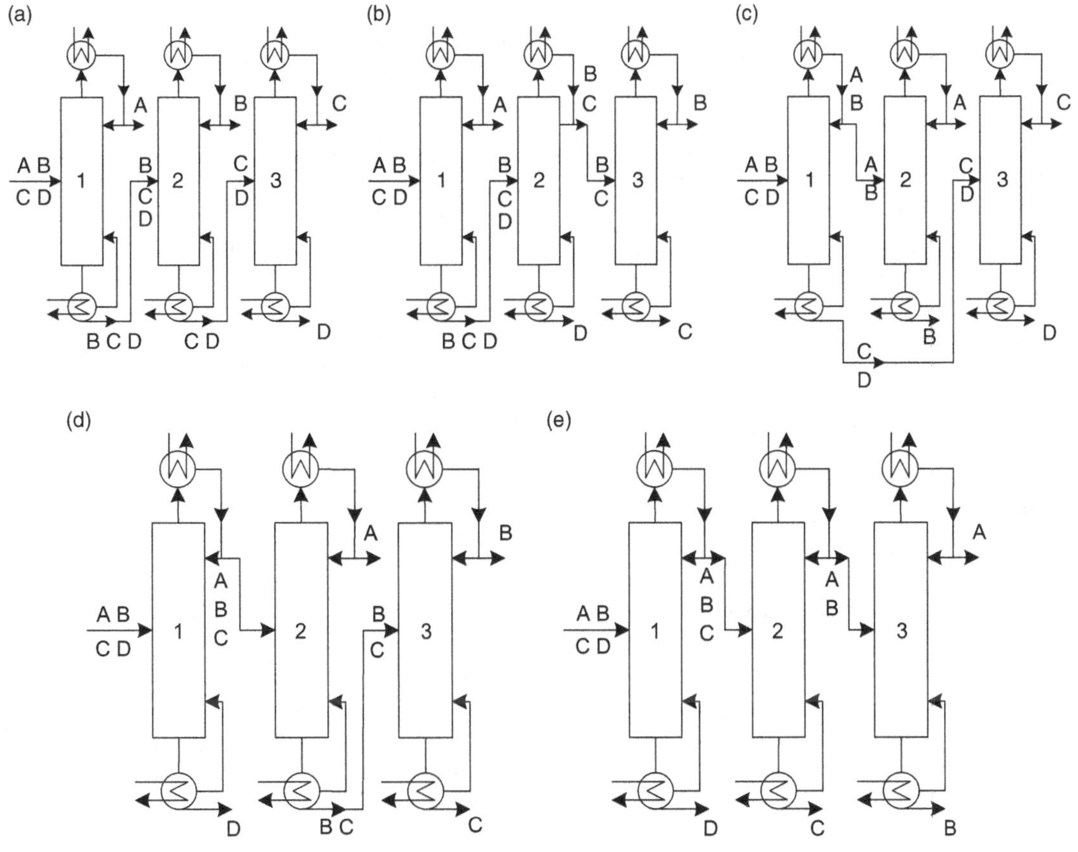

Figure 9.2.2. Four-component mixture of species A, B, C and D, showing five different coloumn sequences.

are available in King (1980, pp. 715–720) and Doherty and Malone (2001, Table 7.2).

(1) Remove reactive or corrosive species as soon as possible; the rest of the column sequence thereby remains free of these problems.
(2) Select a sequence where the overhead of each column removes one species, namely the direct sequence.
(3) Remove a species present in large quantities in the feed as early as possible. This will reduce the total flow rates in the rest of the column sequence, resulting in substantial energy savings (see Section 10.1.4.2).
(4) Separation of components whose separation factors (relative volatilities) are low and close to 1 should be carried out near the end of the cascade where other components are not present (see equation (10.1.47)). Choose a method other than distillation if $1 < \alpha_{ij} \lesssim 1.1$.
(5) Remove the most volatile species in the first column distillate.
(6) If a species is to be obtained in a product with high fractional recovery, implement it near the end of the column sequence.
(7) It is likely to be a safe bet to select the cheapest column operation as the next step in a sequence of columns.

9.3 Cascades for multicomponent mixture separation involving other separation processes

Large-scale utilization of a cascade of separation devices of one particular type for the separation of a multicomponent mixture is primarily found in the case of distillation based separation of a multicomponent mixture of volatile chemicals (see Section 9.2). We refer here to steady state operation of these distillation devices since chromatographic operation separates a multicomponent mixture generally via unsteady operation in one fixed-bed device (see Section 7.1.5.1, Figure 7.1.22) or by having steady state operation in one device having crossflow of two bulk phases (see Section 8.3.1). In general, if there is a multicomponent mixture containing a variety of species consisting of various combinations of molecules, macromolecules, colloids, particles, etc., then separation/purification involves a sequence of separation processes of different types. Such sequences are illustrated in Chapter 11 for a variety of feed streams focused on recovering/purifying/concentrating a particular species or a number of species.

There are examples, however, where a particular type of separation process has been used in a cascade for separating/recovering a variety of molecules in a feed mixture. Consider the case of recovering a variety of heavy metals and acids in a number of effluent streams obtained in hydrometallurgical processes, metal pickling processes, etc. Ritcey and Ashbrook (1979) have described the recovery of metals and strong acids from the effluent of a stainless steel pickling bath. The waste from the HF–HNO_3 pickling bath contained substantial amounts of iron, nickel and molybdenum in addition to fluorides and nitrates. Solvent extraction was primarily employed as follows.

A solvent containing 75% TBP in kerosene (see Section 5.2.2.4, equation (5.2.110)) is used to extract HNO_3, HF and molybdenum from the effluent after adding H_2SO_4; addition of H_2SO_4 also enabled the extraction of HF since the solvent used prefers to extract HNO_3. Using water to back extract/strip HNO_3 and HF from this solvent based extract, one can recycle HNO_3 and HF back to the metal pickling bath. The solvent based extract is left with molybdenum; 1M NaOH solution is used to strip this molybdenum and then precipitated for recovery. Meanwhile, the treated original effluent (the aqueous raffinate from TPB extraction) containing Cr, Fe and Ni is recovered via precipitation in two steps.

Problems

9.1.1 The dimensions and volume of any stage are in general proportional to the flow rate through that stage. Therefore the total volume of a separation plant/device will be proportional to the sum total of the flow rates through all stages in the separation plant/device. Consider expression (9.1.30) for light fraction flow rate W^t_{t1n} leaving stage n and going upward. Obtain the following expression for the total light fraction flow rate (upflow) from stage 1 to stage N in the enriching section of an ideal cascade:

$$\sum_{n=1}^{N} W^t_{t1n} = \frac{W^t_{t1N}}{(\alpha^{hf}_{12}-1)}\left[x_{11t}\,\alpha^{hf}_{12}\left\{N - \frac{1-\left(\alpha^{hf}_{12}\right)^{-N}}{\left(\alpha^{hf}_{12}-1\right)}\right\} + (1-x_{11t})\left\{\frac{\alpha^{hf}_{12}((\alpha^{hf}_{12})^N-1)}{\left(\alpha^{hf}_{12}-1\right)} - N\right\}\right],$$

given

$$\sum_{n=1}^{N} a^{N-n+1} = \frac{a(a^N - 1)}{(a-1)}.$$

Now employ the result (9.1.6) for N to develop the following expression for the total light fraction flow rate (upflow) from stage 1 to N in the enriching section of an ideal cascade:

$$\sum_{n=1}^{N} W_{t1n}^{t} = \frac{W_{t1N}^{t}}{(\alpha_{12}^{hf} - 1)} \left[\frac{\left\{x_{11t}(\alpha_{12}^{hf} + 1) - 1\right\} \ln\left(\frac{x_{11t}(1-x_{1f})}{x_{1f}(1-x_{11t})}\right)}{\ln \alpha_{12}^{hf}} + \frac{\alpha_{12}^{hf}}{\left(\alpha_{12}^{hf} - 1\right)} \frac{(x_{11t} - x_{1f})(1 - 2x_{1f})}{x_{1f}(1 - x_{1f})} \right].$$

Note: $x_{11t} = x_{11d}$; $\alpha_{12}^{hf} = \alpha_{12}^{ft}$.

9.1.2 This problem is focused on the ideal cascade of Figure 9.1.1(b), where the enrichment achieved per stage is quite small, i.e. we have *close separation*.

(1) The first problem is to develop a difference equation for x_{i1n} between two consecutive stages $(n-1)$ and n employing equation (9.1.23). Then convert this equation to a differential equation for a differential change in x_{i1n} for a differential change in the stage number n.

(2) Obtain an expression for the total heavy fraction flow rate (downflow) in the enriching section of the cascade, i.e. develop a continuous version of $\sum_{n=1}^{N} W_{2tn}$ (see Problem 9.1.1 for an illustration of the summation); you do not have to complete the integration.

(3) Find out the condition under which the total heavy fraction flow rate is minimized in the enriching section.

9.1.3 (1) Calculate the values of the total light fraction flow rate (upflow) from stage 1 to stage N in the enriching section of an ideal cascade, as described in Example 9.1.2 for uranium isotope separation using countercurrent gas centrifuges. The overhead top product flow rate is 2 gmol/time; $\alpha_{12}^{hf} = 1.18$; $x_{11t} = 0.9$, $x_{1f} = 0.0072$. Employ the result provided in Problem 9.1.1.

(2) If, instead of a countercurrent gas centrifuge, each stage is a gaseous diffusion stage, determine the corresponding value of the total light fraction flow rate (upflow) from stage 1 to stage N, given $\alpha_{12}^{hf} = 1.0021$.

9.1.4 Consider a variation of the two-dimensional cascade (Figure 9.1.8) for the solvent extraction based purification of an aqueous feed (for example) shown in Figure 9.P.1. Assume that the stages identified by 21, 31, 32, 41, 42

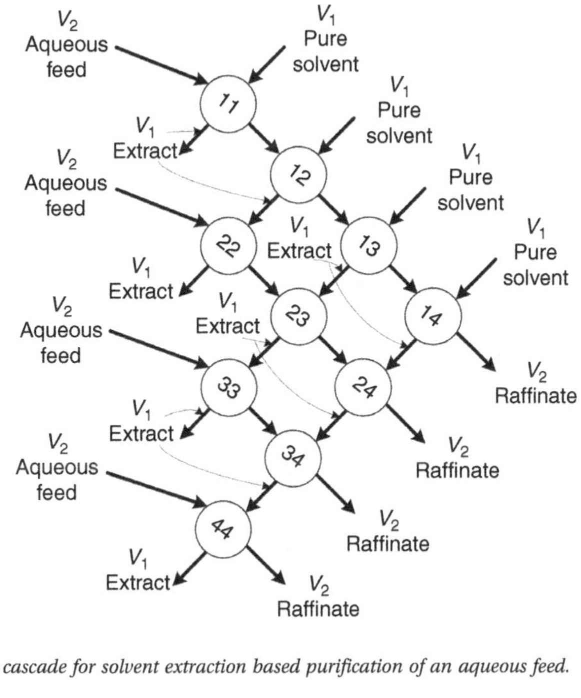

Figure 9.P.1. Two-dimensional cascade for solvent extraction based purification of an aqueous feed.

and 43 in Figure 9.1.8 do not exist. Further, instead of feed crystals, a feed aqueous solution of volume V_2 is introduced into stages 11, 22, 33 and 44. In addition, a volume V_1 of fresh extracting solvent is introduced into each of stages 11, 12, 13 and 14 to extract solute *i*. Instead of crystals, aqueous raffinate of volume V_2 is moved from stages 11 to 12 to 13 to 14 and out of the cascade. Correspondingly, the extract from each of stages 11, 12, 13 and 14 is moved downward: extract from 12 goes to 22; extract from 13 goes to 23 and then to 33; extract from 14 goes to 24 and then to 34 and then to 44. Let the extraction factor for species *i* be defined by $E_i = \kappa_{i1}\ V_1/V_2$. Let the constant fractions *p* and *q* of the amount of solute *i* entering any stage via the feed or the raffinate and the extract go into the extract phase and the raffinate phase, respectively.

(1) Show that

$$p = \frac{E_i}{1+E_i}, \quad q = \frac{1}{1+E_i}$$

if the aqueous-organic system consists of immiscible phases and κ_{i1} is independent of solute concentration.

(2) If the fresh feed stream entering stages 11, 22, 33 and 44 contains 1 gmol of solute, identify the amount of solute in each of the raffinate and extract streams in the cascade containing stages 11, 12, 13, 14, 22, 23, 24, 33, 34 and 44.

10

Energy required for separation

Energy is needed to separate a mixture into pure compounds. Some separation operations are highly energy consuming; others are less so. But all separations require a minimum amount of energy. Section 10.1 introduces the concept of minimum energy required for separation, independent of the process employed. This section then illustrates the minimum energy required in various separation processes such as membrane gas permeation, distillation, extraction and adsorption. How to reduce the amount of energy needed for a given separation by particular separation processes is the general topic of Section 10.2. The specific separation processes and topics covered in this section include: evaporation of water for desalination by multiple-effect evaporation, multistage flash evaporation, vapor-compression evaporation; distillation based separation of mixtures by multieffect and heat pump configuration. The role of free energy of mixing in influencing the energy requirement for separation is briefly considered here. Finally, the special considerations needed for processing dilute solutions of solutes of interest are discussed in the context of reducing the total energy required.

10.1 Minimum energy required for separation

We will briefly calculate here the minimum energy required for separation for a few separation processes for illustrative purposes. It is known (Hougen *et al.*, 1959) that the minimum energy required (or work to be done $(-\underline{W})$) for the isothermal separation of m moles of a mixture present in phase j by a reversible process is given by the change in the total free energy of all molecules, G_t (where G is the Gibbs free energy), namely ΔG_t per mole of mixture multiplied by m moles of mixture:

$$-\underline{W}_{\min} = m\Delta G_t = \sum_{i=1}^{n} m_{if}\,\Delta\overline{G}_{ij} = \sum_{i=1}^{n} m_{if}\,\Delta\mu_{ij}, \quad (10.1.1)$$

where μ_{ij} is the chemical potential of species i in phase j. Employing the definition (3.3.21) for μ_{ij}, we get

$$-\underline{W}_{\min} = \sum_{i=1}^{n} m_{if}\Delta\left(\mu_i^0 + \overline{V}_{ij}\left(P_j - P^0\right) + RT\ \ell n\, a_{ij}\right). \quad (10.1.2)$$

Per mole of the mixture, the minimum energy required is given by

$$-W_{\min}\big|_{\text{mol}} = \sum_{i=1}^{n} x_{if}\Delta\left(\overline{V}_{ij}P_j + RT\ \ell n\left(\gamma_{ij}x_{ij}\right)\right), \quad (10.1.3)$$

since there is no change in the standard state conditions. At constant pressure, this expression is reduced to

$$-W_{\min}\big|_{\text{mol}} = \sum_{i=1}^{n} x_{if}\,RT\left[\ell n\left(\gamma_{ij}x_{ij}\right)_{\text{final}} - \ell n\left(\gamma_{if}x_{if}\right)\right]$$
$$= -RT\sum_{i=1}^{n} x_{if}\,\ell n\left(\gamma_{if}x_{if}\right), \quad (10.1.4)$$

since the final state of each species is pure at unit activity,

$$\left(a_{ij}\right)_{\text{final}} = 1 = \left(\gamma_{ij}x_{ij}\right)_{\text{final}}.$$

King (1980, p. 661) has provided this expression with the appropriate earlier references/sources. For feed mixtures behaving as ideal mixtures at constant pressure and temperature,

$$-W_{\min}\big|_{\text{mol}} = -RT\sum_{i=1}^{n} x_{if}\ \ell n\, x_{if}. \quad (10.1.5)$$

First we will illustrate the minimum energy required to separate a small amount of mixture for the following processes: evaporation of water from a saline solution; recovery of water by reverse osmosis; separation of an ideal binary gas mixture by membrane permeation. Then we will consider the definition of net work consumption for thermally driven processes. Next we will consider a variety of separation processes vis-à-vis their minimum energy requirement for separation.

10.1.1 Evaporation of water from sea water

To illustrate this process, we will employ the approach of Spiegler (1977). Consider two reservoirs (see Figure 10.1.1(a)), one containing sea water and the other containing fresh water, in a thermally insulated container. The vapor spaces in both vessels do not contain any air (it was removed earlier); however, the vapor spaces in each vessel do contain water vapor, which is in equilibrium with the liquid. The objective is to remove some pure water from the sea water container vapor space to the fresh-water container vapor space, which means that the vapor (which is pure water) has to be transferred. At any temperature, the equilibrium water vapor pressure over sea water is less than that over fresh water. This difference is about 1.84% of the pure water vapor pressure at 25 °C, which is 0.0312 atm.

If we now have a hypothetical reversible compressor (Figure 10.1.1(b)) which takes 43.4 liters of water vapor (the volume of 1 g of water at 25 °C and 0.0312 atm) from the sea water vessel and compresses it isothermally over the small pressure difference, namely 0.0184 × 0.0312 atm, we can transfer it to the fresh-water vessel, where it will condense. The work done to separate this amount of water is given by

$$\begin{aligned}\left(\frac{\text{work done}}{\text{g of water}}\right) &= \int_{P_1}^{P_2} V_{ij}\,\mathrm{d}P = V_{ij}(\Delta P)\\ &= 43.4\,\text{liter} \times 0.0184 \times 0.0312\,\text{atm}\\ &= 0.0249\,\text{liter-atm/g of water}\\ &\cong 0.0249\,\text{liter-atm/cm}^3\text{of water}\\ &= 24\,900\,\text{liter-atm/m}^3\text{of water}\\ &= 0.703\ \text{kW-hr/m}^3\\ &= 2.65\ \text{kW-hr/1000 US gal.}\end{aligned} \tag{10.1.6}$$

(Since we are transferring pure water vapor from one vessel to the other, from equation (10.1.3) the work that has to be done for the separation is just

$$-W_{\min} = \Delta(\overline{V}_{wg}P_g) = \overline{V}_{wg}\,\Delta P_g, \tag{10.1.7}$$

where we have assumed that the partial molar volume of water is constant.)

There are a few important issues here. First, as a small amount of water vapor is taken away from the sea water vessel, a small amount of water would immediately be evaporated from the sea water. This would lead to some cooling of the sea water vessel. Correspondingly, the excess vapor transferred to the fresh-water vessel will lead to some condensation and therefore heating up of the vessel. However, the two vessels are in an isothermal container; therefore the excess heat will be exchanged with the fresh-water vessel. Second, to carry out the reversible process, the volume of water vapor to be transferred should be differentially small. We used 1 g of water simply for the purpose of calculation, even though the volume of water vapor used for calculation was high because of the low pressure.

10.1.2 Recovery of water by reverse osmosis

We will now illustrate the method adopted by Spiegler (1977) to determine the amount of energy required to recover water by reverse osmosis in the limit of the minimum amount of energy required. Consider Figure 10.1.2, which shows a chamber containing sea water and a piston separated from a fresh-water chamber by a semipermeable membrane which rejects salt completely. The pressure on the solution exerted by the piston is the osmotic pressure of sea water, $\pi_{\text{sea water}} = \pi_{sw}$. Therefore there is osmotic equilibrium and no water is being transferred in either direction through the membrane. If we now increase the

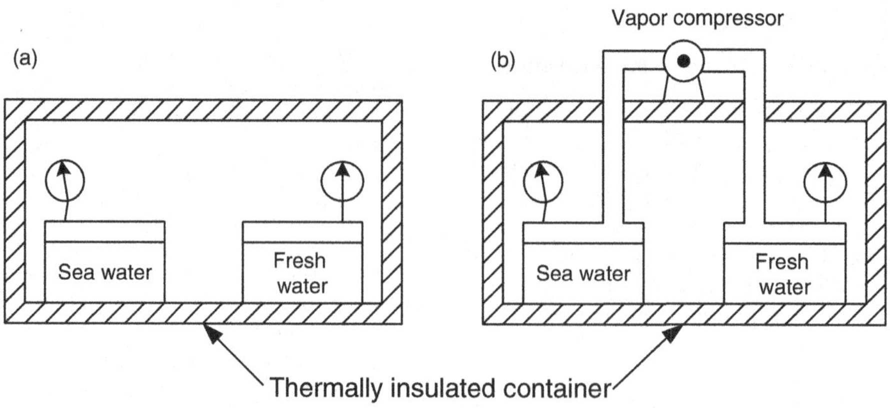

Figure 10.1.1. Thermally insulated container accommodating two separate vessels, one containing sea water, the other containing fresh water. (a) In this configuration, two separate vessels are not connected in any way. (b) In this configuration, a small amount of vapor from the sea water vessel is reversibly compressed and sent to the vapor space of the fresh-water vessel. (After Spiegler (1977).)

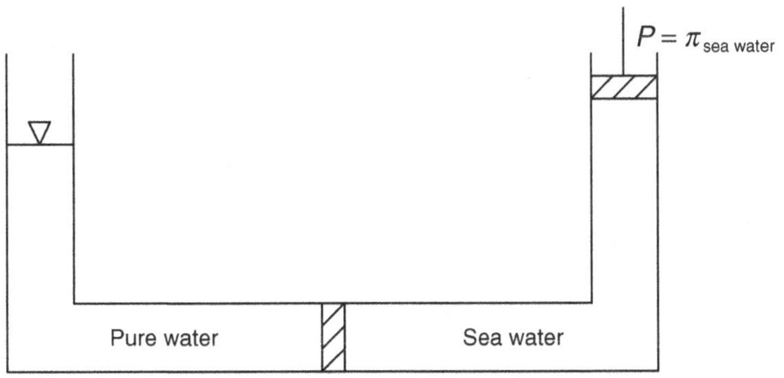

Figure 10.1.2. A pure-water chamber at atmospheric pressure is separated from the sea water chamber (which has a piston and is at a pressure of $\pi_{\text{sea water}}$) by a semipermeable membrane M, which allows only water to pass through.

pressure on the piston by an infinitesimally small amount, $d\pi_{sw}$, a small amount of pure water of volume dV will be removed to the pure-water side. Immediately, the sea water side solution concentration will be increased differentially, leading to a new equilibrium osmotic pressure and a new osmotic equilibrium. To calculate the work done to recover this volume of pure water, we note that

$$-dW = (\pi_{sw} + d\pi_{sw})dV \cong \pi_{sw}\,dV, \qquad (10.1.8)$$

after neglecting second-order terms. If the initial sea water volume was V_1 and we keep carrying out this process slowly till the final volume of concentrated sea water is V_2 ($<V_1$), then the total work done to recover one liter of pure water is

$$-W|_{\text{liter}} = \frac{1}{V_1 - V_2}\int_{V_1}^{V_2} \pi_{sw}\,dV. \qquad (10.1.9)$$

From equation (3.3.86a), we know that the osmotic pressure of saline water is given by

$$\pi_{sw} = \frac{RT}{\overline{V}_{sw}}\ln a_{sw}, \qquad (10.1.10)$$

where a_{sw} is the activity of the solvent, namely water in the saline water. From basic thermodynamics, we know that

$$a_{ij} = \frac{\hat{f}_{ij}}{f^0_{ij}} \Rightarrow a_{sw} = \frac{\hat{f}_{sw}}{f^0_{sw}}. \qquad (10.1.11)$$

However, $f^0_{sw} = f_{sw} = P^{\text{sat}}_s\,\Phi^{\text{sat}}_s$ and

$$\hat{f}_{sw} = \hat{f}_{sg} \cong p_{sg}\,\widehat{\Phi}_{sg} \Rightarrow a_{sw} = \frac{p_{sg}\,\hat{\Phi}_{sg}}{P^{\text{sat}}_s\,\Phi^{\text{sat}}_s}$$

$$= \frac{\text{vapor pressure of water over solution}}{\text{vapor pressure of pure water}}.$$

However, the equilibrium vapor pressure of water over the saline water, $p_{sg}\Phi_{sg}$, has been related to the equilibrium vapor pressure of pure water, $P^{\text{sat}}_s\,\Phi^{\text{sat}}_s$, via the salinity of the saline water, s:

$$p_{sg}\,\hat{\Phi}_{sg} = P^{\text{sat}}_s\,\Phi^{\text{sat}}_s(1 - As), \qquad (10.1.13)$$

where the salinity s is defined as the total amount of solids (in grams) per kilogram of sea water[1] and $A = 0.000537$. We may now rewrite equation (10.1.9) using the other three relations, (10.1.10), (10.1.12) and (10.1.13) as

$$-W|_{\text{liter}} = \frac{RT}{(V_1 - V_2)\,\overline{V}_{sw}}\int_{V_1}^{V_2}[\ln(1 - As)]\,dV. \qquad (10.1.14)$$

Since As is very small compared to 1, we may use the approximation $(1 - As) \cong -As$ to obtain

$$-W|_{\text{liter}} = -\frac{ART}{\overline{V}_{sw}(V_1 - V_2)}\int_{V_1}^{V_2} s\,dV. \qquad (10.1.15)$$

Since the membrane is semipermeable and rejects salt completely, the following salt balance relation holds from initial saline water volume V_1 (salinity s_1) and any other saline water volume V (salinity s) for dilute solutions:

$$sV = s_1 V_1 \Rightarrow s = s_1 V_1/V. \qquad (10.1.16)$$

Therefore

$$\begin{aligned} -W|_{\text{liter}} &= -\frac{ART\,s_1 V_1}{\overline{V}_{sw}(V_1 - V_2)}\int_{V_1}^{V_2}\frac{dV}{V} \\ &= -\frac{ART\,s_1 V_1}{\overline{V}_{sw}(V_1 - V_2)}\ln\left(\frac{V_2}{V_1}\right) \\ &= -\frac{ART\,s_1 V_1}{\overline{V}_{sw}(V_1 - V_2)}\ln\left(1 - \frac{V_1 - V_2}{V_1}\right). \qquad (10.1.17) \end{aligned}$$

[1] When all carbonate has been converted to oxide, bromide and iodide have been replaced by chloride and all organic matter has been completely oxidized.

Since in our process $(V_1 - V_2)$ is small compared to V_1, we obtain

$$-W|_{\text{liter}} = -\frac{ART\,s_1\,V_1}{\overline{V}_{sw}(V_1 - V_2)}\,(-)\,\frac{V_1 - V_2}{V_1} = \frac{ART\,s_1}{\overline{V}_{sw}} \tag{10.1.18}$$

in the limit of $(V_1 - V_2) \to 0$.

For sea water, $s = 35$ (say, 35 g/kg of sea water), R = 2.31×10^{-6} kW-hr/gmol-K, $A = 0.000537$, $T = 298\,\text{K}$ (25 °C), $\overline{V}_{sw} = 0.018$ liter/gmol, so

$$-W|_{\text{liter}} = \frac{0.000537 \times 2.31 \times 10^{-6} \times 298 \times 35}{0.018}\,\frac{\text{kW-hr}}{\text{gmol-K}}\,\frac{\text{K}}{\text{liter/gmol}}; \tag{10.1.19}$$

$$-W|_{\text{liter}} = 7.18 \times 10^{-4}\,\text{kW-hr/liter}$$
$$= 0.718\,\text{kW-hr/m}^3. \tag{10.1.20}$$

Note that this energy requirement to recover 1 liter of pure water is essentially identical to that calculated using the evaporation process in Section 10.1.1. This is as it should be since, by equation (10.1.1), the minimum energy required merely depends on the initial and final values of the Gibbs free energy, which is a state function. It does not matter how one goes from $G_t|_{\text{initial}}$ to $G_t|_{\text{final}}$; therefore the separation technique followed does not influence the minimum energy required. In practice, the actual amount of energy required for separation depends strongly on the separation technique used and the associated irreversibilities.

10.1.3 Net work consumption

In the two examples considered in Sections 10.1.1 and 10.1.2, the minimum energy required for separation in getting pure water from sea water was mechanical/electrical energy supplied through a piston or a compressor (assuming frictionless operation or complete conversion of electrical to mechanical energy). In many separation processes, the energy required for separation is supplied as heat at a high temperature T_h and withdrawn as heat at a low temperature T_ℓ. The maximum amount of work that can be obtained from an ideal heat engine operating between two temperatures T_h and T_0 is, according to the Carnot principle, a fraction $(T_h - T_0)/T_h$ of the total amount of heat supplied. For the amount of heat Q_h supplied, the corresponding amount of work available is $Q_h\ [(T_h - T_0)/T_h]$. If the amount of heat rejected by the separation device is Q_ℓ at temperature T_ℓ, then the amount of work one could get out of this heat, from the Carnot principle, is $Q_\ell\ [(T_\ell - T_0)/T_\ell]$, where T_0 is the lowest temperature heat sink (Figure 10.1.3). Therefore the highest amount of work one can get out from an amount of heat Q_h supplied to the separation device at T_h and an amount of heat Q_ℓ rejected by the separation device at T_ℓ is as follows:

$$W_{\text{net}} = Q_h\left(\frac{T_h - T_0}{T_h}\right) - Q_\ell\left(\frac{T_\ell - T_0}{T_\ell}\right). \tag{10.1.21}$$

If $Q_h = Q_\ell$ and no other energy sources are involved, then the net amount of work required for separation is

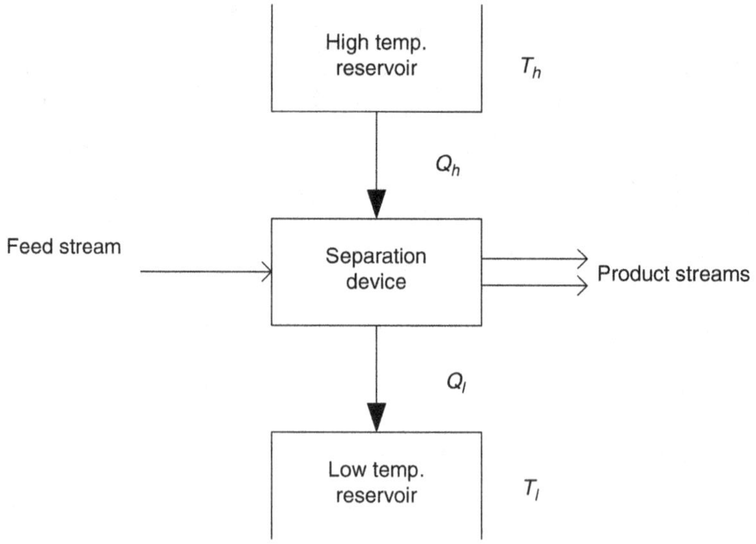

Figure 10.1.3. Heat input to a separation device at temperature T_h *and heat rejection at a lower temperature* T_l *provides the work needed for separation.*

$$W_{net} = Q_h T_0 \left(\frac{1}{T_\ell} - \frac{1}{T_h} \right). \quad (10.1.22)$$

Both expressions (10.1.21) and (10.1.22) represent what is called the *net work consumption* for the separation process driven by heat.

Suppose the separation device is using heat working between 120 °C and 25 °C, $Q_h = Q_\ell$ and $T_0 = (25 + 273)$ K. Then the fraction of the heat actually used for separation is

$$T_0 \left(\frac{T_h - T_\ell}{T_h T_\ell} \right) = 298 \left(\frac{393 - 298}{393 \times 298} \right) = 0.24.$$

In Sections 10.1.1 and 10.1.2, we obtained the minimum energy required for separating 1 m^3 of water as ~0.71 kW-hr/m^3. If the energy supplied for separation is thermal energy between 120 °C and 25 °C, then the minimum amount of heat energy required for separation is (0.71/0.24) = 2.95 kW-hr/m^3 of water.

10.1.4 Minimum energy required for membrane gas permeation, distillation, extraction and other separation processes

In this section, we will illustrate the calculation of the minimum energy required for a variety of separation processes, e.g. membrane gas permeation, distillation, extraction and adsorption.

10.1.4.1 *Membrane gas permeation*

Consider air at 1 atm pressure and 25 °C as an ideal gas mixture of 21% O_2, 79% N_2 (mole %). We are interested in separating 1 gmol of this mixture into pure O_2 and pure N_2 at 1 atm and 25 °C through two semipermeable membranes and two isothermal reversible compressors between the membranes and the final pure gas product streams (Figure 10.1.4). We could obtain an estimate of the energy required for this separation using equation (10.1.2). Instead we will use the basic principles of thermodynamics and calculate the changes in appropriate thermodynamic quantities as we go from location 1 (feed air mixture, x_{O_2f}, x_{N_2f}, 1 atm) to locations A (pure O_2 at a pressure x_{O_2f} atm), C (pure N_2 at a pressure x_{N_2f} atm), B (pure O_2 at 1 atm) and D (pure N_2 at 1 atm).

Consider location 1. The total enthalpy of the mixture is $x_{O_2f}\overline{H}_{O_2} + x_{N_2f}\overline{H}_{N_2}$, where $\overline{H}_i$ is the partial molar enthalpy of species i in the mixture. At location A, the enthalpy of pure oxygen is $x_{O_2f}H_{O_2}$ since there are x_{O_2f} gmols of pure O_2, whose molar enthalpy is H_{O_2}. Since we have ideal gas behavior, $H_{O_2} = \overline{H}_{O_2}$. Similarly, $H_{N_2} = \overline{H}_{N_2}$ at location C, where the enthalpy of pure N_2 is $x_{N_2f}H_{N_2}$. Note that this O_2 semipermeable membrane allows O_2 to come through at the feed partial pressure; similarly for the N_2 semipermeable membrane. Calculate the Gibbs free energy change as follows:

$$\Delta G|_{1\to A/C} = \Delta(H - TS) = (\Delta H - T\Delta S)|_{1\to A+C}; \quad (10.1.23)$$

$$\Delta H|_{1\to A+C} = \Delta H|_{A+C} - \Delta H|_1 = \left(x_{O_2f} H_{O_2} + x_{N_2f} H_{N_2}\right)_{A+C} - \left(x_{O_2f}\overline{H}_{O_2} + x_{N_2f}\overline{H}_{N_2}\right)_1 \Rightarrow \Delta H|_{1\to A+C} = 0. \quad (10.1.24)$$

We know, from basic thermodynamics, $\Delta S|_{1\to A+C} = Q/T$. For an isothermal reversible process, $Q = 0$, since no heat was supplied or withdrawn (see Problem 10.2.1). Therefore

$$\Delta G|_{1\to A+C} = \Delta H|_{1\to A+C} - T\Delta S|_{1\to A+C} = 0 - 0 = 0. \quad (10.1.25)$$

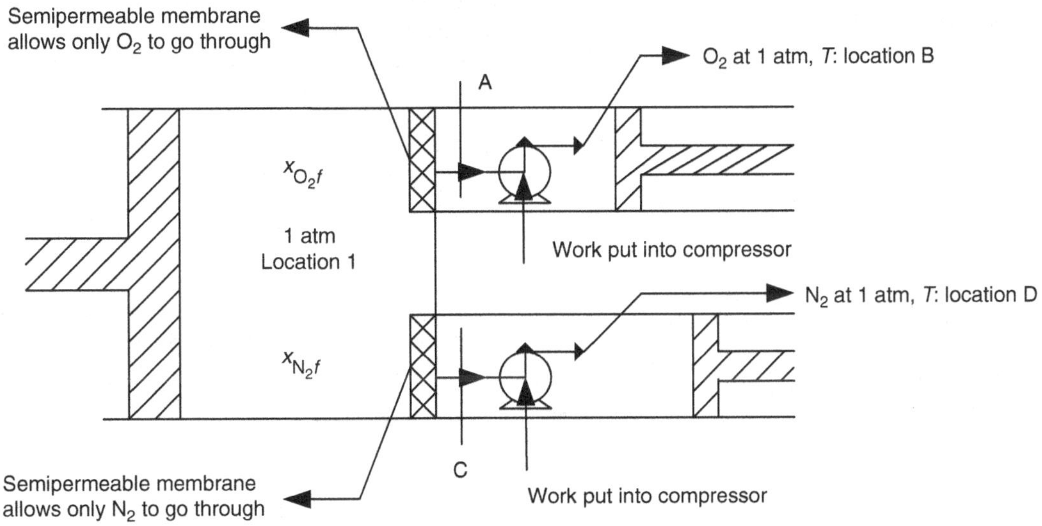

Figure 10.1.4. Reversible isothermal separation process of an O_2–N_2 mixture at 1 atm through two semipermeable membranes and two reversible isothermal compressors to produce two pure streams of O_2 and N_2, each at 1 atm.

Now we will calculate the energy required for isothermal compression in the two compressors for pure O_2 and pure N_2. From standard textbooks on thermodynamics (e.g. Smith and Van Ness (1975, p. 451)), the mechanical energy balance equation for 1 gmol of a pure fluid going from pressure P_1 to P_2 is:

$$\int_{P1}^{P_2} V_i \, dP + \frac{\Delta v^2}{2g_c} + \frac{g}{g_c}\Delta z + F + W_s = 0. \tag{10.1.26}$$

Here, v is the velocity of the fluid, Δz is the change in height during the process, F is an estimate of the frictional energy loss and W_s is the shaft work done by the system on the surroundings. For the O_2 stream between locations A and B, $\Delta v^2 = 0$, $\Delta z = 0$, $F = 0$,

$$W_s = -\int_{P_1}^{P_2} V_{O_2}\, dP$$

per gmol of O_2. Since we have x_{O_2f} gmol of O_2,

$$\begin{aligned} W_s &= -x_{O_2f}\int_{P_1}^{P_2} V_{O_2}\, dP \\ &= -x_{O_2f}\, RT \ln\frac{P_2}{P_1} = -x_{O_2f}\, RT\ln\frac{1}{x_{O_2f}\times 1} \\ &= x_{O_2f}\, RT\ln x_{O_2f}. \end{aligned} \tag{10.1.27}$$

Therefore, the value for the shaft work supplied from outside is $-x_{O_2f}\, RT\ln x_{O_2f}$. The corresponding value for the N_2 compressor is $-x_{N_2f}\, RT\ln x_{N_2f}$. Thus, the total shaft work supplied from outside to the two compressors is given by

$$-x_{O_2f}\, RT\ln x_{O_2f} - x_{N_2f}\, RT\ln x_{N_2f}. \tag{10.1.28}$$

We will now focus on calculating ΔG for the whole process. We need to calculate the values for the step A + C → B + D. For the enthalpy change,

$$\Delta H|_{A+C\to B+D} = Q_R - W_s = 0$$

since, for the ideal gas,

$$H = U + PV = U + RT \Rightarrow \Delta H = \Delta U + R\Delta T = 0, \tag{10.1.29}$$

for the isothermal process. Therefore $Q_R = W_s$ for the isothermal process. For the reversible process,

$$\Delta S|_{A+C\to B+D} = \frac{W_s}{T} = x_{O_2f}\, R\ln x_{O_2f} + x_{N_2f}\, R\ln x_{N_2f}. \tag{10.1.30}$$

Therefore

$$\begin{aligned} &\Delta G|_{total} = \Delta H|_{total} - T\Delta S|_{total} \Rightarrow \Delta G|_{1\to A+C} + \Delta G|_{A+C\to B+D} \\ &= \Delta H|_{1\to A+C} + \Delta H|_{A+C\to B+D} - T[\Delta S_{1\to A+C}] - T[\Delta S_{A+C\to B+D}] \\ &= 0 + 0 - T\times 0 - T x_{O_2f}\, R\ln x_{O_2f} - T x_{N_2f}\, R\ln x_{N_2f}; \end{aligned} \tag{10.1.31}$$

$$\Delta G|_{total} = -RT\{x_{O_2f}\ln x_{O_2f} + x_{N_2f}\ln x_{N_2f}\}, \tag{10.1.32}$$

a result already illustrated in relation (10.1.5).

10.1.4.2 *Distillation and evaporation*

We will illustrate first the thermodynamic minimum energy requirement according to equation (10.1.4) for an example of distillation of a binary mixture of styrene (46.5 mol%) and ethylbenzene (53.5 mol%) producing two pure products at 110 °C (Humphrey and Keller, 1997). Then we will calculate the minimum energy requirement for distillation based on the minimum heat required. We consider evaporation briefly at the end of this section.

According to equation (10.1.4), the thermodynamic minimum energy requirement per gmol of an ideal binary feed mixture of styrene ($i = 2$) and ethylbenzene ($i = 1$) yielding pure products is given by

$$-W_{min}|_{mol} = RT\left[-x_{1f}\ln x_{1f} - x_{2f}\ln x_{2f}\right]. \tag{10.1.33}$$

Here, $x_{1f} = 0.535$ and $x_{2f} = 0.465$, resulting in

$$\begin{aligned} -W_{min}|_{mol} &= -1.987\,\frac{\text{cal}}{\text{gmol-K}}\times 383\,\text{K}\times[0.535\ln 0.535 \\ &\quad + 0.465\ln 0.465] = 523.9\ \text{cal/gmol of feed}. \end{aligned} \tag{10.1.34}$$

If this process is carried out in a distillation column, the minimum energy required may be determined from the heat Q_R supplied in the reboiler/gmol of feed at T_R if we may assume that the total heat supplied at the reboiler is equal to that withdrawn in the condenser (i.e. Q_C) at T_C. Further, this minimum will occur at the minimum reflux ratio, which means that there will be an infinite number of plates. Following Humphrey and Keller (1997), we assume the following: complete separation of feed into two pure products; constant relative volatility α_{12}; constant molar overflow; feed at bubble point; minimum reflux ratio; single reboiler and condenser; liquid feed at bubble point. Consider now the distillation column shown in Figure 10.1.5(a). The overall and component material balance equations are:

$$W_{tf} = W_{td} + W_{tb}; \quad W_{tf}x_{1f} = W_{td}; \quad W_{tf}(1 - x_{1f}) = W_{tb}; \; x_{1f} = x_{1\ell}^{f}. \tag{10.1.35}$$

At any location in the column, the vapor composition is related to the liquid composition via the relative volatility relation (4.1.24):

$$x_{1v} = \frac{\alpha_{12}x_{1\ell}}{1 + (\alpha_{12} - 1)\,x_{1\ell}}. \tag{10.1.36}$$

For the operating conditions, the McCabe–Thiele diagram is shown in Figure 10.1.5(b). It is clear from the figure that the slope of the stripper section operating line for the given conditions is x_{1v}/x_{1f}^{ℓ}' which is also equal to $W_{t\ell}/W_{tv}$. Therefore, using (10.1.36),

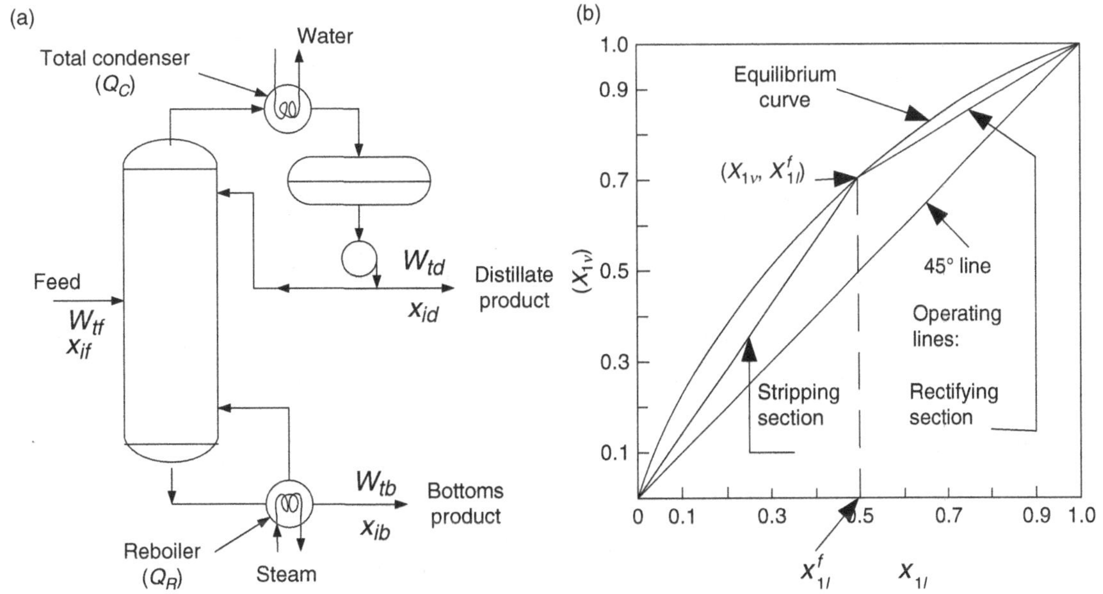

Figure 10.1.5. (a) Distillation column with a reboiler and a total condenser for a binary mixture separation. (b) Separation of a binary mixture into two pure compounds by binary distillation: McCabe–Thiele diagram. (After Humphrey and Keller (1997).)

$$\frac{W_{t\ell}}{W_{tv}} = \frac{\alpha_{12}}{1+(\alpha_{12}-1)\,x^f_{1\ell}} = \frac{W_{tv}+W_{tb}}{W_{tv}} = \frac{W_{tv}+W_{tf}(1-x_{1f})}{W_{tv}}, \tag{10.1.37}$$

which leads to the following result:

$$(W_{tv}/W_{tf}) = (1/(\alpha_{12}-1)) + x^f_{1\ell} = (1/(\alpha_{12}-1)) + x_{1f}. \tag{10.1.38}$$

If ΔH_{vb} is the heat of vaporization of the bottoms product, then the minimum heat supplied at the reboiler per mole of feed is

$$Q_R|_{\min} = \Delta H_{vb}(W_{tv}/W_{tf}) = \Delta H_{vb}\{(1/(\alpha_{12}-1)) + x_{1f}\}. \tag{10.1.39}$$

An alternative way of expressing this result is

$$Q_R|_{\min} = \Delta H_{vb}(R+1)\,(W_{td}/W_{tf}), \tag{10.1.40}$$

where R is the reflux ratio.

For the styrene (2)-ethylbenzene (1) system mentioned earlier, it is known that ΔH_v of styrene (bottom product) is 9350 cal/gmol. Further, $\alpha_{12} \cong 1.37$ (between 1.34 and 1.40 for vacuum operation, see King (1980, p. 645)). Therefore we may obtain a value of $Q_R|_{\min}$ for x_{1f}=0.535 (see the system conditions prior to result (10.1.34)):

$$Q_R|_{\min} = 9350\,\frac{\text{cal}}{\text{gmol}}\left\{\frac{1}{0.37}+0.535\right\} = 30\,265\,\frac{\text{cal}}{\text{gmol of feed}}. \tag{10.1.41}$$

When we compare this result with the thermodynamic minimum work needed (see (10.1.34)) for such a feed liquid, namely 523.9 cal/gmol of feed, the energy efficiency, given by

$$\eta = 100 \times \frac{\text{minimum energy needed}}{\text{minimum energy supplied}}, \tag{10.1.42}$$

is very low, namely (Humphrey and Keller, 1997)

$$\eta = 100 \times \frac{523.9}{30\,265} = 1.73\%. \tag{10.1.43}$$

If we consider the net work consumption basis of the heat supplied at the reboiler (at ~110 °C, $T_R = T_h$) and that rejected at the condenser (at ~50 °C, $T_C = T_\ell$), then, from definition (10.1.22), we may write the net work consumption per mol of feed to be (if T_0 corresponds to 50 °C, say)

$$(-W_{\text{net}}) = Q_R|_{\min}\left(\frac{T_0}{T_0}-\frac{T_0}{T_h}\right) = Q_R|_{\min}\left(1-\frac{323}{383}\right) = Q_R|_{\min} \times 0.157. \tag{10.1.44}$$

Therefore, the energy efficiency is the thermodynamic minimum divided by $(-W_{\text{net}})$:

$$\eta = 100 \times \frac{523.9}{30\,265 \times 0.157} \cong 11\%. \tag{10.1.45}$$

For close-boiling mixtures, where $\ell n\ \alpha_{12}$ may be approximated by $(\alpha_{12}-1)$, and all other conditions of distillation identified in the previous analysis are valid, Hougen *et al.* (1959) and King (1980, p. 666) have provided a compact expression for the net work consumption in distillation between a reboiler temperature T_R and the condenser temperature T_C under the condition of minimum reflux. Since $Q_R = Q_C$,

$$-W_{\text{net}} = Q_R T_0 \left(\frac{1}{T_C} - \frac{1}{T_R} \right) = Q_C T_0 \left(\frac{1}{T_C} - \frac{1}{T_R} \right). \tag{10.1.46}$$

For a relatively pure distillate stream, the minimum reflux liquid flow rate $W_{t\ell}|_{\min}$ (by (8.1.177)) and Q_C are given by

$$W_{t\ell}|_{\min} = \frac{1}{(\alpha_{12}-1)} W_{tf};\ Q_C \cong \frac{\Delta H_{v2}}{(\alpha_{12}-1)} \left(W_{t\ell}|_{\min} + W_{td} \right). \tag{10.1.47}$$

From the Clausius–Clapeyron equation for vapor pressure P_i^{sat} of species *i*,

$$\frac{d \,\ell n P_i^{\text{sat}}}{d(1/T)} = \frac{-\Delta H_{vi}}{R} \tag{10.1.48}$$

$$\Rightarrow \ \ell n P_1^{\text{sat}}|_{T_C} = a_1 - \frac{\Delta H_{v1}}{R} \frac{1}{T_C}; \qquad \ell n P_1^{\text{sat}}|_{T_R} = a_1 - \frac{\Delta H_{v1}}{R} \frac{1}{T_R}. \tag{10.1.49}$$

However, $P_1^{\text{sat}}|_{T_C}$ for complete separation corresponds to column pressure, whereas $P_1^{\text{sat}}|_{T_R}$ is more likely to be α_{12} times the column pressure. Therefore

$$\begin{aligned} &\ell n P_1^{\text{sat}}|_{T_R} - \ell n P_1^{\text{sat}}|_{T_C} \\ &= \frac{\Delta H_{v1}}{R} \left(\frac{1}{T_C} - \frac{1}{T_R} \right) = \ell n \left(P_1^{\text{sat}}|_{T_R} / P_1^{\text{sat}}|_{T_C} \right) \\ &\cong \ell n(\alpha_{12}) \cong (\alpha_{12} - 1) \end{aligned} \tag{10.1.50}$$

for close separation.

Substituting (10.1.47) and (10.1.50) into (10.1.46), we get

$$\begin{aligned} -W_{\text{net}} &= \frac{\Delta H_{v2}(W_{t\ell}|_{\min} + W_{td})\, T_0}{\dfrac{\Delta H_{v1}}{R} \left(\dfrac{1}{T_C} - \dfrac{1}{T_R} \right)} \times \left(\frac{1}{T_C} - \frac{1}{T_R} \right) \\ &= R T_0 W_{tf}, \end{aligned} \tag{10.1.51}$$

where we have assumed $\Delta H_{v2} \cong \Delta H_{v1}$, which is a good assumption for close-boiling mixtures. Therefore, per gmol of feed,

$$-W_{\text{net}}|_{\text{gmol of feed}} = R\, T_0 \tag{10.1.52}$$

for close-boiling mixtures in distillation between a single reboiler and a condenser. The corresponding energy efficiency η vis-à-vis the thermodynamic minimum at T_0 is

$$\eta = \frac{R\, T_0 \left[-x_{1f} \,\ell n\, x_{1f} - x_{2f} \,\ell n\, x_{2f} \right]}{R\, T_0}. \tag{10.1.53}$$

This efficiency is maximum at $x_{1f} = x_{2f} = 0.5$, and is quite high compared to the result given in (10.1.43).

Evaporation is generally carried out in one stage and the vapor is pure solvent, e.g. evaporative desalination of saline water produces pure water vapor. Following relation (10.1.39), the minimum heat required per mole of feed is

$$Q_R|_{\min} = \Delta H_{v1} \left(W_{tv} / W_{tf} \right), \tag{10.1.54}$$

where ΔH_{v1} is the heat of vaporization of the solvent being evaporated. If the evaporative heat is supplied at T_h and the heat rejection is taking place at T_0, then the net work consumption per mole of feed is

$$-W_{\text{net}}|_{\text{mole}} = Q_R|_{\min}\, T_0 \left(\frac{1}{T_0} - \frac{1}{T_h} \right). \tag{10.1.55}$$

10.1.4.3 *Solvent extraction*

The direct expenditure of energy in the separation process of solvent extraction is due to the pumping costs and costs of agitation/dispersion in conventional solvent extraction devices (membrane solvent extraction devices do not have dispersion; only pumping costs are relevant). Let us ignore the small amount of energy required for these steps. Instead, focus on the extract stream from which the extracted product and the solvent have to be separated to reuse the solvent and recover the product. Usually this is achieved by distillation. This cost must be less than that required to separate the product from the feed solution being subjected to extraction by distillation. This is the reason to choose solvent extraction to start with, unless the thermal stability of the product is in question, as is true for many pharmaceutical molecules. Solvent extraction is a necessity in such cases, e.g. penicillins. A small amount of extraction solvent will also be present in the raffinate. This solvent should be removed and recovered also (usually by distillation) (Null, 1980).

If the solvent is high-boiling or essentially nonvolatile, the energy requirement for distillation to recover the volatile product is minimized. For a nonvolatile solvent, a flash distillation will remove the volatile product. Often the solvent is quite volatile when the products being considered are high-boiling. In such a case, while separating the extract for recycling the solvent or recovering the solvent lost in the raffinate, the solvent will appear in the distillate.

Consider now the case where the extraction solvent is high-boiling with respect to the extracted solute ($i = 1$) in extract ($= x_{1E}$). The minimum heat supplied at the reboiler per mole of feed which undergoes the solvent extraction process is obtained, prior to using relation (10.1.39), as

$$Q_R|_{\min}|_{\text{mole of feed to be extracted}} = Q_{R|\text{mole of extract}} \times \frac{\text{moles of extract}}{\text{moles of feed to undergo extraction}} = Q_R|_{\text{mole of extract}} \frac{x_{1f}}{x_{1E}},$$

since

$$\frac{x_{1f}}{x_{1E}} = \frac{\text{moles of solute 1 in feed to undergo extraction/mole of feed to undergo extraction}}{\text{moles of solute 1 in extract } (j{=}E)\text{/mole of extract}}. \quad (10.1.56)$$

However, from relation (10.1.39),

$$\begin{aligned} Q_{R|\text{mole of extract}} &= \Delta H_{vb|\text{extract}}\left(W_{tv}/W_{tf}\right) \\ &= \Delta H_{vb|\text{extract}}\left\{\frac{1}{(\alpha_{12}-1)} + x_{1E}\right\}, \quad (10.1.57) \end{aligned}$$

since here x_{1E} represents the feed mole fraction of the more volatile species, $i = 1$, in the distillation column for the extract. Therefore

$$Q_{R|\min|\text{mole of feed to be extracted}} = \frac{x_{1f}}{x_{1E}} \Delta H_{vb|extract}\left\{\frac{1}{(\alpha_{12}-1)_{\text{extract}}} + x_{1E}\right\}, \quad (10.1.58)$$

whereas ΔH_{vb} corresponds to the ΔH_v for the solvent in the distillation tower bottoms section (Humphrey and Keller, 1997). We could also determine the minimum net work consumption for distillation of the extract at the solvent extract distillation column bottom temperature of $T_{R,sE}$:

$$-W_{\min}\big|_{T_{R,sE}} = Q_{R|\min|\text{mole of feed to be extracted}}\left(\frac{T_0}{T_0} - \frac{T_0}{T_{R,sE}}\right). \quad (10.1.59)$$

Solvent extraction will appear to be useful if the original feed to undergo solvent extraction underwent distillation at a bottoms temperature T_h instead, where

$$\begin{aligned} -W_{\min}\big|_{T_h} &= Q_{R|\min|\text{mole of feed to be extracted}}\left(\frac{T_0}{T_0} - \frac{T_0}{T_h}\right) \\ &= \Delta H_{vb}\big|_f\left\{\left(\frac{1}{\alpha_{12}-1}\right)_{\text{feed}} + x_{1f}\right\}\left(1 - \frac{T_0}{T_h}\right) \end{aligned} \quad (10.1.60)$$

is greater than that given by expression (10.1.59):

$$\begin{aligned} &\Delta H_{vb}\big|_f\left\{\left(\frac{1}{\alpha_{12}-1}\right)_{\text{feed}} + x_{1f}\right\}\left(1 - \frac{T_0}{T_h}\right) \\ &> \frac{x_{1f}}{x_{1E}} \Delta H_{vb}\big|_{\text{extract}}\left\{\left(\frac{1}{\alpha_{12}-1}\right)_{\text{extract}} + x_{1f}\right\}\left(1 - \frac{T_0}{T_{R,sE}}\right). \end{aligned} \quad (10.1.61)$$

Null (1980) has provided an analysis of distillation vs. solvent extraction separation of a feed solution under a variety of conditions.

10.1.4.4 *Adsorption*

The regeneration step in adsorption can be accomplished by pressure-swing (for gaseous species), purge-gas, temperature-swing (TSA processes) or displacement processes (see Section 7.1.1.3). Assuming that the adsorption process does not involve substantial energy (due to large pressure drops, etc., which are unlikely), the regeneration step does require significant energy input. For the case of adsorption from a gaseous stream, pressure reduction or temperature increase needs energy input. For liquid-phase adsorption, the bed may be regenerated by steam. We will consider here very briefly the energy required for regeneration of the bed for a TSA process with a gaseous feed stream (Null, 1980); steam desorption for liquid-phase adsorption is also relevant. The minimum energy required for the thermal-swing step at a temperature T_{des} per unit gmol of species i adsorbed at a temperature T_{ads} for a gaseous feed is

$$Q_{R|\min|\text{gmol}\,i} = \left[\left\{c_{pi} + \left(c_{p\text{bed}}/\overline{q}_{i1}\right)\right\}\left(T_{\text{des}} - T_{\text{ads}}\right) + \Delta H_{\text{ads}}\right], \quad (10.1.62)$$

where ΔH_{ads} is the heat of adsorption of species i in units of J/gmol of i, c_{pi} is the specific heat of the adsorbed species i in J/((gmol of i)-K), $c_{p\text{bed}}$ is the specific heat of the adsorbent bed in J/((g of adsorbent bed)-K), $\overline{q}_{i1}$ is the average adsorbate concentration to be desorbed in (gmol adsorbed of species i)/(g of adsorbent bed). The net work consumption basis for this heat supplied for regeneration (10.1.62) is

$$W_{\text{net}|\text{gmol}\,i} = Q_{R|\min|\text{gmol}\,i}\left(1 - \frac{T_0}{T_{\text{des}}}\right). \quad (10.1.63)$$

The expressions for minimum energy required and the net work consumption per unit time are obtained as follows:

$$\begin{aligned} Q_{R|\min|\text{time}} &= Q_{R|\min|\text{gmol}\,i} \times W_{ti|\text{ads}}; \\ -W_{\text{net}|\text{time}} &= W_{\text{net}|\text{gmol}\,i} \times W_{ti|\text{ads}}, \end{aligned} \quad (10.1.64)$$

where $W_{ti|\text{ads}}$ is the rate of adsorption of species i per unit time, say, per hour. Since, in many such calculations, the thermal energy required for separation is determined, one often wonders how such a process would compare with the process of distillation for liquid feeds. As Null (1980) has shown, when small amounts of high-boiling species are to be removed, adsorption is preferred over distillation. Similarly, when the reflux ratio needed is very large (due to lower α_{12}), adsorption energy requirements may be lower.

In ion exchange processes, the regeneration is achieved by passing a different solution with different composition or pH through the ion exchange resin bed. The energy cost is not directly available.

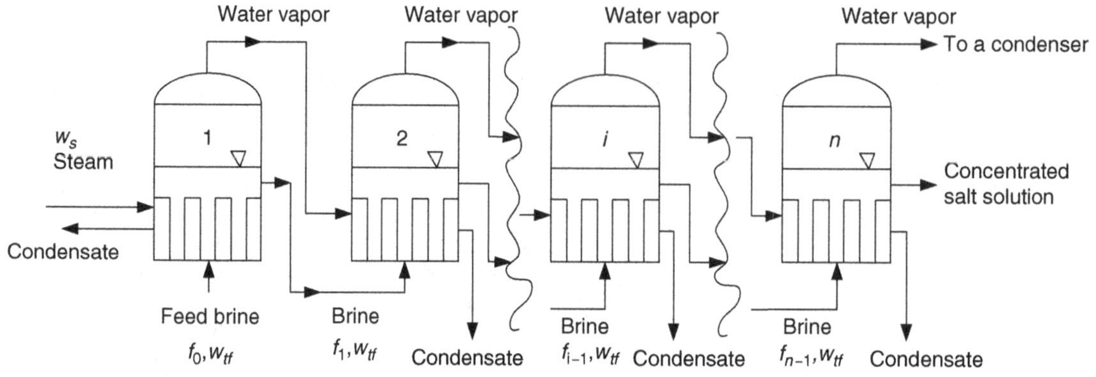

Figure 10.2.1. Multiple-effect evaporation. The numbers in each evaporator, 1, 2,. . ., i, . . ., n correspond to the first effect, second effect, ith effect and the nth effect, respectively.

10.2 Reducing energy required for separation

The total cost of implementing any separation includes the cost of an investment in the separation equipment (the capital cost), the cost of energy needed for separation and the required operating costs. Considerable efforts have been invested over many years to reduce the cost of the separation equipment as well as the cost of energy needed for separation. Here we focus on a few concepts employed to reduce the cost of energy required for separation.[2] The concepts of interest are: multiple-effect evaporation; vapor-recompression evaporation; multiple pressure based operation. We illustrate these concepts first with evaporation of water for desalination. Then we briefly consider their application to distillation. We treat next the free energy lost by mixing different streams and how it may be recovered. At the end of the section, we discuss the nature of challenges posed by a dilute solution. We begin with multiple-effect evaporation, multistage flash evaporation and vapor-recompression evaporation for desalination.

10.2.1 Evaporation of water for desalination

10.2.1.1 *Multiple-effect evaporation*

Let us consider the case of sea water desalination by evaporation. Suppose w_{tf} kg/s is the flow rate of brine into an evaporator (evaporator 1), where w_s kg/s of steam is supplied to evaporate water from brine. Let f_0 be the weight fraction of water in the feed stream (~0.965). Let the weight fraction of water left in the concentrated brine leaving this evaporator be f_1. Recognize that the amount of water vapor produced in this evaporator has considerable thermal energy in it. We can now condense this vapor in another evaporator (evaporator 2) fed with the concentrated brine leaving evaporator 1 (Figure 10.2.1). Let the weight fraction of water left in the concentrated brine leaving this second evaporator be f_2. Assuming that all of the heat obtained by condensing the steam in the first evaporator is also used to evaporate water in the second evaporator by condensing the water vapor from the first evaporator, and that the latent heat of vaporization/condensation λ is the same in both evaporators operating at the same temperature, we may write the heat balance as

$$w_s\lambda = w_{tf}(f_0 - f_1)\lambda = w_{tf}(f_1 - f_2)\lambda. \qquad (10.2.1)$$

If such a process is continued with n evaporators in series, as shown in Figure 10.2.1, and if we may assume that the rate of heat supply by condensation of steam in evaporator 1 is also supplied in each subsequent evaporator, then we obtain the following heat balance relation for n evaporators:

$$w_s\lambda = w_{tf}(f_0 - f_1)\lambda = w_{tf}(f_1 - f_2)\lambda = \cdots = w_{tf}(f_{n-1} - f_n)\lambda, \qquad (10.2.2)$$

which yields the following mass balance relation:

$$w_s = w_{tf}(f_0 - f_1) = w_{tf}(f_1 - f_2) = \cdots = w_{tf}(f_{n-1} - f_n). \qquad (10.2.3)$$

Summing these relations containing weight fractions of water entering or leaving a given evaporator for n evaporators, we get

$$w_{tf}(f_0 - f_n) = nw_s = n(f_{i-1} - f_i)w_{tf}. \qquad (10.2.4)$$

This relation indicates that the total rate of water evaporation in this series of evaporators (total n in number) is n times that in any one evaporator (evaporator i, say), and that the total amount of water evaporated is n times the steam supplied to the first evaporator. Therefore the amount of energy supplied to evaporate 1 kg of water in

[2] Usually, the capital cost goes up when energy costs are reduced.

this assembly of n evaporators in series (Figure 10.2.1) is λ/n, which is considerably smaller than that if we had only one evaporator (where it would be λ, 1 kg of water evaporated per kg of steam condensed). In this multi-evaporator assembly, the first evaporator based system is called the first effect, the next one is called the second effect, and so on.

The situation presented above would suggest that as we keep on increasing the value of n, the net heat required to evaporate 1 kg of water will decrease as λ/n. This is a simplistic picture. There are limits to the achievable value of n. We will consider now these aspects vis-à-vis the evaporation of water in sea water desalination.

There are two basic limitations. First, if steam is supplied into evaporator 1 at 100 °C, the vapor space in the evaporator has to be maintained at a pressure less than 1 atm due to the boiling-point elevation of sea water. (Due to the presence of salts, the vapor pressure of sea water at a given temperature is lower than that of pure water; therefore, at any pressure, the boiling point of sea water is higher than that of pure water by the amount of the boiling-point elevation.) Similarly, the brine in the second evaporator being more concentrated will require a somewhat lower vapor space pressure (than in evaporator 1), and so on till the last stage.

Second, a much more important factor comes into play. For effective heat transfer from the water vapor condensing in evaporator 2, the temperature of the brine in evaporator 2 should be significantly smaller, by, say, 4–5 °C from the temperature of water vapor coming from evaporator 1. Correspondingly, the vapor space pressure in evaporator 2 will be significantly lower than that in evaporator 1. In this fashion, the temperature in the vapor space of the last evaporator approaches a lower limit governed by the environmental temperature and the vacuum level available via a steam ejector and a vacuum pump. In general, the lowest temperature at the last stage (which are here called *effects*) used for practical purposes is ~38 °C. Therefore the total number of effects, is usually limited to somewhat between 10 and 25. Further, as a result of these and other limitations, the number of kilograms of fresh water produced per kilogram of steam condensed in multiple-effect (ME) evaporators decreases with the number n in the following fashion (for illustration only): 0.9 ($n = 1$); 1.75 ($n = 2$); 2.5 ($n = 3$); 3.2 ($n = 4$); 4.0 ($n = 5$), etc. (Spiegler, 1977).

If we, however, continue with the highly simplistic analysis suggesting that the net heat required to evaporate 1 kg of water in an n-effect system is λ/n, then the cost of steam to produce 1 kg of water in an n-effect system will be given by $(C_{\text{steam}} \times w_s)/nw_s$, where the cost of steam per kg is C_{steam}. The other component of the cost is the cost of the evaporators: suppose each evaporator is identical and costs C_{evap} per kg of water produced; then the total cost of the evaporators is $C_{\text{evap}} \times n$. The total cost of producing water is the sum of the capital costs and the operating costs. If we simplistically express the total cost per kg of water evaporated as the following sum (C_{total} is the total cost),

$$C_{\text{total}} = \frac{C_{\text{steam}}}{n} + \left(C_{\text{evap}}\right) n, \tag{10.2.5}$$

we can obtain an optimum value of n as n_{opt} by differentiating the above with respect to n and equating it to 0:

$$\frac{C_{\text{steam}}}{n_{\text{opt}}^2} = C_{\text{evap}} \Rightarrow n_{\text{opt}} = \sqrt{C_{\text{steam}}/C_{\text{evap}}}. \tag{10.2.6}$$

From relations (10.2.3) and (10.2.4) the expression for the total cost $C_{\text{cost}}^{\text{T}}$ of the multi-effect system is

$$C_{\text{cost}}^{\text{T}} = C_{\text{steam}} \frac{w_{tf}(f_0 - f_n)}{n} + w_{tf}(f_0 - f_n)\left(C_{\text{evap}}\right) n. \tag{10.2.7}$$

King (1980, p. 788) has provided the following expression for the evaporator cost:

$$w_{tf}(f_0 - f_n)\left(C_{\text{evap}}\right)n = \text{const}_1 \left[\frac{w_{tf}(f_0 - f_n)\lambda}{U(T_s - T_n)}\right]^m n, \tag{10.2.8}$$

where U is the overall heat transfer coefficient in the evaporator (assumed equal for all evaporators), T_s is the condensation temperature of the steam in the first evaporator (first effect), T_n is the temperature of boiling in effect n (neglecting boiling-point elevation) and const_1 and m are cost-related factors. Just as the expression (10.2.5) for n_{opt} was obtained, similarly we can obtain from expressions (10.2.7) and (10.2.8) the following expression for n_{opt} (King, 1980, p. 789):

$$n_{\text{opt}} = \left[\frac{C_{\text{steam}}}{\text{const}_1}\left\{U\frac{(T_s - T_n)}{\lambda}\right\}^m \left\{(f_0 - f_n)w_{tf}\right\}^{1-m}\right]^{1/2}. \tag{10.2.9}$$

There are several other real-life features which should be considered for a more detailed analysis. From considerations on practical rates of heat transfer without too much heat transfer surface area, the brine temperature in effect 2 should be at a temperature 4–5 °C lower than that of the vapor temperature coming from the first effect. Therefore, the saturated brine coming from the first effect (at a higher pressure) to the second effect where the saturated brine is at a temperature 4–5 °C lower (and at a slightly lower pressure) will undergo a small flash, and a little bit of water vapor would be released in addition to that generated by boiling in effect 2. Details of the calculation procedures incorporating these effects are provided in El-Sayed and Silver (1980). However, the subject of flashing is the basis of another widely used technology in desalination called *multistage-flash* (MSF) evaporation. Just as in multiple-effect (ME) evaporation, the energy required for evaporation of water is substantially reduced in the MSF process considered next.

10.2.1.2 *Multistage-flash evaporation*

In this evaporation process for desalination of say, sea water, there is a series of n stages, each stage being an assembly of a feed brine heater, a water vapor condenser and a flashing evaporator (Figure 10.2.2). Cold brine at feed temperature T_{bf} enters the feed brine heater in stage n. It is heated there by the latent heat of the condensing vapor at a slightly higher temperature. The condensate is collected and becomes the distillate. The condensing vapor is generated by flashing hotter brine coming at the bottom of the stage from stage $n-1$ at a slightly higher pressure. At the upper end of the cascade (Figure 10.2.2), the feed brine which has been heated in n stages reaches T_{b1}, which is the next to highest temperature in the scheme as it comes out of the feed brine heater section of stage 1. This hot brine is then introduced to a steam-heated separate brine heater, where its temperature is raised to say, 100 °C; however, its pressure is also raised, so that boiling is prevented. This hot sea water next enters the flashing evaporator section of stage 1 maintained at a pressure lower than that of the steam-heated brine heater at the top of the cascade. A small fraction of the water is evaporated by flashing; the vapor goes up and condenses on the tubes carrying the feed brine to be heated up. The design of the device allows the condensate to be collected separately from the hot brine, which is then taken to stage 2 where the pressure is somewhat lower than that of stage 1, leading to some more flashing of water vapor. As the flashing takes place, the temperature of the brine is reduced; the sensible heat of the brine generates the latent heat of vaporization.

If the brine is cooled by such a multistage-flash process from 100 °C to, say, 60 °C, the fraction of brine that will be evaporated will be around 0.071. If the lowest temperature of the concentrated brine is around 40 °C, the fraction of brine evaporated will increase slightly from 0.071. However, from an energy cost perspective, the sensible heat of brine over a range of temperature, from T_{b1} exiting stage 1 (and entering the steam-heated brine heater) to the temperature of the cold feed brine T_{bf}, is being continuously recycled. The extra energy supplied from outside in the steam heater is to raise the temperature of the hot brine from T_{b1} to, say, 100 °C. Therefore, from a simplistic calculation point of view, the kilograms of steam produced by the MSF process per kilogram of steam supplied in the brine heater is the ratio $(T_{b1}- T_{bf})/(100 - T_{b1})$. One can see that this ratio can be large if T_{b1} is close to 100 °C, say 95 °C, and T_{bf} is, say, 35 °C; i.e. $(60/5) = 12$. For a detailed analysis, see El-Sayed and Silver (1980).

An added advantage of this process is that the flashing takes place from the liquid without any heat transfer surface being present. Therefore any scales formed float in the liquid, unlike those in the multieffect evaporation process, where the evaporator tube surfaces are exposed to scaling due to evaporation. (Typical scaling salts are $CaSO_4$ and $CaCO_3$; they deposit on the heat transfer surface, reducing the heat transfer coefficient.) The heat transfer surface between the condensing vapor produced by the flash and the feed brine does not encounter scaling since there is no evaporation from a solution on either side of the heat transfer surface.

10.2.1.3 *Vapor-compression evaporation*

In the two processes of evaporation described above, the highest temperature in the process is provided by steam from an external source. In the vapor-compression evaporation process, except for an initial source of heat to start the process, the heat at the highest temperature is provided by the vapor from the evaporation process itself. This is achieved by mechanically compressing the vapor generated by evaporation in a compressor (either a mechanical compressor or a steam-ejector based compressor) from a pressure P_1 in the evaporator to P_2 at the exit of the compressor; the corresponding saturation temperatures of the two vapors are T_1 and T_2 ($>T_1$) (Figure 10.2.3). The difference $(T_2 - T_1)$ should be larger than the temperature difference needed for heat transfer to the brine (in the

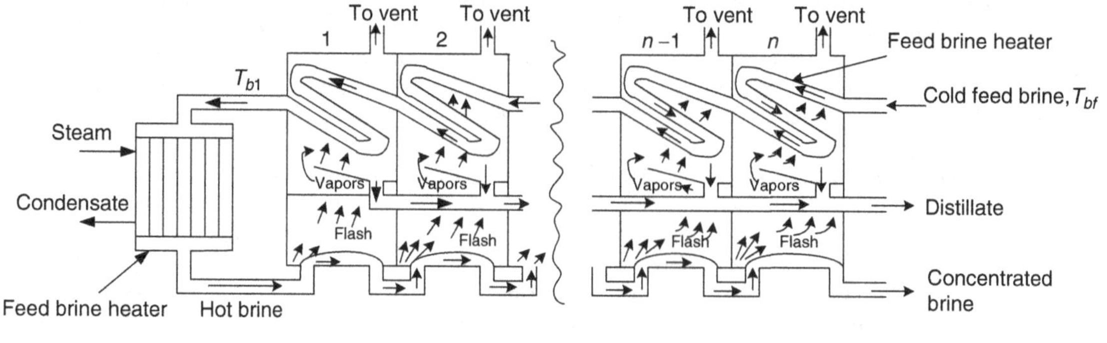

Figure 10.2.2. Multistage-flash evaporation for desalination: n *stages. (After Spiegler (1977).)*

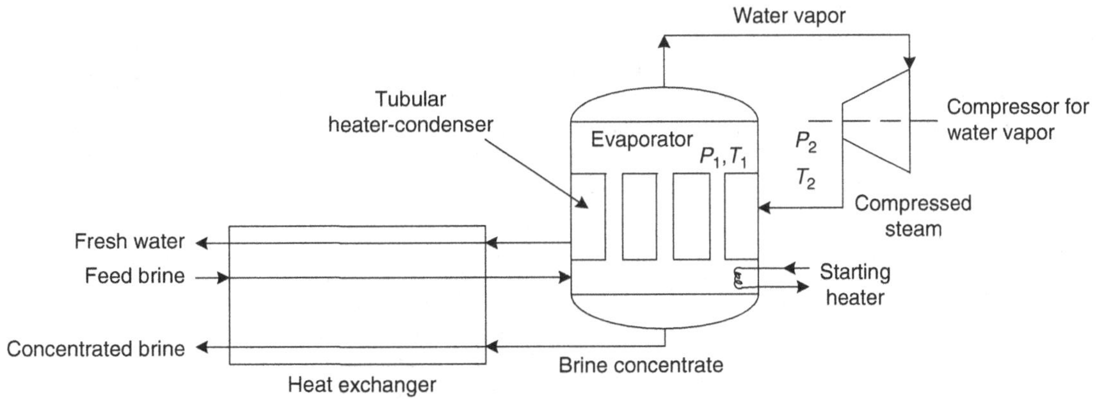

Figure 10.2.3. Vapor-compression evaporation for desalination.

evaporator) from the condenser by the boiling-point elevation, as well as any frictional losses in the vapor conduits (El-Sayed and Silver, 1980).

If the process of water vapor compression is described like that for an ideal gas undergoing an adiabatic process, then the reversible work required for the compression is given by that for an isentropic process:

$$-W_s|_{rev} = \frac{\gamma P_1 V_1}{(\gamma-1)}\left[\left(\frac{P_2}{P_1}\right)^{\frac{\gamma-1}{\gamma}} - 1\right]. \quad (10.2.10)$$

Here, $\gamma = (c_p/c_v)$ and V_1 is the volume of the vapor at temperature T_1. The value of γ for water vapor is 1.32. For processes that are not completely adiabatic, γ will be replaced by another quantity δ, where PV^δ = constant instead of PV^γ = constant for an ideal gas following an adiabatic compression. In vapor-compression evaporation processes, where the ratio of the pressures (P_2/P_1) (= compression ratio = r) is close to 1, one can employ the following simplification:

$$\left(\frac{P_2}{P_1}\right)^{\frac{\gamma-1}{\gamma}} = \left(1 + \frac{P_2-P_1}{P_1}\right)^{\frac{\gamma-1}{\gamma}} \cong 1 + \frac{\gamma-1}{\gamma}\left(\frac{P_2-P_1}{P_1}\right), \quad (10.2.11)$$

yielding

$$-W_s = \frac{\gamma P_1 V_1}{(\gamma-1)}\left[\frac{\gamma-1}{\gamma}\left(\frac{P_2-P_1}{P_1}\right)\right] = (P_2-P_1)V_1$$
$$= (P_1V_1)(r-1) = \frac{(P_1V_1)(r^2-1)}{(r+1)} \cong \frac{P_1V_1(r^2-1)}{2r}, \quad (10.2.12)$$

for $r = (P_2/P_1)$ close to 1 (Spiegler, 1977).

We can now obtain an estimate of the work needed for this compression work by equating P_1V_1 to RT_1. Consider a somewhat concentrated sea water boiling at 101.05 °C; now R = 2.31×10^{-6} kW-hr/gmol-K. Therefore

$$-W_s|_{gmol} = 2.3 \times 10^{-6}\,\frac{\text{kW-hr}}{\text{gmol-K}} \times 374.2\,\text{K} \times \frac{(r^2-1)}{2r}$$
$$= 430.33 \times 10^{-6}\,\frac{(r^2-1)}{r}\,\frac{\text{kW-hr}}{\text{gmol}}.$$

On a volumetric basis, this energy requirement turns out to be

$$-W_s|_{liter} = 430.33 \times 10^{-6}\,\frac{(r^2-1)}{r}\,\frac{\text{kW-hr}}{\text{gmol}}\,\frac{1}{0.018}\,\frac{\text{gmol}}{\text{liter}}$$
$$= 23\,907 \times 10^{-6}\,\frac{(r^2-1)}{r}\,\frac{\text{kW-hr}}{\text{liter}};$$
$$-W_s|_{m^3} = 23.9(r^2-1)/r\,\frac{\text{kW-hr}}{\text{m}^3}. \quad (10.2.13)$$

Let r have the following values: 1.1, 1.2. Then the corresponding values for $-W_s|_{m^3}$ are 4.562, 8.763 kW-hr/m^3. We can compare these values with the minimum energy required per equation (10.1.18) for, say, sea water concentrated twice at 101.5 °C:

$$-W_s|_{liter} = \frac{ART\,s_1}{\overline{V}_s} = \frac{0.000537 \times 2.31 \times 10^{-6} \times 374.2 \times 70}{0.018}\,\frac{\text{kW-hr}}{\text{liter}}$$
$$= 18.03 \times 10^{-4}\,\text{kW-hr/liter} = 1.8\,\text{kW-hr/m}^3.$$

The compressed water vapor condensate is the product of this process. Its sensible heat, as well as that of the brine concentrate, are used to heat the feed brine in a heat-exchanger arrangement (Figure 10.2.3).

10.2.2 Distillation

The three common strategies that may be employed to reduce the energy requirement in the distillation method for separation of liquid solutions involve the principles of multieffect operation, multipressure operation and heat pump. However, unlike that for evaporation in desalination, there exists a considerable variety of approaches under each one of these broad principles. We will briefly introduce the reader to the basics of such principles one by one. All such methods are of interest since distillation continues to be the separation process which uses the largest amount of energy among all separation processes.

10.2.2.1 *Multieffect distillation*

In conventional distillation processes, the main source of external energy input is in the reboiler at the highest column temperature. If, therefore, the amount of this external energy input per mole of feed is reduced, the energy efficiency ought to improve. One way to achieve this involves using the heat from the condensation of the overhead vapor in the reboiler of another column; the two columns become thermally linked. For a binary feed mixture, one method of implementing this is to have two columns connected in the fashion of Figure 10.2.4(a), with the lower column (column 1) operating at a pressure P_H high enough to let the vapors from the top come out and condense in the reboiler of the upper column (column 2) operating a lower pressure P_L. However, since the feed input rate is the same for both columns, the steam input to the bottom column reboiler is only for half the total feed flow rate being processed. This energy expenditure reduction requires an increased capital investment due to two columns. *Note*: The column diameters are reduced due to the lower vapor flow rate. However, the total separation equipment volume is similar to that for a single-column operation. Additional details may be found in Robinson and Gilliland (1950) (as the earliest source for such a scheme), King (1980, pp. 697–698), Henley and Seader (1981) and Wankat (1988). Wankat (1993) has considered 23 different multieffect distillation schemes and has suggested heuristics to reduce the number of options.

Consider the overhead vapor from column 1 in Figure 10.2.4(a). The dew-point temperature of this vapor must be greater than the bubble-point temperature of the bottom liquid of column 2. In addition, there has to be two other considerations for successful column coupling. The total heat available from condensation of the vapor from the top of column 1 should be what is needed to vaporize the required fraction of the bottoms from column 2. Correspondingly, there has to be a sufficient temperature

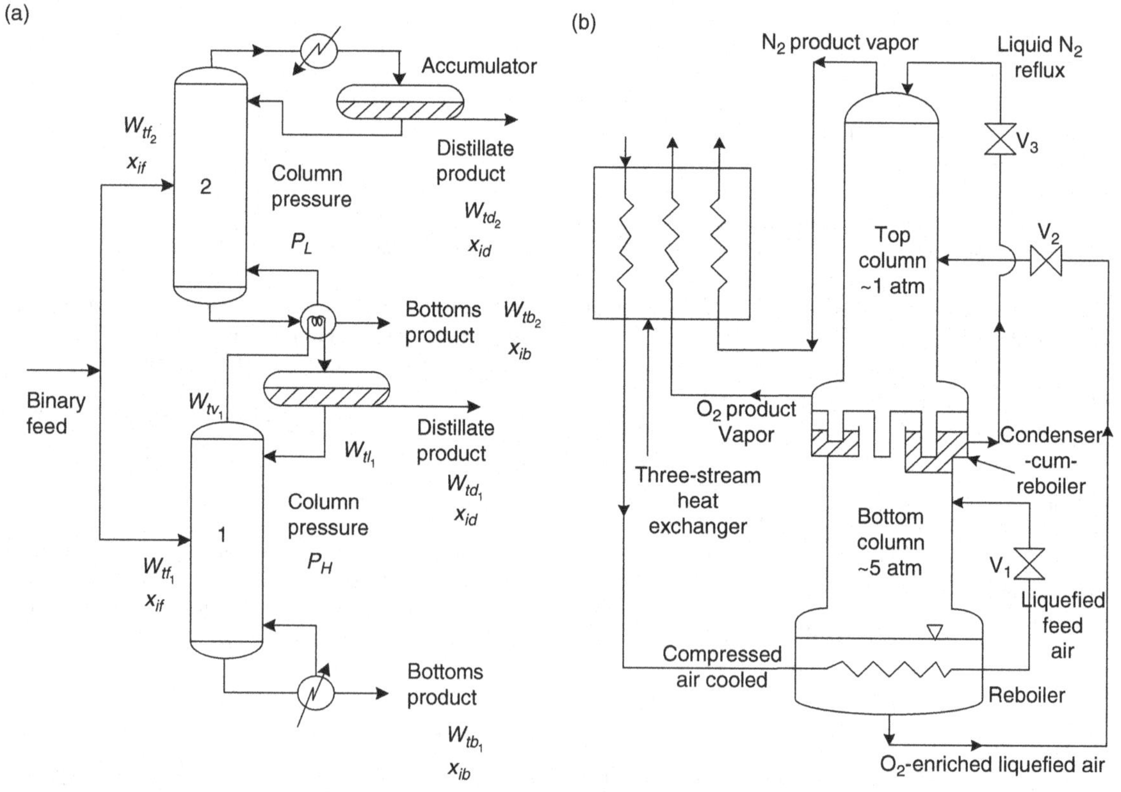

Figure 10.2.4. (a) Multiple-effect distillation with two columns and feed to both columns. (b) Cryogenic air separation in a Linde double-column device: forward feed of liquefied feed air to the bottom column.

difference between these two streams so that the heat-exchanger surface area required is not too large. The balancing of the thermal load is achieved in some multieffect distillation processes by appropriate adjustment of the feed flow rates. Regardless of the details of such balancing procedures, it is clear that, if the condenser of one distillation column is going to be the reboiler of another distillation column, there has to be a difference in pressure between the columns. Therefore multieffect distillation requires operation at multiple pressures: two columns require operation at two different pressures. Further, often fresh feed is introduced to only one column, unlike that shown in Figure 10.2.4(a). Two modes are quite common when the feed is being introduced to only one column: *forward feed*, where the fresh feed is introduced to the bottom column operating at a higher pressure P_H; *backward feed*, where the fresh feed is introduced to the top column operating at a lower pressure P_L. Cryogenic distillation of air provides an example of a forward feed and a Linde double-column arrangement where the low-pressure column is simply physically located above the high-pressure column (Figure 10.2.4(b)).

In such a double-column device, compressed and cooled air[3] is introduced at the bottom of the bottom column into a coil, where it provides a heat source (reboiler) for the evaporation of the liquid on the shell side. The compressed and cooled air then flows through a valve V_1, undergoes Joule–Thompson expansion (and liquefaction) to a pressure of 5 atm. This liquefied air is introduced as feed to the bottom column. Since N_2 is more volatile than O_2, the liquid collecting at the bottom of this column is about 45% O_2: this liquid is introduced now as feed to the middle of the top column operating at 1 atm through a pressure-reducing valve V_2. Rectification in the top column produces almost pure N_2 in the vapor phase at the top and almost pure O_2 as liquid at the bottom of the top column. Correspondingly, the top of the bottom column produces almost pure N_2 at this location. The N_2 condenser-O_2 evaporator at the top of this bottom column is the O_2 evaporator-N_2 condenser of the top column. The almost pure N_2 vapor liquefied here is introduced as reflux at the top of the top column after pressure reduction through valve V_3. Cold purified N_2 vapor at the top of the top column and cold O_2 vapor from the bottom of the top column are taken out as products through a three-channel heat exchanger, which warms up the gaseous O_2 and N_2 streams and cools down the compressed air feed stream going to the reboiler of the bottom column.

There is one potential disadvantage to the multieffect distillation schemes from energy considerations. Although the total heat input into the multieffect distillation is reduced, for example, in the binary distillation of Figure 10.2.4(a), the heat at the bottom of column 1 has to be supplied at a higher temperature due to the higher pressure. Therefore the net work consumption (equation (10.1.22)) goes up due to a higher value of T_h. This has been characterized also as trading "first law heat" for "second law ΔTs" (Westerberg, 1985); the utilities cost goes up as T_h goes up or T_ℓ goes down. An alternative strategy employs the concept of a prefractionator before the main distillation column such that the bottoms utility temperature remains the same in both columns but the total energy consumed is reduced. Although there is a change in the capital costs, the overall costs were reduced (Agrawal *et al.*, 1996).

Figure 10.2.5 illustrates such a concept where an intermediate reboiler is used in the main lower-pressure column below the feed location to condense the overhead vapors from the prefractinator operating at a higher pressure. All of the feed is introduced to the prefractionator (this does not have to be the case). The reboilers of this prefractionator and the main column are running on the same high-temperature utility, and are therefore at the same temperature, but they have different compositions. The prefractionator in this scheme is primarily an enricher, and it produces part of the overall distillate, whereas the bottoms from the prefractionator is introduced as the feed to the main column (Agrawal *et al.*, 1996) after a pressure reduction. The above discussion was focused on a binary mixture separation. These and other configurations require additional considerations when a ternary mixture has to be separated, as has been pointed out already in Chapter 9.

10.2.2.2 *Heat pump*

We have been introduced to the notion of vapor-compression evaporation for desalination in Section 10.2.1.3. Its application to distillation based separation of mixtures is generally identified as a heat pump method of distillation. Figure 10.2.6 illustrates the *open-loop method*, which is one of the two methods of carrying it out for a binary mixture using one distillation column. The overhead vapor from the column top is compressed to a pressure high enough for it to provide boil-up at the bottom of the column as it condenses. Part of the condensate becomes the distillate product, while the pressure of the rest is reduced through a valve V before it is introduced as liquid reflux at the top of the column. The second heat pump method is called the *close-loop heat pump*: here, an external working fluid is used to carry out the boil-up at the column bottom at a high enough pressure. After condensation, the pressure of this liquid is reduced through a Joule–Thompson valve, cooling it substantially, and then it is used to condense the overhead vapor from the column

[3] The air we are referring to here has been supplied, via a compressor, from the ambient atmosphere. The air is then cooled to ~10 °C and purified of water vapor, oil, particles through a series of separators.

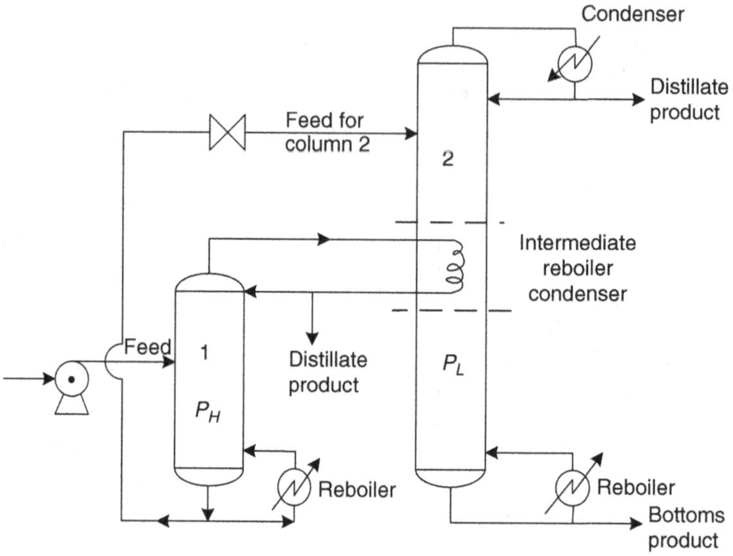

Figure 10.2.5. Prefractionator column at a higher pressure P_H *has been heat integrated with the main column at a lower pressure* P_L *via an intermediate reboiler (Agrawal* et al.*, 1996).*

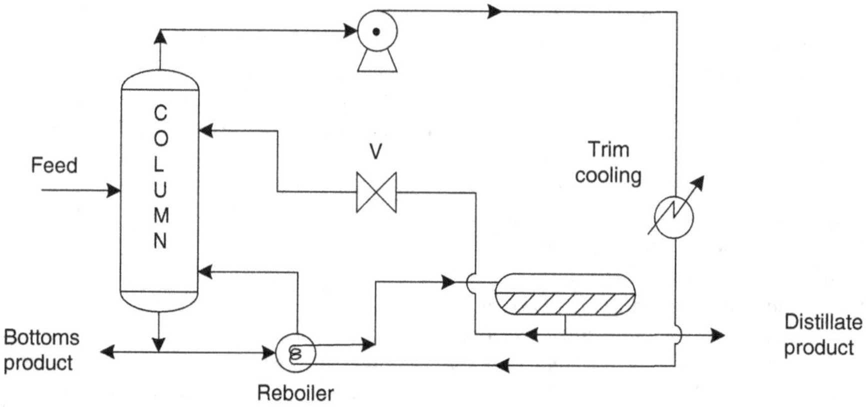

Figure 10.2.6. Open-loop heat-pump arrangement for binary distillation in a single column.

as it is vaporized. The vaporized external working fluid is then prepared to act as a high-temperature utility at the column bottom by compressing the vaporized external working fluid. An introduction to heat-pump schemes for binary distillation using one column has been provided by King (1980, pp. 695–697). For ternary separation heat pump schemes using two thermally linked distillation columns, see Agrawal and Yee (1994).

We will now provide the fundamental expression for the work required to pump the heat from a lower temperature T_2 to a higher temperature T_1 by an ideal reversible Carnot engine. If the amount of heat Q_2 extracted at the low temperature T_2 is pumped by supplying an amount of work $(-W_s)$ to a higher temperature T_1 where the amount of heat rejected is Q_1, then[4]

[4]See the analysis in basic texts on applied thermodynamics, e.g., Smith *et al.* (2001, p. 310). Remember W_s is positive when it is done by the system on the surroundings; it is negative when the surroundings works on the system.

$$Q_1 = Q_2 + (-W_s), \tag{10.2.14}$$

$$\left((-W_s)/Q_1\right) = \frac{T_1 - T_2}{T_1} \tag{10.2.15}$$

and

$$\left((-W_s)/Q_2\right) = \frac{T_1 - T_2}{T_2}. \tag{10.2.16}$$

Result (10.2.16) follows from the other two relations (10.2.14) and 10.2.15). Therefore the amount of work needed to pump one unit of heat ($Q_2 = 1$ calorie) is $((T_1 - T_2)/T_2)$. If this amount of work were to be generated from a thermal source at T_1 rejecting the heat at T_0, then the amount of heat Q_s needed will have to be estimated from the Carnot principle discussed in (10.1.21), namely

$$Q_s\left(\frac{T_1 - T_0}{T_1}\right) = (-W_s) = \frac{T_1 - T_2}{T_2};$$
$$Q_s = \frac{T_1}{T_2}\left(\frac{T_1 - T_2}{T_1 - T_0}\right). \tag{10.2.17}$$

10.2.3 Free energy of mixing

In Chapter 9, we came across the "no mixing condition" during cascade analysis. This condition suggests that the energy spent in developing separation should not be lost by mixing two streams having different compositions since they have different chemical potentials. This criterion is not enough since streams having the same composition but having different phases (e.g. vapor and liquid) will have different chemical potentials; if these two streams are mixed, their chemical potential difference would be lost. Therefore a more general principle guiding the mixing of two streams in cascades is: there should be no mixing of two streams having different chemical potentials.

On the other hand, in real-life conditions, one may have to undertake such mixing. The strategy to be pursued then is: can one recover some energy during the process of mixing? For example, consider reverse osmosis desalination of sea water. It produces a concentrate whose salt concentration is higher than that of the sea water, the concentration level depending on the fractional water recovery. Normally, this concentrate would be dumped into the sea. However, if there was a device which could generate power resulting from the mixing of this concentrate with the sea water, then the power so generated would reduce the power required for desalination.

If we consider now a relation of the type represented in (10.1.1) and (10.1.2), we may write the following expression for the maximum amount of work recoverable from mixing under ideal conditions when a mole of each feed stream (f_1 for concentrate, f_2 for sea water) is mixed with the other to produce a product stream ($j = p$) dumped into the sea:

$$W_{\max} = \sum_{i=1}^{n}\left(\mu_{if_1} + \mu_{if_2} - 2\mu_{ip}\right); \tag{10.2.18}$$

$$W_{\max} = \sum_{i=1}^{n}\Big(\overline{V}_{if}P_{f_1} + RT\,\ell n\,\gamma_{if_1}x_{if_1} + \overline{V}_{if_2}P_{f_2} + RT\,\ell n\,\gamma_{if_2}x_{if_2} - 2\overline{V}_{ip}P_p - 2RT\,\ell n\,\gamma_{ip}x_{ip}\Big). \tag{10.2.19}$$

If $P_{f_1} = P_{f_2} = P_p$ and $\overline{V}_{if_1} = \overline{V}_{if_2} = \overline{V}_{ip}$ then

$$W_{\max} = RT\sum_{i=1}^{n}\left(\ell n\,\gamma_{if_1}x_{if_1} + \ell n\,\gamma_{if_2}x_{if_2} - 2\ell n\,\gamma_{ip}x_{ip}\right). \tag{10.2.20}$$

Generally the reverse osmosis concentrate is at a high pressure (less than that of the entering high-pressure feed brine); this pressure energy is recovered to a large extent in energy recovery devices. The above expression assumes that the concentrate pressure has been reduced to atmospheric.

We have seen in Section 3.4.2.5 that application of an electrical voltage in a stack of ion exchange membranes when appropriately configured can lead to desalination via the process of electrodialysis. Weinstein and Leitz (1976) and Jagur-Grodzinski and Kramer (1986) have shown how an electrodialytic stack of membranes may be operated in reverse, in what is called a *dialytic battery*. Concentrated brine enters one chamber and dilute brine enters the contiguous chamber: the mixing of these streams through the ion exchange membranes generates a voltage at the electrodes, and therefore power. Estimate of this salinity power, which could be potentially generated from the mixing of river water with sea water worldwide, is huge (Wick, 1978). The cost of recovering such an energy source is, however, a strong deterrent at this time.

We will briefly provide an expression for the voltage generated between the electrodes in a dialytic battery. First, when a membrane is placed between two solutions of an electrolyte having mean ionic activities of $a'_{\pm}$ and $a''_{\pm}(< a'_{\pm})$ (see (3.3.119d), where the mean ionic activity $a_{\pm}$ has been related to the individual ionic activity of the electrolyte AY), it develops a potential difference between the two sides called the *membrane potential*; it is also called *concentration potential*. In an electrodialysis cell, there are many cation exchange membranes and generally an equal number of anion exchange membranes. One cation exchange membrane (CEM), the two solutions on two sides of it and an anion exchange membrane (AEM) constitute a cell pair; an electrodialysis cell may have N cell pairs. If each ion exchange membrane behaves ideally, with a cation exchange membrane allowing only cations to go through and an anion exchange membrane allowing only anions to go through, then the total open-circuit voltage (no external load) for N cell pairs containing N

CEMs and N AEMs developed is (Weinstein and Leitz, 1976)

$$V^o_{\text{stack}} = 2N \frac{RT}{\mathcal{F}} \ell n \left(a'_{\pm} / a''_{\pm} \right) \tag{10.2.21}$$

as the individual ionic species moves through the corresponding ion exchange membrane from a higher activity to lower activity. For nonideal ion exchange membranes, the open-circuit voltage developed will be somewhat lower.

Another way to carry out mixing of two saline solutions would be what is practiced in the concept known as *osmotic power plant*. Suppose you have a saline water (1) at a pressure (P_1) less than its osmotic pressure (π_1). Let the aqueous solution (2) on the other side of the reverse osmosis membrane (having a lower osmotic pressure $\pi_2 < \pi_1$) be present at atmospheric pressure, P_0. If the osmotic pressure of this solution 2 is less than the pressure of the saline water 1 (i.e. $\pi_2 < P_1$), then water from solution 2 will go to solution 1 and will dilute it. However, this water will increase the volume of the saline solution present at pressure P_1. Therefore one could use this extra energy through an appropriate hydraulic device/arrangement (Loeb, 1976).

10.2.4 Dilute solutions

Dilute solutions of solutes/macrosolutes are frequently encountered in practice. Examples are: a dilute fermentation broth containing small volatile molecules such as alcohol, small nonvolatile molecules such as amino acids, large protein molecules that are secreted externally or those that are obtained in a cell lysate from a cell culture. In these cases, recovery of these molecules in a substantially pure condition is the separation goal. Fractionation of isotopes where the desired isotope is present in a very dilute solution is another problem of the same type (separation of D_2O from H_2O; separation of U^{235}/U^{238} isotopes present in a gaseous mixture of UF_6, where the $U^{235}F_6$ atom fraction is 0.00711, a very low concentration). Recovery and purification of a synthetic pharmaceutically active organic compound present in a very dilute aqueous/organic solution is essentially another related problem of a similar nature. Dilute solutions of radioactive impurities obtained as byproducts of fission confront us with a similar challenge in that water has to be purified and the radioactive impurity has to be substantially concentrated for safe disposal in a secure environment. Dilute solutions of organic pollutants in water pose a somewhat similar separation problem, even though the organic pollutant may not have much value, unlike the fermentation products/byproducts mentioned earlier.

There are considerable differences in the separation problems discussed above. In the case of purification of proteins produced by fermentation or cell culture, one needs to concentrate and then purify to obtain as pure a protein as possible that is now substantially free from impurities as well as a large amount of water; similarly for many other products, with water sometimes replaced by an organic solvent. In the case of uranium isotope separation, a product enriched to the level of 0.03 atom fraction of the desired isotope $U^{235}F_6$ is enough. However, all such problems have a common feature:[5] separation requires handling a large volume of an essentially inert or undesirable substance, e.g. water, in a number of the problems described above. There is significant literature available on the cost of a particular desired product, Cost_i, as a function of other quantities, such as mass of inert material in the feed per unit mass of the desired product, $(w_{\text{in},f} / w_{i,f})$ (Lightfoot and Cockrem, 1987). One such expression is

$$\text{Cost}_i = \text{a}\left(w_{\text{in},f}/w_{i,f}\right), \tag{10.2.22}$$

where a is a universal constant in the so-called Sherwood plot, independent of the species. Obviously, the energy cost of treating a large volume of a feed containing a small amount of the desired product is significant. It will vary, however, with the separation method adopted.

Consider a distillation column under minimum reflux for a relatively pure distillate stream. From relation (10.1.47), we can write

$$W_{t\ell}\big|_{\min} = \frac{1}{(\alpha_{12}-1)} W_{tf} = \frac{1}{(\alpha_{12}-1)} \frac{W_{td}}{x_{1f}}$$

$$\Rightarrow \frac{W_{t\ell}\big|_{\min}}{W_{td}} = \frac{1}{(\alpha_{12}-1)} \frac{1}{x_{1f}}. \tag{10.2.23}$$

Since $W_{t\ell}$ is related to W_{tv} by the slope of the operating lines, they are not too far apart on either side of 1 for a low α_{12} system. The molar vapor flow rate from the reboiler provides a reasonable estimate of the energy requirement for distillation. Relation (10.2.23) suggests that as x_{1f} decreases, i.e. the feed becomes more and more dilute in the species 1 being recovered in the overhead as an almost pure product, the energy requirement increases enormously since a very large volume flow rate of inerts has to be handled (Lightfoot and Cockrem, 1987) for this dilute feed. For example, in the separation of D_2O from H_2O (and just a bit of HDO), the natural abundance of deuterium in water in atom ppm varies between 119 and 163 (Benedict *et al.*, 1981, p. 709). These authors have indicated that, in a distillation plant, 200 000 gmols of steam are required per gmol of heavy water produced (mole fraction of D_2O in product is 0.998); the very low separation factor of 1.05 used here (see relation (5.2.46) for an expression for α_{12}) contributes to this enormous energy requirement. As a result, alternative pathways to obtain D_2O are followed (Benedict *et al.*, 1981, chap. 13).

Lightfoot and Cockrem (1987) have pointed out the applicability of equation (10.2.22) for a variety of separation

[5]The differences between enrichment, concentration and purification have been illustrated in Section 1.7.

techniques used to recover products from a dilute fermentation broth. Generally the cost of processing for separation, and therefore to some extent the energy needed for separation, is proportional to the volume of the liquid to be processed in processes such as filtration, membrane processes, centrifugation, solvent extraction, precipitation, crystallization, etc. One can argue that in processes such as solvent extraction and adsorption, if one can have highly selective solvents or adsorbents, then one can achieve a rapid reduction in processing volume via the solute transfer between the phases. It has been pointed out, however, that, in conventional dispersive extraction, due to the limitations of processing flow rates, the phase volume reduction can be only 5–10 fold, regardless of how selective the extracting solvent is (Lightfoot and Cockrem, 1987). New nondispersive membrane based solvent extraction techniques (see Section 3.4.3.2) bypass this limitation and can operate with phase volume reductions as much as 100 times or more in porous hollow fiber membrane based devices.

A general problem in separation from dilute solutions may be posed as follows via an example. Suppose you have a nonvolatile solute, say a protein, a small-molecule drug or a crystalline solute, etc., present in a very dilute aqueous solution. The separation goal is to get this in as pure a solid form as possible. Lightfoot and Cockrem (1987) have identified two alternative strategies as follows.

(a) One can first, in step 1, precipitate the solute from the dilute solution; the precipitate phase will have brought the solute to unit activity since we will have pure solid (see Section 3.3.7.5). Then one can, in step 2, concentrate this precipitate from this solution where it is suspended. In both steps, the whole solution volume has to be treated, and consequently the energy expenditure is likely to be significant.
(b) One can first concentrate the solution of the solute (step 1). Then one can bring the activity of the solute to pure solute activity (unit activity) via precipitation in step 2 from a much smaller solution volume.

In both cases, there will be additional steps taken to achieve final purification. In approach (b), however, the volume of solution to be processed in step 2 is much lower; this is due to the concentration step implemented in step 1, which was not present in step 1 of approach (a). The goal in both approaches is to achieve a product whose concentration is as close as possible to 1 in terms of mole fraction (x_{ij}) and where the species activity is also as close as possible to 1 (a_{ij}). Therefore in an x_{ij}–a_{ij} plot, approaches (a) and (b) follow very different pathways. In approach (a), the first step increases a_{ij} then increases x_{ij}. In approach (b), the first step increases x_{ij}, then a_{ij} is increased. One should remember that any such discussion must be qualified by the results of the actual separation step undertaken. For example, in approach (a), the first step may be concentration of the solution via ion exchange adsorption. The desorption step should lead to a much more concentrated solution.

Problems

10.1.1 Using the expression for the minimum energy required to desalinate sea water (salinity, $s = 35$) at 25 °C, calculate the minimum energy required per m^3 of fresh water produced if the final volume of the concentrated sea water ($= V_2$) is 90%, 75%, 50% and 25% of the original sea-water volume V_1.

10.1.2 Calculate the minimum amount of work you have to do to recover 1 cm^3 and 1 liter of fresh water from a brackish water source using a salt-impermeable membrane. Show that the theoretical expression for this work is equal to $RTC_{salt}|_1$, where $C_{salt}|_1$, is the molar concentration of salt. The osmotic pressure of the salt solution is given by Van't Hoff's law. The temperature is 25 °C. The brackish water contains 0.5 g of NaCl per 100 g of water.

10.2.1 To reduce the energy required for a distillation process with almost pure water as overhead vapor, a heat-pump arrangement is proposed. The column is operating at atmospheric pressure. The reboiler needs a vapor condensing at 150 °C. The overhead water vapor should be compressed by a reversible adiabatic compression process such that it can be the condensing vapor in the reboiler. Determine the heat delivered by the condensing vapor per pound of the vapor; calculate the ratio of the heat delivered in the reboiler to the work required in the reversible compressor. Use the steam tables from Smith and Van Ness (1975) or any other suitable source. (Ans. 1020 Btu/ℓb (567 cal/g); 7.254.)

10.2.2 Consider the possibility of combining the vapor-compression evaporation principle for desalination with the multiple-effect evaporation principle for a two-effect system. Draw a schematic of such a plant and identify the possible advantages of such a combination with respect to Figures 10.2.3 and 10.2.1 (for a two-effect system).

10.2.3 This problem is concerned with balancing the two columns, 1 and 2, in the multieffect distillation process of Figure 10.2.4(a) by adjusting the two feed flow rates, W_{tf1} and W_{tf2}. Assume that the feed to both the columns is saturated liquid. Further, the two columns are being operated with $(W_{t\ell}/W_{td})$ ratios related to their minimum values in the following way:

$$(W_{t\ell}/W_{td})_1 = r_1(W_{t\ell}/W_{td})_{1,\min}; \qquad (W_{t\ell}/W_{td})_2 = r_2(W_{t\ell}/W_{td})_{2,\min}.$$

Develop the following expression for the ratio (W_{tf1}/W_{tf2}) in terms of r_1, r_2, the minimum values of $(W_{t\ell}/W_{td})$ in both columns and the latent heats of vaporization/condensation in the two columns where they are thermally coupled at the condenser-reboiler between columns 1 and 2:

$$\frac{W_{tf_1}}{W_{tf_2}} = \frac{\left[r_2\left(W_{t\ell}/W_{td}\right)_{2,\min}+1\right]}{\left(\Delta H_{v1}/\Delta H_{v2}\right)\left\{r_1\left(W_{t\ell}/W_{td}\right)_{1,\min}+1\right\}}.$$

Assume that the bottoms and the distillate compositions are identical for each column.

10.2.4 A waste heat source is available at a temperature lower than the steam temperature used in the reboiler of a distillation column for a binary mixture. There are three possible ways of using this waste heat: use an additional reboiler somewhere in the stripping section; heat the feed stream to the column so that part of it is vapor; the feed is split into two streams of flow rates W_{tf1} and W_{tf2}, where the feed stream f_2 is completely vaporized by the waste heat at an intermediate reboiler and introduced at an appropriate location in the stripping section (feed stream f_1 is introduced higher up). Focus on this third method of split feed, which divides the column into three sections. Assume constant molar overflow with appropriate allowances for two feed input locations. Use minimum reflux methodology and two potential pinch locations (at the pinch location or pinch point the operating line intersects the equilibrium curve). Determine the minimum vapor flow rate from the bottom reboiler for each pinch location and comment on which arrangement will lead to the reduction of energy requirement in the bottom reboiler.

10.2.5 Consider a saturated liquid feed flow rate of W_{tf} mol/s having a composition x_{1f} for a binary mixture of species $i = 1, 2$ for which $\alpha_{12} = \left(P_1^{\text{sat}}/P_2^{\text{sat}}\right) = 5$ over the range of temperatures and pressures of interest. We are interested in the separation achieved by distillation for the following two cases.

(1) We heat this feed liquid sufficiently for it to reach vapor-liquid equilibrium in the flash drum (see Section 6.3.2.1); determine graphically the vapor and liquid compositions $(x_{1v}, x_{1\ell})$ if $x_{1f} = 0.5$, $q = 0.5$ and $\alpha_{12}=5$.

(2) We split the saturated liquid feed into two fractions having flow rates of W_{tf1} and W_{tf2}; introduce one fraction at the top of a distillation column, vaporize the second fraction completely and introduce it at the bottom the column. Plot the q-lines for the two streams entering the column if $x_{1f} = 0.5$. Develop, under conditions of constant molal overflow, the operating line for this distillation column. Plot the line for the situation where $W_{tf1} = W_{tf2}$. Obtain graphically the compositions of the overhead vapor product and the bottoms liquid product for two values of the number of ideal equilibrium plates in column, $N = 1$ and $N = 5$.

(3) Discuss how $N = 5$ leads to higher separation for the same energy input for all other cases.

10.2.6 In Section 2.2.3, the *separative power* δU was introduced for a single-entry separator in an isotope separation plant via equation (2.2.32a). Expression (2.2.42) was developed for the value function in the case of *close separation*. The developments in Chapter 2 were carried out in mole fraction units and molar flow rates. In the market for nuclear fuels, an estimate of the energy requirement for isotope separation is sometimes expressed in units of separative work, SWU, which is essentially identical to the δU expression (2.2.32a) containing expression (2.2.42) for the value function, except mass units are used for the flow rates w_{t1}, w_{t2}, w_{tf}:

$$\delta U = w_{t1}(2x_{11}-1)\,\ell n\left(\frac{x_{11}}{1-x_{11}}\right) + w_{t2}(2x_{12}-1)\,\ell n\left(\frac{x_{12}}{1-x_{12}}\right) - w_{tf}(2x_{1f}-1)\,\ell n\left(\frac{x_{1f}}{1-x_{1f}}\right).$$

If the w_{ij} values are expressed in kilograms (ignore the time dependence for the time being since the w_{ij} terms are mass flow rates) and the x_{ij} values are estimated based on atom fractions, then calculate the value of δU in kilograms for the following:

(1) a feed of 2 kg of natural uranium containing 0.711% U^{235};

(2) an enriched uranium product of 0.33 kg containing 2.8% U^{235}.

Show that the value of δU is essentially 1 kg; this is identified as 1 SWU.

11

Common separation sequences

We have studied a variety of separation processes and techniques. Our focus was on developing an elementary understanding of an individual separation process/technique. In practice, more often than not, a combination of more than one separation process is employed, regardless of the scale of operation involved. Here we introduce very briefly the separation sequences employed in a few specific industries. The separation sequences of interest are considered under the following headings: bioseparations (Section 11.1); water treatment (Section 11.2); chemical and petrochemical industries (Section 11.3); hydrometallurgical processes (Section 11.4). It is to be noted here that often the separation sequences are reinforced by chemical reactions within such a sequence or before/after the separation steps. The intent here is to provide an elementary view of the complexity and demands of practical systems where certain types of separation sequences are crucial/primary/dominant components.

More often than not, we will find that certain types of separation techniques and processes are much more prevalent in certain industries. For example, solvent extraction and back extraction processes are dominant in the recovery and purification of metals and metallic compounds via hydrometallurgical processes. On the other hand, distillation and, to a much lesser extent, absorption/stripping followed by solvent extraction are the primary separation processes in the chemical/petrochemical industries. Water treatment industries/plants are however, focused much more on deactivation/removal of biological contaminants, suspended materials and dissolved impurities from water via oxidation processes, filtration, membrane techniques and ion exchange processes. Biological separations share some of these characteristics of water treatment processes in terms of the separation techniques; however, since the focus is on recovering/purifying the biologically relevant compound, processes such as chromatography are in great demand.

11.1 Bioseparations

Separation processes utilized to recover, concentrate and purify biologically produced molecules, macromolecules or cellular entities in the biotechnology industry (Ladisch, 2001; Shuler and Kargi, 2002) are commonly studied under the general title of "bioseparations." Often these and other related activities are lumped under the title *downstream processing* (Belter *et al.*, 1988). Generally a sequence of separation processes is implemented to obtain the final product. This sequence is governed by the source of the compound as well as the desired final product purity and form. The variety of products from the biotechnology industry is enormous. Table 11.1.1 provides a very brief list of products in various categories, such as smaller molecules, somewhat larger molecules, proteins, enzymes, viruses and whole cells. Compounds identified in the list are/can be/were produced via biotechnology.

In general, such products are present in low concentrations or as dilute solutions, often as a result of a batch process. The process may be one of cell culture, fermentation or may involve biochemical reactions/transformations based on enzymes as such or enzymes present inside whole cells. In a few cases, the starting mixture for separation may be a natural product, e.g. corn, or a biological fluid, e.g. blood plasma. Correspondingly, one could classify the latter as biomedical separations, all others being biochemical separations. For products that are not whole cells, the products may have been secreted by whole cells and are therefore present externally in the solution: these are called *extracellular* products. Those products which are not secreted by whole cells, and therefore have to be recovered from the cell by destroying the cell (the so-called "lysing" of the cell), are identified as *intracellular*. The cost of recovery and purification of biologically produced products can be more than half of the total production cost. The more dilute the initial solution of the product, the higher will be the ultimate cost of production.

Table 11.1.1. Illustrative list of bioproducts in a few categories

Category	Bioproduct example	Molecular weight	Physical dimension
Smaller molecules	Ethanol, acetone, n-butanol, acetic acid, citric acid, lactic acid, butanediol, amino acids, glucose, fructose	Generally less than 200	
Somewhat larger molecules	Antibiotics, steroids, disaccharides, larger fatty acids	Generally less than 600–700 but greater than 200	
Macromolecules	Proteins (enzymes), polysaccharides, nucleic acids	10^3 to 10^{10}	
Particles - not free-living	Virus		100 nm
Particles - free-living	Bacteria, [a]fungi (Baker's yeast), animal cells, plant cells		1-10 μm[b]

[a] See Table 2.3, Shuler and Kargi (2002) for more details about various components of bacteria.
[b] See Tables 7.3.1 for additional details.

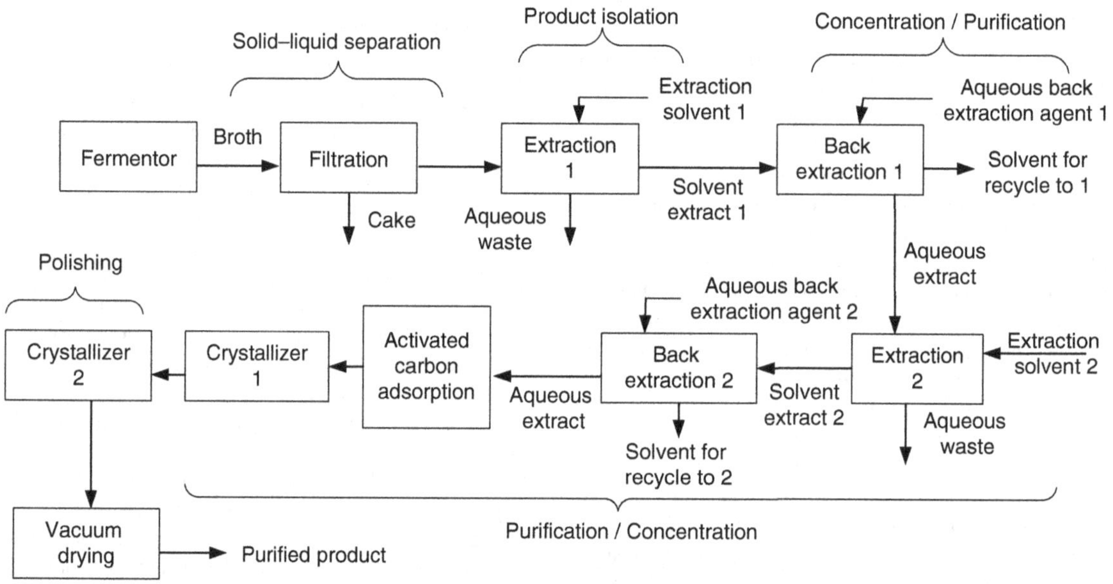

Figure 11.1.1. Block diagram for the production process for a typical antibiotic, penicillin, an extracellular product.

Consider a typical antibiotic production process, as illustrated in Figure 11.1.1. The antibiotic is present in a fermentation broth as a very dilute solution in the presence of considerable soluble impurities, as well as a significant concentration of whole cells and cellular debris. The particulate materials, such as whole cells, cell debris, etc., are removed first by filtration or centrifugation. The filtration method is described under rotary vacuum filtration (Section 7.2.1.5) or microfiltration (Section 7.2.1.4) of the tangential-flow type (TFF). Centrifugation is illustrated in Sections 7.3.2.1 and 7.3.2.2 via different types of centrifuges. This step has been characterized as *removal of insolubles* (Belter *et al.*, 1988) – it is essentially separation of cellular particles from a liquid solution of the product.

In the next step, the filtered fermentation broth is contacted with an extracting solvent in a mixer-settler type of device (Section 6.4.1.2); the solvent extracts the antibiotic from the broth, along with many related and nonrelated compounds. Countercurrent extracting cascades in the form of centrifugal extractors are employed (Section 8.1.4, Figure 8.1.35) to reduce the contact/residence time. This process is often called *product isolation* in that the product has been isolated from the broth; however, the solvent extraction process extracts other compounds as well from the broth. Often, adsorption (Section 7.1.1) as well as membrane processes such as ultrafiltration and reverse osmosis may be used (Sections 7.2.1.3 and 7.2-1.2). Sometimes, such a step results in a significant increase in product concentration.

The next stage in the antibiotic processing in Figure 11.1.1 involves *purification* and *concentration*. In the case of antibiotics, first there occurs a process of back extraction into an aqueous solution, which selectively back extracts the product more than the other impurities. Then additional solvent extraction and aqueous back extraction steps are carried out under appropriate *pH* conditions to achieve further purification and concentration. This process is usually followed by activated carbon adsorption to remove impurities, then crystallization (Section 6.4.1), which provides a very high level of purification. For many soluble products, the purification step is usually implemented via chromatography (Section 7.1.5), in the case of proteins by the highly selective affinity adsorption/chromatography (Section 7.1.5.1.8). Traditionally, the last step in purifying the product in the desired form is often termed the *polishing step* (Belter *et al.*, 1988). It involves purification via crystallization/chromatography. In the case of antibiotics, crystallization is followed by vacuum drying (Section 6.3.2.4).

It is useful now to identify the concentration of the antibiotic product (i.e. the titer) during the different steps identified earlier (Belter *et al.*, 1988). The product concentration level in the fermentation broth may vary between 0.1 and 5 g/liter. Removal of insolubles via microfiltration, rotary vacuum filter and centrifuging increases the product concentration marginally, in the range of 1-5 g/liter. Product isolation via solvent extraction enhances the product concentration to 5-50 g/liter. Purification by chromatography or crystallization leads to a product concentration level between 50 and 200 g/liter. The polishing step may not enhance the concentration in general much, but it certainly improves the purity. If a membrane process such as reverse osmosis is used to concentrate a very dilute solution of the antibiotic, 10-30 times the initial concentration may be achieved.

The concentration of penicillin in the fermentation broth has recently increased to as much as 10-50 g/liter. On the other hand, the concentration of a therapeutic protein in the fermentation broth may be as low as 10^{-3}-10^{-5} g/liter. At the other extreme, small molecule bioproducts, such as citric acid or alcohol, are present in the bioreactor at levels of around 100 g/liter and 70-120 g/liter, respectively.

Antibiotics are not the only extracellular products obtained in this way; alcohol, citric acid, lactic acid, l-lysine are typical small-molecule examples of extracellular products. For biological macromolecules, examples of extracellular products include proteins, such as tissue-type plasminogen activator, monoclonal antibodies (mAbs), such as immunoglobulins, proteases (enzymes which degrade proteins such as subtilisins), used in a variety of applications of food processing, industrial and household applications, and polysaccharides, such as xanthan gum.

We will now briefly illustrate an abstracted stepwise scheme that may be followed for recovery and purification of an intracellular product (Figure 11.1.2). Typical examples of intracellular products are: human insulin (a hormone, MW 5734), bovine growth hormone, human growth hormone (HGH), human leukocyte interferon, glucose isomerase from *Streptomyces* species, hemoglobin from red blood cells.

Amongst many differences in the processing schematics between an extracellular product and an intracellular product are the additional steps to be pursued to get the product into the usual downstream processing steps (Figure 11.1.2). Usually, with intracellular products, the cells have to be harvested first via centrifugation (into pellets) or tangential-flow microfiltration (TFF) (retentate). These cells are then lysed (ruptured, broken open) via mechanical disruption (as in a homogenizer, bead mills, etc.), chemical disruption (via osmotic shock - which happens when cells are immersed in pure water which enters the cell and swells it, leading to rupture - or solubilization by a concentrated solution of detergents, which destroy the cell membrane, releasing the materials from the cell interior; a dilute solution (e.g. 10%) of toluene will essentially achieve the same goal except organic solvents may not be desirable as a contaminant).

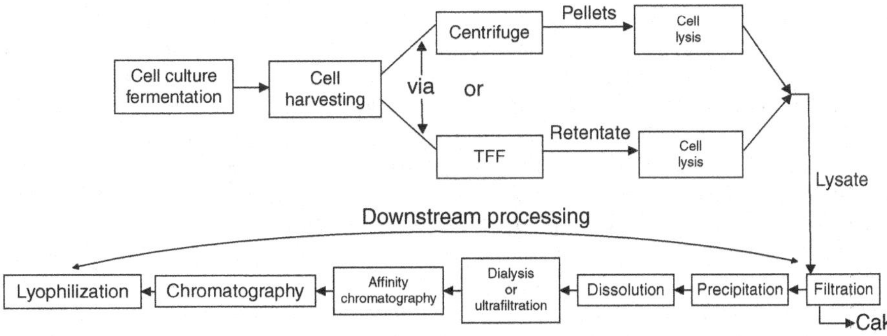

Figure 11.1.2. Block diagram containing additional steps in the production process for a typical intracellular product prior to downstream processing (TFF = tangential-flow filtration).

Once the lysate is obtained, it is subjected to filtration to remove cellular debris; the product is in the clarified solution. Usually, it (e.g. a protein) is precipitated or extracted by aqueous two-phase extraction (Sections 4.1.3 and 8.1.4). The precipitate may be dissolved and then subjected to the usual concentration and purification steps.

Such a scheme has to be preceded by a number of complex steps if the protein (for example) product is present inside the cell as an inclusion body. Inclusion bodies are specific regions in cells containing a large amount of protein, often misfolded; once they are released, after cell disruption, refolding of the protein has to be carried out. An example of such a situation is encountered in the production of human insulin, a polypeptide. An illustrative account of human insulin production is provided in Harrison *et al.* (2003, chap. 11). The same reference provides details of the production of extracellular products, such as citric acid as well as a monoclonal antibody. One of the earliest textbooks on bioseparations is by Belter *et al.* (1988). Among more recent textbooks, that by Ladisch (2001) has a greater emphasis on chromatography since it supposedly consumes about 50% of the bioseparation costs. Additional textbooks on bioseparations are those by Garcia *et al.* (1999) and Harrison *et al.* (2003). Sections on bioseparations in Bailey and Ollis (1986) and Shuler and Kargi (2002) are also useful. These books are to be consulted for additional information.

Two other types of products should be mentioned. When the product is a whole cell (e.g. baker's yeast, single-cell proteins), many of the downstream processing steps considered so far are not needed. Filtration, followed by any other treatment necessary for the cell to be useful as a final product, is sufficient. For baker's yeast, the filtered cake is extruded into particles and dried at ~45–50 °C. The second type of product is of much smaller molecular weight and volatile in nature, e.g. ethanol, acetone, butanol, acetic acid, etc. In all such cases, whether a concentrated or a diluted product (e.g. low-alcohol beer) is needed, often the first step is to go to a beer still, where ethanol and water, for example, are evaporated and taken out through the top whereas the suspended solids are taken out from the bottom of the still as "stillage." The stillage is passed through screens and then through various driers to obtain dried grains, etc., as additional products. The evaporated products are purified. In these cases, evaporation and distillation are used, unlike processes for almost any other bioproduct development.

Distillation is implemented almost universally for the recovery of higher concentrations of volatile bioproducts such as alcohol, acetone and butanol. However, this step consumes almost 50% of the energy cost of the alcohol production plant (Shuler and Kargi, 2002). As a result, lower energy consuming techniques, such as pervaporation (Section 6.3.3.4) (Thongsukmak and Sirkar, 2007), reverse osmosis (Section 7.2.1.2), etc., are being explored. On the other hand, to produce a low-alcohol beer, processes such as dialysis and reverse osmosis are practiced along with distillation to produce simultaneously two products, e.g. low-alcohol beer (~0.5–2.5% alcohol) and a high-alcohol content product (Tilgner and Schmitz, 1981; Moonen and Niefind, 1982; Attenborough, 1988; Ladisch, 2001).

What we have observed so far in the production of bioproducts is that a large number of separation/purification steps have to be implemented consecutively, one step at a time, till the final product is obtained in the desired form and at the required purity level. These steps include:

(1) cell harvesting and lysis for intracellular products;
(2) separation of cells and cellular debris from the product obtained from fermentation or cell culture in solution/suspension via centrifugation, microfiltration, filtration;
(3) isolation of the product;
(4) concentration and purification of the products;
(5) polishing of the product.

An item of considerable concern is the extent of loss of the bioproduct along the train of bioseparation steps. Each step encounters some loss. For example, suppose there are five steps and each step achieves a fractional recovery $Y_{iR}\,(=0.95)$ of species *i*. The overall fractional recovery $(Y_{iR})_{\text{ov}}$ will be (Ladisch, 2001).

$$(Y_{iR})_{\text{ov}} = (Y_{iR})^5 = (0.95)^5 = 0.774. \tag{11.1.1}$$

Therefore the overall process recovery of the product is significantly reduced.

Processes have been developed that combine at least two of the steps in one device to attempt higher recovery and additional device cost reduction. Section 7.1.6 (Figure 7.1.28) illustrates the expanded-bed adsorption process for protein recovery and partial purification directly from a fermentation broth/lysate. Figure 11.1.3 illustrates a combined microfiltration/adsorption chromatography process (Xu *et al.*, 2005). A hollow fiber microfiltration membrane module has its shell side packed with chromatographic resin beads. A cyclic process is carried out. A broth/lysate containing protein products in solution is passed through the bore of the microfiltration hollow fibers for a certain period. The permeate goes through the resin beads on the shell side and the proteins are adsorbed. After this loading period, the feed broth/lysate flow is stopped. Then an eluent solution is passed through the fiber bore into the shell side to elute the proteins in sequence with the fiber bore end closed. Next the resin bed is regenerated by passing the starting buffer to initiate the loading–elution–regeneration cycle again. The advantages of this combined process are: recovery and separation of proteins directly from an unclarified feed; self-cleaning of the membrane fouling via the washing and elution steps; recovery of the proteins adsorbed on the membrane; prevention of product degradation in the holding tank prior to chromatography; elimination of the holding tank.

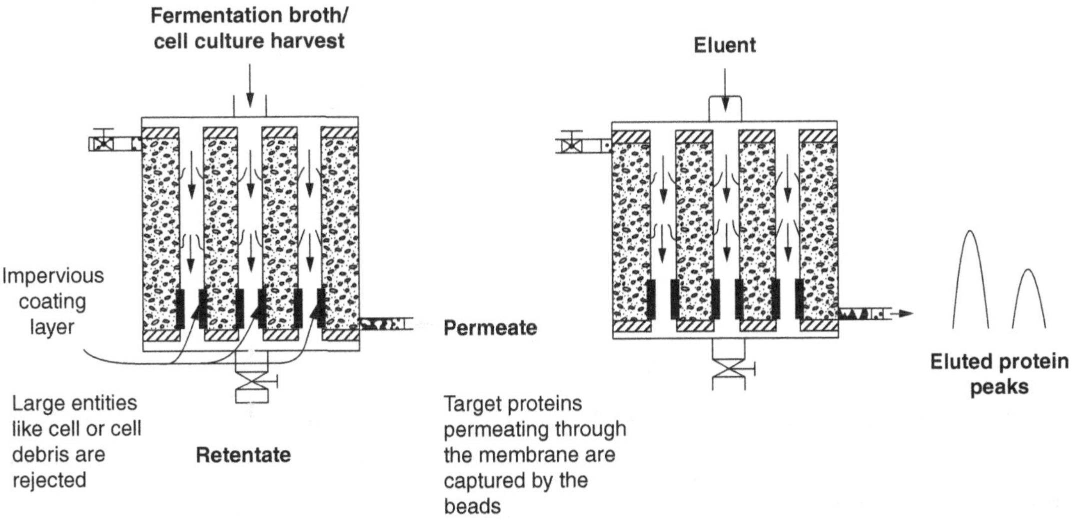

Figure 11.1.3. Integrated membrane filtration and chromatography device. (After Xu et al. *(2005).)*

In the purification of therapeutic monoclonal antibodies (mAbs), produced as an extracellular product from bacteria, yeast and mammalian cells, the isolation and purification steps involve a number of demanding processes. For example, when a mammalian cell line such as Chinese hamster ovary (CHO) is employed, the impurities in the harvested cell culture liquid contain Chinese hamster ovary cell proteins (CHOPs), DNAs, antibody variants, viral particles, endotoxins and small molecules (Follman and Fahrner, 2004). Conventionally, affinity chromatography using protein A (Figure 7.1.27(a)) is employed to bind antibodies from such a solution, leading to >99.5% removal of the impurities. However, the protein A based affinity chromatography step is very costly; it may cost as much as 35% of the total raw material costs for downstream purification. A common process sequence used includes a three-column sequence of a protein A affinity column, a cation exchange column and an anion exchange column. Additional steps include a virus filtration step and an ultrafiltration/diafiltration step. Mehta *et al.* (2008) have described in detail how a two-column nonaffinity process, when integrated with a high-performance tangential-flow filtration (HPTFF) technique, can replace the protein A affinity column based purification sequence for the purification of mAbs. This reduces the purification cost for mAbs substantially; the plant footprint is also much smaller.

11.2 Separation sequences for water treatment

Water is essential not only to human existence, but also for a variety of industrial and agricultural activities. Purification of water is essential for potable water; ultrapurification of water is indispensable to the semiconductor industry, where the purified water used for cleaning and rinsing of wafers is known as *ultrapure water* (UPW). The separation sequence followed to obtain a particular water product depends on the nature of the ultimate product water needed and the quality of the feed water. In the multistep scenario followed, processes involving chemical reactions invariably appear along with a host of different separation processes. Examples of different process sequences are related either to the water source to be treated or to the ultimate product water desired. Sea-water and brackish-water desalination, surface-water, ground-water and waste-water treatment are examples of the first category; potable water (municipal water), pharmaceutical grade water, ultrapure water, etc., are examples of the second category. We will illustrate very briefly the water treatment processes for sea-water and brackish-water desalination, ultrapure water production and municipal water production.

11.2.1 Sea-water and brackish-water desalination

Sea water usually contains total dissolved solids (TDS) at the level of 35 000 ppm; around the Middle East, this can go up to 40 000–50 000 ppm. Brackish waters can have TDS as low as 700–1000 ppm, going all the way up to 10 000 ppm plus. Desalination of sea water and brackish water involves a significant number of pretreatment steps, which are essential to the successful long-term performance of the reverse osmosis (RO) membranes increasingly and invariably employed in new desalination plants. There are also a few post-treatment steps after desalination. The exact nature and details of the pretreatment steps are influenced by the source water for the plant. As shown in Figure 11.2.1, typical pretreatment steps consist of a variety of separation

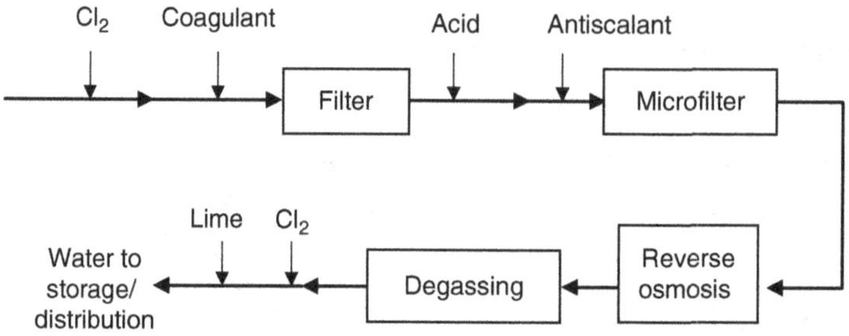

Figure 11.2.1. Schematic for a RO desalination plant.

processes, with some reaction processes taking place before, after or in between. Feed water to be desalinated undergoes the following treatment steps, in various combinations, prior to desalination by a reverse osmosis (RO) membrane: chlorination via sodium hypochlorite addition to control growth of bacteria or algae; addition of coagulants such as ferric chloride and polyelectrolytes to remove colloidal materials and suspended solids in granular filter beds (Section 7.2.2); addition of acid to convert scaling salts, e.g. $CaCO_3$, to nonscaling salts; addition of antiscalants to prevent scale formation from salts such as $CaSO_4$ and $CaCO_3$; microfiltration to remove micron-size particles/precipitates, etc. (Section 7.2.1.4); degassing to remove dissolved gases, e.g. H_2S, O_2 (Sections 8.1.2.2.1 and 8.1.2.4).

Post-treatment of the desalted permeate from the reverse osmosis unit (Section 7.2.1.2) consists of the addition of Cl_2 and lime for disinfection and corrosion protection. If H_2S is present, it is eliminated by air stripping. For brackish-water feeds containing hydrocarbons, an activated carbon adsorber is used to remove dissolved hydrocarbons prior to microfiltration, which is followed by steps needed during the pretreatment and post-treatment processes: dechlorination is required as a pretreatment if the RO membrane for desalination cannot tolerate residual chlorine; dissolved oxygen is often removed to avoid damaging the RO membrane via vacuum based deaeration or addition of sodium bisulfite. An introduction to the pretreatment and post-treatment processes for membrane based sea-water and brackish-water desalination has been provided by Williams *et al.* (2001). The scale of such desalination plants is quite large, as much as 87 million gallons per day at Ashkelon, Israel, for example.

A very significant fraction of desalination is achieved via thermal desalination processes such as multistage flash, multi-effect distillation, etc. (Section 10.2.1) (primarily in older plants). Pretreatment processes include filtration, chlorination, deaeration and scale control via acid addition, antiscalant treatment, etc. Post-treatment processes include building up the dissolved salt concentration to the level of 50 ppm via blending with brackish ground water or addition of lime, etc. (Howe, 1974), since distilled water is highly corrosive.

11.2.2 Ultrapure water production

Ultrapure water (UPW) is routinely used in the microelectronics industry for repeated cleaning and rinsing of wafers; a wafer may contact as much as 100–1000 liter of UPW. It is also used to clean process equipment during such manufacturing. A typical microelectronics plant may use as much as 2 million gallons per day. Such ultrapure water should not have particulate, bacterial, ionic, organic or dissolved gaseous impurities, which would all damage integrated circuits. Some of the standards to be satisfied are as follows: total organic carbon (TOC) ~ 1 ppb; dissolved O_2, 1 ppb; resistivity at 25 °C, 18.20 megohm-cm; metals at ppt (parts per trillion) level; cations, 1–5 ppt; anions, 20–30 ppt; reactive silica, 50 ppt; number of 0.05–0.1 μm particles per liter, 100; endotoxin units/ml, 0.03; bacteria, 1 coliform unit per 1000 ml (Baird and Williams, 2005).

The plant producing UPW obtains the potable feed water from the municipality. Multimedia filters made up of sand, coal, etc., are employed first to eliminate large particles and suspended solids in the feed water (Figure 11.2.2). Microfilters used to remove 1–5 μm+ particles are included next in the loop, before this water is introduced to a reverse osmosis (RO) unit in two stages. The first RO unit, along with the preceding units, are part of the makeup system. The second RO unit is part of the main treatment system - the primary system. Microbes such as bacteria, virus and yeast present in the water (especially introduced by filter beds of carbon or otherwise) and in connecting lines are destroyed/disrupted by ultraviolet disinfection at 254 nm. The permeate from RO units still contains a significant amount of electrolytes, which are almost completely removed by continuous electrodeionization (CEDI)[1] units. Any remaining trace ions are removed by a mixed ion exchange bed to produce water possessing a resistivity of >18 megohm-cm. This water is passed through a 0.2 μm filter. Part of this water is recycled

[1] If the brine feed channels of an electrodialysis (ED) unit (Section 8.1.7.2) are filled with mixed ion exchange resin beads, the deionization of the water can be carried out continuously to limits much greater than that in any ED unit.

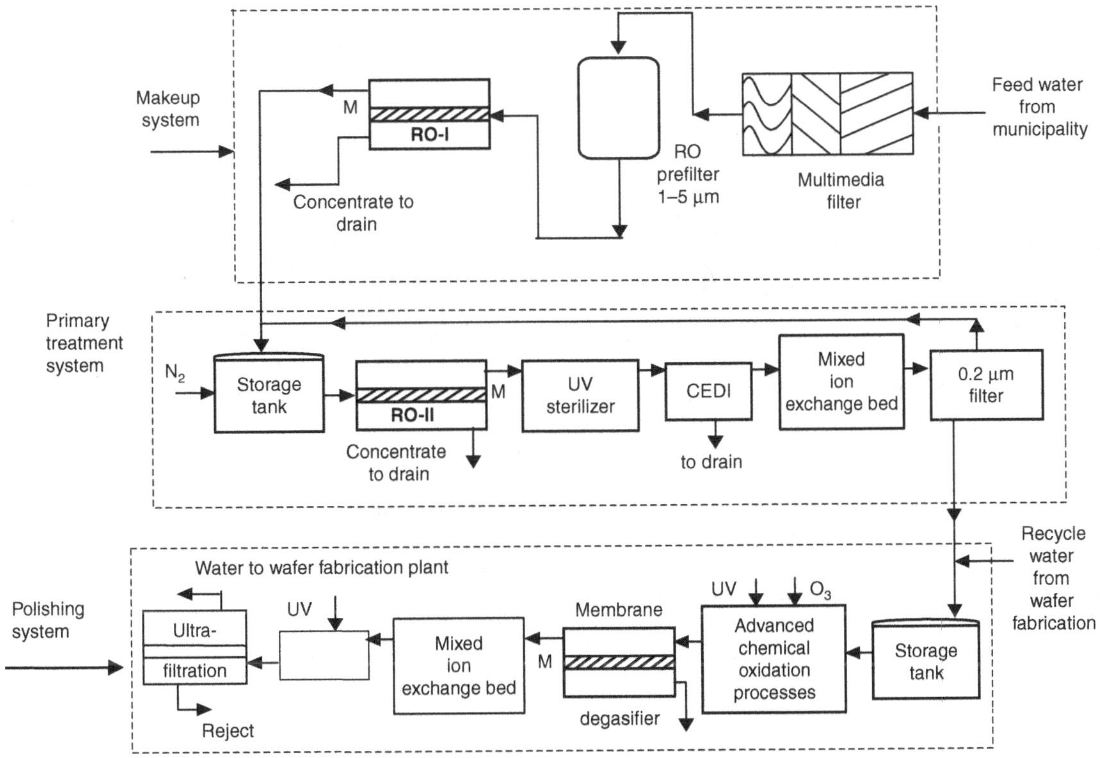

Figure 11.2.2. Process flowsheet for UPW production. (After Baird and Williams (2005).)

to a storage tank (in between the RO-I and RO-II units), where it is mixed with the product of the make-up system. The rest of the water out of the 0.2 μm filter is treated as the output of the primary treatment system.

This primary treatment system output undergoes an additional polishing treatment, in particular because water from the wafer fabrication is recycled to the tank receiving the primary treatment system output. Specifically, advanced chemical oxidation processes using O_3 and UV are employed to destroy microbes and organics, although the latter are reduced to compounds removed in the subsequent treatment steps. These steps include: membrane degassing (Section 8.1.2.2.1) to remove dissolved O_2 and CO_2 (the O_2 level <1–5 ppb); mixed ion exchange resin beds to remove residual ions and organic compounds; a final UV polishing treatment to destroy microbes; an ultrafiltration unit to remove any sub-submicron particles, microbes and residual large organic molecules (MW > 1000 dalton) as the last polishing filter.

11.2.3 Pharmaceutical grade water

Water required for pharmaceutical purposes has to satisfy two different requirements: water for injection (WFI) and purified water (WPU). Water satisfying the requirements for WPU is subjected to the additional step of distillation (evaporation) in order to quality as WFI. However, the requirements for WPU are less stringent than those for UPW. For example, the TOC level may be as much as 500 ppb; there are no requirements for dissolved oxygen as such (in UPW it is ≤ 1 ppb); water resistivity can be as low as 0.77 megohm-cm; the bacteria level in WFI can be 10 coliform units per 100 ml, etc. (Baird and Williams, 2005). Plants producing WPU are much smaller than UPW units and typically have a capacity of 10 000 to 50 000 gallons per day.

The process sequence employed to produce WPU grade water in the pharmaceutical industry typically comprises the following steps: multimedia filtration, water softening, activated carbon filtration, prefiltration, UV disinfection, reverse osmosis, continuous electrodeionization (CEDI). For WFI, as mentioned earlier, the WPU water is subjected to a distillation process, and may be supplied hot if needed.

11.3 Chemical and petrochemical industries

In terms of the physical dimensions of the separation equipment employed, the footprint and the volume of product produced, chemical and petrochemical operations are huge, much larger than the previous two process categories discussed. Further, the number of products and the types of processes used are much larger. It is fruitless to attempt to describe here such variety vis-à-vis the separation processes/techniques employed. We will merely illustrate here a few flowsheets to point out the major differences from those considered in Sections 11.1 and 11.2.

More often than not, the sequence of separations in chemical and petrochemical operations comprises part of a chemical production process where chemical reactions play a crucial part. Separation processes are often used to purify the feed stream entering the reactor. The products of reaction need to be separated from each other and from the residual feed, with the separated unreacted feed components recycled back to the reactor inlet where the feed stream is introduced. Figure 11.3.1 illustrates schematically this basic mode of operation, without reference to any particular process. Figure 11.3.2 provides an example. Naptha fraction (Shreve and Hatch, 1984) from crude oil distillation is taken to a reformer, where the octane number is increased by producing more olefins, compounds having lower molecular weight and achieving more cyclization and aromatization.

The process of aromatization, e.g. cyclohexane to benzene, produces a gaseous stream containing some H_2 (so does cyclization, i.e. linear alkanes converted to cyclic compounds). The H_2 is taken away as a bleed and the other cyclic compounds are recycled back to the feed. The liquid product from this gas–liquid separator goes to a debutanizer column, which concentrates light hydrocarbons at the top. The heavier hydrocarbons go next to a solvent extraction column, where a polar organic solvent such as ethylene glycol or propylene glycol is introduced as a solvent to extract aromatics preferentially (Figure 4.1.9(a)). The extract is taken to a solvent regenerator, which provides volatile aromatics as the top product and the much less volatile polar solvent as the bottom product recycled back to the aromatics extraction column. The volatile aromatics fraction is then taken to a distillation column producing benzene and toluene as the top product; the bottom product stream contains xylenes and heavier aromatics, which are subjected to further separation as well as isomerization.

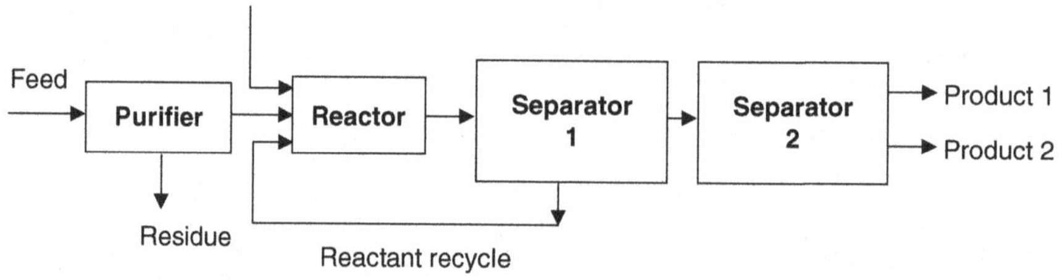

Figure 11.3.1. Hypothetical process schematic for production of chemicals – role of separators.

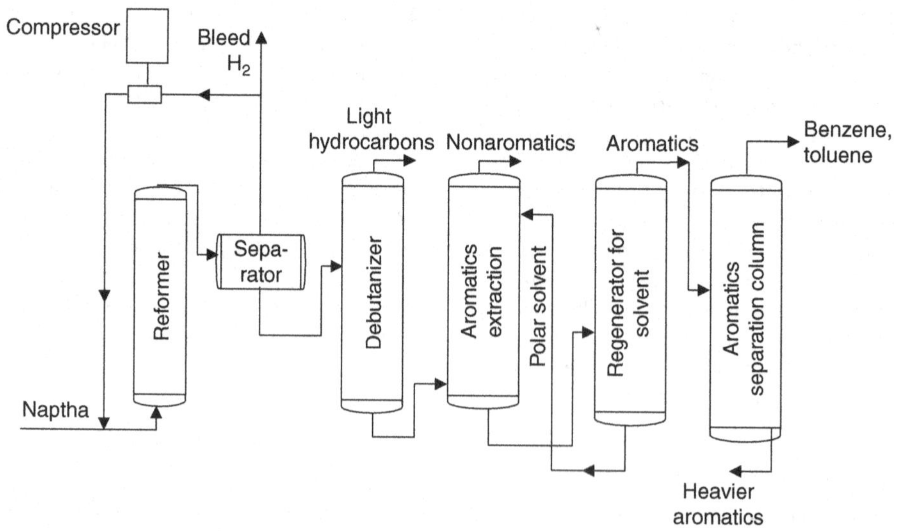

Figure 11.3.2. Separation train after reforming of naptha.

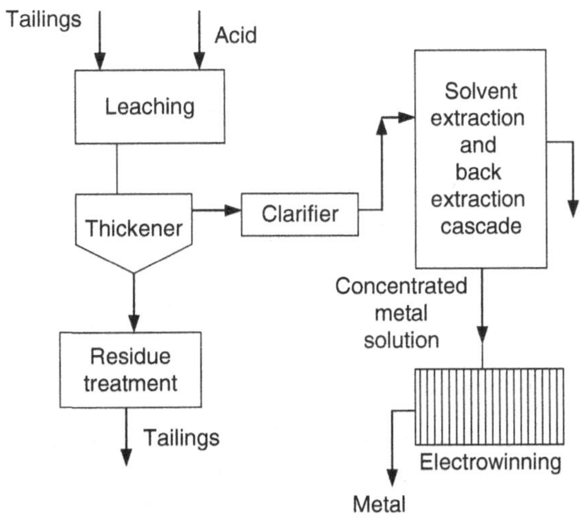

Figure 11.4.1. General schematic of a hydrometallurgical process employing electrowinning to recover the metal.

11.4 Hydrometallurgical processes

Hydrometallurgical processes involve many steps to achieve recovery of metal in the form and purity desired, starting from ores or other sources such as tailings or waste solutions/leachates. In the first step, an ore, which probably contains the metal at a low concentration (e.g. 1%), is ground in a grinding circuit and then sent to a leaching tank, where an acid, often sulphuric acid, is employed to leach the metal into solution (Figure 11.4.1). The slurry is then sent to a thickener, where flocculation[2] is employed with flocculating agents and the floc containing the slurry particles are settled. The liquid overflow from thickeners is filtered/clarified, and the filtered clear liquid is sent to a solvent extraction cascade, where the metal in the liquid containing the metal salt in solution is extracted into a solvent. The solvent extract is then subjected to back extraction into an aqueous solution. The back extract is generally a more concentrated solution of the metal in salt form.

From such a concentrated aqueous solution of the metal, the recovery of the metal is achieved in a variety of the ways. The metal may be precipitated as a hydroxide, an oxide, or calcined as an oxide or recovered as a metal via electrowinning (e.g., Cu, Co, Ni), etc. Illustrations of process flowsheets for the recovery of a variety of metals, e.g. copper, zinc, nickel, uranium, chromium, beryllium, etc., are available in Ritcey and Ashbrook (1984b). Benedict *et al.* (1981) also provide process flowsheets and descriptions of metals relevant to the nuclear power industry.

[2]Certain types of particles, for example precipitates, have a tendency to come together while in suspension. Flocculants, especially large molecular weight polymers with or without positive or negative charges, when added to such a suspension, bring the particles together much more readily and allow the larger assembly of particles, "flocs," to settle rapidly by gravity.

Postface

The area of separations is extremely broad. Most separation processes/techniques have an extensive literature. Each process has its own universe. Yet there are a few common principles and characteristics that are shared by many separation techniques/processes. This book has provided an elementary introduction to such principles and characteristics, even as it provides some details of many of the most commonly used separation processes. It is hoped that this book will help the reader to be prepared to undertake more extensive studies of individual processes/techniques or of a certain class of techniques or certain sequences of separation processes/techniques.

An important goal of this book is to develop an appreciation that separation processes/techniques may be understood on the basis of a few key features/concepts: the nature of the force causing the selective movement of molecules, macromolecules and particles; the nature of the various regions/phases in the separation system; the four types of bulk flow pattern of the regions/phases vis-à-vis the direction(s) of the forces causing selective movement of species/particles to be separated; the method of feed introduction. Further, for relatively simple systems, one could develop the governing equations that describe the separation achievable from the set of basic governing equations provided in Section 6.2 for most systems described in the book. In addition, a useful introduction to the role played by chemical reactions of certain types for a variety of separations has been provided.

Many topics could not be covered in this book. A much abbreviated list includes: the molecular basis of equilibrium partitioning of molecules between different phases; enthalpic and entropic contributions to partitioning/selectivity; the molecular basis of affinity binding in bioseparations; nonisothermal analysis of absorption columns, adsorption beds, distillation columns, etc.; multicomponent multistage separations in distillation columns; numerical methods for multicomponent multistage countercurrent separation processes; experimental methods in separation studies; hybrid separation processes; selection of separation processes for solving a separation problem; reaction–separation/separation–reaction/reaction–separation–reaction processes and devices.

Quite a few separation processes were barely touched upon. A number of separation processes could not be covered: azeotropic distillation; Calutron; cell disruption; coagulation; continuous deionization; expression; extractive distillation; forward osmosis; gaseous diffusion; ion retardation; leaching; membrane distillation; nanofiltration; pressure-swing distillation, etc. Many analytical techniques were also not covered.

Appendix A

Units, various constants and equivalent values of various quantities in different units

The basic units of various quantities and fundamental dimensions are identified below for SI, CGS and FPS systems. Additional common abbreviations are also listed.

SI (Système International d' Unités) system

Current, amp (ampere); length, m (meter); mass, kg (kilogram); mole, mol; kmol (kilogram mole); time, s (second); temperature, K (kelvin); force, N (newton); pressure, N/m^2 (pascal, Pa); energy, N-m (joule, J); power, N-m/s (watt, W).

CGS system

Length, cm (centimeter); mass, g (gram); mole, gmol (gram mole); time, s (second); temperature °C (centigrade); force, dyne; energy, erg (dyne-cm).

FPS system

Length, ft; mass, lbmass;[1] mole, lbmol (pound mole); time, s (second); temperature °F (Fahrenheit); force, lbforce;[1] pressure, psia (pounds per square inch).

Additional common units and abbreviations

atm (atmosphere), C (coulomb), cal (calorie), farad (F), hr (hour), Hz (hertz), km (kilometer), kV (kilovolt), kW (kilowatt), l (liter), ml (milliliter), rad (radian), S (Siemens), tesla (T), townsend (Td), W (watt)

---

[1] Although conventional practice is to indicate lbmass and lbforce by lb_m and lb_f, respectively, we have used the nonsubscripted versions in this book.

Appendix B

Constants

Fundamental constants

Mathematical constants: $\pi = 3.14159...$; $\sqrt{2} = 1.41421...$.

Logarithmic constants: $e = 2.71828...$; $\log_e 10 = \ell n\ 10 = 2.30258...$; $\log_{10} e = 0.43429$.

Physical constants

Gas law constant: R = 8.31451 J/gmol-K = 1.987 cal/(gmol-K) = 1.987 Btu/(lbmol °R) = 8314.3 J/kgmol-K = 82.057×10^{-3} m^3-atm/kgmol-K = 8314.34 m^3-Pa/kgmol-K = 0.0820578 L-atm/gmol-K = 82.0578 atm-cm^3/gmol-K = 10.73 psi-ft^3(lb mol °R) = 1.314 ft^3-atm/lbmol-K = 0.7302 ft^3-atm/lbmol-°R.

Avogadro's number: $\tilde{N} = 6.02214 \times 10^{23}$ molecules/gmol.

Boltzmann constant: $k^B (= R/\tilde{N}) = 1.381 \times 10^{-23}$ J/K $= 1.381 \times 10^{-16}$ erg/K.

Faraday constant: $\mathcal{F} = 96\,485$ coulomb (C)/g-equivalent.

Standard acceleration of gravity: g = 980 cm/s^2 = 9.806 m/s^2 = 32.174 ft/s^2. (Gravitational conversion factor g_c = 32.174 lbmass-ft/lbforce-s^2 = 980.66 gmass-cm/gforce-s^2.)

Speed of light in vacuum: $c = 2.99792 \times 10^{10}$ cm/s.

Speed of sound in dry air at 0 °C = 331.3 m/s.

Charge of an electron: $e = 1.6021 \times 10^{-19}$ coulomb (C).

Electrical permittivity of vacuum: $\varepsilon_0 = 8.8542 \times 10^{-14}$ coulomb/volt-cm (or farad/cm).

Planck's constant: $h = 6.62608 \times 10^{-34}$ J-s.

Joule's constant (mechanical equivalent of heat) = 4.184 J/cal = 4.184×10^7 erg/cal = 778.16 ft-lbforce/Btu.

Atmospheric pressure (at sea level) = 760 mm Hg at 0 °C = 760 torr = 29.92 in Hg at 0 °C = 33.9 ft H_2O at 4 °C = 1.1013 bar = 14.696 psia = 1.0132×10^5 N/m^2 = 1.0132×10^5 Pa = 1 atm = 76 cm Hg at 0 °C.

Appendix C

Various quantities expressed in different units

Acceleration: 1 m/s^2 = 100 cm/s^2 = 3.28 ft/s^2 = 0.425 × 10^8 ft/h^2.

Area: 1 m^2 = 10^4 cm^2 = 10.76 ft^2 = 1550 in^2.

Density: 1 kg/m^3 = 1000 g/cm^3 = 1 g/liter = 1 g/ℓ = 0.0624 $lbmass/ft^3$ = 8.34 × 10^{-3} lbmass/US gal = 1.002 × 10^{-2} lb/UK gal.

Diffusivity: 1 m^2/s = 10^4 cm^2/s = 38 750 ft^2/hr = 10.76 ft^2/s.

Energy: 1 joule = 1 J = 1 N-m = 10^7 dyne-cm = 10^7 erg = 10^7 g-cm^2/s^2 = 1 kg-m^2/s^2 = 0.239 cal = 9.47 × 10^{-4} Btu = 1 W-s = 2.7778 × 10^{-7} kW-hr = 3.725 × 10^{-7} hp-hr = 0.73756 ft-lbforce. These conversions are based on International Steam Table Convention of the definition of calorie.

Energy flux: 1 J/m^2-s = 1 W/m^2 = 0.317 Btu/ft^2-hr = 2.39 × 10^{-5} cal/cm^2-s.

Enthalpy: 1 J/kg = 2.39 × 10^{-4} cal/g = 4.299 × 10^{-4} Btu/lb.

Force: 1 N = 1 kg-m/s^2 = 10^5 dyne = 10^5 g-cm/s^2 = 7.233 lbmass-ft/s^2 = 7.233 poundals[1] = 2.2488 × 10^{-1} lbforce.

Heat capacity (specific heat): 1 J/kg-K = 10^{-3} kJ/kg-K = 2.388 × 10^{-4} Btu/lb-°F = 2.388 × 10^{-4} cal/g-°C.

Heat transfer coefficient: 1 N-m/m^2-s-K = 1 W/m^2-K = 1 kg/s^3-K = 10^{-4} W/cm^2-K = 10^3 g/s^3-K = 2.39 × 10^{-5} cal/cm^2-s-K = 1.761 × 10^{-1} Btu/ft^2-hr-°F.

Interfacial tension (surface tension): 1 N/m = 10^3 dyne/cm = 10^3 erg/cm^2.

Latent heat: 1 kJ/kg = 0.43 Btu/lbmass = 1 J/g = 0.239 cal/g.

Length: 1 m = 100 cm = 10^6 μm = 10^9 nm = 10^{10} Å = 3.28 ft = 39.37 inch = 1.0933 yard.

Mass: 1 kg = 1000g = 2.20462 lbmass = 35.2739 oz = 1.1022 × 10^{-3} ton = 10^{-3} metric ton.

Mass-flow rate: 1 kg/s = 2.2045 lbmass/s = 1000g/s.

Mass flux (mass velocity): 1 kg/m^2-s = 737.46 lbmass/ft^2-hr = 0.1 g/cm^2-s.

Mass transfer coefficient:

Molar concentration based: 1 kgmol/m^2-s-(kgmol/m^3) = 1 m/s = 100 cm/s = 11808 lb mol/ft^2-hr-(lb mol/ft^3) = 11808 ft/hr = 1 gmol/m^2-s-(gmol/m^3) = 100 gmol/cm^2-s-(gmol/cm^3).

Mole fraction based: 1 kgmol/m^2-s-mole fraction = 1000 gmol/m^2-s-mole fraction = 737.46 lb mol/ft^2-hr-mole fraction.

Partial pressure based: 1 kgmol/m^2-s-(N/m^2) = 747 × 10^5 lb mol/ft^2-hr-atm.

Power: 1 W = 1 J/s = 1N-m/s = 0.73756 ft-lbforce/s = 0.239 cal/s = 9.47 × 10^{-4} Btu/s = 3.41 Btu/hr = 1.34 × 10^{-3} hp.

Power/volume: 1 W/m^3 = 1 N-m/s-m^3 = 0.0209 ft-lbforce/s-ft^3.

Pressure: 1 Pa = 1 N/m^2 = 1.45 × 10^{-4} lbforce/$inch^2$ (psi) = 208.8 lbforce/ft^2 = 10 dyne/cm^2 =7.5006 × 10^{-3} mm Hg = 9.8692 × 10^{-6} atm.

[1] 1 poundal = 1(lbmass-ft)/s^2.

Specific heat: *see* "Heat capacity (specific heat)."

Surface area/volume (specific area): 1 m^2/m^3 = 0.01 cm^2/cm^3 = 0.304 ft^2/ft^3.

Temperature: °C = (5/9)(°F - 32); 0 °C = 32 °F = 273.15 K = 491.67R; °R = °F + 459.67; K = °C + 273.15 = (1/1.8)°R; 100 °C = 373.15K = 671.67 °R = 212 °F.

Thermal conductivity: 1 W/m-K = 1 kg-m/s^3-K = 10^5 erg/s-cm-K = 2.3901 × 10^{-3} cal/s-cm-K = 1 J/s-m-K = 5.778 × 10^{-1} Btu/hr-ft-°F.

Velocity: 1 m/s = 3.28 ft/s = 100 cm/s.

Viscosity: 1 Pa-s = 1 kg/m-s = 10 g/cm-s (poise) = 10^3 cp (centipoise) = 2.0886 × 10^{-2} lb force-s/ft^2 = 0.6719 lbmass/ft-s.

Volume: 1 m^3 = 10^6 cm^3 (ml) = 1000 l = 35.287 ft^3 = 264.17 US gallon = 220.83 UK gallon.

Volume flow rate: 1 m^3/s = 35.287 ft^3/s = 10^6 cm^3/s.

Volume flux: 1 m^3/m^2-s = 100 cm^3/cm^2-s = 3.28 ft^3/ft^2-s.

References

Abegg, C.F., J.D. Stevens and M.A. Larson, *AIChE J.*, **14**, 118 (1968).

Abou-Nemeh, I., A. Das, A. Saraf and K.K. Sirkar, *J. Membr. Sci.*, **158**, 187 (1998).

Adamson, A.W., *Physical Chemistry of Surfaces*, 2nd edn., Wiley Interscience, London (1967).

Agrawal, R. and T.F. Yee, *Ind. Eng. Chem. Res.*, **33**, 2717 (1994).

Agrawal R., Z.T. Fidkowski and J. Xu, *AIChE J.*, **42**(8), 2118 (1996).

AIChE, *Bubble-Tray Design Manual*, American Institute of Chemical Engineers, New York (1958).

AIChE, Equipment Testing Procedures Committee, *AIChE Equipment Testing Procedure - Tray Distillation Columns*, 2nd edn., American Institute of Chemical Engineers, New York (1987).

Aiken, R.C., *Chem. Eng. Sci.*, **37**, 1031 (1982).

Albertins, R., W.C. Gates and J.E. Powers, in *Fractional Solidification*, M. Zief and W.R. Wilcox (eds.), Vol. I, Marcel Dekker, New York (1967), chap. 11.

Albertsson, P.-Å., *Partition of Cell Particles and Macromolecules*, 3rd edn., Wiley Interscience, New York (1986).

Alexandratos, S.D and D.W. Crick, *Ind. Eng. Chem. Res.*, **35**, 635 (1996).

Alishusky, J.J. and R.L. Fournier, *AIChE J.*, **36**(10), 1605 (1990).

Allcock, H.R., F.W. Lampe and J.E. Mark, *Contemporary Polymer Chemistry*, 3rd edn., Pearson Education, Upper Saddle River, NJ (2003), chap 15.

Alpers, A., B. Keil, O. Lüdtke and K. Ohlrogge, *Ind. Eng. Chem. Res.*, **38**, 3754 (1999).

Ambler, C.M., *Chem. Eng. Prog.*, **48**, 150 (1952).

Amini, A., U. Paulsen-Sorman and D. Westerlund, *Chromatographia*, **50**, 497 (1999).

Anderson, J.B., *AIChE Symp. Series*, **76**(192), 89 (1980).

Anderson, J.L. and J.A. Quinn, *J. Chem. Soc. Faraday Trans., 1*, **68**(4), 744 (1972).

Anderson, J.L., M.E. Lowell and D.C. Prieve, *J. Fluid Mech.*, **117**, 107 (1982).

Anderson, J.L. and J.A. Quinn, *Biophys. J.*, **14**, 130 (1974).

Andrews, A.T., *Electrophoresis: Theory, Techniques, and Biochemical and Clinical Applications*, Clarendon Press, Oxford (1981), pp. 39, 87.

Annesini, M.C., F. Gironi and L. Marrelli, *Ind. Eng. Chem. Res.*, **27**(7), 1212 (1988).

Anwar, M.M., C. Hanson and M.W.T. Pratt, *Trans. Inst. Chem. Eng., London*, **49**, 95 (1971).

Applegate, L.E. and C.R. Antonson, in *Reverse Osmosis and Membrane Research*, H.K. Lonsdale and H.E. Podall (eds.), Plenum Press, New York (1972).

Aris, R., *Proc. Roy. Soc. London, Ser. A.*, **235**, 67 (1956).

Aris, R., *Ind. Eng. Chem. Fund.*, **8**, 603 (1969).

Aris, R. and N.R. Amundson, *Mathematical Methods in Chemical Engineering*, vol. 2, Prentice Hall, Englewood Cliffs, NJ (1973).

Armstrong, D. and L. He, *Anal. Chem.*, **73**, 4551 (2001).

Armstrong, D.W. and H.L. Jin, *Anal. Chem.*, **59**, 2237 (1987).

Armstrong, D.W., Y. Tang, T. Ward and M. Nichols, *Anal. Chem.*, **65**, 1114 (1993).

Arsalani, V., K. Rostami and A. Kheirolomoom, *Ind. Eng. Chem. Res.*, **44**, 7469 (2005).

Asai, S., *Ind. Eng. Chem. Proc. Des. Dev.*, **24**, 1105 (1985).

Asai, S. and S. Ikegami, *Ind. Chem. Fund.*, **21**, 181 (1982).

Ash, R., R.M. Baker and R. M. Barrer, *Proc. Roy Soc. A*, **299**, 434 (1967).

Ash, R., R.M. Barrer and J.H. Clint, R.J. Dolphin and C.L. Murray, *Phil. Trans. Roy. Soc. London*, **275**(1249), 255 (1973a).

Ash, R., R. M. Barrer, and R.J. Lawson, *J. Chem. Soc. Faraday Trans. 1*, **69**, 2166 (1973b).

Ashkin, A., *Phys. Rev. Lett.*, **24**(4), 156 (1970).

Astarita, G., *Mass Transfer with Chemical Reaction*, Elsevier, Amsterdam (1967).

Astarita, G., D.W. Savage and A. Bisio, *Gas Treating with Chemical Solvents*, John Wiley, New York (1983).

Atkinson, G., in *American Institute of Physics Handbook*, D.E. Gray (ed.), 3rd edn., McGraw-Hill, New York (1972).

Attenborough, W.M., *Ferment*, **1**(2), 40 (1988).

Atwood, J.L., J.E.D. Davies and D.D. MacNicol (eds.), *Inclusion Compounds*, Academic Press, London (1984).

Auvil, S.R. and B.W. Wilkinson, *AIChE J.*, **22**, 564 (1976).

Babcock, W.C., R.W. Baker, E.D. Lachapelle and K.L. Smith, *J. Membr. Sci.*, **7**, 89 (1980).

Bailey, J.E. and D.F. Ollis, *Biochemical Engineering Fundamentals*, 2nd edn., McGraw-Hill, New York (1986).

Bailey, P.D., *An Introduction to Peptide Chemistry*, John Wiley, New York (1990).

Baird, A. and R. Williams, *Chem. Eng.*, **May**, 36–43 (2005).

Baird, M.H.I., "Extraction by chemical reaction," 88th AIChE National Meeting, Philadelphia, PA, June 8–12 (1980), paper 36b.

Bajpai, R.K., A.K. Gupta and M.G. Rao, *AIChE J.*, **20**, 989 (1974).

Baker, B. and R.L. Pigford, *Ind. Eng. Chem. Fund.*, **10**, 283 (1971).

Baker, R.W., *Membrane Technology and Applications*, 2nd edn., John Wiley, Hoboken, NJ (2004).

Baker, R.W. and J.G. Wijmans, "Membrane fractionation process," US Patent 5,032,148, July 16 (1991).

Baker, R.W. and J.G. Wijmans, in *Polymeric Gas Separation Membranes*, D.R. Paul and Y.P. Yampolskii (eds.), CRC Press, Boca Raton, FL (1994), p. 361.

Bard, A.J. and M.V. Mirkin (eds.), *Scanning Electrochemical Microscopy*, Marcel Dekker, New York (2001).

Barker, P.E., *Progr. Separ. Purif.*, **4**, 325 (1971).

Barrer, R.M., *Appl. Mater. Res.*, **2**, 129 (1963).

Basak, S. and M.R. Ladisch, *Anal. Biochem.*, **226**, 51 (1995).

Basmadjian, D., D.K. Ha and C.Y. Pan, *Ind. Eng. Chem. Proc. Des. Dev.*, **14**(3), 328 (1975a).

Basmadjian, D., D.K. Ha and C.Y. Pan, *Ind. Eng. Chem. Proc. Des. Dev.*, **14**(3), 340 (1975b).

Basu, R., R. Prasad and K.K. Sirkar, *AIChE J.*, **36**(3), 450 (1990).

Bates, R.G., *Determination of pH: Theory and Practice*, John Wiley, New York (1964), p. 183.

Batt, B.C., V.M. Yabannavar and V. Singh, *Bioseparations* **5**, 41 (1995).

Beckwith, J.B. and C.F. Ivory, *Chem. Eng. Commun.*, **54**(1–6) 301 (1987).

Begovich, J.M. and W.G. Sisson, *AIChE J.*, **30**(5), 705 (1984).

Beier, P.M. and I. Stahl, "The electrostatic processing of raw salts at Kali and Salz GmbH," *Proc. XXIMPC, Aachen*, Sept. 21–26 (1997), p. 657.

Belfort, G., *Synthetic Membrane Processes*, Academic Press, New York (1984).

Belter, P.A., E.L. Cussler and W-S. Hu, *Bioseparations: Downstream Processing in Biotechnology*, John Wiley, New York (1988).

Bender, M.L. and M. Komiyama, *Cyclodextrin Chemistry*, Springer-Verlag, New York (1978).

Benedict M. and T. H. Pigford, *Nuclear Chemical Engineering*, McGraw-Hill, New York, (1957), p. 500.

Benedict, M., T.H. Pigford and H.W. Levi, *Nuclear Chemical Engineering*, 2nd edn., McGraw-Hill, New York (1981).

Benitez, J., *Process Engineering and Design for Air Pollution Control*, Prentice Hall, Englewood Cliffs, NJ (1993).

Benitez, J., *Principles and Modern Applications of Mass Transfer Operations*, 2nd edn., John Wiley, Hoboken, NJ (2009).

Berglund, K.A., in *Handbook of Industrial Crystallization*, A.S. Myerson (ed.), Butterworths-Heinemann, Boston (1993), p. 89.

Berglund, K.A. and E.J. de Jong, *Separ. Technol.*, **1**, 1 (1990).

Bhattacharyya, D., T. Barranger, M. Jevtitch and S. Greenleaf, *Separation of Dilute Hazardous Organics by Low Pressure Composite Membranes*, NTIS Report no.: PB 87-214870, July (1987).

Bhattacharyya, D., M.E. Williams, R.J. Ray and S.B. McCray, in *Membrane Handbook*, W.S.W. Ho and K.K. Sirkar (eds.), Van Nostrand Reinhold, New York (1992), chap. 23.

Bhattacharyya, D., M.E. Williams, R.J. Ray and S.B. McCray, in *Membrane Handbook*, W.S.W. Ho and K.K. Sirkar (eds.), Kluwer Academic Publishers, Boston, MA (2001).

Bhaumik, S., S. Majumdar and K.K. Sirkar, *AIChE J.*, **42**(2), 409 (1996).

Bhaumik, D., S. Majumdar and K.K. Sirkar, *J. Membr. Sci.*, **138**, 77 (1998).

Bhave, R.R., *Inorganic Membranes, Synthesis, Characteristics and Applications*, Van Nostrand Reinhold, New York (1991), chap. 5.

Bhave, R.R. and M.M. Sharma, *Chem. Eng. Sci.*, **38**, 141 (1983).

Bhave, R.R. and K.K. Sirkar, in *Liquid Membranes: Theory and Applications*, R.D. Noble and J.D. Way (eds.), ACS Symposium Series. 347, ACS, Washington, D.C. (1987), chap. 10.

Biesenberger, J.A., *Devolatilization of Polymers*, Hanser Publishers, Munich (1983).

Bikerman, J.J., *Physical Surfaces*, Academic Press, New York (1970), p. 315.

Bird, R.B., W.E. Stewart and E.N. Lightfoot, *Transport Phenomena*, John Wiley, New York (1960).

Bird, R.B., W.E. Stewart and E.N. Lightfoot, *Transport Phenomena*, 2nd edn., John Wiley, New York (2002).

Birss, R.R. and M.R. Parker, in *Progress in Filtration and Separation*, R.J. Wakeman (ed.), vol. 2, Elsevier Scientific Publishing Co., Amsterdam, (1981), chap. 4.

Blakebrough, N., *Biochemical and Biological Engineering Science*, vol. 1, Academic Press, New York (1967).

Block, B., *Chem. Eng.*, **68**(3), 87 (1961).

Böddeker, K.W., G. Bengston and E. Bode, *"Pervaporation of low volatile aromatics from water,"* IMTEC '88, Sydney, Australia, Nov. 15–17 (1988).

Bogart, M.J.P., *Trans. AIChE.*, **33**, 139 (1937).

Bonmati, R., R. Margulis and G.C. Letourneux, "Gas chromatography, a new industrial process," 88th AIChE National Meeting, Philadelphia, PA, June 8–12 (1980).

Boyd, G.E., A.W. Adamson and L.S. Myers Jr., *J. Am. Chem. Soc.*, **69**, 2836 (1947).

Boyde, T.R.C. *Separ. Sci.*, **6**(6), 771 (1971).

Boyle, M.D.P., E.L. Faulmann and D.W. Metzger, in *Molecular Interactions in Bioseparations*, T.T. Ngo (ed.), Plenum Press, New York (1993), p. 91.

Bradley, D., *The Hydrocyclone*, Pergamon Press, Oxford (1965).

Bramley, A., *Science*, **92**, 427 (1940).

Brandt, S., R.K. Goeffe, S.B. Kessler, J.L. O'Connor and S.E. Zale, *Biotechnol.*, **6**, 779 (1988).

Brant, L.R., K.L. Stellner and J.F. Scamehorn, in *Surfactant-based Separation Processes*, J.F. Scamehorn and J.H. Harwell (eds.), Marcel Dekker, New York (1989), chap. 12.

Breck, D.W., *Zeolite Molecular Sieves*, Wiley-Interscience, New York (1974).

Brian, P.L.T., *Ind. Eng. Chem. Fund.*, **4**, 439 (1965).

Brooks, C.A. and S. M. Cramer, *AIChE J.*, **38**(12), 1969 (1992).

Burgess, J., *Ions in Solution: Basic Principles of Chemical Interactions*, Ellis Horwood Ltd., Chichester; John Wiley, New York (1988), p. 81.

Burton, J.A., R.C. Prim and W.P. Slichter, *J. Chem. Phys.* **21**, 1987 (1953).

Cahn, R.P. and N.N. Li, *Separ. Sci. Technol.*, **9**, 505 (1974).

Cahn, R.P. and N.N. Li, in *Membrane Separation Processes*, P. Meares (ed.), Elsevier, Amsterdam (1976), p. 327.

Cahn, R.P., N.N. Li and R.C. Minday, *Environ. Sci. Technol.*, **12**, 1051 (1978).

Calderbank, P.H. and M.B. Moo-Young, *Chem. Eng. Sci.*, **16**, 39 (1961).

Caldwell, K.D. and Y-S. Gao, *Anal. Chem.*, **65**, 1764 (1993).

Callahan, R.W., *AIChE Symp. Series*, **84**(261), 54 (1988).

Calvert, S., in *Handbook of Air Pollution Technology*, S. Calvert and H.M. Englund (eds.), Wiley, New York (1984), p. 215.

Carlson, A. and R. Nagarajan, *Biotechnol. Prog.*, **8**, 85 (1992).

Carslaw, H.S. and J.C. Jaeger, *Conduction of Heat in Solids*, 2nd edn., Clarendon Press, Oxford (1959).

Catsimpoolas, N. (ed.), *Methods of Cell Separation*, Vol. I, Plenum Press, New York (1977).

Chan, Y.N.I., F.B. Hill and Y.W. Wong, *Chem. Eng. Sci.*, **36**, 243 (1981).

Chang, Y.-K. and H.A. Chase, *Biotechnol. Bioeng.*, **49**, 204 (1996).

Chao, K.C. and J.D. Seader, *AIChE J.*, **7**, 598 (1961).

Chapman, T.W., in *Lectures in Transport Phenomena*, R.B. Bird, W.E. Stewart, E.N. Lightfoot and T.W. Chapman (eds.), AIChE Continuing Education Series 4, AIChE, New York (1969).

Chem. Eng., "Gas chromatography analysis to production," *Chem. Eng.*, March 24 (1980), p. 70.
Chem. Eng., "Separating paraffin isomers using chromatography," *Chem. Eng.*, May 18 (1981), p. 92.
Chen, C-J., *Separ. Sci. Technol*, **36**(3), 499 (2001).
Chen, H., A.S. Kovvali, S. Majumdar and K.K. Sirkar, *Ind. Eng. Chem. Res.*, **38**(9), 3489 (1999a).
Chen, H., A.S. Kovvali and K.K. Sirkar, *Ind. Eng. Chem. Res.*, **39**(7), 2447 (2000).
Chen, H-P., Z-J. Lin, D-C. Liu, X.S. Wang and M.J. Rhodes, *Ind. Eng. Chem. Res.*, **38**(4), 1605 (1999b).
Chen, H.T. and F.B. Hill, *Separ. Sci.*, **6**, 411 (1971).
Chen, H.T., Y.W. Wong and S. Wu, *AIChE J.*, **25**, 320 (1979).
Chen, H.T., W.T. Yang, U. Pancharoen and R. Parisi, *AIChE J.*, **26**, 839 (1980).
Chen, R-R., R.N. Zare, E.C. Peters, F. Svec and J.J. Frechét, *Anal. Chem.*, **73**, 1987 (2001).
Chen, S., Y.K. Kao and S.T. Hwang, *J. Membr. Sci.*, **26**, 143 (1986).
Cherkasov, A.N. and A.E. Polotsky, *J. Membr. Sci.*, **110**, 79 (1996).
Chern, R.T. and N.F. Brown, *Macromolecules*, **23**(8), 2370 (1990).
Chern, R.T., W.J. Koros, H.B. Hopfenberg and V.T. Stannett, in *Material Science of Synthetic Membranes*, D. Lloyd (ed.), ACS Symposium Series 269, ACS, Washington, D.C. (1985), p. 25.
Cheryan, M., *Ultrafiltration Handbook*, Technomic Publishing, Lancaster, PA (1986), p. 60.
Cheryan, M., *Ultrafiltration and Microfiltration Handbook*, Technomic Publishing, Lancaster, PA (1998).
Chimowitz, E.H. and K.J. Pennisi, *AIChE J.*, **32**, 1665 (1986).
Ching, C.B. and D.M. Ruthven, *Can. J. Chem. Eng.*, **62**, 398 (1984).
Ching, C.B. and D.M. Ruthven, *AIChE Symp. Series*, **81**(242), (1985).
Ching, C.B., C. Ho and D.M. Ruthven, *AIChE J.*, **32**(11), 1876 (1986).
Choudhury, A.P.R. and D.A. Dahlstrom, *AIChE J.*, **3**, 433 (1957).
Choy, E.M., D.F. Evans and E.L. Cussler, *J. Am. Chem. Soc.*, **96**, 7085 (1974).
Christian, S.D. and J.F. Scamehorn, in *Surfactant-based Separation Processes*, J.F. Scamehorn and J.H. Harwell (eds.), Marcel Dekker, New York (1989), p. 3.
Christian, S.D., S.N. Bhat, E.E. Tucker, J.F. Scamehorn and D.A. El-Sayed, *AIChE J.*, **34**, 189 (1988).
Chung, J., *Pharmaceut. Technol.*, **June**, 39 (1996).
Ciliberti, D.F. and B.W. Lancaster, *AIChE J.*, **22**(2), 394 (1976a).
Ciliberti, D.F. and B.W. Lancaster, *Chem. Eng. Sci.*, **31**, 499 (1976b).
Cleary, W. and M.F. Doherty, *Ind. Eng. Chem. Proc. Des. Dev.*, **24**, 1071 (1985).
Clift, R., M. Ghadiri and A.C. Hoffman, *AIChE J.*, **37**(2), 285 (1991).
Cohen, K., *The Theory of Isotope Separation*, McGraw-Hill, New York (1951).
Cohn, E.J. and J.D. Ferry, in *Proteins, Amino Acids and Peptides as Ions and Dipolar Ions*, E.J. Cohn and J.T. Edsall (eds.), ACS Monograph Series no. 90, Reinhold, New York (1943), p. 602.
Colburn, A.P., *Trans. AIChE*, **35**, 211 (1939).
Collins, J.P. and J.D. Way, *Ind. Eng. Chem. Res.*, **32**, 3006 (1993).
Colman, D.A. and M.T. Thew, "Cyclone separator," US Patent 4,764,287, August 16 (1988).
Colton, C.K., K.A. Smith, E.W. Merrill and P.C. Farrell, *J. Biomed. Mater. Res.*, **5**, 459 (1971).
Colton, C.K., C.N. Sattersfield and C.J. Lai, *AIChE J.*, **21**, 289 (1975).
Cornish, A.R.H., *Trans. Inst. Chem. Eng., London*, **43**, T332 (1965).
Cotterman, R.L. and J.M. Prausnitz, *Ind. Eng. Chem. Proc. Des. Dev.*, **24**, 434 (1985).
Cotterman, R.L., R. Bender and J.M. Prausnitz, *Ind. Eng. Chem. Proc. Des. Dev.*, **24**, 194 (1985).
Cox, M. and D.S. Flett, in *Handbook of Solvent Extraction*, T.C. Lo, M.H.I. Baird and C. Hanson (eds.), Wiley, New York (1983), chap. 2.2. Reprinted, Krieger Publishing Co., Marlboro, FL (1991).
Craig, H. and R. C. Wiens, *Science*, **271**, 1708 (1996).
Craig, H., Y. Horibe and T. Sowers, *Science*, **242**, 1675 (1988).
Crank, J. and G.S. Park, in *Diffusion in Polymers*, J. Crank and G.S. Park (eds.), Academic Press, London (1968), chap. 1.
Curtiss, C.F and R.B. Bird, *Ind. Eng. Chem. Res.*, **38**, 2515 (1999).
Curtiss, C.F. and J.O. Hirschfelder, *J. Chem. Phys.*, **17**, 550 (1949).
Cussler, E.L., *AIChE J.*, **17**, 405 (1971).
Cussler, E.L., *Diffusion: Mass Transfer in Fluid Systems*, Cambridge University Press, New York, (1984), p. 407.
Cussler, E.L., *Diffusion: Mass Transfer in Fluid Systems*, 2nd edn., Cambridge University Press, Cambridge (1997), pp. 251, 304.
Cussler, E.L., M.R. Stokar and J.E. Varberg, *AIChE J.*, **30**(4), 578 (1984).
Dadyburjor, D.B., *Chem. Eng. Progr.*, **74**(4), 85 (1978).
Dai, X.P., S. Majumdar, R.G. Luo and K.K. Sirkar, *Biotechnol. Bioeng.*, **83**(2), 125 (2003).
Danckwerts, P.V., *Chem. Eng. Sci.*, **2**, 1 (1953).
Danckwerts, P.V., *Gas-Liquid Reactions*, McGraw-Hill, New York (1970).
Danckwerts, P.V. and M.M. Sharma, *The Chem. Engr.*, no. 202, CE 244 (1966).
Danesi, P.R., *Separ. Sci. Technol.*, **19**(11&12), 857 (1984–85).
Danesi, P.R., E.P. Horwitz and P.G. Rickert, *J. Phys. Chem.*, **87**, 4708 (1983).
Danesi, P.R., L. Reichley-Yinger, C. Cianetti and P.G. Rickert, *Solv. Extract. Ion Exchange*, **2**, 781 (1984).
Darken, L.S. and R.W. Gurry, *Physical Chemistry of Metals*, McGraw-Hill, New York (1953).
Datta, R., *Biotechnol. Prog.*, **6**, 485 (1990).
Davies, J.T. and E.K. Rideal, *Interfacial Phenomena*, Academic Press, New York (1963).
Davis, J.C., R.J. Valus, R. Eshraghi and A.E. Velikoff, *Separ. Sci. Technol.*, **28**, 463 (1993).
Davis, R.H., in *Membrane Handbook*, W.S.W. Ho and K.K. Sirkar (eds.), Kluwer Academic, Boston, MA (2001), chaps. 31, 33.
Davis, R.H. and D.C. Grant, in *Membrane Handbook*, W.S.W. Ho and K.K. Sirkar (eds.), Van Nostrand Reinhold, New York, (1992), chap. 32.
Davis, R.H. and D.C. Grant, in *Membrane Handbook*, W.S.W. Ho and K.K. Sirkar (eds.), Kluwer Academic Publishers, Boston, MA (2001).
Davis, R.H. and D.T. Leighton, *Chem. Eng. Sci.*, **42**(2), 275 (1987).
Davis, R.H. and J.D. Sherwood, *Chem. Eng. Sci.*, **45**(11), 3203 (1990).
Davis, R.H., X. Zhang and J.P. Agarwala, *Ind. Eng. Chem. Res.*, **28**, 785 (1989).
de Clerk, K. and C.E. Cloete, *Separ. Sci.*, **6**(5), 627 (1971).
de Groot, S.R. and P. Mazur, *Nonequilibrium Thermodynamics*, North Holland, Amsterdam (1962).
Debenedetti, P.G. and R.C. Reid, *AIChE J.*, **32**, 2034 (1986).
DeHoff, R.T., *Thermodynamics in Material Science*, McGraw-Hill, New York (1993).

Dekker, M., K. Van't Reit, J.J. Van Der Pol, J.W.A. Baltussen, R. Hilhorst and B.J. Bijsterbosch, *Chem. Eng. J.*, **46**, B69 (1991).

Denbigh, K., *The Principles of Chemical Equilibrium*, Cambridge University Press, London (1971), pp. 87, 103.

Deng, Y., J. Zhang, T. Tsuda, P.H. Yu, A.A. Boulton and R.M. Cassidy, *Anal. Chem.*, **70**, 4586 (1998).

DePriester, C.L., *Chem. Eng. Prog. Symp. Ser.*, **49**(7), 1 (1953).

Deutsch, W., *Ann. Phys. (Leipzig)*, **68**, 335 (1922).

DeVault, D., *J. Am. Chem. Soc.*, **65**, 532 (1943).

Dewitt, T.S., "Liquid-Liquid Extraction of COS into Caustic Soda," 88th AIChE National Meeting, Philadelphia, PA June 8-12 (1980), paper 61b.

Dietz, P.W., *AIChE J.*, **27**(6), 888 (1981).

Ding, H.B., M.C. Yang, D. Schisla and E.L. Cussler, *AIChE J.*, **35**(5), 814 (1989).

Diwekar, U.M., *Batch Distillation: Simulation, Optimum Design and Control*, Taylor and Francis, Washington, D.C. (1996).

Dobby, G. and J.A. Finch, *Powder Technol.*, **17**, 73 (1977).

Dodge, B.F. and A.K. Dunbar, *J. Am. Chem. Soc.*, **49**, 591 (1927).

Doherty, M.F. and M.F. Malone, *Conceptual Design of Distillation Systems*, McGraw-Hill, New York (2001), pp. 186-199, 259-265, 507-513.

Doherty, M.F. and J.D. Perkins, *Chem. Eng. Sci.*, **33**, 569 (1978).

Donovan, R.P., *Fabric Filtration for Combustion Sources*, Marcel Dekker, New York (1985).

Doong, S.J. and R.T. Yang, *AIChE J.*, **32**(3), 397 (1986).

Dose, E.V. and G. Guiochon, *Anal. Chem.*, **62**, 174 (1990).

Dresner, L., *Desalination*, **10**, 27 (1972).

Dunn, R.O., J.F. Scamehorn and S.D. Christian, *Separ. Sci. Technol.*, **20**, 257 (1985).

Dunn, R.O., J.F. Scamehorn and S.D. Christian, *Separ. Sci. Technol.*, **22** (2&3), 763 (1987).

Dutta, P. and A. Beskok, *Anal. Chem.*, **73**, 1979 (2001).

Dwivedi, P.N. and S.N. Upadhyay, *Ind. Eng. Chem. Proc. Des. Dev.*, **16**, 157 (1977).

Dye, S.R., J.P. DeCasli, II, and G. Carta, *Ind. Eng. Chem. Res.*, **29**, 849 (1990).

Dyer, P.N., R.E. Richards, S.L. Russek and D.M. Taylor, *Solid State Ionics*, **134**, 21 (2000).

Eckert, J.S., *Chem. Eng. Progr.*, **66**(3), 39 (1970).

Eckstein, E.C., P.G. Bailey and A.H. Shapiro, *J. Fluid Mech.*, **79**, 191 (1977).

Eduljee, H.E., *Hydrocarbon Proc.*, **54**, 120 (1975).

Edwards, T.J., G. Maurer, J. Newman and J.M. Prausnitz, *AIChE J.*, **24**(6), 966 (1978).

Eggerstedt, P.M., J.F. Zievers and E.C. Zievers, *Chem. Eng. Progr.*, **Jan.**, 62 (1993).

Eiceman, G.A. and Z. Karpas, *Ion Mobility Spectrometry*, CRC Press, Boca Raton, FL (1994).

Eiceman, G.A., E.V. Krylov, N.S. Krylova, E.G. Nazarov and R.A. Miller, *Anal. Chem.*, **76**(17), 4397 (2004).

Eisinger, R.S. and G.E. Keller, *Environ. Progr.*, **9**(4), Nov., 235 (1990).

Ellerbe, R.W., *Chem. Eng.*, **80**, 110 (1973).

El-Sayed, Y.M. and R.S. Silver, in *Principles of Desalination*, Part A, K.S. Spiegler and A.D.K. Laird (eds.), 2nd edn., Academic Press, New York (1980), chap. 2.

Englezos, P., *Ind. Eng. Chem. Res.*, **32**, 1251 (1993).

Erbar, J.H. and R.N Maddox, *Petroleum Refiner*, **40**(5), 183 (1961).

Estrin, J., in *Handbook of Industrial Crystallization*, A.S. Myerson (ed.), Butterworth-Heinemann, Boston, MA (1993), chap. 6.

Evangelista, F. and G. Jonsson, *Chem. Eng. Commun*, **72**, 69 (1988).

Everaerts, F.M., F.E.P. Mikkers and Th.P.E. Verheggen, *Separ. Purif. Meth.*, **6**(2), 287 (1977).

Fair, J.R., *Petro/Chem. Engr.*, **33**, 45 (1961).

Fair, J.R., H.R. Null and W.L. Bolles, *Ind. Eng. Chem. Proc. Des. Dev.*, **22**, 53 (1983).

Fair, J.R., D. R. Steinmeyer, W.R. Penney and B.B. Crocker, in *Perry's Chemical Engineers' Handbook*, 6th edn., R. H. Perry and D. Green (eds.), McGraw-Hill, New York (1984), sect. 18.

Fanali, S., in *Capillary Electrophoresis Technology*, N.A. Guzman (ed.), Marcel Dekker, New York (1993), p. 731.

Farmer, J.C., "Method and apparatus for capacitive deionization, electrochemical purification and regeneration of electrodes," US Patent 5,425,859 (1995).

Farmer, J.C., D.V. Fix, G.V. Mack, R.W. Pekola and J.F. Poco, *J. Electrochem. Soc.*, **143**, 159 (1996).

Farrell, P.C. and A.L. Babb, *J. Biomed. Mater. Res.*, **7**, 275 (1973).

Feins, M., "Novel internally-staged ultrafiltration for protein," *Ph. D. Thesis, Department of Chemical Engineering*, New Jersey Institute of Technology, Newark, NJ (2004).

Feins, M. and K.K. Sirkar, *Biotechnol. Bioeng.* **86**(6), 603 (2004).

Feins, M. and K.K. Sirkar, *J. Membr. Sci.*, **248**(1-2), 137 (2005).

Fenske, M.R., *Ind. Eng. Chem.*, **24**, 482 (1932).

Fernandez, G.F. and C.N. Kenney, *Chem. Eng. Sci.*, **38**, 827 (1983).

Ferry, J.D., *J. Gen. Physiol.*, **20**, 95 (1936).

Field, R.W., D. Wu, J.A. Howell and B.B. Gupta, *J. Membr. Sci.*, **100**, 259 (1995).

Figdor, C.G., F. Preijers, R. Huijbens, P. Ruijs, T.J.M. de Witte and W.S. Bont, in *Cell Separation Methods and Applications*, D. Recktenwald and A. Radbruch (eds.), Marcel Dekker, New York (1998), chap. 3.

Findlay, R.A., in *New Chemical Engineering Separation Techniques*, H.A. Schoen (ed.), Interscience Publishers, New York (1962), chap. 4.

Fish, B.B., R.W. Carr and R. Aris, *AIChE J.*, **35**(5), 737 (1989).

Fitz Jr., C.W., J.G. Kunesh and A. Shariat, *Ind. Eng. Chem. Res.*, **38**, 512 (1999).

Flagan, R.C. and J.H. Seinfeld, *Fundamentals of Air Pollution Engineering*, Prentice Hall, Englewood Cliffs, NJ (1988).

Flett, D.S., J. Melling and M. Cox, in *Handbook of Solvent Extraction*, T.C. Lo, M.H.I. Baird and C. Hansen (eds.), Krieger Publishing Co., Malabar, FL, (1991), chap. 24.

Flory, P.J., *Principles of Polymer Chemistry*, Cornell University Press, Ithaca, NY (1953).

Follman, D.K. and R.L. Fahrner, *J. Chromatog. A*, **1024**, 79 (2004).

Foust, A.S., L.A. Wenzel, C.W. Clump, L. Maus and L.B. Andersen, *Principles of Unit Operations*, John Wiley, New York (1960), p. 532.

Freeman, B.D., *Macromolecules*, **32**, 375 (1999).

Frej, A.-K. Barnfield, H.J. Johannson, S. Johansson and P. Leijon, *Bioprocess Eng.*, **16**, 57 (1997).

Frenz, J. and C. Horvath, *AIChE J.*, **31**(3), 400 (1985).

Friedlander, S.K., *Smoke, Dust and Haze: Fundamentals of Aerosol Behavior*, John Wiley, New York (1977).

Froment, G.F. and K.B. Bischoff, *Chemical Reactor Analysis and Design*, John Wiley, New York (1979), chap. 14.

Fu, S. and C.A. Lucy, *Anal. Chem.*, **70**(1), 173 (1998).

Fuchs, N.A., *The Mechanics of Aerosols*, McMillan, New York (1964).

Fuerstenau, D.W. and T.W. Healy, in *Adsorptive Bubble Separation Techniques*, R. Lemlich (ed.), Academic Press, New York (1972).

Fuerstenau, D.W. and R. Herrera-Urbina, in *Surfactant-based Separation Processes*, J.F. Scamehorn and J.A. Harwell (eds.), Marcel Dekker, New York (1989), p. 259.

Fuh, C.B., M.N. Myers and J.C. Giddings, *Anal. Chem.*, **64**, 3125 (1992).

Fujita, Y., H. Nakamura and T. Muto, *J. Appl. Electrochem.*, **16**, 935 (1986).

Fuller, E.J., in *Separation and Purification Methods*, E.S. Perry and C.J. Van Oss (eds.), vol. 1, Marcel Dekker, New York (1972), p. 253.

Gandhi, S.K., *VLSI Fabrication Principles*, John Wiley, New York (1983).

Garcia, A.A., M.R. Bonen, J. Ramirez-Vick, M. Sadaka and A. Vuppu, *Bioseparation Process Science*, Blackwell Science, Malden, MA (1999).

Garg, D.R. and D.M. Ruthven, *Adv. Chem. Ser.*, **121**, 345 (1973).

Geankoplis, C.J., *Mass Transport Phenomena*, Ohio State University Bookstores, Columbus, OH (1972).

Geankoplis, C.J., *Transport Processes and Separation Process Principles*, 4th edn., Prentice Hall, Upper Saddle River, NJ (2003). pp. 487, 659, 681.

Gebauer, K.H., J. Thommes and M.R. Kula, *Chem., Eng. Sci.*, **57**(3), 405 (1997).

Gee, A.P., in *Cell Separation Methods and Applications*, D. Rectenwald and A. Radbruch (eds.), Marcel Dekker, New York (1998), chap. 9.

Geller, D.A. and G.W. Swift, *J. Acoust. Soc. Am.*, **111**(4), 1675 (2002a).

Geller, D.A. and G.W. Swift, *J. Acoust. Soc. Am.*, **112**(2), 504 (2002b).

Gellings, P.J. and H.J.M. Bouwmeester, *Catalysis Today*, **12**(1), Feb. (1992).

Gentilcore, M.J., *Chem. Eng. Progr.*, **Jan.**, 56 (2002).

Gerster, J.A., A.B. Hill, N.N. Hochgraf and D.G. Robinson, "Tray efficiencies in distillation columns," Final Report, University of Delaware, Research Committee, American Institute of Chemical Engineers, New York (1958).

Ghowsi, K., J.P. Foley and R.J. Gale, *Anal. Chem.*, **62**, 2714 (1990).

Gidaspow, D., Y-T. Shih, J. Bouillard and D. Wasan, *AIChE J.*, **35**(5), 714 (1989).

Giddings, J.C., *Dynamics of Chromatography, Part I. Principles and Theory*, Marcel Dekker, New York (1965).

Giddings, J.C., *Separ. Sci.*, **1**, 123 (1966).

Giddings, J.C., *Separ. Sci.*, **4**, 181 (1969).

Giddings, J.C., *Separ. Sci. Technol.*, **13**, 3 (1978).

Giddings, J.C., in *Treatise on Analytical Chemistry, Part I. Theory and Practice*, P.J. Elving, E. Grushka and I.M. Kolthoff (eds.), vol. 5, Wiley-Interscience, New York, (1982), chap. 3.

Giddings, J.C., *Separ. Sci Technol.*, **20** (9&10), 249 (1985).

Giddings, J.C., *Unified Separation Science*, John Wiley, New York (1991).

Giddings, J.C., *Science*, **260**, 1456 (1993).

Giddings, J.C. and K. Dahlgren, *Separ. Sci.*, **6**, 343 (1971).

Giddings, J.C., E. Kucera, C.P. Russel and M.N. Myers, *J. Phys. Chem.*, **72**, 4397 (1968).

Giddings, J.C., F.J. Yang and M.N. Myers, *Science*, **193**, 1244 (1976).

Gilbert, S.W., *AIChE J.*, **37**(8), 1205 (1991).

Gilliland, E.R., *Ind. Eng. Chem.*, **32**, 1220 (1940).

Glatz, C.E., M. Hoare and J. Landa-Vertiz, *AIChE J.*, **32**, 1196 (1967).

Glueckauf, E., *Discuss. Faraday Soc.*, **7**, 12 (1949).

Glueckauf, E., *Trans. Faraday Soc.*, **51**, 34 (1955a).

Glueckauf, E., *Trans. Faraday Soc.*, **51**, 1540 (1955b).

Glueckauf, E. and J.I. Coates, *J. Chem. Soc.*, 1315 (1947).

Gnanasundaram, S., T.E. Degaleesan and G.S. Laddha, *Can. J. Chem. Eng.*, **57**(2), 141 (1979).

Gobie, W.A., J.B. Beckwith and C.F. Ivory, *Biotechnol. Prog.*, **1**(1), 60 (1985).

Göklen, K.E. and T.A. Hatton, *Biotechnol. Prog.*, **1**(1), 69 (1985).

Göklen, K.E. and T.A. Hatton, *Separ. Sci. Techol.*, **22**, 831 (1987).

Goel, V., M.A. Accomazzo, A.J. Dileo, P. Meier, A. Pitt and M. Pluskal, in *Membrane Handbook*, W.S.W. Ho and K.K. Sirkar (eds.), Van Nostrand Reinhold, New York (1992), chap. 34.

Goel, V., M.A. Accomazzo, A.J. Dileo, P. Meier, A. Pitt and M. Pluskai, in *Membrane Handbook*, W.S.W. Ho and K.K. Sirkar (eds.), Kluwer Academic Publishers, Boston, MA (2001), chap 34.

Goettler, L.A. and R.L. Pigford, *AIChE J.*, **17**, 793 (1971).

Gooding, C.H. and R.M. Felder, *AIChE J.*, **27**(2), 193 (1981).

Gray, W.G., *Chem. Eng. Sci.*, **30**, 229 (1975).

Green, G. and G. Belfort, *Desalination*, **35**, 129 (1980).

Greenlaw. F.W., R.A. Sheldon and E.V. Thompson, *J. Membr. Sci.*, **2**, 333 (1977).

Gregory, R.A. and N.H. Sweed, *Chem. Eng. J.*, **1**, 207 (1970).

Grew, K.E. and T.L. Ibbs., *Thermal Diffusion in Gases*, Cambridge University Press, Cambridge (1952).

Grieves, R.B., in *Treatise on Analytical Chemistry, Part I. Theory and Practice*, P.J. Elving, E. Grushka and I.M. Kolthoff (eds.), vol. 5, Wiley-Interscience, New York (1982), p. 371.

Grosjean, P.R.L. and H. Sawistowski, *Trans. Inst. Chem. Eng., London*, **58**, 59 (1980).

Grosser, J.H., M.F. Doherty and M.F. Malone, *Ind. Eng. Chem. Res.*, **26**, 983 (1987).

Grossman, M.W. and T.W. Shepp, *IEEE Trans. Plasma Sci.*, **19**(6), 1114 (1991).

Grubisic, Z., P. Rempp and H. Benoit, *Polymer Lett.*, **5**, 753 (1967).

Grushka, E., K.D. Caldwell, M.N. Myers and J.C. Giddings, *Separ. Purif. Meth.* **2**(1), 127 (1973).

Guggenheim, E.A., *Thermodynamics*, 5th edn., North-Holland Publishing Co., Amsterdam (1967).

Guha, A.K., S. Majumdar and K.K. Sirkar, *Ind. Eng. Chem. Res.*, **31**, 593 (1992).

Gupta, A.S. and G. Thodos, *AIChE J.*, **8**, 608 (1962).

Gupta, R. and N.H Sweed, *Ind. Eng. Chem. Fund.*, **10**, 283 (1971).

Gutierrez, A.P., D. Haase, D.A. Keyworth, D.G. Walker, D.L. Klein, "ESEP, A Process for the recovery of ethylene," 175th ACS National Meeting, Anaheim, CA, March 12-17 (1978).

Guzman, N.A. (ed.), *Capillary Electrophoresis Technology*, Marcel Dekker, New York (1993).

Hagel, L., in *Protein Purification, Principles, High Resolution Methods and Applications*, J.C. Jansen and L. Ryder (eds.), VCH, New York (1989), p. 63.

Haimour, N., A. Bidarian and O.C. Sandall, *Separ. Sci. Technol.*, **22** (2&3), 921 (1987).

Halliday, D. and R. Resnick, *Physics, Part II*, John Wiley, New York (1962), p. 921.

Han, S., F.C. Ferreira and A. Livingston, *J. Membr. Sci.*, **188**, 219 (2001).

Hannig, K., *Z. Anal. Chem.*, **181**, 244 (1961).

Hansen, C.M., *Ind. Eng. Chem. Prod. Res. Dev.*, **8**, 2 (1969).

Hansen, C.M. and A. Beerbower, in *Encyclopedia of Chemical Technology*, 2nd edn., A. Standen (ed.), suppl. vol., Interscience, New York (1971), p. 889.

Happel, J., *AIChE J.*, **4**, 197 (1958).

Happel, J., *AIChE J.*, **5**, 174 (1959).

Happel J. and H. Brenner, *Low Reynolds Number Hydrodynamics*, Prentice Hall, Englewood Cliffs, NJ (1965).

Happel J. and H. Brenner, *Low Reynolds Number Hydrodynamics*, Martinus Nijhoff, The Hague (1983).

Harriott, P., *Separ. Sci.*, **8**, 291 (1973).

Harrison, R.G., P. Todd, S.R. Rudge and D.P. Petrides, *Bioseparations Science and Engineering*, Oxford University Press, New York (2003), chap. 8, pp. 349-362.

Hartland, S. and J.C. Mecklenburgh, *Chem. Eng. Sci.*, **21**, 1209 (1966).

Hatta, S., *Tohoku Imperial Univ. Tech. Rep.* **8**, 1 (1928).

Hatta, S., *Tohoku Imperial Univ. Tech. Rep.* **10**, 119 (1932).

Hatton, T.A., in *Surfactant-based Separation Processes*, J.F. Scamehorn and J.A. Harwell (eds.), Marcel Dekker, New York (1989), p. 55.

Hausmann, M., C. Cremer, R. Hartig, H.G. Liebich, G.H. Luers, A. Saalmüler and R. Teichmann, in *Cell Separation Methods and Applications*, D. Recktenwald and A. Radbruch (eds.), Marcel Dekker, New York (1998), chap. 10.

Haynes, C.A., J. Carson, H.W. Blanch and J.M. Prausnitz, *AIChE J.*, **37**(9), 1401 (1991).

Haynes, C.A., F.J. Benitez, H.W. Blanch and J.M. Prausnitz, *AIChE J.*, **39**(9), 1539 (1993).

Helfferich, F., *Ion Exchange*, McGraw-Hill, New York (1962).

Helfferich, F., *Ion Exchange*, Dover Publications, New York (1995).

Helfferich, F. and G. Klein, *Multicomponent Chromatography*, Marcel Dekker, New York (1970).

Helmbrecht, C., R. Niessner and C. Haisch, *Anal. Chem.*, **79**, 7097 (2007).

Hengstebeck, R.J., *Distillation: Principles and Design Procedures*, Reinhold, New York (1961), chap. 7.

Henis, J.M.S. and M.K. Tripodi, *J. Membr. Sci.*, **8**, 233 (1981).

Henley, E.J. and J.D. Seader, *Equilibrium-Stage Separation Operations in Chemical Engineering*, John Wiley, New York (1981), pp. 690-692, 714.

Henry, J.D. Jr., in *Perry's Chemical Engineers' Handbook*, R.H. Perry and D.W. Green (eds.), 6th edn., McGraw-Hill, New York (1984), sect. 17.

Henry, J.D. Jr. and C.G Moyers, Jr., in *Perry's Chemical Engineers' Handbook*, R.H. Perry and D.W. Green (eds.), 6th edn., McGraw-Hill, New York (1984), pp. 17-6 to 17-12.

Herberhold, M., *Metal π-complexes, Part II: Specific Aspects*, vol. II, Elsevier, New York (1974).

Herzig, J.P., D.M. Leclerc and P. Le Goff, *Ind. Eng. Chem.*, **62**(5), 8 (1970).

Hestekin, J.A., L.G. Bachas and D. Bhattacharyya, *Ind. Eng. Chem. Res.*, **40**, 2668 (2001).

Heyne, L., in *Solid Electrolytes*, S. Geller (ed.), Springer-Verlag, New York (1977), chap. 7.

Higuchi, A., Y. Ishida and T. Nakagawa, *Desalination*, **90**, 127 (1993).

Hill, F.B., *Chem. Eng. Commun.*, **7**, 37 (1980).

Hill, T.L., *J. Chem. Phys.*, **17**, 520 (1949).

Hills, G.J., P.W.M. Jacobs and N. Lakshminarayaniah, *Proc. Roy. Soc. London, Ser. A*, **262**, 246 (1961).

Hines, A.L. and R.M. Maddox, *Mass Transfer: Fundamentals and Applications*, Prentice-Hall PTR, Upper Saddle River, NJ (1985).

Hirschfelder, J.O., C.F. Curtis and R.B. Bird, *Molecular Theory of Gases and Liquids*, John Wiley, New York (1954).

Ho, W.S.W and D.C. Dalrymple, *J. Membr. Sci.*, **91**, 13 (1994).

Ho, W.S.W and N.N. Li, in *Perry's Chemical Engineers' Handbook*, R.H. Perry and D.W. Green (eds.), 6th edn., McGraw-Hill, New York (1984a), pp. 17-14 to 17-35.

Ho, W.S.W and N.N. Li, in *Hydrometallurgical Process Fundamentals*, R.G. Bautista (ed.), Plenum Press, New York (1984b), p. 555.

Ho, W.S.W. and N.N Li, in *Membrane Handbook*, W.S.W. Ho and K.K. Sirkar (eds.), Kluwer Academic Publishers, Boston, MA (2001), chaps. 36, 37.

Ho, W.S.W and K.K. Sirkar (eds.), *Membrane Handbook*, Kluwer Academic Publishers, Boston, MA (2001).

Ho, W.S.W., G. Doyle, D.W. Savage and R.L. Pruett, *Ind. Eng. Chem., Res.*, **27**, 334 (1988).

Hochhauser, A.M. and E.L. Cussler, *AIChE Symp. Ser.* **71**(152), 136 (1975).

Hoffman, R.A. and D.W. Houck, in *Cell Separation Methods and Applications*, D. Recktenwald and A. Radbruch (eds.), Marcel Dekker, New York (1998), chap. 11.

Holusha, J., "From the ashes, a plus for a utility," *New York Times*, Nov. 24 (1993).

Horvath, C.S. and H.J. Lin, *J. Chromatog.*, **149**, 43 (1978).

Hougen, O. A., K. M. Watson and R.A. Ragatz, *Chemical Process Principles, Part I, Material and Energy Balances*, 2nd edn., John Wiley, New York (1954), p. 163.

Hougen, O.A., K.M. Watson and R.A. Ragatz, *Chemical Process Principles, Part II, Thermodynamics*, 2nd edn., John Wiley, New York (1959), pp. 968-969.

Howard, A.J., G. Carta and C.H. Byers, "Novel applications of continuous annular chromatography: separation of sugars," ACS National Meeting, Denver, CO, April, 5-10 (1987).

Howe, E.D., *Fundamentals of Water Desalination*, Marcel Dekker, New York (1974), chap. 5 II.

Howell, J.A., *J. Membr. Sci.*, **107**, 165 (1995).

Hovingh, M.E., G.H. Thompson and J.C. Giddings, *Anal. Chem.*, **42**, 195 (1970).

Hsu, H-W., *Separation by Centrifugal Phenomena, Techniques of Chemistry*, E.S. Perry (ed.), vol. XVI, John Wiley, New York (1981).

Huang, S.H., W.C. Lee and G.T. Tsao, *Chem. Eng. J.*, **38**, 179 (1988a).

Huang, S.H., S. Roy, K.C Hou and G.T. Tsao, *Biotechnol. Prog.*, **4**(3), 159 (1988b).

Huang, X.C., M.A. Quesada and R.A. Mathies, *Anal. Chem.*, **64**(8), 967 (1992).

Huckins, H.E. and K. Kammermeyer, *Chem. Eng. Prog.*, **49**, 180 (1953a).

Huckins, H.E. and K. Kammermeyer, *Chem. Eng. Prog.*, **49**, 295 (1953b).

Hughes, R.D., J.A. Maloney and E.F. Steiglemann, in *Recent Developments in Separation Science*, N.N. Li and J.M. Calo (eds.), vol. 9, CRC Press, Boca Raton, FL (1986), p. 174.

Hulbert, H.M. and S. Katz, *Chem. Eng. Sci.*, **19**, 555 (1964).

Humphrey, J.L. and G.E. Keller, II, *Separation Process Technology*, McGraw-Hill, New York (1997), p. 297.

Hunter, J.B., *Separ. Sci. Technol.*, **23**(8&9), 913 (1988).

Hwang, S.T. and K. Kammermeyer, "Membranes in separations," in *Techniques of Chemistry*, A. Weissberger (ed.), vol. 7, Wiley-Interscience (1975). Reprinted, Kreiger Publishing, Melborne, FL (1984).

Hwang, Y-L., G.E. Keller, II, and J.D. Olson, *Ind. Eng. Chem. Res.*, **31**, 1753 (1992a).

Hwang, Y-L., J.D. Olson and G.E. Keller, II, *Ind. Eng. Chem. Res.*, **31**, 1759 (1992b).

Ibbs, T.L., K.E. Grew and A.A. Hirst, *Proc. Roy. Soc. (London)*, A**173**, 543 (1939).

Inculet, I.I., *Electrostatic Mineral Separation*, Research Studies Press Ltd., John Wiley, New York (1984).

Ivory, C.F., *J. Chromatog.*, **195**, 165 (1980).

Ivory, C.F. and W.A. Gobie, *Biotechnol. Prog.*, **6**, 21 (1990).

Ivory, C.F., M. Gilmartin, W.A. Gobie, C.A. McDonald and R.L. Zollars, *Biotechnol. Prog.*, **11**, 21 (1995).

Jaasund, S.A., *Chem. Eng.*, **Nov. 23, 159** (1987).

Jacobs, D.B. and J. Zimmerman, in *Polymerization Process*, C.E. Schildknecht (ed.), Wiley, New York (1977), chap. 12.

Jacques, J., A. Collet and S.H. Wilen, *Enantiomers, Racemates and Resolutions*, Wiley-Interscience, John Wiley, New York (1981), chap. 5.1.

Jafvert, C.T., W. Chu and P.L. Van Hoof, in *Surfactant-enhanced Subsurface Remediation: Emerging Technologies*, D.A. Sabatini, R.C. Knox and J.H. Harwell (eds.), ACS Symposium Series 594, Amercian Chemical Society, Washington, D.C. (1995), p. 24.

Jagur-Grodzinski, J. and R. Kramer, *Ind. Eng. Chem. Proc. Des. Dev.*, **25**(2), 443 (1986).

James, A.T. and A.J.P. Martin, *Biochem. J.*, **50**, 679 (1952).

Janini, G.M. and H.J. Issaq, in *Capillary Electrophoresis Technology*, N.A. Guzman (ed.), Marcel Dekker, New York (1993), p. 119.

Jassim, M.S., G. Rochelle, D. Eimer and C. Ramshaw, *Ind. Eng. Chem. Res.*, **46**, 2823 (2007).

Jennings, W., *Analytical Gas Chromatography*, Academic Press, New York (1987), p. 79.

Jillavenkatesa, A., S.J. Dapkunas and L.S.H. Lum, *Particle Size Characterization*, NIST Recommended Practice Guide, Special Publication 960-1, January, NIST (2001).

Johnson, J.S., L. Dresner and K.A. Kraus, in *Principles of Desalination*, K.S. Spiegler (ed.), Academic Press, New York (1966), chap. 8.

Johnson, J.S., L. Dresner and K.A. Kraus, in *Principles of Desalination*, K.S. Spiegler and A.D.K. Laird, 2nd edn., Part B, Academic Press, New York (1980), chap. 8.

Johnston, K.P., S.E. Barry, N.K. Read and T.R. Holcomb, *Ind. Eng. Chem. Res.*, **26**, 2372 (1987).

Jolley, J.E. and J.H. Hildebrand, *J. Am. Chem. Soc., 450*, **80**, 1050 (1958).

Jones, J.B., in *Techniques of Chemistry, Vol. X, Applications of Biochemical Systems in Organic Chemistry*, J.B. Jones, C.J. Sih and D. Perlman, (eds.), Part 1, John Wiley, New York (1976), p. 18.

Jones, J.B. and J.F. Beck, in *Techniques of Chemistry, Vol. X, Applications of Biochemical Systems in Organic Chemistry*, J.B. Jones, C.J. Sih and D. Perlman (eds.), Part 1, John Wiley, New York (1976), p. 139.

Jones, T.B., *Electromechanics of Particles*, Cambridge University Press, New York (1995).

Jorgenson, J.W. and K.D. Lukacs, *Anal. Chem.*, **53** 1298 (1981).

Jung, M., S. Mayer and V. Schuring, *LCGC North Am.*, **12**(6), 458 (1994).

Jury, S.H. and W.L. Locke, *AIChE J.*, **3**(4), 480 (1957).

Kammermeyer, K., in *Progress in Separation and Purification*, E.S. Perry (ed.), vol. 1, Interscience, New York (1968), p. 335.

Karger, B.L. and F. Foret, in *Capillary Electrophoresis Technology*, N.A. Guzman (ed.), Marcel Dekker, New York (1993), p. 3.

Karger, B.L., L.R. Snyder and C. Horvath, *An Introduction to Separation Science*, Wiley, New York (1973), chap. 8, p. 269.

Karlson, E.L. and S.P. Edkins, *AIChE Symp. Ser.*, **71**(151), 286 (1975).

Kay, R.L., *Pure Appl. Chem.*, **63**, 1393 (1991).

Keey, R.B., *Drying: Principles and Practice*, Pergamon Press, New York (1972).

Kehlen, H., M.T. Ratzsch and J. Bergmann, *AIChEJ.*, **31**(7), 1136 (1985).

Keller, G.E., A.E. Marcinkowsky, S.K. Verma and K.D. Williamson, in *Separation and Purification Technology*, N.N. Li and J.M. Calo (eds.), Marcel Dekker, New York (1992), chap. 3.

Keller, G.E., II, in *Industrial Gas Separations*, ACS Symposium Series 223, American Chemical Society, Washington, D.C. (1983).

Keller, G.E. II and C.H.A. Kuo, "Enhanced gas separation by selective adsorption," US Patent 4,354,859, Oct. 19 (1982).

Keller, K.H. and T.R. Stein, *Math. Biosci.*, **1**, 421 (1967).

Kelsall, D.F., in *Solid-Liquid Separation*, J.B. Pool and D. Doyle (eds.), HMSO, London (1966).

Kennard, E.H., *Kinetic Theory of Gases*, McGraw Hill, New York (1938).

Kenyon, N.S., C. Ricordi, J.G. Gribben, L.M. Nadler, R.K. Zwerner and T.R. Russell, in *Cell Separation Methods and Applications*, D. Reckenwald and A. Radbruch (eds.) Marcel Dekker, New York (1998), p. 103.

Kertes, A.S. and C.J. King, *Biotechnol. Bioeng.*, **28**, 269 (1986).

Kessler, D.P. and P.C. Wankat, *Chem. Eng.*, Sept. 26, 71 (1988).

Kessler, S.B and E. Klein, in *Membrane Handbook*, W.S. Winston Ho and K.K. Sirkar (eds.) Kluwer Academic Publishers, Boston, MA (2001), chaps. 11-15.

Kheterpal, I. and R.A. Mathies, *Anal. Chem.*, **71**(1), 31A (1999).

Kiani, A., R.R. Bhave and K.K. Sirkar, *J. Membr. Sci.*, **20**, 125 (1984).

King, C.J., *Separation Processes*, 2nd edn., McGraw-Hill, New York (1980).

King, C.J., *Freeze-Drying of Foods*, CRC Press, Cleveland, OH (1971).

King, R.S., H.W. Blanch and J.M Prausnitz, *AIChE J.*, **34**, 1585 (1988).

Kirchner, J.J. and E.F. Hasselbrink, Jr., *Anal. Chem.*, **77**, 1140 (2005).

Kirkland, J.J. and C.H. Dilks, Jr., *Anal. Chem.*, **64**, 2836 (1992).

Kirwan, D.J. and C.J. Orella, in *Handbook of Industrial Crystallization*, A.S. Myerson (ed.), Butterworths-Heinemann, Boston, MA (1993), chap. 11.

Kister, H.Z., *Distillation Operation*, McGraw-Hill, New York (1990).

Kister, H.Z., *Distillation Design*, McGraw-Hill, New York (1992).

Kister, H.Z., *Chem. Eng. Prog.*, June, 39 (2008).

Kister, H.Z. and D.R. Gill, *Chem. Eng. Progr.*, **87**(2), 32, Feb. (1991).

Kister, H.Z., J. Scherffius, K. Afshar and E. Akbar, *Chem. Eng. Prog.*, **28**, July (2007).

Kitiyanan, B., J.H. O'Haver, J.H. Harwell and D. Sabatini in *Surfactant-based Separations*, J.F. Scamehorn and J.H. Harwell (eds.), ACS Symposium. Series, 740, American Chemical Society, Washington, D.C. (1999), p. 76.

Klein, E., J.K. Smith, R.P. Wendt and S.V. Desai, *Separ. Sci.*, **7**, 285 (1972).

Klein, E., J.K. Smith, R.E.C. Weaver, R. P. Wendt and S.V. Desai, *Separ. Sci.*, **8**, 585 (1973).

Klein, E., F.F. Holland and K. Eberle, *J. Membr. Sci.*, **5**, 173 (1979).

Knaebel, K.S. and F.B. Hill, "Analysis of gas purification by heatless adsorption", AIChE National Meeting, Los Angeles, CA, Nov. 14–18 (1982), paper 91d.

Knaebel, K.S. and F.B. Hill, *Chem. Eng. Sci.*, **40**(12), 2351 (1985).

Knaebel, K.S. and R.L. Pigford, *Ind. Eng. Chem. Fund.*, **22**, 336 (1983).

Kohl, A.L. and F.C. Riesenfeld, *Gas Purification*, 3rd edn., Gulf Publishing Co., Houston, TX (1979), p.783.

Komasawa, I and T. Otake, *Ind. Eng. Chem. Fund.*, **22**, 122 (1983)

Koros, W.J., *J. Polym. Sci., Polym. Phys. Ed.*, **18**, 981 (1980).

Koros, W.J., R.T. Chern, V.T. Stannett and H.B. Hopfenberg, *J. Polym. Sci., Polym. Phys. Ed.*, **19**, 1513 (1981).

Kouketsu, T., S. Duan, T. Kai, S. Kazama and K. Yamada, *J. Membr. Sci.*, **287**, 51 (2007).

Koval, C.A., T. Spontarelli and R.D. Noble, *Ind. Eng. Chem. Res.*, **28**, 1020 (1989).

Kovvali, A.S. and K.K. Sirkar, *Ind. Eng. Chem. Res.*, **40**, 2502 (2001).

Kovvali, A.S. and K.K. Sirkar, *Ann. NY Acad. Sci.*, **984**, 279 (2003).

Kovvali, A.S., H. Chen and K.K. Sirkar, *J. Am. Chem. Soc.*, **122**(31), 7594 (2000).

Kovvali, A.S., H. Chen and K.K. Sirkar, *Ind. Eng. Chem. Res.*, **41**(3), 347 (2002).

Kowler, D.E. and R.H. Kadlec, *AIChE J.*, **18**, 1207 (1972).

Kozinski, A.A. and E.N. Lightfoot, *AIChE J.*, **18**(5), 1030 (1972).

Kralj, J.G., M.T.W. Lis, M.A. Schmidt and K.F. Jensen, *Anal. Chem.*, **78**, 5019 (2006).

Kremser, A., *Nat. Petrol. News.*, **22**(21), May 30, 43 (1930).

Krishna, R. and G.L. Standart, *Chem. Eng. Commun.*, **3**, 201 (1979).

Krishna, R., H.F. Martinez, R. Sreedhar and G.L. Standart, *Trans. Inst. Chem. Eng.*, **55**, 178 (1977).

Krishnamurthy, R. and R. Taylor, *AIChE J.*, **31**, 449 (1985).

Kulkarni, S.S., E.W. Funk and N.N Li, in *Membrane Handbook*, W.S. Winston Ho and K.K. Sirkar (eds.), Kluwer Academic, Boston, MA (2001), chaps. 26–30.

Kulov, N.N., *Chem. Eng. Commun.*, **21**, 259 (1983).

Kumar, R. and G.R. Dissinger, *Ind. Eng. Chem. Proc. Des. Dev.*, **25**, 456 (1986).

Kumar, R. and S. Sircar, *Chem. Eng. Commun.*, **26**, 339 (1984).

Kuo, Y. and H.P. Gregor, *Separ. Sci. Technol.*, **18**, 421 (1983).

Kurnik, R.T., A.W. Yu, G.S. Blank, A.R. Burton, D. Smith, A.M. Athalye and R. van Reis, *Biotechnol. Bioeng.*, **45**, 149 (1995).

Kurrelmeyer, B. and W.H. Mais, *Electricity and Magnetism*, Van Nostrand Co., Princeton, NJ (1967).

Kuwabara, S., *J. Phys. Soc. Jpn.*, **14**, 527 (1959).

Kynch, G.J., *Trans. Faraday Soc.*, **48**, 166 (1952).

La Mer, V.K., *Ind. Eng. Chem.*, **44**, 1270 (1952).

Ladisch, M.R., *Bioseparations Engineering: Principles, Practice and Economics*, John Wiley, New York (2001).

Lahiere, R.J., J.L. Humphrey and J.R. Fair, *Separ. Sci. Technol.*, **22**(2&3), 379 (1987).

Landau, L.D. and E.M. Lifshitz, *Fluid Mechanics*, Addison Wesley, Reading, MA (1959), p. 97.

Lane, J.A., in *Chemical Engineers Handbook*, 3rd edn., J. H. Perry (ed.), McGraw-Hill, New York (1950), p. 753.

Lane, J.A. and J.W. Riggle, *Chem. Eng. Progr. Symp. Ser.*, **55**(24), 135 (1959).

Langer, S. H. and R.G. Haldeman, *J. Phys. Chem.*, **68**(4), 962 (1964).

Langhaar, H.L., *Dimensional Analysis and Theory of Models*, Wiley, New York (1951).

Langmuir, I., *Phys. Rev.*, **8**, 149 (1916).

Lapidus, L. and N.R. Amundson, *J. Phys. Colloid Chem.*, **54**, 821 (1950).

Lapidus, L. and N.R. Amundson, *J. Phys. Chem.*, **56**, 984 (1952).

Lapple C., *Air Pollution Engineering Manual*, USEPA, AP-40, 94 (1951).

Larsen, M.U. and N.C. Shapley, *Anal. Chem.*, **79**, 1947 (2007).

Lasic, D., *Am. Scientist*, **80**, 20 (1992).

Lee, C-K., *Ind. Eng. Chem. Res.*, **34**, 2104 (1995).

Lee, H.H., *Fundamentals of Microelectronic Processing*, McGraw-Hill, New York (1990), chap. 3.

Lee, H.L., J.F.G Reis, J. Dohner and E.N. Lightfoot, *AIChE J.*, **20**, 776 (1974).

Lee, H.L., E.N. Lightfoot, J.F.G., Reis and M.D. Waissbluth, in *Recent Developments in Separation Science*, N.N. Li (ed.), vol. III, Part A, CRC Press, Cleveland OH (1977a), p. 1.

Lee, K-H., D.F. Evans and E.L. Cussler, *AIChE J.*, **24**, 860 (1978).

Lee, K.L., H.B. Hopfenberg and V.T. Stannett, *J. Appl. Polym. Sci.*, **21**, 1795 (1977b).

Lehninger, A.L., *Principles of Biochemistry*, 3rd edn., Worth Publishers, New York (1982).

Leighton, D.T. and A. Acrivos, *J. Fluid Mech.*, **177**, 109 (1987).

Leith, D. and W. Licht, *AIChE Symp. Ser.*, **68**(126), 196 (1972).

Lemlich, R., in *Progress in Separation and Purification*, E.S. Perry (ed.), vol. 1, Wiley (Interscience), New York (1968).

Leonard, E.F. and L.W. Bleumle, Jr., *Trans. NY Acad. Sci. Ser.*, **2**(21), 585 (1959).

Leung, P.S., in *Ultrafiltration Membranes and Applications*, A.R. Cooper (ed.), Plenum, New York (1979), p. 415.

LéVêque, J., *Ann. Mines Ser.*, **12**, **13**, 201, 305–362, 381–415 (1928).

Levich, V.G., *Physicochemical Hydrodynamics*, Prentice-Hall, Englewood Cliffs, NJ (1962), sect, 94.

Lewis, G.N. and M. Randall, *Thermodynamics*, 2nd edn., revised by K.S. Pitzer and L. Brewer, McGraw-Hill, New York (1961).

Lewis, W.K., *Ind. Eng. Chem.*, **28**, 399 (1936).

Liddle, C.J., *Chem. Eng.*, **75**(23), Oct. 21, 137 (1968).

Liepmann, H.W., *J. Fluid Mech.*, **10**, 65 (1961).

Lightfoot, E.N., *Transport Phenomena and Living Systems*, John Wiley, New York (1974).

Lightfoot, E.N., *Separ. Sci. Technol.*, **40**, 739 (2005).

Lightfoot, E.N. and M.C.M. Cockrem, *Separ. Sci. Technol.*, **22**(2&3), 165 (1987).

Lightfoot, E.N., R.J. Sanchez-Palma and D.O. Edwards, in *New Chemical Engineering Separation Techniques*, H.A. Schoen (ed.), Interscience Publishers, New York (1962), chap. 2, pp. 99–181.

Lightfoot, E.N., P.T. Noble, A.S. Chiang and T.A. Ugulini, *Separ. Sci. Technol.*, **16**, 619 (1981).

Lin, I.J. and L. Benguigui, *Powder Technol.*, **17**, 95 (1977).

Lin, Y-S., W. Wang and J. Han, *AIChE J.*, **40**(5), 786 (1994).

Lindahl, P.E., *Nature*, **161**(4095), 648 (1948).

Lindahl, P.E., *Biochem. Biophys. Acta*, **21**, 411 (1956).

Linton, W.H. and T. K. Sherwood, *Chem. Eng. Progr.*, **46**, 258 (1950).

Liow, J.L. and C.N. Kenney, *AIChE J.*, **36**(1), 53 (1990).

Liu, Y.A. and M.J. Oak, *AIChE J.*, **29**(5), 771 (1983).

Lo, T.-C., M.H.I. Baird and C. Hanson, *Handbook of Solvent Extraction*, Wiley-Interscience, New York (1983), chaps. 2.1, 2.2, 3.

Lo, T.C., M.H.I. Baird and C. Hanson, *Handbook of Solvent Extraction*, Kreiger Publishing Co., Malabar, FL (1991).

Lockett, M.J., *Distillation Tray Fundamentals*, Cambridge University Press, Cambridge (1986).

Lockhart, F.J. and R.J. McHenry, *Petroleum Refiner*, **37**, 209 (1958).

Loeb, S., *J. Membr. Sci.*, **1**, 49 (1976).

Long, R.B., *Ind. Eng. Chem. Fund.*, **4**(4), 445 (1965).

Lonsdale, H.K., in *Desalination by Reverse Osmosis*, U. Merten (ed.), The MIT Press, Cambridge, MA (1966), p. 117.

Lonsdale, H.K., U. Merten and R.L. Riley, *J. Appl. Polym. Sci.*, **9**, 1341 (1965).

Lopez-Leiva, M., in *Ultrafiltration Membranes and Applications*, A.R. Cooper (ed.), Polymer Science and Technology Series, vol. 13, Plenum Press, New York (1980).

Lotfian, P., M.S. Levy, R.S. Coffin, T. Fearn and P. Ayazi-Shamlou, *Biotechnol. Prog.*, **19**, 209 (2003).

Lowe, A.B. and C.L. McCormick, in *Stimuli-Responsive Water Soluble and Amphiphilic Polymers*, C.L. McCormick (ed.), ACS Symposium. Series 780, American Chemical Society, Washington, D.C. (2001), p. 1.

Lowenheim F. A. and M. K. Moran, *Faith, Keyes and Clark's Industrial Chemicals*, 4th edn., John Wiley, New York (1975), p. 654.

Luborsky F.E and B.J. Drummond, *IEEE Trans. Magnetics*, **11**(6), 1696 (1975).

Lye, G.J., J.A. Asenjo and D.L. Pyle, *AIChE J.*, **42**(3), 713 (1996).

McCabe, W.L. and J.C. Smith, *Unit Operations of Chemical Engineering*, 3rd edn., McGraw-Hill, New York (1976).

McCabe, W.L., J.C. Smith and P. Harriott, *Unit Operations of Chemical Engineering*, 5th edn., McGraw-Hill, New York (1993).

Macias-Salinas, R. and J.R. Fair, *Ind. Eng. Chem. Res.*, **41**, 3429 (2002).

McCoy, B.J., *AIChE J.*, **32**(9), 1570 (1986).

McHugh, M.A. and V.J. Krukonis, *Supercritical Fluid Extraction*, Butterworths, Boston, MA (1986).

MacKenzie, P.D. and C.J. King, *Ind. Eng. Chem. Proc. Des. Dev.*, **24**, 1192 (1985).

McNab, G.S. and A. Meisen, *J. Colloid Interface Sci.*, **44**, 339 (1973).

Maget, H.J.R. "Process for gas purification," US Patent 3,489,670, January 13 (1970).

Majors, R.E., *LCGC North Am.*, **23**(12), 1248 (2005).

Majumdar, S., A.K. Guha and K.K. Sirkar, *AIChE J.*, **34**, 1135 (1988).

Majumdar, S., K.K. Sirkar and A. Sengupta, in *Membrane Handbook*, W.S. Winston Ho and K.K. Sirkar (eds.), Kluwer Academic Publishers, Boston, MA (2001), chap. 42.

Mallinson, R.H. and C. Ramshaw, "Mass transfer apparatus and process," European Patent 0053881 (1982).

Mandal, D.K., A.K. Guha and K.K. Sirkar, *J. Membr. Sci.*, **144**, 13 (1998).

Marin G.B. (ed.), *Multiscale Analysis*, Advances in Chemical Engineering 30, Elsevier Academic Press, Amsterdam (2005).

Markham, E.C. and A.F. Benton, *J. Am. Chem. Soc.*, **53**, 497 (1931).

Marmur, A., *J. Am. Chem. Soc.*, **122**, 2120 (2000).

Marr, R. and A. Kopp, *Int. Chem. Eng.*, **22**(1), 44 (1982).

Martell, A.E. and R.M. Smith, *Critical Stability Constants, Vol. 1*, Plenum Press, New York (1974).

Martell, A.E. and R.M. Smith, *Critical Stability Constants, Vol. 5*, Plenum Press, New York (1982).

Martin, A.J.P. and R.L.M. Synge, *Biochem. J.*, **35**, 1358 (1941).

Martin, H. and W. Kuhn, *Z. Physik. Chem.*, **189A**, 219 (1941).

Martin, J.J., *Chemie Ingenieur Technik*, **44**, 249 (1972).

Martin, J.J., *Chem. Eng. Edu.*, **Summer**, 119 (1983).

Mason, E.A. and W. Juda, *Chem. Eng. Progr. Symp. Ser.*, **24**(55), 155 (1959).

Mason, E.A. and T.A. Kirkham, *Chem. Eng. Progr. Symp. Ser.*, **24** (55), 173 (1959).

Mason, E.A. and E.W. McDaniel, *Transport Properties of Ions in Gases*, John Wiley, New York (1988).

Mass, J.H., in *Handbook of Separation Techniques for Chemical Engineers*, P.A. Schweitzer (ed.), McGraw-Hill, New York (1979), sect. 6.1.

Matson, S.L. and J.A. Quinn, *Ann. NY Acad. Sci.*, **469**, 152 (1986).

Matson, S.L, C.S. Herrick and W.J. Ward, III, *Ind. Eng. Chem. Proc. Des. Dev.*, **16**, 370 (1977).

Mattiasson, B. and T.G.I. Ling, in *Membrane Separations in Biotechnology*, W.C. McGregor (ed.), Marcel Dekker, New York (1986).

Mattock, P., G.F. Aitchison and A.R. Thomson, *Separ. Purif. Meth.*, **9**(1), 1 (1980).

Mayer, S.W. and E.R. Tompkins, *J. Am. Chem. Soc.*, **69**, 2792 (1947).

Meares, P., in *Membrane Separation Processes*, P. Meares (ed.) Elsevier, New York (1976), chap. 1.

Medina, J.C., N. Wu and M.L. Lee, *Anal. Chem.*, **73**(6), 1301 (2001).

Mehta, A., M.L. Tse, J. Fogle *et al.*, *Chem. Eng. Progr.*, special issue on *Purifying Therapeutic Monoclonal Antibodies, Bioprocessing* (2008), pp. S14–S19.

Meister, B.J. and A.E. Platt, *Ind. Eng. Chem. Res.*, **28**, 1659 (1989).

Merten, U., *Desalination by Reverse Osmosis*, MIT Press, Cambridge, MA (1966).

Meselson, M. and F.W. Stahl, *Proc. Natl. Acad. Sci. USA*, **44**, 671 (1958).

Meselson, M., F.W. Stahl and J. Vinograd, *Proc. Natl. Acad. Sci. USA*, **43**, 581 (1957).

Michaels, A.S., *Trans. Am. Soc. Artif. Intern. Organs*, **12**, 387 (1966).

Michaels, A., *Chem. Eng. Progr.*, **64**(12), 31 (1968a).

Michaels, A.S., in *Progress in Separation and Purification*, E.S. Perry (ed.), vol. 1, Wiley-Interscience, New York (1968b), pp. 297–334.

Michaels, A.S. and H.J. Bixler, *J. Polym. Sci.*, **50**, 393 (1961).

Michaels, A.S. and H.J. Bixler, in *Progress in Separation and Purification*, E.S. Perry (ed.), vol. 1, Wiley-Interscience, New York (1968), p. 143.

Middleman, S., *An Introduction to Mass and Heat Transfer, Principles of Analysis and Design*, John Wiley, New York (1998).

Miers, H.A., *J. Inst. Metals*, **37**, 331 (1927).

Miers, H.A. and F. Isaac, *J. Chem. Soc., London*, **89**, 413 (1906).

Migita, H., K. Soga and Y.H. Mori, *AIChE J.*, **51**(8), 2190 (2005).

Mikkers, F.E.P., F.M. Evaeraerts and Th.P.E.M. Verheggen, *J. Chromatog.*, **169**, 11 (1979).

Milnes, A.G., *Deep Impurities in Semiconductors*, John Wiley, New York (1973).

Mittlefehldt, E., *Separ. Times*, **8**(1), 10 (2002).

Miyauchi, T. and T. Vermeulen, *Ind. Eng. Chem. Fund.*, **2**, 113 (1963).

Moates, G.H. and J.K. Kennedy, in *Fractional Solidification*, M. Zief and W.R. Wilcox (eds.), vol. I, Marcel Dekker, New York (1967), chap. 10.

Molinari, J.G.D., in *Fractional Solidification*, M. Zief and W.R. Wilcox (eds.), vol. I, Marcel Dekker, New York (1967), chap. 13.

Moody, HW., *J. Chem. Ed.*, **59**(4), Apr., 290 (1982).

Moonen, H. and N.J. Niefind, *Desalination*, **41**, 327 (1982).

Moore, L.R., S. Milliron, P.S. Williams, J.J. Chalmers, S. Margel and M. Zborowski, *Anal. Chem.*, **76**, 3899 (2004).

Mothes, H. and F. Löffler, *Chem. Eng. Process*, **18**, 323 (1984).

Mulder, M., *Basic Principles of Membrane Technology*, Kluwer Academic Publishers, Dordrecht (1991).

Mullin, J.W., *Crystallization*, Butterworths, London (1961).

Mullin, J.W., *Crystallization*, 2nd edn., Butterworths, London (1972).

Myers, A.L. and J.M. Prausnitz, *AIChE J.*, **11**, 121 (1965).

Myers, A.L., C. Minka and D.Y. Ou, *AIChE J.*, **28**, 97 (1982).

Myerson, A.S., *Handbook of Industrial Crystallization*, Butterworth-Heinemann, Boston, MA (1993), chaps. 1–2.

Nakano, Y., C. Tien and W.N. Gill, *AIChE J.*, **13**(6), 1092 (1967).

National Research Council, *International Critical Tables, Vol. III*, McGraw-Hill, New York (1929).

Naylor, R.W. and P.O. Backer, *AIChE J.*, **1**, 95 (1955).

Neel, J., in *Membrane Separations Technology: Principles and Applications*, R.D. Noble and S.A. Stern (eds.) Elsevier, New York (1995), chap. 5.

Nerenberg, S.T. and G. Pogojeff, *Am. J. Clin. Path.*, **51**(6), 728 (1969).

Neumann, L., E.T. White and T. Howes, "What does a mean size mean?", AIChE Annual Meeting, San Francisco, CA, Nov. 16–21 (2003), paper 39a.

Newman, J., *Electrochemical Systems*, Prentice Hall, Englewood Cliffs, NJ (1973), p. 193.

Newman, J. *Electrochemical Systems*, 2nd edn., Prentice Hall, Englewood Cliffs, NJ (1991), p. 215.

Ng, P.K., J. Lundblad and G. Mitra, *Separ. Sci.*, **11**(5), 499 (1976).

Noble, P.T., *Biotechnol. Prog.*, **1**(4), 237 (1985).

Noble, R.D. and S.A. Stern (eds.), *Membrane Separations Technology: Principles and Applications*, Elsevier, Amsterdam (1995).

Noble, R.D. and P.A. Terry, *Principles of Chemical Separations with Environmental Applications*, Cambridge University Press, Cambridge (2004).

Noble, R.D. and J.D. Way (eds.), *Liquid Membranes: Theory and Applications*, ACS Symposium Series 347, American Chemical Society. Washington, D.C. (1987).

Noble, R.D., J.D. Way and L.A. Powers, *Ind. Eng. Chem. Fund.*, **25**, 450 (1986).

Norman, M.A., C.B. Evans, A.R. Fuoco, R.D. Noble and C.A. Koval, *Anal. Chem.*, **77**(19), 6374 (2005).

Null, H.R., *Chem. Eng. Progr.*, **Aug.**, 42 (1980).

Nyborg, W.L., in *Ultrasound: Its Application in Medicine and Biology*, F.J. Fry (ed.), vol. 1, Elsevier, New York (1978), p. 1.

Nyström, M., P. Aimar, S. Luque, M. Kulovaara and S. Metsämuuronen, *Colloids Surf. A: Physicochem. Eng. Aspects*, **138**, 185 (1998).

Nývlt, J., *Design of Crystallizers*, CRC Press, Boca Raton, FL (1992).

Nývlt, J., O. Sohnel, M. Matuchova and M. Brout, *The Kinetics of Industrial Crystallization*, Elsevier, Amsterdam (1985).

Oak, M.J., "Modeling and experimental study of high gradient magnetic separation with application to coal beneficiation," unpublished M.S. thesis, Auburn University, AL, Dec. 8 (1977).

O'Connell, H.E., *Trans. AIChE.*, **42**, 741 (1946).

O'Farrell, P.H., *Science*, **227**, 1586 (1985).

Ohno, M., O. Ozaki, H. Sato, S. Kimura and T. Miyauchi, *J. Nucl. Sci. Technol.*, **14**(8), 589 (1977).

Ohno, M., T. Morisue, O. Ozaki and T. Miyauchi, *J. Nucl. Sci. Technol.*, **15**(5), 376 (1978).

Ohya, H. and Y. Taniguchi, *Desalination*, **16**, 359 (1975).

Olander, D.R., *Adv. Nuclear Sci. Technol.*, **6**, 105 (1972).

Orcutt, J.C., in *Fractional Solidification*, M. Zief and W.R. Wilcox (eds.), vol. 1, Marcel Dekker, New York (1967), chap. 17.

Ornatski, N.V., E.V. Sergeev and Y.M. Shekhtman, *Investigations of the Process of Clogging of Sands*, University of Moscow, Moscow (1955).

Overdevest, P.E.M., J.T.F. Keurentjes, A. Van der Padt and K. van't Reit, in *Surfactant-based Separation: Science and Technology*, J.F. Scamehorn and J.H. Harwell (eds.), ACS Symposium Series 740, Amercian Chemical Society, Washington, D. C. (2000), chap. 9.

Padin, J., R.T. Yang and C.L. Munson, *Ind. Eng. Chem. Res.*, **38** (10), 3614 (1999).

Pamme, N. and A. Manz, *Anal. Chem.*, **76**, 7250 (2004).

Pan, C.Y., *AIChE J.*, **29**, 545 (1983).

Pan, C.Y., *AIChE J.*, **32**, 2020 (1986).

Pan, C.Y. and H.W. Habgood, *Ind. Eng. Chem. Fund.*, **13**, 323 (1974).

Pan, C.Y. and H.W. Habgood, *Can. J. Chem. Eng.*, **56**, 197 (1978a).

Pan, C.Y. and H.W. Habgood, *Can. J. Chem. Eng.*, **56**, 210 (1978b).

Patel, D.C. and R.G. Luo, in *Adsorption and its Applications in Industry and Environmental Protection*, A. Dabrowski (ed.), Studies in Surface Science and Catalysis 120, Elsevier Science, New York (1998).

Paul, D.R., *Ind. Eng. Chem. Proc. Des. Dev.*, **10**(3), 375 (1971).

Paul, D.R. and W.J. Koros, *J. Polym. Sci., Polym. Phys. Ed.*, **14**, 675 (1976).

Pearce, C.W., in *VLSI Technology*, S.M. Sze (ed.), McGraw-Hill, New York (1983).

Pedersen, C.J., *J. Am. Chem. Soc*, **89**(26), 7017 (1967).

Pedersen, C.J., *Science*, **241**, 536 (1988).

Perkins, T.W., D.S. Mak, T.W. Root and E.N. Lightfoot, *J. Chromatog. A*, **766**, 1 (1997).

Perry, R.H. and D. Green, *Perry's Chemical Engineers' Handbook*, 6th edn., McGraw-Hill, New York (1984).

Petenate, A. and C.E. Glatz, *Biotechnol. Bioeng.*, **25**, 3059 (1983).

Peters, E.A.J.F., B. Breure, P. van den Henvel and P.J.A.M. Kerkhof, *Ind. Eng. Chem. Res.*, **47**, 3937 (2008).

Petersson, F., A. Nilsson, H. Jönsson and T. Laurell, *Anal. Chem.*, **77**, 1216 (2005).

Pfann, W.G. *Zone Melting*, 2nd edn., John Wiley, New York (1966), p. 31.

Pfeffer, R., *Ind. Eng. Chem. Fund.*, **3**, 380 (1964).

Phelps, D.S.C. and D.M. Ruthven, *Ind. Eng. Chem. Res.*, **40**, 2168 (2001).

Philpot, J. St. L., *Trans. Faraday Soc.*, **36**, 38 (1940).

Pigford, R.L., B. Baker, III and D.E. Blum, *Ind. Eng. Chem. Fund.*, **8**, 144 (1969a).

Pigford, R.L., B. Baker, III and D.E. Blum, *Ind. Eng. Chem. Fund.*, **8**, 848 (1969b).

Pitzer, K.S. and L. Brewer, *Thermodynamics by G.N. Lewis and M. Randall*, 2nd edn. McGraw-Hill, New York (1961), p. 666.

Pohl, H.A., in *Methods of Cell Separation*, N. Castimpoolas (ed.), vol. 1, Plenum Press, New York (1977), chap. 3.

Pohl, H.A., *Dielectrophoresis: The Behavior of Matter in Nonuniform Electric Fields*, Cambridge University Press, New York (1978).

Pohl, H.A. and K. Kaler, *Cell Biophys.* **1**, 15 (1979).

Porath, J., *Protein Expression Purif.* **3**, 263 (1992).

Porath, J. and P. Flodin, *Nature*, **83**, 1657 (1959).

Porath, J., J. Carlsson, I. Olsson and G. Belfrage, *Nature*, **258**, 598 (1975).

Porter, H.F., G.A. Schurr, D.F. Wells and K.T. Semrau, in *Perry's Chemical Engineers' Handbook*, R.H. Perry and D. Green (eds.), 6th edn., McGraw-Hill, New York (1984), sect. 20.

Porter, K.E., M.J. Lockett and C.T. Lim, *Trans. Inst. Chem. Eng.*, **50**, 91 (1972).

Powers, J.E., in *New Chemical Engineering Separation Techniques*, H.M. Schoen (ed.), Interscience, New York (1962).

Pozrikidis, C., *J. Fluid Mech.*, **351**, 139 (1997).

Prasad, R. and K.K. Sirkar, *Ind. Eng. Chem. Proc. Des. Dev.*, **24**, 350 (1985).

Prasad, R. and K.K. Sirkar, *AIChE J.*, **33**, 1057 (1987).

Prasad, R. and K.K. Sirkar, *AIChE J.*, **34**, 177 (1988).

Prasad, R. and K.K. Sirkar, in *Membrane Handbook*, W.S. Winston Ho and K.K. Sirkar (eds.), Kluwer Academic, Boston, MA (2001), chap. 41.

Pratt, H.R.C., *Countercurrent Separation Processes*, Elsevier, Amsterdam (1967).

Pratt, H.R.C., *Solv. Extr. Ion Exchange*, **1**(4), 669 (1983).

Pratt, H.R.C. and G.W. Stevens, *Ind. Eng. Chem. Res.*, **30**, 733 (1991).

Prausnitz, J.M., *Molecular Thermodynamics of Fluid-Phase Equilibria*, Prentice-Hall, Englewood Cliffs, NJ (1969), p. 273.

Prausnitz, J.M. and P.L. Chueh, *Computer Calculations for High-Pressure Vapor-Liquid Equilibria*, Prentice-Hall, Englewood Cliffs, NJ (1968).

Prausnitz, J.M., R.N. Lichtenthaler and E.G. de Azevedo, *Molecular Thermodynamics of Fluid-Phase Equilibria*, 2nd edn., Prentice Hall, Englewood Cliffs, NJ (1986).

Prausnitz, J.M., R.N. Lichtenthaler and E.G. Azevedo, *Molecular Thermodynamics of Fluid Phase Equilibria*, 3rd edn., Prentice Hall, Englewood Cliffs, NJ (1999).

Present, R.D., *Kinetic Theory of Gases*, McGraw-Hill, New York (1958), p. 118.

Probstein, R.F., *Physicochemical Hydrodynamics: An Introduction*, Butterworths, Boston, MA (1989), pp. 129–141.

Pujar, N.S. and A.L. Zydney, *Ind. Eng. Chem. Res.*, **33**, 2473 (1994).

Pusch, W., H.G. Burghoff and E. Staude, in *Proc. 5th Int. Symp. Fresh Water From the Sea*, vol. 4, Alghero, Italy May 16–20 (1976), p. 143.

Qin, Y.J., J.P. Sheth and K.K. Sirkar, *Ind. Eng. Chem. Res.*, **42**, 582 (2003).

Rachford, H.H. Jr. and J.D. Rice, *J. Petrol. Technol.* **4**(10), sect. 1, Oct., p. 19 (1952a).

Rachford, H.H. Jr. and J.D. Rice, *J. Petrol. Technol.*, **4**(10), sect. 2, Oct., p. 3 (1952b).

Radke, C.J. and J.M. Prausnitz, *AIChE J.*, **18**(4), 761 (1972).

Raghavan, S. and D.W. Fuerstenau, *AIChE Symp. Ser.*, **71**(150), 59 (1975).

Raghavan, N.S., M.M. Hassan and D.M. Ruthven, *AIChE J.*, **31**(3), 385 (1985).

Ramkrishna, D., *Population Balances: Theory and Applications to Particle Systems in Engineering*, Academic Press, New York (2000).

Ramshaw C. and R.H. Mallinson, "Mass transfer process," US Patent 4,283,255, Aug. 11 (1981).

Randolph, A.D. and M.A. Larson, *Theory of Particulate Processes: Analysis and Techniques of Continuous Crystallization*, 2nd edn., Academic Press, New York (1988).

Rao, G.H. and K.K. Sirkar, *Desalination*, **27**, 99 (1978).

Ratanathanawongs, S.K. and J.C. Giddings, *Anal. Chem.*, **64**, 6 (1992).

Rathore, A.S. and C.S. Horvath, *Anal. Chem.*, **70**, 3069 (1998).

Rautenbach, R. and R. Albrecht, *Membrane Processes*, John Wiley, New York (1989).

Recktenwald, D. and A. Radbruch (eds.), *Cell Separation Methods and Applications*, Marcel Dekker, New York (1998), p. 293.

Reed, B.W., M.L. Semmens and E.L. Cussler, in *Membrane Separations: Principles and Applications*, R.D. Noble and S.A. Stern (eds.), Elsevier, Amsterdam (1995), p. 467.

Rege, S.U. and R.T. Yang, *Separ. Sci. Technol.*, **36**(15), 3355 (2001).

Reid, R.C., J.M. Prausnitz and T.K. Sherwood, *The Properties of Gases and Liquids*, 3rd edn., McGraw-Hill, New York (1977).

Reis, J.F.G., E.N. Lightfoot and H-L. Lee, *AIChE J.*, **20**(2), 362 (1974).

Reiss, L.P., *Ind. Eng. Chem. Proc. Des. Dev.*, **6**, 486 (1967).

Renkin, E.M., *J. Gen. Physiol.*, **38**, 225 (1954).

Reschke, M. and K. Schügerl, *Chem. Eng. J.*, **29**, 825 (1984).

Rhee, H.K. and N.R. Amundson, *Ind. Eng. Chem. Fund.*, **9**, 303 (1970).

Rhee, H.K. and N.R. Amundson, *AIChE J.*, **28**(3), 423 (1982).

Rhee, H.K., R. Aris and N.R. Amundson, *Phil. Trans. Roy. Soc. London.*, **267**A, 419 (1970a).

Rhee, H.K., E.D. Heerdt and N.R. Amundson, *Chem. Eng. J.*, **1**, 279 (1970b).

Rhee, H.K., E.D. Heerdt and N.R. Amundson, *Chem. Eng. J.*, **3**, 22 (1972).

Richman, D., E.A. Wynne and F.D. Rosi, in *Fractional Solidification*, M. Zief and W.R. Wilcox (eds.), vol. I, Marcel Dekker, New York (1967), chap. 9.

Ridgway, K. and E.E. Thrope, in *Handbook of Solvent Extraction*, T.C. Lo, M.H.I. Baird and C. Hanson (eds.), Wiley, New York, Ch. 19 (1983). Reprinted, Krieger Publishing Co., Malabar, FL (1991).

Rietema, K., *Chem. Eng. Sci.*, **7**, 89 (1957).

Rietema, K., *Chem. Eng. Sci.*, **15**, 298 (1969).

Ritcey, G.M. and A.W. Ashbrook, *Solvent Extraction: Principles and Applications to Process Metallurgy*, Part II, Elsevier Scientific Publishing Company, Amsterdam (1979).

Ritcey, G.M. and A.W. Ashbrook, *Solvent Extraction, Part I*, Elsevier, New York (1984a).

Ritcey, G.M. and A.W. Ashbrook, *Solvent Extraction, Part II*, Elsevier, New York (1984b), pp. 607–608.

Ritter, J.A. and R.T. Yang, *Ind. Eng. Chem. Res.*, **28**, 599 (1989).

Rixey, W.G., "Nonwet adsorbents for the selective recovery of polar organic solutes from dilute aqueous solution," unpublished Ph.D. dissertation, Dept. Chem. Eng., University of California, Berkeley, CA (1987).

Roberts, D.L. and S.K. Friedlander, *AIChE J.* **26**, 593 (1980a).

Roberts, D.L. and S.K. Friedlander, *AIChE J.* **26**, 602 (1980b).

Roberts, D.L., J.F. Scamehorn and S.D. Christian, in *Surfactant-based Separations: Science and Technology*, J.F. Scamehorn and J.H. Harwell (eds.), ACS Symposium Series 740, American Chemical Society, Washington, D.C. (2000), chap. 11.
Robeson, L., *J. Membr. Sci.*, **62**, 165 (1991).
Robinson, C.S. and E.R. Gilliland, *Elements of Fractional Distillation*, 4th edn., McGraw-Hill, New York (1950), pp. 162–175.
Robinson, J.P., in *Encyclopedia of Biomaterials and Biomedical Engineering*, G.E. Wnek and G.L. Bowlin (eds.), Marcel Dekker, New York (2004), p. 630.
Robinson, J.S., D. Scott and J. Winnick, *AIChE J.*, **44**(10), 2168 (1998).
Robinson, R.G. and D.Y. Cha, *Biotechnol. Prog.*, **1**, 18 (1985).
Rogers, T.H., F.V. Grim and N.E. Lemmon, *Ind. Eng. Chem.*, **18**, 164 (1926).
Rolchigo, P.M. and D.J. Graves, *AIChE J.*, **34** (3), 483 (1988).
Romero, C.A. and R.H. Davis, *J. Membr. Sci.*, **39**, 157 (1988).
Romero, J. and A.L. Zydney, *Separ. Sci. Technol.*, **36**(7), 1575 (2001).
Rony, P.R., *Separ. Sci.*, **3**(3), 239 (1968a).
Rony, P.R., *Separ. Sci.*, **3**(4), 357 (1968b).
Rony, P.R., *Separ. Sci.*, **4**, 413 (1969a).
Rony, P.R., *Separ. Sci.*, **4**(6), 493 (1969b).
Rony, P.R., *Separ. Sci.*, **5**(1), 1 (1970).
Rony, P.R., in *Recent Advances in Separation Techniques*, N. N. Li (ed.), AIChE Symposium Series 120, no. 68 (1972), pp. 1–58.
Rosen, J.B., *J. Chem. Phys.*, **20**, 387 (1952).
Rosen, J.B., *Ind. Eng. Chem.*, **46**, 1590 (1954).
Rothfeld, L.B., *AIChE J.*, **9**, 19 (1963).
Rousseau, R.W. and J.S. Staton, *Chem. Eng.*, **July**, 91 (1988).
Rubin, E. and E.L. Gaden, in *New Chemical Engineering Separation Techniques*, H.M. Schoen (ed.), Interscience, New York (1962).
Rubin, E. and J. Jorne, *Ind. Eng. Chem. Fund.*, **8**, 474 (1969).
Rubow, K.L., "Submicron aerosol filtration characteristics of membrane filters," unpublished Ph.D. Thesis, Mechanical Engineering Department, University of Minnesota, MN (1981).
Rudge, S.R. and M.R. Ladisch, *Biotechnol. Prog.*, **4**(13), 123 (1988).
Ruhemann, M., *The Separation of Gases*, 2nd edn., Oxford University Press, London (1949).
Rumeau, J., F. Persin, V. Sciers, M. Persin and J. Sarrazin, *J. Membr. Sci.*, **73**, 313 (1992).
Russel, W.B., D.A. Saville and W.K. Schowalter, *Colloidal Dispersions*, Cambridge University Press, Cambridge (1989).
Ruthven, D.M., *Principles of Adsorption and Adsorption Processes*, John Wiley, New York (1984), pp. 381–383.
Sabadell, J.E. and N.H Sweed, *Separ. Sci.*, **5**, 171 (1970).
Safarik, D.J. and R.B. Eldridge, *Ind. Eng. Chem. Res.*, **37**, 2571 (1998).
Saffman, P.G., *J. Fluid Mech.*, **22**, 385 (1965).
Sage, B.H., *Thermodynamics of Multicomponent Systems*, Reinhold Publishing Corp., New York (1965), p. 256.
Said, A.S. (1956), *AIChE J.*, **2**(4), 477 (1956).
Said, A.S., *Separ. Sci. Technol.*, **13**(8), 647 (1978).
Saksena, S. and A.L. Zydney, *Biotechnol. Bioeng.*, **43**, 960 (1994).
Salcudean, M., I. Gartshore and E.C. Statie, *Chem. Eng.*, **Apr.**, 66 (2003).
Sandell, E.B., *Anal. Chem.*, **40**, 834 (1968).
Sanders, E.S., W.J. Koros, H.B. Hopfenberg and V.T. Stannett, *J. Membr. Sci.*, **18**, 53 (1984).
Sanyal, D., N. Vasishtha and D.N. Saraf, *Ind. Eng. Chem. Res.*, **27**, 2149 (1988).
Sarkar, S., C.J. Mumford and C.R. Phillips, *Ind. Eng. Chem. Proc. Des. Dev.*, **19**, 665 (1980).
Sartori, G., W.S. Ho, D.W. Savage, G.R. Chludzinski and S. Wiechert, *Separ. Purif. Meth.*, **16**(2), 171 (1987).
Saskawa, S. and H. Walter, *Biochemistry*, **11**, 2760 (1972).
Satterfield, C.M., C.K. Colton and W.K. Pitcher Jr., *AIChE J.*, **19**, 628 (1973).
Saunders, M.S., J.B. Vierow and G. Carta, *AIChE J.*, **35**(1), 53 (1989).
Savage, D.W., E.W. Funk, W.C. Yu and G. Astarita, *Ind. Eng. Chem. Fund.*, **25**, 326 (1986).
Scamehorn, J.F., S.D. Christian and R.T. Ellington, in *Surfactant-based Separation Processes*, J.F. Scamehorn and J.H. Harwell (eds.), Marcel Dekker, New York (1989), chap. 2.
Schachman, H.K., *J. Phys. Colloid Chem.*, **52**, 1034 (1948).
Schaetzel, P., C. Vauclair, G. Luo and Q.T. Nguyen, *J. Membr. Sci.*, **191**, 103 (2001).
Schaetzel, P., C. Vanclair, Q.T. Nguyen and R. Bouzerar, *J. Membr. Sci.*, **244**, 117 (2004).
Schasfoort, R.B.M., S. Schlautmann, J. Hendrikse and A. van den Berg, *Science*, **286**, 942 (1999).
Schock, G. and A. Miquel, *Desalination*, **64**, 339 (1987).
Schork, J.M., R. Srinivasan and S.R. Auvil, *Ind. Eng. Chem. Res.*, **32**, 2226 (1993).
Schottky, W., *Phys. Z.*, **25**, 635 (1924).
Schultz, J.S., J.S. Goddard and S.R. Suchdeo, *AIChE J.*, **20**, 417 (1974).
Schure, M.R. and A.M. Lenhoff, *Anal. Chem.*, **65**, 3024 (1993).
Schweitz, L., P. Spégel and S. Nilsson, *Analyst*, **125**, 1899 (2000).
Schweitzer, P.A., *Handbook of Separation Techniques for Chemical Engineers*, 3rd edn., McGraw-Hill, New York (1997).
Schwinge, J., D.E. Wiley and D.F. Fletcher, *Ind. Eng. Chem. Res.*, **41**, 4879 (2002).
Schwinge, J., D.E. Wiley and D.F. Fletcher, *Ind. Eng. Chem. Res.*, **42**, 4962 (2003).
Scopes, R.K., *Protein Purification: Principles and Practice*, 2nd edn., Springer Verlag, New York (1987), p. 101.
Scopes, R.K., *Protein Purification: Principles and Practice*, 3rd edn., Springer, New York (1994), chaps. 6, 7.
Scott, C.D., R.D. Spence and W.G. Sisson, *J. Chromatogr.*, **126**, 381 (1976).
Scott, D.S. and F.A.L. Dullien, *Chem. Eng. Sci.*, **17**, 771 (1962).
Seader, J.D. and E.J. Henley, *Separation Process Principles*, John Wiley & Sons, New York (1998).
Seader, J.D. and E.J. Henley, *Separation Process Principles*, 2nd edn., John Wiley, New York (2006).
Sengupta, A. and K.K. Sirkar, in *Progress in Filtration and Separation*, R.J. Wakeman (ed.), vol. 4, Elsevier, Amsterdam (1986), p. 289.
Sengupta, A. and K.K. Sirkar, in *Membrane Separations Technology: Principles and Applications*, R.D. Noble and S.A. Stern (eds.), Elsevier Science, Amsterdam (1995), chap. 11.
Sengupta, A., R. Basu and K.K. Sirkar, *AIChE J.*, **34**, 1698 (1988).
Sengupta, A., B. Raghuraman, and K.K. Sirkar, *J. Membr. Sci.*, **51**, 105 (1990).
Sengupta, A., P.A. Petersen, B.D. Miller, J. Schneider and C.W. Fulk Jr., *Separ. Purif. Technol.* **14**, 189 (1998).

Setchenow, M., *Ann. Chim. Phys.*, **25**, 226 (1892).

Shaffer, L.H. and M.S. Mintz, in *Principles of Desalination*, K.S. Spiegler (ed.), Academic Press, New York, Ch. 6 (1966); 2nd edn., Part A, Ch. 6, K.S. Spiegler and A.D.K. Laird (eds.) (1980).

Shaffer, L.H. and M.S. Mintz, in *Principles of Desalination*, K.S. Spiegler (ed.), Academic Press, New York (1980).

Shamsai, B.M. and H.G. Monobouquette, *J. Membr. Sci.*, **130**, 173 (1997).

Shapiro, H.P., *Practical Flow Cytometry*, 3rd edn., Wiley-Liss, New York (1995).

Sharnez, R. and D. Sammons, *Strategies for Enhancing Performance in Continuous-Flow Electrophoresis, I: Selective Manipulation of Particle Trajectories, Preprints of First Separations Division Topical Conference on Separation Technologies: New Developments and Opportunities*, Miami Beach, FL, Nov. 2-6 (1992), pp. 264–270.

Shelden, R.A. and E.V. Thompson, *J. Membr. Sci.*, **4**, 115 (1978).

Shendalman, L.H. and J.E. Mitchell, *Chem. Eng. Sci.*, **27**, 1449 (1972).

Sheng, H.P., *Separ. Purif. Methods*, **6**(1), 89 (1977).

Sherwood, T.K. and F.A.L. Holloway, *Trans. AIChE J.*, **36**, 39 (1940).

Sherwood, T.K., P.L.T. Brian and R.E. Fisher, *Ind. Eng. Chem. Fund.*, **6**, 2 (1967).

Sherwood, T.K., R.L. Pigford and C.R. Wilke, *Mass Transfer*, McGraw-Hill, New York (1975), chap. 11.

Shiras, R.N., D.N. Hanson and C.H. Gibson, *Ind. Eng. Chem.*, **42**, 871 (1950).

Shreve, R.N. and G.T. Hatch, *Chemical Process Industries*, 5th edn, McGraw-Hill, New York (1984).

Shuler, M.L. and F. Kargi, *Bioprocess Engineering: Basic Concepts*, 2nd edn., Prentice Hall PTR, Upper Saddle River, NJ (2002).

Sidhoum, M., A. Sengupta and K.K. Sirkar, *AIChE J.*, **34**(3), 417 (1988).

Siegell, J.H., G.D. Dupre and J.C. Pirkle Jr., *Chem. Eng. Progr.*, **Nov.**, 57 (1986).

Siirola, J.J. and S.D. Barnicki, in *Perry's Chemical Engineers' Handbook*, R.H. Perry and D.W. Green (eds.), 7th edn., McGraw-Hill, New York (1997), pp. 13–54.

Silver, R., in *Principles of Desalination*, K.S. Spiegler (ed.), Academic Press, New York (1966), p. 77.

Sircar, S and J.R. Hufton, *AIChE J.*, **46** (3), 659 (2000).

Sircar, S. and R. Kumar, *I&E. Chem. Proc. Des. Dev.*, **22**(2), 271 (1983).

Sircar, S. and J.W. Zondlo, *US Patent* **4**, 013, 429 (1977).

Sirkar, K.K., *Separ. Sci.*, **12**(3), 211 (1977).

Sirkar, K.K., *Separ. Sci. Technol.*, **15** (4), 1091 (1980).

Sirkar, K.K., "Immobilized interface solute-transfer process," US Patent 4997569 (1991).

Sirkar, K.K., in *Membrane Handbook*, W.S. Winston Ho and K. K. Sirkar (eds.), Van Nostrand Reinhold, New York (1992), chap. 46.

Sirkar, K.K., "Immobilized interface solute transfer apparatus," US Patent 4,789,468, Dec. 6 (1988); US Patent Re, 34, 828, Jan. 17 (1995).

Sirkar, K.K., in *Membrane Handbook*, W.S. Winston Ho and K.K. Sirkar (eds.), Kluwer Academic, Boston, MA (2001), chap. 46.

Sirkar, K.K., *Ind. Eng. Chem. Res.*, **47**(15), 5250 (2008).

Sirkar, K.K., P.T. Dang and G.H. Rao, *Ind. Eng. Proc. Des. Dev.*, **21**, 517 (1982).

Skarstrom, C.W., "Method and apparatus for fractionating gas mixtures by adsorption," US Patent 2,944,627, July 12 (1960).

Skarstrom, C.W., in *Recent Developments in Separation Science*, N. Li (ed.), vol 2, CRC Press, Cleveland, OH (1975), p. 95.

Skelland, A.H.P., *Diffusional Mass Transfer*, Wiley, New York (1974).

Slattery, J.C., *Momentum, Energy and Mass Transfer in Continua*, McGraw-Hill, New York (1972).

Sleicher, C.A. Jr., *AIChE J.*, **5**(2), 145 (1959).

Sleicher, C.A. Jr., *AIChE J.*, **6**(2), 529 (1960).

Smith, D.R. and J.A. Quinn, *AIChE J.*, **26**, 112 (1980).

Smith, D.R., R.J. Lander and J.A. Quinn, in *Recent Developments in Separation Science*, vol. 3B, N.N. Li (ed.), CRC Press, Cleveland, OH (1977), p. 225.

Smith, J.M. and H.C. Van Ness, *Introduction to Chemical Engineering Thermodynamics*, 3rd edn., McGraw-Hill, New York (1975).

Smith, J.M., H.C. Van Ness and M.M. Abbott, *Introduction to Chemical Engineering Thermodynamics*, 6th edn., McGraw-Hill, New York (2001).

Smith, K.A., J.K. Meldon and C.K. Colton, *AIChE J.*, **19**, 102 (1973).

Snyder, L.R. and J.J. Kirkland, *Introduction to Modern Liquid Chromatography*, 2nd edn., Wiley-Interscience, New York (1979), pp. 260–289.

Solan, A., Y. Winograd and U. Katz, *Desalination*, **9**, 89 (1971).

Soltanieh, M. and W.N. Gill, *Chem. Eng. Commun*, **12**, 279 (1981).

Song, L., Z. Ma, X. Liao, P.B. Kosaraju, J.R. Irish and K.K. Sirkar, *J. Membr. Sci.*, **323**, 257 (2008).

Soo, S.L., *Particulates and Continuum: Multiphase Fluid Dynamics*, Hemisphere Publishing, New York (1989), chap. 2 .

Soo, S.L. and L.W. Rodgers, *Powder Technol.*, **5**, 43 (1971).

Souders, M., G.J. Pierotti and C.L. Dunn, *Chem. Eng. Progr. Symp. Ser.*, **66**(100), 41 (1970).

Sourirajan, S., *Reverse Osmosis*, Logos Press, London (1970).

Spiegler, K.S., *Salt-Water Purification*, 2nd edn., Plenum Press, New York (1977).

Spiegler, K.S. and O. Kedem, *Desalination*, **1**, 311 (1966).

Spielman, L.A. and P.M. Cukor, *J. Colloid Interface Sci.*, **43**, 51 (1973).

Spielman, L.A. and J.A. Fitzpatrick, *J. Colloid Interface Sci.*, **42**, 607 (1972).

Spoor, P.S. and G.W. Swift, *Phys. Rev. Lett.*, **85**(8), 1646 (2000).

Spriggs, H.D. and J.L. Gainer, *Ind. Eng. Chem. Fund.*, **12**, 291 (1973).

Spriggs, H.D. and N.N. Li, in *Membrane Separation Processes*, P. Meares (ed.), Elsevier, Amsterdam (1976), chap. 2 .

Stannett, V.T., in *Diffusion in Polymers*, J. Crank and G.S. Park (eds.), Academic Press, London (1968), p. 41.

Stannett, V. T., W.J. Koros, D.L. Paul, H. Lonsdale and R. Baker, *Adv. Polym. Sci.*, **32**, 69 (1979).

Steed, J.W. and J.L. Atwood, *Supramolecular Chemistry*, John Wiley, New York (2000).

Stella, A., H.R.C. Pratt, K.H. Mensforth, G.W. Stevens and T. Bowser, *Ind. Eng. Chem. Res.*, **45**, 6555 (2006).

Stern, S.A. and H.L. Frisch, *Ann. Rev. Mater. Sci.*, **11**, 523 (1981).

Stern, S.A. and W.P. Walawender Jr., *Separ. Sci.*, **4**(2), 129 (1969).

Sternling, C.V. and L.E. Scriven, *AIChE J.*, **5**, 514 (1959).

Stewart, G.H., *Separ. Sci. Technol.*, **13**(3), 201 (1978).

Strathmann, H., *Separ. Sci. Technol.*, **15**, 1135 (1980).

Strathmann, H., in *Membrane Handbook*, W.S. Winston Ho and K.K. Sirkar (eds.), Kluwer Academic, Boston, MA (2001), chaps. 16–20.

Strickland, A.D., R.G. De Krester and P.J. Scales, *AIChE J.*, **51**(9) 2481 (2005).

Strigle, R.F. Jr., *Packed Tower Design and Applications*, 2nd edn., Gulf Publishing, Houston, TX (1994).

Suen, S.-Y. and M. Etzel, *Chem. Eng. Sci.*, **47**(6), 1355 (1992).

Suh, S.S. and P.C. Wankat, *AIChE J.*, **35**(3), 523 (1987).

Sussman, M.V. and C.C. Huang, *Science*, **156**, 974 (1967).

Sussman, M.V., K.N. Astill, R. Rombach, A. Cerullo and S.S. Chen, *Ind. Eng. Chem. Fund.*, **11**(2), 181 (1972).

Sussman, M.V., K.N. Astill and R.N.S. Rathore, *J. Chromatograph. Sci.*, **12**, 91 (1974).

Sutija, D.P. and J.M. Prausnitz, *Chem. Eng. Ed.*, **Winter**, 20 (1990).

Suzuki, I., H. Yagi, H. Komatsu and M. Hirata, *J. Chem. Eng. Jpn.*, **4**, 26 (1971).

Svarovsky, L., in *Solid-Liquid Separation*, L. Svarovsky (ed.), Butterworths, London (1977).

Svarovsky, L., in *Progress in Separation and Filtration*, R.J. Wakeman (ed.), vol. 1, Elsevier, Amsterdam (1979), p. 251.

Svarovsky, L., *Solid-Gas Separation*, Elsevier Scientific Publishing, Amsterdam (1981).

Svarovsky, L., *Solid-liquid Separation*, Butterworths, London (1982), chap. 6.

Svedberg, T. and K.O Pedersen, *The Ultracentrifuge*, Clarendon Press, Oxford (1940).

Svrcek, W.Y. and W.D. Monnery, *Chem. Eng. Progr.*, **Oct.**, 53 (1993).

Sweed, N.H., in *Progress in Separation and Purification*, E.S Perry and C.J. Van Oss (eds.), vol. 4, John Wiley, New York, (1971).

Sweed, N.H., in *Recent Developments in Separation Science*, N.N. Li (ed.), vol. I, CRC Press, Boca Raton, FL (1972).

Sweeney, M.J. and P.M. Galletti, *Trans. Am. Soc. Artif. Intern. Organs*, **10**, 3 (1964).

Tabatabai, A.J., F. Scamehorn and S.D. Christian, *J. Membr. Sci.*, **100**, 193 (1995).

Talbot, J.B., *Separ. Sci. Technol.*, **15**(3), 277 (1980).

Talbot, L., R.K. Chang, R.W. Schefer and D.R. Willis, *J. Fluid Mech.*, **101**, 737 (1980).

Tamada, J.A., A.S. Kertes and C.J. King, *Ind. Eng. Chem. Res.*, **29**, 1319 (1990).

Tamon, H., H. Mizota, N. Sano, S. Schulze and M. Oakazaki, *AIChE J.*, **41**(7), 1701 (1995).

Tan, H.K.S., *AIChE J.*, **40**(2), 369 (1994).

Tanford, C., *Physical Chemistry of Macromolecules*, John Wiley, New York (1961).

Tanford, C., *The Hydrophobic Effect: Formation of Micelles and Biological Membranes*, 2nd edn., Wiley Interscience, New York (1980).

Tanford, C., Y. Nozaki, J.A. Reynolds and S. Mikano, *Biochemistry*, **13**, 2369 (1974).

Tang, K.E.S. and V. Bloomfield, *Biophys. J.*, **82**, 2876 (2002).

Taylor, G.I., *Proc. Roy. Soc. London Ser. A.*, **219**, 186 (1953).

Taylor, G.I., *Proc. Roy. Soc. London, Ser. A.*, **225**, 473 (1954).

Taylor, R. and R. Krishna, *Multicomponent Mass Transfer*, John Wiley, New York (1993).

Taylor, R., H.A. Kooijman and J.-S Huang, *Comput. Chem. Eng.*, **18**, 205 (1994).

Taylor, R., R. Krishna and H. Kooijman, *Chem. Eng. Progr.*, July, 28 (2003).

Ter Haar, G. and S.J. Wyard, *Ultrasound Med. Biol.*, **4**, 111 (1978).

Ter Linden, A.J., *Proc. Inst. Mech. Eng.*, **160**, 233 (1949).

Terabe, S., in *Capillary Electrophoresis Technology*, N.A. Guzman (ed.), Marcel Dekker, New York (1993), p. 65.

Terabe, S., K. Otsuka, K. Ichikawa, A. Tsuchiya and T. Ando, *Anal. Chem.*, **56**, 111 (1984).

Teraoka, Y., H-M. Zhang, S. Furukawa and N. Yamazoe, *Chem. Lett.*, p. 1743 (1985).

Terrill, D.L., L.F. Sylvestre and M.F. Doherty, *Ind. Eng. Chem, Proc. Des. Dev.*, **24**, 1062 (1985).

Theodore, L., *Chem. Eng. Progr.*, **Sept.**, 16 (2005).

Thijssen, H.A., *Chem. Eng. Progr.*, **75**(7), 21 (1979).

Thomas, H., *J. Am. Chem. Soc.*, **66**, 1664 (1944).

Thomas, H., *Ann. NY Acad. Sci.*, **49**, 161 (1948).

Thongsukmak, A. and K.K. Sirkar, *J. Membr. Sci.*, **302**, 45 (2007).

Thongsukmak, A., and K.K. Sirkar, *J. Membr. Sci.*, **329**, 119 (2009).

Thorman, J.M., H. Rhim and S.T. Hwang, *Chem. Eng. Sci.*, **30**(7), 751 (1975).

Thorman, J.M., S.T. Hwang and K.H. Yuen, *Separ. Sci. Technol.*, **15**(4), 1069 (1980).

Tien, C., *Granular Filtration of Aerosols and Hydrosols*, Butterworths, London (1989), pp. 108–109.

Tilgner, H.G. and F.J. Schmitz, European Patent 36, 175; assigned to Akzo GmbH (1981).

Tomida, T., K. Hamaguchi, S. Tashima, M. Katoh and S. Masuda, *Ind. Eng. Chem. Res.*, **40**, 3557 (2001).

Tompkins, C.J., A.S. Michaels and S.W. Peretti, *J. Membr. Sci.*, **75**, 277 (1992).

Tondre, C., in *Surfactant-based Separations: Science and Technology*, J.F. Scamehorn and J.H. Harwell (eds.), ACS Symposium Series 740, American Chemical Society, Washington, D.C. (2000), chap. 10.

Toor, H.L., *AIChE J.*, **3**, 198 (1957).

Towns, J.K. and F.E. Regnier, *Anal. Chem.*, **64**, 243 (1992).

Treybal, R.E., *Liquid Extraction*, 2nd edn., McGraw-Hill, New York, (1963).

Treybal, R.E., *Mass Transfer Operations*, 3rd edn., McGraw-Hill, New York (1980).

Tsuda, T., in *Capillary Electrophoresis Technology*, N.A. Guzman (ed.), Marcel Dekker, New York (1993), chap. 14.

Tsuda, T. (ed.), *Electric Field Applications in Chromatography, Industrial and Chemical Processes*, VCH, Weinheim (1995).

Tsuji, A., E. Nakashima, S. Hamano and T. Yamana, *J. Pharm. Sci.*, **67**, 1059 (1978).

Turco, R.P., O.B. Toon, T.P. Ackerman, J.B Pollack and C. Sagan, *Science*, **247**, 166 (1990).

Turner, D.B., *Workbook of Atmospheric Dispersion Estimates*, HEW, Washington, D.C. (1969).

Turnock, P.H. and R.H. Kadlec, *AIChE J.*, **17**, 335 (1971).

Uemasu, I., *J. Inclusion Phenom. Mol. Recog. Chem.*, **13**, 1 (1992).

Ulrich, J., in *Handbook of Industrial Crystallization*, A.S. Myerson (ed.), Butterworth-Heinemann, Boston, MA (1993), chap. 7.

Underwood, A.J.V., *Trans. Inst. Chem. Engrs.*, **16**, 112 (1932).

Underwood, A.J.V., *Chem. Eng. Progr.*, **44**, 603 (1948).

Van Deemter, J.K., F.J. Zuiderweg and A. Klinkenberg, *Chem. Eng. Sci.*, **5**, 271 (1956).

Van der Kolk, H., in *Cyclones in Industry*, K. Rietema and C.G. Verver (eds.), Elsevier, Amsterdam (1961), p. 76.

Van Dongen, D.B. and M.F. Doherty, *Chem. Eng. Sci.*, **39**, 883 (1984).

Van Ebbenhorst Tengbergen H.J. and K. Rietema, in *Cyclones in Industry*, K. Rietema and C.G. Verver (eds.), Elsevier, Amsterdam (1961), p. 23.

Van Eijndhoven, R.H., S. Saksena and A.L. Zydney, *Biotechnol. Bioeng.*, **48**, 406 (1995).

Van Ness, H.C., *Ind. Eng. Chem. Fund.*, **8**, 464 (1969).

Van Ness, H.C. and M.M. Abbott, *Classical Thermodynamics of Nonelectrolyte Solutions*, McGraw-Hill, New York (1982).

Van Reis, R. and S. Saksena, *J. Membr. Sci.*, **129**, 19 (1997).

Van Vlasselaer, P., V.C. Palathumpat, G. Strang and M.J. Shapero, in *Cell Separation Methods and Applications*, D. Recktenwald and A. Radbruch (eds.), Marcel Dekker, New York (1998), p. 14.

Vermeulen, T., *Ind. Eng. Chem.*, **45**, 1664 (1953).

Vermeulen, T., L. Nady, J.M. Korchta, E. Ravoo and D. Howery, *Ind. Eng. Chem. Proc. Des. Dev.*, **10**(1), 91 (1971).

Vermuelen, T., G. Klein and N.K. Hiester, in *Chemical Engineers' Handbook*, J.H. Perry (ed.), McGraw-Hill, New York (1973), sect. 16.

Vieth, W.R., J.M. Howell and J.H. Hsieh, *J. Membr. Sci.*, **1**, 177 (1976).

Vilker, V.L., C.K. Colton and K.A. Smith, *AIChE J.*, **27**(4), 637 (1981).

Vink, H., *J. Chromatogr.*, **69**, 237 (1972).

Viovy, J-L., *Rev. Mod. Phys.*, **72**(3), 813 (2000).

Von Hippel, A., *Dielectrics and Waves*, John Wiley, New York (1954).

Von Stockar, U. and X-P. Lu, *Ind. Eng. Chem. Res.*, **30**, 1248 (1991).

Vrentas, J.S., J.L. Duda and S.T. Hsieh, *Ind. Eng. Chem. Prod. Res. Dev.*, **22**, 326 (1983).

Wadekar, V.V. and M.M. Sharma, *J. Separ. Process Technol.*, **1**, 1 (1981).

Wagener, K., H.D. Freyer and B.A. Billal, *Separ. Sci.*, **6**(4), 483 (1971).

Wakao, N. and T. Funazkri, *Chem. Eng. Sci.*, **33**, 1375 (1978).

Walawender, W.P. and S.A. Stern, *Separ. Sci.*, **7**(5), 553 (1972).

Wallace, R.M., *Ind. Eng. Chem. Proc. Des. Dev.*, **6**, 423 (1967).

Walsh, A.J. and H.G. Monobouquette, *J. Membr. Sci.*, **84**, 107 (1993).

Walter, J.E., *J. Chem. Phys.*, **13**(6), 229 (1945).

Wang, M., S. Hou and J.F. Parcher, *Anal. Chem.*, **78**(4), 1242 (2006).

Wankat, P.C., *AIChE J.*, **23**(6), 859 (1977).

Wankat, P.C., *Large Scale Adsorption and Chromatography*, vols. I & II, CRC Press, Boca Raton, FL (1986).

Wankat, P.C., *Ind. Eng. Chem. Res.*, **26**, 1579 (1987).

Wankat, P.C., *Equilibrium-Staged Separations*, Elsevier, New York (1988), chap. 10.

Wankat, P.C., *Rate-Controlled Separations*, Elsevier Applied Science, New York (1990).

Wankat, P.C., *Ind. Eng. Chem. Res.*, **32**, 894 (1993).

Wankat, P.C., *Separation Process Engineering*, 2nd edn., Prentice Hall, Upper Saddle River, NJ (2007), pp. 334–338, 649, 653.

Wankat, P.C., J.C. Dore and W.C. Nelson, *Separ. Purif. Meth.*, **4**(2), 215 (1975).

Ward, W.J., *AIChE J.*, **16**, 405 (1970).

Ward, W.J. and W.L. Robb, *Science*, **156**, 1481 (1967).

Wark, K. and C.F. Warner, *Air Pollution: Its Origin and Control*, Harper and Row Publishers, New York (1976).

Watson, J.H.P., *J. Appl. Phys.*, **44**(9), 4209 (1973).

Wauters, C.N. and J. Winnick, *AIChE J.*, **44**(10), 2144 (1998).

Way, J.D. and R.D. Noble, in *Membrane Handbook*, W.S. Winston Ho and K.K. Sirkar (eds.) Kluwer Academic, Boston, MA (2001), chap. 44.

Way, J.D., R.D. Noble, T.M. Flynn and F.D. Sloan, *J. Membr. Sci.*, **12**, 239 (1982).

Weaver, K. and C.E. Hamrin, *Chem. Eng. Sci.*, **29**, 1873 (1974).

Wei, J. and M.J. Realff, *AIChE J.*, **49**(12), 3138 (2003).

Weinstein, J.N. and F.B. Leitz, *Science*, **191**, 557 (1976).

Weiser, M.A.H., R.E. Apfel and E.A. Neppiras, *Acoustica*, **56**, 114 (1984).

Weller, S. and W.A. Steiner, *J. Appl. Phys.*, **21**, 279 (1950a).

Weller, S. and W.A. Steiner, *Chem. Eng. Progr.*, **46**, 585 (1950b).

Weaver, K. and C.E. Hamrin, *Chem. Eng. Sci.*, **29**, 1873 (1974).

Werezak, G.N., *Chem. Eng. Progr. Symp. Series* **65**(91), 6 (1969).

Westerberg, A.W., *Comput. Chem. Eng.*, **9**(5), 421 (1985).

Whitaker, S., *Chem. Eng. Sci.*, **28**, 139 (1973).

White, D.H. and P.G. Barkley, *Chem. Eng. Progr.*, **Jan.**, 25 (1989).

White, H.J., *Industrial Electrostatic Precipitation*, Addison-Wesley, Reading, MA (1963).

Whitney, R.P. and J.E. Vivian, *Chem. Eng. Progr.*, **45**, 323 (1949).

Wick, G.L., *Energy*, **3**, 95 (1978).

Wickramasinghe, S.R., M.J. Semmens and E.L. Cussler, *J. Membr. Sci.*, **84**, 1 (1993).

Wieme, R.J., in *Chromatography: A Laboratory Handbook of Chromatographic and Electrophoretic Methods*, E. Heftmann (ed.), Van Nostrand Reinhold, New York (1975), pp. 228–281.

Wijmans, J.G. and R.W. Baker, *J. Membr. Sci.*, **79**, 101 (1993).

Wijmans, J.G., A.L. Athayde, R. Daniels, J.H. Ly, H. D. Kamaruddin and I. Pinnau, *J. Membr. Sci.*, **109**, 135 (1996).

Wilcox, W.R., in *Fractional Soldification*, M. Zief and W.R. Wilcox (eds.), vol. I, Marcel Dekker, New York (1967), chap. 3.

Wilcox, W.R. and C.R. Wilke, *AIChE J.*, **10**, 160 (1964).

Wiley, D.E and D.F. Fletcher, *J. Membr. Sci.*, **211**, 127 (2003).

Wilhelm, R.H. and N.H. Sweed, *Science*, **159**, 522 (1968).

Wilhelm, R.H., A.W. Rice and A.R. Bendelius, *Ind. Eng. Chem. Fund.*, **5**, 141 (1966).

Wilhelm, R.H., A.W. Rice, R.W. Rolke and N.H. Sweed, *Ind. Eng. Chem. Fund.*, **7**, 337 (1968).

Wilke, C.R. and P. Chang, *AIChE J.*, **1**, 264 (1975).

Williams, M.E., D. Bhattacharyya, R.J. Ray and S.B. McCray, in *Membrane Handbook*, W.S. Winston Ho and K.K. Sirkar (eds.), Kluwer Academic, Boston, MA (2001), chap. 24.

Williams, P.S., S. Levin, T. Lenczycki and J.C. Giddings, *Ind. Eng. Chem. Res.*, **31**, 2172 (1992).

Wilson, E.J. and C.J. Geankpolis, *Ind. Eng. Chem. Fund.*, **5**, 9 (1966).

Wise, W.S. and D.F. Williams, "The principles of dissociation extraction," *Proc. Symp. Less Common Means of Separation*, Birmingham, UK, Apr. 24–26, 1963, The Institution of Chemical Engineers, London (1963), pp. 112–118.

Wolbert, D., B.-F. Ma, Y. Aurelle and J. Seureau, *AIChE J.*, **41** (6), 1395 (1995).

Wolf, D., J.L. Borowitz, A. Gabor and Y. Shraga, *Ind. Eng. Chem. Fund.*, **15**, 15 (1976).

Wyatt, G.R. and S.S. Cohen, *Nature*, **170**, 1072 (1952).

Wynn, N.P., *Chem. Eng. Progr.*, **Mar.**, 52 (1992).

Xu, Y., K.K. Sirkar, X-P. Dai and R.G. Luo, *Biotechnol. Prog.*, **21**, 590 (2005).

Yanar, D.K. and B.A. Kwetkus, *J. Electrostat.*, **35**, 257 (1995).

Yang, M.C. and E.L. Cussler, *AIChE J.*, **32**, 1910 (1986).

Yang, Q. and N.M. Kocherginsky, *J. Membr. Sci.*, **286**, 301 (2006).

Yang, R.T., *Gas Separation by Adsorption Processes*, Butterworths, Boston, MA (1987).

Yang, R.T., *Adsorbents: Fundamentals and Applications*, Wiley-Interscience, Hoboken, NJ (2003).

Yang, Z.F., A.K. Guha and K.K. Sirkar, *Ind. Eng. Chem. Res.*, **35**(4), 1383 (1996a).

Yang, Z.F., A.K. Guha and K.K. Sirkar, *Ind. Eng. Chem. Res.*, **35**(11), 4214 (1996b).

Yasuda, H., A. Peterlin, C.K. Colton, K.A. Smith and E.W. Merrill, *Makromol. Chem.*, **126**, 177 (1969).

Yih, C.S., *Phys. Fluids*, **11**, 477 (1968).

Yun, C.H., R. Prasad, A.K. Guha and K.K. Sirkar, *Ind. Eng. Chem. Res.*, **32**(6), 1186 (1993).

Zabasajja, J. and R.F. Savinell, *AIChE J.*, **35**(5), 755 (1989).

Zeman, L.J. and A.L. Zydney, *Microfiltration and Ultrafiltration: Principles and Applications*, Marcel Dekker, New York (1996).

Zief M. and W.R. Wilcox, in *Fractional Solidification*, M. Zief and W.R. Wilcox (eds.), vol. I, Marcel Dekker, New York (1967), chap. 1.

Zolandz, R.R. and G.K. Fleming, in *Membrane Handbook*, W.S.W. Ho and K.K. Sirkar (eds.), Kluwer Academic, Boston, MA (2001), p. 44.

Zubritsky, E., *Anal. Chem.*, **74**, 23A–26A (2002).

Zuiderweg, F.J. and A. Harmens, *Chem. Eng. Sci.*, **9**, 89 (1958).

Zumstein, R.C. and R.W. Rousseau, *Ind. Eng. Chem. Res.*, **28**, 1226 (1989).

Zwiebel, I., R.L. Gariepy and J.L. Schnitzer, *AIChE J.*, **18**(6), 1139 (1972).

Index

Printed in the United States
by Baker & Taylor Publisher Services